BRONC … ER
1983-91 REPAIR MANUAL

Sr. Vice President	Ronald A. Hoxter
Publisher and Editor-In-Chief	Kerry A. Freeman, S.A.E.
Managing Editors	Peter M. Conti, Jr. □ W. Calvin Settle, Jr., S.A.E.
Assistant Managing Editor	Nick D'Andrea
Senior Editors	Richard J. Rivele, S.A.E. □ Ron Webb
Director of Manufacturing	Mike D'Imperio
Manager of Manufacturing	John F. Butler
Editors	Peter M. Conti, Jr.
	Nick D'Andrea
	Neil Lenard, A.S.E.
	Steven Morgan
	Anthony Tortorici, A.S.E., S.A.E.

CHILTON BOOK COMPANY

ONE OF THE ***DIVERSIFIED PUBLISHING COMPANIES,***
A PART OF ***CAPITAL CITIES/ABC, INC.***

Manufactured in USA

Chilton Way Radnor, Pa. 19089
ISBN 0-8019-8159-X
Library of Congress Catalog Card No. 91-055479

2 3 4 5 6 7 8 9 0 1 1 0 9 8 7 6 5 4 3 2

Contents

Contents

SAFETY NOTICE

Proper service and repair procedures are vital to the safe, reliable operation of all motor vehicles, as well as the personal safety of those performing repairs. This manual outlines procedures for servicing and repairing vehicles using safe, effective methods. The procedures contain many NOTES, CAUTIONS and WARNINGS which should be followed along with standard safety procedures to eliminate the possibility of personal injury or improper service which could damage the vehicle or compromise its safety.

It is important to note that the repair procedures and techniques, tools and parts for servicing motor vehicles, as well as the skill and experience of the individual performing the work vary widely. It is not possible to anticipate all of the conceivable ways or conditions under which vehicles may be serviced, or to provide cautions as to all of the possible hazards that may result. Standard and accepted safety precautions and equipment should be used when handling toxic or flammable fluids, and safety goggles or other protection should be used during cutting, grinding, chiseling, prying, or any other process that can cause material removal or projectiles.

Some procedures require the use of tools specially designed for a specific purpose. Before substituting another tool or procedure, you must be completely satisfied that neither your personal safety, nor the performance of the vehicle will be endangered

Although information in this manual is based on industry sources and is complete as possible at the time of publication, the possibility exists that some car manufacturers made later changes which could not be included here. While striving for total accuracy, Chilton Book Company cannot assume responsibility for any errors, changes or omissions that may occur in the compilation of this data.

PART NUMBERS

Part numbers listed in this reference are not recommendations by Chilton for any product by brand name. They are references that can be used with interchange manuals and aftermarket supplier catalogs to locate each brand supplier's discrete part number.

SPECIAL TOOLS

Special tools are recommended by the vehicle manufacturer to perform their specific job. Use has been kept to a minimum, but where absolutely necessary, they are referred to in the text by the part number of the tool manufacturer. These tools can be purchased under the appropriate part number, from Ford dealer or regional distributor or an equivalent tool can be purchased locally from a tool supplier or parts outlet. Before substituting any tool for the recommended one, read the SAFETY NOTICE at the top of this page.

ACKNOWLEDGMENTS

The Chilton Book Company expresses its appreciation to Ford Motor Co., Detroit, Michigan for their generous assistance.

1 General Information and Maintenance

QUICK REFERENCE INDEX

GENERAL INDEX

HOW TO USE THIS BOOK

Chilton's Total Car Care Manual for Ford Ranger/Bronco II/ Explorer models from 1983 through 1991 is intended to teach you about the inner workings of your vehicle and save you money on its upkeep.

The first two Sections will be the most used, since they contain maintenance and tune-up information and procedures. Studies have shown that a properly tuned and maintained engine can get at least 10% better gas mileage (which translates into lower operating costs) and periodic maintenance will catch minor problems before they turn into major repair bills. The other Sections deal with the more complex systems of your vehicle. Operating systems from engine through brakes are covered. This book will give you detailed instructions to help you change your own brake pads and shoes, tune-up the engine, replace spark plugs and filters, and do many more jobs that will save you money, give you personal satisfaction and help you avoid expensive problems.

A secondary purpose of this book is a reference guide for owners who want to understand their vehicle and/or their mechanics better. In this case, no tools at all are required. Knowing just what a particular repair job requires in parts and labor time will allow you to evaluate whether or not you're getting a fair price quote and help decipher itemized bills from a repair shop.

Before attempting any repairs or service on your vehicle, read through the entire procedure outlined in the appropriate Section. This will give you the overall view of what tools and supplies will be required. There is nothing more frustrating than having to walk to the bus stop on Monday morning because you were short one gasket on Sunday afternoon. So read ahead and plan ahead. Each operation should be approached logically and all procedures thoroughly understood before attempting any work. Some special tools that may be required can often be rented from local automotive jobbers or places specializing in renting tools and equipment. Check the yellow pages of your phone book.

All Sections contain adjustments, maintenance, removal and installation procedures, and overhaul procedures. When overhaul is not considered practical, we tell you how to remove the failed part and then how to install the new or rebuilt replacement. In this way, you at least save the labor costs. Backyard overhaul of some components (such as the alternator or water pump) is just not practical, but the removal and installation procedure is often simple and well within the capabilities of the average owner.

Two basic mechanic's rules should be mentioned here. First, whenever the LEFT side of the vehicle or engine is referred to, it is meant to specify the DRIVER'S side of the vehicle. Conversely, the RIGHT side of the vehicle means the PASSENGER'S side. Second, all screws and bolts are removed by turning counterclockwise, and tightened by turning clockwise (left loosen, right tighten).

Safety is always the MOST important rule. Constantly be aware of the dangers involved in working on or around any vehicle and take proper precautions to avoid the risk of personal injury or damage to the vehicle. See the section in this section, Servicing Your Vehicle Safely, and the SAFETY NOTICE on the acknowledgment page before attempting any service procedures and pay attention to the instructions provided. There are 3 common mistakes in mechanical work:

1. Incorrect order of assembly, disassembly or adjustment. When taking something apart or putting it together, doing things in the wrong order usually just costs you extra time; however it CAN break something. Read the entire procedure before beginning disassembly. Do everything in the order in which the instructions say you should do it, even if you can't immediately see a reason for it. When you're taking apart something that is very intricate (for example a carburetor), you might want to draw a picture of how it looks when assembled at one point in order to make sure you get everything back in its proper position. We will supply exploded views whenever possible, but sometimes the job requires more attention to detail than an illustration provides. When making adjustments (especially tune-up adjustments), do them in order. One adjustment often affects another and you cannot expect satisfactory results unless each adjustment is made only when it cannot be changed by any other.

2. Overtorquing (or undertorquing) nuts and bolts. While it is more common for overtorquing to cause damage, undertorquing can cause a fastener to vibrate loose and cause serious damage, especially when dealing with aluminum parts. Pay attention to torque specifications and utilize a torque wrench in assembly. If a torque figure is not available remember that, if you are using the right tool to do the job, you will probably not have to strain yourself to get a fastener tight enough. The pitch of most threads is so slight that the tension you put on the wrench will be multiplied many times in actual force on what you are tightening. A good example of how critical torque is can be seen in the case of spark plug installation, especially where you are putting the plug into an aluminum cylinder head. Too little torque can fail to crush the gasket, causing leakage of combustion gases and consequent overheating of the plug and engine parts. Too much torque can damage the threads or distort the plug, which changes the spark gap at the electrode. Since more and more manufacturers are using aluminum in their engine and chassis parts to save weight, a torque wrench should be in any serious do-it-yourselfer's tool box.

There are many commercial chemical products available for ensuring that fasteners won't come loose, even if they are not torqued just right (a very common brand is Loctite®). If you're worried about getting something together tight enough to hold, but loose enough to avoid mechanical damage during assembly, one of these products might offer substantial insurance. Read the label on the package and make sure the product is compatible with the materials, fluids, etc. involved before choosing one.

3. Crossthreading. This occurs when a part such as a bolt is screwed into a nut or casting at the wrong angle and forced, causing the threads to become damaged. Crossthreading is more likely to occur if access is difficult. It helps to clean and lubricate fasteners, and to start threading with the part to be installed going straight in, using your fingers. If you encounter resistance, unscrew the part and start over again at a different angle until it can be inserted and turned several times without much effort. Keep in mind that many parts, especially spark plugs, use tapered threads so that gentle turning will automatically bring the part you're threading to the proper angle if you don't force it or resist a change in angle. Don't put a wrench on the part until it's been turned in a couple of times by hand. If you suddenly encounter resistance and the part has not seated fully, don't force it. Pull it back out and make sure it's clean and threading properly.

Always take your time and be patient; once you have some experience, working on your vehicle will become an enjoyable hobby.

TOOLS AND EQUIPMENT

Naturally, without the proper tools and equipment it is impossible to properly service your vehicle. It would be impossible to catalog each tool that you would need to perform each or every operation in this book. It would also be unwise for the amateur to rush out and buy an expensive set of tools an the theory that he may need one or more of them at sometime.

The best approach is to proceed slowly, gathering together a good quality set of those tools that are used most frequently. Don't be misled by the low cost of bargain tools. It is far better to spend a little more for better quality. Forged wrenches, 6- or 12-point sockets and fine tooth ratchets are by far preferable to their less expensive counterparts. As any good mechanic can tell you, there are few worse experiences than trying to work on any vehicle with bad tools. Your monetary savings will be far outweighed by frustration and mangled knuckles.

Certain tools, plus a basic ability to handle them, are required to get started. A basic mechanics tool set, a torque wrench and a Torx® bits set. Torx® bits are hexlobular drivers which fit both inside and outside on special Torx® head fasteners used in various places on modern vehicles. Begin accumulating those tools that are used most frequently; those associated with routine maintenance and tune-up.

In addition to the normal assortment of screwdrivers and pliers you should have the following tools for routine maintenance jobs (your Ranger/Bronco II and Explorer uses both SAE and metric fasteners):

1. SAE/Metric wrenches, sockets and combination open end/box end wrenches in sizes from ⅛ in. (3mm) to ¾ in. (19mm) and a spark plug socket (13/16 in. or ⅝ in.). If possible, buy various length socket drive extensions. One break in this department is that the metric sockets available in the U.S. will all fit the ratchet handles and extensions you may already have (¼ in., ⅜ in., and ½ in. drive).
2. Jackstands for support.
3. Oil filter wrench.
4. Oil filter spout for pouring oil.
5. Grease gun for chassis lubrication.
6. Hydrometer for checking the battery.
7. A container for draining oil.
8. Many rags (paper or cloth) for wiping up the inevitable mess.

In addition to the above items there are several others that are not absolutely necessary, but handy to have around. These include a hydraulic floor jack, oil-dry, a transmission funnel and the usual supply of lubricants, antifreeze and fluids, although these can be purchased as needed. This is a basic list for routine maintenance, but only your personal needs and desires can accurately determine your list of necessary tools.

The second list of tools is for tune-ups. While the tools involved here are slightly more sophisticated, they need not be outrageously expensive. There are several inexpensive tach/dwell meters on the market that are every bit as good for the average mechanic as an expensive professional model. Just be sure that it goes to at least 1,200–1,500 rpm on the tach scale and that it works on 4, 6 and 8 cylinder engines. A basic list of tune-up equipment could include:

1. Tach/dwell meter.
2. Spark plug wrench.
3. Timing light (a DC light that works from the vehicle's battery is best, although an AC light that plugs into 110V house current will suffice at some sacrifice in brightness).
4. Wire spark plug gauge/adjusting tools.

Here again, be guided by your own needs. While not absolutely necessary, an ohmmeter can be useful in determining whether or not a spark plug wire is any good by measuring its resistance. In addition to these basic tools, there are several other tools and gauges you may find useful. These include:

1. A compression gauge. The screw-in type is slower to use, but eliminates the possibility of a faulty reading due to escaping pressure.
2. A manifold vacuum gauge.
3. A test light.
4. An induction meter. This is used for determining whether or not there is current in a wire. These are handy for use if a wire is broken somewhere in a wiring harness.

As a final note, you will probably find a torque wrench neces-

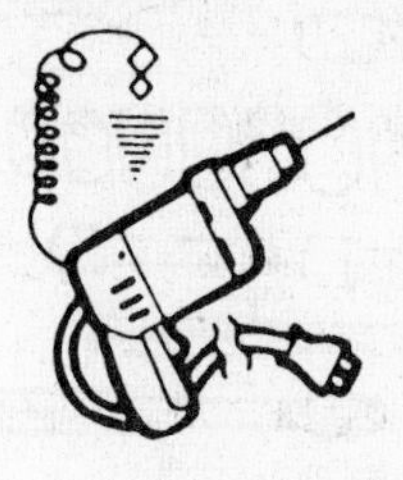

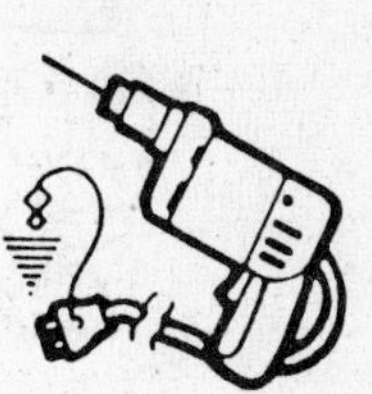

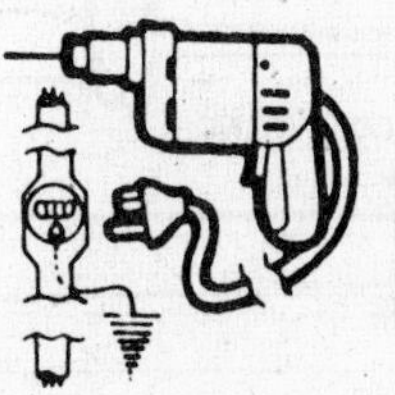

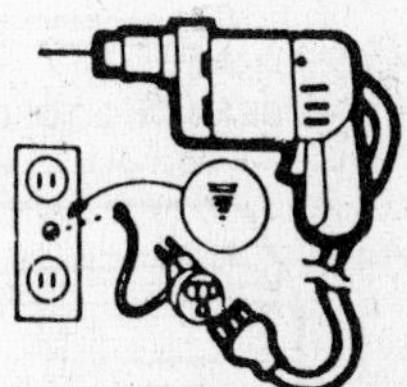

When using electric tools, make sure they are properly grounded.

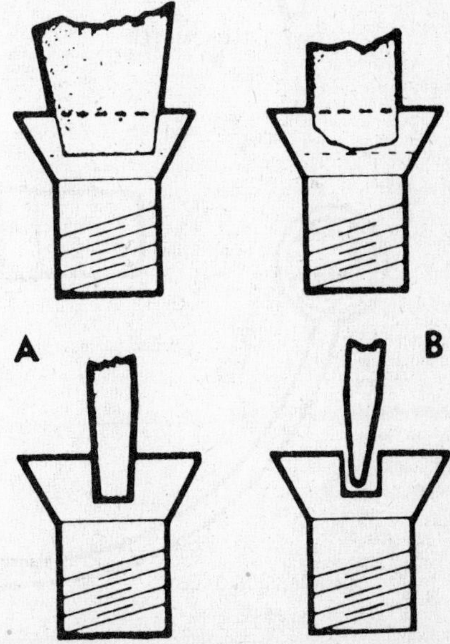

Keep screwdriver tips in good shape. They should fit in the screw head slots in the manner shown in A. If they look like the tip shown in B, they need grinding or replacing.

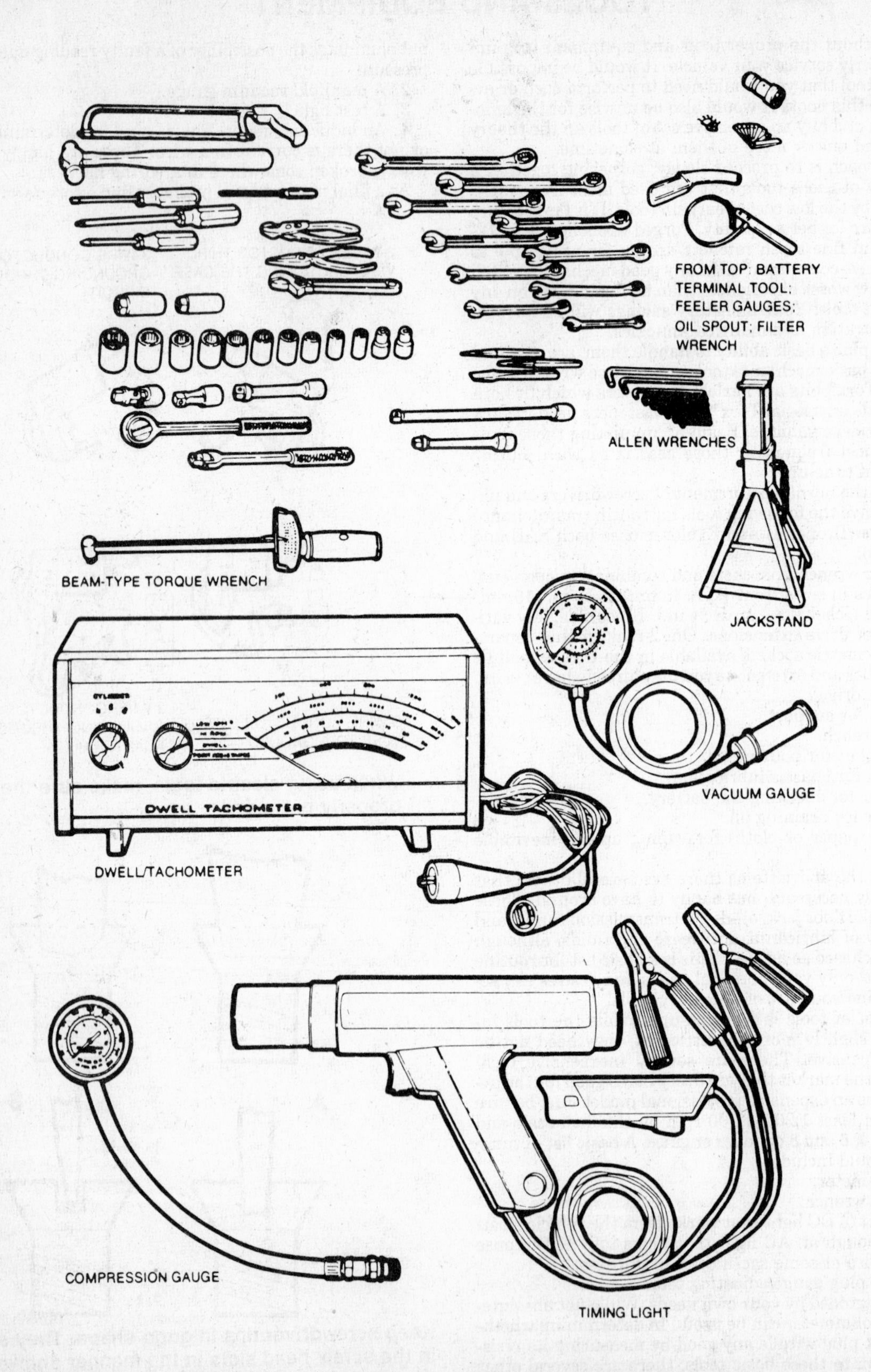

You need only a basic assortment of hand tools and test instruments for most maintenance and repair jobs.

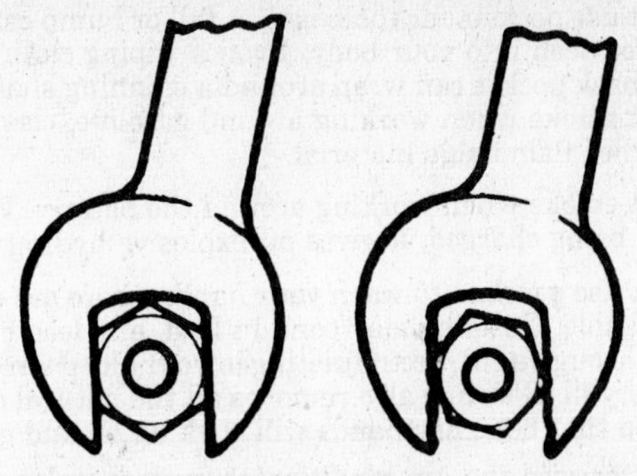

When using an open-end wrench, use the exact size needed and position it squarely on the flats of the bolt or nut.

sary for all but the most basic work. The beam type models are perfectly adequate, although the newer click (breakaway) type are more precise, and you don't have to crane your neck to see a torque reading in awkward situations. The breakaway torque wrenches are more expensive and should be recalibrated periodically.

Torque specification for each fastener will be given in the procedure in any case that a specific torque value is required. If no torque specifications are given, use the following values as a guide, based upon fastener size:

Bolts marked 6T
6mm bolt/nut – 5–7 ft. lbs.
8mm bolt/nut – 12–17 ft. lbs.
10mm bolt/nut – 23–34 ft. lbs.
12mm bolt/nut – 41–59 ft. lbs.
14mm bolt/nut – 56–76 ft. lbs.

Bolts marked 8T
6mm bolt/nut – 6–9 ft. lbs.
8mm bolt/nut – 13–20 ft. lbs.
10mm bolt/nut – 27–40 ft. lbs.
12mm bolt/nut – 46–69 ft. lbs.
14mm bolt/nut – 75–101 ft. lbs.

Special Tools

Normally, the use of special factory tools is avoided for repair procedures, since these are not readily available for the do-it-yourself mechanic. When it is possible to perform the job with more commonly available tools, it will be pointed out, but occasionally a special tool was designed to perform a specific function and should be used. Before substituting another tool, you should be convinced that neither your safety nor the performance of the vehicle will be compromised. Where possible, an illustration of the special tool will be provided so that an equivalent tool may be used.

Some special tools are available commercially from Owatonna Tool Co., Owatonna, MN 55060. Others can be purchased through your Ford dealer or local parts supplier.

SERVICING YOUR VEHICLE SAFELY

It is virtually impossible to anticipate all of the hazards involved with maintenance and service work, but care and common sense will prevent most accidents.

The rules of safety for mechanics range from "don't smoke around gasoline," to "use the proper tool for the job." The trick to avoiding injuries is to develop safe work habits and take every possible precaution.

Dos

- Do keep a fire extinguisher and first aid kit within easy reach.
- Do wear safety glasses or goggles when cutting, drilling or prying, even if you have 20–20 vision. If you wear glasses for the sake of vision, they should be made of hardened glass that can also serve as safety glasses, or wear safety goggles over your regular glasses.
- Do shield your eyes whenever you work around the battery. Batteries contain sulphuric acid. In case of contact with the eyes or skin, flush the area with water or a mixture of water and baking soda and get medical attention immediately.
- Do use safety stands for any under vehicle service. Jacks are for raising vehicles; safety stands are for making sure the vehicle stays raised until you want it to come down. Whenever the vehicle is raised, block the wheels remaining on the ground and set the parking brake.
- Do use adequate ventilation when working with any chemicals. Like carbon monoxide, the asbestos dust resulting from brake lining wear can be poisonous in sufficient quantities.
- Do disconnect the negative battery cable when working on the electrical system. The primary ignition system can contain up to 40,000 volts.
- Do follow manufacturer's directions whenever working with potentially hazardous materials. Both brake fluid and antifreeze are poisonous if taken internally.
- Do properly maintain your tools. Loose hammerheads, mushroomed punches and chisels, frayed or poorly grounded electrical cords, excessively worn screwdrivers, spread wrenches (open end), cracked sockets, slipping ratchets, or faulty droplight sockets can cause accidents.
- Do use the proper size and type of tool for the job being done.
- Do when possible, pull on a wrench handle rather than push on it, and adjust your stance to prevent a fall.
- Do be sure that adjustable wrenches are tightly adjusted on the nut or bolt and pulled so that the face is on the side of the fixed jaw.

• Do select a wrench or socket that fits the nut or bolt. The wrench or socket should sit straight, not cocked.

• Do strike squarely with a hammer—avoid glancing blows.

• Do set the parking brake and block the drive wheels if the work requires that the engine be running.

Don'ts

• Don't run an engine in a garage or anywhere else without proper ventilation—EVER! Carbon monoxide is poisonous. It takes a long time to leave the human body and you can build up a deadly supply of it in your blood stream by simply breathing in a little every day. You may not realize you are slowly poisoning yourself. Always use power vents, windows, fans or open the garage doors.

• Don't work around moving parts while wearing a necktie or other loose clothing. Short sleeves are much safer than long, loose sleeves and hard-toed shoes with neoprene soles protect your toes and give a better grip on slippery surfaces. Jewelry such as watches, fancy belt buckles, beads or body adornment of any kind is not safe working around any vehicle. Long hair should be hidden under a hat or cap.

• Don't use pockets for toolboxes. A fall or bump can drive a screwdriver deep into your body. Even a wiping cloth hanging from the back pocket can wrap around a spinning shaft or fan.

• Don't smoke when working around gasoline, cleaning solvent or other flammable material.

• Don't smoke when working around the battery. When the battery is being charged, it gives off explosive hydrogen gas.

• Don't use gasoline to wash your hands; there are excellent soaps available. Gasoline may contain lead, and lead can enter the body through a cut, accumulating in your blood stream until you are very ill. Gasoline also removes all the natural oils from the skin so that bone dry hands will suck up oil and grease.

• Don't service the air conditioning system unless you are equipped with the necessary tools and training. The refrigerant, R-12, is extremely cold and when exposed to the air, will instantly freeze any surface it comes in contact with, including your eyes. Although the refrigerant is normally non-toxic, R-12 becomes a deadly poisonous gas (phosgene) in the presence of an open flame. One good whiff of the vapors from burning refrigerant can be fatal.

SERIAL NUMBER IDENTIFICATION

Vehicle Identification (VIN) Number

A 17 digit combination of numbers and letters forms the vehicle identification number (VIN). The VIN is stamped on a metal tab that is riveted to the instrument panel close to the windshield. The VIN plate is visible by looking through the windshield on the driver's side. The VIN number is also found on the Safety Compliance Certification Label which is described below.

By looking at the 17 digit VIN number, a variety of information about the vehicle can be determined.

• The 1st digit identifies the country of origin. 1 = USA; 2 = Canada.

• The 2nd digit identifies the manufacturer. F = Ford.

• The 3rd digit identifies the type of vehicle.

C = Basic (stripped) chassis
D = Incomplete vehicle
M = Multi-purpose vehicle
T = Truck (complete vehicle)

• The 4th digit identifies the gross vehicle weight rating (GVWR Class) and brake system. For incomplete vehicles, the 4th digit determines the brake system only. All brake systems are hydraulic.

A = up to 3,000 lbs.
B = 3,001–4,000 lbs.
C = 4,001–5,000 lbs.
D = 5,001–6,000 lbs.
E = 6,001–7,000 lbs.
F = 7,001–8,000 lbs.
G = 8,001–8,500 lbs.
H = 8,500–9,000 lbs.
J = 9,001–10,000 lbs.

• The 5th digits identifies the model or line. R = Ranger U = Bronco II and Explorer.

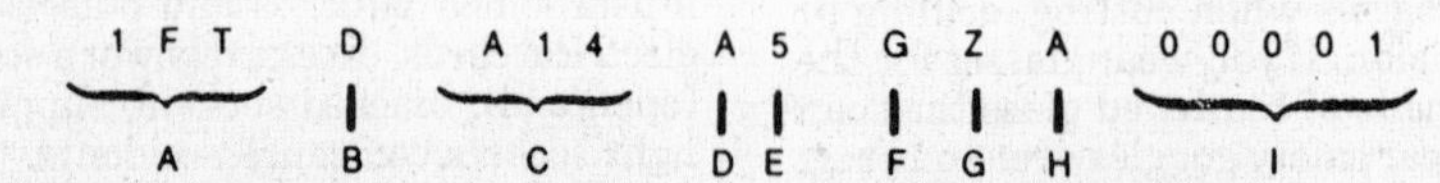

A. Position 1, 2, and 3—Manufacturer, Make and Type (World Manufacturer Identifier)
B. Position 4—Brake System/GVWR
C. Position 5, 6, and 7—Model or Line, Series, Chassis, Cab Type
D. Position 8—Engine Type
E. Position 9—Check Digit
F. Position 10—Model Year
G. Position 11—Assembly Plant
H. Position 12—Constant "A" until sequence number of 99,999 is reached, then changes to a constant "B" and so on
I. Position 13 through 17—Sequence number—begins at 00001

Sample VIN number

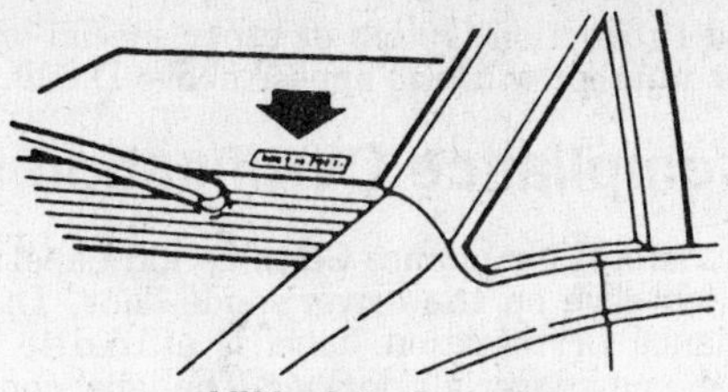

VIN plate location

- The 6th and 7th digits identify chassis and body type.
  - 10 = 4x2 pickup regular cab
  - 11 = 4x4 pickup regular cab
  - 14 = 4x2 pickup super cab
  - 15 = 4x4 pickup super cab
  - 12 = 4x2 standard Bronco II
  - 14 = 4x4 standard Bronco II
  - 22 = 4x2 standard Explorer 2 dr.
  - 24 = 4x4 standard Explorer 2 dr.
  - 32 = 4x2 standard Explorer 4 dr.
  - 34 = 4x4 standard Explorer 4 dr.
- The 8th digit identifies the engine.
  - C = 2.0L 4-cylinder
  - P = 2.2L 4-cylinder
  - A = 2.3L 4-cylinder
  - E = 2.3L 4-cylinder
  - S = 2.8L 4-cylinder
  - T = 2.9L 6-cylinder
  - U = 3.0L 6-cylinder
  - X = 4.0L 6-cylinder
- The 9th digit is a check digit.
- The 10th digit identifies the model year.
  - D = 1983
  - E = 1984
  - F = 1985
  - G = 1986
  - H = 1987
  - J = 1988
  - K = 1989
  - L = 1990
  - M = 1991
- The 11th digit identifies the assembly plant.
  - C = Ontario, Canada
  - H = Lorain, OH
  - K = Claycomo, MO
  - L = Wayne, MI
  - N = Norfolk, VA
  - P = St. Paul, MN
  - U = Louisville, KY
  - Z = Hazlewood, MO
- Digits twelve through seventeen make up the sequential serial and warranty number. Digit twelve uses the letter A until the production or sequence of 99,999 units (digits thirteen through seventeen) is reached. Letter A then becomes B for the next production sequence of vehicles.

Build Date Stamp Location

The vehicle build date is stamped on the front surface of the radiator support on the passenger side of the vehicle. Yellow ink is used for the date stamp. When the marking surface is painted the body color, the date stamp will be marked in red ink. Units from the Ontario truck plant (code C) will be marked with silver ink.

Safety Standard Certification Labels Complete Vehicles

(UNITED STATES)

MFD. BY FORD MOTOR CO. IN U.S.A.
DATE: 2/83 GVWR: 3740 LBS/1696 KG
FRONT GAWR: 1910 LBS REAR GAWR: 2012 LBS
866 KG WITH 866 KG WITH
P195/75R14SL TIRES P195/75R14SL TIRES
14x5.0JJ RIMS 14x5.0JJ RIMS
AT 35 PSI COLD AT 35 PSI COLD
THIS VEHICLE CONFORMS TO ALL APPLICABLE FEDERAL MOTOR VEHICLE SAFETY STANDARDS IN EFFECT ON THE DATE OF MANUFACTURE SHOWN ABOVE
VEHICLE IDENTIFICATION NO. 1FTCR10Z 5DUA00001
TYPE TRUCK
EXTERIOR PAINT COLORS DSO
WB | TYPE GVW | BODY | TRANS | AXLE

(QUEBEC)

FABR. AUX E-U PAR LA FORD MOTOR CO.
DATE: PNBV:
PNBE AVANT: PNBE ARRIERE:
AVEC
<PNEUS>
<JANTES>
A LB/PO² A FROID A LB/PO² A FROID
CE VEHICULE EST CONFORME A TOUTES LES NORMES FEDERALES DE SECURITE DES V.A. EN VIGUEUR A LA DATE DE FABR. INDIQUEE CI-DESSUS.
N° D'IDENT. DU VEHICULE
TYPE

FOR VEHICLES MFD IN U.S.A. FOR QUEBEC, CANADA.

AUX. LABEL
MFD. BY FORD MOTOR CO. IN U.S.A.
SNOW PLOW PREP OR MAX. FRONT GAWR OPTION
FRONT GAWR: 2750 LB/1247 KG
ONLY WHEN AIR CYLINDER ASSISTED FRONT SPRINGS ARE PRESSURIZED TO
55 PSI
FOR COMPLETE INFORMATION, SEE MAIN LABEL
E37A-1020472-AA

INCOMPLETE VEHICLES

THE INCOMPLETE VEHICLE RATING DECAL IS INSTALLED ON THE DRIVER'S DOOR LOCK PILLAR IN PLACE OF THE SAFETY STANDARD CERTIFICATION LABEL.

VEHICLE RATING DECAL

INCOMPLETE VEHICLE MANUFACTURED BY
GVWR: 4220 LBS/1914 KG
VEHICLE IDENTIFICATION NUMBER 1FTCR10Z 5DUA00001
EXTERIOR PAINT COLORS 2H 48 DSO

WB	TYPE-GVW	BODY	TRANS	AXLE	
114	R106	CARS	W	84	2 D

Vehicle certification labels

Vehicle Data

The vehicle data appears on the Safety Compliance Certification Label on the second and third lines following the identification number. The code set (two numbers or a number and letter) above COLOR identify the exterior paint color, with two sets of codes designating two tone paint. The three digits under W.N. designate the wheelbase in inches. The letter and three digits under TYPE/G.V.W. designate the truck model within a series and the gross vehicle weight rating. The letters and/or numbers under BODY designate the interior trim, seat and body type. The transmission installed in the vehicle is identified under TRANS by an alphabetical code.

A letter and a number or two numbers under AXLE identify the rear axle ratio and, when required, a letter or number is also stamped after the rear axle code to identify the front axle. The letters and/or numerals under TAPE designate the external body side tape stripe code. The spring usage codes for the vehicle are identified under SPRING.

A two digit number is stamped above D.S.O. to identify the district which originally ordered the vehicle. If the vehicle is built to special order (Domestic Special Order, Foreign Special Order, Limited Production Option or other special order), the complete order number will also appear above D.S.O.

Safety Compliance Certification Label

The English Safety Compliance Certification Label is affixed to the door latch edge on the driver's side door. The French Safety Compliance Certification Label is affixed to the door latch edge on the passenger's side door. The label contains the following information: name of manufacturer, the month and year of manufacture, the certification statement, and the Vehicle Identification number. The label also contains information on Gross Vehicle weight ratings, Wheel and tire data, and additional vehicle data information codes.

Emission Calibration Label

The emission calibration number label is attached to the left side door or the left door post pillar. This label plate identifies the engine calibration number, engine code number and the revision level. These numbers are used to determine if parts are unique to specific engines. The engine codes and calibration are

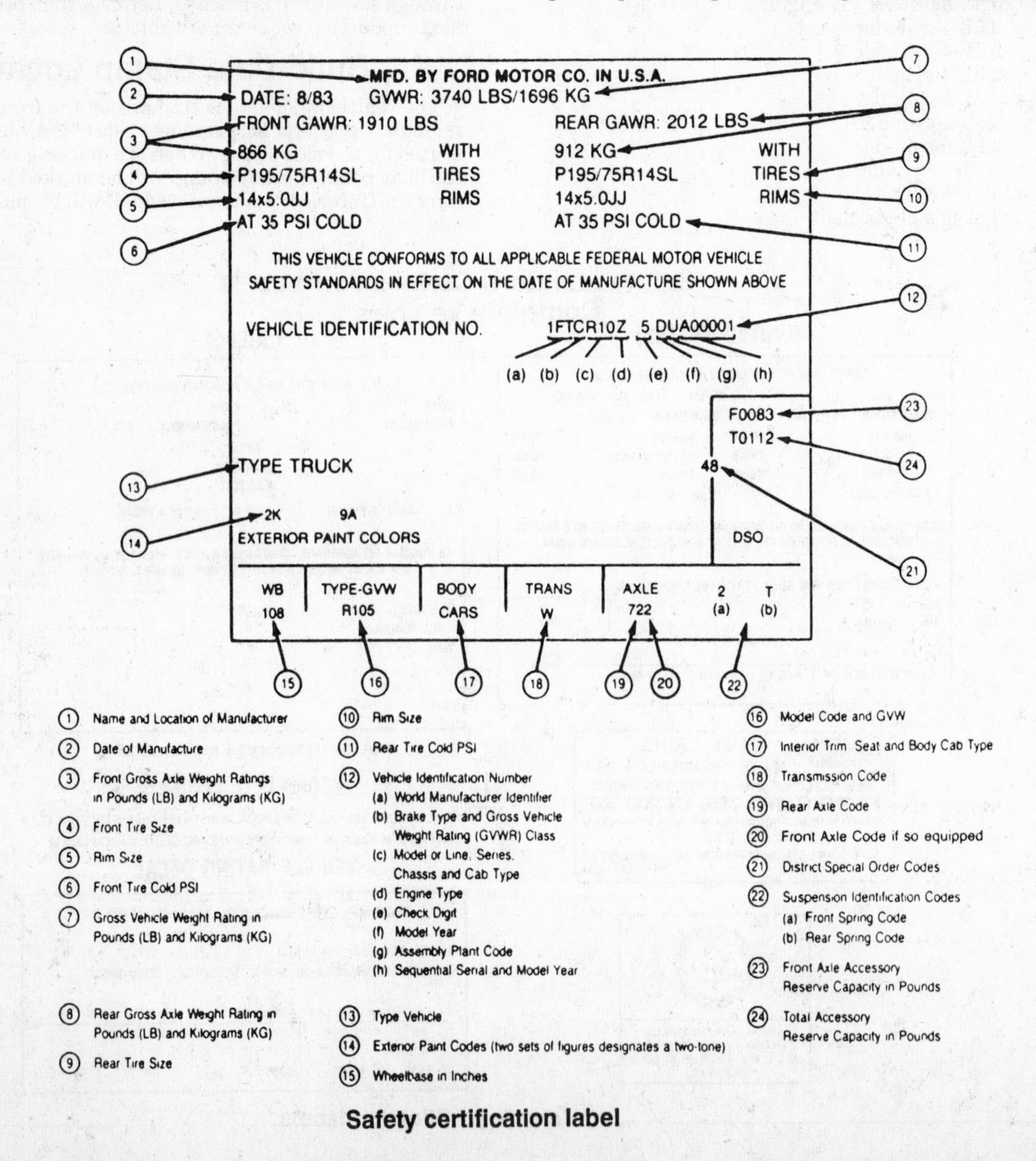

Safety certification label

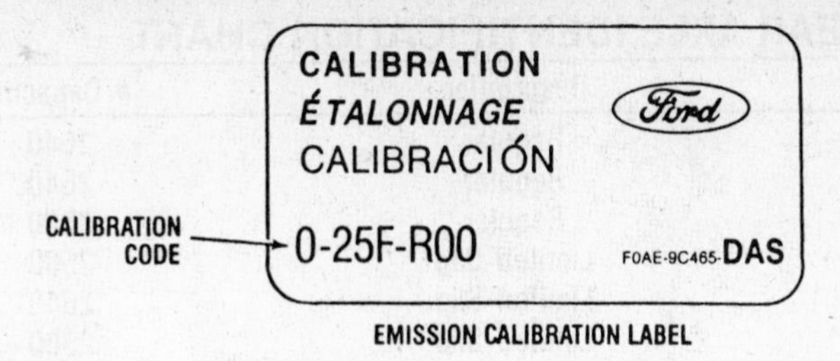

EMISSION CALIBRATION LABEL

CALIBRATION CODE

Use the following to interpret the calibration code from the Emission Calibration Label:

1. MODEL YEAR—This number represents the model year in which the Calibration was first introduced. As shown below, the model year is 1990. (Represented by the Number 0)
2. CALIBRATION DESIGN LEVEL—Represents the design level assigned to the engine (25F).
3. CALIBRATION REVISION LEVEL—Represents the revision level of the calibration (R00). These numbers will advance as revisions occur.

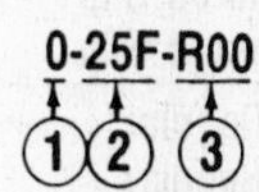

CALIBRATION CODE

Emission calibration label

necessary for ordering parts and asking questions related to the engine.

Engine

The engine identification code is a letter located in the eigth digit of the Vehicle Identification Number stamped on a metal tab that is riveted to the instrument panel close to the windshield. Specific engine data is located on a label attached to the timing cover.

Transmission

The transmission code may be found in two places on the vehicle. One is on the Safety Standard Certification label attached to the left driver's side door lock post. The code appears as a letter in the "Trans" column of the label. M = 5-speed manual transmission; T = 4-speed automatic transmision.

The other location is on the transmission body itself. On manual transmissions, the identification number is located on a plate attached to the main transmission case. On the plate you find Ford's assigned part number, the serial number and a bar code used for inventory purposes. On automatic transmissions, the identification number is stamped on plate that hangs from the lower left extension housing bolt. The plate identifies when the transmission was built, it's code letter and model number.

Transfer Case

All vehicles can be equipped with a mechanical or electronic shift transfer case. The identification number is stamped on a plate on the side of the case.

TRANSMISSION IDENTIFICATION CHART

Code	Transmission Type
V	Automatic—C3
W	Automatic—C5
T	Automatic—4 speed OD (A4LD)
X	Manual—4 Speed
5	Manual—5 Speed Overdrive
M/D	Manual—5 Speed Overdrive

ENGINE IDENTIFICATION CHART

VIN Code	Engine Displacement Liter/CID	No. of Cylinders	Fuel System	Manufacturer
C	2.0/122	4	Gas	Ford
P	2.2/134	4	Diesel	Ford
A	2.3/140	4	Gas	Ford
E	2.3/140	4	Diesel	Mitsubishi
S	2.8/173	6	Gas	Ford
T	2.9/177	6	Gas	Ford
U	3.0/183	6	Gas	Ford
X	4.0/241	6	Gas	Ford

REAR AXLE IDENTIFICATION CHART

Model	Code	Description	# Capacity	Ratio
Bronco II	42	Regular	2640	3.45
	44	Regular	2640	3.73
	47	Regular	2500	4.10
	D2	Limited Slip	2500	3.45
	D4	Limited Slip	2640	3.73
	D7	Limited Slip	2500	4.10
Ranger	72	Regular	2200	3.08
	74	Regular	2200	3.45
	82	Regular	2700	3.08
	84	Regular	2700	3.45
	85	Regular	2750	3.55
	86	Regular	2700	3.73
	87	Regular	2700	4.10
	96	Regular	3200	3.73
	F4	Limited Slip	2700	3.45
	F5	Limited Slip	2750	3.55
	F6	Limited Slip	2700	3.73
	F7	Limited Slip	2700	4.10
	K6	Limited Slip	3200	4.10
Explorer	43	Regular	3200	3.08
	41	Regular	3200	3.27
	45	Regular	3200	3.55
	04	Limited Slip	3200	3.73

Front Drive Axle

The identification number is stamped on a plate on the differential housing.

Rear Axle

The rear axle code may be found in two places on the vehicle. One is on the Safety Standard Certification label attached to the left driver's side door lock post. The code appears as a number or letter/number combination in the "Axle" column of the label. The rear axle identification code is also stamped on a metal tag hanging from the axle cover-to-carrier bolt at the 2 o'clock position in the cover bolt circle.

ROUTINE MAINTENANCE

Air Cleaner

The air cleaner is a paper element type. The paper cartridge should be replaced according to the Preventive Maintenance Schedule at the end of this Section.

NOTE: Check the air filter more often if the vehicle is operated under severe dusty conditions and replace or clean it as necessary.

REPLACEMENT

Carbureted Engines

1. Open the engine compartment hood.

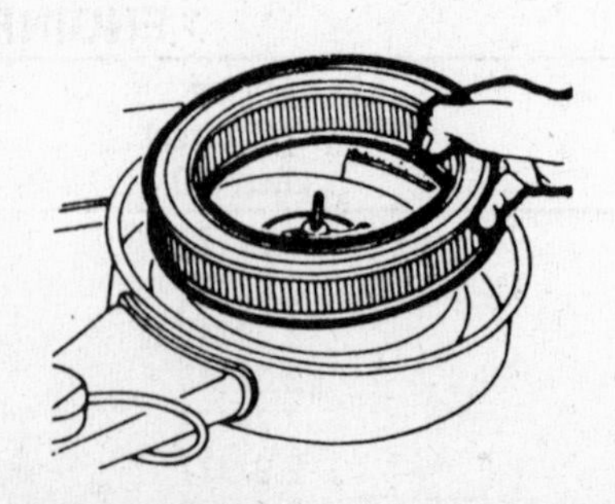

Removing the carburetor air cleaner element

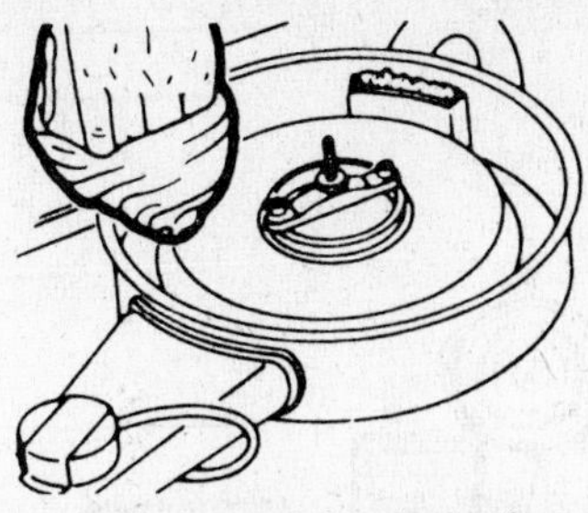

Clean out the air cleaner body before installing the new filter

2. Remove the wing nut holding the air cleaner assembly to the top of the carburetor.
3. Disconnect the crankcase ventilation hose at the air cleaner and remove the entire air cleaner assembly from the carburetor.
4. Remove and discard the old filter element, and inspect the condition of the air cleaner mounting gasket. Replace the gasket as necessary.

NOTE: A crankcase ventilation filter is located in the side of the air cleaner body. The filter should be replaced rather than cleaned. Simply pull the old filter out of the body every 20,000 miles (or more frequently if the vehicle has been used in extremely dusty conditions) and push a new filter into place.

5. Install the air cleaner body on the carburetor so that the word **FRONT** faces toward the front of the vehicle.
6. Place the new filter element in the air cleaner body and install the cover and tighten the wing nut. If the word **TOP** appears on the element, make sure that the side that the word appears on is facing up when the element is in place.
7. Connect the crankcase ventilation hose to the air cleaner.

Fuel Injected Engines

1. Loosen the clamps that secure the hose assembly to the air cleaner.
2. Remove the screws that attach the air cleaner to the bracket.
3. Disconnect the hose and inlet tube from the air cleaner.
4. Remove the screws attaching the air cleaner cover.
5. Remove the air filter and tubes.
6. Installation is the reverse of removal. Don't overtighten the hose clamps! A torque of 12–15 inch lbs. is sufficient.

Diesel Engines

2.2L ENGINE

1. Open the engine compartment hood.
2. Remove the wing nut holding the air cleaner assembly.
3. Remove and discard the old filter element, and inspect the condition of the air cleaner mounting gasket. Replace the gasket as necessary.
4. Place the new filter element in the air cleaner body and install the cover and tighten the wing nut.

2.3L ENGINE

1. Open the engine compartment hood.
2. Remove the clips holding the air cleaner assembly.
3. Remove and discard the old filter element, and inspect the condition of the air cleaner mounting gasket. Replace the gasket as necessary.
4. Place the new filter element in the air cleaner body and install the cover and tighten the wing nut.

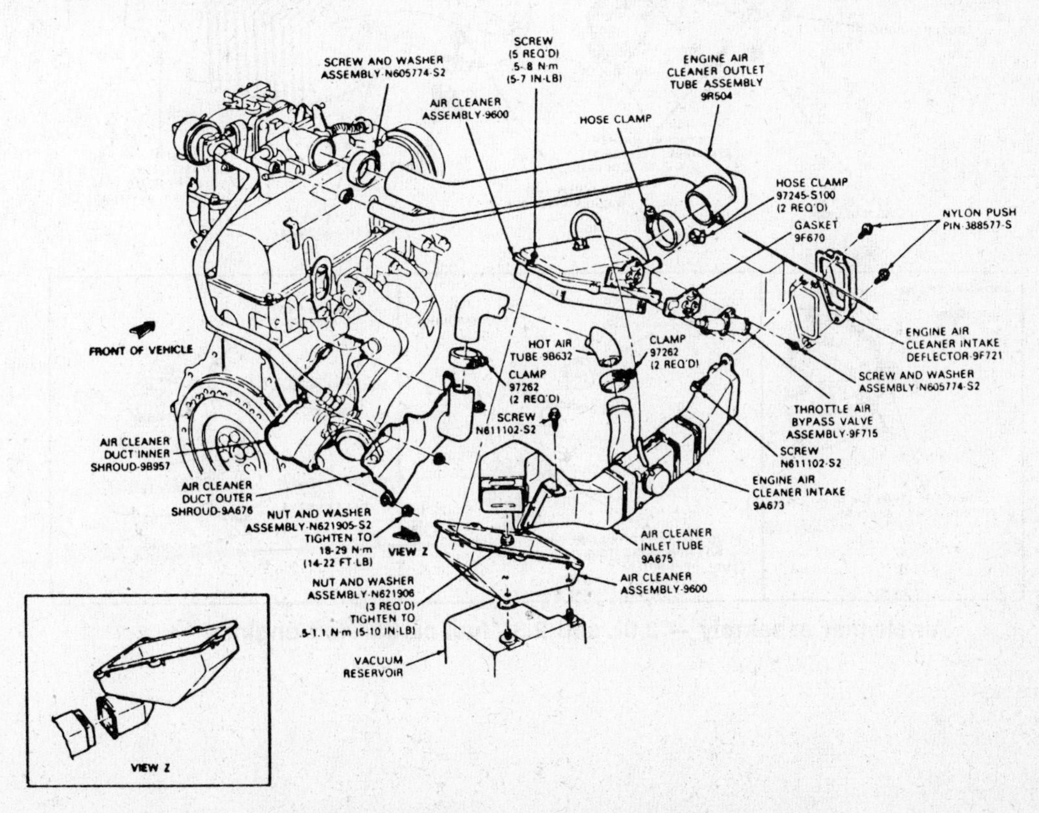

Air cleaner assembly — 2.3L fuel injected engine

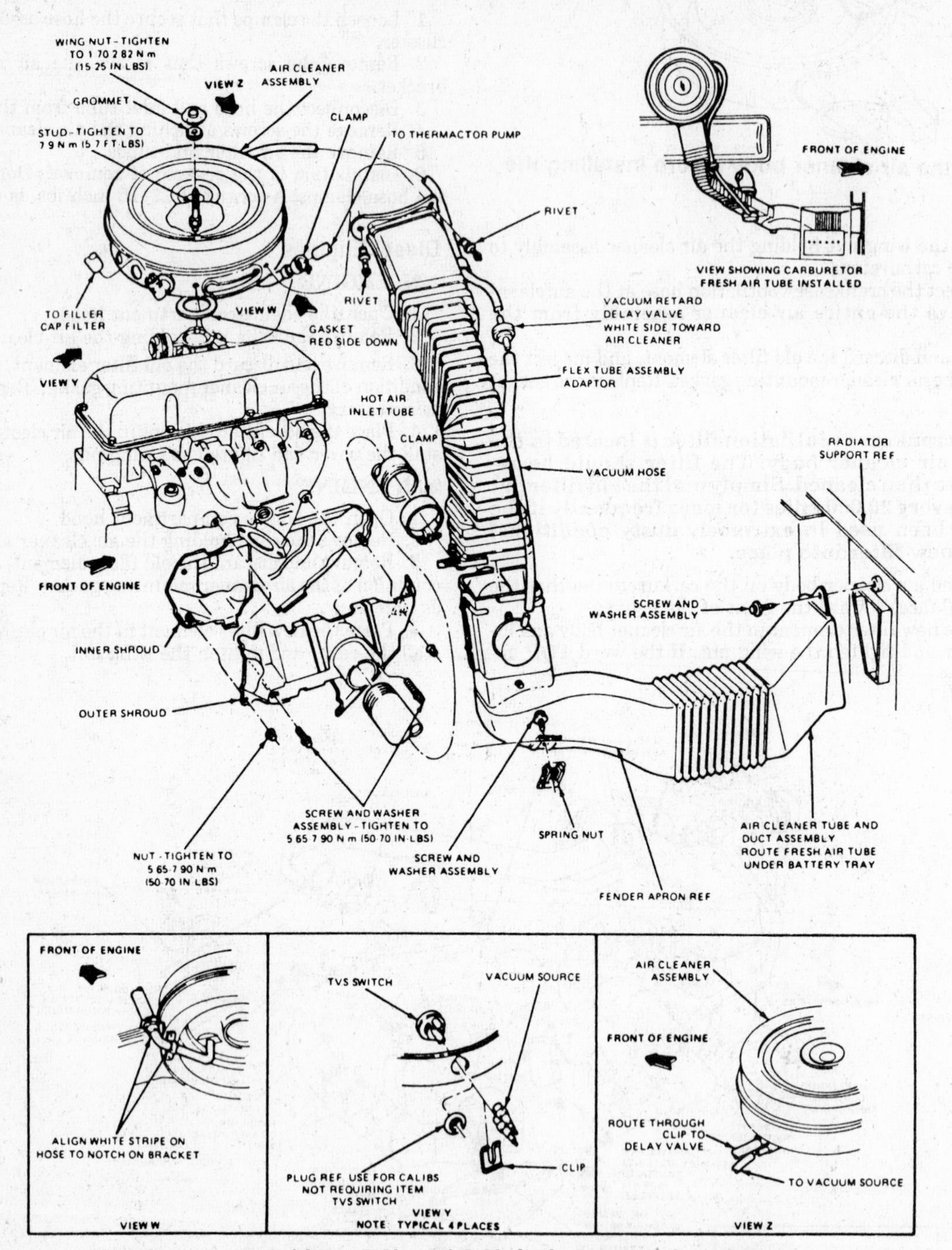

Air cleaner assembly – 2.0L and 2.3L fuel carbureted engines

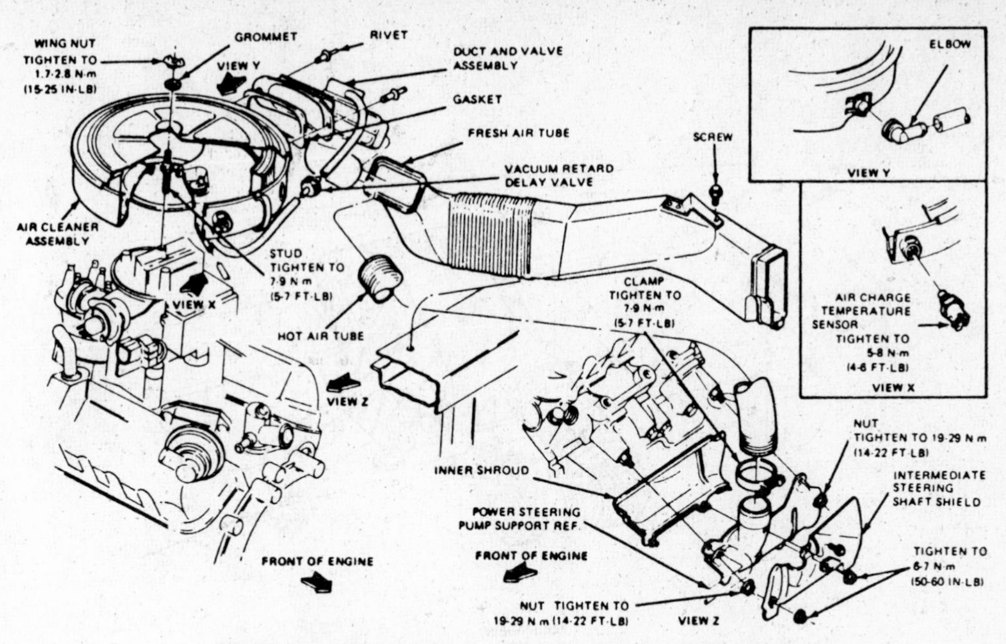

Air cleaner assembly — 2.8L carbureted engine

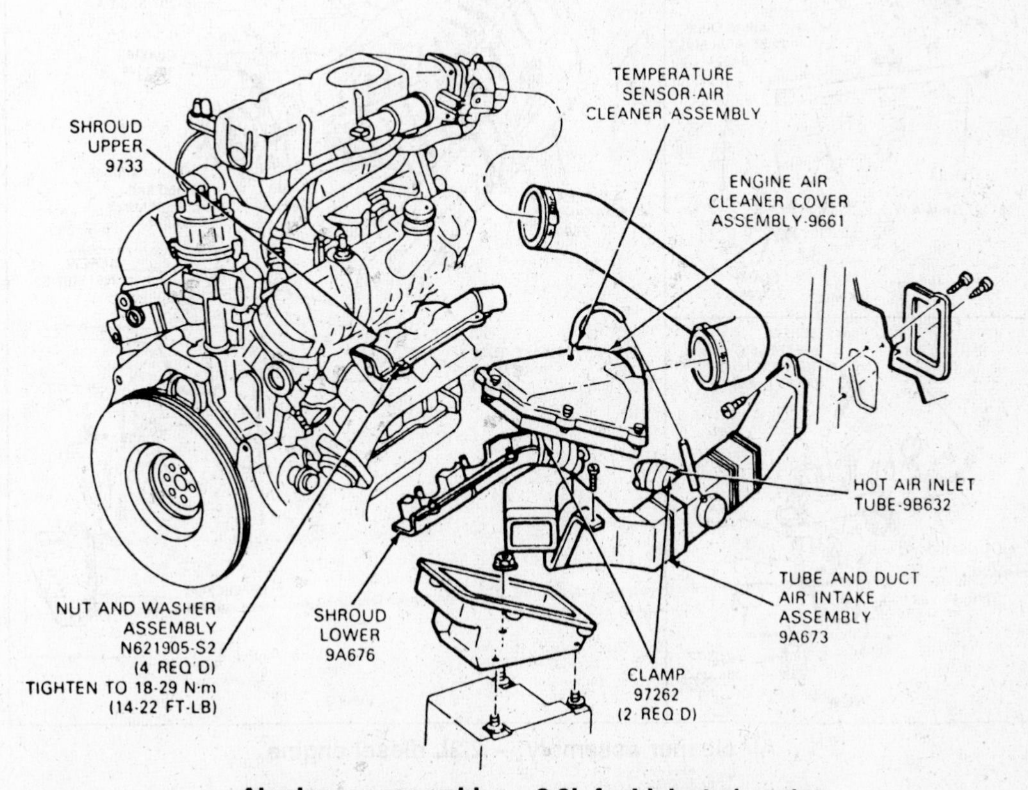

Air cleaner assembly — 2.9L fuel injected engine

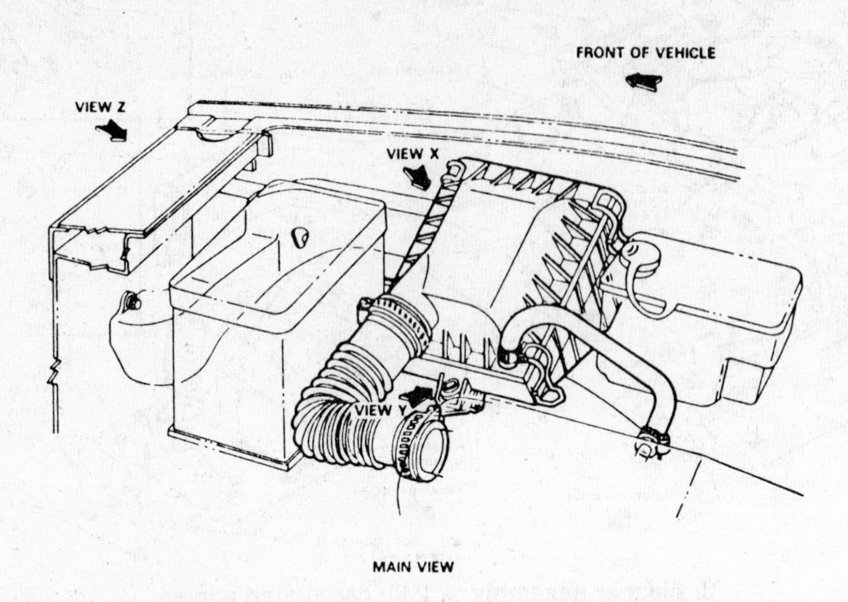

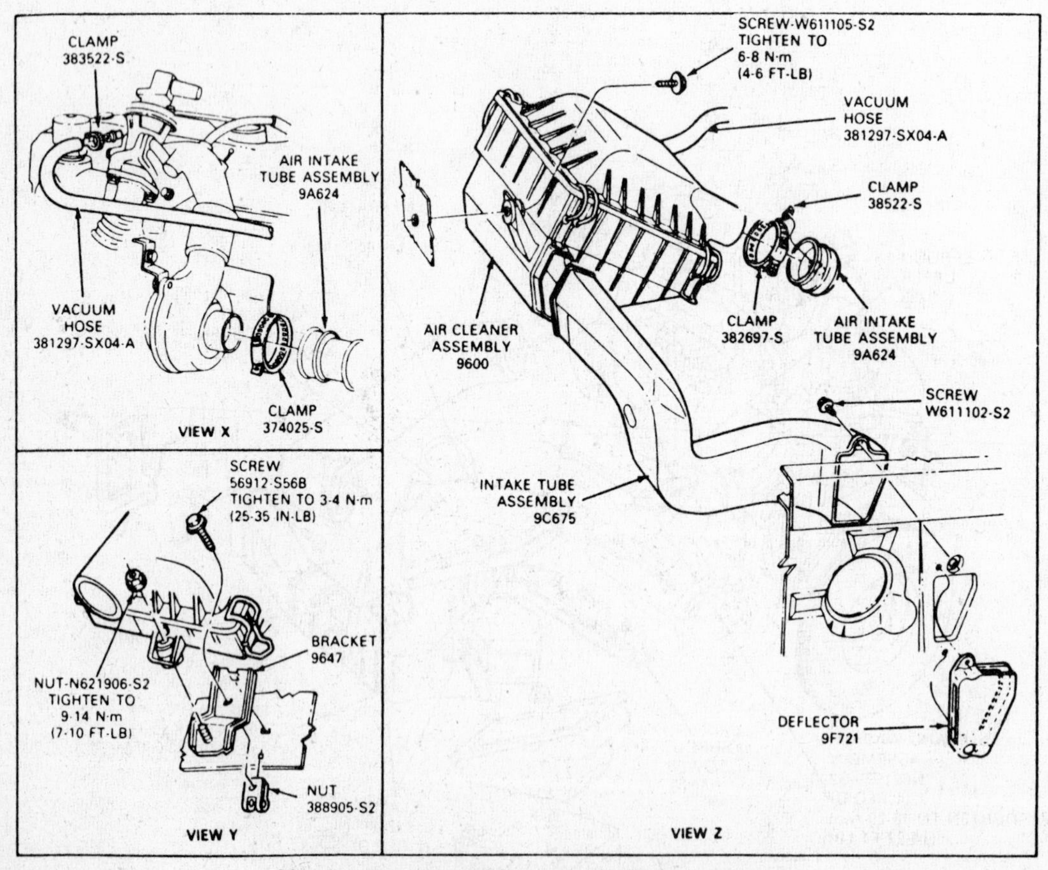

Air cleaner assembly — 2.3L diesel engine

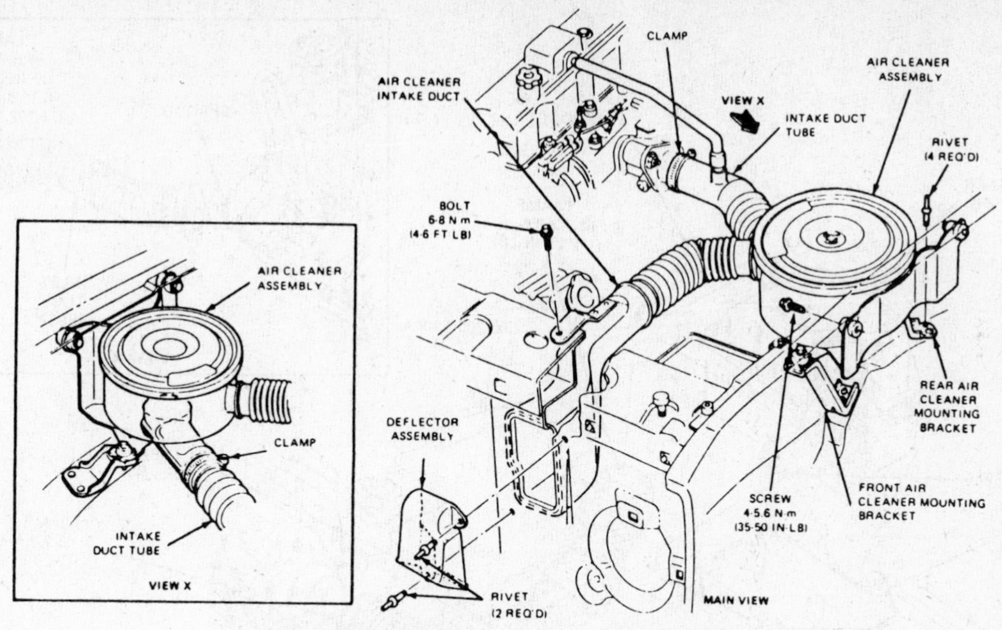

Air cleaner assembly — 2.2L diesel engine

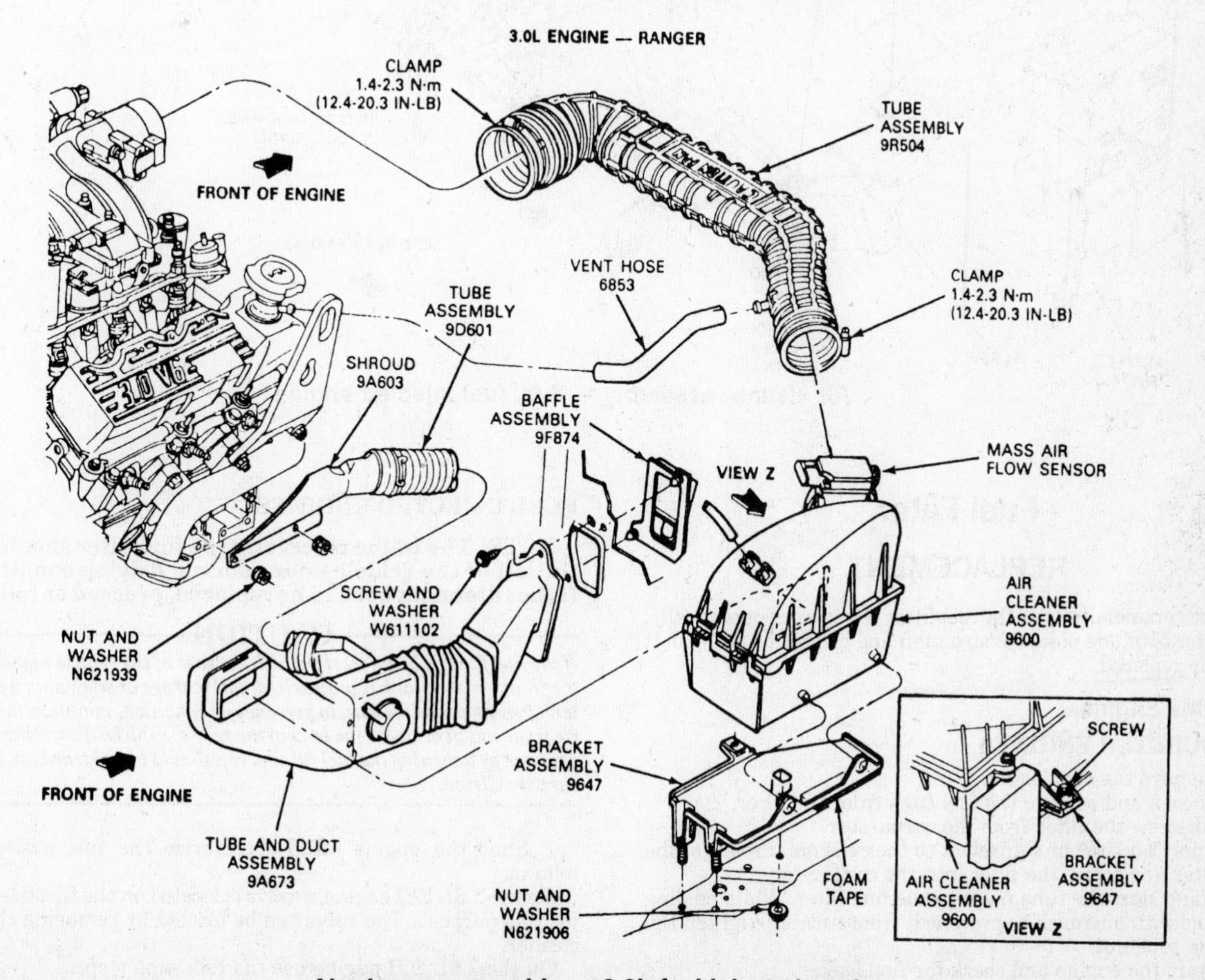

Air cleaner assembly — 3.0L fuel injected engine

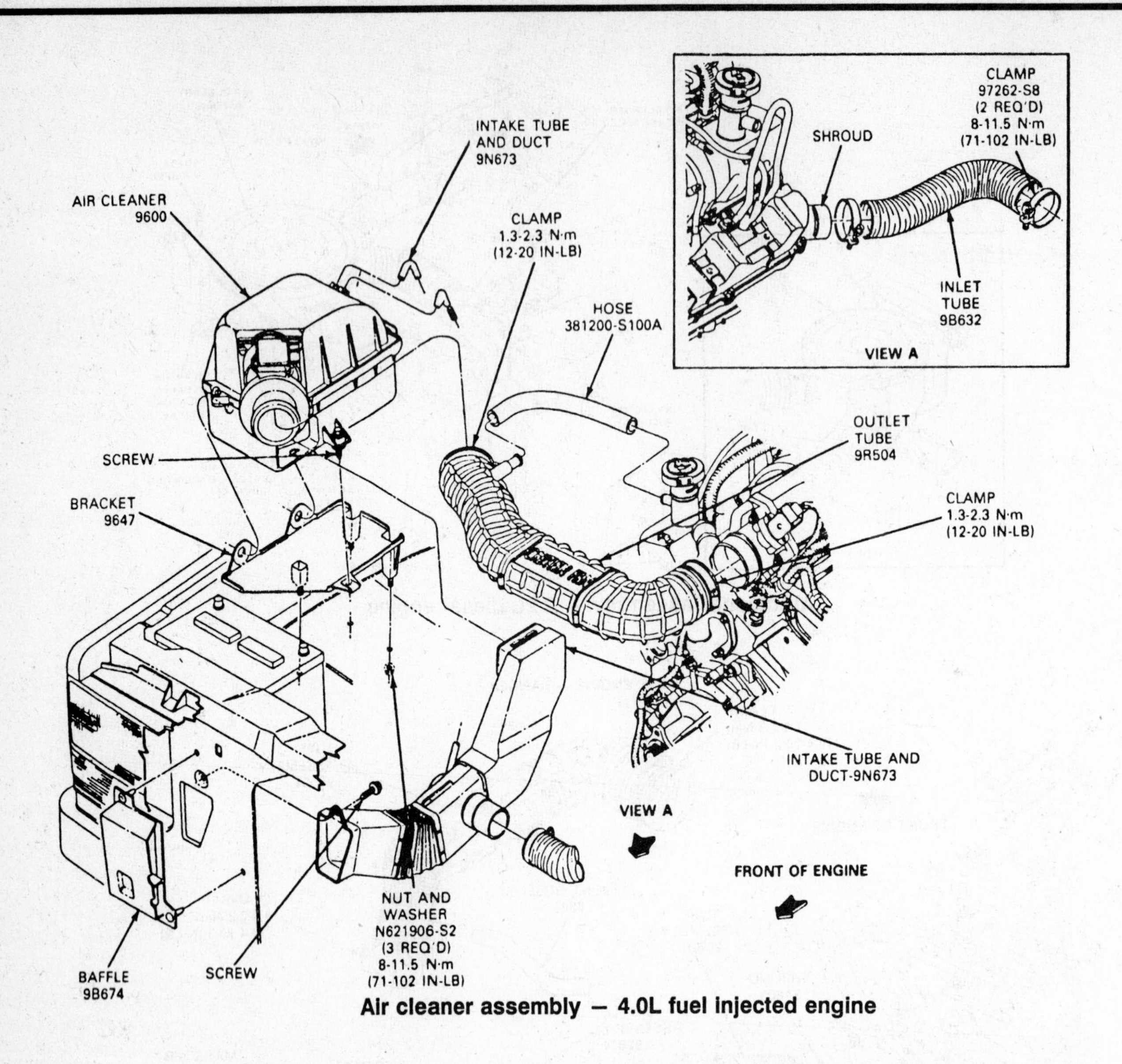

Air cleaner assembly — 4.0L fuel injected engine

Fuel Filter

REPLACEMENT

It is recommended that the fuel filter be replaced periodically. The filter is of one piece construction and cannot be cleaned, it must be replaced.

Gasoline Engines

CARBURETED ENGINES

1. Remove the air cleaner.
2. Loosen and remove the fuel tube from the filter.
3. Unscrew the filter from the carburetor.
4. Apply Loctite® or equivalent to the external threads of the new filter and screw the filter into the carburetor.
5. Hand start the tube nut into the fuel filter. While holding the filter with a wrench to prevent it from turning, tighten the fuel line tube nut.
6. Start the engine and check for fuel leaks.
7. Replace the air cleaner.

FUEL INJECTED ENGINES

NOTE: The inline reservoir type fuel filter should last the left of the vehicle under normal driving conditions. If the filter does need to be replaced, proceed as follows:

CAUTION

If the fuel filter is being serviced with the rear of the vehicle higher than the front, or if the tank is pressurized, fuel leakage or siphoning from the tank fuel lines could occur. to prevent this condition, maintain the vehicle front end at or above the level of the rear of vehicle. also, relieve tank pressure by loosening the fuel fill cap. cap should be tightened after pressure is relieved.

1. Shut the engine off. Depressurize the fuel system as follows:

On the 2.3L EFI engine, a valve is located on the throttle body for this purpose. The valve can be located by removing the air cleaner.

On the 2.9L EFI engine use the following steps:

- Remove the gas cap.

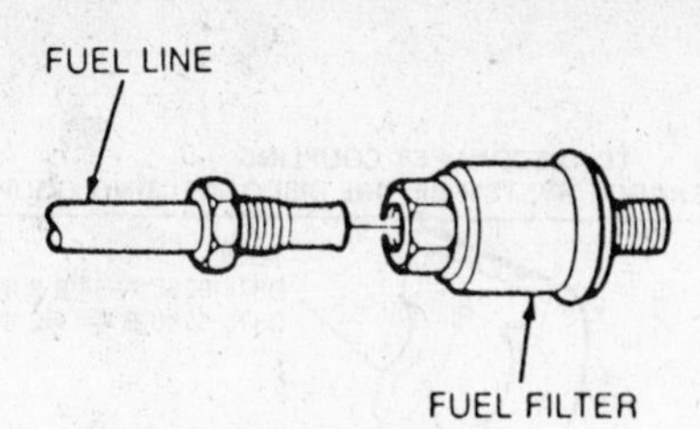

Carbureted engine fuel filter

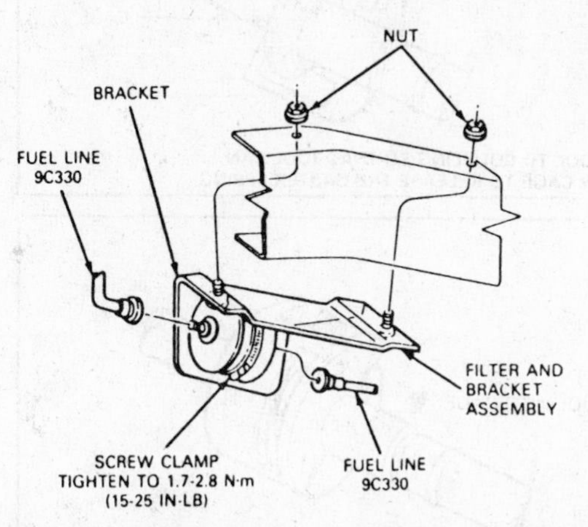

Inline reservoir fuel filter — fuel injected engines

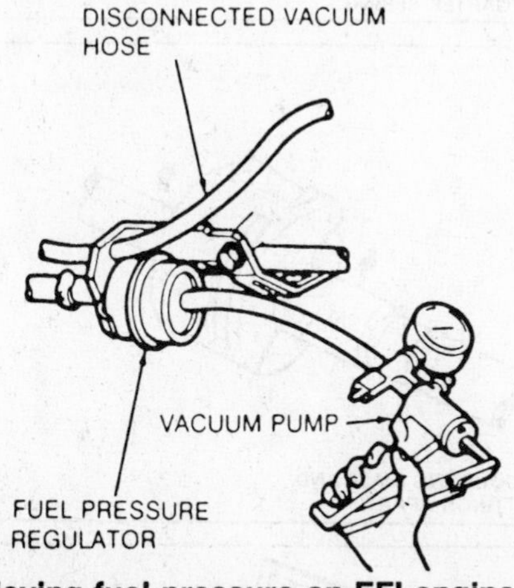

Relieving fuel pressure on EFI engines

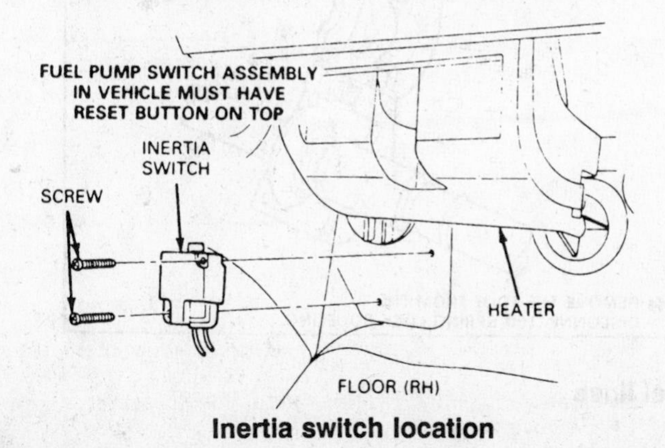

Inertia switch location

- Disconnect the vacuum hose from the pressure regulator located on the engine fuel rail.
- Using a hand vacuum pump, apply approximately 25 in.Hg pressure to the pressure regulator.

Fuel pressure will be released into the fuel tank through the fuel return hose.

On most EFI engines the fuel pressure can be relieved as follows. Use special tool T80L-9974-B or equivalent attached to the pressure measuring port on the engine fuel rail. Direct the drain hose to a suitable container and depress the pressure relief button.

An alternate method of relieving the fuel system pressure is to disconnect the electrical connector from the inertia switch and crank the engine for about 15–30 seconds until it runs out of fuel and dies.

2. Raise and support the vehicle safely.
3. Remove the push connect fittings at both ends of fuel filter. Install new retainer clips in each push connect fitting.
4. Remove the filter from the bracket by loosening the filter retaining clamp enough to allow the filter to pass through.

NOTE: The flow arrow direction should be positioned as installed in the bracket to ensure proper flow of fuel through the replacement filter.

5. Install the filter in the bracket, ensuring proper direction of flow as noted by arrow. Tighten clamp to 15–25 inch lbs.
6. Install the push connect fittings at both ends of the filter.
7. Lower the vehicle.
8. Start the engine and check for leaks.

Clean all dirt and/or grease from the fuel filter fittings. "Quick Connect" fittings are used on models equipped with a pressurized fuel system. These fittings must be disconnected using the proper procedure or the fittings may be damaged. The fuel filter uses a "hairpin" clip retainer. Spread the two hairpin clip legs about 1/8 in. (3mm) each to disengage it from the fitting, then pull the clip outward. Use finger pressure only; do not use any tools. Push the quick connect fittings onto the filter ends. Ford recommends that the retaining clips be replaced whenever removed. The fuel tubes used on these fuel systems are manufactured in 5/16 in. and 3/8 in. diameters. Each fuel tube takes a different size harpin clip, so keep this in mind when purchasing new clips. A click will be heard when the hairpin clip snaps into its proper position. Pull on the lines with moderate pressure to ensure proper connection. Start the engine and check for fuel leaks. If the inertia switch (reset switch) was disconnected to relieve the fuel system pressure, cycle the ignition switch from the **OFF** to **ON** position several times to re-charge the fuel system before attempting to start the engine.

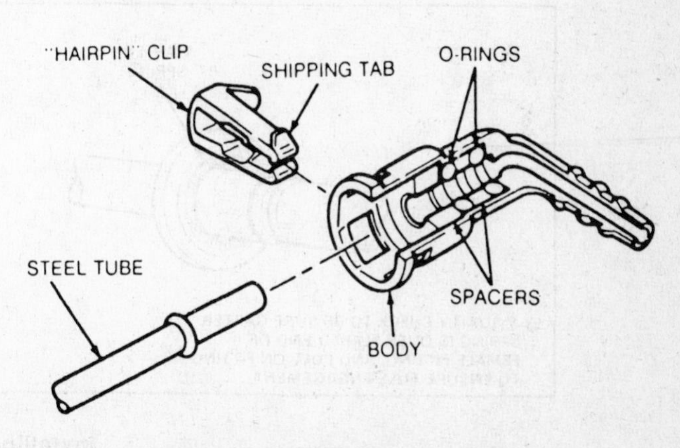

5/16 in. quick connect fuel fitting

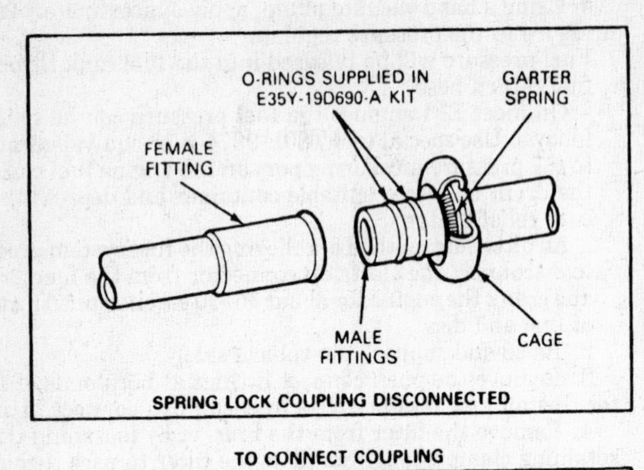

SPRING LOCK COUPLING DISCONNECTED

TO DISCONNECT COUPLING

CAUTION — DISCHARGE SYSTEM BEFORE DISCONNECTING COUPLING

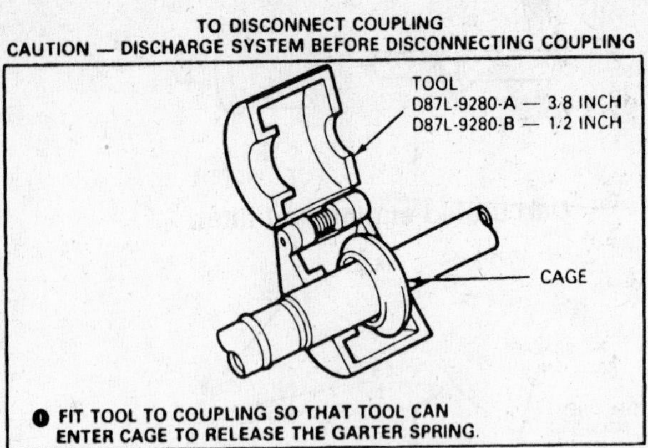

❶ FIT TOOL TO COUPLING SO THAT TOOL CAN ENTER CAGE TO RELEASE THE GARTER SPRING.

TO CONNECT COUPLING

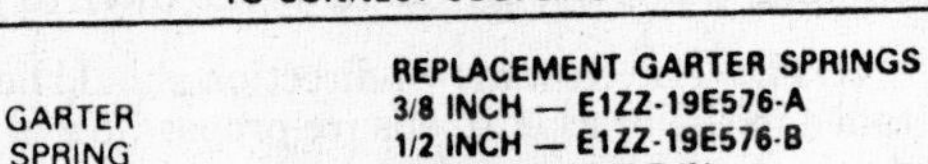

❶ CHECK FOR MISSING OR DAMAGED GARTER SPRING — REMOVE DAMAGED SPRING WITH SMALL HOOKED WIRE — INSTALL NEW SPRING IF DAMAGED OR MISSING.

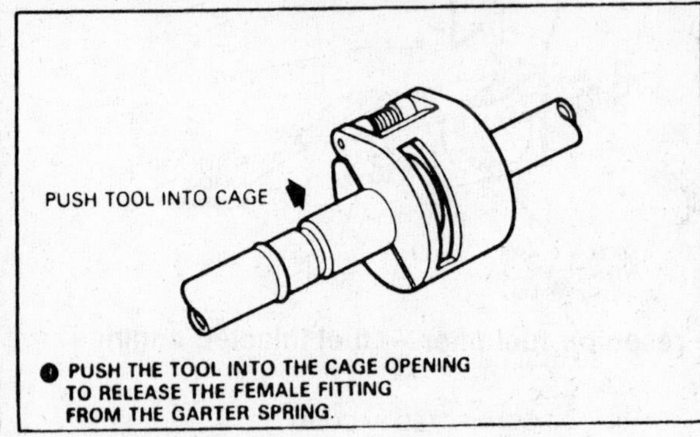

❷ PUSH THE TOOL INTO THE CAGE OPENING TO RELEASE THE FEMALE FITTING FROM THE GARTER SPRING.

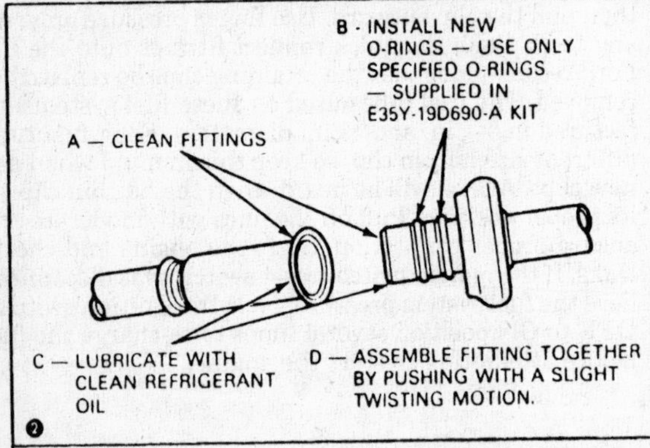

❷

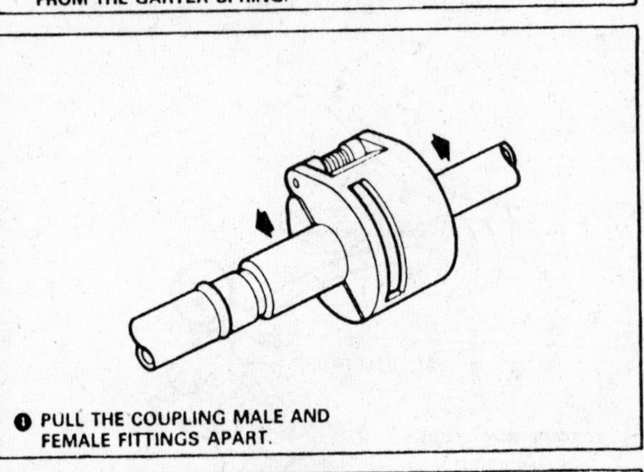

❸ PULL THE COUPLING MALE AND FEMALE FITTINGS APART.

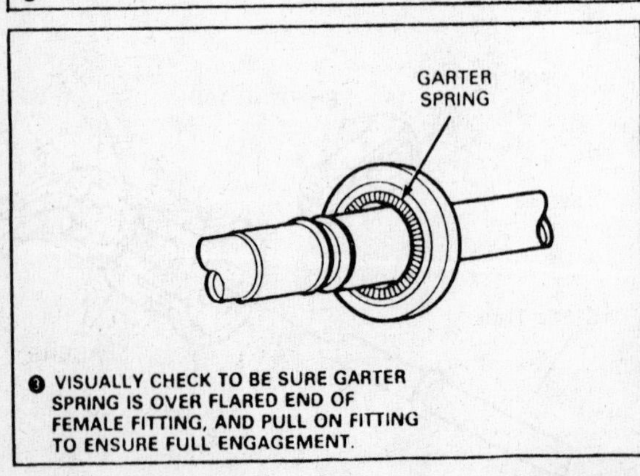

❸ VISUALLY CHECK TO BE SURE GARTER SPRING IS OVER FLARED END OF FEMALE FITTING, AND PULL ON FITTING TO ENSURE FULL ENGAGEMENT.

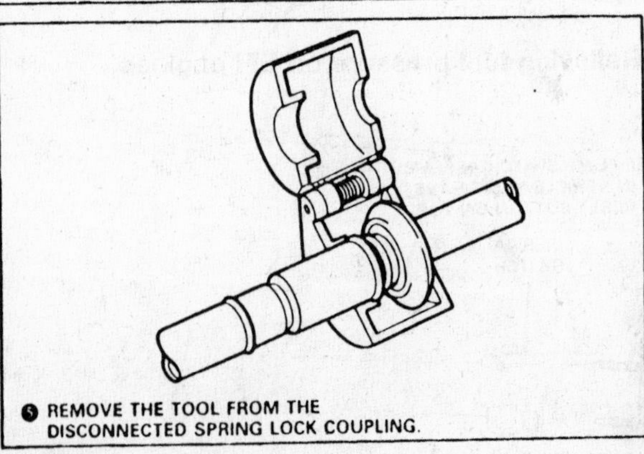

❺ REMOVE THE TOOL FROM THE DISCONNECTED SPRING LOCK COUPLING.

Installing fuel lines

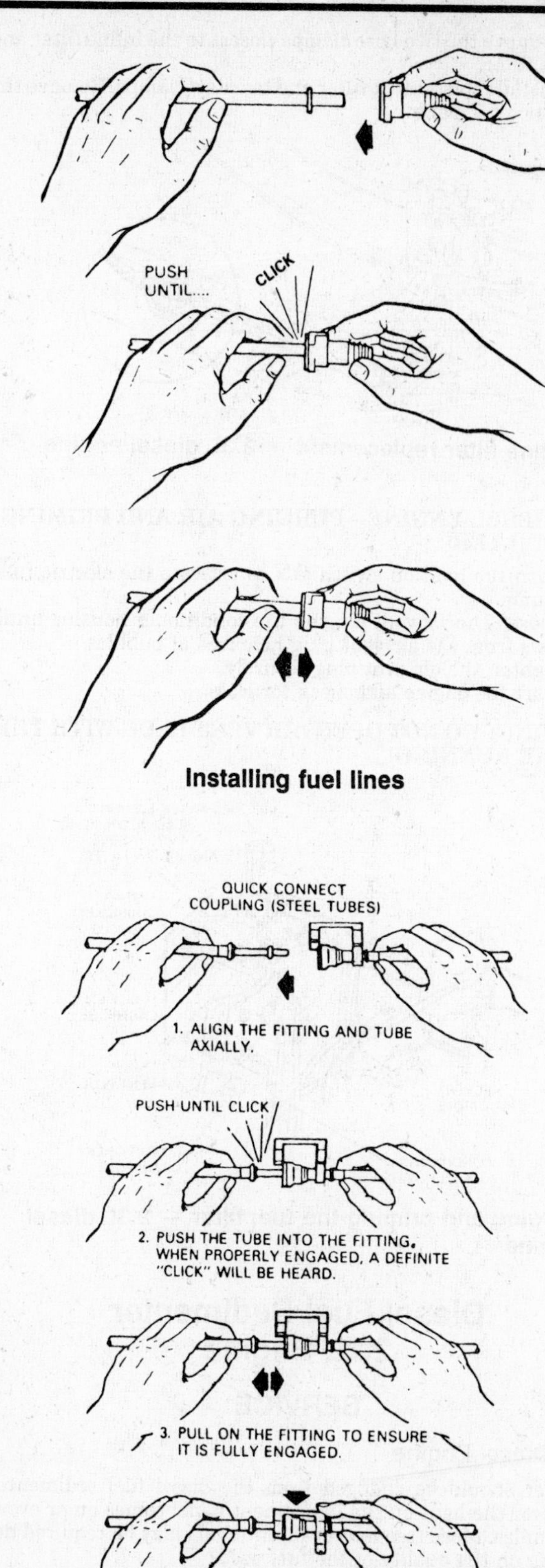

Installing fuel lines

Installing fuel lines

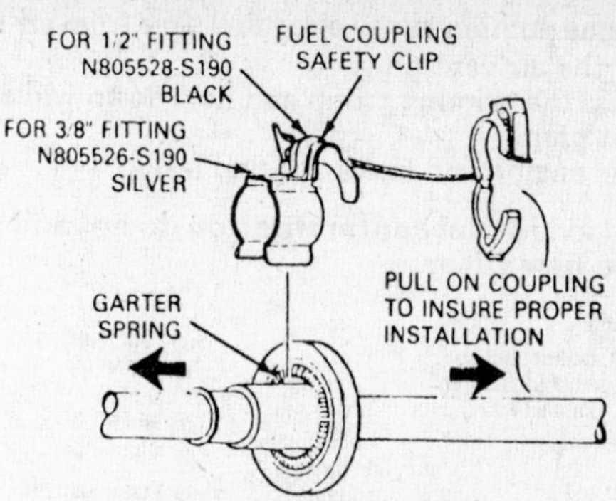

SPRING LOCK COUPLING – FOR FUEL LINE TO ENGINE FUEL RAIL CONNECTIONS

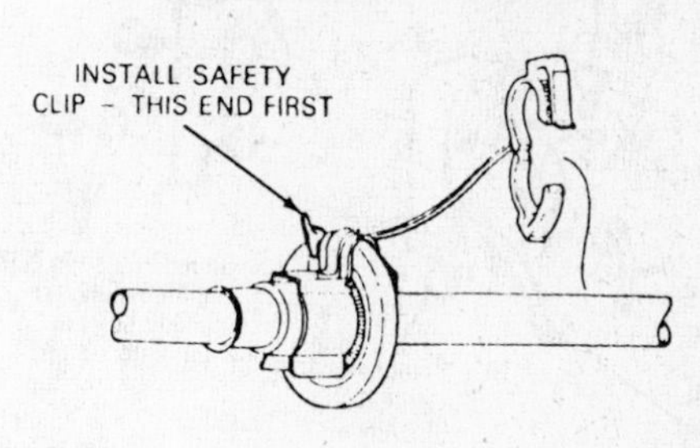

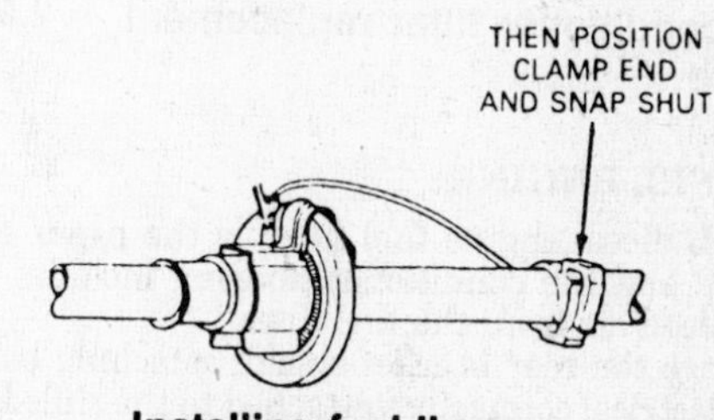

Installing fuel lines

Diesel Engines

2.2L DIESEL ENGINE

1. Remove the spin on filter by turning counterclockwise with hands or suitable tool, and discard filter.
2. Clean the filter mounting surfaces.
3. Coat the gasket of the new filter with clean diesel fuel.
4. Tighten the filter until the gasket touches the filter header, then tighten an additional ½ turn.
5. Air Bleed the fuel system using the following procedure:
   a. Loosen the fuel filter air vent plug.
   b. Pump the priming pump on the top of the filter adapter.

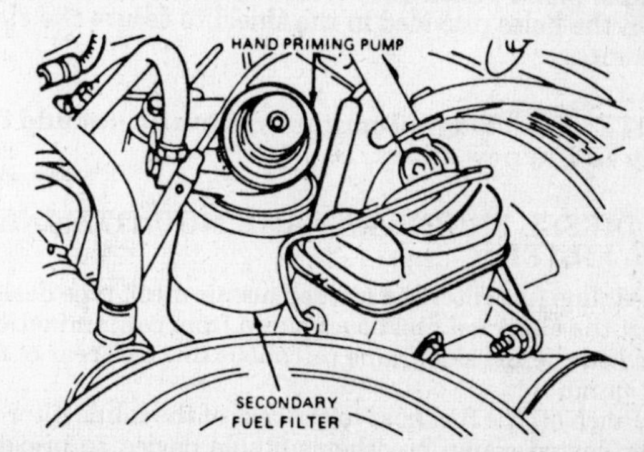

Priming the pump and fuel filter assembly – 2.2L diesel engine

c. Continue pumping until clear fuel, free from air bubbles, flows from the air vent plug.

d. Depress the priming pump and hold down while closing the air vent plug.

6. Start the engine and check for fuel leaks.

NOTE: To avoid fuel contamination do not add fuel directly to the new filter.

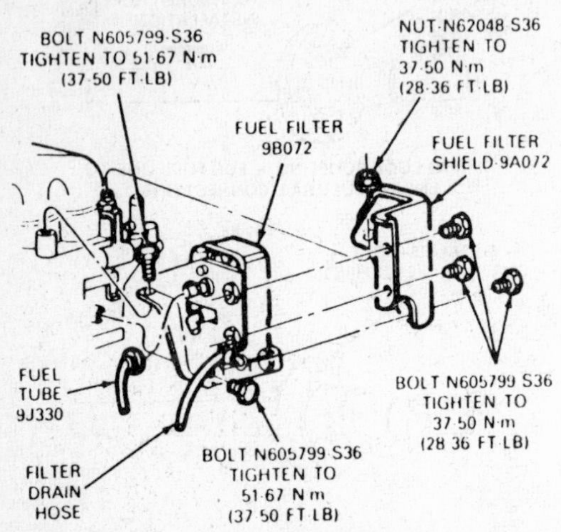

Fuel conditioner filter replacement – 2.3L diesel engine

2.3L DIESEL ENGINE

The 2.3L diesel engine fuel filter is the paper element cartridge type and the conditioner housing includes an air vent screw to bleed air from the fuel lines.

1. Remove the rear bracket shield, attaching bolts, and unplug the electrical connectors attached to the shiled. These connectors pull apart by pulling on the wire bundle on each side. Be resting the shield on the engine valve cover, the electrical connector halves leading to the fuel conditioner can be left attached to the shield.
2. Remove the rectangular filter element cartridge by unlatching holddown clamps by hand or with suitable tool, and pull element cartridge rear ward until it clears the base.
3. Clean the filter mounting pad.
4. Install the new element by pushing straight on after lining up filter element grommet holes with corresponding inlet/outlet tubes on the base.
5. Snap on the clamps.
6. Install the rear bracket shield, tighten the bolts to specification and plug the electrical connections back together. If the connectors were pulled away from the shield, push the locators back in the holes provided in the shield to secure the electrical connectors.

NOTE: To avoid fuel contamination do not add fuel directly to the new filter.

2.3L DIESEL ENGINE w/FRAME MOUNTED INLINE FUEL FILTER

The inline fuel filter is a molded plastic mesh type designed to protect the electrical fuel boost pump from contamination. It is located on the LH side frame rail about tow feet rear of the fuel boost pump.

1. Pinch off the fuel hose to the rear of the inline filter using a rubber coated clamp or other suitable device to prevent fuel from siphoning from the tank. Care must be taken not to damage the fuel hose.
2. Remove the two hose clamps closest to the inline filter and remove filter.
3. Install replacement filter and two new clamps. Remove the hose pinch-off clamp.

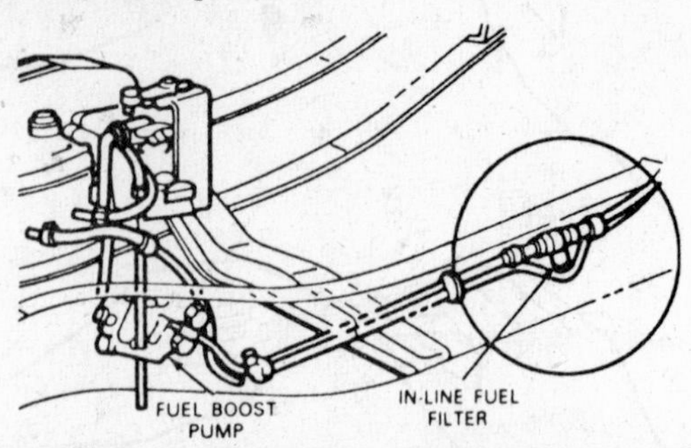

Inline filter replacement – 2.3L diesel engine

2.3L DIESEL ENGINE – PURGING AIR AND PRIMING FUEL FILTER

1. Turn the ignition switch **ON** to activate the electric fuel boost pump.
2. Loosen the air vent plug on the conditioner housing until fuel flows from the air vent plug hole free of bubbles.
3. Tighten the air vent plug securely.
4. Start the engine and check for leaks.

WARNING: DO NOT OPEN AIR VENT PLUG WITH THE ENGINE RUNNING!

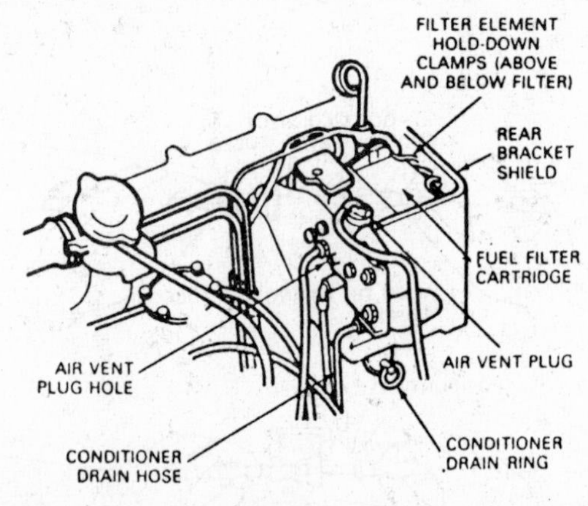

Purging and priming the fuel filter – 2.3L diesel engine

Diesel Fuel Sedimenter 2.2L Engine

SERVICE

2.2L Diesel Engine

Water should be changed from the diesel fuel sedimenter whenever the light on the instrument panel comes on or every 5,000 miles. More frequent drain intervals may be required depending on the quality of the fuel used.

CAUTION

The vehicle must be stopped with the engine off when draining the sedimenter. Fuel may ignite if sedimenter is drained while the engine is running or the vehicle is moving.

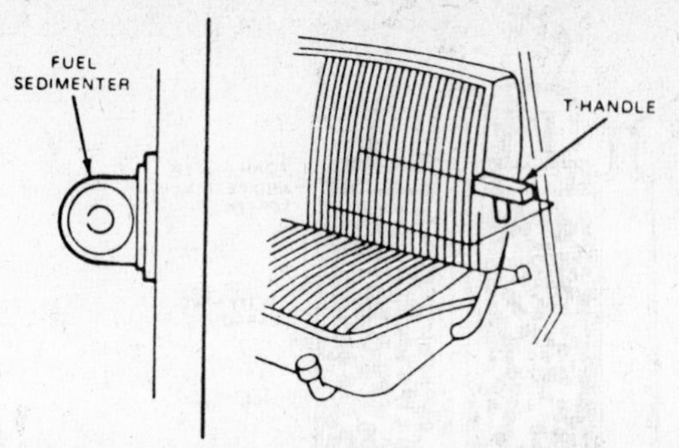

Diesel fuel sedimenter draining — 2.2L diesel engine

The instrument panel warning light **(WATER IN FUEL)** will glow when approximately ½ liter of water has accumulated in the sedimenter. When the warning light glows, shut off the engine as soon as safely possible. A suitable drain pan or container should be placed under the sedimenter, which is mounted inside the frame rail, underneath the driver's side of the cab. To drain the fuel sedimenter, pull up on the T-handle (located on the cab floor behind the driver's seat) until resistance is felt. Turn the ignition switch to the **ON** position so the warning light glows and hold T-handle up for approximately 45 seconds after light goes out.

To stop draining fuel, release T-handle and inspect sedimenter to verify that draining has stopped. Discard drained fluid suitably.

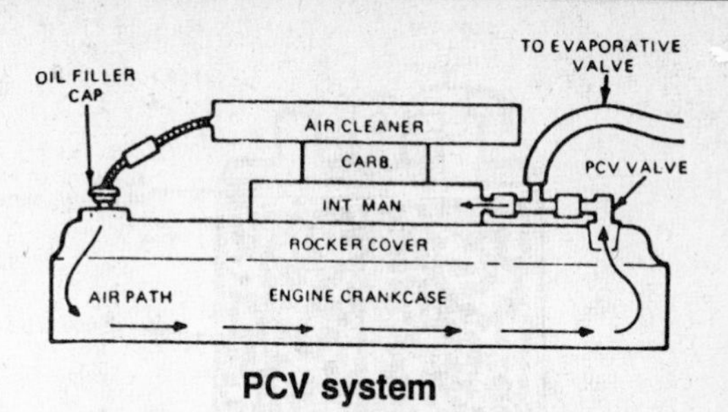

PCV system

PCV Valve

REMOVAL AND INSTALLATION

Check the PCV valve according to the Preventive Maintenance Schedule at the end of this Section to see if it is free and not gummed up, stuck or blocked. To check the valve, remove it from the engine and work the valve by sticking a screwdriver in the crankcase side of the valve. It should move. It is possible to clean the PCV valve by soaking it in a solvent and blowing it out with compressed air. This can restore the valve to some level of operating order.

This should be used only in emergency situations. Otherwise, the valve should be replaced. Always check PCV valve hose for wear or cracks during sevice procedure.

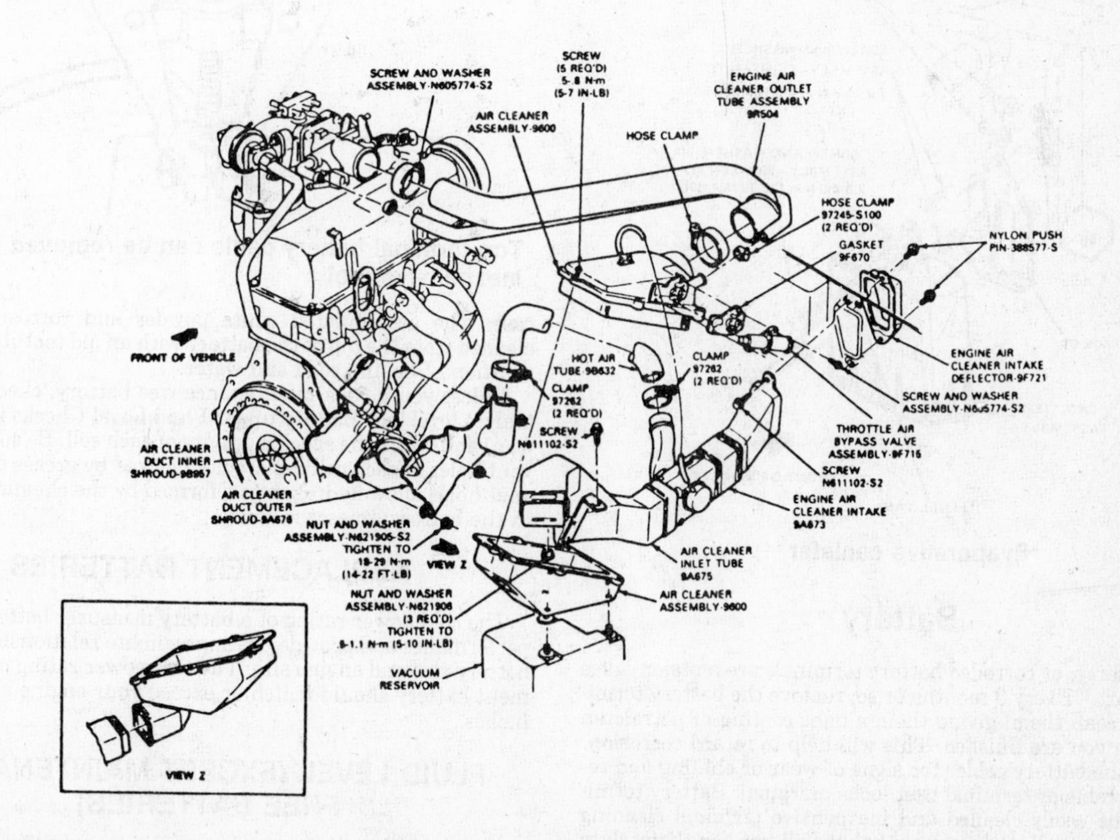

Air cleaner assembly — 2.3L fuel injected engine

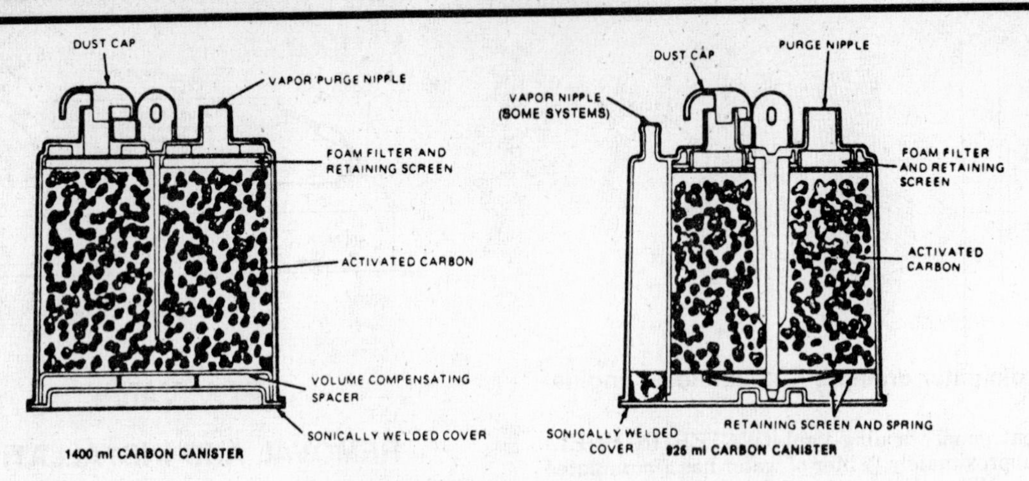

Evaporative canister cross-sections

Evaporative Canister

The fuel evaporative emission control canister should be inspected for damage or leaks at the hose fittings. Repair or replace any old or cracked hoses. Replace the canister if it is damaged in any way. The canister is located on the left side radiator support, under the hood.

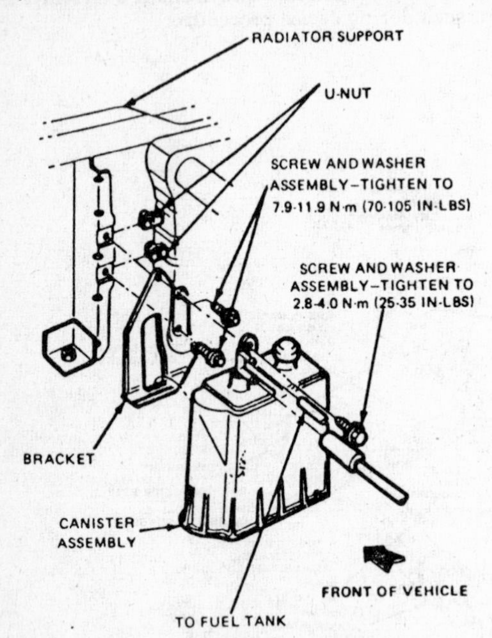

Evaporative canister

Battery

Loose, dirty, or corroded battery terminals are a major cause of "no-start." Every 3 months or so, remove the battery terminals and clean them, giving them a light coating of petroleum jelly when you are finished. This will help to retard corrosion.

Check the battery cables for signs of wear or chafing and replace any cable or terminal that looks marginal. Battery terminals can be easily cleaned and inexpensive terminal cleaning tools are an excellent investment that will pay for themselves many times over. They can usually be purchased from any well equipped auto store or parts department. Side terminal batteries require a different tool to clean the threads in the battery case. The accumulated white powder and corrosion can be cleaned from the top of the battery with an old toothbrush and a solution of baking soda and water.

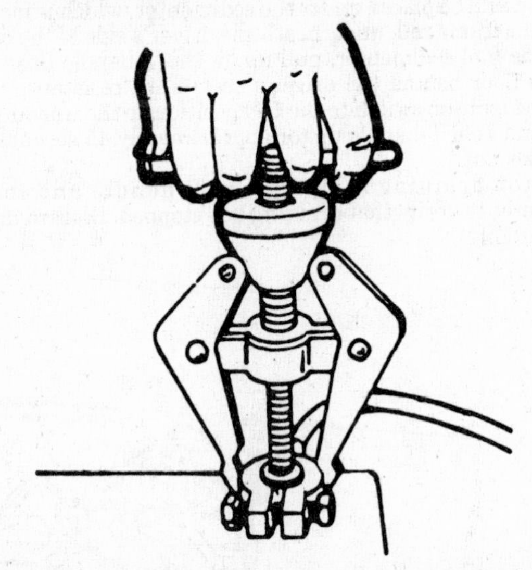

Top terminal battery cable can be removed with this inexpensive tool

Unless you have a maintenance free battery, check the electrolyte level (see Battery under Fluid Level Checks in this Section) and check the specific gravity of each cell. Be sure that the vent holes in each cell cap are not blocked by grease or dirt. The vent holes allow hydrogen gas, formed by the chemical reaction in the battery, to escape safely.

REPLACEMENT BATTERIES

The cold power rating of a battery measures battery starting performance and provides an approximate relationship between battery size and engine size. The cold power rating of a replacement battery should match or exceed your engine size in cubic inches.

FLUID LEVEL (EXCEPT MAINTENANCE FREE BATTERIES)

Check the battery electrolyte level at least once a month, or more often in hot weather or during periods of extended vehicle operation. The level can be checked through the case on translu-

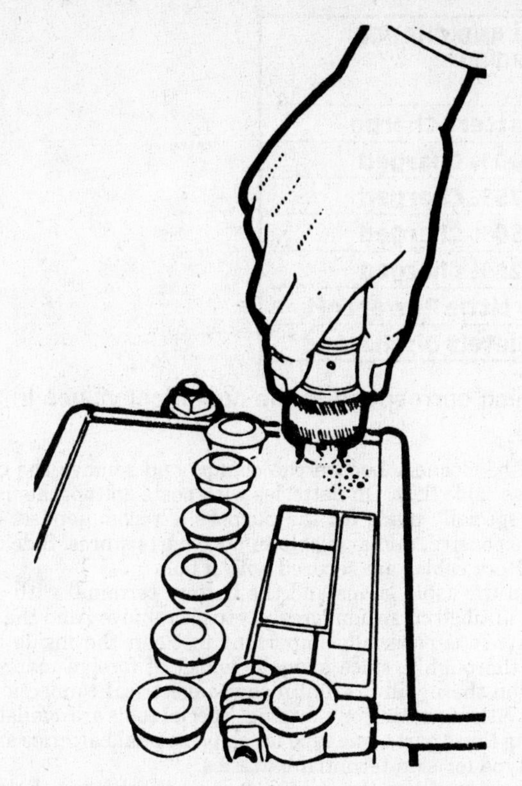

Clean the battery posts with a wire terminal cleaner

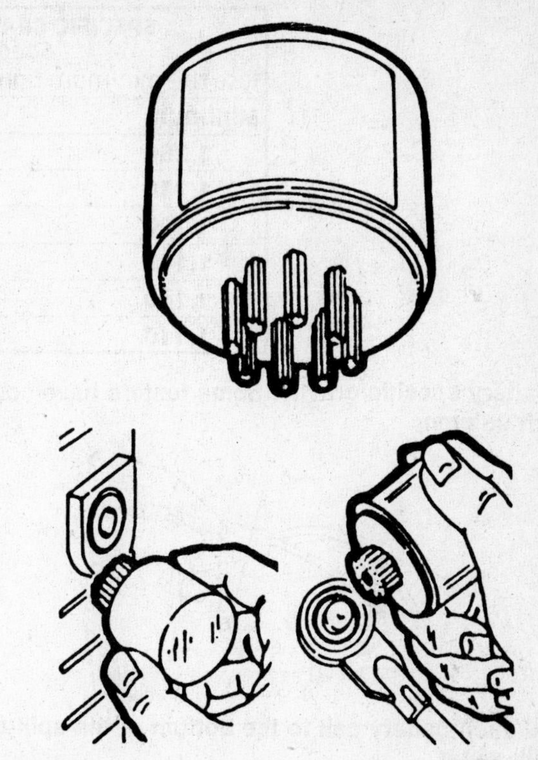

Side terminal batteries require a special wire brush for cleaning

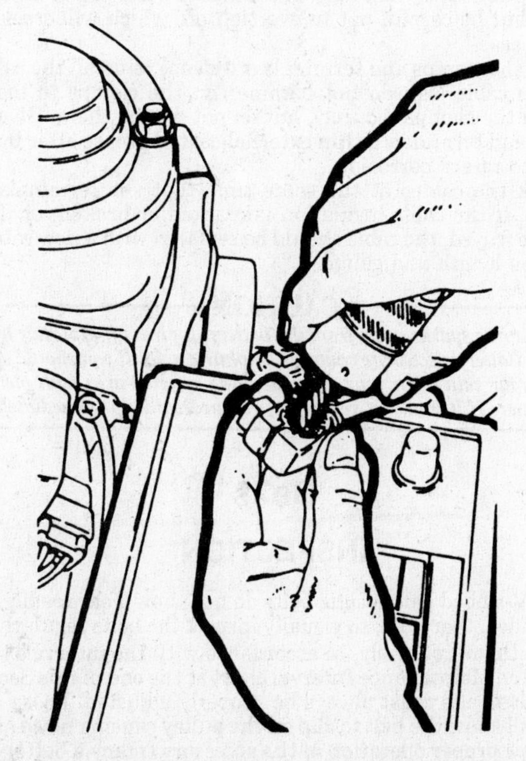

Clean the cable ends with a stiff wire cleaning tool

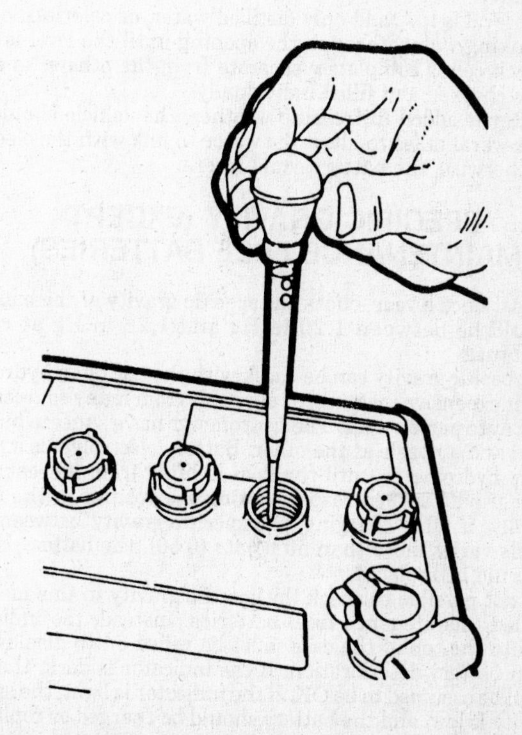

An inexpensive hydrometer will quickly test the battery's state of charge

SPECIFIC GRAVITY (@ 80°F.) AND CHARGE Specific Gravity Reading (use the minimum figure for testing)	
Minimum	Battery Charge
1.260	100% Charged
1.230	75% Charged
1.200	50% Charged
1.170	25% Charged
1.140	Very Little Power Left
1.110	Completely Discharged

Battery specific gravity. Some testers have colored balls which correspond to the numerical values in the left column

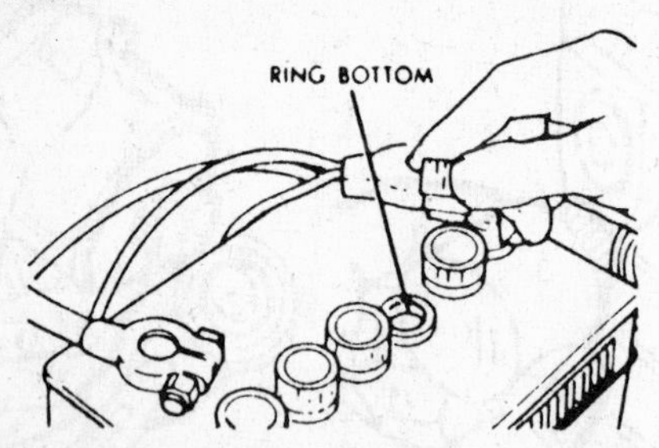

Fill each battery cell to the bottom of the split ring with water

cent polypropylene batteries; the cell caps must be removed on other models. The electrolyte level in each cell should be kept filled to the split ring inside, or the line marked on the outside of the case.

If the level is low, add only distilled water, or colorless, odorless drinking water, through the opening until the level is correct. Each cell is completely separate from the others, so each must be checked and filled individually.

If water is added in freezing weather, the vehicle should be driven several miles to allow the water to mix with the electrolyte. Otherwise, the battery could freeze.

SPECIFIC GRAVITY (EXCEPT MAINTENANCE FREE BATTERIES)

At least once a year, check the specific gravity of the battery. It should be between 1.20 in.Hg and 1.26 in.Hg at room temperature.

The specific gravity can be check with the use of an hydrometer, an inexpensive instrument available from many sources, including auto parts stores. The hydrometer has a squeeze bulb at one end and a nozzle at the other. Battery electrolyte is sucked into the hydrometer until the float is lifted from its seat. The specific gravity is then read by noting the position of the float. Generally, if after charging, the specific gravity between any two cells varies more than 50 points (0.50), the battery is bad and should be replaced.

It is not possible to check the specific gravity in this manner on sealed (maintenance free) batteries. Instead, the indicator built into the top of the case must be relied on to display any signs of battery deterioration. If the indicator is dark, the battery can be assumed to be OK. If the indicator is light, the specific gravity is low, and the battery should be charged or replaced.

CABLES AND CLAMPS

Once a year, the battery terminals and the cable clamps should be cleaned. Loosen the clamps and remove the cables, negative cable first. On batteries with posts on top, the use of a puller specially made for the purpose is recommended. These are inexpensive, and available in auto parts stores. Side terminal battery cables are secured with a bolt.

Clean the cable lamps and the battery terminal with a wire brush, until all corrosion, grease, etc., is removed and the metal is shiny. It is especially important to clean the inside of the clamp thoroughly, since a small deposit of foreign material or oxidation there will prevent a sound electrical connection and inhibit either starting or charging. Special tools are available for cleaning these parts, one type for conventional batteries and another type for side terminal batteries.

Before installing the cables, loosen the battery holddown clamp or strap, remove the battery and check the battery tray. Clear it of any debris, and check it for soundness. Rust should be wire brushed away, and the metal given a coat of anti-rust paint. Replace the battery and tighten the holddown clamp or strap securely, but be careful not to overtighten, which will crack the battery case.

After the clamps and terminals are clean, reinstall the cables, negative cable last; do not hammer on the clamps to install. Tighten the clamps securely, but do not distort them. Give the clamps and terminals a thin external coat of grease after installation, to retard corrosion.

Check the cables at the same time that the terminals are cleaned. If the cable insulation is cracked or broken, or if the ends are frayed, the cable should be replaced with a new cable of the same length and gauge.

CAUTION

Keep flame or sparks away from the battery; it gives off explosive hydrogen gas. Battery electrolyte contains sulphuric acid. If you should splash any on your skin or in your eyes, flush the affected area with plenty of clear water. If it lands in your eyes, get medical help immediately.

Belts

INSPECTION

The V-ribbed belt design, belts do not show wear readily. It is a good idea, therefore, to visually inspect the belts regularly and replace them, routinely, in accordance with the intervals specified in the Maintenance Interval chart at the end of this Section. The drive belts must always be properly adjusted. Loose drive belts will allow the belt to slip on the pulley causing noise or will not allow proper operation of the accessory (many a battery has died due to a loose alternator belt). On the other hand, overly tight belts will place strain on accessory bearings causing them to fail prematurely.

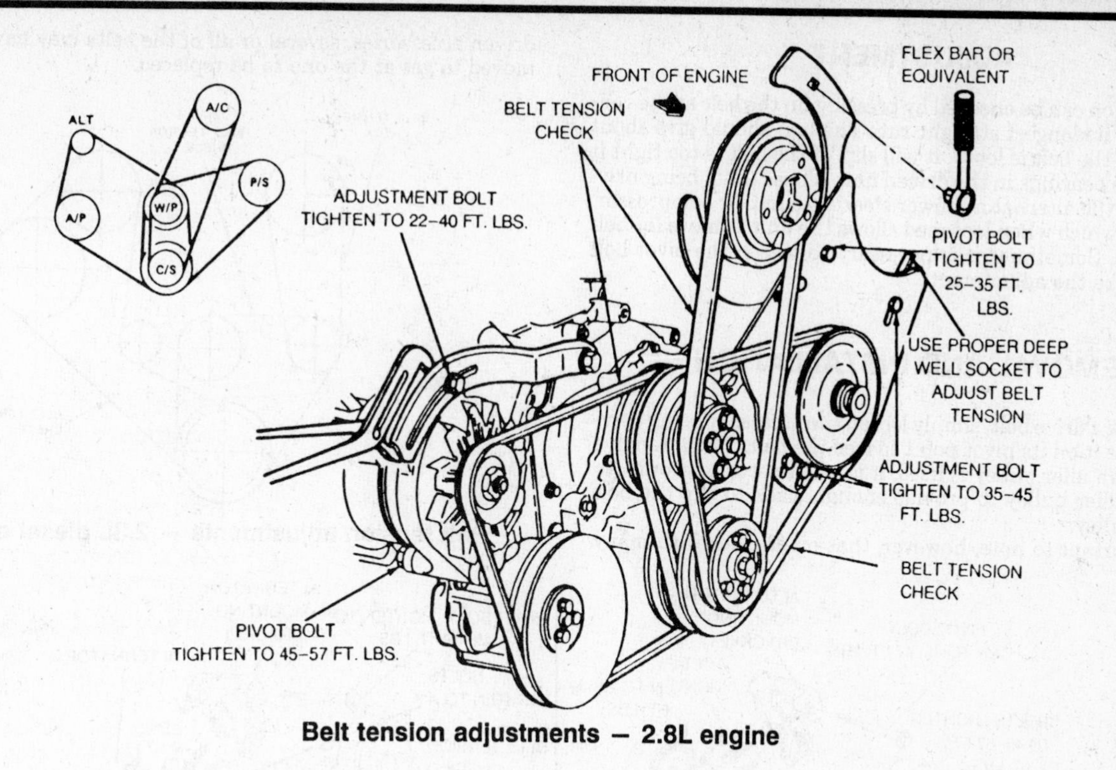

Belt tension adjustments – 2.8L engine

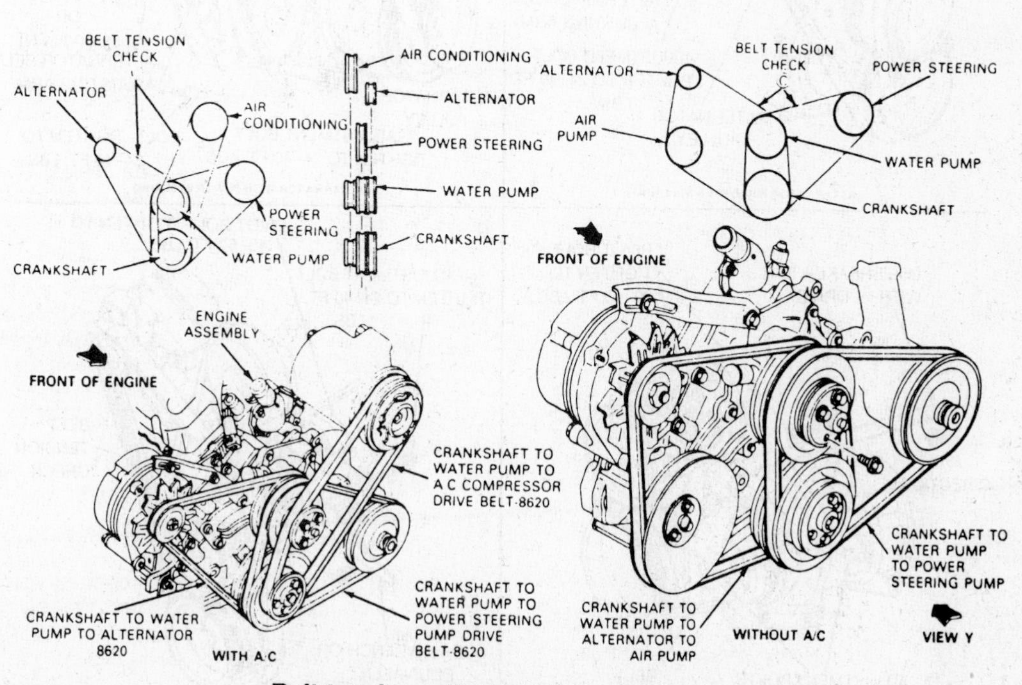

Belt tension adjustments – 2.9L engine

ADJUSTMENT

Belt tension can be checked by pressing on the belt at the center point of its longest straight run. The belt should give about 1/4–1/2 in.. If the belt is loose, it will slip. If the belt is too tight it will damage bearings in the driven unit. Those units being driven, such as the alternator, power steering pump or compressor, have a bolt which when loosened allows the unit to move for belt adjustment. Sometimes it is necessary to loosen the pivot bolt also, to make the adjustment.

REMOVAL AND INSTALLATION

To remove a drive belt, simply loosen the accessory being driven and move it on its pivot point to free the belt. Then, remove the belt. If an idler pulley is used, it is often necessary, only, to loosen the idler pulley to provide enough slack to slip the belt from the pulley.

It is important to note, however, that on engines with many driven accessories, several or all of the belts may have to be removed to get at the one to be replaced.

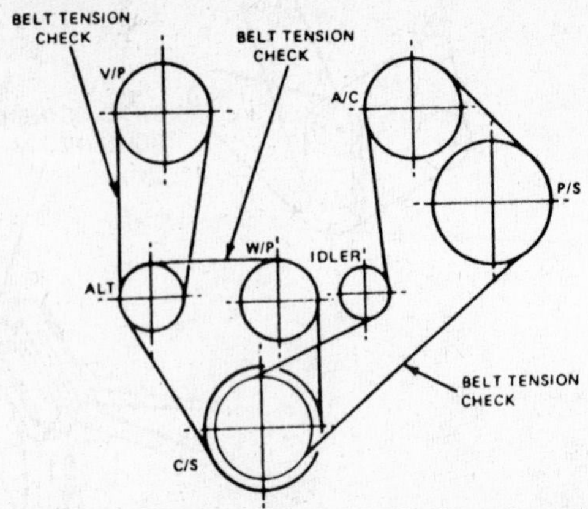

Belt tension adjustments – 2.3L diesel engine

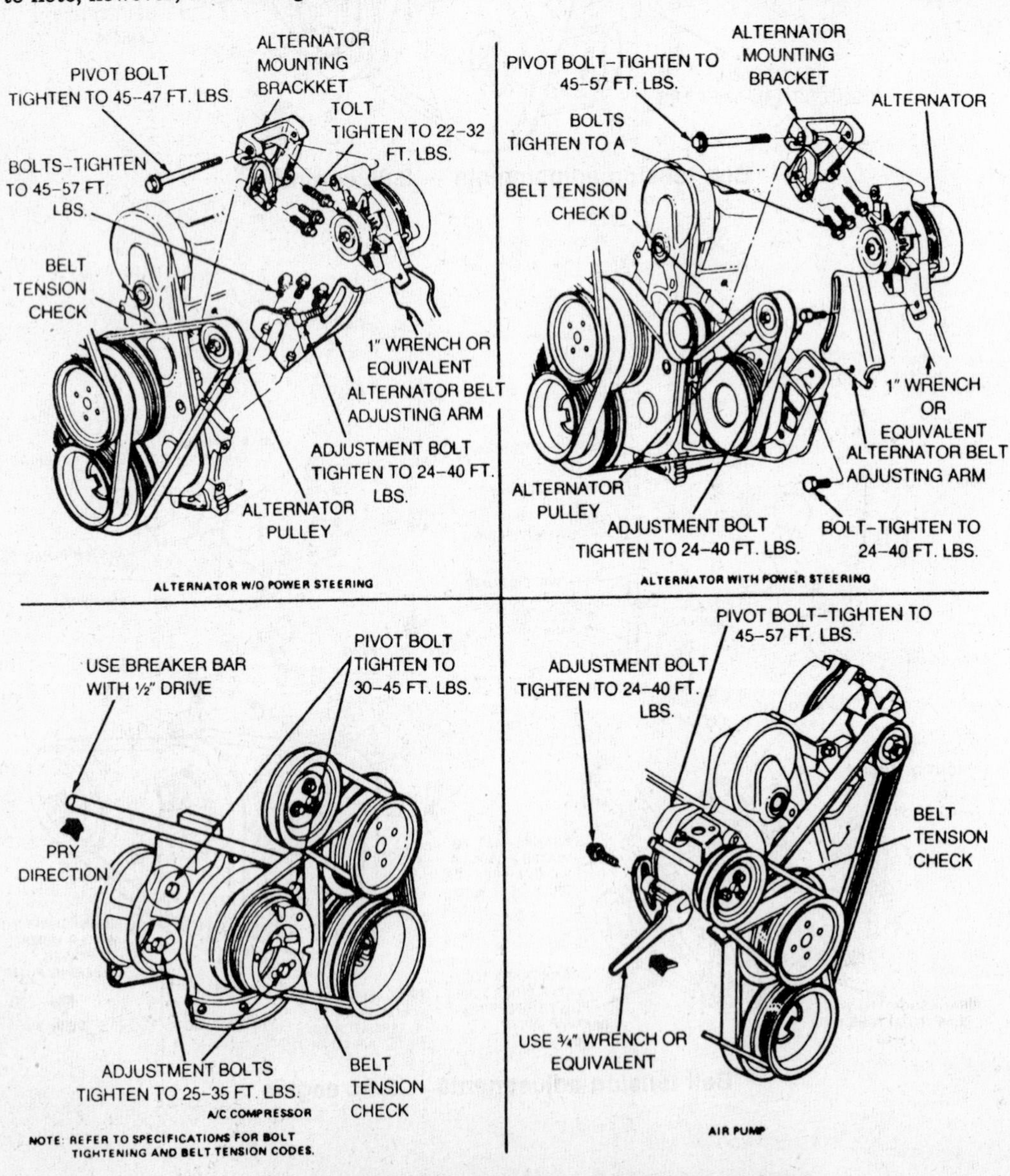

Belt tension adjustments – 2.0L, 2.3L engines

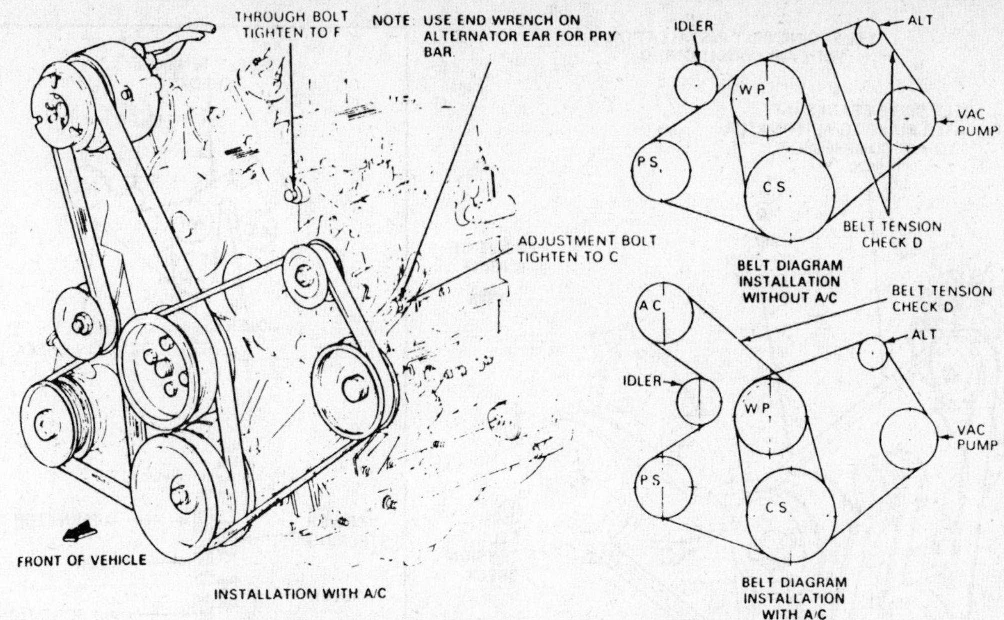

Belt tension adjustments – 2.3L diesel engine

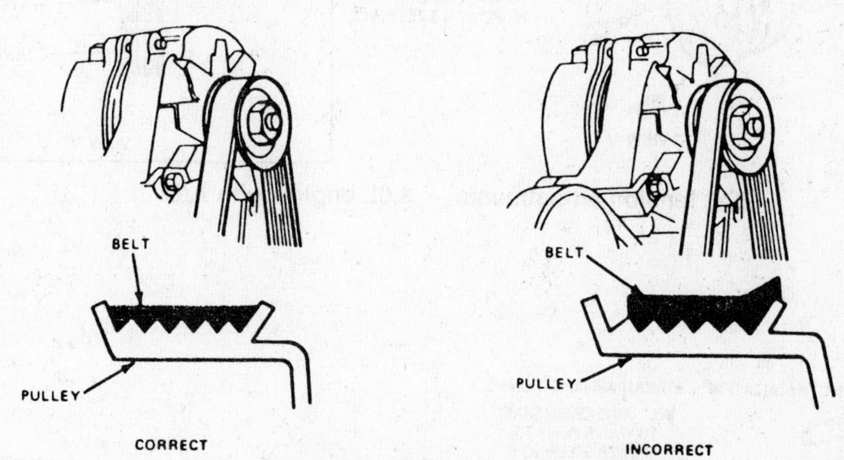

Checking ribbed belt alignment

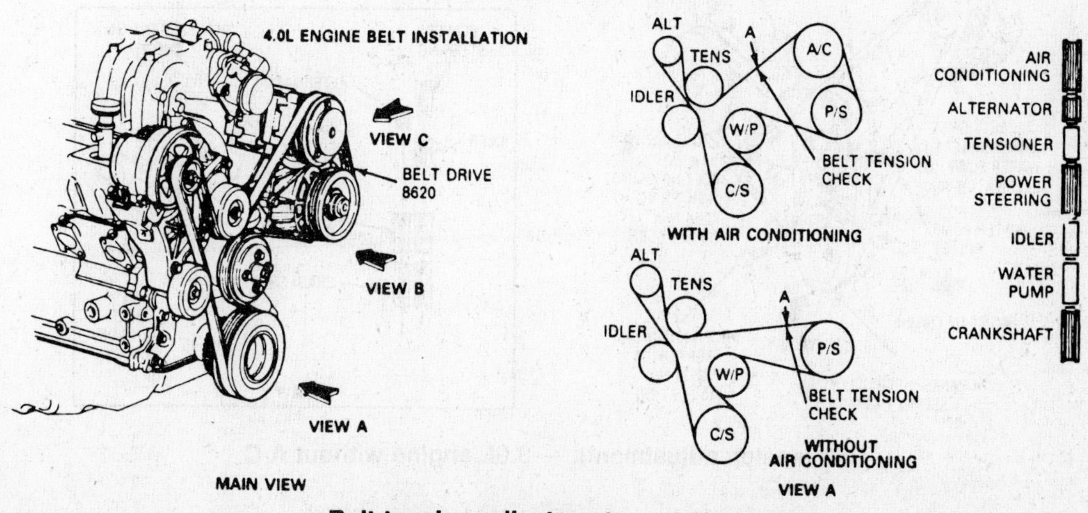

Belt tension adjustments – 4.0L engine

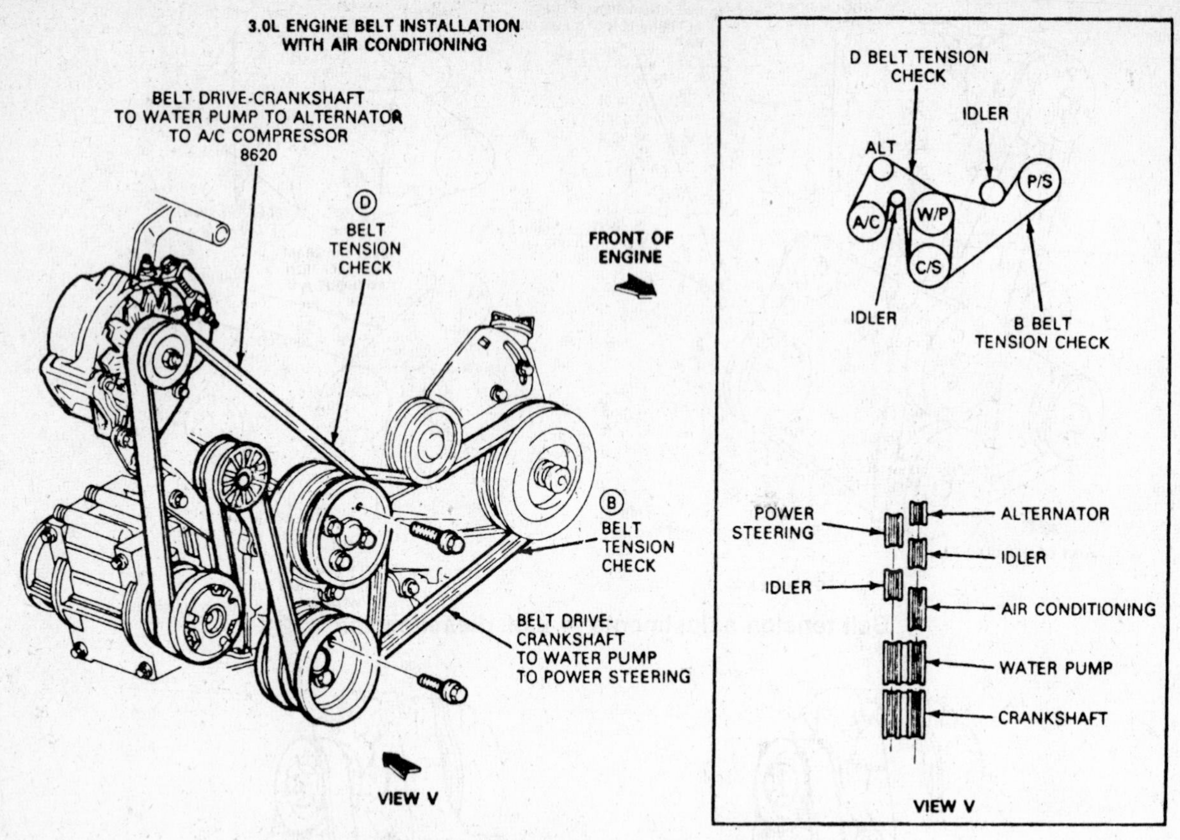

Belt tension adjustments — 3.0L engine with A/C

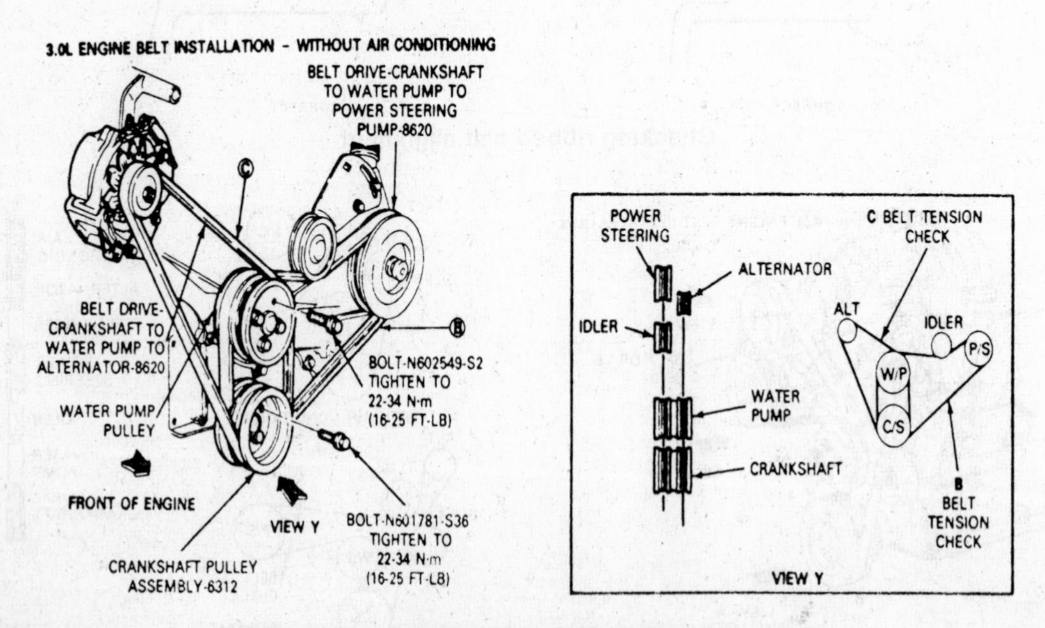

Belt tension adjustments — 3.0L engine without A/C

HOW TO SPOT WORN V-BELTS

V–Belts are vital to efficient engine operation – they drive the fan, water pump and other accessories. They require little maintenance (occasional tightening) but they will not last forever. Slipping or failure of the V–belt will lead to overheating. If your V–belt looks like any of these, it should be replaced.

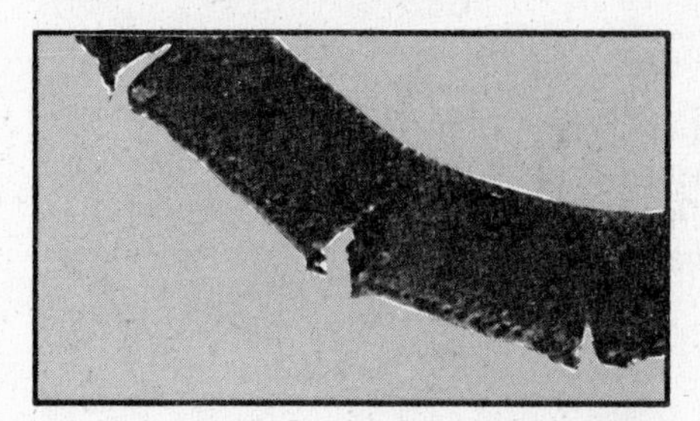

Cracking or Weathering

This belt has deep cracks, which cause it to flex. Too much flexing leads to heat build–up and premature failure. These cracks can be caused by using the belt on a pulley that is too small. Notched belts are available for small diameter pulleys.

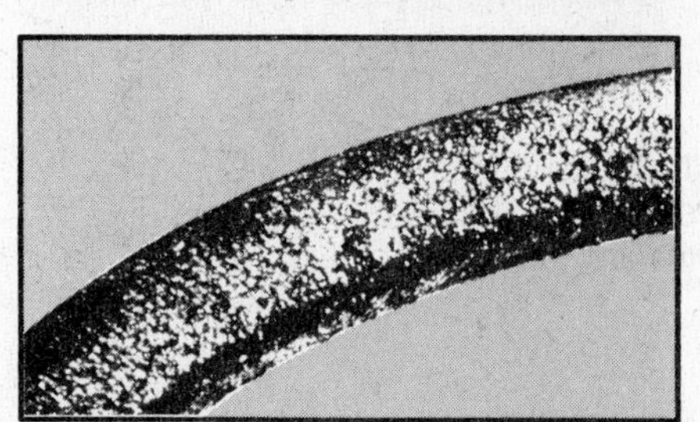

Softening (Grease and Oil)

Oil and grease on a belt can cause the belt's rubber compounds to soften and separate from the reinforcing cords that hold the belt together. The belt will first slip, then finally fail altogether.

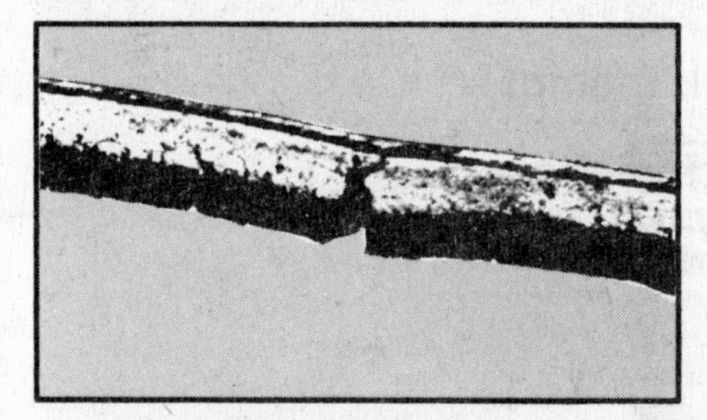

Glazing

Glazing is caused by a belt that is slipping. A slipping belt can cause a run-down battery, erratic power steering, overheating or poor accessory performance. The more the belt slips, the more glazing will be built up on the surface of the belt. The more the belt is glazed, the more it will slip. If the glazing is light, tighten the belt.

Worn Cover

The cover of this belt is worn off and is peeling away. The reinforcing cords will begin to wear and the belt will shortly break. When the belt cover wears in spots or has a rough jagged appearance, check the pulley grooves for roughness.

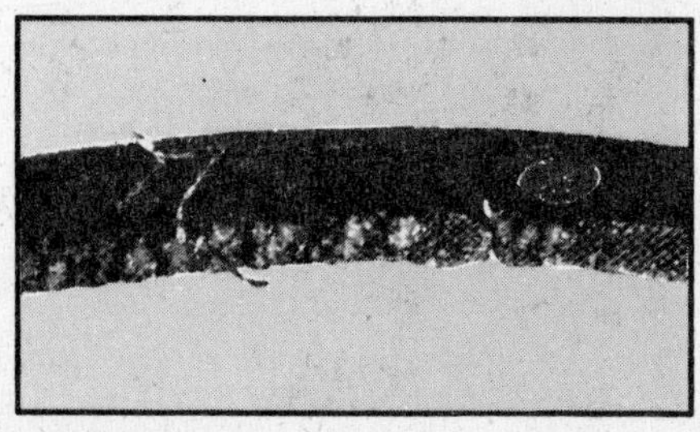

Separation

This belt is on the verge of breaking and leaving you stranded. The layers of the belt are separating and the reinforcing cords are exposed. It's just a matter of time before it breaks completely.

HOW TO SPOT BAD HOSES

Both the upper and lower radiator hoses are called upon to perform difficult jobs in an inhospitable environment. They are subject to nearly 18 psi at under hood temperatures often over 280°F, and must circulate nearly 7500 gallons of coolant an hour—3 good reasons to have good hoses.

Swollen Hose

A good test for any hose is to feel it for soft or spongy spots. Frequently these will appear as swollen areas of the hose. The most likely cause is oil soaking. This hose could burst at any time, when hot or under pressure.

Cracked Hose

Cracked hoses can usually be seen but feel the hoses to be sure they have not hardened; a prime cause of cracking. This hose has cracked down to the reinforcing cords and could split at any of the cracks.

Frayed Hose End (Due to Weak Clamp)

Weakened clamps frequently are the cause of hose and cooling system failure. The connection between the pipe and hose has deteriorated enough to allow coolant to escape when the engine is hot.

Debris in Cooling System

Debris, rust and scale in the cooling system can cause the inside of a hose to weaken. This can usually be felt on the outside of the hose as soft or thinner areas.

Hoses

REPLACEMENT

1. Drain the existing antifreeze and coolant. Open the radiator and engine drain petcocks, or disconnect the bottom radiator hose, at the radiator outlet.

CAUTION

When draining the coolant, keep in mind that cats and dogs are attracted by the ethylene glycol antifreeze, and are quite likely to drink any that is left in an uncovered container or in puddles on the ground. This will prove fatal in sufficient quantity. Always drain the coolant into a sealable container. Coolant should be reused unless it is contaminated or several years old.

NOTE: Before opening the radiator petcock, spray it with some penetrating lubricant.

2. Loosen the clamp on each end of the hose to be removed.
3. Slide the hose off the connections.
4. Position the clamps on each end of the new hose.
5. Slide the hose onto the connections, then tighten the clamps. If the connections have a bead around the edges, make sure the clamps are located beyond the beads.
6. Refill the cooling system with coolant. Run the engine for several minutes, then check the hose connections for leaks.

Cooling System

CAUTION

Never remove the radiator cap under any conditions while the engine is running! Failure to follow these instructions could result in damage to the cooling system or engine and/or personal injury. To avoid having scalding hot coolant or steam blow out of the radiator, use extreme care when removing the radiator cap from a hot radiator. Wait until the engine has cooled, then wrap a thick cloth around the radiator cap and turn it slowly to the first stop. Step back while the pressure is released from the cooling system. When you are sure the pressure has been released, press down on the radiator cap (still have the cloth in position) turn and remove the radiator cap.

At least once every 2 years, the engine cooling system should be inspected, flushed, and refilled with fresh coolant. If the coolant is left in the system too long, it loses its ability to prevent rust and corrosion. If the coolant has too much water, it won't protect against freezing.

The pressure cap should be looked at for signs of age or deterioration. Fan belt and other drive belts should be inspected and adjusted to the proper tension. (See checking belt tension).

Hose clamps should be tightened, and soft or cracked hoses replaced. Damp spots, or accumulations of rust or dye near hoses, water pump or other areas, indicate possible leakage, which must be corrected before filling the system with fresh coolant.

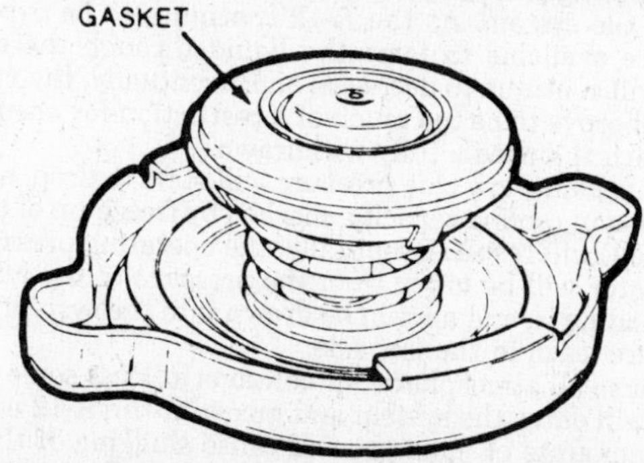

Check the radiator cap gasket for cracks or wear

CHECK THE RADIATOR CAP

While you are checking the coolant level, check the radiator cap for a worn or cracked gasket. It the cap doesn't seal properly, fluid will be lost and the engine will overheat.

Worn caps should be replaced with a new one.

CLEAN RADIATOR OF DEBRIS

Periodically clean any debris – leaves, paper, insects, etc. – from the radiator fins. Pick the large pieces off by hand. The smaller pieces can be washed away with water pressure from a hose.

Carefully straighten any bent radiator fins with a pair of needle nose pliers. Be careful – the fins are very soft. Don't wiggle the fins back and forth too much. Straighten them once and try not to move them again.

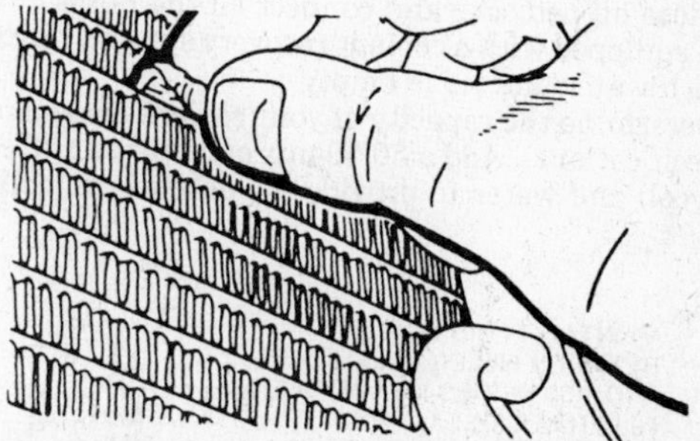

Keep the radiator fins clear of debris for maximum cooling

DRAIN AND REFILL THE COOLING SYSTEM

Completely draining and refilling the cooling system every two years at least will remove accumulated rust, scale and other deposits. Coolant in late model vehicles is a 50/50 mixture of ethylene glycol and water for year round use. Use a good quality antifreeze with water pump lubricants, rust inhibitors and other corrosion inhibitors along with acid neutralizers.

1. Drain the existing antifreeze and coolant. Open the radiator and engine drain petcocks, or disconnect the bottom radiator hose, at the radiator outlet.

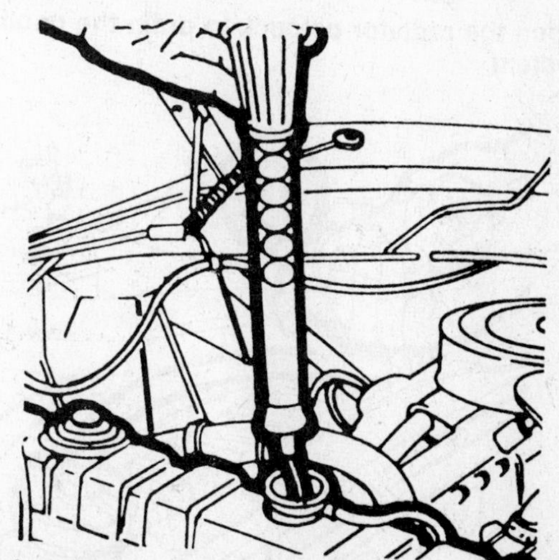

Check antifreeze protection with an inexpensive tester

CAUTION

When draining the coolant, keep in mind that cats and dogs are attracted by the ethylene glycol antifreeze, and are quite likely to drink any that is left in an uncovered container or in puddles on the ground. This will prove fatal in sufficient quantity. Always drain the coolant into a sealable container. Coolant should be reused unless it is contaminated or several years old.

NOTE: Before opening the radiator petcock, spray it with some penetrating lubricant.

2. Close the petcock or reconnect the lower hose and fill the system with water.
3. Add a can of quality radiator flush.
4. Idle the engine until the upper radiator hose gets hot.
5. Drain the system again.
6. Repeat this process until the drained water is clear and free of scale.
7. Close all petcocks and connect all the hoses.
8. If equipped with a coolant recovery system, flush the reservoir with water and leave empty.
9. Determine the capacity of your coolant system (see capacities specifications). Add a 50/50 mix of quality antifreeze (ethylene glycol) and water to provide the desired protection.

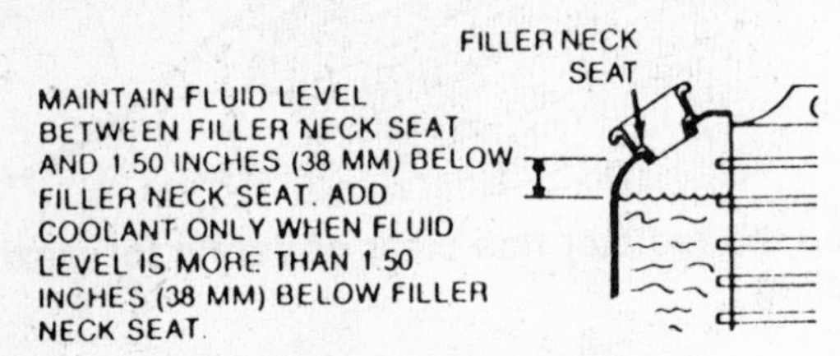

Coolant level check

Open the radiator petcock to drain the cooling system

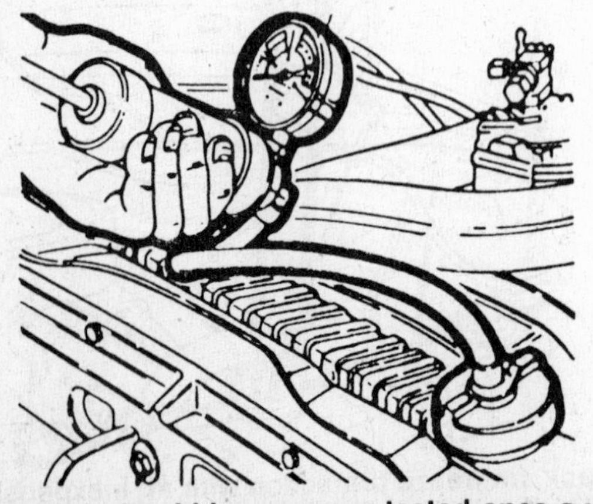

The system should be pressure tested once a year

10. Run the engine to operating temperature.
11. Stop the engine and check the coolant level.
12. Check the level of protection with an antifreeze tester, replace the cap and check for leaks.

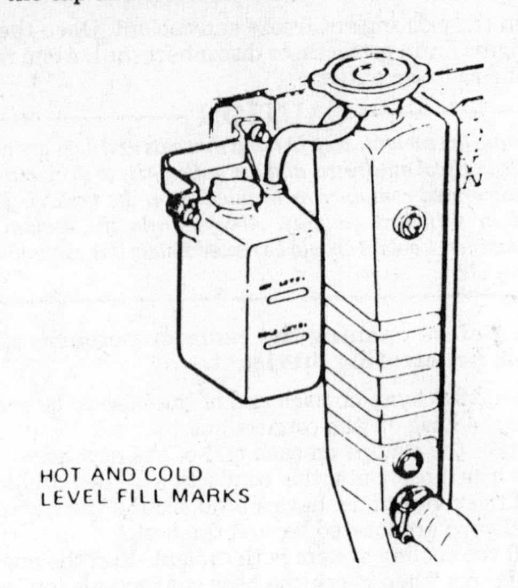

Coolant recovery system

Air Conditioning

GENERAL SERVICING PROCEDURES

The most important aspect of air conditioning service is the maintenance of pure and adequate charge of refrigerant in the system. A refrigeration system cannot function properly if a significant percentage of the charge is lost. Leaks are common because the severe vibration encountered in an automobile can easily cause a sufficient cracking or loosening of the air conditioning fittings. As a result, the extreme operating pressures of the system force refrigerant out.

The problem can be understood by considering what happens to the system as it is operated with a continuous leak. Because the expansion valve regulates the flow of refrigerant to the evaporator, the level of refrigerant there is fairly constant. The receiver/drier stores any excess of refrigerant, and so a loss will first appear there as a reduction in the level of liquid. As this level nears the bottom of the vessel, some refrigerant vapor bubbles will begin to;appear in the stream of liquid supplied to the expansion valve. This vapor decreases the capacity of the expansion valve very little as the valve opens to compensate for its presence. As the quantity of liquid in the condenser decreases, the operating pressure will drop there and throughout the high side of the system. As the R-12 continues to be expelled, the pressure available to force the liquid through the expansion valve will continue to decrease, and, eventually, the valve's orifice will prove to be too much of a restriction for adequate flow even with the needle fully withdrawn.

At this point, low side pressure will start to drop, and severe reduction in cooling capacity, marked by freeze-up of the evaporator coil, will result. Eventually, the operating pressure of the evaporator will be lower than the pressure of the atmosphere surrounding it, and air will be drawn into the system wherever there are leaks in the low side.

Because all atmospheric air contains at least some moisture, water will enter the system and mix with the R-12 and the oil. Trace amounts of moisture will cause sludging of the oil, and corrosion of the system. Saturation and clogging of the filter/drier, and freezing of the expansion valve orifice will even-

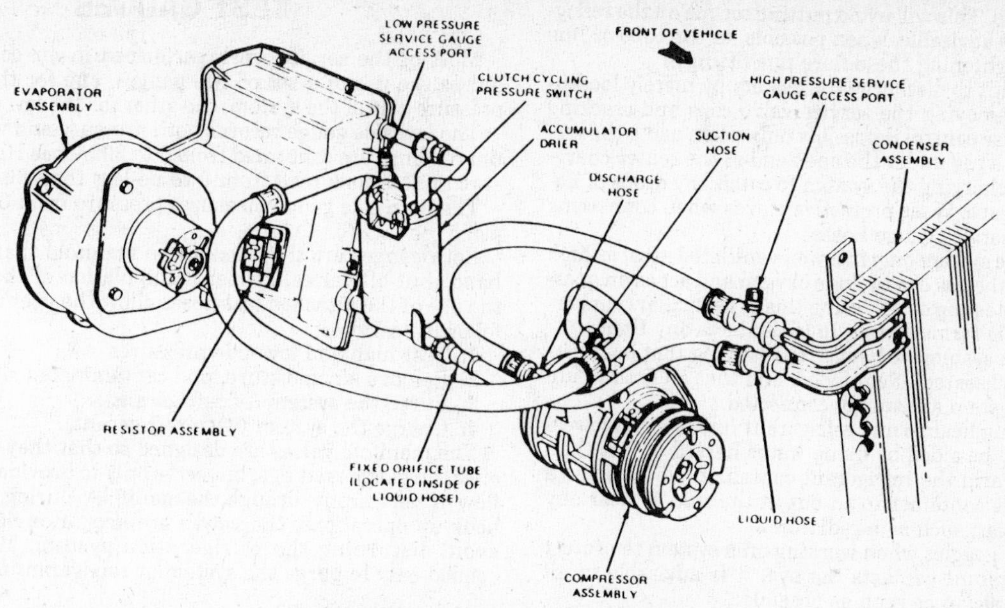

A/C system, showing all components

tually result. As air fills the system to a greater and greater extend, it will interfere more and more with the normal flows of refrigerant and heat.

A list of general precautions that should be observed while doing this follows:

1. Keep all tools as clean and dry as possible.
2. Thoroughly purge the service gauges and hoses of air and moisture before connecting them to the system. Keep them capped when not in use.
3. Thoroughly clean any refrigerant fitting before disconnecting it, in order to minimize the entrance of dirt into the system.
4. Plan any operation that requires opening the system beforehand in order to minimize the length of time it will be exposed to open air. Cap or seal the open ends to minimize the entrance of foreign material.
5. When adding oil, pour it through an extremely clean and dry tube or funnel. Keep the oil capped whenever possible. Do not use oil that has not been kept tightly sealed.
6. Use only refrigerant 12. Purchase refrigerant intended for use in only automotive air conditioning system. Avoid the use of refrigerant 12 that may be packaged for another use, such as cleaning, or powering a horn, as it is impure.
7. Completely evacuate any system that has been opened to replace a component, other than when isolating the compressor, or that has leaked sufficiently to draw in moisture and air. This requires evacuating air and moisture with a good vacuum pump for at least one hour.

If a system has been open for a considerable length of time it may be advisable to evacuate the system for up to 12 hours (overnight).

8. Use a wrench on both halves of a fitting that is to be disconnected, so as to avoid placing torque on any of the refrigerant lines.

ADDITIONAL PREVENTIVE MAINTENANCE CHECKS

Antifreeze

In order to prevent heater core freeze-up during A/C operation, it is necessary to maintain permanent type antifreeze protection of +15°F (−9°C) or lower. A reading of −15°F (−26°C) is ideal since this protection also supplies sufficient corrosion inhibitors for the protection of the engine cooling system.

WARNING: Do not use antifreeze longer than specified by the manufacturer.

Radiator Cap

For efficient operation of an air conditioned cooling system, the radiator cap should have a holding pressure which meets manufacturer's specifications. A cap which fails to hold these pressure should be replaced.

Condenser

Any obstruction of or damage to the condenser configuration will restrict the air flow which is essential to its efficient operation. It is therefore, a good rule to keep this unit clean and in proper physical shape.

NOTE: Bug screens are regarded as obstructions.

Condensation Drain Tube

This single molded drain tube expels the condensation, which accumulates on the bottom of the evaporator housing, into the engine compartment.

If this tube is obstructed, the air conditioning performance can be restricted and condensation buildup can spill over onto the vehicle's floor.

SAFETY PRECAUTIONS

Because of the importance of the necessary safety precautions that must be exercised when working with air conditioning systems and R-12 refrigerant, a recap of the safety precautions are outlined.

1. Avoid contact with a charged refrigeration system, even when working on another part of the air conditioning system or vehicle. If a heavy tool comes into contact with a section of copper tubing or a heat exchanger, it can easily cause the relatively soft material to rupture.
2. When it is necessary to apply force to a fitting which contains refrigerant, as when checking that all system couplings are securely tightened, use a wrench on both parts of the fitting

involved, if possible. This will avoid putting torque on the refrigerant tubing. (It is advisable, when possible, to use tube or line wrenches when tightening these flare nut fittings.)

3. Do not attempt to discharge the system by merely loosening a fitting, or removing the service valve caps and cracking these valves. Precise control is possibly only when using the service gauges. Place a rag under the open end of the center charging hose while discharging the system to catch any drops of liquid that might escape. Wear protective gloves when connecting or disconnecting service gauge hoses.

4. Discharge the system only in a well ventilated area, as high concentrations of the gas can exclude oxygen and act as an anesthetic. When leak testing or soldering this is particularly important, as toxic gas is formed when R-12 contacts any flame.

5. Never start a system without first verifying that both service valves are backseated, if equipped, and that all fittings are throughout the system are snugly connected.

6. Avoid applying heat to any refrigerant line or storage vessel. Charging may be aided by using water heated to less than 125°F (52°C) to warm the refrigerant container. Never allow a refrigerant storage container to sit out in the sun, or near any other source of heat, such as a radiator.

7. Always wear goggles when working on a system to protect the eyes. If refrigerant contacts the eye, it is advisable in all cases to see a physician as soon as possible.

8. Frostbite from liquid refrigerant should be treated by first gradually warming the area with cool water, and then gently applying petroleum jelly. A physician should be consulted.

9. Always keep refrigerant can fittings capped when not in use. Avoid sudden shock to the can which might occur from dropping it, or from banging a heavy tool against it. Never carry a refrigerant can in the passenger compartment of a vehicle.

10. Always completely discharge the system before painting the vehicle (if the paint is to be baked on), or before welding anywhere near the refrigerant lines.

TEST GAUGES

Most of the service work performed in air conditioning requires the use of a set of two gauges, one for the high (head) pressure side of the system, the other for the low (suction) side.

The low side gauge records both pressure and vacuum. Vacuum readings are calibrated from 0 to 30 inches Hg and the pressure graduations read from 0 to no less than 60 psi.

The high side gauge measures pressure from 0 to at last 600 psi.

Both gauges are threaded into a manifold that contains two hand shut-off valves. Proper manipulation of these valves and the use of the attached test hoses allow the user to perform the following services:

1. Test high and low side pressures.
2. Remove air, moisture, and contaminated refrigerant.
3. Purge the system (of refrigerant).
4. Charge the system (with refrigerant).

The manifold valves are designed so that they have no direct effect on gauge readings, but serve only to provide for, or cut off, flow of refrigerant through the manifold. During all testing and hook-up operations, the valves are kept in a close position to avoid disturbing the refrigeration system. The valves are opened only to purge the system or refrigerant or to charge it.

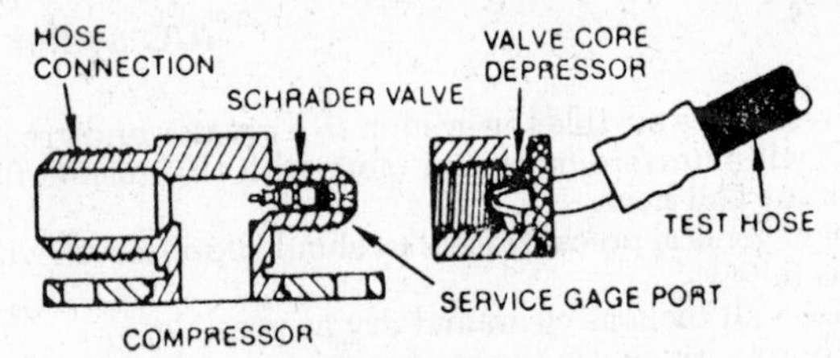

Schrader valve

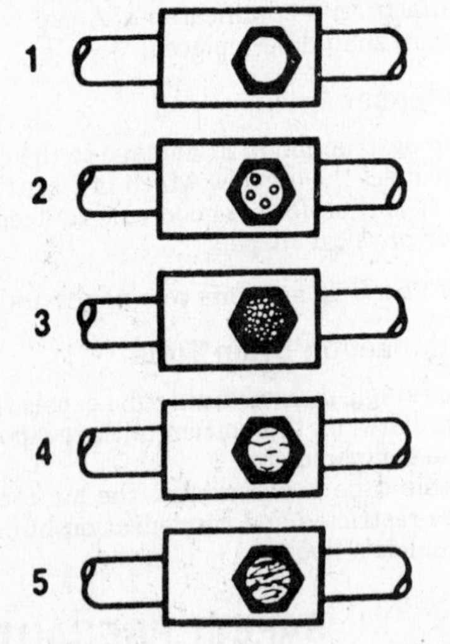

1 Clear sight glass – system correctly charged or overcharged

2 Occasional bubbles – refrigerant charge slightly low

3 Oil streaks on sight glass – total lack of refrigerant

4 Heavy stream of bubbles – serious shortage of refrigerant

5 Dark or clouded sight glass – contaminent present

Sight glass inspection

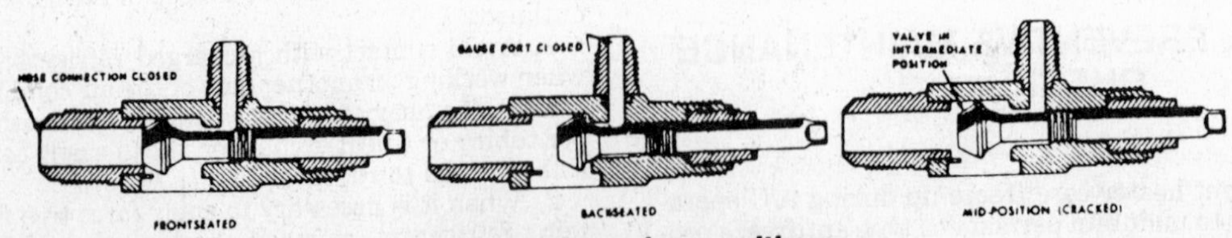

Manual service positions

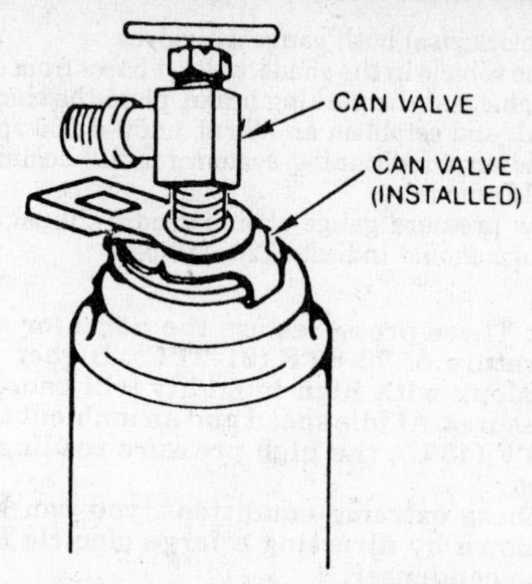

One pound R-12 can with opener valve connected

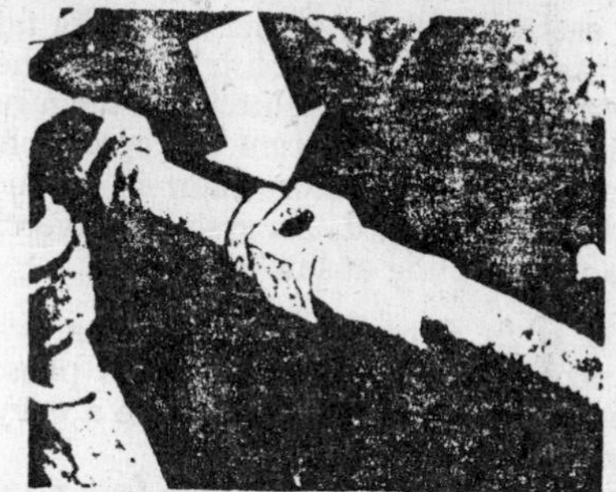

Some sight glasses are in the A/C lines

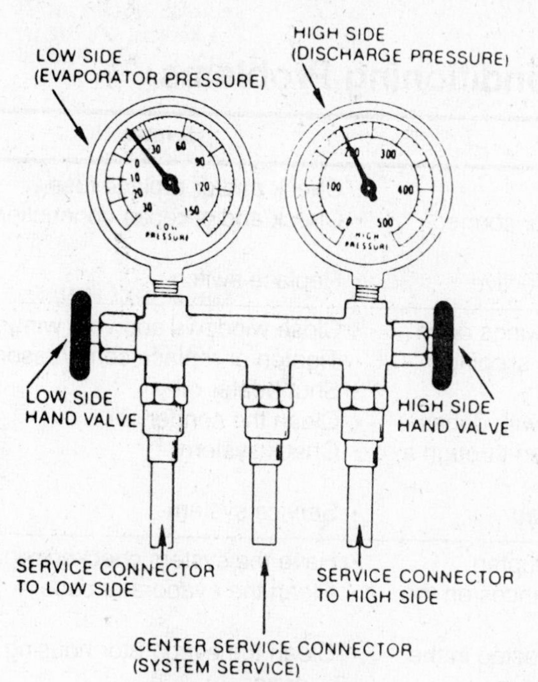

Manifold gauge set

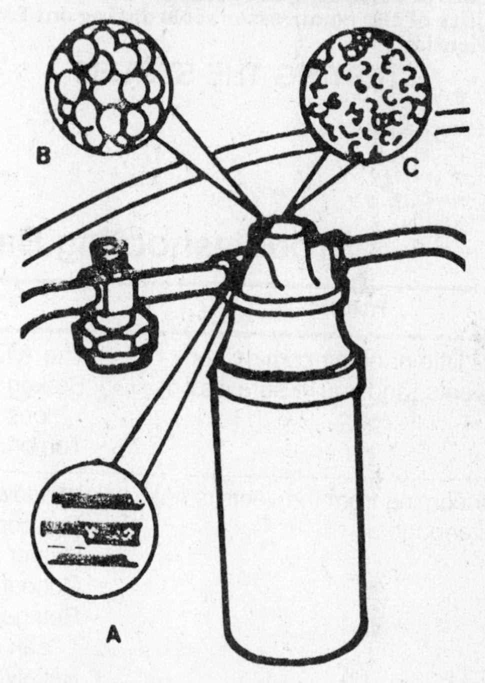

Oil streaks (A), constant bubbles (B) or foam (C) indicate there is not enough refrigerant in the system. Occasional bubbles during initial operation is normal. A clear sight glass indicates a proper charge of refrigerant or no refrigerant at all, which can be determined by the presence of cold air at the outlets in the car. If the glass is clouded with a milky white substance, have the receiver/drier checked professionally

INSPECTION

CAUTION

The compressed refrigerant used in the air conditioning system expands into the atmosphere at a temperature of −21.7°F (−30°C) or lower. This will freeze any surface, including your eyes, that it contacts. In addition, the refrigerant decomposes into a poisonous gas in the presence of a flame. Do not open or disconnect any part of the air conditioning system.

Sight Glass Check

You can safely make a few simple checks to determine if your air conditioning system needs service. The tests work best if the temperature is warm (about 70°F [21.1°C]).

NOTE: If your vehicle is equipped with an aftermarket air conditioner, the following system check may not apply. You should contact the manufacturer of the unit for instructions on systems checks.

1. Place the automatic transmission in Park or the manual transmission in Neutral. Set the parking brake.
2. Run the engine at a fast idle (about 1,500 rpm) either with the help of a friend or by temporarily readjusting the idle speed screw.
3. Set the controls for maximum cold with the blower on High.
4. Locate the sight glass in one of the system lines. Usually it is on the left alongside the top of the radiator.
5. If you see bubbles, the system must be recharged. Very likely there is a leak at some point.
6. If there are no bubbles, there is either no refrigerant at all or the system is fully charged. Feel the two hoses going to the belt driven compressor. If they are both at the same temperature, the system is empty and must be recharged.
7. If one hose (high pressure) is warm and the other (low pressure) is cold, the system may be all right. However, you are probably making these tests because you think there is something wrong, so proceed to the next step.

8. Have an assistant in the vehicle turn the fan control on and off to operate the compressor clutch. Watch the sight glass.
9. If bubbles appear when the clutch is disengaged and disappear when it is engaged, the system is properly charged.
10. If the refrigerant takes more than 45 seconds to bubble when the clutch is disengaged, the system is overcharged. This usually causes poor cooling at low speeds.

WARNING: If it is determined that the system has a leak, it should be corrected as soon as possible. Leaks may allow moisture to enter and cause a very expensive rust problem.

Exercise the air conditioner for a few minutes, every two weeks or so, during the cold months. This avoids the possibility of the compressor seals drying out from lack of lubrication.

TESTING THE SYSTEM

1. Connect a gauge set.
2. Close (clockwise) both gauge set valves.
4. Park the vehicle in the shade, at least 5 feet from any walls. Start the engine, set the parking brake, place the transmission in NEUTRAL and establish an idle of 1,100–1,300 rpm.
5. Run the air conditioning system for full cooling, in the MAX or COLD mode.
6. The low pressure gauge should read 5–20 psi; the high pressure gauge should indicate 120–180 psi.

WARNING: These pressures are the norm for an ambient temperature of 70–80°F (21–27°C). Higher air temperatures along with high humidity will cause higher system pressures. At idle speed and an ambient temperature of 110°F (43°C), the high pressure reading can exceed 300 psi.

Under these extreme conditions, you can keep the pressures down by directing a large electric floor fan through the condenser.

Troubleshooting Basic Air Conditioning Problems

Problem	Cause	Solution
There's little or no air coming from the vents (and you're sure it's on)	· The A/C fuse is blown · Broken or loose wires or connections · The on/off switch is defective	· Check and/or replace fuse · Check and/or repair connections · Replace switch
The air coming from the vents is not cool enough	· Windows and air vent wings open · The compressor belt is slipping · Heater is on · Condenser is clogged with debris · Refrigerant has escaped through a leak in the system · Receiver/drier is plugged	· Close windows and vent wings · Tighten or replace compressor belt · Shut heater off · Clean the condenser · Check system · Service system
The air has an odor	· Vacuum system is disrupted · Odor producing substances on the evaporator case · Condensation has collected in the bottom of the evaporator housing	· Have the system checked/repaired · Clean the evaporator case · Clean the evaporator housing drains
System is noisy or vibrating	· Compressor belt or mountings loose · Air in the system	· Tighten or replace belt; tighten mounting bolts · Have the system serviced
Sight glass condition		
Constant bubbles, foam or oil streaks	· Undercharged system	· Charge the system
Clear sight glass, but no cold air	· No refrigerant at all	· Check and charge the system
Clear sight glass, but air is cold	· System is OK	
Clouded with milky fluid	· Receiver drier is leaking dessicant	· Have system checked
Large difference in temperature of lines	· System undercharged	· Charge and leak test the system
Compressor noise	· Broken valves · Overcharged	· Replace the valve plate · Discharge, evacuate and install the correct charge

Troubleshooting Basic Air Conditioning Problems (cont.)

Problem	Cause	Solution
Compressor noise	· Incorrect oil level	· Isolate the compressor and check the oil level. Correct as necessary.
	· Piston slap	· Replace the compressor
	· Broken rings	· Replace the compressor
	· Drive belt pulley bolts are loose	· Tighten with the correct torque specification
Excessive vibration	· Incorrect belt tension	· Adjust the belt tension
	· Clutch loose	· Tighten the clutch
	· Overcharged	· Discharge, evacuate and install the correct charge
	· Pulley is misaligned	· Align the pulley
Condensation dripping in the passenger compartment	· Drain hose plugged or improperly positioned	· Clean the drain hose and check for proper installation
	· Insulation removed or improperly installed	· Replace the insulation on the expansion valve and hoses
Frozen evaporator coil	· Faulty thermostat	· Replace the thermostat
	· Thermostat capillary tube improperly installed	· Install the capillary tube correctly
	· Thermostat not adjusted properly	· Adjust the thermostat
Low side low—high side low	· System refrigerant is low	· Evacuate, leak test and charge the system
	· Expansion valve is restricted	· Replace the expansion valve
Low side high—high side low	· Internal leak in the compressor—worn	· Remove the compressor cylinder head and inspect the compressor. Replace the valve plate assembly if necessary. If the compressor pistons, rings or
Low side high—high side low (cont.)		cylinders are excessively worn or scored replace the compressor
	· Cylinder head gasket is leaking	· Install a replacement cylinder head gasket
	· Expansion valve is defective	· Replace the expansion valve
	· Drive belt slipping	· Adjust the belt tension
Low side high—high side high	· Condenser fins obstructed	· Clean the condenser fins
	· Air in the system	· Evacuate, leak test and charge the system
	· Expansion valve is defective	· Replace the expansion valve
	· Loose or worn fan belts	· Adjust or replace the belts as necessary
Low side low—high side high	· Expansion valve is defective	· Replace the expansion valve
	· Restriction in the refrigerant hose	· Check the hose for kinks—replace if necessary
Low side low—high side high	· Restriction in the receiver/drier	· Replace the receiver/drier
	· Restriction in the condenser	· Replace the condenser

Troubleshooting Basic Air Conditioning Problems (cont.)

Problem	Cause	Solution
Low side and high normal (inadequate cooling)	• Air in the system	• Evacuate, leak test and charge the system
	• Moisture in the system	• Evacuate, leak test and charge the system

DISCHARGING THE SYSTEM

1. Remove the caps from the high and low pressure charging valves in the high and low pressure lines.
2. Turn both manifold gauge set hand valves to the fully closed (clockwise) position.
3. Connect the manifold gauge set.
4. If the gauge set hoses do not have the gauge port actuating pins, install fitting adapters T71P-19703-S and R on the manifold gauge set hoses. If the vehicle does not have a service access gauge port valve, connect the gauge set low pressure hose to the evaporator service access gauge port valve. A special adapter, T77L-19703-A, is required to attach the manifold gauge set to the high pressure service access gauge port valve.
5. Place the end of the center hose away from you and the vehicle.
6. Open the low pressure gauge valve slightly and allow the system pressure to bleed off.
7. When the system is just about empty, open the high pressure valve very slowly to avoid losing an excessive amount of refrigerant oil. Allow any remaining refrigerant to escape.

EVACUATING THE SYSTEM

NOTE: This procedure requires the use of a vacuum pump.

1. Connect the manifold gauge set.
2. Discharge the system.
3. Make sure that the low pressure gauge set hose is connected to the low pressure service gauge port on the top center of the accumulator/drier assembly and the high pressure hose connected to the high pressure service gauge port on the compressor discharge line.
4. Connect the center service hose to the inlet fitting of the vacuum pump.
5. Turn both gauge set valves to the wide open position.
6. Start the pump and note the low side gauge reading.
7. Operate the pump until the low pressure gauge reads 25–30 in.Hg. Continue running the vacuum pump for 10 minutes more. If you've replaced some component in the system, run the pump for an additional 20–30 minutes.
8. Leak test the system. Close both gauge set valves. Turn off the pump. The needle should remain stationary at the point at which the pump was turned off. If the needle drops to zero rapidly, there is a leak in the system which must be repaired.

LEAK TESTING

Some leak tests can be performed with a soapy water solution. There must be at least a ½ lb. charge in the system for a leak to be detected. The most extensive leak tests are performed with either a Halide flame type leak tester or the more preferable electronic leak tester.

In either case, the equipment is expensive, and, the use of a Halide detector can be **extremely** hazardous!

CHARGING THE SYSTEM

CAUTION

NEVER OPEN THE HIGH PRESSURE SIDE WITH A CAN OF REFRIGERANT CONNECTED TO THE SYSTEM! OPENING THE HIGH PRESSURE SIDE WILL OVER PRESSURIZE THE CAN, CAUSING IT TO EXPLODE!

1. Connect the gauge set.
2. Close (clockwise) both gauge set valves.
3. Connect the center hose to the refrigerant can opener valve.
4. Make sure the can opener valve is closed, that is, the needle is raised, and connect the valve to the can. Open the valve, puncturing the can with the needle.
5. Loosen the center hose fitting at the pressure gauge, allowing refrigerant to purge the hose of air. When the air is bled, tighten the fitting.

CAUTION

IF THE LOW PRESSURE GAUGE SET HOSE IS NOT CONNECTED TO THE ACCUMULATOR/DRIER, KEEP THE CAN IN AN UPRIGHT POSITION!

6. Disconnect the wire harness snap-lock connector from the clutch cycling pressure switch and install a jumper wire across the two terminals of the connector.
7. Open the low side gauge set valve and the can valve.
8. Allow refrigerant to be drawn into the system.
9. When no more refrigerant is drawn into the system, start the engine and run it at about 1,500 rpm. Turn on the system and operate it at the full high position. The compressor will operate and pull refrigerant gas into the system.

NOTE: To help speed the process, the can may be placed, upright, in a pan of warm water, not exceeding 125°F (52°C).

10. If more than one can of refrigerant is needed, close the can valve and gauge set low side valve when the can is empty and connect a new can to the opener. Repeat the charging process until the sight glass indicates a full charge. The frost line on the outside of the can will indicate what portion of the can has been used.

CAUTION

NEVER ALLOW THE HIGH PRESSURE SIDE READING TO EXCEED 240 psi!

11. When the charging process has been completed, close the gauge set valve and can valve. Remove the jumper wire and reconnect the cycling clutch wire. Run the system for at least five minutes to allow it to normalize. Low pressure side reading should be 4–25 psi; high pressure reading should be 120–210 psi at an ambient temperature of 70–90°F (21–32°C).
12. Loosen both service hoses at the gauges to allow any refrigerant to escape. Remove the gauge set and install the dust caps on the service valves.

NOTE: Multi-can dispensers are available which allow a simultaneous hook-up of up to four 1 lb. cans of R-12.

CAUTION

Never exceed the recommended maximum charge for the system! The maximum charge for systems is 3 lb.

Windshield Wipers

Intense heat from the sun, snow and ice, road oils and the chemicals used in windshield washer solvents combine to deteriorate the rubber wiper refills. The refills should be replaced about twice a year or whenever the blades begin to streak or chatter.

WIPER REFILL REPLACEMENT

Normally, if the wipers are not cleaning the windshield properly, only the refill has to be replaced. The blade and arm usually require replacement only in the event of damage. It is not necessary (except on new Tridon refills) to remove the arm or the blade to replace the refill (rubber part), though you may have to position the arm higher on the glass. You can do this turning the ignition switch **ON** and operating the wipers. When they are positioned where they are accessible, turn the ignition switch **OFF**.

There are several types of refills and your vehicle could have any kind, since aftermarket blades and arms may not use exactly the same refill as the original equipment.

Most Anco styles use a release button that is pushed down to allow the refill to slide out of the yoke jaws. The new refills slide in and locks in place. Some Anco refills are removed by locating where the metal backing strip or the refill is wider. Insert a small screwdriver blade between the frame and metal backing strip. Press down to release the refill from the retaining tab.

The Trico style is unlocked at one end by squeezing 2 metal tabs, and the refill is slid out of the frame jaws. When the new refill is installed, the tabs will click into place, locking the refill.

The polycarbonate type is held in place by a locking lever that is pushed downward out of the groove in the arm to free the refill. When the new refill is installed, it will lock in place automatically.

The Tridon refill has a plastic backing strip with a notch about 1 in. from the end. Hold the blade (frame) on a hard surface so that the frame is tightly bowed. Grip the tip of the backing strip and pull up while twisting counterclockwise. The backing strip will snap out of the retaining tab. Do this for the remaining tabs until the refill is free of the arm. The length of these refills is molded into the end and they should be replaced with identical types.

No matter which type of refill you use, be sure that all of the frame claws engage the refill. Before operating the wipers, be sure that no part of the metal frame is contacting the windshield.

WIPER ARM REPLACEMENT

To remove the arm and blade assembly, raise the blade end of the arm off of the windshield and move the slide latch away from the pivot shaft. The wiper arm can now be removed from the shaft without the use of any tools.

To install, push the main head over the pivot shaft. Be sure the wipers are in the pared position, and the blade assembly is in its correct position. Hold the main arm head onto the pivot shaft while raising the blade end of the wiper arm and push the slide latch into the lock under the pivot shaft head. Then, lower the blade to the windshield. If the blade does not lower to the windshield, the slide latch is not completely in place.

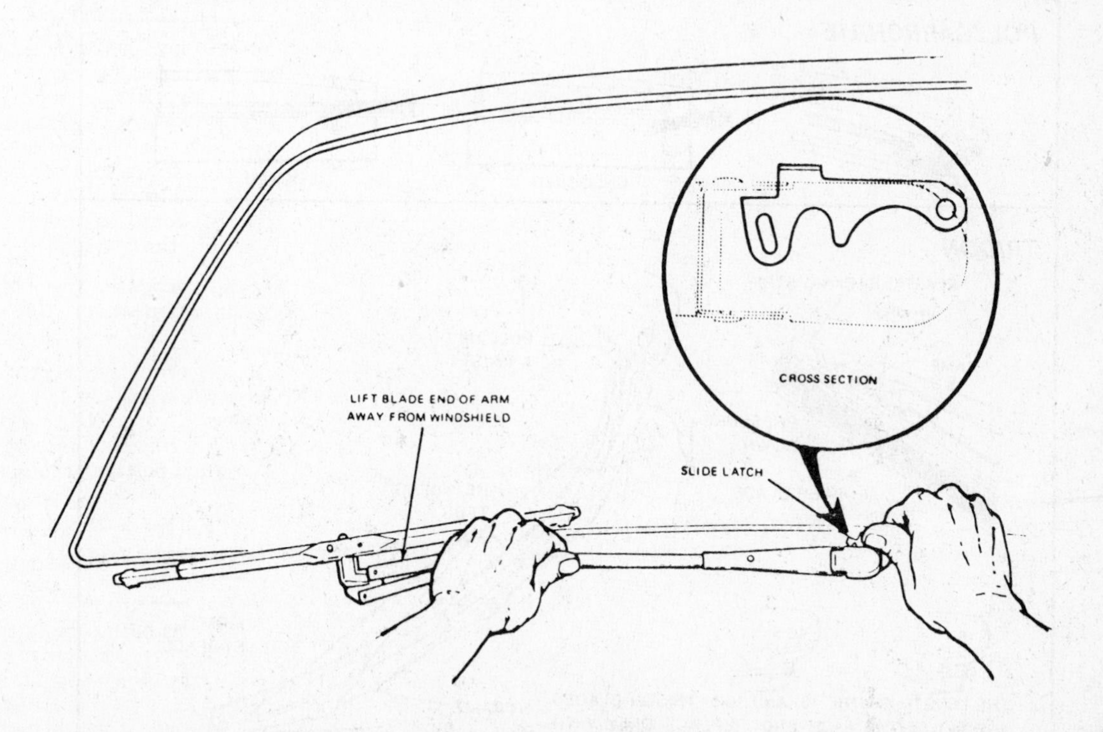

Wiper arm replacement

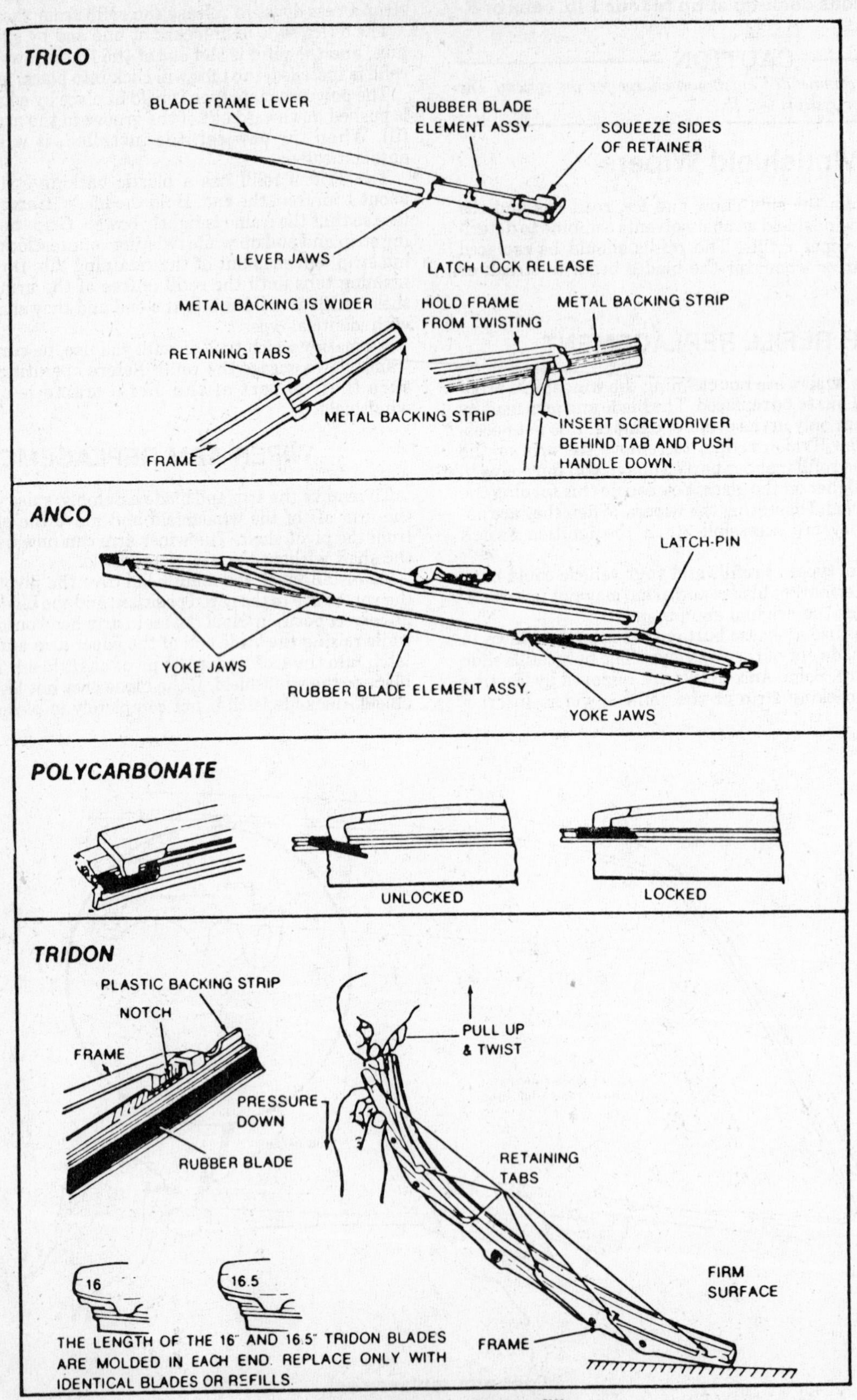

Popular styles of wiper refills

Tires and Wheels

The tires should be rotated as specified in the Maintenance Intervals Chart. Refer to the accompanying illustrations for the recommended rotation patterns.

The tires on your vehicle should have built-in tread wear indicators, which appear as ½ in. (12.7mm) bands when the tread depth gets as low as $^{1}/_{16}$ in. (1.5mm). When the indicators appear in 2 or more adjacent grooves, it's time for new tires.

For optimum tire life, you should keep the tires properly inflated, rotate them often and have the wheel alignment checked periodically.

Some late models have the maximum load pressures listed in the V.I.N. plate on the left door frame. In general, pressure of 28–32 psi would be suitable for highway use with moderate loads and passenger truck type tires (load range B, non-flotation) of original equipment size. Pressures should be checked before driving, since pressure can increase as much as 6 psi due to heat. It is a good idea to have an accurate gauge and to check pressures weekly. Not all gauges on service station air pumps are to be trusted. In general, truck type tires require higher pressures and flotation type tires, lower pressures.

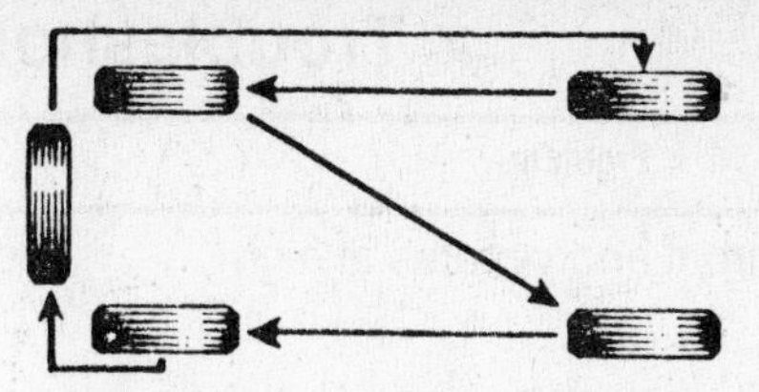

Bias-ply tire rotation diagram

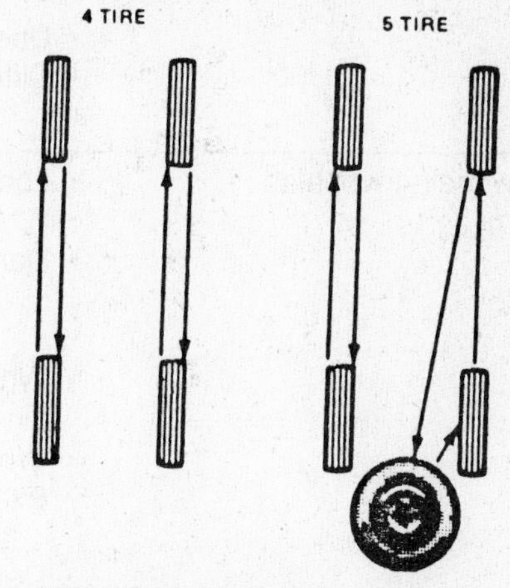

Radial tire rotation diagram

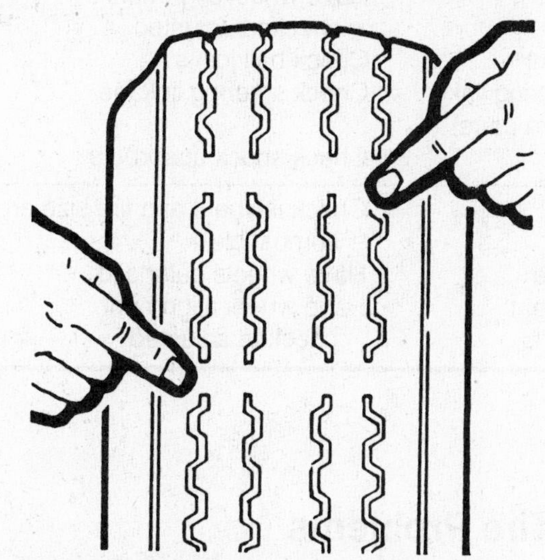

Tread wear indicators are built into tires

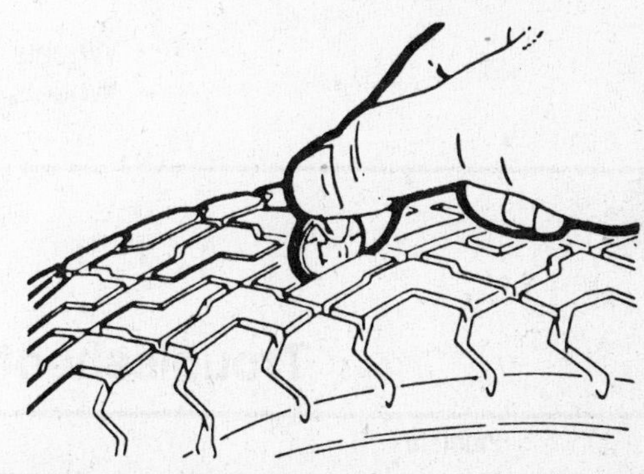

Tread depth can be checked with a penny; when the top of Lincoln's head is visible, it's time for new tires

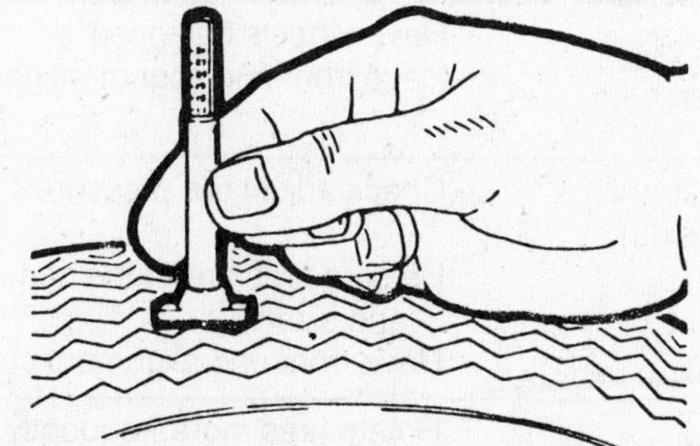

Checking tread depth with an inexpensive depth gauge

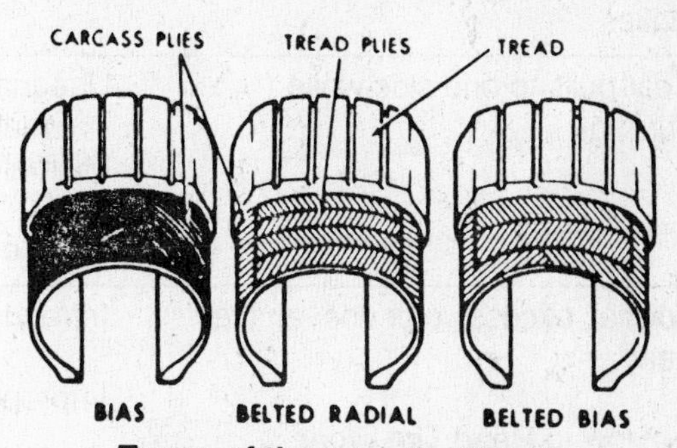

Types of tire construction

TIRE ROTATION

It is recommended that you have the tires rotated every 6,000 miles. There is no way to give a tire rotation diagram for every combination of tires and vehicles, but the accompanying diagrams are a general rule to follow. Radial tires should not be cross-switched; they last longer if their direction of rotation is not changed. Truck tires sometimes have directional tread, indicated by arrows on the sidewalls; the arrow shows the direction of rotation. They will wear very rapidly if reversed. Studded snow tires will lose their studs if their direction of rotation is reversed.

Troubleshooting Basic Wheel Problems

Problem	Cause	Solution
The car's front end vibrates at high speed	• The wheels are out of balance • Wheels are out of alignment	• Have wheels balanced • Have wheel alignment checked/adjusted
Car pulls to either side	• Wheels are out of alignment • Unequal tire pressure • Different size tires or wheels	• Have wheel alignment checked/adjusted • Check/adjust tire pressure • Change tires or wheels to same size
The car's wheel(s) wobbles	• Loose wheel lug nuts • Wheels out of balance • Damaged wheel • Wheels are out of alignment • Worn or damaged ball joint • Excessive play in the steering linkage (usually due to worn parts) • Defective shock absorber	• Tighten wheel lug nuts • Have tires balanced • Raise car and spin the wheel. If the wheel is bent, it should be replaced • Have wheel alignment checked/adjusted • Check ball joints • Check steering linkage • Check shock absorbers
Tires wear unevenly or prematurely	• Incorrect wheel size • Wheels are out of balance • Wheels are out of alignment	• Check if wheel and tire size are compatible • Have wheels balanced • Have wheel alignment checked/adjusted

Troubleshooting Basic Tire Problems

Problem	Cause	Solution
The car's front end vibrates at high speeds and the steering wheel shakes	• Wheels out of balance • Front end needs aligning	• Have wheels balanced • Have front end alignment checked
The car pulls to one side while cruising	• Unequal tire pressure (car will usually pull to the low side) • Mismatched tires • Front end needs aligning	• Check/adjust tire pressure • Be sure tires are of the same type and size • Have front end alignment checked
Abnormal, excessive or uneven tire wear See "How to Read Tire Wear"	• Infrequent tire rotation • Improper tire pressure • Sudden stops/starts or high speed on curves	• Rotate tires more frequently to equalize wear • Check/adjust pressure • Correct driving habits
Tire squeals	• Improper tire pressure • Front end needs aligning	• Check/adjust tire pressure • Have front end alignment checked

Tire Size Comparison Chart

"Letter" sizes			Inch Sizes	Metric-inch Sizes		
"60 Series"	"70 Series"	"78 Series"	1965–77	"60 Series"	"70 Series"	"80 Series"
			5.50-12, 5.60-12	165/60-12	165/70-12	155-12
		Y78-12	6.00-12			
		W78-13	5.20-13	165/60-13	145/70-13	135-13
		Y78-13	5.60-13	175/60-13	155/70-13	145-13
			6.15-13	185/60-13	165/70-13	155-13, P155/80-13
A60-13	A70-13	A78-13	6.40-13	195/60-13	175/70-13	165-13
B60-13	B70-13	B78-13	6.70-13	205/60-13	185/70-13	175-13
			6.90-13			
C60-13	C70-13	C78-13	7.00-13	215/60-13	195/70-13	185-13
D60-13	D70-13	D78-13	7.25-13			
E60-13	E70-13	E78-13	7.75-13			195-13
			5.20-14	165/60-14	145/70-14	135-14
			5.60-14	175/60-14	155/70-14	145-14
			5.90-14			
A60-14	A70-14	A78-14	6.15-14	185/60-14	165/70-14	155-14
	B70-14	B78-14	6.45-14	195/60-14	175/70-14	165-14
	C70-14	C78-14	6.95-14	205/60-14	185/70-14	175-14
D60-14	D70-14	D78-14				
E60-14	E70-14	E78-14	7.35-14	215/60-14	195/70-14	185-14
F60-14	F70-14	F78-14, F83-14	7.75-14	225/60-14	200/70-14	195-14
G60-14	G70-14	G77-14, G78-14	8.25-14	235/60-14	205/70-14	205-14
H60-14	H70-14	H78-14	8.55-14	245/60-14	215/70-14	215-14
J60-14	J70-14	J78-14	8.85-14	255/60-14	225/70-14	225-14
L60-14	L70-14		9.15-14	265/60-14	235/70-14	
	A70-15	A78-15	5.60-15	185/60-15	165/70-15	155-15
B60-15	B70-15	B78-15	6.35-15	195/60-15	175/70-15	165-15
C60-15	C70-15	C78-15	6.85-15	205/60-15	185/70-15	175-15
	D70-15	D78-15				
E60-15	E70-15	E78-15	7.35-15	215/60-15	195/70-15	185-15
F60-15	F70-15	F78-15	7.75-15	225/60-15	205/70-15	195-15
G60-15	G70-15	G78-15	8.15-15/8.25-15	235/60-15	215/70-15	205-15
H60-15	H70-15	H78-15	8.45-15/8.55-15	245/60-15	225/70-15	215-15
J60-15	J70-15	J78-15	8.85-15/8.90-15	255/60-15	235/70-15	225-15
	K70-15		9.00-15	265/60-15	245/70-15	230-15
L60-15	L70-15	L78-15, L84-15	9.15-15			235-15
	M70-15	M78-15				255-15
		N78-15				

NOTE: Every size tire is not listed and many size comaprisons are approximate, based on load ratings. Wider tires than those supplied new with the vehicle should always be checked for clearance

NOTE: Mark the wheel position or direction of rotation on radial tires or studded snow tires before removing them.

If your vehicle is equipped with tires having different load ratings on the front and the rear, the tires should not be rotated front to rear. Rotating these tires could affect tire life (the tires with the lower rating will wear faster, and could become overloaded), and upset the handling of the vehicle.

TIRE USAGE

The tires on your vehicle were selected to provide the best all around performance for normal operation when inflated as specified. Oversize tires (Load Range D) will not increase the maximum carrying capacity of the vehicle, although they will provide an extra margin of tread life. Be sure to check overall height before using larger size tires which may cause interference with suspension components or wheel wells. When replacing conventional tire sizes with other tire size designations, be sure to check the manufacturer's recommendations. Interchangeability is not always possible because of differences in load ratings, tire dimensions, wheel well clearances, and rim size. Also due to differences in handling characteristics, 70 Series and 60 Series tires should be used only in pairs on the same axle; radial tires should be used only in sets of four.

The wheels must be the correct width for the tire. Tire dealers have charts of tire and rim compatibility. A mismatch can cause sloppy handling and rapid tread wear. The old rule of thumb is that the tread width should match the rim width (inside bead to inside bead) within 1 in.. For radial tires, the rim width should be 80% or less of the tire (not tread) width.

The height (mounted diameter) of the new tires can greatly change speedometer accuracy, engine speed at a given road speed, fuel mileage, acceleration, and ground clearance. Tire manufacturers furnish full measurement specifications. Speedometer drive gears are available for correction.

NOTE: Dimensions of tires marked the same size may vary significantly, even among tires from the same manufacturer.

The spare tire should be usable, at least for low speed operation, with the new tires.

TIRE DESIGN

For maximum satisfaction, tires should be used in sets of five. Mixing or different types (radial, bias-belted, fiberglass belted) should be avoided. Conventional bias tires are constructed so that the cords run bead-to-bead at an angle. Alternate plies run at an opposite angle. This type of construction gives rigidity to both tread and sidewall. Bias-belted tires are similar in construction to conventional bias ply tires. Belts run at an angle and also at a 90° angle to the bead, as in the radial tire. Tread life is improved considerably over the conventional bias tire. The radial tire differs in construction, but instead of the carcass plies running at an angle of 90° to each other, they run at an angle of 90° to the bead. This gives the tread a great deal of rigidity and the sidewall a great deal of flexibility and accounts for the characteristic bulge associated with radial tires.

Radial tire are recommended for use on all Ford trucks. If they are used, tire sizes and wheel diameters should be selected to maintain ground clearance and tire load capacity equivalent to the minimum specified tire. Radial tires should always be used in sets of five, but in an emergency radial tires can be used with caution on the rear axle only. If this is done, both tires on the rear should be of radial design.

NOTE: Radial tires should never be used on only the front axle.

FLUIDS AND LUBRICANTS

Oil and Fuel Recommendations

Gasoline Engines

All Ford Ranger/Bronco II/Explorer must use lead-free gasoline.

The recommended oil viscosities for sustained temperatures ranging from below 0°F (−18°C) to above 32°F (0°C) are listed in this Section. They are broken down into multiviscosities and single viscosities. Multiviscosity oils are recommended because of their wider range of acceptable temperatures and driving conditions.

When adding oil to the crankcase or changing the oil or filter, it is important that oil of an equal quality to original equipment be used in your vehicle. The use of inferior oils may void the warranty, damage your engine, or both.

The SAE (Society of Automotive Engineers) grade number of

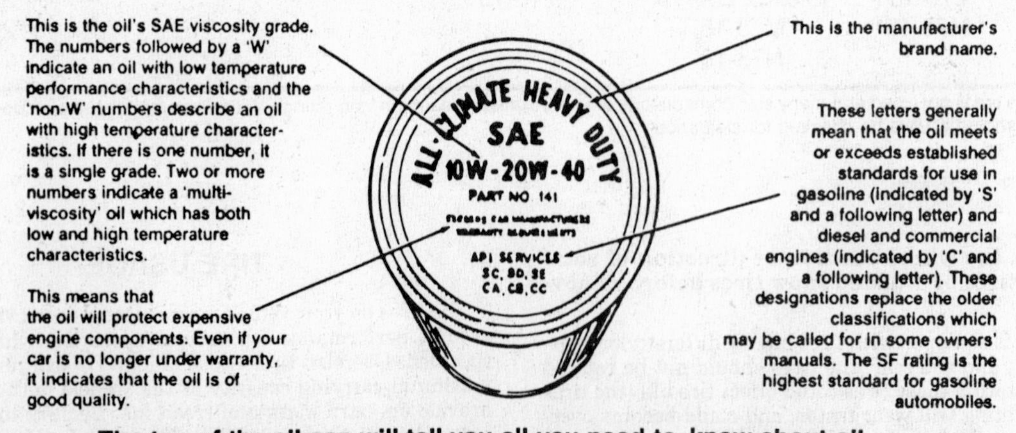

The top of the oil can will tell you all you need to know about oil

oil indicates the viscosity of the oil (its ability to lubricate at a given temperature). The lower the SAE number, the lighter the oil; the lower the viscosity, the easier it is to crank the engine in cold weather but the less the oil will lubricate and protect the engine in high temperatures. This number is marked on every oil container.

Oil viscosities should be chosen from those oils recommended for the lowest anticipated temperatures during the oil change interval. Due to the need for an oil that embodies both good lubrication at high temperatures and easy cranking in cold weather, multigrade oils have been developed. Basically, a multigrade oil is thinner at low temperatures and thicker at high temperatures. For example, a 10W–40 oil (the W stands for winter) exhibits the characteristics of a 10 weight (SAE 10) oil when the vehicle is first started and the oil is cold. Its lighter weight allows it to travel to the lubricating surfaces quicker and offer less resistance to starter motor cranking than, say, a straight 30 weight (SAE 30) oil. But after the engine reaches operating temperature, the 10W–40 oil begins acting like straight 40 weight (SAE 40) oil, its heavier weight providing greater lubrication with less chance of foaming than a straight 30 weight oil.

The API (American Petroleum Institute) designations, also found on the oil container, indicates the classification of engine oil used under certain given operating conditions. Only oils designated for use Service SG heavy duty detergent should be used in your vehicle. Oils of the SG type perform may functions inside the engine besides their basic lubrication. Through a balanced system of metallic detergents and polymeric dispersants, the oil prevents high and low temperature deposits and also keeps sludge and dirt particles in suspension. Acids, particularly sulphuric acid, as well as other by-products of engine combustion are neutralized by the oil. If these acids are allowed to concentrate, they can cause corrosion and rapid wear of the internal engine parts.

CAUTION

Non-detergent motor oils or straight mineral oils should not be used in your Ford gasoline engine.

Diesel Engines

Diesel engines require different engine oil from those used in gasoline engines. Besides doing the things gasoline engine oil does, diesel oil must also deal with increased engine heat and the diesel blow-by gases, which create sulphuric acid, a high corrosive.

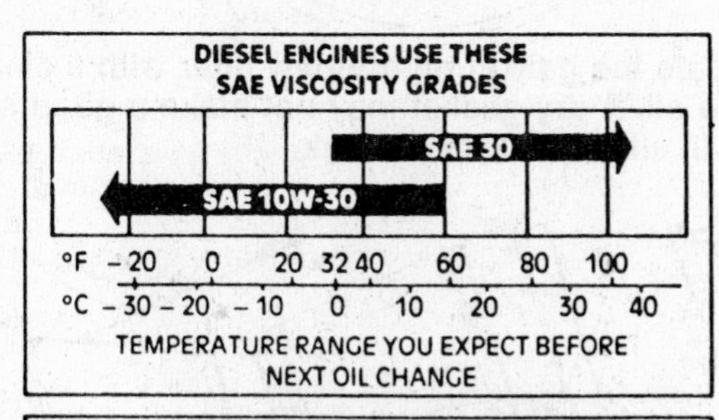

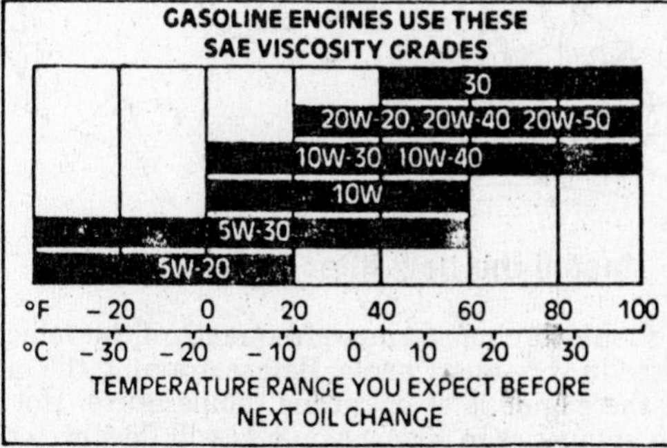

Engine oil viscosities

Under the American Petroleum Institute (API) classifications, gasoline engine oil codes begin with an **S**, and diesel engine oil codes begin with a **C**. This first letter designation is followed by a second letter code which explains what type of service (heavy, moderate, light) the oil is meant for. For example, the top of a typical oil can will include: API SERVICES SG, CD. This means the oil in the can is a superior, heavy duty engine oil when used in a diesel engine.

Many diesel manufacturers recommend an oil with both gasoline and diesel engine API classifications.

NOTE: Ford specifies the use of an engine oil conforming to API service categories of both SG and CD. DO NOT use oils labeled as only SG or only CD as they could cause engine damage.

FUEL

Fuel makers produce two grades of diesel fuel, No. 1 and No. 2, for use in automotive diesel engines. Generally speaking, No. 2 fuel is recommended over No. 1 for driving in temperatures above 20°F (−7°C). In fact, in many areas, No. 2 diesel is the only fuel available. By comparison, No. 2 diesel fuel is less volatile than No. 1 fuel, and gives better fuel economy. No. 2 fuel is also a better injection pump lubricant.

Two important characteristics of diesel fuel are its cetane number and its viscosity.

The cetane number of a diesel fuel refers to the ease with which a diesel fuel ignites. High cetane numbers mean that the fuel will ignite with relative ease or that it ignites well at low temperatures. Naturally, the lower the cetane number, the higher the temperature must be to ignite the fuel. Most commercial fuels have cetane numbers that range from 35 to 65. No. 1 diesel fuel generally has a higher cetane rating than No. 2 fuel.

Viscosity is the ability of a liquid, in this case diesel fuel, to flow. Using straight No. 2 diesel fuel below 20°F (−7°C) can cause problems, because this fuel tends to become cloudy, meaning wax crystals begin forming in the fuel. 20°F (−7°C) is often call the cloud point for No. 2 fuel. In extremely cold weather, No. 2 fuel can stop flowing altogether. In either case, fuel flow is restricted, which can result in no start condition or poor engine performance. Fuel manufacturers often winterize No. 2 diesel fuel by using various fuel additives and blends (no. 1 diesel fuel, kerosene, etc.) to lower its winter time viscosity. Generally speaking, though, No. 1 diesel fuel is more satisfactory in extremely cold weather.

NOTE: No. 1 and No. 2 diesel fuels will mix and burn with no ill effects, although the engine manufacturer recommends one or the other. Consult the owner's manual for information.

Depending on local climate, most fuel manufacturers make winterized No. 2 fuel available seasonally.

Many automobile manufacturers publish pamphlets giving the locations of diesel fuel stations nationwide. Contact the local dealer for information.

Do not substitute home heating oil for automotive diesel fuel. While in some cases, home heating oil refinement levels equal those of diesel fuel, many times they are far below diesel engine requirements. The result of using dirty home heating oil will be a clogged fuel system, in which case the entire system may have to be dismantled and cleaned.

One more word on diesel fuels. Don't thin diesel fuel with gasoline in cold weather. The lighter gasoline, which is more explosive, will cause rough running at the very least, and may cause extensive damage to the fuel system if enough is used.

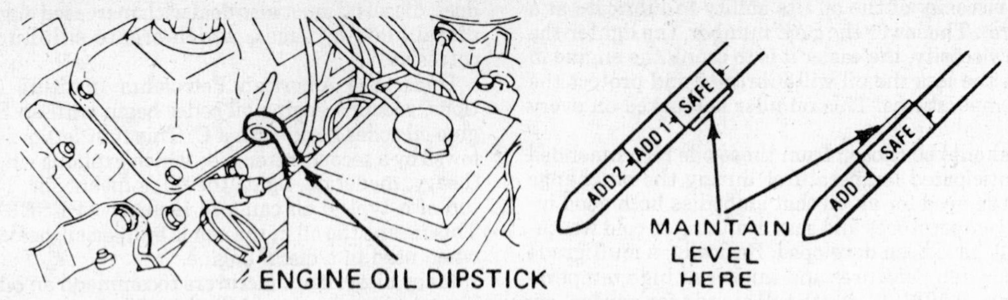

Checking engine oil level

OIL LEVEL CHECK

Check the engine oil level every time you fill the gas tank. The oil level should be above the ADD mark and not above the FULL mark on the dipstick. Make sure that the dipstick is inserted into the crankcase as far as possible and that the vehicle is resting on level ground. Also, allow a few minutes after turning off the engine for the oil to drain into the pan or an inaccurate reading will result.

1. Open the hood and remove the engine oil dipstick.
2. Wipe the dipstick with a clean, lint-free rag and reinsert it. Be sure to insert it all the way.
3. Pull out the dipstick and note the oil level. It should be between the **SAFE** (MAX) mark and the **ADD** (MIN) mark.
4. If the level is below the lower mark, replace the dipstick and add fresh oil to bring the level within the proper range. Do not overfill.
5. Recheck the oil level and close the hood.

NOTE: Use a multi-grade oil with API classification SG.

OIL AND FILTER CHANGE

NOTE: The engine oil and oil filter should be changed at the same time, at the recommended intervals on the maintenance schedule chart.

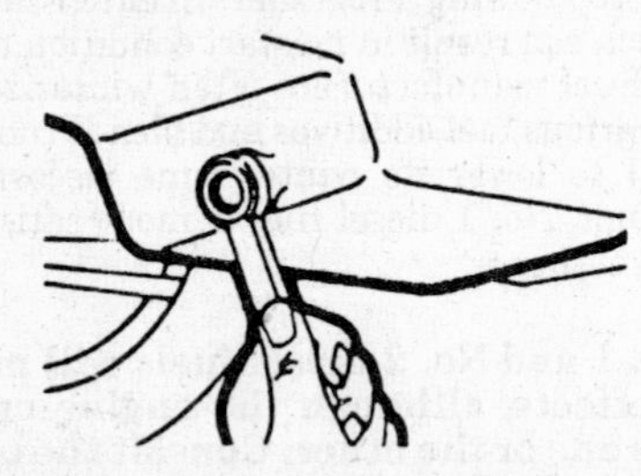

Loosen, but do not remove, the drain plug on the bottom of the oil pan. Get your drain plug ready

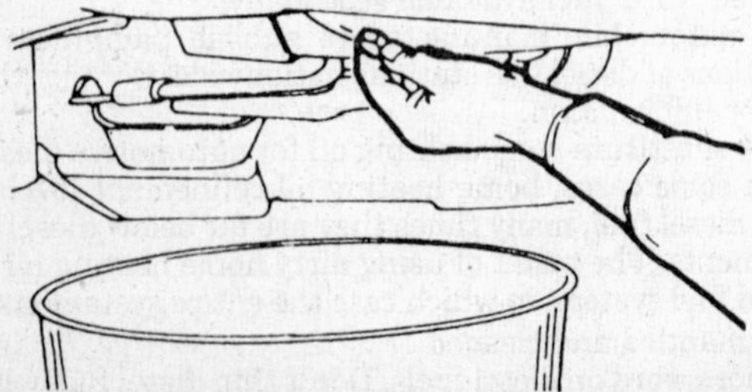

Unscrew the plug by hand. Keep an inward pressure on the plug as you unscrew it, so the oil won't escape until you pull the plug away

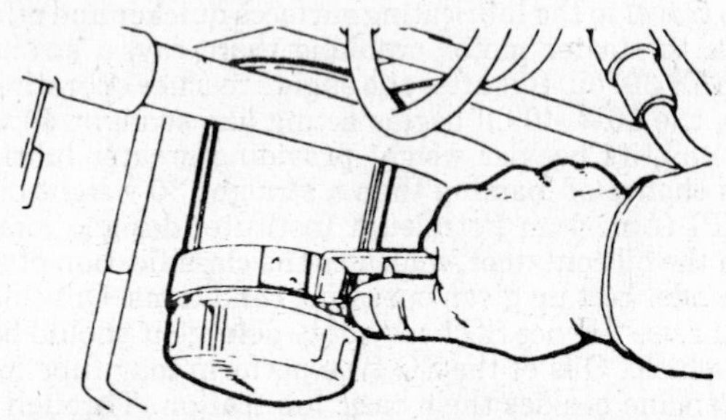

Move the drain pan underneath the oil filter. Use a strap wrench to remove the oil filter – remember it is still filled with about a quart of hot, dirty engine oil

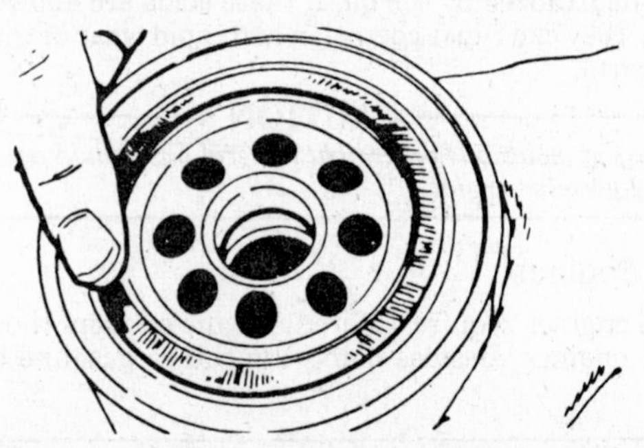

Lubricate the gasket on the new filter with a clean engine oil. A dry gasket may not make a good seal and will allow the filter to leak

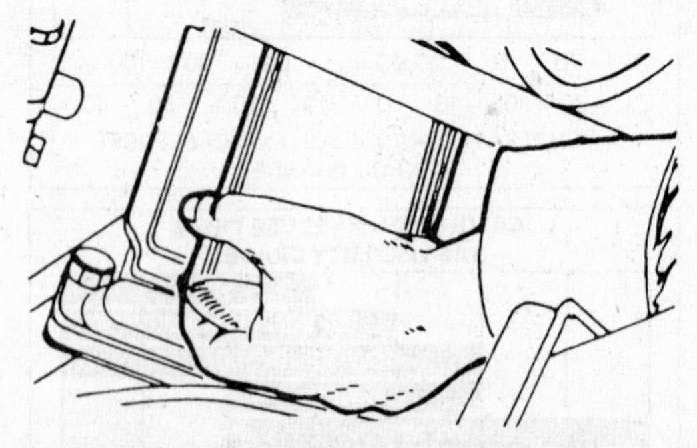

Install the new filter by hand only

The oil should be changed more frequently if the vehicle is being operated in very dusty areas. Before draining the oil, make sure that the engine is at operating temperature. Hot oil will hold more impurities in suspension and will flow better, allowing the removal of more oil and dirt.

Loosen the drain plug with a wrench, then, unscrew the plug

with your fingers, using a rag to shield your fingers from the heat. Push in on the plug as you unscrew it so you can feel when all of the screw threads are out of the hole. You can then remove the plug quickly with the minimum amount of oil running down your arm and you will also have the plug in your hand and not in the bottom of a pan of hot oil. Drain the oil into a suitable receptacle. Be careful of the oil. If it is at operating temperatures it is hot enough to burn you.

CAUTION

The EPA warns that prolonged contact with used engine oil may cause a number of skin disorders, including cancer! You should make every effort to minimize your exposure to used engine oil. Protective gloves should be worn when changing the oil. Wash your hands and any other exposed skin areas as soon as possible after exposure to used engine oil. Soap and water, or waterless hand cleaner should be used.

The oil filter SHOULD BE changed every time the oil is changed. To remove the filter, you may need an oil filter wrench since the filter may have been fitted too tightly and the heat from the engine may have made it even tighter. A filter wrench can be obtained at an auto parts store and is well worth the investment, since it will save you a lot of grief. Loosen the filter with the filter wrench. With a rag wrapped around the filter, unscrew the filter from the boss on the side of the engine. Be careful of hot oil that will run down the side of the filter. Make sure that you have a pan under the filter before you start to remove it from the engine; should some of the hot oil happen to get on you, you will have a place to dump the filter in a hurry. Wipe the base of the mounting boss with a clean, dry cloth. When you install the new filter, smear a small amount of oil on the gasket with your finger, just enough to coat the entire surface, where it comes in contact with the mounting plate. When you tighten the filter, rotate if only a half turn after it comes in contact with the mounting boss.

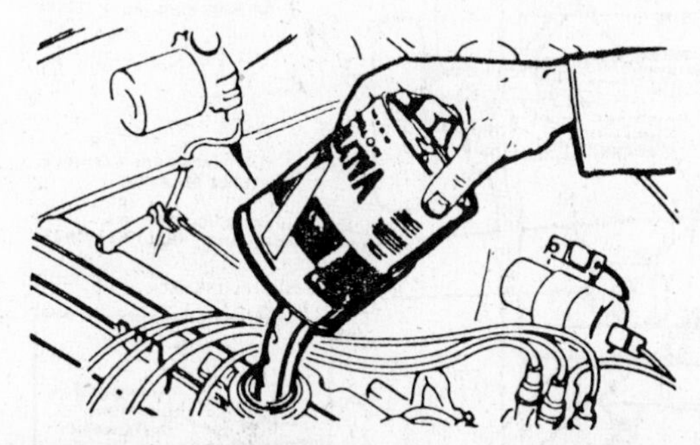

Don't forget to install the drain plug before refilling the engine with fresh oil

Manual Transmission

FLUID RECOMMENDATION

The lubricant in the transmission should be checked and changed periodically, except when the vehicle has been operated in deep water and water has entered the transmission. When this happens, change the lubricant in the transmission as soon as possible. Use Standard Transmission Lube SAE 80 in the manual transmission.

LEVEL CHECK

Before checking the lubricant level in the transmission, make sure that the vehicle is on level ground. Remove the fill plug from the transmission. Remove the plug slowly when it starts to reach the end of the threads on the plug. Hold the plug up against the hole and move it away slowly. This is to minimize the loss of lubricant through the fill hole. The level of the lubricant should be up to the bottom of the fill hole. If lubricant is not present at the bottom of the fill hole, add Standard Transmission Lube SAE 80 until it reaches the proper level. A squeeze bottle or siphon gun is used to fill a manual transmission with lubricant.

DRAIN AND REFILL

Drain and refill the transmission daily if the vehicle has been operating in water. All you have to do is remove the drain plug which is located at the bottom of the transmission. Allow all the lubricant to run out before replacing the plug. Replace the oil with the correct fluid. If you are experiencing hard shifting and the weather is very cold, use a lighter weight fluid in the transmission. If you don't have a pressure gun to install the oil, use a suction gun.

Automatic Transmission

FLUID RECOMMENDATION

Refer to the dipstick to confirm automatic transmission fluid specifications. With a C5 automatic transmission, add only Type H automatic transmission fluid. With the C3 and A4LD automatic transmissions, use only Dexron® II automatic transmission fluid.

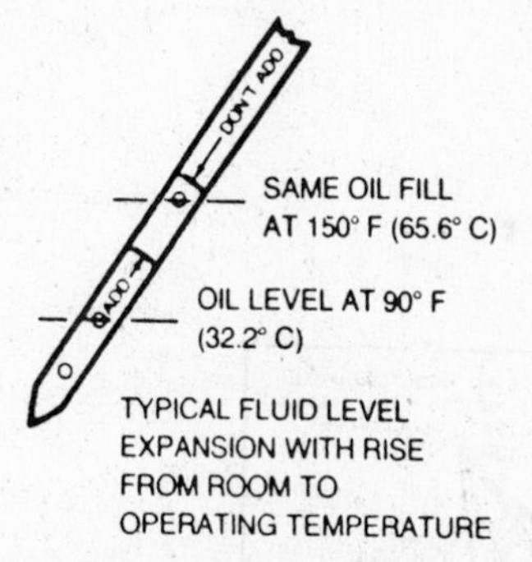

Checking automatic transmission fluid

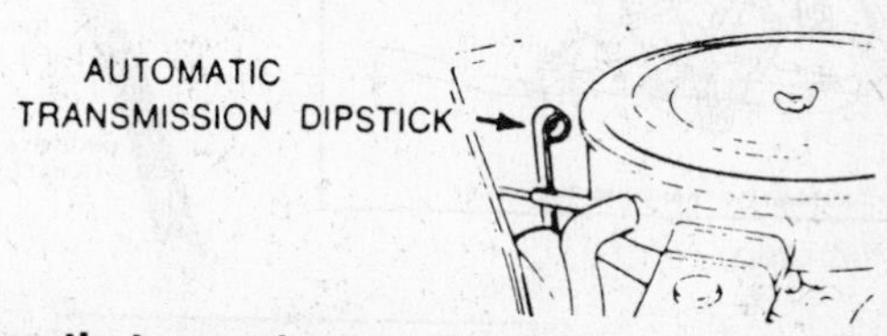

Automatic transmission dipstick is located towards the rear of the engine

LEVEL CHECK

The fluid level in an automatic transmission is checked when the transmission is at operating temperatures. If the vehicle has been sitting and is cold, drive it at highway speeds for at least 20 minutes to warm up the transmission. The transmission dipstick is located under the hood, against the firewall, on the right side.

1. With the transmission in Park, the engine running at idle speed, the foot brakes applied and the vehicle resting on level ground, move the transmission gear selector through each of

the gear positions, including Reverse, allowing time for the transmission to engage. Return the shift selector to the Park position and apply the parking brake. Do not turn the engine off, but leave it running at idle speed.

2. Clean all dirt from around the transmission dipstick cap and the end of the filler tube.

3. Pull the dipstick out of the tube, wipe it off with a clean cloth, and push it back into the tube all the way, making sure that it seats completely.

4. Pull the dipstick out of the tube again and read the level of the fluid on the stick. The level should be between the ADD and FULL marks. If fluid must be added, add enough fluid through the tube to raise the level up to between the ADD and FULL marks. Do not overfill the transmission because this will cause foaming and loss of fluid through the vent and malfunctioning of the transmission.

DRAIN AND REFILL

The transmission is filled at the factory with a high quality fluid that both transmits power and lubricates and will last a long time. In most cases, the need to change the fluid in the automatic transmission will never arise under normal use. But since this is a truck and most likely will be subjected to more severe operating conditions than a conventional vehicle, the fluid may have to be replaced. An internal leak in the radiator could develop and contaminate the fluid, necessitating fluid replacement.

The extra load of operating the vehicle in deep sand, towing a heavy trailer, etc., causes the transmission to create more heat due to increased friction. This extra heat is transferred to the transmission fluid and, if the oil is allowed to become too hot, it will change its chemical composition or become scorched. When this occurs, valve bodies become clogged and the transmission doesn't operate as efficiently as it should. Serious damage to the transmission can result.

You can tell if the transmission fluid is scorched by noting a distinctive **burned** smell and discoloration. Scorched transmission fluid is dark brown or black as opposed to its normal bright, clear red color. Since transmission fluid **cooks** in stages, it may develop forms of sludge or varnish. Pull the dipstick out and place the end on a tissue or paper towel. Particles of sludge can be seen more easily this way. If any of the above conditions do exist, the transmission fluid should be completely drained, the filtering screens cleaned, the transmission inspected for possible damage and new fluid installed. Refer to Section 7 under Automatic Transmission for Pan Removal and Filter Service Procedures.

CAUTION

Use of a fluid other than those specified could result in transmission malfunction and/or failure.

NOTE: If it is necessary to completely drain and refill the transmission, it will be necessary to remove the residual fluid from the torque converter and the cooler lines.

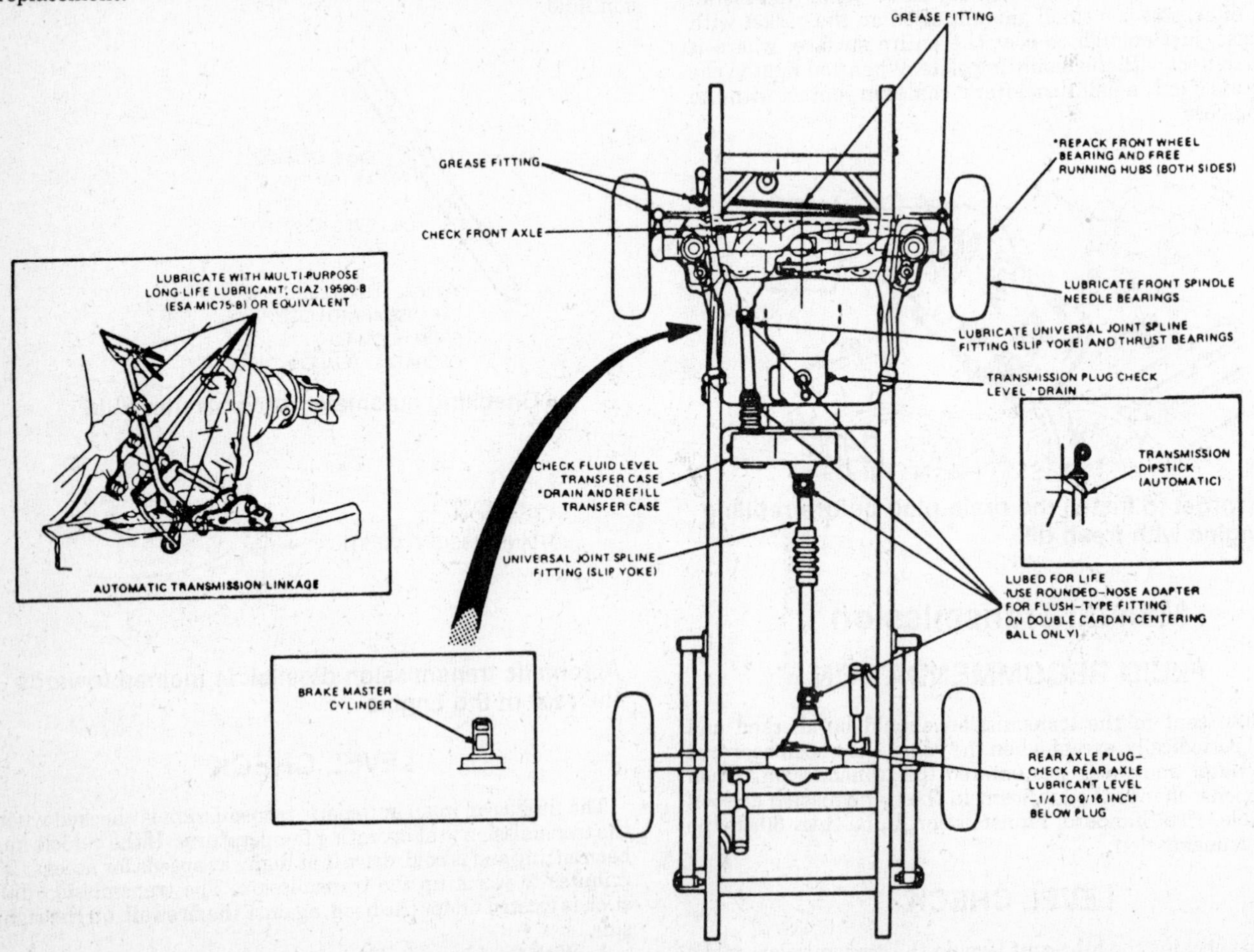

Lubrication chart — 4WD models

The procedure for partial drain and refill, for a vehicle that is in service, is as follows:

1. Place a drain pan under the transmission. Loosen the pan bolts and pull one corner down to start the fluid draining. Remove and empty the pan.
2. When all the fluid has drained from the transmission, remove and clean the pan and screen. Make sure not to leave any solvent residue or lint from the rags in the pan.
3. Install the pan with a new gasket and tighten the bolts in a crisscross pattern.
4. Add three quarts of fluid through the dipstick tube.
   With a C5 automatic transmission, add only fluid meeting Ford Specification ESP-MZC166-H, Type H automatic transmission fluid. With a C3 and A4LD automatic transmission, use only DEXRON®II automatic transmission fluid. The level should be at or below the ADD mark.
5. Check the fluid level as soon as the transmission reaches operating temperature for the first time. Make sure that the level is between ADD and FULL.

To drain the torque converter:

1. Remove the converter housing lower cover.
2. Rotate the torque converter until the drain plug comes into view.
3. Remove the drain plug and allow the transmission fluid to drain.
4. Flush the cooler lines completely.

Front and Rear Axle

FLUID RECOMMENDATION

Use hypoid gear lubricant SAE 80 or 90.

NOTE: On models with the front locking differential, add 2 oz. of friction modifier Ford part No. EST-M2C118-A. On models with the rear locking differential, use only locking differential fluid Ford part No. ESP-M2C154-A or its equivalent, and add 4 oz. of friction modifier Ford part No. EST-M2C118-A.

LEVEL CHECK

Clean the area around the fill plug, which is located in the housing cover, before removing the plug. The lubricant level should be maintained to the bottom of the fill hole with the axle in its normal running position. If lubricant does not appear at

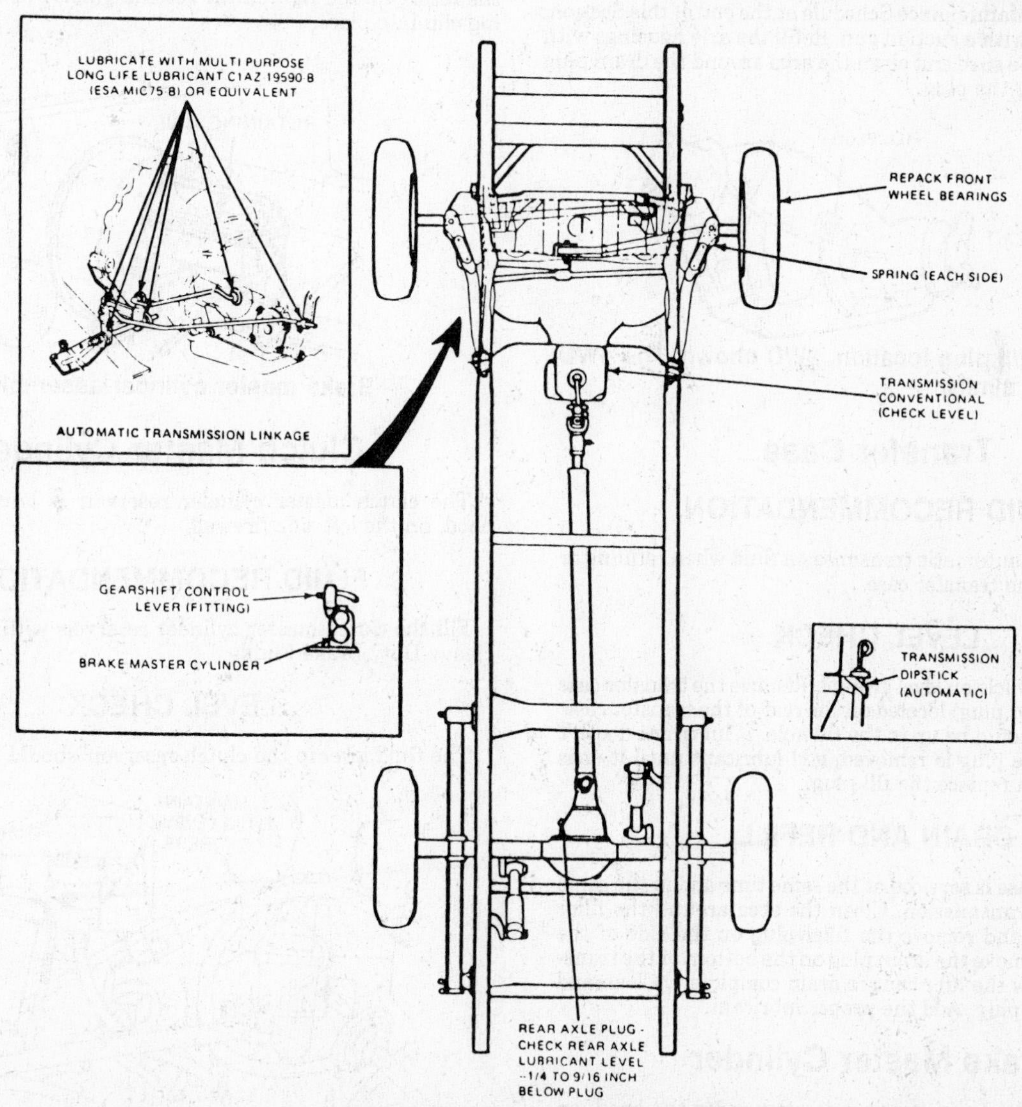

Lubrication chart – 2WD models

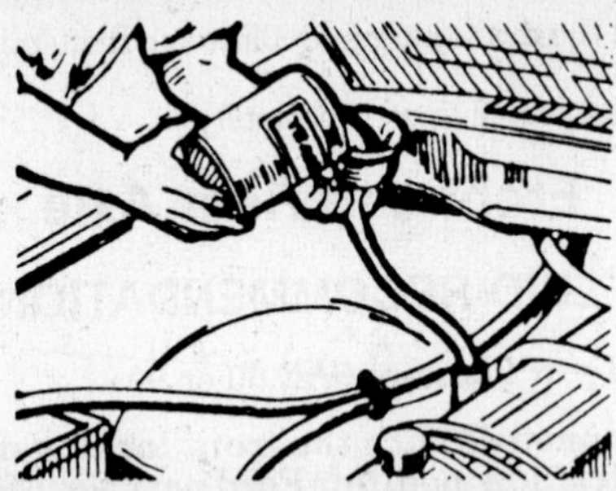

Fill the transmission with the required amount of fluid. Do not overfill. Start the engine and run the selector through all shift points. Check the fluid and add as necessary.

the hole when the plug is removed additional lubricant should be added.

DRAIN AND REFILL

Drain and refill the front and rear axle housings according to the Preventive Maintenance Schedule at the end of this Section. Remove the oil with a suction gun. Refill the axle housings with the proper oil. Be sure and clean the area around the drain plug before removing the plug.

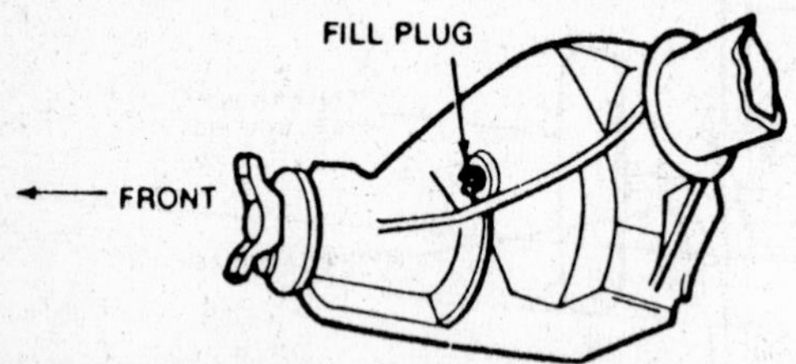

Differential fill plug location, 2WD shown. The 4WD front axle is similar

Transfer Case

FLUID RECOMMENDATION

Use Dexron®II automatic transmission fluid when refilling or adding fluid to the transfer case.

LEVEL CHECK

Position the vehicle on level ground. Remove the transfer case fill plug (the upper plug) located on the rear of the transfer case. The fluid level should be up to the fill hole. If lubricant doesn't run out when the plug is removed, add lubricant until it does run out and then replace the fill plug.

DRAIN AND REFILL

The transfer case is serviced at the same time and in the same manner as the transmission. Clean the area around the filler and drain plugs and remove the filler plug on the side of the transfer case. Remove the drain plug on the bottom of the transfer case and allow the lubricant to drain completely. Clean and install the drain plug. Add the proper lubricant.

Brake Master Cylinder

The master cylinder reservoir is located under the hood, on the left side firewall.

FLUID RECOMMENDATION

Fill the master cylinder with a good quality Heavy-Duty Dot 3 Brake Fluid.

LEVEL CHECK

Before removing the master cylinder reservoir cap, make sure the vehicle is resting on level ground and clean all dirt away from the top of the master cylinder. Pry off the retaining clip and remove the cap. The brake fluid level should be within ¼ in. of the top of the reservoir.

If the level of the brake fluid is less than half the volume of the reservoir, it is advised that you check the brake system for leaks. Leaks in a hydraulic brake system most commonly occur at the wheel cylinder.

There is a rubber diaphragm in the top of the master cylinder cap. As the fluid level lowers in the reservoir due to normal brake shoe wear or leakage, the diaphragm takes up the space. This is to prevent the loss of brake fluid out the vented cap and contamination by dirt. After filling the master cylinder to the proper level with brake fluid, but before replacing the cap, fold the rubber diaphragm up into the cap, then replace the cap on the reservoir and tighten the retaining bolt or snap the retaining clip into place.

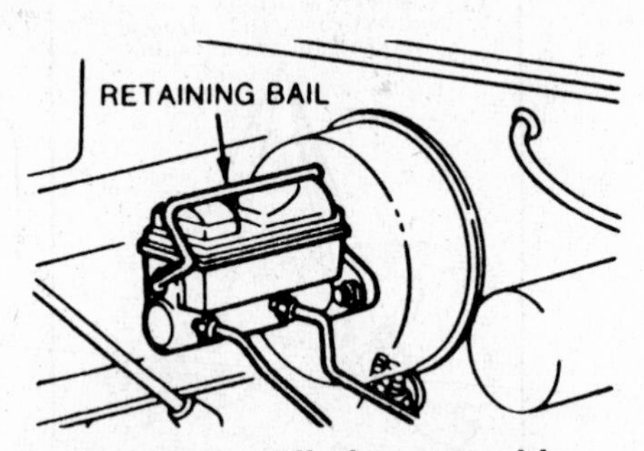

Brake master cylinder assembly

Clutch Master Cylinder

The clutch master cylinder reservoir is located under the hood, on the left side firewall.

FLUID RECOMMENDATION

Fill the clutch master cylinder reservoir with a good quality Heavy-Duty Brake Fluid.

LEVEL CHECK

The fluid level in the clutch reservoir should be visible at or

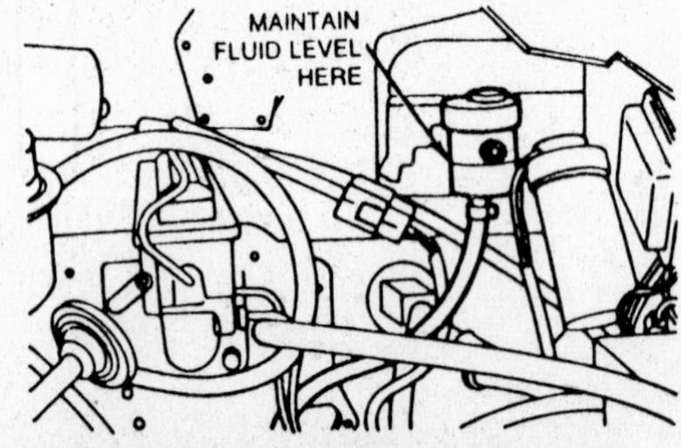

Clutch master cylinder reservoir

above the step in the translucent reservoir body, filling above this point is not necessary.

NOTE: The fluid level in the clutch reservoir will slowly increase as the clutch wears.

Before removing the clutch master cylinder reservoir cap, make sure the vehicle is resting on level ground and clean all dirt away from the top of the reservoir.

Power Steering Reservoir

FLUID RECOMMENDATION

Fill the power steering reservoir with a good quality power steering fluid or Auto. Trans. Fluid-Type F.

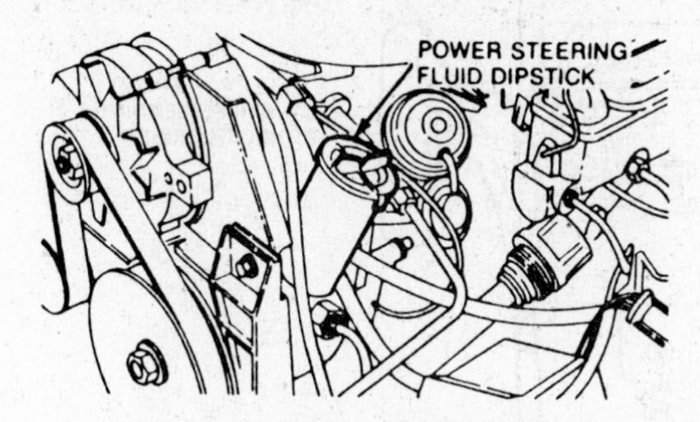

Power steering pump reservoir

LEVEL CHECK

Position the vehicle on level ground. Run the engine until the fluid is at normal operating temperature. Turn the steering wheel all the way to the left and right several times. Position the wheels in the straight ahead position, then shut off the engine. Check the fluid level on the dipstick which is attached to the reservoir cap. The level should be between the ADD and FULL marks on the dipstick. Add fluid accordingly. Do not overfill.

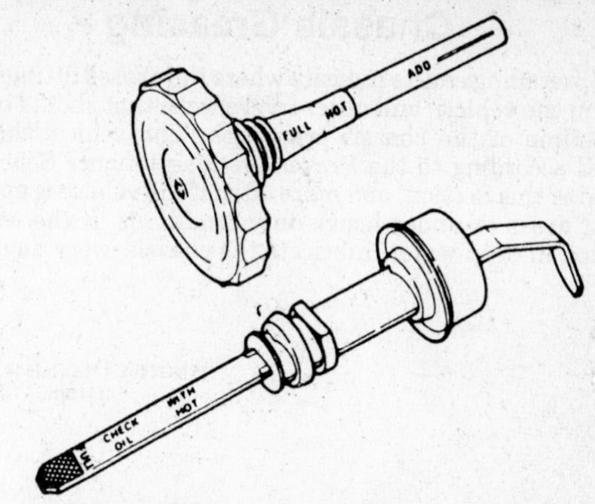

Power steering pump reservoir dipsticks

Steering Gear

The steering gear is factory-filled with steering gear grease. Changing of this lubricant should not be performed and the housing should not be drained, lubricant is not required for the life of the steering gear.

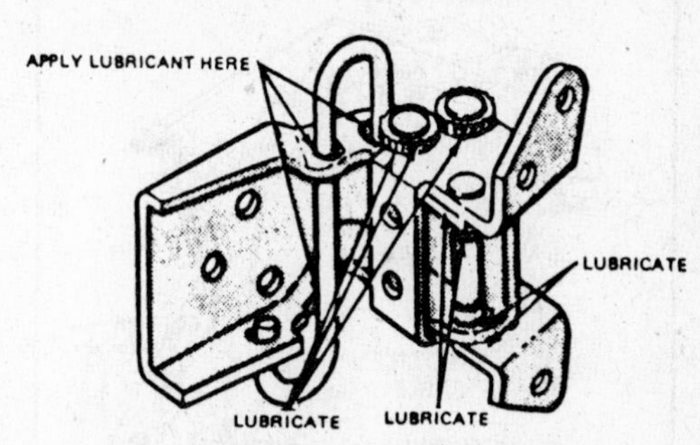

Door hinge lubrication points

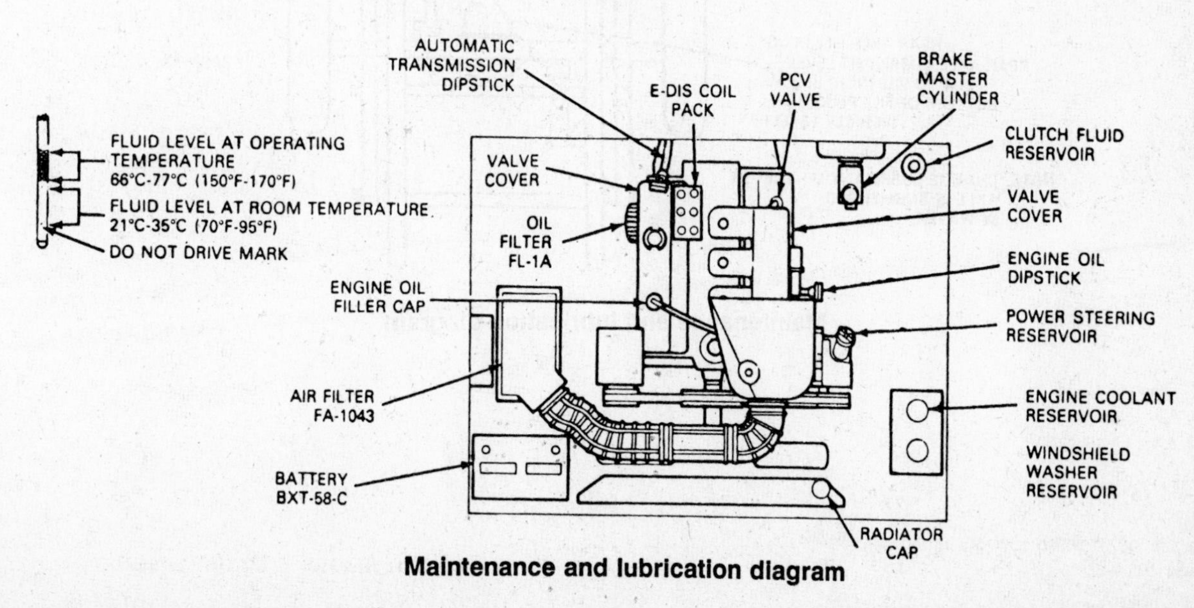

Maintenance and lubrication diagram

Chassis Greasing

The preceding charts indicate where the grease fittings are located on the vehicle, and other level checks that should be made at the time of the chassis grease job. The vehicle should be greased according to the Preventive Maintenance Schedule at the end of this Section, and more often if the vehicle is operating in dust areas or under heavy-duty conditions. If the vehicle is operated in deep water, lubricate the chassis every day.

Body Lubrication

Lubricate the door and tailgate hinges, door locks, door latches, and the hood latch when they become noisy or difficult to operate. A high quality Polyethylene Grease should be used as a lubricant.

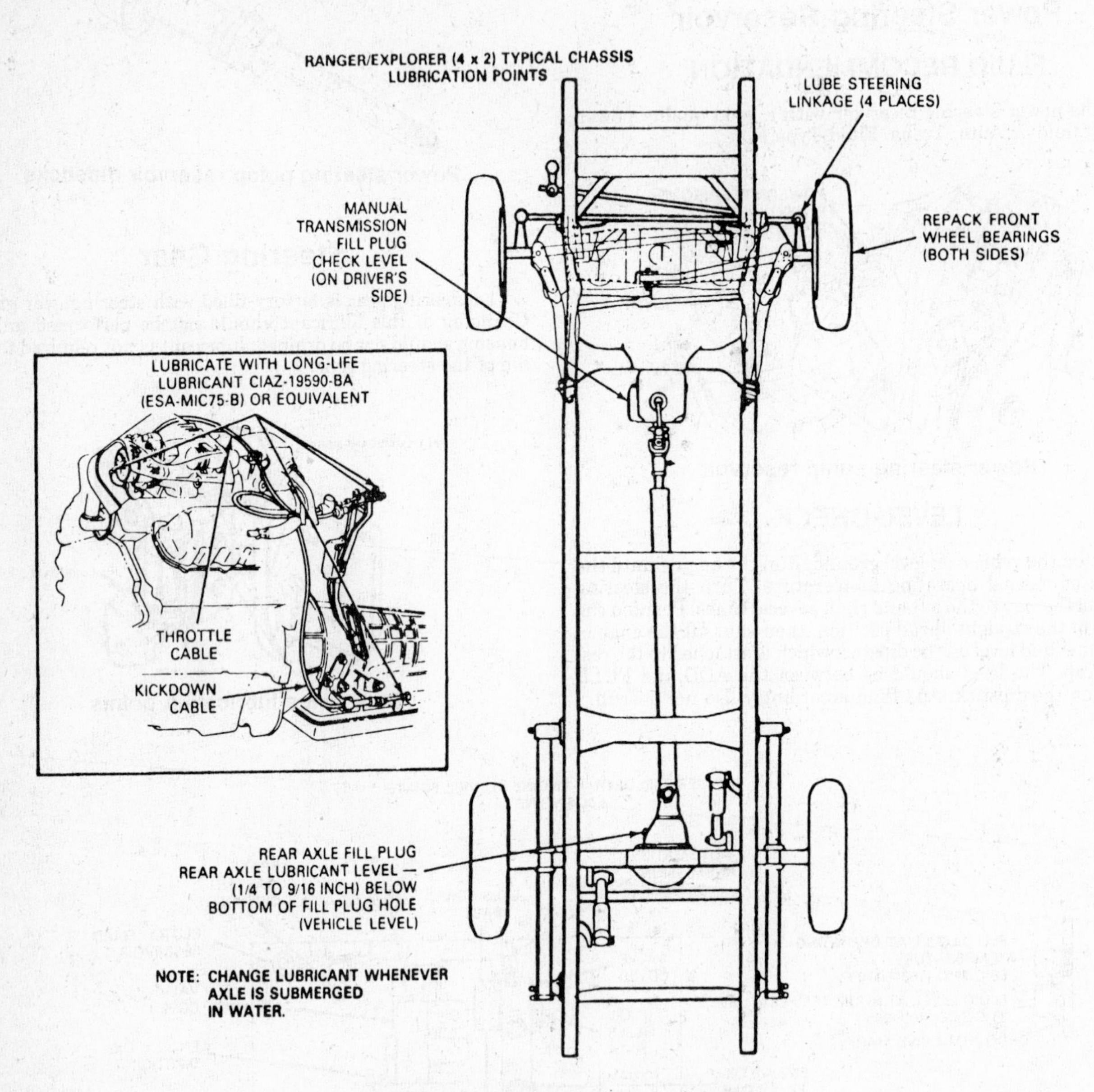

Maintenance and lubrication diagram

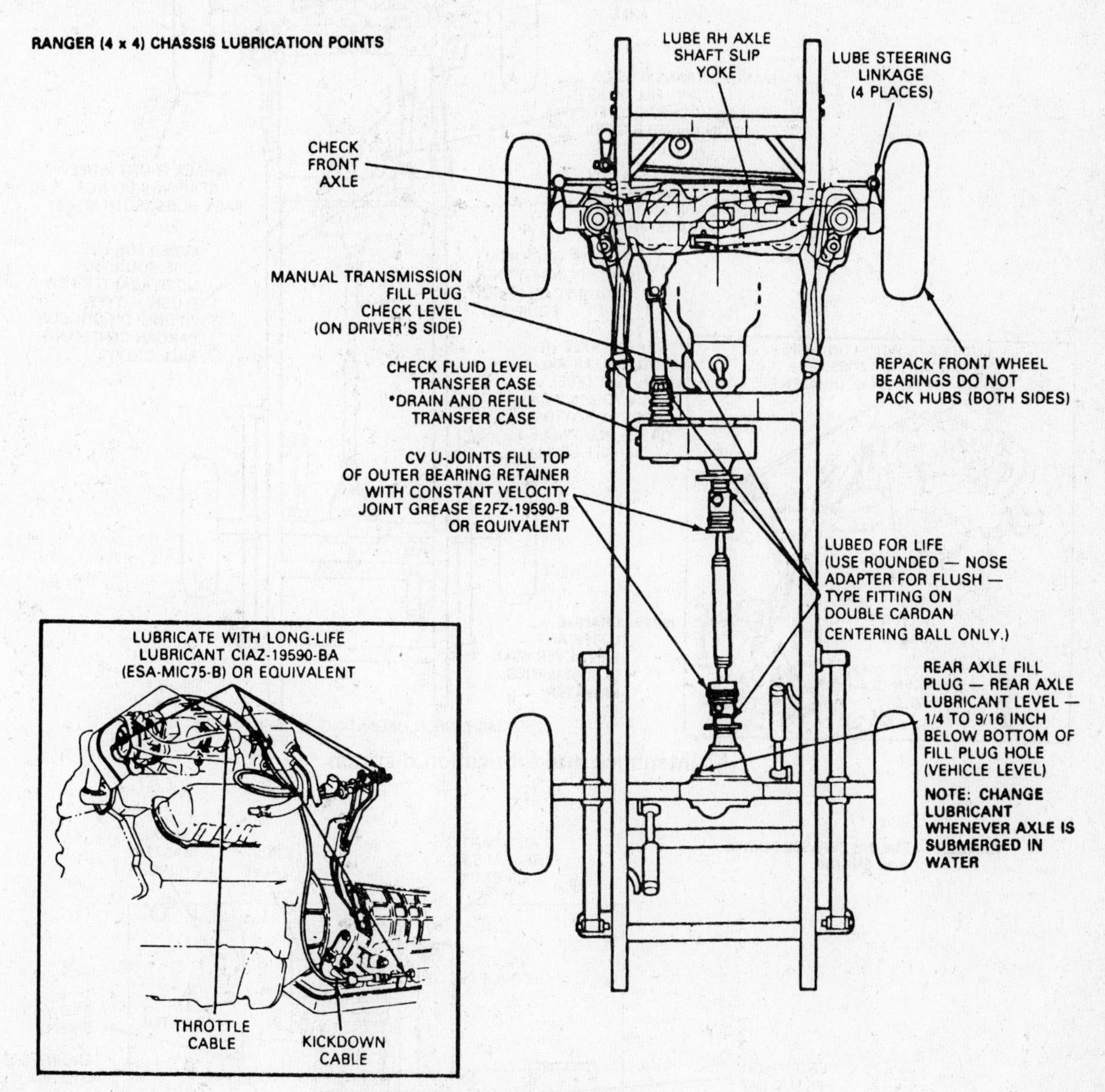

Maintenance and lubrication diagram

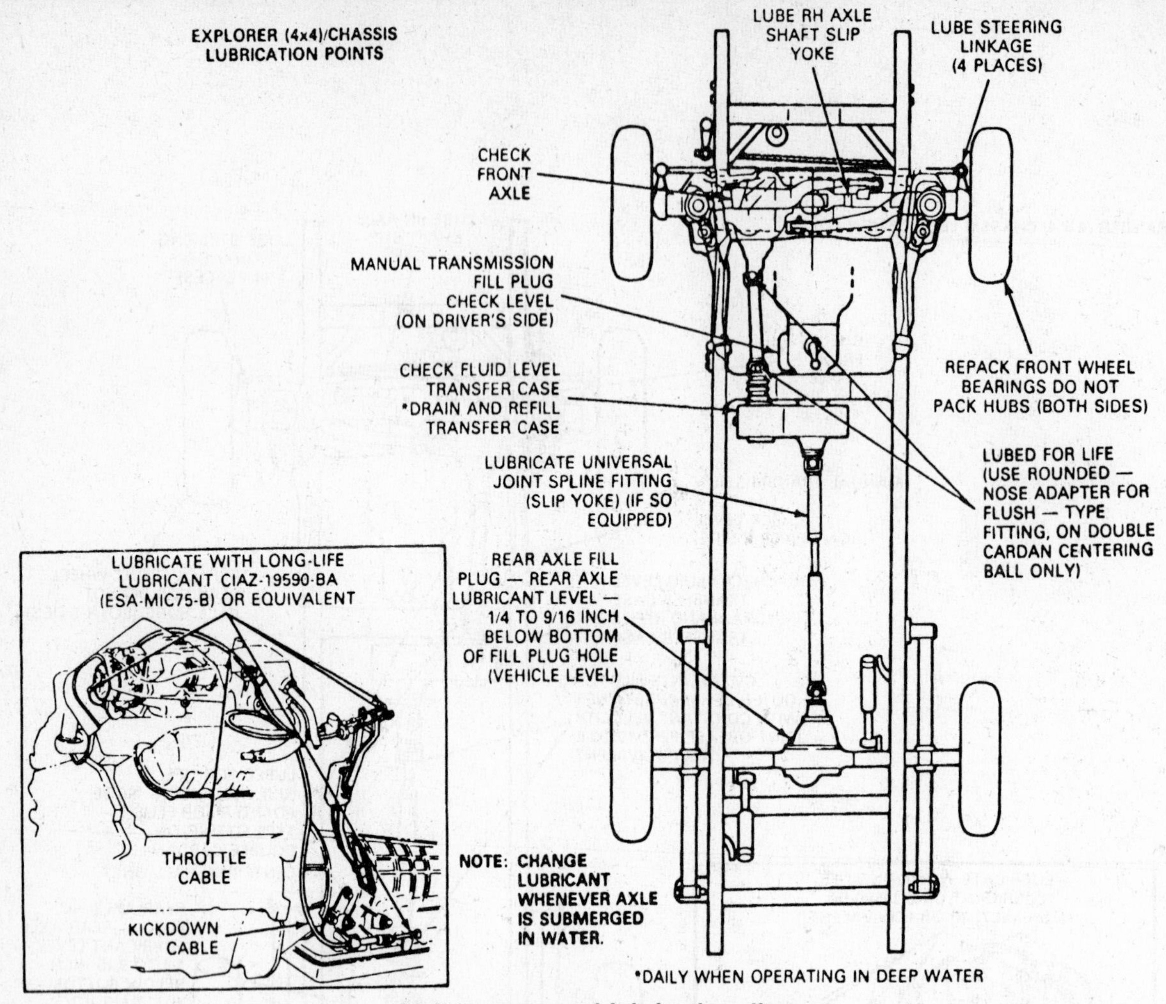

Maintenance and lubrication diagram

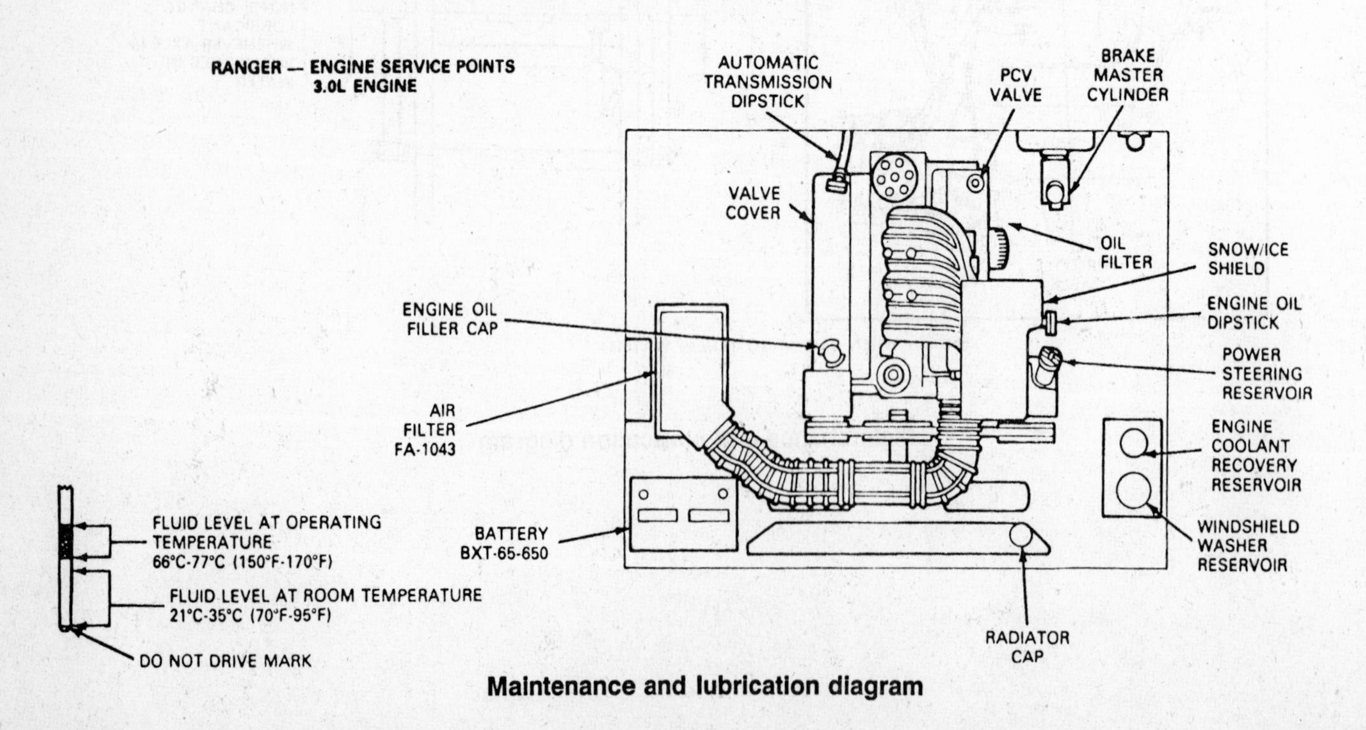

Maintenance and lubrication diagram

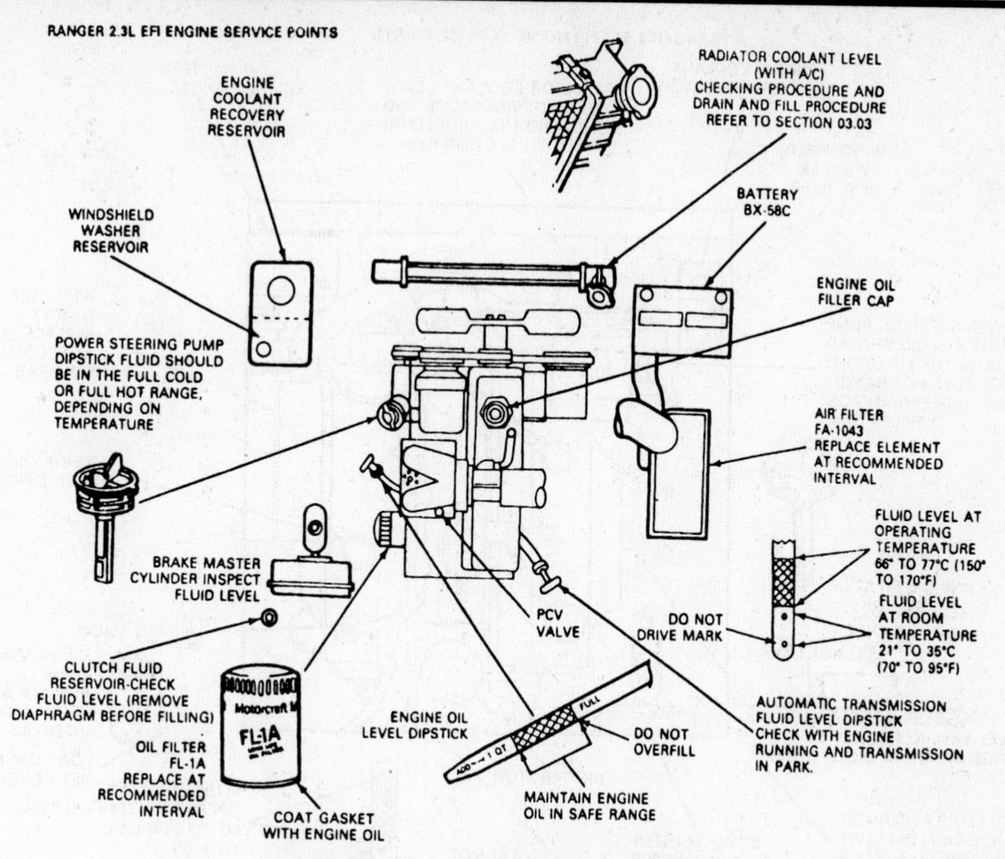

Maintenance and lubrication diagram

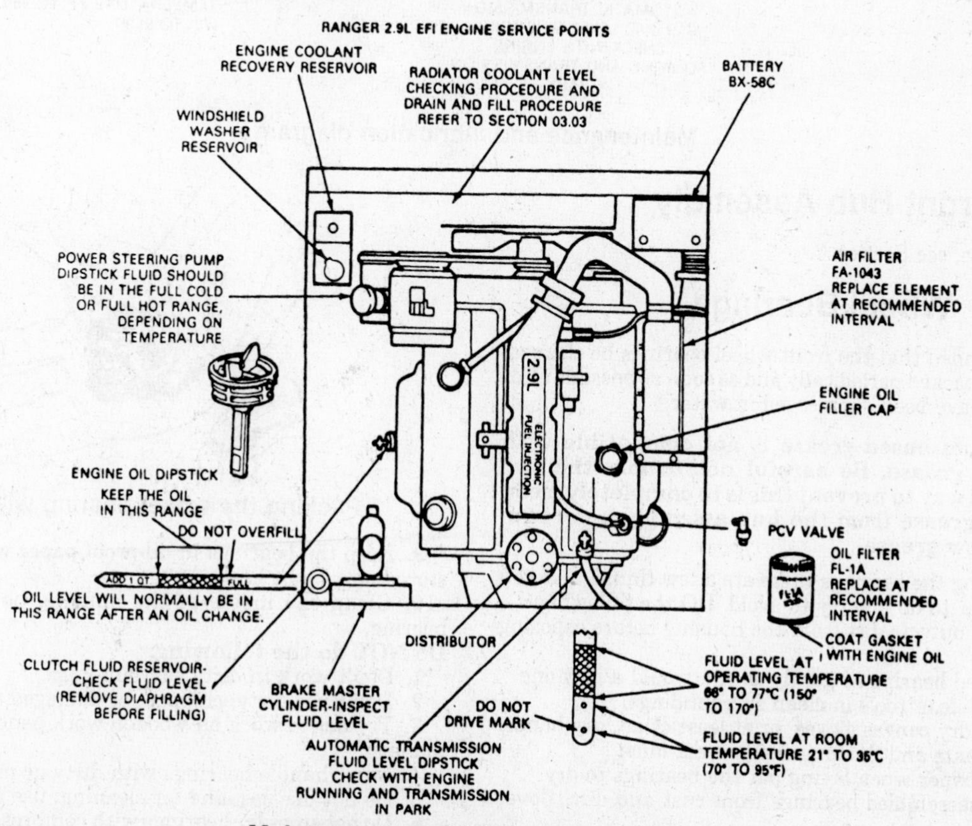

Maintenance and lubrication diagram

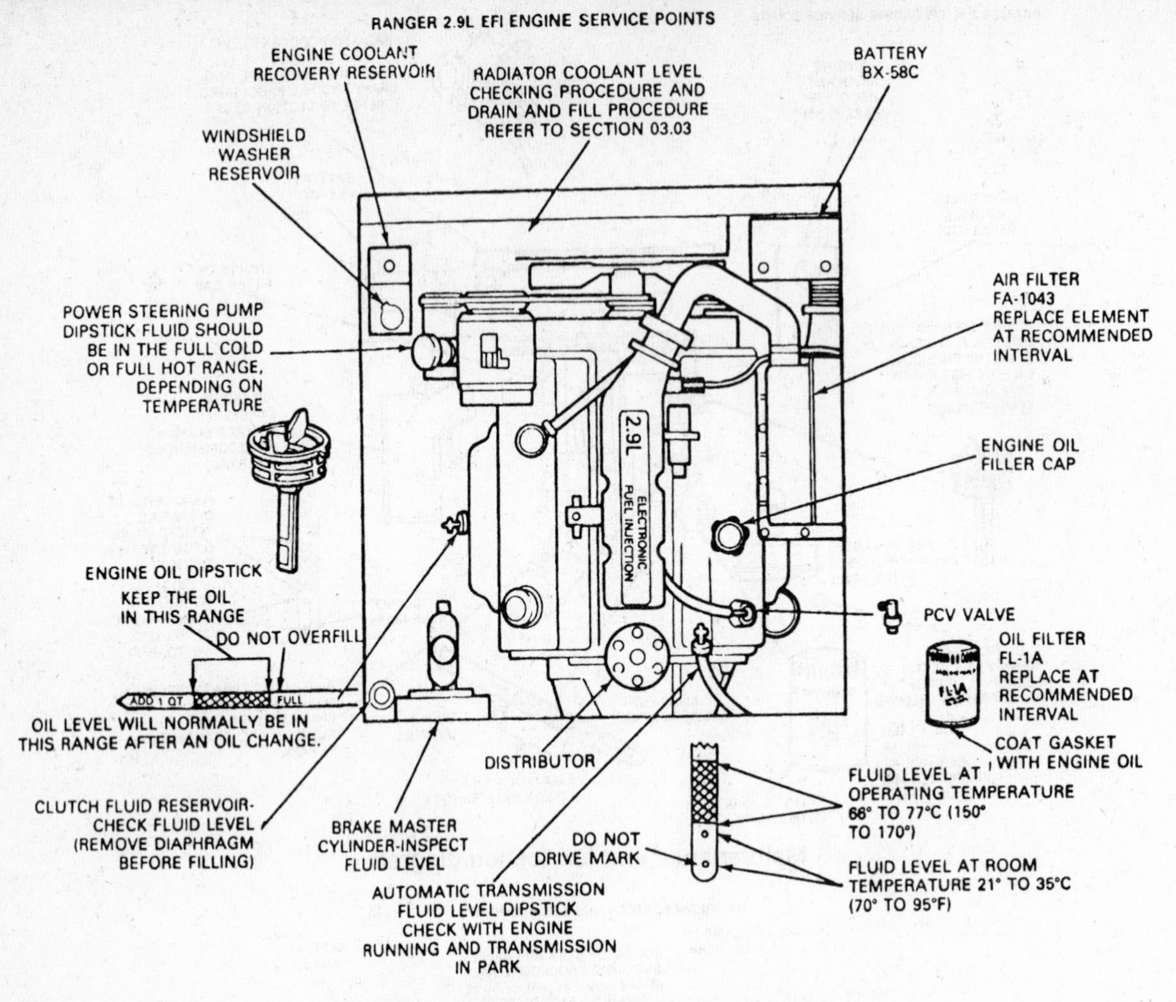

Maintenance and lubrication diagram

Front Hub Assembly

For 4WD vehicles, see Section 7.

Wheel Bearings

It is recommended that the front wheel bearings be cleaned, inspected and repacked periodically and as soon as possible after the front hubs have been submerged in water.

NOTE: Sodium based grease is not compatible with lithium based grease. Be careful not to mix the two types. The best way to prevent this is to completely clean all of the old grease from the hub assembly before installing any new grease.

Before handling the bearings there are a few things that you should remember to do and try to avoid. DO the following:

1. Remove all outside dirt from the housing before exposing the bearing.
2. Treat a used bearing as gently as you would a new one.
3. Work with clean tools in clean surroundings.
4. Use clean, dry canvas gloves, or at least clean, dry hands.
5. Clean solvents and flushing fluids are a must.
6. Use clean paper when laying out the bearings to dry.
7. Protect disassembled bearings from rust and dirt. Cover them up.
8. Use clean rags to wipe bearings.

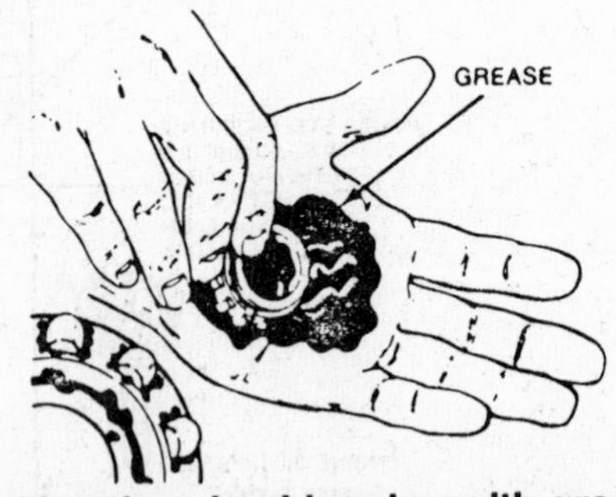

Packing the wheel bearing with grease

9. Keep the bearings in oil-proof paper when they are to be stored or are not in use.
10. Clean the inside of the housing before replacing the bearing.

Do NOT do the following:

1. Don't work in dirty surroundings.
2. Don't use dirty, chipped, or damaged tools.
3. Try not to work on wooden work benches, or use wooden mallets.
4. Don't handle bearings with dirty or moist hands.
5. Do not use gasoline for cleaning; use a safe solvent.
6. Do not spin-dry bearings with compressed air. They will be damaged.

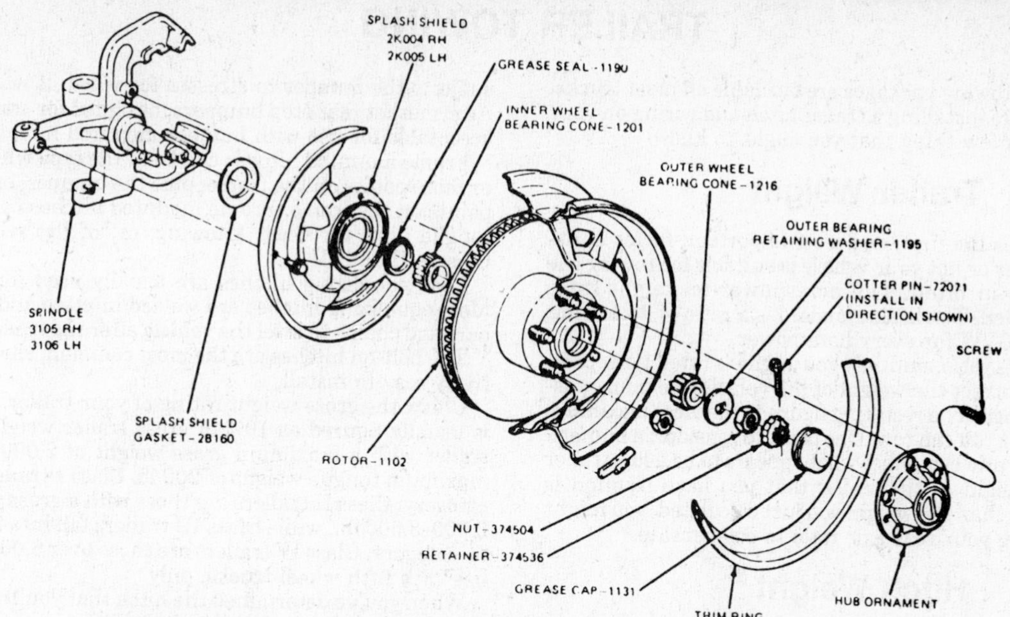

Exploded view of the wheel bearings, grease seal and front hub

7. Do not spin unclean bearings.
8. Avoid using cotton waste or dirty clothes to wipe bearings.
9. Try not to scratch or nick bearing surfaces.
10. Do not allow the bearing to come in contact with dirt or rust at any time.

REMOVAL AND INSTALLATION

NOTE: For 4WD vehicles, see Section 7.

1. Raise the vehicle and support with jackstands. Remove the wheel from the rotor.
2. Remove the caliper and support it from the underbody with a piece of wire.
3. Remove the grease cap from the hub and the cotter pin, nut lock, adjusting nut, and the flat washer from the spindle. Remove the outer bearing assembly from the hub.
4. Carefully pull the hub and rotor assembly off the spindle.
5. Carefully drive out the inner bearing cone and grease seal from the hub.
6. Clean the inner and outer bearing cups with solvent. Inspect the cups for scratches, pits, excessive wear, and other damage. The cups are removed from the hub by driving them out with a drift pin. They are installed in the same manner.
7. If it is determined that the cups are in satisfactory condition and are to remain in the hub, clean and inspect the cones (bearings). Replace the bearings if necessary. When replacing either the cone or the cup, both parts should be replaced as a unit.
8. Thoroughly clean all components in a suitable solvent and blow them dry with compressed air or allow them to dry while resting on clean paper.

NOTE: Do not spin the bearings with compressed air while drying them.

9. Cover the spindle with a clean cloth, and brush all loose dirt from the dust shield. Carefully remove the cloth to prevent dirt from falling from it.
10. Install the inner or outer bearing cups if they were removed. Thoroughly clean the old grease from the surrounding surfaces.
11. Pack the inside of the hub with wheel bearing grease. Add grease to the hub until the grease is flush with the inside diameter of the bearing cup.

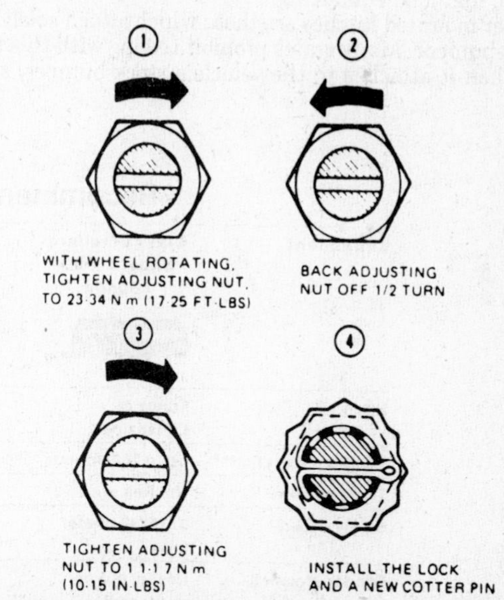

Front wheel bearing adjustment procedures

12. Pack the bearing cone and roller assembly with wheel bearing grease. A bearing packer is desirable for this operation. If a packer is not available, place a large portion of grease into the palm of your hand and sliding the edge of the roller cage through the grease with your other hand, work as much grease in between the rollers as possible.
13. Position the inner bearing cone and roller assembly in the inner cup. Apply a light film of grease to the lips of a new grease seal and install the seal into the hub.
14. Carefully position the hub and rotor assembly onto the spindle. Be careful not to damage the grease seal.
15. Place the outer bearing into position on the spindle and into the bearing cup. Install the adjusting nut finger tight.
16. Adjust the wheel bearings as shown in the illustration. Install the grease cap.
17. Install the caliper to the spindle and the wheel to the hub.
18. Remove the jackstands and lower the vehicle. Torque the lug nuts to 85–115 ft. lbs.

TRAILER TOWING

Factory trailer towing packages are available on most trucks. However, if you are installing a trailer hitch and wiring on your truck, there are a few thing that you ought to know.

Trailer Weight

Trailer weight is the first, and most important, factor in determining whether or not your vehicle is suitable for towing the trailer you have in mind. The horsepower-to-weight ratio should be calculated. The basic standard is a ratio of 35:1. That is, 35 pounds of GVW for every horsepower.

To calculate this ratio, multiply you engine's rated horsepower by 35, then subtract the weight of the vehicle, including passengers and luggage. The resulting figure is the ideal maximum trailer weight that you can tow. One point to consider: a numerically higher axle ratio can offset what appears to be a low trailer weight. If the weight of the trailer that you have in mind is somewhat higher than the weight you just calculated, you might consider changing your rear axle ratio to compensate.

Hitch Weight

There are three kinds of hitches: bumper mounted, frame mounted, and load equalizing.

Bumper mounted hitches are those which attach solely to the vehicle's bumper. Many states prohibit towing with this type of hitch, when it attaches to the vehicle's stock bumper, since it subjects the bumper to stresses for which it was not designed. Aftermarket rear step bumpers, designed for trailer towing, are acceptable for use with bumper mounted hitches.

Frame mounted hitches can be of the type which bolts to two or more points on the frame, plus the bumper, or just to several points on the frame. Frame mounted hitches can also be of the tongue type, for Class I towing, or, of the receiver type, for Classes II and III.

Load equalizing hitches are usually used for large trailers. Most equalizing hitches are welded in place and use equalizing bars and chains to level the vehicle after the trailer is hooked up.

The bolt-on hitches are the most common, since they are relatively easy to install.

Check the gross weight rating of your trailer. Tongue weight is usually figured as 10% of gross trailer weight. Therefore, a trailer with a maximum gross weight of 2,000 lb. will have a maximum tongue weight of 200 lb. Class I trailers fall into this category. Class II trailers are those with a gross weight rating of 2,000–3,500 lb., while Class III trailers fall into the 3,500–6,000 lb. category. Class IV trailers are those over 6,000 lb. and are for use with fifth wheel trucks, only.

When you've determined the hitch that you'll need, follow the manufacturer's installation instructions, exactly, especially when it comes to fastener torques. The hitch will subjected to a lot of stress and good hitches come with hardened bolts. Never substitute an inferior bolt for a hardened bolt.

Recommended Equipment Checklist

Equipment	Class I Trailers Under 2,000 pounds	Class II Trailers 2,000-3,500 pounds	Class III Trailers 3,500-6,000 pounds	Class IV Trailers 6,000 pounds and up
Hitch	Frame or Equalizing	Equalizing	Equalizing	Fifth wheel Pick-up truck only
Tongue Load Limit**	Up to 200 pounds	200-350 pounds	350-600 pounds	600 pounds and up
Trailer Brakes	Not Required	Required	Required	Required
Safety Chain	3/16" diameter links	1/4" diameter links	5/16" diameter links	—
Fender Mounted Mirrors	Useful, but not necessary	Recommended	Recommended	Recommended
Turn Signal Flasher	Standard	Constant Rate or heavy duty	Constant Rate or heavy duty	Constant Rate or heavy duty
Coolant Recovery System	Recommended	Required	Required	Required
Transmission Oil Cooler	Recommended	Recommended	Recommended	Recommended
Engine Oil Cooler	Recommended	Recommended	Recommended	Recommended
Air Adjustable Shock Absorbers	Recommended	Recommended	Recommended	Recommended
Flex or Clutch Fan	Recommended	Recommended	Recommended	Recommended
Tires	***	***	***	***

NOTE The information in this chart is a guide Check the manufacturer's recommendations for your car if in doubt

*Local laws may require specific equipment such as trailer brakes or fender mounted mirrors Check your local laws Hitch weight is usually 10-15% of trailer gross weight and should be measured with trailer loaded

**Most manufacturer's do not recommend towing trailers of over 1.000 pounds with compacts Some intermediates cannot tow Class III trailers

***Check manufacturer's recommendations for your specific car trailer combination

—Does not apply

Wiring

Wiring the vehicle for towing is fairly easy. There are a number of good wiring kits available and these should be used, rather than trying to design your own. All trailers will need brake lights and turn signals as well as tail lights and side marker lights. Most states require extra marker lights for overly wide trailers. Also, most states have recently required back-up lights for trailers, and most trailer manufacturers have been building trailers with back-up lights for several years.

Additionally, some Class I, most Class II and just about all Class III trailers will have electric brakes.

Add to this number an accessories wire, to operate trailer internal equipment or to charge the trailer's battery, and you can have as many as seven wires in the harness.

Determine the equipment on your trailer and buy the wiring kit necessary. The kit will contain all the wires needed, plus a plug adapter set which included the female plug, mounted on the bumper or hitch, and the male plug, wired into, or plugged into the trailer harness.

When installing the kit, follow the manufacturer's instructions. The color coding of the wires is standard throughout the industry.

One point to note, some domestic vehicles, and most imported vehicles, have separate turn signals. On most domestic vehicles, the brake lights and rear turn signals operate with the same bulb. For those vehicles with separate turn signals, you can purchase an isolation unit so that the brake lights won't blink whenever the turn signals are operated, or, you can go to your local electronics supply house and buy four diodes to wire in series with the brake and turn signal bulbs. Diodes will isolate the brake and turn signals. The choice is yours. The isolation units are simple and quick to install, but far more expensive than the diodes. The diodes, however, require more work to install properly, since they require the cutting of each bulb's wire and soldering in place of the diode.

One final point, the best kits are those with a spring loaded cover on the vehicle mounted socket. This cover prevents dirt and moisture from corroding the terminals. Never let the vehicle socket hang loosely. Always mount it securely to the bumper or hitch.

Cooling

ENGINE

One of the most common, if not THE most common, problem associated with trailer towing is engine overheating.

With factory installed trailer towing packages, a heavy duty cooling system is usually included. Heavy duty cooling systems are available as optional equipment on most trucks, with or without a trailer package. If you have one of these extra-capacity systems, you shouldn't have any overheating problems.

If you have a standard cooling system, without an expansion tank, you'll definitely need to get an aftermarket expansion tank kit, preferably one with at least a 2 quart capacity. These kits are easily installed on the radiator's overflow hose, and come with a pressure cap designed for expansion tanks.

Another helpful accessory is a Flex Fan. These fan are large diameter units are designed to provide more airflow at low speeds, with blades that have deeply cupped surfaces. The blades then flex, or flatten out, at high speed, when less cooling air is needed. These fans are far lighter in weight than stock fans, requiring less horsepower to drive them. Also, they are far quieter than stock fans.

If you do decide to replace your stock fan with a flex fan, note that if your truck has a fan clutch, a spacer between the flex fan and water pump hub will be needed.

Aftermarket engine oil coolers are helpful for prolonging engine oil life and reducing overall engine temperatures. Both of these factors increase engine life.

While not absolutely necessary in towing Class I and some Class II trailers, they are recommended for heavier Class II and all Class III towing.

Engine oil cooler systems consist of an adapter, screwed on in place of the oil filter, a remote filter mounting and a multi-tube, finned heat exchanger, which is mounted in front of the radiator or air conditioning condenser.

TRANSMISSION

An automatic transmission is usually recommended for trailer towing. Modern automatics have proven reliable and, of course, easy to operate, in trailer towing.

The increased load of a trailer, however, causes an increase in the temperature of the automatic transmission fluid. Heat is the worst enemy of an automatic transmission. As the temperature of the fluid increases, the life of the fluid decreases.

It is essential, therefore, that you install an automatic transmission cooler.

The cooler, which consists of a multi-tube, finned heat exchanger, is usually installed in front of the radiator or air conditioning compressor, and hooked inline with the transmission cooler tank inlet line. Follow the cooler manufacturer's installation instructions.

Select a cooler of at least adequate capacity, based upon the combined gross weights of the truck and trailer.

Cooler manufacturers recommend that you use an aftermarket cooler in addition to, and not instead of, the present cooling tank in your truck's radiator. If you do want to use it in place of the radiator cooling tank, get a cooler at least two sizes larger than normally necessary.

NOTE: A transmission cooler can, sometimes, cause slow or harsh shifting in the transmission during cold weather, until the fluid has a chance to come up to normal operating temperature. Some coolers can be purchased with or retrofitted with a temperature bypass valve which will allow fluid flow through the cooler only when the fluid has reached operating temperature, or above.

PUSHING AND TOWING

To push-start your vehicle, (manual transmissions only) follow the procedures below. Check to make sure that the bumpers of both vehicles are aligned so neither will be damaged. Be sure that all electrical system components are turned off (headlights, heater, blower, etc.). Turn on the ignition switch. Place the shift lever in Third or Fourth and push in the clutch pedal. At about 15 mph, signal the driver of the pushing vehicle to fall back, depress the accelerator pedal, and release the clutch pedal slowly. The engine should start.

When you are doing the pushing or pulling, make sure that the two bumpers match so you won't damage the vehicle you are to push. Another good idea is to put an old tire between the two vehicles. If the bumpers don't match, perhaps you should tow the other vehicle. If the other vehicle is just stuck, use First gear to slowly push it out. Tell the driver of the other vehicle to go slowly too. try to keep your truck right up against the other vehicle while you are pushing. If the two vehicles do separate, stop and start over again instead of trying to catch up and ramming the other vehicle. Also try, as much as possible, to avoid riding or slipping the clutch. When the other vehicle gains enough traction, it should pull away from your vehicle.

If you have to tow the other vehicle, make sure that the two chain or rope is sufficiently long and strong, and that it is attached securely to both vehicles at a strong place. Attach the chain at a point on the frame or as close to it as possible. Once again, go slowly and tell the other driver to do the same. Warn the other driver not to allow too much slack in the line when he gains traction and can move under his own power. Otherwise he may run over the tow line and damage both vehicles. If your truck has to be towed by a tow truck, it can be towed forward for any distance with the driveshaft connected as long as it is dine fairly slowly. If your truck has to be towed backward and is a 4WD model, unlock the front axle driving hubs, to prevent the front differential from rotating and place the transfer case in neutral. Also clamp the steering wheel on all models, in the straight ahead position with a clamping device designed for towing service.

JACKING AND HOSTING

It is very important to be careful about running the engine, on vehicles equipped with limited slip differentials, while the vehicle is up on a jack. This is because if the drive train is engaged, power is transmitted to the wheel with the best traction and the vehicle will drive off the jack, resulting in possible damage or injury.

Jack the truck from under the axles, radius arms, or spring hangers and the frame. Be sure and block the diagonally opposite wheel to prevent the vehicle from moving. Place jackstands under the vehicle at the points mentioned above when you are going to work under the vehicle.

CAUTION

On models equipped with an under chassis mounted spare tire, remove the tire, wheel or tire carrier from the vehicle before it is placed in a high lift position in order to avoid sudden weight release from the chassis.

When raising the vehicle on a hoist, position the front end adapters under the center of the lower suspension arm or the spring supports as near to the wheels as practical. The rear hoist adapters should be placed under the spring mounting pads or the rear axle housing. Be careful not to touch the rear shock absorber mounting brackets.

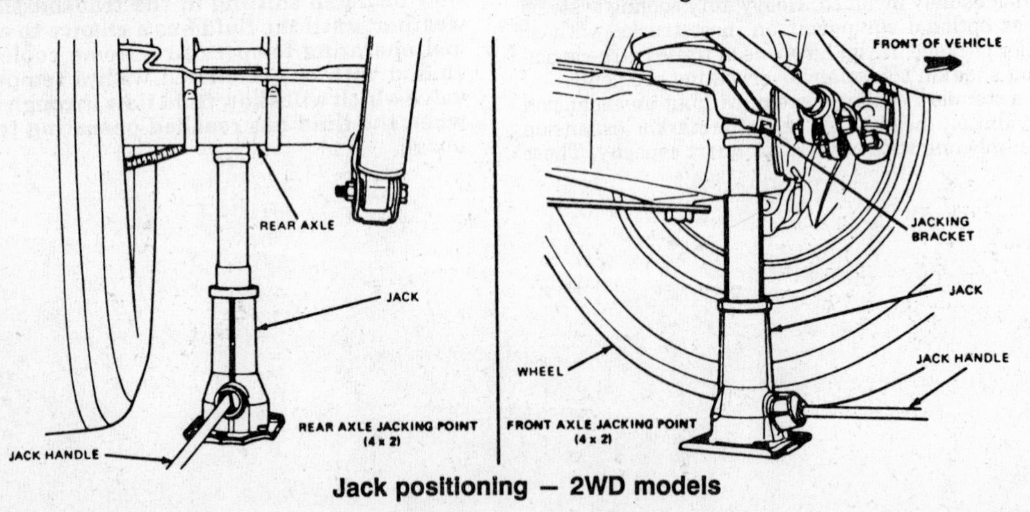

Jack positioning – 2WD models

JUMP STARTING A DEAD BATTERY

The chemical reaction in a battery produces explosive hydrogen gas. This is the safe way to jump start a dead battery, reducing the chances of an accidental spark that could cause an explosion.

Jump Starting Precautions

1. Be sure both batteries are of the same voltage.
2. Be sure both batteries are of the same polarity (have the same grounded terminal).
3. Be sure the vehicles are not touching.
4. Be sure the vent cap holes are not obstructed.
5. Do not smoke or allow sparks around the battery.
6. In cold weather, check for frozen electrolyte in the battery. Do not jump start a frozen battery.
7. Do not allow electrolyte on your skin or clothing.
8. Be sure the electrolyte is not frozen.

CAUTION: Make certin that the ignition key, in the vehicle with the dead battery, is in the OFF position. Connecting cables to vehicles with on-board computers will result in computer destruction if the key is not in the OFF position.

Jump Starting Procedure

1. Determine voltages of the two batteries; they must be the same.
2. Bring the starting vehicle close (they must not touch) so that the batteries can be reached easily.
3. Turn off all accessories and both engines. Put both vehicles in Neutral or Park and set the handbrake.
4. Cover the cell caps with a rag—do not cover terminals.
5. If the terminals on the run-down battery are heavily corroded, clean them.
6. Identify the positive and negative posts on both batteries and connect the cables in the order shown.
7. Start the engine of the starting vehicle and run it at fast idle. Try to start the car with the dead battery. Crank it for no more than 10 seconds at a time and let it cool for 20 seconds in between tries.
8. If it doesn't start in 3 tries, there is something else wrong.
9. Disconnect the cables in the reverse order.
10. Replace the cell covers and dispose of the rags.

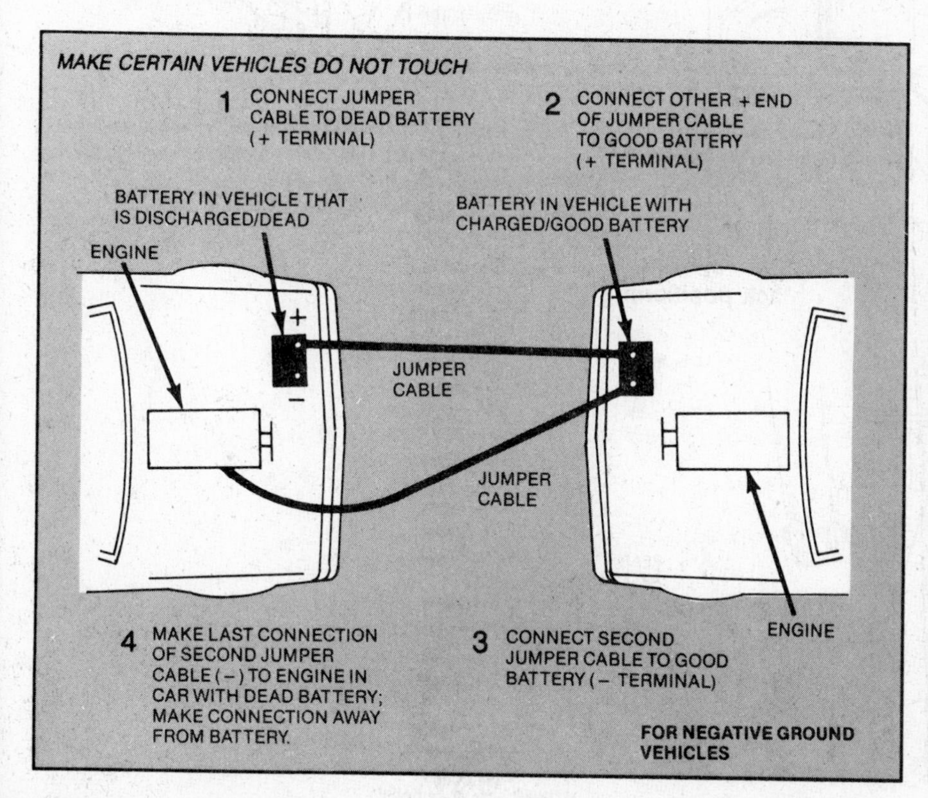

Side terminal batteries occasionally pose a problem when connecting jumper cables. There frequently isn't enough room to clamp the cables without touching sheet metal. Side terminal adaptors are available to alleviate this problem and should be removed after use

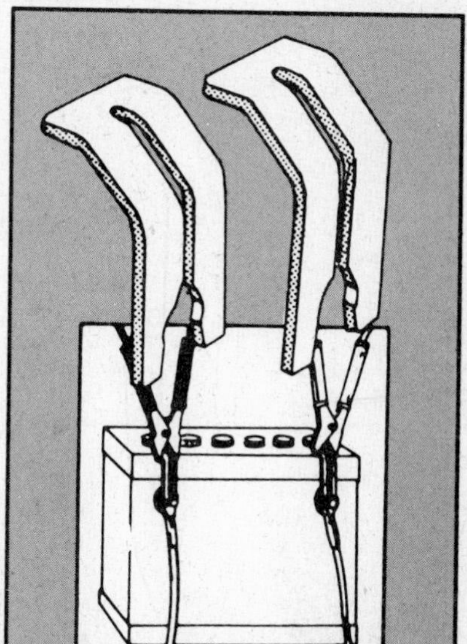

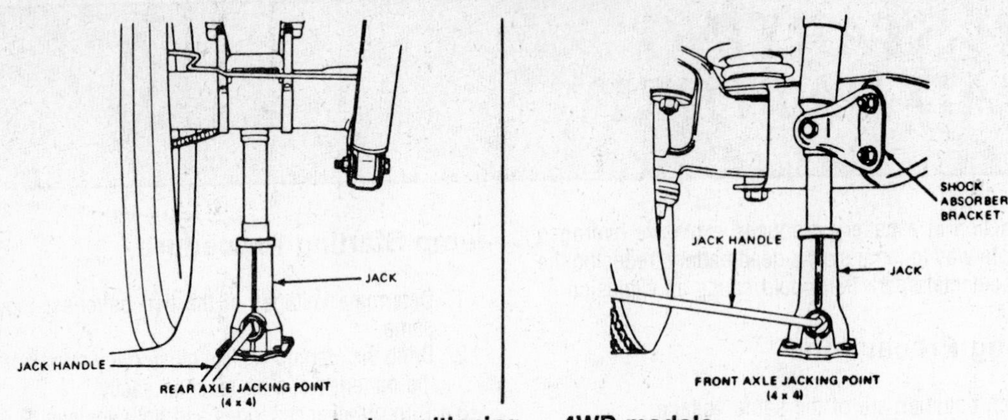

Jack positioning — 4WD models

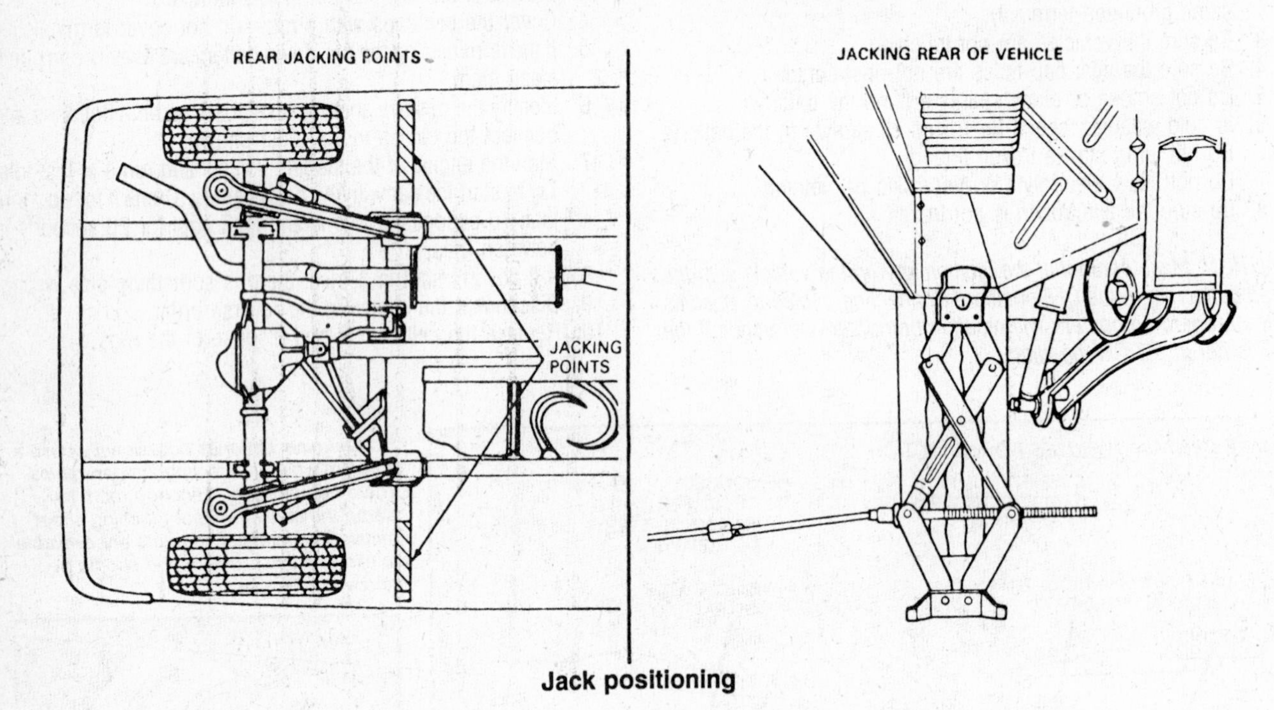

Jack positioning

REAR
AXLE
TUBE
JACK
JACK
HANDLE
REAR AXLE JACKING POINT
RANGER

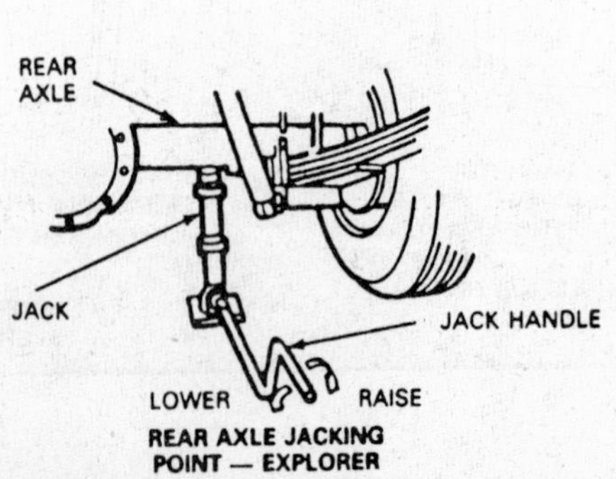

Jack positioning

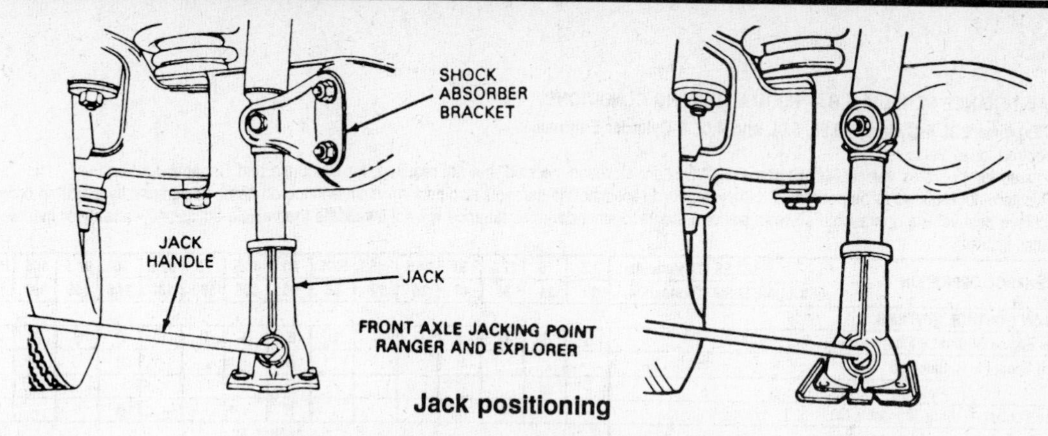

Jack positioning

Preventive Maintenance Schedule

Interval	Item	Service
Perform at each month or distance shown, whichever comes first.		
Gasoline Engine		
7½ mos/7,500 miles	engine oil & filter	change
	drive belts	check tension & condition
	idle speed (2.0L & 2.3L only)	check
	valve clearance (2.8L only)	check
	wheel lug nuts	torque
	u-joints	lubricate
	double cardin joint	lubricate
12 mos/12,000 miles	cooling system	check & inspect
30 mos/30,000 miles	spark plugs	replace
	air cleaner filter	replace
	PCV valve	replace
	crankcase emission filter	replace
	choke linkage	clean
	clutch reservoir	check fluid level
	disc brake system	inspect
	front wheel bearings (4x2)	inspect & lubricate
	spindle needle bearings (4x4)	inspect & lubricate
	thrust bearings (4x4)	inspect & lubricate
	hub lock (4x4)	inspect & lubricate
	driveshaft output slip yokes	lubricate
	exhaust system	inspect
	drive axle R.H. axle shaft slip yoke	lubricate
	steering gear	inspect
Diesel Engines		
5 mos/5,000 miles	engine oil & filter	change
	fuel sedimenter	drain water
	wheel lug nuts	torque
12 mos/12,000 miles	bypass oil filter	change
	cooling system	check & inspect
15 mos/15,000 miles	drive belts	check tension & condition
	valve clearance	check
30 mos/30,000 miles	air cleaner filter	replace
	air cleaner hoses	inspect
	secondary fuel filter	replace
	disc brake system	inspect
	clutch reservoir	check fluid level
	front wheel bearings	inspect & lubricate
	driveshaft slip yoke	lubricate
	u-joints	lubricate
	exhaust system	inspect
	steering gear	inspect
36 mos/36,000 miles	coolant	replace

NOTE: *When operating under severe conditions, cut maintenance schedules in half. When operating daily in water, repack hubs daily and change axle, transmission and transfer case fluids every 1,000 miles.*

1991 MAINTENANCE SCHEDULE B — NORMAL DRIVING CONDITIONS

Ranger/Explorer 2.3L 4-Cylinder, 2.9L, 3.0L and 4.0L 6-Cylinder Engines

B — Required for all vehicles.
b — Required for 49 States vehicles (all States except California); recommended, but not required, for California and Canada vehicles.
(b) — This item not required to be performed. However, Ford recommends that you also perform maintenance on items designated by a "(b)" in order to achieve best vehicle operation. Failure to perform this recommended maintenance will not invalidate the vehicle emissions warranty or manufacturer recall liability.

MAINTENANCE OPERATION — MILES (Thousands)	7.5	15	22.5	30	37.5	45	52.5	60	67.5	75	82.5	90	97.5	105	112.5	120
KILOMETERS (Thousands)	12	24	36	48	60	72	84	96	108	120	132	144	156	168	181	193
EMISSION CONTROL SYSTEMS																
Change Engine Oil and Oil Filter — every 6 months OR	B	B	B	B	B	B	B	B	B	B	B	B	B	B	B	b
Replace Spark Plugs: Standard				B				B				B				b
Platinum Type 3.0L/4.0L								B								b
Replace Coolant — every 36 months OR				B				B				B				b
*Check Cooling System, Hoses and Clamps	ANNUALLY															
Replace Air Cleaner Filter				B				b				b				b
*Check/Clean Idle Speed Control Air Bypass Valve (2.3L)								(b)								(b)
*Check/Clean Throttle Body								(b)								(b)
Replace PCV Valve (1)								b/1								b
Replace Ignition Wires								b								b
Inspect Drive Belt Condition and Tension — 2.3L								b								b
OTHER SYSTEMS																
Check Wheel Lug Nut Torque§	B	B	B	B	B	B	B	B	Beyond 60,000 miles/96 000 km, continue recommended maintenance operations at intervals indicated for 0-60,000 miles/96 000 km.							
Rotate Tires	B		B		B		B									
Check Clutch Reservoir Fluid Level	B	B	B	B	B	B	B	B								
Inspect and Lubricate Automatic Transmission Shift Linkage (Cable System)	B	B	B	B	B	B	B	B								
Inspect and Lubricate Front Wheel Bearings				B				B								
Inspect Disc Brake System and Lubricate Caliper Slide Rails		B		B		B		B								
Inspect Drum Brake Linings, Lines and Hoses		B		B		B		B								
Inspect Exhaust System for Leaks, Damage or Loose Parts				B				B								
Inspect and Remove any Foreign Material Trapped by Exhaust System Shielding	B	B	B	B	B	B	B	B								
Lubricate Driveshaft U-Joints If Equipped with Grease Fitting	B	B	B	B	B	B	B	B								
Inspect Parking Brake System for Damage and Operation		B		B		B		B								
Lubricate Throttle/Kick Down Cable Ball Stud				B				B								
Lubricate Rear Driveshaft Double Cardan Joint Centering Ball (Ranger SWB 4x4)	B	B	B	B	B	B	B	B								
Lubricate Front Drive Axle R.H. Axle — Shaft Slip Yoke (4x4)				B				B								
Inspect Spindle Needle Bearing Spindle Thrust Bearing Lubrication (4x4)				B				B								
Inspect Hub Lock Lubrication (4x4)				B				B								
Change Transfer Case Oil (4x4)								B								
Lubricate Steering Linkage Joints If equipped with Grease Fittings	B	B	B	B	B	B	B	B								

* Check means a function measurement of system's operation (performance, leaks or conditions of parts). Correct as required.

§ Wheel lug nuts must be retightened to proper torque specifications at 500 miles/800 km of new vehicle operation (100 miles/160 km and 500 miles/800 km for vehicles equipped for snowplowing). See your Owner Guide for proper torque specifications. Also retighten to proper torque specification at 500 miles/800 km after (1) any wheel change or (2) any other time the wheel lug nuts have been loosened.

/1 At 60,000 miles/96 000 km, your dealer will replace the PCV Valve at no cost on 2.3L, 2.9L, 3.0L and 4.0L engines except California and Canada vehicles. **NOTE: Refer to page 2 of the Maintenance Schedule Record Book for "NO COST PCV VALVE REPLACEMENT."**

NOTE: Change rear axle lubrication at 100,000 miles.

NOTES:

Unique Driving Conditions

If your driving habits **FREQUENTLY** include one or more of the following conditions:

- Operating when outside temperatures remain **below freezing** and most trips are less than 16 km (10 miles).
- Operating during **HOT WEATHER** in stop-and-go "rush hour" traffic.
- Towing a trailer, using a camper or roof-top carrier, or carrying maximum loads.
- Operating in severe dust conditions.
- Extensive idling, such as police, taxi or door-to-door delivery service.
- High speed operation with a fully loaded vehicle (MAX. GVW).
- Snow plowing on low speed operation.

Change ENGINE OIL and OIL FILTER every 3 months or 4 800 km (3,000 miles) whichever occurs first.

Check/Regap SPARK PLUGS every 15,000 miles (24 000 km).

AIR CLEANER AND CRANKCASE VENTILATION FILTERS –
If operating in severe dust conditions, ask your dealer for proper replacement intervals.

AUTOMATIC TRANSMISSION FLUID — Change each 30,000 miles (48 000 km) — If your driving habits FREQUENTLY include one or more of the following conditions:

- Operating during hot weather (above 90°F, 32°C) and carrying heavy loads and driving in hilly terrain.
- Towing a trailer or slide-in camper.
- Door-to-door delivery, police or taxi.

Extreme Service Items

If vehicle is operated off-highway, perform the following items every 1,000 miles (1 600 km).
If vehicle is operated in mud and/or water perform the following items daily:

- Inspect disc brake system, lube caliper slide rails.
- Inspect drum brake system, hoses, and lines.
- Lubricate automatic transmission external controls (Cable system).
- Inspect front wheel bearings and lubrication.
- Inspect exhaust system for leaks, damage or loose parts and remove any foreign material trapped by shielding.
- Lubricate front axle, steering and clutch linkages, axle and driveshaft U-joints and slip yoke if equipped with fittings.

CAPACITIES

Year	Model	Engine Displacement Liter/CID	Engine Crankcase with Filter (qts)	Transmission Manual (pts)	Transmission Automatic (qts)	Transfer Case (pts)	Drive Axle Front (pts)	Drive Axle Rear (pts)	Fuel Tank (gal)	Cooling System wo/AC (qts)	Cooling System w/AC (qts)
1983	Ranger	2.0/122	5	3①	8②	2	1	5	15	6.5	—
		2.2/134 Diesel	7	3①	8②	2	1	5	15	10.0	10.7
		2.3/140	6	3①	8②	2	1	5	15	6.5	7.2
1984	Bronco II	2.8/173	5	3①	8②	2	1	5.5	23	7.2	7.8
	Ranger	2.0/122	5	3①	8②	2	1	5	15	6.5	—
		2.2/134 Diesel	7	3①	8②	2	1	5	15	10.0	10.7
		2.3/140	6	3①	8②	2	1	5	15	6.5	7.2
		2.8/173	5	3①	8②	2	1	5	15	7.2	7.8
1985	Bronco II	2.8/173	5	3①	8②	2	1	5.5	23	7.2	7.8
	Ranger	2.0/122	5	3①	8②	2	1	5	15	6.5	—
		2.3/140	6	3①	8②	2	1	5	15	6.5	7.2
		2.3/144 Diesel	7	3①	8②	2	1	5	15	13	13
		2.8/173	5	3①	8②	2	1	5.5	15	7.2	7.8
1986	Bronco II	2.9/177	5	3①	8②	2	1	5.5	23	7.2	7.8
	Ranger	2.0/122	5	3①	8②	2	1	5	15	6.5	—
		2.3/140	6	3①	8②	2	1	5	15	6.5	7.2
		2.3/144 Diesel	7	3①	8②	2	1	5	15	13	13
		2.9/177	5	3①	8②	2	1	5.5	15	7.2	7.8
1987	Bronco II	2.9/177	5	3①	8②	2	1	5.5	23	7.2	7.8
	Ranger	2.0/122	5	3①	8②	2	1	5	15	6.5	—
		2.3/130	6	3①	8②	2	1	5	15	6.5	7.2
		2.3/144 Diesel	7	3①	8②	2	1	5	15	13	13
		2.9/177	5	3①	8②	2	1	5.5	15	7.2	7.8
1988	Bronco II	2.9/177	5	3①	8②	2	1	5.5	23	7.2	7.8
	Ranger	2.0/122	5	3①	8②	2	1	5	15	6.5	—
		2.3/140	6	3①	8②	2	1	5	15	6.5	7.2
		2.9/177	5	3①	8②	2	1	5.5	15	7.2	7.8
1989	Bronco II	2.9/177	5	3①	8②	2	3	5.5	15	7.2	7.8
	Ranger	2.3/140	5	3①	8②	2	3	5	15	6.5	7.2
		2.9/177	5	3①	8②	2	3	5.5	15	7.2	7.8
1990	Bronco II	2.9/177	5	3①	8②	2	3	5.5	15	7.2	7.8
	Ranger	2.3/140	5	3①	8②	2	3	5	15	6.5	7.2
		2.9/177	5	3①	8②	2	3	5.5	15	7.2	7.8
		4.0/241	5	5	10	2	3	5	16.3③	7.8	8.6

CAPACITIES

Year	Model	Engine Displacement Liter/CID	Engine Crankcase with Filter (qts)	Transmission		Transfer Case (pts)	Drive Axle		Fuel Tank (gal)	Cooling System	
				Manual (pts)	Automatic (qts)		Front (pts)	Rear (pts)		wo/AC (qts)	w/AC (qts)
1991	Explorer	4.0/241	5	5	10	2	3	5	19.3	7.8	8.6
	Ranger	2.3/140	5	5	10	2	3	5	16.3③	6.5	7.2
		2.9/177	5	5	10	2	3	5	16.3③	7.2	7.8
		3.0/183	5	5	10	2	3	5	16.3③	9.5	10.2
		4.0/241	5	5	10	2	3	5	16.3③	7.8	8.6

wo–A/C: without Air Conditioning
w–A/C: with Air Conditioning
① 5-speed overdrive transmission: 3.6 pints
② C-5 transmission (2WD): 7.5 quarts
③ 19.6 optional

2 Engine Performance and Tune-Up

QUICK REFERENCE INDEX

GENERAL INDEX

Diagnosis of Spark Plugs

Problem	Possible Cause	Correction
Brown to grayish-tan deposits and slight electrode wear.	• Normal wear.	• Clean, regap, reinstall.
Dry, fluffy black carbon deposits.	• Poor ignition output.	• Check distributor to coil connections.
Wet, oily deposits with very little electrode wear.	• "Break-in" of new or recently overhauled engine. • Excessive valve stem guide clearances. • Worn intake valve seals.	• Degrease, clean and reinstall the plugs. • Refer to Section 3. • Replace the seals.
Red, brown, yellow and white colored coatings on the insulator. Engine misses intermittently under severe operating conditions.	• By-products of combustion.	• Clean, regap, and reinstall. If heavily coated, replace.
Colored coatings heavily deposited on the portion of the plug projecting into the chamber and on the side facing the intake valve.	• Leaking seals if condition is found in only one or two cylinders.	• Check the seals. Replace if necessary. Clean, regap, and reinstall the plugs.
Shiny yellow glaze coating on the insulator.	• Melted by-products of combustion.	• Avoid sudden acceleration with wide-open throttle after long periods of low speed driving. Replace the plugs.
Burned or blistered insulator tips and badly eroded electrodes.	• Overheating.	• Check the cooling system. • Check for sticking heat riser valves. Refer to Section 1. • Lean air-fuel mixture. • Check the heat range of the plugs. May be too hot. • Check ignition timing. May be over-advanced. • Check the torque value of the plugs to ensure good plug-engine seat contact.
Broken or cracked insulator tips.	• Heat shock from sudden rise in tip temperature under severe operating conditions. Improper gapping of plugs.	• Replace the plugs. Gap correctly.

GASOLINE ENGINE TUNE-UP SPECIFICATIONS

Year	Model	Engine Displacement Liter/CID	Spark Plugs		Ignition Timing (deg.)		Fuel Pump Pressure (psi)	Compression Pressure (psi)	Idle Speed (rpm)		Valve Clearance	
			Type*	Gap (in.)	MT	AT			MT	AT	Intake	Exhaust
1983	Ranger	2.0/122	①	①	①	①	5–7	②	①	①	Hyd.	Hyd.
		2.3/140	①	①	①	①	5–7	②	①	①	Hyd.	Hyd.
1984	Bronco II	2.8/173	AWSF-42	0.044	10B	10B	4.5–6.5	②	①	①	0.014	0.016
	Ranger	2.0/122	AWSF-42	0.044	8B	8B	5–7	②	800	800	Hyd.	Hyd.
		2.3/140	AWSF-44	0.044	①	①	5–7	②	800	800	Hyd.	Hyd.
		2.8/173	AWSF-42	0.044	10B	10B	4.5–6.5	②	①	①	0.014	0.016
1985	Bronco II	2.8/173	AWSF-42C	0.044	10B	10B	4.5–6.5	②	①	①	0.014	0.016
	Ranger	2.0/122	AWSF-52C	0.044	6B	6B	5–7	②	800	800	Hyd.	Hyd.
		2.3/140	AWSF-44C	0.044	10B	10B	30–40	②	①	①	Hyd.	Hyd.
		2.8/173	AWSF-42C	0.044	10B	10B	4.5–6.5	②	①	①	0.014	0.016
1986	Bronco II	2.9/177	AWSF-42C	0.044	10B	10B	30–40	②	①	①	Hyd.	Hyd.
	Ranger	2.0/122	AWSF-52C	0.044	6B	6B	5–7	②	800	800	Hyd.	Hyd.
		2.3/140	AWSF-44C	0.044	10B	10B	30–40	②	①	①	Hyd.	Hyd.
		2.9/177	AWSF-42C	0.044	10B	10B	30–40	②	①	①	Hyd.	Hyd.
1987	Bronco II	2.9/177	AWSF-42C	0.044	10B	10B	30–40	②	①	①	Hyd.	Hyd.
	Ranger	2.0/122	AWSF-52C	0.044	6B	6B	5–7	②	800	800	Hyd.	Hyd.
		2.3/140	AWSF-44C	0.044	10B	10B	30–40	②	①	①	Hyd.	Hyd.
		2.9/177	AWSF-42C	0.044	10B	10B	30–40	②	①	①	Hyd.	Hyd.
1988	Bronco II	2.9/177	AWSF-42C	0.044	10B	10B	30–40	②	①	①	Hyd.	Hyd.
	Ranger	2.0/122	AWSF-52C	0.044	6B	6B	5–7	②	800	800	Hyd.	Hyd.
		2.3/140	AWSF-44C	0.044	10B	10B	30–40	②	①	①	Hyd.	Hyd.
		2.9/177	AWSF-42C	0.044	10B	10B	30–40	②	①	①	Hyd.	Hyd.
1989	Bronco II	2.9/177	AWSF-42C	0.044	10B	10B	30–40	②	①	①	Hyd.	Hyd.
	Ranger	2.3/140	AWSF-44C	0.044	10B	10B	30–40	②	①	①	Hyd.	Hyd.
		2,8/177	AWSF-42C	0.044	10B	10B	30–40	②	①	①	Hyd.	Hyd.
1990	Bronco II	2.9/177	AWSF-42C	0.044	10B	10B	30–40	②	①	①	Hyd.	Hyd.
	Ranger	2.3/140	AWSF-44C	0.044	10B	10B	30–40	②	①	①	Hyd.	Hyd.
		2.9/177	AWSF-42C	0.044	10B	10B	30–40	②	①	①	Hyd.	Hyd.
		4.0/241	AWSF-42C	①	10B	10B	30–40	②	①	①	Hyd.	Hyd.
1991	Ranger	2.3	①	①	10B	10B	30–40	②	①	①	Hyd.	Hyd.
		2.9	AWSF-42C	0.044	10B	10B	30–40	②	①	①	Hyd.	Hyd.
		3.0	AWSF-32P	0.044	10B	10B	30–40	②	①	①	Hyd.	Hyd.
		4.0	AWSF-42C	①	10B	10B	30–40	②	①	①	Hyd.	Hyd.
	Explorer	4.0	AWSF-42C	①	10B	10B	30–40	②	①	①	Hyd.	Hyd.

NOTE: The underhood specification sticker often reflects tune-up specification changes made in production. Sticker figures must be used if they conflict with those in this chart.

*Spark plugs shown are original equipment. Part numbers in this reference are not recommendations by Chilton for any product by brand name.

① See underhood specification sticker.

② The lowest compression reading should be within 75 percent of the highest reading.

ALWAYS REFER TO VEHICLE EMISSION CONTROL INFORMATION DECAL FOR SPECIFICATIONS AND TIMING PROCEDURES.

DIESEL TUNE-UP SPECIFICATIONS

Year	Model	Engine Displacement Liter/CID	Valve Clearance (warm) Intake (in.)	Valve Clearance (warm) Exhaust (in.)	Injection Timing (ATDC)	Injection Nozzle Pressure (psi)	Idle Speed (rpm)	Cranking Compression Pressure (psi)
1983	Ranger	2.2/134 Diesel	0.012	0.012	2	1957	700	427 ②
1984	Ranger	2.2/134 Diesel	0.012	0.012	2	1957	700	427 ②
1985	Ranger	2.2/140 Diesel	0.010	0.010	①	1707	①	384 ③
1986	Ranger	2.2/140 Diesel	0.010	0.010	①	1707	①	384 ③
1987	Ranger	2.2/140 Diesel	0.010	0.010	①	1707	①	384 ③

① See underhood specification sticker
② At 200 rpm
③ At 250 rpm

GASOLINE ENGINE TUNE-UP PROCEDURES

Spark Plugs

A typical spark plug consists of a metal shell surrounding a ceramic insulator. A metal electrode extends downward through the center of the insulator and protrudes a small distance. Located at the end of the plug and attached to the side of the outer metal shell is the side electrode. The side electrode bends in at a 90° angle so that its tip is even with, and parallel to, the tip of the center electrode. The distance between these two electrodes (measured in thousandths of an inch) is called the spark plug gap. The spark plug in no way produces a spark but merely provides a gap across which the current can arc. The coil produces anywhere from 20,000 to 40,000 volts which travels to the distributor where it is distributed through the spark plug wires to the spark plugs. The current passes along the center electrode and jumps the gap to the side electrode, and, in do doing, ignites the air/fuel mixture in the combustion chamber.

Spark plugs ignite the air and fuel mixture in the cylinder as the piston reaches the top of the compression stroke. The controlled explosion that results forces the piston down, turning the crankshaft and the rest of the drive train.

The average life of a spark plug is dependent on a number of factors: the mechanical condition of the engine; the type of engine; the type of fuel; driving conditions; and the driver.

When you remove the spark plugs, check their condition. They are a good indicator of the condition of the engine. It it a good idea to remove the spark plugs at regular intervals, such as every 2,000 or 3,000 miles, just so you can keep an eye on the mechanical state of your engine.

A small deposit of light tan or gray material on a spark plug that has been used for any period of time is to be considered normal.

The gap between the center electrode and the side or ground electrode can be expected to increase not more than 0.001 in. every 1,000 miles under normal conditions.

When a spark plug is functioning normally or, more accurately, when the plug is installed in an engine that is functioning properly, the plugs can be taken out, cleaned, regapped, and reinstalled in the engine without doing the engine any harm.

When, and if, a plug fouls and beings to misfire, you will have to investigate, correct the cause of the fouling, and either clean or replace the plug.

There are several reasons why a spark plug will foul and you can learn which is at fault by just looking at the plug. A few of the most common reasons for plug fouling, and a description of the fouled plug's appearance, are listed in the Color Section, which also offers solutions to the problems.

SPARK PLUG HEAT RANGE

Spark plug heat range is the ability of the plug to dissipate heat. The longer the insulator (or the farther it extends into the engine), the hotter the plug will operate; the shorter the insulator the cooler it will operate. A plug that absorbs little heat and remains too cool will quickly accumulate deposits of oil and carbon since it is not hot enough to burn them off. This leads to plug fouling and consequently to misfiring. A plug that absorbs too much heat will have no deposits, but, due to the excessive heat, the electrodes will burn away quickly and in some instances, preignition may result. Preignition takes place when plug tips get so hot that they glow sufficiently to ignite the fuel/air mixture before the actual spark occurs. This early ignition will usually cause a pinging during low speeds and heavy loads.

The general rule of thumb for choosing the correct heat range when picking a spark plug is: if most of your driving is long distance, high speed travel, use a colder plug; if most of your driving is stop and to, use a hotter plug. Original equipment plugs are compromise plugs, but most people never have occasion to change their plugs from the factory-recommended heat range.

REPLACING SPARK PLUGS

Ford recommends that spark plugs be changed every 30,000 miles with electronic ignition systems. Under severe driving conditions, those intervals should be halved. Severe driving conditions are:

1. Extended periods of idling or low speed operation, such as off-road or door-to-door delivery.
2. Driving short distances (less than 10 miles) when the average temperature is below 10° for 60 days or more.
3. Excessive dust or blowing dirt conditions.

When you're removing spark plugs, you should work on one at a time. Don't start by removing the plug wires all at once, because unless you number them, they may become mixed up. Take a minute before you begin and number the wires with tape. The best location for numbering is near where the wires come out of the cap.

1. Twist the spark plug boot and remove the boot and wire form the plug. Do not pull on the wire itself as this will ruin the wire.
2. If possible, use a brush or rag to clean the area around the spark plug. Make sure that all the dirt is removed so that none will enter the cylinder after the plug is removed.
3. Remove the spark plug using the proper size socket. Turn the socket counterclockwise to remove the plug. Be sure to hold the socket straight on the plug to avoid breaking the plug, or rounding off the hex on the plug.
4. Once the plug is out, check it against the plugs shown in this Section to determine engine condition. This is crucial since plug readings are vital signs of engine condition.
5. Use a round wire feeler gauge to check the plug gap. The correct size gauge should pass through the electrode gap with a slight drag. If you're in doubt, try one size smaller and one larger. The smaller gauge should go through easily while the larger one shouldn't go through at all. If the gap is incorrect, use the electrode bending tool on the end of the gauge to adjust the gap. When adjusting the gap, always bend the side electrode. The center electrode is non-adjustable.
6. Squirt a drop of penetrating oil on the threads of the new plug and install it. Don't oil the treads too heavily. Turn the plug in clockwise by hand until it is snug.
7. When the plug is finger tight, tighten it with a wrench.

NOTE: Whenever a high tension wire is removed for any reason from a spark plug, coil or distributor terminal housing, silicone grease must be applied to the boot before it is reconnected. Using a small clean tool, coat the entire interior surface of the boot with Ford silicone grease D7AZ 19A331-A or equivalent.

8. Install the plug boot firmly over the plug. Proceed to the next plug.

Spark Plug Wires

INSPECTION

Visually inspect the spark plug cables for burns, cuts, or breaks in the insulation. Check the spark plug boots and the nipples on the distributor cap and coil. Replace any damaged wiring. If no physical damage is obvious, the wire can be checked with an ohmmeter for excessive resistance.

When installing a new set of spark plug cables, replace the cables one at a time so there will be no mixup. Start by replacing the longest cables first. Install the boot firmly over the spark plug. Route the wire exactly the same as the original. Insert the nipple firmly into the tower on the distributor cap. Repeat the process for each cable.

REMOVAL

When removing spark plug wires, use great care. Grasp and twist the insulator back and forth on the spark plug to free the insulator. Do not pull on the wire directly as it may become separated from the connector inside the insulator.

INSTALLATION

NOTE: Whenever a high tension wire is removed for any reason form a spark plug, coil or distributor terminal housing, silicone grease must be applied to the boot before it is reconnected. Using a small clean tool, coat the entire interior surface of the boot with Ford silicone grease D7AZ 19A331-A or equivalent.

1. Install each wire in or on the proper terminal of the distributor cap. Be sure the terminal connector inside the insulator is fully seated. The No. 1 terminal is identified on the cap.
2. Remove wire separators from old wire set and install them on new set in approximately same position.
3. Connect wires to proper spark plugs. Install ignition coil wire. Be certain all wires are fully seated on terminals.

Always use a wire gauge to check the electrode gap; a flat feeler gauge may not give the proper reading

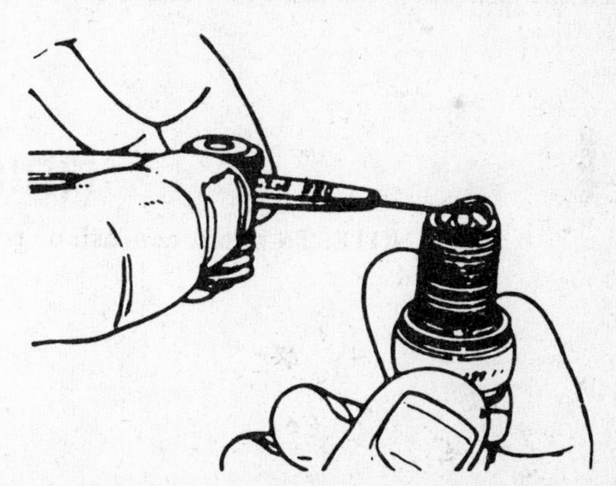

Always use a wire gauge to check the electrode gap; a flat feeler gauge may not give the proper reading

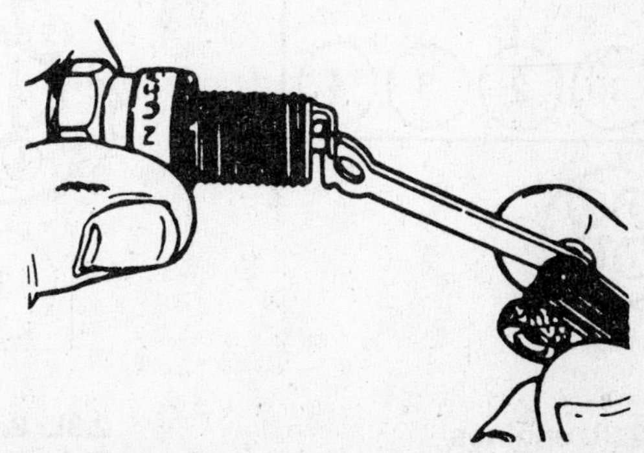

Adjust the electrode gap by bending the side electrode

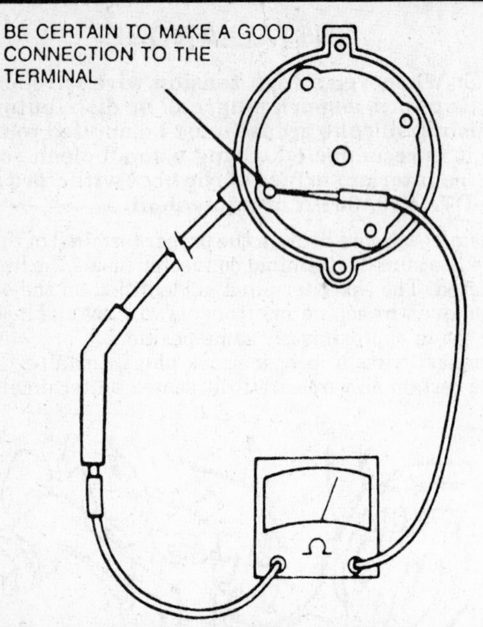

Testing spark plug wire resistance with an ohmmeter

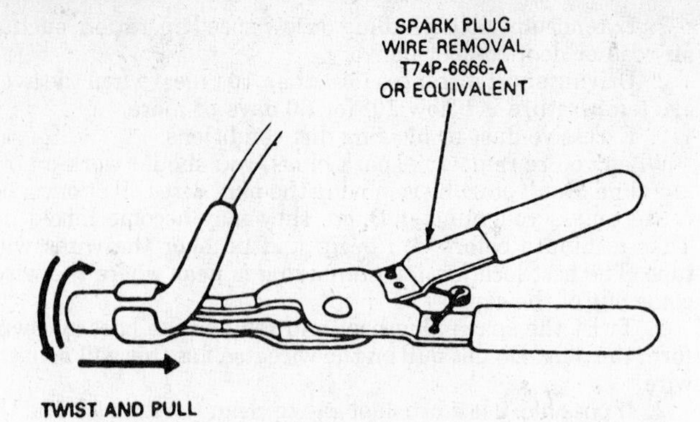

Spark plug removal tool

TESTING

1. Remove the distributor cap from the distributor assembly.
2. Visually inspect the spark plug wires for burns, cuts or breaks in the insulation. Check the spark plug boots and the nipples on the distributor cap and coil. Replace any damaged wiring.
3. Inspect the spark plug wires to insure that they are firmly seated on the distributor cap.
4. Disconnect the spark plug wire(s) thought to be defective at the spark plug.
5. Using an ohmmeter, measure the resistance between the distributor cap terminal and the spark plug terminal.

NOTE: Make certain that a good connection exists between the distributor cap and the spark terminal. Never, under any circumstances, measure resistance by puncturing the spark plug wire.

6. If the measured resistance is less than 7000Ω per foot of wire, the wire is good. If the measured resistance is greater than 7000Ω per foot, the wire is defective and should be replaced.

FIRING ORDERS

NOTE: To avoid confusion, replace spark plugs and wires one at a time.

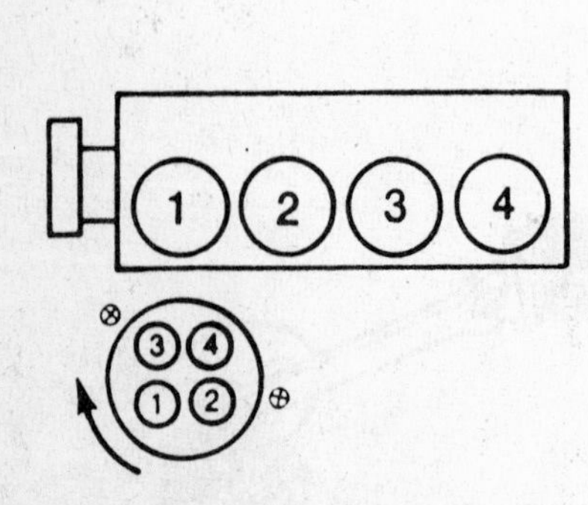

2.0L, 2.3L engines
Firing order: 1-3-4-2
Distributor rotation: clockwise

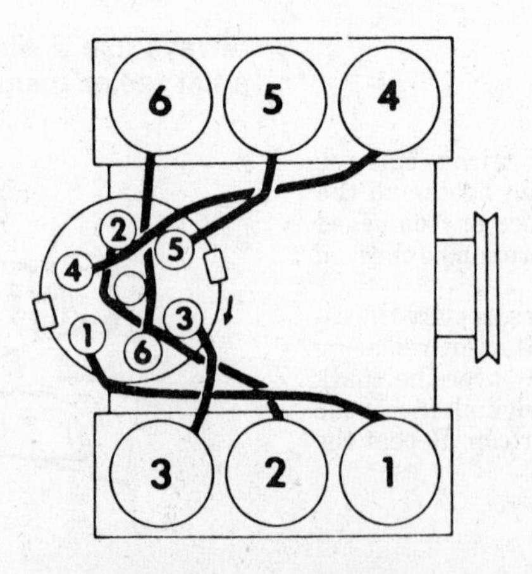

2.8L, 2.9L, 3.0L engines
Firing order: 1-4-2-5-3-6
Distributor rotation: clockwise

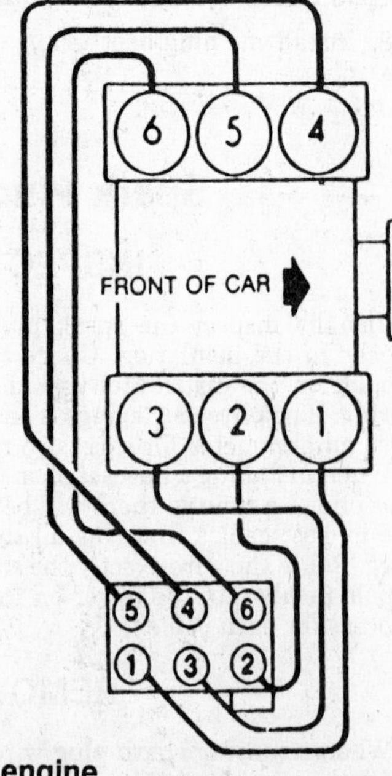

4.0L engine
Firing order: 1-4-2-5-3-6
Distributor rotation: no distributor

Electronic Ignition

The 2.0L and the 2.3L (1983-84) engines are equipped with Dura Spark II ignition system. However, depending on engine calibration, the Dura Spark II system may use a standard module or a **Universal Ignition Module.** The Universal Ignition Module (UIM) is capable of providing spark timing retard in response to barometric or engine sensors, or MCU signal.

The Ranger/Bronco II equipped with the 2.3L (1985-88), 2.8L, 2.9L or 3.0L engines uses a universal distributor design which incorporates an integrally mounted TFI-IV module.

NOTE: On some models, 2.3L (twin plug) and 4.0L engine applications use a distributorless ignition system. This system consists of the following components: Crankshaft timing sensor, EDIS or DIS ignition module, one or two Ignition coil packs, EEC-IV module and related wiring. Initial timing is PRESET at 10°BTDC and is NOT ADJUSTABLE.

The distributor uses a **Hall Effect** vane switch stator assembly and has provision for fixed octane adjustment. A new cap, adapter and rotor are designed for use on the Universal Distributor. The Thick Film Integrated (TFI) module is contained in molded thermo-plastic and is mounted on the distributor base. The TFI-IV features a **push start** mode which allows push starting of the vehicle, if necessary. The TFI-IV system uses an **E-Core** ignition coil, which replaces the oil-filled coil found on other systems.

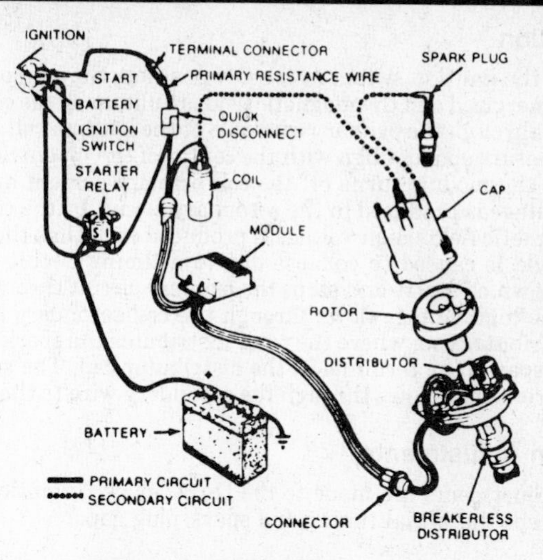

Dura Spark II Ignition system

DUAL MODE TIMING IGNITION MODULE

On some applications, a special Dura Spark II ignition module is used with altitude compensation. This special module plus the barometric pressure switch, allows the base engine timing to be modified to suit altitude conditions. All other elements and performance characteristics of this module are identical in both modes of operation to the basic Dura Spark II system. All Dura Spark II modules equipped with altitude features, have three connectors instead of the normal two. A barometric switch provides an automatic retard signal to the module at different altitudes, giving appropriate advanced timing at higher altitude and retard mode for spark knock control at lower altitudes.

DISTRIBUTOR

The distributor are equipped with both vacuum and centrifugal spark advances which operate the same regardless of the type of ignition system used. A dual vacuum advance is used on certain engines to provide ignition retard during engine closed throttle operation, to help control engine exhaust emissions.

CIRCUIT OPERATION

All systems consist of a primary (low voltage) and secondary (high voltage) circuit.

The Primary Circuit:
The components involved in the primary circuit are:

1. Battery
2. Ignition switch
3. Integral primary circuit resistance wire
4. Primary windings of the ignition coil
5. Magnetic pickup coil assembly in the distributor
6. Ignition module

The Secondary Circuit:
The components of the secondary circuit are:

1. Secondary windings of the ignition coil
2. Distributor rotor
3. Distributor cap and adapter
4. Secondary spark plug wires
5. Spark plugs

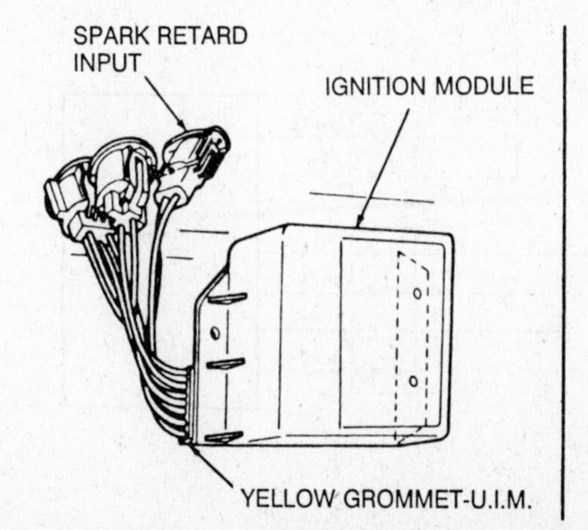

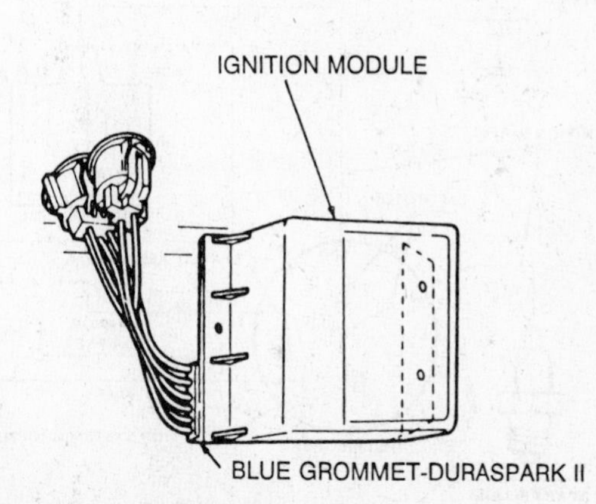

Two types of Dura Spark II ignition modules

Operation

With the ignition switch in the **ON** position, the primary circuit is energized and the magnetic field is built up by the current flowing through the primary windings of the ignition coil. When the armature spokes align with the center of the magnetic pick-up coil, the module turns off the coil primary current and the high voltage is produced in the secondary circuit by the collapsing magnetic field. High voltage is produced each time the magnetic field is caused to collapse due to a timing circuit in the module, which starts and stops the primary circuit through the coil. The high voltage flows through the coil secondary lead to the distributor cap, where the rotor distributes the spark to the proper spark plug terminal in the distributor cap. The secondary current then flows through the secondary wire to the spark plug.

System Adjustments

No adjustments are made to the Dura Spark II ignition system except the initial timing and spark plug gap.

SECONDARY WIRE USAGE

Spark plug wires that are used with the Dura Spark II system are 8mm in size, to contain the higher output voltage. Two types of wires are used in this system and some engines will have both types. It is important to identify the type of wire to a cylinder before a replacement is obtained and installed. Both types are blue in color and have silicone jacketing. The insulation material underneath the jacketing can be a EPDM or have another silicone layer, separated by glass braid. EPDM wires are used where the engine temperatures are cooler and are identified by the letters **SE**. The silicone jacket type are used where the engine temperatures are high and are identified by the letters **SS**.

NOTE: Whenever a Dura Spark II high tension wire is removed for any purpose from a spark plug, coil or distributor cap, silicone grease must be applied to the boot before it is reconnected.

The spark plug wires are marked with the cylinder number, model year and date of cable manufacture (quarter and year). Service replacement wires do not have this information.

TROUBLESHOOTING

To properly diagnose the ignition system, a starting place must be established and an order of inspection followed until the fault is found and repaired. A recheck should be made, again in its order of inspection, to verify the repairs and to assure trouble-free operation.

Run Mode Test

1. If no spark is available at the spark plug, remove the coil high tension lead at the distributor and either place it ¼ in. from the engine block or place a modified spark plug into the coil wire and ground the spark plug body.
2. Turn the ignition switch to the **RUN** position and tap the distributor body with a screwdriver type tool handle. Check for spark while tapping.
3. If spark is available, crank the engine with the starter and check for spark. If spark occurs, the primary ignition system is OK.
4. If no spark occurs, turn the key to the **OFF** position and crank the engine to align the engine timing pointer with the initial timing degree line on the damper pulley. Turn the key to the **RUN** position and again tap the distributor and check for spark.
5. If no spark occurs, measure battery voltage and measure the battery voltage on the module's red wire without disconnecting any connectors. The voltage in the red wire should equal battery voltage.
6. If battery voltage is not present in the module red wire, repair the circuit between the battery and the module connector. Recheck the voltage supply.
7. With the voltage present in the modules's red wire, cycle the ignition switch between the **RUN** and **OFF** position. A spark should be seen each time the switch is turned to the **OFF** position.
8. If no spark occurs, measure the voltage on the battery side of the coil.
   a. Less than 6 volts – Repair the wire carrying current to the battery terminal of the coil and repeat test.
   b. If voltage is 6–8 volts – Substitute, but do not install, a known good module and repeat the test. If spark then occurs, reconnect the original module to verify its being defective. Replace as required. Refer to step 10 if the battery voltage is present.

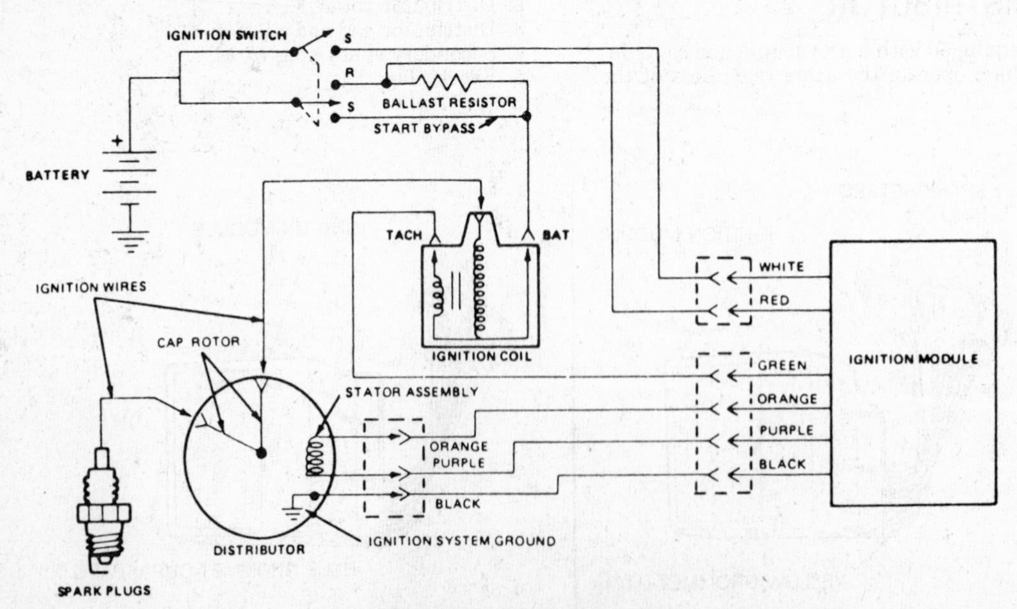

Dura Spark II circuit operation

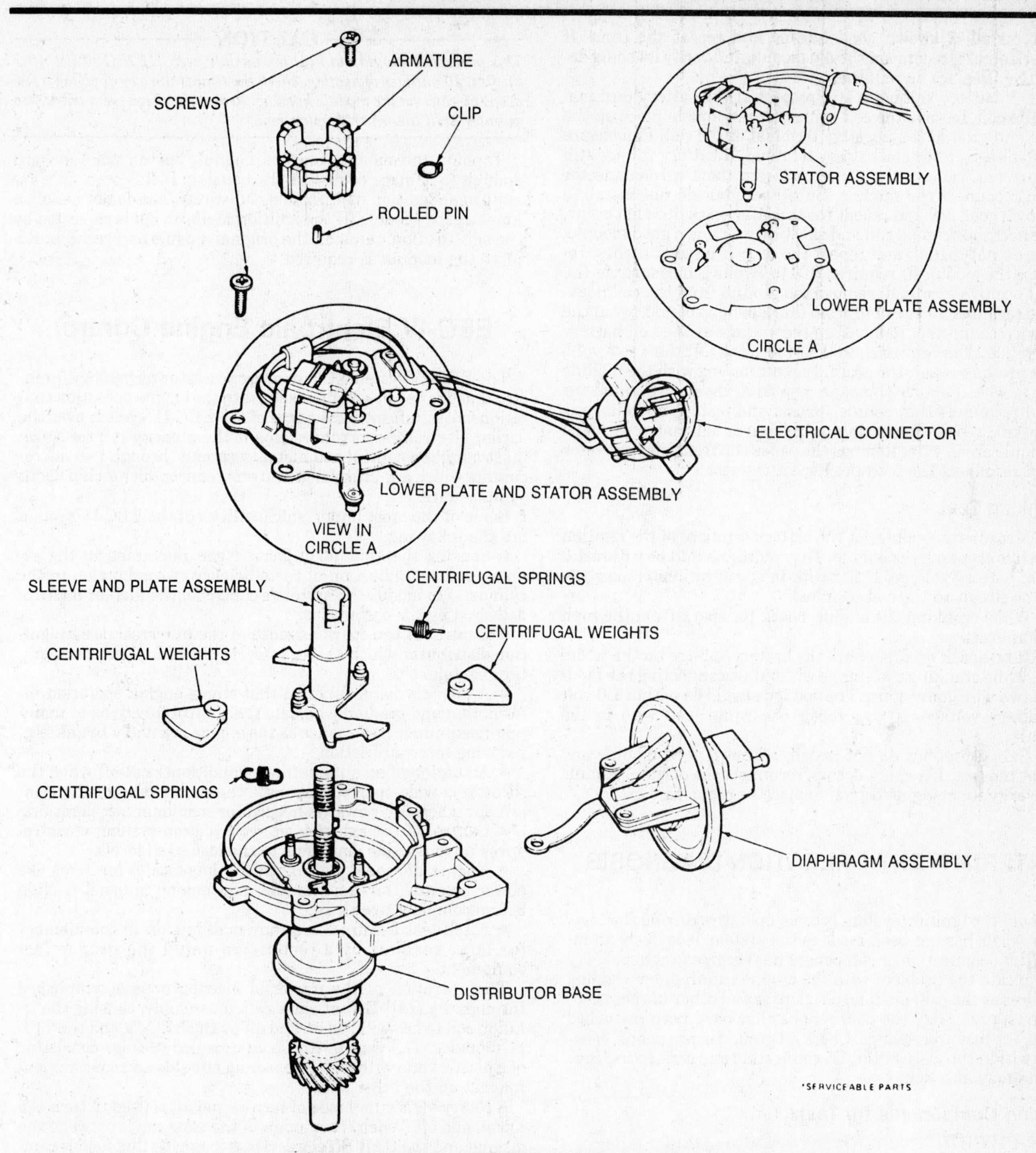

Dura Spark II distributor assembly – 4 cylinder 122, 140 engines

9. If a spark occurs from step 7, substitute, but do not install, and ground a good distributor of any calibration 4-, 6- or 8-cylinder. Spin the distributor shaft and check for high tension spark.
   a. If a spark occurs, reconnect the old distributor and verify its being defective. Replace as required.
   b. If no spark occurs, disconnect the distributor connector and 4 post connector at the module. Check the harness wires that mate with the module and distributor orange and purple wires for continuity between the module and distributor end of the harness. Check to be sure there is no short between the two wires and there is an open circuit to ground. If not OK, repair the wiring and repeat the test to verify repairs.
   c. If no spark occurs after completing step 9b, reconnect the distributor connector and substitute, but do not install, a known good module and repeat the test. If a spark occurs, reconnect the original module and verify it is defective. Repair as required.
10. If battery voltage is present at the terminal of the coil:
   a. Disconnect the 4 wire connector at the module. Insert a paper clip between the green and black wires of the module and remeasure the voltage at the battery terminal of the coil.
   b. If the voltage is between 6 to 8 volts, substitute, but do

not install, a known good module and repeat the tests. If spark occurs, reconnect the old module and verify its being defective. Replace as required.

c. If battery voltage is still present at the battery terminal of the coil, be sure the coil connector remains in place on the coil and ground the negative terminal of the coil. Remeasure the voltage on the coil battery terminal. If battery voltage still is present, remove the paper clip from the 4 wire connector and reconnect the module. Substitute, but do not install, a known good coil and repeat the test. If a spark does not occur, connect the original coil and substitute a known good module, but do not install, and repeat the test. If a spark occurs, replace the module as required. If 4 to 7 volts is measured at the coil positive terminal, remove the ground from the coil negative terminal and ground from the paper clip connector in the 4 wire connector. Remeasure the voltage at the coil battery terminal. The voltage should be 4 to 7 volts. If the 4 to 7 volts are present, repair the ground circuit mating with the module black wire. Remove the paper clip from the 4 wire connector and reconnect the module. Repeat the test. If no voltage is present, repair the module to coil wire that mates with the module green wire. Remove the paper clip from the connector and reconnect the module. Repeat the test.

Cranking Test

1. Measure the voltage at the battery terminal of the ignition coil while cranking the engine. The reading should be within 1.0 volt of battery voltage. If not with-in specifications, repair the wire or circuit to the coil terminal.
2. While cranking the engine, check for spark from the high tension leads.
3. If no spark occurs, check the battery voltage on the white wire, while cranking the engine without disconnecting the module's two wire connectors. The voltage should be within 1.0 volt of battery voltage. If not, repair the white feed wire to the module.
4. Substitute, but do not install, a known good module and repeat the test. If a spark occurs, reconnect the original module and verify its being defective. Replace as required.

INTERMITTENT OPERATION DIAGNOSIS

Should the ignition system become operative during the tests and a repair has not been made to the system, it is likely an intermittent connection or component has become functional. Try to duplicate the problem with the engine running, by wiggling the wires at the coil, module, distributor and other harness connections, preferably the connections that have been disturbed during the test proceedings. Check all ground connections, especially within the distributor. Disconnecting and connecting connectors may also help.

Heating Components for Tests

PICK-UP COIL

Using a 250 watt heat lamp, approximately 1–2 in. from the pick-up coil, apply heat for 4 to 6 minutes while monitoring the pick-up coil continuity between the parallel blades of the disconnected distributor connector. The resistance should be 400 to 1000 ohms. Tapping with a screwdriver type handle may also be helpful to locate problem. If specifications cannot be met or held, replace the pick-up coil.

IGNITION MODULE

With the engine running, heat the module by placing a 250 watt heat lamp bulb approximately 1–2 in. from the module top surface.

CAUTION

This procedure should not heat the module over 212°F (100°C). After the first 10 minutes of heating, check the temperature by applying a few drops of water on the module housing. Repeat the check every one to two minutes until the water droplets boil.

Tapping the module may be helpful, but do not tap hard enough to damage or distort the housing. If this procedure results in an ignition malfunction, substitute, but do not install, a known good module. If the ignition malfunction is corrected by the substitution, reinstall the original module and recheck. Replace the module as required.

EEC-IV Electronic Engine Control

The EEC-IV system utilizes microprocessor technology for instantaneous detection and response to the engine operation conditions. A significant advantage of the EEC-IV system over the earlier electronic-control systems is the capacity to process almost a million control commands a second through two microcircuits which are integrated into one semiconductor chip about ¼ in. square.

Some of the most important functions of the EEC-IV system are the following:

- Sensing the amount of free oxygen remaining in the exhaust gas (an indication of the efficiency of combustion in the cylinders) to modify the system's control of the fuel-air mixture delivered by the carburetor.
- Control of the spark advance of the universal design ignition distributor which has an integral thick film electronic ignition module.
- A memory bank function that stores engine operation information and gradually adjusts the control functions to maintain maximum performance as the engine gradually **breaks in**, reducing internal friction.
- Actuation of an automatic air-conditioner cut-off when the throttle is wide-open to eliminate the power drag of the compressor when full power is needed for maximum acceleration.
- Control of the exhaust-gas recirculation system which reduces nitrous oxide emissions to the legislated levels.
- Engine idle speed control which compensates for items like power steering, air-conditioning, and engine internal friction before complete **break-in**.
- A self-test feature stores any malfunction in the memory for later readout by a technician using the proper test equipment.

The operation of the universal distributor is accomplished through the **Hall Effect** vane switch assembly, causing the ignition coil to be switched on and off by the EEC-IV and the TFI-IV modules. The vane switch is an capsuled package consisting of a plastic vane with a Hall sensor on one side and a permanent magnet on the other side.

A rotary vane cup, made of ferrous metal, is used to turn the coil on and off. When the window of the vane cup is between the magnet and the **Hall Effect** device, a magnetic flux field is completed from the magnet through the **Hall Effect** device and back to the magnet. As the vane passes through this opening, the flux lines are shunted through the vane and back to the magnet. During this time, the **Hall Effect** device is turned on and a voltage pulse is produced as the vane passes through the opening. When the vane clears the opening, the window edge shuts the device down and the signal is turned off. The signal is then used by the EEC-IV system for crankshaft position sensing and the computation of the proper spark advance based on the engine demand. The conditioned spark advance and the voltage distribution is accomplished through a conventional rotor, cap and ignition wires.

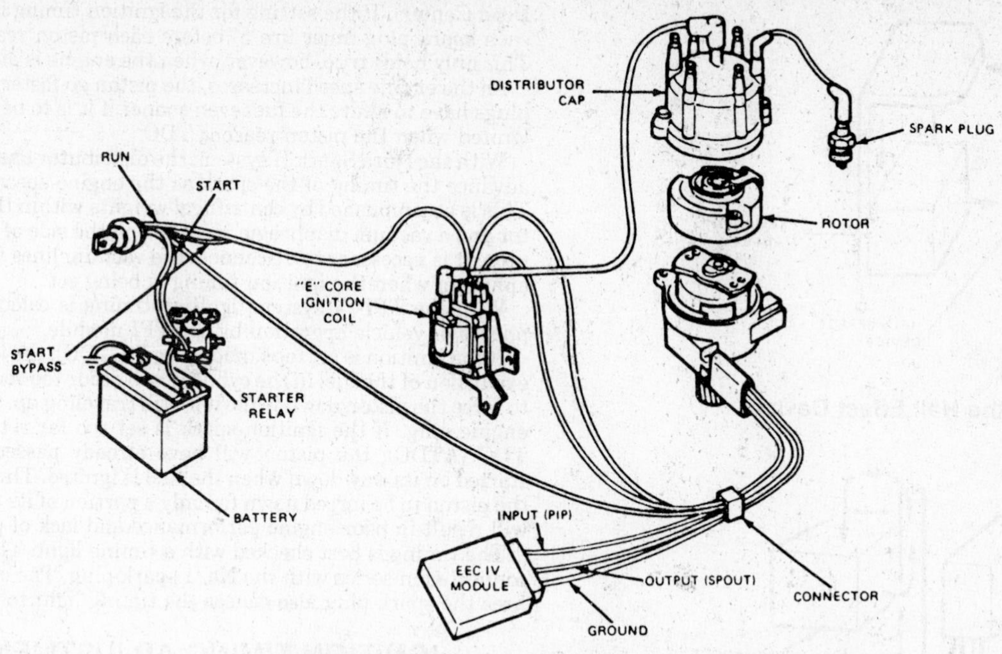

EEC-IV Thick Film Integrated (TFI) Ignition system

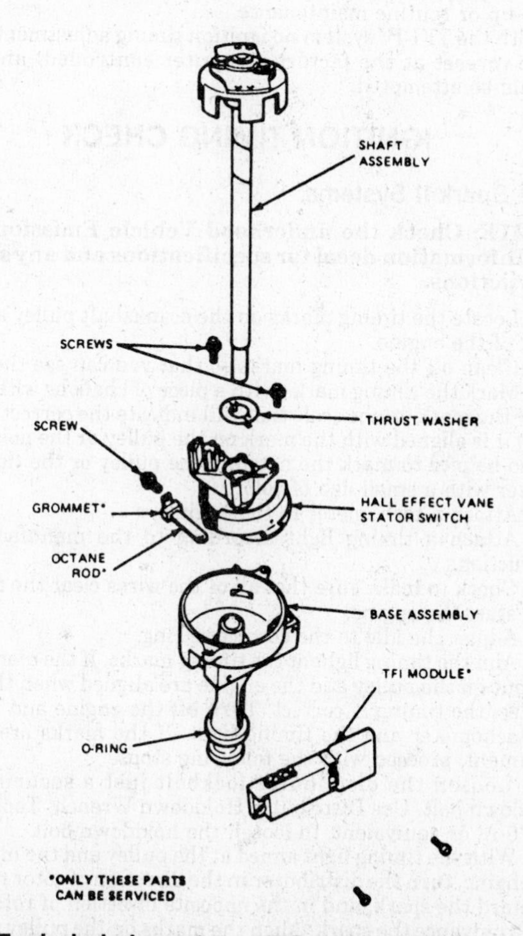

Exploded view of the EEC-IV distributor

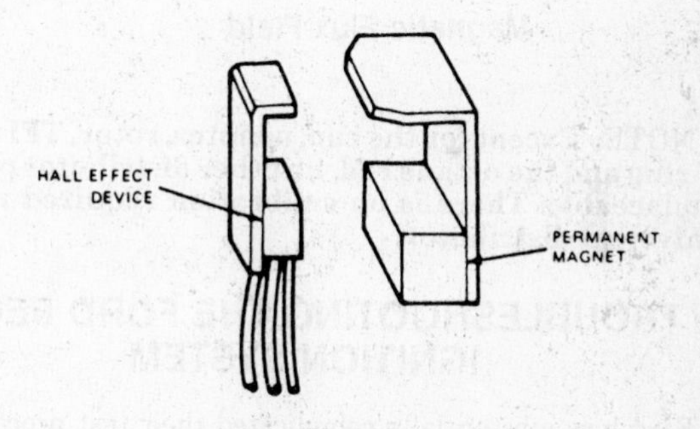

Hall effect Device

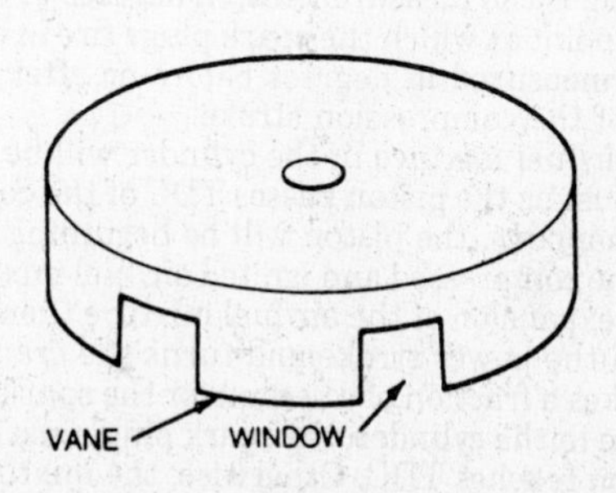

Rotary Vane cap

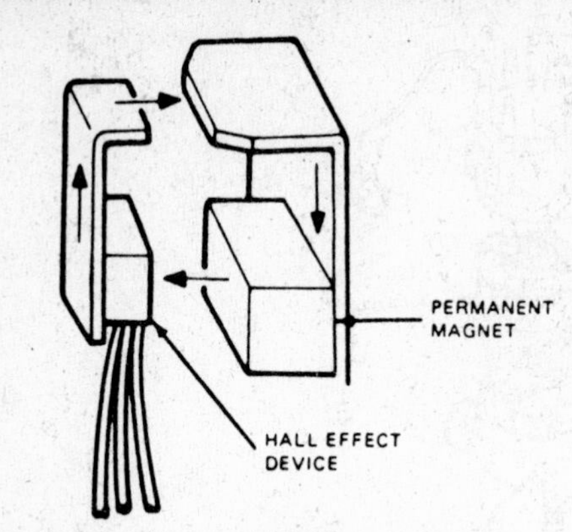

Activating the Hall Effect Device

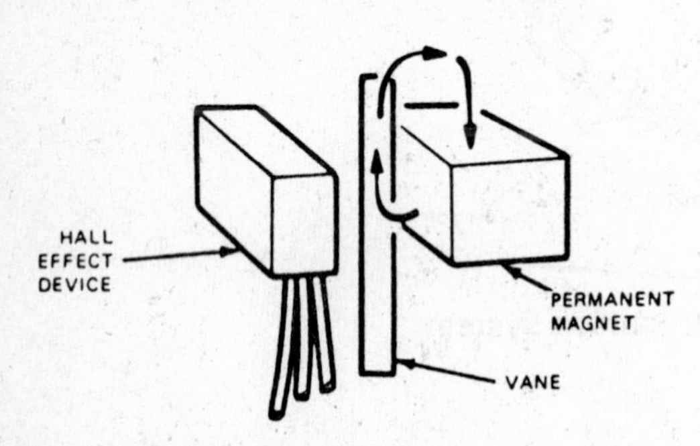

Magnetic Flux Field

NOTE: Except for the cap, adapter, rotor, TFI module, O-ring and the octane rod, no other distributor parts are replaceable. There is no calibration required with the universal distributor.

TROUBLESHOOTING THE FORD EEC-IV IGNITION SYSTEM

Ford has substantially complicated their test procedure for the EEC-IV electronic ignition system. Due to the sensitive nature of the system and the complexity of the test procedures, it is recommended that you refer to your dealer if you suspect a problem in your electronic ignition system.

For all repair and adjustment procedures, please refer to Section 3, Engine Electrical Section.

Ignition Timing

Ignition timing is the measurement, in degrees of crankshaft rotation, of the point at which the spark plugs fire in each of the cylinders. It is measured in degrees before or after Top Dead Center (TDC) of the compression stroke.

Ideally, the air/fuel mixture in the cylinder will be ignited by the spark plug just as the piston passes TDC of the compression stroke. If this happens, the piston will be beginning the power stroke just as the compressed and ignited air/fuel mixture starts to expand. The expansion of the air/fuel mixture then forces the piston down on the power stroke and turns the crankshaft.

Because it takes a fraction of a second for the spark plug to ignite the mixture in the cylinder, the spark plug must fire a little before the piston reaches TDC. Otherwise, the mixture will not be completely ignited as the piston passes TDC and the full power of the explosion will not be used by the engine.

The timing measurement is given in degrees of crankshaft rotation before the piston reaches TDC (BTDC, or Before Top Dead Center). If the setting for the ignition timing is 5° BTDC, each spark plug must fire 5° before each piston reaches TDC. This only holds true, however, when the engine is at idle speed.

As the engine speed increases, the piston go faster. The spark plugs have to ignite the fuel even sooner if it is to be completely ignited when the piston reaches TDC.

With the Dura Spark II system, the distributor has a means to advance the timing of the spark as the engine speed increases. This is accomplished by centrifugal weights within the distributor and a vacuum diaphragm mounted on the side of the distributor. It is necessary to disconnect the vacuum lines from the diaphragm when the ignition timing is being set.

With the TFI-IV system, ignition timing is calculated at all phases of vehicle operation by the TFI module.

If the ignition is set too far advanced (BTDC), the ignition and expansion of the fuel in the cylinder will occur too soon and tend to force the piston down while it is still traveling up. This causes engine ping. If the ignition spark is set too far retarded after TDC (ATDC), the piston will have already passed TDC and started on its way down when the fuel is ignited. This will cause the piston to be forced down for only a portion of its travel. This will result in poor engine performance and lack of power.

The timing is best checked with a timing light. This device is connected in series with the No. 1 spark plug. The current that fires the spark plug also causes the timing light to flash.

IGNITION TIMING ADJUSTMENT

With the Dura Spark II system, only an initial timing adjustment is possible. Ignition timing is not considered to be a part of tune-up or routine maintenance.

With the TFI-IV system no ignition timing adjustment is possible (preset at the factory-computer controlled) and none should be attempted.

IGNITION TIMING CHECK

Dura Spark II Systems

NOTE: Check the underhood Vehicle Emission Control Information decal for specifications and any special instructions.

1. Locate the timing marks on the crankshaft pulley and the front of the engine.
2. Clean off the timing marks so that you can see them.
3. Mark the timing marks with a piece of chalk or with paint. Color the mark on the scale that will indicate the correct timing when it is aligned with the mark on the pulley or the pointer. It is also helpful to mark the notch in the pulley or the tip of the pointer with a small dab of color.
4. Attach a tachometer to the engine.
5. Attach a timing light according to the manufacturer's instructions.
6. Check to make sure that all of the wires clear the fan and then start the engine.
7. Adjust the idle to the correct setting.
8. Aim the timing light at the timing marks. If the marks that you put on the pulley and the engine are aligned when the light flashes, the timing is correct. Turn off the engine and remove the tachometer and the timing light. If the marks are not in alignment, proceed with the following stops.
9. Loosen the distributor lockbolt just a security-type holddown bolt. Use Distributor Holddown Wrench, Tool T82L-12270-A, or equivalent, to loosen the holddown bolt.
10. With the timing light aimed at the pulley and the marks on the engine, turn the distributor in the direction of rotor rotation to retard the spark, and in the opposite direction of rotor rotation to advance the spark. Align the marks on the pulley and the engine with the flashes of the timing light.
11. When the marks are aligned, tighten the distributor

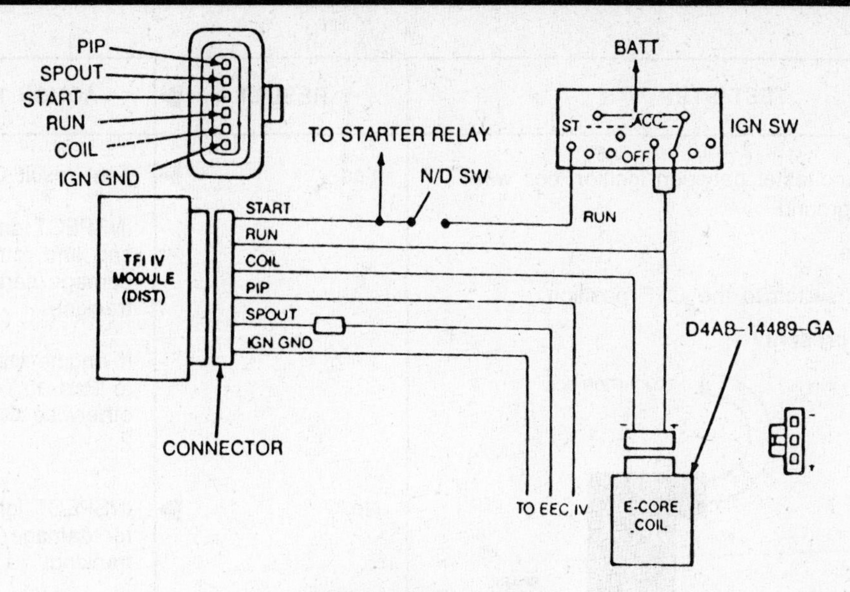

TFI-IV Ignition system

lockbolt and recheck the timing with the timing light to make sure that the distributor did not move when you tightened the lockbolt.

12. Turn off the engine and remove the timing light.

TFI-IV Electronic Ignition System

This system has a universal distributor design which is gear driven and has a die cast metal base that incorporates an integrally mounted TFI-IV ignition module.

The distributor uses a "Hall Effect" stator assembly and eliminates the conventional centrifugal and vacuum advance mechanisms. No distributor calibration is required and it is not normally necessary to adjust initial timing. The cap, adapter and rotor are designed for use with the universal distributor and the ignition module is a Thick Film Integrated (TFI) design. The module is contained in molded thermoplastic and is mounted on the distributor base retained by two screws. The distributor assembly can be identified by the part number information printed on a decal attached to the side of the distributor base.

In addition to the Hall Effect switch, the ignition module contains the Profile Ignition Pick-up (PIP) sensor, which sends an electronically oriented crankshaft position signal to the ECA and the TFI module circuitry. The ECA, after taking all the sensors information, produces a new signal called the Spout. This Spout signal is then sent back to the TFI module for comparison with the PIP signal. The TFI-IV module then uses both of these signals to fire the ignition coil at the proper timing interval. A modification to the circuitry allows for a Push-Start mode for manual transmission equipped vehicles.

The operation of the universal distributor is accomplished through the Hall Effect vane switch assembly, causing the ignition coil to be switched on and off by the EEC-IV and TFI-IV modules. The vane switch is an encapsulated package consisting of a Hall sensor on one side and a permanent magnet on the other side. A rotary vane cup, made of ferrous (magnetic) metal is used to trigger the Hall Effect switch.

When the window of the vane cup is between the magnet and Hall Effect device, a magnetic flux field is completed from the magnet through the Hall Effect device and back to the magnet. As the vane passes through this opening, the flux lines are shunted through the vane and back to the magnet. As the vane passes through this opening, the flux lines are shunted through the vane and back to the magnet. A voltage is produced while the vane passes through the opening. When the vane clears the opening, the window causes the signal to go to zero volts. The signal is then used by the EEC-IV system for crankshaft position sensing and the computation of the desired spark advance based on engine demand and calibration. The conditioned spark advance and voltage distribution is accomplished through a conventional rotor, cap and ignition wires.

NOTE: The ignition timing is preset at the factory and computer controlled. No attempt should be made to alter the ignition timing from the factory specifications.

TROUBLESHOOTING

Many times a quick test can locate the cause of a problem without going into full system checkout. Included are tests which may isolate the cause of the problem. The first step is to verify that a problem exists and then to make some preliminary tests to determine if the problem is in the ignition system, a related system or a completely unrelated system. The following procedures are intended to provide tests to identify and locate some of the more frequently encountered problems.

Intermittant faults are the hardest faults to identify simply because they alternately appear and go away. Intermittant faults may be the result of corroded terminals, cracked or broken wires, voltage leakage, heat related failures, etc and they will drive evn the most initiated technician crazy. Verify the mode of the ignition system and engine when the malfunction occurs and relate to this mode for failure indications. (examples = engine hot or cold, acceleration or deceleration, etc).

TEST STEP	RESULT ▶	ACTION TO TAKE
1. Connect spark tester between ignition coil wire and engine ground. 2. Crank engine. 3. Turn ignition switch to the OFF position. 4. **Was spark present?**	Yes ▶	Test result OK. INSPECT distributor cap and rotor for damage/carbon tracking. If engine starts, GO to Part 1, Test 2, otherwise GO to Test 2.
	No ▶	INSPECT ignition coil for damage/carbon tracking. CRANK engine to verify distributor rotation. GO to Test 4.

Test 1 – secondary coil voltage

TEST STEP	RESULT ▶	ACTION TO TAKE
1. Place the transmission shift lever in the PARK (A/T) or NEUTRAL (M/T) position and set the parking brake. CAUTION **Failure to perform this step may result in the vehicle moving when the starter is subsequently engaged during the test.** 2. Disconnect wire at S terminal of starter relay. 3. Attach remote starter switch. 4. Turn ignition switch to the RUN position. 5. Crank the engine using remote starter switch. 6. Turn ignition switch to the OFF position. 7. Remove remote starter switch. 8. Reconnect wire to S terminal of starter relay. 9. **Was spark present?** TO IGNITION COIL SPARK TESTER ENGINE GROUND	Yes ▶ No ▶	Test result OK. Problem is not in the ignition system. GO to Test 3.

Test 2 — secondary coil voltage

TEST STEP	RESULT	ACTION TO TAKE
1. Separate wiring harness connector from ignition module. Inspect for dirt, corrosion, and damage. **NOTE: Push connector tabs to separate.** 2. Verify that the wire to the S terminal of starter relay is disconnected. 3. Attach negative (−) VOM lead to distributor base. 4. Measure battery voltage. 5. Following the appropriate table below, measure connector terminal voltage by attaching VOM to small straight pin inserted into connector terminal and turning ignition switch to position shown. **CAUTION** **Do not allow straight pin to contact electrical ground.** (see tables below) 6. Turn ignition switch to OFF position. 7. Remove straight pin. 8. Reconnect wire to S terminal of starter relay. 9. **Was the value at least 90 percent of battery voltage in each case?**	Yes ▶	REPLACE TFI module.
	No ▶	INSPECT for faults in wiring harness and connectors. REFER to vehicle wiring diagram for appropriate circuit. Damaged or worn ignition switch.

TFI without CCD

Connector Terminal	Wire/Circuit	Ignition Switch Test Position
#3	Run Circuit	Run and Start
#4	Start Circuit	Start

TFI with CCD

Connector Terminal	Wire/Circuit	Ignition Switch Test Position
#3	Run Circuit	Run and Start

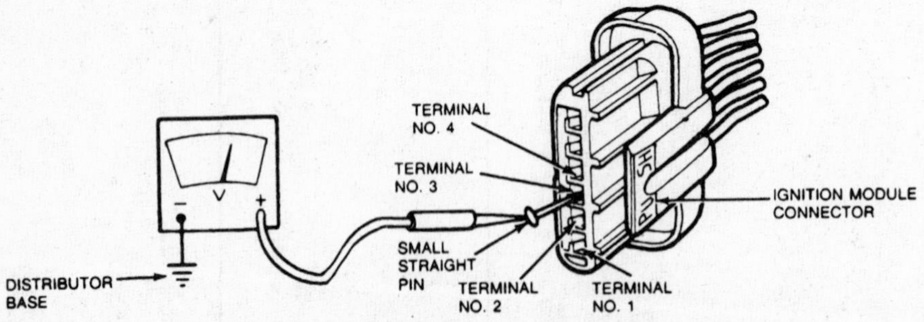

Test 3 — wiring harness

TEST STEP	RESULT ▶	ACTION TO TAKE
1. Place the transmission shift lever in the PARK (A/T) or NEUTRAL (M/T) position and set the parking brake. **CAUTION** **Failure to perform this step may result in the vehicle moving when the starter is subsequently engaged during the test.** 2. Disconnect the harness connector from the TFI module and connect the TFI tester. 3. Connect the red lead from the tester to the positive (+) side of the battery. 4. Disconnect the wire at the S terminal of the starter relay, and attach remote starter switch. 5. Crank the engine using the remote starter switch and note the status of the two LED lamps. 6. Remove the tester and remote starter switch. 7. Reconnect the wire to the starter relay and the connector to the TFI. 8. **Did the PIP light blink?**	Yes ▶ No ▶	GO to Test 6. REMOVE distributor cap and VERIFY rotation. If OK, GO to Test 5.

Test 4 – stator

TEST STEP	RESULT ▶	ACTION TO TAKE
1. Remove the distributor from the engine and the TFI module from the distributor. 2. Measure resistance between TFI module terminals as shown below. **Measure Between These Terminals** / **Resistance Should Be**: GND — PIP In: Greater than 500 Ohms PIP PWR — PIP IN: Less than 2K Ohms PIP PWR — TFI PWR: Less than 200 Ohms GND — IGN GND: Less than 2 Ohms PIP In — PIP: Less than 200 Ohms 3. **Are all these readings as specified?**	Yes ▶ No ▶	Replace stator. Replace TFI.

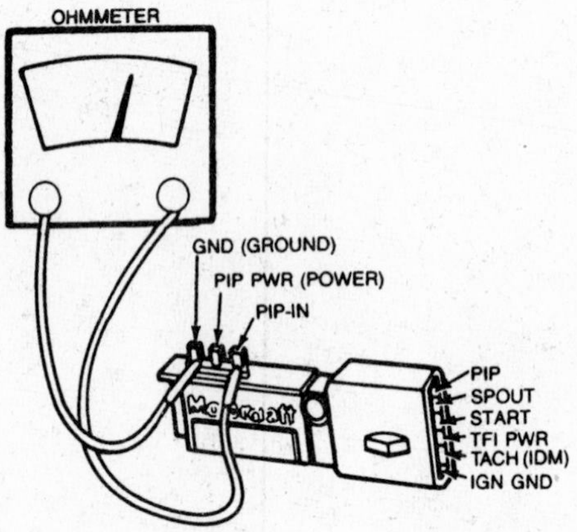

Measure Between These Terminals	Resistance Should Be
GND — PIP In	Greater than 500 Ohms
PIP PWR — PIP IN	Less than 2K Ohms
PIP PWR — TFI PWR	Less than 200 Ohms
GND — IGN GND	Less than 2 Ohms
PIP In — PIP	Less than 200 Ohms

Test 5 – stator

TEST STEP	RESULT ▶	ACTION TO TAKE
1. Use status of Tach light from Test 4. 2. **Did the Tach light blink?**	Yes ▶	GO to Test 7.
	No ▶	REPLACE TFI module and CHECK for spark using the method described in Test 1. If spark was not present REPLACE the coil also.

Test 6 -- module

TEST STEP	RESULT ▶	ACTION TO TAKE
1. Disconnect ignition coil connector. Inspect for dirt, corrosion and damage. 2. Connect the ignition coil connector to a known good ignition coil. 3. Connect one end of a known good secondary wire to the spark tester. Connect the other end to the known good ignition coil. CAUTION **DO NOT HOLD THE COIL while performing this test. Dangerous voltages may be present on the metal laminations as well as the high voltage tower.** 4. Crank engine. 5. Turn ignition switch to OFF position. 6. **Was spark present?**	Yes ▶	MEASURE resistance of the ignition coil wire (from vehicle). REPLACE if greater than 7,000 ohms per foot. If OK, REPLACE ignition coil.
	No ▶	RECONNECT coil connector to the vehicle coil and spark tester to vehicle secondary wire and GO to Test 8.

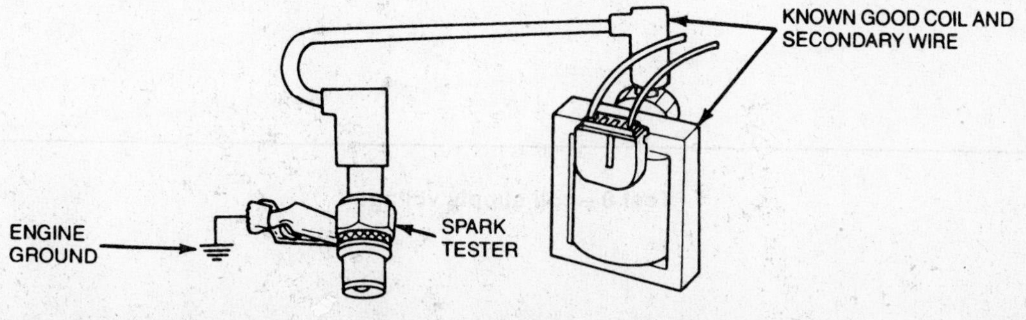

Test 7 – secondary coil wire

TEST STEP	RESULT ▶	ACTION TO TAKE
1. Attach negative (−) VOM lead to distributor base. 2. Measure battery voltage. 3. Turn ignition switch to RUN position. 4. Measure voltage at POSITIVE (+) terminal of ignition coil. 5. Turn ignition switch to OFF position. 6. **Was the value 90 percent of battery voltage or more?**	Yes ▶	GO to Test 10.
	No ▶	INSPECT and SERVICE wiring between ignition coil and ignition switch. REFER to vehicle wiring diagram. Worn or damaged ignition switch. REFER to Charging/Electrical Shop Manual Group (Electrical Power Supply System Group for Compact Truck).

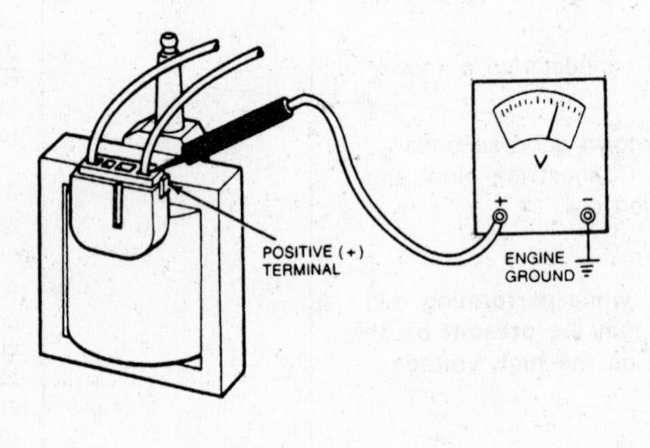

Test 8 – coil supply voltage

Preliminary Checks

Before beginning any organized test procedures on the electronic ignition system, first perform these simple checks:

1. Visually inspect the engine compartment to make sure all vacuum hoses and spark plug wires are properly routed and securely connected.
2. Examine all wiring harnesses and connectors for insulation damage, burned, overheated, loose or broken connections. *DO NOT* remove the lubricant compound from inside the connectors.
3. Check that the TFI module is securely fastened to the distributor base.
4. Make sure the battery is fully charged and the cables are clean and tight.
5. Make sure all accessories are switch to the **OFF** position during diagnosis.
6. Check for loose or damaged spark plug or coil wires. If boots or nipples are removed on 8mm ignition wires, reline inside of each with new silicone dielectric compound.

Test Equipment

The following test equipment, or equivalents, are necessary to diagnose the ignition system:

- Spark Tester D81P–6666–A, which resembles a spark plug with the side electrode removed. A spark plug with the side electrode removed IS NOT sufficient to check for spark and may lead to incorrect results.
- Digital Volt/Ohmmeter (Rotunda 014–00407 or any good digital volt/ohmmeter)
- 12 volt test light.
- Small straight pin.

When instructed to inspect a wiring harness, both a visual inspection and a continuity test should be performed. When making measurements on a wiring harness or connector, it is good practice to wiggle the wires while measuring. The following tests are designed to be performed in order to gradually narrow down the cause of a problem in the ignition system.

COMPONENT TESTS

Ignition Coil Secondary Voltage Test

Connect the spark tester (D81P–6666–A or equivalent) between the ignition coil wire and engine ground, then crank the engine. If spark is present, the secondary voltage is OK. If there is no spark, measure the resistance of the ignition coil wire and replace the wire if the resistance is greater than 7,000Ω per foot. Inspect the ignition coil for damage or carbon tracking and crank the engine with the distributor cap removed to verify distributor rotation. All distributors rotate in the **CLOCKWISE** direction.

Ignition Coil Primary Resistance Test

Turn the ignition switch off, then disconnect the ignition coil connector. Check for dirt, corrosion or damage. Use an ohmmeter to measure the resistance from the positive (+) to negative (–) terminals of the ignition coil. If the reading is between 0.3–1.0Ω the ignition coil is OK; continue on to the Ignition Coil Secondary Resistance Test. If the reading is less than 0.3Ω or greater than 1.0Ω, replace the ignition coil.

Ignition Coil Secondary Resistance Test

Use an ohmmeter to measure the resistance between the negative (–) terminal to the high voltage terminal of the ignition coil. If the reading is between 6,500–11,500Ω, the ignition coil is OK; go on to the Wiring Harness Test. If the reading is less than 6,500Ω or more than 11,500Ω, replace the ignition coil.

Wiring Harness Test

1. Separate the wiring harness connector from the ignition module and check for dirt, corrosion and damage. Push the connector tabs to separate the connector.
2. Disconnect the wire at the **S** terminal of the starter relay.
3. Touch the negative (–) lead from a volt/ohmmeter to the distributor base.
4. Measure battery voltage.
5. Measure the connector terminal voltage by attaching the positive volt/ohmmeter lead to a small straight pin inserted into the connector terminal and turning the ignition switch to the position shown.

CAUTION

Do not allow the straight pin to touch any electrical ground

6. If, in all cases, the reading is at least 90% of battery voltage, the harness is OK; continue on to the Stator Test for 1986–87 vehicles and on 1988–89 vehicles, replace the TFI module. If any reading is less than 90% of battery voltage, check for faults in the wiring harness and connectors or for a damaged or worn ignition switch.
7. After all tests are complete, turn the ignition switch off. Remove the straight pin and reconnect the wire to the **S** terminal of the starter relay.

Stator Test

1986–87

1. Turn the ignition switch to the **OFF** position.
2. Remove the coil wire and ground it.
3. Touch the negative (–) lead from a volt/ohmmeter to the distributor base.
4. Disconnect the pin-in-line connector near the distributor and attach the positive (+) volt/ohmmeter lead to the TFI module side of the connector.
5. Turn the ignition switch to the **ON** position.
6. Bump the starter and measure the voltage levels with the engine not moving. Allow sufficient time for the digital voltage reading to stabilize before taking the measurement. Record all values for possible use in additional tests.
7. If the highest reading is greater than 90 percent 90% of battery voltage , go to Step 8. If the highest value is less than 90% of battery voltage, replace the stator assembly.
8. If the lowest value is greater than 0.5 volts, remove the distributor from the engine. Remove the TFI module from the distributor and check the stator connector terminals and TFI terminals for misalignment; service as necessary. If OK, replace the stator assembly. If the lowest value is less than 0.5 volts, go to Step 9.
9. If all values are between 0.5 volts and 90% of battery voltage, replace the stator assembly. If no values are between 0.5 volts and 90% of battery voltage, go on to the EEC-IV/TFI-IV Test.

1988

1. Turn the ignition switch to the **OFF** position.
2. Remove the coil wire and ground it.
3. Touch the negative (–) lead from a volt/ohmmeter to the distributor base.
4. Disconnect the pin-in-line connector near the distributor and attach the positive (+) volt/ohmmeter lead to the TFI module side of the connector.
5. Turn the ignition switch to the **ON** position.
6. Bump the starter and measure the voltage levels with the engine not moving. Allow sufficient time for the digital voltage reading to stabilize before taking the measurement. Record all values for possible use in additional tests.
7. If the highest reading is greater than 70% of battery voltage , go to Step 8. If the highest value is less than 70% of battery voltage, replace the stator assembly.

8. If the lowest value is greater than 0.5 volts, remove the distributor from the engine. Remove the TFI module from the distributor and check the stator connector terminals and TFI terminals for misalignment; service as necessary. If OK, go to Step 9. If the lowest value is less than 0.5 volts, go to Step 10.

9. If the lowest value was less than 0.5 volts, measure the resistance between the TFI module terminals and record the readings. Compare the readings to those shown in the accompanying table. If all the resistance readings are within specification, replace the stator. If any of the readings are off, replace the TFI module.

10. If there are any voltage readings between 0.5 volts and 70% of battery voltage, replace the stator. If there are no readings bewteen 0.5 volts and battry voltage, go on to the EEC-IV/TFI-IV Test.

11. Using the above illustration, measure the resistance between the "PIP" and "Spout" terminals of the TFI module with an ohmmeter. The resistance should be less than 7 kΩ. If the resistance is not less than 7 kΩ, replace the TFI module.

1989–90

NOTE: This test requires the use of special Rotunda TFI module tester tool 105-0002 or its equivalent.

1. Place the transmission shift lever in the **PARK** (AT) or **NEUTRAL** (MT) position. Block the wheels.

2. Disconnect the harness connector from the TFI module and connect the special TFI tester tool in accordance with the manufacturer's instructions.

3. Touch the red lead of the tester to the positive (+) battery terminal.

4. Disconnect the wire located at the "S" terminal of the starter relay.

5. Connect a remote starter to the starter relay.

6. Crank the engine using the remote starter switch and observe the two LED "PIP" (Profile Ignition Pickup) and "Tach" lights on the tester display.

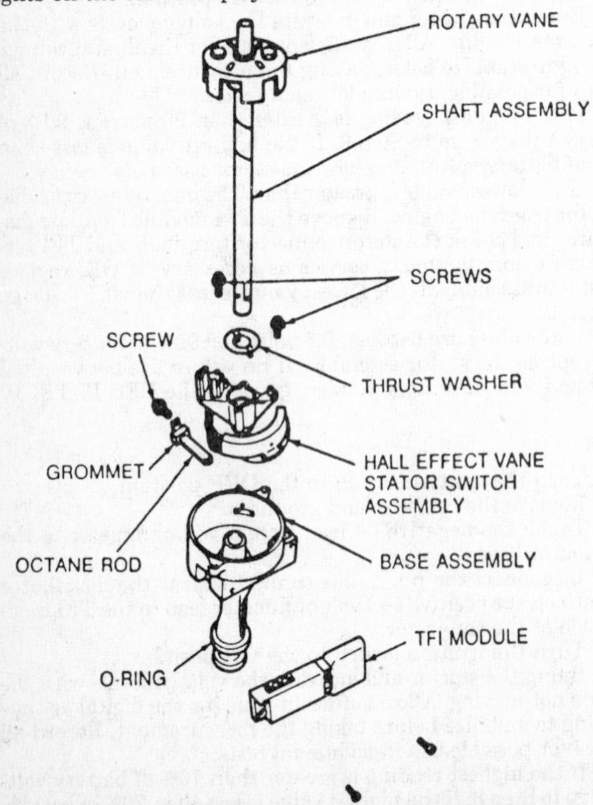

Exploded view of Hall Effect distributor

7. Did the "PIP" light blink? If it did, go on to Step 8. If it didn't, remove the distributor cap and crank the engine over to verify that the distributor is rotating in the CLOCKWISE direction. If the distributor rotation is O.K., continue on to Step 9.

8. Did the "Tach" light blink? If it did, go on to the EEC-IV/TFI-IV Test. If it didn't, replace the TFI module and then check for spark. If there is no spark, replace the coil also.

9. Remove the distibutor from the engine and then separate the TFI module from the distributor. Using an ohmmeter, measure the resistance between the TFI module terminals and record the readings. Compare the readings to those shown in the table above. If all the resistance readings are within specification, replace the stator. If any of the readings are off, replace the TFI module.

10. Disconnect the tester and remote starter switch.

EEC-IV/TFI-IV TEST

Connect a spark tester between the ignition coil wire and engine ground. Crank the engine and check for spark. If no spark is present, replace the TFI-IV module. On 1989–90 models, after replacing the TFI module, measure the coil primary resistance. If the resistance is less than 3Ω, replace the coil also. If spark is present, check the PIP (Profile Ignition Pickup) and ground wires for continuity and repair as necessary. If OK, the EEC-IV system will have to be diagnosed. See Section 4 for details.

Primary Circuit Continuity Test

1. Push the connector tabs and separate the wiring harness connector from the ignition module. Check for dirt, corrosion and damage and repair as necessary.

2. Touch the negative (-) lead from a volt/ohmmeter to the distributor base.

3. Measure battery voltage.

4. Touch the positive (+) lead of the volt/ohmmeter to a small straight pin inserted into connector terminal No. 2. Be careful not to let the pin touch any electrical ground.

5. Turn the ignition switch to the **RUN** position and measure the voltage at terminal No. 2. If the reading is at least 90% of battery voltage, go back to the Wiring Harness Test. If the reading is less than 90% of battery voltage, continue on to the Ignition Coil Primary Voltage Test.

Ignition Coil Primary Voltage Test

1. Attach the negative (-) lead of the volt/ohmmeter to the distributor base.

2. Measure the battery voltage.

3. Turn the ignition switch to the **RUN** position and measure the voltage at the negative (-) terminal of the ignition coil.

4. If the reading is at least 90% of battery voltage, check the wiring harness between the ignition module and coil negative terminal for dirt corrosion and other damage. If the reading is less than 90% of battery voltage, check the wiring harness between the ignition coil and coil negative terminal, then go on to the next test.

Ignition Coil Supply Voltage Test

1. Remove the coil connector.

2. Attach the negative (-) lead from a volt/ohmmeter to the distributor base.

3. Measure the battery voltage.

4. Turn the ignition switch to the RUN position.

5. Measure the voltage at the positive (+) terminal of the ignition coil. If the reading is at least 90% of battery voltage, check the ignition coil connector for dirt, corrosion and damage. Check the ignition coil terminals for dirt, corrosion and damage; if no problem is found, replace the ignition coil. If the reading is less than 90% of battery voltage, check the wiring between the ignition coil and ignition switch, or for a worn or damaged ignition switch.

Ignition Coil and Secondary Wire Test

NOTE: This test applies to 1989 vehicles only. You will a known good coil and secondary wire to perform the test.

1. Disconnect the connector from the ignition coil. Inspect the coil terminals for dirt and corrosion.
2. Connect the ignition coil connector to a known good coil.
3. Connect one end of a known good secondary wire to a spark tester. Connect the other end of the wire to the good coil.

CAUTION

DO NOT hold the coil while performing this test, as it presents a serious shock hazard!

4. Crank the engine and check for spark.
5. Turn the ignition switch to the **OFF** position.
6. If spark was present, measure the ignition coil wire resistance. If the resistance is greater than 7,000Ω, replace the coil wire. If the resistance was O.K., replace the ignition coil. If there was no spark, reconnect coil connector to the coil and the spark tester to the secondary wire and go on to the EEC-IV/TFI-IV Test.

Initial Timing

ADJUSTMENT

1. Place transmission in Park (automatic) or Neutral (manual), with the A/C and heater off.
2. Connect an inductive timing light (Rotunda 059–00006 or equivalent) according to the manufacturer's instructions.
3. Disconnect the single wire inline spout connector near the distributor or remove the shorting bar from the double wire spout connector.
4. Start the engine and allow it to reach normal operating temperature.
5. With the engine at the timing rpm specified, illuminate the timing marks with the timing light and check and/or adjust the initial timing to the specification listed on the underhood emission control label by loosening the distributor holddown bolt and rotating the distributor slightly. A Torx® bit may be necessary to loosen the security type distributor holddown, if equipped.
6. Reconnect the single wire inline spout connector and check the timing advance to verify the distributor is advancing beyond the initial setting.
7. Once the timing is set, shut off the engine and disconnect the timing light.

EEC-IV EDIS Electronic Ignition System

In the distributorless ignition system (DIS) used on the 4.0L V6, all engine timing and spark distribution is handled electronically with no moving parts. This system has fewer parts that require replacement and provides a more accurately timed spark. During basic operation, the EEC-IV determines the ignition timing required by the engine and a DIS module determines which ignition coil to fire.

The purpose of the DIS module is to deliver a full energy spark at a crank angle targeted by the EEC-IV and to provide the EEC-IV module with speed and position information. An Ignition Diagnostic Monitor (IDM) Clean Tach Out (CTO) line is also provided. The DIS inputs and outputs are listed below;

a. Inputs – Variable Reluctance Sensor Input (VRS) and Spark Advance Word (SAW).

b. Outputs – Coil drivers, one for every 2 cylinders. Profile Ignition Pick-up (PIP) and Ignition Diagnostic Monitor (IDM) Clean Tach Out (CTO) line.

The Variable Reluctance Sensor (VRSA) input is derived by sensing the passage of the teeth from a 36 minus one tooth crankshaft mounted wheel. The signal is processed by the DIS in order to identify the missing tooth. Once the missing tooth is found, the module is said to be syncronized. Synchronization is essential to the DIS in order to ascertain and track the angular position of the crankshaft relative to a fixed reference. Since the missing tooth is indexed from top dead center (TDC) of the No. 1 cylinder, the cylinder pair of No. 1 and No. 4 is identified for 4 cylinder DIS. Cylinder pair indentification is a requirement for the Distributorless Ignition Systems (DIS) which fires a pair of simultaneous sparks (one on compression and one on exhaust).

A Profile Ignition Pick-Up (PIP) output is synthesized from the High Data Rate (HDR) VRTS signal and sent to the EEC-IV control module. This signal is a 50 percent duty cycle square wave with its rising edge at 10° before top dead center of each cylinder event. The PIP signal is required by the EEC-IV control module to determine engine speed and position. Once the EEC-IV control module reconizes the the PIP signal, fuel and spark functions are enable. The Calculated spark target is sent to the DIS module as a pulse width modulated digital signal (SAW). This signal is sent once per cylinder event. It is then up to the DIS module to decode the SAW signal and to fire the next spark at the command spark target.

Coil firing is initiated by energizing the DIS coil in sequence referenced to the missing tooth (example 4 cylinder A-B) and firing at the command spark target. (default spark advance is 10° BTDC). By energizing the primary side of the coils in proper sequence and by connecting the secondary wires in accordance with the engine firing order, a power stroke is achieved on each cylinder event with a wasted spark fired in the exhausting cylinder. In addition, an IDM/CTO signal is transmitted on each spark firing. This signal is intended to communicate diagnostic information by pulse width modulation and provide a clean buffered signal, proportional to engine speed, for tachometer operation. A brief description of the system components and their operation follows:

Variable Reluctance Sensor (VRS) – is a magnetic transducer with a pole piece wrapped with fine wire which , when exposed to a change in flux linkage will induce a differential voltage across the terminals of the wire windings. The output differential voltage of the transducer is a function of the sensor to tooth air gap (voltage increase with decreasing air gap), tooth and sensor pole piece width (fixed for a given design) and engine angular velocity (the voltage increases with increasing rpm).

When the VRS encounters a rotating timing wheel, the passing ferromagnetic teeth causes a change in the sensor reluctance. The varying reluctance alters the amount of magnetic flux linkage passing through the wire windings and induces a voltage proportional to both the rate of change in the magnetic flux and the number of coil windings. As a tooth approaches the pole piece of the sensor, a positive differential voltage is induced. This voltage is positive as long as the flux change is increasing. When the pole piece is in the center of the tooth there is no net change in the flux across the coil windings and the output voltage is zero. As the tooth moves away from the pole piece (increasing reluctance, decreasing flux), the output voltage swings negative.

The negative zero crossing of the differential VRS signal occurs when the pole piece of the sensor corresponds to the center of each physical tooth. The DIS module is sensitive only to this negative zero crossing and uses this transition to establish crankshaft position.

At normal engine running speeds, the VRS signal waveform is an approximation of a sine wave with increased amplitude adjacent to the missing tooth region. The variable reluctance sensor and wheel are designed to provide a nearly symetrical signal and meet the following specifications:

a. Minimum engine speed 30 rpm with a VRS output voltage of 150 milli-volts peak to peak.

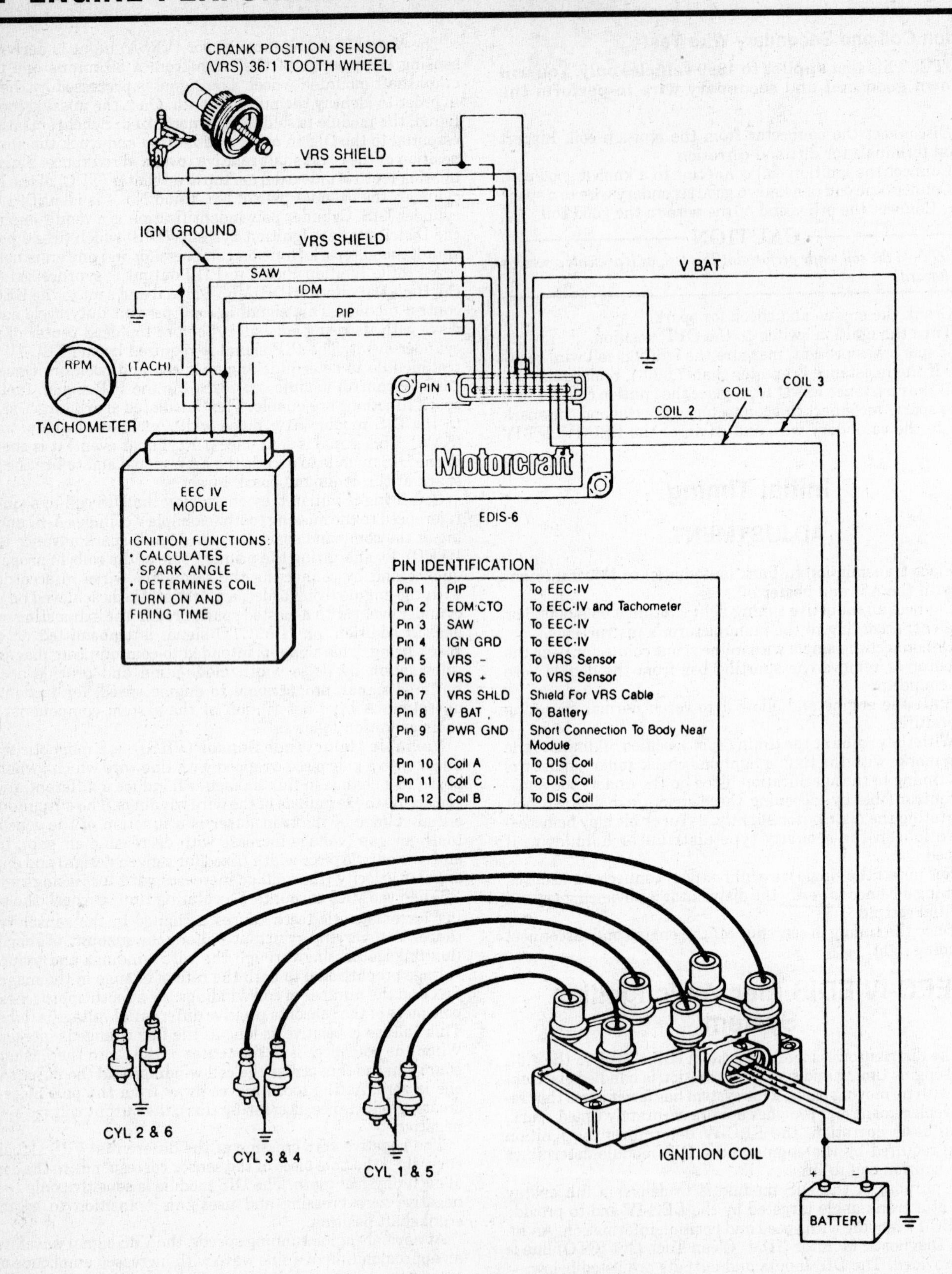

Pin	Name	Connection
Pin 1	PIP	To EEC-IV
Pin 2	EDM-CTO	To EEC-IV and Tachometer
Pin 3	SAW	To EEC-IV
Pin 4	IGN GRD	To EEC-IV
Pin 5	VRS −	To VRS Sensor
Pin 6	VRS +	To VRS Sensor
Pin 7	VRS SHLD	Shield For VRS Cable
Pin 8	V BAT	To Battery
Pin 9	PWR GND	Short Connection To Body Near Module
Pin 10	Coil A	To DIS Coil
Pin 11	Coil C	To DIS Coil
Pin 12	Coil B	To DIS Coil

EDIS (Electronic Distributorless Ignition System) – 4.0L engine

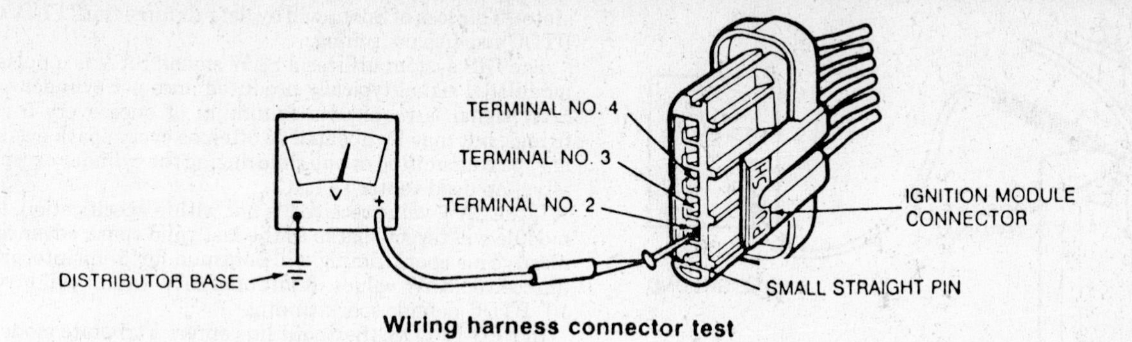

Wiring harness connector test

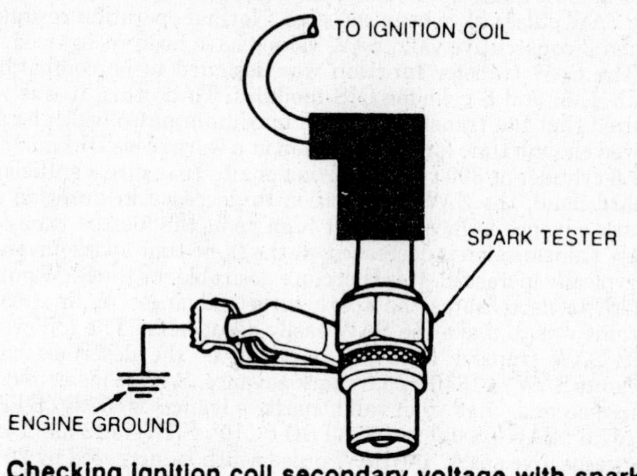

Checking ignition coil secondary voltage with spark tester

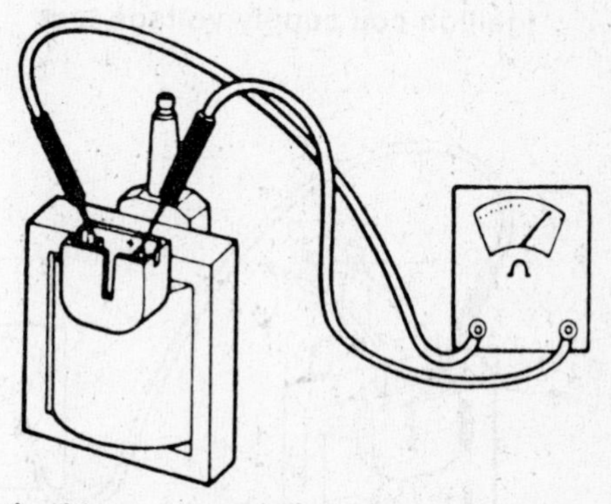

Ignition coil primary resistance test

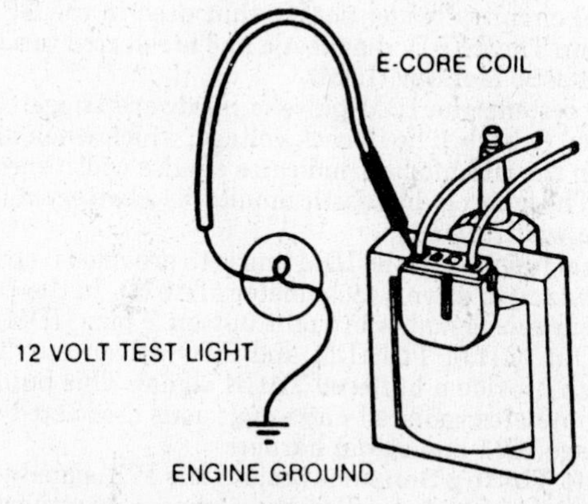

Ignition coil primary circuit switching test

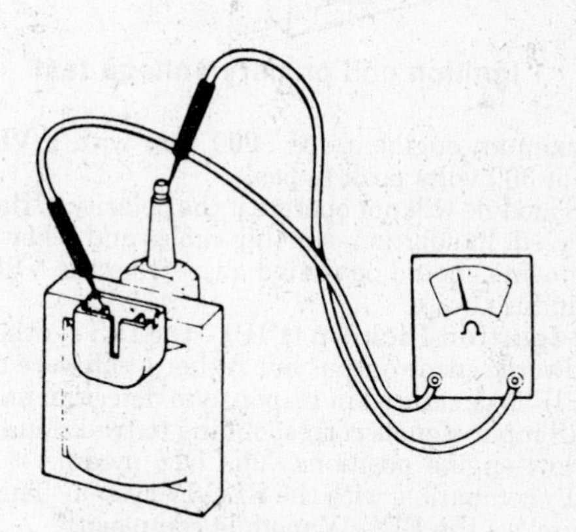

Ignition coil secondary test

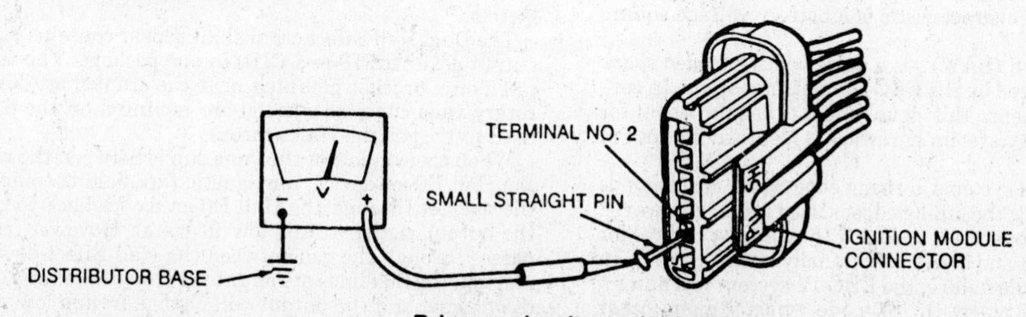

Primary circuit continuity test

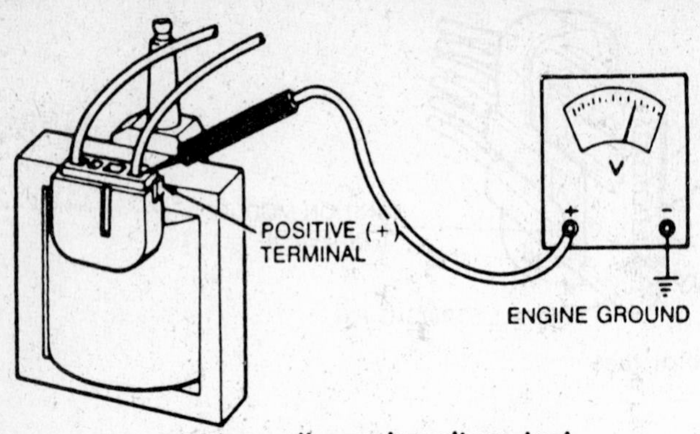

Ignition coil supply voltage test

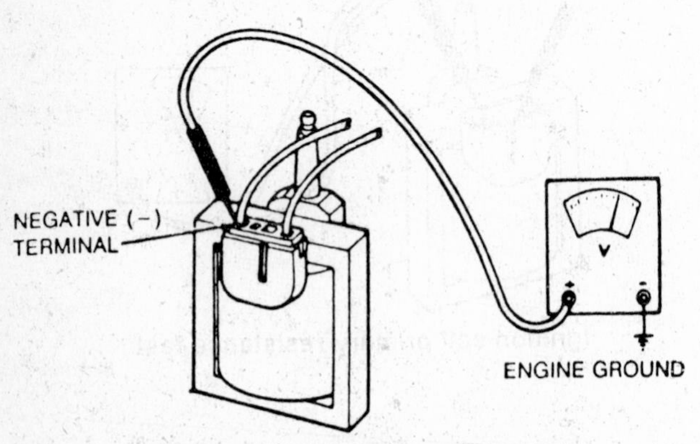

Ignition coil primary voltage test

b. Maximum engine speed 8000 rpm with a VRS output voltage of 300 volts peak to peak.

The DIS module will not operate if the polarity of the VRS signal is reversed. In addition, starting cables and other noise producing elements should be routed away from the VRS wires so as not to induce noise.

Profile Ignition Pick-Up (PIP) – the DIS synthetic PIP is not a hardware sensor output but rather a software driven signal. The PIP is generated in response to detection and recognition of VRS input signals corresponding to traditional PIP high and PIP low engine positions. The DIS module is hardware functionally compatible with the EEC-IV system. This compatibility minimizes the EEC-IV module complexity.

The PIP interface signal, historically has been generated by a magnetic Hall pick up sensor located in the distributor. The PIP output signal is synthesized from the VRS sensor and is the basic engine timing signal sent to the EEC-IV control module. The signal profile is a rising edge at 10° BTDC of every cylinder event with a 50 percent duty cycle, i.e. (the falling edge is 90° later on a 4-cylinder; 60 degrees later on a 6-cylinder and 45° later on an 8 cylinder). This signal characteristic is a battery voltage square wave.

Spark Angle Word (SAW) – is a pulse width encoded spark advance word generated by the EEC-IV control module. In conventional TFI-IV systems, this signal line is called the Spout for spark out. The EEC-IV system hardware is identical for both of these systems.

In the conventional systems, a rising edge Spout corresponds to a spark firing event; the falling edge is used for dwell control. The EEC-IV control module is tasked with real time control for both spark placement and dwell. Additionally in the event of an EEC-IV control module failure, the EEC-IV system will enter a Limited Operation Strategy (LOS) mode which, among other things will put the Spout in a tri-state (i.e. high impedance state). This loss of Spout will by default force the TFI-IV to a 10° BTDC fixed spark timing.

The DIS system utilizes a SAW signal. SAW is a pulse width modulated signal typically produced once per cylinder event. A SAW signal is required a minimum of once every 5 cylinder firings, but may by updated as often as every spark event. SAW may be transmitted at any time during the cylinder except at 10° after top dead center (ATDC).

If the SAW value received is not within specification, the DIS module will target sparks to the last valid spark advance recognized. This spark timing will continue for 5 misinterpreted or unreceived SAW values. Additional SAW faults will invoke the 10° BTDC default spark timing.

In EEC-IV LOS, the Spout line enters a tri-state mode, which results in a no SAW signal to the DIS module. This forces the DIS module to the 10° BTDC timing after the criteria for 5 missing SAW pulses have been satisfied. Normal operation resumes when 3 consecutive valid SAW values have been recognized.

The SAW transfer function was designed to be compatible with 4, 6, and 8 cylinder DIS modules. To do this, it was required that the transfer function maximum pulse width be allowed enough time for transmission at a worse case DIS module for 8 cylinder at 800 rpm operating point. To insure a sufficient guard band, the SAW is specified to decrease in duration as spark advance is increased. At high rpm, the alotted time for SAW transmission is decreased; at the same time spark advance is typically increased. It is therefore desirable for the SAW pulse width to decrease as the spark target advances. A final constraint was to make the SAW easily decodeable. The following DIS SAW transfer function meets all of the described constraints SAW = 1540 – 25.6 spark advance. SAW is measured in micro-seconds and the valid spark advance is 57.5° BTDC (+57.5°: SAW-68 us) to 10° ATDC (–10°: SAW-1792 us). During respective spark, the SAW pulse width is increased by 2048 micro-seconds. This longer pulse width indicates to the DIS module that repetitive spark is desired.

Ignition Diagnostic Monitor (IDM) – is a signal line used to convey the DIS operating status to the EEC-IV system. The EEC-IV signal circuitry is identical to that used in the TFI-IV ignition system. The TFI-IV diagnostic line is referred to as the Ignition Diganostic Monitor (IDM).

In the DIS system, the IDM pulse is hardware triggered by the ignition coil flyback. The flyback voltage, which when internally sensed in the DIS module, indicates a pulse width encoded IDM output. The ignition diagnostic monitor is a battery voltage variable pulse width signal.

An additional function of the IDM line is to provide an optional method in which to drive a tachometer (TACH). In the DIS 4 cylinder system this signal is brought out on 2 pins, IDM and Clean Tach Out (CTO). The IDM and CTO driven tach lines may be used to provide a buffered TACH signal. This buffered tach signal eliminates radiated noise emissions associated with the high voltage IDM line signal harness.

Crankshaft Timing Sensor – besides the PIP signal there is another signal generated called the Cylinder Identification (CID). The CID signal is used to synchronize the ignition coils, due to the fact that the Ranger uses a 2 ignition coil pack DIS system.

The Dual Hall Effect crankshaft sensor contains 2 hall digital output devices (PIP and CID) in one package. The sensor is located on a bracket mounted near the crankshaft damper. Two rotary vane cups (or wheels) are mounted on the damper and used to trigger the Hall sensors.

When the window of the vane cup is between the magnet and the Hall Effect device, a magnetic flux field is completed from the magnet through the Hall Effect device back to the magnet the output signal will be low (0 volts). However, as the vane passes through the gap between the Hall Effect device and the magnet, the flux lines are shunted through the vane and back to the magnet and the output will change from a low voltage to a high voltage.

The PIP cup has 2 teeth resulting in a 2 positive going edges each revolution of the crankshaft, where as the CID cup has 1 tooth and generates a signal that is high half of the crankshaft revolution and low the other half. The CID is used by the DIS module to enable it to select the proper coils to fire. When the CID is high, coils 2 and 3 are enabled and when the CID is low 1 and 4 are enabled. The EEC-IV control module tells the DIS module when to fire but the DIS module has to select one of the 2 coils based on the CID (which 2 of 4 if in the DPI mode).

DIS module—the main function of the DIS module is to switch between the ignition coils and trigger the coils to spark. The DIS ignition module receives the PIP and the CID signals from the crankshaft timing sensor and the Spout (spark out) signal from the EEC-IV control module. During normal operation, the PIP signal is passed onto the EEC-IV control module and provides base timing and rpm information. The CID signal provides the DIS module with information required to switch between the coils for cylinders No. 1 and No. 4 and the coils for cylinders No. 2 and No. 3. The Spout signal (from the EEC-IV control module) contains the optimum spark timing and dwell time information. The spark angle is determined by the rising edge of the Spout while the falling edge of the Spout controls the coil current on or dwell time. The dwell time is controlled or varied by varying the duty cycle of the Spout signal.

This feature is called Computer Controlled Dwell (CCD). With the proper inputs of the PIP, CID and Spout the DIS module turns the ignition coils on and off in the proper sequence for spark control.

TROUBLESHOOTING

Crankshaft Timing Sensor

REMOVAL AND INSTALLATION

1. Disconnect the negative battery cable.
2. Disconnect the sensor electrical connectors from the engine wiring harness.
3. Remove the large electrical connector from the crankshaft timing sensor assembly by prying out the red retaining clip and removing the 4 wires.
4. Remove the crankshaft pulley assembly by removing the accessory drive belts and then the 4 bolts that retaining the crankshaft pulley hub assembly. Remove the timing belt outer cover.
5. Rotate the crankshaft so that the keyway is at the 10 o'clock position. This will place the vane window of both the inner and outer vane cups over the crankshaft sensor timing assembly.

NOTE: The vane cups are attached to the crankshaft pulley hub assembly.

6. Remove the 2 crankshaft timing sensor retaining bolts and the plastic wire harness retainer which secures the crankshaft sensor to its mounting bracket.
7. Remove the crankshaft timing sensor assembly, sliding the wires out from behind the inner timing belt cover.

To install:

8. Remove the large electrical connector from the new crankshaft timing sensor assembly.
9. Position the crankshaft timing sensor assembly. First slide the electrical wires behind the inner timing belt cover, now hold the sensor assembly loosely in place with the retaining bolts but do not tighten the bolts at this time.
10. Install the large electrical connector onto the crankshaft timing sensor assembly.

NOTE: Be sure that the 4 wires to the large electrical connector are installed in the proper locations. The sensor will not function properly if the wires are installed incorrectly.

11. Reconnect both of the crankshaft timing sensor electrical connectors to the engine harness.
12. Rotate the crankshaft so that the outer vane on the crankshaft pulley hub assembly engages both sides of the crankshaft Hall Effect sensor positioner tool T89P–6316–A or equivalent and tighten the sensor assembly retaining bolts.
13. Rotate the crankshaft so that the vane on the crankshaft pulley hub is no longer engaged in the crankshaft sensor positioner tool and remove the tool.
14. Install a new plastic wire harness retainer to secure the crankshaft timing sensor harness to its mounting bracket and trim off the excess.
15. Install the timing belt outer cover.
16. Install the crankshaft pulley assembly and tighten the 4 attaching bolts to specifications.
17. Install the drive belts and adjust as necessary. Reconnect the negative battery cable and perform a vehicle road test.

TROUBLESHOOTING

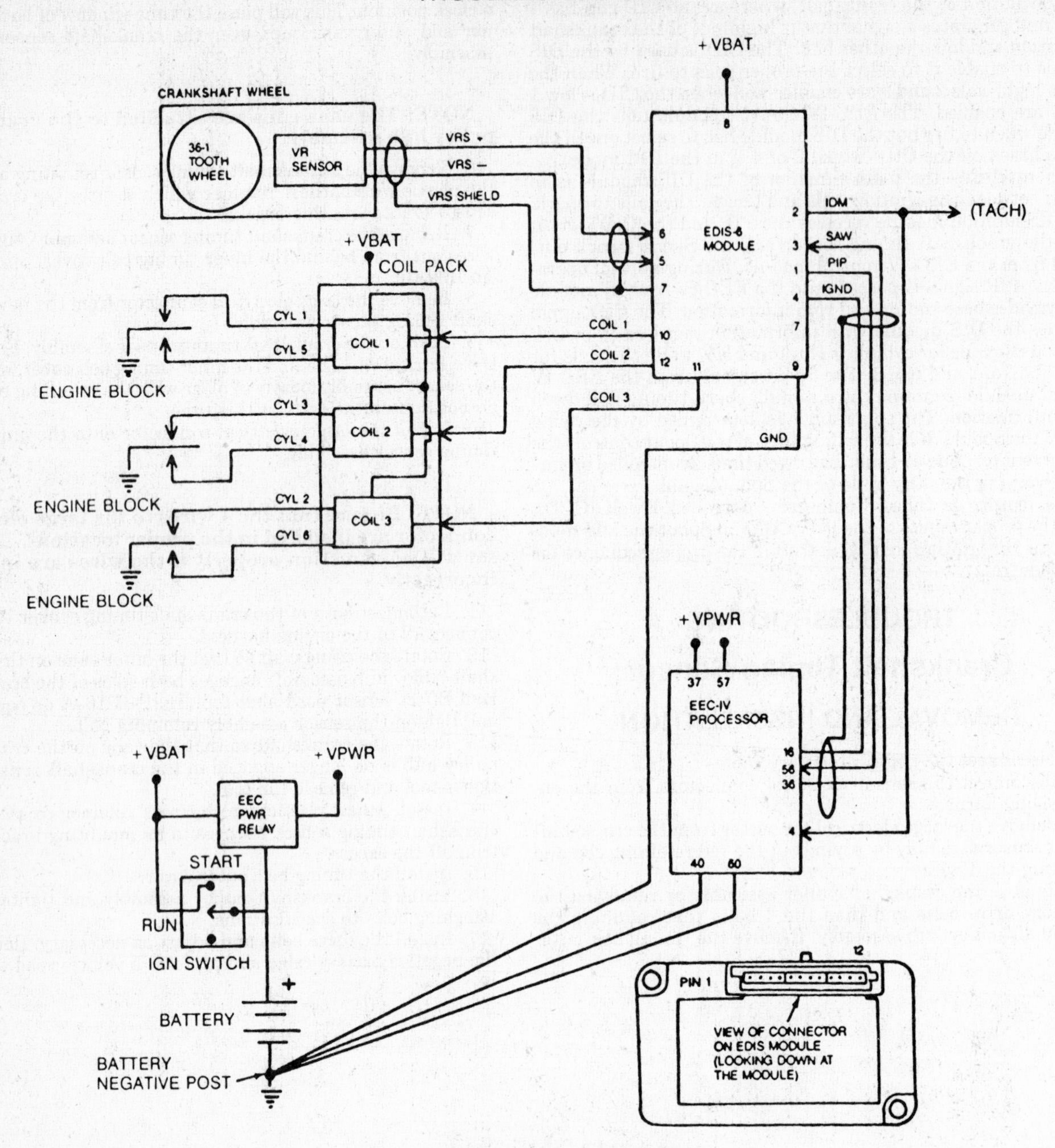

EDIS wiring schematic

TEST STEP	RESULT ▶	ACTION TO TAKE
A1 PERFORM EEC-IV QUICK TEST		
• perform EEC-IV Quick Test. • **Has Quick Test been performed?**	Yes ▶ No ▶	GO to A2 . GO to Section 14 first and perform Quick Test.
A2 CHECK FOR SPARK DURING CRANK		
• Using a Neon Bulb Spark Tester (OTC D 89P-6666-A), check for spark at all spark plug wires while cranking. • **Was spark present on ALL spark plug wires and consistent (one spark per crankshaft revolution)?**	Yes ▶ No ▶	GO to A3 . GO to A10 .
A3 CHECK PIP AT EDIS MODULE		
• Key off. • Install EDIS diagnostic cable to breakout box. • Use EDIS 6 overlay. • Connect LED test lamp between J43 (PIP E) and J7 (GND). • **Does the LED test lamp blink continuously during crank?**	Yes ▶ No ▶	GO to A4 . GO to A7 .
A4 CHECK PIP CONTINUITY TO EEC		
• Key off. • Disconnect the EEC processor. • **Is the resistance between J43 (PIP E) and Pin 56 (PIP) of the EEC connector less than 5 ohms?**	Yes ▶ No ▶	GO to A5 . SERVICE open circuit. RECONNECT the EEC processor and RERUN Quick Test.
A5 CHECK IGND AT EDIS		
• Key off. • **Is the resistance between J47 (IGND E) and J7 (BAT −) less than 5 ohms?**	Yes ▶ No ▶	GO to A6 . REPLACE EDIS module and RERUN Quick Test.

Test A — no start, part 1

TEST STEP	RESULT ▶	ACTION TO TAKE
A6 CHECK IGND CONTINUITY TO EEC		
• Key off. • **Is the resistance between J47 (IGND E) and Pin 16 (IGN GND) of the EEC connector less than 5 ohms?**	Yes ▶	REPLACE EDIS module. RUN Quick Test.
	No ▶	SERVICE open circuit. RECONNECT EEC and RERUN Quick Test.
A7 CHECK PIP AT EDIS MODULE/ISOLATE EEC		
• Key off. • Disconnect the EEC processor. • Crank the engine. • **Does the LED test lamp blink continuously during crank?**	Yes ▶	REPLACE EEC processor. RERUN Quick Test.
	No ▶	GO to A8.
A8 CHECK PIP CIRCUIT FOR SHORT TO GROUND		
• Key off. • Disconnect the EDIS module from the EDIS diagnostic cable. • Disconnect the HEGO sensor. • **Is the resistance between J43 (PIP E) and J7 (BAT –) less than 5 ohms?**	Yes ▶	PIP is shorted to ground in harness. CHECK connectors, SERVICE or REPLACE harness. RERUN Quick Test.
	No ▶	GO to A9.
A9 CHECK PIP CIRCUIT FOR SHORT TO POWER		
• **Is the resistance between J43 (PIP E) and J51 (VBAT E) less than 5 ohms?**	Yes ▶	PIP is shorted to power in harness. CHECK connectors, SERVICE or REPLACE harness. RUN Quick Test.
	No ▶	REPLACE EDIS module. RERUN Quick Test.

Test A – no start, part 2

TEST STEP	RESULT ▶	ACTION TO TAKE
A10 DETERMINE MISSING SPARK PATTERN		
• **Is spark missing from BOTH cylinder 1 and 5 plug wires?** or • **Is spark missing from BOTH cylinder 2 and 6 plug wires?** or • **Is spark missing from BOTH cylinder 3 and 4 plug wires?**	Yes ▶ No ▶	GO to A12. GO to A11.
A11 CHECK PLUGS AND WIRES		
• Check spark plug wires for insulation damage, looseness, shorting or other damage. • Remove and check spark plugs for damage, wear, carbon deposits and proper plug gap. • **Are spark plugs and wires OK?**	Yes ▶ No ▶	REINSTALL plugs and wires. GO to A12. SERVICE or REPLACE damaged component. RERUN Quick Test.
A12 CHECK VBAT C AT COIL		
• Key off. • Install EDIS diagnostic cable to breakout box. • Use EDIS 6 overlay. • Connect the LED test lamp between J5 (VBAT C) and J7 (BAT-). • Key On. • **Is the LED test lamp on and bright?**	Yes ▶ No ▶	GO to A13. SERVICE open circuit. RECONNECT the EEC processor and RERUN Quick Test.
A13 CHECK COIL FOR FAILURE		
• Connect LED test lamp between J7 (BAT-) to each coil - J3, J6 and J10, one at a time. • Key on. • **Is the LED test lamp on and bright?**	Yes ▶ No ▶	GO to A14. REPLACE Coil pack. RERUN Quick Test.
A14 CHECK GND CONTINUITY TO EDIS MODULE		
• Key off. • **Is the resistance between J27 (GND E) and J7 (BAT-) less than 5 ohms?**	Yes ▶ No ▶	GO to A15. SERVICE open circuit. RECONNECT EEC and RERUN Quick Test.

Test A – no start, part 3

TEST STEP	RESULT ▶	ACTION TO TAKE
A15 CHECK GND IN EDIS MODULE		
• Key off. • **Is the resistance between J27 (GND E) and J47 (IGND E) less than 5 ohms?**	Yes ▶ No ▶	GO to **A16**. REPLACE EDIS module. RERUN Quick Test.
A16 CHECK VBAT E AT EDIS MODULE		
• Connect the LED test lamp between J51 (VBAT E) and J7 (BAT-). • Key on. • **Is the LED test lamp on and bright?**	Yes ▶ No ▶	GO to **A17**. SERVICE open circuit. RECONNECT the EEC processor and RERUN Quick Test.
A17 CHECK VRS AMPLITUDE AT EDIS MODULE		
• **Is the A/C voltage between J35 (VRS+ E) and J48 (VRS- E) more than 1 volt when the engine is cranked?**	Yes ▶ No ▶	GO to **A18**. GO to **A19**.
A18 CHECK VRS BIAS AT EDIS MODULE		
• Key on. • **Is the DC voltage between J35 (VRS+ E) and J27 (GND E) between 1.0 and 2.0 volts?**	Yes ▶ No ▶	REPLACE EDIS module. RERUN Quick Test. GO to **A28**.
A19 CHECK VRS CRANKING VOLTAGE AT EDIS		
• Key off. • Disconnect the EDIS module from the EDIS diagnostic cable. • **Is the A/C voltage more than 1 volt when the engine is cranked?**	Yes ▶ No ▶	REPLACE EDIS module. RERUN Quick Test. GO to **A20**.
A20 CHECK VRS CRANKING VOLTAGE AT SENSOR		
• Key off. • Disconnect the VRS engine harness from the EDIS diagnostic cable. • **Is the A/C voltage between J32 (VRS- S) and J31 (VRS+ S) more than 1 volt when the engine is cranked?**	Yes ▶ No ▶	GO to **A22**. GO to **A21**.

Test A – no start, part 4

TEST STEP	RESULT ▶	ACTION TO TAKE
A21 CHECK VR SENSOR AND CRANKSHAFT WHEEL		
• Check the crankshaft data wheel for damage. • Check the air gap. • **Is the air gap correct and is the crankshaft data wheel OK?**	Yes ▶	REPLACE VR sensor. RERUN Quick Test.
	No ▶	REPAIR or REPLACE bad parts. SET air gap.
A22 CHECK VRS+ CIRCUIT FOR OPEN		
• Key off. • Reconnect the EDIS diagnostic cable. • **Is the resistance between J35 (VRS+ E) and J31 (VRS+ S) less than 5 ohms?**	Yes ▶	GO to A23 .
	No ▶	SERVICE open circuit in VRS+ harness. RERUN Quick Test.
A23 CHECK VRS- CIRCUIT FOR OPEN		
• Key off. • **Is the resistance between J32 (VRS- S) and J48 (VRS – E) less than 5 ohms?**	Yes ▶	GO to A24 .
	No ▶	SERVICE open circuit in VRS- harness. RERUN Quick Test.
A24 CHECK VRS- CIRCUIT FOR SHORT TO POWER		
• Key off. • Disconnect the HEGO sensor. • **Is the resistance between J48 (VRS- E) and J51 (VBAT E) greater than 10K ohms?**	Yes ▶	GO to A25 .
	No ▶	SERVICE short circuit in VRS- harness. RERUN Quick Test.
A25 CHECK VRS- CIRCUIT FOR SHORT TO GROUND		
• **Is the resistance between J48 (VRS- E) and J7 (BAT-) greater than 10K ohms?**	Yes ▶	GO to A26 .
	No ▶	SERVICE short circuit in VRS- harness. RERUN Quick Test.

Test A — no start, part 5

TEST STEP	RESULT ▶	ACTION TO TAKE
A26 CHECK VRS+ CIRCUIT FOR SHORT TO GROUND		
• **Is the resistance between J35 (VRS+ E) and J7 (BAT-) greater than 10K ohms?**	Yes ▶	GO to A27.
	No ▶	SERVICE short circuit in VRS+ harness. RERUN Quick Test.
A27 CHECK VRS+ CIRCUIT FOR SHORT TO POWER		
• **Is the resistance between J35 (VRS+ E) and J51 (VBAT E) greater than 10K ohms?**	Yes ▶	VRS+ is shorted to VRS-. SERVICE short in harness. RERUN Quick Test.
	No ▶	SERVICE short circuit in VRS+ harness. RERUN Quick Test.
A28 CHECK VRS BIAS AT EDIS MODULE		
• Key off. • Disconnect the VR sensor from the EDIS diagnostic cable. • Key on. • **Is the DC voltage between J35 (VRS+ E) and J7 (BAT-) between 1.0 and 2.0 volts?**	Yes ▶	REPLACE VR sensor. RERUN Quick Test.
	No ▶	GO to A29.
A29 CHECK VRS+ CIRCUIT FOR SHORT TO POWER		
• Key off. • Disconnect the EDIS module from the EDIS diagnostic cable. • Disconnect the HEGO sensor. • **Is the resistance between J35 (VRS+ E) and J51 (VBAT E) greater than 10K ohms?**	Yes ▶	GO to A30.
	No ▶	VRS+ is shorted to power in harness. CHECK connectors, SERVICE or REPLACE harness. RERUN Quick Test.
A30 CHECK VRS+ CIRCUIT FOR SHORT TO GROUND		
• **Is the resistance between J35 (VRS+ E) and J7 (BAT-) greater than 10K ohms?**	Yes ▶	GO to A31.
	No ▶	SERVICE short circuit in VRS+ harness. RERUN Quick Test.

Test A — no start, part 6

TEST STEP		RESULT	▶ ACTION TO TAKE
A31	CHECK VRS− CIRCUIT FOR SHORT TO POWER		
	• **Is the resistance between J48 (VRS− E) and J51 (VBAT E) greater than 10K ohms?**	Yes	▶ GO to A32 .
		No	▶ SERVICE short circuit in VRS− harness. RERUN Quick Test.
A32	CHECK VRS− CIRCUIT FOR SHORT TO GROUND		
	• **Is the resistance between J48 (VRS− E) and J7 (BAT−) less than 5 ohms?**	Yes	▶ SERVICE short circuit in VRS− harness. RERUN Quick Test.
		No	▶ REPLACE EDIS module. RERUN Quick Test.

Test A — no start, part 7

TEST STEP	RESULT ▶	ACTION TO TAKE
B1 CHECK IDM CONTINUITY TO EEC		
• Key off. • Install EDIS diagnostic cable to breakout box. • Use EDIS 6 overlay. • Disconnect the EEC processor. • **Is the resistance between J41 (IDM E) and Pin 4 (IDM) of the EEC connector less than 5 ohms?**	Yes ▶ No ▶	GO to **B2**. SERVICE open in IDM circuit. RERUN Quick Test.
B2 CHECK IDM CIRCUIT FOR SHORT TO GROUND		
• Key off. • Disconnect the EDIS module from the EDIS diagnostic cable. • Disconnect the HEGO sensor. • **Is the resistance between J41 (IDM E) and J7 (BAT-) greater than 10K ohms?**	Yes ▶ No ▶	GO to **B3**. IDM is shorted to ground in harness. CHECK connectors, SERVICE or REPLACE harness. RERUN Quick Test.
B3 CHECK IDM CIRCUIT FOR SHORT TO POWER		
• **Is the resistance between J41 (IDM E) and J51 (VBAT E) greater than 10K ohms?**	Yes ▶ No ▶	REPLACE EDIS module. RECONNECT EEC processor. RERUN Quick Test. If fault persists, REPLACE EEC processor and RERUN Quick Test. IDM is shorted to power in harness. CHECK connectors, SERVICE or REPLACE harness. RERUN Quick Test.

Test B – code 16, memory code 18 and/or "Check Engine Light On" IDM failure

TEST STEP	RESULT	ACTION TO TAKE
C1 CHECK FOR COIL SIGNALS AT EDIS		
• Key off. • Install EDIS diagnostic cable to breakout box. • Use the EDIS-6 overlay. • Key on. • Connect LED test lamp from J51 (VBAT E) to each coil - J53, J54 and J55, one at a time. • **Does the lamp turn on brightly for each coil?**	Yes ▶ No ▶	GO to C6 . GO to C2 .
C2 CHECK EDIS COIL AT CRANK		
• Crank engine. • **Does lamp blink at each coil during crank?**	Yes ▶ No ▶	GO to C3 . GO to C4 .
C3 CHECK COIL CONTINUITY TO EDIS MODULE		
• Key off. • Disconnect EDIS module from the EDIS diagnostic cable. • Disconnect the coil pack from the EDIS diagnostic cable. • **Is the resistance between J3 and J53, J6 and J55, J10 and J54 less than 5 ohms?**	Yes ▶ No ▶	REPLACE coil pack. RERUN Quick Test. SERVICE open circuit in coil harness. RERUN Quick Test.
C4 CHECK COIL FOR SHORT TO POWER		
• Key off. • Disconnect the coil pack from the EDIS diagnostic cable. • Crank the engine. • **Does lamp blink at any coil during crank?**	Yes ▶ No ▶	REPLACE coil pack. RERUN Quick Test. GO to C5 .
C5 CHECK COIL CIRCUITS FOR SHORT TO POWER		
• Key off. • HEGO disconnected. • Disconnect EDIS module from EDIS diagnostic cable. • **Is the resistance between J51 (VBAT E) and each coil - J3, J6 and J10 - greater than 10K ohms?**	Yes ▶ No ▶	REPLACE EDIS module. RERUN Quick Test. SERVICE short to power in coil harness RERUN Quick Test.

Test C – memory code 45 and/or "Check Engine Light On" coil failure

TEST STEP	RESULT ▶	ACTION TO TAKE
C6 CHECK COIL FOR SHORT TO GROUND		
• Key off. • Disconnect the coil pack from the EDIS diagnostic cable. • Key on. • Connect LED test lamp from J7 (BAT-) to each coil - J53, J54 and J55, one at a time. • **Does the lamp turn on for each coil?**	Yes ▶ No ▶	GO to C7 . REPLACE coil pack. RERUN Quick Test.
C7 CHECK COIL CIRCUITS FOR SHORT TO GROUND		
• Key off. • Disconnect the EDIS module from the EDIS diagnostic cable. • Key on. • Connect LED test lamp from J7 (BAT-) to each coil - J53, J54 and J55, one at a time. • **Does the lamp turn on for each coil?**	Yes ▶ No ▶	SERVICE short circuit in coil harness. RERUN Quick Test. REPLACE EDIS module. RERUN Quick Test.

Test C – part 2

TEST STEP	RESULT	ACTION TO TAKE
D1 CHECK FOR SPARK AT WIRES		
• Engine on. • Using a Neon Bulb Spark Tester (OTC D 89P-6666-A), check for spark at all spark plug wires. • **Was spark present on ALL spark plug wires and consistent (one spark per crankshaft revolution)?**	Yes ▶ No ▶	GO to D2. GO to D6.
D2 CHECK TIMING WITH TIMING LIGHT		
• Install timing light. • Start engine and warm to normal engine temperature. • Transmission out of gear. • **Is the spark angle greater than 19 degrees BTDC?**	Yes ▶ No ▶	REFER to engine diagnostics in Shop Manual. GO to D3.
D3 CHECK SAW CONTINUITY TO EEC		
• Key off. • Install EDIS diagnostic cable but do not connect the EDIS module or coil pack to the cable. • Connect the cable to the breakout box. • Use the EDIS 6 overlay. • Disconnect the EEC processor. • **Is the resistance between J45 (SAW E) and Pin 36 (SAW) of the EEC connector less than 5 ohms?**	Yes ▶ No ▶	GO to D4. SERVICE open in SAW circuit. RECONNECT the EEC processor and RERUN Quick Test.
D4 CHECK SAW CIRCUIT FOR SHORT TO GROUND		
• Key off. • HEGO disconnected. • **Is the resistance between J45 (SAW E) and J7 (BAT -) greater than 10K ohms?**	Yes ▶ No ▶	GO to D5. SAW is shorted to ground in harness. CHECK connectors, SERVICE or REPLACE harness. RERUN Quick Test.

Test D — engine running with code 18, SAW failure

TEST STEP	RESULT ▶	ACTION TO TAKE
D5 CHECK SAW CIRCUIT FOR SHORT-TO-POWER		
• **Is the resistance between J45 (SAW E) and J51 (VBAT E) greater than 10K ohms?**	Yes ▶	REPLACE EDIS module. RECONNECT EEC processor. RERUN Quick Test. If fault persists, REPLACE EEC processor and RERUN Quick Test.
	No ▶	SAW is shorted to ground in harness. CHECK connectors, SERVICE or REPLACE harness. RERUN Quick Test.
D6 CHECK PLUGS AND WIRES		
• Check spark plug wires for insulation damage, looseness, shorting or other damage. • Remove and check spark plugs for damage, wear, carbon deposits and proper plug gap. • **Are spark plugs and wires OK?**	Yes ▶	REPLACE coil pack.
	No ▶	SERVICE or REPLACE damaged component. RERUN Quick Test.

Test D – part 2

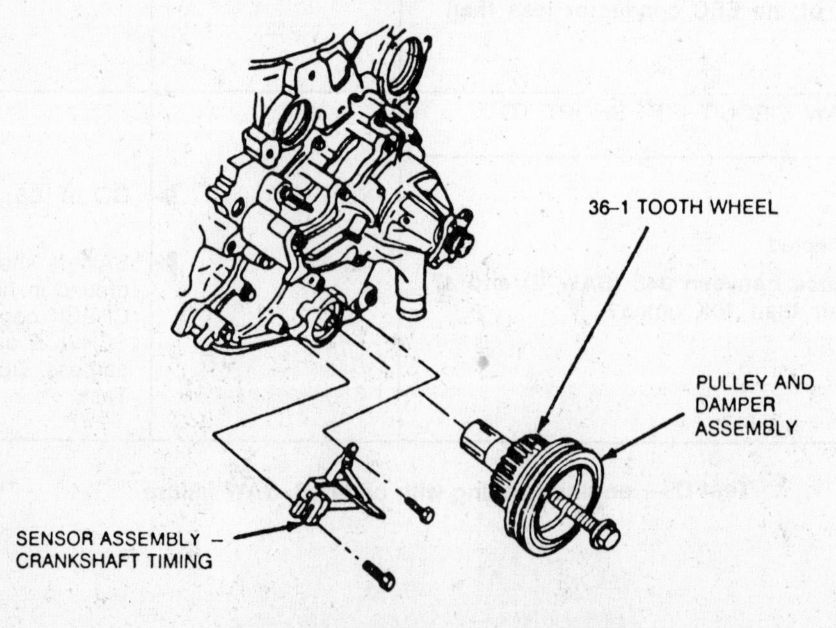

Crankshaft sensor -- 4.0L engine

EDIS Module

REMOVAL AND INSTALLATION

NOTE: Use this general service procedure as guide. Modify service steps as neceesary.

1. Disconnect the negative battery cable.
2. Disconnect each electrical connector of the DIS ignition module assembly by pushing down the connector locking tabs where it is stamped **PUSH** and then pull it away from the module.
3. Remove the 3 retaining screws, remove the ignition module assembly from the lower intake manifold.

To install:

4. Apply an even coat (approximately $\frac{1}{32}$ in.) of a suitable silicone dielectric compound to the mounting surface of the DIS module.
5. Mount the DIS module assembly onto the intake assembly and install the retaining screws. Torque the screws to 22–31 inch lbs.
6. Install the electrical connectors to the DIS ignition module assembly. Reconnect the negative battery cable.

Ignition Coil Pack

REMOVAL AND INSTALLATION

1. Disconnect the negative battery cable.
2. Disconnect the electrical harness connector from the ignition coil pack.
3. Remove the spark plug wires by squeezing the locking tabs to release the coil boot retainers.
4. Remove the coil pack mounting screws and remove the coil pack.

NOTE: On vehicle equipped with power steering it may be necessary to remove the intake (left hand) coil and bracket as an assembly.

To install:

5. Install the coil pack and the retaining screws. Torque the retaining screws to 40–62 inch lbs.
6. Connect the spark plug wires and connect the electrical connector to the coil pack.
7. Reconnect the negative battery cable.

NOTE: Be sure to place some dielectric compound into each spark plug boot prior to installation of the spark plug wire.

Ignition Timing

Timing is preset at 10° BTC and is not adjustable.

Valve Lash

Valve adjustment determines how far the valves enter the cylinder and how long they stay open and closed.

If the valve clearance is too large, part of the life of the camshaft will be used to removing the excessive clearance. Consequently, the valve will not be opening as far as it should. This condition has two effects: the valve train components will emit a tapping sound as they take up the excessive clearance and the engine will perform poorly because the valves don't open fully and allow the proper amount of gases to flow into and out of the engine.

If the valve clearance is too small, the intake valve and the exhaust valves will open too far and they will not fully seat on the cylinder head when they close. When a seat on the cylinder head when they close. When a valve seats itself on the cylinder head, it does two things: it seals the combustion chamber so that none of the gases in the cylinder escape and it cools itself by transferring some of the heat it absorbs from the combustion in the cylinder to the cylinder head and to the engine's cooling system. If the valve clearance is too small, the engine will run poorly because of the gases escaping from the combustion chamber. The valves will also become overheated and will warp, since they cannot transfer heat unless they are touching the valve seat in the cylinder head.

NOTE: While all valve adjustments must be made as accurately as possible, it is better to have the valve adjustment slightly loose than slightly tight as a burned valve may result from overly tight adjustments.

ADJUSTMENT

2.0L, 2.3L Engines

NOTE: The 4-cylinder gasoline engines in this vehicle are equipped with hydraulic valve lash adjusters. Adjustment is not necessary as a tune up procedure. To check the valve lash use the following procedure.

THE ALLOWABLE COLLAPSED TAPPET GAP IS 0.035–0.055 AT THE CAMSHAFT.

THE DESIRED COLLAPSED TAPPET GAP IS 0.040–0.050 AT THE CAMSHAFT.

1. Disconnect the battery ground cable.
2. Remove the rocker arm cover following the procedure in Section 3.
3. Position the camshaft so that the base circle of the lobe is facing the cam follower of the valve to be checked.
4. Using tool T74P-6565-A, slowly apply pressure to the cam follower until the lash adjuster is completely collapsed. Hold the follower in this position and insert the proper size feeler gauge between the base circle of the cam and the follower.
5. If the clearance is excessive, remove the cam follower and inspect for damage.
6. If the cam follower appears to be intact, and not excessively worn, measure the valve spring damper assembly assembled height to be sure the valve is not sticking.
7. If the valve spring damper spring assembled height is correct, check the dimensions of the camshaft following the procedure in Section 3.
8. If the camshaft dimensions are to specifications, remove, clean and test the lash adjuster.
9. Reinstall the lash adjuster and check the clearance. Replace damaged or worn parts as necessary.

2.8L Engine

NOTE: The following procedure should be performed on a cold engine.

1. Disconnect the battery ground cable.
2. Remove the rocker arm cover following the procedure in Section 3.
3. Place your finger on the adjusting screw of the intake valve rocker arm for the number 5 cylinder. You should be able to feel any movement in the rocker arm.
4. Using a remote starter switch, **bump** the engine over until the intake valve for the number 5 cylinder just begins to open. The valves on the number 1 cylinder may now be adjusted.
5. Adjust the number 1 intake valve so that a 0.35mm feeler gauge has a light drag and a 0.38mm feeler gauge is very tight. Turn the adjusting screw clockwise to decrease the gap and counterclockwise to increase the gap. The adjusting screws are self-locking and will stay in position once they are set.

NOTE: When checking the valve lash, be sure to insert the feeler gauge between the rocker arm and the valve tip at the front or (rear) edge of the valve and move it to-

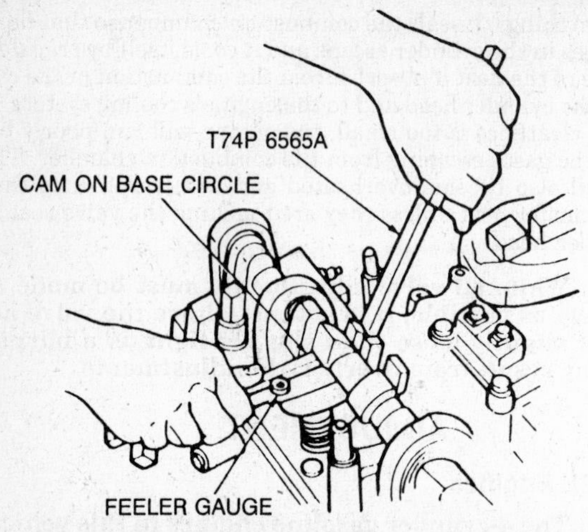

Checking valve lash — 2.0L, 2.3L engines

ward the opposite edge with a rearward or (forward) motion. DO NOT insert the feeler gauge at the outer edge and move toward the inner edge (inward toward the carburetor), this will produce an incorrect reading which will result in overly tight valves.

6. Using the same method, adjust the number 1 exhaust valve lash so that a 0.40mm feeler gauge has a light drag and a 0.43mm feeler gauge is very tight.
7. Adjust the remaining valves in the same manner, in the firing order (1-4-2-5-3-6) by positioning the camshaft according to the chart below.
8. Install the rocker arm covers following the procedure in Section 3 under **Engine Mechanical**.
9. Reconnect the battery ground cable.
10. Start the engine and check for oil and vacuum leaks.

2.9L Engine

NOTE: The following procedure should be performed on a cold engine.

1. On the cylinder to be adjusted, position the cams so that the tappets are in the base circle area.
2. Loosen the adjusting screws until a distinct lash between the roller arm pad and the valve tip can be noticed.

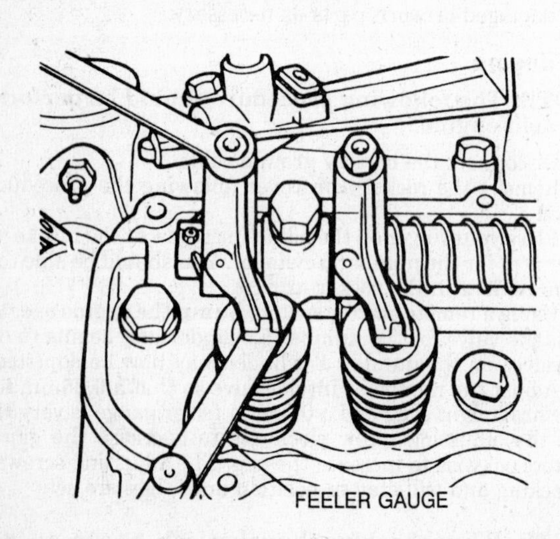

Valve adjustment — 2.8L engine

NOTE: The plunger of the hydraulic tappet should now be fully extended under the load of the internal spring.

3. Carefully screw in the adjustment screws until the roller arms slightly touch the valves.
4. To achieve the normal working position of the plunger, screw in the adjusting screw 1½ turns, equivalent to 2mm.

To adjust both valves for cylinder number	1	4	2	5	3	6
The intake valve must be opening for cylinder number	5	3	6	1	4	2

Valve adjusting arrangement — 2.8L engine

Fuel System

This Section contains only tune-up adjustment procedures for the carburetor. Descriptions, adjustments and overhaul procedures for the carburetor can be found in Section 5.

When the engine in your vehicle is running, the air/fuel mixture from the carburetor is being drown into the engine by a partial vacuum created by the downward movement of the pistons on the intake stroke. The amount of air/fuel mixture that enters the engine is controlled by the throttle plate(s) in the bottom of the carburetor. When the engine is not running, the throttle plates are closed, completely blocking off the air/fuel passage(s) at the bottom of the carburetor. The throttle plates are connected by the throttle linkage to the accelerator pedal in the passenger compartment of the truck. When you depress the pedal, you open the throttle plates in the carburetor to admit more air/fuel mixture to the engine.

When the engine is idling, it is necessary to have the throttle plates open slightly. To prevent having to hold your foot on the pedal, an idle speed adjusting screw is located on the carburetor linkage.

The idle adjusting screw contacts a lever (throttle lever) on the outside of the carburetor. When the screw is turned, it opens or closes the throttle plates of the carburetor, raising or lowering the idle speed of the engine. This screw is called the curb idle adjusting screw. There are three different types of carburetors used on the Ford Ranger/Bronco II. The 4-2.0L, 2.3L engines use the Carter model YFA 1-bbl. carburetor except on the California and High Altitude models which are equipped with the Carter model YFA 1-bbl. Feedback carburetor. The 6-2.8L engine is equipped with a Motorcraft model 2150 2-bbl. carburetor.

IDLE SPEED ADJUSTMENTS

2.0L, 2.3L Engines with Carter YFA-1V & YFA-1V Feedback Carburetor

1. Block the wheels and apply the parking brake.
2. Place the transmission in Neutral or Park.
3. Bring engine to normal operating temperature.
4. Place the air conditioning selector in the Off position.
5. Place transmission in specified position as referred to on the emission decal.
6. Check/adjust curb idle RPM. If adjustment is required, turn the hex head adjustment at the rear of the TSP (throttle solenoid positioner) housing.
7. Place the transmission in Neutral or Park. Rev the engine momentarily. Place transmission in specified position and recheck curb idle RPM. Readjust if required.
8. Turn the ignition key to the Off position.

9. If a curb idle RPM adjustment was required and the carburetor is equipped with a dashpot, adjust the dashpot clearance to specification as follows:
 a. Turn key to On position. Open throttle to allow TSP solenoid plunger to extend to the curb idle position.
 b. Collapse dashpot plunger to maximum extent. Measure clearance between tip of plunger and extension pad on throttle vent lever. If required, adjust to specification. Tighten dashpot locknut. Recheck clearance. Turn key to Off position.
10. If curb idle adjustment was required:
 a. Turn ignition key to the On position to activate the TSP (engine not running). Open throttle to allow the TSP solenoid plunger to extend to the curb idle position.
 b. Secure the choke plate in the wide open position.
 c. Open throttle so that the throttle vent lever does not touch the bowl vent rod. Close the throttle to the idle set position and measure the travel of the fuel bowl vent rod from the open throttle position.
 d. Travel of the bowl vent rod should be 2.5-3.5mm.
 e. If out of specification, bend the throttle vent lever at notch to obtain required travel.
11. Remove all test equipment and reinstall air cleaner assembly. Tighten the holddown bolt.

2.8L Engine with Motorcraft 2150A-2V Carburetor

NOTE: On models equipped with air conditioning, the air conditioner-On RPM speed must be set prior to setting the Curb Idle Speed adjustment. This adjustment is made with the vacuum operated throttle modulator (VOTM) on.

AIR CONDITIONER-ON RPM ADJUSTMENT

1. Remove the air cleaner and disconnect and plug the vacuum lines.
2. Block the wheels, apply the parking brake, turn off all accessories, start the engine and run it to normalize underhood temperatures.
3. Check that the chile plate is fully open and connect a tachometer according to the manufacturer's instructions.
4. Disconnect the air conditioner clutch wire at the compressor.
5. Place the heater control selector to maximum cooling and set the blower switch in the high position.
6. Place the manual transmission in neutral; the automatic transmission in drive.
7. Using the saddle bracket adjusting screw, adjust the air conditioner-ON RPM to the specifications shown on the under hood emission sticker.
8. Reconnect the air conditioner compressor clutch wire.
9. Proceed to step 4 below and set the Curb Idle Speed adjustment.

CURB IDLE SPEED ADJUSTMENT

1. Remove the air cleaner and disconnect and plug the vacuum lines.
2. Block the wheels, apply the parking brake, turn off all accessories, start the engine and run it to normalize underhood temperatures.
3. Check that the choke plate is fully open and connect a tachometer according to the manufacturer's instructions.
4. Place the manual transmission in neutral; the automatic transmission in drive and make certain the (TSP) throttle stop positioner plunger is extended.
5. Turn the saddle bracket adjustment screw (non-air conditioned), or the hex head protruding from the rear of the TSP diaphragm assembly (air conditioned models) until the specified idle speed is obtained.
6. Check the TSP-off speed as follows:
 a. disconnect the TSP wire.
 b. place the transmission in neutral and check the RPM. If necessary, adjust to the specified TSP-off speed with the throttle adjusting screw. Check the underhood sticker for specifications.
7. Install the air cleaner and connect the vacuum lines. Recheck the idle speed. Adjust, if necessary, with the air cleaner on.

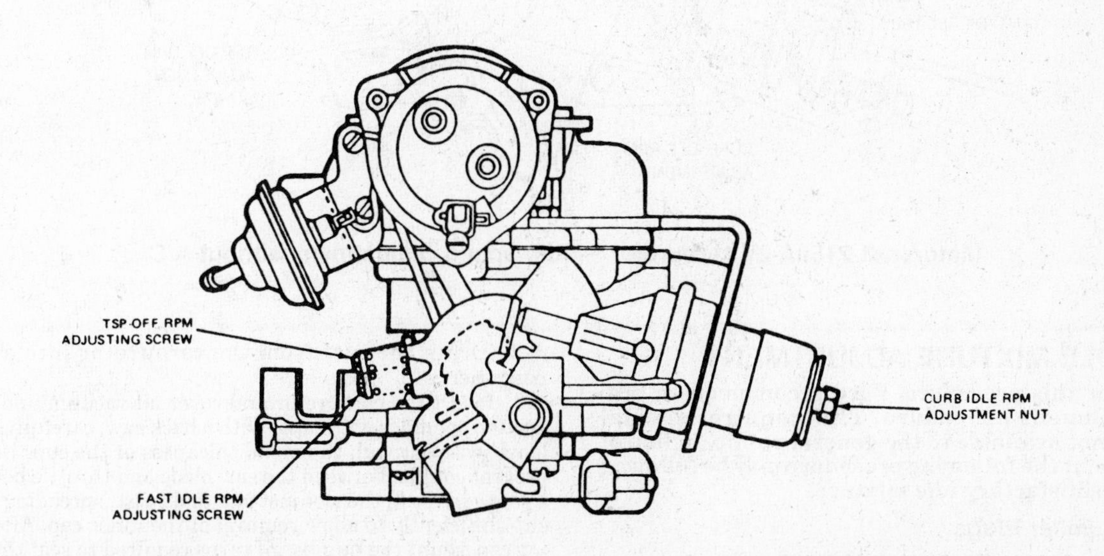

Carter YFA-1V & YFA-1V Feedback carburetor – idle speed adjustment

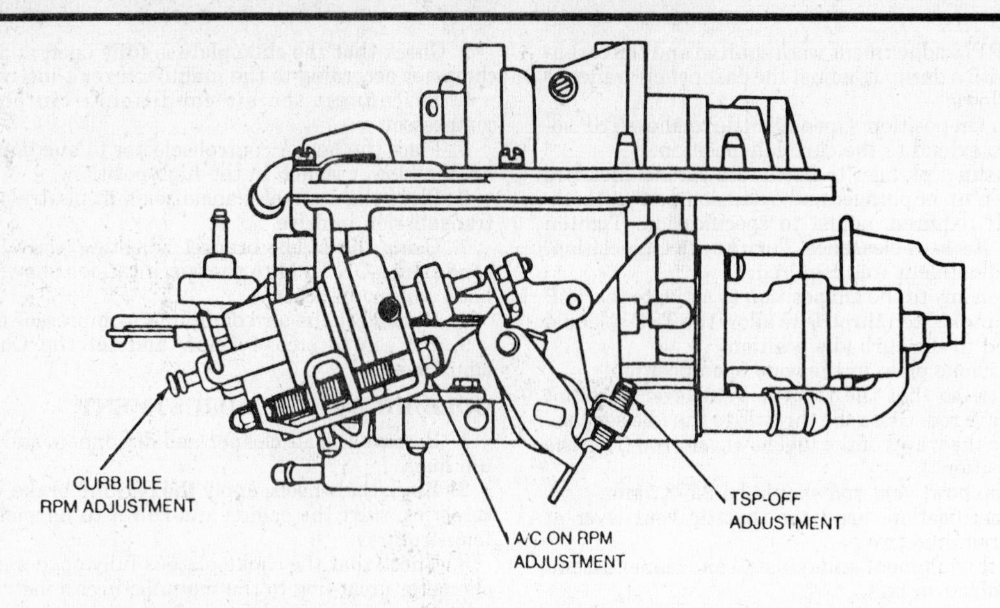

Motorcraft 2150A-2V carburetor – idle speed adjustment with A/C

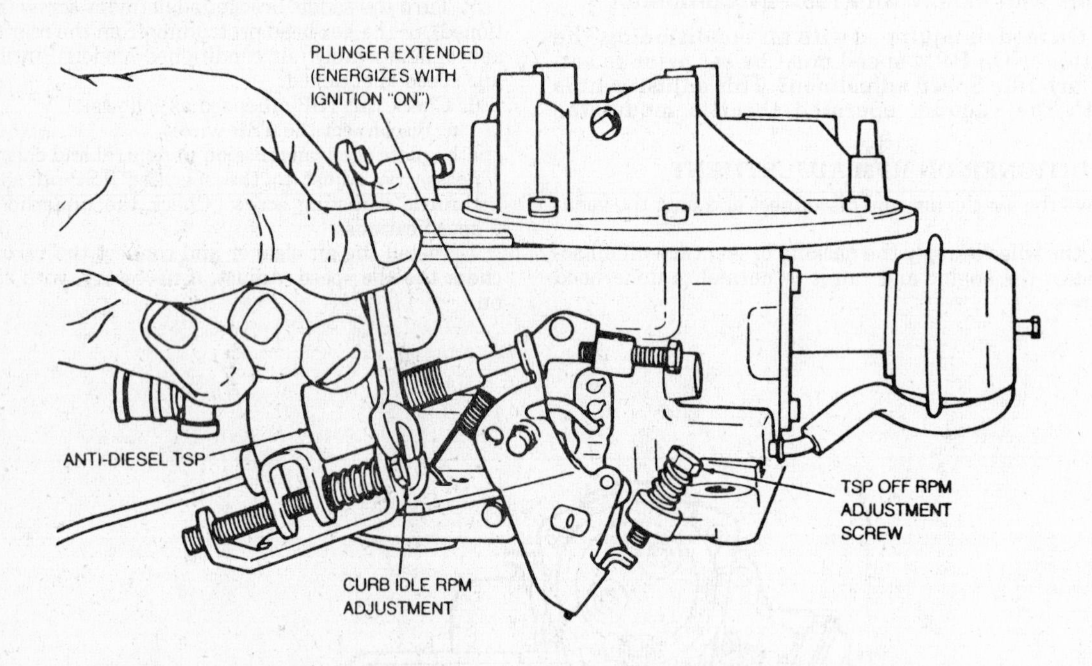

Motorcraft 2150A-2V carburetor – idle speed adjustment without A/C

IDLE MIXTURE ADJUSTMENT

NOTE: For this procedure, Ford recommends a propane enrichment procedure. This requires special equipment not available to the general public. In lieu of this equipment the following procedure may be followed to obtain a satisfactory idle mixture.

Removing Limiter Plugs

2.0L, 2.3L ENGINES WITH CARTER YFA-1V FEEDBACK CARBURETORS

1. Remove the carburetor from the engine as described in Section 5.

2. Drain the fuel from the carburetor into a suitable container.

3. Invert the carburetor and cover all vacuum and fuel connection openings with tape. With a hack saw, carefully saw a slot lengthwise through the metal thickness of the cup. Use care to prevent contact between the saw blade and throttle body. Insert a screwdriver in the slot just cut, and twist, spreading the outer cup sufficiently to allow removal of the inner cap. After removing cap, count the number of turns required to seat the mixture screw needle lightly. This information will be used in assembly. Remove the screw and cap. After cleaning the metal shavings from the carburetor, remove the tape from the openings.

4. Install the idle mixture screw and spring, and a new adjust-

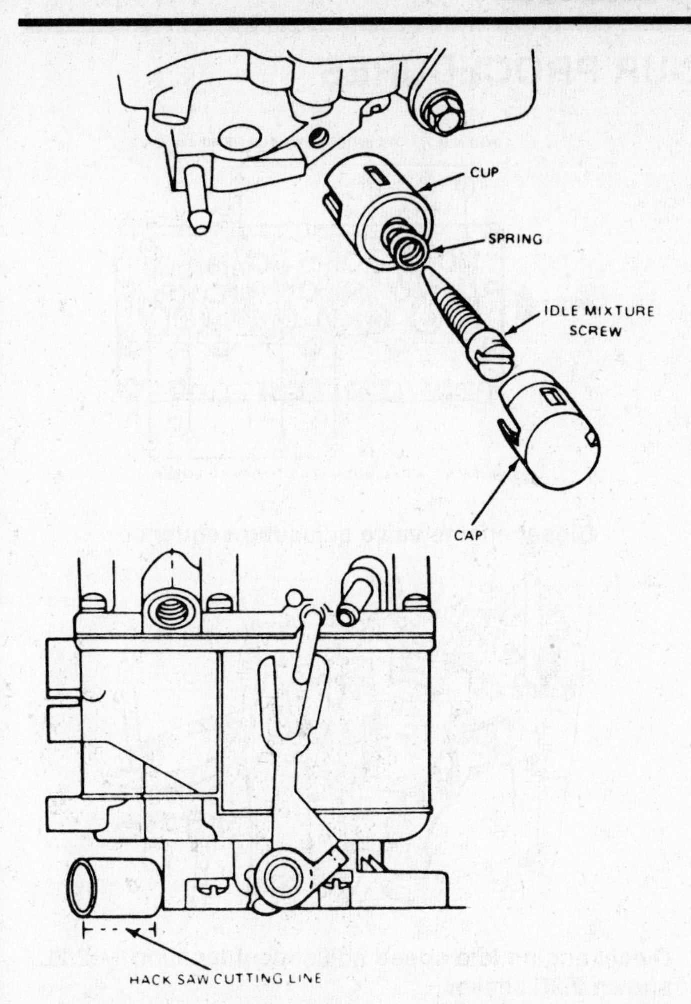

Mixture screw locking caps — Carter YFA-1V & YFA-1V Feedback carburetor

ment limiting cup. Set the screw to the same number of turns out from the lightly seated position as noted during disassembly.

5. Install the carburetor on the vehicle and perform the idle mixture setting procedure.

6. After making the mixture adjustment, install the mixture limiting cap.

2.8L ENGINE WITH MOTORCRAFT 2150A-2V CARBURETOR

The idle mixture adjusting screws are covered with a two-piece tamper-resistant limiter plugs. To adjust the idle mixture the plugs must be removed using the following procedure:

1. Remove and drain the carburetor using the procedure in Section 5 under **Fuel System**.

2. Turn the carburetor over and locate the locking tab on each locking cap.

3. Using a blunt punch and a light hammer, tap the locking cap until the locking tab has cleared the detent in the locking plug.

NOTE: Support the area under the limiter plug when removing it to prevent the adjusting screw from bending.

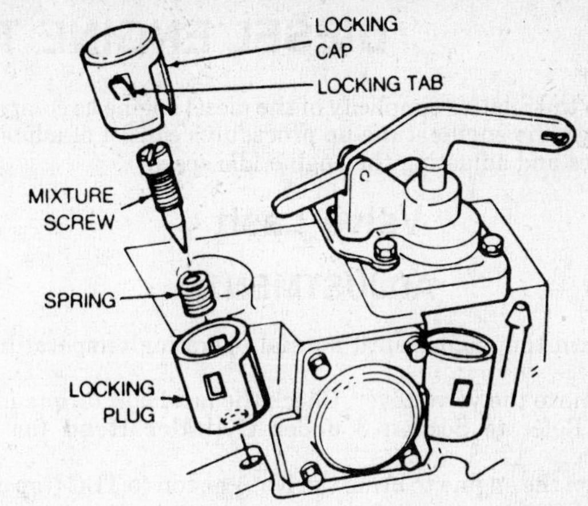

Mixture screw locking caps — Motorcraft 2150-2V

4. Remove the locking cap from the locking plug and remove the support form under the locking cap.

5. Repeat steps 3 and 4 for the other locking cap.

6. Install the carburetor on the engine and perform the idle mixture adjustment.

7. To install the cap, align with the detent in the plug and press the cap into the plug.

Mixture Adjustment

1. Block the wheels, set the parking brake and run the engine to bring it to normal operating temperature.

2. Disconnect the hose between the emission canister and the air cleaner.

3. On engines equipped with the Thermactor air injection system, the routing of the vacuum lines connected to the dump valve will have to be temporarily changed. Mark them for reconnection before switching them.

4. For valves with one or two vacuum lines at the side, disconnect and plug the lines.

5. For valves with one vacuum line at the top, check the line to see if it is connected to the intake manifold or an intake manifold source such as the carburetor or distributor vacuum line. If not, remove and plug the line at the dump valve and connect a temporary length of vacuum hose from the dump valve fitting to a source of intake manifold vacuum.

6. Remove the limiter caps form the mixture screws by CAREFULLY cutting them with a sharp knife.

7. Place the transmission in neutral and run the engine at 2500 rpm for 15 seconds.

8. Place the automatic transmission in Drive; the manual in neutral.

9. Adjust the idle speed to the higher of the two figures given on the underhood sticker.

10. Turn the idle mixture screws to obtain the highest possible rpm, leaving the screws in the leanest position that will maintain this rpm.

11. Repeat steps 7 thru 10 until further adjustment of the mixture screws does not increase the rpm.

12. Turn the screws in until the lower of the two idle speed figures is reached. Turn the screws in ¼ turn increments each to insure a balance.

13. Turn the engine off and remove the tachometer. Reinstall all equipment.

Fuel Injected Engines

These engines have idle speed controlled by the TFI-IV/EEC-IV system and no adjustment is possible.

DIESEL ENGINE TUNE-UP PROCEDURES

Due to the relative simplicity of the diesel engine as compared to the gasoline engine, tune-up procedures consist of adjusting the valves and adjusting the engine idle speed.

Valve Lash

ADJUSTMENT

1. Warm the engine until normal operating temperature is reached.
2. Remove the valve cover. Check the head bolt torque in sequence. Refer to Section 3 under **Cylinder Head** for this procedure.
3. Turn the engine to bring the No. 1 piston to TDC (top dead center) of the compression stroke.
4. Adjust the following valves:
- No. 1 Intake
- No. 1 Exhaust
- No. 2 Intake
- No. 3 Exhaust
5. Rotate the crankshaft 360° and bring No. 4 piston to TDC of the compression stroke.
6. Adjust the following valves:
- No. 2 Exhaust
- No. 3 Intake
- No. 4 Intake
- No. 4 Exhaust
7. To adjust the valves, loosen the locknut on the rocker arm. Rotate the adjusting screw clockwise to reduce clearance, counterclockwise to increase clearance. Clearance is checked with a flat feeler gauge (0.012 inch gauge intake and 0.012 gauge exhaust-engine must be hot on the 2.2L diesel engine) (0.010 inch gauge intake and 0.010 gauge exhaust-engine must be hot on the 2.3L Turbo diesel engine) that is passed between the rocker arm and valve stem.
8. After adjustments are made, be sure the locknuts are tight. Be sure mounting surfaces are clean. Install the valve cover and new valve cover gasket.

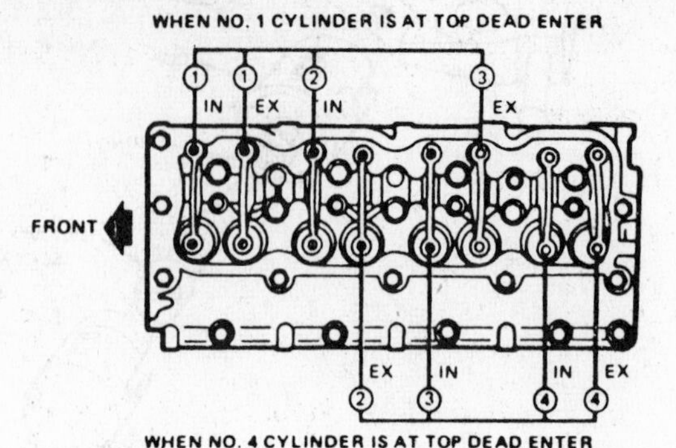

Diesel engine valve adjusting sequence

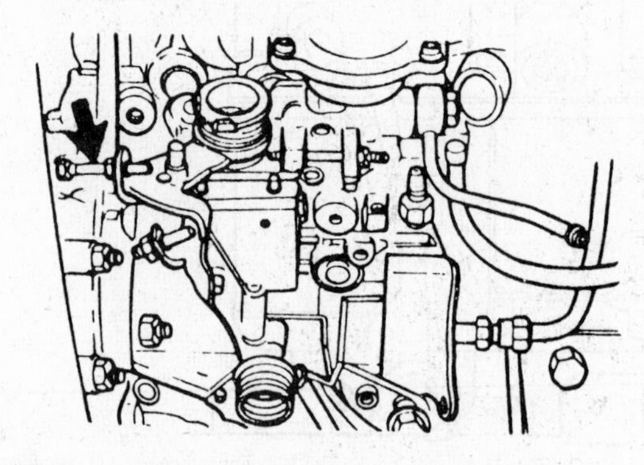

Diesel engine idle speed adjustment location -- 2.2L shown 2.3L similar

Setting the Idle Speed

NOTE: A special tachometer is required to check engine RPM on a diesel engine.

1. Block the wheels and apply the parking brake.
2. Start and run engine until the normal operating temperature is reached. Shut off engine.
3. Connect diesel engine tachometer.
4. Start engine and check RPM. Refer to emissions decal for latest specifications. RPM is usually adjusted in Neutral for manual transmissions and Drive for automatic models.
5. The adjustment bolt is located on the bell crank at the top of the injector pump. The upper bolt is for curb idle, the lower for max speed.
6. Loosen the locknut. Turn the adjustment screw clockwise to increase RPM, counterclockwise to lower the RPM.
7. Tighten the locknut, Increase engine speed several times and recheck idle. Readjust if necessary

Injection Timing

NOTE: For injection pump timing please refer to Section 5 under Diesel Fuel Systems.

3

Engine and Engine Overhaul

QUICK REFERENCE INDEX

GENERAL INDEX

ENGINE ELECTRICAL

Understanding the Engine Electrical System

The engine electrical system can be broken down into three separate and distinct systems:

1. The starting system.
2. The charging system.
3. The ignition system.

BATTERY AND STARTING SYSTEM

Basic Operating Principles

The battery is the first link in the chain of mechanisms which work together to provide cranking of the automobile engine. In most modern cars, the battery is a lead/acid electrochemical device consisting of six 2v subsections connected in series so the unit is capable of producing approximately 12v of electrical pressure. Each subsection, or cell, consists of a series of positive and negative plates held a short distance apart in a solution of sulfuric acid and water. The two types of plates are of dissimilar metals. This causes a chemical reaction to be set up, and it is this reaction which produces current flow from the battery when its positive and negative terminals are connected to an electrical appliance such as a lamp or motor. The continued transfer of electrons would eventually convert the sulfuric acid in the electrolyte to water, and make the two plates identical in chemical composition. As electrical energy is removed from the battery, its voltage output tends to drop. Thus, measuring battery voltage and battery electrolyte composition are two ways of checking the ability of the unit to supply power. During the starting of the engine, electrical energy is removed from the battery. However, if the charging circuit is in good condition and the operating conditions are normal, the power removed from the battery will be replaced by the generator (or alternator) which will force electrons back through the battery, reversing the normal flow, and restoring the battery to its original chemical state.

The battery and starting motor are linked by very heavy electrical cables designed to minimize resistance to the flow of current. Generally, the major power supply cable that leaves the battery goes directly to the starter, while other electrical system needs are supplied by a smaller cable. During starter operation, power flows from the battery to the starter and is grounded through the car's frame and the battery's negative ground strap.

The starting motor is a specially designed, direct current electric motor capable of producing a very great amount of power for its size. One thing that allows the motor to produce a great deal of power is its tremendous rotating speed. It drives the engine through a tiny pinion gear (attached to the starter's armature), which drives the very large flywheel ring gear at a greatly reduced speed. Another factor allowing it to produce so much power is that only intermittent operation is required of it. This, little allowance for air circulation is required, and the windings can be built into a very small space.

The starter solenoid is a magnetic device which employs the small current supplied by the starting switch circuit of the ignition switch. This magnetic action moves a plunger which mechanically engages the starter and electrically closes the heavy switch which connects it to the battery. The starting switch circuit consists of the starting switch contained within the ignition switch, a transmission neutral safety switch or clutch pedal switch, and the wiring necessary to connect these in series with the starter solenoid or relay.

A pinion, which is a small gear, is mounted to a one-way drive clutch. This clutch is splined to the starter armature shaft. When the ignition switch is moved to the **START** position, the solenoid plunger slides the pinion toward the flywheel ring gear via a collar and spring. If the teeth on the pinion and flywheel match properly, the pinion will engage the flywheel immediately. If the gear teeth butt one another, the spring will be compressed and will force the gears to mesh as soon as the starter turns far enough to allow them to do so. As the solenoid plunger reaches the end of its travel, it closes the contacts that connect the battery and starter and then the engine is cranked.

As soon as the engine starts, the flywheel ring gear begins turning fast enough to drive the pinion at an extremely high rate of speed. At this point, the one-way clutch begins allowing the pinion to spin faster than the starter shaft so that the starter will not operate at excessive speed. When the ignition switch is released from the starter position, the solenoid is de-energized, and a spring contained within the solenoid assembly pulls the gear out of mesh and interrupts the current flow to the starter.

Some starter employ a separate relay, mounted away from the starter, to switch the motor and solenoid current on and off. The relay thus replaces the solenoid electrical switch, buy does not eliminate the need for a solenoid mounted on the starter used to mechanically engage the starter drive gears. The relay is used to reduce the amount of current the starting switch must carry.

THE CHARGING SYSTEM

Basic Operating Principles

The automobile charging system provides electrical power for operation of the vehicle's ignition and starting systems and all the electrical accessories. The battery services as an electrical surge or storage tank, storing (in chemical form) the energy originally produced by the engine driven generator. The system also provides a means of regulating generator output to protect the battery from being overcharged and to avoid excessive voltage to the accessories.

The storage battery is a chemical device incorporating parallel lead plates in a tank containing a sulfuric acid/water solution. Adjacent plates are slightly dissimilar, and the chemical reaction of the two dissimilar plates produces electrical energy when the battery is connected to a load such as the starter motor. The chemical reaction is reversible, so that when the generator is producing a voltage (electrical pressure) greater than that produced by the battery, electricity is forced into the battery, and the battery is returned to its fully charged state.

The vehicle's generator is driven mechanically, through V-belts, by the engine crankshaft. It consists of two coils of fine wire, one stationary (the stator), and one movable (the rotor). The rotor may also be known as the armature, and consists of fine wire wrapped around an iron core which is mounted on a shaft. The electricity which flows through the two coils of wire (provided initially by the battery in some cases) creates an intense magnetic field around both rotor and stator, and the interaction between the two fields creates voltage, allowing the generator to power the accessories and charge the battery.

There are two types of generators: the earlier is the direct current (DC) type. The current produced by the DC generator is generated in the armature and carried off the spinning armature by stationary brushes contacting the commutator. The commutator is a series of smooth metal contact plates on the end of the armature. The commutator is a series of smooth metal contact plates on the end of the armature. The commutator plates, which are separated from one another by a very short gap, are connected to the armature circuits so that current will flow in one directions only in the wires carrying the generator output. The generator stator consists of two stationary coils of wire which draw some of the output current of the generator to form a powerful magnetic field and create the interaction of fields which generates the voltage. The generator field is wired in series with the regulator.

Newer automobiles use alternating current generators or alternators, because they are more efficient, can be rotated at higher speeds, and have fewer brush problems. In an alternator, the field rotates while all the current produced passes only through the stator winding. The brushes bear against continuous slip rings rather than a commutator. This causes the current produced to periodically reverse the direction of its flow. Diodes (electrical one-way switches) block the flow of current from traveling in the wrong direction. A series of diodes is wired together to permit the alternating flow of the stator to be converted to a pulsating, but unidirectional flow at the alternator output. The alternator's field is wired in series with the voltage regulator.

The regulator consists of several circuits. Each circuit has a core, or magnetic coil of wire, which operates a switch. Each switch is connected to ground through one or more resistors. The coil of wire responds directly to system voltage. When the voltage reaches the required level, the magnetic field created by the winding of wire closes the switch and inserts a resistance into the generator field circuit, thus reducing the output. The contacts of the switch cycle open and close many times each second to precisely control voltage.

While alternators are self-limiting as far as maximum current is concerned, DC generators employ a current regulating circuit which responds directly to the total amount of current flowing through the generator circuit rather than to the output voltage. The current regulator is similar to the voltage regulator except that all system current must flow through the energizing coil on its way to the various accessories.

Ignition Coil

REMOVAL AND INSTALLATION

1. Disconnect the battery ground.
2. Disconnect the two small and one large wires from the coil.
3. Disconnect the condenser connector from the coil, if equipped.
4. Unbolt and remove the coil.
5. Installation is the reverse of service removal procedure.

Ignition Module

REMOVAL AND INSTALLATION

Removing the module, on all models, is a matter of simply removing the fasteners that attach it to the fender firewall and pulling apart the connectors. When unplugging the connectors, pull them apart with a firm, straight pull. NEVER PRY THEM APART! To pry them will cause damage. When reconnecting them, coat the mating ends with silicone dielectric grease to waterproof the connection. Press the connectors together firmly to overcome any vacuum lock caused by the grease.

NOTE: If the locking tabs weaken or break, don't replace the unit. Just secure the connection with electrical tape or tie straps.

Distributor Cap, Adapter and Rotor

REMOVAL AND INSTALLATION

1. Tag all spark plug wires with a piece of tape according to cylinder number for reference when installing the wires, then remove them from the distributor cap. Note the position of No. 1 spark plug tower.
2. Unclip the distributor cap and lift it straight up and off the distributor.
3. Using a screwdriver, loosen the adapter attaching screws and remove the adapter.
4. Loosen the screws attaching the rotor to the distributor and remove the rotor.
5. Wipe the distributor cap and rotor with a clean, damp cloth. Inspect the cap for cracks, broken carbon button, carbon tracks, dirt or corrosion on the terminals and replace the cap if questionable. Replace the rotor if cracks, carbon tracks, burns, damaged blade or a damaged spring is noted.
6. Position the distributor rotor with the square and round locator pins matched to the rotor mounting plate. Tighten the screws to 24–36 inch lbs. (2–4 Nm).
7. Install the adapter and tighten the attaching screws to 18–23 inch lbs. (2–3 Nm).
8. Install the cap, noting the square alignment locator, then tighten the holddown screws to 18–23 inch lbs. (2–3 Nm).
9. Install the spark plug wires in firing order, starting from No. 1 tower and working in sequence around the cap. Refer to the firing order illustrations in Section 2, if necessary. Make sure the ignition wires are installed correctly and are firmly seated in the distributor cap towers.

TFI Ignition Module

REMOVAL AND INSTALLATION

1. Remove the distributor cap with the ignition wires attached and position it out of the way. Remove the adapter.
2. Disconnect the TFI harness connector.
3. Remove the distributor from the engine as previously described.
4. Place the distributor on a clean workbench and remove the two TFI module attaching screws.
5. Pull the right side of the module down toward the distributor mounting flange and then back up to disengage the module terminals from the connector in the distributor base. The module may then be pulled toward the flange and away from the distributor.

CAUTION

Do not attempt to lift the module from its mounting surface prior to moving the entire TFI module toward the distributor flange or you will break the pins at the distributor/module connector.

6. Coat the metal base of the new TFI module with a $\frac{1}{32}$ in. (0.8mm) thick film of Silicone Dielectric Compound D7AZ–19A331–A or equivalent. This is extremely important to help dissipate the heat when the module is operating.
7. Place the TFI module on the distributor base mounting flange.
8. Carefully position the TFI module assembly toward the distributor bowl and engage the three distributor connector pins securely. Be careful when performing this step. It is very easy to bend one of the connector pins when installing.

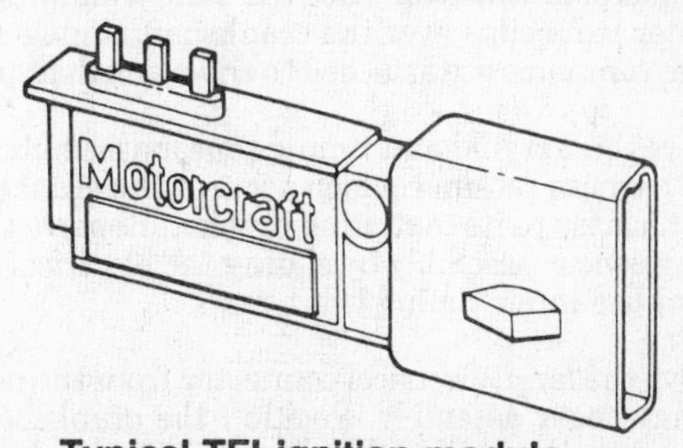

Typical TFI ignition module

9. Install the two TFI mounting screws and tighten them to 15–35 inch lbs. (1–4 Nm).
10. Install the distributor on the engine as previously described.
11. Install the distributor cap and tighten the mounting screws to 18–23 inch lbs. (2–3 Nm).
12. Reconnect the TFI wiring harness connector.
13. Attach a timing light according to the manufacturer's instructions and set the initial timing.

Octane Rod

REMOVAL AND INSTALLATION

1. Remove the distributor cap, adapter and rotor as previously described.
2. Remove the octane rod 4mm retaining screw carefully. Don't drop it.
3. Slide the octane rod grommet out to a point where the rod can be disengaged from the stator retaining post and remove the octane rod. Retain the grommet for use with the new octane rod.
4. Install the grommet on the new octane rod.
5. Install the octane rod into the distributor, making sure it engages the stator retaining post.
6. Install the retaining screw and tighten it to 15–35 inch lbs. (2–4 Nm).
7. Install the rotor, adapter and cap as described above.

NOTE: The 4.0L engine is equipped with the EDIS distributorless ignition system. For all ignition system component procedures, see Section 2 of this book. The 2.3L (twin plug) and 4.0L engine have ignition systems which do not use a distributor assembly.

The 2.3L (twin plug) engine uses a distributorless ignition system. This system consists of the following components: Crankshaft timing sensor, DIS ignition module, two ignition coil packs, EEC-IV module and related wiring. Initial timing is PRESET at 10°BTDC and is NOT ADJUSTABLE. Refer to service procedures below.

Crankshaft Timing Sensor Assembly

REMOVAL AND INSTALLATION

1. Disconnect the negative battery cable. Disconnect the crankshaft timing sensor electrical connectors from the engine harness.
2. Remove the large electrical connector (mark location of wires for correct installation) from the crankshaft timing sensor assembly by prying out the red retaining clip and removing the four wires.
3. Remove the crankshaft pulley assembly by removing the accessory drive belts and then the four bolts that retain it the crankshaft pulley hub assembly.
4. Remove the timimng belt cover outer cover.
5. Rotate the crankshaft so that the keyway is at 10 O'CLOCK position. This will place the vane window of both inner and outer vane cups over the crankshaft timing sensor assembly. The vane cups are attached to the crankshaft pulley hub assembly.
6. Remove the 2 crankshaft timing sensor assembly retaining bolts, wire harness retainer which secures the crankshaft timing sensor harness to its mounting bracket. Remove the crankshaft timing sensor assembly by sliding the electrical wires out from behind the inner timing belt cover.

To install:

7. Remove the large electrical connector from the new crankshaft timing sensor assembly. Position the crankshaft timing sensor assembly. First, slide the electrical wires behind the inner timing belt cover. Then, hold the sensor assembly loosely in place with the retaining bolts, but do not tighten at this time.

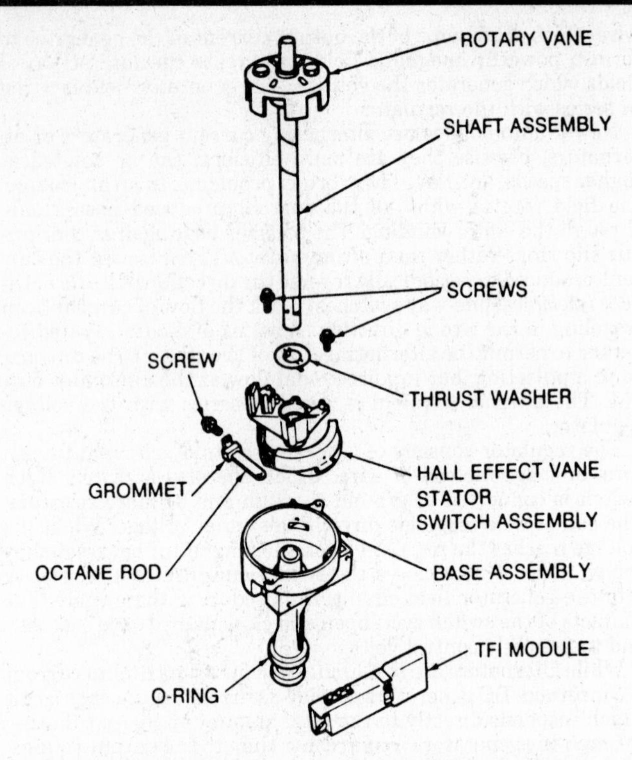

Exploded view of the universal distributor assembly

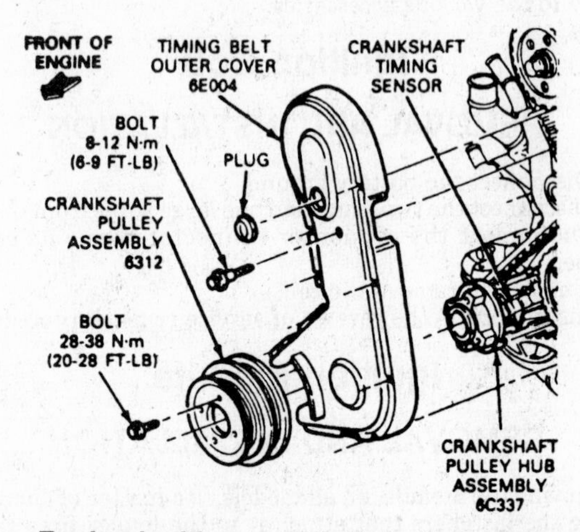

Engine ignition — 2.3L Twin Plug Engine

8. Install the large electrical connector (electrical connector must be installed in the proper location) to the crankshaft timing sensor assembly.
9. Reconnect both of the crankshaft timing sensor electrical connectors to the engine harness.
10. Rotate the crankshaft such that the outer vane on the crankshaft pulley hub assembly engages both sides of the Crankshaft Hall Sensor Positioner Tool 89P–6316–A or equivalent and tighten sensor bolts to 22–31 inch lbs.
11. Rotate the crankshaft, then remove the Special Tool. Install new wire harness retainer and secure the crankshaft timing sensor harness to its mounting bracket. Trim off the excess as necessary.
12. Install the timing belt outer cover. Install the crankshaft

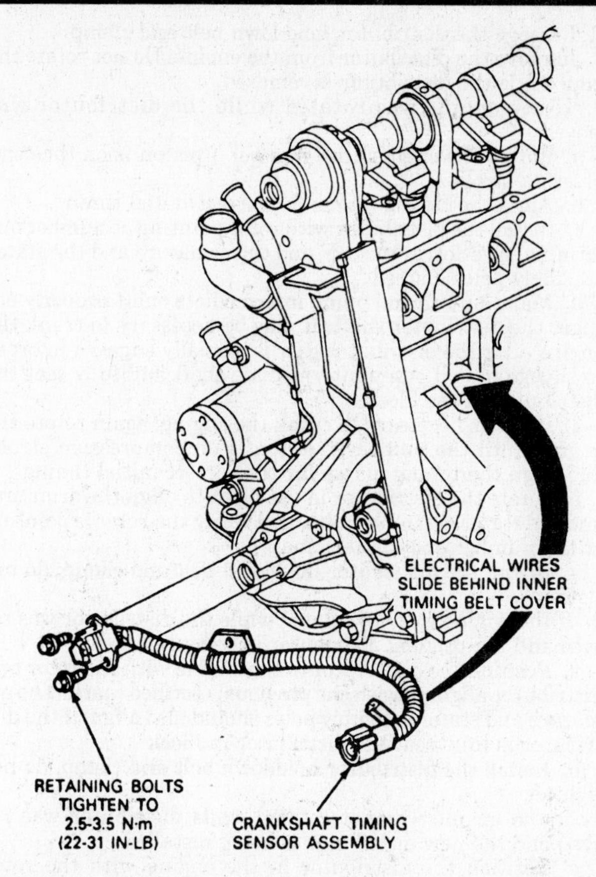

Crankshaft timing sensor assembly – 2.3L Twin Plug Engine

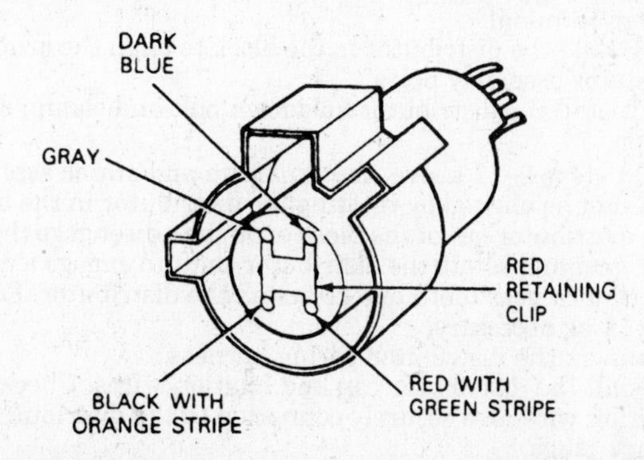

Electrical connector crankshaft timing sensor assembly

pulley assembly and tighten retaining bolts to 20–28 ft. lbs. Install drive belts. Reconnect battery, start engine and check for proper operation.

DIS Module Assembly

REMOVAL AND INSTALLATION

1. Disconnect the negative battery cable.
2. Disconnect each electrical connector of the DIS ignition module assembly by pushing down the connector locking tabs where it is stamped **PUSH** and then pull it away from the module.

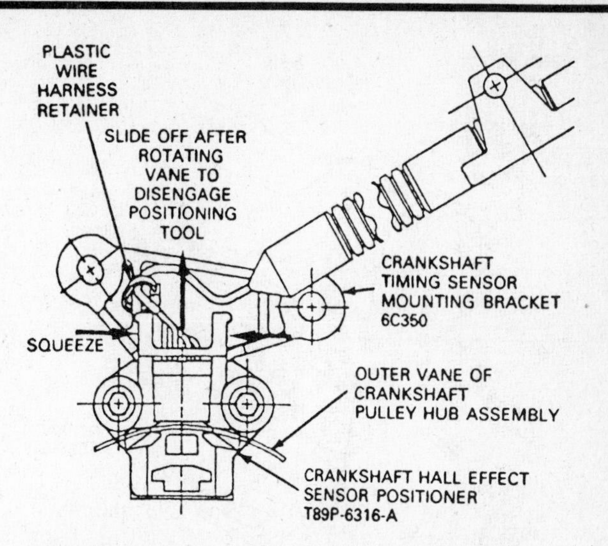

Installation of Crankshaft Hall Effect Sensor Positioner

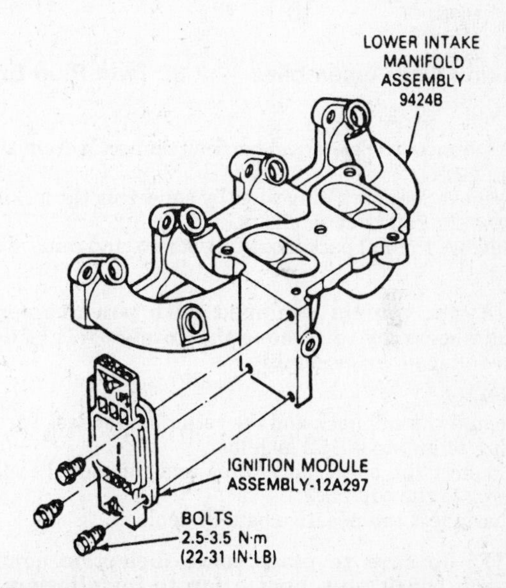

DIS ignition module – 2.3L Twin Plug Engine

3. Remove the 3 retaining screws, remove the ignition module assembly from the lower intake manifold.

To install:

4. Apply an even coat (approximately $\frac{1}{32}$ in.) of a suitable silicone dielectric compound to the mounting surface of the DIS module.
5. Mount the DIS module assembly onto the intake assembly and install the retaining screws. Torque the screws to 22–31 inch lbs.
6. Install the electrical connectors to the DIS ignition module assembly. Reconnect the negative battery cable.

Ignition Coil Pack

REMOVAL AND INSTALLATION

1. Disconnect the negative battery cable.

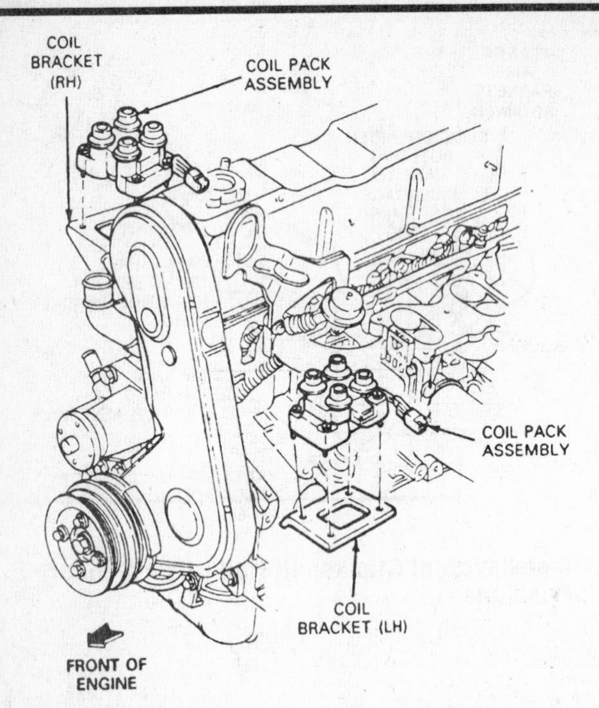

Ignition Coil Assemblies – 2.3L Twin Plug Engine

2. Disconnect the electrical harness connector from the ignition coil pack.
3. Remove the spark plug wires by squeezing the locking tabs to release the coil boot retainers.
4. Remove the coil pack mounting screws and remove the coil pack.

NOTE: On vehicle equipped with power steering it may be necessary to remove the intake (left hand) coil and bracket as an assembly.

To install:
5. Install the coil pack and the retaining screws. Torque the retaining screws to 40–62 inch lbs.
6. Connect the spark plug wires and connect the electrical connector to the coil pack.
7. Reconnect the negative battery cable.

NOTE: Be sure to place some dielectric compound into each spark plug boot prior to installation of the spark plug wire.

Distributor

REMOVAL AND INSTALLATION

2.0L & 1983–84 2.3L Engines
Dura Spark II Ignition System

1. Remove one alternator mounting bolt and drive belt. Swing the alternator to one side.
2. Remove the distributor cap. Position it and ignition wires to one side.
3. Disconnect and plug the vacuum advance hose.
4. Separate the distributor connector from the wiring harness.
5. Rotate the engine to align the stator assembly pole and any armature pole.
6. Scribe a mark on the distributor body and engine block to indicate the position of distributor in the engine, and the position of the rotor.
7. Remove the distributor holddown bolt and clamp.
8. Remove the distributor from the engine. Do not rotate the engine while the distributor is removed.
9. If the engine was rotated while the distributor was removed:
   a. Rotate the engine until number 1 piston is on the compression stroke.
   b. Align the timing marks for correct initial timing.
   c. Install the distributor with rotor pointing at number one terminal position in the cap, and the armature and the stator assembly poles aligned.
   d. Make sure the oil pump intermediate shaft properly engages the distributor shaft. It may be necessary to crank the engine after the distributor gear is partially engaged in order to engage the oil pump intermediate shaft and fully seat the distributor in the block.
   e. If it was necessary to crank the engine, again rotate the engine until the number 1 piston is on compression stroke and align the timing marks for the correct initial timing.
   f. Rotate the distributor in the block to align the armature and the stator assembly poles, and verify the rotor is pointing at the number one cap terminal.
   g. Install the distributor holddown bolt and clamp; do not tighten.
10. If the engine was not rotated while the distributor was removed and the original distributor is being replaced:
   a. Position the distributor in the engine with the rotor and distributor aligning with the previously scribed mark. The armature and stator assembly poles should also align, if the distributor is fully seat the distributor in block.
   b. Install the distributor holddown bolt and clamp; do not tighten.
11. If the engine was not rotated while distributor was removed and the new distributor is being installed:
   a. Position the distributor in the engine with the rotor aligned with the previously scribed mark. If necessary, crank the engine to fully seat the distributor.
   b. Rotate the engine until the timing marks for the correct initial timing align and the rotor is pointing at the number one cap terminal.
   c. Rotate the distributor in the block to align the armature and stator assembly poles.
   d. Install the distributor holddown bolt and clamp; do not tighten.
12. If in steps 9–11 above, the armature and stator assembly poles cannot be aligned by rotating the distributor in the block, pull the distributor out of the block enough to disengage the distributor gear and rotate the distributor shaft to engage a different distributor gear tooth and re-install the distributor. Repeat steps 9–11 as necessary.
13. Connect the distributor wiring harness.
14. Install the distributor cap and ignition wires. Check that the ignition wires are securely connected to the distributor cap and spark plugs.
15. Reinstall the alternator mounting bolt and drive belt. Adjust to specification. Refer to Belt Tension Adjustment, described later in this Section.
16. Set the initial timing per specification on the Vehicle Emission Control Information Decal.
17. Tighten the distributor holddown bolt to 17–25 ft.lbs.
18. Recheck the initial timing. Readjust if necessary.
19. Connect the vacuum advance hose.

2.3L 1985–88, 2.8L, 2.9L and 3.0L Engines
TFI-IV Ignition

1. Remove the air cleaner assembly, taking note of the hose locations.
2. Disconnect the primary wiring connector from the distributor.

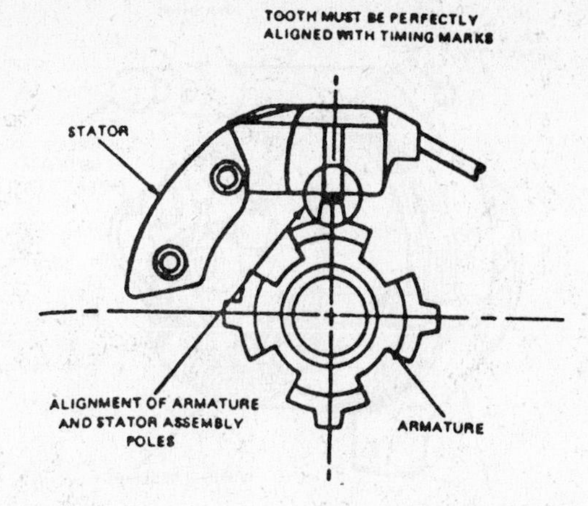

Dura Spark II armature-stator assembly alignment

NOTE: Before removing the distributor cap, mark the position of the No. 1 wire tower on the distributor base for future reference.

3. Using a screwdriver, remove the distributor cap and adapter and position it and the attached wires out of the way.
4. Remove the rotor and place it out of the way to avoid damage.
5. Remove the Thick Film Integrated module connector.
6. Remove the distributor holddown bolt and clamp and remove the distributor.

NOTE: Some engines may be equipped with a security-type holddown bolt. Use Distributor Holddown Wrench, Tool T82L–12270–A, or equivalent, to remove the holddown bolt.

7. Rotate the engine until the No. 1 piston is on the compression stroke.
8. Align the timing marks for the correct initial timing.
9. Rotate the distributor shaft so that the rotor tip is pointing toward the mark previously made on the distributor base.
10. Continue rotating slightly so that the leading edge of the vane is centered in the vane switch stator assembly.
11. Rotate the distributor in the engine block to align the leading edge of the vane and the vane switch and verify that the rotor is pointing at No. 1 cap terminal.

NOTE: If the vane and vane switch stator cannot be aligned by rotating the distributor out of the block, pull the distributor out of the block enough to disengage the distributor and rotate the distributor to engage a different distributor gear tooth. Repeat Steps 8, 9 and 10 as necessary.

12. Install the distributor holddown bolt and clamp. Do not tighten at this time.
13. Connect the distributor Thick Film Integrated (TFI) and the primary wiring harnesses.
14. Install the distributor rotor and tighten the attaching screws.
15. Install the distributor cap adapter and tighten the attaching screws.
16. Install the distributor cap and wires. Check that the ignition wires are securely attached to the cap and spark plugs.

NOTE: Before installing the plug wires, coat the inside of each boot with silicone lubricant.

17. Set the initial timing, with a timing light, to specification. Refer to the underhood Vehicle Emission Control Information Decal.

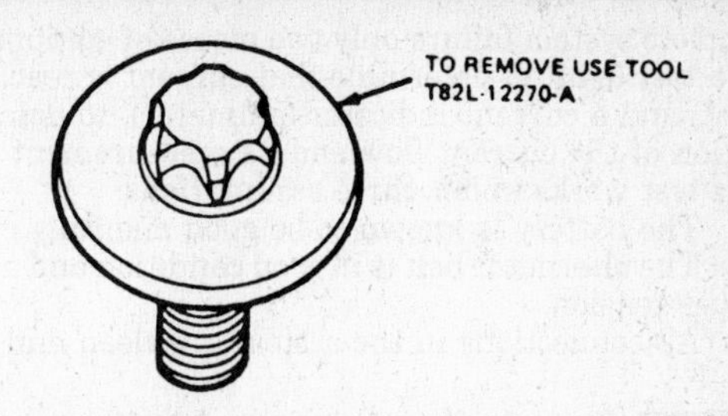

Security type hold-down bolt

18. Tighten the distributor hold-down bolt to 17–25 ft. lbs.
19. Recheck and adjust the timing, if necessary.

Alternator

The alternator charging system is a negative (–) ground system which consists of an alternator, a regulator, a charge indicator, a storage battery and wiring connecting the components.

The alternator is belt-driven from the engine. Energy is supplied from the alternator regulator system to the rotating field through two brushes to two slip rings. The slip rings. The slip rings are mounted on the rotor shaft and are connected to the field coil. This energy supplied to the rotating field from the battery is called excitation current and is used to initially energize the field to begin the generation electricity. Once the alternator starts to generate electricity, the excitation current comes from its own output rather than the battery.

The alternator produces power in the form of alternating current. The alternating current is rectified to direct current by 6 diodes. The direct current is used to charge the battery and power the rest of the electrical system.

ALTERNATOR PRECAUTIONS

To prevent damage to the alternator and regulator, the following precautionary measures must be taken when working with the electrical system.

1. Never reverse battery connections. Always check the battery polarity visually. This should be done before any connections are made to be sure that all of the connections correspond to the battery ground polarity of the truck.
2. Booster batteries for starting must be connected properly. Make sure that the positive cable of the booster battery is connected to the positive terminal of the battery that is getting the boost. The same applies to the negative cables.
3. Disconnect the battery cables before using a fast charger; the charger has a tendency to force current through the diodes in the opposite direction for which they were designed. This burns out the diodes.
4. Never use a fast charger as a booster for starting the vehicle.
5. Never disconnect the voltage regulator while the engine is running.
6. Do not ground the alternator output terminal.
7. Do not operate the alternator on an open circuit with the field energized.
8. Do not attempt to polarize an alternator.

CHARGING SYSTEM TROUBLESHOOTING

There are many possible ways in which the charging system can malfunction. Often the source of a problem is difficult to diagnose, requiring special equipment and a good deal of experience. This is usually not the case, however, where the charging system fails completely and causes the dash board warning light to come on or the battery to become dead. To troubleshoot a complete system failure only two pieces of equipment are needed: a test light, to determine that current is reaching a certain point; and a current indicator (ammeter), to determine the direction of the current flow and its measurement in amps.

This test works under three assumptions:

1. The battery is known to be good and fully charged.
2. The alternator belt is in good condition and adjusted to the proper tension.
3. All connections in the system are clean and tight.

NOTE: In order for the current indicator to give a valid reading, the car must be equipped with battery cables which are of the same gauge size and quality as original equipment battery cables.

1. Turn off all electrical components on the car. Make sure the doors of the car are closed. If the car is equipped with a clock, disconnect the clock by removing the lead wire from the rear of the clock. Disconnect the positive battery cable from the battery and connect the ground wire on a test light to the disconnected positive battery cable. Touch the probe end of the test light to the positive battery post. The test light should not light. If the light does light, there is a short or open circuit on the car.
2. Disconnect the voltage regulator wiring harness connector at the voltage regulator. Turn on the ignition key. Connect the wire on a test light to a good ground (engine bolt). Touch the probe end of a test light to the ignition wire connector into the voltage regulator wiring connector. This wire corresponds to the **I** terminal on the regulator. If the test light goes on, the charging system warning light circuit is complete. If the test light does not come on and the warning light on the instrument panel is on, either the resistor wire, which is parallel with the warning light, or the wiring to the voltage regulator, is defective. If the test light does not come on and the warning light is not on, either the bulb is defective or the power supply wire form the battery through the ignition switch to the bulb has an open circuit. Connect the wiring harness to the regulator.
3. Examine the fuse link wire at the wiring harness from the starter relay to the alternator. If the insulation on the wire is cracked or split, the fuse link may be melted. Connect a test light to the fuse link by attaching the ground wire on the test light to an engine bolt and touching the probe end of the light to the bottom of the fuse link wire where it splices into the alternator output wire. If the bulb in the test light does not light, the fuse link is melted.
4. Start the engine and place a current indicator on the positive battery cable. Turn off all electrical accessories and make sure the doors are closed. If the charging system is working properly, the gauge will show a draw of less than 5 amps. If the system is not working properly, the gauge will show a draw of more than 5 amps. A charge moves the needle toward the battery, a draw moves the needle away from the battery. Turn the engine off.
5. Disconnect the wiring harness from the voltage regulator at the regulator at the regulator connector. Connect a male (solderless connector) to each end of a jumper wire. Insert one end of the wire into the wiring harness connector which corresponds to the **A** terminal on the regulator. Insert the other end of the wire into the wiring harness connector which corresponds to the **F** terminal on the regulator. Position the connector with the jumper wire installed so that it cannot contact any metal surface under the hood. Position a current indicator gauge on the positive battery cable. Have an assistant start the engine. Observe the reading on the current indicator. Have your assistant slowly raise the speed of the engine to about 2,000 rpm or until the current indicator needle stops moving, whichever comes first. Do not run the engine for more than a short period of time in this condition. If the wiring harness connector or jumper wire becomes excessively hot during this test, turn off the engine and check for a grounded wire in the regulator wiring harness. If the current indicator charge of about three ampsee amps less than the output of the alternator, the alternator is working properly. If the previous tests showed a draw, the voltage regulator is defective. If the gauge does not show the proper charging rate, the alternator is defective.

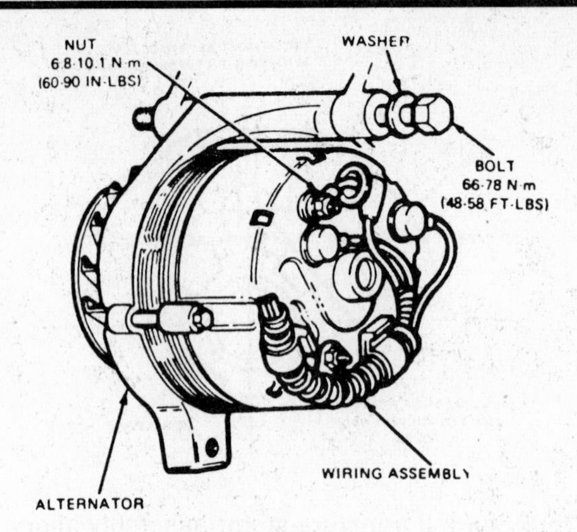

Rear terminal alternator contact locations

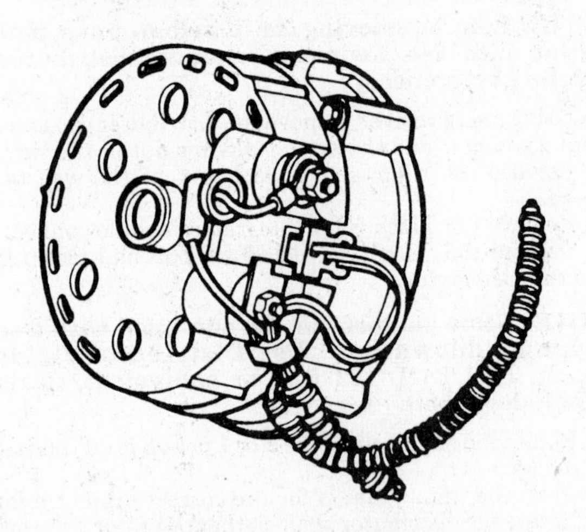

Side terminal alternator

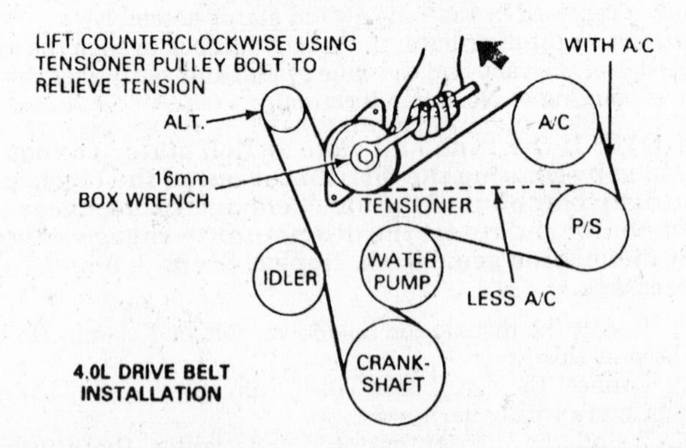

Drive belt installation and adjustment — 4.0L engine

REMOVAL AND INSTALLATION

1. Open the hood and disconnect the battery ground cable.

2. Remove the adjusting arm bolt and loosen the pivot bolt.
3. Remove the drive belt from the alternator pulley.
4. Label all the leads to the alternator so that they can be re-installed correctly and remove the leads from the alternator.
5. Remove the alternator pivot bolt and remove the alternator from the truck.
6. To install, reverse the above service procedure.

BELT TENSION ADJUSTMENT

The fan belt drives the alternator and water pump. If the belt is too loose, it will slip and the alternator will not be able to produce its rated current.

Also, the water pump will not operate efficiently and the engine could overheat. Check the tension of the fan belt by pushing your thumb down on the longest span of the belt, midway between the pulleys. Belt deflection should be approximately ½ in.

1. Loosen the alternator mounting bolt and the adjusting arm bolts.
2. Apply pressure on the alternator front housing only, moving the alternator away from the engine to tighten the belt. Do not apply pressure to the rear of the cast aluminum housing of an alternator; damage to the housing could result.
3. Tighten the alternator mounting bolt and the adjusting arm bolts when the correct tension is reached.

Regulator

The alternator regulator has been designed to control the charging system's rate of charge and to compensate for seasonal temperature changes. This regulator is 100 percent solid state, consisting of transistors, diodes, and resistors. The operating functions are achieved in basically four circuits: The output stage, the voltage control stage, the solid state relay, and the field circuit overload protection stage. There are two different regulators used on your Ford vehicle. The units both look alike, but are not interchangeable due to the different wiring connector plugs. One unit is used on trucks equipped with an ammeter and the other is used on alternator warning light equipped trucks. The regulators are calibrated by the manufacturer and no adjustment is required or possible on these units.

REMOVAL AND INSTALLATION

1. Disconnect the positive terminal of the battery.
2. Disconnect all of the electrical leads at the regulator. Label them as removed, so you can replace them in the correct order on the replacement unit.
3. Remove all of the hold-down screws, then remove the unit from the vehicle.
4. Install the new voltage regulator using the hold-down screws from the old one, or new ones if they are provided with the replacement regulator. Tighten the hold-down screws.
5. Connect all the leads to the new regulator.

Battery

REMOVAL AND INSTALLATION

1. Disconnect the negative cable then positive cable. (ALWAYS DISCONNECT THE NEGATIVE BATTERY CABLE

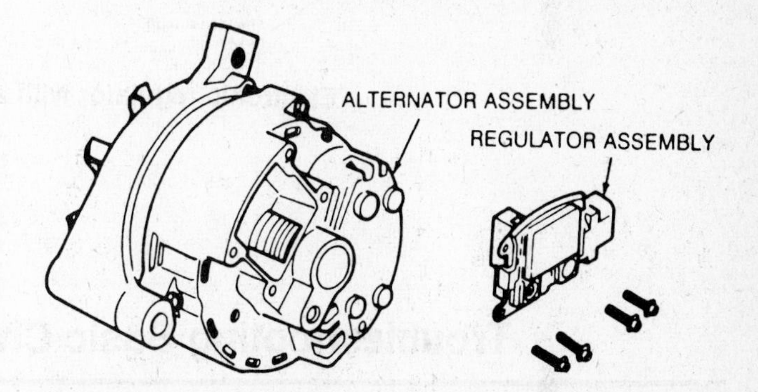

Removing the voltage regulator from the alternator

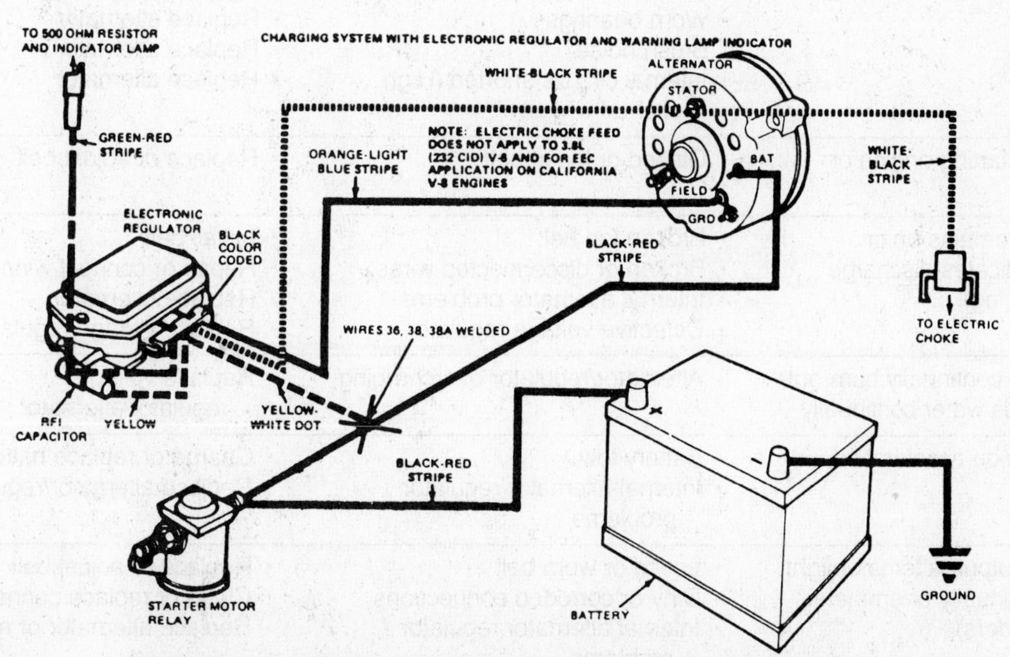

Electronic regulator with warning lamp charging system

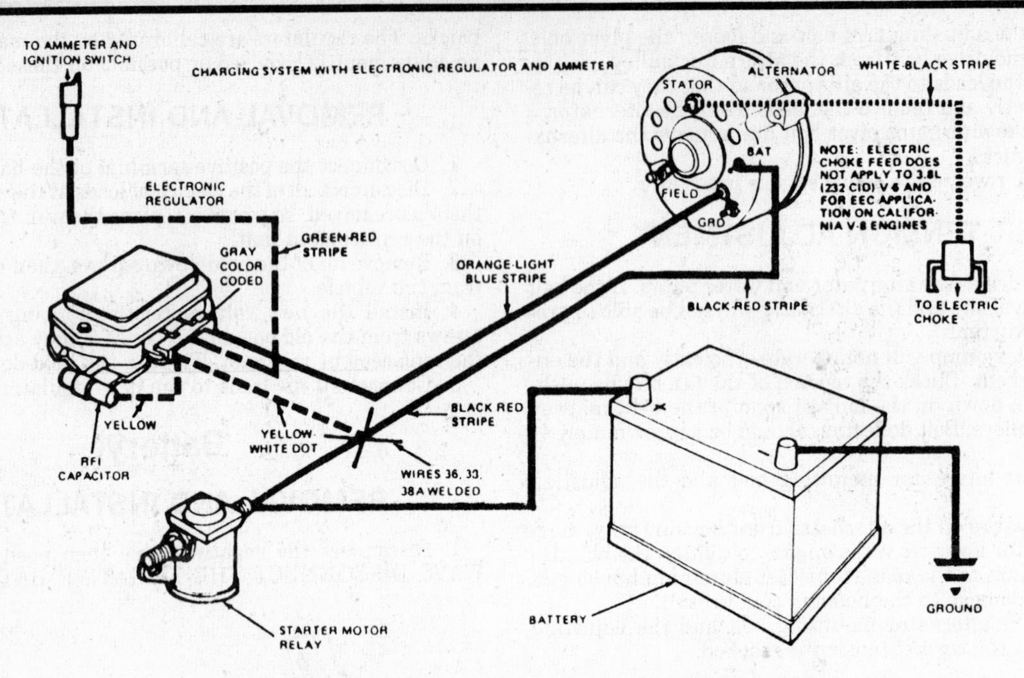

Electronic regulator with ammeter charging system

Troubleshooting Basic Charging System Problems

Problem	Cause	Solution
Noisy alternator	• Loose mountings • Loose drive pulley • Worn bearings • Brush noise • Internal circuits shorted (High pitched whine)	• Tighten mounting bolts • Tighten pulley • Replace alternator • Replace alternator • Replace alternator
Squeal when starting engine or accelerating	• Glazed or loose belt	• Replace or adjust belt
Indicator light remains on or ammeter indicates discharge (engine running)	• Broken fan belt • Broken or disconnected wires • Internal alternator problems • Defective voltage regulator	• Install belt • Repair or connect wiring • Replace alternator • Replace voltage regulator
Car light bulbs continually burn out—battery needs water continually	• Alternator/regulator overcharging	• Replace voltage regulator/alternator
Car lights flare on acceleration	• Battery low • Internal alternator/regulator problems	• Charge or replace battery • Replace alternator/regulator
Low voltage output (alternator light flickers continually or ammeter needle wanders)	• Loose or worn belt • Dirty or corroded connections • Internal alternator/regulator problems	• Replace or adjust belt • Clean or replace connections • Replace alternator or regulator

ALTERNATOR SPECIFICATION CHART

Supplier	Rating: Amperes @ 15V	Rating: Watts @ 15V	Slip-Ring Turning: Min. Diameter MM	Slip-Ring Turning: Min. Diameter Inches	Slip-Ring Turning: Max. Runout MM	Slip-Ring Turning: Max. Runout Inches	Brush Length: New MM	Brush Length: New Inches	Brush Length: Wear Limit MM	Brush Length: Wear Limit Inches	Pulley Nut Lb-Ft
Ford	40A	600W	31	1.22	.013	0.0005	12.19	.480	6.35	.25	60–100
Ford	40A HE	600W	31	1.22	.013	0.0005	12.19	.480	6.35	.25	60–100
Ford	60A	900W	31	1.22	.013	0.0005	12.19	.480	6.35	.25	60–100
Ford	65A	975W	31	1.22	.013	0.0005	12.19	.480	6.35	.25	60–100
Ford	75A	1125W	31	1.22	.013	0.0005	12.19	.480	6.35	.25	60–100
Ford	80A	1200W	31	1.22	.013	0.0005	12.19	.480	6.35	.25	60–100

FIRST) To do this, loosen the cable end-clamp bolts and twist the end-clamps until they are free. You can also buy an inexpensive clamp puller which makes the job easier.

2. Remove the battery hold-down clamps.
3. Using a battery lifting strap, lift the battery from the vehicle.
4. Installation is the reverse of service removal procedure.

NOTE: Keeping the battery top clean and dry reduces the need for service and extends battery life. Baking soda and water is excellent for this procedure also it neutralize corrosion. Always lubricate each battery post to help prevent corrosion with long-life lubricant or equivalent.

Before installing the battery in the vehicle, make sure that the battery terminals are clean and free from corrosion. Use a battery terminal cleaner on the terminals and on the inside of the battery cable ends. If a cleaner is not available, use coarse grade sandpaper to remove the corrosion. A mixture of baking soda and water poured over the terminals and cable ends will help remove and neutralize any acid build up. Before installing the cables onto the terminals, cut a piece of felt cloth or something similar into a circle about the size of the battery terminals at their base. Push the cloth pieces over the terminals so they lie flat on the top of the battery. Soak the pieces of cloth with oil or equivalent. This will keep the formation of oxidized acid to a minimum. Place the battery in the vehicle. Install the cables onto the terminals. Tighten the nuts on the cable ends. Smear a light coating of grease on the cable ends and tops of the terminals. This will further prevent the build up of oxidized acid on the terminals and the cable ends. Install and tighten the nuts of the battery hold bracket.

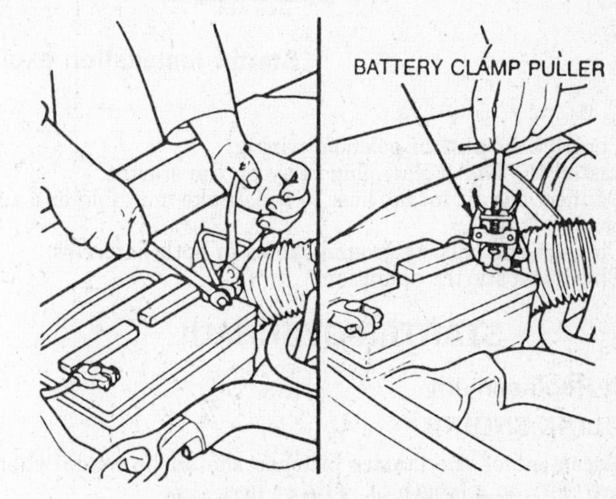

Disconnecting the battery

Starter Motor

REMOVAL AND INSTALLATION

All Gasoline Engines Except 4.0L Engine

1. Disconnect the negative battery cable.
2. Raise the vehicle and support it safely on jackstands.
3. Disconnect the relay-to-starter cable at the starter terminal.
4. Remove the starter mounting bolts and lower the starter from the engine.

To install:

5. Position the new starter assembly to the flywheel housing and start the mounting bolts in by hand.
6. Snug all bolts while holding the starter squarely against its mounting surface and fully inserted into the pilot hole. Tighten the mounting bolts to 15–20 ft. lbs.
7. Reconnect the relay-to-starter cable assembly to the starter motor. Tighten the screw and washer assemblies to 70–130 inch lbs..
8. Lower the vehicle, then connect the negative battery cable.

4.0L Engine

1. Disconnect the negative battery cable.
2. Raise the vehicle and support it safely on jackstands.
3. Matchmark and disconnect the wires at the starter terminals.
4. Remove the starter mounting bolts and lower the starter from the engine.

To install:

5. Position the new starter assembly to the flywheel housing and start the mounting bolts in by hand.
6. Snug all bolts while holding the starter squarely against its mounting surface. Starting with the topmost bolt, tighten the mounting bolts to 15–20 ft. lbs. (21–27 Nm).
7. Reconnect the wires.
8. Lower the vehicle, then connect the negative battery cable.

Diesel Engines

1. Disconnect the battery ground cables from both batteries.
2. Remove the air intake hose between the air cleaner and the intake manifold.
3. Remove the No. 1 glow plug relay from the starter and position it out of the way.
4. Disconnect the starter solenoid wiring.
5. Remove the three starter mounting bolts, and remove the starter.

To install:

6. Position the starter on the engine and install the mounting bolts. Tighten bolts to 48–65 ft. lbs.

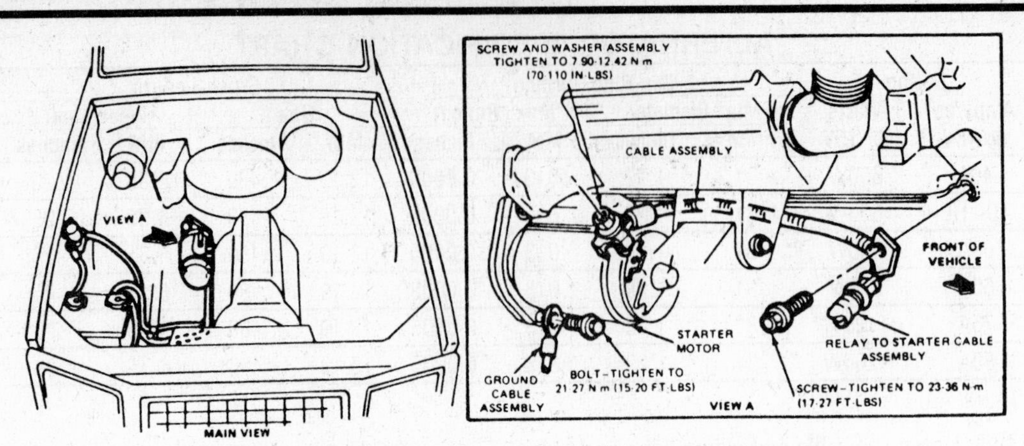

Starter installation exploded view – gasoline engines

7. Connect the starter solenoid wiring.
8. Install the No. 1 glow plug relay on the starter.
9. Connect the air intake hose to the intake manifold and air cleaner.
10. Connect the battery ground cables to both batteries.
11. Check the starter operation.

STARTER OVERHAUL

Brush Replacement

GASOLINE ENGINE

Replacement of the starter brushes should be made when they are worn to a length of ¼ in. or less.

NOTE: This procedure requires the use of a 300 watt soldering iron and rosin core solder. If you are unfamiliar with the use of this tool, do not attempt this procedure.

1. Remove the two through bolts from the starter frame.
2. Remove the brush end plate, brush springs and brushes from the holder.
3. Remove the ground brush attaching screws from the frame and remove the brushes.
4. Cut the insulated brush leads from the field coils, as close to the field connection point as possible.
5. Check the plastic brush holder for cracks or broken mounting pads. Replace it if necessary.
6. Position the new insulated field brushes lead on the field coil connection. Position and crimp the clip provided with the brushes to hold the brush lead to the connection. Solder the lead, clip, and connection together.
7. Install the ground brush leads to the frame with the attaching screws.
8. Install the brush holder and insert the brushes in the holder and install the brush springs. Positive brush leads should be positioned in their respective slots in the brush holder to prevent potential grounding.
9. Install the brush end plate. Be sure end plate insulator is positioned properly on the end plate.
10. Install the two through bolts to the starter frame and tighten.
11. Connect the starter to a battery to check its operation.

DIESEL ENGINE

Replacement or the starter brushes should be made when they are worn to a length of $^7/_{16}$ in. or less.

NOTE: This procedure requires the use of a 300 watt soldering iron and rosin core solder. If you are unfamiliar with the use of this tool, do not attempt this procedure.

1. Remove the two through bolts from the starter frame.
2. Remove the brush end plate, brush lead wires and brushes from the holder.
3. Break the brush to remove it from the lead wire.
4. Clean the brush lead wire and position the new brush on the brush lead wire, through the small taper side.
5. Solder the brush and brush lead wire using rosin core solder and a 300 watt iron.
6. Install the brush holder and insert the brushes in the holder and install the brush springs. Positive brush leads should be positioned in their respective slots in the brush holder to prevent potential grounding.
7. Install the brush end plate. Be sure end plate insulator is positioned properly on the end plate.
8. Install the two through bolts to the starter frame and tighten.
9. Connect the starter to a battery to check its operation.

Starter Drive Replacement

GASOLINE ENGINES

1. Remove the cover of the starter drive's plunger lever arm. Remove the through-bolts, starter drive gear housing and the return spring of the driver gear's actuating lever.
2. Remove the pivot pin which retains the starter gear plunger level and remove the lever.
3. Remove the stop-ring which holds the drive gear to the armature shaft and then remove the drive gear assembly.

To install the drive gear assembly:

4. Lightly Lubriplate® the armature shaft splines and install the starter drive gear assembly on the shaft. Install a new stop-ring and stop-ring retainer.
5. Position the starter drive gear plunger lever to the frame and starter drive assembly.
6. Install the pivot pin.
7. Position the drive plunger lever return spring and the drive gear housing to the frame, then install and tighten the throughbolts. Be sure that the stop-ring retainer is properly seated in the drive housing.
8. Position the starter drive plunger lever cover with its gasket, on the starter. Tighten the cover retaining screw.

DIESEL ENGINES

NOTE: The starter used on diesel engines requires disassembly to repair or replace the starter drive. The procedure is as follows.

1. Remove the starter solenoid.

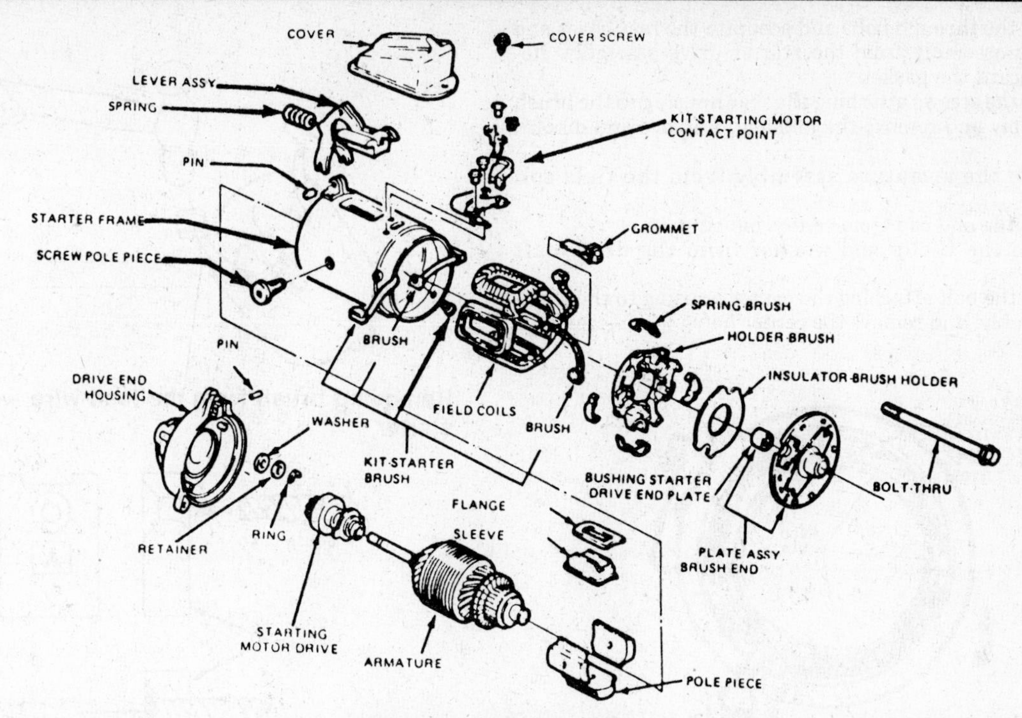

Starter, exploded view

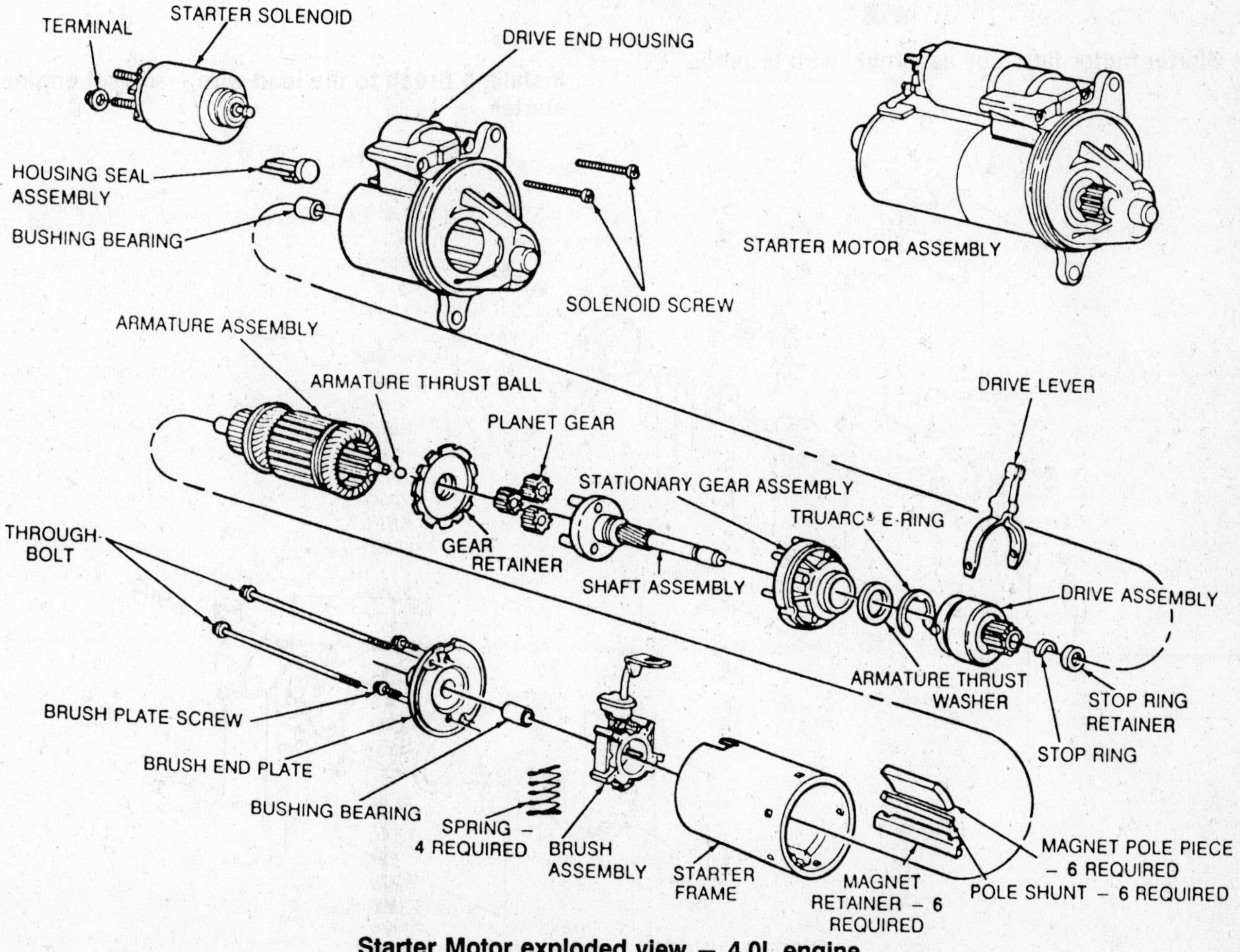

Starter Motor exploded view – 4.0L engine

2. Remove the through bolts and separate the rear cover and the armature assembly from the starter drive assembly. Remove and discard the gasket.
3. Remove the screws attaching the rear housing to the brush holder assembly and remove the housing. Remove and discard the gasket.
4. Remove the armature assembly from the field coil assembly.
5. Remove the end cap from center housing.
6. Remove the C-clip and washer from the driveshaft assembly.
7. Remove the bolt attaching the center housing to the starter drive assembly, and remove the center housing, if necessary.

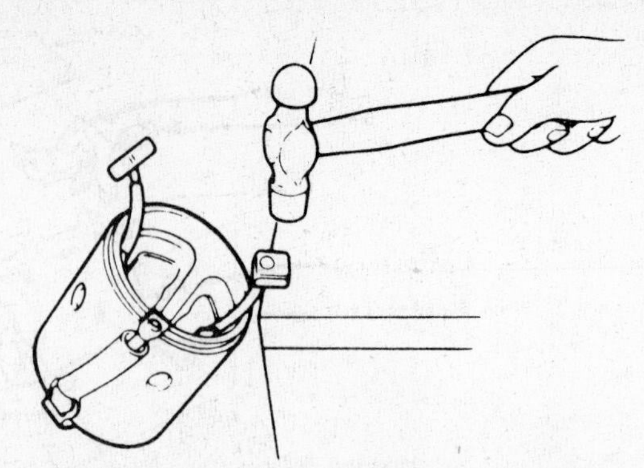

Removing brush from the lead wire — diesel engine starter

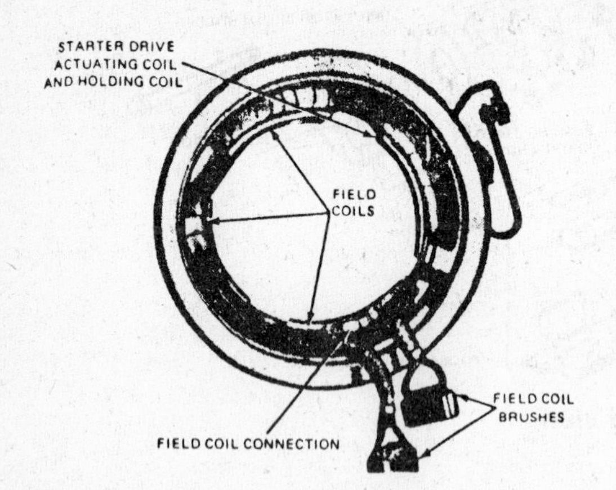

Starter motor field coil assembly with brushes

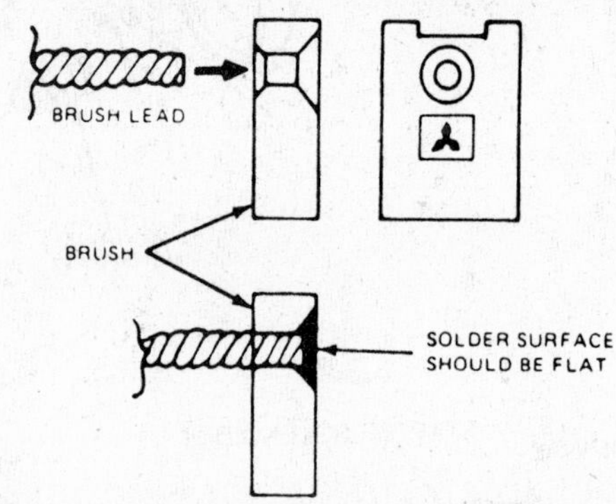

Installing brush to the lead wire — diesel engine starter

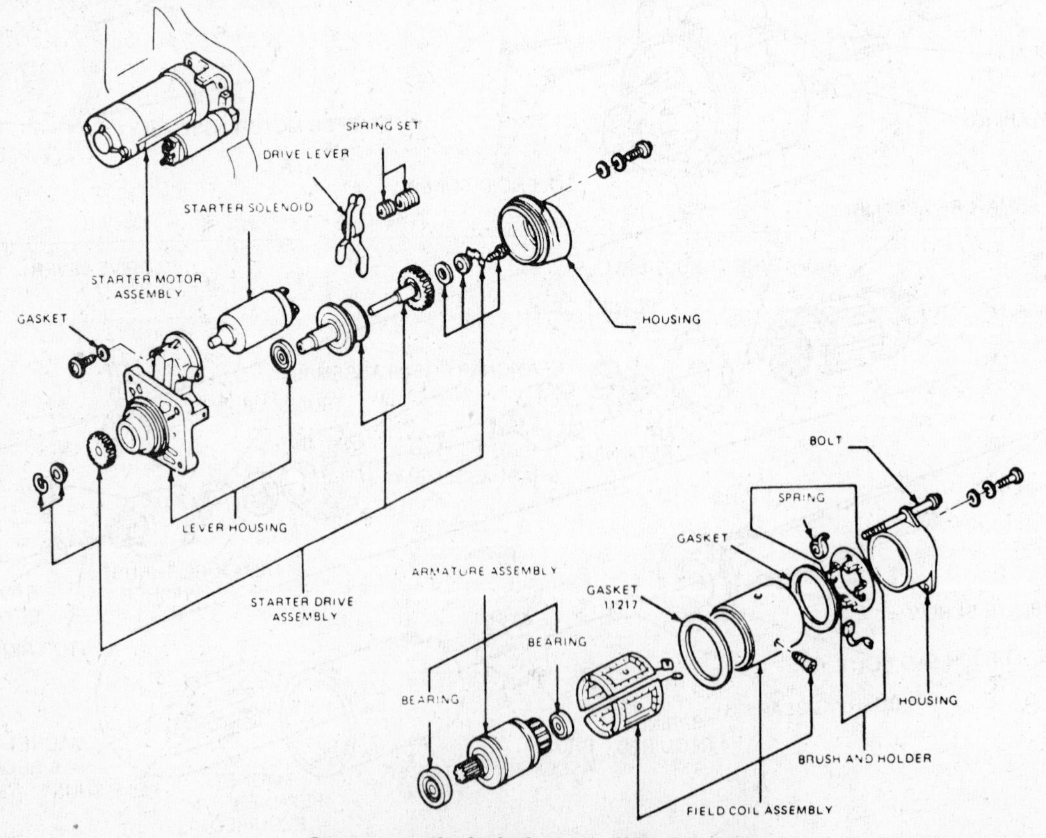

Starter exploded view — diesel engine

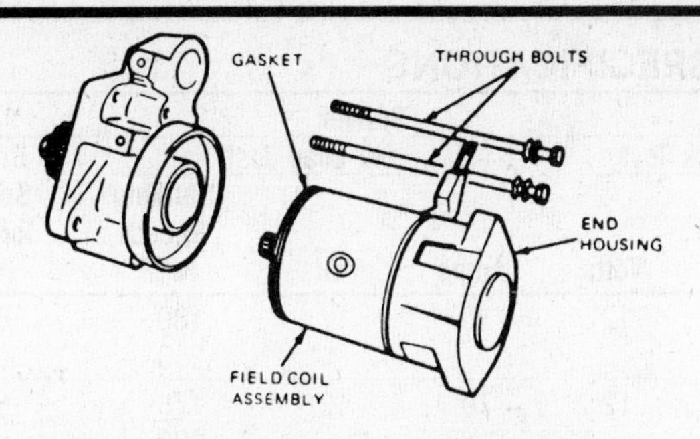

Removing the starter through bolts

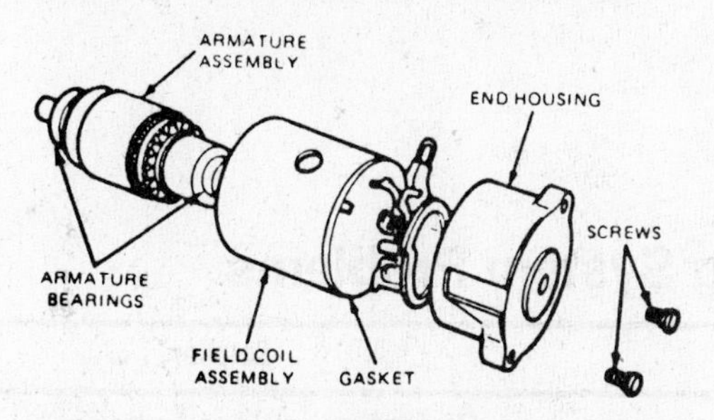

Removing the armature from the starter

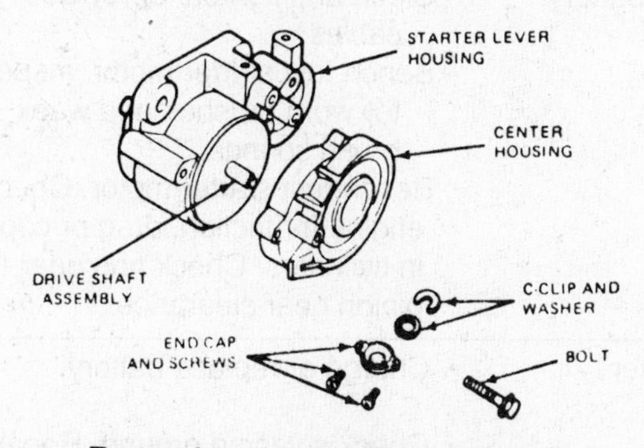

Removing the center housing

8. Using a suitable tool and arbor press, remove bearings from armature assembly, if necessary.
9. Clean, inspect, repair and replace parts as necessary.

To assemble:

10. Using a suitable tool and arbor press, install new armature assembly bearings, if they were removed.

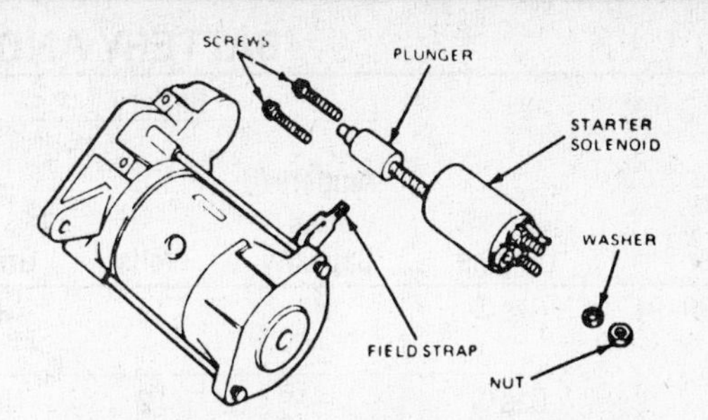

Removing the starter solenoid

11. Position the center housing on the starter drive assembly and install the attaching bolt. Tighten bolt to 5–7 ft. lbs.
12. Install the washer and C-clip on the driveshaft assembly.
13. Install end cap on the center housing.
14. Install armature assembly in the field coil assembly, making sure the brushes are positioned correctly on the commutator.
15. Using a new gasket, position the rear housing and install the bolts attaching the housing to the brush holder. Tighten bolts to 5–7 ft. lbs.
16. Using a new gasket, position the rear housing and armature assembly on the starter drive housing. Install the through bolts and tighten to 5–7 ft. lbs.
17. Install the starter solenoid.

STARTER RELAY REPLACEMENT

Gasoline Engine

The starter relay is mounted on the inside of the right wheel well. To replace it, disconnect the positive battery cable from the battery, disconnect all of the electrical leads from the relay and remove the relay from the fender well. Replace in the reverse order of removal procedure.

STARTER SOLENOID REPLACEMENT

Diesel Engine

1. Remove the starter as described in this Section.
2. Remove the nut and washer from the M terminal of the starter solenoid and position the field strap out of the way.
3. Position the solenoid on the starter and install the attaching screws. Tighten to 5–7 ft. lbs.
4. Position the strap to the M terminal on the solenoid and install the nut and washer. Tighten to 80–120 inch lbs.
5. Install the starter as described in this Section.

BATTERY AND STARTER SPECIFICATIONS

		Battery			Starter					
					Lock Test		No Load Test			Brush
Year	Engine	Ampere/Hour Capacity	Volts	Ground	Amps	Volts	Amps	Volts	Cranking Speed RPM	Spring Tension (oz)
1983-88	Gas	45	12	Neg.	200	12	70	12	180–250	40
	Gas	63	12	Neg.	180	12	80	12	150–290	80
	Diesel	54	12	Neg.	500	12	180	12	150–220	

BATTERY AND STARTER SPECIFICATIONS

		Battery			Starter					
					Lock Test		No Load Test			
Year	Engine	Ampere/ Hour Capacity	Volts	Ground	Amps	Volts	Amps	Volts	Cranking Speed RPM	Brush Spring Tension (oz)
1989-91	Gas ①	—	12	Neg.	150-200	12	80	12	180-250	80
	Gas ②	—	12	Neg.	140-200	12	70	12	170-220	64

NOTE: Maximum starting circuit voltage drop between battery positive terminal to starter assembly terminal at normal engine temperature is 0.5 volt.

① 4 inch diameter starter motor

② 3 inch diameter starter motor

Troubleshooting Basic Starting System Problems

Problem	Cause	Solution
Starter motor rotates engine slowly	• Battery charge low or battery defective	• Charge or replace battery
	• Defective circuit between battery and starter motor	• Clean and tighten, or replace cables
	• Low load current	• Bench-test starter motor. Inspect for worn brushes and weak brush springs.
	• High load current	• Bench-test starter motor. Check engine for friction, drag or coolant in cylinders. Check ring gear-to-pinion gear clearance.
Starter motor will not rotate engine	• Battery charge low or battery defective	• Charge or replace battery
	• Faulty solenoid	• Check solenoid ground. Repair or replace as necessary.
	• Damage drive pinion gear or ring gear	• Replace damaged gear(s)
	• Starter motor engagement weak	• Bench-test starter motor
	• Starter motor rotates slowly with high load current	• Inspect drive yoke pull-down and point gap, check for worn end bushings, check ring gear clearance
	• Engine seized	• Repair engine
Starter motor drive will not engage (solenoid known to be good)	• Defective contact point assembly	• Repair or replace contact point assembly
	• Inadequate contact point assembly ground	• Repair connection at ground screw
	• Defective hold-in coil	• Replace field winding assembly
Starter motor drive will not disengage	• Starter motor loose on flywheel housing	• Tighten mounting bolts
	• Worn drive end busing	• Replace bushing
	• Damaged ring gear teeth	• Replace ring gear or driveplate
	• Drive yoke return spring broken or missing	• Replace spring

Troubleshooting Basic Starting System Problems

Problem	Cause	Solution
Starter motor drive disengages prematurely	• Weak drive assembly thrust spring • Hold-in coil defective	• Replace drive mechanism • Replace field winding assembly
Low load current	• Worn brushes • Weak brush springs	• Replace brushes • Replace springs

ENGINE MECHANICAL

Design

4-CYLINDER GASOLINE ENGINES

The 2.0L,2.3L overhead cam engines are of lightweight iron construction. The crankshaft is supported on five main bearings and the camshaft by four. Main, connecting rod, camshaft and auxiliary shaft bearings are all replaceable.

The camshaft is driven from the crankshaft by a cogged belt, which also operates the auxiliary shaft, and through this shaft, the oil pump, fuel pump and distributor. Tension on the cam drive belt is maintained by a locked idler pulley bearing on the outside of the belt.

Water pump and fan are separately driven from the crankshaft by a 6-ribbed belt which also drives the alternator.

Hydraulic valve lash adjusters are used in the valve train. These units are placed at the fulcrum point of the cam followers (or rocker arms). Their action is similar to the hydraulic tappets used in push-rod engines and they are constructed and serviced in the same manner. The cylinder head has drilled oil passages to provide engine oil pressure to the lash adjusters.

A set of metric wrenches is required to service the 4–2.0L and 2.3L engines.

V6 GASOLINE ENGINE

The V6 engine is of the standard, two-bank, V-design with the banks of cylinders opposed to each other at a 60° angle.

The crankshaft is supported by 4 main bearings, with crankshaft end thrust controlled by the flanged No. 3 bearing.

The camshaft, which is located in the center of the V design of the engine, is mounted on 4 bearings and is gear driven by the crankshaft. An eccentric on the front of the camshaft operates the fuel pump. A gear on the rear of the camshaft drives the distributor, which drives the oil pump through an intermediate shaft. The oil pump is located in the rear of the oil pan.

The engine is equipped with solid valve lifters.

The engine is equipped with a closed positive crankcase ventilation system which directs crankcase fumes to the intake manifold.

The engine is equipped with the Thermactor exhaust emission control system, otherwise known as the air injection system.

On the 3.0L V6 engine The camshaft, which is located in the center of the V design of the engine, is mounted on 4 bearings and is chain driven by the crankshaft. The cast iron cylinder heads feature central plugs, in dual quench type combustion chambers. The hydraulic lifters are activated by fulcrum-mounted rocker arms.

On the 4.0L V6 engine The camshaft, which is located in the center of the V design of the engine, is mounted on 4 bearings and is chain driven by the crankshaft. The cast iron cylinder heads feature central plugs, in dual quench type combustion chambers. The hydraulic, roller type lifters are activated by shaft mounted, non-adjustable rocker arms. The ignition system is EDIS distributorless system.

4-CYLINDER DIESEL ENGINES

2.2L Diesel Engine

The 2.2L Diesel engine is a 4-cylinder, 4-cycle, water cooled, overhead valve engine. The cylinders are numbered 1–2–3–4 from the front of the engine. The injection order is 1–3–4–2.

The valve mechanism is of the overhead valve type. The rocker arm shaft is prevented from rotating by a taper pin installed through the first rocker arm support into the shaft. The valve tappers are of the solid type, which require periodic valve adjustment. Valve caps are mounted on the end of valve stems. The caps provide a large area for contacting the rocker arm, increased durability and reduced thrust pressure. Both intake and exhaust valves are offset by 1mm from the center of the rocker arm. This allows the valves to rotate to prevent carbon buildup on the valve face and seat, and to prevent uneven wear of the valves.

The cylinder head is designed with the intake ports on the left side and exhaust ports on the right side. This crossflow design provides high intake/exhaust efficiency.

The cylinder block is of thin, ductile, cast iron, with dry-type , pressed in, cylinder liners made of heat-resistant cast iron.

The crankshaft is supported by five main bearings. The bear-

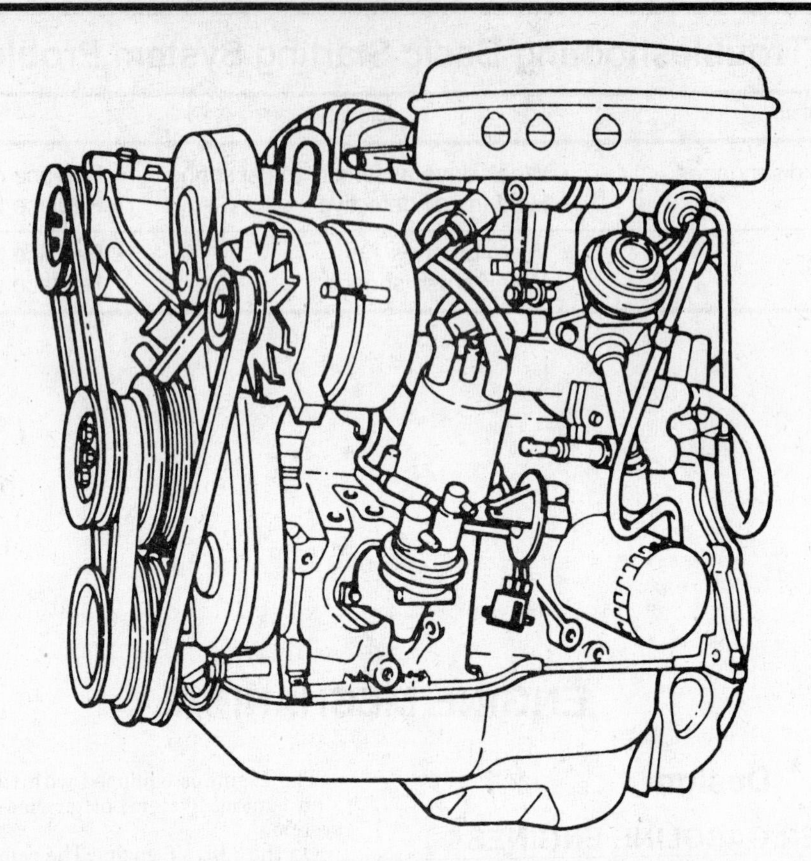

2.8L engine assembly — 2.3L carbureted engine similar

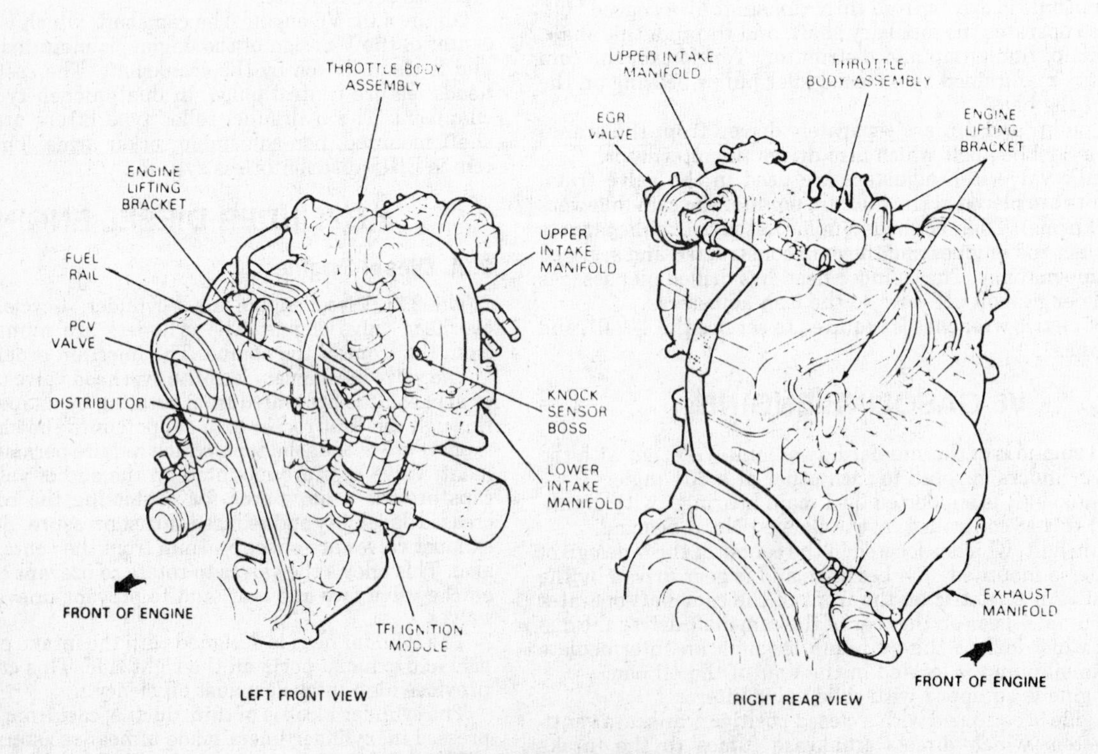

2.3L EFI engine assembly

ing caps are not interchangeable with each other. The No. 1 and No. 2 main caps are numbered, and the No. 4 cap is identical to the No. 1 and No. 2 caps, but is not marked. The No. 3 and No. 5 caps are unique and not marked. The crank pins and main journals are hardened for high wear resistance.

The camshaft is supported by bores in the cylinder block, and is held in place by a thrust plate. There are no bearing inserts in the camshaft bore.

The aluminum alloy pistons are fitted with two compression rings and one oil ring. The top compression ring and the oil ring are plated. A nickel alloy insert is cast in the top compression ring groove to reduce wear due to heat. To reduce piston slap, a steel strut is cast into the piston skirt to help control expansion rates. There are no oversize pistons available.

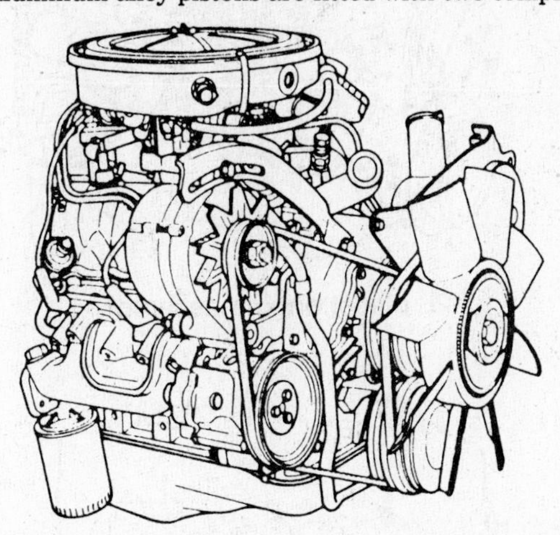

2.8L engine with Thermactor® emission system

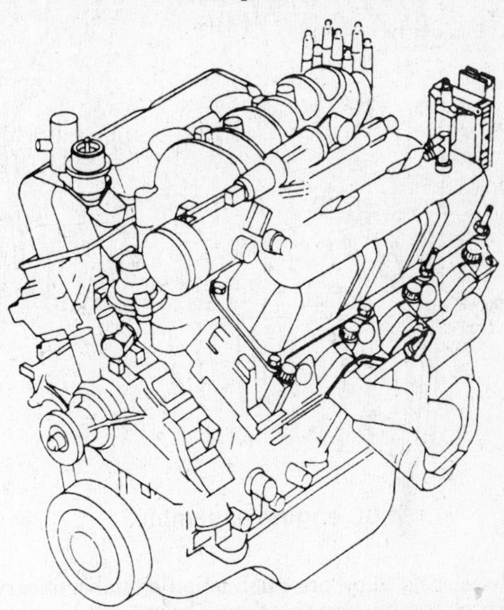

3.0L engine assembly

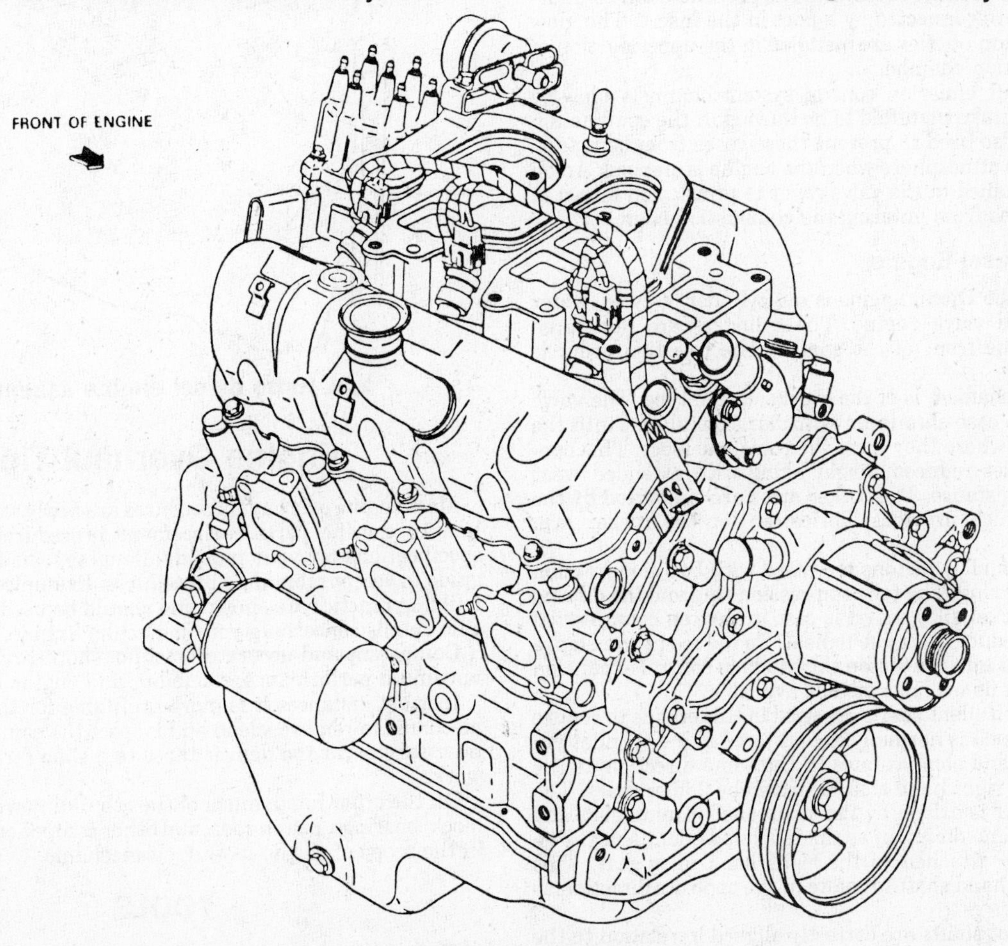

2.9L EFI engine assembly

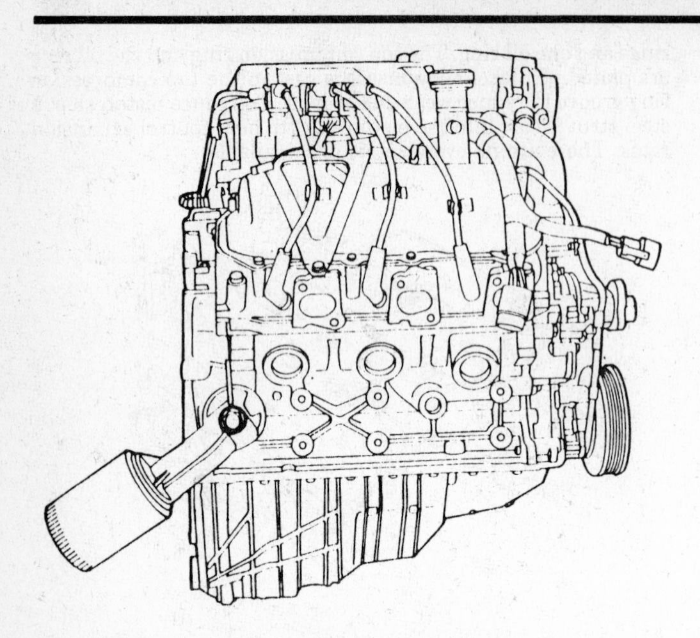

4.0L engine assembly

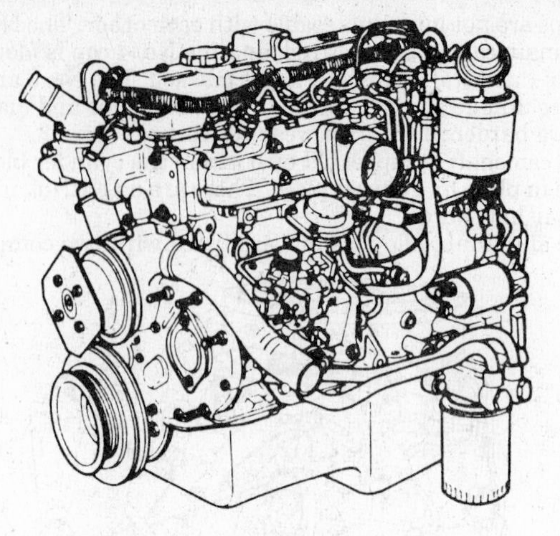

2.2L diesel engine assembly

A heat-resisting alloy pre-combustion chamber insert is installed in each combustion chamber. The pre-combustion chamber is designed to produce fuel swirl. The pre- and main combustion chambers are connected by a port in the insert. The glow plugs and injection nozzles are installed in the upper portion of the pre-combustion chamber.

The crankshaft emission control system channels blow-by gases into the intake manifold to be burned in the combustion chamber. It is also used to prevent these gases from being discharged into the atmosphere when the engine is stopped. An oil separator is installed in the valve cover to prevent oil mixed in the blow-by gases from entering the combustion chamber.

2.3L Turbo Diesel Engine

The 2.3L Turbo Diesel engine is a 4-cylinder, 4-cycle, water cooled, overhead valve engine. The cylinders are numbered 1–2–3–4 from the front of the engine. The injection order is 1–3–4–2.

The valve mechanism is of the overhead cam type. The valve rocker arms are cast aluminum with hard metal cast into the sliding surfaces where they meet the camshaft lobes. This construction provides reduced weight along with extended wear and abrasion resistance. The rocker arms are supported by the rocker shaft which provides lubrication for the rocker arm bores.

The aluminum alloy pistons are fitted with two compression rings and one oil ring. The top compression ring and the oil ring are plated. A nickel alloy insert is cast in the top compression ring groove to reduce wear due to heat. To reduce piston slap, a steel strut is cast into the piston skirt to help control expansion rates. There are no oversize pistons available.

Two belt driven silent shafts are used in this engine to reduce the vibration which is normally encountered in a 4 cylinder engine. The left hand shaft rotates in the same direction as the crankshaft. The right hand shaft is driven by 2 idler gears. The bottom idler gear is driven by the silent shaft timing belt, and rotates in the same direction as the left hand silent shaft. The top idler gear is attached to the right hand idler gear. This causes the right hand shaft to rotate in the opposite direction as the crankshaft.

When the silent shafts are correctly aligned in relation to the crankshaft the counter weights on the silent shaft counter the vibrations of the crankshaft, reducing the engine vibration

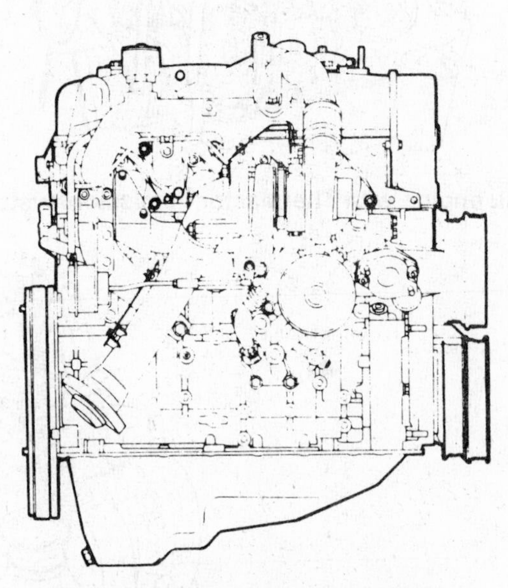

2.3L turbo diesel engine assembly

Engine Overhaul Tips

Most engine overhaul procedures are fairly standard. In addition to specific parts replacement procedures and complete specifications for your individual engine, this section also is a guide to accept rebuilding procedures. Examples of standard rebuilding practice are shown and should be used along with specific details concerning your particular engine.

Competent and accurate machine shop services will ensure maximum performance, reliability and engine life.

In most instances it is more profitable for the do-it-yourself mechanic to remove, clean and inspect the component, buy the necessary parts and deliver these to a shop for actual machine work.

On the other hand, much of the rebuilding work (crankshaft, block, bearings, piston rods, and other components) is well within the scope of the do-it-yourself mechanic.

TOOLS

The tools required for an engine overhaul or parts replacement will depend on the depth of your involvement. With a few exceptions, they will be the tools found in a mechanic's tool kit. More in-depth work will require any or all of the following:

- a dial indicator (reading in thousandths) mounted on a universal base
- micrometers and telescope gauges
- jaw and screw-type pullers
- scraper
- valve spring compressor
- ring groove cleaner
- piston ring expander and compressor
- ridge reamer
- cylinder hone or glaze breaker
- Plastigage®
- engine stand

The use of most of these tools is illustrated in this Section. Many can be rented for a one-time use from a local parts jobber or tool supply house specializing in automotive work.

Occasionally, the use of special tools is called for. See the information on Special Tools and Safety Notice in the front of this book before substituting another tool.

INSPECTION TECHNIQUES

Procedures and specifications are given in this Section for inspecting, cleaning and assessing the wear limits of most major components. Other procedures such as Magnaflux® and Zyglo® can be used to locate material flaws and stress cracks. Magnaflux® is a magnetic process applicable only to ferrous materials. The Zyglo® process coats the material with a fluorescent dye penetrant and can be used on any material Check for suspected surface cracks can be more readily made using spot check dye. The dye is sprayed onto the suspected area, wiped off and the area sprayed with a developer. Cracks will show up brightly.

OVERHAUL TIPS

Aluminum has become extremely popular for use in engines, due to its low weight. Observe the following precautions when handling aluminum parts:

- Never hot tank aluminum parts (the caustic hot tank solution will eat the aluminum.
- Remove all aluminum parts (identification tag, etc.) from engine parts prior to the tanking.
- Always coat threads lightly with engine oil or anti-seize compounds before installation, to prevent seizure.
- Never overtorque bolts or spark plugs especially in aluminum threads.

Stripped threads in any component can be repaired using any of several commercial repair kits (Heli-Coil®, Microdot®, Keenserts®, etc.).

When assembling the engine, any parts that will be frictional contact must be prelubed to provide lubrication at initial start-up. Any product specifically formulated for this purpose can be used, but engine oil is not recommended as a prelube.

When semi-permanent (locked, but removable) installation of bolts or nuts is desired, threads should be cleaned and coated with Loctite® or other similar, commercial non-hardening sealant.

REPAIRING DAMAGED THREADS

Several methods of repairing damaged threads are available. Heli-Coil® (shown here), Keenserts® and Microdot® are among the most widely used. All involve basically the same principle—drilling out stripped threads, tapping the hole and installing a prewound insert—making welding, plugging and oversize fasteners unnecessary.

Two types of thread repair inserts are usually supplied: a standard type for most Inch Coarse, Inch Fine, Metric Course and Metric Fine thread sizes and a spark lug type to fit most spark plug port sizes. Consult the individual manufacturer's catalog to determine exact applications. Typical thread repair kits will contain a selection of prewound threaded inserts, a tap (corresponding to the outside diameter threads of the insert) and an installation tool. Spark plug inserts usually differ be-

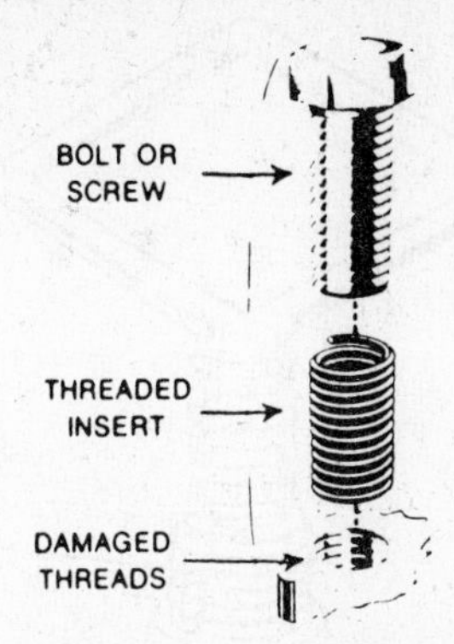

Damaged bolt holes can be repaired with thread repair inserts

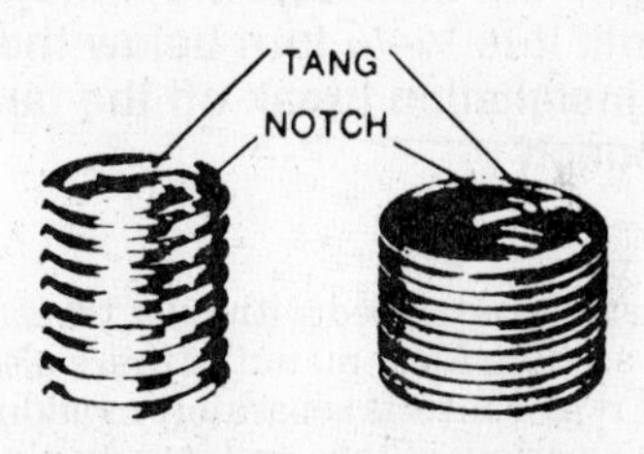

Standard thread repair insert (left) and spark plug thread insert (right)

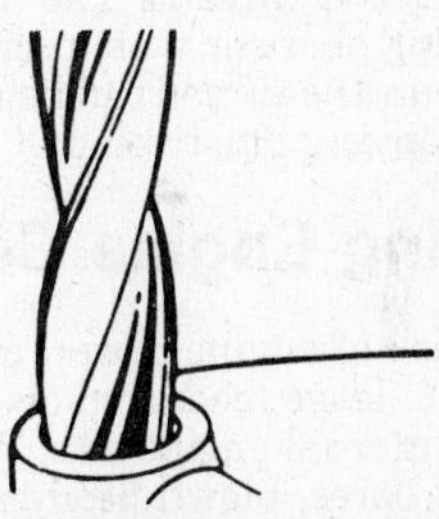

Drill out the damaged threads with the specified drill – drill completely through the hole or to the bottom of a blind hole

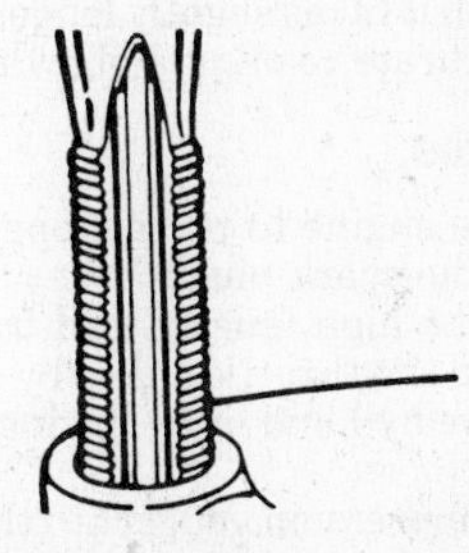

With the tap supplied, tap the hole to receive the thread insert – keep the tap well oiled and back it out frequently to avoid clogging the threads

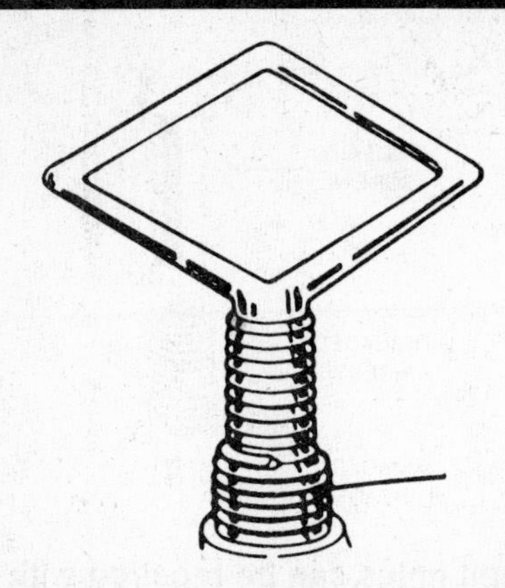

Screw the thread insert onto the installation tool until the tang engages the slot. Screw the insert into the tapped hole until it is ¼–½ turn below the top surface. After installation break off the tang with a hammer and punch

cause they require a tap equipped with pilot threads and a combined reamer/tap section. Most manufacturers also supply blister-packed thread repair inserts separately in addition to a master kit containing a variety of taps and inserts plus installation tools.

Before effecting a repair to a threaded hole, remove any snapped, broken or damaged bolts or studs. Penetrating oil can be used to free frozen threads. The offending item can be removed with locking pliers or with a screw or stud extractor. After the hole is clear, the thread can be repaired, as shown in the series of accompanying illustrations.

Checking Engine Compression

A noticeable lack of engine power, excessive oil consumption and/or poor fuel mileage measured over an extended period are all indicators of internal engine war. Worn piston rings, scored or worn cylinder bores, blown head gaskets, sticking or burnt valves and worn valve seats are all possible culprits here. A check of each cylinder's compression will help you locate the problems.

As mentioned earlier, a screw-in type compression gauge is more accurate that the type you simply hold against the spark plug hole, although it takes slightly longer to use. It's worth it to obtain a more accurate reading. Follow the procedures below.

Gasoline Engines

1. Warm up the engine to normal operating temperature.
2. Remove all the spark plugs.
3. Disconnect the high tension lead from the ignition coil.
4. On fully open the throttle either by operating the carburetor throttle linkage by hand or by having an assistant floor the accelerator pedal.
5. Screw the compression gauge into the no.1 spark plug hole until the fitting is snug.

WARNING: Be careful not to crossthread the plug hole. On aluminum cylinder heads use extra care, as the threads in these heads are easily ruined.

6. Ask an assistant to depress the accelerator pedal fully on both carbureted and fuel injected vehicles. Then, while you read the compression gauge, ask the assistant to crank the engine two or three times in short bursts using the ignition switch.
7. Read the compression gauge at the end of each series of cranks, and record the highest of these readings. Repeat this procedure for each of the engine's cylinders. Compare the highest reading of each cylinder to the compression pressure specification in the Tune-Up Specifications chart. The specs in this chart are maximum values.

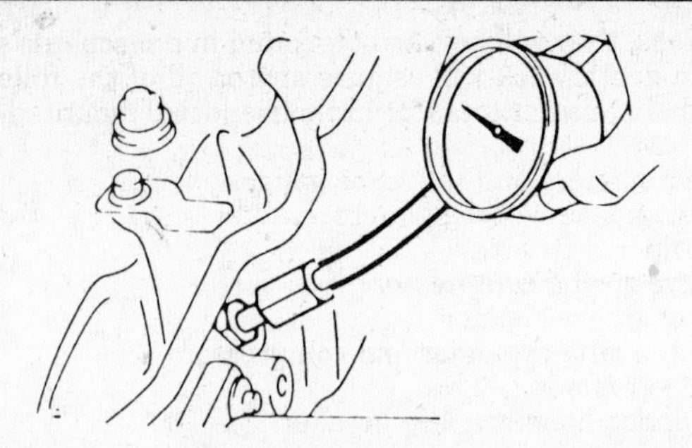

The screw-in type compression gauge is more accurate

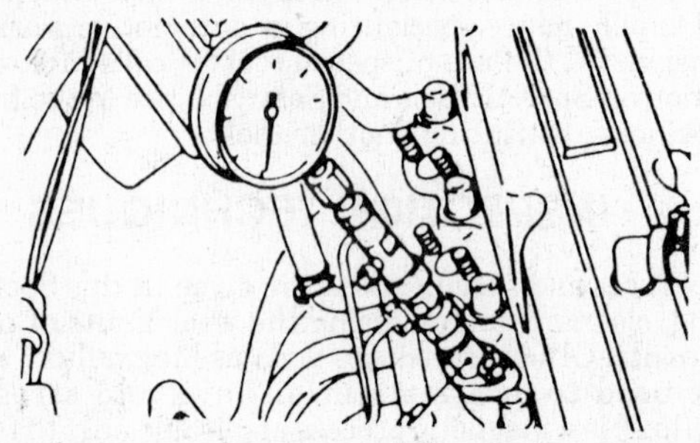

Diesel engines require a special compression gauge adapter

A cylinder's compression pressure is usually acceptable if it is not less than 80% of maximum. The difference between any two cylinders should be no more than 12–14 pounds.

8. If a cylinder is unusually low, pour a tablespoon of clean engine oil into the cylinder through the spark plug hole and repeat the compression test. If engine compression comes up after adding the oil, it appears that the cylinder's piston rings or bore are damaged or worn. If the pressure remains low, the valves may not be seating properly (a valve job is needed), or the head gasket may be blown near that cylinder. If compression in any two adjacent cylinders is low, and if the addition of oil doesn't help the compression, there is leakage past the head gasket. Oil and coolant water in the combustion chamber can result from this problem. There may be evidence of water droplets on the engine dipstick when a head gasket has blown.

Diesel Engines

Checking cylinder compression on diesel engines is basically the same procedure as on gasoline engines except for the following:

1. A special compression gauge adaptor suitable for diesel engines (because these engines have much greater compression pressures) must be used.
2. Remove the injector tubes and remove the injectors from each cylinder.

WARNING: Do not forget to remove the washer underneath each injector. Otherwise, it may get lost when the engine is cranked.

3. When fitting the compression gauge adaptor to the cylinder head, make sure the bleeder of the gauge (if equipped) is closed.
4. When reinstalling the injector assemblies, install new washers underneath each injector.

Troubleshooting Engine Mechanical Problems

Problem	Cause	Solution
External oil leaks	• Fuel pump gasket broken or improperly seated	• Replace gasket
	• Cylinder head cover RTV sealant broken or improperly seated	• Replace sealant; inspect cylinder head cover sealant flange and cylinder head sealant surface for distortion and cracks
	• Oil filler cap leaking or missing	• Replace cap
External oil leaks	• Oil filter gasket broken or improperly seated	• Replace oil filter
	• Oil pan side gasket broken, improperly seated or opening in RTV sealant	• Replace gasket or repair opening in sealant; inspect oil pan gasket flange for distortion
	• Oil pan front oil seal broken or improperly seated	• Replace seal; inspect timing case cover and oil pan seal flange for distortion
	• Oil pan rear oil seal broken or improperly seated	• Replace seal; inspect oil pan rear oil seal flange; inspect rear main bearing cap for cracks, plugged oil return channels, or distortion in seal groove
	• Timing case cover oil seal broken or improperly seated	• Replace seal
	• Excess oil pressure because of restricted PCV valve	• Replace PCV valve
	• Oil pan drain plug loose or has stripped threads	• Repair as necessary and tighten
	• Rear oil gallery plug loose	• Use appropriate sealant on gallery plug and tighten
	• Rear camshaft plug loose or improperly seated	• Seat camshaft plug or replace and seal, as necessary
	• Distributor base gasket damaged	• Replace gasket
Excessive oil consumption	• Oil level too high	• Drain oil to specified level
	• Oil with wrong viscosity being used	• Replace with specified oil
	• PCV valve stuck closed	• Replace PCV valve
	• Valve stem oil deflectors (or seals) are damaged, missing, or incorrect type	• Replace valve stem oil deflectors
	• Valve stems or valve guides worn	• Measure stem-to-guide clearance and repair as necessary
	• Poorly fitted or missing valve cover baffles	• Replace valve cover
	• Piston rings broken or missing	• Replace broken or missing rings
	• Scuffed piston	• Replace piston
	• Incorrect piston ring gap	• Measure ring gap, repair as necessary
	• Piston rings sticking or excessively loose in grooves	• Measure ring side clearance, repair as necessary
	• Compression rings installed upside down	• Repair as necessary
	• Cylinder walls worn, scored, or glazed	• Repair as necessary

Troubleshooting Engine Mechanical Problems (cont.)

Problem	Cause	Solution
	• Piston ring gaps not properly staggered	• Repair as necessary
	• Excessive main or connecting rod bearing clearance	• Measure bearing clearance, repair as necessary
No oil pressure	• Low oil level	• Add oil to correct level
	• Oil pressure gauge, warning lamp or sending unit inaccurate	• Replace oil pressure gauge or warning lamp
	• Oil pump malfunction	• Replace oil pump
	• Oil pressure relief valve sticking	• Remove and inspect oil pressure relief valve assembly
	• Oil passages on pressure side of pump obstructed	• Inspect oil passages for obstruction
	• Oil pickup screen or tube obstructed	• Inspect oil pickup for obstruction
	• Loose oil inlet tube	• Tighten or seal inlet tube
Low oil pressure	• Low oil level	• Add oil to correct level
	• Inaccurate gauge, warning lamp or sending unit	• Replace oil pressure gauge or warning lamp
	• Oil excessively thin because of dilution, poor quality, or improper grade	• Drain and refill crankcase with recommended oil
	• Excessive oil temperature	• Correct cause of overheating engine
	• Oil pressure relief spring weak or sticking	• Remove and inspect oil pressure relief valve assembly
	• Oil inlet tube and screen assembly has restriction or air leak	• Remove and inspect oil inlet tube and screen assembly. (Fill inlet tube with lacquer thinner to locate leaks.)
	• Excessive oil pump clearance	• Measure clearances
	• Excessive main, rod, or camshaft bearing clearance	• Measure bearing clearances, repair as necessary
High oil pressure	• Improper oil viscosity	• Drain and refill crankcase with correct viscosity oil
	• Oil pressure gauge or sending unit inaccurate	• Replace oil pressure gauge
	• Oil pressure relief valve sticking closed	• Remove and inspect oil pressure relief valve assembly
Main bearing noise	• Insufficient oil supply	• Inspect for low oil level and low oil pressure
	• Main bearing clearance excessive	• Measure main bearing clearance, repair as necessary
	• Bearing insert missing	• Replace missing insert
	• Crankshaft end play excessive	• Measure end play, repair as necessary
	• Improperly tightened main bearing cap bolts	• Tighten bolts with specified torque
	• Loose flywheel or drive plate	• Tighten flywheel or drive plate attaching bolts
	• Loose or damaged vibration damper	• Repair as necessary

Troubleshooting Engine Mechanical Problems (cont.)

Problem	Cause	Solution
Connecting rod bearing noise	• Insufficient oil supply	• Inspect for low oil level and low oil pressure
	• Carbon build-up on piston	• Remove carbon from piston crown
	• Bearing clearance excessive or bearing missing	• Measure clearance, repair as necessary
	• Crankshaft connecting rod journal out-of-round	• Measure journal dimensions, repair or replace as necessary
	• Misaligned connecting rod or cap	• Repair as necessary
	• Connecting rod bolts tightened improperly	• Tighten bolts with specified torque
Piston noise	• Piston-to-cylinder wall clearance excessive (scuffed piston)	• Measure clearance and examine piston
	• Cylinder walls excessively tapered or out-of-round	• Measure cylinder wall dimensions, rebore cylinder
	• Piston ring broken	• Replace all rings on piston
	• Loose or seized piston pin	• Measure piston-to-pin clearance, repair as necessary
	• Connecting rods misaligned	• Measure rod alignment, straighten or replace
	• Piston ring side clearance excessively loose or tight	• Measure ring side clearance, repair as necessary
	• Carbon build-up on piston is excessive	• Remove carbon from piston
Valve actuating component noise	• Insufficient oil supply	• Check for: (a) Low oil level (b) Low oil pressure (c) Plugged push rods (d) Wrong hydraulic tappets (e) Restricted oil gallery (f) Excessive tappet to bore clearance
	• Push rods worn or bent	• Replace worn or bent push rods
	• Rocker arms or pivots worn	• Replace worn rocker arms or pivots
	• Foreign objects or chips in hydraulic tappets	• Clean tappets
	• Excessive tappet leak-down	• Replace valve tappet
	• Tappet face worn	• Replace tappet; inspect corresponding cam lobe for wear
	• Broken or cocked valve springs	• Properly seat cocked springs; replace broken springs
	• Stem-to-guide clearance excessive	• Measure stem-to-guide clearance, repair as required
	• Valve bent	• Replace valve
	• Loose rocker arms	• Tighten bolts with specified torque
	• Valve seat runout excessive	• Regrind valve seat/valves
	• Missing valve lock	• Install valve lock
	• Push rod rubbing or contacting cylinder head	• Remove cylinder head and remove obstruction in head
	• Excessive engine oil (four-cylinder engine)	• Correct oil level

Troubleshooting the Cooling System

Problem	Cause	Solution
High temperature gauge indication—overheating	• Coolant level low	• Replenish coolant
	• Fan belt loose	• Adjust fan belt tension
	• Radiator hose(s) collapsed	• Replace hose(s)
	• Radiator airflow blocked	• Remove restriction (bug screen, fog lamps, etc.)
	• Faulty radiator cap	• Replace radiator cap
	• Ignition timing incorrect	• Adjust ignition timing
	• Idle speed low	• Adjust idle speed
	• Air trapped in cooling system	• Purge air
	• Heavy traffic driving	• Operate at fast idle in neutral intermittently to cool engine
	• Incorrect cooling system component(s) installed	• Install proper component(s)
	• Faulty thermostat	• Replace thermostat
	• Water pump shaft broken or impeller loose	• Replace water pump
	• Radiator tubes clogged	• Flush radiator
	• Cooling system clogged	• Flush system
	• Casting flash in cooling passages	• Repair or replace as necessary. Flash may be visible by removing cooling system components or removing core plugs.
	• Brakes dragging	• Repair brakes
	• Excessive engine friction	• Repair engine
	• Antifreeze concentration over 68%	• Lower antifreeze concentration percentage
	• Missing air seals	• Replace air seals
	• Faulty gauge or sending unit	• Repair or replace faulty component
	• Loss of coolant flow caused by leakage or foaming	• Repair or replace leaking component, replace coolant
	• Viscous fan drive failed	• Replace unit
Low temperature indication—undercooling	• Thermostat stuck open	• Replace thermostat
	• Faulty gauge or sending unit	• Repair or replace faulty component
Coolant loss—boilover	• Overfilled cooling system	• Reduce coolant level to proper specification
	• Quick shutdown after hard (hot) run	• Allow engine to run at fast idle prior to shutdown
	• Air in system resulting in occasional "burping" of coolant	• Purge system
	• Insufficient antifreeze allowing coolant boiling point to be too low	• Add antifreeze to raise boiling point
	• Antifreeze deteriorated because of age or contamination	• Replace coolant
	• Leaks due to loose hose clamps, loose nuts, bolts, drain plugs, faulty hoses, or defective radiator	• Pressure test system to locate source of leak(s) then repair as necessary

Troubleshooting the Cooling System (cont.)

Problem	Cause	Solution
Coolant loss—boilover	• Faulty head gasket • Cracked head, manifold, or block • Faulty radiator cap	• Replace head gasket • Replace as necessary • Replace cap
Coolant entry into crankcase or cylinder(s)	• Faulty head gasket • Crack in head, manifold or block	• Replace head gasket • Replace as necessary
Coolant recovery system inoperative	• Coolant level low • Leak in system • Pressure cap not tight or seal missing, or leaking • Pressure cap defective • Overflow tube clogged or leaking • Recovery bottle vent restricted	• Replenish coolant to FULL mark • Pressure test to isolate leak and repair as necessary • Repair as necessary • Replace cap • Repair as necessary • Remove restriction
Noise	• Fan contacting shroud • Loose water pump impeller • Glazed fan belt • Loose fan belt • Rough surface on drive pulley • Water pump bearing worn • Belt alignment	• Reposition shroud and inspect engine mounts • Replace pump • Apply silicone or replace belt • Adjust fan belt tension • Replace pulley • Remove belt to isolate. Replace pump. • Check pulley alignment. Repair as necessary.
No coolant flow through heater core	• Restricted return inlet in water pump • Heater hose collapsed or restricted • Restricted heater core • Restricted outlet in thermostat housing • Intake manifold bypass hole in cylinder head restricted • Faulty heater control valve • Intake manifold coolant passage restricted	• Remove restriction • Remove restriction or replace hose • Remove restriction or replace core • Remove flash or restriction • Remove restriction • Replace valve • Remove restriction or replace intake manifold

NOTE: *Immediately after shutdown, the engine enters a condition known as heat soak. This is caused by the cooling system being inoperative while engine temperature is still high. If coolant temperature rises above boiling point, expansion and pressure may push some coolant out of the radiator overflow tube. If this does not occur frequently it is considered normal.*

Troubleshooting the Serpentine Drive Belt

Problem	Cause	Solution
Tension sheeting fabric failure (woven fabric on outside circumference of belt has cracked or separated from body of belt)	• Grooved or backside idler pulley diameters are less than minimum recommended	• Replace pulley(s) not conforming to specification
	• Tension sheeting contacting (rubbing) stationary object	• Correct rubbing condition
	• Excessive heat causing woven fabric to age	• Replace belt
	• Tension sheeting splice has fractured	• Replace belt
Noise (objectional squeal, squeak, or rumble is heard or felt while drive belt is in operation)	• Belt slippage	• Adjust belt
	• Bearing noise	• Locate and repair
	• Belt misalignment	• Align belt/pulley(s)
	• Belt-to-pulley mismatch	• Install correct belt
	• Driven component inducing vibration	• Locate defective driven component and repair
	• System resonant frequency inducing vibration	• Vary belt tension within specifications. Replace belt.
Rib chunking (one or more ribs has separated from belt body)	• Foreign objects imbedded in pulley grooves	• Remove foreign objects from pulley grooves
	• Installation damage	• Replace belt
	• Drive loads in excess of design specifications	• Adjust belt tension
	• Insufficient internal belt adhesion	• Replace belt
Rib or belt wear (belt ribs contact bottom of pulley grooves)	• Pulley(s) misaligned	• Align pulley(s)
	• Mismatch of belt and pulley groove widths	• Replace belt
	• Abrasive environment	• Replace belt
	• Rusted pulley(s)	• Clean rust from pulley(s)
	• Sharp or jagged pulley groove tips	• Replace pulley
	• Rubber deteriorated	• Replace belt
Longitudinal belt cracking (cracks between two ribs)	• Belt has mistracked from pulley groove	• Replace belt
	• Pulley groove tip has worn away rubber-to-tensile member	• Replace belt
Belt slips	• Belt slipping because of insufficient tension	• Adjust tension
	• Belt or pulley subjected to substance (belt dressing, oil, ethylene glycol) that has reduced friction	• Replace belt and clean pulleys
	• Driven component bearing failure	• Replace faulty component bearing
	• Belt glazed and hardened from heat and excessive slippage	• Replace belt
"Groove jumping" (belt does not maintain correct position on pulley, or turns over and/or runs off pulleys)	• Insufficient belt tension	• Adjust belt tension
	• Pulley(s) not within design tolerance	• Replace pulley(s)
	• Foreign object(s) in grooves	• Remove foreign objects from grooves

Troubleshooting the Serpentine Drive Belt (cont.)

Problem	Cause	Solution
"Groove jumping" (belt does not maintain correct position on pulley, or turns over and/or runs off pulleys)	• Excessive belt speed • Pulley misalignment • Belt-to-pulley profile mismatched • Belt cordline is distorted	• Avoid excessive engine acceleration • Align pulley(s) • Install correct belt • Replace belt
Belt broken (Note: identify and correct problem before replacement belt is installed)	• Excessive tension • Tensile members damaged during belt installation • Belt turnover • Severe pulley misalignment • Bracket, pulley, or bearing failure	• Replace belt and adjust tension to specification • Replace belt • Replace belt • Align pulley(s) • Replace defective component and belt
Cord edge failure (tensile member exposed at edges of belt or separated from belt body)	• Excessive tension • Drive pulley misalignment • Belt contacting stationary object • Pulley irregularities • Improper pulley construction • Insufficient adhesion between tensile member and rubber matrix	• Adjust belt tension • Align pulley • Correct as necessary • Replace pulley • Replace pulley • Replace belt and adjust tension to specifications
Sporadic rib cracking (multiple cracks in belt ribs at random intervals)	• Ribbed pulley(s) diameter less than minimum specification • Backside bend flat pulley(s) diameter less than minimum • Excessive heat condition causing rubber to harden • Excessive belt thickness • Belt overcured • Excessive tension	• Replace pulley(s) • Replace pulley(s) • Correct heat condition as necessary • Replace belt • Replace belt • Adjust belt tension

GENERAL ENGINE SPECIFICATIONS

Year	Model	Engine Displacement Liter/CID	Net Horsepower (@ rpm)	Net Torque (@ rpm)	Bore × Stroke (in.)	Compression Ratio	Oil Pressure (@ 2000 rpm)
1983	Ranger	2.0/122	73 @ 4000	107 @ 2400	3.52 × 3.13	9.0:1	40–60
		2.2/134 Diesel	59 @ 4000	90 @ 2500	3.50 × 3.50	22:1	51 ③
		2.3/140	79 @ 3800 ①	124 @ 2200 ②	3.78 × 3.13	9.0:1	40–60
1984	Bronco II	2.8/173	115 @ 4600	150 @ 2600	3.66 × 2.70	8.7:1	40–60
	Ranger	2.0/122	73 @ 4000	107 @ 2400	3.52 × 3.13	9.0:1	40–60
		2.2/134 Diesel	59 @ 4000	90 @ 2500	3.50 × 3.50	22:1	51 ③
		2.3/140	79 @ 3800 ①	124 @ 2200 ②	3.78 × 3.13	9.0:1	40–60
		2.8/173	115 @ 4600	150 @ 2600	3.66 × 2.70	8.7:1	40–60
1985	Bronco II	2.8/173	115 @ 4600	150 @ 2600	3.66 × 2.70	8.7:1	40–60
	Ranger	2.0/122	73 @ 4000	107 @ 2400	3.52 × 3.13	9.0:1	40–60
		2.3/140	79 @ 3800 ①	124 @ 2200 ②	3.78 × 3.13	9.0:1	40–60
		2.3/140 Diesel	86 @ 4200	134 @ 2000	3.59 × 3.54	21:1	11.4 ④
		2.8/173	115 @ 4600	150 @ 2600	3.66 × 2.70	8.7:1	40–60

GENERAL ENGINE SPECIFICATIONS

Year	Model	Engine Displacement Liter/CID	Net Horsepower (@ rpm)	Net Torque (@ rpm)	Bore × Stroke (in.)	Compression Ratio	Oil Pressure (@ 2000 rpm)
1986	Bronco II	2.9/177	140 @ 4600	170 @ 2600	3.66 × 2.83	9.0:1	40–60
	Ranger	2.0/122	73 @ 4000	107 @ 2400	3.52 × 3.13	9.0:1	40–60
		2.3/140	90 @ 4000	130 @ 1800	3.78 × 3.13	9.0:1	40–60
		2.3/140 Diesel	86 @ 4200	134 @ 2000	3.59 × 3.54	21:1	11.4 ④
		2.9/177	140 @ 4600	170 @ 2600	3.66 × 2.83	9.0:1	40–60
1987	Bronco II	2.9/177	140 @ 4600	170 @ 2600	3.66 × 2.83	9.0:1	40–60
	Ranger	2.0/122	80 @ 4200	106 @ 2600	3.52 × 3.13	9.0:1	40–60
		2.3/140	90 @ 4000	134 @ 2000	3.78 × 3.13	9.0:1	40–60
		2.3/140 Diesel	86 @ 4200	134 @ 2000	3.59 × 3.54	21:1	11.4 ④
		2.9/177	140 @ 4600	170 @ 2600	3.66 × 2.83	9.0:1	40–60
1988	Bronco II	2.9/177	140 @ 4600	170 @ 2600	3.66 × 2.83	9.0:1	40–60
	Ranger	2.0/122	80 @ 4200	106 @ 2600	3.52 × 3.13	9.0:1	40–60
		2.3/140	90 @ 4000	134 @ 2000	3.78 × 3.13	9.0:1	40–60
		2.9/177	140 @ 4600	170 @ 2600	3.66 × 2.83	9.0:1	40–60
1989	Bronco II	2.9/177	140 @ 4800	170 @ 2600	3.66 × 2.83	9.0:1	40–60
	Ranger	2.3/140	100 @ 4600	133 @ 2600	3.78 × 3.13	9.2:1	40–60
		2.9/177	140 @ 4600	170 @ 2600	3.66 × 2.83	9.0:1	40–60
1990	Bronco II	2.9/177	140 @ 4600	170 @ 2600	3.66 × 2.83	9.0:1	40–60
	Ranger	2.3/140	100 @ 4600	133 @ 2600	3.78 × 3.13	9.2:1	40–60
		2.9/177	140 @ 4600	170 @ 2600	3.66 × 2.83	9.0:1	40–60
		4.0/241	155 @ 4200	220 @ 2400	3.94 × 3.31		40–60
1991	Ranger	2.3/140	100 @ 4600	133 @ 2600	3.78 × 3.13	9.2:1	40–60
		2.9/177	140 @ 4600	170 @ 2600	3.66 × 2.83	9.0:1	40–60
		3.0/183	145 @ 4800	165 @ 3600	3.50 × 3.14	9.3:1	40–60
		4.0/241	155 @ 4200	220 @ 2400	3.94 × 3.31	9.0:1	40–60
	Explorer	4.0/241	155 @ 4200	220 @ 2400	3.94 × 3.31	9.0:1	40–60

① Auto. trans.: 82 @ 4200
② Auto. trans.: 126 @ 2200
③ @ 3600 rpm
④ @ Idle

VALVE SPECIFICATIONS

Year	Model	Engine Displacement Liter/CID	Seat Angle (deg)	Face Angle (deg)	Spring Test Pressure (lbs @ in.)	Spring Installed Height (in.)	Stem-to-Guide Clearance (in.)		Stem Diameter (in.)	
							Intake	Exhaust	Intake	Exhaust
1983	Ranger	2.0/122	45	44	149 @ 1.12	1.49–1.55	0.0010–0.0027	0.0015–0.0032	0.3416–0.3423	0.3411–0.3418
		2.2/134 Diesel	①	①	②	③	0.0015–0.0046	0.0020–0.0051	0.3150	0.3150
		2.3/140	45	44	149 @ 1.12	1.53–1.59	0.0010–0.0027	0.0015–0.0032	0.3416–0.3423	0.3411–0.3418

VALVE SPECIFICATIONS

Year	Model	Engine Displacement Liter/CID	Seat Angle (deg)	Face Angle (deg)	Spring Test Pressure (lbs @ in.)	Spring Installed Height (in.)	Stem-to-Guide Clearance (in.)		Stem Diameter (in.)	
							Intake	Exhaust	Intake	Exhaust
1984	Bronco II	2.8/173	45	44	143 @ 1.22	1.58–1.61	0.0008–0.0025	0.0018–0.0035	0.3159–0.3167	0.3149–0.3156
	Ranger	2.0/122	45	44	149 @ 1.12	1.49–1.55	0.0010–0.0027	0.0015–0.0032	0.3416–0.3423	0.3411–0.3418
		2.2/134 Diesel	①	①	②	③	0.0015–0.0046	0.0020–0.0051	0.3150	0.3150
		2.3/140	45	44	149 @ 1.12	1.53–1.59	0.0010–0.0027	0.0015–0.0032	0.3416–0.3423	0.3411–0.3418
		2.8/173	45	44	143 @ 1.22	1.58–1.61	0.0008–0.0025	0.0018–0.0035	0.3159–0.3167	0.3149–0.3156
1985	Bronco II	2.8/173	45	44	143 @ 1.22	1.58–1.61	0.0008–0.0025	0.0018–0.0035	0.3159–0.3167	0.3149–0.3156
	Ranger	2.0/122	45	44	149 @ 1.12	1.49–1.55	0.0010–0.0027	0.0015–0.0032	0.3416–0.3423	0.3411–0.3418
		2.3/140	45	44	149 @ 1.12	1.53–1.59	0.0010–0.0027	0.0015–0.0032	0.3416–0.3423	0.3411–0.3418
		2.3/140 Diesel	45	44	61 @ 1.591	1.591	0.0012–0.0024	0.0020–0.0035	0.3150	0.3150
		2.8/173	45	44	143 @ 1.22	1.58–1.61	0.0008–0.0025	0.0018–0.0035	0.3159–0.3167	0.3149–0.3156
1986	Bronco II	2.9/177	45	44	143 @ 1.22	1.58–1.61	0.0008–0.0025	0.0018–0.0035	0.3159–0.3167	0.3149–0.3156
	Ranger	2.0/122	45	44	149 @ 1.12	1.49–1.55	0.0010–0.0027	0.0015–0.0032	0.3416–0.3423	0.3411–0.3418
		2.3/140	45	44	149 @ 1.12	1.53–1.59	0.0010–0.0027	0.0015–0.0032	0.3416–0.3423	0.3411–0.3418
		2.3/140 Diesel	45	44	61 @ 1.591	1.591	0.0012–0.0024	0.0020–0.0035	0.3150	0.3150
		2.9/177	45	44	143 @ 1.22	1.58–1.61	0.0008–0.0025	0.0018–0.0035	0.3159–0.3167	0.3149–0.3156
1987	Bronco II	2.9/177	45	44	143 @ 1.22	1.58–1.61	0.0008–0.0025	0.0018–0.0035	0.3159–0.3167	0.3149–0.3156
	Ranger	2.0/122	45	44	149 @ 1.12	1.49–1.55	0.0010–0.0027	0.0015–0.0032	0.3416–0.3423	0.3411–0.3418
		2.3/140	45	44	149 @ 1.12	1.53–1.59	0.0010–0.0027	0.0015–0.0032	0.3416–0.3423	0.3411–0.3418
		2.3/140 Diesel	45	44	61 @ 1.591	1.591	0.0012–0.0024	0.0020–0.0035	0.3150	0.3150
		2.9/177	45	44	143 @ 1.22	1.58–1.61	0.0008–0.0025	0.0018–0.0035	0.3159–0.3167	0.3149–0.3156
1988	Bronco II	2.9/177	45	44	143 @ 1.22	1.58–1.61	0.0008–0.0025	0.0018–0.0035	0.3159–0.3167	0.3149–0.3156
	Ranger	2.0/122	45	44	149 @ 1.12	1.49–1.55	0.0010–0.0027	0.0015–0.0032	0.3416–0.3423	0.3411–0.3418
		2.3/140	45	44	149 @ 1.12	1.53–1.59	0.0010–0.0027	0.0015–0.0032	0.3416–0.3423	0.3411–0.3418
		2.9/177	45	44	143 @ 1.22	1.58–1.61	0.0008–0.0025	0.0018–0.0035	0.3159–0.3167	0.3149–0.3156

VALVE SPECIFICATIONS

Year	Model	Engine Displacement Liter/CID	Seat Angle (deg)	Face Angle (deg)	Spring Test Pressure (lbs @ in.)	Spring Installed Height (in.)	Stem-to-Guide Clearance (in.)		Stem Diameter (in.)	
							Intake	Exhaust	Intake	Exhaust
1989	Bronco II	2.9/177	45	44	143 @ 1.22	1.58–1.61	0.0008–0.0025	0.0018–0.0035	0.3159–0.3167	0.3149–0.3156
	Ranger	2.3/140	45	44	149 @ 1.12	1.49–1.55	0.0010–0.0027	0.0015–0.0032	0.3416–0.3423	0.3411–0.3418
		2.9/177	45	44	143 @ 1.22	1.58–1.61	0.0008–0.0025	0.0018–0.0035	0.3159–0.3167	0.3149–0.3156
1990	Bronco II	2.9/177	45	44	143 @ 1.22	1.58–1.61	0.0008–0.0025	0.0018–0.0035	0.3159–0.3167	0.3149–0.3156
	Ranger	2.3/140	45	44	149 @ 1.12	1.49–1.55	0.0010–0.0027	0.0015–0.0032	0.3416–0.3423	0.3411–0.3418
		2.9/177	45	44	143 @ 1.22	1.58–1.61	0.0008–0.0025	0.0018–0.0035	0.3159–0.3167	0.3149–0.3156
		4.0/241	45	44	138 @ 1.22	1.58–1.61	0.0008–0.0025	0.0018–0.0035	0.3159–0.3167	0.3149–0.3156
1991	Ranger	2.3/140	45	44	149 @ 1.12	1.49–1.55	0.0010–0.0027	0.0015–0.0032	0.3416–0.3423	0.3411–0.3418
		2.9/177	45	44	143 @ 1.22	1.58–1.61	0.0008–0.0025	0.0018–0.0035	0.3159–0.3167	0.3149–0.3156
		3.0/183	45	44	180 @ 1.16	1.85	0.0010–0.0027	0.0015–0.0032	0.3134–0.3126	0.3129–0.3121
		4.0/241	45	44	138 @ 1.22	1.58–1.61	0.0008–0.0025	0.0018–0.0035	0.3159–0.3167	0.3149–0.3156
	Explorer	4.0/241	45	44	138 @ 1.22	1.58–1.61	0.0008–0.0025	0.0018–0.0035	0.3159–0.3167	0.3149–0.3156

① Intake: 45 degrees Exhaust: 30 degrees
② Outer: 40 @ 1.59 in.
Inner: 28 @ 1.49 in.
③ Outer: 1.587 in.
Inner: 1.488 in.

CRANKSHAFT AND CONNECTING ROD SPECIFICATIONS

All measurements are given in inches.

Year	Model	Engine Displacement Liter/CID	Crankshaft				Connecting Rod		
			Main Brg Journal Dia.	Main Brg Oil Clearance	Shaft End-play	Thrust on No.	Journal Diameter	Oil Clearance	Side Clearance
1983	Ranger	2.0/122	2.3982–2.3990	0.0008–0.0015	0.004–0.008	3	2.0462–2.0472	0.0008–0.0015	0.0035–0.0105
		2.2/134 Diesel	2.5591	0.0016–0.0036	0.0055–0.0154	3	2.0866	0.0014–0.0030	0.0094–0.0134
		2.3/140	2.3982–2.3990	0.0008–0.0015	0.004–0.008	3	2.0462–2.0472	0.0008–0.0015	0.0035–0.0105
1984	Bronco II	2.8/173	2.2433–2.2441	0.0008–0.0015	0.004–0.008	3	2.1252–2.1260	0.0006–0.0016	0.004–0.011
	Ranger	2.0/122	2.3982–2.3990	0.0008–0.0015	0.004–0.008	3	2.0462–2.0472	0.0008–0.0015	0.0035–0.0105
		2.2/134 Diesel	2.5591	0.0016–0.0036	0.0055–0.0154	3	2.0866	0.0014–0.0030	0.0094–0.0134
		2.3/140	2.3982–2.3990	0.0008–0.0015	0.004–0.008	3	2.0462–2.0472	0.0008–0.0015	0.0035–0.0105
		2.8/173	2.2433–2.2441	0.0008–0.0015	0.004–0.008	3	2.1252–2.1260	0.0006–0.0016	0.004–0.011

CRANKSHAFT AND CONNECTING ROD SPECIFICATIONS

All measurements are given in inches.

Year	Model	Engine Displacement Liter/CID	Crankshaft				Connecting Rod		
			Main Brg Journal Dia.	Main Brg Oil Clearance	Shaft End-play	Thrust on No.	Journal Diameter	Oil Clearance	Side Clearance
1985	Bronco II	2.8/173	2.2433–2.2441	0.0008–0.0015	0.004–0.008	3	2.1252–2.1260	0.0006–0.0016	0.004–0.011
	Ranger	2.0/122	2.3982–2.3990	0.0008–0.0015	0.004–0.008	3	2.0462–2.0472	0.0008–0.0015	0.0035–0.0105
		2.2/134 Diesel	2.5591	0.0016–0.0036	0.0055–0.0154	3	2.0866	0.0014–0.0030	0.0094–0.0134
		2.3/140	2.3982–2.3990	0.0008–0.0015	0.004–0.008	3	2.0462–2.0472	0.0008–0.0015	0.0035–0.0105
		2.8/173	2.2433–2.2441	0.0008–0.0015	0.004–0.008	3	2.1252–2.1260	0.0006–0.0016	0.004–0.011
1986	Bronco II	2.9/177	2.2433–2.2441	0.0008–0.0015	0.004–0.008	3	2.1252–2.1260	0.0006–0.0016	0.004 0.011
	Ranger	2.0/122	2.3982–2.3990	0.0008–0.0015	0.004–0.008	3	2.0462–2.0472	0.0008–0.0015	0.0035–0.0105
		2.3/140	2.3982–2.3990	0.0008–0.0015	0.004–0.008	3	2.0462–2.0472	0.0008–0.0015	0.0035–0.0105
		2.3/140 Diesel	2.5980	0.0008–0.0015	0.0008–0.0020	3	2.087	0.0008–0.0024	0.004–0.010
		2.9/177	2.2433–2.2441	0.0008–0.0015	0.004–0.008	3	2.1252–2.1260	0.0006–0.0016	0.004–0.011
1987	Bronco II	2.9/177	2.2433–2.2441	0.0008–0.0015	0.004–0.008	3	2.1252–2.1260	0.0006–0.0016	0.004–0.011
	Ranger	2.0/122	2.3982–2.3990	0.0008–0.0015	0.004–0.008	3	2.0462–2.0472	0.0008–0.0015	0.0035–0.0105
		2.3/140	2.3982–2.3990	0.0008–0.0015	0.004–0.008	3	2.0462–2.0472	0.0008–0.0015	0.0035–0.0105
		2.3/140 Diesel	2.5980	0.0008–0.0015	0.0008–0.0020	3	2.087	0.0008–0.0024	0.004–0.010
		2.9/177	2.2433–2.2441	0.0008–0.0015	0.004–0.008	3	2.1252–2.1260	0.0006–0.0016	0.004–0.011
1988	Bronco II	2.9/177	2.2433–2.2441	0.0008–0.0015	0.004–0.008	3	2.1252–2.1260	0.0006–0.0016	0.004–0.011
	Ranger	2.0/122	2.3982–2.3990	0.0008–0.0015	0.004–0.008	3	2.0462–2.0472	0.0008–0.0015	0.0035–0.0105
		2.3/140	2.3982–2.3990	0.0008–0.0015	0.004–0.008	3	2.0462–2.0472	0.0008–0.0015	0.0035–0.0105
		2.9/177	2.2433–2.2441	0.0008–0.0015	0.004–0.008	3	2.1252–2.1260	0.0006–0.0016	0.004–0.011
1989	Bronco II	2.9/177	2.2433–2.2441	0.0008–0.0015	0.004–0.008	3	2.1252–2.1260	0.0006–0.0016	0.004–0.011
	Ranger	2.3/140	2.3982–2.3990	0.0008–0.0015	0.004–0.008	3	2.0462–2.0472	0.0008–0.0015	0.0035–0.0105
		2.9/177	2.2433–2.2441	0.0008–0.0015	0.004–0.008	3	2.1252–8.1260	0.0006–0.0016	0.004–0.011
1990	Bronco II	2.9/177	2.2433–2.2441	0.0008–0.0015	0.004–0.008	3	2.1252–2.1260	0.0006–0.0016	0.004–0.011
	Ranger	2.3/140	2.3982–2.3990	0.0008–0.0015	0.004–0.008	3	2.0462–2.0472	0.0008–0.0015	0.0035–0.0105

CRANKSHAFT AND CONNECTING ROD SPECIFICATIONS

All measurements are given in inches.

Year	Model	Engine Displacement Liter/CID	Crankshaft				Connecting Rod		
			Main Brg Journal Dia.	Main Brg Oil Clearance	Shaft End-play	Thrust on No.	Journal Diameter	Oil Clearance	Side Clearance
1990		2.9/177	2.2433–2.2441	0.0008–0.0015	0.004–0.008	3	2.1252–2.1260	0.0006–0.0016	0.004–0.011
		4.0/241	2.2433–2.2441	0.0008–0.0015	0.0160–0.0126	—	2.1252–2.1260	0.0003–0.0024	0.0002–0.0025
1991	Ranger	2.3/140	2.3982–2.3990	0.0008–0.0015	0.004–0.008	3	2.0462–2.0472	0.0008–0.0015	0.0035–0.0105
		2.9/177	2.2433–2.2441	0.0008–0.0015	0.004–0.008	3	2.1252–2.1260	0.0006–0.0016	0.004–0.011
		3.0/183	2.5190–2.5198	0.0010–0.0014	0.004 0.008	3	2.1253–2.1261	0.0010–0.0014	0.0060–0.0140
		4.0/241	2.2433–2.2441	0.0008–0.0015	0.0160–0.0126	—	2.1252–2.1260	0.0003–0.0024	0.0002–0.0025
	Explorer	4.0/241	2.2433–2.2441	0.0008–0.0015	0.0160–0.0126	—	2.1252–2.1260	0.0003–0.0024	0.0002–0.0025

PISTON AND RING SPECIFICATIONS

All measurements are given in inches.

Year	Model	Engine Displacement Liter/CID	Piston to Bore Clearance	Ring Side Clearance			Ring Gap		
				Top Compression	Bottom Compression	Oil Control	Top Compression	Bottom Compression	Oil Control
1983	Ranger	2.0/122	0.0014–0.0022	0.0020–0.0040	0.0020–0.0040	Snug	0.010–0.020	0.010–0.020	0.015–0.055
		2.2/134 Diesel	0.0021–0.0031	0.0020–0.0035	0.0016–0.0031	.0012–.0028	0.0157–0.0217	0.0118–0.0157	0.0138–0.0217
		2.3/140	0.0014–0.0022	0.0020–0.0040	0.0020–0.0040	Snug	0.010–0.020	0.010–0.020	0.015–0.055
1984	Bronco II	2.8/173	0.0011–0.0019	0.0020–0.0033	0.0020–0.0033	Snug	0.015–0.023	0.015–0.023	0.015–0.023
	Ranger	2.0/122	0.0014–0.0022	0.0020–0.0040	0.0020–0.0040	Snug	0.010–0.020	0.010–0.020	0.015–0.055
		2.2/134 Diesel	0.0021–0.0031	0.0020–0.0035	0.0016–0.0031	.0012–.0028	0.0157–0.0217	0.0118–0.0157	0.0138–0.0217
		2.3/140	0.0014–0.0022	0.0020–0.0040	0.0020–0.0040	Snug	0.010–0.020	0.010–0.020	0.015–0.055
		2.8/173	0.0011–0.0019	0.0020–0.0033	0.0020–0.0033	Snug	0.015–0.023	0.015–0.023	0.015–0.023
1985	Bronco II	2.8/173	0.0011–0.0019	0.0020–0.0033	0.0020–0.0033	Snug	0.015–0.023	0.015–0.023	0.015–0.023
	Ranger	2.0/122	0.0014–0.0022	0.0020–0.0040	0.0020–0.0040	Snug	0.010–0.020	0.010–0.020	0.015–0.055
		2.3/140	0.0014–0.0022	0.0020–0.0040	0.0020–0.0040	Snug	0.010–0.020	0.010–0.020	0.015–0.055
		2.3/140 Diesel	0.0021–0.0031	0.0020–0.0035	0.0016–0.0031	.0012–.0028	0.0157–0.0217	0.0118–0.0157	0.0138–0.0217
		2.8/173	0.0011–0.0019	0.0020–0.0033	0.0020–0.0033	Snug	0.015–0.023	0.015–0.023	0.015–0.023

PISTON AND RING SPECIFICATIONS

All measurements are given in inches.

Year	Model	Engine Displacement Liter/CID	Piston to Bore Clearance	Ring Side Clearance Top Compression	Ring Side Clearance Bottom Compression	Ring Side Clearance Oil Control	Ring Gap Top Compression	Ring Gap Bottom Compression	Ring Gap Oil Control
1986	Bronco II	2.9/177	0.0011–0.0019	0.0020–0.0033	0.0020–0.0033	Snug	0.015–0.023	0.015–0.023	0.015–0.023
	Ranger	2.0/122	0.0014–0.0022	0.0020–0.0040	0.0020–0.0040	Snug	0.010–0.020	0.010–0.020	0.015–0.055
		2.2/134 Diesel	0.0021–0.0031	0.0020–0.0035	0.0016–0.0031	.0012–.0028	0.0157–0.0217	0.0118–0.0157	0.0138–0.0217
		2.3/140	0.0014–0.0022	0.0020–0.0040	0.0020–0.0040	Snug	0.010–0.020	0.010–0.020	0.015–0.055
		2.9/177	0.0011–0.0019	0.0020–0.0033	0.0020–0.0033	Snug	0.015–0.023	0.015–0.023	0.015–0.023
1987	Bronco II	2.9/177	0.0011–0.0019	0.0020–0.0033	0.0020–0.0033	Snug	0.015–0.023	0.015–0.023	0.015–0.023
	Ranger	2.0/122	0.0014–0.0022	0.0020–0.0040	0.0020–0.0040	Snug	0.010–0.020	0.010–0.020	0.015–0.055
		2.3/140	0.0014–0.0022	0.0020–0.0040	0.0002–0.0040	Snug	0.010–0.020	0.010–0.020	0.015–0.055
		2.3/140 Diesel	0.0021–0.0031	0.0020–0.0035	0.0016–0.0031	.0012–.0028	0.0157–0.0217	0.0118–0.0157	0.0138–0.0217
		2.9/177	0.0011–0.0019	0.0020–0.0033	0.0020–0.0033	Snug	0.015–0.023	0.015–0.023	0.015–0.023
1988	Bronco II	2.9/177	0.0011–0.0019	0.0020–0.0033	0.0020–0.0033	Snug	0.015–0.023	0.015–0.023	0.015–0.023
	Ranger	2.0/122	0.0014–0.0022	0.0020–0.0040	0.0020–0.0040	Snug	0.010–0.020	0.010–0.020	0.015–0.055
		2.3/140	0.0014–0.0022	0.0020–0.0040	0.0020–0.0040	Snug	0.010–0.020	0.010–0.020	0.015–0.055
		2.9/177	0.0011–0.0019	0.0020–0.0033	0.0020–0.0033	Snug	0.015–0.023	0.015–0.023	0.015–0.023
1989	Bronco II	2.9/177	0.0011–0.0019	0.0020–0.0033	0.0020–0.0033	Snug	0.015–0.023	0.015–0.023	0.015–0.055
	Ranger	2.3/140	0.0014–0.0022	0.0020–0.0040	0.0020–0.0040	Snug	0.010–0.020	0.010–0.020	0.010–0.049
		2.9/177	0.0011–0.0019	0.0020–0.0033	0.0020–0.0033	Snug	0.015–0.023	0.015–0.023	0.015–0.055
1990	Bronco II	2.9/177	0.0011–0.0019	0.0020–0.0033	0.0020–0.0033	Snug	0.015–0.023	0.015–0.023	0.015–0.055
	Ranger	2.3/140	0.0014–0.0022	0.0020–0.0040	0.0020–0.0040	Snug	0.010–0.020	0.010–0.020	0.010–0.049
		2.9/177	0.0011–0.0019	0.0020–0.0033	0.0020–0.0033	Snug	0.015–0.023	0.015–0.023	0.015–0.055
		4.0/241	0.0008–0.0019	0.0020–0.0033	0.0020–0.0033	Snug	0.015–0.023	0.015–0.023	0.015–0.055
1991	Ranger	2.3/140	0.0014–0.0022	0.0020–0.0040	0.0020–0.0040	Snug	0.010–0.020	0.010–0.020	0.010–0.049
		2.9/177	0.0011–0.0019	0.0020–0.0033	0.0020–0.0033	Snug	0.015–0.023	0.015–0.023	0.015–0.055
		3.0/183	0.0012–0.0023	0.0016–0.0037	0.0016–0.0037	Snug	0.010–0.020	0.010–0.020	0.010–0.049

PISTON AND RING SPECIFICATIONS

All measurements are given in inches.

Year	Model	Engine Displacement Liter/CID	Piston to Bore Clearance	Ring Side Clearance Top Compression	Ring Side Clearance Bottom Compression	Ring Side Clearance Oil Control	Ring Gap Top Compression	Ring Gap Bottom Compression	Ring Gap Oil Control
		4.0/241	0.0008–0.0019	0.0020–0.0033	0.0020–0.0033	Snug	0.015–0.023	0.015–0.023	0.015–0.055
	Explorer	4.0/241	0.0008–0.0019	0.0020–0.0033	0.0020–0.0033	Snug	0.015–0.023	0.015–0.023	0.015–0.055

TORQUE SPECIFICATIONS

All readings in ft. lbs.

Year	Model	Engine Displacement Liter/CID	Cylinder Head Bolts	Main Bearing Bolts	Rod Bearing Bolts	Crankshaft Pulley Bolts	Flywheel to Crankshaft Bolts	Manifold Intake	Manifold Exhaust
1983	Ranger	2.0/122	80-90 ①	80-90 ①	30-36 ②	100–120	56-64	14–21 ③	16–23 ④
		2.2/134 Diesel	80-85	80-85	50-54	253-289	95–137	12-17	17–20 ⑦
		2.3/140	80-90 ①	80-90 ①	30-36 ②	100–120	56-64	14–21 ③	16–23 ④
1984	Bronco II	2.8/173	70-85 ⑤	65-75	19-24	85-96	47-52	15–18 ⑥	20-30
	Ranger	2.0/122	80-90 ①	80-90 ①	30-36 ②	100–120	56-64	14–21 ③	16–23 ④
		2.2/134 Diesel	80-85	80-85	50-54	253-289	95–137	12-17	17–20 ⑦
		2.3/140	80-90 ①	80-90 ①	30-36 ②	100–120	56-64	14–21 ③	16–23 ④
		2.8/173	70-85 ⑤	65-75	19-24	85-96	47-52	15–18 ⑥	20-30
1985	Bronco II	2.8/173	70-85 ⑤	65-75	19-24	85-96	47-52	15–18 ⑥	20-30
	Ranger	2.0/122	80-90 ①	80-90 ①	30-36 ②	100–120	56-64	14–21 ③	16–23 ④
		2.3/140	80-90 ①	80-90 ①	30-36 ②	100–120	56-64	14–21 ③	16–23 ④
		2.3/140 Diesel	76-83 ⑧	55-61	33-34	123-137	94-101	11-14	11-14
		2.8/173	70-85 ⑤	65-75	19-24	85-96	47-52	15–18 ⑥	20-30
1986	Bronco II	2.9/177	⑨	65-75	19-24	85-96	47-52	15–18 ⑥	20-30
	Ranger	2.0/122	80-90 ①	80-90 ①	30-36 ②	100–120	56-64	14–21 ③	16–23 ④
		2.3/140	80-90 ①	80-90 ①	30-36 ②	100–120	56-64	14–21 ③	16–23 ④
		2.3/140 Diesel	76-83 ⑧	55-61	33-34	123-137	94-101	11-14	11-14
		2.9/177	⑨	65-75	19-24	85-96	47-52	15–18 ⑥	20-30
1987	Bronco II	2.9/177	⑨	65-75	19-24	85-96	47-52	15–18 ⑥	20-30
	Ranger	2.0/122	80-90 ①	80-90 ①	30-36 ②	100–120	56-64	14–21 ③	16–23 ④
		2.3/140	80-90 ①	80-90 ①	30-36 ②	100–120	56-64	14–21 ③	16–23 ④
		2.3/140 Diesel	76-83 ⑧	55-61	33-34	123-137	94-101	11-14	11-14
		2.9/177	⑨	65-75	19-24	85-96	47-52	15–18 ⑥	20-30
1988	Bronco II	2.9/177	⑨	65-75	19-24	85-96	47-52	15–18 ⑥	20-30
	Ranger	2.0/122	80-90 ①	80-90 ①	30-36 ②	100–120	56-64	14–21 ③	16–23 ④
		2.3/140	80-90 ①	80-90 ①	30-36 ②	100–120	56-64	14–21 ③	16–23 ④
		2.9/177	⑨	65-75	19-24	85-96	47-52	15–18 ⑥	20-30

TORQUE SPECIFICATIONS

All readings in ft. lbs.

Year	Model	Engine Displacement Liter/CID	Cylinder Head Bolts	Main Bearing Bolts	Rod Bearing Bolts	Crankshaft Pulley Bolts	Flywheel to Crankshaft Bolts	Manifold	
								Intake	Exhaust
1989	Bronco II	2.9/177	⑨	65–75	19–24	85–96	47–52	15–18 ⑥	20–30
	Ranger	2.3/140	80–90 ①	75–85 ⑩	30–36 ②	103–133	56–64	14–21 ③	16–23 ④
		2.9/177	⑨	65–75	19–24	85–96	47–52	15–18 ⑥	20–30
1990	Bronco II	2.9/177	⑨	65–75	19–24	85–96	47–52	15–18 ⑥	20–30
	Ranger	2.3/140	80–90 ①	75–85 ⑩	30–36 ②	103–133	56–64	14–21 ③	20–30
		2.9/177	⑨	65–75	19–24	85–96	47–52	15–18 ⑥	20–30
		4.0/241	⑪	66–77	18–24	N/A	N/A	⑪	20
1991	Ranger	2.3/140	80–90 ①	75–85 ⑩	30–36 ②	103–133	56–64	14–21 ③	20–30
		2.9/177	⑨	65–75	19–24	85–96	47–52	15–18 ⑥	20–30
		3.0/183	68 ⑫	60	26	107	59	18 ⑫	19
		4.0/241	⑪	66–77	18–24	N/A	N/A	⑪	20
	Explorer	4.0/241	⑪	66–77	18–24	N/A	N/A	⑪	20

① Torque in two steps: 1st 50–60 ft. lbs., 2nd 80–90 ft. lbs.
② Torque in two steps: 1st 25–30 ft. lbs., 2nd 30–36 ft. lbs.
③ Torque in two steps: 1st 5–7 ft. lbs., 2nd 14–21 ft. lbs.
④ Torque in two steps: 1st 5–7 ft. lbs., 2nd 16–23 ft. lbs.
⑤ Torque in three steps: 1st 29–40 ft. lbs., 2nd 40–51 ft. lbs., 3rd 70–85 ft. lbs.
⑥ Torque in five steps: 1st Hand start and snug nuts, 2nd 3–6 ft. lbs., 3rd 6–11 ft. lbs., 4th 11–15 ft. lbs., 5th 15–18 ft. lbs., Repeat Step 5 after warm up.
⑦ Retorque after warm up.
⑧ Cold, 84–90 ft. lbs. Hot
⑨ Torque in three steps: 1st 22 ft. lbs., 2nd 51–55 ft. lbs., 3rd Turn 90 degrees
⑩ Torque in 2 steps:
Step 1: 50–60 ft. lbs.
Step 2: 75–85 ft. lbs.
⑪ Refer to text for procedure
⑫ Torque in 2 steps
N/A—Not available at time of publication
NOTE: Always refer to text for sequence before starting torque procedure.

Engine

REMOVAL AND INSTALLATION

2.0L, 2.3L Engines

1. Raise the hood and install protective fender covers. Drain the coolant from the radiator. On the carbureted engines remove the air cleaner and duct assembly. On the 2.3L EFI engine, disconnect the air cleaner outlet tube at the throttle body, idle speed control hose and the heat riser tube.

CAUTION

When draining the coolant, keep in mind that cats and dogs are attracted by the ethylene glycol antifreeze, and are quite likely to drink any that is left in an uncovered container or in puddles on the ground. This will prove fatal in sufficient quantity. Always drain the coolant into a sealable container. Coolant should be reused unless it is contaminated or several years old.

2. Disconnect the battery ground cable at the engine and disconnect the battery positive cable at the battery and set aside.
3. Mark the location of the hood hinges and remove the hood.
4. Disconnect the upper and lower radiator hoses from the engine. Remove the radiator shroud screws. Remove the radiator upper supports.
5. Remove engine fan and shroud assembly. Then remove the radiator. Remove the oil fill cap.
6. Disconnect the coil primary wire at the coil. Disconnect the oil pressure and the water temperature sending unit wires from the sending units.
7. Disconnect the alternator wire from the alternator, the starter cable from the starter and the accelerator cable from the carburetor. If so equipped, disconnect the transmission kickdown rod.
8. If so equipped, remove the A/C compressor from the mounting bracket and position it out of the way, leaving the refrigerant lines attached.
9. Disconnect the power brake vacuum hose. Disconnect the heater hoses from the engine.
10. On the carbureted engines, disconnect the fuel line from the fuel pump. On the 2.3L EFI engine depressurize the fuel system, then disconnect the 2 push connect fittings at the engine fuel rail.
11. Remove the engine mount nuts. Raise the vehicle and safely support on jackstands. Drain the engine oil from the crankcase. Remove the starter motor.

CAUTION

The EPA warns that prolonged contact with used engine oil may cause a number of skin disorders, including cancer! You should make every effort to minimize your exposure to used engine oil. Protective gloves should be worn when changing the oil. Wash your hands and any other exposed skin areas as soon as possible after exposure to used engine oil. Soap and water, or waterless hand cleaner should be used.

12. Disconnect the muffler exhaust inlet pipe at the exhaust manifold.
13. Remove the dust cover (manual transmission) or converter inspection plate (automatic transmission).
14. On vehicles with a manual transmission, remove the flywheel housing cover lower attaching bolts. On vehicles with automatic transmissions, remove the converter-to-flywheel bolts, then remove the converter housing lower attaching bolts.
15. Remove clutch slave cylinder (manual transmission). Lower the vehicle.
16. Support the transmission and flywheel or converter housing with a jack.
17. Remove the flywheel housing or converter housing upper attaching bolts.
18. Attach the engine lifting hooks to the existing lifting brackets. Carefully, so as not to damage any components, lift the engine out of the vehicle.

To install the engine:

19. If clutch was removed, reinstall. Carefully lower the engine into the engine compartment. On a vehicle with automatic transmission, start the converter pilot into the crankshaft. On a vehicle with a manual transmission, start the transmission main drive gear into the clutch disc. It may be necessary to adjust the position of the transmission in relation to the engine if the input shaft will not enter the clutch disc. If the engine hangs up after the shaft enters, turn the crankshaft in the clockwise direction slowly (transmission in gear), until the shaft splines mesh with the clutch disc splines.
20. Install the flywheel or converter housing upper attaching bolts. Remove the engine lifting hooks from the lifting brackets.
21. Remove the jack from under the transmission. Raise the vehicle and safely support on jackstands.
22. On a vehicle with a manual transmission, install the flywheel lower housing bolts and tighten to specifications. On a vehicle with an automatic transmission, attach the converter to the flywheel bolts and tighten to specifications. Install the converter housing-to-engine bolts and tighten to specifications.
23. Install clutch slave cylinder.
24. Install the dust cover (manual transmission). Correct the exhaust inlet pipe to the exhaust manifold.
25. Install the starter motor and connect the starter cables.
26. Lower the vehicle. Install the engine mounting nuts and tighten to 65–85 ft. lbs.
27. Connect the heater hoses to the engine. On the carbureted engines, connect the fuel line to the fuel pump. On the 2.3L EFI engine connect the 2 push connect fittings at the engine fuel rail. Connect the power brake vacuum hose.
28. Connect the alternator wire to the alternator, connect the accelerator cable to the carburetor. If so equipped, connect the transmission kickdown rod. If so equipped, install the A/C compressor to the mounting bracket.
29. Connect the coil primary wire at the coil. Connect the oil pressure and water temperature sending unit wires. Install oil fill cap.
30. Install the radiator and secure with upper support brackets. Install the fan and shroud assembly. Connect upper and lower radiator hoses.
31. Install the hood and align.
32. On the carbureted engines install the air cleaner and duct assembly. On the 2.3L EFI engine, connect the air cleaner outlet tube, the idle speed control hose, and the heat riser tube at the throttle body. Fill and bleed the cooling system.
33. Fill the crankcase with specified oil. Connect battery ground cable to engine and battery positive cable to battery.
34. Start the engine and check for leaks.

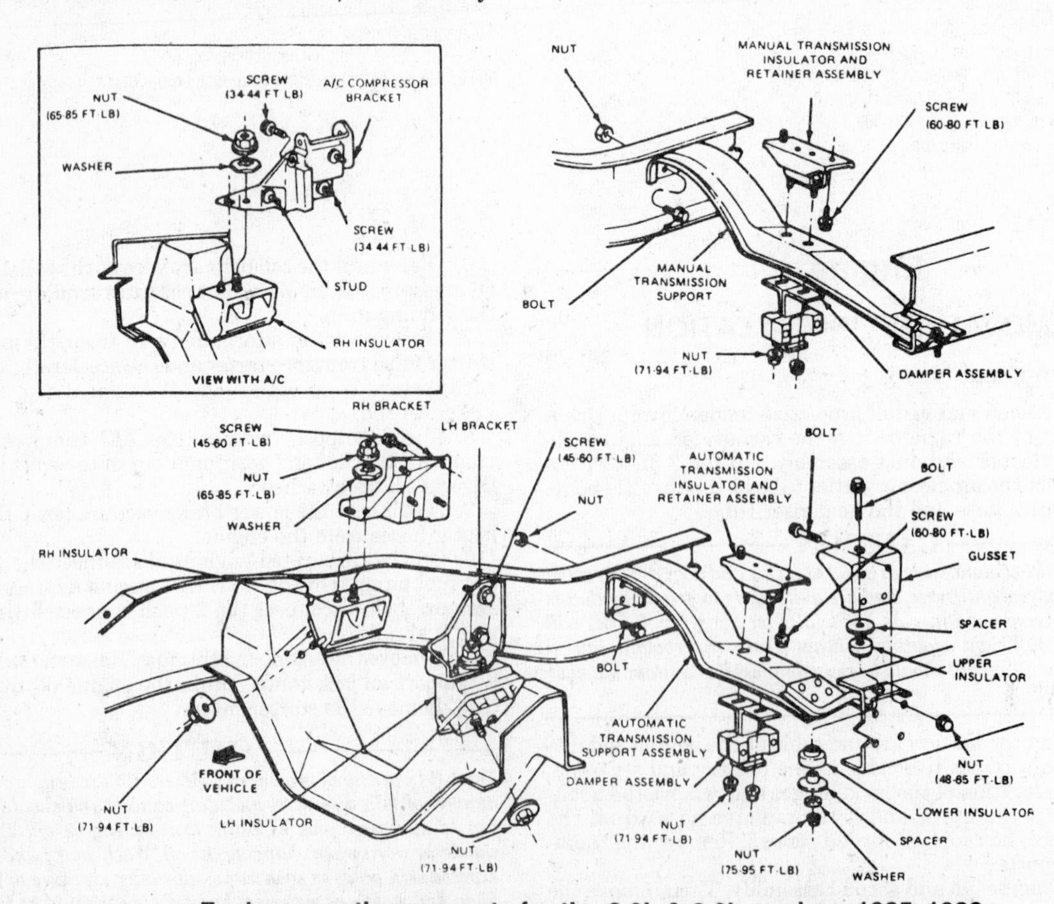

Engine mounting supports for the 2.0L & 2.3L engines 1985–1988

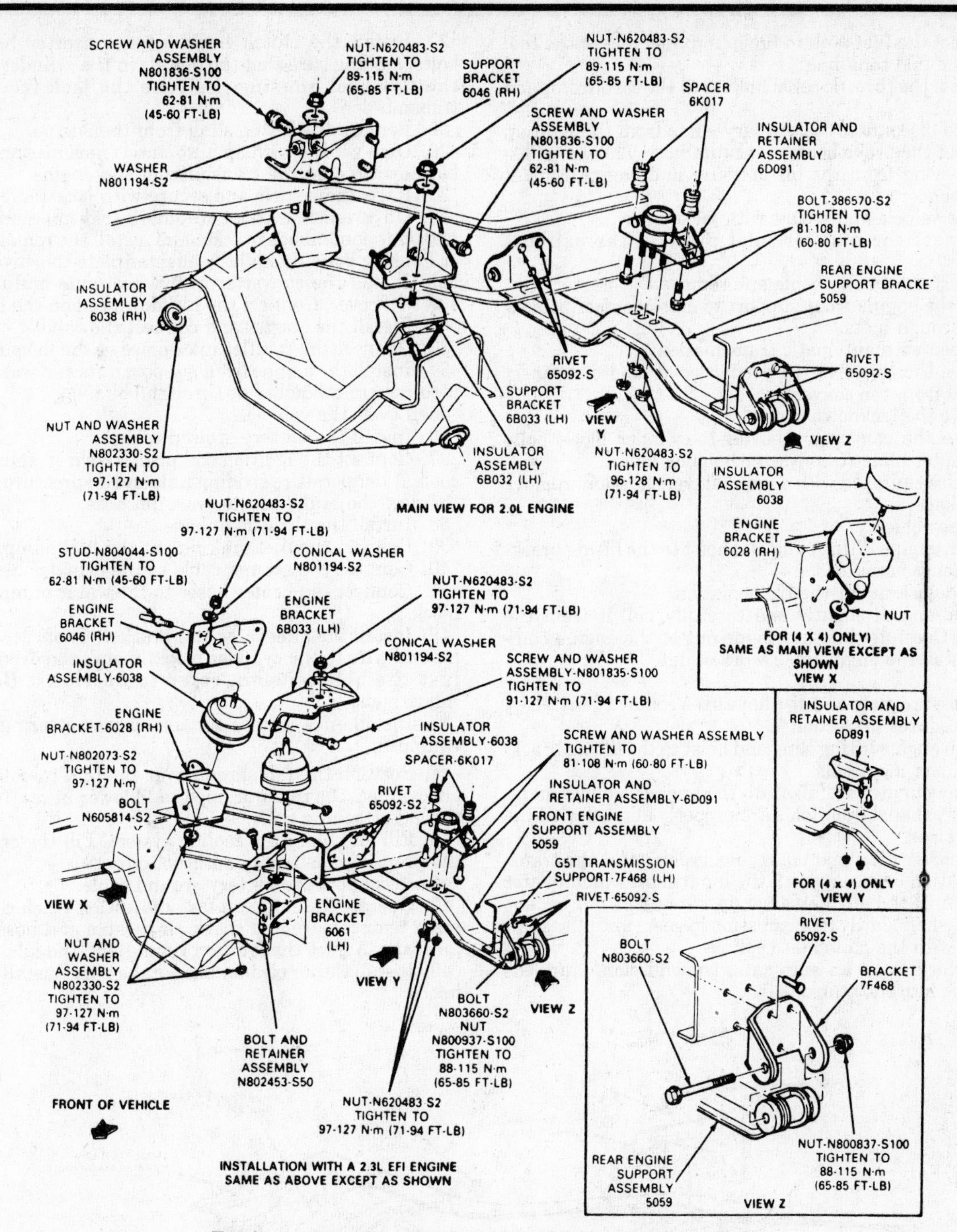

Engine mounting supports for the 2.3L EFI engine

2.8L Engine

Remove or disconnect the Thermactor system parts that will interfere with the removal or installation of the engine.

1. Disconnect the battery ground cable and drain the cooling system.

CAUTION

When draining the coolant, keep in mind that cats and dogs are attracted by the ethylene glycol antifreeze, and are quite likely to drink any that is left in an uncovered container or in puddles on the ground. This will prove fatal in sufficient quantity. Always drain the coolant into a sealable container. Coolant should be reused unless it is contaminated or several years old.

2. Remove the hood.
3. Remove the air cleaner and intake duct assembly.
4. Disconnect the radiator upper and lower hoses at the radiator.
5. Remove the fan shroud attaching bolts and position the shroud over the fan. Remove the radiator and shroud.
6. Remove the alternator and bracket. Position the alternator out of the way. Disconnect the alternator ground wire from the cylinder block.
7. Remove A/C compressor and power steering and position them out of the way, if so equipped. DO NOT disconnect the A/C refrigerant lines.
8. Disconnect the heater hoses at the block and water pump.
9. Remove the ground wires from the cylinder block.

10. Disconnect the fuel tank to fuel pump fuel line at the fuel pump. Plug the fuel tank line.
11. Disconnect the throttle cable linkage at the carburetor and intake manifold.
12. Label and disconnect the primary wires from the ignition coil. Disconnect the brake booster vacuum hose. Label and disconnect the wiring from the oil pressure and engine coolant temperature senders.
13. Raise the vehicle and secure with jackstands.
14. Disconnect the muffler inlet pipes at the exhaust manifolds.
15. Disconnect the starter cable and remove the starter.
16. Remove the engine front support to crossmember attaching nuts or through bolts.
17. If equipped with automatic transmission:
  a. Remove the converter inspection cover and disconnect the flywheel from the converter.
  b. Remove the kickdown rod.
  c. Remove the converter housing-to-cylinder block bolts and the adapter plate-to-converter housing bolt.
18. On vehicles equipped with a manual transmission, remove the clutch linkage.
19. Lower the vehicle.
20. Attach an engine lifting sling and hoist to the lifting brackets at the exhaust manifolds.
21. Position a jack under the transmission.
22. Raise the engine slightly and carefully pull it from the transmission. Carefully lift the engine out of the engine compartment. Install the engine on a work stand.

To install:

If clutch pressure plate and disc have been removed, install by following procedures in Section 7.
23. Attach an engine lifting sling and hoist to the lifting brackets at the exhaust manifolds.
24. Lower the engine carefully into the engine compartment. Make sure the exhaust manifolds are properly aligned with the muffler inlet pipes.
25. On a vehicle with a manual transmission, start the transmission in relation to the engine if the input shaft will not enter the clutch disc. If the engine hangs up after the shaft enters, turn the crankshaft slowly (transmission in gear) until the shaft splines mesh with the clutch disc splines.
26. On a vehicle with an automatic transmission, start the converter pilot into the crankshaft.
27. Install the clutch housing or converter housing upper bolts, making sure that the dowels in the cylinder block engage the flywheel housing. Remove the jack from under the transmission.
28. Remove the lifting sling from the engine.
29. On a vehicle with an automatic transmission, position the kickdown rod on the transmission and engine.
30. Raise the vehicle and secure with jackstands.
31. On a vehicle with an automatic transmission, position the transmission linkage bracket and install the remaining converter housing bolts. Install the adapter plate-to-converter housing bolt. Install the converter-to-flywheel nuts and install the inspection cover. Connect the kickdown rod on the transmission.
32. Install the starter and connect the cable.
33. Connect the muffler inlet pipes at the exhaust manifolds.
34. Install the engine front support nuts and washer attaching it to the crossmember or through bolts.
35. Lower the vehicle.
36. Install the battery ground cable.
37. Connect the ignition coil primary wires, then connect the coolant temperature sending unit and oil pressure sending unit. Connect the brake booster vacuum hose.
38. Install the throttle linkage.
39. Connect the fuel tank line at the fuel pump.
40. Connect the ground cable at the cylinder block.
41. Connect the heater hoses to the water pump and cylinder block.
42. Install the alternator and bracket. Connect the alternator ground wire to the cylinder block. Install the drive belt and adjust the belt tension, refer to Section 1 Belt Tension Adjustment.
43. Install A/C compressor and power steering pump, if so equipped.
44. Position the fan shroud over the fan. Install the radiator and connect the radiator upper and lower hoses. Install the fan shroud attaching bolts.
45. Fill and bleed the cooling system. Fill the crankcase with the proper grade and quantity of oil.
46. Reconnect the battery ground cable.
47. Operate the engine at fast idle until it reaches normal operating temperature and check all gaskets and hose connections for leaks. Adjust the ignition timing and the idle speed.
48. Install the air cleaner and intake duct. Install and align the hood.

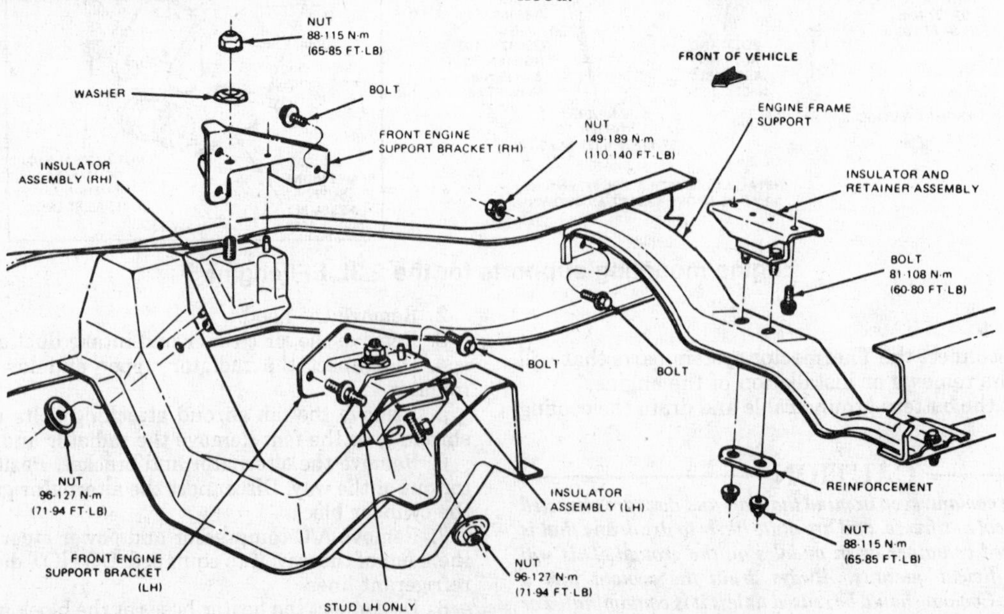

Engine mounting supports for the 2.8L engine

2.9L Engine

Remove or disconnect the Thermactor system parts that will interfere with the removal or installation of the engine.

1. Disconnect the battery ground cable and drain the cooling system.

CAUTION

When draining the coolant, keep in mind that cats and dogs are attracted by the ethylene glycol antifreeze, and are quite likely to drink any that is left in an uncovered container or in puddles on the ground. This will prove fatal in sufficient quantity. Always drain the coolant into a sealable container. Coolant should be reused unless it is contaminated or several years old.

2. Remove the hood.
3. Remove the air cleaner and intake hose.
4. Disconnect the radiator upper and lower hoses at the radiator.
5. Remove the fan shroud attaching bolts and position the shroud over the fan. Remove the radiator and shroud.
6. Remove the alternator and bracket. Position the alternator out of the way. Disconnect the alternator ground wire from the cylinder block.
7. Remove A/C compressor and power steering and position them out of the way, if so equipped. DO NOT disconnect the A/C refrigerant lines.
8. Disconnect the heater hoses at the block and water pump.
9. Remove the ground wires from the cylinder block.
10. Disconnect the fuel tank to fuel pump fuel line at the fuel pump. Plug the fuel tank line.
11. Disconnect the throttle cable shield and linkage at the throttle body and intake manifold. Disconnect the vacuum hoses from the EGR valve, the front fitting in the upper intake manifold and position them out of the way. Also, disconnect all vacuum connectors from the rear vacuum fitting in the upper intake manifold.
12. Label and disconnect the primary wires from the ignition coil. Disconnect the brake booster vacuum hose. Label and disconnect the wiring from the oil pressure and engine coolant temperature senders. Also the injector harness, air charge temperature sensor, throttle position sensor and the EGR pressure sensor.
13. Raise the vehicle and secure with jackstands.
14. Disconnect the muffler inlet pipes at the exhaust manifolds. Disconnect the starter cable and remove the starter.
15. If equipped with manual transmission, remove the clutch housing attaching bolts. Remove the hydraulic clutch hose.
16. Remove the engine front support to crossmember attaching nuts or through bolts.

17a. If equipped with automatic transmission:

a. Remove the converter inspection cover and disconnect the flywheel from the converter.

b. Remove the cable.

c. Remove the converter housing-to-cylinder block bolts and the adapter plate-to-converter housing bolt.

17b. On vehicles equipped with a manual transmission, remove the clutch linkage.

18. Lower the vehicle.
19. Attach an engine lifting sling and hoist to the lifting brackets at the exhaust manifolds.
20. Position a jack under the transmission.
21. Raise the engine slightly and carefully pull it from the transmission. Carefully lift the engine out of the engine compartment. Install the engine on a work stand.

To install:

If clutch pressure plate and disc have been removed, install by following procedures in Section 7.

22. Attach an engine lifting sling and hoist to the lifting brackets at the exhaust manifolds.
23. Lower the engine carefully into the engine compartment. Make sure the exhaust manifolds are properly aligned with the muffler inlet pipes.

On a vehicle with a manual transmission, start the transmission in relation to the engine if the input shaft will not enter the clutch disc. If the engine hangs up after the shaft enters, turn the crankshaft slowly (transmission in gear) until the shaft splines mesh with the clutch disc splines. On a vehicle with an automatic transmission, start the converter pilot into the crankshaft.

24. Install the clutch housing or converter housing upper bolts, making sure that the dowels in the cylinder block engage the flywheel housing. Remove the jack from under the transmission.
25. Remove the lifting sling from the engine.
26. On a vehicle with an automatic transmission, position the kickdown rod on the transmission and engine.
27. Raise the vehicle and secure with jackstands.
28. On a vehicle with an automatic transmission, position the transmission linkage bracket and install the remaining converter housing bolts. Install the adapter plate-to-converter housing bolt. Install the converter-to-flywheel nuts and install the inspection cover. Connect the kickdown rod on the transmission.
29. Install the starter and connect the cable.
30. Connect the muffler inlet pipes at the exhaust manifolds.
31. Install the engine front support nuts and washer attaching it to the crossmember or through bolts.
32. Lower the vehicle.
33. Install the battery ground cable.
34. Connect the ignition coil primary wires, then connect the coolant temperature sending unit and oil pressure sending unit. Connect the brake booster vacuum hose.
35. Install the throttle linkage.
36. Connect the fuel tank line at the fuel rail.
37. Connect the ground cable at the cylinder block.
38. Connect the heater hoses to the water pump and cylinder block.
39. Install the alternator and bracket. Connect the alternator ground wire to the cylinder block. Install the drive belt and adjust the belt tension, refer to Section 1 Belt Tension Adjustment.
40. Install A/C compressor and power steering pump, if so equipped.
41. Position the fan shroud over the fan. Install the radiator and connect the radiator upper and lower hoses. Install the fan shroud attaching bolts.
42. Fill and bleed the cooling system. Fill the crankcase with the proper grade and quantity of oil.
43. Reconnect the battery ground cable.
44. Operate the engine at fast idle until it reaches normal operating temperature and check all gaskets and hose connections for leaks. Adjust the ignition timing and the idle speed.
45. Install the intake hose. Install and align the hood.

3.0L and 4.0L Engines

1. Disconnect the negative battery cable. Drain the cooling system. Relieve fuel system pressure.

CAUTION

When draining the coolant, keep in mind that cats and dogs are attracted by the ethylene glycol antifreeze, and are quite likely to drink any that is left in an uncovered container or in puddles on the ground. This will prove fatal in sufficient quantity. Always drain the coolant into a sealable container. Coolant should be reused unless it is contaminated or several years old.

2. Remove the hood (matchmark hinges for correct installation).
3. Remove the air cleaner and intake hose.
4. Disconnect the radiator upper and lower hoses at the radiator.

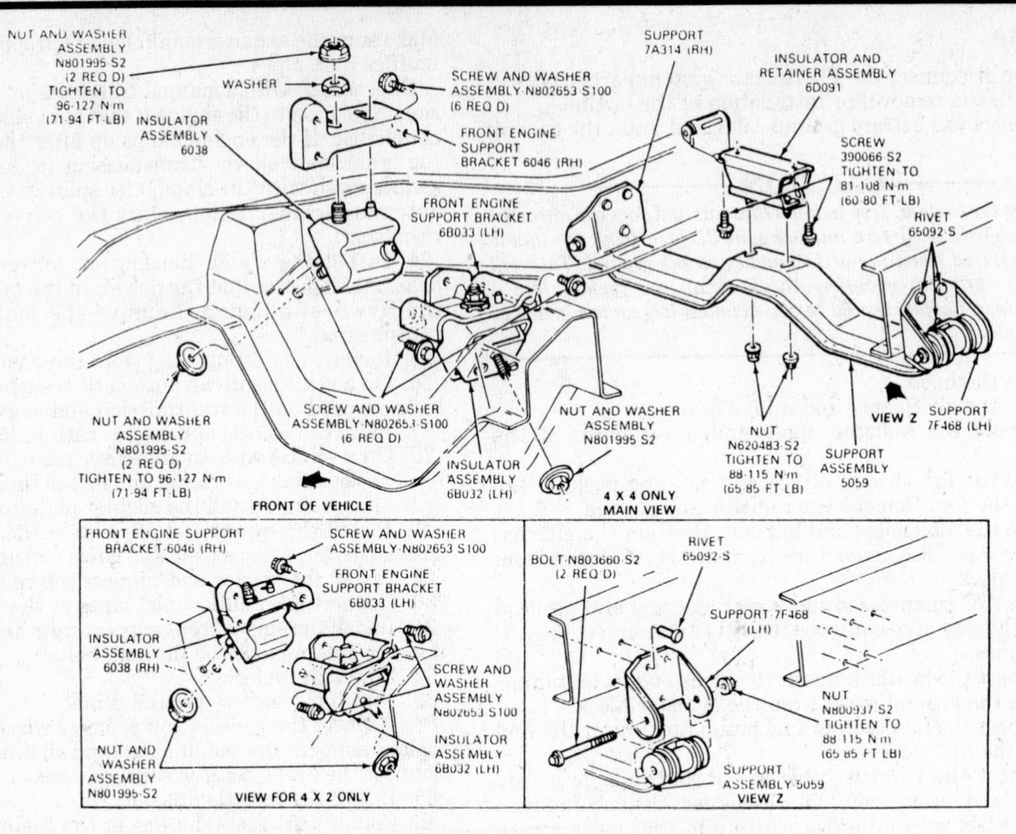

Engine mounting supports for the 2.9L engine

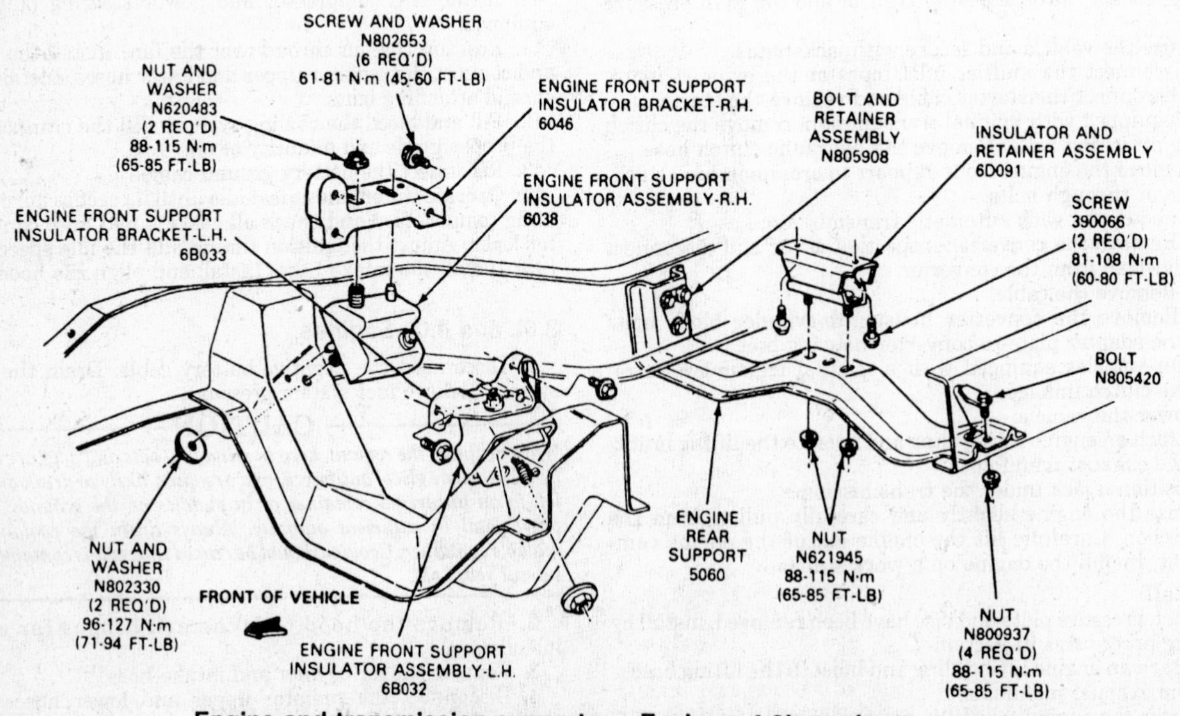

Engine and transmission supports – Explorer 4.0L engine

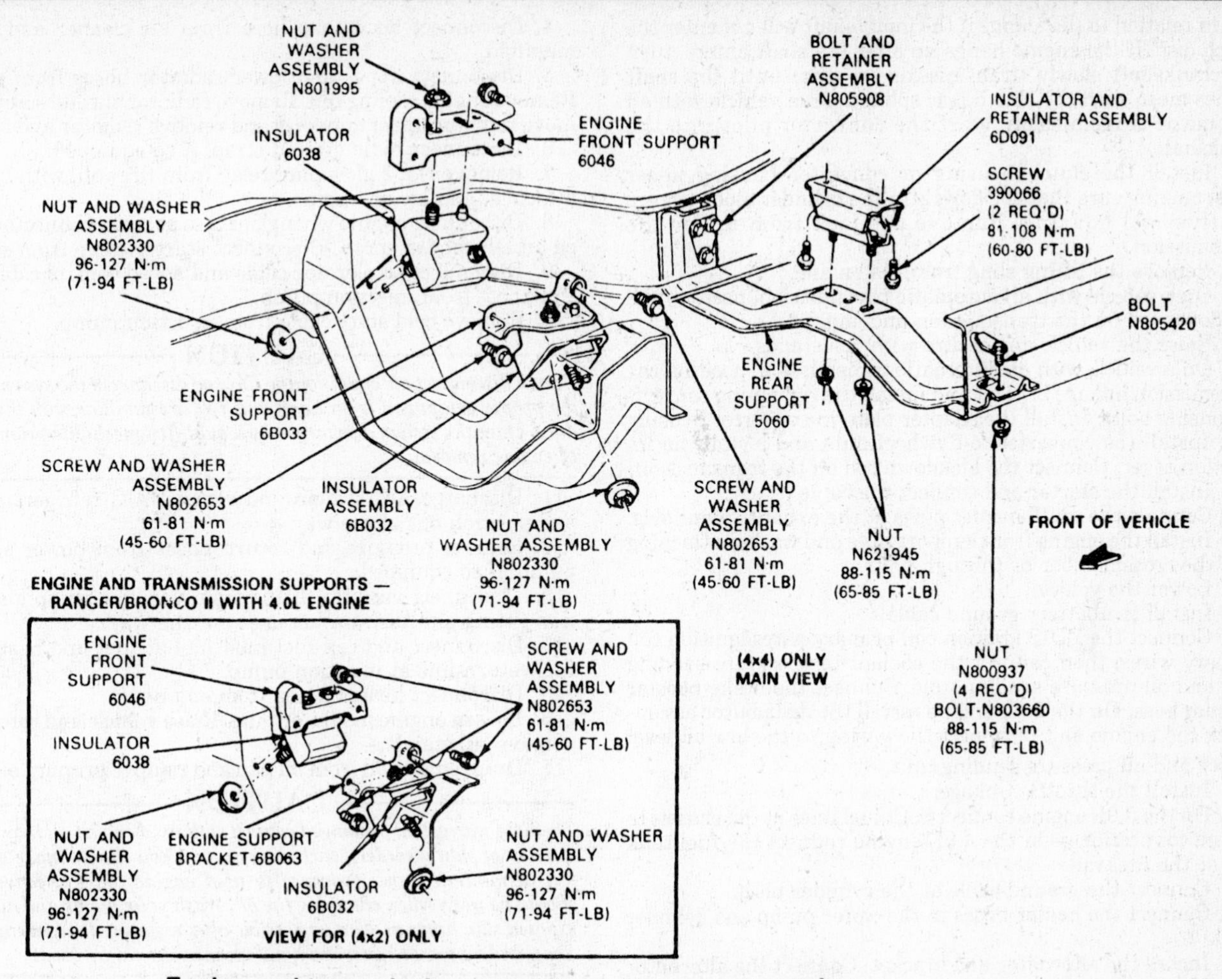

Engine and transmission supports — Ranger 4.0L engine

5. Remove the fan shroud attaching bolts and position the shroud over the fan. Remove the radiator and shroud.
6. Remove the alternator and bracket. Position the alternator out of the way. Disconnect the alternator ground wire from the cylinder block.
7. Remove A/C compressor and power steering and position them out of the way, if so equipped. DO NOT disconnect the A/C refrigerant lines.
8. Disconnect the heater hoses at the block and water pump.
9. Remove the ground wires from the cylinder block.
10. On the 3.0L engine disconnect both fuel lines at the chassis to engine connections. On 4.0L engine disconnect the fuel tank to fuel rail lines at the fuel rail.
11. Disconnect the throttle cable shield and linkage at the throttle body and intake manifold. Disconnect all vacuum connections from the rear vacuum fitting in upper intake manifold.
12. Label and disconnect the primary wires from the EDIS ignition coil/primary wires from the ignition coil. Disconnect the brake booster vacuum hose. Label and disconnect the wiring from the oil pressure and engine coolant temperature senders. Also the injector harness, air charge temperature sensor, throttle position sensor. On the 3.0L engine remove the distributor assembly (mark assembly before removal) from the engine and disconnect the wiring from the low oil level sensor and oil pressure sending unit.
13. Raise the vehicle and secure with jackstands.
14. Disconnect the muffler inlet pipes at the exhaust manifolds. Disconnect the starter cable and remove the starter.
15. If equipped with manual transmission, remove the clutch housing attaching bolts. Remove the hydraulic clutch hose.
16. Remove the engine front support to crossmember attaching nuts or through bolts.
17a. If equipped with automatic transmission:
 a. Remove the converter inspection cover and disconnect the flywheel from the converter.
 b. Remove the cable.
 c. Remove the converter housing-to-cylinder block bolts and the adapter plate-to-converter housing bolt.
17b. On vehicles equipped with a manual transmission, remove the clutch linkage.
18. Lower the vehicle.
19. Attach an engine lifting sling and hoist to the lifting brackets at the exhaust manifolds.
20. Position a jack under the transmission.
21. Raise the engine slightly and carefully pull it from the transmission. Carefully lift the engine out of the engine compartment. Install the engine on a work stand.

To install:

If clutch pressure plate and disc have been removed, install by following procedures in Section 7.
22. Attach an engine lifting sling and hoist to the lifting brackets at the exhaust manifolds.
23. Lower the engine carefully into the engine compartment. Make sure the exhaust manifolds are properly aligned with the muffler inlet pipes.

On a vehicle with a manual transmission, start the transmission in relation to the engine if the input shaft will not enter the clutch disc. If the engine hangs up after the shaft enters, turn the crankshaft slowly (transmission in gear) until the shaft splines mesh with the clutch disc splines. On a vehicle with an automatic transmission, start the converter pilot into the crankshaft.

24. Install the clutch housing or converter housing upper bolts, making sure that the dowels in the cylinder block engage the flywheel housing. Remove the jack from under the transmission.
25. Remove the lifting sling from the engine.
26. On a vehicle with an automatic transmission, position the kickdown rod on the transmission and engine.
27. Raise the vehicle and secure with jackstands.
28. On a vehicle with an automatic transmission, position the transmission linkage bracket and install the remaining converter housing bolts. Install the adapter plate-to-converter housing bolt. Install the converter-to-flywheel nuts and install the inspection cover. Connect the kickdown rod on the transmission.
29. Install the starter and connect the cable.
30. Connect the muffler inlet pipes at the exhaust manifolds.
31. Install the engine front support nuts and washer attaching it to the crossmember or through bolts.
32. Lower the vehicle.
33. Install the battery ground cable.
34. Connect the EDIS ignition coil primary wires/ignition coil primary wires, then connect the coolant temperature sending unit and oil pressure sending unit. Connect the brake booster vacuum hose. On the 3.0L engine install the distributor assembly to the engine and reconnect the wiring to the low oil level sensor and oil pressure sending unit.
35. Install the throttle linkage.
36. On the 3.0L engine connect both fuel lines at the chassis to engine connections. On the 4.0L engine connect the fuel tank line at the fuel rail.
37. Connect the ground cable at the cylinder block.
38. Connect the heater hoses to the water pump and cylinder block.
39. Install the alternator and bracket. Connect the alternator ground wire to the cylinder block. Install the drive belt and adjust the belt tension, refer to Section 1 Belt Tension Adjustment.
40. Install A/C compressor and power steering pump, if so equipped.
41. Position the fan shroud over the fan. Install the radiator and connect the radiator upper and lower hoses. Install the fan shroud attaching bolts.
42. Fill and bleed the cooling system. Fill the crankcase with the proper grade and quantity of oil.
43. Reconnect the battery ground cable.
44. Operate the engine at fast idle until it reaches normal operating temperature and check all gaskets and hose connections for leaks. On the 3.0L engine adjust ignition timing and idle speed.
45. Install the intake hose. Install and adjust the hood.

2.2L Diesel Engine

1. Open hood and install protective fender covers. Mark location of hood hinges and remove hood.
2. Disconnect battery ground cables from both batteries. Disconnect battery ground cables at engine.
3. Drain coolant from radiator.

CAUTION

When draining the coolant, keep in mind that cats and dogs are attracted by the ethylene glycol antifreeze, and are quite likely to drink any that is left in an uncovered container or in puddles on the ground. This will prove fatal in sufficient quantity. Always drain the coolant into a sealable container. Coolant should be reused unless it is contaminated or several years old.

4. Disconnect air intake hose from air cleaner and intake manifold.
5. Disconnect upper and lower radiator hoses from engine. Remove engine cooling fan. Remove radiator shroud screws. Remove radiator upper supports and remove radiator and shroud.
6. Disconnect radio ground strap, if so equipped.
7. Remove No. 2 glow plug relay from firewall, with harness attached, and lay on engine.
8. Disconnect engine wiring harness at main connector located on left fender apron. Disconnect starter cable from starter.
9. Disconnect accelerator cable and speed control cable, if so equipped, from injection pump.
10. Remove cold start cable from injection pump.

CAUTION

Do not disconnect air conditioning lines or discharger the system unless the proper equipment is on hand and you are familiar with the procedure. Have the system discharged by a qualified mechanic prior to start of engine removal.

11. Discharge A/C system and remove A/C refrigerant lines and position out of the way.
12. Remove pressure and return hoses from power steering pump, if so equipped.
13. Disconnect vacuum fitting from vacuum pump and position fitting and vacuum hoses out of the way.
14. Disconnect and cap fuel inlet line at fuel line heater and fuel return line at injection pump.
15. Disconnect heater hoses from engine.
16. Loosen engine insulator nuts. Raise vehicle and safely support on jackstands.
17. Drain engine oil from oil pan and remove primary oil filter.

CAUTION

The EPA warns that prolonged contact with used engine oil may cause a number of skin disorders, including cancer! You should make every effort to minimize your exposure to used engine oil. Protective gloves should be worn when changing the oil. Wash your hands and any other exposed skin areas as soon as possible after exposure to used engine oil. Soap and water, or waterless hand cleaner should be used.

18. Disconnect oil pressure sender hose from oil filter mounting adapter.
19. Disconnect muffler inlet pipe at exhaust manifold.
20. Remove bottom engine insulator nuts. Remove transmission bolts. Lower vehicle. Attach engine lifting sling and chain hoist.
21. Carefully lift engine out of vehicle to avoid damage to components.
22. Install engine on work stand, if necessary.

To install:

23. When installing the engine; Carefully lower engine into engine compartment to avoid damage to components.
24. Install two top transmission-to-engine attaching bolts. Remove engine lifting sling.
25. Raise vehicle and safely support on jackstands.
26. Install engine insulator nuts and tighten to specification.
27. Install remaining transmission-to-engine attaching bolts and tighten all bolts to specification.
28. Connect muffler inlet pipe to exhaust manifold and tighten to specification.
29. Install oil pressure sender hose and install new oil filter as described in Section 1.
30. Lower vehicle.
31. Tighten upper engine insulator nuts to specification.
32. Connect heater hoses to engine. Connect fuel inlet line to fuel line heater and fuel return line to injection pump. Connect vacuum fitting and hoses to vacuum pump. Connect pressure and return hoses to power steering pump, if so equipped. Check and add power steering fluid.
33. Install A/C refrigerant lines and charge system, if so equipped.

NOTE: System can be charged after engine installation is complete.

34. Install A/C drive belt, and tighten to specification.
35. Connect cold start cable to injection pump. Connect accelerator cable and speed control cable, if so equipped, to injection pump.

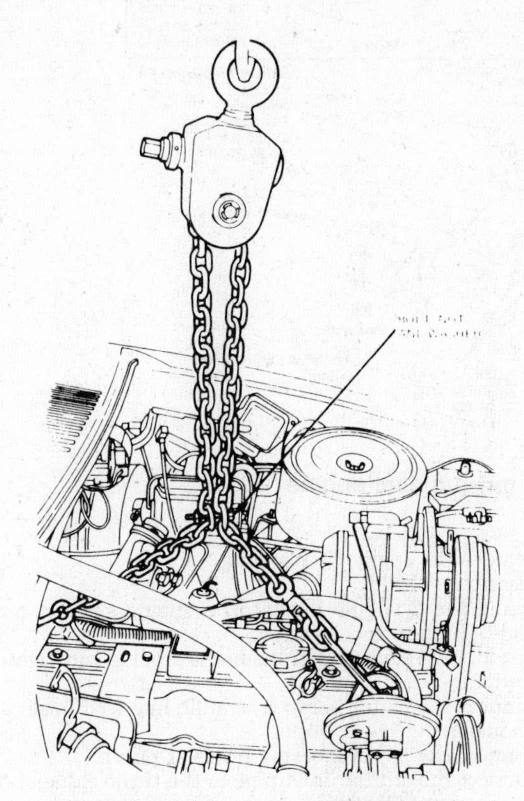

Removing the 2.2L diesel engine

36. Connect engine wiring harness to main wiring harness at left fender apron. Connect radio ground strap, if so equipped.
37. Position radiator in vehicle, install radiator upper support brackets and tighten to specification. Install radiator fan shroud and tighten to specification. Install radiator fan and tighten to specification.
38. Connect upper and lower radiator hoses to engine and tighten clamps to specification. Connect air intake hose to air cleaner and intake manifold.
39. Fill and bleed cooling system.
40. Fill crankcase with specified quantity and quality of oil.
41. Connect battery ground cables to engine. Connect battery ground cables to both batteries.
42. Run engine and check for oil, fuel and coolant leaks. Close hood.

2.3L Diesel Engine

1. Mark the location of the hood hinges and remove the hood.
2. Disconnect the battery ground cables from both batteries.
3. Disconnect the battery ground cables at engine.
4. Drain the engine coolant.

CAUTION

When draining the coolant, keep in mind that cats and dogs are attracted by the ethylene glycol antifreeze, and are quite likely to drink any that is left in an uncovered container or in puddles on the ground. This will prove fatal in sufficient quantity. Always drain the coolant into a sealable container. Coolant should be reused unless it is contaminated or several years old.

5. Remove the crankcase breather hose at the rocker cover.
6. Remove the intake hose between the air cleaner and turbocharger.

NOTE: Cap the turbocharger inlet.

7. Remove the A/C compressor and position out of the way.

NOTE: It is not necessary to disconnect A/C hoses from compressor.

8. Disconnect the heater hoses from heater core inlet and outlet.

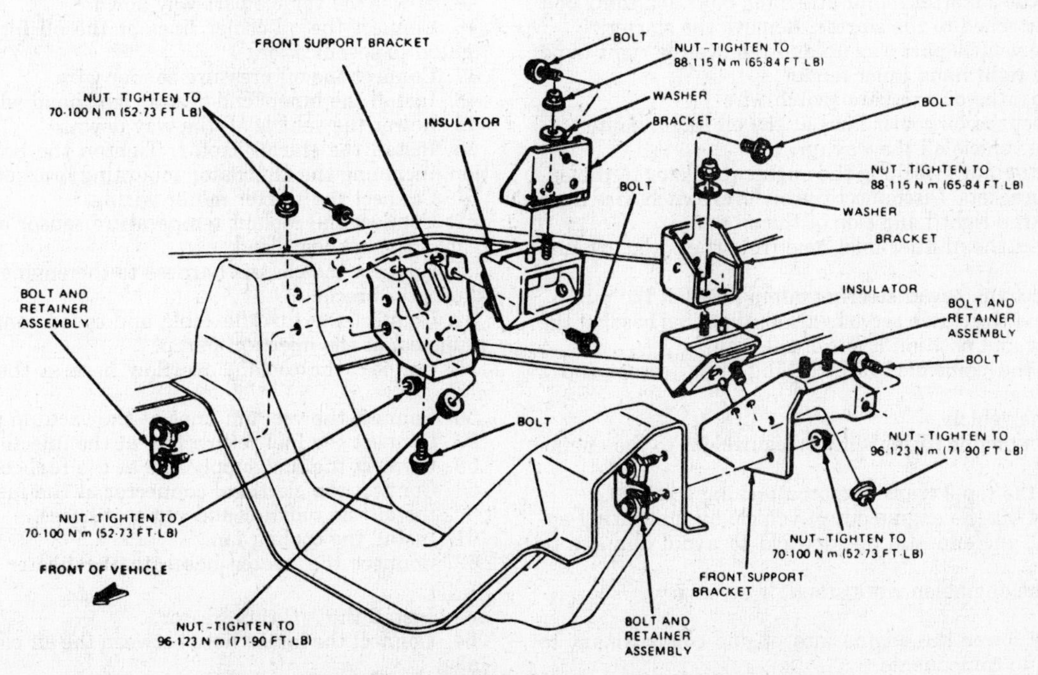

Engine mounting supports for the 2.2L diesel engine

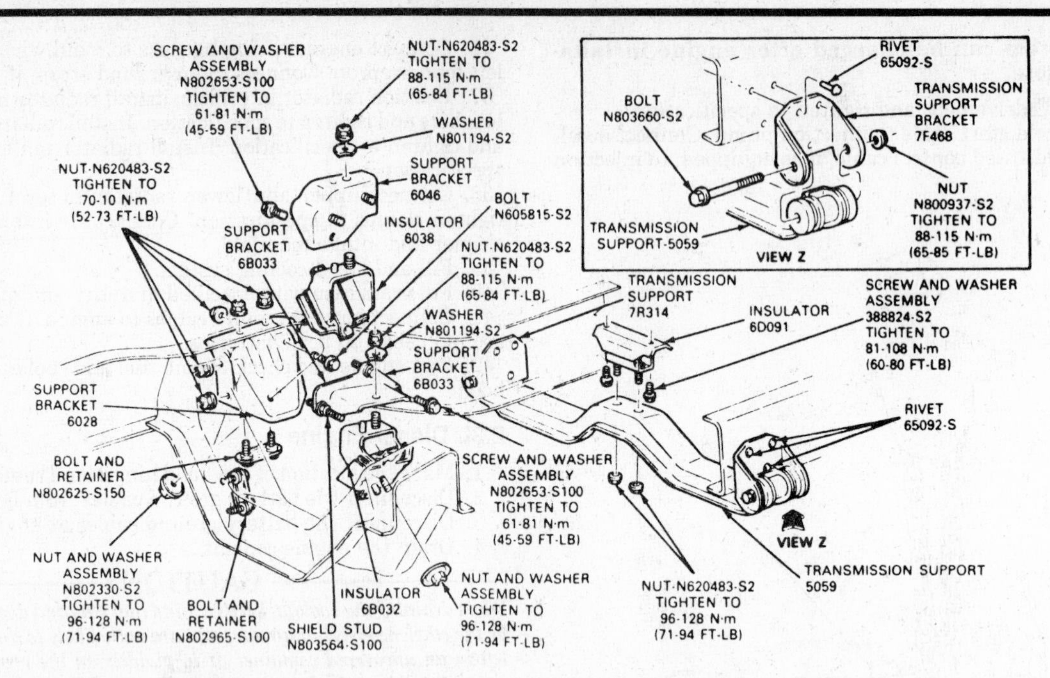

Engine mounting supports for the 2.3L turbo diesel engine

9. Remove the cooling fan.
10. Disconnect the radiator hoses and remove the radiator.
11. Disconnect the electrical connector at the fuel conditioner.
12. Disconnect the fuel supply line at the fuel conditioner.
13. Disconnect the fuel return line at the injection pump.
14. Disconnect the vacuum lines at the vacuum pump fitting.
15. Disconnect the coolant overflow hose from the filler neck.
16. Disconnect the throttle cable at the injection pump.
17. Disconnect the engine harness from the chassis harness at the alternator bracket.
18. Disconnect the wires from the glow plug bus bar.
19. Disconnect the starter motor wiring.
20. Remove the 2 starter motor attaching bolts and the 1 bolt at the brace attached to the starter. Remove the starter.
21. Raise the vehicle part way up and remove the right hand wheel and the right hand inner fender.
22. Disconnect the oil pressure switch wire.
23. Disconnect the oil cooler lines at the oil filter adapter.
24. Raise the vehicle all the way up.
25. Remove the nut attaching the engine insulator to the engine support brackets. Disconnect the oil level switch wire from the switch on the right hand side of the oil pan.
26. Disconnect the muffler inlet pipe from the turbo exhaust outlet pipe.
27. Disconnect the power steering pump hoses at the pump.
28. Disconnect the clutch servo hydraulic line (red hose) at the clutch housing and position it out of the way.
29. Remove the transmission attaching bolts (except top 2 bolts).
30. Lower the vehicle.
31. Attach an engine lifting sling or equivalent to the engine lifting brackets.
32. Remove the top 2 transmission attaching bolts.
33. Carefully lift the engine out of vehicle by first lifting approximately 3 in. and sliding forward to avoid damage to components.
34. Install the engine on work stand, if necessary.

To install:

35. Carefully lower the engine into engine compartment to avoid damage to components.
36. Install the 2 top transmission bolts.
37. Remove the engine lifting sling.
38. Raise the vehicle.
39. Install the 2 engine insulator-to-engine support bracket bolts and tighten to specification.
40. Install the remaining transmission bolts and tighten to specification.
41. Connect the clutch servo hydraulic line (red hose) at the clutch housing.
42. Connect the power steering hoses at pump.
43. Connect the muffler inlet pipe to the turbo exhaust outlet and tighten to specification.
44. Connect the oil level switch wire.
45. Lower the vehicle part way down.
46. Connect the oil cooler lines at the oil filter adapter and tighten to specification.
47. Connect the oil pressure sender wire.
48. Install the inner fender and right hand wheel assembly.
49. Lower the vehicle all the way down.
50. Install the starter motor. Tighten the bolts to specification, including the alternator mounting bracket brace.
51. Connect the starter motor wiring.
52. Connect the coolant temperature sensor at the left hand rear of the cylinder head.
53. Connect the chassis harness to the engine harness at the alternator bracket.
54. Connect the throttle cable and speed control cable, if so equipped, at the injection pump.
55. Connect the coolant overflow hose at the radiator filler neck.
56. Connect the vacuum lines at the vacuum pump fitting.
57. Connect the fuel return line at the injection pump.
58. Connect the fuel supply line at the fuel conditioner.
59. Connect the electrical connector at the fuel conditioner.
60. Install the radiator and radiator hoses.
61. Install the cooling fan.
62. Connect the heater hoses to the heater core inlet and outlet.
63. Install the A/C compressor.
64. Connect the intake hose between the air cleaner and turbo inlet.
65. Connect the breather hose to rocker cover.

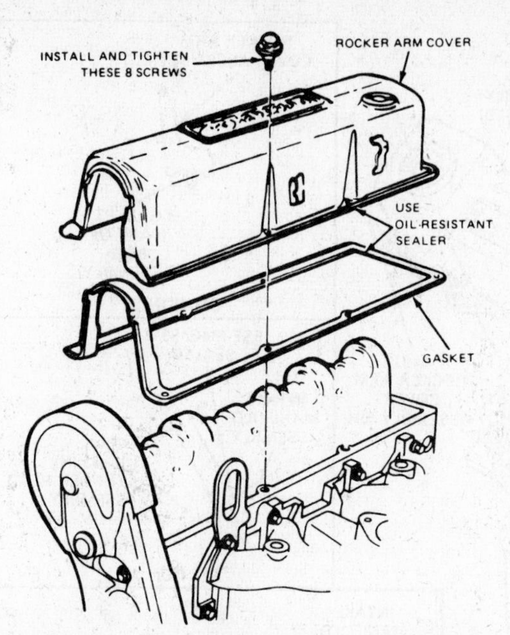

Rocker arm cover installation – 2.0L, 2.3L carbureted engines

66. Fill and bleed the cooling system.
67. Fill the engine with specified quantity and quality of oil.
68. Connect the battery ground cables to engine.
69. Start the engine and check for oil, fuel and coolant leaks.
70. Stop the engine.
Check oil and coolant levels. Add if necessary.
71. Fill and bleed the power steering pump, if necessary.
72. Install the hood.

Valve Rocker Arm Cover

REMOVAL AND INSTALLATION

2.0L, 2.3L, 2.8L and 2.9L Engines

NOTE: To service the valve rocker cover on the 2.3L EFI engine the throttle body assembly and EGR supply tube must first be removed. Refer to the necessary service procedures.

1. Disconnect the negative battery cable. Remove the air cleaner and attaching parts. Label each spark plug wire prior to its removal in order to ease the installation of the wires on the correct spark plugs.
2. Remove the spark plug wires.
3. Remove the PCV valve and hose.
4. Remove the carburetor choke air deflector plate (shield).
5. Remove the rocker arm cover attaching screws and the load distribution washers (patch pieces). Be sure the washers are installed in their original position.
6. Remove the transmission fluid level indicator tube and bracket, which is attached to rocker arm cover.
7. Disconnect the kickdown linkage and the carburetor (automatic transmission only).
8. Position the thermactor air hose and the wiring harness away from the right hand rocker arm cover.
9. Remove the engine oil fill cap.
10. Disconnect the vacuum line at the canister purge solenoid and disconnect the line routed from the canister to the purge solenoid (disconnect the power brake booster hose, if so equipped).
11. With a light plastic hammer, tap the rocker arm covers to break the seal.

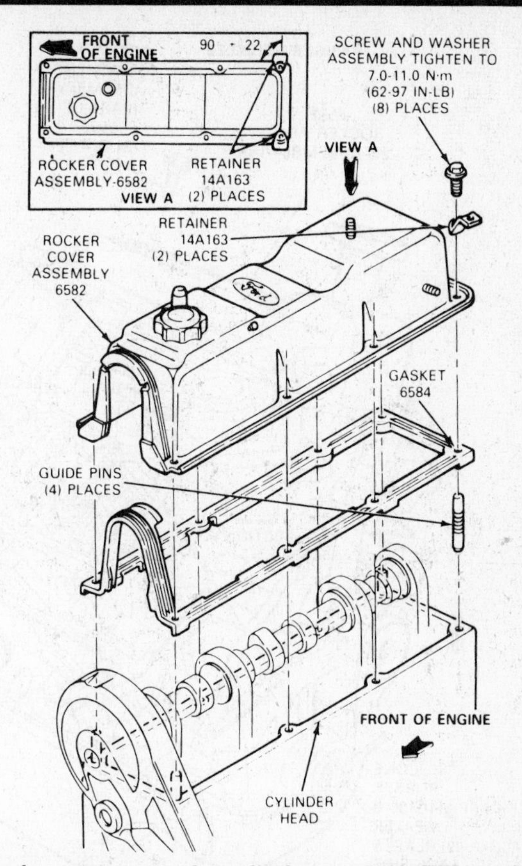

Rocker arm cover installation – 2.3L EFI engine

12. Remove the rocker arm covers.

To install:

13. Clean all gasket material from the cylinder heads and rocker arm cover gasket surfaces.
14. Install the rocker arm covers, using new gaskets and install the attaching screw and rocker arm cover reinforcement pieces.
15. Install the transmission fluid level indicator tube and the bracket (attaches to rocker arm cover).
16. Connect the kickdown linkage (automatic transmission only).
17. After ensuring all rocker arm cover reinforcement washers are installed in their original position, tighten the rocker arm cover screws.
18. Install the spark plug. wires.
19. Install the PCV valve and hose.
20. Install the carburetor choke air deflector plate (shield).
21. Reposition the thermactor air hose and the wiring harness in their original places.
22. Install the engine oil fill cap.
23. Connect the vacuum line at the canister purge (connect power brake hose, if so equipped) solenoid and connect the line routed from canister to the purge solenoid.
24. Install the air cleaner and the attaching parts.
25. Reconnect the negative battery cable. Start the engine and check for oil leaks.

3.0L Engine

NOTE: The rocker covers installed on the 3.0L engine incorporate integral (built in) gaskets which should last the life of the vehicle. Replacement gaskets are available if required.

1. Disconnect the negative battery cable. Disconnect the igni-

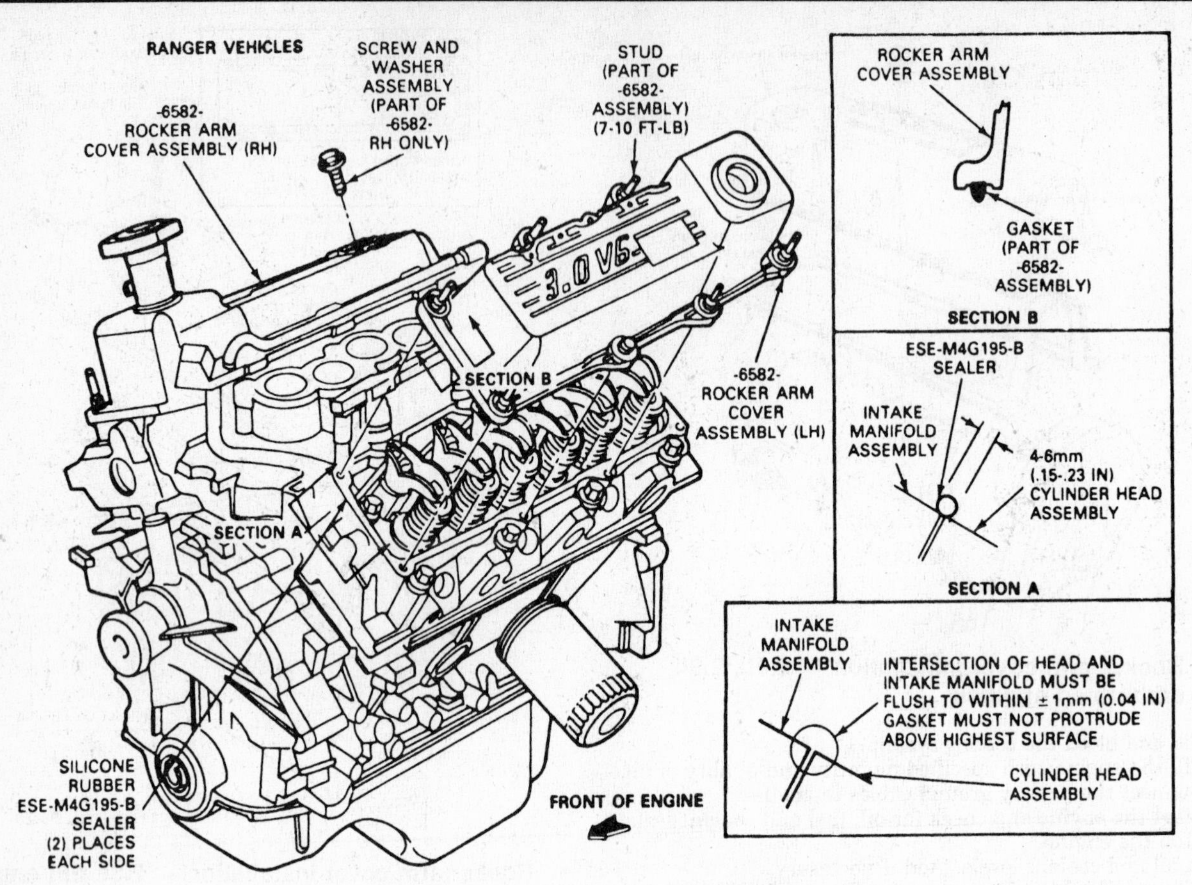

Rocker arm installation – 3.0L engine

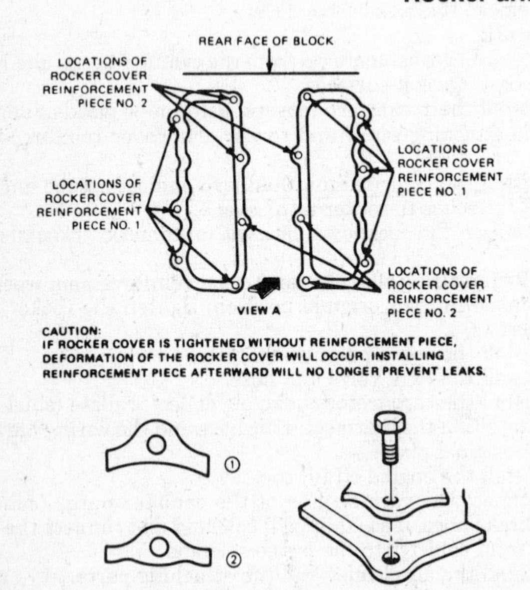

Rocker arm cover reinforcement washer locations – 2.8L and 2.9L engines

tion wires from the spark plugs, but leave them attached to their wire looms.

2. Remove the ignition wire separators from the rocker arm cover attaching bolt studs with the wires attached, then lay the wires out of the way.

3. If the left hand cover is being removed, remove the throttle body assembly remove the PCV valve and fuel injector harness stand-offs. If the right hand cover is being removed, remove the engine harness connectors, fuel injector harness stand-offs and air cleaner closure hose from oil fill adapter.

4. Using caution, slide a sharp thin blade knife between cylinder head gasket surface and rocker cover gasket at four RTV junctions. CUT ONLY THE RTV SEALER AND AVOID CUTTING INTEGRAL GASKET.

5. Remove the integral gasket from the rocker cover gasket channel. Note bolt/stud fasteners locations before removing gasket for correct installation. Clean gasket channel and remove any traces of RTV sealant.

6. Align fastener holes, lay new gasket onto channel and install by hand. Install gasket to each fastener, seat fastener against cover and at the same time roll gasket around fastener collar. If installeed correctly all fasteners will be be secured by gasket and not fall out.

7. To install valve cover to the engine, lightly oil all bolts and stud threads. Apply a bead of RTV sealant at the cylinder head to intake manifold rail step (two places per rail).

8. Place the rocker cover on the cylinder head and instal attaching bolts and studs. Tighten the attaching bolts to 9 ft. lbs.

9. Install all remaining components in reverse order of removal procedure. Connect the ignition wires to the spark plugs and reconnect the negative battery cable. Start the engine and run to normal operating temperature and check for oil and vacuum leaks.

4.0L Engine

NOTE: Failure to install new rocker cover gaskets and rocker cover reinforcement pieces will result in oil leaks.

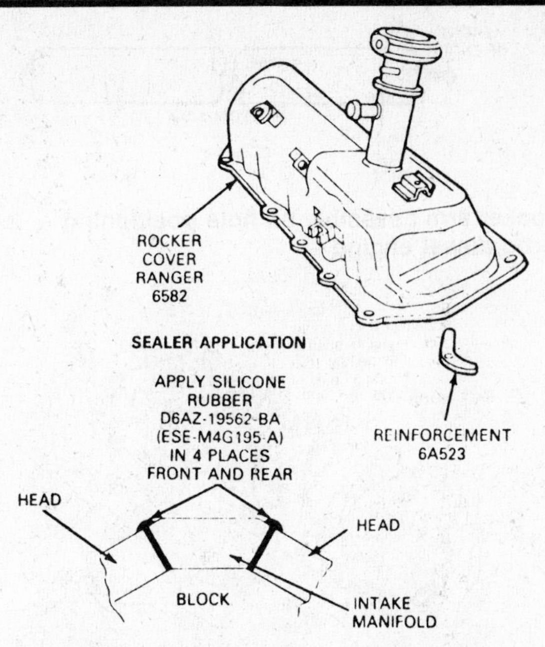

Rocker arm installation – 4.0L engine

1. Disconnect the negative battery cable. Tag and remove the spark plug wires.
2. Disconnect and remove the fuel supply and return lines. See Section 5.
3. For left rocker cover removal, remove the upper intake manifold.
4. For right rocker cover removal, remove air inlet duct and hose to oil fill tube, drive belt from alternator, alternator. Drain cooling system remove the upper radiator hose from the engine. Remove the EDIS ignition coil and bracket assembly. Remove the A/C low pressure hose bracket if so equipped. Remove the PCV valve hose and breather.

CAUTION

When draining the coolant, keep in mind that cats and dogs are attracted by the ethylene glycol antifreeze, and are quite likely to drink any that is left in an uncovered container or in puddles on the ground. This will prove fatal in sufficient quantity. Always drain the coolant into a sealable container. Coolant should be reused unless it is contaminated or several years old.

5. Remove the rocker cover bolts and load distribution pieces. The washers must be installed in their original positions, so keep track of them.
6. Remove the rocker cover. It will probably be necessary to tap the cover loose with a plastic or rubber mallet.
7. Remove the rocker covers.
8. Clean all gasket material from the cover and head.
9. Installation is the reverse of removal. Always use a new gasket coated with sealer. If any of the RTV silicone gasket material was removed from the mating area of the head(s) and intake manifold, replace it. Torque the bolts to 3–5 ft. lbs.
10. Reconnect the negative battery cable. Start the engine and run to normal operating temperature and check for oil and fuel leaks.

Rocker Arms

REMOVAL AND INSTALLATION

2.0L, 2.3L Engines

NOTE: A special tool is required to compress the lash adjuster.

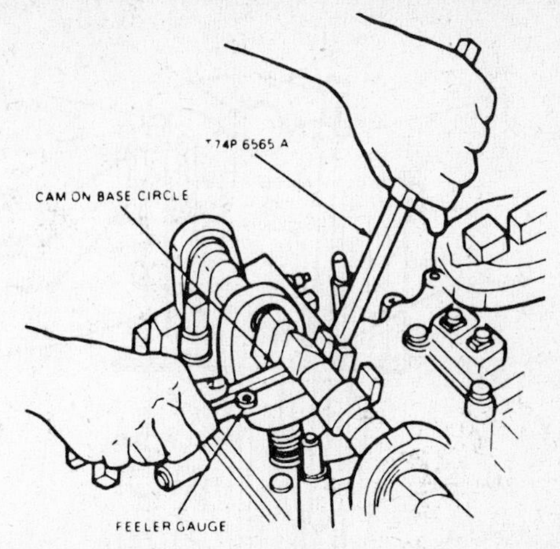

Using the special tool to collapse the lash adjuster

1. Remove the valve cover and associated parts as required.
2. Rotate the camshaft so that the base circle of the cam is against the cam follower you intend to remove.
3. Remove the retaining spring from the cam follower, if so equipped.
4. Using special tool T74P–6565–B or a valve spring compressor tool, collapse the lash adjuster and/or depress the valve spring, as necessary, and slide the cam follower over the lash adjuster and out from under the camshaft.
5. Install the cam follower in the reverse order of removal. Make sure that the lash adjuster is collapsed and released before rotating the camshaft.

Rocker Arm Shaft Assembly

REMOVAL AND INSTALLATION

2.2L Diesel Engine

1. Remove the valve cover.
2. Remove the bolts attaching the rocker shaft to the cylinder head and remove the rocker shaft.
3. Remove the push rods, if necessary.

NOTE: Note position of push rods so they may be returned to their original positions.

To install:

4. Install the push rods, if removed.

NOTE: Ball end goes toward tappet.

5. Position the rocker shaft on the cylinder head and install the retaining bolts.

NOTE: Before tightening rocker shaft bolts, make sure the rocker arms are fully seated in the push rod cups.

6. Tighten the rocker shaft bolts alternately, two to three turns at a time, working from the center to the ends until all rocker shaft brackets are seated.
7. Adjust valve rocker arms as described in Section 2 under Valve Lash Adjustment.
8. Install the valve cover.
9. Connect the heater tube assembly to valve cover.
10. Run the engine and check for oil leaks.

2.3L Diesel Engine

1. Remove the valve cover.

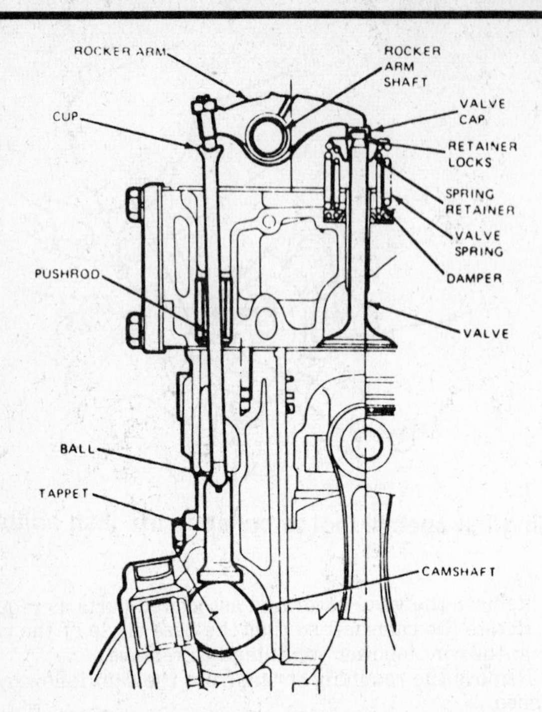

2.2L diesel engine valve train assembly

2. Remove the bolts attaching the rocker shaft to the cylinder head, 1 turn at a time from the front to the rear and remove the rocker shaft.

To install:

3. Position the rocker shaft on the cylinder head.

NOTE: Be certain the end of the rocker shaft with the single oil hole is toward the front of the engine. Also, ensure the rocker shaft is installed with the oil holes DOWN.

4. Install the rocker shaft retaining bolts and tighten one turn at a time from front to rear, repeating the sequence until all the bolts are seated. Tighten the bolts to 25–28 ft. lbs.
5. Adjust valve rocker arms as described in Section 2 under Valve Lash Adjustment.
6. Install the valve cover.
7. Run the engine and check for oil leaks.

2.8L, 2.9L Engines

1. Remove the valve rocker arm covers following the procedure given above.
2. Remove the rocker arm shaft stand attaching bolts by loosening the bolts two turns at a time, in sequence (from the end of shaft to middle shaft).
3. Lift off the rocker arm and shaft assembly and the oil baffle.

To install:

4. Loosen the valve lash adjusting screws a few turns. Apply the engine oil to the assembly to provide the initial lubrication.
5. Install the oil baffle and rocker arm shaft assembly to the cylinder head and guide adjusting screws on to the push rods.
6. Install and tighten rocker arm stand attaching bolts to 43–50 ft. lbs., two turns at a time, in sequence (from middle of shaft to the end of shaft). Adjust valve lash to cold specified setting.
7. Adjust the valve lash to the cold specified setting. Refer to Section 2 under Valve Lash Adjustment for adjustment procedures.
8. Install the valve rocker arm covers following the procedure given above.

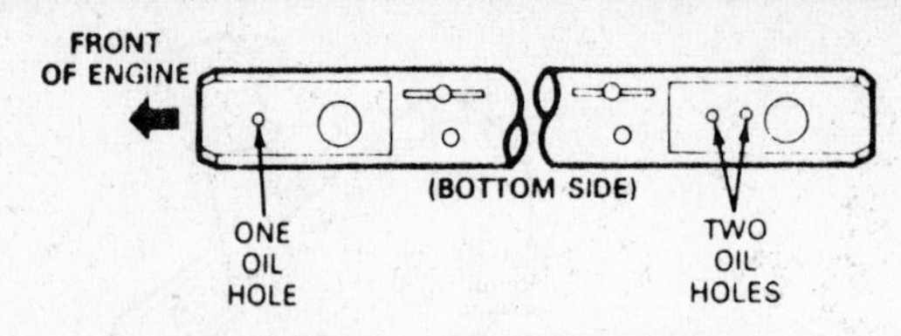

Rocker arm assembly oil hole positioning – 2.3L turbo diesel engine

Rocker arm assembly – 2.3L turbo diesel engine

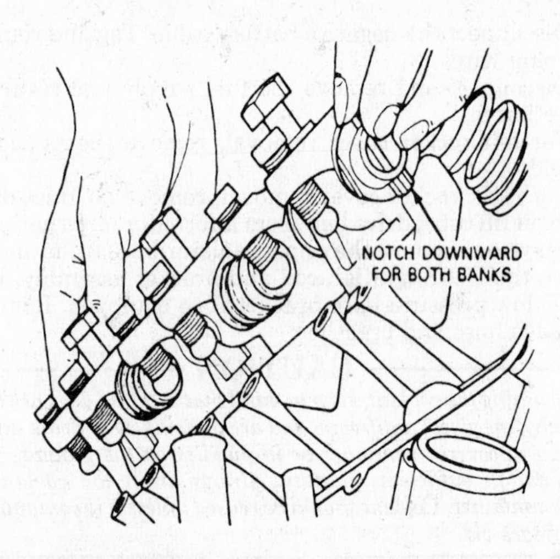

Rocker arm shaft assembly – 2.8L and 2.9L engines

DISASSEMBLY AND REASSEMBLY

1. Remove the spring washer and pin from each end of the valve rocker arm shaft.
2. Slide the rocker arms, springs and rocker arm shaft supports off the shaft. Be sure to mark the parts for re-assembly in the same locations.
3. If it is necessary to remove the plugs from each end of the shaft, drill or pierce the plug on one end. Use a steel rod to knock out the plug on the opposite end. Working from the open end, knock out the remaining plug.
4. The oil holes in the rocker arm shaft must point down when the shaft is installed. This position of the shaft can be recognized by a notch on the front face of the shaft.
5. If the plugs were removed from the shaft, use a blunt tool and install a plug, cup side out, in each end of the shaft.
6. Install a spring washer and pin on one end of the shaft, coat the rocker arm shaft with heavy engine oil and install the parts in the same sequence they were removed.

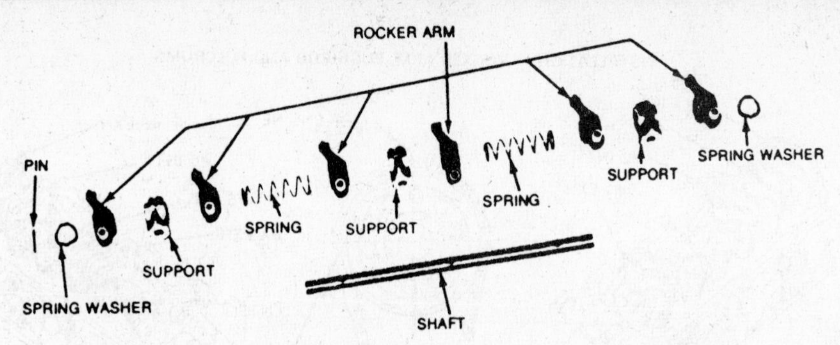

Rocker arm shaft assembly – 2.8L and 2.9L engines

3.0L Engine

1. Remove the rocker arm covers.
2. Removing the single retaining bolt at each rocker arm.
3. The rocker arm and pushrod may then be removed from the engine. Keep all rocker arms and pushrods in order so they may be installed in their original locations.
4. Installation is the reverse of removal. Tighten the rocker arm fulcrum bolts in two stages, first to 5–11 ft. lbs. (7–15 Nm), then to 18–26 ft. lbs. (25–35 Nm). Refer to the illustration for initial valve adjustment.

4.0L Engine

1. Remove the rocker cover.
2. Loosen the rocker arm shaft stand bolts, 2 turns at a time, front-to-rear, until they are free.
3. Lift off the rocker arm shaft assembly. If the pushrods are to be removed, tag them, since they have to be installed in their original positions.
4. Installation is the reverse of removal procedure. Tighten the rocker arm shaft bolts, 2 turns at a time, front to rear, until the specified torque of 52 ft. lbs. is reached. Install the rocker cover.

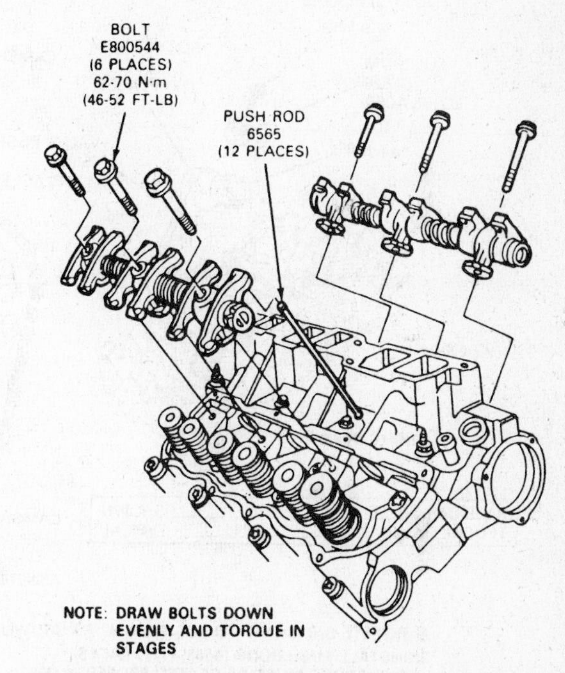

Rocker arm shaft assembly – 4.0L engine

Thermostat

REMOVAL AND INSTALLATION

2.0L and 2.2L Engines

1. Drain the cooling system below the level of the coolant outlet housing.

CAUTION

When draining the coolant, keep in mind that cats and dogs are attracted by the ethylene glycol antifreeze, and are quite likely to drink any that is left in an uncovered container or in puddles on the ground. This will prove fatal in sufficient quantity. Always drain the coolant into a sealable container. Coolant should be reused unless it is contaminated or several years old.

2. Disconnect the heater return hose at the thermostat housing located on the left front lower side of engine.
3. Remove the coolant outlet housing retaining bolts and slide the housing with the hose attached to one side.
4. Turn the thermostat counterclockwise to unlock it from the outlet.
5. Remove the gasket from the engine block and clean both mating surfaces.

NOTE: It is good practice to check the operation of a new thermostat before it is installed in an engine. Place the thermostat in a pan of boiling water. If it does not open more than 1/4 in., do not install it in the engine.

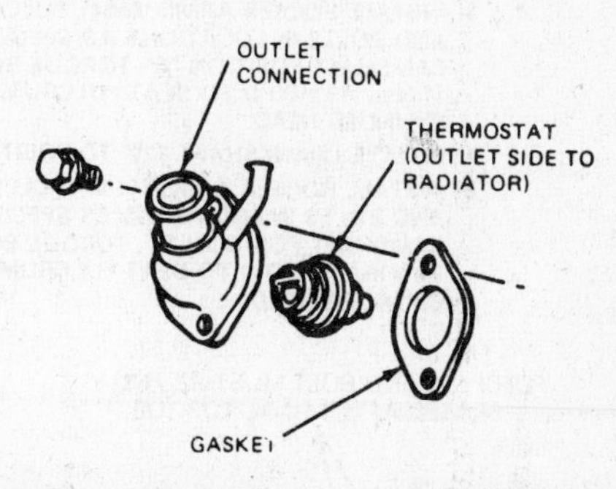

Thermostat and housing – exploded view

6. To install the thermostat, coat a new gasket with water resistant sealer and position it on the outlet of the engine. The gasket must be in place before the thermostat is installed.
7. Install the thermostat with the bridge (opposite end from the spring) inside the elbow connection and turn it clockwise to lock it in position, with the bridge against the flats cast into the elbow connection.
8. Position the elbow connection onto the mounting surface

INSTALLATION – ROCKER ARM, PUSH ROD AND FULCRUMS

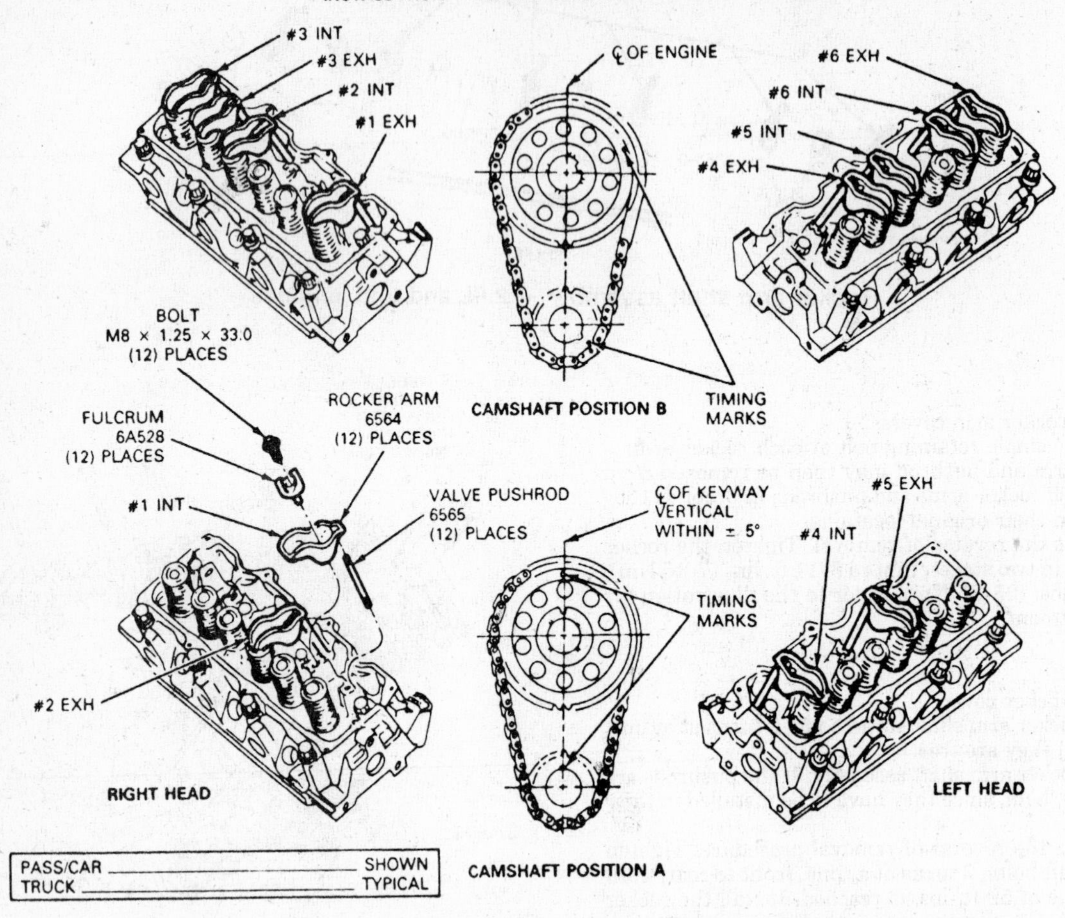

ASSEMBLY PROCEDURE

1. ROTATE CAMSHAFT TO POSITION "A" AS SHOWN.
2. INSTALL PUSH RODS (6565) (12) PLACES-PUSH RODS MUST BE SEATED PROPERLY ON TAPPET ASSEMBLY
3. INSTALL ROCKER ARMS (6564), FULCRUMS (6A528) AND BOLTS IN LOCATIONS AS SPECIFIED IN CAMSHAFT POSITION "A", TORQUE BOLTS TO 11 N·m AS REQ'D TO SEAT FULCRUMS IN CYLINDER HEAD
4. ROTATE CRANKSHAFT 120° TO POSITION "B"
5. INSTALL ROCKER ARMS (6564), FULCRUMS (6A528), AND BOLTS IN LOCATIONS AS SPECIFIED IN CAMSHAFT POSITION "B", TORQUE BOLTS TO 11 N·m AS REQ'D TO SEAT FULCRUMS IN CYLINDER HEAD

NOTE: FULCRUMS MUST BE FULLY SEATED IN CYLINDER HEADS AND PUSH RODS MUST BE FULLY SEATED IN ROCKER ARM SOCKETS PRIOR TO FINAL TORQUE.

6. APPLY ESE-M2C39-F OIL TO ROCKER ARM ASSEMBLIES.
7. FINAL TORQUE BOLTS TO 32.0 N·m (CAMSHAFT MAY BE IN ANY POSITION).

NOTE: CAMSHAFT POSITIONS "A" AND "B" ARE REQUIRED TO PLACE TAPPET ASSEMBLY ON BASE CIRCLE OF CAMSHAFT LOBE TO CHECK COLLAPSED TAPPET GAP.

FULCRUM AND BOLT MUST BE FULLY SEATED AFTER FINAL TORQUE

4.69-2.15 WITH TAPPET FULLY COLLAPSED ON BASE CIRCLE OF CAM LOBE AFTER ASSEMBLY REF. QUALITY AUDIT ONLY.

CYL. NO.	CAMSHAFT POSITION	
	A	B
	SET GAP OF VALVES NOTED	
1	INT.	EXH.
2	EXH.	INT.
3	NONE	INT.-EXH.
4	INT.	EXH.
5	EXH.	INT.
6	NONE	INT.-EXH.

Initial valve adjustment — 3.0L engine

of the outlet, so that the thermostat flange is resting on the gasket and install the retaining bolts.

9. Connect the heater hose to the thermostat housing.

10. Fill the radiator and operate the engine until it reaches operating temperature. Check the coolant level and adjust as necessary.

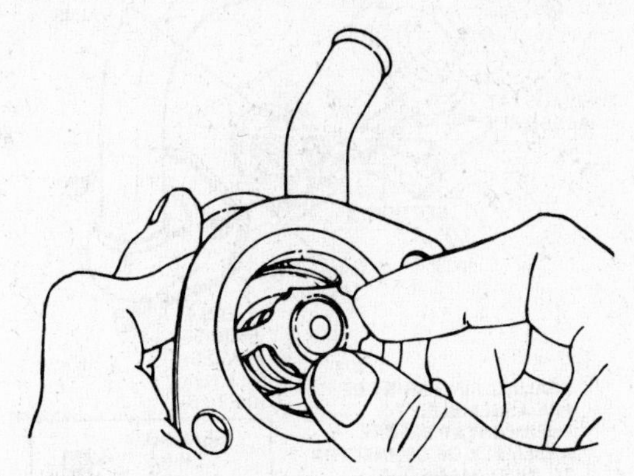

Installing the thermostat in the thermostat housing

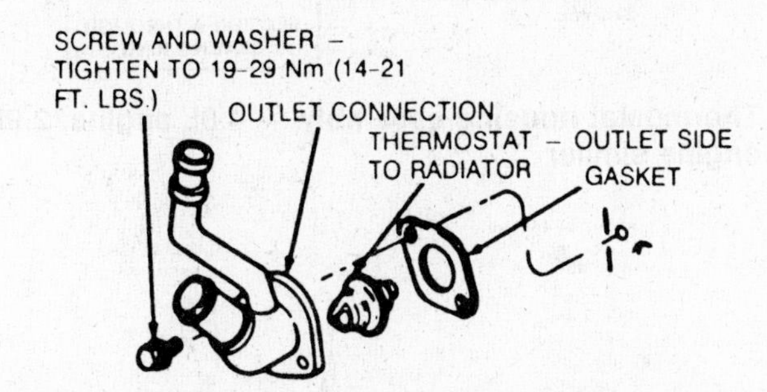

Thermostat housing assembly – 2.3L engine

2.3L Engine

1. Drain the cooling system (engine cold).

CAUTION

When draining the coolant, keep in mind that cats and dogs are attracted by the ethylene glycol antifreeze, and are quite likely to drink any that is left in an uncovered container or in puddles on the ground. This will prove fatal in sufficient quantity. Always drain the coolant into a sealable container. Coolant should be reused unless it is contaminated or several years old.

2. Remove the retaining bolts for the thermostat housing. Lift the housing clear and remove the thermostat. It may be easier to clean the gasket mating surfaces with the heater and radiator hoses removed from the thermostat housing.

3. Clean all gasket mating surfaces and make sure the thermostat is in the housing properly. Always use a new gasket. Tighten the thermostat housing retaining bolts to 14–21 ft. lbs. Refill the cooling system, start the engine and check for leaks.

2.8L Engine

1. Drain the cooling system (engine cold).

CAUTION

When draining the coolant, keep in mind that cats and dogs are attracted by the ethylene glycol antifreeze, and are quite likely to drink any that is left in an uncovered container or in puddles on the ground. This will prove fatal in sufficient quantity. Always drain the coolant into a sealable container. Coolant should be reused unless it is contaminated or several years old.

2. Remove the retaining bolts from the thermostat housing. Move the housing out of the way, then lift out the thermostat.

3. Clean all gasket mating surfaces and make sure the new thermostat is installed correctly in the housing. Always use a new gasket. Tighten the thermostat housing bolts to 12–15 ft. lbs.

4. Refill the cooling system, start the engine and check for leaks.

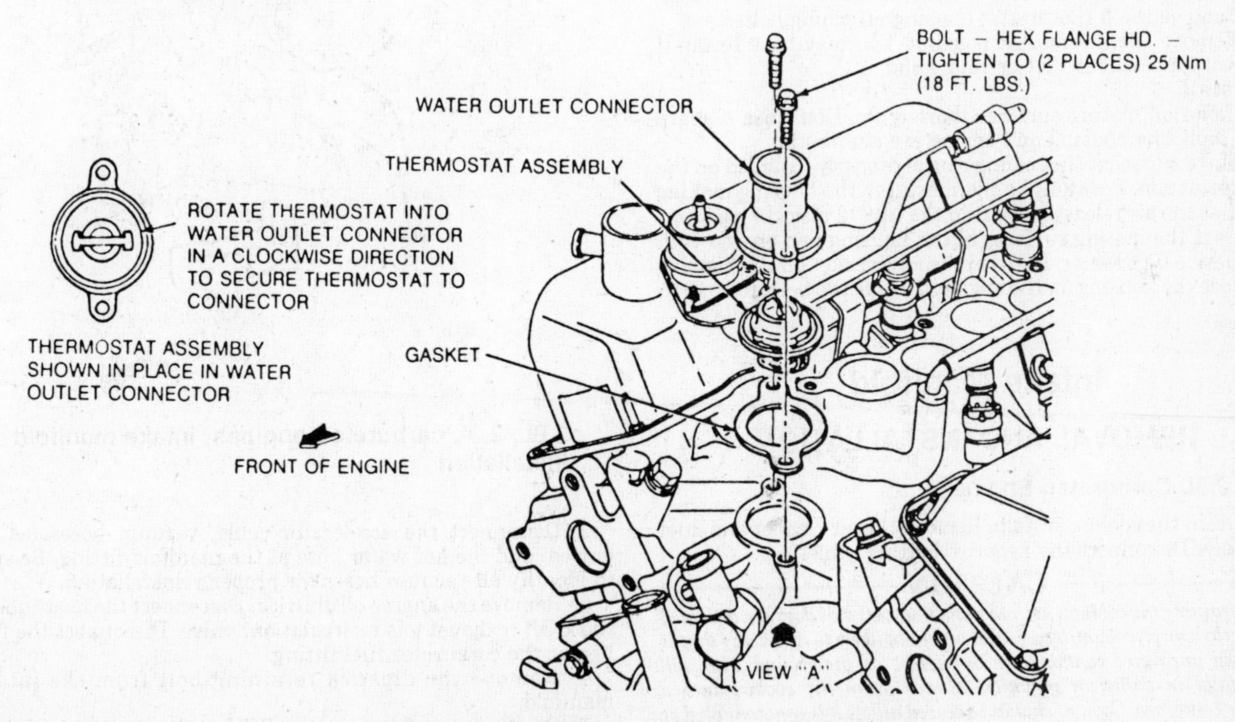

Thermostat housing assembly – 3.0L engine

3.0L Engine

1. Drain the cooling system.

CAUTION

When draining the coolant, keep in mind that cats and dogs are attracted by the ethylene glycol antifreeze, and are quite likely to drink any that is left in an uncovered container or in puddles on the ground. This will prove fatal in sufficient quantity. Always drain the coolant into a sealable container. Coolant should be reused unless it is contaminated or several years old.

2. Disconnect the battery ground cable.
3. Remove the upper radiator hose.
4. Remove the thermostat housing bolts.
5. Remove the housing and thermostat as an assembly.
6. Turn the thermostat counterclockwise to remove it from the housing.
7. Clean all gasket material from the housing and engine.
8. Turn the thermostat clockwise into the housing until the thermostat bridge is perpendicular to the mounting holes.
9. Position the housing on the engine, using a new gasket coated with sealer. Tighten the bolts to 19 ft. lbs.
10. Instal the hose, fill and bleed the cooling system, install the air cleaner and resonator, connect the battery ground cable and start the engine. Check for leaks.

2.9L and 4.0L Engines

1. Drain the cooling system.

CAUTION

When draining the coolant, keep in mind that cats and dogs are attracted by the ethylene glycol antifreeze, and are quite likely to drink any that is left in an uncovered container or in puddles on the ground. This will prove fatal in sufficient quantity. Always drain the coolant into a sealable container. Coolant should be reused unless it is contaminated or several years old.

2. Disconnect the battery ground.
3. Remove the air cleaner duct assembly.
4. Remove the upper radiator hose.
5. Remove the 3 thermostat housing attaching bolts.
6. Remove the thermostat housing. You may have to tap it loose with a plastic mallet or your hand.

To install:

7. Clean all mating surfaces thoroughly. Don't use a sharp metal tool! The housing and engine are aluminum.
8. Make sure that the sealing ring is properly installed on the thermostat rim. Position the thermostat in the housing making sure that the air release valve is in the **up** (12 o'clock) position.
9. Coat the mating surfaces of the housing and engine with an adhesive type sealer. Position the new gasket on the housing and place the housing on the engine. Torque the bolts to 7–10 ft. lbs.

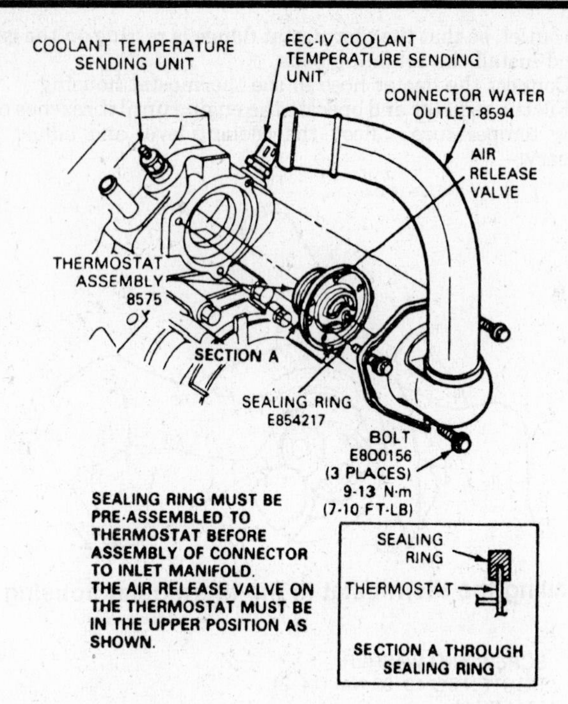

Thermostat housing assembly — 4.0L engine, 2.9L engine similar

Intake Manifold

REMOVAL AND INSTALLATION

2.0L, 2.3L Carbureted Engines

1. Drain the cooling system. Remove the air cleaner and duct assembly. Disconnect the negative battery cable.

CAUTION

When draining the coolant, keep in mind that cats and dogs are attracted by the ethylene glycol antifreeze, and are quite likely to drink any that is left in an uncovered container or in puddles on the ground. This will prove fatal in sufficient quantity. Always drain the coolant into a sealable container. Coolant should be reused unless it is contaminated or several years old.

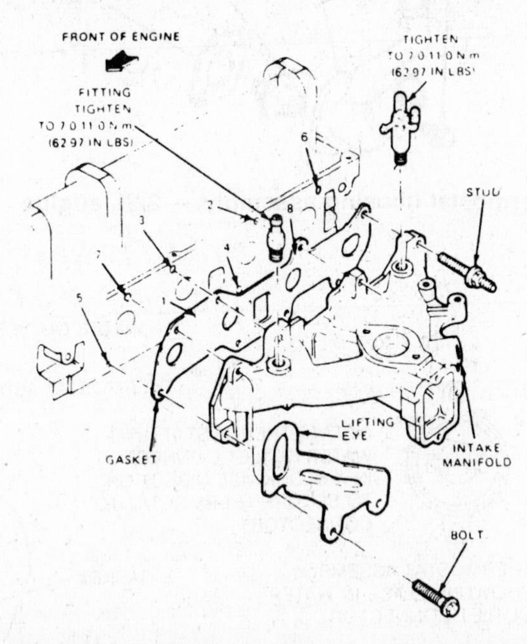

2.0L, 2.3L carbureted engines, intake manifold installation

2. Disconnect the accelerator cable, vacuum hoses (as required) and the hot water hose at the manifold fitting. Be sure to identify all vacuum hoses for proper reinstallation.
3. Remove the engine oil dipstick. Disconnect the heat tube at the EGR (exhaust gas recirculation) valve. Disconnect the fuel line at the carburetor fuel fitting.
4. Remove the dipstick retaining bolt from the intake manifold.
5. Disconnect and remove the PCV at the engine and intake manifold.

6. Remove the distributor cap and position the cap and wires out of the way, after removing the plastic plug connector from the valve cover.
7. Remove the intake manifold retaining bolts. Remove the manifold from the engine.
8. Clean all gasket mounting surfaces.
9. Install a new mounting gasket and intake manifold on the engine. Torque the bolts in proper sequence in two steps, first 5–7 ft. lbs., then 14–21 ft. lbs. The rest of the installation is in the reverse order of removal.

2.3L EFI And 2.3L Twin Plug Engines

The intake manifold is a two-piece (upper and lower) aluminum casting. Runner lengths are tuned to optimize engine torque and power output. The manifold provides mounting flanges for the air throttle body assembly, fuel supply manifold, accelerator control bracket and the EGR valve and supply tube. A vacuum fitting is installed to provide vacuum to various engine accessories. Pockets for the fuel injectors are machined to prevent both air and fuel leakage. The following procedure is for the removal of the intake manifold with the fuel charging assembly attached.

1. Make sure the ignition is OFF, then drain the coolant from the radiator (engine cold).

CAUTION

When draining the coolant, keep in mind that cats and dogs are attracted by the ethylene glycol antifreeze, and are quite likely to drink any that is left in an uncovered container or in puddles on the ground. This will prove fatal in sufficient quantity. Always drain the coolant into a sealable container. Coolant should be reused unless it is contaminated or several years old.

2. Disconnect the negative battery cable and secure it out of the way.
3. Remove the fuel filler cap to vent tank pressure. Release the pressure from the fuel system at the fuel pressure relief valve using EFI pressure gauge T80L–9974–A or equivalent. The fuel pressure relief valve is located on the fuel line in the upper right hand corner of the engine compartment. Remove the valve cap to gain access to the valve.
4. Disconnect the electrical connectors at the throttle position sensor, knock sensor, injector wiring harness, air charge temperature sensor and engine coolant temperature sensor. On the 2.3L twin plug engine disconnect TPS sensor, air bypass valve, injector wiring harness from main harness and water temperature indicator sensor, engine coolant temperature sensor, ignition control assembly.
5. Tag and disconnect the vacuum lines at the upper intake manifold vacuum tree, at the EGR valve and at the fuel pressure regulator and canister purge line as necessary.
6. Remove the throttle linkage shield and disconnect the throttle linkage and speed control cable (if equipped). Unbolt the accelerator cable from the bracket and position the cable out of the way.
7. Disconnect the air intake hose, air bypass hose and crankcase vent hose.
8. Disconnect the PCV hose from the fitting on the underside of the upper intake manifold.
9. Loosen the clamp on the coolant bypass line at the lower intake manifold and disconnect the hose.
10. Disconnect the EGR tube from the EGR valve by removing the flange nut.
11. Remove the upper intake manifold retaining nuts. Remove the upper intake manifold and air throttle body assembly.
12. Disconnect the push connect fitting at the fuel supply manifold and fuel return lines. Disconnect the fuel return line from the fuel supply manifold.
13. Remove the engine oil dipstick bracket retaining bolt.

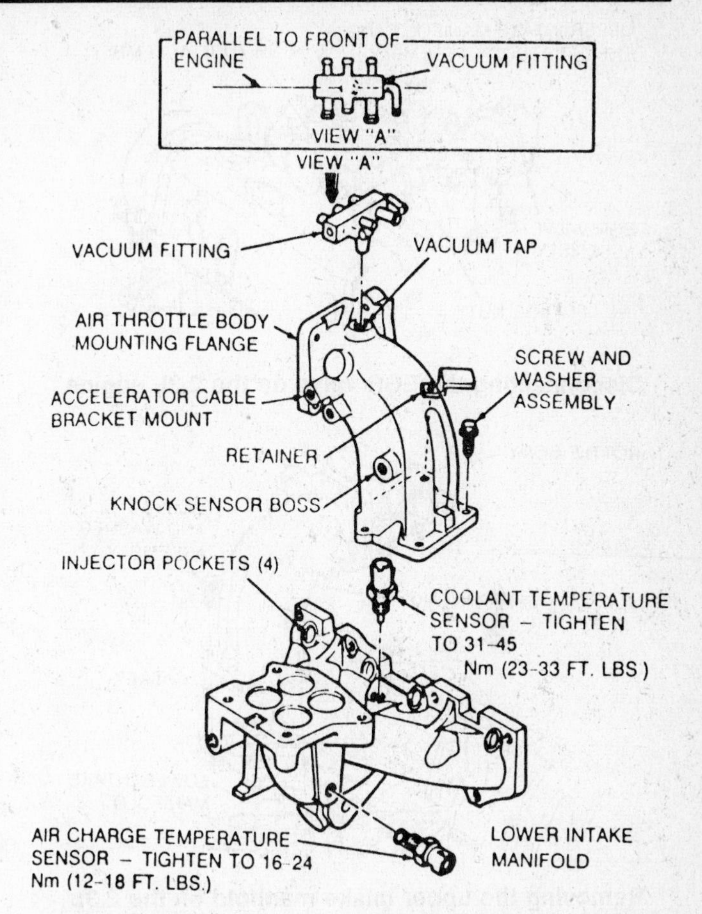

Upper and lower intake manifold assemblies on the 2.3L engine

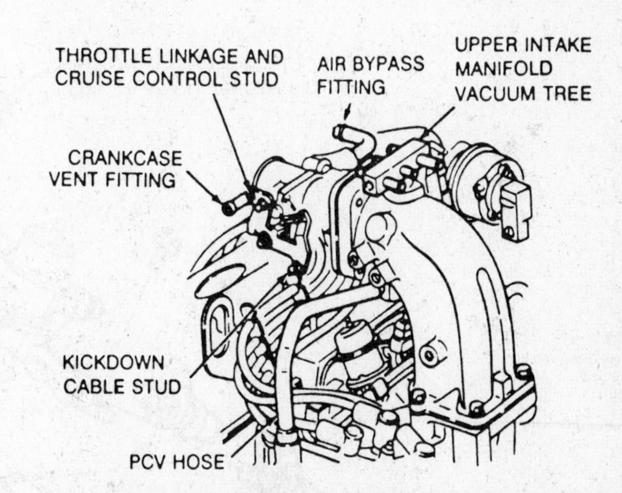

Linkage and hose location on the 2.3L engine

14. Disconnect the electrical connectors from all four fuel injectors and move the harness aside.
15. Remove the two fuel supply manifold retaining bolts, then carefully remove the fuel supply manifold and injectors. Remove the injectors by exerting a slight twisting/pulling motion.
16. Remove the four bottom retaining bolts from the lower manifold. The front two bolts also secure an engine lifting bracket. Once the bolts are removed, remove the lower intake manifold.

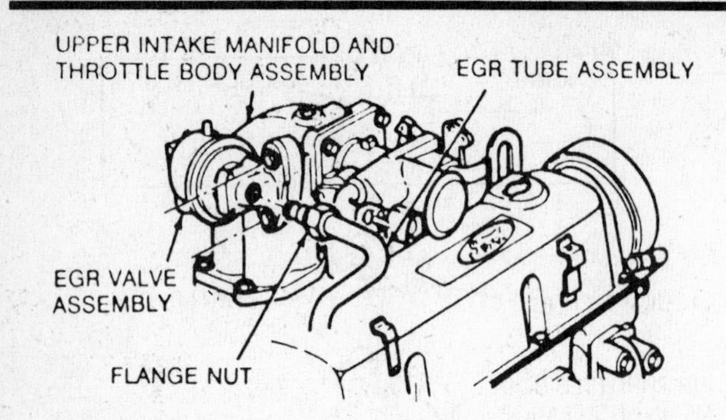

Disconnecting the EGR valve on the 2.3L engine

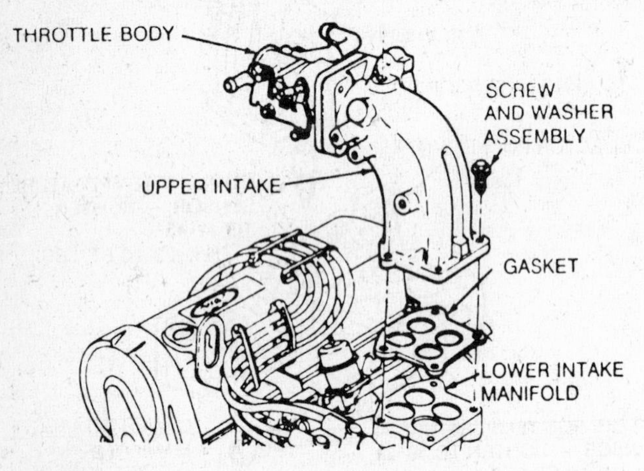

Removing the upper intake manifold on the 2.3L engine

17. Clean and inspect the mounting faces of the lower intake manifold and cylinder head. Both surfaces must be clean and flat. If the intake manifold upper or lower section is being replaced, it will be necessary to transfer components from the old to the new part.

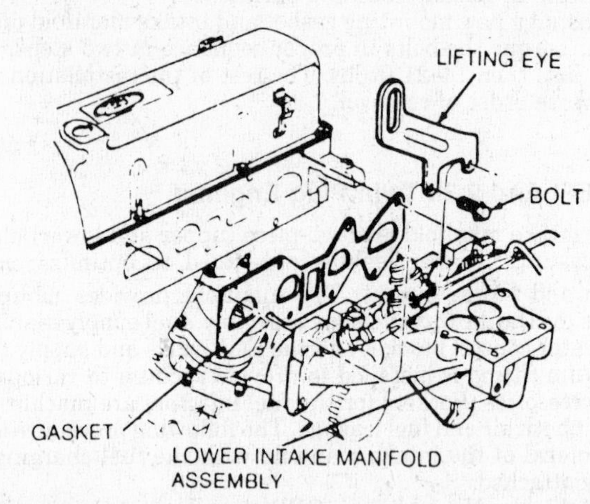

Removing the lower intake manifold on the 2.3L engine

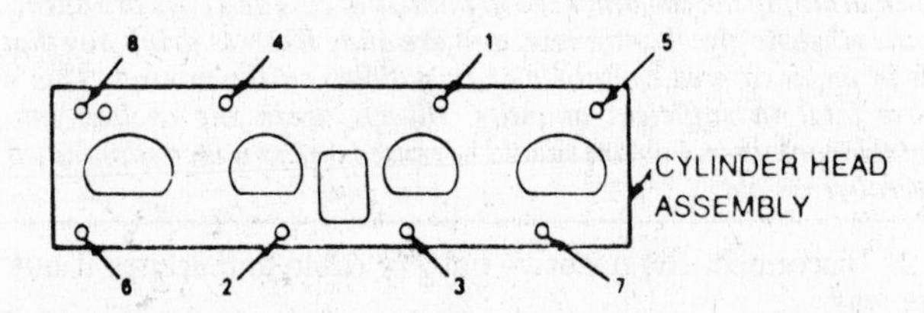

Torque sequence for the lower intake manifold – 2.3L engines

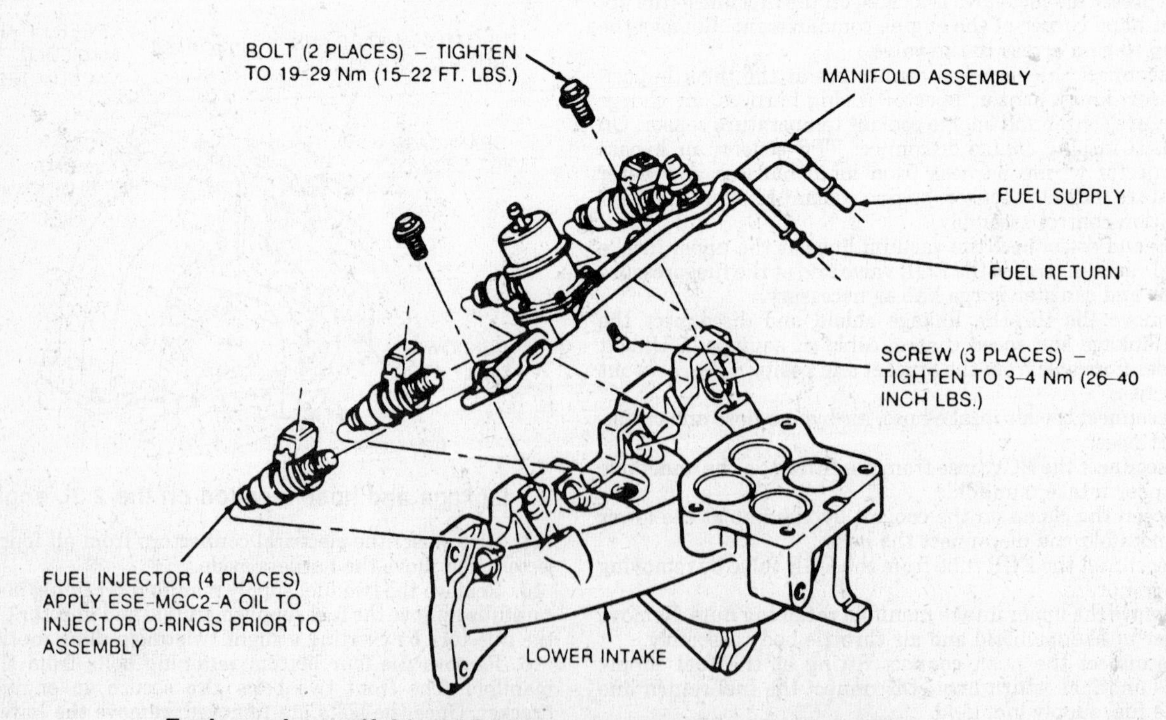

Fuel supply manifold and injector mounting on the 2.3L engine

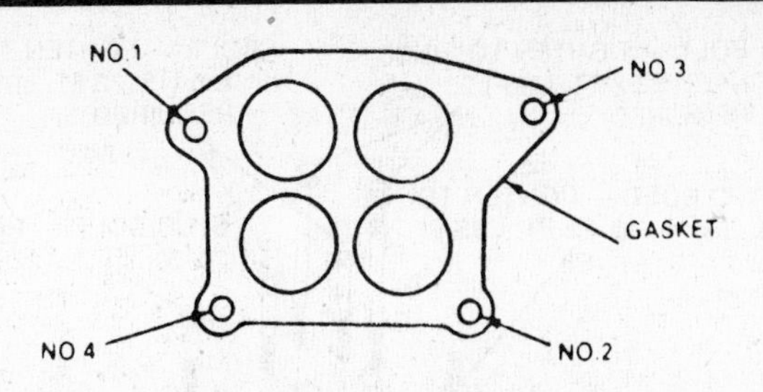

Upper intake manifold torque sequence – 2.3L engines

To install:

18. To install, first clean and oil the manifold bolt threads. Install a new lower manifold gasket.
19. Position the lower manifold assembly to the head and install the engine lifting bracket. Install the four top manifold retaining bolts finger tight. Install the four remaining manifold bolts and tighten all bolts to 12–15 ft. lbs. and 15–22 ft. lbs. on the 2.3L twin plug engine, following the sequence illustrated.
20. Install the fuel supply manifold and injectors with two retaining bolts. Tighten the retaining bolts to 12–15 ft. lbs. and 15–22 ft. lbs. on the 2.3L twin plug engine.
21. Connect the four electrical connectors to the injectors.
22. Make sure the gasket surfaces of the upper and lower intake manifolds are clean. Place a gasket on the lower intake manifold assembly, then place the upper intake manifold in position.
23. Install the retaining bolts and tighten in sequence to 15–22 ft. lbs.
24. Install the engine oil dipstick, then connect the fuel return and supply lines to the fuel supply manifold.
25. Connect the EGR tube to the EGR valve and tighten it to 6–9 ft. lbs. and 18 ft. lbs. on the 2.3L twin plug engine.
26. Connect the coolant bypass line and tighten the clamp. Connect the PCV system hose to the fitting on the underside of the upper intake manifold.
27. If removed, install the vacuum tee on the upper intake manifold. Use Teflon® tape on the threads and tighten to 12–18 ft. lbs. Reconnect the vacuum lines to the tee, the EGR valve and the fuel pressure regulator and canister purge line as necessary.
28. Hold the accelerator cable bracket in position on the upper intake manifold and install the retaining bolt. Tighten the bolt to 10–15 ft. lbs.
29. Install the accelerator cable to the bracket.
30. Position a new gasket on the fuel charging assembly air throttle body mounting flange. Install the air throttle body to the fuel charging assembly. Install two retaining nuts and two bolts and tighten to 12–15 ft. lbs. and 15–25 ft. lbs. on the 2.3L twin plug engine.
31. Connect the accelerator and speed control cable (if equipped), then install the throttle linkage shield.
32. Reconnect the throttle position sensor, injector wiring harness, knock sensor, air charge temperature sensor and engine coolant temperature sensor. On the 2.3L twin plug engine reconnect TPS sensor, air bypass valve, injector wiring harness to main harness and water temperature indicator sensor, engine coolant temperature sensor, ignition control assembly.
33. Connect the air intake hose, air bypass hose and crankcase ventilation hose.
34. Reconnect the negative battery cable. Refill the cooling system to specifications and pressurize the fuel system by turning the ignition switch on and off (without starting the engine) at least six times, leaving the ignition on for at least five seconds each time.
35. Start the engine and let it idle while checking for fuel, coolant and vacuum leaks. Correct as necessary. Road test the vehicle for proper operation.

2.8L Engine

1. Disconnect the negative battery cable.
2. Remove the air cleaner.
3. Disconnect the throttle cable.
4. Drain the coolant. Disconnect and remove the hose from the water outlet to the radiator and bypass hose from the intake manifold to the thermostat housing rear cover.

CAUTION

When draining the coolant, keep in mind that cats and dogs are attracted by the ethylene glycol antifreeze, and are quite likely to drink any that is left in an uncovered container or in puddles on the ground. This will prove fatal in sufficient quantity. Always drain the coolant into a sealable container. Coolant should be reused unless it is contaminated or several years old.

5. Remove the distributor cap and spark plug wires as an assembly.

NOTE: Mark each plug wire with its cylinder number.

6. Disconnect distributor wiring harness. Observe and mark the location of the distributor rotor and housing so ignition timing can be maintained at reassembly. Remove distributor holddown screw and clamp and lift out distributor.

NOTE: Some engines may be equipped with a security-type hold-down bolt. Use Distributor Hold-Down Wrench, Tool T82L-12270-A, or equivalent, to remove the hold-down bolt.

7. Remove the rocker arm covers.
8. Remove the fuel line and filter.
9. Remove the intake manifold attaching bolts and nuts. Tap manifold lightly with a plastic mallet to break the gasket seal. Lift off the manifold.
10. Remove all the old gasket material and sealing compound.

To install:

11. Apply silicone sealer to the joining surfaces. Place the intake manifold gasket in position. Make sure that the tab on the right bank cylinder head gasket fits into the cutout of the manifold gasket.
12. Apply silicone sealer to the attaching bolt bosses on the intake manifold and position the intake manifold. Follow the torque sequence and torque the bolts to specifications.
13. Install the distributor so that the rotor and housing are in the same position marked at removal.
14. Install the distributor clamp and attaching bolt. Connect distributor wires.
15. Install the fuel line.
16. Replace the rocker arm cover gaskets, and reinstall the rocker arm valve covers using the procedure given under Rocker Arm Cover removal and installation.
17. Install the distributor cap. Coat the inside of each spark plug wire connector with silicone grease with a small screwdriver, and install the wires. Connect the distributor wiring harness.
18. Install and adjust the throttle linkage.
19. Install the air cleaner and the air cleaner tube at carburetor.
20. Connect the negative battery cable.
21. Connect the hoses from the water outlet to the radiator and the bypass hose from the thermostat housing rear cover to the intake manifold.
22. Refill and bleed the cooling system.
23. Recheck the ignition timing and reset the engine idle speed to specification.
24. Run the engine at fast idle and check for coolant and oil leaks.

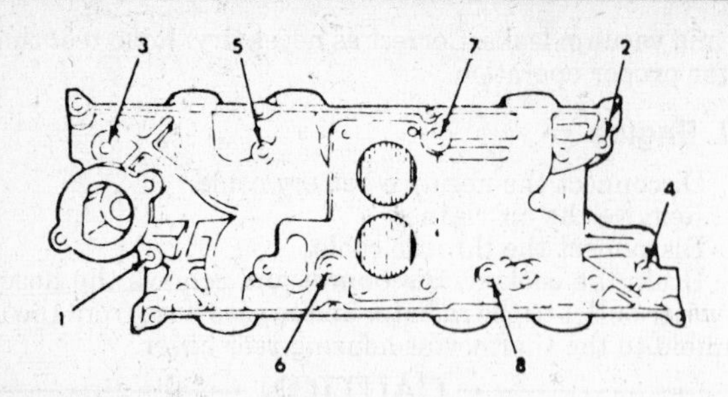

Intake manifold torque sequence on the 2.8L and 2.9L engines

2.9L Engine

1. Disconnect battery negative cable.
2. Remove air cleaner air intake duct from throttle body.
3. Disconnect throttle cable and bracket assembly.
4. Disconnect EGR tube at EGR valve.
5. Disconnect all vacuum hoses from fittings on upper intake manifold.
6. Disconnect electrical connections at throttle body, EGR pressure sensor, intake manifold upper and lower and distributor. Also disconnect fuel injector subharness from main EEC harness.
7. Remove upper intake manifold (plenum) assembly.
8. Drain coolant. Disconnect and remove hose from water outlet to radiator and heat supply.

CAUTION

When draining the coolant, keep in mind that cats and dogs are attracted by the ethylene glycol antifreeze, and are quite likely to drink any that is left in an uncovered container or in puddles on the ground. This will prove fatal in sufficient quantity. Always drain the coolant into a sealable container. Coolant should be reused unless it is contaminated or several years old.

9. Remove distributor cap and spark plug wires as an assembly.
10. Observe and mark the location of the distributor rotor and housing so ignition timing can be maintained as reassembly. Remove distributor hold-down screw and clamp and lift out distributor.
11. Remove rocker arm covers.
12. Remove intake manifold attaching bolts and nuts. Note length of manifold attaching bolts during removal so that they may be installed in their original positions. Tap manifold lightly with a plastic mallet to break gasket seal. Lift off manifold.
13. Remove all old gasket material and sealing compound.

To install:

14. Apply sealing compound to the joining surfaces. Place the intake manifold gasket in position. Ensure the tab on the right hand bank cylinder head gasket fits into the cutout of the manifold gasket.
15. Apply sealing compound to the attaching bolt bosses on the intake manifold and position the intake manifold. Follow the tightening sequence and tighten the bolts to specifications.
16. Install distributor so that rotor and housing are in the same position marked at removal.
17. Install distributor clamp and attaching bolts.
18. Replace rocker arm cover gasket, and install rocker arm valve covers using the procedure under Rocker Arm Cover and Rocker Arm.
19. Install distributor cap. Coat the inside of each spark plug wire connector with silicone grease with a small screwdriver, and install the wires. Connect distributor wiring harness.
20. Apply sealing compound, joining surfaces of upper and lower intake manifold. Install upper intake manifold gaskets.

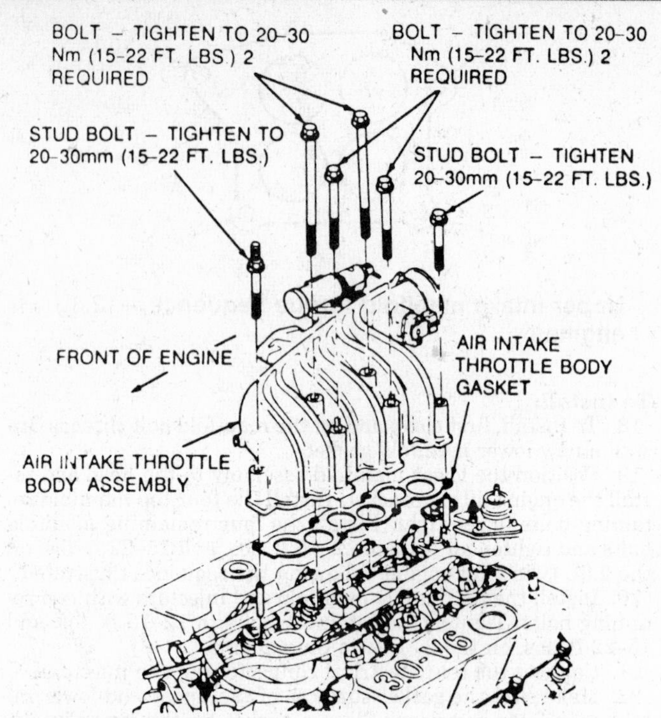

Removing the air intake throttle body on the 3.0L engine

21. Install upper intake manifold (plenum) assembly.
22. Connect all vacuum hoses to fittings on upper intake manifold.
23. Connect electrical connections at throttle body, EGR pressure sensor, intake manifolds sub harness to EEC main harness.
24. Install and adjust throttle linkage bracket assembly and cover.
25. Connect hoses from water outlet to radiator, and bypass hose from thermostat housing rear cover to intake manifold.
26. Connect battery negative cable.
27. Refill and bleed the cooling system.
28. Recheck ignition timing and reset engine idle speed to specification.
29. Run engine at fast idle and check for coolant and oil leaks.

3.0L Engine

1. Drain the cooling system (engine cold). Review complete service procedure before starting this repair.

CAUTION

When draining the coolant, keep in mind that cats and dogs are attracted by the ethylene glycol antifreeze, and are quite likely to drink any that is left in an uncovered container or in puddles on the ground. This will prove fatal in sufficient quantity. Always drain the coolant into a sealable container. Coolant should be reused unless it is contaminated or several years old.

2. Disconnect the battery ground cable.
3. Depressurize the fuel system and remove the air intake throttle body as outlined in Section 5.
4. Disconnect the fuel return and supply lines.
5. Remove the fuel injector wiring harness from the engine.
6. Disconnect the upper radiator hose.
7. Disconnect the water outlet heater hose.
8. Disconnect the distributor cap with the spark plug wires attached. Matchmark and remove the distributor assembly.

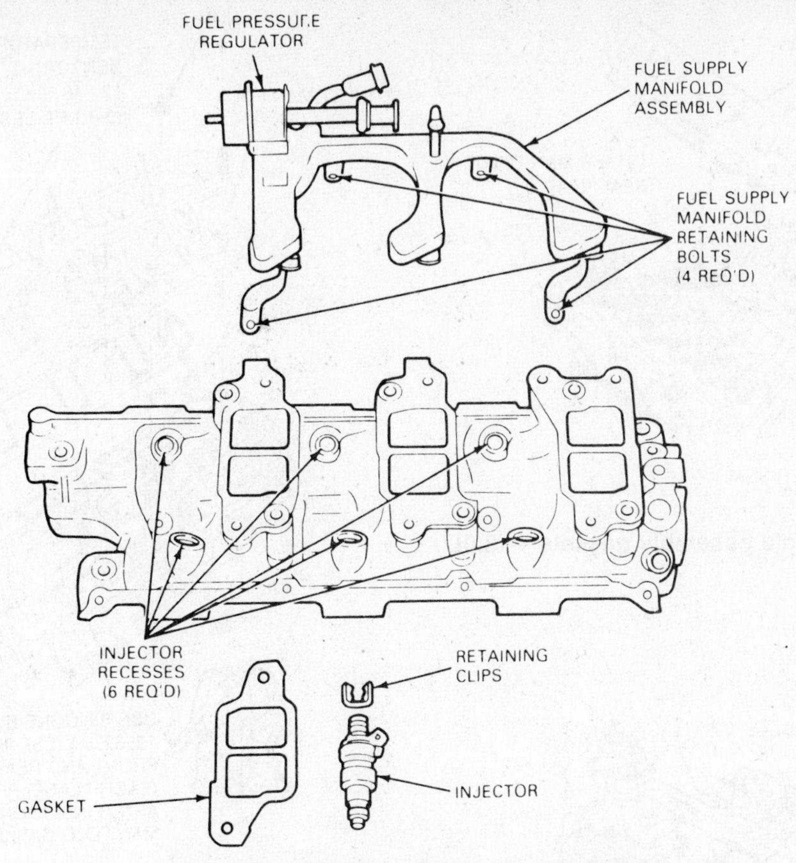

Intake manifold assembly – 2.9L engine

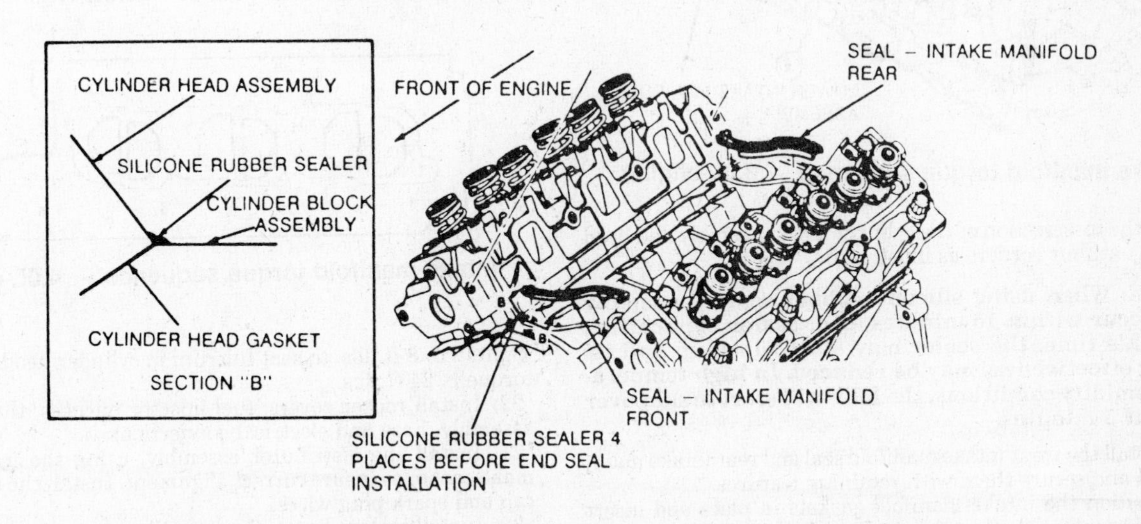

Apply silicone sealer as shown intake manifold – 3.0L engine

9. Remove the rocker covers. Loosen retaining nut from cylinder #3 intake valve and rotate rocker arm fulcrum away from valve retainer. Remove the pushrod
10. Remove the intake manifold attaching bolts and studs (Torx type socket needed).
11. Lift the intake manifold off the engine. Use a plastic mallet to tap lightly around the intake manifold to break it loose, if necessary. Do not pry between the manifold and cylinder head with any sharp instrument. The manifold can be removed with the fuel rails and injectors in place.
12. Remove the manifold side gaskets and end seals and discard. If the manifold is being replaced, transfer the fuel injector and fuel rail components to the new manifold on a clean workbench. Clean all gasket mating surfaces.

To install:

13. First lightly oil all attaching bolts and stud threads. The intake manifold, cylinder head and cylinder block mating surfaces should be clean and free of old silicone rubber sealer. Use a suitable solvent to clean these areas.
14. Apply silicone rubber sealer (D6AZ–19562–A or equiva-

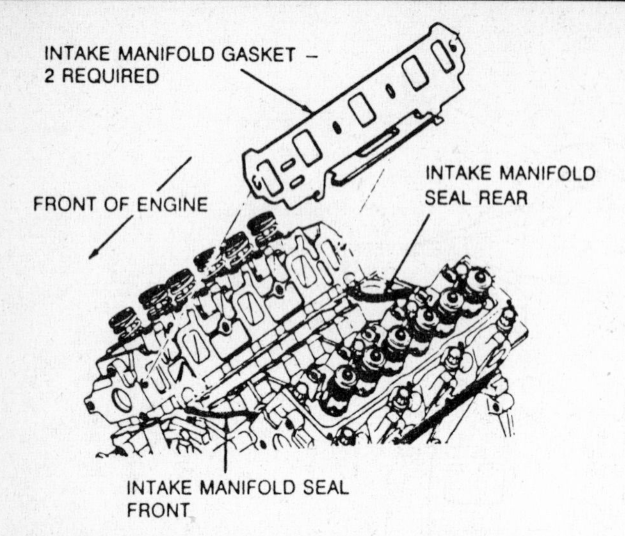

Installing intake manifold assembly gaskets – 3.0L engine

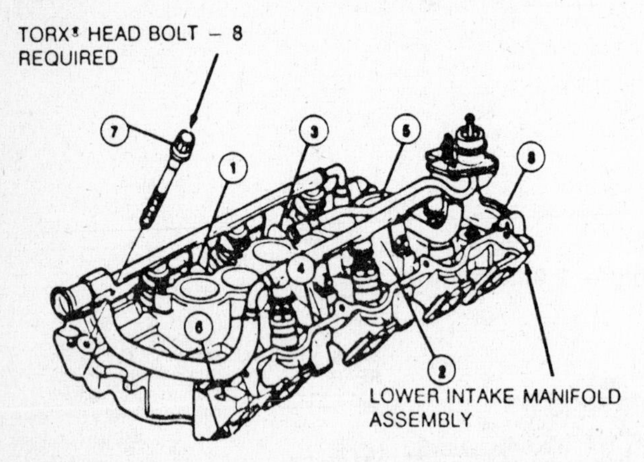

Intake manifold torque sequence – 3.0L engine

lent) to the intersection of the cylinder block assembly and head assembly at four corners as illustrated.

NOTE: When using silicone rubber sealer, assembly must occur within 15 minutes after sealer application. After this time, the sealer may start to set-up and its sealing effectiveness may be reduced. In high temperature/humidity conditions, the RTV will start to skin over in about 5 minutes.

15. Install the front intake manifold seal and rear intake manifold seal and secure them with retaining features.
16. Position the intake manifold gaskets in place and insert the locking tabs over the tabs on the cylinder head gaskets.
17. Apply silicone rubber sealer over the gasket in the same places as in Step 14.
18. Carefully lower the intake manifold into position on the cylinder block and cylinder heads to prevent smearing the silicone sealer and causing gasketing voids.
19. Install the retaining bolts and tighten in two stages, in the sequence illustrated, first to 11 ft. lbs. and then to 18 ft. lbs.
20. If installing a new manifold, install fuel supply rail and injectors. Refer to the necessary service procedures.
21. Install #3 cylinder intake valve pushrod. Apply oil to pushrod and fulcrum prior to installation. Rotate the crankshaft to place the lifter on the heel position or base circle of camshaft.

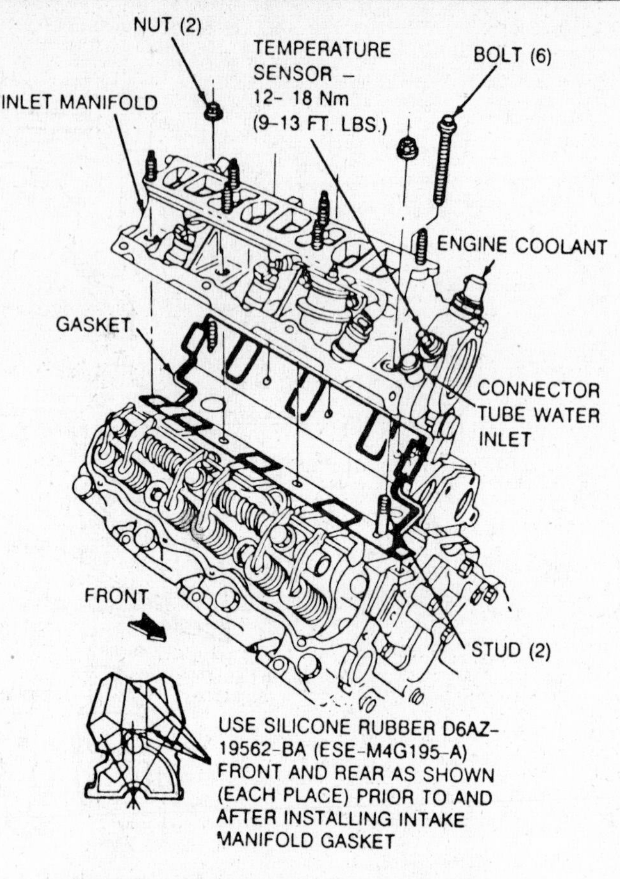

Intake manifold gasket – 4.0L engine

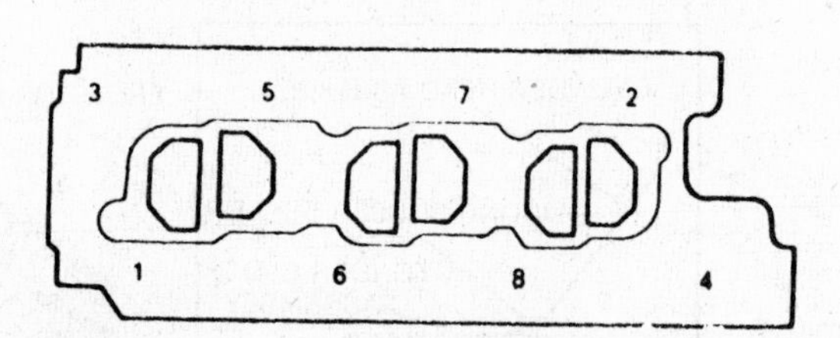

Intake manifold torque sequence – 4.0L engine

Tighten to 8 ft. lbs. to seat fulcrum in cylinder head. Final bolt torque is 24 ft. lbs.
22. Install rocker covers, fuel injector harness, throttle body assembly, hose and electrical connections.
21. Install the distributor assembly, using the matchmarks make earlier to insure correct alignment. Install the distributor cap and spark plug wires.
22. Install coolant hoses. Connect all vacuum lines. Reconnect fuel lines. Install fuel ine safety clips.
23. Fiil and bleed cooling system. Replace crankcase oil.
24. Install air cleaner hose. Connect battery ground cable. Start engine and check for coolant, oil, fuel and vacuum leaks.
25. Verify distributor base initial timing as outlined. Check and adjust engine idle as necessary.

4.0L Engine

The intake manifold is a 4-piece assembly, consisting of the upper intake manifold, the throttle body, the fuel supply manifold, and the lower intake manifold.
1. Disconnect the battery ground cable.

2. Remove the air cleaner and intake duct.
3. Remove the weather shield.
4. Disconnect the throttle cable and bracket.
5. Tag and disconnect all vacuum lines connected to the manifold.
6. Tag and disconnect all electrical wires connected to the manifold assemblies.
7. Relieve the fuel system pressure. See Section 1 Fuel Filter.
8. Tag and remove the spark plug wires.
9. Remove the EDIS ignition coil and bracket.
10. Remove the throttle body. See Section 5.
11. Remove the 6 attaching nuts and lift off the upper manifold.
12. Remove the rocker covers.
13. Remove the lower intake manifold bolts. Tap the manifold lightly with a plastic mallet and remove it.
14. Clean all surfaces of old gasket material.

To install:

15. Apply RTV silicone gasket material at the junction points of the heads and manifold.

NOTE: This material will set within 15 minutes, so work quickly!

16. Install new manifold gaskets and again apply the RTV material.
17. Position the manifold and install the nuts hand tight. Torque the nuts, in 4 stages, in the sequence shown, to 18 ft. lbs.
18. Once again, apply RTV material to the manifold/head joints.
19. Install the rocker covers.
20. Install the upper manifold. Tighten the nuts to 18 ft. lbs.
21. Install the EDIS coil.
22. Connect the fuel and return lines.
23. Install the throttle body.
24. Connect all wires.
25. Connect all vacuum lines.
26. Connect the throttle linkage.
27. Install the weather shield.
28. Install the air cleaner and duct.
29. Fill and bleed the cooling system.
30. Connect the battery ground.
31. Run the engine and check for leaks.

2.2L Diesel Engine

1. Disconnect the battery ground cables from both batteries.
2. Disconnect the air inlet hose from the air cleaner and intake manifold. Disconnect and remove the fuel injection lines from the nozzles and injection pump. Cap all lines and fittings to prevent dirt pickup.
3. Remove the nut attaching the lower fuel return line brace to the intake manifold.
4. Disconnect and remove the lower fuel line from the injector pump and upper fuel return line.
5. Remove the air conditioner compressor with the lines attached and position out of the way. Remove the power steering pump and rear support with the lines still attached and position out of the way.
6. Remove the air inlet adapter, dropping resistor (electrical measuring device) and the gaskets.
7. Disconnect the fuel filter inlet line, remove the fuel filter mounting bracket from the cylinder head and position the filter assembly out of the way.
8. Remove the mounting nuts for the fuel line heater assembly to intake manifold and position the heater out of the way.
9. Remove the nuts that attach the intake manifold to the cylinder head. Remove the intake manifold and gasket.

To install:

10. Position intake manifold on cylinder head and install two nuts.
11. Do not tighten the mounting nuts until No. 3 lower nut that holds the fuel return line bracket is installed. After installation of the No. 3 nut, tighten all of the mounting nuts to 12–17 ft. lbs.
12. Install inlet fitting on intake manifold. Tighten bolts to specification.
13. Install A/C compressor and bracket brace.
14. Connect battery ground cables to both batteries.
15. Run engine and check for oil and intake air leaks.

2.3L Turbo Diesel Engine

1. Disconnect battery ground cables from both batteries.
2. Remove support braces from A/C compressor bracket, inlet fitting, and intake manifold.
3. Remove A/C compressor from mounting bracket and position out of the way.
4. Remove inlet fitting from intake manifold.
5. Remove turbo oil feed line from cylinder head and turbo center housing. Remove oil line clamp bolt.
6. Loosen bolts attaching turbo heat shield to exhaust manifold. Remove top bolt and position shield out of the way.
7. Remove wastegate actuator from turbo and mounting bracket.
8. Remove two top actuator mounting bracket bolts. Loosen bottom bracket bolts and position bracket out of the way.
9. Remove five remaining intake manifold bolts and two nuts and remove intake manifold.

To install:

10. Position intake manifold on cylinder head and install two nuts.
11. Rotate wastegate actuator mounting bracket and install two top mounting bolts. Install five remaining intake manifold bolts. Tighten all intake manifold and wastegate actuator mounting bracket hardware to specification.
12. Install wastegate actuator. Tighten mounting bolts to specification.
13. Install turbo heat shield on exhaust manifold.
14. Install turbo oil feed line. Tighten fittings and clamp bolt to specification.
15. Install inlet fitting on intake manifold. Tighten bolts to specification.
16. Install A/C compressor and bracket brace.
17. Connect battery ground cables to both batteries.
18. Run engine and check for oil and intake air leaks.

Exhaust Manifold

REMOVAL AND INSTALLATION

2.0L, 2.3L Engines

1. Remove the air cleaner and dust assembly. Disconnect the negative battery cable.
2. Remove the EGR line at the exhaust manifold. Loosen the EGR tube. Remove the check valve at the exhaust manifold and disconnect the hose at the end of the air by-pass valve.
3. Remove the bracket attaching the heater hoses to the valve cover. Disconnect the exhaust pipe from the exhaust manifold.
4. Remove the exhaust manifold mounting bolts/nuts and remove the manifold.
5. Install the exhaust manifold in the reverse order. Torque the manifold in sequence in two steps, first 5–7 ft. lbs. and then 16–23 ft. lbs.

2.8L, 2.9L Engines

1. Remove the air cleaner.
2. Remove the attaching nuts from the exhaust manifold shroud (left side).
3. Disconnect the attaching nuts from the muffler inlet pipe. Remove thermactor components as necessary to allow the removal of the exhaust manifold(s).

4. Disconnect the choke heat tubes at the carburetor.
5. Remove the manifold attaching bolts.
6. Lift the manifold from the cylinder head.

To install:

7. Position the manifold on the heads and install and tighten the attaching bolts to specification.
8. Install a new inlet pipe gasket. Install and tighten the inlet pipe attaching nuts.
9. Position the exhaust manifold shroud on the manifold and install and tighten the attaching nuts to specification (left side). Install the thermactor components that had been removed.
10. Install the air cleaner. Reinstall the choke heat tube (if so equipped).

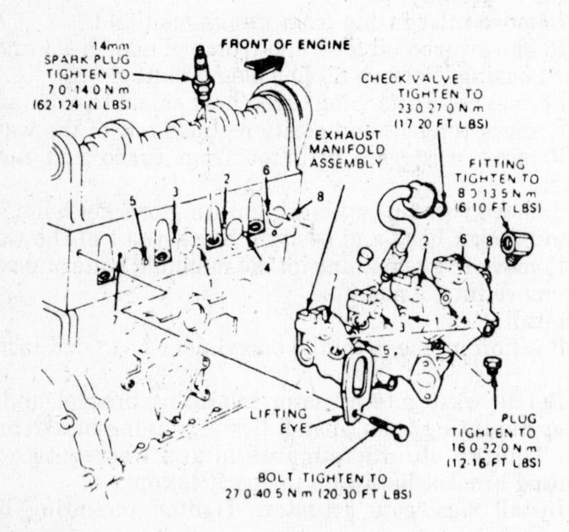

2.0L and 2.3L carbureted engines, exhaust manifold installation

3.0L Engine

1. Disconnect the negative battery cable. Raise and safely support the vehicle as necessary.
2. Remove the spark plugs.
3. If removing the left side exhaust manifold remove the oil level indicator tube retaining nut, rotate the dipstick assembly out of the way.

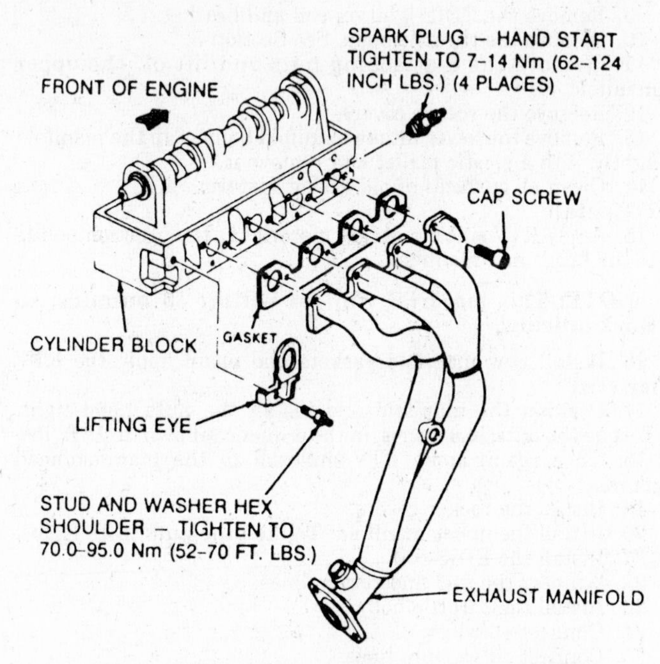

Exhaust manifold assembly – 2.3L engine

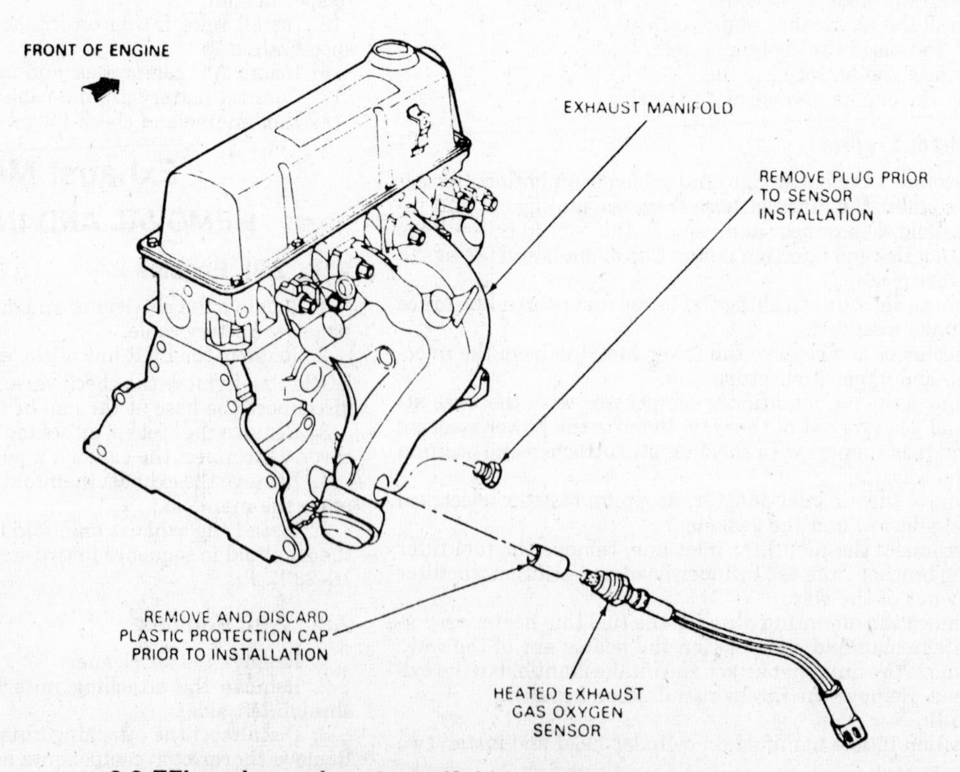

2.3 EFI engine, exhaust manifold installation

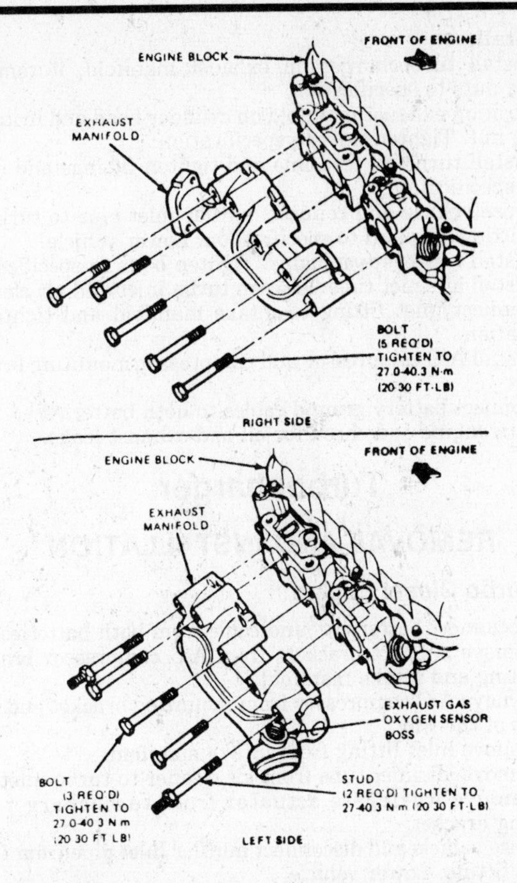

2.8L and 2.9L engine exhaust manifold torque sequence

4. Remove the manifold to exhaust pipe attaching nuts, then separate the exhaust pipe from the manifold.
5. Remove the exhaust manifold attaching bolts and the manifold.

To install:

6. Clean all gasket mating surfaces.
7. Lightly oil all bolt and stud threads before installation. If a new manifold is being installed, the oxygen sensor will have to be transferred to the new part.
8. Position the exhaust manifold on the cylinder head and install the manifold attaching bolts. Tighten them to 18 ft. lbs.
9. Connect (replace gasket if so equipped) the exhaust pipe to the manifold, then tighten the attaching nuts to 30 ft. lbs. TIGHTEN BOTH NUTS IN EQUAL AMOUNTS TO CORRECTLY SEAT INLET PIPE FLANGE.
10. Install oil tube dipstick assembly (apply sealer to tube-if removed) as necessary. Install spark plugs.
11. Connect the negative battery cable. Start the engine and check for oil and exhaust leaks.

4.0L Engine

LEFT SIDE

1. Disconnect the negative battery cable. Remove the oil level indicator tube bracket.
2. Remove the power steering pump hoses.
3. Remove the exhaust pipe-to-manifold bolts.
4. Unbolt and remove the manifold.
5. Clean and lightly oil all fastener threads.
6. Installation is the reverse of removal. Replace all gaskets if so equipped. Torque the manifold bolts to 19 ft. lbs.; the exhaust pipe nuts to 20 ft. lbs. TIGHTEN BOTH EXHAUST PIPE RETAINING NUTS IN EQUAL AMOUNTS TO CORRECTLY SEAT INLET PIPE FLANGE.

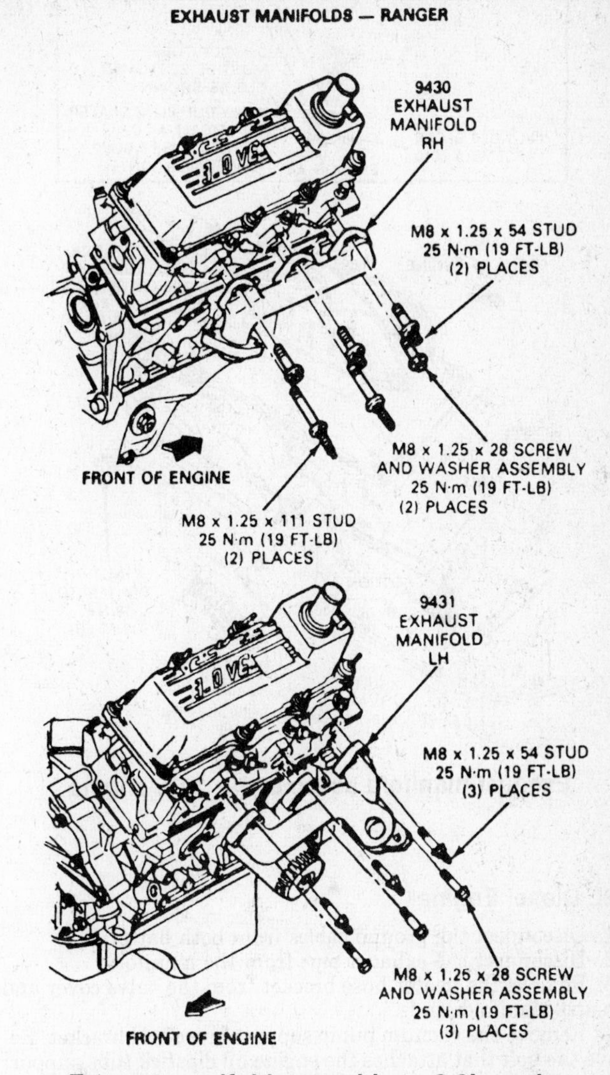

Exhaust manifold assembly — 3.0L engine

RIGHT SIDE

1. Drain the cooling system.

CAUTION

When draining the coolant, keep in mind that cats and dogs are attracted by the ethylene glycol antifreeze, and are quite likely to drink any that is left in an uncovered container or in puddles on the ground. This will prove fatal in sufficient quantity. Always drain the coolant into a sealable container. Coolant should be reused unless it is contaminated or several years old.

2. Remove the heater hose support bracket.
3. Disconnect the heater hoses.
4. Remove the exhaust pipe-to-manifold nuts.
5. Unbolt and remove the manifold.
6. Installation is the reverse of removal. Replace all gaskets if so equipped. Torque the manifold bolts to 19 ft. lbs.; the exhaust pipe nuts to 20 ft. lbs. TIGHTEN BOTH EXHAUST PIPE RETAINING NUTS IN EQUAL AMOUNTS TO CORRECTLY SEAT INLET PIPE FLANGE.

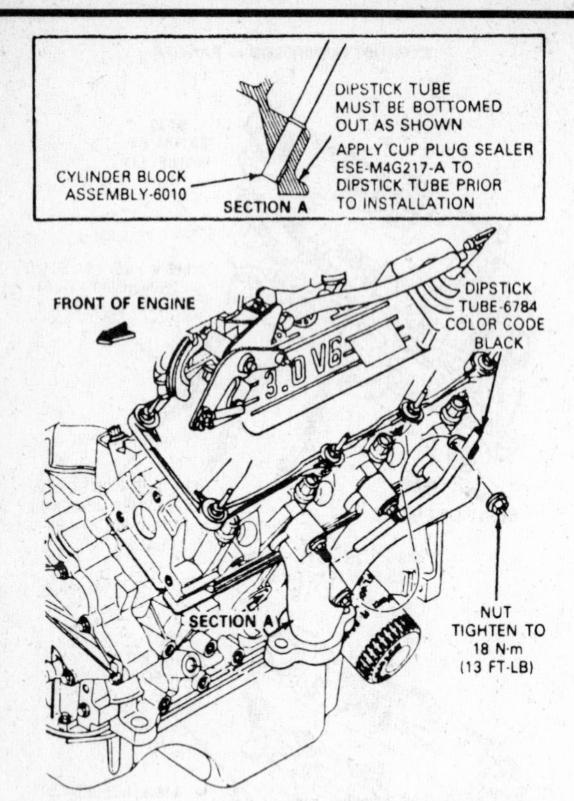

Exhaust manifold assembly — 3.0L engine

2.2L Diesel Engine

1. Disconnect the ground cables from both batteries.
2. Disconnect the exhaust pipe from the manifold.
3. Remove the heater hose bracket from the valve cover and exhaust manifold studs.
4. Remove the vacuum pump support brace and bracket. Remove the bolt that attaches the engine oil dipstick tube support bracket to the exhaust manifold.
5. Remove the nuts that attach the exhaust manifold to the engine and remove the manifold.
6. Clean all gasket mounting surfaces. Install a new mounting gasket and install the exhaust manifold and components in the reverse order of removal. Torque the mounting bolts to 17–20 ft. lbs. After warning up engine retighten to the same torque specification.

2.3L Turbo Diesel Engine

1. Disconnect battery ground cable from both batteries.
2. Remove support brackets from A/C compressor bracket, inlet fitting and intake manifold.
3. Remove A/C compressor from mounting bracket and position out of the way.
4. Remove inlet fitting from intake manifold.
5. Remove air inlet tube from air cleaner-to-turbo inlet.
6. Remove wastegate actuator from turbocharger and mounting bracket.
7. Raise vehicle and disconnect muffler inlet pipe from turbo exhaust fitting. Lower vehicle.
8. Disconnect turbo oil feed line from cylinder head and turbo center housing.
9. Remove nuts attaching exhaust manifold to cylinder head and remove exhaust manifold and turbocharger as an assembly.
10. Remove turbocharger from exhaust manifold if necessary.

To install:

11. Install turbocharger on exhaust manifold, if removed. Tighten nuts to specification.
12. Position exhaust manifold on cylinder head and install attaching nut. Tighten nuts to specification.
13. Install turbo oil feed line and tighten fittings and clamp bolt to specification.
14. Raise vehicle and connect muffler inlet pipe to turbo exhaust fitting. Tighten to specification. Lower vehicle.
15. Install wastegate actuator. Tighten bolts to specification.
16. Install air inlet tube between turbo inlet and air cleaner.
17. Connect inlet fitting to intake manifold and tighten to specification.
18. Install A/C compressor and compressor mounting bracket braces.
19. Connect battery ground cables to both batteries.
20. Run engine and check for oil and exhaust leaks.

Turbocharger

REMOVAL AND INSTALLATION

2.3L Turbo Diesel Engine

1. Disconnect battery ground cable from both batteries.
2. Remove support brackets from A/C compressor bracket, inlet fitting and intake manifold.
3. Remove A/C compressor from mounting bracket and position out of the way.
4. Remove inlet fitting from intake manifold.
5. Remove air inlet tube from air cleaner-to-turbo inlet.
6. Remove wastegate actuator from turbocharger and mounting bracket.
7. Raise vehicle and disconnect muffler inlet pipe from turbo exhaust fitting. Lower vehicle.
8. Disconnect turbo oil feed line from cylinder head and turbo center housing.
9. Remove nuts attaching exhaust manifold to cylinder head and remove exhaust manifold and turbocharger as an assembly.
10. Remove turbocharger from exhaust manifold.

To install:

11. Install turbocharger on exhaust manifold. Tighten nuts to specification.
12. Position exhaust manifold on cylinder head and install attaching nut. Tighten nuts to specification.
13. Install turbo oil feed line and tighten fittings and clamp bolt to specification.
14. Raise vehicle and connect muffler inlet pipe to turbo exhaust fitting. Tighten to specification. Lower vehicle.
15. Install wastegate actuator. Tighten bolts to specification.
16. Install air inlet tube between turbo inlet and air cleaner.
17. Connect inlet fitting to intake manifold and tighten to specification.
18. Install A/C compressor and compressor mounting bracket braces.
19. Connect battery ground cables to both batteries.
20. Run engine and check for oil and exhaust leaks.

Air Conditioning Compressor

REMOVAL AND INSTALLATION

1. Discharge the refrigerant system. See Section 1 for the proper procedure.
2. Disconnect the two refrigerant lines from the compressor. Cap the openings immediately!
3. Remove tension from the drive belt. Remove the belt.
4. Disconnect the clutch wire at the connector.
5. Remove the bolt attaching the support brace to the front

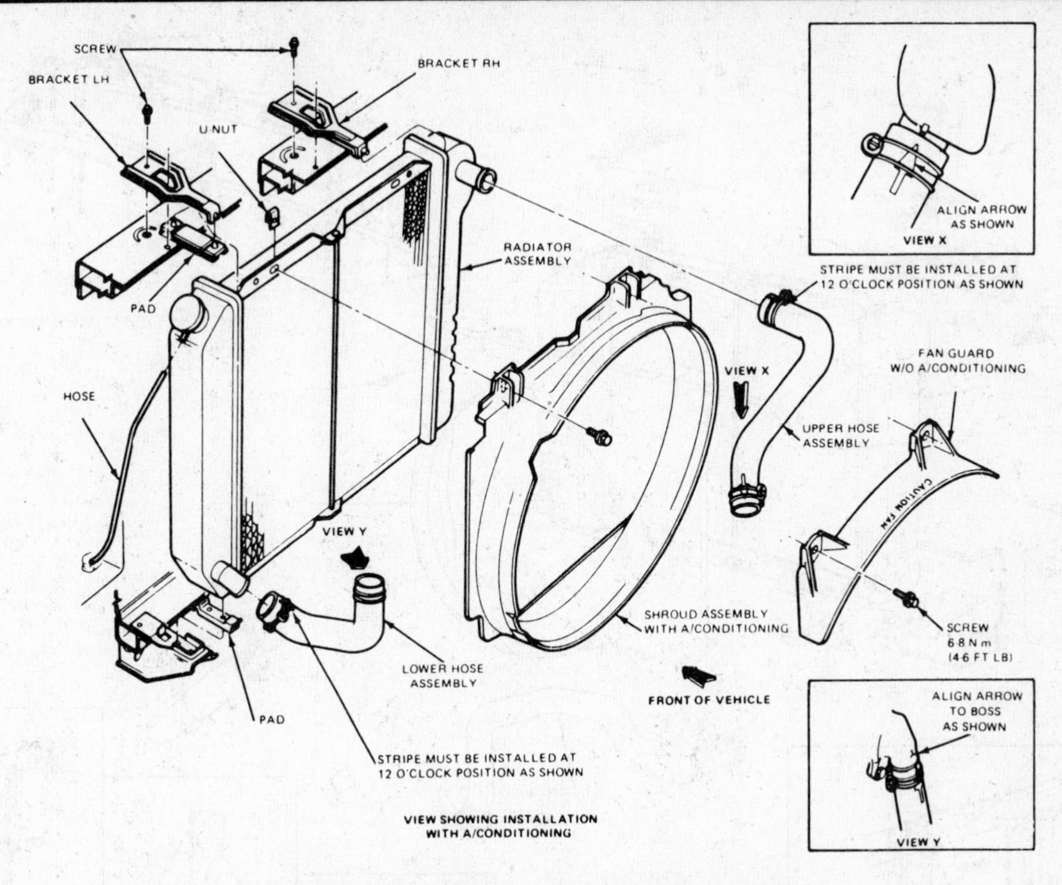

Radiator assembly — 2.8L and 2.9L engines

brace and the nut attaching the support brace to the intake manifold. Remove the support brace.

6. Remove the two bolts attaching the rear support to the bracket.

7. Remove the bolt attaching the compressor tab to the front brace and the two bolts attaching the compressor front legs to the bracket.

8. Remove the compressor.

9. Installation is the reverse of removal service procedure. Use new O-rings coated with clean refrigerant oil at all fittings. Evacuate, charge and leak test the system.

Radiator

REMOVAL AND INSTALLATION

1. Drain the cooling system. Remove the overflow tube from the coolant recovery bottle and from the radiator.

CAUTION

When draining the coolant, keep in mind that cats and dogs are attracted by the ethylene glycol antifreeze, and are quite likely to drink any that is left in an uncovered container or in puddles on the ground. This will prove fatal in sufficient quantity. Always drain the coolant into a sealable container. Coolant should be reused unless it is contaminated or several years old.

2. Disconnect the transmission cooling lines from the bottom of the radiator, if so equipped.

3. Remove the retaining bolts at the top of the shroud, and position the shroud over the fan, clear of the radiator.

4. Disconnect the upper and lower hoses from the radiator.

5. Remove the radiator retaining bolts or the upper supports and lift the radiator from the vehicle.

6. Install the radiator in the reverse order of removal. Fill the cooling system and check for leaks.

Air Conditioning Condenser

REMOVAL AND INSTALLATION

1. Discharge the refrigerant system. See Section 1.

2. Disconnect the refrigerant lines from the condenser using the proper spring lock tool. Cap all opening immediately!

NOTE: The fittings are spring-lock couplings and a special tool, T81P-19623-G, should be used. The larger opening end of the tool is for ½ in. discharge lines; the smaller end for ⅜ in. liquid lines.

To operate the tool, close the tool and push the tool into the open side of the cage to expand the garter spring and release the female fitting. If the tool is not inserted straight, the garter spring will cock and not release.

After the garter spring is released, pull the fittings apart.

3. Drain the cooling system.

CAUTION

When draining the coolant, keep in mind that cats and dogs are attracted by the ethylene glycol antifreeze, and are quite likely to drink any that is left in an uncovered container or in puddles on the ground. This will prove fatal in sufficient quantity. Always drain the coolant into a sealable container. Coolant should be reused unless it is contaminated or several years old.

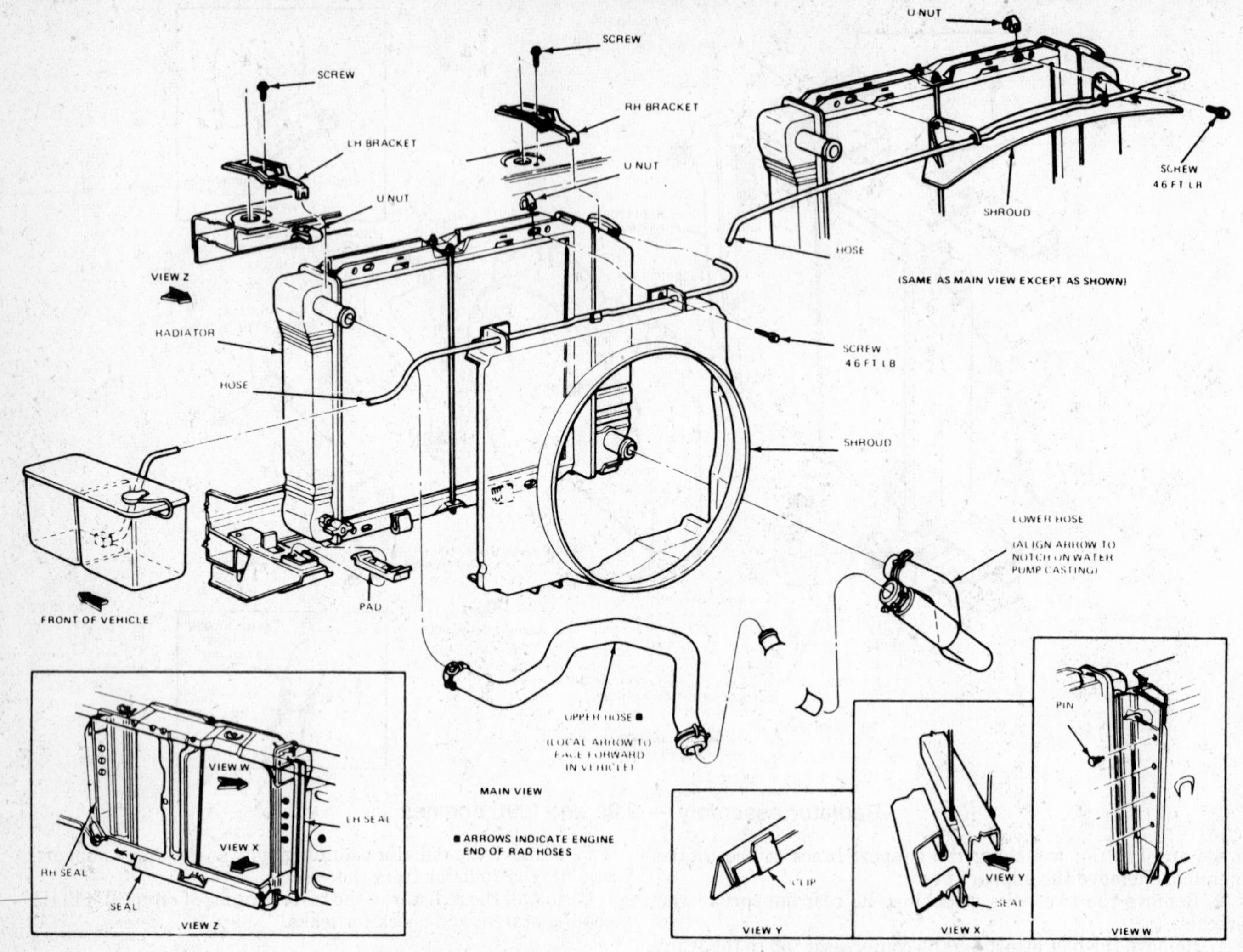

Radiator installation — 2.0L and 2.3L engines

4. Disconnect the upper radiator hose.
5. Remove the bolts retaining the ends of the radiator upper support to the side supports.
6. Carefully pull the top edge of the radiator rearward and remove the condenser upper support.
7. Lift out the condenser.
8. If a new condenser is being installed, add 1 fluid.oz. of new refrigerant oil to the new condenser. Installation is the reverse of removal. Always use new O-rings coated with clean refrigerant oil on the line fittings. Evacuate, charge and leak test the system.

Engine Fan

REMOVAL AND INSTALLATION

NOTE: Refer to to exploded view illustration before starting this sevice procedure.

2.0L and 2.3L Gas Engine
2.2L Diesel Engine

1. Unbolt and remove the fan finger guard.
2. Disconnect the overflow tube from the shroud, remove the mounting screws and lift the shroud off the brackets.
3. Place the shroud behind the fan.
4. Remove the 4 clutch/fan assembly-to-pulley screws and remove the clutch/fan assembly.
5. Remove the 4 fan-to-clutch screws.
6. Installation is the reverse of removal. Torque the fan-to-clutch screws to 55–70 inch lbs.; the fan/clutch assembly-to-pulley bolts to 12–18 ft. lbs.

2.8L Engine
2.9L Engine
3.0L Engine
2.3 Diesel Engines

1. Remove the fan shroud.
2. Using Strap Wrench D79L–6731–A and Fan Clutch Nut Wrench T83T–6312–B, or their equivalents, loosen the large nut attaching the clutch to the water pump hub.

NOTE: The nut is loosened clockwise.

3. Remove the fan/clutch assembly.
4. Remove the fan-to-clutch bolts.

To install:

5. Installation is the reverse of removal. Torque the fan-to-clutch bolts to 55–70 inch lbs.; the hub nut to 50–100 ft. lbs. Don't forget, the hub is tightened counterclockwise.

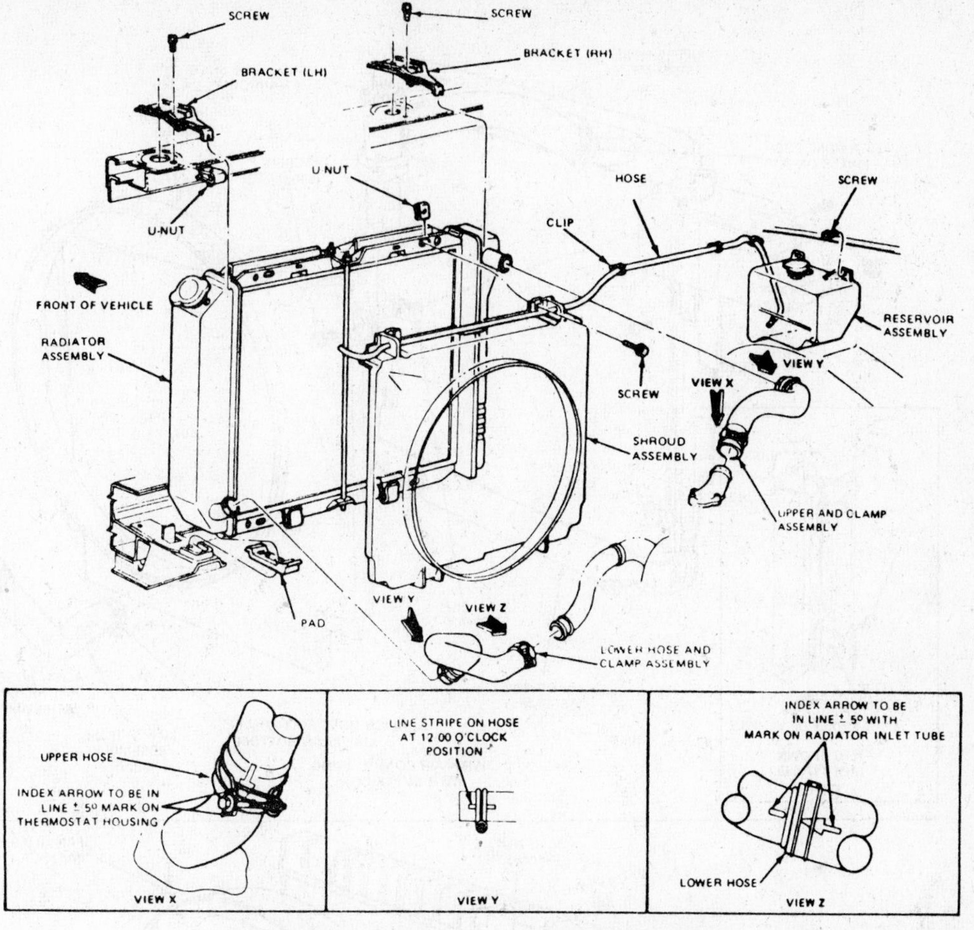

Radiator installation – 2.2L diesel engine

4.0L Engine

1. Remove the fan shroud.
2. Using Fan Clutch Pulley Holder T84T–6312–C and Fan Clutch Nut Wrench T84T–6312–D, or their equivalents, loosen the large nut attaching the clutch to the water pump hub.

NOTE: The nut is loosened clockwise.

3. Remove the fan/clutch assembly.
4. Remove the fan-to-clutch bolts.

To install:

5. Installation is the reverse of removal. Torque the fan-to-clutch bolts to 55–70 inch lbs.; the hub nut to 50–100 ft. lbs. Don't forget, the hub is tightened counterclockwise.

Water Pump

REMOVAL AND INSTALLATION

2.0L, 2.3L Engines

1. Disconnect the negative battery cable.
2. Remove the two bolts that retain the fan shroud and position the shroud back over the fan.
3. Remove the four bolts that retain the cooling fan. Remove the fan and shroud.
4. Loosen and remove the power steering and A/C compressor drive belts.
5. Remove the water pump pulley and the vent hose to the emissions canister.
6. Remove the heater hose at the water pump.
7. Remove the cam belt cover. Remove the lower radiator hose from the water pump.
8. Remove the water pump mounting bolts and the water pump. Clean all gasket mounting surfaces.
9. Install the water pump in the reverse order of removal. Coat the threads of the mounting bolts with sealer before installation.

2.8L, 2.9L Engines

1. Drain the coolant from the radiator and remove the lower hose and the return hose from the water inlet housing.

CAUTION

When draining the coolant, keep in mind that cats and dogs are attracted by the ethylene glycol antifreeze, and are quite likely to drink any that is left in an uncovered container or in puddles on the ground. This will prove fatal in sufficient quantity. Always drain the coolant into a sealable container. Coolant should be reused unless it is contaminated or several years old.

2. Using Tools T83T–6312–A and B remove the fan and clutch assembly from the front of the water pump.

NOTE: The fan clutch assembly uses a left hand thread, remove by turning the nut counterclockwise.

3. Loosen the alternator mounting bolts and remove the alternator belt.

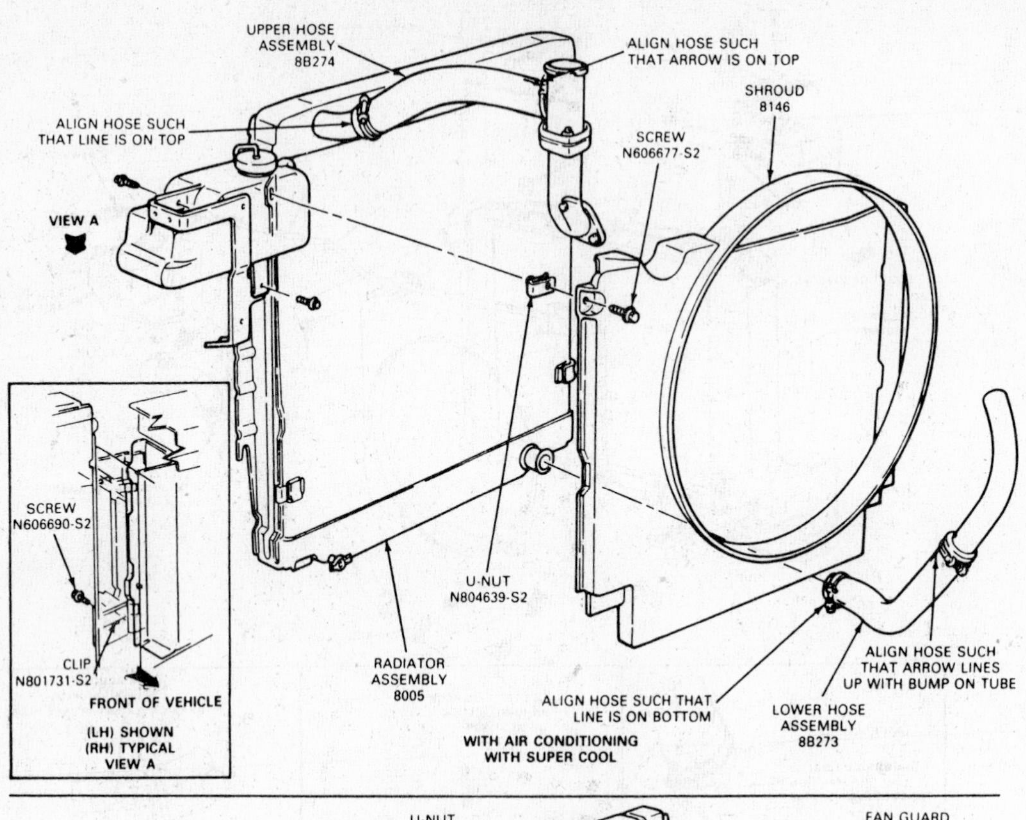

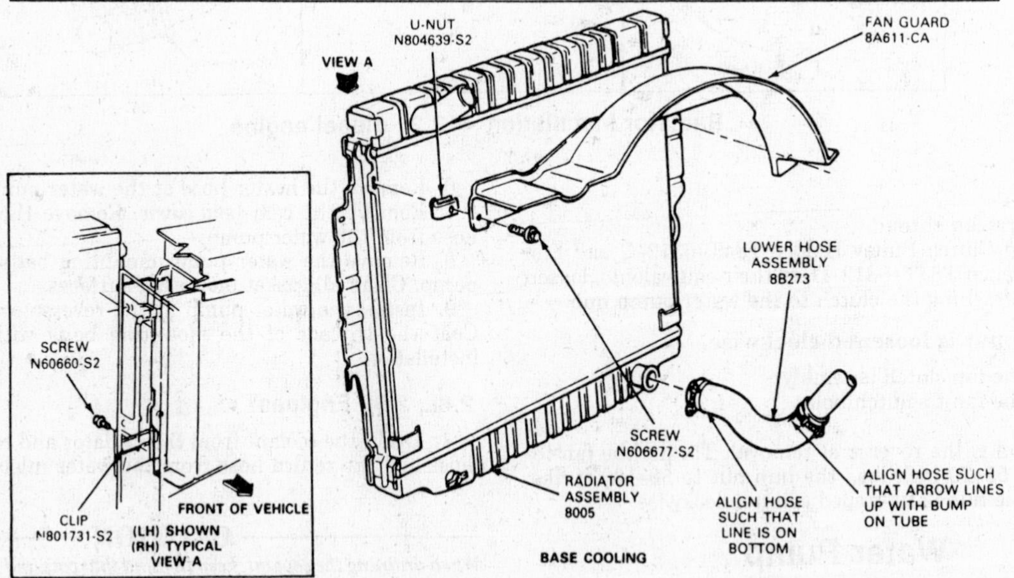

Radiator installation – 2.3L turbo diesel engine

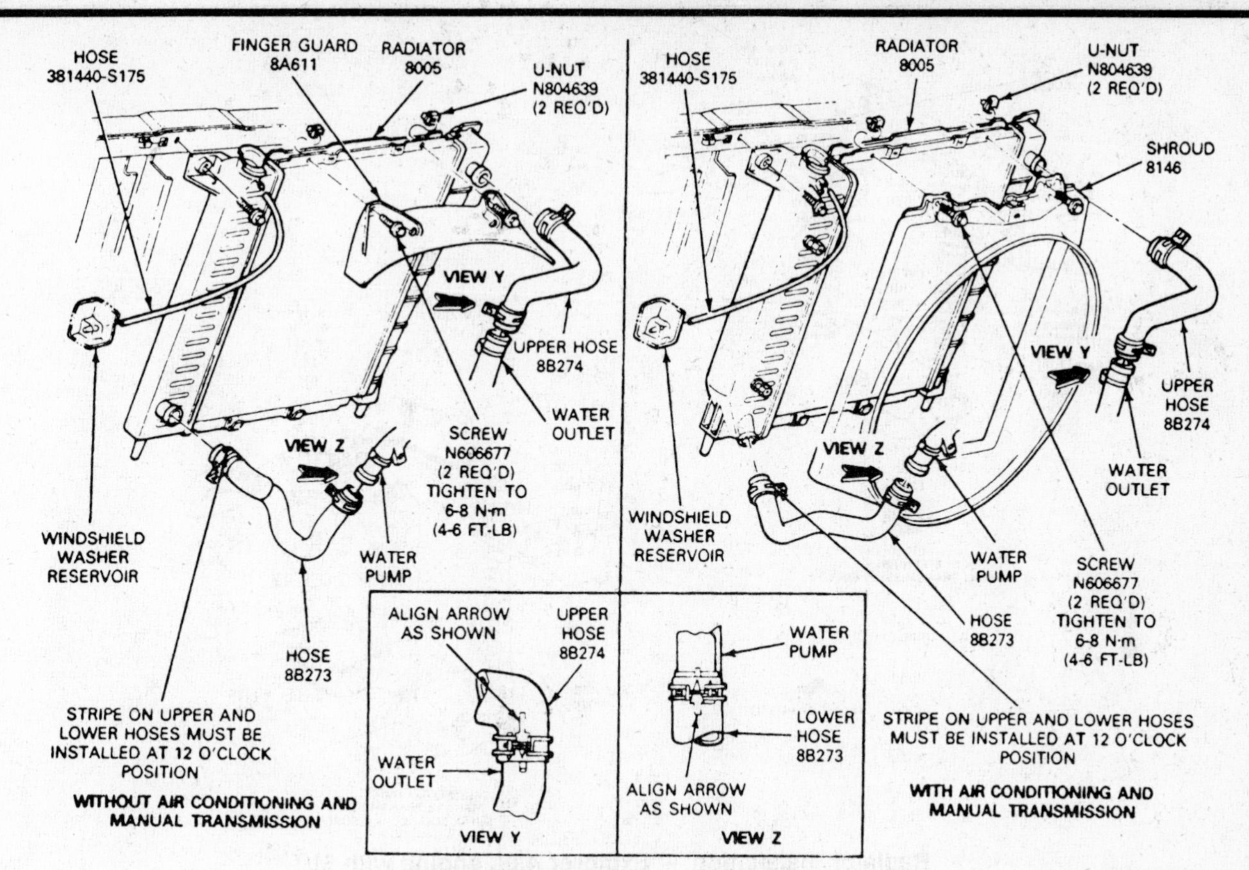

Radiator installation – 2.9L engine

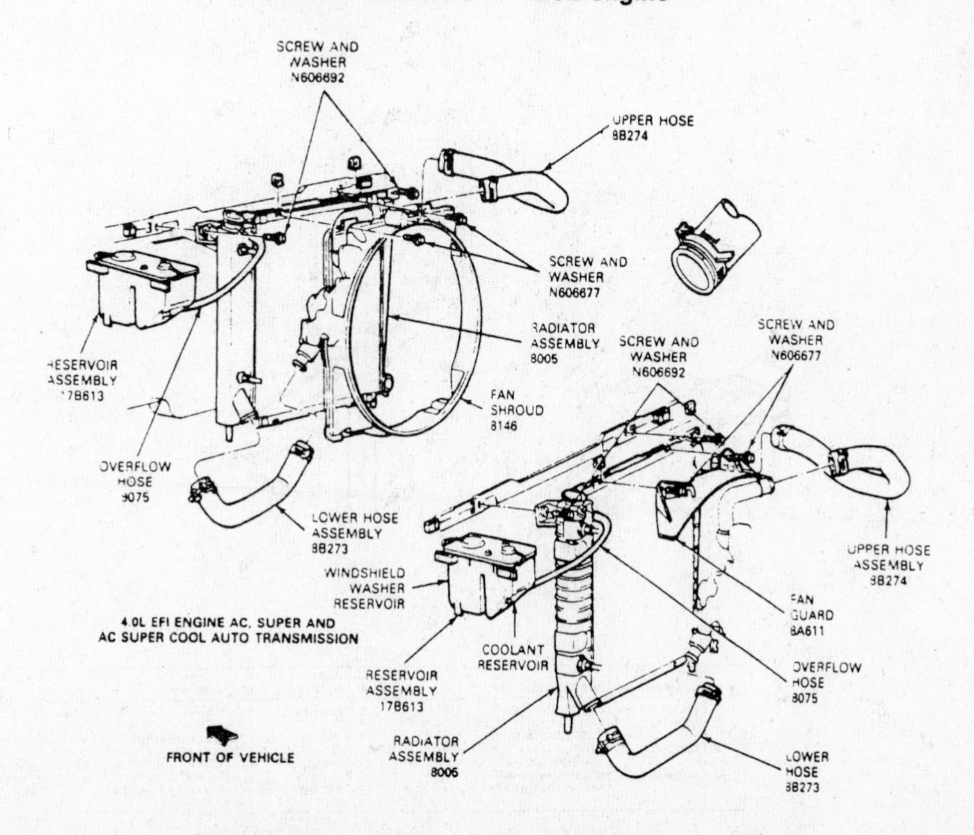

Radiator installation – Explorer 4.0L engine with A/T

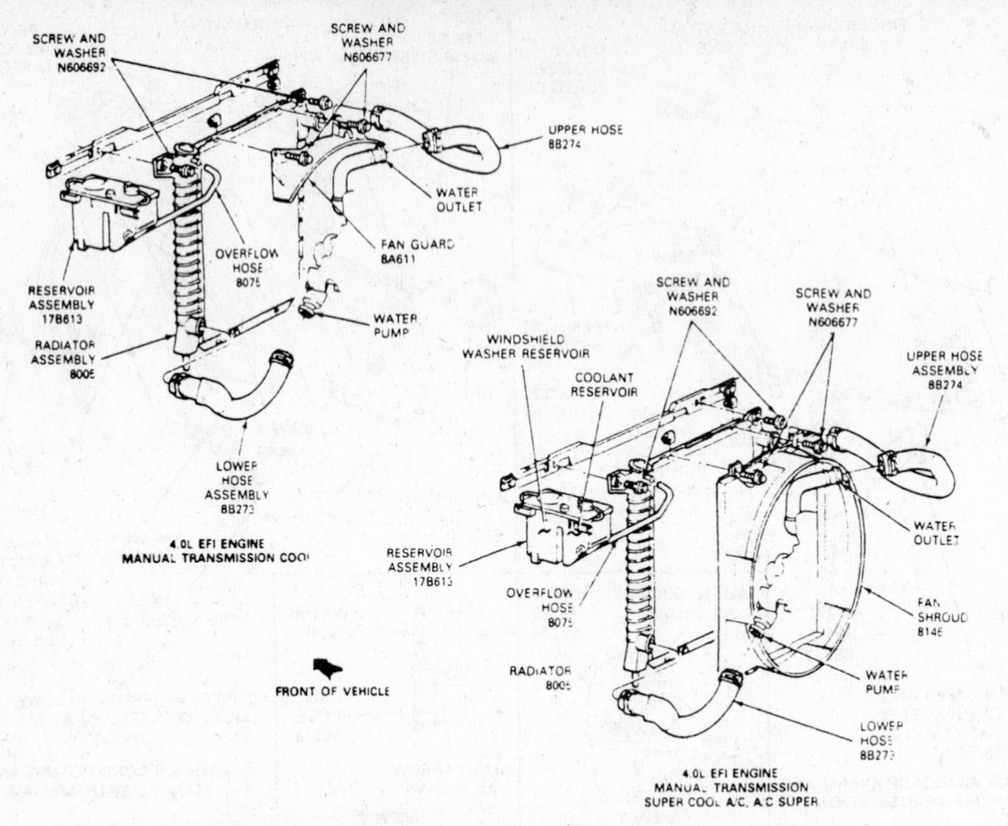

Radiator installation – Explorer 4.0L engine with M/T

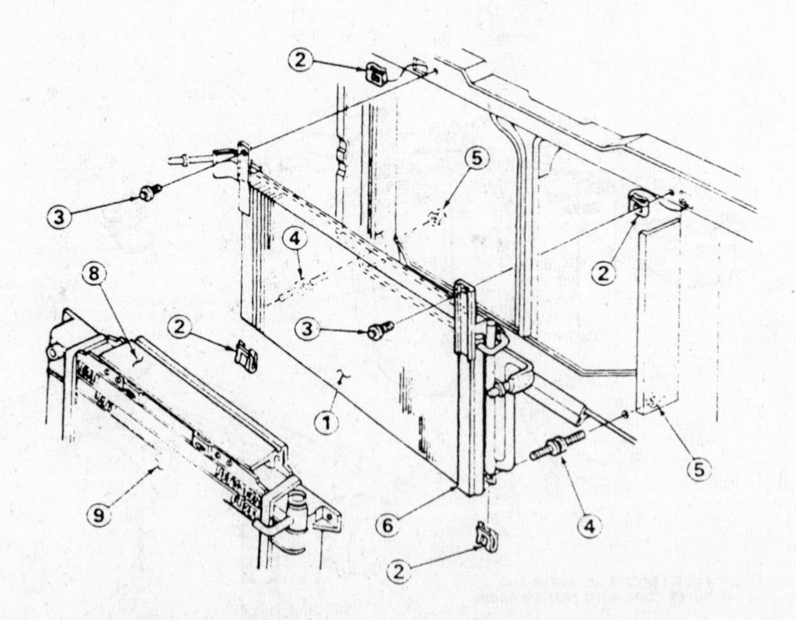

ITEM	PART NUMBER	DESCRIPTION	ITEM	PART NUMBER	DESCRIPTION
1.	19710	CONDENSER ASSY	5	N806046-S2	NUT & WASHER (19710 TO RADIATOR SUPPORT) (2 REQ'D)
2.	N623342-S2	"U" NUT (19710 TO RADIATOR SUPPORT) (4 REQ'D)	6	19E572	SEAL (2 REQ'D ON SOME AUTO TRANS. APPLICATIONS)
3	N605892-S2	BOLT — (19710 TO RADIATOR SUPPORT) (2 REQ'D)	7.	19E572	CONDENSER BOTTOM SEAL (WITH AUTO TRANS. ONLY)
4	N806047-S2	STUD & WASHER (TO LOWER MOUNTING BRACKET) (2 REQ'D)	8.	19E572	CONDENSER TOP SEAL
			9.	(REF)	RADIATOR ASSY

Condenser assembly

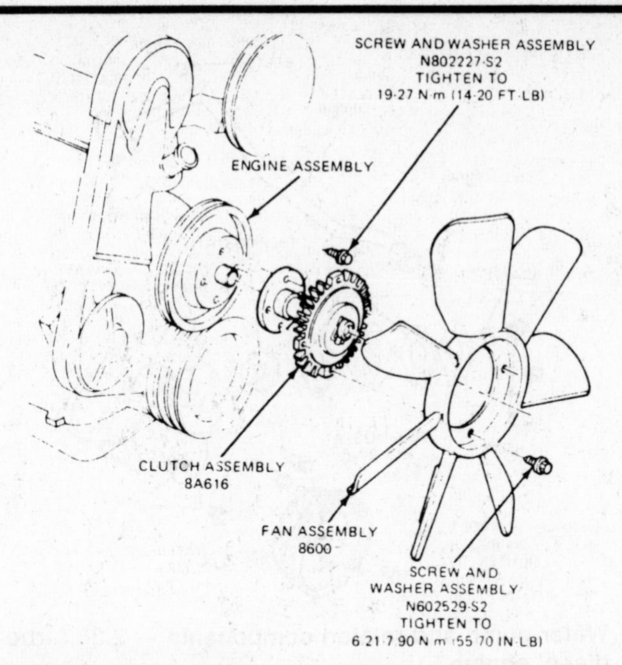

Radiator fan assembly – 2.0L and 2.3L gas engines

CAUTION: This nut has a LH thread and must be rotated clockwise for removal.

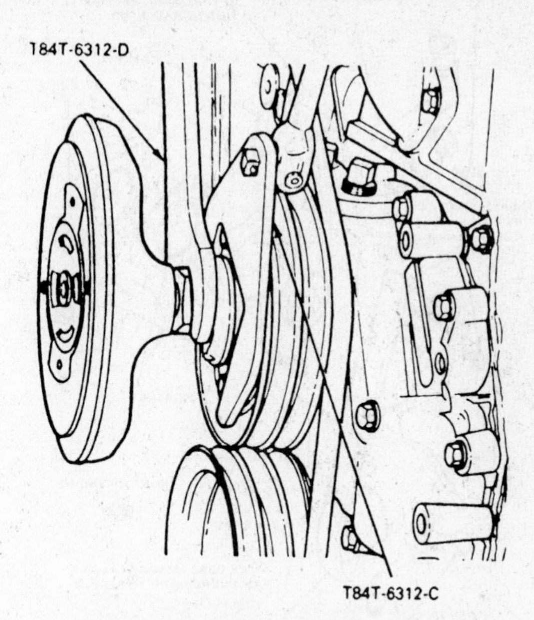

Removing the fan/clutch assembly

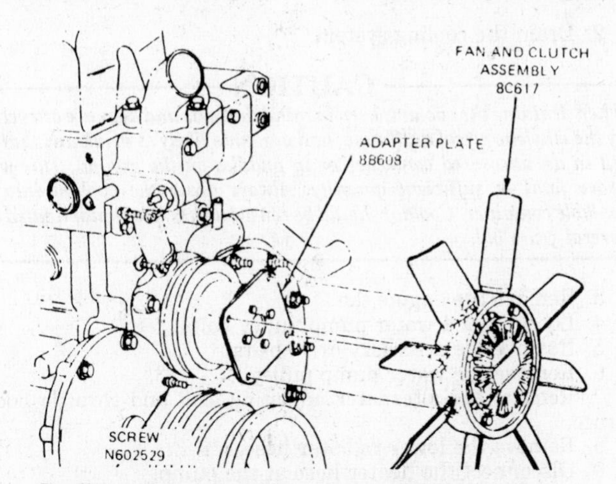

Radiator fan assembly – 2.2 diesel engine

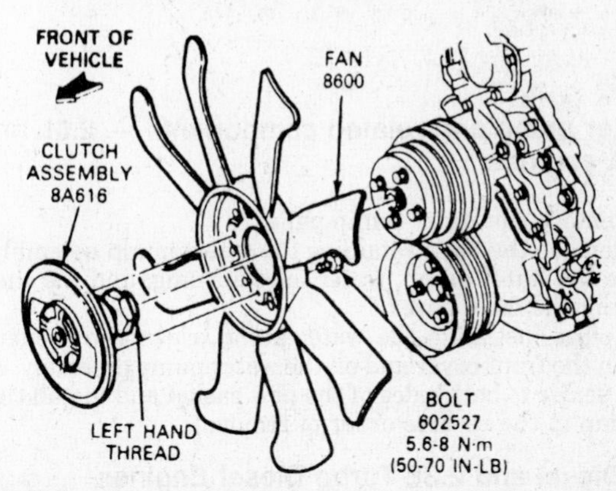

Radiator fan assembly – 2.9L and 3.0L engines

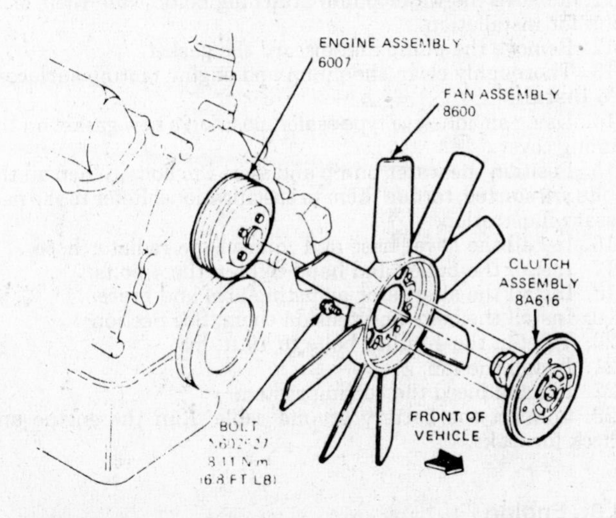

Radiator fan assembly – 2.8L gas engine

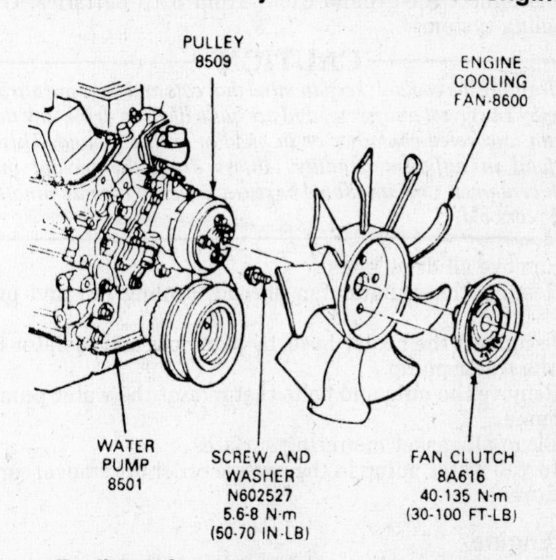

Radiator fan assembly – 4.0L engine

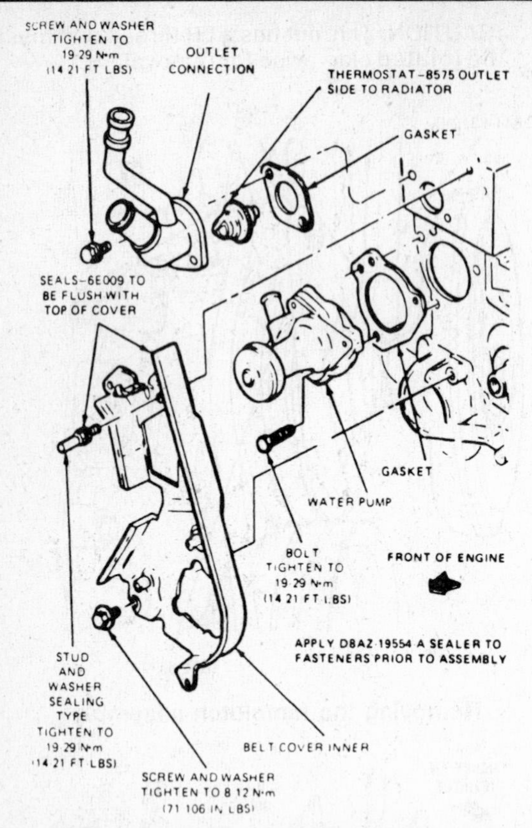

Water pump and related components – 2.0L and 2.3L engines

4. Remove the water pump pulley.
5. Remove the bolts retaining the water pump assembly and remove the water pump, water inlet housing, and the thermostat from the front cover.
6. Before installing the water pump, clean the gasket surfaces on the front cover and on the water pump assembly. Apply gasket sealer to both sides of the new gasket and install the water pump in the reverse order of removal.

2.2L Diesel and 2.3L Turbo Diesel Engines

1. Disconnect the ground cables from both batteries. Drain the cooling system.

CAUTION

When draining the coolant, keep in mind that cats and dogs are attracted by the ethylene glycol antifreeze, and are quite likely to drink any that is left in an uncovered container or in puddles on the ground. This will prove fatal in sufficient quantity. Always drain the coolant into a sealable container. Coolant should be reused unless it is contaminated or several years old.

2. Remove all drive belts.
3. Remove the radiator fan shroud, cooling fan and pump pulley.

Disconnect the heater hose, by-pass hose and radiator hose from the water pump.

4. Remove the nuts and bolts that mount the water pump to the engine.
5. Clean all gasket mounting surfaces.
6. Install water pump in the reverse order of removal service procedure.

3.0L Engine

1. Disconnect the battery ground cable.

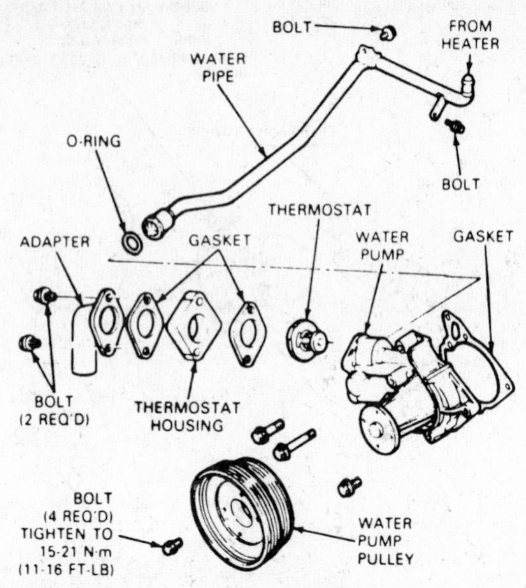

Water pump and related components – 2.3L turbo diesel engine

2. Drain the cooling system.

CAUTION

When draining the coolant, keep in mind that cats and dogs are attracted by the ethylene glycol antifreeze, and are quite likely to drink any that is left in an uncovered container or in puddles on the ground. This will prove fatal in sufficient quantity. Always drain the coolant into a sealable container. Coolant should be reused unless it is contaminated or several years old.

3. Remove the engine fan.
4. Loosen the 4 water pump pulley bolts.
5. Remove the accessory drive belts.
6. Remove the water pump pulley.
7. Remove the alternator adjusting arm and throttle body brace.
8. Remove the lower radiator hose.
9. Disconnect the heater hose at the pump.
10. Rotate the belt adjuster out of the way.
11. Remove the water pump attaching bolts. Note their location for installation.
12. Remove the pump and discard the gasket.
13. Thoroughly clean the pump and engine mating surfaces.

To install:

14. Using an adhesive type sealer, position a new gasket on the timing cover.
15. Position the water pump and start the bolts. When all the bolts are started, torque them to specifications. Refer to the necessary illustration.
16. Install the lower hose and connect the radiator hose.
17. Install the pulley and hand tighten the 4 bolts.
18. Install the alternator adjusting arm and brace.
19. Install the belts and tension them. See Section 1.
20. Tighten the 4 pulley bolts to 19 ft. lbs.
21. Install the fan.
22. Fill and bleed the cooling system.
23. Connect the battery ground cable. Run the engine and check for leaks.

4.0L Engine

1. Drain the cooling system.

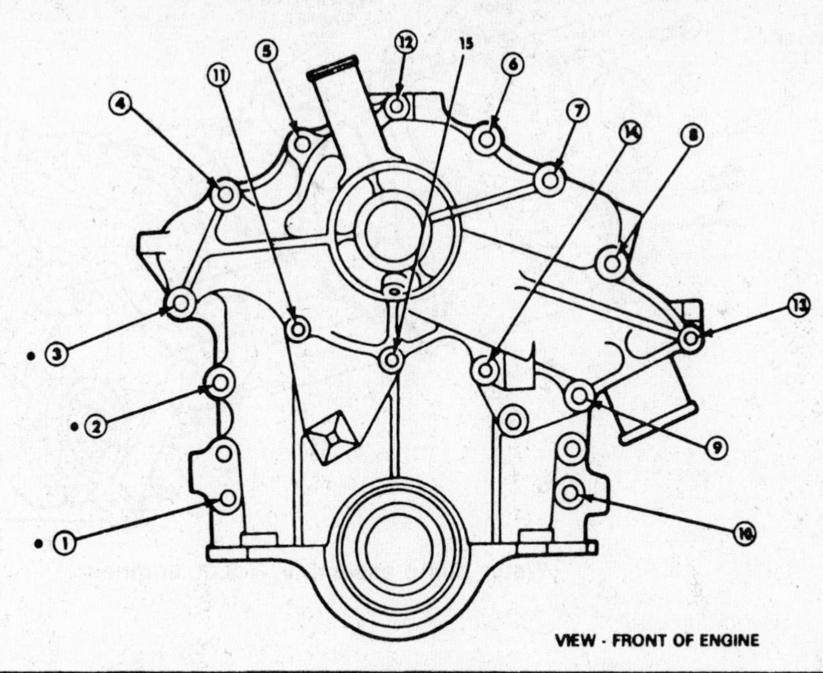

Fastener and Hole No.	Fasteners			
	Part No.	Size	N·m	Ft-Lb
● 1	N804215	M8 x 1.25 x 72.25	25	19
● 2	N804215	M8 x 1.25 x 72.25	25	19
● 3	N606547-S8	M8 x 1.25 x 70.0	25	19
4	N606547-S8	M8 x 1.25 x 70.0	25	19
5	N605909-S8	M8 x 1.25 x 42.0	25	19
6	N804154-S8	M8 x 1.25 x 99.3	25	19
7	N606547-S8	M8 x 1.25 x 70.0	25	19
8	N606547-S8	M8 x 1.25 x 70.0	25	19
9	N606547-S8	M8 x 1.25 x 70.0	25	19
10	N605909-S8	M8 x 1.25 x 42.0	25	19
11	N804168-S8	M6 x 1.0 x 25.0	10	7
12	N804168-S8	M6 x 1.0 x 25.0	10	7
13	N804168-S8	M6 x 1.0 x 25.0	10	7
14	N804168-S8	M6 x 1.0 x 25.0	10	7
15	N804168-S8	M6 x 1.0 x 25.0	10	7

NOTE: ●Apply Pipe Sealant with Teflon D8AZ-19554-A (ESG-M4G194-A) Sealer to Fastener Threads

Water pump front cover fastener chart — 3.0L engine

CAUTION

When draining the coolant, keep in mind that cats and dogs are attracted by the ethylene glycol antifreeze, and are quite likely to drink any that is left in an uncovered container or in puddles on the ground. This will prove fatal in sufficient quantity. Always drain the coolant into a sealable container. Coolant should be reused unless it is contaminated or several years old.

2. Remove the lower radiator hose.
3. Disconnect the heater hose at the pump.
4. Remove the fan and fan clutch assembly. You'll have to hold the pulley while loosening the fan clutch nut. There is a tool made for this purpose which will make the job easier, Ford tool No. T84T–6312–C.

NOTE: The nut has left-hand threads. It is removed by turning it clockwise.

5. Loosen the alternator mounting bolts and remove the belt. On vehicles with air conditioning, remove the alternator and bracket.
6. Remove the water pump pulley.
7. Remove the attaching bolts and remove the water pump.

To install:

8. Clean the mounting surfaces of the pump and front cover thoroughly. Remove all traces of gasket material.
9. Apply adhesive gasket sealer to both sides of a new gasket and place the gasket on the pump.
10. Position the pump on the cover and install the bolts finger-

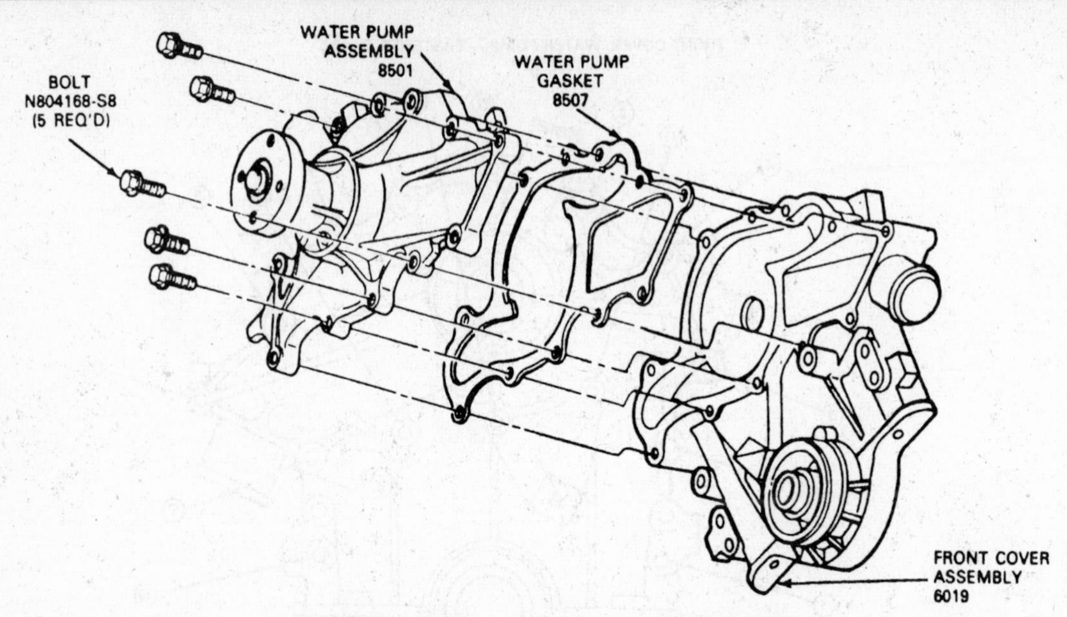

Water pump assembly — 3.0L engine

tight. When all bolts are in place, torque them to 72–108 inch lbs. (6–9 ft. lbs.).

11. Install the pulley.
12. On vehicles with air conditioning, install the alternator and bracket.
13. Install and adjust the drive belt.
14. Connect the hoses and tighten the clamps.
15. Install the fan and clutch assembly. Tighten the nut to 50–100 ft. lbs.

NOTE: The nut is tightened counterclockwise.

16. Fill and bleed the cooling system. Start the engine and check for leaks.

Cylinder Head

REMOVAL AND INSTALLATION

2.0L, 2.3L Carbureted Engines

1. Drain the cooling system. Disconnect the negative battery cables.

CAUTION

When draining the coolant, keep in mind that cats and dogs are attracted by the ethylene glycol antifreeze, and are quite likely to drink any that is left in an uncovered container or in puddles on the ground. This will prove fatal in sufficient quantity. Always drain the coolant into a sealable container. Coolant should be reused unless it is contaminated or several years old.

2. Remove the air cleaner.
3. Remove the valve cover.

NOTE: On models with air conditioning, remove the mounting bolts and the drive belt, and position the compressor out of the way. Remove the compressor upper mounting bracket from the cylinder head.

CAUTION

If the compressor refrigerant lines do not have enough slack to permit repositioning of the compressor without first disconnecting the refrigerant lines, the air conditioning system will have to be evacuated by a trained air conditioning serviceman. Under no circumstances should an untrained person attempt to disconnect the air conditioning refrigerant lines.

4. Remove the intake and exhaust manifolds from the head.
5. Remove the camshaft drive belt cover. Note the location of the belt cover attaching screws that have rubber grommets.
6. Loosen the drive belt tensioner and remove the belt.
7. Remove the water outlet elbow from the cylinder head with the hose attached.
8. Remove the cylinder head attaching bolts.
9. Remove the cylinder head from the engine.

To install:

10. Clean all gasket material and carbon from the top of the cylinder block and pistons and from the bottom of the cylinder head.
11. Position a new cylinder head gasket on the engine and place the head on the engine.

NOTE: If you encounter difficulty in positioning the cylinder head on the engine block, it may be necessary to install guide studs in the block to correctly align the head and the block. To fabricate guide studs, obtain two new cylinder head bolts and cut their heads off with a hack saw. Install the bolts in the holes in the engine block which correspond with cylinder head bolt holes Nos. 3 and 4, as identified in the cylinder head bolt tightening sequence illustration. Then, install the head gasket and head over the bolts. Install the cylinder head attaching bolts, replacing the studs with the original head bolts.

12. Using a torque wrench, tighten the head bolts in the sequence in two steps, first 50–60 ft. lbs., then 80–90 ft. lbs.
13. Install the camshaft drive belt.
14. Install the camshaft drive belt cover and its attaching bolts. Make sure the rubber grommets are installed on the bolts. Tighten the bolts to 6–13 ft. lbs.

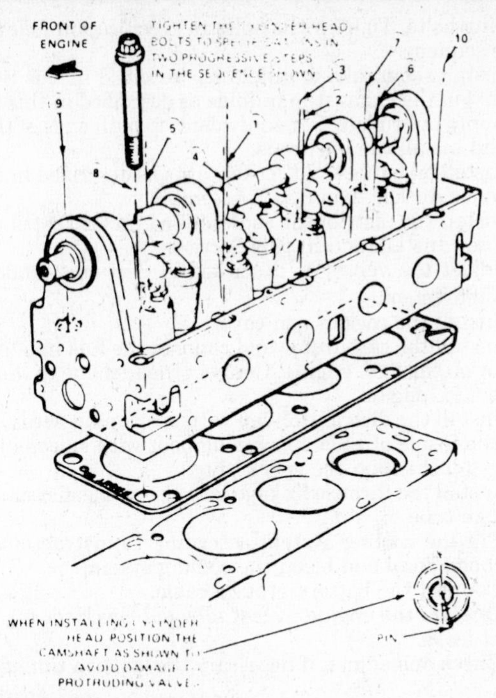

Cylinder head installation – 2.0L and 2.3L carbureted engines

15. Install the water outlet elbow and a new gasket on the engine and tighten the attaching bolts to 12–15 ft. lbs.
16. Install the intake and exhaust manifolds.
17. Assemble the rest of the components in reverse order of removal procedure.

2.3L EFI Engine

1. Drain cooling system.

CAUTION

When draining the coolant, keep in mind that cats and dogs are attracted by the ethylene glycol antifreeze, and are quite likely to drink any that is left in an uncovered container or in puddles on the ground. This will prove fatal in sufficient quantity. Always drain the coolant into a sealable container. Coolant should be reused unless it is contaminated or several years old.

2. Remove air cleaner assembly.
3. Remove one heater hose retaining screw to rocker cover.
4. Disconnect distributor cap and spark plug wire and remove assembly.
5. Remove spark plugs.
6. Disconnect required vacuum hoses.
7. Remove dipstick.
8. Remove rocker retaining bolts and remove cover.
9. Remove intake manifold retaining bolts.
10. Loosen alternator retaining bolts and remove belt from the pulley. Remove mounting bracket retaining bolts to the head.
11. Disconnect upper radiator hose at both ends and remove from vehicle.
12. Remove cam belt cover four bolts and remove cover. For power steering-equipped vehicles, move power steering pump bracket.
13. Loosen cam idler retaining bolts. Position idler in the unloaded position and tighten the retaining bolts.
14. Remove cam belt from the cam pulley and auxiliary pulley.
15. Remove four nuts and/or stud bolts retaining heat stove to exhaust manifold.
16. Remove the eight exhaust manifold retaining bolts.
17. Remove the cam belt idler and two bracket bolts.
18. Remove cam belt idler spring stop from the cylinder head.
19. Disconnect oil sending unit lead wire.
20. Remove cylinder head retaining bolts.
21. Remove the cylinder head.
22. Clean cylinder head gasket surface at the block.
23. Clean intake manifold gasket surface at the intake manifold.
24. Clean exhaust manifold gasket surface at the exhaust manifold.
25. Clean exhaust manifold gasket surface at the cylinder head.
26. Clean cylinder head gasket surface at the cylinder head.
27. Clean intake manifold gasket surface at the cylinder head.
28. Blow oil out of the cylinder head bolt block hoses.
29. Clean rocker cover gasket surface on the head.
30. Check cylinder head for flatness.

To install:

31. Position head gasket on the block.
32. Clean rocker arm cover (cam cover).
33. Install rocker cover gasket to the rocker cover.
34. Position cylinder head to block.
35. Install cylinder head retaining bolts and tighten to specifications.
36. Connect oil sending unit lead wires.
37. Install cam belt idler spring stop to the cylinder head.
38. Position cam belt idler to cylinder head, and install retaining bolts.
39. Install the eight exhaust manifold retaining bolts and/or stud bolts.
40. Install four nuts and/or stud bolts retaining heat stove to exhaust manifold.
41. Align distributor rotor with No. one plug location in the distributor cap.
42. Align cam gear with pointer.
43. Align crank pulley (TDC) with pointer on cam belt cover.
44. Position cam belt to pulleys (cam and auxiliary).
45. Loosen idler retaining bolts, rotate engine and check timing alignment.
46. Adjust belt tensioner and tighten retaining bolts.
47. Install four cam belt cover and four retaining bolts.
48. Connect upper radiator hose to engine and raditor and tighten retaining clamps.
49. Position alternator bracket to cylinder head and install retainers.
50. Position drive belt to pulley and adjust belt tension using Belt Tension Gauge Rotunda 021–00045 or equivalent.
51. Position intake manifold to head, and install retaining bolts.
52. Install rocker arm covers and retaining bolts.
53. Install spark plugs.
54. Install dipstick.
55. Connect appropriate vacuum hoses.
56. Position and connect spark plug wires and distributor.
57. Install one retaining heater/hose screw to the rocker cover.
58. Fill cooling system.
59. Start engine and check for leaks.
60. Adjust ignition timing and connect distributor vacuum line.
61. Adjust carburetor idle speed in mixture.
62. Install air cleaner.

2.8L, 2.9L Engines

1. Disconnect the battery ground cable.
2. Drain the radiator coolant.

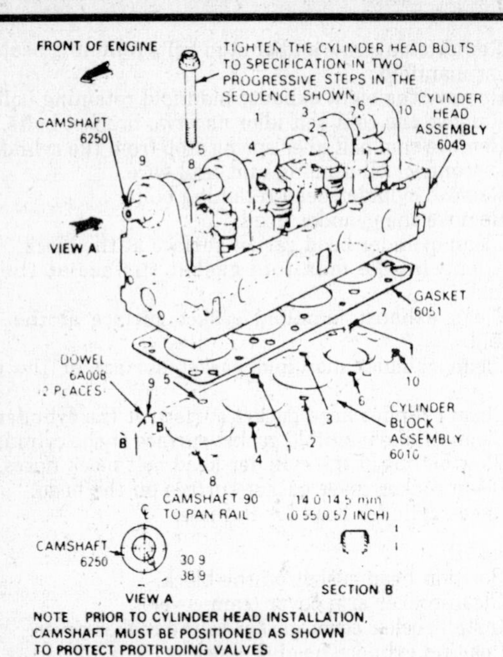

Cylinder head installation – 2.3L EFI engine

CAUTION

When draining the coolant, keep in mind that cats and dogs are attracted by the ethylene glycol antifreeze, and are quite likely to drink any that is left in an uncovered container or in puddles on the ground. This will prove fatal in sufficient quantity. Always drain the coolant into a sealable container. Coolant should be reused unless it is contaminated or several years old.

3. Remove the air cleaner from the carburetor and disconnect the throttle linkage on the 2.8L. Remove the intake tube from the throttle body and disconnect the throttle linkage and cover on the 2.9L.
4. Remove the distributor. Refer to Distributor, Removal and Installation as described earlier in this Section.
5. Remove the radiator hose and the by-pass hose from the thermostat housing and intake manifold.
6. Remove the rocker arm covers and the rocker arm shafts as described in this Section.
7. Remove the fuel line from carburetor and remove the carburetor on the 2.8L. Depressureize the fuel system and remove the fuel line from the fuel rail on the 2.9L.
8. Remove the intake manifold as described in this Section.
9. Remove and label the pushrods in order to keep them in sequence for proper assembly.
10. Remove the exhaust manifolds as described in this Section.
11. Remove the cylinder head attaching bolts. Remove the cylinder heads and discard the head gaskets.

To install:

12. Clean the cylinder heads, intake manifold, valve rocker arm cover and cylinder block gasket surfaces.
13. Place the cylinder head gaskets in position on the cylinder block.

NOTE: Gaskets are marked with the words FRONT and TOP for correct positioning. Left and right cylinder head gaskets are not interchangeable. Use new cylinder head bolts.

14. Install the fabricated alignment dowels in the cylinder block and install the cylinder head assemblies on the cylinder block, one at a time.
15. Remove the alignment dowels and install the cylinder head attaching bolts. Tighten the bolts to specification following the torque sequence.
16. Install the intake manifold as described in this Section.
17. Install the exhaust manifolds as described in this Section.
18. Apply Lubriplate®, or equivalent, to both ends of the pushrods and install the pushrods.
19. Install oil baffles and rocker arms as described in this Section under Rocker Arm Shaft Assembly.
20. Install the distributor as described under Distributor Removal and Installation in this Section.
21. Adjust the valves as described in Section 2 under Valve Lash adjustment.
22. Install the rocker arm covers.
23. Install the carburetor and connect the fuel line to the carburetor on the 2.8L engine. Connect the fuel line to the fuel rail on the 2.9L engine.
24. Install the distributor cap with spark plug wires attached. Coat the inside of each spark plug boot with silicone lubricant and install them on the spark plug.
25. Install the throttle linkage and, the air cleaner or air cleaner intake tube.
26. Fill the cooling system according to instructions on the underhood decal and bleed the cooling system.
27. Connect the battery ground cable.
28. Operate the engine at fast idle and check for oil, fuel and coolant leaks.
29. Check and adjust, if necessary the ignition timing and idle speed.

3.0L Engine

NOTE: Review the complete service procedure before starting this repair.

1. Drain the cooling system (engine cold) into a clean container and save the coolant for reuse.

CAUTION

When draining the coolant, keep in mind that cats and dogs are attracted by the ethylene glycol antifreeze, and are quite likely to drink any that is left in an uncovered container or in puddles on the ground. This will prove fatal in sufficient quantity. Always drain the coolant into a sealable container. Coolant should be reused unless it is contaminated or several years old.

2. Disconnect the battery ground cable.
3. Remove the air cleaner.
4. Relieve fuel pressure. Disconnect fuel lines as necessary. Mark vacuum line location and remove lines.
5. Disconnect upper and lower radiator hoses-position out of the way.
6. Remove coil assembly. Remove the throttle body. See Section 5.
7. Mark distributor housing to block and note rotor position. Remove the ignition wires from the spark plugs and locating studs. Remove the distributor.
8. If the left hand cylinder head is being removed:
   a. Remove the accessory drive belt.
   b. Remove the power steering pump and bracket assembly. DO NOT disconnect the hoses. Tie the assembly out of the way.
   c. Remove the engine oil dipstick and tube. Rotate or remove tube assembly.
   d. Remove the fuel line retaining bracket bolt from the front of cylinder head.
9. If the right hand head is being removed:
   a. Remove the accessory drive belt.
   b. Disconnect alternator electrical harnesses.
   c. Remove belt tensioner assembly.
   d. Remove the alternator and bracket.
   e. Remove hose from rocker cover to oil fill adapter.
10. Remove the spark plugs.

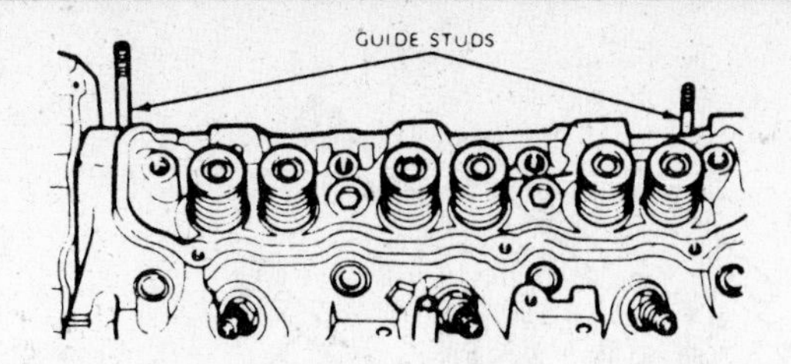

Cylinder head alignment studs — 2.8L and 2.9L engines

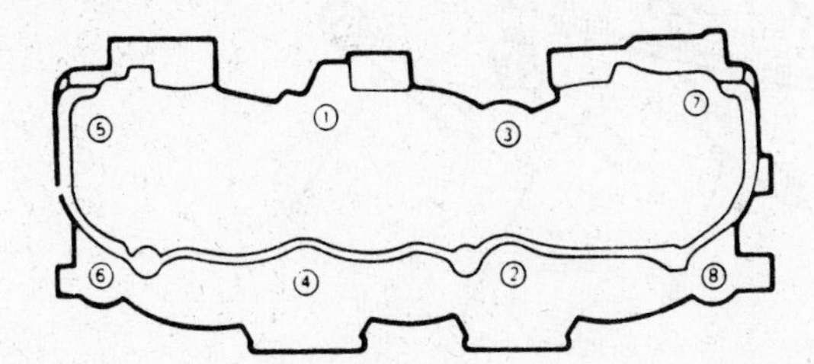

Cylinder head torque sequence — 2.8L engine

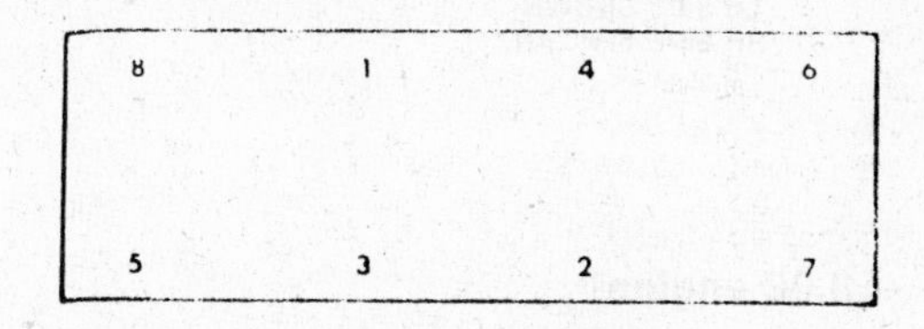

STEP 1 TIGHTEN IN SEQUENCE TO 30 N·m (22 FT-LB)
STEP 2 TIGHTEN IN SEQUENCE TO 70-75 N·m (51-55 FT-LB
STEP 3 WAIT 5 MINUTES
STEP 4 IN SEQUENCE, TURN ALL BOLTS 90 DEGREES

Cylinder head torque sequence — 2.9L engine

11. Remove the exhaust manifold(s).
12. Remove the rocker arm covers as previously described.
13. Loosen rocker arm fulcrum retaining bolts enough to allow the rocker arm to be lifted off the pushrod and rotate to one side.

NOTE: Regardless of which cylinder head is being removed #3 cylinder intake valve pushrod must be removed to allow removal of the intake manifold.

14. Remove the pushrods, keeping them in order so they may be installed in their original locations.
15. Remove the intake manifold as outlined. Refer to the necessary service procedure.
16. Loosen the cylinder head attaching bolts in reverse of the torque sequence, then remove the bolts and lift off the cylinder head(s). Remove and discard the old cylinder head gasket(s).

To install:

17. Clean the cylinder heads, intake manifold, valve rocker arm cover and cylinder block gasket surfaces of all traces of old gasket material and/or sealer. Refer to the following overhaul procedures for cylinder head component removal, valve replacement, resurfacing, etc.
18. Lightly oil all bolt and stud bolt threads except those specifying special sealant. Position the new head gasket(s) on the cylinder block, using the dowels for alignment. The dowels should be replaced if damaged.
19. Position the cylinder head(s) on the block and install the attaching bolts. Tighten the head bolts in sequence to 59 ft. lbs. Back off all bolts one full turn (360 DEGREES). Retighten the

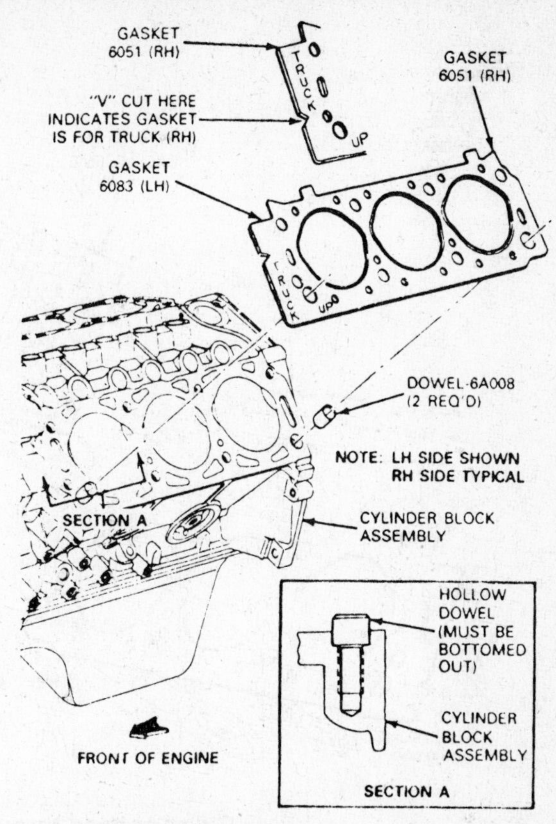

Cylinder head installation — 3.0L engine

cylinder head bolts in sequence in two service steps 37 ft. lbs. and final torque specification 68 ft. lbs.

20. Install intake manifold as outlined.
21. Dip each pushrod in heavy engine oil then install the pushrods in their original locations.
22. For each valve, rotate the crankshaft until the tappet rests on the heel (base circle) of the camshaft lobe before tightening the fulcrum attaching bolts. Position the rocker arms over the pushrods, install the fulcrums (TORQUE 8 FT. LBS. TO SEAT FULCRUM) and then tighten the fulcrum attaching bolts (FINAL TORQUE) to 24 ft. lbs. (32 Nm). REFER to the necessary illustration for details if necessary.

CAUTION

The fulcrums must be fully seated in the cylinder head and pushrods must be seated in the rocker arm sockets prior to final tightening.

23. Lubricate all rocker arm assemblies with heavy engine oil. If the original valve train components are being installed, a valve clearance check is not required. If, however, a component has been replaced, the valve clearance should be checked.
24. Install the exhaust manifold(s).
25. Install the dipstick tube and spark plugs.
26. Position the rocker arm cover with a new gasket on the cylinder head and install the retaining bolts. Note the location of the spark plug wire routing clip stud bolts.
27. Install the injector harness.
28. Install the distributor and wires.
29. Install the throttle body and new gasket. Refer to the necessary service procedures.
30. If the left hand cylinder head was removed, perform the following:

   a. Install the fuel line retaining bracket bolt to the front of cylinder head. Torque to 26 ft. lbs.

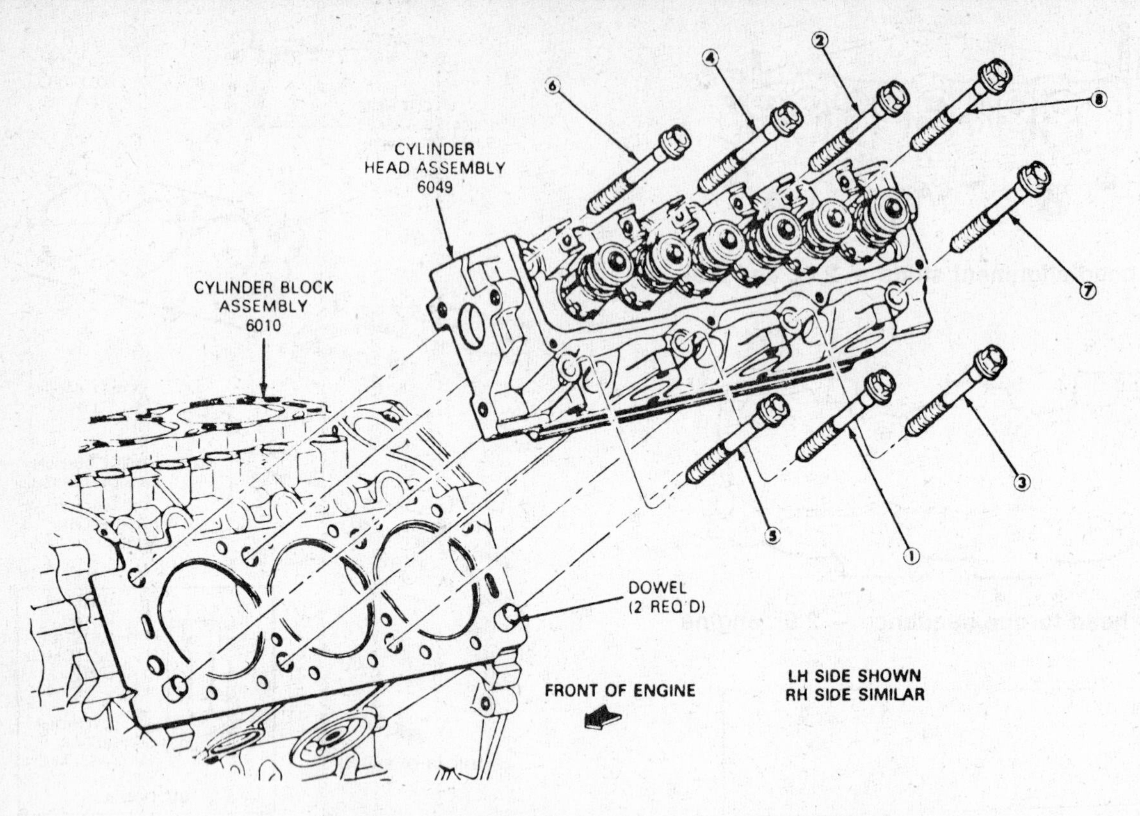

Cylinder head torque sequence — 3.0L engine

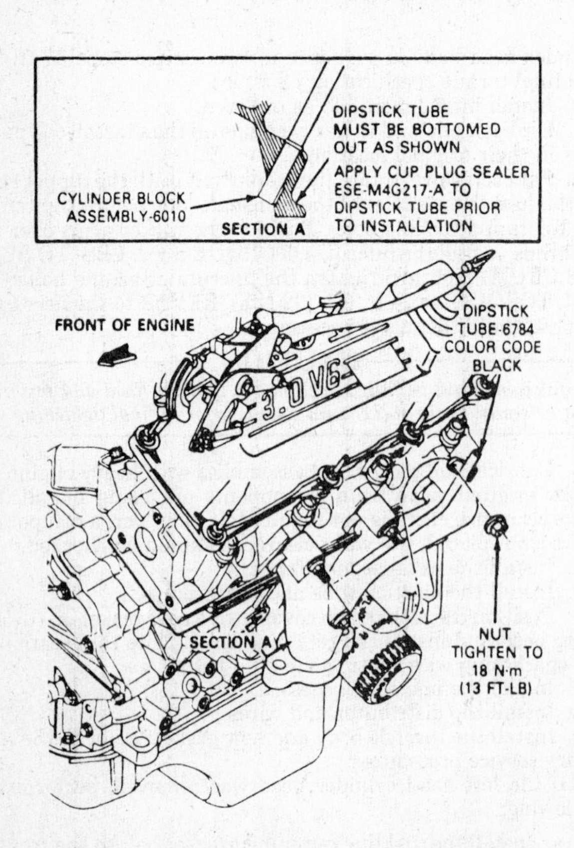

Cylinder head installation — 3.0L engine

b. Install the engine oil dipstick and tube assembly.
c. Install the power steering pump and bracket assembly.
d. Install the accessory drive belt.

31. If the right hand cylinder head was removed, perform the following:
    a. Install hose from rocker cover to oil fill adapter.
    b. Install the alternator and bracket.
    c. Install belt tensioner assembly.
    d. Reconnect alternator electrical harnesses.
    e. Install the accessory drive belt.
32. Connect fuel lines. Install fuel line saftey clips.
33. Install all radiator hoses. Connect vacuum lines.
34. Drain and change engine oil.

CAUTION

The EPA warns that prolonged contact with used engine oil may cause a number of skin disorders, including cancer! You should make every effort to minimize your exposure to used engine oil. Protective gloves should be worn when changing the oil. Wash your hands and any other exposed skin areas as soon as possible after exposure to used engine oil. Soap and water, or waterless hand cleaner should be used.

35. Install the air cleaner.
36. Fill and bleed the cooling system.
37. Connect the battery ground cable.
38. Start the engine and check for leaks. Verify base ignition timing. Adjust the speed control linkage as necessary.

4.0L Engine

1. Drain the cooling system (engine cold) into a clean container and save the coolant for reuse.

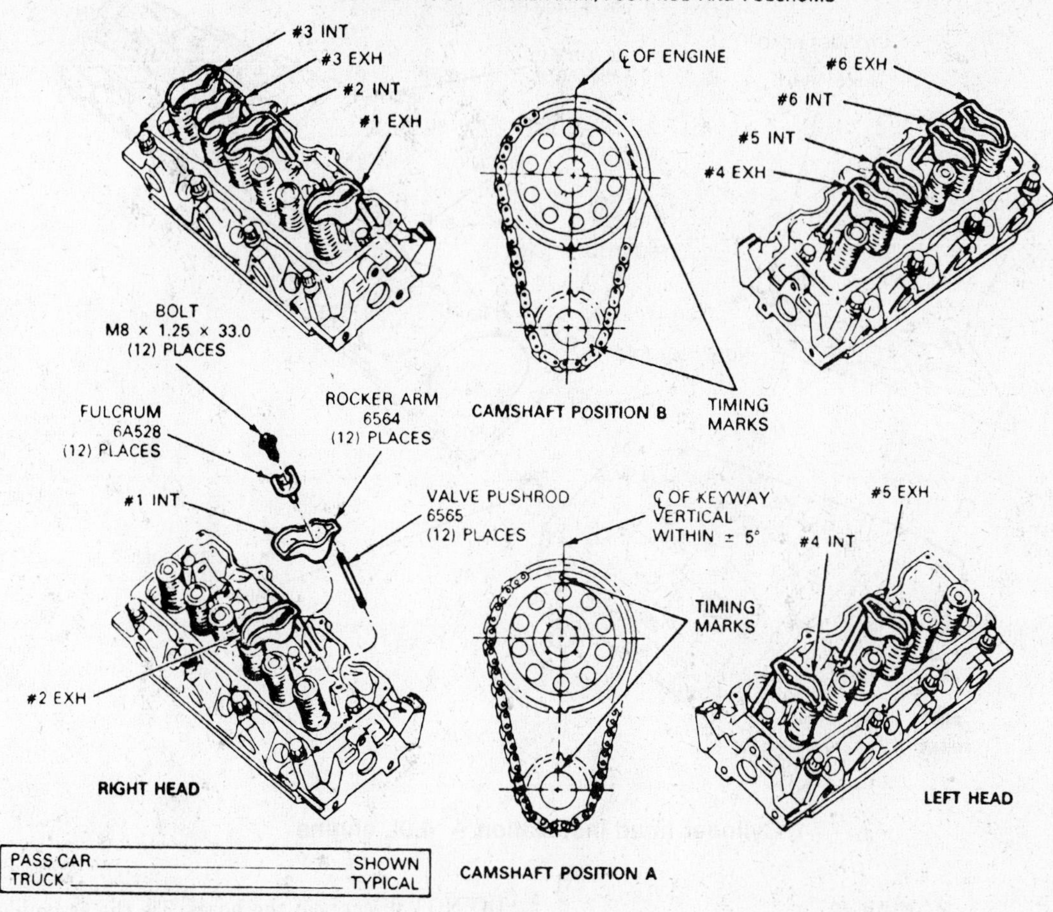

ASSEMBLY PROCEDURE

1. ROTATE CAMSHAFT TO POSITION "A" AS SHOWN.
2. INSTALL PUSH RODS (6565) (12) PLACES-PUSH RODS MUST BE SEATED PROPERLY ON TAPPET ASSEMBLY
3. INSTALL ROCKER ARMS (6564), FULCRUMS (6A528) AND BOLTS IN LOCATIONS AS SPECIFIED IN CAMSHAFT POSITION "A", TORQUE BOLTS TO 11 N·m AS REQ'D TO SEAT FULCRUMS IN CYLINDER HEAD
4. ROTATE CRANKSHAFT 120° TO POSITION "B"
5. INSTALL ROCKER ARMS (6564), FULCRUMS (6A528), AND BOLTS IN LOCATIONS AS SPECIFIED IN CAMSHAFT POSITION "B", TORQUE BOLTS TO 11 N·m AS REQ'D TO SEAT FULCRUMS IN CYLINDER HEAD

NOTE: FULCRUMS MUST BE FULLY SEATED IN CYLINDER HEADS AND PUSH RODS MUST BE FULLY SEATED IN ROCKER ARM SOCKETS PRIOR TO FINAL TORQUE.

6. APPLY ESE-M2C39-F OIL TO ROCKER ARM ASSEMBLIES.
7. FINAL TORQUE BOLTS TO 32.0 N·m (CAMSHAFT MAY BE IN ANY POSITION).

NOTE: CAMSHAFT POSITIONS "A" AND "B" ARE REQUIRED TO PLACE TAPPET ASSEMBLY ON BASE CIRCLE OF CAMSHAFT LOBE TO CHECK COLLAPSED TAPPET GAP.

FULCRUM AND BOLT MUST BE FULLY SEATED AFTER FINAL TORQUE

4.69-2.15 WITH TAPPET FULLY COLLAPSED ON BASE CIRCLE OF CAM LOBE AFTER ASSEMBLY REF. QUALITY AUDIT ONLY.

CYL. NO.	CAMSHAFT POSITION A	B
	SET GAP OF VALVES NOTED	
1	INT.	EXH.
2	EXH.	INT.
3	NONE	INT.-EXH.
4	INT.	EXH.
5	EXH.	INT.
6	NONE	INT.-EXH.

Initial valve adjustment — 3.0L engine

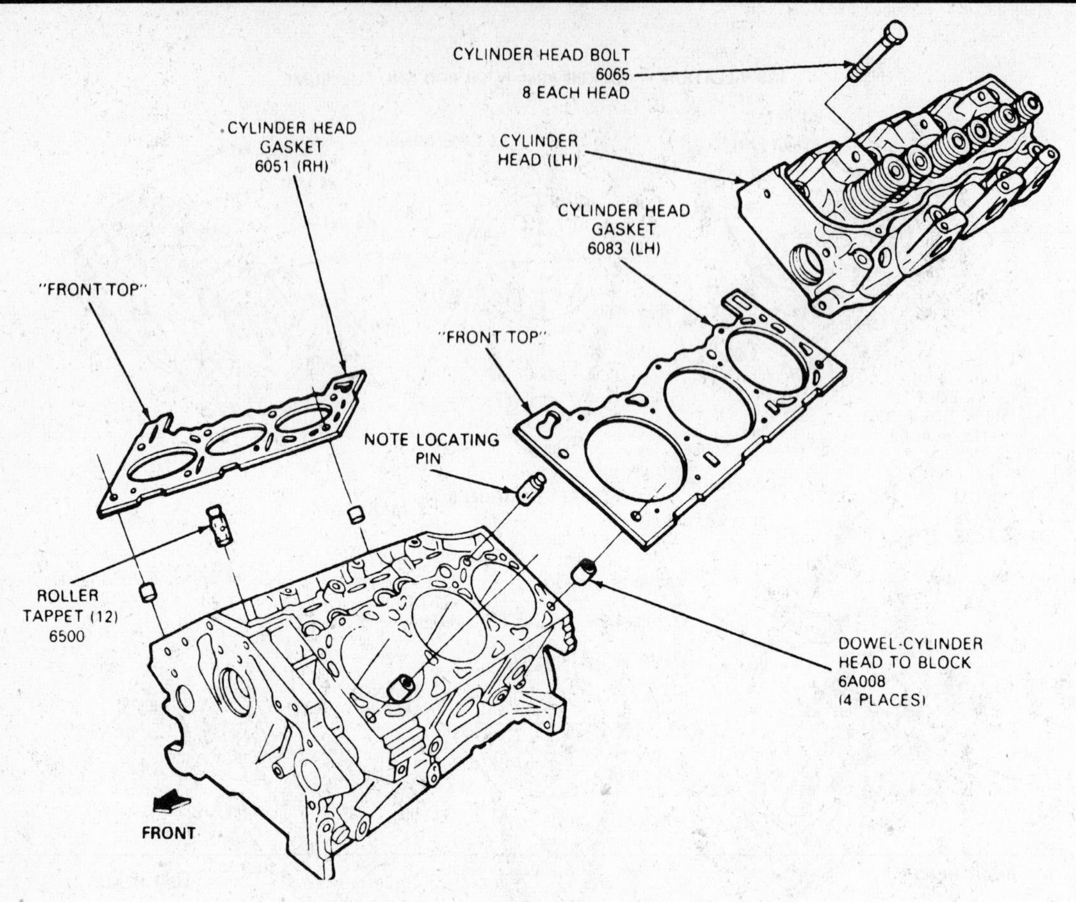

Cylinder head installation — 4.0L engine

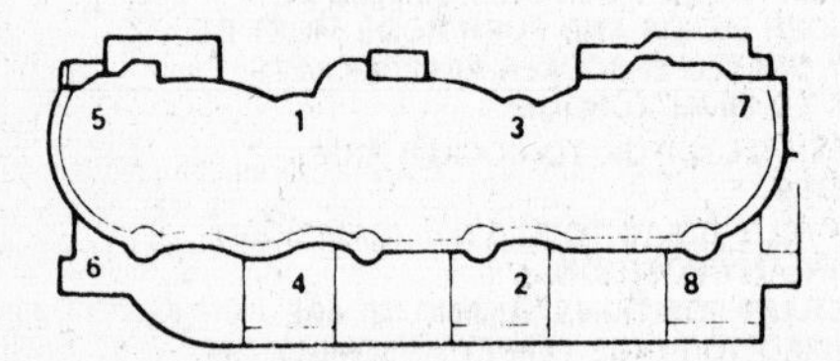

Cylinder head torque sequence — 4.0L engine

CAUTION

When draining the coolant, keep in mind that cats and dogs are attracted by the ethylene glycol antifreeze, and are quite likely to drink any that is left in an uncovered container or in puddles on the ground. This will prove fatal in sufficient quantity. Always drain the coolant into a sealable container. Coolant should be reused unless it is contaminated or several years old.

2. Disconnect the battery ground cable.
3. Remove the air cleaner.
4. Remove the upper and lower intake manifolds as described earlier.
5. If the left cylinder head is being removed:
   a. Remove the accessory drive belt.
   b. Remove the air conditioning compressor.
   c. Remove the power steering pump and bracket assembly. DO NOT disconnect the hoses. Tie the assembly out of the way.
   d. Remove the spark plugs.
6. If the right head is being removed:
   a. Remove the accessory drive belt.
   b. Remove the alternator and bracket.
   c. Remove the EDIS ignition coil and bracket.
   d. Remove the spark plugs.
7. Remove the exhaust manifold(s).
8. Remove the rocker arm covers as previously described.
9. Remove the rocker shaft assembly.
10. Remove the pushrods, keeping them in order so they may be installed in their original locations.
11. Loosen the cylinder head attaching bolts in reverse of the torque sequence, then remove the bolts and discard them. They cannot be re-used.
12. Lift off the cylinder head(s).
13. Remove and discard the old cylinder head gasket(s).

To install:

14. Clean the cylinder heads, intake manifolds, valve rocker arm cover and cylinder block gasket surfaces of all traces of old gasket material and/or sealer. Refer to the following overhaul procedures for cylinder head component removal, valve replacement, resurfacing, etc.
15. Lightly oil all bolt and stud bolt threads except those specifying special sealant. Position the new head gasket(s) on the cylinder block, using the dowels for alignment. The dowels should be replaced if damaged.

NOTE: The cylinder head(s) and intake manifold are torqued alternately and in sequence, to assure a correct fit and gasket crush.

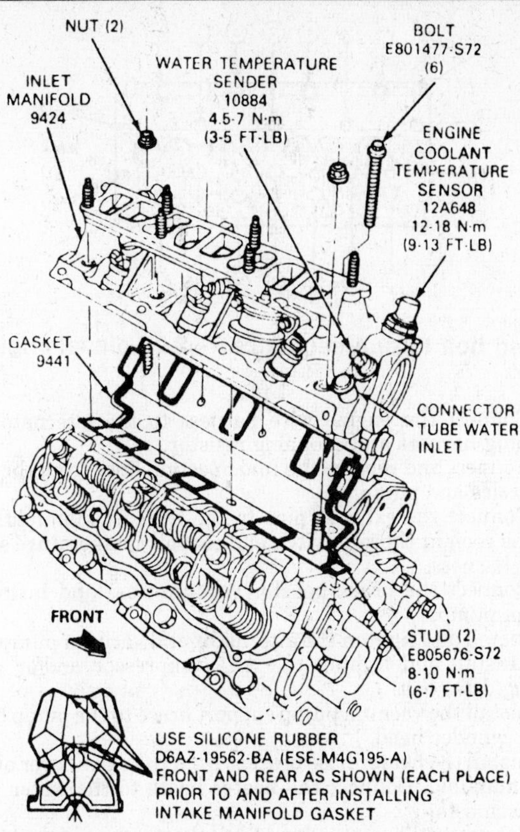

Cylinder head installation – 4.0L engine

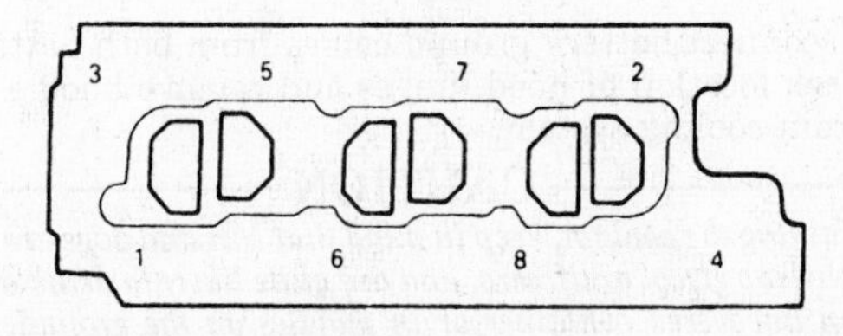

Lower intake manifold torque sequence – 4.0L engine

16. Position the cylinder head(s) on the block.
17. Apply a bead of RTV silicone gasket material to the mating joints of the head and block at the 4 corners. Install the intake manifold gasket and again apply the sealer.

NOTE: This sealer sets within 15 minutes, so work quickly!

18. Install the lower intake manifold and intsall the bolts and nuts for the manifold and head(s). Tighten all fasteners finger-tight.
19. Tighten the intake manifold fasteners, in sequence, to 36–72 inch lbs.

WARNING: Do not re-use the old head bolts. ALWAYS use new head bolts!

20. Torque the head bolts, in sequence, to 59 ft. lbs.

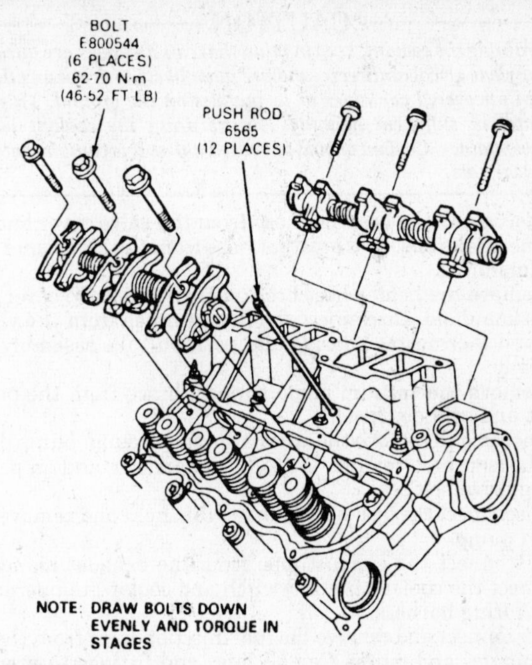

Installation rocker arm shaft assemblies – 4.0L engine

21. Tighten the intake manifold fasteners, in sequence, to 6–11 ft. lbs.
22. Tighten the head bolts, in sequence, an additional 80–85 DEGREES tighter. 85 degrees is a little less than 1/4 turn. 1/4 turn would equal 90 degrees.
23. Torque the intake manifold fasteners, in sequence, to 11–15 ft. lbs.; then, in sequence, to 15–18 ft. lbs.
24. Dip each pushrod in heavy engine oil then install the pushrods in their original locations.
25. Install the rocker shaft assembly(ies) and tighten the bolts to 46–52 ft. lbs., front to rear, in several equal stages.
26. Apply another bead of RTV sealer at the 4 corners where the intake manifold and heads meet.
27. Install the rocker covers, using new gaskets coated with sealer. Torque the bolts to 36–60 inch lbs. After 2 minutes, retorque the cover bolts.
28. Install the upper intake manifold. Torque the nuts to 15–18 ft. lbs.
29. Install the exhaust manifold(s).
30. Install the spark plugs and wires.
31. If the left head was removed, install the power steering pump, compressor and drive belt.
32. If the right head was removed, install the EDIS coil and bracket, alternator and bracket, and the drive belt.
33. Install the air cleaner.
34. Fill the cooling system. See Chapter 1.

NOTE: At this point, it's a good idea to change the engine oil. Coolant contamination of the engine oil often occurs during cylinder head removal.

35. Connect the battery ground cable.
36. Start the engine and check for leaks.

2.2L Diesel Engine

1. Disconnect the ground cables from both batteries.
2. Mark the hood hinges for realignment on installation and remove the hood. Drain the cooling system.

CAUTION

When draining the coolant, keep in mind that cats and dogs are attracted by the ethylene glycol antifreeze, and are quite likely to drink any that is left in an uncovered container or in puddles on the ground. This will prove fatal in sufficient quantity. Always drain the coolant into a sealable container. Coolant should be reused unless it is contaminated or several years old.

3. Disconnect the breather hose from the valve cover and remove the intake hose and breather hose from the air cleaner and intake manifold.
4. Remove the heater hose bracket from the valve cover and exhaust manifold. Disconnect the heater hoses from the water pump and thermostat housing and position tube assembly out of the way.
5. Remove the vacuum pump support brace from the pump bracket and cylinder head.
6. Loosen and remove the alternator and vacuum pump drive belts. Loosen and remove the A/C compressor and/or power steering drive belt.
7. Disconnect the brake booster vacuum hose and remove the vacuum pump.
8. Disconnect the exhaust pipe from the exhaust manifold. Disconnect the coolant thermoswitch and coolant temperature sender wiring harness.
9. Disconnect and remove the fuel injection lines from the injector nozzles and pump. Cap all lines and fittings to prevent dirt from entering the system.
10. Disconnect the engine wire harness from the alternator, the glow plug harness and dropping resistor and position the harness out of the way.
11. Disconnect the fuel lines from both sides of the fuel heater. Remove the fuel filter assembly from the mounting bracket and position out of the way with the fuel line attached.
12. Loosen the lower No. 3 intake port nut and the bolt on the injection pump; disconnect the lower fuel return line from the intake manifold stud and the upper fuel return line.
13. If equipped with power steering, remove the bolt that attaches the pump rear support bracket to the cylinder head.
14. Remove the upper radiator hose. Disconnect the by-pass hose from the thermostat housing.
15. Remove the A/C compressor and position out of the way with the lines still attached.

CAUTION

Do not disconnect the compressor lines unless the proper tools are on hand to discharge the system and you are familiar with the procedure.

16. Remove the valve cover, rocker arm shaft assembly and the pushrods. Identify the pushrods and keep them in order for return to their original position.
17. Remove the cylinder head attaching bolts, starting at the ends of the head, working alternately toward the center. Remove the cylinder head from the truck.
18. Clean all gasket mounting surfaces.

To install:

19. Position a new cylinder head gasket into position. Install the cylinder head attaching bolts, torque the cylinder head bolts in the proper sequence to 80–85 ft. lbs.
20. Install the pushrods, rocker arm shaft assembly and the valve cover.
21. Install the A/C compressor.
22. Install the upper radiator hose and the by-pass hose to the thermostat housing.
23. If equipped with power steering, install the bolt that attaches the pump rear support bracket to the cylinder head.
24. Connect the lower fuel return line to the intake manifold stud and the upper fuel return line. Tighten the lower No. 3 intake port nut and the bolt on the injection pump;
25. Install the fuel filter assembly. Connect the fuel lines from both sides of the fuel heater.

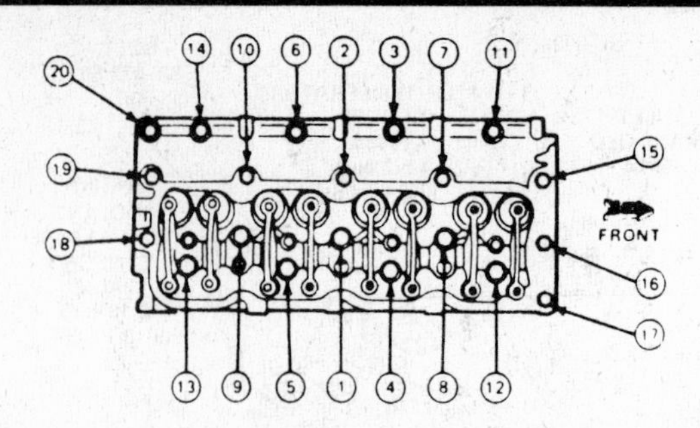

Head bolt torque sequence – 2.2L diesel engine

26. Connect the engine wire harness to the alternator, the glow plug harness and dropping resistor.
27. Connect and remove the fuel injection lines from the injector nozzles and pump.
28. Connect the exhaust pipe to the exhaust manifold. Connect the coolant thermoswitch and coolant temperature sender wiring harness.
29. Connect the brake booster vacuum hose and install the vacuum pump.
30. Install and tighten the alternator and vacuum pump drive belts. Install and tighten the A/C compressor and/or power steering drive belt.
31. Install the vacuum pump support brace to the pump bracket and cylinder head.
32. Install the heater hose bracket from the valve cover and exhaust manifold. Connect the heater hoses to the water pump and thermostat.
33. Connect the breather hose from the valve cover and install the intake hose and breather hose to the air cleaner and intake manifold.
34. Install the hood.
35. Connect the ground cables from both batteries. Start the engine and check for leaks.

2.3L Turbo Diesel Engine

1. Disconnect battery ground cables from both batteries.
2. Mark location of hood hinges and remove hood.
3. Drain cooling system.

CAUTION

When draining the coolant, keep in mind that cats and dogs are attracted by the ethylene glycol antifreeze, and are quite likely to drink any that is left in an uncovered container or in puddles on the ground. This will prove fatal in sufficient quantity. Always drain the coolant into a sealable container. Coolant should be reused unless it is contaminated or several years old.

4. Disconnect breather hose from rocker cover.
5. Remove heater hose clamp from rocker cover and position hoses out of the way.
6. Remove cooling fan and shroud.
7. Remove accessory drive belts.
8. Remove upper front timing belt cover.
9. Loosen and remove camshaft/injection pump timing belt from camshaft sprocket.
10. Remove inlet hose between air cleaner and turbo inlet.
11. Raise vehicle and disconnect muffler inlet pipe from turbo exhaust fitting. Lower vehicle.
12. Remove fuel conditioner and bracket and position assembly out of the way.
13. Disconnect and remove fuel lines between injection pump and nozzles. Cap all lines and fittings using Protective Cap Set T85L–9395–A or equivalent.
14. Disconnect heater hose from fitting on LH rear of cylinder head.

15. Remove A/C compressor and mounting bracket.
16. disconnect glow plug electrical leads from No. 2 and No. 3 glow plugs.
17. Disconnect coolant temperature switch wire.
18. Remove intake and exhaust manifold.
19. Remove rocker cover.
20. Loosen cylinder head bolts using a 10mm hex-head socket. Loosen bolts in sequence as shown.
21. Remove cylinder head. Remove old head gasket.
22. Remove components as necessary.

To install:

23. Assemble components to head as necessary.
24. Clean gasket mating surfaces on cylinder head and engine block.
25. Position new cylinder head gasket on engine block.
26. Position cylinder head on engine block and install cylinder head bolts. Tighten bolts in sequence shown, as follows:
    a. Tighten bolts in sequence to 38–42 ft. lbs.
    b. Tighten bolts in sequence to 76–83 ft. lbs.
27. Install rocker cover.

NOTE: Ensure half-moon gasket is installed in rear of cylinder.

28. Install intake and exhaust manifolds.
29. Connect coolant temperature switch connector.
30. Connect glow plug connectors to No. 3 and No. 4 glow plugs.
31. Install A/C compressor bracket and A/C compressor.
32. Connect heater hose to fitting on LH rear of cylinder head.
33. Connect fuel lines to injection pump and nozzles. Tighten line nuts to specification.
34. Install fuel conditioner and bracket.
35. Raise vehicle and connect muffler inlet pipe to turbo exhaust fitting. Lower vehicle.
36. Connect inlet hose between air cleaner and turbo inlet.
37. Install and adjust injection pump/camshaft timing belt.
38. Install upper timing belt cover.
39. Install accessory drive belts.
40. Position heater hoses on rocker cover and install clamp.
41. Connect breather hose to rocker cover.
42. Change engine oil and filter. Fill engine with specified quantity and quality of oil.
43. Fill and bleed cooling system.
44. Connect battery ground cables to both batteries.
45. Run engine and check for oil, fuel and coolant leaks.
46. Install hood.

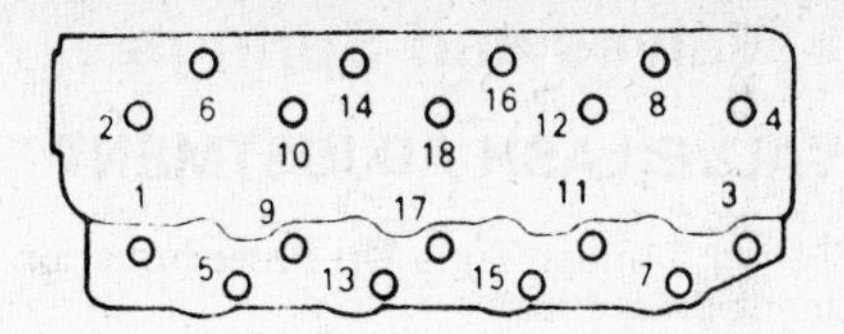

Cylinder head loosening sequence – 2.3L turbo diesel engine

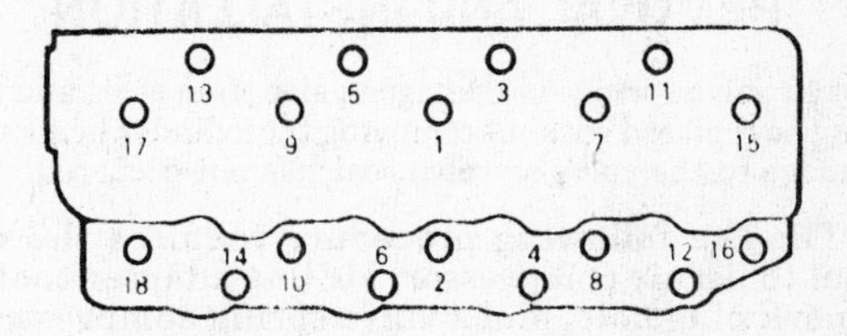

Cylinder head torque sequence – 2.3L turbo diesel engine

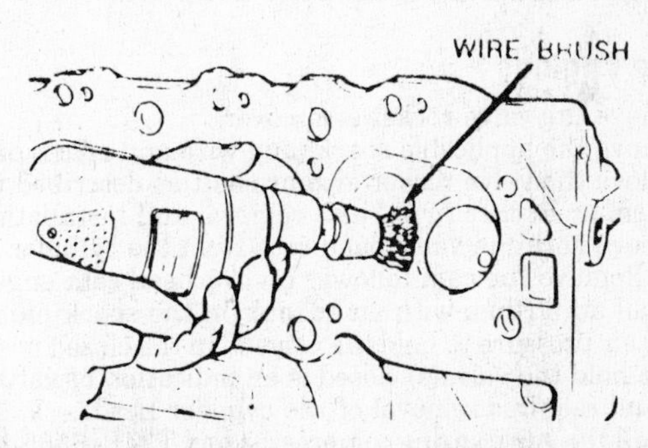

remove the carbon from the cylinder head with a wire brush and electric drill

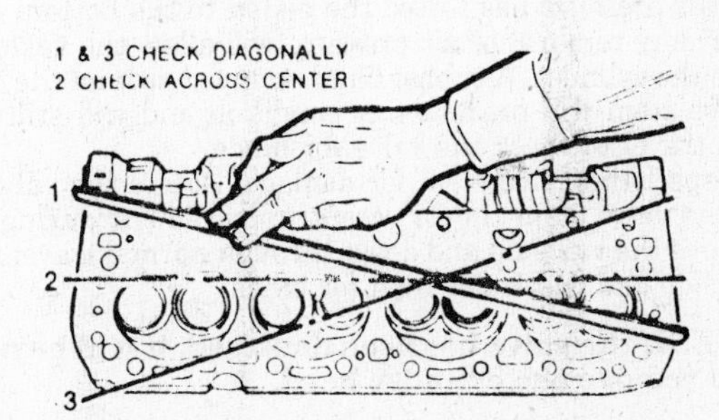

Check the cylinder head for warpage

CLEANING AND INSPECTION

1. With the valves installed to protect the valve seats, remove deposits from the combustion chambers and valve heads with a scraper and a wire brush. Be careful not to damage the cylinder head gasket surface. After the valves are removed, clean the valve guide bores with a valve guide cleaning tool. Using cleaning solvent to remove dirt, grease and other deposits, clean all bolt holes; be sure the oil passage is clean.
2. Remove all deposits from the valves with a fine wire brush or buffing wheel.
3. Inspect the cylinder heads for cracks or excessively burned areas in the exhaust outlet ports.
4. Check the cylinder head for cracks and inspect the gasket surface for burrs and nicks. Replace the head if it is cracked.
5. On cylinder heads that incorporate valve seat inserts, check the inserts for excessive wear, cracks or looseness.

RESURFACING

Cylinder Head Flatness

When a cylinder head is removed because of gasket leaks, check the flatness of the cylinder head gasket surface.

1. Place a straight-edge across the gasket surface of the cylinder head. Using feeler gauges, determine the clearance at the center of the straight-edge.
2. If warpage exceeds 0.08mm in a 152mm span, or 0.15mm over the total length, the cylinder head must be resurfaced.
3. If necessary to refinish the cylinder head gasket surface, do not plane or grind off more than 0.25mm from the original gasket surface.

NOTE: When milling the cylinder heads of V-6 engines, the intake manifold mounting position is altered, and must be corrected by milling the manifold flange a proportionate amount. Refer to a Machine shop as necessary.

Valves and Springs

VALVE LASH ADJUSTMENT

Refer to Section 2 under Tune-Up Procedures for the Valve Lash Adjustment procedure.

Valve Spring, Retainer and Seal

REMOVAL AND INSTALLATION

Broken valve springs or damaged valve stem seals and retainers may be replaced without removing the cylinder head, provided damage to the valve or valve seat has not occurred.

NOTE: The following procedure requires the use of special tools: air compressor, air line adapter tool to fit the spark plug hole, and a valve spring compressor tool designed to be used with the head on the engine. If the head has been removed from the engine the procedure will only require the use of a valve spring compressor tool designed to be used with the head off.

Gasoline Engines

1. Remove the valve rocker arm cover.
2. Remove the applicable spark plug wire and spark plug.
3. Remove the valve rocker arm or shaft as described under Rocker Arm or Rocker Arm Shaft removal and installation.
4. Remove both the valve push rods from the cylinder being serviced. Remove the cam follower on overhead cam engines.
5. Install an air line with an adapter in the spark plug hole and apply air pressure to hold the valve(s) in the closed position. Failure to hold the valve(s) closed is an indication of valve seat damage and requires removal of the cylinder head.
6. Install the valve spring compressor tool T74P–6565–A & B or equivalent. Compress the valve spring and remove the retainer locks, spring retainer and valve spring.
7. Remove and discard the valve stem seal.
8. If air pressure has forced the piston to the bottom of the cylinder, any removal of air pressure will allow the valve(s) to fall into the cylinder. A rubber band wrapped around the end of the valve stem will prevent this condition and will still allow enough travel to check the valve for binds.
9. Inspect the valve stem for damage. Rotate the valve and check the valve stem tip for eccentric movement during rotation. Move the valve up and down through normal travel in the valve guide and check the stem for binds.

NOTE: If the valve has been damaged, it will be necessary to remove the cylinder head.

10. If the valve condition proves satisfactory, lubricate the valve stem with engine oil. Hold the valve in the closed position and apply air pressure within the cylinder.
11. Install a new valve stem seal. Place the spring in position over the valve and install the valve spring retainer. Compress the valve spring and install the valve spring retainer locks. Remove the valve spring compressor tools.
12. Lubricate the push rod ends with Lubriplate® or equivalent and install the push rod. Apply Lubriplate® or equivalent to the tip of the valve stem and to both ends of the rocker arm.
13. Turn off the air and remove the air line and adapter. Install the spark plug and connect the spark plug wire.
14. Clean and install the remaining valve train components and the rocker arm cover.

Diesel Engines

1. Remove valve rocker arm cover.
2. Remove the rocker arm shaft as described in this Section.

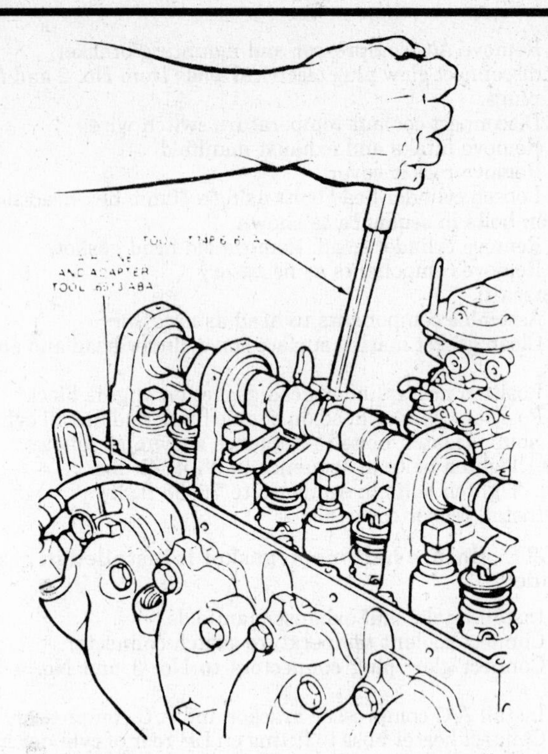

Compressing the valve spring – 2.0L and 2.3L engines

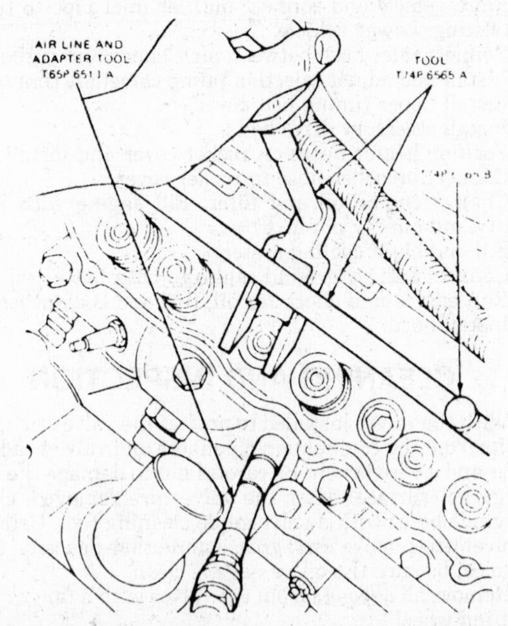

Compressing the valve spring on the cylinder head – 2.8L and 2.9L engines

3. Remove the glow plug harness from glow plug and remove glow plugs.
4. Install the adapter from the compression Test kit, Rotunda 19–6001 or equivalent, in the glow plug hole, attach an air-line, and turn on the air supply.

NOTE: An alternate method of holding the valves up, is to rotate the crankshaft until the affected piston is at TDC.

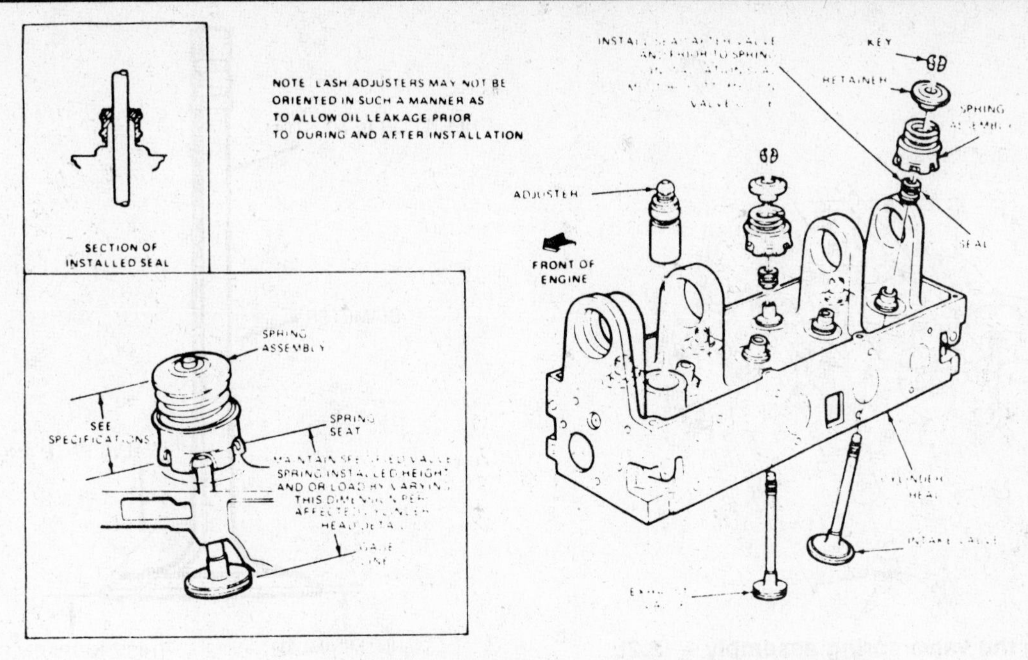

Valve train components – 2.0L and 2.3L engines

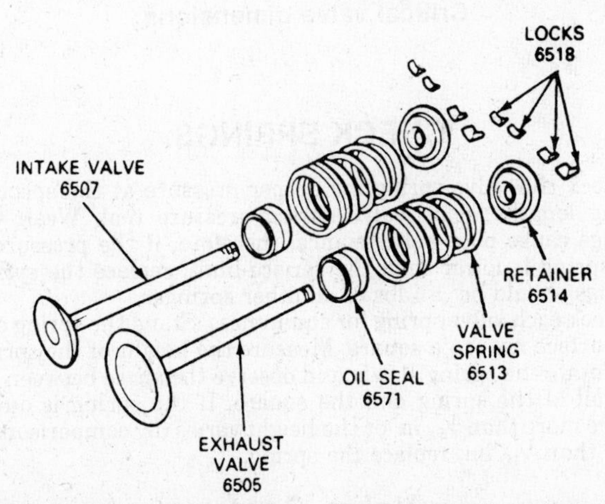

Valve train components – 2.8L and 2.9L engines

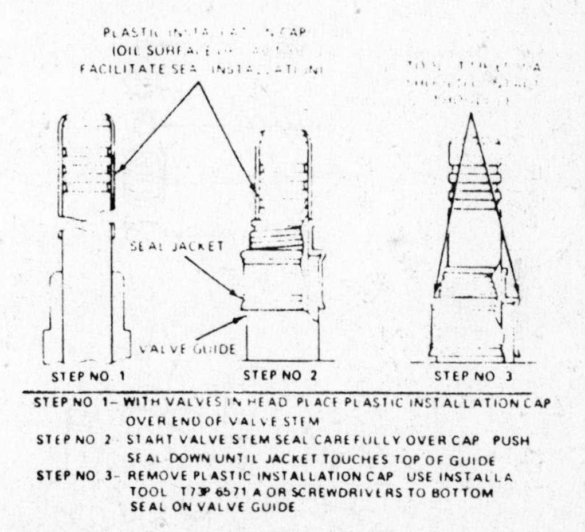

Installing the valve stem seals – 2.0L and 2.3L engines

5. Install the valve spring compressor bar, Tool T83T–6513–B, and using Valve Spring compressor Tool T74P–6565–A or equivalent, compress valve the spring and remove retainer locks, spring retainer, valve spring and damper.
6. Remove valve stem seal(s).

To Install:

7. Install new valve stem seal(s).
8. Install damper, vale spring and valve spring retainer over the vale stem. Using Tool T74P–6565–A or equivalent, compress the valve spring assembly and install the retainer locks.

CAUTION

Make sure retainer locks are fully seated in groove on valve stem.

9. Repeat the procedure for each vale spring assembly as necessary.
10. Disconnect the air supply line and remove the adapter from glow plug hole.
11. Install the glow plugs and tighten to 11–15 ft. lbs.
12. Install glow plug harness.
13. Install the rocker arm shaft as described in this Section.
14. Adjust the valves as described in Section 2 under Diesel Engine Tune Up Procedures.
15. Install valve cover, with a new gasket.
16. Connect the heater hose tube assembly to the valve cover.
17. Run the engine and check for oil leaks.

INSPECTION

Valves

Minor pits, grooves, etc., may be removed. Discard valves that are severely damaged, or if the face runout cannot be corrected by refinishing or if the stem clearance exceeds specifications. Discard any worn or damaged valve train parts.

REFACING VALVES

NOTE: The valve seat refacing operation should be coordinated with the valve refacing operations so that the finished angles of the valve seat and valve face will be to specifications and provide a compression tight fit.

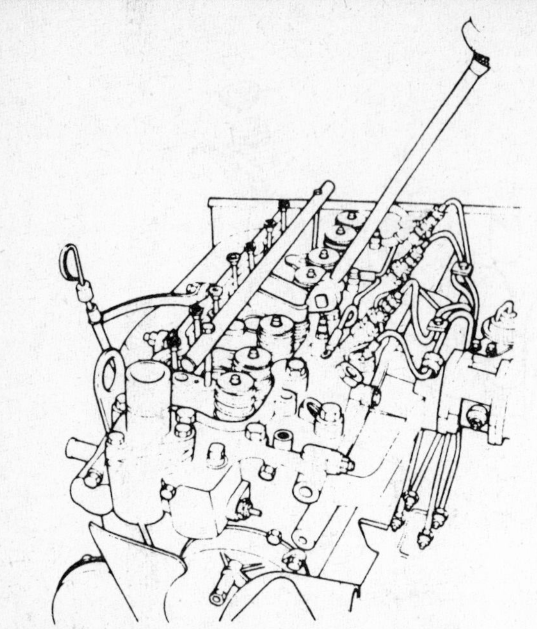

Compressing the valve spring assembly – 2.2L diesel engine

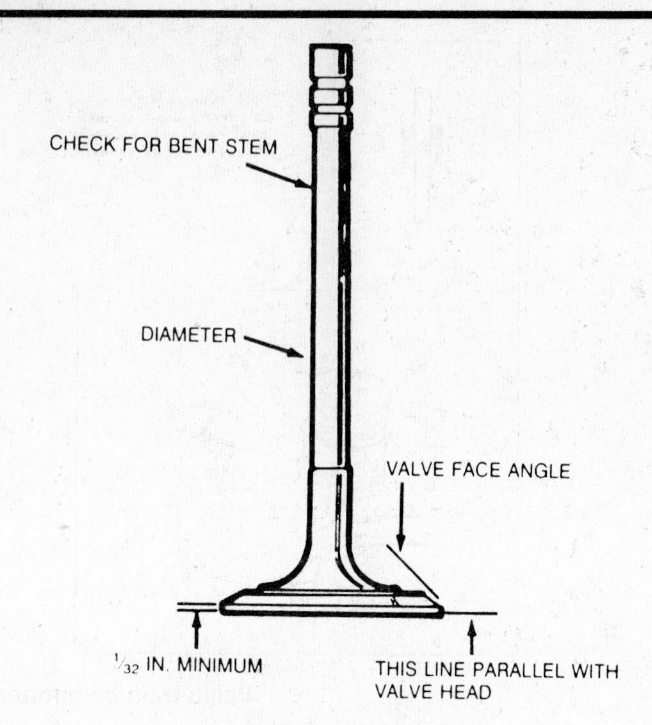

Critical valve dimensions

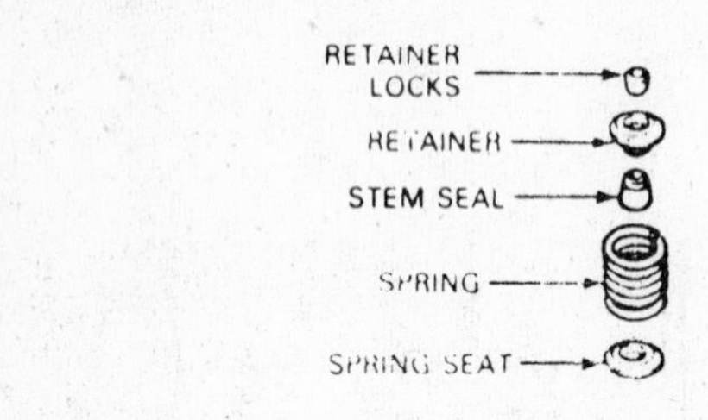

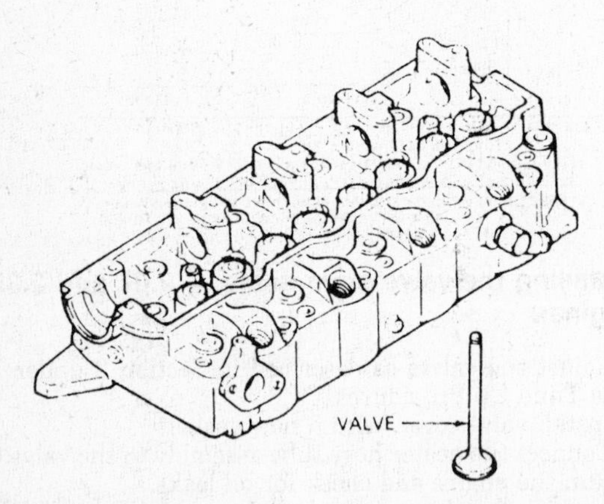

Valve train components – 2.3L turbo diesel engine

If the valve face runout is excessive and/or to remove pits and grooves, reface the valves to a true 44 degree angle. Remove only enough stock to correct the runout or to clean up the grooves and pits.

If the edge of the head is less than 0.8mm from the end of the valve stem.

If the valve and/or valve seat has been refaced, it will be necessary to check the clearance between the rocker arm pad and the valve stem tip with the valve train assembly installed in the engine.

CHECK SPRINGS

Check the valve spring for proper pressure at the specified spring lengths using valve spring pressure tool. Weak vale springs cause poor performance; therefore, if the pressure of any spring is lower than the service limit, replace the spring. Springs should be ±5 lbs of all other springs.

Check each valve spring for squareness. Stand the spring on a flat surface next to a square. Measure the height of the spring, and rotate the spring slowly and observe the space between the top coil of the spring and the square. If the spring is out of square more than $^5/_{64}$ in. or the height varies (by comparison) by more than $^1/_{16}$ in., replace the spring.

Valve Seats

CUTTING THE SEATS

NOTE: The valve refacing operation should be coordinated with the refacing of the valve seats so that the finished angles of the valve seat and valve face will be to specifications and provide a compression tight fit.

Grind the valve seats of all engines to a true 45 degree angle. Remove only enough stock to clean up pits and grooves or to correct the valve seat runout.

The finished valve seat should contact the approximate center of the valve face. It is good practice to determine where the valve seat contacts the valve face. To do this, coat the seat with Prussian blue and set the valve in place. Rotate the valve with light pressure. If the blue is transferred to the top edge of the valve face, lower the valve seat. If the blue is transferred to the bottom edge of the valve face, raise the valve seat.

LAPPING THE VALVES

When valve faces and seats have been refaced and recut, or if they are determined to be in good condition, the valves must be

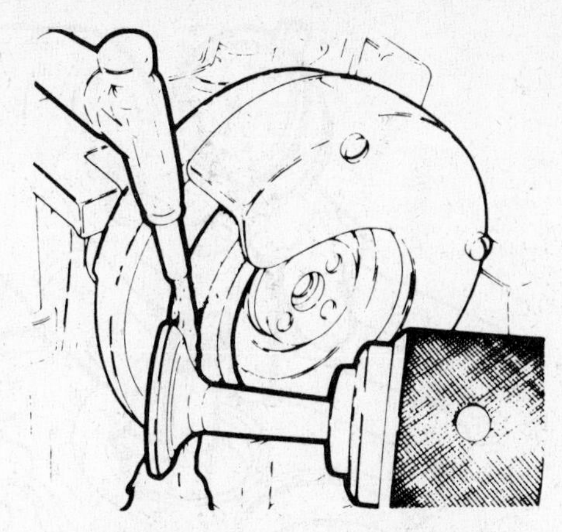

Valve grinding by machine

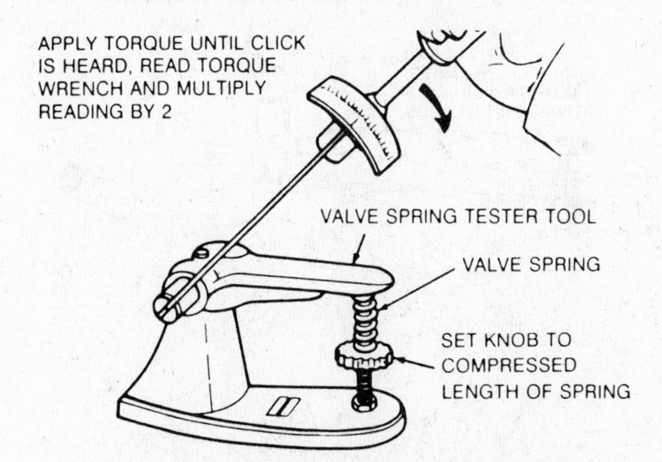

Checking the valve spring tension

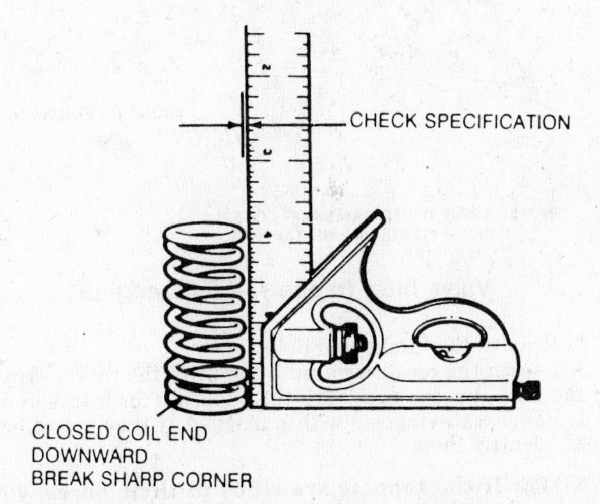

Checking the valve spring squareness

lapped in to ensure efficient sealing when the valve closes against the seat.

1. Invert the cylinder head so that the combustion chambers are facing up.
2. Lightly lubricate the valve stems with clean oil, and coat the valve seats with valve grinding compound. Install the valves in the head as numbered.

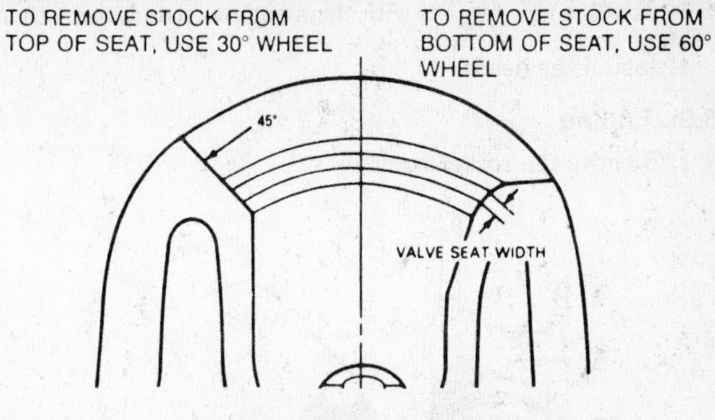

Refacing the valve seats

3. Attach the suction cup of a valve lapping tool to a valve head. You'll probably have to moisten the cup to securely attach the tool to the valve.
4. Rotate the tool between the palms. Changing position and lifting the tool often to prevent grooving. Lap the valve until a smooth, polished seat is evident (you may have to add a bit more compound after some lapping is done).
5. Remove the valve and tool, and remove ALL traces of grinding compound with solvent soaked rag, or rinse the head with solvent.

NOTE: Valve lapping can also be done by fastening a suction cup to a piece of drill rod in a hand eggbeater type drill. Proceed as above, using the drill as a lapping tool. Due to the higher speeds involved when using the hand drill, care must be exercised to avoid grooving the seat. Lift the tool and change direction of rotation often.

Valve Guides

REAMING VALVE GUIDES

If it becomes necessary to ream a valve guide to install a valve with an oversize stem, a reaming kit is available which contains oversize reamers and pilot tools.

When replacing a standard size valve with an oversize valve always use the reamer in sequence (smallest oversize first, then next smallest, etc.) so as not to overload the reamers. Always reface the valve seat after the valve guide has been reamed, and use a suitable scraper to break the sharp corner at the top of the valve guide.

Knurling is a process in which the metal on the valve guide bore is displaced and raised, thereby reducing clearance. Knurling also provides excellent oil control. The option of knurling rather than reaming valve guides should be discussed with a reputable machinist or engine specialist.

Valve Lifter (Tappets)

REMOVAL AND INSTALLATION

2.8L and 2.9L Engines

1. Remove the cylinder heads.
2. Lift out the tappets with a magnet. If they are to be re-used, mark them for installation. They must be inserted in their original locations!

NOTE: If the tappets are stuck in their bores, you'll need a claw-type removal tool.

3. Coat the new tappets with clean engine oil and insert them in their bores.
4. Install the heads.

3.0L Engine

1. Remove the rocker covers.

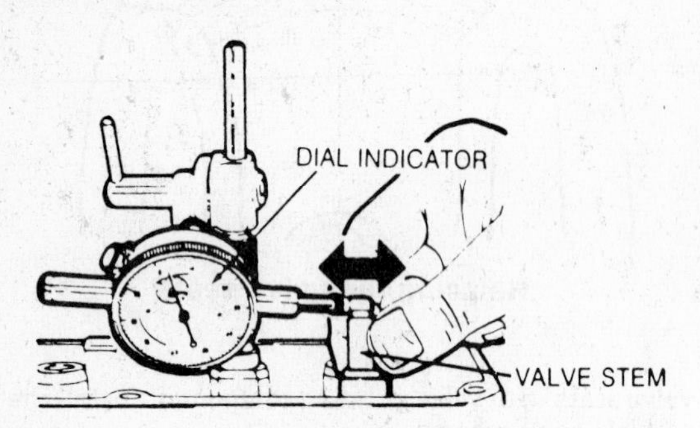

Measuring the valve stem-to-guide clearance

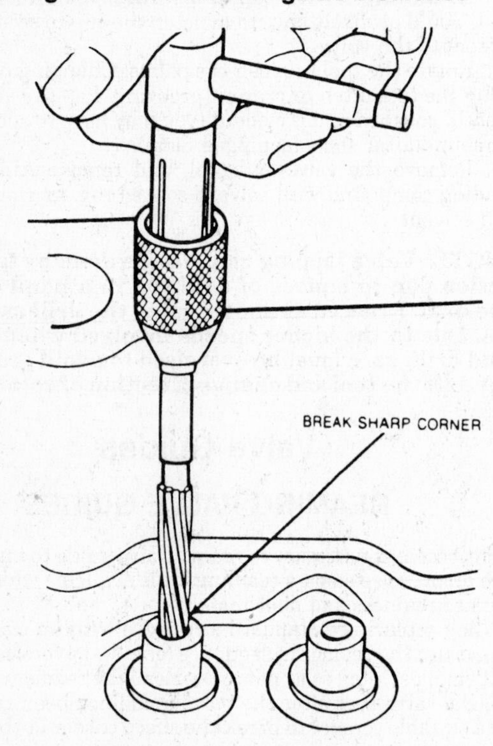

Reaming the valve guide

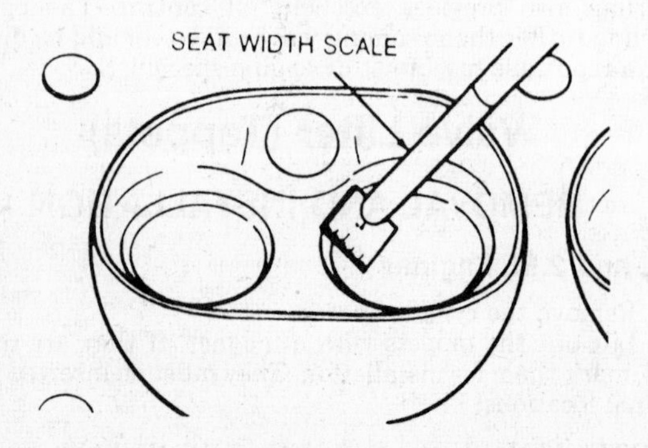

Measuring the valve seat width

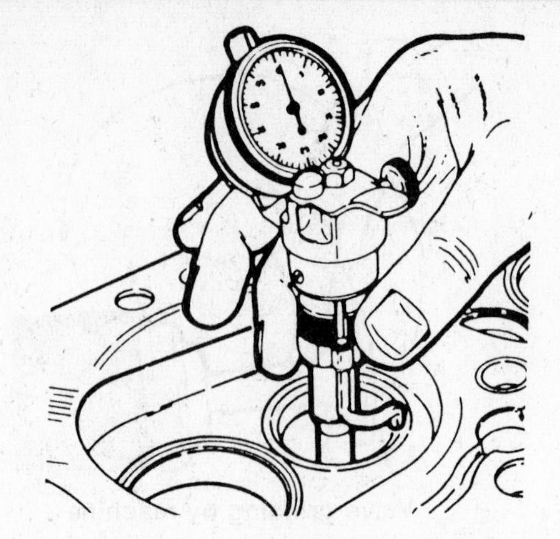

Measuring valve seat runout with a dial indicator

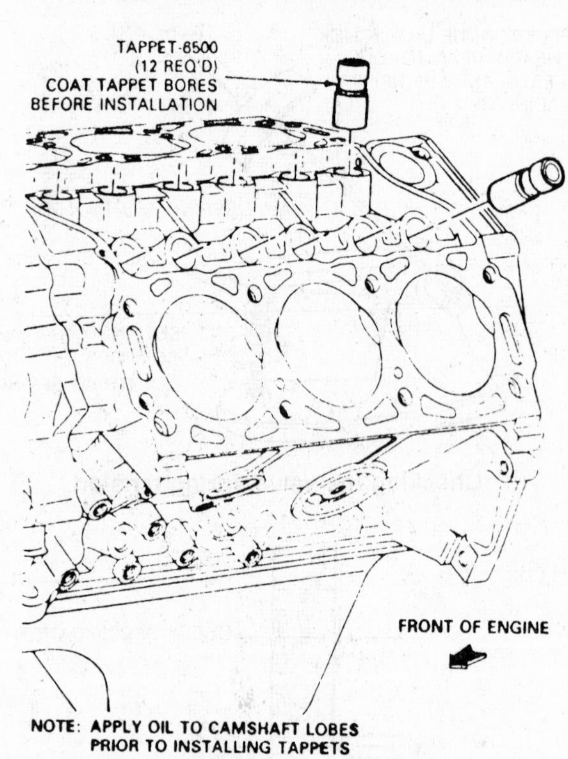

Valve lifter (tappet) – 3.0L engine

2. Remove the intake manifold.
3. Loosen the rocker arm nuts and pivot the rocker arm out of the way. Remove and mark the pushrods for installation.
4. Remove the tappets with a magnet. If they are to be re-used, identify them.

NOTE: If the tappets are stuck in their bores, you'll need a claw-type removal tool.

5. Coat the new tappets with clean engine oil and insert them in their bores.
6. Coat the pushrods with heavy engine oil and insert them into the bores from which they came.
7. Replace the rocker arms. See the procedure above.
8. Install the manifold.
9. Install the rocker covers.

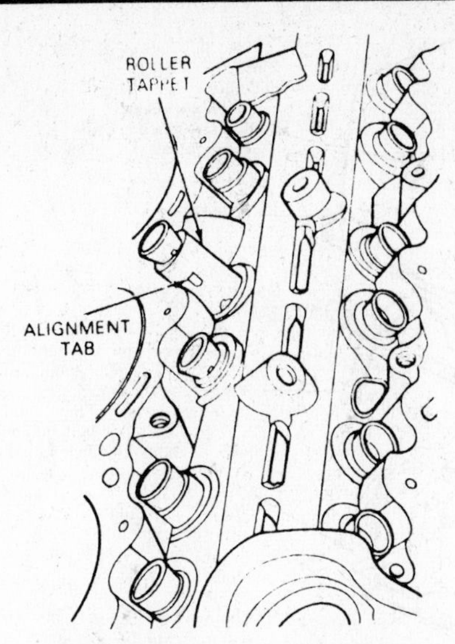

Roller tappet – 4.0L engine

4.0L Engine

1. Remove the upper and lower intake manifolds.
2. Remove the rocker covers.
3. Remove the rocker shaft assembly.
4. Remove and mark the pushrods for installation.
5. Remove the tappets with a magnet. If they are to be reused, identify them.

NOTE: If the tappets are stuck in their bores, you'll need a claw-type removal tool.

6. Coat the new tappets with clean engine oil and insert them in their bores.
7. Coat the pushrods with heavy engine oil and insert them into the bores from which they came.
8. Install the rocker shaft assembly.
9. Install the rocker covers.
10. Install the upper and lower manifold.

Oil Pan

REMOVAL AND INSTALLATION

2.0L, 2.3L Engines

NOTE: Before starting this repair procedure, view necessary illustrations for additional details.

1. Disconnect the negative battery cable.
2. Remove the air cleaner assembly. Remove the oil dipstick. Remove the engine mount retaining nuts.
3. Remove the oil cooler lines at the radiator, if so equipped. Remove the (2) bolts retaining the fan shroud to the radiator and remove shroud.
4. Remove the radiator retaining bolts (automatic only). Position radiator upward and wire to the hood (automatic only).
5. Raise the vehicle and safely support on jackstands.
6. Drain the oil from crankcase.

CAUTION

The EPA warns that prolonged contact with used engine oil may cause a number of skin disorders, including cancer! You should make every effort to minimize your exposure to used engine oil. Protective gloves should be worn when changing the oil. Wash your hands and any other exposed skin areas as soon as possible after exposure to used engine oil. Soap and water, or waterless hand cleaner should be used.

7. Remove the starter cable from starter and remove the starter.
8. Disconnect the exhaust manifold tube to the inlet pipe bracket at the thermactor check valve.
9. Remove the transmission mount retaining nuts to the crossmember.
10. Remove the bellcrank from the converter housing (automatic only).
11. Remove the oil cooler lines from retainer at the block (automatic only).
12. Remove the front crossmember (automatic only).
13. Disconnect the right front lower shock absorber mount (manual only).
14. Position the jack under the engine, raise and block with a piece of wood approximately 2½ in. high. Remove the jack.
15. Position the jack under the transmission and raise slightly (automatic only).
16. Remove the oil pan retaining bolts, lower the pan to the chassis. Remove the oil pump drive and pick-up tube assembly.
17. Remove the oil pan (out the front on automatics) (out the rear on manuals).

To install:

18. Clean the oil pan and inspect for damage. Clean the oil pan gasket surface at the cylinder block. Clean the oil pump exterior and oil pump pick-up tube screen.
19. Position the oil pan gasket and end seals to the cylinder block (use contact cement to retain).
20. Position the oil pan to the crossmember.
21. Install the oil pump and pick-up tube assembly. Install the oil pan to cylinder block with retaining bolts.
22. Lower the jack under transmission (automatic only).
23. Position the jack under the engine, raise slightly, and remove the wood spacer block.
24. Replace the oil filter.
25. Connect the exhaust manifold tube to the inlet pipe bracket at the thermactor check valve.
26. Install the transmission mount to the crossmember.
27. Install the oil cooler lines to the retainer at the block (automatic only).
28. Install the bellcrank to the converter housing (automatic only).
29. Install the right front lower shock absorber mount (manual only). Install the front crossmember (automatic only).
30. Install the starter and connect the cable. Lower vehicle.
31. Install the engine mount bolts.
32. Locate the radiator to the supports and install the (2) retaining bracket bolts (automatic only). Install the fan shroud on the radiator.
33. Connect the oil cooler lines to the radiator (automatic only).
34. Install the air cleaner assembly.
35. Install the oil dipstick. Fill the crankcase with oil.
36. Start the engine and check for leaks.

2.8L, 2.9L Engines

1. Disconnect the battery ground cable.
2. Remove the carburetor air cleaner assembly.
3. Remove the fan shroud and position it over the fan.
4. Remove the distributor cap with the wires still attached, and position it forward of the dash panel.
5. Remove the distributor and cover the opening with a clean rag.
6. Remove the nuts attaching the front engine mounts to the cross member.
7. Remove the engine oil dipstick tube.
8. Raise the truck on a hoist and support with jackstands.
9. Drain the engine crankcase and remove the oil filter.

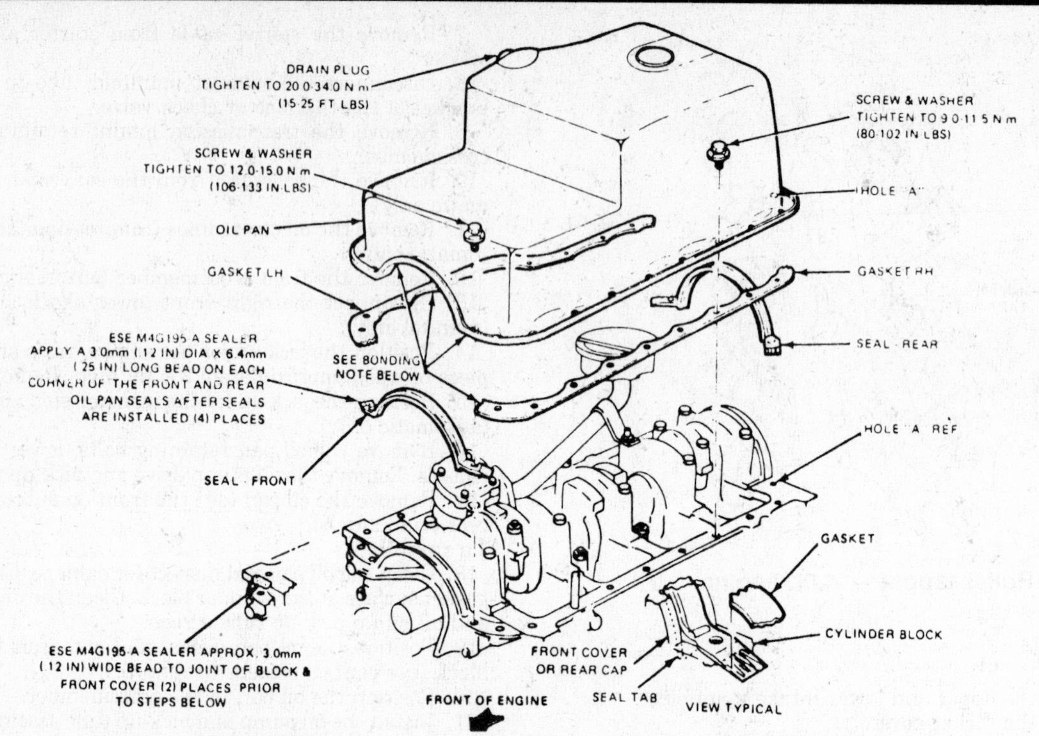

THERMAL BONDING INSTRUCTIONS · OIL PAN GASKETS TO BE BONDED SECURELY TO OIL PAN USING A THERMAL PROCESS MEETING THE REQUIREMENTS OF THE (ES-DOAE 6584 A OR EQUIVALENT) ADHESIVE COATING SPECIFICATION. IF NECESSARY IN PLACE OF THERMAL BONDING USE ADHESIVE (ESE M2G52 A OR B OR EQUIVALENT) APPLY EVENLY TO OIL PAN FLANGE & TO PAN SIDE OF GASKETS ALLOW ADHESIVE TO DRY PAST "WET" STAGE THEN INSTALL GASKETS TO OIL PAN.

1. APPLY SEALER AS NOTED ABOVE
2. INSTALL SEALS TO FRONT COVER & REAR BEARING CAP · PRESS SEAL TABS FIRMLY INTO BLOCK
3. INSTALL (2) GUIDE PINS
4. INSTALL OIL PAN OVER GUIDE PINS & SECURE WITH (4) BOLTS
5. INSTALL (18) BOLTS
6. TORQUE ALL BOLTS IN SEQUENCE CLOCKWISE FROM HOLE "A" AS NOTED ABOVE

Oil pan installation – 2.0L and 2.3L engines

OIL PAN
ASSEMBLY
6675

SPACERS-6C629
(2 REQ'D)

GASKET
6710

SEALER-ESE-MG195-C
(FORD SILICONE GASKET
AND SEALANT E3AZ-19562-A)
(6 PLACES)

SEALER

FRONT OF ENGINE

ASSEMBLY PROCEDURE

1. APPLY SILICONE GASKET AND SEALANT E3AZ-19562-A, SIX PLACES AS SHOWN.
2. INSTALL OIL PAN GASKET IN THE OIL PAN.
3. INSTALL OIL PAN ASSEMBLY.
4. INSTALL OIL PAN FLANGE BOLTS TIGHT ENOUGH TO COMPRESS THE CORK/RUBBER OIL PAN GASKET TO THE POINT THAT THE TWO TRANSMISSION HOLES ARE ALIGNED WITH THE TWO TAPPED HOLES IN THE OIL PAN, BUT LOOSE ENOUGH TO ALLOW MOVEMENT OF THE PAN, RELATIVE TO THE BLOCK.
5. INSTALL THE TWO OIL PAN/ TRANSMISSION BOLTS AND TIGHTEN TO 40-50 N·m (30-39 FT-LB) TO ALIGN THE OIL PAN WITH THE TRANSMISSION. THEN LOOSEN BOLTS 1/2 TURN.
6. TIGHTEN ALL OIL PAN FLANGE BOLTS TO 10-13.5 N·m (90-120 IN-LB).
7. TIGHTEN THE TWO OIL PAN/ TRANSMISSION BOLTS TO 40-54 N·m (30-39 FT-LB).

Oil pan installation – 2.3L Twin Plug engine

CAUTION

The EPA warns that prolonged contact with used engine oil may cause a number of skin disorders, including cancer! You should make every effort to minimize your exposure to used engine oil. Protective gloves should be worn when changing the oil. Wash your hands and any other exposed skin areas as soon as possible after exposure to used engine oil. Soap and water, or waterless hand cleaner should be used.

10. Remove the transmission fluid filler tube and plug the hole in the pan, On automatic transmission models.
11. Disconnect the muffler inlet pipes.
12. If equipped with an oil cooler, disconnect the bracket and lower the cooler.
13. Remove the starter motor.
14. On automatic transmission models, position the cooler lines out of the way.
15. Disconnect the front stabilizer bar and position it forward.
16. Position a jack under the engine and raise the engine until it touches the dash panel. Install wooden blocks between the front motor mounts and the no. 2 crossmember.
17. Lower the engine onto the blocks and remove the jack.
18. Remove the oil pan attaching bolts and lower the pan assembly.
19. Installation is the reverse of the removal procedure

3.0L Engine

1. Disconnect the negative battery cable.
2. Remove the oil level dipstick.
3. Remove the fan shroud-leave the fan shroud over the fan assembly.
4. Remove the motor mount nuts from the frame.
5. Mark and remove the distributor assembly from the engine.
6. Raise and support the vehicle safely. Remove the oil level sensor wire.
7. Drain the engine oil from the crankcase into a suitable container and dispose of it properly.

CAUTION

The EPA warns that prolonged contact with used engine oil may cause a number of skin disorders, including cancer! You should make every effort to minimize your exposure to used engine oil. Protective gloves should be worn when changing the oil. Wash your hands and any other exposed skin areas as soon as possible after exposure to used engine oil. Soap and water, or waterless hand cleaner should be used.

8. Remove the starter motor from the engine.
9. Remove the transmission inspection cover.
10. Remove the right hand axle I-Beam. The brake caliper must be removed and wired out of the way. Refer to the necessary service procedures.
11. Remove the oil pan attaching bolts, usingf a suitable lifting device, raise the engine about 2 inches. Remove the oil pan from the engine block.

NOTE: Oil pan fits tightly between the transmission spacer plate and oil pump pickup tube. Use care when removing the oil pan from the engine.

12. Clean all gasket surfaces on the engine and oil pan. Remove all traces of old gasket and/or sealer.

To install:

13. Apply a 4mm bead of RTV sealer to the junctions of the rear main bearing cap and block, and the front cover and block. The sealer sets in 15 minutes, so work quickly!
14. Apply adhesive to the gasket mating surfaces and install oil pan gasket.
15. Install the oil pan on the engine block.
16. Torque the pan bolts EVENLY to 9 ft. lbs. working from the center to the end position on the oil pan.

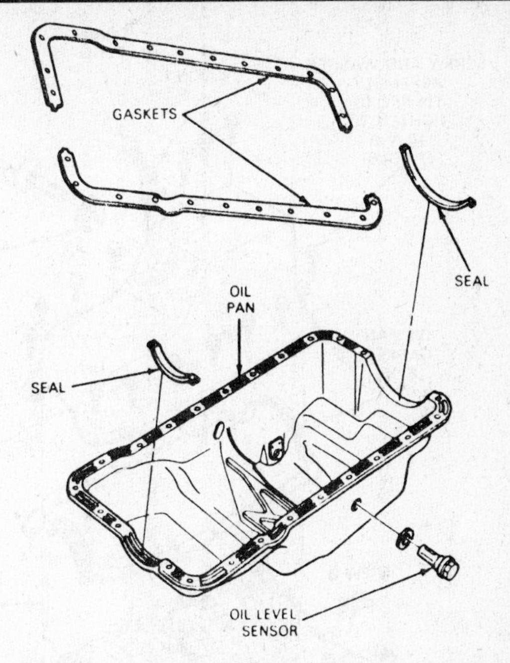

Oil pan assembly – 2.8L and 2.9L engines

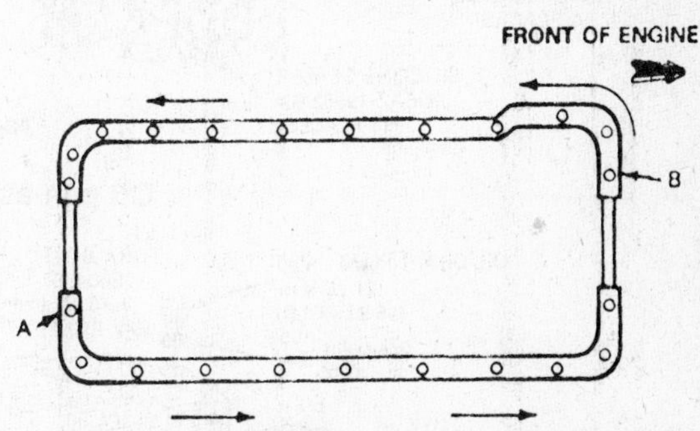

Oil pan torque sequence – 2.8L and 2.9L engines

17. Install low-oil level sensor connector. Lower engine assembly to original position.
18. Install right hand axle I-Beam. Install the brake caliper. Refer to the necessary service procedures.
19. Install transmission inspection cover. Install starter motor.
20. Lower the vehicle and install the fan shroud.
21. Install motor mount retaining nuts. Install distributor assembly.
22. Replace the oil level dipstick. Connect the battery ground. Fill crankcase with the correct amount of new engine oil. Start engine and check for leaks.

4.0L Engine

NOTE: Review the complete service procedure before starting this repair.

1. Disconnect the negative battery cable. Remove the complete engine assembly from the vehicle. Refer to the necessary service procedures in this Section.

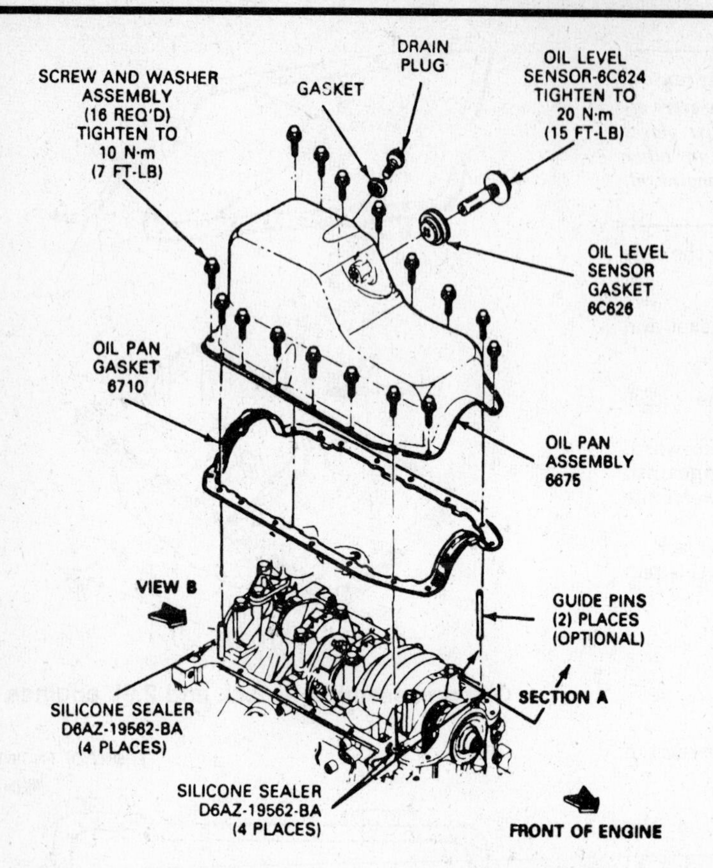

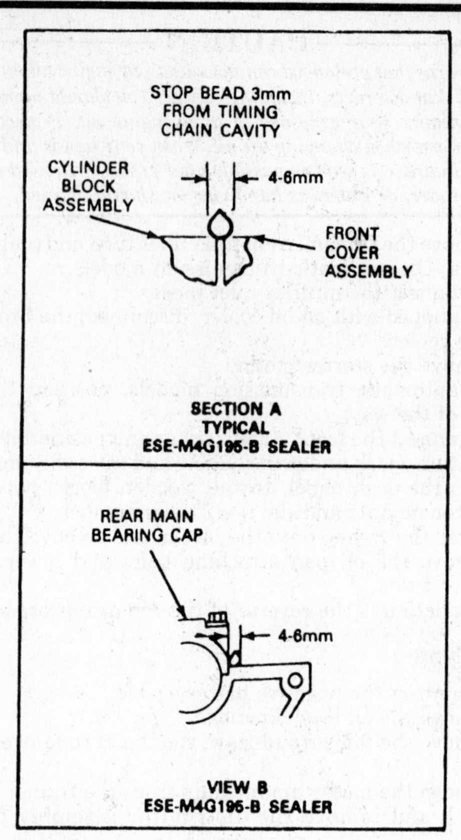

Oil pan assembly – 3.0L engine

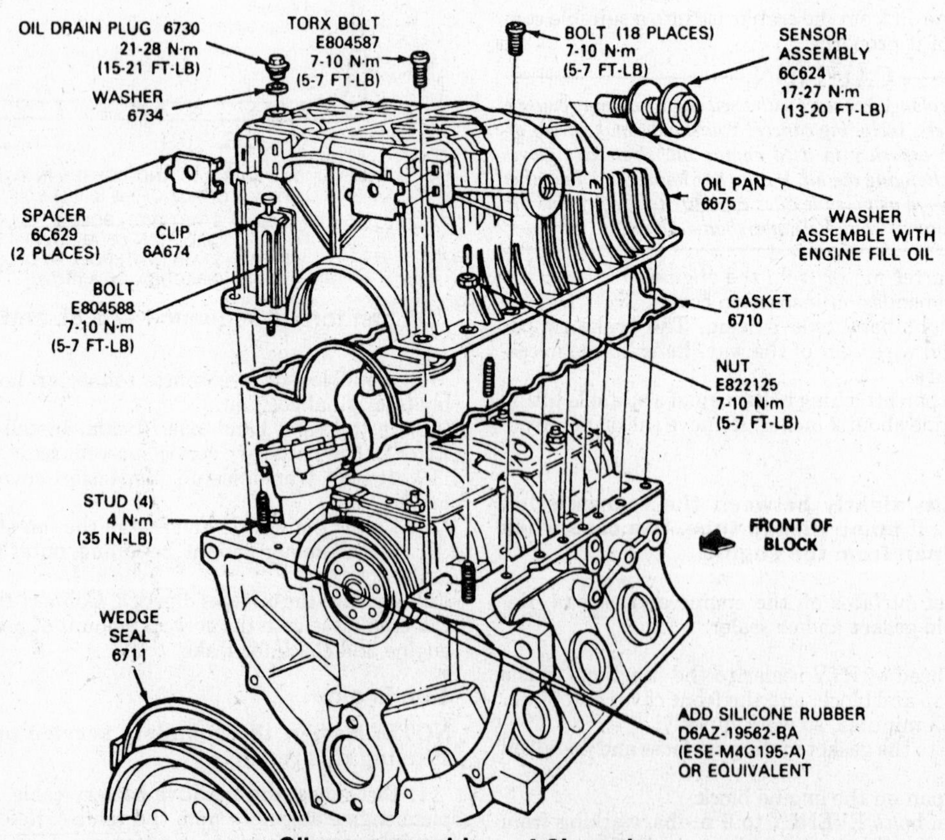

Oil pan assembly – 4.0L engine

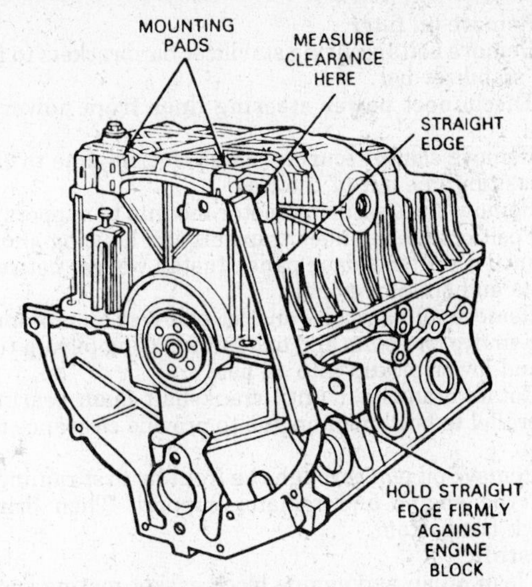

Oil pan installation – 4.0L engine

2. Mount the engine on a suitable engine stan with oil pan facing up.
3. Remove the oil pan attaching bolts (note location of 2 spacers) and remove the pan from the engine block.
4. Remove the oil pan gasket and crankshaft rear main bearing cap wedge seal.
5. Clean all gasket surfaces on the engine and oil pan. Remove all traces of old gasket and/or sealer.

To install:

6. Install a new crankshaft rear main bearing cap wedge seal. The seal should fit snugly into the sides of the rear main bearing cap.
7. Position the oil pan gasket to the engine block and place the oil pan in correct position on the 4 locating studs.
8. Torque the oil pan retaining bolts EVENLY to 5–7 ft. lbs.

NOTE: The transmission bolts to the engine and oil pan. There are 2 spacers on the rear of the oil pan to allow proper mating of the transmission and oil pan. If these spacers were lost, or the oil pan was replaced, you must determine the proper spacers to install. To do this:

a. With the oil pan installed, place a straightedge across the machined mating surface of the rear of the block, extending over the oil pan-to-transmission mounting surface.

b. Using a feeler gauge, measure the gap bewteen the oil pan mounting pad and the straightedge.

c. Repeat the procedure for the other side.

d. Select the spacers as follows:

Gap = 0.011–0.020 in.; spacer = 0.010 in.
Gap = 0.021–0.029 in.; spacer = 0.020 in.
Gap = 0.030–0.039 in.; spacer = 0.030 in.

Failure to use the correct spacers will result in damage to the oil pan and oil leakage.

9. Install the selected spacers to the mounting pads on the rear of the oil pan before bolting the engine and transmission together. Install the engine assembly in the vehicle.
10. Connect the negative battery cable. Start the engine and check for leaks.

2.2L Diesel Engine

1. Disconnect the ground battery cables from both batteries.
2. Remove the engine oil dipstick. Disconnect the air intake hose from the air cleaner and the intake manifold.
3. Drain the coolant and remove the fan and fan shroud.

CAUTION

When draining the coolant, keep in mind that cats and dogs are attracted by the ethylene glycol antifreeze, and are quite likely to drink any that is left in an uncovered container or in puddles on the ground. This will prove fatal in sufficient quantity. Always drain the coolant into a sealable container. Coolant should be reused unless it is contaminated or several years old.

4. Disconnect the radiator hoses. Remove the radiator upper support brackets and remove radiator and fan shroud.
5. Disconnect and cap the fuel inlet and outlet lines at the fuel filter and the return line at the injection pump.
6. Remove the fuel filter assembly from the mounting bracket. Remove the fuel filter mounting bracket from the cylinder head.
7. Remove the nuts and washers attaching the engine brackets to the insulators.
8. Raise the vehicle and safely support on jackstands.
9. Loosen the transmission insulator bolts at the rear of the transmission. Remove the bottom engine insulator bolts.
10. Drain the engine oil from the crankcase. Remove the primary oil filter from the left side of engine.

CAUTION

The EPA warns that prolonged contact with used engine oil may cause a number of skin disorders, including cancer! You should make every effort to minimize your exposure to used engine oil. Protective gloves should be worn when changing the oil. Wash your hands and any other exposed skin areas as soon as possible after exposure to used engine oil. Soap and water, or waterless hand cleaner should be used.

11. Remove the by-pass filter mounting bracket and hoses.
12. Lower the vehicle.
13. Attach an engine lifting sling and hoist. Raise the engine until the insulator studs clear the insulator. Slide the engine forward, then raise the engine approximately 3 in.
14. Install a wooden block 3 in. high between the left mount and bracket. Install a wooden block 4¼ in. high between the right mount and bracket. Lower the engine.
15. Remove the lifting sling and raise the vehicle.
16. Remove the oil pan attaching bolts, and lower the oil pan onto the cross member.
17. Disconnect the oil pickup from the oil pump and bearing cap, and lay in the oil pan.
18. Move the oil pan forward and up between the front of engine and the front body sheet metal. If additional clearance is needed, move the A/C condenser forward.

To install:

19. Clean the gasket mating surfaces of the oil pan and engine block with a suitable solvent and dry thoroughly. Apply 3mm bead of Silicone Sealer on the split line between the engine block and the engine front cover and along the side rails.
20. Locate the oil pan gaskets in position with Gasket Cement and make sure that the gasket tabs are seated in seal cap grooves.
21. Press the front and rear oil pan seals in the seal cap grooves with both ends of the seals contacting oil pan gaskets.
22. Apply the 3mm bead of sealer at the ends of the oil pan seals where they meet the oil pan gaskets.
23. Position the oil pan with the pickup tube on the No. 1 crossmember.
24. Install the oil pickup tube, with a new gasket, and tighten bolts to 6–9 ft. lbs. Install the oil pan with attaching bolts and plates. Tighten bolts to 7–12 ft. lbs.
25. Lower the vehicle.
26. Install a lifting sling, raise the engine and remove the wooden blocks.
27. Lower the engine onto the insulators and install and tighten the nuts and washers.
28. Raise the vehicle and safely support on jackstands.

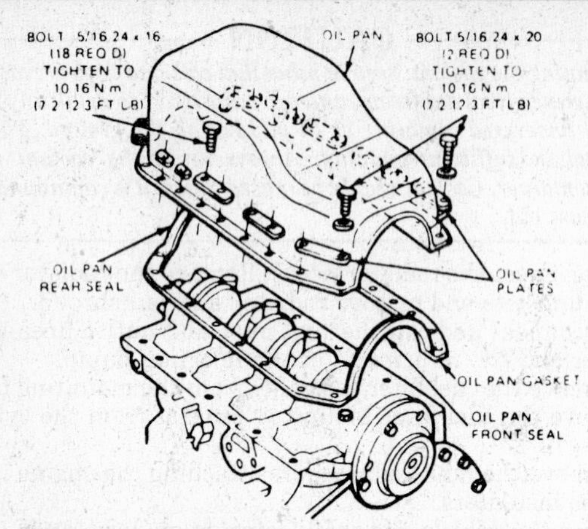

Oil pan installation – 2.2L diesel engine

29. Install the transmission mount nuts.
30. Install the by-pass filter bracket and hoses. Install the by-pass oil filter.
31. Install the oil pan drain plug. Install the new primary oil filter.
32. Lower the vehicle.
33. Install the fuel filter bracket on engine.
34. Install the fuel filter and adapter on mounting bracket.
35. Install the fuel return line on the injection pump and the fuel lines on fuel filter.
36. Position the radiator in the vehicle, install the radiator hoses and upper support brackets.
37. Install the radiator fan shroud. Install the radiator fan and tighten.
38. Fill and bleed the cooling system.
39. Fill the crankcase with the specified quantity and quality of oil.
40. Install the engine oil dipstick.
41. Install the air intake hose on the air cleaner and intake manifold.
42. Connect the battery ground cables to both batteries.
43. Run the engine and check for oil, fuel and coolant leaks.

2.3L Turbo Diesel Engine

1. Disconnect battery ground cable from both batteries.
2. Remove engine oil dipstick.
3. Remove cooling fan and fan shroud.
4. Drain cooling system and remove radiator.

CAUTION

When draining the coolant, keep in mind that cats and dogs are attracted by the ethylene glycol antifreeze, and are quite likely to drink any that is left in an uncovered container or in puddles on the ground. This will prove fatal in sufficient quantity. Always drain the coolant into a sealable container. Coolant should be reused unless it is contaminated or several years old.

5. Remove alternator belt.
6. Remove bolts securing A/C condenser to radiator support and position condenser up and out of the way.
7. Raise vehicle.
8. Disconnect oil level switch wire.
9. Drain engine oil.

CAUTION

The EPA warns that prolonged contact with used engine oil may cause a number of skin disorders, including cancer! You should make every effort to minimize your exposure to used engine oil. Protective gloves should be worn when changing the oil. Wash your hands and any other exposed skin areas as soon as possible after exposure to used engine oil. Soap and water, or waterless hand cleaner should be used.

10. Remove oil filter.
11. Remove bolts securing stabilizer bar brackets to frame and lower stabilizer bar.
12. Disconnect power steering lines from power steering pump.
13. Remove clamp securing power steering line to crossmember; position line out of the way.
14. Remove nuts securing motor mounts to support brackets.
15. Position a jack under transmission housing and raise engine until it contacts dash panel. Install wedges between motor mounts and crossmembers.
16. Remove oil pan bolts and let pan rest on crossmember.
17. Remove two bolts and one nut securing pickup tube to engine and lower pickup into oil pan.
18. Rotate crankshaft until crankshaft main bearing throws are parallel to bottom of engine to provide clearance to remove oil pan.
19. Remove oil pan through the front by first raining it up between the engine and radiator support. Then bring it out through the bottom.

To Install:

20. Clean oil an and engine block gasket mating surfaces.
21. Place oil pickup tube in oil pan and install oil pan with new gasket in position under engine.
22. Install oil pickup tube and tighten nuts and bolts to specification.
23. Apply a 3mm bead of Silicone Sealer D6AZ–19562–B or equivalent, along split line between cylinder block and front lower case and rear oil seal retainer.
24. Install oil pan and tighten bolts to specification.
25. Remove wedges and lower engine.
26. Install motor mount retaining nuts and tighten to specification.
27. Connect power steering lines to power steering pump and install line clamp.
28. Raise stabilizer bar into position and install retaining bolts. Tighten to specification.
29. Install No. 1A crossmember and tighten to specification.
30. Connect wire to oil level switch.
31. Lower vehicle.
32. Install A/C condenser.
33. Install alternator belt.
34. Install radiator, cooling fan and shroud.
35. Fill and bleed cooling system.
36. Fill engine with specified quantity and quality of oil.
37. install engine oil dipstick.
38. Connect battery ground cables to both batteries.
39. Run engine and check for oil, coolant and power steering fluid brakes.

Oil Pump

REMOVAL AND INSTALLATION

All Gasoline Engines

Follow the service procedures under Oil Pan Removal and remove the oil pan assembly. Remove the oil pump retainer bolts and remove the oil pump. Prime the oil pump with clean engine oil by filling either the inlet or outlet port with clean engine oil. Rotate the pump shaft to distribute the oil within the pump body. Install the pump and tighten the mounting bolts to 14–21 ft. lbs. on 2.3L engines; 6–10 ft. lbs. on 2.8L & 2.9L engines; 30–40 ft. lbs. on 3.0L engines; 13–15 ft. lbs on 4.0L engines. Install the oil pan as previously described. The oil pumps are not serviceable. If defective, they must be replaced.

Do not force the oil pump if it does not seat readily. The oil pump driveshaft may be misaligned with the distributor shaft assembly. To align, rotate the intermediate driveshaft into a new position.

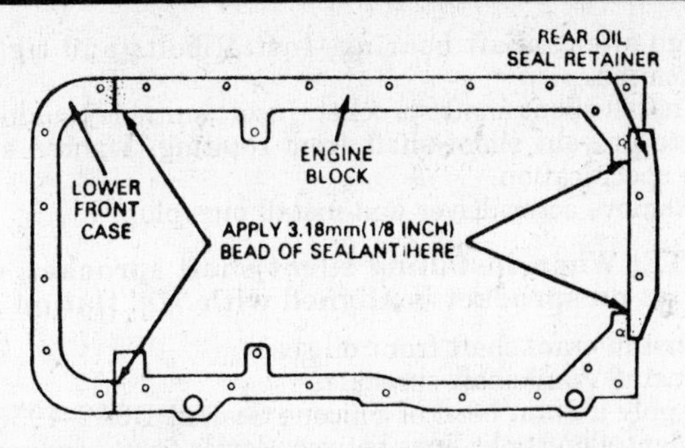

Oil pan gasket mounting – 2.3L turbo diesel engine

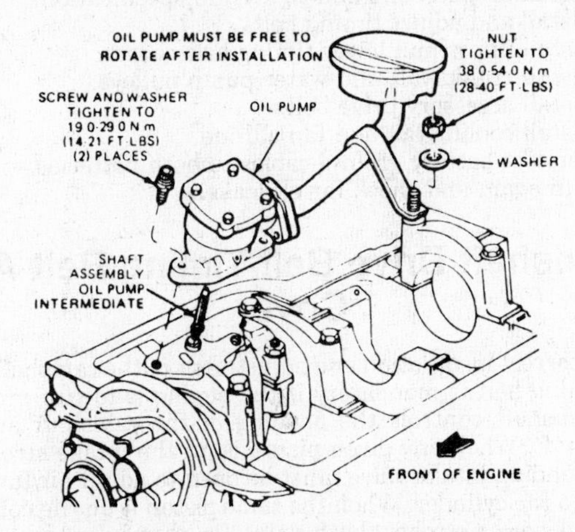

Oil pump installation – 2.0L and 2.3L engines

2.2L Diesel Engine

1. Disconnect the battery ground cables from both batteries.
2. Remove the oil pan.
3. Disconnect the oil pump outlet tube from cylinder block.
4. Remove the oil pump set screw and remove oil pump.

To install:

5. Install the oil pump.
6. Apply teflon tape, or equivalent, to the threads of the oil pump set screws. Install the set screw and tighten.
7. Install a new gasket on the oil pump outlet tube and tighten the bolts to 6–9 ft. lbs.
8. Install the oil pan.
9. Fill the crankcase with specified quantity and quality of oil.
10. Fill and bleed the cooling system.
11. Connect the battery ground cables to both batteries.

2.3L Turbo Diesel Engine

1. Disconnect battery ground cables from both batteries.
2. Remove cooling fan and fan shroud.
3. Remove water pump pulley, crankshaft pulley, upper and lower timing belt covers, timing belts, and crankshaft sprockets.
4. Loosen oil pan bolts and remove six front oil pan-to-front case bolts.
5. Remove pipe plug in right hand side of engine block, Insert cross point screwdriver into hole to prevent right hand silent shaft from rotating. Remove nut attaching silent shaft sprocket to drive gear and remove sprocket.
6. Remove bolts attaching front case to engine block and remove front case.
7. Remove front case and gasket.
8. Remove silent shaft reverse rotation gear cover and remove silent shaft and gears.

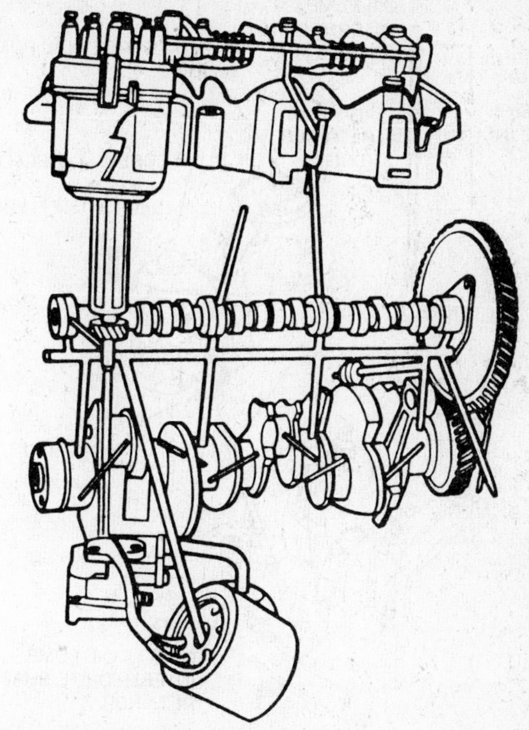

Oil pump and oil pressure lubrication system – 2.8L and 2.9L engines

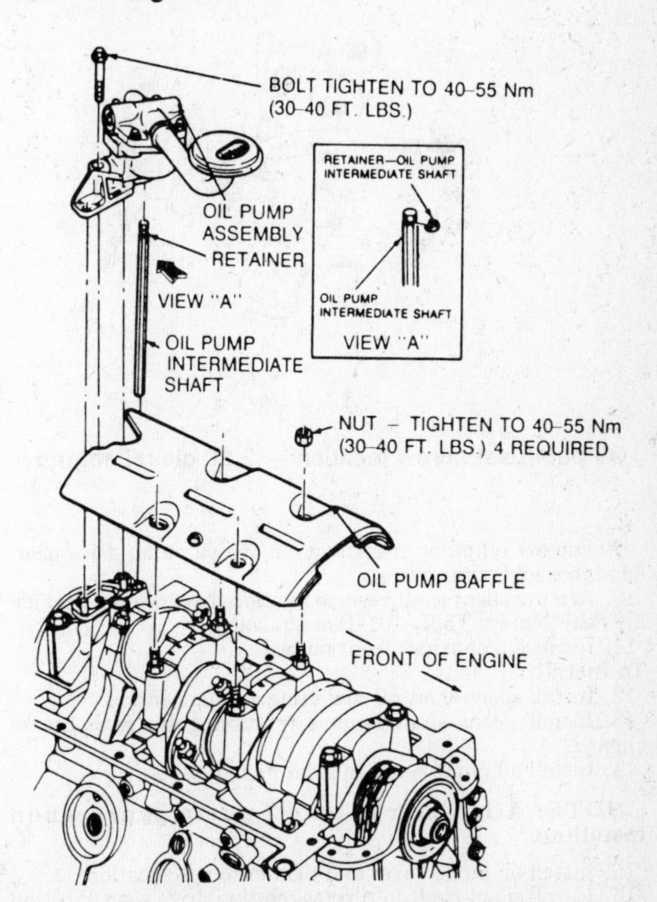

Oil pump installation – 3.0L engine

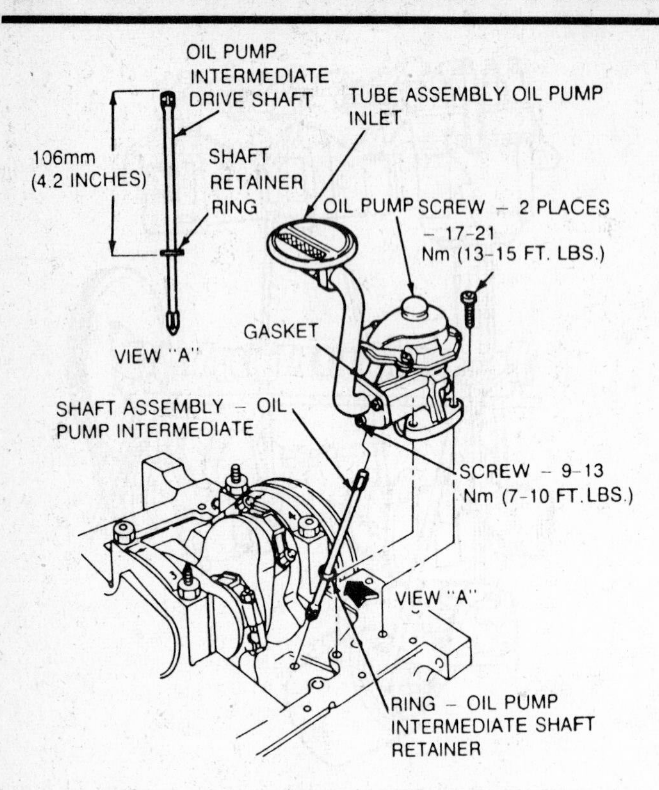

Oil pump installation – 4.0L engine

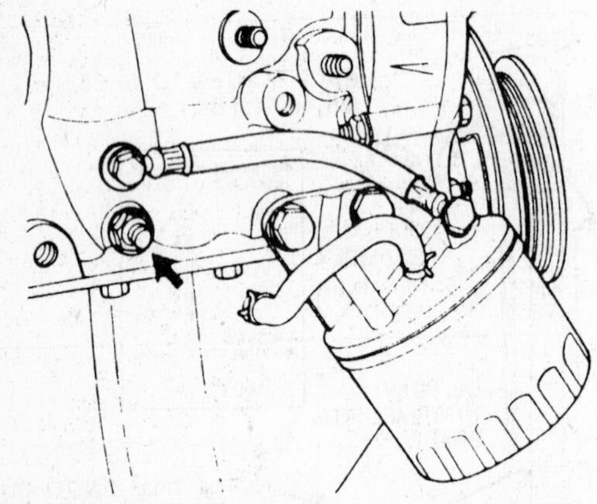

Oil pump set screw location – 2.2L diesel engine

9. Remove oil pump cover and remove oil pump drive gear, and inner and outer gears.
10. Remove silent shaft reverse rotation drive gear oil seal using Seal Remove T58L–101–B or equivalent.
11. Remove crankshaft front oil seal.

To install:

12. Install silent shaft oil seal using a 21mm socket.
13. Install silent shaft reverse rotation gears with marks aligned.
14. Install oil pump gears in front housing.

NOTE: Align marks on oil pump gears when installing.

15. Install oil pump cover and tighten to specification.
16. Install silent shaft in reverse rotation drive gear. Position front cover and new gasket on engine block using care not to damage silent shaft bearing. Install bolts and tighten to specification.
17. Install silent shaft sprocket. Insert suitable tool in hole in block to prevent silent shaft from rotating. Tighten sprocket nut to specification.
18. Remove screwdriver and install pipe plug.

NOTE: When installing silent shaft sprocket, ensure "D" flat on sprocket is aligned with "D" flat on shaft.

19. Install crankshaft front oil seal.
20. Install crankshaft sprocket.
21. Apply a 3mm bead of Silicone Sealant D6A2–19562–B or equivalent along split lines between lower front case cover and rear oil seal retainer and engine block.
22. Install oil pan bolts and tighten to specification.
23. Install and adjust timing belts.
24. Install upper and lower timing belt covers.
25. Install crankshaft and water pump pulleys.
26. Install accessory drive belts.
27. Install cooling fan and fan shroud.
28. Connect battery ground cables to both batteries.
29. Run again and check for oil leaks.

Camshaft Drive Belt/Timing Belt And Cover

The correct installation and adjustment of the camshaft drive belt/timing belt is mandatory if the engine is to run properly. The camshaft controls the opening of the camshaft and the crankshaft. When any given piston is on the intake stroke the corresponding intake valve must be open to admit air/fuel mixture into the cylinder. When the same piston is on the compression and power strokes, both valves in that cylinder must be closed. When the piston is on the exhaust stroke, the exhaust valve for that cylinder must be open. If the opening and closing of the valves is not coordinated with the movements of the pistons, the engine will run very poorly, if at all.

The camshaft drive belt/timing belt also turns the engine auxiliary shaft. The distributor is driven by the engine auxiliary shaft. Since the distributor controls ignition timing, the auxiliary shaft must be coordinated with the camshaft and the crankshaft, since both valves in any given cylinder must be closed and the piston in that cylinder near the top of the compression stroke when the spark plug fires.

Due to this complex interrelationship between the camshaft, the crankshaft and the auxiliary shaft, the cogged pulleys on each component must be aligned when the camshaft drive belt/timing belt is installed.

TROUBLESHOOTING

Should the camshaft drive belt/timing belt jump timing by a tooth or two, the engine could still run; but very poorly. To visually check for correct timing of the crankshaft, auxiliary shaft, and the camshaft follow this procedure:

NOTE: There is an access plug provided in the cam drive belt cover so that the camshaft timing cam be checked without moving the drive belt cover.

1. Remove the access plug.
2. Turn the crankshaft until the timing marks on the crankshaft indicate TDC.
3. Make sure that the timing mark on the camshaft drive sprocket is aligned with the pointer on the inner belt cover. Also, the rotor of the distributor must align with the No. 1 cylinder firing position.

NOTE: Never turn the crankshaft of any of the overhead cam engines in the opposite direction of normal ro-

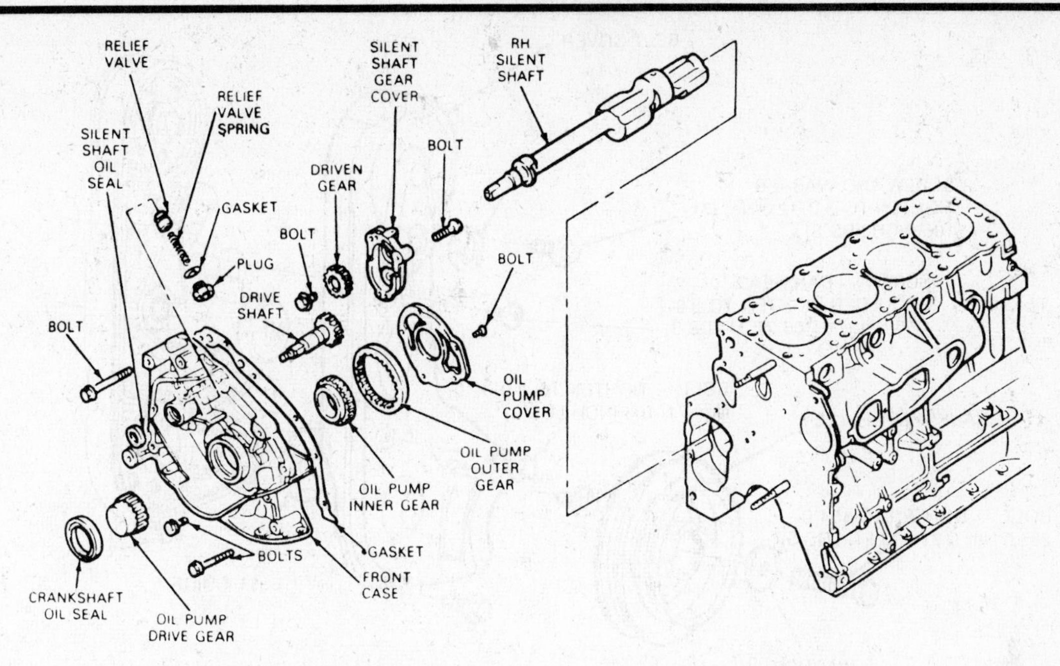

Oil pump assembly and front case – 2.3L turbo diesel engine

tation. Backward rotation of the crankshaft may cause the timing belt to slip and alter the timing.

REMOVAL AND INSTALLATION

2.0L and 2.3L Engines

1. Rotate the engine so that No. 1 cylinder is at TDC on the compression stroke. Check that the timing marks are aligned on the camshaft and crankshaft pulleys. An access plug is provided in the cam belt cover so that the camshaft timing can be checked without removal of the cover or any other parts. Set the crankshaft to TDC by aligning the timing mark on the crank pulley with the TC mark on the belt cover. Look through the access hole in the belt cover to make sure that the timing mark on the cam drive sprocket is lined up with the pointer on the inner belt cover.

NOTE: Always turn the engine in the normal direction of rotation. Backward rotation may cause the timing belt to jump time, due to the arrangement of the belt tensioner.

2. Drain cooling system. Remove the upper radiator hose as necessary. Remove the fan blade and water pump pulley bolts.

CAUTION

When draining the coolant, keep in mind that cats and dogs are attracted by the ethylene glycol antifreeze, and are quite likely to drink any that is left in an uncovered container or in puddles on the ground. This will prove fatal in sufficient quantity. Always drain the coolant into a sealable container. Coolant should be reused unless it is contaminated or several years old.

3. Loosen the alternator retaining bolts and remove the drive belt from the pulleys. Remove the water pump pulley.
4. Loosen and position the power steering pump mounting bracket and position it aside.
5. Remove the four timing belt outer cover retaining bolts and remove the cover. Remove the crankshaft pulley and belt guide.
6. Loosen the belt tensioner pulley assembly, then position a camshaft belt adjuster tool (T74P–6254–A or equivalent) on the tension spring rollpin and retract the belt tensioner away from the timing belt. Tighten the adjustment bolt to lock the tensioner in the retracted position.
7. Remove the timing belt.

To install:

8. Install the new belt over the crankshaft sprocket and then counterclockwise over the auxiliary and camshaft sprockets, making sure the lugs on the belt properly engage the sprocket teeth on the pulleys. Be careful not to rotate the pulleys when installing the belt.
9. Release the timing belt tensioner pulley, allowing the tensioner to take up the belt slack. If the spring does not have enough tension to move the roller against the belt (belt hangs loose), it might be necessary to manually push the roller against the belt and tighten the bolt.

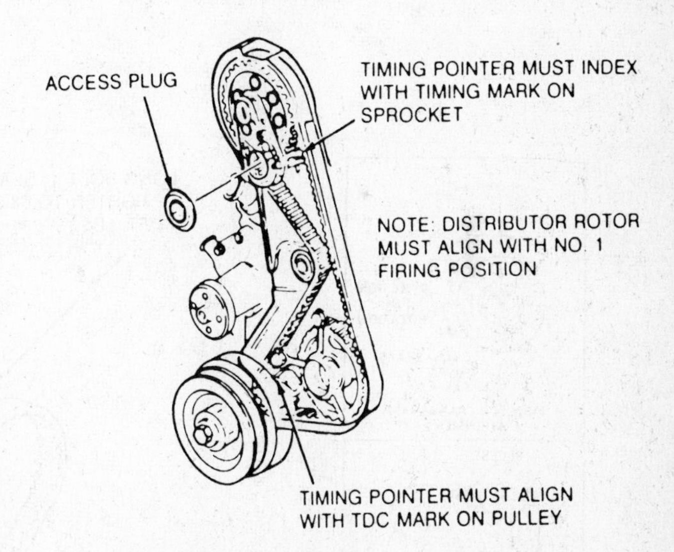

Timing mark alignment – 2.3L engine

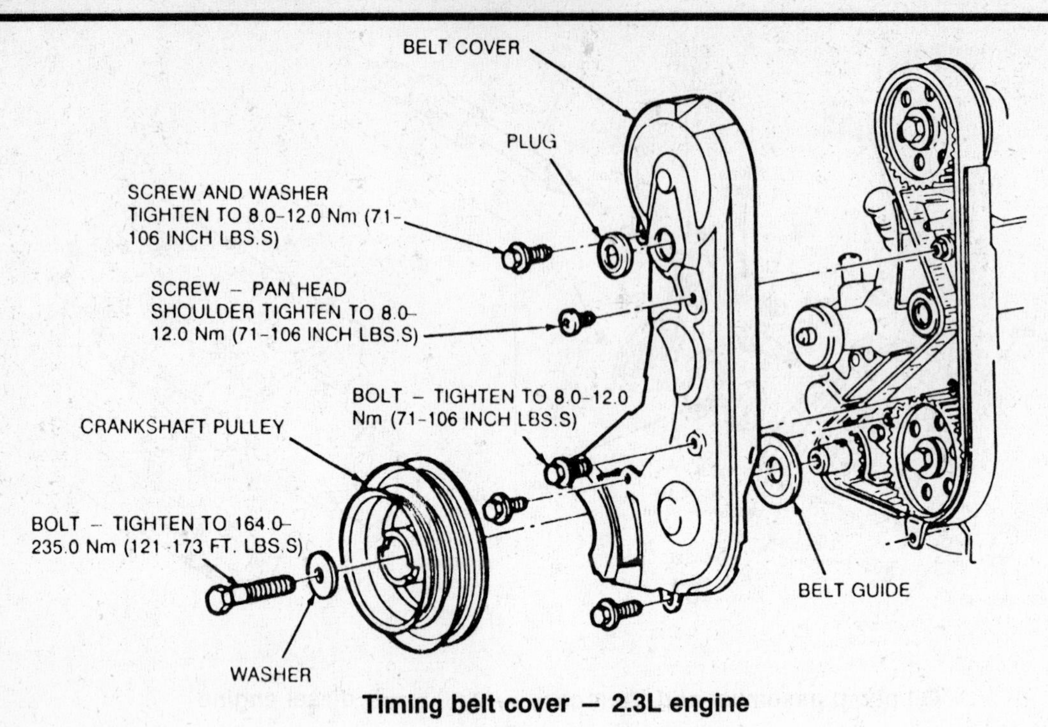

Timing belt cover — 2.3L engine

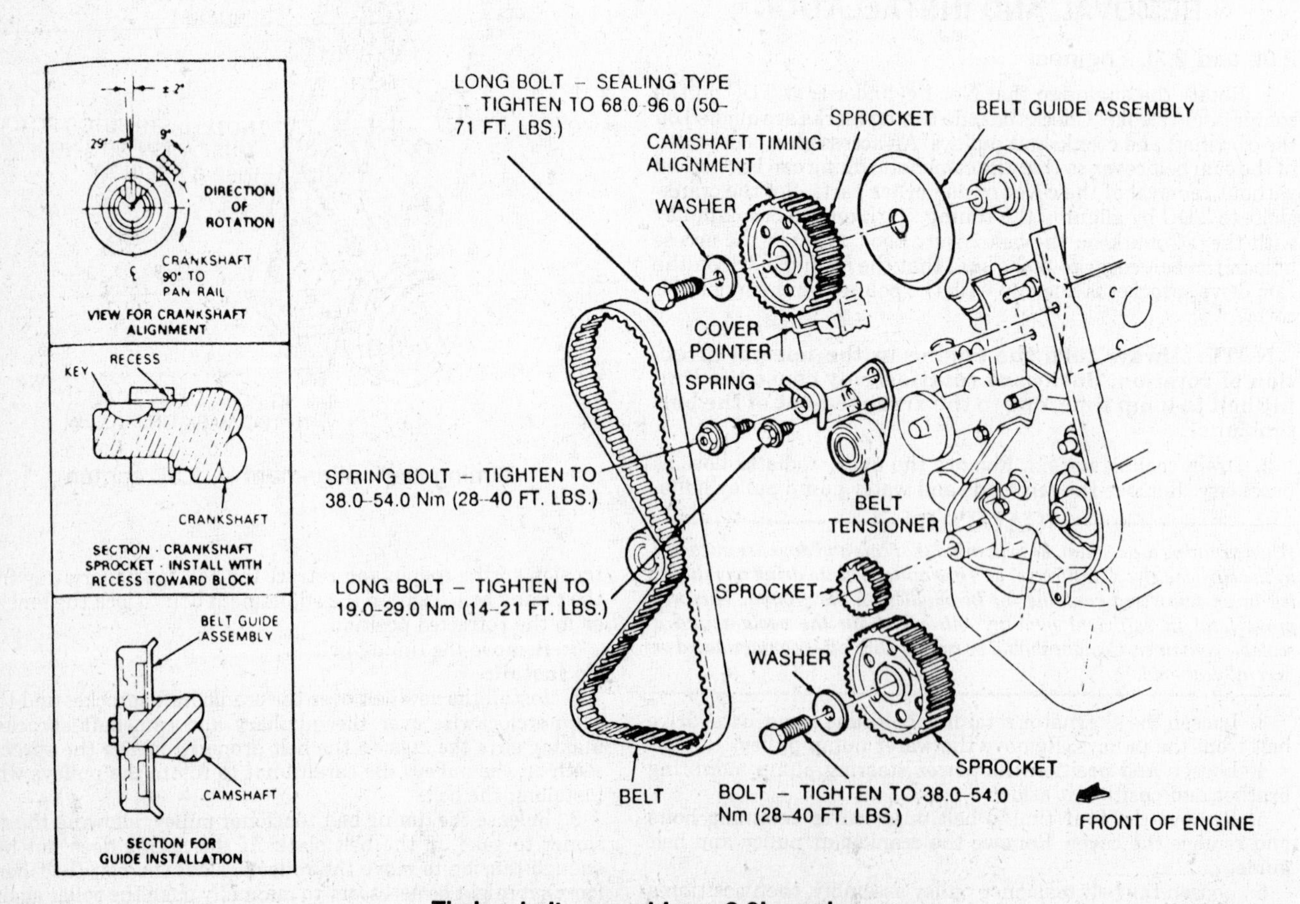

Timing belt assembly — 2.3L engine

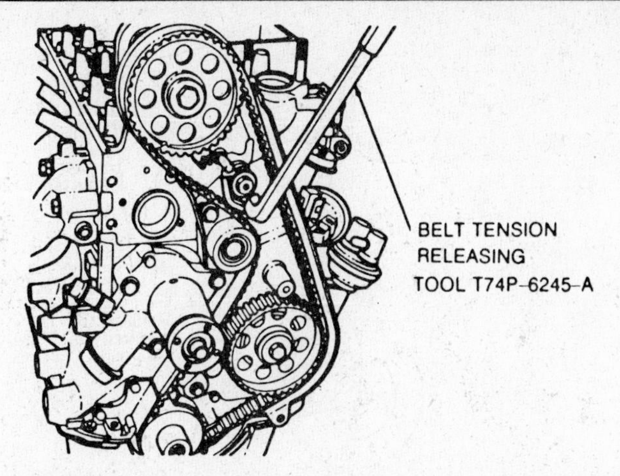

Releasing the timing belt tensioner with special tool

NOTE: The spring cannot be used to set belt tension; a wrench must be used on the tensioner assembly.

10. Rotate the crankshaft two complete turns by hand (in the normal direction of rotation) to remove the slack from the belt, then tighten the tensioner adjustment and pivot bolts to specifications. Refer to the necessary illustrations. Make sure the belt is seated properly on the pulleys and that the timing marks are still in alignment when No. 1 cylinder is again at TDC/compression.
11. Install the crankshaft pulley and belt guide.
12. Install the timing belt cover.
13. Install the water pump pulley and fan blades. Install upper radiator hose if necessary. Refill the cooling system.
14. Position the alternator and drive belts, then adjust and tighten it to specifications.
15. Start the engine and check the ignition timing. Adjust the timing, if necessary.

2.3L Turbo Diesel Engine

1. Disconnect battery ground cables from both batteries.
2. Remove cooling fan and fan shroud.
3. Remove accessory drive belts.
4. Rotate crankshaft in direction of engine rotation to bring No. 1 piston to TDC of compression stroke.
5. Remove crankshaft pulley.
6. Remove upper and lower front covers.
7. Loosen belt tensioners from timing belts. Remove belt(s), as necessary.

To install:

8. Align crankshaft timing marks.
9. Align LH and right hand silent shaft timing marks.
10. Install silent shaft belt.

NOTE: Install belt in original direction of rotation. For ease of installation, pry on tensioner spring to reduce load on tensioner.

11. Align camshaft timing marks.
12. Install injection pump/camshaft timing belt as follows.

NOTE: To ease installation, locally manufacture the following tool. Using a pivot nut from a rear brake shoe adjuster, weld a piece of bar stock on the end.

a. Using the tensioner spring tool, release the tension on the tension spring for the injection pump/camshaft timing belt.
b. After releasing spring tension, rotate tensioner toward water pump and tighten top bolts.

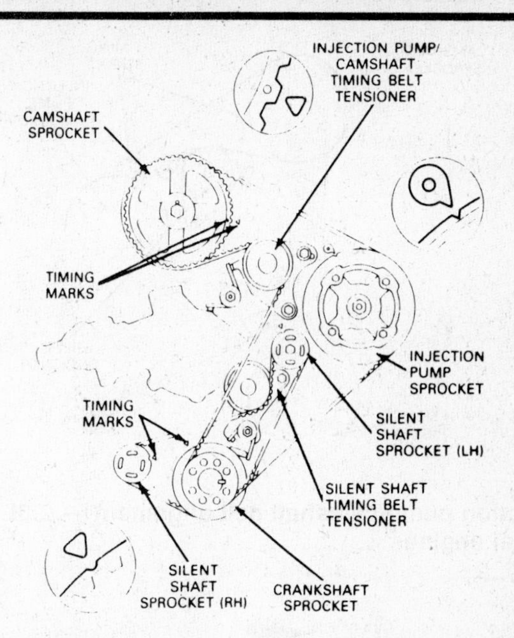

Timing belt alignment – 2.3L turbo diesel engine

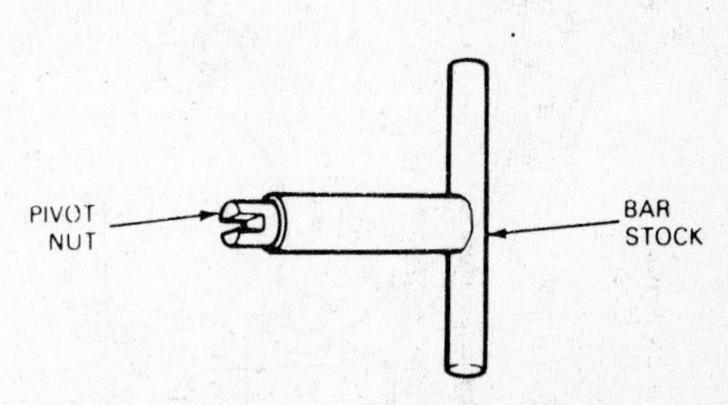

Timing belt tension tool – 2.3L turbo diesel engine

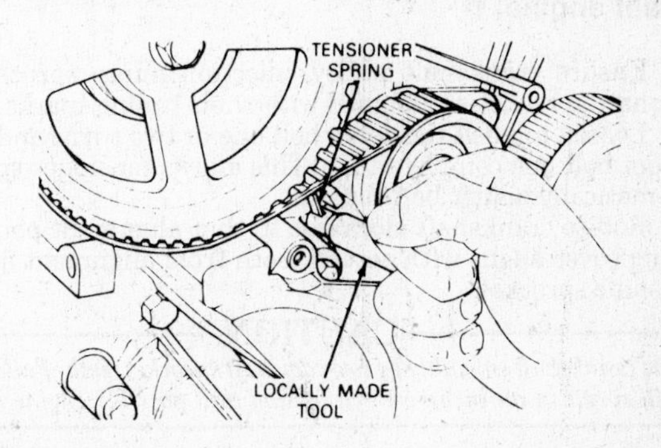

Timing belt tension adjustment – 2.3L turbo diesel engine

c. Install the belt.

13. Adjust silent shaft belt and camshaft/injection pump belt tensions.
14. Install upper and lower front timing covers.
15. Install crankshaft pulley.
16. Install accessory drive belts.
17. Install cooling fan and fan shroud.
18. Connect battery ground cables to both batteries.

INJECTION PUMP/CAMSHAFT BELT ADJUSTMENT

1. Remove timing belt upper cover.
2. Rotate engine until No. 1 piston is at TDC on compression stroke.

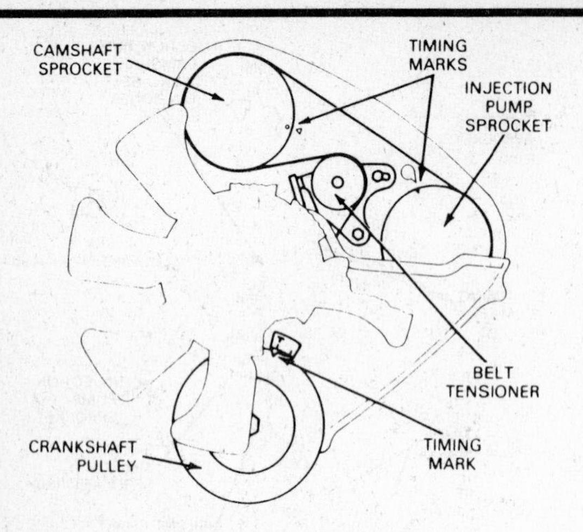

Injection pump/camshaft belt alignment — 2.3L turbo diesel engine

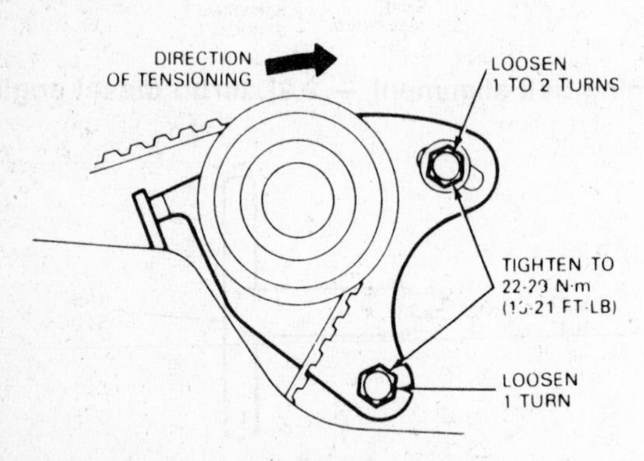

Injection pump/camshaft belt tensioner — 2.3L turbo diesel engine

3. Ensure crankshaft pulley, injection pump sprocket and camshaft sprocket are aligned with their timing marks.
4. Loosen top belt tensioner bolt one or two turns and loosen bottom bolt one complete turn. This allows tensioner spring to automatically adjust belt tension.
5. Rotate crankshaft clockwise to that alignment pointer on timing cover aligns with second tooth from alignment mark on camshaft sprocket.

CAUTION

Rotate crankshaft smoothly by two camshaft sprocket teeth. Failure to do so will result in an incorrect belt tension and possible engine damage.

6. Tighten top belt tensioner mounting bolt to specification, then tighten bottom bolt to specification.
7. Rotate crankshaft counterclockwise until timing marks are aligned. Push belt down halfway between injection pump sprocket and camshaft sprocket and check deflection. If properly tensioned, belt should deflect 4–5mm.

SILENT SHAFT BELT ADJUSTMENT

1. Rotate crankshaft until No. 1 piston is at TDC on compression stroke.
2. Remove access cover for top belt tensioner bolt by inserting a suitable tool in slot shown and prying out.
3. Loosen top belt tension mounting bolt one complete turn. Then, loosen bottom bolt one to two turns. This allows tensioner spring to automatically adjust belt tension.
4. Tighten bottom tensioner bolt to specification, then tighten top bolt to specification.

Camshaft belt alignment — 2.3L turbo diesel engine

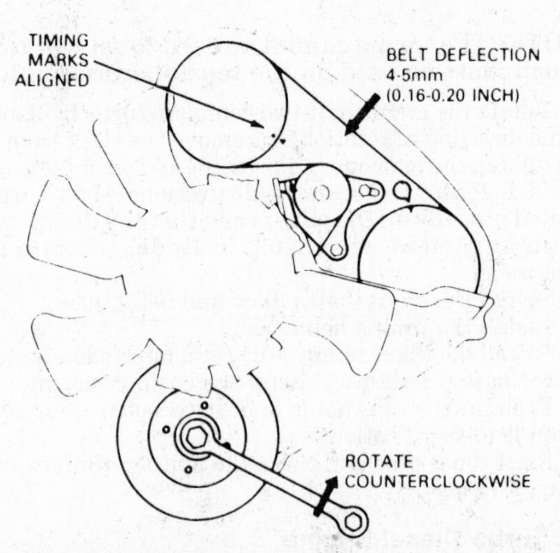

Camshaft belt deflection — 2.3L turbo diesel engine

CAUTION

Tighten bottom bolt first. Tighten top bolt first can cause tensioner to rotate and over-tension the timing belt.

5. Install access cover for top tensioner bolt by sliding down along two guide lines embossed on front lower cover.

Timing Gear Cover & Oil Seal

REMOVAL AND INSTALLATION

2.8L, 2.9L Engines

1. Remove the oil pan as described under Oil Pan removal and installation.
2. Drain the coolant. Remove the radiator.

CAUTION

When draining the coolant, keep in mind that cats and dogs are attracted by the ethylene glycol antifreeze, and are quite likely to drink any that is left in an uncovered container or in puddles on the ground. This will prove fatal in sufficient quantity. Always drain the coolant into a sealable container. Coolant should be reused unless it is contaminated or several years old.

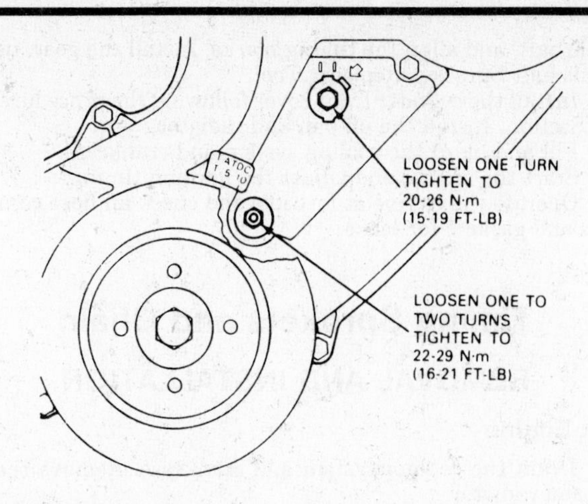

Silent shaft belt tensioner – 2.3L turbo diesel engine

3. Remove the A/C compressor and the power steering bracket, if so equipped, and position then out of the way. DO NOT disconnect the A/C refrigerant lines.
4. Remove the alternator, thermactor pump and drive belt(s).
5. Remove the fan.
6. Remove the water pump and the heater and radiator hoses.
7. Remove the drive pulley from the crankshaft.
8. Remove the front cover retaining bolts. If necessary, tap the cover lightly with a plastic hammer to break the gasket seal. Remove the front cover. If the front cover plate gasket needs replacement, remove the two screws and remove the plate. If necessary, remove the guide sleeves from the cylinder block.
9. If the front cover oil seal needs replacement use the following procedure:
   a. Support the front cover to prevent damage while driving out the seal.
   b. Drive out the seal from front cover with Front Cover Aligner, T74P–6019–A, or equivalent.
   c. Support the front cover to prevent damage while installing the seal.
   d. Coat the new front cover oil seal with Lubriplate® or equivalent. Install the new seal in the front cover.

To install:

10. Clean the front cover mating surfaces of gasket material. Apply sealer to the gasket surfaces on the cylinder block and back side of the front cover plate. Install the guide sleeves, with new seal rings, with the chamfered end toward the front cover, if removed. Position the gasket and the front cover plate on the cylinder block. Temporarily install the four front cover screws to position the gasket and cover plate in plate. Install and tighten the two cover plate attaching bolts, then remove the four screws that were temporarily installed.
11. Apply gasket sealer to the front cover gasket surface. Place the gasket in position on the front cover.
12. Place the front cover on the engine and start all the retaining screws two or three turns. Center the cover by inserting Front Cover Aligner, Tool T74P–6019–A, or equivalent in oil seal.
13. Torque the front cover attaching screws to 13–16 ft. lbs.
14. Install the belt drive pulley.
15. Install the oil pan as described under Oil Pan removal and installation.
16. Install the water pump, heater hose, A/C compressor, alternator, thermactor pump and drive belt(s). Adjust drive belt tension.
17. Fill and bleed the cooling system.
18. Operate the engine at fast idle speed and check for coolant and oil leaks.

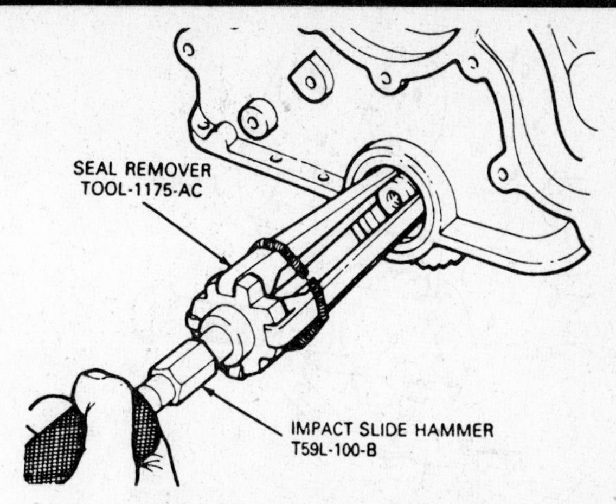

Removing the front oil seal

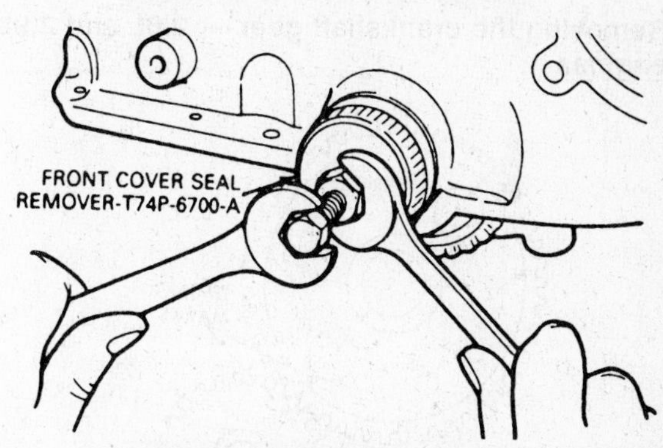

Installing the front oil seal

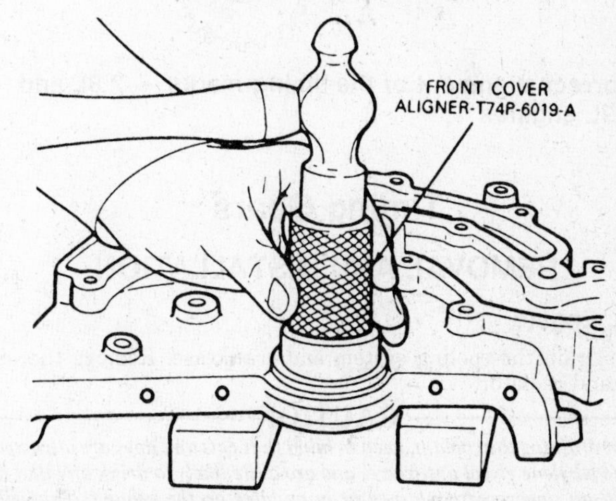

Removing and installing front oil seal with front cover assembly removed

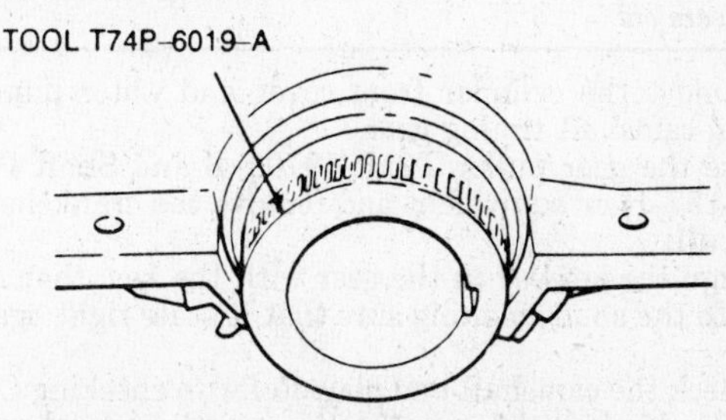

Front cover alignment special tool

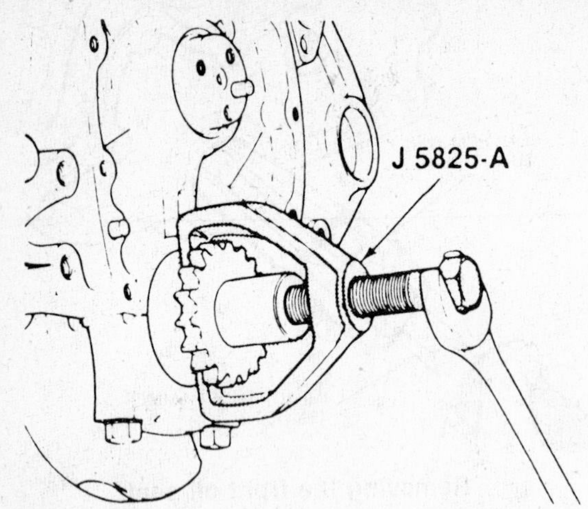

Removing the crankshaft gear – 2.8L and 2.9L engines

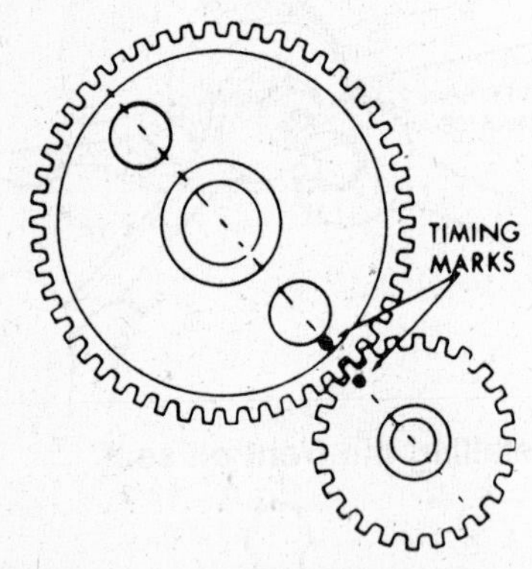

Correct alignment of the timing marks – 2.8L and 2.9L engines

Timing Gears

REMOVAL AND INSTALLATION

2.8L Engine

1. Drain the cooling system and crankcase. Remove the oil pan and radiator.

CAUTION

When draining the coolant, keep in mind that cats and dogs are attracted by the ethylene glycol antifreeze, and are quite likely to drink any that is left in an uncovered container or in puddles on the ground. This will prove fatal in sufficient quantity. Always drain the coolant into a sealable container. Coolant should be reused unless it is contaminated or several years old.

2. Remove the cylinder front cover and water pump, drive belt, and camshaft timing gear.
3. Use the gear puller T71P–19703–B and Shaft Protector T71P–7137–H or equivalent and remove the crankshaft gear.

To install:

4. Align the keyway in the gear with the key, then slide the gear onto the shaft, making sure that it seats tight against the spacer.
5. Check the camshaft end play. Refer to checking Camshaft procedure. If not within specifications, replace the thrust plate.
6. Align the keyway in the crankshaft gear with key in the crankshaft, and align the timing marks. Install the gear, using Crankshaft Sprocket Replacer Tool.
7. Install the cylinder front cover following the procedures in this Section. Install the oil pan and radiator.
8. Fill and bleed the cooling system and crankcase.
9. Start the engine and adjust the ignition timing.
10. Operate the engine at fast idle and check all hose connections and gaskets for leaks.

Timing Sprokets and Chain

REMOVAL AND INSTALLATION

2.9L Engine

1. Drain the cooling system and crankcase. Remove the oil pan and radiator.

CAUTION

When draining the coolant, keep in mind that cats and dogs are attracted by the ethylene glycol antifreeze, and are quite likely to drink any that is left in an uncovered container or in puddles on the ground. This will prove fatal in sufficient quantity. Always drain the coolant into a sealable container. Coolant should be reused unless it is contaminated or several years old.

2. Remove the cylinder front cover and water pump, drive belt, and camshaft timing gear.
3. Use the clutch aligner T71P–P–7137–H or equivalent and remove the crankshaft sproket.

To install:

4. Align the keyway in the gear with the key, then slide the gear onto the shaft, making sure that it seats tight against the spacer.

NOTE: Feed the timing chain around the crankshaft sprocket and then around the camshaft sprocket. Install the camshaft sprocket together with timing chain as an assembly. Align all necessary timing marks.

5. Check the camshaft end play. Refer to checking Camshaft. If not within specifications, replace the thrust plate.
6. Align the keyway in the crankshaft gear with key in the crankshaft, and align the timing marks. Install the gear, using Crankshaft Sprocket Replacer Tool.
7. Install the cylinder front cover following the procedures in this Section. Install the oil pan and radiator.
8. Fill and bleed the cooling system and crankcase.
9. Start the engine and adjust the ignition timing.
10. Operate the engine at fast idle and check all hose connections and gaskets for leaks.

TIMING CHAIN DEFLECTION

2.9L Engine

1. Remove the timing chain tensioner.
2. Rotate the crankshaft counterclockwise (as viewed from the front of the engine) to take up the slack on the left hand side of the chain.
3. Mark a reference point on a block approximately at midpoint of the chain. Measure from this point to the chain.
4. Rotate the crankshaft in the opposite direction to take up the slack on the right hand side of the chain. Force the left hand side of the chain out with your fingers and measure the distance between the reference point and the chain. The deflection is the difference between the two measurements.
5. If the deflection measurement exceeds specification, replace the timing chain and sprockets.
6. If the wear on the tensioner face exceeds 1.5mm, replace the tensioner.

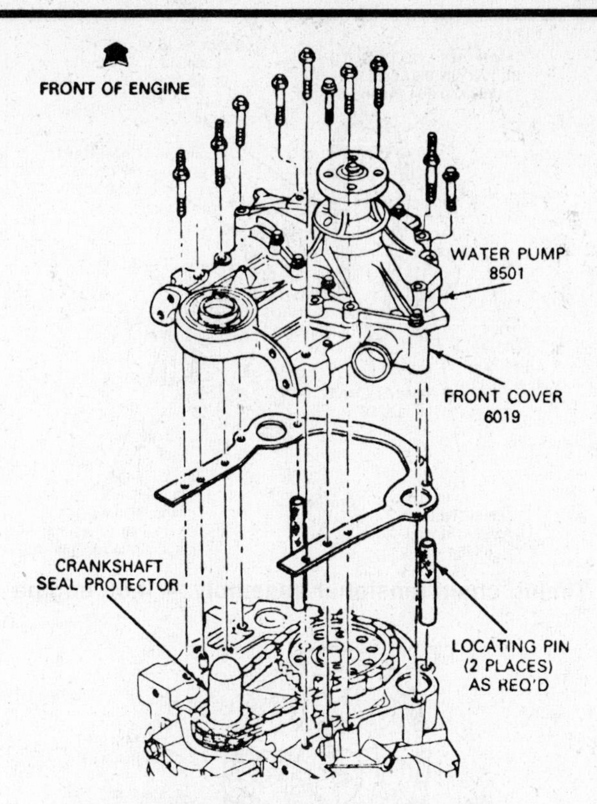

Front cover assembly — 3.0L engine

7. When installing the crankshaft sprocket, fill the keyway chamfer cavity with EOAZ–19554–AA Threadlock and Sealer or equivalent, flush with the front face of the sprocket.

Timing Cover and Chain

REMOVAL AND INSTALLATION

3.0L Engine

1. Disconnect the negative battery cable.
2. Drain the cooling system and crankcase.

CAUTION

When draining the coolant, keep in mind that cats and dogs are attracted by the ethylene glycol antifreeze, and are quite likely to drink any that is left in an uncovered container or in puddles on the ground. This will prove fatal in sufficient quantity. Always drain the coolant into a sealable container. Coolant should be reused unless it is contaminated or several years old.

3. Remove the cooling fan.
4. Loosen the water pump pulley bolts. Loosen and remove the accessory drive belts. Remove the water pump pulley.
5. Remove the alternator adjusting arm and brace assembly. Remove the heated air intake duct from the engine.
6. Remove the upper motor mount retaining nuts. Remove the A/C compressor upper bolts, then remove the front cover front nuts on vehicles with automatic transmission and A/C.
7. Mark and remove the distributor assembly from the vehicle.
8. Raise the vehicle. Remove the A/C compressor bolts and braket if so equipped and position the assembly aside.
9. Remove the crankshaft pulley and damper assembly. Remove the oil pan. Refer to the necessary service procedures in this Section.

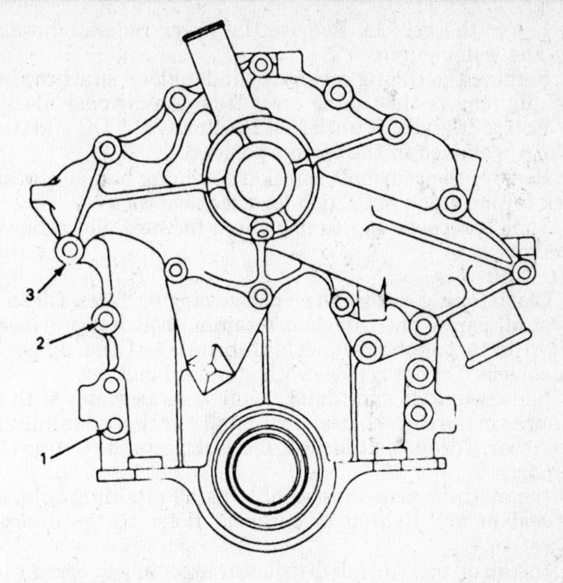

FASTENERS

Ref. No.	Part No.	Size	N·m	Ft-Lb
1	N804215-S100	M8 x 1.25 x 72.25	25	19
2	N804215-S100	M8 x 1.25 x 72.25	25	19
3	N804811-S100	M8 x 1.25 x 70.0	25	19

NOTE: Apply Pipe Sealant ESG-M4G194-A to Bolt Threads A10107-1C

Front cover bolt installation chart — 3.0L engine

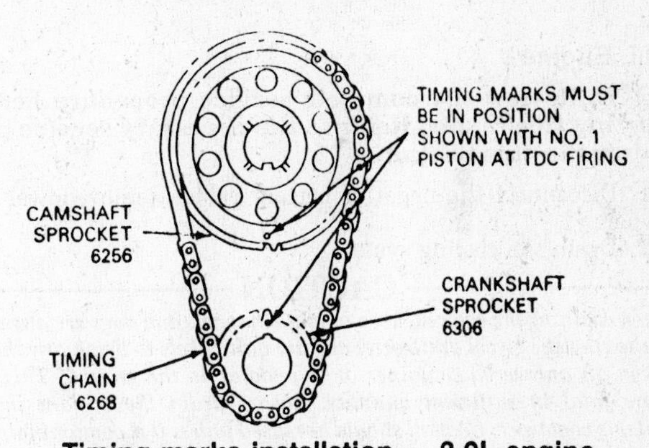

Timing marks installation — 3.0L engine

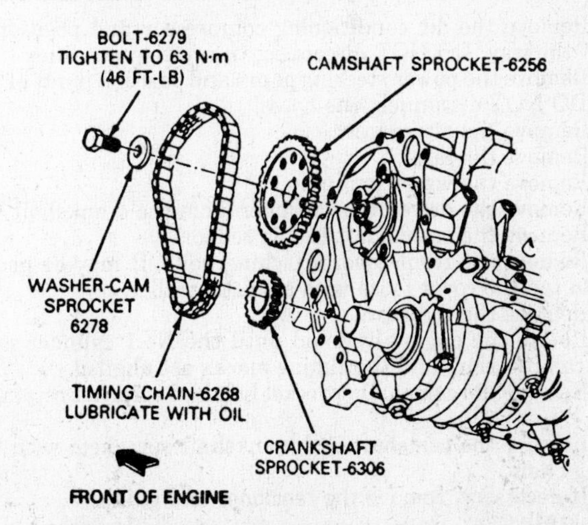

Timing chain assembly — 3.0L engine

10. Lower the vehicle. Remove the lower radiator hose. Remove the water pump.
11. Remove the timing cover to cylinder block attaching bolts. Carefully remove the timing cover from the cylinder block.
12. Rotate crankshaft until NO. 1 piston is at TDC and timing marks are aligned in the correct position.
13. Remove the camshaft sprocket retaining bolt and washer. Check timing chain deflection for excessive wear.
14. Slide sprockets and timing chain forward and remove as assembly.

To install:

15. Clean timing cover and oil pan sealing surfaces. Clean and inspect all parts. The camshaft retaining bolt has a drilled oil passage for timing chain assembly lubrication. Clean oil passage with solvent. Do not replace with standard bolt.
16. Slide sprockets and timing chain on as assembly with timing marks in the correct location. Install camshaft retaining bolt and washer. Torque bolt to 46 ft. lbs. Lubricate the timing chain assembly.
17. Install timing cover assembly. Install retaining bolts with pipe sealant and tighten as outlined. Refer to the necessary illustration.
18. Install oil pan (install distributor assembly in correct position) and water pump. Refer to the necessary procedures in this Section.
19. Install crankshaft damper and pulley assembly.
20. Install drive belt components. Install drive belts and adjust.
21. Fill crankcase. Refill and bleed coling system. Connect the negative battery cable. Start engine check for coolant, oil and exhaust leaks.

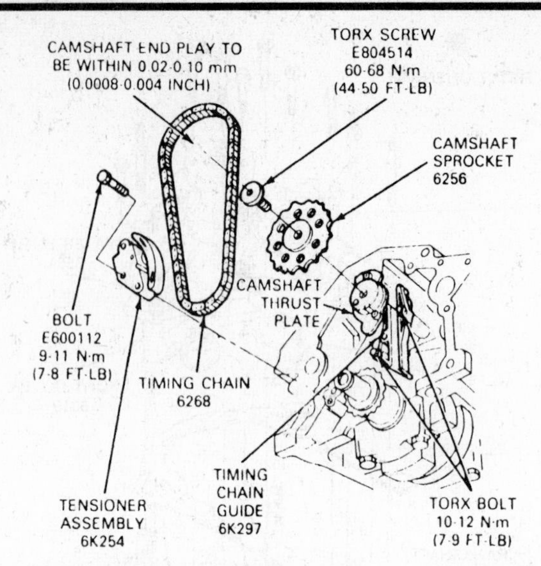

Timing chain tensioner assembly — 4.0L engine

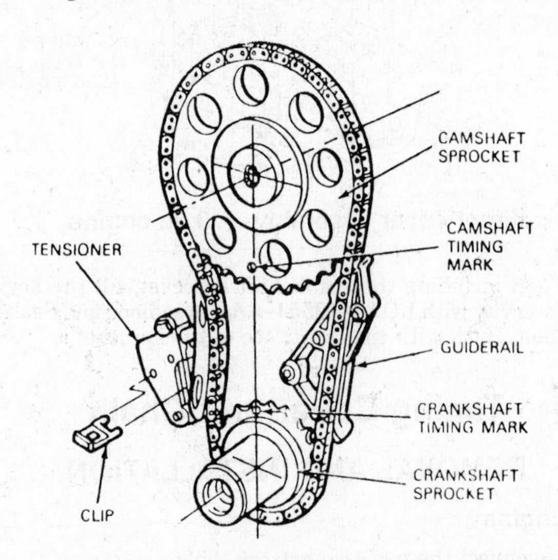

Timing marks installation — 4.0L engine

4.0L Engine

NOTE: Review the complete service procedure before starting this repair. Refer to the necessary service procedures in this Section.

1. Disconnect the negative battery cable. Remove/lower the oil pan.
2. Drain the cooling system.

CAUTION

When draining the coolant, keep in mind that cats and dogs are attracted by the ethylene glycol antifreeze, and are quite likely to drink any that is left in an uncovered container or in puddles on the ground. This will prove fatal in sufficient quantity. Always drain the coolant into a sealable container. Coolant should be reused unless it is contaminated or several years old.

3. Remove the air conditioning compressor and position it out of the way. DO NOT disconnect the refrigerant lines!
4. Remove the power steering pump and position it out of the way. DO NOT disconnect the hoses!
5. Remove the alternator.
6. Remove the fan.
7. Remove the water pump.
8. Remove the drive pulley/damper from the crankshaft.
9. Remove the crankshaft timing sensor.
10. Remove the front cover attaching bolts. It may be necessary to tap the cover loose with a plastic mallet.
11. Remove the radiator.
12. Rotate the engine by hand until the No.1 cylinder is at TDC compression, and the timing marks are aligned.
13. Remove the camshaft sprocket bolt and sprocket retaining key.
14. Remove the camshaft and crankshaft sprockets with the timing chain.
15. If necessary, remove the tensioner and guide.

To install:

16. Install the timing chain guide. Make sure the pin of the guide is in the hole in the block. Tighten the bolts to 84–96 inch lbs.
17. Align the timing marks on the crankshaft and camshaft sprockets and install the sprockets and chain.
18. Install the camshaft sprocket bolt and sprocket retaining key. Make sure that the timing marks are still aligned.
19. Install the tensioner with the clip in place to keep it retracted.
20. Install the crankshaft key. Make sure the timing marks are still aligned.
21. Make sure the tensioner side of the chain is held inward and the other side is straight and tight.
22. Install the camshaft sprocket bolt and tighten it to 50 ft. lbs.
23. Remove the tensioner clip.
24. Check camshaft endplay.
25. Install the front cover and attaching bolts. Refer to the necessary illustration.
26. Install the radiator.
27. Install the crankshaft timing sensor.
28. Install the drive pulley/damper.
29. Install the water pump.

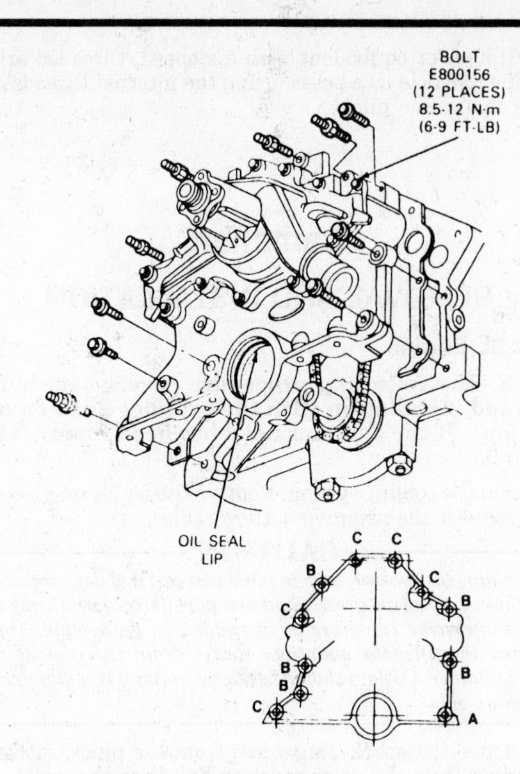

FRONT COVER BOLTS				
FASTENER	LOCATION	SIZE	QUANTITY	TORQUE N·M (FT-LB)
BOLT	A	M8 × 47	(1)	17-21 (13-15)
BOLT	B	M8 × 25	(5)	17-21 (13-15)
STUD	C	M8 × 25	(5)	17-21 (13-15)

Front cover assembly/bolt location — 4.0L engine

30. Install the fan.
31. Install the alternator.
32. Install the power steering pump.
33. Install the air conditioning compressor.
34. Fill the cooling system.
35. Install the oil pan.
36. Fill the crankcase to the proper level. Connect the negative battery cable. Start engine check for leaks and roadtest the vehicle for proper operation.

Timing Cover/Seal and Timing Chain/Gears

REMOVAL AND INSTALLATION

2.2L Diesel Engine

1. Bring the engine to No. 1 piston at TDC on the compression stroke.
2. Disconnect the ground cables from the batteries. Drain the cooling system.

CAUTION

When draining the coolant, keep in mind that cats and dogs are attracted by the ethylene glycol antifreeze, and are quite likely to drink any that is left in an uncovered container or in puddles on the ground. This will prove fatal in sufficient quantity. Always drain the coolant into a sealable container. Coolant should be reused unless it is contaminated or several years old.

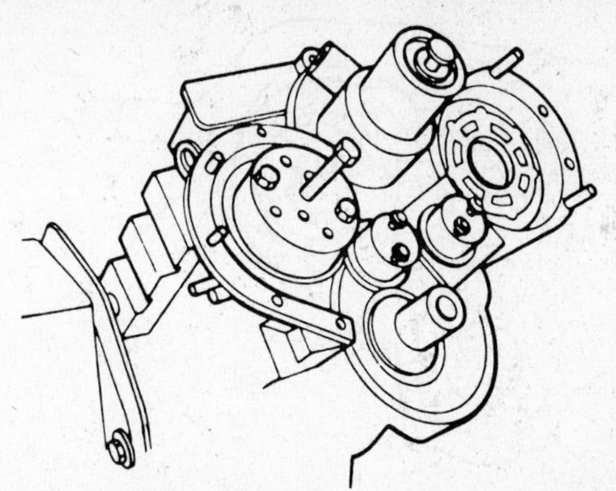

Remove the camshaft gear — 2.2L diesel engine

3. Remove the radiator fan shroud and cooling fan. Drain the engine oil from the crankcase.

CAUTION

The EPA warns that prolonged contact with used engine oil may cause a number of skin disorders, including cancer! You should make every effort to minimize your exposure to used engine oil. Protective gloves should be worn when changing the oil. Wash your hands and any other exposed skin areas as soon as possible after exposure to used engine oil. Soap and water, or waterless hand cleaner should be used.

4. Loosen the idler pulley and remove the A/C compressor belt. Remove the power steering belt. Remove the power steering pump and mounting bracket, position out of the way with the hoses attached.
5. Loosen and remove the alternator and vacuum pump drive belts.
6. Remove the water pump. Using a suitable puller, remove the crankshaft pulley.
7. Remove the nuts and bolts retaining the timing case cover to the engine block. Remove the timing case cover.
8. Remove the engine oil pan.
9. Verify that all timing marks are aligned. Rotate the engine, if necessary, to align marks.
10. Remove the bolt attaching the camshaft gear and remove the washer and friction gear.
11. Remove the bolt attaching the injection pump gear and remove the washer and friction gear.
12. Install Ford tool T83T6306A or equivalent on to the camshaft drive gear and remove the gear. Attach the puller to the injection pump drive gear and remove the gear.
13. Remove the nuts attaching the idler gears after marking reference points on the idler gears for reinstallation position. Remove the idler gear assemblies.
14. Remove the nuts attaching the injection pump to the timing gear case. Support the injection pump in position.
15. Remove the bolts that attaching the timing gear case to the engine block and remove the case if necessary.
16. Clean all gasket mounting surfaces. Clean all parts, replace as necessary.
17. Remove the old oil seal from the front cover and replace.
18. Position the timing gear cover case with a new mounting gasket and install.
19. Install the timing gears as follows:
    a. Verify that the crankshaft and right idler pulley timing marks align and install the right idler gear assembly.
    b. Install the camshaft gear so that the timing marks align with the timing mark on the right idler gear.
    c. Install the left idler gear assembly so that the timing marks align with the timing marks align with the timing mark on the right idler gear.

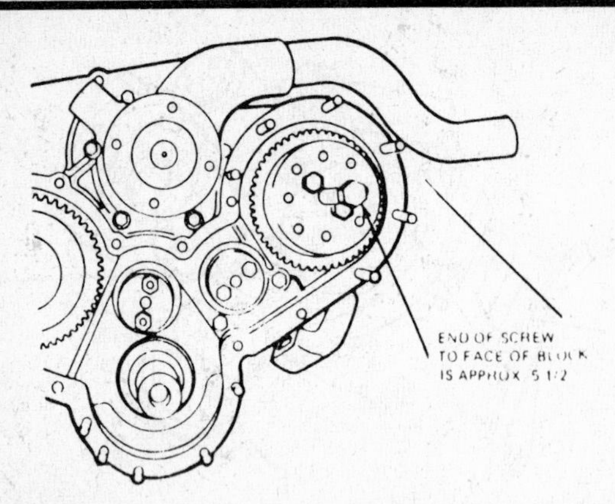

Remove the injection pump gear — 2.2L diesel engine

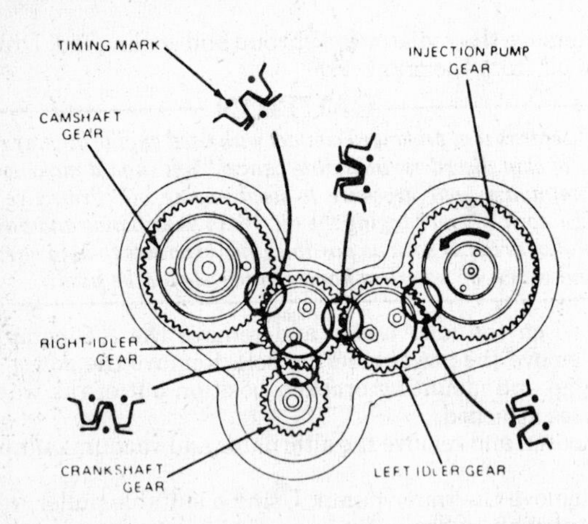

Timing gear alignment — 2.2L diesel engine

d. Install the injection pump gear so that the timing marks align with the timing mark on the left idler gear.

e. Install all friction gears, washers, nuts and bolts on the gears.

20. Install the timing case covers using a new mounting gasket.

21. Install the remaining components in the reverse order of removal.

Camshaft and Auxiliary Shaft Sprockets and Seals

REMOVAL AND INSTALLATION

The cylinder front cover, camshaft and auxiliary shaft seals are replaced in the same manner with the same tools after the respective gear has been removed. Always use a new attaching bolt when replacing the camshaft sprocket or use new Teflon® sealing tape on the threads of the old bolt. To remove the sprockets, first remove the timing cover and belt, then use tool T74P–6256–B, or equivalent to pull the cam drive sprocket. The same tool is used in exactly the same manner to remove the auxiliary shaft sprocket, as well as to hold the sprockets while the attaching bolts are installed and tightened.

A front cover seal remover tool T74P–6700–B or equivalent is used to remove all the seals. When positioning this tool, make sure that the jaws are gripping the thin edge of the seal very tightly before operating the jack-screw portion of the tool.

To install the seals, a cam and auxiliary shaft seal replacer T74P–6150–A or equivalent with a stepped, threaded arbor is used. The tool acts as a press, using the internal threads of the various shafts as a pilot.

Camshaft

REMOVAL AND INSTALLATION

2.0L, 2.3L Engine

NOTE: The following procedure covers camshaft removal and installation with the cylinder head on or off the engine. If the cylinder head has been removed start at Step 9.

1. Drain the cooling system. Remove the air cleaner assembly and disconnect the negative battery cable.

CAUTION

When draining the coolant, keep in mind that cats and dogs are attracted by the ethylene glycol antifreeze, and are quite likely to drink any that is left in an uncovered container or in puddles on the ground. This will prove fatal in sufficient quantity. Always drain the coolant into a sealable container. Coolant should be reused unless it is contaminated or several years old.

2. Remove the spark plug wires from the plugs, disconnect the retainer from the valve cover and position the wires out of the way. Disconnect rubber vacuum lines as necessary.
3. Remove all drive belts. Remove the alternator mounting bracket-to-cylinder head mounting bolts, position bracket and alternator out of the way.
4. Disconnect and remove the upper radiator hose. Disconnect the radiator shroud.
5. Remove the fan blades and water pump pulley and fan shroud. Remove cam belt and valve covers.
6. Align engine timing marks at TDC for No. 1 cylinder. Remove cam drive belt.
7. Jack up the front of the vehicle and support on jackstands. Remove the front motor mount bolts. Disconnect the lower radiator hose from the radiator. Disconnect and plug the automatic transmission cooler lines.
8. Position a piece of wood on a floor jack and raise the engine carefully as far as it will go. Place blocks of wood between the engine mounts and crossmember pedestals.
9. Remove the rocker arms (camshaft followers).
10. Remove the camshaft drive gear and belt guide using a suitable puller. Remove the front oil seal with a sheet metal screw and slide hammer.
11. Remove the camshaft retainer located on the rear mounting stand by unbolting the two bolts.
12. Remove the camshaft by carefully withdrawing toward the front of the engine. Caution should be used to prevent damage to cam bearings, lobes and journals.
13. Check the camshaft journals and lobes for wear. Inspect the cam bearings, if worn (unless the proper bearing installing tool is on hand), the cylinder head must be removed for new bearings to be installed by a machine shop.
14. Camshaft installation is in the reverse order of service removal procedure. See following notes.

NOTE: Coat the camshaft with heavy SF oil before sliding it into the cylinder head. Install a new front seal. Apply a coat of sealer or teflon tape to the cam drive gear bolt before installation.

After any procedure requiring removal of the rocker arms, each lash adjuster must be fully collapsed after assembly, then released. This must be done before the camshaft is turned.

15. Refill cooling system. Start engine and check for leaks. Roadtest the vehicle for proper operation.

2.8L, 2.9L Engines

1. Disconnect the battery ground cable from the battery.
2. Drain the oil from the crankcase.

CAUTION

The EPA warns that prolonged contact with used engine oil may cause a number of skin disorders, including cancer! You should make every effort to minimize your exposure to used engine oil. Protective gloves should be worn when changing the oil. Wash your hands and any other exposed skin areas as soon as possible after exposure to used engine oil. Soap and water, or waterless hand cleaner should be used.

3. Remove the radiator, fan and spacer, drive belt and pulley.
4. Label and remove the spark plug wires from the spark plugs.
5. Remove the distributor cap with spark plug wires as an assembly.
6. Disconnect the distributor wiring harness and remove the distributor.
7. Remove the alternator.
8. Remove the thermactor pump.
9. Remove the fuel lines, fuel filter and carburetor.
10. Remove the intake manifold as described earlier.
11. Remove the rocker arm covers and rocker arm and shaft assemblies as described in this Section. Label and remove the push rods and the tappets, so they can be reinstalled in the same location.

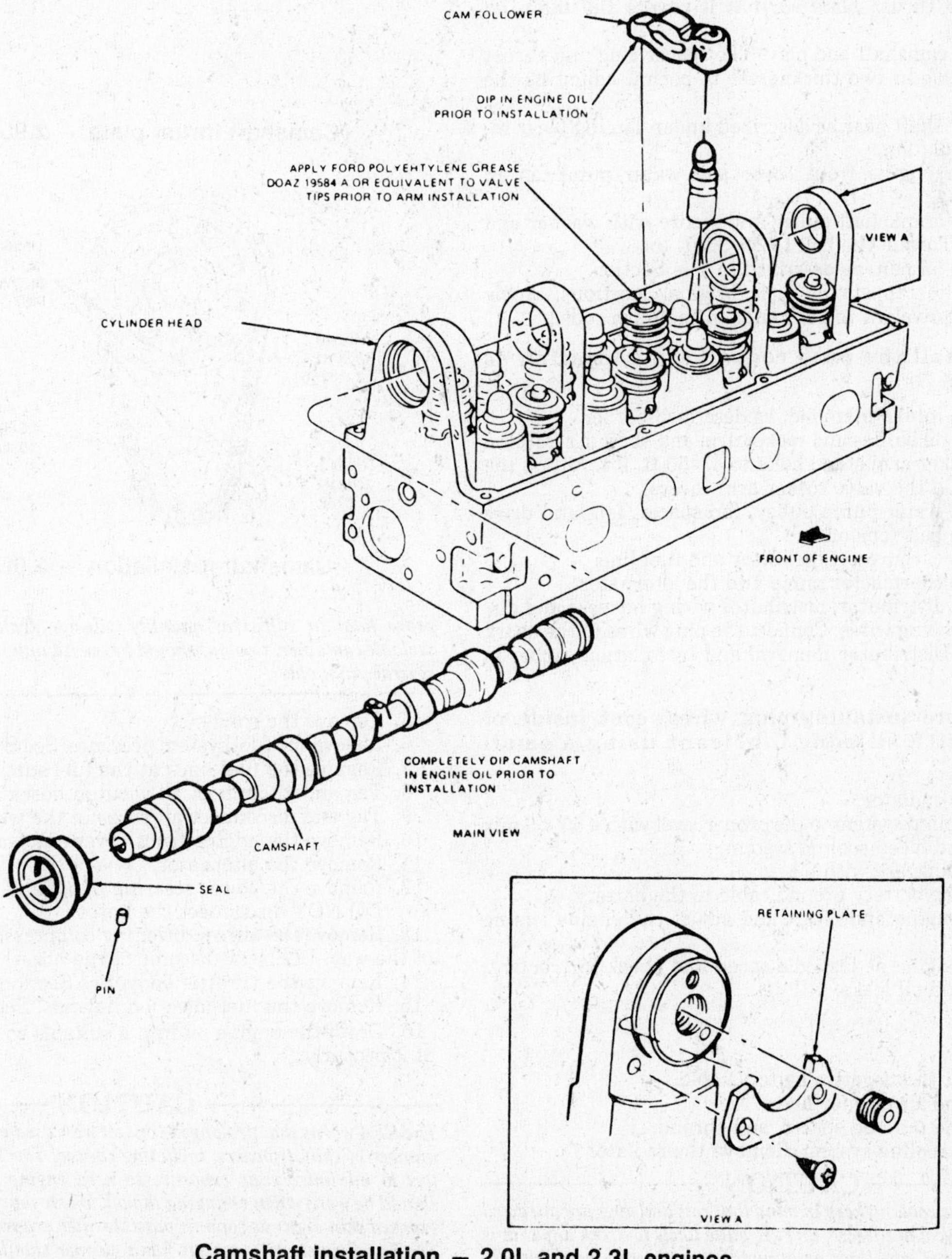

Camshaft installation – 2.0L and 2.3L engines

12. Remove the oil pan as described in this Section.
13. Remove the crankshaft damper.
14. Remove the engine front cover and water pump as an assembly.
15. Remove the camshaft gear attaching bolt and washer, and slide gear off camshaft.
16. Remove the camshaft thrust plate.
17. Carefully remove the camshaft from the block, avoiding any damage to the camshaft bearings.
18. Remove the camshaft drive gear and spacer ring.

To install:

19. Oil the camshaft journals with heavy SF grade engine oil and apply Lubriplate® or equivalent lubricant to the cam lobes. Install the spacer ring with the chamfered side toward the camshaft. Insert the camshaft key.
20. Install the camshaft in the block, carefully avoiding damage to the bearing surfaces.
21. Install the thrust plate so that it covers the main oil gallery.
22. Check the camshaft end play. The spacer ring and thrust plate are available in two thicknesses to permit adjusting the end play.
23. Install camshaft gear as described under Timing Gear removal and installation.
24. Install the engine front cover and water pump as an assembly.
25. Install the crankshaft pulley and secure with washer and attaching bolt. Torque the bolt to 85–96 ft. lbs.
26. Install the oil pan, as described in this Section.
27. Position the tappets in their original locations. Apply Lubriplate or equivalent to both ends of the push rods.

NOTE: Install the push rods in same location as removed.

28. Install the intake manifold, as described earlier.
29. Install the oil baffles and rocker arm and shaft assemblies. Tighten the rocker arm stand bolts to 43–50 ft. lbs. Adjust the valves and install the valve rocker arm covers.
30. Install the water pump pulley, fan spacer, fan, and drive belt. Adjust the belt tension.
31. Install the carburetor, fuel filter and fuel line.
32. Install the thermactor pump and the alternator.
33. Install the distributor, distributor wiring harness and distributor cap and plug wires. Connect the plug wires to the spark plugs. Refer to Distributor Removal and Installation earlier in this Section.

NOTE: Before installing plug wires, coat inside of each boot with silicone lubricant using a small screwdriver.

34. Install the radiator.
35. Fill the cooling system to the proper level with a 50–50 mix of antifreeze and bleed cooling system.
36. Fill the crankcase with oil.
37. Connect the battery ground cable to the battery.
38. Run the engine and check and adjust the engine timing and idle speed.
39. Run the engine at fast idle speed and check for coolant, fuel, vacuum and oil leaks.

3.0L Engine

1. Disconnect the negative battery cable.
2. Remove the air cleaner hoses.
3. Remove the fan and spacer, and shroud.
4. Drain the cooling system. Remove the radiator.

CAUTION

When draining the coolant, keep in mind that cats and dogs are attracted by the ethylene glycol antifreeze, and are quite likely to drink any that is left in an uncovered container or in puddles on the ground. This will prove fatal in sufficient quantity. Always drain the coolant into a sealable container. Coolant should be reused unless it is contaminated or several years old.

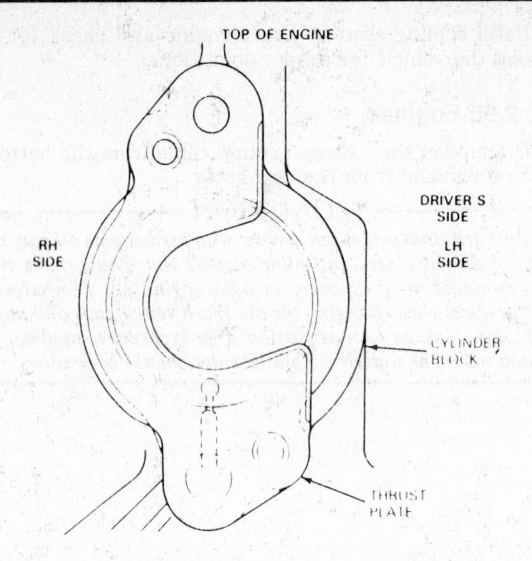

Camshaft thrust plate – 2.9L engine

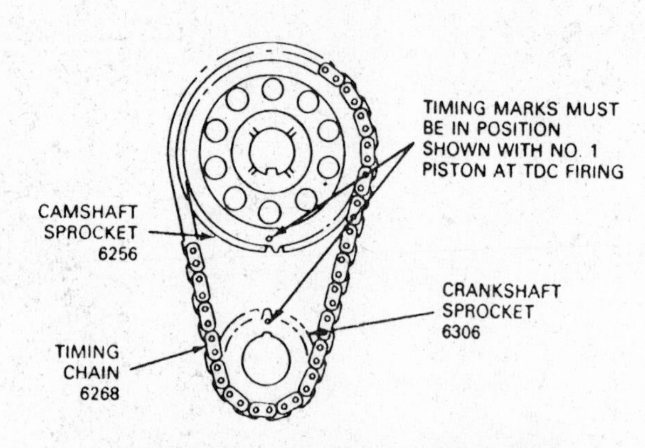

Camshaft installation – 3.0L engine

5. Remove the condenser.
6. Relieve the fuel system pressure. See Section 1 Fuel Filter.
7. Remove the fuel lines at the fuel supply manifold.
8. Tag and disconnect all vacuum hoses in the way.
9. Tag and disconnect all wires in the way.
10. Remove the engine front cover and water pump.
11. Remove the alternator.
12. Remove the power steering pump and secure it out of the way. DO NOT disconnect the hoses!
13. Remove the air conditioning compressor and secure it out of the way. DO NOT disconnect the hoses!
14. Remove the throttle body. See Section 5.
15. Remove the fuel injection harness. See Section 5.
16. Drain the engine oil into a suitable container and dispose of it properly.

CAUTION

The EPA warns that prolonged contact with used engine oil may cause a number of skin disorders, including cancer! You should make every effort to minimize your exposure to used engine oil. Protective gloves should be worn when changing the oil. Wash your hands and any other exposed skin areas as soon as possible after exposure to used engine oil. Soap and water, or waterless hand cleaner should be used.

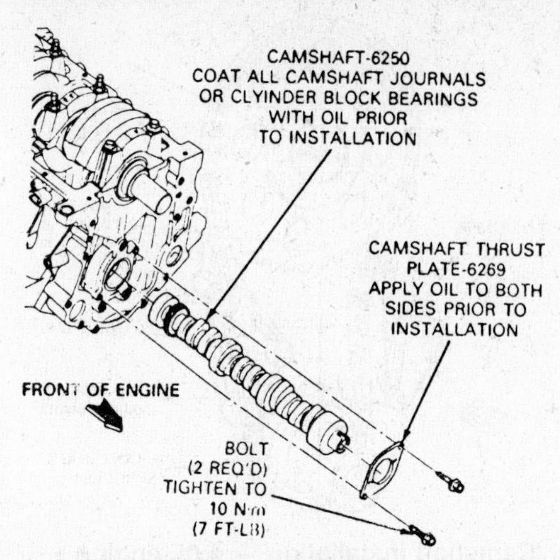

Camshaft assembly installation — 3.0L engine

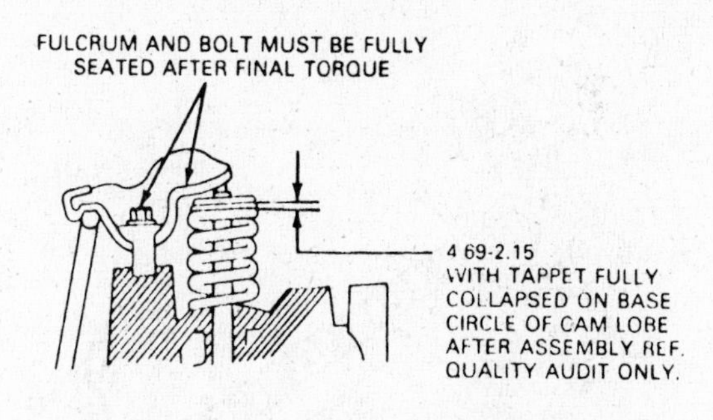

CYL. NO.	CAMSHAFT POSITION A	CAMSHAFT POSITION B
	SET GAP OF VALVES NOTED	
1	INT.	EXH.
2	EXH.	INT.
3	NONE	INT.-EXH.
4	INT.	EXH.
5	EXH.	INT.
6	NONE	INT.-EXH.

Camshaft assembly installation — 3.0L engine

17. Turn the engine by hand to 0 BTDC of the power stroke on No. 1 cylinder.
18. Disconnect the spark plug wires from the plugs.
19. Remove the distributor cap with the spark plug wires as an assembly.
20. Matchmark the rotor, distributor body and engine. Disconnect the distributor wiring harness and remove the distributor.
21. Remove the rocker arm covers.
22. Remove the intake manifold as previously described.
23. Loosen the rocker arm bolts enough to pivot the rocker arms out of the way and remove the pushrods. Identify them for installation. They must be installed in their original positions!
24. Remove the tappets. Identify them for installation.
25. Remove the crankshaft pulley/damper.
26. Remove the starter.
27. Remove the oil pan as previously described.
28. Turn the engine by hand until the timing marks align at TDC of the power stroke on No.1 piston.
29. Check the camshaft endplay. If excessive, you'll have to replace the thrust plate.
30. Remove the camshaft gear attaching bolt and washer, then slide the gear off the camshaft.
31. Remove the camshaft thrust plate.
32. Carefully slide the camshaft out of the engine block, using caution to avoid any damage to the camshaft bearings.

To install:

33. Oil the camshaft journals and cam lobes with heavy SG engine oil (50W). Install the spacer ring with the chamfered side toward the camshaft, then insert the camshaft key.
34. Install the camshaft in the block, using caution to avoid any damage to the camshaft bearings.
35. Install the thrust plate. Tighten the attaching screws to 84 inch lbs.
36. Rotate the camshaft and crankshaft as necessary to align the timing marks. Install the camshaft gear and chain. Tighten the attaching bolt to 46 ft. lbs.
30. Coat the tappets with 50W engine oil and place them in their original locations.
31. Apply 50W engine oil to both ends of the pushrods. Install the pushrods in their original locations.
32. Pivot the rocker arms into position. Tighten the fulcrum bolts to 8 ft. lbs.
33. Rotate the engine until both timing marks are at the tops of their sprockets and aligned. Tighten the following fulcrum bolts to 18 ft. lbs.:
- No.1 intake
- No.2 exhaust
- No.4 intake
- No.5 exhaust

34. Rotate the engine until the camshaft timing mark is at the bottom of the sprocket and the crankshaft timing mark is at the top of the sprocket, and both are aligned. Tighten the following fulcrum bolts to 18 ft. lbs.:
- No.1 exhaust
- No.2 intake
- No.3 intake and exhaust
- No.4 exhaust
- No.5 intake
- No.6 intake and exhaust

35. Now, tighten all the bolts to 24 ft. lbs.
36. Turn the engine by hand to 0 BTDC of the power stroke on No. 1 cylinder.
37. Install the engine front cover and water pump assembly.
38. Install the oil pan.
39. Install the crankshaft damper/pulley and tighten the retaining bolt to 107 ft. lbs.
40. Install the intake manifold and tighten the mounting bolts to the specifications and in the sequence described under Intake Manifold Removal And Installation.
41. Install the rocker covers.
42. Install the injector harness.
43. Install the distributor.
44. Install the cap and wires.
45. Install the throttle body.
46. Install the alternator.
47. Install the power steering pump.
48. Install the compressor.
49. Connect all wires.
50. Connect all vacuum lines.
51. Install the radiator and condenser.
52. Install the fan and clutch.
53. Install the fuel lines.
54. Install the starter.
55. Refill the cooling system.
56. Replace the oil filter and refill the crankcase with the specified amount of engine oil.
57. Reconnect the battery ground cable.
58. Start the engine and check the ignition timing and idle speed. Adjust if necessary. Run the engine at fast idle and check for coolant, fuel, vacuum or oil leaks.

4.0L Engine

NOTE: Review the complete service procedure before starting this repair. Refer to the necessary service procedures in this Section.

1. Disconnect the negative battery cable.
2. Drain the engine oil into a suitable container and dispose of it properly.

CAUTION

The EPA warns that prolonged contact with used engine oil may cause a number of skin disorders, including cancer! You should make every effort to minimize your exposure to used engine oil. Protective gloves should be worn when changing the oil. Wash your hands and any other exposed skin areas as soon as possible after exposure to used engine oil. Soap and water, or waterless hand cleaner should be used.

3. Drain the cooling system.

CAUTION

When draining the coolant, keep in mind that cats and dogs are attracted by the ethylene glycol antifreeze, and are quite likely to drink any that is left in an uncovered container or in puddles on the ground. This will prove fatal in sufficient quantity. Always drain the coolant into a sealable container. Coolant should be reused unless it is contaminated or several years old.

4. Remove the radiator.
5. Remove the condenser.
6. Remove the fan and spacer, and shroud.
7. Remove the air cleaner hoses.
8. Tag and remove the spark plug wires.
9. Remove the EDIS ignition coil and bracket.
10. Remove the crankshaft pulley/damper.
11. Remove the clamp. bolt and oil pump drive from the rear of the block.
12. Remove the alternator.
13. Relieve the fuel system pressure. See Section 1 Fuel Filter.
14. Remove the fuel lines at the fuel supply manifold.
15. Remove the upper and lower intake manifolds as previously described.
16. Remove the rocker arm covers.
17. Remove the rocker shaft assemblies.
18. Remove the pushrods. Identify them for installation. They must be installed in their original positions!
19. Remove the tappets. Identify them for installation.
20. Remove the oil pan as previously described.
21. Remove the engine front cover and water pump.
22. Place the timing chain tensioner in the retracted position and install the retaining clip.
23. Turn the engine by hand until the timing marks align at TDC of the power stroke on No.1 piston.
24. Check the camshaft endplay. If excessive, you'll have to replace the thrust plate.
25. Remove the camshaft gear attaching bolt and washer, then slide the gear off the camshaft.
26. Remove the camshaft thrust plate.
27. Carefully slide the camshaft out of the engine block, using caution to avoid any damage to the camshaft bearings.

To install:

28. Oil the camshaft journals and cam lobes with heavy SG engine oil (50W).
29. Install the camshaft in the block, using caution to avoid any damage to the camshaft bearings.
30. Install the thrust plate. Make sure that it covers the main oil gallery. Tighten the attaching screws to 96 inch lbs.
31. Rotate the camshaft and crankshaft as necessary to align the timing marks. Install the camshaft gear and chain. Tighten the attaching bolt to 50 ft. lbs.
32. Remove the clip from the chain tensioner.

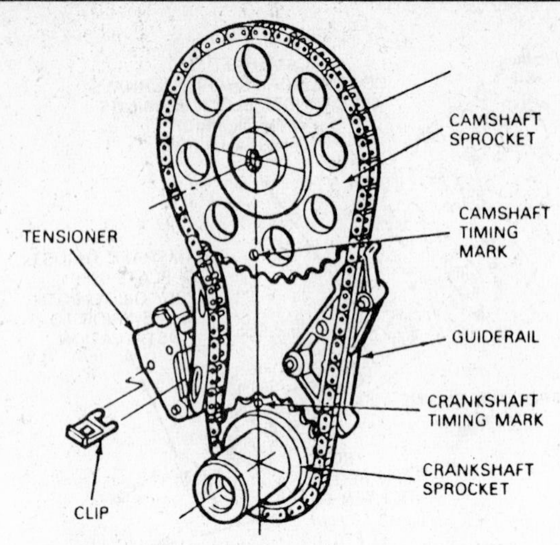

Camshaft installation – 4.0L engine

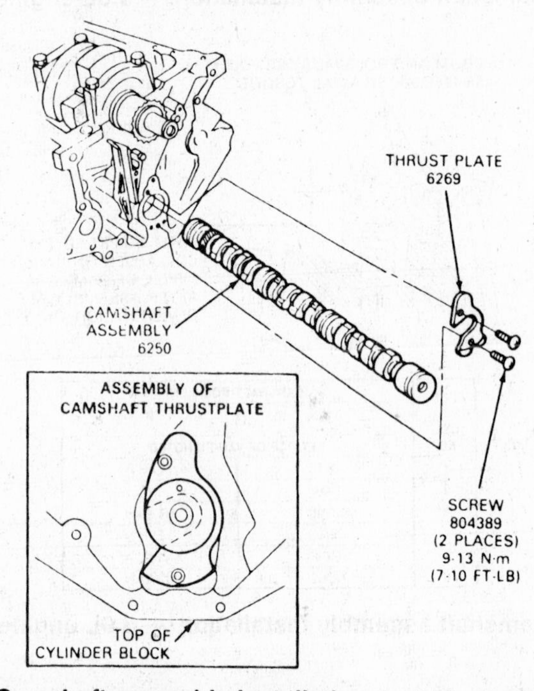

Camshaft assembly installation – 4.0L engine

33. Install the engine front cover and water pump assembly. Refer to the necessary service procedures in this Section.
34. Install the crankshaft damper/pulley and tighten the retaining bolt to 107 ft. lbs.
35. Install the oil pan.
36. Coat the tappets with 50W engine oil and place them in their original locations.
37. Apply 50W engine oil to both ends of the pushrods. Install the pushrods in their original locations.
38. Install the upper and lower intake manifolds.
39. Install the rocker shaft assemblies.
40. Install the rocker covers.
41. Install the fan and clutch.
42. Install the fuel lines.
43. Install the oil pump drive.
44. Install the alternator.
45. Install the EDIS coil and plug wires. Coat the inside of each wire boot with silicone lubricant.

46. Install the radiator and condenser.
47. Refill the cooling system.
48. Replace the oil filter and refill the crankcase with the specified amount of engine oil.
49. Reconnect the battery ground cable.
50. Start the engine and check the ignition timing and idle speed. Adjust if necessary. Run the engine at fast idle and check for coolant, fuel, vacuum or oil leaks.

2.2L Diesel Engine

1. Ford recommends that the engine be removed from the vehicle when camshaft replacement is necessary.
2. With the engine removed; remove the valve cover, rocker arms and shaft assembly and the pushrods. Remove the lifters, identify and keep in order if they are to be reused.
3. Remove the front timing case cover and camshaft gear.
4. Remove the engine oil pan and oil pump.
5. Remove the camshaft thrust plate and the camshaft. Take care when removing the camshaft not to damage lobes or bearings.
6. Apply oil to the camshaft bearings and bearing journals. Apply Polyethylene grease to the camshaft lobes and install the camshaft into the engine.
7. Reinstall components in the reverse order of removal.

2.3L Turbo Diesel Engine

1. Disconnect battery ground cables from both batteries.
2. Remove rocker cover.
3. Remove upper front timing cover.
4. Rotate crankshaft until No. 1 piston is at TDC on compression stroke.
5. Loosen camshaft/injection pump drive belt tensioner and remove timing belt from camshaft pulley.
6. Remove camshaft pulley bolt and remove pulley.
7. Remove rocker arm shaft.
8. Remove camshaft bearing caps and remove camshaft.
9. Remove and discard camshaft oil seal.
10. Inspect camshaft and bearings.

To install:

11. Position camshaft on cylinder head and install bearing caps. Tighten bolts to specification.

CAUTION

Ensure bearing caps are installed in their original positions.

12. Coat sealing lip of new camshaft seal with engine oil and install seal using Camshaft Oil Seal Replacer T85T-6250-A or equivalent.

CAUTION

Ensure seal installer is positioned with hole over spring pin on camshaft.

13. Install rocker arm.
14. Install camshaft pulley and tighten bolt to specification.
15. Install camshaft/injection pump drive belt. Adjust drive belt.
16. Adjust valves.
17. Install upper front timing cover.
18. Install rocker cover.
19. Fill cooling system.
20. Connect battery ground cable to both batteries.
21. Run engine and check for oil leaks.

CAMSHAFT INSPECTION

Camshaft Lobe Lift

2.0L, 2.3L ENGINES

Check the lift of each lobe in consecutive order and make a note of the readings. Camshaft assembly specifications are sometimes modify by Ford after production. Refer to a local reputable machine shop as necessary.

1. Remove the air cleaner and the valve rocker arm cover.
2. Measure the distance between the major (A-A) and minor (B-B) diameters of each cam lobe with a Vernier caliper and record the readings. The difference in the readings on each cam diameter is the lobe lift.
3. If the readings do not meet specifications, replace the camshaft and all rocker arms.
4. Install the valve rocker arm cover and the air cleaner.

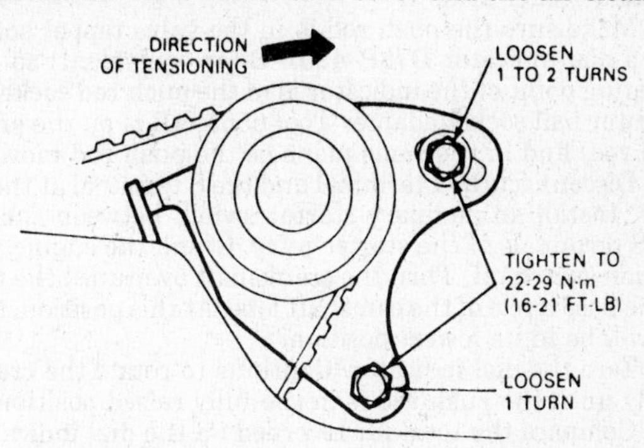

Camshaft belt tensioner – 2.3L turbo diesel engine

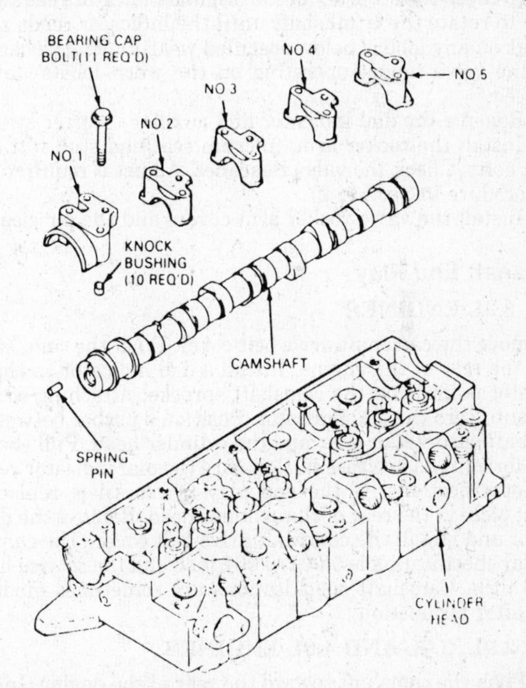

Camshaft assembly – 2.3L turbo diesel engine

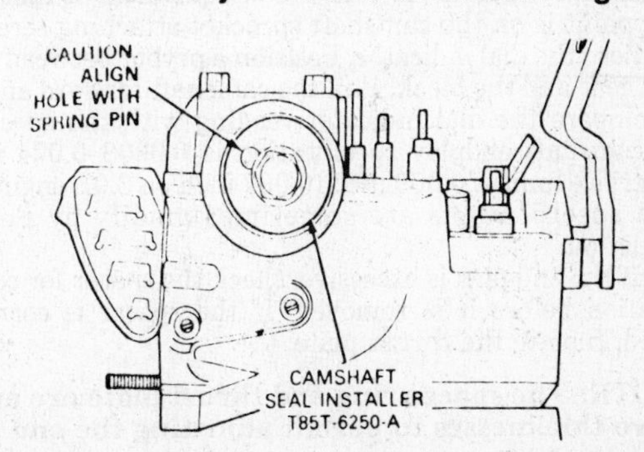

Camshaft seal installation – 2.3L turbo diesel engine

2.8L, 2.9L, 3.0L, 4.0L AND 2.2L DIESEL ENGINES

Check the lift of each lobe in consecutive order and make a note of the reading. Camshaft assembly specifications are sometimes modify by Ford after production. Refer to a local reputable machine shop as necessary.

1. Remove the fresh air inlet tube and the air cleaner. Remove the heater hose and crankcase ventilation hoses. Remove valve rocker arm cover(s).
2. Remove the rocker arm stud nut or fulcrum bolts, fulcrum seat and rocker arm.
3. Make sure the push rod is in the valve tappet socket. Install a dial indicator D78P-4201-B (or equivalent) so that the actuating point of the indicator is in the push rod socket (or the indicator ball socket adapter Tool 6565-AB is on the end of the push rod) and in the same plane as the push rod movement.
4. Disconnect the I terminal and the S terminal at the starter relay. Install an auxiliary starter switch between the battery and S terminals of the starter relay. Crank the engine with the ignition switch off. Turn the crankshaft over until the tappet is on the base circle of the camshaft lobe. At this position, the push rod will be in its lowest position.
5. Zero the dial indicator. Continue to rotate the crankshaft slowly until the push rod is in the fully raised position.
6. Compare the total lift recorded on the dial indicator with the specification.

   To check the accuracy of the original indicator reading, continue to rotate the crankshaft until the indicator reads zero. If the lift on any lobe is below specified wear limits, the camshaft and the valve tappet operating on the worn lobe(s) must be replaced.
7. Remove the dial indicator and auxiliary starter switch.
8. Install the rocker arm, fulcrum seat and stud nut or fulcrum bolts. Check the valve clearance. Adjust if required (refer to procedure in Section 2).
9. Install the valve rocker arm covers and the air cleaner.

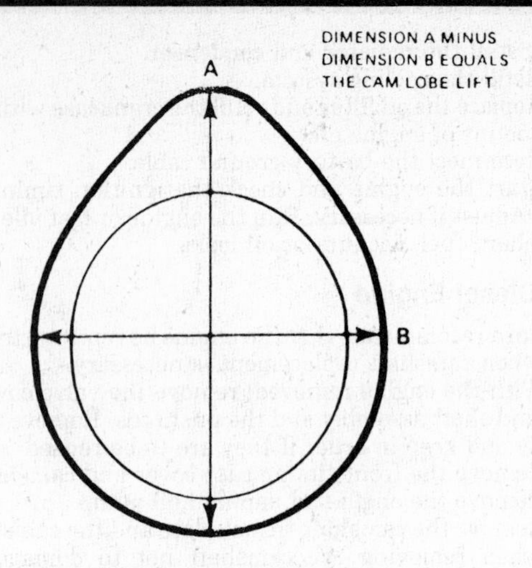

Checking OHC camshaft lobe lift

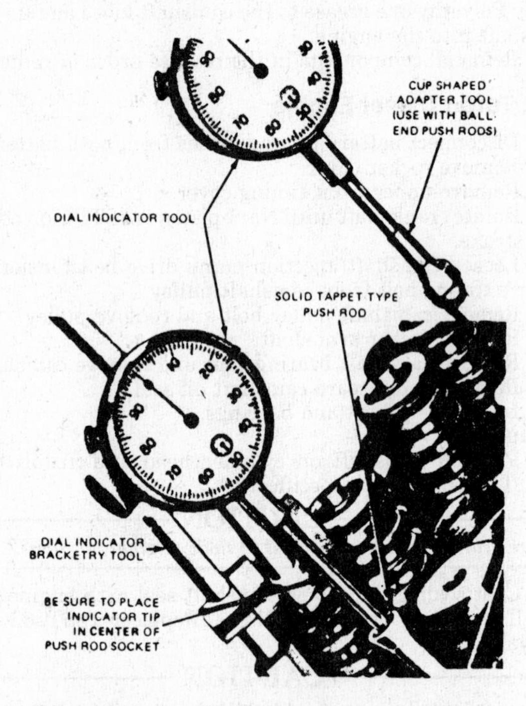

Checking the OHV camshaft lobe lift

Camshaft End Play

2.0L, 2.3L ENGINES

Remove the camshaft drive belt cover. Push the camshaft toward the rear of the engine. Install a dial indicator so that the indicator point is on the camshaft sprocket attaching screw or gear hub. Zero the dial indicator. Position a prybar between the camshaft sprocket or gear and the cylinder head. Pull the camshaft forward and release it. Compare the dial indicator reading with specifications. If the end play is excessive, replace the thrust plate at the rear of the cylinder head. Remove the dial indicator and install the camshaft drive belt cover. The camshaft endplay specification is 0.001-0.007 inch and the service limit is 0.003 inch. Camshaft specifications are sometimes modify by Ford after production.

2.8L, 2.9L, 3.0L AND 4.0L ENGINES

1. Push the camshaft toward the rear of the engine. Install a dial indicator (Tool D78P-4201-C or equivalent so that the indicator point is on the camshaft sprocket attaching screw.
2. Zero the dial indicator. Position a prybar between the camshaft gear and the block. Pull the camshaft forward and release it. Compare the dial indicator reading with the specification. The camshaft endplay specification is 0.0008-0.004 inch and the service limit is 0.009 inch (0.007 inch on 3.0L engine). Camshaft specifications are sometimes modify by Ford after production.
3. If the end play is excessive, check the spacer for correct installation before it is removed. If the spacer is correctly installed, replace the thrust plate.

NOTE: The spacer ring and thrust plate are available in two thicknesses to permit adjusting the end play.

4. Remove the dial indicator.

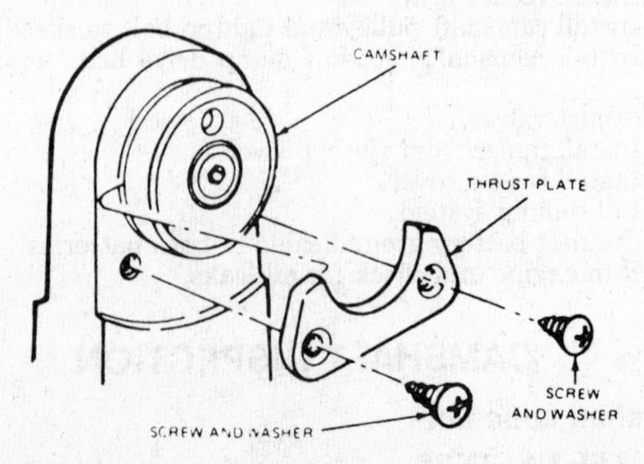

Camshaft thrust plate installation — 2.0L and 2.3L engines

2.2L DIESEL ENGINE

1. Remove the camshaft as described earlier in this Section.
2. Mount the thrust plate. Camshaft gear and the friction gear on the camshaft.
3. Install and tighten the lock bolt.
4. Measure the end play by inserting a feeler gauge between the thrust plate and the cam gear.
5. If the end play exceeds specification replace the thrust plate. Refer to a local reputable machine shop as necessary. Camshaft specifications are sometimes modify by Ford after production.

Camshaft Bearings

REMOVAL AND INSTALLATION

If excessive camshaft wear is found, or if the engine is completely rebuilt, the camshaft bearings should be replaced. Use these service repair procedures as a guide-modify as necessary for Diesel engines.

2.0L And 2.3L Engines

1. Remove the head and place it on a work stand.
2. Remove the camshaft.
3. Using a tool such as Bearing Replacer T71P–6250–A, remove the bearings.
4. Coat the new bearings with clean 50W engine oil and install them with the tool.

2.8L And 2.9L Engines

1. Remove the engine and place it on a work stand.
2. Remove the flywheel.

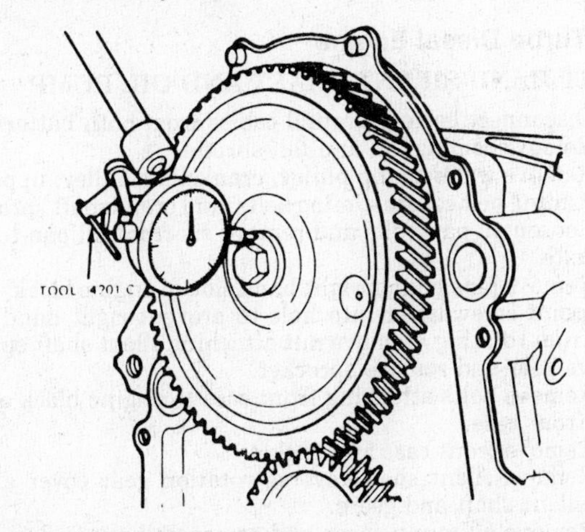

Checking camshaft endplay – 2.8L and 2.9L engines

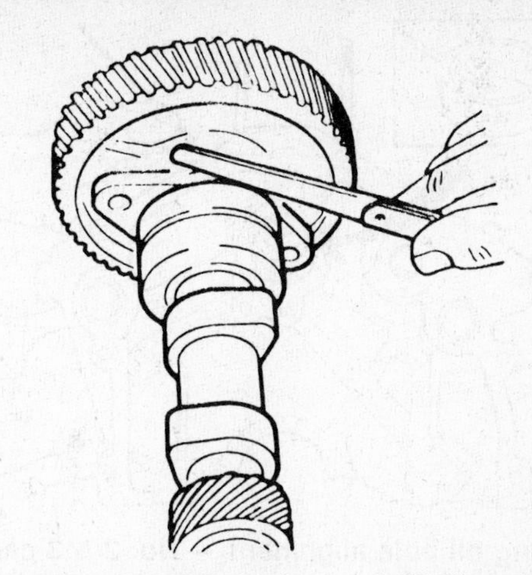

Checking camshaft endplay – 2.2L diesel engine

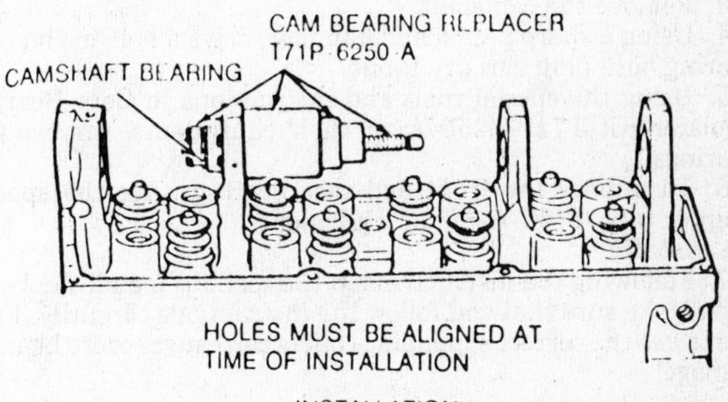

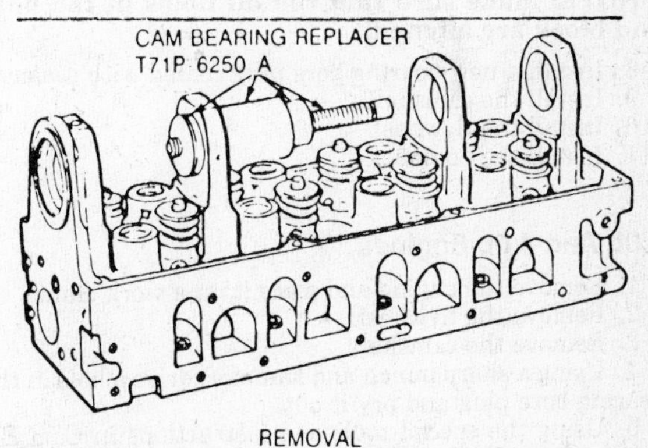

Camshaft bearing procedures – 2.3L engine

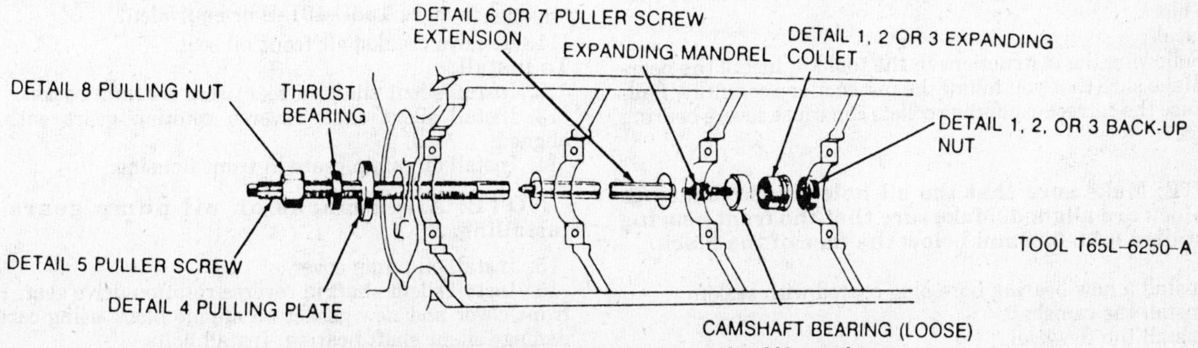

Camshaft bearing procedures on the V6 engines

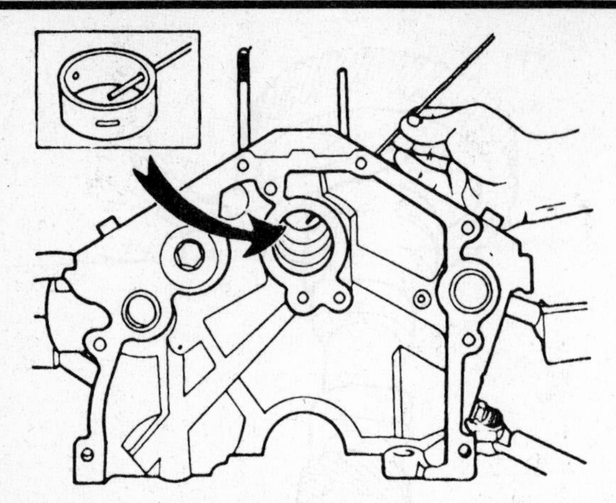

Checking oil hole alignment – No. 2 & 3 camshaft bearings – 2.9L engine

3. Remove the camshaft.
4. Using a sharp punch and hammer, drive a hole in the rear bearing bore plug and pry it out.
5. Using the special tools and instructions in Cam Bearing Replacer Kit T71P–6250–A, or their equivalents, remove the bearings.
6. To remove the front and rear bearings, use the special adapter tube T72C–6250, or equivalent.

To install:

7. Following the instructions in the tool kit, install the bearings. Make sure that you follow the instructions carefully. Failure to use the correct expanding collets can cause severe bearing damage!

NOTE: Make sure that the oil holes in the bearings and block are aligned!

8. Install a new bearing bore plug coated with sealer.
9. Install the camshaft.
10. Install the flywheel.
11. Install the engine.

3.0L And 4.0L Engines

1. Remove the engine and place it on a work stand.
2. Remove the flywheel.
3. Remove the camshaft.
4. Using a sharp punch and hammer, drive a hole in the rear bearing bore plug and pry it out.
5. Using the special tools and instructions in Cam Bearing Replacer Kit T65L–6250–A, or their equivalents, remove the bearings.
6. To remove the front bearing, install the tool from the rear of the block.

To install:

7. Following the instructions in the tool kit, install the bearings. Make sure that you follow the instructions carefully. Failure to use the correct expanding collets can cause severe bearing damage!

NOTE: Make sure that the oil holes in the bearings and block are aligned! Make sure that the front bearing is installed 0.51–0.89mm below the face of the block.

8. Install a new bearing bore plug coated with sealer.
9. Install the camshaft.
10. Install the flywheel.
11. Install the engine.

Auxiliary Shaft

REMOVAL AND INSTALLATION

2.0L, 2.3L Engines

1. Remove the camshaft drive belt cover.
2. Remove the drive belt. Remove the auxiliary shaft sprocket. A puller may be necessary to remove the sprocket.
3. Remove the distributor and fuel pump.
4. Remove the auxiliary shaft cover and thrust plate.
5. Withdraw the auxiliary shaft from block.

NOTE: The distributor drive gear and the fuel pump eccentric on the auxiliary shaft must not be allowed to touch the auxiliary shaft bearings during removal and installation. Completely coat the shaft with oil before sliding it into place.

6. Slide the auxiliary shaft into the housing and insert the thrust plate to hold the shaft.
7. Install a new gasket and auxiliary shaft cover.

NOTE: The auxiliary shaft cover and cylinder front cover share a gasket. Cut off the old gasket around the cylinder cover and use half of the new gasket on the auxiliary shaft cover.

8. Fit a new gasket into the fuel pump and install the pump.
9. Insert the distributor and install the auxiliary shaft sprocket.
10. Align the timing marks and install the drive belt.
11. Install the drive belt cover.
12. Check the ignition timing.

2.3L Turbo Diesel Engine

RIGHT HAND SILENT SHAFT AND OIL PUMP

1. Disconnect battery ground cables from both batteries.
2. Remove cooling fan and fan shroud.
3. Remove water pump pulley, crankshaft pulley, upper and lower timing belt covers, timing belts, and crankshaft sprockets.
4. Loosen oil pan bolts and remove six front oil pan-to-front case bolts.
5. Remove pipe plug in right hand side of engine block, Insert cross point screwdriver into hole to prevent right hand silent shaft from rotating. Remove nut attaching silent shaft sprocket to drive gear and remove sprocket.
6. Remove bolts attaching front case to engine block and remove front case.
7. Remove front case and gasket.
8. Remove silent shaft reverse rotation gear cover and remove silent shaft and gears.
9. Remove oil pump cover and remove oil pump drive gear, and inner and outer gears.
10. Remove silent shaft reverse rotation drive gear oil seal using Seal Remove T58L–101–B or equivalent.
11. Remove crankshaft front oil seal.

To install:

12. Install silent shaft oil seal using a 21mm socket.
13. Install silent shaft reverse rotation gears with marks aligned.
14. Install oil pump gears in front housing.

NOTE: Align marks on oil pump gears when installing.

15. Install oil pump cover.
16. Install silent shaft in reverse rotation drive gear. Position front cover and new gasket on engine block using care not to damage silent shaft bearing. Install bolts.
17. Install silent shaft sprocket. Insert suitable tool in hole in

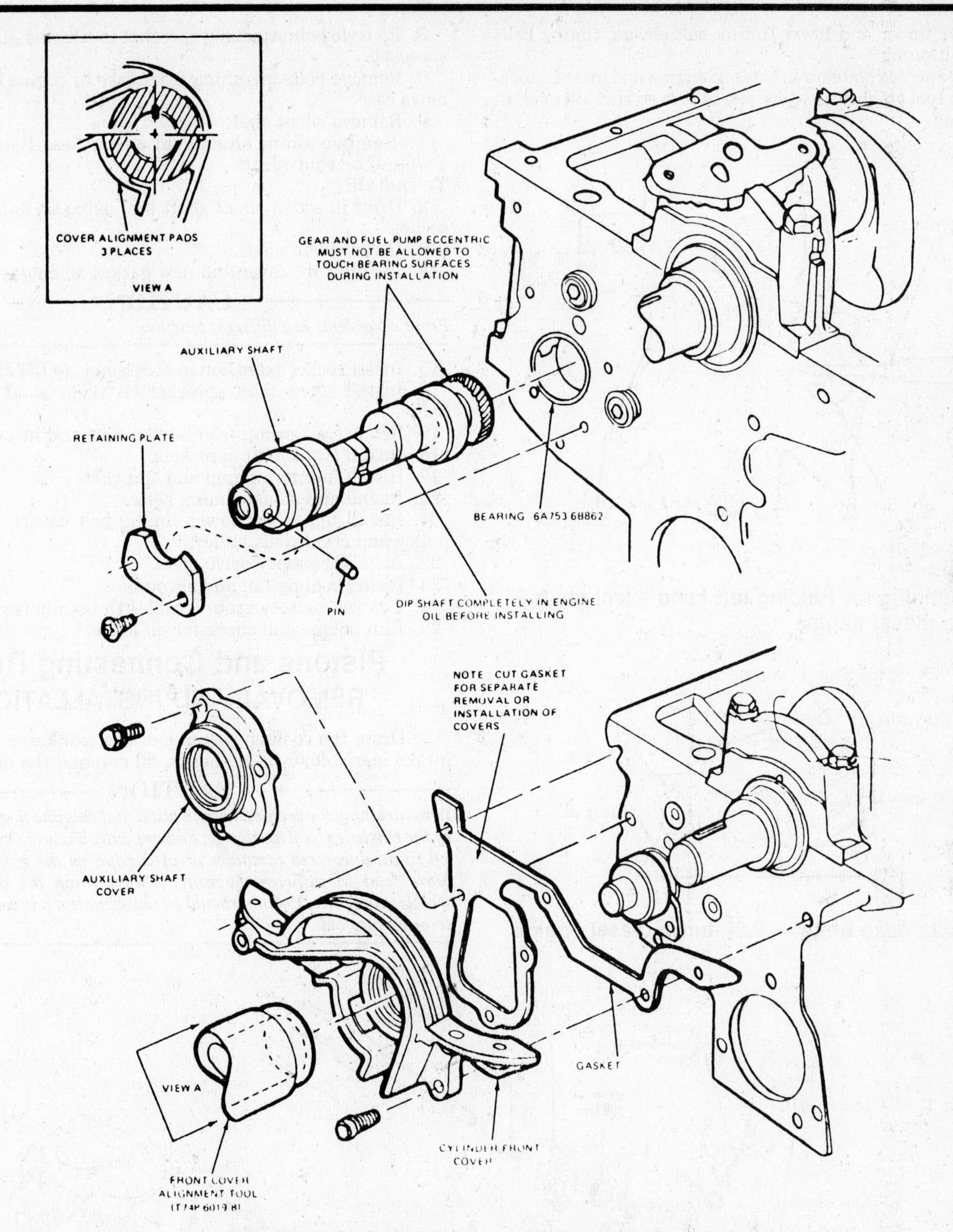

Auxiliary shaft installation – 2.0L and 2.3L engines

block to prevent silent shaft from rotating. Tighten sprocket nut.

18. Remove screwdriver and install pipe plug.

NOTE: When installing silent shaft sprocket, ensure "D" flat on sprocket is aligned with "D" flat on shaft.

19. Install crankshaft front oil seal.
20. Install crankshaft sprocket.
21. Apply a 3mm bead of Silicone Sealant D6A2–19562–B or equivalent along split lines between lower front case cover and rear oil seal retainer and engine block.
22. Install oil pan bolts.
23. Install and adjust timing belts.
24. Install upper and lower timing belt covers.
25. Install crankshaft and water pump pulleys.
26. Install accessory drive belts.
27. Install cooling fan and fan shroud.
28. Connect battery ground cables to both batteries.
29. Run again and check for oil leaks.

LEFT HAND SILENT SHAFT

1. Disconnect battery ground cables from both batteries.
2. Remove cooling fan and fan shroud.
3. Remove accessory drive belts.
4. Remove alternator and bracket.
5. Remove water pump and crankshaft pulleys.

6. Remove upper and lower timing belt covers, timing belts and injection pump.
7. Remove access plate on LH side of engine and insert a socket extension tool or equivalent in hole to prevent LH silent shaft from rotating.

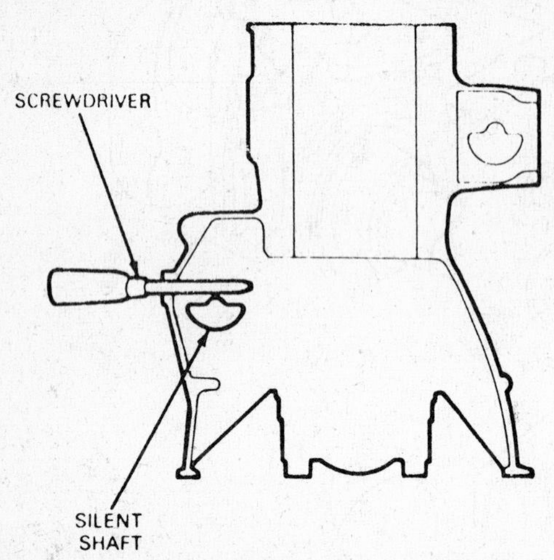

Tool positioning for holding left hand silent shaft – 2.3L turbo diesel engine

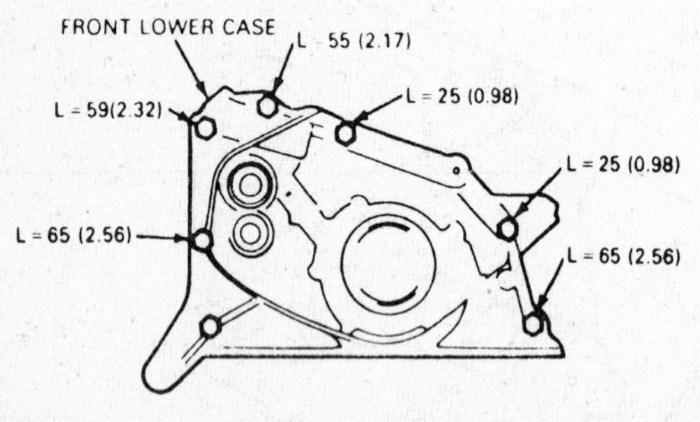

Front lower case bolts – 2.3L turbo diesel engine

8. Remove bolt attaching sprocket to silent shaft and remove sprocket.
9. Remove bolts attaching front case to engine block and remove case.
10. Remove silent shaft.
11. Remove silent shaft seal using Seal Remover Tool-1175–AC or equivalent.

To install:

12. Drive in a new silent shaft seal using an appropriate size socket.
13. Install silent shaft.
14. Install front cover and new gasket on engine block.

CAUTION

Front cover bolts are different lengths.

15. Insert socket extension in access hole in LH side of engine.
16. Install silent shaft sprocket on silent shaft and tighten bolt.
17. Remove extension from access hole and install cover.
18. Install crankshaft sprockets.
19. Install injection pump and sprocket.
20. Install and adjust timing belts.
21. Install upper and lower timing belt covers, water pump pulley and crankshaft pulley.
22. Install accessory drive belts.
23. Install cooling fan and shroud.
24. Connect battery ground cables to both batteries.
25. Run engine and check for oil leaks.

Pistons and Connecting Rods

REMOVAL AND INSTALLATION

1. Drain the cooling system and the crankcase. Remove the intake manifold, cylinder heads, oil pan and the oil pump.

CAUTION

When draining the coolant, keep in mind that cats and dogs are attracted by the ethylene glycol antifreeze, and are quite likely to drink any that is left in an uncovered container or in puddles on the ground. This will prove fatal in sufficient quantity. Always drain the coolant into a sealable container. Coolant should be reused unless it is contaminated or several years old.

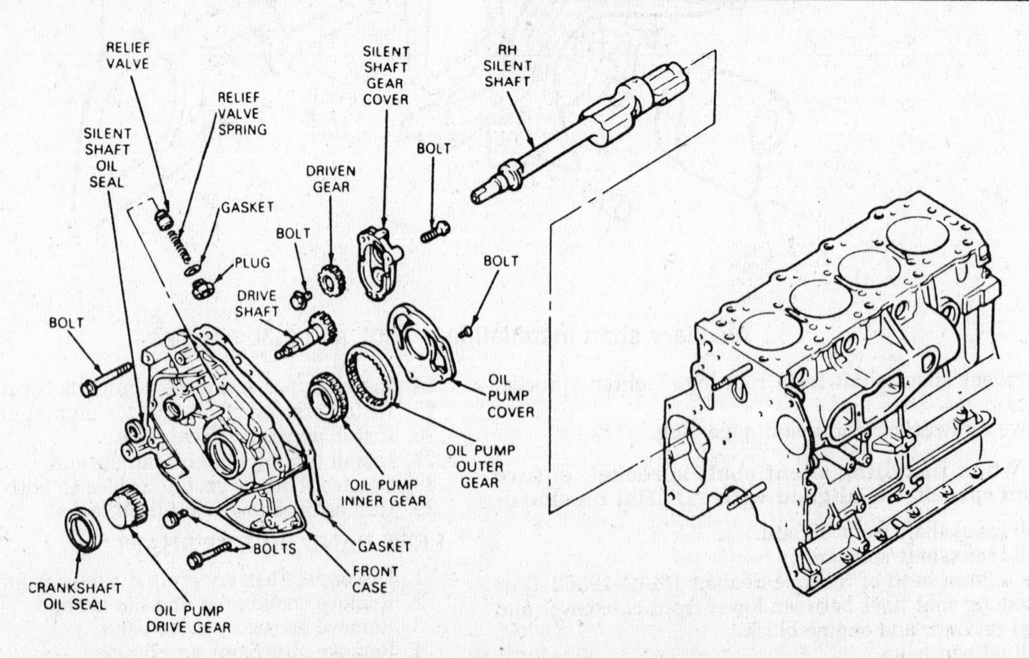

Lower front case cover and right hand silent shaft – 2.3L turbo diesel engine

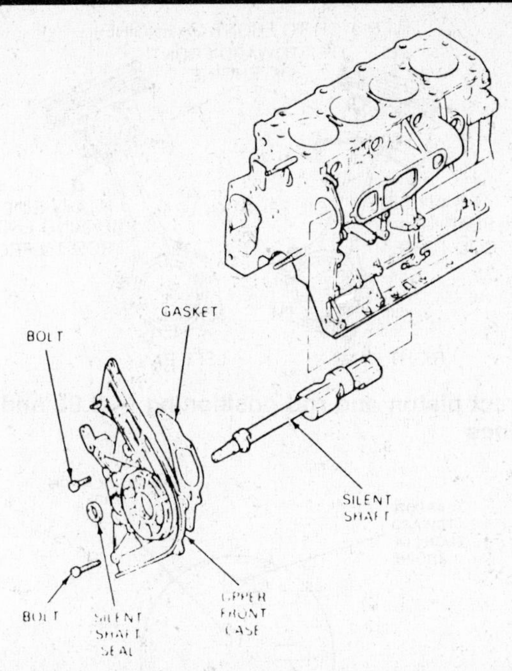

Upper front case cover and left hand silent shaft – 2.3L turbo diesel engine

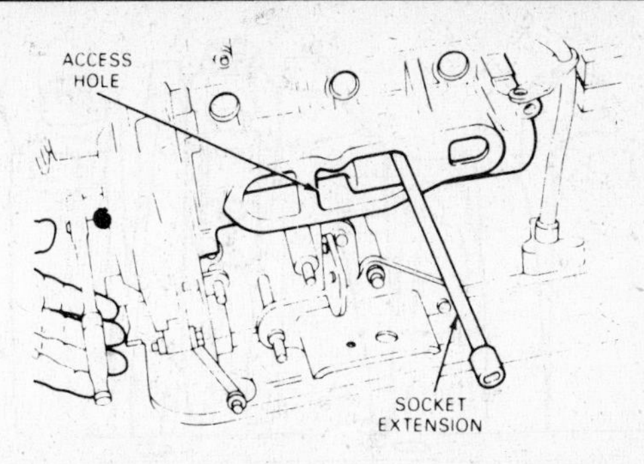

Tool positioning for holding right hand silent shaft – 2.3L turbo diesel engine

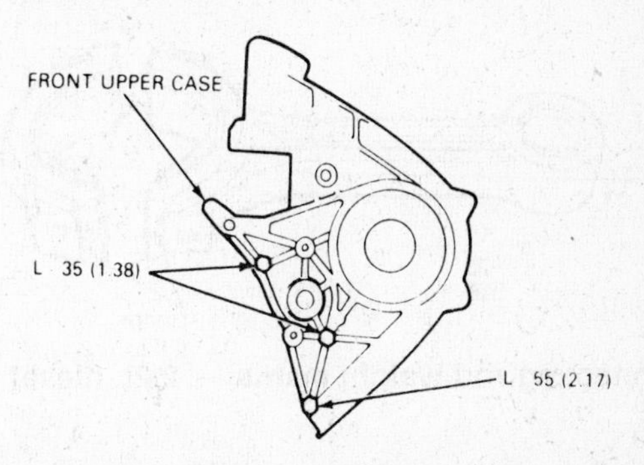

Front upper case bolts – 2.3L turbo diesel engine

2. Turn the crankshaft until the piston to be removed is at the bottom of its travel, then place a cloth on the piston head to collect filings. Remove any ridge of deposits at the end of the piston travel from the upper cylinder bore, using a ridge reaming tool. Do not cut into the piston ring travel area more than $1/32$ in. when removing the ridge.

3. Make sure that all of the connecting rod bearing caps can be identified, so they will be reinstalled in their original positions.

4. Turn the crankshaft until the connecting rod that is to be removed is at the bottom of its stroke and remove the connecting rod nuts and bearing cap.

5. Push the connecting rod and piston assembly out the top of the cylinder bore with the wooden end of a hammer handle. Be careful not to damage the crankshaft bearing journal or the cylinder wall when removing the piston and rod assembly.

6. Remove the bearing inserts from the connecting rod and cap if the bearings are to be replaced, and place the cap onto the piston/rod assembly from which it was removed.

NOTE: On the diesel engines, be sure to install pistons in same cylinders from which they were removed or to which they were fitted. Connecting rod and bearing caps have weight marks stamped on one side of main bearing bore boss. If rod replacement is necessary, all rods should be the same weight to maintain proper balance. Numbers on connecting rod and bearing cap must be on same side when installed in cylinder bore. If a connecting rod is ever transposed from one block or cylinder to another, new bearings should be fitted and connecting rod should be numbered to correspond with new cylinder number.

7. Before installing the piston/connecting rod assembly, be sure to clean all gasket mating surfaces, oil the pistons, piston rings and the cylinder walls with light engine oil.

8. Be sure to install the pistons in the cylinders from which they were removed. The connecting rod and bearing caps are numbered from 1 to 3 in the right bank and from 4 to 6 in the left back on the V6 engine, beginning at the front of the engine. The numbers on the connecting rod and bearing cap must be on the same side when installed in the cylinder bore. If a connecting rod is ever transposed from one engine or cylinder to another, new bearings should be fitted and the connecting rod should be numbered to correspond with the new cylinder number. The notch on the piston head goes toward the front of the engine.

9. Make sure the ring gaps are properly spaced around the circumference of the piston. Fit a piston ring compressor around the piston and slide the piston and connecting rod assembly down into the cylinder bore, pushing it in with the wooden hammer handle. Push the piston down until it is only slightly below the top of the cylinder bore. Guide the connecting rods onto the crankshaft bearing journals carefully, to avoid damaging the crankshaft.

10. Check the bearing clearance of all the rod bearings, fitting them to the crankshaft bearing journals.

11. After the bearings have been fitted, apply a light coating of engine oil to the journals and bearings.

12. Turn the crankshaft until the appropriate bearing journal is at the bottom of its stroke, then push the piston assembly all the way down until the connecting rod bearing seats on the crankshaft journal. Be careful not to allow the bearing cap screws to strike the crankshaft bearing journals and damage them.

13. After the piston and connecting rod assemblies have been installed, check the connecting rod side clearance on each crankshaft journal.

14. Prime and install the oil pump and the oil pump intake tube, then install the oil pan.

15. Reassemble the rest of the engine in the reverse order of disassembly.

PISTON RING REPLACEMENT

1. Select the proper ring set for the size cylinder bore.
2. Position the ring in the bore in which it is going to be used.

Connecting rod and bearing cap numbering

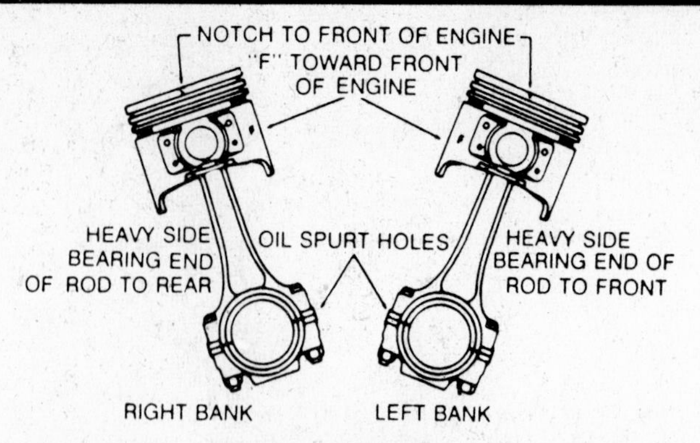

Correct piston and rod positioning – 2.8L and 2.9L engines

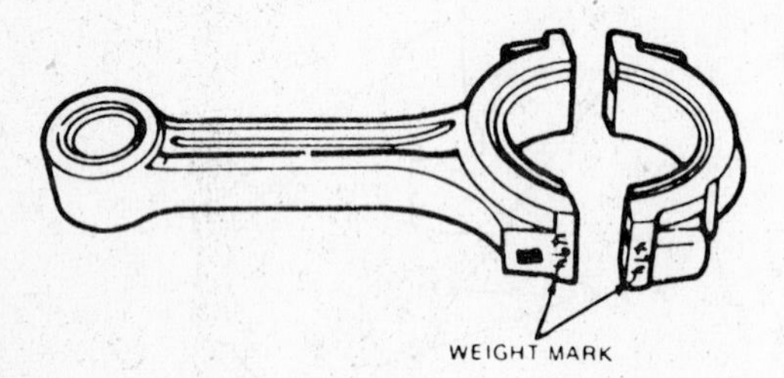

Connecting rod weight marks – 2.2L diesel engine

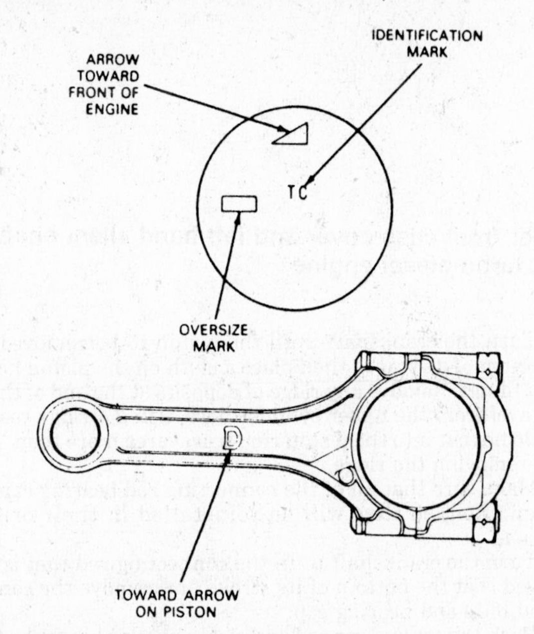

Correct piston and rod positioning – 2.3L turbo diesel engine

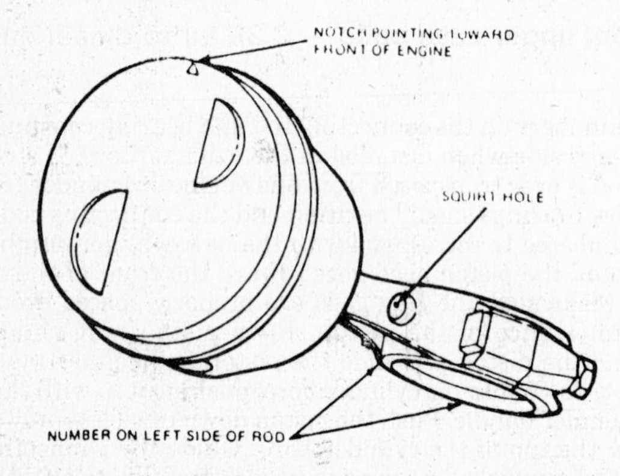

Correct piston and rod positioning – 2.0L and 2.3L engines

3. Push the ring down into the bore area where normal ring wear is not encountered.
4. Use the head of the piston to position the ring in the bore so that the ring is square with the cylinder wall. Use caution to avoid damage to the ring or cylinder bore.
5. Measure the gap between the ends of the ring with a feeler gauge. Ring gap in a worn cylinder is normally greater than specification. If the ring gap is greater than the specified limits, try an oversize ring set.
6. Check the ring side clearance of the compression rings with a feeler gauge inserted between the ring and its lower land according to specification. The gauge should slide freely around the entire ring circumference without binding. Any wear that occurs will form a step at the inner portion of the lower land. If the lower lands have high steps, the piston should be replaced.

CLEANING AND INSPECTION

Connecting Rods

1. Remove the bearings from the rod and cap. Identify the

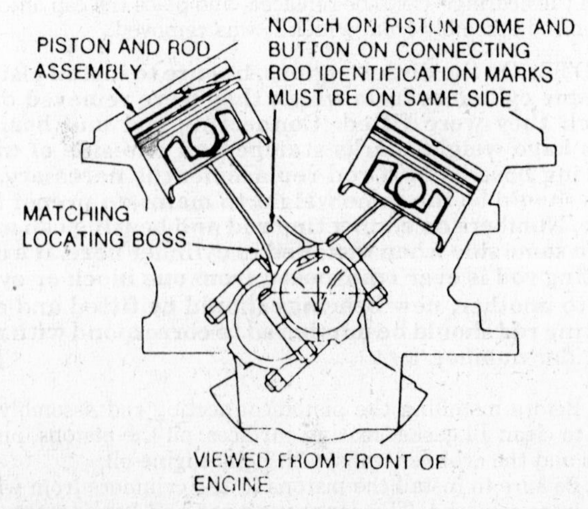

Correct piston and rod positioning – 3.0L & 4.0L engines

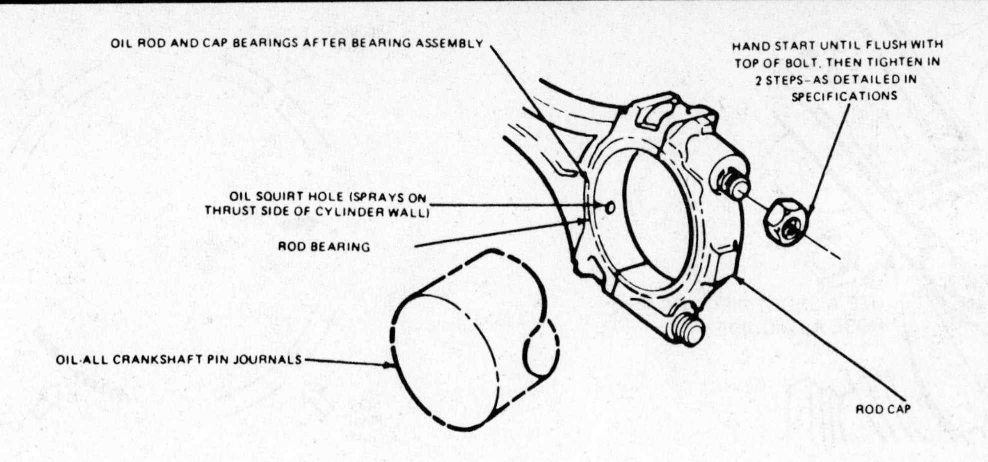

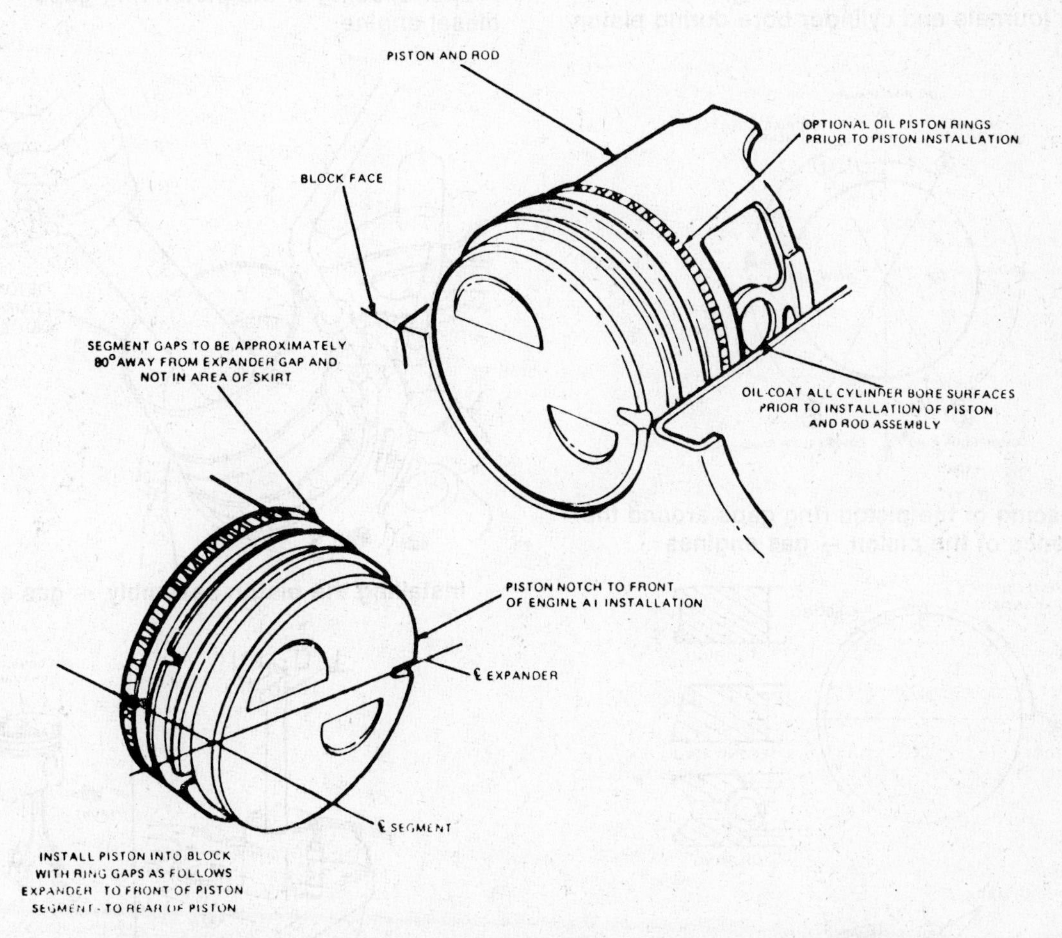

Connecting rod, piston and ring installation – 2.0L and 2.3L engines

bearings if they are to be used again. Clean the connecting rod in solvent, including the rod bore and the back of the bearing inserts. Do not use a caustic cleaning solution. Blow out all passages with compressed air.

2. The connecting rods and related parts should be carefully inspected and checked for conformance to specifications. Various forms of engine wear caused by these parts can be readily identified.

3. A shiny surface on the pin boss side of the piston usually indicates that a connecting rod is bent or the piston pin hole is not in proper relation to the piston skirt and ring grooves.

4. Abnormal connecting rod bearing wear can be caused by either a bent connecting rod, an improperly machined journal, or a tapered connecting rod bore.

5. Twisted connect rods will not create an easily identifiable wear pattern, but badly twisted rods will disturb the action of the entire piston, rings, and connecting rod assembly and may be the cause of excessive oil consumption.

6. Inspect the connecting rods for signs of fractures and the bearing bores for out-of-round and taper. If the bore exceeds the maximum limit and/or if the rod is fractured, it should be replaced.

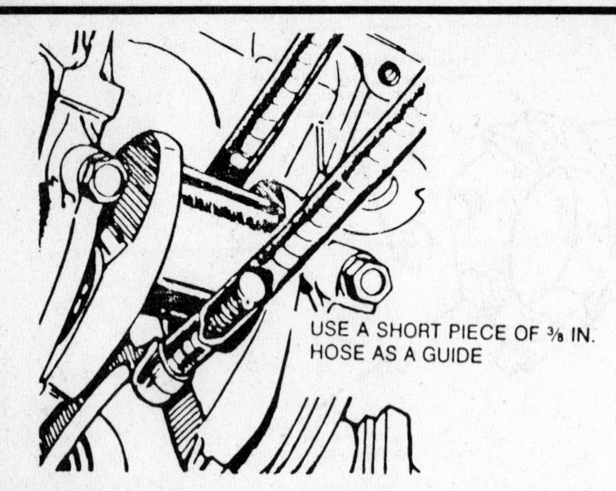

Use a length of hose or rubber tubing to protect the crankshaft journals and cylinder bore during piston installation

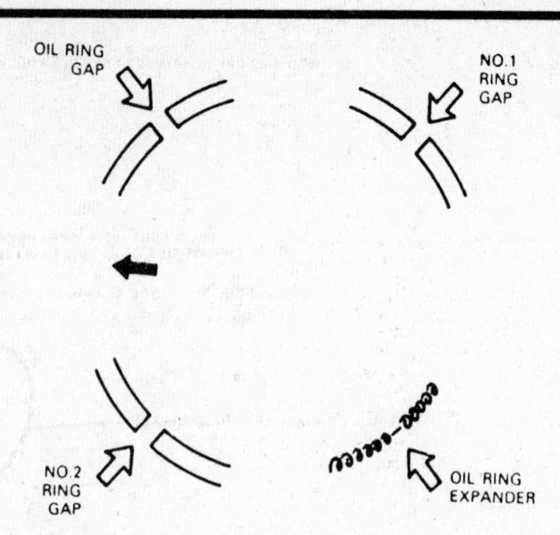

Proper spacing of the piston ring gaps – 2.3L turbo diesel engine

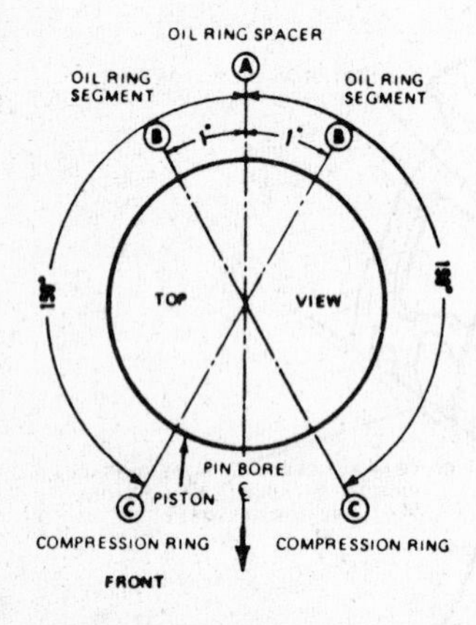

Proper spacing of the piston ring gaps around the circumference of the piston – gas engines

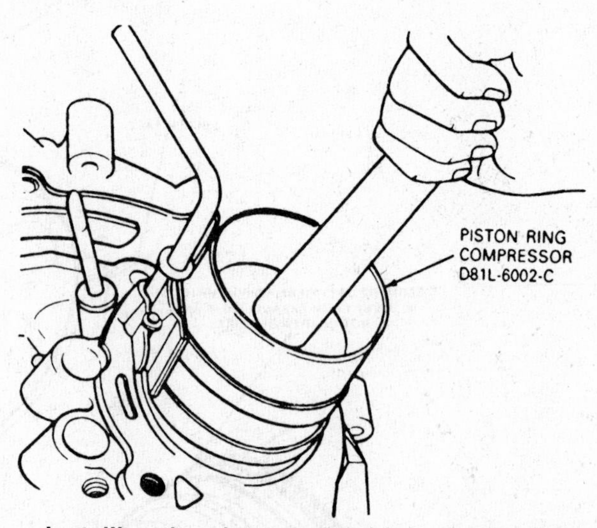

Installing the piston assembly – gas engines

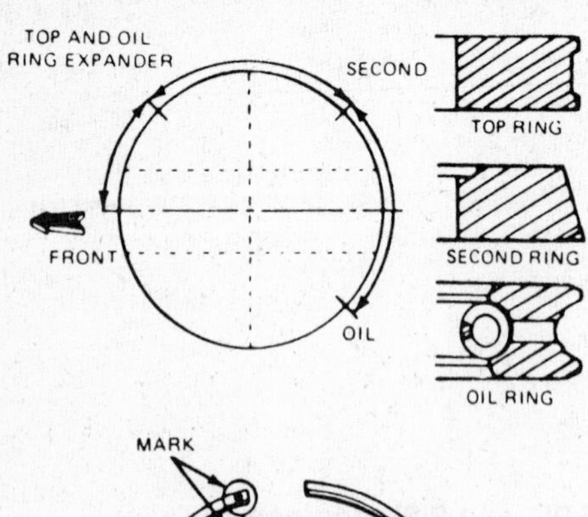

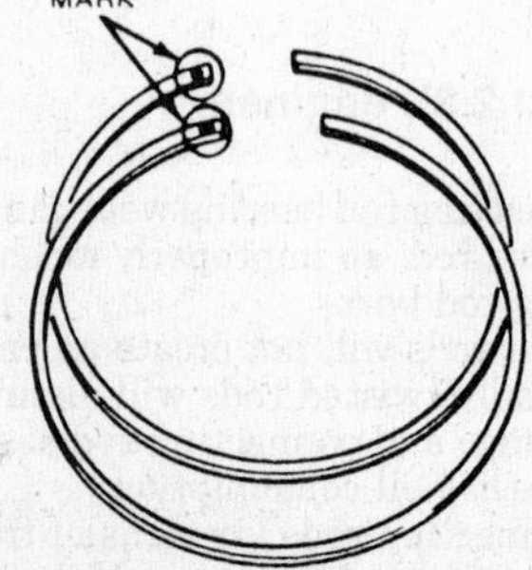

Proper spacing of the piston ring gaps around the circumference of the piston – 2.2L diesel engine

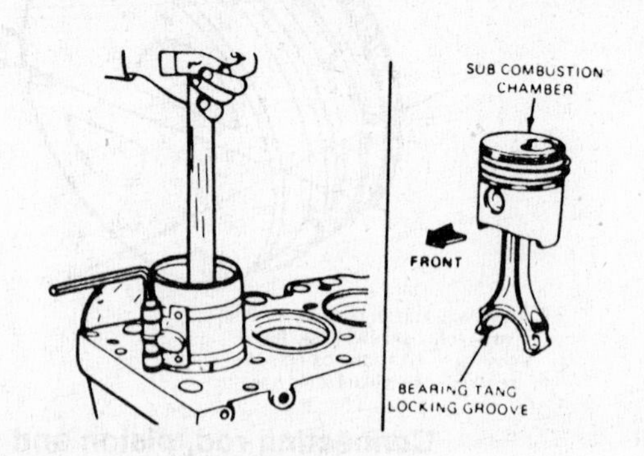

Installing the piston assembly – 2.2L engine

7. Check the ID of the connecting rod piston pin bore. Install oversize piston pin if the pin bore is not within specifications. Replace worn or damaged connecting rod nuts and bolts.
8. After the connecting rods are assembled to the piston, check the rods for bends or twists on a suitable alignment fixture. Follow the instructions of the fixture manufacturer. If the bend and/or twist exceeds specifications, the rod must be straightened or replaced.

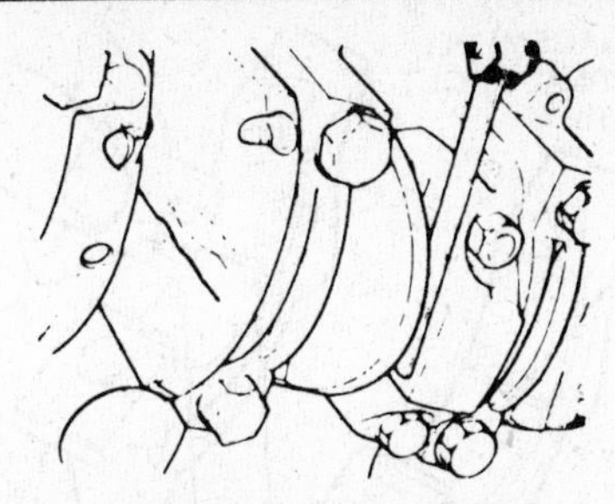

Checking the connecting rod side clearance on the crankshaft bearing journal

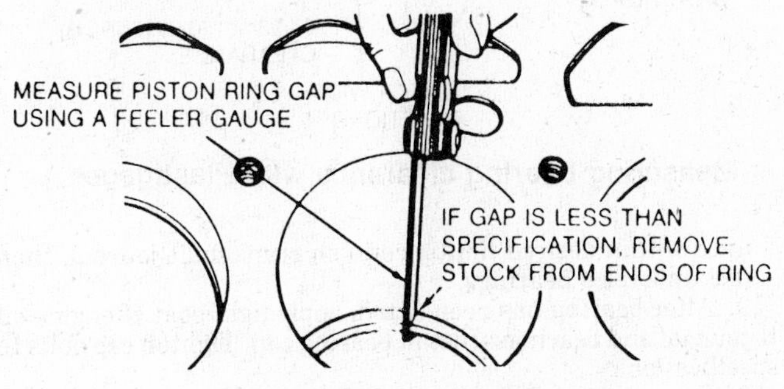

Measuring the ring end gap

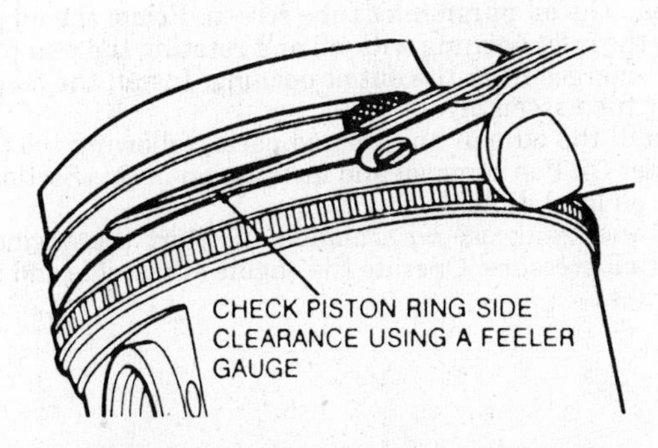

Measuring the ring side clearance

Pistons, Pins and Rings

1. Remove deposits from the piston surfaces. Clean gum or varnish from the piston skirt, piston pins and rings with solvent. Do not use a caustic cleaning solution or a wire brush to clean pistons. Clean the ring groove with a ring groove cleaner. Make sure the oil ring slots (or holes) are clean.
2. Carefully inspect the pistons for fractures at the ring lands, skirts, and pin bosses, and for scuffed, rough, or scored skirts. If the lower inner portion of the ring grooves have high steps, replace the piston. The step will interfere with ring operation and cause excessive ring side clearance.
3. Spongy, eroded ares near the edge of the piston top are usually caused by detonation or pre-ignition. A shiny surface on the thrust surface of the piston, offset from the centerline between the piston pin holes, can be caused by a bent connecting rod. Replace pistons that show signs of excessive wear, wavy ring lands or fractures, or damage from detonation or pre-ignition.
4. Check the piston to cylinder bore clearance by measuring the piston and bore diameters. Measure the OD of the piston with micrometers at the centerline of the piston pin bore and at 90 degrees to the pin bore axis. Check the ring side clearance following the procedure under Piston Ring Replacement in this Section.
5. Replace piston pins showing signs of fracture, etching or wear. Check the piston pin fit in the piston and rod.
6. Check the OD of the piston pin and the ID of the pin bore in the piston. Replace any piston pin or piston that is not within specifications.
7. Replace all rings that are scored, chipped or cracked. Check the end gap and side clearance. It is good practice to always install new rings when overhauling an engine. Rings should not be transferred from one piston to another regardless of mileage.

PISTON PIN REPLACEMENT

1. Remove the bearing inserts from the connecting rod and cap.
2. Mark the pistons to assure assembly with the same rod, rod position and installation in the same cylinders from which they were removed.
3. Using an Arbor press and tool T68P–6135–A or equivalent, press the piston pin from the piston and connecting rod. Remove the piston rings if they are to be replaced.

NOTE: Check the fit of a new piston in the cylinder bore before assembling the piston and piston pin to the connecting rod.

4. Apply a light coat of engine oil to all parts. Assemble the piston to the connecting rod with the indentation or notch in the original position.
5. Start the piston pin in the connecting rod (this may require a very light tap with a mallet). Using an Arbor press and tool T68P–6135–A or equivalent, press the piston pin through the piston and connecting rod until the pin is centered in the piston.
6. install the piston rings using a piston ring installation tool of the proper size (refer to Piston Ring Replacement in this Section).
7. Be sure the bearing inserts and the bearing bore in the connect rod and cap are clean. Foreign material under the inserts will distort the bearing and cause it to fail.

ROD BEARING REPLACEMENT

1. Drain the crankcase. Remove the oil level dipstick. Remove the oil pan and related parts, following the procedure under Oil Pan removal and installation in this Section.

CAUTION

The EPA warns that prolonged contact with used engine oil may cause a number of skin disorders, including cancer! You should make every effort to minimize your exposure to used engine oil. Protective gloves should be worn when changing the oil. Wash your hands and any other exposed skin areas as soon as possible after exposure to used engine oil. Soap and water, or waterless hand cleaner should be used.

2. Remove the oil pump inlet tube assembly and the oil pump.
3. Turn the crankshaft until the connecting rod to which new bearings are to be fitted is down. Remove the connecting rod cap. Remove the bearing inserts from the rod and cap.
4. Be sure the bearing inserts and the bearing bore in the connecting rod and cap are clean. Foreign material under the inserts will distort the bearing and cause failure.
5. Clean the crankshaft journal. Inspect journals for nicks, burrs or bearing pick-up that would cause premature bearing wear.
6. Install the bearing inserts in the connect god and cap with the tangs fitting in slots provided.
7. Pull the connecting rod assembly down firmly on the crankshaft journal.
8. Select fit the bearing using the following procedures:
   a. Place a piece of Plastigage® or it's equivalent, or the bearing surface across the full width of the bearing cap and about 6mm off center.

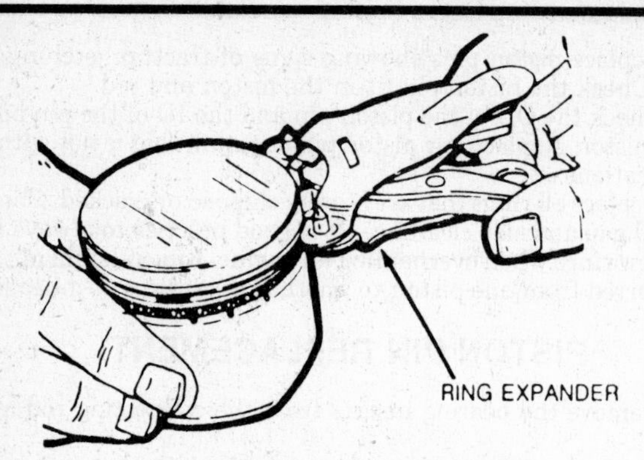

Removing the piston rings

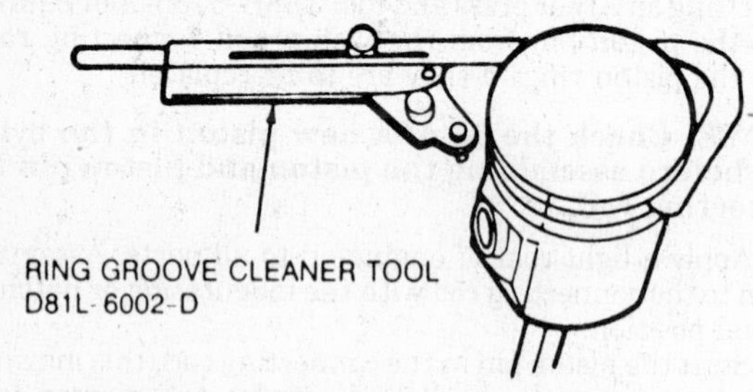

Cleaning the piston grooves

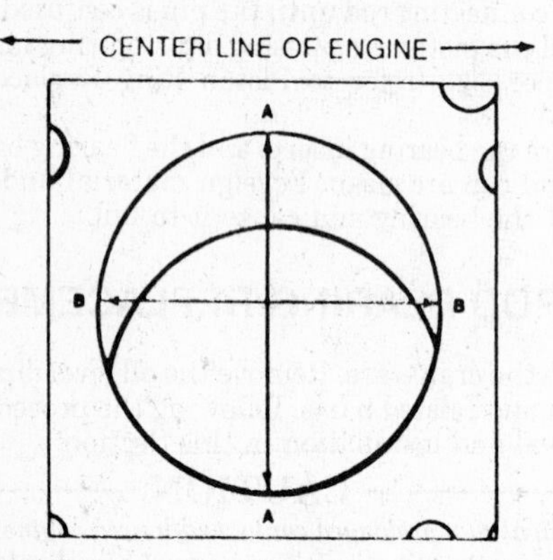

1. OUT OF ROUND = DIFFERENCE BETWEEN "A" AND "B"
2. TAPER = DIFFERENCE BETWEEN THE "A" MEASUREMENT AT TOP OF CYLINDER BORE AND THE "A" MEASUREMENT OF CYLINDER BORE

"A" AT RIGHT ANGLE TO CENTER LINE OF ENGINE
"B" PARALLEL TO CENTER LINE OF ENGINE

Measure the cylinder bore at the points indicated

b. Install cap and tighten bolts to specifications. Do not turn crankshaft while Plastigage® is in place.

c. Remove cap. Using Plastigage® scale, check width of Plastigage® at widest point to get minimum clearance. Check at narrowest point to get maximum clearance. Difference between readings is taper of journal.

d. If clearance exceeds specified limits, try a 0.001 in. or 0.002 in. undersize bearing in combination with the standard bearing. Bearing clearance must be within specified limits. If standard and 0.002 in. undersize bearing does not bring clearance within desired limits, refinish crankshaft journal, then use undersize bearings.

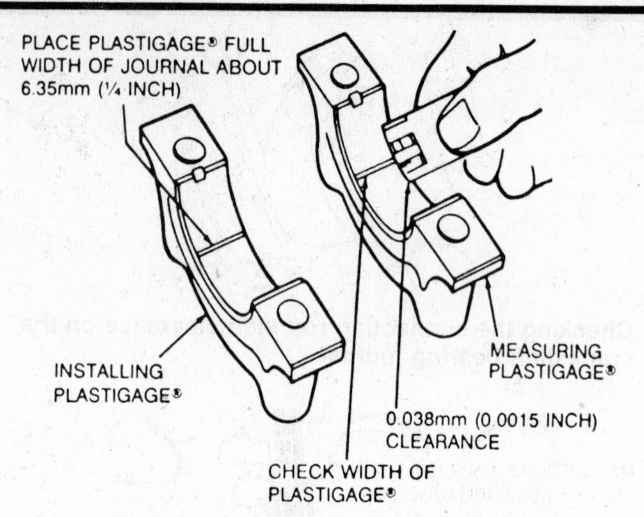

Measuring bearing clearance with Plastigage®

9. After bearing has been fitted, apply light coat of engine oil to journal and bearings. Install bearing cap. Tighten cap bolts to specifications.

10. Repeat procedures for remaining bearings that require replacement.

11. Clean the oil pump inlet tube screen. Prime the oil pump by filing the inlet opening with oil and rotating the pump shaft until oil emerges from the outlet opening. Install the oil pump and inlet tube assembly.

12. Install the oil pan and related parts, following the procedure under Oil Pan removal and installation in this Section. Install the oil level dipstick.

13. Fill the crankcase with engine oil. Start the engine and check for oil pressure. Operate the engine at fast idle and check for oil leaks.

Freeze (Core) Plugs

REMOVAL AND INSTALLATION

All Engines

1. Drain the the complete cooling system and engine block.

CAUTION

When draining the coolant, keep in mind that cats and dogs are attracted by the ethylene glycol antifreeze, and are quite likely to drink any that is left in an uncovered container or in puddles on the ground. This will prove fatal in sufficient quantity. Always drain the coolant into a sealable container. Coolant should be reused unless it is contaminated or several years old.

2. Drill or punch a hole in the center of the freeze plug and pull it out from the engine bolck with a slide hammer or equivalent. Note stay away from the freeze plug location when working, as coolant will flow from the engine block when the plug is removed.

3. Check the bore for roughness or burrs. If the bore is damaged, hone it and use an oversized freeze plug.

4. Coat the new freeze plug with sealer and drive into the correct position.

5. Refill the cooling system. Start engine and check for coolant leaks.

Rear Main Oil Seal

REMOVAL AND INSTALLATION

Gasoline Engines

If the crankshaft rear oil seal replacement is the only operation being performed, it can be done in the vehicle as detailed in the following procedure. If the oil seal is being replaced in conjunction with a rear main bearing replacement, the engine must be removed from the vehicle and installed on a work stand.

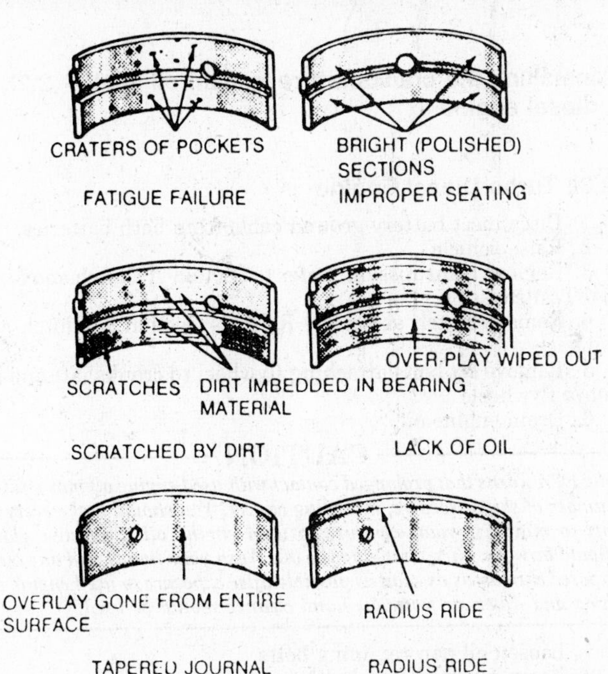

Inspecting the bearings for damage

1. Remove the starter.
2. Remove the transmission from the vehicle, following the procedures in Section 7.
3. On a manual shift transmission, remove the pressure plate and cover assembly and the clutch disc following the procedure in Section 7.
4. Remove the flywheel attaching bolts and remove the flywheel and engine rear cover plate.
5. Use an awl to punch two holes in the crankshaft rear oil seal. Punch the holes on opposite sides of the crankshaft and just above the bearing cap to cylinder block split line. Install a sheet metal screw in each hole. Use two large screwdrivers or small pry bars and pry against both screws at the same time to remove the crankshaft rear oil seal. It may be necessary to place small blocks of wood against the cylinder block to provide a fulcrum point for the pry bars. Use caution throughout this procedure to avoid scratching or otherwise damaging the crankshaft oil seal surface.

To install:

6. Clean the oil seal recess in the cylinder block and main bearing cap.
7. Clean, inspect and polish the rear oil seal rubbing surface on the crankshaft. Coat a new oil seal and the crankshaft with a light film of engine oil. Start the seal in the recess with the seal lip facing forward and install it with a seal driver. Keep the tool, T82L–6701–A (4–cyl. engines) or T72C–6165 (6–cyl. engine) straight with the centerline of the crankshaft and install the seal until the tool contacts the cylinder block surface. Remove the tool and inspect the seal to be sure it was not damaged during installation.
8. Install the engine rear cover plate. Position the flywheel on the crankshaft flange. Coat the threads of the flywheel attaching bolts with oil-resistant sealer and install the bolts. Tighten the bolts in sequence across from each other to the specifications listed in the Torque chart at the beginning of this Section.
9. On a manual shift transmission, install the clutch disc and the pressure plate assembly following the procedure in Section 7.
10. Install the transmission, following the procedure in Section 7.

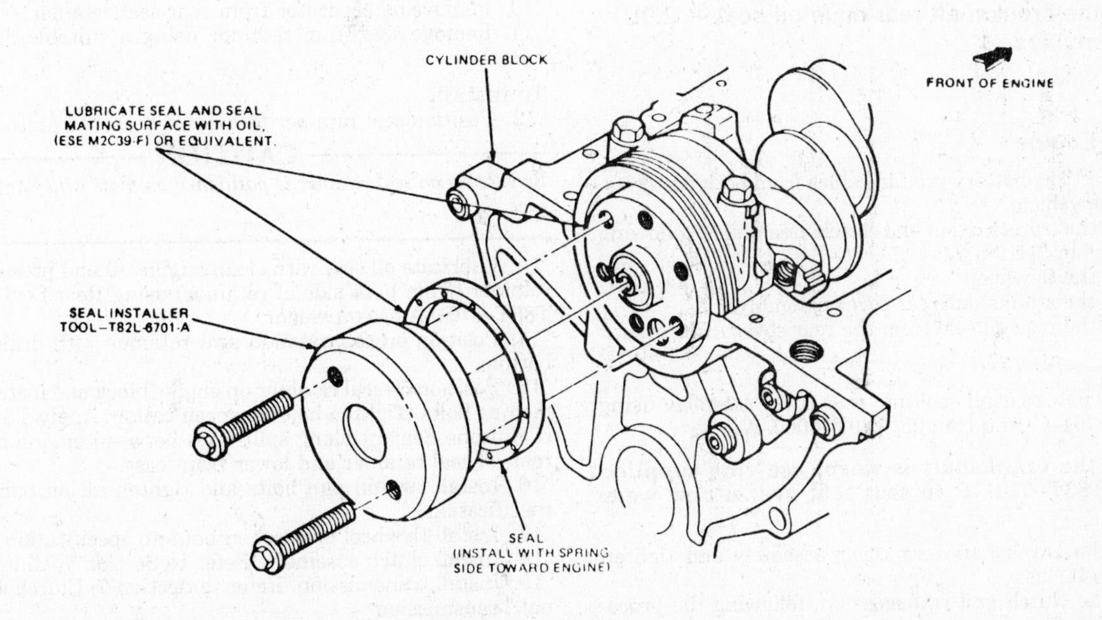

Installing the crankshaft rear oil seal – 2.0L and 2.3L engines other engines similar

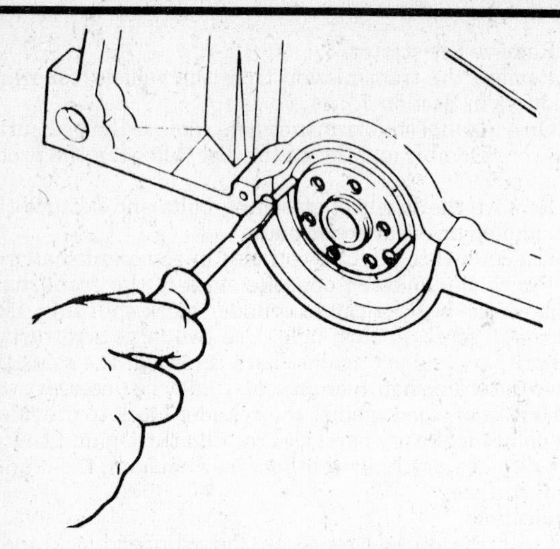

Removing the rear oil seal

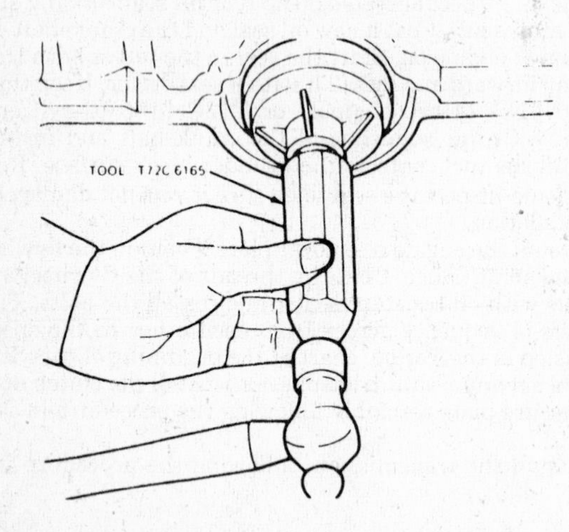

Installing the crankshaft rear main oil seal — 2.8L and 2.9L engines

2.2L Diesel Engine

1. Disconnect the battery ground cables from both batteries.
2. Raise the vehicle.
3. Remove the transmission and clutch assemblies, following the procedures in Section 7.
4. Remove the flywheel.
5. Remove the crankshaft rear cover assembly.
6. Remove the rear oil seal from the rear cover using a suitable tool.

To install:

7. Install a new rear oil seal into rear cover assembly using Tool, T83T–6701–C, and Handle, T80T–4000–W.

NOTE: If the crankshaft is worn, use ring supplied with Tool, T83T–6701–C to seat seal over a new wear area.

8. Install the crankshaft rear cover assembly and tighten bolts to 11–15 ft. lbs.
9. Install the clutch and transmission, following the procedures in Section 7.
10. Lower the vehicle.
11. Connect the battery ground cables to both batteries.
12. Start the engine and check for oil leaks.

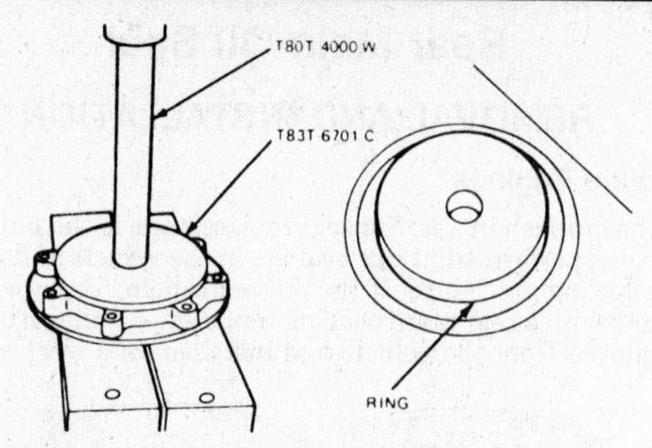

Installing the crankshaft rear main oil seal — 2.2L diesel engine

2.3L Turbo Diesel Engine

1. Disconnect battery ground cable from both batteries.
2. Raise vehicle.
3. Remove transmission. Refer to Section 7, Clutch and Manual Transmission.
4. Remove clutch assembly. Refer to Section 7, Clutch and Linkage.
5. Remove six bolts attaching flywheel to crankshaft and remove flywheel.
6. Drain engine oil.

CAUTION

The EPA warns that prolonged contact with used engine oil may cause a number of skin disorders, including cancer! You should make every effort to minimize your exposure to used engine oil. Protective gloves should be worn when changing the oil. Wash your hands and any other exposed skin areas as soon as possible after exposure to used engine oil. Soap and water, or waterless hand cleaner should be used.

7. Loosen oil pan retaining bolts.
8. Remove two oil pan bolts from rear seal retainer.
9. Remove five bolts attaching rear seal retainer to engine block and remove retainer and gasket.
10. Remove oil separator from rear seal retainer.
11. Remove seal from retainer using a suitable drift and a hammer.

To install:

12. Position seal retainer face down on arbor press plate.

CAUTION

Be sure lip on seal retainer is positioned on plate with flange over the edge.

13. Lubricate oil seal with clean engine oil and press seal into retainer (from back side of retainer) using Rear Seal Replacer T85T–6701–A or equivalent.
14. Position oil separator in seal retainer with drain hole at bottom.
15. Position oil seal retainer on engine block and install five attaching bolts. Tighten bolts to specification. Apply a 3mm bead of Silicone Sealant along split lines between engine block and rear oil seal retainer and lower front case.
16. Install two oil pan bolts and tighten all oil pan bolts to specification.
17. Install flywheel and tighen bolts to specification.
18. Install clutch assembly. Refer to Section 7, Clutch.
19. Install transmission. Refer to Section 7, Clutch and Manual Transmission.
20. Lower vehicle.
21. Fill crankcase with specified quantity and quality of oil.
22. Connect battery ground cables to both batteries.
23. Run engine and check for oil leaks.

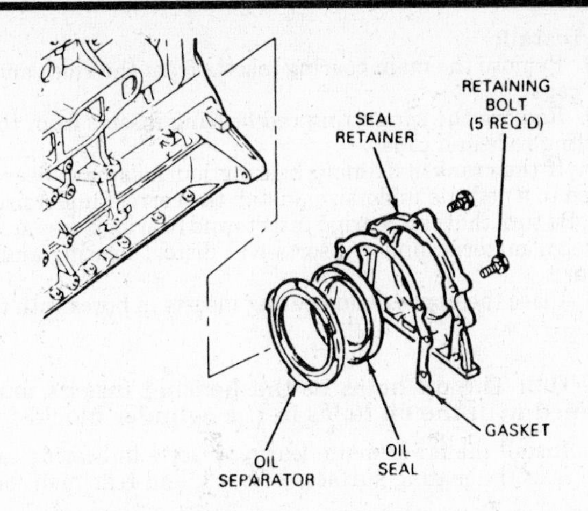

Rear main oil seal — 2.3L turbo diesel engine

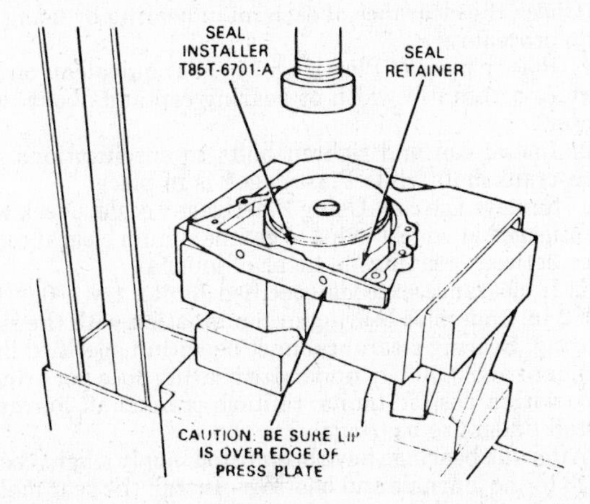

Rear main oil seal replacement — 2.3L turbo diesel engine

Crankshaft and Main Bearings

REMOVAL AND INSTALLATION

Gasoline Engines

1. Remove the engine from the vehicle as previously described, then place it on a work stand.
2. Remove the transmission (if attached), bell housing, flywheel or flex plate and rear plate.
3. Drain the crankcase and remove the oil pan with the engine in a normal upright position.

CAUTION

The EPA warns that prolonged contact with used engine oil may cause a number of skin disorders, including cancer! You should make every effort to minimize your exposure to used engine oil. Protective gloves should be worn when changing the oil. Wash your hands and any other exposed skin areas as soon as possible after exposure to used engine oil. Soap and water, or waterless hand cleaner should be used.

4. Remove the components from the front of the engine and the front cover.
5. Invert the engine and remove the oil pump, pickup tube and baffle, if equipped.
6. Make sure all main and connecting rod bearing caps are marked so they can be installed in their original locations.
7. Remove the connecting rod nuts and lift off the cap with its bearing insert. Install short pieces of rubber hose over the connecting rod studs to protect the crankshaft journals, then carefully push the piston and rod assemblies down into the cylinder bores.
8. Remove the main bearing caps with their bearing inserts. Inspect the crankshaft journals for nicks, burrs or bearing pickup that would cause premature bearing wear. When replacing standard bearings with new bearings, it is good practice to fit the bearing to minimum specified clearance. If the desired clearance cannot be obtained with a standard bearing, try one half of a 0.001 in. (0.025mm) or 0.002 in. (0.050mm) undersize in combination with a standard bearing to obtain the proper clearance.
9. Place a piece of Plastigage® on the bearing surface across the full width of the bearing cap, about ¼ in. (6mm) off center. Install the cap and tighten the bolts to the specified torque given in the General Engine Specifications Chart. Do not rotate the crankshaft with the Plastigage® in place. Remove the cap and use the scale provided with the kit to check the Plastigage width at its widest and narrowest points. Widest point is minimum clearance, narrowest point is maximum clearance; the difference between the two is the taper reading of the journal.
10. Bearing clearance must be within specified limits. If standard 0.002 in. (0.050mm) undersize bearings do not bring the clearance within desired limits, the crankshaft will have to be refinished or replaced. Remove the remaining main bearing caps and lift out the crankshaft, being careful not to damage the thrust bearing surfaces. Discard the rear main oil seal. The crankshaft should be refinished at a machine shop to give the proper clearance with the next undersize bearing. If the journal will not clean up to the maximum undersize bearing, the crankshaft will have to be replaced.
11. Clean the bearing bores in the block and caps. Foreign material under the inserts or on the bearing surfaces may distort the insert and cause bearing failure.
12. Assemble the main bearing inserts in their correct location in the bearing caps and cylinder block. Check the oil hole alignment between the bearing inserts and block. Apply a liberal coating of clean heavy SG engine oil to the bearing surfaces, then carefully lower the crankshaft into position.
13. Insert the remaining bearing shells into the main bearing caps and coat the bearings with clean heavy SG engine oil, then install the caps with the arrows pointing toward the front of the engine. Apply a thin even coating of sealing compound to the rear sealing surface of the rear main bearing cap before installing. Install and tighten all main bearing cap bolts finger tight after lightly oiling the threads.
14. Tighten all bearing cap bolts, except for the thrust bearing cap, to the specifications given in the Torque Specifications Chart.
15. Align the thrust bearing surfaces by forcing the crankshaft forward and the thrust bearing cap rearward. While holding in this position, tighten the thrust bearing cap to specifications.
16. Install a new rear main oil seal as previously described.
17. On some engines, use a flat tool such as a large blunt end screwdriver to push the two wedge shaped seals between the cylinder block and rear main bearing cap. Position the seals with the round side facing the main bearing cap.
18. Pull the connecting rods up one at a time and install rod caps after applying a liberal coating of heavy SG engine oil to the bearings. Tighten all bearing caps to specifications. On V6 engines, check the connecting rod side clearance as previously described. Check the crankshaft end play with a dial indicator.
19. Install the oil pump, pickup tube and baffle, if equipped. Prime the oil pump before installation as described under Oil Pump Removal and Installation.
20. Install the front cover and timing chain, belt or gears. Replace the front cover oil seal.
21. Install the rear cover plate (if equipped) and the flywheel or flex plate. Tighten the mounting bolts to specifications.

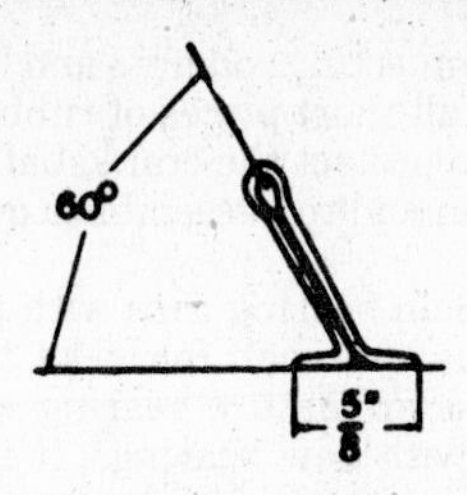

Installing the upper main bearing inserts

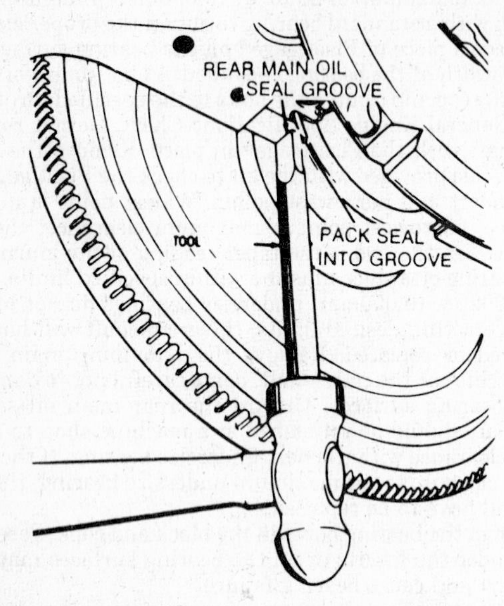

Applying RTV sealer to the rear main bearing cap – 2.8L and 2.9L engines

22. Install the clutch disc, pressure plate and bell housing on manual transmission models.
23. Install the oil pan and tighten the bolts to specifications. See Oil Pan Removal and Installation for gasket and sealer placement.
24. Invert the engine to its normal, upright position and fill the crankcase with the specified amount and type of engine oil. Replace the oil filter.
25. Install the transmission, if removed with the engine.
26. Install the engine in the vehicle as previously described. Roadtest the vehicle for proper operation.

2.2L, 2.3L Diesel Engines

1. Remove the engine assembly from the vehicle as described in this Section, and install in work stand.
2. Remove the water pump pulley, crankshaft pulley, timing gear case cover, timing gears, fuel injection pump and timing gear case.
3. Remove the starter, flywheel, and engine-to-transmission adapter plate.
4. Remove the oil pan and oil pump.
5. Remove the rear oil seal assembly as described in this Section under Rear Main Oil Seal.
6. Remove the crankshaft main bearing caps.

NOTE: Main bearing caps are numbered 1 through 5, front to rear, and must be returned to their original positions.

7. Carefully lift the crankshaft out of block so the thrust bearing surfaces are not damaged.

To install:

8. Remove the main bearing inserts from the block and bearing caps.
9. Remove the connecting rod bearing inserts from the connecting rods and caps.
10. If the crankshaft main bearing journals have been refinished to a definite undersize, install the correct undersize bearing. Be sure that the bearing inserts and bearing bores are clean. Foreign material under inserts will distort bearing and cause failure.
11. Place the upper main bearing inserts in bores with tang in slot.

NOTE: The oil holes in the bearing inserts must be aligned with the oil holes in the cylinder block.

12. Install the lower main bearing inserts in bearing caps.
13. lean the mating surfaces of block and rear main bearing cap.
14. Carefully lower the crankshaft into place. Be careful not to damage bearing surfaces.
15. Check the clearance of each main bearing by using the following procedure:
    a. Place a piece of Plastigage® or it's equivalent, on bearing surface across full width of bearing cap and about ¼ in. off center.
    b. Install cap and tighten bolts to specifications. Do not turn crankshaft while Plastigage® is in place.
    c. Remove the cap. Using Plastigage® scale, check width of Plastigage® at widest point to get maximum clearance. Difference between readings is taper of journal.
    d. If clearance exceeds specified limits, try a 0.001 in. or 0.002 in. undersize bearing in combination with the standard bearing. Bearing clearance must be within specified limits. If standard and 0.002 in. undersize bearing does not bring clearance within desired limits, refinish crankshaft journal, then install undersize bearings.
16. After the bearings have been fitted, apply a light coat of engine oil to the journals and bearings. Install the rear main bearing cap. Install all bearing caps except the thrust bearing cap (No. 3 bearing). Be sure that main bearing caps are installed in original locations. Tighten the bearing cap bolts to; 80–85 ft. lbs. on the 2.2L engine, 55–61 ft. lbs. on the 2.3L engine.
17. Install the thrust bearing cap with bolts fingertight.
18. Pry the crankshaft forward against the thrust surface of upper half of bearing.
19. Hold the crankshaft forward and pry the thrust bearing cap to the rear. This aligns the thrust surfaces of both halves of the bearing.
20. Retain the forward pressure on the crankshaft. Tighten the cap bolts to; 80–85 ft. lbs. on the 2.2L engine, 55–61 ft. lbs. on the 2.3L engine.
21. Force the crankshaft toward the rear of engine.
22. Install new bearing inserts in the connecting rods and caps. Check the clearance of each bearing, following procedure described under step 15.
23. After the connecting rod bearings have been fitted, apply light coat of engine oil to the journals and bearings.
24. Turn the crankshaft throw to bottom of its stroke. Push the piston all the way down until the rod bearing seats on crankshaft journal.
25. Install the connecting rod cap. Be sure that the connecting rod bolt heads are properly seated in the connecting rod. Tighten nuts to 50–54 ft. lbs.
26. After piston and connecting rod assemblies have been installed, check side clearance (refer to Crankshaft and Connecting Rod Specifications) between connecting rods on each connecting rod crankshaft journal.
27. Install a new rear main oil seal in oil seal adapter as described in this Section under Rear Main Oil Seal.
28. Install rear main oil seal adapter.

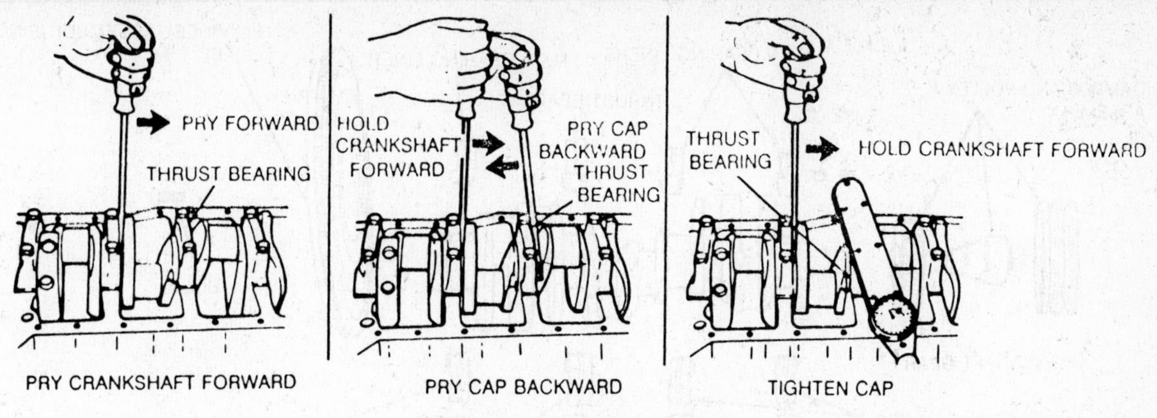

Aligning the thrust bearing

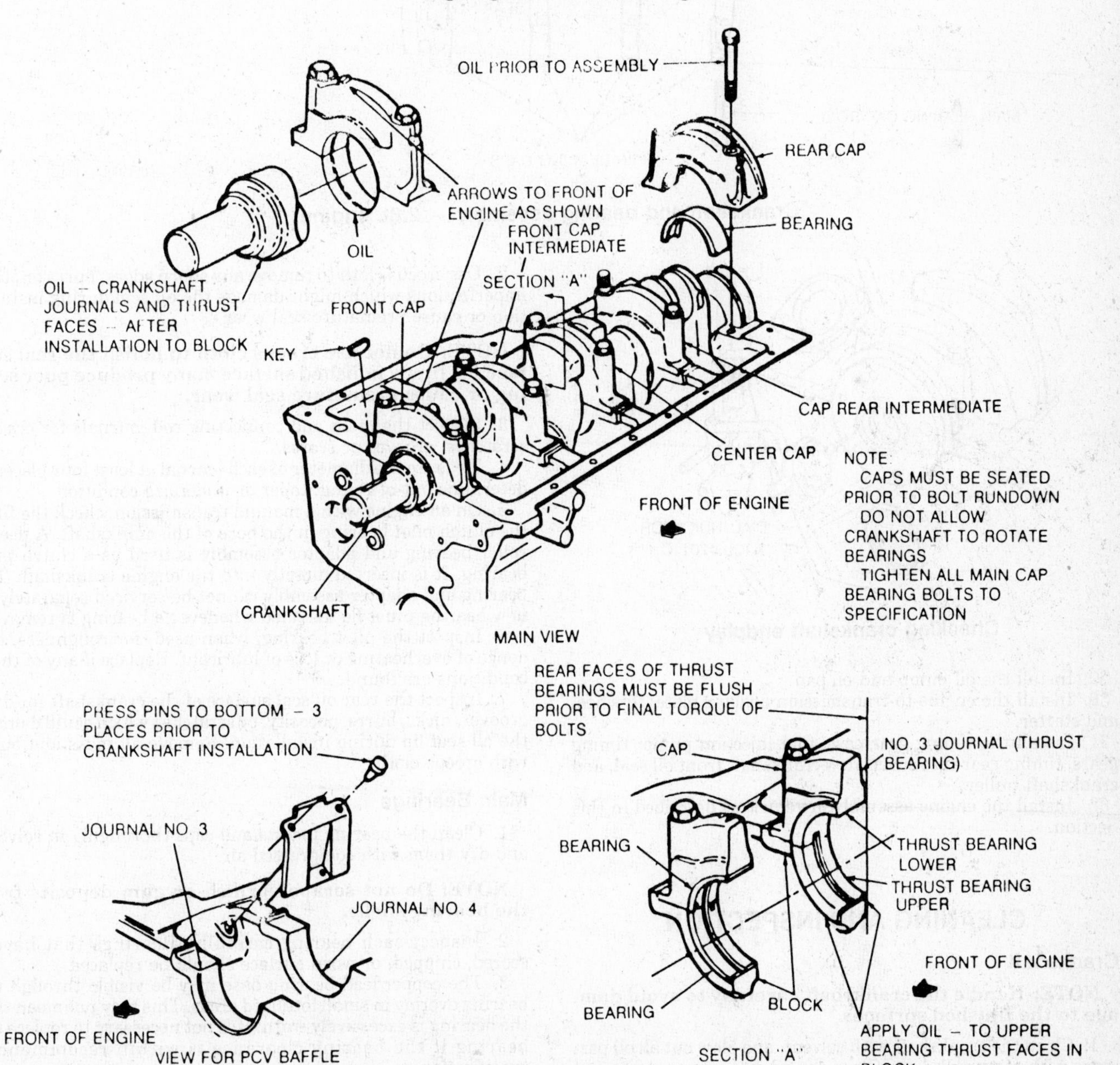

Crankshaft and bearing installation — 2.3L engine

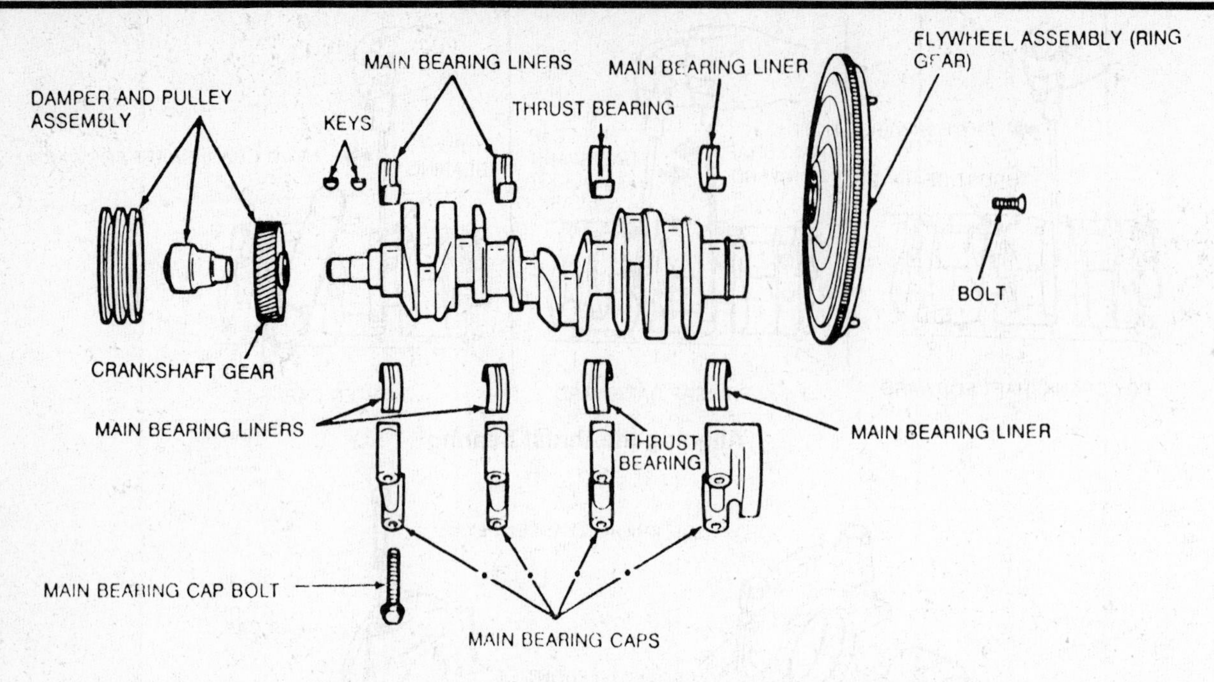

Crankshaft and bearing assembly – 2.8L engine

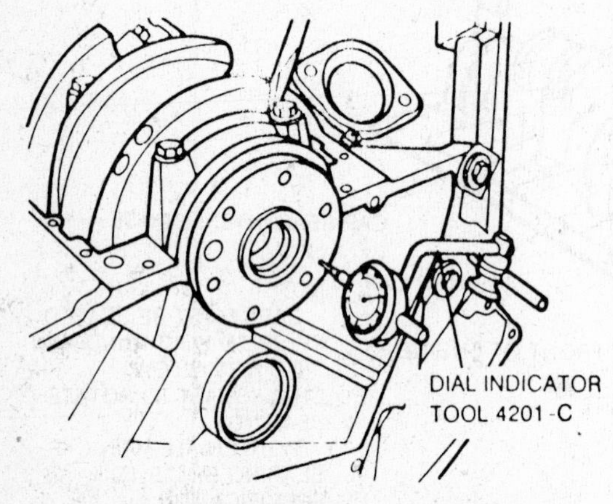

Checking crankshaft endplay

29. Install the oil pump and oil pan.
30. Install the engine-to-transmission adapter plate, flywheel and starter.
31. Install the timing gear case, fuel injection pump, timing gears, timing gear case cover, new crankshaft front oil seal, and crankshaft pulley.
32. Install the engine assembly in vehicle as described in this Section.

CLEANING AND INSPECTION

Crankshaft

NOTE: Handle the crankshaft carefully to avoid damage to the finished surfaces.

1. Clean the crankshaft with solvent, and blow out all oil passages with compressed air. Clean the oil seal contact surface at the rear of the crankshaft with solvent to remove any corrosion, sludge or varnish deposits.
2. Use crocus cloth to remove any sharp edges, burrs or other imperfections which might damage the oil seal during installation or cause premature seal wear.

NOTE: Do not use crocus cloth to polish the seal surfaces. A finely polished surface many produce poor sealing or cause premature seal wear.

3. Inspect the main and connecting rod journals for cracks, scratches, grooves or scores.
4. Measure the diameter of each journal at least four places to determine out-of-round, taper or undersize condition.
5. On an engine with a manual transmission, check the fit of the clutch pilot bearing in the bore of the crankshaft. A needle roller bearing and adapter assembly is used as a clutch pilot bearing. It is inserted directly into the engine crankshaft. The bearing and adapter assembly cannot be serviced separately. A new bearing must be installed whenever a bearing is removed.
6. Inspect the pilot bearing, when used, for roughness, evidence of overheating or loss of lubricant. Replace if any of these conditions are found.
7. Inspect the rear oil seal surface of the crankshaft for deep grooves, nicks, burrs, porosity, or scratches which could damage the oil seal lip during installation. Remove all nicks and burrs with crocus cloth.

Main Bearings

1. Clean the bearing inserts and caps thoroughly in solvent, and dry them with compressed air.

NOTE: Do not scrape varnish or gum deposits from the bearing shells.

2. Inspect each bearing carefully. Bearings that have a scored, chipped, or worn surface should be replaced.
3. The copper/lead bearing base may be visible through the bearing overlay in small localized areas. This may not mean that the bearing is excessively worn. It is not necessary to replace the bearing if the bearing clearance is within recommended specifications.
4. Check the clearance of bearings that appear to be satisfactory with Plastigage® or it's equivalent. Fit the new bearings fol-

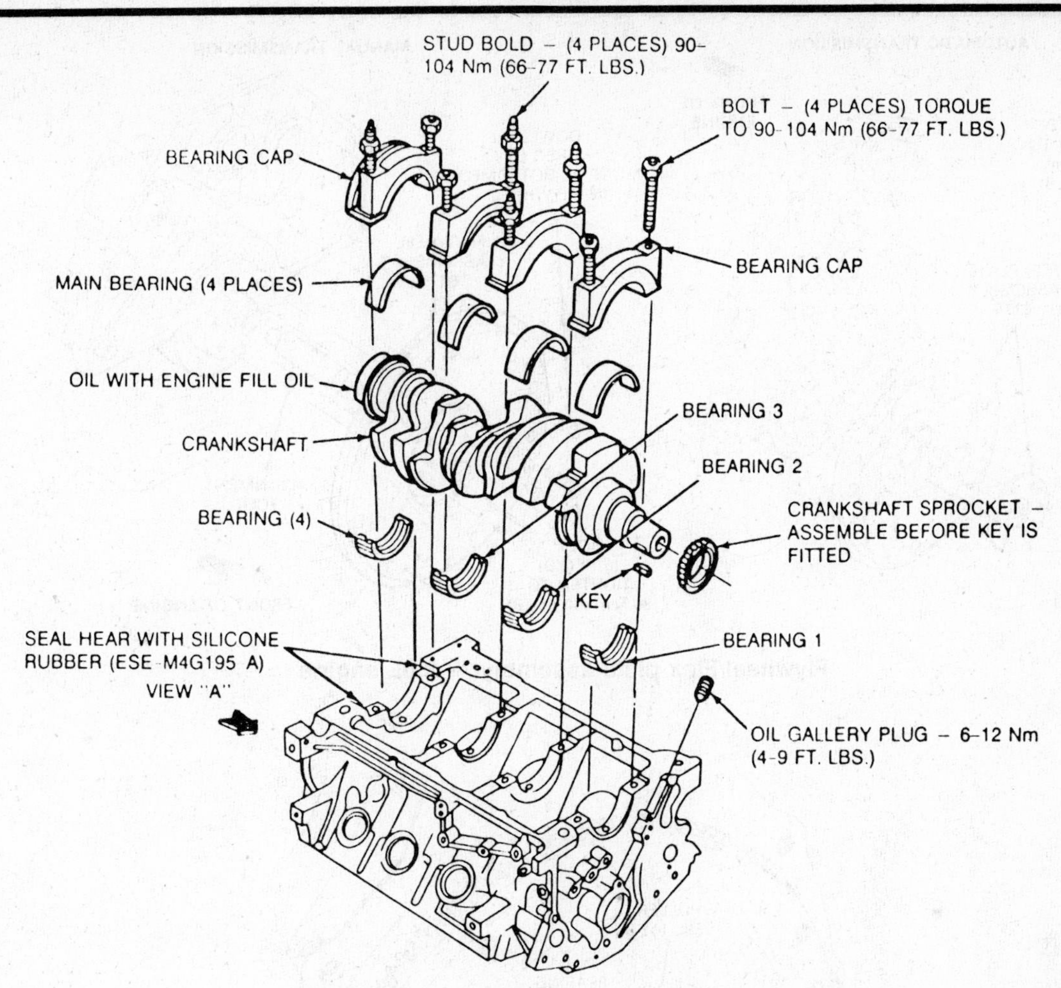

Crankshaft and bearing assembly – 4.0L engine

lowing the procedure Crankshaft and Main Bearing removal and installation in this Section.

REGRINDING JOURNALS

1. Dress minor scores with an oil stone. If the journals are severely marred or exceed the for the next undersize bearing.
2. Regrind the journals to give the proper clearance with the next undersize bearing. If the journal will not clean up to maximum undersize bearing available, replace the crankshaft.
3. Always reproduce the same journal shoulder radius that existed originally. Too small a radius will result in fatigue failure of the crankshaft. Too large a radius will result in bearing failure due to radius ride of the bearing.
4. After regrinding the journals, chamfer the oil holes; then polish the journals with a No. 320 grit polishing cloth and engine oil. Crocus cloth may also be used as a polishing agent.

Flywheel and Ring Gear

REMOVAL AND INSTALLATION

1. Remove the transmission, following procedures in Section 7, Clutch and Transmission.
2. On a manual shift transmission, remove the clutch pressure plate and cover assembly and clutch disc, following the procedures in Section 7, Clutch and Transmission.
3. Remove the flywheel attaching bolts and remove the flywheel.

To install:

4. Position the flywheel on the crankshaft flange. Coat the threads of the flywheel attaching bolts with Loctite® or equivalent and install the bolts. Tighten the bolts in sequence across from each other to specifications.
5. On a manual shift transmission, install the clutch disc and pressure plate and cover assembly following the procedures in Section 7, Clutch and Transmission.
6. Install the transmission following the procedure in Section 7, Clutch and Transmission.

RING GEAR REPLACEMENT

NOTE: This procedure is for manual shift transmission only. On automatic transmission if the ring gear has worn, chipped or cracked teeth, replace the flywheel assembly.

1. Heat the ring gear with a blow torch on the engine side of the gear, and knock it off the flywheel. Do not hit the flywheel when removing the ring gear.
2. Heat the new ring gear evenly until the gear expands enough to slip onto the flywheel. Make sure the gear is properly seated against the shoulder. Do not heat any part of the gear more than 500 degrees F. If this limit is exceeded, the hardness will be removed from the ring gear teeth.

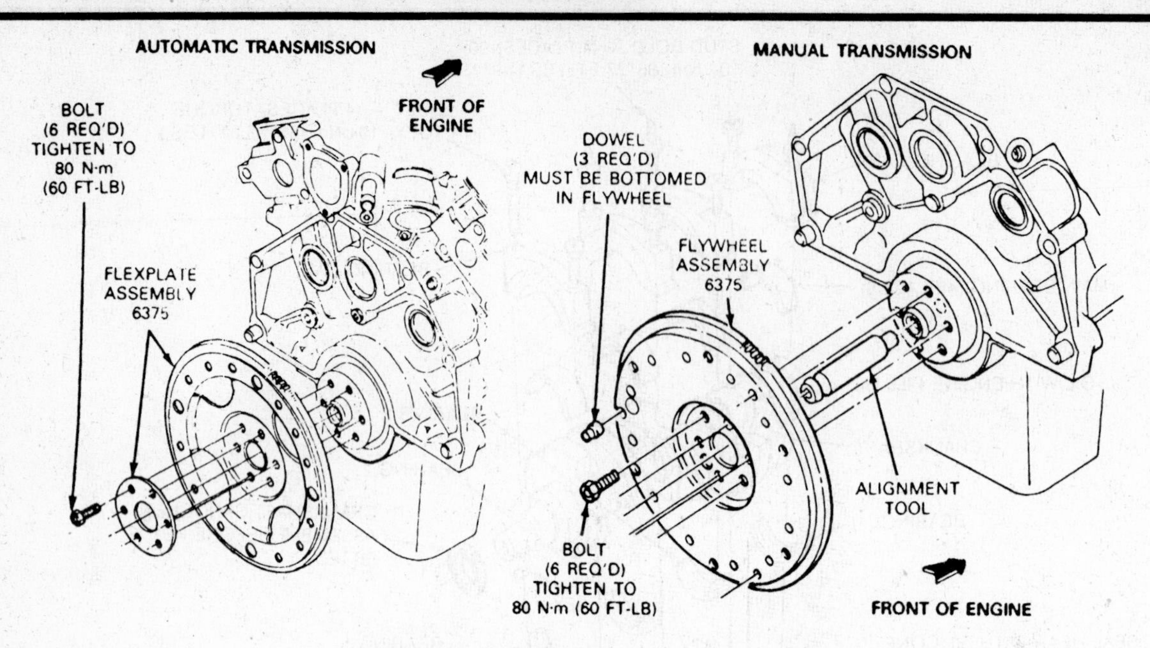

Flywheel/Flex plate assembly — 3.0L engine

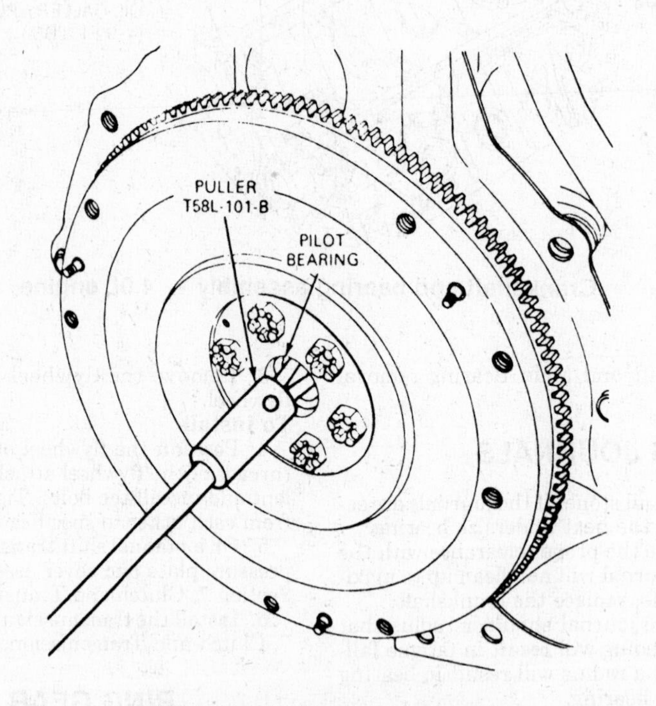

Clutch pilot bearing removal

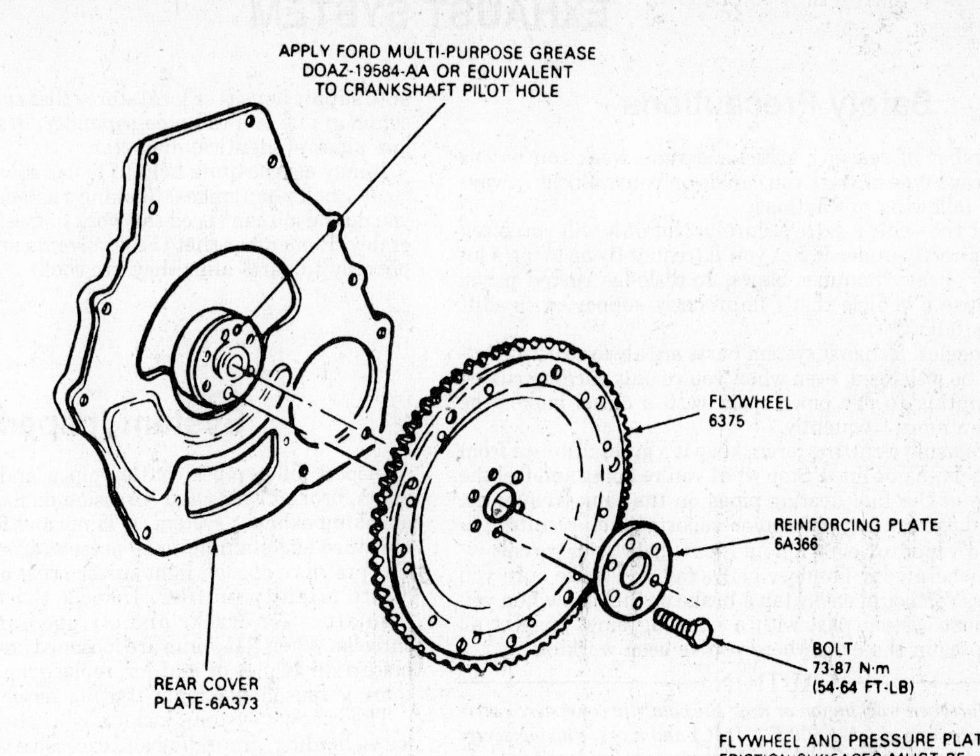

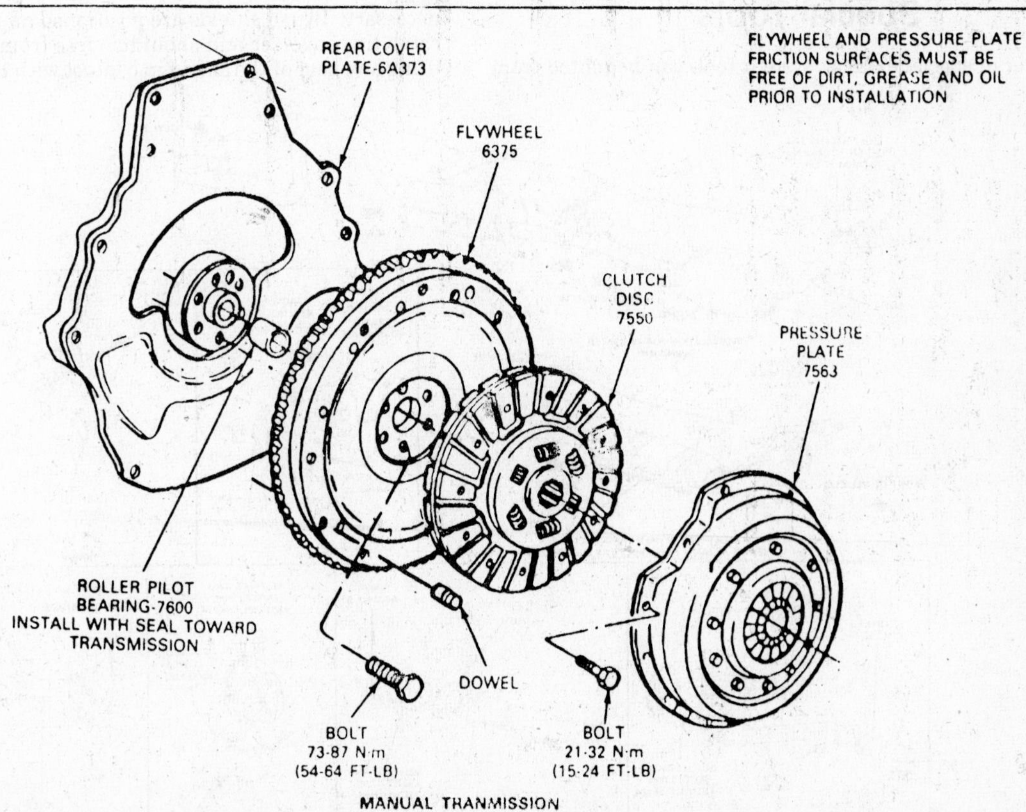

Flywheel/Flex plate assembly — 2.3L engine

EXHAUST SYSTEM

Safety Precautions

For a number of reasons, exhaust system work can be the most dangerous type of work you can do on your vehicle. Always observe the following precautions:

- Support the vehicle extra securely. Not only will you often be working directly under it, but you'll frequently be using a lot of force, say, heavy hammer blows, to dislodge rusted parts. This can cause a vehicle that's improperly supported to shift and possibly fall.
- Wear goggles. Exhaust system parts are always rusty. Metal chips can be dislodged, even when you're only turning rusted bolts. Attempting to pry pipes apart with a chisel makes the chips fly even more frequently.
- If you're using a cutting torch, keep it a great distance from either the fuel tank or lines. Stop what you're doing and feel the temperature of the fuel bearing pipes on the tank frequently. Even slight heat can expand and/or vaporize fuel, resulting in accumulated vapor, or even a liquid leak, near your torch.
- Watch where your hammer blows fall and make sure you hit squarely. You could easily tap a brake or fuel line when you hit an exhaust system part with a glancing blow. Inspect all lines and hoses in the area where you've been working.

CAUTION

Be very careful when working on or near the catalytic converter. External temperatures can reach 1,500°F (816°C) and more, causing severe burns. Removal or installation should be performed only on a cold exhaust system.

Special Tools

A number of special exhaust system tools can be rented from auto supply houses or local stores that rent special equipment. A common one is a tail pipe expander, designed to enable you to join pipes of identical diameter.

It may also be quite helpful to use solvents designed to loosen rusted bolts or flanges. Soaking rusted parts the night before you do the job can speed the work of freeing rusted parts considerably. Remember that these solvents are often flammable. Apply only to parts after they are cool!

System Inspection

Inspect inlet pipes, outlet pipes and mufflers for cracked joints, broken welds and corrosion damage that would result in a leaking exhaust system. It is normal for a certain amount of moisture and staining to be present around the muffler seams. The presence of soot, light surface rust or moisture does not indicate a faulty muffler. Inspect the clamps, brackets and insulators for cracks and stripped or badly corroded bolt threads. When flat joints are loosened and/or disconnected to replace a shield pipe or muffler, replace the bolts and flange nuts if there is reasonable doubt that its service life is limited.

The exhaust system, including brush shields, must be free of leaks, binding, grounding and excessive vibrations. These conditions are usually caused by loose or broken flange bolts, shields, brackets or pipes. If any of these conditions exist, check the exhaust system components and alignment. Align or replace as necessary. Brush shields are positioned on the underside of the catalytic converter and should be free from bends which would bring any part of the shield in contact with the catalytic convert-

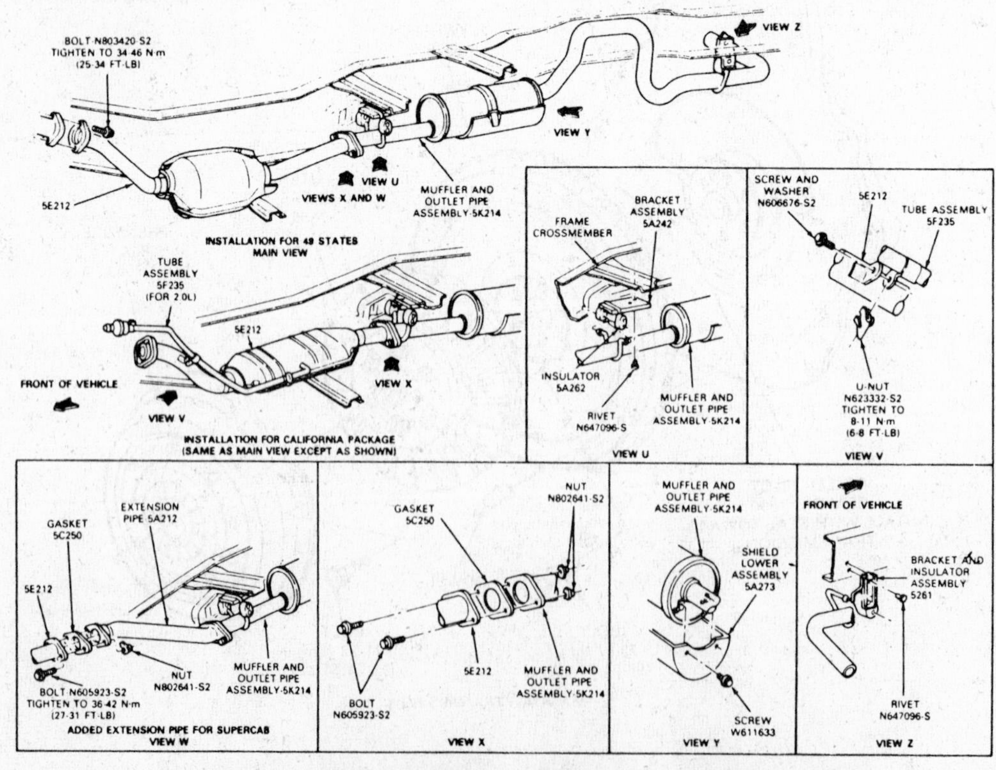

Exhaust system assembly – Ranger (4 cylinder engine)

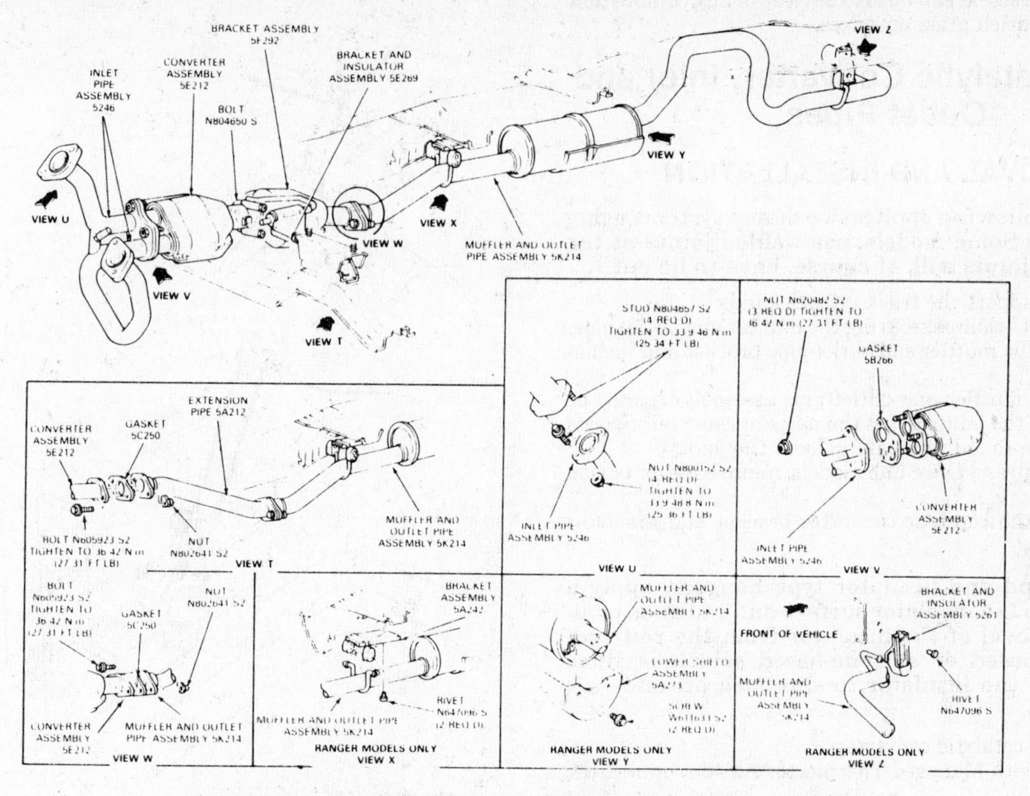

Exhaust system assembly – Ranger (6 cylinder engine)

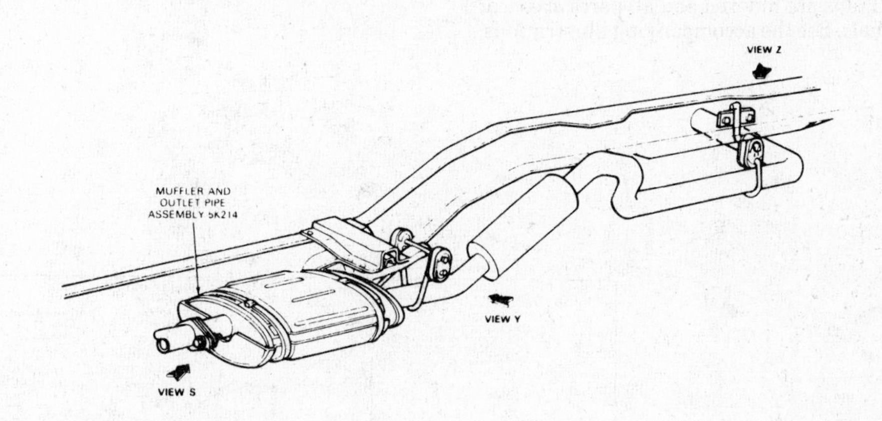

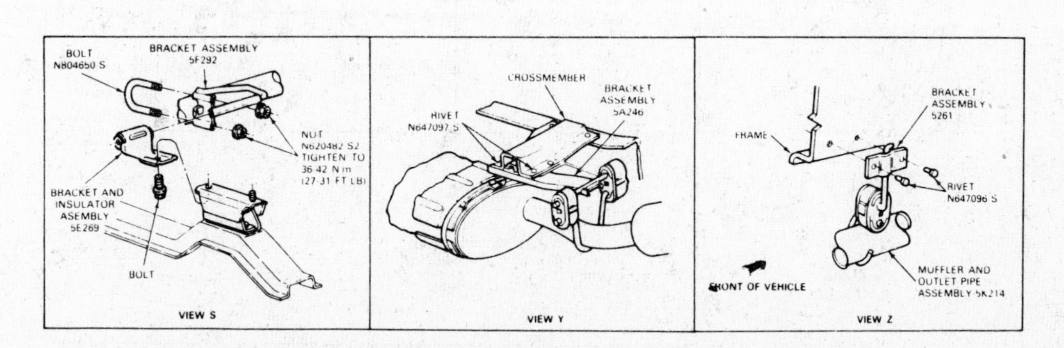

Exhaust system assembly – Bronco II (6 cylinder engine)

er or muffler. The shield should also be clear of any combustible material such as dried grass or leaves.

Muffler, Catalytic Converter, Inlet and Outlet Pipes

REMOVAL AND INSTALLATION

NOTE: The following applies to exhaust systems using clamped joints. Some models, use welded joints at the muffler. These joints will, of course, have to be cut.

1. Raise and support the truck on jackstands.
2. Remove the U-clamps securing the muffler and outlet pipe.
3. Disconnect the muffler and outlet pipe bracket and insulator assemblies.
4. Remove the muffler and outlet pipe assembly. It may be necessary to heat the joints to get the parts to come off. Special tools are available to aid in breaking loose the joints.
5. On Super Cab and Crew Cab models, remove the extension pipe.
6. Disconnect the catalytic converter bracket and insulator assembly.

NOTE: For rod and insulator type hangers, apply a soap solution to the insulator surface and rod ends to allow easier removal of the insulator from the rod end. Don't use oil-based or silicone-based solutions since they will allow the insulator to slip back off once it's installed.

7. Remove the catalytic converter.
8. On models with Managed Thermactor Air, disconnect the MTA tube assembly.
9. Remove the inlet pipe assembly.
10. Install the components making sure that all the components in the system are properly aligned before tightening any fasteners. Make sure all tabs are indexed and all parts are clear of surrounding body panels. See the accompanying illustrations.

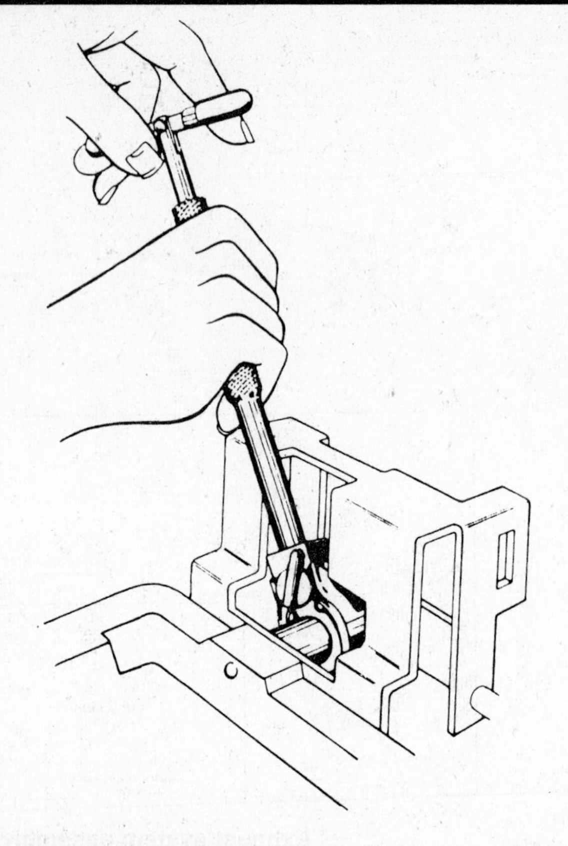

Special tool to remove exhaust mounting brackets retaining bolts in confine areas

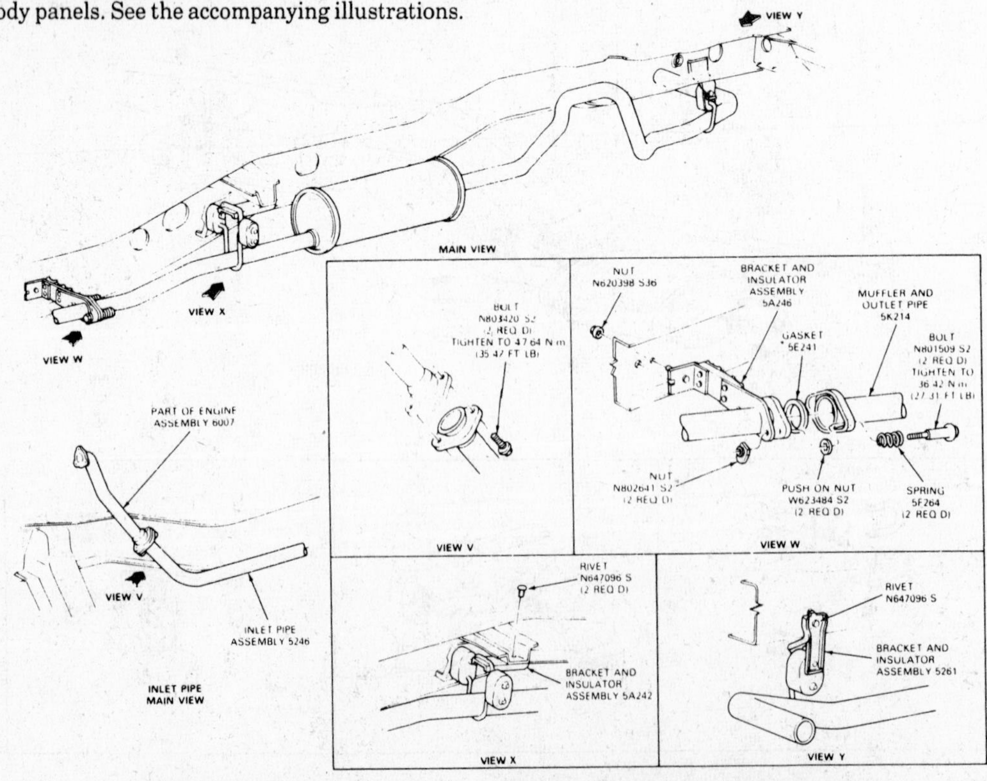

Exhaust system assembly – Ranger (4 cylinder diesel engine)

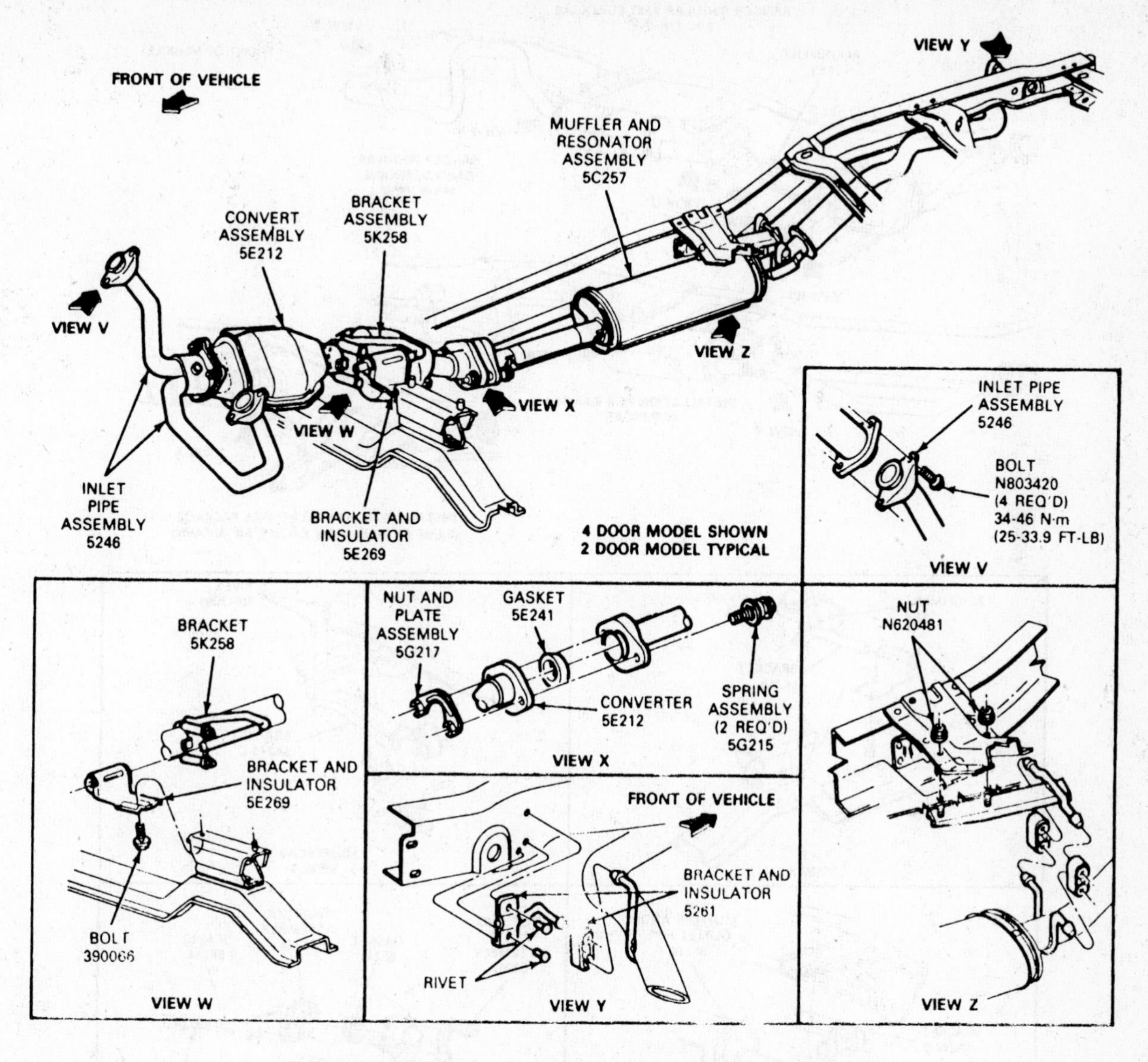

Exhaust system assembly – Explorer (4.0L engine)

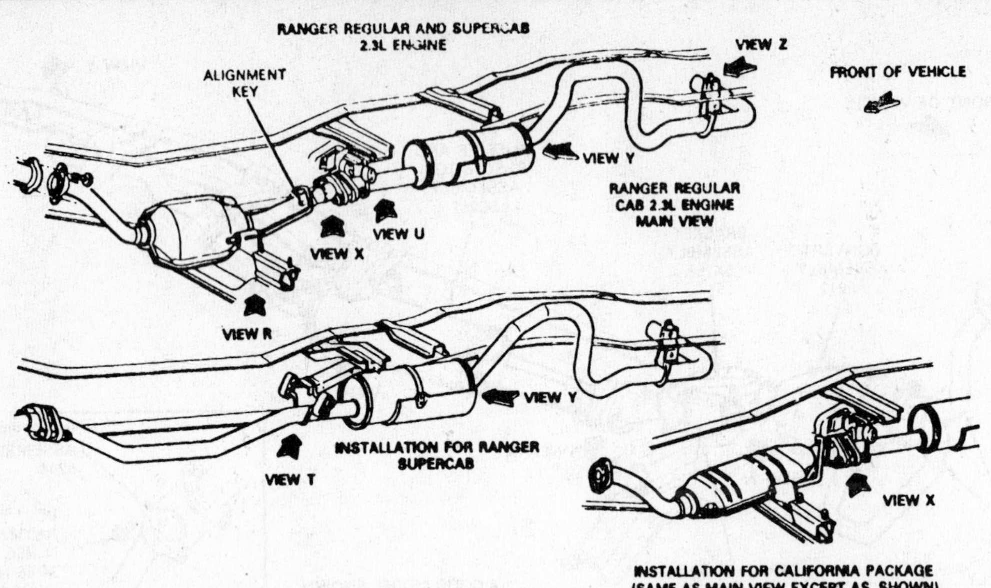

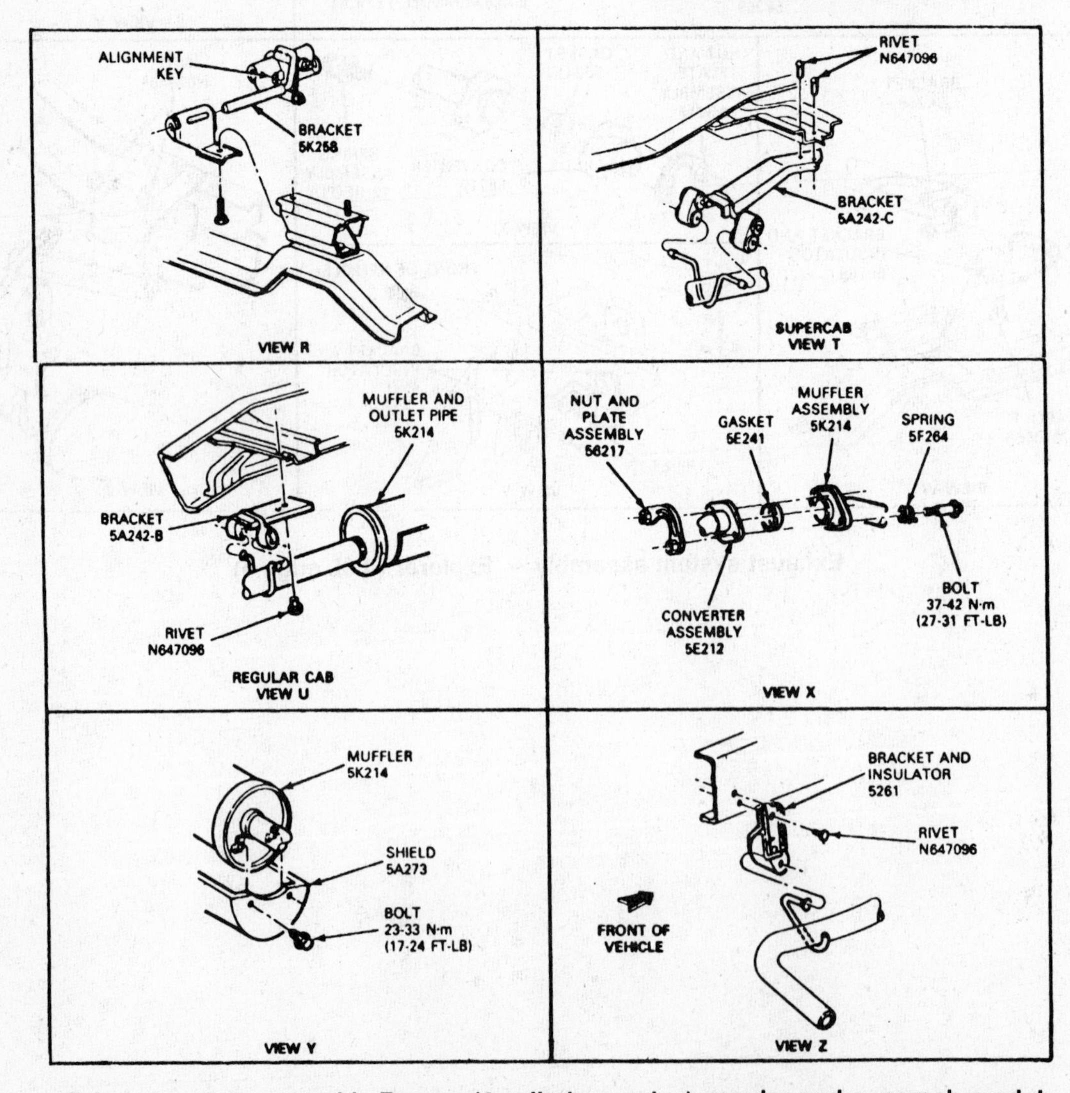

Exhuast system assembly-Ranger (4 cylinder engine) regular and supercab models

4 Emission Controls

QUICK REFERENCE INDEX

GENERAL INDEX

EMISSION CONTROLS

NOTE: Some emission components may be covered under Ford Dealer Emission Warranty (free of charge) see your local dealership.

There are three types of automobile pollutants that concern automotive engineers: crankcase fumes, exhaust gases and gasoline vapors from evaporation. The devices and systems used to limit these pollutants are commonly called emission control equipment.

Crankcase Emission Controls

The crankcase emission control equipment consists of a positive crankcase ventilation (PCV) valve, a closed oil filler cap and the hoses that connect this equipment.

When the engine is running, a small portion of the gases which are formed in the combustion chamber leak by the piston rings and enter the crankcase. Since these gases are under pressure they tend to escape from the crankcase and enter into the atmosphere. If these gases were allowed to remain in the crankcase for any length of time, they would contaminate the engine oil and cause sludge to build up. If the gases are allowed to escape into the atmosphere, they would pollute the air, as they contain unburned hydrocarbons. The crankcase emission control equipment recycles these gases back into the engine combustion chamber, where they are burned.

Crankcase gases are recycled in the following manner. While the engine is running, clean filtered air is drawn into the crankcase either directly through the oil filler cap or through the carburetor air filter and then through a hose leading to the oil filler cap. As the air passes through the crankcase it picks up the combustion gases and carries them out of the crankcase up through the PCV valve and into the intake manifold. After they enter the

LOW SPEED OPERATION—HIGH MANIFOLD VACUUM

HIGH SPEED OPERATION—LOW MANIFOLD VACUUM

A cutaway section of a PCV valve showing its operation

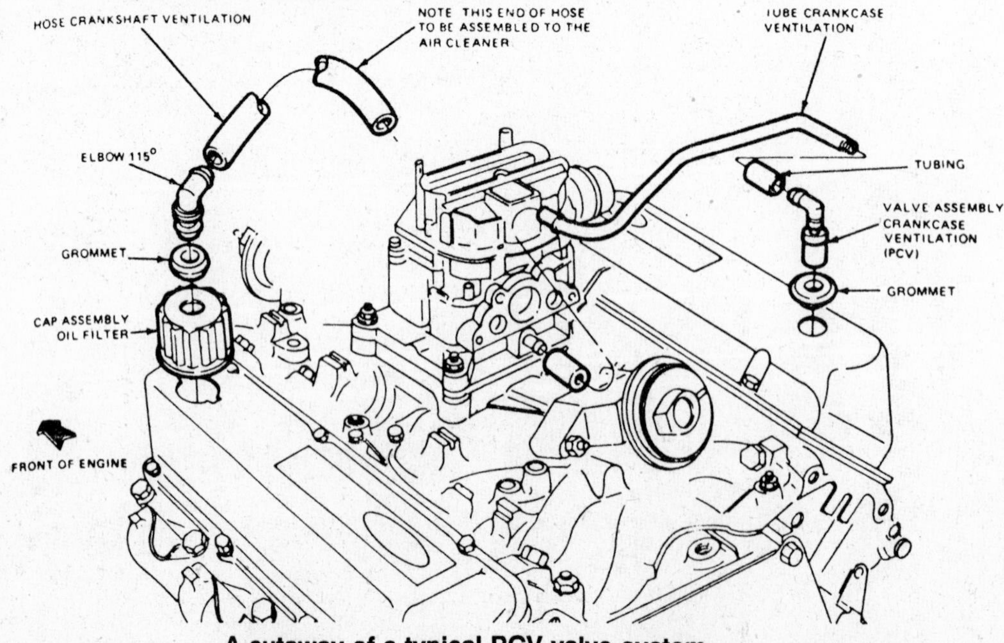

A cutaway of a typical PCV valve system

intake manifold they are drawn into the combustion chamber and are burned.

The most critical component of the system is the PCV valve. This vacuum-controlled valve regulates the amount of gases which are recycled into combustion chamber. At low engine speeds the valve is partially closed, limiting the flow of gases into the intake manifold. As engine speed increases, the valve opens to admit greater quantities of the gasses into the intake manifold. If the valve should become blocked or plugged, the gases will bee prevented from escaping the crankcase by the normal route. Since these gases are under pressure, they will find their own way out of the crankcase. This alternate route is usually a weak oil seal or gasket in the engine. As the gas escapes by the gasket, it also creates an oil leak. Besides causing oil leaks, a clogged PCV valve also allows these gases to remain in the crankcase for an extended period of time, promoting the formation of sludge in the engine.

TROUBLESHOOTING

With the engine running, pull the PCV valve and hose from the valve rocker cover rubber grommet. Block off the end of the valve with your finger. A strong vacuum should be felt. Shake the valve; a clicking noise indicates it is free. Replace the valve if it is suspected of being blocked. REPLACE THE PCV VALVE INSTEAD OF TRYING TO CLEAN IT-ALSO CHECK PCV VALVE HOSE FOR CRACKS OR WEAR.

REMOVAL AND INSTALLATION

1. Pull the PCV valve and hose from the rubber grommet in the rocker cover.
2. Remove the PCV valve from the hose. Inspect the inside of the PCV valve. If it is dirty, disconnect it from the intake manifold and clean it in a suitable, safe solvent.

To install, proceed as follows:

1. If the PCV valve hose was removed, connect it to the intake manifold.
2. Connect the PCV valve to its hose.
3. Install the PCV valve into the rubber grommet in the valve rocker cover.

Evaporative Emission Controls

All gasoline powered vehicles are equipped with fuel evaporative emission control. The system is designed to limit fuel vapors released into the atmosphere.

Changes in atmospheric temperature cause fuel tanks to breathe that is, the air within the tank expands and condenses with outside temperature changes. As the temperature rises, air escapes through the tank vent tube or the vent in the tank cap. The air which escapes contains gasoline vapors. In a similar manner, the gasoline which fills the carburetor float bowl expands when the engine is stopped. Engine heat causes this expansion. The vapors escape through the carburetor and air cleaner.

The Evaporative Emission Control System provides a sealed fuel system with the capability to store and condense fuel vapors. The system has three parts: a fill control vent system; a vapor vent and storage system; and a pressure and vacuum relief system (special fill cap).

The fill control vent system is a modification to the fuel tank. It uses an air space within the tank which is 10–12% of the tank's volume. The air space is sufficient to provide for the thermal expansion of the fuel. The space also serves as part of the in-tank vapor vent system.

The in-tank vent system consists of the air space previously described and a vapor separator assembly. The separator assembly is mounted to the top of the fuel tank and is secured by a cam-lockring, similar to the one which secures the fuel sending unit. Foam material fills the vapor separator assembly. The foam material separates raw fuel and vapors, thus retarding the entrance of fuel into the vapor line.

The sealed filler cap has a pressure vacuum relief valve. Under normal operating conditions, the filler cap operates as a check valve, allowing air to enter the tank to replace the fuel consumed. At the same time, it prevents vapors from escaping through the cap. In case of excessive pressure within the tank, the filler cap valve opens to relieve the pressure.

Because the filler cap is sealed, fuel vapors have but one place through which they may escape-the vapor separator assembly at the top of the fuel tank. The vapors pass through the foam material and continue through a single vapor line which leads to a canister in the engine compartment. The canister is filled with activated charcoal.

Another vapor line runs from the top of the carburetor float chamber to the charcoal canister.

As the fuel vapors (hydrocarbons), enter the charcoal canister, they are absorbed by the charcoal. The air is dispelled through the open bottom of the charcoal canister, leaving the hydrocarbons trapped within the charcoal. When the engine is started, vacuum causes fresh air to be drawn into the canister from its open bottom. The fresh air passes through the charcoal picking up the hydrocarbons which are trapped there and feeding them into the carburetor for burning with the fuel mixture.

EVAPORATIVE EMISSION CONTROL SYSTEM CHECK/SERVICE

Other than a visual check to determine that none of the vapor lines are broken, there is no test for this equipment.

The only maintenance on the evaporative system is to periodically check all hoses and connections for leaks and deterioration. Replace any hoses which are found to be damaged in any way. Under normal circumstances, the charcoal canister is expected to last the life of the vehicle, but it should be periodically inspected for any damage or contamination by raw gasoline. Replace any gasoline soaked canister found. Refer to the illustrations for canister mounting and evaporative hose routing on the various engines. Filler cap damage or contamination that clogs the pressure/vacuum valve may result in deformation of the fuel tank.

NOTE: Evaporative emission components are designed (warrantied) and tested to exceed 120,000 miles/10 years of vehicle use. Some components may be covered under Ford Dealer Emission Warranty see your local dealership for details.

Carbon Canister/Vapor Hoses

REMOVAL AND INSTALLATION

All Engines

1. Disconnect the negative battery cable.
2. Mark and disconnect the vapor hoses from the canister assembly.
3. Remove the screw securing the canister to the bracket or fender apron.
4. Lift up on the canister assembly to disengage the tab on the back side and remove the canister.
5. Installation is the reverse of the service removal procedure. Always install vapor hose in correct location. Refer to the necessary illustrations in this Section.

To disconnect a vapor hose from any component securely grip component with one hand and vapor hose with the other hand

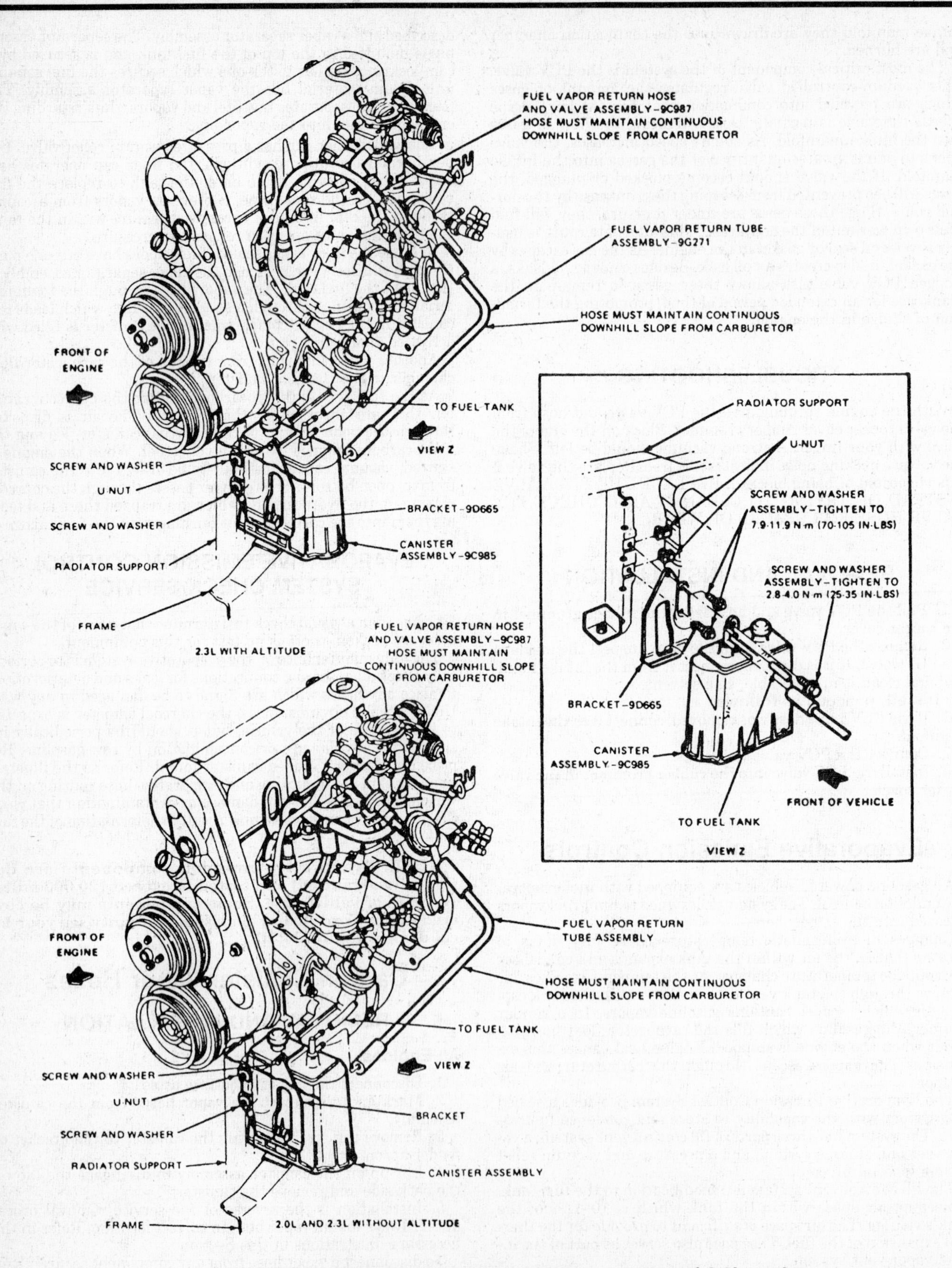

Carburetor and evaporative canister venting – 2.0L and 2.3L engines

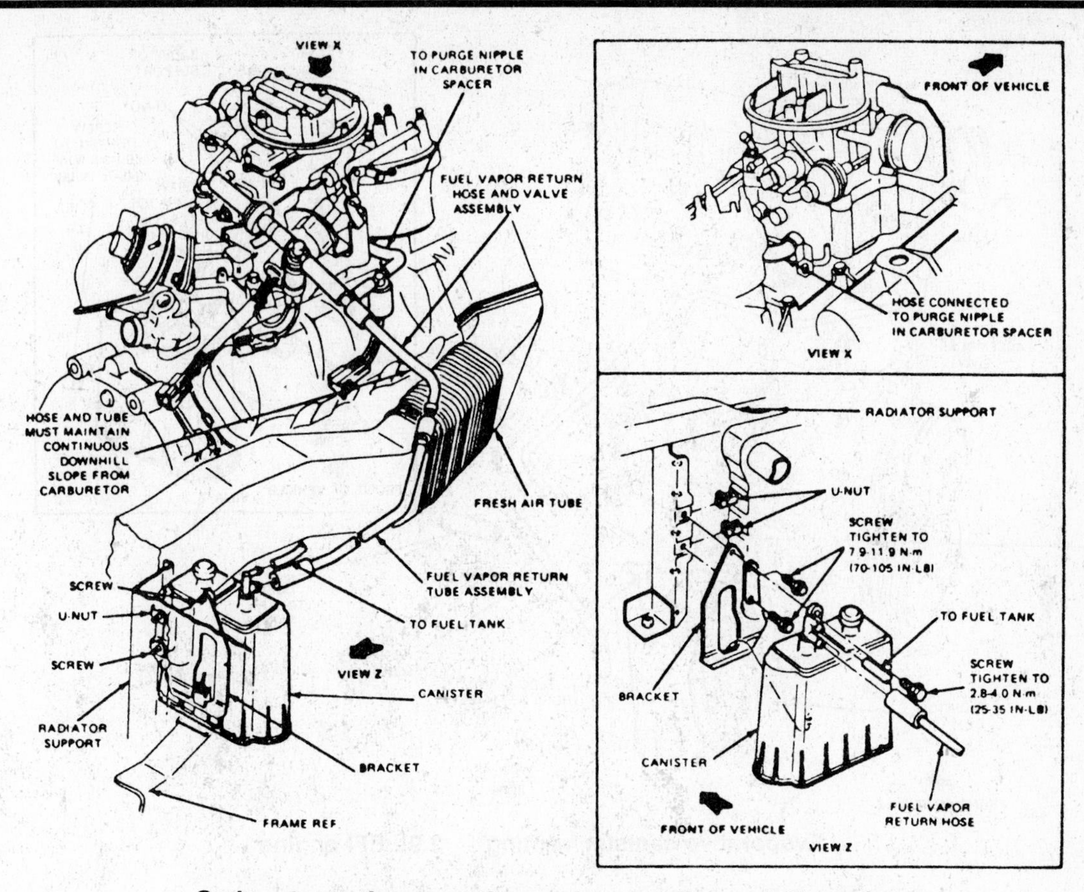

Carburetor and evaporative canister venting – 2.8L engine

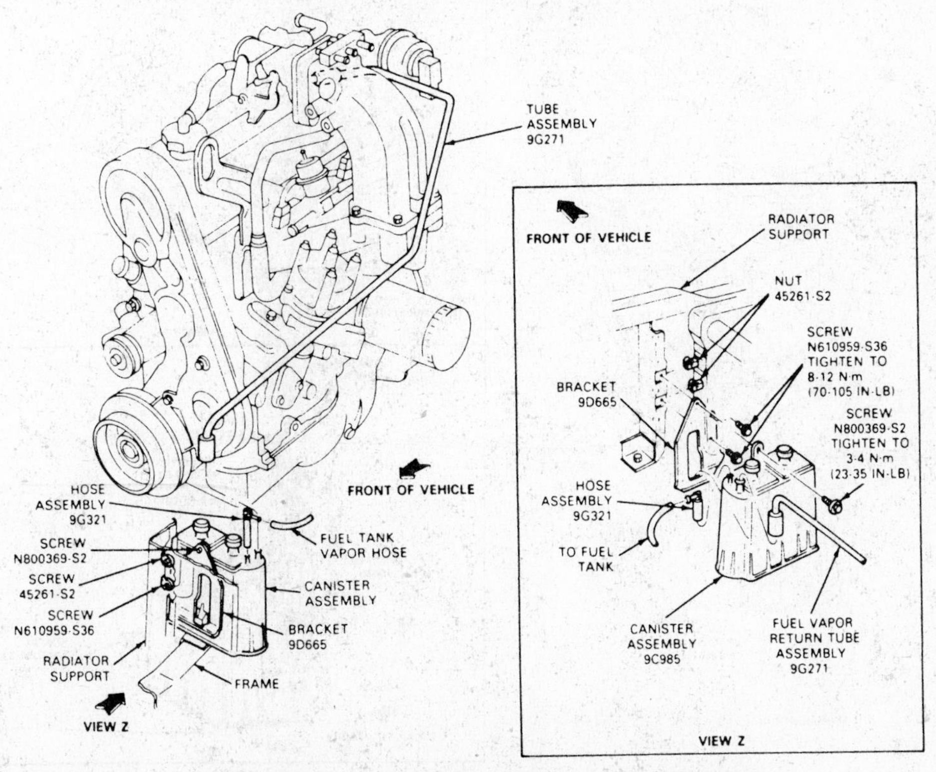

Evaporative canister venting – 2.3L EFI engine

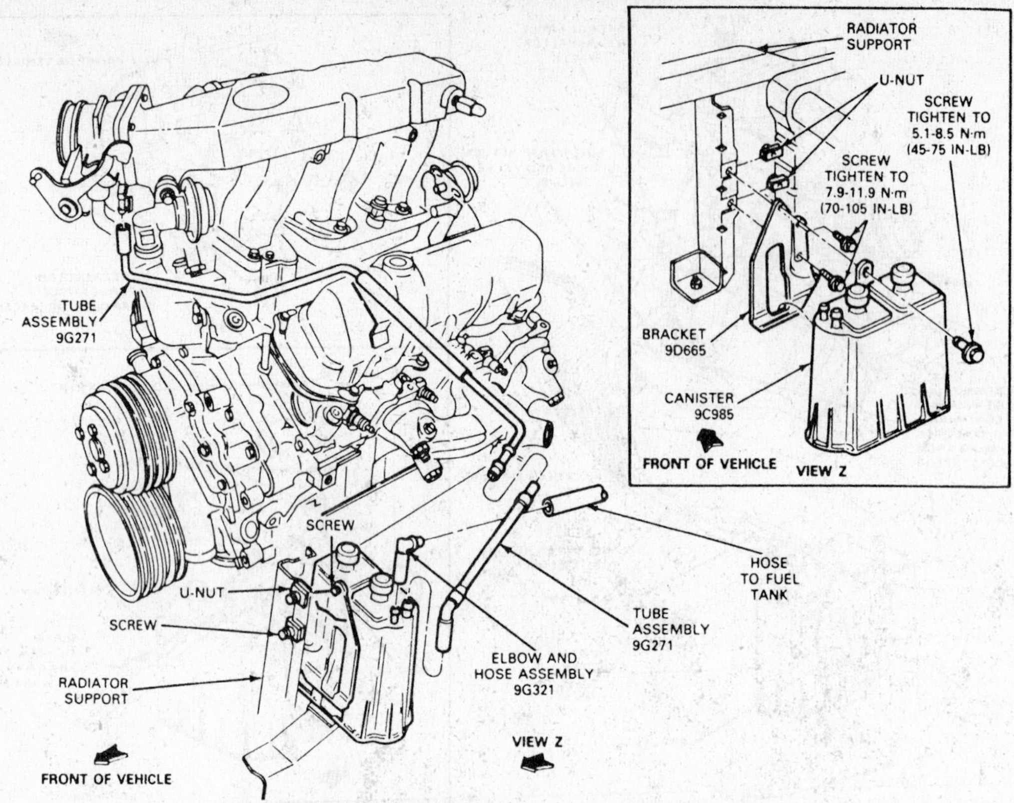

Evaporative canister venting – 2.9L EFI engine

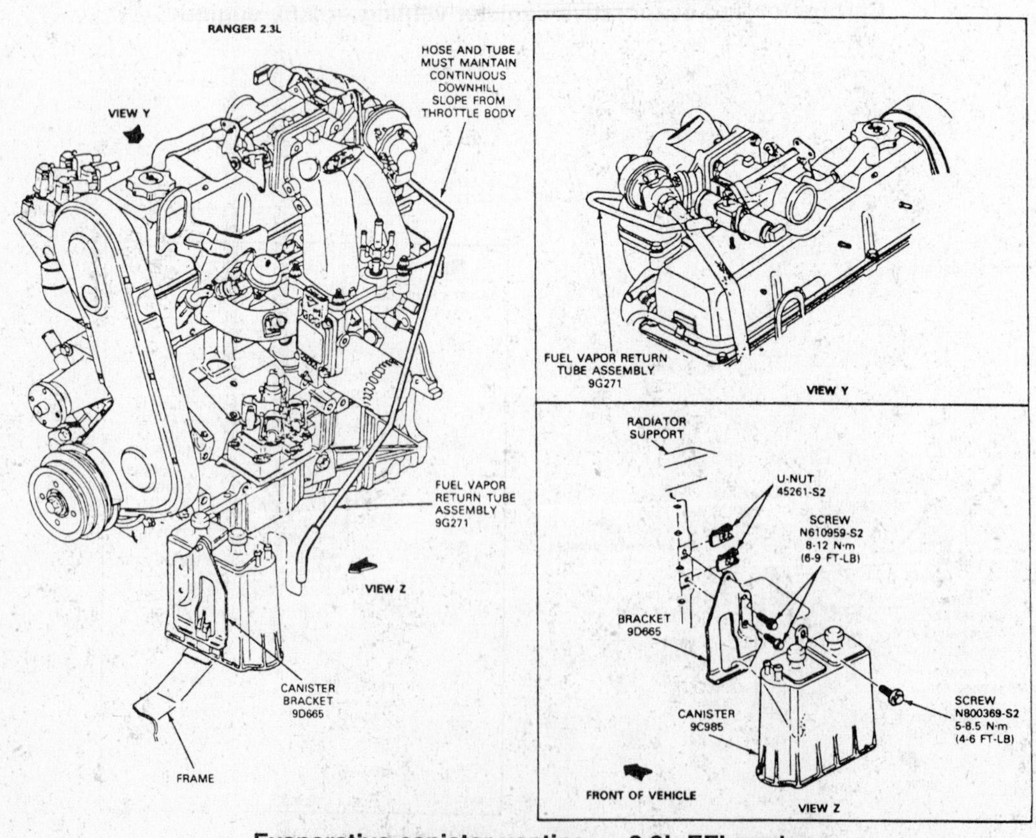

Evaporative canister venting – 2.3L EFI engine

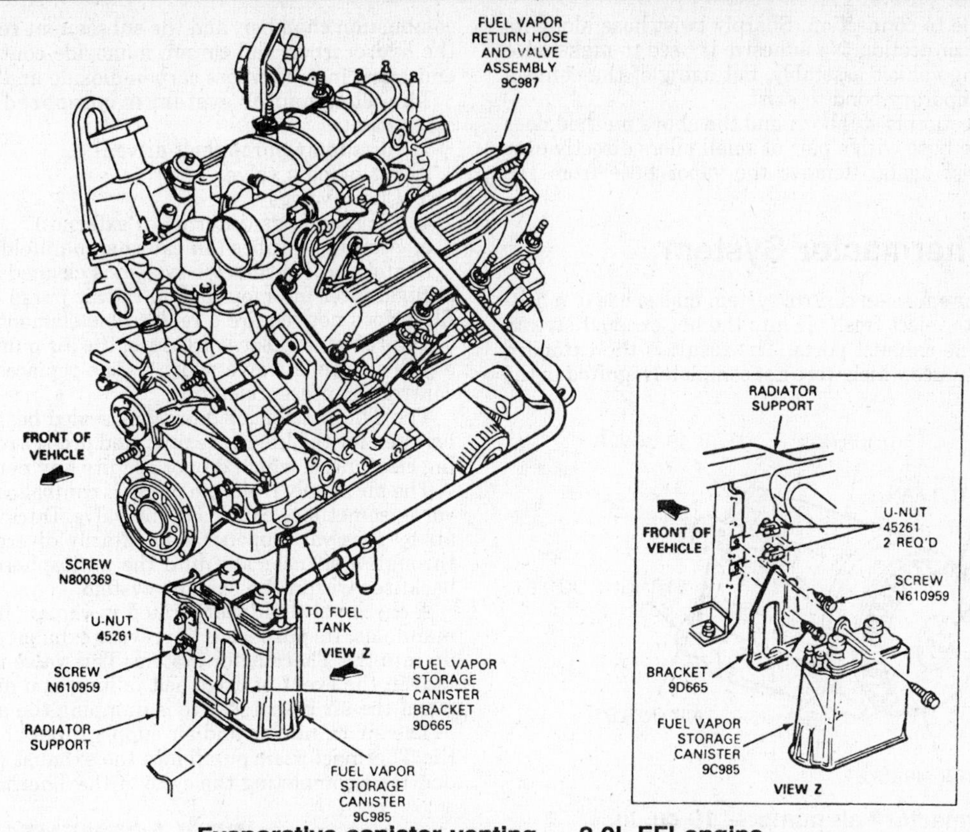

Evaporative canister venting — 3.0L EFI engine

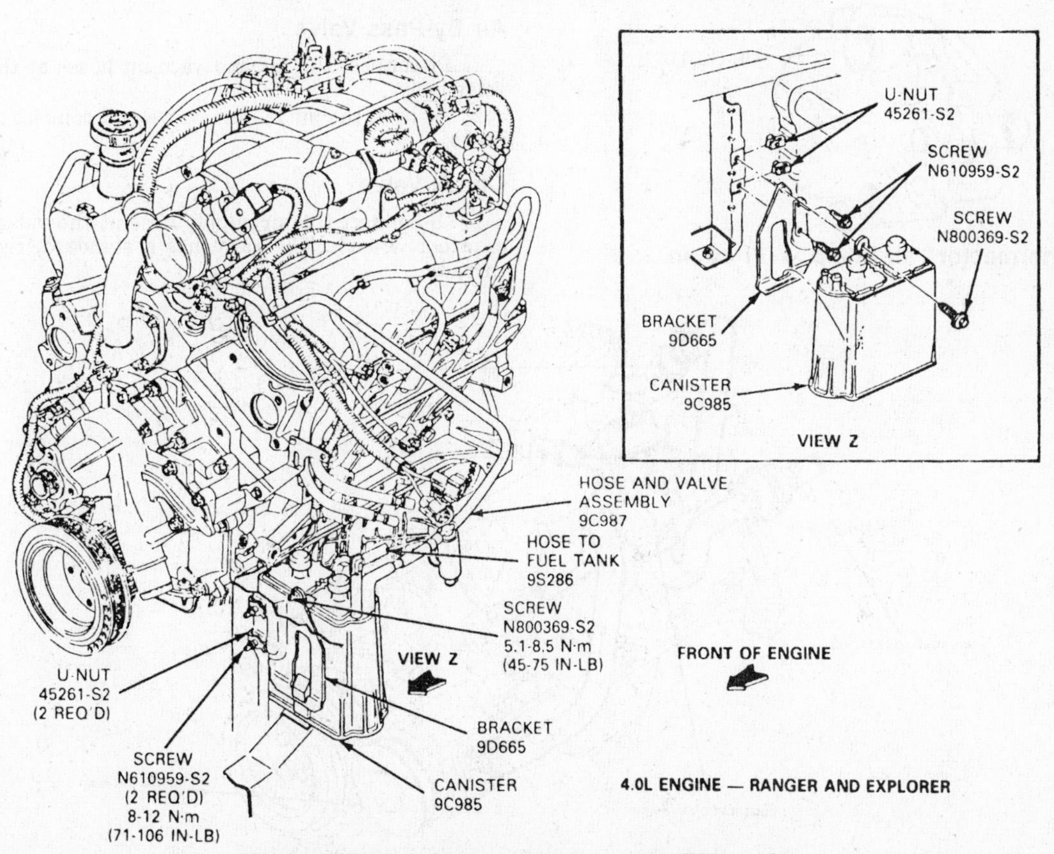

Evaporative canister venting — 4.0L EFI engine

as close as possible to connection. Sharply twist hose along its axis to break the connection. No adhesive is used to make hose connections during vehicle assembly, but aging of the connections causes a temporary bond to exist.

If the the connection is stubborn and the above method does not work, grip the hose with a pair of small pilers directly over the joint and twist again. Remove the vapor hose from the component.

Thermactor System

The Thermactor emission control system makes use of a belt-driven air pump to inject fresh air into the hot exhaust stream through the engine exhaust ports. The result is the extended burning of those fumes which were not completely ignited in the

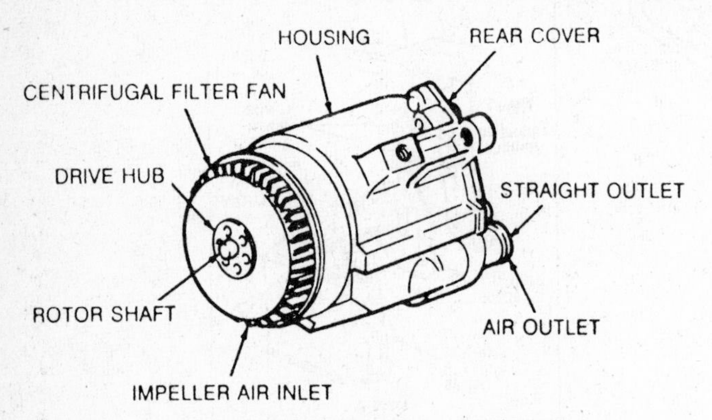

Thermactor® air pump — 19 cu. in.

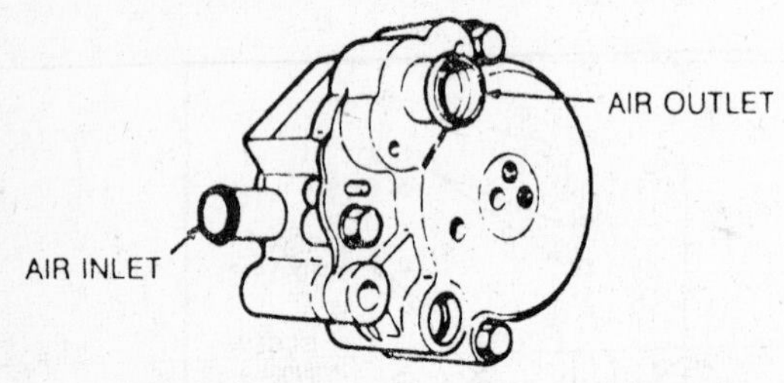

Thermactor® air pump — 11 cu. in.

combustion chamber, and the subsequent reduction of some of the hydrocarbon and carbon monoxide content of the exhaust emissions into harmless carbon dioxide and water.

The Thermactor system is composed of the following components:

1. Air supply pump (belt-driven)
2. Air by pass valve
3. Check valves
4. Air manifolds (internal or external)
5. Air supply tubes (on external manifolds only)

Air for the Thermactor system is cleaned by means of a centrifugal filter fan mounted on the air pump driveshaft. The air filter does not require a replaceable element.

To prevent excessive pressure, the air pump is equipped with a pressure relief valve which uses a replaceable plastic plug to control the pressure setting.

The Thermactor air pump has sealed bearings which are lubricated for the life of the unit, and pre-set rotor vane and bearing clearances, which do not require any periodic adjustments.

The air supply from the pump is controlled by the air by-pass valve, sometimes called a dump valve. During deceleration, the air by-pass valve opens, momentarily diverting the air supply through a silencer and into the atmosphere, thus preventing backfires within the exhaust system.

A check valve is incorporated in the air inlet side of the air manifolds. Its purpose is to prevent exhaust gases from backing up into the Thermactor system. This valve is especially important in the event of drive belt failure, and during deceleration, when the air by-pass valve is dumping the air supply.

The air manifolds and air supply tubes channel the air from the Thermactor air pump into the exhaust ports of each cylinder, thus completing the cycle of the Thermactor system.

REPLACEMENT

Air By-Pass Valve

1. Disconnect the air and vacuum hoses at the air by-pass valve body.
2. Position the air by-pass valve and connect the respective hoses.

Check Valve

1. Disconnect the air supply hose at the valve. Use 1¼ in. crowfoot wrench. The valve has a standard, right-hand pipe thread.

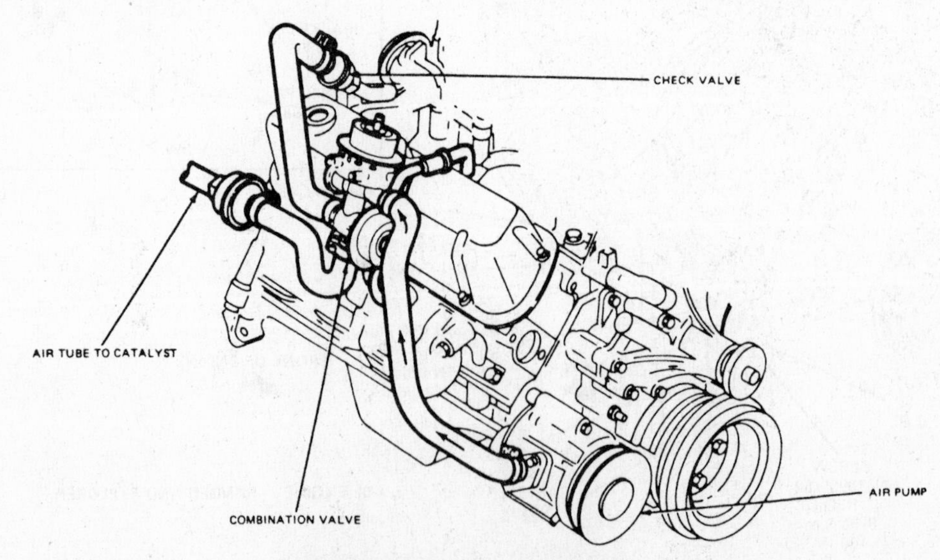

Typical Thermactor® System

2. Clean the threads on the air supply tube with a wire brush. Do not blow compressed air through the check valve in either direction.
3. Install the check valve and tighten.
4. Connect the air supply hose.

Air Pump and Filter Fan

1. Loosen the air pump attaching bolts.
2. Remove the drive pulley attaching bolts and pull the pulley off the air pump shaft.
3. Pry the outer disc loose, then remove the centrifugal filter fan. Care must be used to prevent foreign matter from entering the air intake hole, especially if the fan breaks during removal. Do not attempt to remove the metal drive hub.
4. Install the new filter fan by drawing it into position with the pulley bolts.

Air Pump

1. Disconnect the air outlet hose at the air pump.
2. Loosen the pump belt tension adjuster.
3. Disengage the drive belt.
4. Remove the mounting bolt and air pump.
5. Position the air pump on the mounting bracket and install the mounting bolt.
6. Place the drive belt in the pulley and attach the adjusting arm to the air pump.
7. Adjust the drive belt tension and tighten the adjusting arm and mounting bolts.
8. Connect the air outlet hose to the air pump.

Relief Valve

Do not disassemble the air pump on the truck to replace the relief valve, but remove the pump from the engine.

1. Remove the relief valve on the pump housing and hold it in position with a block of wood.
2. Use a hammer to lightly tap the wood block until the relief valve is seated.

Relief Valve Pressure-Setting Plug

1. Compress the locking tabs inward (together) and remove the plastic pressure-setting plug.
2. Before installing the new plug, be sure that the plug is the correct one. The plugs are color-coded.
3. Insert the plug in the relief vale hose and push in until it snaps into place.

SYSTEM TESTING

Air Pump Functional Check

Check the air pump belt tension and adjust it, if necessary. Disconnect the air supply hose from the bypass control valve. The pump is operating properly if air flow is felt at the pump outlet and the flow increases as the engine speed is increased. Do not pry on the pump to adjust the belt as the aluminum housing is likely to collapse.

Normally Closed Bypass Valve Check

1. Disconnect the air supply hose at the valve outlet.
2. Remove the vacuum line to check to see that a vacuum signal is present at the vacuum nipple. Remove or bypass any restrictors or delay valves in the vacuum line. There must be a vacuum present at the nipple before proceeding.
3. With the engine at 1,500 rpm and the vacuum line connected to the vacuum nipple, air pump supply air should be heard and felt at the air bypass valve outlet.
4. With the engine at 1,500 rpm, disconnect the vacuum line. Air at the outlet should be significantly decreased or shut off. Air pump supply air should be heard or felt at the silencer ports.
5. If the normally closed air bypass valve does not successful-

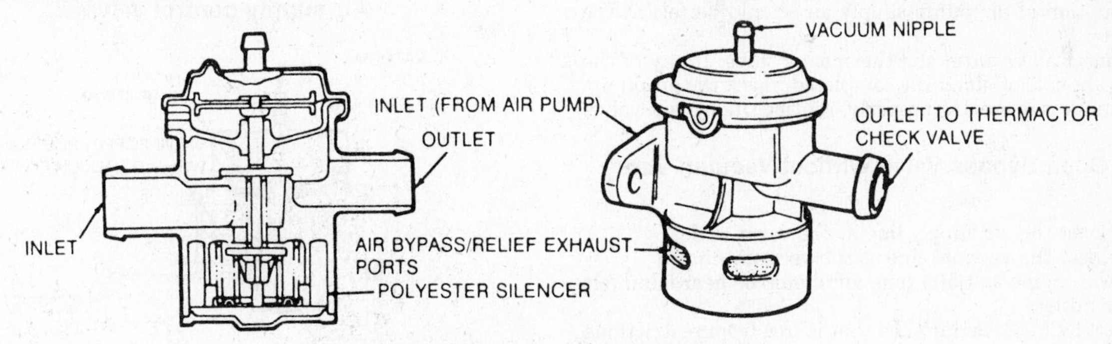

Normally closed air bypass valve

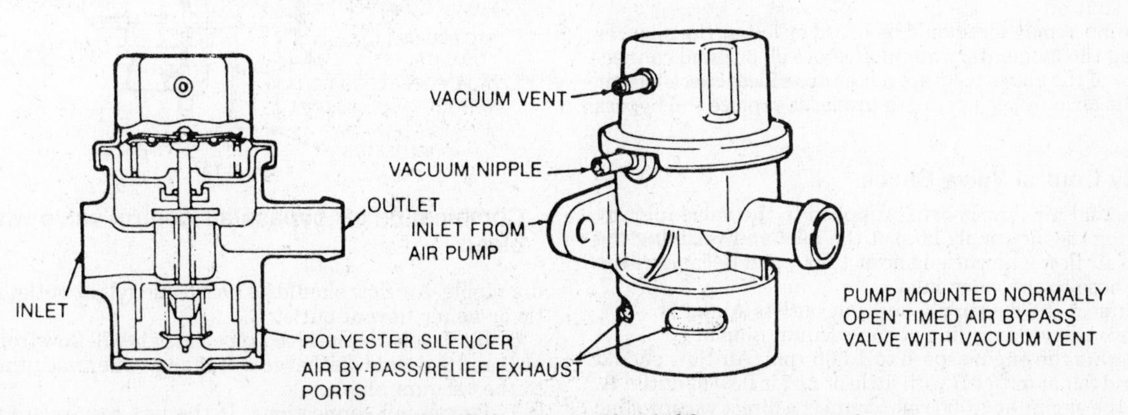

Normally open air bypass valve with vacuum vents

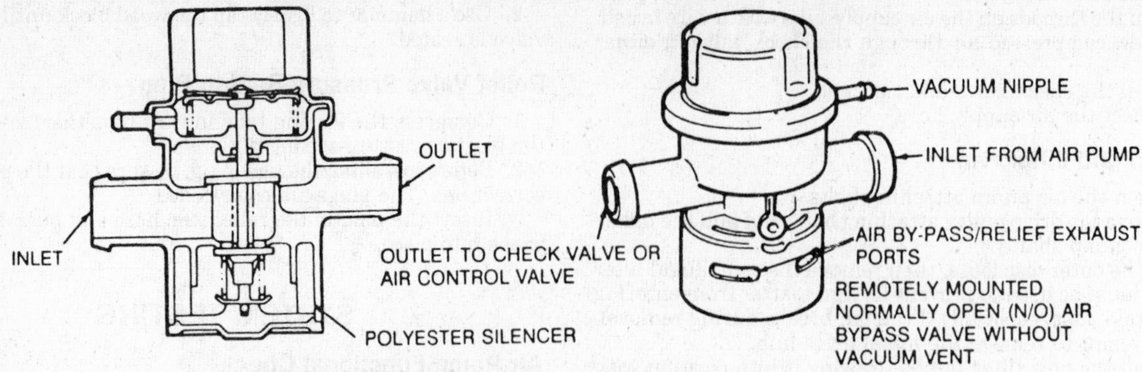

Normally open air bypass valve without vacuum vents

ly complete the above tests, check the air pump. If the pump is operating properly, replace the air bypass valve.

Normally Open Bypass Valve Check

1. Disconnect the air pump supply line at the outlet.
2. Disconnect all vacuum lines from the vacuum nipple and the vacuum vent.
3. Start the engine and raise the engine speed to 1,500 rpm. The air pump supply air should be heard and felt at the outlet.
4. Using a length of vacuum hose with no restrictors or devices, connect the vacuum nipple to one of the manifold vacuum fittings on the intake manifold. With the vacuum vent open to the atmosphere and the engine at 1,500 rpm, virtually no air should be felt at the valve outlet and virtually all air should be bypassed through the silencer ports.
5. Using the same direct vacuum line to an intake manifold vacuum source, cap the vacuum vent. Accelerate the engine speed to 2,000 rpm and suddenly release the throttle. A momentary interruption of air pump supply air should be felt at the valve outlet.
6. Reconnect all vacuum and thermactor lines. If any of the above tests are not satisfactorily completed, check the air pump. If the air pump is operating properly, replace the bypass valve.

Normally Open Bypass Valve Without Vacuum Vent Check

1. Disconnect the air supply line at the valve outlet.
2. Disconnect the vacuum line at the vacuum nipple.
3. With the engine at 1,500 rpm, air should be heard and felt at the valve outlet.
4. Connect a direct vacuum line that is free from restrictions from any manifold vacuum source to the vacuum nipple on the air bypass valve. Air at the outlet should be momentarily decreased or shut off.
5. Air pump supply air should be heard or felt at the silencer ports during the momentary dump. Restore all original connections. If any of the above tests are not as described, check the air pump. If the air pump is operating properly, replace the bypass valve.

Air Supply Control Valve Check

1. Verify that air flow is being supplied to the valve inlet by disconnecting the air supply hose at the inlet and verifying the presence of air flow with the engine at 1,500 rpm. Reconnect the air supply hose to the valve inlet.
2. Disconnect the air supply hoses at outlets **A** and **B**.
3. Remove the vacuum line at the vacuum nipple.
4. Accelerate the engine speed to 1,500 rpm. Air flow should be heard and felt at outlet **B** with little or no air flow at outlet **A**.
5. With the engine at 1,500 rpm, connect a direct vacuum line from any manifold vacuum fitting to the air control valve vacuum nipple. Air flow should be heard and felt at outlet **A** with little or no air flow at outlet **B**.
6. If the valve is the bleed type, less air will flow from outlet **A** or **B** and the main discharge will change when vacuum is applied to the vacuum nipple.
7. Restore all connections. If the test results are not as described, replace the air control valve.

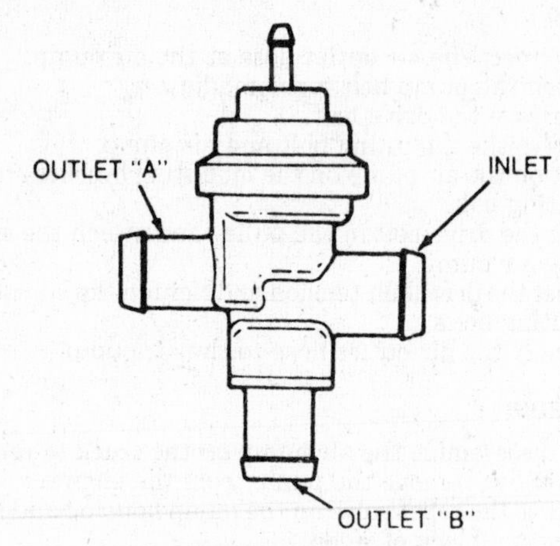

Air supply control valve

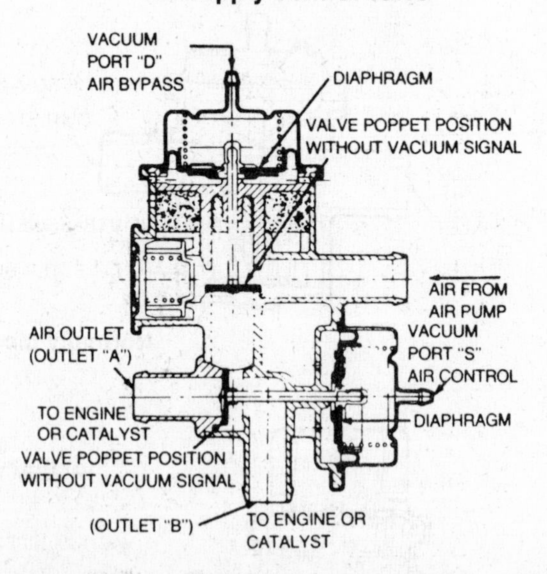

Combination air bypass/air control valve without bleed

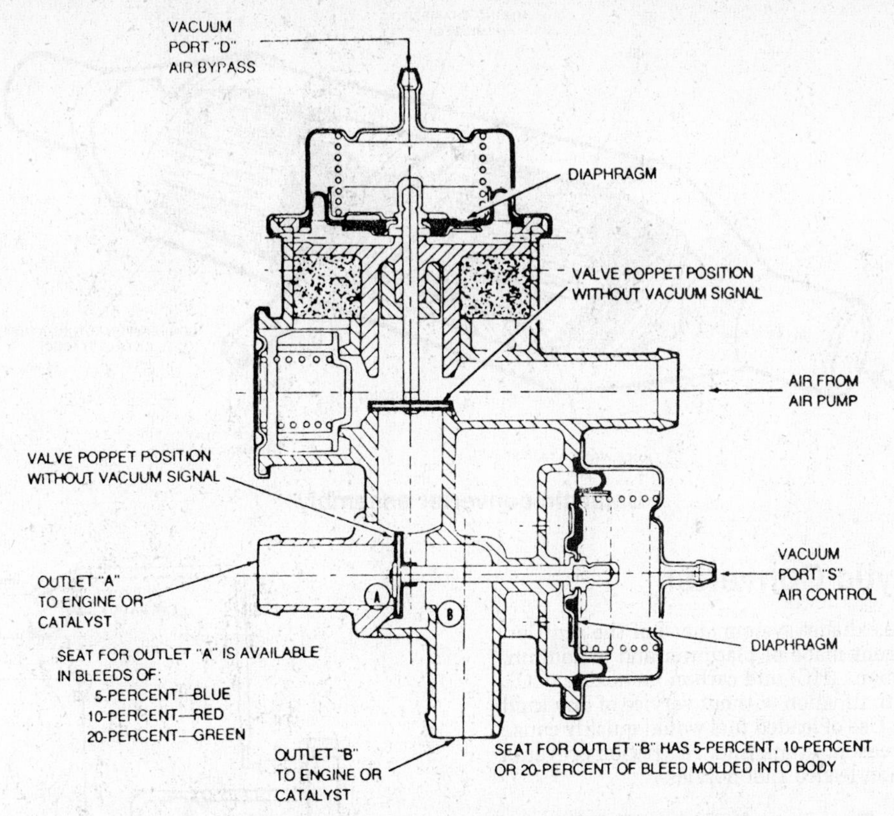

Combination air bypass/air control valve bleed

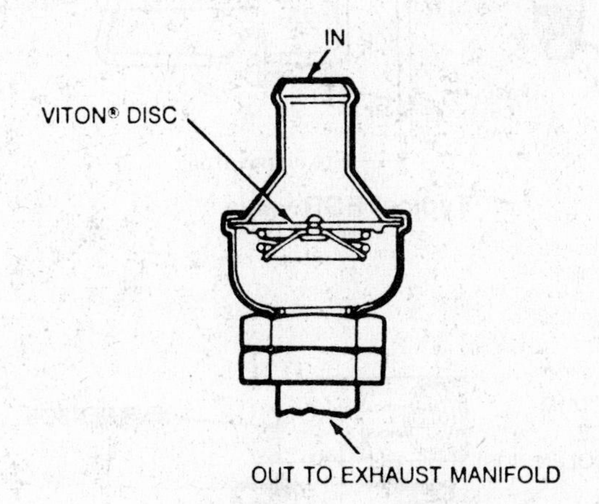

Typical air check valve

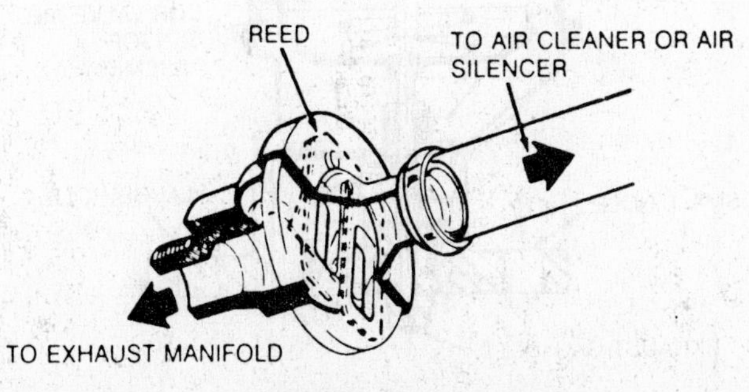

Typical pulse air valve

Combination Air Bypass/Air Control Valve Check

The combination air bypass/air control valve combines the functions of the air bypass and air control valve into a single unit. There are two normally closed valves; the non-bleed and bleed type, both of which look alike. One distinguishing feature will be that the bleed type will have the percent of bleed molded into the plastic case.

1. Disconnect the hoses from outlets **A** and **B**.
2. Disconnect and plug the vacuum line to port **D**.
3. With the engine operating at 1,500 rpm, air flow should be noted coming out of the bypass vents.
4. Reconnect the vacuum line to port **D** and disconnect and plug the vacuum line to port **S**. Make sure vacuum is present in the line to vacuum port **D**.
5. With the engine operating at 1,500 rpm, air flow should be noted coming out of outlet **B** and no air flow should be coming from outlet **A**.
6. With the engine at 1,500 rpm, apply 8–10 in.Hg of vacuum to port **S**. Air should now flow from outlet **A**.
7. If the valve is the bleed type, some lesser amount of air will flow from outlet **A** or **B** and the main discharge will change when vacuum is applied to port **S**.

NOTE: If there is a small air tap attached to the inlet tube from the air pump, air flow should be present during engine operation.

Air Check Valve/Pulse Air Valve Test

1. Inspect all hoses, tubes and the air valve for leaks.
2. Disconnect the hose on the inlet side if the air valve and attempt to blow through the valve. Air should pass freely.
3. Repeat the test, only this time attempt to suck air through the valve. No air should pass.
4. If any other results are obtained, replace the check valve.

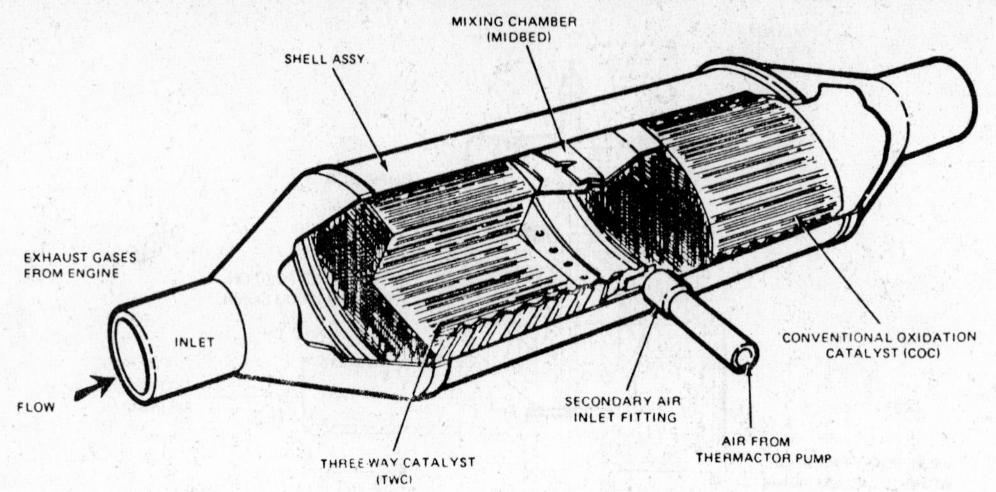

Catalytic converter assembly

Catalytic Converter

The converter is in the exhaust system ahead of the muffler. It contains a catalytic agent made of platinum and palladium, used to oxidize hydrocarbons (HC) and carbon monoxide (CO). The catalyst is expected to function without service of any kind for at least 50,000 miles. Use of leaded fuel would quickly cause catalyst failure; for this reason, a tank filler restriction prevents the entry of service station leaded fuel nozzles.

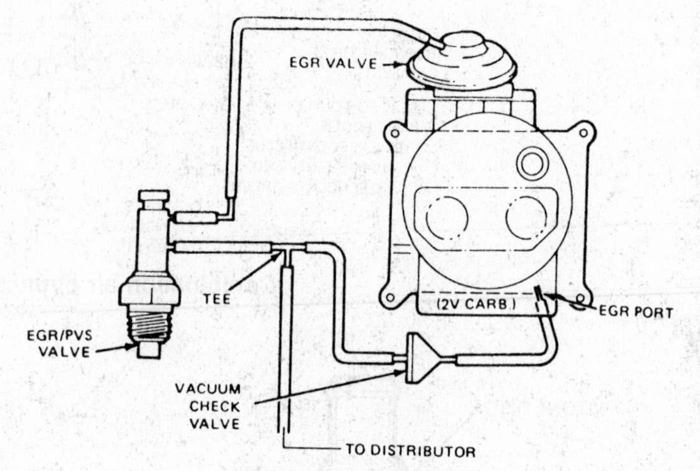

Typical EGR valve

Exhaust Gas Recirculation (EGR) System

The exhaust gas recirculation (EGR) system is designed to reintroduce inert exhaust gas into the combustion chamber, thereby lowering peak combustion temperatures and reducing the formation of Nitrous Oxide (NOx). The amount of exhaust gas recirculated and the timing of the cycle varies by calibration and is controlled by various factors, such as engine speed, engine vacuum, exhaust system backpressure, coolant temperature and throttle angle depending on the calibration. All EGR valves are vacuum actuated, but controlled by the EEC-IV onboard computer. The electronic EGR valve is not serviceable, however the EGR valve position (EVP) sensor and EGR valve can be replaced as individual components.

SYSTEM SERVICE

The EGR valve assembly (including the EVP sensor) should be replaced/serviced every 60,000 miles. Disconnect the vacuum hose, electrical connector and EGR line (if equipped), then remove the mounting bolts and lift off the EGR valve assembly. When replacing the EGR valve, the exhaust gas passages should be cleaned of carbon deposits. Excessive carbon deposits may require the removal of the mounting plate or intake manifold for cleaning. Excessive carbon deposits should not be pushed into the intake manifold where they can be drawn into the combustion chambers when the engine is started.

EGR SUPPLY PASSAGES AND CARBURETOR SPACE CLEANING

Remove the carburetor and carburetor spacer on engines so equipped. Clean the supply tube with a small power-driven rotary type wire brush or blast cleaning equipment. Clean the exhaust gas passages in the spacer using a suitable wire brush

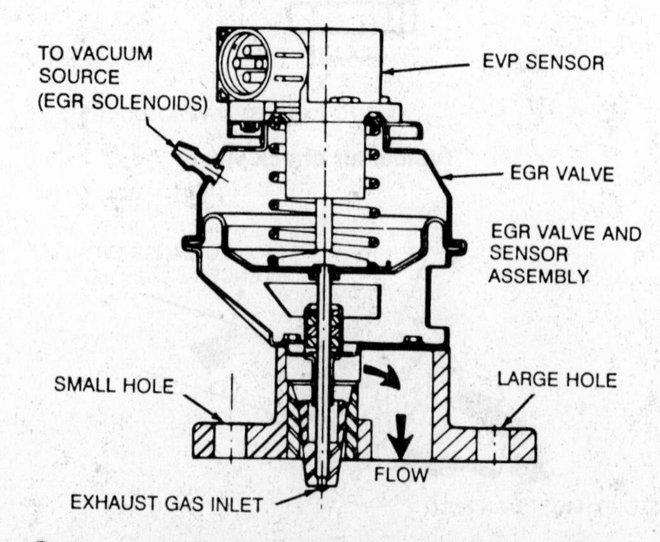

Cross section of the base entry type EGR valve

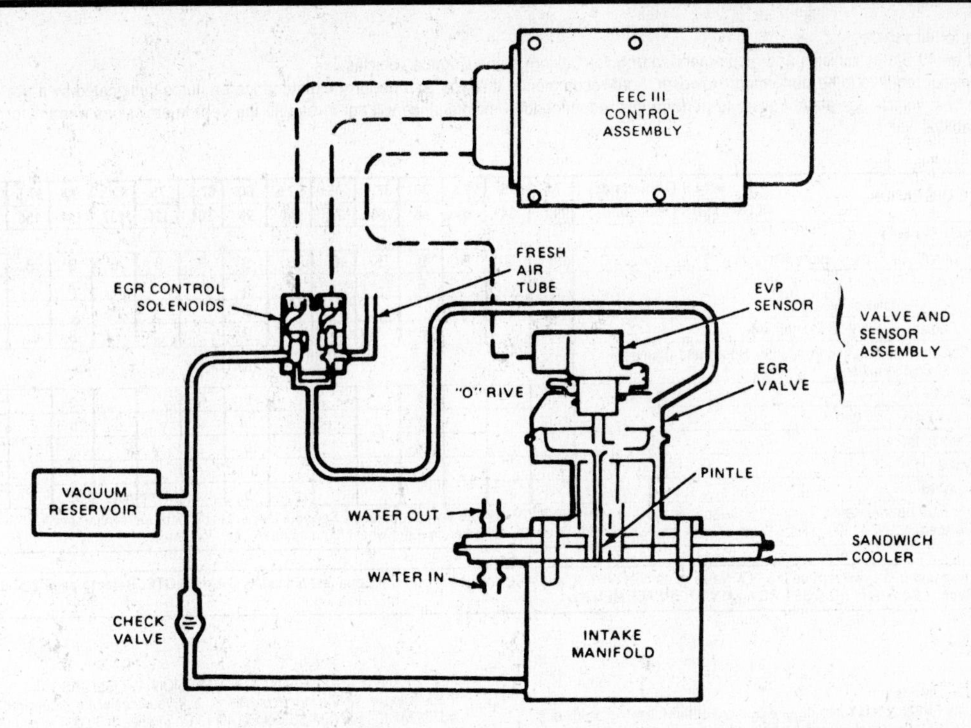

EGR system on vehicles equipped with electronic engine control

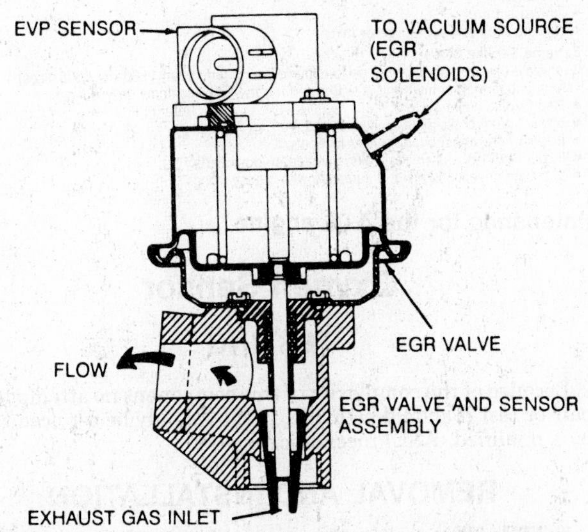

Cross section of the side entry type EGR valve

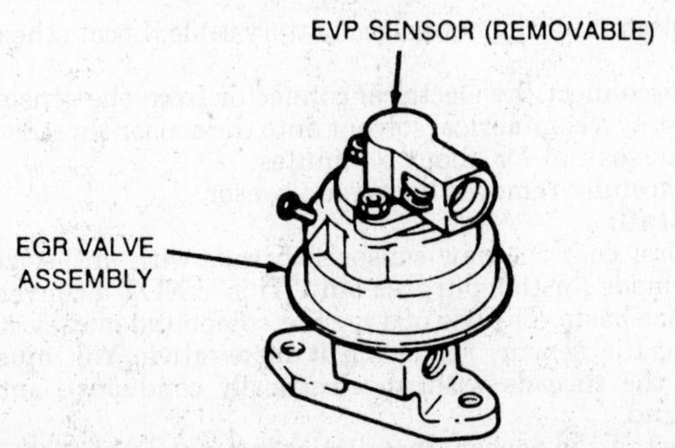

Location of the EGR valve position (EVP) sensor on the EGR valve

and/or scraper. The machined holes in the spacer can be cleaned by using a suitable round wire brush. Hard encrusted material should be probed loose first, then brushed out.

EGR EXHAUST GAS CHANNEL CLEANING

Clean the exhaust has channel, where applicable, in the intake manifold, using a suitable carbon scraper. Clean the exhaust gas entry port in the intake manifold by hand passing a suitable drill bit through the holes to auger out the deposits. Do not use a wire brush. The manifold riser bore(s) should be suitably plugged during the above action to prevent any of the residue from entering the induction system.

Emission Maintenance Warning Light

The emission maintenance warning light system (starting 1985 model year) consists of an instrument panel mounted amber lens (with EGR or EMISS printed on it that is electrically connected to a sensor module located under the instrument panel. The purpose of the system is to alert the driver that emission system maintenance is required. Specific maintenance requirements are listed in the Emission System Scheduled Maintenance Charts. Note for your vehicle use these charts as a guide for recommended service.

The system actually measures accumulated vehicle ignition key on-time and is designed to continuously close an electrical circuit to the amber lens after 2000 hours of vehicle operation. Assuming an average vehicle speed of 30 mph, the 2000 hours equates to approximately 60,000 miles of vehicle operation. Actual vehicle mileage intervals will vary considerably as individual driving habits vary.

Every time the ignition is switched on, the warning light will glow for 2–5 seconds as a bulb check and to verify that the system is operating properly. When approximately 60,000 miles is reached, the warning light will remain on continuously to indicate that service is required. After the required maintenance is performed, the sensor must be reset for another 60,000 mile pe-

B — Required for all vehicles.
b — Required for 49 States vehicles and recommended only for California and Canada vehicles.
(b) — This item not required to be performed. However, Ford recommends that you also perform maintenance on items designated by a "(b)" in order to achieve best vehicle operation. Failure to perform this recommended maintenance will not invalidate the vehicle emissions warranty or manufacturer recall liability.

MAINTENANCE OPERATION — MILES (Thousands)	7.5	15	22.5	30	37.5	45	52.5	60	67.5	75	82.5	90	97.5	105	112.5	120
KILOMETERS (Thousands)	12	24	36	48	60	72	84	96	108	120	132	144	156	168	181	193
Emission Control Systems																
Change Engine Oil and Oil Filter — every 6 months OR	B	B	B	B	B	B	B	B	B	B	B	B	B	B	B	b
Replace Spark Plugs — Standard				B				B				B				b
— Platinum Type 3.0L								B								b
Replace Engine Coolant — every 36 months OR				B				B				B				b
Check Engine Coolant Condition and Protection, Hoses and Clamps Annually — Prior to Cold Weather	ANNUALLY															
Replace Air Cleaner Filter				B				b				b				b
Replace Crankcase Emission Filter				B				b				b				b
*Check Clean Throttle Body								(b)								(b)
Replace PCV Valve (1)								b/1								b
Replace Ignition Wires								b								b

* Wheel lug nuts must be retightened to proper torque specifications at 500 miles/800 km of new vehicle operation. See your Owner Guide for proper torque specification. Also retighten to proper torque specification at 500 miles/800 km after (1) any wheel change or (2) any other time the wheel lug nuts have been loosened.

/1 At 60,000 miles your dealer will replace the PCV Valve at no cost on 3.0L and 4.0L engines except California and Canada vehicles. **NOTE: Refer to page 2 of the Maintenance Schedule Record Book for "NO COST PCV VALVE REPLACEMENT".**

NOTES:
Unique Driving Conditions
If your driving habits **FREQUENTLY** include one or more of the following conditions:
- Operating when outside temperatures remain **below freezing** and most trips are less than 16 km (10 miles).
- Operating during **HOT WEATHER** in stop-and-go "rush hour" traffic.
- Towing a trailer, using a camper or roof-top carrier, or carrying maximum loads.
- Operating in severe dust conditions.
- Extensive idling, such as police, taxi or door-to-door delivery use.
- High speed operation with a fully loaded vehicle (MAX. GVW).

Change ENGINE OIL and OIL FILTER every 3 months or 3,000 miles (4 800 km) whichever occurs first.
Check/Regap SPARK PLUGS every 30,000 miles (48 000 km).

AIR CLEANER AND CRANKCASE EMISSION AIR FILTERS.
If operating in severe dust conditions, ask your dealer for proper replacement intervals.
AUTOMATIC TRANSMISSION FLUID — Change each 30,000 miles (48 000 km) — if your driving habits FREQUENTLY include one or more of the following conditions.
- Operating during hot weather (above 90°F, 32°C) and carrying heavy loads and driving hilly terrain.
- Towing a trailer.
- Door-to-door delivery, police or taxi.

Extreme Service Items
If vehicle is operated off-highway, perform the following items every 1 600 km (1,000 miles). If vehicle is operated in mud and/or water perform the following items daily:
- Inspect disc brake system.
- Inspect drum brake system, hoses, and lines.
- Inspect front wheel bearings and lubrication.
- Inspect exhaust system for leaks, damage or loose parts.
- Lubricate driveshaft U-joint if equipped with grease fittings.

Emission system scheduled maintenance for the 4.0L engine

riod. The sensor module is located above the right front corner of the glove box assembly.

EMISSION WARNING LIGHT RESET PROCEDURE

1. Make sure the ignition key is OFF.
2. Locate the sensor (above the right front corner of the glove box), and lightly push a phillips screwdriver or small rod tool through the 0.2 in. (5mm) diameter hole with the sticker labeled "RESET" and lightly press down and hold.
3. While lightly holding the screwdriver or tool down, turn the ignition switch to the RUN position. The emission warning light will then light and should remain on for as long as the screwdriver is held down. Hold the screwdriver down for approximately 5 seconds.
4. Remove the screwdriver or tool. The lamp should go out within 2–5 seconds, indicating that a reset has occurred. If the light remains on, begin again at Step 1. If the light goes out, turn the ignition off and go to the next Step.
5. Turn the ignition to the RUN position. The warning light should illuminate for 2–5 seconds and then go out. This verifies that a proper reset of the module has been accomplished. If the light remains on, repeat the reset procedure.

NOTE: Some models (Non-EEC 2.0L engine) use a non-resettable control unit. When reset has occurred, replace it with a resettable type if available.

Oxygen Sensor

TESTING

Because of the complexity of this component no attempt to repair or test it should be made. It should only be serviced/tested by a qualified (ASE) mechanic.

REMOVAL AND INSTALLATION

NOTE: This service is a general procedure modify service steps as necessary.

1. Disconnect the newgative battery cable. Locate the oxygen sensor.
2. Disconnect the electrical connector from the sensor.
3. Spray a commerical solvent onto the sensor threads and allow it to soak in for about 5 minutes.
4. Carefully remove the oxygen sensor.

To install:

5. First coat the new sensor's threads with anti-seize compound made for this purpose only. This is NOT a conventional anti-seize paste. The use of a regular compound may electrically insulate the sensor, rendering it inoperative. You must coat ONLY the threads with an electrically conductive anti-seize compound.
6. Install the sensor (installation torque is about 30 ft. lbs. most vehicles) reconnect the electrical connector. Be careful not to damage the electrical connector.

2.3L EFI, 3.0L EFI and 2.8L Engines Emission System Scheduled Maintenance

All items designated with a B code are required to be performed in all states and Canada. (B) coded items are required for Canada and all states except California, and recommended only for California: b coded items are required in all states except California, and recommended only for California and Canada vehicles: (B) coded items are recommended for all vehicles. However, Ford recommends that you perform maintenance on all designated items to achieve best vehicle operation.

NORMAL DRIVING SERVICE INTERVALS
Perform at the months or distances shown, whichever comes first.

MILES (Thousands)	7.5	15	22.5	30	37.5	45	52.5	60	67.5	75	82.5	90	97.5	105	112.5	120
MAINTENANCE OPERATION KILOMETERS (Thousands)	12	24	36	48	60	72	84	96	108	120	132	144	156	168	181	193
Emission Control Systems																
Change engine oil — every 12 months OR	B	B	B	B	B	B	b	b	b	b	b	b	b	b	b	b
Change engine oil filter — every 12 months OR	B	B	B	B	B	B	b	b	b	b	b	b	b	b	b	b
Replace spark plugs				B				b				b				b
Replace engine coolant — every 36 months OR				B				b				b				b
Check engine coolant condition & protection, hoses and clamps annually — prior to cold weather	ANNUALLY															
Inspect drive belt condition and tension				B				b				b				b
Replace air cleaner filter				B				b				b				b
Replace crankcase emission filter — if equipped				B				b				b				b
Inspect and clean injector tips — (2.3L EFI)								(b)								b
Replace PCV valves								b								b
Replace ignition wires								b								b
Check thermactor hoses and clamps								b								b
Clean choke linkages and external controls and inspect function of carburetor "hang on" devices — (2.8L only)				B				b				b				b
Replace EGR valve assembly (including EVP sensor on electronic EGR valve)*								b								b
Replace EGR vacuum solenoid(s) and filter (2.3L and 2.8L)								b								b
Replace EGO/HEGO sensor*								b								b
Check engine valve clearance (2.8L)	B			B				b				b				b

*This vehicle may be equipped with an Emissions Maintenance Warning Light. If so equipped, these parts are to be replaced either at 60,000 miles or when the Emissions Maintenance Warning Light remains on continuously with the key in the "On" position, whichever occurs first.

NOTES:

Unique Driving Conditions

If your driving habits **FREQUENTLY** include:

- Operating when outside temperatures remain **below freezing** and most trips are less than 8 km (5 miles).
- Operating during **HOT WEATHER** (above +90°F or +32°C) and
  - — Driving continuously in excess of normal highway speeds;
  - — Driving in stop-and-go "rush hour" traffic.
- Towing a trailer, using a camper or car-top carrier, or carrying maximum loads.
- Operating in severe dust conditions.
- Extensive idling, such as police, taxi or door-to-door delivery use.
- High speed operation with a fully loaded vehicle.

Change ENGINE OIL and OIL FILTER every 3 months or 4 800 km (3,000 miles) whichever occurs first.

Check/Regap SPARK PLUGS every 9 600 km (6,000 miles).

AIR CLEANER and CRANKCASE EMISSION AIR FILTERS — If operating in severe dust conditions.

Replace EGR SOLENOID FILTER(S) at 48 000 km (30,000 miles) and 144 000 km (90,000 miles) if operating in severe dust conditions. (2.3L and 2.8L)

Extreme Service Items

If vehicle is operated off-highway, perform the following items every 1 600 km (1,000 miles). If vehicle is operated in mud and or water perform the following items daily:

- Inspect disc brake system.
- Inspect front wheel bearings and lubrication.
- Inspect exhaust system for leaks, damage or loose parts.

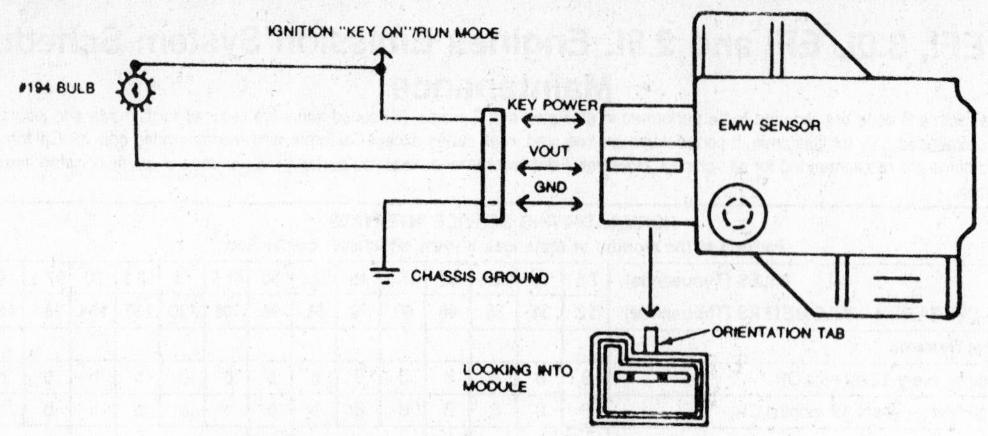

Ford emission maintenance light system schematic

ELECTRONIC ENGINE CONTROLS

MCU System

The MCU system is used on the 2.0L and 2.3L carbureted in-line 4–cylinder engines. The heart of the system is the fuel control system. This is necessary to keep the air-fuel ratio at proper chemical balance (14.7) to obtain maximum catalyst efficiency. The fuel control **Loop** consists of an Exhaust Gas Oxygen (EGO) sensor, Micro-processor Control Unit (MCU), and Fuel Control Solenoid.

The EGO sensor senses whether the exhaust gas is rich or lean of proper chemical balance. This signal is sent to the MCU module, which sends a varying signal to the fuel control solenoid to move the air-fuel ratio back to the proper chemical balance. The operation is called **closed loop** operation.

The other mode of operation is called **open loop.** In this mode, the MCU module sends out a fixed signal to the fuel control solenoid. During this time the input from the EGO sensor is ignored, thus opening the loop.

The determining factor when the system goes into open or closed loop is based upon information from the switch inputs, which sense coolant temperature, manifold vacuum, and throttle position. Generally, the vehicle will be in closed loop when the vehicle is at operating temperature and at a steady part throttle cruise.

Other functions controlled by the MCU module, by means of vacuum solenoids, are the Thermactor Air Bypass (TAB) valve and the Thermactor Air Diverter (TAD) valve. Also controlled, but not on all calibrations, are Canister Purge and the Spark Retard Solenoid.

NOTE: Because of the complexity of this system no attempt to repair it should be made. It should only be serviced by a qualified (ASE) mechanic.

EEC-IV System

This system is a controlling system which serves to improve emission control, fuel economy, driveability, and engine performance. This is achieved by the means of an on-board Electronic Control Assembly which reads the inputs from various sensors. The Electronic Control Assembly makes computations based on these inputs and then sends controlling outputs to various engine components in order to provide the optimum air/fuel ratio.

Engine Controls

The EEC-IV system is similar to other Ford engine control systems in that the center of the system is a microprocessor called an electronic control assembly. This assembly receives data from the sensors, switches, relays and other electronic components, then issues commands to control engine functions. The electronic control assembly is calibrated to optimize emissions, fuel economy and driveability. The electronic control assembly in the EEC-IV system is a microprocessor like other EEC systems, but the calibration modules are located within the electronic control assembly instead of being attached to the outside, as in previous models. The harness connectors are edge type connectors which provide a more positive connection and allow probing from the rear while connected. The electronic control assembly is usually mounted in the passenger compartment under the front section of the center console. The EEC-IV system does not control the pulse air injection or the upshift lamp. The system does control the fuel injectors for air fuel mixture, spark timing, deceleration fuel cut-off, EGR function, curb and fast idle speed, evaporative emission purge, air condition cut-off during wide open throttle, cold engine start and enrichment, electric fuel pump and self test engine diagnostics.

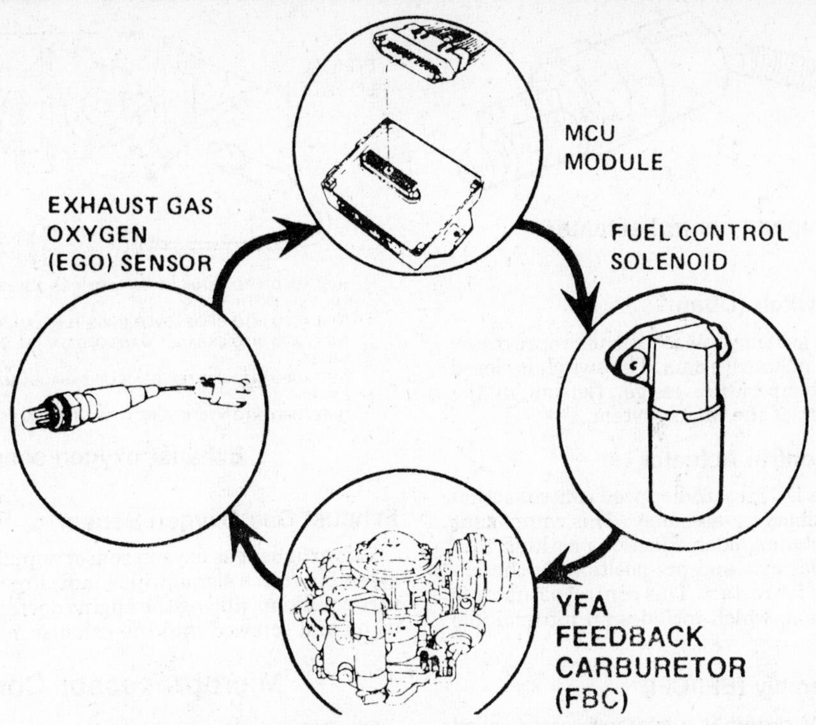

2.0L and 2.3L carbureted engines, fuel control loop — MCU system

SYSTEM COMPONENTS

Air Bypass Solenoid Valve

The air bypass solenoid is used to control engine idle speed. This component is operated by the electronic control module. The function of the valve is to allow air to pass around the throttle plates in order to control cold engine fast idle, no touch start, dashpot operation, over temperature idle boost and engine idle load correction. The electrical signal to the solenoid should be one volt or less, with all accessories off. Applying twelve volts to the solenoid will cause a neutral idle speed change greater than 1,000 rpm.

AIR CHARGE TEMPERATURE SENSOR

The air charge temperature sensor provides the electronic fuel injection system with the correct air/fuel mixture information. This sensor is used as both a density corrector to calculate air flow and to proportion cold enrichment fuel flow. This sensor is similar in construction to the engine coolant temperature sensor. The sensor is located in the engine cylinder runner of the intake manifold or attached to the air cleaner.

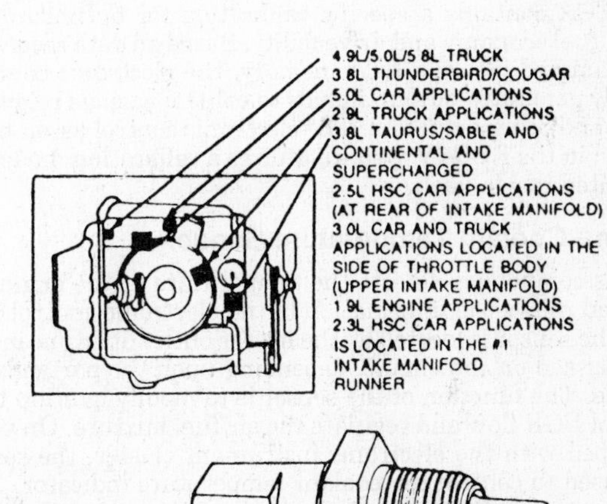

Air charge temperature sensor

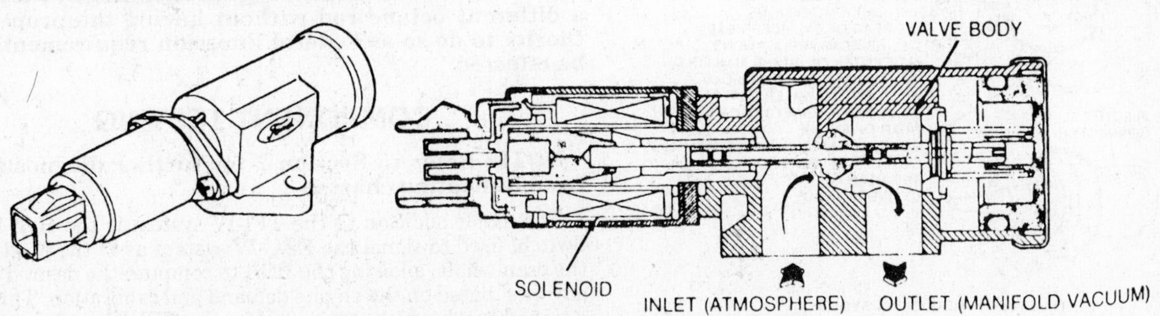

Idle air bypass valve — EFI engines

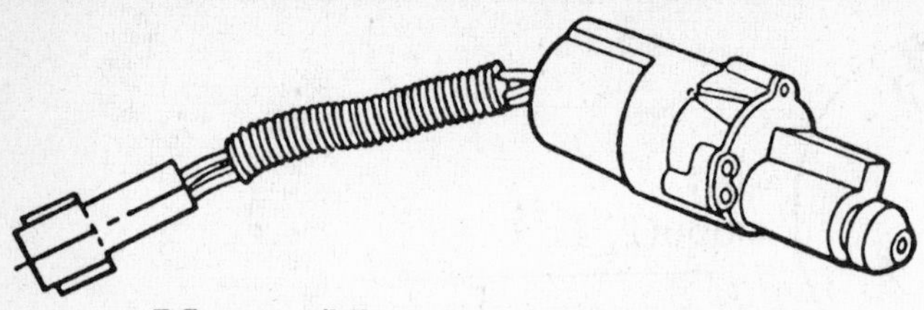

DC motor/idle speed control actuator

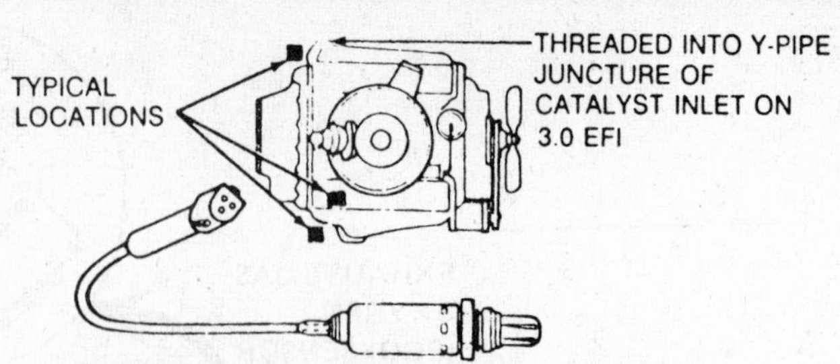

Exhaust oxygen sensor

Coolant Temperature Switch (Dual)

The coolant temperature switch provides the microprocessor control unit with coolant temperature data. The switch is closed in the normal operating temperature range. Defects in the switch are diagnosed as part of the MCU system.

DC Motor/Idle Speed Control Actuator

On vehicles equipped, the DC motor/idle speed control actuator is mounted to the fuel charging assembly. This component controls the idle speed, including such functions as high cam rpm, anti-diesel shut off, dashpot and pre-positioning the engine fuel charging assembly for restart. This control actuator is driven by the EEC-IV system, which includes an integral idle tracking switch.

Electronic Control Assembly (EFI/CFI)

The center of the EEC-IV system is a microprocessor called the electronic control assembly (ECA). This component receives data from a number of sensors and other electronic components. The ECA contains a specific calibration for optimizing emissions, fuel economy and driveability. Based on data received and programmed into the ECA memory, the electronic control assembly generates output signals to control various relays, solenoids and other actuators. The electronic control assembly that is used in the EEC-IV system, utilizes a calibration module that is located inside itself.

Engine Coolant Temperature Sensor

This component detects the temperature of the engine coolant and relays the information to the electronic control assembly. The sensor is located by the heater outlet fitting or in a cooling passage on the engine, depending upon the particular type vehicle. The function of the sensor is to modify ignition timing, control EGR flow and regulate the air/fuel mixture. On vehicles equipped with the electronic instrument cluster, the sensor is also used to control the coolant temperature indicator.

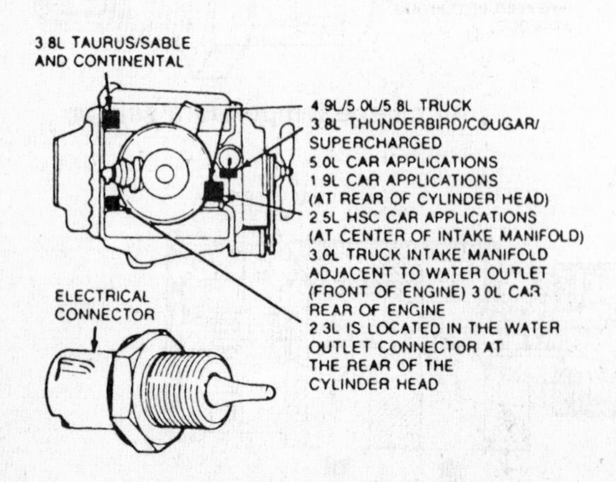

Engine coolant temperature switch

Exhaust Gas Oxygen Sensor

The exhaust gas oxygen sensor supplies the electronic control assembly with a signal which indicates either a rich or lean mixture condition, during the engine operation. This sensor is located on and screwed into the exhaust manifold.

Microprocessor Control Unit

The Microprocessor Control Unit (MCU) is a microprocessor based module that is programmed to interface with various types of sensors, switches and actuators and to perform the engine control functions. The control unit is a major component in the complete MCU system. This system has the capability to diagnose a malfunction within its own system.

Thick Film Integrated (TFI) System

GENERAL

The Thick Film Integrated (TFI) ignition system is used for all EEC-IV/EFI models. The TFI ignition system uses a new style distributor, called the Universal distributor and is equipped with no centrifugal and vacuum advance units. It has a distributor base mounted TFI ignition module, which is self-contained in a moulded thermo-plastic unit. The TFI-IV system module can be identified by having 6 pins. The system uses an E-Core ignition coil, named after the shape of the laminations making up the core.

The Universal Distributor used with the TFI-IV system, contains a provision to change the basic distributor calibration with the use of a replaceable "Octane" rod, from the standard of zero degrees to either three or six degree retard rods. No other calibration changes are possible.

NOTE: Do not change the ignition timing by the use of a different octane rod without having the proper authority to do so as Federal Emission requirements will be effected.

COMPONENT TESTING

NOTE: Refer to Section 2 for further diagnosis and troubleshooting charts.

The major addition to the TFI-IV system is the Hall Effect Switch, used to signal the EEC-IV system as to the position of the crankshaft, allowing the ECA to compute the desired spark advance, based on the engine demand and calibration. This conditioned spark advance then pulses the TFI-IV module to turn the ignition coil current OFF and ON, generating the high volt-

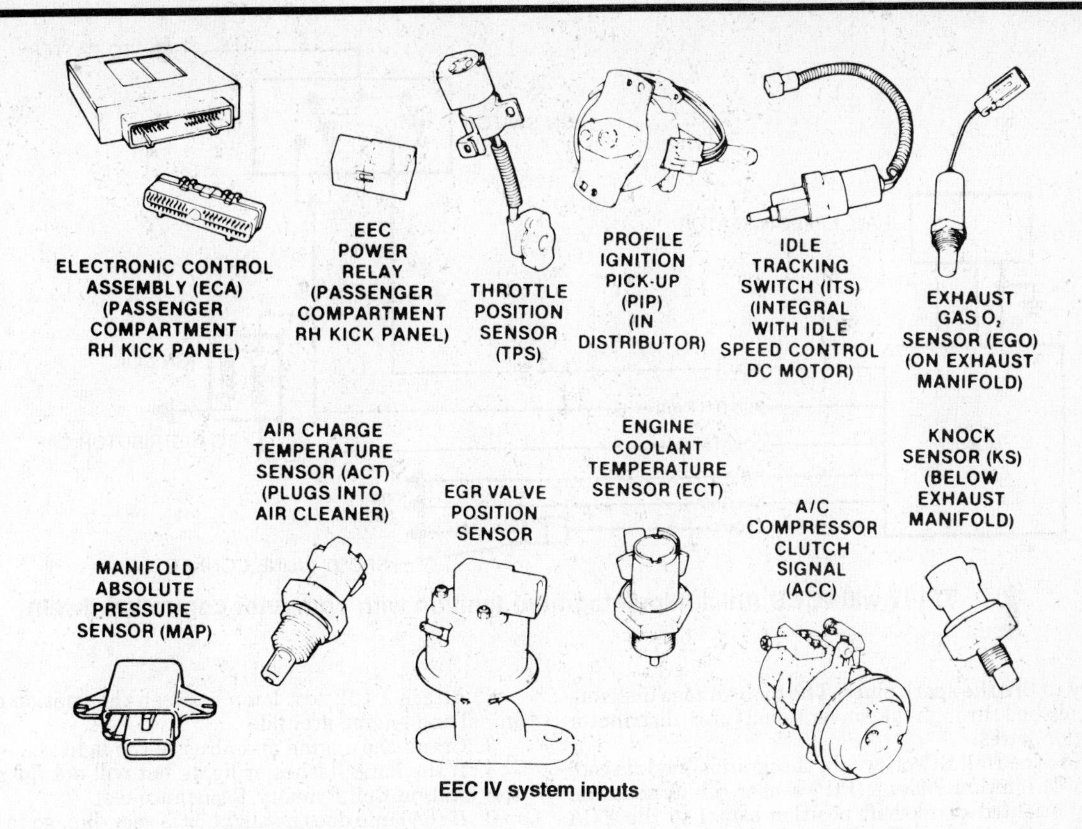

EEC IV system inputs

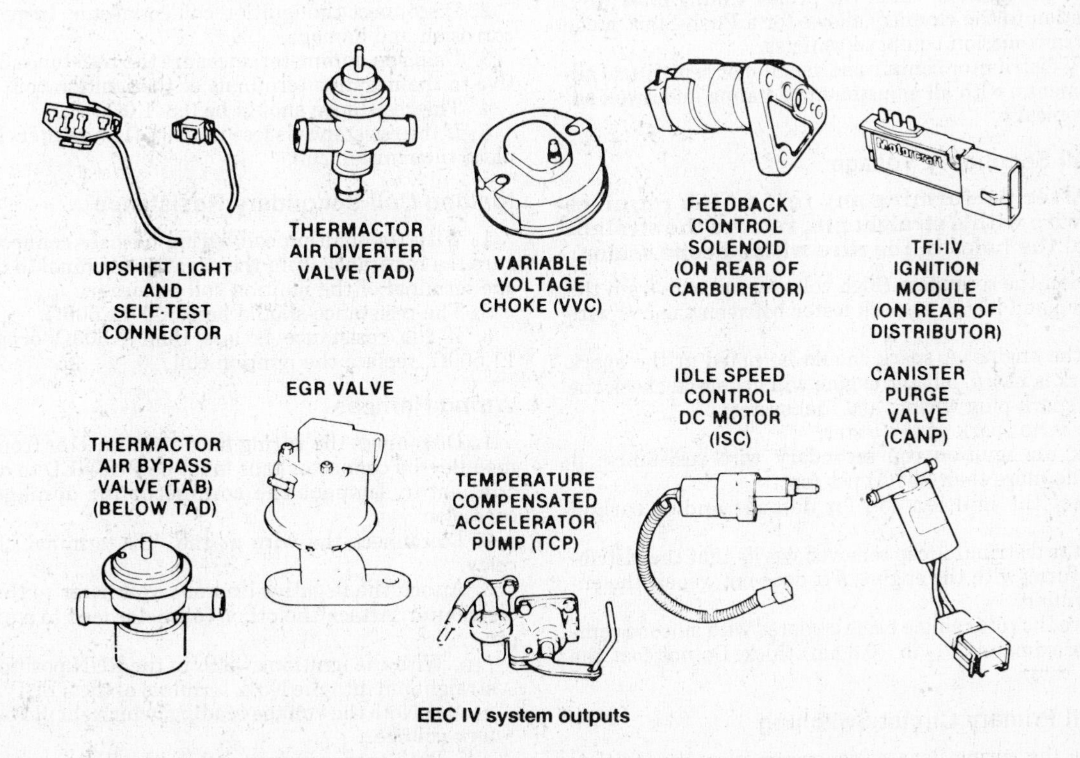

EEC IV system outputs

EEC-IV system components

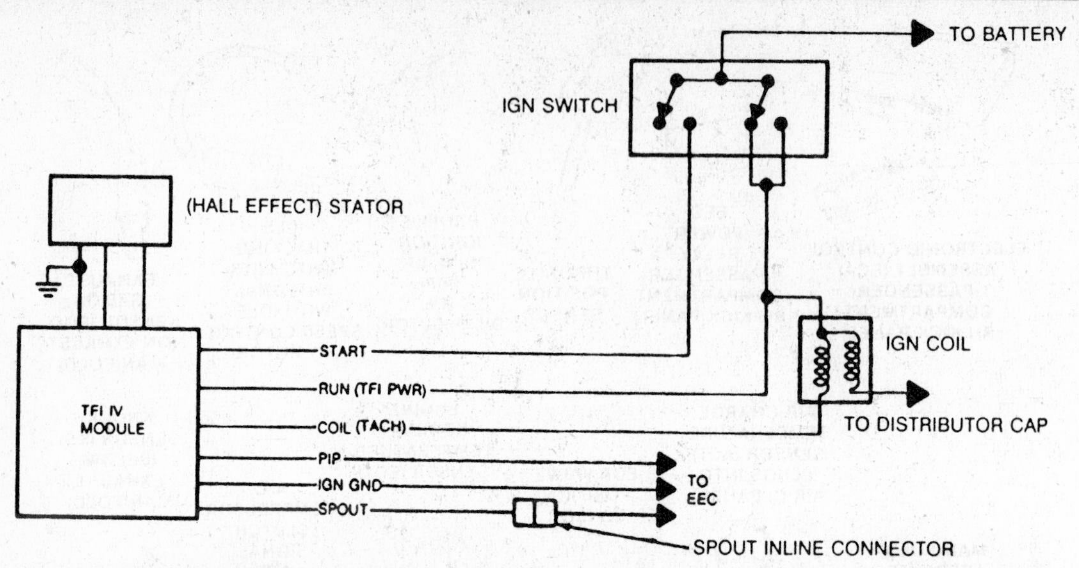

TFI-IV with CCD (thick film integrated ignition with computer controlled dwell)

age necessary to fire the spark plugs. The high voltage distribution is accomplished through the conventional rotor, distributor cap and ignition wires.

In addition to the Hall Effect switch, the ignition module contains the Profile Ignition Pick-up (PIP) sensor, which sends an electronically oriented crankshaft position signal to the ECA and the TFI module circuitry. The ECA, after taking all the sensors information, produces a new signal called the Spout. This Spout signal is then sent back to the TFI module for comparison with the PIP signal. The TFI-IV module then uses both of these signals to fire the ignition coil at the proper timing interval.

A modification to the circuitry allows for a Push-Start mode for manual transmission equipped vehicles.

The TFI-IV distributor contains no vacuum or centrifugal advance components, with all adjustment of timing and dwell adjusted electronically.

Ignition Coil Secondary Voltage

NOTE: After performing any test which requires piercing a wire with a straight pin, remove the straight pin and seal the holes in the wire with silicone sealer.

1. Disconnect the secondary (high voltage) coil wire from the distributor cap and install a spark tester between the coil wire and ground.
2. Crank the engine. A spark should be noted at the spark tester. If spark is noted, but the engine will not start, check the spark plugs, spark plug wiring, and fuel system.
3. If there is no spark at the tester:
   a. Check the ignition coil secondary wire resistance; it should be no more than 7000Ω per foot.
   b. Inspect the ignition coil for damage and/or carbon tracking.
   c. With the distributor cap removed, verify that the distributor shaft turns with the engine; if it does not, repair the engine as required.
   d. Be sure the rotor single blade is coated with silicone compound, approximately $\frac{1}{32}$ in. (0.8mm) thick. Do not coat the multipoint rotor.

Ignition Coil Primary Circuit Switching

1. Separate the wiring harness connector from the ignition module. Inspect for dirt, corrosion and damage. Re-connect the harness connector.
2. Attach a 12V test lamp between the ignition coil tach terminal and engine ground.
3. Crank the engine and observe the light.
4. If the lamp flashes or lights but will not flash, proceed to the Ignition Coil Primary Resistance test.
5. If the lamp does not light or is very dim, go to the Primary Circuit Continuity test.

Ignition Coil Primary Resistance

1. Have the ignition switch in the OFF position.
2. Disconnect the ignition coil connector. Inspect it for dirt, corrosion and damage.
3. Using an ohmmeter, measure the resistance from the positive to the negative terminals of the ignition coil.
4. The resistance should be 0.3–1.0Ω.
5. If the resistance is less than 0.3Ω or greater than 1.0Ω, replace the ignition coil.

Ignition Coil Secondary Resistance

1. With the ignition coil wiring harness connector off, measure the resistance from the negative terminal to the high voltage terminal of the ignition coil connector.
2. The resistance should be 6,500–11,500Ω.
3. If the resistance is less than 6,500Ω or greater than 11,500Ω, replace the ignition coil.

Wiring Harness

1. Disconnect the wiring harness connector from the TFI-IV module; the connector tabs must be PUSHED to disengage the connector. Inspect the connector for damage, dirt, and corrosion.
2. Disconnect the wire at the "S" terminal of the starter relay.
3. Attach the negative lead of a voltmeter to the base of the distributor. Attach the other voltmeter lead to a small straight pin.
   a. With the ignition switch in the RUN position, insert the straight pin into the No. 2 terminal of the TFI-IV module connector. Note the voltage reading, which should be 90% of battery voltage.
   b. With the ignition switch in the RUN position, move the straight pin to the No. 3 connector terminal. Again, note the voltage reading, which should be 90% of battery voltage.

c. Move the straight pin to the No. 4 connector terminal, then turn the ignition switch to the START position. Note the voltage reading, which should be 90% of battery voltage.

4. If any reading is less than 90% of the battery voltage, inspect the wiring, connectors, and/or ignition switch for defects.

5. Turn the ignition switch to the OFF position. Remove the straight pin and reconnect the wire at the starter relay.

Stator

1. Turn the ignition switch to the OFF position.
2. Remove the coil wire and ground it.
3. Attach the negative voltmeter lead to the distributor base.
4. Disconnect the pin-in-line connector near the distributor and attach the positive voltmeter lead to the TFI-IV module side of the connector.
5. Turn the ignition switch to the ON position.
6. "Bump" the starter with the ignition switch and measure the voltage levels with the engine not operating. Record all measurements.
7. If the highest value is less than 90% of battery voltage, replace the stator assembly.
8. If the lowest value is greater than 0.5 volts, remove the distributor from the engine, remove the TFI-IV module from the distributor and inspect the stator connector terminals and the TFI terminals for misalignment. If OK, replace the stator.
9. If the values are between 0.5 volts and 90% of battery voltage, replace the stator assembly.
10. If there are no values between 0.5 volts and 90% of battery voltage, connect a spark tester between the ignition coil wire and the engine ground.
11. Crank the engine. If a spark occurs, check the PIP and ignition ground wires for continuity. Repair as required. If no fault is found, refer to EEC-IV diagnostics.
12. If no spark occurs, replace the TFI-IV module.

Primary Circuit Continuity

1. Separate the wiring harness connector from the ignition module. Inspect for dirt, corrosion and damage.
2. Attach the negative voltmeter lead to the distributor base.
3. Measure the battery voltage.
4. Attach the voltmeter lead to a straight pin and insert the pin into the connector terminal No. 2.
5. Turn the ignition switch to the RUN position and measure the voltage at terminal No. 2.
6. The voltage should be no less than 90% of battery voltage.
7. If less than 90% of battery voltage, perform the Ignition Coil Primary Voltage test.
8. Turn the ignition switch to the OFF position and remove the pin.

Ignition Coil Primary Voltage

1. Attach the negative voltmeter lead to the distributor base.
2. Measure the battery voltage.
3. Turn the ignition switch to the RUN position.
4. Measure the voltage at the negative terminal of the ignition coil.
5. If the voltage is 90% of battery voltage, inspect the wiring harness between the ignition module and the coil negative terminal.
6. If the voltage is less the 90% of battery voltage, inspect the wiring harness between the ignition module and the coil negative terminal.
7. Turn the ignition switch to the OFF position.

Ignition Coil Supply Voltage

1. Remove the coil connector.
2. Attach the negative voltmeter lead to the distributor base.
3. Measure the battery voltage.
4. Turn the ignition switch to the RUN psoition.
5. Measure the voltage at the positive terminal of the ignition coil.
6. The voltage should be 90% of battery voltage. Inspect the ignition coil and connector for dirt, corrosion and damage. If required, replace the ignition coil.
7. If the voltage is less than 90% of battery voltage, inspect and repair, as required, the wiring between the ignition coil and the ignition switch. Check for a worn or damaged ignition switch.
8. Turn the ignition switch to the OFF position. Reconnect the ignition module connector.

FORD MOTOR COMPANY DIAGNOSTIC TROUBLE CODES

Trouble Codes	Quick Test Mode	Car Engines														Truck Engines								
		1.9L EFI	1.9L CFI	2.3L OHC EFI	2.3L EFI Turbo	2.3L HSC EFI	2.5L CLC CFI	2.5L MTX CFI	3.0L EFI	3.0L SHO SEFI	3.8L SEFI AXOD	3.8L SEFI RWD	3.8L SEFI SC	5.0L SEFI	5.0L MA SEFI	2.3L TPH EFI	2.9L EFI	3.0L EFI	4.9L EFI	5.0L EFI	5.8L EFI	5.8L E40D EFI	7.5L EFI	7.5L E40D EFI
11—System pass	O/R/C	X	X	X	X	X	X	X	X	X	X	X	X	X	X	X	X	X	X	X	X	X	X	X
12—Rpm unable to reach upper test limit	R	X	X	X	X	X	X	X	X	X	X	X	X	X	X	X	X	X	X	X	X	X	X	X
13—Rpm unable to reach lower test limit	R	X	X	X	X	X	X	X	X	X	X	X	X	X	X	X	X	X	X	X	X	X	X	X
13—DC motor did not move	O	—	—	—	—	—	X	X	—	—	—	—	—	—	—	—	—	—	—	—	—	—	—	—
13—DC motor did follow dashpot	O	—	X	—	—	—	X	X	—	—	—	—	—	—	—	—	—	—	—	—	—	—	—	—
14—PIP circuit failure	C	X	X	X	X	X	X	X	X	X	X	X	X	X	X	X	X	X	X	X	X	X	X	X
15—ROM test failure	O	X	X	X	X	X	X	X	X	X	X	X	X	X	X	X	X	X	X	X	X	X	X	X
15—Power interrupted to keep alive memory	C	X	X	X	X	X	X	X	X	X	X	X	X	X	X	X	X	X	X	X	X	X	X	X
16—Rpm above self-test limit, set too high	R	(1)	—	—	—	—	X	X	—	—	—	—	—	—	—	—	—	—	—	—	—	—	—	—
16—Rpm too low to perform test	R	—	—	X	—	—	—	—	—	—	—	—	—	X	—	—	—	—	—	—	—	—	—	—
17—Rpm below self-test limit, set too low	R	(1)	X	—	—	—	X	X	—	—	—	—	—	—	—	—	—	—	—	—	—	—	—	—
18—Loss of tach input to ECU, spout grounded	C	X	X	X	X	X	X	X	X	X	X	X	X	X	X	(14)	X	X	X	X	X	X	X	X
18—Spout circuit open	R	X	X	—	—	X	X	X	X	X	X	X	X	X	X	X	X	X	X	X	X	X	X	X
19—Erratic rpm during test or rpm too low	R	(1)	(7)	—	—	—	—	—	—	—	—	—	—	—	—	—	—	—	—	—	—	—	—	—
19—Failure of EEC power supply	O	—	—	X	—	X	—	—	X	—	X	X	—	X	X	X	X	X	X	X	X	X	X	X
19—CID sensor input failed	C	—	—	—	—	—	—	—	—	X	—	—	X	—	—	—	—	—	—	—	—	—	—	—
21—ECT sensor input out of test range	O/R	X	X	X	X	X	X	X	X	X	X	X	X	X	X	X	X	X	X	X	X	X	X	X
22—BP sensor input out of test range	O/R/C	X	—	—	X	—	—	—	—	(7)	—	—	(7)	—	(7)	—	—	—	—	—	—	—	—	—
22—MAP sensor input out of test range	O/R/C	—	X	X	—	X	X	X	X	—	X	X	—	X	—	X	X	X	X	X	X	X	X	X
23—TP sensor input out of test range	O/R	X	(3)	X	X	X	(3)	(3)	X	X	X	X	X	X	X	X	X	X	X	X	X	X	X	X
24—ACT sensor input out of test range	O/R	—	X	X	—	X	X	X	X	X	X	X	X	X	X	X	X	X	X	X	X	X	X	X
24—VAT sensor input out of test range	O/R	—	—	—	X	—	—	—	—	—	—	—	—	—	—	—	—	—	—	—	—	—	—	—
25—KS sensor not detected during test	R	—	—	X	X	—	—	—	X	X	—	—	X	—	—	—	—	—	X	X	—	—	—	—
26—VAF sensor input out of test range	O/R	X	—	—	X	—	—	—	—	—	—	—	—	—	—	—	—	—	—	—	—	—	—	—
26—MAF sensor input out of test range	O/R	—	—	—	—	—	—	—	—	X	—	—	X	—	X	—	—	—	—	—	—	—	—	—
26—TOT sensor input out of test range	O/R	—	—	—	—	—	—	—	—	—	—	—	—	—	—	—	—	—	—	—	—	X	—	X
28—VAT sensor input out of test range	O/R	X	—	—	—	—	—	—	—	—	—	—	—	—	—	—	—	—	—	—	—	—	—	—
28—Loss of primary tach, right side	C	—	—	—	—	—	—	—	—	—	—	—	—	—	—	X	—	—	—	—	—	—	—	—
29—Insufficient input from vehicle speed sensor	C	—	—	—	—	X	X	X	X	X	X	X	X	X	X	X	X	X	X	X	X	X	—	X
31—PFE circuit below minimum voltage	O/R/C	—	X	—	—	X	—	—	X	X	X	X	X	—	—	—	—	—	—	—	—	—	—	—
31—EVP circuit below minimum voltage	O/R/C	—	—	X	—	—	X	X	—	—	—	—	—	X	X	X	—	—	X	X	X	X	X	X
32—EGR valve not seated	R/C	—	X	(6)	—	X	—	—	X	X	X	X	X	—	—	—	—	—	—	—	—	—	—	—
32—EVP voltage below closed limit	O/R/C	—	—	—	—	—	X	X	—	—	—	—	—	X	X	X	—	—	X	X	X	X	X	X
33—EGR valve not opening	R/C	—	X	—	—	X	X	X	X	X	X	X	X	X	X	X	—	—	X	X	X	X	X	X

Ford Motor Company diagnostic trouble codes

Trouble Codes	Quick Test Mode	Car Engines														Truck Engines								
		1.9L EFI	1.9L CFI	2.3L OHC EFI	2.3L EFI Turbo	2.3L HSC EFI	2.5L CLC CFI	2.5L MTX CFI	3.0L EFI	3.0L SHO SEFI	3.8L SEFI AXOD	3.8L SEFI RWD	3.8L SEFI SC	5.0L SEFI	5.0L MA SEFI	2.3L TPH EFI	2.9L EFI	3.0L EFI	4.9L EFI	5.0L EFI	5.8L EFI	5.8L E40D EFI	7.5L EFI	7.5L E40D EFI
33—EVP not closing in limits	R	—	—	X	—	—	—	—	—	—	—	—	—	—	—	—	—	—	—	—	—	—	—	—
34—Insufficient EGR flow	R	—	—	X	X	—	—	—	—	—	—	—	—	—	—	—	—	—	—	—	—	—	—	—
34—Defective PFE sensor	O	—	X	—	—	X	—	—	X	X	X	X	X	—	—	—	—	—	—	—	—	—	—	—
34—Excess exhaust back pressure	R/C	—	X	—	—	X	—	—	X	X	X	X	X	—	—	—	—	—	—	—	—	—	—	—
34—EVP voltage above closed limit	O/R/C	—	—	—	—	—	X	X	—	—	—	—	—	X	X	X	—	—	X	X	X	X	X	X
35—EVP circuit above maximum voltage	O/R/C	—	—	—	—	—	X	X	—	—	—	—	—	X	X	X	—	—	X	X	X	X	X	X
35—PFE circuit above maximum voltage	O/R/C	—	X	—	—	X	—	—	X	X	X	X	X	—	—	—	—	—	—	—	—	—	—	—
35—Rpm too low for EGR test	R	—	—	X	—	—	—	—	—	—	—	—	—	—	—	—	—	—	—	—	—	—	—	—
38—Idle track switch circuit open	C	—	X	—	—	—	X	X	—	—	—	—	—	—	—	—	—	—	—	—	—	—	—	—
39—AXOD by-pass clutch not applying properly	C	—	—	—	—	—	—	—	X	—	X	—	—	—	—	—	—	—	—	—	—	—	—	—
41—EGO/HEGO circuit shows system lean	R	X	X	X	X	X	X	X	X	⑩	⑩	⑩	⑩	⑩	⑩	X	X	X	X	X	X	X	X	X
41—No EGO/HEGO switching detected, system lean	C	X	X	X	X	X	X	X	X	⑩	⑩	⑩	⑩	⑩	⑩	X	X	X	X	X	X	X	X	X
42—EGO/HEGO shows system rich	R	X	X	X	X	X	X	X	X	⑩	⑩	⑩	⑩	⑩	⑩	X	X	X	X	X	X	X	X	X
42—No EGO/HEGO switching detected, system rich	C	X	—	—	X	—	—	—	—	—	—	—	—	—	—	—	—	—	—	—	—	—	—	—
43—EGO/HEGO lean at wide open throttle	C	X	—	—	—	—	—	—	—	—	—	—	—	—	—	—	—	—	—	—	—	—	—	—
44—Thermactor air system inoperative (cyl. 1-4)	R	—	—	—	—	—	—	—	—	—	—	—	—	X	X	—	—	—	X	X	X	X	X	X
45—DIS coil Pack 3 circuit failure	C	—	—	—	—	—	—	—	—	X	—	—	X	—	—	—	—	—	—	—	—	—	—	—
45—Thermactor air upstream during self test	R	—	—	—	—	—	—	—	—	—	—	—	—	X	X	—	—	—	X	X	X	X	—	—
46—DIS coil Pack 1 circuit failure	C	—	—	—	—	—	—	—	—	X	—	—	X	—	—	—	—	—	—	—	—	—	—	—
46—Thermactor air not bypassed during self-test	R	—	—	—	—	—	—	—	—	—	—	—	—	X	X	—	—	—	X	X	X	X	—	—
47—Airflow at base idle	R	X	—	—	—	—	—	—	—	—	—	—	—	—	—	—	—	—	—	—	—	—	—	—
47—4x4 switch is closed	O	—	—	—	—	—	—	—	—	—	—	—	—	—	—	—	—	—	—	—	—	X	—	X
48—Airflow high at base idle	R	X	—	—	—	—	—	—	—	—	—	—	—	—	—	—	—	—	—	—	—	—	—	—
48—DIS coil pack 2 circuit failure	C	—	—	—	—	—	—	—	—	X	—	—	X	—	—	—	—	—	—	—	—	—	—	—
48—Loss of secondary tach, left side	C	—	—	—	—	—	—	—	—	—	—	—	—	—	—	X	—	—	—	—	—	—	—	—
49—1-2 shift error	C	—	—	—	—	—	—	—	—	—	—	—	—	—	—	—	—	—	—	—	—	X	—	X
49—Spout signal defaulted to 10 degrees BTDC	C	—	—	—	—	—	—	—	—	X	—	—	X	—	—	—	—	—	—	—	—	—	—	—
51—ECT sensor input exceeds test max.	O/C	X	X	X	X	X	X	X	X	X	X	X	X	X	X	X	X	X	X	X	X	X	X	X
52—PSPS circuit open	O	—	—	X	—	X	X	X	X	X	X	—	X	—	—	X	—	X	X	X	—	—	—	—
52—PSPS always open or always closed	R	—	—	—	—	X	X	X	X	X	X	—	X	—	—	X	—	X	X	X	—	—	—	—
53—TP sensor input exceeds test maximum	O/C	X	X	X	X	X	X	X	X	X	X	X	X	X	X	X	X	X	X	X	X	X	X	X
54—ACT sensor input exceeds test maximum	O/C	—	X	X	—	X	X	X	X	X	X	X	X	X	X	X	X	X	X	X	X	X	X	X
54—VAT sensor input exceed test maximum	O/C	—	—	—	X	—	—	—	—	—	—	—	—	—	—	—	—	—	—	—	—	—	—	—
55—Key power input to processor is open	R	—	X	—	—	—	X	X	—	—	—	—	—	—	—	—	—	—	—	—	—	—	—	—

Ford Motor Company diagnostic trouble codes

Trouble Codes	Quick Test Mode	Car Engines 1.9L EFI	1.9L CFI	2.3L OHC EFI	2.3L EFI Turbo	2.3L HSC EFI	2.5L CLC CFI	2.5L MTX CFI	3.0L EFI	3.0L SHO SEFI	3.8L SEFI AXOD	3.8L SEFI RWD	3.8L SEFI SC	5.0L SEFI	5.0L MA SEFI	Truck Engines 2.3L TPH EFI	2.9L EFI	3.0L EFI	4.9L EFI	5.0L EFI	5.8L EFI	5.8L E40D EFI	7.5L EFI	7.5L E40D EFI
56—VAF sensor input exceeds test maximum	O/C	X	—	—	—	—	—	—	—	—	—	—	—	—	—	—	—	—	—	—	—	—	—	—
56—MAF sensor input exceeds test maximum	O/C	—	—	—	—	—	—	—	—	X	—	—	X	—	X	—	—	—	—	—	—	—	—	—
56—TOT sensor input exceeds test maximum	O/C	—	—	—	—	—	—	—	—	—	—	—	—	—	—	—	—	—	—	—	—	X	—	X
57—AXOD neutral pressure switch failed open	C	—	—	—	—	—	—	—	X	—	X	—	—	—	—	—	—	—	—	—	—	—	—	—
58—VAT sensor input exceeds test maximum	O/C	X	—	—	—	—	—	—	—	—	—	—	—	—	—	—	—	—	—	—	—	—	—	—
58—Idle tracking switch circuit closed	R	—	X	—	—	—	X	X	—	—	—	—	—	—	—	—	—	—	—	—	—	—	—	—
58—Idle tracking switch circuit open	O	—	X	—	—	—	X	X	—	—	—	—	—	—	—	—	—	—	—	—	—	—	—	—
59—AXOD 4/3 pressure switch failed open	C	—	—	—	—	—	—	—	X	—	X	—	—	—	—	—	—	—	—	—	—	—	—	—
59—Low speed fuel pump circuit failure	O/C	—	—	—	—	—	—	—	—	X	—	—	—	—	—	—	—	—	—	—	—	—	—	—
59—AXOD 4/3 pressure switch failed closed	O	—	—	—	—	—	—	—	—	—	X	—	—	—	—	—	—	—	—	—	—	—	—	—
59—2-3 shift error	C	—	—	—	—	—	—	—	—	—	—	—	—	—	—	—	—	—	—	—	—	X	—	X
61—ECT test sensor input below test minimum	O/C	X	X	X	X	X	X	X	X	X	X	X	X	X	X	X	X	X	X	X	X	X	X	X
62—AXOD 4/3 or 3/2 pressure switch failed closed	O	—	—	—	—	—	—	—	X	—	—	—	—	—	—	—	—	—	—	—	—	—	—	—
63—Converter clutch failure	C	—	—	—	—	—	—	—	—	—	—	—	—	—	—	—	—	—	—	—	—	—	—	—
63—TP sensor below test minimum	O/C	X	X	X	X	X	X	X	X	X	X	X	X	X	X	X	X	X	X	X	X	X	X	X
64—ACT sensor input below test minimum	O/C	—	X	X	—	X	X	X	X	X	X	X	X	X	X	X	X	X	X	X	X	X	X	X
64—VAT sensor input below test minimum	O/C	—	—	—	X	—	—	—	—	—	—	—	—	—	—	—	—	—	—	—	—	—	—	—
65—Failed to enter closed loop mode	C	X	—	—	—	—	—	—	—	—	—	—	—	—	—	—	—	—	—	—	—	—	—	—
65—Overdrive cancel switch not changing state	R	—	—	—	—	—	—	—	—	—	—	—	—	—	—	—	—	—	—	—	—	X	—	X
66—VAF sensor input below test minimum	O/C	X	—	—	X	—	—	—	—	—	—	—	—	—	—	—	—	—	—	—	—	—	—	—
66—MAF sensor input below test minimum	C	—	—	—	—	—	—	—	—	X	—	—	X	—	X	—	—	—	—	—	—	—	—	—
66—TOT sensor input below test minimum	O/C	—	—	—	—	—	—	—	—	—	—	—	—	—	—	—	—	—	—	—	—	X	—	X
67—Neutral drive switch open, A/C input high	O	X	④	X	X	X	X	⑧	X	⑨	⑫	—	X	⑫	X	X	X	X	X	X	X	—	X	—
67—Clutch switch circuit failure	C	X	—	—	X	—	X	X	—	—	—	—	X	—	—	—	X	—	—	—	—	—	—	—
67—AXOD neutral pressure switch failed closed	O	—	—	—	—	—	—	—	—	—	X	—	—	—	—	—	—	—	—	—	—	—	—	—
67—MLP sensor out of range, A/C input high	O/C	—	—	—	—	—	—	—	—	—	—	—	—	—	—	—	—	—	—	—	—	X	—	X
68—VAT sensor input below test minimum	O/C	X	—	—	—	—	—	—	—	—	—	—	—	—	—	—	—	—	—	—	—	—	—	—
68—Idle tracking switch circuit open	R	—	X	—	—	—	X	X	—	—	—	—	—	—	—	—	—	—	—	—	—	—	—	—
68—Idle tracking switch closed	O	—	X	—	—	—	X	X	—	—	—	—	—	—	—	—	—	—	—	—	—	—	—	—
68—AXOD temperature switch failed open	O/R/C	—	—	—	—	—	—	—	—	—	X	—	—	—	—	—	—	—	—	—	—	—	—	—
69—AXOD 3/4 pressure switch failed open	C	—	—	—	—	—	—	—	X	—	X	—	—	—	—	—	—	—	—	—	—	—	—	—
69—AXOD 3/2 pressure switch failed closed	O	—	—	—	—	—	—	—	—	—	X	—	—	—	—	—	—	—	—	—	—	—	—	—
69—3-4 shift error	C	—	—	—	—	—	—	—	—	—	—	—	—	—	—	—	—	—	—	—	—	X	—	X

Ford Motor Company diagnostic trouble codes

Trouble Codes	Quick Test Mode	Car Engines 1.9L EFI	1.9L CFI	2.3L OHC EFI	2.3L EFI Turbo	2.3L HSC EFI	2.5L CLC CFI	2.5L MTX CFI	3.0L EFI	3.0L SHO SEFI	3.8L SEFI AXOD	3.8L SEFI RWD	3.8L SEFI SC	5.0L SEFI	5.0L MA SEFI	Truck Engines 2.3L TPH EFI	2.9L EFI	3.0L EFI	4.9L EFI	5.0L EFI	5.8L EFI	5.8L E40D EFI	7.5L EFI	7.5L E40D EFI
70—EEC-IV data transmission link failed	C	—	—	—	—	—	—	—	—	—	X	—	—	—	—	—	—	—	—	—	—	—	—	—
71—Software re-initialization detected	C	②	—	—	X	—	—	—	—	—	—	—	—	—	—	—	—	—	—	—	—	—	—	—
71—Idle tracking switch closed on pre-position	C	—	X	—	—	—	X	X	—	—	—	—	—	—	—	—	—	—	—	—	—	—	—	—
71—Cluster control assembly circuit failed	C	—	—	—	—	—	—	—	—	—	X	—	—	—	—	—	—	—	—	—	—	—	—	—
72—Power interrupt detected	C	X	—	—	X	—	X	X	—	—	—	—	—	—	—	—	—	—	—	—	—	—	—	—
72—Insufficient BP change during test	R	—	—	—	—	—	—	—	—	X	—	—	X	—	—	—	—	—	—	—	—	—	—	—
72—Insufficient MAP output change during test	R	—	—	X	—	X	—	—	X	—	—	—	—	—	—	X	X	X	X	X	X	X	X	X
72—Message center control circuit failed	C	—	—	—	—	—	—	—	—	—	X	—	—	—	—	—	—	—	—	—	—	—	—	—
73—Insufficient TP output change during test	R	X	—	X	X	X	—	—	X	X	—	—	X	—	—	X	X	X	X	X	X	X	X	X
73—Insufficient TP change	0	—	X	—	—	—	X	X	—	—	—	—	—	—	—	—	—	—	—	—	—	—	—	—
74—Brake on/off circuit open, not on during test	R	—	—	X	—	—	X	X	X	X	X	X	X	X	—	X	X	X	—	—	—	X	—	—
75—Brake on/off circuit closed, always high	R	—	—	X	—	—	—	—	—	—	—	—	—	X	—	—	—	—	—	—	—	—	—	—
76—Insufficient VAF output change during test	R	X	—	—	X	—	—	—	—	—	—	—	—	—	—	—	—	—	—	—	—	—	—	—
77—Wide open throttle not sensed during test	R	X	—	X	X	X	—	—	X	X	—	—	X	—	X	X	X	X	X	X	X	X	X	X
79—A/C on during self test	0	—	—	—	—	—	—	—	—	X	X	X	X	X	X	—	—	—	—	—	—	—	—	—
81—Insufficient IAS output during test	0	—	—	—	—	—	—	—	—	X	—	—	—	—	—	—	—	—	—	—	—	—	—	—
81—Air management 2 circuit failure	0	—	—	—	—	—	—	—	—	—	—	—	—	X	X	—	—	—	X	X	X	X	—	—
82—Supercharger bypass circuit failure	0	—	—	—	—	—	—	—	—	—	—	—	X	—	—	—	—	—	—	—	—	—	—	—
82—Air management 1 circuit failure	0	—	—	—	—	—	—	—	—	—	—	—	—	X	X	—	—	—	X	X	X	X	X	X
83—EGRC solenoid circuit failure	0	—	—	X	—	—	—	—	—	—	—	—	—	—	—	—	—	—	—	—	—	—	—	—
83—High speed electro drive fan circuit failure	0	—	—	—	—	—	X	X	X	—	X	—	X	—	—	—	—	—	—	—	—	—	—	—
83—Low speed fuel pump relay circuit open	0/C	—	—	—	—	—	—	—	—	X	—	—	—	—	—	—	—	—	—	—	—	—	—	—
84—EGR VAC regulator circuit failure	0	—	④	—	—	X	X	X	X	X	X	X	X	X	X	X	—	—	X	X	X	X	X	X
84—EGRV solenoid circuit failure	0	—	—	X	—	—	—	—	—	—	—	—	—	—	—	—	—	—	—	—	—	—	—	—
85—Adaptive lean limit reached	C	X	—	—	—	—	—	—	—	—	—	—	—	—	—	—	—	—	—	—	—	—	—	—
85—Canister purge circuit failure	0	—	④	—	—	X	X	X	X	X	X	X	X	X	X	—	—	X	—	—	—	X	—	X
86—Adaptive rich limit reached	C	X	—	—	—	—	—	—	—	—	—	—	—	—	—	—	—	—	—	—	—	—	—	—
86—3-4 shift solenoid circuit failure	0	—	—	—	—	—	—	—	—	—	—	—	—	—	—	X	X	X	—	—	—	—	—	—
87—Fuel pump primary circuit failure	0/C	—	④	⑤	—	X	X	X	X	X	X	X	X	X	X	X	X	X	X	X	X	X	X	X
88—Electro drive fan circuit failure	0	—	—	—	—	—	X	X	X	X	X	—	X	—	—	—	—	—	—	—	—	—	—	—
88—Loss of dual plug input input control	C	—	—	—	—	—	—	—	—	—	—	—	—	—	—	X	—	—	—	—	—	—	—	—
89—Clutch converter overdrive circuit failure	0	—	—	X	—	—	—	—	—	—	—	—	—	—	—	X	X	X	—	—	—	—	—	—
89—AXOD lock-up solenoid circuit failure	0	—	—	—	—	—	—	—	X	—	X	—	—	—	—	—	—	—	—	—	—	—	—	—
91—HEGO sensor circuit shows system lean	R	—	—	—	—	—	—	—	—	⑪	⑪	⑪	⑪	⑪	⑪	—	—	—	—	—	—	—	—	—

Ford Motor Company diagnostic trouble codes

Trouble Codes	Quick Test Mode	Car Engines 1.9L EFI	1.9L CFI	2.3L OHC EFI	2.3L EFI Turbo	2.3L HSC EFI	2.5L CLC CFI	2.5L MTX CFI	3.0L EFI	3.0L SHO SEFI	3.8L SEFI AXOD	3.8L SEFI RWD	3.8L SEFI SC	5.0L SEFI	5.0L MA SEFI	Truck Engines 2.3L TPH EFI	2.9L EFI	3.0L EFI	4.9L EFI	5.0L EFI	5.8L EFI	5.8L E40D EFI	7.5L EFI	7.5L E40D EFI
91—Shift solenoid 1 circuit failure	0	—	—	—	—	—	—	—	—	—	—	—	—	—	—	—	—	—	—	—	—	X	—	X
91—No HEGO switching sensed	C	—	—	—	—	—	—	—	—	⑪	⑪	⑪	⑪	⑪	⑪	—	—	—	—	—	—	—	—	—
92—HEGO sensor circuit shows system rich	R	—	—	—	—	—	—	—	—	⑪	⑪	⑪	⑪	⑪	⑪	—	—	—	—	—	—	—	—	—
92—Shift solenoid 2 circuit failure	0	—	—	—	—	—	—	—	—	—	—	—	—	—	—	—	—	—	—	—	—	X	—	X
93—TP sensor input low at max DC motor extension	0	—	X	—	—	—	X	X	—	—	—	—	—	—	—	—	—	—	—	—	—	—	—	—
94—Thermactor air system inoperative (cyl. 5–8)	R	—	—	—	—	—	—	—	—	—	—	—	—	X	X	—	—	—	—	—	—	—	—	—
94—Converter clutch solenoid circuit failure	0	—	—	—	—	—	—	—	—	—	—	—	—	—	—	—	—	—	—	—	—	X	—	X
95—Fuel pump secondary circuit failure	O/C	X	X	—	—	X	X	X	X	X	X	X	—	—	X	X	X	X	X	X	X	X	X	X
96—Fuel pump secondary circuit failure	O/C	X	X	—	—	X	X	X	X	—	X	X	X	—	X	X	X	X	X	X	X	X	X	X
96—High speed fuel pump relay circuit open	O/C	—	—	—	—	—	—	—	—	X	—	—	—	—	—	—	—	—	—	—	—	—	—	—
97—Overdrive cancel indicator circuit failure	0	—	—	—	—	—	—	—	—	—	—	—	—	—	—	—	—	—	—	—	—	—	—	X
98—Hard fault is present	R	—	X	X	—	X	—	—	X	X	X	X	X	X	X	X	X	X	X	X	X	X	X	X
98—Electronic pressure control driver failure	0	—	—	—	—	—	—	—	—	—	—	—	—	—	—	—	—	—	—	—	—	X	—	X
99—EEC system has not learned to control idle	R	—	X	—	—	—	X	X	—	—	—	—	—	—	—	—	—	—	—	—	—	—	—	—
99—Electronic pressure control circuit failure	O/C	—	—	—	—	—	—	—	—	—	—	—	—	—	—	—	—	—	—	—	—	X	—	X
No code—unable to run self-test or output codes	⑬	X	X	X	X	X	X	X	X	X	X	X	X	X	X	X	X	X	X	X	X	X	X	X
Code not listed—does not apply to vehicle tested	⑬	X	X	X	X	X	X	X	X	X	X	X	X	X	X	X	X	X	X	X	X	X	X	X

0—Key on, engine off
R—Engine running
C—Continuous memory

① ISC off
② 1989 only
③ Occurs also in continuous memory (C)
④ Occurs also with the engine running (R)
⑤ Occurs only in key on, engine off (O)
⑥ Occurs only with the engine running (R)
⑦ Occurs only in key on, engine off (O) or continuous memory (C)
⑧ A/C input high only
⑨ Neutral pressure switch open
⑩ Right side HEGO
⑪ Left side HEGO
⑫ N-D switch only (no A/C signal)
⑬ Refer to system diagnostics
⑭ Erratic input to processor

Ford Motor Company diagnostic trouble codes

VACUUM DIAGRAMS

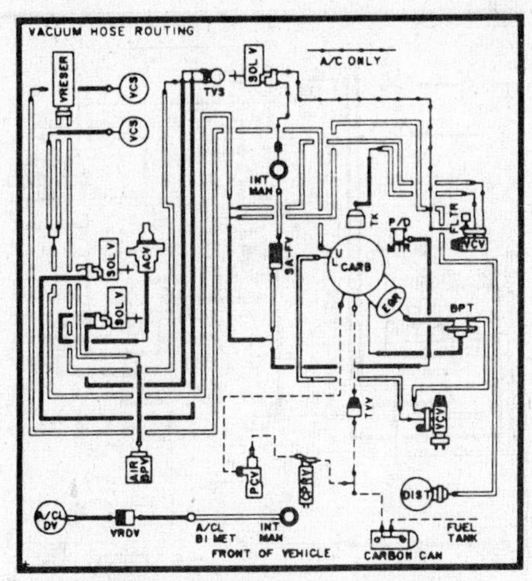

Calibration 3-49S-R10—2.3L Ranger 1983

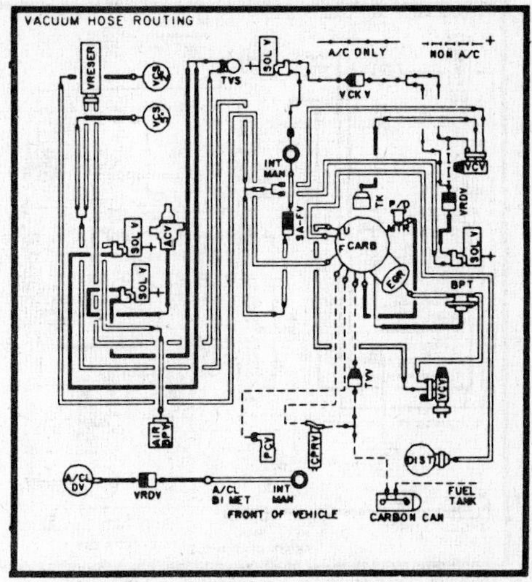

Calibration 3-49T-R17—2.3L Ranger 1983

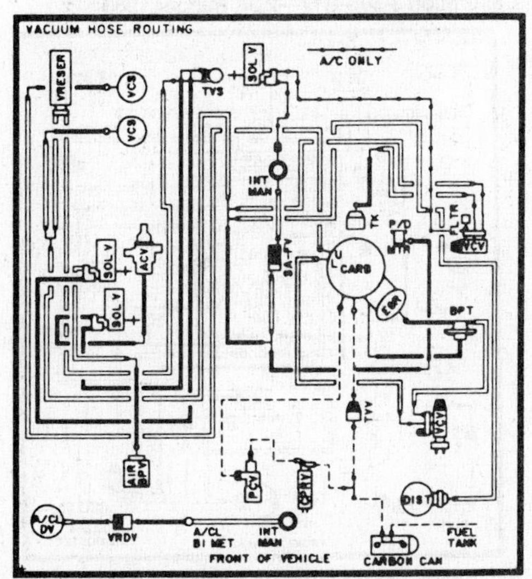

Calibration 3-49S-R11—2.3L Ranger 1983

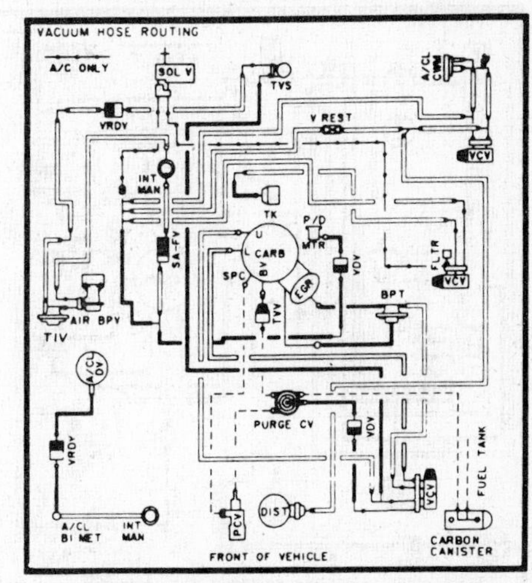

Calibration 3-49X-R11—2.3L Ranger 1983

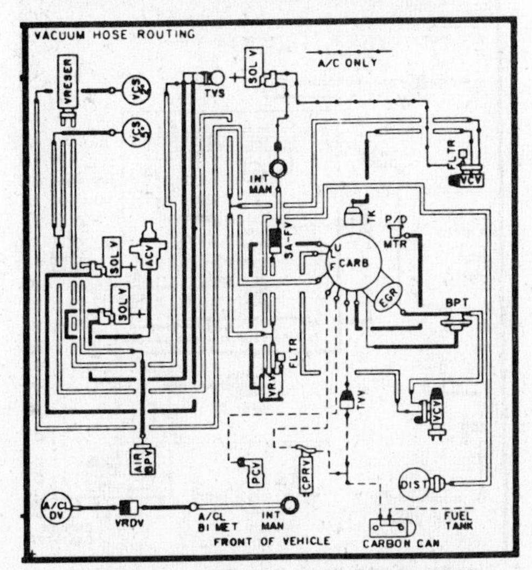

Calibration 3-49S-R15—2.3L Ranger 1983

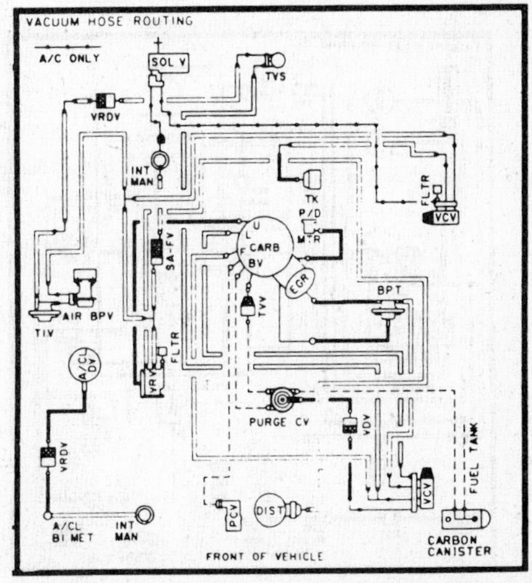

Calibration 3-49Y-R17—2.3L Ranger 1983

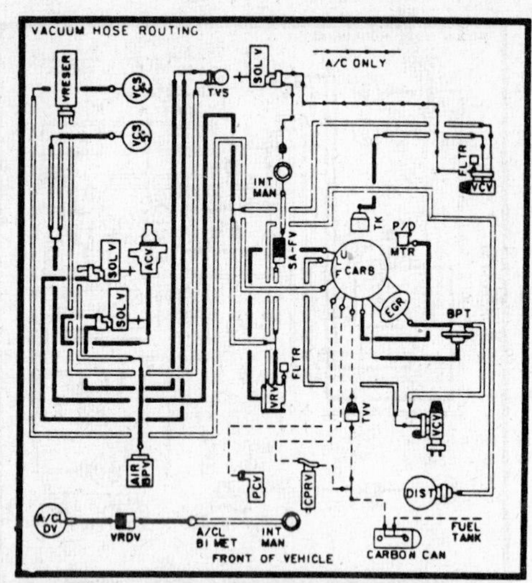

Calibration 3-50H-R17—2.3L Ranger 1983

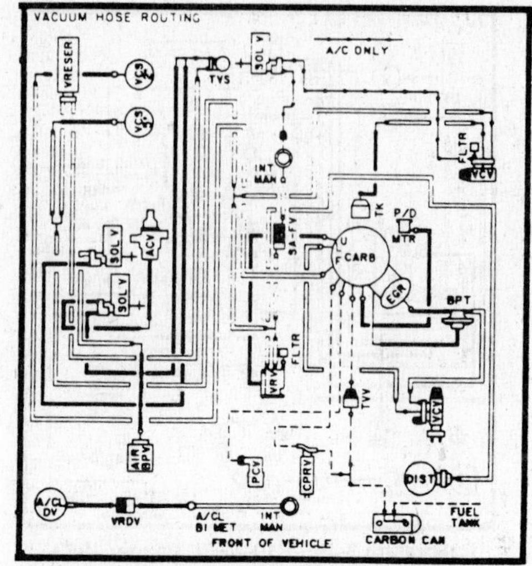

Calibration 3-50S-R17—2.3L Ranger 1983

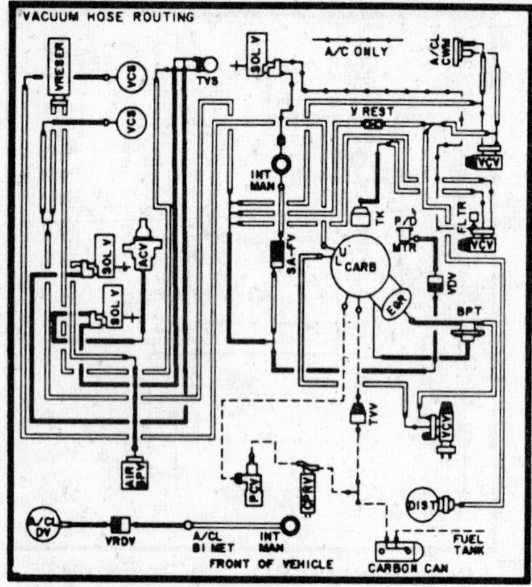

Calibration 3-50S-R10—2.3L Ranger 1983

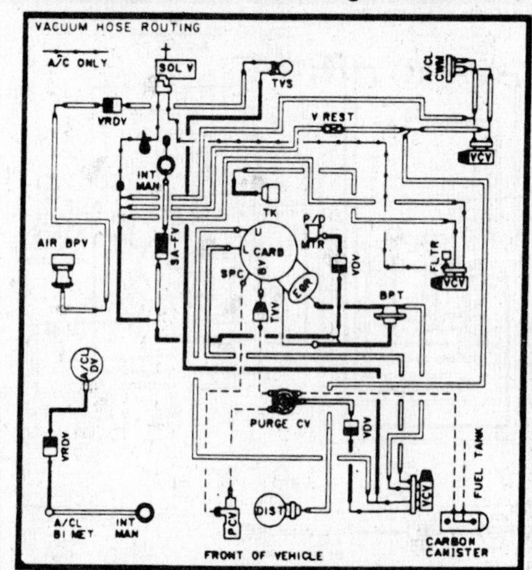

Calibration 3-50X-R10—2.3L Ranger 1983

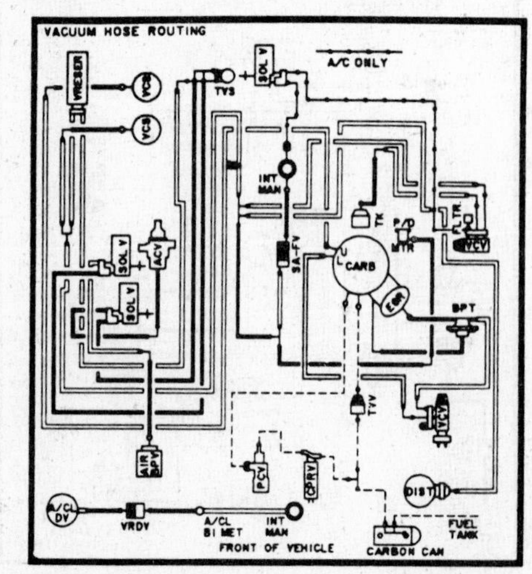

Calibration 3-50S-R11—2.3L Ranger 1983

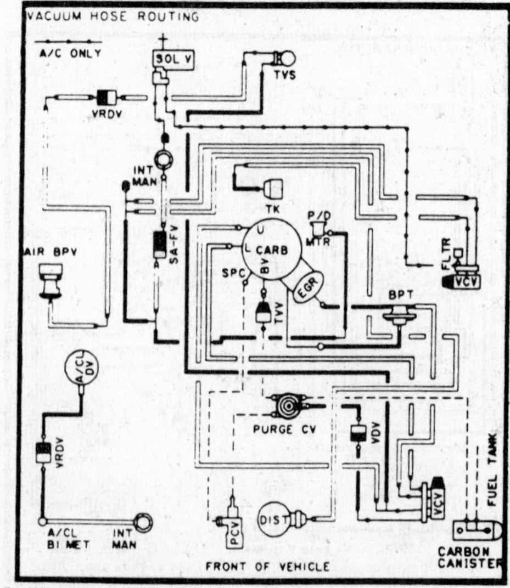

Calibration 3-50X-R11—2.3L Ranger 1983

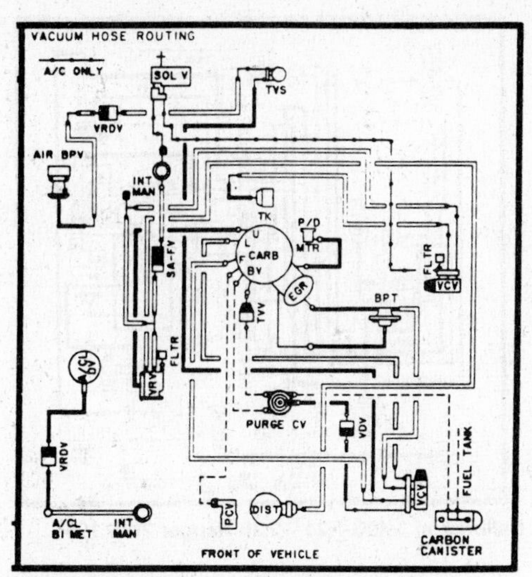

Calibration 3-50Y-R17—2.3L Ranger 1983

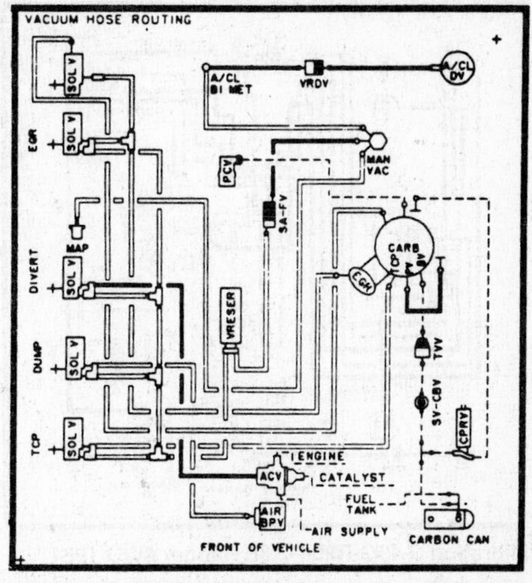

Calibration 3-62D-R00—2.8L Ranger/Bronco II 1983

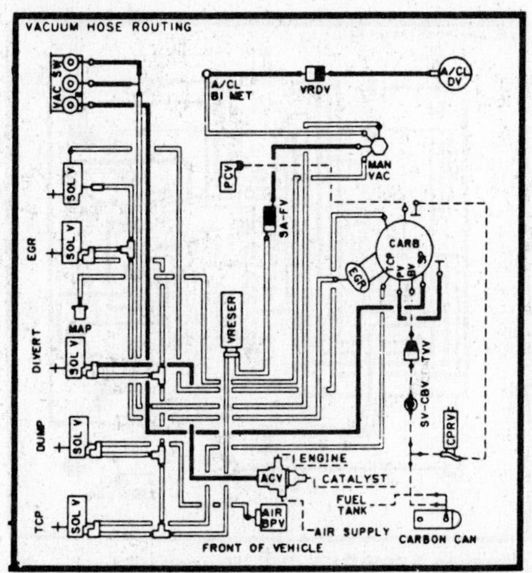

Calibration 3-61F-R00—2.8L Ranger/Bronco II 1983

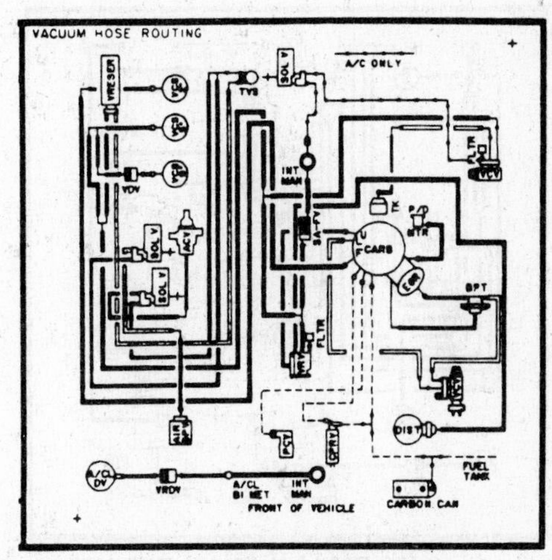

Calibration 3-41P-R15—2.0L Ranger 1984

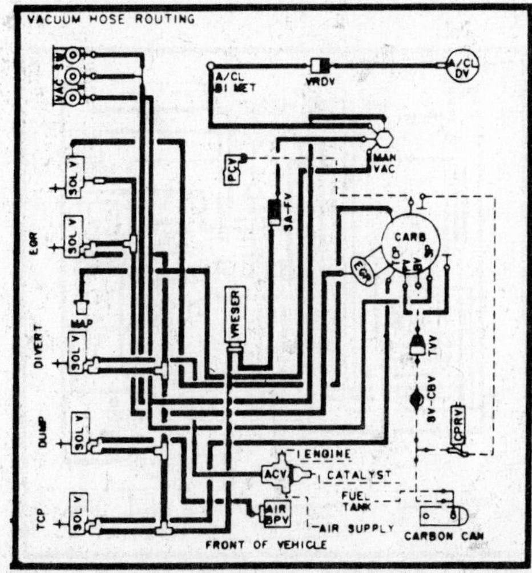

Calibration 3-61S-R00—2.8L Ranger/Bronco II 1983

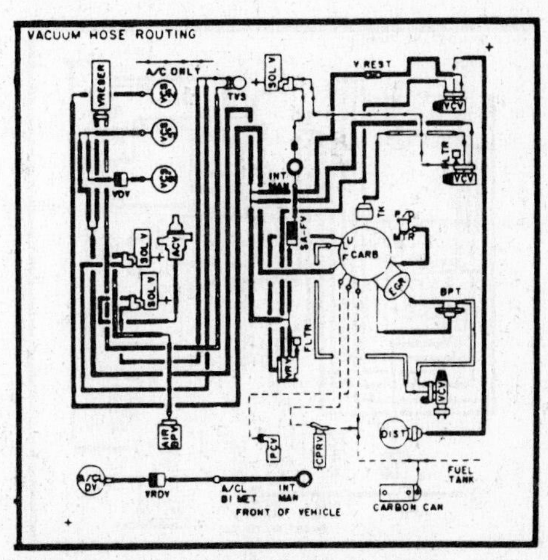

Calibration 3-41S-R18—2.0L Ranger 1984

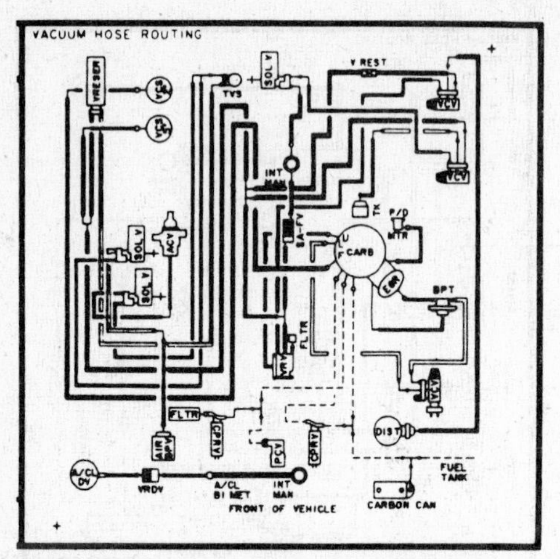

Calibration 3-49G-R20—2.3L Ranger 4WD 1984

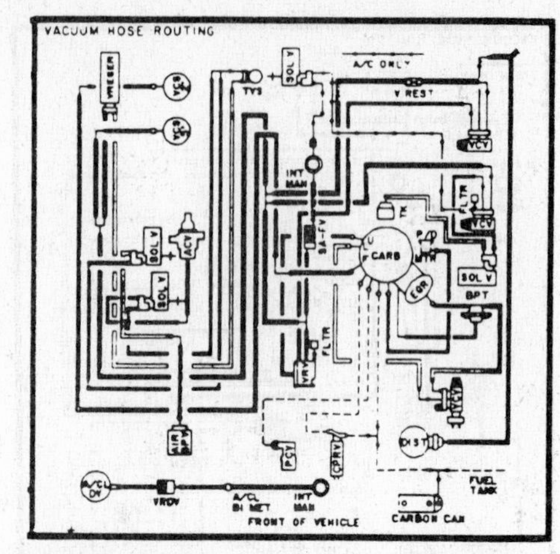

Calibration 3-49G-R21—2.3L Ranger 2WD 1984

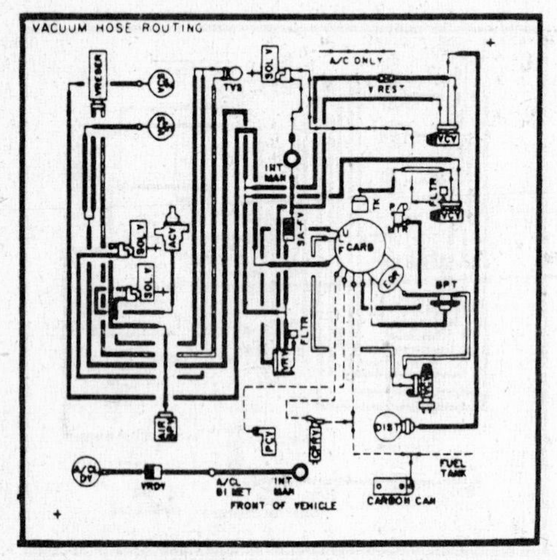

Calibration 3-49G-R20—2.3L Ranger 2WD 1984

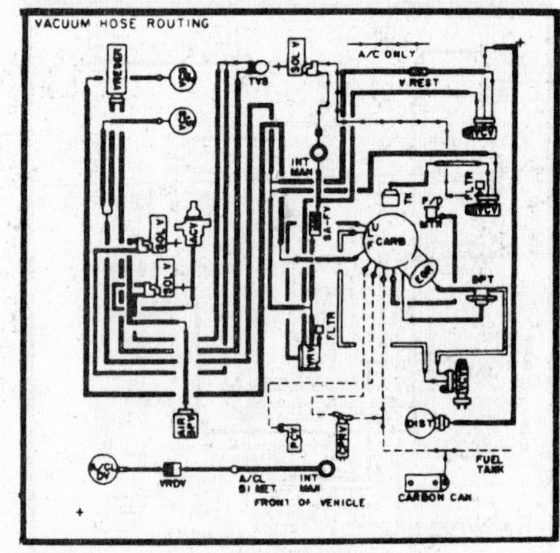

Calibration 3-49H-R17—2.3L Ranger 1984

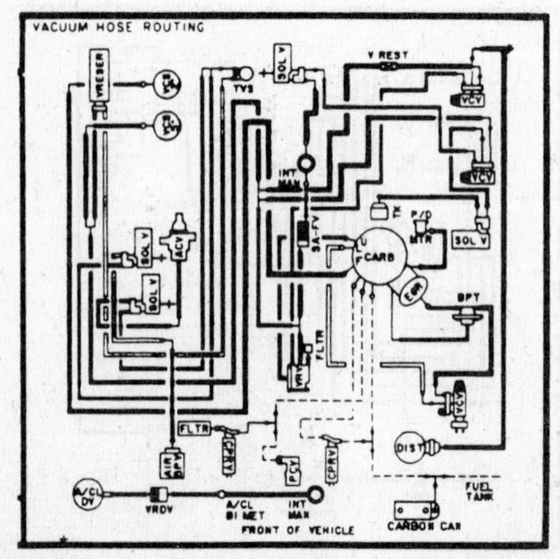

Calibration 3-49G-R21—2.3L Ranger 4WD 1984

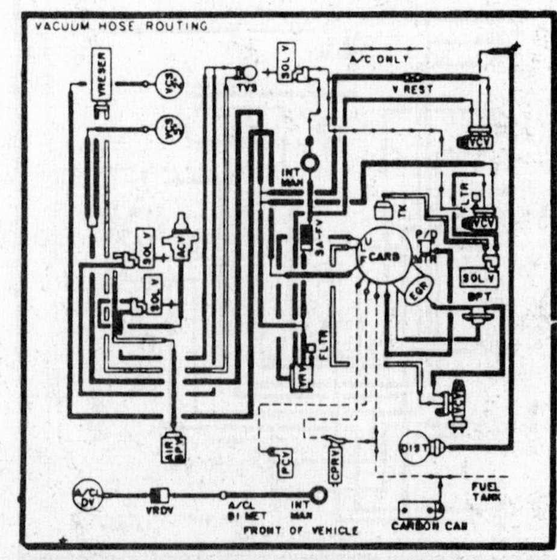

Calibration 3-49H-R18—2.3L Ranger 1984

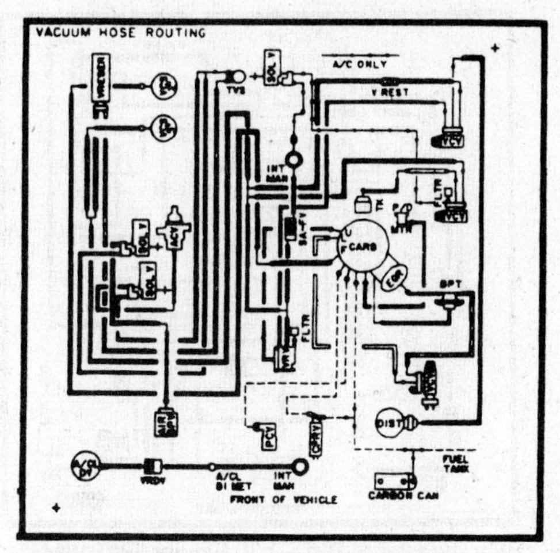

Calibration 3-49S-R16—2.3L Ranger 1984

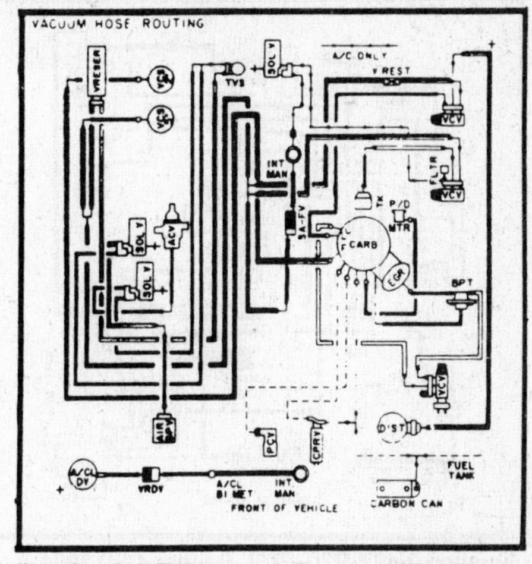

Calibration 3-49T-R20—2.3L Ranger 2WD 1984

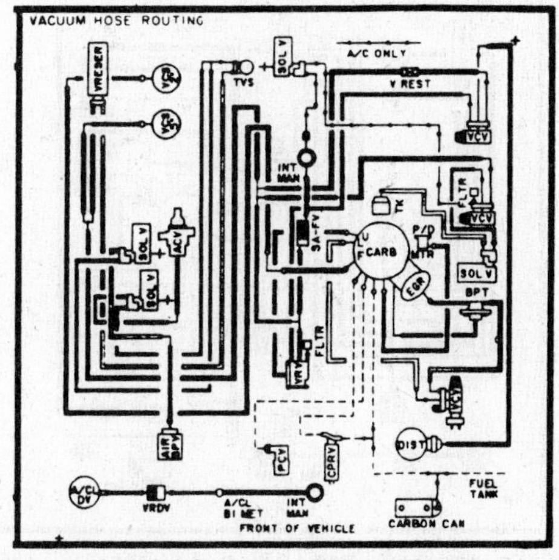

Calibration 3-49S-R17—2.3L Ranger 1984

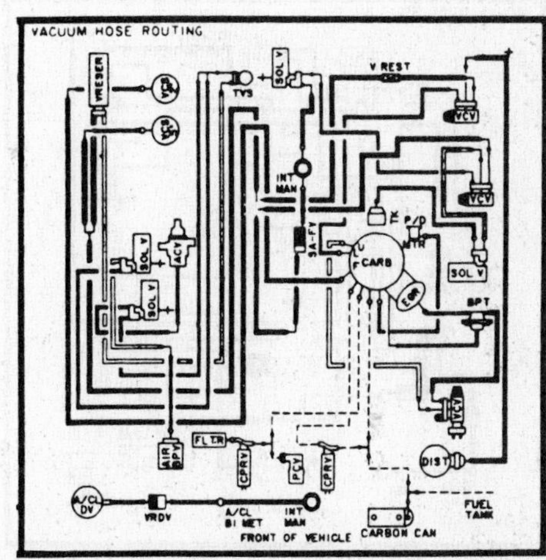

Calibration 3-49T-R21—2.3L Ranger 4WD 1984

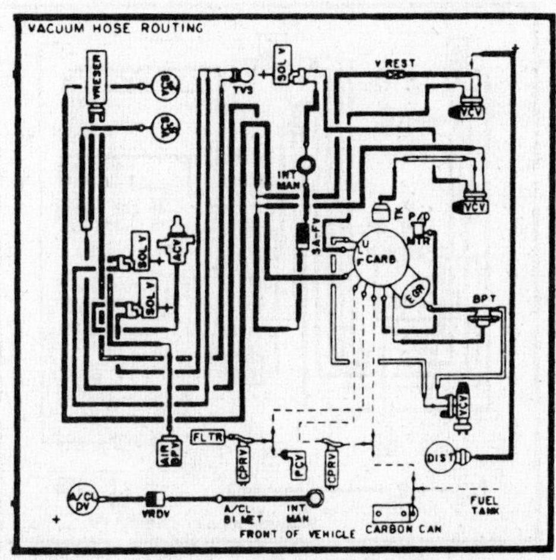

Calibration 3-49T-R20—2.3L Ranger 4WD 1984

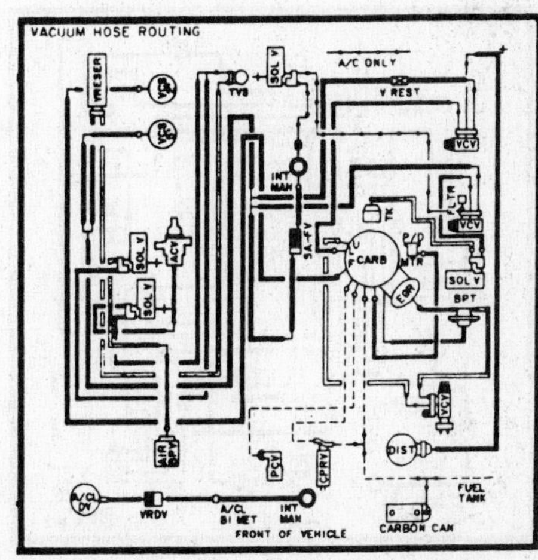

Calibration 3-49T-R21—2.3L Ranger 2WD 1984

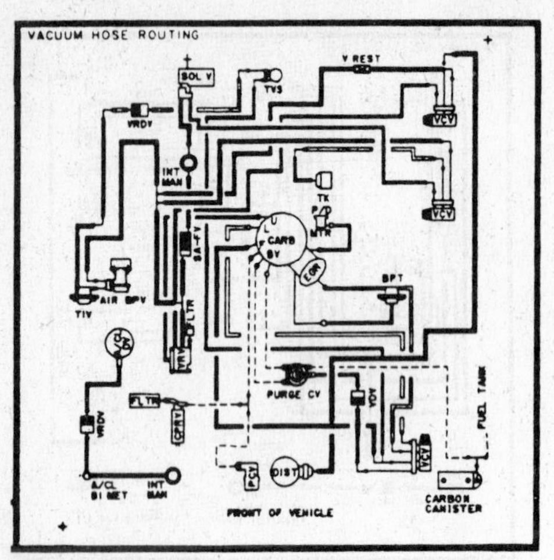

Calibration 3-49Y-R20—2.3L Ranger 4WD 1984

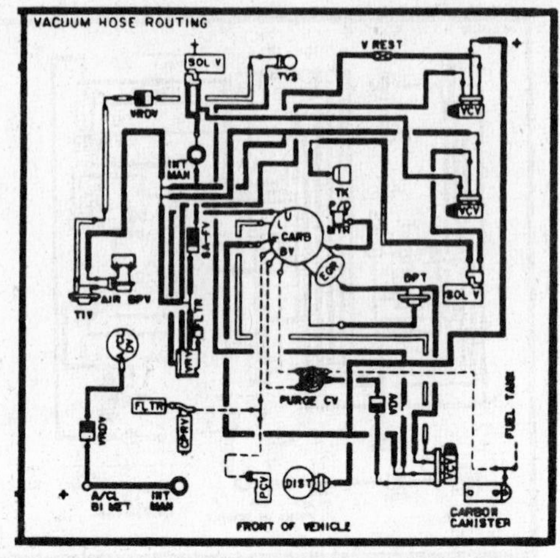

Calibration 3-49Y-R21—2.3L Ranger 2WD 1984

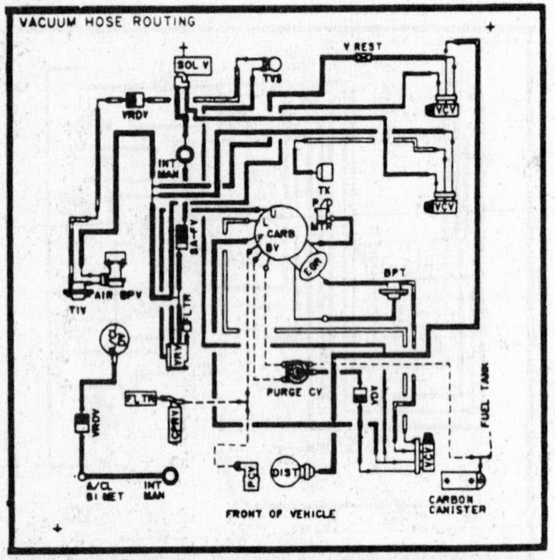

Calibration 3-49Y-R20—2.3L Ranger 2WD 1984

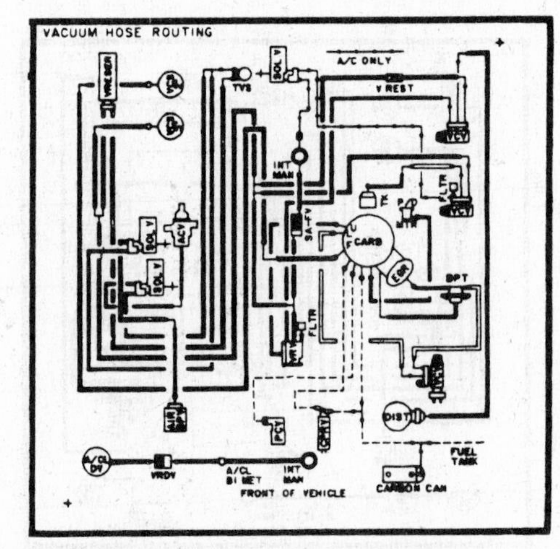

Calibration 3-50H-R18—2.3L Ranger 1984

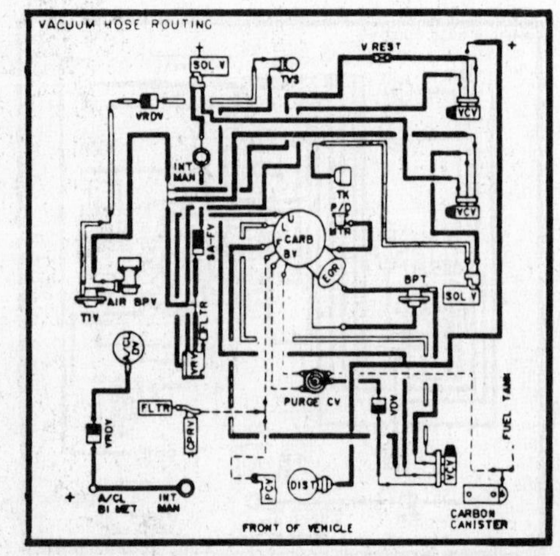

Calibration 3-49Y-R21—2.3L Ranger 4WD 1984

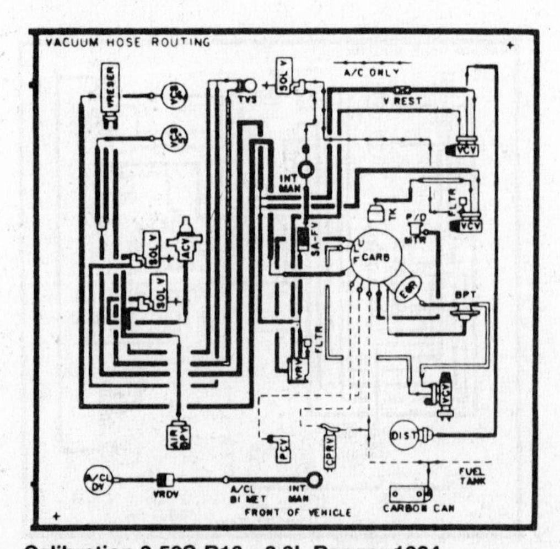

Calibration 3-50S-R18—2.3L Ranger 1984

Calibration 3-61F-R10—2.8L Ranger/Bronco II 1984

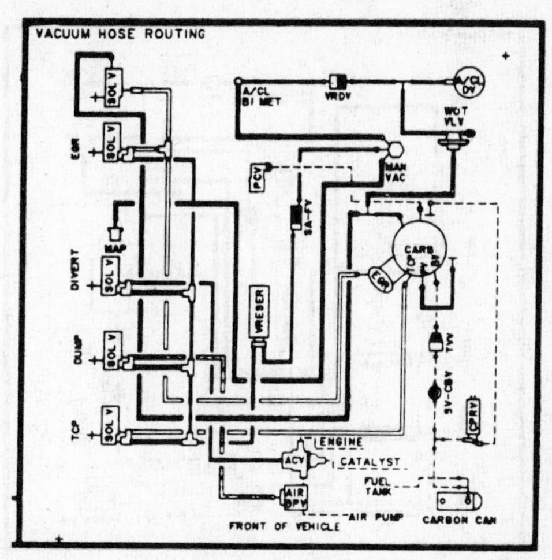

Calibration 3-61G-R10—2.8L Ranger/Bronco II 1984

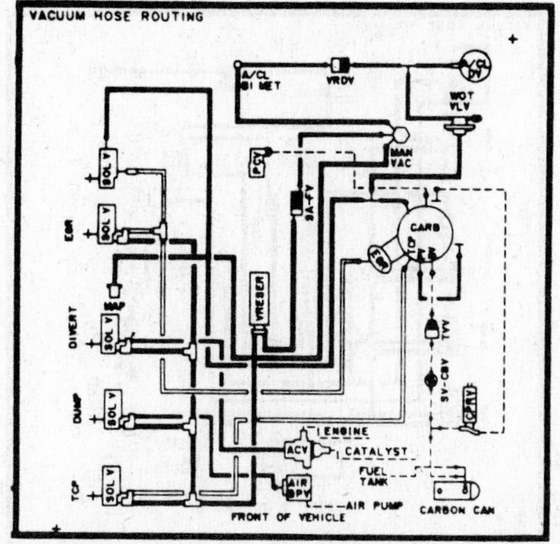

Calibration 3-61F-R11—2.8L Ranger/Bronco II 1984

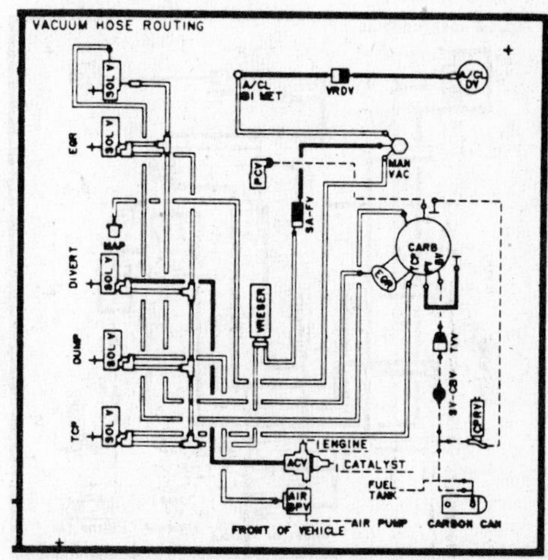

Calibration 3-61K-R01—2.8L Ranger/Bronco II 1984

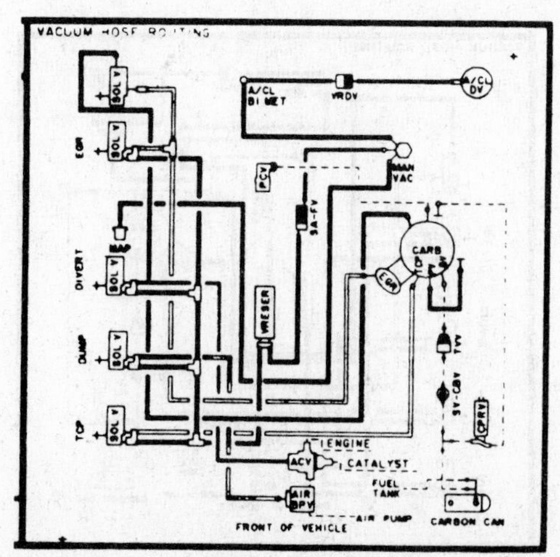

Calibration 3-61G-R00—2.8L Ranger/Bronco II 1984

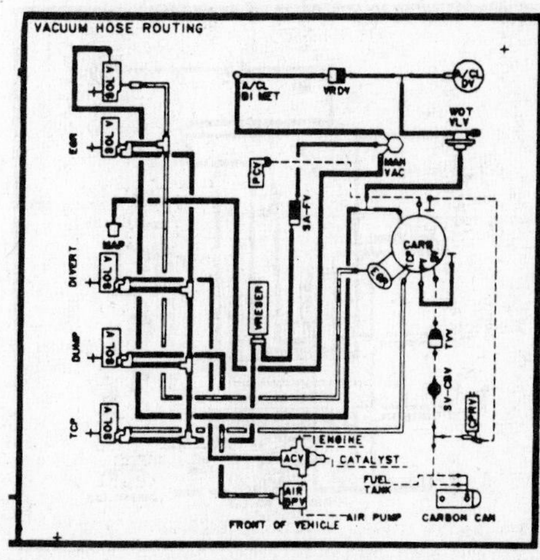

Calibration 3-61K-R10—2.8L Ranger/Bronco II 1984

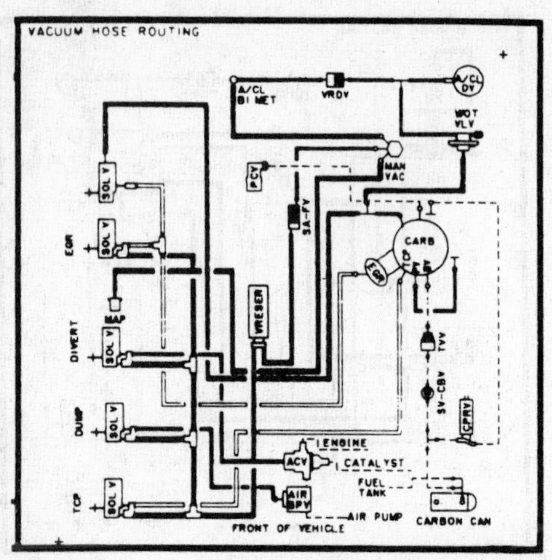

Calibration 3-61S-R12—2.8L Ranger/Bronco II 1984

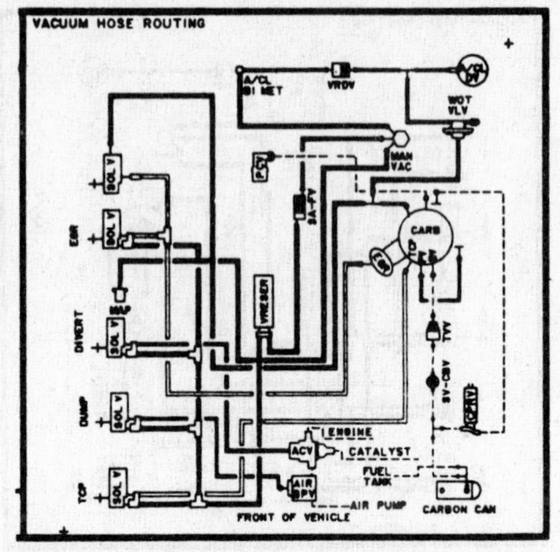

Calibration 3-62D-R12—2.8L Ranger/Bronco II 1984

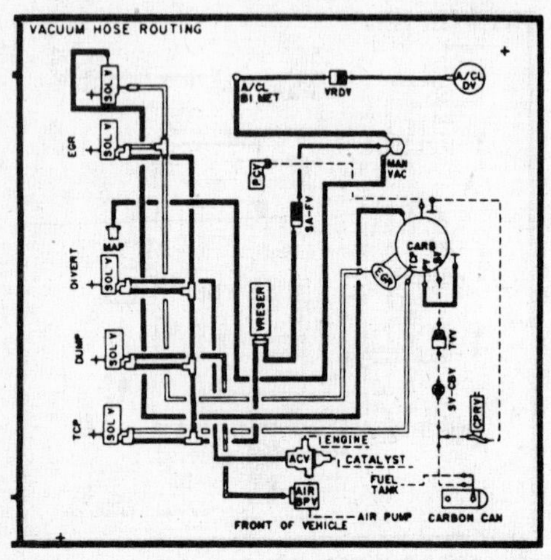

Calibration 3-62D-R00—2.8L Ranger/Bronco II 1984

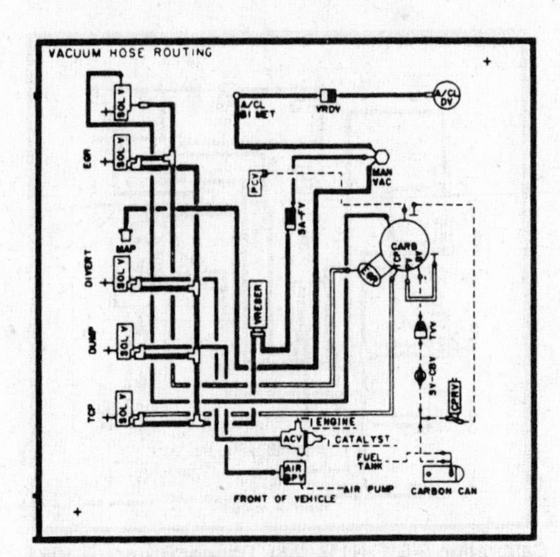

Calibration 3-62S-R14—2.8L Ranger/Bronco II 1984

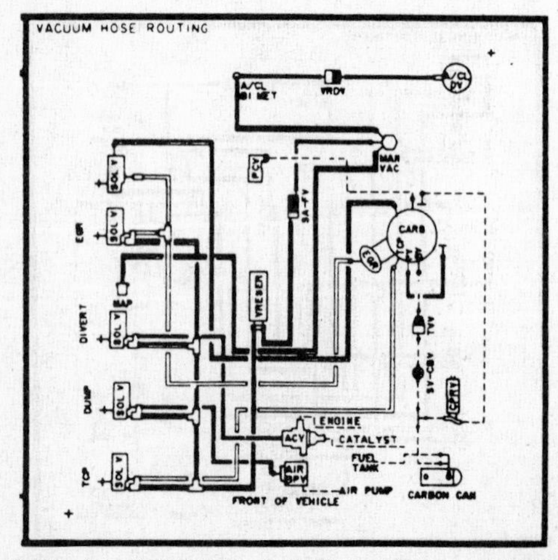

Calibration 3-62D-R10—2.8L Ranger/Bronco II 1984

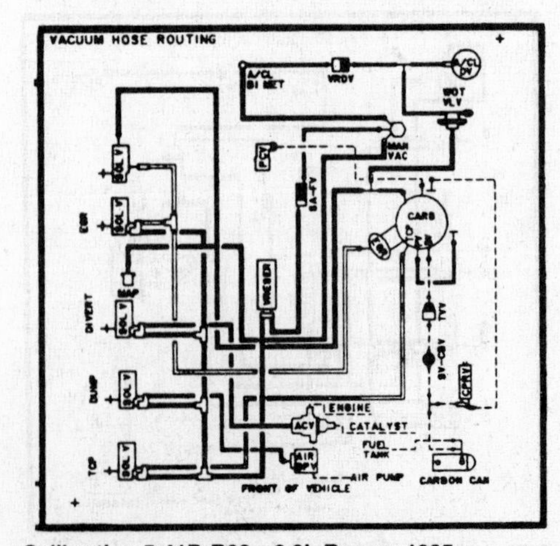

Calibration 5-41D-R00—2.0L Ranger 1985

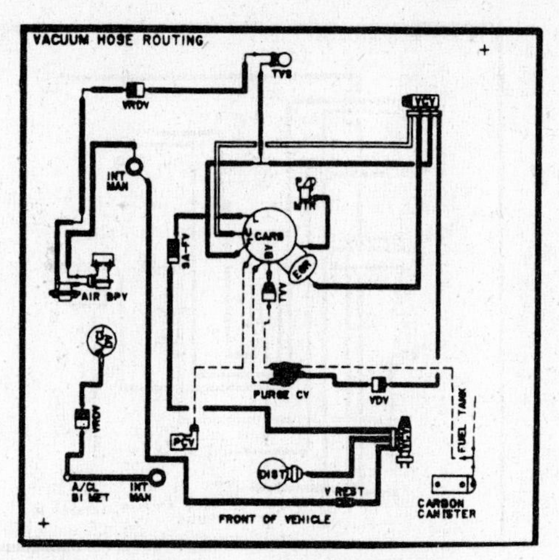

Calibration 5-41D-R10—2.0L Ranger 1985

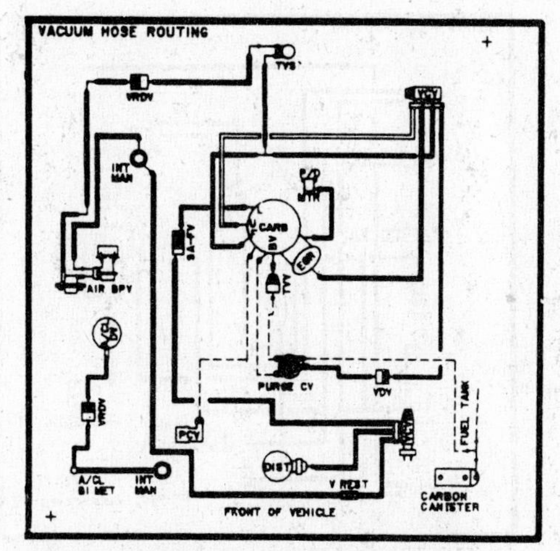

Calibration 5-41D-R10—2.0L Ranger 1985

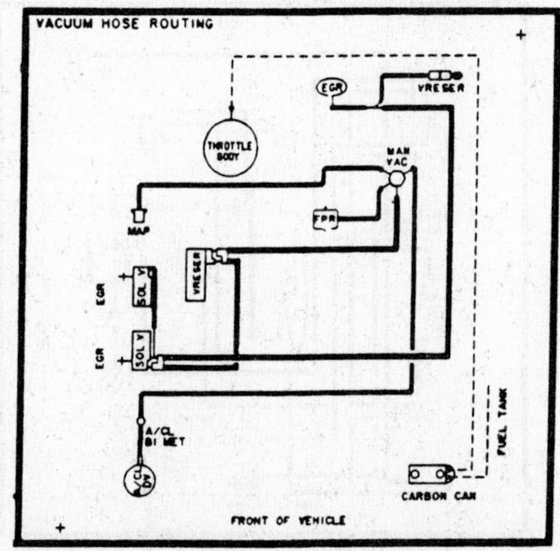

Calibration 5-49F-R01—2.3L Ranger 1985

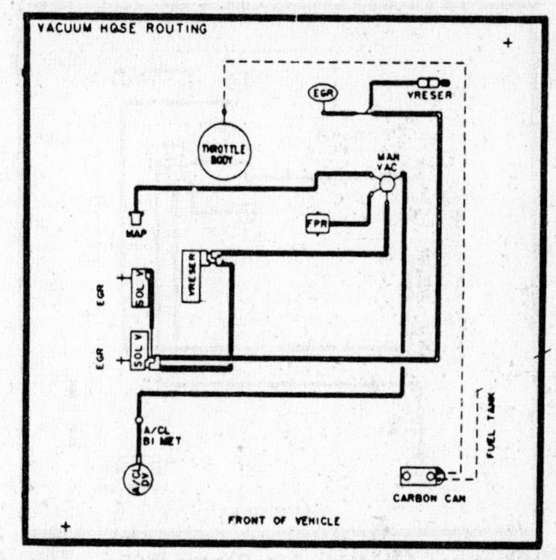

Calibration 5-49F-R01—2.3L Ranger (Chassis Cab) 1985

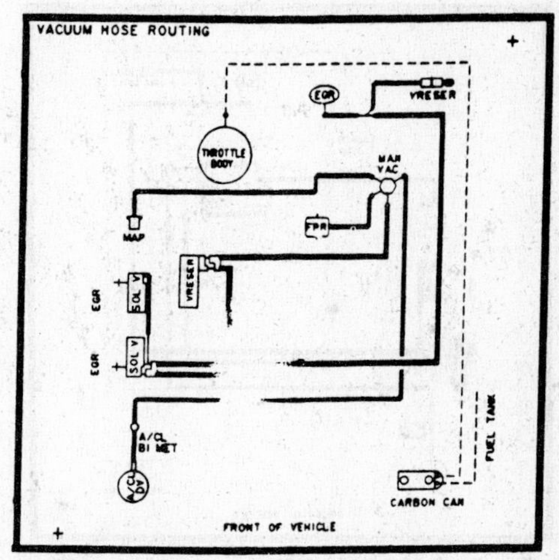

Calibration 5-49S-R01—2.3L Ranger 1985

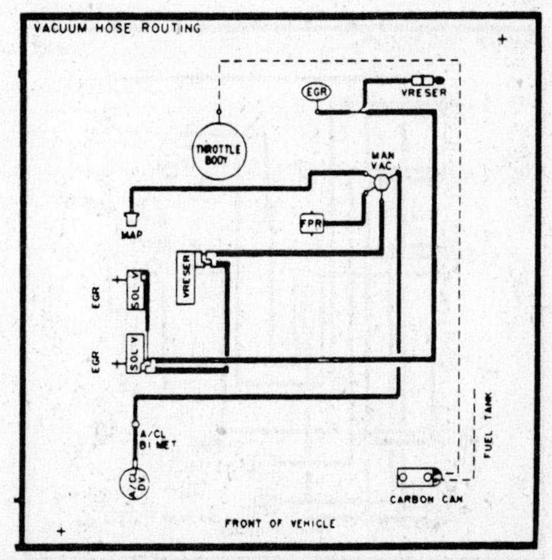

Calibration 5-49S-R01—2.3L Ranger (Chassis Cab) 1985

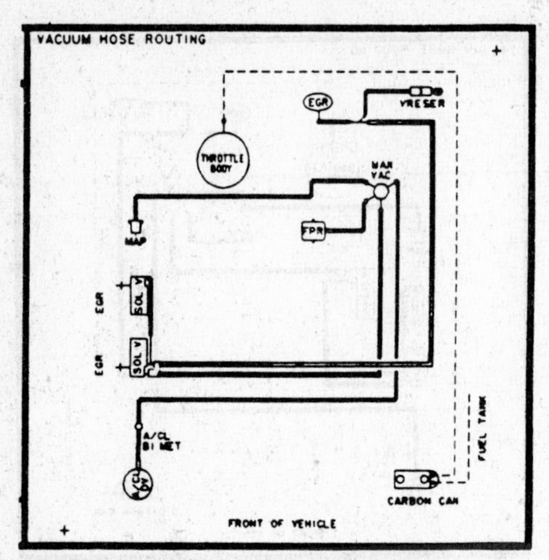

Calibration 5-50H-R02—2.3L Ranger 1985

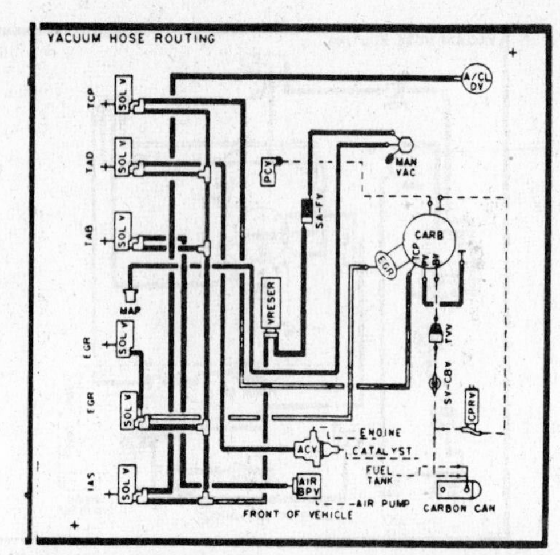

Calibration 5-61F-R01—2.8L Ranger (Chassis Cab HD) 1985

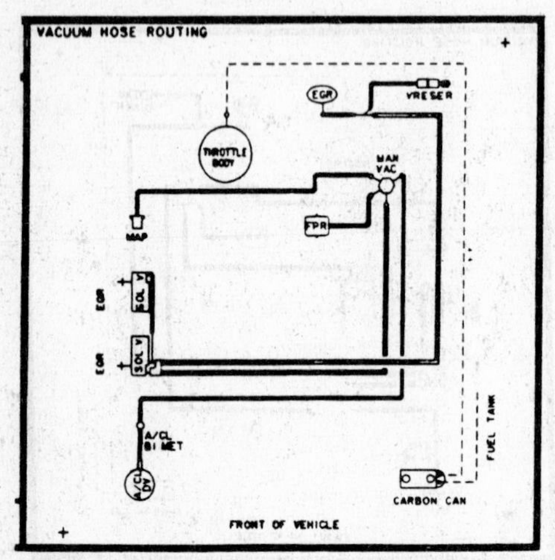

Calibration 5-50S-R02—2.3L Ranger 1985

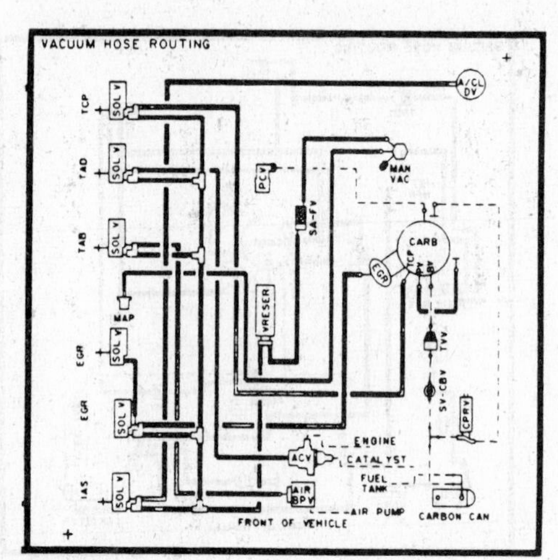

Calibration 5-61F-R01—2.8L Ranger (Chassis Cab LD) 1985

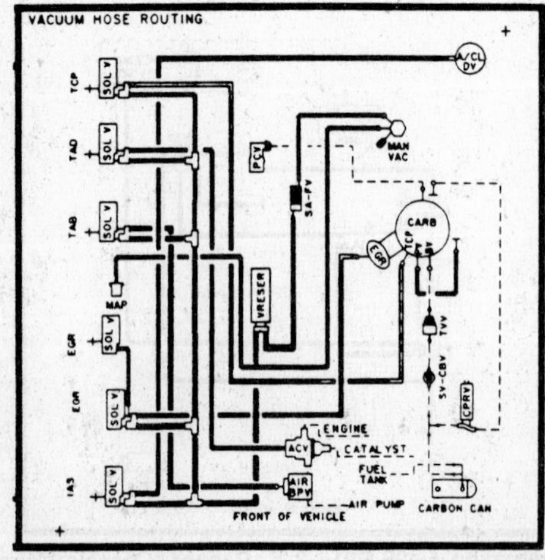

Calibration 5-61F-R01—2.8L Ranger/Bronco II 1985

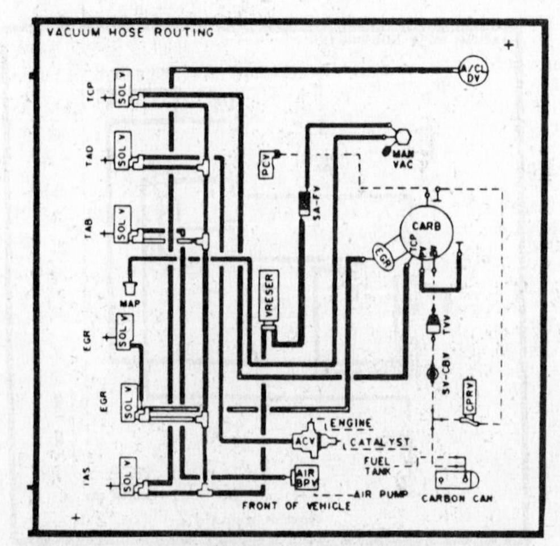

Calibration 5-61S-R02—2.8L Ranger/Bronco II 1985

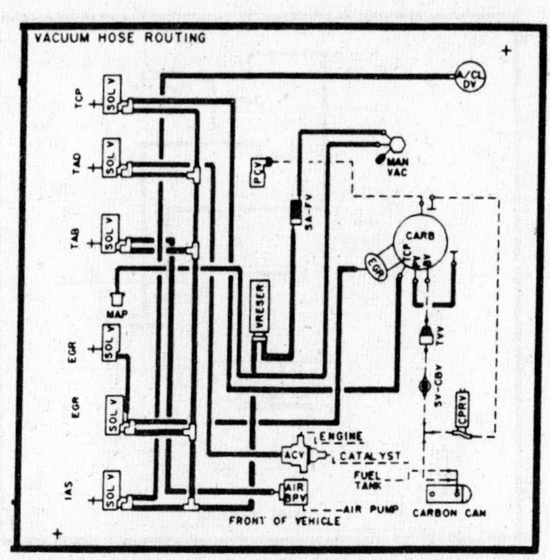

Calibration 5-62E-R01—2.8L Ranger/Bronco II 1985

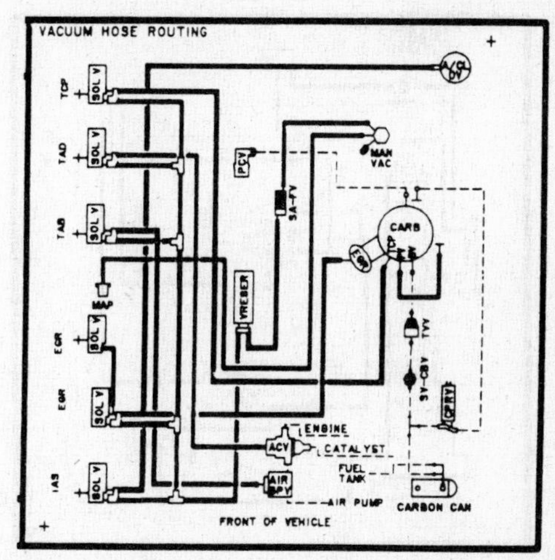

Calibration 5-62E-R01—2.8L Ranger (Chassis Cab) 1985

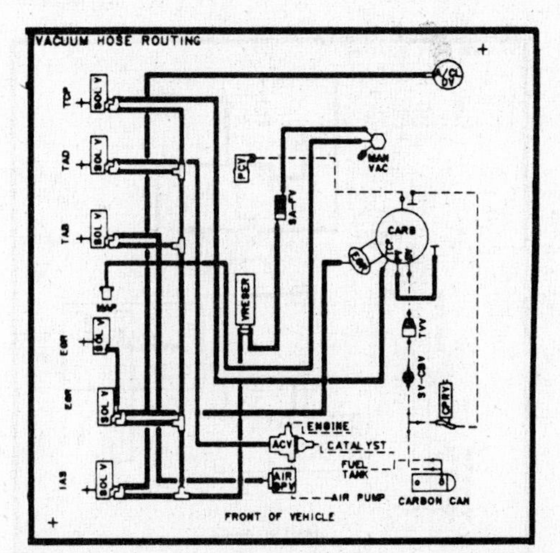

Calibration 5-62E-R01—2.8L Ranger (Chassis Cab) 1985

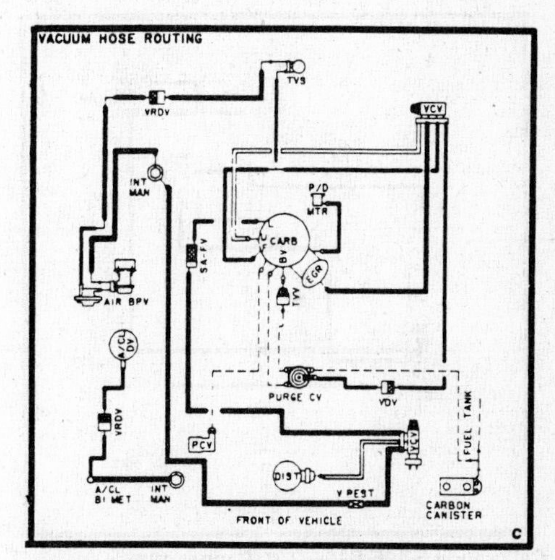

Calibration 6-41D-R10—2.0L Ranger 1986

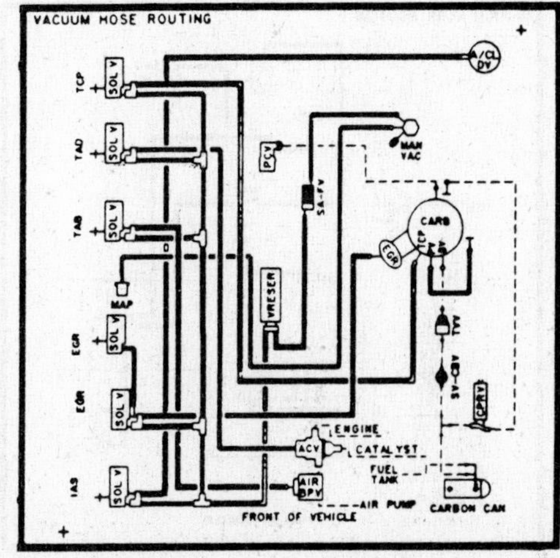

Calibration 5-62E-R01—2.8L Ranger/Bronco II 1985

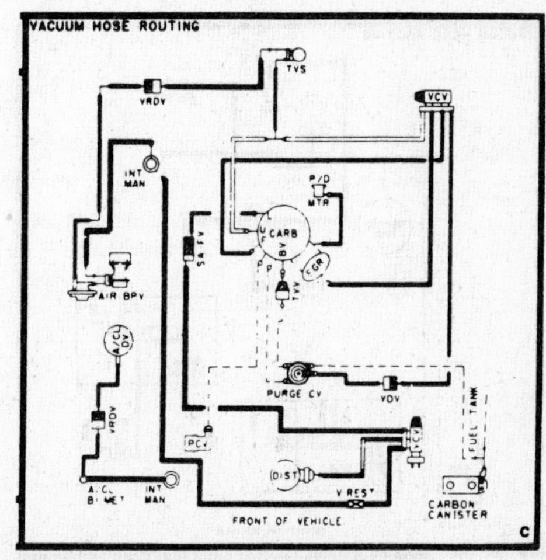

Calibration 6-41D-R00—2.0L Ranger 1986

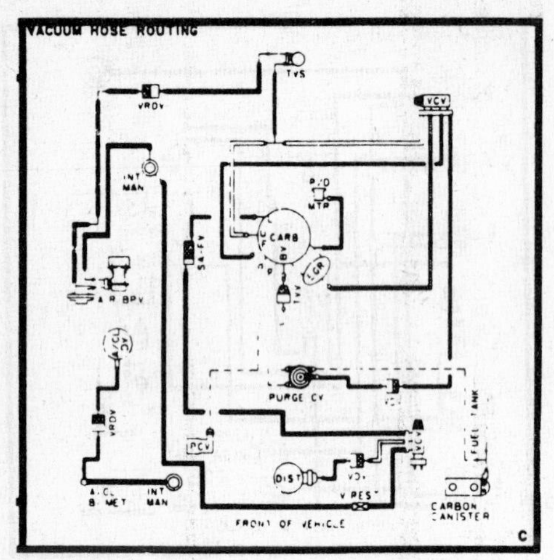

Calibration 6-41D-R01—2.0L Ranger 1986

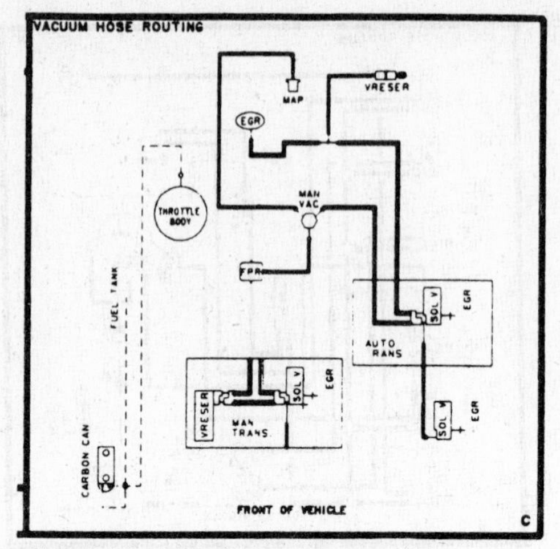

Calibration 5-49J-R10—2.3L Ranger 1986

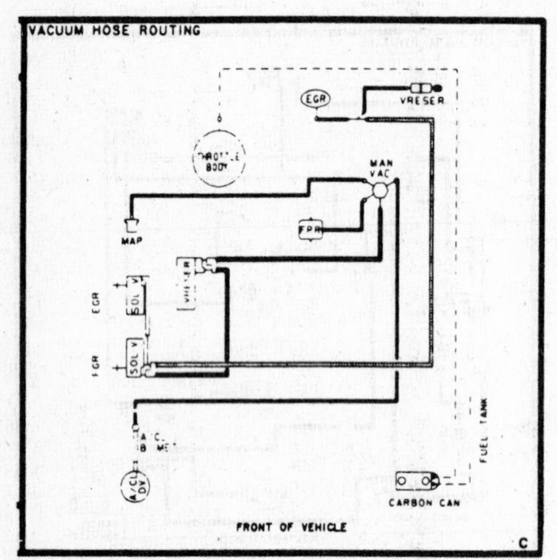

Calibration 5-49F-R10—2.3L Ranger 1986

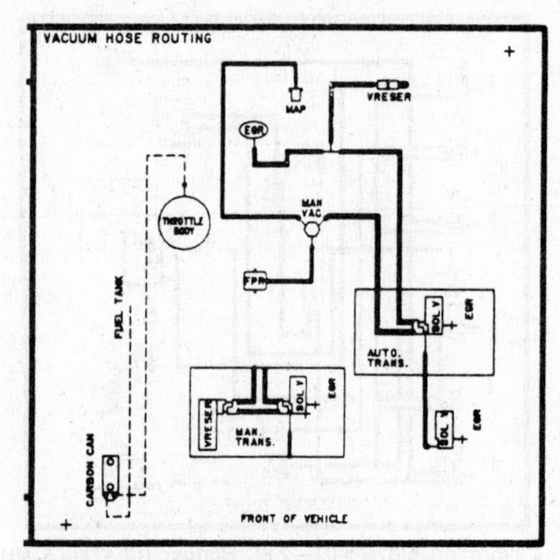

Calibration 5-49R-R00—2.3L Ranger 1986

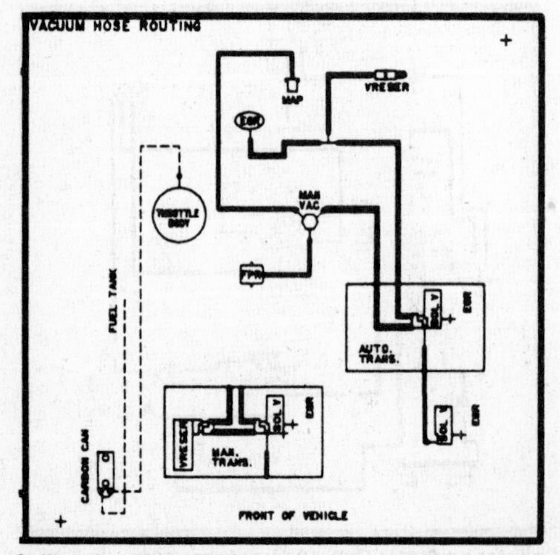

Calibration 5-49J-R00—2.3L Ranger 1986

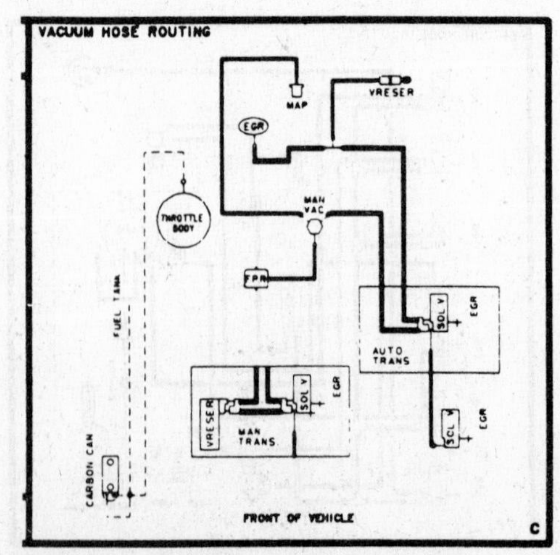

Calibration 5-49R-R10—2.3L Ranger 1986

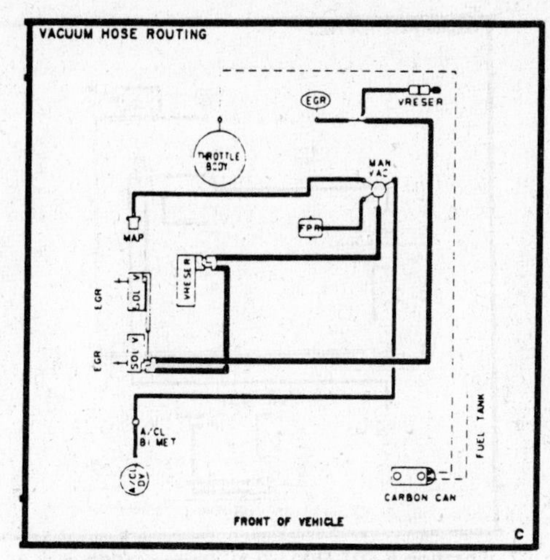

Calibration 5-49S-R10—2.3L Ranger 1986

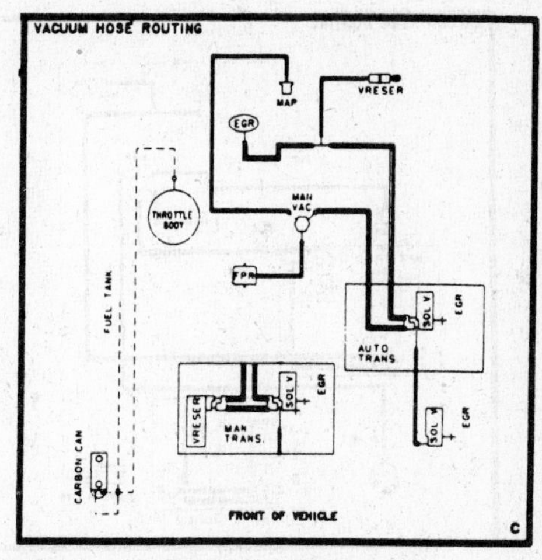

Calibration 5-50R-R00—2.3L Ranger 1986

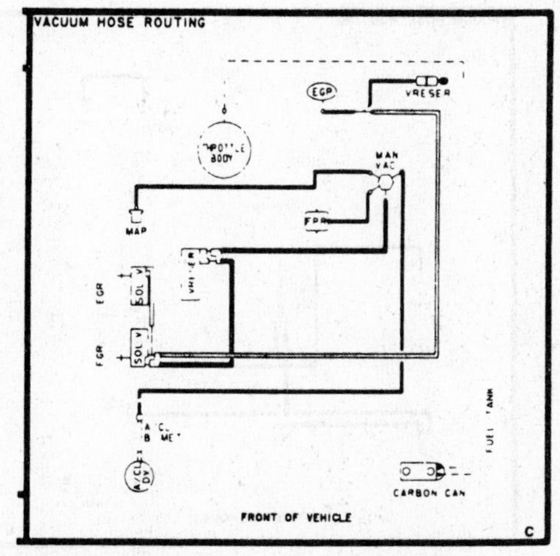

Calibration 5-50H-R10—2.3L Ranger 1986

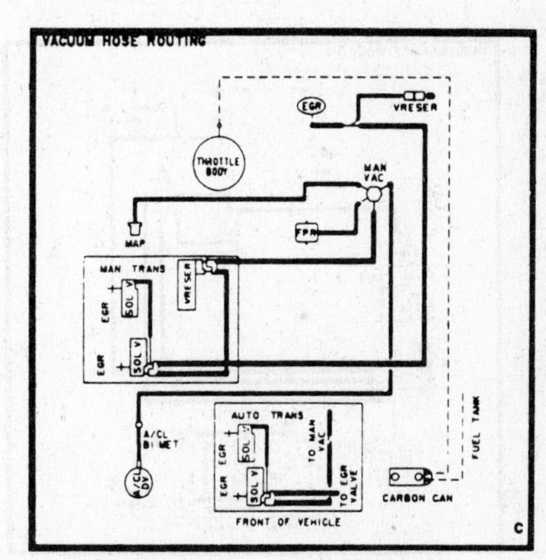

Calibration 5-50S-R10—2.3L Ranger 1986

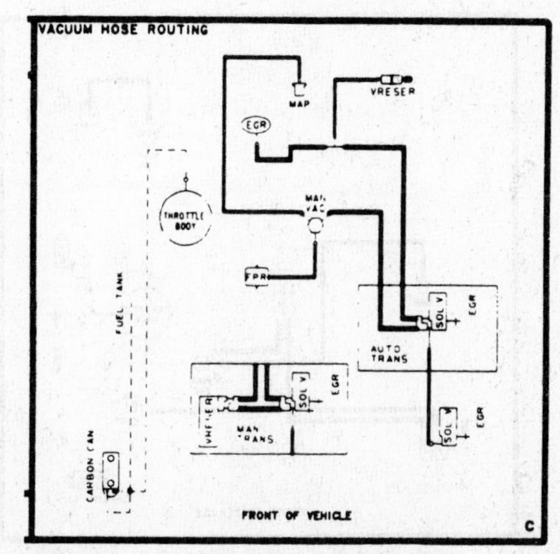

Calibration 5-50J-R00—2.3L Ranger 1986

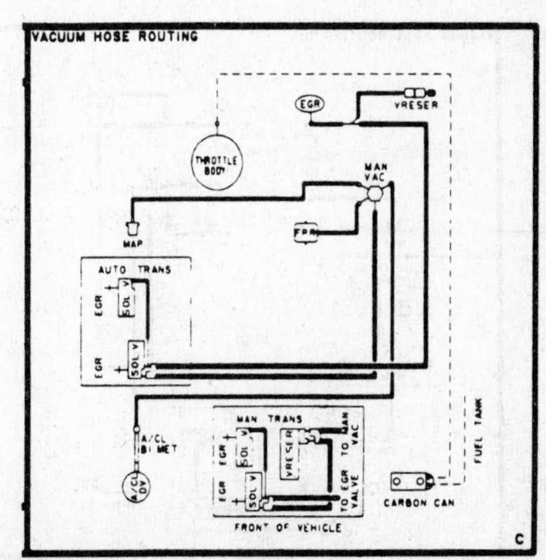

Calibration 6-49F-R00—2.3L Ranger 1986

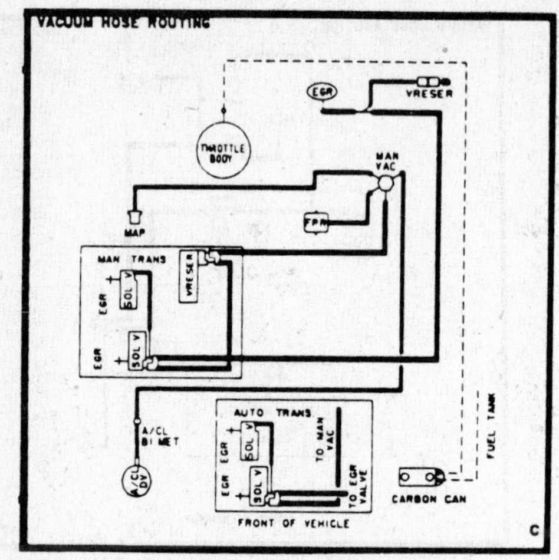

Calibration 6-49S-R00—2.3L Ranger 1986

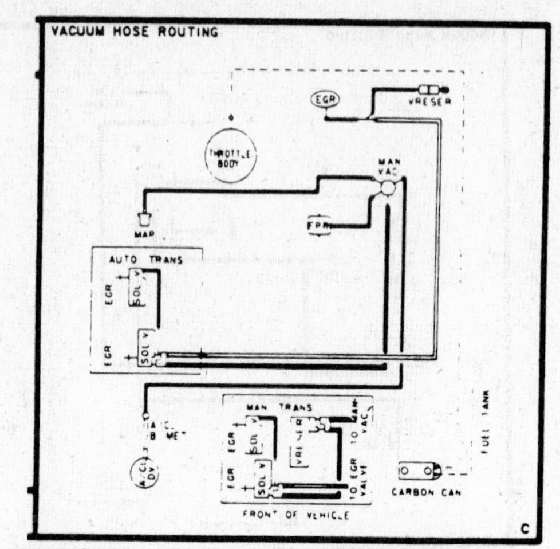

Calibration 6-50T-R00—2.3L Ranger 1986

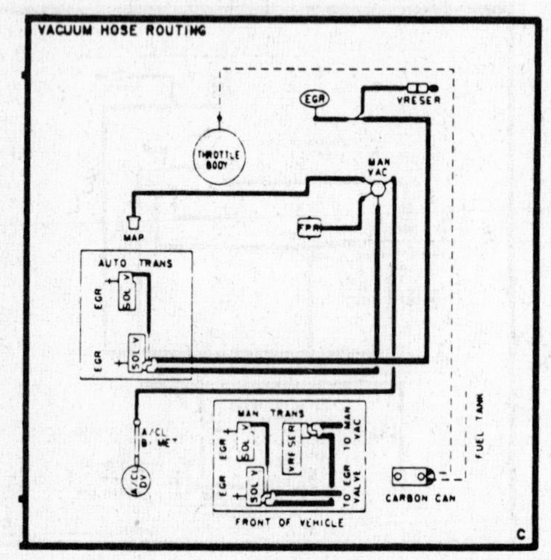

Calibration 6-49S-R00—2.3L Ranger (Super Cab) 1986

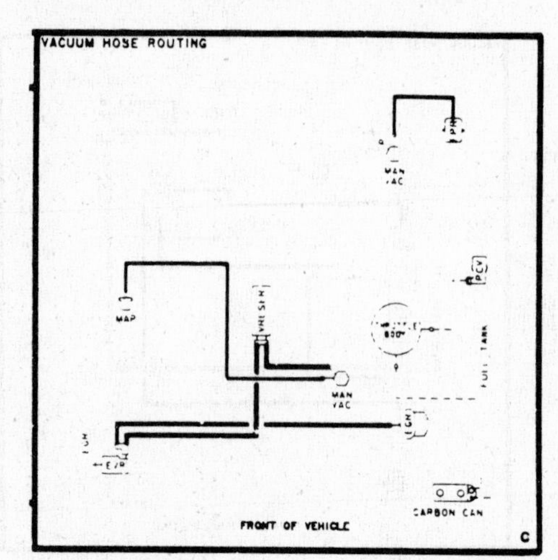

Calibration 6-65F-R01—2.9L Ranger/Bronco II 1986

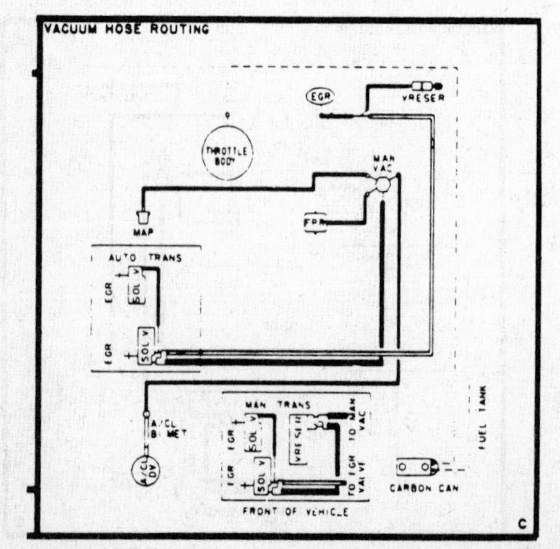

Calibration 6-50L-R00—2.3L Ranger (Super Cab) 1986

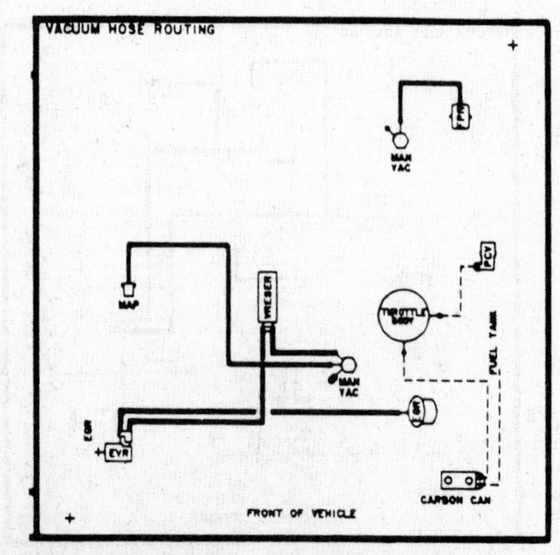

Calibration 6-65F-R01—2.9L Ranger 1986

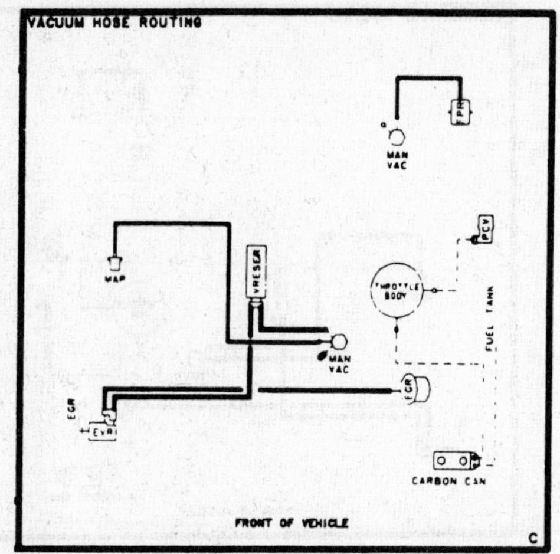

Calibration 6-65H-R00—2.9L Ranger (Chassis Cab HD) 1986

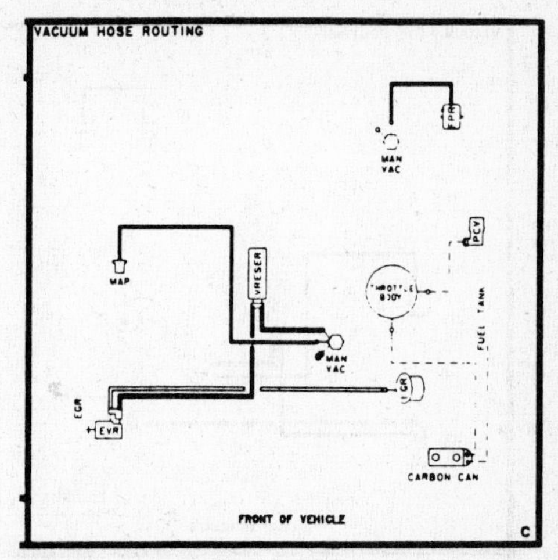

Calibration 6-65H-R10—2.9L Ranger (Chassis Cab LD) 1986

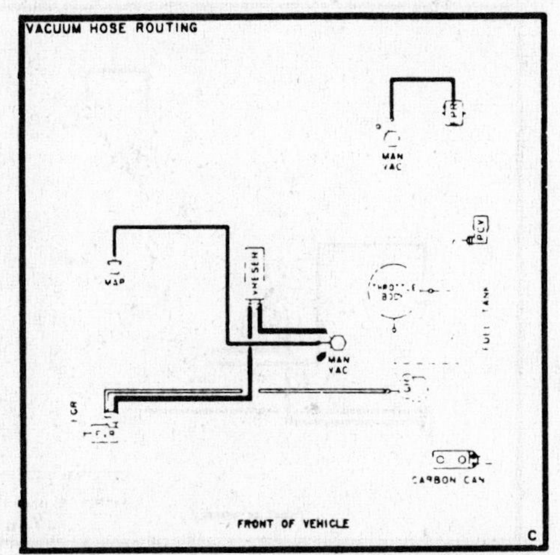

Calibration 6-65H-R00—2.9L Ranger (Chassis Cab LD) 1986

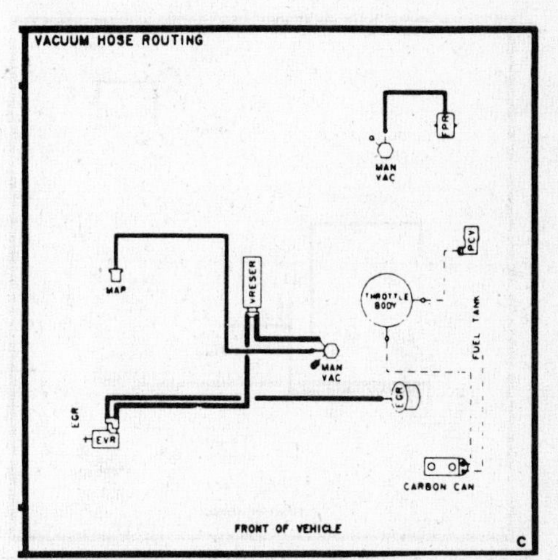

Calibration 6-65T-R00—2.9L Ranger (chassis Cab HD) 1986

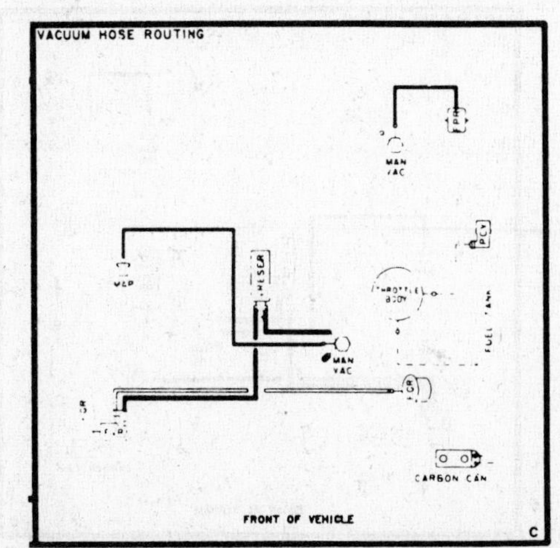

Calibration 6-65H-R10—2.9L Ranger (Chassis Cab HD) 1986

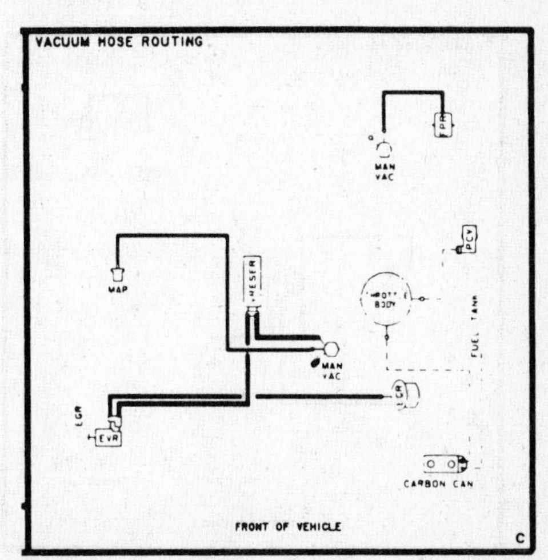

Calibration 6-65T-R00—2.9L Ranger (Chassis Cab LD) 1986

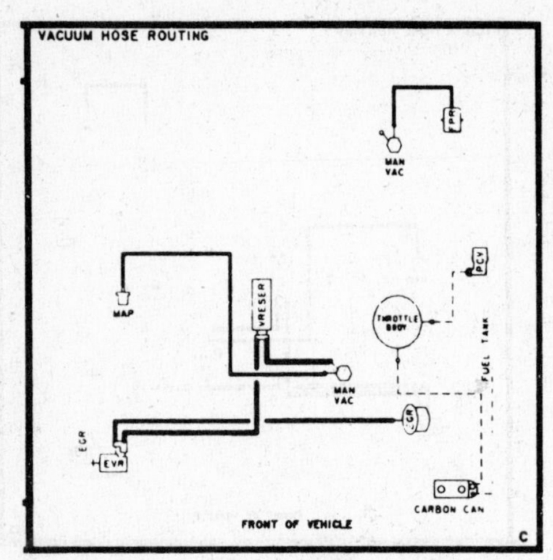

Calibration 6-65T-R00—2.9L Ranger 1986

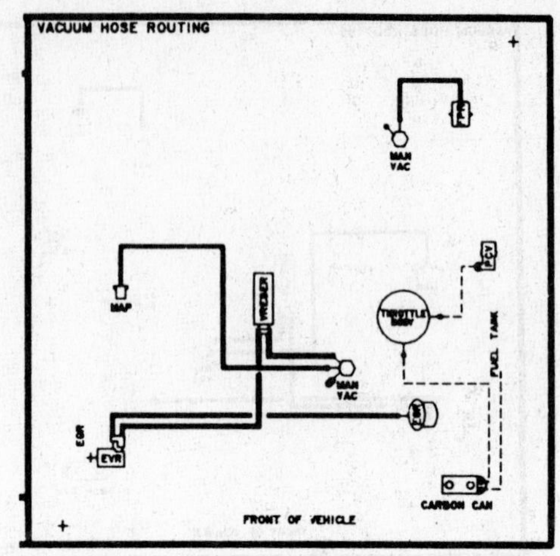

Calibration 6-65T-R10—2.9L Ranger/Bronco II 1986

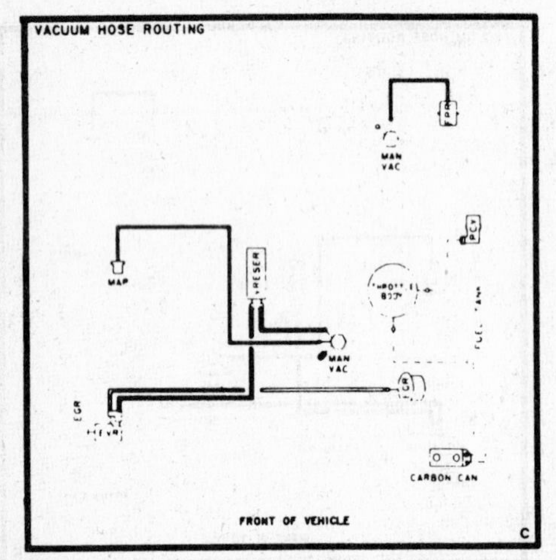

Calibration 6-65T-R10—2.9L Ranger (Chassis Cab HD) 1986

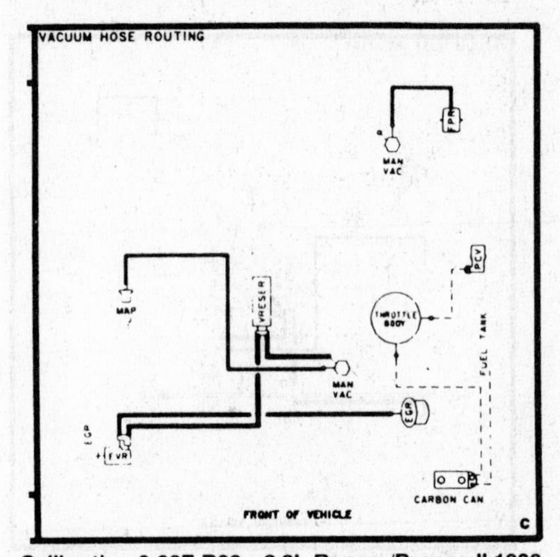

Calibration 6-66F-R00—2.9L Ranger/Bronco II 1986

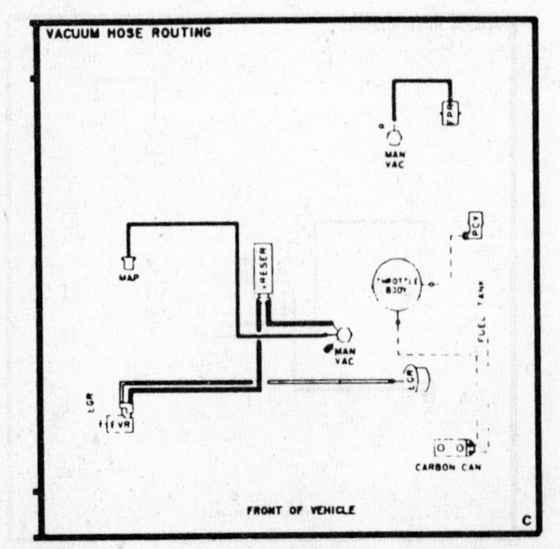

Calibration 6-65T-R10—2.9L Ranger (Chassis Cab LD) 1986

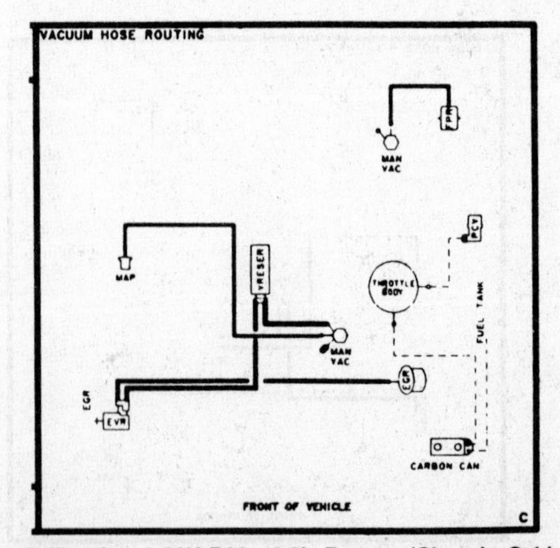

Calibration 6-66H-R00—2.9L Ranger (Chassis Cab) 1986

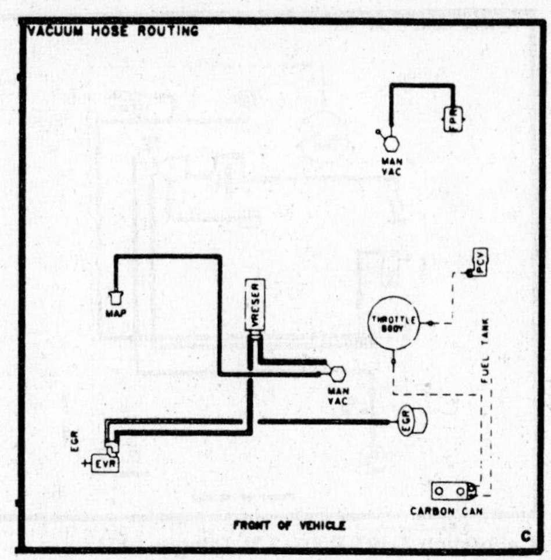

Calibration 6-66H-R00—2.9L Ranger/Bronco II 1986

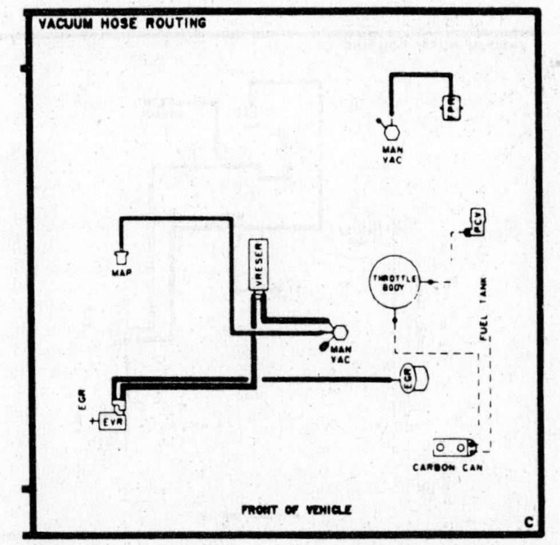

Calibration 6-66S-R00—2.9L Ranger (Chassis Cab) 1986

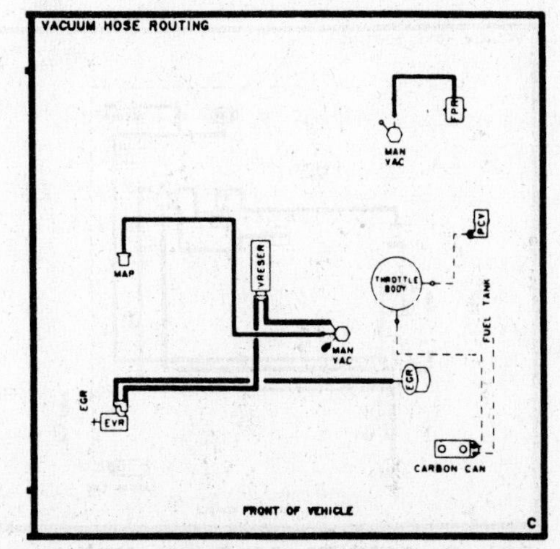

Calibration 6-66S-R00—2.9L Ranger/Bronco II 1986

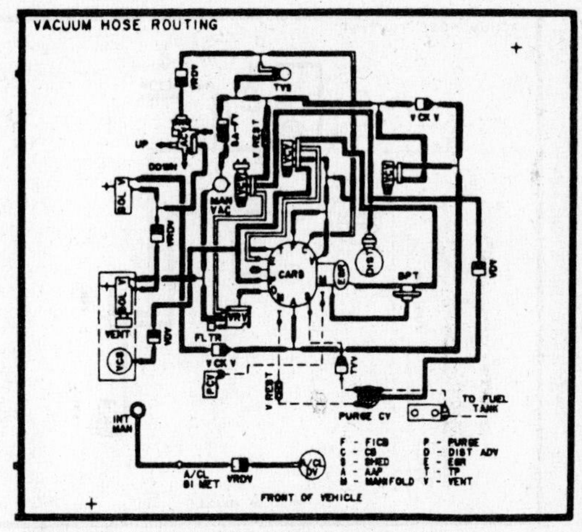

Calibration 7-41D-R00—2.0L Ranger 1987

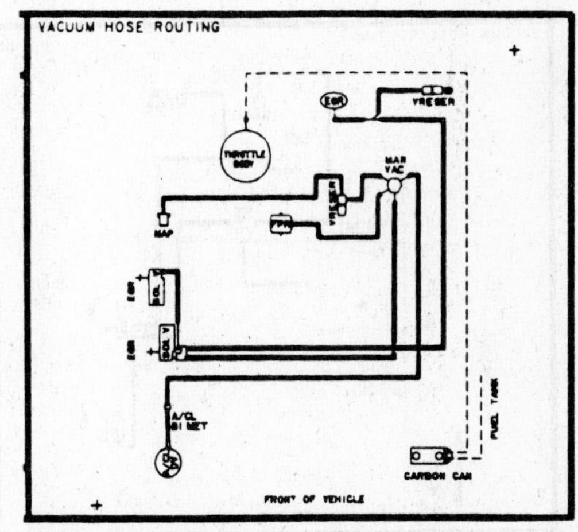

Calibration 7-49F-R00—2.3L Ranger 1987

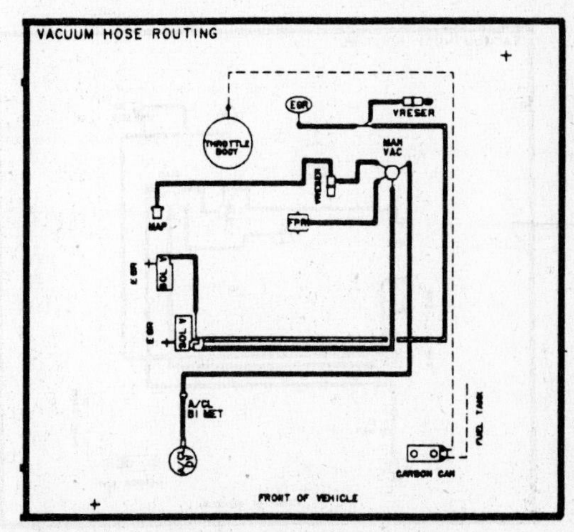

Calibration 7-49H-R00—2.3L Ranger 1987

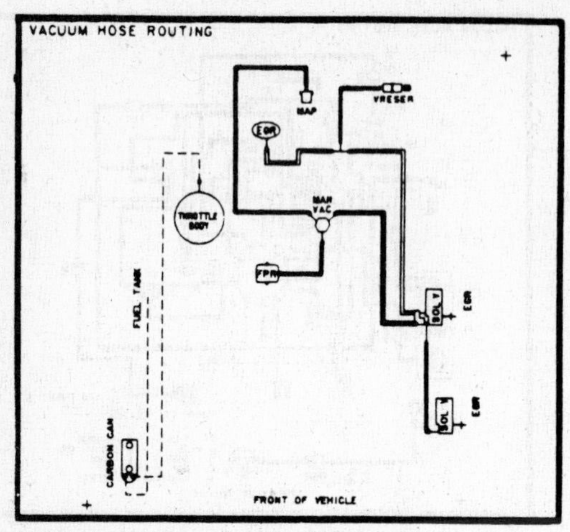

Calibration 7-49J-R00—2.3L Ranger 1987

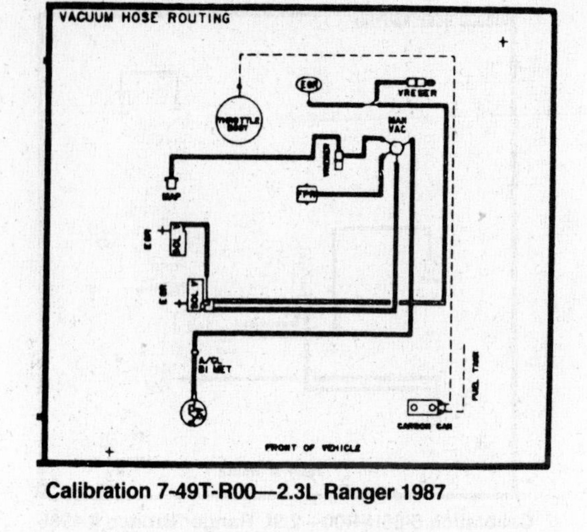

Calibration 7-49T-R00—2.3L Ranger 1987

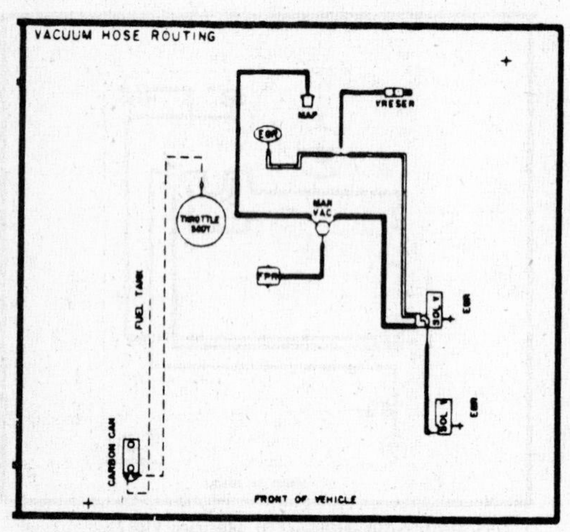

Calibration 7-49R-R00—2.3L Ranger 1987

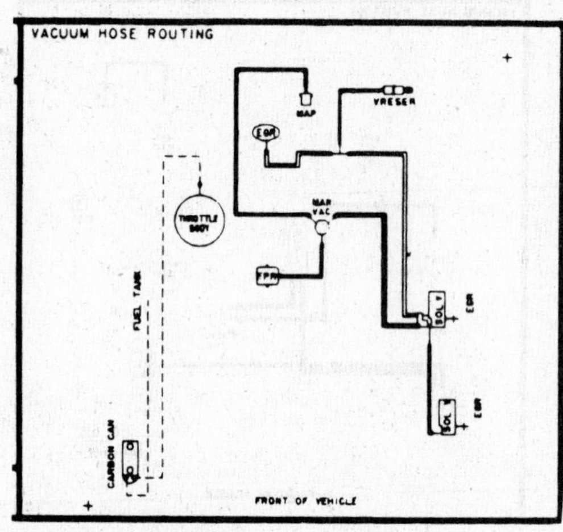

Calibration 7-50J-R00—2.3L Ranger 1987

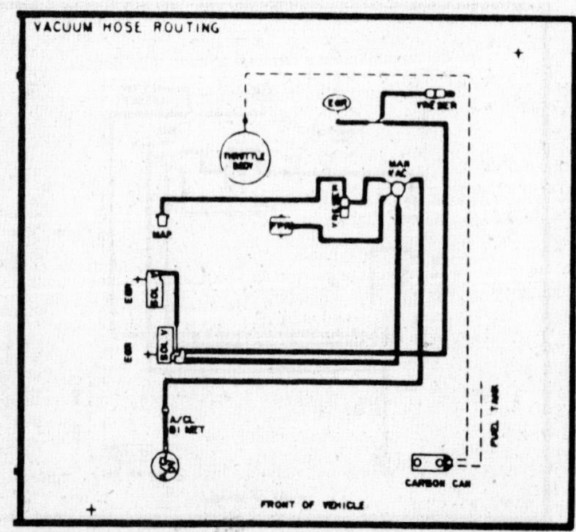

Calibration 7-49S-R00—2.3L Ranger 1987

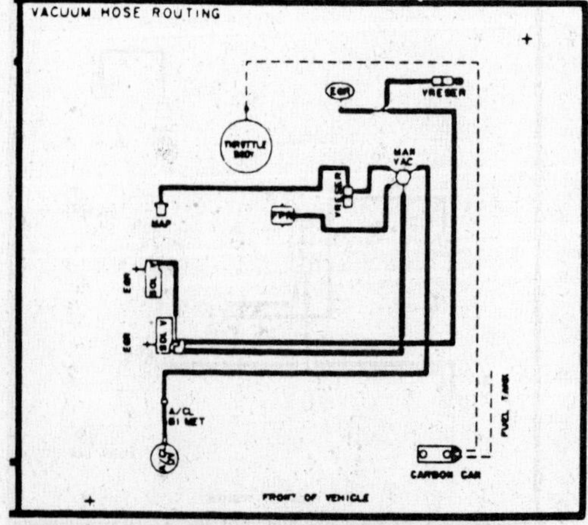

Calibration 7-50K-R00—2.3L Ranger 1987

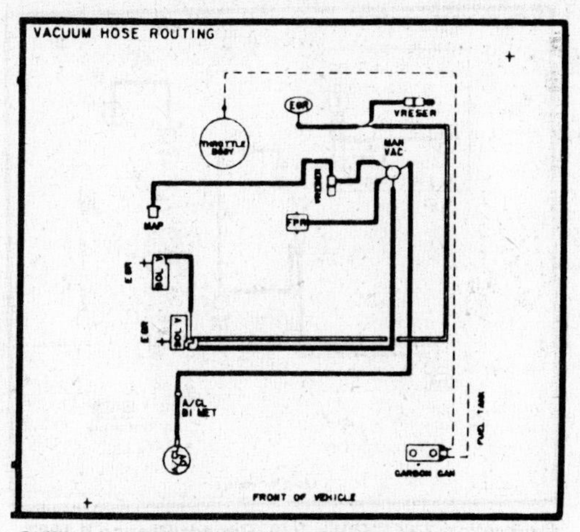

Calibration 7-50L-R00—2.3L Ranger 1987

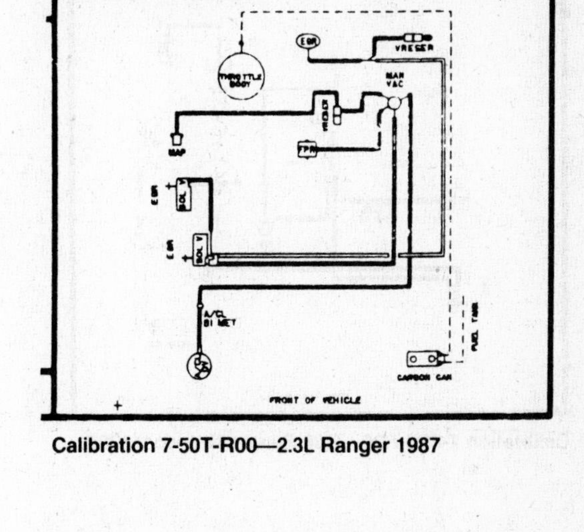

Calibration 7-50T-R00—2.3L Ranger 1987

Calibration 7-50R-R00—2.3L Ranger 1987

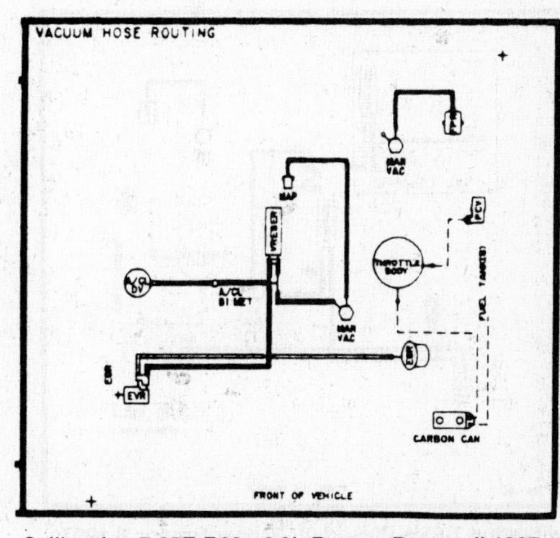

Calibration 7-65F-R00—2.3L Ranger/Bronco II 1987

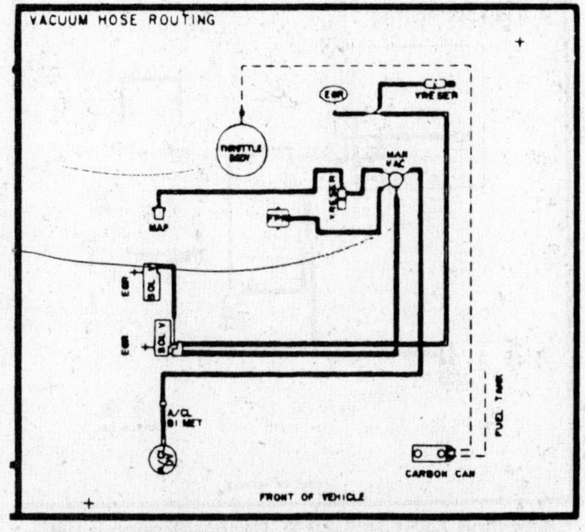

Calibration 7-50S-R00—2.3L Ranger 1987

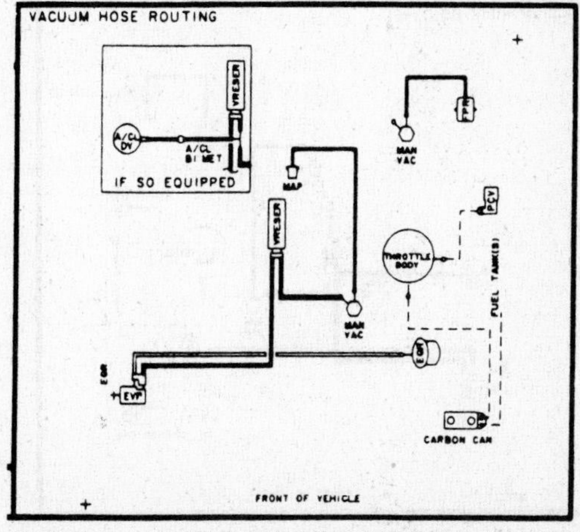

Calibration 7-65F-R05—2.9L Ranger/Bronco II 1987

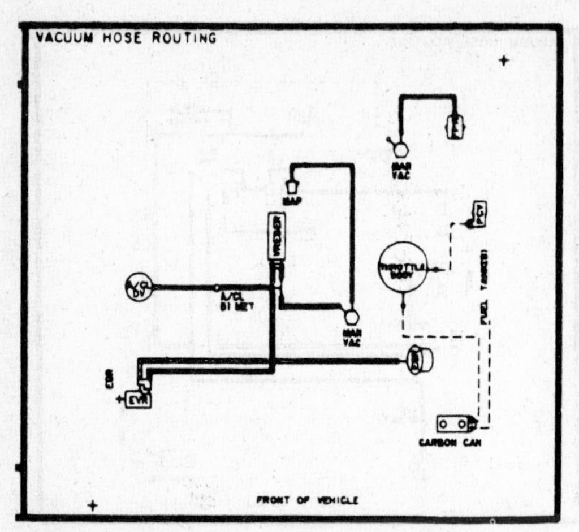

Calibration 7-65H-R00—2.9L Ranger/Bronco II 1987

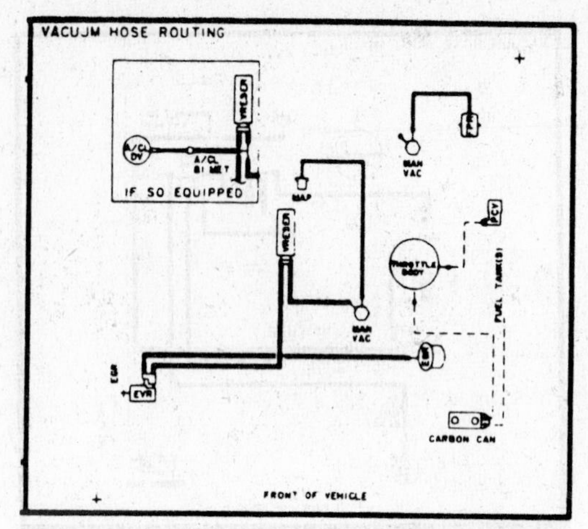

Calibration 7-65S-R05—2.9L Ranger/Bronco II 1987

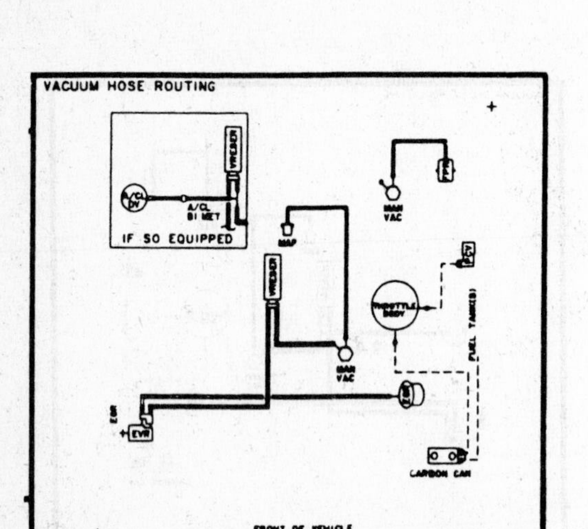

Calibration 7-65H-R05—2.9L Ranger/Bronco II 1987

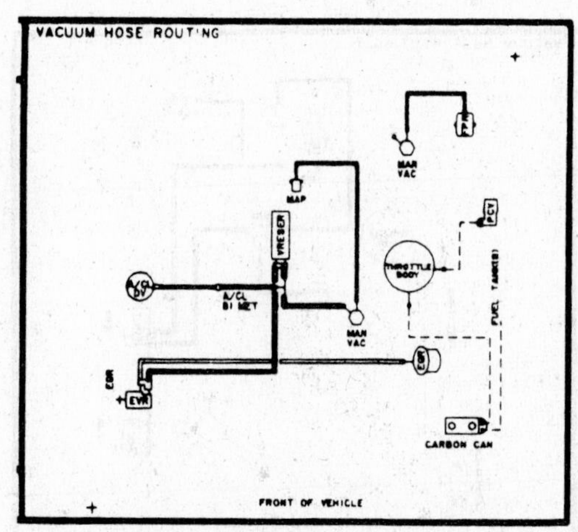

Calibration 7-65T-R00—2.9L Ranger/Bronco II 1987

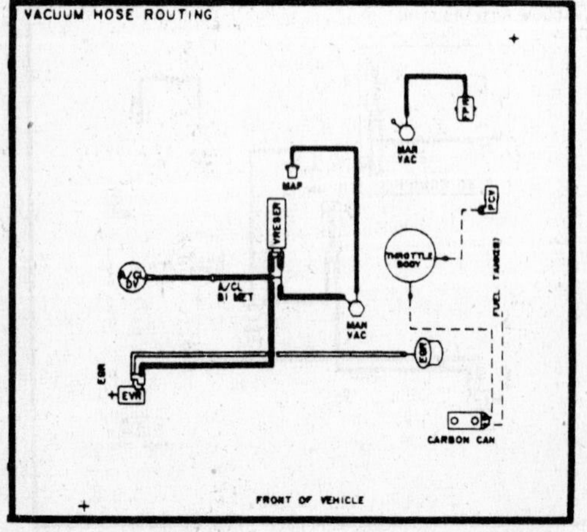

Calibration 7-65S-R00—2.9L Ranger/Bronco II 1987

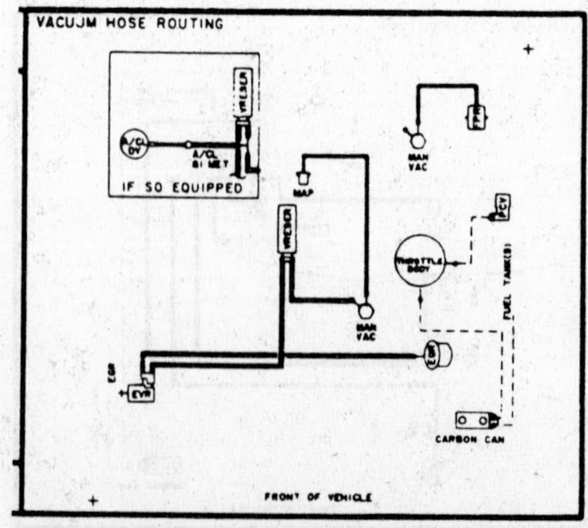

Calibration 7-65T-R05—2.9L Ranger/Bronco II 1987

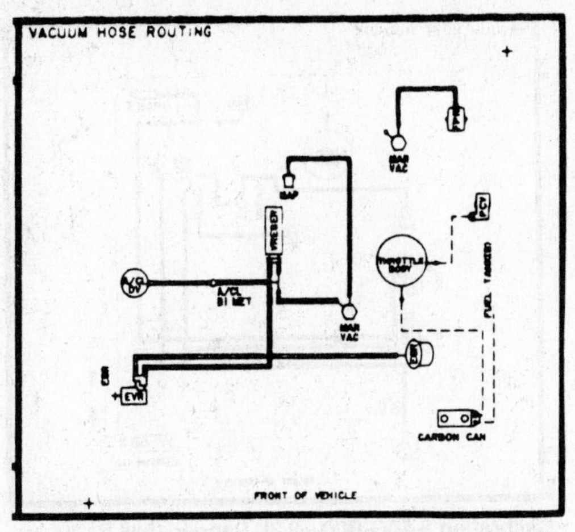

Calibration 7-66F-R00—2.9L Ranger/Bronco II 1987

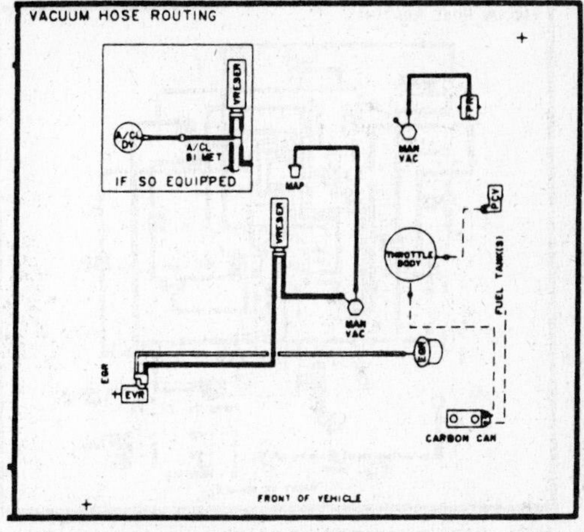

Calibration 7-66H-R05—2.9L Ranger/Bronco II 1987

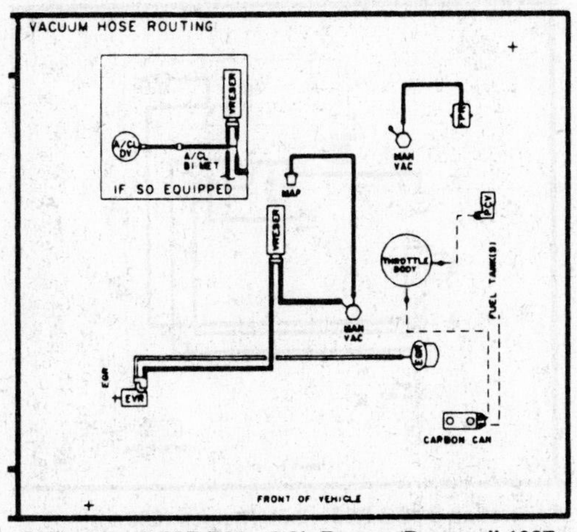

Calibration 7-66F-R05—2.9L Ranger/Bronco II 1987

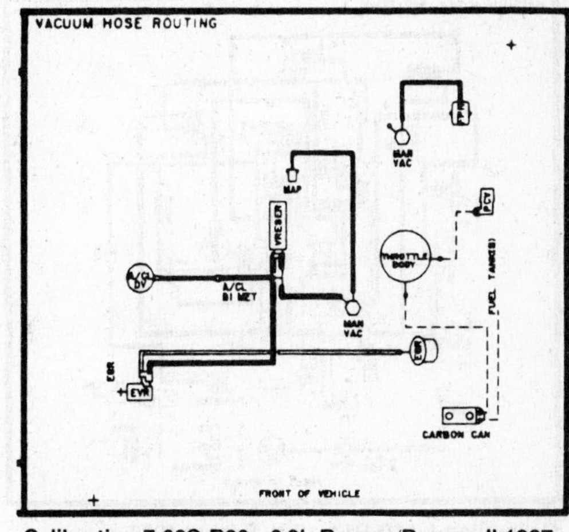

Calibration 7-66S-R00—2.9L Ranger/Bronco II 1987

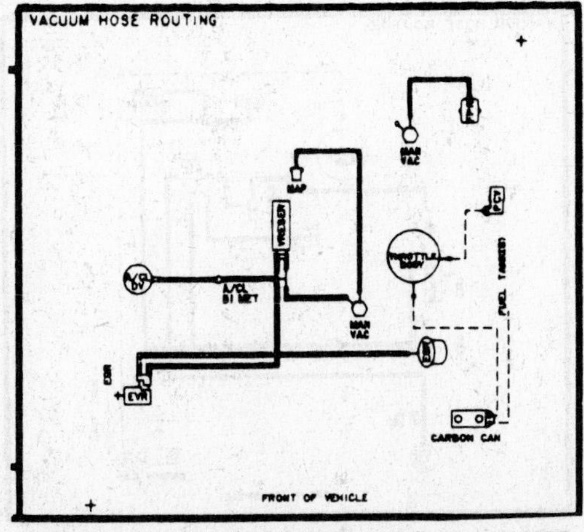

Calibration 7-66H-R00—2.9L Ranger/Bronco II 1987

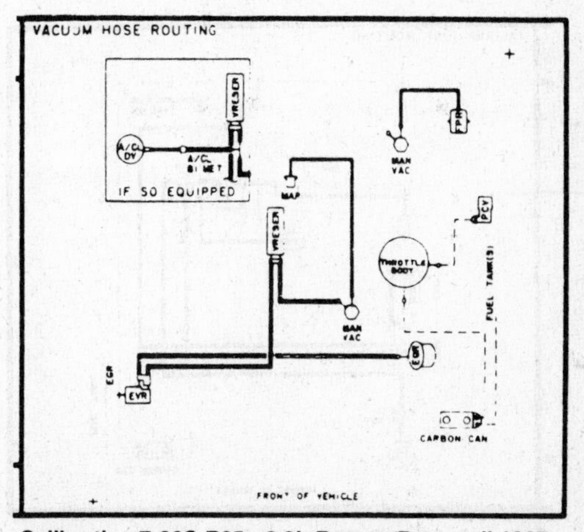

Calibration 7-66S-R05—2.9L Ranger/Bronco II 1987

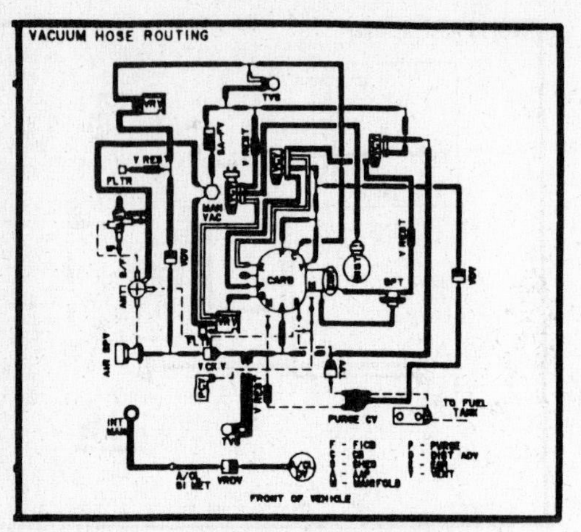

Calibration 8-41D-R04—2.0L Ranger 1988

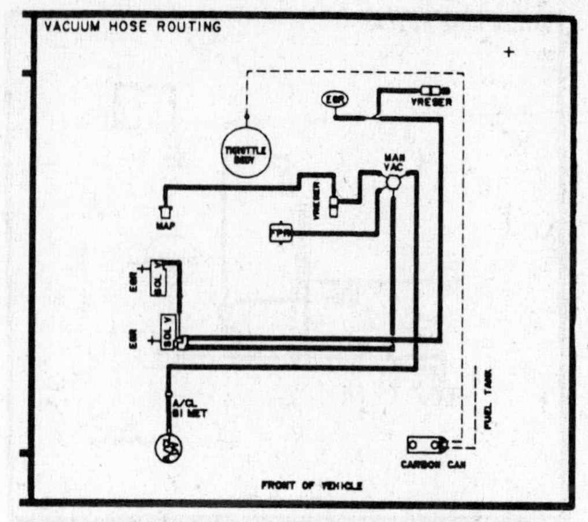

Calibration 8-49G-R00—2.3L Ranger 1988

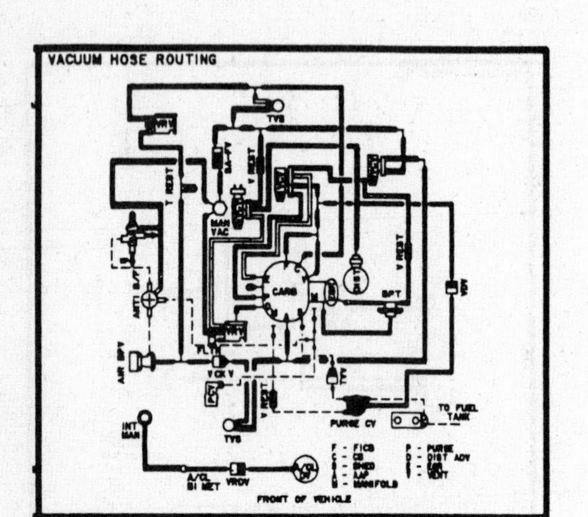

Calibration 8-41D-R11—2.0L Ranger 1988

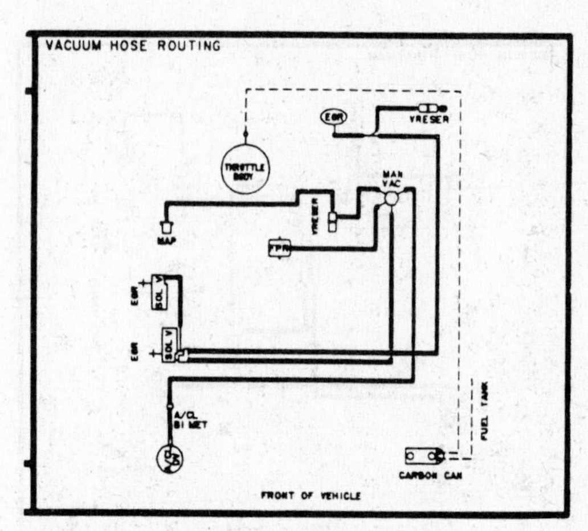

Calibration 8-49G-R00—2.3L Ranger 1988

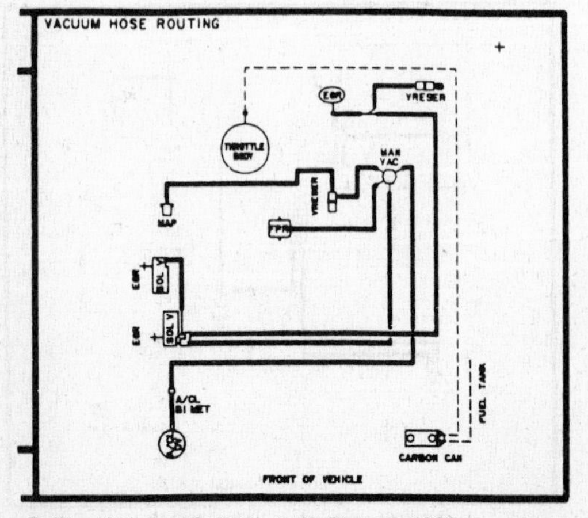

Calibration 8-49G-R00—2.3L Ranger 1988

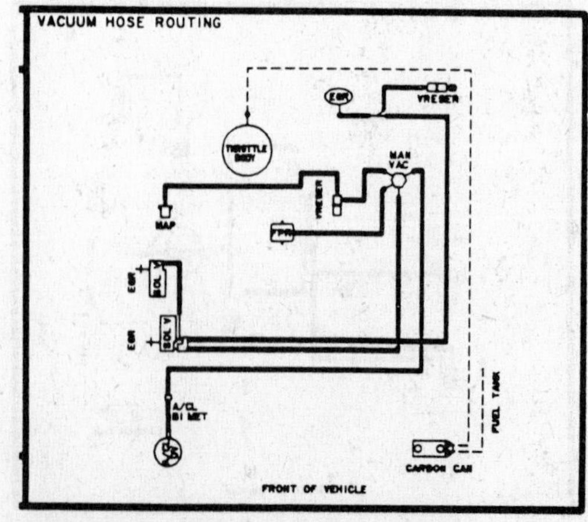

Calibration 8-49T-R10—2.3L Ranger 1988

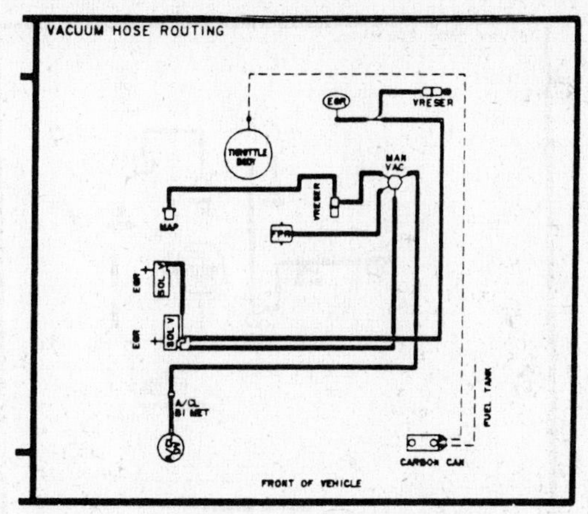

Calibration 8-50T-R00—2.3L Ranger 1988

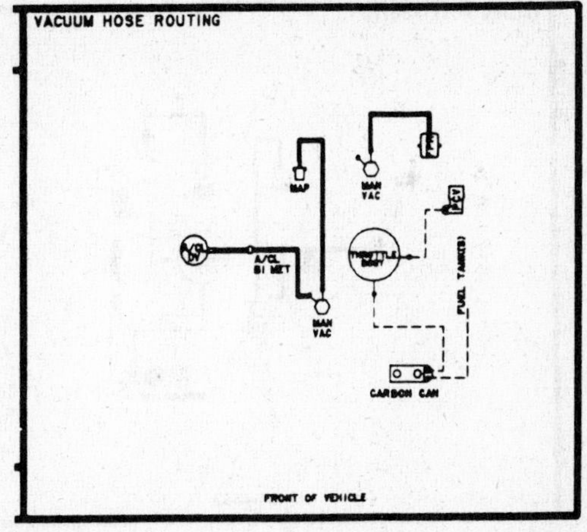

Calibration 8-65T-R00—2.9L Ranger/Bronco II 1988

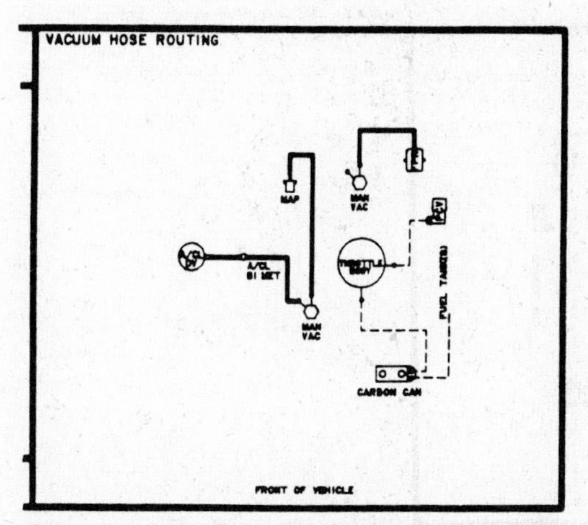

Calibration 8-65H-R00—2.9L Ranger/Bronco II 1988

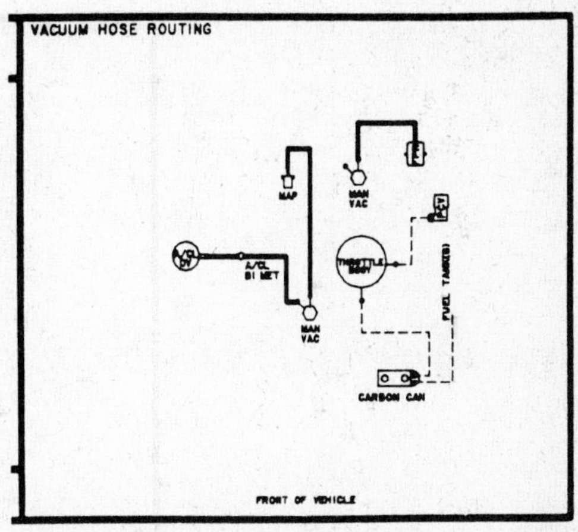

Calibration 8-66F-R00—2.9L Ranger/Bronco II 1988

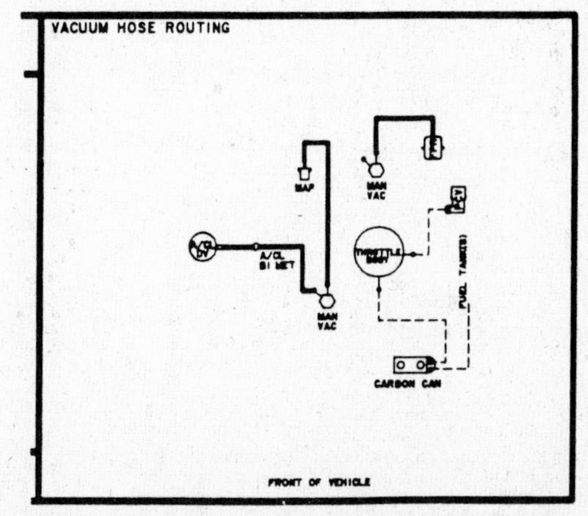

Calibration 8-65S-R00—2.9L Ranger/Bronco II 1988

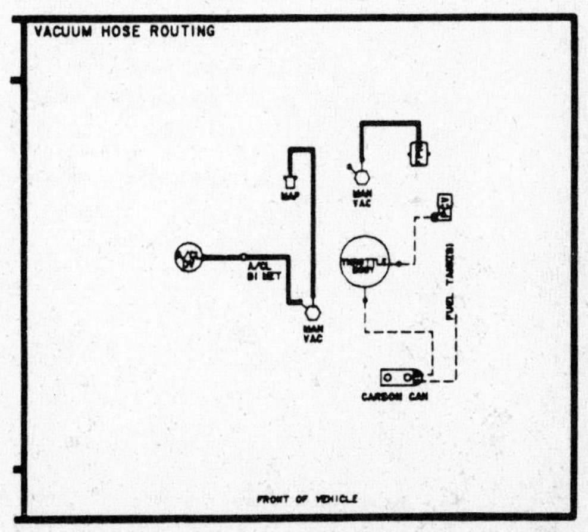

Calibration 8-66F-R11—2.9L Ranger/Bronco II 1988

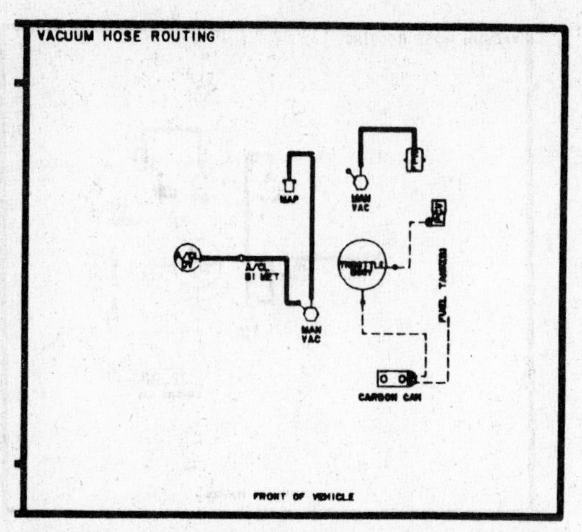

Calibration 8-66H-R00—2.9L Ranger/Bronco II 1988

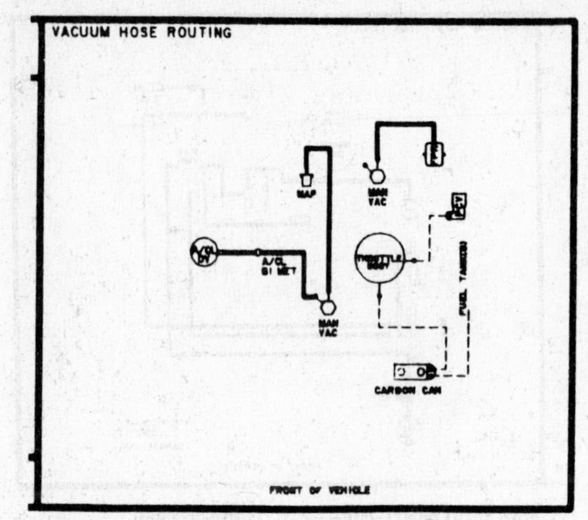

Calibration 8-66S-R00—2.9L Ranger/Bronco II 1988

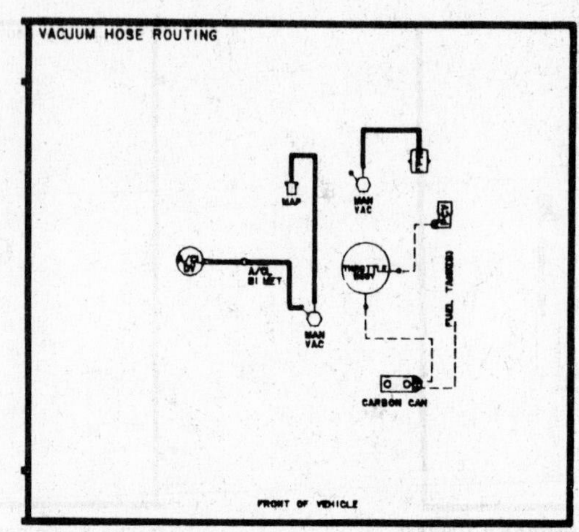

Calibration 8-66H-R11—2.9L Ranger/Bronco II 1988

CALIBRATION: 9–49F–R00

2.3L EFI

FORD MOTOR COMPANY
IMPORTANT VEHICLE INFORMATION

THIS VEHICLE IS EQUIPPED WITH ELECTRONIC FUEL INJECTION. IDLE MIXTURE, COLD ENGINE IDLE SPEED AND COLD ENGINE FUEL ENRICHMENT ARE NOT ADJUSTABLE.

SET PARKING BRAKE AND BLOCK WHEELS. MAKE ALL ADJUSTMENTS WITH ENGINE AT NORMAL OPERATING TEMPERATURE, TRANSMISSION IN NEUTRAL AND ACCESSORIES OFF.

IGNITION TIMING–
(1) TURN OFF ENGINE.
(2) DISCONNECT THE IN-LINE SPOUT CONNECTOR (⇒⊏⊐).
(3) RE-START PREVIOUSLY WARMED-UP ENGINE.
(4) ADJUST IGNITION TIMING TO 10° BTDC.
(5) TURN OFF ENGINE AND RESTORE ELECTRICAL CONNECTION.

THIS ENGINE IS EQUIPPED WITH AUTOMATIC IDLE SPEED CONTROL. IDLE RPM IS NOT ADJUSTABLE. IF NOT WITHIN SPECIFIED RPM RANGE, SEE SHOP MANUAL:
MANUAL TRANS. IN NEUTRAL: 645–795 RPM

USE SAE 5W-30 OIL – API CATEGORY SG, SG/CC OR SG/CD.

THIS VEHICLE CONFORMS TO U.S. EPA REGULATIONS APPLICABLE TO 1989 MODEL YEAR NEW LIGHT-DUTY TRUCKS.

E9AE-9C485- | CATALYST | SPARK PLUG: AWSF-32C GAP- .042-.046 2.3L-9HM KFM2.3T5FNFX - EGS/TWC/FI/EGR

VACUUM HOSE ROUTING

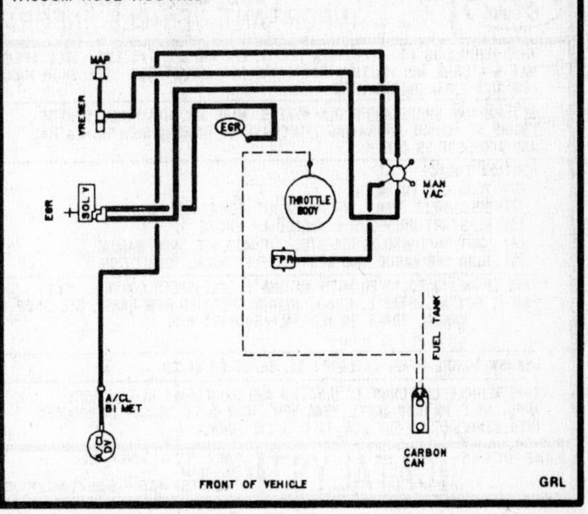

CALIBRATION PARTS LIST

CODE: BFB

NAME/DESCRIPTION	ENGINEERING NO.	SERVICE NO.
SENSOR, Exhaust Gas Recirculation Valve	E7TF 9G428-AA	E7TZ 9G428-A
CONTROL ASSY., EGR Vacuum Regulator	E8TE 9J459-AA	E8TZ 9J459-A
SENSOR ASSY., Exhaust Gas Oxygen	E77F 9F472-AA, BA, E7TF 9F472-AA	E7TZ 9F472-A
SENSOR ASSY., Manifold Absolute Pressure	E7EF 9F479-A1A, A2A	E7FZ 9F479-A
VALVE ASSY., E.G.R. Pressure External	E8TE 9F483-D2A	E8TZ 9F483-D
INJECTOR ASSY., Fuel	E59E 9F593-A2B	E5TZ 9F593-A
SENSOR ASSY., Air Cleaner Air Temperature Control		D7FZ 9E607-A
MOTOR ASSY., Air Cleaner Vacuum		D4FZ 9D612-A
VALVE ASSY., Throttle Air Bypass	E9TE 9F715-A1A	E9TZ 9F715-A
SENSOR ASSY., Speed	E3AF 9E731-AB	E3AZ 9E731-A
REGULATOR ASSY., Fuel Charging Pressure	E7EE 9C968-AA, E77E 9C968-AA	E7FZ 9C968-A
POTENTIOMETER ASSY., Throttle Position Sensor	E7DF 9B989-AA	E7DZ 9B989-A
SENSOR ASSY., E.E.C. Coolant Temperature	E1AF 12A648-AA, E4AF 12A648-AA	E1AZ 12A648-A
PROCESSOR AND CALIBRATOR ASSY., E.E.C. IV	E9TF 12A650-A1A, A2A	E9TZ 12A650-AA
SENSOR ASSY., Air Charge Temperature	E4AF 12A697-AA	E1AZ 12A697-A

CALIBRATION: 9–49S–R00 — 2.3L EFI

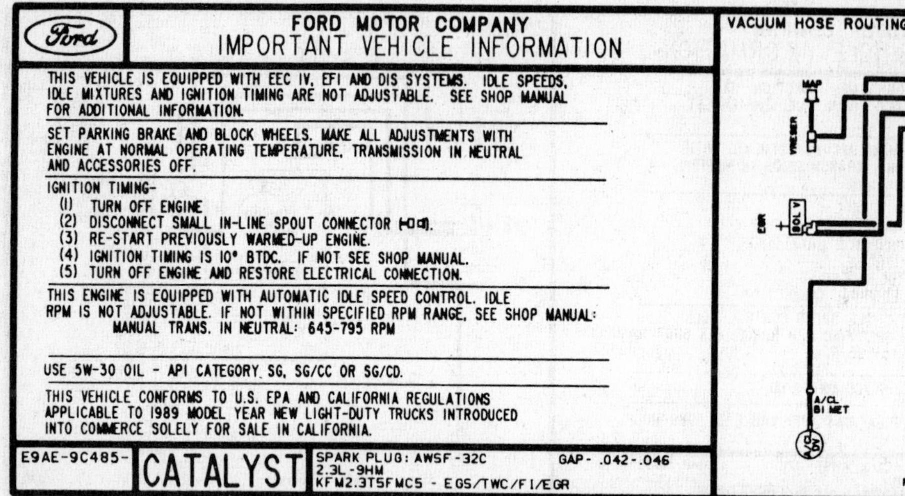
FORD MOTOR COMPANY
IMPORTANT VEHICLE INFORMATION

THIS VEHICLE IS EQUIPPED WITH EEC IV, EFI AND DIS SYSTEMS. IDLE SPEEDS, IDLE MIXTURES AND IGNITION TIMING ARE NOT ADJUSTABLE. SEE SHOP MANUAL FOR ADDITIONAL INFORMATION.

SET PARKING BRAKE AND BLOCK WHEELS. MAKE ALL ADJUSTMENTS WITH ENGINE AT NORMAL OPERATING TEMPERATURE, TRANSMISSION IN NEUTRAL AND ACCESSORIES OFF.

IGNITION TIMING-
(1) TURN OFF ENGINE
(2) DISCONNECT SMALL IN-LINE SPOUT CONNECTOR.
(3) RE-START PREVIOUSLY WARMED-UP ENGINE.
(4) IGNITION TIMING IS 10° BTDC. IF NOT SEE SHOP MANUAL.
(5) TURN OFF ENGINE AND RESTORE ELECTRICAL CONNECTION.

THIS ENGINE IS EQUIPPED WITH AUTOMATIC IDLE SPEED CONTROL. IDLE RPM IS NOT ADJUSTABLE. IF NOT WITHIN SPECIFIED RPM RANGE, SEE SHOP MANUAL:
MANUAL TRANS. IN NEUTRAL: 645-795 RPM

USE 5W-30 OIL - API CATEGORY SG, SG/CC OR SG/CD.

THIS VEHICLE CONFORMS TO U.S. EPA AND CALIFORNIA REGULATIONS APPLICABLE TO 1989 MODEL YEAR NEW LIGHT-DUTY TRUCKS INTRODUCED INTO COMMERCE SOLELY FOR SALE IN CALIFORNIA.

E9AE-9C485- | CATALYST | SPARK PLUG: AWSF-32C GAP- .042-.046
2.3L-9HM
KFM2.3T5FMC5 - EGS/TWC/FI/EGR

CALIBRATION PARTS LIST — CODE: BHH

NAME/DESCRIPTION	ENGINEERING NO.	SERVICE NO.
SENSOR, Exhaust Gas Recirculation Valve	E7TF 9G428-AA	E7TZ 9G428-A
CONTROL ASSY., E.G.R. Vacuum Regulator	E8TE 9J459-AA	E8TZ 9J459-A
SENSOR ASSY., Exhaust Gas Oxygen	E77F 9F472-AA, BA, E7TF 9F472-AA	E7TZ 9F472-A
SENSOR ASSY., Manifold Absolute Pressure	E7EF 9F479-A1A, A2A	E7FZ 9F479-A
VALVE ASSY., E.G.R. Pressure External	E8TE 9F483-D2A	E8TZ 9F483-D
INJECTOR ASSY., Fuel	E59E 9F593-A2B	E5TZ 9F593-A
SENSOR ASSY., Air Cleaner Air Temperature Control		D7FZ 9E607-A
MOTOR ASSY., Air Cleaner Vacuum		D4FZ 9D612-A
VALVE ASSY., Throttle Air Bypass	E9TE 9F715-A1A	E9TZ 9F715-A
SENSOR ASSY., Speed	E3AF 9E731-AB	E3AZ 9E731-A
REGULATOR ASSY., Fuel Charging Pressure	E7EE 9C968-AA, E77E 9C968-AA	E7FZ 9C968-A
POTENTIOMETER ASSY., Throttle Position Sensor	E7DF 9B989-AA	E7DZ 9B989-A
SENSOR ASSY., E.E.C. Coolant Temperature	E1AF 12A648-AA, E4AF 12A648-AA	E1AZ 12A648-A
PROCESSOR AND CALIBRATOR ASSY., E.E.C. IV	E9TF 12A650-A1A, A2A	E9TZ 12A650-AA
SENSOR ASSY., Air Charge Temperature	E4AF 12A697-AA	E1AZ 12A697-A

CALIBRATION: 9–50K–R00 — 2.3L EFI

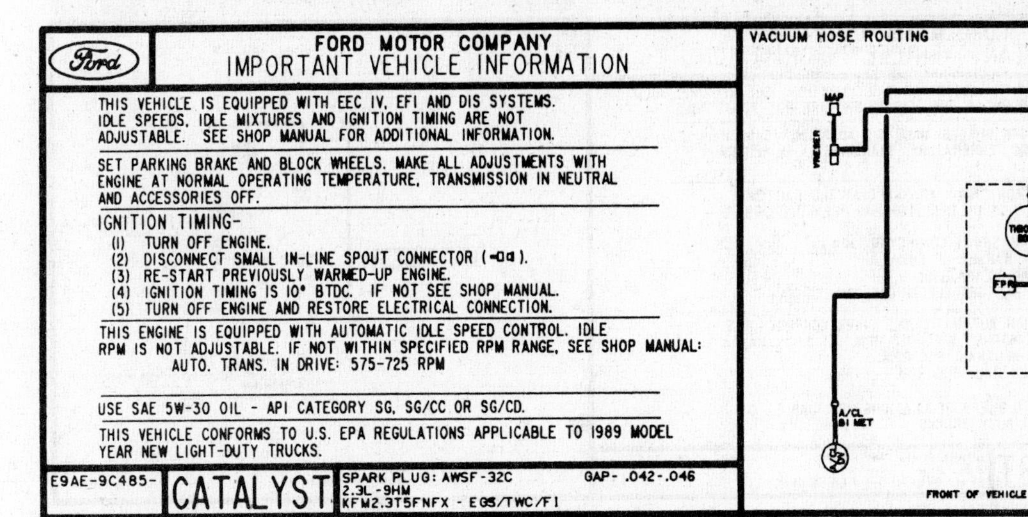
FORD MOTOR COMPANY
IMPORTANT VEHICLE INFORMATION

THIS VEHICLE IS EQUIPPED WITH EEC IV, EFI AND DIS SYSTEMS. IDLE SPEEDS, IDLE MIXTURES AND IGNITION TIMING ARE NOT ADJUSTABLE. SEE SHOP MANUAL FOR ADDITIONAL INFORMATION.

SET PARKING BRAKE AND BLOCK WHEELS. MAKE ALL ADJUSTMENTS WITH ENGINE AT NORMAL OPERATING TEMPERATURE, TRANSMISSION IN NEUTRAL AND ACCESSORIES OFF.

IGNITION TIMING–
(1) TURN OFF ENGINE.
(2) DISCONNECT SMALL IN-LINE SPOUT CONNECTOR.
(3) RE-START PREVIOUSLY WARMED-UP ENGINE.
(4) IGNITION TIMING IS 10° BTDC. IF NOT SEE SHOP MANUAL.
(5) TURN OFF ENGINE AND RESTORE ELECTRICAL CONNECTION.

THIS ENGINE IS EQUIPPED WITH AUTOMATIC IDLE SPEED CONTROL. IDLE RPM IS NOT ADJUSTABLE. IF NOT WITHIN SPECIFIED RPM RANGE, SEE SHOP MANUAL:
AUTO. TRANS. IN DRIVE: 575-725 RPM

USE SAE 5W-30 OIL - API CATEGORY SG, SG/CC OR SG/CD.

THIS VEHICLE CONFORMS TO U.S. EPA REGULATIONS APPLICABLE TO 1989 MODEL YEAR NEW LIGHT-DUTY TRUCKS.

E9AE-9C485- CATALYST SPARK PLUG: AWSF-32C GAP-.042-.046
2.3L-9HM
KFM2.3T5FNFX - EGS/TWC/FI

CALIBRATION PARTS LIST — CODE: BHJ

NAME/DESCRIPTION	ENGINEERING NO.	SERVICE NO.
SENSOR ASSY., Exhaust Gas Oxygen	E7FF 9F472-AA, BA, E7TF 9F472-AA	E7TZ 9F472-A
SENSOR ASSY., Manifold Absolute Pressure	E7EF 9F479-A1A, A2A	E7FZ 9F479-A
INJECTOR ASSY., Fuel	E59E 9F593-A2B	E5TZ 9F593-A
SENSOR ASSY., Air Cleaner Air Temperature Control		D7FZ 9E607-A
MOTOR ASSY., Air Cleaner Vacuum		D4FZ 9D612-A
VALVE ASSY., Throttle Air Bypass	E9TE 9F715-A1A	E9TZ 9F715-A
SENSOR ASSY., Speed	E3AF 9E731-AB	E3AZ 9E731-A
REGULATOR ASSY., Fuel Charging Pressure	E7EE 9C968-AA, E77E 9C968-AA	E7FZ 9C968-A
POTENTIOMETER ASSY., Throttle Position Sensor	E7DF 9B989-AA	E7DZ 9B989-A
SENSOR ASSY., E.E.C. Coolant Temperature	E1AF 12A648-AA, E4AF 12A648-AA	E1AZ 12A648-A
PROCESSOR AND CALIBRATOR ASSY., E.E.C. IV	E9TF 12A650-C1A, C2A	E9TZ 12A650-CA
SENSOR ASSY., Air Charge Temperature	E4AF 12A697-AA	E1AZ 12A697-A

CALIBRATION: 9-65F-R00 — 2.9L EFI

FORD MOTOR COMPANY
IMPORTANT VEHICLE INFORMATION

THIS VEHICLE IS EQUIPPED WITH ELECTRONIC FUEL INJECTION. IDLE MIXTURE, COLD ENGINE IDLE SPEED AND COLD ENGINE FUEL ENRICHMENT ARE NOT ADJUSTABLE.

SET PARKING BRAKE AND BLOCK WHEELS. MAKE ALL ADJUSTMENTS WITH ENGINE AT NORMAL OPERATING TEMPERATURE, TRANSMISSION IN NEUTRAL AND ACCESSORIES OFF.

IGNITION TIMING- ELECTRONICALLY ADJUSTED AT THE FACTORY-ADJUSTMENT REQUIRED ONLY IF DISTRIBUTOR HAS BEEN DISTURBED.
(1) TURN OFF ENGINE.
(2) DISCONNECT THE IN-LINE SPOUT CONNECTOR (=0⊏).
(3) RE-START PREVIOUSLY WARMED-UP ENGINE.
(4) ADJUST IGNITION TIMING TO 10° BTDC.
(5) TURN OFF ENGINE AND RESTORE ELECTRICAL CONNECTION.

THIS ENGINE IS EQUIPPED WITH AUTOMATIC IDLE SPEED CONTROL. IDLE RPM IS NOT ADJUSTABLE. IF NOT AT SPECIFIED RPM, SEE SHOP MANUAL:
MANUAL TRANS. IN NEUTRAL: 850 RPM
AUTO TRANS. IN NEUTRAL: 800 RPM

THIS VEHICLE CONFORMS TO U.S. EPA REGULATIONS APPLICABLE TO 1989 MODEL YEAR NEW LIGHT-DUTY TRUCKS.

E9AE-9C485- | CATALYST | SPARK PLUG: AWSF-42C GAP- .042-.046
2.9L-9HM
KFM2.9T5FMFO - FI/EGS/TWC

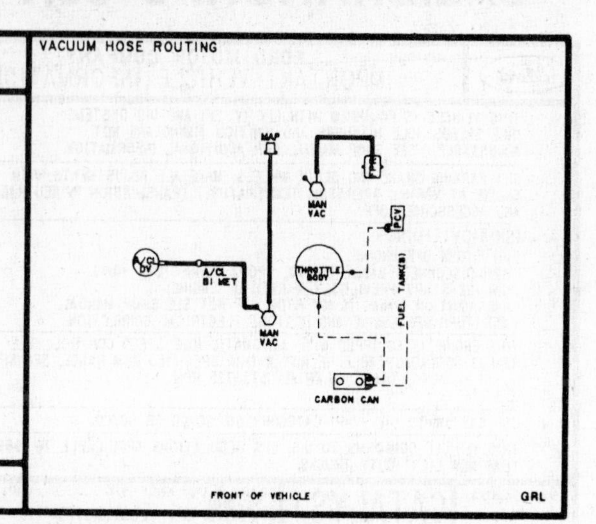

CALIBRATION PARTS LIST — CODE: BDF

NAME/DESCRIPTION	ENGINEERING NO.	SERVICE NO.
SENSOR ASSY., Exhaust Gas Oxygen	E77F 9F472-AA, BA, E7TF 9F472-AA	E7TZ 9F472-A
SENSOR ASSY., Exhaust Gas Oxygen	E7TF 9F472-CA, DA, EA	E7TZ 9F472-B
SENSOR ASSY., Manifold Absolute Pressure	E7EF 9F479-A1A, A2A	E7FZ 9F479-A
INJECTOR ASSY., Fuel	E67E 9F593-B1B, B4B	E6SZ 9F593-A
SENSOR ASSY., Air Cleaner Air Temperature Control		D7FZ 9E607-A
MOTOR ASSY., Air Cleaner Vacuum		D4FZ 9D612-A
VALVE ASSY., Throttle Air Bypass	E9AE 9F715-A1A, A2A	E9AZ 9F715-A
SENSOR ASSY., Speed	E3AF 9E731-AB	E3AZ 9E731-A
REGULATOR ASSY., Fuel Charging Pressure	88TF 9C968-EA	E6TZ 9C968-B
POTENTIOMETER ASSY., Throttle Position Sensor	E9TF 9B989-AA	E9TZ 9B989-A
DISTRIBUTOR ASSY.	E67E 12127-AA	E6TZ 12127-A
SENSOR ASSY., E.E.C. Coolant Temperature	E4AF 12A648-AA	E1AZ 12A648-A
PROCESSOR AND CALIBRATOR ASSY., E.E.C. IV	E9TF 12A650-D1A, D2A	E9TZ 12A650-DA
SENSOR ASSY., Air Charge Temperature	E4AF 12A697-AA	E1AZ 12A697-A

CALIBRATION: 9–65H–R00

2.9L EFI

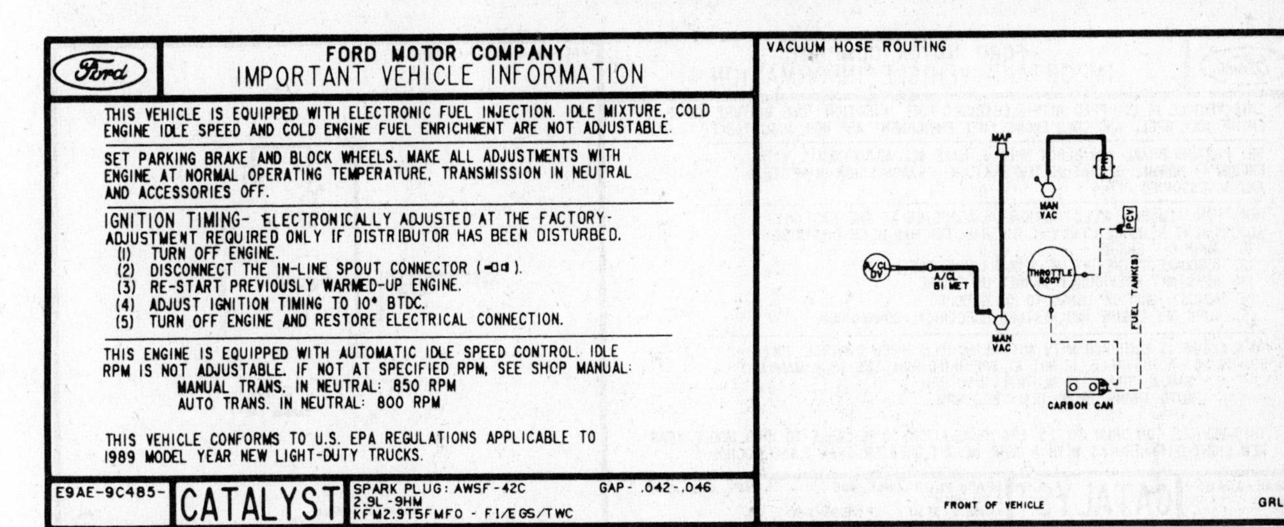

FORD MOTOR COMPANY
IMPORTANT VEHICLE INFORMATION

THIS VEHICLE IS EQUIPPED WITH ELECTRONIC FUEL INJECTION. IDLE MIXTURE, COLD ENGINE IDLE SPEED AND COLD ENGINE FUEL ENRICHMENT ARE NOT ADJUSTABLE.

SET PARKING BRAKE AND BLOCK WHEELS. MAKE ALL ADJUSTMENTS WITH ENGINE AT NORMAL OPERATING TEMPERATURE, TRANSMISSION IN NEUTRAL AND ACCESSORIES OFF.

IGNITION TIMING- ELECTRONICALLY ADJUSTED AT THE FACTORY- ADJUSTMENT REQUIRED ONLY IF DISTRIBUTOR HAS BEEN DISTURBED.
(1) TURN OFF ENGINE.
(2) DISCONNECT THE IN-LINE SPOUT CONNECTOR (-⊏⊐-).
(3) RE-START PREVIOUSLY WARMED-UP ENGINE.
(4) ADJUST IGNITION TIMING TO 10° BTDC.
(5) TURN OFF ENGINE AND RESTORE ELECTRICAL CONNECTION.

THIS ENGINE IS EQUIPPED WITH AUTOMATIC IDLE SPEED CONTROL. IDLE RPM IS NOT ADJUSTABLE. IF NOT AT SPECIFIED RPM, SEE SHOP MANUAL:
MANUAL TRANS. IN NEUTRAL: 850 RPM
AUTO TRANS. IN NEUTRAL: 800 RPM

THIS VEHICLE CONFORMS TO U.S. EPA REGULATIONS APPLICABLE TO 1989 MODEL YEAR NEW LIGHT-DUTY TRUCKS.

E9AE-9C485- CATALYST SPARK PLUG: AWSF-42C GAP- .042-.046
2.9L-9HM
KFM2.9T5FMFO - FI/EGS/TWC

CALIBRATION PARTS LIST

CODE: BDF

NAME/DESCRIPTION	ENGINEERING NO.	SERVICE NO.
SENSOR ASSY., Exhaust Gas Oxygen	E77F 9F472-AA, BA, E7TF 9F472-AA, Alt., E7TF 9F472-CA, DA, EA, Alt.	E7TZ 9F472-A, E7TZ 9F472-B, Alt.
SENSOR ASSY., Manifold Absolute Pressure	E7EF 9F479-A1A, A2A	E7FZ 9F479-A
INJECTOR ASSY., Fuel	E67E 9F593-B1B, B4B	E6SZ 9F593-A
SENSOR ASSY., Air Cleaner Air Temperature Control		D7FZ 9E607-A
MOTOR ASSY., Air Cleaner Vacuum		D4FZ 9D612-A
VALVE ASSY., Throttle Air Bypass	E9AE 9F715-A1A, A2A	E9AZ 9F715-A
SENSOR ASSY., Speed	E3AF 9E731-AB	E3AZ 9E731-A
REGULATOR ASSY., Fuel Charging Pressure	88TF 9C968-EA	E6TZ 9C968-B
POTENTIOMETER ASSY., Throttle Position Sensor	E9TF 9B989-AA	E9TZ 9B989-A
DISTRIBUTOR ASSY.	E67E 12127-AA	E6TZ 12127-A
SENSOR ASSY., Engine Electronic Control Coolant Temperature	E4AF 12A648-AA	E1AZ 12A648-A
PROCESSOR AND CALIBRATOR ASSY., E.E.C. IV	E9TF 12A650-G1A, G2A	E9TZ 12A650-GA
SENSOR ASSY., Air Charge Temperature	E4AF 12A697-AA	E1AZ 12A697-A

CALIBRATION: 9–65H–R00 — 2.9L EFI O/3450 LB GVW

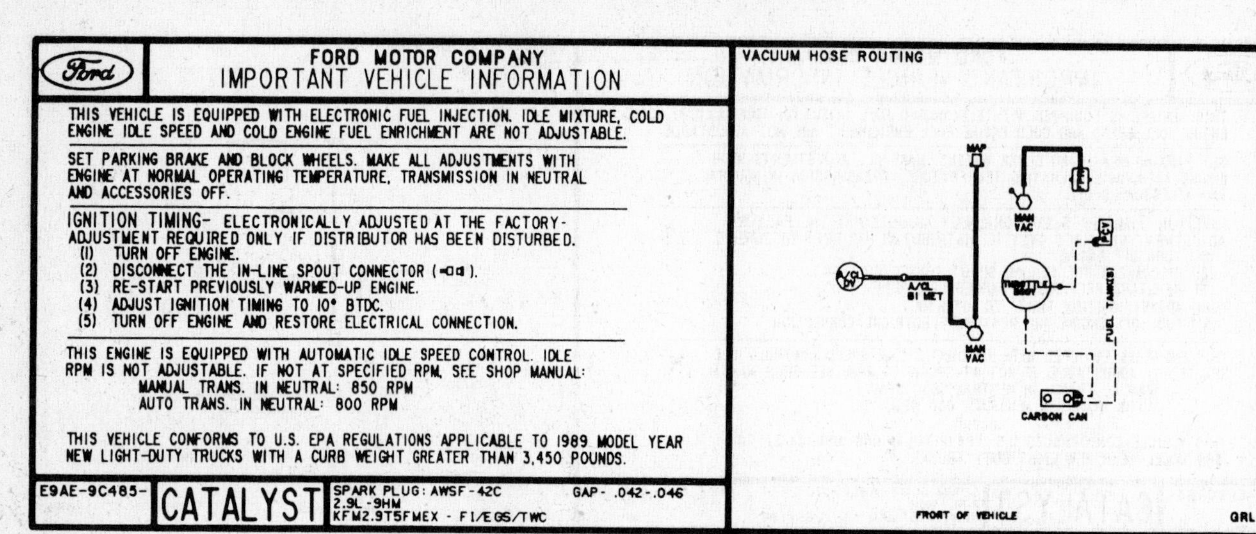
FORD MOTOR COMPANY
IMPORTANT VEHICLE INFORMATION

THIS VEHICLE IS EQUIPPED WITH ELECTRONIC FUEL INJECTION. IDLE MIXTURE, COLD ENGINE IDLE SPEED AND COLD ENGINE FUEL ENRICHMENT ARE NOT ADJUSTABLE.

SET PARKING BRAKE AND BLOCK WHEELS. MAKE ALL ADJUSTMENTS WITH ENGINE AT NORMAL OPERATING TEMPERATURE, TRANSMISSION IN NEUTRAL AND ACCESSORIES OFF.

IGNITION TIMING- ELECTRONICALLY ADJUSTED AT THE FACTORY-ADJUSTMENT REQUIRED ONLY IF DISTRIBUTOR HAS BEEN DISTURBED.
(1) TURN OFF ENGINE.
(2) DISCONNECT THE IN-LINE SPOUT CONNECTOR (=▭▯).
(3) RE-START PREVIOUSLY WARMED-UP ENGINE.
(4) ADJUST IGNITION TIMING TO 10° BTDC.
(5) TURN OFF ENGINE AND RESTORE ELECTRICAL CONNECTION.

THIS ENGINE IS EQUIPPED WITH AUTOMATIC IDLE SPEED CONTROL. IDLE RPM IS NOT ADJUSTABLE. IF NOT AT SPECIFIED RPM, SEE SHOP MANUAL:
MANUAL TRANS. IN NEUTRAL: 850 RPM
AUTO TRANS. IN NEUTRAL: 800 RPM

THIS VEHICLE CONFORMS TO U.S. EPA REGULATIONS APPLICABLE TO 1989 MODEL YEAR NEW LIGHT-DUTY TRUCKS WITH A CURB WEIGHT GREATER THAN 3,450 POUNDS.

E9AE-9C485- CATALYST SPARK PLUG: AWSF-42C GAP-.042-.046
2.9L-9HM
KFM2.9T5FMEX - FI/EGS/TWC

VACUUM HOSE ROUTING

CALIBRATION PARTS LIST — CODE: BDR

NAME/DESCRIPTION	ENGINEERING NO.	SERVICE NO.
SENSOR ASSY., Exhaust Gas Oxygen	E77F 9F472-AA, BA, E7TF 9F472-AA, Alt., E7TF 9F472-CA, DA, EA, Alt.	E7TZ 9F472-A, E7TZ 9F472-B, Alt.
SENSOR ASSY., Manifold Absolute Pressure	E7EF 9F479-A1A, A2A	E7FZ 9F479-A
INJECTOR ASSY., Fuel	E67E 9F593-B1B, B4B	E6SZ 9F593-A
SENSOR ASSY., Air Cleaner Air Temperature Control		D7FZ 9E607-A
MOTOR ASSY., Air Cleaner Vacuum		D4FZ 9D612-A
VALVE ASSY., Throttle Air Bypass	E9AE 9F715-A1A, A2A	E9AZ 9F715-A
SENSOR ASSY., Speed	E3AF 9E731-AB	E3AZ 9E731-A
REGULATOR ASSY., Fuel Charging Pressure	88TF 9C968-EA	E6TZ 9C968-B
POTENTIOMETER ASSY., Throttle Position Sensor	E9TF 9B989-AA	E9TZ 9B989-A
DISTRIBUTOR ASSY.	E67E 12127-AA	E6TZ 12127-A
SENSOR ASSY., Engine Electronic Control Coolant Temperature	E4AF 12A648-AA	E1AZ 12A648-A
PROCESSOR AND CALIBRATOR ASSY., E.E.C. IV	E9TF 12A650-G1A, G2A	E9TZ 12A650-GA
SENSOR ASSY., Air Charge Temperature	E4AF 12A697-AA	E1AZ 12A697-A

CALIBRATION: 9–65S–R00 — 2.9L EFI

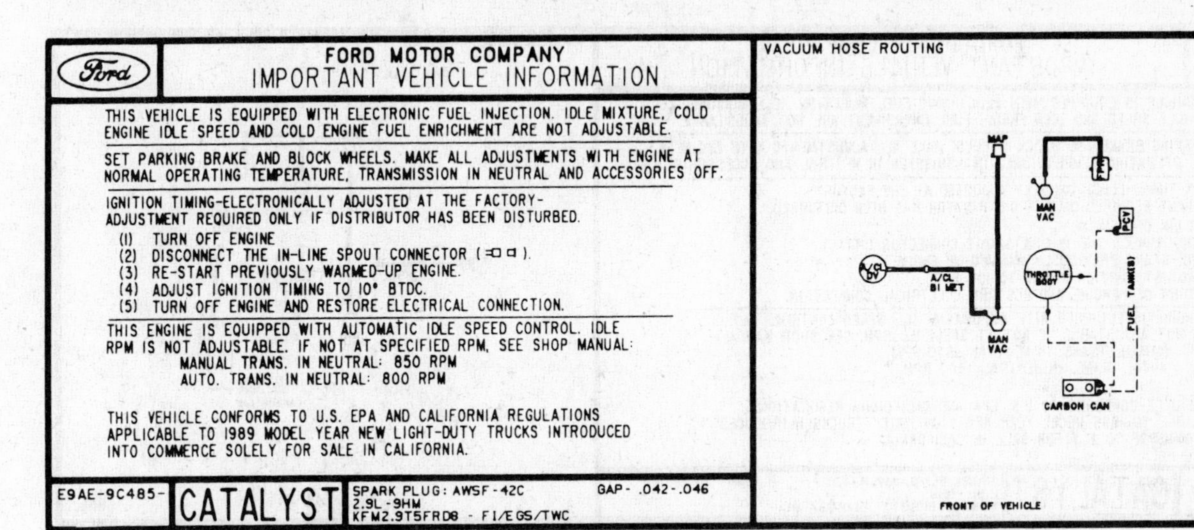

Ford

FORD MOTOR COMPANY
IMPORTANT VEHICLE INFORMATION

THIS VEHICLE IS EQUIPPED WITH ELECTRONIC FUEL INJECTION. IDLE MIXTURE, COLD ENGINE IDLE SPEED AND COLD ENGINE FUEL ENRICHMENT ARE NOT ADJUSTABLE.

SET PARKING BRAKE AND BLOCK WHEELS. MAKE ALL ADJUSTMENTS WITH ENGINE AT NORMAL OPERATING TEMPERATURE, TRANSMISSION IN NEUTRAL AND ACCESSORIES OFF.

IGNITION TIMING-ELECTRONICALLY ADJUSTED AT THE FACTORY-ADJUSTMENT REQUIRED ONLY IF DISTRIBUTOR HAS BEEN DISTURBED.

(1) TURN OFF ENGINE
(2) DISCONNECT THE IN-LINE SPOUT CONNECTOR (=□ □).
(3) RE-START PREVIOUSLY WARMED-UP ENGINE.
(4) ADJUST IGNITION TIMING TO 10° BTDC.
(5) TURN OFF ENGINE AND RESTORE ELECTRICAL CONNECTION.

THIS ENGINE IS EQUIPPED WITH AUTOMATIC IDLE SPEED CONTROL. IDLE RPM IS NOT ADJUSTABLE. IF NOT AT SPECIFIED RPM, SEE SHOP MANUAL:
MANUAL TRANS. IN NEUTRAL: 850 RPM
AUTO. TRANS. IN NEUTRAL: 800 RPM

THIS VEHICLE CONFORMS TO U.S. EPA AND CALIFORNIA REGULATIONS APPLICABLE TO 1989 MODEL YEAR NEW LIGHT-DUTY TRUCKS INTRODUCED INTO COMMERCE SOLELY FOR SALE IN CALIFORNIA.

E9AE-9C485- | CATALYST | SPARK PLUG: AWSF-42C GAP- .042-.046
2.9L-9HM
KFM2.9T5FRD8 - FI/EGS/TWC

VACUUM HOSE ROUTING

CALIBRATION PARTS LIST — CODE: BDT

NAME/DESCRIPTION	ENGINEERING NO.	SERVICE NO.
SENSOR ASSY., Exhaust Gas Oxygen	E7TF 9F472-AA, BA, E7TF 9F472-AA, Alt., E7TF 9F472-CA, DA, EA, Alt.	E7TZ 9F472-A, E7TZ 9F472-B, Alt.
SENSOR ASSY., Manifold Absolute Pressure	E7EF 9F479-A1A, A2A	E7FZ 9F479-A
INJECTOR ASSY., Fuel	E67E 9F593-B1B, B4B	E6SZ 9F593-A
SENSOR ASSY., Air Cleaner Air Temperature Control		D7FZ 9E607-A
MOTOR ASSY., Air Cleaner Vacuum		D4FZ 9D612-A
VALVE ASSY., Throttle Air Bypass	E9AE 9F715-A1A, A2A	E9AZ 9F715-A
SENSOR ASSY., Speed	E3AF 9E731-AB	E3AZ 9E731-A
REGULATOR ASSY., Fuel Charging Pressure	88TF 9C968-EA	E6TZ 9C968-B
POTENTIOMETER ASSY., Throttle Position Sensor	E9TF 9B989-AA	E9TZ 9B989-A
DISTRIBUTOR ASSY.	E67E 12127-AA	E6TZ 12127-A
SENSOR ASSY., Engine Electronic Control Coolant Temperature	E4AF 12A648-AA	E1AZ 12A648-A
PROCESSOR AND CALIBRATOR ASSY., E.E.C. IV	E9TF 12A650-E1A, E2A	E9TZ 12A650-EA
SENSOR ASSY., Air Charge Temperature	E4AF 12A697-AA	E1AZ 12A697-A

CALIBRATION: 9–65S–R00 — 2.9L EFI

FORD MOTOR COMPANY
IMPORTANT VEHICLE INFORMATION

THIS VEHICLE IS EQUIPPED WITH ELECTRONIC FUEL INJECTION. IDLE MIXTURE, COLD ENGINE IDLE SPEED AND COLD ENGINE FUEL ENRICHMENT ARE NOT ADJUSTABLE.

SET PARKING BRAKE AND BLOCK WHEELS. MAKE ALL ADJUSTMENTS WITH ENGINE AT NORMAL OPERATING TEMPERATURE, TRANSMISSION IN NEUTRAL AND ACCESSORIES OFF.

IGNITION TIMING-ELECTRONICALLY ADJUSTED AT THE FACTORY-ADJUSTMENT REQUIRED ONLY IF DISTRIBUTOR HAS BEEN DISTURBED.

(1) TURN OFF ENGINE
(2) DISCONNECT THE IN-LINE SPOUT CONNECTOR ( ).
(3) RE-START PREVIOUSLY WARMED-UP ENGINE.
(4) ADJUST IGNITION TIMING TO 10° BTDC.
(5) TURN OFF ENGINE AND RESTORE ELECTRICAL CONNECTION.

THIS ENGINE IS EQUIPPED WITH AUTOMATIC IDLE SPEED CONTROL. IDLE RPM IS NOT ADJUSTABLE. IF NOT AT SPECIFIED RPM, SEE SHOP MANUAL:
MANUAL TRANS. IN NEUTRAL: 850 RPM
AUTO. TRANS. IN NEUTRAL: 800 RPM

THIS VEHICLE CONFORMS TO U.S. EPA AND CALIFORNIA REGULATIONS APPLICABLE TO 1989 MODEL YEAR NEW LIGHT-DUTY TRUCKS INTRODUCED INTO COMMERCE SOLELY FOR SALE IN CALIFORNIA.

E9AE-9C485- | CATALYST | SPARK PLUG: AWSF-42C GAP-.042-.046 2.9L-9HM KFM2.9T5FRC7 - FI/EGS/TWC

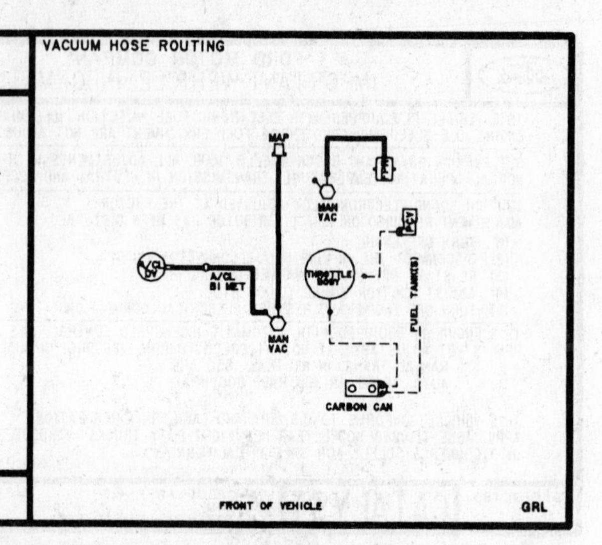

CALIBRATION PARTS LIST — CODE: BDU

NAME/DESCRIPTION	ENGINEERING NO.	SERVICE NO.
SENSOR ASSY., Exhaust Gas Oxygen	E77F 9F472-AA, BA, E7TF 9F472-AA, Alt., E7TF 9F472-CA, DA, EA, Alt.	E7TZ 9F472-A, E7TZ 9F472-B, Alt.
SENSOR ASSY., Manifold Absolute Pressure	E7EF 9F479-A1A, A2A	E7FZ 9F479-A
INJECTOR ASSY., Fuel	E67E 9F593-B1B, B4B	E6SZ 9F593-A
SENSOR ASSY., Air Cleaner Air Temperature Control		D7FZ 9E607-A
MOTOR ASSY., Air Cleaner Vacuum		D4FZ 9D612-A
VALVE ASSY., Throttle Air Bypass	E9AE 9F715-A1A, A2A	E9AZ 9F715-A
SENSOR ASSY., Speed	E3AF 9E731-AB	E3AZ 9E731-A
REGULATOR ASSY., Fuel Charging Pressure	88TF 9C968-EA	E6TZ 9C968-B
POTENTIOMETER ASSY., Throttle Position Sensor	E9TF 9B989-AA	E9TZ 9B989-A
DISTRIBUTOR ASSY.	E67E 12127-AA	E6TZ 12127-A
SENSOR ASSY., Engine Electronic Control Coolant Temperature	E4AF 12A648-AA	E1AZ 12A648-A
PROCESSOR AND CALIBRATOR ASSY., E.E.C. IV	E9TF 12A650-E1A, E2A	E9TZ 12A650-EA
SENSOR ASSY., Air Charge Temperature	E4AF 12A697-AA	E1AZ 12A697-A

CALIBRATION: 9–65T–R00 — 2.9L EFI

FORD MOTOR COMPANY
IMPORTANT VEHICLE INFORMATION

THIS VEHICLE IS EQUIPPED WITH ELECTRONIC FUEL INJECTION. IDLE MIXTURE, COLD ENGINE IDLE SPEED AND COLD ENGINE FUEL ENRICHMENT ARE NOT ADJUSTABLE.

SET PARKING BRAKE AND BLOCK WHEELS. MAKE ALL ADJUSTMENTS WITH ENGINE AT NORMAL OPERATING TEMPERATURE, TRANSMISSION IN NEUTRAL AND ACCESSORIES OFF.

IGNITION TIMING-ELECTRONICALLY ADJUSTED AT THE FACTORY-ADJUSTMENT REQUIRED ONLY IF DISTRIBUTOR HAS BEEN DISTURBED.
(1) TURN OFF ENGINE
(2) DISCONNECT THE IN-LINE SPOUT CONNECTOR ( ).
(3) RE-START PREVIOUSLY WARMED-UP ENGINE.
(4) ADJUST IGNITION TIMING TO 10° BTDC.
(5) TURN OFF ENGINE AND RESTORE ELECTRICAL CONNECTION.

THIS ENGINE IS EQUIPPED WITH AUTOMATIC IDLE SPEED CONTROL. IDLE RPM IS NOT ADJUSTABLE. IF NOT AT SPECIFIED RPM, SEE SHOP MANUAL:
MANUAL TRANS. IN NEUTRAL: 850 RPM
AUTO. TRANS. IN NEUTRAL: 800 RPM

THIS VEHICLE CONFORMS TO U.S. EPA AND CALIFORNIA REGULATIONS APPLICABLE TO 1989 MODEL YEAR NEW LIGHT-DUTY TRUCKS INTRODUCED INTO COMMERCE SOLELY FOR SALE IN CALIFORNIA.

E9AE-9C485- CATALYST SPARK PLUG: AWSF-42C GAP-.042-.046
2.9L-9HM
KFM2.9T5FRD8 - FI/EGS/TWC

VACUUM HOSE ROUTING

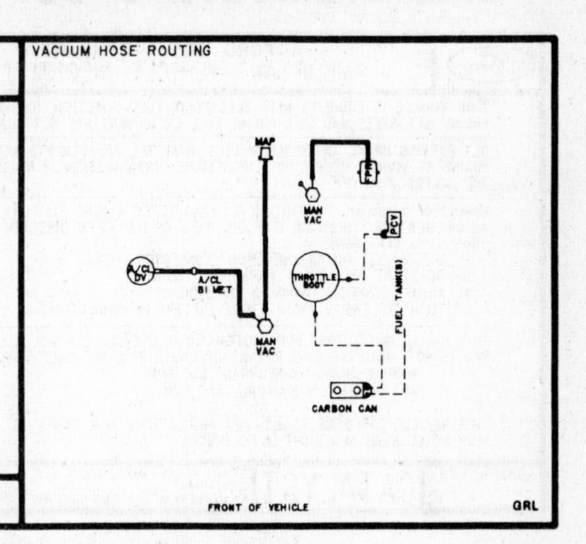

CALIBRATION PARTS LIST — CODE: BDF

NAME/DESCRIPTION	ENGINEERING NO.	SERVICE NO.
SENSOR ASSY., Exhaust Gas Oxygen	E77F 9F472-AA, BA, E7TF 9F472-AA	E7TZ 9F472-A
SENSOR ASSY., Exhaust Gas Oxygen	E7TF 9F472-CA, DA, EA	E7TZ 9F472-B
SENSOR ASSY., Manifold Absolute Pressure	E7EF 9F479-A1A, A2A	E7FZ 9F479-A
INJECTOR ASSY., Fuel	E67E 9F593-B1B, B4B	E6SZ 9F593-A
SENSOR ASSY., Carburetor Air Cleaner Air Temperature Control		D7FZ 9E607-A
MOTOR ASSY., Carburetor Air Cleaner Vacuum		D4FZ 9D612-A
VALVE ASSY., Throttle Air Bypass	E9AE 9F715-A1A, A2A	E9AZ 9F715-A
SENSOR ASSY., Speed	E3AF 9E731-AB	E3AZ 9E731-A
REGULATOR ASSY., Fuel Charging Pressure	88TF 9C968-EA	E6TZ 9C968-B
POTENTIOMETER ASSY., Throttle Position Sensor	E9TF 9B989-AA	E9TZ 9B989-A
DISTRIBUTOR ASSY.	E67E 12127-AA	E6TZ 12127-A
SENSOR ASSY., E.E.C. Coolant Temperature	E4AF 12A648-AA	E1AZ 12A648-A
PROCESSOR AND CALIBRATOR ASSY., E.E.C. IV	E9TF 12A650-U1A, U2A	E9TZ 12A650-UA
SENSOR ASSY., Air Charge Temperature	E4AF 12A697-AA	E1AZ 12A697-A

CALIBRATION: 9-66F-R10 — 2.9L EFI

FORD MOTOR COMPANY
IMPORTANT VEHICLE INFORMATION

THIS VEHICLE IS EQUIPPED WITH ELECTRONIC FUEL INJECTION. IDLE MIXTURE, COLD ENGINE IDLE SPEED AND COLD ENGINE FUEL ENRICHMENT ARE NOT ADJUSTABLE.

SET PARKING BRAKE AND BLOCK WHEELS. MAKE ALL ADJUSTMENTS WITH ENGINE AT NORMAL OPERATING TEMPERATURE, TRANSMISSION IN NEUTRAL AND ACCESSORIES OFF.

IGNITION TIMING- ELECTRONICALLY ADJUSTED AT THE FACTORY-ADJUSTMENT REQUIRED ONLY IF DISTRIBUTOR HAS BEEN DISTURBED.
(1) TURN OFF ENGINE.
(2) DISCONNECT THE IN-LINE SPOUT CONNECTOR ( ).
(3) RE-START PREVIOUSLY WARMED-UP ENGINE.
(4) ADJUST IGNITION TIMING TO 10° BTDC.
(5) TURN OFF ENGINE AND RESTORE ELECTRICAL CONNECTION.

THIS ENGINE IS EQUIPPED WITH AUTOMATIC IDLE SPEED CONTROL. IDLE RPM IS NOT ADJUSTABLE. IF NOT AT SPECIFIED RPM, SEE SHOP MANUAL:
MANUAL TRANS. IN NEUTRAL: 850 RPM
AUTO TRANS. IN NEUTRAL: 800 RPM

THIS VEHICLE CONFORMS TO U.S. EPA REGULATIONS APPLICABLE TO 1989 MODEL YEAR NEW LIGHT-DUTY TRUCKS.

E9AE-9C485- CATALYST SPARK PLUG: AWSF-42C GAP- .042-.046
2.9L-9HM
KFM2.9T5FMFO - FI/EGS/TWC

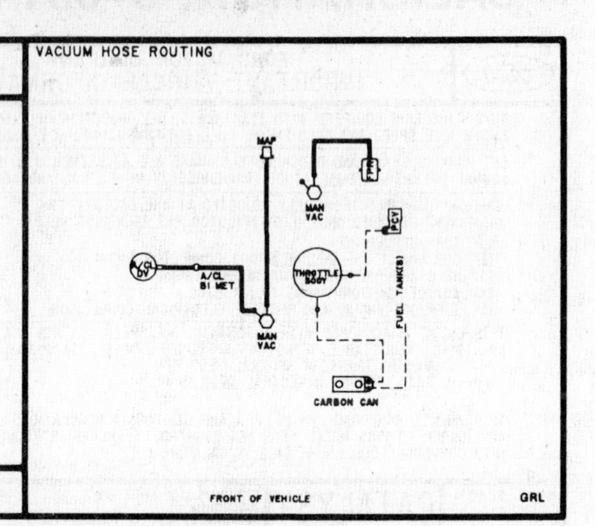

CALIBRATION PARTS LIST — CODE: BDT

NAME/DESCRIPTION	ENGINEERING NO.	SERVICE NO.
SENSOR ASSY., Exhaust Gas Oxygen	E77F 9F472-AA, BA, E7TF 9F472-AA	E7TZ 9F472-A
SENSOR ASSY., Exhaust Gas Oxygen	E7TF 9F472-CA, DA, EA	E7TZ 9F472-B
SENSOR ASSY., Manifold Absolute Pressure	E7EF 9F479-A1A, A2A	E7FZ 9F479-A
INJECTOR ASSY., Fuel	E67E 9F593-B1B, B4B	E6SZ 9F593-A
SENSOR ASSY., Carburetor Air Cleaner Air Temperature Control		D7FZ 9E607-A
MOTOR ASSY., Carburetor Air Cleaner Vacuum		D4FZ 9D612-A
VALVE ASSY., Throttle Air Bypass	E9AE 9F715-A1A, A2A	E9AZ 9F715-A
SENSOR ASSY., Speed	E3AF 9E731-AB	E3AZ 9E731-A
REGULATOR ASSY., Fuel Charging Pressure	88TF 9C968-EA	E6TZ 9C968-B
POTENTIOMETER ASSY., Throttle Position Sensor	E9TF 9B989-AA	E9TZ 9B989-A
DISTRIBUTOR ASSY.	E67E 12127-AA	E6TZ 12127-A
SENSOR ASSY., E.E.C. Coolant Temperature	E4AF 12A648-AA	E1AZ 12A648-A
PROCESSOR AND CALIBRATOR ASSY., E.E.C. IV	E9TF 12A650-H1B, H2B	E9TZ 12A650-HB
SENSOR ASSY., Air Charge Temperature	E4AF 12A697-AA	E1AZ 12A697-A

CALIBRATION: 9–66H–R10 — 2.9L EFI

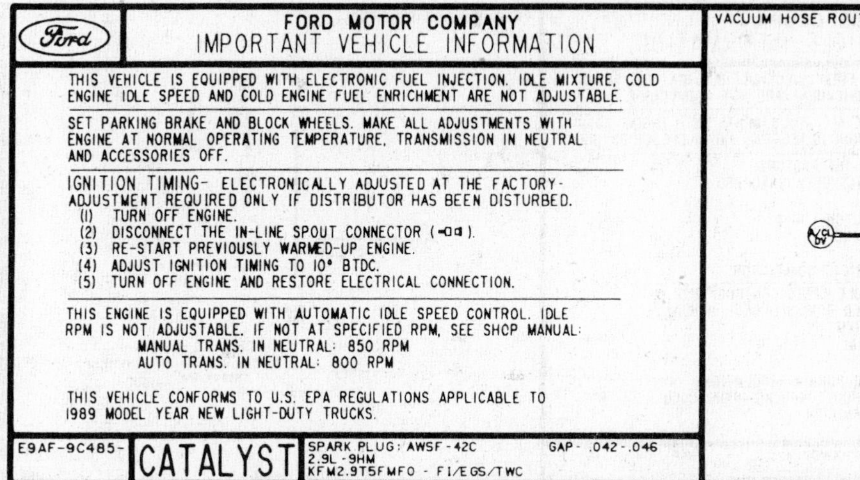

Ford

FORD MOTOR COMPANY
IMPORTANT VEHICLE INFORMATION

THIS VEHICLE IS EQUIPPED WITH ELECTRONIC FUEL INJECTION. IDLE MIXTURE, COLD ENGINE IDLE SPEED AND COLD ENGINE FUEL ENRICHMENT ARE NOT ADJUSTABLE.

SET PARKING BRAKE AND BLOCK WHEELS. MAKE ALL ADJUSTMENTS WITH ENGINE AT NORMAL OPERATING TEMPERATURE, TRANSMISSION IN NEUTRAL AND ACCESSORIES OFF.

IGNITION TIMING- ELECTRONICALLY ADJUSTED AT THE FACTORY-ADJUSTMENT REQUIRED ONLY IF DISTRIBUTOR HAS BEEN DISTURBED.
(1) TURN OFF ENGINE.
(2) DISCONNECT THE IN-LINE SPOUT CONNECTOR ( ).
(3) RE-START PREVIOUSLY WARMED-UP ENGINE.
(4) ADJUST IGNITION TIMING TO 10° BTDC.
(5) TURN OFF ENGINE AND RESTORE ELECTRICAL CONNECTION.

THIS ENGINE IS EQUIPPED WITH AUTOMATIC IDLE SPEED CONTROL. IDLE RPM IS NOT ADJUSTABLE. IF NOT AT SPECIFIED RPM, SEE SHOP MANUAL:
MANUAL TRANS. IN NEUTRAL: 850 RPM
AUTO TRANS. IN NEUTRAL: 800 RPM

THIS VEHICLE CONFORMS TO U.S. EPA REGULATIONS APPLICABLE TO 1989 MODEL YEAR NEW LIGHT-DUTY TRUCKS.

E9AF-9C485- CATALYST SPARK PLUG: AWSF-42C GAP-.042-.046
2.9L-9HM
KFM2.9T5FMFO - FI/EGS/TWC

VACUUM HOSE ROUTING

CALIBRATION PARTS LIST — CODE: BDF

NAME/DESCRIPTION	ENGINEERING NO.	SERVICE NO.
SENSOR ASSY., Exhaust Gas Oxygen	E77F 9F472-AA, BA, E7TF 9F472-AA	E7TZ 9F472-A
SENSOR ASSY., Exhaust Gas Oxygen	E7TF 9F472-CA, DA, EA	E7TZ 9F472-B
SENSOR ASSY., Manifold Absolute Pressure	E7EF 9F479-A1A, A2A	E7FZ 9F479-A
INJECTOR ASSY., Fuel	E67E 9F593-B1B, B4B	E6SZ 9F593-A
SENSOR ASSY., Carburetor Air Cleaner Air Temperature Control		D7FZ 9E607-A
MOTOR ASSY., Carburetor Air Cleaner Vacuum		D4FZ 9D612-A
VALVE ASSY., Throttle Air Bypass	E9AE 9F715-A1A, A2A	E9AZ 9F715-A
SENSOR ASSY., Speed	E3AF 9E731-AB	E3AZ 9E731-A
REGULATOR ASSY., Fuel Charging Pressure	88TF 9C968-EA	E6TZ 9C968-B
POTENTIOMETER ASSY., Throttle Position Sensor	E9TF 9B989-AA	E9TZ 9B989-A
DISTRIBUTOR ASSY.	E67E 12127-AA	E6TZ 12127-A
SENSOR ASSY., E.E.C. Coolant Temperature	E4AF 12A648-AA	E1AZ 12A648-A
PROCESSOR AND CALIBRATOR ASSY., E.E.C. IV	E9TF 12A650-K1B, K2B	E9TZ 12A650-KB
SENSOR ASSY., Air Charge Temperature	E4AF 12A697-AA	E1AZ 12A697-A

CALIBRATION: 9–66S–R02 — 2.9L EFI

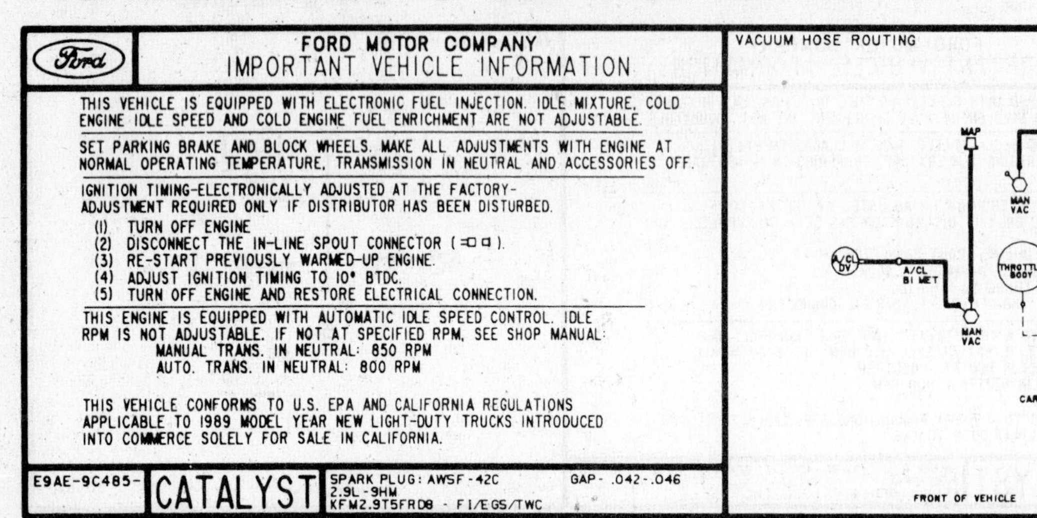

Ford

FORD MOTOR COMPANY
IMPORTANT VEHICLE INFORMATION

THIS VEHICLE IS EQUIPPED WITH ELECTRONIC FUEL INJECTION. IDLE MIXTURE, COLD ENGINE IDLE SPEED AND COLD ENGINE FUEL ENRICHMENT ARE NOT ADJUSTABLE.

SET PARKING BRAKE AND BLOCK WHEELS. MAKE ALL ADJUSTMENTS WITH ENGINE AT NORMAL OPERATING TEMPERATURE, TRANSMISSION IN NEUTRAL AND ACCESSORIES OFF.

IGNITION TIMING-ELECTRONICALLY ADJUSTED AT THE FACTORY-ADJUSTMENT REQUIRED ONLY IF DISTRIBUTOR HAS BEEN DISTURBED.
(1) TURN OFF ENGINE
(2) DISCONNECT THE IN-LINE SPOUT CONNECTOR ( ).
(3) RE-START PREVIOUSLY WARMED-UP ENGINE.
(4) ADJUST IGNITION TIMING TO 10° BTDC.
(5) TURN OFF ENGINE AND RESTORE ELECTRICAL CONNECTION.

THIS ENGINE IS EQUIPPED WITH AUTOMATIC IDLE SPEED CONTROL. IDLE RPM IS NOT ADJUSTABLE. IF NOT AT SPECIFIED RPM, SEE SHOP MANUAL:
MANUAL TRANS. IN NEUTRAL: 850 RPM
AUTO. TRANS. IN NEUTRAL: 800 RPM

THIS VEHICLE CONFORMS TO U.S. EPA AND CALIFORNIA REGULATIONS APPLICABLE TO 1989 MODEL YEAR NEW LIGHT-DUTY TRUCKS INTRODUCED INTO COMMERCE SOLELY FOR SALE IN CALIFORNIA.

E9AE-9C485- CATALYST SPARK PLUG: AWSF-42C GAP-.042-.046
2.9L-9HM
KFM2.9T5FRD8 - FI/EGS/TWC

VACUUM HOSE ROUTING

CALIBRATION PARTS LIST — CODE: BDT

NAME/DESCRIPTION	ENGINEERING NO.	SERVICE NO.
SENSOR ASSY., Exhaust Gas Oxygen	E77F 9F472-AA, BA, E7TF 9F472-AA	E7TZ 9F472-A
SENSOR ASSY., Exhaust Gas Oxygen	E7TF 9F472-CA, DA, EA	E7TZ 9F472-B
SENSOR ASSY., Manifold Absolute Pressure	E7EF 9F479-A1A, A2A	E7FZ 9F479-A
INJECTOR ASSY., Fuel	E67E 9F593-B1B, B4B	E6SZ 9F593-A
SENSOR ASSY., Carburetor Air Cleaner Air Temperature Control		D7FZ 9E607-A
MOTOR ASSY., Carburetor Air Cleaner Vacuum		D4FZ 9D612-A
VALVE ASSY., Throttle Air Bypass	E9AE 9F715-A1A, A2A	E9AZ 9F715-A
SENSOR ASSY., Speed	E3AF 9E731-AB	E3AZ 9E731-A
REGULATOR ASSY., Fuel Charging Pressure	88TF 9C968-EA	E6TZ 9C968-B
POTENTIOMETER ASSY., Throttle Position Sensor	E9TF 9B989-AA	E9TZ 9B989-A
DISTRIBUTOR ASSY.	E67E 12127-AA	E6TZ 12127-A
SENSOR ASSY., E.E.C. Coolant Temperature	E4AF 12A648-AA	E1AZ 12A648-A

NAME/DESCRIPTION	ENGINEERING NO.	SERVICE NO.
PROCESSOR AND CALIBRATOR ASSY., E.E.C. IV	E9TF 12A650-J1B, J2B	E9TZ 12A650-JB
SENSOR ASSY., Air Charge Temperature	E4AF 12A697-AA	E1AZ 12A697-A

CALIBRATION: 9–66T–R02 — 2.9L EFI

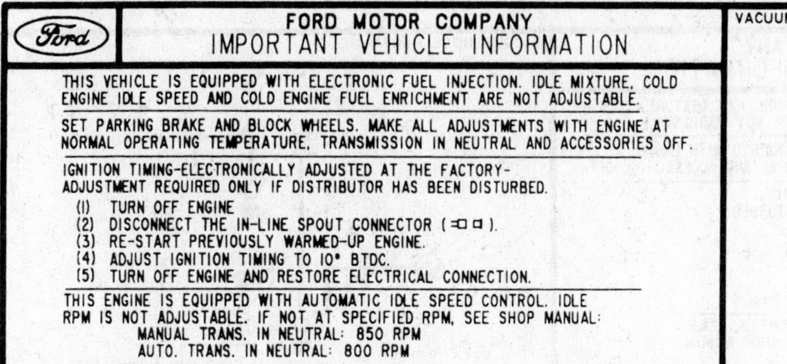
Ford

FORD MOTOR COMPANY
IMPORTANT VEHICLE INFORMATION

THIS VEHICLE IS EQUIPPED WITH ELECTRONIC FUEL INJECTION. IDLE MIXTURE, COLD ENGINE IDLE SPEED AND COLD ENGINE FUEL ENRICHMENT ARE NOT ADJUSTABLE.

SET PARKING BRAKE AND BLOCK WHEELS. MAKE ALL ADJUSTMENTS WITH ENGINE AT NORMAL OPERATING TEMPERATURE, TRANSMISSION IN NEUTRAL AND ACCESSORIES OFF.

IGNITION TIMING-ELECTRONICALLY ADJUSTED AT THE FACTORY-ADJUSTMENT REQUIRED ONLY IF DISTRIBUTOR HAS BEEN DISTURBED.

(1) TURN OFF ENGINE
(2) DISCONNECT THE IN-LINE SPOUT CONNECTOR ( ).
(3) RE-START PREVIOUSLY WARMED-UP ENGINE.
(4) ADJUST IGNITION TIMING TO 10° BTDC.
(5) TURN OFF ENGINE AND RESTORE ELECTRICAL CONNECTION.

THIS ENGINE IS EQUIPPED WITH AUTOMATIC IDLE SPEED CONTROL. IDLE RPM IS NOT ADJUSTABLE. IF NOT AT SPECIFIED RPM, SEE SHOP MANUAL:
MANUAL TRANS. IN NEUTRAL: 850 RPM
AUTO. TRANS. IN NEUTRAL: 800 RPM

THIS VEHICLE CONFORMS TO U.S. EPA AND CALIFORNIA REGULATIONS APPLICABLE TO 1989 MODEL YEAR NEW LIGHT-DUTY TRUCKS INTRODUCED INTO COMMERCE SOLELY FOR SALE IN CALIFORNIA.

E9AE-9C485- CATALYST SPARK PLUG: AWSF-42C GAP-.042-.046
2.9L-9HM
KFM2.9T5FRD8 - FI/EGS/TWC

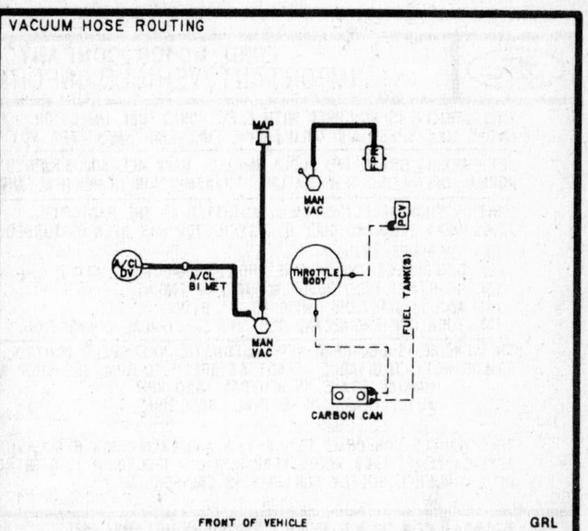

CALIBRATION PARTS LIST — CODE: BDT

NAME/DESCRIPTION	ENGINEERING NO.	SERVICE NO.
SENSOR ASSY., Exhaust Gas Oxygen	E77F 9F472-AA, BA, E7TF 9F472-AA	E7TZ 9F472-A
SENSOR ASSY., Exhaust Gas Oxygen	E7TF 9F472-CA, DA, EA	E7TZ 9F472-B
SENSOR ASSY., Manifold Absolute Pressure	E7EF 9F479-A1A, A2A	E7FZ 9F479-A
INJECTOR ASSY., Fuel	E67E 9F593-B1B, B4B	E6SZ 9F593-A
SENSOR ASSY., Air Cleaner Air Temperature Control		D7FZ 9E607-A
MOTOR ASSY., Air Cleaner Vacuum		D4FZ 9D612-A
VALVE ASSY., Throttle Air Bypass	E9AE 9F715-A1A, A2A	E9AZ 9F715-A
SENSOR ASSY., Speed	E3AF 9E731-AB	E3AZ 9E731-A
REGULATOR ASSY., Fuel Charging Pressure	88TF 9C968-EA	E6TZ 9C968-B
POTENTIOMETER ASSY., Throttle Position Sensor	E9TF 9B989-AA	E9TZ 9B989-A
DISTRIBUTOR ASSY.	E67E 12127-AA	E6TZ 12127-A
SENSOR ASSY., E.E.C. Coolant Temperature	E4AF 12A648-AA	E1AZ 12A648-A
PROCESSOR AND CALIBRATOR ASSY., E.E.C. IV	E9TF 12A650-V1B, V2B	E9TZ 12A650-VB
SENSOR ASSY., Air Charge Temperature	E4AF 12A697-AA	E1AZ 12A697-A

CALIBRATION: 9–66T–R02 — 2.9L EFI

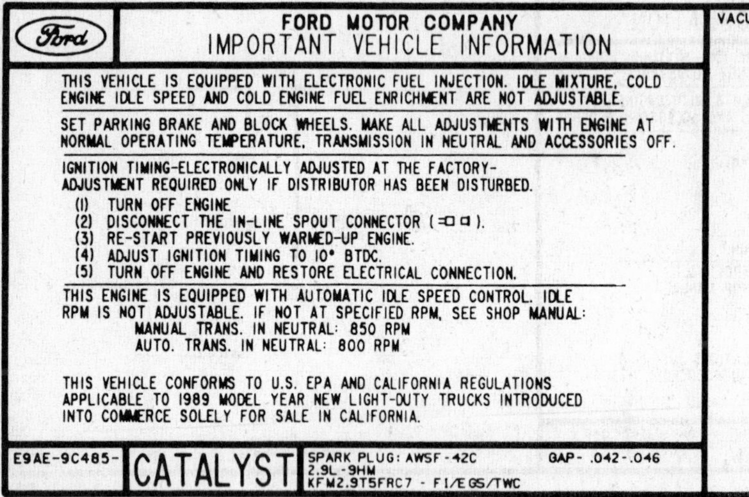

FORD MOTOR COMPANY
IMPORTANT VEHICLE INFORMATION

THIS VEHICLE IS EQUIPPED WITH ELECTRONIC FUEL INJECTION. IDLE MIXTURE, COLD ENGINE IDLE SPEED AND COLD ENGINE FUEL ENRICHMENT ARE NOT ADJUSTABLE.

SET PARKING BRAKE AND BLOCK WHEELS. MAKE ALL ADJUSTMENTS WITH ENGINE AT NORMAL OPERATING TEMPERATURE, TRANSMISSION IN NEUTRAL AND ACCESSORIES OFF.

IGNITION TIMING-ELECTRONICALLY ADJUSTED AT THE FACTORY-ADJUSTMENT REQUIRED ONLY IF DISTRIBUTOR HAS BEEN DISTURBED.

(1) TURN OFF ENGINE
(2) DISCONNECT THE IN-LINE SPOUT CONNECTOR (=◻ ◻).
(3) RE-START PREVIOUSLY WARMED-UP ENGINE.
(4) ADJUST IGNITION TIMING TO 10° BTDC.
(5) TURN OFF ENGINE AND RESTORE ELECTRICAL CONNECTION.

THIS ENGINE IS EQUIPPED WITH AUTOMATIC IDLE SPEED CONTROL. IDLE RPM IS NOT ADJUSTABLE. IF NOT AT SPECIFIED RPM, SEE SHOP MANUAL:
MANUAL TRANS. IN NEUTRAL: 850 RPM
AUTO. TRANS. IN NEUTRAL: 800 RPM

THIS VEHICLE CONFORMS TO U.S. EPA AND CALIFORNIA REGULATIONS APPLICABLE TO 1989 MODEL YEAR NEW LIGHT-DUTY TRUCKS INTRODUCED INTO COMMERCE SOLELY FOR SALE IN CALIFORNIA.

E9AE-9C485- | CATALYST | SPARK PLUG: AWSF-42C GAP- .042-.046 | 2.9L-9HM | KFM2.9T5FRC7 - FI/EGS/TWC

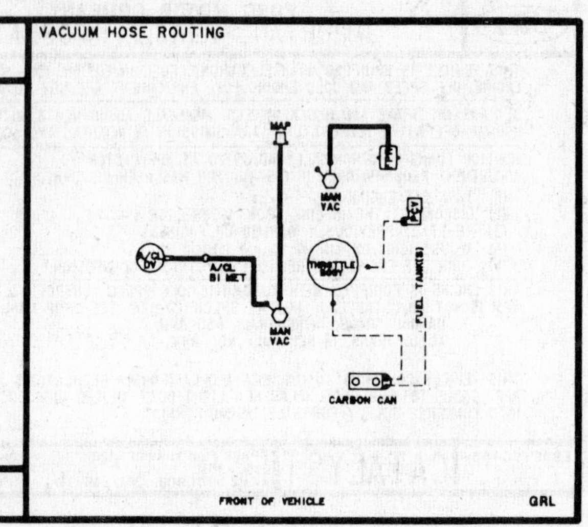

CALIBRATION PARTS LIST — CODE: BDU

NAME/DESCRIPTION	ENGINEERING NO.	SERVICE NO.
SENSOR ASSY., Exhaust Gas Oxygen	E77F 9F472-AA, BA, E7TF 9F472-AA, Alt., E7TF 9F472-CA, DA, EA, Alt.	E7TZ 9F472-A, E7TZ 9F472-B, Alt.
SENSOR ASSY., Manifold Absolute Pressure	E7EF 9F479-A1A, A2A	E7FZ 9F479-A
INJECTOR ASSY., Fuel	E67E 9F593-B1B, B4B	E6SZ 9F593-A
SENSOR ASSY., Air Cleaner Air Temperature Control		D7FZ 9E607-A
MOTOR ASSY., Air Cleaner Vacuum		D4FZ 9D612-A
VALVE ASSY., Throttle Air Bypass	E9AE 9F715-A1A, A2A	E9AZ 9F715-A
SENSOR ASSY., Speed	E3AF 9E731-AB	E3AZ 9E731-A
REGULATOR ASSY., Fuel Charging Pressure	88TF 9C968-EA	E6TZ 9C968-B
POTENTIOMETER ASSY., Throttle Position Sensor	E9TF 9B989-AA	E9TZ 9B989-A
DISTRIBUTOR ASSY.	E67E 12127-AA	E6TZ 12127-A
SENSOR ASSY., Engine Electronic Control Coolant Temperature	E4AF 12A648-AA	E1AZ 12A648-A
PROCESSOR AND CALIBRATOR ASSY., E.E.C. IV	E9TF 12A650-V1B, V2B	E9TZ 12A650-VB
SENSOR ASSY., Air Charge Temperature	E4AF 12A697-AA	E1AZ 12A697-A

CALIBRATION: 0-49S-R00 — 2.3L EFI

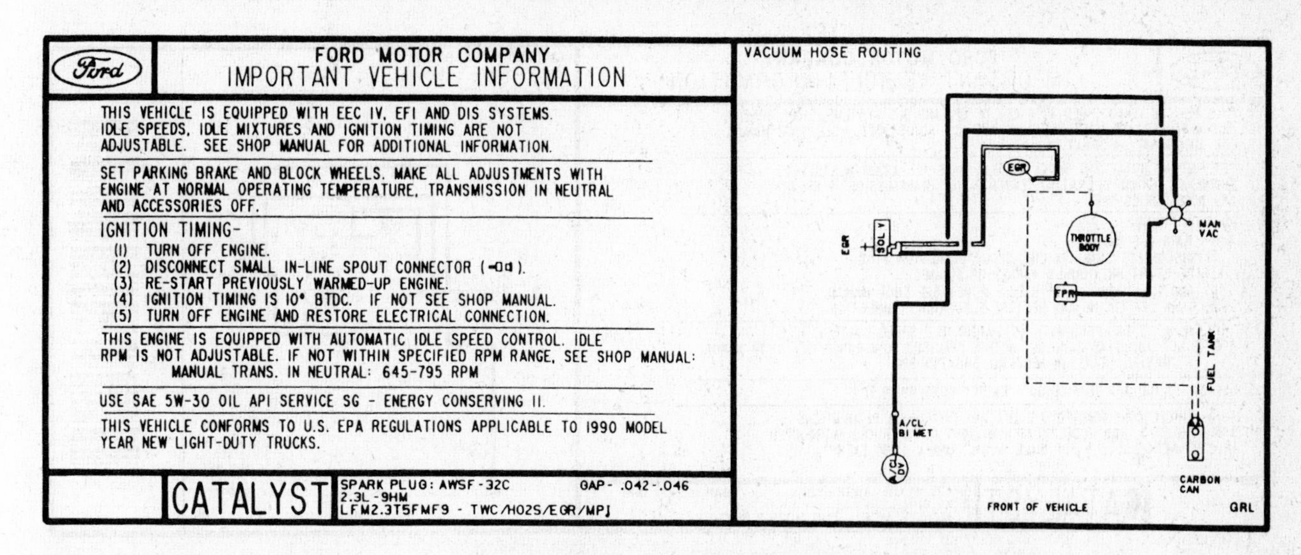
FORD MOTOR COMPANY
IMPORTANT VEHICLE INFORMATION

THIS VEHICLE IS EQUIPPED WITH EEC IV, EFI AND DIS SYSTEMS. IDLE SPEEDS, IDLE MIXTURES AND IGNITION TIMING ARE NOT ADJUSTABLE. SEE SHOP MANUAL FOR ADDITIONAL INFORMATION.

SET PARKING BRAKE AND BLOCK WHEELS. MAKE ALL ADJUSTMENTS WITH ENGINE AT NORMAL OPERATING TEMPERATURE, TRANSMISSION IN NEUTRAL AND ACCESSORIES OFF.

IGNITION TIMING-
(1) TURN OFF ENGINE.
(2) DISCONNECT SMALL IN-LINE SPOUT CONNECTOR ( ).
(3) RE-START PREVIOUSLY WARMED-UP ENGINE.
(4) IGNITION TIMING IS 10° BTDC. IF NOT SEE SHOP MANUAL.
(5) TURN OFF ENGINE AND RESTORE ELECTRICAL CONNECTION.

THIS ENGINE IS EQUIPPED WITH AUTOMATIC IDLE SPEED CONTROL. IDLE RPM IS NOT ADJUSTABLE. IF NOT WITHIN SPECIFIED RPM RANGE, SEE SHOP MANUAL:
MANUAL TRANS. IN NEUTRAL: 645-795 RPM

USE SAE 5W-30 OIL API SERVICE SG - ENERGY CONSERVING II.

THIS VEHICLE CONFORMS TO U.S. EPA REGULATIONS APPLICABLE TO 1990 MODEL YEAR NEW LIGHT-DUTY TRUCKS.

CATALYST
SPARK PLUG: AWSF-32C GAP- .042-.046
2.3L-9HM
LFM2.3T5FMF9 - TWC/HO2S/EGR/MPI

CALIBRATION PARTS LIST — CODE: DHK

NAME/DESCRIPTION	ENGINEERING NO.	SERVICE NO.
SENSOR, Exhaust Gas Recirculation Valve	E7TF 9G428-AA	E7TZ 9G428-A
CONTROL ASSY, EGR Vacuum Regulator	E8TE 9J459-AA	E8TZ 9J459-A
SENSOR ASSY, Exhaust Gas Oxygen	F0WF 9F472-AA	F0WZ 9F472-B
VALVE ASSY, EGR Pressure External	E97E 9F483-A2A, A4A	E9TZ 9F483-A
INJECTOR ASSY, Fuel	E59E 9F593-A2B	E5TZ 9F593-A
SENSOR ASSY, Carburetor Air Cleaner Air Temperature Control		D7FZ 9E607-A
MOTOR ASSY, Carburetor Air Cleaner Vacuum		D4FZ 9D612-A
VALVE ASSY, Throttle Air By-Pass	E9TE 9F715-A1A F0TE 9F715-AA	E9TZ 9F715-A
SENSOR ASSY, Speed	E3AF 9E731-AB	E3AZ 9E731-A
REGULATOR ASSY, Fuel Charging Pressure	E7EE 9C968-AA E77E 9C968-AA	E7FZ 9C968-A
POTENTIOMETER ASSY, Throttle Position Sensor	E7DF 9B989-AA	E7DZ 9B989-A
SENSOR ASSY, Mass Airflow	F0VF 12B579-AA	F0VY 12B579-A
SENSOR ASSY, Engine Electronic Control Coolant Temperature	E1AF 12A648-AA E4AF 12A648-AA	E1AZ 12A648-A
PROCESSOR AND CALIBRATOR ASSY, EEC-IV	F07F 12A650-AB	F0TZ 12A650-RAB
SENSOR ASSY, Air Charge Temperature	E4AF 12A697-AA	E1AZ 12A697-A

CALIBRATION: 0-49S-R00 — 2.3L EFI

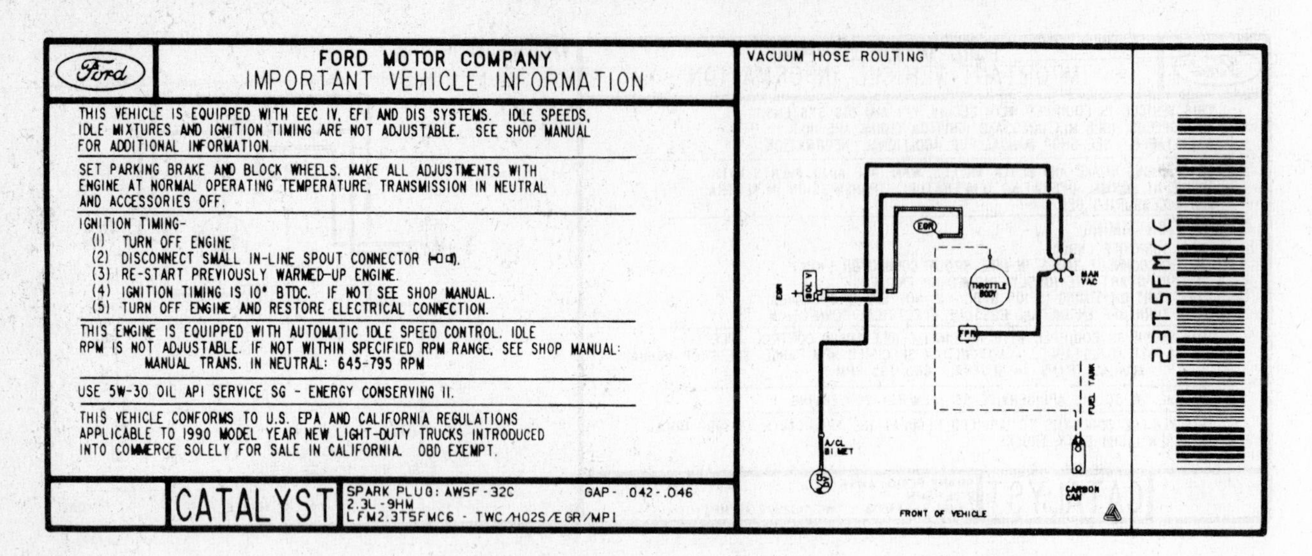

Ford

FORD MOTOR COMPANY
IMPORTANT VEHICLE INFORMATION

THIS VEHICLE IS EQUIPPED WITH EEC IV, EFI AND DIS SYSTEMS. IDLE SPEEDS, IDLE MIXTURES AND IGNITION TIMING ARE NOT ADJUSTABLE. SEE SHOP MANUAL FOR ADDITIONAL INFORMATION.

SET PARKING BRAKE AND BLOCK WHEELS. MAKE ALL ADJUSTMENTS WITH ENGINE AT NORMAL OPERATING TEMPERATURE, TRANSMISSION IN NEUTRAL AND ACCESSORIES OFF.

IGNITION TIMING-
(1) TURN OFF ENGINE
(2) DISCONNECT SMALL IN-LINE SPOUT CONNECTOR.
(3) RE-START PREVIOUSLY WARMED-UP ENGINE.
(4) IGNITION TIMING IS 10° BTDC. IF NOT SEE SHOP MANUAL.
(5) TURN OFF ENGINE AND RESTORE ELECTRICAL CONNECTION.

THIS ENGINE IS EQUIPPED WITH AUTOMATIC IDLE SPEED CONTROL. IDLE RPM IS NOT ADJUSTABLE. IF NOT WITHIN SPECIFIED RPM RANGE, SEE SHOP MANUAL:
MANUAL TRANS. IN NEUTRAL: 645-795 RPM

USE 5W-30 OIL API SERVICE SG - ENERGY CONSERVING II.

THIS VEHICLE CONFORMS TO U.S. EPA AND CALIFORNIA REGULATIONS APPLICABLE TO 1990 MODEL YEAR NEW LIGHT-DUTY TRUCKS INTRODUCED INTO COMMERCE SOLELY FOR SALE IN CALIFORNIA. OBD EXEMPT.

CATALYST

SPARK PLUG: AWSF-32C GAP- .042-.046
2.3L-9HM
LFM2.3T5FMC6 - TWC/HO2S/EGR/MPI

VACUUM HOSE ROUTING

CALIBRATION PARTS LIST — CODE: DDV

NAME/DESCRIPTION	ENGINEERING NO.	SERVICE NO.
SENSOR, Exhaust Gas Recirculation Valve	E7TF 9G428-AA	E7TZ 9G428-A
CONTROL ASSY, EGR Vacuum Regulator	E8TE 9J459-AA	E8TZ 9J459-A
SENSOR ASSY, Exhaust Gas Oxygen	F0WF 9F472-AA	F0WZ 9F472-B
VALVE ASSY, EGR Pressure External	E97E 9F483-A2A, A4A	E9TZ 9F483-A
INJECTOR ASSY, Fuel	E59E 9F593-A2B	E5TZ 9F593-A
SENSOR ASSY, Carburetor Air Cleaner Air Temperature Control		D7FZ 9E607-A
MOTOR ASSY, Carburetor Air Cleaner Vacuum		D4FZ 9D612-A
VALVE ASSY, Throttle Air By-Pass	E9TE 9F715-A1A F0TE 9F715-AA	E9TZ 9F715-A
SENSOR ASSY, Speed	E3AF 9E731-AB	E3AZ 9E731-A
REGULATOR ASSY, Fuel Charging Pressure	E7EE 9C968-AA E77E 9C968-AA	E7FZ 9C968-A
POTENTIOMETER ASSY, Throttle Position Sensor	E7DF 9B989-AA	E7DZ 9B989-A
SENSOR ASSY, Mass Airflow	F0VF 12B579-AA	F0VY 12B579-A
SENSOR ASSY, Engine Electronic Control Coolant Temperature	E1AF 12A648-AA E4AF 12A648-AA	E1AZ 12A648-A
PROCESSOR AND CALIBRATOR ASSY, EEC-IV	F07F 12A650-AB	F0TZ 12A650-RAB
SENSOR ASSY, Air Charge Temperature	E4AF 12A697-AA	E1AZ 12A697-A

CALIBRATION: 0-49S-R00 — 2.3L EFI

IMPORTANT VEHICLE INFORMATION — FORD MOTOR COMPANY — RENSEIGNEMENTS IMPORTANTS SUR LE VÉHICULE

THIS VEHICLE IS EQUIPPED WITH EEC IV, EFI AND DIS SYSTEMS. IDLE SPEEDS, IDLE MIXTURES AND IGNITION TIMING ARE NOT ADJUSTABLE. SEE SHOP MANUAL FOR ADDITIONAL INFORMATION.

SET PARKING BRAKE AND BLOCK WHEELS. MAKE ALL ADJUSTMENTS WITH ENGINE AT NORMAL OPERATING TEMPERATURE, TRANSMISSION IN NEUTRAL, ACCESSORIES AND HEADLIGHTS OFF.

IGNITION TIMING-
(1) TURN OFF ENGINE
(2) DISCONNECT SMALL IN-LINE SPOUT CONNECTOR (=◻◅).
(3) RE-START PREVIOUSLY WARMED-UP ENGINE.
(4) IGNITION TIMING IS 10° BTDC. IF NOT SEE SHOP MANUAL.
(5) TURN OFF ENGINE AND RESTORE ELECTRICAL CONNECTION.

THIS ENGINE IS EQUIPPED WITH AUTOMATIC IDLE SPEED CONTROL. IDLE RPM IS NOT ADJUSTABLE. IF NOT WITHIN SPECIFIED RPM RANGE, SEE SHOP MANUAL:
MANUAL TRANS. IN NEUTRAL: 645-795 RPM

USE SAE 5W-30 OIL API SERVICE SG - ENERGY CONSERVING II.

CE VÉHICULE EST MUNI DES SYSTÈMES EEC IV, EFI ET DIS. • LES RÉGIMES DE RALENTI, LES MÉLANGES DE RALENTI ET L'ALLUMAGE NE SONT PAS RÉGLABLES. POUR PLUS DE DÉTAILS, CONSULTER LE MANUEL DE RÉPARATION.

SERRER LE FREIN DE STATIONNEMENT, BLOQUER LES ROUES. EFFECTUER TOUT RÉGLAGE SUR MOTEUR NORMALEMENT CHAUD, B.V. AU POINT MORT, CONTACT DES ACCESSOIRES ET DES PHARES COUPÉ.

CALAGE DE L'ALLUMAGE :
(1) ARRÊTER LE MOTEUR.
(2) DÉBRANCHER LE PETIT CONNECTEUR (=◻◅) INTERCALÉ DANS LE CIRCUIT DE DÉCLENCHEMENT DE L'ÉTINCELLE.
(3) REDÉMARRER LE MOTEUR PRÉALABLEMENT RÉCHAUFFÉ.
(4) L'ALLUMAGE DOIT ÊTRE CALÉ À 10° AVANT PMH. SINON, VOIR LE MANUEL DE RÉPARATION.
(5) ARRÊTER LE MOTEUR ET REBRANCHER LE CONNECTEUR.

CE MOTEUR EST À COMMANDE DE RALENTI AUTOMATIQUE. LE RÉGIME DE RALENTI N'EST PAS RÉGLABLE. S'IL N'EST PAS DANS LES LIMITES PRESCRITES, CONSULTER LE MANUEL DE RÉPARATION:
B.V.M. AU POINT MORT : 645-795 TR/MIN

EEC IV= COMMANDE ÉLECTRONIQUE DU MOTEUR, VERSION IV
EFI = INJECTION ÉLECTRONIQUE
DIS = ALLUMAGE SANS DISTRIBUTEUR
ALLUMAGE : 10° AVANT PMH.SINON,V.MAN.DE RÉPARATION.

HUILE PRÉCONISÉE : *SAE* 5W-30, CLASSE *API* « SG » - « ÉCONOMIE D'ÉNERGIE II »

	SPARK PLUG / *BOUGIES*	GAP / *ÉLECTRODES*
2.3 L	AWSF-32C	.042-.046

CALIBRATION PARTS LIST — CODE: DHL

NAME/DESCRIPTION	ENGINEERING NO.	SERVICE NO.
SENSOR, Exhaust Gas Recirculation Valve	E7TF 9G428-AA	E7TZ 9G428-A
CONTROL ASSY, EGR Vacuum Regulator	E8TE 9J459-AA	E8TZ 9J459-A
SENSOR ASSY, Exhaust Gas Oxygen	F0WF 9F472-AA	F0WZ 9F472-B
VALVE ASSY, EGR Pressure External	E97E 9F483-A2A, A4A	E9TZ 9F483-A
INJECTOR ASSY, Fuel	E59E 9F593-A2B	E5TZ 9F593-A
SENSOR ASSY, Carburetor Air Cleaner Air Temperature Control		D7FZ 9E607-A
MOTOR ASSY, Carburetor Air Cleaner Vacuum		D4FZ 9D612-A
VALVE ASSY, Throttle Air By-Pass	E9TE 9F715-A1A F0TE 9F715-AA	E9TZ 9F715-A
SENSOR ASSY, Speed	E3AF 9E731-AB	E3AZ 9E731-A
REGULATOR ASSY, Fuel Charging Pressure	E7EE 9C968-AA E77E 9C968-AA	E7FZ 9C968-A
POTENTIOMETER ASSY, Throttle Position Sensor	E7DF 9B989-AA	E7DZ 9B989-A
SENSOR ASSY, Mass Airflow	F0VF 12B579-AA	F0VY 12B579-A
SENSOR ASSY, Engine Electronic Control Coolant Temperature	E1AF 12A648-AA E4AF 12A648-AA	E1AZ 12A648-A
PROCESSOR AND CALIBRATOR ASSY, EEC-IV	F07F 12A650-AB	F0TZ 12A650-RAB
SENSOR ASSY, Air Charge Temperature	E4AF 12A697-AA	E1AZ 12A697-A

CALIBRATION: 0-49T-R10 — 2.3L EFI

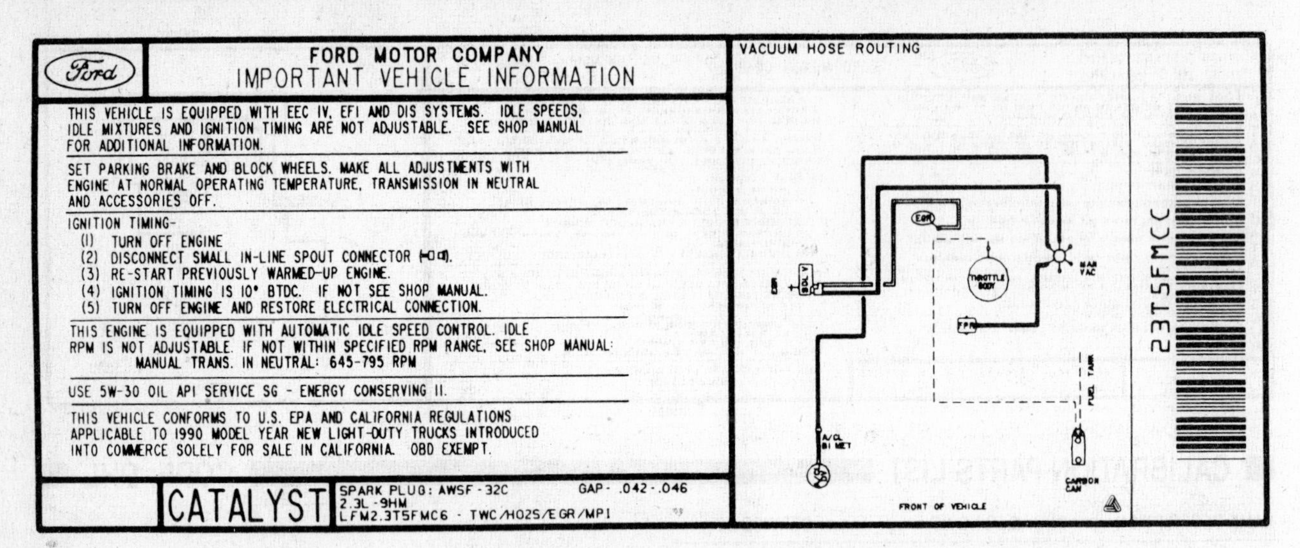
FORD MOTOR COMPANY
IMPORTANT VEHICLE INFORMATION

THIS VEHICLE IS EQUIPPED WITH EEC IV, EFI AND DIS SYSTEMS. IDLE SPEEDS, IDLE MIXTURES AND IGNITION TIMING ARE NOT ADJUSTABLE. SEE SHOP MANUAL FOR ADDITIONAL INFORMATION.

SET PARKING BRAKE AND BLOCK WHEELS. MAKE ALL ADJUSTMENTS WITH ENGINE AT NORMAL OPERATING TEMPERATURE, TRANSMISSION IN NEUTRAL AND ACCESSORIES OFF.

IGNITION TIMING-
(1) TURN OFF ENGINE
(2) DISCONNECT SMALL IN-LINE SPOUT CONNECTOR.
(3) RE-START PREVIOUSLY WARMED-UP ENGINE.
(4) IGNITION TIMING IS 10° BTDC. IF NOT SEE SHOP MANUAL.
(5) TURN OFF ENGINE AND RESTORE ELECTRICAL CONNECTION.

THIS ENGINE IS EQUIPPED WITH AUTOMATIC IDLE SPEED CONTROL. IDLE RPM IS NOT ADJUSTABLE. IF NOT WITHIN SPECIFIED RPM RANGE, SEE SHOP MANUAL:
MANUAL TRANS. IN NEUTRAL: 645-795 RPM

USE 5W-30 OIL API SERVICE SG - ENERGY CONSERVING II.

THIS VEHICLE CONFORMS TO U.S. EPA AND CALIFORNIA REGULATIONS APPLICABLE TO 1990 MODEL YEAR NEW LIGHT-DUTY TRUCKS INTRODUCED INTO COMMERCE SOLELY FOR SALE IN CALIFORNIA. OBD EXEMPT.

CATALYST
SPARK PLUG: AWSF-32C GAP-.042-.046
2.3L-9HM
LFM2.3T5FMC6 - TWC/HO2S/EGR/MPI

CALIBRATION PARTS LIST — CODE: DDV

NAME/DESCRIPTION	ENGINEERING NO.	SERVICE NO.
SENSOR, Exhaust Gas Recirculation Valve	E7TF 9G428-AA	E7TZ 9G428-A
CONTROL ASSY, EGR Vacuum Regulator	E8TE 9J459-AA	E8TZ 9J459-A
SENSOR ASSY, Exhaust Gas Oxygen	F0WF 9F472-AA	F0WZ 9F472-B
VALVE ASSY, EGR Pressure External	E97E 9F483-A2A, A4A	E9TZ 9F483-A
INJECTOR ASSY, Fuel	E59E 9F593-A2B	E5TZ 9F593-A
SENSOR ASSY, Carburetor Air Cleaner Air Temperature Control		D7FZ 9E607-A
MOTOR ASSY, Carburetor Air Cleaner Vacuum		D4FZ 9D612-A
VALVE ASSY, Throttle Air By-Pass	E9TE 9F715-A1A F0TE 9F715-AA	E9TZ 9F715-A
SENSOR ASSY, Speed	E3AF 9E731-AB	E3AZ 9E731-A
REGULATOR ASSY, Fuel Charging Pressure	E7EE 9C968-AA E77E 9C968-AA	E7FZ 9C968-A
POTENTIOMETER ASSY, Throttle Position Sensor	E7DF 9B989-AA	E7DZ 9B989-A
SENSOR ASSY, Mass Airflow	F0VF 12B579-AA	F0VY 12B579-A
SENSOR ASSY, Engine Electronic Control Coolant Temperature	E1AF 12A648-AA E4AF 12A648-AA	E1AZ 12A648-A
PROCESSOR AND CALIBRATOR ASSY, EEC-IV	F07F 12A650-ADA	F0TZ 12A650-RADA
SENSOR ASSY, Air Charge Temperature	E4AF 12A697-AA	E1AZ 12A697-A

■ CALIBRATION: 0-50S-R00 ■ 2.3L EFI ■

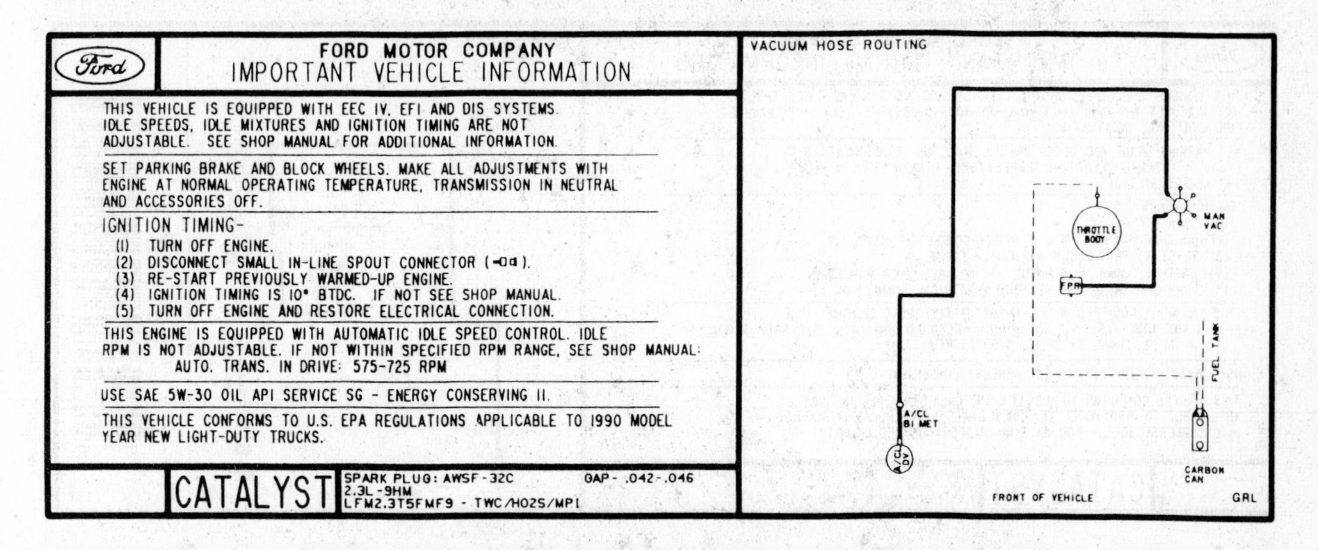
FORD MOTOR COMPANY
IMPORTANT VEHICLE INFORMATION

THIS VEHICLE IS EQUIPPED WITH EEC IV, EFI AND DIS SYSTEMS. IDLE SPEEDS, IDLE MIXTURES AND IGNITION TIMING ARE NOT ADJUSTABLE. SEE SHOP MANUAL FOR ADDITIONAL INFORMATION.

SET PARKING BRAKE AND BLOCK WHEELS. MAKE ALL ADJUSTMENTS WITH ENGINE AT NORMAL OPERATING TEMPERATURE, TRANSMISSION IN NEUTRAL AND ACCESSORIES OFF.

IGNITION TIMING-
(1) TURN OFF ENGINE.
(2) DISCONNECT SMALL IN-LINE SPOUT CONNECTOR (-▭▭).
(3) RE-START PREVIOUSLY WARMED-UP ENGINE.
(4) IGNITION TIMING IS 10° BTDC. IF NOT SEE SHOP MANUAL.
(5) TURN OFF ENGINE AND RESTORE ELECTRICAL CONNECTION.

THIS ENGINE IS EQUIPPED WITH AUTOMATIC IDLE SPEED CONTROL. IDLE RPM IS NOT ADJUSTABLE. IF NOT WITHIN SPECIFIED RPM RANGE, SEE SHOP MANUAL:
AUTO. TRANS. IN DRIVE: 575-725 RPM

USE SAE 5W-30 OIL API SERVICE SG - ENERGY CONSERVING II.

THIS VEHICLE CONFORMS TO U.S. EPA REGULATIONS APPLICABLE TO 1990 MODEL YEAR NEW LIGHT-DUTY TRUCKS.

CATALYST
SPARK PLUG: AWSF-32C GAP-.042-.046
2.3L-9HM
LFM2.3T5FMF9 - TWC/HO2S/MPI

VACUUM HOSE ROUTING

■ CALIBRATION PARTS LIST ■ CODE: DHM ■

NAME/DESCRIPTION	ENGINEERING NO.	SERVICE NO.
SENSOR ASSY, Exhaust Gas Oxygen	F0WF 9F472-AA	F0WZ 9F472-B
INJECTOR ASSY, Fuel	E59E 9F593-A2B	E5TZ 9F593-A
SENSOR ASSY, Carburetor Air Cleaner Air Temperature Control		D7FZ 9E607-A
MOTOR ASSY, Carburetor Air Cleaner Vacuum		D4FZ 9D612-A
VALVE ASSY, Throttle Air By-Pass	E9TE 9F715-A1A F0TE 9F715-AA	E9TZ 9F715-A
SENSOR ASSY, Speed	E3AF 9E731-AB	E3AZ 9E731-A
REGULATOR ASSY, Fuel Charging Pressure	E7EE 9C968-AA E77E 9C968-AA	E7FZ 9C968-A
POTENTIOMETER ASSY, Throttle Position Sensor	E7DF 9B989-AA	E7DZ 9B989-A
SENSOR ASSY, Mass Airflow	F0VF 12B579-AA	F0VY 12B579-A
SENSOR ASSY, Engine Electronic Control Coolant Temperature	E1AF 12A648-AA E4AF 12A648-AA	E1AZ 12A648-A
PROCESSOR AND CALIBRATOR ASSY, EEC-IV	F07F 12A650-BB	F0TZ 12A650-RBB
SENSOR ASSY, Air Charge Temperature	E4AF 12A697-AA	E1AZ 12A697-A

CALIBRATION: 0-50S-R00 — 2.3L EFI

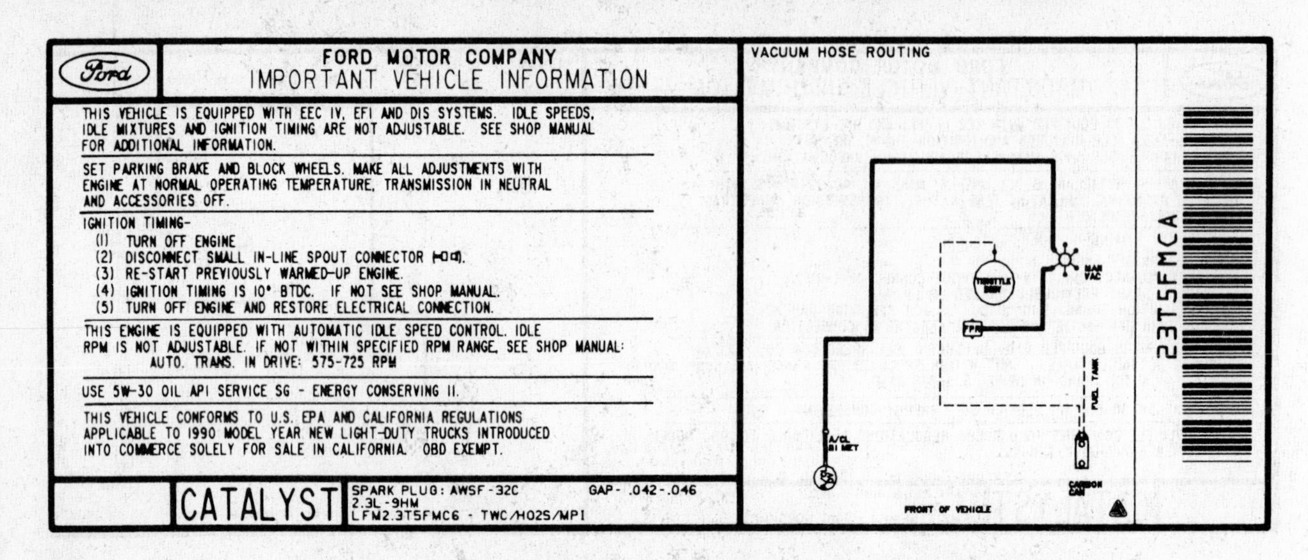
FORD MOTOR COMPANY
IMPORTANT VEHICLE INFORMATION

THIS VEHICLE IS EQUIPPED WITH EEC IV, EFI AND DIS SYSTEMS. IDLE SPEEDS, IDLE MIXTURES AND IGNITION TIMING ARE NOT ADJUSTABLE. SEE SHOP MANUAL FOR ADDITIONAL INFORMATION.

SET PARKING BRAKE AND BLOCK WHEELS. MAKE ALL ADJUSTMENTS WITH ENGINE AT NORMAL OPERATING TEMPERATURE, TRANSMISSION IN NEUTRAL AND ACCESSORIES OFF.

IGNITION TIMING-
(1) TURN OFF ENGINE
(2) DISCONNECT SMALL IN-LINE SPOUT CONNECTOR.
(3) RE-START PREVIOUSLY WARMED-UP ENGINE.
(4) IGNITION TIMING IS 10° BTDC. IF NOT SEE SHOP MANUAL.
(5) TURN OFF ENGINE AND RESTORE ELECTRICAL CONNECTION.

THIS ENGINE IS EQUIPPED WITH AUTOMATIC IDLE SPEED CONTROL. IDLE RPM IS NOT ADJUSTABLE. IF NOT WITHIN SPECIFIED RPM RANGE, SEE SHOP MANUAL:
AUTO. TRANS. IN DRIVE: 575-725 RPM

USE 5W-30 OIL API SERVICE SG - ENERGY CONSERVING II.

THIS VEHICLE CONFORMS TO U.S. EPA AND CALIFORNIA REGULATIONS APPLICABLE TO 1990 MODEL YEAR NEW LIGHT-DUTY TRUCKS INTRODUCED INTO COMMERCE SOLELY FOR SALE IN CALIFORNIA. OBD EXEMPT.

CATALYST

SPARK PLUG: AWSF-32C GAP-.042-.046
2.3L-9HM
LFM2.3T5FMC6 - TWC/HO2S/MPI

CALIBRATION PARTS LIST — CODE: DDY

NAME/DESCRIPTION	ENGINEERING NO.	SERVICE NO.
SENSOR ASSY, Exhaust Gas Oxygen	F0WF 9F472-AA	F0WZ 9F472-B
INJECTOR ASSY, Fuel	E59E 9F593-A2B	E5TZ 9F593-A
SENSOR ASSY, Carburetor Air Cleaner Air Temperature Control		D7FZ 9E607-A
MOTOR ASSY, Carburetor Air Cleaner Vacuum		D4FZ 9D612-A
VALVE ASSY, Throttle Air By-Pass	E9TE 9F715-A1A F0TE 9F715-AA	E9TZ 9F715-A
SENSOR ASSY, Speed	E3AF 9E731-AB	E3AZ 9E731-A
REGULATOR ASSY, Fuel Charging Pressure	E7EE 9C968-AA E77E 9C968-AA	E7FZ 9C968-A
POTENTIOMETER ASSY, Throttle Position Sensor	E7DF 9B989-AA	E7DZ 9B989-A
SENSOR ASSY, Mass Airflow	F0VF 12B579-AA	F0VY 12B579-A
SENSOR ASSY, Engine Electronic Control Coolant Temperature	E1AF 12A648-AA E4AF 12A648-AA	E1AZ 12A648-A
PROCESSOR AND CALIBRATOR ASSY, EEC-IV	F07F 12A650-BB	F0TZ 12A650-RBB
SENSOR ASSY, Air Charge Temperature	E4AF 12A697-AA	E1AZ 12A697-A

CALIBRATION: 0-50S-R00 — 2.3L EFI

IMPORTANT VEHICLE INFORMATION — FORD MOTOR COMPANY — RENSEIGNEMENTS IMPORTANTS SUR LE VÉHICULE

CATALYST CATALYSEUR

THIS VEHICLE IS EQUIPPED WITH EEC IV, EFI AND DIS SYSTEMS. IDLE SPEEDS, IDLE MIXTURES AND IGNITION TIMING ARE NOT ADJUSTABLE. SEE SHOP MANUAL FOR ADDITIONAL INFORMATION.

SET PARKING BRAKE AND BLOCK WHEELS. MAKE ALL ADJUSTMENTS WITH ENGINE AT NORMAL OPERATING TEMPERATURE, TRANSMISSION IN NEUTRAL, ACCESSORIES AND HEADLIGHTS OFF.

IGNITION TIMING-
(1) TURN OFF ENGINE
(2) DISCONNECT SMALL IN-LINE SPOUT CONNECTOR (=Dd).
(3) RE-START PREVIOUSLY WARMED-UP ENGINE.
(4) IGNITION TIMING IS 10° BTDC. IF NOT SEE SHOP MANUAL.
(5) TURN OFF ENGINE AND RESTORE ELECTRICAL CONNECTION.

THIS ENGINE IS EQUIPPED WITH AUTOMATIC IDLE SPEED CONTROL. IDLE RPM IS NOT ADJUSTABLE. IF NOT WITHIN SPECIFIED RPM RANGE, SEE SHOP MANUAL:
AUTO. TRANS. IN DRIVE: 575-725 RPM

USE SAE 5W-30 OIL API SERVICE SG - ENERGY CONSERVING II.

CE VÉHICULE EST MUNI DES SYSTÈMES EEC IV, EFI ET DIS. • LES RÉGIMES DE RALENTI, LES MÉLANGES DE RALENTI ET L'ALLUMAGE NE SONT PAS RÉGLABLES. POUR PLUS DE DÉTAILS, CONSULTER LE MANUEL DE RÉPARATION.

SERRER LE FREIN DE STATIONNEMENT, BLOQUER LES ROUES. EFFECTUER TOUT RÉGLAGE SUR MOTEUR NORMALEMENT CHAUD, B.V. AU POINT MORT, CONTACT DES ACCESSOIRES ET DES PHARES COUPÉ.

CALAGE DE L'ALLUMAGE :
(1) ARRÊTER LE MOTEUR.
(2) DÉBRANCHER LE PETIT CONNECTEUR (=Dd) INTERCALÉ DANS LE CIRCUIT DE DÉCLENCHEMENT DE L'ÉTINCELLE.
(3) REDÉMARRER LE MOTEUR PRÉALABLEMENT RÉCHAUFFÉ.
(4) L'ALLUMAGE DOIT ÊTRE CALÉ À 10° AVANT PMH. SINON, VOIR LE MANUEL DE RÉPARATION.
(5) ARRÊTER LE MOTEUR ET REBRANCHER LE CONNECTEUR.

CE MOTEUR EST À COMMANDE DE RALENTI AUTOMATIQUE. LE RÉGIME DE RALENTI N'EST PAS RÉGLABLE. S'IL N'EST PAS DANS LES LIMITES PRESCRITES, CONSULTER LE MANUEL DE RÉPARATION:
B.V.A. EN POSITION "D" : 575-725 TR/MIN

EEC IV= COMMANDE ÉLECTRONIQUE DU MOTEUR, VERSION IV
EFI = INJECTION ÉLECTRONIQUE
DIS = ALLUMAGE SANS DISTRIBUTEUR
ALLUMAGE : 10° AVANT PMH.SINON,V.MAN.DE RÉPARATION.

HUILE PRÉCONISÉE : SAE 5W-30, CLASSE API « SG » - « ÉCONOMIE D'ÉNERGIE II »

	SPARK PLUG / BOUGIES	GAP / ÉLECTRODES
2.3 L	AWSF-32C	.042-.046

MAN VAC
THROTTLE BODY
FPR
FUEL TANK
A/CL BI MET
CARBON CAN
FRONT OF VEHICLE
GRL

CALIBRATION PARTS LIST — CODE: DHR

NAME/DESCRIPTION	ENGINEERING NO.	SERVICE NO.
SENSOR ASSY, Exhaust Gas Oxygen	F0WF 9F472-AA	F0WZ 9F472-B
INJECTOR ASSY, Fuel	E59E 9F593-A2B	E5TZ 9F593-A
SENSOR ASSY, Carburetor Air Cleaner Air Temperature Control		D7FZ 9E607-A
MOTOR ASSY, Carburetor Air Cleaner Vacuum		D4FZ 9D612-A
VALVE ASSY, Throttle Air By-Pass	E9TE 9F715-A1A F0TE 9F715-AA	E9TZ 9F715-A
SENSOR ASSY, Speed	E3AF 9E731-AB	E3AZ 9E731-A
REGULATOR ASSY, Fuel Charging Pressure	E7EE 9C968-AA E77E 9C968-AA	E7FZ 9C968-A
POTENTIOMETER ASSY, Throttle Position Sensor	E7DF 9B989-AA	E7DZ 9B989-A
SENSOR ASSY, Mass Airflow	F0VF 12B579-AA	F0VY 12B579-A
SENSOR ASSY, Engine Electronic Control Coolant Temperature	E1AF 12A648-AA E4AF 12A648-AA	E1AZ 12A648-A
PROCESSOR AND CALIBRATOR ASSY, EEC-IV	F07F 12A650-BB	F0TZ 12A650-RBB
SENSOR ASSY, Air Charge Temperature	E4AF 12A697-AA	E1AZ 12A697-A

CALIBRATION: 0-50T-R10 — 2.3L EFI

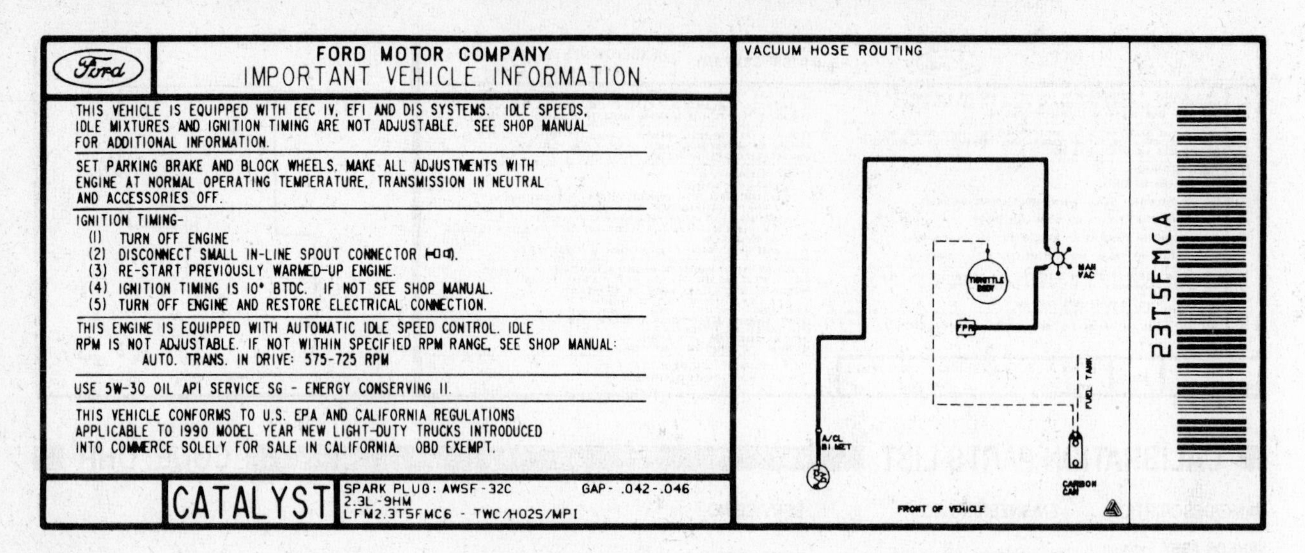

Ford

FORD MOTOR COMPANY
IMPORTANT VEHICLE INFORMATION

THIS VEHICLE IS EQUIPPED WITH EEC IV, EFI AND DIS SYSTEMS. IDLE SPEEDS, IDLE MIXTURES AND IGNITION TIMING ARE NOT ADJUSTABLE. SEE SHOP MANUAL FOR ADDITIONAL INFORMATION.

SET PARKING BRAKE AND BLOCK WHEELS. MAKE ALL ADJUSTMENTS WITH ENGINE AT NORMAL OPERATING TEMPERATURE, TRANSMISSION IN NEUTRAL AND ACCESSORIES OFF.

IGNITION TIMING-
(1) TURN OFF ENGINE
(2) DISCONNECT SMALL IN-LINE SPOUT CONNECTOR.
(3) RE-START PREVIOUSLY WARMED-UP ENGINE.
(4) IGNITION TIMING IS 10° BTDC. IF NOT SEE SHOP MANUAL.
(5) TURN OFF ENGINE AND RESTORE ELECTRICAL CONNECTION.

THIS ENGINE IS EQUIPPED WITH AUTOMATIC IDLE SPEED CONTROL. IDLE RPM IS NOT ADJUSTABLE. IF NOT WITHIN SPECIFIED RPM RANGE, SEE SHOP MANUAL:
AUTO. TRANS. IN DRIVE: 575-725 RPM

USE 5W-30 OIL API SERVICE SG - ENERGY CONSERVING II.

THIS VEHICLE CONFORMS TO U.S. EPA AND CALIFORNIA REGULATIONS APPLICABLE TO 1990 MODEL YEAR NEW LIGHT-DUTY TRUCKS INTRODUCED INTO COMMERCE SOLELY FOR SALE IN CALIFORNIA. OBD EXEMPT.

CATALYST

SPARK PLUG: AWSF-32C GAP-.042-.046
2.3L-9HM
LFM2.3T5FMC6 - TWC/HO2S/MPI

VACUUM HOSE ROUTING

CALIBRATION PARTS LIST — CODE: DDY

NAME/DESCRIPTION	ENGINEERING NO.	SERVICE NO.
SENSOR ASSY, Exhaust Gas Oxygen	F0WF 9F472-AA	F0WZ 9F472-B
INJECTOR ASSY, Fuel	E59E 9F593-A2B	E5TZ 9F593-A
SENSOR ASSY, Carburetor Air Cleaner Air Temperature Control		D7FZ 9E607-A
MOTOR ASSY, Carburetor Air Cleaner Vacuum		D4FZ 9D612-A
VALVE ASSY, Throttle Air By-Pass	E9TE 9F715-A1A F0TE 9F715-AA	E9TZ 9F715-A
SENSOR ASSY, Speed	E3AF 9E731-AB	E3AZ 9E731-A
REGULATOR ASSY, Fuel Charging Pressure	E7EE 9C968-AA E77E 9C968-AA	E7FZ 9C968-A
POTENTIOMETER ASSY, Throttle Position Sensor	E7DF 9B989-AA	E7DZ 9B989-A
SENSOR ASSY, Mass Airflow	F0VF 12B579-AA	F0VY 12B579-A
SENSOR ASSY, Engine Electronic Control Coolant Temperature	E1AF 12A648-AA E4AF 12A648-AA	E1AZ 12A648-A
PROCESSOR AND CALIBRATOR ASSY, EEC-IV	F07F 12A650-AEA	F0TZ 12A650-RAEA
SENSOR ASSY, Air Charge Temperature	E4AF 12A697-AA	E1AZ 12A697-A

CALIBRATION: 9-49F-R00 — 2.3L EFI

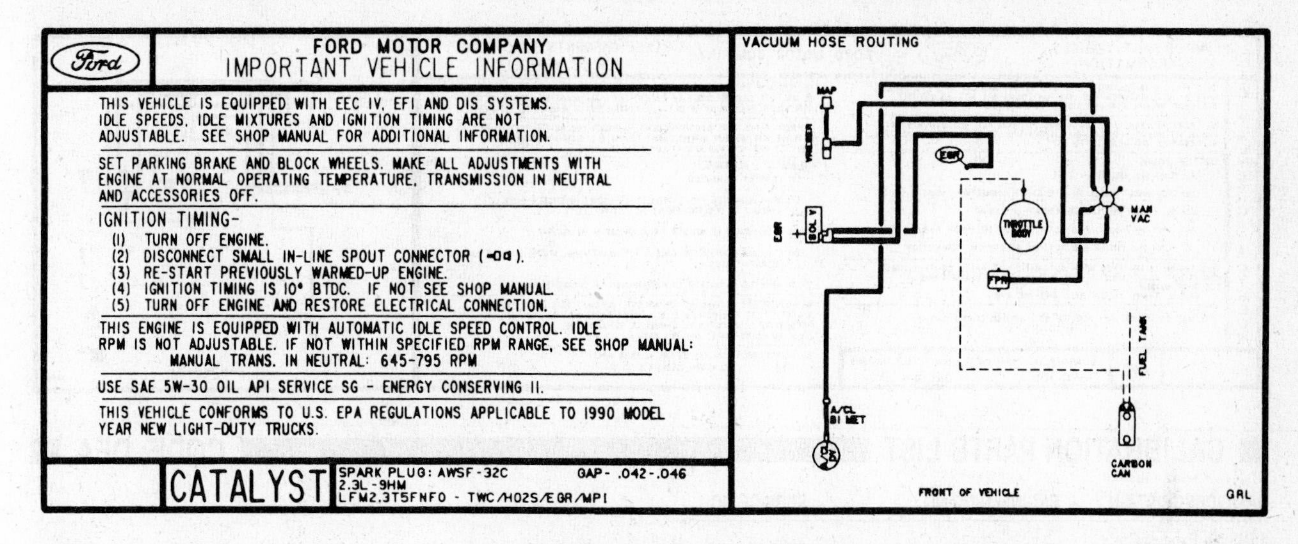
FORD MOTOR COMPANY
IMPORTANT VEHICLE INFORMATION

THIS VEHICLE IS EQUIPPED WITH EEC IV, EFI AND DIS SYSTEMS. IDLE SPEEDS, IDLE MIXTURES AND IGNITION TIMING ARE NOT ADJUSTABLE. SEE SHOP MANUAL FOR ADDITIONAL INFORMATION.

SET PARKING BRAKE AND BLOCK WHEELS. MAKE ALL ADJUSTMENTS WITH ENGINE AT NORMAL OPERATING TEMPERATURE, TRANSMISSION IN NEUTRAL AND ACCESSORIES OFF.

IGNITION TIMING-
(1) TURN OFF ENGINE.
(2) DISCONNECT SMALL IN-LINE SPOUT CONNECTOR.
(3) RE-START PREVIOUSLY WARMED-UP ENGINE.
(4) IGNITION TIMING IS 10° BTDC. IF NOT SEE SHOP MANUAL.
(5) TURN OFF ENGINE AND RESTORE ELECTRICAL CONNECTION.

THIS ENGINE IS EQUIPPED WITH AUTOMATIC IDLE SPEED CONTROL. IDLE RPM IS NOT ADJUSTABLE. IF NOT WITHIN SPECIFIED RPM RANGE, SEE SHOP MANUAL:
MANUAL TRANS. IN NEUTRAL: 645-795 RPM

USE SAE 5W-30 OIL API SERVICE SG - ENERGY CONSERVING II.

THIS VEHICLE CONFORMS TO U.S. EPA REGULATIONS APPLICABLE TO 1990 MODEL YEAR NEW LIGHT-DUTY TRUCKS.

CATALYST
SPARK PLUG: AWSF-32C GAP- .042-.046
2.3L-9HM
LFM2.3T5FNFO - TWC/HO2S/EGR/MPI

CALIBRATION PARTS LIST — CODE: DDZ

NAME/DESCRIPTION	ENGINEERING NO.	SERVICE NO.
SENSOR, Exhaust Gas Recirculation Valve	E7TF 9G428-AA	E7TZ 9G428-A
CONTROL ASSY, EGR Vacuum Regulator	E8TE 9J459-AA	E8TZ 9J459-A
SENSOR ASSY, Exhaust Gas Oxygen	E77F 9F472-AA, BA E7TF 9F472-AA	E7TZ 9F472-A
SENSOR ASSY, Exhaust Gas Oxygen	E9SF 9F472-AA E97F 9F472-AA	F0SZ 9F472-A
SENSOR ASSY, Exhaust Gas Oxygen	E93F 9F472-AA, DA, EA	F03Z 9F472-A
SENSOR ASSY, Manifold Absolute Pressure	E7EF 9F479-A1A, A2A	E7FZ 9F479-A
VALVE ASSY, EGR Pressure External	E97E 9F483-A2A, A4A	E9TZ 9F483-A
INJECTOR ASSY, Fuel	E59E 9F593-A2B	E5TZ 9F593-A
SENSOR ASSY, Carburetor Air Cleaner Air Temperature Control		D7FZ 9E607-A
MOTOR ASSY, Carburetor Air Cleaner Vacuum		D4FZ 9D612-A
VALVE ASSY, Throttle Air By-Pass	E9TE 9F715-A1A F0TE 9F715-AA	E9TZ 9F715-A
SENSOR ASSY, Speed	E3AF 9E731-AB	E3AZ 9E731-A
REGULATOR ASSY, Fuel Charging Pressure	E7EE 9C968-AA E77E 9C968-AA	E7FZ 9C968-A
POTENTIOMETER ASSY, Throttle Position Sensor	E7DF 9B989-AA	E7DZ 9B989-A
SENSOR ASSY, Engine Electronic Control Coolant Temperature	E1AF 12A648-AA E4AF 12A648-AA	E1AZ 12A648-A
PROCESSOR AND CALIBRATOR ASSY, EEC-IV	E9TF 12A650-A1A, A2A	E9TZ 12A650-AA
SENSOR ASSY, Air Charge Temperature	E4AF 12A697-AA	E1AZ 12A697-A

CALIBRATION: 9-49F-R00 — 2.3L EFI

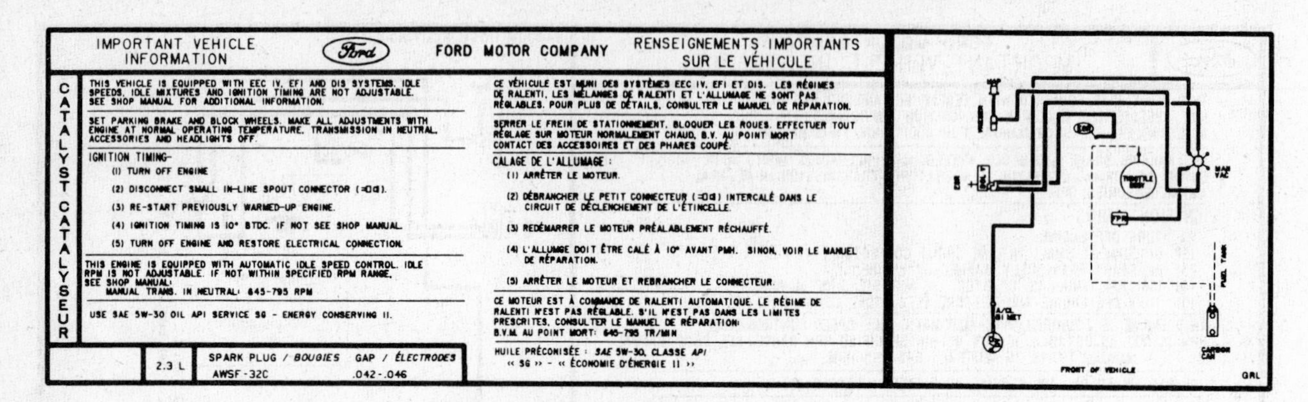

IMPORTANT VEHICLE INFORMATION — FORD MOTOR COMPANY — RENSEIGNEMENTS IMPORTANTS SUR LE VÉHICULE

CATALYST CATALYSEUR

THIS VEHICLE IS EQUIPPED WITH EEC IV, EFI AND DIS SYSTEMS. IDLE SPEEDS, IDLE MIXTURES AND IGNITION TIMING ARE NOT ADJUSTABLE. SEE SHOP MANUAL FOR ADDITIONAL INFORMATION.

SET PARKING BRAKE AND BLOCK WHEELS. MAKE ALL ADJUSTMENTS WITH ENGINE AT NORMAL OPERATING TEMPERATURE, TRANSMISSION IN NEUTRAL, ACCESSORIES AND HEADLIGHTS OFF.

IGNITION TIMING–
(1) TURN OFF ENGINE
(2) DISCONNECT SMALL IN-LINE SPOUT CONNECTOR (=Dd).
(3) RE-START PREVIOUSLY WARMED-UP ENGINE.
(4) IGNITION TIMING IS 10° BTDC. IF NOT SEE SHOP MANUAL.
(5) TURN OFF ENGINE AND RESTORE ELECTRICAL CONNECTION.

THIS ENGINE IS EQUIPPED WITH AUTOMATIC IDLE SPEED CONTROL. IDLE RPM IS NOT ADJUSTABLE. IF NOT WITHIN SPECIFIED RPM RANGE, SEE SHOP MANUAL:
MANUAL TRANS. IN NEUTRAL: 645-795 RPM

USE SAE 5W-30 OIL API SERVICE SG - ENERGY CONSERVING II.

CE VÉHICULE EST MUNI DES SYSTÈMES EEC IV, EFI ET DIS. LES RÉGIMES DE RALENTI, LES MÉLANGES DE RALENTI ET L'ALLUMAGE NE SONT PAS RÉGLABLES. POUR PLUS DE DÉTAILS, CONSULTER LE MANUEL DE RÉPARATION.

SERRER LE FREIN DE STATIONNEMENT, BLOQUER LES ROUES. EFFECTUER TOUT RÉGLAGE SUR MOTEUR NORMALEMENT CHAUD, B.V. AU POINT MORT CONTACT DES ACCESSOIRES ET DES PHARES COUPÉ.

CALAGE DE L'ALLUMAGE :
(1) ARRÊTER LE MOTEUR.
(2) DÉBRANCHER LE PETIT CONNECTEUR (=Dd) INTERCALÉ DANS LE CIRCUIT DE DÉCLENCHEMENT DE L'ÉTINCELLE.
(3) REDÉMARRER LE MOTEUR PRÉALABLEMENT RÉCHAUFFÉ.
(4) L'ALLUMGE DOIT ÊTRE CALÉ À 10° AVANT PMH. SINON, VOIR LE MANUEL DE RÉPARATION.
(5) ARRÊTER LE MOTEUR ET REBRANCHER LE CONNECTEUR.

CE MOTEUR EST À COMMANDE DE RALENTI AUTOMATIQUE. LE RÉGIME DE RALENTI N'EST PAS RÉGLABLE. S'IL N'EST PAS DANS LES LIMITES PRESCRITES, CONSULTER LE MANUEL DE RÉPARATION:
B.V.M. AU POINT MORT: 645-795 TR/MIN

HUILE PRÉCONISÉE : *SAE* 5W-30, CLASSE *API* « SG » - « ÉCONOMIE D'ÉNERGIE II »

	2.3 L	SPARK PLUG / *BOUGIES*	GAP / *ÉLECTRODES*
		AWSF-32C	.042-.046

CALIBRATION PARTS LIST — CODE: DEA

NAME/DESCRIPTION	ENGINEERING NO.	SERVICE NO.
SENSOR, Exhaust Gas Recirculation Valve	E7TF 9G428-AA	E7TZ 9G428-A
CONTROL ASSY, EGR Vacuum Regulator	E8TE 9J459-AA	E8TZ 9J459-A
SENSOR ASSY, Exhaust Gas Oxygen	E77F 9F472-AA, BA E7TF 9F472-AA	E7TZ 9F472-A
SENSOR ASSY, Exhaust Gas Oxygen	E9SF 9F472-AA E97F 9F472-AA	F0SZ 9F472-A
SENSOR ASSY, Exhaust Gas Oxygen	E93F 9F472-AA, DA, EA	F03Z 9F472-A
SENSOR ASSY, Manifold Absolute Pressure	E7EF 9F479-A1A, A2A	E7FZ 9F479-A
VALVE ASSY, EGR Pressure External	E97E 9F483-A2A, A4A	E9TZ 9F483-A
INJECTOR ASSY, Fuel	E59E 9F593-A2B	E5TZ 9F593-A
SENSOR ASSY, Carburetor Air Cleaner Air Temperature Control		D7FZ 9E607-A
MOTOR ASSY, Carburetor Air Cleaner Vacuum		D4FZ 9D612-A
VALVE ASSY, Throttle Air By-Pass	E9TE 9F715-A1A F0TE 9F715-AA	E9TZ 9F715-A
SENSOR ASSY, Speed	E3AF 9E731-AB	E3AZ 9E731-A
REGULATOR ASSY, Fuel Charging Pressure	E7EE 9C968-AA E77E 9C968-AA	E7FZ 9C968-A
POTENTIOMETER ASSY, Throttle Position Sensor	E7DF 9B989-AA	E7DZ 9B989-A
SENSOR ASSY, Engine Electronic Control Coolant Temperature	E1AF 12A648-AA E4AF 12A648-AA	E1AZ 12A648-A
PROCESSOR AND CALIBRATOR ASSY, EEC-IV	E9TF 12A650-A1A, A2A	E9TZ 12A650-AA
SENSOR ASSY, Air Charge Temperature	E4AF 12A697-AA	E1AZ 12A697-A

■ CALIBRATION: 9-50K-R00 — 2.3L EFI ■

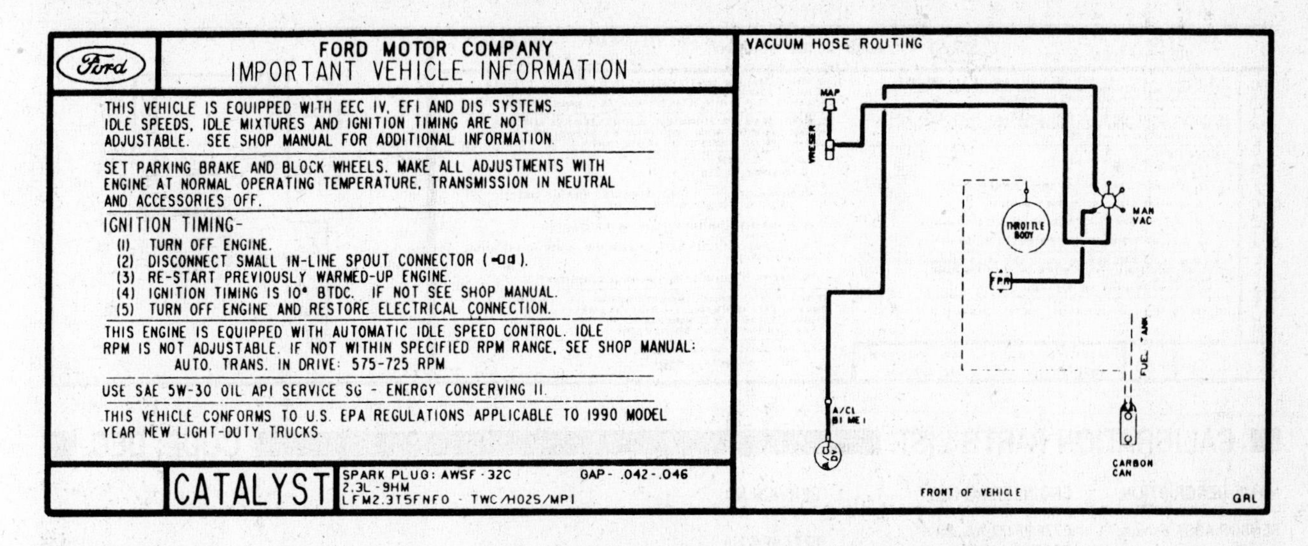

FORD MOTOR COMPANY
IMPORTANT VEHICLE INFORMATION

THIS VEHICLE IS EQUIPPED WITH EEC IV, EFI AND DIS SYSTEMS. IDLE SPEEDS, IDLE MIXTURES AND IGNITION TIMING ARE NOT ADJUSTABLE. SEE SHOP MANUAL FOR ADDITIONAL INFORMATION.

SET PARKING BRAKE AND BLOCK WHEELS. MAKE ALL ADJUSTMENTS WITH ENGINE AT NORMAL OPERATING TEMPERATURE, TRANSMISSION IN NEUTRAL AND ACCESSORIES OFF.

IGNITION TIMING-
(1) TURN OFF ENGINE.
(2) DISCONNECT SMALL IN-LINE SPOUT CONNECTOR (⊸ɑ).
(3) RE-START PREVIOUSLY WARMED-UP ENGINE.
(4) IGNITION TIMING IS 10° BTDC. IF NOT SEE SHOP MANUAL.
(5) TURN OFF ENGINE AND RESTORE ELECTRICAL CONNECTION.

THIS ENGINE IS EQUIPPED WITH AUTOMATIC IDLE SPEED CONTROL. IDLE RPM IS NOT ADJUSTABLE. IF NOT WITHIN SPECIFIED RPM RANGE, SEE SHOP MANUAL:
AUTO. TRANS. IN DRIVE: 575-725 RPM

USE SAE 5W-30 OIL API SERVICE SG - ENERGY CONSERVING II.

THIS VEHICLE CONFORMS TO U.S. EPA REGULATIONS APPLICABLE TO 1990 MODEL YEAR NEW LIGHT-DUTY TRUCKS.

CATALYST

SPARK PLUG: AWSF-32C GAP-.042-.046
2.3L-9HM
LFM2.3T5FNFO - TWC/HO2S/MPI

VACUUM HOSE ROUTING

■ CALIBRATION PARTS LIST — CODE: DGE ■

NAME/DESCRIPTION	ENGINEERING NO.	SERVICE NO.
SENSOR ASSY, Exhaust Gas Oxygen	E77F 9F472-AA, BA E7TF 9F472-AA	E7TZ 9F472-A
SENSOR ASSY, Exhaust Gas Oxygen	E9SF 9F472-AA E97F 9F472-AA	F0SZ 9F472-A
SENSOR ASSY, Exhaust Gas Oxygen	E93F 9F472-AA, DA, EA	F03Z 9F472-A
SENSOR ASSY, Manifold Absolute Pressure	E7EF 9F479-A1A, A2A	E7FZ 9F479-A
INJECTOR ASSY, Fuel	E59E 9F593-A2B	E5TZ 9F593-A
SENSOR ASSY, Carburetor Air Cleaner Air Temperature Control		D7FZ 9E607-A
MOTOR ASSY, Carburetor Air Cleaner Vacuum		D4FZ 9D612-A
VALVE ASSY, Throttle Air By-Pass	E9TE 9F715-A1A F0TE 9F715-AA	E9TZ 9F715-A
SENSOR ASSY, Speed	E3AF 9E731-AB	E3AZ 9E731-A
REGULATOR ASSY, Fuel Charging Pressure	E7EE 9C968-AA E77E 9C968-AA	E7FZ 9C968-A
POTENTIOMETER ASSY, Throttle Position Sensor	E7DF 9B989-AA	E7DZ 9B989-A
SENSOR ASSY, Engine Electronic Control Coolant Temperature	E1AF 12A648-AA E4AF 12A648-AA	E1AZ 12A648-A
PROCESSOR AND CALIBRATOR ASSY, EEC-IV	E9TF 12A650-C1A, C2A	E9TZ 12A650-CA
SENSOR ASSY, Air Charge Temperature	E4AF 12A697-AA	E1AZ 12A697-A

CALIBRATION: 9-50K-R00 — 2.3L EFI

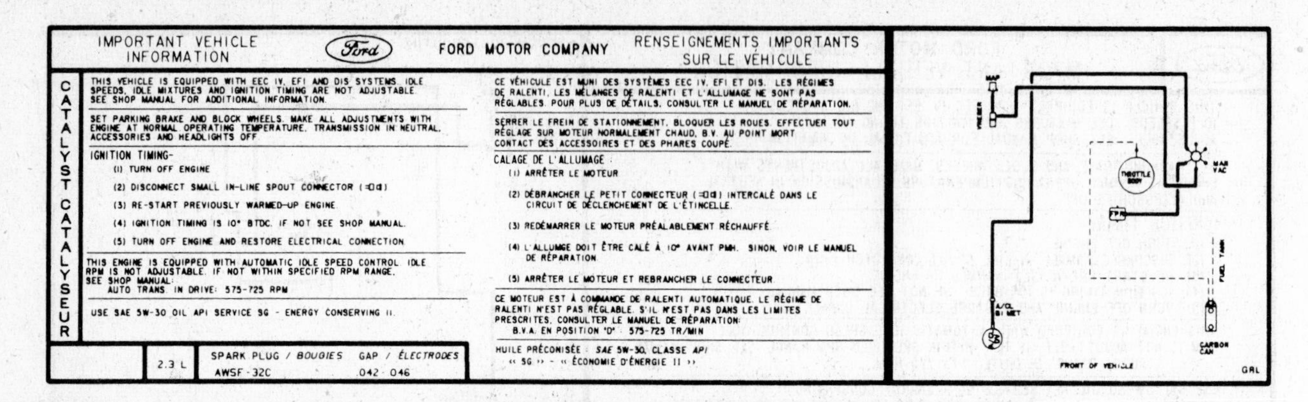

IMPORTANT VEHICLE INFORMATION — FORD MOTOR COMPANY — RENSEIGNEMENTS IMPORTANTS SUR LE VÉHICULE

CATALYST / CATALYSEUR

THIS VEHICLE IS EQUIPPED WITH EEC IV, EFI AND DIS SYSTEMS. IDLE SPEEDS, IDLE MIXTURES AND IGNITION TIMING ARE NOT ADJUSTABLE. SEE SHOP MANUAL FOR ADDITIONAL INFORMATION.

SET PARKING BRAKE AND BLOCK WHEELS. MAKE ALL ADJUSTMENTS WITH ENGINE AT NORMAL OPERATING TEMPERATURE, TRANSMISSION IN NEUTRAL, ACCESSORIES AND HEADLIGHTS OFF.

IGNITION TIMING-
(1) TURN OFF ENGINE
(2) DISCONNECT SMALL IN-LINE SPOUT CONNECTOR
(3) RE-START PREVIOUSLY WARMED-UP ENGINE
(4) IGNITION TIMING IS 10° BTDC. IF NOT SEE SHOP MANUAL.
(5) TURN OFF ENGINE AND RESTORE ELECTRICAL CONNECTION

THIS ENGINE IS EQUIPPED WITH AUTOMATIC IDLE SPEED CONTROL. IDLE RPM IS NOT ADJUSTABLE. IF NOT WITHIN SPECIFIED RPM RANGE, SEE SHOP MANUAL:
AUTO TRANS. IN DRIVE: 575-725 RPM

USE SAE 5W-30 OIL API SERVICE SG - ENERGY CONSERVING II.

CE VÉHICULE EST MUNI DES SYSTÈMES EEC IV, EFI ET DIS. LES RÉGIMES DE RALENTI, LES MÉLANGES DE RALENTI ET L'ALLUMAGE NE SONT PAS RÉGLABLES. POUR PLUS DE DÉTAILS, CONSULTER LE MANUEL DE RÉPARATION.

SERRER LE FREIN DE STATIONNEMENT, BLOQUER LES ROUES. EFFECTUER TOUT RÉGLAGE SUR MOTEUR NORMALEMENT CHAUD, B.V. AU POINT MORT CONTACT DES ACCESSOIRES ET DES PHARES COUPÉ.

CALAGE DE L'ALLUMAGE :
(1) ARRÊTER LE MOTEUR
(2) DÉBRANCHER LE PETIT CONNECTEUR INTERCALÉ DANS LE CIRCUIT DE DÉCLENCHEMENT DE L'ÉTINCELLE.
(3) REDÉMARRER LE MOTEUR PRÉALABLEMENT RÉCHAUFFÉ.
(4) L'ALLUMGE DOIT ÊTRE CALÉ À 10° AVANT PMH. SINON, VOIR LE MANUEL DE RÉPARATION.
(5) ARRÊTER LE MOTEUR ET REBRANCHER LE CONNECTEUR.

CE MOTEUR EST À COMMANDE DE RALENTI AUTOMATIQUE. LE RÉGIME DE RALENTI N'EST PAS RÉGLABLE. S'IL N'EST PAS DANS LES LIMITES PRESCRITES, CONSULTER LE MANUEL DE RÉPARATION:
B.V.A. EN POSITION "D" : 575-725 TR/MIN

HUILE PRÉCONISÉE : SAE 5W-30, CLASSE API « SG » - « ÉCONOMIE D'ÉNERGIE II »

	SPARK PLUG / BOUGIES	GAP / ÉLECTRODES
2.3 L	AWSF-32C	.042-.046

CALIBRATION PARTS LIST — CODE: DEC

NAME/DESCRIPTION	ENGINEERING NO.	SERVICE NO.
SENSOR ASSY, Exhaust Gas Oxygen	**E77F 9F472-AA, BA** **E7TF 9F472-AA**	**E7TZ 9F472-A**
SENSOR ASSY, Exhaust Gas Oxygen	**E9SF 9F472-AA** **E97F 9F472-AA**	**F0SZ 9F472-A**
SENSOR ASSY, Exhaust Gas Oxygen	**E93F 9F472-AA, DA, EA**	**F03Z 9F472-A**
SENSOR ASSY, Manifold Absolute Pressure	**E7EF 9F479-A1A, A2A**	**E7FZ 9F479-A**
INJECTOR ASSY, Fuel	**E59E 9F593-A2B**	**E5TZ 9F593-A**
SENSOR ASSY, Carburetor Air Cleaner Air Temperature Control		**D7FZ 9E607-A**
MOTOR ASSY, Carburetor Air Cleaner Vacuum		**D4FZ 9D612-A**
VALVE ASSY, Throttle Air By-Pass	**E9TE 9F715-A1A** **F0TE 9F715-AA**	**E9TZ 9F715-A**
SENSOR ASSY, Speed	**E3AF 9E731-AB**	**E3AZ 9E731-A**
REGULATOR ASSY, Fuel Charging Pressure	**E7EE 9C968-AA** **E77E 9C968-AA**	**E7FZ 9C968-A**
POTENTIOMETER ASSY, Throttle Position Sensor	**E7DF 9B989-AA**	**E7DZ 9B989-A**
SENSOR ASSY, Engine Electronic Control Coolant Temperature	**E1AF 12A648-AA** **E4AF 12A648-AA**	**E1AZ 12A648-A**
PROCESSOR AND CALIBRATOR ASSY, EEC-IV	**E9TF 12A650-C1A, C2A**	**E9TZ 12A650-CA**
SENSOR ASSY, Air Charge Temperature	**E4AF 12A697-AA**	**E1AZ 12A697-A**

■ CALIBRATION: 0-65T-R00 — 2.9L EFI ■

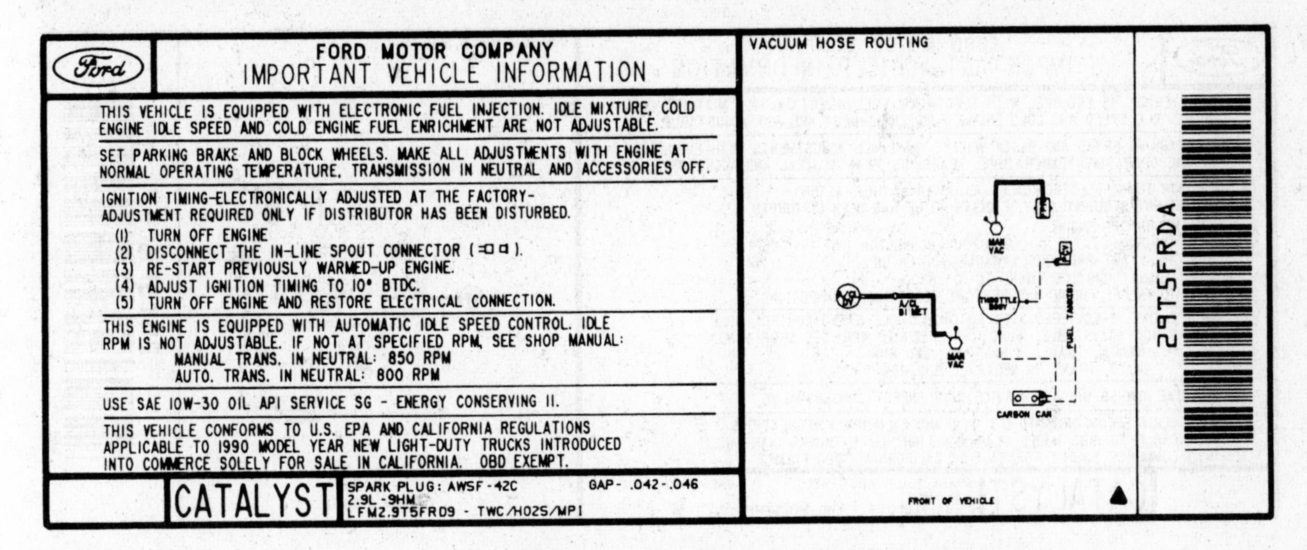

FORD MOTOR COMPANY
IMPORTANT VEHICLE INFORMATION

THIS VEHICLE IS EQUIPPED WITH ELECTRONIC FUEL INJECTION. IDLE MIXTURE, COLD ENGINE IDLE SPEED AND COLD ENGINE FUEL ENRICHMENT ARE NOT ADJUSTABLE.

SET PARKING BRAKE AND BLOCK WHEELS. MAKE ALL ADJUSTMENTS WITH ENGINE AT NORMAL OPERATING TEMPERATURE, TRANSMISSION IN NEUTRAL AND ACCESSORIES OFF.

IGNITION TIMING-ELECTRONICALLY ADJUSTED AT THE FACTORY-ADJUSTMENT REQUIRED ONLY IF DISTRIBUTOR HAS BEEN DISTURBED.
(1) TURN OFF ENGINE
(2) DISCONNECT THE IN-LINE SPOUT CONNECTOR.
(3) RE-START PREVIOUSLY WARMED-UP ENGINE.
(4) ADJUST IGNITION TIMING TO 10° BTDC.
(5) TURN OFF ENGINE AND RESTORE ELECTRICAL CONNECTION.

THIS ENGINE IS EQUIPPED WITH AUTOMATIC IDLE SPEED CONTROL. IDLE RPM IS NOT ADJUSTABLE. IF NOT AT SPECIFIED RPM, SEE SHOP MANUAL:
MANUAL TRANS. IN NEUTRAL: 850 RPM
AUTO. TRANS. IN NEUTRAL: 800 RPM

USE SAE 10W-30 OIL API SERVICE SG - ENERGY CONSERVING II.

THIS VEHICLE CONFORMS TO U.S. EPA AND CALIFORNIA REGULATIONS APPLICABLE TO 1990 MODEL YEAR NEW LIGHT-DUTY TRUCKS INTRODUCED INTO COMMERCE SOLELY FOR SALE IN CALIFORNIA. OBD EXEMPT.

CATALYST

SPARK PLUG: AWSF-42C GAP-.042-.046
2.9L-9HM
LFM2.9T5FRD9 - TWC/HO2S/MPI

VACUUM HOSE ROUTING

■ CALIBRATION PARTS LIST — CODE: DCN ■

NAME/DESCRIPTION	ENGINEERING NO.	SERVICE NO.
SENSOR ASSY, Exhaust Gas Oxygen	F07F 9F472-CA, DA	F0TZ 9F472-C
INJECTOR ASSY, Fuel	E67E 9F593-B1B, B4B	E6SZ 9F593-A
SENSOR ASSY, Carburetor Air Cleaner Air Temperature Control		D7FZ 9E607-A
MOTOR ASSY, Carburetor Air Cleaner Vacuum		D4FZ 9D612-A
VALVE ASSY, Throttle Air By-Pass	E9AE 9F715-A1A, A2A F0AE 9F715-A1A, A2A	E9AZ 9F715-A
SENSOR ASSY, Speed	E3AF 9E731-AB	E3AZ 9E731-A
REGULATOR ASSY, Fuel Charging Pressure	90TF 9C968-AA	F0TZ 9C968-A
POTENTIOMETER ASSY, Throttle Position Sensor	E9TF 9B989-AA	E9TZ 9B989-A
DISTRIBUTOR ASSY	E67E 12127-AA	E6TZ 12127-A
SENSOR ASSY, Mass Airflow	F0VF 12B579-AA	F0VY 12B579-A
SENSOR ASSY, Engine Electronic Control Coolant Temperature	E4AF 12A648-AA	E1AZ 12A648-A
PROCESSOR AND CALIBRATOR ASSY, EEC-IV	F07F 12A650-DA	F0TZ 12A650-DA
SENSOR ASSY, Air Charge Temperature	E4AF 12A697-AA	E1AZ 12A697-A

CALIBRATION: 0-65T-R00 — 2.9L EFI

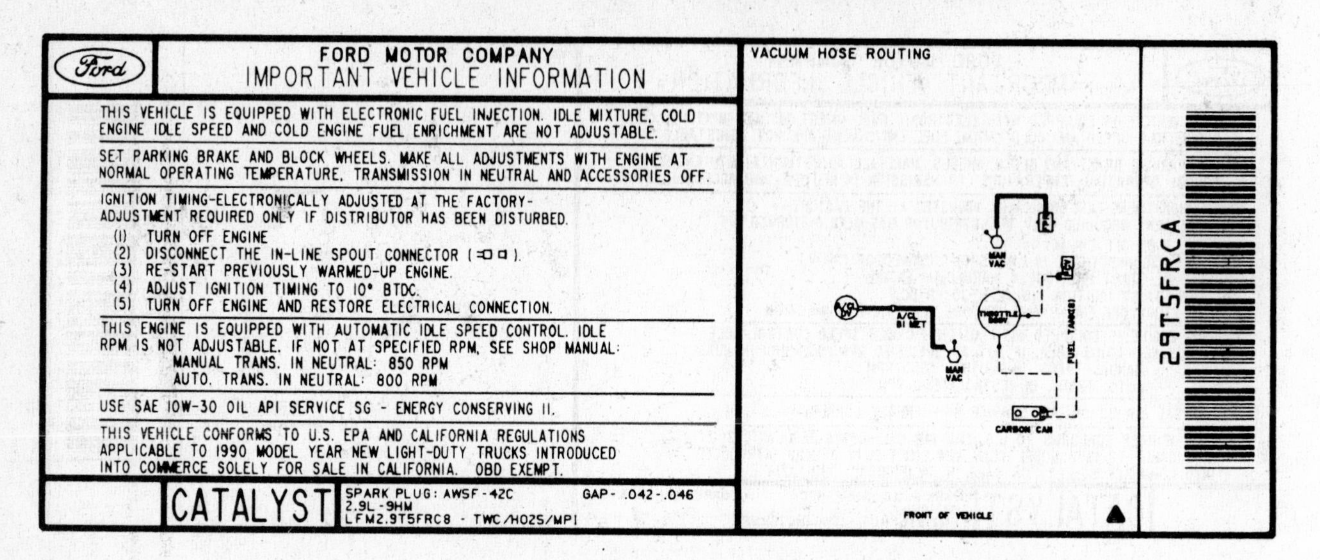

Ford

FORD MOTOR COMPANY
IMPORTANT VEHICLE INFORMATION

THIS VEHICLE IS EQUIPPED WITH ELECTRONIC FUEL INJECTION. IDLE MIXTURE, COLD ENGINE IDLE SPEED AND COLD ENGINE FUEL ENRICHMENT ARE NOT ADJUSTABLE.

SET PARKING BRAKE AND BLOCK WHEELS. MAKE ALL ADJUSTMENTS WITH ENGINE AT NORMAL OPERATING TEMPERATURE, TRANSMISSION IN NEUTRAL AND ACCESSORIES OFF.

IGNITION TIMING-ELECTRONICALLY ADJUSTED AT THE FACTORY-ADJUSTMENT REQUIRED ONLY IF DISTRIBUTOR HAS BEEN DISTURBED.
(1) TURN OFF ENGINE
(2) DISCONNECT THE IN-LINE SPOUT CONNECTOR ( ).
(3) RE-START PREVIOUSLY WARMED-UP ENGINE.
(4) ADJUST IGNITION TIMING TO 10° BTDC.
(5) TURN OFF ENGINE AND RESTORE ELECTRICAL CONNECTION.

THIS ENGINE IS EQUIPPED WITH AUTOMATIC IDLE SPEED CONTROL. IDLE RPM IS NOT ADJUSTABLE. IF NOT AT SPECIFIED RPM, SEE SHOP MANUAL:
MANUAL TRANS. IN NEUTRAL: 850 RPM
AUTO. TRANS. IN NEUTRAL: 800 RPM

USE SAE 10W-30 OIL API SERVICE SG - ENERGY CONSERVING II.

THIS VEHICLE CONFORMS TO U.S. EPA AND CALIFORNIA REGULATIONS APPLICABLE TO 1990 MODEL YEAR NEW LIGHT-DUTY TRUCKS INTRODUCED INTO COMMERCE SOLELY FOR SALE IN CALIFORNIA. OBD EXEMPT.

CATALYST

SPARK PLUG: AWSF-42C GAP-.042-.046
2.9L-9HM
LFM2.9T5FRC8 - TWC/HO2S/MPI

CALIBRATION PARTS LIST — CODE: DCR

NAME/DESCRIPTION	ENGINEERING NO.	SERVICE NO.
SENSOR ASSY, Exhaust Gas Oxygen	F07F 9F472-CA, DA	F0TZ 9F472-C
INJECTOR ASSY, Fuel	E67E 9F593-B1B, B4B	E6SZ 9F593-A
SENSOR ASSY, Carburetor Air Cleaner Air Temperature Control		D7FZ 9E607-A
MOTOR ASSY, Carburetor Air Cleaner Vacuum		D4FZ 9D612-A
VALVE ASSY, Throttle Air By-Pass	E9AE 9F715-A1A, A2A F0AE 9F715-A1A, A2A	E9AZ 9F715-A
SENSOR ASSY, Speed	E3AF 9E731-AB	E3AZ 9E731-A
REGULATOR ASSY, Fuel Charging Pressure	90TF 9C968-AA	F0TZ 9C968-A
POTENTIOMETER ASSY, Throttle Position Sensor	E9TF 9B989-AA	E9TZ 9B989-A
DISTRIBUTOR ASSY	E67E 12127-AA	E6TZ 12127-A
SENSOR ASSY, Mass Airflow	F0VF 12B579-AA	F0VY 12B579-A
SENSOR ASSY, Engine Electronic Control Coolant Temperature	E4AF 12A648-AA	E1AZ 12A648-A
PROCESSOR AND CALIBRATOR ASSY, EEC-IV	F07F 12A650-DA	F0TZ 12A650-DA
SENSOR ASSY, Air Charge Temperature	E4AF 12A697-AA	E1AZ 12A697-A

CALIBRATION: 0-66F-R10 — 2.9L EFI

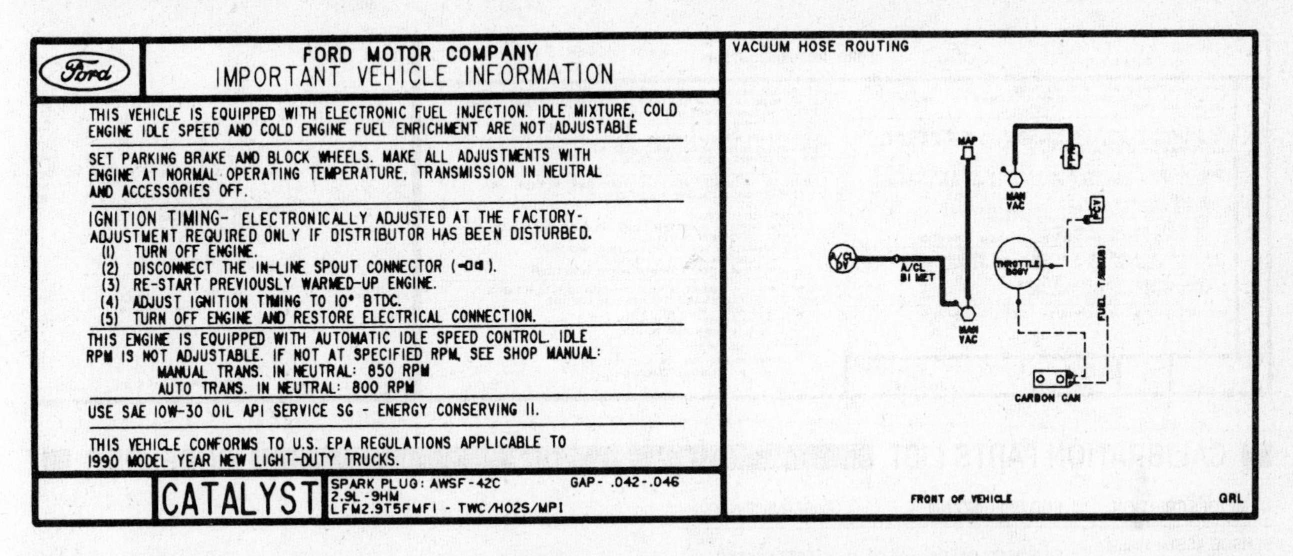
FORD MOTOR COMPANY
IMPORTANT VEHICLE INFORMATION

THIS VEHICLE IS EQUIPPED WITH ELECTRONIC FUEL INJECTION. IDLE MIXTURE, COLD ENGINE IDLE SPEED AND COLD ENGINE FUEL ENRICHMENT ARE NOT ADJUSTABLE

SET PARKING BRAKE AND BLOCK WHEELS. MAKE ALL ADJUSTMENTS WITH ENGINE AT NORMAL OPERATING TEMPERATURE, TRANSMISSION IN NEUTRAL AND ACCESSORIES OFF.

IGNITION TIMING- ELECTRONICALLY ADJUSTED AT THE FACTORY-ADJUSTMENT REQUIRED ONLY IF DISTRIBUTOR HAS BEEN DISTURBED.
(1) TURN OFF ENGINE.
(2) DISCONNECT THE IN-LINE SPOUT CONNECTOR (-□□).
(3) RE-START PREVIOUSLY WARMED-UP ENGINE.
(4) ADJUST IGNITION TIMING TO 10° BTDC.
(5) TURN OFF ENGINE AND RESTORE ELECTRICAL CONNECTION.

THIS ENGINE IS EQUIPPED WITH AUTOMATIC IDLE SPEED CONTROL. IDLE RPM IS NOT ADJUSTABLE. IF NOT AT SPECIFIED RPM, SEE SHOP MANUAL:
MANUAL TRANS. IN NEUTRAL: 850 RPM
AUTO TRANS. IN NEUTRAL: 800 RPM

USE SAE 10W-30 OIL API SERVICE SG - ENERGY CONSERVING II.

THIS VEHICLE CONFORMS TO U.S. EPA REGULATIONS APPLICABLE TO 1990 MODEL YEAR NEW LIGHT-DUTY TRUCKS.

CATALYST | SPARK PLUG: AWSF-42C GAP- .042-.046
2.9L-9HM
LFM2.9T5FMFI - TWC/HO2S/MPI

VACUUM HOSE ROUTING

CALIBRATION PARTS LIST — CODE: DDB

NAME/DESCRIPTION	ENGINEERING NO.	SERVICE NO.
SENSOR ASSY, Exhaust Gas Oxygen	F07F 9F472-CA, DA	F0TZ 9F472-C
SENSOR ASSY, Manifold Absolute Pressure	E7EF 9F479-A1A, A2A	E7FZ 9F479-A
INJECTOR ASSY, Fuel	E67E 9F593-B1B, B4B	E6SZ 9F593-A
SENSOR ASSY, Carburetor Air Cleaner Air Temperature Control		D7FZ 9E607-A
MOTOR ASSY, Carburetor Air Cleaner Vacuum		D4FZ 9D612-A
VALVE ASSY, Throttle Air By-Pass	E9AE 9F715-A1A, A2A F0AE 9F715-A1A, A2A	E9AZ 9F715-A
SENSOR ASSY, Speed	E3AF 9E731-AB	E3AZ 9E731-A
REGULATOR ASSY, Fuel Charging Pressure	88TF 9C968-EA	E6TZ 9C968-B
REGULATOR ASSY, Fuel Charging Pressure	90TF 9C968-AA	F0TZ 9C968-A
POTENTIOMETER ASSY, Throttle Position Sensor	E9TF 9B989-AA	E9TZ 9B989-A
DISTRIBUTOR ASSY	E67E 12127-AA	E6TZ 12127-A
SENSOR ASSY, Engine Electronic Control Coolant Temperature	E4AF 12A648-AA	E1AZ 12A648-A
PROCESSOR AND CALIBRATOR ASSY, EEC-IV	F0TF 12A650-BHA	F0TZ 12A650-BHA
SENSOR ASSY, Air Charge Temperature	E4AF 12A697-AA	E1AZ 12A697-A

CALIBRATION: 0-66F-R10 — 2.9L EFI

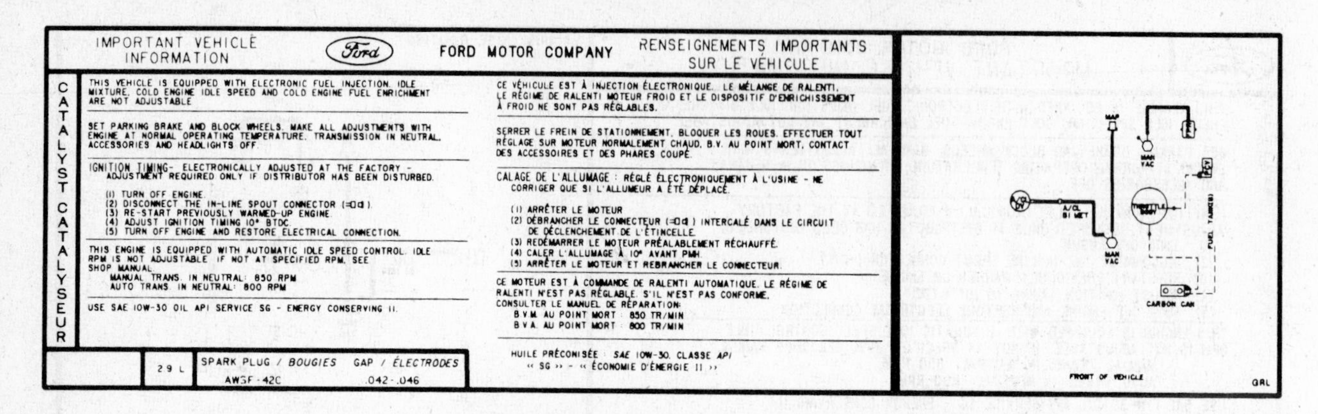

IMPORTANT VEHICLE INFORMATION — FORD MOTOR COMPANY — RENSEIGNEMENTS IMPORTANTS SUR LE VÉHICULE

CATALYST CATALYSEUR

THIS VEHICLE IS EQUIPPED WITH ELECTRONIC FUEL INJECTION. IDLE MIXTURE, COLD ENGINE IDLE SPEED AND COLD ENGINE FUEL ENRICHMENT ARE NOT ADJUSTABLE.

SET PARKING BRAKE AND BLOCK WHEELS. MAKE ALL ADJUSTMENTS WITH ENGINE AT NORMAL OPERATING TEMPERATURE, TRANSMISSION IN NEUTRAL, ACCESSORIES AND HEADLIGHTS OFF.

IGNITION TIMING- ELECTRONICALLY ADJUSTED AT THE FACTORY - ADJUSTMENT REQUIRED ONLY IF DISTRIBUTOR HAS BEEN DISTURBED.

(1) TURN OFF ENGINE.
(2) DISCONNECT THE IN-LINE SPOUT CONNECTOR (=▢◁).
(3) RE-START PREVIOUSLY WARMED-UP ENGINE.
(4) ADJUST IGNITION TIMING 10° BTDC.
(5) TURN OFF ENGINE AND RESTORE ELECTRICAL CONNECTION.

THIS ENGINE IS EQUIPPED WITH AUTOMATIC IDLE SPEED CONTROL. IDLE RPM IS NOT ADJUSTABLE. IF NOT AT SPECIFIED RPM, SEE SHOP MANUAL.
MANUAL TRANS. IN NEUTRAL: 850 RPM
AUTO TRANS. IN NEUTRAL: 800 RPM

USE SAE 10W-30 OIL API SERVICE SG - ENERGY CONSERVING II.

CE VÉHICULE EST À INJECTION ÉLECTRONIQUE. LE MÉLANGE DE RALENTI, LE RÉGIME DE RALENTI MOTEUR FROID ET LE DISPOSITIF D'ENRICHISSEMENT À FROID NE SONT PAS RÉGLABLES.

SERRER LE FREIN DE STATIONNEMENT, BLOQUER LES ROUES. EFFECTUER TOUT RÉGLAGE SUR MOTEUR NORMALEMENT CHAUD, B.V. AU POINT MORT, CONTACT DES ACCESSOIRES ET DES PHARES COUPÉ.

CALAGE DE L'ALLUMAGE : RÉGLÉ ÉLECTRONIQUEMENT À L'USINE - NE CORRIGER QUE SI L'ALLUMEUR A ÉTÉ DÉPLACÉ.

(1) ARRÊTER LE MOTEUR
(2) DÉBRANCHER LE CONNECTEUR (=▢◁) INTERCALÉ DANS LE CIRCUIT DE DÉCLENCHEMENT DE L'ÉTINCELLE.
(3) REDÉMARRER LE MOTEUR PRÉALABLEMENT RÉCHAUFFÉ.
(4) CALER L'ALLUMAGE À 10° AVANT PMH.
(5) ARRÊTER LE MOTEUR ET REBRANCHER LE CONNECTEUR.

CE MOTEUR EST À COMMANDE DE RALENTI AUTOMATIQUE. LE RÉGIME DE RALENTI N'EST PAS RÉGLABLE. S'IL N'EST PAS CONFORME, CONSULTER LE MANUEL DE RÉPARATION:
B.V.M. AU POINT MORT : 850 TR/MIN
B.V.A. AU POINT MORT : 800 TR/MIN

HUILE PRÉCONISÉE : SAE 10W-30, CLASSE API « SG » - « ÉCONOMIE D'ÉNERGIE II »

2.9 L	SPARK PLUG / BOUGIES	GAP / ÉLECTRODES
	AWSF-42C	.042-.046

CALIBRATION PARTS LIST — CODE: DDC

NAME/DESCRIPTION	ENGINEERING NO.	SERVICE NO.
SENSOR ASSY, Exhaust Gas Oxygen	F07F 9F472-CA, DA	F0TZ 9F472-C
SENSOR ASSY, Manifold Absolute Pressure	E7EF 9F479-A1A, A2A	E7FZ 9F479-A
INJECTOR ASSY, Fuel	E67E 9F593-B1B, B4B	E6SZ 9F593-A
SENSOR ASSY, Carburetor Air Cleaner Air Temperature Control		D7FZ 9E607-A
MOTOR ASSY, Carburetor Air Cleaner Vacuum		D4FZ 9D612-A
VALVE ASSY, Throttle Air By-Pass	E9AE 9F715-A1A, A2A F0AE 9F715-A1A, A2A	E9AZ 9F715-A
SENSOR ASSY, Speed	E3AF 9E731-AB	E3AZ 9E731-A
REGULATOR ASSY, Fuel Charging Pressure	88TF 9C968-EA	E6TZ 9C968-B
REGULATOR ASSY, Fuel Charging Pressure	90TF 9C968-AA	F0TZ 9C968-A
POTENTIOMETER ASSY, Throttle Position Sensor	E9TF 9B989-AA	E9TZ 9B989-A
DISTRIBUTOR ASSY	E67E 12127-AA	E6TZ 12127-A
SENSOR ASSY, Engine Electronic Control Coolant Temperature	E4AF 12A648-AA	E1AZ 12A648-A
PROCESSOR AND CALIBRATOR ASSY, EEC-IV	F0TF 12A650-BHA	F0TZ 12A650-BHA
SENSOR ASSY, Air Charge Temperature	E4AF 12A697-AA	E1AZ 12A697-A

■ CALIBRATION: 0-66H-R10 — 2.9L EFI ■

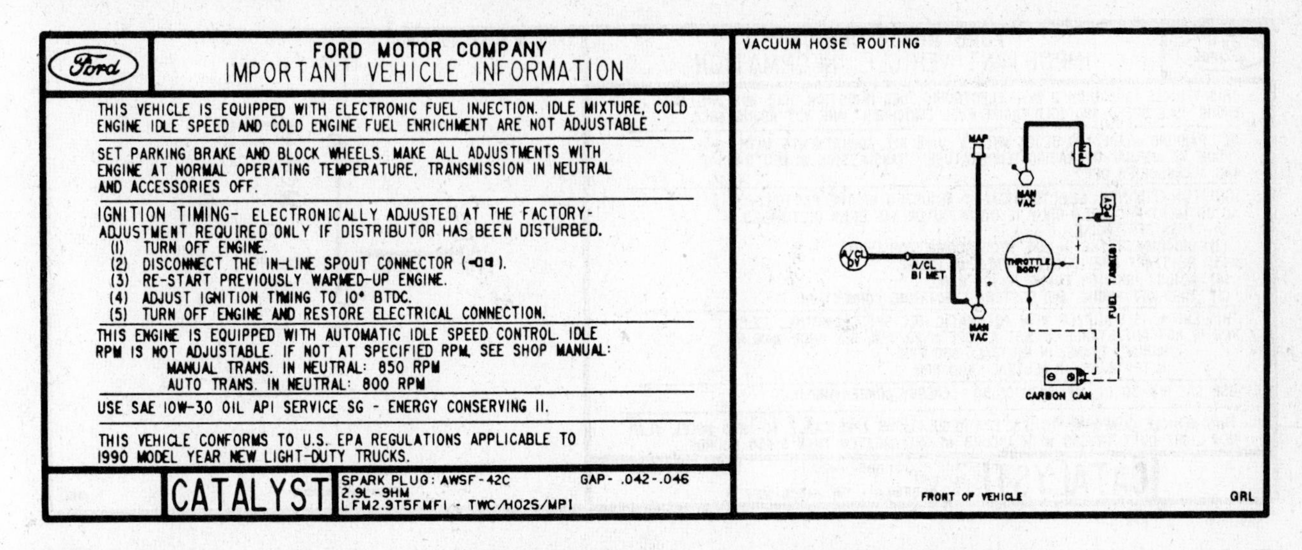
Ford

FORD MOTOR COMPANY
IMPORTANT VEHICLE INFORMATION

THIS VEHICLE IS EQUIPPED WITH ELECTRONIC FUEL INJECTION. IDLE MIXTURE, COLD ENGINE IDLE SPEED AND COLD ENGINE FUEL ENRICHMENT ARE NOT ADJUSTABLE

SET PARKING BRAKE AND BLOCK WHEELS. MAKE ALL ADJUSTMENTS WITH ENGINE AT NORMAL OPERATING TEMPERATURE, TRANSMISSION IN NEUTRAL AND ACCESSORIES OFF.

IGNITION TIMING- ELECTRONICALLY ADJUSTED AT THE FACTORY-ADJUSTMENT REQUIRED ONLY IF DISTRIBUTOR HAS BEEN DISTURBED.
(1) TURN OFF ENGINE.
(2) DISCONNECT THE IN-LINE SPOUT CONNECTOR (-◘◘).
(3) RE-START PREVIOUSLY WARMED-UP ENGINE.
(4) ADJUST IGNITION TIMING TO 10° BTDC.
(5) TURN OFF ENGINE AND RESTORE ELECTRICAL CONNECTION.

THIS ENGINE IS EQUIPPED WITH AUTOMATIC IDLE SPEED CONTROL. IDLE RPM IS NOT ADJUSTABLE. IF NOT AT SPECIFIED RPM, SEE SHOP MANUAL:
MANUAL TRANS. IN NEUTRAL: 850 RPM
AUTO TRANS. IN NEUTRAL: 800 RPM

USE SAE 10W-30 OIL API SERVICE SG - ENERGY CONSERVING II.

THIS VEHICLE CONFORMS TO U.S. EPA REGULATIONS APPLICABLE TO 1990 MODEL YEAR NEW LIGHT-DUTY TRUCKS.

CATALYST
SPARK PLUG: AWSF-42C GAP- .042-.046
2.9L-9HM
LFM2.9T5FMFI - TWC/HO2S/MPI

VACUUM HOSE ROUTING

■ CALIBRATION PARTS LIST — CODE: DDB ■

NAME/DESCRIPTION	ENGINEERING NO.	SERVICE NO.
SENSOR ASSY, Exhaust Gas Oxygen	F07F 9F472-CA, DA	F0TZ 9F472-C
SENSOR ASSY, Manifold Absolute Pressure	E7EF 9F479-A1A, A2A	E7FZ 9F479-A
INJECTOR ASSY, Fuel	E67E 9F593-B1B, B4B	E6SZ 9F593-A
SENSOR ASSY, Carburetor Air Cleaner Air Temperature Control		D7FZ 9E607-A
MOTOR ASSY, Carburetor Air Cleaner Vacuum		D4FZ 9D612-A
VALVE ASSY, Throttle Air By-Pass	E9AE 9F715-A1A, A2A F0AE 9F715-A1A, A2A	E9AZ 9F715-A
SENSOR ASSY, Speed	E3AF 9E731-AB	E3AZ 9E731-A
REGULATOR ASSY, Fuel Charging Pressure	88TF 9C968-EA	E6TZ 9C968-B
REGULATOR ASSY, Fuel Charging Pressure	90TF 9C968-AA	F0TZ 9C968-A
POTENTIOMETER ASSY, Throttle Position Sensor	E9TF 9B989-AA	E9TZ 9B989-A
DISTRIBUTOR ASSY	E67E 12127-AA	E6TZ 12127-A
SENSOR ASSY, Engine Electronic Control Coolant Temperature	E4AF 12A648-AA	E1AZ 12A648-A
PROCESSOR AND CALIBRATOR ASSY, EEC-IV	F0TF 12A650-BGA	F0TZ 12A650-BGA
SENSOR ASSY, Air Charge Temperature	E4AF 12A697-AA	E1AZ 12A697-A

CALIBRATION: 0-66H-R10 — 2.9L EFI

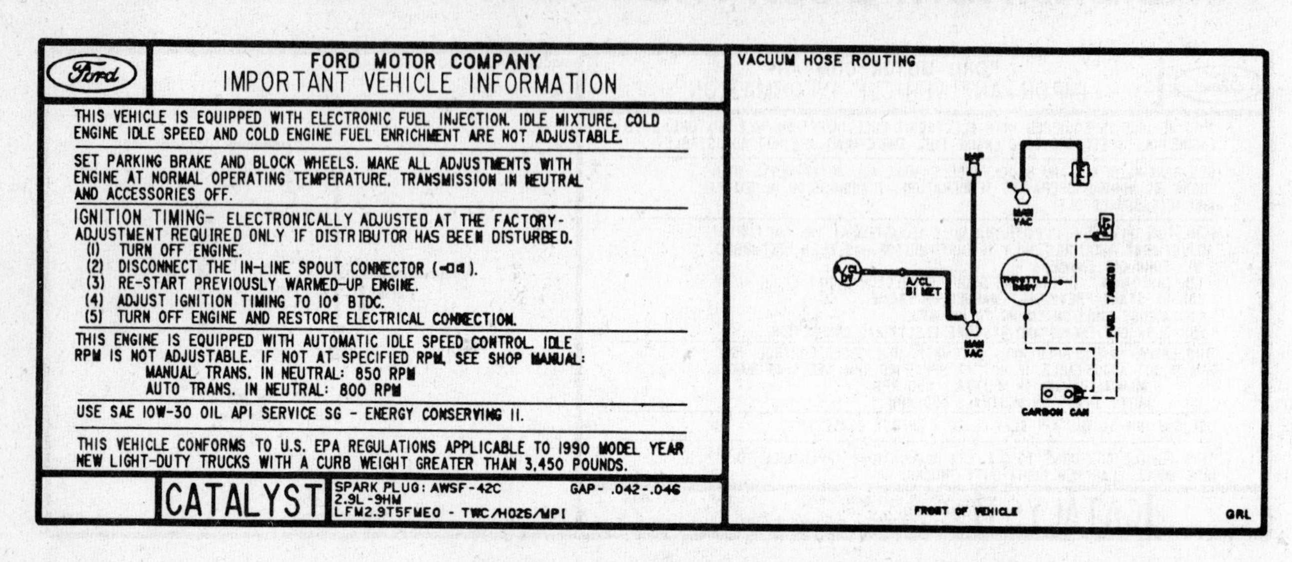
Ford
FORD MOTOR COMPANY
IMPORTANT VEHICLE INFORMATION

THIS VEHICLE IS EQUIPPED WITH ELECTRONIC FUEL INJECTION. IDLE MIXTURE, COLD ENGINE IDLE SPEED AND COLD ENGINE FUEL ENRICHMENT ARE NOT ADJUSTABLE.

SET PARKING BRAKE AND BLOCK WHEELS. MAKE ALL ADJUSTMENTS WITH ENGINE AT NORMAL OPERATING TEMPERATURE, TRANSMISSION IN NEUTRAL AND ACCESSORIES OFF.

IGNITION TIMING– ELECTRONICALLY ADJUSTED AT THE FACTORY- ADJUSTMENT REQUIRED ONLY IF DISTRIBUTOR HAS BEEN DISTURBED.
(1) TURN OFF ENGINE.
(2) DISCONNECT THE IN-LINE SPOUT CONNECTOR (-□◁).
(3) RE-START PREVIOUSLY WARMED-UP ENGINE.
(4) ADJUST IGNITION TIMING TO 10° BTDC.
(5) TURN OFF ENGINE AND RESTORE ELECTRICAL CONNECTION.

THIS ENGINE IS EQUIPPED WITH AUTOMATIC IDLE SPEED CONTROL. IDLE RPM IS NOT ADJUSTABLE. IF NOT AT SPECIFIED RPM, SEE SHOP MANUAL:
MANUAL TRANS. IN NEUTRAL: 850 RPM
AUTO TRANS. IN NEUTRAL: 800 RPM

USE SAE 10W-30 OIL API SERVICE SG - ENERGY CONSERVING II.

THIS VEHICLE CONFORMS TO U.S. EPA REGULATIONS APPLICABLE TO 1990 MODEL YEAR NEW LIGHT-DUTY TRUCKS WITH A CURB WEIGHT GREATER THAN 3,450 POUNDS.

CATALYST
SPARK PLUG: AWSF-42C GAP- .042-.046
2.9L-9HM
LFM2.9T5FMEO - TWC/HO2S/MPI

VACUUM HOSE ROUTING

CALIBRATION PARTS LIST — CODE: DDD

NAME/DESCRIPTION	ENGINEERING NO.	SERVICE NO.
SENSOR ASSY, Exhaust Gas Oxygen	F07F 9F472-CA, DA	F0TZ 9F472-C
SENSOR ASSY, Manifold Absolute Pressure	E7EF 9F479-A1A, A2A	E7FZ 9F479-A
INJECTOR ASSY, Fuel	E67E 9F593-B1B, B4B	E6SZ 9F593-A
SENSOR ASSY, Carburetor Air Cleaner Air Temperature Control		D7FZ 9E607-A
MOTOR ASSY, Carburetor Air Cleaner Vacuum		D4FZ 9D612-A
VALVE ASSY, Throttle Air By-Pass	E9AE 9F715-A1A, A2A F0AE 9F715-A1A, A2A	E9AZ 9F715-A
SENSOR ASSY, Speed	E3AF 9E731-AB	E3AZ 9E731-A
REGULATOR ASSY, Fuel Charging Pressure	88TF 9C968-EA	E6TZ 9C968-B
REGULATOR ASSY, Fuel Charging Pressure	90TF 9C968-AA	F0TZ 9C968-A
POTENTIOMETER ASSY, Throttle Position Sensor	E9TF 9B989-AA	E9TZ 9B989-A
DISTRIBUTOR ASSY	E67E 12127-AA	E6TZ 12127-A
SENSOR ASSY, Engine Electronic Control Coolant Temperature	E4AF 12A648-AA	E1AZ 12A648-A
PROCESSOR AND CALIBRATOR ASSY, EEC-IV	F0TF 12A650-BGA	F0TZ 12A650-BGA
SENSOR ASSY, Air Charge Temperature	E4AF 12A697-AA	E1AZ 12A697-A

CALIBRATION: 0-66H-R10 — 2.9L EFI

IMPORTANT VEHICLE INFORMATION — FORD MOTOR COMPANY — RENSEIGNEMENTS IMPORTANTS SUR LE VÉHICULE

CATALYST CATALYSEUR

THIS VEHICLE IS EQUIPPED WITH ELECTRONIC FUEL INJECTION. IDLE MIXTURE, COLD ENGINE IDLE SPEED AND COLD ENGINE FUEL ENRICHMENT ARE NOT ADJUSTABLE.

SET PARKING BRAKE AND BLOCK WHEELS. MAKE ALL ADJUSTMENTS WITH ENGINE AT NORMAL OPERATING TEMPERATURE. TRANSMISSION IN NEUTRAL. ACCESSORIES AND HEADLIGHTS OFF.

IGNITION TIMING- ELECTRONICALLY ADJUSTED AT THE FACTORY - ADJUSTMENT REQUIRED ONLY IF DISTRIBUTOR HAS BEEN DISTURBED.

(1) TURN OFF ENGINE.
(2) DISCONNECT THE IN-LINE SPOUT CONNECTOR (=□◘).
(3) RE-START PREVIOUSLY WARMED-UP ENGINE.
(4) ADJUST IGNITION TIMING 10° BTDC.
(5) TURN OFF ENGINE AND RESTORE ELECTRICAL CONNECTION.

THIS ENGINE IS EQUIPPED WITH AUTOMATIC IDLE SPEED CONTROL. IDLE RPM IS NOT ADJUSTABLE. IF NOT AT SPECIFIED RPM, SEE SHOP MANUAL.
MANUAL TRANS. IN NEUTRAL: 850 RPM
AUTO TRANS. IN NEUTRAL: 800 RPM

USE SAE 10W-30 OIL API SERVICE SG - ENERGY CONSERVING II.

CE VÉHICULE EST À INJECTION ÉLECTRONIQUE. LE MÉLANGE DE RALENTI, LE RÉGIME DE RALENTI MOTEUR FROID ET LE DISPOSITIF D'ENRICHISSEMENT À FROID NE SONT PAS RÉGLABLES.

SERRER LE FREIN DE STATIONNEMENT, BLOQUER LES ROUES. EFFECTUER TOUT RÉGLAGE SUR MOTEUR NORMALEMENT CHAUD, B.V. AU POINT MORT, CONTACT DES ACCESSOIRES ET DES PHARES COUPÉ.

CALAGE DE L'ALLUMAGE : RÉGLÉ ÉLECTRONIQUEMENT À L'USINE - NE CORRIGER QUE SI L'ALLUMEUR A ÉTÉ DÉPLACÉ.

(1) ARRÊTER LE MOTEUR
(2) DÉBRANCHER LE CONNECTEUR (=□◘) INTERCALÉ DANS LE CIRCUIT DE DÉCLENCHEMENT DE L'ÉTINCELLE.
(3) REDÉMARRER LE MOTEUR PRÉALABLEMENT RÉCHAUFFÉ.
(4) CALER L'ALLUMAGE À 10° AVANT PMH.
(5) ARRÊTER LE MOTEUR ET REBRANCHER LE CONNECTEUR.

CE MOTEUR EST À COMMANDE DE RALENTI AUTOMATIQUE. LE RÉGIME DE RALENTI N'EST PAS RÉGLABLE. S'IL N'EST PAS CONFORME, CONSULTER LE MANUEL DE RÉPARATION:
B.V.M. AU POINT MORT : 850 TR/MIN
B.V.A. AU POINT MORT : 800 TR/MIN

HUILE PRÉCONISÉE : *SAE* 10W-30, CLASSE *API* « SG » - « ÉCONOMIE D'ÉNERGIE II »

	SPARK PLUG / *BOUGIES*	GAP / *ÉLECTRODES*
2.9 L	AWSF-42C	.042-.046

MAP
MAN VAC
A/CL DV
A/CL BI MET
THROTTLE BODY
PCV
FUEL TANK(S)
CARBON CAN
FRONT OF VEHICLE
GRL

CALIBRATION PARTS LIST — CODE: DDC

NAME/DESCRIPTION	ENGINEERING NO.	SERVICE NO.
SENSOR ASSY, Exhaust Gas Oxygen	F07F 9F472-CA, DA	F0TZ 9F472-C
SENSOR ASSY, Manifold Absolute Pressure	E7EF 9F479-A1A, A2A	E7FZ 9F479-A
INJECTOR ASSY, Fuel	E67E 9F593-B1B, B4B	E6SZ 9F593-A
SENSOR ASSY, Carburetor Air Cleaner Air Temperature Control		D7FZ 9E607-A
MOTOR ASSY, Carburetor Air Cleaner Vacuum		D4FZ 9D612-A
VALVE ASSY, Throttle Air By-Pass	E9AE 9F715-A1A, A2A F0AE 9F715-A1A, A2A	E9AZ 9F715-A
SENSOR ASSY, Speed	E3AF 9E731-AB	E3AZ 9E731-A
REGULATOR ASSY, Fuel Charging Pressure	88TF 9C968-EA	E6TZ 9C968-B
REGULATOR ASSY, Fuel Charging Pressure	90TF 9C968-AA	F0TZ 9C968-A
POTENTIOMETER ASSY, Throttle Position Sensor	E9TF 9B989-AA	E9TZ 9B989-A
DISTRIBUTOR ASSY	E67E 12127-AA	E6TZ 12127-A
SENSOR ASSY, Engine Electronic Control Coolant Temperature	E4AF 12A648-AA	E1AZ 12A648-A
PROCESSOR AND CALIBRATOR ASSY, EEC-IV	F0TF 12A650-BGA	F0TZ 12A650-BGA
SENSOR ASSY, Air Charge Temperature	E4AF 12A697-AA	E1AZ 12A697-A

CALIBRATION: 0-66T-R00 — 2.9L EFI

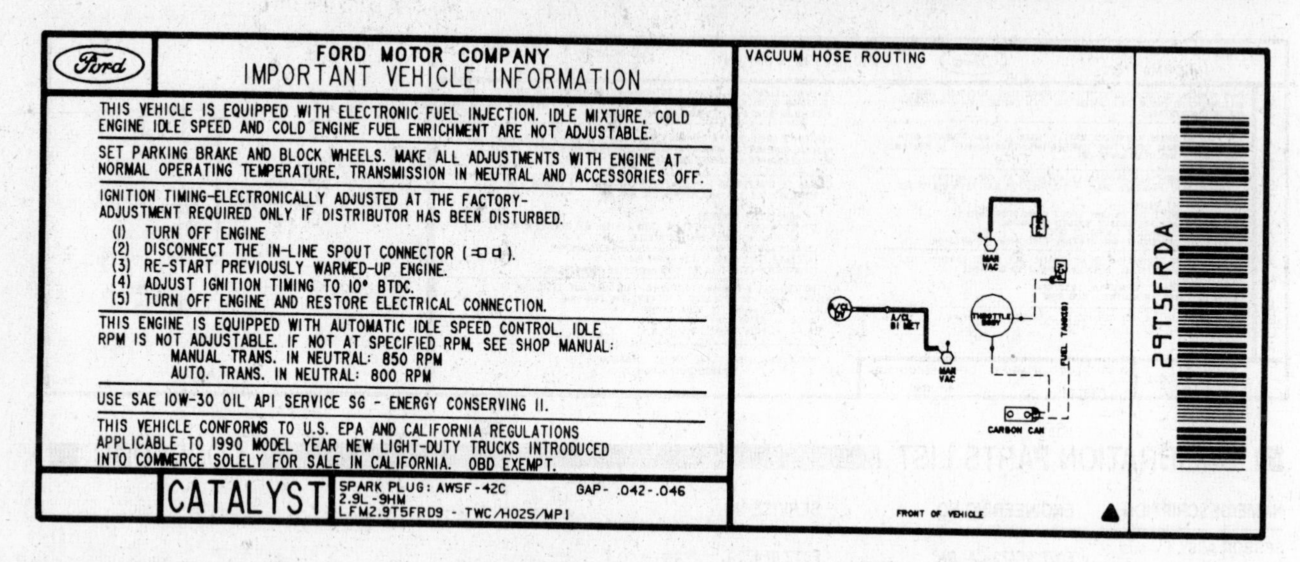
Ford

FORD MOTOR COMPANY
IMPORTANT VEHICLE INFORMATION

THIS VEHICLE IS EQUIPPED WITH ELECTRONIC FUEL INJECTION. IDLE MIXTURE, COLD ENGINE IDLE SPEED AND COLD ENGINE FUEL ENRICHMENT ARE NOT ADJUSTABLE.

SET PARKING BRAKE AND BLOCK WHEELS. MAKE ALL ADJUSTMENTS WITH ENGINE AT NORMAL OPERATING TEMPERATURE, TRANSMISSION IN NEUTRAL AND ACCESSORIES OFF.

IGNITION TIMING-ELECTRONICALLY ADJUSTED AT THE FACTORY-ADJUSTMENT REQUIRED ONLY IF DISTRIBUTOR HAS BEEN DISTURBED.
(1) TURN OFF ENGINE
(2) DISCONNECT THE IN-LINE SPOUT CONNECTOR (=□ □).
(3) RE-START PREVIOUSLY WARMED-UP ENGINE.
(4) ADJUST IGNITION TIMING TO 10° BTDC.
(5) TURN OFF ENGINE AND RESTORE ELECTRICAL CONNECTION.

THIS ENGINE IS EQUIPPED WITH AUTOMATIC IDLE SPEED CONTROL. IDLE RPM IS NOT ADJUSTABLE. IF NOT AT SPECIFIED RPM, SEE SHOP MANUAL:
MANUAL TRANS. IN NEUTRAL: 850 RPM
AUTO. TRANS. IN NEUTRAL: 800 RPM

USE SAE 10W-30 OIL API SERVICE SG - ENERGY CONSERVING II.

THIS VEHICLE CONFORMS TO U.S. EPA AND CALIFORNIA REGULATIONS APPLICABLE TO 1990 MODEL YEAR NEW LIGHT-DUTY TRUCKS INTRODUCED INTO COMMERCE SOLELY FOR SALE IN CALIFORNIA. OBD EXEMPT.

CATALYST
SPARK PLUG: AWSF-42C GAP- .042-.046
2.9L-9HM
LFM2.9T5FRD9 - TWC/HO2S/MPI

VACUUM HOSE ROUTING

CALIBRATION PARTS LIST — CODE: DCN

NAME/DESCRIPTION	ENGINEERING NO.	SERVICE NO.
SENSOR ASSY, Exhaust Gas Oxygen	F07F 9F472-CA, DA	F0TZ 9F472-C
INJECTOR ASSY, Fuel	E67E 9F593-B1B, B4B	E6SZ 9F593-A
SENSOR ASSY, Carburetor Air Cleaner Air Temperature Control		D7FZ 9E607-A
MOTOR ASSY, Carburetor Air Cleaner Vacuum		D4FZ 9D612-A
VALVE ASSY, Throttle Air By-Pass	E9AE 9F715-A1A, A2A F0AE 9F715-A1A, A2A	E9AZ 9F715-A
SENSOR ASSY, Speed	E3AF 9E731-AB	E3AZ 9E731-A
REGULATOR ASSY, Fuel Charging Pressure	90TF 9C968-AA	F0TZ 9C968-A
POTENTIOMETER ASSY, Throttle Position Sensor	E9TF 9B989-AA	E9TZ 9B989-A
DISTRIBUTOR ASSY	E67E 12127-AA	E6TZ 12127-A
SENSOR ASSY, Mass Airflow	F0VF 12B579-AA	F0VY 12B579-A
SENSOR ASSY, Engine Electronic Control Coolant Temperature	E4AF 12A648-AA	E1AZ 12A648-A
PROCESSOR AND CALIBRATOR ASSY, EEC-IV	F07F 12A650-FA	F0TZ 12A650-FA
SENSOR ASSY, Air Charge Temperature	E4AF 12A697-AA	E1AZ 12A697-A

CALIBRATION: 0-66T-R00 — 2.9L EFI

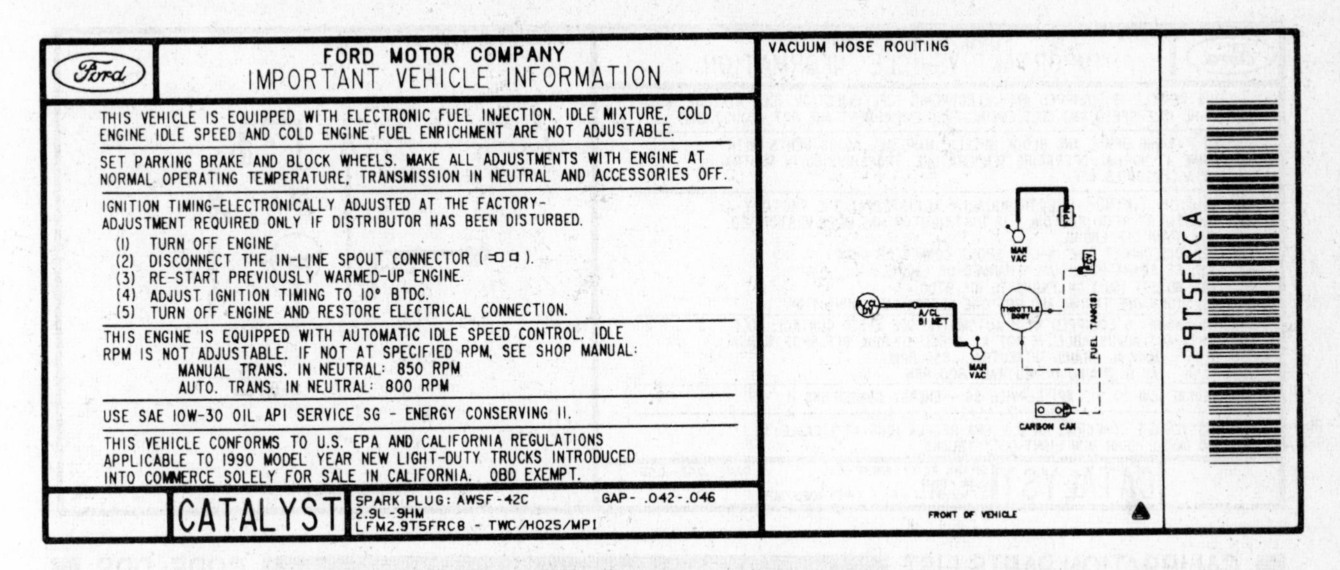
Ford

FORD MOTOR COMPANY
IMPORTANT VEHICLE INFORMATION

THIS VEHICLE IS EQUIPPED WITH ELECTRONIC FUEL INJECTION. IDLE MIXTURE, COLD ENGINE IDLE SPEED AND COLD ENGINE FUEL ENRICHMENT ARE NOT ADJUSTABLE.

SET PARKING BRAKE AND BLOCK WHEELS. MAKE ALL ADJUSTMENTS WITH ENGINE AT NORMAL OPERATING TEMPERATURE, TRANSMISSION IN NEUTRAL AND ACCESSORIES OFF.

IGNITION TIMING-ELECTRONICALLY ADJUSTED AT THE FACTORY-ADJUSTMENT REQUIRED ONLY IF DISTRIBUTOR HAS BEEN DISTURBED.
(1) TURN OFF ENGINE
(2) DISCONNECT THE IN-LINE SPOUT CONNECTOR (⊏□ □).
(3) RE-START PREVIOUSLY WARMED-UP ENGINE.
(4) ADJUST IGNITION TIMING TO 10° BTDC.
(5) TURN OFF ENGINE AND RESTORE ELECTRICAL CONNECTION.

THIS ENGINE IS EQUIPPED WITH AUTOMATIC IDLE SPEED CONTROL. IDLE RPM IS NOT ADJUSTABLE. IF NOT AT SPECIFIED RPM, SEE SHOP MANUAL:
MANUAL TRANS. IN NEUTRAL: 850 RPM
AUTO. TRANS. IN NEUTRAL: 800 RPM

USE SAE 10W-30 OIL API SERVICE SG - ENERGY CONSERVING II.

THIS VEHICLE CONFORMS TO U.S. EPA AND CALIFORNIA REGULATIONS APPLICABLE TO 1990 MODEL YEAR NEW LIGHT-DUTY TRUCKS INTRODUCED INTO COMMERCE SOLELY FOR SALE IN CALIFORNIA. OBD EXEMPT.

CATALYST

SPARK PLUG: AWSF-42C GAP-.042-.046
2.9L-9HM
LFM2.9T5FRC8 - TWC/HO2S/MPI

VACUUM HOSE ROUTING

CALIBRATION PARTS LIST — CODE: DCR

NAME/DESCRIPTION	ENGINEERING NO.	SERVICE NO.
SENSOR ASSY, Exhaust Gas Oxygen	F07F 9F472-CA, DA	F0TZ 9F472-C
INJECTOR ASSY, Fuel	E67E 9F593-B1B, B4B	E6SZ 9F593-A
SENSOR ASSY, Carburetor Air Cleaner Air Temperature Control		D7FZ 9E607-A
MOTOR ASSY, Carburetor Air Cleaner Vacuum		D4FZ 9D612-A
VALVE ASSY, Throttle Air By-Pass	E9AE 9F715-A1A, A2A F0AE 9F715-A1A, A2A	E9AZ 9F715-A
SENSOR ASSY, Speed	E3AF 9E731-AB	E3AZ 9E731-A
REGULATOR ASSY, Fuel Charging Pressure	90TF 9C968-AA	F0TZ 9C968-A
POTENTIOMETER ASSY, Throttle Position Sensor	E9TF 9B989-AA	E9TZ 9B989-A
DISTRIBUTOR ASSY	E67E 12127-AA	E6TZ 12127-A
SENSOR ASSY, Mass Airflow	F0VF 12B579-AA	F0VY 12B579-A
SENSOR ASSY, Engine Electronic Control Coolant Temperature	E4AF 12A648-AA	E1AZ 12A648-A
PROCESSOR AND CALIBRATOR ASSY, EEC-IV	F07F 12A650-FA	F0TZ 12A650-FA
SENSOR ASSY, Air Charge Temperature	E4AF 12A697-AA	E1AZ 12A697-A

CALIBRATION: 9-65F-R00 — 2.9L EFI

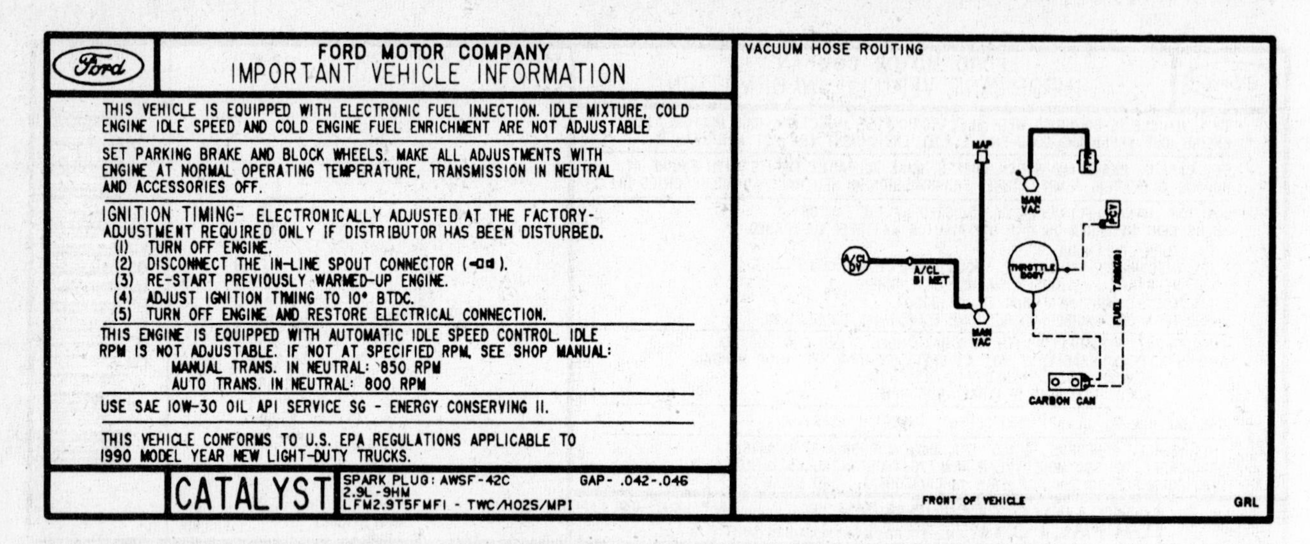

FORD MOTOR COMPANY
IMPORTANT VEHICLE INFORMATION

THIS VEHICLE IS EQUIPPED WITH ELECTRONIC FUEL INJECTION. IDLE MIXTURE, COLD ENGINE IDLE SPEED AND COLD ENGINE FUEL ENRICHMENT ARE NOT ADJUSTABLE

SET PARKING BRAKE AND BLOCK WHEELS. MAKE ALL ADJUSTMENTS WITH ENGINE AT NORMAL OPERATING TEMPERATURE, TRANSMISSION IN NEUTRAL AND ACCESSORIES OFF.

IGNITION TIMING- ELECTRONICALLY ADJUSTED AT THE FACTORY- ADJUSTMENT REQUIRED ONLY IF DISTRIBUTOR HAS BEEN DISTURBED.
(1) TURN OFF ENGINE.
(2) DISCONNECT THE IN-LINE SPOUT CONNECTOR.
(3) RE-START PREVIOUSLY WARMED-UP ENGINE.
(4) ADJUST IGNITION TIMING TO 10° BTDC.
(5) TURN OFF ENGINE AND RESTORE ELECTRICAL CONNECTION.

THIS ENGINE IS EQUIPPED WITH AUTOMATIC IDLE SPEED CONTROL. IDLE RPM IS NOT ADJUSTABLE. IF NOT AT SPECIFIED RPM, SEE SHOP MANUAL:
MANUAL TRANS. IN NEUTRAL: 850 RPM
AUTO TRANS. IN NEUTRAL: 800 RPM

USE SAE 10W-30 OIL API SERVICE SG - ENERGY CONSERVING II.

THIS VEHICLE CONFORMS TO U.S. EPA REGULATIONS APPLICABLE TO 1990 MODEL YEAR NEW LIGHT-DUTY TRUCKS.

CATALYST | SPARK PLUG: AWSF-42C GAP- .042-.046 | 2.9L-9HM | LFM2.9T5FMFI - TWC/HO2S/MPI

CALIBRATION PARTS LIST — CODE: DDB

NAME/DESCRIPTION	ENGINEERING NO.	SERVICE NO.
SENSOR ASSY, Exhaust Gas Oxygen	F07F 9F472-CA, DA	F0TZ 9F472-C
SENSOR ASSY, Exhaust Gas Oxygen	E77F 9F472-AA, BA E7TF 9F472-AA	E7TZ 9F472-A
SENSOR ASSY, Exhaust Gas Oxygen	E7TF 9F472-CA, DA, EA	E7TZ 9F472-B
SENSOR ASSY, Manifold Absolute Pressure	E7EF 9F479-A1A, A2A	E7FZ 9F479-A
INJECTOR ASSY, Fuel	E67E 9F593-B1B, B4B	E6SZ 9F593-A
SENSOR ASSY, Carburetor Air Cleaner Air Temperature Control		D7FZ 9E607-A
MOTOR ASSY, Carburetor Air Cleaner Vacuum		D4FZ 9D612-A
VALVE ASSY, Throttle Air By-Pass	E9AE 9F715-A1A, A2A F0AE 9F715-A1A, A2A	E9AZ 9F715-A
SENSOR ASSY, Speed	E3AF 9E731-AB	E3AZ 9E731-A
REGULATOR ASSY, Fuel Charging Pressure	88TF 9C968-EA	E6TZ 9C968-B
REGULATOR ASSY, Fuel Charging Pressure	90TF 9C968-AA	F0TZ 9C968-A
POTENTIOMETER ASSY, Throttle Position Sensor	E9TF 9B989-AA	E9TZ 9B989-A
DISTRIBUTOR ASSY	E67E 12127-AA	E6TZ 12127-A
SENSOR ASSY, Engine Electronic Control Coolant Temperature	E4AF 12A648-AA	E1AZ 12A648-A
PROCESSOR AND CALIBRATOR ASSY, EEC-IV	E9TF 12A650-D1A, D2A	E9TZ 12A650-DA
SENSOR ASSY, Air Charge Temperature	E4AF 12A697-AA	E1AZ 12A697-A

CALIBRATION: 9-65F-R00 — 2.9L EFI

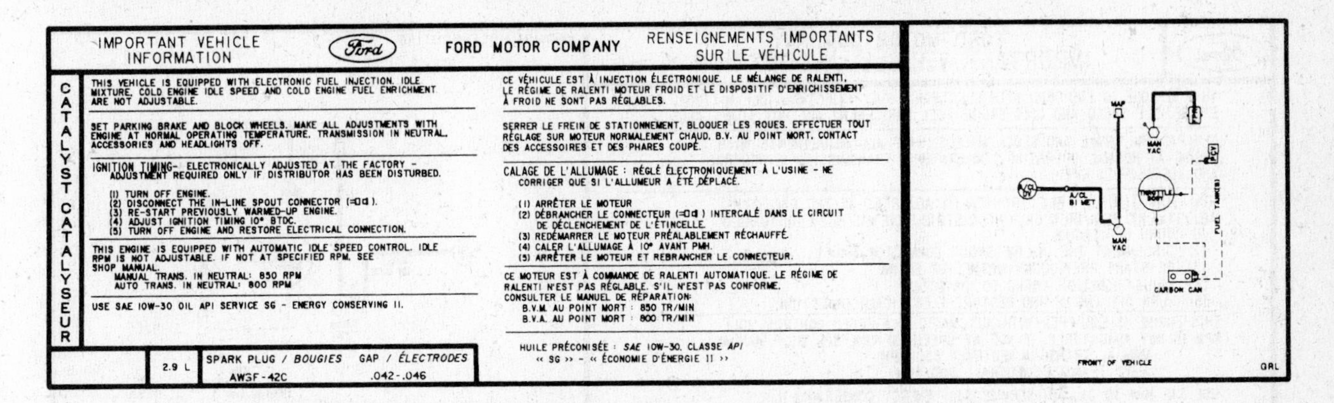
IMPORTANT VEHICLE INFORMATION — FORD MOTOR COMPANY — RENSEIGNEMENTS IMPORTANTS SUR LE VÉHICULE

CATALYST CATALYSEUR

THIS VEHICLE IS EQUIPPED WITH ELECTRONIC FUEL INJECTION. IDLE MIXTURE, COLD ENGINE IDLE SPEED AND COLD ENGINE FUEL ENRICHMENT ARE NOT ADJUSTABLE.

SET PARKING BRAKE AND BLOCK WHEELS. MAKE ALL ADJUSTMENTS WITH ENGINE AT NORMAL OPERATING TEMPERATURE, TRANSMISSION IN NEUTRAL, ACCESSORIES AND HEADLIGHTS OFF.

IGNITION TIMING- ELECTRONICALLY ADJUSTED AT THE FACTORY - ADJUSTMENT REQUIRED ONLY IF DISTRIBUTOR HAS BEEN DISTURBED.

(1) TURN OFF ENGINE.
(2) DISCONNECT THE IN-LINE SPOUT CONNECTOR.
(3) RE-START PREVIOUSLY WARMED-UP ENGINE.
(4) ADJUST IGNITION TIMING 10° BTDC.
(5) TURN OFF ENGINE AND RESTORE ELECTRICAL CONNECTION.

THIS ENGINE IS EQUIPPED WITH AUTOMATIC IDLE SPEED CONTROL. IDLE RPM IS NOT ADJUSTABLE. IF NOT AT SPECIFIED RPM, SEE SHOP MANUAL.
MANUAL TRANS. IN NEUTRAL: 850 RPM
AUTO TRANS. IN NEUTRAL: 800 RPM

USE SAE 10W-30 OIL API SERVICE SG - ENERGY CONSERVING II.

CE VÉHICULE EST À INJECTION ÉLECTRONIQUE. LE MÉLANGE DE RALENTI, LE RÉGIME DE RALENTI MOTEUR FROID ET LE DISPOSITIF D'ENRICHISSEMENT À FROID NE SONT PAS RÉGLABLES.

SERRER LE FREIN DE STATIONNEMENT, BLOQUER LES ROUES. EFFECTUER TOUT RÉGLAGE SUR MOTEUR NORMALEMENT CHAUD, B.V. AU POINT MORT, CONTACT DES ACCESSOIRES ET DES PHARES COUPÉ.

CALAGE DE L'ALLUMAGE : RÉGLÉ ÉLECTRONIQUEMENT À L'USINE - NE CORRIGER QUE SI L'ALLUMEUR A ÉTÉ DÉPLACÉ.

(1) ARRÊTER LE MOTEUR
(2) DÉBRANCHER LE CONNECTEUR INTERCALÉ DANS LE CIRCUIT DE DÉCLENCHEMENT DE L'ÉTINCELLE.
(3) REDÉMARRER LE MOTEUR PRÉALABLEMENT RÉCHAUFFÉ.
(4) CALER L'ALLUMAGE À 10° AVANT PMH.
(5) ARRÊTER LE MOTEUR ET REBRANCHER LE CONNECTEUR.

CE MOTEUR EST À COMMANDE DE RALENTI AUTOMATIQUE. LE RÉGIME DE RALENTI N'EST PAS RÉGLABLE. S'IL N'EST PAS CONFORME, CONSULTER LE MANUEL DE RÉPARATION:
B.V.M. AU POINT MORT : 850 TR/MIN
B.V.A. AU POINT MORT : 800 TR/MIN

HUILE PRÉCONISÉE : *SAE* 10W-30, CLASSE *API* « SG » - « ÉCONOMIE D'ÉNERGIE II »

	SPARK PLUG / *BOUGIES*	GAP / *ÉLECTRODES*
2.9 L	AWSF-42C	.042-.046

CALIBRATION PARTS LIST — CODE: DDC

NAME/DESCRIPTION	ENGINEERING NO.	SERVICE NO.
SENSOR ASSY, Exhaust Gas Oxygen	F07F 9F472-CA, DA	F0TZ 9F472-C
SENSOR ASSY, Exhaust Gas Oxygen	E77F 9F472-AA, BA E7TF 9F472-AA	E7TZ 9F472-A
SENSOR ASSY, Exhaust Gas Oxygen	E7TF 9F472-CA, DA, EA	E7TZ 9F472-B
SENSOR ASSY, Manifold Absolute Pressure	E7EF 9F479-A1A, A2A	E7FZ 9F479-A
INJECTOR ASSY, Fuel	E67E 9F593-B1B, B4B	E6SZ 9F593-A
SENSOR ASSY, Carburetor Air Cleaner Air Temperature Control		D7FZ 9E607-A
MOTOR ASSY, Carburetor Air Cleaner Vacuum		D4FZ 9D612-A
VALVE ASSY, Throttle Air By-Pass	E9AE 9F715-A1A, A2A F0AE 9F715-A1A, A2A	E9AZ 9F715-A
SENSOR ASSY, Speed	E3AF 9E731-AB	E3AZ 9E731-A
REGULATOR ASSY, Fuel Charging Pressure	88TF 9C968-EA	E6TZ 9C968-B
REGULATOR ASSY, Fuel Charging Pressure	90TF 9C968-AA	F0TZ 9C968-A
POTENTIOMETER ASSY, Throttle Position Sensor	E9TF 9B989-AA	E9TZ 9B989-A
DISTRIBUTOR ASSY	E67E 12127-AA	E6TZ 12127-A
SENSOR ASSY, Engine Electronic Control Coolant Temperature	E4AF 12A648-AA	E1AZ 12A648-A
PROCESSOR AND CALIBRATOR ASSY, EEC-IV	E9TF 12A650-D1A, D2A	E9TZ 12A650-DA
SENSOR ASSY, Air Charge Temperature	E4AF 12A697-AA	E1AZ 12A697-A

CALIBRATION: 9-65H-R00 — 2.9L EFI

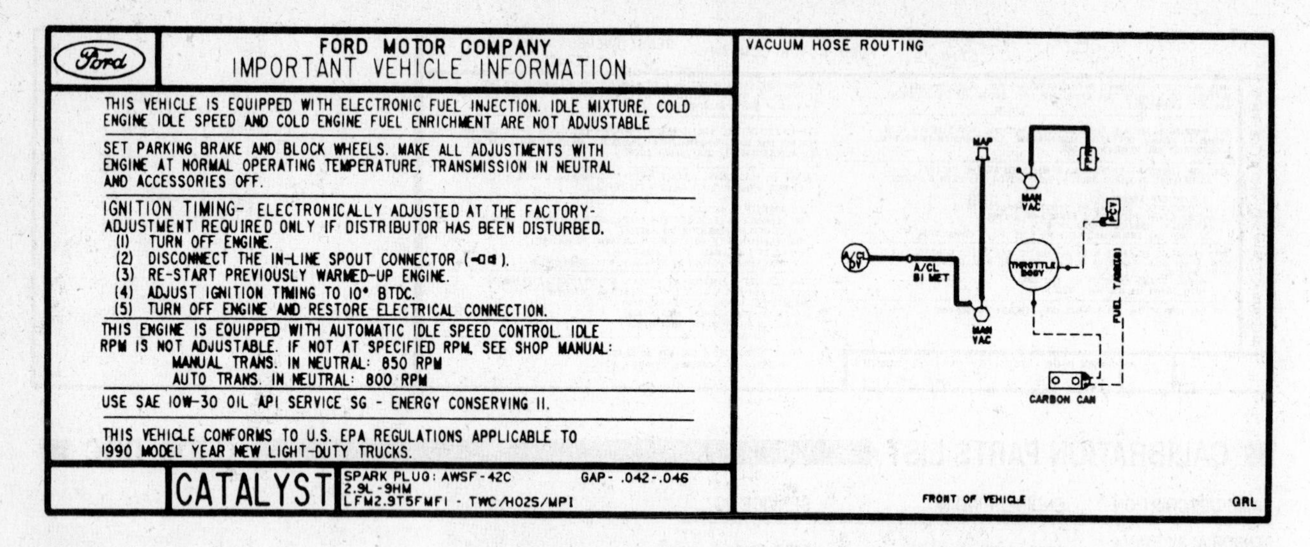
Ford

FORD MOTOR COMPANY
IMPORTANT VEHICLE INFORMATION

THIS VEHICLE IS EQUIPPED WITH ELECTRONIC FUEL INJECTION. IDLE MIXTURE, COLD ENGINE IDLE SPEED AND COLD ENGINE FUEL ENRICHMENT ARE NOT ADJUSTABLE

SET PARKING BRAKE AND BLOCK WHEELS. MAKE ALL ADJUSTMENTS WITH ENGINE AT NORMAL OPERATING TEMPERATURE, TRANSMISSION IN NEUTRAL AND ACCESSORIES OFF.

IGNITION TIMING- ELECTRONICALLY ADJUSTED AT THE FACTORY-ADJUSTMENT REQUIRED ONLY IF DISTRIBUTOR HAS BEEN DISTURBED.
(1) TURN OFF ENGINE.
(2) DISCONNECT THE IN-LINE SPOUT CONNECTOR.
(3) RE-START PREVIOUSLY WARMED-UP ENGINE.
(4) ADJUST IGNITION TIMING TO 10° BTDC.
(5) TURN OFF ENGINE AND RESTORE ELECTRICAL CONNECTION.

THIS ENGINE IS EQUIPPED WITH AUTOMATIC IDLE SPEED CONTROL. IDLE RPM IS NOT ADJUSTABLE. IF NOT AT SPECIFIED RPM, SEE SHOP MANUAL:
MANUAL TRANS. IN NEUTRAL: 850 RPM
AUTO TRANS. IN NEUTRAL: 800 RPM

USE SAE 10W-30 OIL API SERVICE SG - ENERGY CONSERVING II.

THIS VEHICLE CONFORMS TO U.S. EPA REGULATIONS APPLICABLE TO 1990 MODEL YEAR NEW LIGHT-DUTY TRUCKS.

CATALYST — SPARK PLUG: AWSF-42C GAP- .042-.046
2.9L -9HM
LFM2.9T5FMFI - TWC/HO2S/MPI

VACUUM HOSE ROUTING

CALIBRATION PARTS LIST — CODE: DDB

NAME/DESCRIPTION	ENGINEERING NO.	SERVICE NO.
SENSOR ASSY, Exhaust Gas Oxygen	F07F 9F472-CA, DA	F0TZ 9F472-C
SENSOR ASSY, Exhaust Gas Oxygen	E77F 9F472-AA, BA E7TF 9F472-AA	E7TZ 9F472-A
SENSOR ASSY, Exhaust Gas Oxygen	E7TF 9F472-CA, DA, EA	E7TZ 9F472-B
SENSOR ASSY, Manifold Absolute Pressure	E7EF 9F479-A1A, A2A	E7FZ 9F479-A
INJECTOR ASSY, Fuel	E67E 9F593-B1B, B4B	E6SZ 9F593-A
SENSOR ASSY, Carburetor Air Cleaner Air Temperature Control		D7FZ 9E607-A
MOTOR ASSY, Carburetor Air Cleaner Vacuum		D4FZ 9D612-A
VALVE ASSY, Throttle Air By-Pass	E9AE 9F715-A1A, A2A F0AE 9F715-A1A, A2A	E9AZ 9F715-A
SENSOR ASSY, Speed	E3AF 9E731-AB	E3AZ 9E731-A
REGULATOR ASSY, Fuel Charging Pressure	88TF 9C968-EA	E6TZ 9C968-B
REGULATOR ASSY, Fuel Charging Pressure	90TF 9C968-AA	F0TZ 9C968-A
POTENTIOMETER ASSY, Throttle Position Sensor	E9TF 9B989-AA	E9TZ 9B989-A
DISTRIBUTOR ASSY	E67E 12127-AA	E6TZ 12127-A
SENSOR ASSY, Engine Electronic Control Coolant Temperature	E4AF 12A648-AA	E1AZ 12A648-A
PROCESSOR AND CALIBRATOR ASSY, EEC-IV	E9TF 12A650-G1A, G2A	E9TZ 12A650-GA
SENSOR ASSY, Air Charge Temperature	E4AF 12A697-AA	E1AZ 12A697-A

CALIBRATION: 9-65H-R00 — 2.9L EFI

IMPORTANT VEHICLE INFORMATION — FORD MOTOR COMPANY — RENSEIGNEMENTS IMPORTANTS SUR LE VÉHICULE

CATALYST CATALYSEUR

THIS VEHICLE IS EQUIPPED WITH ELECTRONIC FUEL INJECTION. IDLE MIXTURE, COLD ENGINE IDLE SPEED AND COLD ENGINE FUEL ENRICHMENT ARE NOT ADJUSTABLE.

SET PARKING BRAKE AND BLOCK WHEELS. MAKE ALL ADJUSTMENTS WITH ENGINE AT NORMAL OPERATING TEMPERATURE, TRANSMISSION IN NEUTRAL, ACCESSORIES AND HEADLIGHTS OFF.

IGNITION TIMING- ELECTRONICALLY ADJUSTED AT THE FACTORY - ADJUSTMENT REQUIRED ONLY IF DISTRIBUTOR HAS BEEN DISTURBED.

(1) TURN OFF ENGINE.
(2) DISCONNECT THE IN-LINE SPOUT CONNECTOR (=[]d).
(3) RE-START PREVIOUSLY WARMED-UP ENGINE.
(4) ADJUST IGNITION TIMING 10° BTDC.
(5) TURN OFF ENGINE AND RESTORE ELECTRICAL CONNECTION.

THIS ENGINE IS EQUIPPED WITH AUTOMATIC IDLE SPEED CONTROL. IDLE RPM IS NOT ADJUSTABLE. IF NOT AT SPECIFIED RPM, SEE SHOP MANUAL.
MANUAL TRANS. IN NEUTRAL: 850 RPM
AUTO TRANS. IN NEUTRAL: 800 RPM

USE SAE 10W-30 OIL API SERVICE SG - ENERGY CONSERVING II.

CE VÉHICULE EST À INJECTION ÉLECTRONIQUE. LE MÉLANGE DE RALENTI, LE RÉGIME DE RALENTI MOTEUR FROID ET LE DISPOSITIF D'ENRICHISSEMENT À FROID NE SONT PAS RÉGLABLES.

SERRER LE FREIN DE STATIONNEMENT, BLOQUER LES ROUES. EFFECTUER TOUT RÉGLAGE SUR MOTEUR NORMALEMENT CHAUD, B.V. AU POINT MORT, CONTACT DES ACCESSOIRES ET DES PHARES COUPÉ.

CALAGE DE L'ALLUMAGE : RÉGLÉ ÉLECTRONIQUEMENT À L'USINE - NE CORRIGER QUE SI L'ALLUMEUR A ÉTÉ DÉPLACÉ.

(1) ARRÊTER LE MOTEUR
(2) DÉBRANCHER LE CONNECTEUR (=[]d) INTERCALÉ DANS LE CIRCUIT DE DÉCLENCHEMENT DE L'ÉTINCELLE.
(3) REDÉMARRER LE MOTEUR PRÉALABLEMENT RÉCHAUFFÉ.
(4) CALER L'ALLUMAGE À 10° AVANT PMH.
(5) ARRÊTER LE MOTEUR ET REBRANCHER LE CONNECTEUR.

CE MOTEUR EST À COMMANDE DE RALENTI AUTOMATIQUE. LE RÉGIME DE RALENTI N'EST PAS RÉGLABLE. S'IL N'EST PAS CONFORME, CONSULTER LE MANUEL DE RÉPARATION:
B.V.M. AU POINT MORT : 850 TR/MIN
B.V.A. AU POINT MORT : 800 TR/MIN

HUILE PRÉCONISÉE : *SAE* 10W-30, CLASSE *API* « SG » - « ÉCONOMIE D'ÉNERGIE II »

2.9 L	SPARK PLUG / *BOUGIES*	GAP / *ÉLECTRODES*
	AWSF-42C	.042-.046

MAP
MAN VAC
A/CL BI MET
MAN VAC
THROTTLE BODY
FUEL TANK(S)
CARBON CAN
FRONT OF VEHICLE
GRL

CALIBRATION PARTS LIST — CODE: DDC

NAME/DESCRIPTION	ENGINEERING NO.	SERVICE NO.
SENSOR ASSY, Exhaust Gas Oxygen	F07F 9F472-CA, DA	F0TZ 9F472-C
SENSOR ASSY, Exhaust Gas Oxygen	E77F 9F472-AA, BA E7TF 9F472-AA	E7TZ 9F472-A
SENSOR ASSY, Exhaust Gas Oxygen	E7TF 9F472-CA, DA, EA	E7TZ 9F472-B
SENSOR ASSY, Manifold Absolute Pressure	E7EF 9F479-A1A, A2A	E7FZ 9F479-A
INJECTOR ASSY, Fuel	E67E 9F593-B1B, B4B	E6SZ 9F593-A
SENSOR ASSY, Carburetor Air Cleaner Air Temperature Control		D7FZ 9E607-A
MOTOR ASSY, Carburetor Air Cleaner Vacuum		D4FZ 9D612-A
VALVE ASSY, Throttle Air By-Pass	E9AE 9F715-A1A, A2A F0AE 9F715-A1A, A2A	E9AZ 9F715-A
SENSOR ASSY, Speed	E3AF 9E731-AB	E3AZ 9E731-A
REGULATOR ASSY, Fuel Charging Pressure	88TF 9C968-EA	E6TZ 9C968-B
REGULATOR ASSY, Fuel Charging Pressure	90TF 9C968-AA	F0TZ 9C968-A
POTENTIOMETER ASSY, Throttle Position Sensor	E9TF 9B989-AA	E9TZ 9B989-A
DISTRIBUTOR ASSY	E67E 12127-AA	E6TZ 12127-A
SENSOR ASSY, Engine Electronic Control Coolant Temperature	E4AF 12A648-AA	E1AZ 12A648-A
PROCESSOR AND CALIBRATOR ASSY, EEC-IV	E9TF 12A650-G1A, G2A	E9TZ 12A650-GA
SENSOR ASSY, Air Charge Temperature	E4AF 12A697-AA	E1AZ 12A697-A

CALIBRATION: 9-65H-R00 — 2.9L EFI

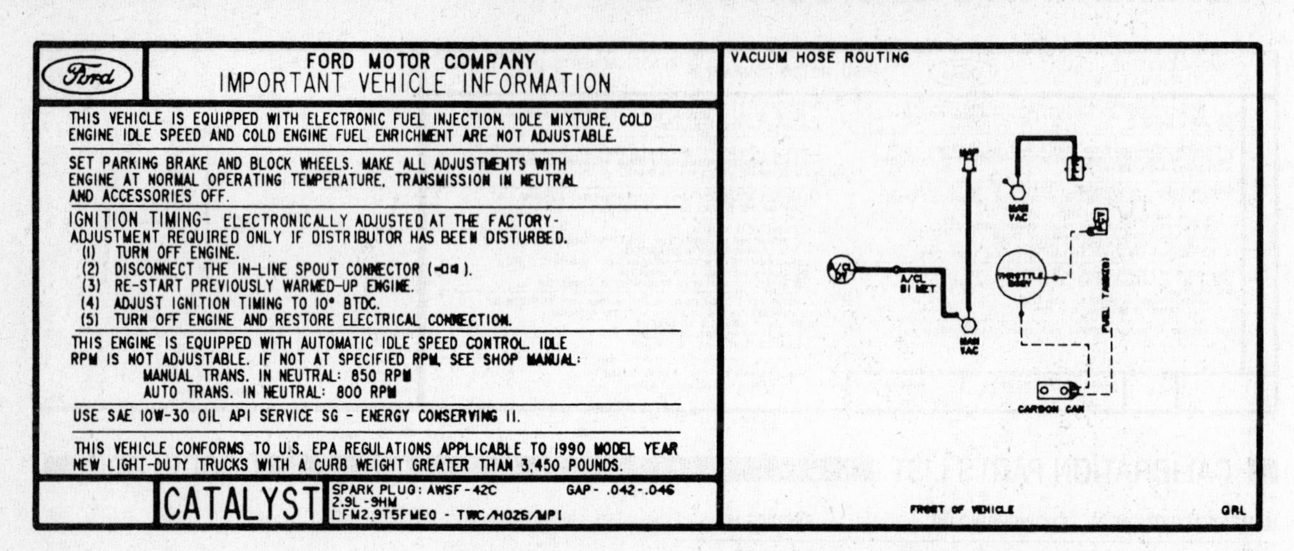

FORD MOTOR COMPANY
IMPORTANT VEHICLE INFORMATION

THIS VEHICLE IS EQUIPPED WITH ELECTRONIC FUEL INJECTION. IDLE MIXTURE, COLD ENGINE IDLE SPEED AND COLD ENGINE FUEL ENRICHMENT ARE NOT ADJUSTABLE.

SET PARKING BRAKE AND BLOCK WHEELS. MAKE ALL ADJUSTMENTS WITH ENGINE AT NORMAL OPERATING TEMPERATURE, TRANSMISSION IN NEUTRAL AND ACCESSORIES OFF.

IGNITION TIMING- ELECTRONICALLY ADJUSTED AT THE FACTORY- ADJUSTMENT REQUIRED ONLY IF DISTRIBUTOR HAS BEEN DISTURBED.
(1) TURN OFF ENGINE.
(2) DISCONNECT THE IN-LINE SPOUT CONNECTOR (-◻◅).
(3) RE-START PREVIOUSLY WARMED-UP ENGINE.
(4) ADJUST IGNITION TIMING TO 10° BTDC.
(5) TURN OFF ENGINE AND RESTORE ELECTRICAL CONNECTION.

THIS ENGINE IS EQUIPPED WITH AUTOMATIC IDLE SPEED CONTROL. IDLE RPM IS NOT ADJUSTABLE. IF NOT AT SPECIFIED RPM, SEE SHOP MANUAL:
MANUAL TRANS. IN NEUTRAL: 850 RPM
AUTO TRANS. IN NEUTRAL: 800 RPM

USE SAE 10W-30 OIL API SERVICE SG - ENERGY CONSERVING II.

THIS VEHICLE CONFORMS TO U.S. EPA REGULATIONS APPLICABLE TO 1990 MODEL YEAR NEW LIGHT-DUTY TRUCKS WITH A CURB WEIGHT GREATER THAN 3,450 POUNDS.

CATALYST
SPARK PLUG: AWSF-42C GAP- .042-.046
2.9L-9HM
LFM2.9T5FMEO - TWC/HO2S/MPI

CALIBRATION PARTS LIST — CODE: DDD

NAME/DESCRIPTION	ENGINEERING NO.	SERVICE NO.
SENSOR ASSY, Exhaust Gas Oxygen	F07F 9F472-CA, DA	F0TZ 9F472-C
SENSOR ASSY, Exhaust Gas Oxygen	E77F 9F472-AA, BA E7TF 9F472-AA	E7TZ 9F472-A
SENSOR ASSY, Exhaust Gas Oxygen	E7TF 9F472-CA, DA, EA	E7TZ 9F472-B
SENSOR ASSY, Manifold Absolute Pressure	E7EF 9F479-A1A, A2A	E7FZ 9F479-A
INJECTOR ASSY, Fuel	E67E 9F593-B1B, B4B	E6SZ 9F593-A
SENSOR ASSY, Carburetor Air Cleaner Air Temperature Control		D7FZ 9E607-A
MOTOR ASSY, Carburetor Air Cleaner Vacuum		D4FZ 9D612-A
VALVE ASSY, Throttle Air By-Pass	E9AE 9F715-A1A, A2A F0AE 9F715-A1A, A2A	E9AZ 9F715-A
SENSOR ASSY, Speed	E3AF 9E731-AB	E3AZ 9E731-A
REGULATOR ASSY, Fuel Charging Pressure	88TF 9C968-EA	E6TZ 9C968-B
REGULATOR ASSY, Fuel Charging Pressure	90TF 9C968-AA	F0TZ 9C968-A
POTENTIOMETER ASSY, Throttle Position Sensor	E9TF 9B989-AA	E9TZ 9B989-A
DISTRIBUTOR ASSY	E67E 12127-AA	E6TZ 12127-A
SENSOR ASSY, Engine Electronic Control Coolant Temperature	E4AF 12A648-AA	E1AZ 12A648-A
PROCESSOR AND CALIBRATOR ASSY, EEC-IV	E9TF 12A650-G1A, G2A	E9TZ 12A650-GA
SENSOR ASSY, Air Charge Temperature	E4AF 12A697-AA	E1AZ 12A697-A

■ CALIBRATION: 9-55J-R11 — 3.0L EFI ■

FORD MOTOR COMPANY
IMPORTANT VEHICLE INFORMATION

THIS VEHICLE IS EQUIPPED WITH EEC IV/EFI SYSTEMS. IDLE SPEEDS AND IDLE MIXTURES ARE NOT ADJUSTABLE. SEE SHOP MANUAL FOR ADDITIONAL INFORMATION.

ADJUST IGNITION TIMING WITH THE TRANSMISSION IN NEUTRAL, PARKING BRAKE SET AND THE WHEELS BLOCKED. ENGINE MUST BE AT NORMAL OPERATING TEMPERATURE.
(1) TURN OFF ENGINE.
(2) DISCONNECT SMALL IN-LINE SPOUT CONNECTOR (=◻◻) LOCATED NEAR THE DISTRIBUTOR.
(3) RE-START PREVIOUSLY WARMED-UP ENGINE.
(4) ADJUST IGNITION TIMING TO 10° BTDC.
(5) TURN OFF ENGINE AND RESTORE ELECTRICAL CONNECTION.

THIS VEHICLE CONFORMS TO U.S. EPA REGULATIONS APPLICABLE TO 1990 MODEL YEAR NEW LIGHT-DUTY TRUCKS WITH A CURB WEIGHT GREATER THAN 3,450 POUNDS.

USE SAE 5W-30 OIL API SERVICE SG - ENERGY CONSERVING II.

CATALYST — SPARK PLUG: AWSF-32P GAP- .042-.046
3.0L-9HM
LFM3.0T5FYK3 - TWC/HO2S/MPI

VACUUM HOSE ROUTING

■ CALIBRATION PARTS LIST — CODE: DDE ■

NAME/DESCRIPTION	ENGINEERING NO.	SERVICE NO.
SENSOR ASSY, Exhaust Gas Oxygen	E77F 9F472-AA, BA E7TF 9F472-AA	E7TZ 9F472-A
SENSOR ASSY, Exhaust Gas Oxygen	E9SF 9F472-AA E97F 9F472-AA	F0SZ 9F472-A
SENSOR ASSY, Manifold Absolute Pressure	E7EF 9F479-A1A, A2A	E7FZ 9F479-A
INJECTOR ASSY, Fuel	E59E 9F593-A1A, A2A, AB	E5TZ 9F593-A
INJECTOR ASSY, Fuel	E67E 9F593-B1A, B1B	E6SZ 9F593-A
VALVE ASSY, Throttle Air By-Pass	E6AE 9F715-D1A, D1B	E6AZ 9F715-D
VALVE ASSY, Throttle Air By-Pass	E9DE 9F715-A1A E9EE 9F715-A1A	E9DZ 9F715-A
SENSOR ASSY, Speed	E3AF 9E731-AB	E3AZ 9E731-A
SENSOR ASSY, Speed	E45F 9E731-AA	E45Y 9E731-A
REGULATOR ASSY, Fuel Charging Pressure	E6AE 9C968-AA, AB E7DE 9C968-BA	E6AZ 9C968-A
REGULATOR ASSY, Fuel Charging Pressure	E7EE 9C968-AA E77E 9C968-AA	E7FZ 9C968-A
POTENTIOMETER ASSY, Throttle Position Sensor	E7DF 9B989-AA	E7DZ 9B989-A
DISTRIBUTOR ASSY	E6AE 12127-EA	E6AZ 12127-C
SENSOR ASSY, Engine Electronic Control Coolant Temperature	E1AF 12A648-AA E4AF 12A648-AA	E1AZ 12A648-A
PROCESSOR AND CALIBRATOR ASSY, EEC-IV	E9TF 12A650-L1C, L2C	E9TZ 12A650-LC
SENSOR ASSY, Air Charge Temperature	E1AF 12A697-AA E4AF 12A697-AA	E1AZ 12A697-A

CALIBRATION: 9-55J-R11 — 3.0L EFI

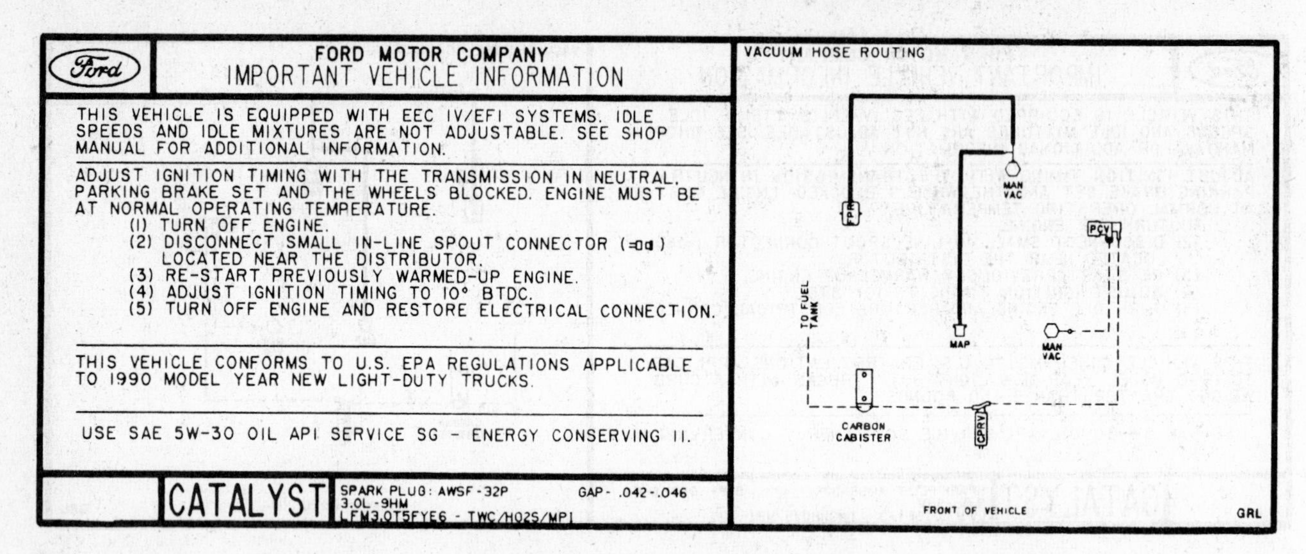
Ford

FORD MOTOR COMPANY
IMPORTANT VEHICLE INFORMATION

THIS VEHICLE IS EQUIPPED WITH EEC IV/EFI SYSTEMS. IDLE SPEEDS AND IDLE MIXTURES ARE NOT ADJUSTABLE. SEE SHOP MANUAL FOR ADDITIONAL INFORMATION.

ADJUST IGNITION TIMING WITH THE TRANSMISSION IN NEUTRAL, PARKING BRAKE SET AND THE WHEELS BLOCKED. ENGINE MUST BE AT NORMAL OPERATING TEMPERATURE.
(1) TURN OFF ENGINE.
(2) DISCONNECT SMALL IN-LINE SPOUT CONNECTOR (=⊏⊐) LOCATED NEAR THE DISTRIBUTOR.
(3) RE-START PREVIOUSLY WARMED-UP ENGINE.
(4) ADJUST IGNITION TIMING TO 10° BTDC.
(5) TURN OFF ENGINE AND RESTORE ELECTRICAL CONNECTION.

THIS VEHICLE CONFORMS TO U.S. EPA REGULATIONS APPLICABLE TO 1990 MODEL YEAR NEW LIGHT-DUTY TRUCKS.

USE SAE 5W-30 OIL API SERVICE SG - ENERGY CONSERVING II.

CATALYST
SPARK PLUG: AWSF-32P GAP-.042-.046
3.0L-9HM
LFM3.0T5FYE6 - TWC/HO2S/MPI

CALIBRATION PARTS LIST — CODE: DDH

NAME/DESCRIPTION	ENGINEERING NO.	SERVICE NO.
SENSOR ASSY, Exhaust Gas Oxygen	E77F 9F472-AA, BA E7TF 9F472-AA	E7TZ 9F472-A
SENSOR ASSY, Exhaust Gas Oxygen	E9SF 9F472-AA E97F 9F472-AA	F0SZ 9F472-A
SENSOR ASSY, Manifold Absolute Pressure	E7EF 9F479-A1A, A2A	E7FZ 9F479-A
INJECTOR ASSY, Fuel	E59E 9F593-A1A, A2A, AB	E5TZ 9F593-A
INJECTOR ASSY, Fuel	E67E 9F593-B1A, B1B	E6SZ 9F593-A
VALVE ASSY, Throttle Air By-Pass	E6AE 9F715-D1A, D1B	E6AZ 9F715-D
VALVE ASSY, Throttle Air By-Pass	E9DE 9F715-A1A E9EE 9F715-A1A	E9DZ 9F715-A
SENSOR ASSY, Speed	E3AF 9E731-AB	E3AZ 9E731-A
SENSOR ASSY, Speed	E45F 9E731-AA	E45Y 9E731-A
REGULATOR ASSY, Fuel Charging Pressure	E6AE 9C968-AA, AB E7DE 9C968-BA	E6AZ 9C968-A
REGULATOR ASSY, Fuel Charging Pressure	E7EE 9C968-AA E77E 9C968-AA	E7FZ 9C968-A
POTENTIOMETER ASSY, Throttle Position Sensor	E7DF 9B989-AA	E7DZ 9B989-A
DISTRIBUTOR ASSY	E6AE 12127-EA	E6AZ 12127-C
SENSOR ASSY, Engine Electronic Control Coolant Temperature	E1AF 12A648-AA E4AF 12A648-AA	E1AZ 12A648-A
PROCESSOR AND CALIBRATOR ASSY, EEC-IV	E9TF 12A650-L1C, L2C	E9TZ 12A650-LC
SENSOR ASSY, Air Charge Temperature	E1AF 12A697-AA E4AF 12A697-AA	E1AZ 12A697-A

CALIBRATION: 9-55J-R11 — 3.0L EFI

FORD MOTOR COMPANY
IMPORTANT VEHICLE INFORMATION

THIS VEHICLE IS EQUIPPED WITH EEC IV/EFI SYSTEMS. IDLE SPEEDS AND IDLE MIXTURES ARE NOT ADJUSTABLE. SEE SHOP MANUAL FOR ADDITIONAL INFORMATION.

ADJUST IGNITION TIMING WITH THE TRANSMISSION IN NEUTRAL, PARKING BRAKE SET AND THE WHEELS BLOCKED. ENGINE MUST BE AT NORMAL OPERATING TEMPERATURE.

(1) TURN OFF ENGINE.
(2) DISCONNECT SMALL IN-LINE SPOUT CONNECTOR (=▭◨) LOCATED NEAR THE DISTRIBUTOR.
(3) RE-START PREVIOUSLY WARMED-UP ENGINE.
(4) ADJUST IGNITION TIMING TO 10° BTDC.
(5) TURN OFF ENGINE AND RESTORE ELECTRICAL CONNECTION.

THIS VEHICLE CONFORMS TO U.S. EPA AND CALIFORNIA REGULATIONS APPLICABLE TO 1990 MODEL YEAR NEW LIGHT DUTY TRUCKS INTRODUCED INTO COMMERCE SOLELY FOR SALE IN CALIFORNIA. OBD EXEMPT.

USE SAE 5W-30 OIL
API SERVICE SG - ENERGY CONSERVING II.

CATALYST

SPARK PLUG: AWSF-32P GAP-.042-.046
3.0L-9HM
LFM3.0T5FEDX - TWC/HO2S/MPI

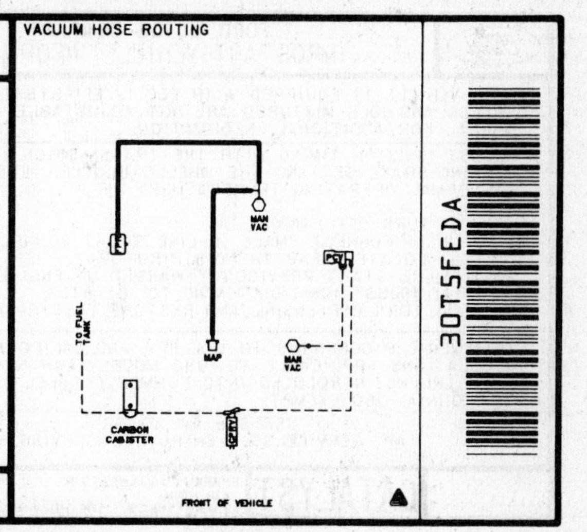

CALIBRATION PARTS LIST — CODE: DDJ

NAME/DESCRIPTION	ENGINEERING NO.	SERVICE NO.
SENSOR ASSY, Exhaust Gas Oxygen	E77F 9F472-AA, BA E7TF 9F472-AA	E7TZ 9F472-A
SENSOR ASSY, Exhaust Gas Oxygen	E9SF 9F472-AA E97F 9F472-AA	F0SZ 9F472-A
SENSOR ASSY, Manifold Absolute Pressure	E7EF 9F479-A1A, A2A	E7FZ 9F479-A
INJECTOR ASSY, Fuel	E59E 9F593-A1A, A2A, AB	E5TZ 9F593-A
INJECTOR ASSY, Fuel	E67E 9F593-B1A, B1B	E6SZ 9F593-A
VALVE ASSY, Throttle Air By-Pass	E6AE 9F715-D1A, D1B	E6AZ 9F715-D
VALVE ASSY, Throttle Air By-Pass	E9DE 9F715-A1A E9EE 9F715-A1A	E9DZ 9F715-A
SENSOR ASSY, Speed	E3AF 9E731-AB	E3AZ 9E731-A
SENSOR ASSY, Speed	E45F 9E731-AA	E45Y 9E731-A
REGULATOR ASSY, Fuel Charging Pressure	E6AE 9C968-AA, AB E7DE 9C968-BA	E6AZ 9C968-A
REGULATOR ASSY, Fuel Charging Pressure	E7EE 9C968-AA E77E 9C968-AA	E7FZ 9C968-A
POTENTIOMETER ASSY, Throttle Position Sensor	E7DF 9B989-AA	E7DZ 9B989-A
DISTRIBUTOR ASSY	E6AE 12127-EA	E6AZ 12127-C
SENSOR ASSY, Engine Electronic Control Coolant Temperature	E1AF 12A648-AA E4AF 12A648-AA	E1AZ 12A648-A
PROCESSOR AND CALIBRATOR ASSY, EEC-IV	E9TF 12A650-L1C, L2C	E9TZ 12A650-LC
SENSOR ASSY, Air Charge Temperature	E1AF 12A697-AA E4AF 12A697-AA	E1AZ 12A697-A

CALIBRATION: 9-55J-R11 — 3.0L EFI

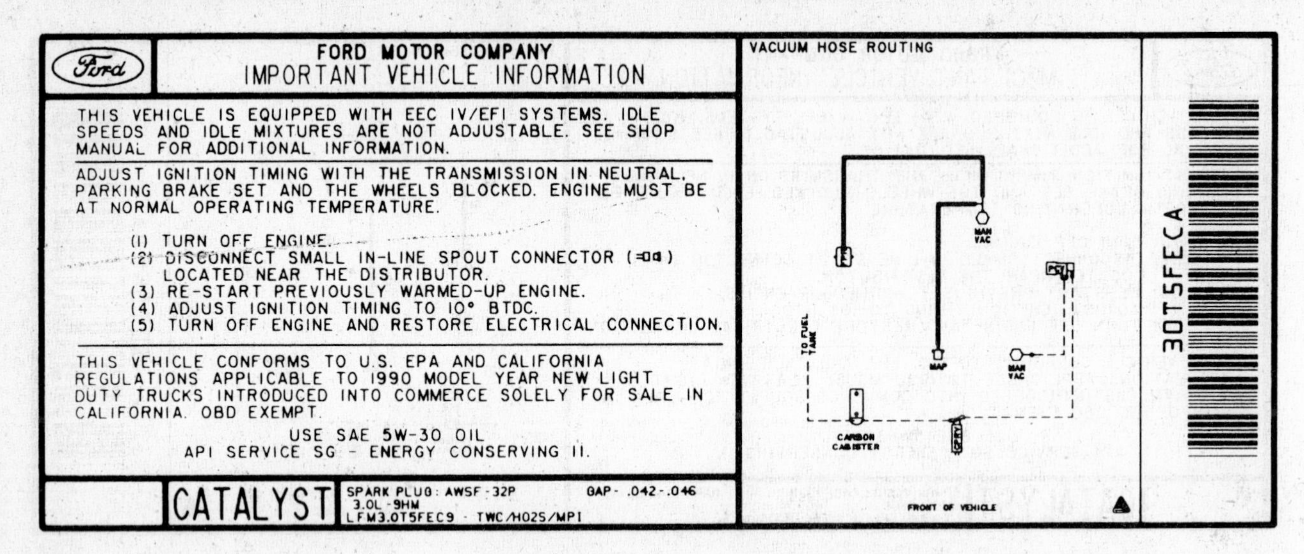
FORD MOTOR COMPANY
IMPORTANT VEHICLE INFORMATION

THIS VEHICLE IS EQUIPPED WITH EEC IV/EFI SYSTEMS. IDLE SPEEDS AND IDLE MIXTURES ARE NOT ADJUSTABLE. SEE SHOP MANUAL FOR ADDITIONAL INFORMATION.

ADJUST IGNITION TIMING WITH THE TRANSMISSION IN NEUTRAL, PARKING BRAKE SET AND THE WHEELS BLOCKED. ENGINE MUST BE AT NORMAL OPERATING TEMPERATURE.

(1) TURN OFF ENGINE.
(2) DISCONNECT SMALL IN-LINE SPOUT CONNECTOR (=▭◁) LOCATED NEAR THE DISTRIBUTOR.
(3) RE-START PREVIOUSLY WARMED-UP ENGINE.
(4) ADJUST IGNITION TIMING TO 10° BTDC.
(5) TURN OFF ENGINE AND RESTORE ELECTRICAL CONNECTION.

THIS VEHICLE CONFORMS TO U.S. EPA AND CALIFORNIA REGULATIONS APPLICABLE TO 1990 MODEL YEAR NEW LIGHT DUTY TRUCKS INTRODUCED INTO COMMERCE SOLELY FOR SALE IN CALIFORNIA. OBD EXEMPT.

USE SAE 5W-30 OIL
API SERVICE SG - ENERGY CONSERVING II.

CATALYST

SPARK PLUG: AWSF-32P GAP- .042-.046
3.0L-9HM
LFM3.0T5FEC9 - TWC/HO2S/MPI

VACUUM HOSE ROUTING

CALIBRATION PARTS LIST — CODE: DDK

NAME/DESCRIPTION	ENGINEERING NO.	SERVICE NO.
SENSOR ASSY, Exhaust Gas Oxygen	E77F 9F472-AA, BA E7TF 9F472-AA	E7TZ 9F472-A
SENSOR ASSY, Exhaust Gas Oxygen	E9SF 9F472-AA E97F 9F472-AA	F0SZ 9F472-A
SENSOR ASSY, Manifold Absolute Pressure	E7EF 9F479-A1A, A2A	E7FZ 9F479-A
INJECTOR ASSY, Fuel	E59E 9F593-A1A, A2A, AB	E5TZ 9F593-A
INJECTOR ASSY, Fuel	E67E 9F593-B1A, B1B	E6SZ 9F593-A
VALVE ASSY, Throttle Air By-Pass	E6AE 9F715-D1A, D1B	E6AZ 9F715-D
VALVE ASSY, Throttle Air By-Pass	E9DE 9F715-A1A E9EE 9F715-A1A	E9DZ 9F715-A
SENSOR ASSY, Speed	E3AF 9E731-AB	E3AZ 9E731-A
SENSOR ASSY, Speed	E45F 9E731-AA	E45Y 9E731-A
REGULATOR ASSY, Fuel Charging Pressure	E6AE 9C968-AA, AB E7DE 9C968-BA	E6AZ 9C968-A
REGULATOR ASSY, Fuel Charging Pressure	E7EE 9C968-AA E77E 9C968-AA	E7FZ 9C968-A
POTENTIOMETER ASSY, Throttle Position Sensor	E7DF 9B989-AA	E7DZ 9B989-A
DISTRIBUTOR ASSY	E6AE 12127-EA	E6AZ 12127-C
SENSOR ASSY, Engine Electronic Control Coolant Temperature	E1AF 12A648-AA E4AF 12A648-AA	E1AZ 12A648-A
PROCESSOR AND CALIBRATOR ASSY, EEC-IV	E9TF 12A650-L1C, L2C	E9TZ 12A650-LC
SENSOR ASSY, Air Charge Temperature	E1AF 12A697-AA E4AF 12A697-AA	E1AZ 12A697-A

CALIBRATION: 9-55J-R11 — 3.0L EFI

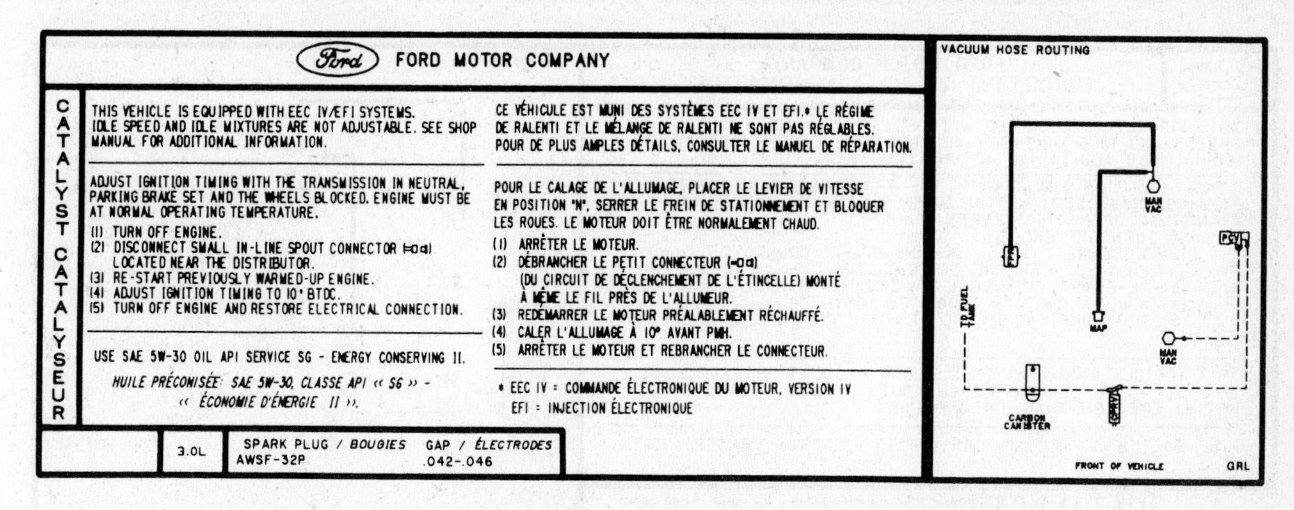
FORD MOTOR COMPANY

CATALYST CATALYSEUR

THIS VEHICLE IS EQUIPPED WITH EEC IV/EFI SYSTEMS.
IDLE SPEED AND IDLE MIXTURES ARE NOT ADJUSTABLE. SEE SHOP MANUAL FOR ADDITIONAL INFORMATION.

ADJUST IGNITION TIMING WITH THE TRANSMISSION IN NEUTRAL, PARKING BRAKE SET AND THE WHEELS BLOCKED. ENGINE MUST BE AT NORMAL OPERATING TEMPERATURE.
(1) TURN OFF ENGINE.
(2) DISCONNECT SMALL IN-LINE SPOUT CONNECTOR LOCATED NEAR THE DISTRIBUTOR.
(3) RE-START PREVIOUSLY WARMED-UP ENGINE.
(4) ADJUST IGNITION TIMING TO 10° BTDC.
(5) TURN OFF ENGINE AND RESTORE ELECTRICAL CONNECTION.

USE SAE 5W-30 OIL API SERVICE SG - ENERGY CONSERVING II.

HUILE PRÉCONISÉE: SAE 5W-30, CLASSE API « SG » - « ÉCONOMIE D'ÉNERGIE II ».

CE VÉHICULE EST MUNI DES SYSTÈMES EEC IV ET EFI.* LE RÉGIME DE RALENTI ET LE MÉLANGE DE RALENTI NE SONT PAS RÉGLABLES. POUR DE PLUS AMPLES DÉTAILS, CONSULTER LE MANUEL DE RÉPARATION.

POUR LE CALAGE DE L'ALLUMAGE, PLACER LE LEVIER DE VITESSE EN POSITION "N", SERRER LE FREIN DE STATIONNEMENT ET BLOQUER LES ROUES. LE MOTEUR DOIT ÊTRE NORMALEMENT CHAUD.
(1) ARRÊTER LE MOTEUR.
(2) DÉBRANCHER LE PETIT CONNECTEUR (DU CIRCUIT DE DÉCLENCHEMENT DE L'ÉTINCELLE) MONTÉ À MÊME LE FIL PRÈS DE L'ALLUMEUR.
(3) REDÉMARRER LE MOTEUR PRÉALABLEMENT RÉCHAUFFÉ.
(4) CALER L'ALLUMAGE À 10° AVANT PMH.
(5) ARRÊTER LE MOTEUR ET REBRANCHER LE CONNECTEUR.

* EEC IV = COMMANDE ÉLECTRONIQUE DU MOTEUR, VERSION IV
EFI = INJECTION ÉLECTRONIQUE

3.0L	SPARK PLUG / *BOUGIES*	GAP / *ÉLECTRODES*
	AWSF-32P	.042-.046

CALIBRATION PARTS LIST — CODE: DNR

NAME/DESCRIPTION	ENGINEERING NO.	SERVICE NO.
SENSOR ASSY, Exhaust Gas Oxygen	E77F 9F472-AA, BA E7TF 9F472-AA	E7TZ 9F472-A
SENSOR ASSY, Exhaust Gas Oxygen	E9SF 9F472-AA E97F 9F472-AA	F0SZ 9F472-A
SENSOR ASSY, Manifold Absolute Pressure	E7EF 9F479-A1A, A2A	E7FZ 9F479-A
INJECTOR ASSY, Fuel	E59E 9F593-A1A, A2A, AB	E5TZ 9F593-A
INJECTOR ASSY, Fuel	E67E 9F593-B1A, B1B	E6SZ 9F593-A
VALVE ASSY, Throttle Air By-Pass	E6AE 9F715-D1A, D1B	E6AZ 9F715-D
VALVE ASSY, Throttle Air By-Pass	E9DE 9F715-A1A E9EE 9F715-A1A	E9DZ 9F715-A
SENSOR ASSY, Speed	E3AF 9E731-AB	E3AZ 9E731-A
SENSOR ASSY, Speed	E45F 9E731-AA	E45Y 9E731-A
REGULATOR ASSY, Fuel Charging Pressure	E6AE 9C968-AA, AB E7DE 9C968-BA	E6AZ 9C968-A
REGULATOR ASSY, Fuel Charging Pressure	E7EE 9C968-AA E77E 9C968-AA	E7FZ 9C968-A
POTENTIOMETER ASSY, Throttle Position Sensor	E7DF 9B989-AA	E7DZ 9B989-A
DISTRIBUTOR ASSY	E6AE 12127-EA	E6AZ 12127-C
SENSOR ASSY, Engine Electronic Control Coolant Temperature	E1AF 12A648-AA E4AF 12A648-AA	E1AZ 12A648-A
PROCESSOR AND CALIBRATOR ASSY, EEC-IV	E9TF 12A650-L1C, L2C	E9TZ 12A650-LC
SENSOR ASSY, Air Charge Temperature	E1AF 12A697-AA E4AF 12A697-AA	E1AZ 12A697-A

CALIBRATION: 9-56J-R11 — 3.0L EFI

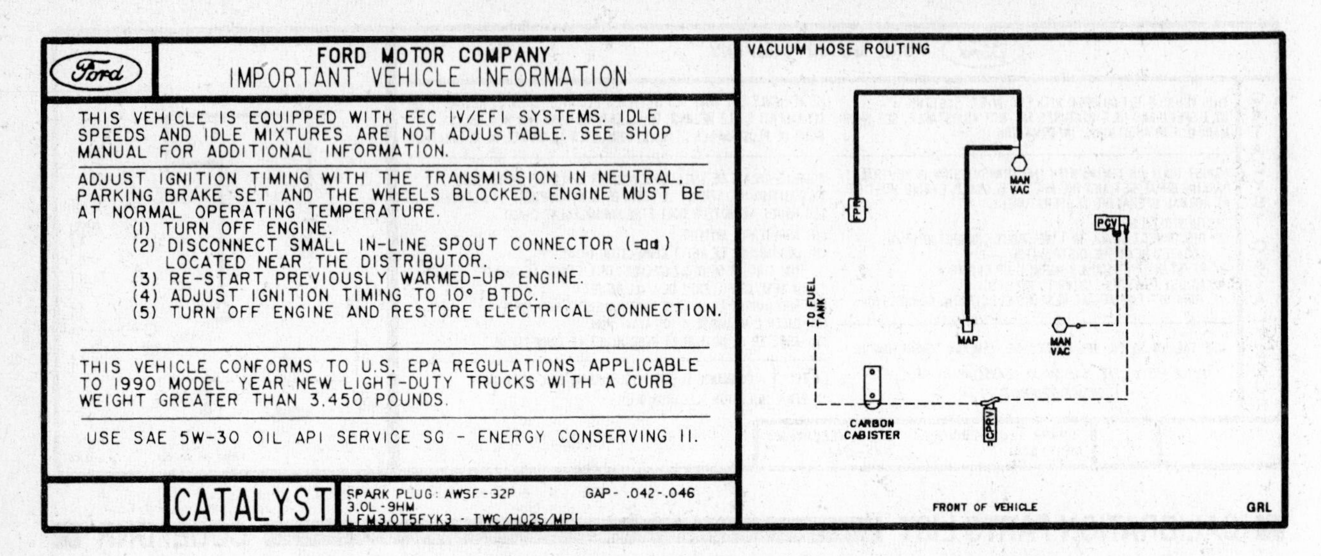

FORD MOTOR COMPANY
IMPORTANT VEHICLE INFORMATION

THIS VEHICLE IS EQUIPPED WITH EEC IV/EFI SYSTEMS. IDLE SPEEDS AND IDLE MIXTURES ARE NOT ADJUSTABLE. SEE SHOP MANUAL FOR ADDITIONAL INFORMATION.

ADJUST IGNITION TIMING WITH THE TRANSMISSION IN NEUTRAL, PARKING BRAKE SET AND THE WHEELS BLOCKED. ENGINE MUST BE AT NORMAL OPERATING TEMPERATURE.
(1) TURN OFF ENGINE.
(2) DISCONNECT SMALL IN-LINE SPOUT CONNECTOR (=□□) LOCATED NEAR THE DISTRIBUTOR.
(3) RE-START PREVIOUSLY WARMED-UP ENGINE.
(4) ADJUST IGNITION TIMING TO 10° BTDC.
(5) TURN OFF ENGINE AND RESTORE ELECTRICAL CONNECTION.

THIS VEHICLE CONFORMS TO U.S. EPA REGULATIONS APPLICABLE TO 1990 MODEL YEAR NEW LIGHT-DUTY TRUCKS WITH A CURB WEIGHT GREATER THAN 3,450 POUNDS.

USE SAE 5W-30 OIL API SERVICE SG - ENERGY CONSERVING II.

CATALYST

SPARK PLUG: AWSF-32P GAP-.042-.046
3.0L-9HM
LFM3.0T5FYK3 - TWC/HO2S/MPI

CALIBRATION PARTS LIST — CODE: DDE

NAME/DESCRIPTION	ENGINEERING NO.	SERVICE NO.
SENSOR ASSY, Exhaust Gas Oxygen	E77F 9F472-AA, BA E7TF 9F472-AA	E7TZ 9F472-A
SENSOR ASSY, Exhaust Gas Oxygen	E9SF 9F472-AA E97F 9F472-AA	F0SZ 9F472-A
SENSOR ASSY, Manifold Absolute Pressure	E7EF 9F479-A1A, A2A	E7FZ 9F479-A
INJECTOR ASSY, Fuel	E59E 9F593-A1A, A2A, AB	E5TZ 9F593-A
INJECTOR ASSY, Fuel	E67E 9F593-B1A, B1B	E6SZ 9F593-A
VALVE ASSY, Throttle Air By-Pass	E6AE 9F715-D1A, D1B	E6AZ 9F715-D
VALVE ASSY, Throttle Air By-Pass	E9DE 9F715-A1A E9EE 9F715-A1A	E9DZ 9F715-A
SENSOR ASSY, Speed	E3AF 9E731-AB	E3AZ 9E731-A
SENSOR ASSY, Speed	E45F 9E731-AA	E45Y 9E731-A
REGULATOR ASSY, Fuel Charging Pressure	E6AE 9C968-AA, AB E7DE 9C968-BA	E6AZ 9C968-A
REGULATOR ASSY, Fuel Charging Pressure	E7EE 9C968-AA E77E 9C968-AA	E7FZ 9C968-A
POTENTIOMETER ASSY, Throttle Position Sensor	E7DF 9B989-AA	E7DZ 9B989-A
DISTRIBUTOR ASSY	E6AE 12127-EA	E6AZ 12127-C
SENSOR ASSY, Engine Electronic Control Coolant Temperature	E1AF 12A648-AA E4AF 12A648-AA	E1AZ 12A648-A
PROCESSOR AND CALIBRATOR ASSY, EEC-IV	E9TF 12A650-M1C, M2C	E9TZ 12A650-MC
SENSOR ASSY, Air Charge Temperature	E1AF 12A697-AA E4AF 12A697-AA	E1AZ 12A697-A

CALIBRATION: 9-56J-R11 — 3.0L EFI

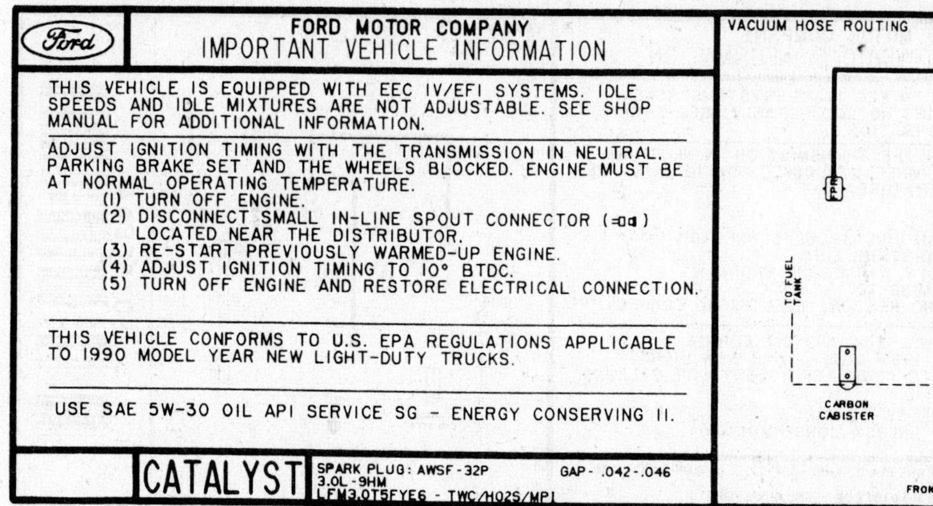
FORD MOTOR COMPANY
IMPORTANT VEHICLE INFORMATION

THIS VEHICLE IS EQUIPPED WITH EEC IV/EFI SYSTEMS. IDLE SPEEDS AND IDLE MIXTURES ARE NOT ADJUSTABLE. SEE SHOP MANUAL FOR ADDITIONAL INFORMATION.

ADJUST IGNITION TIMING WITH THE TRANSMISSION IN NEUTRAL, PARKING BRAKE SET AND THE WHEELS BLOCKED. ENGINE MUST BE AT NORMAL OPERATING TEMPERATURE.
(1) TURN OFF ENGINE.
(2) DISCONNECT SMALL IN-LINE SPOUT CONNECTOR LOCATED NEAR THE DISTRIBUTOR.
(3) RE-START PREVIOUSLY WARMED-UP ENGINE.
(4) ADJUST IGNITION TIMING TO 10° BTDC.
(5) TURN OFF ENGINE AND RESTORE ELECTRICAL CONNECTION.

THIS VEHICLE CONFORMS TO U.S. EPA REGULATIONS APPLICABLE TO 1990 MODEL YEAR NEW LIGHT-DUTY TRUCKS.

USE SAE 5W-30 OIL API SERVICE SG - ENERGY CONSERVING II.

CATALYST

SPARK PLUG: AWSF-32P GAP-.042-.046
3.0L-9HM
LFM3.0T5FYE6 - TWC/HO2S/MPI

CALIBRATION PARTS LIST — CODE: DDH

NAME/DESCRIPTION	ENGINEERING NO.	SERVICE NO.
SENSOR ASSY, Exhaust Gas Oxygen	E77F 9F472-AA, BA E7TF 9F472-AA	E7TZ 9F472-A
SENSOR ASSY, Exhaust Gas Oxygen	E9SF 9F472-AA E97F 9F472-AA	F0SZ 9F472-A
SENSOR ASSY, Manifold Absolute Pressure	E7EF 9F479-A1A, A2A	E7FZ 9F479-A
INJECTOR ASSY, Fuel	E59E 9F593-A1A, A2A, AB	E5TZ 9F593-A
INJECTOR ASSY, Fuel	E67E 9F593-B1A, B1B	E6SZ 9F593-A
VALVE ASSY, Throttle Air By-Pass	E6AE 9F715-D1A, D1B	E6AZ 9F715-D
VALVE ASSY, Throttle Air By-Pass	E9DE 9F715-A1A E9EE 9F715-A1A	E9DZ 9F715-A
SENSOR ASSY, Speed	E3AF 9E731-AB	E3AZ 9E731-A
SENSOR ASSY, Speed	E45F 9E731-AA	E45Y 9E731-A
REGULATOR ASSY, Fuel Charging Pressure	E6AE 9C968-AA, AB E7DE 9C968-BA	E6AZ 9C968-A
REGULATOR ASSY, Fuel Charging Pressure	E7EE 9C968-AA E77E 9C968-AA	E7FZ 9C968-A
POTENTIOMETER ASSY, Throttle Position Sensor	E7DF 9B989-AA	E7DZ 9B989-A
DISTRIBUTOR ASSY	E6AE 12127-EA	E6AZ 12127-C
SENSOR ASSY, Engine Electronic Control Coolant Temperature	E1AF 12A648-AA E4AF 12A648-AA	E1AZ 12A648-A
PROCESSOR AND CALIBRATOR ASSY, EEC-IV	E9TF 12A650-M1C, M2C	E9TZ 12A650-MC
SENSOR ASSY, Air Charge Temperature	E1AF 12A697-AA E4AF 12A697-AA	E1AZ 12A697-A

CALIBRATION: 9-56J-R11 — 3.0L EFI

FORD MOTOR COMPANY
IMPORTANT VEHICLE INFORMATION

THIS VEHICLE IS EQUIPPED WITH EEC IV/EFI SYSTEMS. IDLE SPEEDS AND IDLE MIXTURES ARE NOT ADJUSTABLE. SEE SHOP MANUAL FOR ADDITIONAL INFORMATION.

ADJUST IGNITION TIMING WITH THE TRANSMISSION IN NEUTRAL, PARKING BRAKE SET AND THE WHEELS BLOCKED. ENGINE MUST BE AT NORMAL OPERATING TEMPERATURE.

(1) TURN OFF ENGINE.
(2) DISCONNECT SMALL IN-LINE SPOUT CONNECTOR LOCATED NEAR THE DISTRIBUTOR.
(3) RE-START PREVIOUSLY WARMED-UP ENGINE.
(4) ADJUST IGNITION TIMING TO 10° BTDC.
(5) TURN OFF ENGINE AND RESTORE ELECTRICAL CONNECTION.

THIS VEHICLE CONFORMS TO U.S. EPA AND CALIFORNIA REGULATIONS APPLICABLE TO 1990 MODEL YEAR NEW LIGHT DUTY TRUCKS INTRODUCED INTO COMMERCE SOLELY FOR SALE IN CALIFORNIA. OBD EXEMPT.

USE SAE 5W-30 OIL
API SERVICE SG - ENERGY CONSERVING II.

CATALYST

SPARK PLUG: AWSF-32P GAP-.042-.046
3.0L-9HM
LFM3.0T5FEDX - TWC/HO2S/MPI

VACUUM HOSE ROUTING

TO FUEL TANK
MAN VAC
MAP
MAN VAC
CARBON CANISTER
FRONT OF VEHICLE

30T5FEDA

CALIBRATION PARTS LIST — CODE: DDJ

NAME/DESCRIPTION	ENGINEERING NO.	SERVICE NO.
SENSOR ASSY, Exhaust Gas Oxygen	E77F 9F472-AA, BA E7TF 9F472-AA	E7TZ 9F472-A
SENSOR ASSY, Exhaust Gas Oxygen	E9SF 9F472-AA E97F 9F472-AA	F0SZ 9F472-A
SENSOR ASSY, Manifold Absolute Pressure	E7EF 9F479-A1A, A2A	E7FZ 9F479-A
INJECTOR ASSY, Fuel	E59E 9F593-A1A, A2A, AB	E5TZ 9F593-A
INJECTOR ASSY, Fuel	E67E 9F593-B1A, B1B	E6SZ 9F593-A
VALVE ASSY, Throttle Air By-Pass	E6AE 9F715-D1A, D1B	E6AZ 9F715-D
VALVE ASSY, Throttle Air By-Pass	E9DE 9F715-A1A E9EE 9F715-A1A	E9DZ 9F715-A
SENSOR ASSY, Speed	E3AF 9E731-AB	E3AZ 9E731-A
SENSOR ASSY, Speed	E45F 9E731-AA	E45Y 9E731-A
REGULATOR ASSY, Fuel Charging Pressure	E6AE 9C968-AA, AB E7DE 9C968-BA	E6AZ 9C968-A
REGULATOR ASSY, Fuel Charging Pressure	E7EE 9C968-AA E77E 9C968-AA	E7FZ 9C968-A
POTENTIOMETER ASSY, Throttle Position Sensor	E7DF 9B989-AA	E7DZ 9B989-A
DISTRIBUTOR ASSY	E6AE 12127-EA	E6AZ 12127-C
SENSOR ASSY, Engine Electronic Control Coolant Temperature	E1AF 12A648-AA E4AF 12A648-AA	E1AZ 12A648-A
PROCESSOR AND CALIBRATOR ASSY, EEC-IV	E9TF 12A650-M1C, M2C	E9TZ 12A650-MC
SENSOR ASSY, Air Charge Temperature	E1AF 12A697-AA E4AF 12A697-AA	E1AZ 12A697-A

CALIBRATION: 9-56J-R11 — 3.0L EFI

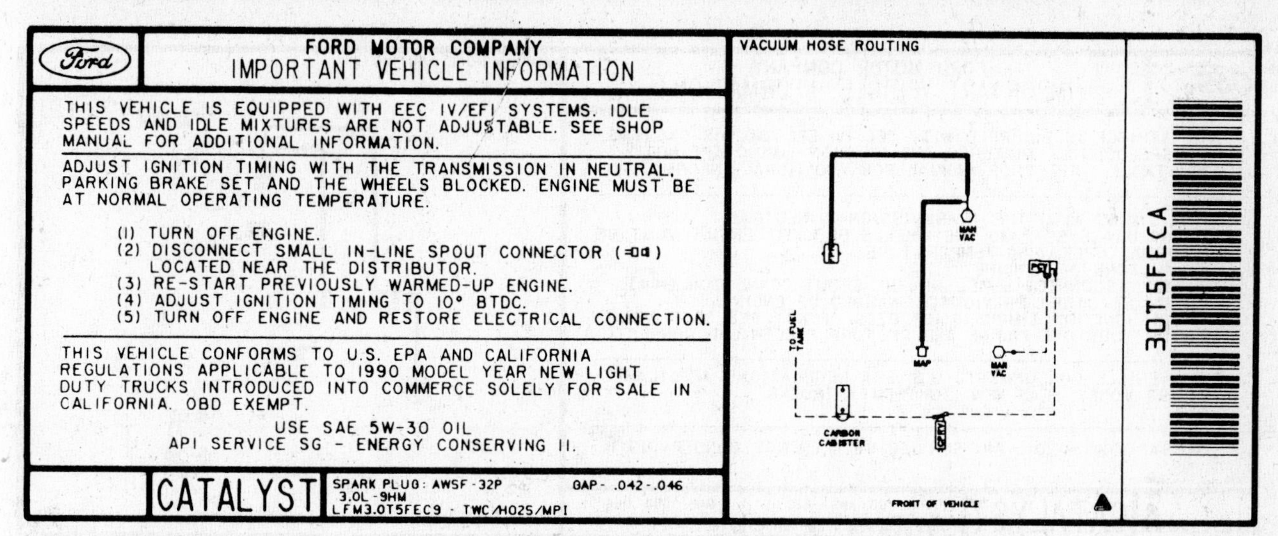
FORD MOTOR COMPANY
IMPORTANT VEHICLE INFORMATION

THIS VEHICLE IS EQUIPPED WITH EEC IV/EFI SYSTEMS. IDLE SPEEDS AND IDLE MIXTURES ARE NOT ADJUSTABLE. SEE SHOP MANUAL FOR ADDITIONAL INFORMATION.

ADJUST IGNITION TIMING WITH THE TRANSMISSION IN NEUTRAL, PARKING BRAKE SET AND THE WHEELS BLOCKED. ENGINE MUST BE AT NORMAL OPERATING TEMPERATURE.

(1) TURN OFF ENGINE.
(2) DISCONNECT SMALL IN-LINE SPOUT CONNECTOR LOCATED NEAR THE DISTRIBUTOR.
(3) RE-START PREVIOUSLY WARMED-UP ENGINE.
(4) ADJUST IGNITION TIMING TO 10° BTDC.
(5) TURN OFF ENGINE AND RESTORE ELECTRICAL CONNECTION.

THIS VEHICLE CONFORMS TO U.S. EPA AND CALIFORNIA REGULATIONS APPLICABLE TO 1990 MODEL YEAR NEW LIGHT DUTY TRUCKS INTRODUCED INTO COMMERCE SOLELY FOR SALE IN CALIFORNIA. OBD EXEMPT.

USE SAE 5W-30 OIL
API SERVICE SG - ENERGY CONSERVING II.

CATALYST
SPARK PLUG: AWSF-32P GAP-.042-.046
3.0L-9HM
LFM3.0T5FEC9 - TWC/HO2S/MPI

VACUUM HOSE ROUTING

CALIBRATION PARTS LIST — CODE: DDK

NAME/DESCRIPTION	ENGINEERING NO.	SERVICE NO.
SENSOR ASSY, Exhaust Gas Oxygen	E77F 9F472-AA, BA E7TF 9F472-AA	E7TZ 9F472-A
SENSOR ASSY, Exhaust Gas Oxygen	E9SF 9F472-AA E97F 9F472-AA	F0SZ 9F472-A
SENSOR ASSY, Manifold Absolute Pressure	E7EF 9F479-A1A, A2A	E7FZ 9F479-A
INJECTOR ASSY, Fuel	E59E 9F593-A1A, A2A, AB	E5TZ 9F593-A
INJECTOR ASSY, Fuel	E67E 9F593-B1A, B1B	E6SZ 9F593-A
VALVE ASSY, Throttle Air By-Pass	E6AE 9F715-D1A, D1B	E6AZ 9F715-D
VALVE ASSY, Throttle Air By-Pass	E9DE 9F715-A1A E9EE 9F715-A1A	E9DZ 9F715-A
SENSOR ASSY, Speed	E3AF 9E731-AB	E3AZ 9E731-A
SENSOR ASSY, Speed	E45F 9E731-AA	E45Y 9E731-A
REGULATOR ASSY, Fuel Charging Pressure	E6AE 9C968-AA, AB E7DE 9C968-BA	E6AZ 9C968-A
REGULATOR ASSY, Fuel Charging Pressure	E7EE 9C968-AA E77E 9C968-AA	E7FZ 9C968-A
POTENTIOMETER ASSY, Throttle Position Sensor	E7DF 9B989-AA	E7DZ 9B989-A
DISTRIBUTOR ASSY	E6AE 12127-EA	E6AZ 12127-C
SENSOR ASSY, Engine Electronic Control Coolant Temperature	E1AF 12A648-AA E4AF 12A648-AA	E1AZ 12A648-A
PROCESSOR AND CALIBRATOR ASSY, EEC-IV	E9TF 12A650-M1C, M2C	E9TZ 12A650-MC
SENSOR ASSY, Air Charge Temperature	E1AF 12A697-AA E4AF 12A697-AA	E1AZ 12A697-A

CALIBRATION: 0-58F-R00 — 4.0L EFI

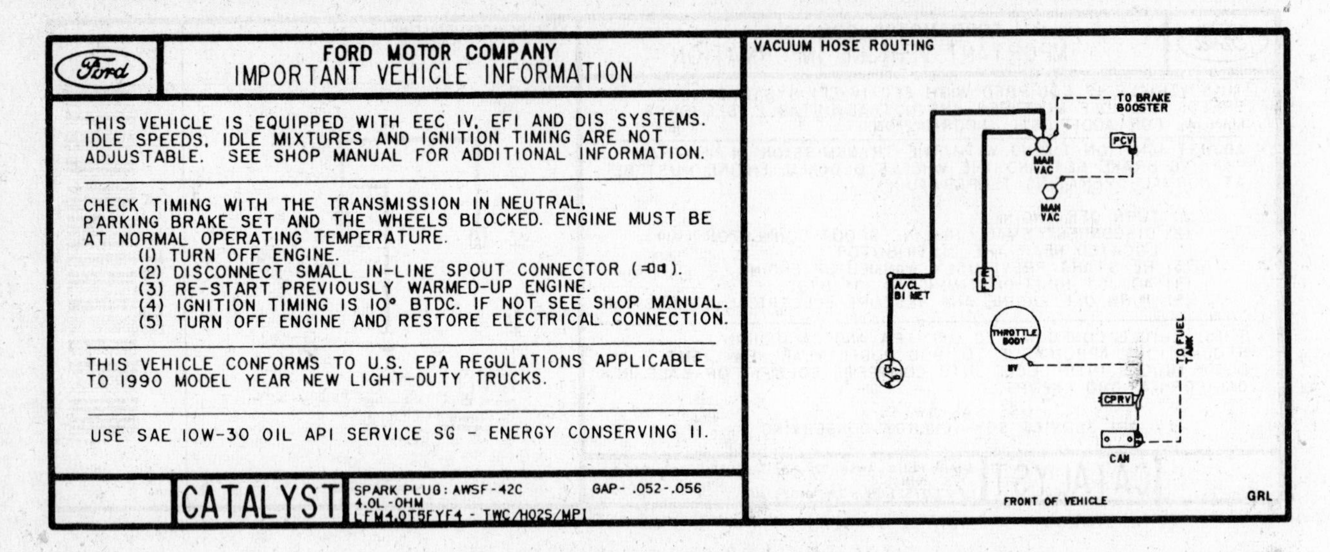
Ford

FORD MOTOR COMPANY
IMPORTANT VEHICLE INFORMATION

THIS VEHICLE IS EQUIPPED WITH EEC IV, EFI AND DIS SYSTEMS. IDLE SPEEDS, IDLE MIXTURES AND IGNITION TIMING ARE NOT ADJUSTABLE. SEE SHOP MANUAL FOR ADDITIONAL INFORMATION.

CHECK TIMING WITH THE TRANSMISSION IN NEUTRAL, PARKING BRAKE SET AND THE WHEELS BLOCKED. ENGINE MUST BE AT NORMAL OPERATING TEMPERATURE.
(1) TURN OFF ENGINE.
(2) DISCONNECT SMALL IN-LINE SPOUT CONNECTOR (=▯◁).
(3) RE-START PREVIOUSLY WARMED-UP ENGINE.
(4) IGNITION TIMING IS 10° BTDC. IF NOT SEE SHOP MANUAL.
(5) TURN OFF ENGINE AND RESTORE ELECTRICAL CONNECTION.

THIS VEHICLE CONFORMS TO U.S. EPA REGULATIONS APPLICABLE TO 1990 MODEL YEAR NEW LIGHT-DUTY TRUCKS.

USE SAE 10W-30 OIL API SERVICE SG – ENERGY CONSERVING II.

CATALYST
SPARK PLUG: AWSF-42C GAP- .052-.056
4.0L-OHM
LFM4.0T5FYF4 - TWC/HO2S/MPI

VACUUM HOSE ROUTING

CALIBRATION PARTS LIST — CODE: DLA

NAME/DESCRIPTION	ENGINEERING NO.	SERVICE NO.
SENSOR ASSY, Exhaust Gas Oxygen	F07F 9F472-CA, DA	F0TZ 9F472-C
INJECTOR ASSY, Fuel	90TF 9F593-AA	F0TZ 9F593-B
SENSOR ASSY, Carburetor Air Cleaner Air Temperature Control		D7FZ 9E607-A
MOTOR ASSY, Carburetor Air Cleaner Vacuum		D4FZ 9D612-A
VALVE ASSY, Throttle Air By-Pass	E9TE 9F715-BA, F0TE 9F715-B1A	E9TZ 9F715-B
SENSOR ASSY, Speed	E3AF 9E731-AB	E3AZ 9E731-A
REGULATOR ASSY, Fuel Charging Pressure	90TF 9C968-AA	F0TZ 9C968-A
POTENTIOMETER ASSY, Throttle Position Sensor	F07F 9B989-BA	F07Z 9B989-B
SENSOR ASSY, Mass Airflow	F07F 12B579-AA	F0TZ 12B579-A
SENSOR ASSY, Engine Control Barometric Pressure	E7EF 12A644-A1A, A2A	E7FZ 12A644-A
SENSOR ASSY, Engine Electronic Control Coolant Temperature	E4AF 12A648-AA	E1AZ 12A648-A
PROCESSOR AND CALIBRATOR ASSY, EEC-IV	F07F 12A650-L1A	F0TZ 12A650-LAA
SENSOR ASSY, Air Charge Temperature	E4AF 12A697-AA	E1AZ 12A697-A

CALIBRATION: 0-58F-R00 — 4.0L EFI

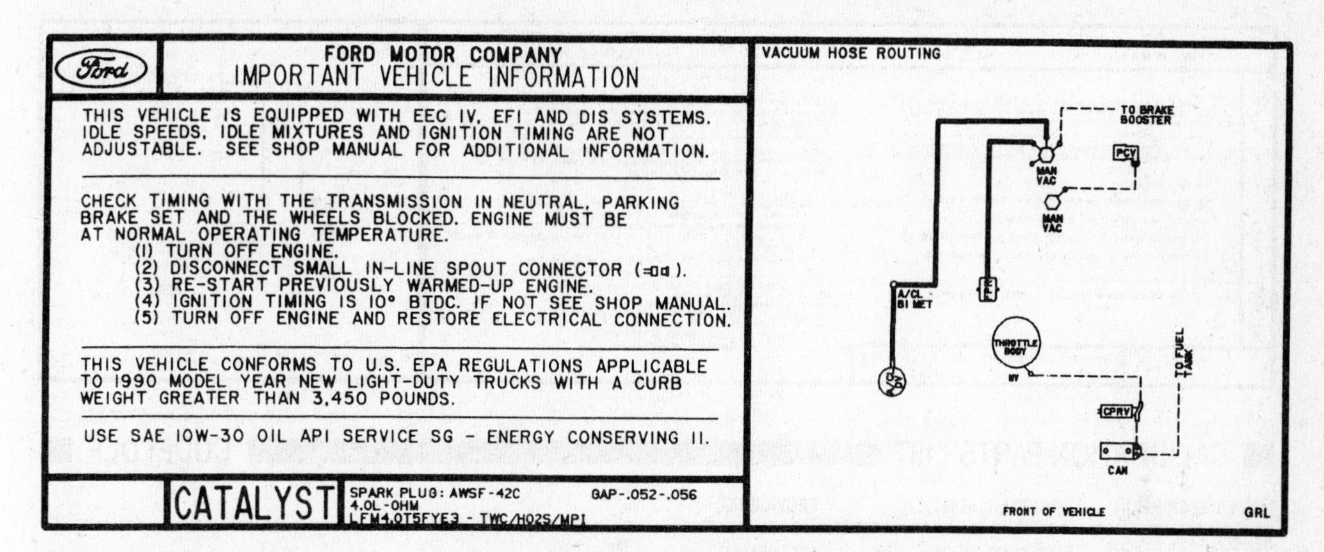

Ford

FORD MOTOR COMPANY
IMPORTANT VEHICLE INFORMATION

THIS VEHICLE IS EQUIPPED WITH EEC IV, EFI AND DIS SYSTEMS. IDLE SPEEDS, IDLE MIXTURES AND IGNITION TIMING ARE NOT ADJUSTABLE. SEE SHOP MANUAL FOR ADDITIONAL INFORMATION.

CHECK TIMING WITH THE TRANSMISSION IN NEUTRAL, PARKING BRAKE SET AND THE WHEELS BLOCKED. ENGINE MUST BE AT NORMAL OPERATING TEMPERATURE.

(1) TURN OFF ENGINE.
(2) DISCONNECT SMALL IN-LINE SPOUT CONNECTOR (=□□).
(3) RE-START PREVIOUSLY WARMED-UP ENGINE.
(4) IGNITION TIMING IS 10° BTDC. IF NOT SEE SHOP MANUAL.
(5) TURN OFF ENGINE AND RESTORE ELECTRICAL CONNECTION.

THIS VEHICLE CONFORMS TO U.S. EPA REGULATIONS APPLICABLE TO 1990 MODEL YEAR NEW LIGHT-DUTY TRUCKS WITH A CURB WEIGHT GREATER THAN 3,450 POUNDS.

USE SAE 10W-30 OIL API SERVICE SG - ENERGY CONSERVING II.

CATALYST

SPARK PLUG: AWSF-42C GAP-.052-.056
4.0L-0HM
LFM4.0T5FYE3 - TWC/HO2S/MPI

VACUUM HOSE ROUTING

CALIBRATION PARTS LIST — CODE: DLB

NAME/DESCRIPTION	ENGINEERING NO.	SERVICE NO.
SENSOR ASSY, Exhaust Gas Oxygen	F07F 9F472-CA, DA	F0TZ 9F472-C
INJECTOR ASSY, Fuel	90TF 9F593-AA	F0TZ 9F593-B
SENSOR ASSY, Carburetor Air Cleaner Air Temperature Control		D7FZ 9E607-A
MOTOR ASSY, Carburetor Air Cleaner Vacuum		D4FZ 9D612-A
VALVE ASSY, Throttle Air By-Pass	E9TE 9F715-BA, F0TE 9F715-B1A	E9TZ 9F715-B
SENSOR ASSY, Speed	E3AF 9E731-AB	E3AZ 9E731-A
REGULATOR ASSY, Fuel Charging Pressure	90TF 9C968-AA	F0TZ 9C968-A
POTENTIOMETER ASSY, Throttle Position Sensor	F07F 9B989-BA	F07Z 9B989-B
SENSOR ASSY, Mass Airflow	F07F 12B579-AA	F0TZ 12B579-A
SENSOR ASSY, Engine Control Barometric Pressure	E7EF 12A644-A1A, A2A	E7FZ 12A644-A
SENSOR ASSY, Engine Electronic Control Coolant Temperature	E4AF 12A648-AA	E1AZ 12A648-A
PROCESSOR AND CALIBRATOR ASSY, EEC-IV	F07F 12A650-L1A	F0TZ 12A650-LAA
SENSOR ASSY, Air Charge Temperature	E4AF 12A697-AA	E1AZ 12A697-A

CALIBRATION: 0-58F-R00 — 4.0L EFI

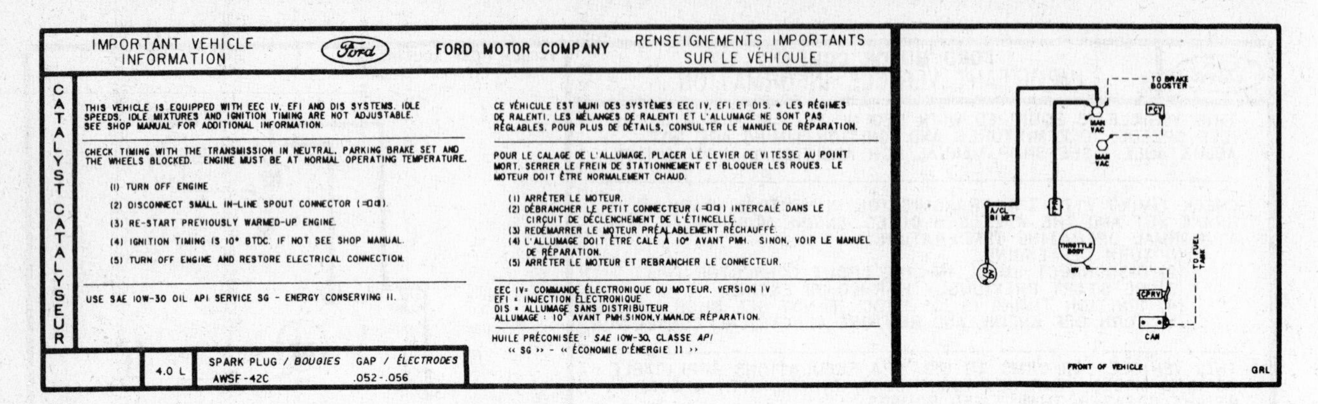
IMPORTANT VEHICLE INFORMATION — FORD MOTOR COMPANY — RENSEIGNEMENTS IMPORTANTS SUR LE VÉHICULE

CATALYST CATALYSEUR

THIS VEHICLE IS EQUIPPED WITH EEC IV, EFI AND DIS SYSTEMS. IDLE SPEEDS, IDLE MIXTURES AND IGNITION TIMING ARE NOT ADJUSTABLE. SEE SHOP MANUAL FOR ADDITIONAL INFORMATION.

CHECK TIMING WITH THE TRANSMISSION IN NEUTRAL, PARKING BRAKE SET AND THE WHEELS BLOCKED. ENGINE MUST BE AT NORMAL OPERATING TEMPERATURE.

(1) TURN OFF ENGINE
(2) DISCONNECT SMALL IN-LINE SPOUT CONNECTOR (=⊐⊏).
(3) RE-START PREVIOUSLY WARMED-UP ENGINE.
(4) IGNITION TIMING IS 10° BTDC. IF NOT SEE SHOP MANUAL.
(5) TURN OFF ENGINE AND RESTORE ELECTRICAL CONNECTION.

USE SAE 10W-30 OIL API SERVICE SG - ENERGY CONSERVING II.

CE VÉHICULE EST MUNI DES SYSTÈMES EEC IV, EFI ET DIS. • LES RÉGIMES DE RALENTI, LES MÉLANGES DE RALENTI ET L'ALLUMAGE NE SONT PAS RÉGLABLES. POUR PLUS DE DÉTAILS, CONSULTER LE MANUEL DE RÉPARATION.

POUR LE CALAGE DE L'ALLUMAGE, PLACER LE LEVIER DE VITESSE AU POINT MORT, SERRER LE FREIN DE STATIONNEMENT ET BLOQUER LES ROUES. LE MOTEUR DOIT ÊTRE NORMALEMENT CHAUD.

(1) ARRÊTER LE MOTEUR.
(2) DÉBRANCHER LE PETIT CONNECTEUR (=⊐⊏) INTERCALÉ DANS LE CIRCUIT DE DÉCLENCHEMENT DE L'ÉTINCELLE.
(3) REDÉMARRER LE MOTEUR PRÉALABLEMENT RÉCHAUFFÉ.
(4) L'ALLUMAGE DOIT ÊTRE CALÉ À 10° AVANT PMH. SINON, VOIR LE MANUEL DE RÉPARATION.
(5) ARRÊTER LE MOTEUR ET REBRANCHER LE CONNECTEUR.

EEC IV= COMMANDE ÉLECTRONIQUE DU MOTEUR, VERSION IV
EFI = INJECTION ÉLECTRONIQUE
DIS = ALLUMAGE SANS DISTRIBUTEUR
ALLUMAGE : 10° AVANT PMH.SINON,V.MAN.DE RÉPARATION.

HUILE PRÉCONISÉE : *SAE* 10W-30, CLASSE *API* « SG » - « ÉCONOMIE D'ÉNERGIE II »

	SPARK PLUG / *BOUGIES*	GAP / *ÉLECTRODES*
4.0 L	AWSF-42C	.052-.056

CALIBRATION PARTS LIST — CODE: DLF

NAME/DESCRIPTION	ENGINEERING NO.	SERVICE NO.
SENSOR ASSY, Exhaust Gas Oxygen	F07F 9F472-CA, DA	F0TZ 9F472-C
INJECTOR ASSY, Fuel	90TF 9F593-AA	F0TZ 9F593-B
SENSOR ASSY, Carburetor Air Cleaner Air Temperature Control		D7FZ 9E607-A
MOTOR ASSY, Carburetor Air Cleaner Vacuum		D4FZ 9D612-A
VALVE ASSY, Throttle Air By-Pass	E9TE 9F715-BA, F0TE 9F715-B1A	E9TZ 9F715-B
SENSOR ASSY, Speed	E3AF 9E731-AB	E3AZ 9E731-A
REGULATOR ASSY, Fuel Charging Pressure	90TF 9C968-AA	F0TZ 9C968-A
POTENTIOMETER ASSY, Throttle Position Sensor	F07F 9B989-BA	F07Z 9B989-B
SENSOR ASSY, Mass Airflow	F07F 12B579-AA	F0TZ 12B579-A
SENSOR ASSY, Engine Control Barometric Pressure	E7EF 12A644-A1A, A2A	E7FZ 12A644-A
SENSOR ASSY, Engine Electronic Control Coolant Temperature	E4AF 12A648-AA	E1AZ 12A648-A
PROCESSOR AND CALIBRATOR ASSY, EEC-IV	F07F 12A650-L1A	F0TZ 12A650-LAA
SENSOR ASSY, Air Charge Temperature	E4AF 12A697-AA	E1AZ 12A697-A

■ CALIBRATION: 0-58H-R00 ■ 4.0L EFI ■

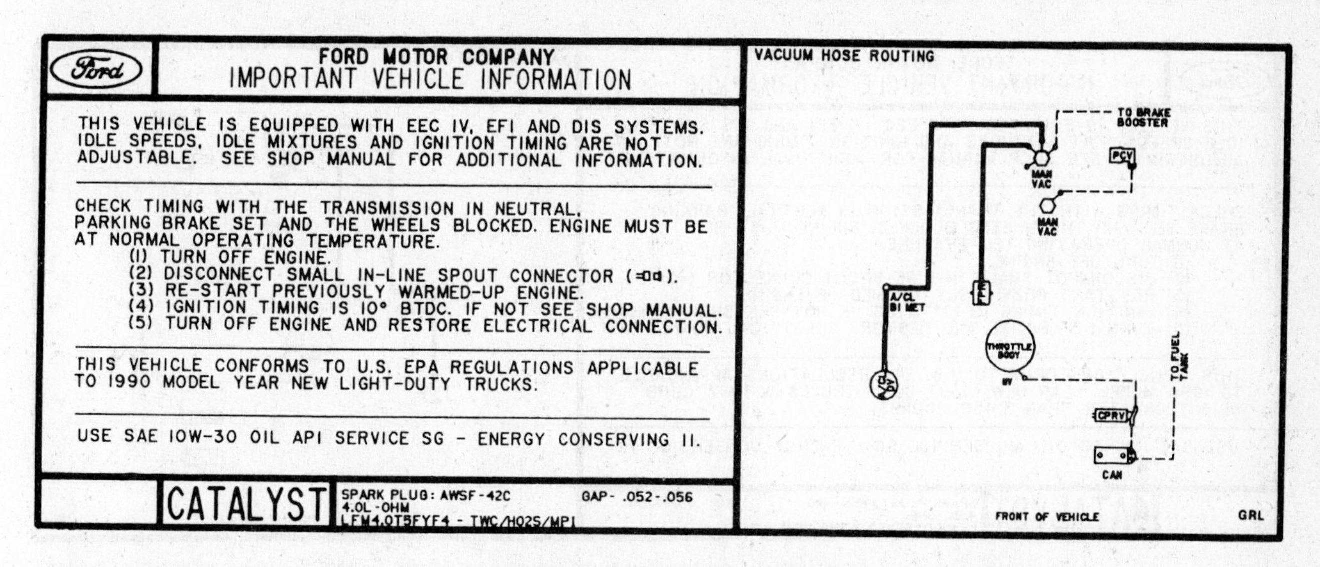

Ford
FORD MOTOR COMPANY
IMPORTANT VEHICLE INFORMATION

THIS VEHICLE IS EQUIPPED WITH EEC IV, EFI AND DIS SYSTEMS. IDLE SPEEDS, IDLE MIXTURES AND IGNITION TIMING ARE NOT ADJUSTABLE. SEE SHOP MANUAL FOR ADDITIONAL INFORMATION.

CHECK TIMING WITH THE TRANSMISSION IN NEUTRAL, PARKING BRAKE SET AND THE WHEELS BLOCKED. ENGINE MUST BE AT NORMAL OPERATING TEMPERATURE.
(1) TURN OFF ENGINE.
(2) DISCONNECT SMALL IN-LINE SPOUT CONNECTOR (=⊐⊏).
(3) RE-START PREVIOUSLY WARMED-UP ENGINE.
(4) IGNITION TIMING IS 10° BTDC. IF NOT SEE SHOP MANUAL.
(5) TURN OFF ENGINE AND RESTORE ELECTRICAL CONNECTION.

THIS VEHICLE CONFORMS TO U.S. EPA REGULATIONS APPLICABLE TO 1990 MODEL YEAR NEW LIGHT-DUTY TRUCKS.

USE SAE 10W-30 OIL API SERVICE SG - ENERGY CONSERVING II.

CATALYST
SPARK PLUG: AWSF-42C GAP-.052-.056
4.0L-OHM
LFM4.0T5FYF4 - TWC/HO2S/MPI

VACUUM HOSE ROUTING

■ CALIBRATION PARTS LIST ■ CODE: DLA ■

NAME/DESCRIPTION	ENGINEERING NO.	SERVICE NO.
SENSOR ASSY, Exhaust Gas Oxygen	F07F 9F472-CA, DA	F0TZ 9F472-C
INJECTOR ASSY, Fuel	90TF 9F593-AA	F0TZ 9F593-B
SENSOR ASSY, Carburetor Air Cleaner Air Temperature Control		D7FZ 9E607-A
MOTOR ASSY, Carburetor Air Cleaner Vacuum		D4FZ 9D612-A
VALVE ASSY, Throttle Air By-Pass	E9TE 9F715-BA, F0TE 9F715-B1A	E9TZ 9F715-B
SENSOR ASSY, Speed	E3AF 9E731-AB	E3AZ 9E731-A
REGULATOR ASSY, Fuel Charging Pressure	90TF 9C968-AA	F0TZ 9C968-A
POTENTIOMETER ASSY, Throttle Position Sensor	F07F 9B989-BA	F07Z 9B989-B
SENSOR ASSY, Mass Airflow	F07F 12B579-AA	F0TZ 12B579-A
SENSOR ASSY, Engine Control Barometric Pressure	E7EF 12A644-A1A, A2A	E7FZ 12A644-A
SENSOR ASSY, Engine Electronic Control Coolant Temperature	E4AF 12A648-AA	E1AZ 12A648-A
PROCESSOR AND CALIBRATOR ASSY, EEC-IV	F07F 12A650-M1A	F0TZ 12A650-MAA
SENSOR ASSY, Air Charge Temperature	E4AF 12A697-AA	E1AZ 12A697-A

CALIBRATION: 0-58H-R00 — 4.0L EFI

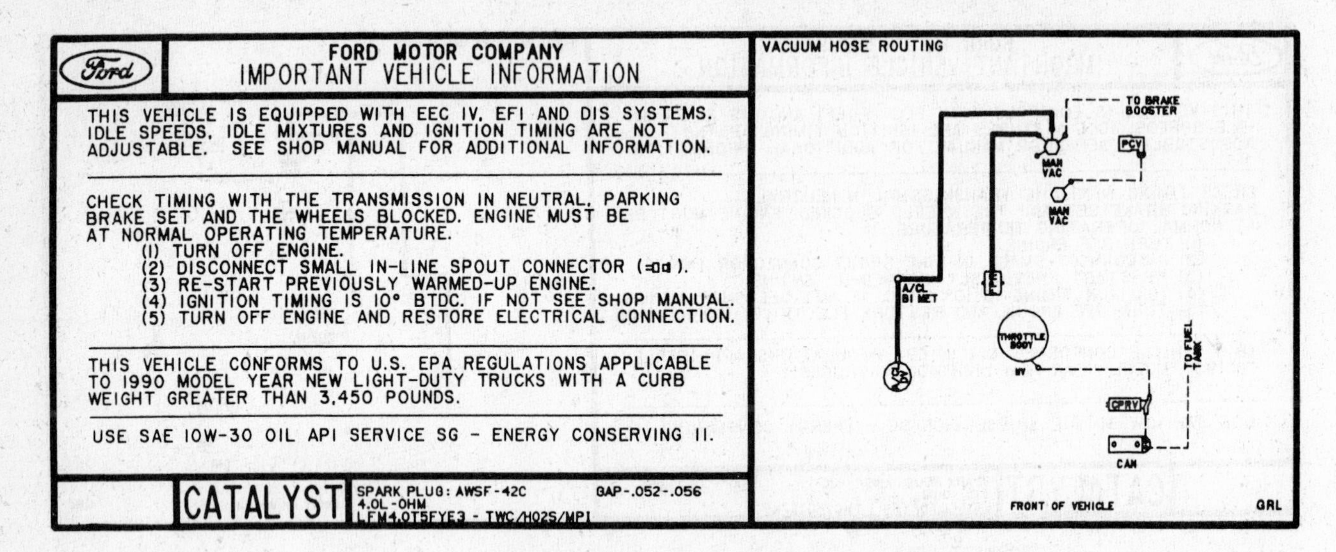
Ford

FORD MOTOR COMPANY
IMPORTANT VEHICLE INFORMATION

THIS VEHICLE IS EQUIPPED WITH EEC IV, EFI AND DIS SYSTEMS. IDLE SPEEDS, IDLE MIXTURES AND IGNITION TIMING ARE NOT ADJUSTABLE. SEE SHOP MANUAL FOR ADDITIONAL INFORMATION.

CHECK TIMING WITH THE TRANSMISSION IN NEUTRAL, PARKING BRAKE SET AND THE WHEELS BLOCKED. ENGINE MUST BE AT NORMAL OPERATING TEMPERATURE.
(1) TURN OFF ENGINE.
(2) DISCONNECT SMALL IN-LINE SPOUT CONNECTOR (=▢◨).
(3) RE-START PREVIOUSLY WARMED-UP ENGINE.
(4) IGNITION TIMING IS 10° BTDC. IF NOT SEE SHOP MANUAL.
(5) TURN OFF ENGINE AND RESTORE ELECTRICAL CONNECTION.

THIS VEHICLE CONFORMS TO U.S. EPA REGULATIONS APPLICABLE TO 1990 MODEL YEAR NEW LIGHT-DUTY TRUCKS WITH A CURB WEIGHT GREATER THAN 3,450 POUNDS.

USE SAE 10W-30 OIL API SERVICE SG - ENERGY CONSERVING II.

CATALYST

SPARK PLUG: AWSF-42C GAP-.052-.056
4.0L-OHM
LFM4.0T5FYE3 - TWC/HO2S/MPI

VACUUM HOSE ROUTING

CALIBRATION PARTS LIST — CODE: DLB

NAME/DESCRIPTION	ENGINEERING NO.	SERVICE NO.
SENSOR ASSY, Exhaust Gas Oxygen	F07F 9F472-CA, DA	F0TZ 9F472-C
INJECTOR ASSY, Fuel	90TF 9F593-AA	F0TZ 9F593-B
SENSOR ASSY, Carburetor Air Cleaner Air Temperature Control		D7FZ 9E607-A
MOTOR ASSY, Carburetor Air Cleaner Vacuum		D4FZ 9D612-A
VALVE ASSY, Throttle Air By-Pass	E9TE 9F715-BA, F0TE 9F715-B1A	E9TZ 9F715-B
SENSOR ASSY, Speed	E3AF 9E731-AB	E3AZ 9E731-A
REGULATOR ASSY, Fuel Charging Pressure	90TF 9C968-AA	F0TZ 9C968-A
POTENTIOMETER ASSY, Throttle Position Sensor	F07F 9B989-BA	F07Z 9B989-B
SENSOR ASSY, Mass Airflow	F07F 12B579-AA	F0TZ 12B579-A
SENSOR ASSY, Engine Control Barometric Pressure	E7EF 12A644-A1A, A2A	E7FZ 12A644-A
SENSOR ASSY, Engine Electronic Control Coolant Temperature	E4AF 12A648-AA	E1AZ 12A648-A
PROCESSOR AND CALIBRATOR ASSY, EEC-IV	F07F 12A650-M1A	F0TZ 12A650-MAA
SENSOR ASSY, Air Charge Temperature	E4AF 12A697-AA	E1AZ 12A697-A

CALIBRATION: 0-58H-R00 — 4.0L EFI

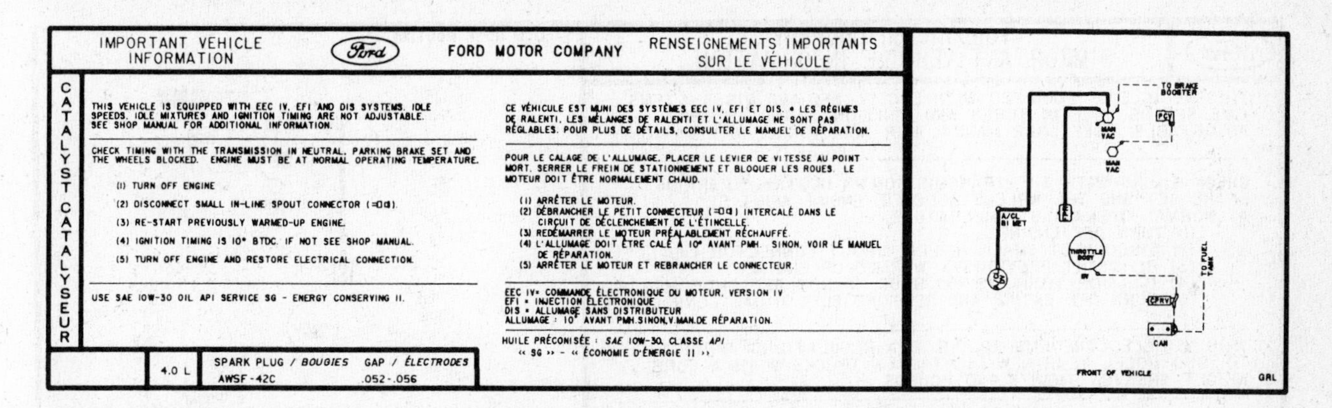
IMPORTANT VEHICLE INFORMATION

FORD MOTOR COMPANY

RENSEIGNEMENTS IMPORTANTS SUR LE VÉHICULE

CATALYST CATALYSEUR

THIS VEHICLE IS EQUIPPED WITH EEC IV, EFI AND DIS SYSTEMS. IDLE SPEEDS, IDLE MIXTURES AND IGNITION TIMING ARE NOT ADJUSTABLE. SEE SHOP MANUAL FOR ADDITIONAL INFORMATION.

CHECK TIMING WITH THE TRANSMISSION IN NEUTRAL, PARKING BRAKE SET AND THE WHEELS BLOCKED. ENGINE MUST BE AT NORMAL OPERATING TEMPERATURE.

(1) TURN OFF ENGINE
(2) DISCONNECT SMALL IN-LINE SPOUT CONNECTOR (=▢◁).
(3) RE-START PREVIOUSLY WARMED-UP ENGINE.
(4) IGNITION TIMING IS 10° BTDC. IF NOT SEE SHOP MANUAL.
(5) TURN OFF ENGINE AND RESTORE ELECTRICAL CONNECTION.

USE SAE 10W-30 OIL API SERVICE SG - ENERGY CONSERVING II.

CE VÉHICULE EST MUNI DES SYSTÈMES EEC IV, EFI ET DIS. • LES RÉGIMES DE RALENTI, LES MÉLANGES DE RALENTI ET L'ALLUMAGE NE SONT PAS RÉGLABLES. POUR PLUS DE DÉTAILS, CONSULTER LE MANUEL DE RÉPARATION.

POUR LE CALAGE DE L'ALLUMAGE, PLACER LE LEVIER DE VITESSE AU POINT MORT, SERRER LE FREIN DE STATIONNEMENT ET BLOQUER LES ROUES. LE MOTEUR DOIT ÊTRE NORMALEMENT CHAUD.

(1) ARRÊTER LE MOTEUR.
(2) DÉBRANCHER LE PETIT CONNECTEUR (=▢◁) INTERCALÉ DANS LE CIRCUIT DE DÉCLENCHEMENT DE L'ÉTINCELLE.
(3) REDÉMARRER LE MOTEUR PRÉALABLEMENT RÉCHAUFFÉ.
(4) L'ALLUMAGE DOIT ÊTRE CALÉ À 10° AVANT PMH. SINON, VOIR LE MANUEL DE RÉPARATION.
(5) ARRÊTER LE MOTEUR ET REBRANCHER LE CONNECTEUR.

EEC IV = COMMANDE ÉLECTRONIQUE DU MOTEUR, VERSION IV
EFI = INJECTION ÉLECTRONIQUE
DIS = ALLUMAGE SANS DISTRIBUTEUR
ALLUMAGE : 10° AVANT PMH.SINON,V.MAN.DE RÉPARATION.

HUILE PRÉCONISÉE : *SAE* 10W-30, CLASSE *API* « SG » - « ÉCONOMIE D'ÉNERGIE II »

	SPARK PLUG / *BOUGIES*	GAP / *ÉLECTRODES*
4.0 L	AWSF-42C	.052-.056

CALIBRATION PARTS LIST — CODE: DLF

NAME/DESCRIPTION	ENGINEERING NO.	SERVICE NO.
SENSOR ASSY, Exhaust Gas Oxygen	F07F 9F472-CA, DA	F0TZ 9F472-C
INJECTOR ASSY, Fuel	90TF 9F593-AA	F0TZ 9F593-B
SENSOR ASSY, Carburetor Air Cleaner Air Temperature Control		D7FZ 9E607-A
MOTOR ASSY, Carburetor Air Cleaner Vacuum		D4FZ 9D612-A
VALVE ASSY, Throttle Air By-Pass	E9TE 9F715-BA, F0TE 9F715-B1A	E9TZ 9F715-B
SENSOR ASSY, Speed	E3AF 9E731-AB	E3AZ 9E731-A
REGULATOR ASSY, Fuel Charging Pressure	90TF 9C968-AA	F0TZ 9C968-A
POTENTIOMETER ASSY, Throttle Position Sensor	F07F 9B989-BA	F07Z 9B989-B
SENSOR ASSY, Mass Airflow	F07F 12B579-AA	F0TZ 12B579-A
SENSOR ASSY, Engine Control Barometric Pressure	E7EF 12A644-A1A, A2A	E7FZ 12A644-A
SENSOR ASSY, Engine Electronic Control Coolant Temperature	E4AF 12A648-AA	E1AZ 12A648-A
PROCESSOR AND CALIBRATOR ASSY, EEC-IV	F07F 12A650-M1A	F0TZ 12A650-MAA
SENSOR ASSY, Air Charge Temperature	E4AF 12A697-AA	E1AZ 12A697-A

■ CALIBRATION: 0-58K-R00 4.0L EFI ■

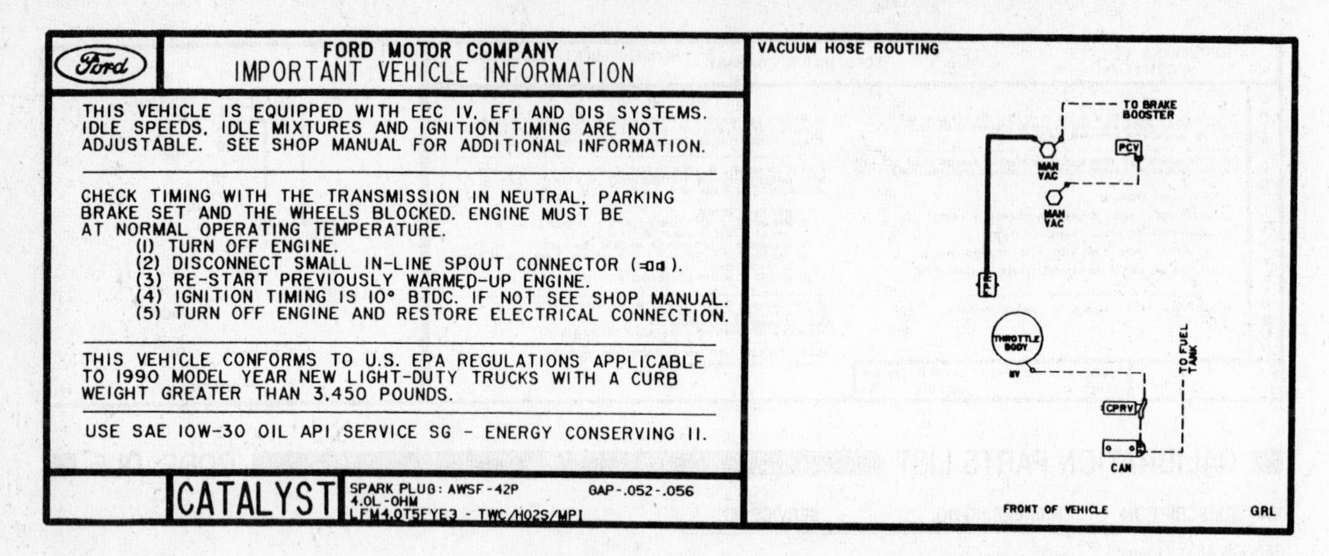

FORD MOTOR COMPANY
IMPORTANT VEHICLE INFORMATION

THIS VEHICLE IS EQUIPPED WITH EEC IV, EFI AND DIS SYSTEMS. IDLE SPEEDS, IDLE MIXTURES AND IGNITION TIMING ARE NOT ADJUSTABLE. SEE SHOP MANUAL FOR ADDITIONAL INFORMATION.

CHECK TIMING WITH THE TRANSMISSION IN NEUTRAL, PARKING BRAKE SET AND THE WHEELS BLOCKED. ENGINE MUST BE AT NORMAL OPERATING TEMPERATURE.
(1) TURN OFF ENGINE.
(2) DISCONNECT SMALL IN-LINE SPOUT CONNECTOR (=□□).
(3) RE-START PREVIOUSLY WARMED-UP ENGINE.
(4) IGNITION TIMING IS 10° BTDC. IF NOT SEE SHOP MANUAL.
(5) TURN OFF ENGINE AND RESTORE ELECTRICAL CONNECTION.

THIS VEHICLE CONFORMS TO U.S. EPA REGULATIONS APPLICABLE TO 1990 MODEL YEAR NEW LIGHT-DUTY TRUCKS WITH A CURB WEIGHT GREATER THAN 3,450 POUNDS.

USE SAE 10W-30 OIL API SERVICE SG - ENERGY CONSERVING II.

CATALYST

SPARK PLUG: AWSF-42P GAP-.052-.056
4.0L-0HM
LFM4.0T5FYE3 - TWC/HO2S/MPI

VACUUM HOSE ROUTING

■ CALIBRATION PARTS LIST CODE: DKZ ■

NAME/DESCRIPTION	ENGINEERING NO.	SERVICE NO.
SENSOR ASSY, Exhaust Gas Oxygen	F07F 9F472-CA, DA	F0TZ 9F472-C
INJECTOR ASSY, Fuel	90TF 9F593-AA	F0TZ 9F593-B
VALVE ASSY, Throttle Air By-Pass	E9TE 9F715-BA, F0TE 9F715-B1A	E9TZ 9F715-B
SENSOR ASSY, Speed	E3AF 9E731-AB	E3AZ 9E731-A
REGULATOR ASSY, Fuel Charging Pressure	90TF 9C968-AA	F0TZ 9C968-A
POTENTIOMETER ASSY, Throttle Position Sensor	F07F 9B989-BA	F07Z 9B989-B
SENSOR ASSY, Mass Airflow	F07F 12B579-AA	F0TZ 12B579-A
SENSOR ASSY, Engine Control Barometric Pressure	E7EF 12A644-A1A, A2A	E7FZ 12A644-A
SENSOR ASSY, Engine Electronic Control Coolant Temperature	E4AF 12A648-AA	E1AZ 12A648-A
PROCESSOR AND CALIBRATOR ASSY, EEC-IV	F07F 12A650-SA	F0TZ 12A650-SA
SENSOR ASSY, Air Charge Temperature	E4AF 12A697-AA	E1AZ 12A697-A

CALIBRATION: 0-58K-R00 — 4.0L EFI

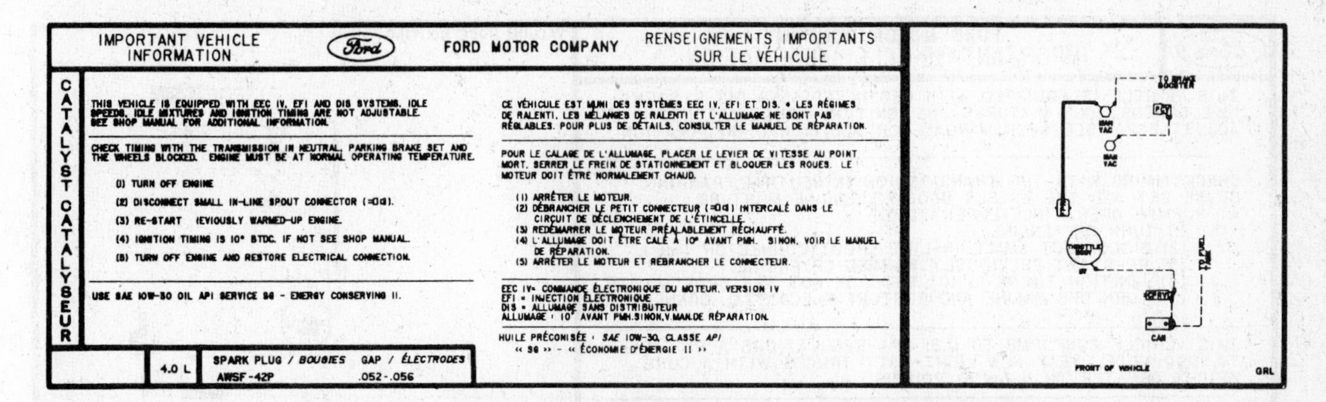

IMPORTANT VEHICLE INFORMATION — Ford — FORD MOTOR COMPANY — RENSEIGNEMENTS IMPORTANTS SUR LE VÉHICULE

CATALYST CATALYSEUR

THIS VEHICLE IS EQUIPPED WITH EEC IV, EFI AND DIS SYSTEMS. IDLE SPEEDS, IDLE MIXTURES AND IGNITION TIMING ARE NOT ADJUSTABLE. SEE SHOP MANUAL FOR ADDITIONAL INFORMATION.

CHECK TIMING WITH THE TRANSMISSION IN NEUTRAL, PARKING BRAKE SET AND THE WHEELS BLOCKED. ENGINE MUST BE AT NORMAL OPERATING TEMPERATURE.

(1) TURN OFF ENGINE
(2) DISCONNECT SMALL IN-LINE SPOUT CONNECTOR (=Dd).
(3) RE-START IEVIOUSLY WARMED-UP ENGINE.
(4) IGNITION TIMING IS 10° BTDC. IF NOT SEE SHOP MANUAL.
(5) TURN OFF ENGINE AND RESTORE ELECTRICAL CONNECTION.

USE SAE 10W-30 OIL API SERVICE SG - ENERGY CONSERVING II.

CE VÉHICULE EST MUNI DES SYSTÈMES EEC IV, EFI ET DIS. • LES RÉGIMES DE RALENTI, LES MÉLANGES DE RALENTI ET L'ALLUMAGE NE SONT PAS RÉGLABLES. POUR PLUS DE DÉTAILS, CONSULTER LE MANUEL DE RÉPARATION.

POUR LE CALAGE DE L'ALLUMAGE, PLACER LE LEVIER DE VITESSE AU POINT MORT, SERRER LE FREIN DE STATIONNEMENT ET BLOQUER LES ROUES. LE MOTEUR DOIT ÊTRE NORMALEMENT CHAUD.

(1) ARRÊTER LE MOTEUR.
(2) DÉBRANCHER LE PETIT CONNECTEUR (=Dd) INTERCALÉ DANS LE CIRCUIT DE DÉCLENCHEMENT DE L'ÉTINCELLE.
(3) REDÉMARRER LE MOTEUR PRÉALABLEMENT RÉCHAUFFÉ.
(4) L'ALLUMAGE DOIT ÊTRE CALÉ À 10° AVANT PMH. SINON, VOIR LE MANUEL DE RÉPARATION.
(5) ARRÊTER LE MOTEUR ET REBRANCHER LE CONNECTEUR.

EEC IV= COMMANDE ÉLECTRONIQUE DU MOTEUR, VERSION IV
EFI = INJECTION ÉLECTRONIQUE
DIS = ALLUMAGE SANS DISTRIBUTEUR
ALLUMAGE : 10° AVANT PMH.SINON,V.MAN.DE RÉPARATION.

HUILE PRÉCONISÉE : *SAE* 10W-30, CLASSE *API* « SG » - « ÉCONOMIE D'ÉNERGIE II »

	SPARK PLUG / *BOUGIES*	GAP / *ÉLECTRODES*
4.0 L	AWSF-42P	.052-.056

CALIBRATION PARTS LIST — CODE: DLC

NAME/DESCRIPTION	ENGINEERING NO.	SERVICE NO.
SENSOR ASSY, Exhaust Gas Oxygen	F07F 9F472-CA, DA	F0TZ 9F472-C
INJECTOR ASSY, Fuel	90TF 9F593-AA	F0TZ 9F593-B
VALVE ASSY, Throttle Air By-Pass	E9TE 9F715-BA F0TE 9F715-B1A	E9TZ 9F715-B
SENSOR ASSY, Speed	E3AF 9E731-AB	E3AZ 9E731-A
REGULATOR ASSY, Fuel Charging Pressure	90TF 9C968-AA	F0TZ 9C968-A
POTENTIOMETER ASSY, Throttle Position Sensor	F07F 9B989-BA	F07Z 9B989-B
SENSOR ASSY, Mass Airflow	F07F 12B579-AA	F0TZ 12B579-A
SENSOR ASSY, Engine Control Barometric Pressure	E7EF 12A644-A1A, A2A	E7FZ 12A644-A
SENSOR ASSY, Engine Electronic Control Coolant Temperature	E4AF 12A648-AA	E1AZ 12A648-A
PROCESSOR AND CALIBRATOR ASSY, EEC-IV	F07F 12A650-SA	F0TZ 12A650-SA
SENSOR ASSY, Air Charge Temperature	E4AF 12A697-AA	E1AZ 12A697-A

CALIBRATION: 0-58J-R00 — 4.0L EFI

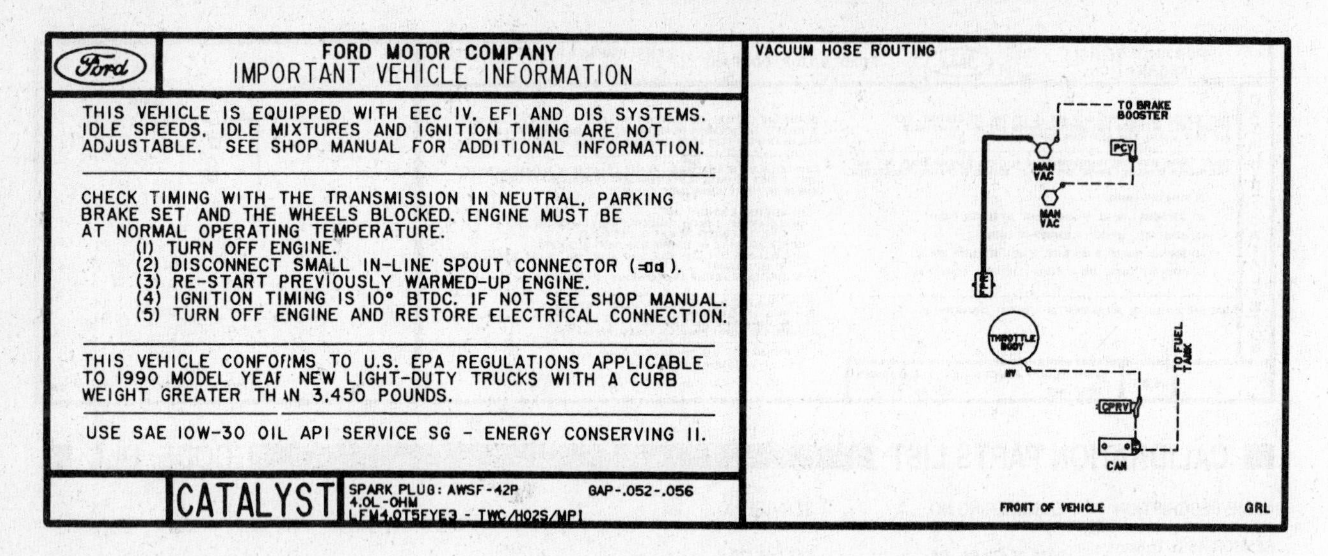
FORD MOTOR COMPANY
IMPORTANT VEHICLE INFORMATION

THIS VEHICLE IS EQUIPPED WITH EEC IV, EFI AND DIS SYSTEMS. IDLE SPEEDS, IDLE MIXTURES AND IGNITION TIMING ARE NOT ADJUSTABLE. SEE SHOP MANUAL FOR ADDITIONAL INFORMATION.

CHECK TIMING WITH THE TRANSMISSION IN NEUTRAL, PARKING BRAKE SET AND THE WHEELS BLOCKED. ENGINE MUST BE AT NORMAL OPERATING TEMPERATURE.
(1) TURN OFF ENGINE.
(2) DISCONNECT SMALL IN-LINE SPOUT CONNECTOR (=0ɑ).
(3) RE-START PREVIOUSLY WARMED-UP ENGINE.
(4) IGNITION TIMING IS 10° BTDC. IF NOT SEE SHOP MANUAL.
(5) TURN OFF ENGINE AND RESTORE ELECTRICAL CONNECTION.

THIS VEHICLE CONFORMS TO U.S. EPA REGULATIONS APPLICABLE TO 1990 MODEL YEAR NEW LIGHT-DUTY TRUCKS WITH A CURB WEIGHT GREATER THAN 3,450 POUNDS.

USE SAE 10W-30 OIL API SERVICE SG - ENERGY CONSERVING II.

CATALYST
SPARK PLUG: AWSF-42P GAP-.052-.056
4.0L-OHM
LFM4.0T5FYE3 - TWC/HO2S/MPI

VACUUM HOSE ROUTING

CALIBRATION PARTS LIST — CODE: DKZ

NAME/DESCRIPTION	ENGINEERING NO.	SERVICE NO.
SENSOR ASSY, Exhaust Gas Oxygen	F07F 9F472-CA, DA	F0TZ 9F472-C
INJECTOR ASSY, Fuel	90TF 9F593-AA	F0TZ 9F593-B
VALVE ASSY, Throttle Air By-Pass	E9TE 9F715-BA, F0TE 9F715-B1A	E9TZ 9F715-B
SENSOR ASSY, Speed	E3AF 9E731-AB	E3AZ 9E731-A
REGULATOR ASSY, Fuel Charging Pressure	90TF 9C968-AA	F0TZ 9C968-A
POTENTIOMETER ASSY, Throttle Position Sensor	F07F 9B989-BA	F07Z 9B989-B
SENSOR ASSY, Mass Airflow	F07F 12B579-AA	F0TZ 12B579-A
SENSOR ASSY, Engine Control Barometric Pressure	E7EF 12A644-A1A, A2A	E7FZ 12A644-A
SENSOR ASSY, Engine Electronic Control Coolant Temperature	E4AF 12A648-AA	E1AZ 12A648-A
PROCESSOR AND CALIBRATOR ASSY, EEC-IV	F07F 12A650-R1A	F0TZ 12A650-RAA
SENSOR ASSY, Air Charge Temperature	E4AF 12A697-AA	E1AZ 12A697-A

CALIBRATION: 0-58J-R00 — 4.0L EFI

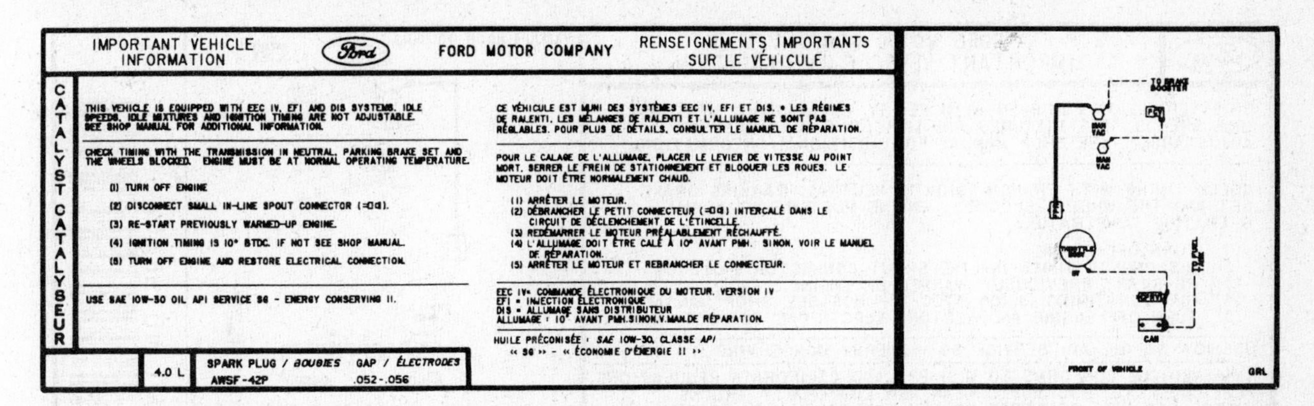
IMPORTANT VEHICLE INFORMATION — FORD MOTOR COMPANY — RENSEIGNEMENTS IMPORTANTS SUR LE VÉHICULE

CATALYST CATALYSEUR

THIS VEHICLE IS EQUIPPED WITH EEC IV, EFI AND DIS SYSTEMS. IDLE SPEEDS, IDLE MIXTURES AND IGNITION TIMING ARE NOT ADJUSTABLE. SEE SHOP MANUAL FOR ADDITIONAL INFORMATION.

CHECK TIMING WITH THE TRANSMISSION IN NEUTRAL, PARKING BRAKE SET AND THE WHEELS BLOCKED. ENGINE MUST BE AT NORMAL OPERATING TEMPERATURE.

(1) TURN OFF ENGINE
(2) DISCONNECT SMALL IN-LINE SPOUT CONNECTOR (=⊐⊲).
(3) RE-START PREVIOUSLY WARMED-UP ENGINE.
(4) IGNITION TIMING IS 10° BTDC. IF NOT SEE SHOP MANUAL.
(5) TURN OFF ENGINE AND RESTORE ELECTRICAL CONNECTION.

USE SAE 10W-30 OIL API SERVICE SG - ENERGY CONSERVING II.

CE VÉHICULE EST MUNI DES SYSTÈMES EEC IV, EFI ET DIS. • LES RÉGIMES DE RALENTI, LES MÉLANGES DE RALENTI ET L'ALLUMAGE NE SONT PAS RÉGLABLES. POUR PLUS DE DÉTAILS, CONSULTER LE MANUEL DE RÉPARATION.

POUR LE CALAGE DE L'ALLUMAGE, PLACER LE LEVIER DE VITESSE AU POINT MORT, SERRER LE FREIN DE STATIONNEMENT ET BLOQUER LES ROUES. LE MOTEUR DOIT ÊTRE NORMALEMENT CHAUD.

(1) ARRÊTER LE MOTEUR.
(2) DÉBRANCHER LE PETIT CONNECTEUR (=⊐⊲) INTERCALÉ DANS LE CIRCUIT DE DÉCLENCHEMENT DE L'ÉTINCELLE.
(3) REDÉMARRER LE MOTEUR PRÉALABLEMENT RÉCHAUFFÉ.
(4) L'ALLUMAGE DOIT ÊTRE CALÉ À 10° AVANT PMH. SINON, VOIR LE MANUEL DE RÉPARATION.
(5) ARRÊTER LE MOTEUR ET REBRANCHER LE CONNECTEUR.

EEC IV= COMMANDE ÉLECTRONIQUE DU MOTEUR, VERSION IV
EFI = INJECTION ÉLECTRONIQUE
DIS = ALLUMAGE SANS DISTRIBUTEUR
ALLUMAGE : 10° AVANT PMH.SINON,V.MAN.DE RÉPARATION.

HUILE PRÉCONISÉE : SAE 10W-30, CLASSE API « SG » - « ÉCONOMIE D'ÉNERGIE II »

	SPARK PLUG / BOUGIES	GAP / ÉLECTRODES
4.0 L	AWSF-42P	.052-.056

CALIBRATION PARTS LIST — CODE: DLC

NAME/DESCRIPTION	ENGINEERING NO.	SERVICE NO.
SENSOR ASSY, Exhaust Gas Oxygen	F07F 9F472-CA, DA	F0TZ 9F472-C
INJECTOR ASSY, Fuel	90TF 9F693-AA	F0TZ 9F593-B
VALVE ASSY, Throttle Air By-Pass	E9TE 9F715-BA F0TE 9F715-B1A	E9TZ 9F715-B
SENSOR ASSY, Speed	E3AF 9E731-AB	E3AZ 9E731-A
REGULATOR ASSY, Fuel Charging Pressure	90TF 9C968-AA	F0TZ 9C968-A
POTENTIOMETER ASSY, Throttle Position Sensor	F07F 9B989-BA	F07Z 9B989-B
SENSOR ASSY, Mass Airflow	F07F 12B579-AA	F0TZ 12B579-A
SENSOR ASSY, Engine Control Barometric Pressure	E7EF 12A644-A1A, A2A	E7FZ 12A644-A
SENSOR ASSY, Engine Electronic Control Coolant Temperature	E4AF 12A648-AA	E1AZ 12A648-A
PROCESSOR AND CALIBRATOR ASSY, EEC-IV	F07F 12A650-R1A	F0TZ 12A650-RAA
SENSOR ASSY, Air Charge Temperature	E4AF 12A697-AA	E1AZ 12A697-A

CALIBRATION: 0-58L-R00 — 4.0L EFI

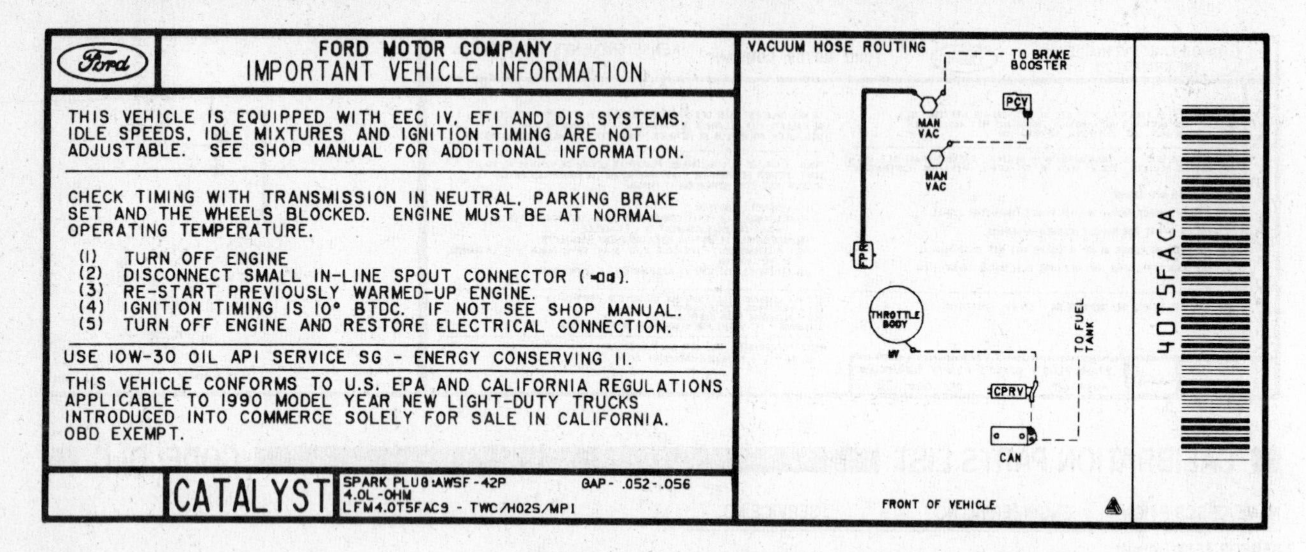

FORD MOTOR COMPANY
IMPORTANT VEHICLE INFORMATION

THIS VEHICLE IS EQUIPPED WITH EEC IV, EFI AND DIS SYSTEMS. IDLE SPEEDS, IDLE MIXTURES AND IGNITION TIMING ARE NOT ADJUSTABLE. SEE SHOP MANUAL FOR ADDITIONAL INFORMATION.

CHECK TIMING WITH TRANSMISSION IN NEUTRAL, PARKING BRAKE SET AND THE WHEELS BLOCKED. ENGINE MUST BE AT NORMAL OPERATING TEMPERATURE.

(1) TURN OFF ENGINE
(2) DISCONNECT SMALL IN-LINE SPOUT CONNECTOR (-dd).
(3) RE-START PREVIOUSLY WARMED-UP ENGINE.
(4) IGNITION TIMING IS 10° BTDC. IF NOT SEE SHOP MANUAL.
(5) TURN OFF ENGINE AND RESTORE ELECTRICAL CONNECTION.

USE 10W-30 OIL API SERVICE SG - ENERGY CONSERVING II.

THIS VEHICLE CONFORMS TO U.S. EPA AND CALIFORNIA REGULATIONS APPLICABLE TO 1990 MODEL YEAR NEW LIGHT-DUTY TRUCKS INTRODUCED INTO COMMERCE SOLELY FOR SALE IN CALIFORNIA. OBD EXEMPT.

CATALYST
SPARK PLUG:AWSF-42P GAP-.052-.056
4.0L-OHM
LFM4.0T5FAC9 - TWC/HO2S/MPI

CALIBRATION PARTS LIST — CODE: DKT

NAME/DESCRIPTION	ENGINEERING NO.	SERVICE NO.
SENSOR ASSY, Exhaust Gas Oxygen	F07F 9F472-CA, DA	F0TZ 9F472-C
INJECTOR ASSY, Fuel	90TF 9F593-AA	F0TZ 9F593-B
VALVE ASSY, Throttle Air By-Pass	E9TE 9F715-BA, F0TE 9F715-B1A	E9TZ 9F715-B
SENSOR ASSY, Speed	E3AF 9E731-AB	E3AZ 9E731-A
REGULATOR ASSY, Fuel Charging Pressure	90TF 9C968-AA	F0TZ 9C968-A
POTENTIOMETER ASSY, Throttle Position Sensor	F07F 9B989-BA	F07Z 9B989-B
SENSOR ASSY, Mass Airflow	F0TZ 12B579-AA	F0TZ 12B579-A
SENSOR ASSY, Engine Electronic Control Barometric Pressure	E7EF 12A644-A1A, A2A	E7FZ 12A644-A
SENSOR ASSY, Engine Electronic Control Coolant Temperature	E4AF 12A648-AA	E1AZ 12A648-A
PROCESSOR AND CALIBRATOR ASSY, EEC-IV	F07F 12A650-T1A	F0TZ 12A650-TAA
SENSOR ASSY, Air Charge Temperature	E4AF 12A697-AA	E1AZ 12A697-A

CALIBRATION: 0-58Q-R00 — 4.0L EFI

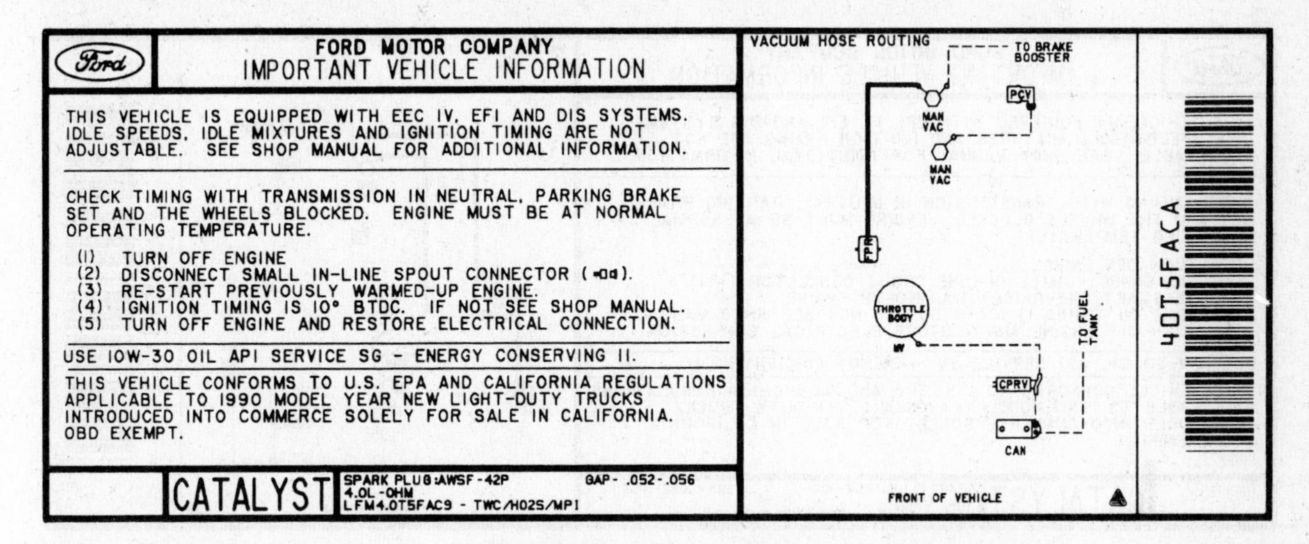
FORD MOTOR COMPANY
IMPORTANT VEHICLE INFORMATION

THIS VEHICLE IS EQUIPPED WITH EEC IV, EFI AND DIS SYSTEMS. IDLE SPEEDS, IDLE MIXTURES AND IGNITION TIMING ARE NOT ADJUSTABLE. SEE SHOP MANUAL FOR ADDITIONAL INFORMATION.

CHECK TIMING WITH TRANSMISSION IN NEUTRAL, PARKING BRAKE SET AND THE WHEELS BLOCKED. ENGINE MUST BE AT NORMAL OPERATING TEMPERATURE.

(1) TURN OFF ENGINE
(2) DISCONNECT SMALL IN-LINE SPOUT CONNECTOR ( ).
(3) RE-START PREVIOUSLY WARMED-UP ENGINE.
(4) IGNITION TIMING IS 10° BTDC. IF NOT SEE SHOP MANUAL.
(5) TURN OFF ENGINE AND RESTORE ELECTRICAL CONNECTION.

USE 10W-30 OIL API SERVICE SG - ENERGY CONSERVING II.

THIS VEHICLE CONFORMS TO U.S. EPA AND CALIFORNIA REGULATIONS APPLICABLE TO 1990 MODEL YEAR NEW LIGHT-DUTY TRUCKS INTRODUCED INTO COMMERCE SOLELY FOR SALE IN CALIFORNIA. OBD EXEMPT.

CATALYST

SPARK PLUG:AWSF-42P GAP-.052-.056
4.0L-OHM
LFM4.0T5FAC9 - TWC/HO2S/MPI

VACUUM HOSE ROUTING

CALIBRATION PARTS LIST — CODE: DKT

NAME/DESCRIPTION	ENGINEERING NO.	SERVICE NO.
SENSOR ASSY, Exhaust Gas Oxygen	F07F 9F472-CA, DA	F0TZ 9F472-C
INJECTOR ASSY, Fuel	90TF 9F593-AA	F0TZ 9F593-B
VALVE ASSY, Throttle Air By-Pass	E9TE 9F715-BA, F0TE 9F715-B1A	E9TZ 9F715-B
SENSOR ASSY, Speed	E3AF 9E731-AB	E3AZ 9E731-A
REGULATOR ASSY, Fuel Charging Pressure	90TF 9C968-AA	F0TZ 9C968-A
POTENTIOMETER ASSY, Throttle Position Sensor	F07F 9B989-BA	F07Z 9B989-B
SENSOR ASSY, Mass Airflow	F07F 12B579-AA	F0TZ 12B579-A
SENSOR ASSY, Engine Control Barometric Pressure	E7EF 12A644-A1A, A2A	E7FZ 12A644-A
SENSOR ASSY, Engine Electronic Control Coolant Temperature	E4AF 12A648-AA	E1AZ 12A648-A
PROCESSOR AND CALIBRATOR ASSY, EEC-IV	F07F 12A650-U1A	F0TZ 12A650-UAA
SENSOR ASSY, Air Charge Temperature	E4AF 12A697-AA	E1AZ 12A697-A

CALIBRATION: 0-58S-R00 — 4.0L EFI

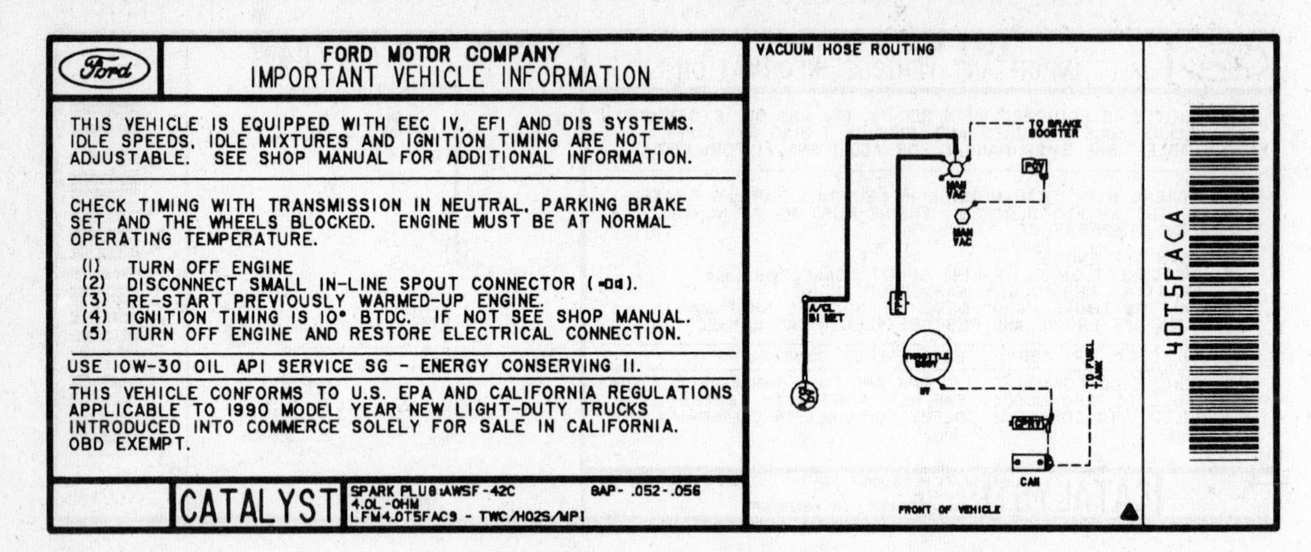
Ford

FORD MOTOR COMPANY
IMPORTANT VEHICLE INFORMATION

THIS VEHICLE IS EQUIPPED WITH EEC IV, EFI AND DIS SYSTEMS. IDLE SPEEDS, IDLE MIXTURES AND IGNITION TIMING ARE NOT ADJUSTABLE. SEE SHOP MANUAL FOR ADDITIONAL INFORMATION.

CHECK TIMING WITH TRANSMISSION IN NEUTRAL, PARKING BRAKE SET AND THE WHEELS BLOCKED. ENGINE MUST BE AT NORMAL OPERATING TEMPERATURE.

(1) TURN OFF ENGINE
(2) DISCONNECT SMALL IN-LINE SPOUT CONNECTOR (=◘).
(3) RE-START PREVIOUSLY WARMED-UP ENGINE.
(4) IGNITION TIMING IS 10° BTDC. IF NOT SEE SHOP MANUAL.
(5) TURN OFF ENGINE AND RESTORE ELECTRICAL CONNECTION.

USE 10W-30 OIL API SERVICE SG - ENERGY CONSERVING II.

THIS VEHICLE CONFORMS TO U.S. EPA AND CALIFORNIA REGULATIONS APPLICABLE TO 1990 MODEL YEAR NEW LIGHT-DUTY TRUCKS INTRODUCED INTO COMMERCE SOLELY FOR SALE IN CALIFORNIA. OBD EXEMPT.

CATALYST

SPARK PLUG:AWSF-42C GAP-.052-.056
4.0L-OHM
LFM4.0T5FAC9 - TWC/HO2S/MPI

CALIBRATION PARTS LIST — CODE: DKU

NAME/DESCRIPTION	ENGINEERING NO.	SERVICE NO.
SENSOR ASSY, Exhaust Gas Oxygen	F07F 9F472-CA, DA	F0TZ 9F472-C
INJECTOR ASSY, Fuel	90TF 9F593-AA	F0TZ 9F593-B
SENSOR ASSY, Carburetor Air Cleaner Air Temperature Control		D7FZ 9E607-A
MOTOR ASSY, Carburetor Air Cleaner Vacuum		D4FZ 9D612-A
VALVE ASSY, Throttle Air By-Pass	E9TE 9F715-BA F0TE 9F715-B1A	E9TZ 9F715-B
SENSOR ASSY, Speed	E3AF 9E731-AB	E3AZ 9E731-A
REGULATOR ASSY, Fuel Charging Pressure	90TF 9C968-AA	F0TZ 9C968-A
POTENTIOMETER ASSY, Throttle Position Sensor	F07F 9B989-BA	F07Z 9B989-B
SENSOR ASSY, Mass Airflow	F07F 12B579-AA	F0TZ 12B579-A
SENSOR ASSY, Engine Control Barometric Pressure	E7EF 12A644-A1A, A2A	E7FZ 12A644-A
SENSOR ASSY, Engine Electronic Control Coolant Temperature	E4AF 12A648-AA	E1AZ 12A648-A
PROCESSOR AND CALIBRATOR ASSY, EEC-IV	F07F 12A650-N1A	F0TZ 12A650-NAA
SENSOR ASSY, Air Charge Temperature	E4AF 12A697-AA	E1AZ 12A697-A

■ CALIBRATION: 0-58S-R00 ■ 4.0L EFI ■

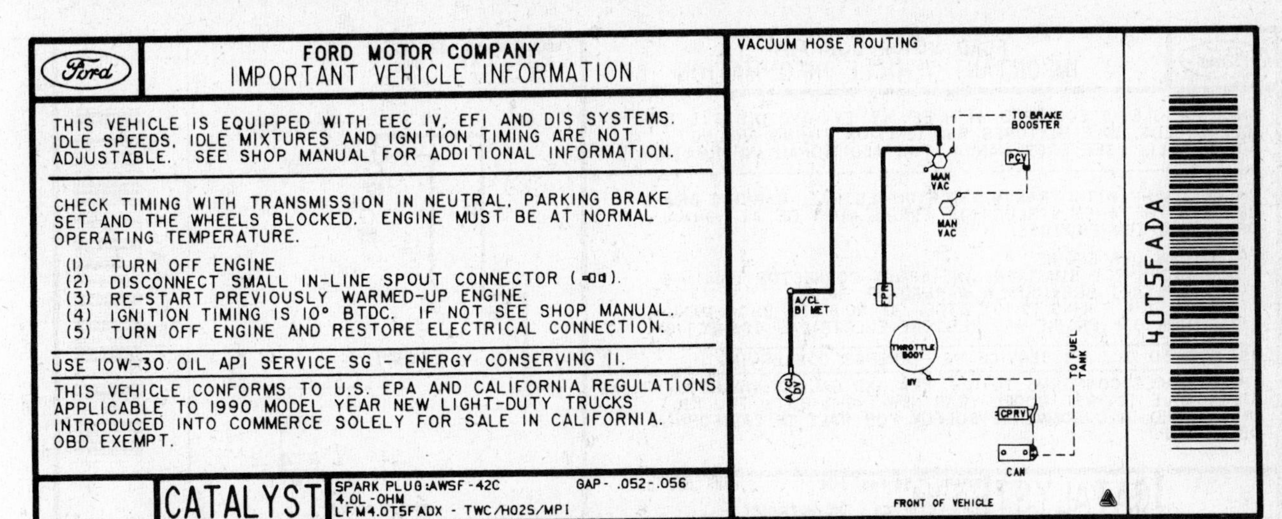
Ford

FORD MOTOR COMPANY
IMPORTANT VEHICLE INFORMATION

THIS VEHICLE IS EQUIPPED WITH EEC IV, EFI AND DIS SYSTEMS. IDLE SPEEDS, IDLE MIXTURES AND IGNITION TIMING ARE NOT ADJUSTABLE. SEE SHOP MANUAL FOR ADDITIONAL INFORMATION.

CHECK TIMING WITH TRANSMISSION IN NEUTRAL, PARKING BRAKE SET AND THE WHEELS BLOCKED. ENGINE MUST BE AT NORMAL OPERATING TEMPERATURE.

(1) TURN OFF ENGINE
(2) DISCONNECT SMALL IN-LINE SPOUT CONNECTOR ( ).
(3) RE-START PREVIOUSLY WARMED-UP ENGINE.
(4) IGNITION TIMING IS 10° BTDC. IF NOT SEE SHOP MANUAL.
(5) TURN OFF ENGINE AND RESTORE ELECTRICAL CONNECTION.

USE 10W-30 OIL API SERVICE SG - ENERGY CONSERVING II.

THIS VEHICLE CONFORMS TO U.S. EPA AND CALIFORNIA REGULATIONS APPLICABLE TO 1990 MODEL YEAR NEW LIGHT-DUTY TRUCKS INTRODUCED INTO COMMERCE SOLELY FOR SALE IN CALIFORNIA. OBD EXEMPT.

CATALYST

SPARK PLUG:AWSF-42C GAP-.052-.056
4.0L-OHM
LFM4.0T5FADX - TWC/HO2S/MPI

■ CALIBRATION PARTS LIST ■ CODE: DKY ■

NAME/DESCRIPTION	ENGINEERING NO.	SERVICE NO.
SENSOR ASSY, Exhaust Gas Oxygen	F07F 9F472-CA, DA	F0TZ 9F472-C
INJECTOR ASSY, Fuel	90TF 9F593-AA	F0TZ 9F593-B
SENSOR ASSY, Carburetor Air Cleaner Air Temperature Control		D7FZ 9E607-A
MOTOR ASSY, Carburetor Air Cleaner Vacuum		D4FZ 9D612-A
VALVE ASSY, Throttle Air By-Pass	E9TE 9F715-BA F0TE 9F715-B1A	E9TZ 9F715-B
SENSOR ASSY, Speed	E3AF 9E731-AB	E3AZ 9E731-A
REGULATOR ASSY, Fuel Charging Pressure	90TF 9C968-AA	F0TZ 9C968-A
POTENTIOMETER ASSY, Throttle Position Sensor	F07F 9B989-BA	F07Z 9B989-B
SENSOR ASSY, Mass Airflow	F07F 12B579-AA	F0TZ 12B579-A
SENSOR ASSY, Engine Control Barometric Pressure	E7EF 12A644-A1A, A2A	E7FZ 12A644-A
SENSOR ASSY, Engine Electronic Control Coolant Temperature	E4AF 12A648-AA	E1AZ 12A648-A
PROCESSOR AND CALIBRATOR ASSY, EEC-IV	F07F 12A650-N1A	F0TZ 12A650-NAA
SENSOR ASSY, Air Charge Temperature	E4AF 12A697-AA	E1AZ 12A697-A

CALIBRATION: 0-58T-R00 — 4.0L EFI

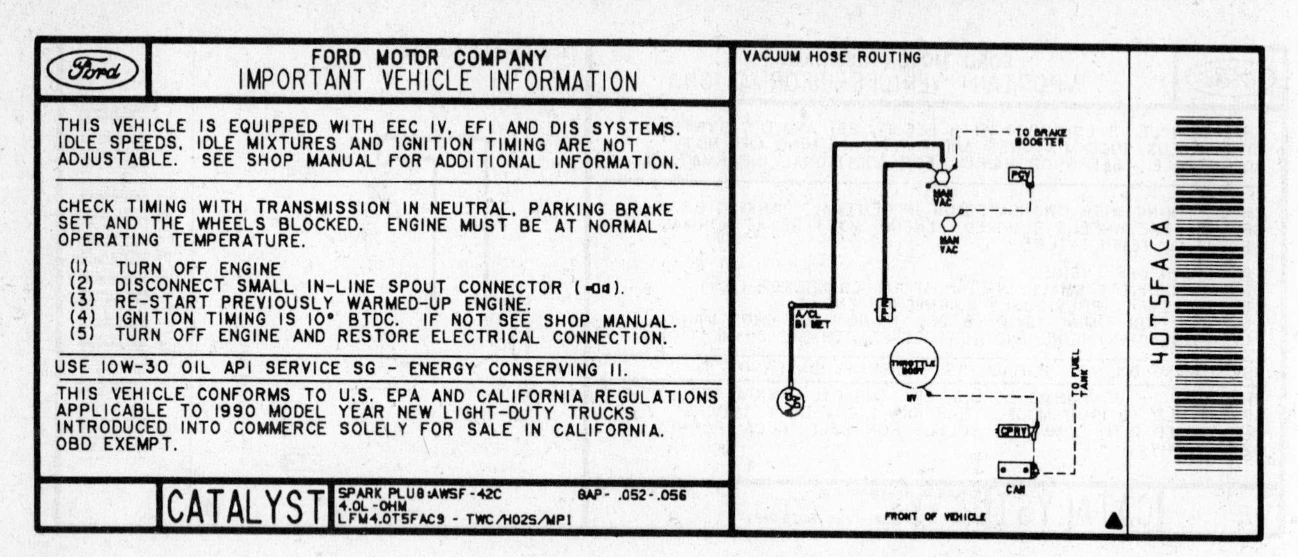

Ford

FORD MOTOR COMPANY
IMPORTANT VEHICLE INFORMATION

THIS VEHICLE IS EQUIPPED WITH EEC IV, EFI AND DIS SYSTEMS. IDLE SPEEDS, IDLE MIXTURES AND IGNITION TIMING ARE NOT ADJUSTABLE. SEE SHOP MANUAL FOR ADDITIONAL INFORMATION.

CHECK TIMING WITH TRANSMISSION IN NEUTRAL, PARKING BRAKE SET AND THE WHEELS BLOCKED. ENGINE MUST BE AT NORMAL OPERATING TEMPERATURE.

(1) TURN OFF ENGINE
(2) DISCONNECT SMALL IN-LINE SPOUT CONNECTOR (-◻◻).
(3) RE-START PREVIOUSLY WARMED-UP ENGINE.
(4) IGNITION TIMING IS 10° BTDC. IF NOT SEE SHOP MANUAL.
(5) TURN OFF ENGINE AND RESTORE ELECTRICAL CONNECTION.

USE 10W-30 OIL API SERVICE SG - ENERGY CONSERVING II.

THIS VEHICLE CONFORMS TO U.S. EPA AND CALIFORNIA REGULATIONS APPLICABLE TO 1990 MODEL YEAR NEW LIGHT-DUTY TRUCKS INTRODUCED INTO COMMERCE SOLELY FOR SALE IN CALIFORNIA. OBD EXEMPT.

CATALYST

SPARK PLUG :AWSF-42C GAP- .052-.056
4.0L-OHM
LFM4.0T5FAC9 - TWC/HO2S/MPI

VACUUM HOSE ROUTING

CALIBRATION PARTS LIST — CODE: DKU

NAME/DESCRIPTION	ENGINEERING NO.	SERVICE NO.
SENSOR ASSY, Exhaust Gas Oxygen	F07F 9F472-CA, DA	F0TZ 9F472-C
INJECTOR ASSY, Fuel	90TF 9F593-AA	F0TZ 9F593-B
SENSOR ASSY, Carburetor Air Cleaner Air Temperature Control		D7FZ 9E607-A
MOTOR ASSY, Carburetor Air Cleaner Vacuum		D4FZ 9D612-A
VALVE ASSY, Throttle Air By-Pass	E9TE 9F715-BA F0TE 9F715-B1A	E9TZ 9F715-B
SENSOR ASSY, Speed	E3AF 9E731-AB	E3AZ 9E731-A
REGULATOR ASSY, Fuel Charging Pressure	90TF 9C968-AA	F0TZ 9C968-A
POTENTIOMETER ASSY, Throttle Position Sensor	F07F 9B989-BA	F07Z 9B989-B
SENSOR ASSY, Mass Airflow	F07F 12B579-AA	F0TZ 12B579-A
SENSOR ASSY, Engine Control Barometric Pressure	E7EF 12A644-A1A, A2A	E7FZ 12A644-A
SENSOR ASSY, Engine Electronic Control Coolant Temperature	E4AF 12A648-AA	E1AZ 12A648-A
PROCESSOR AND CALIBRATOR ASSY, EEC-IV	F07F 12A650-P1A	F0TZ 12A650-PAA
SENSOR ASSY, Air Charge Temperature	E4AF 12A697-AA	E1AZ 12A697-A

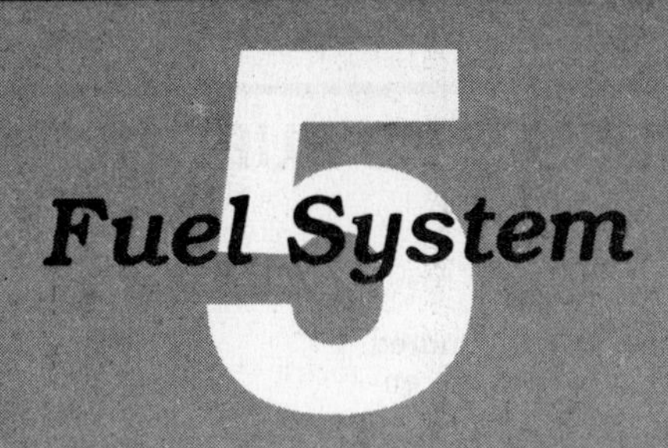

5 Fuel System

QUICK REFERENCE INDEX

GENERAL INDEX

CARBURETED FUEL SYSTEM

Mechanical Fuel Pump

The carbureted engines use a camshaft eccentric-actuated combination fuel pump located on the lower left side of the engine block.

REMOVAL

1. Disconnect the fuel inlet and outlet lines at the fuel pump. Discard the fuel inlet retaining clamp.
2. Remove the pump retaining bolts then remove the pump assembly and gasket from the engine. Discard the gasket.

INSTALLATION

1. If a new pump is to be installed, remove the fuel line connector fitting from the old pump and install it in the new pump (if so equipped).
2. Remove all gasket material from the mounting pad and pump flange. Apply oil resistant sealer to both sides of a new gasket.
3. Position the new gasket on the pump flange and hold the pump in position against the mounting pad. Make sure that the rocker arm is riding on the camshaft eccentric.
4. Press the pump tight against the pad, install the retaining bolts and alternately torque them to 14–21 ft. lbs. Connect the fuel lines. Use a new clamp on the fuel inlet line.
5. Operate the engine and check for leaks.

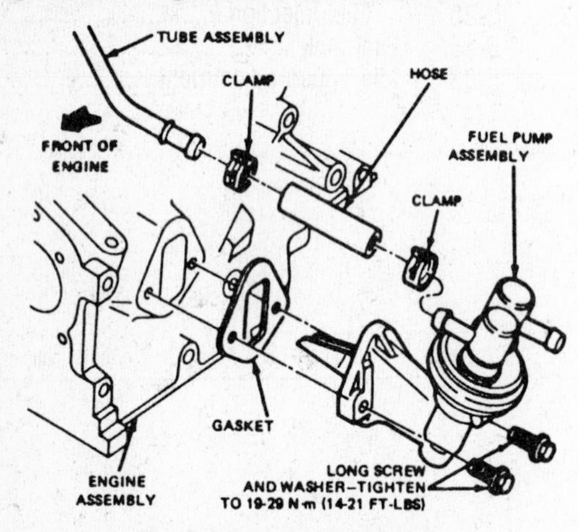

Fuel pump installation

TESTING

Incorrect fuel pump pressure and low volume (flow rate) are the two most likely fuel pump troubles that will affect engine performance. Low pressure will cause a lean mixture and fuel starvation at high speeds and excessive pressure will cause high fuel consumption and carburetor flooding.

To determine that the fuel pump is in satisfactory operating condition, tests for both fuel pump pressure and volume should be performed.

The tests are performed with the fuel pump installed on the engine and the engine at normal operating temperature and at idle speed.

Before the test, make sure that the replaceable fuel filter has been changed at the proper mileage interval. If in doubt, install a new filter.

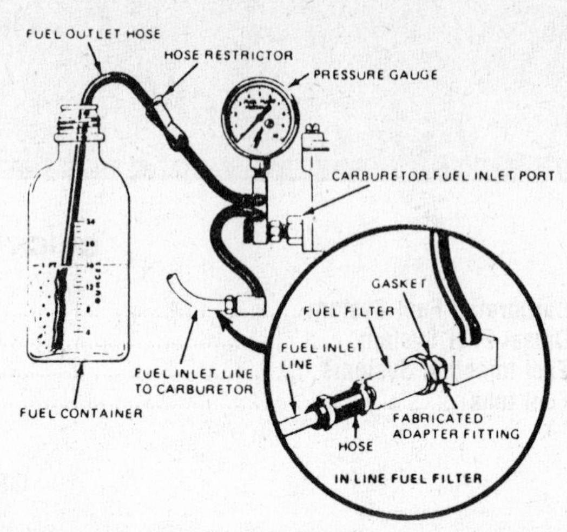

Fuel pump volume and pressure test equipment

Pressure Test

1. Remove the air cleaner assembly. Disconnect the fuel inlet line of the fuel filter at the carburetor. Use care to prevent fire, due to fuel spillage. Place an absorbent cloth under the connection before removing the line to catch any fuel that might flow out of the line.
2. Connect a pressure gauge, a restrictor and a flexible hose between the fuel filter and the carburetor.
3. Position the flexible hose and the restrictor so that the fuel can be discharged into a suitable, graduated container.
4. Before taking a pressure reading, operate the engine at the specified idle rpm and vent the system into the container by opening the hose restrictor momentarily.
5. Close the hose restrictor, allow the pressure to stabilize and note the reading.

If the pump pressure is not within 4.5–6.5 psi and the fuel lines and filter are in satisfactory condition, the pump is defective and should be replaced.

If the pump pressure is within the proper range, perform the test for fuel volume.

Volume Test

1. Operate the engine at the specified idle rpm.
2. Open the hose restrictor and catch the fuel in the container while observing the time it takes to pump 1 pint. It should take 30 seconds for 1 pint to be expelled. If the pump does not pump to specifications, check for proper fuel tank venting or a restriction in the fuel line leading from the fuel tank to the carburetor before replacing the fuel pump.

Carburetor

The carburetor identification tag is attached to the carburetor. To obtain replacement parts, it is necessary to know the part number prefix, suffix and, in some cases, the design change code. If the carburetor is ever replaced by a new unit, make sure that the identification tag stays with the new carburetor and the vehicle.

REMOVAL AND INSTALLATION

1. Remove the air cleaner.
2. Remove the throttle cable and transmission linkage from the throttle lever. Disconnect all vacuum lines, emission hoses, the fuel line and electrical connections.

3. Remove the carburetor retaining nuts then remove the carburetor. Remove the carburetor mounting gasket, spacer (if so equipped), and the lower gasket from the intake manifold.
4. Before installing the carburetor, clean the gasket mounting surfaces of the spacer and carburetor. Place the spacer between two new gaskets and position the spacer and the gaskets on the intake manifold. Position the carburetor on the spacer and gasket and secure it with the retaining nuts. To prevent leakage, distortion or damage to the carburetor body flange, snug the nuts, then alternately tighten each nut in a criss-cross pattern.
5. Connect the inline fuel line, throttle cable, transmission linkage and all electrical connections and vacuum lines on the carburetor.
6. Adjust the engine idle speed, the idle fuel mixture and install the air cleaner.

FLOAT AND FUEL LEVEL ADJUSTMENTS

Aisan Model Y Feedback 1-bbl (Dry Adjustment)

Stabilize engine temperature. With vehicle parked on a level surface and running at curb idle, check that fuel level is within the limits on sight glass as shown, If not, proceed as follows.
1. Remove carburetor air horn from carburetor. Remove and discard air horn gasket.
2. Remove power valve piston and spring.
3. Invert air horn assembly. Using a drill of the specified diameter, check the clearance between the top of the float and bottom surface of air horn.

NOTE: The float lever should be resting on needle pin when checking clearance.

4. If required, bend float air as shown to adjust float level.

CAUTION

Do not load the needle when adjusting the float. Also, do not bend the tab at the end of the float arm. This tab prevents the float from striking the bottom of the fuel bowl when empty. Refer to float drop adjustments.

5. Install power valve piston and spring.
6. install air horn on carburetor, using a new air horn gasket.
7. Install carburetor on engine.
8. check ISC motor maximum extension rpm, adjust as necessary.

Carter Model YFA & YFA Feedback 1-bbl (Dry Adjustment)

1. Remove the air cleaner.
2. Disconnect the choke heat tube at the carburetor air horn. Disconnect the fuel inlet line at the filter.

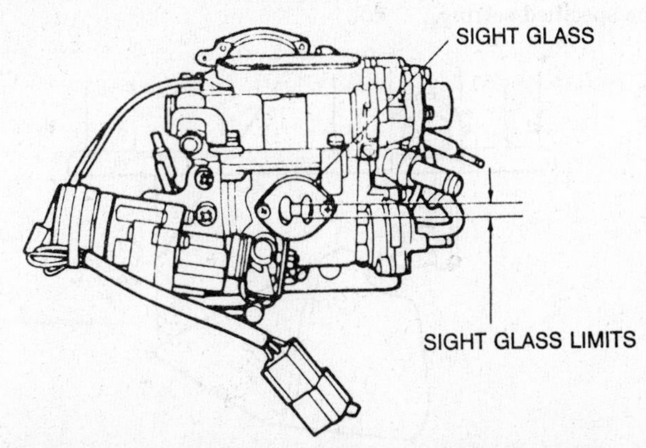

Float level sight glass check – Aisan model Y carburetor

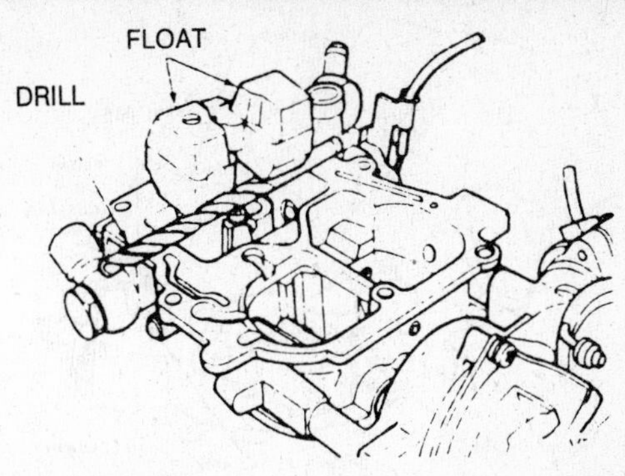

Float level adjustment – Aisan model Y carburetor

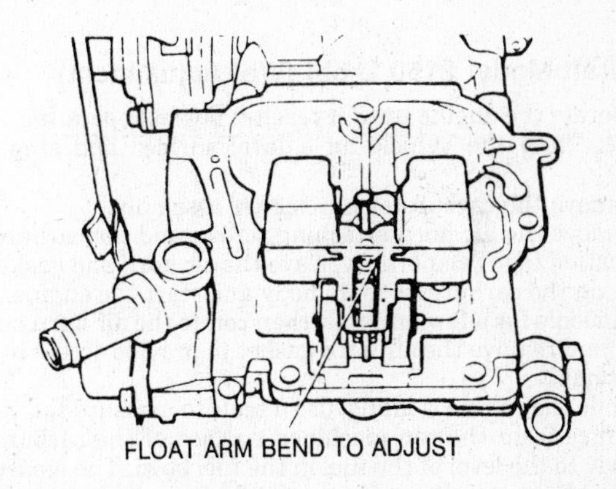

Float arm adjustment – Aisan model Y carburetor

3. Disconnect the electric choke wire at the connector.
4. Remove the wire clip retaining the link joining the fast idle choke lever to the fast idle cam and remove the link. Remove the air horn assembly attaching screws, dashpot and bracket assembly and air horn gasket. Discard the gasket.
5. Fabricate a float level gauge to the specified float level dimension. Refer to the Carburetor Specification chart for dimensions.
6. Invert the air horn assembly, and check the clearance from the float indentation on the top of the float to the bottom of the air horn with the float level gauge. Hold the air horn at eye level when gauging the float level. The float arm (lever) should be resting on the needle pin. Do not load the needle when adjusting the float. Bend the float arm as necessary to adjust the float level (clearance). Do not bend the tab at the end of the float arm. It prevents the float from striking the bottom of the fuel bowl when empty.
7. Install a new air horn to main body gasket. Make sure all holes in the new gasket have been properly punched and that no foreign material has adhered to the gasket. Install the air horn assembly, connect vent line to canister (if so equipped), and bracket assembly and air horn attaching screws and tighten to 27–37 in. lbs. Position the link and plastic bushing joining the fast idle cam to the fast idle choke lever and retain in place on the fast idle cam with the plastic bushing and wire clip. Make sure the mechanical fuel bowl vent rod is engaged with the forked actuating lever (if so equipped).
8. Connect the fuel inlet line to the fuel filter.
9. Connect the electric choke wire.
10. Install the air cleaner. Starter the engine and run it until normal operating temperature is reached. Adjust the idle fuel mixture and idle speed.

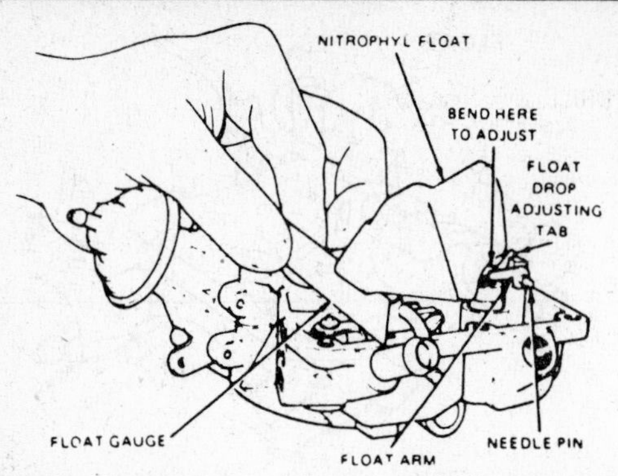

Float level adjustment – Carter YFA & YFA Feedback carburetor

Motorcraft Model 2150 2-bbl (Wet Adjustment)

1. Operate the engine until it reaches normal operating temperature. Place the vehicle on a level surface and stop the engine.
2. Remove the carburetor air cleaner assembly.
3. Remove the air horn attaching screws and the carburetor identification tag. Temporarily, leave the air horn and gasket in position on the carburetor main body and start the engine. Let the engine idle for a few minutes, then rotate the air horn out of the way and remove the air horn gasket to provide access to the float assembly.
4. While the engine is idling, use a scale to measure the vertical distance from the top machined surface of the carburetor main body to the level of the fuel in the fuel bowl. The measurement must be made at least ¼ in. away from any vertical surface to assure an accurate reading, because the surface of the fuel is concave, being higher at the edges than the center. Care must be exercised to measure the fuel level at the point of contact with the float.
5. If any adjustment is required, stop the engine to minimize the hazard of fire due to spilled gasoline. To adjust the fuel level, bend the float tab contacting the fuel inlet valve upward in relation to the original position to raise the fuel level, and downward to lower it. Each time the float is adjusted, the engine must be started and permitted to idle for a few minutes to stabilize the fuel level. Check the fuel level after each adjustment, until the specified level is obtained.
6. Assemble the carburetor in the reverse order of disassembly, using a new gasket between the air horn and the main carburetor body.

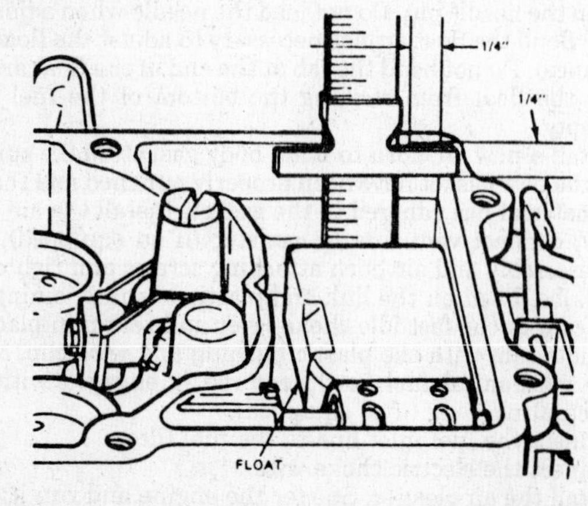

Float level adjustment – Motorcraft 2150A 2-bbl. carburetor

FLOAT DROP ADJUSTMENT

Aisan Model Y Feedback 1-bbl Carburetor

1. Remove carburetor air horn from carburetor. Remove and discard air horn gasket.
2. Hold air horn upright and let float hang free. Using vernier calipers, measure the maximum dimension from toe end of float to casting surface. Hold air horn at eye level when gauging the dimension.
3. The float drop dimension should be as shown in the Carburetor Chart. Adjust to specification by bending tab as shown.
4. Install air horn on carburetor, using a new air horn gasket.
5. Check ISC motor maximum extension rpm and adjust as necessary.

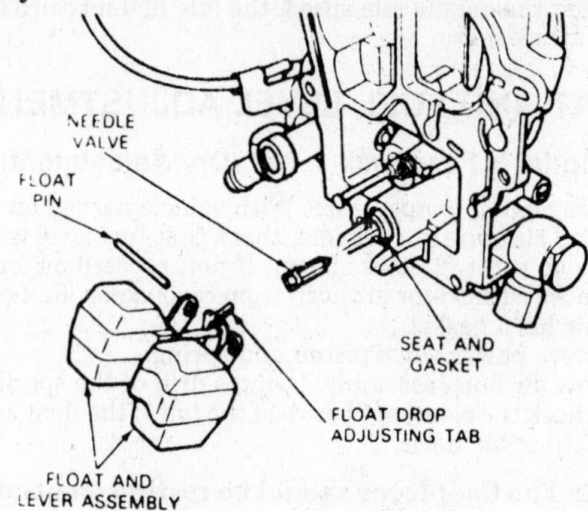

Float drop adjustment – Aisan model Y carburetor

Carter Model YFA & YFA Feedback 1-bbl

1. Remove the air cleaner.
2. Disconnect the choke heat tube at the carburetor air horn. Disconnect the fuel inlet line at the filter.
3. Disconnect the electric choke wire at the connector.
4. Remove the wire clip retaining the link joining the fast idle choke lever to the fast idle cam and remove the link. Remove the air horn assembly attaching screws, dashpot and bracket assembly and air horn gasket. Discard the gasket.
5. Fabricate a float drop gauge to the specified dimension 38mm minimum.
6. Hold the air horn upright and let the float hang free. Measure the maximum clearance from the foe end of the float to the casting surface. Hold the air horn at eye level when gauging the dimension.
7. To adjust, bend the tab at the end of the float arm to obtain the specified setting.

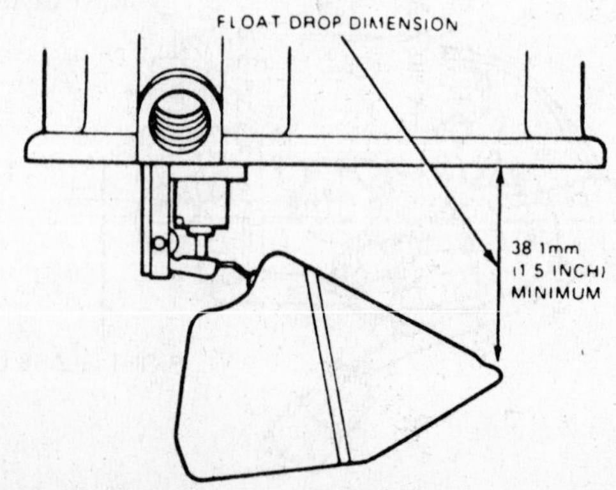

Float level adjustment – Carter YFA & YFA Feedback carburetor

8. Install a new air horn to main body gasket. Make sure all holes in the new gasket have been properly punched and that no foreign material has adhered to the gasket. Install the air horn assembly, connect vent line to canister (if so equipped, and bracket assembly and air horn attaching screws and tighten to 27–37 in. lbs. Position the link and plastic bushing joining the fast idle cam to the fast idle choke lever and retain in place on the fast idle cam with the plastic bushing and wire clip. Make sure the mechanical fuel bowl vent rod is engaged with the forked actuating lever (if so equipped).
9. Connect the fuel inlet line to the fuel filter.
10. Connect the electric choke wire.
11. Install the air cleaner. Start the engine and run it until normal operating temperature is reached. Adjust the idle fuel mixture and idle speed.

FAST IDLE SPEED ADJUSTMENT

Carter YFA & YFA Feedback 1-bbl

1. Place the transmission in Neutral or Park.
2. Bring the engine to normal operating temperature.
3. Turn the ignition key to the Off position.
4. Put the air conditioner selector in the Off position.
5. Disconnect the vacuum hose at the EGR valve and plug.
6. Place the fast idle RPM adjusting screw on the specified step of the fast idle cam.
7. Start the engine without touching the accelerator pedal: Check/adjust fast idle RPM to specification. Refer to the under hood sticker for specifications.
8. Rev the engine momentarily, allowing the engine to return to idle and turn the ignition key to the Off position.
9. Remove the plug from the EGR vacuum hose and reconnect it.

Motorcraft Model 2150A 2-bbl

1. Place the transmission in park or neutral.
2. Bring the engine to normal operating temperature.
3. Disconnect the plug the vacuum hose at the EGR and purge valves.

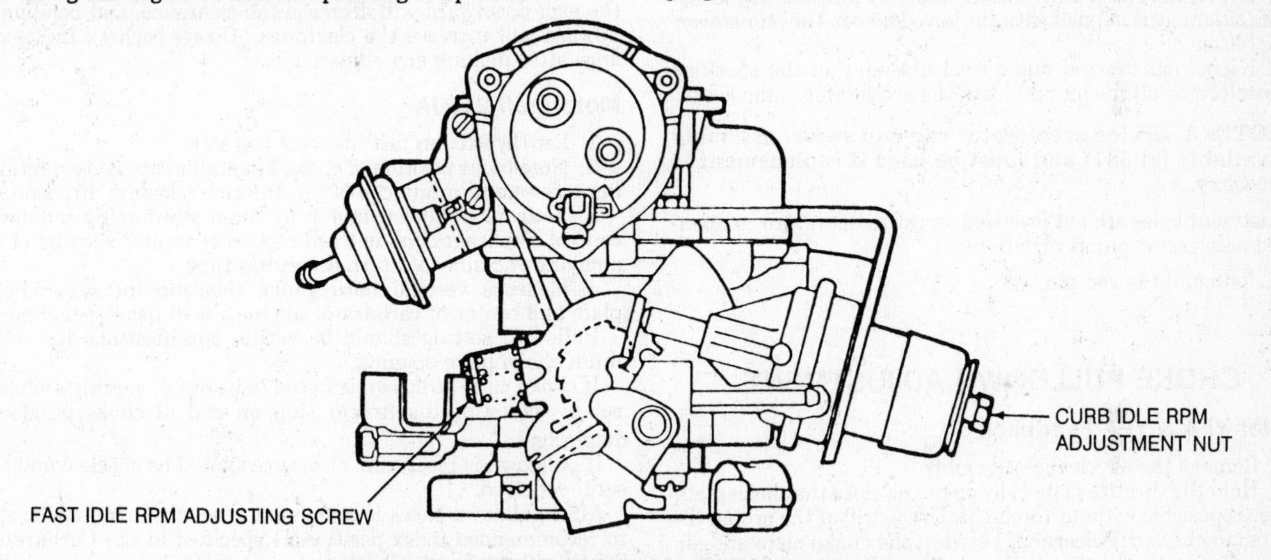

Carter YFA & YFA Feedback fast idle speed control

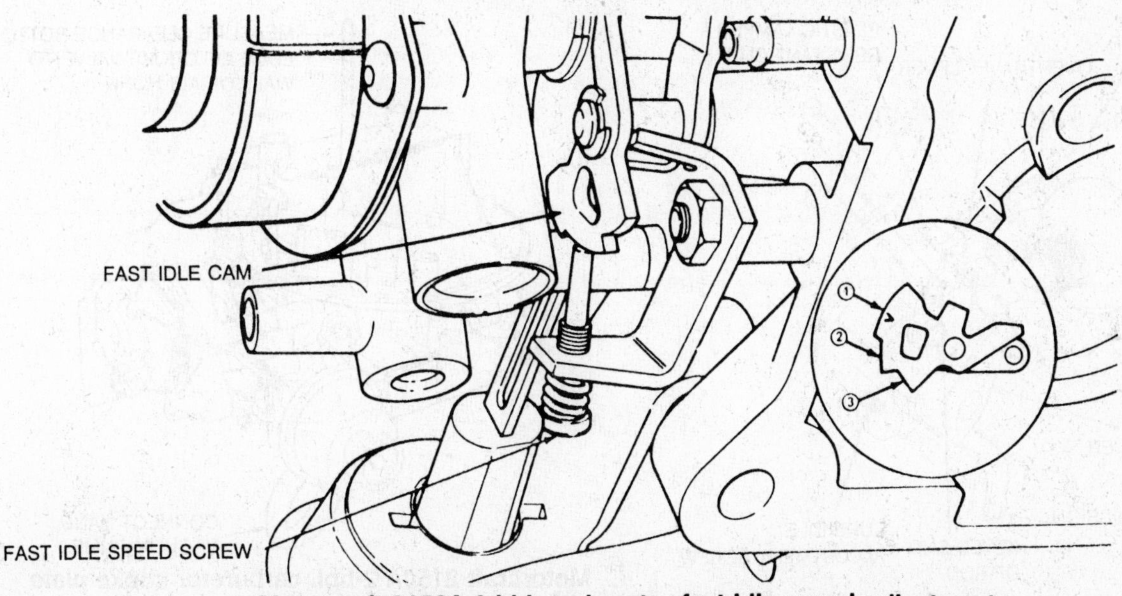

Motorcraft 2150A 2-bbl. carburetor fast idle speed adjustment

4. Place the fast idle lever on the **V** step of the fast idle cam.
5. Adjust the fast idle rpm to specifications.
6. Reconnect the EGR and purge vacuum hoses.

ACCELERATING PUMP STROKE ADJUSTMENT

Motorcraft Model 2150A 2-bbl

The accelerating pump stroke has been factory set for a particular engine application and should not be readjusted. If the stroke has been changed from the specified hole reset to specifications by following these procedures

1. Using a blunt-tipped punch, remove and retain the roll pin from the accelerator pump cover.

NOTE: Support the area under the roll pin when removing the pin.

2. Rotate the pump link and rod assembly until the keyed end of the assembly is aligned with the keyed hole in the pump overtravel lever.
3. Reposition the rod and swivel assembly in the specified hole and reinstall the pump link in the accelerator pump cover.

NOTE: A service accelerator rod and swivel assembly is available (9F687) and must be used if replacement is necessary.

Adjustment holes are not provided on the temperature compensated accelerator pump carburetors.

4. Reinstall the rod pin.

CHOKE PULLDOWN ADJUSTMENT

Carter YFA & YFA Feedback

1. Remove the air cleaner assembly.
2. Hold the throttle plate fully open and close the choke plate as far as possible without forcing it. Use a drill of the proper diameter to check the clearance between the choke plate and air horn. Refer to the Carburetor Specification chart for specifications.

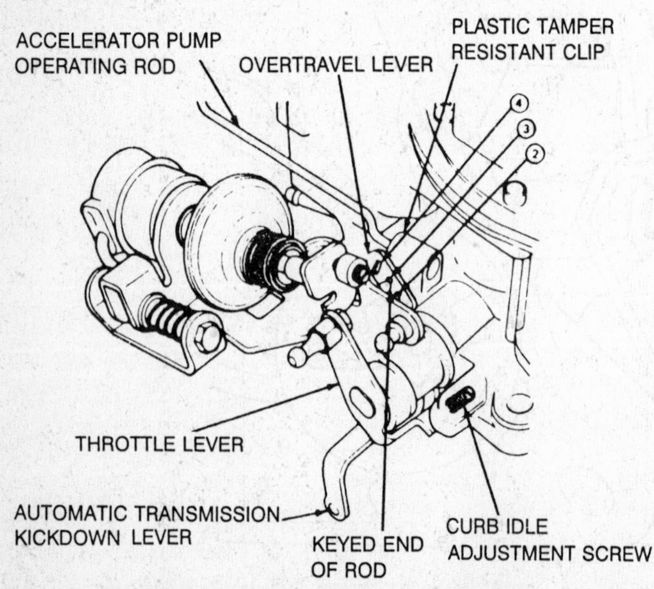

Accelerator pump stroke adjustment

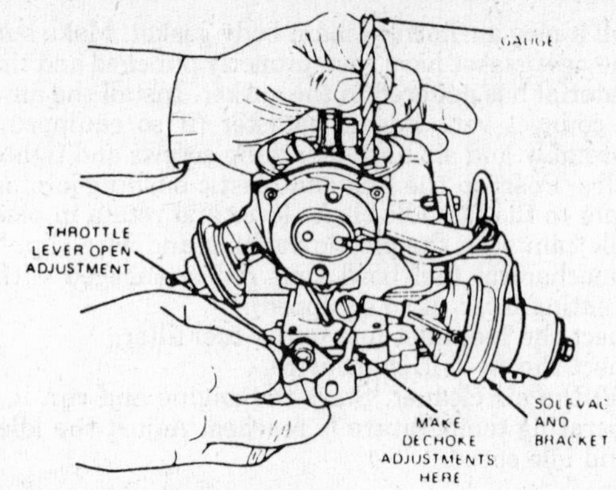

Choke plate pulldown adjustment — Carter YFA & YFA Feedback carburetor

3. If the clearance is not within specification, adjust by bending the arm on the choke trip lever of the throttle lever. Bending the arm downward will decrease the clearance, and bending it upward will increase the clearance. Always recheck the clearance after making any adjustment.

Motorcraft 2150A

1. Set throttle on fast idle cam top step.
2. Note index position of choke bimetallic cap. Loosen retaining screws and rotate cap 90° in the rich (closing) direction.
3. Activate pulldown motor by manually forcing pulldown control diaphragm link in the direction of applied vacuum or by applying vacuum to external vacuum tube.
4. Measure vertical hard gauge clearance between choke plate and center of carburetor air horn wall nearest fuel bowl.

Pulldown setting should be within specifications for minimum choke plate opening.

If choke plate pulldown is found to be out of specification, reset by adjusting diaphragm stop on end of choke pulldown diaphragm.

If pulldown is reset, cam clearance should be checked and reset if required.

After pulldown check is completed, reset choke bimetallic cap to recommended index position as specified in the Carburetor Specifications Chart. Check and reset fast idle speed to specifications if necessary.

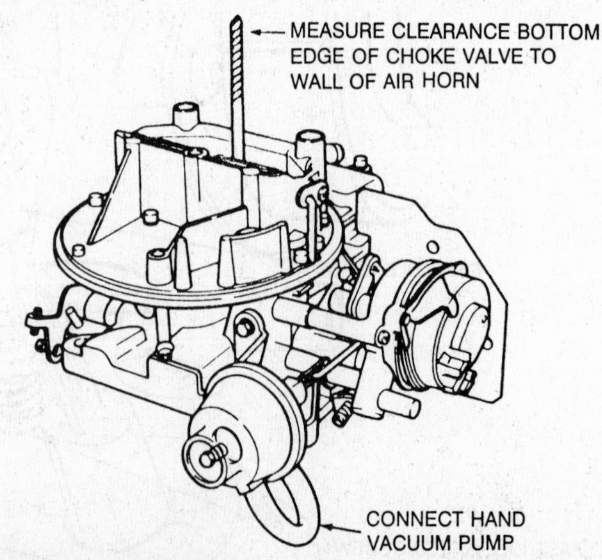

Motorcraft 2150A 2-bbl. carburetor choke plate pulldown adjustment

SECONDARY TOUCH CLEARANCE

Aisan Model Y Carburetor

1. Remove carburetor from vehicle, as described.
2. Check and adjust secondary touch clearance as follows:
   a. Open throttle until secondary touch adjustment tang just touches secondary kicker lever.
   b. Using a drill of the specified diameter, check the clearance between the primary throttle plate and the wall of the bore.
   c. If required, adjustment is made by bending secondary touch adjustment tang located on the primary throttle lever.
3. Install carburetor onto engine.

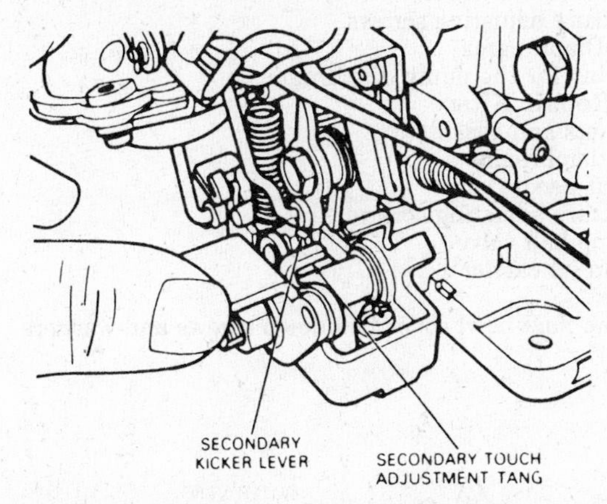

Secondary touch clearance adjustment – Aisan model Y carburetor

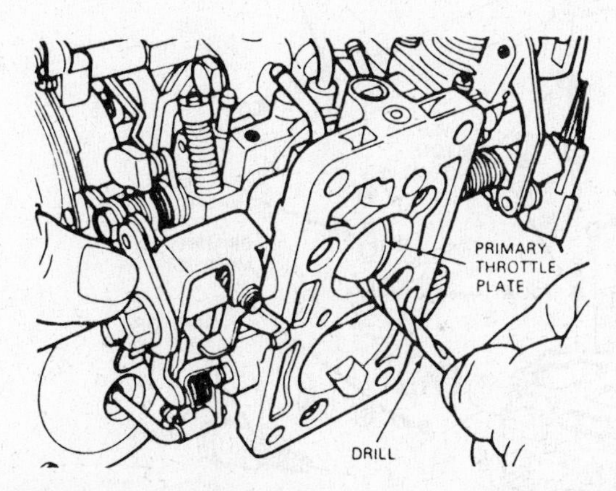

Checking secondary touch clearance adjustment Aisan model Y carburetor

SECONDARY KICK CLEARANCE

Aisan Model Y Carburetor

1. Remove carburetor from vehicle as described.
2. Check and adjust secondary kick clearance as follows: Open primary throttle to wide-open throttle. Using a drill of the specified diameter, check the clearance between the secondary throttle plate and the wall of the bore. If required, adjustment is made by bending the secondary kick tang on the secondary throttle lever.
3. Install carburetor onto engine.

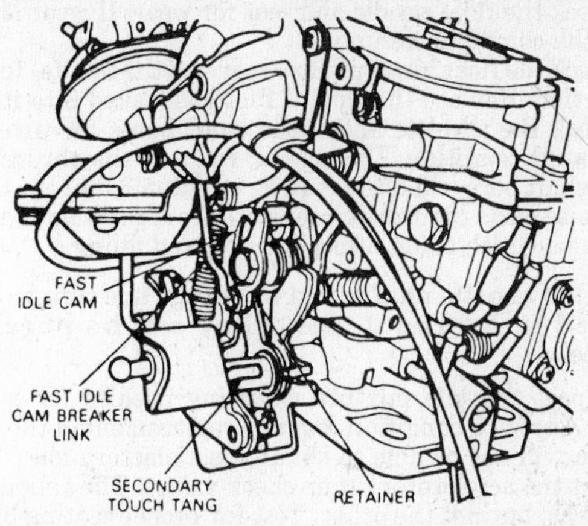

Adjusting secondary touch clearance adjustment – Aisan model Y carburetor

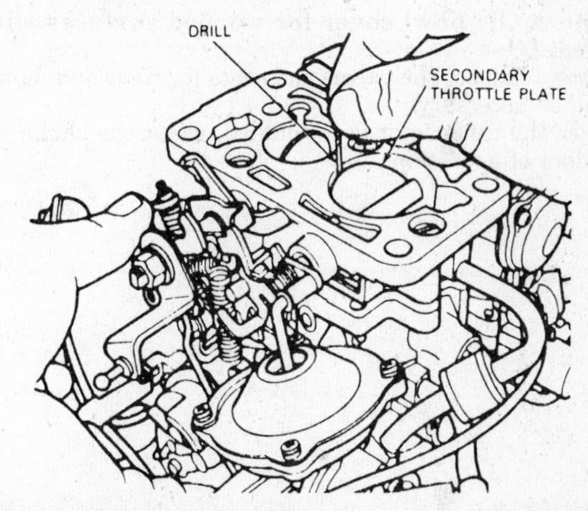

Checking secondary kick clearance adjustment – Aisan model Y carburetor

OVERHAUL

Efficient carburetion depends greatly on careful cleaning and inspection during overhaul since dirt, gum, water or varnish in or on the carburetor parts are often responsible for poor performance.

Overhaul the carburetor in a clean, dust free area. Carefully disassemble the carburetor, referring often to the exploded views. Keep all similar and look-alike parts segregated during disassembly and cleaning to avoid accidental interchange during assembly. Make a note of all jet sizes.

When the carburetor is disassembled, wash all parts (except diaphragms, electric choke unit, pump plunger and any other plastic, leather, fiber, or rubber parts) in clean carburetor solvent. Do not leave the parts in the solvent any longer than is necessary to sufficiently loosen the dirt and deposits. Excessive cleaning may remove the special finish from the float bowl and choke valve bodies, leaving these parts unfit for service. Rinse all parts in clean solvent and blow them dry with compressed air or allow them to air dry, while resting on clean, lint less paper. Wipe clean all cork, plastic, leather and fiber parts with a clean, lint-free cloth.

Blow out all passages and jets with compressed air and be sure that there are no restrictions or blockages. Never use wire or similar tools to clean jets, fuel passages or air bleeds. Clean all jets and valves separately to avoid accidental interchange.

Examine all parts for wear or damage. If wear or damage is found, replace the defective parts. Especially, inspect the following:

1. Check the float needle and seat for wear. If wear is found, replace the complete assembly.
2. Check the float hinge pin for wear and the float(s) for dents or distortion. Replace the float if fuel has leaked into it.
3. Check the throttle and choke shaft bores for wear or an out-of-round condition. Damage or wear to the throttle arm, shaft or shaft bore will often require replacement of the throttle body. These parts require a close tolerance of fit; wear may allow air leakage, which could affect starting and idling.

NOTE: Throttle shafts and bushings are not normally included in overhaul kits. They can be purchased separately.

4. Inspect the idle mixture adjusting needles for burrs or grooves. Any such condition requires replacement of the needle, since you will not be able to obtain a satisfactory idle.
5. Test the accelerator pump check valves. They should pass air one way, but not the other. Test for proper seating by blowing and sucking on the valve. Replace the valve as necessary. If the valve is satisfactory, wash the valve again to remove moisture.
6. Check the bowl cover for warped surfaces with a straightedge.
7. Closely inspect the valves and seats for wear and damage, replacing as necessary.
8. After the carburetor is assembled, check the choke valve for freedom of operation.

Carburetor overhaul kits are recommended for each overhaul. These kits contain all gaskets and new pars to replace those which deteriorate most rapidly. Failure to replace all of the parts supplied with the kit (especially gaskets) can result in poor performance later.

NOTE: Most carburetor rebuilding kits include specific procedures which should be followed during overhaul.

Most carburetor manufacturers supply overhaul kits of these basic types: minor repair; major repair; and gasket kits. Basically, they contain the following:

Minor Repair Kits:
- All gaskets
- Float needle valve
- Mixture adjusting screws
- All Diaphragms
- Spring for the pump diaphragm

Major Repair Kits:
- All jets and gaskets
- All diaphragms
- Float needle valve
- Mixture adjusting screws
- Pump ball valve
- Main jet carrier
- Float
- Some float bowl cover holddown screws and washers

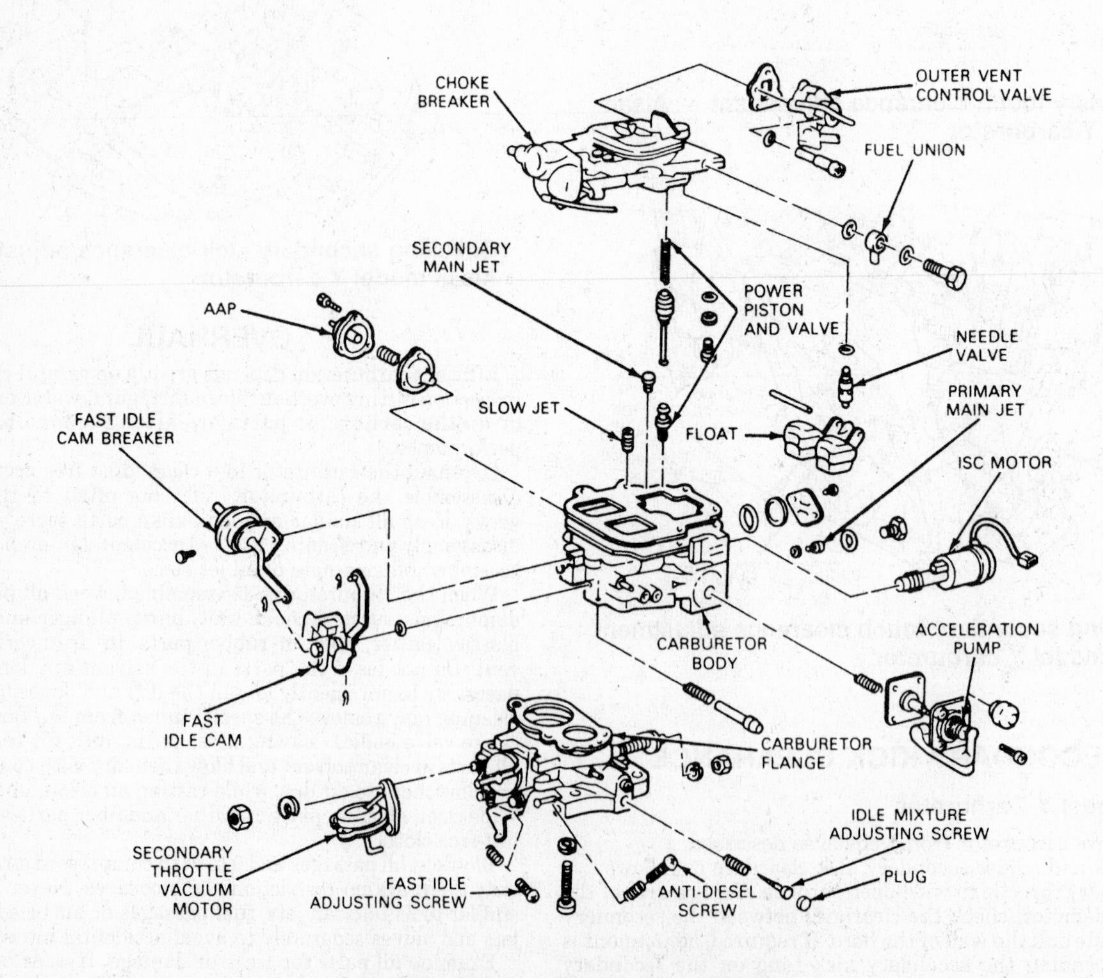

Aisan model Y 1-bbl. carburetor

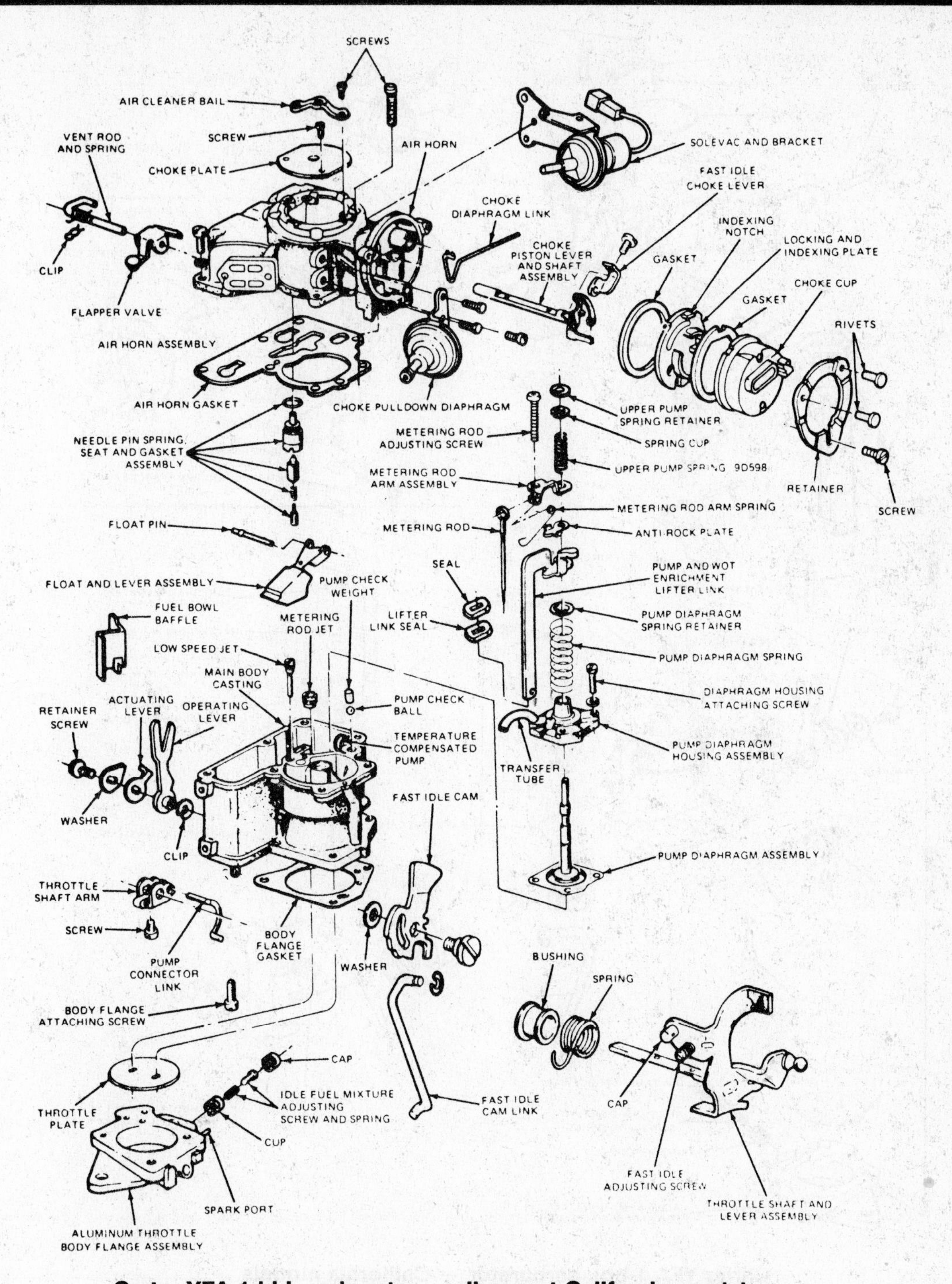

Carter YFA 1-bbl. carburetor — all except California models

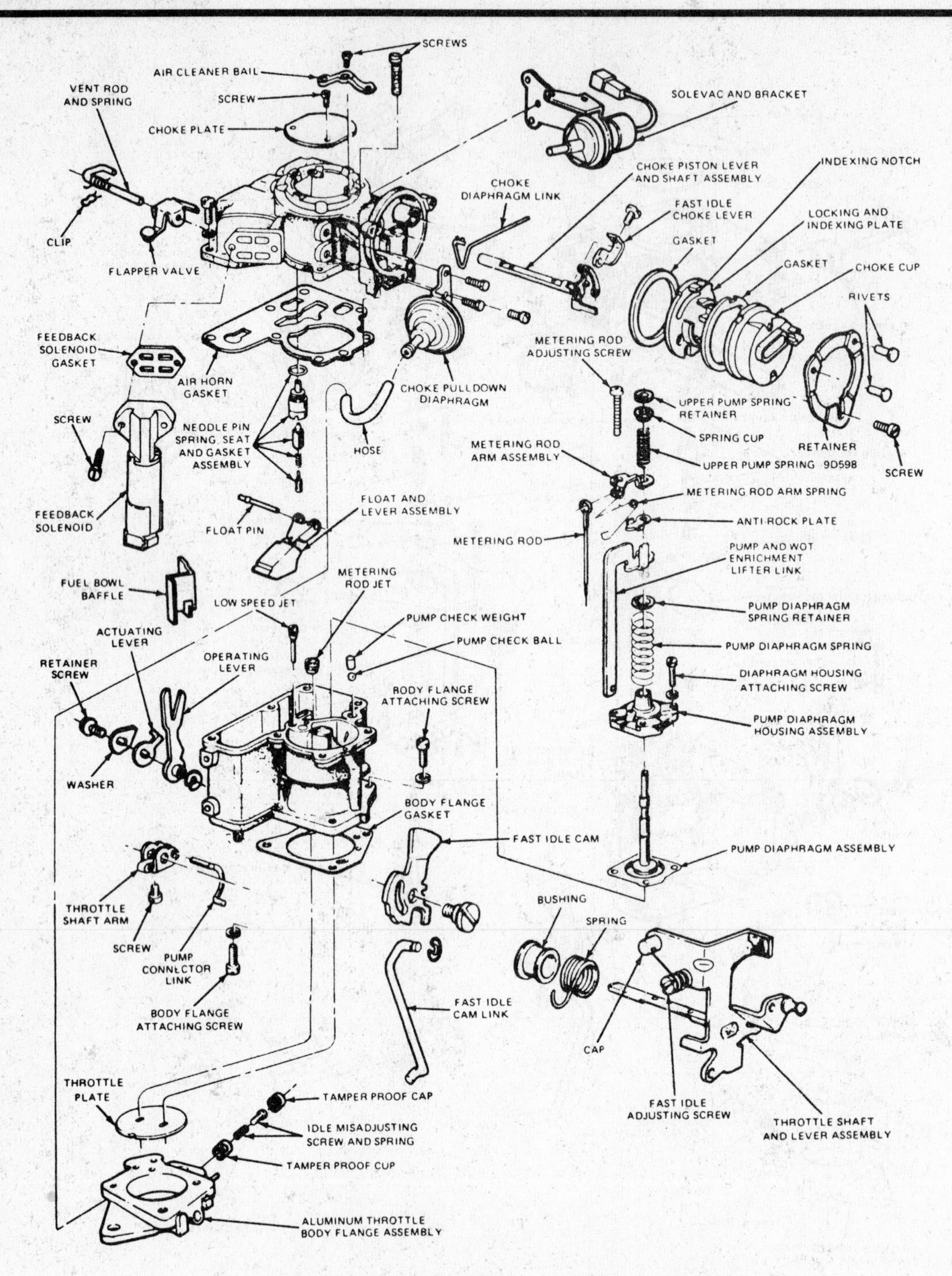

Carter YFA 1-bbl. carburetor — California models

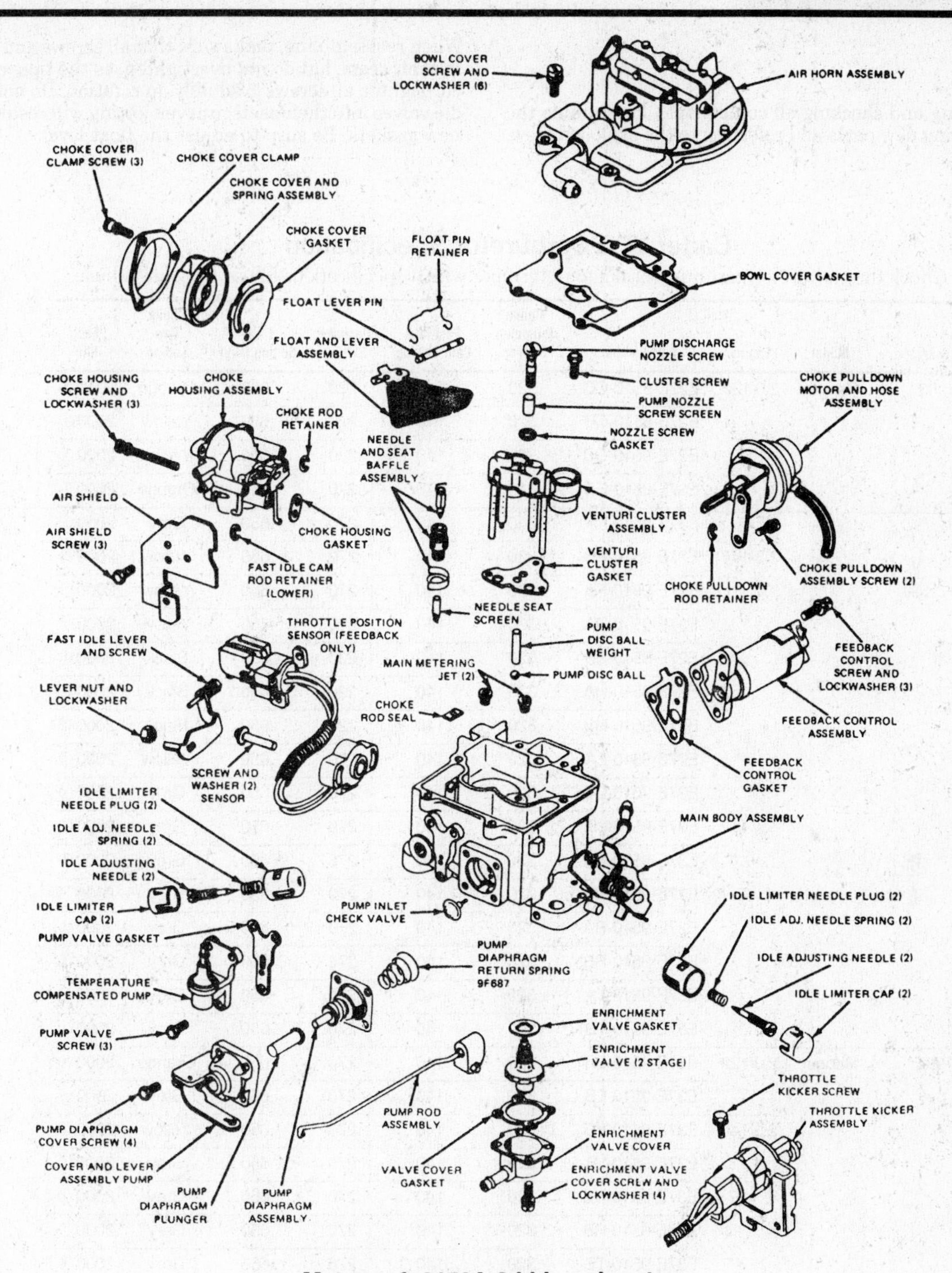

Motorcraft 2150A 2-bbl. carburetor

Gasket Kits:

- All gaskets

After cleaning and checking all components, reassemble the carburetor, using new parts and referring to the exploded view. When reassembling, make sure that all screws and jets are tight in their seats, but do not overtighten, as the tips will be distorted. Tighten all screws gradually, in rotation. Do not tighten needle valves into their seats; uneven jetting will result. Always use new gaskets. Be sure to adjust the float level.

Carter YFA Carburetor Specification

Check the carburetor part number tag to determine which specifications to use for your vehicle.

Year	Model	Engine	Part Number	Choke Pulldown Setting	Fast Idle Cam Setting	Dechoke Setting	Float Setting (Dry)	Choke Cap Setting	Fast Idle
1983	Ranger	2.0/122	E27E-9510-CC	.320	.140	.220	.650	Orange	2000 ①
			E27E-9510-CB	.320	.140	.220	.650	Yellow	2000 ①
			E27E-9510-GB	.320	.140	.220	.650	Yellow	2000 ①
			E37E-9510-EA	.320	.140	.270	.650	Orange	2000 ①
			E37E-9510-FA	.320	.140	.270	.650	Gray	2000 ①
		2.3/140	E27E-9510-BB	.320	.140	.270	.650	Yellow	2000 ①
			E27E-9510-FB	.320	.140	.270	.650	Yellow	2000 ①
			E37E-9510-BA	.320	.140	.270	.650	Yellow	2000 ①
			E27E-9510-EB	.320	.140	.220	.650	Black	2000 ①
			E27E-9510-HA	.320	.140	.220	.650	Black	2000 ①
			E27E-9510-HB	.320	.140	.220	.650	Black	2000 ①
			E27E-9510-FA	.320	.140	.270	.650	Yellow	2000 ①
			E37E-9510-LA	.320	.140	.270	.650	Grey	2000 ①
			E37E-9510-LB	.320	.140	.270	.650	Grey	2000 ①
			E37E-9510-NA	.320	.140	.270	.650	Grey	2000 ①
			E37E-9510-NB	.320	.140	.270	.650	Grey	2000 ①
			E37E-9510-RA	.320	.140	.270	.650	Grey	2000 ①
			E37E-9510-RB	.320	.140	.270	.650	Grey	2000 ①
			E37E-9510-TA	.320	.140	.270	.650	Grey	2000 ①
			E37E-9510-TB	.320	.140	.270	.650	Grey	2000 ①
1984	Ranger	2.0/122	E37E-9510-EB	.320	.140	.270	.650	Orange	2000 ①
			E37E-9510-FB	.320	.140	.270	.650	Grey	2000 ①
		2.3/140	E37E-9510-BB	.320	.140	.270	.650	Yellow	2000 ①
			E37E-9510-LB	.320	.140	.270	.650	Grey	2000 ①
			E37E-9510-NB	.320	.140	.270	.650	Grey	2000 ①
			E37E-9510-RB	.320	.140	.270	.650	Grey	2000 ①
			E37E-9510-TB	.320	.140	.270	.650	Grey	2000 ①
1985	Ranger	2.0/122	E57E-9510-DA	.320	.140	.270	.650	Grey	1700 ②
1986	Ranger	2.0/122	E57E-9510-DA	.320	.140	.270	.650	Grey	1700 ②
			E57E-9510-DB	.320	.140	.270	.650	Grey	1700 ②

① 1900 rpm for vehicles with less than 100 miles.

② 1600 rpm for vehicles with less than 100 miles.

Motorcraft 2150A Carburetor Specification

Check the carburetor part number tag to determine which specifications to use for your vehicle.

Year	Model	Engine	Part Number	Choke Pulldown Setting	Fast Idle Cam Setting	Dechoke Setting	Float Setting (Wet)	Float Setting (Dry)	Accelerator Pump Lever Location	Choke Cap Setting	Fast Idle
1984	Bronco II	2.8/173	E37E-9510-AAA	.136	V-notch	.250	.810	7/16"	#4	V-notch	3000 ①
	Ranger		E37E-9510-ABA	.136	V-notch	.250	.810	7/16"	#4	V-notch	3000 ①
			E37E-9510-ADA	.136	V-notch	.250	.810	7/16"	#4	V-notch	3000 ①
			E37E-9510-AEA	.136	V-notch	.250	.810	7/16"	#4	V-notch	3000 ①
			E47E-9510-TA	.136	V-notch	.250	.810	7/16"	#4	V-notch	3000 ①
			E47E-9510-VA	.136	V-notch	.250	.810	7/16"	#4	V-notch	3000 ①
1985	Bronco II	2.8/173	E57E-9510-BA	.136	Hi-Cam	.250	.810	1/16"	#4	3NR	3000 ①
	Ranger		E57E-9510-CA	.136	Hi-Cam	.250	.810	1/16"	#4	3NR	3000 ①

① 2800 rpm for vehicles with less than 100 miles.

Aisan Model Y Carburetor Specification

Check the carburetor part number tag to determine which specifications to use for your vehicle.

Year	Model	Engine	Part Number	Choke Pulldown Setting	Fast Idle Cam Setting	Dechoke Setting	Float Setting (Dry)	Choke Cap Setting	Fast Idle
1987	Ranger	2.0/122	E77E-9510-AA	18°	14.5° ①	195 sec.	47.1 mm	20°C	3200 ②
1988	Ranger	2.0/122	E87E-9510-AA	18°	14.5° ①	195 sec.	47.1 mm	20°C	3200 ②

① On step #1
② 3100 rpm for vehicles with less than 100 miles.

GASOLINE FUEL INJECTION SYSTEM

NOTE: This book contains testing and service procedures for your vehicle's fuel injection system. More comprehensive testing and diagnostic procedures may be found in CHILTONS'S GUIDE TO FUEL INJECTION AND FEEDBACK CARBURETORS, available at your local retailer.

CAUTION

BEFORE SERVICING ANY COMPONENTS OF THE FUEL SYSTEM, PERFORM THE FOLLOWING PROCEDURE FOR RELEASING THE FUEL SYSTEM PRESSURE. THE FUEL INJECTION SYSTEMS MAY BE UNDER PRESSURE EVEN WHEN THE ENGINE IS NOT RUNNING.

BLEEDING THE FUEL SYSTEM/RELIEVE FUEL PRESSURE

1. Remove the fuel tank cap and the air filter.
2. Disconnect the negative battery cable.
3. Using the Fuel Pressure Gauge tool No. T80L–9974–A, connect it to the pressure relief valve (remove the valve cap) on the fuel injection manifold/fuel rail assembly.

NOTE: Some 2.3L EFI engines have a pressure relief valve located on the throttle body. On later model 2.3L EFI engines the fuel pressure relief valve is located on the fuel injection supply manifold assembly in the upper RH corner of the engine compartment. To gain access to the pressure relief valve, the valve cap must be removed.

4. Open the pressure relief valve and reduce the fuel pressure.
5. To pressurize the fuel system, perform the following:
   a. Tighten the pressure relief valve and remove the pressure gauge.
   b. Install fuel tank cap and the air filter. Reinstall the negative battery cable.
   c. Turn ignition ON/OFF several times (leave key ON position---15 seconds each time) without starting the engine to pressurize the fuel system. Check for fuel leaks at pressure regulator, fuel injectors and fuel connect fittings.
   d. Start the engine.

Inertia Switch

A safety inertia switch is installed to shut off the electric fuel pump in case of collision. The switch is located on the toe-board to the right of the transmission hump. If the pump shuts off, or if the vehicle has been hit and will not start, check for leaks-then reset the switch. The switch is reset by pushing down on the button provided. To relieve the fuel system pressure disconnect the electrical connection to inertia switch crank the engine for about 20 seconds.

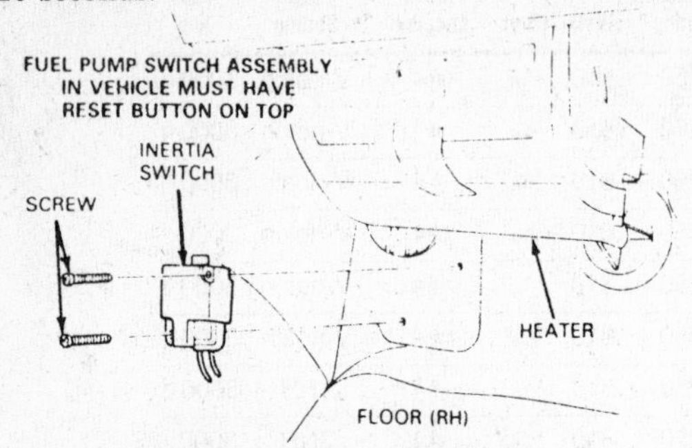

Inertia switch assembly

Electric Fuel Pump

2.3L AND 2.9L ENGINES/TILL 1988

On the 2.3L and the 2.9L engines to model year 1988 the electric fuel pump system used on these EFI engines consisted of two fuel pumps: a low pressure boost pump mounted in the fuel tank, and a high pressure fuel pump mounted on the frame rail. The low pressure electric fuel pump is located in the fuel tank and is a part of the fuel gauge sending unit.

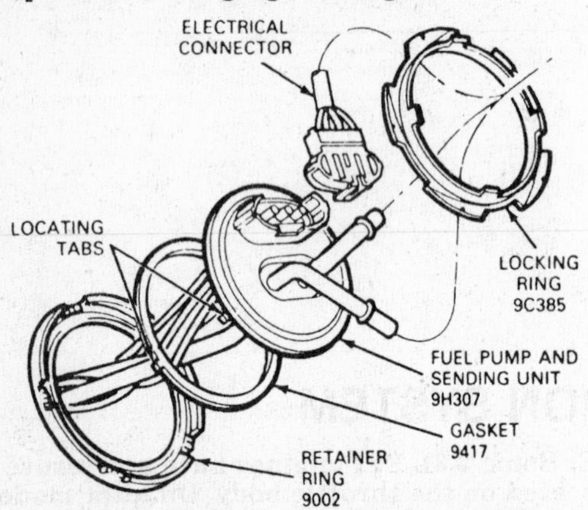

Low pressure fuel pump assembly

REMOVAL AND INSTALLATION

External Frame Mounted Pump Assembly

The high pressure fuel pump is frame mounted and can be accessed from under the vehicle. The fuel pump assembly is retained to the frame with 3 bolts. Remove the fuel pump and bracket assembly (disconnect electrical connector) from the the frame of the vehicle. Before removing the fuel pump assembly, disconnect the negative battery cable and relieve the system pressure. Refer to "BLEEDING THE FUEL SYSTEM/RELIEVE FUEL PRESSURE" procedure in this Section.

Clean all dirt and/or grease from the fuel line fittings. "Quick Connect" fittings are used on all models equipped with a pressurized fuel system. These fittings must be disconnected using the proper procedure or the fittings may be damaged. Remove the fuel pump assembly from the metal mounting bracket. Remove the fuel pump. Install the fuel pump to mounting bracket and install the assembly to frame of the vehicle. Connect all electrical and fuel line connections, connect battery cable, pressurize the fuel system. Start engine and check for leaks.

Internal Pump In Fuel Tank

The low pressure fuel pump is located in the fuel tank and may require removal of the fuel tank assembly. Refer to Fuel Tank Removal and Installation procedures in This Section as necessary.

1. Disconnect the negative battery cable. Remove the fuel from the tank or remove the tank assembly.
2. Disconnect all electrical connections to the fuel system sender unit.
3. Remove any dirt around the fuel sender assembly.
4. Loosen the Quick Connect Fittings. Turn the fuel sender locking ring counterclockwise with special tool or equivalent (brass drift pin) remove the locking ring, sender assembly and sealing gasket. Remove the fuel pump from the fuel sender assembly.

To install:

5. Install the fuel pump to sender assembly. Place a new sealing gasket in the groove of the fuel tank. Install the fuel sender assembly in the fuel tank so that the tabs of the sender are positioned into the slots of the fuel tank. The sealing gasket must remain in place during and after the fuel sender installation.
6. Holding the fuel sender and sealing gasket in place, install and rotate the locking ring clockwise until the stop is against the retainer ring tab.
7. Install the tank assembly as necessary. Connect all electrical and fuel line connections.
8. Refill the tank, connect battery cable, pressurize the fuel system. Start engine and check for leaks.

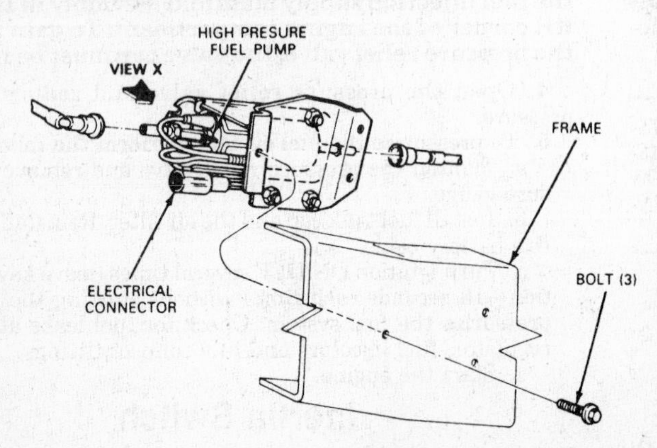

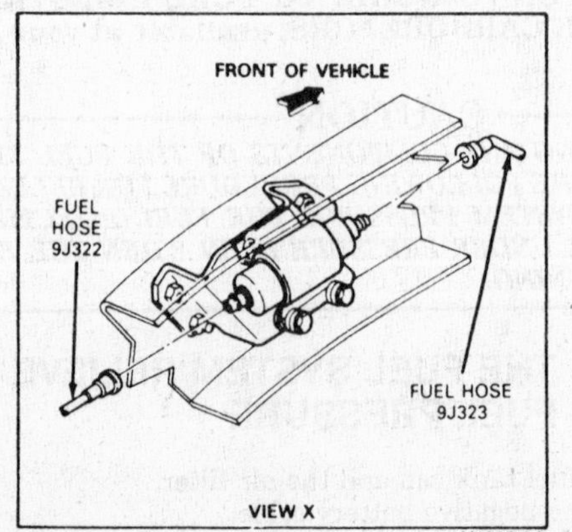

High pressure fuel pump assembly

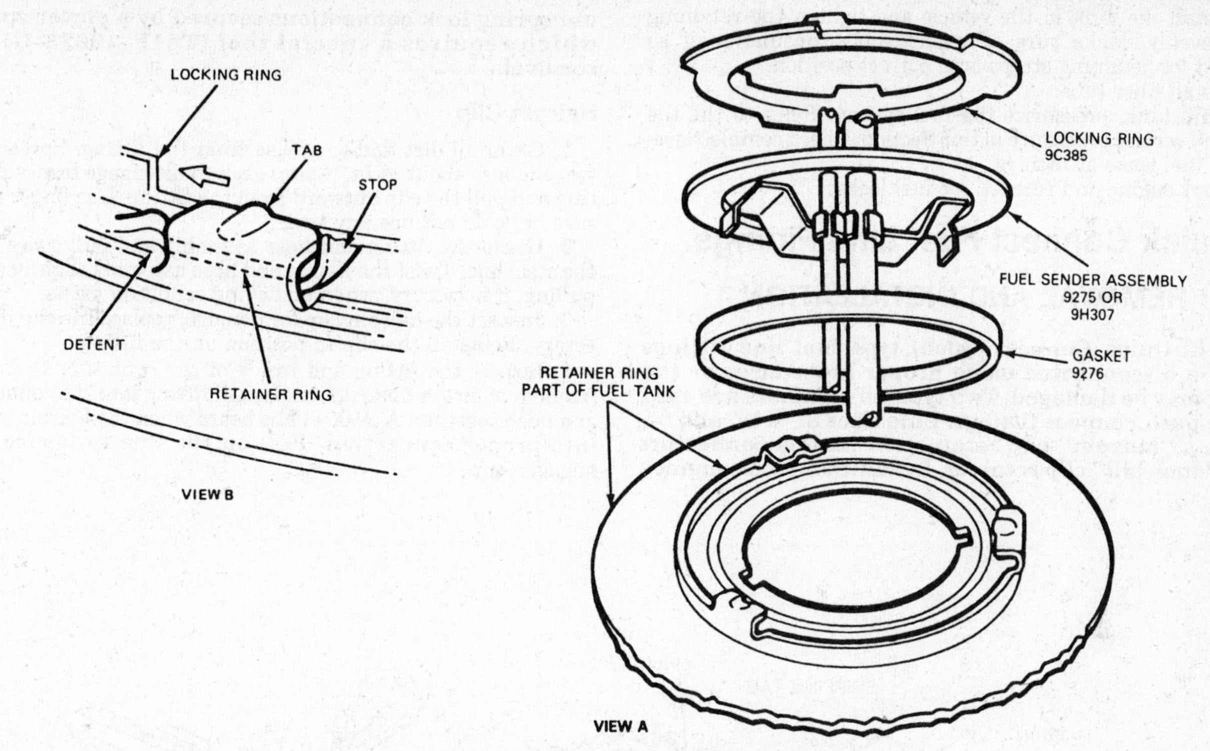

Fuel tank Sender installation (metal lockring type)

2.3L TWIN PLUG 2.9L, 3.0L AND 4.0L ENGINES

On the 2.3L twin plug engine and the 2.9L, 3.0L and 4.0L engines starting in model year 1989 the electric fuel injection system uses a fuel tank and fuel pump/sender assembly.

REMOVAL AND INSTALLATION

1. Disconnect the negative battery cable. Depressurize the fuel system. Refer to the necessary service procedure in this Section.

NOTE: On Ranger vehicles the fuel pump/sender unit may be serviced by removing the pick-up box from the chassis, instead of removing the fuel tank assembly.

CAUTION

The fuel system is under pressure. Release pressure slowly and contain spillage. Observe no smoking/no open flame precautions. Have a Class B–C (dry powder) fire extinguisher within arm's reach at all times.

2. Drain the fuel tank. Drain the gasoline into a suitable safety container and take precautions to avoid the risk of fire. Remove the fuel filler tube.
3. Support the fuel tank assembly and remove the fuel tank supports straps. Lower fuel tank partially and remove all other connections. Remove the bolt from the front strap and remove the front strap. Remove the bolt from the rear strap and remove the rear strap.
4. Remove the fuel feed hose at the fuel gauge sender push connector.
5. Remove the fuel hose from the sender unit push connector.
6. Remove the fuel vapor hose from the vapor valve.
7. Lower the fuel tank from the chassis.
8. Remove the shield from the fuel tank.
9. Remove any dirt from the fuel pump flange.
10. Turn the fuel pump locking ring counterclockwise to remove it. There is a special wrench for this purpose, but you can loosen the ring by tapping it around with a wood dowel and plastic or rubber mallet.

CAUTION

Never hammer on or near the fuel tank with metal tools! The risk of spark and explosion is always present!

11. Remove the fuel pump and bracket assembly.
12. Remove and discard the seal ring gasket.

WARNING: Do not attempt to apply battery voltage to the pump to check its operation while removed from the vehicle, as running the pump dry will destroy it!

To install:

13. Clean the area thoroughly.
14. Coat the new seal ring with multi-purpose grease to hold it in place and install it in the fuel ring groove.
15. Position the pump/sender assembly in the tank making sure that all keyways align and the seal ring stays in place.
16. Hold the fuel sender and sealing gasket in place, install and rotate the locking ring. Find the fuel tank part number on the front bottom of the tank, for vehicles equipped with E59A–9002–CAE tank: tighten the ring to 60–85 ft. lbs. wait 5 minutes and tighten again to 60–85 ft. lbs. For this tank assembly use the same ring that was removed from the tank. Do not replace the ring with a ring from another tank.
17. For tanks equipped with plastic retaining rings E99A–9A307–D, tighten nuts to 40–55 ft. lbs. use the same ring that was removed from the tank. If a new tank is installed, use a new ring.
18. For E69A–9002–PA tanks: tighten the locknut ring once to 80–113 ft. lbs. and for some vehicles tighten the polyethylene lock ring to 40–45 ft. lbs.
19. Support the tank assembly while connecting the fuel lines, vent line and electrical connectors.

20. Install the tank in the vehicle and tighten the retaining straps evenly. Make sure all gasket/rubber insulation if so equipped for retaining straps is in correct position.
21. Install filler tube.
22. Refill tank, pressurize the fuel system. Refer to the the necessary service procedure in this Section. Check complete system for fuel leaks at fittings.
23. Start engine and recheck for fuel leaks.

Quick Connect Fuel Line Fittings

REMOVAL AND INSTALLATION

NOTE: Quick Connect (push) type fuel line fittings must be disconnected using proper procedures or the fitting may be damaged. Two types of retainers are used on the push connect fittings. Line sizes of 3/8 in. and 5/16 in. use a "hairpin" clip retainer. 1/4 in. line connectors use a "duck bill" clip retainer. In addition, some engines use spring lock connections secured by a garter spring which requires a special tool (T81P-19623-G) for removal.

Hairpin Clip

1. Clean all dirt and/or grease from the fitting. Spread the two clip legs about 1/8 in. (3mm) each to disengage from the fitting and pull the clip outward from the fitting. Use finger pressure only, do not use any tools.
2. Grasp the fitting and hose assembly and pull away from the steel line. Twist the fitting and hose assembly slightly while pulling, if necessary, when a sticking condition exists.
3. Inspect the hairpin clip for damage, replace the clip if necessary. Reinstall the clip in position on the fitting.
4. Inspect the fitting and inside of the connector to insure freedom of dirt or obstruction. Install fitting into the connector and push together. A click will be heard when the hairpin snaps into proper connection. Pull on the line to insure full engagement.

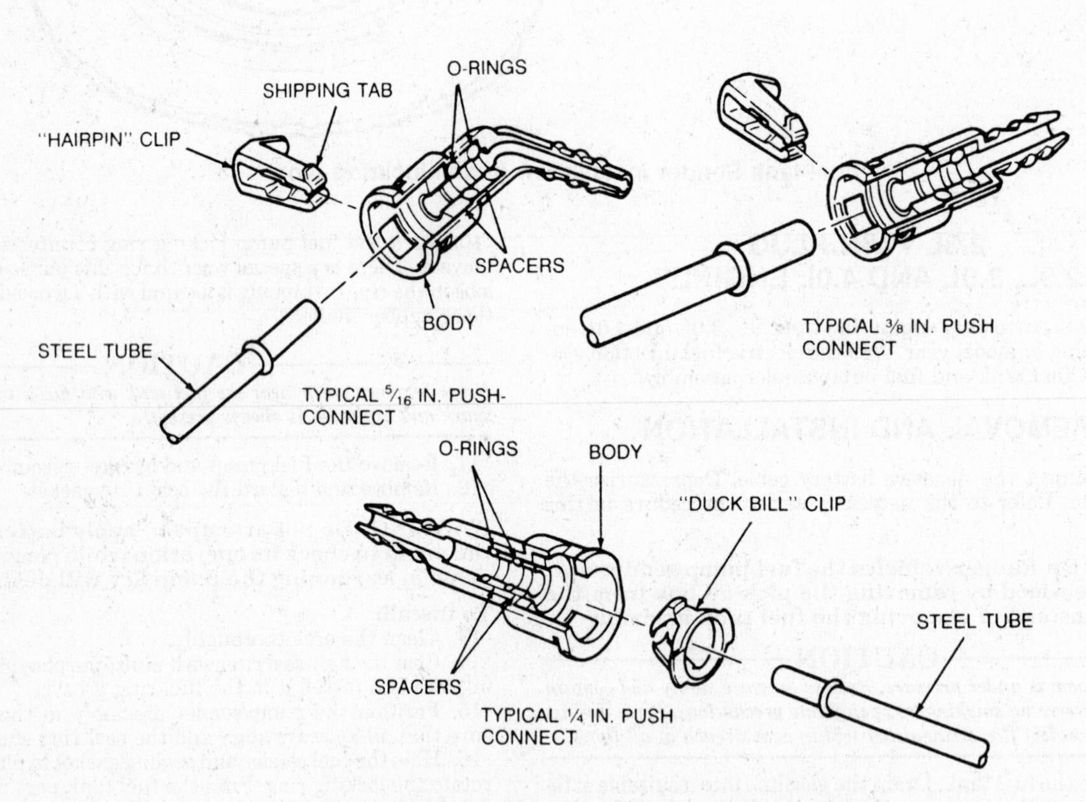

Various types of quick connect fuel line fittings

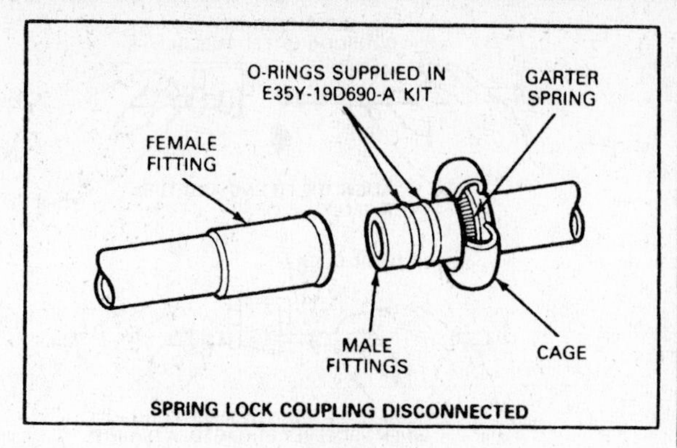

TO DISCONNECT COUPLING
CAUTION — DISCHARGE SYSTEM BEFORE DISCONNECTING COUPLING

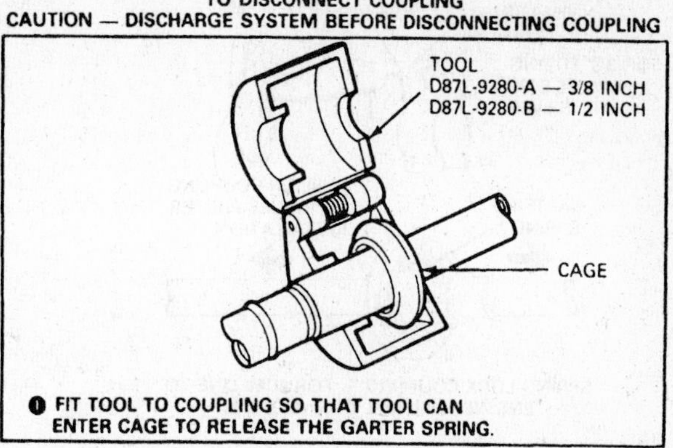

❶ FIT TOOL TO COUPLING SO THAT TOOL CAN ENTER CAGE TO RELEASE THE GARTER SPRING.

TO CONNECT COUPLING

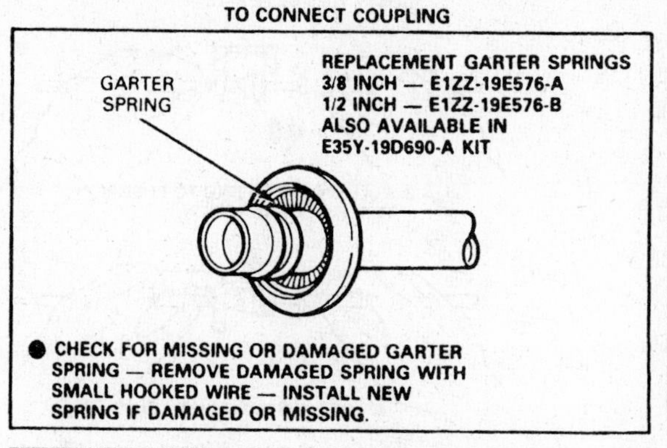

● CHECK FOR MISSING OR DAMAGED GARTER SPRING — REMOVE DAMAGED SPRING WITH SMALL HOOKED WIRE — INSTALL NEW SPRING IF DAMAGED OR MISSING.

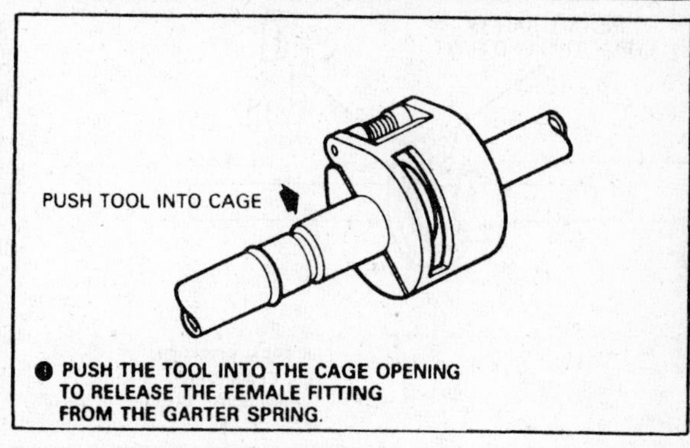

❷ PUSH THE TOOL INTO THE CAGE OPENING TO RELEASE THE FEMALE FITTING FROM THE GARTER SPRING.

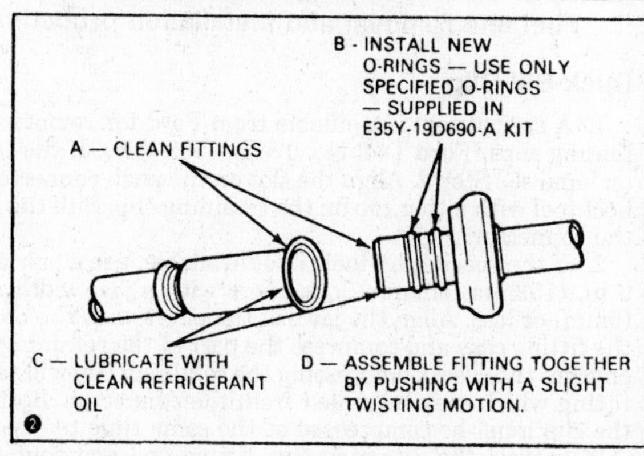

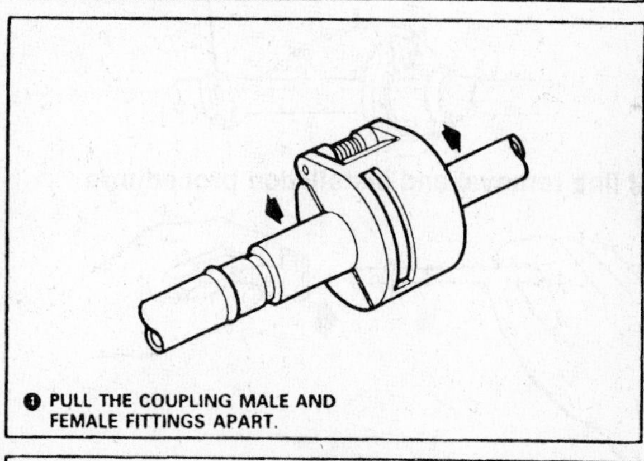

❸ PULL THE COUPLING MALE AND FEMALE FITTINGS APART.

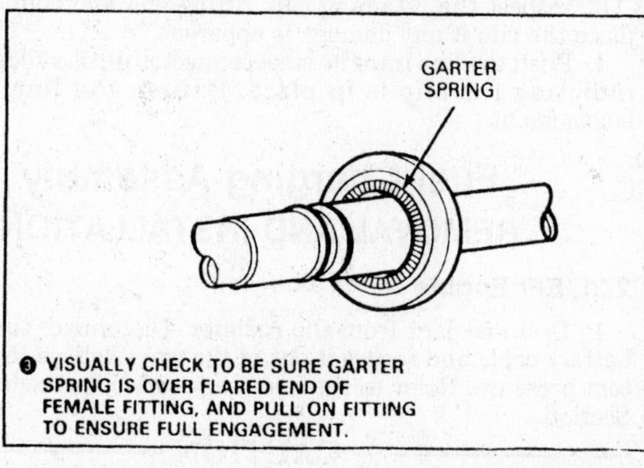

❸ VISUALLY CHECK TO BE SURE GARTER SPRING IS OVER FLARED END OF FEMALE FITTING, AND PULL ON FITTING TO ENSURE FULL ENGAGEMENT.

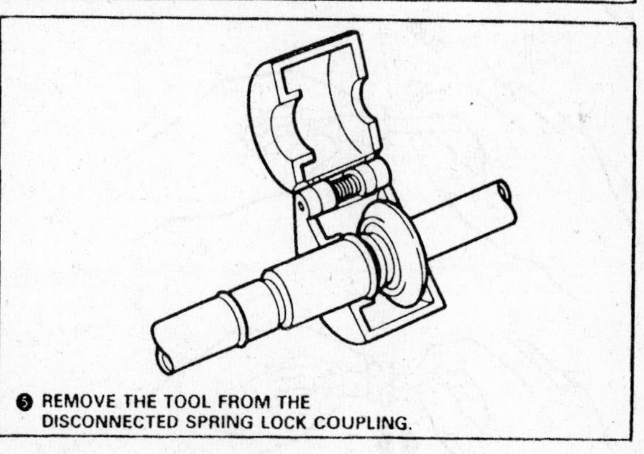

❹ REMOVE THE TOOL FROM THE DISCONNECTED SPRING LOCK COUPLING.

Fuel line removal and installation procedures

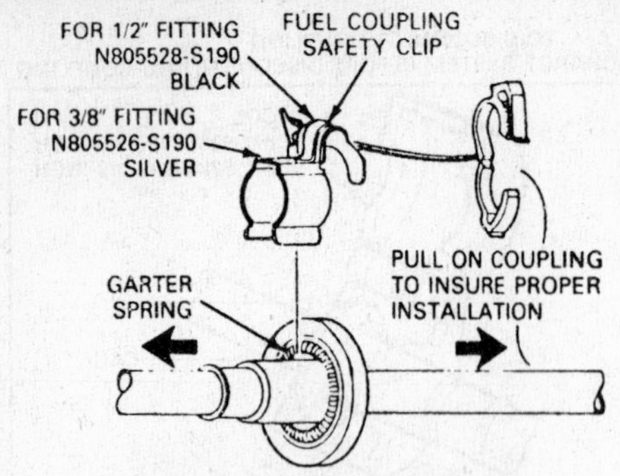

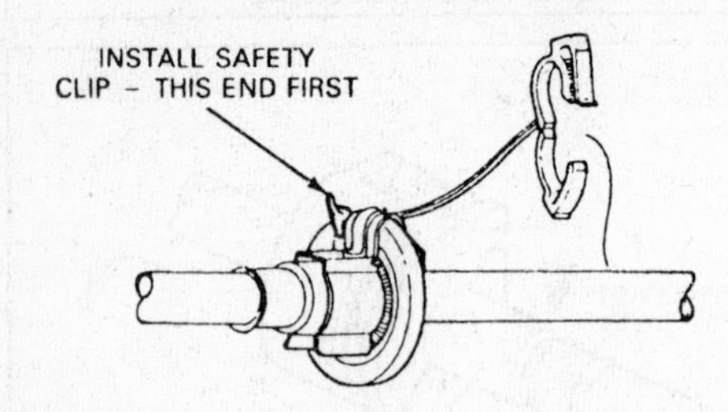

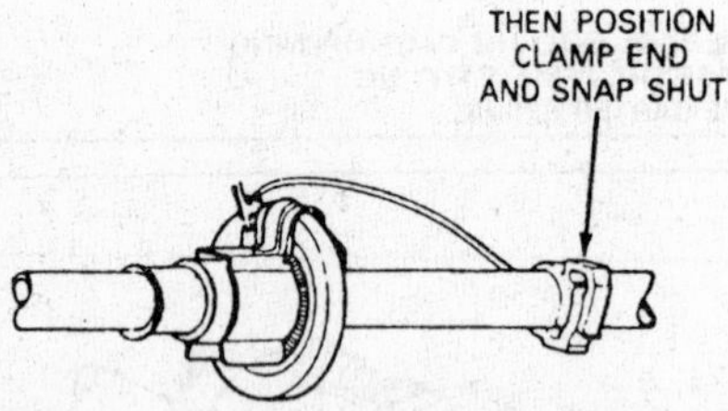

Fuel line removal and installation procedures

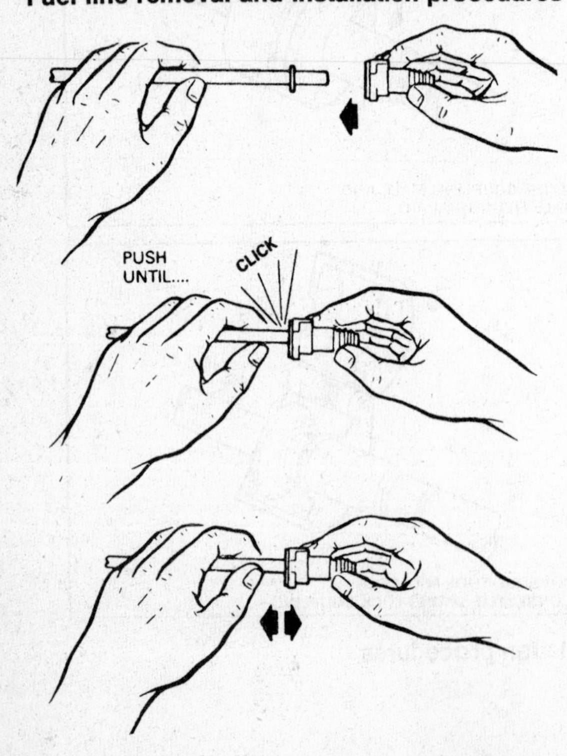

Fuel line removal and installation procedures

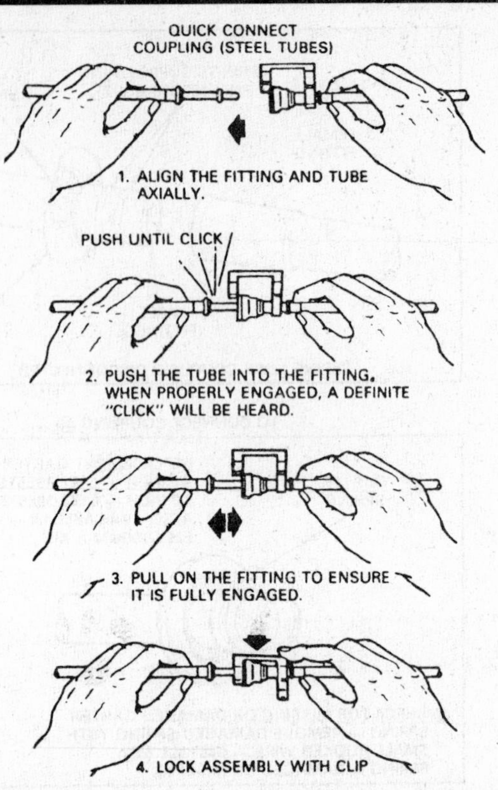

Fuel line removal and installation procedures

Duck Bill Clip

1. A special tool is available from Ford for removing the retaining clips (Ford Tool No. T82L–9500–AH). If the tool is not on hand see Step 2. Align the slot on the push connector disconnect tool with either tab on the retaining clip. Pull the line from the connector.
2. If the special clip tool is not available, use a pair of narrow 6 in. (152mm) channel lock pliers with a jaw width of 0.2 in. (5mm) or less. Align the jaws of the pliers with the openings of the fitting case and compress the part of the retaining clip that engages the case. Compressing the retaining clip will release the fitting which may be pulled from the connector. Both sides of the clip must be compressed at the same time to disengage.
3. Inspect the retaining clip, fitting end and connector. Replace the clip if any damage is apparent.
4. Push the line into the steel connector until a click is heard, indicting the clip is in place. Pull on the line to check engagement.

Fuel Charging Assembly

REMOVAL AND INSTALLATION

2.3L EFI Engine

1. Drain coolant from the radiator. Disconnect the negative battery cable and secure it out of the way. Relieve the fuel system pressure. Refer to the necessary service procedure in this Section.

CAUTION

When draining the coolant, keep in mind that cats and dogs are attracted by the ethylene glycol antifreeze, and are quite likely to drink any that is left in an uncovered container or in puddles on the ground. This will prove fatal in sufficient quantity. Always drain the coolant into a sealable container. Coolant should be reused unless it is contaminated or several years old.

2. Disconnect electrical connectors at:
a. Throttle Position Sensor
b. Injector Wiring Harness at main engine harness and at water temperature indicator sensor.
c. Knock Sensor
d. Air Charge Temperature Sensor
e. Engine Coolant Temperature Sensor
f. Disconnect air bypass valve and ignition control assembly as equipped.
3. Disconnect upper intake manifold vacuum fitting connections by disconnecting:
a. Vacuum lines at upper intake manifold vacuum tree. Labeling the hose locations with tape is recommended to aid reinstallation.
b. Vacuum line to EGR valve.
c. Vacuum line to fuel pressure regulator.
d. Canister purge line as equipped.
4. Remove throttle linkage shield, and disconnect throttle linkage, cruise control, and kickdown cable. Unbolt accelerator cable from bracket and position cable out of the way.
5. Disconnect air intake hose, air bypass hose, and crankcase vent hose. Disconnect PCV system by disconnecting hose from fitting on underside of upper intake manifold.
6. Loosen hose clamp on water bypass line at lower intake manifold, and disconnect hose.
7. Disconnect EGR tube from EGR valve by removing flange nut.
8. Remove upper intake manifold retaining nuts.
9. Remove upper intake manifold and air throttle body assembly.
10. Remove engine oil dipstick bracket retaining bolt.
11. Disconnect the push connect fitting at the fuel supply manifold and fuel return lines.
12. Disconnect the electrical connectors from all four fuel injectors and move harness aside.
13. Remove fuel supply manifold retaining bolts. Carefully remove fuel supply manifold and injectors.

NOTE: Injectors can be removed from the fuel supply manifold at this time by exerting a slight twisting/pulling motion.

14. Remove four bottom retaining bolts from lower manifold.
15. Remove four upper retaining bolts from lower manifold.

NOTE: The front two bolts also secure an engine lift bracket. Remove lower intake manifold assembly.

To install:
16. Clean and inspect the mounting faces of the fuel charging manifold assembly and the cylinder head. Both surfaces must be clean and flat.
17. Clean and oil manifold bolt threads.
18. Install a new gasket.
19. Position the lower manifold assembly to head and install engine lift bracket. Install four upper manifold retaining bolts finger tight.
20. Install four remaining manifold bolts. Tighten all bolts to 15–22 ft. lbs. following the tightening sequence.
21. Install the fuel supply manifold and injectors with two retaining bolts. Tighten bolts to 15–22 ft. lbs.

NOTE: Refer to Fuel Injector Installation procedure in this Section.

22. Connect four electrical connectors to injectors.
23. Ensure that gasket surfaces of upper and lower intake manifolds are clean.
24. Place a new service gasket on the lower intake manifold assembly and place the upper intake manifold in position.
25. Install retaining bolts and tighten in sequence to 15–22 ft. lbs. Refer to the necessary illustration.
26. Install engine oil dipstick.
27. Connect the fuel supply and fuel return lines to fuel supply manifold.
28. Connect EGR tube to the EGR valve. Tighten to 18–28 ft. lbs.
29. Connect the water bypass line.
30. Connect the PCV system hose to the fitting on the underside of the upper intake manifold.
31. Connect upper intake manifold vacuum fitting connections by reconnecting:
a. Vacuum lines at upper intake manifold vacuum tree.
b. Vacuum line to EGR valve.
c. Vacuum line to fuel pressure regulator.
d. Canister purge line if equipped.
32. Hold accelerator cable bracket in position on upper manifold and install retaining bolts. Tighten bolt to 10–15 ft. lbs.
33. Install accelerator cable to bracket.
34. Connect accelerator cable and cruise control. Install throttle linkage shield.
35. Connect electrical connectors to:
a. Throttle position sensor
b. Injector wiring harness
c. Injector wiring harness at water temperature indicator sensor.
d. Knock sensor
e. Air charge temperature sensor
f. Engine coolant temperature sensor
e. Air bypass valve and ignition control assembly if equipped.
36. Connect air intake hose, air bypass hose, and crankcase vent hose.
37. Connect negative battery cable.
38. Install engine coolant.
39. Replace fuel pressure relief cap, then build up fuel pressure as follows: without starting the engine, turn key back and forth at least six times from ON to OFF position, leaving key ON for 15 seconds each time; then check for fuel leaks.
40. Use EEC self-test connector to check proper EEC-IV system functioning.
41. Start engine and allow to run at idle until engine temperature stabilizes. Check for cooling system leaks.
42. Verify correct engine idle.

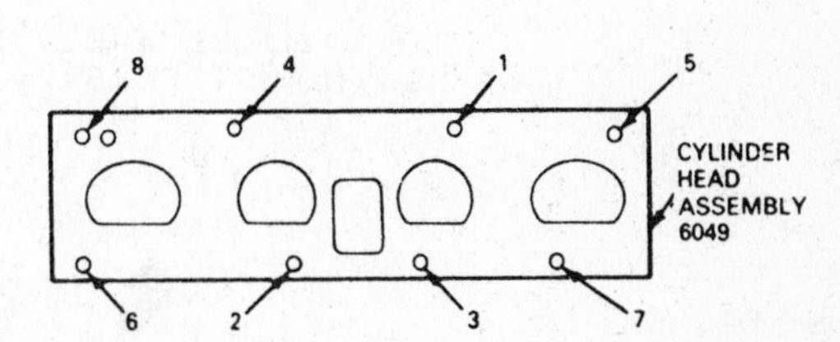

Lower manifold tightening sequence – 2.3L EFI engine

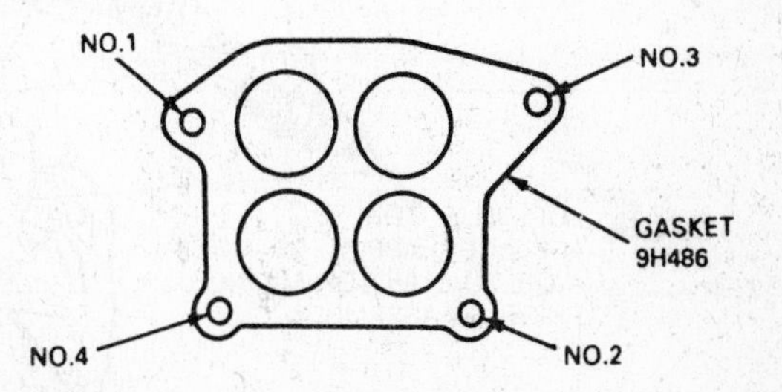

Upper manifold tightening sequence – 2.3L EFI engine

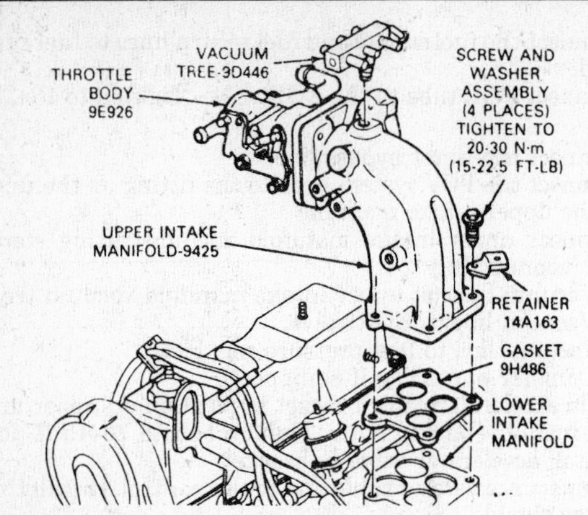

Upper manifold assembly installation — 2.3L EFI engine

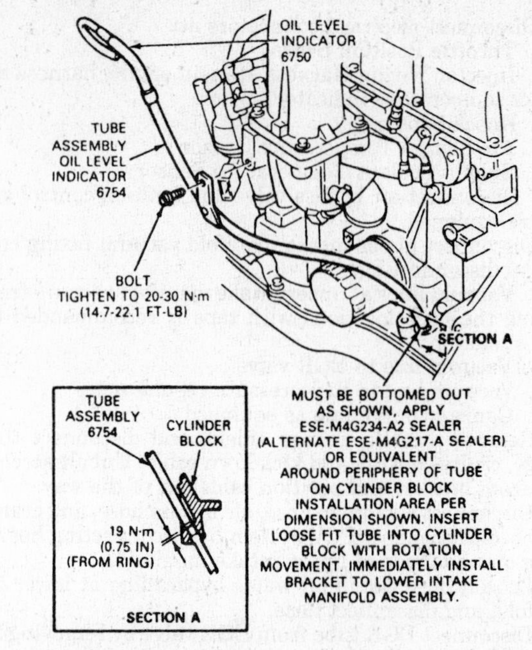

Fuel charging assembly installation — 2.3L EFI

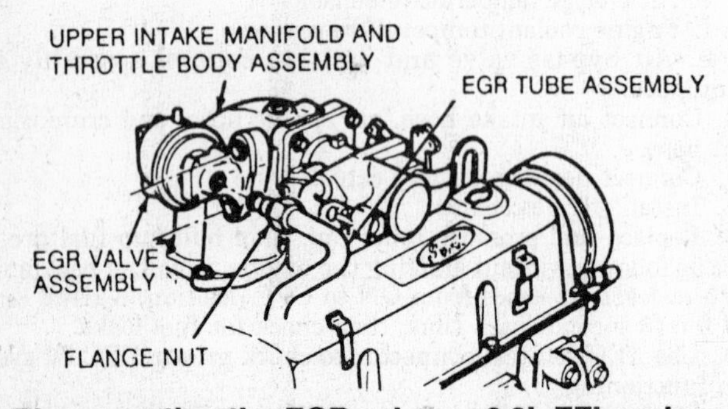

Disconnecting the EGR valve — 2.3L EFI engine

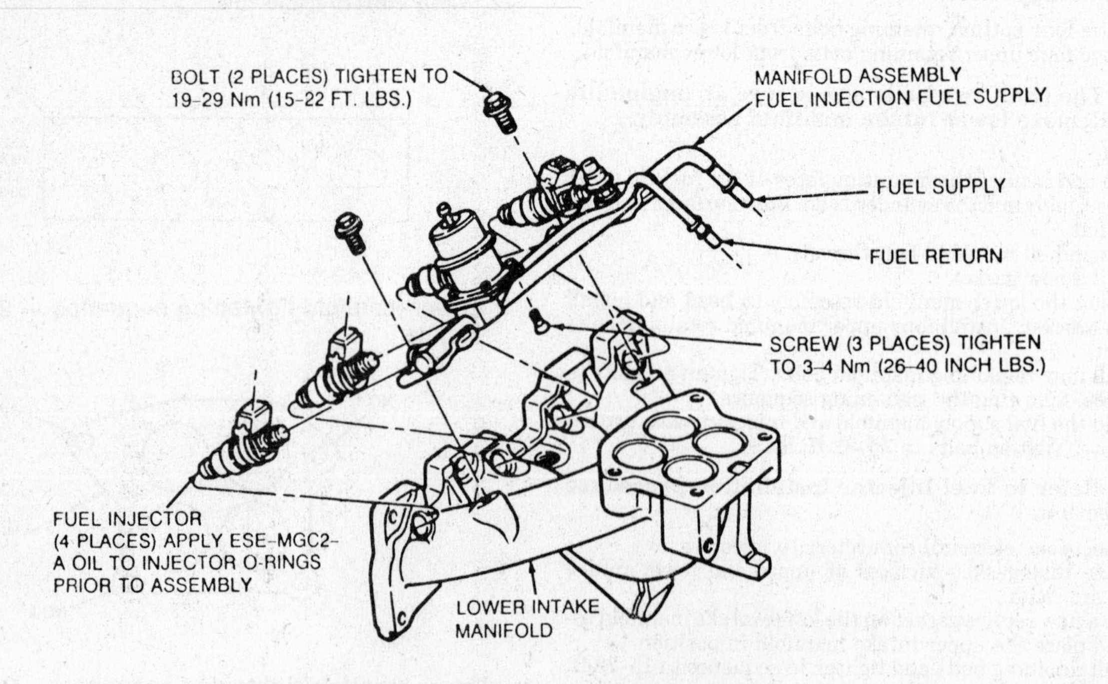

Location of the fuel lines — 2.3L EFI engine

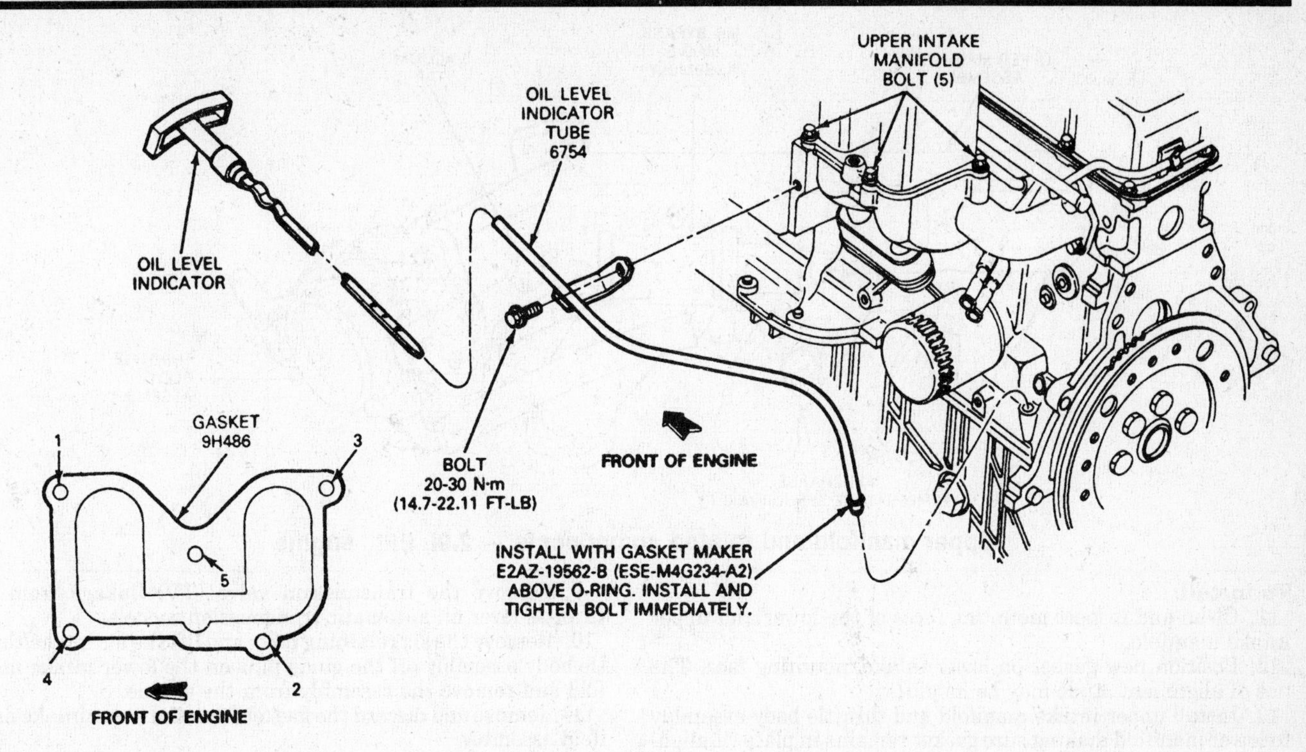

NOTE: The three bolts with stud heads go in hole positions 2, 3 and 4.

Dipstick installation — 2.3L EFI engine

2.9L Engine

1. Disconnect electrical connectors at air bypass valve, throttle position sensor, EGR sensor and air charge temperature sensor (ACT).
2. Remove air inlet tube from air cleaner to throttle body.
3. Remove snow/ice shield to expose throttle linkage. Disconnect throttle cable from ball stud.
4. Disconnect upper intake manifold vacuum connectors; both the front and rear fittings including the EGR valve and vacuum line to fuel pressure regulator.
5. Disconnect PCV closure tube from under throttle body and disconnect PCV vacuum tube from under the manifold.
6. Remove canister purge line from fitting near power steering pump.
7. Disconnect EGR tube from EGR valve by removing flange nut.
8. Loosen bolt which retains A/C line at the upper rear of the upper manifold and disengage retainer.
9. Remove six upper intake manifold retaining bolts.
10. Remove upper intake and throttle body as an assembly from lower intake manifold.

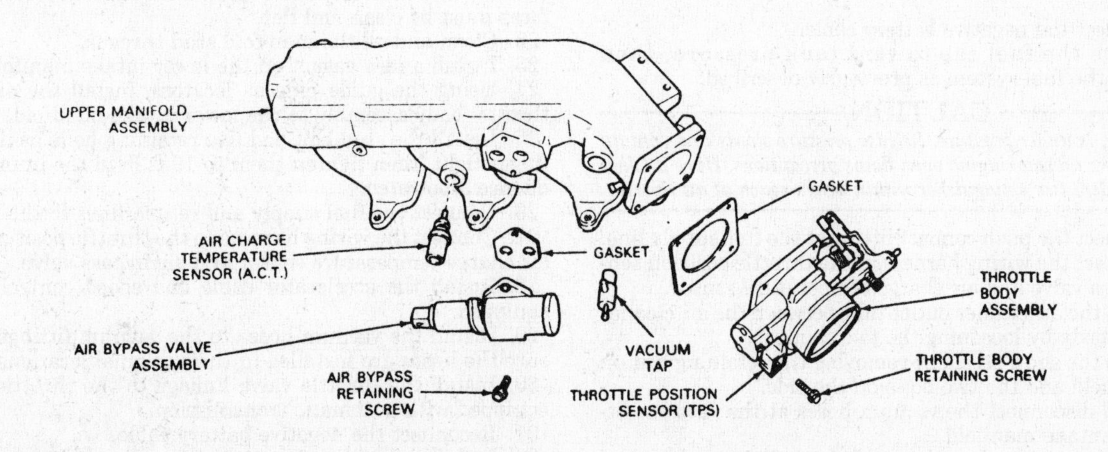

Upper manifold and related components — 2.9L EFI engine

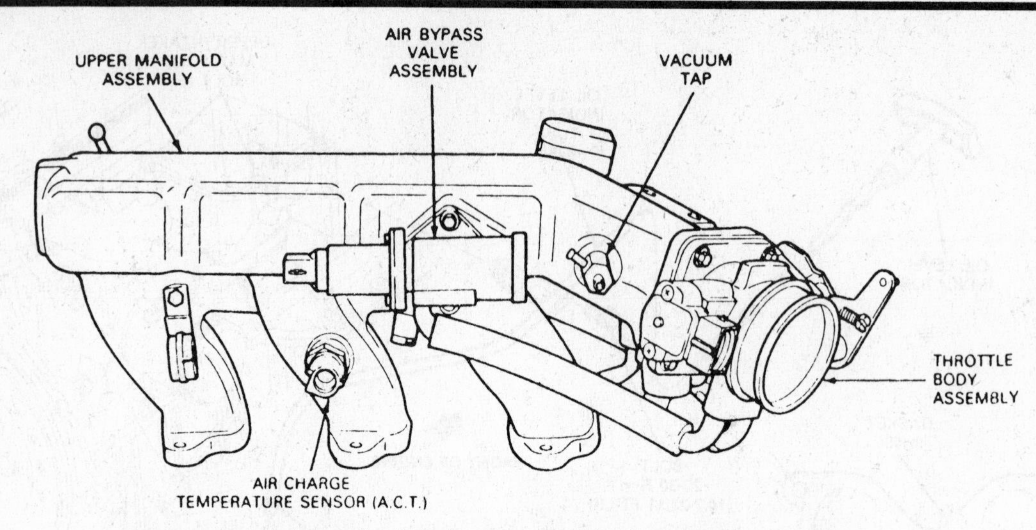

Upper manifold and related components — 2.9L EFI engine

To install:

11. Clean and inspect mounting faces of the lower and upper intake manifold.
12. Position new gasket on lower intake mounting face. The use of alignment studs may be helpful.
13. Install upper intake manifold and throttle body assembly to lower manifold making sure gasket remains in place (if alignment studs aren't used). Align EGR tube in valve.
14. Install six upper intake manifold retaining bolts. Tighten to 11–15 ft. lbs.
15. Engage A/C line retainer cup and tighten bolt to specification.
16. Tighten EGR tube and flare fitting. Tighten lower retainer nut at the exhaust manifold.
17. Install canister purge line to fitting.
18. Connect PCV vacuum hose to bottom of upper manifold and PCV closure hose to throttle body.
19. Connect vacuum lines to vacuum tree, EGR valve, and fuel pressure regulator.
20. Connect throttle cable to throttle body and install snow/ice shield.
21. Connect electrical connector at air bypass valve, TPS sensor, EGR sensor, and ACT sensor.

NOTE: If lower intake manifold assembly was removed, fill and bleed the cooling system.

3.0L Engine

1. Disconnect the negative battery cable.
2. Remove the fuel cap to vent tank pressure, then depressurize the fuel system as previously described.

CAUTION

The fuel system is under pressure. Release pressure slowly and contain spillage. Observe no smoking/no open flame precautions. Have a Class B–C (dry powder) fire extinguisher within arm's reach at all times.

3. Disconnect the push connect fitting at the fuel supply line.
4. Disconnect the wiring harness at the throttle position sensor, air bypass valve and air charge temperature sensor.
5. Remove the air cleaner outlet tube between the air cleaner and throttle body by loosening the two clamps.
6. Remove the snow shield by removing the retaining nut on top of the shield and the two bolts on the side.
7. Tag and disconnect the vacuum hoses at the vacuum fittings on the intake manifold.
8. Disconnect and remove the accelerator and speed control cables (if equipped) from the accelerator mounting bracket and throttle lever.
9. Remove the transmission valve (TV) linkage from the throttle lever on automatic transmission models.
10. Remove the six retaining bolts and lift the air intake/throttle body assembly off the guide pins on the lower intake manifold and remove the assembly from the engine.
11. Remove and discard the gasket from the lower intake manifold assembly.
12. Carefully disconnect the wiring harness from the fuel injectors.
13. Disconnect the vacuum line from the fuel pressure regulator.
14. Remove the four fuel injector manifold retaining bolts, two on each side.
15. Carefully disengage the fuel rail assembly from the fuel injectors by lifting and gently rocking the rail.

To install:

16. Carefully install the fuel rail assembly and injectors into the lower intake manifold, one side at a time, pushing down on the fuel rail to make sure the O-rings are seated.
17. Hold the fuel rail assembly in place and install the retaining bolts finger tight. Tighten the retaining bolts to 6–8 ft. lbs.
18. Connect the fuel supply and return lines.
19. Connect the fuel injector wiring harness at the injectors.
20. Connect the vacuum line to the fuel pressure regulator.
21. Clean and inspect the mounting faces of the air intake/throttle body assembly and the lower intake manifold. Both surfaces must be clean and flat.
22. Clean and oil the manifold stud threads.
23. Install a new gasket on the lower intake manifold.
24. Using the guide pins as locators, install the air intake/throttle body assembly to the lower intake manifold.
25. Install the stud bolt and five retaining bolts as illustrated finger tight, then tighten them to 19 ft.lb in the numbered sequence illustrated.
26. Connect the fuel supply and return lines to the fuel rail.
27. Connect the wiring harness to the throttle position sensor, air charge temperature sensor and air bypass valve.
28. Install the accelerator cable and speed control cable, if equipped.
29. Install the vacuum hoses to the vacuum fittings, making sure the hoses are installed in their original locations.
30. Install the throttle valve linkage to the throttle lever, if equipped with automatic transmission.
31. Reconnect the negative battery cable.
32. Install the fuel tank cap.
33. Install the snow shield and air cleaner outlet tube.
34. Build up fuel pressure by turning the ignition switch ON

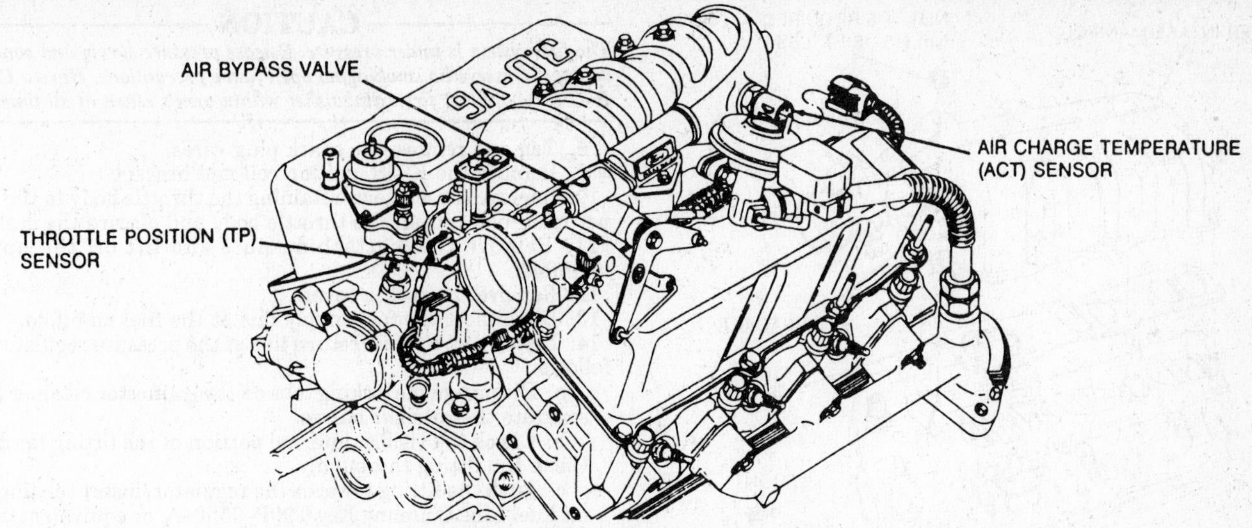

Wiring connections – 3.0L engine

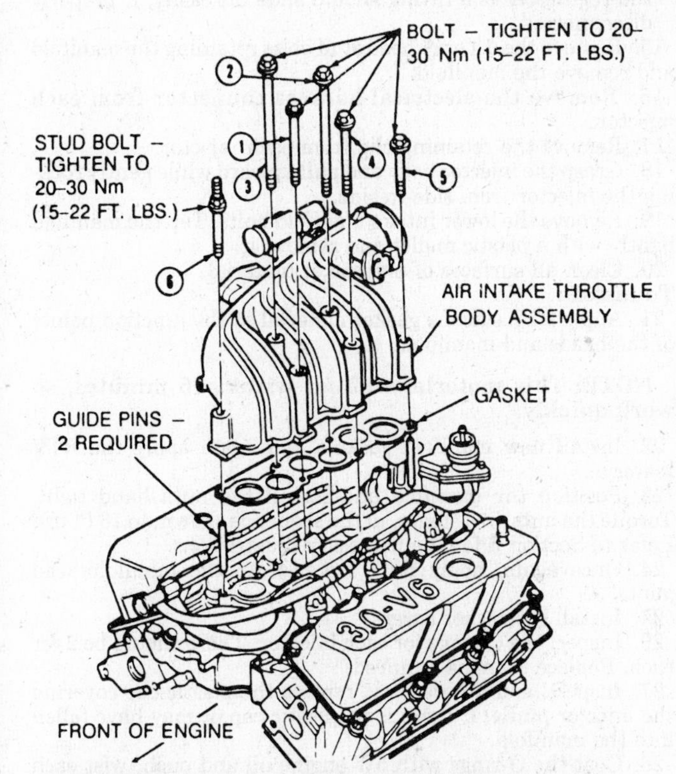

Removing the air intake/throttle body assembly. Tighten the bolts in the numbered sequence when installing

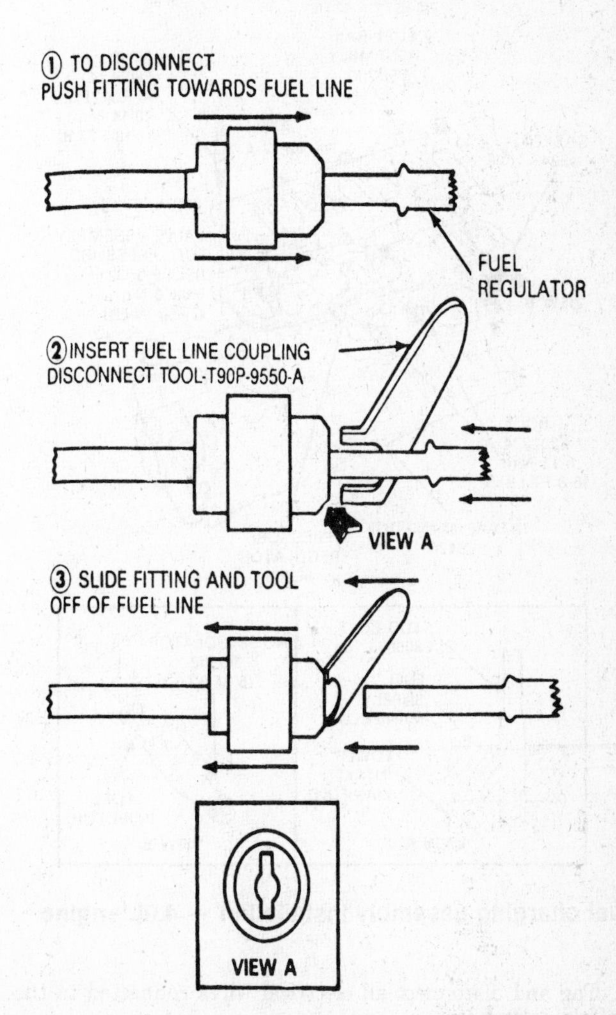

Fuel line removal and installation procedures

and OFF at least 6 times, leaving the ignition on for at least five seconds each time. Check for fuel leaks.

35. Start the engine and adjust the idle speed, if necessary.

4.0L Engine

1. Disconnect the battery ground cable.
2. Remove the air cleaner and intake duct.
3. Remove the weather shield.
4. Disconnect the throttle cable and bracket.
5. Tag and disconnect all vacuum lines connected to the manifold.

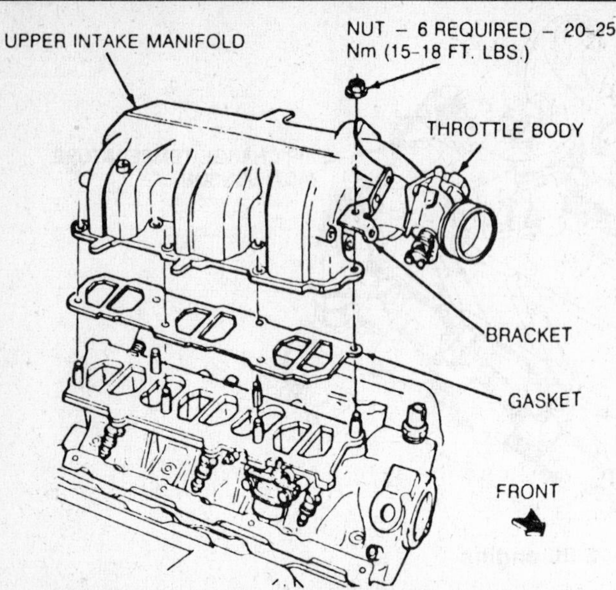

Removing the air intake/throttle body assembly

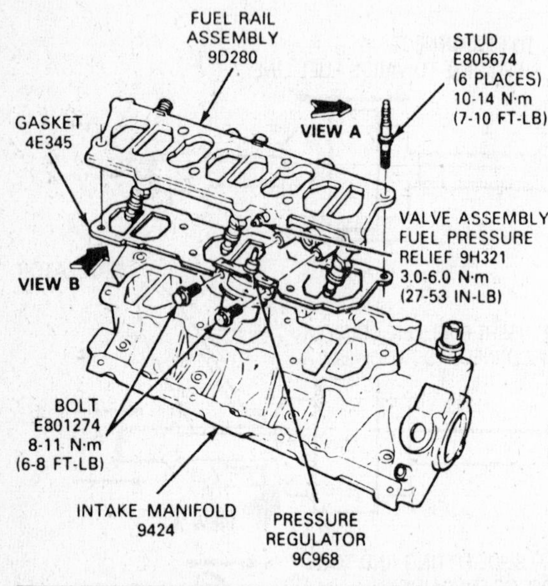

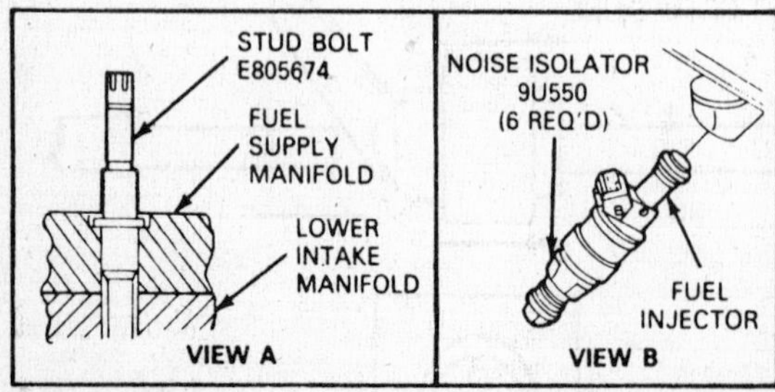

Fuel charging assembly installation – 4.0L engine

6. Tag and disconnect all electrical wires connected to the manifold assemblies.
7. Relieve the fuel system pressure. Refer to the necessary service procedure in this Section.

CAUTION

The fuel system is under pressure. Release pressure slowly and contain spillage. Observe no smoking/no open flame precautions. Have a Class B–C (dry powder) fire extinguisher within arm's reach at all times.

8. Tag and remove the spark plug wires.
9. Remove the EDIS ignition coil and bracket.
10. Remove the 4 screws retaining the throttle body to the upper manifold. Lift off the throttle body and discard the gasket.
11. Remove the 6 attaching nuts and lift off the upper manifold.
12. Remove the rocker covers.
13. Disconnect the fuel supply line at the fuel manifold.
14. Disconnect the fuel return line at the pressure regulator as follows:
    a. Disengage the locking tabs on the connector retainer and separate the retainer halves.
    b. Check the visible, internal portion of the fitting for dirt. Clean the fitting thoroughly.
    c. Push the fitting towards the regulator, insert the fingers on Fuel Line Coupling Key T90P–9550–A, or equivalent, into the slots in the coupling. Using the tool, pull the fitting from the regulator. The fitting should slide off easily, if properly disconnected.
15. Remove the 6 Torx® head stud bolts retaining the manifold and remove the manifold.
16. Remove the electrical harness connector from each injector.
17. Remove the retaining clip from each injector.
18. Grasp the injector body and pull upward while gently rocking the injector from side-to-side.
19. Remove the lower intake manifold bolts. Tap the manifold lightly with a plastic mallet and remove it.
20. Clean all surfaces of old gasket material.

To install:

21. Apply RTV silicone gasket material at the junction points of the heads and manifold.

NOTE: This material will set within 15 minutes, so work quickly!

22. Install new manifold gaskets and again apply the RTV material.
23. Position the manifold and install the nuts hand tight. Torque the nuts, in 4 stages, in the sequence shown, to 18 ft. lbs. Refer to Section 3 for the necessary illustrations.
24. Once again, apply RTV material to the manifold/head joints.
25. Install the rocker covers.
26. Inspect the O-rings for each injector. There should be 2 for each. Replace them as required.
27. Inspect, and if necessary, replace the plastic cap covering the injector pintle. If there is no plastic cap, it may have fallen into the manifold.
28. Coat the O-rings with 5W engine oil and push/twist each injector into the fuel manifold.
29. Install the retainers and electrical harness connectors.
30. Position the fuel supply manifold and press it down firmly until the injectors are fully seated in the fuel supply manifold and lower intake manifold.
31. Install the 6 Torx® head bolts and tighten them to 7–10 ft. lbs.
32. Install the fuel supply line and tighten the fitting to 15–18 ft. lbs.
33. Install the fuel return line on the regulator by pushing it onto the fuel pressure regulator line of to the shoulder.

WARNING: The connector should grip the line securely!

34. Install the connector retainer and snap the two halves of the retainer together.
35. Install the upper manifold. Tighten the nuts to 18 ft. lbs.

36. Install the EDIS coil.
37. Connect the fuel and return lines.
38. Ensure that the mating surfaces of the throttle body and upper manifold are clean and free of gasket material.
39. Install a new gasket on the manifold and position the throttle body on the manifold. Tighten the bolts to 76–106 inch lbs.
40. Connect all wires.
41. Connect all vacuum lines.
42. Connect the throttle linkage.
43. Install the weather shield.
44. Install the air cleaner and duct.
45. Fill and bleed the cooling system.
46. Connect the battery ground.
47. Run the engine and check for leaks.

Air Throttle Body

REMOVAL AND INSTALLATION

2.3L EFI Engine

1. Disconnect the negative battery cable. Remove throttle linkage shield and throttle cable and cruise control.
2. Remove throttle position sensor electrical connector.
3. Remove air intake hose, air bypass hose, and crankcase vent hose.
4. Remove two throttle body nuts (lower) and two bolts (upper).
5. Carefully separate air throttle body from upper intake manifold.
6. Remove and discard gasket between throttle body and upper intake manifold.

WARNING: If scraping is necessary, be careful not to damage gasket surfaces of throttle body and upper manifold assemblies, or allow material to drop into manifold.

To install:

7. Ensure that both throttle body and upper intake manifold gasket surfaces are clean.
9. Install throttle body gasket on the two lower studs of the upper intake manifold.
9. Install throttle body to upper intake manifold and secure with two retaining nuts and two bolts (hand start all first-then evenly tighten) Tighten to 12–25 ft. lbs.
10. Connect throttle position sensor electrical connectors, throttle cable, cruise control, air intake hose, air bypass hose, and crankcase vent hose.
11. Connect throttle linkage shield. Connect battery cable start engine and check for proper operation.

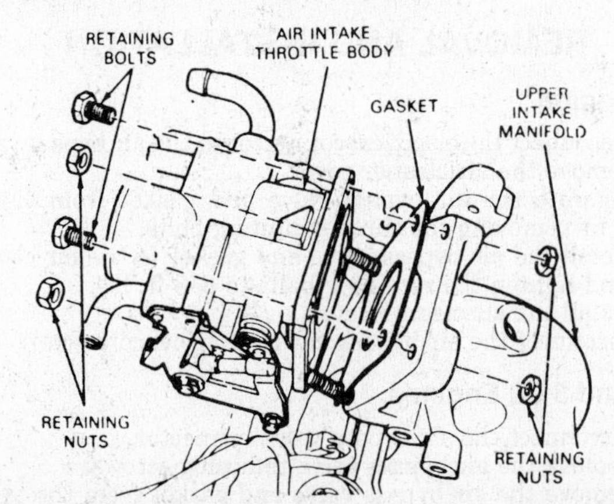

Throttle body replacement – 2.3L EFI engine

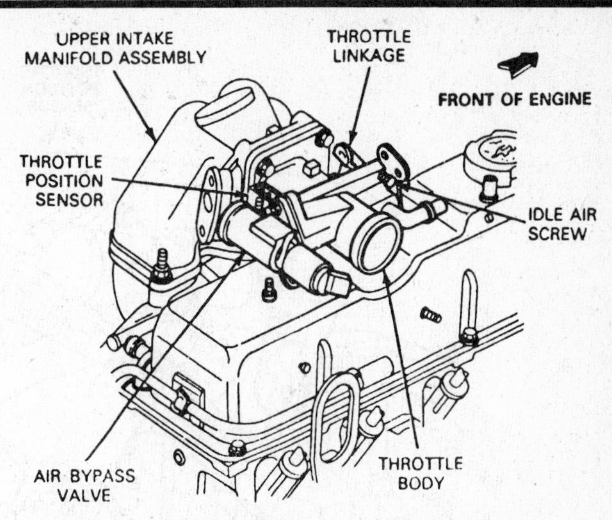

Throttle body assembly – 2.3L EFI engine

2.9L EFI Engine

1. Disconnect the negative battery cable. Disconnect throttle position sensor electrical connector.
2. Remove air inlet duct.
3. Remove snow/ice shield and disconnect throttle cable from ball stud.
4. Disconnect air bypass hose, PCV closure hose, and canister purge hose from the fittings beneath throttle body.
5. Remove four screws retaining throttle body to upper intake manifold.
6. Carefully separate air throttle body from upper intake manifold.
7. Remove and discard gasket between throttle body and upper intake manifold.

To install:

8. Ensure that both throttle body and upper intake manifold gasket surfaces are clean.
9. Install throttle body and gasket to the upper intake manifold. Tighten retaining screws to 6–9 ft. lbs.
10. Reconnect all electrical, vacuum and battery connections. Start engine and check for proper operation.

3.0L Engine

1. Disconnect the negative battery cable. Remove the air cleaner outlet tube.
2. Remove the snow shield by removing the three plastic retainers.
3. Matchmark and disconnect the vacuum and PCV hoses.
4. Disconnect and remove the accelerator and speed control cables (if equipped) from the accelerator mounting bracket and throttle lever.
5. Disconnect the wiring harness at the throttle position sensor, air bypass valve and air charge temperature sensor.
6. Remove the 4 retaining bolts and 2 stud bolts and lift the air intake/throttle body assembly off the lower intake manifold.
7. Remove and discard the gasket.

To install:

8. Clean and inspect the mounting faces of the throttle body assembly and the lower intake manifold. Both surfaces must be clean and flat.
9. Install a new gasket on the lower intake manifold.
10. Install the air throttle body assembly on the lower intake manifold.
11. Install the bolts finger tight, then tighten them evenly in sequence fronm the center outward (see illustration in this Section) to 19 ft. lbs.

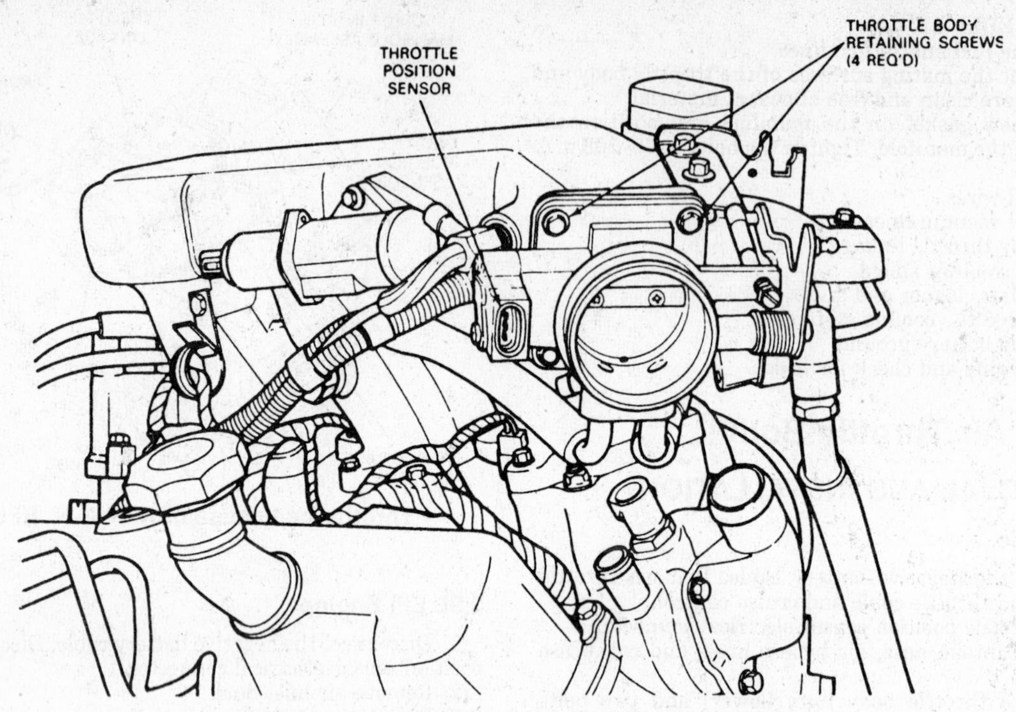

Throttle body replacement — 2.9L EFI engine

12. Connect the wiring harness to the throttle position sensor, air charge temperature sensor and air bypass valve.
13. Install the accelerator cable and speed control cable, if equipped.
14. Install the vacuum hoses to the vacuum fittings, making sure the hoses are installed in their original locations.
15. Install the snow shield and air cleaner outlet tube. Reconnect the battery cable, start engine and check for proper operation.

4.0L Engine

1. Disconnect the negative battery cable. Remove the air cleaner inlet tube.
2. Remove the snow shield.
3. Disconnect the throttle cable at the ball stud.
4. Disconnect the canister purge hose from under the throttle body.

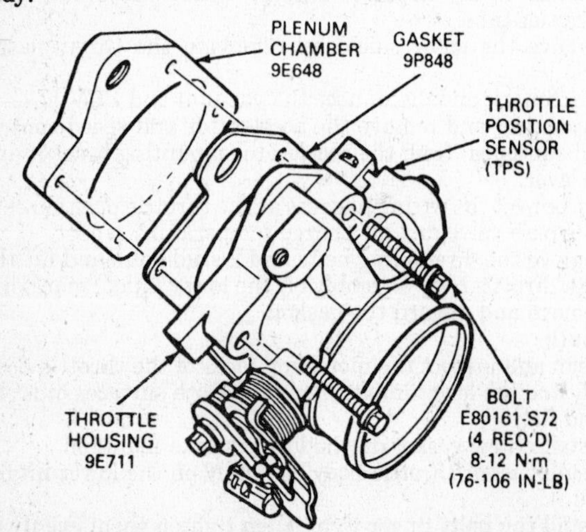

Throttle body assembly — 4.0L EFI engine

5. Disconnect the wiring harness at the throttle position sensor.
6. Remove the 4 retaining bolts and lift the throttle body assembly off the upper intake manifold.
7. Remove and discard the gasket.

To install:

8. Clean and inspect the mounting faces of the throttle body assembly and the upper intake manifold. Both surfaces must be clean and flat.
9. Install a new gasket on the manifold.
10. Install the air throttle body assembly on the intake manifold.
11. Install the bolts finger tight, then tighten them evenly to 76–106 inch lbs.
12. Connect the wiring harness to the throttle position sensor.
13. Install the canister purge hose.
14. Install the snow shield and air cleaner outlet tube. Reconnect the battery cable. Start engine, check for proper operation.

Air Bypass Valve

REMOVAL AND INSTALLATION

2.3L Engine

1. Disconnect the electrical connector at the air bypass valve.
2. Remove the air cleaner cover.
3. Separate the air bypass valve and gasket from the air cleaner by removing the three mounting bolts.
4. Install the air bypass valve and gasket to the air cleaner cover and tighten the retaining bolts to 6–8 ft. lbs.
5. Install the air cleaner cover.
6. Reconnect the air bypass valve electrical connector.

2.9L And 3.0L Engines

1. Disconnect the air bypass valve connector.
2. Remove the air bypass valve retaining screws.
3. Remove the air bypass valve and gasket from the air intake/throttle body assembly. If scraping is necessary to remove

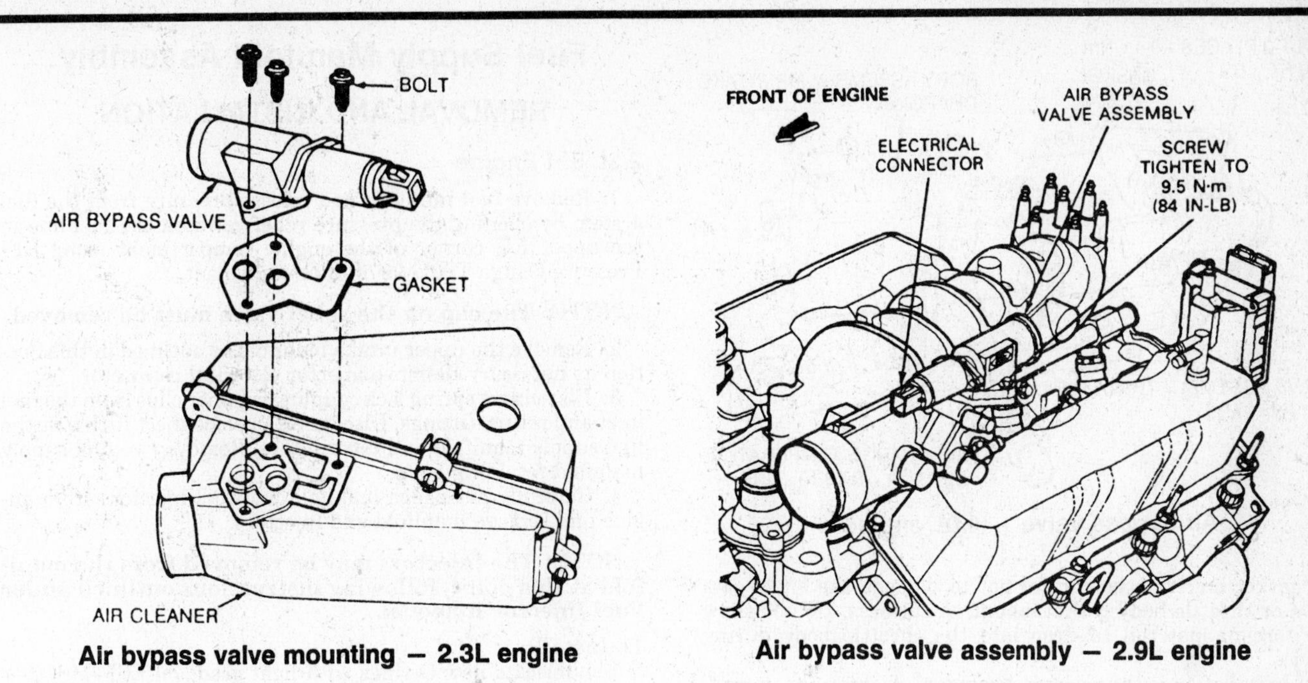

Air bypass valve mounting – 2.3L engine

Air bypass valve assembly – 2.9L engine

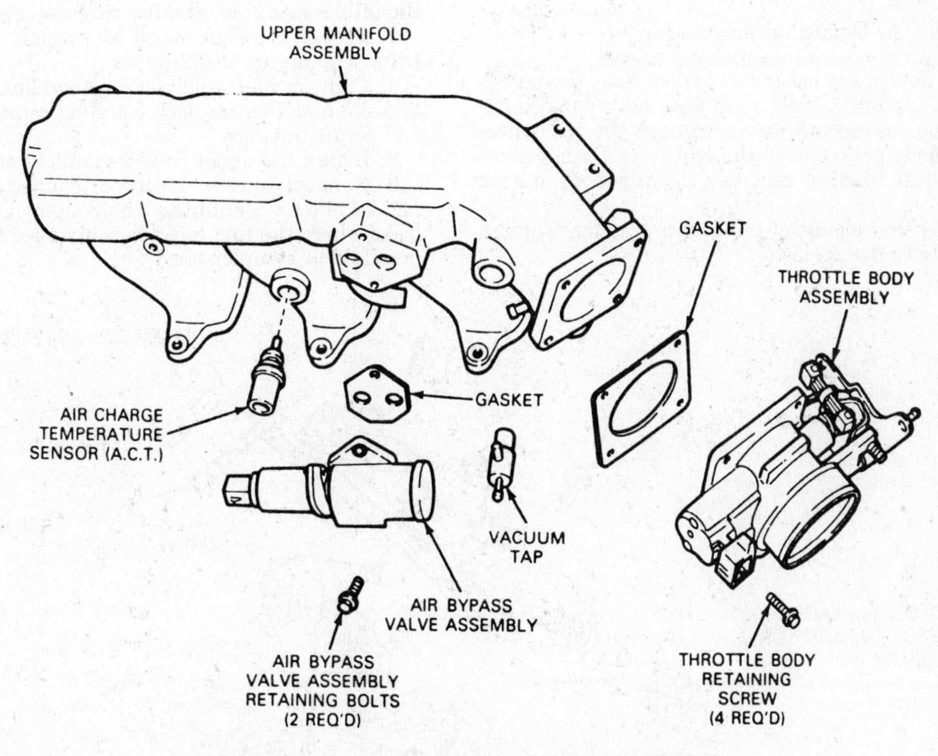

Air bypass valve assembly – 3.0L engine

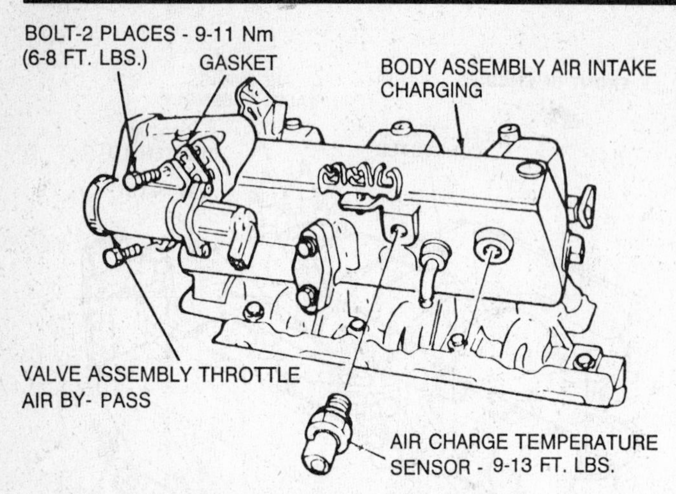

Air bypass valve — 4.0L engine

old gasket material, be careful not to damage the air bypass valve or throttle body gasket mounting surfaces. Do not allow any foreign material to drop into the throttle body during service.

4. Installation is the reverse of removal. Tighten the mounting bolts to 6–8 ft. lbs.

6-4.0L

1. Disconnect the air bypass valve connector.
2. Remove the air bypass valve retaining screws.
3. Remove the air bypass valve and gasket from the air intake/throttle body assembly. If scraping is necessary to remove old gasket material, be careful not to damage the air bypass valve or throttle body gasket mounting surfaces. Do not allow any foreign material to drop into the throttle body during service.
4. Installation is the reverse of removal procedure. Tighten the mounting bolts to 6–8 ft. lbs.

Fuel Supply Manifold Assembly

REMOVAL AND INSTALLATION

2.3L EFI Engine

1. Remove fuel tank cap and release pressure from the fuel system by opening the pressure relief valve on the fuel line in the upper RH corner of the engine compartment using EFI Pressure Gauge T80L–9974–B or equivalent.

NOTE: The cap on the relief valve must be removed.

2. Remove the upper intake manifold as outlined in this Section as necessary. Remove injector electrical connector.
3. Disconnect spring lock coupling retainer clips from the fuel inlet and return fittings. Disconnect push connect fitting at the fuel supply manifold and return lines. Remove two fuel supply manifold retaining bolts.
4. Carefully disengage manifold and fuel injectors from engine and remove manifold and injectors.

NOTE: The injectors may be removed from the manifold at this point, following instructions outlined under Fuel Injector Removal.

To install:

5. Lubricate new O-rings with light grade oil ESF–M6C2–A or equivalent and install two on each injector (one per injector if injectors were not removed from fuel supply manifold).
6. Install the fuel injector supply manifold and injectors into the intake manifold, making sure the injectors are well seated.
7. Secure the fuel manifold assembly using two retaining bolts. Tighten to 15–22 ft. lbs.
8. Connect fuel supply and return lines to fuel supply manifold. Reinstall spring lock coupling retainer clips on fuel inlet and return fittings.
9. Install the upper intake manifold assembly as outlined.
10. Connect injector electrical connectors. Start the engine and let idle for 2 minutes. Turn engine OFF and check for fuel leaks where the fuel injector is installed in the fuel rail, intake manifold or cylinder head.

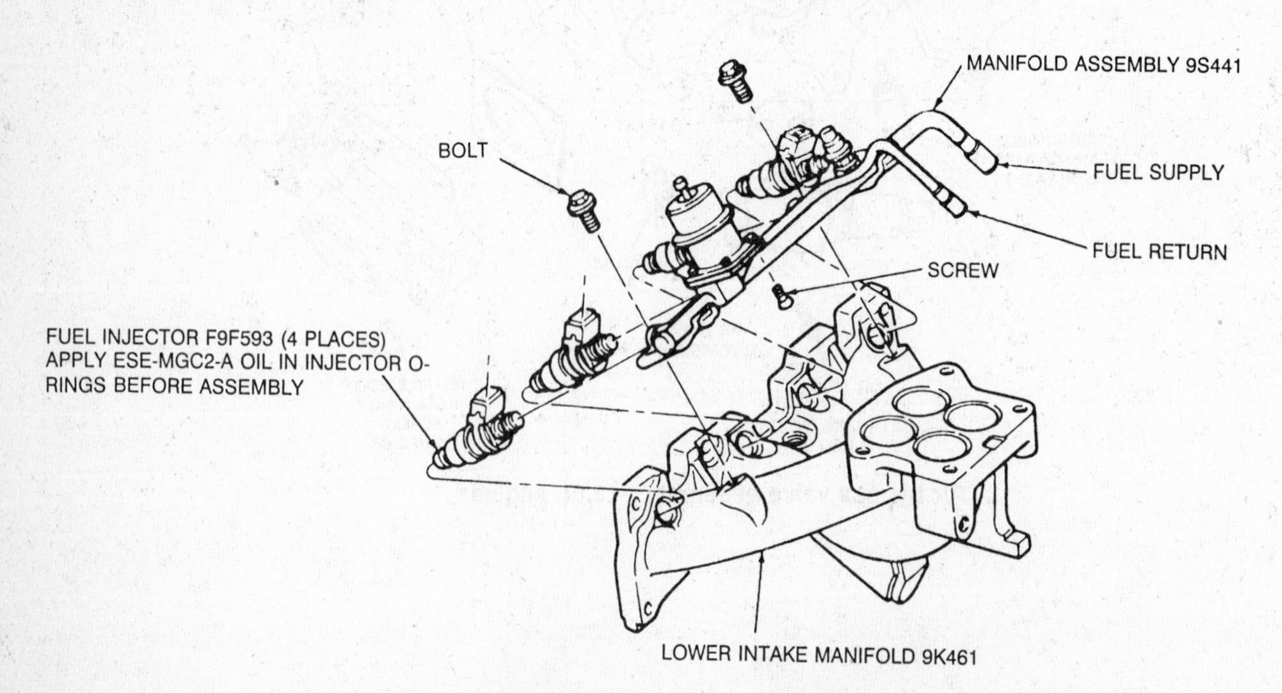

Fuel supply manifold — 2.3L EFI engine

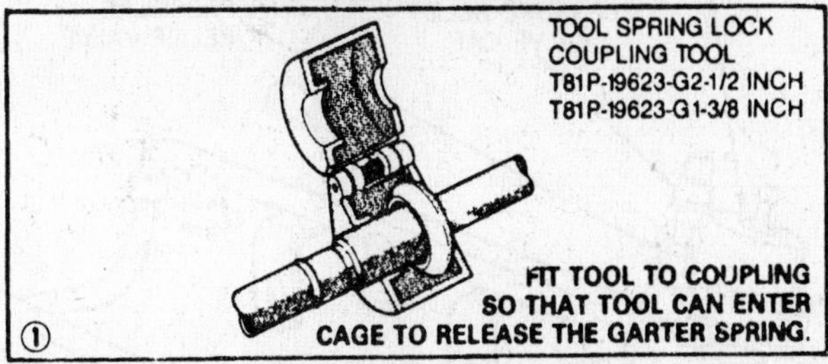

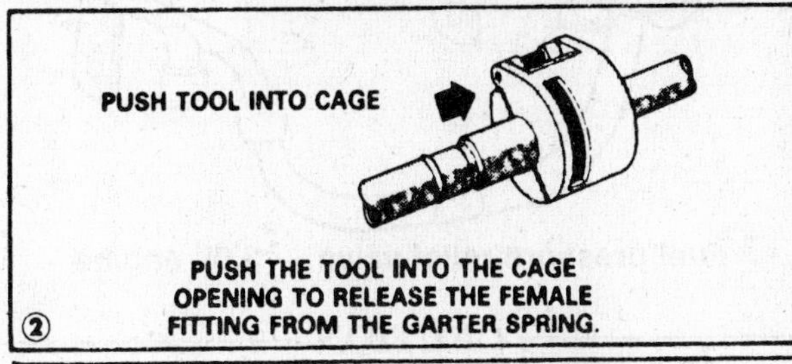

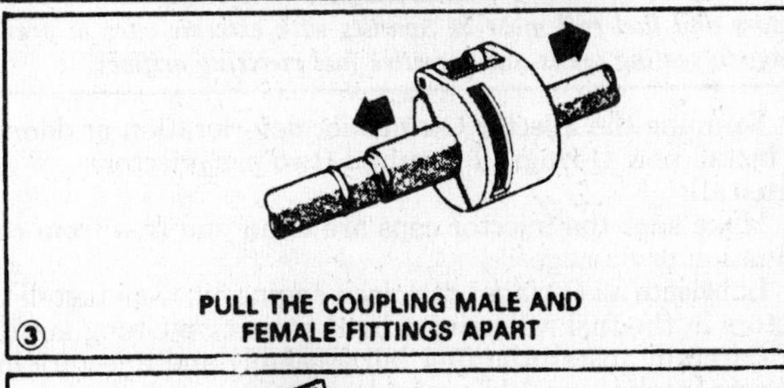

Fuel line removal and installation procedures

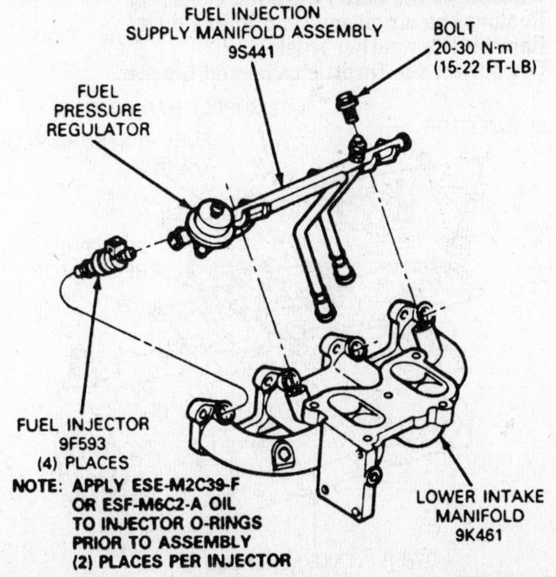

Fuel supply manifold installation – 2.3L EFI engine

2.9L EFI Engine

1. Thoroughly clean the engine.
2. Depressurize the fuel system.
3. Remove upper manifold assembly.
4. Disconnect the fuel supply and return line connections at the fuel supply manifold.
5. Remove four fuel supply manifold retaining bolts.
6. Using Spring Lock coupling Tools D87L–9280–A or B or equivalent, disconnect the crossover fuel hose from the fuel supply manifold.
7. Carefully disengage fuel supply manifold from lower intake manifold. The fuel injectors are retained in the fuel supply manifold with clips.
8. Installation is the reverse of the service removal procedure. Make sure the injectors are fully seated in the fuel rail cups and intake manifold all fuel line connections are connected properly. Start the engine and let idle for 2 minutes. Turn engine OFF and check for fuel leaks where the fuel injector is installed in the fuel rail and intake manifold.

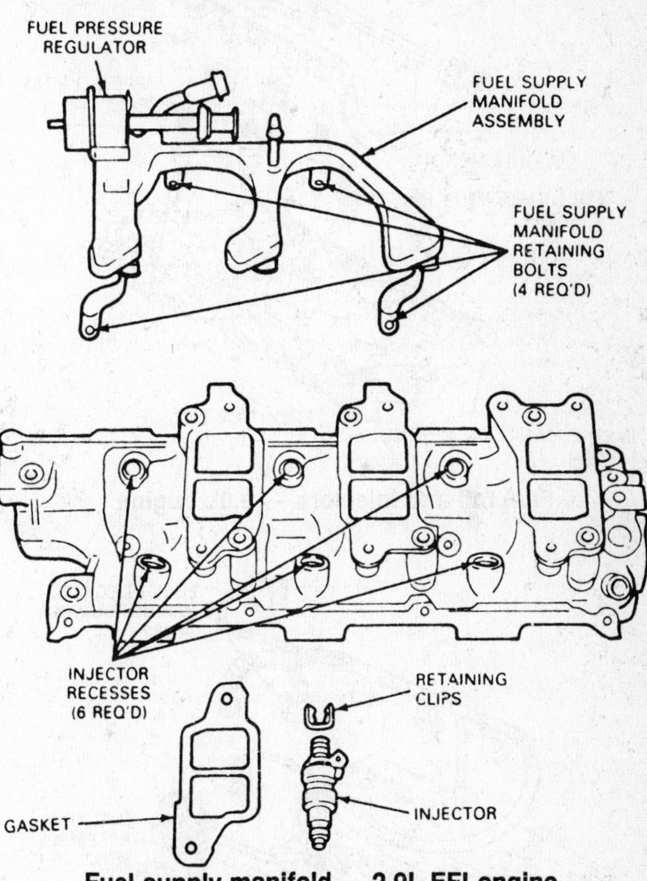

Fuel supply manifold – 2.9L EFI engine

6-3.0L Engine

1. Remove the air intake/throttle body assembly as previously described. Be sure to depressurize the fuel system before disconnecting any fuel lines.

CAUTION

The fuel system is under pressure. Release pressure slowly and contain spillage. Observe no smoking/no open flame precautions. Have a Class B–C (dry powder) fire extinguisher within arm's reach at all times.

2. Carefully disconnect the wiring harness from the fuel injectors.
3. Disconnect the vacuum line from the fuel pressure regulator.
4. Remove the four fuel injector manifold retaining bolts, two on each side.
5. Carefully disengage the fuel rail assembly from the fuel injectors by lifting and gently rocking the rail.
6. Remove the fuel injectors from the intake manifold by lifting while gently rocking from side to side. Place all removed components on a clean surface to prevent contamination by dirt or grease.

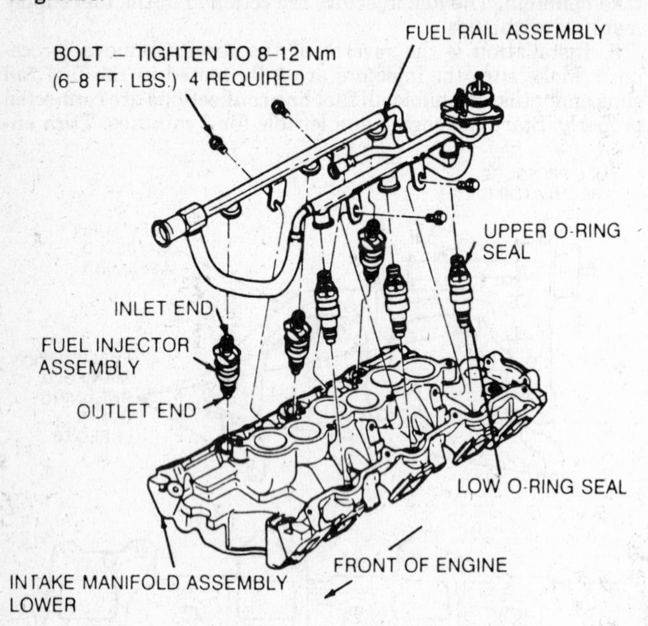

Fuel rail and injectors – 3.0L engine

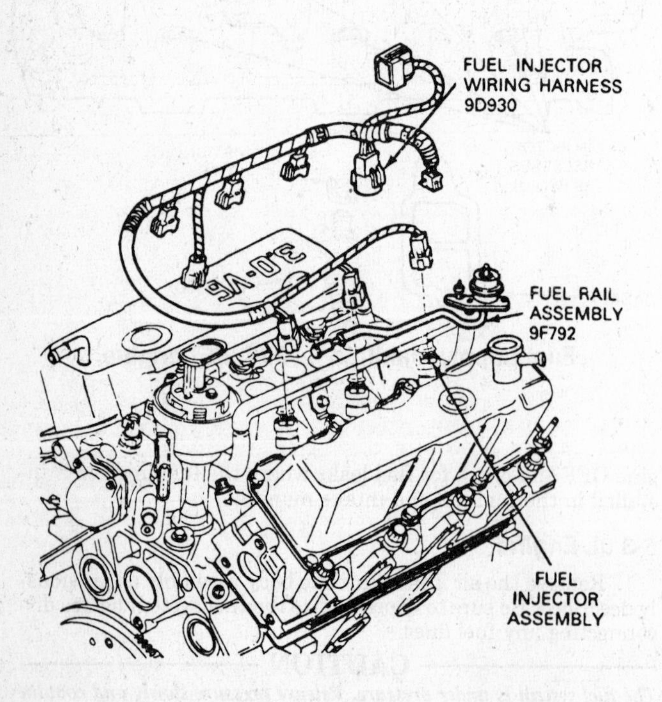

Fuel rail assembly – 3.0L engine

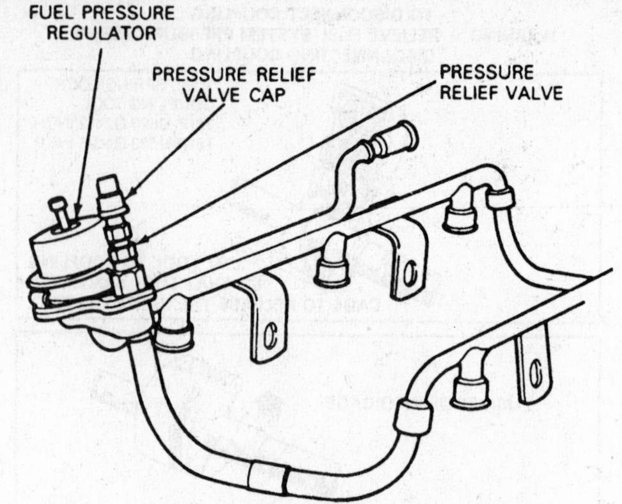

Fuel pressure relief valve – 3.0L engine

CAUTION

Injectors and fuel rail must be handles with extreme care to prevent damage to sealing areas and sensitive fuel metering orifices.

7. Examine the injector O-rings for deterioration or damage and install new O-rings, if required (two per injector).

To install:

8. Make sure the injector caps are clean and free from contamination or damage.
9. Lubricate all O-rings with clean engine oil, then install the injectors in the fuel rail using a light twisting/pushing motion.
10. Carefully install the fuel rail assembly and injectors into the lower intake manifold, one side at a time, pushing down on the fuel rail to make sure the O-rings are seated.
11. Hold the fuel rail assembly in place and install the retaining bolts finger tight. Tighten the retaining bolts to 6–8 ft. lbs.
12. Connect the fuel supply and return lines.
13. Connect the fuel injector wiring harness at the injectors.
14. Connect the vacuum line to the fuel pressure regulator.
15. Install the air intake/throttle body as previously described.

6-4.0L Engine

1. Disconnect the battery ground cable.
2. Remove the air cleaner and intake duct.
3. Remove the weather shield.
4. Disconnect the throttle cable and bracket.

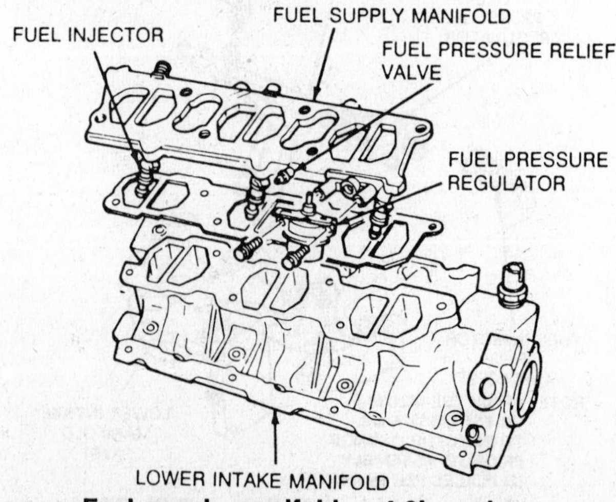

Fuel supply manifold – 4.0L engine

5. Tag and disconnect all vacuum lines connected to the manifold.
6. Tag and disconnect all electrical wires connected to the manifold assemblies.
7. Relieve the fuel system pressure.

CAUTION

The fuel system is under pressure. Release pressure slowly and contain spillage. Observe no smoking/no open flame precautions. Have a Class B–C (dry powder) fire extinguisher within arm's reach at all times.

8. Tag and remove the spark plug wires.
9. Remove the EDIS ignition coil and bracket.
10. Remove the 4 screws retaining the throttle body to the upper manifold. Lift off the throttle body and discard the gasket.
11. Remove the 6 attaching nuts and lift off the upper manifold.
12. Disconnect the fuel supply line at the fuel manifold.
13. Disconnect the fuel return line at the pressure regulator as follows:
   a. Disengage the locking tabs on the connector retainer and separate the retainer halves.
   b. Check the visible, internal portion of the fitting for dirt. Clean the fitting thoroughly.
   c. Push the fitting towards the regulator, insert the fingers on Fuel Line Coupling Key T90P–9550–A, or equivalent, into the slots in the coupling. Using the tool, pull the fitting from the regulator. The fitting should slide off easily, if properly disconnected.
14. Remove the 6 Torx® head stud bolts retaining the manifold and remove the manifold.

To install:

15. Position the fuel supply manifold and press it down firmly until the injectors are fully seated in the fuel supply manifold and lower intake manifold.
16. Install the 6 Torx® head bolts and tighten them to 7–10 ft. lbs.
17. Install the fuel supply line and tighten the fitting to 15–18 ft. lbs.
18. Install the fuel return line on the regulator by pushing it onto the fuel pressure regulator line of to the shoulder.

WARNING: The connector should grip the line securely!

19. Install the connector retainer and snap the two halves of the retainer together.
20. Install the upper manifold. Tighten the nuts to 18 ft. lbs.
21. Install the EDIS coil.
22. Connect the fuel and return lines.
23. Ensure that the mating surfaces of the throttle body and upper manifold are clean and free of gasket material.
24. Install a new gasket on the manifold and position the throttle body on the manifold. Tighten the bolts to 76–106 inch lbs.
25. Connect all wires.
26. Connect all vacuum lines.
27. Connect the throttle linkage.
28. Install the weather shield.
29. Install the air cleaner and duct.
30. Fill and bleed the cooling system.
31. Connect the battery ground.
32. Run the engine and check for leaks.

Fuel Pressure Regulator

REMOVAL AND INSTALLATION

4-2.3L EFI Engine

1. Ensure assembly is depressurized by removing fuel tank cap and releasing pressure from fuel system by opening the pressure relief valve on the fuel line in the upper RH corner of the engine compartment. Use EFI Pressure Gauge T80L–9974–B or equivalent.
2. Remove vacuum line at pressure regulator.
3. Remove three Allen retaining screws from regulator housing.
4. Remove pressure regulator assembly, gasket and O-ring. Discard gasket and inspect O-ring for signs of cracks or deterioration.
5. If scraping is necessary, be careful not to damage fuel pressure regulator or fuel supply line gasket surfaces.

To install:

6. Lubricate fuel pressure regulator O-ring with light oil ESF–M6C2–A or equivalent.

NOTE: Never use silicone grease. It will clog the injectors.

7. Make sure gasket surfaces of fuel pressure regulator and fuel injection manifold are clean.
8. Install O-ring and new gasket on regulator.
9. Install fuel pressure regulator on injector manifold. Tighten three retaining screws to 26–40 in. lbs.
10. Install vacuum line at pressure regulator.

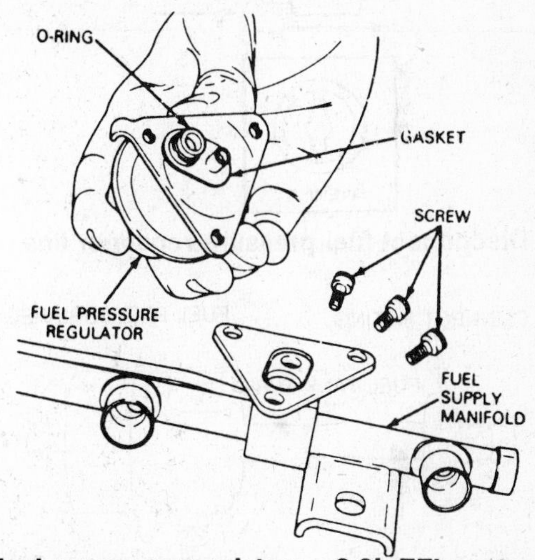

Fuel pressure regulator – 2.3L EFI engine

2.9L EFI Engine

1. Depressurize the fuel system as described earlier.
2. Remove the vacuum line at the pressure regulator.
3. Disconnect the fuel return line from the fuel pressure regulator. See illustration.
4. Remove the 2 retaining screws from the regulator housing.
5. Remove the pressure regulator assembly, washer and O-ring. Discard the washer and check the O-ring for signs of cracks or deterioration.

To install:

6. Clean the gasket mating surfaces. If scraping is necessary, be careful not to damage the fuel pressure regulator or supply line gasket mating surfaces.
7. Lubricate the pressure regulator O-ring with with light engine oil. Do not use silicone grease; it will clog the injectors.
8. Install the O-ring and a new gasket on the pressure regulator.
9. Install the pressure regulator on the fuel manifold and tighten the retaining screws to 71–102 in. lbs. (8–11 Nm).
10. Install all fuel line connections. Make sure all fuel line connections are properly connected.
11. Install the vacuum line at the pressure regulator. Build up fuel pressure by turning the ignition switch ON and OFF at least 6 times, leaving the ignition on for at least 5 seconds each time. Check for fuel leaks.

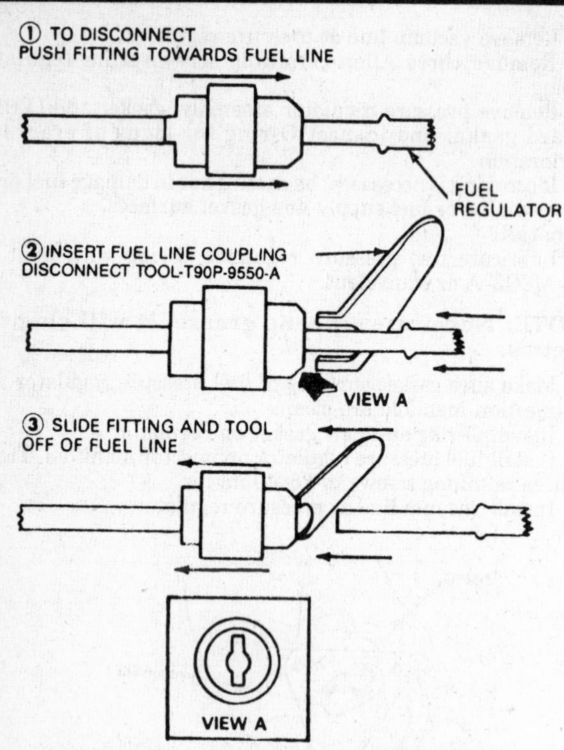

Disconnect fuel pressure regulator line

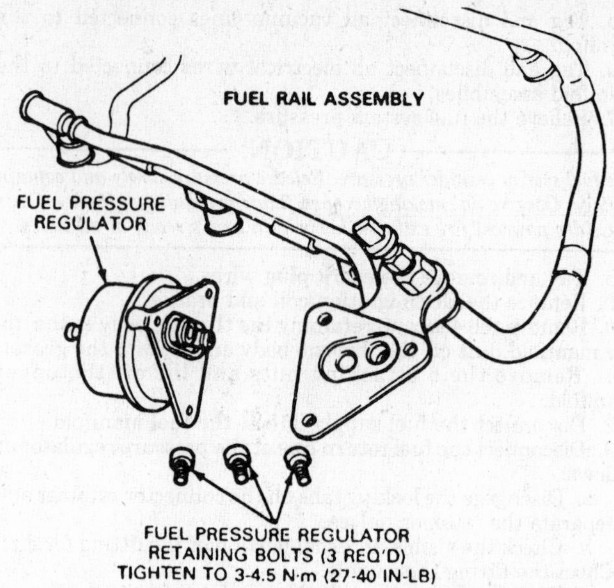

Fuel pressure regulator – 3.0L EFI engine

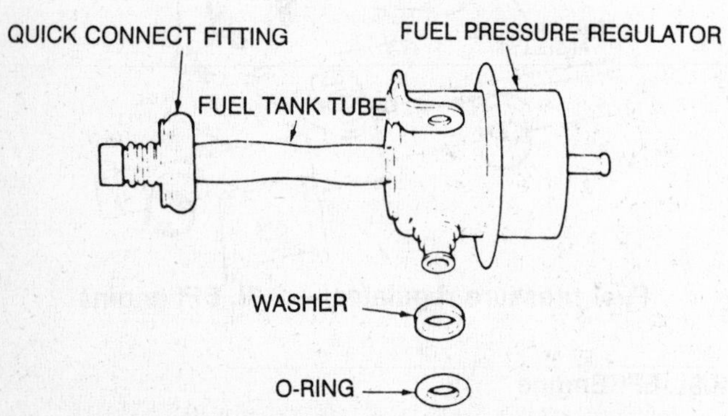

Fuel pressure regulator – 2.9L EFI engine

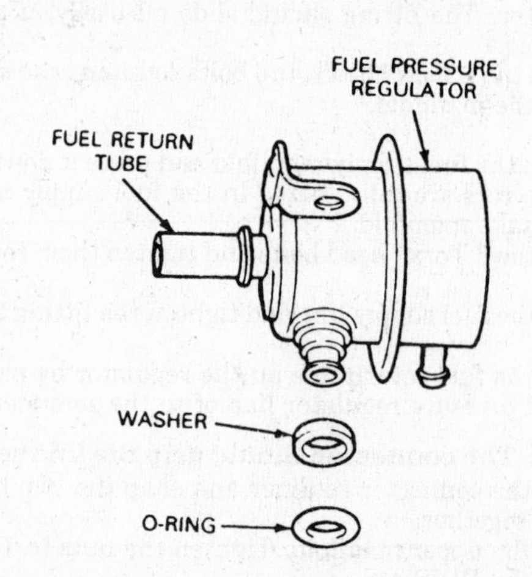

Fuel pressure regulator – 4.0L EFI engine

3.0L And 4.0L Engines

1. Depressurize the fuel system as described earlier in this Section.

CAUTION

The fuel system is under pressure. Release pressure slowly and contain spillage. Observe no smoking/no open flame precautions. Have a Class B–C (dry powder) fire extinguisher within arm's reach at all times.

2. Remove the vacuum and fuel lines at the pressure regulator. See illustration in this Section.

NOTE: On 3.0L engine in order to gain access to pressure regulator screws, loosen or remove fuel rail to intake manifold mounting bolt(s) as required and carefully lift pressure regulator side of the fuel rail.

3. Remove the 2 or 3 allen retaining screws from the regulator housing.
4. Remove the pressure regulator assembly, gasket and O-ring. Discard the gasket and check the O-ring for signs of cracks or deterioration.

To install:

5. Clean the gasket mating surfaces. If scraping is necessary, be careful not to damage the fuel pressure regulator or supply line gasket mating surfaces.
6. Lubricate the pressure regulator O-ring with with light engine oil. Do not use silicone grease; it will clog the injectors.
7. Install the O-ring and a new gasket on the pressure regulator.
8. Install the pressure regulator on the fuel manifold and tighten the retaining screws to 27–40 inch lbs. on the 3.0L engine and 6–8 ft. lbs. on the 4.0L engine.

NOTE: On 3.0L engine install pressure regulator side of the fuel rail to injectors and intake manifold. Ensure that the fuel rail is fully seated and injectors are installed correctly. Tighten fuel rail to manifold bolts to 7 ft. lbs.

9. Install the vacuum and fuel lines at the pressure regulator. Build up fuel pressure by turning the ignition switch ON and OFF at least 6 times, leaving the ignition on for at least 5 seconds each time. Check for fuel leaks.

Fuel Injector

REMOVAL AND INSTALLATION

2.3L EFI Engine

1. Remove fuel tank cap and release pressure from fuel system by opening the fuel pressure relief valve using EFI Pressure Gauge T80L–9974–B or equivalent.

NOTE: The cap on the relief valve must be removed.

2. Carefully remove electrical connectors from individual injectors.
3. Remove fuel supply manifold as outlined.
4. Grasping injector body, pull up while gently rocking injector from side-to-side.
5. Inspect injector O-rings (two per injector) for signs of deterioration. Replace as required.

WARNING: Do not attempt to clean the injector pintle or metering orifice with tools or brushes. Use Rotunda Injector Cleaner/Tester 113–80001 or equivalent.

To install:

6. Lubricate new O-rings with light grade oil ESF–M6C2–A or equivalent and install two on each injector.

NOTE: Never use silicone grease. It will clog the injectors.

7. Install injector(s) into fuel supply manifold. Use a light, twisting, pushing motion to install them.
8. Carefully seat fuel supply manifold assembly and injectors. Secure manifold with two attaching bolts. Tighten to 15–22 ft. lbs.
9. Connect fuel supply and return lines to fuel supply manifold. Reinstall spring lock coupling retainer clips on fuel inlet and return fittings.
10. Install the upper intake manifold assembly as outlined. Install cap on relief valve as necessary.
11. Connect injector electrical connectors. Start the engine and let idle for 2 minutes. Turn engine OFF and check for fuel leaks where the fuel injector is installed in the fuel rail and intake manifold.

6-2.9L EFI Engine

1. Depressurize the fuel system. Refer to the necessary service procedure in this Section.

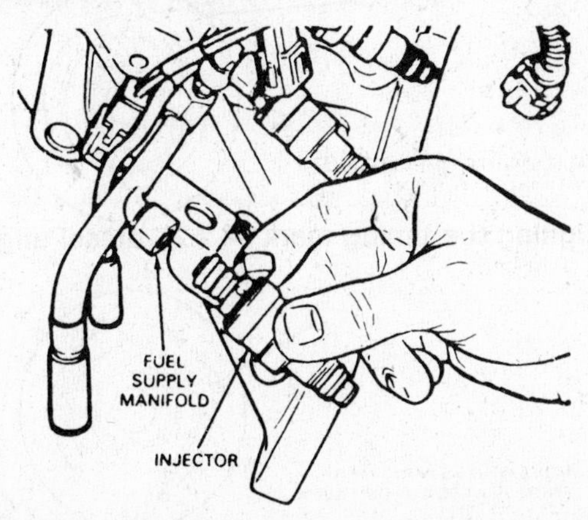

Fuel injector replacement — 2.3 EFI engine

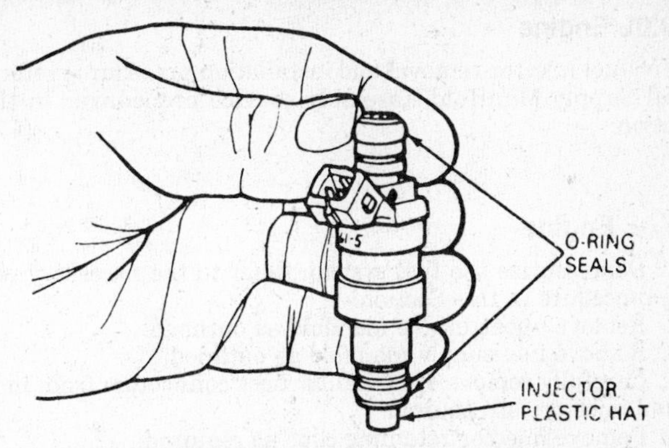

Fuel injector assembly — 2.3 EFI engine

2. Remove upper intake manifold as outlined.
3. Remove fuel supply manifold as outlined.
4. Carefully remove electrical harness connectors from individual injectors as required.
5. Remove injector retaining clips as required.
6. Grasping injector body, pull up while gently rocking injector from side-to-side.
7. Inspect injector O-rings (two per injector) for signs of deterioration. Replace as required.
8. Inspect injector "plastic hat" (covering the injector pintle) and washer for signs of deterioration. Replace as required. If hat is missing, look for it in intake manifold.

To install:

9. Lubricate new O-rings with light grade oil ESP–M6C2–A or equivalent and install two on each injector.

NOTE: Never use silicone grease. It will clog the injector(s).

10. Install injector(s), using a light, twisting, pushing motion.
11. Install injector retainer clip noting engagement with groove injector and flared edge of the cup. Install fuel supply manifold as outlined. Replace retainer clips if distorted.
12. Install electrical harness connectors to injectors.
13. Install upper intake manifold.
14. Make sure the injectors are fully seated in the fuel rail cups and intake manifold all fuel line connections are connected properly. Start the engine and let idle for 2 minutes. Turn engine OFF and check for fuel leaks where the fuel injector is installed in the fuel rail and intake manifold.

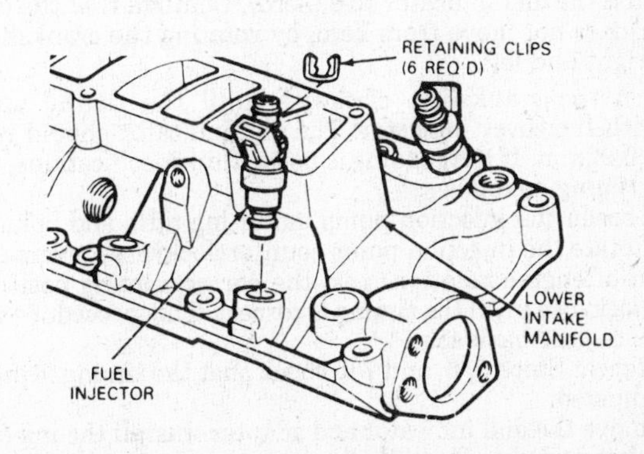

Fuel injector replacement — 2.9 EFI engine

6-3.0L Engine

For fuel injector removal and installation procedures refer to Fuel Supply Manifold Assembly service prcoedures in this Section.

6-4.0L Engine

1. Depressurize the fuel system. Refer to the necessary service procedure in this Section.
2. Remove upper intake manifold as outlined.
3. Remove fuel supply manifold as outlined.
4. Carefully remove electrical harness connectors from individual injectors as required.
5. Remove injector retaining clips as required.
6. Grasping injector body, pull up while gently rocking injector from side-to-side.
7. Inspect injector O-rings (two per injector) for signs of deterioration. Replace as required.
8. Inspect injector "plastic hat" (covering the injector pintle) and washer for signs of deterioration. Replace as required. If hat is missing, look for it in intake manifold.

To install:

9. Lubricate new O-rings with light grade oil ESP-M6C2-A or equivalent and install two on each injector.

NOTE: Never use silicone grease. It will clog the injector(s).

10. Install injector(s), using a light, twisting, pushing motion.
11. Install injector retainer clip noting engagement with groove injector and flared edge of the cup. Install fuel supply manifold as outlined. Replace retainer clips if distorted.
12. Install electrical harness connectors to injectors.
13. Install upper intake manifold assembly.
14. Make sure the injectors are fully seated in the fuel rail cups and intake manifold all fuel line connections are connected properly. Start the engine and let idle for 2 minutes. Turn engine OFF and check for fuel leaks where the fuel injector is installed in the fuel rail and intake manifold.

DIESEL ENGINE FUEL SYSTEM

Injector Timing

2.2L Diesel Engine

NOTE: Special tools Ford 14-0303, Static Timing Gauge Adapter and D82L4201A, Metric Dial Indicator, or the equivalents are necessary to set or check the injector timing.

1. Disconnect both battery ground cables. Remove the air inlet hose from the air cleaner and intake manifold.
2. Remove the distributor head plug bolt and washer from the injector pump.
3. Install the Timing Gauge Adapter and Metric Dial Indicator so that the indicator pointer is in contact with the injector pump plunger and gauge reads approximately 2.0mm.
4. Align the 2°ATDC (after top dead center) on the crankshaft pulley with the indicator on the timing case cover.
5. Slowly turn the engine counterclockwise until the dial indicator pointer stops moving (approximately 30°–50°).
6. Adjust the dial indicator to 0 (Zero). Confirm that the dial indicator does not move from Zero, by rotating the crankshaft slightly right and left.
7. Turn the crankshaft clockwise until the timing mark aligns with the cover indicator. The dial indicator should read 1mm ± 0.02mm. If the reading is not within specifications, adjust the timing as follows:
   a. Loosen the injection pump mounting nuts and bolts.
   b. Rotate the injection pump counterclockwise (reverse direction of engine rotation) past the correct timing position, then clockwise until the timing is correct. This procedure will eliminate gear backlash.
   c. Repeat Steps 5, 6, and 7 to check that the timing is properly adjusted.
8. Remove the dial indicator and adapter. Install the injector head gasket and plug. Install all removed parts.
9. Run engine, check and adjust idle RPM. Check for fuel leaks.

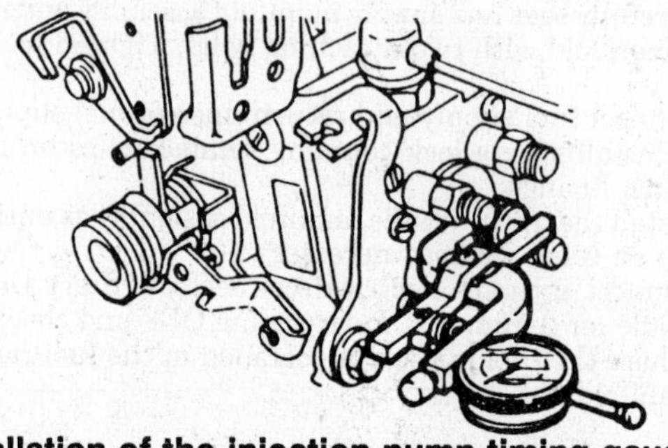

Installation of the injection pump timing gauge – 2.2L diesel engine

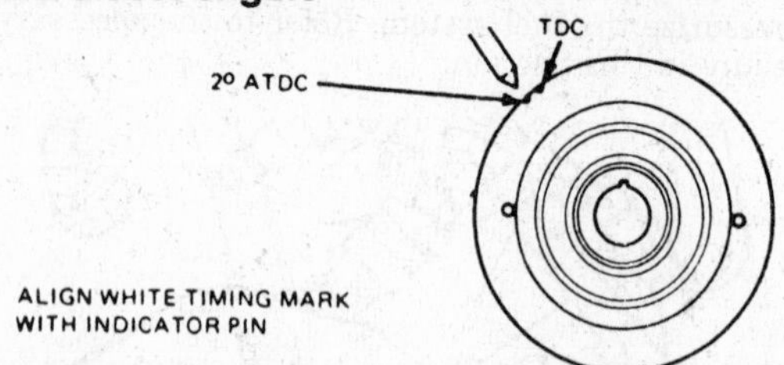

Aligning the timing mark – 2.2L diesel engine

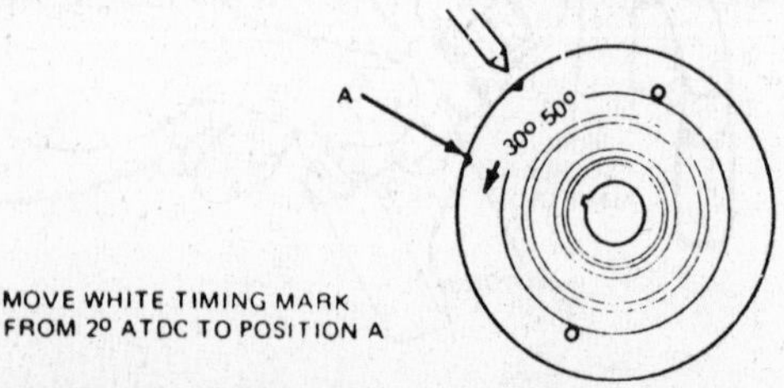

Moving the crankshaft pulley timing mark – 2.2L diesel engine

2.3L Engine

1. Remove top timing belt cover.
2. Rotate engine until No. 1 piston is at TDC on the compression stroke. Verify this by checking timing marks.
3. If engine temperature is below 122°F (50°C), bypass the cold start mechanism as follows:
 - Insert a suitable tool and rotate the fast idle lever.
 - Insert a spacer or a wrench at least 7mm thick between cold start advance lever and the cold start device.
4. Loosen, but do not remove, the two mounting bolts and two nuts attaching the injection pump to the mounting bracket and front cover.
5. To prevent the delivery valve holders from turning with the fuel line nuts, loosen, but do not remove, the nuts securing the fuel lines to the injection pump using a backup wrench.
6. Remove the timing plug bolt from the center of the fuel injection pump hydraulic head.
7. Install adapter Rotunda model 014–00303 or equivalent, into the port in the hydraulic head.
8. Mount Tool D82L–1201–A or equivalent dial indicator, in timing adapter with a minimum preload of 0.010 in. (0.25mm).
9. Rotate the crankshaft approximately 30° counterclockwise. The, set the dial indicator to zero.
10. Rotate the crankshaft clockwise to 5° ATDC and check that the dial indicator indicates 1mm ± 0.03mm.
11. If the timing is out of specification, rotate the injection pump body until the dial indicator indicates specification. Rotate the pump clockwise if the reading is more than specification, or counterclockwise if less than specification. Tighten the injection pump mounting nuts to 11–15 ft. lbs. and tighten the bolts to 15–19 ft. lbs.
12. After tightening injection pump mounting bolts and nuts, repeat Steps 9 and 10 to ensure that the injection timing has not changed.
13. Tighten the fuel line nuts at the injection pump to 17–26 ft. lbs.
14. Using a new copper gasket, install the injection timing plug bolt and tighten to 10–14 ft. lbs.
15. Install the upper timing belt cover.
16. Run the engine and check for fuel leaks at the injection pump.

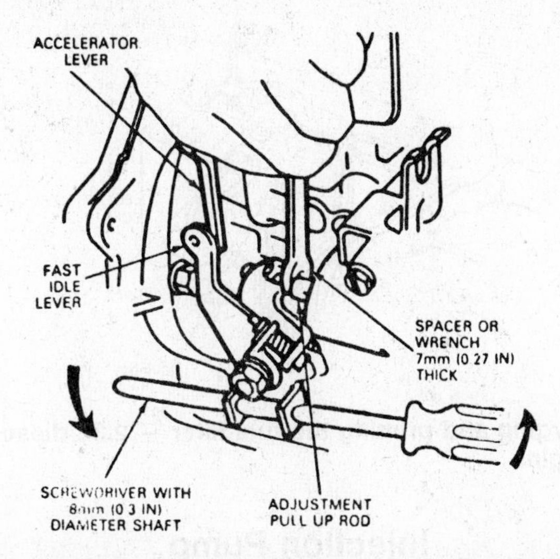

Bypassing the cold start device – 2.3L diesel engine

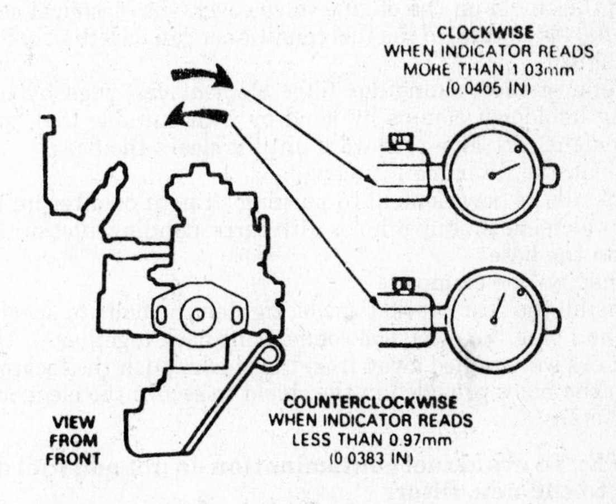

Injection timing adjustment – 2.3L diesel engine

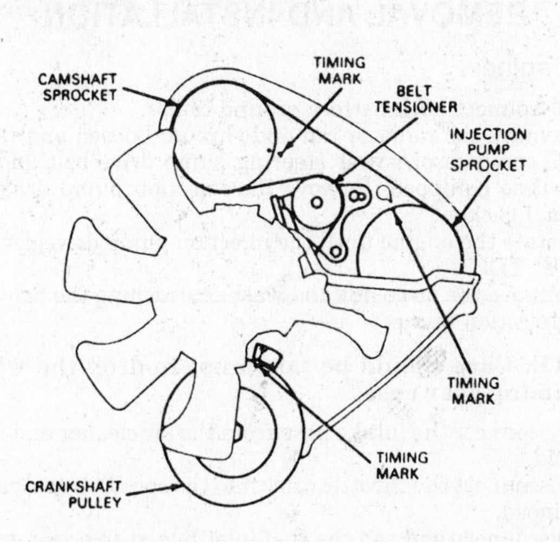

Timing mark alignment – 2.3L diesel engine

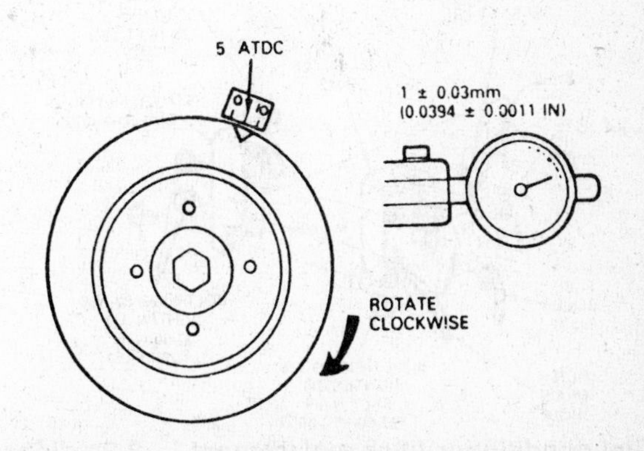

Injection timing check – 2.3L diesel engine

Fuel Filter

REMOVAL AND INSTALLATION

2.2L Engine

1. Disconnect both battery ground cables.
2. Disconnect and cap the fuel filter inlet and outlet lines.
3. Remove the two bolts/nuts attaching the priming pump to the bracket an remove the pump.
4. Remove the two bolts attaching the bracket to the engine and remove the bracket.
5. Install the fuel filter mounting bracket on the engine.
6. Install the priming pump assembly on the bracket.
7. Install the fuel filter lines and clamps.
8. Connect both battery ground cables.
9. Run the engine and check for fuel leaks.

2.3L Engine

The 2.3L diesel engine fuel filter is the paper element cartridge type and the conditioner housing includes an air vent screw to bleed air from the fuel lines.

1. Remove the rear bracket shield, attaching bolts, and unplug the electrical connectors attached to the shield. These connectors pull apart by pulling on the wire bundle on each side. Be resting the shield on the engine valve cover, the electrical connector halves leading to the fuel conditioner can be left attached to the shield.
2. Remove the rectangular filter element cartridge by unlatching holddown clamps by hand or with suitable tool, and pull element cartridge rear ward until it clears the base.
3. Clean the filter mounting pad.
4. Install the new element by pushing straight on after lining up filter element grommet holes with corresponding inlet/outlet tubes on the base.
5. Snap on the clamps.
6. Install the rear bracket shield, tighten the bolts to specification and plug the electrical connections back together. If the connectors were pulled away from the shield, push the locators back in the holes provided in the shield to secure the electrical connectors.

NOTE: To avoid fuel contamination do not add fuel directly to the new filter.

**2.3L ENGINE
FRAME-MOUNTED INLINE FUEL FILTER**

The inline fuel filter is a molded plastic mesh type designed to protect the electrical fuel boost pump from contamination. It is located on the left hand side frame rail about tow feet rear of the fuel boost pump.

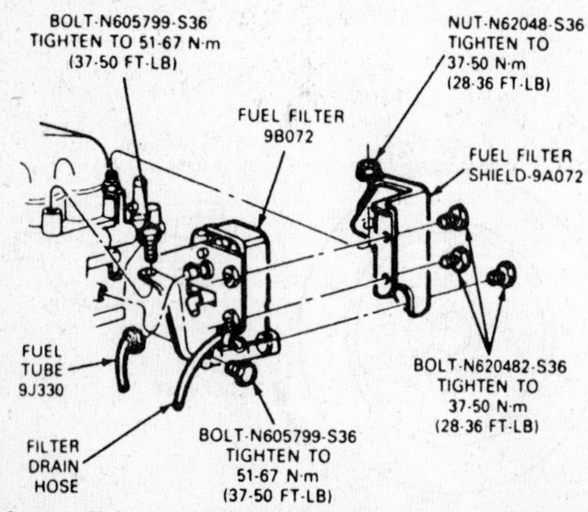

Fuel conditioner filter replacement – 2.3L diesel engine

1. Pinch off the fuel hose to the rear of the inline filter using a rubber coated clamp or other suitable device to prevent fuel from siphoning from the tank. Care must be taken not to damage the fuel hose.
2. Remove the two hose clamps closest to the inline filter and remove filter.
3. Install replacement filter and two new clamps. Remove the hose pinch-off clamp.

2.3L ENGINE – PURGING AIR AND PRIMING FUEL FILTER

1. Turn the ignition switch **ON** to activate the electric fuel boost pump.
2. Loosen the air vent plug on the conditioner housing until fuel flows from the air vent plug hole free of bubbles.
3. Tighten the air vent plug securely.
4. Start the engine and check for leaks.

WARNING: DO NOT OPEN AIR VENT PLUG WITH THE ENGINE RUNNING!

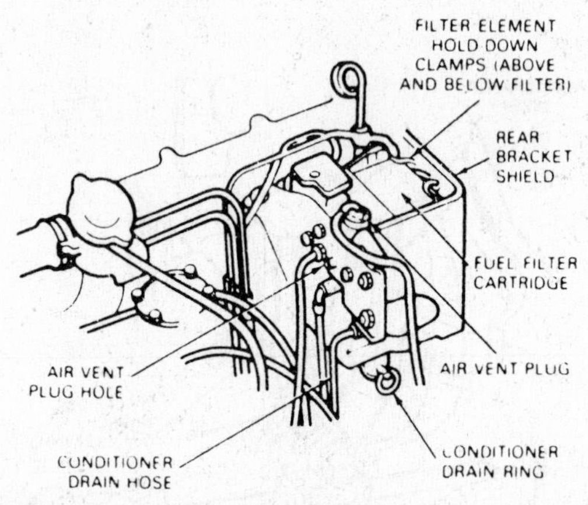

Purging and priming the fuel filter – 2.3L diesel engine

Injection Pump

REMOVAL AND INSTALLATION

2.2L Engine

1. Disconnect both battery ground cables.
2. Remove the radiator fan and shroud. Loosen and remove the A/C compressor/power steering pump drive belt and idler pulley, if so equipped. Remove the injection pump drive gear cover and gasket.
3. Rotate the engine until the injection pump drive gear keyway is at TDC.
4. Remove the large nut and washer attaching the drive gear to the injection pump.

NOTE: Care should be taken not to drop the washer into timing gear case.

5. Disconnect the intake hose from the air cleaner and intake manifold.
6. Disconnect the throttle cable and the speed control cable, if so equipped.
7. Disconnect and cap the fuel inlet line at injection pump.
8. Disconnect the fuel shut-off solenoid lead at the injection pump.
9. Disconnect and remove the fuel injection lines from the nozzles and injection pump. Cap all the fuel lines and fittings.

10. Disconnect the lower fuel return line from the injection pump and the fuel hoses. Loosen the lower No. 3 intake port nut and remove the fuel return line.
11. Remove the two nuts attaching the injection pump to the front timing gear cover and one bolt attaching the pump to the rear support bracket.
12. Install a Gear and Hub Remover, Tool T83T–6306–A or equivalent, in the drive gear cover and attach to the injection pump drive gear. Rotate the screw clockwise until the injection pump disengages from the drive gear. Remove the injection pump.

NOTE: Carefully remove the injection pump to avoid dropping the key into the timing gear case. Disconnect the cold start cable before removing the injection pump from the vehicle.

Connect the cold start cable to pump before positioning the injection pump in the timing gear case.
13. Install the injection pump in position in the timing gear case aligning the key with keyway in the drive gear in the TDC position.

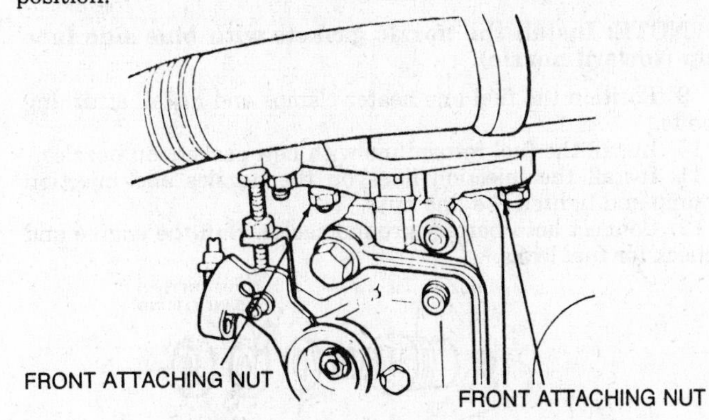

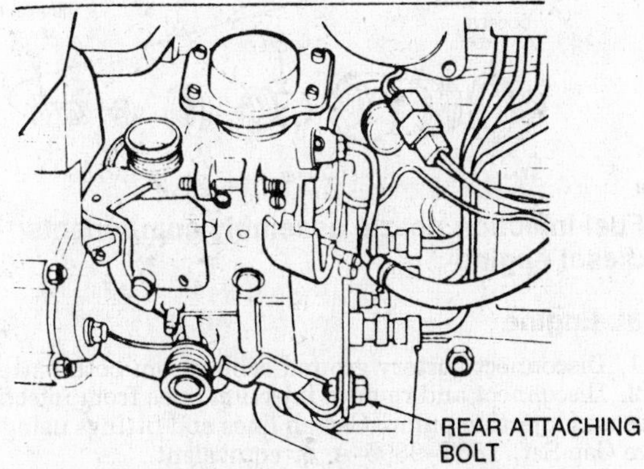

Injection pump attaching locations

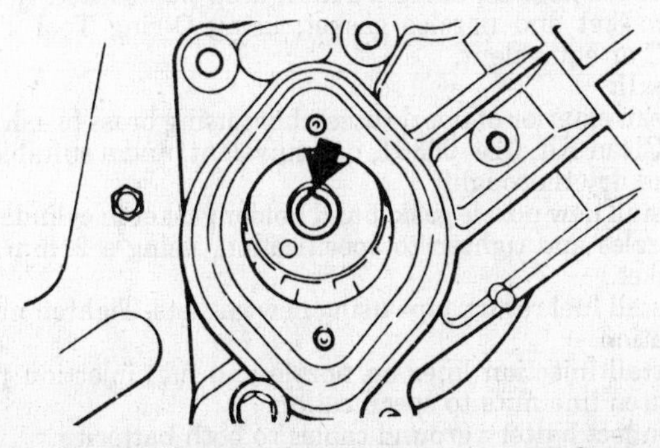

Aligning the key and keyway in the TDC position

NOTE: Use care to avoid dropping the key in the timing gear case.

14. Install the nuts and washers attaching the injection pump to the timing gear case and tighten to draw the injection pump into position.

NOTE: Do not tighten at this time.

15. Install the bolt attaching the injection pump to the rear support. Install the washer and nut attaching the injection drive gear to the injection pump and tighten.
16. Install the injection pump drive gear cover, with a new gasket, on the timing gear case cover and tighten.
17. Adjust the injection timing at this time.
18. Install the lower fuel return line to the injection pump and intake manifold stud. Tighten the Banjo bolt on the injection pump and nut on the intake manifold. Install the connecting fuel hoses and clamps. Install the fuel injection lines to the injection pump an nozzles.
19. Connect the lead to fuel shut-off solenoid on the injection pump Connect the fuel inlet line to the injection pump and install the hose clamp.
20. Install the throttle cable and speed control cable, if so equipped.
21. Air bleed the fuel system.
22. Install the intake hose on the air cleaner and intake manifold.
23. Install the A/C compressor/power steering pump drive belt and idler pulley, if so equipped.
24. Install the radiator shroud and radiator fan.
25. Connect both battery ground cables.
26. Run the engine and check for oil and fuel leaks.

2.3L Engine

1. Disconnect battery ground bales from both batteries.
2. Remove radiator fan and shroud.
3. Loosen and remove accessory drive belts.
4. Rotate crankshaft in direction of engine rotation to bring No. 1 piston to TDC on compression stroke.
5. Remove upper front timing cover as outlined.
6. Loosen and remove timing belt from injection pump.
7. Remove nut attaching sprocket to injection pump.
8. Install puller T77F–4220–B1, or equivalent and remove sprocket.
9. Disconnect throttle cable and speed control cables, if so equipped.
10. Disconnect coolant hoses from injection pump wax element.
11. Disconnect hoses from boost compensator and A/C throttle kicker.
12. Disconnect fuel return line at injection pump from injection return pipe.
13. Disconnect chassis fuel return line from injection pump.
14. Disconnect and cap fuel supply line from fuel conditioner.
15. Disconnect and remove fuel lines at injection pump and nozzles. Cap all lines and fittings using Protective Cap Set, T85T–9395–A or equivalent.
16. Remove two nuts attaching injection pump to rear front case. Then remove two injection pump bracket-to-engine bracket bolts and two engine bracket-to-engine block bolts, and remove pump.
17. Position injection pump on engine and install two nuts attaching injection pump to engine rear front cover. Position injection pump bracket-to-engine bracket on engine block and install two bolts.
18. Install two injection pump bracket-to-engine bracket bolts. Tighten all nuts and bolts to specification.
19. Install injection pump sprocket. Tighten nut to 40–50 ft. lbs.
20. Install and adjust timing belt as outlined in Chapter 3.
21. Adjust injection pump timing.

22. Tighten the two injection pump retaining nuts and four bolts to specification.
23. Install injection lines on injection pump and nozzles. Tighten line nuts to specification.
24. Install fuel supply line and fuel return line on injection pump.
25. Connect nozzle returning line to injection pump.
26. Connect hoses to boost compensator and A/C throttle kicker.
27. Connect coolant hoses to injection pump wax element.
28. Connect throttle cable and speed control cable, if so equipped.
29. Install upper front cover as outlined.
30. Install accessory drive belts.
31. Install radiator fan and shroud.
32. Connect battery ground cables to both batteries.
33. Run engine and check for fuel and coolant leaks.

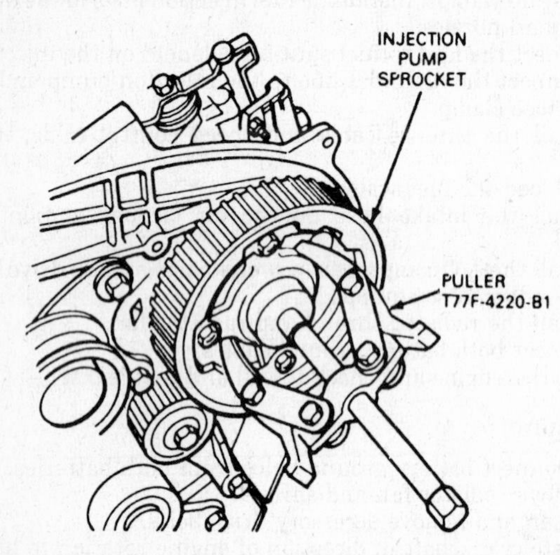

Removing the injection pump sprocket — 2.3L diesel engine

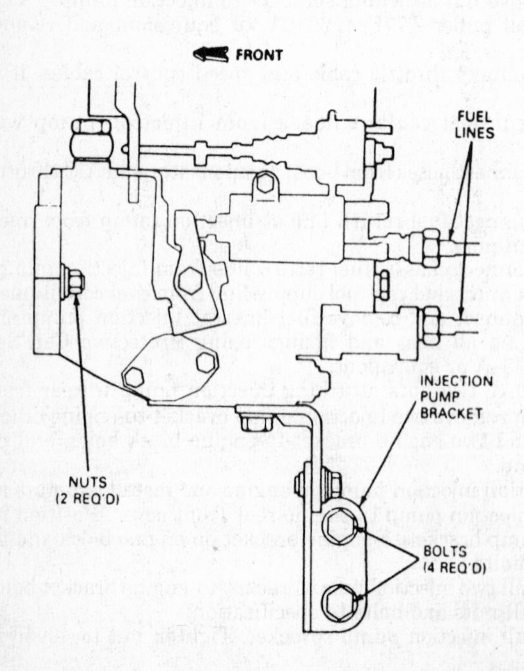

Installing the injection pump — 2.3L diesel engine

Fuel Injectors

REMOVAL AND INSTALLATION

2.2L Engine

1. Disconnect both battery ground cables.
2. Disconnect and remove the injection lines from the nozzles and injection pump. Cap all lines and fittings.
3. Remove the fuel return line and gaskets.
4. Remove the bolts attaching the fuel line heater clamp to the cylinder head and position the heater out of the way.
5. Remove the nozzles, using a 27mm deepwell socket.
6. Remove the nozzle washer (copper) and nozzle gasket (steel), using Tool T71P–19703–C or equivalent.
7. Clean the nozzle assemblies with Nozzle Cleaning Kit, Rotunda 14–0301 or equivalent, and a suitable solvent, and dry thoroughly. Clean the nozzle seats in the cylinder head with Nozzle Seat Cleaner, T83T–9527–B or equivalent.
8. Position the new nozzle washers and gaskets in the nozzle seats, install the nozzles and tighten.

NOTE: Install the nozzle gaskets with blue side face up (toward nozzle).

9. Position the fuel line heater clamps and install attaching bolts.
10. Install the fuel return line with new gaskets on nozzles.
11. Install the injection lines on the nozzles and injection pump and tighten the line nuts.
12. Connect both battery ground cables. Run the engine and check for fuel leaks.

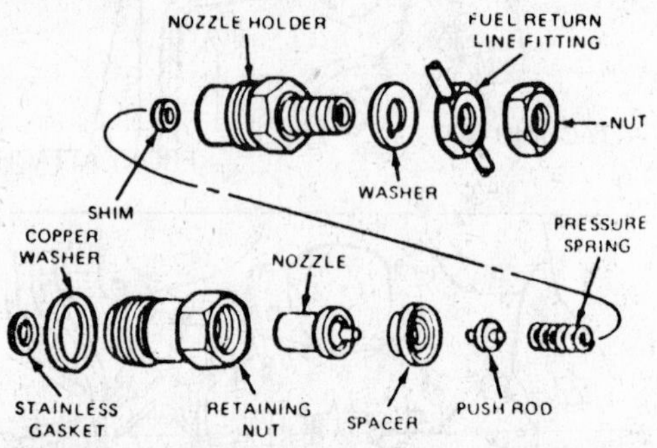

Fuel injection nozzle assembly components — 2.2L diesel engine

2.3L Engine

1. Disconnect battery ground cables from both batteries.
2. Disconnect and remove injection lines from injection nozzles and injection pump. Cap all lines and fittings using Protective Cap Set, T85T–9395–A, or equivalent.
3. Remove fuel return pipe and gaskets.
4. Remove nozzles, using a 21mm deep well socket. Remove holder gasket and nozzles gasket, using O-ring Tool T71P–19703–C or equivalent.

To install:

5. Clean exterior of nozzle assemblies using brass brush from Nozzle Clean Kit, 014–00301, or equivalent, and a suitable solvent, and dry thoroughly.
6. Install new nozzle gasket and holder gasket in cylinder. Install nozzles and tighten to specification, using a 21mm deep well socket.
7. Install fuel return pipe using new gaskets. Tighten nuts to specification.
8. Install injection lines on nozzle and fuel injection pump and tighten line nuts to specification.
9. Connect battery ground cables to both batteries.
10. Run engine and check for fuel leaks.

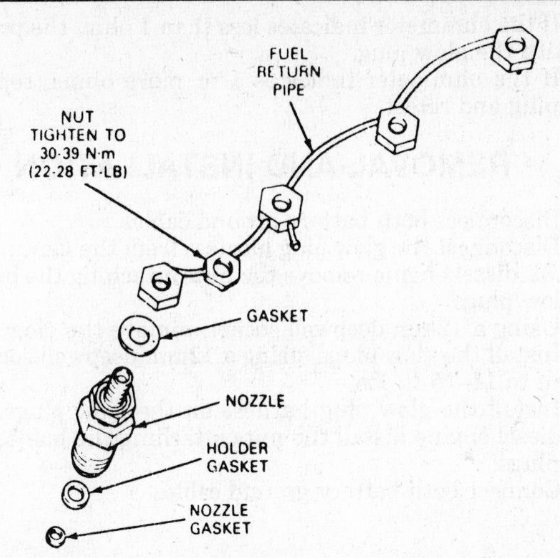

Fuel injection nozzle assembly – 2.3L diesel engine

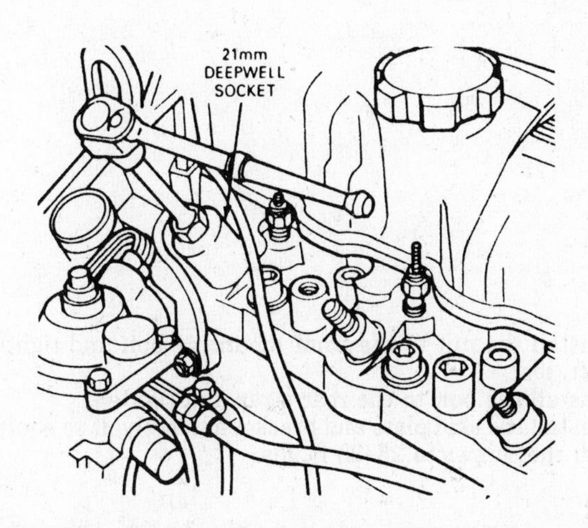

Removing the fuel injector nozzle – 2.3L diesel engine

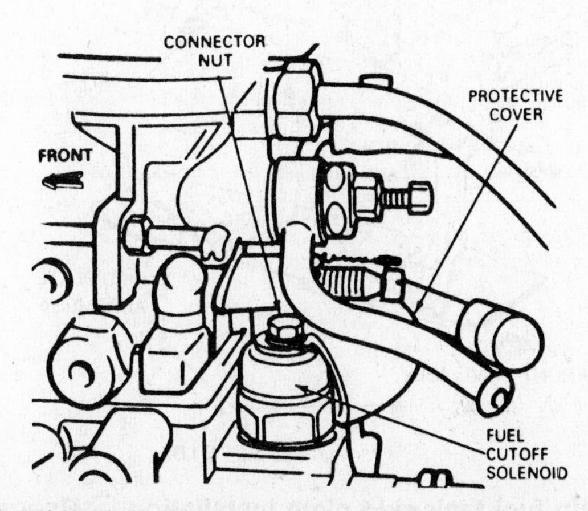

Fuel cutoff solenoid – 2.3L diesel engine

Fuel Cut-Off Solenoid

REMOVAL AND INSTALLATION

1. Disconnect both battery ground cables.
2. Remove the connector from the fuel cut-off solenoid.
3. Remove the fuel cut-off solenoid assembly.
4. Install the fuel cut-off solenoid, with a new O-ring, and tighten.
5. Install the connector on the fuel cut-off solenoid.
6. Connect both battery ground cables. Run the engine and check for fuel leaks.

Glow Plug System

The **quick start; afterglow** system is used to enable the engine to start more quickly when the engine is cold. It consists of the four glow plugs, the control module, two relays, a glow plug resistor assembly, coolant temperature switch, clutch and neutral switches and connecting wiring. Relay power and feedback circuits are protected by fuse links in the wiring harness. The control module is protected by a separate 10A fuse in the fuse panel.

When the ignition switch is turned to the ON position, a Wait-to-Start signal appears near the cold-start knob on the panel. When the signal appears, relay No. 1 also closes and full system voltage is applied to the glow plugs. If engine coolant temperature is below 30°C (86°F), relay No. 2 also closes at this time. After three seconds, the control module turns off the Wait-to-Start light indicating that the engine is ready for starting. If the ignition switch is left in the ON position about three seconds more without cranking, the control opens relay No. 1 and current to the plugs stops to prevent overheating. However, if coolant temperature is below 30°C (86°F) when relay No. 1 opens, relay no. 2 remains closed to apply reduced voltage to the plugs through the glow plug resistor until the ignition switch is turned off.

When the engine is cranked, the control module cycles relay No. 1 intermittently. Thus, glow plug voltage will alternate between 12 and four volts, during cranking, with relay No. 2 closed, or between 12 and zero volts with relay No. 2 open. After

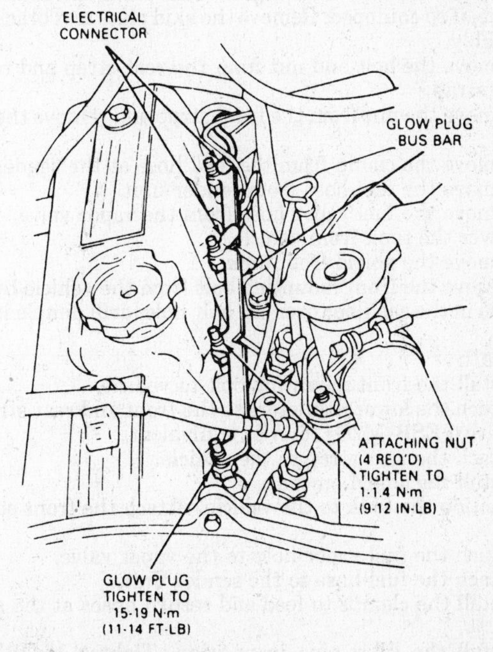

Glow plug harness and buss bar – 2.3L diesel engine

the engine starts, alternator output signals the control module to stop the No. 1 relay cycling and the afterglow function takes over.

If the engine coolant temperature is below 30°C (86°F), the No. 2 relay remains closed. This applies reduced (4.2 to 5.3) voltage to the glow plugs through the glow plug resistor. When the vehicle is under way (clutch and neutral switches closed), or coolant temperature is above 30°C (86°F), the control module opens relay No. 2, cutting off all current to the glow plugs.

TESTING THE GLOW PLUGS

1. Disconnect the leads from each glow plug. Connect one lead of the ohmmeter to the glow plug terminal and the other lead to a good ground. Set the ohmmeter on the ×1 scale. Test each glow plug in the like manner.
2. If the ohmmeter indicates less than 1 ohm, the problem is not with the glow plug.
3. If the ohmmeter indicates 1 or more ohms, replace the glow plug and retest.

REMOVAL AND INSTALLATION

1. Disconnect both battery ground cables.
2. Disconnect the glow plug harness from the glow plugs. On the 2.3L diesel engine remove the nuts attaching the bus bar to the glow plugs.
3. Using a 12mm deepwell socket, remove the glow plugs.
4. Install the glow plugs, using a 12mm deepwell socket, and tighten to 11–15 ft. lbs.
5. Install the glow plug harness on the glow plugs. On the 2.3L diesel engine install the nuts attaching the bus bar to the glow plugs.
6. Connect both battery ground cables.

FUEL TANK

REMOVAL AND INSTALLATION

Ranger Models

MIDSHIP FUEL TANK

1. Drain the fuel from the fuel tank.
2. Loosen the fill pipe clamp.
3. Remove the bolts securing the skid plate and brackets to the frame, if so equipped. Remove the skid plate and brackets as an assembly.
4. Remove the bolt and nut from the rear strap and remove the rear strap.
5. Remove the nut from the front strap and remove the front strap.
6. Remove the clamp from the feed hose at the sender unit.
7. Remove the fuel hose from sender unit.
8. Remove the fuel vapor hose from the vapor valve.
9. Lower the tank from vehicle.
10. Remove the shield from tank.
11. Remove the front mounting bolt from the vehicle by drilling a hole in the cab floor over the bolt hole (drill dimple in floor pan).

To install:

12. Install the front mounting bolt to vehicle.
13. Attach the lower insulators to the front and rear strap using adhesive ESB–M2G115–A or equivalent.
14. Attach the rear strap to the vehicle.
15. Install the shield on tank.
16. Position the tank to the vehicle. Attach the front strap to vehicle.
17. Attach the fuel vapor hose to the vapor valve.
18. Attach the fuel hose to the sender unit.
19. Install the clamps to feed and return hoses at the sender unit.
20. Install the filler pipe in position. Tighten the fill pipe clamp.
21. Install the nut to the front mounting bolt and tighten to 18–20 ft. lbs.
22. Install the bolt to the rear strap and tighten.
23. Install the skid plate and bracket assembly, if so equipped. Tighten the screws to 25–30 ft. lbs.

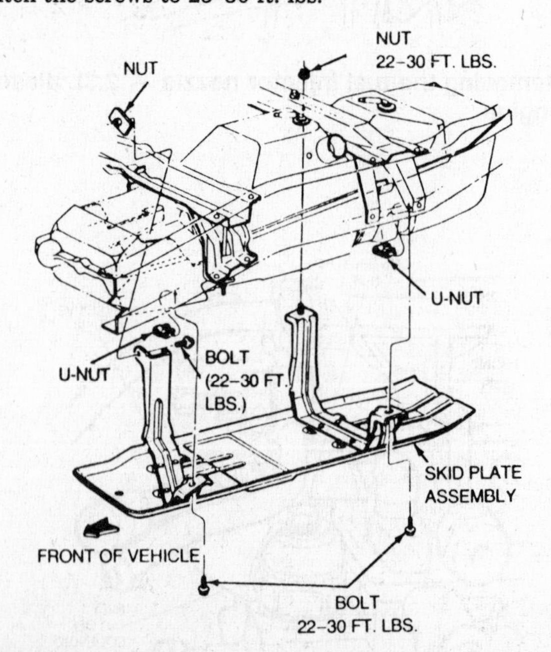

Midship fuel tank skid plate installation – Ranger models

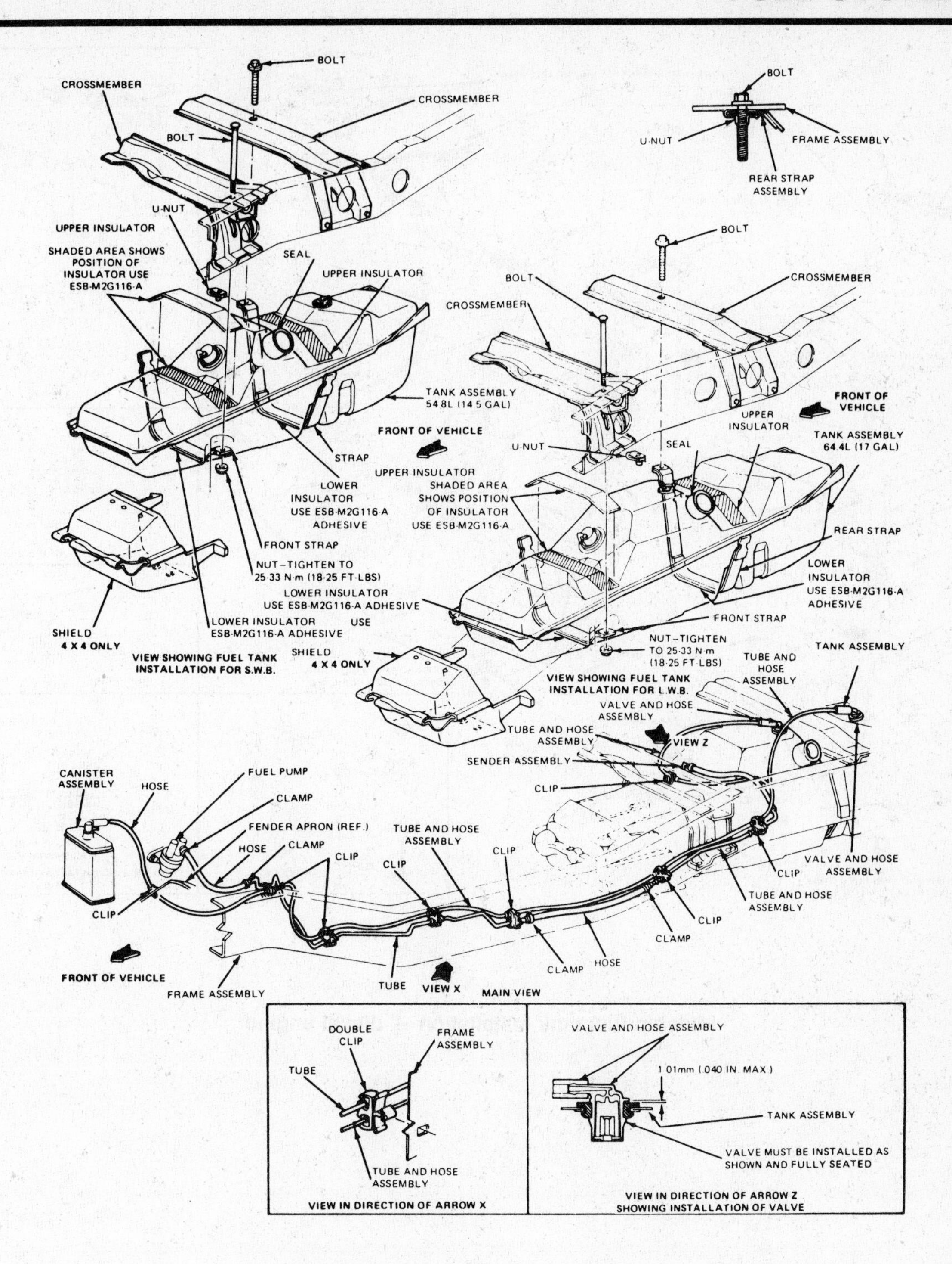

Midship fuel tank installation – Ranger models

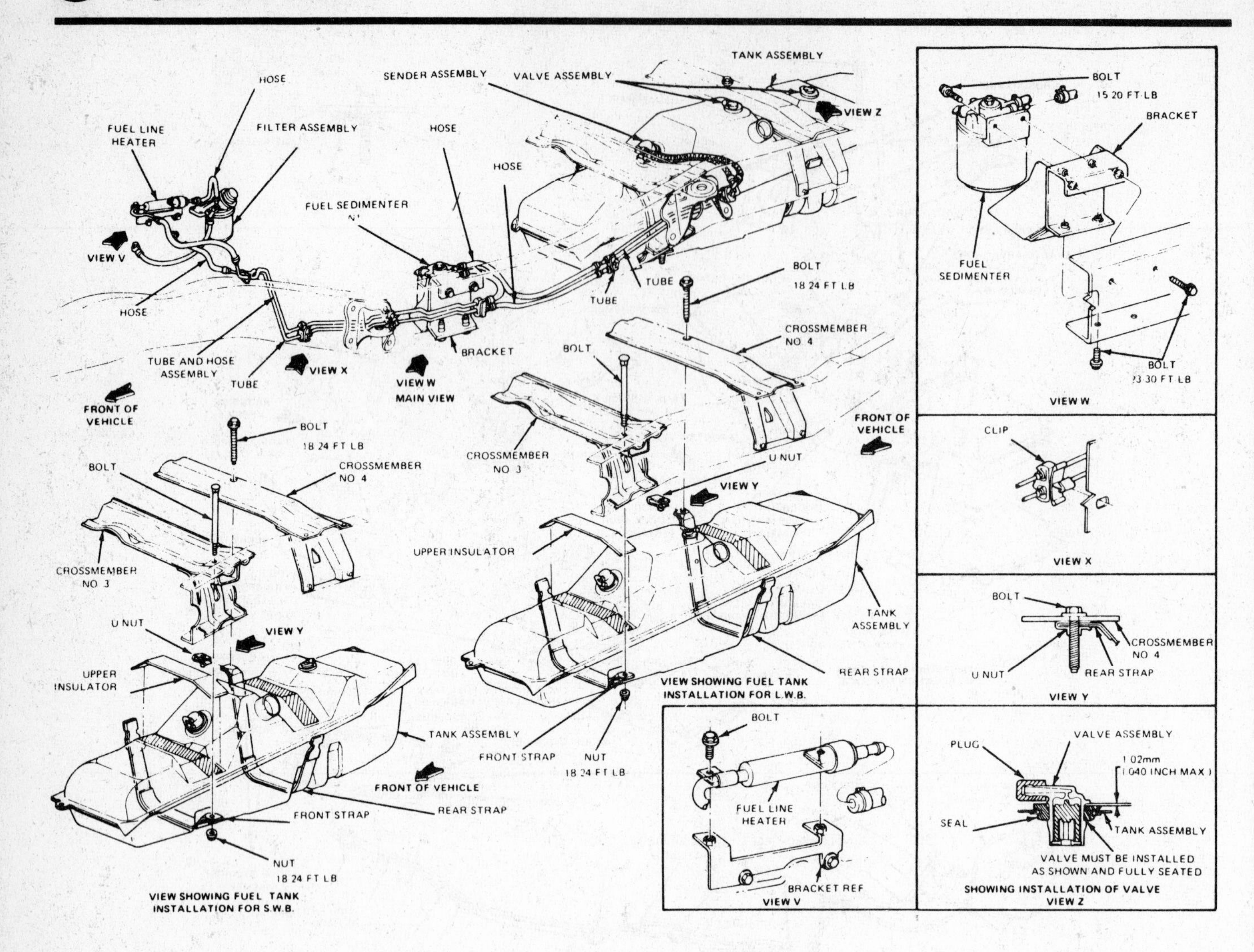

Midship fuel tank installation – diesel engine

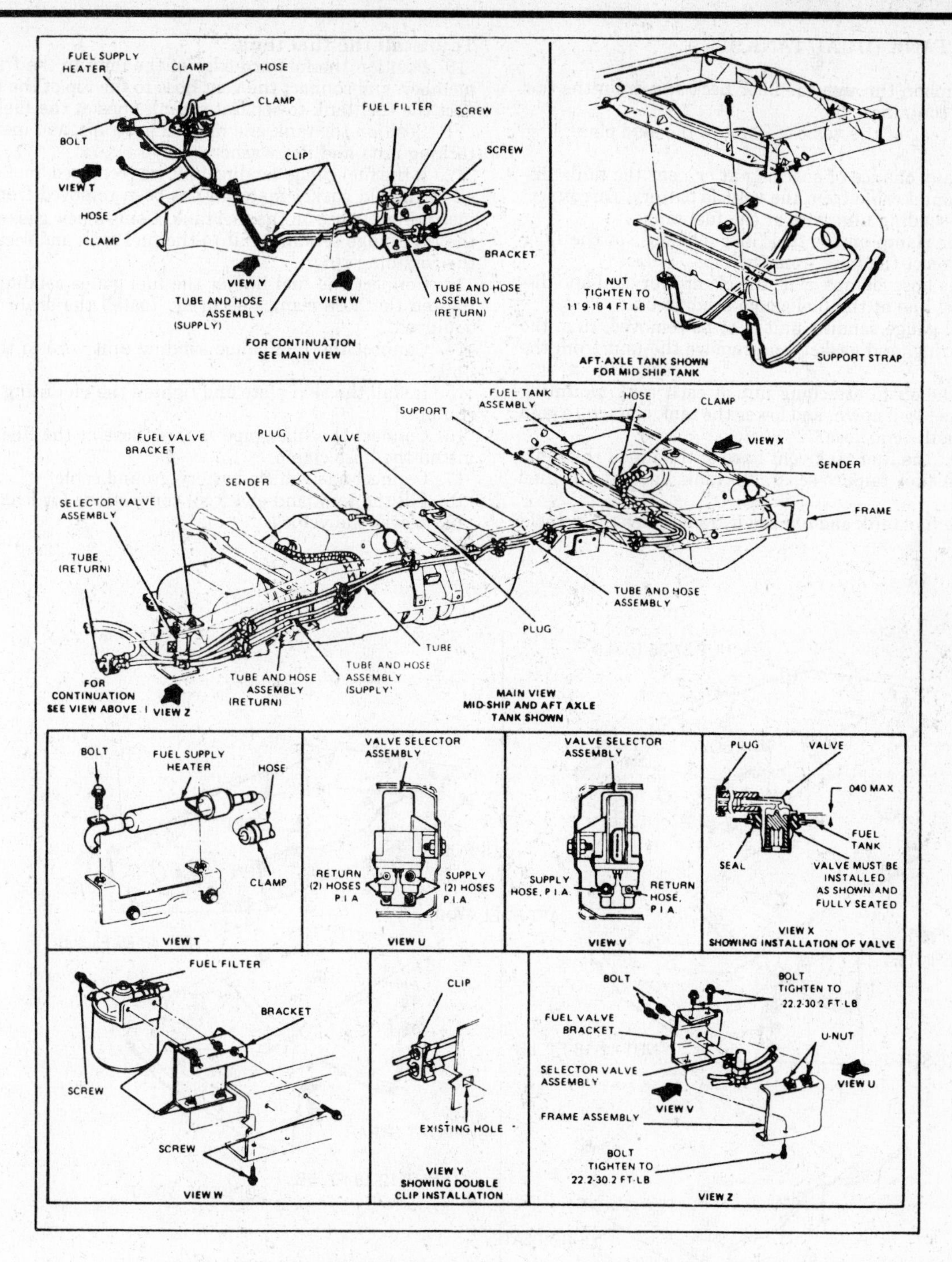

Dual fuel tank installation — diesel engine

REAR FUEL TANK (DUAL TANKS)

1. Insert a siphon through the filler neck and drain the fuel into a suitable container.
2. Raise the rear of the vehicle. Remove the skid plate, if so equipped.
3. To avoid any chance of sparking at or near the tank, disconnect the ground cable from the vehicle battery. Disconnect the fuel gauge sending unit wire at the fuel tank.
4. Loosen the clamp on the fuel filler pipe hose at the filler pipe and disconnect the hose from the pipe.
5. Loosen the hose clamps, slide the clamps forward and disconnect the fuel line at the fuel gauge sending unit.
6. If the fuel gauge sending unit is to be removed, turn the unit retaining ring, and gasket, and remove the unit from the tank.
7. Remove the strap attaching nut at each tank mounting strap, swing the strap down, and lower the tank enough to gain access to the tank vent hose.
8. Disconnect the fuel tank vent hose at the top of the tank. Disconnect the fuel tank-to-separator tank lines at the fuel tank.
9. Lower the fuel tank and remove it from under the vehicle

To install the fuel tank:

10. Position the forward edge of the tank to the frame crossmember, and connect the vent hose to the top of the tank. Connect the fuel tank-to-separator tank lines at the fuel tank.
11. Position the tank and mounting straps, and install the attaching nuts and flat washers.
12. If the fuel gauge sending unit was removed, make sure that all of the old gasket material has been removed from the unit mounting surface on the fuel tank. Using a new gasket, position the fuel gauge sending unit to the fuel tank and secure it with the retaining ring.
13. Connect the fuel line at the fuel gauge sending unit and tighten the hose clamps securely. Install the drain plug, if so equipped.
14. Connect the fuel gauge sending unit wire to the sending unit.
15. Install the skid plate and tighten the mounting nuts, if so equipped.
16. Connect the filler pipe-to-tank hose at the filler pipe and install the hose clamp.
17. Connect the vehicle battery ground cable.
18. Fill the tank and check all connections for leaks.
19. Lower the vehicle.

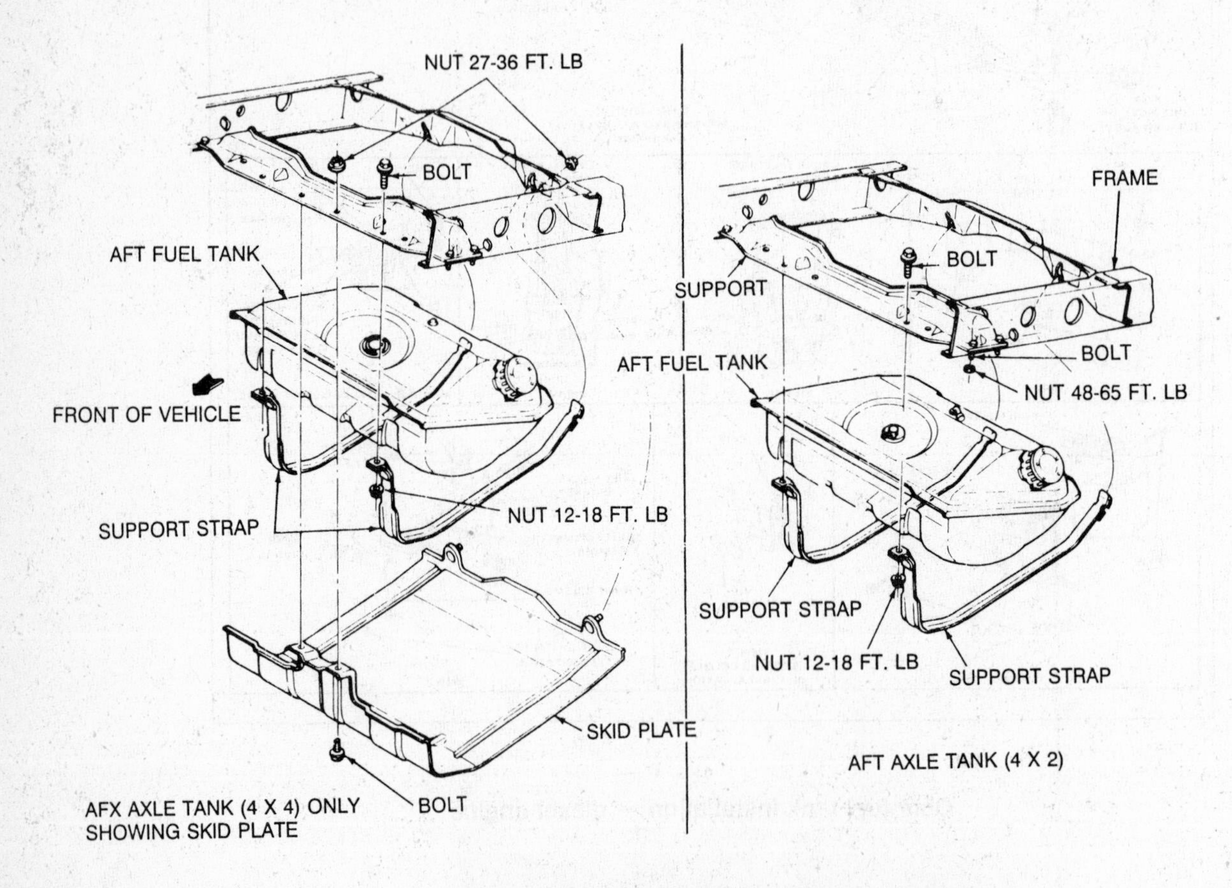

Rear fuel tank installation — Ranger models

Bronco II And Explorer Models

NOTE: Read the entire service procedure and refer to the illustrations before starting repair--modify the service steps as necessary. Always replace all gas line hoses and retaining clamps if necessary.

1. Insert a siphon through the filler neck and drain the fuel into a suitable container. Drain the fuel tank by removing drain plug if so equipped.
2. Raise the rear of the vehicle and safely support. Remove the skid plate, if so equipped.
3. To avoid any chance of sparking at or near the tank, disconnect the ground cable from the vehicle battery. Disconnect the fuel gauge sending unit wire at the fuel tank.
4. Loosen the clamp on the fuel filler pipe hose at the filler pipe and disconnect the hose from the pipe.
5. Loosen the hose clamps, slide the clamps forward and disconnect the fuel line at the fuel gauge sending unit.
6. Support the fuel tank assembly. Remove the strap attaching nut at each tank mounting strap.
7. Swing the strap down, and lower the tank enough to gain access to the tank vent hose.
8. Disconnect the fuel tank vent hose at the top of the tank. Disconnect the fuel tank-to-separator tank lines at the fuel tank.
9. Lower the fuel tank and remove it from under the vehicle

To install:

10. Position the forward edge of the tank to the frame crossmember, and connect the vent hose to the top of the tank. Connect the fuel tank-to-separator tank lines at the fuel tank.
11. Position the tank and mounting straps, and install the attaching nuts and flat washers. Tighten mounting strap retaining nuts evenly.
12. Connect the fuel line at the fuel gauge sending unit and tighten the hose clamps securely. Install the drain plug, if so equipped.
13. Connect the fuel gauge sending unit wire to the sending unit.
14. Install the skid plate and tighten the mounting nuts, if so equipped.
15. Connect the filler pipe-to-tank hose at the filler pipe and install the hose clamp.
16. Connect the vehicle battery ground cable.
17. Fill the tank and check all connections for leaks.
18. Lower the vehicle.

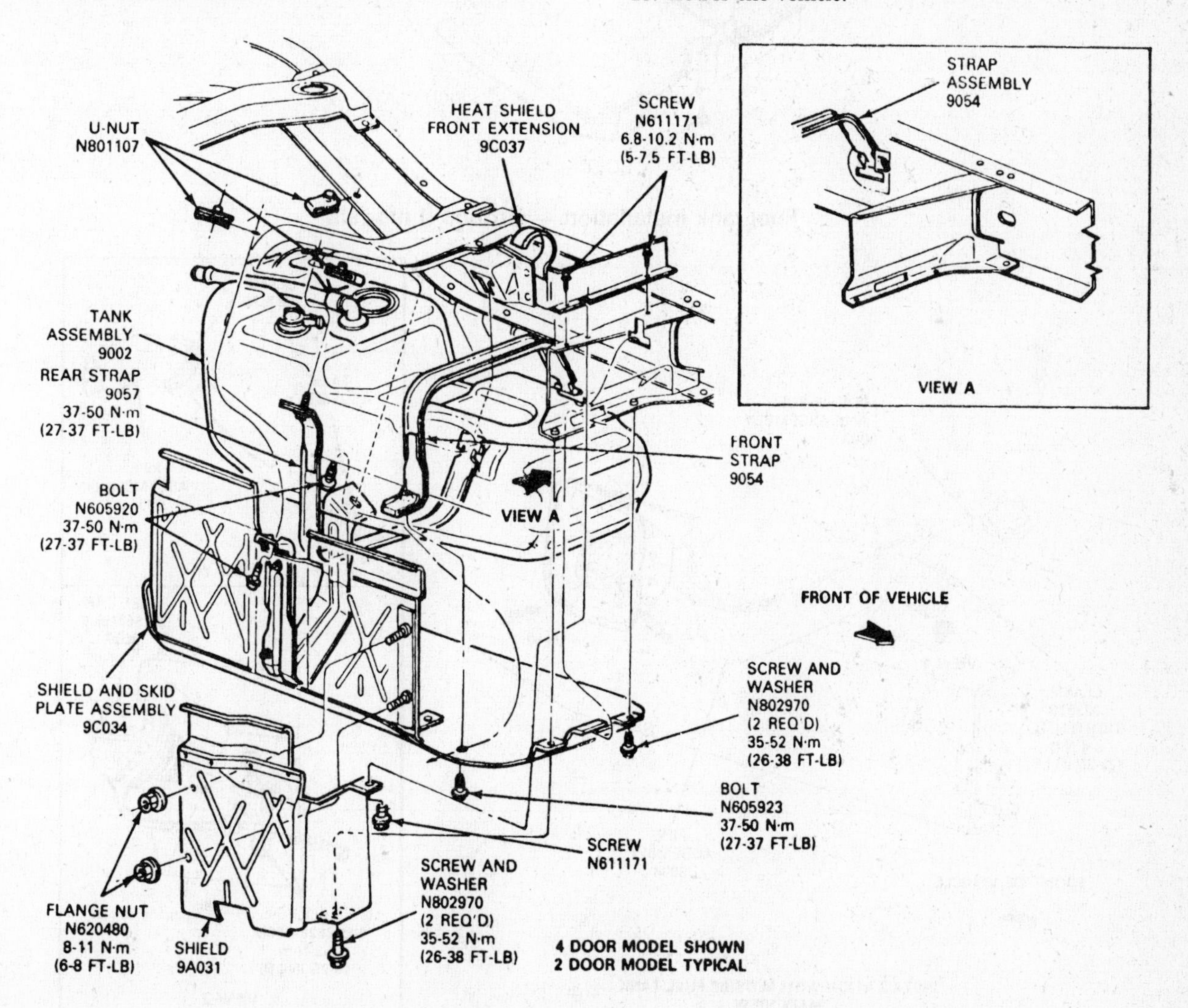

Fuel tank installation — Explorer

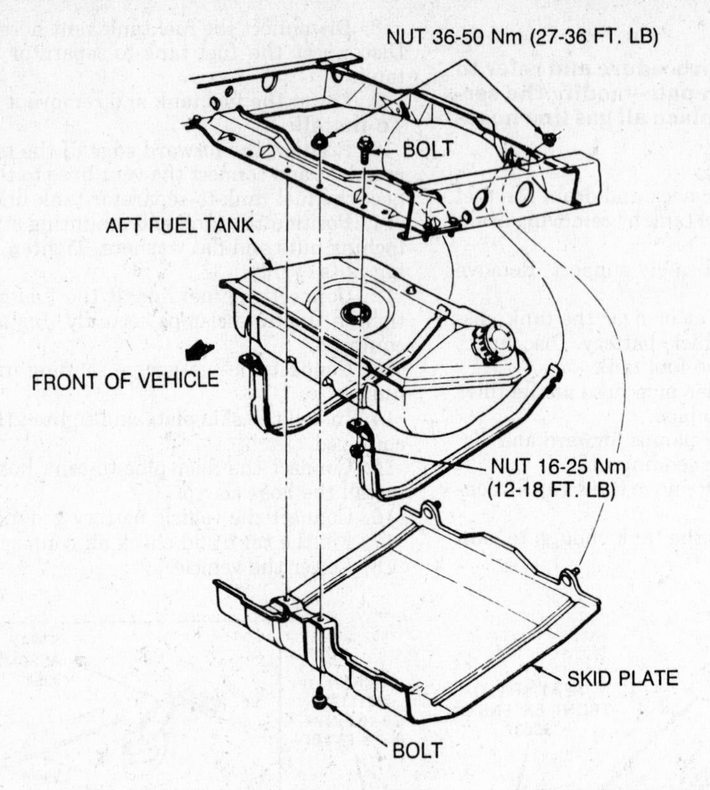

Fuel tank installation – Bronco II models

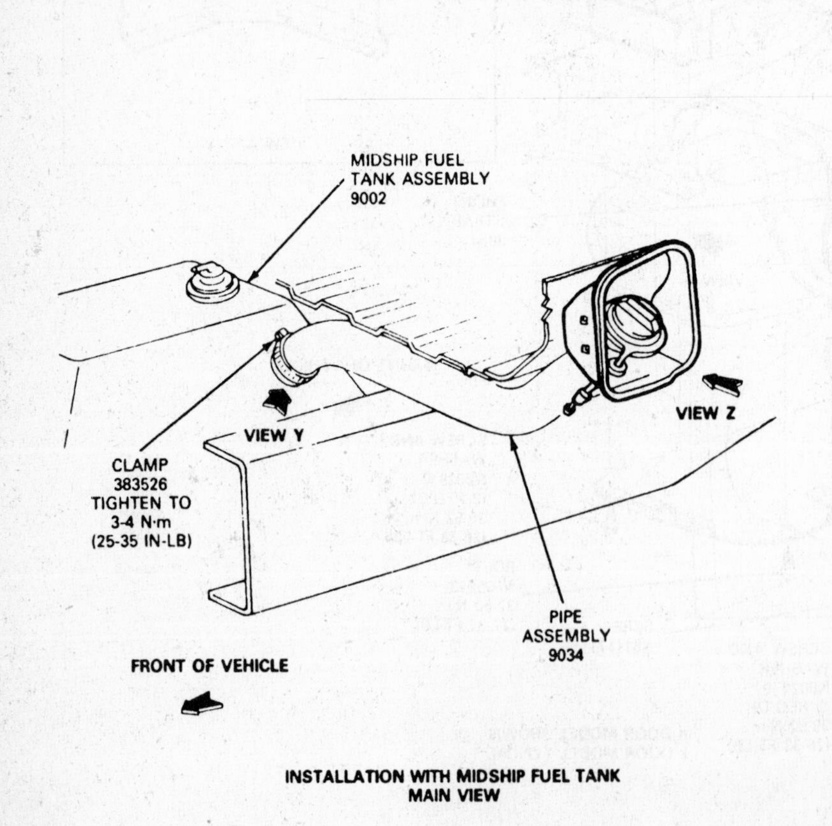

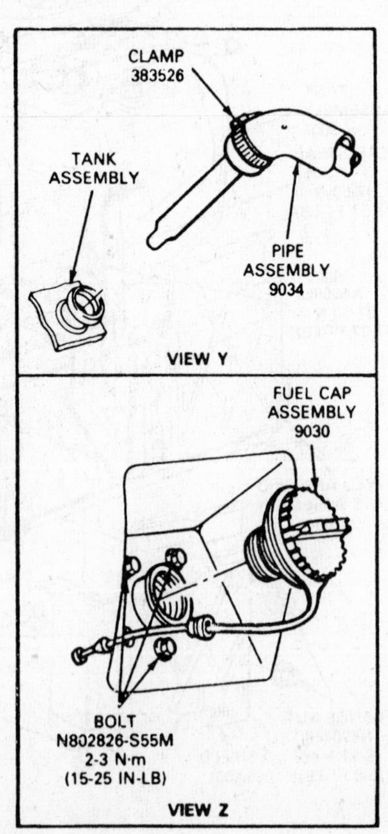

Fuel filler system – Ranger/Bronco II

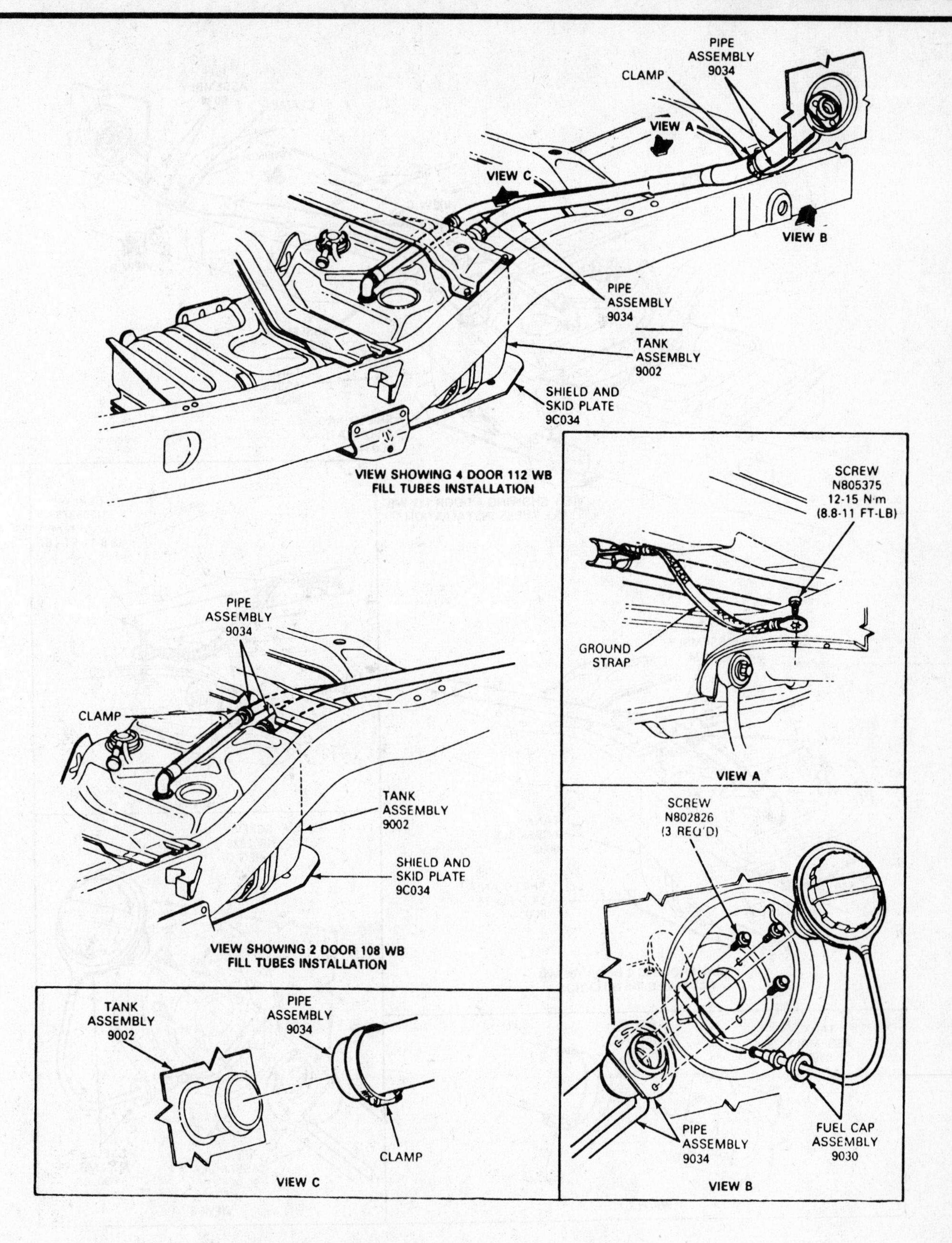

Fuel filler system – Explorer

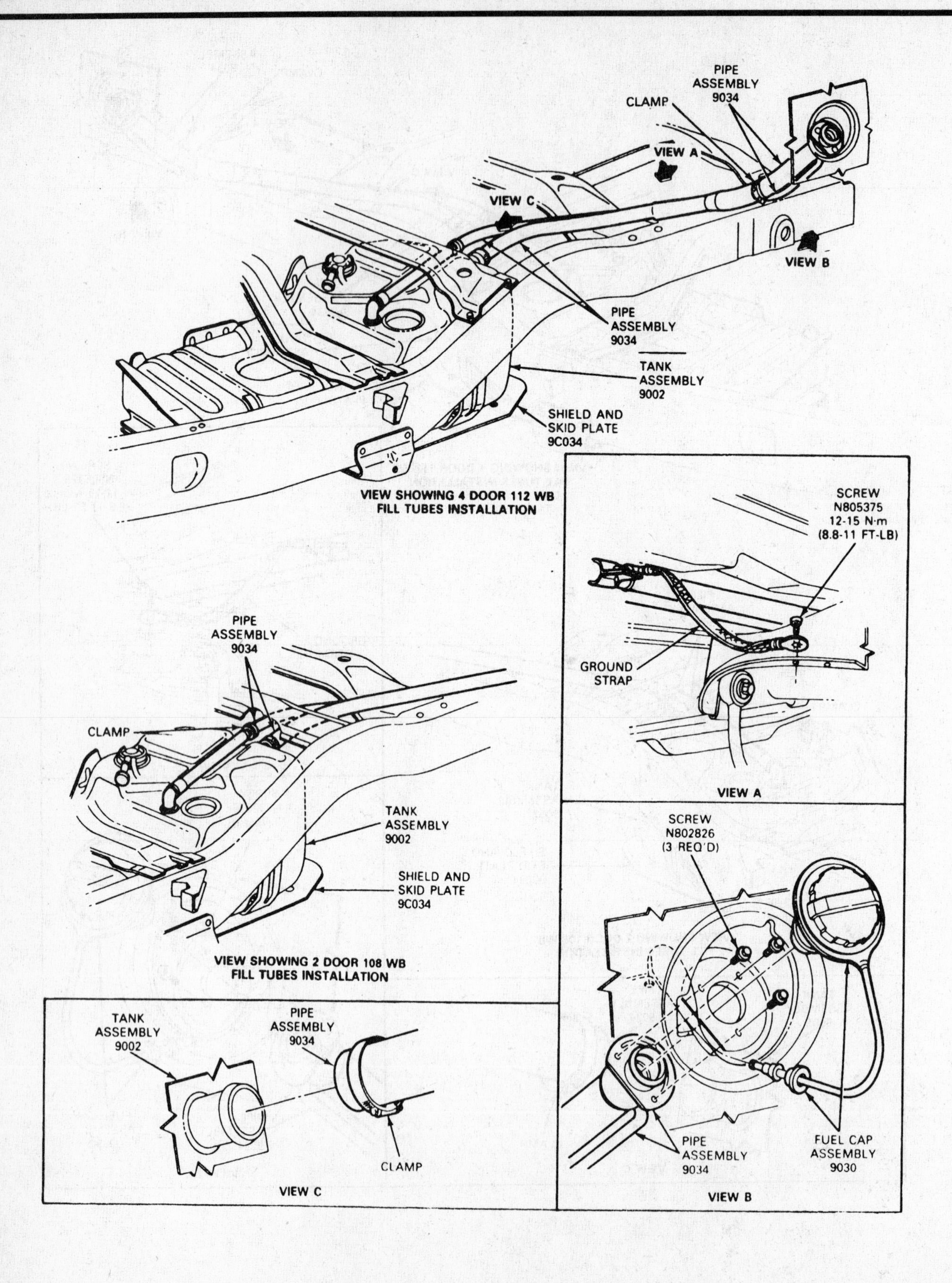
PIPE
ASSEMBLY
9034
CLAMP
VIEW A
VIEW C
VIEW B
PIPE
ASSEMBLY
9034
TANK
ASSEMBLY
9002
SHIELD AND
SKID PLATE
9C034
VIEW SHOWING 4 DOOR 112 WB
FILL TUBES INSTALLATION
SCREW
N805375
12-15 N·m
(8.8-11 FT-LB)
GROUND
STRAP
VIEW A
PIPE
ASSEMBLY
9034
CLAMP
TANK
ASSEMBLY
9002
SHIELD AND
SKID PLATE
9C034
VIEW SHOWING 2 DOOR 108 WB
FILL TUBES INSTALLATION
SCREW
N802826
(3 REQ'D)
TANK
ASSEMBLY
9002
PIPE
ASSEMBLY
9034
CLAMP
VIEW C
PIPE
ASSEMBLY
9034
FUEL CAP
ASSEMBLY
9030
VIEW B

6 Chassis Electrical

QUICK REFERENCE INDEX

GENERAL INDEX

UNDERSTANDING AND TROUBLESHOOTING ELECTRICAL SYSTEMS

At the rate which both import and domestic manufacturers are incorporating electronic control systems into their production lines, it won't be long before every new vehicle is equipped with one or more on-board computer, like the EEC-IV unit installed on the Ranger/Bronco II. These electronic components (with no moving parts) should theoretically last the life of the vehicle, provided nothing external happens to damage the circuits or memory chips.

While it is true that electronic components should never wear out, in the real world malfunctions do occur. It is also true that any computer-based system is extremely sensitive to electrical voltages and cannot tolerate careless or haphazard testing or service procedures. An inexperienced individual can literally do major damage looking for a minor problem by using the wrong kind of test equipment or connecting test leads or connectors with the ignition switch ON. When selecting test equipment, make sure the manufacturers instructions state that the tester is compatible with whatever type of electronic control system is being serviced. Read all instructions carefully and double check all test points before installing probes or making any test connections.

The following section outlines basic diagnosis techniques for dealing with computerized automotive control systems. Along with a general explanation of the various types of test equipment available to aid in servicing modern electronic automotive systems, basic repair techniques for wiring harnesses and connectors is given. Read the basic information before attempting any repairs or testing on any computerized system, to provide the background of information necessary to avoid the most common and obvious mistakes that can cost both time and money. Although the replacement and testing procedures are simple in themselves, the systems are not, and unless one has a thorough understanding of all components and their function within a particular computerized control system, the logical test sequence these systems demand cannot be followed. Minor malfunctions can make a big difference, so it is important to know how each component affects the operation of the overall electronic system to find the ultimate cause of a problem without replacing good components unnecessarily. It is not enough to use the correct test equipment; the test equipment must be used correctly.

Safety Precautions

CAUTION

Whenever working on or around any computer based microprocessor control system, always observe these general precautions to prevent the possibility of personal injury or damage to electronic components.

- Never install or remove battery cables with the key ON or the engine running. Jumper cables should be connected with the key OFF to avoid power surges that can damage electronic control units. Engines equipped with computer controlled systems should avoid both giving and getting jump starts due to the possibility of serious damage to components from arcing in the engine compartment when connections are made with the ignition ON.
- Always remove the battery cables before charging the battery. Never use a high output charger on an installed battery or attempt to use any type of "hot shot" (24 volt) starting aid.
- Exercise care when inserting test probes into connectors to insure good connections without damaging the connector or spreading the pins. Always probe connectors from the rear (wire) side, NOT the pin side, to avoid accidental shorting of terminals during test procedures.
- Never remove or attach wiring harness connectors with the ignition switch ON, especially to an electronic control unit.
- Do not drop any components during service procedures and never apply 12 volts directly to any component (like a solenoid or relay) unless instructed specifically to do so. Some component electrical windings are designed to safely handle only 4 or 5 volts and can be destroyed in seconds if 12 volts are applied directly to the connector.
- Remove the electronic control unit if the vehicle is to be placed in an environment where temperatures exceed approximately 176°F (80°C), such as a paint spray booth or when arc or gas welding near the control unit location in the car.

ORGANIZED TROUBLESHOOTING

When diagnosing a specific problem, organized troubleshooting is a must. The complexity of a modern automobile demands that you approach any problem in a logical, organized manner. There are certain troubleshooting techniques that are standard:

1. Establish when the problem occurs. Does the problem appear only under certain conditions? Were there any noises, odors, or other unusual symptoms?
2. Isolate the problem area. To do this, make some simple tests and observations; then eliminate the systems that are working properly. Check for obvious problems such as broken wires, dirty connections or split or disconnected vacuum hoses. Always check the obvious before assuming something complicated is the cause.
3. Test for problems systematically to determine the cause once the problem area is isolated. Are all the components functioning properly? Is there power going to electrical switches and motors? Is there vacuum at vacuum switches and/or actuators? Is there a mechanical problem such as bent linkage or loose mounting screws? Doing careful, systematic checks will often turn up most causes on the first inspection without wasting time checking components that have little or no relationship to the problem.
4. Test all repairs after the work is done to make sure that the problem is fixed. Some causes can be traced to more than one component, so a careful verification of repair work is important to pick up additional malfunctions that may cause a problem to reappear or a different problem to arise. A blown fuse, for example, is a simple problem that may require more than another fuse to repair. If you don't look for a problem that caused a fuse to blow, for example, a shorted wire may go undetected.

Experience has shown that most problems tend to be the result of a fairly simple and obvious cause, such as loose or corroded connectors or air leaks in the intake system; making careful inspection of components during testing essential to quick and accurate troubleshooting. Special, hand held computerized testers designed specifically for diagnosing the EEC-IV system are available from a variety of after market sources, as well as from the vehicle manufacturer, but care should be taken that any test equipment being used is designed to diagnose that particular computer controlled system accurately without damaging the control unit (ECU) or components being tested.

NOTE: Pinpointing the exact cause of trouble in an electrical system can sometimes only be accomplished by the use of special test equipment. The following describes commonly used test equipment and explains how to put it to best use in diagnosis. In addition to the information covered below, the manufacturer's instructions booklet provided with the tester should be read and clearly understood before attempting any test procedures.

TEST EQUIPMENT

Jumper Wires

Jumper wires are simple, yet extremely valuable, pieces of test equipment. Jumper wires are merely wires that are used to bypass sections of a circuit. The simplest type of jumper wire is merely a length of multistrand wire with an alligator clip at each end. Jumper wires are usually fabricated from lengths of standard automotive wire and whatever type of connector (alligator clip, spade connector or pin connector) that is required for the particular vehicle being tested. The well equipped tool box will have several different styles of jumper wires in several different lengths. Some jumper wires are made with three or more terminals coming from a common splice for special purpose testing. In cramped, hard-to-reach areas it is advisable to have insulated boots over the jumper wire terminals in order to prevent accidental grounding, sparks, and possible fire, especially when testing fuel system components.

Jumper wires are used primarily to locate open electrical circuits, on either the ground (−) side of the circuit or on the hot (+) side. If an electrical component fails to operate, connect the jumper wire between the component and a good ground. If the component operates only with the jumper installed, the ground circuit is open. If the ground circuit is good, but the component does not operate, the circuit between the power feed and component is open. You can sometimes connect the jumper wire directly from the battery to the hot terminal of the component, but first make sure the component uses 12 volts in operation. Some electrical components, such as fuel injectors, are designed to operate on about 4 volts and running 12 volts directly to the injector terminals can burn out the wiring. By inserting an inline fuse holder between a set of test leads, a fused jumper wire can be used for bypassing open circuits. Use a 5 amp fuse to provide protection against voltage spikes. When in doubt, use a voltmeter to check the voltage input to the component and measure how much voltage is being applied normally. By moving the jumper wire successively back from the lamp toward the power source, you can isolate the area of the circuit where the open is located. When the component stops functioning, or the power is cut off, the open is in the segment of wire between the jumper and the point previously tested.

CAUTION

Never use jumpers made from wire that is of lighter gauge than used in the circuit under test. If the jumper wire is of too small gauge, it may overheat and possibly melt. Never use jumpers to bypass high resistance loads (such as motors) in a circuit. Bypassing resistances, in effect, creates a short circuit which may, in turn, cause damage and fire. Never use a jumper for anything other than temporary bypassing of components in a circuit.

12 Volt Test Light

The 12 volt test light is used to check circuits and components while electrical current is flowing through them. It is used for voltage and ground tests. Twelve volt test lights come in different styles but all have three main parts; a ground clip, a probe, and a light. The most commonly used 12 volt test lights have pick-type probes. To use a 12 volt test light, connect the ground clip to a good ground and probe wherever necessary with the pick. The pick should be sharp so that it can penetrate wire insulation to make contact with the wire, without making a large hole in the insulation. The wrap-around light is handy in hard to reach areas or where it is difficult to support a wire to push a probe pick into it. To use the wrap around light, hook the wire to probed with the hook and pull the trigger. A small pick will be forced through the wire insulation into the wire core.

CAUTION

Do not use a test light to probe electronic ignition spark plug or coil wires. Never use a pick-type test light to probe wiring on computer controlled systems unless specifically instructed to do so. Any wire insulation that is pierced by the test light probe should be taped and sealed with silicone after testing.

Like the jumper wire, the 12 volt test light is used to isolate opens in circuits. But, whereas the jumper wire is used to bypass the open to operate the load, the 12 volt test light is used to locate the presence of voltage in a circuit. If the test light glows, you know that there is power up to that point; if the 12 volt test light does not glow when its probe is inserted into the wire or connector, you know that there is an open circuit (no power). Move the test light in successive steps back toward the power source until the light in the handle does glow. When it does glow, the open is between the probe and point previously probed.

NOTE: The test light does not detect that 12 volts (or any particular amount of voltage) is present; it only detects that some voltage is present. It is advisable before using the test light to touch its terminals across the battery posts to make sure the light is operating properly.

Self-Powered Test Light

The self-powered test light usually contains a 1.5 volt penlight battery. One type of self-powered test light is similar in design to the 12 volt test light. This type has both the battery and the light in the handle and pick-type probe tip. The second type has the light toward the open tip, so that the light illuminates the contact point. The self-powered test light is dual purpose piece of test equipment. It can be used to test for either open or short circuits when power is isolated from the circuit (continuity test). A powered test light should not be used on any computer controlled system or component unless specifically instructed to do so. Many engine sensors can be destroyed by even this small amount of voltage applied directly to the terminals.

Open Circuit Testing

To use the self-powered test light to check for open circuits, first isolate the circuit from the vehicle's 12 volt power source by disconnecting the battery or wiring harness connector. Connect the test light ground clip to a good ground and probe sections of the circuit sequentially with the test light. (start from either end of the circuit). If the light is out, the open is between the probe and the circuit ground. If the light is on, the open is between the probe and end of the circuit toward the power source.

Short Circuit Testing

By isolating the circuit both from power and from ground, and using a self-powered test light, you can check for shorts to ground in the circuit. Isolate the circuit from power and ground. Connect the test light ground clip to a good ground and probe any easy-to-reach test point in the circuit. If the light comes on, there is a short somewhere in the circuit. To isolate the short, probe a test point at either end of the isolated circuit (the light should be on). Leave the test light probe connected and open connectors, switches, remove parts, etc., sequentially, until the light goes out. When the light goes out, the short is between the last circuit component opened and the previous circuit opened.

NOTE: The 1.5 volt battery in the test light does not provide much current. A weak battery may not provide enough power to illuminate the test light even when a complete circuit is made (especially if there are high resistances in the circuit). Always make sure that the test battery is strong. To check the battery, briefly touch the ground clip to the probe; if the light glows brightly the

battery is strong enough for testing. Never use a self-powered test light to perform checks for opens or shorts when power is applied to the electrical system under test. The 12 volt vehicle power will quickly burn out the 1.5 volt light bulb in the test light.

Voltmeter

A voltmeter is used to measure voltage at any point in a circuit, or to measure the voltage drop across any part of a circuit. It can also be used to check continuity in a wire or circuit by indicating current flow from one end to the other. Voltmeters usually have various scales on the meter dial and a selector switch to allow the selection of different voltages. The voltmeter has a positive and a negative lead. To avoid damage to the meter, always connect the negative lead to the negative (–) side of circuit (to ground or nearest the ground side of the circuit) and connect the positive lead to the positive (+) side of the circuit (to the power source or the nearest power source). Note that the negative voltmeter lead will always be black and that the positive voltmeter will always be some color other than black (usually red). Depending on how the voltmeter is connected into the circuit, it has several uses.

A voltmeter can be connected either in parallel or in series with a circuit and it has a very high resistance to current flow. When connected in parallel, only a small amount of current will flow through the voltmeter current path; the rest will flow through the normal circuit current path and the circuit will work normally. When the voltmeter is connected in series with a circuit, only a small amount of current can flow through the circuit. The circuit will not work properly, but the voltmeter reading will show if the circuit is complete or not.

Available Voltage Measurement

Set the voltmeter selector switch to the 20V position and connect the meter negative lead to the negative post of the battery. Connect the positive meter lead to the positive post of the battery and turn the ignition switch ON to provide a load. Read the voltage on the meter or digital display. A well charged battery should register over 12 volts. If the meter reads below 11.5 volts, the battery power may be insufficient to operate the electrical system properly. This test determines voltage available from the battery and should be the first step in any electrical trouble diagnosis procedure. Many electrical problems, especially on computer controlled systems, can be caused by a low state of charge in the battery. Excessive corrosion at the battery cable terminals can cause a poor contact that will prevent proper charging and full battery current flow.

Normal battery voltage is 12 volts when fully charged. When the battery is supplying current to one or more circuits it is said to be "under load". When everything is off the electrical system is under a "no-load" condition. A fully charged battery may show about 12.5 volts at no load; will drop to 12 volts under medium load; and will drop even lower under heavy load. If the battery is partially discharged the voltage decrease under heavy load may be excessive, even though the battery shows 12 volts or more at no load. When allowed to discharge further, the battery's available voltage under load will decrease more severely. For this reason, it is important that the battery be fully charged during all testing procedures to avoid errors in diagnosis and incorrect test results.

Voltage Drop

When current flows through a resistance, the voltage beyond the resistance is reduced (the larger the current, the greater the reduction in voltage). When no current is flowing, there is no voltage drop because there is no current flow. All points in the circuit which are connected to the power source are at the same voltage as the power source. The total voltage drop always equals the total source voltage. In a long circuit with many connectors, a series of small, unwanted voltage drops due to corrosion at the connectors can add up to a total loss of voltage which impairs the operation of the normal loads in the circuit.

INDIRECT COMPUTATION OF VOLTAGE DROPS

1. Set the voltmeter selector switch to the 20 volt position.
2. Connect the meter negative lead to a good ground.
3. Probe all resistances in the circuit with the positive meter lead.
4. Operate the circuit in all modes and observe the voltage readings.

DIRECT MEASUREMENT OF VOLTAGE DROPS

1. Set the voltmeter switch to the 20 volt position.
2. Connect the voltmeter negative lead to the ground side of the resistance load to be measured.
3. Connect the positive lead to the positive side of the resistance or load to be measured.
4. Read the voltage drop directly on the 20 volt scale.

Too high a voltage indicates too high a resistance. If, for example, a blower motor runs too slowly, you can determine if there is too high a resistance in the resistor pack. By taking voltage drop readings in all parts of the circuit, you can isolate the problem. Too low a voltage drop indicates too low a resistance. If, for example, a blower motor runs too fast in the MED and/or LOW position, the problem can be isolated in the resistor pack by taking voltage drop readings in all parts of the circuit to locate a possibly shorted resistor. The maximum allowable voltage drop under load is critical, especially if there is more than one high resistance problem in a circuit because all voltage drops are cumulative. A small drop is normal due to the resistance of the conductors.

HIGH RESISTANCE TESTING

1. Set the voltmeter selector switch to the 4 volt position.
2. Connect the voltmeter positive lead to the positive post of the battery.
3. Turn on the headlights and heater blower to provide a load.
4. Probe various points in the circuit with the negative voltmeter lead.
5. Read the voltage drop on the 4 volt scale. Some average maximum allowable voltage drops are:

FUSE PANEL – 7 volts
IGNITION SWITCH – 5 volts
HEADLIGHT SWITCH – 7 volts
IGNITION COIL (+) – 5 volts
ANY OTHER LOAD – 1.3 volts

NOTE: Voltage drops are all measured while a load is operating; without current flow, there will be no voltage drop.

Ohmmeter

The ohmmeter is designed to read resistance (ohms) in a circuit or component. Although there are several different styles of ohmmeters, all will usually have a selector switch which permits the measurement of different ranges of resistance (usually the selector switch allows the multiplication of the meter reading by 10, 100, 1000, and 10,000). A calibration knob allows the meter to be set at zero for accurate measurement. Since all ohmmeters are powered by an internal battery (usually 9 volts), the ohmmeter can be used as a self-powered test light. When the ohmmeter is connected, current from the ohmmeter flows through the circuit or component being tested. Since the ohmmeter's internal resistance and voltage are known values, the amount of current flow through the meter depends on the resistance of the circuit or component being tested.

The ohmmeter can be used to perform continuity test for opens or shorts (either by observation of the meter needle or as a self-powered test light), and to read actual resistance in a circuit. It should be noted that the ohmmeter is used to check the resistance of a component or wire while there is no voltage applied to the circuit. Current flow from an outside voltage source (such as the vehicle battery) can damage the ohmmeter, so the circuit or component should be isolated from the vehicle electrical system before any testing is done. Since the ohmmeter uses its own voltage source, either lead can be connected to any test point.

NOTE: When checking diodes or other solid state components, the ohmmeter leads can only be connected one way in order to measure current flow in a single direction. Make sure the positive (+) and negative (–) terminal connections are as described in the test procedures to verify the one-way diode operation.

In using the meter for making continuity checks, do not be concerned with the actual resistance readings. Zero resistance, or any resistance readings, indicate continuity in the circuit. Infinite resistance indicates an open in the circuit. A high resistance reading where there should be none indicates a problem in the circuit. Checks for short circuits are made in the same manner as checks for open circuits except that the circuit must be isolated from both power and normal ground. Infinite resistance indicates no continuity to ground, while zero resistance indicates a dead short to ground.

RESISTANCE MEASUREMENT

The batteries in an ohmmeter will weaken with age and temperature, so the ohmmeter must be calibrated or "zeroed" before taking measurements. To zero the meter, place the selector switch in its lowest range and touch the two ohmmeter leads together. Turn the calibration knob until the meter needle is exactly on zero.

NOTE: All analog (needle) type ohmmeters must be zeroed before use, but some digital ohmmeter models are automatically calibrated when the switch is turned on. Self-calibrating digital ohmmeters do not have an adjusting knob, but its a good idea to check for a zero readout before use by touching the leads together. All computer controlled systems require the use of a digital ohmmeter with at least 10 megohms impedance for testing. Before any test procedures are attempted, make sure the ohmmeter used is compatible with the electrical system or damage to the on-board computer could result.

To measure resistance, first isolate the circuit from the vehicle power source by disconnecting the battery cables or the harness connector. Make sure the key is OFF when disconnecting any components or the battery. Where necessary, also isolate at least one side of the circuit to be checked to avoid reading parallel resistances. Parallel circuit resistances will always give a lower reading than the actual resistance of either of the branches. When measuring the resistance of parallel circuits, the total resistance will always be lower than the smallest resistance in the circuit. Connect the meter leads to both sides of the circuit (wire or component) and read the actual measured ohms on the meter scale. Make sure the selector switch is set to the proper ohm scale for the circuit being tested to avoid misreading the ohmmeter test value.

CAUTION

Never use an ohmmeter with power applied to the circuit. Like the self-powered test light, the ohmmeter is designed to operate on its own power supply. The normal 12 volt automotive electrical system current could damage the meter.

Ammeters

An ammeter measures the amount of current flowing through a circuit in units called amperes or amps. Amperes are units of electron flow which indicate how fast the electrons are flowing through the circuit. Since Ohms Law dictates that current flow in a circuit is equal to the circuit voltage divided by the total circuit resistance, increasing voltage also increases the current level (amps). Likewise, any decrease in resistance will increase the amount of amps in a circuit. At normal operating voltage, most circuits have a characteristic amount of amperes, called "current draw" which can be measured using an ammeter. By referring to a specified current draw rating, measuring the amperes, and comparing the two values, one can determine what is happening within the circuit to aid in diagnosis. An open circuit, for example, will not allow any current to flow so the ammeter reading will be zero. More current flows through a heavily loaded circuit or when the charging system is operating.

An ammeter is always connected in series with the circuit being tested. All of the current that normally flows through the circuit must also flow through the ammeter; if there is any other path for the current to follow, the ammeter reading will not be accurate. The ammeter itself has very little resistance to current flow and therefore will not affect the circuit, but it will measure current draw only when the circuit is closed and electricity is flowing. Excessive current draw can blow fuses and drain the battery, while a reduced current draw can cause motors to run slowly, lights to dim and other components to not operate properly. The ammeter can help diagnose these conditions by locating the cause of the high or low reading.

Multimeters

Different combinations of test meters can be built into a single unit designed for specific tests. Some of the more common combination test devices are known as Volt/Amp testers, Tach/Dwell meters, or Digital Multimeters. The Volt/Amp tester is used for charging system, starting system or battery tests and consists of a voltmeter, an ammeter and a variable resistance carbon pile. The voltmeter will usually have at least two ranges for use with 6, 12 and 24 volt systems. The ammeter also has more than one range for testing various levels of battery loads and starter current draw and the carbon pile can be adjusted to offer different amounts of resistance. The Volt/Amp tester has heavy leads to carry large amounts of current and many later models have an inductive ammeter pickup that clamps around the wire to simplify test connections. On some models, the ammeter also has a zero-center scale to allow testing of charging and starting systems without switching leads or polarity. A digital multimeter is a voltmeter, ammeter and ohmmeter combined in an instrument which gives a digital readout. These are often used when testing solid state circuits because of their high input impedance (usually 10 megohms or more).

The tach/dwell meter combines a tachometer and a dwell (cam angle) meter and is a specialized kind of voltmeter. The tachometer scale is marked to show engine speed in rpm and the dwell scale is marked to show degrees of distributor shaft rotation. In most electronic ignition systems, dwell is determined by the control unit, but the dwell meter can also be used to check the duty cycle (operation) of some electronic engine control systems. Some tach/dwell meters are powered by an internal battery, while others take their power from the car battery in use. The battery powered testers usually require calibration much like an ohmmeter before testing.

Special Test Equipment

A variety of diagnostic tools are available to help troubleshoot and repair computerized engine control systems. The most sophisticated of these devices are the console type engine analyzers that usually occupy a garage service bay, but there are several types of aftermarket electronic testers available that will allow quick circuit tests of the engine control system by plugging directly into a special connector located in the engine compartment or under the dashboard. Several tool and equipment manufacturers offer simple, hand held testers that measure various circuit voltage levels on command to check all system components for proper operation. Although these testers usually cost about $300–500, consider that the average computer control unit (or ECM) can cost just as much and the money saved by not replacing perfectly good sensors or components in an attempt to correct a problem could justify the purchase price of a special diagnostic tester the first time it's used.

These computerized testers can allow quick and easy test measurements while the engine is operating or while the car is being driven. In addition, the on-board computer memory can be read to access any stored trouble codes; in effect allowing the computer to tell you where it hurts and aid trouble diagnosis by pinpointing exactly which circuit or component is malfunctioning. In the same manner, repairs can be tested to make sure the problem has been corrected. The biggest advantage these special testers have is their relatively easy hookups that minimize or eliminate the chances of making the wrong connections and getting false voltage readings or damaging the computer accidentally.

NOTE: It should be remembered that these testers check voltage levels in circuits; they don't detect mechanical problems or failed components if the circuit voltage falls within the preprogrammed limits stored in the tester PROM unit. Also, most of the hand held testers are designed to work only on one or two systems made by a specific manufacturer.

A variety of after market testers are available to help diagnose different computerized control systems. Owatonna Tool Company (OTC), for example, markets a device called the OTC Monitor which plugs directly into the assembly line diagnostic link (ALDL). The OTC tester makes diagnosis a simple matter of pressing the correct buttons and, by changing the internal PROM or inserting a different diagnosis cartridge, it will work on any model from full size to subcompact, over a wide range of years. An adapter is supplied with the tester to allow connection to all types of ALDL links, regardless of the number of pin terminals used. By inserting an updated PROM into the OTC tester, it can be easily updated to diagnose any new modifications of computerized control systems.

Wiring Harnesses

The average automobile contains about ½ mile of wiring, with hundreds of individual connections. To protect the many wires from damage and to keep them from becoming a confusing tangle, they are organized into bundles, enclosed in plastic or taped together and called wire harnesses. Different wiring harnesses serve different parts of the vehicle. Individual wires are color coded to help trace them through a harness where sections are hidden from view.

A loose or corroded connection or a replacement wire that is too small for the circuit will add extra resistance and an additional voltage drop to the circuit. A ten percent voltage drop can result in slow or erratic motor operation, for example, even though the circuit is complete. Automotive wiring or circuit conductors can be in any one of three forms:

1. Single strand wire
2. Multistrand wire
3. Printed circuitry

Single strand wire has a solid metal core and is usually used inside such components as alternators, motors, relays and other devices. Multistrand wire has a core made of many small strands of wire twisted together into a single conductor. Most of the wiring in an automotive electrical system is made up of multistrand wire, either as a single conductor or grouped together in a harness. All wiring is color coded on the insulator, either as a solid color or as a colored wire with an identification stripe. A printed circuit is a thin film of copper or other conductor that is printed on an insulator backing. Occasionally, a printed circuit is sandwiched between two sheets of plastic for more protection and flexibility. A complete printed circuit, consisting of conductors, insulating material and connectors for lamps or other components is called a printed circuit board. Printed circuitry is used in place of individual wires or harnesses in places where space is limited, such as behind instrument panels.

Wire Gauge

Since computer controlled automotive electrical systems are very sensitive to changes in resistance, the selection of properly sized wires is critical when systems are repaired. The wire gauge number is an expression of the cross section area of the conductor. The most common system for expressing wire size is the American Wire Gauge (AWG) system.

Wire cross section area is measured in circular mils. A mil is 1/000″ (0.001″); a circular mil is the area of a circle one mil in diameter. For example, a conductor ¼″ in diameter is 0.250 in. or 250 mils. The circular mil cross section area of the wire is 250 squared (250″) or 62,500 circular mils. Imported car models usually use metric wire gauge designations, which is simply the cross section area of the conductor in square millimeters (mm).

Gauge numbers are assigned to conductors of various cross section areas. As gauge number increases, area decreases and the conductor becomes smaller. A 5 gauge conductor is smaller than a 1 gauge conductor and a 10 gauge is smaller than a 5 gauge. As the cross section area of a conductor decreases, resistance increases and so does the gauge number. A conductor with a higher gauge number will carry less current than a conductor with a lower gauge number.

NOTE: Gauge wire size refers to the size of the conductor, not the size of the complete wire. It is possible to have two wires of the same gauge with different diameters because one may have thicker insulation than the other.

12 volt automotive electrical systems generally use 10, 12, 14, 16 and 18 gauge wire. Main power distribution circuits and larger accessories usually use 10 and 12 gauge wire. Battery cables are usually 4 or 6 gauge, although 1 and 2 gauge wires are occasionally used. Wire length must also be considered when making repairs to a circuit. As conductor length increases, so does resistance. An 18 gauge wire, for example, can carry a 10 amp load for 10 feet without excessive voltage drop; however if a 15 foot wire is required for the same 10 amp load, it must be a 16 gauge wire.

An electrical schematic shows the electrical current paths when a circuit is operating properly. It is essential to understand how a circuit works before trying to figure out why it doesn't. Schematics break the entire electrical system down into individual circuits and show only one particular circuit. In a schematic, no attempt is made to represent wiring and components as they physically appear on the vehicle; switches and other components are shown as simply as possible. Face views of harness connectors show the cavity or terminal locations in all multi-pin connectors to help locate test points.

If you need to backprobe a connector while it is on the component, the order of the terminals must be mentally reversed. The wire color code can help in this situation, as well as a keyway, lock tab or other reference mark.

NOTE: Wiring diagrams are not included in this book. As trucks have become more complex and available with longer option lists, wiring diagrams have grown in size and complexity. It has become almost impossible to provide a readable reproduction of a wiring diagram in a book this size. Information on ordering wiring diagrams from the vehicle manufacturer can be found in the owner's manual.

WIRING REPAIR

Soldering is a quick, efficient method of joining metals permanently. Everyone who has the occasion to make wiring repairs should know how to solder. Electrical connections that are soldered are far less likely to come apart and will conduct electricity much better than connections that are only "pig-tailed" together. The most popular (and preferred) method of soldering is with an electrical soldering gun. Soldering irons are available in many sizes and wattage ratings. Irons with higher wattage ratings deliver higher temperatures and recover lost heat faster. A small soldering iron rated for no more than 50 watts is recommended, especially on electrical systems where excess heat can damage the components being soldered.

There are three ingredients necessary for successful soldering; proper flux, good solder and sufficient heat. A soldering flux is necessary to clean the metal of tarnish, prepare it for soldering and to enable the solder to spread into tiny crevices. When soldering, always use a resin flux or resin core solder which is non-corrosive and will not attract moisture once the job is finished. Other types of flux (acid core) will leave a residue that will attract moisture and cause the wires to corrode. Tin is a unique metal with a low melting point. In a molten state, it dissolves and alloys easily with many metals. Solder is made by mixing tin with lead. The most common proportions are 40/60, 50/50 and 60/40, with the percentage of tin listed first. Low priced solders usually contain less tin, making them very difficult for a beginner to use because more heat is required to melt the solder. A common solder is 40/60 which is well suited for all-around general use, but 60/40 melts easier, has more tin for a better joint and is preferred for electrical work.

Soldering Techniques

Successful soldering requires that the metals to be joined be heated to a temperature that will melt the solder—usually 360–460°F (182–238°C). Contrary to popular belief, the purpose of the soldering iron is not to melt the solder itself, but to heat the parts being soldered to a temperature high enough to melt the solder when it is touched to the work. Melting flux-cored solder on the soldering iron will usually destroy the effectiveness of the flux.

NOTE: Soldering tips are made of copper for good heat conductivity, but must be "tinned" regularly for quick transference of heat to the project and to prevent the solder from sticking to the iron. To "tin" the iron, simply heat it and touch the flux-cored solder to the tip; the solder will flow over the hot tip. Wipe the excess off with a clean rag, but be careful as the iron will be hot.

After some use, the tip may become pitted. If so, simply dress the tip smooth with a smooth file and "tin" the tip again. An old saying holds that "metals well cleaned are half soldered." Flux-cored solder will remove oxides but rust, bits of insulation and oil or grease must be removed with a wire brush or emery cloth. For maximum strength in soldered parts, the joint must start off clean and tight. Weak joints will result in gaps too wide for the solder to bridge.

If a separate soldering flux is used, it should be brushed or swabbed on only those areas that are to be soldered. Most solders contain a core of flux and separate fluxing is unnecessary. Hold the work to be soldered firmly. It is best to solder on a wooden board, because a metal vise will only rob the piece to be soldered of heat and make it difficult to melt the solder. Hold the soldering tip with the broadest face against the work to be soldered. Apply solder under the tip close to the work, using enough solder to give a heavy film between the iron and the piece being soldered, while moving slowly and making sure the solder melts properly. Keep the work level or the solder will run to the lowest part and favor the thicker parts, because these require more heat to melt the solder. If the soldering tip overheats (the solder coating on the face of the tip burns up), it should be retinned. Once the soldering is completed, let the soldered joint stand until cool. Tape and seal all soldered wire splices after the repair has cooled.

Wire Harness and Connectors

The on-board computer (ECM) wire harness electrically connects the control unit to the various solenoids, switches and sensors used by the control system. Most connectors in the engine compartment or otherwise exposed to the elements are protected against moisture and dirt which could create oxidation and deposits on the terminals. This protection is important because of the very low voltage and current levels used by the computer and sensors. All connectors have a lock which secures the male and female terminals together, with a secondary lock holding the seal and terminal into the connector. Both terminal locks must be released when disconnecting ECM connectors.

These special connectors are weather-proof and all repairs require the use of a special terminal and the tool required to service it. This tool is used to remove the pin and sleeve terminals. If removal is attempted with an ordinary pick, there is a good chance that the terminal will be bent or deformed. Unlike standard blade type terminals, these terminals cannot be straightened once they are bent. Make certain that the connectors are properly seated and all of the sealing rings in place when connecting leads. On some models, a hinge-type flap provides a backup or secondary locking feature for the terminals. Most secondary locks are used to improve the connector reliability by retaining the terminals if the small terminal lock tangs are not positioned properly.

Molded-on connectors require complete replacement of the connection. This means splicing a new connector assembly into the harness. All splices in on-board computer systems should be soldered to insure proper contact. Use care when probing the connections or replacing terminals in them as it is possible to short between opposite terminals. If this happens to the wrong terminal pair, it is possible to damage certain components. Always use jumper wires between connectors for circuit checking and never probe through weatherproof seals.

Open circuits are often difficult to locate by sight because corrosion or terminal misalignment are hidden by the connectors. Merely wiggling a connector on a sensor or in the wiring harness may correct the open circuit condition. This should always be considered when an open circuit or a failed sensor is indicated. Intermittent problems may also be caused by oxidized or loose connections. When using a circuit tester for diagnosis, always probe connections from the wire side. Be careful not to damage sealed connectors with test probes.

All wiring harnesses should be replaced with identical parts, using the same gauge wire and connectors. When signal wires are spliced into a harness, use wire with high temperature insulation only. With the low voltage and current levels found in the system, it is important that the best possible connection at all wire splices be made by soldering the splices together. It is seldom necessary to replace a complete harness. If replacement is necessary, pay close attention to insure proper harness routing. Secure the harness with suitable plastic wire clamps to prevent vibrations from causing the harness to wear in spots or contact any hot components.

NOTE: Weatherproof connectors cannot be replaced with standard connectors. Instructions are provided with replacement connector and terminal packages. Some wire harnesses have mounting indicators (usually pieces of colored tape) to mark where the harness is to be secured.

In making wiring repairs, it's important that you always replace damaged wires with wires that are the same gauge as the wire being replaced. The heavier the wire, the smaller the gauge number. Wires are color-coded to aid in identification and whenever possible the same color coded wire should be used for replacement. A wire stripping and crimping tool is necessary to install solderless terminal connectors. Test all crimps by pulling on the wires; it should not be possible to pull the wires out of a good crimp.

Wires which are open, exposed or otherwise damaged are repaired by simple splicing. Where possible, if the wiring harness is accessible and the damaged place in the wire can be located, it is best to open the harness and check for all possible damage. In an inaccessible harness, the wire must be bypassed with a new insert, usually taped to the outside of the old harness.

When replacing fusible links, be sure to use fusible link wire, NOT ordinary automotive wire. Make sure the fusible segment is of the same gauge and construction as the one being replaced and double the stripped end when crimping the terminal connector for a good contact. The melted (open) fusible link segment of the wiring harness should be cut off as close to the harness as possible, then a new segment spliced in as described. In the case of a damaged fusible link that feeds two harness wires, the harness connections should be replaced with two fusible link wires so that each circuit will have its own separate protection.

NOTE: Most of the problems caused in the wiring harness are due to bad ground connections. Always check all vehicle ground connections for corrosion or looseness before performing any power feed checks to eliminate the chance of a bad ground affecting the circuit.

Repairing Hard Shell Connectors

Unlike molded connectors, the terminal contacts in hard shell connectors can be replaced. Weatherproof hard-shell connectors with the leads molded into the shell have non-replaceable terminal ends. Replacement usually involves the use of a special terminal removal tool that depress the locking tangs (barbs) on the connector terminal and allow the connector to be removed from the rear of the shell. The connector shell should be replaced if it shows any evidence of burning, melting, cracks, or breaks. Replace individual terminals that are burnt, corroded, distorted or loose.

NOTE: The insulation crimp must be tight to prevent the insulation from sliding back on the wire when the wire is pulled. The insulation must be visibly compressed under the crimp tabs, and the ends of the crimp should be turned in for a firm grip on the insulation.

The wire crimp must be made with all wire strands inside the crimp. The terminal must be fully compressed on the wire strands with the ends of the crimp tabs turned in to make a firm grip on the wire. Check all connections with an ohmmeter to insure a good contact. There should be no measurable resistance between the wire and the terminal when connected.

Mechanical Test Equipment

Vacuum Gauge

Most gauges are graduated in inches of mercury (in.Hg), although a device called a manometer reads vacuum in inches of water (in. H2O). The normal vacuum reading usually varies between 18 and 22 in.Hg at sea level. To test engine vacuum, the vacuum gauge must be connected to a source of manifold vacuum. Many engines have a plug in the intake manifold which can be removed and replaced with an adapter fitting. Connect the vacuum gauge to the fitting with a suitable rubber hose or, if no manifold plug is available, connect the vacuum gauge to any device using manifold vacuum, such as EGR valves, etc. The vacuum gauge can be used to determine if enough vacuum is reaching a component to allow its actuation.

Hand Vacuum Pump

Small, hand-held vacuum pumps come in a variety of designs. Most have a built-in vacuum gauge and allow the component to be tested without removing it from the vehicle. Operate the pump lever or plunger to apply the correct amount of vacuum required for the test specified in the diagnosis routines. The level of vacuum in inches of Mercury (in.Hg) is indicated on the pump gauge. For some testing, an additional vacuum gauge may be necessary.

Intake manifold vacuum is used to operate various systems and devices on late model vehicles. To correctly diagnose and solve problems in vacuum control systems, a vacuum source is necessary for testing. In some cases, vacuum can be taken from the intake manifold when the engine is running, but vacuum is normally provided by a hand vacuum pump. These hand vacuum pumps have a built-in vacuum gauge that allow testing while the device is still attached to the component. For some tests, an additional vacuum gauge may be necessary.

HEATING AND AIR CONDITIONING

Blower Motor

REMOVAL AND INSTALLATION

Without Air Conditioning

1. Disconnect the negative battery cable.
2. Remove the air cleaner or air inlet duct, as necessary.
3. Remove the 2 screws attaching the vacuum reservoir to the blower assembly and remove the reservoir.
4. Disconnect the wire harness connector from the blower motor by pushing down on the connector tabs and pulling the connector off of the motor.
5. Disconnect the blower motor cooling tube at the blower motor.
6. Remove the 3 screws attaching the blower motor and wheel to the heater blower assembly.
7. Holding the cooling tube aside, pull the blower motor and wheel from the heater blower assembly and remove it from the vehicle.
8. Remove the blower wheel push-nut or clamp from the motor shaft and pull the blower wheel from the motor shaft.

To install:

9. Install the blower wheel on the blower motor shaft.
10. Install the hub clamp or push-nut.
11. Holding the cooling tube aside, position the blower motor and wheel on the heater blower assembly and install the 3 attaching screws.
12. Connect the blower motor cooling tube and the wire harness connector.

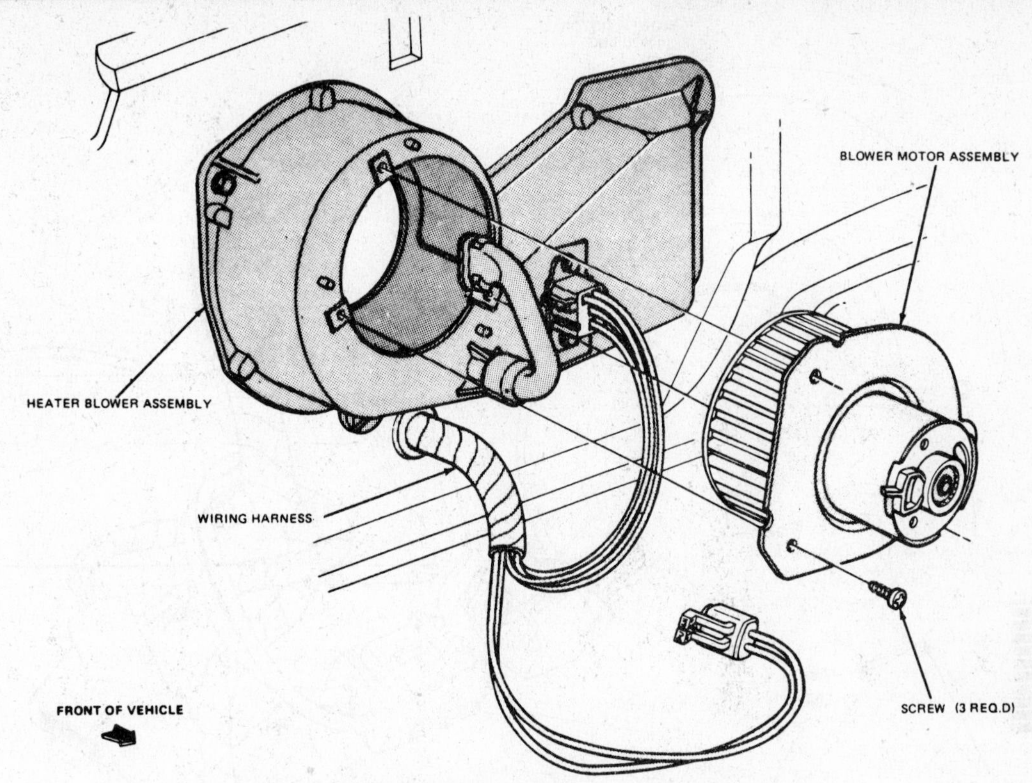

Blower motor assembly removal and installation

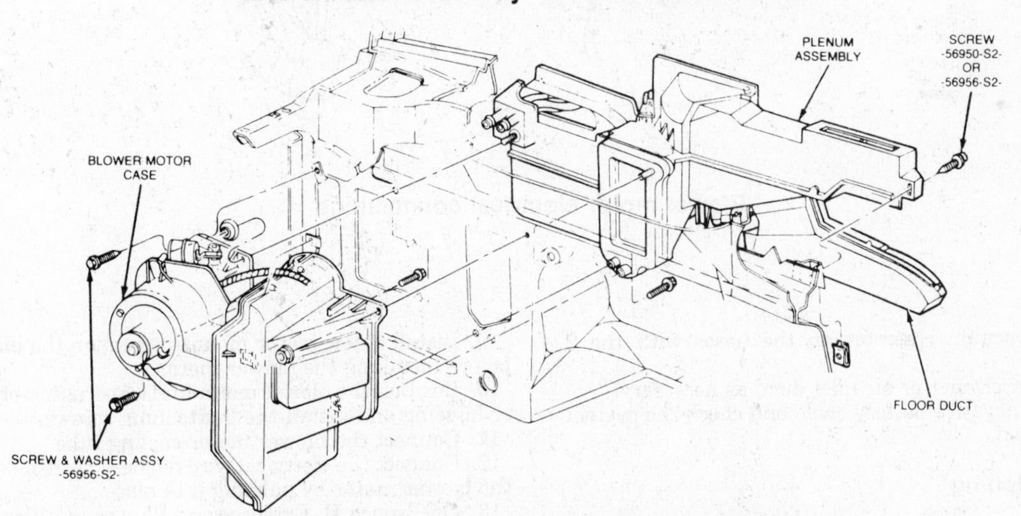

Heater blower motor case and plenum assembly

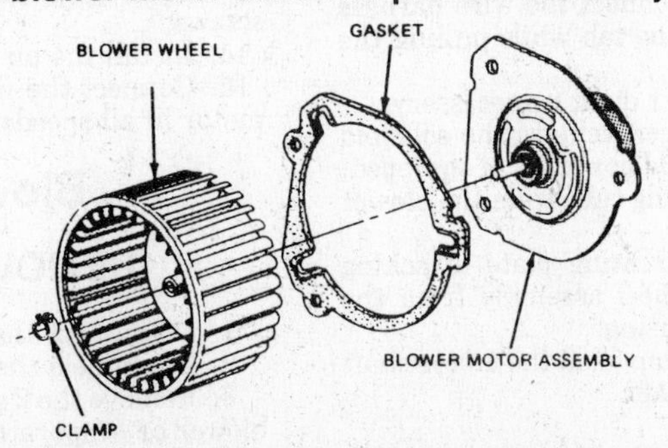

Blower wheel removed from motor

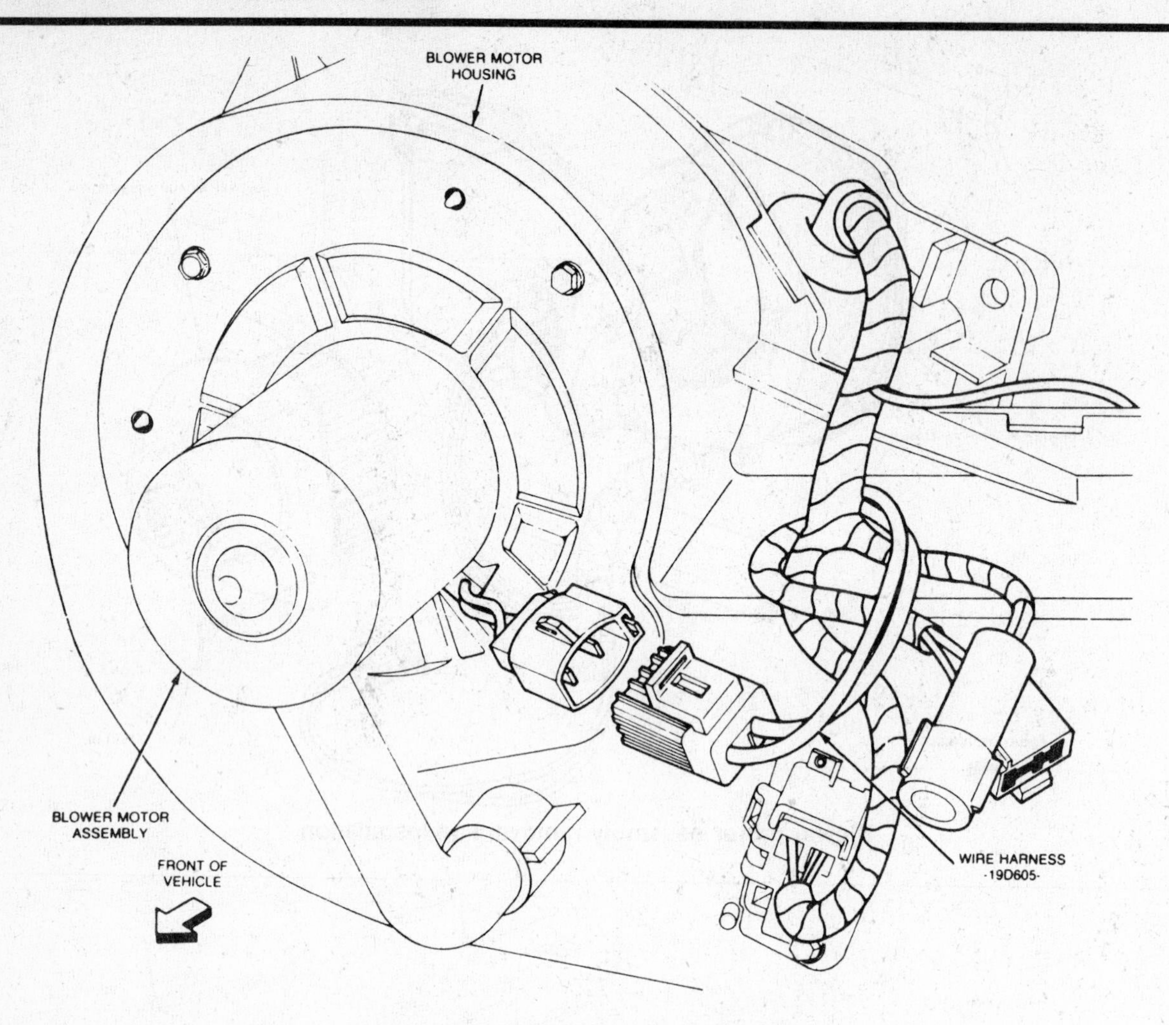

Blower motor electrical connections

13. Install the vacuum reservoir on the hoses with the 2 screws.
14. Install the air cleaner or air inlet duct, as necessary.
15. Connect the negative battery cable and check the system for proper operation.

With Air Conditioning

1. Disconnect the negative battery cable.
2. In the engine compartment, disconnect the wire harness from the motor by pushing down on the tab while pulling the connection off at the motor.
3. Remove the air cleaner or air inlet duct, as necessary.
4. On Bronco II, Explorer and Ranger, remove the solenoid box cover retaining bolts and the solenoid box cover, if equipped.
5. Disconnect the blower motor cooling tube from the blower motor.
6. Remove the 3 blower motor mounting plate attaching screws and remove the motor and wheel assembly from the evaporator assembly blower motor housing.
7. Remove the blower motor hub clamp from the motor shaft and pull the blower wheel from the shaft.

To install:

8. Install the blower motor wheel on the blower motor shaft and install a new hub clamp.
9. Install a new motor mounting seal on the blower housing before installing the blower motor.
10. Position the blower motor and wheel assembly in the blower housing and install the 3 attaching screws.
11. Connect the blower motor cooling tube.
12. Connect the electrical wire harness hardshell connector to the blower motor by pushing into place.
13. On Bronco II, Explorer and Ranger, position the solenoid box cover, if equipped, into place and install the 3 retaining screws.
14. Install the air cleaner or air inlet duct, as necessary.
15. Connect the negative battery cable and check the blower motor in all speeds for proper operation.

Blower Motor Resistor

REMOVAL AND INSTALLATION

1. Disconnect the negative battery cable.
2. Disconnect the wire connector from the resistor assembly.
3. Remove the 2 screws attaching the resistor assembly to the blower or evaporator case and remove the resistor.
4. Installation is the reverse of the removal procedure. Check the blower motor for proper operation in all blower speeds.

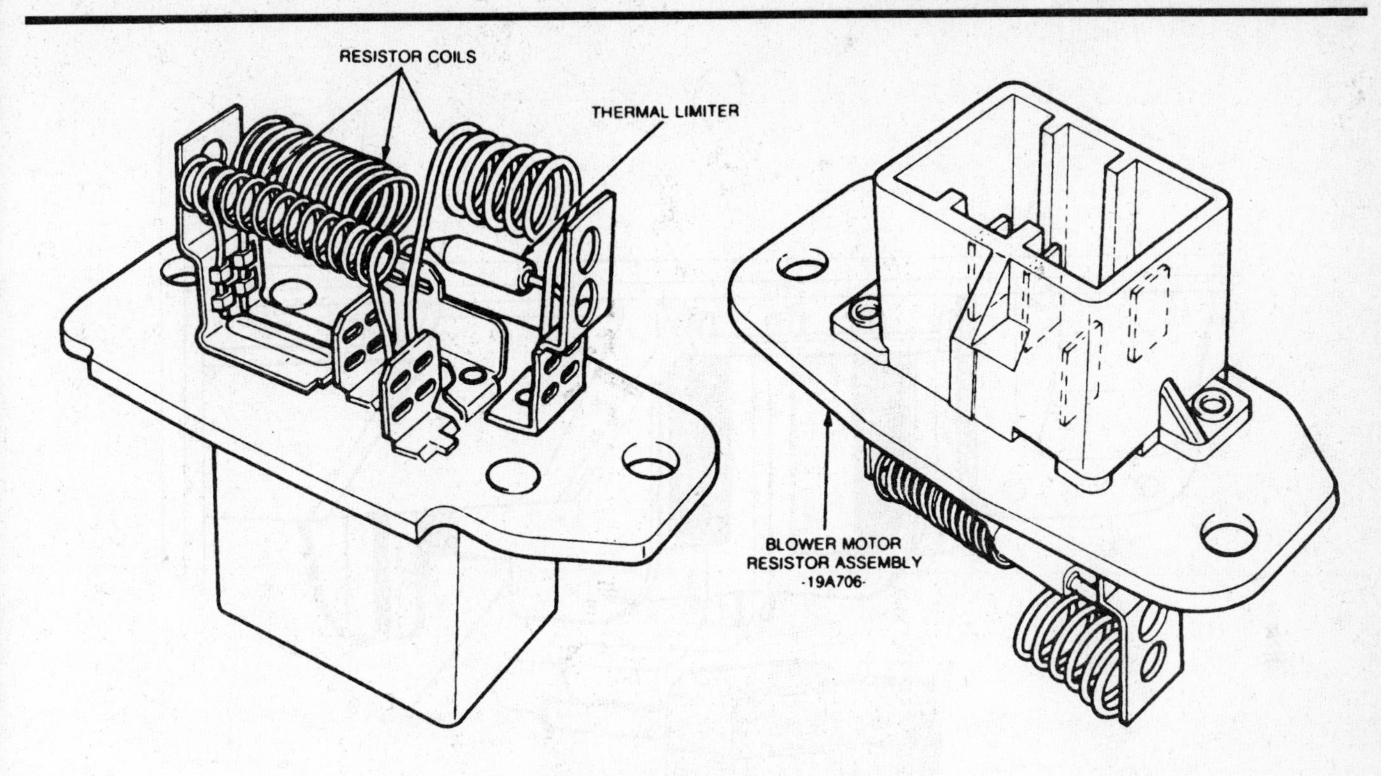

Blower motor resistor assembly

Heater Core

REMOVAL AND INSTALLATION

1. Disconnect the negative battery cable. Allow the engine to cool down. Drain the cooling system.

CAUTION

When draining the coolant, keep in mind that cats and dogs are attracted by the ethylene glycol antifreeze, and are quite likely to drink any that is left in an uncovered container or in puddles on the ground. This will prove fatal in sufficient quantity. Always drain the coolant into a sealable container. Coolant should be reused unless it is contaminated or several years old.

2. Disconnect the heater hoses from the heater core tubes and plug hoses.
3. In the passenger compartment, remove the five screws attaching the heater core access cover to the plenum assembly and remove the access cover.
4. Pull the heater core rearward and down, removing it from the plenum assembly.

To install:

5. Position the heater core and seal in the plenum assembly.
6. Install the heater core access cover to the plenum assembly and secure with five screws.
7. Install the heater hoses to the heater core tubes at the dash panel in the engine compartment. Do not over-tighten hose clamps.
8. Check the coolant level and add coolant as required. Connect the negative battery cable.
9. Start the engine and check the system for coolant leaks.

Control Head

REMOVAL AND INSTALLATION

1. Disconnect the negative battery cable.
2. Open the ash tray and remove the 2 screws that hold the ash tray drawer slide to the instrument panel. Remove the ash tray and drawer slide bracket from the instrument panel.
3. Gently pull the finish panel away from the instrument panel and the cluster. The finish panel pops straight back for approximately 1 in., then up to remove. Be careful not to trap the finish panel around the steering column.

NOTE: If equipped with the electronic 4×4 shift-on-the-fly module, disconnect the wire from the rear of the 4×4 transfer switch before trying to remove the finish panel from the instrument panel.

4. Remove the 4 screws attaching the control assembly to the instrument panel.
5. Pull the control through the instrument panel opening far enough to allow removal of the electrical connections from the blower switch and control assembly illumination lamp. Using a suitable tool, remove the 2 hose vacuum harness from the vacuum switch on the side of the control.
6. At the rear of the control, using a suitable tool, release the temperature and function cable snap-in flags from the white control bracket.
7. On the bottom side of the control, remove the temperature cable from the control by rotating the cable until the T-pin releases the cable. The temperature cable is black with a blue snap-in flag.

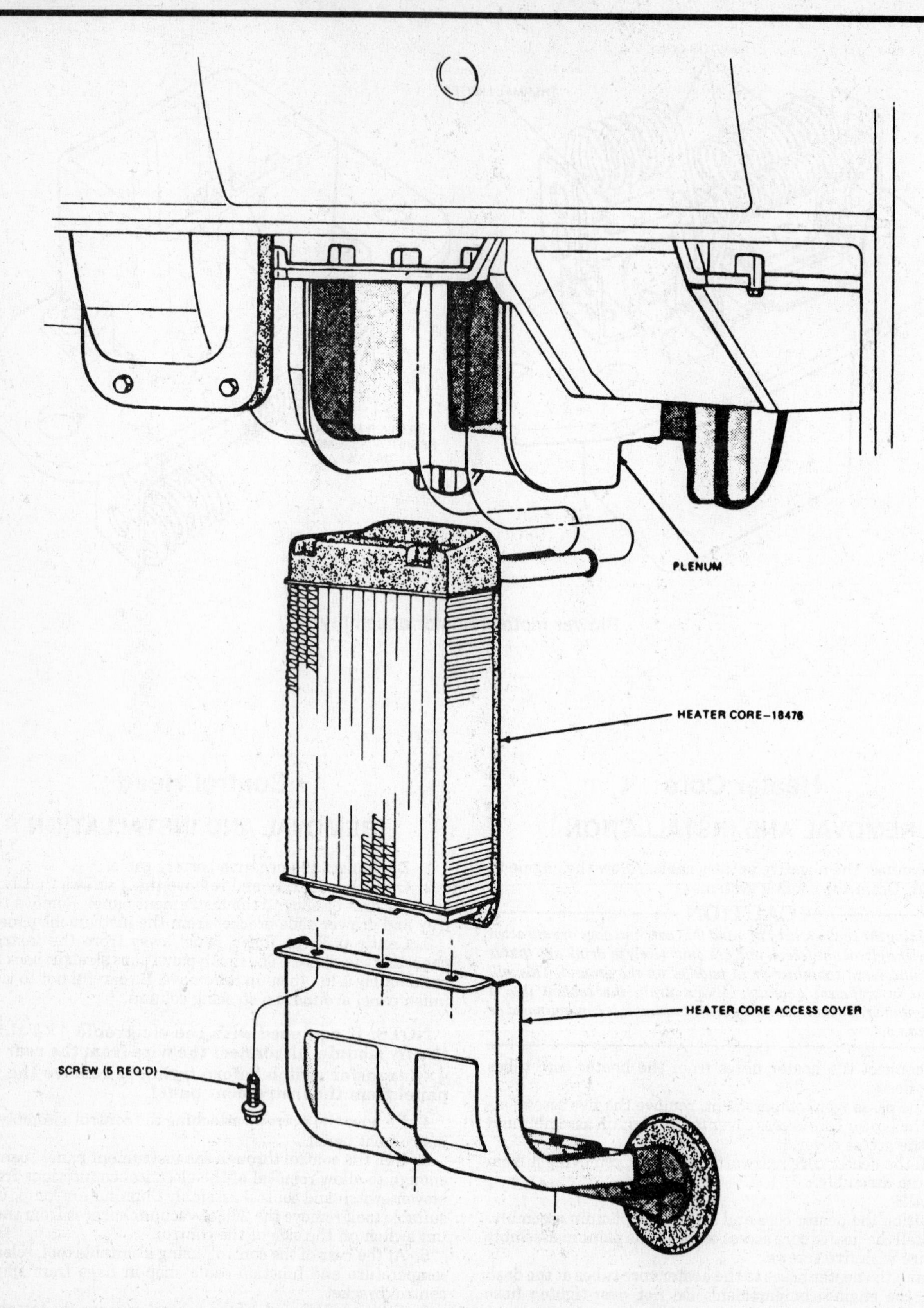

Heater core removal

8. Pull enough cable through the instrument panel opening until the function cable can be held vertical to the control, then remove the control cable from the function lever. The function cable is white with a black snap-in flag.

9. Remove the control assembly from the instrument panel.

To install:

10. Pull the control cables through the control assembly opening in the instrument panel for a distance of approximately 8 in. (203mm).

11. Hold the control assembly up to the instrument panel with it's face directed toward the floor of the vehicle. This will locate the face of the control in a position that is 90 degrees out of it's installed position.

12. Carefully bend and attach the function cable that has a white color code and a black snap-in terminal to the white plastic lever on the control assembly. Rotate the control assembly back to it's normal position for installation, then snap the black cable flag into the control assembly bracket.

13. On the opposite side of the control assembly, attach the black temperature control cable with the blue plastic snap-in flag to the blue plastic lever on the control. Make sure the end of the cable is seated securely with the T-top pin on the control. Rotate the cable to it's operating position and snap the blue cable flag into the control assembly bracket.

14. Connect the wiring harness to the blower switch and the illumination lamp to it's receptacle on the control assembly. Connect the dual terminal on the vacuum hose to the vacuum switch on the control assembly.

15. Position the control assembly into the instrument panel opening and install the 4 mounting screws.

16. If equipped, reconnect the 4×4 electric shift harness on the rear of the cluster finish panel.

17. Install the cluster finish panel with integral push-pins. Make sure that all pins are fully seated around the rim of the panel.

18. Reinsert the ash tray slide bracket and reconnect the illumination connection circuit. Reinstall the 2 screws that retain the ash tray retainer bracket and the finish panel.

19. Replace the ash tray and reconnect the cigarette lighter.

20. Connect the negative battery cable and check the heater system for proper control assembly operation.

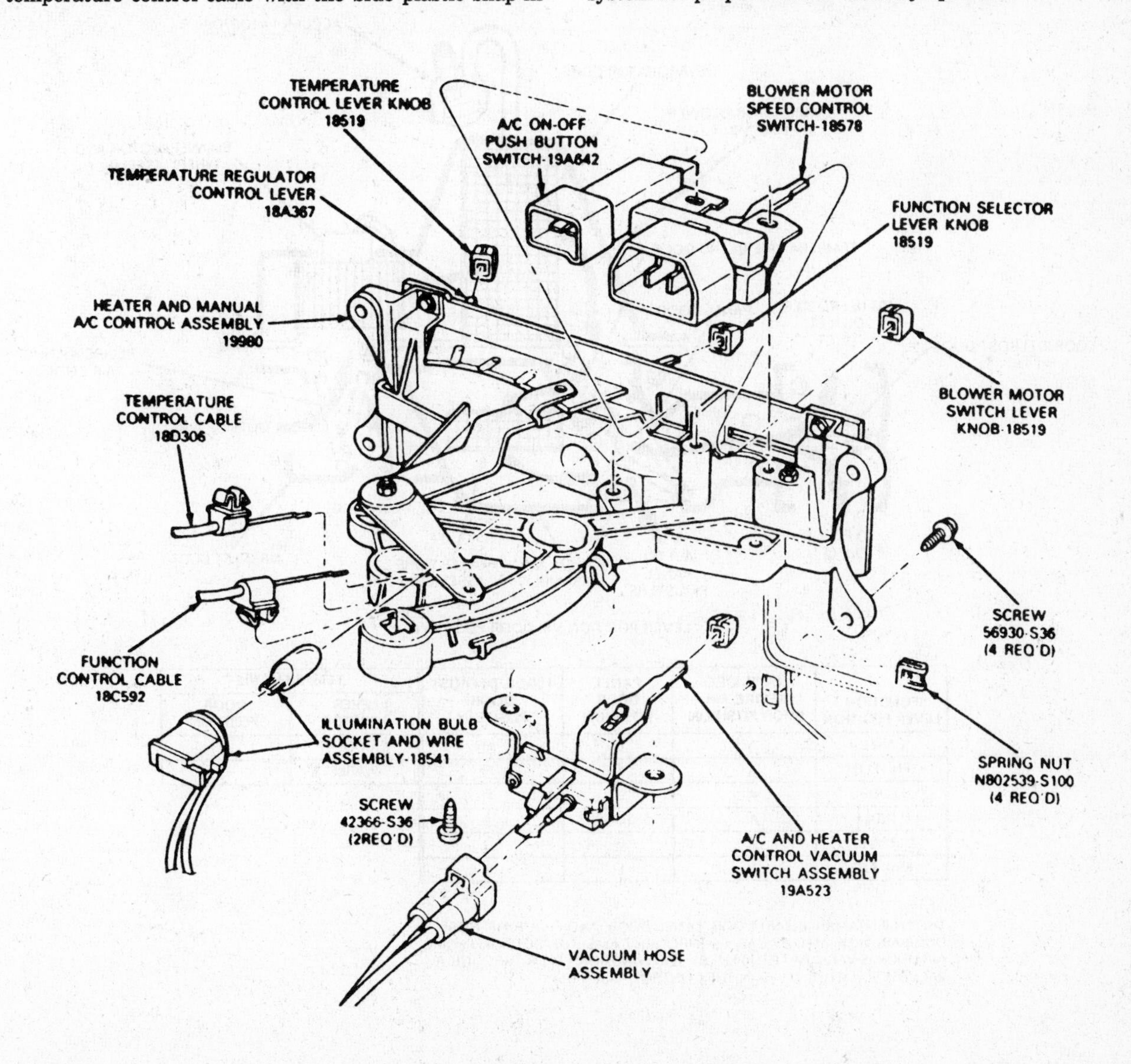

Control head — exploded view

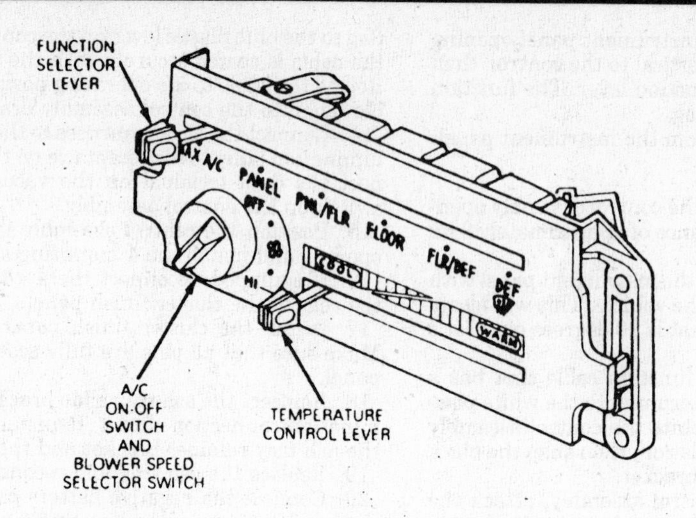

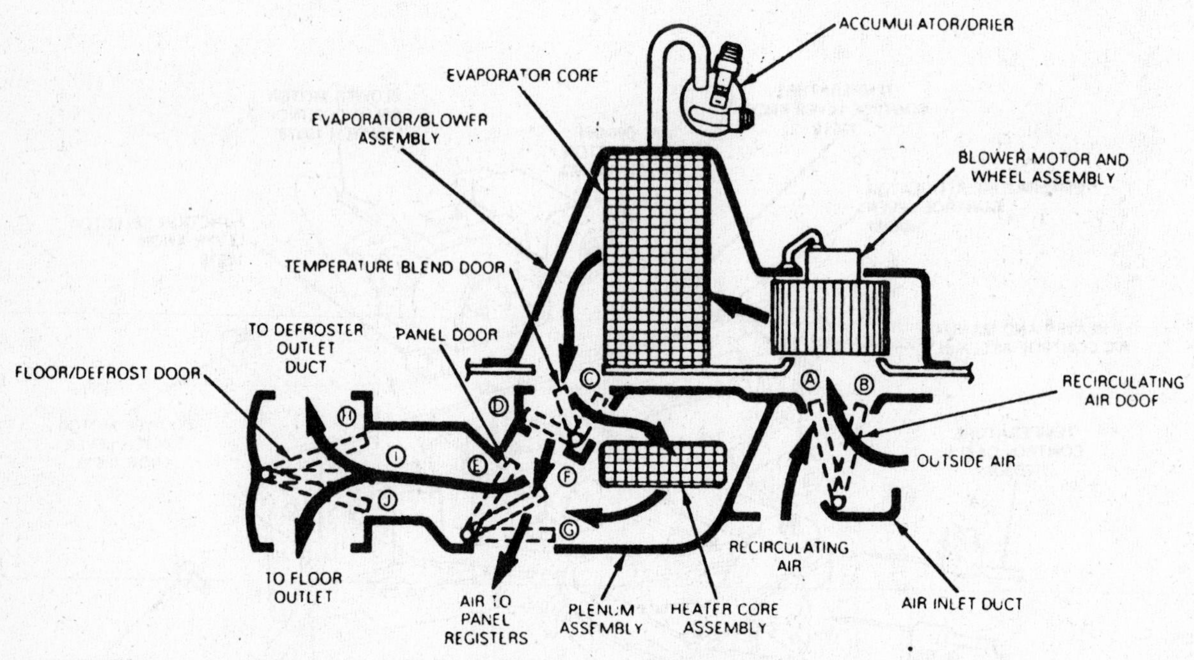

LEVER POSITION VS. DOOR POSITION

FUNCTION LEVER POSITION	OUTSIDE/ RECIRC AIR DOOR POSITION	PANEL DOOR POSITION	FLOOR/DEFROST DOOR POSITION
PANEL	A	E	H
PNL/FLR	A	F	H
FLOOR	A	G	H
FLR/DEF	A	G	I
DEFROST	A	G	J
MAX A/C	B	E	H

TEMPERATURE	
LEVER POSITION	DOOR POSITION
COOL	C
WARM	D

NOTE
THE TEMPERATURE BLEND DOOR, PANEL DOOR, AND FLOOR/DEFROST DOOR ARE ACTUATED BY CABLES THROUGH CAMS. THE OUTSIDE/RECIRC AIR DOOR IS VACUUM ACTUATED BY THE FUNCTION LEVER THROUGH A VACUUM SELECTOR VALVE IN THE CONTROL ASSEMBLY

Control head modes of operation

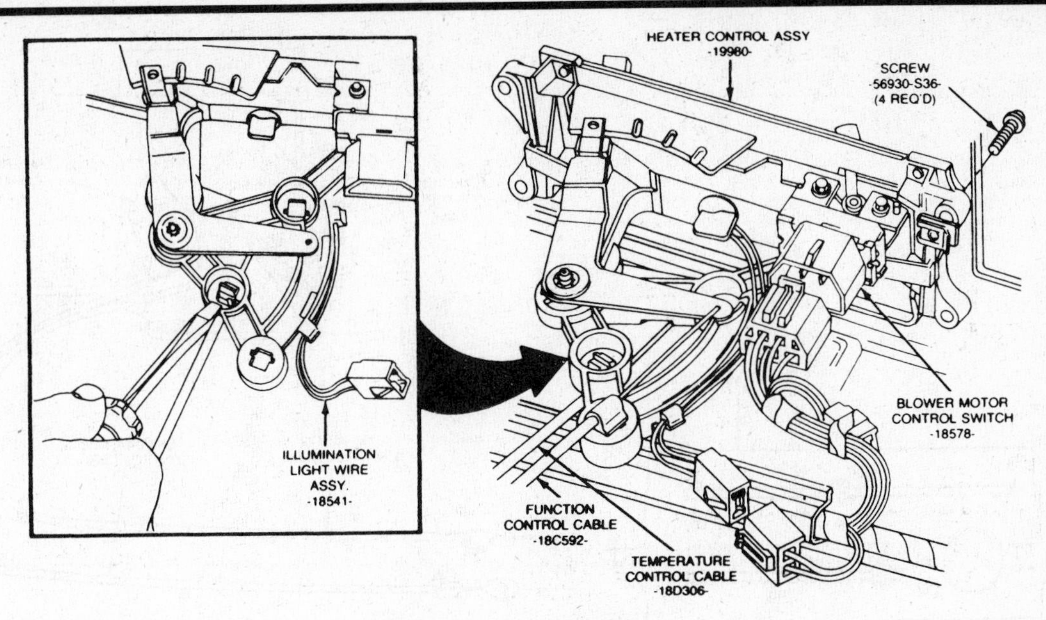

Control head removal

Control Cables

ADJUSTMENT

Function Selector and Temperature Selector Control Cable

To check the temperature cable adjustment, move the temperature control lever all the way to the left, then move it all the way to the right. At the extreme ends of lever travel, the door should be heard to firmly seat, indicated by a loud thumping sound, allowing either maximum or no air flow through the heater core. To check the function cable adjustment, see that the function lever will reach the detents at the far left and right of it's travel. In addition, check that the air flow is correct when the function lever is moved through each detent provided in the control assembly. If cable adjustment is needed, proceed as follows:

1. Disengage the glove compartment door by squeezing it's sides together. Allow the door to hang free.
2. Working through the glove compartment opening, remove the cable jacket from the metal attaching clip on the top of the plenum by depressing the clip tab and pulling the cable out of the clip.

NOTE: The cable end should remain attached to the door cams.

3. To adjust the temperature control cable, set the temperature lever at **COOL** and hold. With the cable end attached to the temperature door cam, push gently on the cable jacket to seat the blend door. Push until resistance is felt. Reinstall the cable to the clip by pushing the cable jacket into the clip from the top until it snaps in.
4. To adjust the function control cable, set the function selector lever in the **DEFROST** detent and hold. With the cable end attached to the function cam, pull on the cam jacket until cam travel stops. Reinstall the cable to the clip by pushing the cable jacket into the clip from the top until it snaps in place.
5. Install the glove compartment.
6. Run the system blower on **HIGH** and actuate the levers, checking for proper adjustment.

REMOVAL AND INSTALLATION

Temperature and Function Cables

1. Disconnect the negative battery cable.
2. Remove the control assembly from the instrument panel.
3. Disengage the glove compartment door by squeezing the sides together and allowing the door to hang free.
4. Working through the glove compartment and/or control opening, remove the temperature and function cable jackets from their clips on top of the plenum by compressing the clip tans and pulling the cables upward.
5. Reach through the glove compartment opening and disconnect the function and temperature cables from their separate cams. The cable ends are secured to the cams under a retention finger.
6. The cables are routed inside the instrument panel with 2 routing aids. Remove the cables from these devices. Reaching through the control opening, pull the cables upward out of the wiring shield cut-out. Reaching through the glove box opening, pull the cables out of the plastic clip up inside the instrument panel.
7. Pull the cables from the instrument panel through the control assembly opening.

To install:

8. Working through the glove compartment opening and the control opening in the instrument panel, feed the end of the cables to the cam area. Feed the cables in from the glove compartment opening, making sure the coiled end of the white function cable and the round hole diecast end of the temperature cable go in first.
9. Attach the coiled end of the function cable to the function cam, making sure the cable is routed under the cable hold-down feature on the cam assembly. The pigtail coil may be facing either up or down.
10. Attach the diecast end of the temperature cable to the temperature cam making sure the cable is routed under the cable hold-down feature on the cam assembly.
11. Route the control end of the cable through the instrument panel until the ends stick out of the control opening. It is not necessary to insert the cable into any routing devices previously

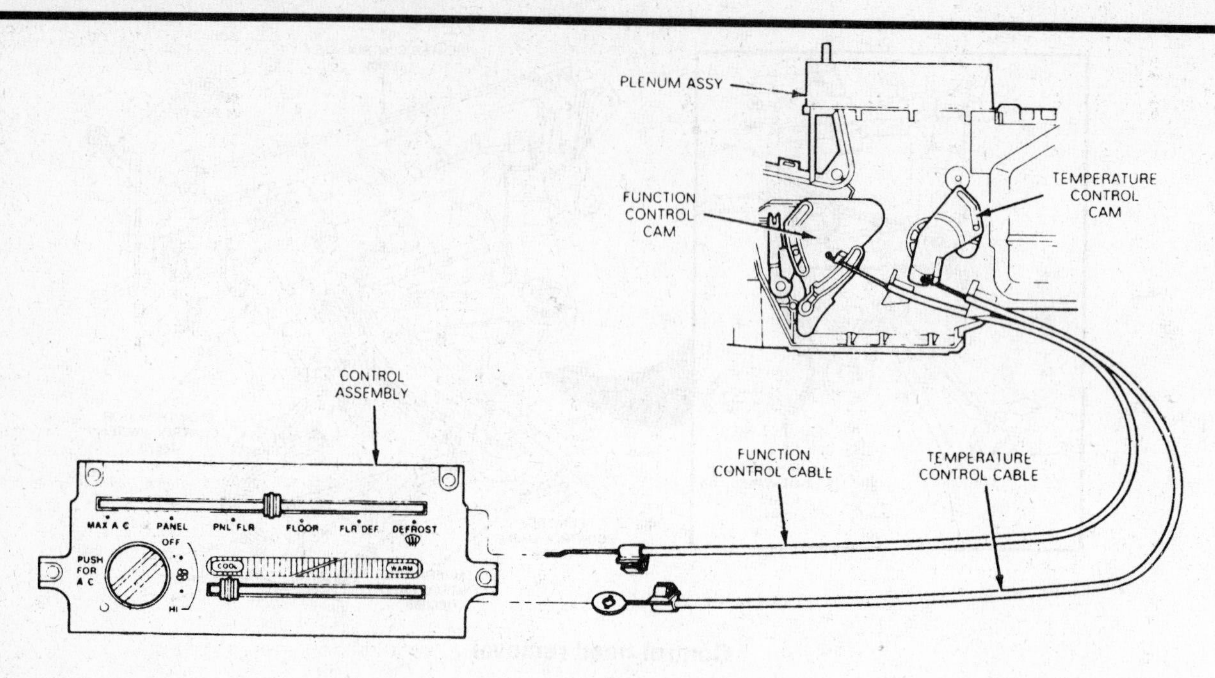

Control cable removal and installation

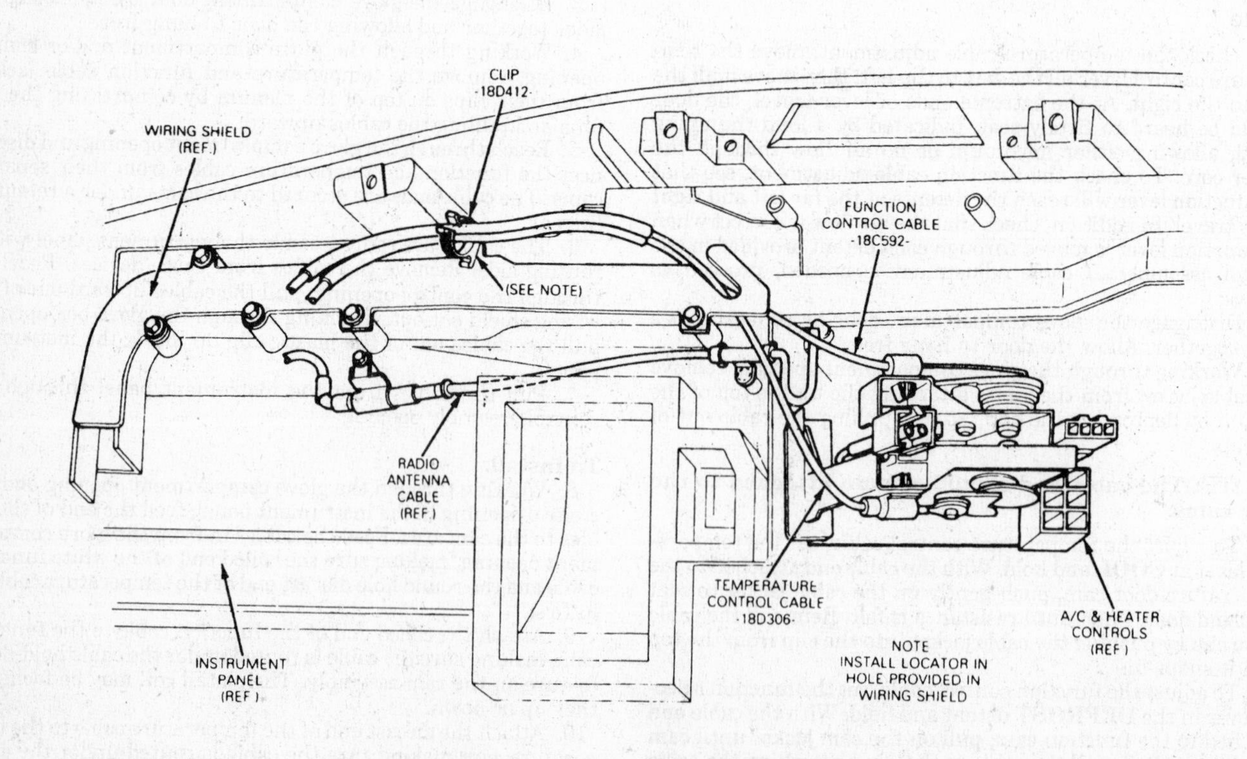

Control cable routing – behind instrument panel

used. The routing aids are only necessary when the entire instrument panel is removed and reinstalled.

12. Attach the function and temperature cables to the control. Install the control assembly in the instrument panel.

13. Adjust the cables in their clips on top of the plenum.

NOTE: Make sure the radio antenna cable does not become disengaged from it's mounting and fall into the plenum cam area where it could cause an increase in control assembly operating effort or a faulty selection of system functions.

14. Connect the negative battery cable and make a final check of the system for proper control cable operation.

Evaporator Core

REMOVAL AND INSTALLATION

1. Disconnect the negative battery cable.
2. Discharge the refrigerant from the air conditioning system according to the proper procedure. See Section 1, for the procedure.
3. Disconnect the electrical connector from the pressure switch located on top of the accumulator/drier. Remove the pressure switch.
4. Disconnect the suction hose from the accumulator/drier using the spring-lock coupling disconnect procedure. Cap the openings to prevent the entrance of dirt and moisture.
5. Disconnect the liquid line from the evaporator core inlet tube using a backup wrench to loosen the fitting. Cap the openings to prevent the entrance of dirt and moisture.
6. Remove the screws holding the evaporator case service cover and vacuum reservoir to the evaporator case assembly.
7. Store the vacuum reservoir in a secure position to avoid vacuum line damage.
8. Remove the 2 dash panel mounting nuts.
9. Remove the evaporator case service cover from the evaporator case assembly.
10. Remove the evaporator core and accumulator/drier assembly from the vehicle.

To install:

NOTE: Add 3 oz. (90ml) of clean refrigerant oil to a new replacement evaporator core to maintain the total system oil charge.

11. Position the evaporator core and accumulator/drier assembly into the evaporator case out-board half.
12. Position the evaporator case service cover into place on the evaporator case assembly.
13. Install the 2 dash panel mounting nuts.
14. Install the screws holding the evaporator service case half to the evaporator case assembly.
15. Place the vacuum reservoir in it's installed position. Attach the reservoir to the case with 2 screws.
16. Connect the liquid line to the evaporator inlet tube using a backup wrench to tighten the fitting. Use a new O-ring lubricated with clean refrigerant oil.

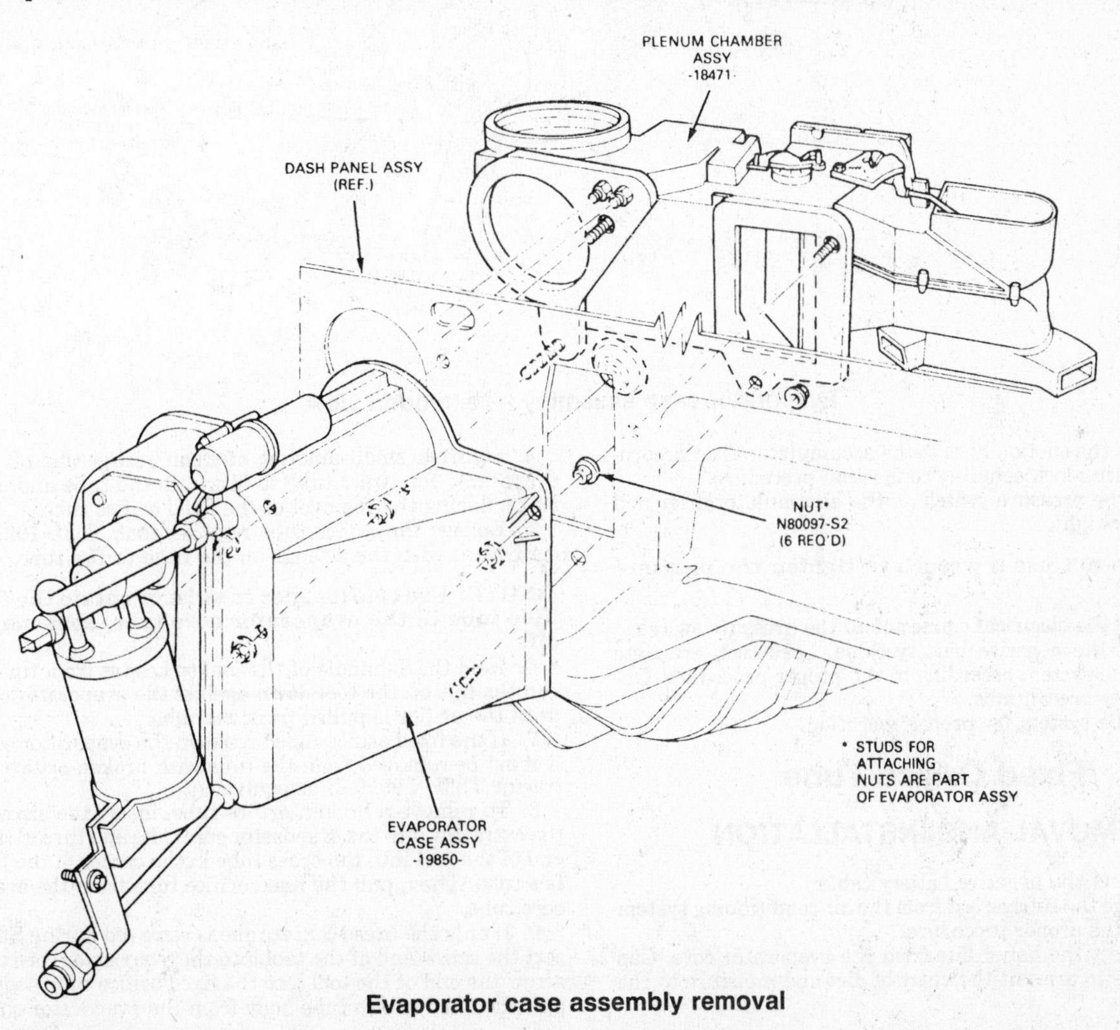

Evaporator case assembly removal

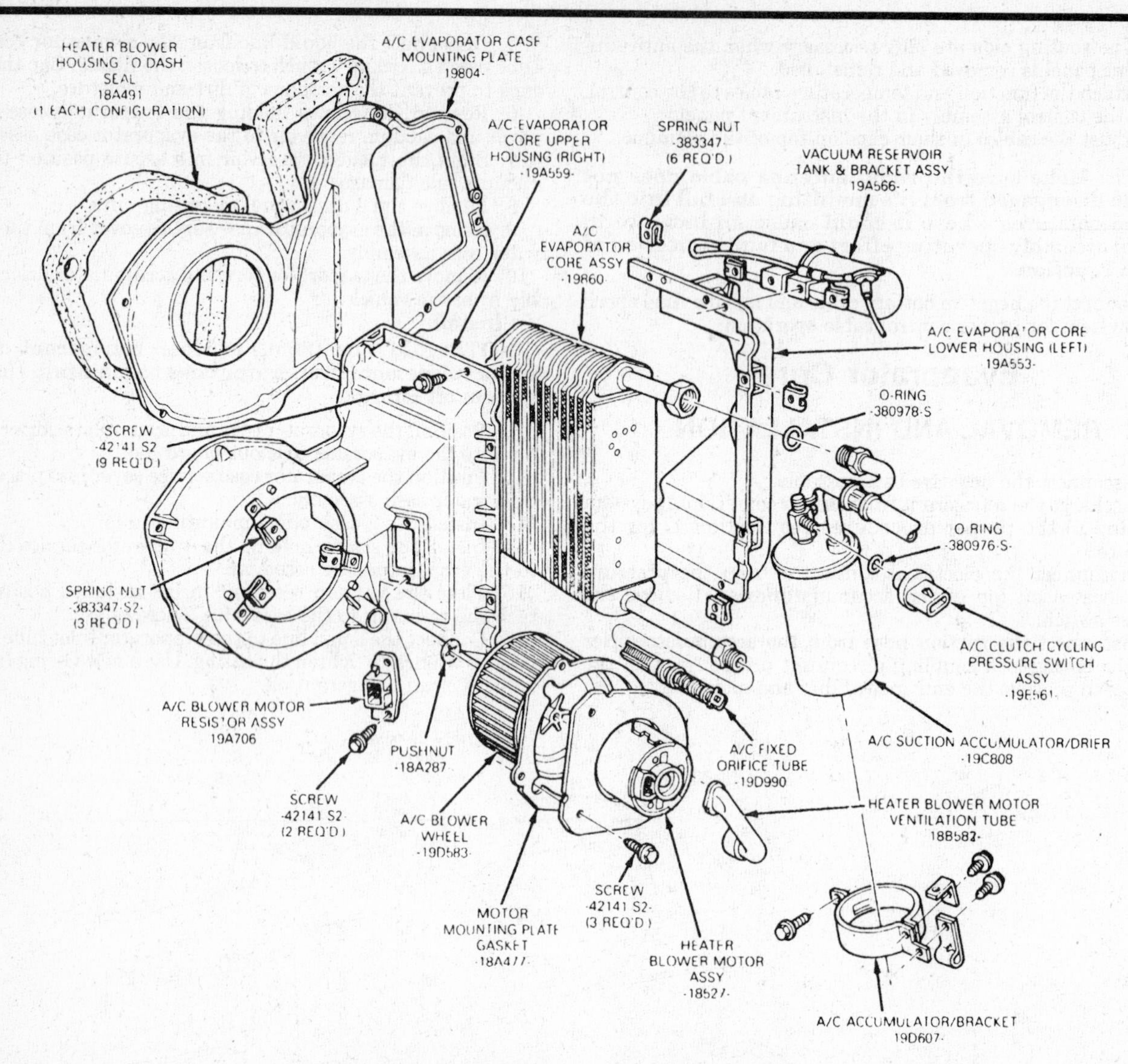

Evaporator case assembly — exploded view

17. Connect the suction hose to the accumulator/drier according to the spring-lock coupling connection procedure.
18. Install the pressure switch on the accumulator/drier and tighten finger-tight.

NOTE: Do not use a wrench to tighten the pressure switch.

19. Connect the electrical connector to the pressure switch.
20. Connect the negative battery cable. Leak test, evacuate and charge the system according to the proper procedure. Observe all safety precautions.
21. Check the system for proper operation.

Fixed Orifice Tube

REMOVAL AND INSTALLATION

1. Disconnect the negative battery cable.
2. Discharge the refrigerant from the air conditioning system according to the proper procedure.
3. Disconnect the liquid line from the evaporator core. Cap the liquid line to prevent the entry of dirt and moisture to the system.
4. Squirt a small amount of clean refrigerant oil into the evaporator core inlet tube to lubricate the tube and orifice O-rings, during the removal of the fixed orifice tube.
5. Engage the orifice tube remover tool, T83L-199990-A or equivalent with the 2 tangs on the fixed orifice tube.

NOTE: Use caution, not to twist or rotate the fixed orifice tube in the evaporator core tube, as it may break off.

6. Hold the T-handle of the tool to keep it from turning and run the nut on the tool down against the evaporator core tube until the orifice is pulled from the tube.
7. If the fixed orifice tube breaks in the evaporator core tube, it must be removed from the tube with broken orifice tube extractor T83L–19990–B or equivalent.
8. To remove a broken orifice tube, insert the screw end of the extractor into the evaporator core tube and thread the screw end of the tool into the brass tube in the center of the fixed orifice tube. Then, pull the fixed orifice tube from the evaporator core tube.
9. If only the brass center tube is removed during Step 8, insert the screw end of the tool into the evaporator core tube and screw the end of the tool into the fixed orifice tube body. Then, pull the fixed orifice tube body from the evaporator core tube.

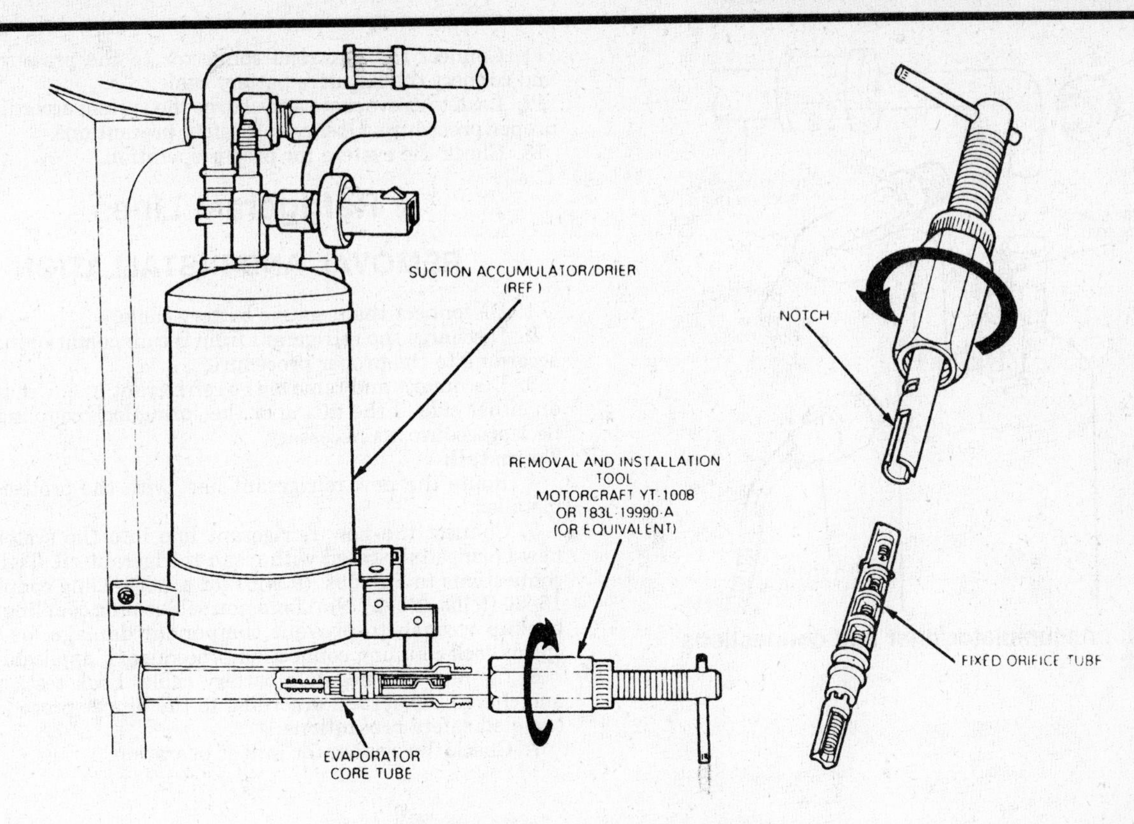

Removing the fixed orifice tube

To install:

10. Lubricate the O-rings on the fixed orifice tube body liberally with clean refrigerant oil.
11. Place the fixed orifice tube in the fixed orifice tube remover/replacer T83L–19990–A or equivalent, and insert the fixed orifice tube into the evaporator core tube until the orifice is seated at the top.
12. Remove the tool from the fixed orifice tube.
13. Connect the liquid line to the evaporator core using the spring-lock coupling connection procedure.
14. Leak test, evacuate and charge the system according to the proper procedure. Observe all safety precautions.
15. Check the system for proper operation.

Accumulator/Drier

REMOVAL AND INSTALLATION

The accumulator/drier should be replaced under the following conditions:

- The accumulator/drier is restricted, plugged or perforated.
- The system has been left open for more than 24 hours.
- There is evidence of moisture in the system: internal corrosion of metal lines or the refrigerant oil is thick and dad compressor or damage to some other major component requires opening of the refrigerant system in order to service the difficulty.

Do not replace the accumulator/drier every time if the following conditions exist:

- There is a loss of refrigerant charge.
- A component, except as described above, is changed.
- A dent is found in the outer shell of the accumulator/drier.

1. Disconnect the negative battery cable.
2. Discharge the refrigerant from the air conditioning system according to the proper procedure.
3. Disconnect the electrical connector from the pressure switch.
4. Remove the pressure switch by unscrewing it from the accumulator/drier.
5. Disconnect the suction hose from the accumulator/drier. Cap the openings to prevent the entrance of dirt and moisture.
6. Loosen the fitting connecting the accumulator/drier to the evaporator core. Use 2 wrenches to prevent component damage.
7. Remove the lower forward screw holding the flanges of the case and bracket together and the screw holding the evaporator inlet tube to the accumulator bracket.
8. Disconnect the accumulator/drier from the evaporator core and remove the bracket from the accumulator/drier.

To install:

9. Perform the following procedure:
   a. Position the bracket on the replacement accumulator/drier loosely.
   b. Using a new O-ring lubricated with clean refrigerant oil, connect the accumulator/drier to the evaporator core tube while aligning the bracket to the slot between the case flanges.
   c. Tighten the accumulator/drier to the evaporator core fitting to 26–31 ft. lbs. (32–42 Nm) using a backup wrench. Install the lower forward screw which retains the bracket between the case flanges. Tighten the bracket on the accumulator and reinstall the clip that holds the evaporator inlet tube to the accumulator bracket.
   d. Connect the suction hose to the accumulator/drier using the spring-lock coupling connection procedure.
10. Install a new O-ring lubricated with clean refrigerant oil on the pressure switch nipple of the accumulator/drier. Install the pressure switch. Tighten the switch to 5–10 ft. lbs. (7–13 Nm) if the switch has a metal base and hand tighten only if the switch has a plastic base.

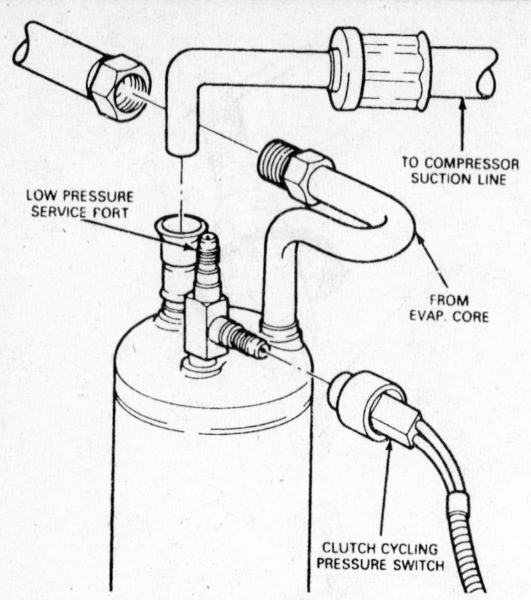

Accumulator/drier and connections

11. Connect the electrical connector to the pressure switch and connect the negative battery cable.
12. Leak test, evacuate and charge the system according to the proper procedure. Observe all safety precautions.
13. Check the system for proper operation.

Refrigerant Lines

REMOVAL AND INSTALLATION

1. Disconnect the negative battery cable.
2. Discharge the refrigerant from the air conditioning system according to the proper procedure.
3. Disconnect and remove the refrigerant line. Use a wrench on either side of the fitting or the spring-lock coupling disconnect procedure, as necessary.

To install:

4. Route the new refrigerant line, with the protective caps installed.
5. Connect the new refrigerant line into the system using new O-rings lubricated with clean refrigerant oil. Tighten the connections to 7 ft. lbs. (9 Nm) for a self-sealing coupling and 15–20 ft. lbs. (21–27 Nm) for a non self-sealing coupling, using a backup wrench to prevent component damage, or use the spring-lock coupling connection procedure, if applicable.
6. Connect the negative battery cable. Leak test, evacuate and charge the system according to the proper procedure. Observe all safety precautions.
7. Check the system for proper operation.

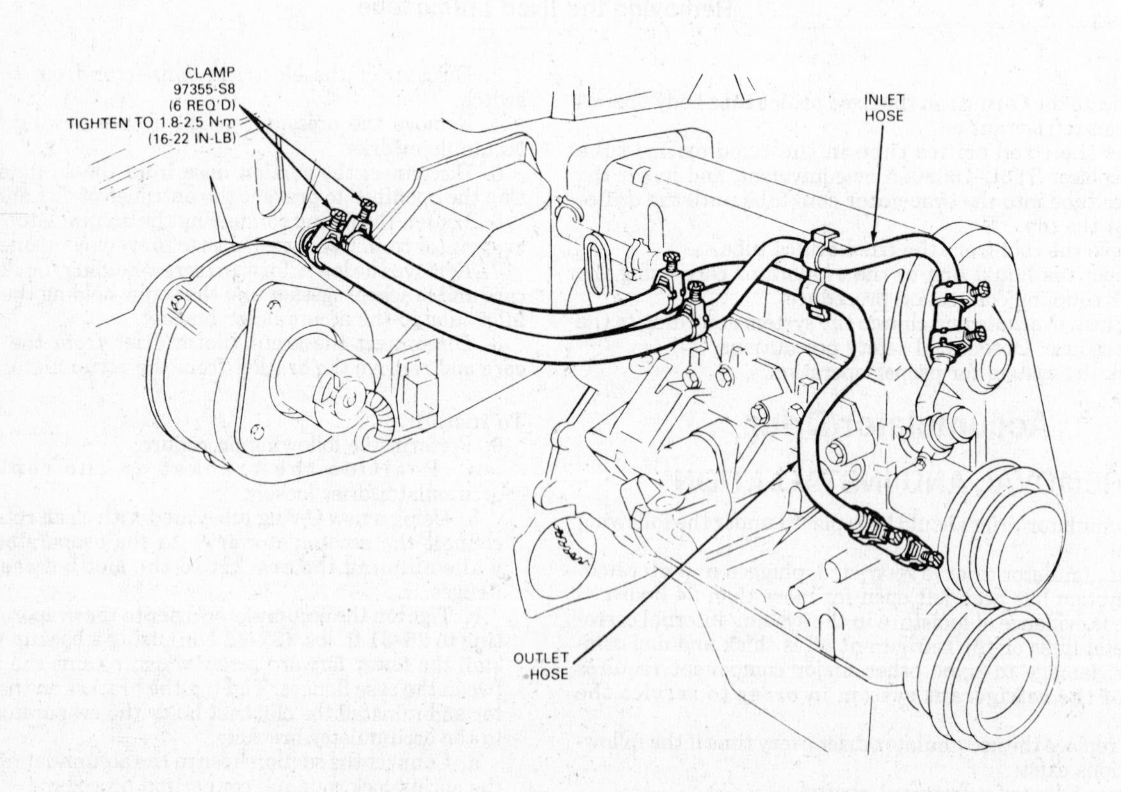

Heater hose routing — 2.0L engine

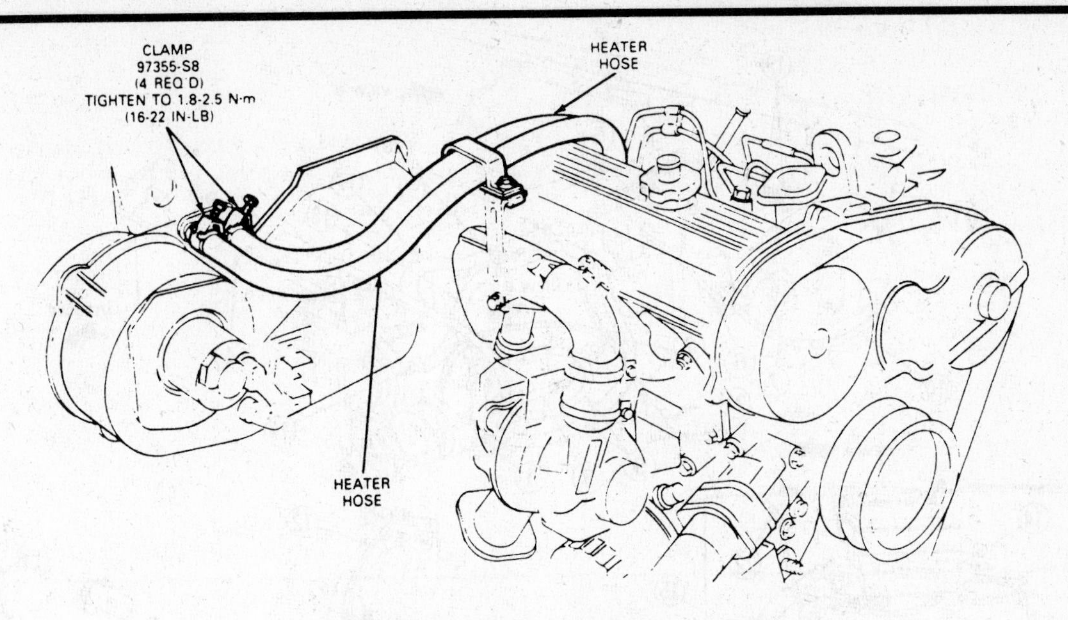

Heater hose routing — 2.3L diesel engine

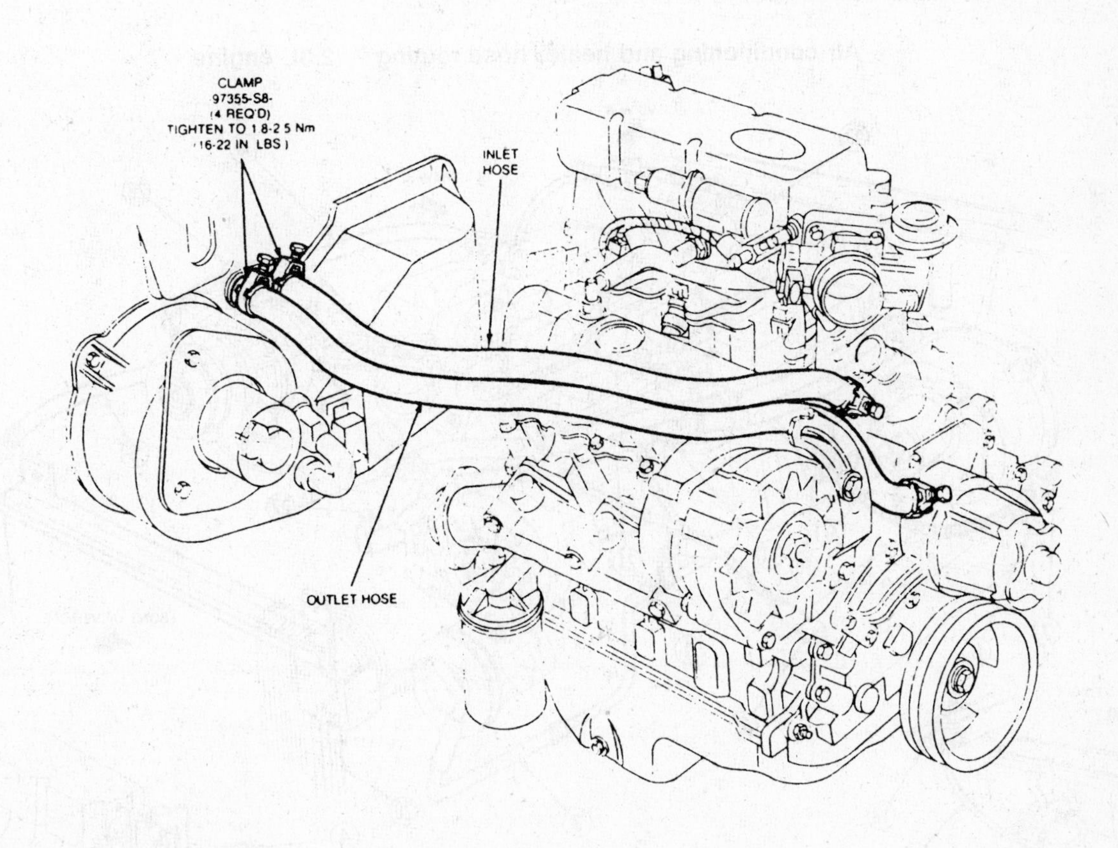

Heater hose routing — 2.8L and 2.9L engines

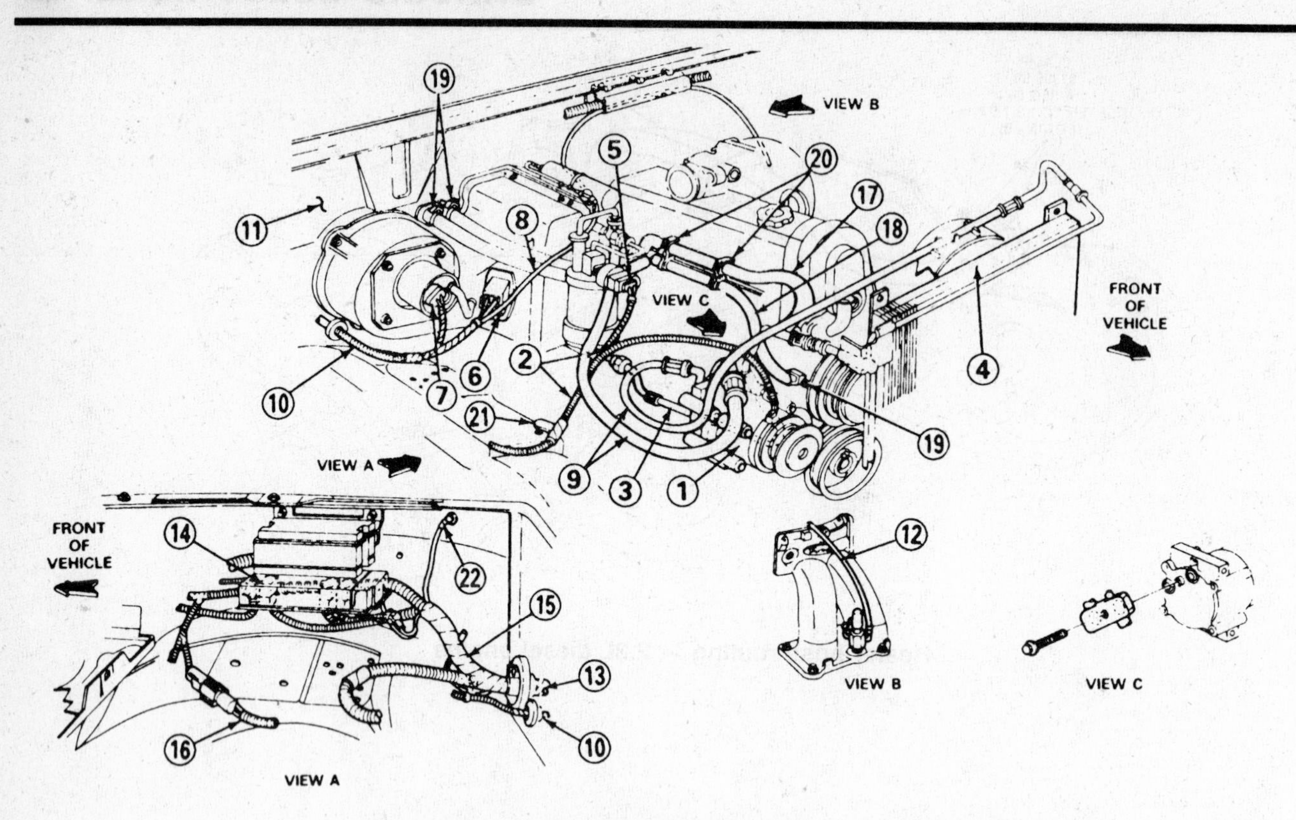

Air conditioning and heater hose routing – 2.3L engine

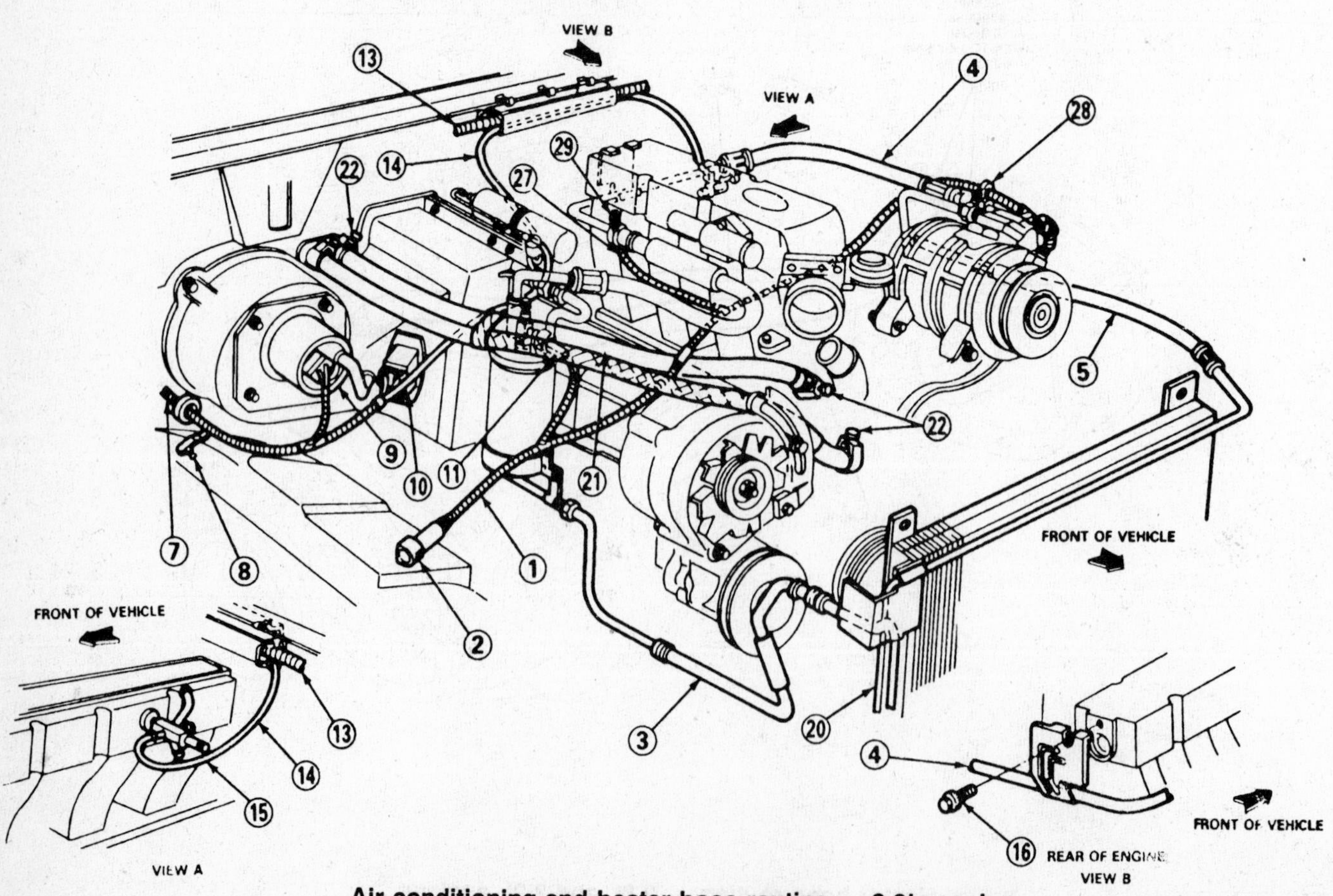

Air conditioning and heater hose routing – 2.9L engine

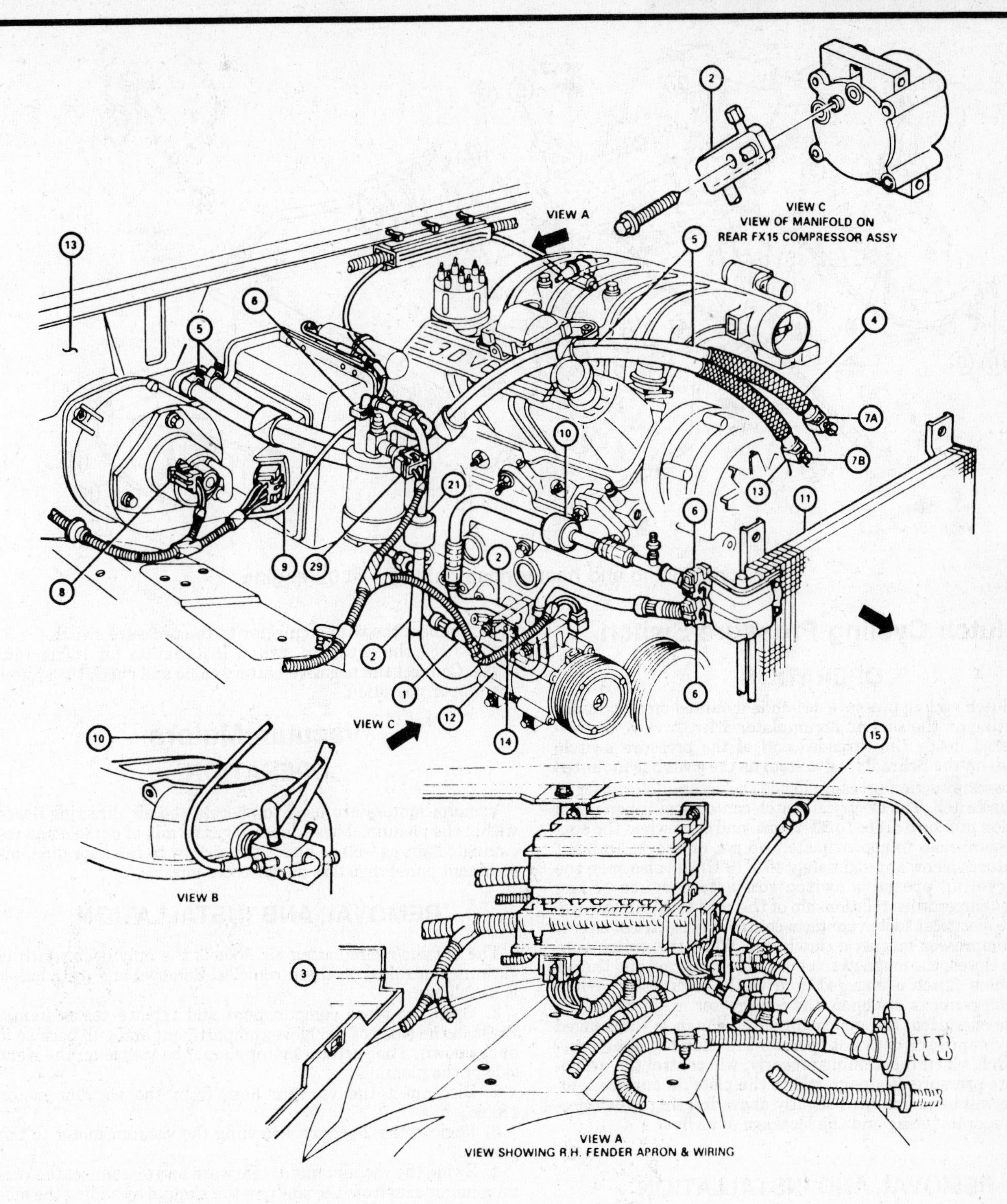

Air conditioning and heater hose routing — 3.0L engine

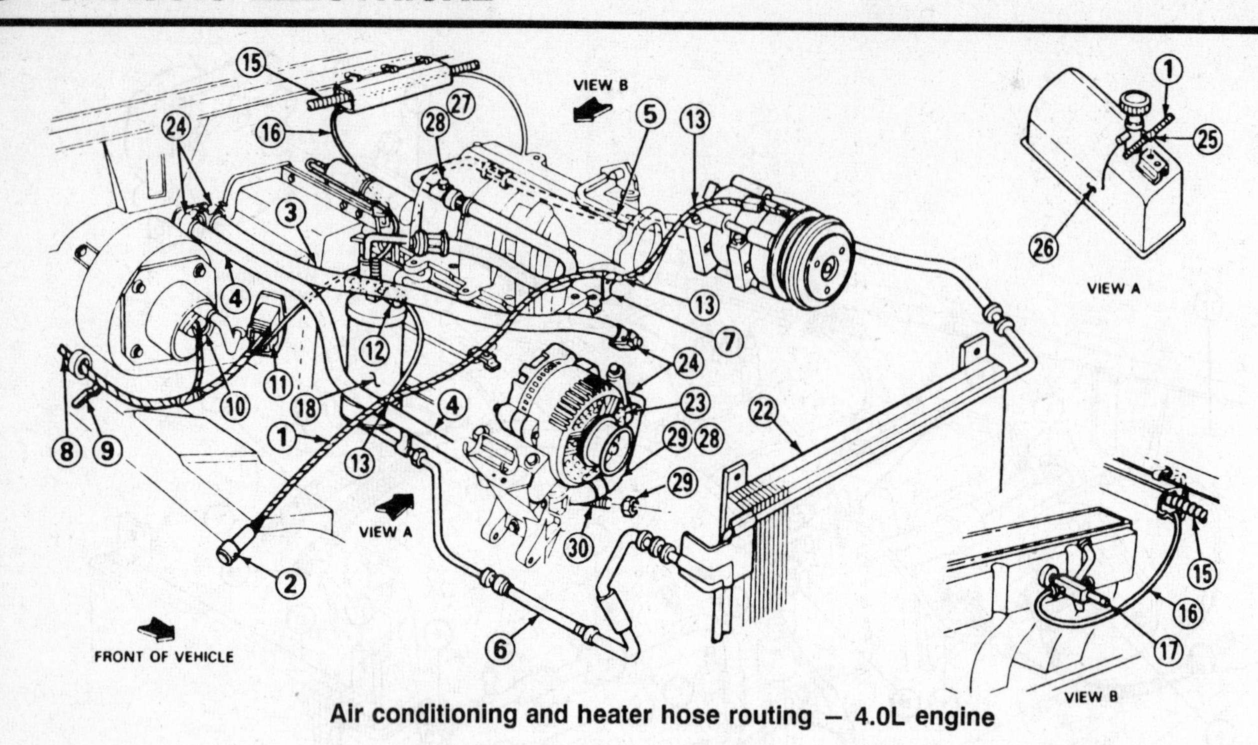

Air conditioning and heater hose routing — 4.0L engine

Clutch Cycling Pressure Switch

OPERATION

The clutch cycling pressure switch is mounted on a Schrader valve fitting on the suction accumulator/drier. A valve depressor, located inside the threaded end of the pressure switch, presses in on the Schrader valve stem as the switch is mounted and allows the suction pressure inside the accumulator/drier to act on the switch. The electrical switch contacts will open when the suction pressure drops to 23–26 psi. and close when the suction pressure rises to approximately 45 psi. or above. Ambient temperatures below approximately 45°F (9°C) will also open the clutch cycling pressure switch contacts because of the pressure/temperature relationship of the refrigerant in the system. The electrical switch contacts control the electrical circuit to the compressor magnetic clutch coil. When the switch contacts are closed, the magnetic clutch coil is energized and the air conditioning clutch is engaged to drive the compressor. When the switch contacts are open, the compressor magnetic clutch coil is de-energized, the air conditioning clutch is disengaged and the compressor does not operate. The clutch cycling pressure switch, when functioning properly, will control the evaporator core pressure at a point where the plate/fin surface temperature will be maintained slightly above freezing which prevents evaporator icing and the blockage of airflow.

REMOVAL AND INSTALLATION

1. Disconnect the negative battery cable.
2. Disconnect the wire harness connector from the pressure switch.
3. Unscrew the pressure switch from the suction accumulator/drier.

To install:

4. Lubricate the O-ring on the accumulator nipple with clean refrigerant oil.
5. Screw the pressure switch on the accumulator nipple. Tighten the switch to 5–10 ft. lbs. (7–13 Nm) if the switch has a metal base. Hand tighten only if the switch has a plastic base.
6. Connect the wire connector to the pressure switch.
7. Check the pressure switch installation for refrigerant leaks. Connect the negative battery cable and check the system for proper operation.

Vacuum Motors

OPERATION

Vacuum motors are used to operate the air directing doors within the plenum. These doors vary the mix of outside and recirculated air, as well as direct the airflow to the floor duct, instrument panel registers or defroster nozzles.

REMOVAL AND INSTALLATION

The outside/recirculating air door is the only door which is vacuum controlled on these vehicles. Removal and installation is as follows:

1. Open the glove compartment and remove the contents. Press in the sides of the glove compartment and pull back so it hangs down. The vacuum motor should be visible on the right side of the plenum.
2. Disconnect the vacuum hose from the vacuum motor nipple.
3. Remove the 2 screws attaching the vacuum motor to the plenum.
4. Swing the vacuum motor rearward and disconnect the vacuum motor arm from the shaft on the plenum by sliding the motor arm to the left.

To install:

5. Position the vacuum motor arm so the shaft on the plenum protrudes through the hole in the vacuum motor.
6. Swing the vacuum motor forward and install 2 screws attaching the vacuum motor to the plenum.
7. Connect the vacuum hose to the vacuum motor nipple.
8. Push the sides of the glove compartment and install to the latched position.
9. Start the engine and move the function lever forward in the control assembly to verify that the vacuum motor functions properly.

AUDIO SYSTEMS

Knob Type Radio

REMOVAL AND INSTALLATION

1. Disconnect the battery ground cable.
2. Remove the knobs and discs from the radio control shafts.
3. Remove the two steering column shroud-to-panel retaining screws and remove the shroud.
4. Detach the cluster trim cover or appliques from the instrument panel by removing the eight screws.
6. Remove the four screws securing the mounting plate assembly to the instrument panel and remove the radio with the mounting plate and rear bracket.
7. Disconnect the antenna lead-in cable, speaker wires and the radio (power) wire.
8. Remove the nut and washer assembly attaching the radio rear support.
9. Remove the nuts and washers from the radio control shafts and remove the mounting plate from the radio.

To install:

10. Install the radio rear support using the nut and washer assembly.
11. Install the mounting plate to the radio using the two lock washers and two nuts.
12. Connect the wiring connectors to the radio and position the radio with the mounting plate to the instrument panel.

NOTE: Make sure that the hair pin area of the rear bracket is engaged to the instrument panel support.

13. Secure the mounting plate to the instrument panel with the four screws.

NOTE: Make sure the mounting plate is fully seated on the instrument panel.

14. Install the panel trim covers and steering column shroud.
15. Install the panel knobs and discs to the radio control shafts.
16. Connect the battery ground cable.

Electronically Tuned Radio (ETR)

REMOVAL AND INSTALLATION

1. Disconnect the negative battery cable.
2. Remove the finish panel from around the radio assembly.
3. Insert the radio removal tool T87P-19061-A or equivalent, into the radio face.
4. Pull the radio out of the instrument panel.
5. Disconnect the wiring connectors and antenna from the radio.
6. Connect the wiring and slide the radio into the instrument panel.
7. Install the finish panel and connect the battery cable.
8. Check the operation of the radio.

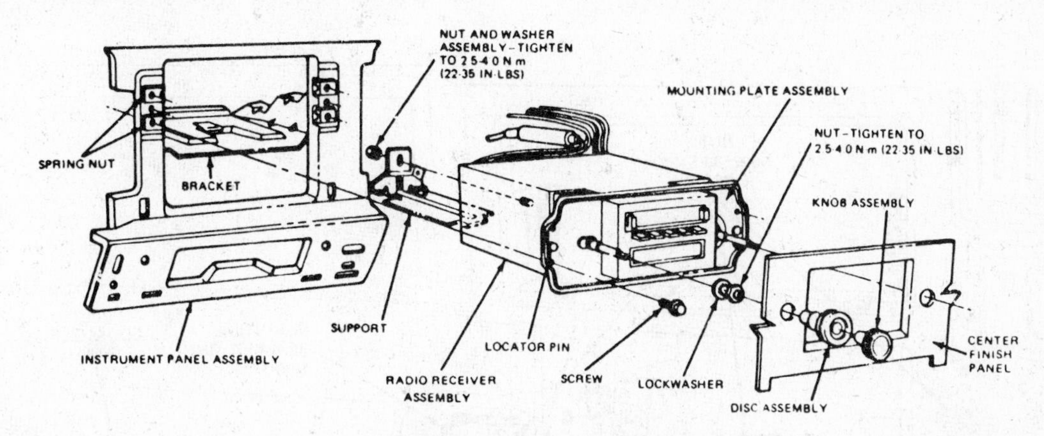

Radio removal – knob style

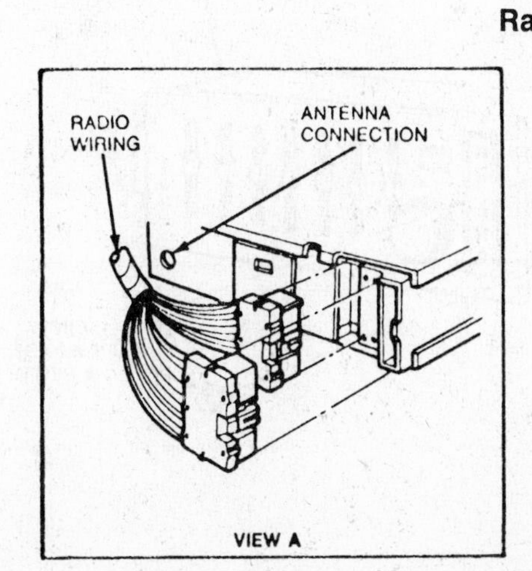

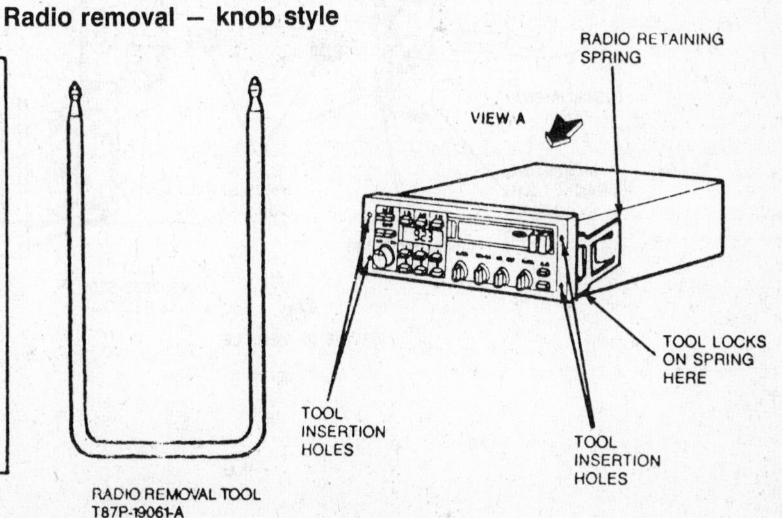

Radio removal tool and use – ETR type

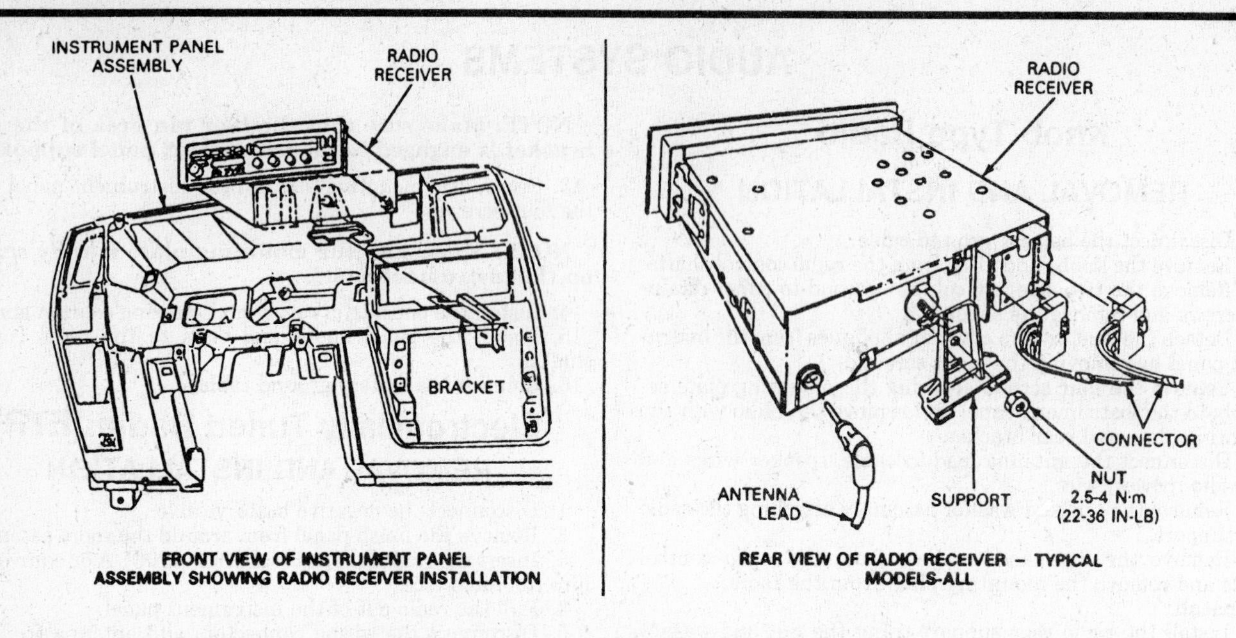

Radio removal and installation – ETR type

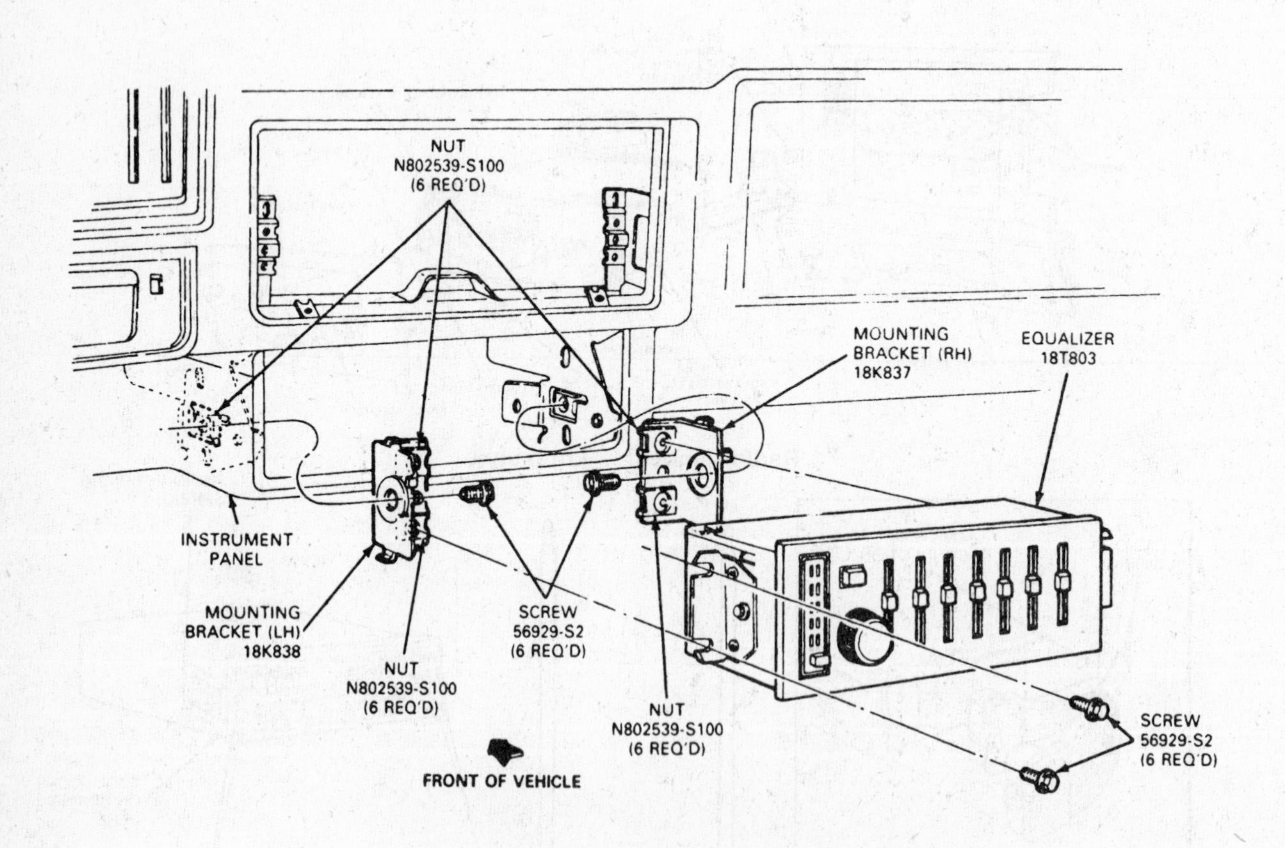

Equalizer removal and installation

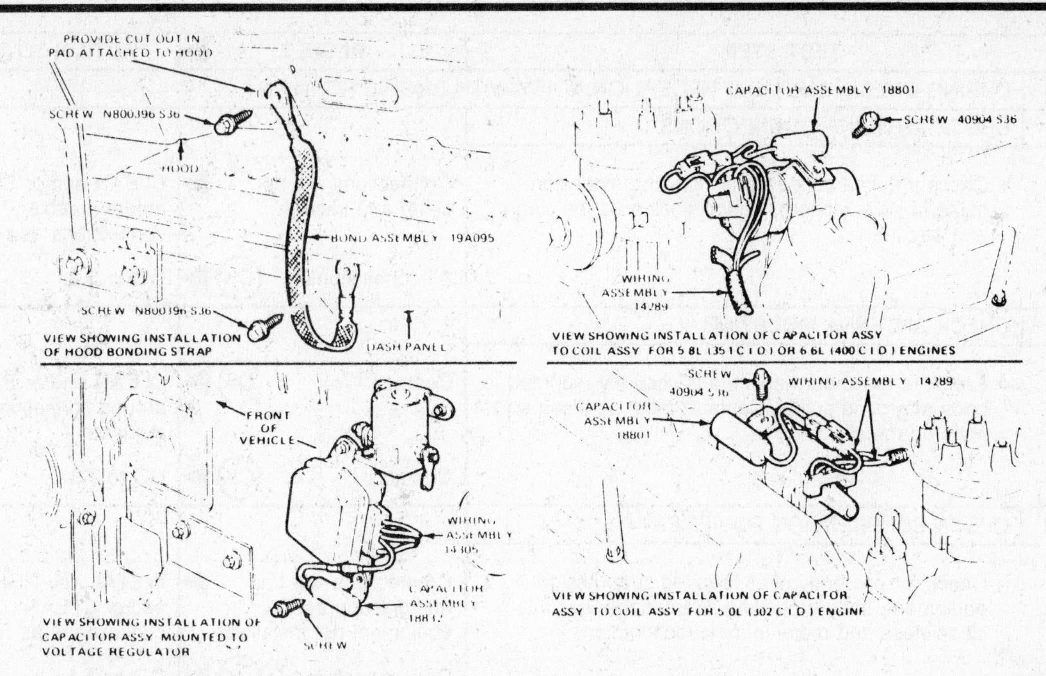

Typical radio suppression equipment, pickups. These, along with spark plug wires and the radio itself, are the areas to check when the radio reception becomes poor

Equalizer

REMOVAL AND INSTALLATION

1. Disconnect the negative battery cable.
2. Remove the finish panel from around the equalizer assembly.
3. Remove the 4 equalizer mounting screws.
4. Pull the equalizer from the instrument panel and disconnect the electrical connector.
5. Connect the electrical lead and install the equalizer into the instrument panel. Install the mounting screws.
6. Install the finish panel and connect the negative battery cable.
7. Check the operation of the stereo system.

Antenna

REMOVAL AND INSTALLATION

1. Disconnect the negative battery cable.
2. Remove the radio assembly. Disconnect the antenna lead from the radio.
3. Remove the antenna mast and lift the cover from the antenna base.
4. Remove the antenna base mounting screws or bolt. Remove the mounting gasket.
5. Pull the antenna wire through the opening.
6. Install the antenna in position, be sure to place the gasket properly.
7. Route antenna cable so as to avoid interference with heater operation.
8. Install the radio assembly. Connect the battery cable and check the operation of the radio.

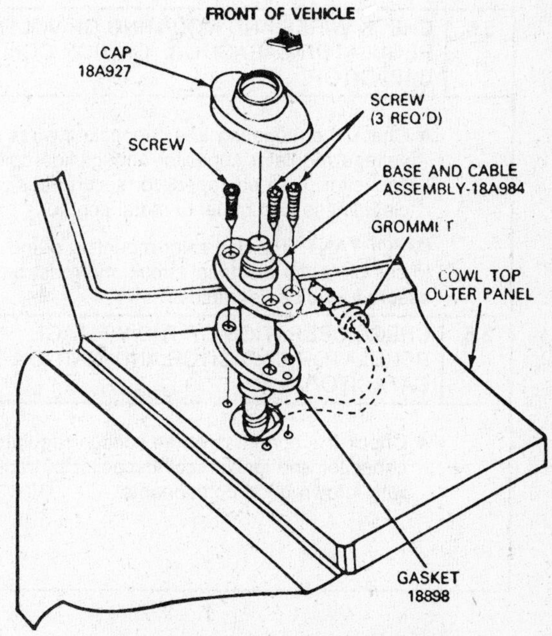

Antenna removal and installation

	TEST STEP	RESULT	▶ ACTION TO TAKE
3.0	DURING CHECK, AM RADIO RECEPTION NOISY WITH ENGINE RUNNING		
3.1	CHECK ANTENNA CONNECTIONS		
	• Check antenna connections including extension cable (if so equipped). Connections must be clean and secure.	Connections are not clean and secure ▶	CLEAN and/or SECURE antenna cable connections as required.
		All connections (OK) ▶	GO to **3.2**.
3.2	CHECK ANTENNA MOUNTING		
	• Check to make sure antenna is securely mounted to body at ground points. Contacts must be clean and metal-to-metal.	Contacts (~~OK~~) ▶	CLEAN and/or SECURE ground connections as required.
		Contacts (OK) ▶	GO to **3.3**.
3.3	CHECK SUPPRESSION EQUIPMENT		
	• Check for presence of all required suppression equipment, body grounding strap usage, security, cleanliness and metal-to-metal connections.	Connections are bad and/or suppression equipment not installed ▶	INSTALL or TIGHTEN and/or CLEAN connections as required.
		Connections are secure and suppression equipment installed correctly (OK) ▶	GO to **3.4**.
3.4	CHECK HOOD BONDING STRAP		
	• Check hood bonding strap for excessive usage, secureness of mounting to sheet metal and contact with hood. Hood bonding strap must scratch hood paint.	Strap (~~OK~~) ▶	INSTALL, SECURE, or FORM hood bonding strap as required.
		Strap (OK) ▶	GO to **3.5**.
3.5	CHECK WIRES AND MOUNTING OF VOLTAGE REGULATOR CAPACITOR, IGNITION COIL CAPACITOR		
	• Check the mounting and connecting wires of the voltage regulator capacitor and ignition coil capacitor (if so equipped) for secureness, cleanliness and metal-to-metal contact. IMPORTANT: The capacitor mounting points are used to complete the electrical circuit and must be mounted securely to clean surfaces.	Connections (~~OK~~) ▶	CLEAN and/or SECURE connections as required.
		Connections (OK) ▶	GO to **3.6**.
3.6	CHECK OPERATION OF THE VOLTAGE REGULATOR CAPACITOR AND IGNITION COIL CAPACITOR		
	• Check the operation of the voltage regulator capacitor and ignition coil capacitor by replacing with known good components.	Voltage regulator capacitor and/or ignition coil capacitor bad ▶	REPLACE capacitor(s) with known good component(s).
		Capacitor(s) are (OK) ▶	GO to **3.7**.

TEST STEP		RESULT	ACTION TO TAKE
3.7	CHECK ALTERNATOR		
	• Check alternator by disconnecting wiring harness from voltage regulator.	Noise eliminated ▶	CHECK alternator
		Noise still present ▶	GO to **3.8**.
3.8	CHECK SPARK PLUG WIRES		
	• Check spark plug wires for proper routing, grounding and secureness of connections.	Spark plug wires not routed, grounded or secured ▶	REROUTE or REPLACE spark plug wires or SECURE connections as required.
		Spark plug wires (OK) ▶	GO to **3.9**.
3.9	CHECK IGNITION SYSTEM		
	• Check ignition system for proper operation. (Use ignition system analyzer or check for open spark plug wires using ohmmeter.) Also check spark plug for cracked insulators.	Ignition system and/or spark plugs not OK ▶	REPAIR or REPLACE components as required.
		Ignition system and spark plugs OK ▶	GO to **3.10**.
3.10	CHECK RADIO CHASSIS MOUNTING		
	• Check all radio chassis mounting points for secureness, cleanliness and metal-to-metal contact.	Mounting (not OK) ▶	CLEAN and/or SECURE as required.
		Mountings (OK) ▶	GO to **3.11**.
3.11	SUBSTITUTE A KNOWN GOOD SPEAKER AND ANTENNA		
	• Substitute a known good speaker, antenna and antenna extension cable (if so equipped). Be sure to ground antenna to an unpainted metal surface.	Noise eliminated ▶	REPAIR or REPLACE antenna, speaker or antenna extension cable.
		Noise not eliminated ▶	GO to **3.12**.
3.12	SUBSTITUTE KNOWN GOOD RADIO		
	• Substitute known good radio.	Noise eliminated ▶	Have radio unit REPAIRED by authorized service center.
		Noise not eliminated ▶	GO to **3.13**.
3.13	REPOSITION ANTENNA, SPEAKER OR RADIO POWER FEED		
	• Check to see if noise can be eliminated by repositioning antenna, speaker or radio power feed wires.	Noise eliminated ▶	REPOSITION permanently by taping.
		Noise not eliminated ▶	GO to **3.14**.
3.14	GROUND VARIOUS PARTS OF TRUCK		
	• Ground various parts of the truck to the frame using a jumper cable. For example: engine, fenders, quarter panel, stone deflectors, air cleaner, body sheet metal.	Noise eliminated ▶	PROVIDE permanent ground where required.

TEST STEP	RESULT ▶	ACTION TO TAKE
4.0 DURING CHECK, FM RADIO RECEPTION IS NOISY WHILE VEHICLE IS NOT IN MOTION		
4.1 NOISE IS ONLY ON FM STERO		
• Check to see if noise is only on FM stereo. Determine if customer concern is due to FM stereo reception limitation. Refer to normal operation description.	Noise only on FM stereo ▶	EXPLAIN and DEMOSTRATE to customer. Inform customer of methods for obtaining best reception.
	Noise is on both FM stereo and FM mono ▶	GO to **4.2**.
4.2 CHECK ANTENNA CABLE CONNECTIONS		
• Check antenna cable connections including extension cable (if so equipped). Connections must be clean and secure.	Connections (OK crossed out) ▶	CLEAN and/or SECURE as required.
	Connections (OK) ▶	GO to **4.3**.
4.3 CHECK ANTENNA MOUNTING		
• Check to ensure antenna is securely mounted to body at mounting nut above antenna; and also, ensure that prongs of grounding collar, at fender underside, are contacting metal. Contact must be clean and metal-to-metal.	Connections (OK crossed out) ▶	CLEAN and/or SECURE as required.
	Connections are (OK) ▶	GO to **4.4**.
4.4 CHECK DISTRIBUTOR ROTOR		
• Check for adequate distributor rotor contact spring tension. Height of spring should be 8.9 mm (0.35 in.) from top of rotor (not applicable to recreational vehicles).	Spring tension (OK crossed out) ▶	REPLACE rotor.
	Spring tension (OK) ▶	GO to **4.5**.
4.5 CHECK DISTRIBUTOR CAP		
• Check to see if carbon center insert in distributor cap is secure.	Carbon center not secure ▶	REPLACE distributor cap.
	Carbon center (OK) ▶	GO to **4.6**.
4.6 CHECK SPARK PLUG WIRES		
Check spark plug wires for proper routing and secureness of connections.	Routing and/or connections (OK crossed out) ▶	REROUTE or SECURE connections as required.
	Routings and connections good ▶	GO to **4.7**.

TEST STEP		RESULT	ACTION TO TAKE
4.7	CHECK IGNITION SYSTEM		
	• Check ignition system for proper operation. (Use ignition system analyzer or check for open spark plug wires using ohmmeter). Also check spark plug for cracked insulators.	Ignition system and/or spark plug wires not working properly, and/ or spark plug insulators cracked ▶	REPAIR or REPLACE components as required.
		Ignition system, spark plug wires and spark plugs in good condition ▶	GO to **4.8**.
4.8	CHECK RADIO CHASSIS MOUNTING		
	• Check all radio chassis mounting points for secureness, cleanliness and metal-to-metal contact.	Contacts are not secure or clean ▶	CLEAN and/or SECURE as required.
		Contacts are (OK) ▶	GO to **4.9**.
4.9	SUBSTITUTE A GOOD SPEAKER AND ANTENNA		
	• Substitute a known good speaker and antenna being sure to ground antenna base to unpainted metal surface.	Noise eliminated ▶	REPAIR or REPLACE antenna or speaker.
		Noise not eliminated ▶	GO to **4.10**.
4.10	SUBSTITUTE EXTENSION CABLE		
	• If equipped with antenna extension cable, substitute with a known good cable. (If not equipped with extension cable GO to **4.11**.)	Noise eliminated ▶	REPLACE antenna extension cable.
		Noise not eliminated ▶	GO to **4.11**.
4.11	SUBSTITUTE KNOWN GOOD RADIO		
	• Remove radio and substitute with a known good radio.	Noise eliminated ▶	Have radio unit REPAIRED by authorized service center.
		Noise not eliminated ▶	GO to **4.12**.
4.12	REPOSITION ANTENNA, SPEAKER, OR RADIO FEED WIRES		
	• Check to see if noise can be eliminated by repositioning antenna, speaker or radio power feed wires.	Noise eliminated ▶	REPOSITION permanently by taping.
		Noise not eliminated ▶	REPLACE distributor cap and rotor with new ungreased cap and rotor.

TEST STEP		RESULT ▶	ACTION TO TAKE
5.0	DURING CHECK, RECEPTION ON RADIO IS NOISY WITH ENGINE RUNNING AND VEHICLE IN MOTION		
5.1	VISUALLY INSPECT CONNECTIONS		
	• Inspect all connections to battery; antenna leads, speaker leads, and radio receiver.	Connections ~~OK~~ ▶	REPAIR or REPLACE wires as necessary.
		Connections (OK) ▶	GO to **5.2**.
5.2	CHECK ANTENNA LEAD IN CABLE		
	• Check for loose antenna lead in cable or loose antenna.	Antenna and/or cable are loose ▶	CONNECT antenna and/or lead in cable securely.
		Antenna and cable secure ▶	Have radio repaired by a qualified radio technician.

TEST STEP		RESULT ▶	ACTION TO TAKE
6.0	DURING CHECK, RADIO RECEPTION IS NOISY WHILE ENGINE IS NOT RUNNING		
6.1	VISUALLY CHECK FOR LOOSE CONNECTIONS		
	• Check all connections to battery, antenna leads, speaker lead and radio receiver for proper connection.	Connections ~~OK~~ ▶	REPAIR or REPLACE connections or wires as required.
		Connections (OK) ▶	GO to **6.2**.
6.2	CHECK ANTENNA LEAD IN CABLE		
	• Check all antenna lead-in cables for bent or missing male pins. Also check the female connectors for position of receptacle with respect to the insulator. The receptacle should be visible.	Connectors and/or connections ~~OK~~ ▶	REPLACE antenna lead in cables.
		Connections (OK) ▶	Have radio REPAIRED by a qualified radio technician.

WINDSHIELD WIPERS

Wiper Arm And Blade

REPLACEMENT

To remove the arm and blade assembly, raise the blade end of the arm off of the windshield and move the slide latch away from the pivot shaft. The wiper arm can now be removed from the shaft without the use of any tools. To install, push the main head over the pivot shaft. Be sure the wipers are in the parked position, and the blade assembly is in its correct position. Hold the main arm head onto the pivot shaft while raising the blade end of the wiper arm and push the slide latch into the lock under the pivot shaft head. Then, lower the blade to the windshield. If the blade does not lower to the windshield, the slide latch is not completely in place.

Windshield Wiper Motor

REMOVAL AND INSTALLATION

1. Turn the wiper switch on. Turn the ignition switch on until the blades are straight up and then turn ignition off to keep them there.
2. Remove the right wiper arm and blade.
3. Remove the negative battery cable.
4. Remove the right pivot nut and allow the linkage to drop into the cowl.
5. Remove the linkage access cover, located on the right side of the dash panel near the wiper motor.
6. Reach through the access cover opening and unsnap the wiper motor clip.
7. Push the clip away from the linkage until it clears the nib on the crank pin. Then, push the clip off the linkage.
8. Remove the wiper linkage from motor crank pin.
9. Disconnect the wiper motor's wiring connector.
10. remove the wiper motor's three attaching screws and remove the motor.

To install:

11. Install the motor and attach the three attaching screws. Tighten to 60-65 inch lbs. (6.7-7.3 Nm).
12. Connect the wiper motor's wiring connector.
13. Install the clip completely on the right linkage. Make sure the clip is completely on.
14. Install the left linkage on the wiper motor crank pin.
15. Install the right linkage on the wiper motor crank pin and pull the linkage on to the crank pin until it snaps.

NOTE: The clip is properly installed if the nib is protruding through the center of the clip.

16. Reinstall the right wiper pivot shaft and nut.
17. Reconnect the battery and turn the ignition **ON**. Turn the wiper switch off so the wiper motor will park, then turn the ignition **OFF**. Replace the right linkage access cover.
18. Install the right wiper blade and arm.
19. Check the system for proper operation.

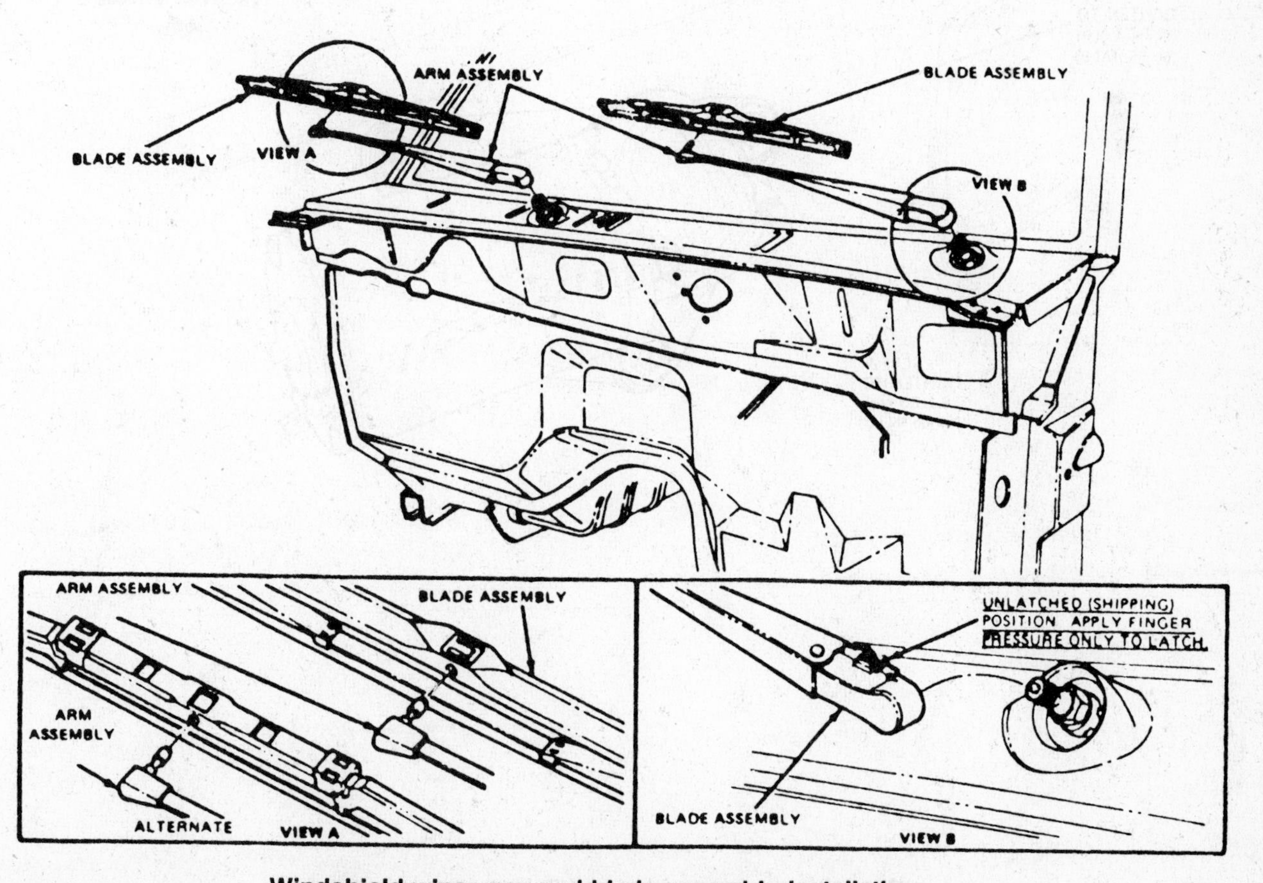

Windshield wiper arm and blade assembly installation

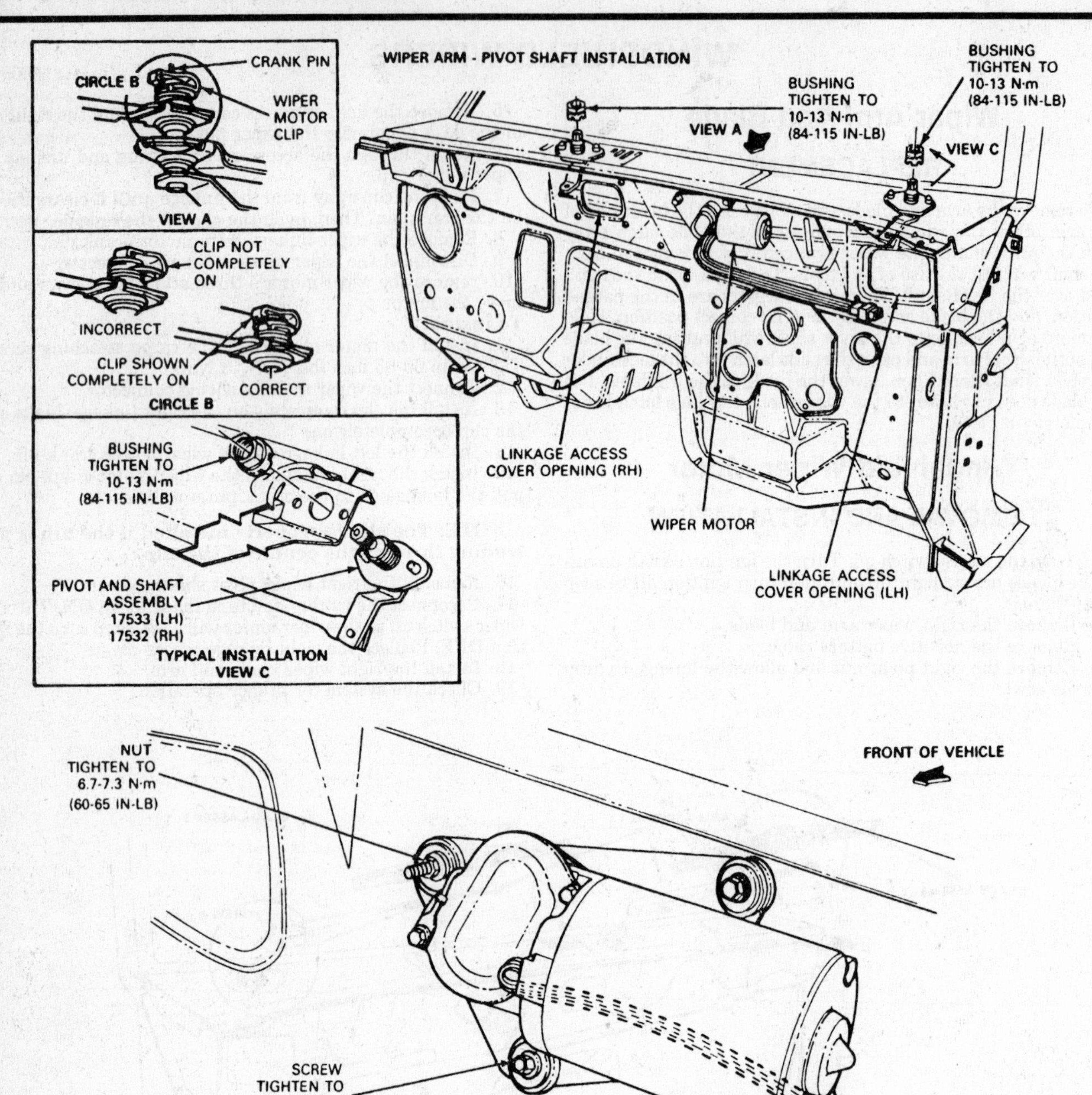

Windshield wiper motor and linkage mounting

Pivot Shaft and Linkage

REMOVAL AND INSTALLATION

1. Perform steps 1 through 8 of the wiper motor removal procedure.
2. Slide the right pivot shaft and linkage assembly out through the R.H. access opening.
3. If the left linkage is to be serviced, remove the L.H. wiper arm and blade assembly.
4. Remove the left linkage access cover.
5. Remove the left pivot nut, lower the linkage and slide it out through the L.H. access opening.

NOTE: The left and right pivot and linkage assemblies are serviced separately.

6. Installation is the reverse of the removal procedure above.

Rear Window Wiper Motor

REMOVAL AND INSTALLATION

1. Disconnect the negative battery cable.
2. Remove the wiper arm and blade.
3. Remove the liftgate interior trim.
4. Remove the motor attaching bolts (3). Disconnect the electrical leads.
5. Remove the wiper motor from the vehicle.
6. Install the wiper motor in position and connect the electrical leads.
7. Install the liftgate trim. Connect the negative battery cable.

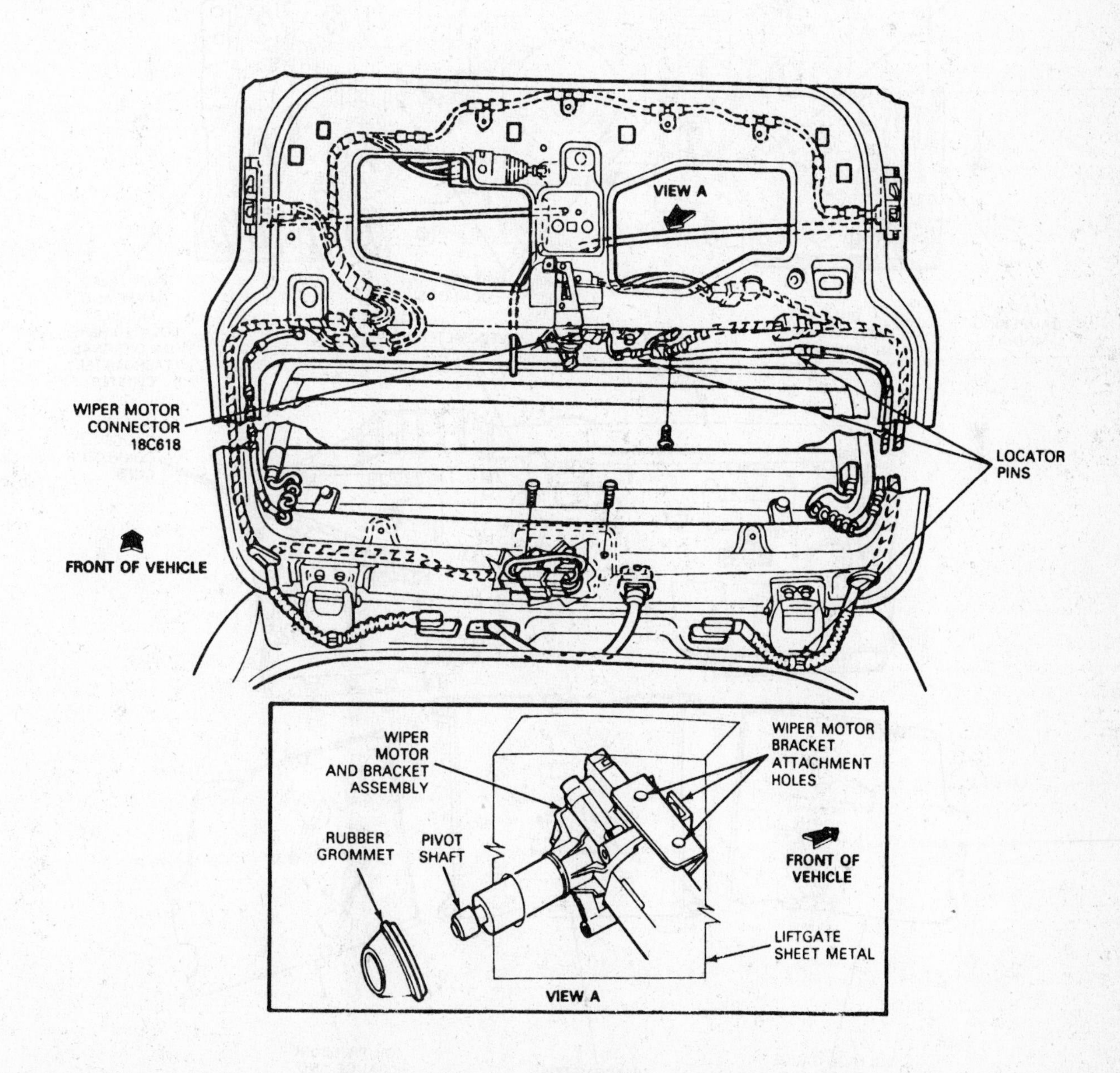

Rear wiper motor mounting – Explorer

INSTRUMENTS AND SWITCHES

Instrument Cluster

REMOVAL AND INSTALLATION

1. Disconnect the battery ground cable.
2. Remove the two steering column shroud-to-panel retaining screws and remove the shroud.
3. Remove the lower instrument panel trim.
4. Remove the cluster trim cover from the instrument panel by removing the eight screws.
5. Remove the four instrument cluster to panel retaining screws.
6. Position the cluster slightly away from the panel for access to the back of the cluster to disconnect the speedometer.

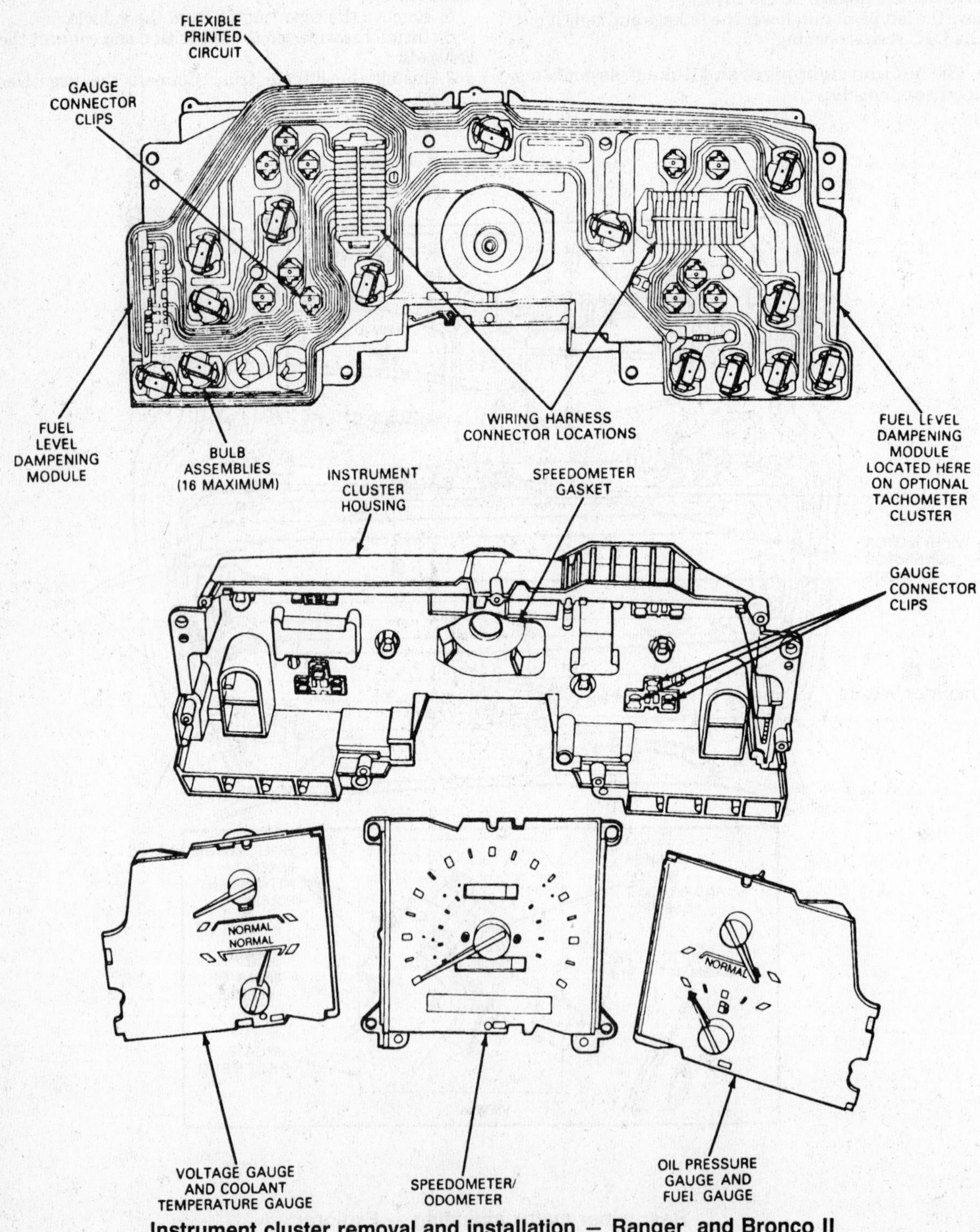

Instrument cluster removal and installation – Ranger and Bronco II

NOTE: If there is not sufficient access to disengage the speedometer cable from the speedometer, it may be necessary to remove the speedometer cable at the transmission and pull cable through cowl, to allow room to reach the speedometer quick disconnect.

7. Disconnect the wiring harness connector from the printed circuit, and any bulb-and-socket assemblies from the wiring harness to the cluster assembly and remove the cluster assembly from the instrument panel.

To install:

8. Apply approximately 1/8 in. diameter ball of D7AZ-19A331-A Silicone Dielectric compound or equivalent in the drive hole of the speedometer head.
9. Position the cluster near its opening in the instrument panel.
10. Connect the wiring harness connector to the printed circuit, and any bulb-and-socket assemblies from the wiring harness to the cluster assembly.
11. Position the cluster to the instrument panel and install the four cluster to panel retaining screws.
13. Install the panel trim covers and the steering column shroud.
14. Connect the battery ground cable.
15. Check operation of all gauges, lamps and signals.

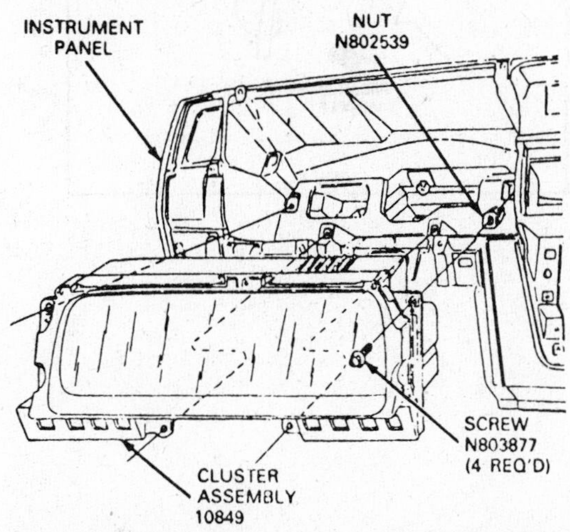

Instrument cluster removal and installation — Explorer

Fuel, Oil Pressure, Voltage and Coolant Temperature Gauges

Each of the gauges can be removed in the same manner, once the instrument cluster is removed.

1. Disconnect the negative battery cable.
2. Remove the instrument cluster assembly.
3. Remove the lens from the instrument cluster.
4. Pull the gauge from the cluster.
5. Install the gauge, by pushing it firmly into position.
6. Install the cluster lens and install the cluster into the instrument panel.

Printed Circuit Board

REMOVAL AND INSTALLATION

1. Disconnect the negative battery cable.
2. Remove the instrument cluster assembly.
3. Remove the lens from the instrument cluster.
4. Pull the gauges from the cluster.
5. Remove the screws that retain the circuit board and carefully remove the circuit board.
6. Install the circuit board and retaining screws.
7. Install the gauges. Install the cluster lens and install the cluster assembly in the instrument panel.
8. Connect the negative battery cable.

Speedometer Cable

REMOVAL AND INSTALLATION

1. Raise and safely support the vehicle.
2. Disengage the cable from the transmission.

NOTE: Disconnect the cable by pulling it out of the speed sensor. Do not attempt to remove the spring retainer clip with the cable and the sensor.

3. Disconnect the cable from its retaining clips.
4. Push the cable through the grommet in the floor and pull it into the passenger compartment.
5. Disconnect the cable from the speedometer, cluster removal may be necessary.
6. Install the speedometer in position and route it through the floor.
7. Connect it at the speedometer and at the transmission.

Speedometer Cable Core

REMOVAL AND INSTALLATION

1. Reach up behind the cluster and disconnect the cable by depressing the quick disconnect tab and pulling the cable away.
2. Remove the cable from the casing. If the cable is broken, raise the vehicle on a hoist and disconnect the cable from the transmission.
3. Remove the cable from the casing.
4. To remove the casing from the vehicle, pull it through the floor pan.
5. To replace the cable, slide the new cable into the casing and connect it at the transmission.
6. Route the cable through the floor pan and position the grommet in its groove in the floor.
7. Push the cable onto the speedometer head.

Instrument Panel

REMOVAL AND INSTALLATION

1. Disconnect the negative battery cable.
2. Disconnect the instrument panel wiring connectors in the engine compartment.
3. Remove the 2 screws retaining the lower steering column cover and remove the cover.
4. Remove the ashtray and retainer.
5. Remove the upper and lower steering column shrouds.
6. Remove the instrument cluster finish panel. Remove the radio assembly and equalizer, if equipped.
7. Remove the screws retaining the instrument cluster. Remove the cluster, making sure to disconnect the electrical leads.
8. Remove the screw that attaches the instrument panel to the brake and clutch pedal support.
9. Disconnect the wiring from the switches on the steering column.
10. Remove the front inside pillar mouldings.
11. Remove the right side cowl trim cover.
12. Remove the lower right insulator from under the instrument panel.

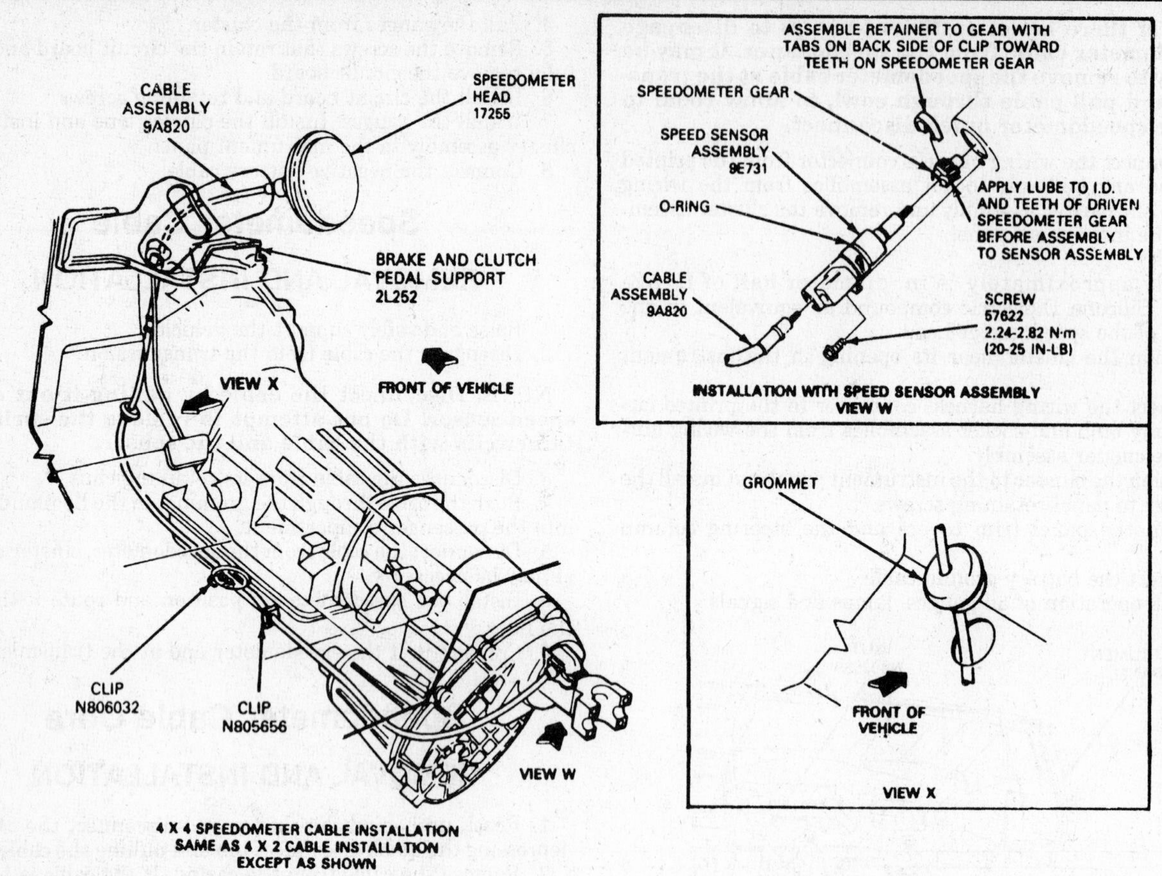

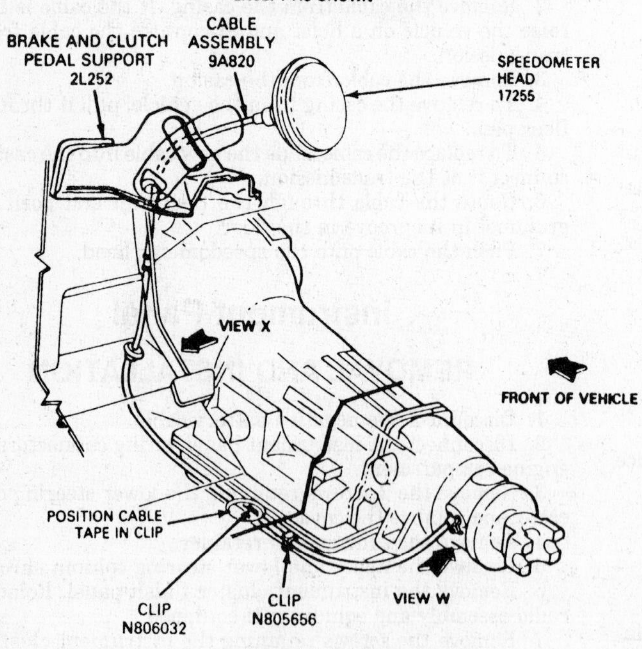

Speedometer cable routing and mounting

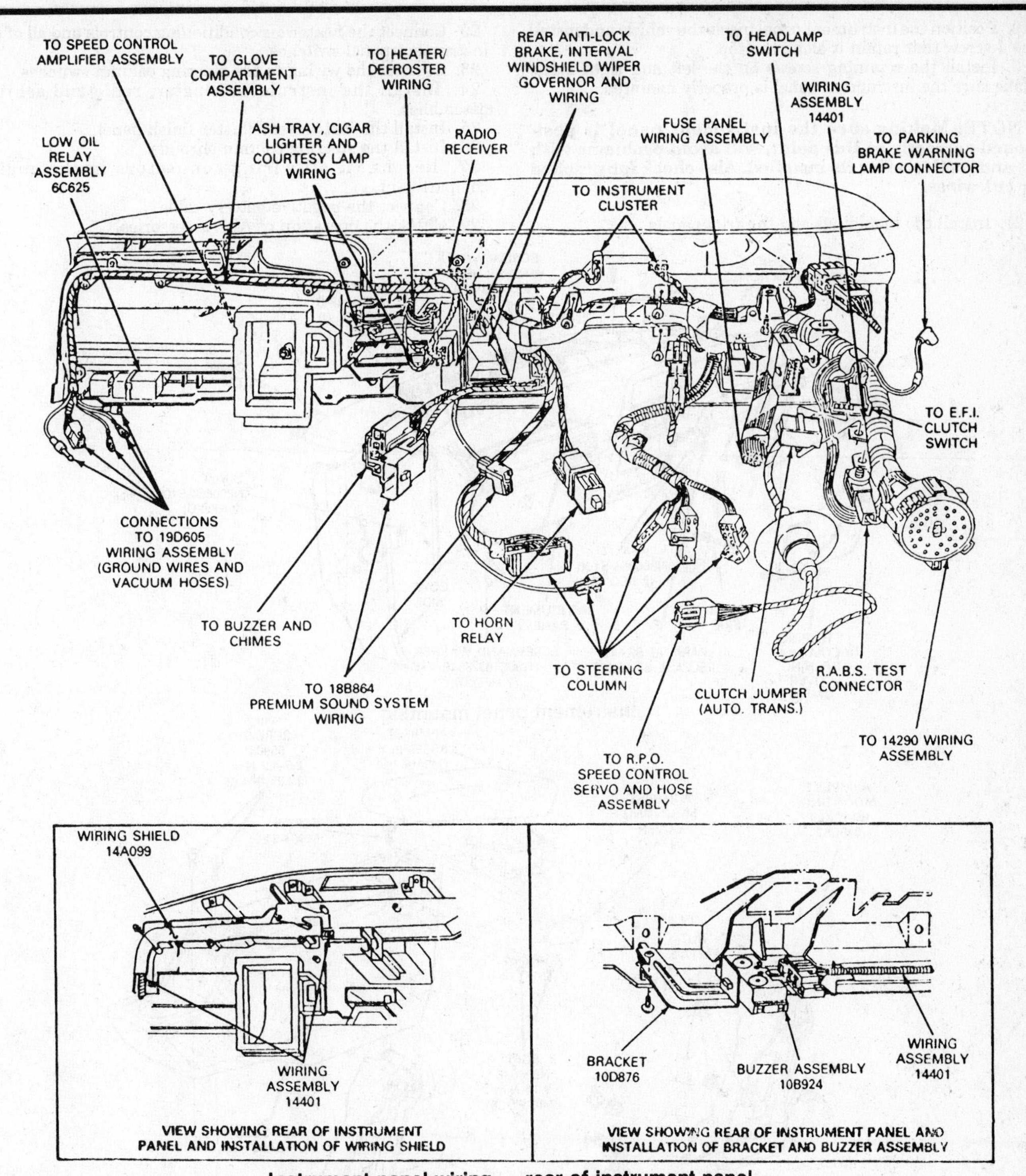

Instrument panel wiring – rear of instrument panel

13. Remove the 2 bolts retaining the instrument panel to the lower right side of the cowl.
14. Remove the 2 screw retaining the instrument panel to the parking brake bracket, on the drivers side.
15. Remove the 4 screw retaining the top of the instrument panel.
16. Reach through the openings in the instrument panel and disconnect any remaining electrical connectors. Disconnect the heater/air conditioning controls.

NOTE: Removing the instrument panel will be much easier with the help of an assistant, as it is extremely bulky and difficult to maneuver.

17. Carefully tilt the instrument panel forward and remove it from the vehicle. Work the instrument panel around the steering wheel.

To install:

18. If the instrument panel is being replaced, transfer all mounting brackets and switches to the new panel.

19. Position the instrument panel inside the vehicle and install the 4 screw that retain it along the top.
20. Install the retaining screws on the left and right sides. Make sure the instrument panel is properly mounted.

NOTE: Making sure the instrument panel is positioned correctly at this point, will avoid problems with fit and rattles, after its installed. Also check for pinched or cut wires.

21. Install the mouldings and the trim panels.
22. Connect the heater/air conditioning controls and all of the instrument panel switches.
23. Connect the wiring to the steering column switches.
24. Install the instrument cluster, radio and ashtray assemblies.
25. Install the instrument cluster finish panel.
26. Install the steering column shrouds.
27. Reconnect all wiring connectors in the engine compartment.
28. Connect the negative battery cable.
29. Check the operation of ALL accessories.

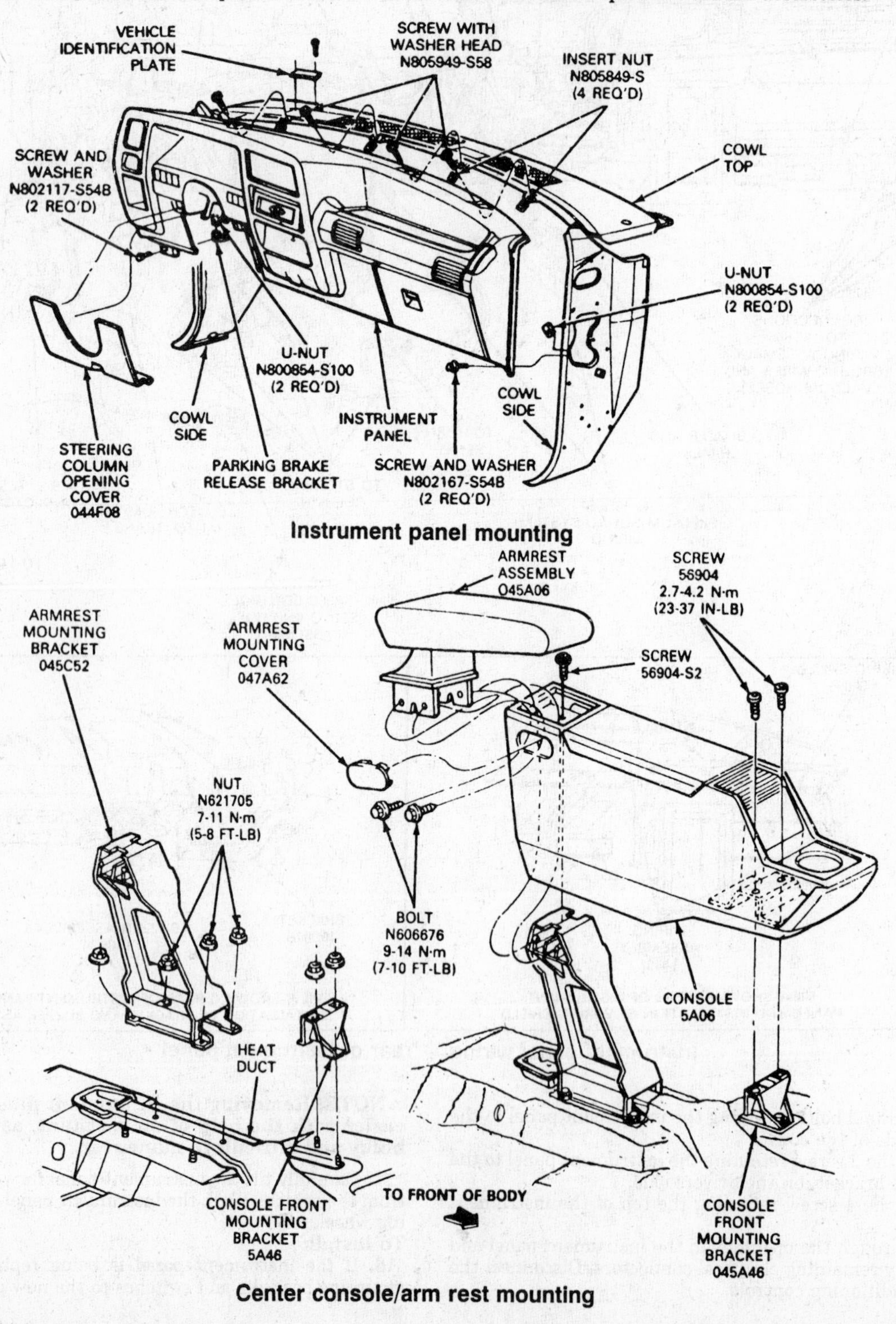

Instrument panel mounting

Center console/arm rest mounting

Center Console

REMOVAL AND INSTALLATION

1. Remove the small arm rest screw covers.
2. Remove the 4 arm rest retaining bolts.
3. Remove the 2 rear arm rest retaining screws and the 2 screws in the front utility tray.
4. Remove the entire assembly, by lifting it from its mounting bracket.
5. Install the console in position and install all mounting screws.

Windshield Wiper Switch

REMOVAL AND INSTALLATION

NOTE: The switch handle is an integral part of the switch and cannot be replaced separately.

1. Disconnect the negative battery cable from the battery.
2. Remove the steering column trim shrouds.
3. Disconnect the quick connect electrical connector.
4. Peel back the foam sight shield, remove the two hex-head screws holding the switch, and remove the windshield wiper switch.
5. Position the switch on the column and install the two hex head screws. Replace the foam sight shield over the witch.
6. Connect the quick connect electrical connector.
7. Install the upper and lower trim shrouds.

Rear Wiper Switch

REMOVAL AND INSTALLATION

1. Disconnect the negative battery cable.
2. Remove the instrument cluster finish panel.
3. Remove the switch from the instrument panel, by carefully prying upward.
4. Disconnect the electrical lead from the switch.
5. Connect the wiring and install the switch in the instrument panel.
6. Install the cluster finish panel.
7. Connect the negative battery cable. Check the operation of the switch.

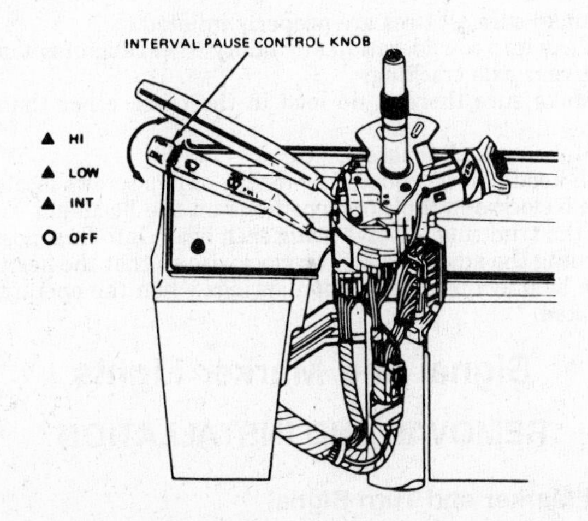

Windshield wiper switch mounting

Headlight Switch

REMOVAL AND INSTALLATION

1. Disconnect the battery ground cable.
2. Pull the headlight switch knob to the headlight on position.
3. Depress the shaft release button and remove the knob and shaft assembly.
4. Remove the instrument panel finish panel.
5. Unscrew the mounting nut and remove the switch from the instrument panel, then remove the wiring connector from the switch.

To install:

6. Connect the wiring connector to the headlamp switch, position the switch in the instrument panel and install the mounting nut.
7. Install the instrument panel finish panel.
8. Install the headlamp switch knob and shaft assembly by pushing the shaft into the switch until it locks into position.
9. Connect the battery ground cable, and check the operation of the headlight switch.

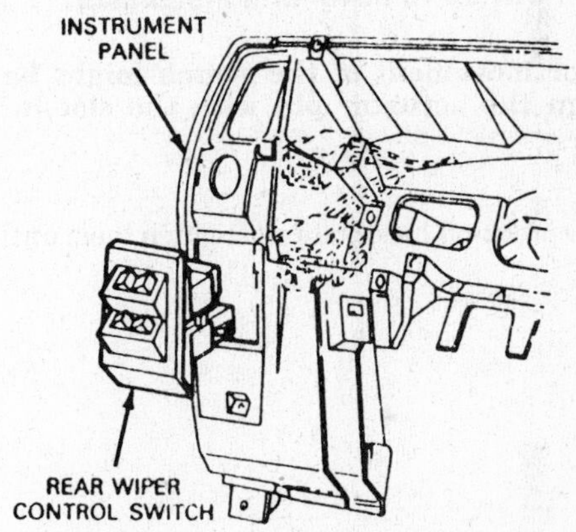

Rear window wiper control switch removal

Ignition Switch

REMOVAL AND INSTALLATION

1. Rotate the ignition key to the **LOCK** position.
2. Disconnect the negative battery cable.
3. Remove the upper and lower steering column trim panels.
4. On models equipped with tilt steering, remove the trim ring from around the steering column.
5. Disconnect the ignition switch electrical connector.
6. Drill out the break-off head bolts connecting the ignition switch to the lock cylinder housing. This can be done with an ⅛ in. drill bit.
7. Remove the remaining pieces of the bolt with and Easy-Out® tool or equivalent, bolt extractor.
8. Disengage the switch from the actuator pin.
9. Remove the switch from the column.

To install:

10. Rotate the ignition switch to the **RUN** position.
11. Install the replacement switch, by aligning the holes on the switch casting base with the holes in the lock cylinder housing. Note that the replacement switch is supplied with the switch in the **RUN** position.

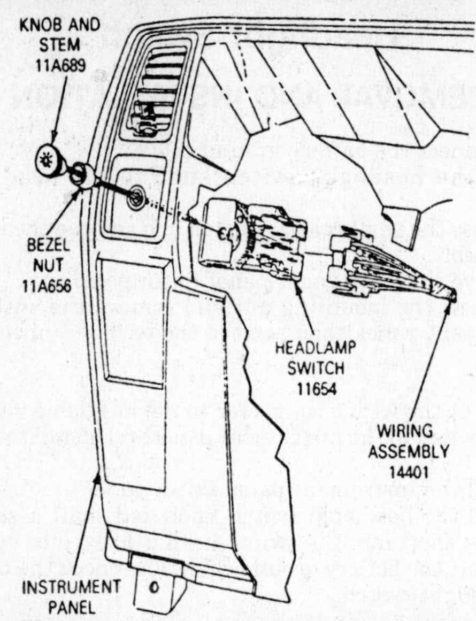

Headlight switch removal and installation

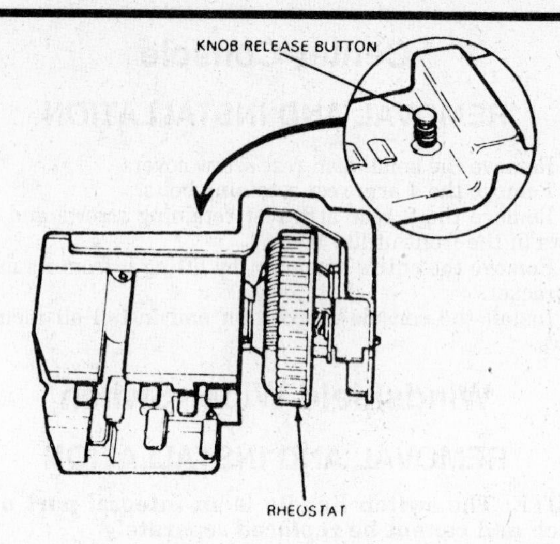

Headlight switch knob — release button location

NOTE: Minor movement of the switch might be required to align the actuator pin with the slot in the carrier.

12. Install new break-off head bolts and tighten them until the heads shear off. This will take approximately 35-50 inch lbs. of torque.
13. Connect the electrical lead to the switch.
14. Connect the negative battery cable and check the operation of the switch in all positions.
15. Install the steering column trim.

LIGHTING

Headlights

REMOVAL AND INSTALLATION

1. Remove the headlamp door attaching screws and remove the headlamp door. On the Explorer, remove the grille.
2. Remove the headlight retaining ring screws, and remove the retaining ring. Do not disturb the adjusting screw settings.
3. Pull the headlight bulb forward and disconnect the wiring assembly plug from the bulb.
4. Connect the wiring assembly plug to the new bulb. Place the bulb in position, making sure that the locating tabs of the bulb are fitted in the positioning slots.
5. Install the headlight retaining ring.
6. Place the headlight trim ring or door into position, and install the retaining screws. Install the grille on Explorer.

HEADLIGHT ADJUSTMENT

NOTE: Before making any headlight adjustments, preform the following steps for preparation:

1. Make sure all tires are properly inflated.
2. Take into consideration any faulty wheel alignment or improper rear axle tracking.
3. Make sure there is no load in the truck other than the driver.
4. Make sure all lenses are clean.

Each headlight is adjusted by means of two screws located at the 12 o'clock and 9 o'clock positions on the headlight underneath the trim ring. Always bring each beam into final position by turning the adjusting screws clockwise so that the headlight will be held against the tension springs when the operation is completed.

Signal and Marker Lights

REMOVAL AND INSTALLATION

Front Marker and Turn Signal

1. Remove the screws retaining the headlight door. On Explorer, remove the grille.

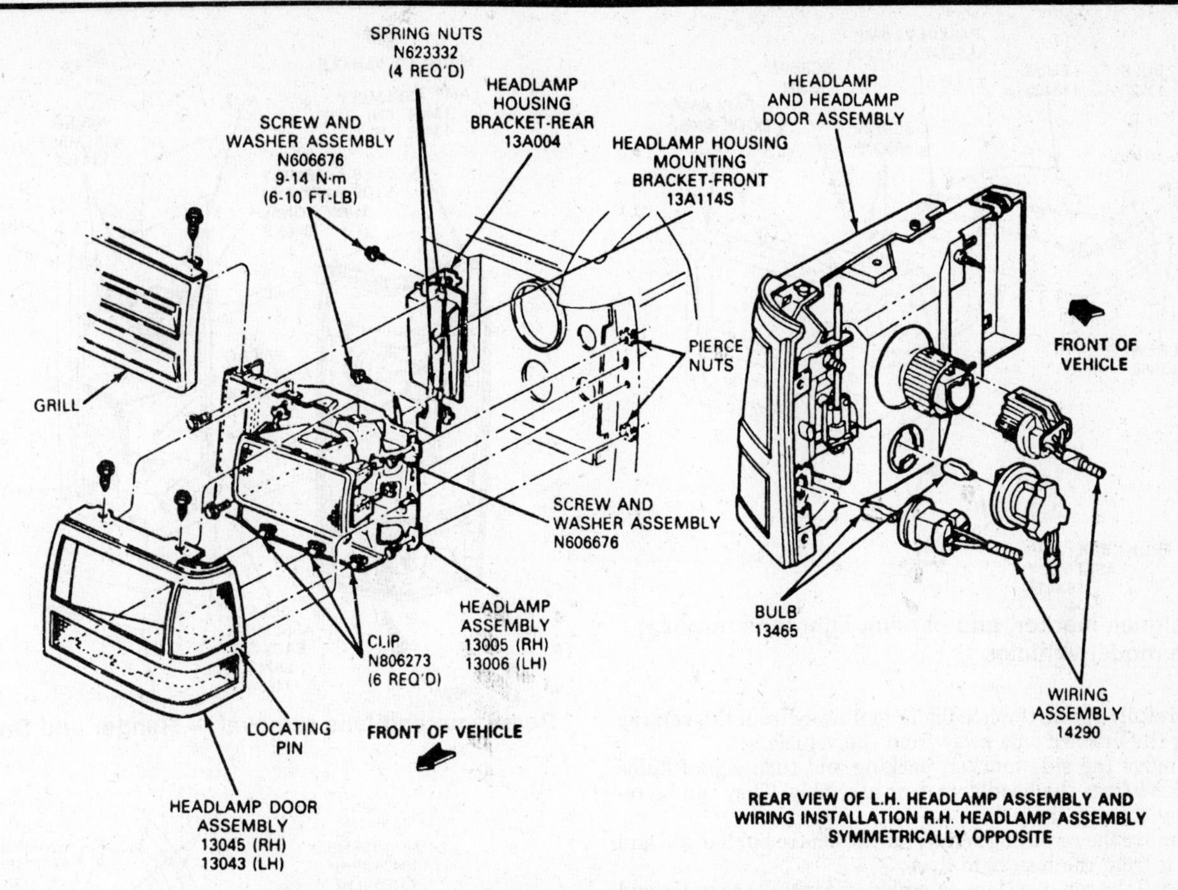

Headlight door and bulb removal – late model vehicles

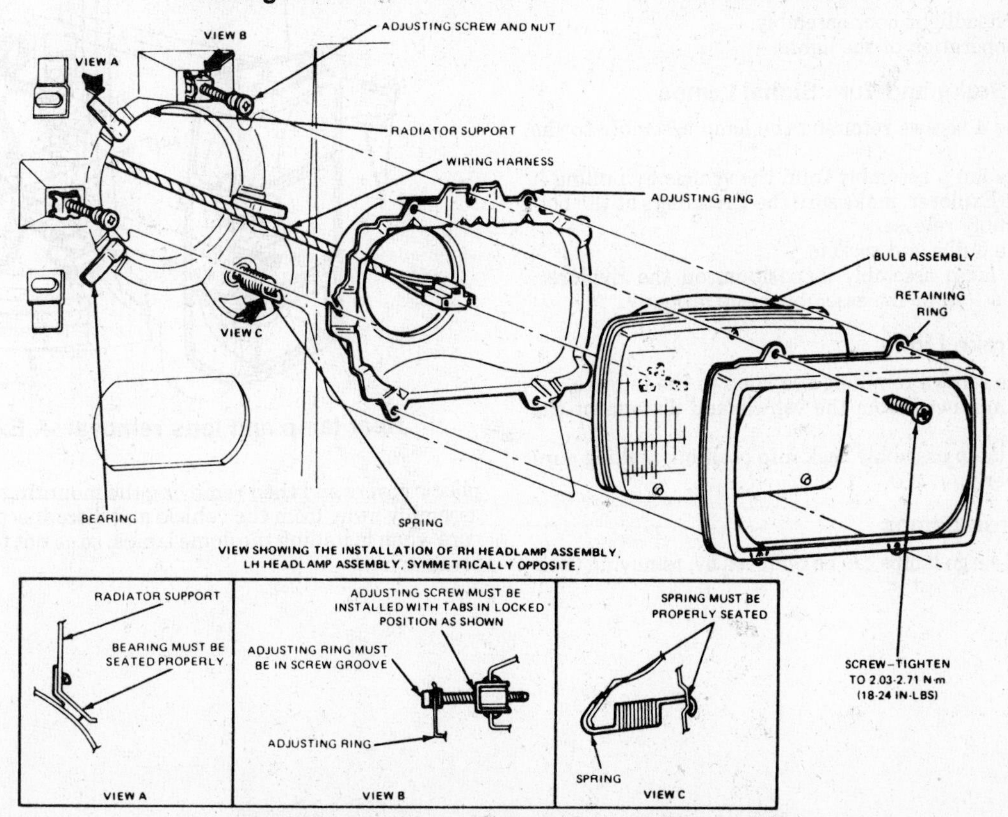

Headlight and bezel removal – earlier models

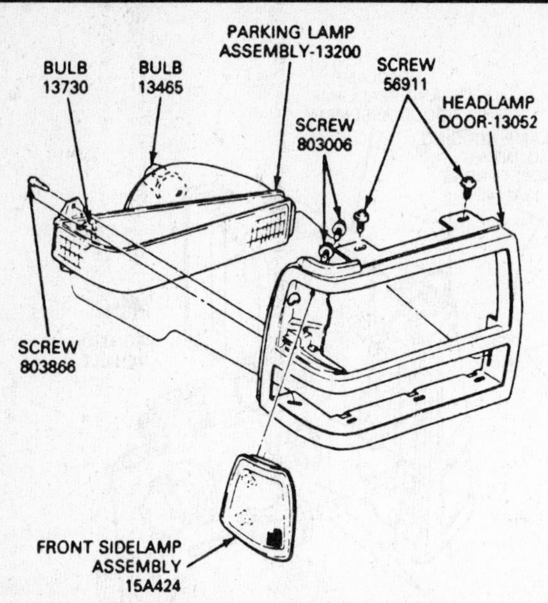

Turn signal, marker, and parking light lens removal – late model vehicles

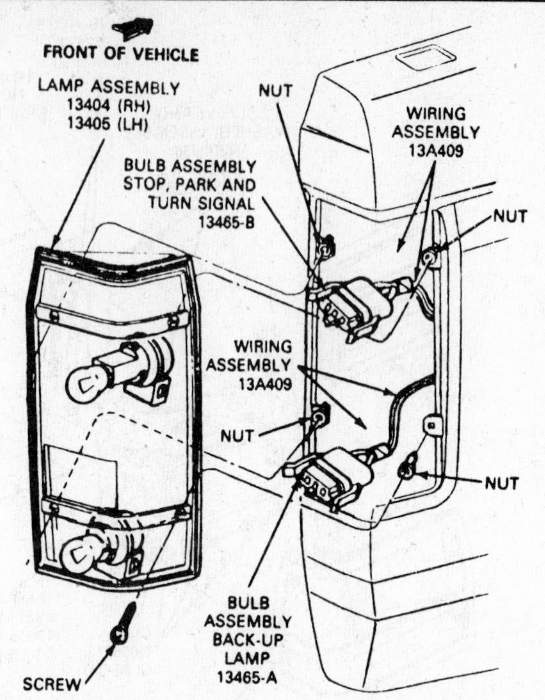

Rear lamp and lens removal – Ranger and Bronco II

2. Carefully rotate the headlight door away from the vehicle, rotating the inboard side away from the vehicle.
3. Remove the side marker, parking and turn signal bulbs and sockets from the headlamp door assembly. They can be removed by turning them.
4. Remove the retaining screws for the lens to be changed and remove it from the headlight door.
5. Install the removed lens assembly and install the bulbs and sockets.
6. Install the headlight door assembly.
7. Check the operation of the lights.

Rear Marker, Brake and Turn Signal Lamps

1. Remove the 4 screws retaining the lamp assembly to the vehicle.
2. Remove the lamp assembly from the vehicle by pulling it outward. On the Explorer, make sure the 2 retainers at the bottom of the assembly release.
3. Remove the bulbs and sockets.
4. Install the lamp assembly in position, on the Explorer, make sure the bottom of the assembly seats properly.

High Mount Brake Light

1. Remove the screws retaining the lamp to the tailgate.
2. Pull the lamp away from the vehicle and disconnect the wiring connector.
3. Install the lamp assembly back into position, making sure it is seated properly.

Dome and Cargo Lamps

The dome and cargo lamps can be removed by, removing their plastic covers and then removing the mounting screws. Pull the assembly away from the vehicle and disconnect the wiring. Use care when installing the dome lamps, so as not to damage the interior trim.

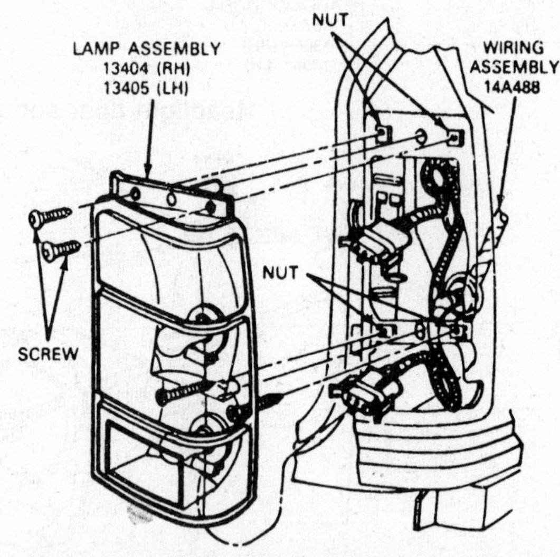

Rear lamp and lens removal – Explorer

TRAILER WIRING

Wiring the truck for towing is fairly easy. There are a number of good wiring kits available and these should be used, rather than trying to design your own. All trailers will need brake lights and turn signals as well as tail lights and side marker lights. Most states require extra marker lights for overly wide trailers. Also, most states have recently required back-up lights for trailers, and most trailer manufacturers have been building trailers with back-up lights for several years.

Additionally, some Class I, most Class II and just about all Class III trailers will have electric brakes.

Add to this number an accessories wire, to operate trailer internal equipment or to charge the trailer's battery, and you can have as many as seven wires in the harness.

Determine the equipment on your trailer and buy the wiring kit necessary. The kit will contain all the wires needed, plus a plug adapter set which included the female plug, mounted on the bumper or hitch, and the male plug, wired into, or plugged into the trailer harness.

When installing the kit, follow the manufacturer's instructions. The color coding of the wires is standard throughout the industry.

One point to note, some domestic vehicles, and most imported vehicles, have separate turn signals. On most domestic vehicles, the brake lights and rear turn signals operate with the same bulb. For those vehicles with separate turn signals, you can purchase an isolation unit so that the brake lights won't blink whenever the turn signals are operated, or, you can go to your local electronics supply house and buy four diodes to wire in series with the brake and turn signal bulbs. Diodes will isolate the brake and turn signals. The choice is yours. The isolation units are simple and quick to install, but far more expensive than the diodes. The diodes, however, require more work to install properly, since they require the cutting of each bulb's wire and soldering in place of the diode.

One final point, the best kits are those with a spring loaded cover on the vehicle mounted socket. This cover prevents dirt and moisture from corroding the terminals. Never let the vehicle socket hang loosely. Always mount it securely to the bumper or hitch.

CIRCUIT PROTECTION

Fusible Links

The fusible link is a short length of special, Hypalon (high temperature) insulated wire, integral with the engine compartment wiring harness and should not be confused with standard wire. It is several wire gauges smaller than the circuit which it protects. Under no circumstances should a fuse link replacement repair be made using a length of standard wire cut from bulk stock or from another wiring harness.

REPLACEMENT

To repair any blown fusible link use the following procedure:

1. Determine which circuit is damaged, its location and the cause of the open fusible link. If the damaged fuse link is one of three fed by a common No. 10 or 12 gauge feed wire, determine the specific affected circuit.
2. Disconnect the negative battery cable.
3. Cut the damaged fusible link from the wiring harness and discard it. If the fusible link is one of three circuits fed by a single feed wire, cut it out of the harness at each splice end and discard it.
4. Identify and procure the proper fusible link and butt connectors for attaching the fusible link to the harness.
5. To repair any fusible link in a 3–link group with one feed:

   a. After cutting the open link out of the harness, cut each of the remaining undamaged fusible links close to the feed wire weld.

   b. Strip approximately ½ in. of insulation from the detached ends of the two good fusible links. Then insert two wire ends into one end of a butt connector and carefully push one stripped end of the replacement fuse link into the same end of the butt connector and crimp all three firmly together.

NOTE: Care must be taken when fitting the three fusible links into the butt connector as the internal diameter is a snug fit for three wires. Make sure to use a proper crimping tool. Pliers, side cutters, etc. will not apply the proper crimp to retain the wires and withstand a pull test.

   c. After crimping the butt connector to the three fusible links, cut the weld portion from the feed wire and strip approximately ½ in. of insulation from the cut end. Insert the stripped end into the open end of the butt connector and crimp very firmly.

   d. To attach the remaining end of the replacement fusible link, strip approximately ½ in. of insulation from the wire end of the circuit from which the blow fusible link was removed, and firmly crimp a butt connector or equivalent to the stripped wire. Then, insert the end of the replacement link into the other end of the connector and crimp firmly.

   e. Using rosin core solder with a consistency of 60 percent tin and 40 percent lead, solder the connectors and the wires at the repairs and insulate with electrical tape.

7. To repair any fusible link which has an eyelet terminal on one end such as the charging circuit, cut off the open fusible link behind the weld, strip approximately ½ in. of insulation from the cut end and attach the appropriate new eyelet fusible link to

the cut stripped wire with an appropriate size butt connector. Solder the connectors and wires at the repair and insulate with tape.

8. Connect the negative battery cable to the battery and test the system for proper operation.

NOTE: Do not mistake a resistor wire for a fusible link. The resistor wire is generally longer and has print stating, "Resistor–don't cut or splice".

When attaching a single No. 16, 17, 18 or 20 gauge fusible link to a heavy gauge wire, always double the stripped wire end of the fusible link before inserting and crimping it into the butt connector for positive wire retention.

Turn Signal and Hazard Flasher Locations

Both the turn signal flasher and the hazard warning flasher are mounted on the fuse panel on the truck. To gain access to the fuse panel,remove the cover from the lower edge of the instrument panel below the steering column. First remove the two fasteners from the lower edge of the cover. Then pull the cover downward until the spring clips disengage from the instrument panel.

The turn signal flasher unit is mounted on the front of the fuse panel, and the hazard warning flasher is mounted on the rear of the fuse panel.

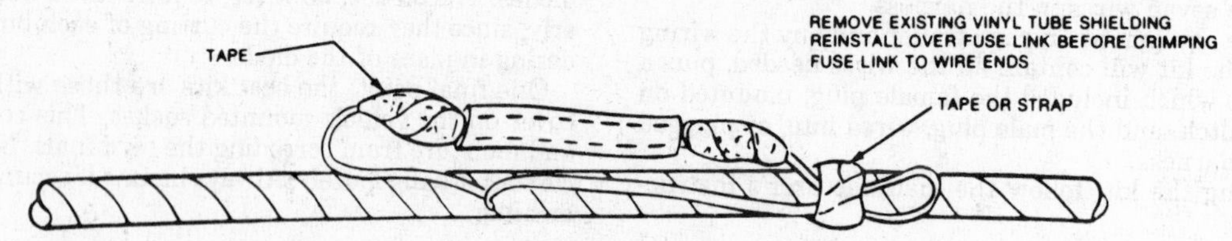

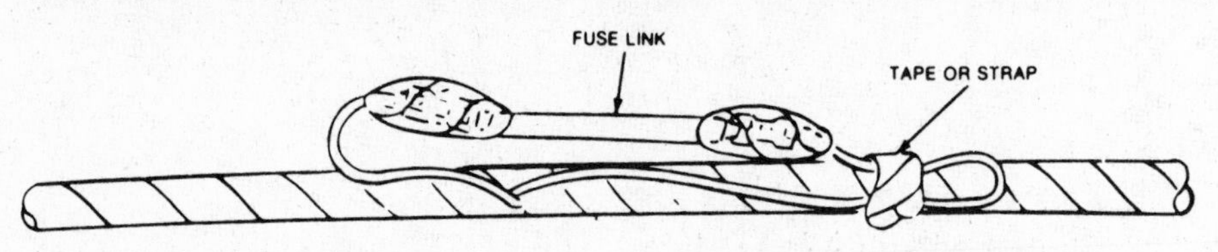

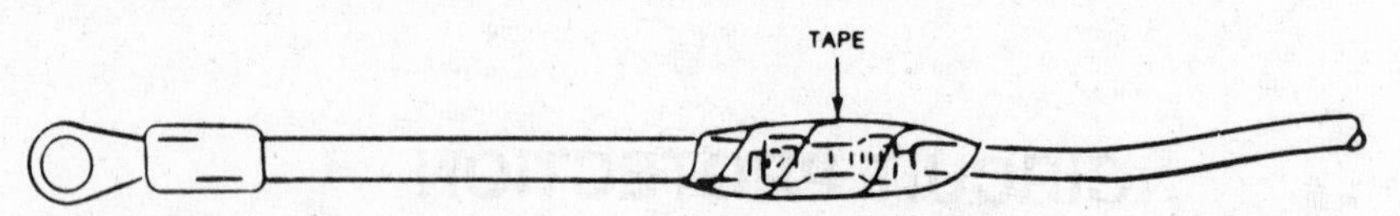

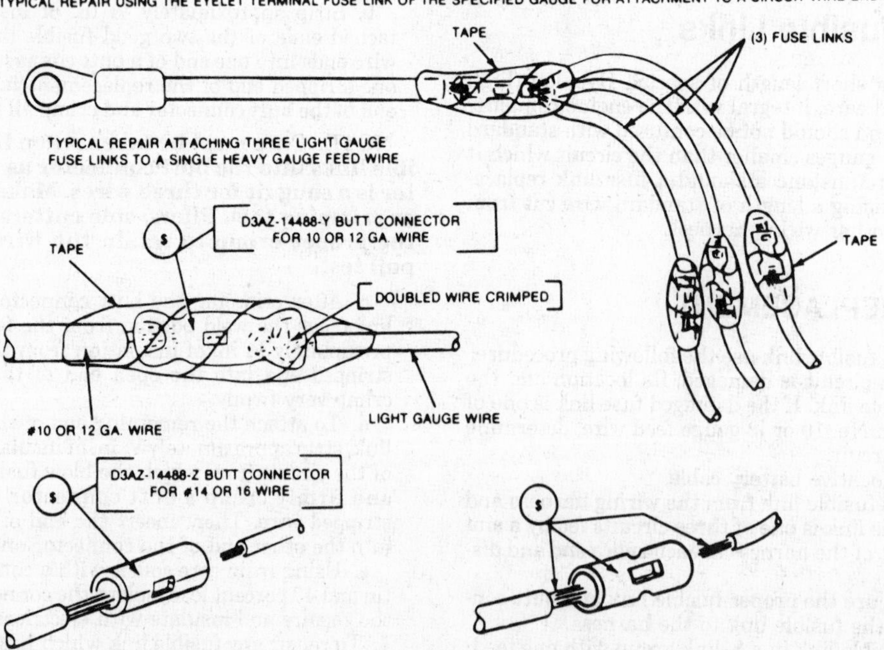

Fusible link repair

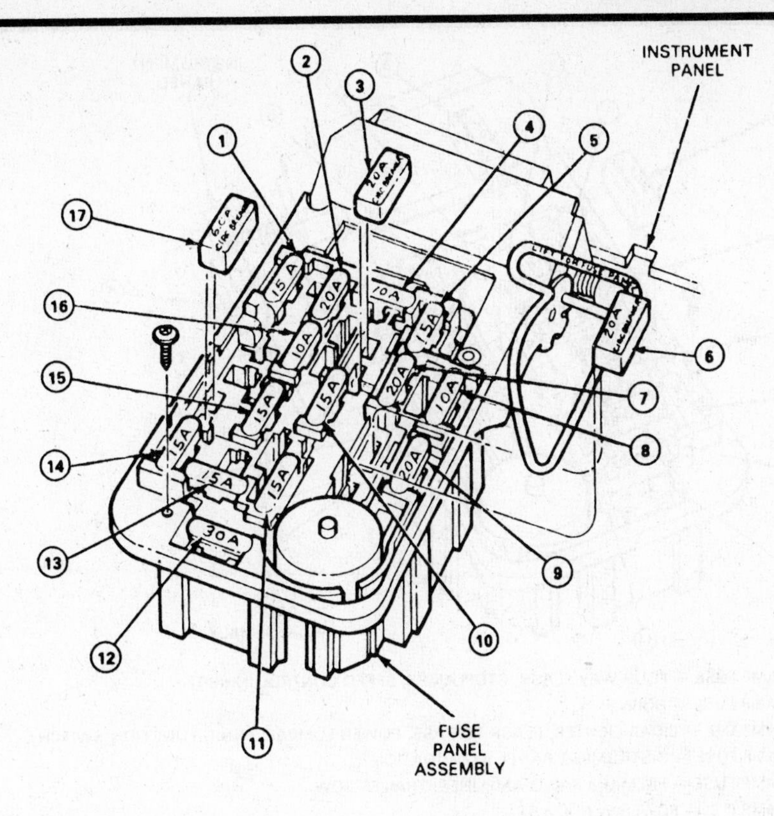

1. 15 Amp Fuse — Four-Way Flash, Stoplamps, Speed Control Inhibit
2. 20 Amp Fuse — Horns
3. 20 Amp C.B. — Cigar Lighter, Flash to Pass
4. 10 Amp Fuse — Instrument Panel Illumination, Park/H Lamp Hot
5. 20 Amp Fuse — Premium Radio Amplifier
6. 30 Amp C.B. — Power Windows, Lumbar
7. 20 Amp Fuse — R.A.B.S. Module
8. 10 Amp Fuse — Hego Heater
9. 20 Amp Fuse — Cluster Warning Lamps, Electronic All Wheel Drive
10. 15 Amp Fuse — Speed Control Amplifier, Radio
11. 15 Amp Fuse — Park/License Lamps
12. 30 Amp Fuse — Blower Motor
13. 15 Amp Fuse — Turn Lamps, B/U Lamps, Turn Indicator, R. Def. Control
14. 15 Amp Fuse — Dome/Courtesy Lamp
15. 15 Amp Fuse — Rear Window Wash/Wipe
16. 10 Amp Fuse — Air Conditioning Switches, Clutch Coil
17. 6 Amp C.B. — Front Wash/Wipe

Fuse panel and fuse identification – Ranger and Bronco II. The fuse location can vary with year and model

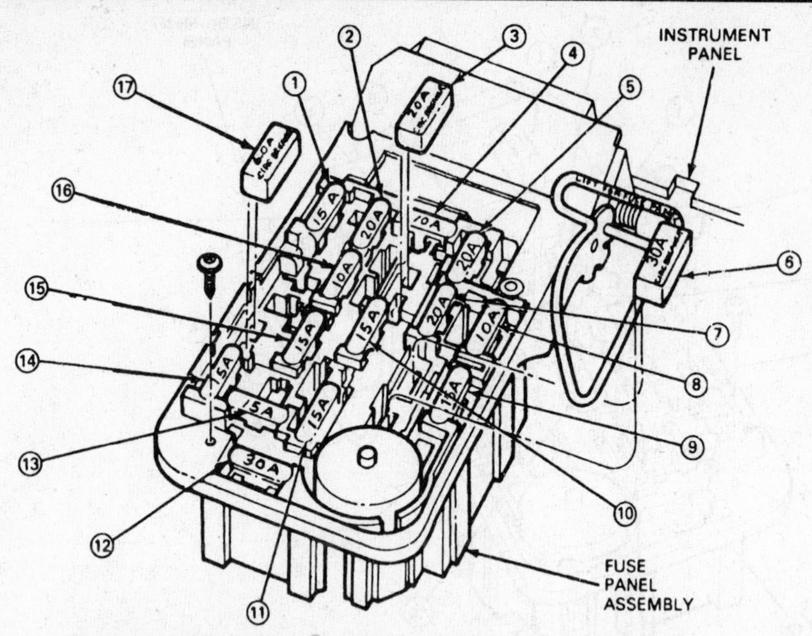

1. 15 AMP FUSE — FOUR-WAY FLASH, STOPLAMPS, SPEED CONTROL INHIBIT
2. 20 AMP FUSE — HORNS
3. 20 AMP C.B. — CIGAR LIGHTER, FLASH TO PASS, POWER LUMBAR, MULTI-FUNCTION SWITCH
4. 10 AMP FUSE — INSTRUMENT PANEL ILLUMINATION
5. 20 AMP FUSE — PREMIUM RADIO AMPLIFIER, TRAILER TOW
6. 30 AMP C.B. — POWER WINDOWS
7. 20 AMP FUSE — R.A.B.S. MODULE
8. 10 AMP FUSE — HEGO HEATER
9. 15 AMP FUSE — CLUSTER WARNING LAMPS, ELECTRONIC ALL WHEEL DRIVE
10. 15 AMP FUSE — SPEED CONTROL AMPLIFIER, RADIO
11. 15 AMP FUSE — PARK/LICENSE LAMPS
12. 30 AMP FUSE — BLOWER MOTOR
13. 15 AMP FUSE — TURN LAMPS, B/U LAMPS, TURN INDICATOR, R. DEF. CONTROL
14. 15 AMP FUSE — DOME/COURTESY LAMP
15. 15 AMP FUSE — REAR WINDOW WASH/WIPE
16. 10 AMP FUSE — AIR CONDITIONING SWITCHES, CLUTCH COIL
17. 6 AMP C.B. — FRONT WASH/WIPE

Fuse panel and fuse identification – Explorer

Troubleshooting Basic Turn Signal and Flasher Problems

Most problems in the turn signals or flasher system can be reduced to defective flashers or bulbs, which are easily replaced. Occasionally, problems in the turn signals are traced to the switch in the steering column, which will require professional service.

F = Front R = Rear • = Lights off o = Lights on

Problem	Solution
Turn signals light, but do not flash	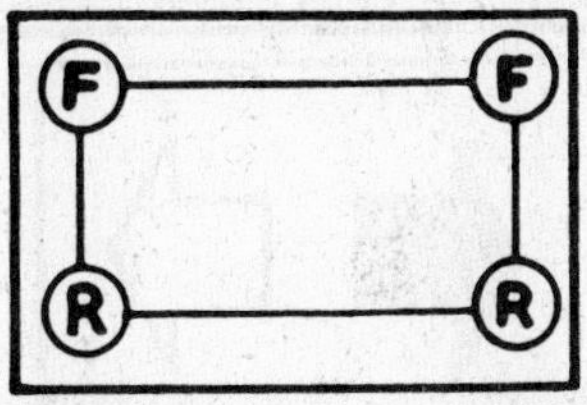 • Replace the flasher
No turn signals light on either side	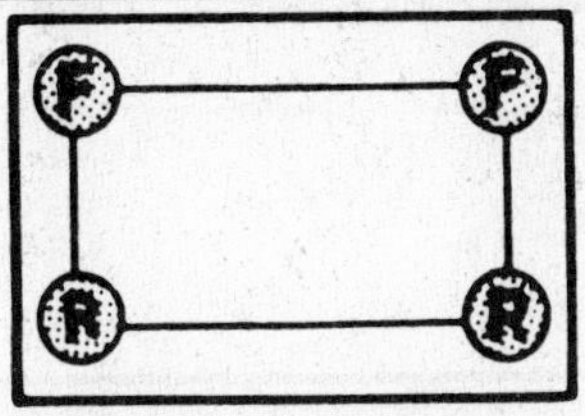 • Check the fuse. Replace if defective. • Check the flasher by substitution • Check for open circuit, short circuit or poor ground
Both turn signals on one side don't work	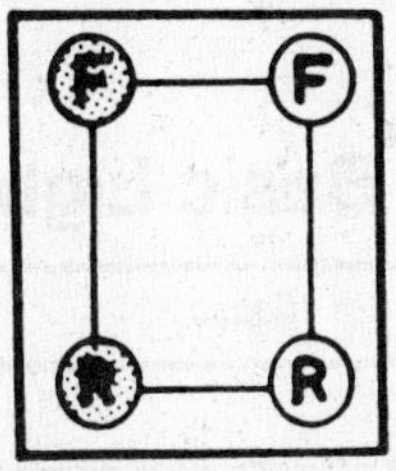 • Check for bad bulbs • Check for bad ground in both housings
One turn signal light on one side doesn't work	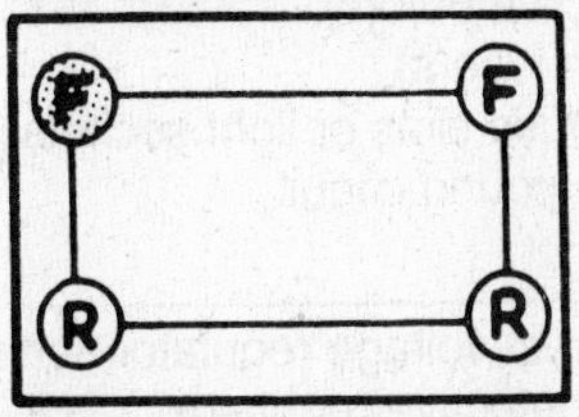 • Check and/or replace bulb • Check for corrosion in socket. Clean contacts. • Check for poor ground at socket
Turn signal flashes too fast or too slow	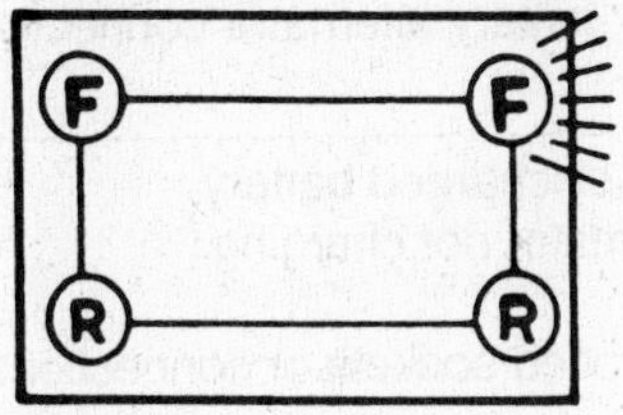 • Check any bulb on the side flashing too fast. A heavy-duty bulb is probably installed in place of a regular bulb. • Check the bulb flashing too slow. A standard bulb was probably installed in place of a heavy-duty bulb. • Check for loose connections or corrosion at the bulb socket
Indicator lights don't work in either direction	 • Check if the turn signals are working • Check the dash indicator lights • Check the flasher by substitution

Troubleshooting Basic Turn Signal and Flasher Problems

Most problems in the turn signals or flasher system can be reduced to defective flashers or bulbs, which are easily replaced. Occasionally, problems in the turn signals are traced to the switch in the steering column, which will require professional service.

F = Front R = Rear ● = Lights off ○ = Lights on

Problem		Solution
One indicator light doesn't light	(diagram: left arrow ●, right arrow ○)	• On systems with 1 dash indicator: See if the lights work on the same side. Often the filaments have been reversed in systems combining stoplights with taillights and turn signals. Check the flasher by substitution • On systems with 2 indicators: Check the bulbs on the same side Check the indicator light bulb Check the flasher by substitution

Troubleshooting Basic Lighting Problems

Problem	Cause	Solution
Lights		
One or more lights don't work, but others do	• Defective bulb(s) • Blown fuse(s) • Dirty fuse clips or light sockets • Poor ground circuit	• Replace bulb(s) • Replace fuse(s) • Clean connections • Run ground wire from light socket housing to car frame
Lights burn out quickly	• Incorrect voltage regulator setting or defective regulator • Poor battery/alternator connections	• Replace voltage regulator • Check battery/alternator connections
Lights go dim	• Low/discharged battery • Alternator not charging • Corroded sockets or connections • Low voltage output	• Check battery • Check drive belt tension; repair or replace alternator • Clean bulb and socket contacts and connections • Replace voltage regulator
Lights flicker	• Loose connection • Poor ground • Circuit breaker operating (short circuit)	• Tighten all connections • Run ground wire from light housing to car frame • Check connections and look for bare wires
Lights "flare"—Some flare is normal on acceleration—if excessive, see "Lights Burn Out Quickly"	• High voltage setting	• Replace voltage regulator

Troubleshooting Basic Lighting Problems

Problem	Cause	Solution
Lights glare—approaching drivers are blinded	• Lights adjusted too high • Rear springs or shocks sagging • Rear tires soft	• Have headlights aimed • Check rear springs/shocks • Check/correct rear tire pressure
Turn Signals		
Turn signals don't work in either direction	• Blown fuse • Defective flasher • Loose connection	• Replace fuse • Replace flasher • Check/tighten all connections
Right (or left) turn signal only won't work	• Bulb burned out • Right (or left) indicator bulb burned out • Short circuit	• Replace bulb • Check/replace indicator bulb • Check/repair wiring
Flasher rate too slow or too fast	• Incorrect wattage bulb • Incorrect flasher	• Flasher bulb • Replace flasher (use a variable load flasher if you pull a trailer)
Indicator lights do not flash (burn steadily)	• Burned out bulb • Defective flasher	• Replace bulb • Replace flasher
Indicator lights do not light at all	• Burned out indicator bulb • Defective flasher	• Replace indicator bulb • Replace flasher

Troubleshooting Basic Dash Gauge Problems

Problem	Cause	Solution
Coolant Temperature Gauge		
Gauge reads erratically or not at all	• Loose or dirty connections • Defective sending unit • Defective gauge	• Clean/tighten connections • Bi-metal gauge: remove the wire from the sending unit. Ground the wire for an instant. If the gauge registers, replace the sending unit. • Magnetic gauge: disconnect the wire at the sending unit. With ignition ON gauge should register COLD. Ground the wire; gauge should register HOT.
Ammeter Gauge—Turn Headlights ON (do not start engine). Note reaction		
Ammeter shows charge	• Connections reversed on gauge	• Reinstall connections
Ammeter shows discharge	• Ammeter is OK	• Nothing
Ammeter does not move	• Loose connections or faulty wiring • Defective gauge	• Check/correct wiring • Replace gauge

Troubleshooting Basic Dash Gauge Problems

Problem	Cause	Solution
Oil Pressure Gauge		
Gauge does not register or is inaccurate	• On mechanical gauge, Bourdon tube may be bent or kinked	• Check tube for kinks or bends preventing oil from reaching the gauge
	• Low oil pressure	• Remove sending unit. Idle the engine briefly. If no oil flows from sending unit hole, problem is in engine.
	• Defective gauge	• Remove the wire from the sending unit and ground it for an instant with the ignition ON. A good gauge will go to the top of the scale.
	• Defective wiring	• Check the wiring to the gauge. If it's OK and the gauge doesn't register when grounded, replace the gauge.
	• Defective sending unit	• If the wiring is OK and the gauge functions when grounded, replace the sending unit
All Gauges		
All gauges do not operate	• Blown fuse • Defective instrument regulator	• Replace fuse • Replace instrument voltage regulator
All gauges read low or erratically	• Defective or dirty instrument voltage regulator	• Clean contacts or replace
All gauges pegged	• Loss of ground between instrument voltage regulator and car • Defective instrument regulator	• Check ground • Replace regulator
Warning Lights		
Light(s) do not come on when ignition is ON, but engine is not started	• Defective bulb • Defective wire • Defective sending unit	• Replace bulb • Check wire from light to sending unit • Disconnect the wire from the sending unit and ground it. Replace the sending unit if the light comes on with the ignition ON.
Light comes on with engine running	• Problem in individual system • Defective sending unit	• Check system • Check sending unit (see above)

Troubleshooting the Heater

Problem	Cause	Solution
Blower motor will not turn at any speed	• Blown fuse • Loose connection • Defective ground • Faulty switch • Faulty motor • Faulty resistor	• Replace fuse • Inspect and tighten • Clean and tighten • Replace switch • Replace motor • Replace resistor
Blower motor turns at one speed only	• Faulty switch • Faulty resistor	• Replace switch • Replace resistor
Blower motor turns but does not circulate air	• Intake blocked • Fan not secured to the motor shaft	• Clean intake • Tighten security
Heater will not heat	• Coolant does not reach proper temperature • Heater core blocked internally • Heater core air-bound • Blend-air door not in proper position	• Check and replace thermostat if necessary • Flush or replace core if necessary • Purge air from core • Adjust cable
Heater will not defrost	• Control cable adjustment incorrect • Defroster hose damaged	• Adjust control cable • Replace defroster hose

Troubleshooting Basic Windshield Wiper Problems

Problem	Cause	Solution
Electric Wipers		
Wipers do not operate—Wiper motor heats up or hums	• Internal motor defect • Bent or damaged linkage • Arms improperly installed on linking pivots	• Replace motor • Repair or replace linkage • Position linkage in park and reinstall wiper arms
Electric Wipers		
Wipers do not operate—No current to motor	• Fuse or circuit breaker blown • Loose, open or broken wiring • Defective switch • Defective or corroded terminals • No ground circuit for motor or switch	• Replace fuse or circuit breaker • Repair wiring and connections • Replace switch • Replace or clean terminals • Repair ground circuits
Wipers do not operate—Motor runs	• Linkage disconnected or broken	• Connect wiper linkage or replace broken linkage
Vacuum Wipers		

Troubleshooting Basic Windshield Wiper Problems

Problem	Cause	Solution
Wipers do not operate	• Control switch or cable inoperative • Loss of engine vacuum to wiper motor (broken hoses, low engine vacuum, defective vacuum/fuel pump) • Linkage broken or disconnected • Defective wiper motor	• Repair or replace switch or cable • Check vacuum lines, engine vacuum and fuel pump • Repair linkage • Replace wiper motor
Wipers stop on engine acceleration	• Leaking vacuum hoses • Dry windshield • Oversize wiper blades • Defective vacuum/fuel pump	• Repair or replace hoses • Wet windshield with washers • Replace with proper size wiper blades • Replace pump

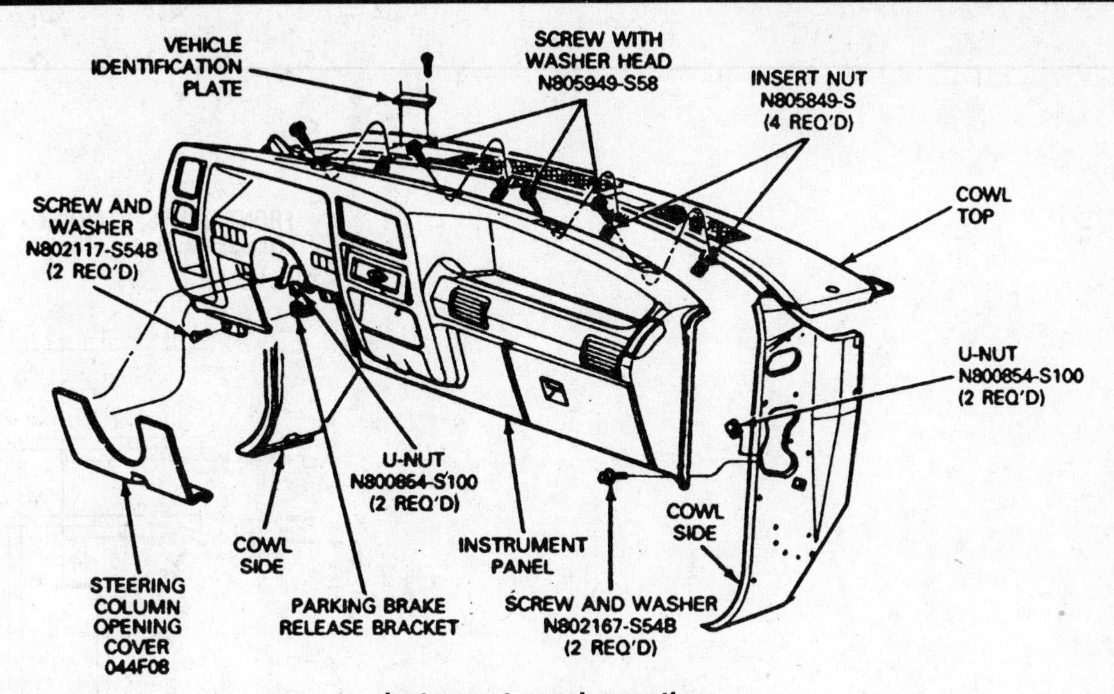

Instrument panel mounting

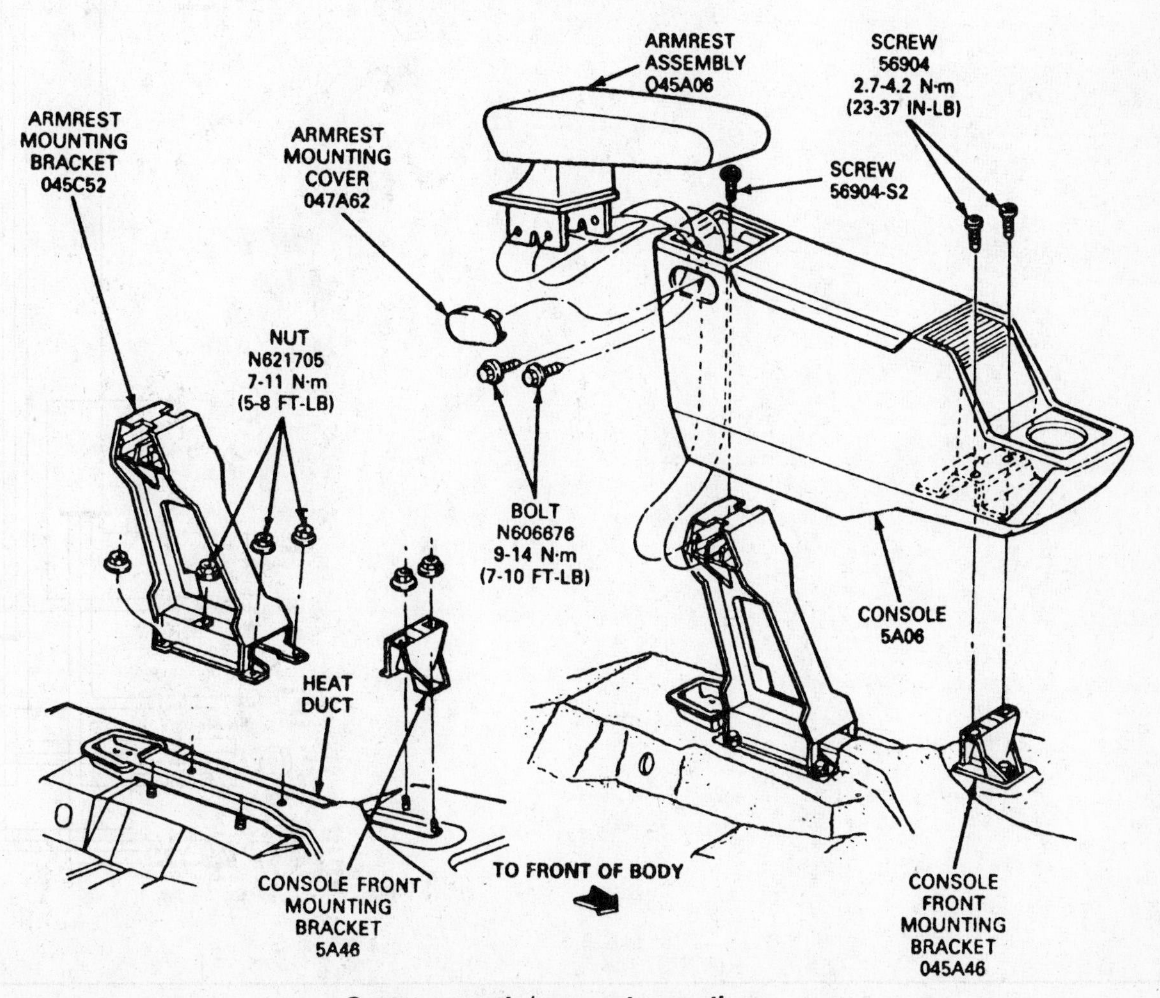

Center console/arm rest mounting

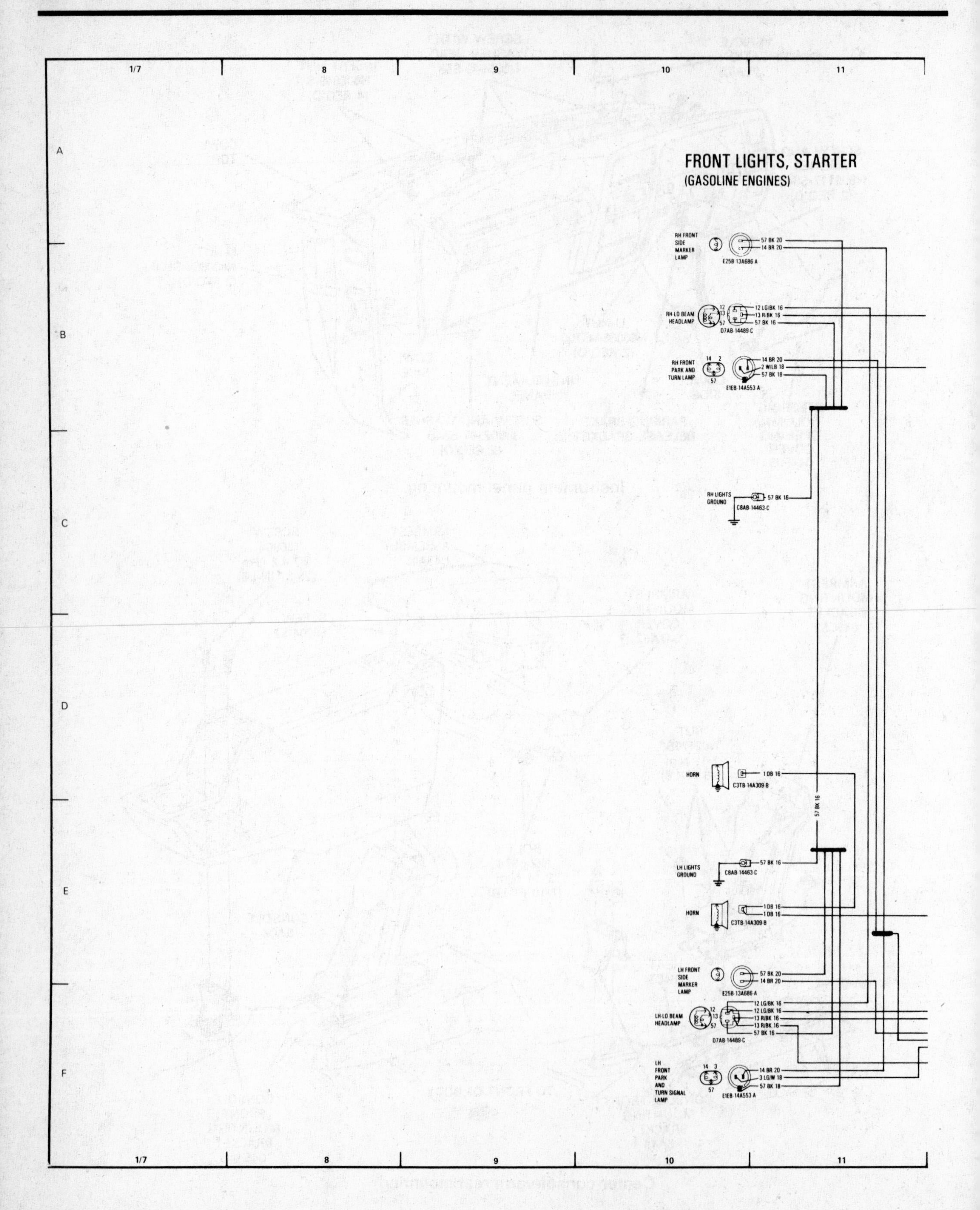
1/7
8
9
10
11
A
B
C
D
E
F
FRONT LIGHTS, STARTER
(GASOLINE ENGINES)
RH FRONT SIDE MARKER LAMP
57 BK 20
14 BR 20
E25B 13A686 A
RH LO BEAM HEADLAMP
12
13
57
12 LG/BK 16
13 R/BK 16
57 BK 16
D7AB 14489 C
RH FRONT PARK AND TURN LAMP
14
2
57
14 BR 20
2 W/LB 18
57 BK 18
E1EB 14A553 A
RH LIGHTS GROUND
57 BK 16
C8AB 14463 C
HORN
1 DB 16
C3TB 14A309 B
57 BK 16
LH LIGHTS GROUND
57 BK 16
C8AB 14463 C
HORN
1 DB 16
1 DB 16
C3TB 14A309 B
LH FRONT SIDE MARKER LAMP
57 BK 20
14 BR 20
E25B 13A686 A
LH LO BEAM HEADLAMP
12
13
57
12 LG/BK 16
12 LG/BK 16
13 R/BK 16
13 R/BK 16
57 BK 16
D7AB 14489 C
LH FRONT PARK AND TURN SIGNAL LAMP
14
3
57
14 BR 20
3 LG/W 18
57 BK 18
E1EB 14A553 A
1/7
8
9
10
11

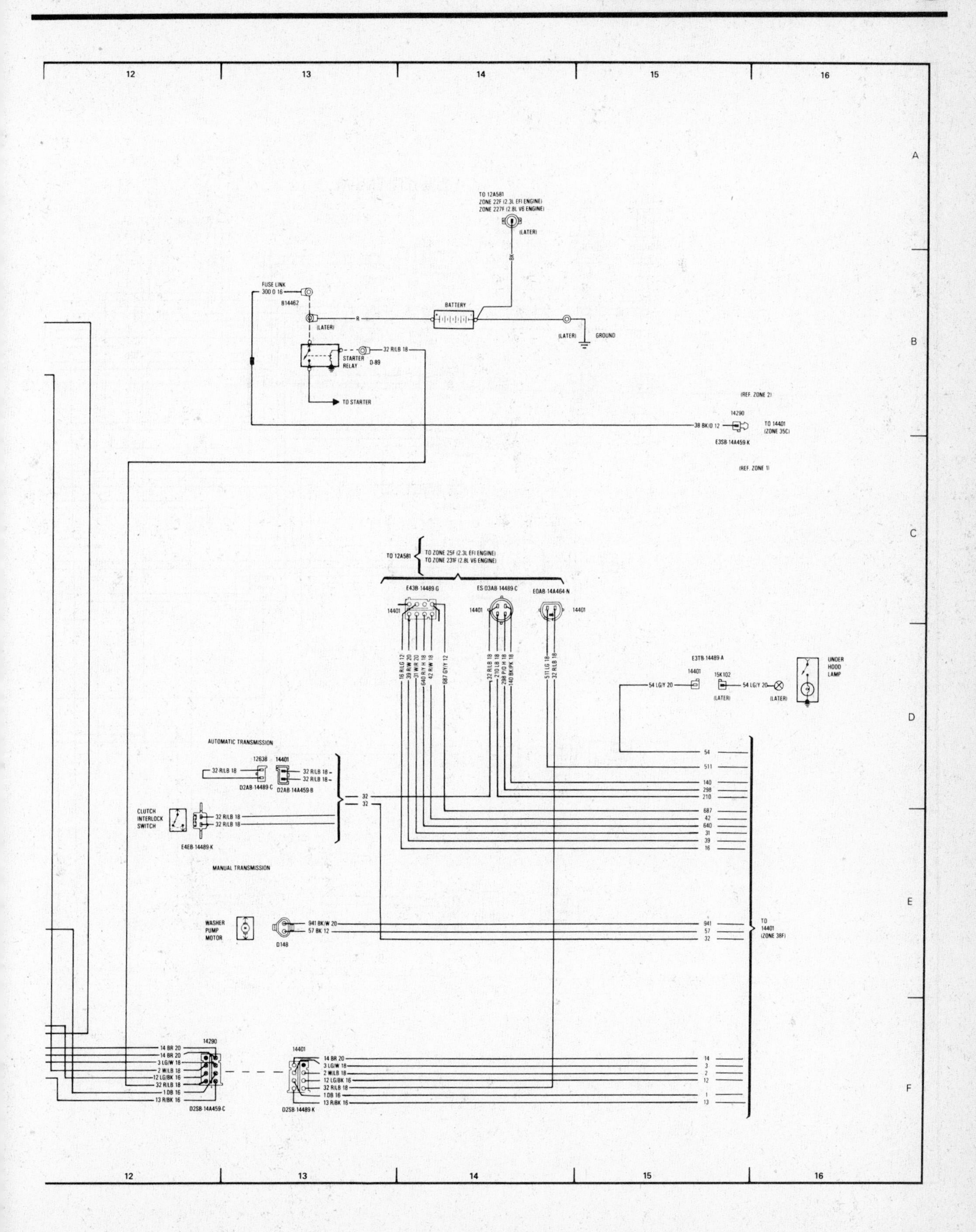
12
13
14
15
16
A
B
C
D
E
F
TO 12A581
ZONE 22F (2.3L EFI ENGINE)
ZONE 227F (2.8L V6 ENGINE)
(LATER)
BK
FUSE LINK
300 O 16
B14462
BATTERY
R
(LATER)
(LATER)
GROUND
32 R/LB 18
STARTER
RELAY
D-89
TO STARTER
(REF. ZONE 2)
14290
38 BK/O 12
TO 14401
(ZONE 35C)
E3SB-14A459-K
(REF. ZONE 1)
TO 12A581
TO ZONE 25F (2.3L EFI ENGINE)
TO ZONE 231F (2.8L V6 ENGINE)
E43B-14489-G
ES-D3AB-14489-C
E0AB-14A464-N
14401
14401
14401
16 R/LG 12
39 R/W 20
31 W/R 20
640 R/Y H 18
42 R/W 18
687 GY/Y 12
32 R/LB 18
210 LB 18
298 P/O H 18
140 BK/PK 18
511 LG 18
32 R/LB 18
E3TB-14489-A
14401
15K102
54 LG/Y 20
54 LG/Y 20
(LATER)
(LATER)
UNDER
HOOD
LAMP
AUTOMATIC TRANSMISSION
12638
14401
32 R/LB 18
32 R/LB 18
32 R/LB 18
D2AB-14489-C
D2AB-14A459-B
32
32
CLUTCH
INTERLOCK
SWITCH
32 R/LB 18
32 R/LB 18
E4EB-14489-K
MANUAL TRANSMISSION
54
511
140
298
210
687
42
640
31
39
16
WASHER
PUMP
MOTOR
941 BK/W 20
57 BK 12
D148
941
57
32
TO
14401
(ZONE 38F)
14290
14 BR 20
14 BR 20
3 LG/W 18
2 W/LB 18
12 LG/BK 16
32 R/LB 18
1 DB 16
13 R/BK 16
D2SB-14A459-C
14401
14 BR 20
3 LG/W 18
2 W/LB 18
12 LG/BK 16
32 R/LB 18
1 DB 16
13 R/BK 16
D2SB-14489-K
14
3
2
12
1
13

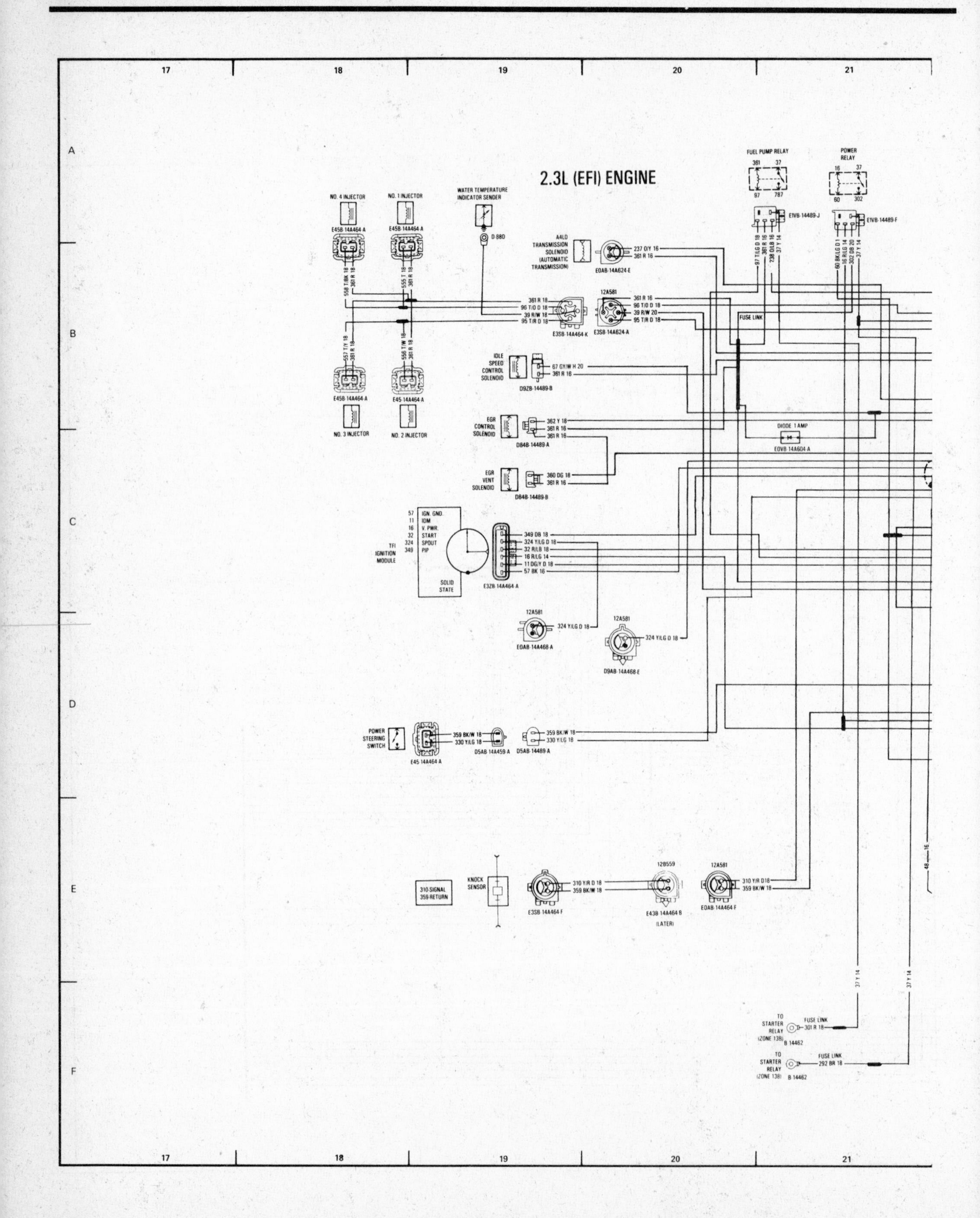
2.3L (EFI) ENGINE
FUEL PUMP RELAY
POWER RELAY
NO. 4 INJECTOR
NO. 1 INJECTOR
WATER TEMPERATURE INDICATOR SENDER
A4LD TRANSMISSION SOLENOID (AUTOMATIC TRANSMISSION)
FUSE LINK
IDLE SPEED CONTROL SOLENOID
NO. 3 INJECTOR
NO. 2 INJECTOR
EGR CONTROL SOLENOID
DIODE 1 AMP
EGR VENT SOLENOID
TFI IGNITION MODULE
IGN. GND.
IDM
V. PWR.
START
SPOUT
PIP
SOLID STATE
POWER STEERING SWITCH
KNOCK SENSOR
310-SIGNAL
359-RETURN
(LATER)
TO STARTER RELAY (ZONE 13B)

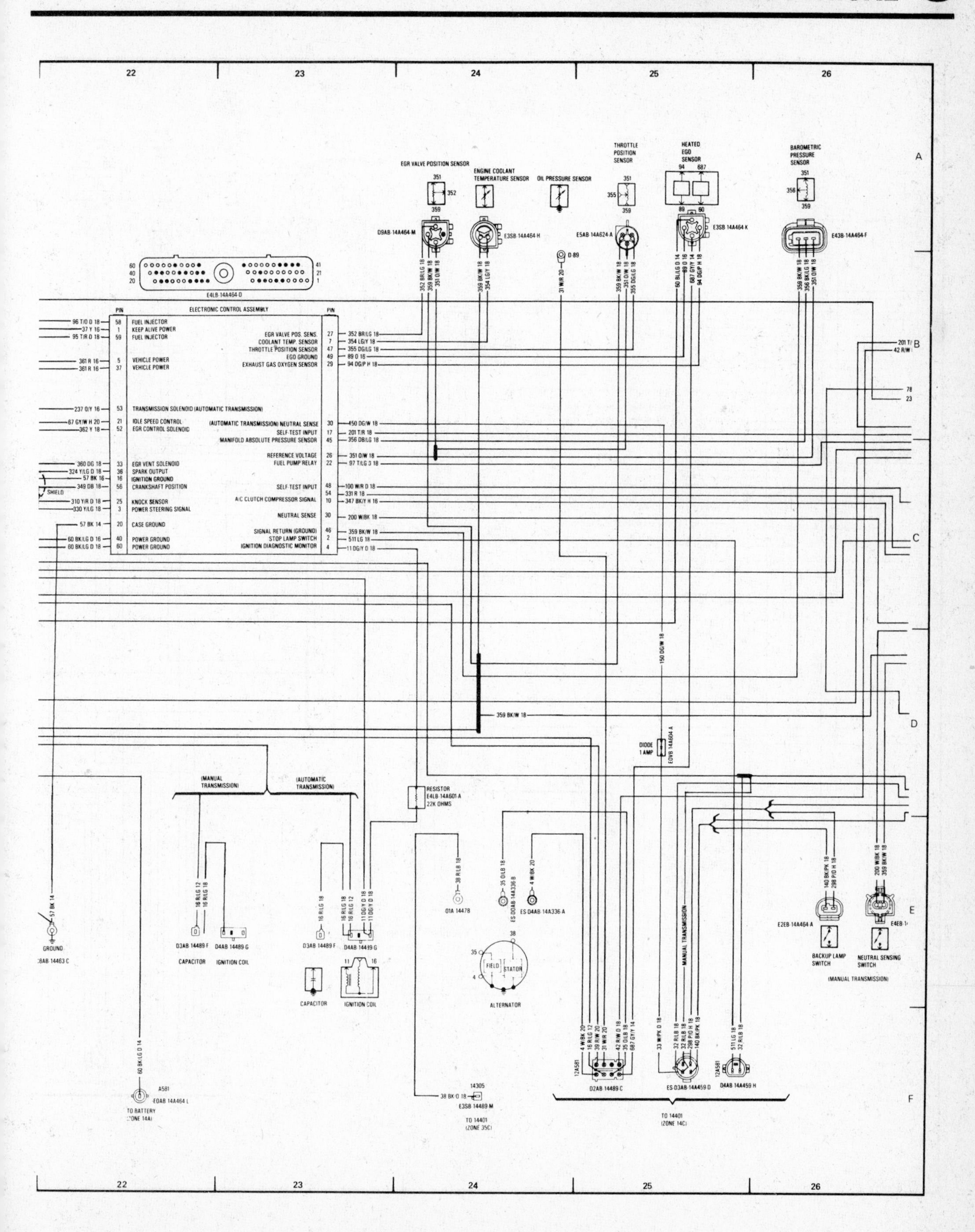
EGR VALVE POSITION SENSOR
ENGINE COOLANT TEMPERATURE SENSOR
OIL PRESSURE SENSOR
THROTTLE POSITION SENSOR
HEATED EGO SENSOR
BAROMETRIC PRESSURE SENSOR
ELECTRONIC CONTROL ASSEMBLY
FUEL INJECTOR
KEEP ALIVE POWER
VEHICLE POWER
TRANSMISSION SOLENOID (AUTOMATIC TRANSMISSION)
IDLE SPEED CONTROL
EGR CONTROL SOLENOID
EGR VENT SOLENOID
SPARK OUTPUT
IGNITION GROUND
CRANKSHAFT POSITION
KNOCK SENSOR
POWER STEERING SIGNAL
CASE GROUND
POWER GROUND
EGR VALVE POS. SENS.
COOLANT TEMP. SENSOR
THROTTLE POSITION SENSOR
EGO GROUND
EXHAUST GAS OXYGEN SENSOR
(AUTOMATIC TRANSMISSION) NEUTRAL SENSE
SELF-TEST INPUT
MANIFOLD ABSOLUTE PRESSURE SENSOR
REFERENCE VOLTAGE
FUEL PUMP RELAY
A/C CLUTCH COMPRESSOR SIGNAL
NEUTRAL SENSE
SIGNAL RETURN (GROUND)
STOP LAMP SWITCH
IGNITION DIAGNOSTIC MONITOR
SHIELD
GROUND
(MANUAL TRANSMISSION)
(AUTOMATIC TRANSMISSION)
CAPACITOR
IGNITION COIL
RESISTOR E4LB-14A601-A 22K OHMS
DIODE 1 AMP
FIELD
STATOR
ALTERNATOR
MANUAL TRANSMISSION
BACKUP LAMP SWITCH
NEUTRAL SENSING SWITCH
(MANUAL TRANSMISSION)
TO BATTERY (ZONE 14A)
TO 14401 (ZONE 35C)
TO 14401 (ZONE 14C)
359 BK/W 18
150 DG/W 18

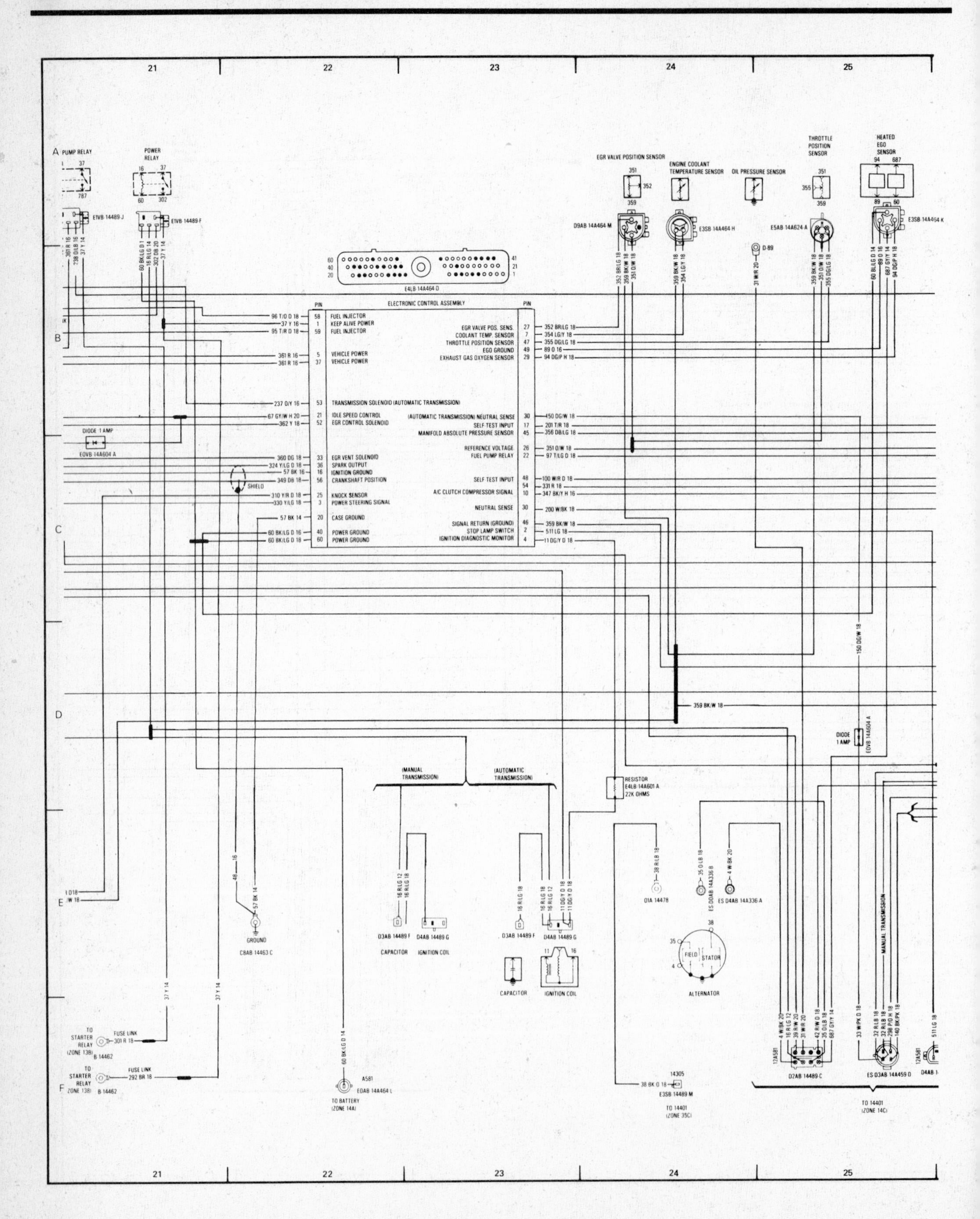

PUMP RELAY
POWER RELAY
EGR VALVE POSITION SENSOR
ENGINE COOLANT TEMPERATURE SENSOR
OIL PRESSURE SENSOR
THROTTLE POSITION SENSOR
HEATED EGO SENSOR
ELECTRONIC CONTROL ASSEMBLY
E4LB 14A464 D
FUEL INJECTOR
KEEP ALIVE POWER
FUEL INJECTOR
VEHICLE POWER
VEHICLE POWER
TRANSMISSION SOLENOID (AUTOMATIC TRANSMISSION)
IDLE SPEED CONTROL
EGR CONTROL SOLENOID
EGR VENT SOLENOID
SPARK OUTPUT
IGNITION GROUND
CRANKSHAFT POSITION
KNOCK SENSOR
POWER STEERING SIGNAL
CASE GROUND
POWER GROUND
POWER GROUND
EGR VALVE POS. SENS.
COOLANT TEMP. SENSOR
THROTTLE POSITION SENSOR
EGO GROUND
EXHAUST GAS OXYGEN SENSOR
(AUTOMATIC TRANSMISSION) NEUTRAL SENSE
SELF-TEST INPUT
MANIFOLD ABSOLUTE PRESSURE SENSOR
REFERENCE VOLTAGE
FUEL PUMP RELAY
SELF TEST INPUT
A/C CLUTCH COMPRESSOR SIGNAL
NEUTRAL SENSE
SIGNAL RETURN (GROUND)
STOP LAMP SWITCH
IGNITION DIAGNOSTIC MONITOR
DIODE 1 AMP
SHIELD
GROUND
C8AB 14463 C
(MANUAL TRANSMISSION)
(AUTOMATIC TRANSMISSION)
CAPACITOR
IGNITION COIL
RESISTOR E4LB-14A601 A 22K OHMS
ALTERNATOR
FIELD
STATOR
TO STARTER RELAY (ZONE 13B)
FUSE LINK
TO BATTERY (ZONE 14A)
TO 14401 (ZONE 35C)
TO 14401 (ZONE 14C)
MANUAL TRANSMISSION
359 BK/W 18
150 DG/W 18

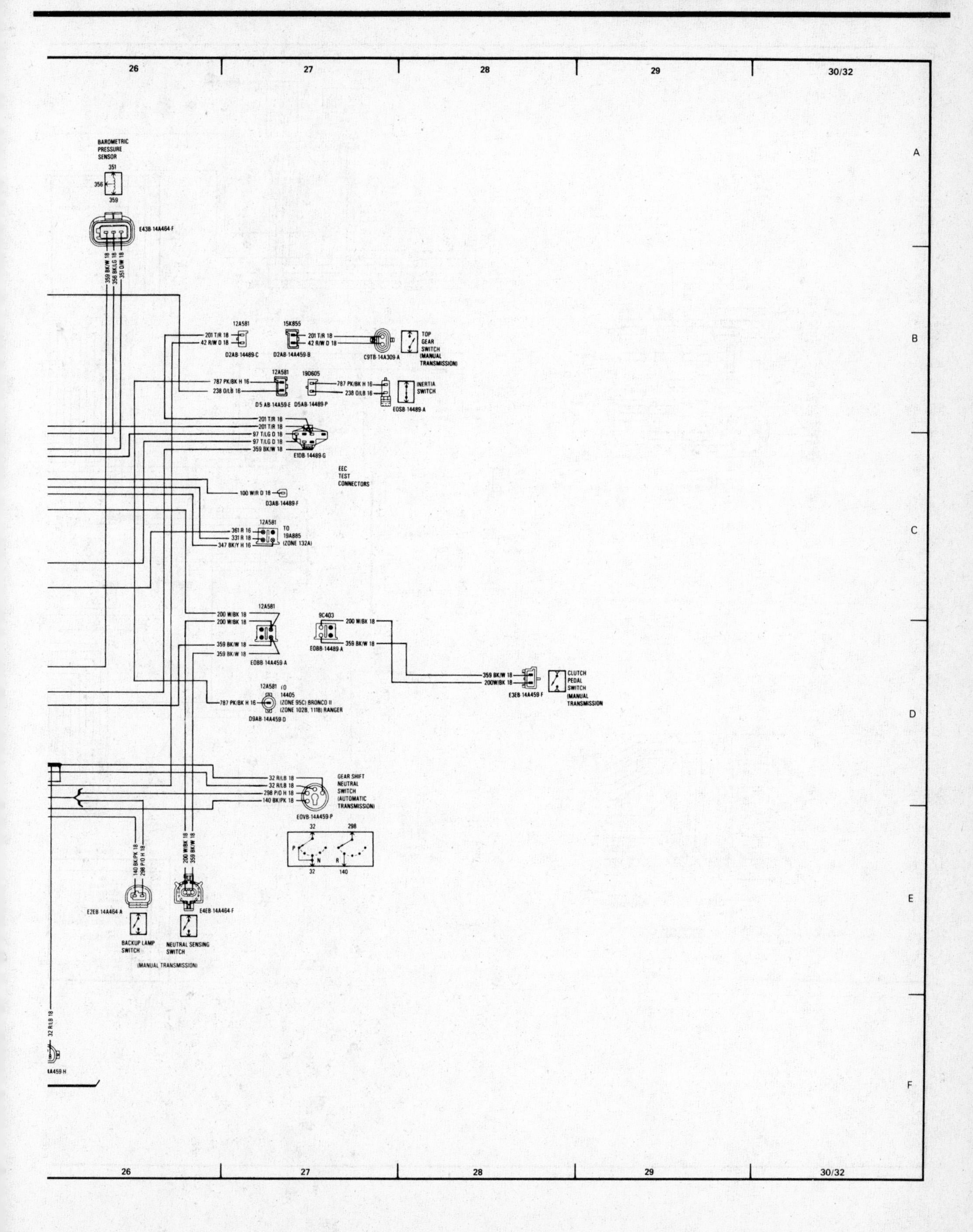

26
27
28
29
30/32
A
B
C
D
E
F
BAROMETRIC PRESSURE SENSOR
E43B-14A464-F
TOP GEAR SWITCH (MANUAL TRANSMISSION)
INERTIA SWITCH
EEC TEST CONNECTORS
CLUTCH PEDAL SWITCH (MANUAL TRANSMISSION)
GEAR SHIFT NEUTRAL SWITCH (AUTOMATIC TRANSMISSION)
E0VB-14A459-P
BACKUP LAMP SWITCH
NEUTRAL SENSING SWITCH
(MANUAL TRANSMISSION)
E2EB-14A464-A
E4EB-14A464-F

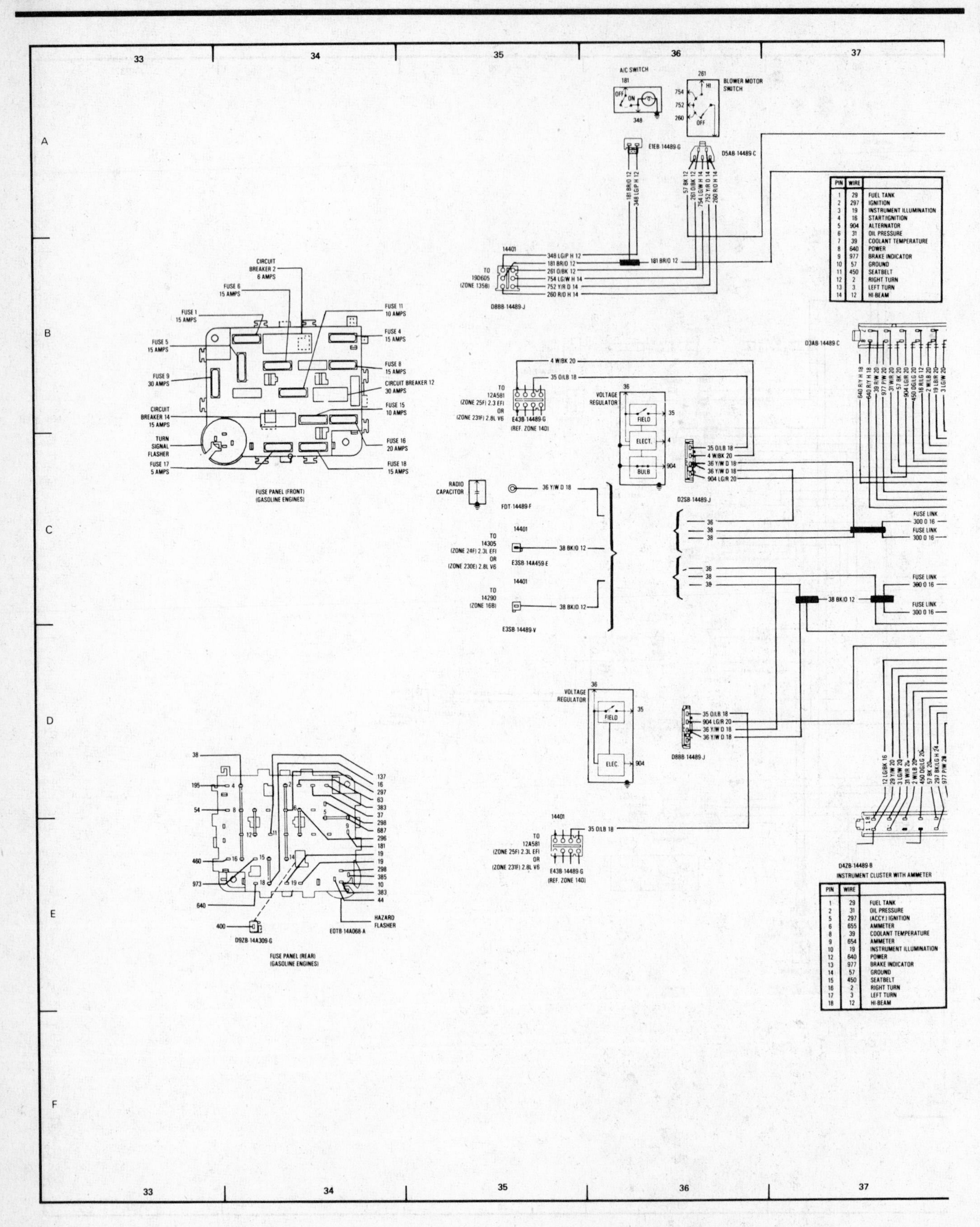

PIN	WIRE	
1	29	FUEL TANK
2	297	IGNITION
3	19	INSTRUMENT ILLUMINATION
4	16	START/IGNITION
5	904	ALTERNATOR
6	31	OIL PRESSURE
7	39	COOLANT TEMPERATURE
8	640	POWER
9	977	BRAKE INDICATOR
10	57	GROUND
11	450	SEATBELT
12	2	RIGHT TURN
13	3	LEFT TURN
14	12	HI-BEAM

D4ZB-14489-B
INSTRUMENT CLUSTER WITH AMMETER

PIN	WIRE	
1	29	FUEL TANK
2	31	OIL PRESSURE
5	297	(ACCY.) IGNITION
6	655	AMMETER
8	39	COOLANT TEMPERATURE
9	654	AMMETER
10	19	INSTRUMENT ILLUMINATION
12	640	POWER
13	977	BRAKE INDICATOR
14	57	GROUND
15	450	SEATBELT
16	2	RIGHT TURN
17	3	LEFT TURN
18	12	HI-BEAM

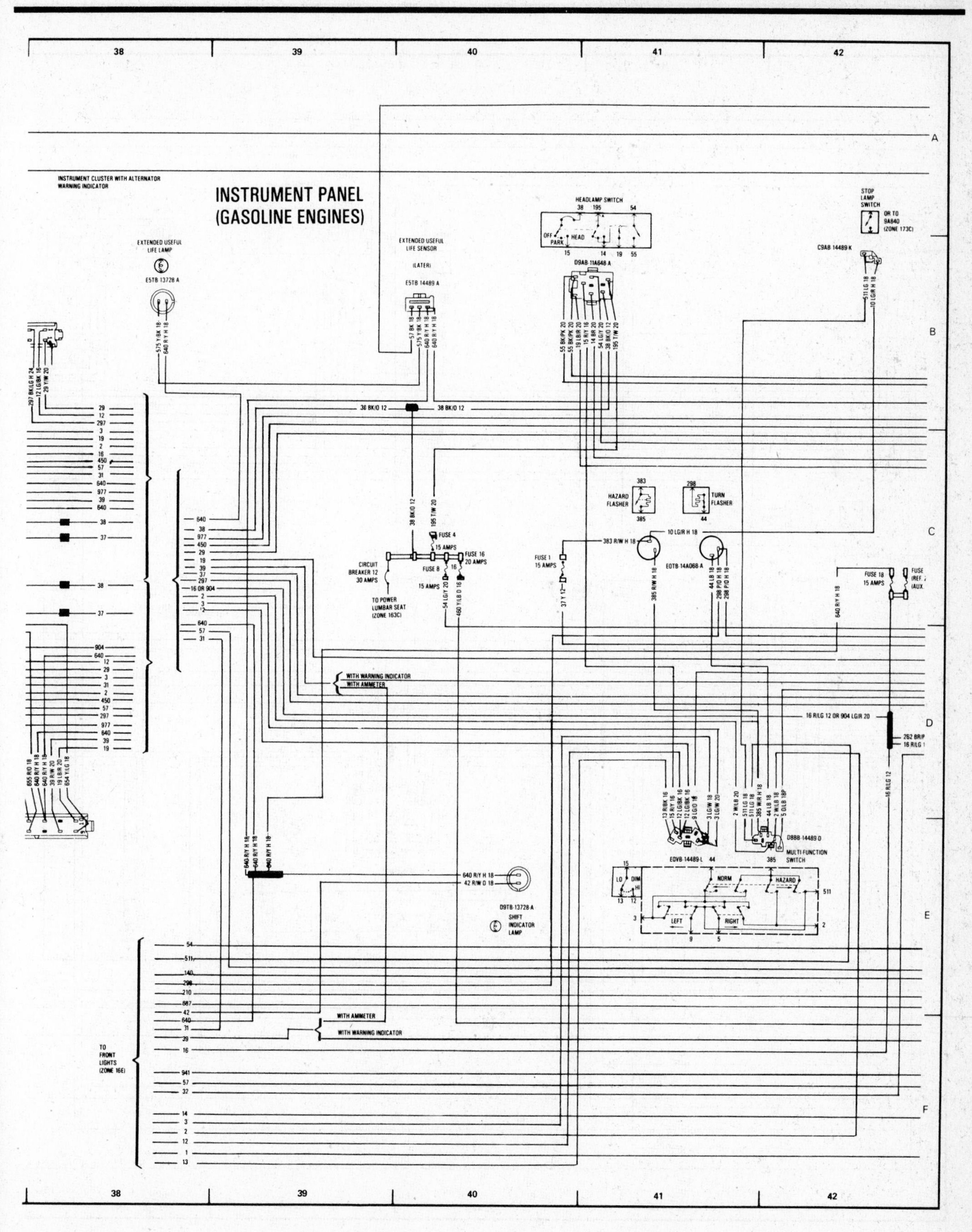
INSTRUMENT PANEL
(GASOLINE ENGINES)
INSTRUMENT CLUSTER WITH ALTERNATOR WARNING INDICATOR
EXTENDED USEFUL LIFE LAMP
E5TB 13728 A
EXTENDED USEFUL LIFE SENSOR
(LATER)
E5TB 14489 A
HEADLAMP SWITCH
OFF
PARK
HEAD
D9AB 11A648 A
STOP LAMP SWITCH
OR TO 9A840 (ZONE 173C)
C9AB 14489 K
HAZARD FLASHER
TURN FLASHER
FUSE 4
15 AMPS
FUSE 16
20 AMPS
FUSE 8
15 AMPS
CIRCUIT BREAKER 12
30 AMPS
TO POWER LUMBAR SEAT (ZONE 163C)
FUSE 1
15 AMPS
E0TB 14A068 A
FUSE 18
15 AMPS
WITH WARNING INDICATOR
WITH AMMETER
D8BB 14489 D
MULTI FUNCTION SWITCH
E0VB 14489 L
D9TB 13728 A
SHIFT INDICATOR LAMP
DIM
HI
LO
NORM
HAZARD
LEFT
RIGHT
TO FRONT LIGHTS (ZONE 16E)
38
39
40
41
42
A
B
C
D
E
F

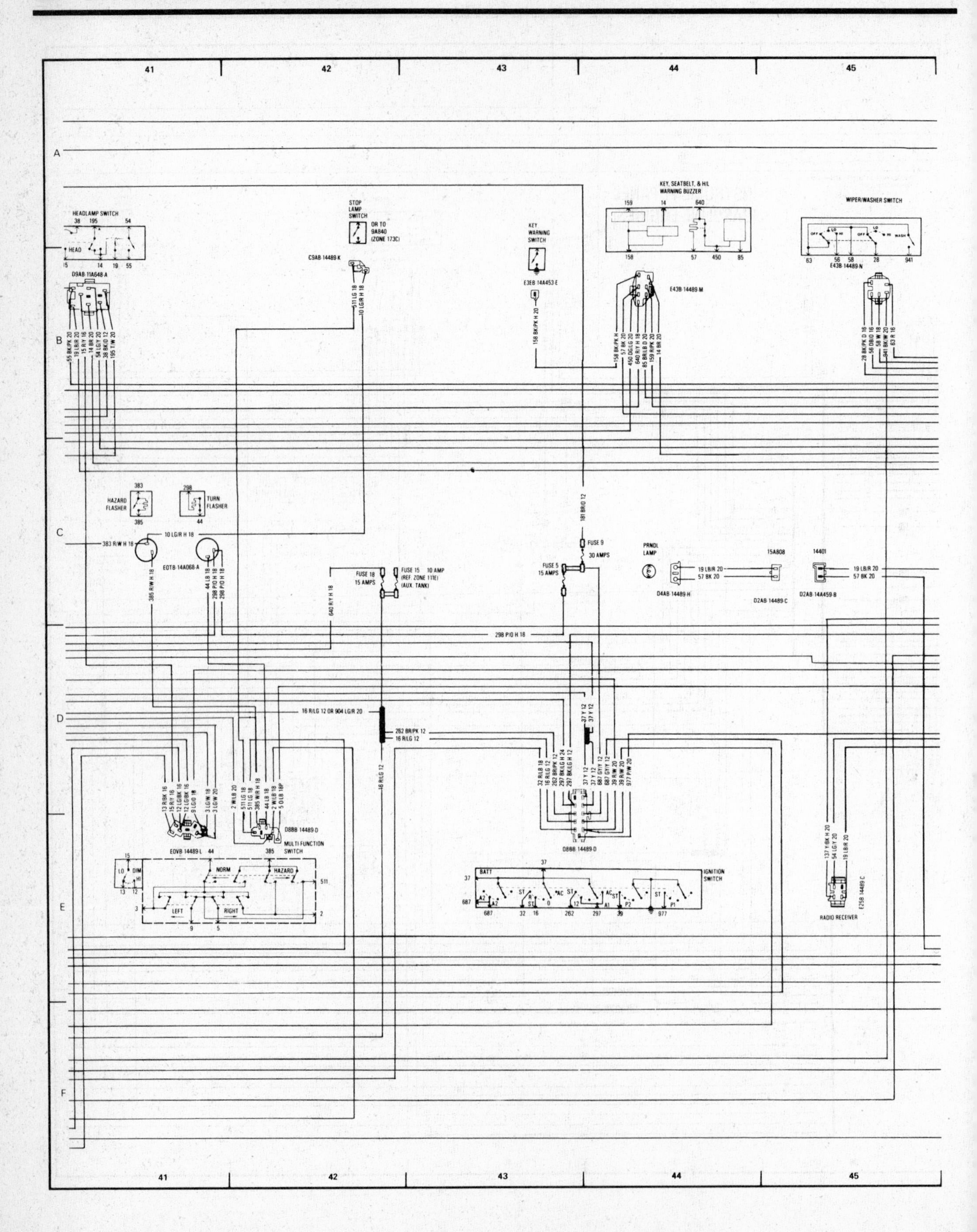
HEADLAMP SWITCH
STOP LAMP SWITCH
KEY WARNING SWITCH
KEY, SEATBELT, & H/L WARNING BUZZER
WIPER/WASHER SWITCH
HAZARD FLASHER
TURN FLASHER
FUSE 18 15 AMPS
FUSE 15 10 AMP
FUSE 9 30 AMPS
FUSE 5 15 AMPS
PRNDL LAMP
MULTI-FUNCTION SWITCH
IGNITION SWITCH
RADIO RECEIVER

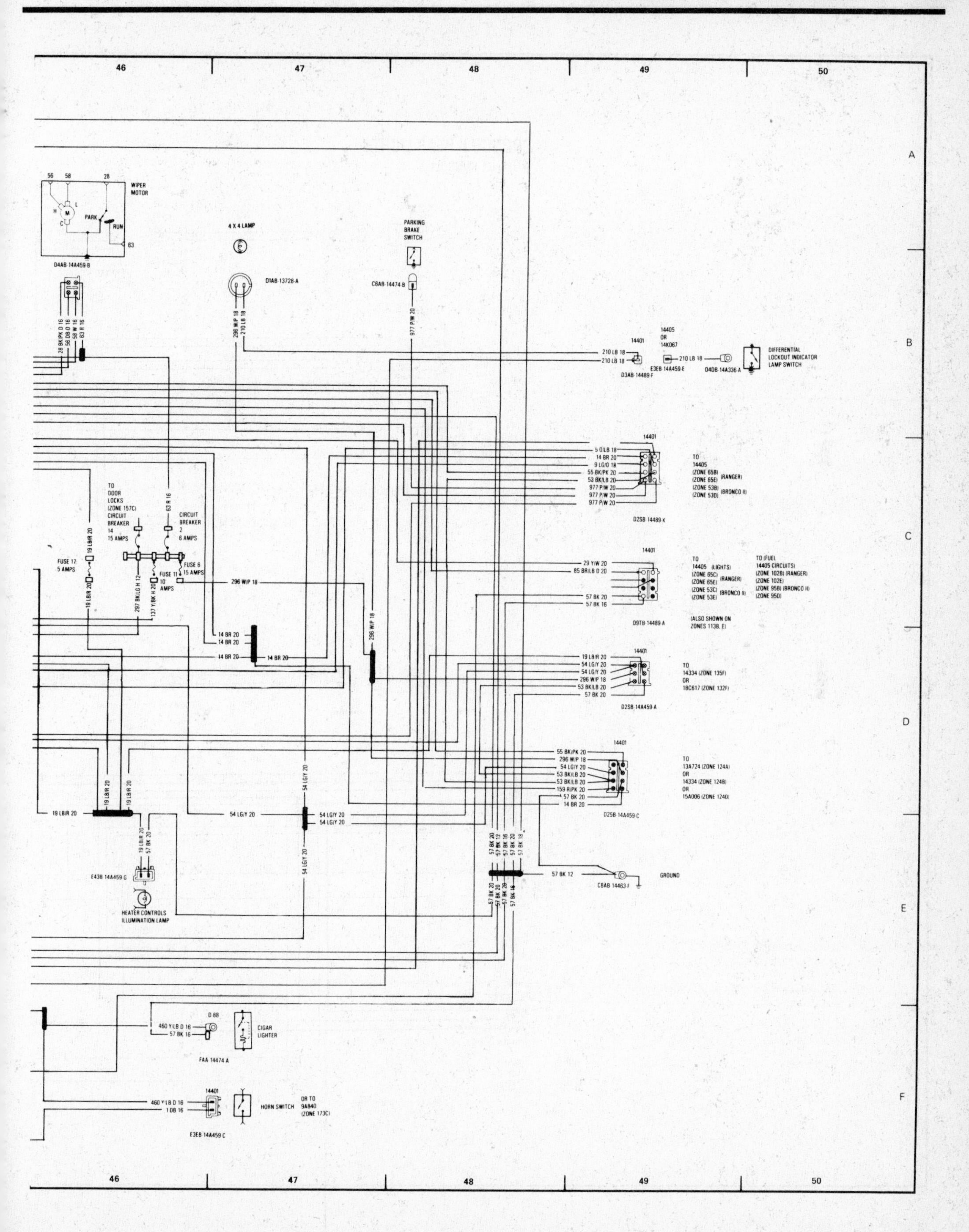

WIPER MOTOR
4 X 4 LAMP
PARKING BRAKE SWITCH
DIFFERENTIAL LOCKOUT INDICATOR LAMP SWITCH
CIRCUIT BREAKER 14 15 AMPS
CIRCUIT BREAKER 2 6 AMPS
FUSE 12 5 AMPS
FUSE 11 10 AMPS
FUSE 6 15 AMPS
HEATER CONTROLS ILLUMINATION LAMP
GROUND
CIGAR LIGHTER
HORN SWITCH

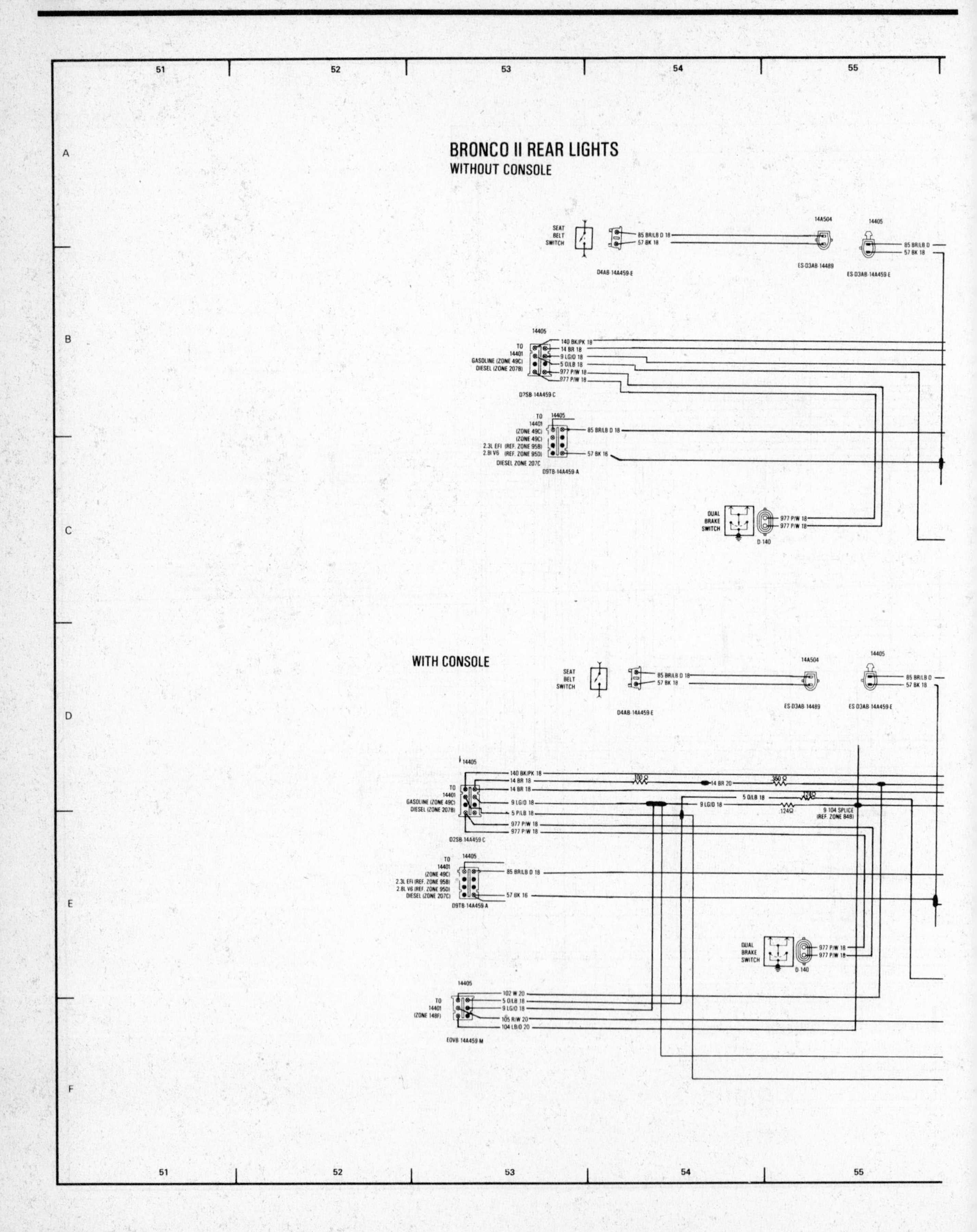
51
52
53
54
55
A
B
C
D
E
F
BRONCO II REAR LIGHTS
WITHOUT CONSOLE
SEAT BELT SWITCH
85 BR/LB D 18
57 BK 18
D4AB-14A459-E
14A504
ES-D3AB-14489
14405
85 BR/LB D
57 BK 18
ES-D3AB-14A459-E
14405
TO 14401 GASOLINE (ZONE 49C) DIESEL (ZONE 207B)
140 BK/PK 18
14 BR 18
9 LG/O 18
5 O/LB 18
977 P/W 18
977 P/W 18
D7SB-14A459-C
TO 14401 (ZONE 49C) (ZONE 49C) 2.3L EFI (REF. ZONE 95B) 2.8I V6 (REF. ZONE 95D) DIESEL ZONE 207C
85 BR/LB D 18
57 BK 16
D9TB-14A459-A
DUAL BRAKE SWITCH
977 P/W 18
977 P/W 18
D-140
WITH CONSOLE
SEAT BELT SWITCH
85 BR/LB D 18
57 BK 18
D4AB-14A459-E
14A504
ES-D3AB-14489
14405
85 BR/LB D
57 BK 18
ES-D3AB-14A459-E
14405
TO 14401 GASOLINE (ZONE 49C) DIESEL (ZONE 207B)
140 BK/PK 18
14 BR 18
14 BR 18
9 LG/O 18
5 P/LB 18
977 P/W 18
977 P/W 18
D2SB-14A459-C
100 Ω
14 BR 20
350 Ω
5 O/LB 18
9 LG/O 18
124Ω
124Ω
9-104 SPLICE (REF. ZONE 84B)
TO 14401 (ZONE 49C) 2.3L EFI (REF. ZONE 95B) 2.8L V6 (REF. ZONE 95D) DIESEL (ZONE 207C)
85 BR/LB D 18
57 BK 16
D9TB-14A459-A
DUAL BRAKE SWITCH
977 P/W 18
977 P/W 18
D-140
14405
TO 14401 (ZONE 148F)
102 W 20
5 O/LB 18
9 LG/O 18
105 R/W 20
104 LB/O 20
EOVB-14A459-M

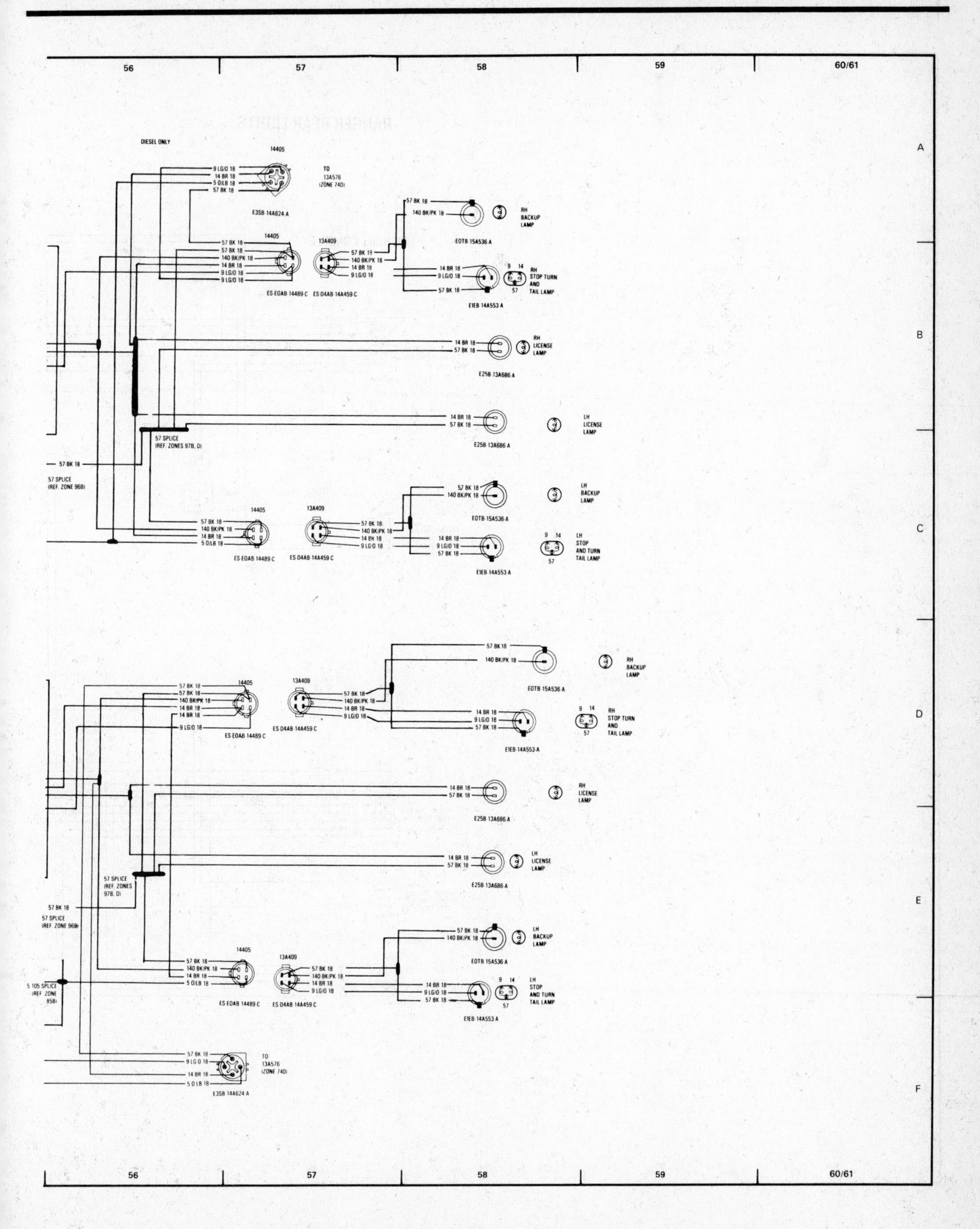
56
57
58
59
60/61
A
B
C
D
E
F
DIESEL ONLY
14405
9 LG/O 18
14 BR 18
5 O/LB 18
57 BK 18
TO
13A576
(ZONE 74D)
E3SB-14A624-A
57 BK 18
140 BK/PK 18
RH
BACKUP
LAMP
E0TB-15A536-A
13A409
14 BR 18
9 LG/O 18
ES-E0AB-14489-C
ES-D4AB-14A459-C
RH
STOP TURN
AND
TAIL LAMP
E1EB-14A553-A
RH
LICENSE
LAMP
E25B-13A686-A
LH
LICENSE
LAMP
57 SPLICE
(REF. ZONES 97B, D)
57 SPLICE
(REF. ZONE 96B)
LH
BACKUP
LAMP
LH
STOP
AND TURN
TAIL LAMP
5 105 SPLICE
(REF. ZONE
85B)

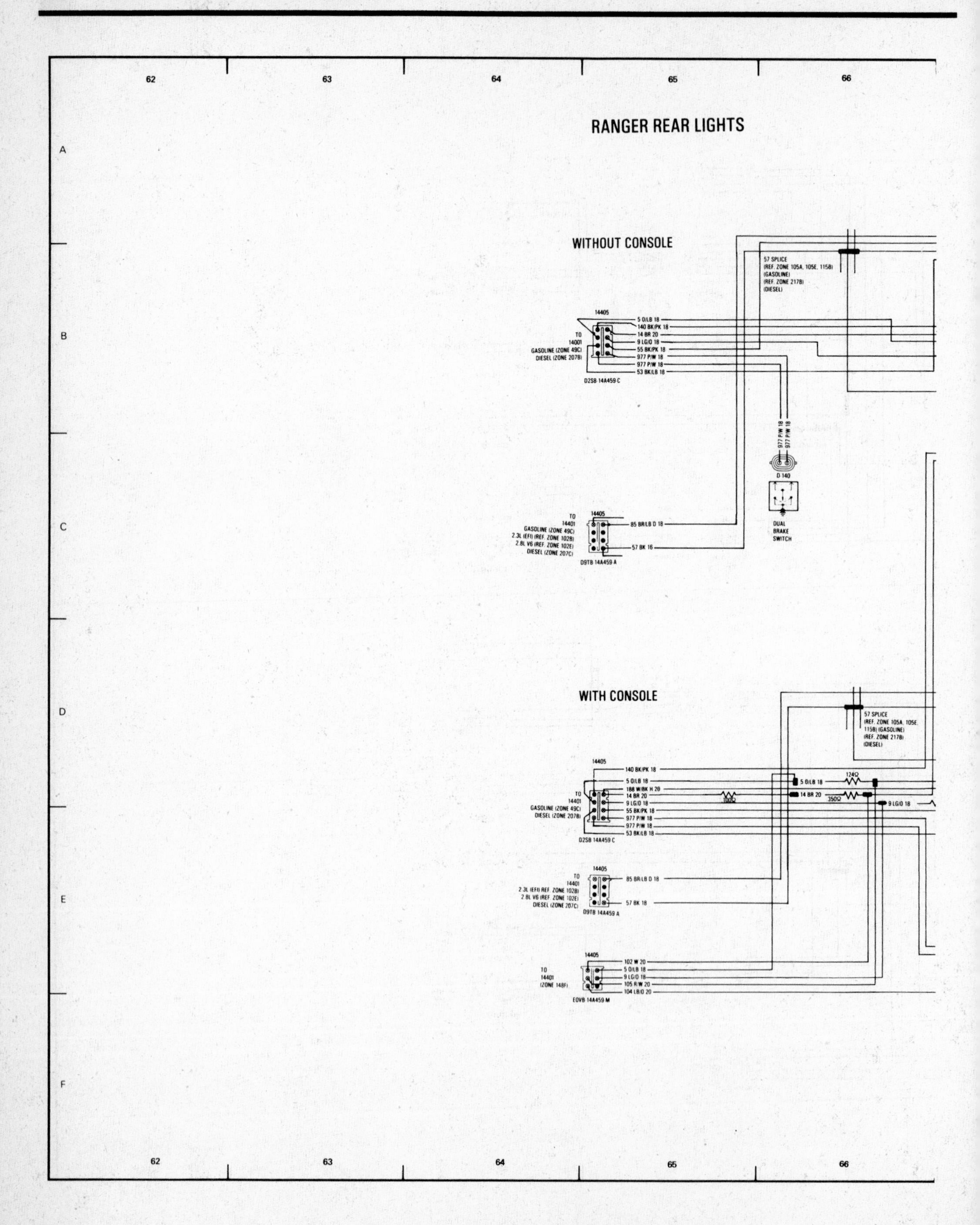
62
63
64
65
66
A
B
C
D
E
F
RANGER REAR LIGHTS
WITHOUT CONSOLE
57 SPLICE
(REF. ZONE 105A, 105E, 115B)
(GASOLINE)
(REF. ZONE 217B)
(DIESEL)
14405
TO
14001
GASOLINE (ZONE 49C)
DIESEL (ZONE 207B)
5 O/LB 18
140 BK/PK 18
14 BR 20
9 LG/O 18
55 BK/PK 18
977 P/W 18
977 P/W 18
53 BK/LB 18
D2S8 14A459 C
977 P/W 18
977 P/W 18
D-140
DUAL
BRAKE
SWITCH
TO
14401
GASOLINE (ZONE 49C)
2.3L (EFI) (REF. ZONE 102B)
2.8L V6 (REF. ZONE 102E)
DIESEL (ZONE 207C)
85 BR/LB D 18
57 BK 16
D9TB 14A459 A
WITH CONSOLE
57 SPLICE
(REF. ZONE 105A, 105E,
115B) (GASOLINE)
(REF. ZONE 217B)
(DIESEL)
14405
140 BK/PK 18
5 O/LB 18
188 W/BK H 20
14 BR 20
9 LG/O 18
55 BK/PK 18
977 P/W 18
977 P/W 18
53 BK/LB 18
TO
14401
GASOLINE (ZONE 49C)
DIESEL (ZONE 207B)
D2SB 14A459 C
100Ω
5 O/LB 18
14 BR 20
124Ω
350Ω
9 LG/O 18
14405
TO
14401
2.3L (EFI) REF. ZONE 102B)
2.8L V6 (REF. ZONE 102E)
DIESEL (ZONE 207C)
85 BR/LB D 18
57 BK 18
D9TB 14A459 A
14405
TO
14401
(ZONE 148F)
102 W 20
5 O/LB 18
9 LG/O 18
105 R/W 20
104 LB/O 20
E0VB 14A459 M

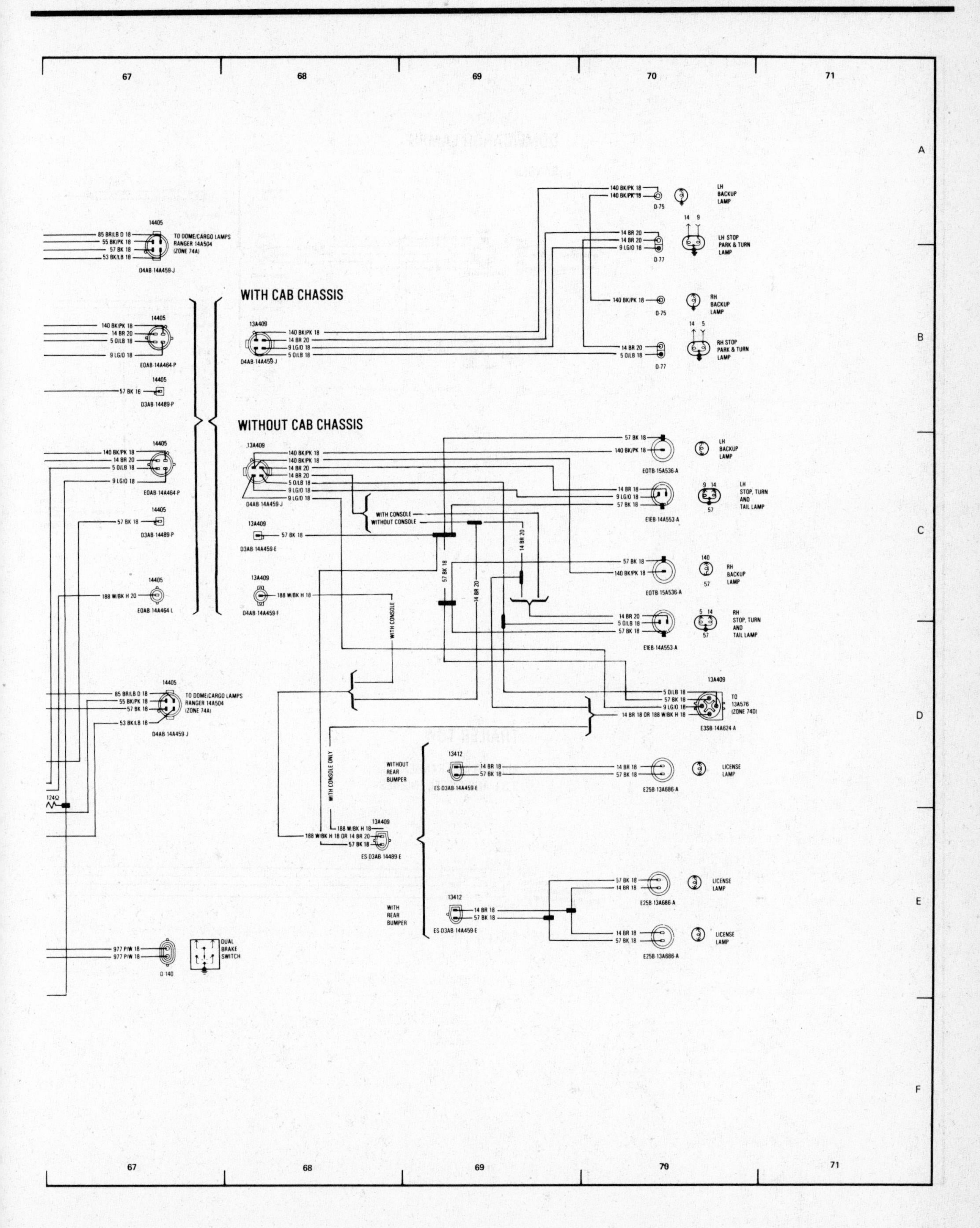
67
68
69
70
71
A
B
C
D
E
F
WITH CAB CHASSIS
WITHOUT CAB CHASSIS
TO DOME/CARGO LAMPS RANGER 14A504 (ZONE 74A)
LH BACKUP LAMP
LH STOP PARK & TURN LAMP
RH BACKUP LAMP
RH STOP PARK & TURN LAMP
LH STOP, TURN AND TAIL LAMP
RH STOP, TURN AND TAIL LAMP
WITH CONSOLE
WITHOUT CONSOLE
WITH CONSOLE ONLY
TO 13A576 (ZONE 74D)
WITHOUT REAR BUMPER
WITH REAR BUMPER
LICENSE LAMP
DUAL BRAKE SWITCH

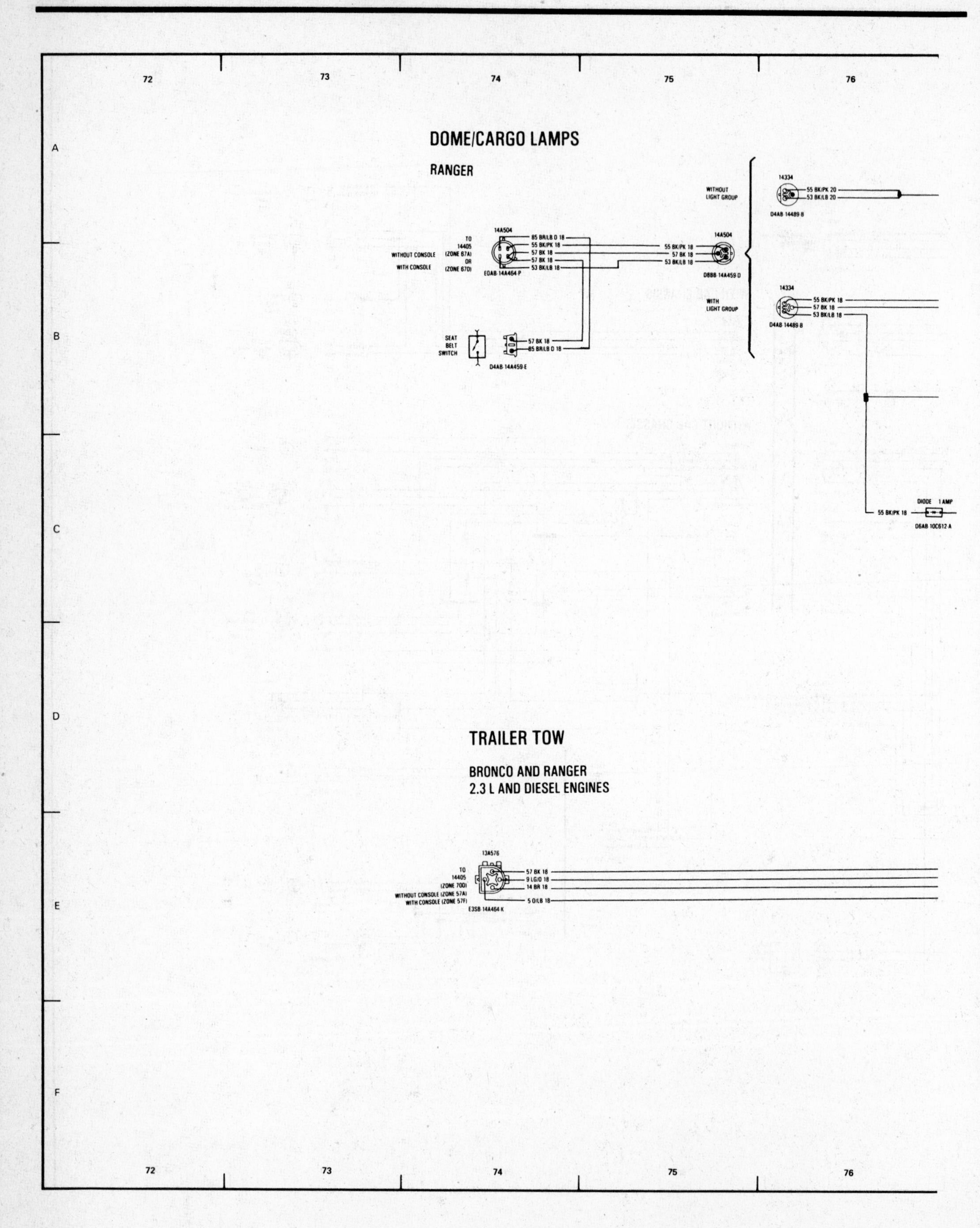
72
73
74
75
76
A
B
C
D
E
F
DOME/CARGO LAMPS
RANGER
TO
14405
WITHOUT CONSOLE (ZONE 67A)
OR
WITH CONSOLE (ZONE 67D)
14A504
85 BR/LB D 18
55 BK/PK 18
57 BK 18
57 BK 18
53 BK/LB 18
E0AB-14A464-P
SEAT
BELT
SWITCH
57 BK 18
85 BR/LB D 18
D4AB-14A459-E
14A504
55 BK/PK 18
57 BK 18
53 BK/LB 18
D8BB-14A459-D
WITHOUT
LIGHT GROUP
14334
55 BK/PK 20
53 BK/LB 20
D4AB-14489-B
WITH
LIGHT GROUP
14334
55 BK/PK 18
57 BK 18
53 BK/LB 18
D4AB-14489-B
DIODE 1 AMP
55 BK/PK 18
D6AB-10C612-A
TRAILER TOW
BRONCO AND RANGER
2.3 L AND DIESEL ENGINES
13A576
TO
14405
(ZONE 70D)
WITHOUT CONSOLE (ZONE 57A)
WITH CONSOLE (ZONE 57F)
57 BK 18
9 LG/O 18
14 BR 18
5 O/LB 18
E3SB-14A464-K

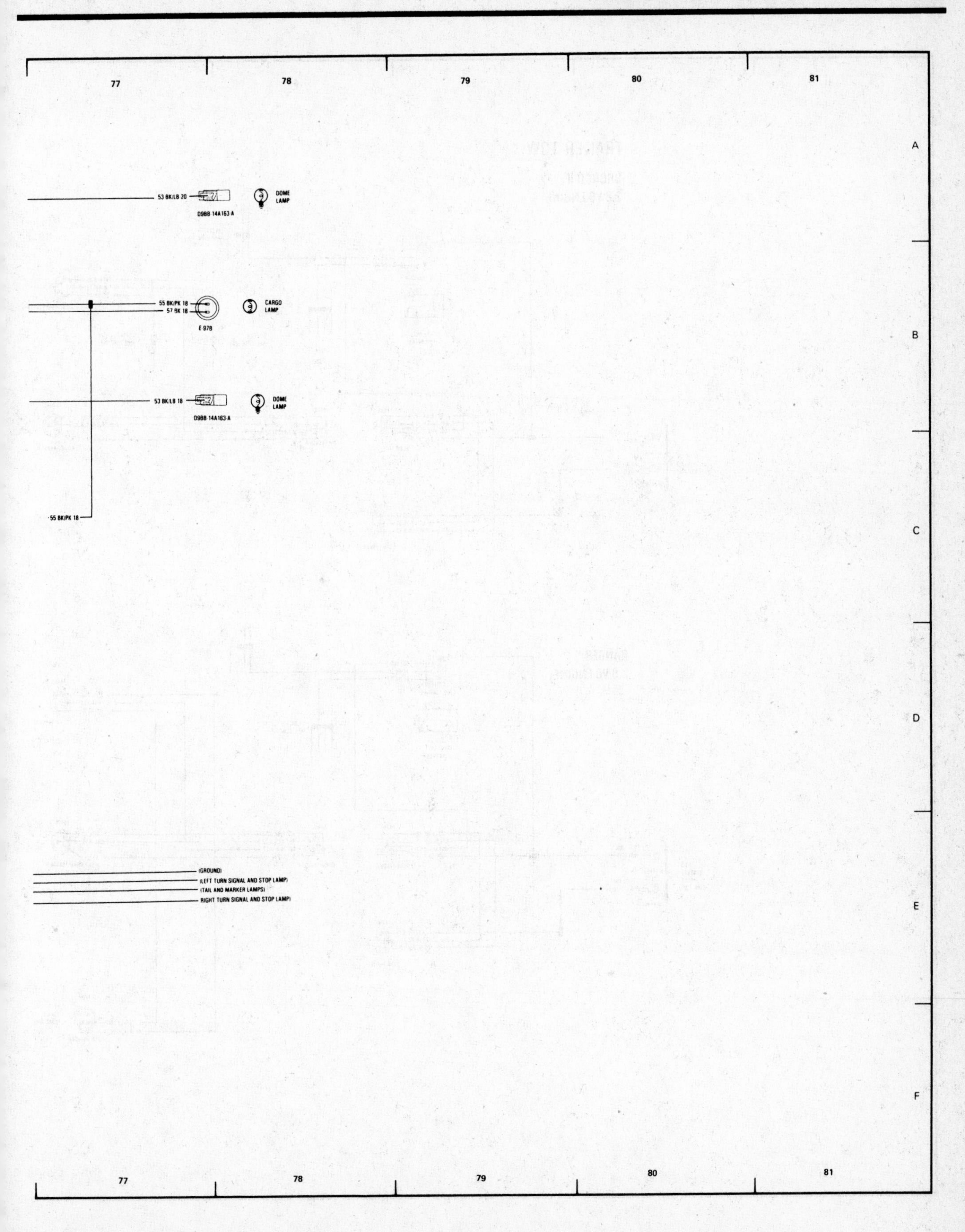
77
78
79
80
81
A
B
C
D
E
F
53 BK/LB 20
D9BB 14A163 A
DOME LAMP
55 BK/PK 18
57 BK 18
E 978
CARGO LAMP
53 BK/LB 18
D9BB 14A163 A
DOME LAMP
55 BK/PK 18
(GROUND)
(LEFT TURN SIGNAL AND STOP LAMP)
(TAIL AND MARKER LAMPS)
RIGHT TURN SIGNAL AND STOP LAMP)

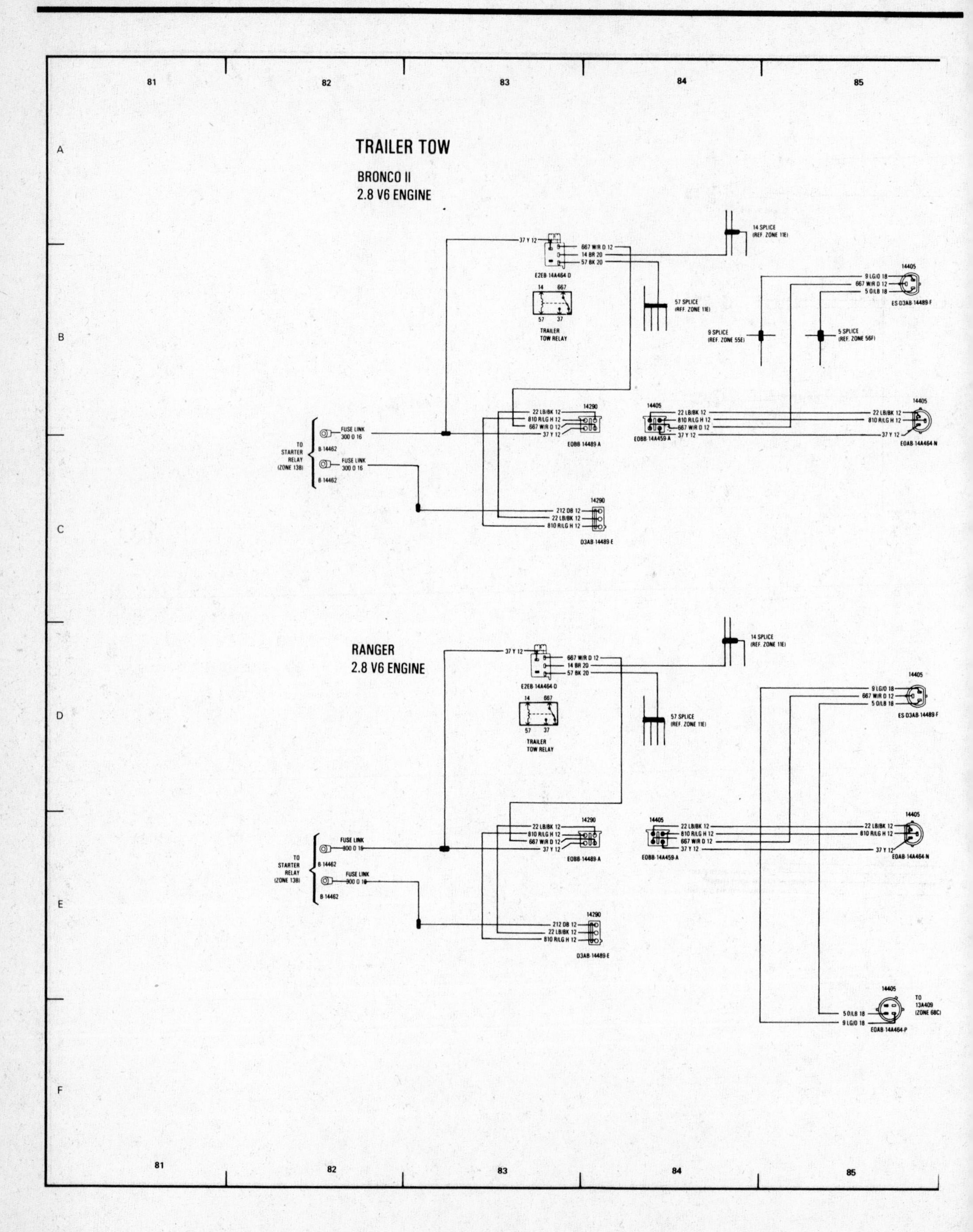

81
82
83
84
85
A
B
C
D
E
F
TRAILER TOW
BRONCO II
2.8 V6 ENGINE
RANGER
2.8 V6 ENGINE
TRAILER
TOW RELAY
14 SPLICE
(REF. ZONE 11E)
57 SPLICE
(REF. ZONE 11E)
9 SPLICE
(REF. ZONE 55E)
5 SPLICE
(REF. ZONE 56F)
FUSE LINK
TO
STARTER
RELAY
(ZONE 13B)
B-14462
14290
14405
E2EB-14A464-D
E0BB-14489-A
D3AB-14489-E
E0BB-14A459-A
E0AB-14A464-N
ES-D3AB-14489-F
E0AB-14A464-P

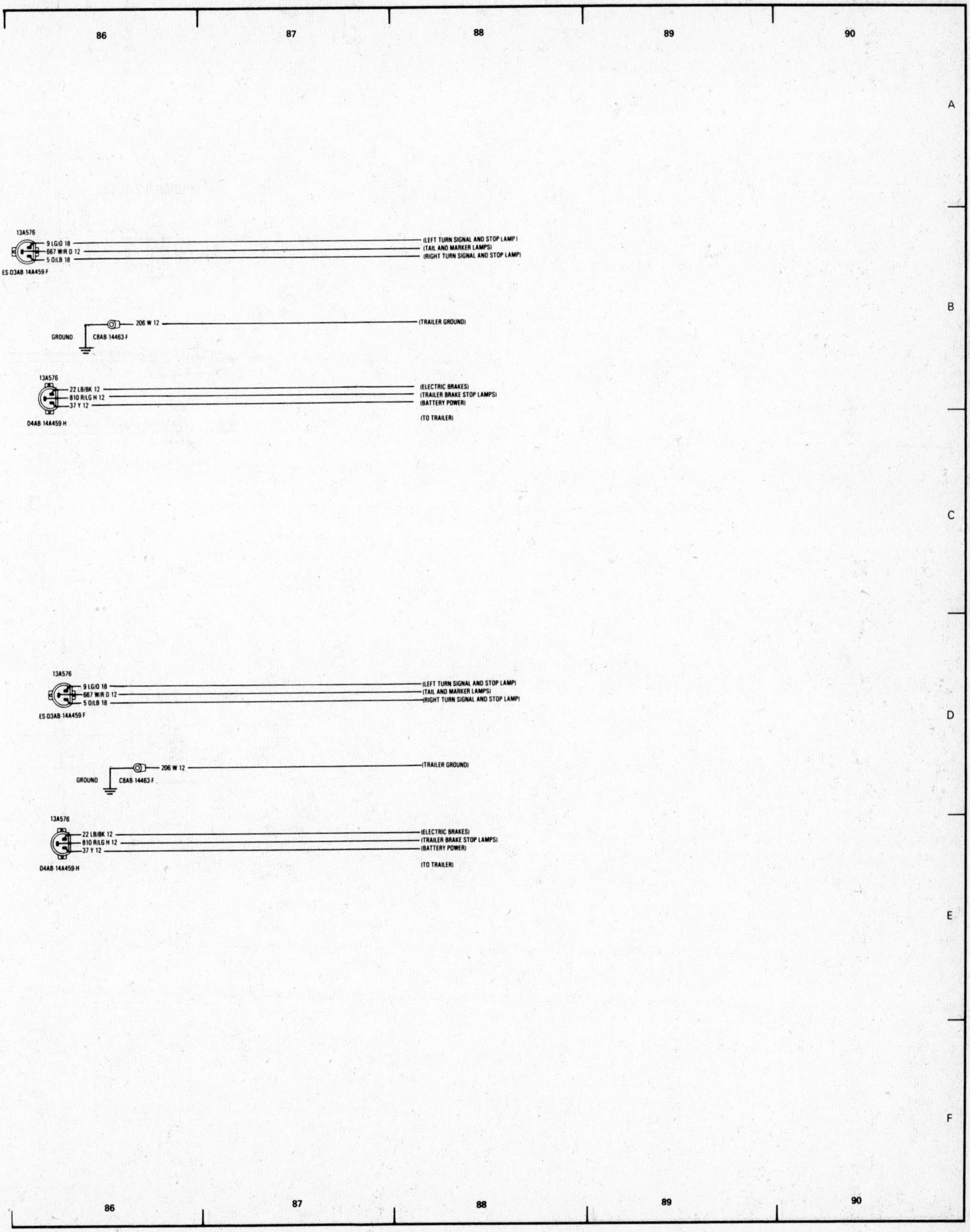

86 87 88 89 90

A

13A576
9 LG/O 18 — (LEFT TURN SIGNAL AND STOP LAMP)
667 W/R D 12 — (TAIL AND MARKER LAMPS)
5 O/LB 18 — (RIGHT TURN SIGNAL AND STOP LAMP)
ES-D3AB-14A459-F

B

206 W 12 — (TRAILER GROUND)
GROUND
C8AB-14463-F

13A576
22 LB/BK 12 — (ELECTRIC BRAKES)
810 R/LG H 12 — (TRAILER BRAKE STOP LAMPS)
37 Y 12 — (BATTERY POWER)
(TO TRAILER)
D4AB-14A459-H

C

13A576
9 LG/O 18 — (LEFT TURN SIGNAL AND STOP LAMP)
667 W/R D 12 — (TAIL AND MARKER LAMPS)
5 O/LB 18 — (RIGHT TURN SIGNAL AND STOP LAMP)
ES-D3AB-14A459-F

D

206 W 12 — (TRAILER GROUND)
GROUND
C8AB-14463-F

13A576
22 LB/BK 12 — (ELECTRIC BRAKES)
810 R/LG H 12 — (TRAILER BRAKE STOP LAMPS)
37 Y 12 — (BATTERY POWER)
(TO TRAILER)
D4AB-14A459-H

E

F

86 87 88 89 90

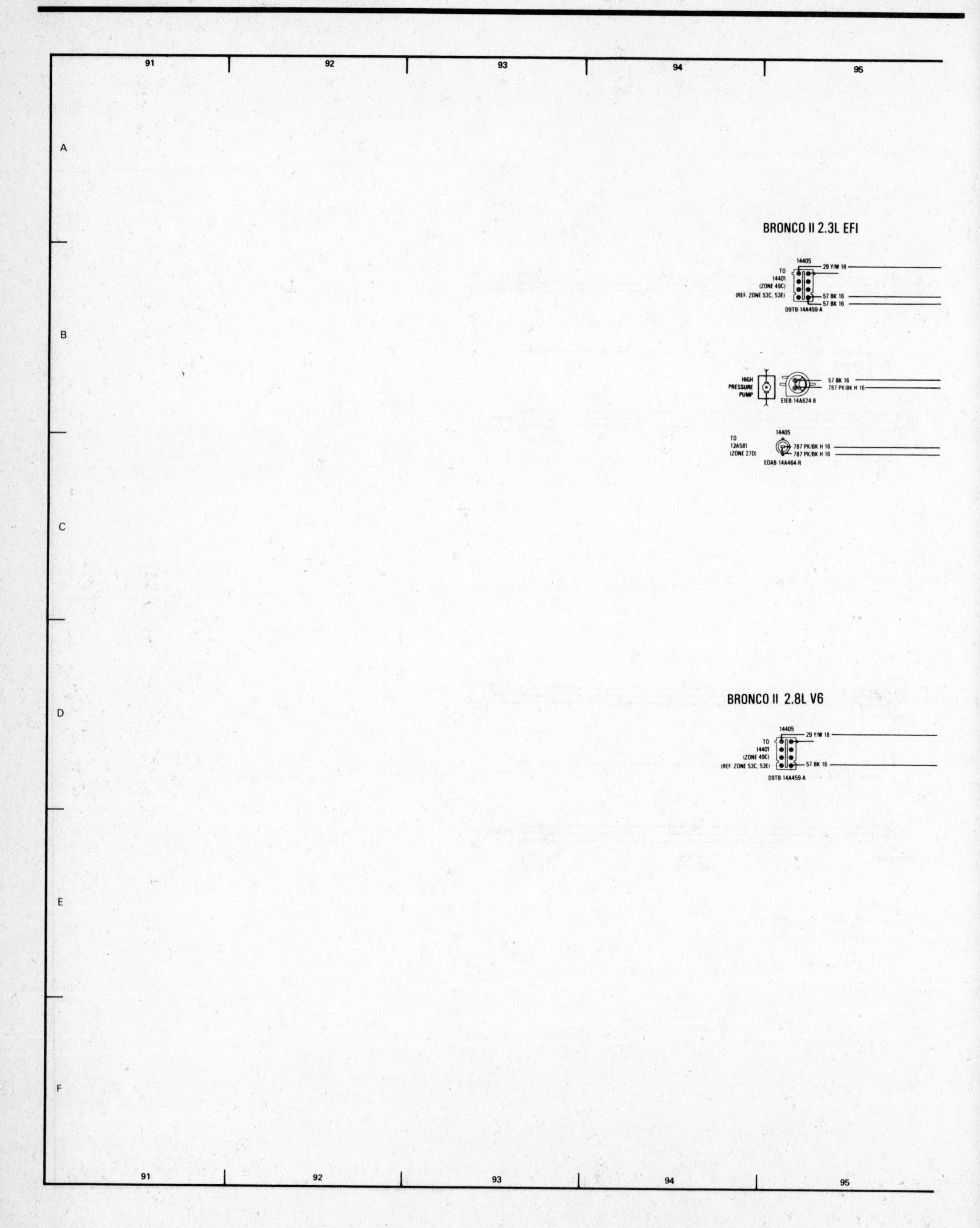
91
92
93
94
95
A
B
C
D
E
F
BRONCO II 2.3L EFI
14405
TO
14401
(ZONE 49C)
(REF. ZONE 53C, 53E)
29 Y/W 18
57 BK 16
57 BK 16
D9TB-14A459-A
HIGH
PRESSURE
PUMP
57 BK 16
787 PK/BK H 16
E1EB-14A624-B
14405
TO
12A581
(ZONE 27D)
787 PK/BK H 16
787 PK/BK H 16
E0AB-14A464-R
BRONCO II 2.8L V6
14405
TO
14401
(ZONE 49C)
(REF. ZONE 53C, 53E)
29 Y/W 18
57 BK 16
D9TB-14A459-A

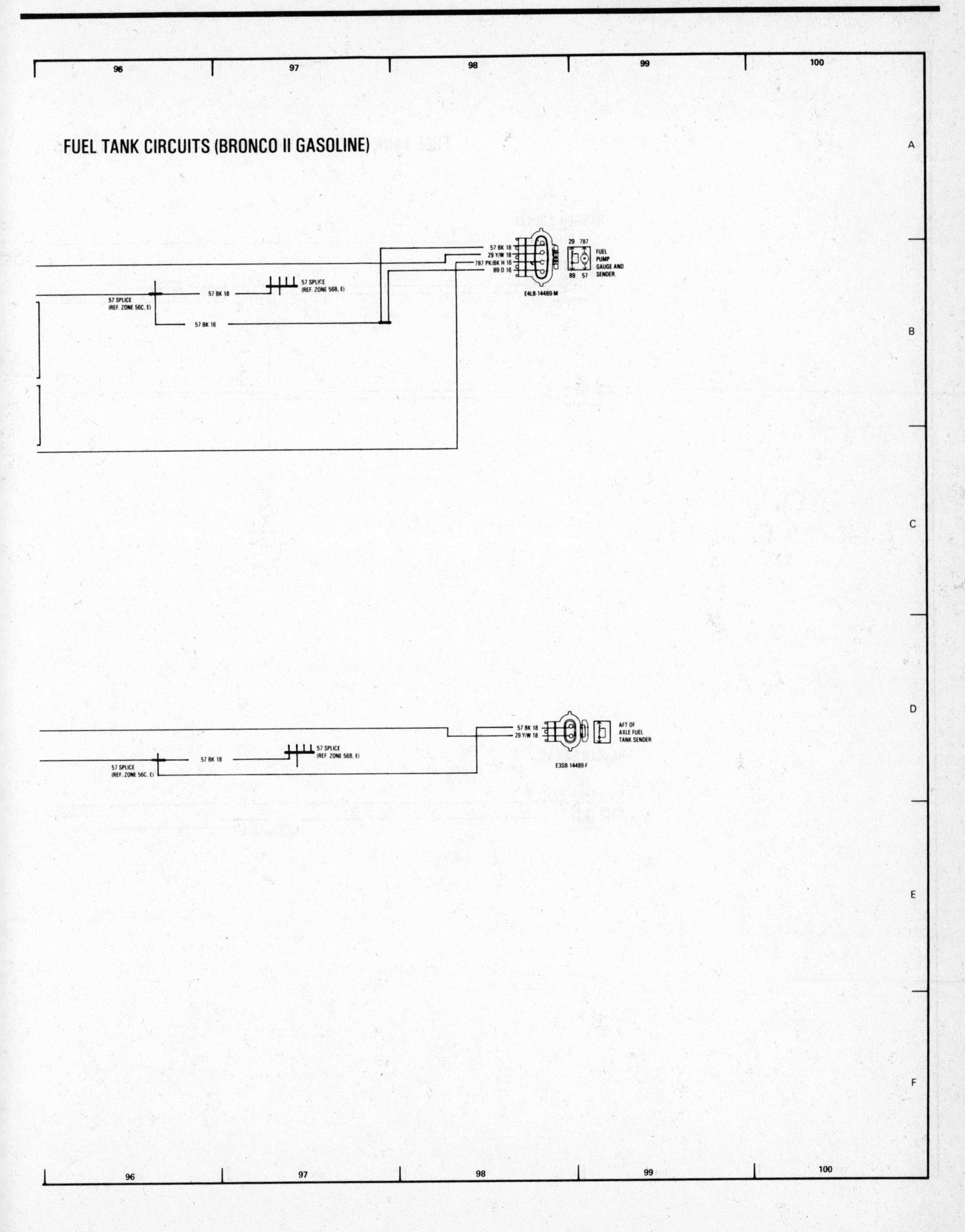
96
97
98
99
100
FUEL TANK CIRCUITS (BRONCO II GASOLINE)
A
57 BK 18
29 Y/W 18
787 PK/BK H 16
89 O 16
29 787
89 57
FUEL PUMP GAUGE AND SENDER
E4LB-14489-M
57 SPLICE (REF. ZONE 56B, E)
57 BK 18
57 SPLICE (REF. ZONE 56C, E)
57 BK 16
B
C
D
57 BK 18
29 Y/W 18
AFT OF AXLE FUEL TANK SENDER
E3SB-14489-F
57 SPLICE (REF. ZONE 56B, E)
57 BK 18
57 SPLICE (REF. ZONE 56C, E)
E
F
96
97
98
99
100

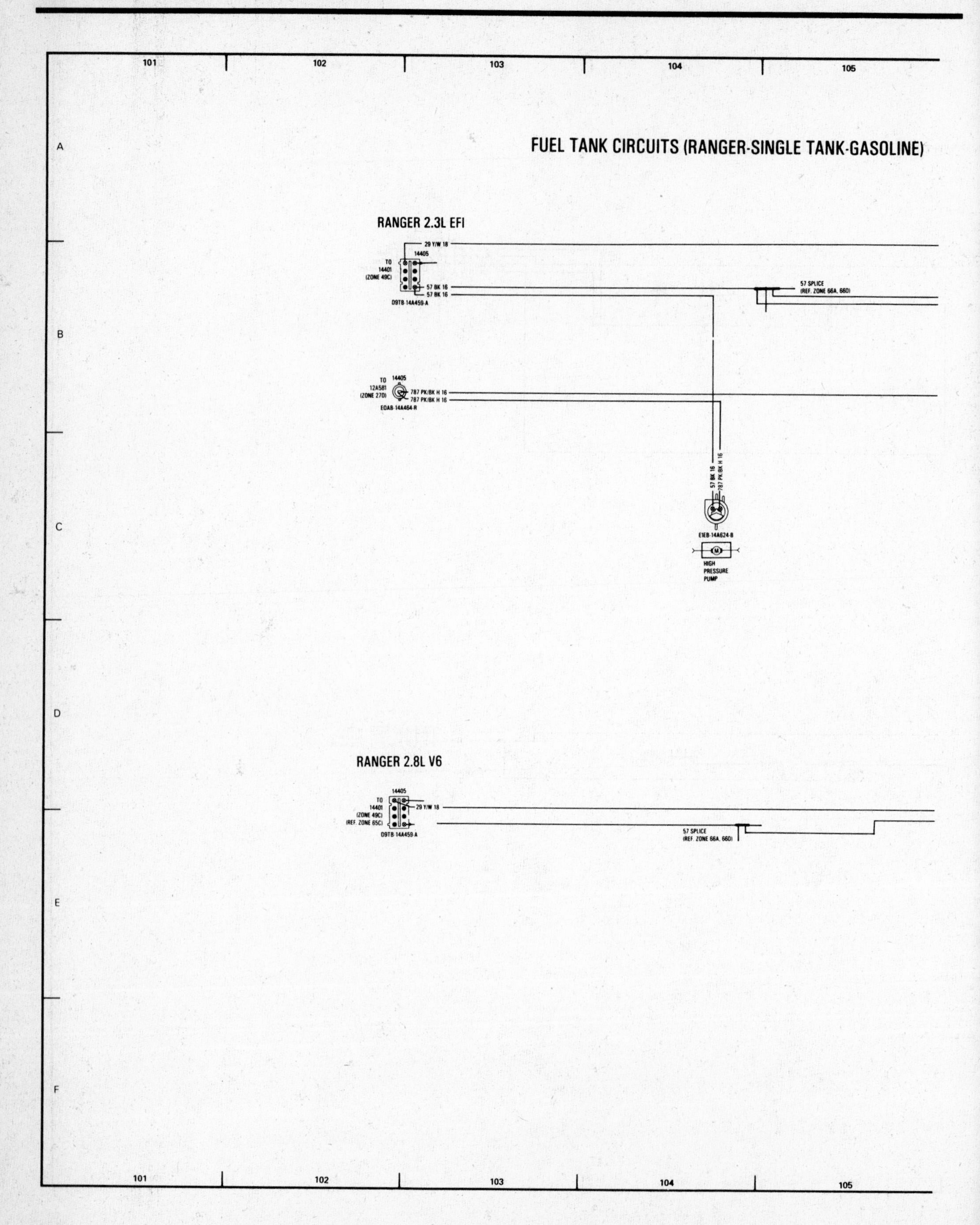
101
102
103
104
105
A
B
C
D
E
F
FUEL TANK CIRCUITS (RANGER-SINGLE TANK-GASOLINE)
RANGER 2.3L EFI
29 Y/W 18
14405
TO
14401
(ZONE 49C)
57 BK 16
57 BK 16
D9TB-14A459-A
57 SPLICE
(REF. ZONE 66A, 66D)
TO
12A581
(ZONE 27D)
14405
787 PK/BK H 16
787 PK/BK H 16
E0AB-14A464-R
57 BK 16
787 PK/BK H 16
E1EB-14A624-B
M
HIGH
PRESSURE
PUMP
RANGER 2.8L V6
14405
TO
14401
(ZONE 49C)
(REF. ZONE 65C)
29 Y/W 18
D9TB-14A459-A
57 SPLICE
(REF. ZONE 66A, 66D)
101
102
103
104
105

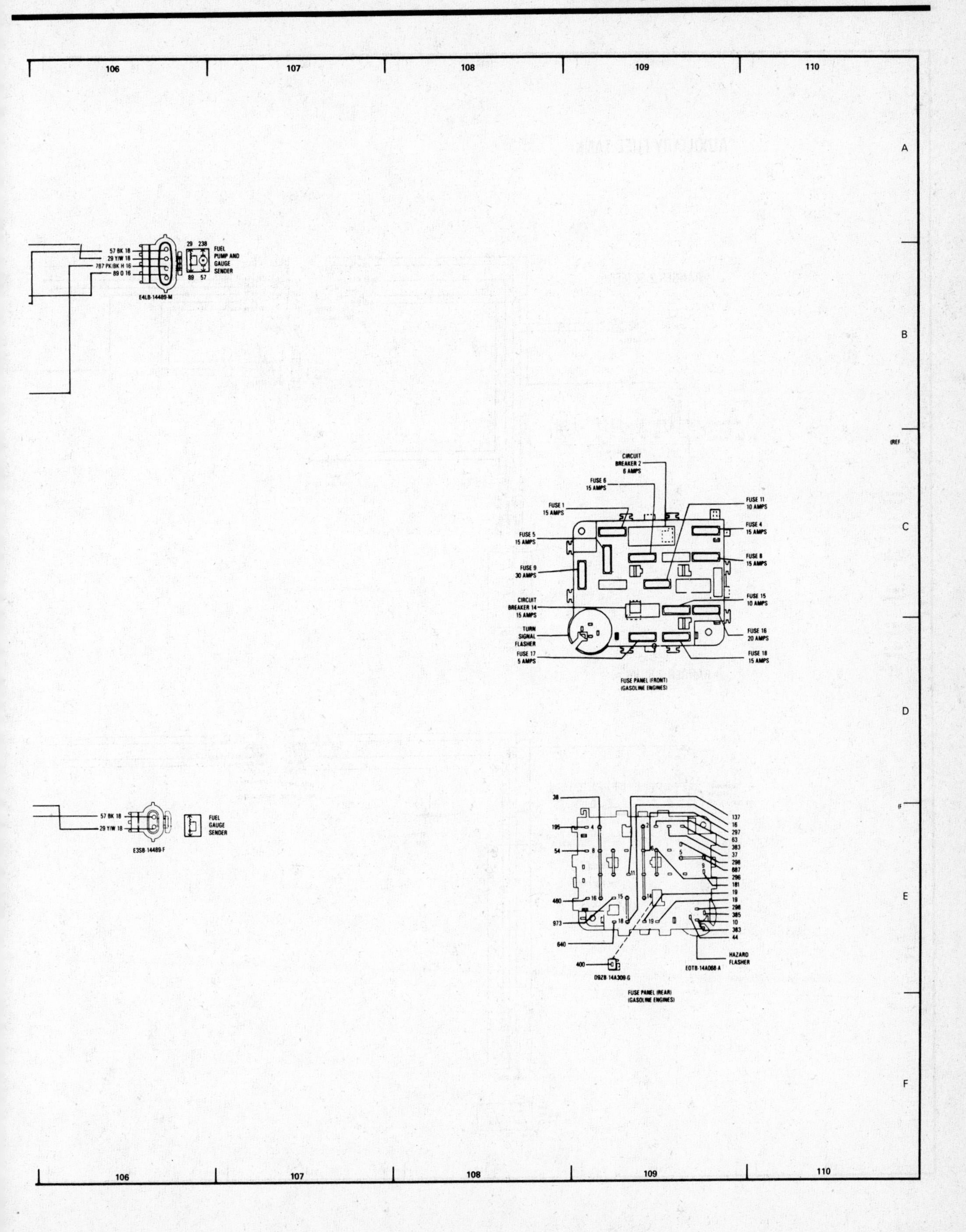
106
107
108
109
110
A
B
C
D
E
F
57 BK 18
29 Y/W 18
787 PK/BK H 16
89 O 16
29 238
89 57
FUEL
PUMP AND
GAUGE
SENDER
E4LB-14489-M
CIRCUIT
BREAKER 2
6 AMPS
FUSE 6
15 AMPS
FUSE 1
15 AMPS
FUSE 11
10 AMPS
FUSE 4
15 AMPS
FUSE 5
15 AMPS
FUSE 8
15 AMPS
FUSE 9
30 AMPS
CIRCUIT
BREAKER 14
15 AMPS
FUSE 15
10 AMPS
TURN
SIGNAL
FLASHER
FUSE 16
20 AMPS
FUSE 17
5 AMPS
FUSE 18
15 AMPS
FUSE PANEL (FRONT)
(GASOLINE ENGINES)
57 BK 18
29 Y/W 18
FUEL
GAUGE
SENDER
E3SB-14489-F
HAZARD
FLASHER
E0TB-14A068-A
D9ZB-14A309-G
FUSE PANEL (REAR)
(GASOLINE ENGINES)

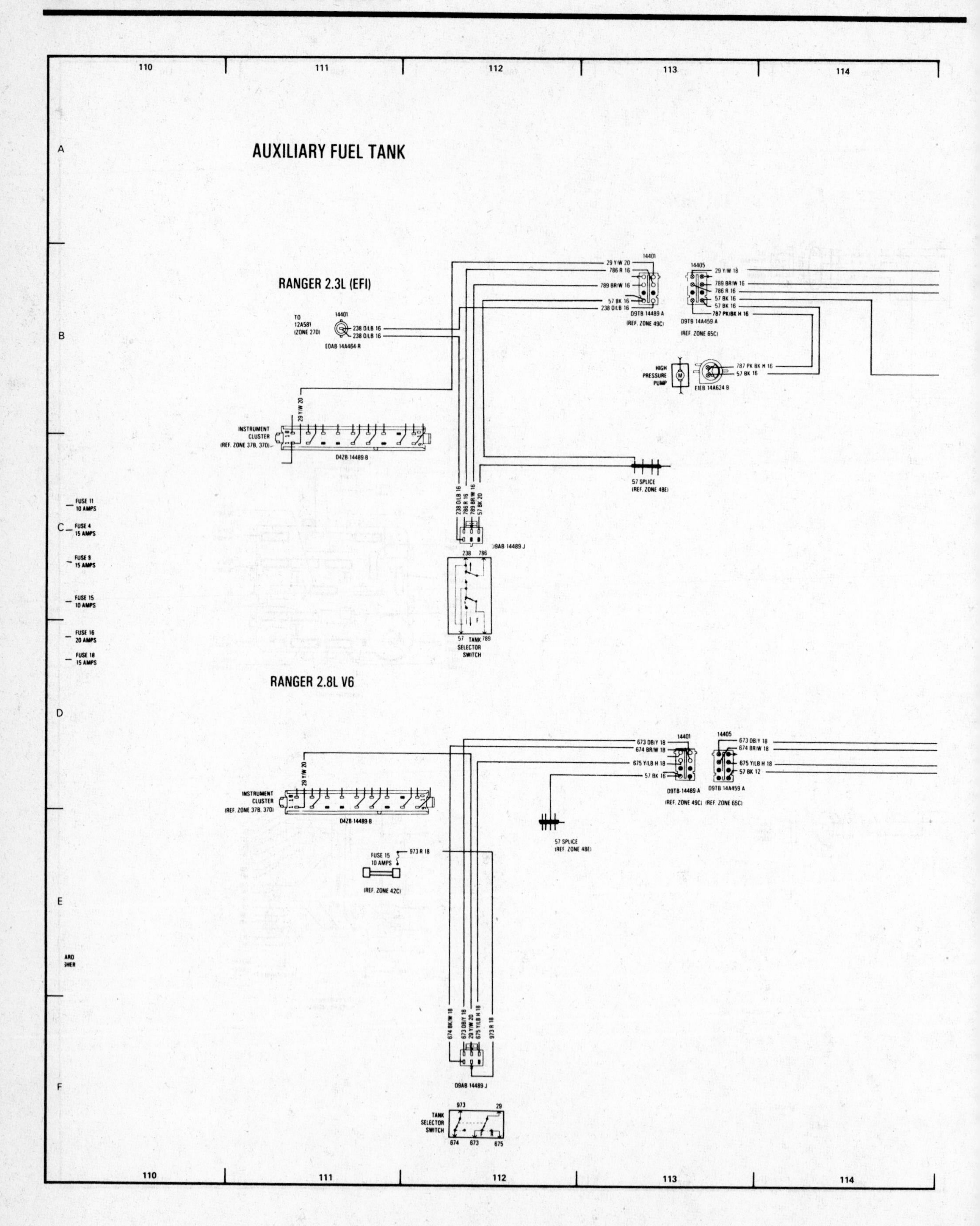
AUXILIARY FUEL TANK
RANGER 2.3L (EFI)
29 Y/W 20
786 R 16
789 BR/W 16
57 BK 16
238 O/LB 16
14401
D9TB 14489 A
(REF. ZONE 49C)
14405
29 Y/W 18
789 BR/W 16
786 R 16
57 BK 16
787 PK/BK H 16
D9TB 14A459 A
(REF. ZONE 65C)
TO
12A581
(ZONE 27D)
E0AB 14A464 R
HIGH
PRESSURE
PUMP
787 PK BK H 16
E1EB 14A624 B
INSTRUMENT
CLUSTER
(REF. ZONE 37B, 37D)
D4ZB 14489 B
57 SPLICE
(REF. ZONE 48E)
57 BK 20
D9AB 14489 J
TANK
SELECTOR
SWITCH
FUSE 11
10 AMPS
FUSE 4
15 AMPS
FUSE 9
15 AMPS
FUSE 15
10 AMPS
FUSE 16
20 AMPS
FUSE 18
15 AMPS
RANGER 2.8L V6
673 DB/Y 18
674 BR/W 18
675 Y/LB H 18
57 BK 16
57 BK 12
973 R 18
(REF. ZONE 42C)
674 BK/W 18
29 Y/W 20

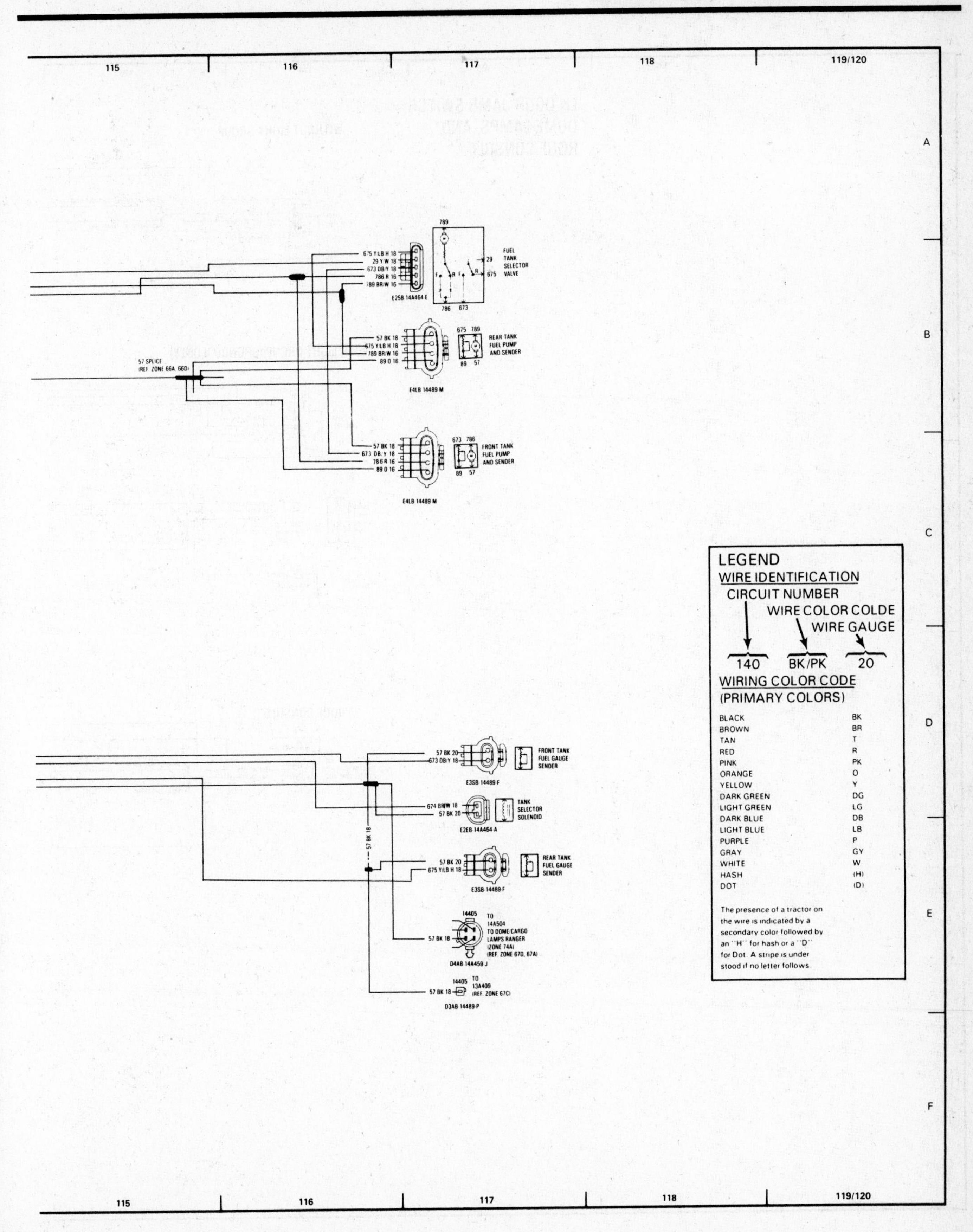
115
116
117
118
119/120
A
B
C
D
E
F
789
6'5 Y/LB H 18
29 Y/W 18
673 DB/Y 18
786 R 16
789 BR/W 16
E25B 14A464 E
29
675
FUEL TANK SELECTOR VALVE
786 673
57 BK 18
675 Y/LB H 18
789 BR/W 16
89 O 16
675 789
89 57
REAR TANK FUEL PUMP AND SENDER
57 SPLICE
(REF. ZONE 66A, 66D)
E4LB 14489 M
57 BK 18
673 DB/Y 18
786 R 16
89 O 16
673 786
89 57
FRONT TANK FUEL PUMP AND SENDER
E4LB 14489 M
LEGEND
WIRE IDENTIFICATION
CIRCUIT NUMBER
WIRE COLOR COLDE
WIRE GAUGE
140 BK/PK 20
WIRING COLOR CODE
(PRIMARY COLORS)
BLACK BK
BROWN BR
TAN T
RED R
PINK PK
ORANGE O
YELLOW Y
DARK GREEN DG
LIGHT GREEN LG
DARK BLUE DB
LIGHT BLUE LB
PURPLE P
GRAY GY
WHITE W
HASH (H)
DOT (D)
The presence of a tractor on the wire is indicated by a secondary color followed by an "H" for hash or a "D" for Dot. A stripe is under stood if no letter follows.
57 BK 20
673 DB/Y 18
FRONT TANK FUEL GAUGE SENDER
E3SB 14489 F
674 BR/W 18
57 BK 20
TANK SELECTOR SOLENOID
E2EB 14A464 A
57 BK 18
57 BK 20
675 Y/LB H 18
REAR TANK FUEL GAUGE SENDER
E3SB 14489 F
14405
TO
14A504
TO DOME/CARGO
LAMPS RANGER
(ZONE 74A)
(REF. ZONE 67D, 67A)
57 BK 18
D4AB 14A459 J
14405
TO
13A409
(REF. ZONE 67C)
57 BK 18
D3AB 14489 P

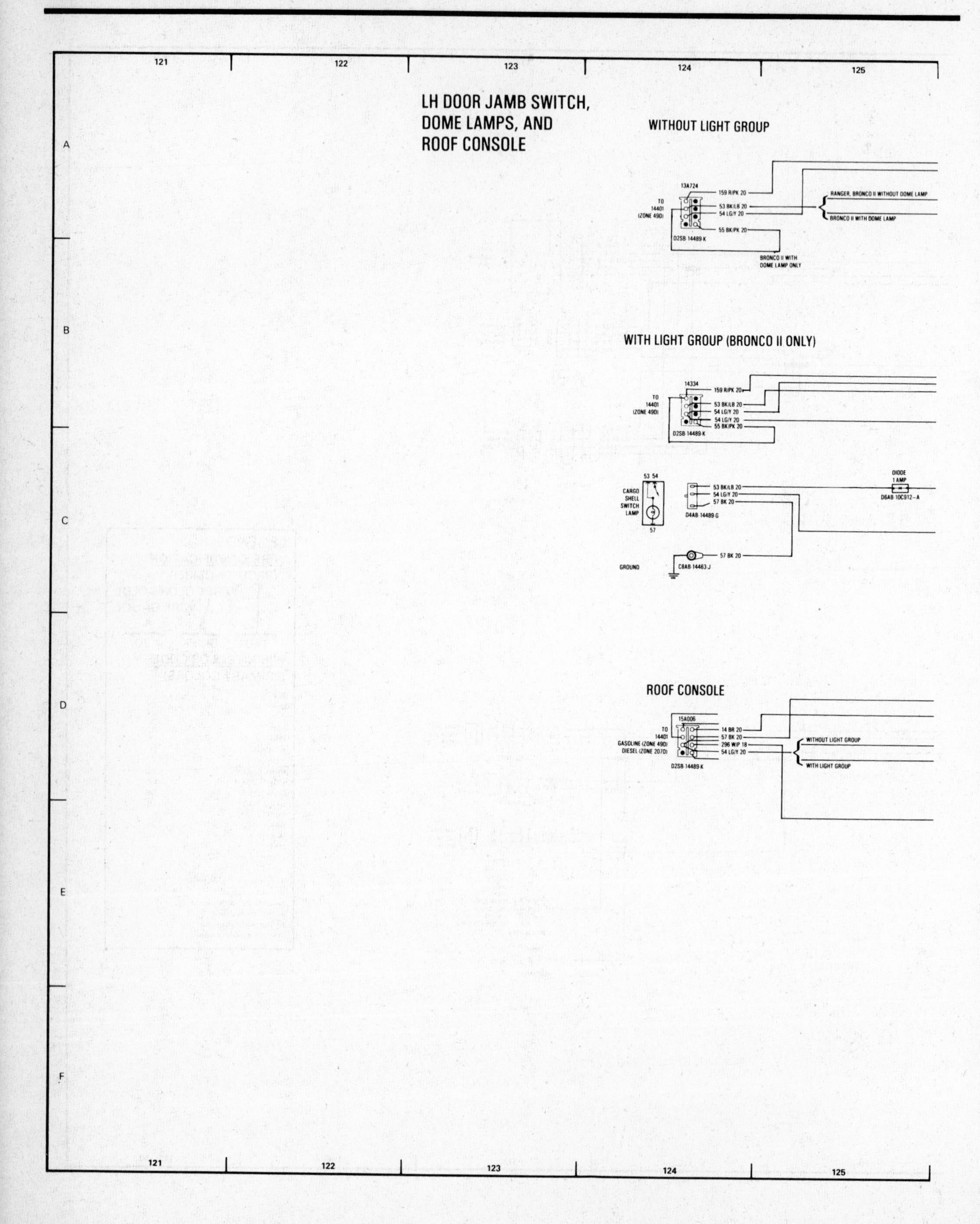
121
122
123
124
125
A
B
C
D
E
F
LH DOOR JAMB SWITCH, DOME LAMPS, AND ROOF CONSOLE
WITHOUT LIGHT GROUP
13A724
TO 14401 (ZONE 49D)
159 R/PK 20
53 BK/LB 20
54 LG/Y 20
55 BK/PK 20
D2SB-14489-K
RANGER, BRONCO II WITHOUT DOME LAMP
BRONCO II WITH DOME LAMP
BRONCO II WITH DOME LAMP ONLY
WITH LIGHT GROUP (BRONCO II ONLY)
14334
TO 14401 (ZONE 49D)
159 R/PK 20
53 BK/LB 20
54 LG/Y 20
54 LG/Y 20
55 BK/PK 20
D2SB-14489-K
53 54
CARGO SHELL SWITCH LAMP
57
53 BK/LB 20
54 LG/Y 20
57 BK 20
D4AB-14489-G
DIODE 1 AMP
D6AB-10C912–A
57 BK 20
GROUND
C8AB-14463-J
ROOF CONSOLE
15A006
TO 14401 GASOLINE-(ZONE 49D) DIESEL-(ZONE 207D)
14 BR 20
57 BK 20
296 W/P 18
54 LG/Y 20
D2SB-14489-K
WITHOUT LIGHT GROUP
WITH LIGHT GROUP

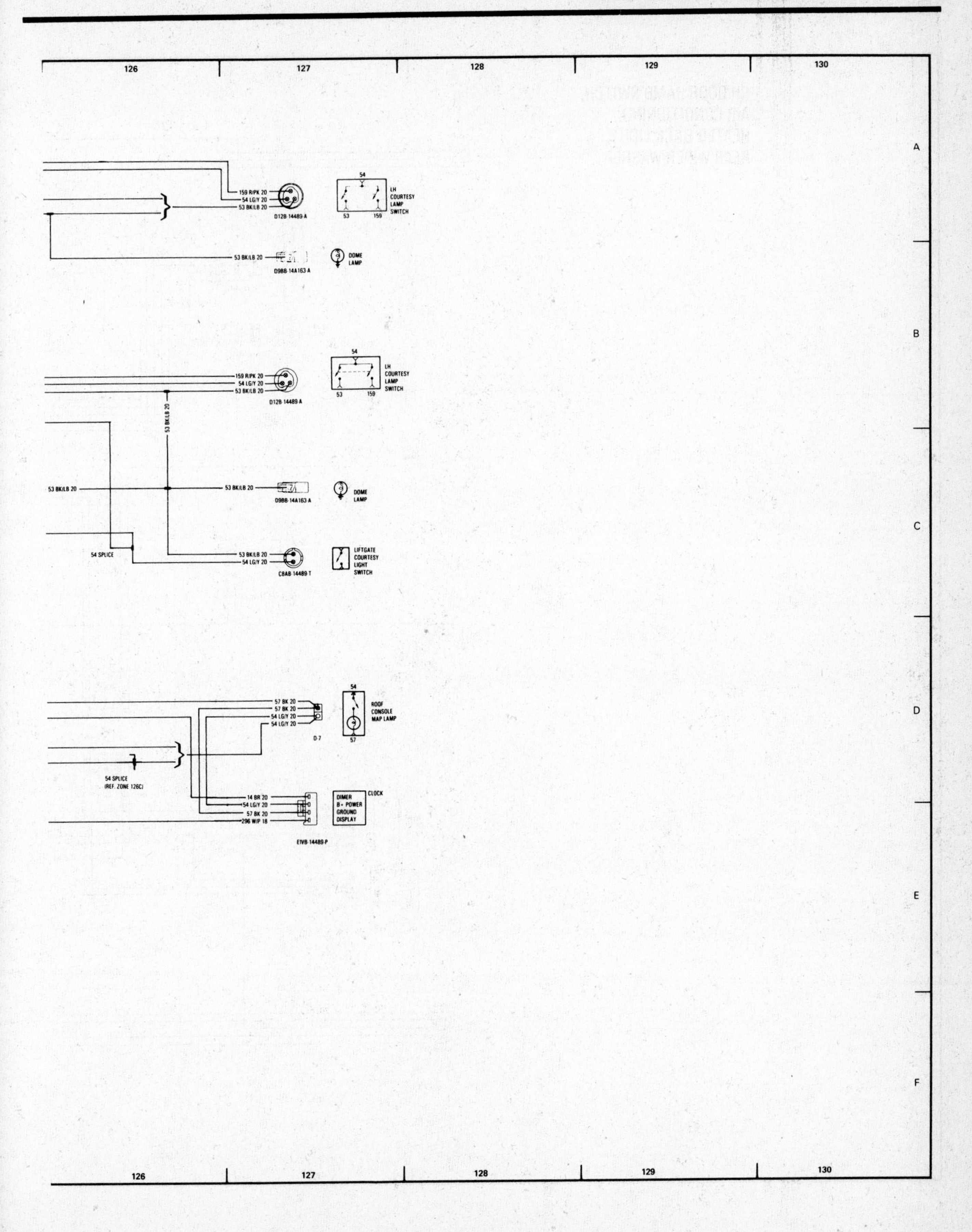
126
127
128
129
130
A
B
C
D
E
F
159 R/PK 20
54 LG/Y 20
53 BK/LB 20
D12B-14489-A
54
LH
COURTESY
LAMP
SWITCH
53
159
53 BK/LB 20
D9BB-14A163-A
DOME
LAMP
159 R/PK 20
54 LG/Y 20
53 BK/LB 20
D12B-14489-A
54
LH
COURTESY
LAMP
SWITCH
53
159
53 BK/LB 20
53 BK/LB 20
53 BK/LB 20
D9BB-14A163-A
DOME
LAMP
54 SPLICE
53 BK/LB 20
54 LG/Y 20
C8AB-14489-T
LIFTGATE
COURTESY
LIGHT
SWITCH
57 BK 20
57 BK 20
54 LG/Y 20
54 LG/Y 20
D-7
54
ROOF
CONSOLE
MAP LAMP
57
54 SPLICE
(REF. ZONE 126C)
14 BR 20
54 LG/Y 20
57 BK 20
296 W/P 18
E1VB-14489-P
DIMER
B+ POWER
GROUND
DISPLAY
CLOCK

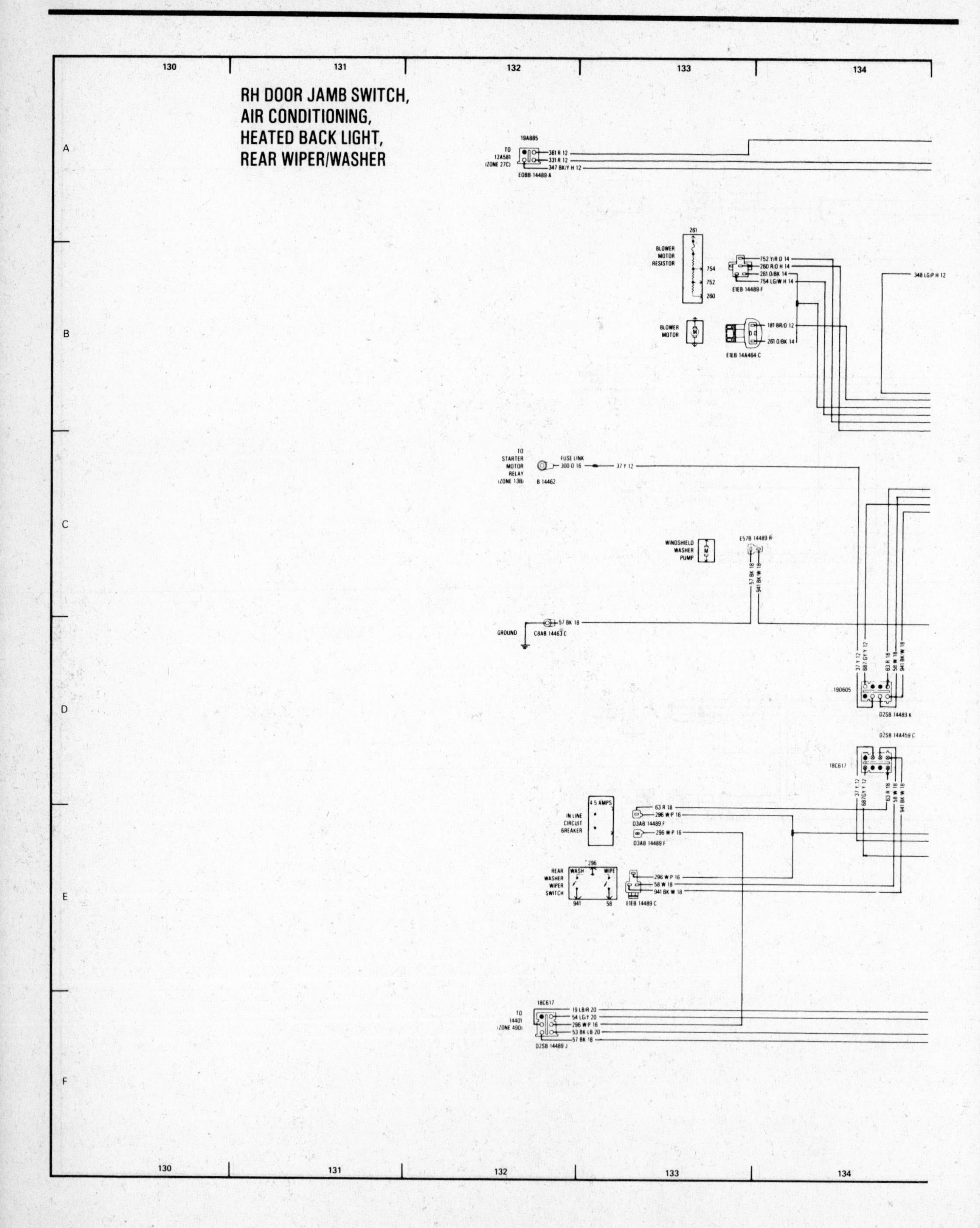

130
131
132
133
134
RH DOOR JAMB SWITCH,
AIR CONDITIONING,
HEATED BACK LIGHT,
REAR WIPER/WASHER
A
B
C
D
E
F
BLOWER MOTOR RESISTOR
BLOWER MOTOR
WINDSHIELD WASHER PUMP
GROUND
IN LINE CIRCUIT BREAKER
REAR WASHER WIPER SWITCH
FUSE LINK

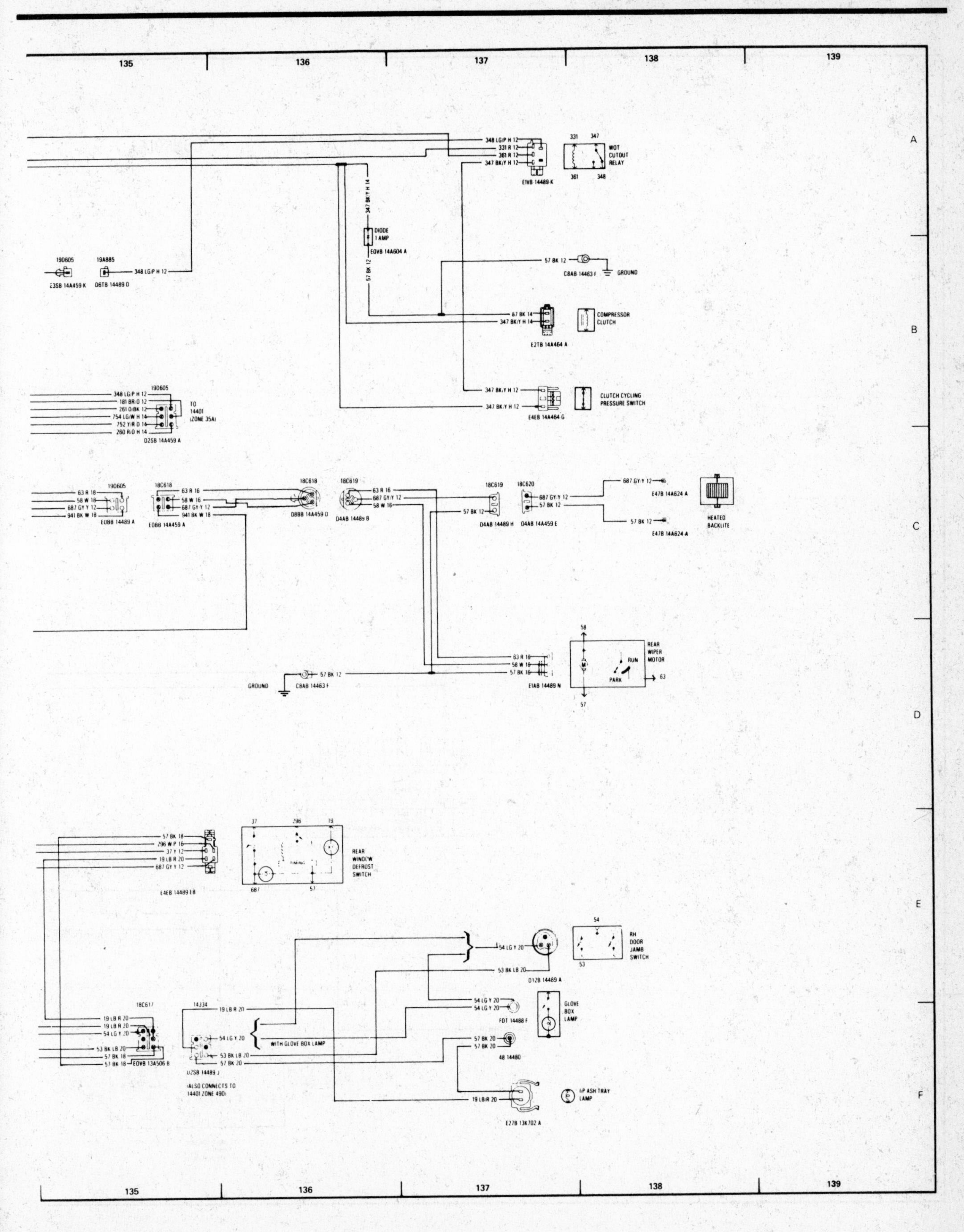
135
136
137
138
139
A
B
C
D
E
F
WOT CUTOUT RELAY
DIODE 1 AMP
GROUND
COMPRESSOR CLUTCH
CLUTCH CYCLING PRESSURE SWITCH
HEATED BACKLITE
REAR WIPER MOTOR
RUN
PARK
REAR WINDOW DEFROST SWITCH
RH DOOR JAMB SWITCH
GLOVE BOX LAMP
WITH GLOVE BOX LAMP
I/P ASH TRAY LAMP

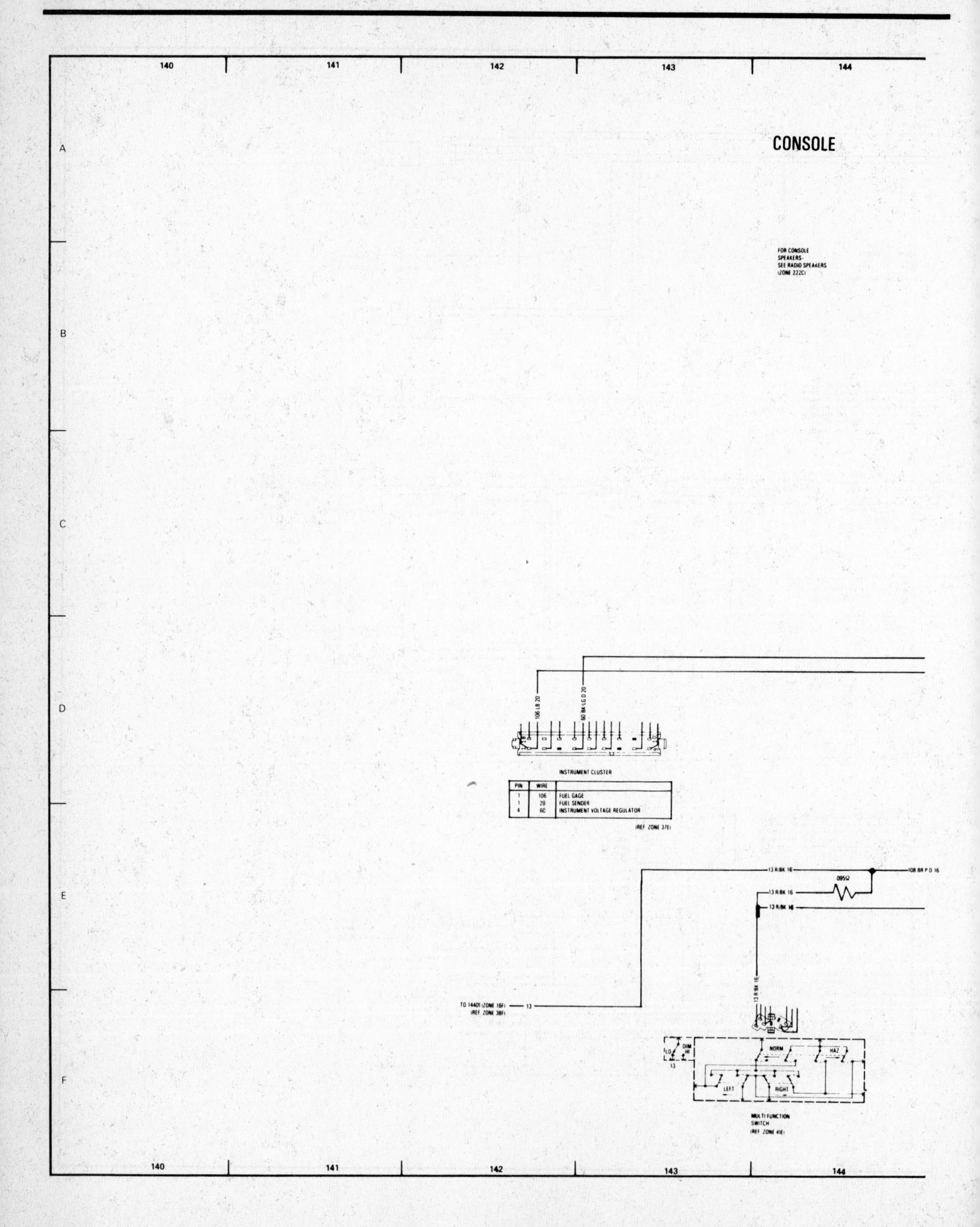

PIN	WIRE	
1	106	FUEL GAGE
1	29	FUEL SENDER
4	60	INSTRUMENT VOLTAGE REGULATOR

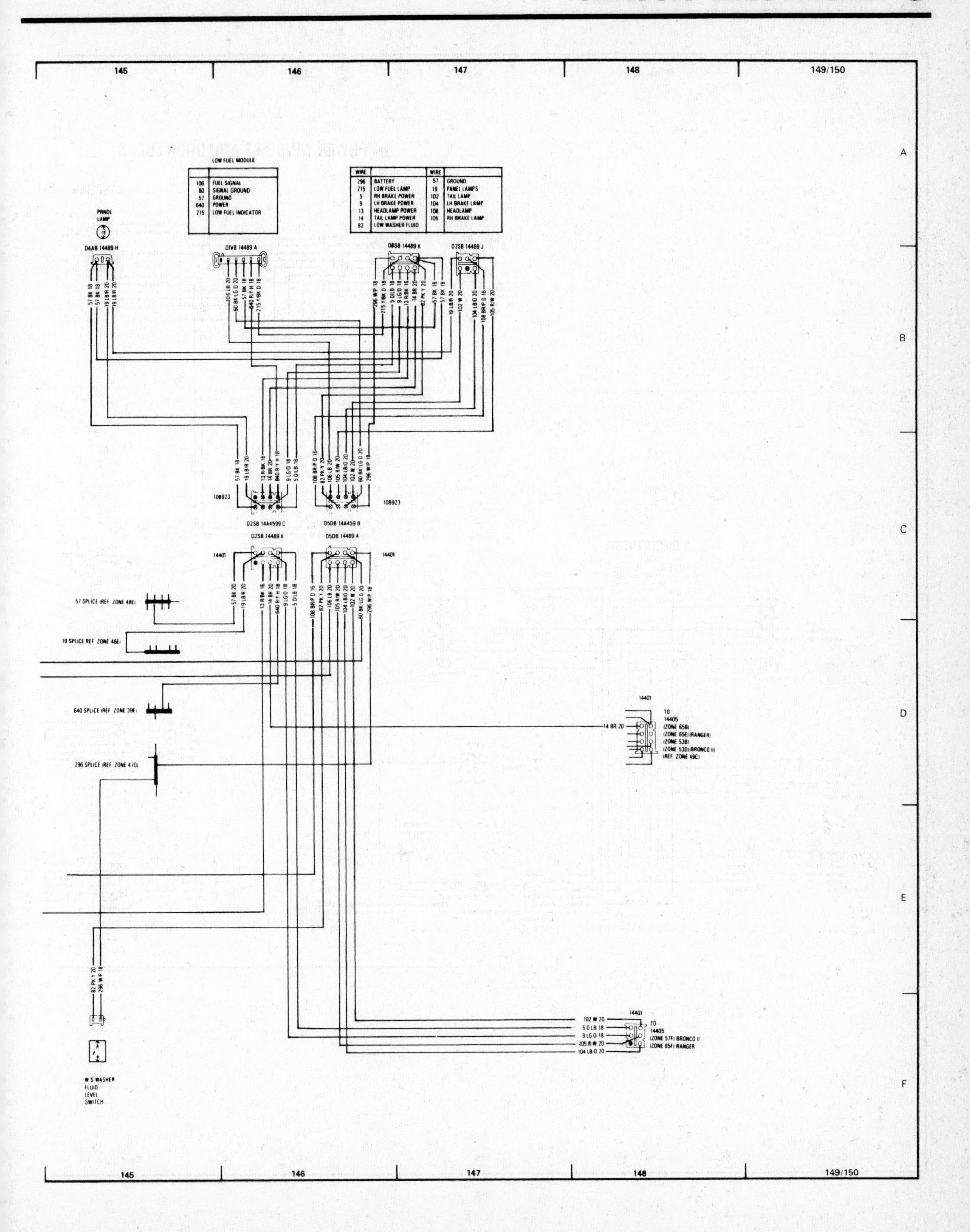
145
146
147
148
149/150
A
B
C
D
E
F
LOW FUEL MODULE
106 FUEL SIGNAL
60 SIGNAL GROUND
57 GROUND
640 POWER
215 LOW FUEL INDICATOR
WIRE
296 BATTERY
215 LOW FUEL LAMP
5 RH BRAKE POWER
9 LH BRAKE POWER
13 HEADLAMP POWER
14 TAIL LAMP POWER
82 LOW WASHER FLUID
WIRE
57 GROUND
19 PANEL LAMPS
102 TAIL LAMP
104 LH BRAKE LAMP
108 HEADLAMP
105 RH BRAKE LAMP
PRNDL LAMP
W S WASHER FLUID LEVEL SWITCH

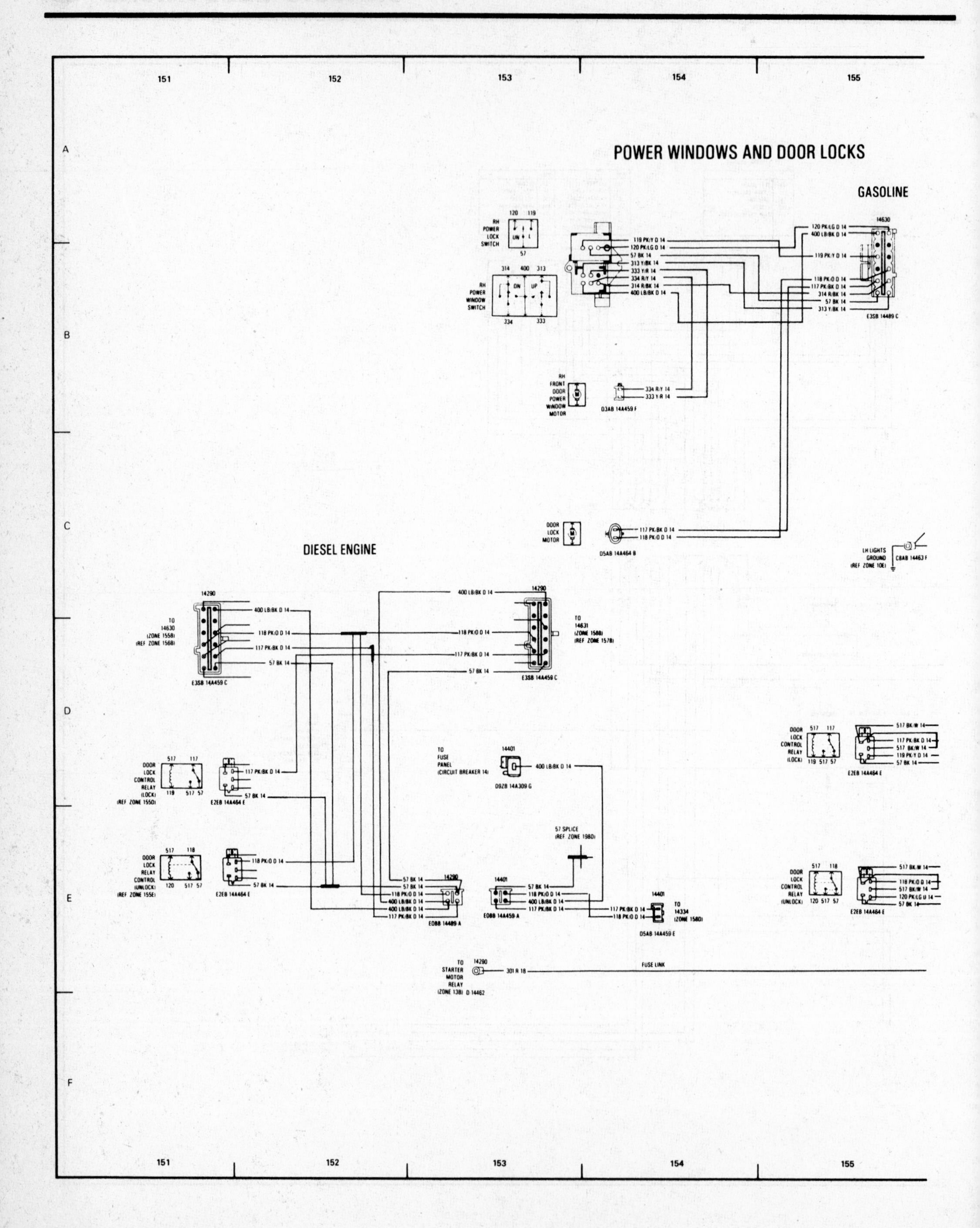

151
152
153
154
155
A
B
C
D
E
F
POWER WINDOWS AND DOOR LOCKS
GASOLINE
RH POWER LOCK SWITCH
RH POWER WINDOW SWITCH
RH FRONT DOOR POWER WINDOW MOTOR
DOOR LOCK MOTOR
LH LIGHTS GROUND (REF ZONE 10E)
DIESEL ENGINE
TO 14630 (ZONE 155B) (REF ZONE 156B)
TO 14631 (ZONE 158B) (REF ZONE 157B)
DOOR LOCK CONTROL RELAY (LOCK) (REF ZONE 155D)
DOOR LOCK CONTROL RELAY (UNLOCK) (REF ZONE 155E)
TO FUSE PANEL (CIRCUIT BREAKER 14)
57 SPLICE (REF ZONE 198D)
TO 14334 (ZONE 158D)
TO STARTER MOTOR RELAY (ZONE 13B)
FUSE LINK
DOOR LOCK CONTROL RELAY (LOCK)
DOOR LOCK CONTROL RELAY (UNLOCK)

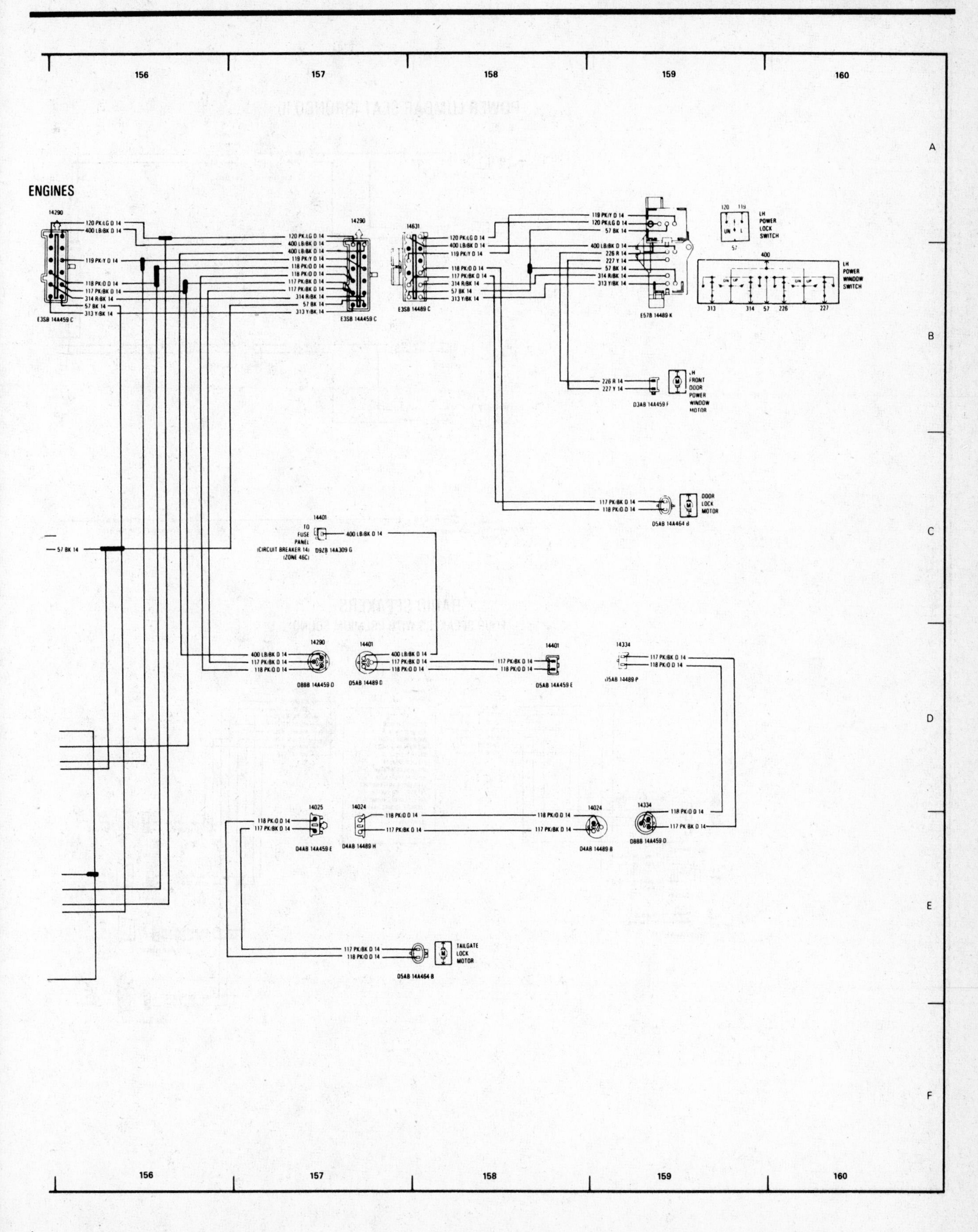

ENGINES
LH POWER LOCK SWITCH
LH POWER WINDOW SWITCH
LH FRONT DOOR POWER WINDOW MOTOR
DOOR LOCK MOTOR
TO FUSE PANEL (CIRCUIT BREAKER 14) (ZONE 46C)
TAILGATE LOCK MOTOR

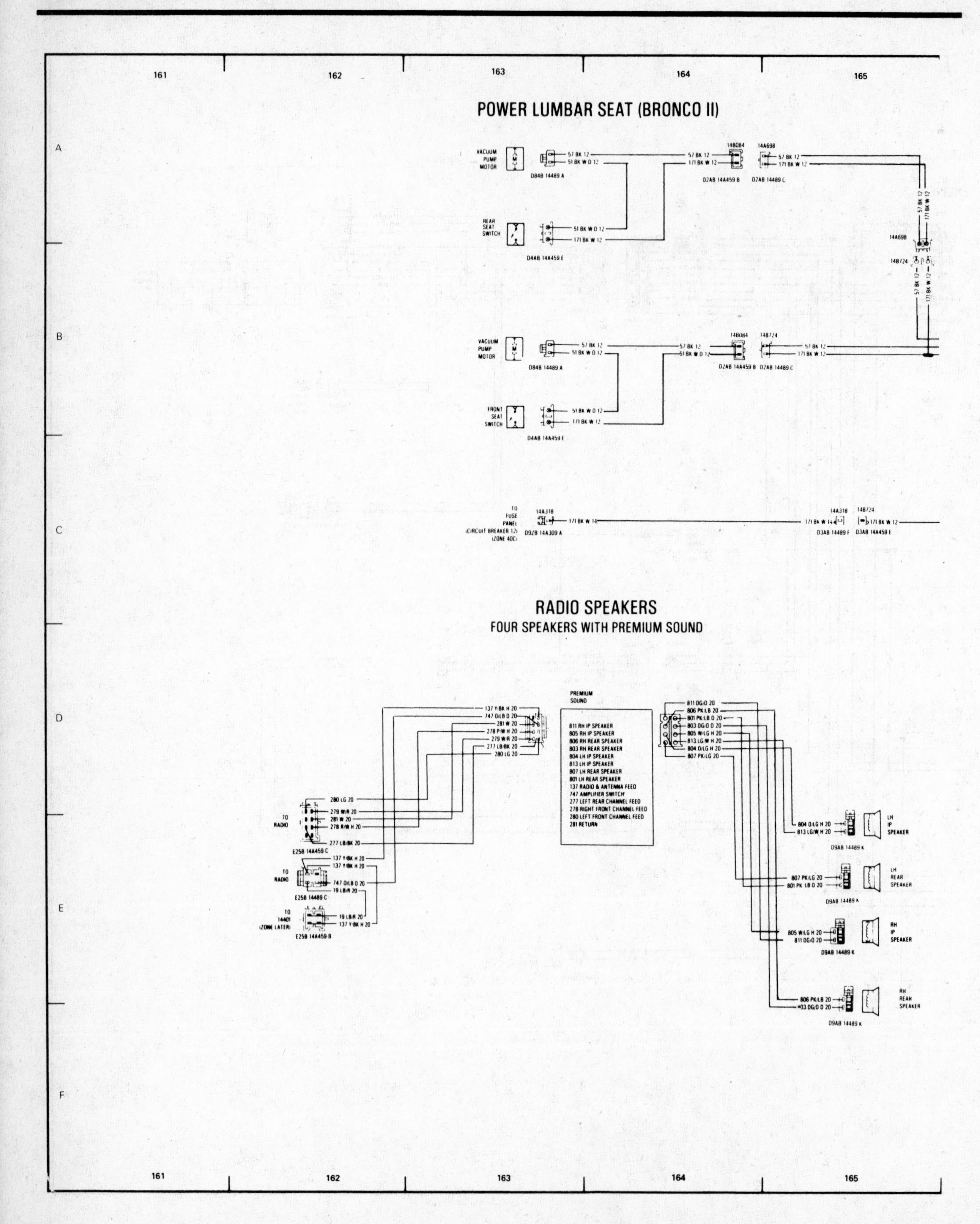
161
162
163
164
165
A
B
C
D
E
F
POWER LUMBAR SEAT (BRONCO II)
VACUUM PUMP MOTOR
REAR SEAT SWITCH
FRONT SEAT SWITCH
TO FUSE PANEL (CIRCUIT BREAKER 12) (ZONE 40C)
RADIO SPEAKERS
FOUR SPEAKERS WITH PREMIUM SOUND
PREMIUM SOUND
811 RH IP SPEAKER
805 RH IP SPEAKER
806 RH REAR SPEAKER
803 RH REAR SPEAKER
804 LH IP SPEAKER
813 LH IP SPEAKER
807 LH REAR SPEAKER
801 LH REAR SPEAKER
137 RADIO & ANTENNA FEED
747 AMPLIFIER SWITCH
277 LEFT REAR CHANNEL FEED
278 RIGHT FRONT CHANNEL FEED
280 LEFT FRONT CHANNEL FEED
281 RETURN
TO RADIO
LH IP SPEAKER
LH REAR SPEAKER
RH IP SPEAKER
RH REAR SPEAKER

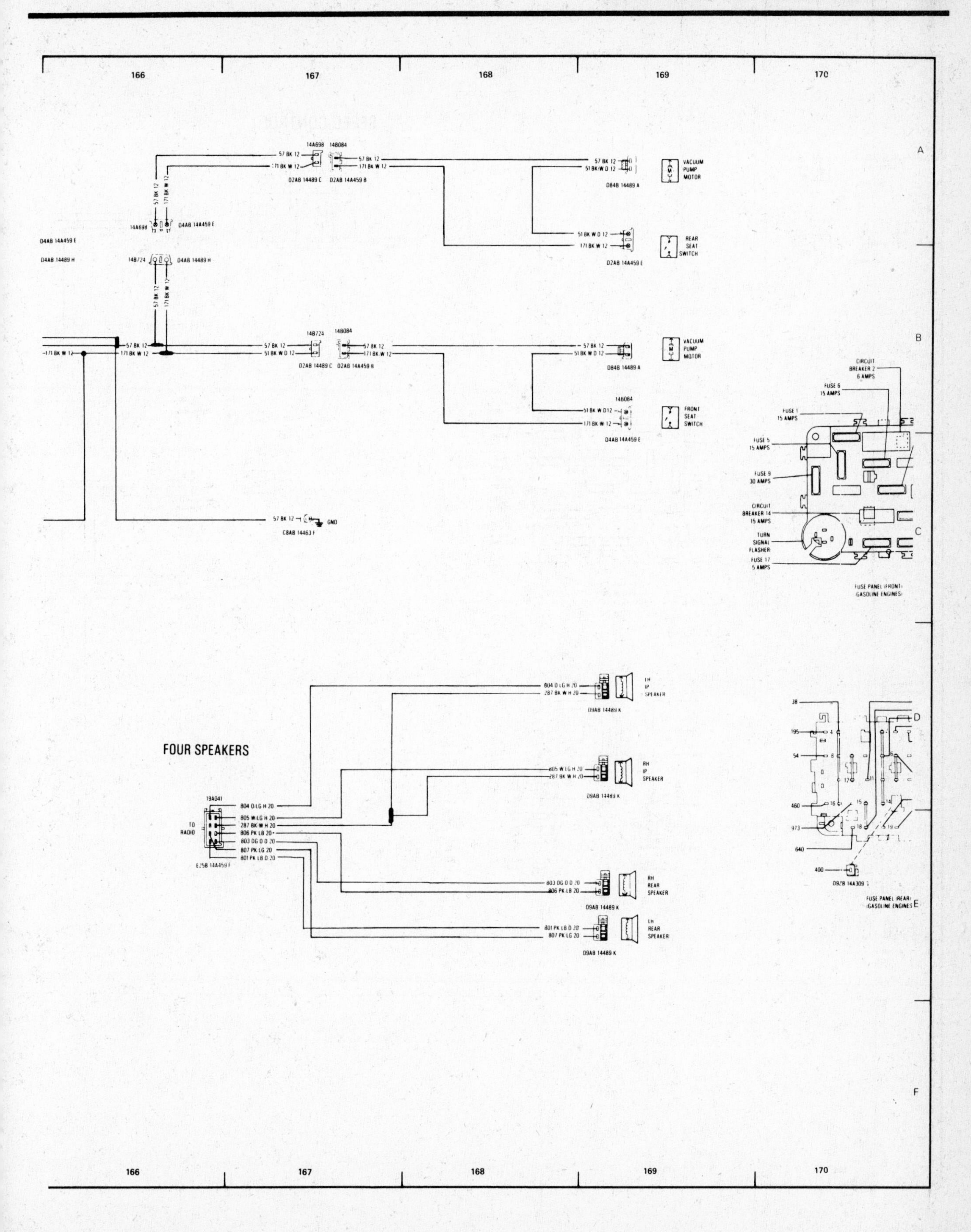

FOUR SPEAKERS
VACUUM PUMP MOTOR
REAR SEAT SWITCH
FRONT SEAT SWITCH
TO RADIO
LH IP SPEAKER
RH IP SPEAKER
RH REAR SPEAKER
LH REAR SPEAKER
FUSE PANEL (FRONT) GASOLINE ENGINES
FUSE PANEL (REAR) GASOLINE ENGINES

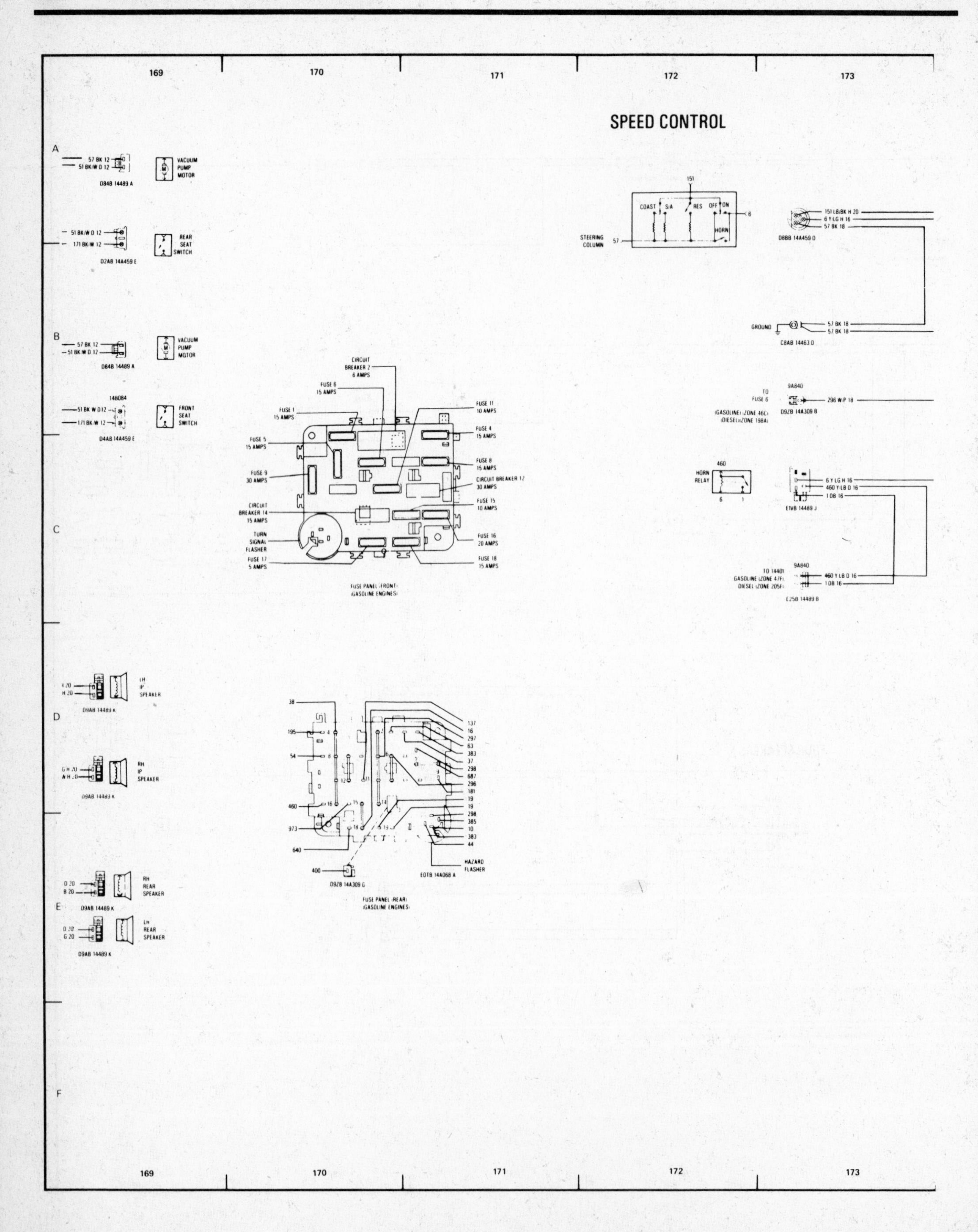
169
170
171
172
173
SPEED CONTROL
VACUUM PUMP MOTOR
REAR SEAT SWITCH
FRONT SEAT SWITCH
STEERING COLUMN
COAST
S/A
RES
OFF
ON
HORN
GROUND
HORN RELAY
CIRCUIT BREAKER 2 6 AMPS
FUSE 6 15 AMPS
FUSE 1 15 AMPS
FUSE 5 15 AMPS
FUSE 9 30 AMPS
CIRCUIT BREAKER 14 15 AMPS
TURN SIGNAL FLASHER
FUSE 17 5 AMPS
FUSE 11 10 AMPS
FUSE 4 15 AMPS
FUSE 8 15 AMPS
CIRCUIT BREAKER 12 30 AMPS
FUSE 15 10 AMPS
FUSE 16 20 AMPS
FUSE 18 15 AMPS
FUSE PANEL (FRONT) (GASOLINE ENGINES)
LH IP SPEAKER
RH IP SPEAKER
RH REAR SPEAKER
LH REAR SPEAKER
HAZARD FLASHER
FUSE PANEL (REAR) (GASOLINE ENGINES)

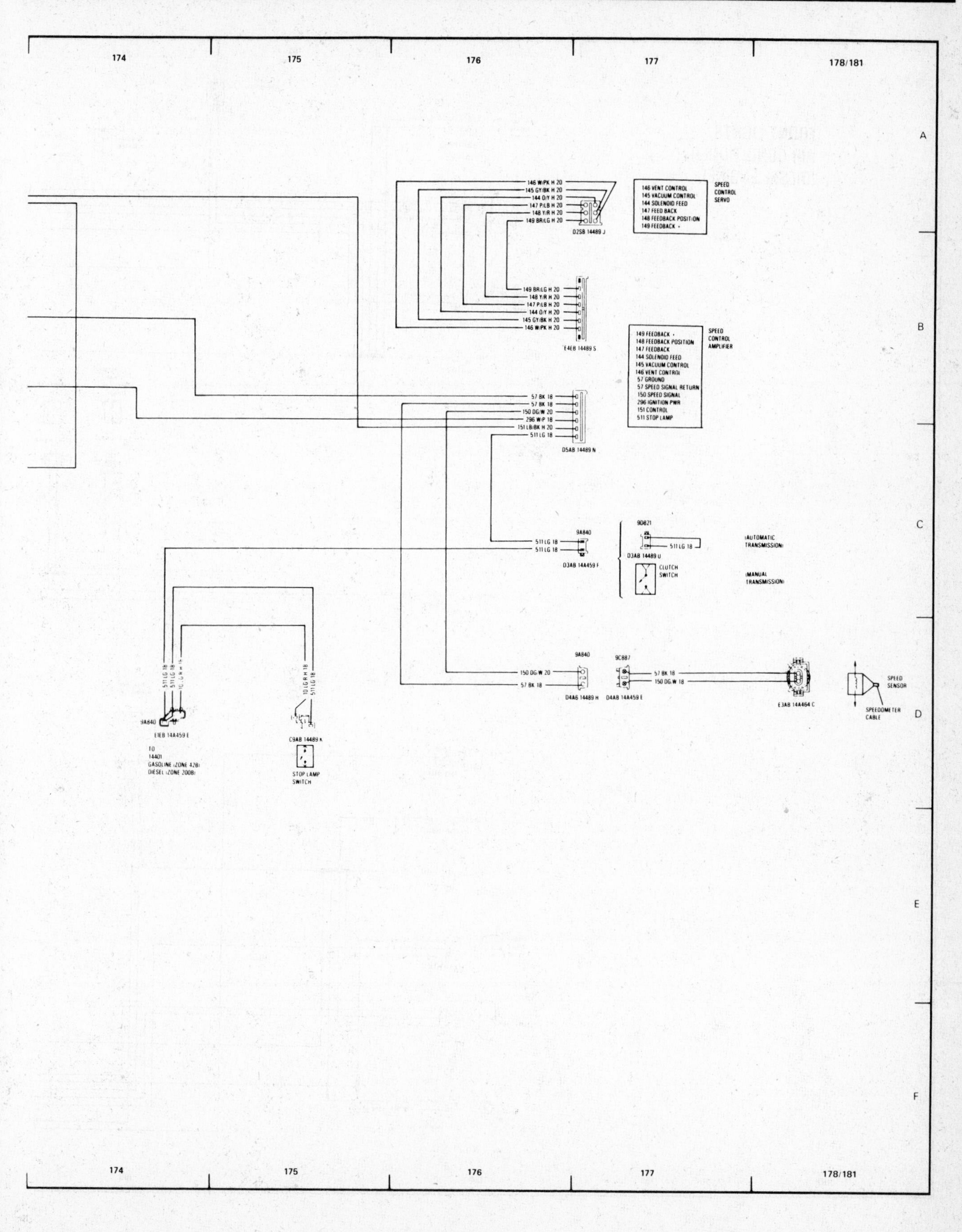
174
175
176
177
178/181
A
B
C
D
E
F
146 W/PK H 20
145 GY/BK H 20
144 O/Y H 20
147 P/LB H 20
148 Y/R H 20
149 BR/LG H 20
D2SB 14489 J
146 VENT CONTROL
145 VACUUM CONTROL
144 SOLENOID FEED
147 FEED BACK
148 FEEDBACK POSITION
149 FEEDBACK +
SPEED CONTROL SERVO
E4EB 14489 S
149 FEEDBACK +
148 FEEDBACK POSITION
147 FEEDBACK
144 SOLENOID FEED
145 VACUUM CONTROL
146 VENT CONTROL
57 GROUND
57 SPEED SIGNAL RETURN
150 SPEED SIGNAL
296 IGNITION PWR
151 CONTROL
511 STOP LAMP
SPEED CONTROL AMPLIFIER
57 BK 18
150 DG/W 20
296 W/P 18
151 LB/BK H 20
511 LG 18
D5AB 14489 N
9A840
D3AB 14A459 F
9D821
D3AB 14489 U
(AUTOMATIC TRANSMISSION)
CLUTCH SWITCH
(MANUAL TRANSMISSION)
9C887
D4A6 14489 H
D4AB 14A459 E
E3AB 14A464 C
SPEED SENSOR
SPEEDOMETER CABLE
E1EB 14A459 E
TO 14401
GASOLINE (ZONE 42B)
DIESEL (ZONE 200B)
C9AB 14489 K
STOP LAMP SWITCH

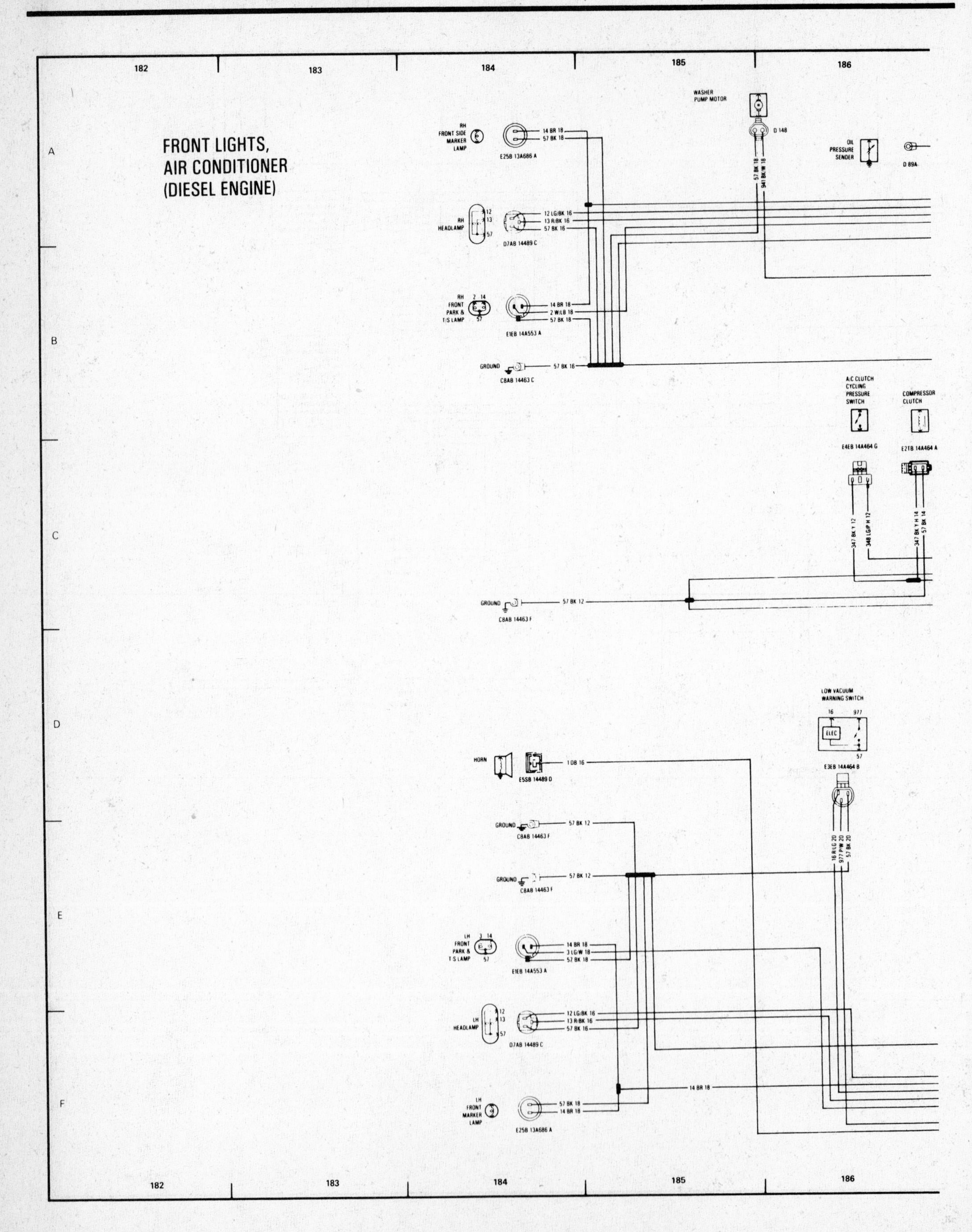
182
183
184
185
186
A
B
C
D
E
F
FRONT LIGHTS,
AIR CONDITIONER
(DIESEL ENGINE)
WASHER PUMP MOTOR
D 148
OIL PRESSURE SENDER
D 89A
RH FRONT SIDE MARKER LAMP
14 BR 18
57 BK 18
E25B 13A686 A
RH HEADLAMP
12 LG/BK 16
13 R/BK 16
57 BK 16
D7AB 14489 C
RH FRONT PARK & T/S LAMP
14 BR 18
2 W/LB 18
57 BK 18
E1EB 14A553 A
GROUND
57 BK 16
C8AB 14463 C
A/C CLUTCH CYCLING PRESSURE SWITCH
COMPRESSOR CLUTCH
E4EB 14A464 G
E2TB 14A464 A
347 BK/Y 12
348 LG/P H 12
57 BK 14
GROUND
57 BK 12
C8AB 14463 F
LOW VACUUM WARNING SWITCH
ELEC
E3EB 14A464 B
HORN
E5SB 14489 D
1 DB 16
GROUND
57 BK 12
C8AB 14463 F
GROUND
57 BK 12
C8AB 14463 F
16 R/LG 20
977 P/W 20
57 BK 20
LH FRONT PARK & T/S LAMP
14 BR 18
3 LG/W 18
57 BK 18
E1EB 14A553 A
LH HEADLAMP
12 LG/BK 16
13 R/BK 16
57 BK 16
D7AB 14489 C
14 BR 18
LH FRONT MARKER LAMP
57 BK 18
14 BR 18
E25B 13A686 A

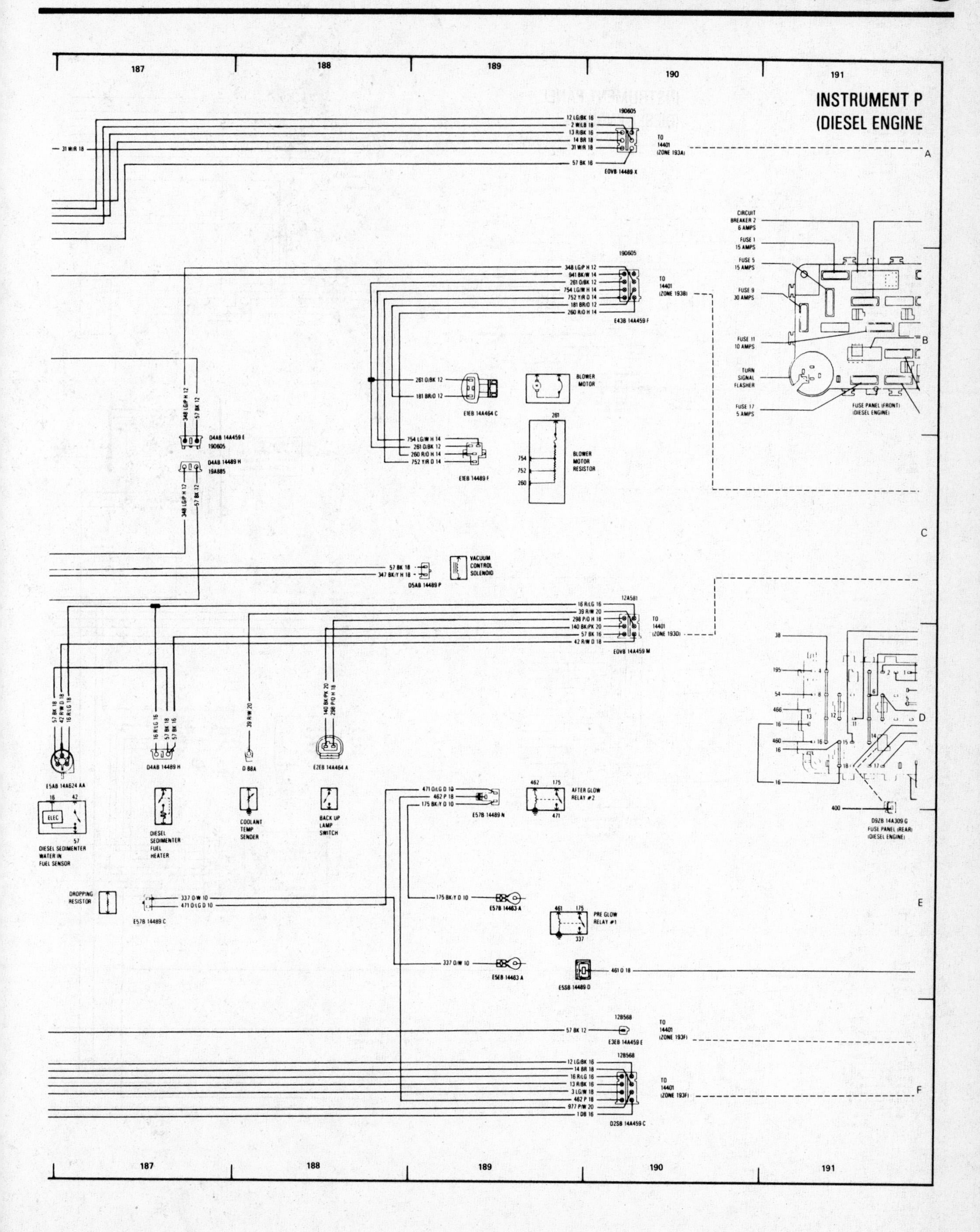

187
188
189
190
191
INSTRUMENT P
(DIESEL ENGINE
BLOWER MOTOR
BLOWER MOTOR RESISTOR
VACUUM CONTROL SOLENOID
DIESEL SEDIMENTER WATER IN FUEL SENSOR
DIESEL SEDIMENTER FUEL HEATER
COOLANT TEMP SENDER
BACK UP LAMP SWITCH
AFTER GLOW RELAY #2
DROPPING RESISTOR
PRE GLOW RELAY #1
FUSE PANEL (FRONT) (DIESEL ENGINE)
FUSE PANEL (REAR) (DIESEL ENGINE)

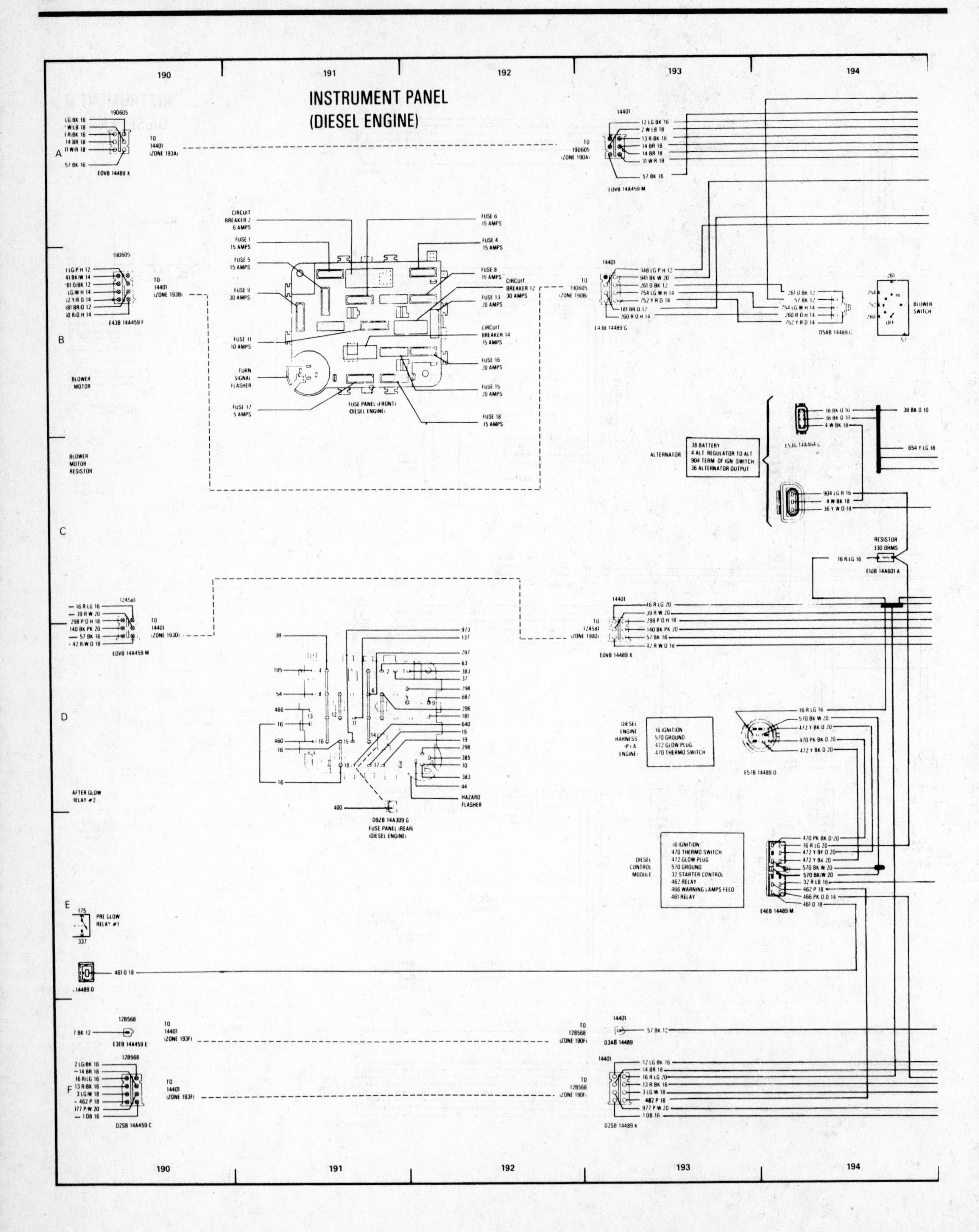
INSTRUMENT PANEL
(DIESEL ENGINE)
190
191
192
193
194
CIRCUIT BREAKER 2 6 AMPS
FUSE 1 15 AMPS
FUSE 5 15 AMPS
FUSE 9 30 AMPS
FUSE 11 10 AMPS
TURN SIGNAL FLASHER
FUSE 17 5 AMPS
FUSE 6 15 AMPS
FUSE 4 15 AMPS
FUSE 8 15 AMPS
CIRCUIT BREAKER 12 30 AMPS
FUSE 13 20 AMPS
CIRCUIT BREAKER 14 15 AMPS
FUSE 16 20 AMPS
FUSE 15 20 AMPS
FUSE 18 15 AMPS
FUSE PANEL (FRONT) (DIESEL ENGINE)
BLOWER MOTOR
BLOWER MOTOR RESISTOR
BLOWER SWITCH
ALTERNATOR
38 BATTERY
4 ALT. REGULATOR TO ALT
904 TERM OF IGN SWITCH
36 ALTERNATOR OUTPUT
RESISTOR 330 OHMS
HAZARD FLASHER
D9ZB 14A309 G
FUSE PANEL (REAR) (DIESEL ENGINE)
DIESEL ENGINE HARNESS (P+A ENGINE)
16 IGNITION
570 GROUND
472 GLOW PLUG
470 THERMO SWITCH
AFTER GLOW RELAY #2
DIESEL CONTROL MODULE
16 IGNITION
470 THERMO SWITCH
472 GLOW PLUG
570 GROUND
32 STARTER CONTROL
462 RELAY
466 WARNING LAMPS FEED
461 RELAY
PRE GLOW RELAY #1

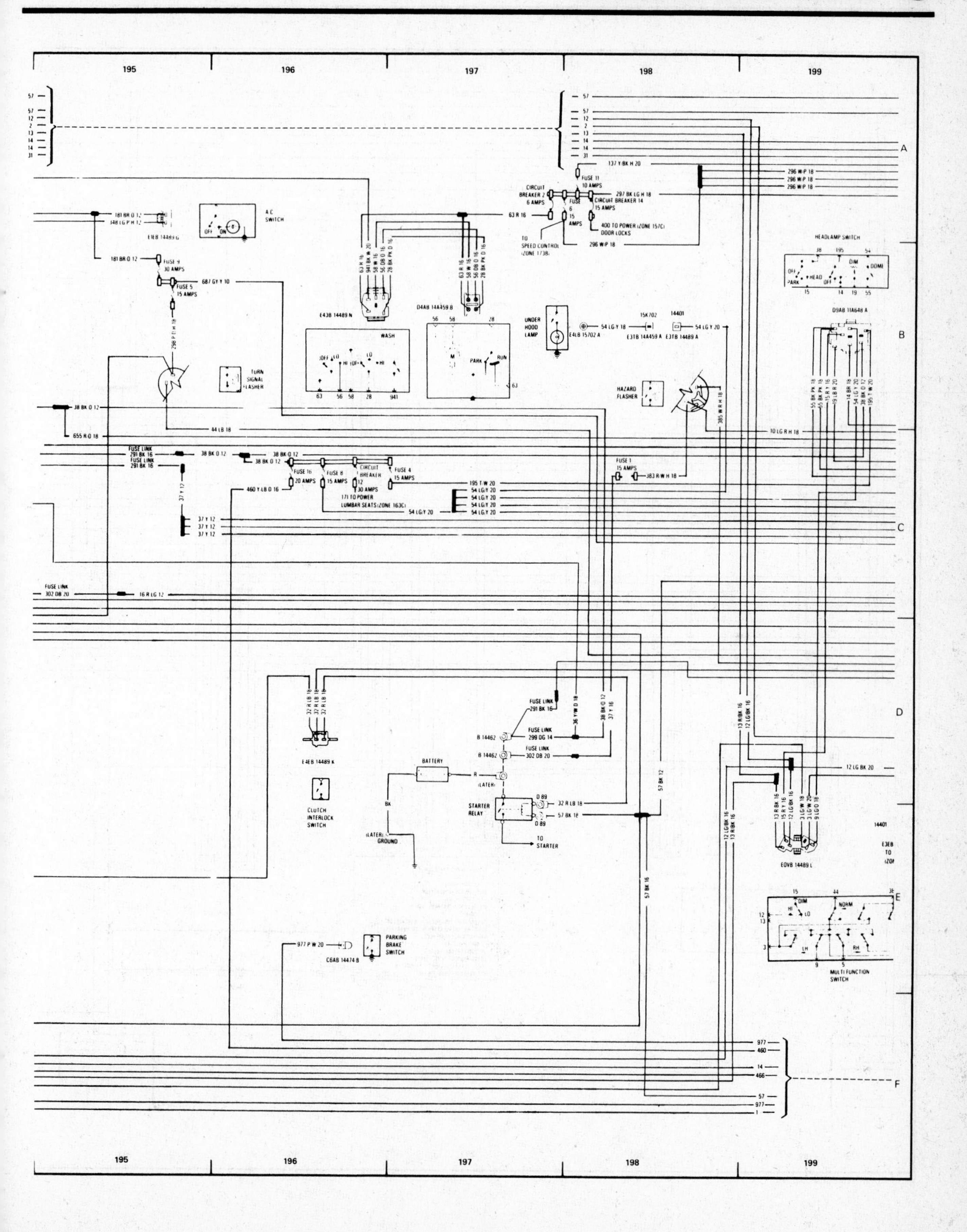
195
196
197
198
199
A
B
C
D
E
F
CIRCUIT BREAKER 2 6 AMPS
FUSE 11 10 AMPS
CIRCUIT BREAKER 14 15 AMPS
400 TO POWER (ZONE 157C) DOOR LOCKS
TO SPEED CONTROL (ZONE 173B)
A C SWITCH
FUSE 9 30 AMPS
FUSE 5 15 AMPS
HEADLAMP SWITCH
WASH
UNDER HOOD LAMP
TURN SIGNAL FLASHER
HAZARD FLASHER
FUSE 16 20 AMPS
FUSE 8 15 AMPS
CIRCUIT BREAKER 12 30 AMPS
FUSE 4 15 AMPS
171 TO POWER LUMBAR SEATS (ZONE 163C)
FUSE 1 15 AMPS
BATTERY
STARTER RELAY
TO STARTER
GROUND
CLUTCH INTERLOCK SWITCH
PARKING BRAKE SWITCH
MULTI FUNCTION SWITCH

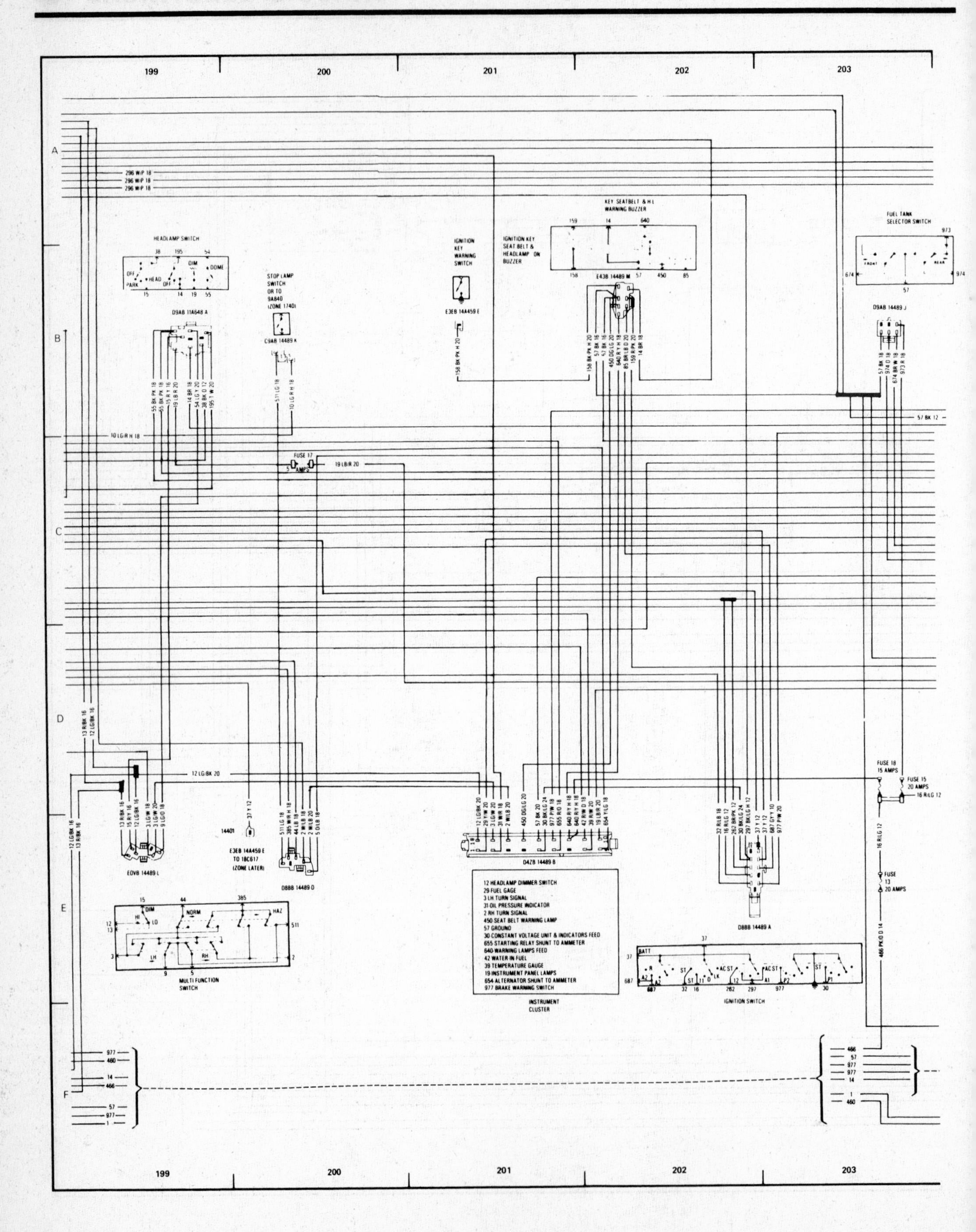

199
200
201
202
203
HEADLAMP SWITCH
D9AB 11A648 A
STOP LAMP SWITCH OR TO 9A840 (ZONE 1740)
C9AB 14489 K
IGNITION KEY WARNING SWITCH
E3EB 14A459 E
IGNITION KEY SEAT BELT & HEADLAMP ON BUZZER
KEY SEATBELT & H L WARNING BUZZER
E43B 14489 M
FUEL TANK SELECTOR SWITCH
D9AB 14489 J
FUSE 17 5 AMPS
10 LG/R H 18
19 LB/R 20
12 LG/BK 20
57 BK 12
E0VB 14489 L
E3EB 14A459 E TO 18C617 (ZONE LATER)
D8BB 14489 D
D4ZB 14489 B
12 HEADLAMP DIMMER SWITCH
29 FUEL GAGE
3 LH TURN SIGNAL
31 OIL PRESSURE INDICATOR
2 RH TURN SIGNAL
450 SEAT BELT WARNING LAMP
57 GROUND
30 CONSTANT VOLTAGE UNIT & INDICATORS FEED
655 STARTING RELAY SHUNT TO AMMETER
640 WARNING LAMPS FEED
42 WATER IN FUEL
39 TEMPERATURE GAUGE
19 INSTRUMENT PANEL LAMPS
654 ALTERNATOR SHUNT TO AMMETER
977 BRAKE WARNING SWITCH
INSTRUMENT CLUSTER
MULTI FUNCTION SWITCH
D8BB 14489 A
IGNITION SWITCH
FUSE 18 15 AMPS
FUSE 15 20 AMPS
FUSE 13 20 AMPS

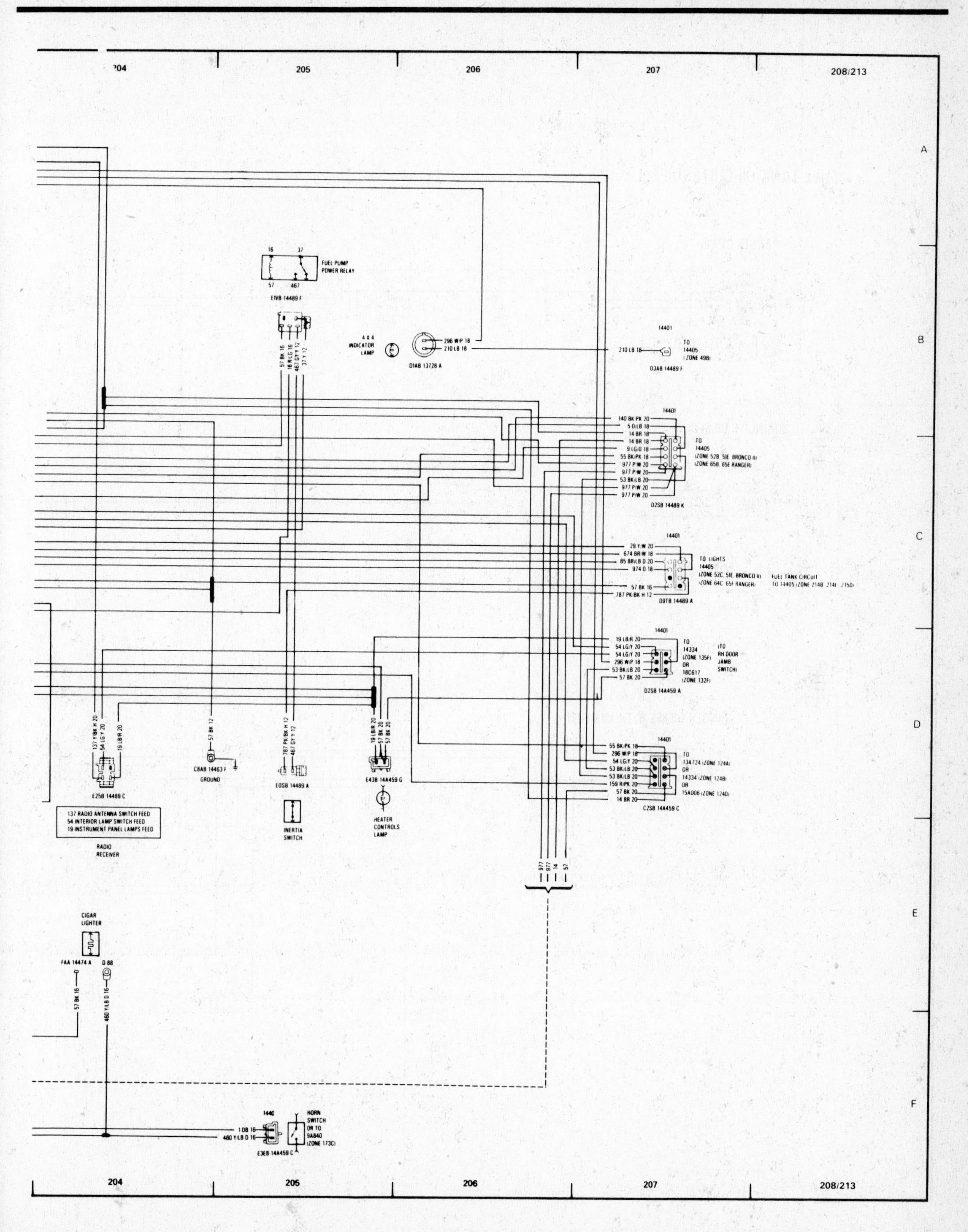

204
205
206
207
208/213
FUEL PUMP POWER RELAY
4 X 4 INDICATOR LAMP
INERTIA SWITCH
HEATER CONTROLS LAMP
GROUND
137 RADIO ANTENNA SWITCH FEED
54 INTERIOR LAMP SWITCH FEED
19 INSTRUMENT PANEL LAMPS FEED
RADIO RECEIVER
CIGAR LIGHTER
HORN SWITCH OR TO 9A840 (ZONE 173C)
FUEL TANK CIRCUIT TO 14405 (ZONE 214B, 214C, 215D)

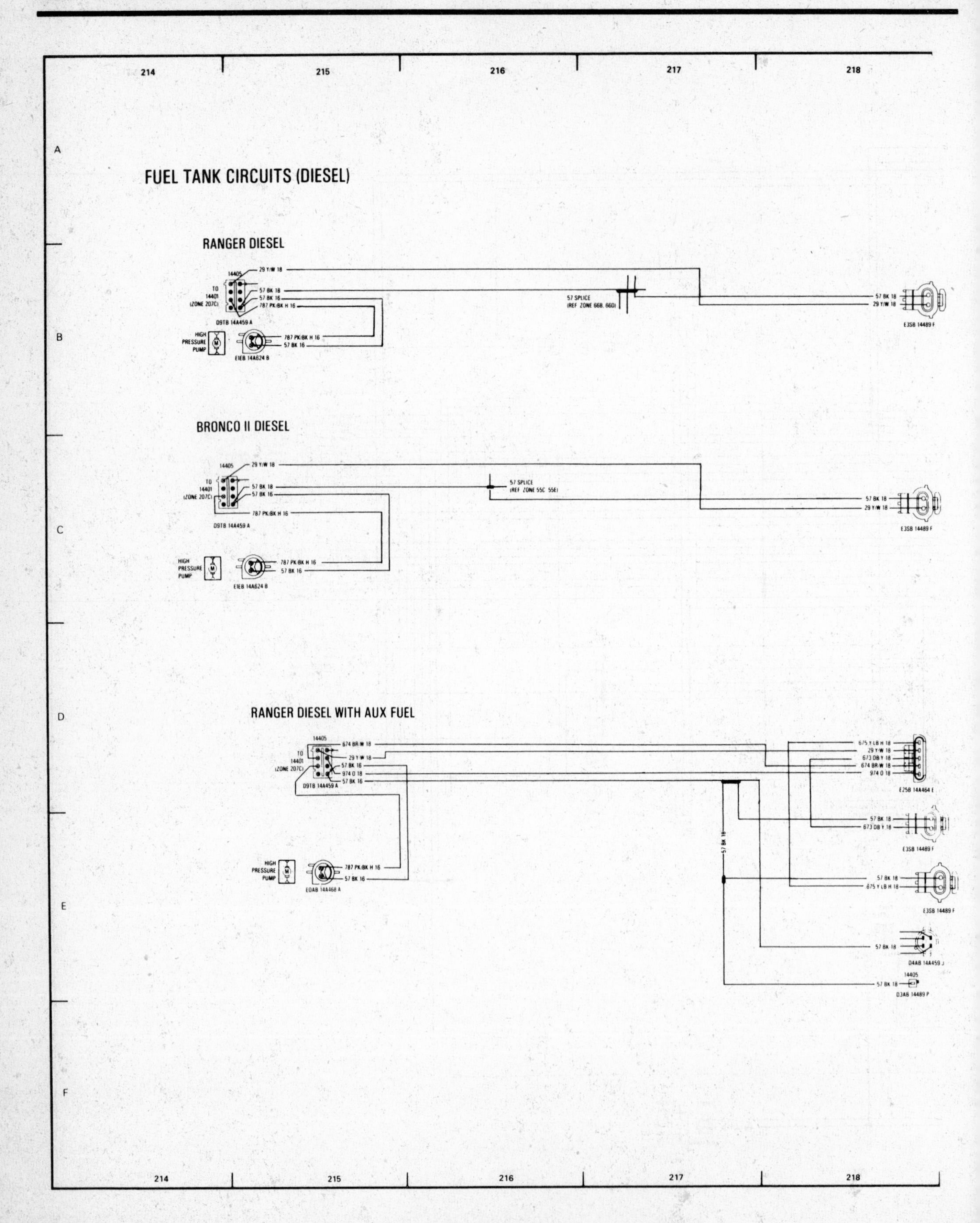
214
215
216
217
218
A
B
C
D
E
F
FUEL TANK CIRCUITS (DIESEL)
RANGER DIESEL
BRONCO II DIESEL
RANGER DIESEL WITH AUX FUEL

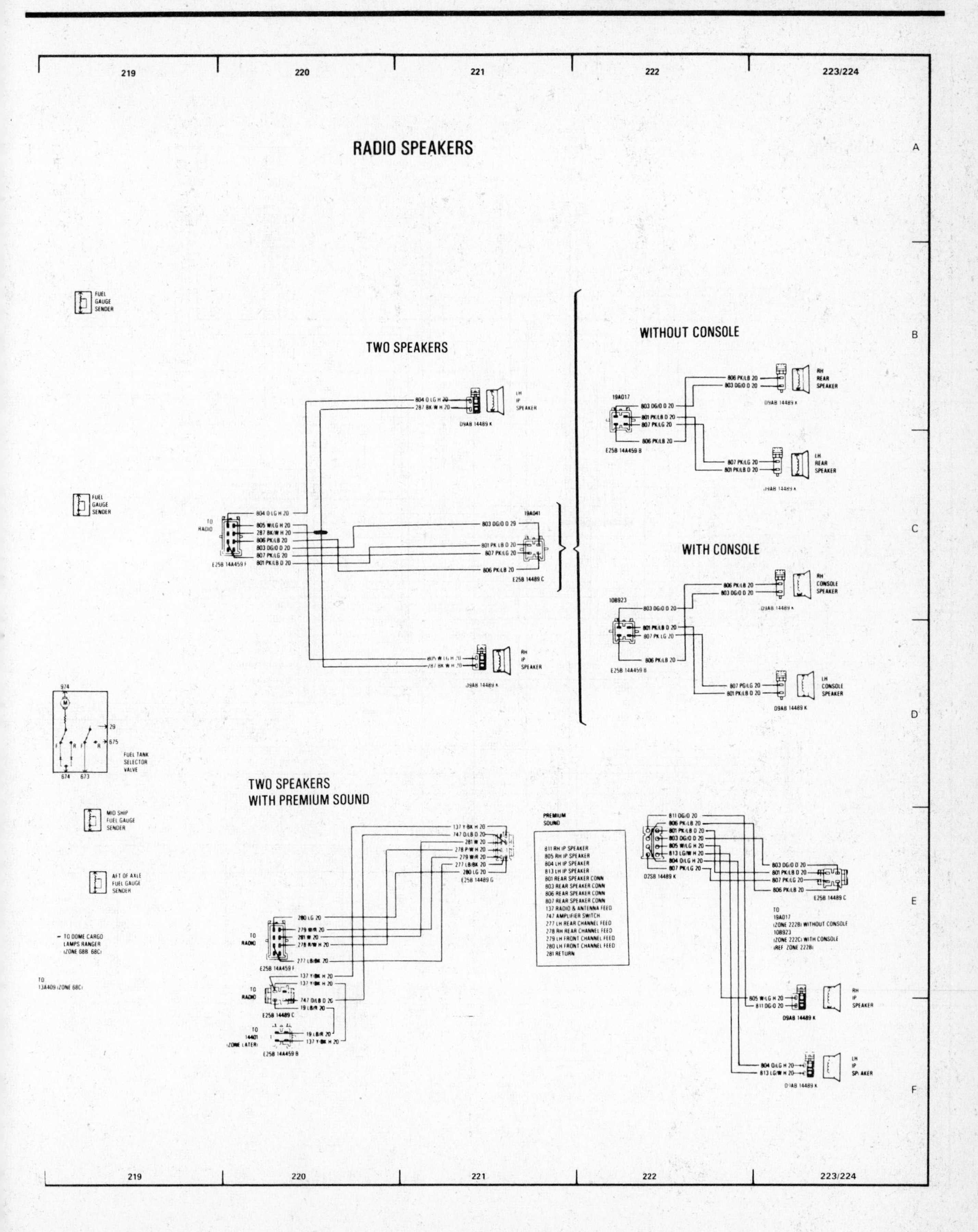
RADIO SPEAKERS
TWO SPEAKERS
WITHOUT CONSOLE
WITH CONSOLE
TWO SPEAKERS
WITH PREMIUM SOUND
219
220
221
222
223/224
A
B
C
D
E
F

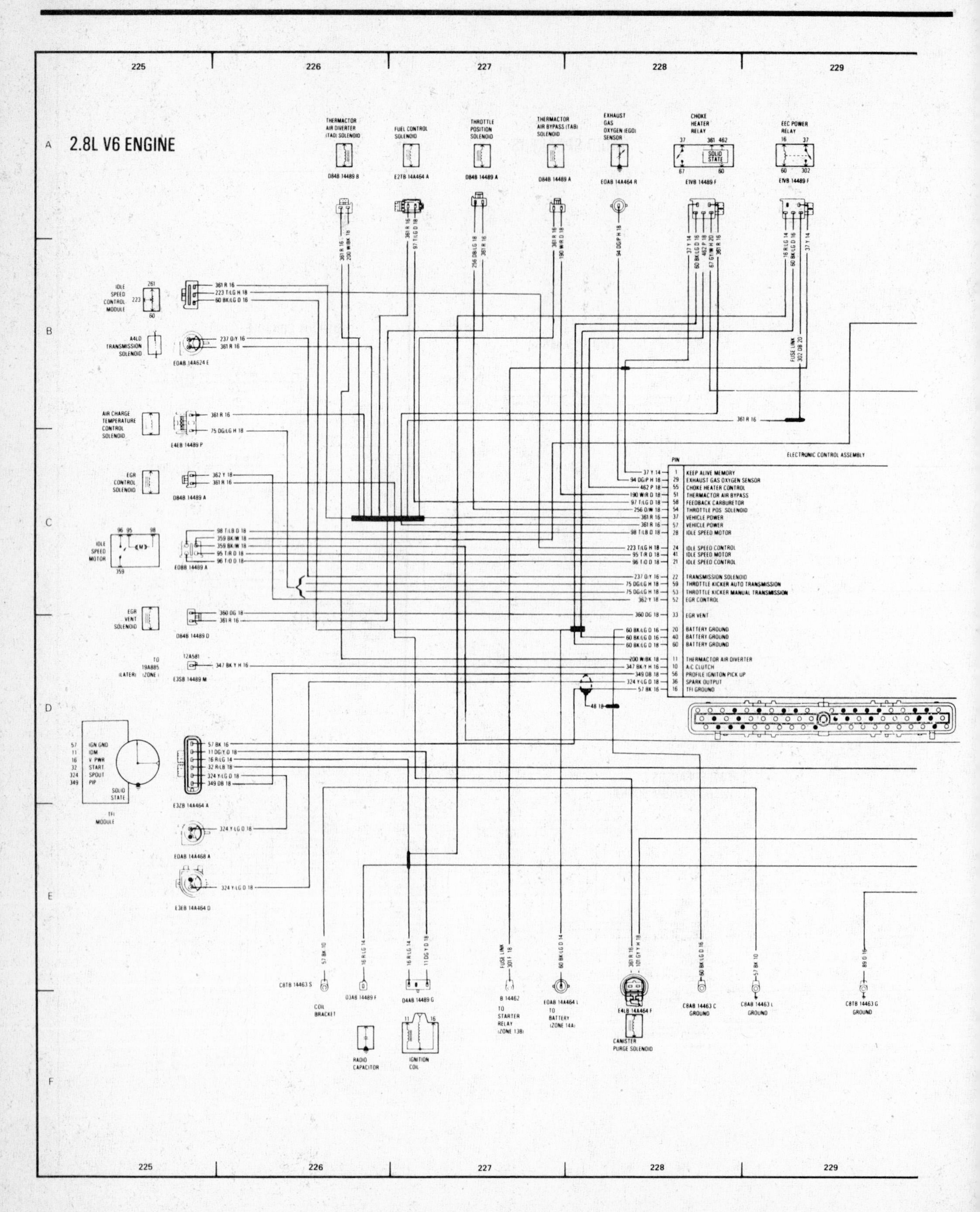
2.8L V6 ENGINE
225
226
227
228
229
THERMACTOR AIR DIVERTER (TAD) SOLENOID
FUEL CONTROL SOLENOID
THROTTLE POSITION SOLENOID
THERMACTOR AIR BYPASS (TAB) SOLENOID
EXHAUST GAS OXYGEN (EGO) SENSOR
CHOKE HEATER RELAY
EEC POWER RELAY
IDLE SPEED CONTROL MODULE
A4LD TRANSMISSION SOLENOID
AIR CHARGE TEMPERATURE CONTROL SOLENOID
EGR CONTROL SOLENOID
IDLE SPEED MOTOR
EGR VENT SOLENOID
TFI MODULE
ELECTRONIC CONTROL ASSEMBLY
PIN
1 KEEP ALIVE MEMORY
29 EXHAUST GAS OXYGEN SENSOR
55 CHOKE HEATER CONTROL
51 THERMACTOR AIR BYPASS
58 FEEDBACK CARBURETOR
54 THROTTLE POS. SOLENOID
37 VEHICLE POWER
57 VEHICLE POWER
28 IDLE SPEED MOTOR
24 IDLE SPEED CONTROL
41 IDLE SPEED MOTOR
21 IDLE SPEED CONTROL
22 TRANSMISSION SOLENOID
59 THROTTLE KICKER AUTO TRANSMISSION
53 THROTTLE KICKER MANUAL TRANSMISSION
52 EGR CONTROL
33 EGR VENT
20 BATTERY GROUND
40 BATTERY GROUND
60 BATTERY GROUND
11 THERMACTOR AIR DIVERTER
10 A/C CLUTCH
56 PROFILE IGNITON PICK UP
36 SPARK OUTPUT
16 TFI GROUND
COIL BRACKET
RADIO CAPACITOR
IGNITION COIL
TO STARTER RELAY (ZONE 13B)
TO BATTERY (ZONE 14A)
CANISTER PURGE SOLENOID
GROUND

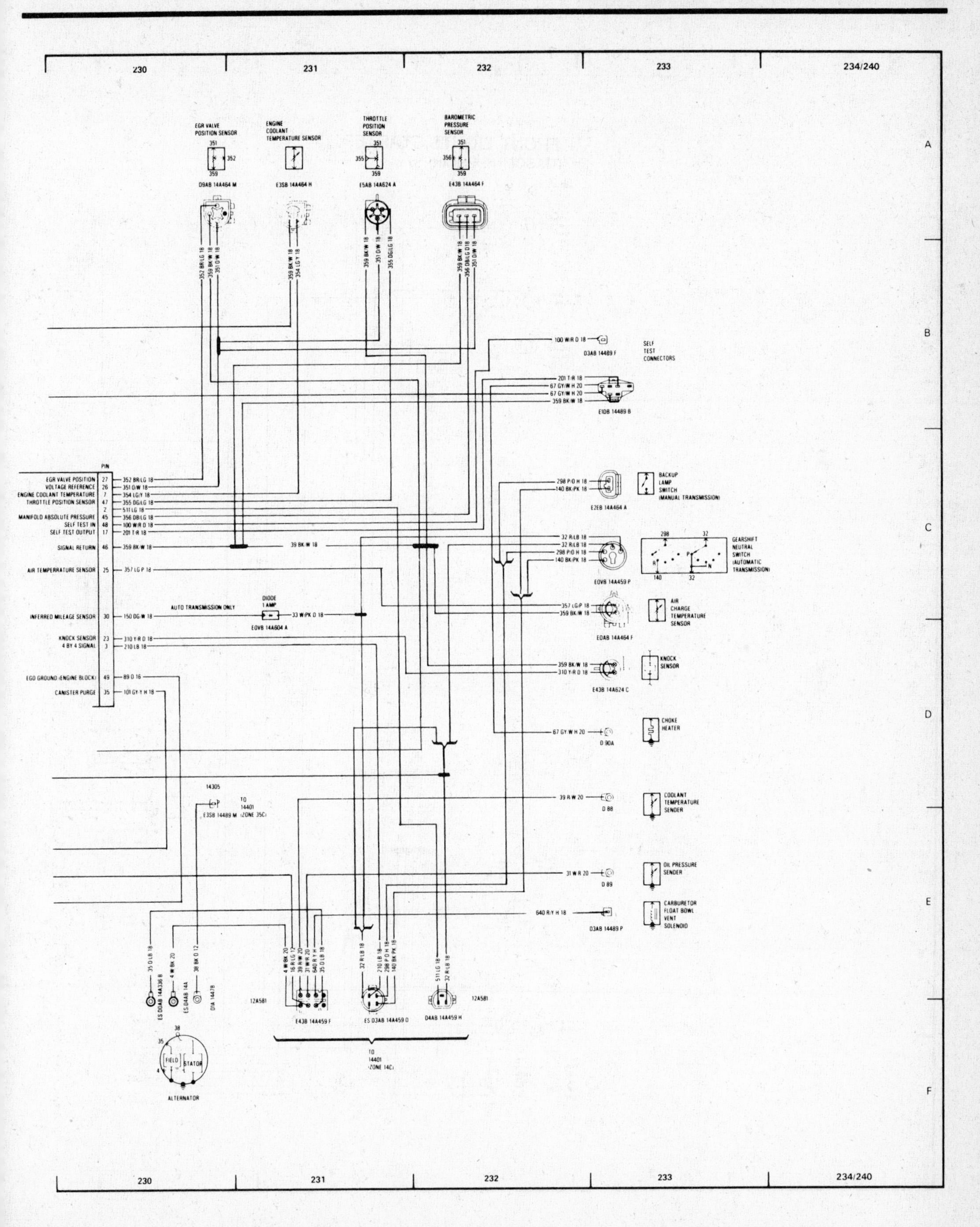

EGR VALVE POSITION SENSOR
ENGINE COOLANT TEMPERATURE SENSOR
THROTTLE POSITION SENSOR
BAROMETRIC PRESSURE SENSOR
SELF TEST CONNECTORS
BACKUP LAMP SWITCH (MANUAL TRANSMISSION)
GEARSHIFT NEUTRAL SWITCH (AUTOMATIC TRANSMISSION)
AIR CHARGE TEMPERATURE SENSOR
KNOCK SENSOR
CHOKE HEATER
COOLANT TEMPERATURE SENDER
OIL PRESSURE SENDER
CARBURETOR FLOAT BOWL VENT SOLENOID
EGR VALVE POSITION
VOLTAGE REFERENCE
ENGINE COOLANT TEMPERATURE
THROTTLE POSITION SENSOR
MANIFOLD ABSOLUTE PRESSURE
SELF TEST IN
SELF TEST OUTPUT
SIGNAL RETURN
AIR TEMPERATURE SENSOR
INFERRED MILEAGE SENSOR
KNOCK SENSOR
4 BY 4 SIGNAL
EGO GROUND (ENGINE BLOCK)
CANISTER PURGE
AUTO TRANSMISSION ONLY
DIODE 1 AMP
ALTERNATOR
FIELD
STATOR
TO 14401 (ZONE 14C)
TO 14401 (ZONE 35C)

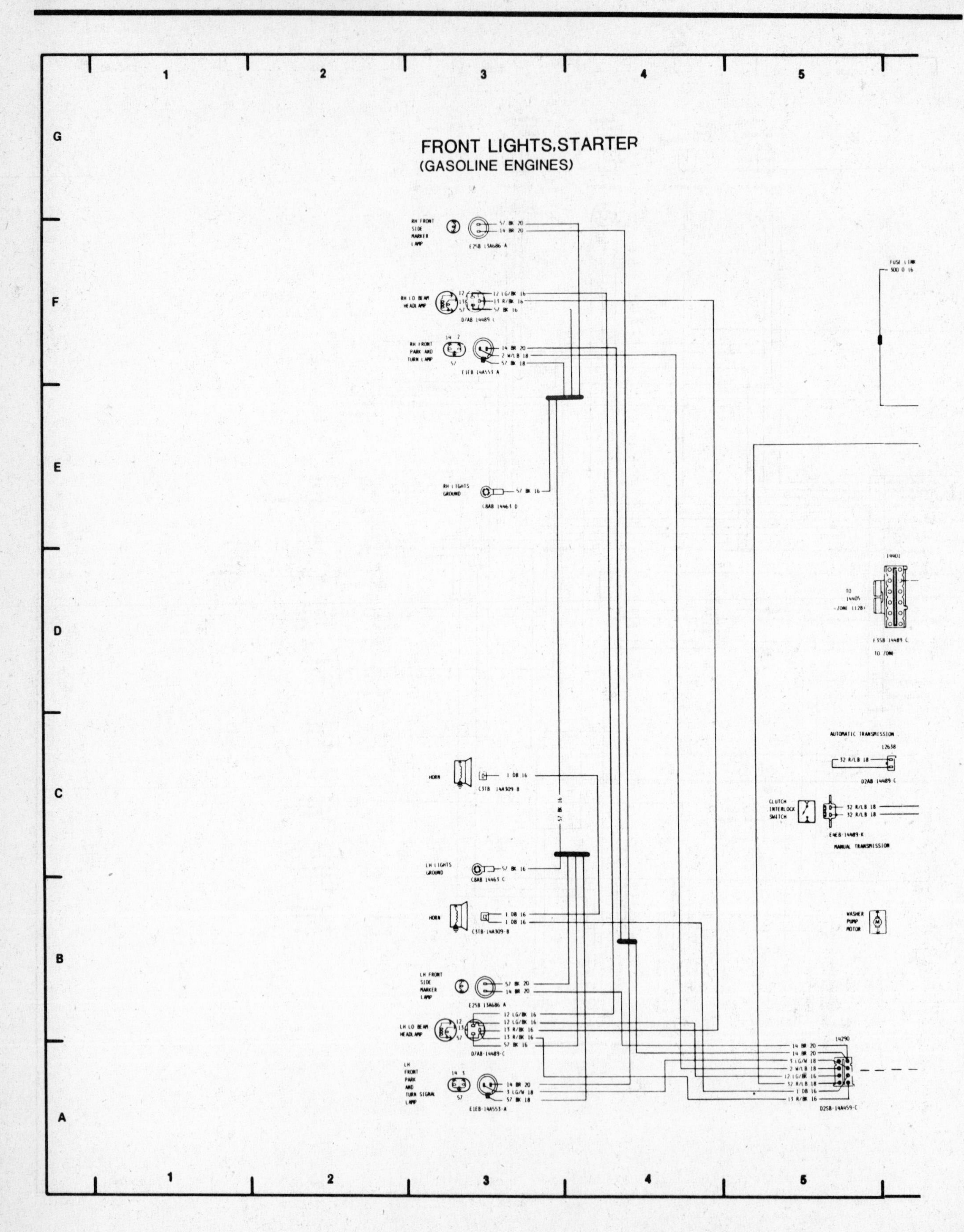
FRONT LIGHTS,STARTER
(GASOLINE ENGINES)
1
2
3
4
5
G
F
E
D
C
B
A
HORN
AUTOMATIC TRANSMISSION
MANUAL TRANSMISSION
CLUTCH INTERLOCK SWITCH
WASHER PUMP MOTOR

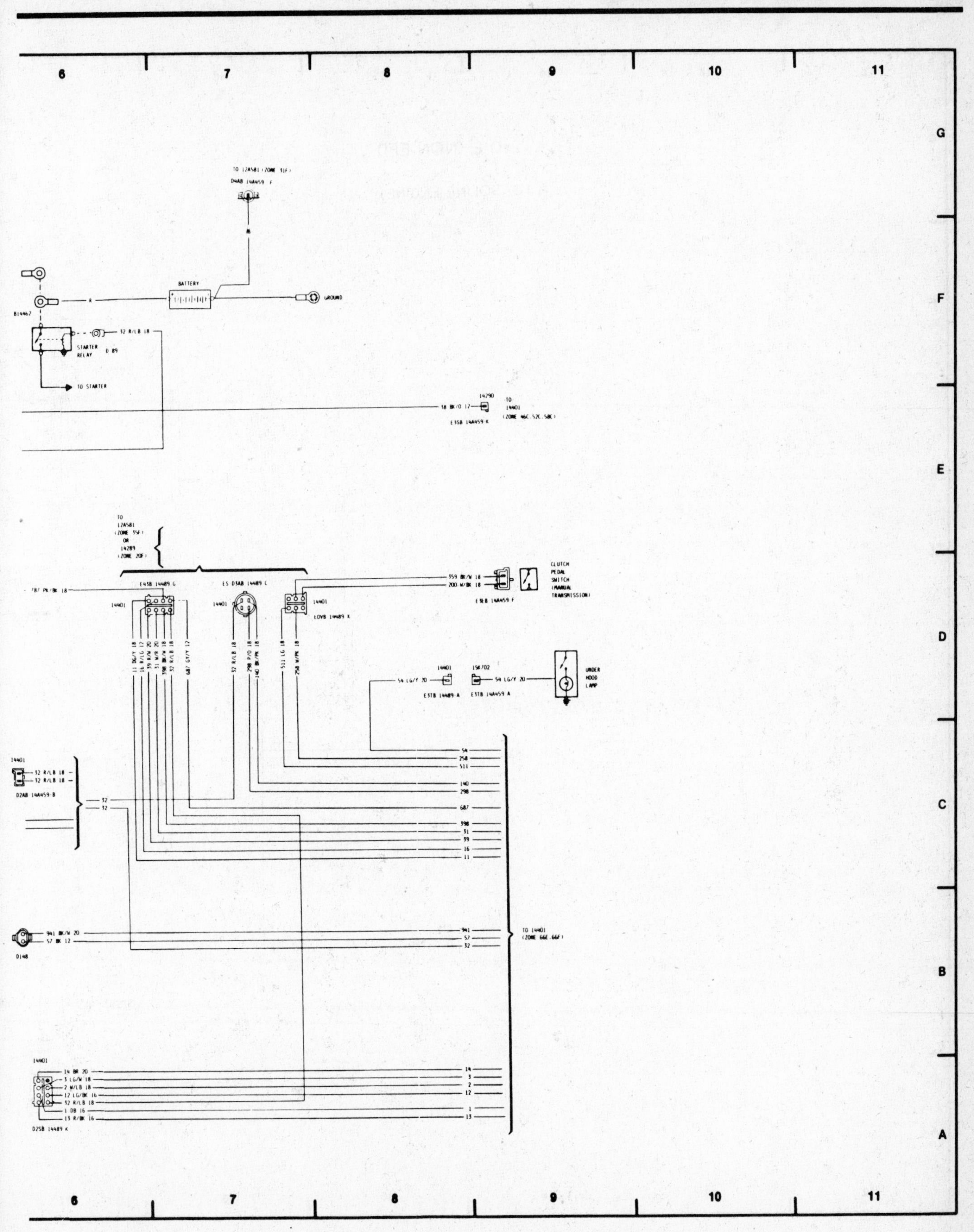
6
7
8
9
10
11
G
F
E
D
C
B
A
BATTERY
GROUND
STARTER RELAY
TO STARTER
CLUTCH PEDAL SWITCH (MANUAL TRANSMISSION)
UNDER HOOD LAMP

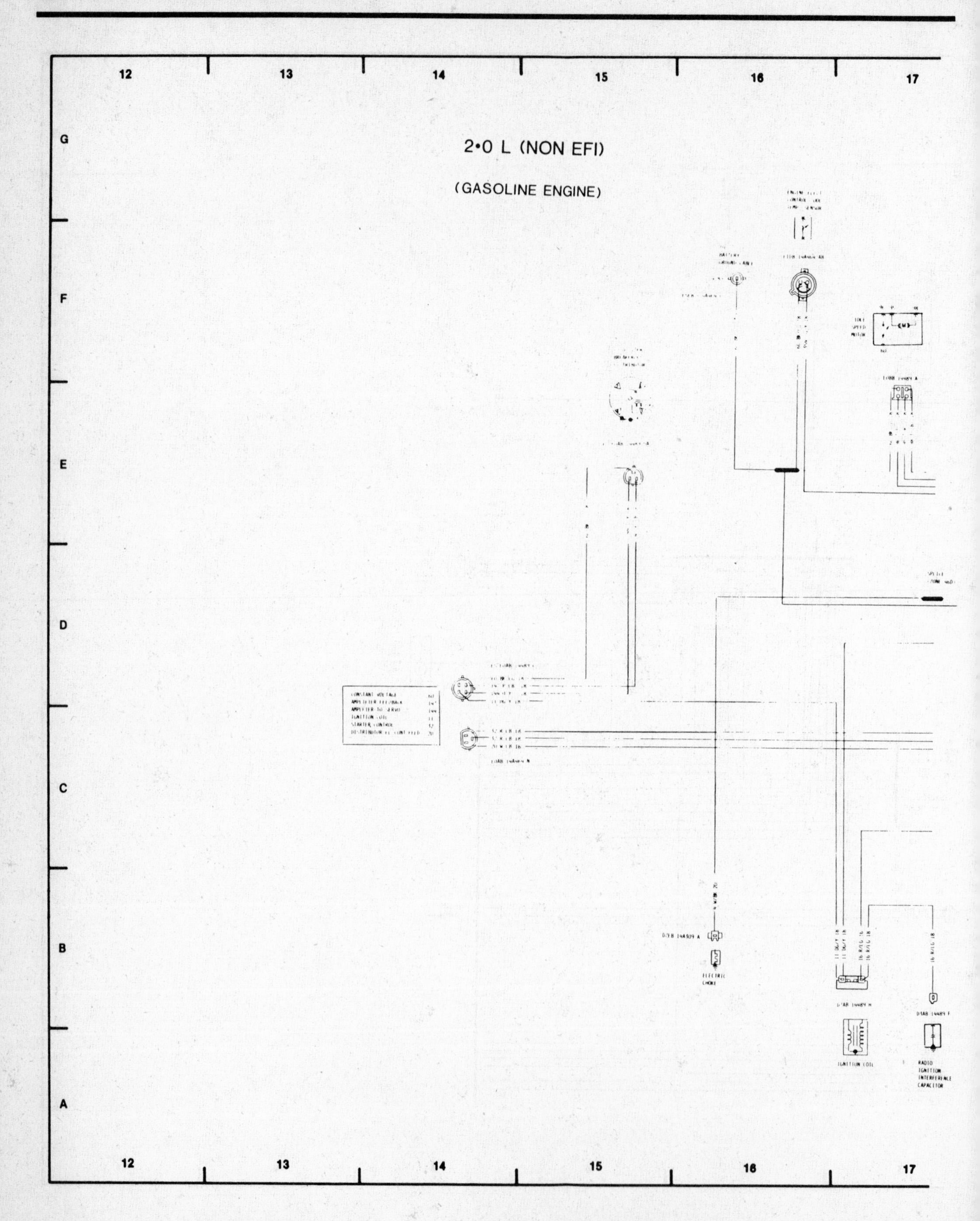
2•0 L (NON EFI)
(GASOLINE ENGINE)
12
13
14
15
16
17
G
F
E
D
C
B
A
ELECTRIC CHOKE
IGNITION COIL
RADIO IGNITION INTERFERENCE CAPACITOR

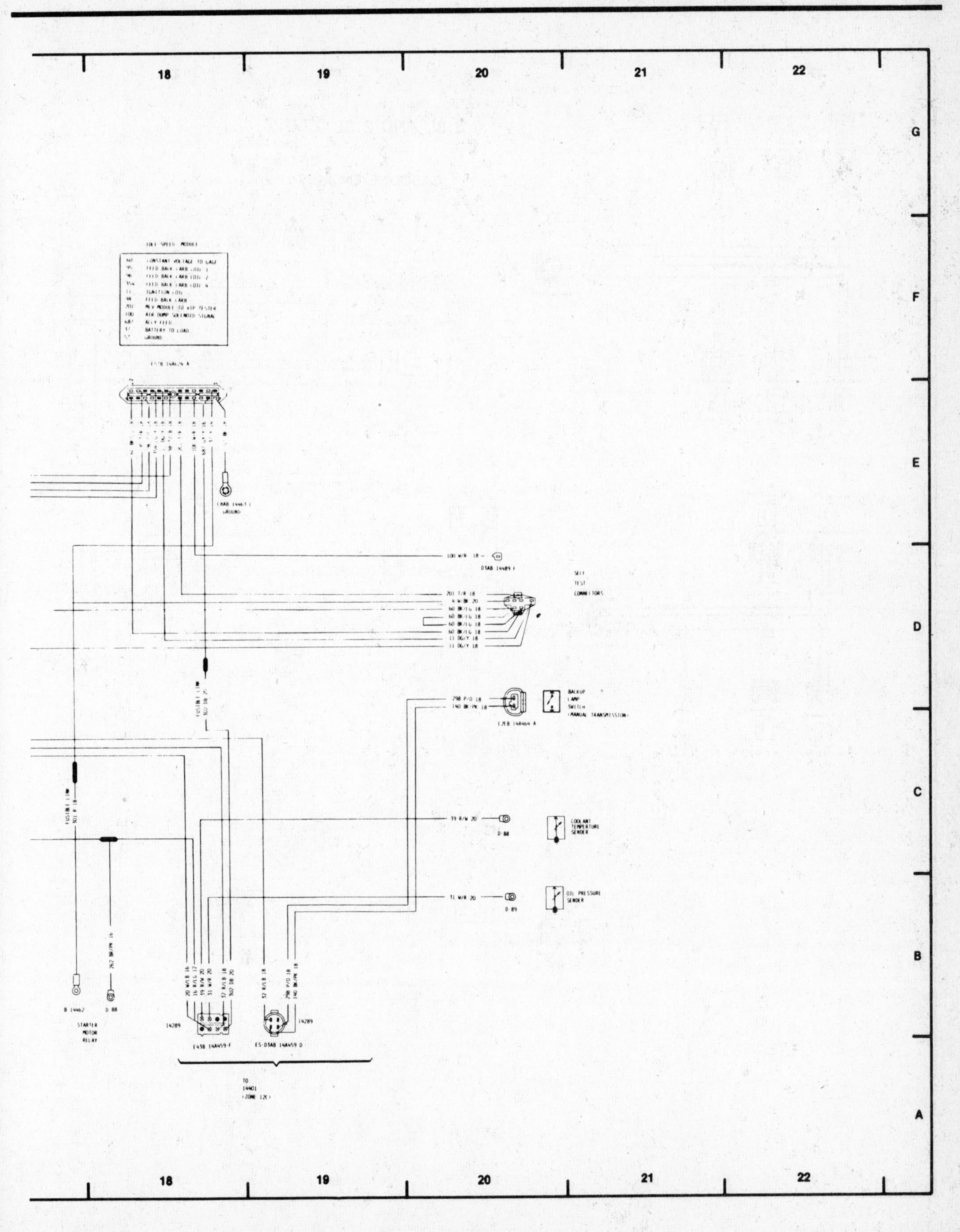
18
19
20
21
22
G
F
E
D
C
B
A
IDLE SPEED MODULE
BATTERY TO LOAD
GROUND
100 W/R 18
D3AB 14489 F
SELF
TEST
CONNECTORS
201 T/R 18
4 W/BK 20
60 BK/LG 18
11 DG/Y 18
298 P/O 18
140 BK/PK 18
E2EB 14A464 A
BACKUP
LAMP
SWITCH
(MANUAL TRANSMISSION)
FUSIBLE LINK
39 R/W 20
D 88
COOLANT
TEMPERTURE
SENDER
31 W/R 20
D 89
OIL PRESSURE
SENDER
762 BK/PK 16
20 W/LB 16
16 R/LG 12
39 R/W 20
31 W/R 20
32 R/LB 18
302 DB 20
B 14462
STARTER
MOTOR
RELAY
14289
E43B 14A459-F
ES-D3AB 14A459 D
TO
14401
(ZONE 12C)

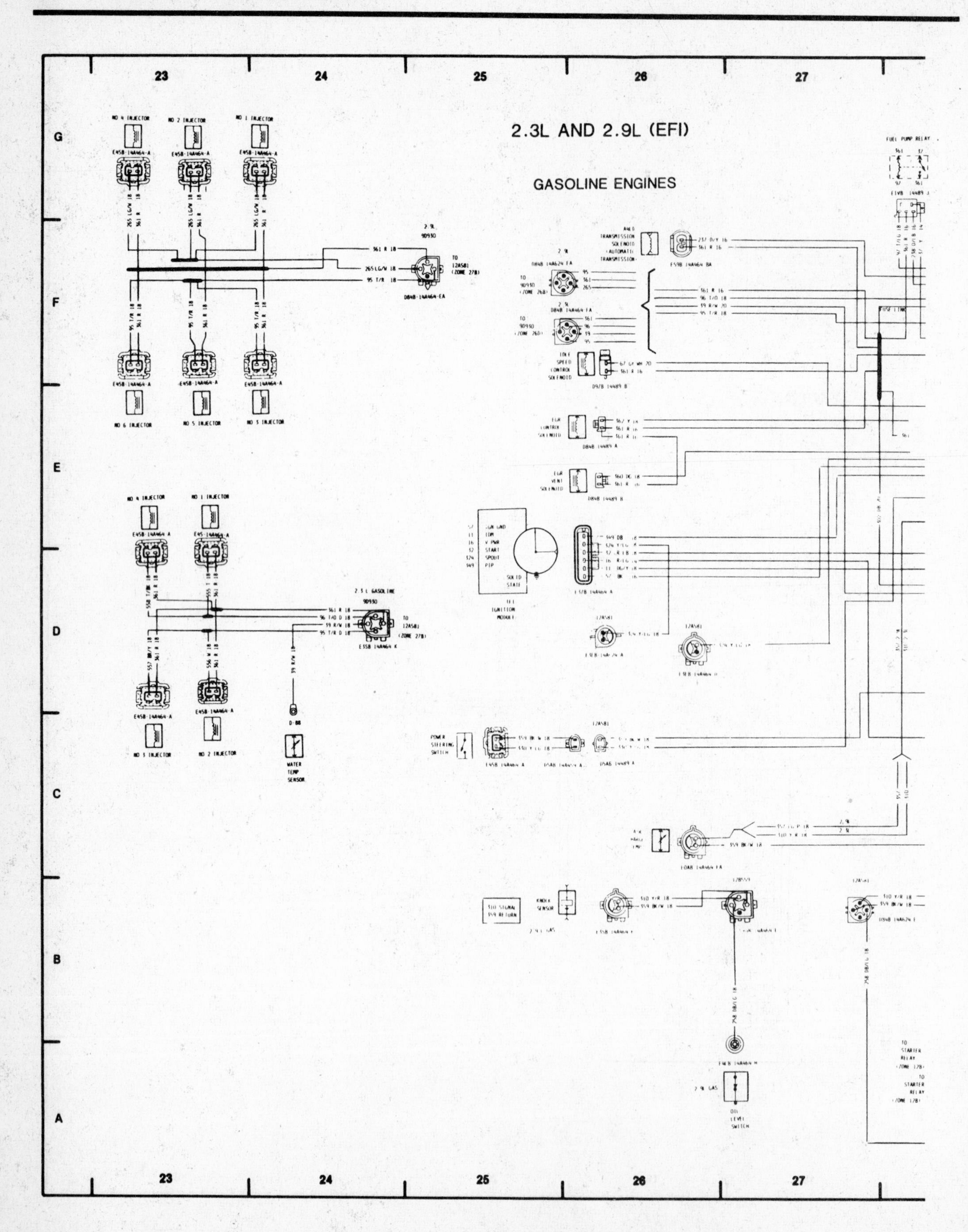
2.3L AND 2.9L (EFI)
GASOLINE ENGINES
NO 4 INJECTOR
NO 2 INJECTOR
NO 1 INJECTOR
NO 6 INJECTOR
NO 5 INJECTOR
NO 3 INJECTOR
POWER STEERING SWITCH
WATER TEMP SENSOR
KNOCK SENSOR
OIL LEVEL SWITCH
FUEL PUMP RELAY
TO STARTER RELAY (ZONE 17B)

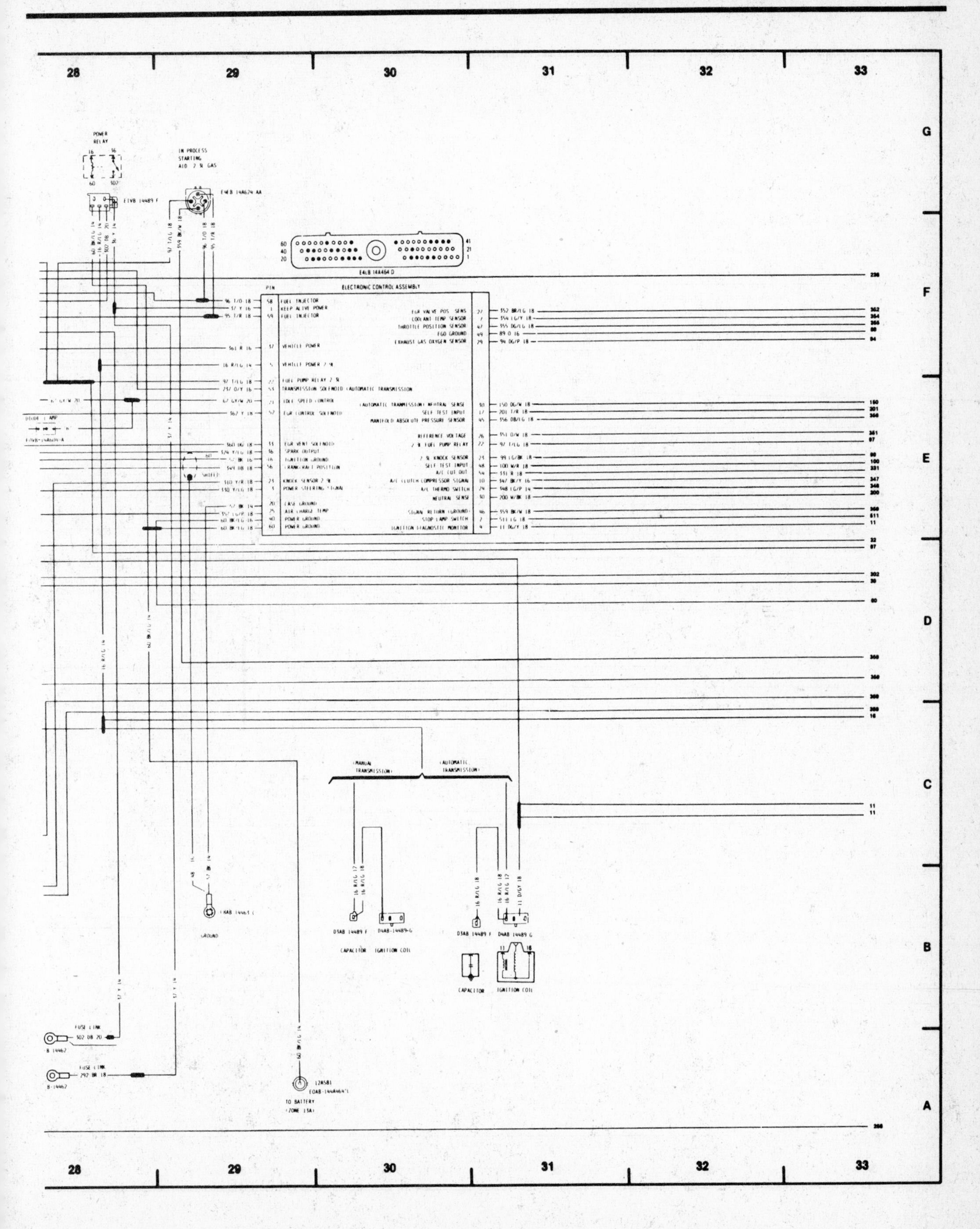
28
29
30
31
32
33
G
F
E
D
C
B
A
POWER RELAY
E1VB 14489 F
IN PROCESS STARTING AID 2.3L GAS
E4EB 14A624 AA
E4LB 14A464 D
PIN
ELECTRONIC CONTROL ASSEMBLY
96 T/O 18
58
FUEL INJECTOR
1
KEEP ALIVE POWER
95 T/R 18
59
FUEL INJECTOR
361 R 16
37
VEHICLE POWER
16 R/LG 14
5
VEHICLE POWER 2.3L
97 T/LG 18
FUEL PUMP RELAY 2.3L
237 O/Y 16
53
TRANSMISSION SOLENOID (AUTOMATIC TRANSMISSION
67 GY/W 20
IDLE SPEED CONTROL
362 Y 18
52
EGR CONTROL SOLENOID
360 DG 18
33
EGR VENT SOLENOID
324 Y/LG 18
36
SPARK OUTPUT
57 BK 16
16
IGNITION GROUND
349 DB 18
56
CRANKCRAFT POSITION
SHIELD
310 Y/R 18
23
KNOCK SENSOR 2.3L
330 Y/LG 18
3
POWER STEERING SIGNAL
57 BK 14
20
CASE GROUND
25
AIR CHARGE TEMP
60 BK/LG 16
40
POWER GROUND
60
POWER GROUND
EGR VALVE POS SENS
27
352 BR/LG 18
COOLANT TEMP SENSOR
7
354 LG/Y 18
THROTTLE POSITION SENSOR
47
355 DG/LG 18
EGO GROUND
49
89 O 16
EXHAUST GAS OXYGEN SENSOR
29
94 DG/P 18
(AUTOMATIC TRANMISSION) NEUTRAL SENSE
30
150 DG/W 18
SELF TEST INPUT
17
201 T/R 18
MANIFOLD ABSOLUTE PRESSURE SENSOR
45
356 DB/LG 18
REFERENCE VOLTAGE
26
351 O/W 18
2.3L FUEL PUMP RELAY
22
97 T/LG 18
2.3L KNOCK SENSOR
23
99 LG/BK 18
SELF TEST INPUT
48
100 W/R 18
A/C CUT OUT
54
331 R 18
A/C CLUTCH COMPRESSOR SIGNAL
10
347 BK/Y 16
A/C THERMO SWITCH
24
348 LG/P 14
NEUTRAL SENSE
30
200 W/BK 18
SIGNAL RETURN (GROUND)
46
359 BK/W 18
STOP LAMP SWITCH
2
511 LG 18
IGNITION DIAGNOSTIC MONITOR
4
11 DG/Y 18
(MANUAL TRANSMISSION)
(AUTOMATIC TRANSMISSION)
D3AB 14489 F
D4AB-14489-G
CAPACITOR
IGNITION COIL
GROUND
FUSE LINK
302 DB 20
B 14462
297 BK 18
12A581
TO BATTERY (ZONE 13A)

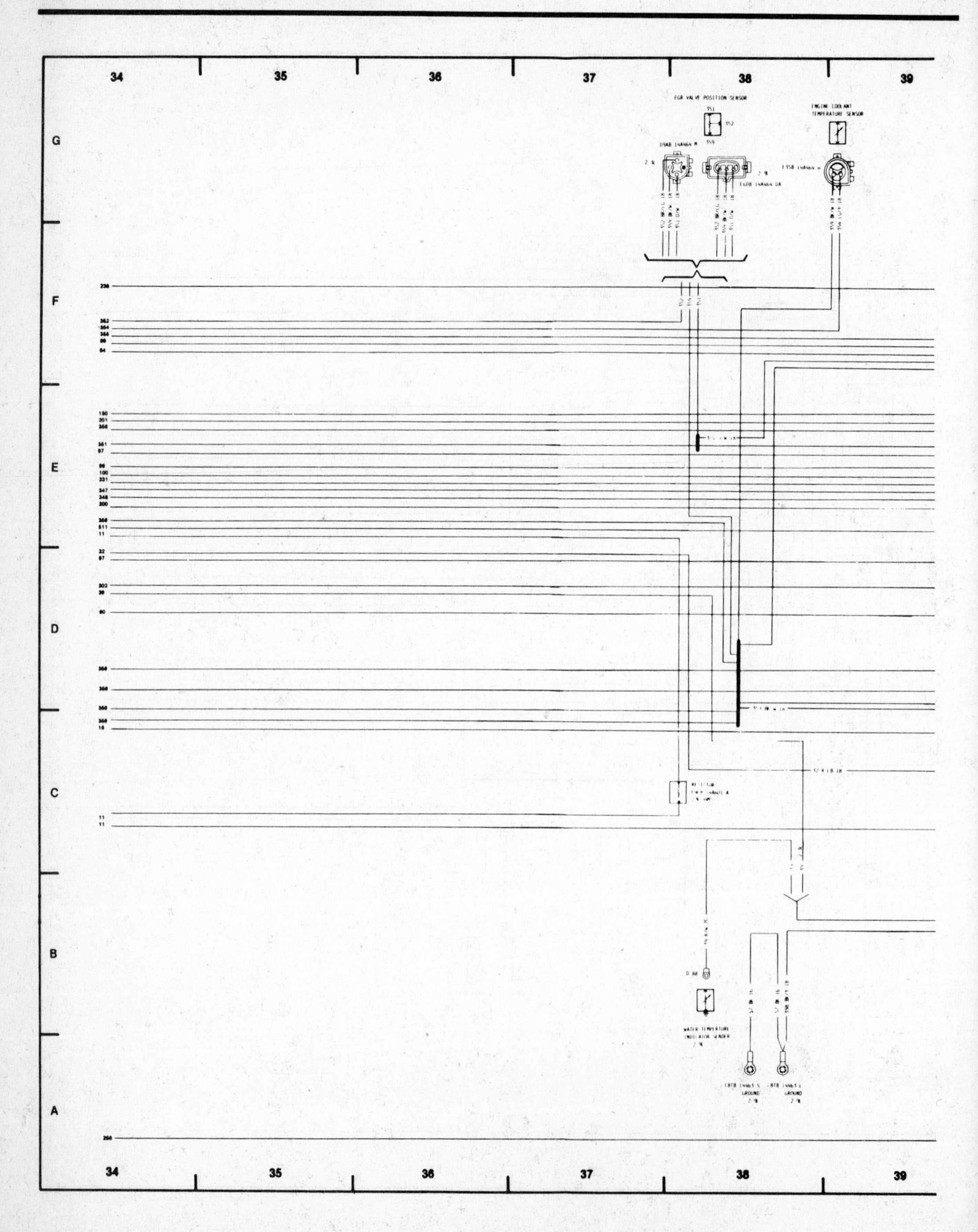
34
35
36
37
38
39
G
F
E
D
C
B
A
EGR VALVE POSITION SENSOR
ENGINE COOLANT TEMPERATURE SENSOR
WATER TEMPERATURE INDICATOR SENDER
GROUND
GROUND

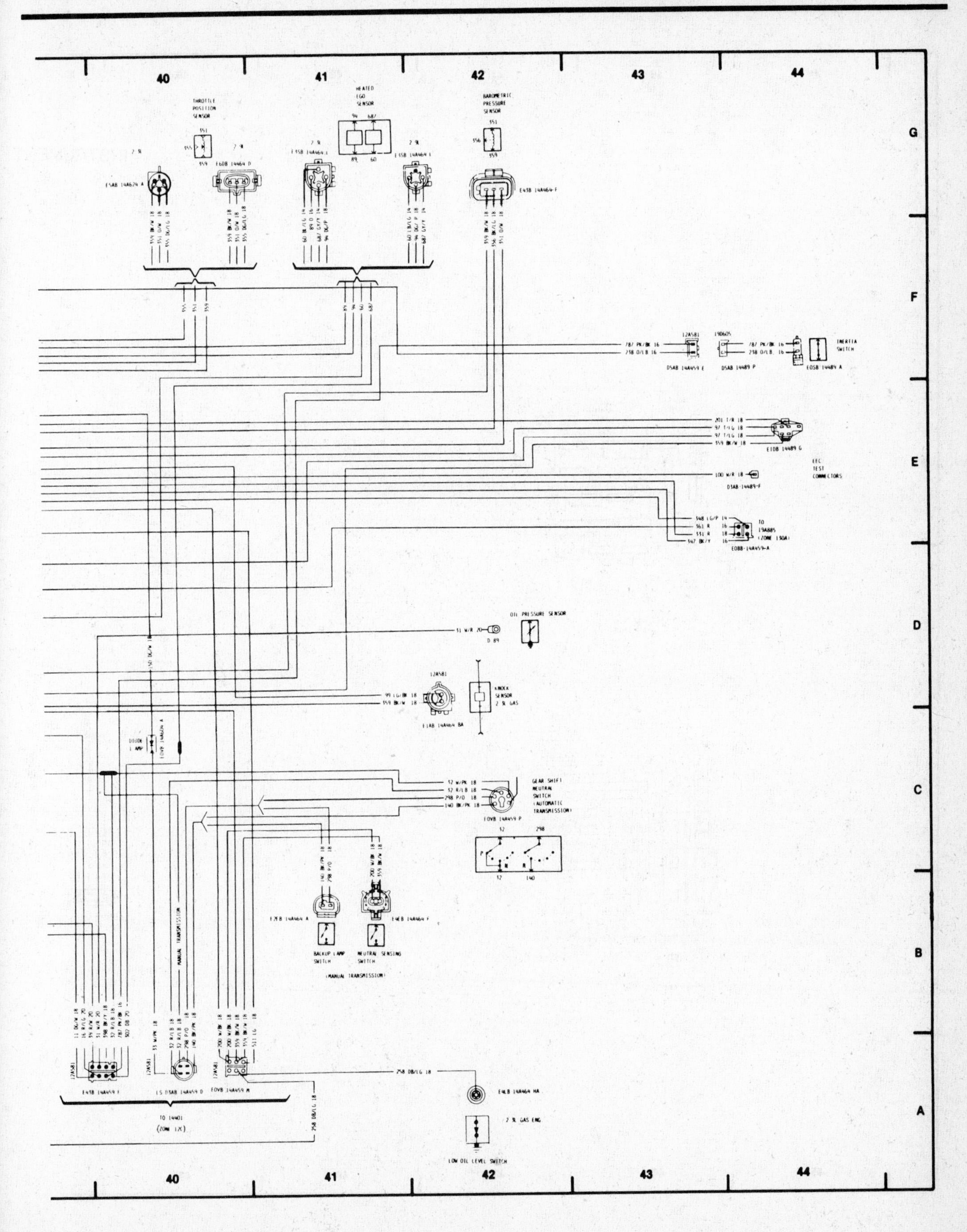
THROTTLE POSITION SENSOR
HEATED EGO SENSOR
BAROMETRIC PRESSURE SENSOR
INERTIA SWITCH
EEC TEST CONNECTORS
OIL PRESSURE SENSOR
KNOCK SENSOR 2.3L GAS
DIODE 1 AMP
GEAR SHIFT NEUTRAL SWITCH (AUTOMATIC TRANSMISSION)
MANUAL TRANSMISSION
BACKUP LAMP SWITCH
NEUTRAL SENSING SWITCH
(MANUAL TRANSMISSION)
TO 14401 (ZONE 12C)
2.3L GAS ENG
LOW OIL LEVEL SWITCH

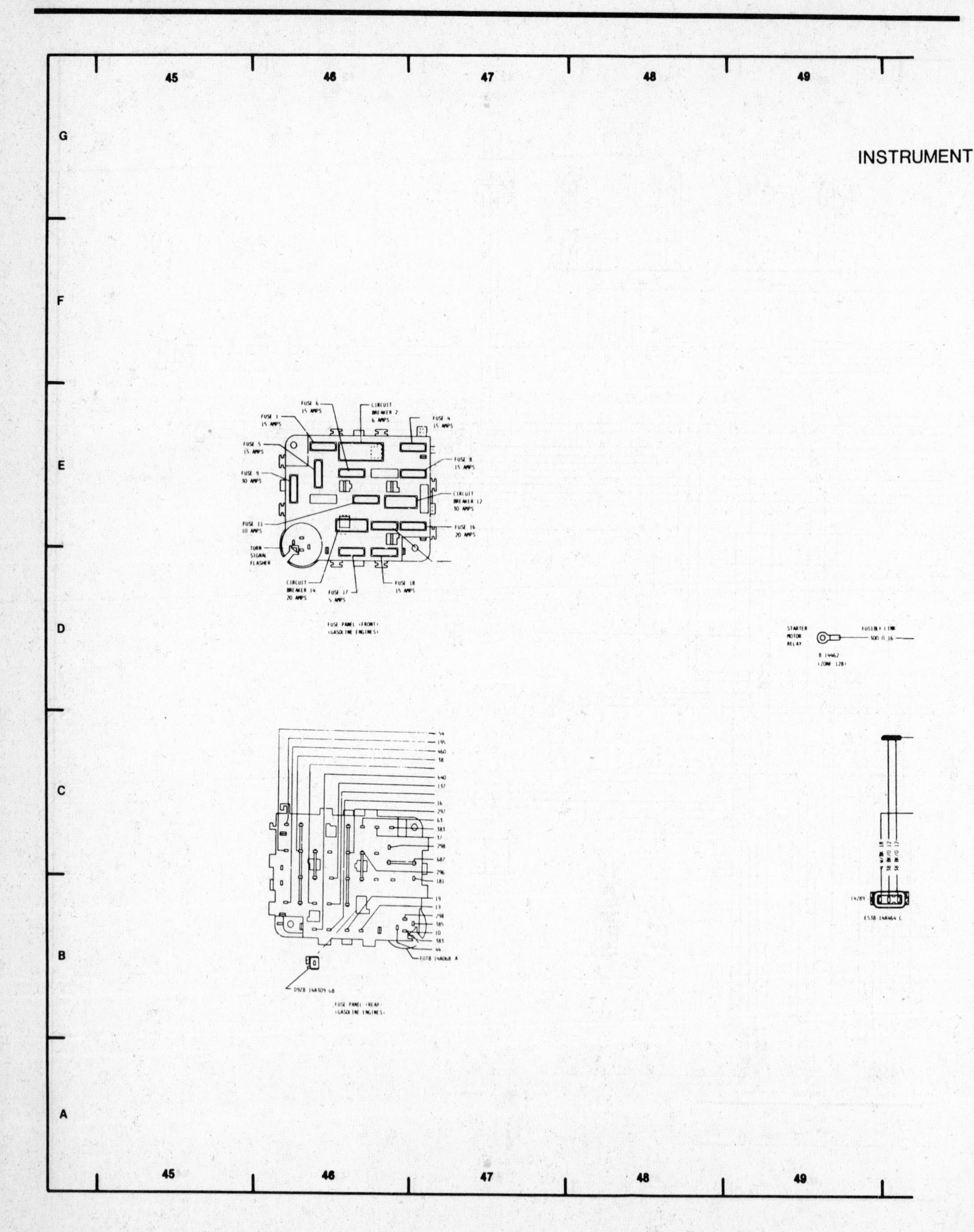
45
46
47
48
49
G
F
E
D
C
B
A
INSTRUMENT
FUSE 6
15 AMPS
CIRCUIT
BREAKER 2
6 AMPS
FUSE 1
15 AMPS
FUSE 4
15 AMPS
FUSE 5
15 AMPS
FUSE 8
15 AMPS
FUSE 9
30 AMPS
CIRCUIT
BREAKER 12
30 AMPS
FUSE 11
10 AMPS
FUSE 16
20 AMPS
TURN
SIGNAL
FLASHER
CIRCUIT
BREAKER 14
20 AMPS
FUSE 17
5 AMPS
FUSE 18
15 AMPS
FUSE PANEL (FRONT)
(GASOLINE ENGINES)
STARTER
MOTOR
RELAY
FUSIBLE LINK
300 0 16
B 14462
(ZONE 12B)
54
195
460
38
640
137
16
297
63
383
37
298
687
296
181
19
13
298
385
10
383
44
E0TB 14A068 A
D9ZB 14A309 GB
FUSE PANEL (REAR)
(GASOLINE ENGINES)
14289
E5TB 14A464 C

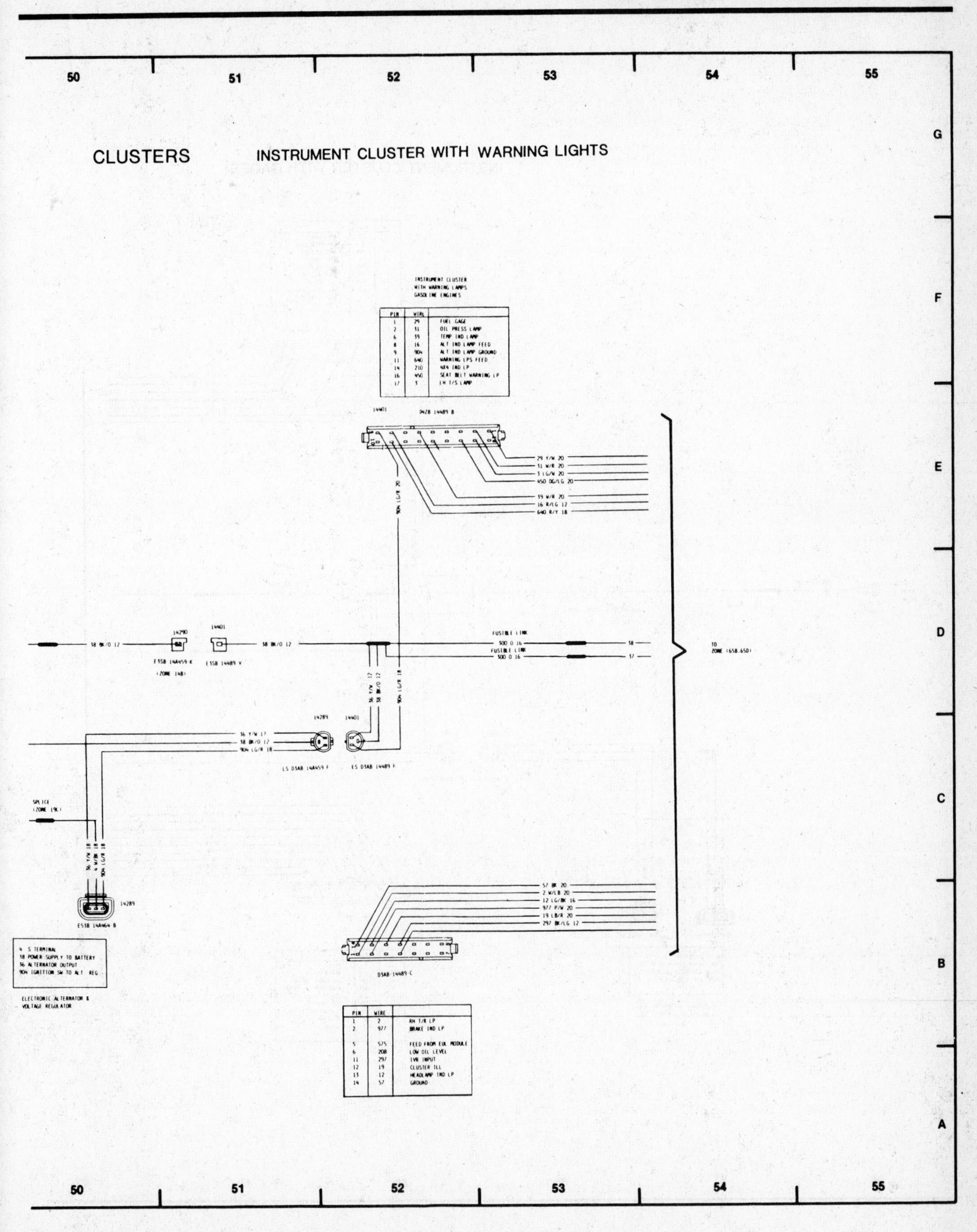

PIN	WIRE	
1	29	FUEL GAGE
2	31	OIL PRESS LAMP
6	39	TEMP IND LAMP
8	16	ALT IND LAMP FEED
9	904	ALT IND LAMP GROUND
11	640	WARNING LPS FEED
14	210	4X4 IND LP
16	450	SEAT BELT WARNING LP
17	3	LH T/S LAMP

PIN	WIRE	
1	2	RH T/R LP
2	977	BRAKE IND LP
5	575	FEED FROM EUL MODULE
6	208	LOW OIL LEVEL
11	297	IVR INPUT
12	19	CLUSTER ILL
13	12	HEADLAMP IND LP
14	57	GROUND

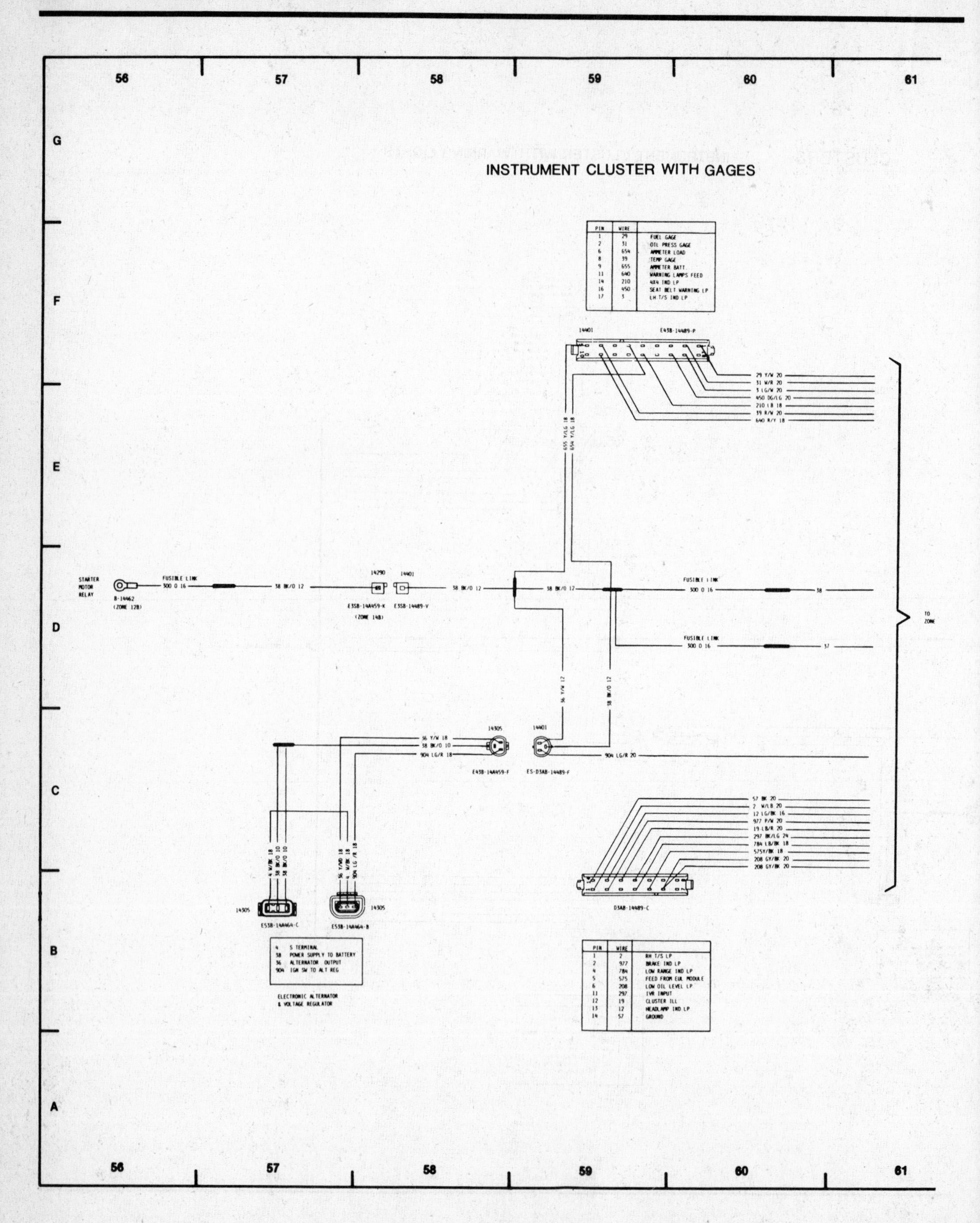
56
57
58
59
60
61
G
F
E
D
C
B
A
INSTRUMENT CLUSTER WITH GAGES
PIN WIRE
1 29 FUEL GAGE
2 31 OIL PRESS GAGE
6 654 AMMETER LOAD
8 39 TEMP GAGE
9 655 AMMETER BATT.
11 640 WARNING LAMPS FEED
14 210 4X4 IND LP
16 450 SEAT BELT WARNING LP
17 3 LH T/S IND LP
14401
E43B-14489-P
29 Y/W 20
31 W/R 20
3 LG/W 20
450 DG/LG 20
210 LB 18
39 R/W 20
640 R/Y 18
655 Y/LG 18
654 Y/LG 18
STARTER MOTOR RELAY
B-14462 (ZONE 12B)
FUSIBLE LINK 300 O 16
38 BK/O 12
14290
14401
E35B-14A459-K (ZONE 14B)
E35B-14489-V
FUSIBLE LINK 300 O 16
38
FUSIBLE LINK 300 O 16
37
TO ZONE
36 Y/W 12
38 BK/O 12
14305
14401
36 Y/W 18
38 BK/O 10
904 LG/R 18
E43B-14A459-F
E5-D3AB-14489-F
904 LG/R 20
57 BK 20
2 W/LB 20
12 LG/BK 16
977 P/W 20
19 LB/R 20
297 BK/LG 24
784 LB/BK 18
575Y/BK 18
208 GY/BK 20
208 GY/BK 20
4 W/BK 18
38 BK/O 10
38 BK/O 10
36 Y/WO 18
4 W/BK 18
904 LG/R 18
14305
E53B-14A464-C
14305
E53B-14A464-B
D3AB-14489-C
4 S TERMINAL
38 POWER SUPPLY TO BATTERY
36 ALTERNATOR OUTPUT
904 IGN SW TO ALT REG
ELECTRONIC ALTERNATOR
& VOLTAGE REGULATOR
PIN WIRE
1 2 RH T/S LP
2 977 BRAKE IND LP
4 784 LOW RANGE IND LP
5 575 FEED FROM EUL MODULE
6 208 LOW OIL LEVEL LP
11 297 IVR INPUT
12 19 CLUSTER ILL
13 12 HEADLAMP IND LP
14 57 GROUND

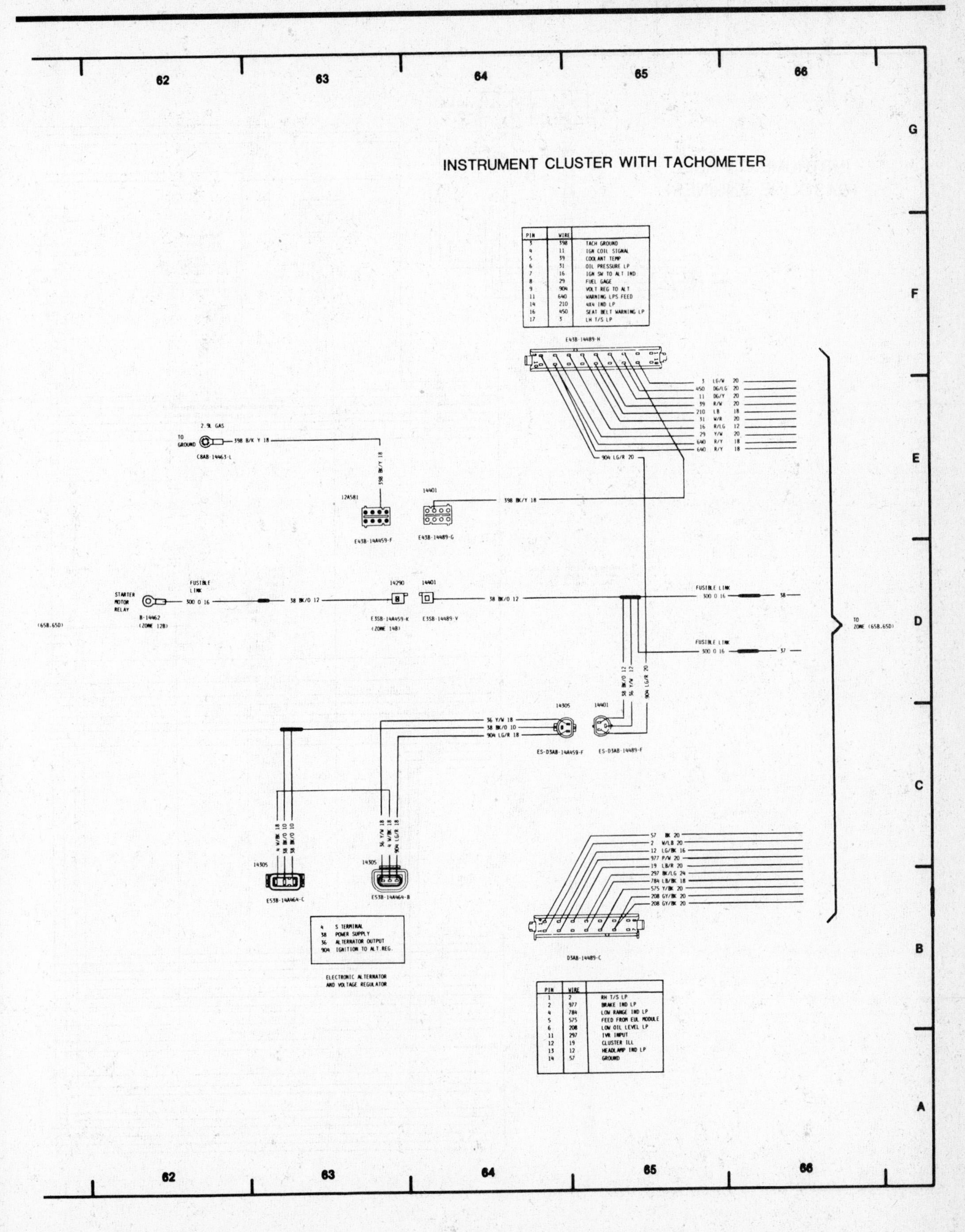

PIN	WIRE	
3	398	TACH GROUND
4	11	IGN COIL SIGNAL
5	39	COOLANT TEMP
6	31	OIL PRESSURE LP
7	16	IGN SW TO ALT IND
8	29	FUEL GAGE
9	904	VOLT REG TO ALT
11	640	WARNING LPS FEED
14	210	4X4 IND LP
16	450	SEAT BELT WARNING LP
17	3	LH T/S LP

PIN	WIRE	
1	2	RH T/S LP
2	977	BRAKE IND LP
4	784	LOW RANGE IND LP
5	575	FEED FROM EUL MODULE
6	208	LOW OIL LEVEL LP
11	297	IVR INPUT
12	19	CLUSTER ILL
13	12	HEADLAMP IND LP
14	57	GROUND

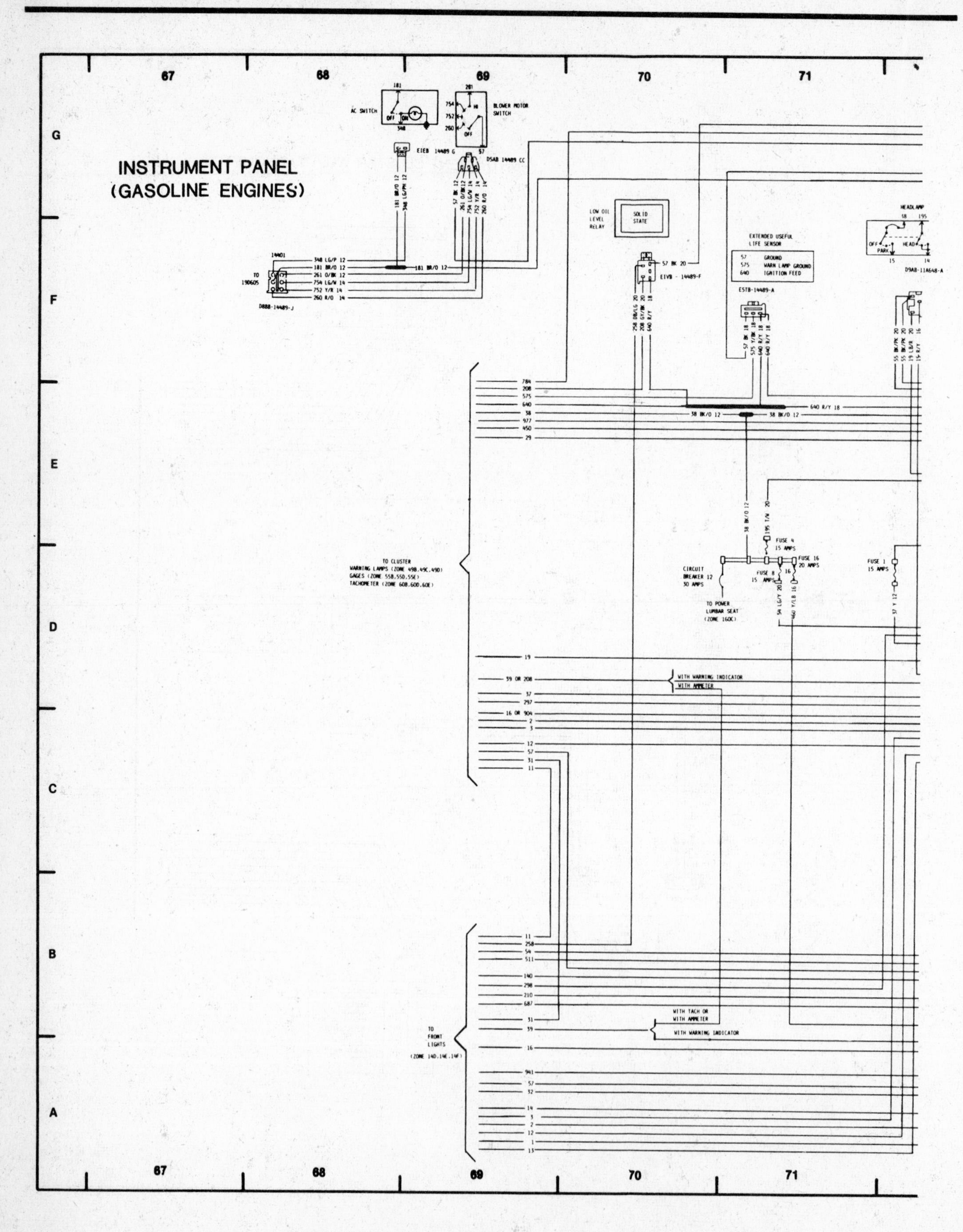

INSTRUMENT PANEL
(GASOLINE ENGINES)
67
68
69
70
71
G
F
E
D
C
B
A
AC SWITCH
BLOWER MOTOR SWITCH
LOW OIL LEVEL RELAY
SOLID STATE
EXTENDED USEFUL LIFE SENSOR
57 GROUND
575 WARN LAMP GROUND
640 IGNITION FEED
HEADLAMP
FUSE 4 15 AMPS
FUSE 16 20 AMPS
FUSE 8 15 AMPS
FUSE 1 15 AMPS
CIRCUIT BREAKER 12 30 AMPS
TO POWER LUMBAR SEAT (ZONE 160C)
TO CLUSTER
WARNING LAMPS (ZONE 49B,49C,49D)
GAGES (ZONE 55B,55D,55E)
TACHOMETER (ZONE 60B,60D,60E)
WITH WARNING INDICATOR
WITH AMMETER
WITH TACH OR WITH AMMETER
TO FRONT LIGHTS (ZONE 14D,14E,14F)
640 R/Y 18
38 BK/O 12

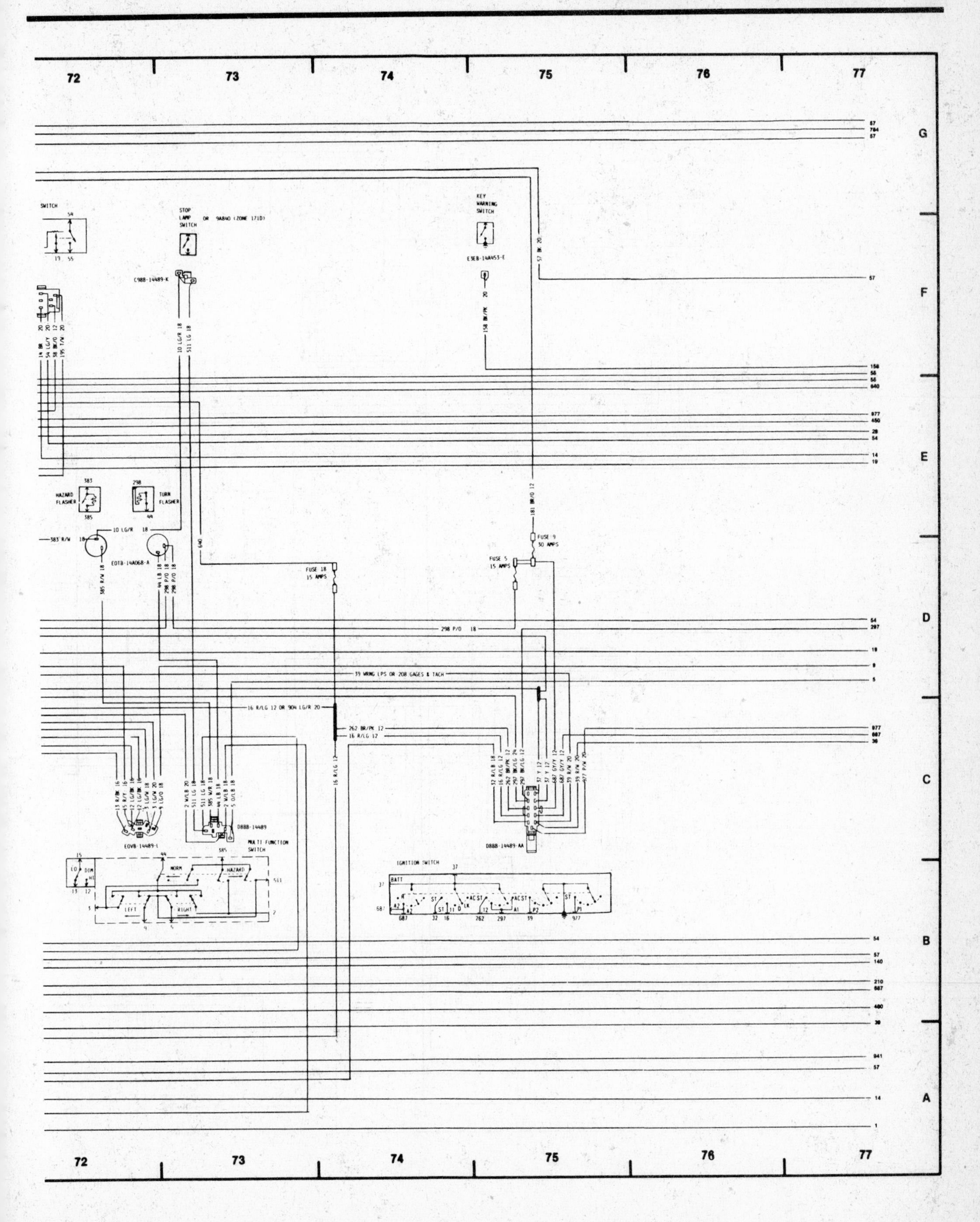

STOP LAMP SWITCH OR 9A840 (ZONE 171D)
KEY WARNING SWITCH
E3EB-14A453-E
C9BB-14489-K
HAZARD FLASHER
TURN FLASHER
E0TB-14A068-A
FUSE 18 15 AMPS
FUSE 9 30 AMPS
FUSE 5 15 AMPS
39 WRNG LPS OR 208 GAGES & TACH
16 R/LG 12 OR 904 LG/R 20
262 BR/PK 12
16 R/LG 12
298 P/O 18
D8BB-14489
MULTI FUNCTION SWITCH
E0VB-14489-L
D8BB-14489-AA
IGNITION SWITCH
BATT
NORM
HAZARD
LEFT
RIGHT

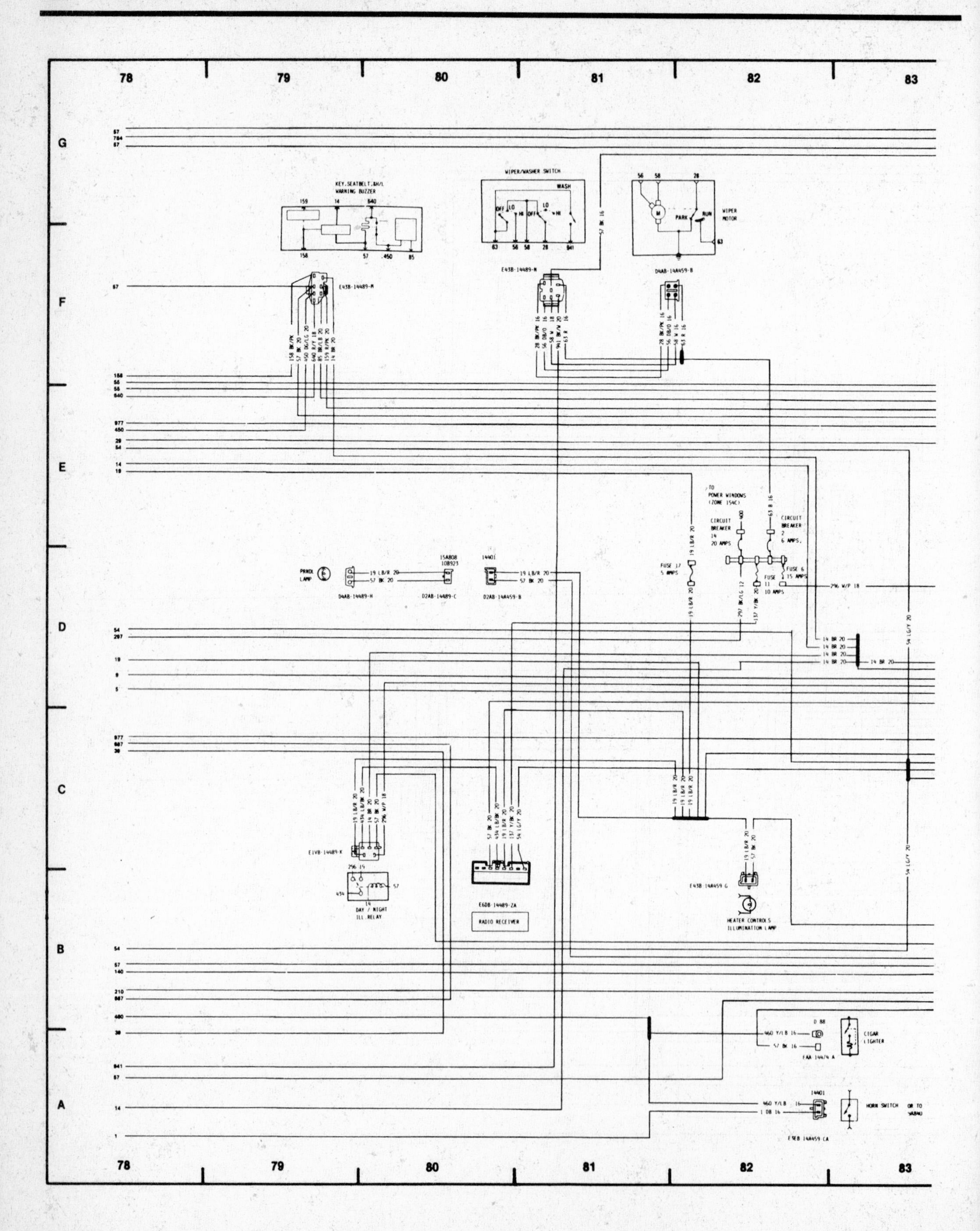
KEY, SEATBELT, &H/L WARNING BUZZER
WIPER/WASHER SWITCH
WASH
OFF
LO
HI
WIPER MOTOR
PARK
RUN
E43B-14489-M
D4AB-14A459-B
TO POWER WINDOWS (ZONE 154C)
CIRCUIT BREAKER 14 20 AMPS
CIRCUIT BREAKER 2 6 AMPS.
FUSE 17 5 AMPS
FUSE 6 15 AMPS
FUSE 11 10 AMPS
PRNDL LAMP
D4AB-14489-H
D2AB-14489-C
D2AB-14A459-B
E1VB-14489-K
DAY / NIGHT ILL RELAY
E60B-14489-ZA
RADIO RECEIVER
E43B-14A459-G
HEATER CONTROLS ILLUMINATION LAMP
CIGAR LIGHTER
HORN SWITCH
78
79
80
81
82
83
G
F
E
D
C
B
A

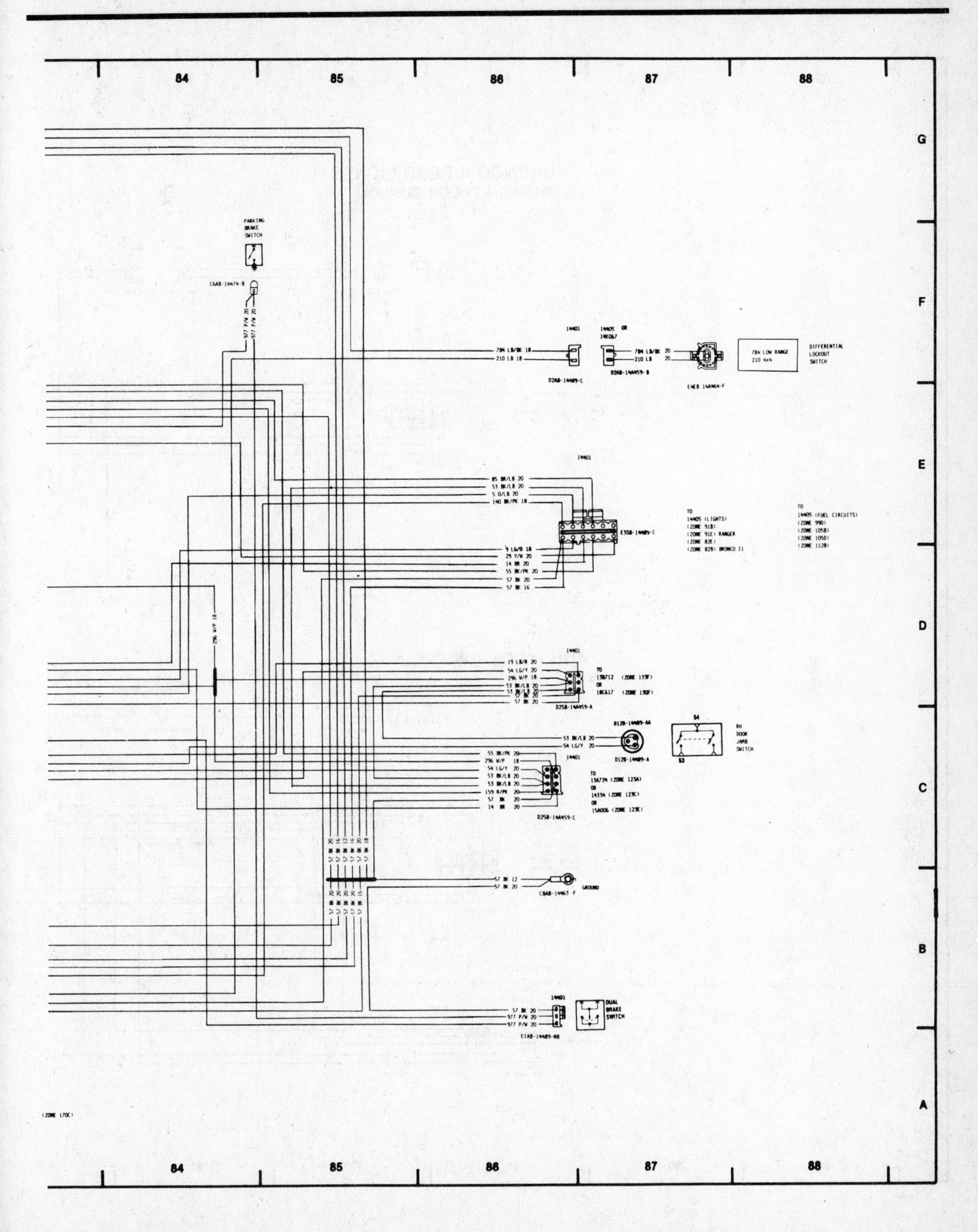
84
85
86
87
88
G
F
E
D
C
B
A
PARKING
BRAKE
SWITCH
C6AB-14474-B
977 P/W 20
977 P/W 20
14401
14405 OR
14K067
784 LB/BK 18
210 LB 18
D2AB-14489-C
784 LB/BK 20
210 LB 20
D2AB-14A459-B
E4EB 14A464-F
784 LOW RANGE
210 4x4
DIFFERENTIAL
LOCKOUT
SWITCH
14401
85 BR/LB 20
53 BK/LB 20
5 O/LB 20
140 BK/PK 18
E3SB-14489-C
9 LG/O 18
29 Y/W 20
14 BR 20
55 BK/PK 20
57 BK 20
57 BK 16
TO
14405 (LIGHTS)
(ZONE 91B)
(ZONE 91E) RANGER
(ZONE 82E)
(ZONE 82B) BRONCO II
TO
14405 (FUEL CIRCUITS)
(ZONE 99D)
(ZONE 105B)
(ZONE 105D)
(ZONE 112B)
296 W/P 18
14401
19 LB/R 20
54 LG/Y 20
296 W/P 18
53 BK/LB 20
53 BK/LB 20
57 BK 20
57 BK 20
D2SB-14A459-A
TO
13B712 (ZONE 133F)
OR
10C617 (ZONE 130F)
D12B-14489-AA
53 BK/LB 20
54 LG/Y 20
D12B-14489-A
54
53
RH
DOOR
JAMB
SWITCH
55 BK/PK 20
296 W/P 18
54 LG/Y 20
53 BK/LB 20
53 BK/LB 20
159 R/PK 20
57 BK 20
14 BR 20
14401
D2SB-14A459-C
TO
13A724 (ZONE 123A)
OR
14334 (ZONE 123C)
OR
15A006 (ZONE 123E)
57 BK 20
57 BK 16
57 BK 12
57 BK 16
57 BK 20
57 BK 18
57 BK 20
57 BK 20
57 BK 20
57 BK 20
57 BK 16
57 BK 12
57 BK 20
C8AB-14463 F
GROUND
14401
57 BK 20
977 P/W 20
977 P/W 20
E1AB-14489-NB
DUAL
BRAKE
SWITCH
(ZONE 170C)

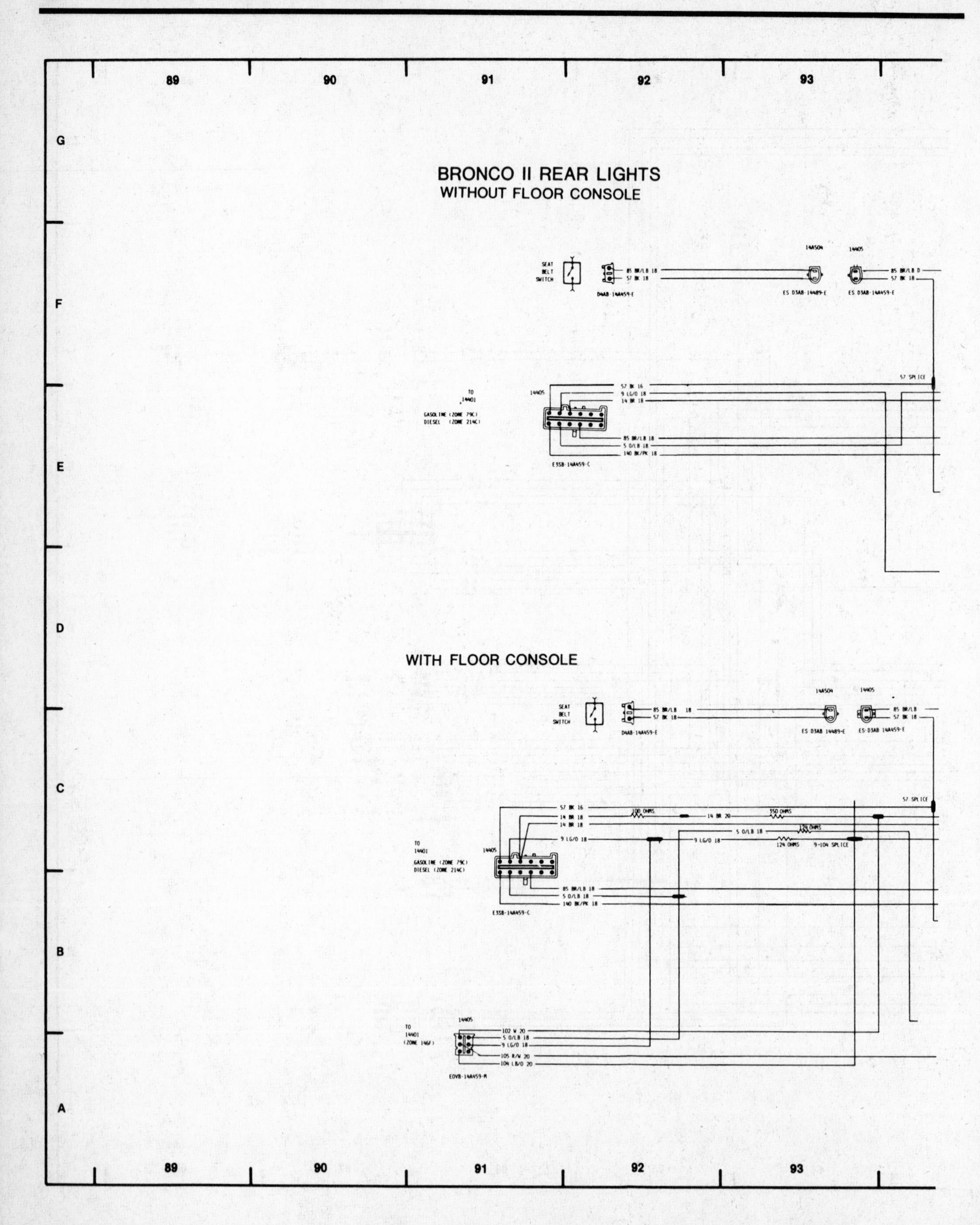

89
90
91
92
93
G
F
E
D
C
B
A
BRONCO II REAR LIGHTS
WITHOUT FLOOR CONSOLE
SEAT BELT SWITCH
85 BR/LB 18
57 BK 18
14A504
14405
57 SPLICE
57 BK 16
9 LG/O 18
14 BR 18
5 O/LB 18
140 BK/PK 18
GASOLINE (ZONE 79C)
DIESEL (ZONE 214C)
WITH FLOOR CONSOLE
100 OHMS
350 OHMS
124 OHMS
9-104 SPLICE
102 W 20
105 R/W 20
104 LB/O 20

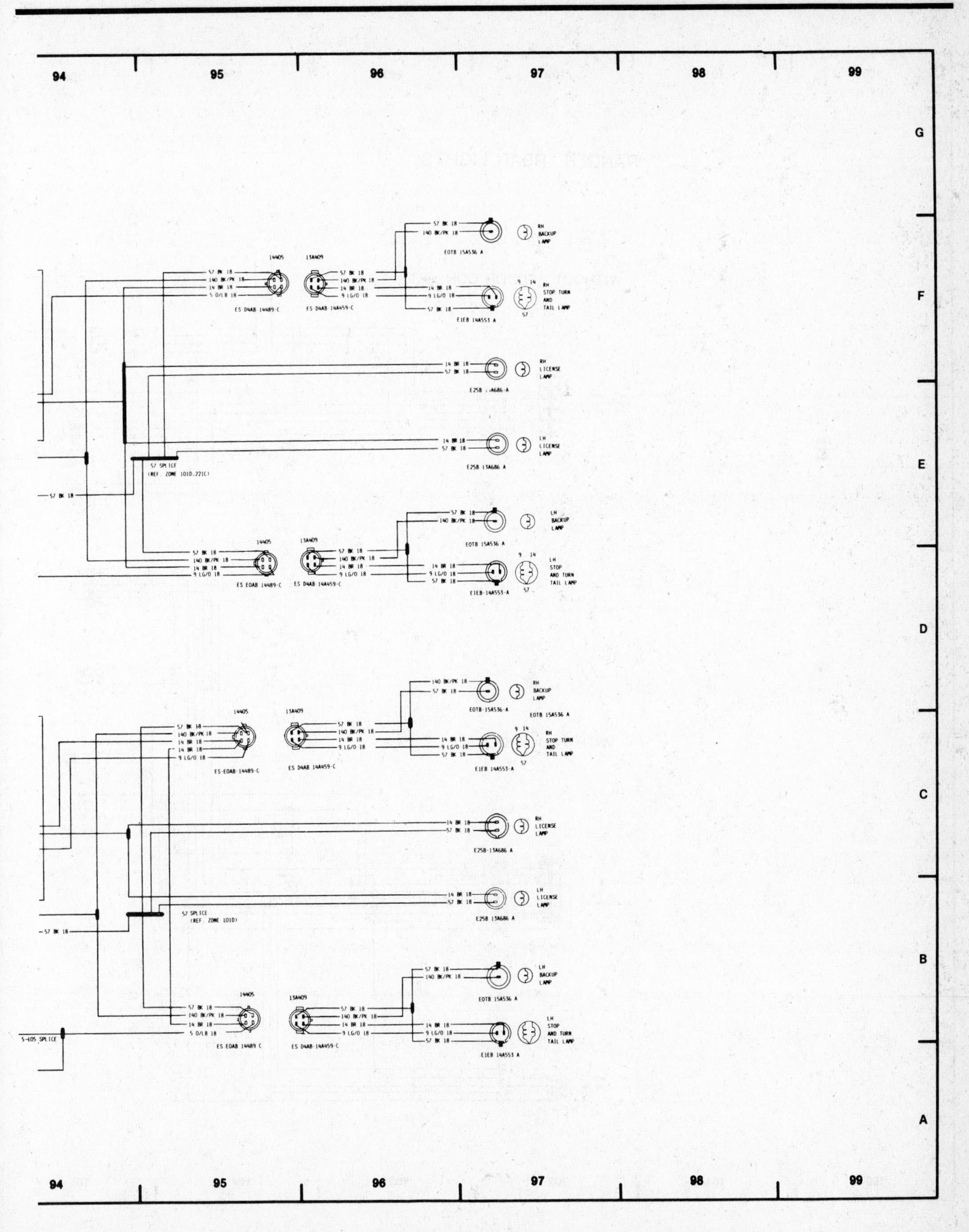

RH BACKUP LAMP
EOTB 15A536 A
RH STOP TURN AND TAIL LAMP
E1EB 14A553 A
RH LICENSE LAMP
LH LICENSE LAMP
E25B 13A686 A
57 SPLICE (REF. ZONE 101D,221C)
LH BACKUP LAMP
LH STOP AND TURN TAIL LAMP
ES EOAB 14489-C
ES D4AB 14A459-C
57 SPLICE (REF. ZONE 101D)
S-105 SPLICE

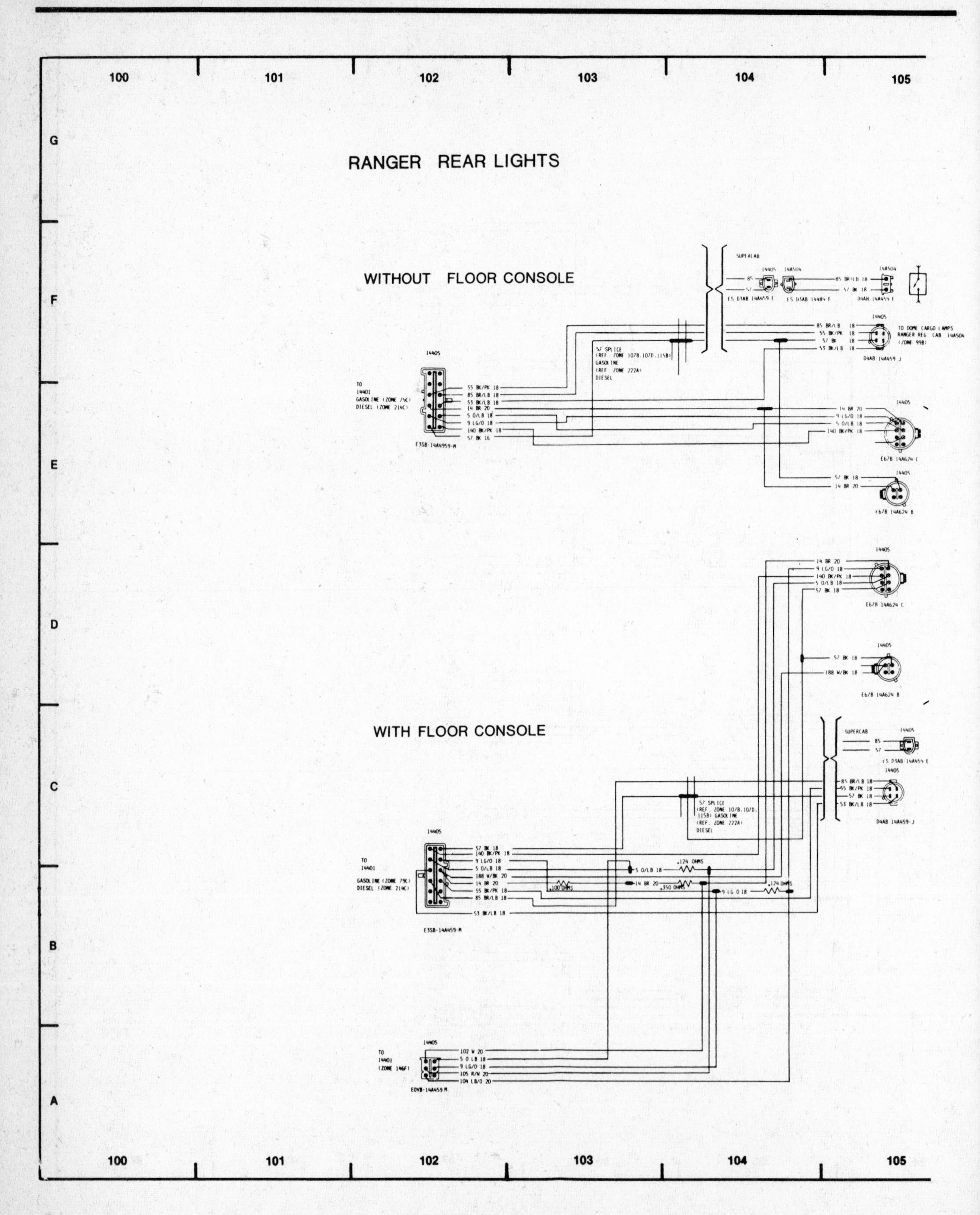
100
101
102
103
104
105
G
F
E
D
C
B
A
RANGER REAR LIGHTS
WITHOUT FLOOR CONSOLE
WITH FLOOR CONSOLE
SUPERCAB
14405
E3SB-14A459-M
E67B 14A624 C
E67B 14A624 B
D4AB 14A459 J
E0VB-14A459 M
TO DOME CARGO LAMPS RANGER REG CAB 14A504 (ZONE 99B)
TO 14401 GASOLINE (ZONE 79C) DIESEL (ZONE 214C)
57 SPLICE (REF. ZONE 107B, 107D, 115B) GASOLINE (REF. ZONE 222A) DIESEL
55 BK/PK 18
85 BR/LB 18
53 BK/LB 18
14 BR 20
5 O/LB 18
9 LG/O 18
140 BK/PK 18
57 BK 16
57 BK 18
188 W/BK 20
102 W 20
105 R/W 20
104 LB/O 20
.124 OHMS
.350 OHMS
.100 OHMS

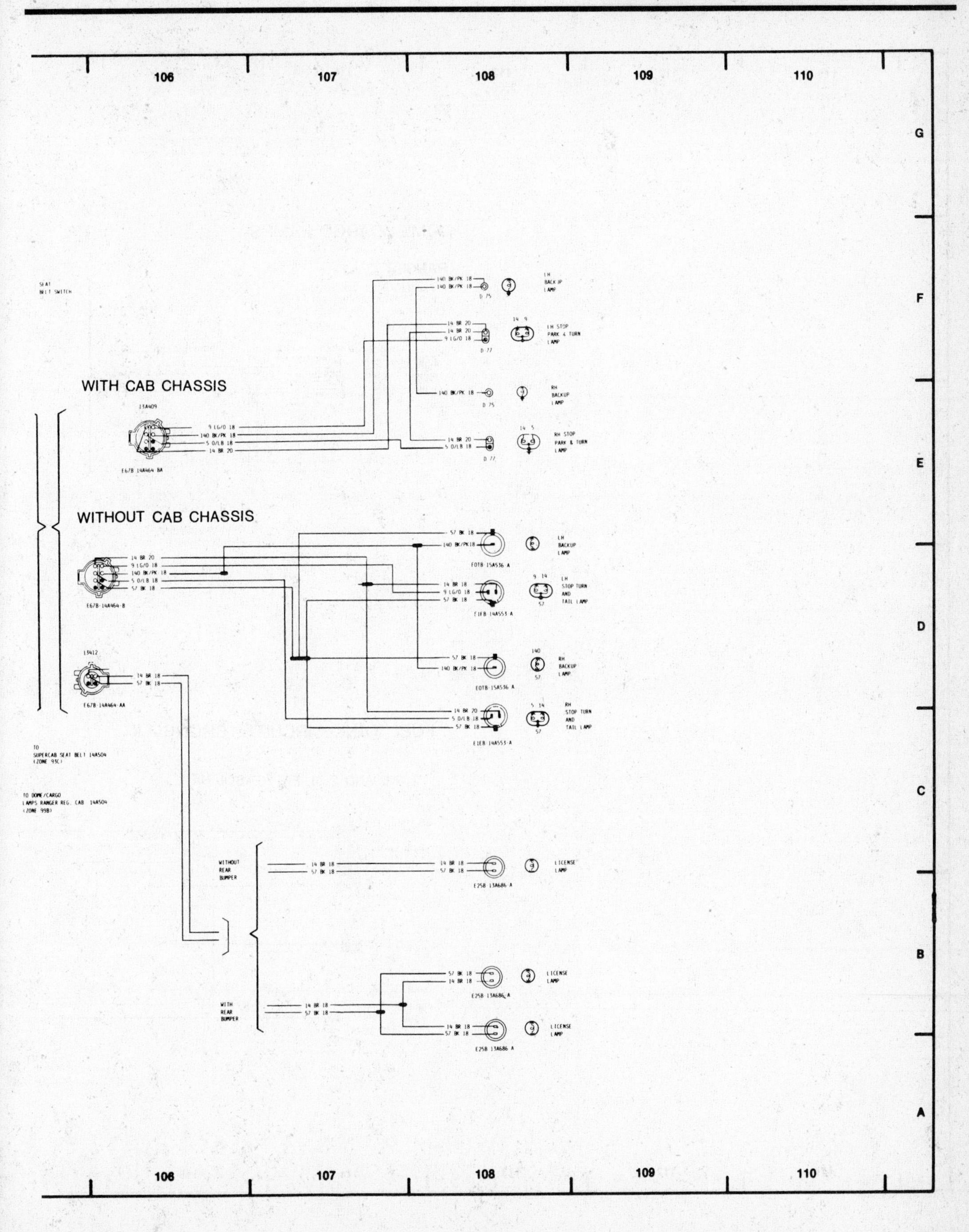
106
107
108
109
110
G
F
E
D
C
B
A
SEAT BELT SWITCH
WITH CAB CHASSIS
13A409
E67B-14A464-BA
WITHOUT CAB CHASSIS
E67B-14A464-B
13412
E67B-14A464-AA
LH BACKUP LAMP
LH STOP PARK & TURN LAMP
RH BACKUP LAMP
RH STOP PARK & TURN LAMP
LH STOP TURN AND TAIL LAMP
RH STOP TURN AND TAIL LAMP
E0TB-15A536-A
E1EB-14A553-A
TO SUPERCAB SEAT BELT 14A504 (ZONE 93C)
TO DOME/CARGO LAMPS RANGER REG. CAB 14A504 (ZONE 99B)
WITHOUT REAR BUMPER
WITH REAR BUMPER
LICENSE LAMP
E25B-13A686-A

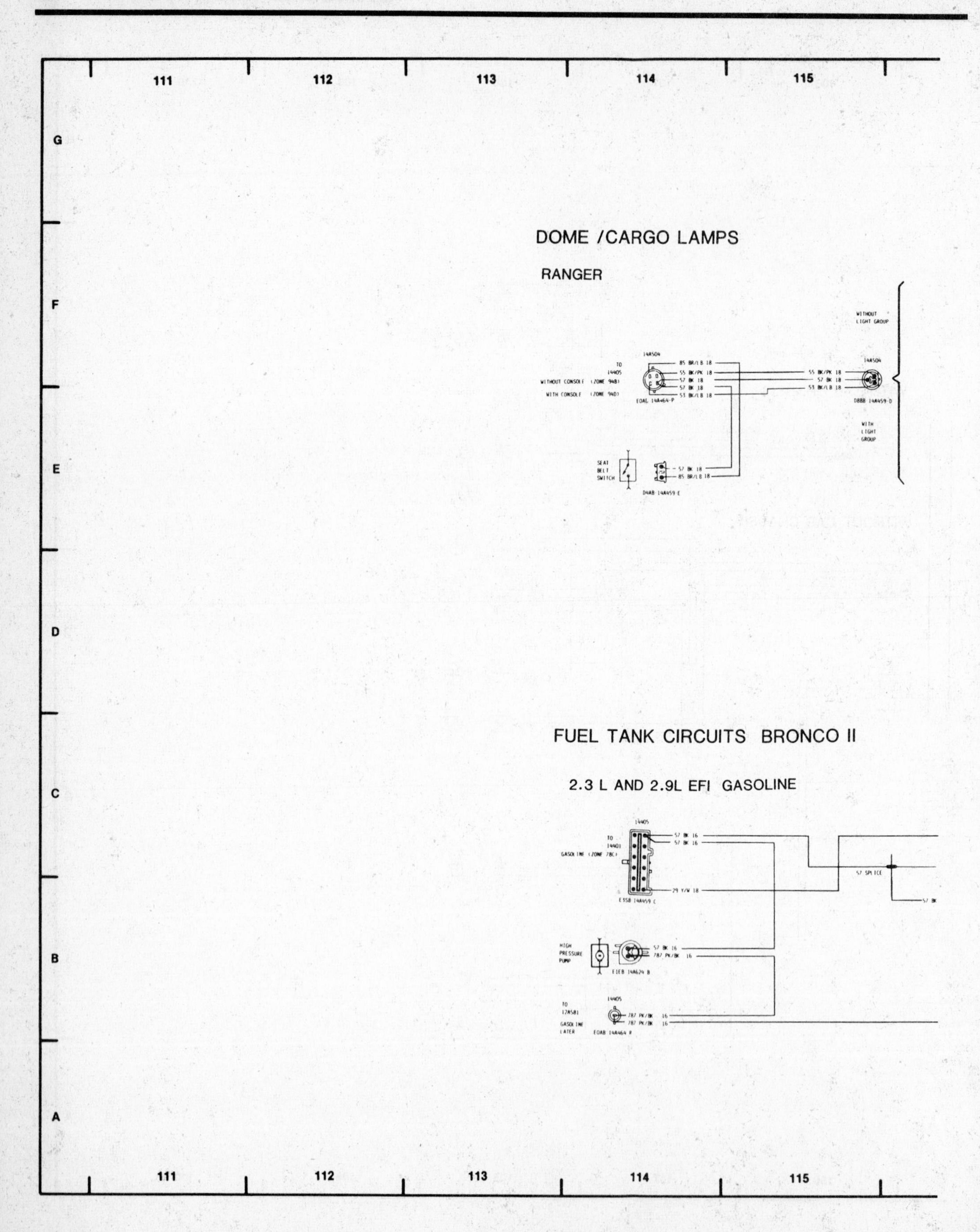

111
112
113
114
115
G
F
E
D
C
B
A
DOME /CARGO LAMPS
RANGER
WITHOUT LIGHT GROUP
14A504
TO 14405
WITHOUT CONSOLE (ZONE 94B)
WITH CONSOLE (ZONE 94D)
85 BR/LB 18
55 BK/PK 18
57 BK 18
57 BK 18
53 BK/LB 18
E0AG 14A464-P
14A504
55 BK/PK 18
57 BK 18
53 BK/LB 18
D8BB 14A459-D
WITH LIGHT GROUP
SEAT BELT SWITCH
57 BK 18
85 BR/LB 18
D4AB 14A459-E
FUEL TANK CIRCUITS BRONCO II
2.3 L AND 2.9L EFI GASOLINE
14405
TO 14401
GASOLINE (ZONE 78C)
57 BK 16
57 BK 16
29 Y/W 18
E35B 14A459 C
57 SPLICE
57 BK
HIGH PRESSURE PUMP
57 BK 16
787 PK/BK 16
E1EB 14A624 B
14405
TO 12A581
GASOLINE LATER
787 PK/BK 16
787 PK/BK 16
E0AB 14A464 R

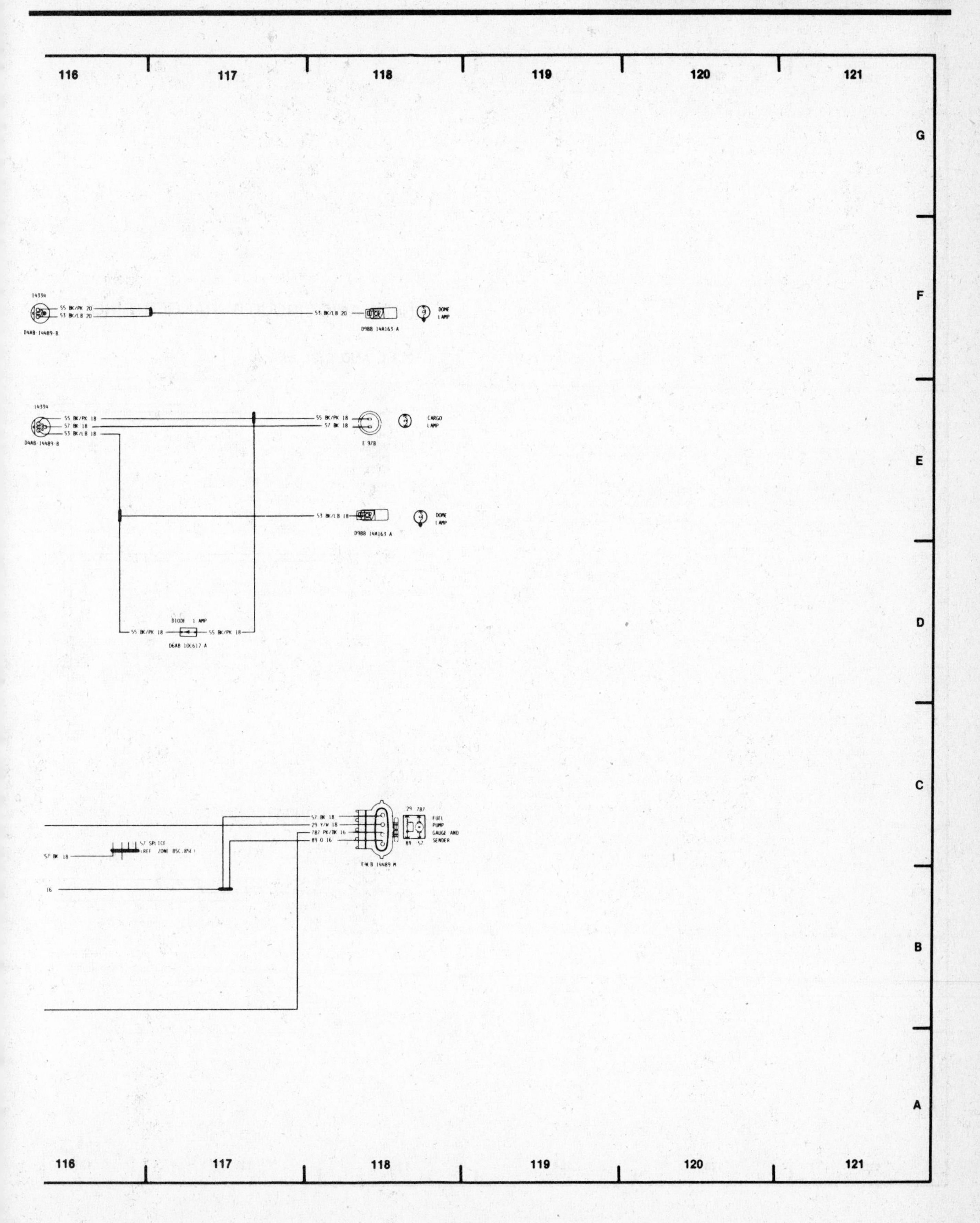
116
117
118
119
120
121
G
F
E
D
C
B
A
14334
55 BK/PK 20
53 BK/LB 20
D4AB-14489-B
53 BK/LB 20
D9BB 14A163-A
DOME LAMP
14334
55 BK/PK 18
57 BK 18
53 BK/LB 18
D4AB-14489-B
55 BK/PK 18
57 BK 18
E 97B
CARGO LAMP
53 BK/LB 18
D9BB 14A163 A
DOME LAMP
DIODE 1 AMP
55 BK/PK 18
55 BK/PK 18
D6AB 10C617 A
57 BK 18
29 Y/W 18
787 PK/BK 16
89 O 16
29 787
89 57
FUEL PUMP GAUGE AND SENDER
E4KB 14489 M
57 SPLICE
(REF ZONE 85C,85E)
57 BK 18
16

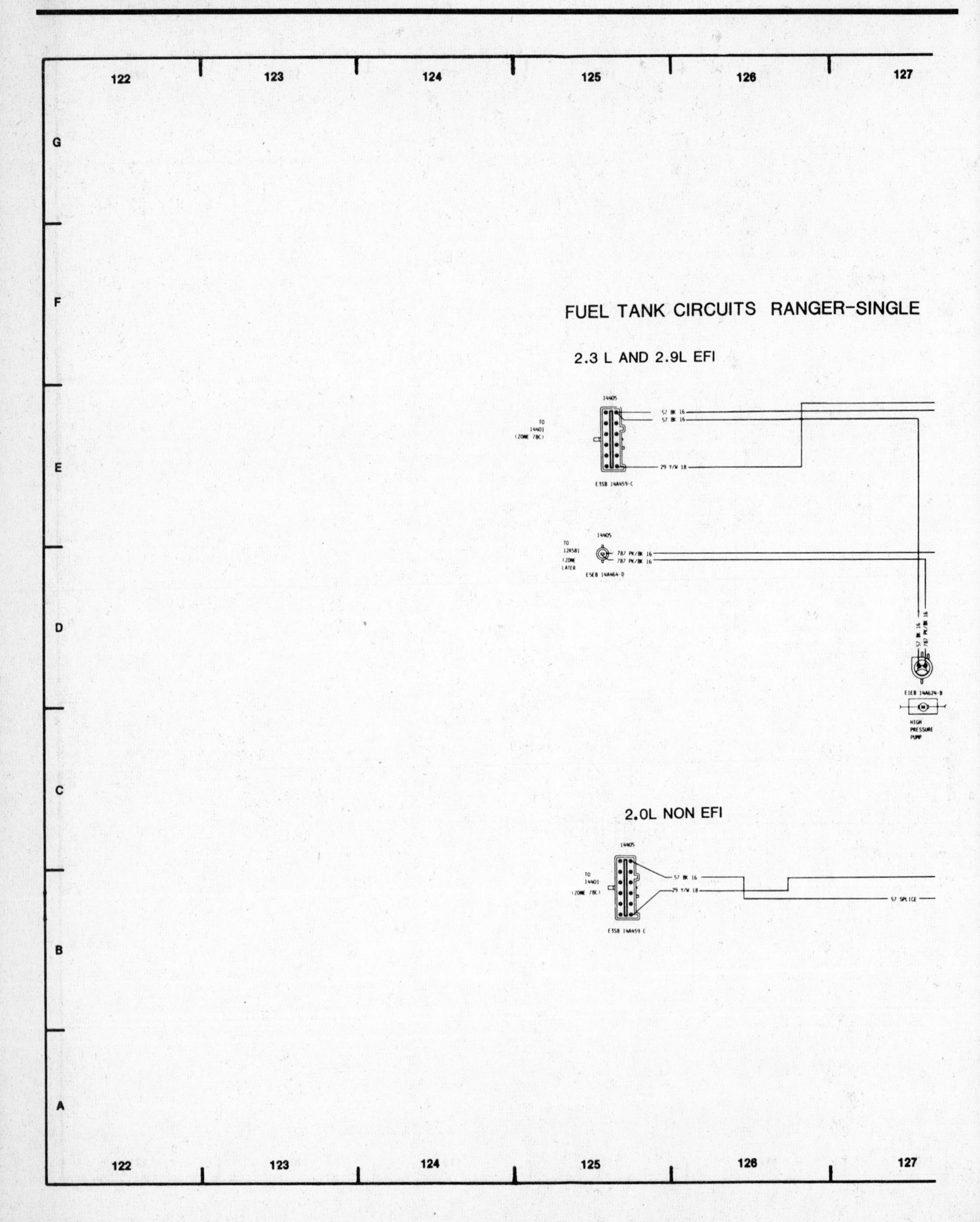
122
123
124
125
126
127
G
F
E
D
C
B
A
FUEL TANK CIRCUITS RANGER-SINGLE
2.3 L AND 2.9L EFI
14405
TO
14401
(ZONE 78C)
57 BK 16
57 BK 16
29 Y/W 18
E3SB 14A459-C
14405
TO
12A581
(ZONE
LATER
787 PK/BK 16
787 PK/BK 16
E5EB 14A464-D
57 BK 16
787 PK/BK 16
E1EB 14A624-B
HIGH
PRESSURE
PUMP
2.0L NON EFI
14405
TO
14401
(ZONE 78C)
57 BK 16
29 Y/W 18
57 SPLICE
E3SB 14A459 C

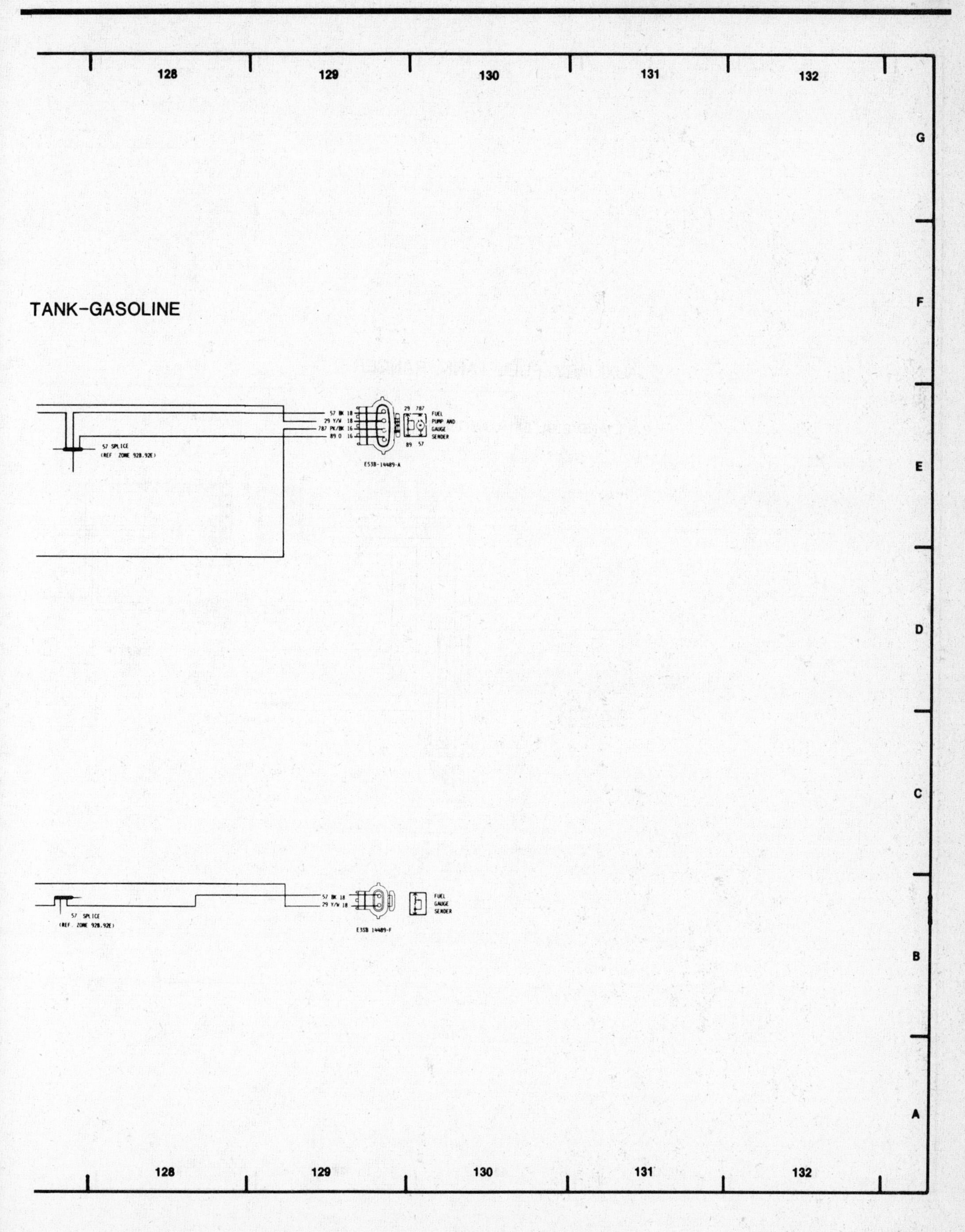
128
129
130
131
132
G
F
E
D
C
B
A
TANK-GASOLINE
57 BK 18
29 Y/W 18
787 PK/BK 16
89 O 16
29 787
FUEL
PUMP AND
GAUGE
SENDER
89 57
57 SPLICE
(REF ZONE 92B.92E)
E53B-14489-A
57 BK 18
29 Y/W 18
FUEL
GAUGE
SENDER
57 SPLICE
(REF. ZONE 92B.92E)
E3SB 14489-F

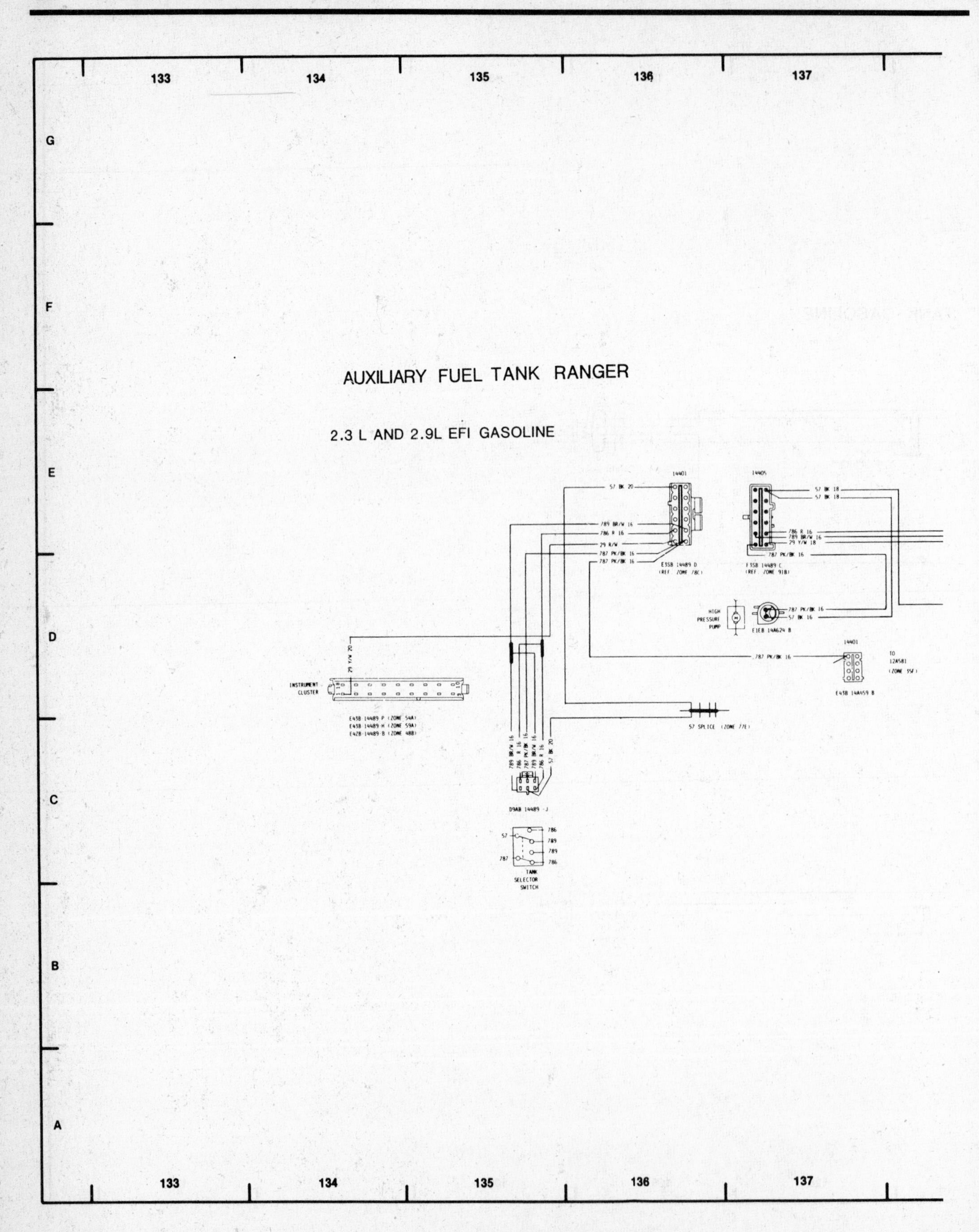

133
134
135
136
137
G
F
E
D
C
B
A
AUXILIARY FUEL TANK RANGER
2.3 L AND 2.9L EFI GASOLINE
HIGH PRESSURE PUMP
INSTRUMENT CLUSTER
TANK SELECTOR SWITCH
57 SPLICE (ZONE 77E)

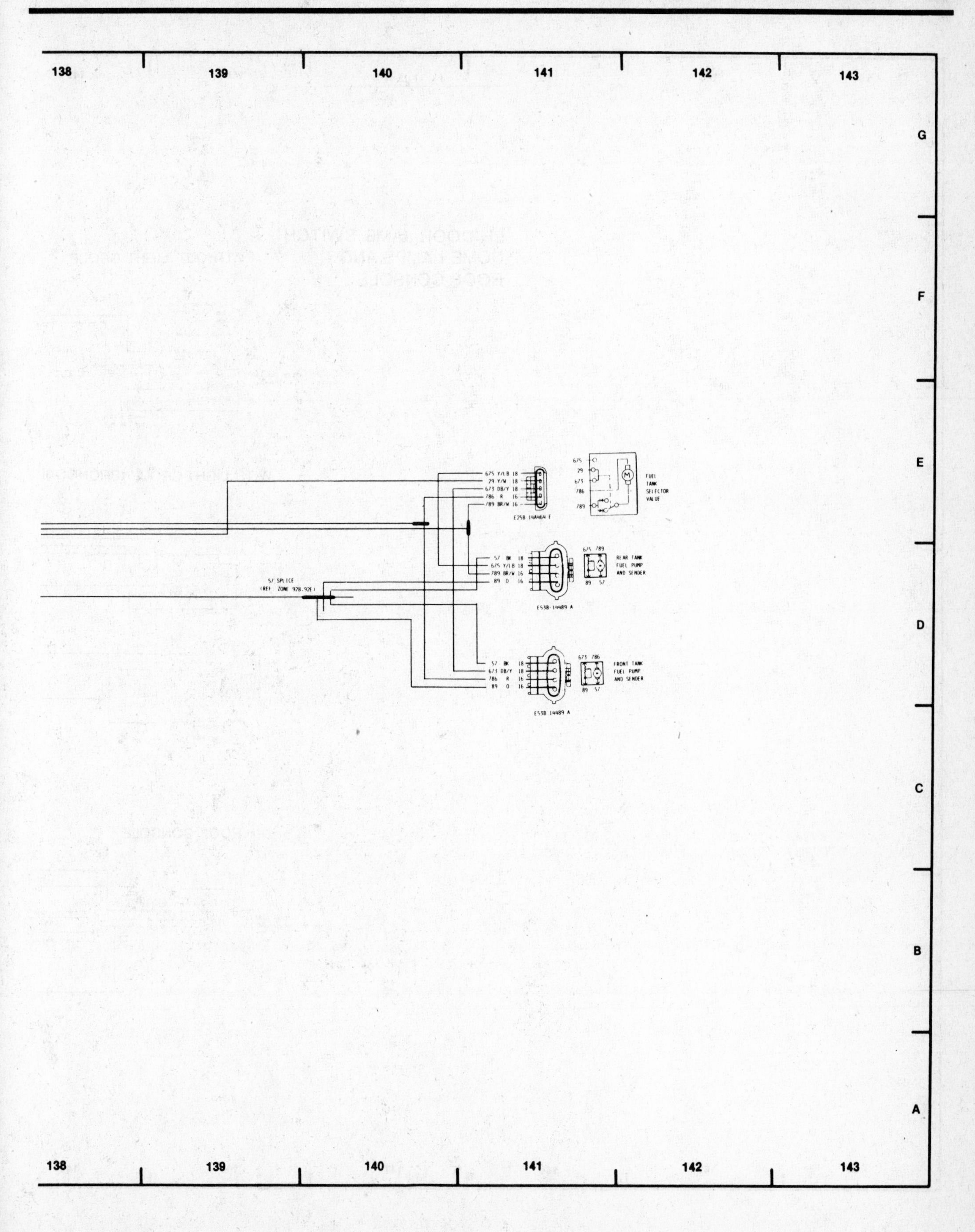
138
139
140
141
142
143
G
F
E
D
C
B
A
675 Y/LB 18
29 Y/W 18
673 DB/Y 18
786 R 16
789 BR/W 16
E25B 14A464 F
675
29
673
786
789
M
FUEL TANK SELECTOR VALUE
57 BK 18
675 Y/LB 18
789 BR/W 16
89 O 16
E53B 14489 A
675 789
89 57
REAR TANK FUEL PUMP AND SENDER
57 SPLICE
(REF ZONE 92B.92E)
57 BK 18
673 DB/Y 18
786 R 16
89 O 16
E53B 14489 A
673 786
89 57
FRONT TANK FUEL PUMP AND SENDER

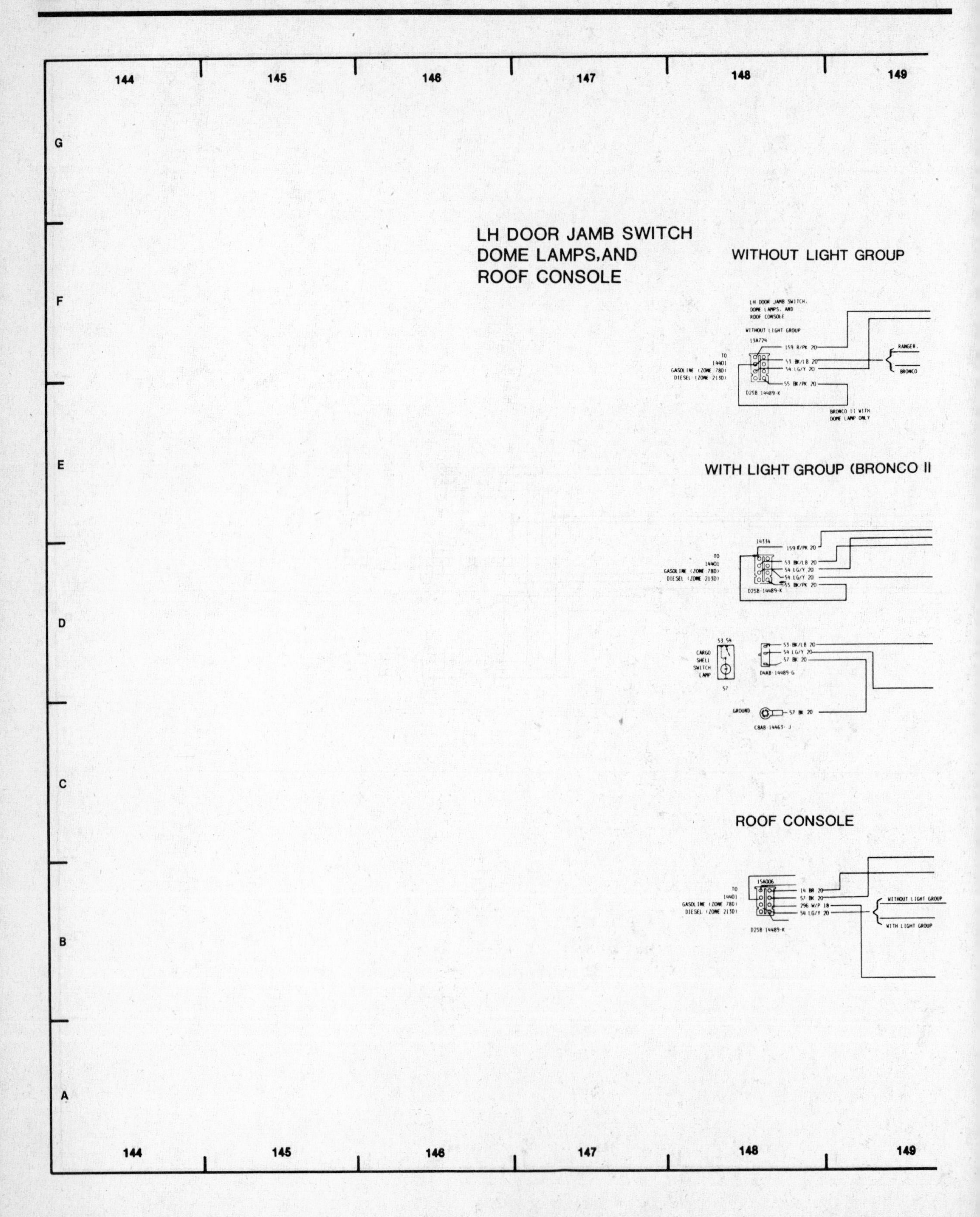
144
145
146
147
148
149
G
F
E
D
C
B
A
LH DOOR JAMB SWITCH
DOME LAMPS,AND
ROOF CONSOLE
WITHOUT LIGHT GROUP
LH DOOR JAMB SWITCH,
DOME LAMPS, AND
ROOF CONSOLE
WITHOUT LIGHT GROUP
13A724
TO
14401
GASOLINE (ZONE 78D)
DIESEL (ZONE 213D)
159 R/PK 20
53 BK/LB 20
54 LG/Y 20
55 BK/PK 20
D2SB 14489-K
RANGER,
BRONCO
BRONCO II WITH
DOME LAMP ONLY
WITH LIGHT GROUP (BRONCO II
14334
TO
14401
GASOLINE (ZONE 78D)
DIESEL (ZONE 213D)
159 R/PK 20
53 BK/LB 20
54 LG/Y 20
54 LG/Y 20
55 BK/PK 20
D2SB-14489-K
53 54
CARGO
SHELL
SWITCH
LAMP
57
53 BK/LB 20
54 LG/Y 20
57 BK 20
D4AB-14489-G
GROUND
57 BK 20
C8AB 14463-J
ROOF CONSOLE
15A006
TO
14401
GASOLINE (ZONE 78D)
DIESEL (ZONE 213D)
14 BR 20
57 BK 20
296 W/P 18
54 LG/Y 20
D2SB 14489-K
WITHOUT LIGHT GROUP
WITH LIGHT GROUP
144
145
146
147
148
149

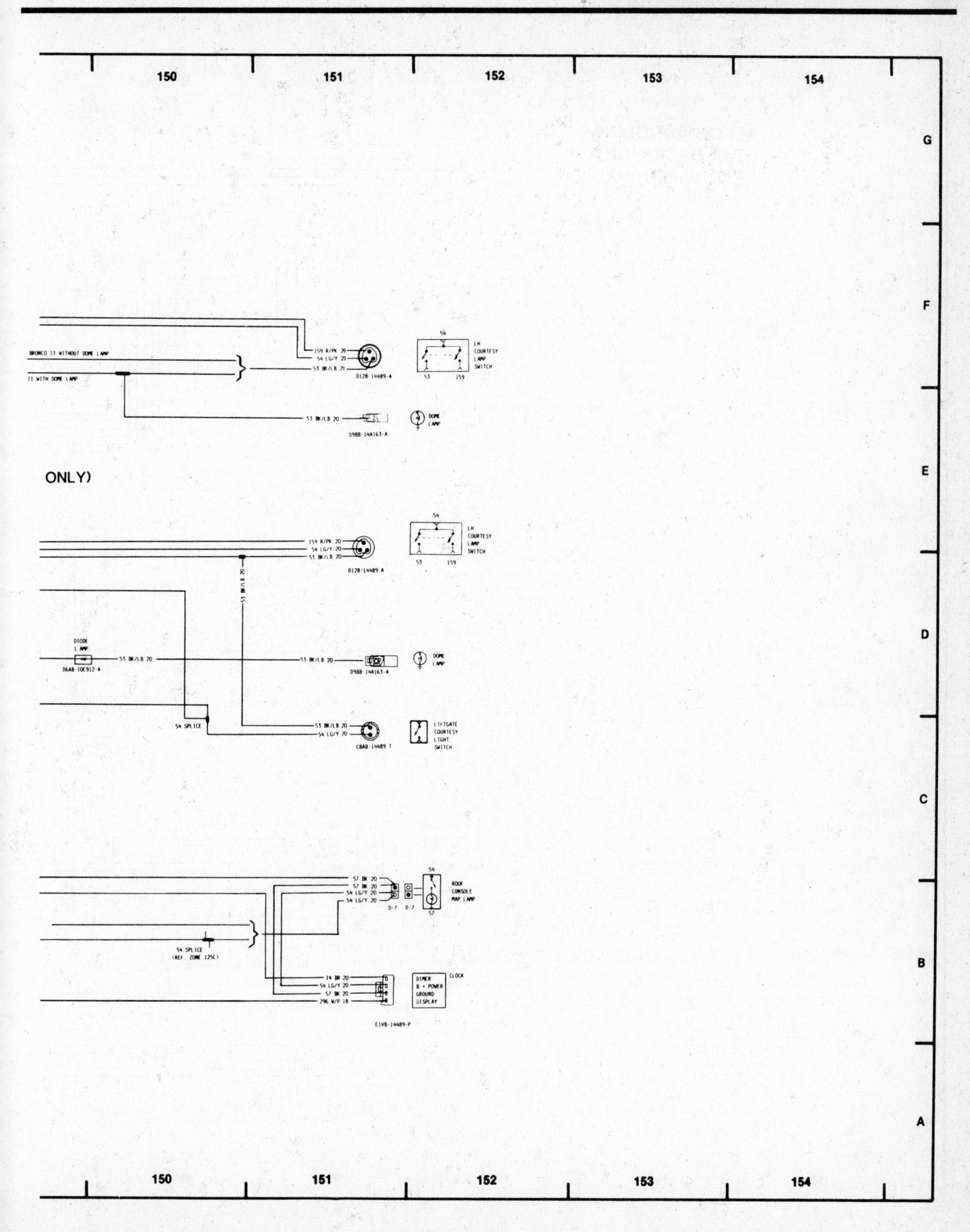
150
151
152
153
154
G
F
E
D
C
B
A
BRONCO II WITHOUT DOME LAMP
II WITH DOME LAMP
159 R/PK 20
54 LG/Y 20
53 BK/LB 20
D12B-14489-A
54
LH
COURTESY
LAMP
SWITCH
53
159
53 BK/LB 20
D9BB-14A163-A
DOME
LAMP
ONLY)
159 R/PK 20
54 LG/Y 20
53 BK/LB 20
D12B-14489-A
54
LH
COURTESY
LAMP
SWITCH
53
159
53 BK/LB 20
DIODE
1 AMP
D6AB-10C912-A
53 BK/LB 20
53 BK/LB 20
D9BB-14A163-A
DOME
LAMP
54 SPLICE
53 BK/LB 20
54 LG/Y 20
C8AB-14489 T
LIFTGATE
COURTESY
LIGHT
SWITCH
57 BK 20
57 BK 20
54 LG/Y 20
54 LG/Y 20
D-7
D-7
54
57
ROOF
CONSOLE
MAP LAMP
54 SPLICE
(REF. ZONE 125C)
14 BR 20
54 LG/Y 20
57 BK 20
296 W/P 18
DIMMER
B + POWER
GROUND
DISPLAY
CLOCK
E1VB-14489-P
150
151
152
153
154

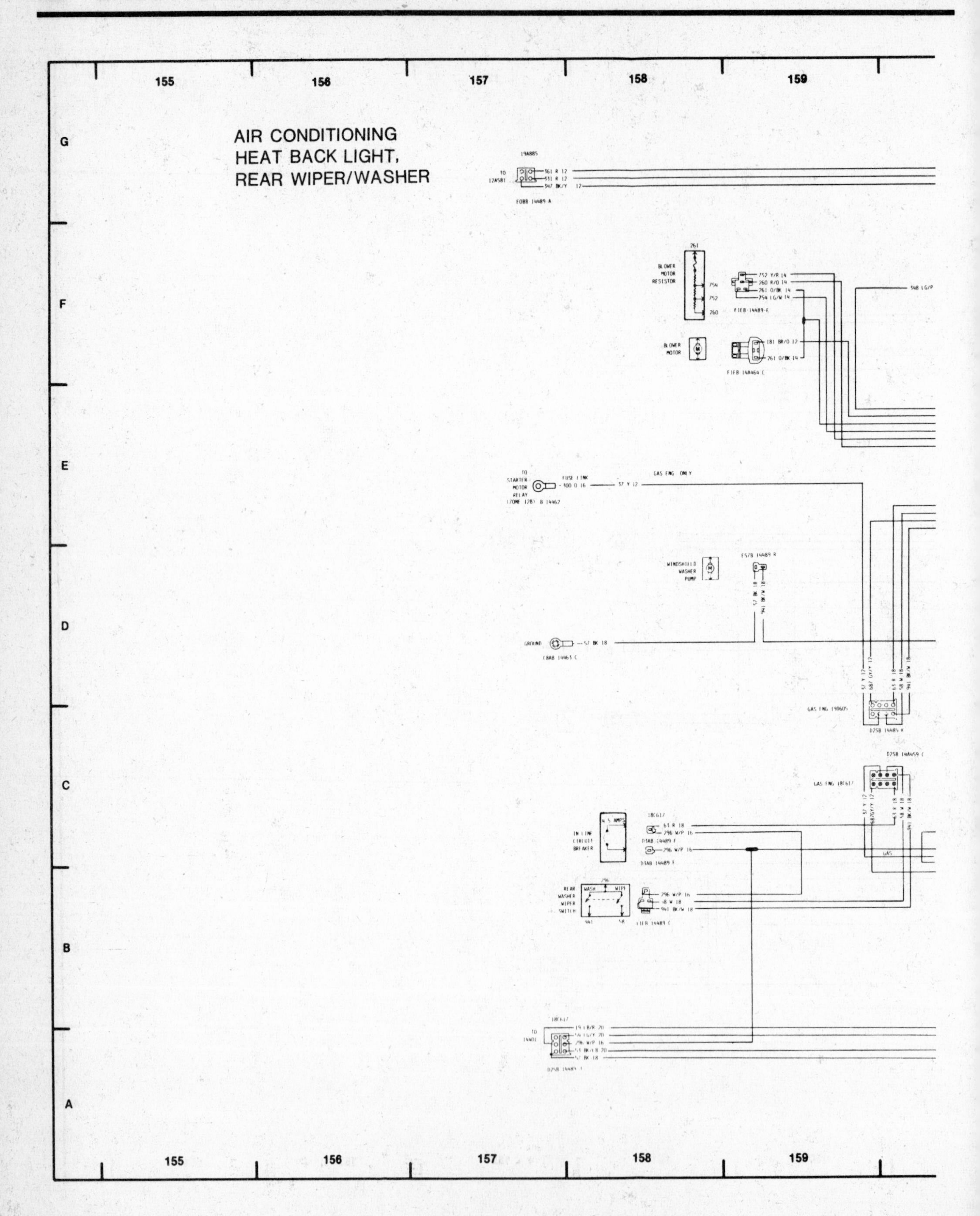
155
156
157
158
159
G
F
E
D
C
B
A
AIR CONDITIONING
HEAT BACK LIGHT,
REAR WIPER/WASHER
BLOWER
MOTOR
RESISTOR
BLOWER
MOTOR
WINDSHIELD
WASHER
PUMP
GROUND
IN LINE
CIRCUIT
BREAKER
REAR
WASHER
WIPER
SWITCH
WASH
WIPE
GAS ENG ONLY
FUSE LINK
GAS

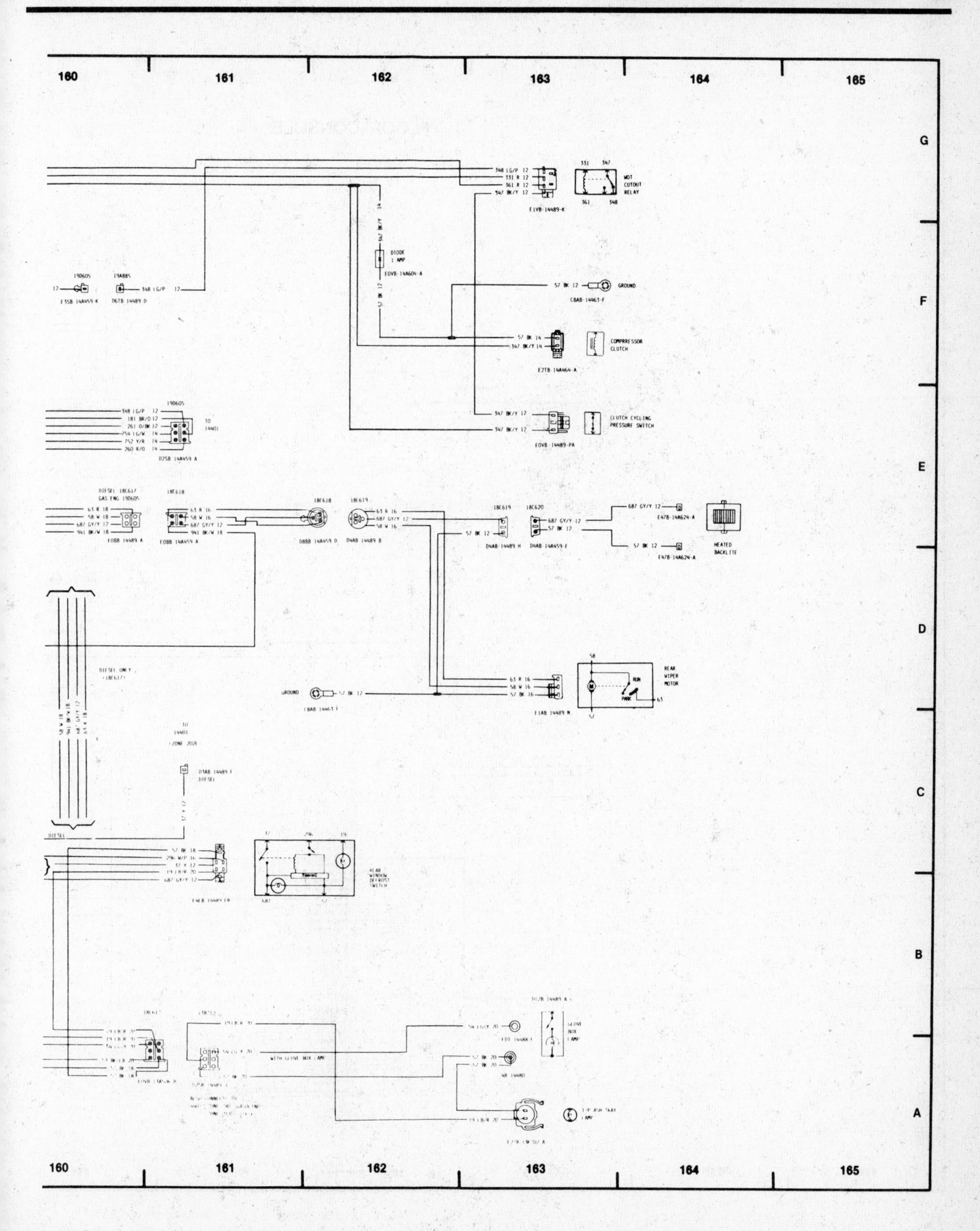

MDT CUTOUT RELAY
E1VB-14489-K
DIODE 1 AMP
E0VB-14A604-A
GROUND
C8AB-14463-F
COMPRESSOR CLUTCH
E2TB-14A464-A
CLUTCH CYCLING PRESSURE SWITCH
E0VB-14489-PA
D25B 14A459 A
HEATED BACKLITE
E47B-14A624-A
D4AB-14489 H
D4AB 14A459 F
REAR WIPER MOTOR
E1AB 14489 N
RUN
PARK
DIESEL ONLY (18C617)
REAR WINDOW DEFROST SWITCH
GLOVE BOX LAMP
WITH GLOVE BOX LAMP

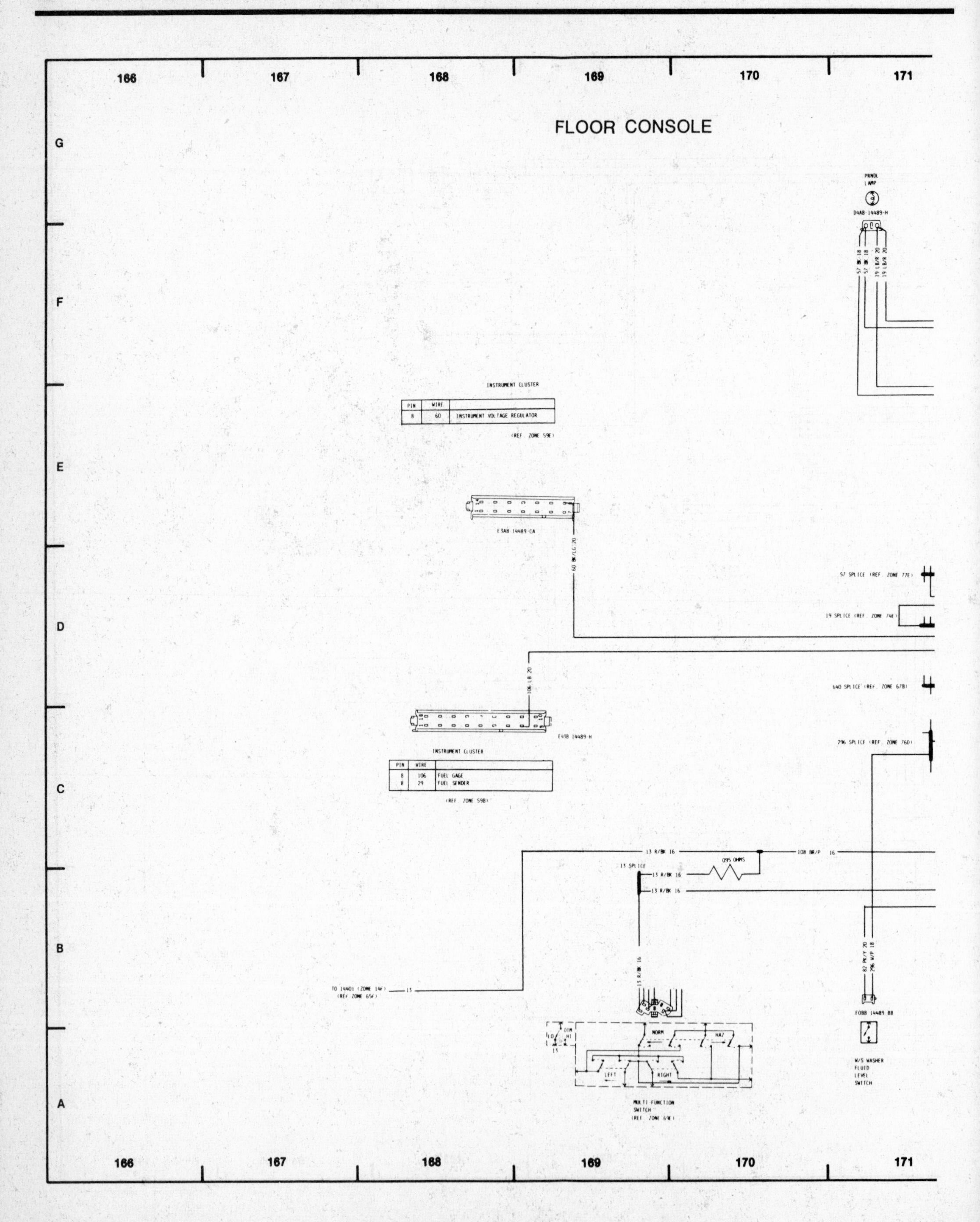
166
167
168
169
170
171
G
F
E
D
C
B
A
FLOOR CONSOLE
PRNDL LAMP
D4AB-14489-H
57 BK 18
57 BK 18
19 LB/R 20
19 LB/R 20
INSTRUMENT CLUSTER
PIN | WIRE
8 | 60 | INSTRUMENT VOLTAGE REGULATOR
(REF. ZONE 59E)
E3AB-14489-CA
60 BK/LG 20
57 SPLICE (REF. ZONE 77E)
19 SPLICE (REF. ZONE 74E)
106 LB 20
640 SPLICE (REF. ZONE 67B)
E43B 14489-H
296 SPLICE (REF. ZONE 76D)
INSTRUMENT CLUSTER
PIN | WIRE
8 | 106 | FUEL GAGE
8 | 29 | FUEL SENDER
(REF. ZONE 59B)
13 R/BK 16
108 BR/P 16
095 OHMS
13 SPLICE
13 R/BK 16
13 R/BK 16
82 PK/Y 20
296 W/P 18
TO 14401 (ZONE 14F)
(REF. ZONE 65F)
13
13 R/BK 16
F0BB 14489 BB
DIM
LO
HI
13
NORM
HAZ
LEFT
RIGHT
W/S WASHER
FLUID
LEVEL
SWITCH
MULTI-FUNCTION
SWITCH
(REF. ZONE 69E)

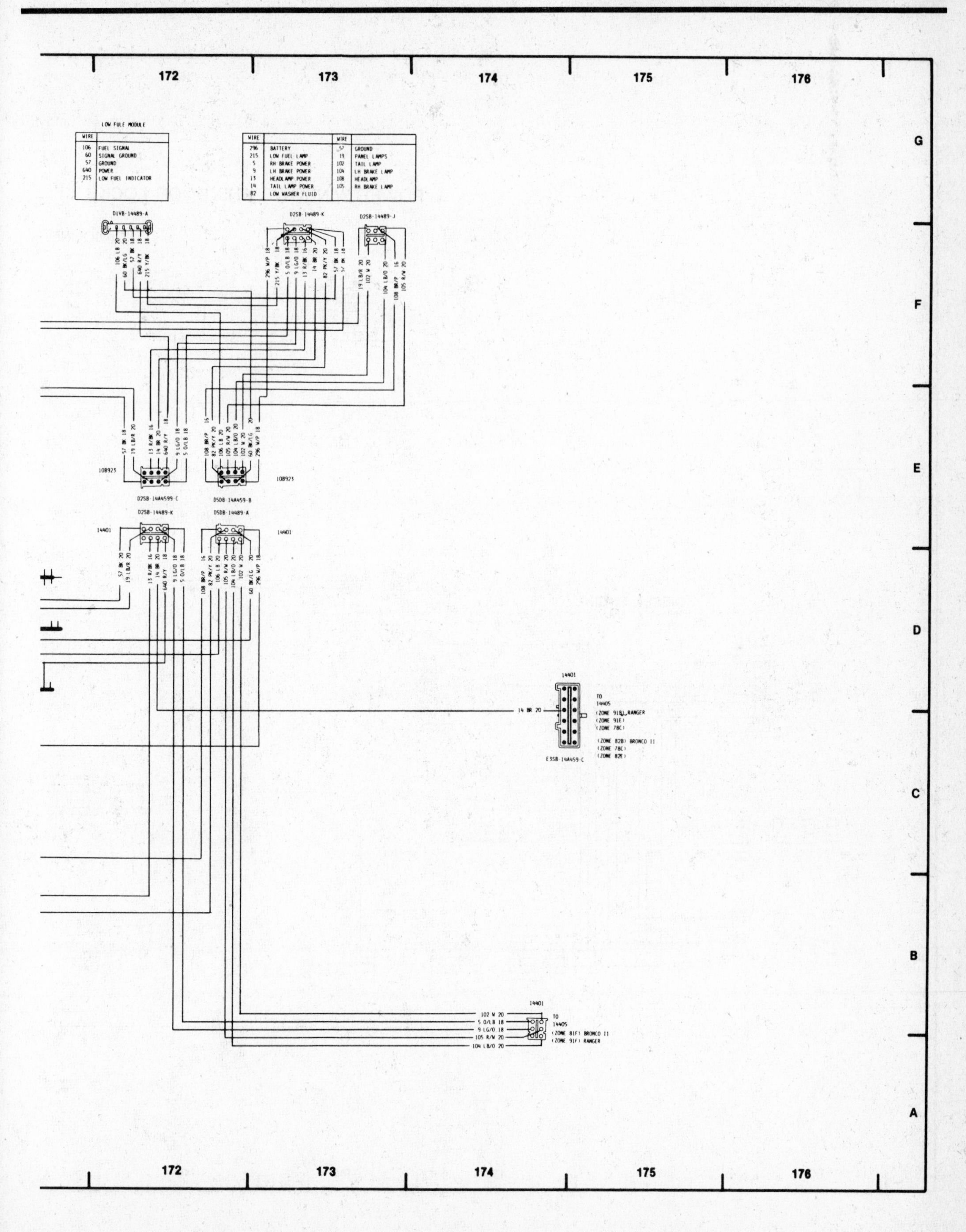
172
173
174
175
176
G
F
E
D
C
B
A
LOW FULE MODULE
WIRE
106 FUEL SIGNAL
60 SIGNAL GROUND
57 GROUND
640 POWER
215 LOW FUEL INDICATOR
WIRE
296 BATTERY
215 LOW FUEL LAMP
5 RH BRAKE POWER
9 LH BRAKE POWER
13 HEADLAMP POWER
14 TAIL LAMP POWER
82 LOW WASHER FLUID
WIRE
57 GROUND
19 PANEL LAMPS
102 TAIL LAMP
104 LH BRAKE LAMP
108 HEADLAMP
105 RH BRAKE LAMP
D1VB-14489-A
D2SB-14489-K
D2SB-14489-J
10B923
D2SB-14A599-C
D5DB-14A459-B
D2SB-14489-K
D5DB-14489-A
14401
14 BR 20
TO 14405
(ZONE 91B) RANGER
(ZONE 91E)
(ZONE 78C)
(ZONE 82B) BRONCO II
(ZONE 78C)
(ZONE 82E)
E3SB-14A459-C
102 W 20
5 O/LB 18
9 LG/O 18
105 R/W 20
104 LB/O 20
(ZONE 81F) BRONCO II
(ZONE 91F) RANGER

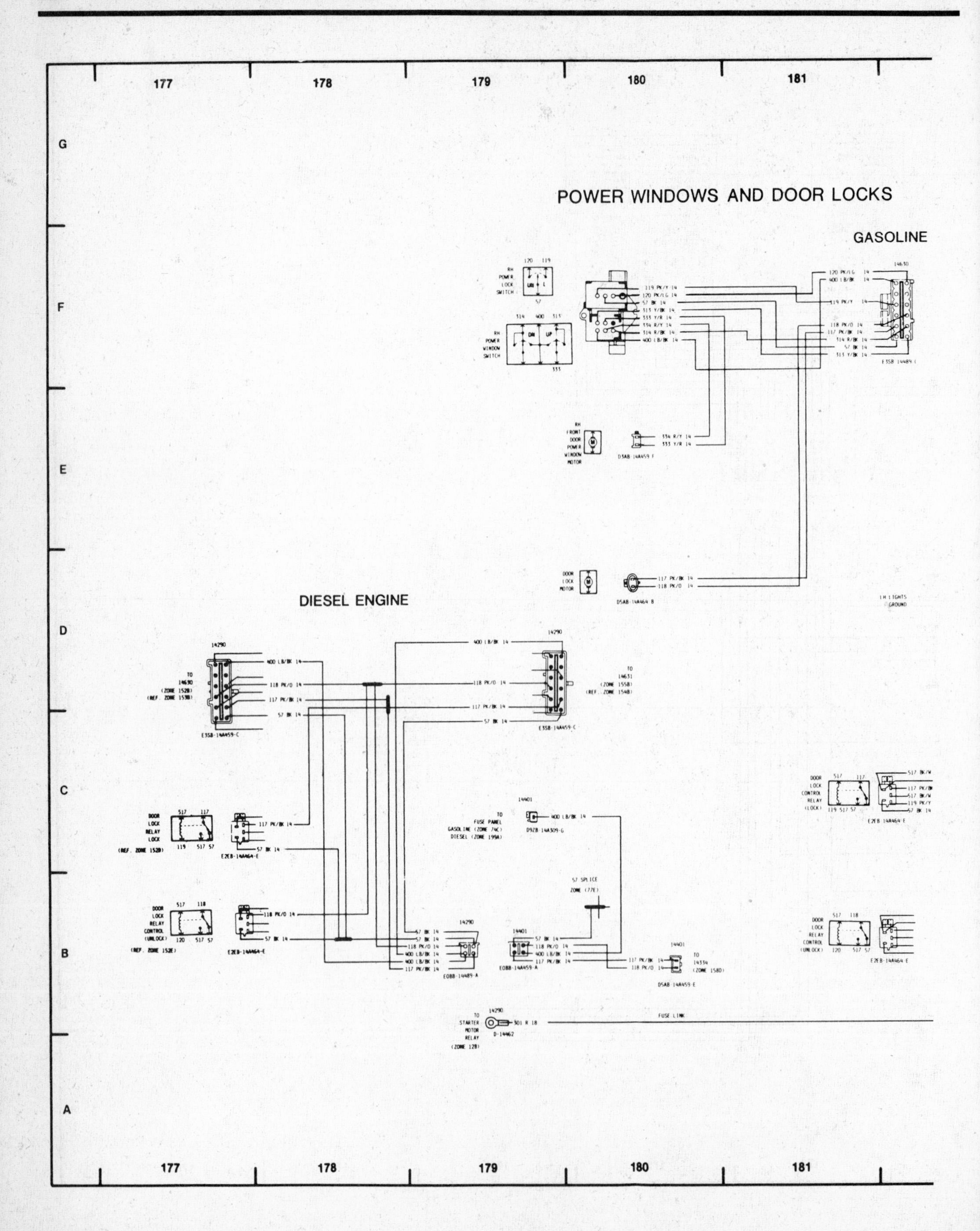

177
178
179
180
181
G
F
E
D
C
B
A
POWER WINDOWS AND DOOR LOCKS
GASOLINE
DIESEL ENGINE
RH POWER LOCK SWITCH
RH POWER WINDOW SWITCH
RH FRONT DOOR POWER WINDOW MOTOR
DOOR LOCK MOTOR
DOOR LOCK RELAY LOCK
DOOR LOCK RELAY CONTROL (UNLOCK)
DOOR LOCK CONTROL RELAY (LOCK)
TO FUSE PANEL GASOLINE (ZONE 74C) DIESEL (ZONE 199A)
TO STARTER MOTOR RELAY (ZONE 12B)
FUSE LINK
57 SPLICE ZONE (77E)
LH LIGHTS GROUND

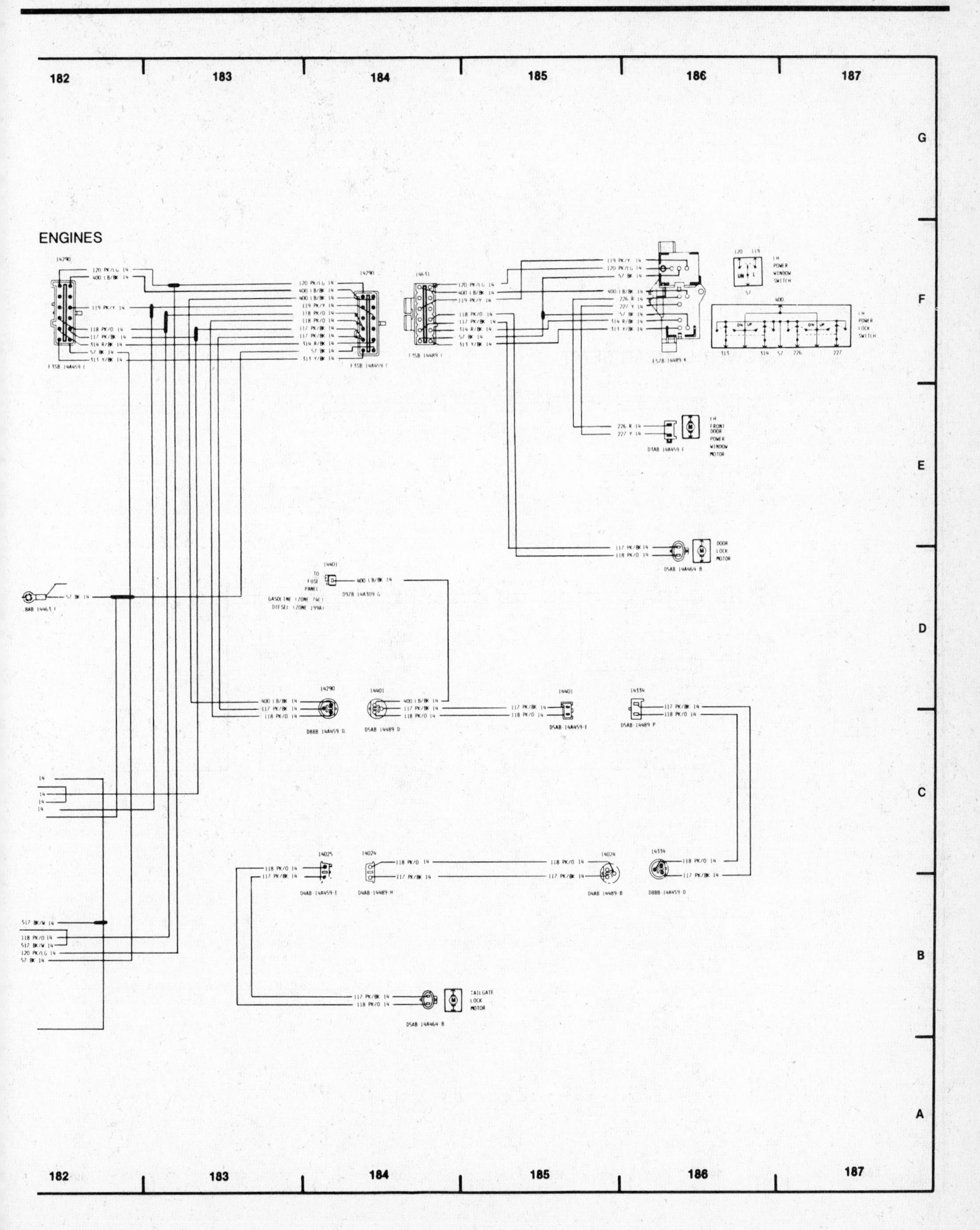
182
183
184
185
186
187
G
F
E
D
C
B
A
ENGINES
14290
120 PK/LG 14
400 LB/BK 14
119 PK/Y 14
118 PK/O 14
117 PK/BK 14
314 R/BK 14
57 BK 14
313 Y/BK 14
14631
LH POWER WINDOW SWITCH
LH POWER LOCK SWITCH
226 R 14
227 Y 14
LH FRONT DOOR POWER WINDOW MOTOR
DOOR LOCK MOTOR
D5AB 14A464 B
14401
TO FUSE PANEL
GASOLINE (ZONE 74C)
DIESEL (ZONE 199A)
D97B 14A309 G
D8BB 14A459 D
D5AB 14489 D
D5AB 14A459 E
14334
D5AB 14489 P
14025
14024
D4AB 14A459 E
D4AB 14489 H
D4AB 14489 B
D8BB 14A459 D
517 BK/W 14
TAILGATE LOCK MOTOR
D5AB 14A464 B

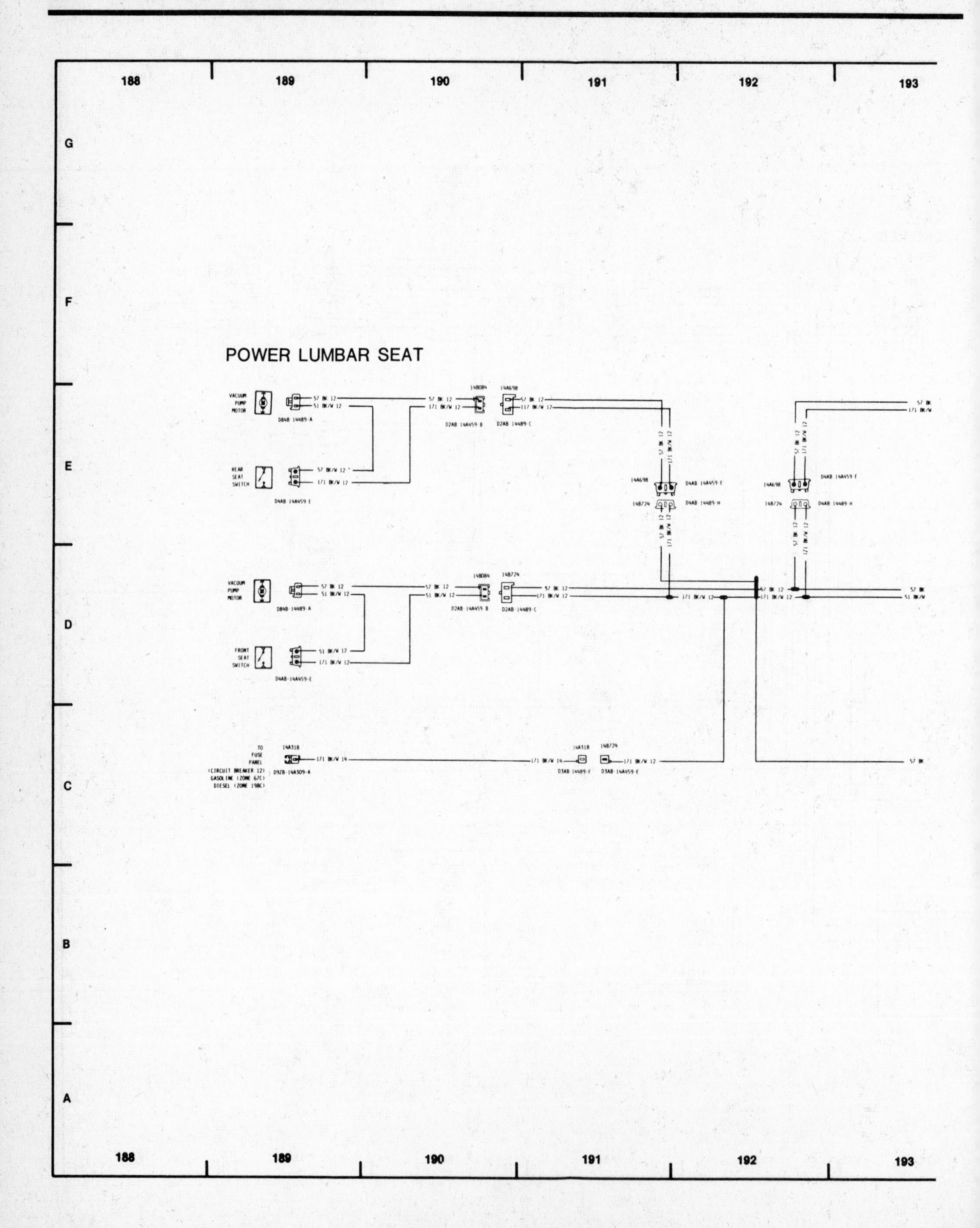
188
189
190
191
192
193
G
F
E
D
C
B
A
POWER LUMBAR SEAT
VACUUM PUMP MOTOR
D84B 14489-A
57 BK 12
51 BK/W 12
14B084
14A698
57 BK 12
171 BK/W 12
D2AB 14A459-B
D2AB 14489-C
57 BK 12
117 BK/W 12
REAR SEAT SWITCH
57 BK/W 12
171 BK/W 12
D4AB 14A459-E
57 BK 12
171 BK/W 12
14A698
D4AB 14A459-E
14B724
D4AB 14489-H
57 BK
171 BK/W
D4AB 14A459-F
VACUUM PUMP MOTOR
D84B 14489-A
57 BK 12
51 BK/W 12
14B084
14B724
D2AB 14A459-B
D2AB 14489-C
FRONT SEAT SWITCH
51 BK/W 12
171 BK/W 12
D4AB 14A459-E
57 BK 12
171 BK/W 12
57 BK
51 BK/W
TO FUSE PANEL
(CIRCUIT BREAKER 12)
GASOLINE (ZONE 67C)
DIESEL (ZONE 198C)
14A318
D92B-14A309-A
171 BK/W 14
14A318
14B724
D3AB 14489-F
D3AB 14A459-E
171 BK/W 12
57 BK

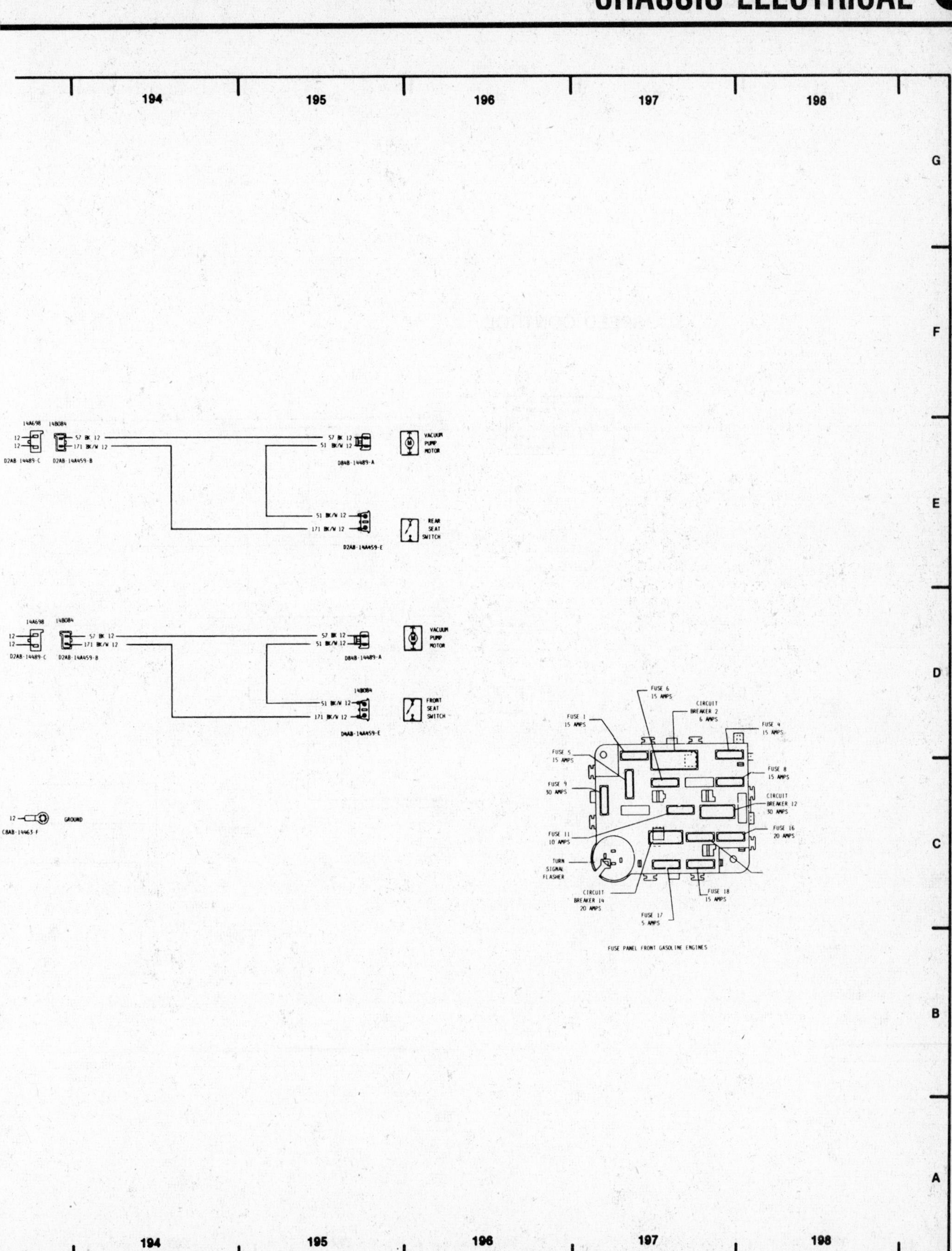
194
195
196
197
198
G
F
E
D
C
B
A
14A698
14B084
12
12
57 BK 12
171 BK/W 12
D2AB-14489-C
D2AB-14A459-B
57 BK 12
51 BK/W 12
D84B-14489-A
VACUUM
PUMP
MOTOR
51 BK/W 12
171 BK/W 12
D2AB-14A459-E
REAR
SEAT
SWITCH
14A698
14B084
12
12
57 BK 12
171 BK/W 12
D2AB-14489-C
D2AB-14A459-B
57 BK 12
51 BK/W 12
D84B-14489-A
VACUUM
PUMP
MOTOR
14B084
51 BK/W 12
171 BK/W 12
D4AB-14A459-E
FRONT
SEAT
SWITCH
12
GROUND
C8AB-14463-F
FUSE 6
15 AMPS
CIRCUIT
BREAKER 2
6 AMPS
FUSE 1
15 AMPS
FUSE 4
15 AMPS
FUSE 5
15 AMPS
FUSE 8
15 AMPS
FUSE 9
30 AMPS
CIRCUIT
BREAKER 12
30 AMPS
FUSE 16
20 AMPS
FUSE 11
10 AMPS
TURN
SIGNAL
FLASHER
CIRCUIT
BREAKER 14
20 AMPS
FUSE 18
15 AMPS
FUSE 17
5 AMPS
FUSE PANEL FRONT GASOLINE ENGINES
194
195
196
197
198

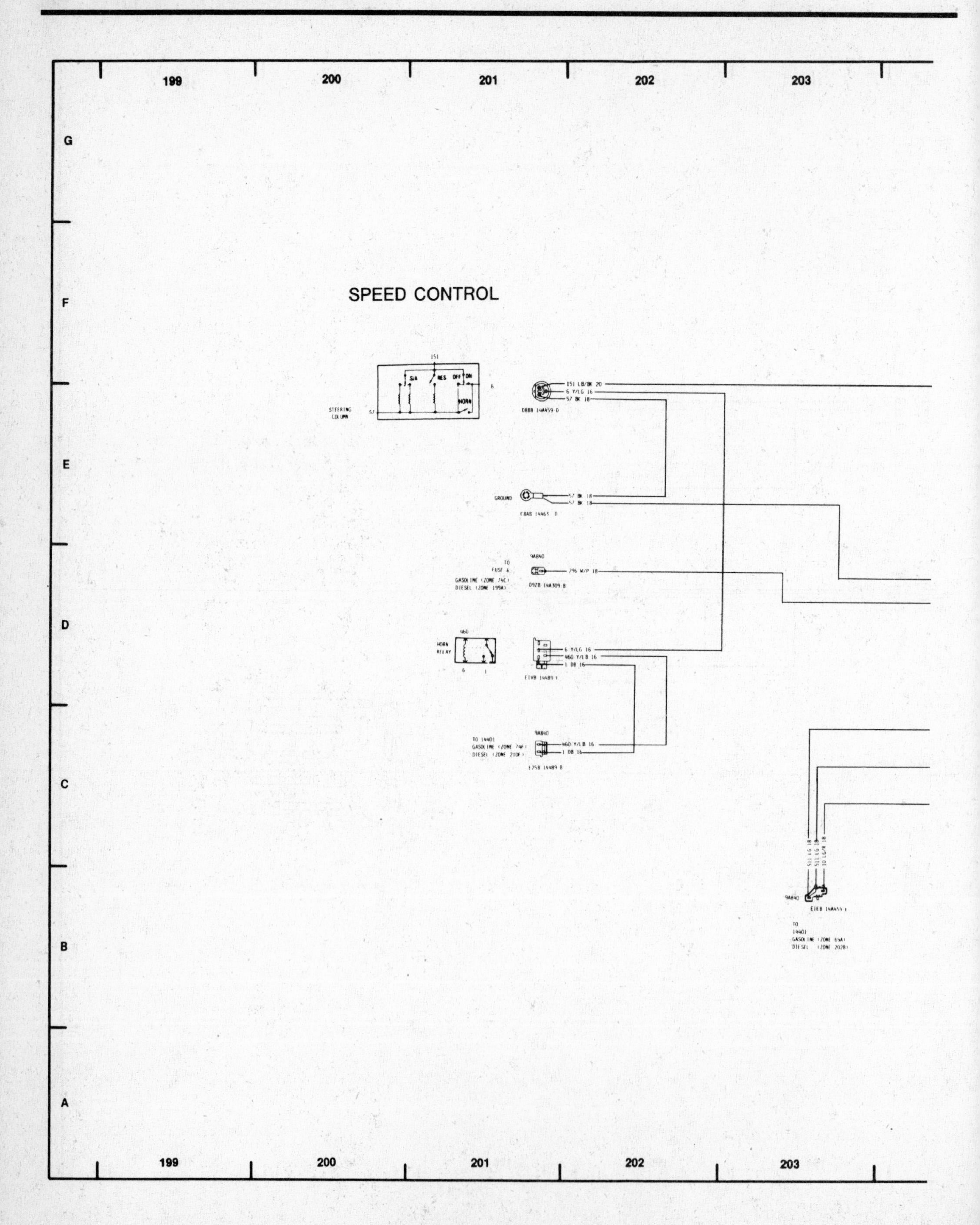
199
200
201
202
203
G
F
E
D
C
B
A
SPEED CONTROL
151
S/A
RES
OFF
ON
HORN
6
STEERING COLUMN
57
151 LB/BK 20
6 Y/LG 16
57 BK 18
D8BB 14A459 D
GROUND
57 BK 18
57 BK 18
C8AB 14463 D
9A840
TO
FUSE 6
GASOLINE (ZONE 74C)
DIESEL (ZONE 199A)
296 W/P 18
D9ZB 14A309 B
460
HORN
RELAY
6
1
6 Y/LG 16
460 Y/LB 16
1 DB 16
E1VB 14489 F
9A840
TO 14401
GASOLINE (ZONE 74F)
DIESEL (ZONE 210F)
460 Y/LB 16
1 DB 16
E25B 14489 B
511 LG 18
511 LG 18
10 LG/R 18
9A840
E1EB 14A455 E
TO
14401
GASOLINE (ZONE 69A)
DIESEL (ZONE 202B)
199
200
201
202
203

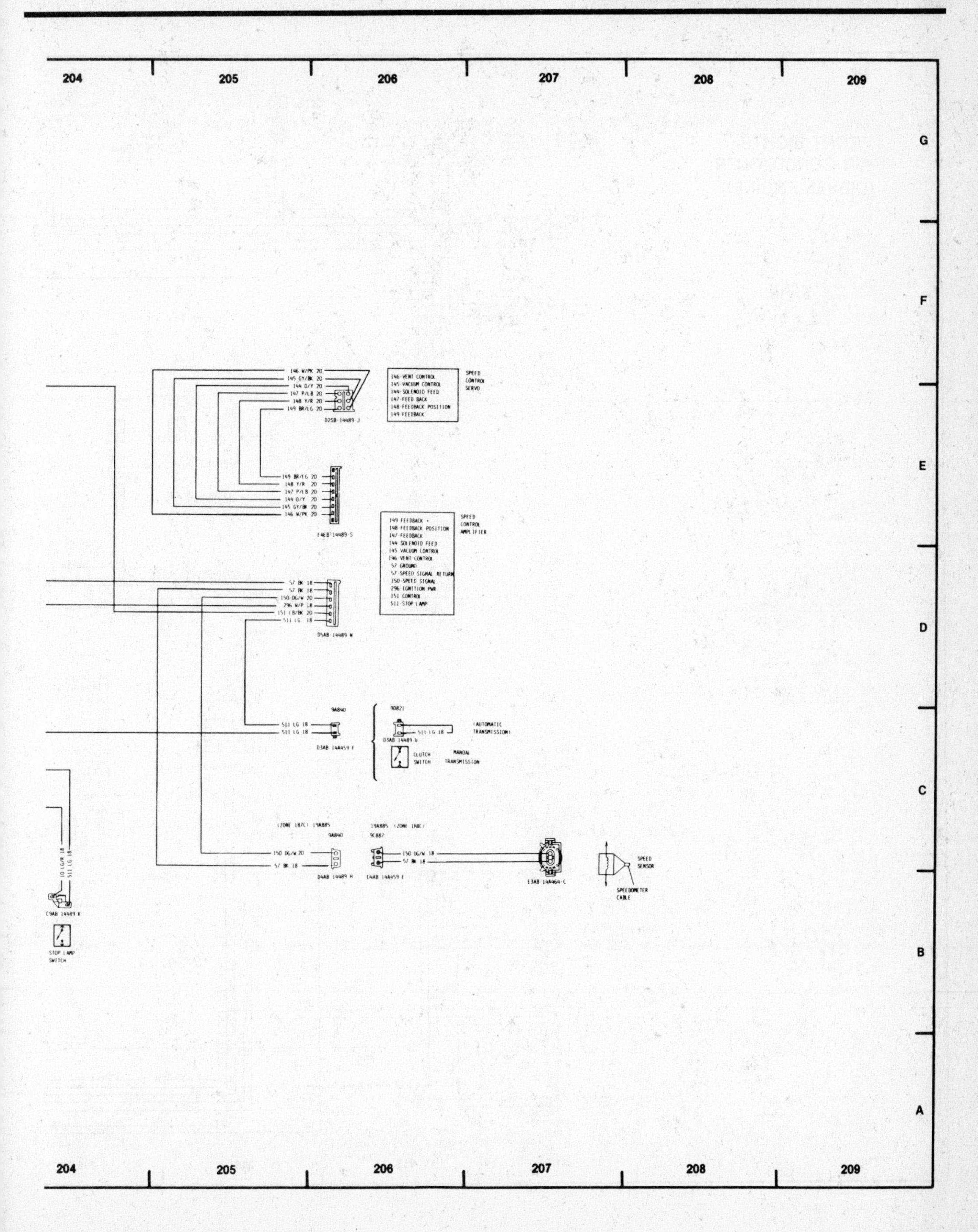
204
205
206
207
208
209
G
F
E
D
C
B
A
146 W/PK 20
145 GY/BK 20
144 O/Y 20
147 P/LB 20
148 Y/R 20
149 BR/LG 20
D2SB-14489-J
146-VENT CONTROL
145-VACUUM CONTROL
144-SOLENOID FEED
147-FEED BACK
148-FEEDBACK POSITION
149-FEEDBACK
SPEED CONTROL SERVO
149 BR/LG 20
148 Y/R 20
147 P/LB 20
144 O/Y 20
145 GY/BK 20
146 W/PK 20
F4EB-14489-S
149-FEEDBACK +
148-FEEDBACK POSITION
147-FEEDBACK
144 SOLENOID FEED
145 VACUUM CONTROL
146 VENT CONTROL
57 GROUND
57-SPEED SIGNAL RETURN
150-SPEED SIGNAL
296-IGNITION PWR
151 CONTROL
511-STOP LAMP
SPEED CONTROL AMPLIFIER
57 BK 18
57 BK 18
150 DG/W 20
296 W/P 18
151 LB/BK 20
511 LG 18
D5AB-14489-N
9A840
511 LG 18
511 LG 18
D3AB-14A459-F
9D821
511 LG 18
D3AB-14489-U
(AUTOMATIC TRANSMISSION)
CLUTCH SWITCH
MANUAL TRANSMISSION
(ZONE 187C) 19A885
9A840
150 DG/W 20
57 BK 18
D4AB-14489-H
19A885 (ZONE 188C)
9C887
150 DG/W 18
57 BK 18
D4AB-14A459-E
E3AB-14A464-C
SPEED SENSOR
SPEEDOMETER CABLE
10 LG/R 18
511 LG 18
C9AB-14489-K
STOP LAMP SWITCH

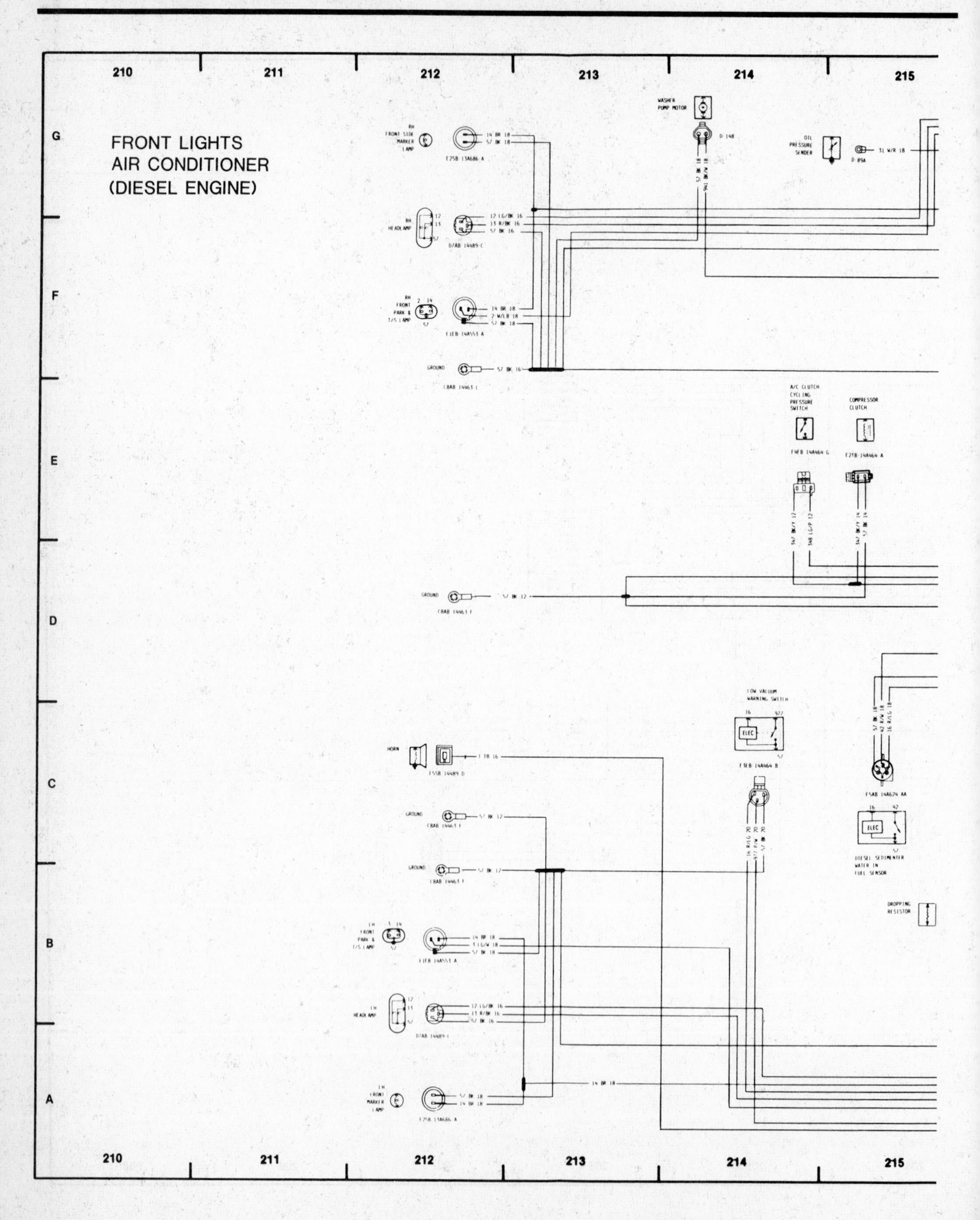
210
211
212
213
214
215
G
F
E
D
C
B
A
FRONT LIGHTS
AIR CONDITIONER
(DIESEL ENGINE)
WASHER PUMP MOTOR
OIL PRESSURE SENDER
RH FRONT SIDE MARKER LAMP
E25B 13A686 A
RH HEADLAMP
D7AB 14489 C
RH FRONT PARK & T/S LAMP
E1EB 14A553 A
GROUND
C8AB 14463 C
A/C CLUTCH CYCLING PRESSURE SWITCH
E4EB 14A464 G
COMPRESSOR CLUTCH
E2TB 14A464 A
C8AB 14463 F
LOW VACUUM WARNING SWITCH
ELEC
E5EB 14A464 B
HORN
E5SB 14489 D
E5AB 14A624 AA
DIESEL SEDIMENTER WATER IN FUEL SENSOR
DROPPING RESISTOR
LH FRONT PARK & T/S LAMP
LH HEADLAMP
LH FRONT MARKER LAMP
E25B 13A686 A

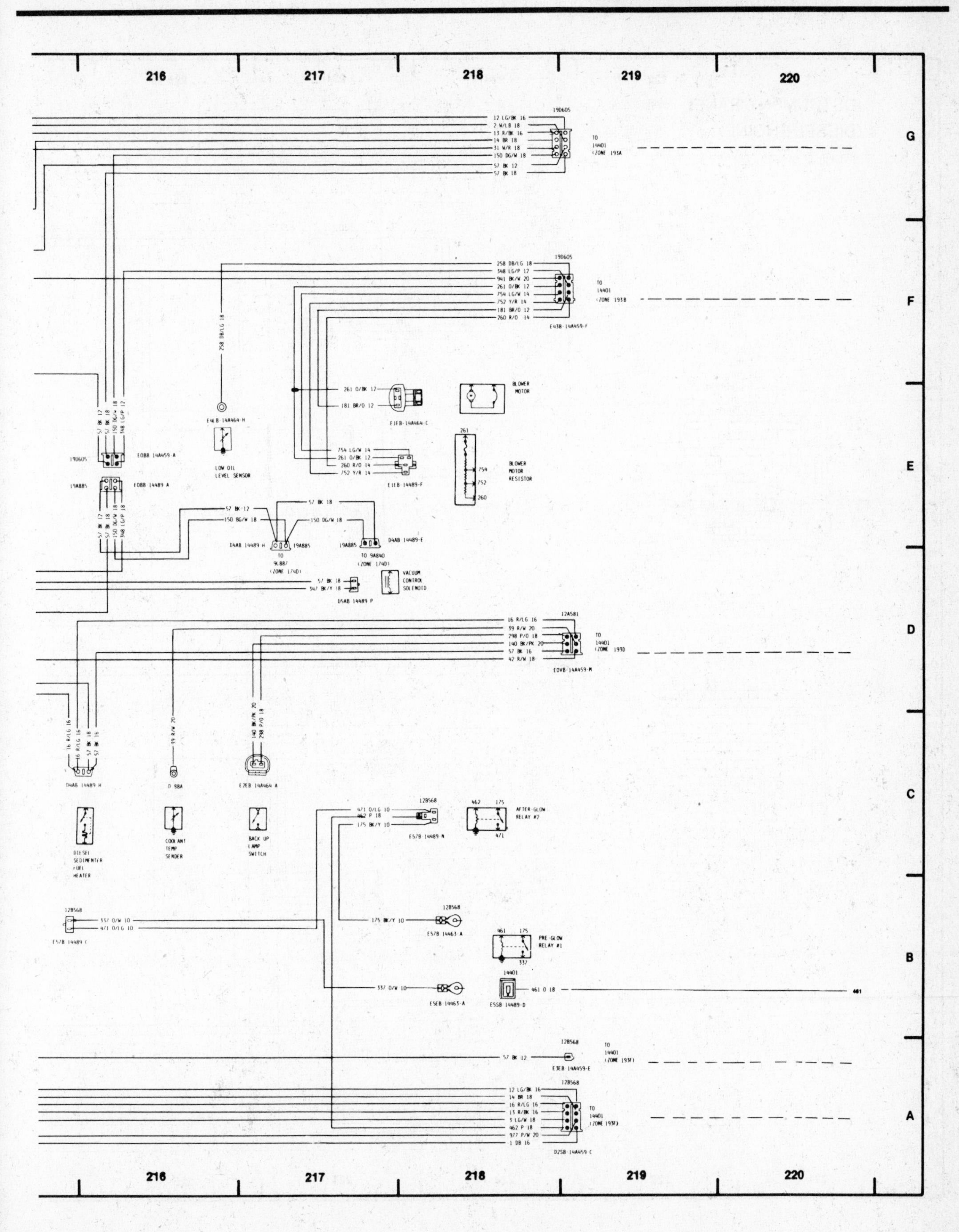

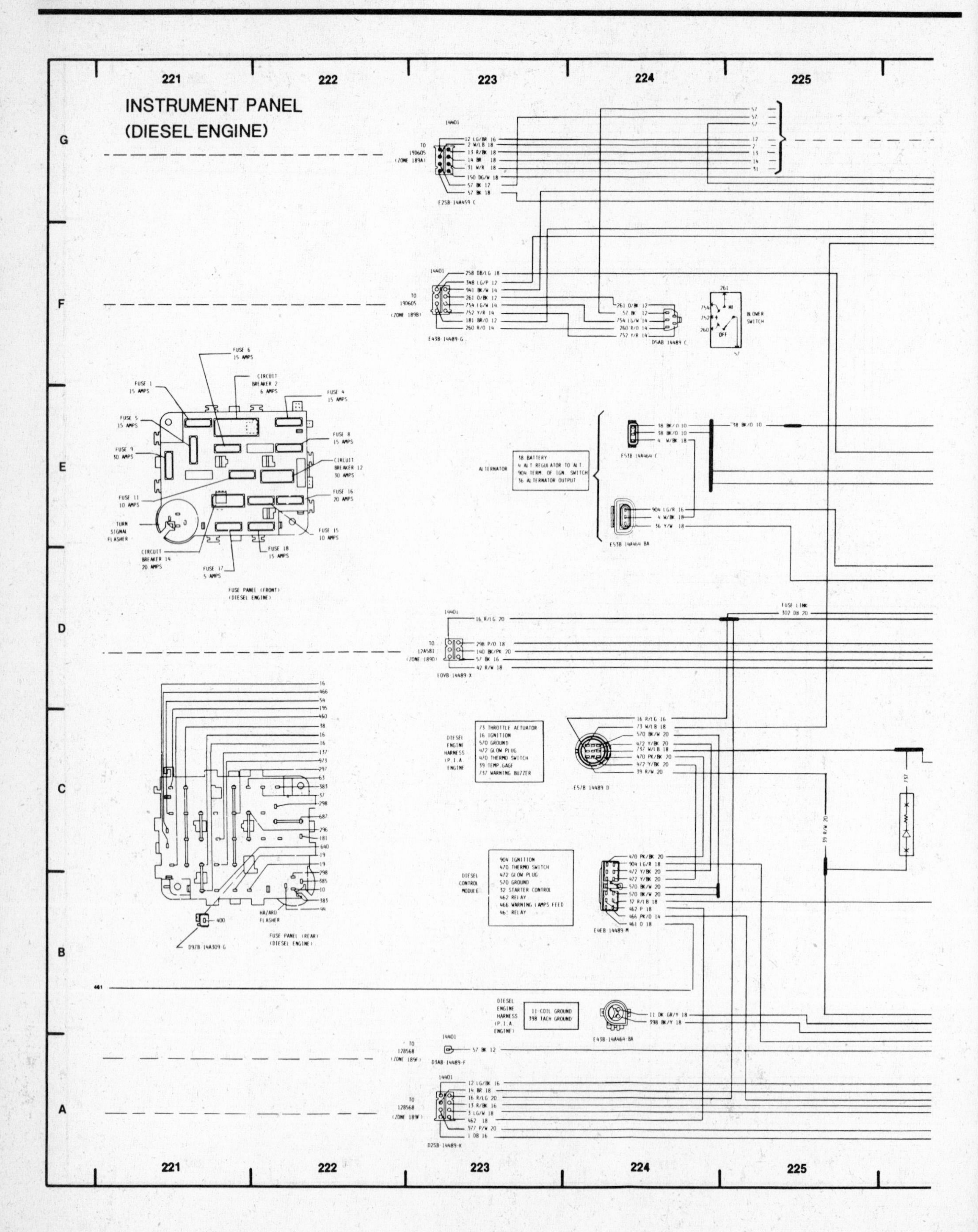

221
222
223
224
225
INSTRUMENT PANEL
(DIESEL ENGINE)
G
F
E
D
C
B
A
BLOWER SWITCH
ALTERNATOR
FUSE PANEL (FRONT) (DIESEL ENGINE)
FUSE PANEL (REAR) (DIESEL ENGINE)
DIESEL ENGINE HARNESS (P.I.A. ENGINE)
DIESEL CONTROL MODULE

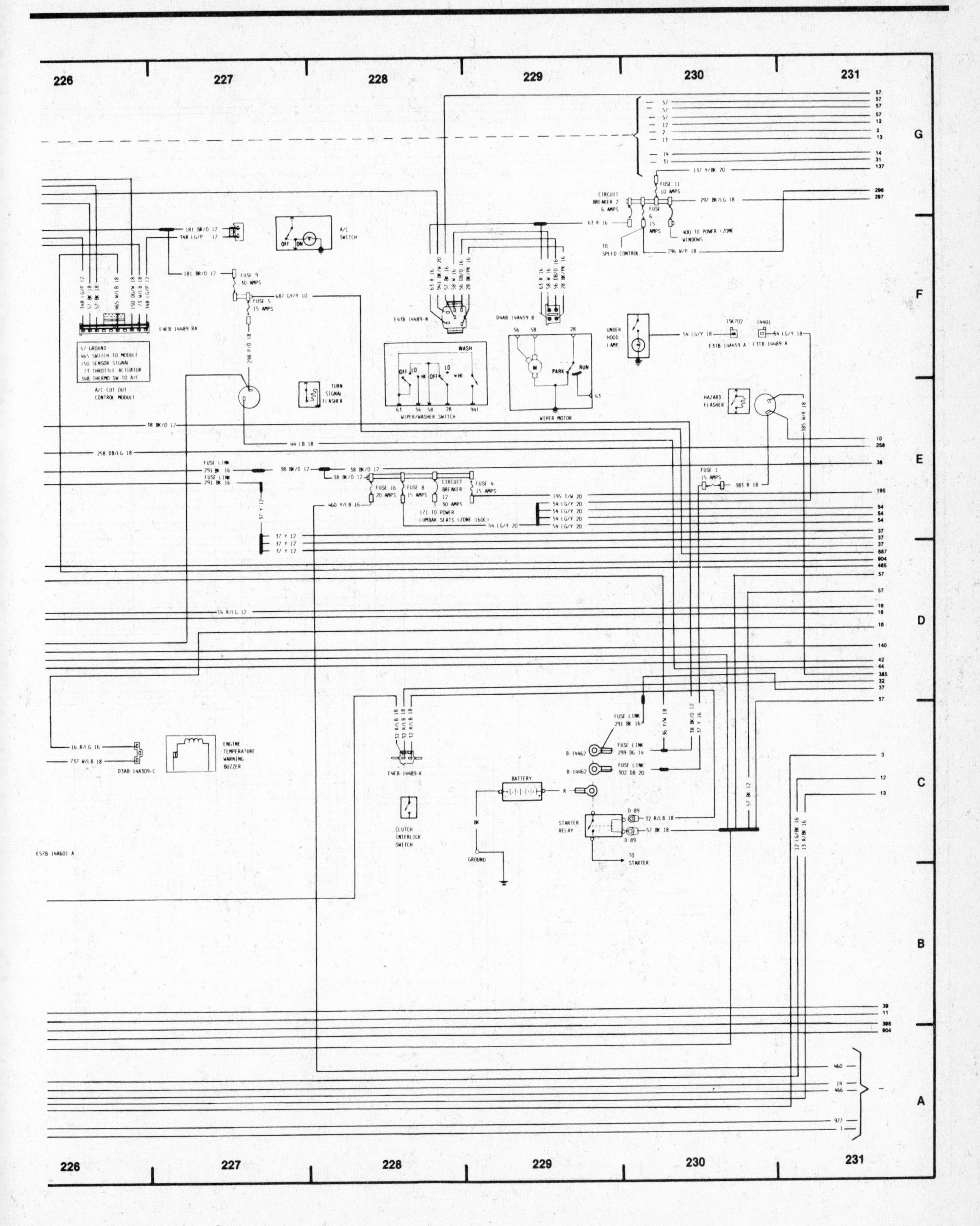

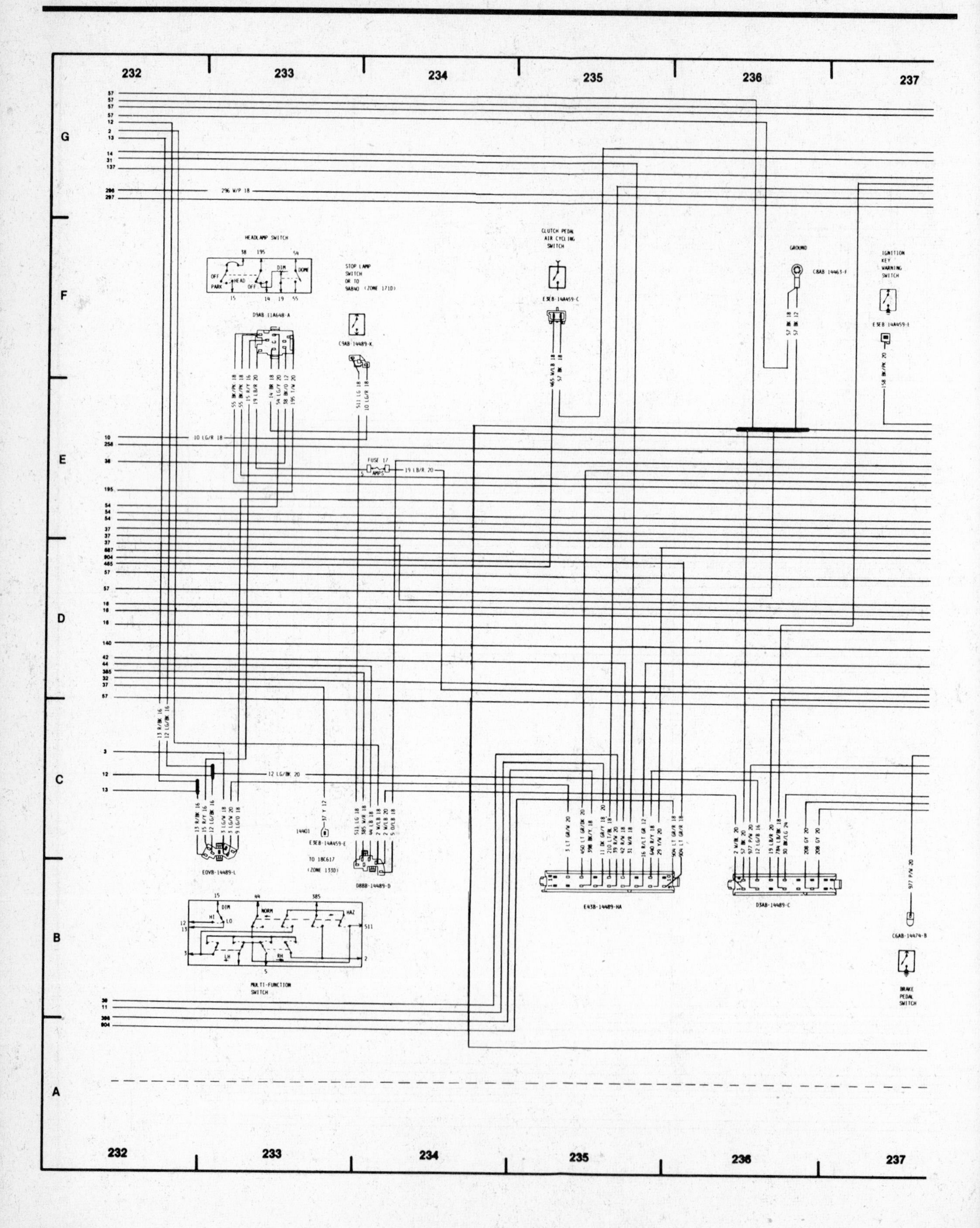
232
233
234
235
236
237
G
F
E
D
C
B
A
HEADLAMP SWITCH
STOP LAMP SWITCH
CLUTCH PEDAL AIR CYCLING SWITCH
GROUND
IGNITION KEY WARNING SWITCH
FUSE 17
MULTI-FUNCTION SWITCH
BRAKE PEDAL SWITCH

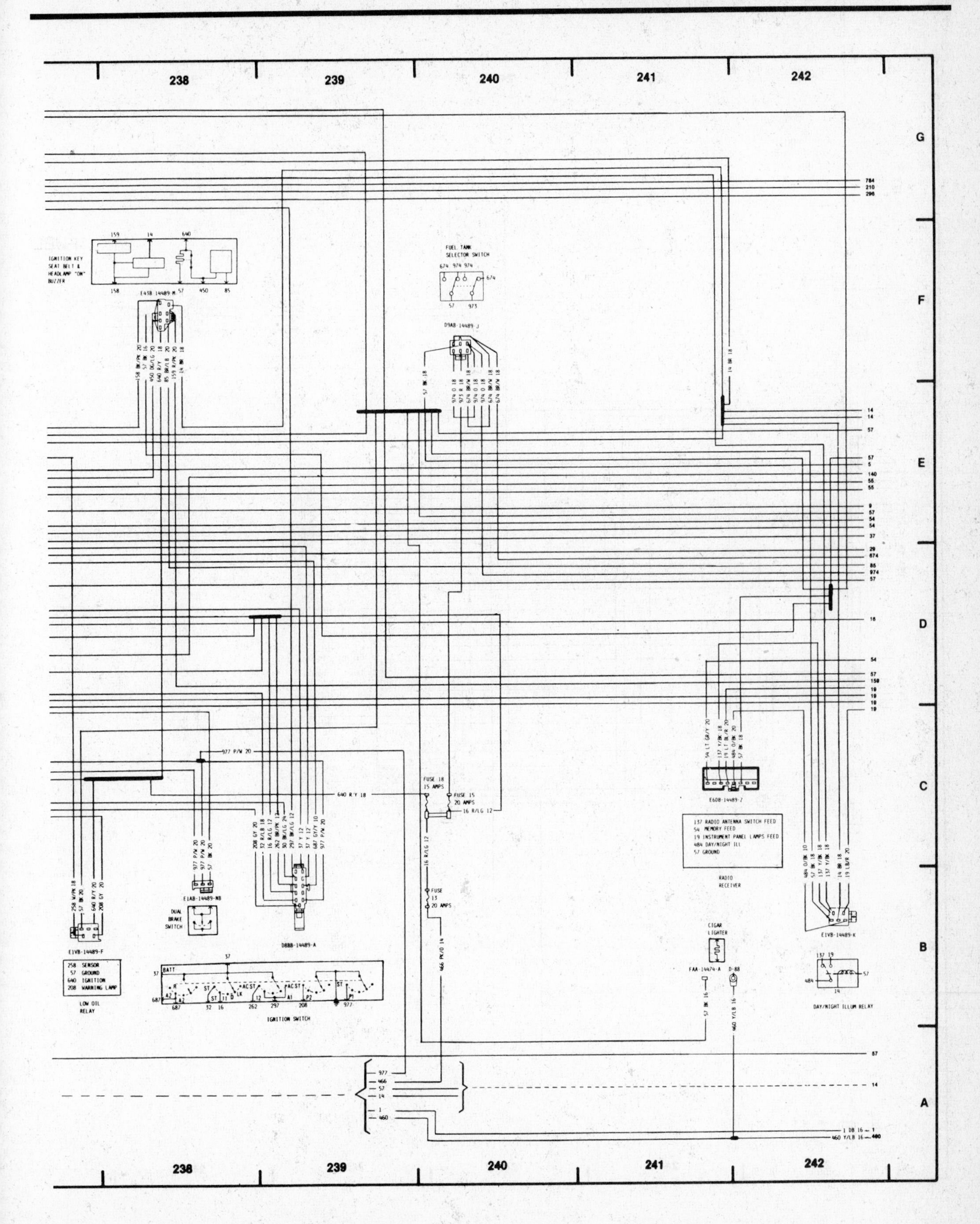
238
239
240
241
242
G
F
E
D
C
B
A
IGNITION KEY SEAT BELT & HEADLAMP "ON" BUZZER
E43B-14489-M
FUEL TANK SELECTOR SWITCH
D9AB-14489-J
FUSE 18 15 AMPS
FUSE 15 20 AMPS
FUSE 13 20 AMPS
E1VB-14489-F
258 SENSOR
57 GROUND
640 IGNITION
208 WARNING LAMP
LOW OIL RELAY
E1AB-14489-NB
DUAL BRAKE SWITCH
D8BB-14489-A
IGNITION SWITCH
E6DB-14489-Z
137 RADIO ANTENNA SWITCH FEED
54 MEMORY FEED
19 INSTRUMENT PANEL LAMPS FEED
484 DAY/NIGHT ILL
57 GROUND
RADIO RECEIVER
CIGAR LIGHTER
FAA-14474-A
D-88
E1VB-14489-K
DAY/NIGHT ILLUM RELAY
238
239
240
241
242

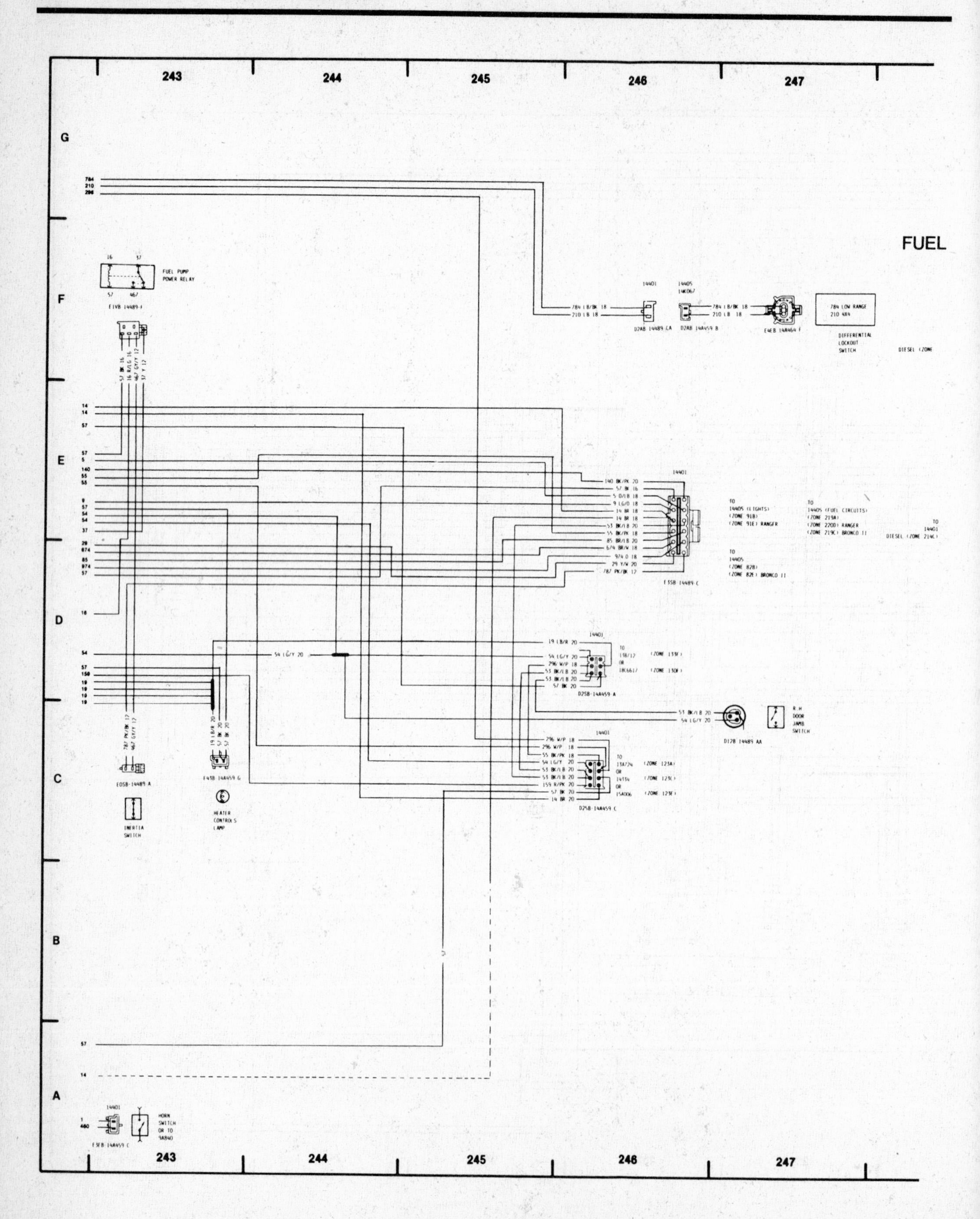
FUEL
FUEL PUMP POWER RELAY
DIFFERENTIAL LOCKOUT SWITCH
INERTIA SWITCH
HEATER CONTROLS LAMP
R.H. DOOR JAMB SWITCH
HORN SWITCH OR TO 9A840

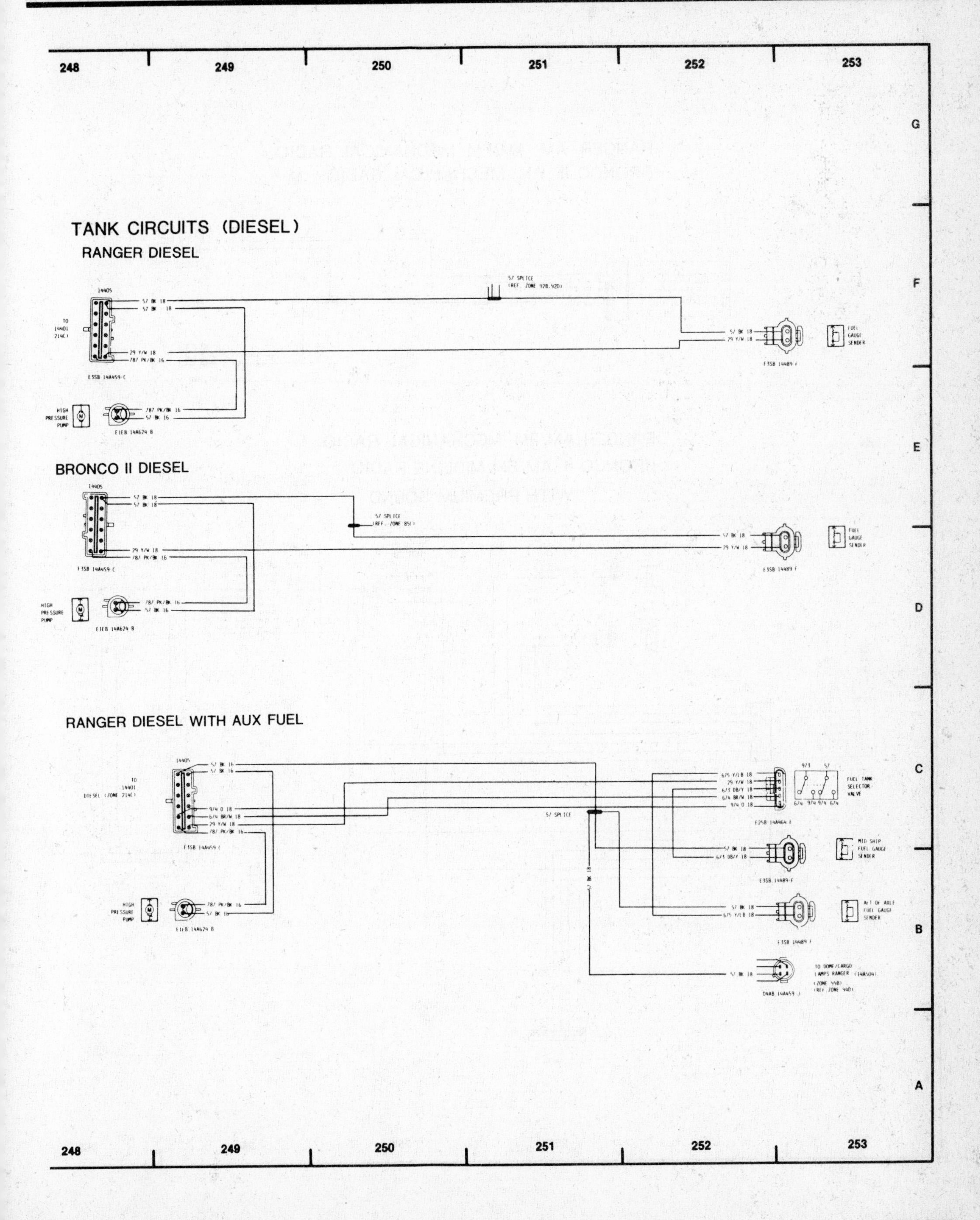
248
249
250
251
252
253
G
F
E
D
C
B
A
TANK CIRCUITS (DIESEL)
RANGER DIESEL
BRONCO II DIESEL
RANGER DIESEL WITH AUX FUEL

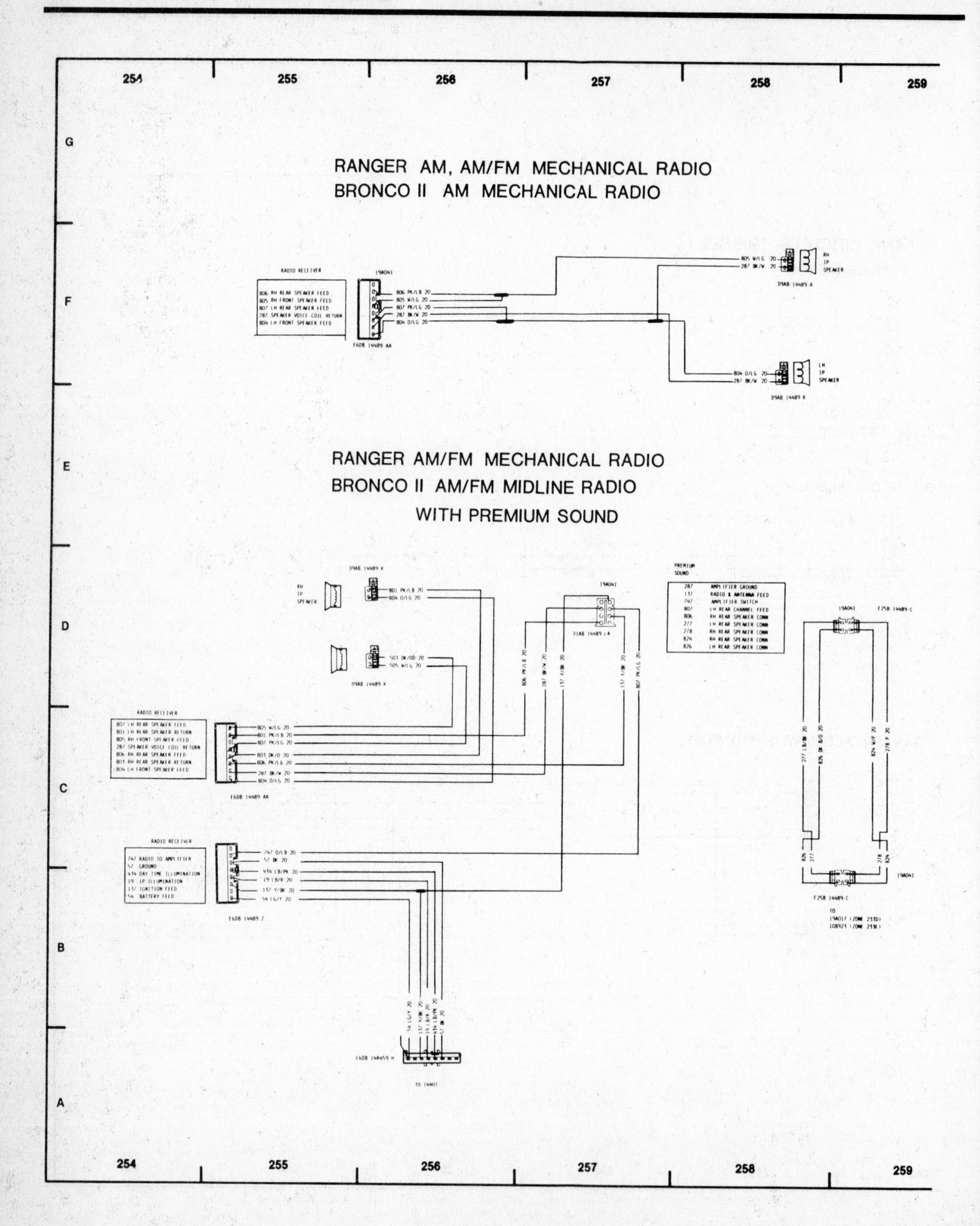
254
255
256
257
258
259
G
F
E
D
C
B
A
RANGER AM, AM/FM MECHANICAL RADIO
BRONCO II AM MECHANICAL RADIO
RADIO RECEIVER
806 RH REAR SPEAKER FEED
805 RH FRONT SPEAKER FEED
807 LH REAR SPEAKER FEED
287 SPEAKER VOICE COIL RETURN
804 LH FRONT SPEAKER FEED
19A041
E6DB 14489-AA
806 PK/LB 20
805 W/LG 20
807 PK/LG 20
287 BK/W 20
804 O/LG 20
RH
IP
SPEAKER
D9AB-14489-A
LH
IP
SPEAKER
D9AB 14489 K
RANGER AM/FM MECHANICAL RADIO
BRONCO II AM/FM MIDLINE RADIO
WITH PREMIUM SOUND
801 PK/LB 20
804 O/LG 20
503 DK/OD 20
505 W/LG 20
PREMIUM
SOUND
287 AMPLIFIER GROUND
137 RADIO & ANTENNA FEED
747 AMPLIFIER SWITCH
807 LH REAR CHANNEL FEED
806 RH REAR SPEAKER CONN
277 LH REAR SPEAKER CONN
278 RH REAR SPEAKER CONN
824 RH REAR SPEAKER CONN
826 LH REAR SPEAKER CONN
D1AB 14489 LA
806 PK/LB 20
287 BK/W 20
137 Y/BK 20
807 PK/LG 20
RADIO RECEIVER
807 LH REAR SPEAKER FEED
801 LH REAR SPEAKER RETURN
805 RH FRONT SPEAKER FEED
287 SPEAKER VOICE COIL RETURN
806 RH REAR SPEAKER FEED
803 RH REAR SPEAKER RETURN
804 LH FRONT SPEAKER FEED
805 W/LG 20
801 PK/LB 20
807 PK/LG 20
803 DK/O 20
806 PK/LG 20
287 BK/W 20
804 O/LG 20
E6DB 14489 AA
RADIO RECEIVER
747 RADIO TO AMPLIFIER
57 GROUND
434 DAY TIME ILLUMINATION
19 IP ILLUMINATION
137 IGNITION FEED
54 BATTERY FEED
747 O/LB 20
57 BK 20
434 LB/PK 20
19 LB/R 20
137 Y/BK 20
54 LG/Y 20
E6DB 14489 Z
F25B 14489-C
277 LB/BK 20
826 DK B/O 20
824 W/P 20
278 P 20
TO
19A017 (ZONE 233D)
10B923 (ZONE 233E)
E6DB 14A459 H
TO 14401

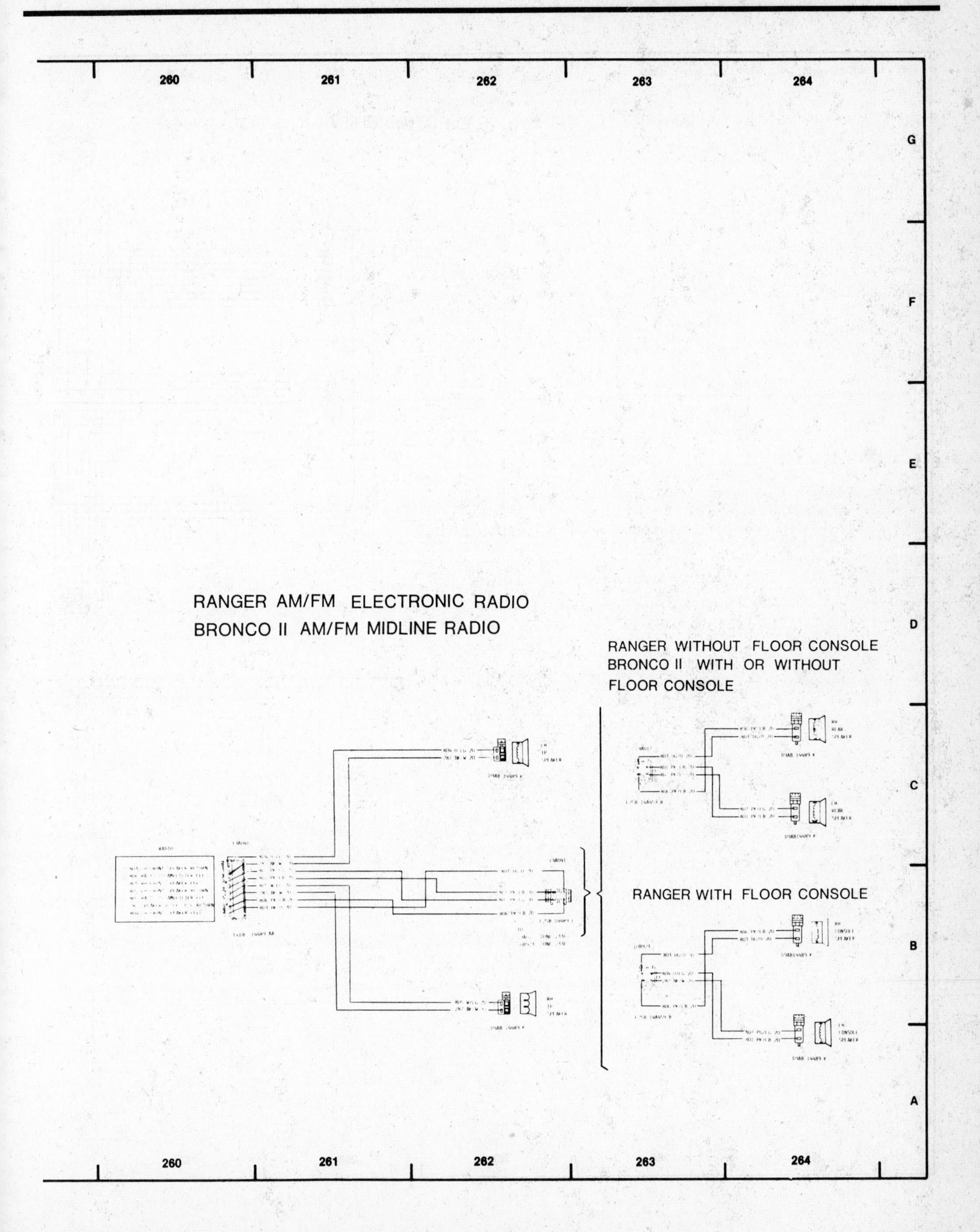
260
261
262
263
264
G
F
E
D
C
B
A
RANGER AM/FM ELECTRONIC RADIO
BRONCO II AM/FM MIDLINE RADIO
RANGER WITHOUT FLOOR CONSOLE
BRONCO II WITH OR WITHOUT
FLOOR CONSOLE
RANGER WITH FLOOR CONSOLE

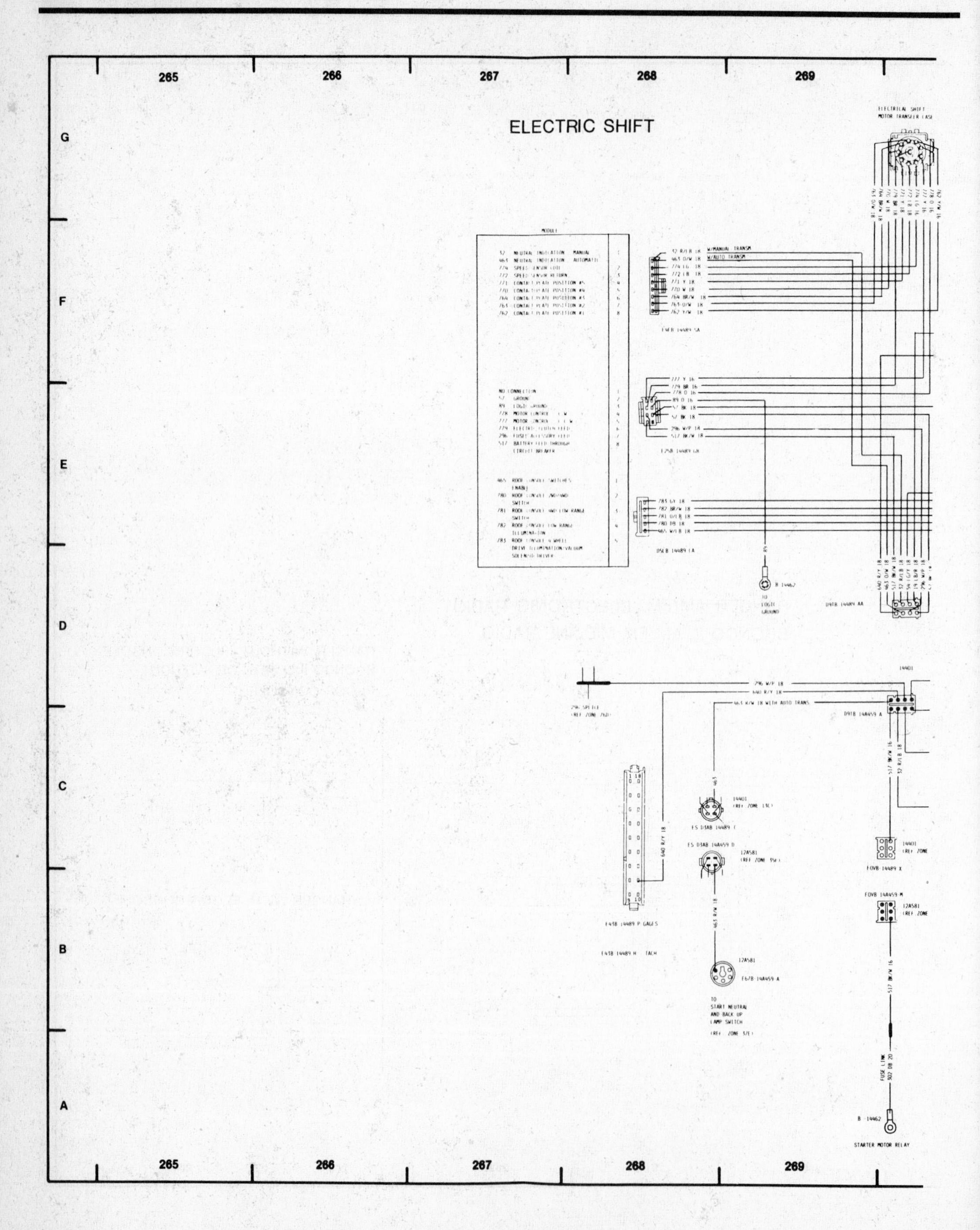
ELECTRIC SHIFT
265
266
267
268
269
G
F
E
D
C
B
A
MODULE
STARTER MOTOR RELAY

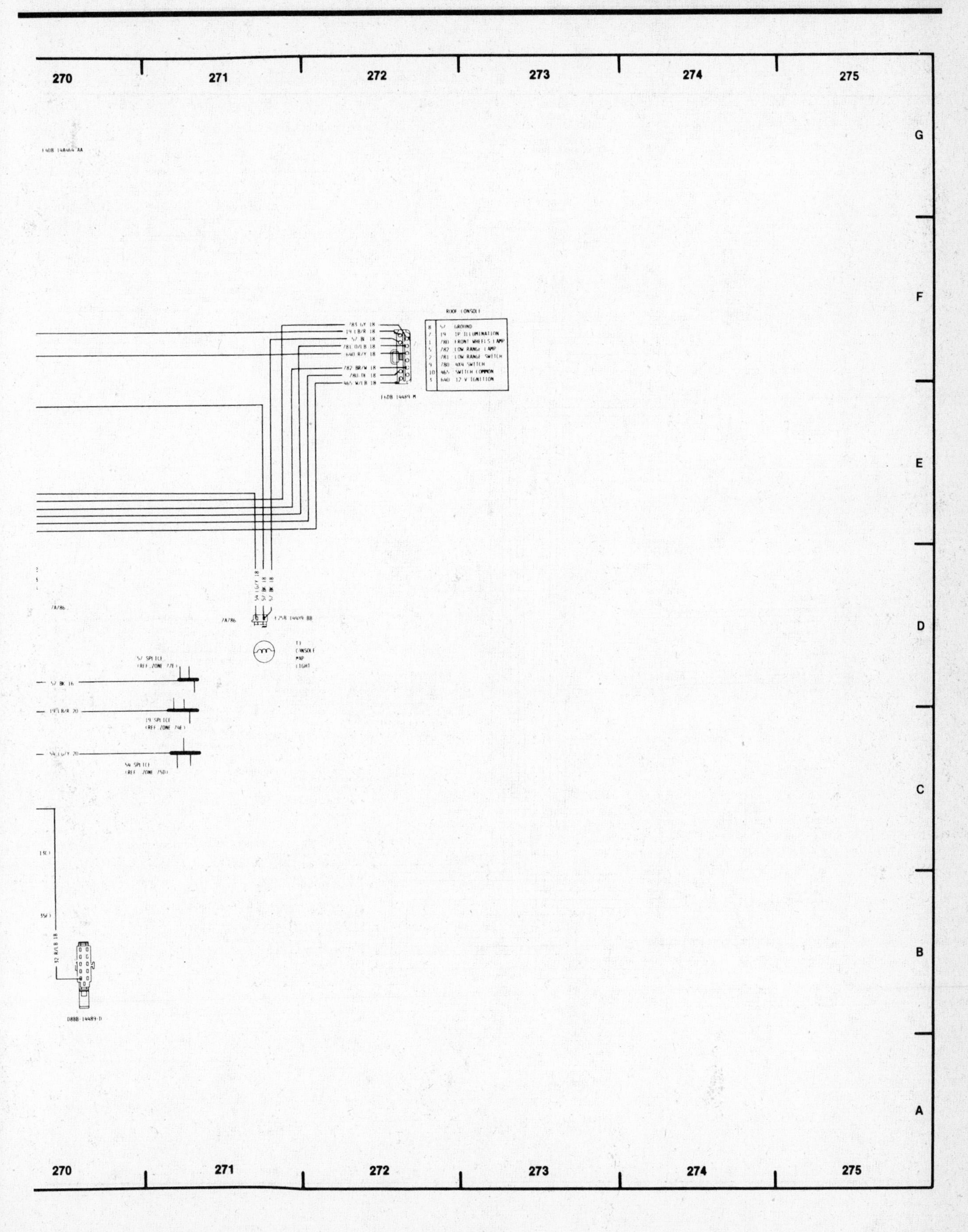
270
271
272
273
274
275
G
F
E
D
C
B
A
ROOF CONSOLE
8 57 GROUND
7 19 IP ILLUMINATION
1 780 FRONT WHEELS LAMP
5 782 LOW RANGE LAMP
2 781 LOW RANGE SWITCH
9 780 4X4 SWITCH
10 465 SWITCH COMMON
3 640 12 V IGNITION
783 GY 18
19 LB/R 18
57 BK 18
781 O/LB 18
640 R/Y 18
782 BR/W 18
780 DB 18
465 W/LB 18
54 LG/Y 18
57 BK 18
57 BK 18
7A786
CONSOLE
MAP
LIGHT
57 SPLICE
(REF. ZONE 77E)
57 BK 16
19 LB/R 20
19 SPLICE
(REF. ZONE 74E)
54 LG/Y 20
54 SPLICE
(REF. ZONE 75D)
52 W/LB 18

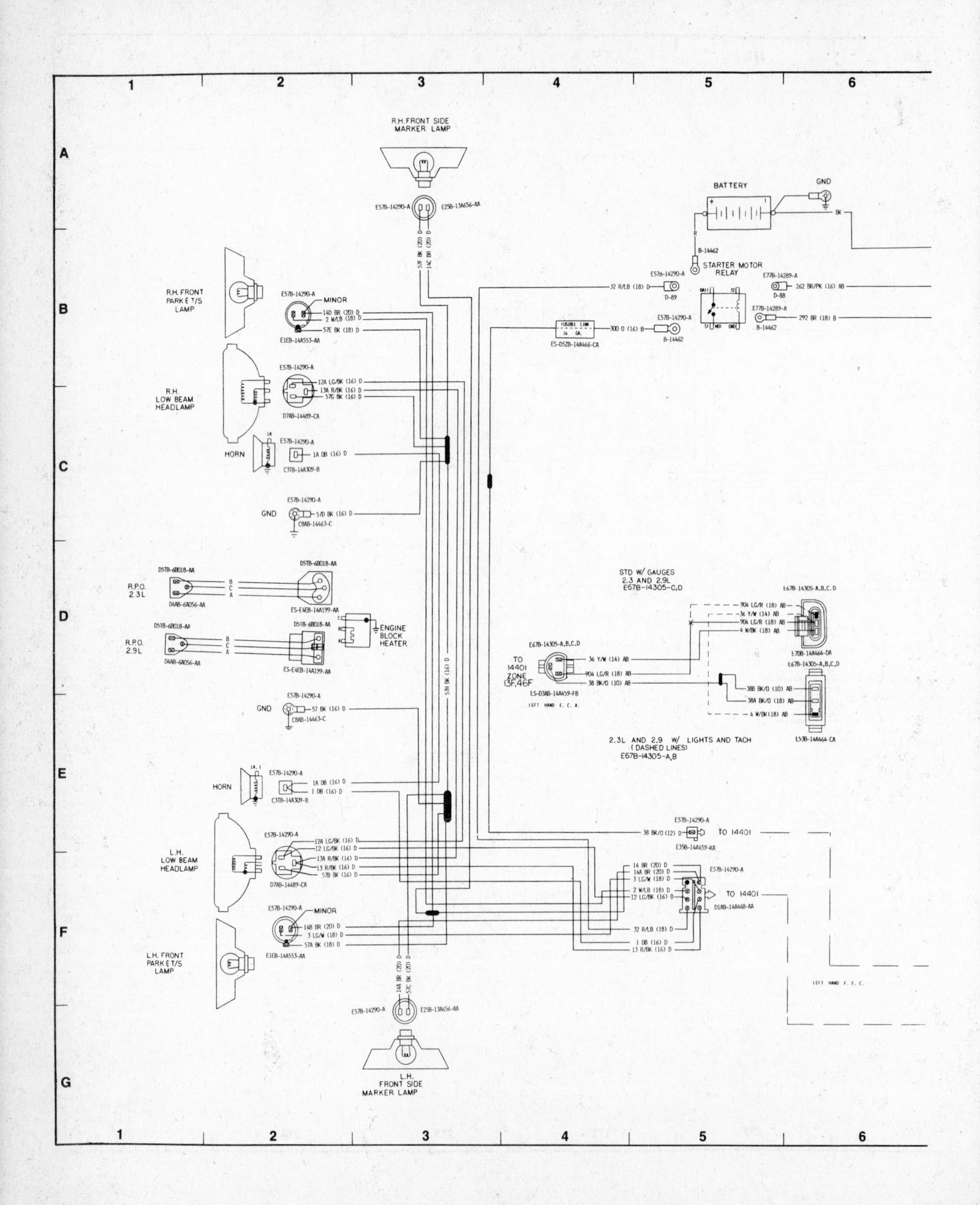

R.H. FRONT SIDE MARKER LAMP
BATTERY
GND
STARTER MOTOR RELAY
R.H. FRONT PARK & T/S LAMP
MINOR
R.H. LOW BEAM HEADLAMP
HORN
GND
R.P.O. 2.3L
R.P.O. 2.9L
ENGINE BLOCK HEATER
GND
STD W/ GAUGES 2.3 AND 2.9L E67B-14305-C,D
TO 14401 ZONE 13F,46F
2.3L AND 2.9 W/ LIGHTS AND TACH (DASHED LINES) E67B-14305-A,B
HORN
L.H. LOW BEAM HEADLAMP
TO 14401
MINOR
L.H. FRONT PARK & T/S LAMP
L.H. FRONT SIDE MARKER LAMP

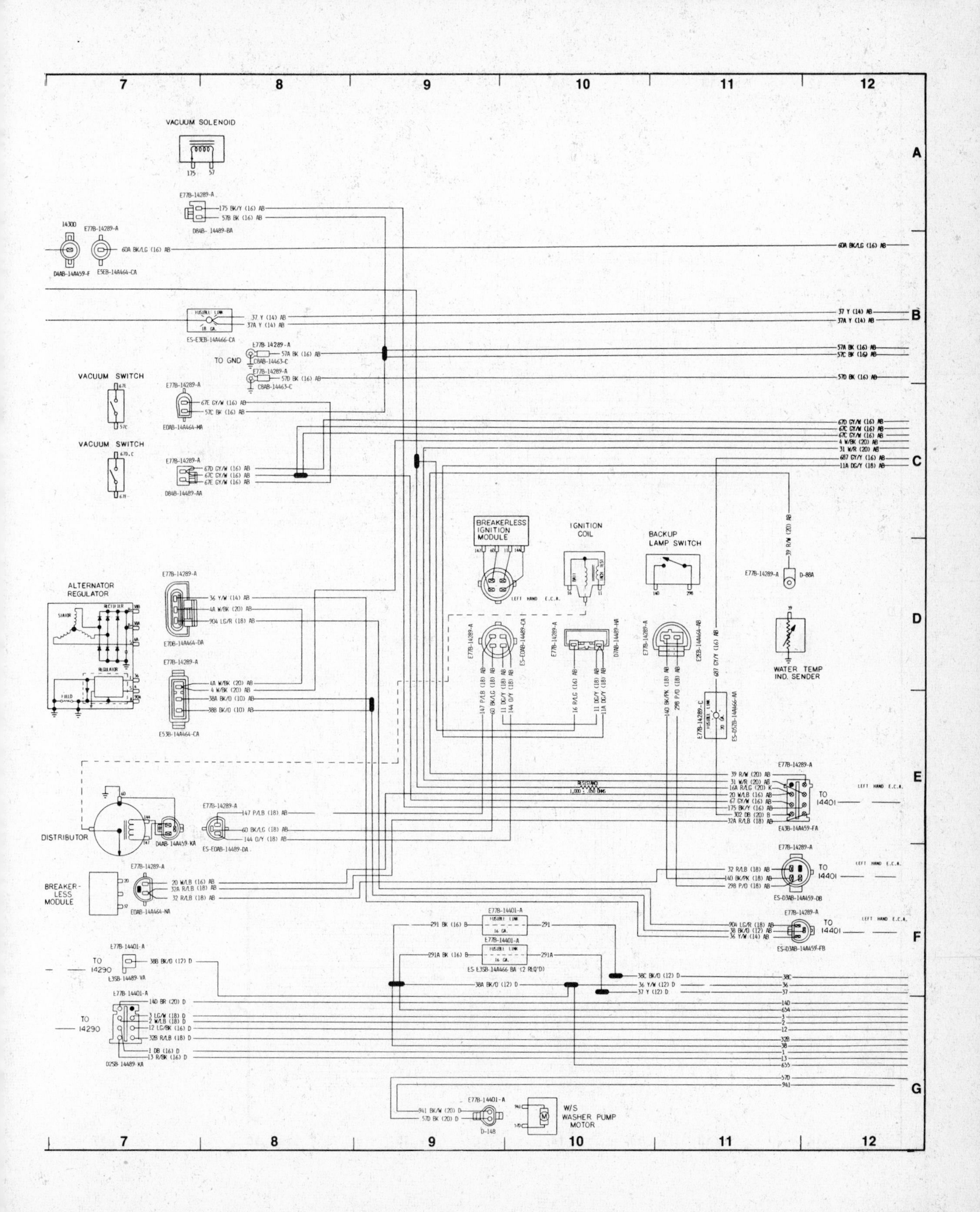

VACUUM SOLENOID
VACUUM SWITCH
VACUUM SWITCH
TO GND
ALTERNATOR REGULATOR
BREAKERLESS IGNITION MODULE
IGNITION COIL
BACKUP LAMP SWITCH
WATER TEMP IND. SENDER
DISTRIBUTOR
BREAKER-LESS MODULE
W/S WASHER PUMP MOTOR
TO 14401
TO 14290
LEFT HAND E.C.A.

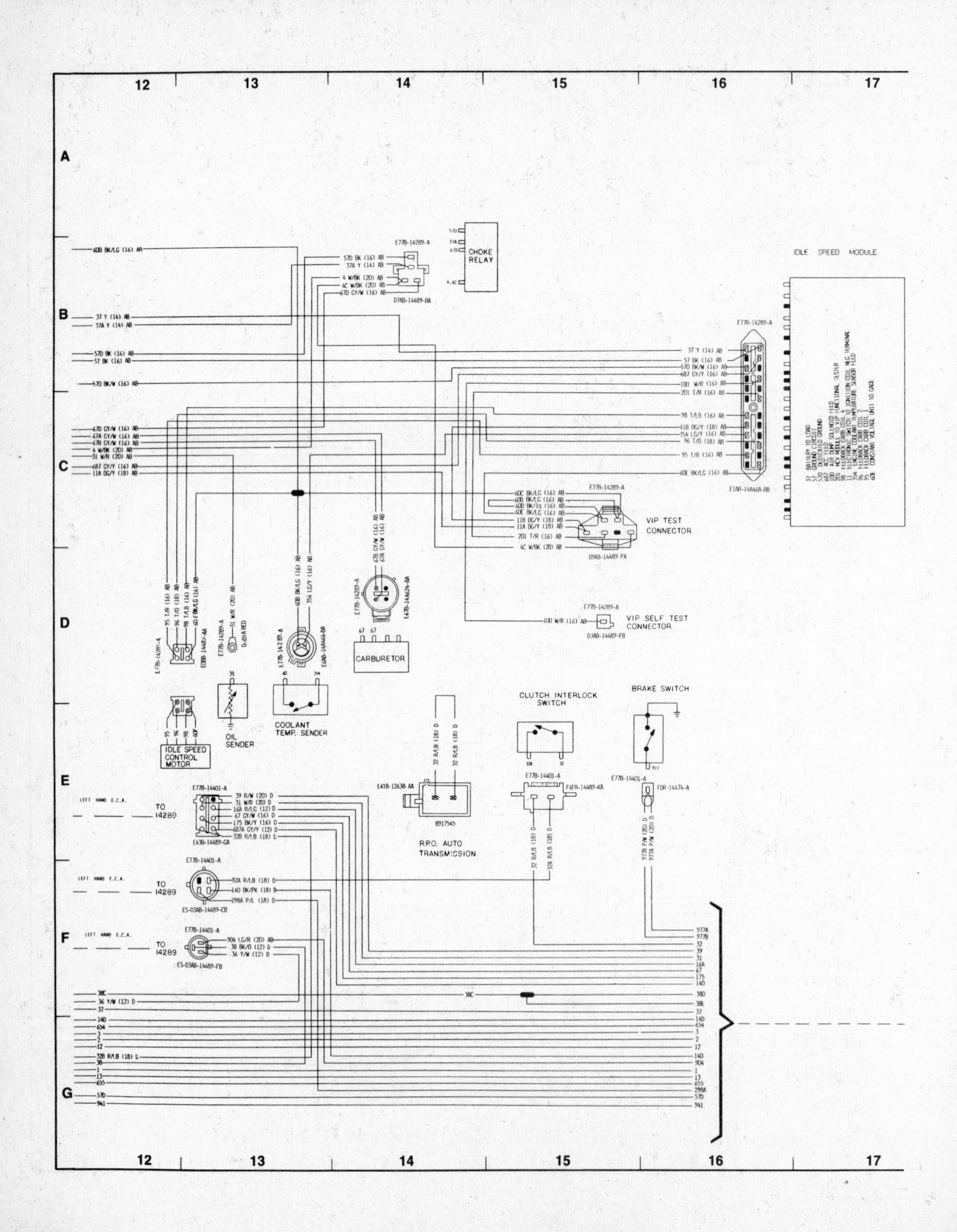
CHOKE RELAY
IDLE SPEED MODULE
VIP TEST CONNECTOR
VIP SELF TEST CONNECTOR
CARBURETOR
IDLE SPEED CONTROL MOTOR
OIL SENDER
COOLANT TEMP. SENDER
CLUTCH INTERLOCK SWITCH
BRAKE SWITCH
R.P.O. AUTO TRANSMISSION
LEFT HAND E.C.A.
TO 14289

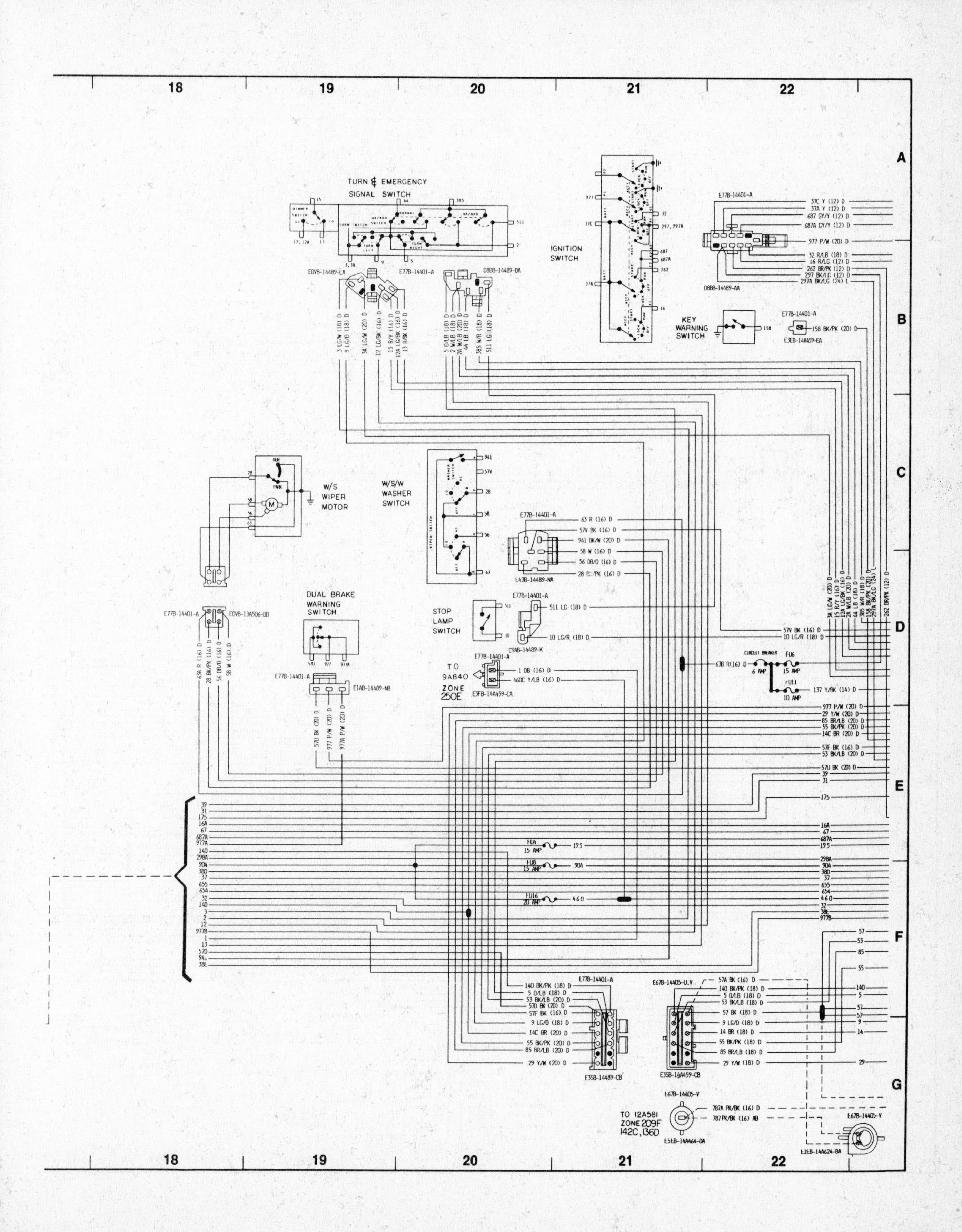
TURN & EMERGENCY SIGNAL SWITCH
IGNITION SWITCH
KEY WARNING SWITCH
W/S WIPER MOTOR
W/S/W WASHER SWITCH
DUAL BRAKE WARNING SWITCH
STOP LAMP SWITCH
TO 9A840 ZONE 250E
TO 12A581 ZONE 209F 142C, 136D
18
19
20
21
22
A
B
C
D
E
F
G

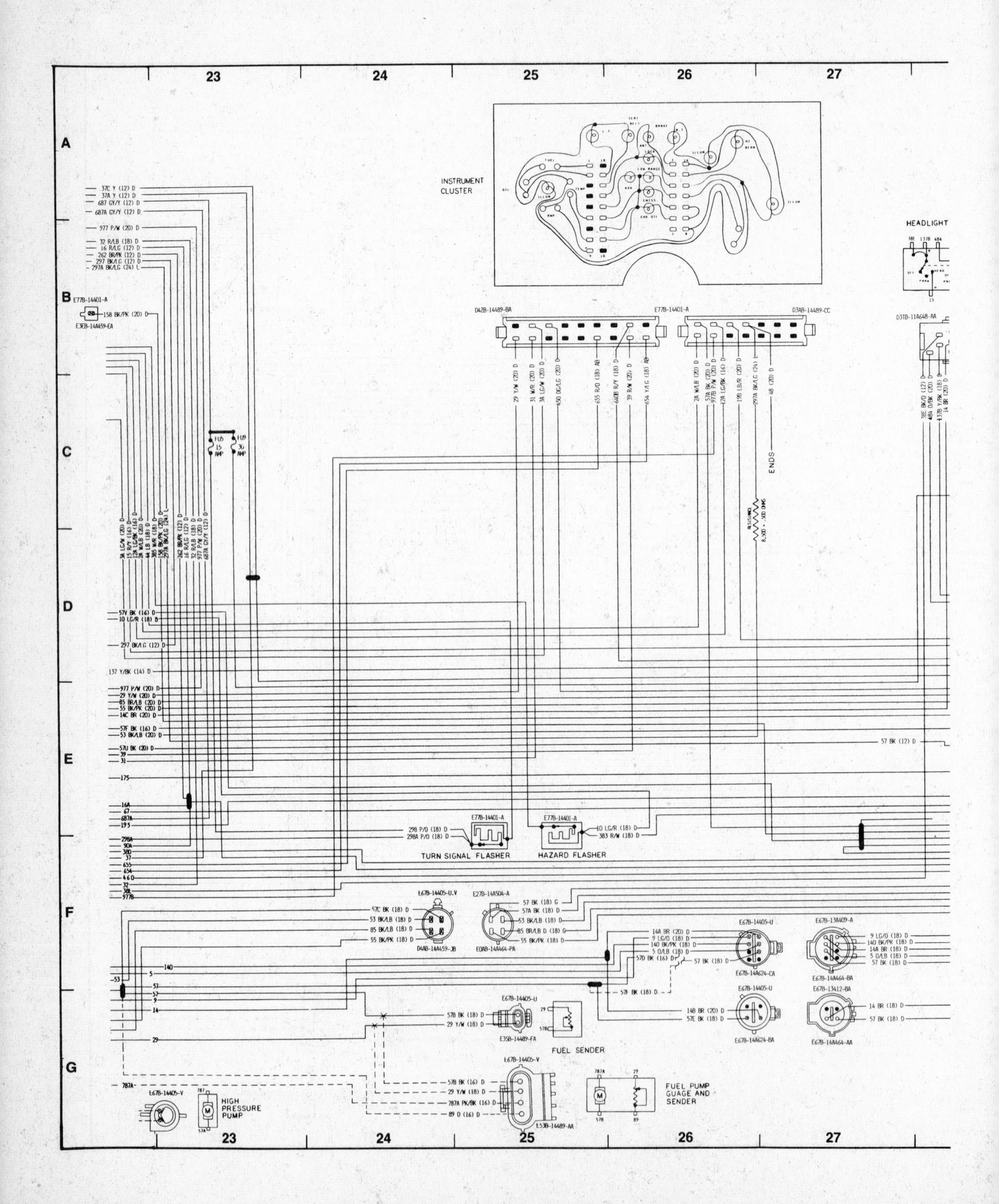
INSTRUMENT CLUSTER
HEADLIGHT
TURN SIGNAL FLASHER
HAZARD FLASHER
FUEL SENDER
HIGH PRESSURE PUMP
FUEL PUMP GUAGE AND SENDER
ENDS

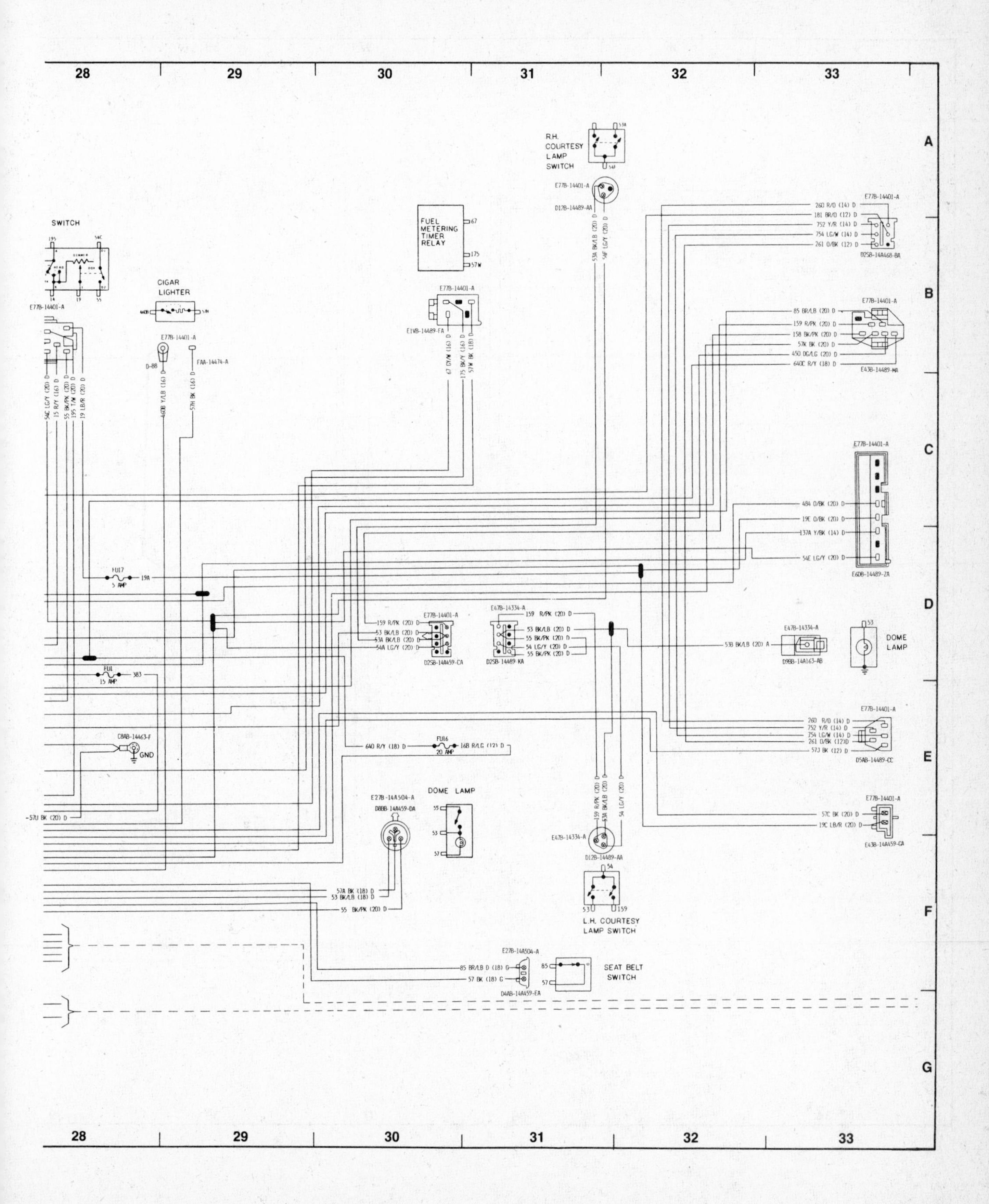
R.H. COURTESY LAMP SWITCH
FUEL METERING TIMER RELAY
SWITCH
CIGAR LIGHTER
DOME LAMP
DOME LAMP
L.H. COURTESY LAMP SWITCH
SEAT BELT SWITCH
GND
FU17 5 AMP
FU1 15 AMP
FU16 20 AMP

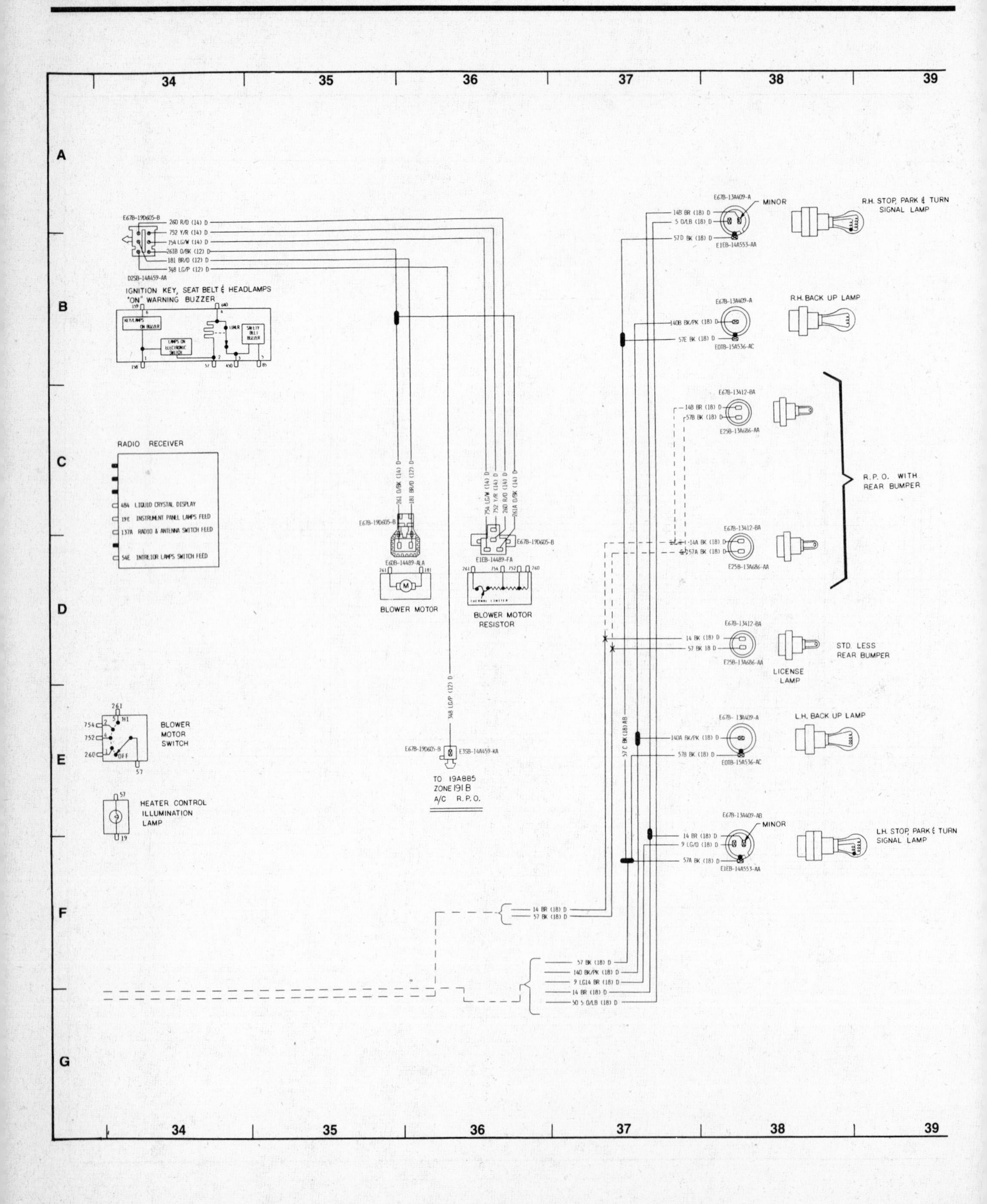
IGNITION KEY, SEAT BELT & HEADLAMPS "ON" WARNING BUZZER
RADIO RECEIVER
484 LIQUID CRYSTAL DISPLAY
19E INSTRUMENT PANEL LAMPS FEED
137A RADIO & ANTENNA SWITCH FEED
54E INTERIOR LAMPS SWITCH FEED
BLOWER MOTOR
BLOWER MOTOR RESISTOR
BLOWER MOTOR SWITCH
HEATER CONTROL ILLUMINATION LAMP
TO 19A885 ZONE 191 B A/C R.P.O.
MINOR
R.H. STOP, PARK & TURN SIGNAL LAMP
R.H. BACK UP LAMP
R.P.O. WITH REAR BUMPER
STD. LESS REAR BUMPER
LICENSE LAMP
L.H. BACK UP LAMP
L.H. STOP, PARK & TURN SIGNAL LAMP

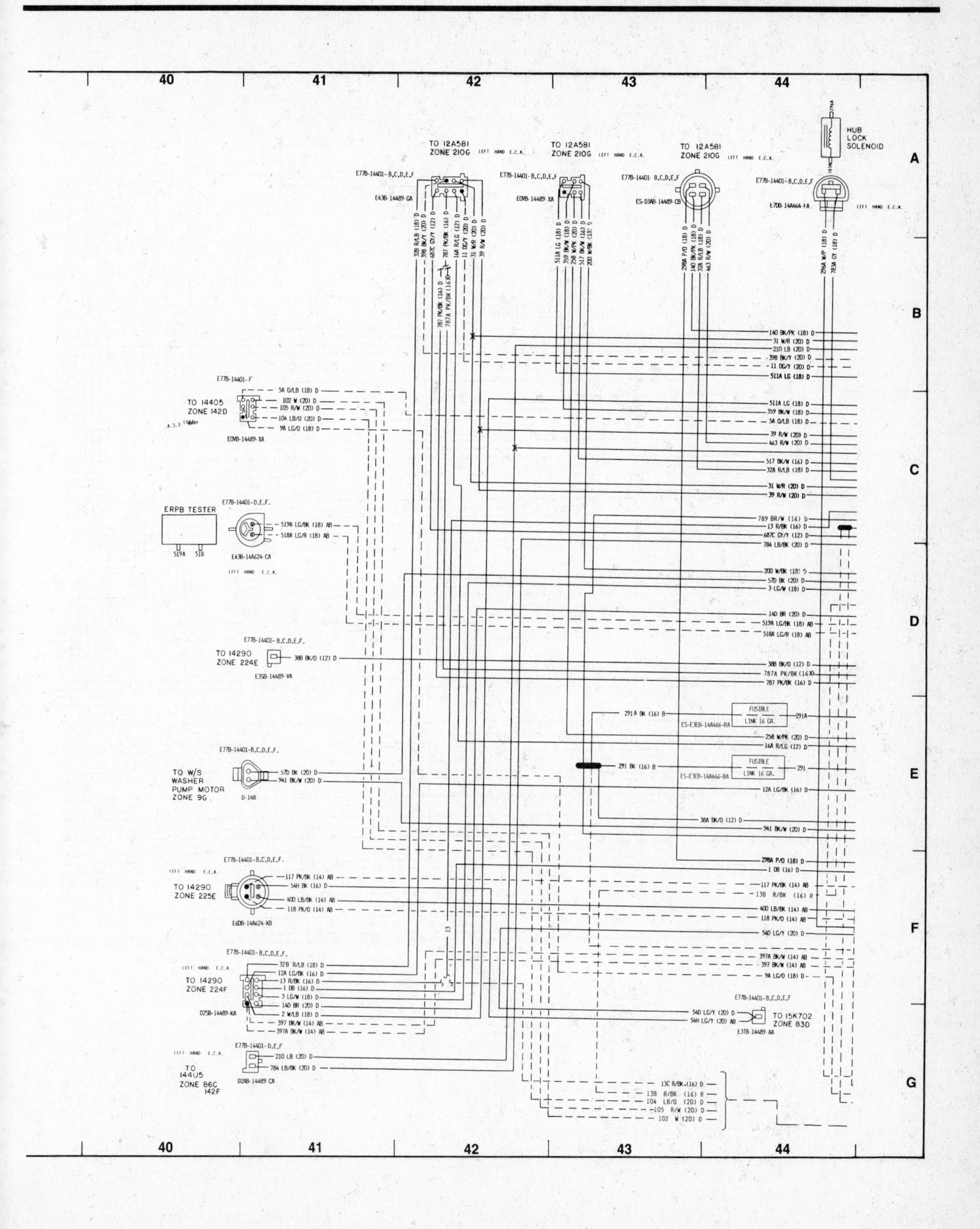
40
41
42
43
44
A
B
C
D
E
F
G
HUB LOCK SOLENOID
TO 12A581 ZONE 210G
LEFT HAND E.C.A.
TO 14405 ZONE 142D
ERPB TESTER
519A
518
TO 14290 ZONE 224E
TO W/S WASHER PUMP MOTOR ZONE 9G
TO 14290 ZONE 225E
TO 14290 ZONE 224F
TO 144U5 ZONE 86C 142F
TO 15K702 ZONE 83D
FUSIBLE LINK 16 GA.
140 BK/PK (18) D
31 W/R (20) D
210 LB (20) D
398 BK/Y (20) D
11 DG/Y (20) D
511A LG (18) D
359 BK/W (18) D
5A O/LB (18) D
39 R/W (20) D
463 R/W (20) D
517 BK/W (16) D
32A R/LB (18) D
789 BR/W (16) D
13 R/BK (16) D
687C GY/Y (12) D
784 LB/BK (20) D
200 W/BK (18) D
57D BK (20) D
3 LG/W (18) D
140 BR (20) D
519A LG/BK (18) AB
518A LG/R (18) AB
38B BK/O (12) D
787A PK/BK (16) D
787 PK/BK (16) D
291A BK (16) B
291 BK (16) B
258 W/PK (20) D
16A R/LG (12) D
12A LG/BK (16) D
38A BK/O (12) D
941 BK/W (20) D
298A P/O (18) D
1 DB (16) D
117 PK/BK (14) AB
13B R/BK (16) R
400 LB/BK (14) AB
118 PK/O (14) AB
54D LG/Y (20) D
54H LG/Y (20) AB
397A BK/W (14) AB
397 BK/W (14) AB
9A LG/O (18) D
13C R/BK (16) D
104 LB/O (20) D
105 R/W (20) D
102 W (20) D
2 W/LB (18) D
32B R/LB (18) D
296A W/P (18) D
783A GY (18) D

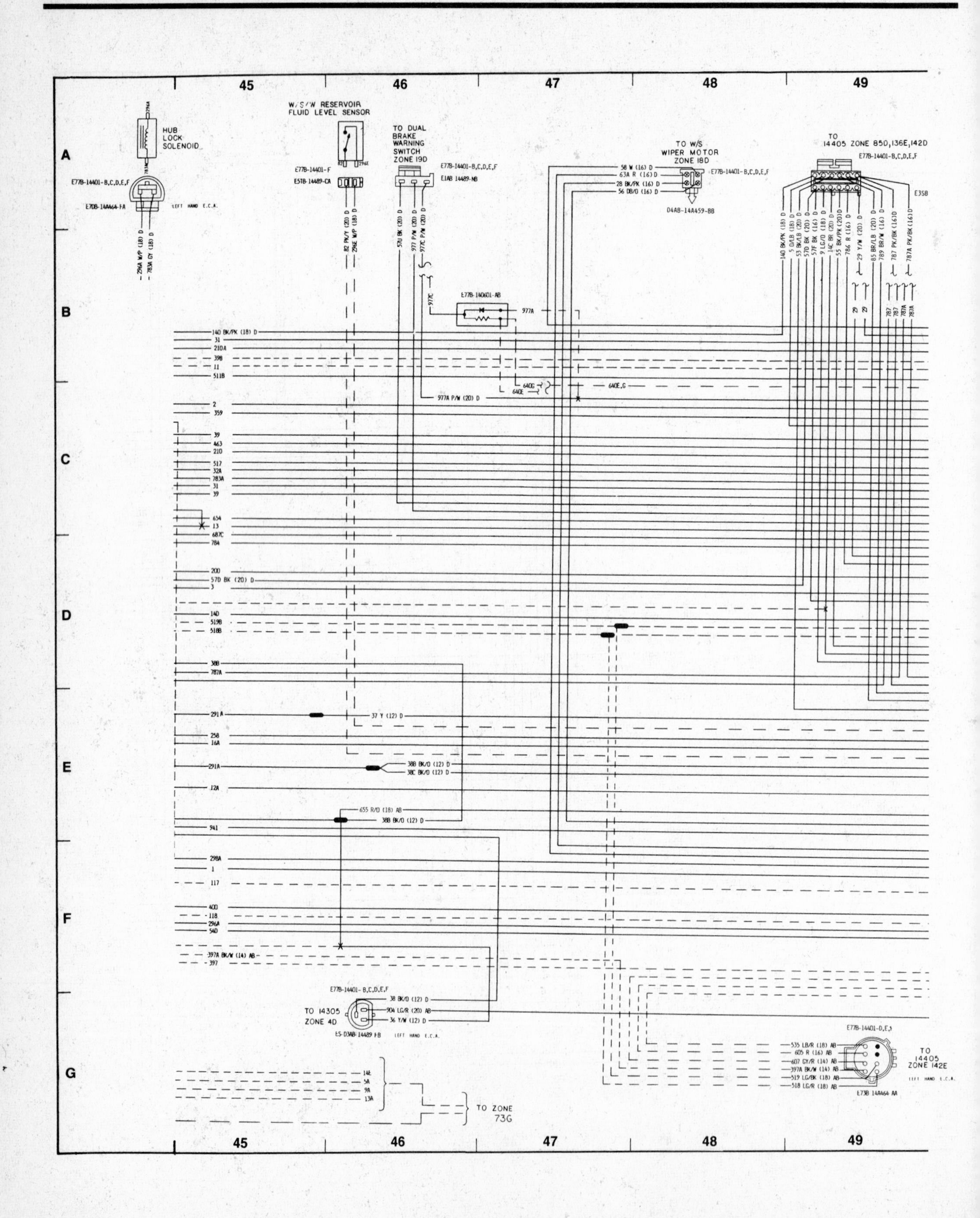
45
46
47
48
49
A
B
C
D
E
F
G
HUB LOCK SOLENOID
W/S/W RESERVOIR FLUID LEVEL SENSOR
TO DUAL BRAKE WARNING SWITCH ZONE 19D
TO W/S WIPER MOTOR ZONE 18D
TO 14405 ZONE 850,136E,142D
TO 14305 ZONE 4D
TO 14405 ZONE 142E
TO ZONE 73G

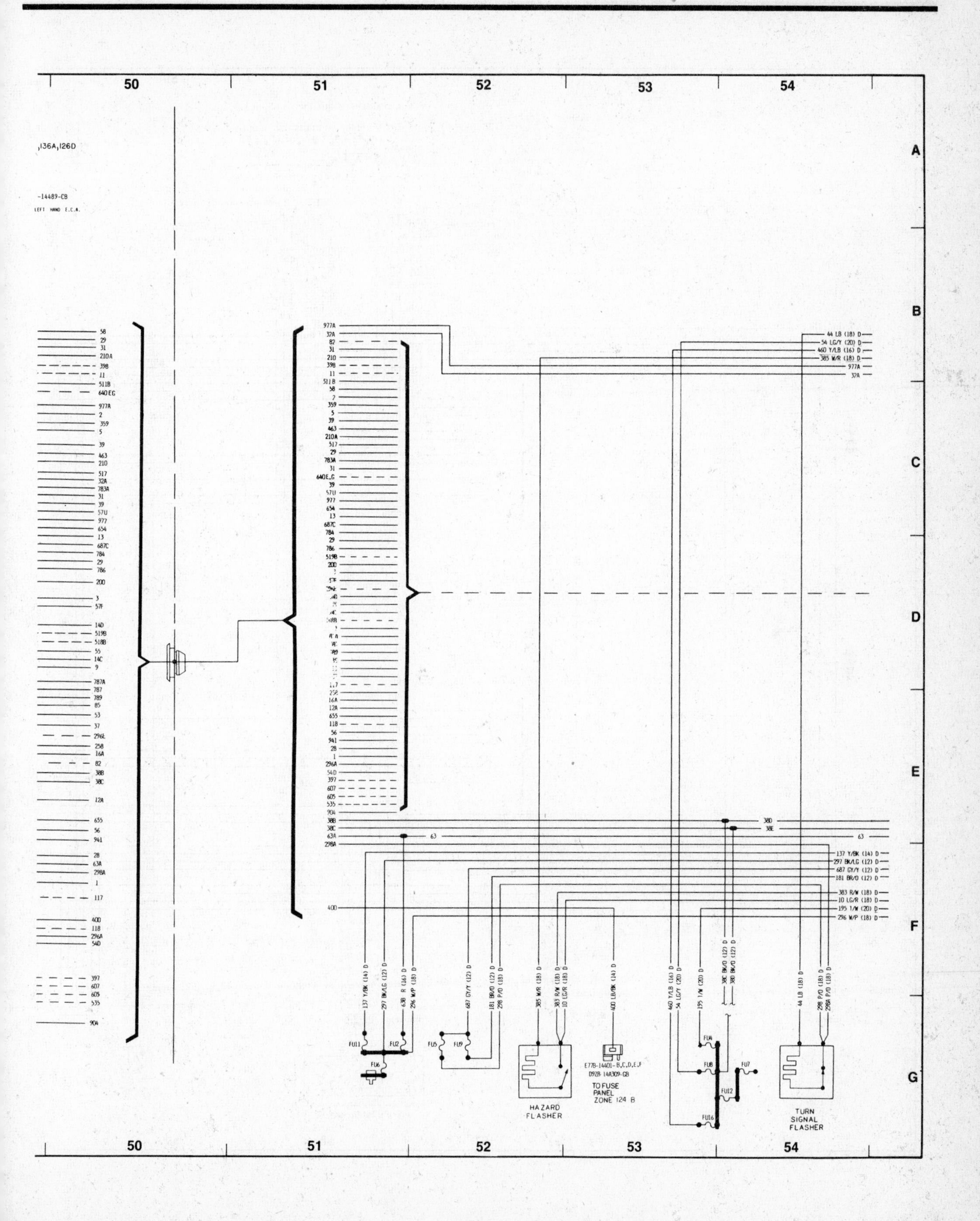

HAZARD FLASHER
TURN SIGNAL FLASHER
TO FUSE PANEL ZONE 124 B

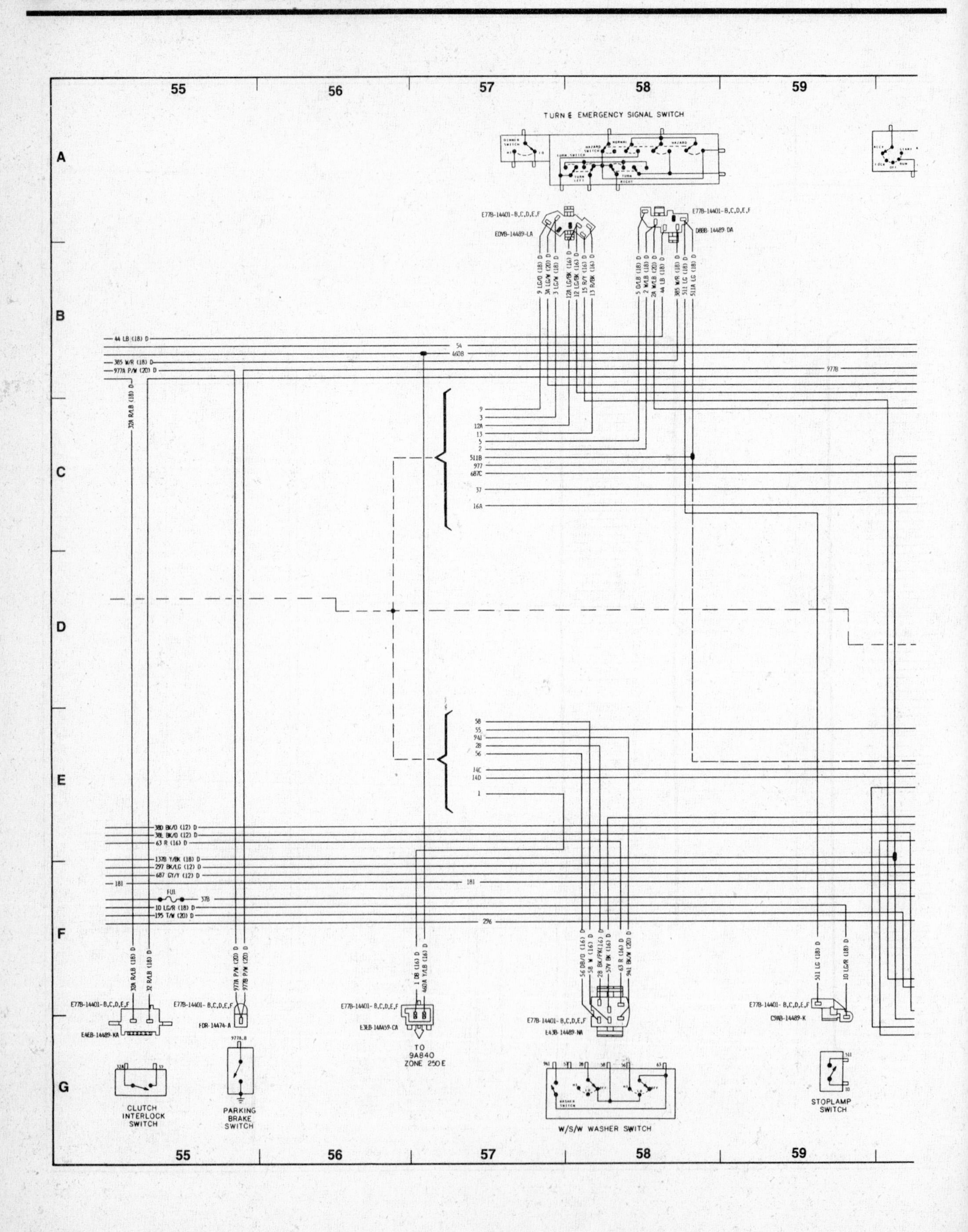

55
56
57
58
59
A
B
C
D
E
F
G
TURN & EMERGENCY SIGNAL SWITCH
E77B-14401-B,C,D,E,F
D8BB-14489-DA
44 LB (18) D
385 W/R (18) D
977A P/W (20) D
32A R/LB (18) D
977B
380 BK/O (12) D
38L BK/O (12) D
63 R (16) D
137B Y/BK (18) D
297 BK/LG (12) D
687 GY/Y (12) D
181
FUI
37B
10 LG/R (18) D
195 T/W (20) D
296
CLUTCH INTERLOCK SWITCH
PARKING BRAKE SWITCH
TO 9A840 ZONE 250E
W/S/W WASHER SWITCH
STOPLAMP SWITCH
511 LG (18) D
10 LG/R (18) D
C9AB-14489-K

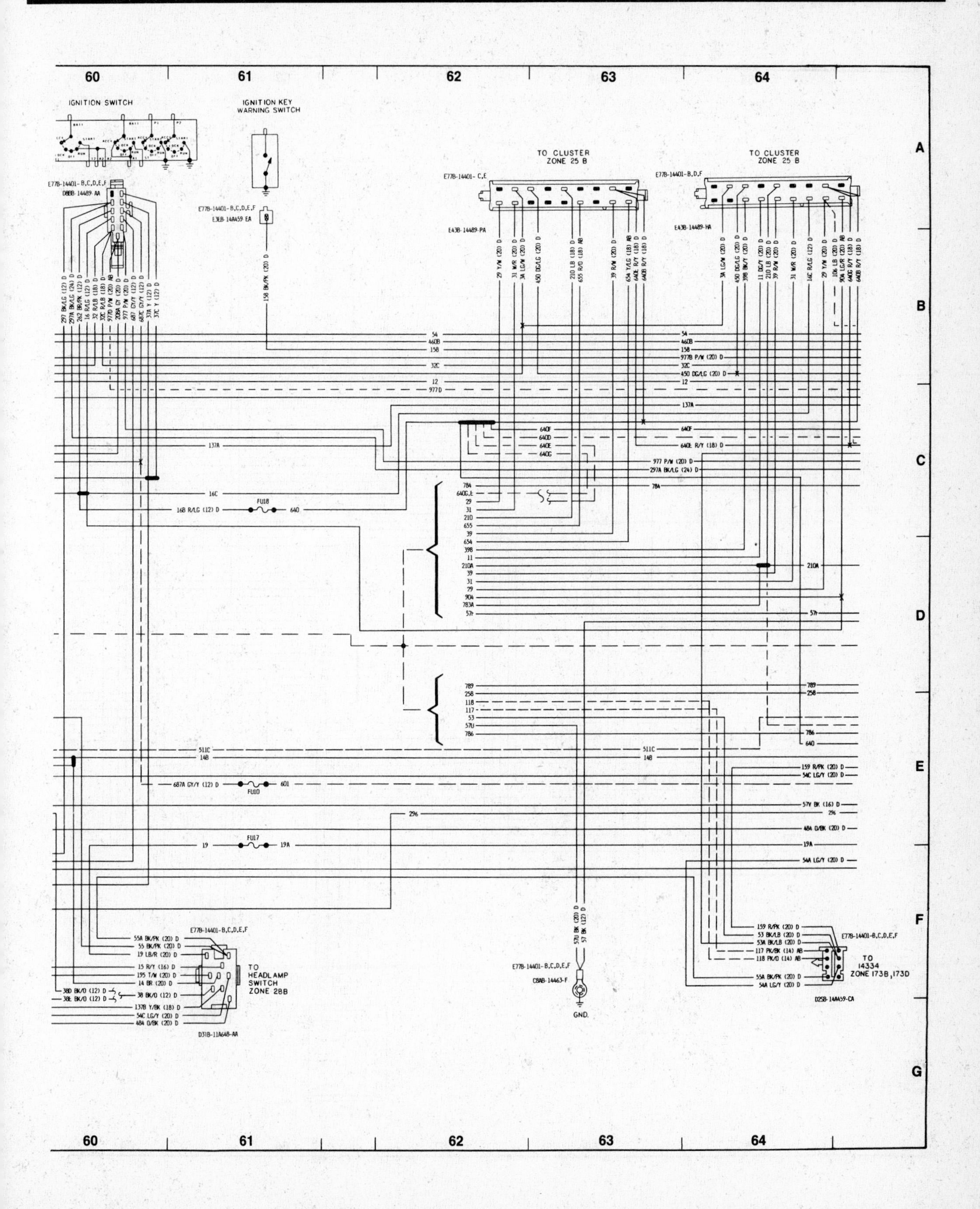
60
61
62
63
64
IGNITION SWITCH
IGNITION KEY WARNING SWITCH
TO CLUSTER ZONE 25 B
TO CLUSTER ZONE 25 B
E77B-14401-B,C,D,E,F
D8BB-14489-AA
E3EB-14A459 EA
E77B-14401-C,E
E43B-14489-PA
E77B-14401-B,D,F
E43B-14489-HA
FU18
FU10
FU17
TO HEADLAMP SWITCH ZONE 28B
D31B-11A648-AA
E77B-14401-B,C,D,E,F
C8AB-14463-F
GND.
TO 14334 ZONE 173B,173D
D25B-14A459-CA
A
B
C
D
E
F
G

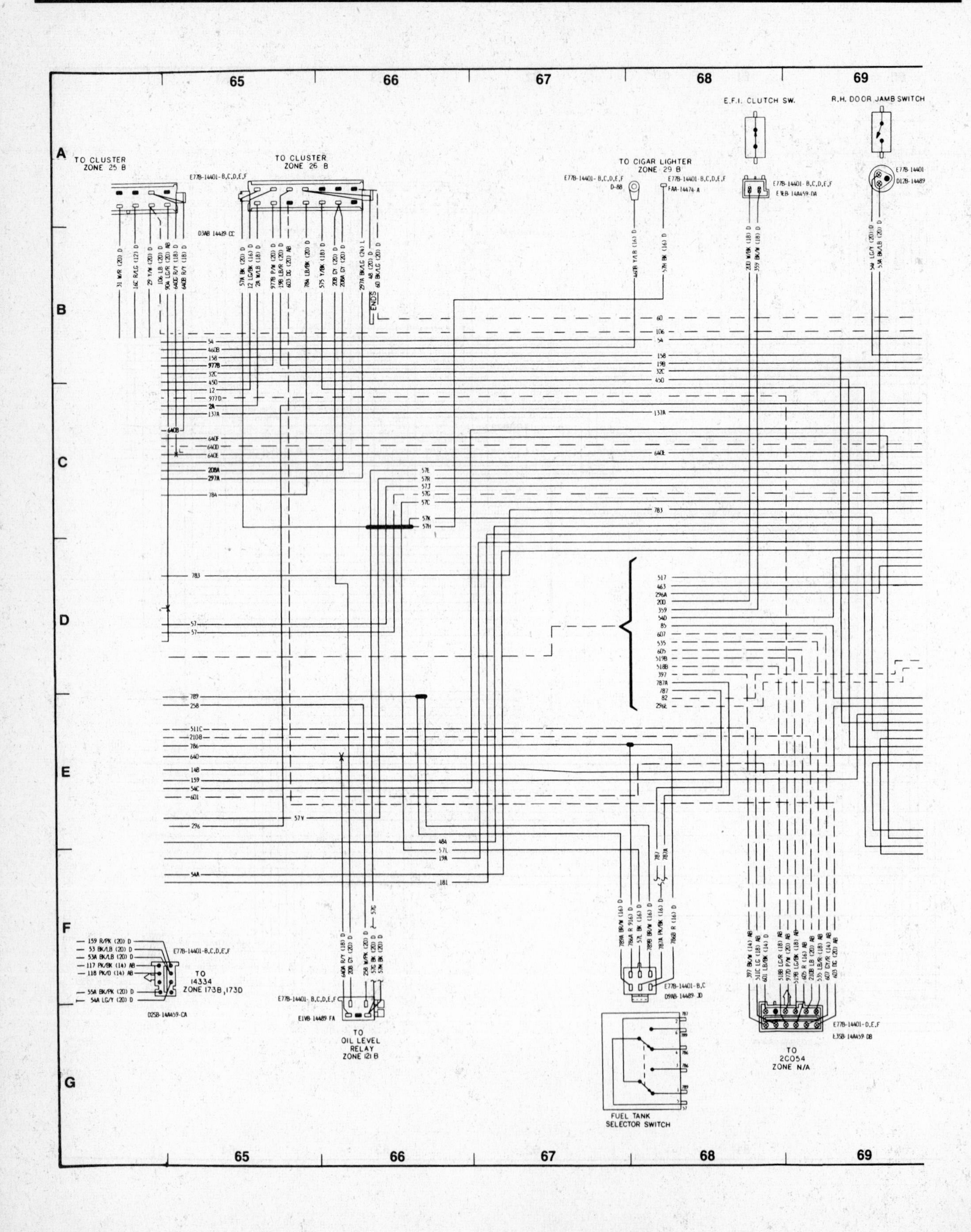

65
66
67
68
69
E.F.I. CLUTCH SW.
R.H. DOOR JAMB SWITCH
TO CLUSTER ZONE 25 B
TO CLUSTER ZONE 26 B
TO CIGAR LIGHTER ZONE 29 B
TO 14334 ZONE 173B, 173D
TO OIL LEVEL RELAY ZONE 121 B
FUEL TANK SELECTOR SWITCH
TO 2C054 ZONE N/A

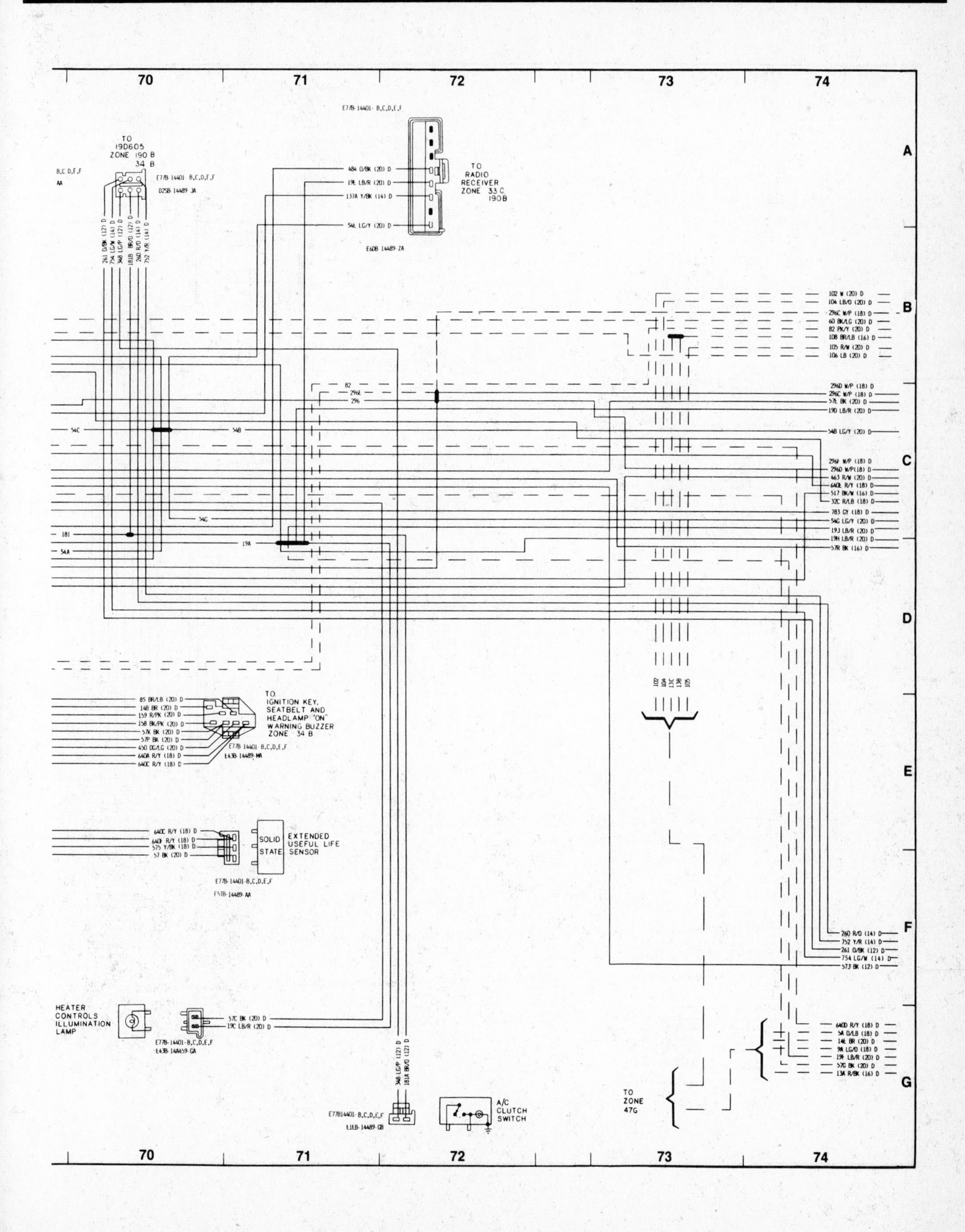
TO RADIO RECEIVER ZONE 33 C 190B
TO 19D605 ZONE 190 B 34 B
TO IGNITION KEY, SEATBELT AND HEADLAMP "ON" WARNING BUZZER ZONE 34 B
SOLID STATE
EXTENDED USEFUL LIFE SENSOR
HEATER CONTROLS ILLUMINATION LAMP
A/C CLUTCH SWITCH
TO ZONE 47G

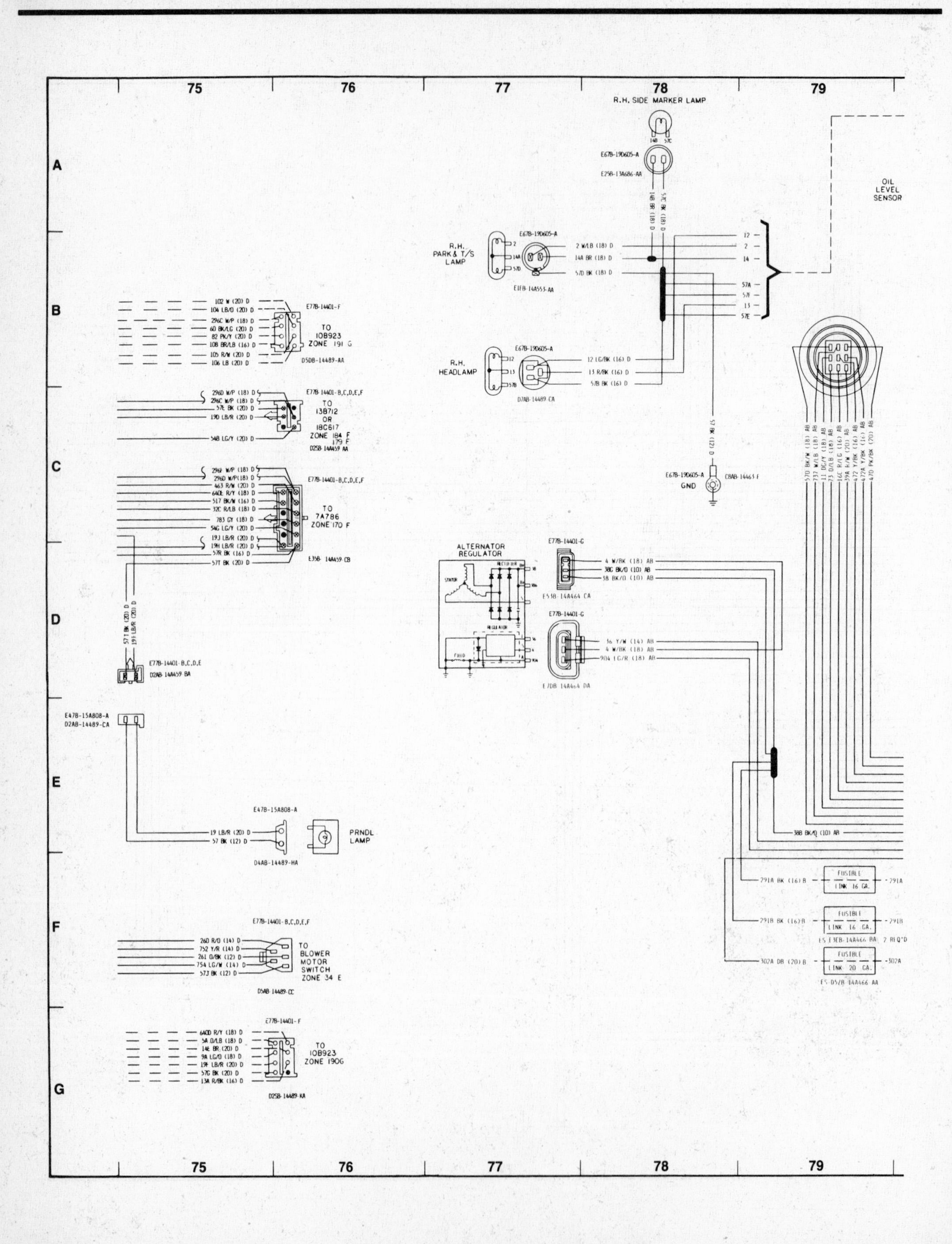

R.H. SIDE MARKER LAMP
OIL LEVEL SENSOR
R.H. PARK & T/S LAMP
R.H. HEADLAMP
TO 10B923 ZONE 191 G
TO 13B712 OR 18C617 ZONE 184 F 179 F
TO 7A786 ZONE 170 F
ALTERNATOR REGULATOR
GND
PRNDL LAMP
TO BLOWER MOTOR SWITCH ZONE 34 E
TO 10B923 ZONE 190G
FUSIBLE LINK 16 GA.
FUSIBLE LINK 20 GA.

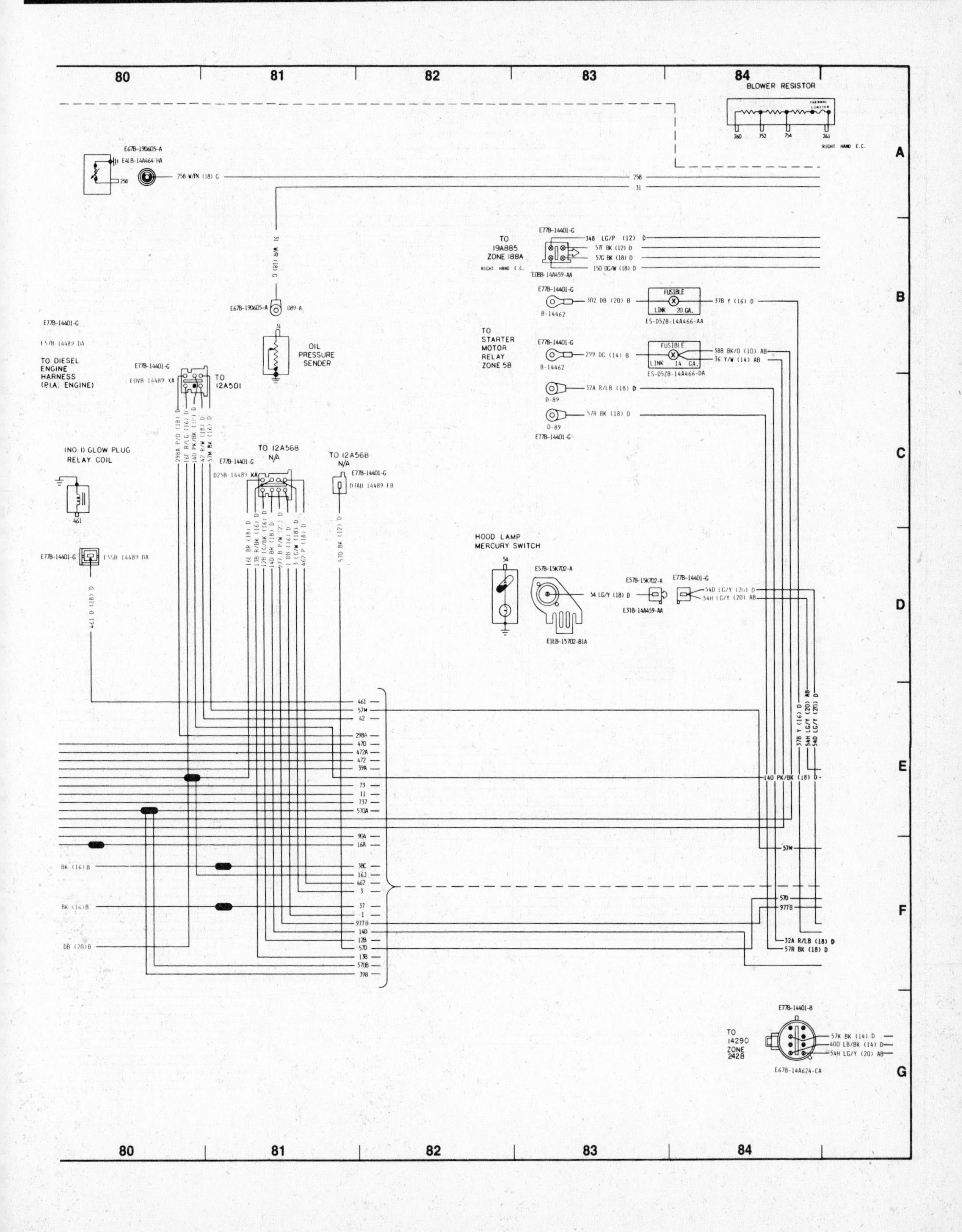
80
81
82
83
84
BLOWER RESISTOR
TO 19A885 ZONE 188A
TO STARTER MOTOR RELAY ZONE 5B
OIL PRESSURE SENDER
TO DIESEL ENGINE HARNESS (P.I.A. ENGINE)
TO 12A501
(NO. 1) GLOW PLUG RELAY COIL
TO 12A568 N/A
TO 12A568 N/A
HOOD LAMP MERCURY SWITCH
FUSIBLE LINK 20 GA.
FUSIBLE LINK 14 GA.
TO 14290 ZONE 242B
A
B
C
D
E
F
G

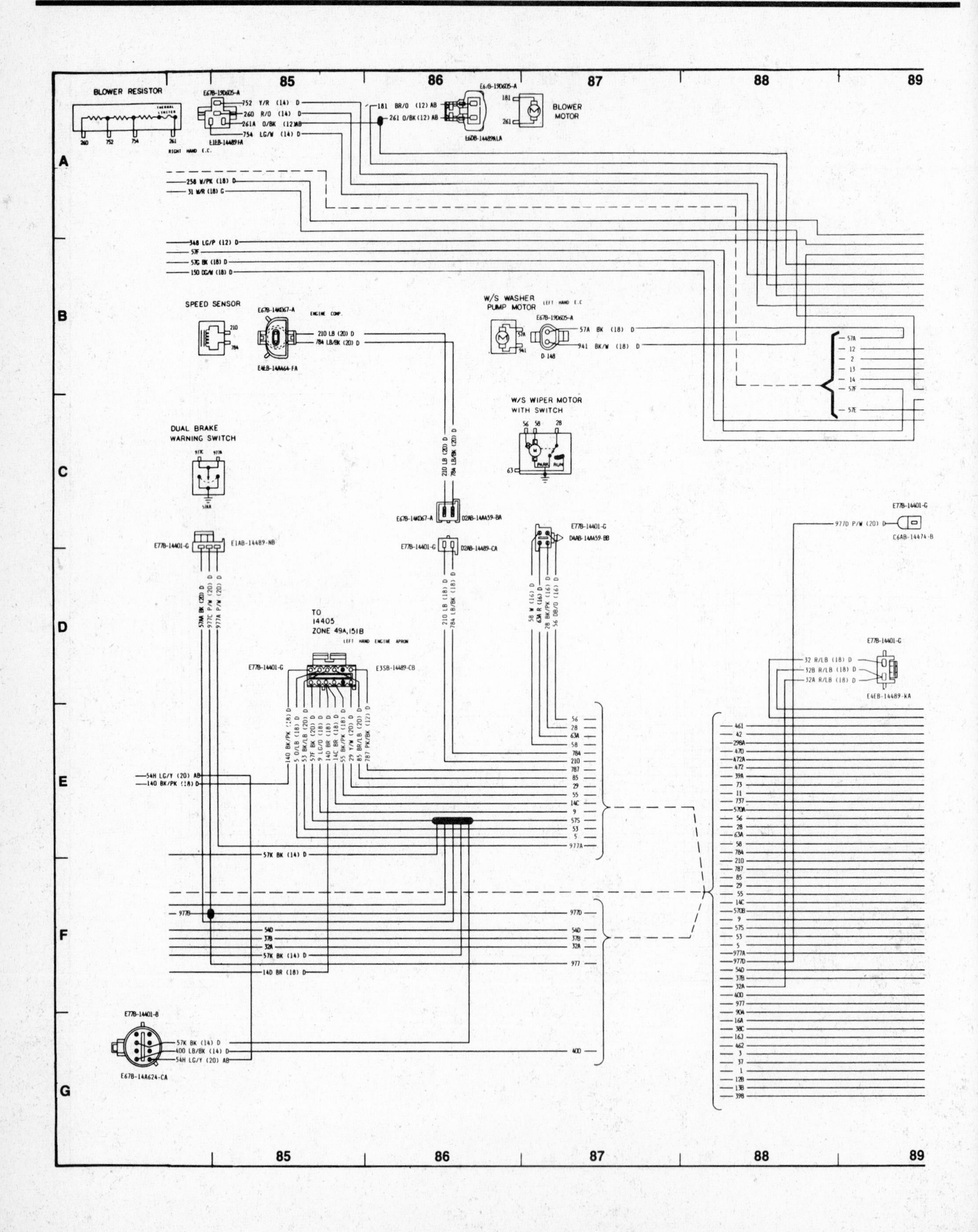

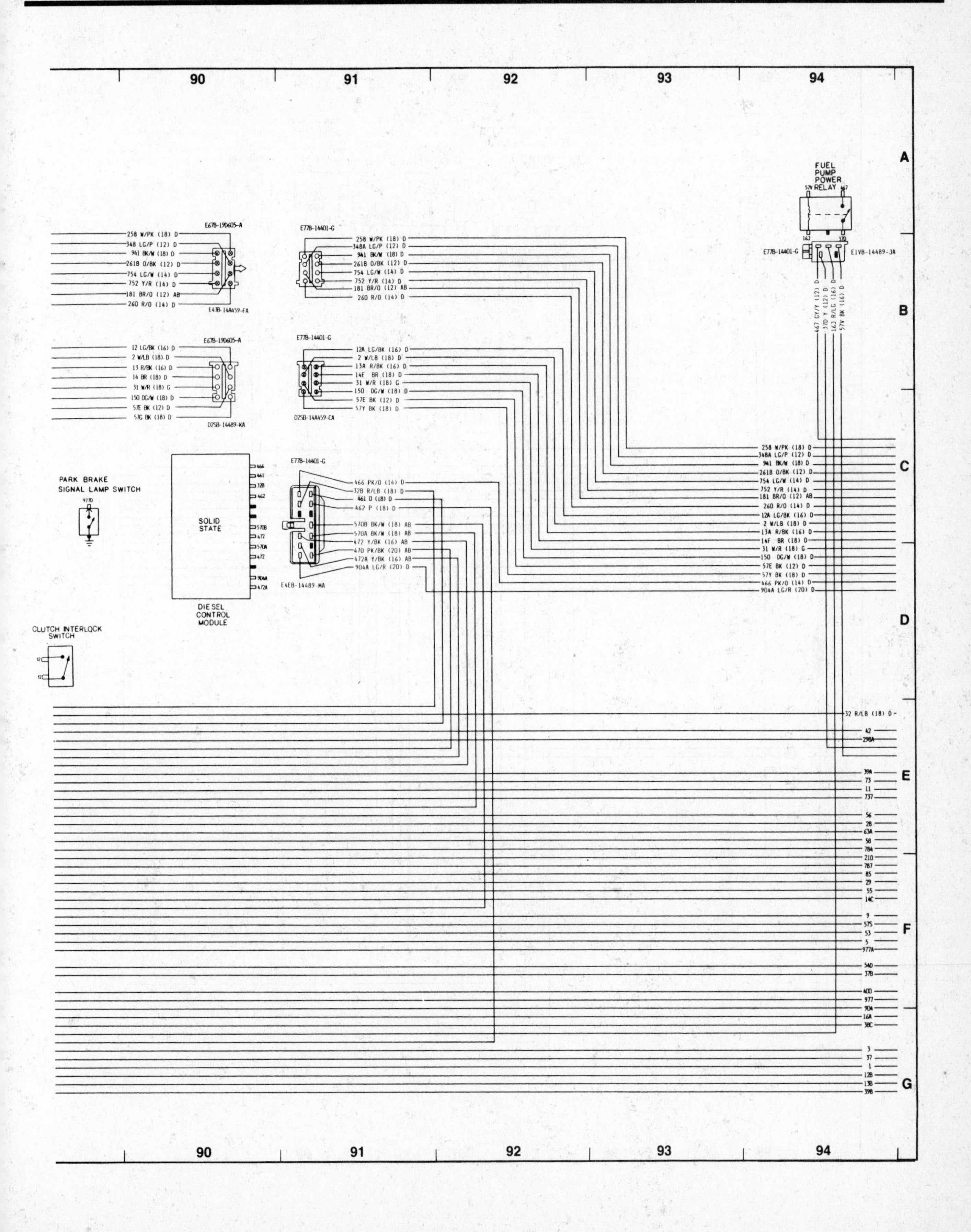

90
91
92
93
94
FUEL PUMP POWER RELAY
PARK BRAKE SIGNAL LAMP SWITCH
SOLID STATE
DIESEL CONTROL MODULE
CLUTCH INTERLOCK SWITCH
E67B-19D605-A
E77B-14401-G
E43B-14A459-FA
D25B-14489-KA
D25B-14A459-CA
E4EB-14489-MA
E1VB-14489-JA
258 W/PK (18) D
348 LG/P (12) D
941 BK/W (18) D
261B O/BK (12) D
754 LG/W (14) D
752 Y/R (14) D
181 BR/O (12) AB
260 R/O (14) D
12 LG/BK (16) D
2 W/LB (18) D
13 R/BK (16) D
14 BR (18) D
31 W/R (18) G
150 DG/W (18) D
57E BK (12) D
57G BK (18) D
466 PK/O (14) D
32B R/LB (18) D
461 O (18) D
462 P (18) D
570B BK/W (18) AB
570A BK/W (18) AB
472 Y/BK (16) AB
470 PK/BK (20) AB
472A Y/BK (16) AB
904A LG/R (20) D
467 GY/Y (12) D
370 Y (12) D
163 R/LG (16) D
57Y BK (16) D
32 R/LB (18) D

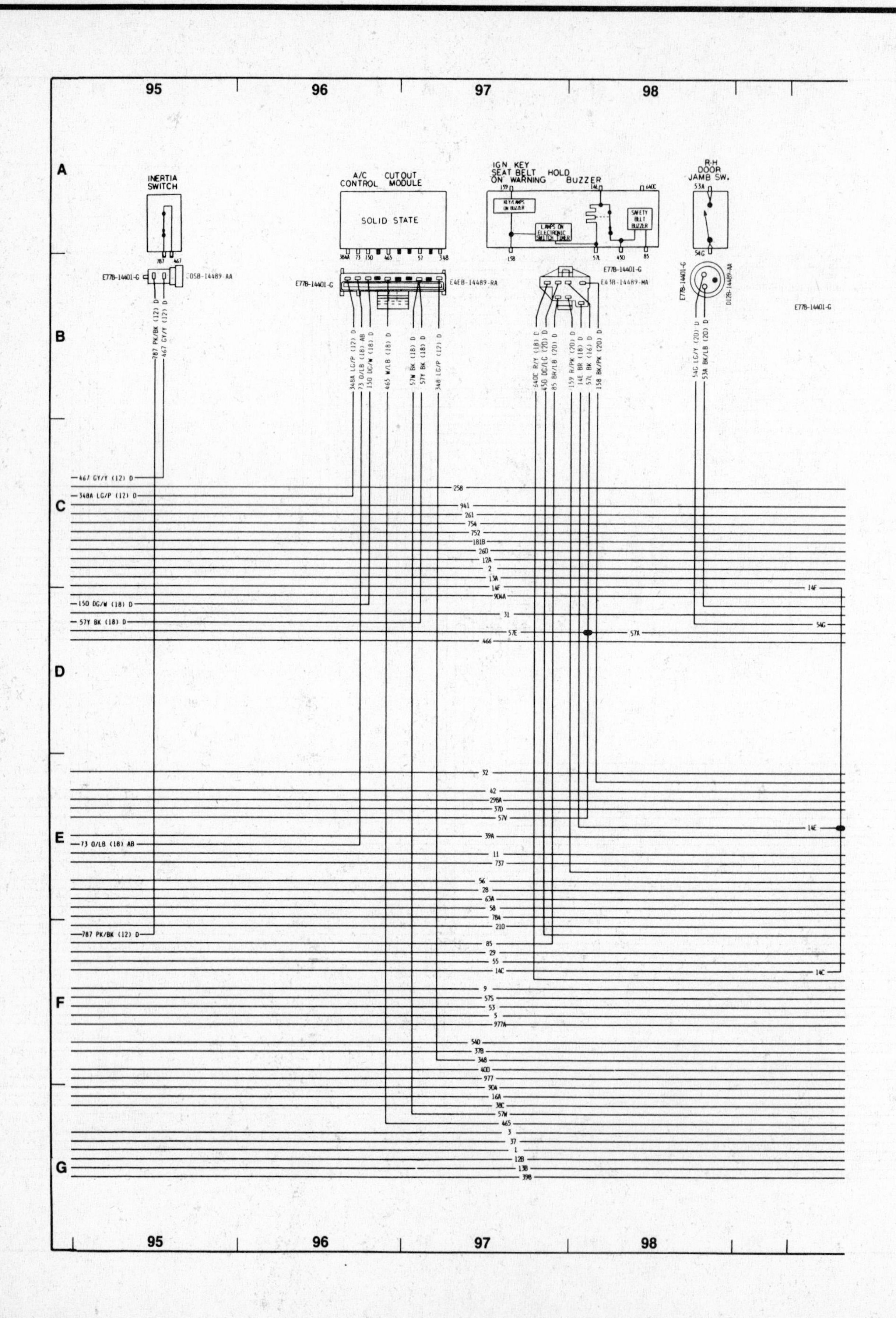
95
96
97
98
A
B
C
D
E
F
G
INERTIA SWITCH
A/C CUTOUT CONTROL MODULE
SOLID STATE
IGN KEY SEAT BELT ON WARNING
HOLD BUZZER
R-H DOOR JAMB SW.
E77B-14401-G
E4EB-14489-RA
E43B-14489-MA
467 GY/Y (12) D
348A LG/P (12) D
150 DG/W (18) D
57Y BK (18) D
73 O/LB (18) AB
787 PK/BK (12) D
14F
54G
57X
14E
14C

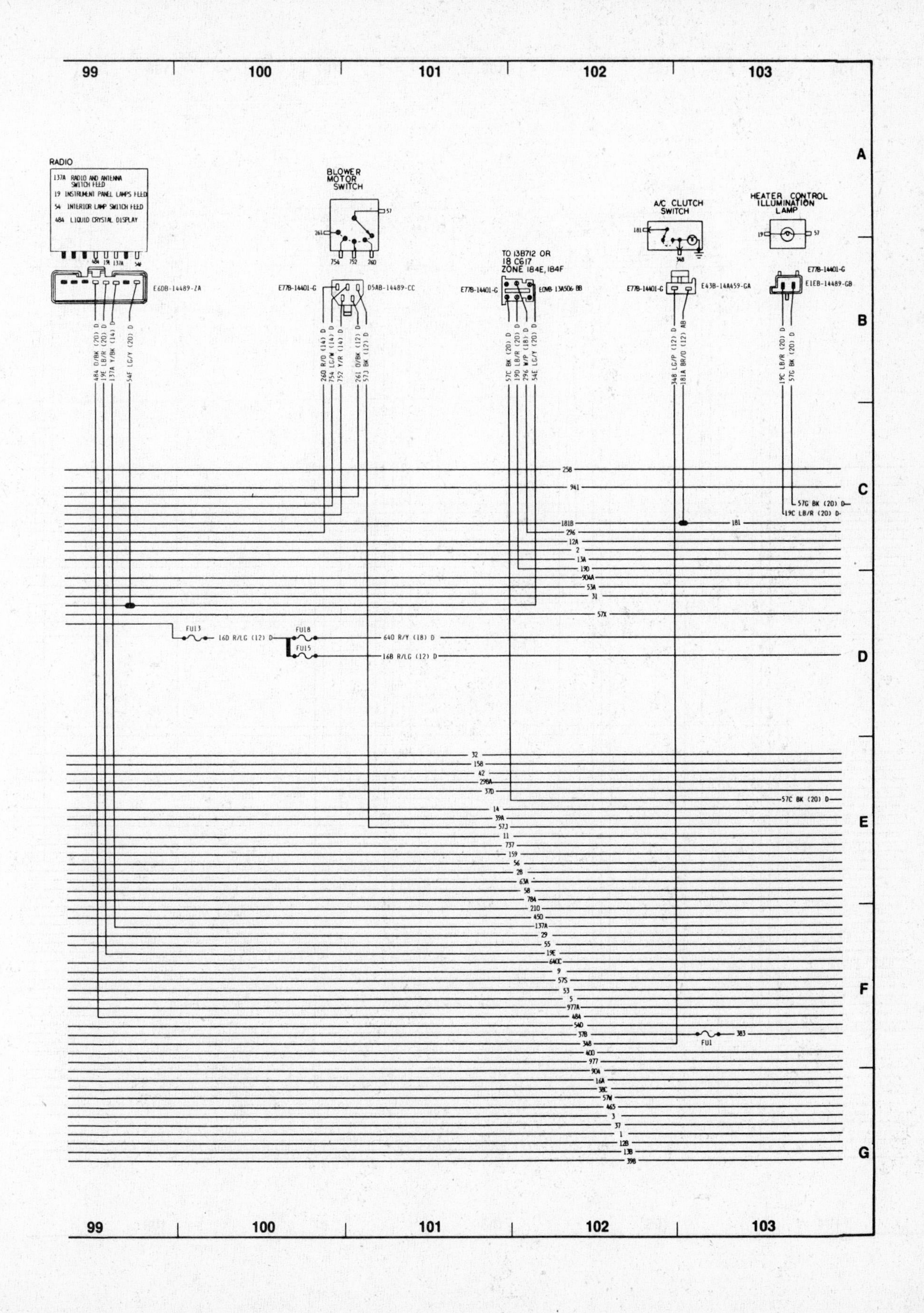
RADIO
137A RADIO AND ANTENNA SWITCH FEED
19 INSTRUMENT PANEL LAMPS FEED
54 INTERIOR LAMP SWITCH FEED
484 LIQUID CRYSTAL DISPLAY
BLOWER MOTOR SWITCH
A/C CLUTCH SWITCH
HEATER CONTROL ILLUMINATION LAMP
TO 13B712 OR 18 C617 ZONE 184E, 184F
E6DB-14489-ZA
E77B-14401-G
D5AB-14489-CC
E0VB-13A506 BB
E43B-14A459-GA
E1EB-14489-GB
FU13
FU18
FU15
FU1
16D R/LG (12) D
640 R/Y (18) D
16B R/LG (12) D
57C BK (20) D
57G BK (20) D
19C LB/R (20) D
99
100
101
102
103
A
B
C
D
E
F
G

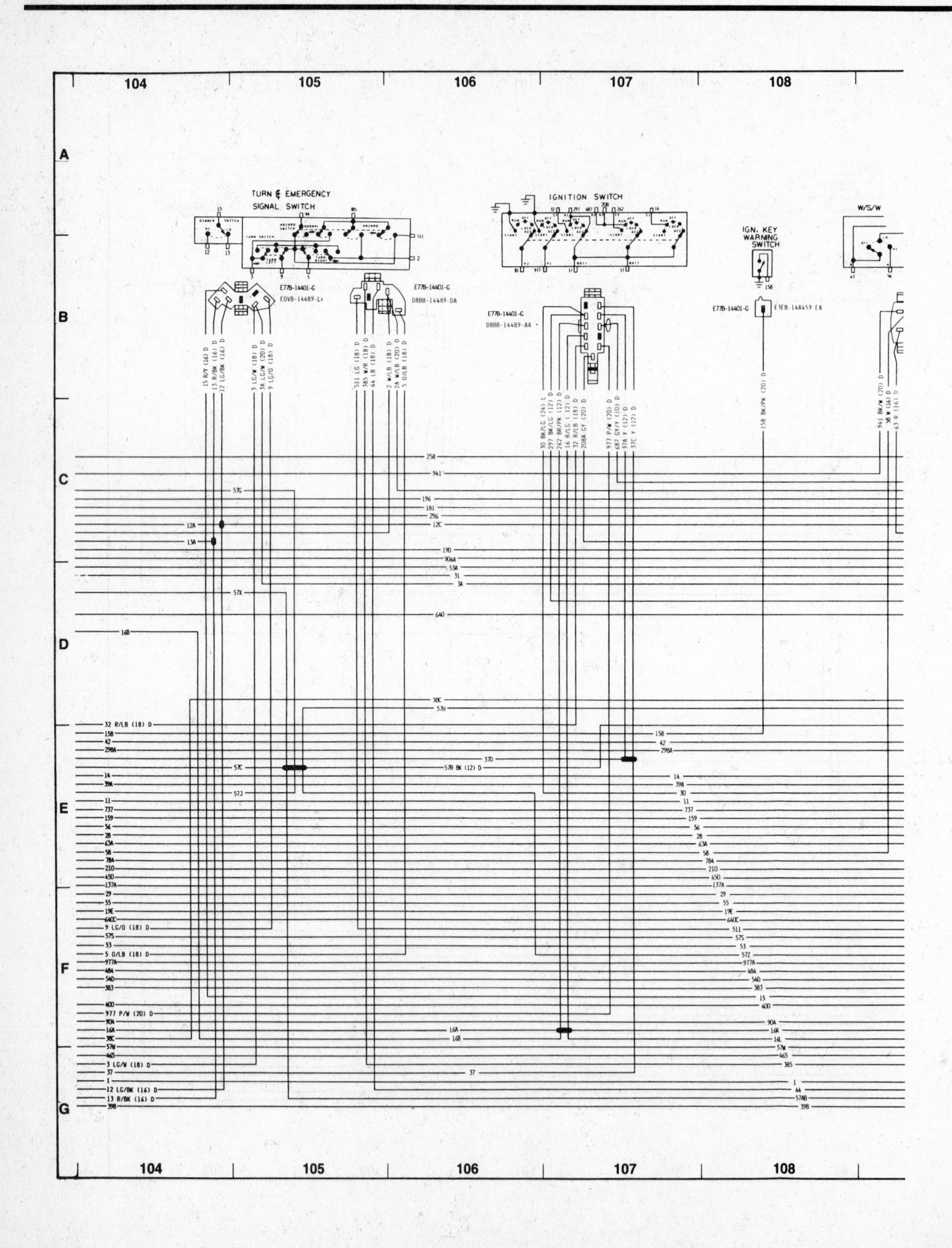

104
105
106
107
108
TURN & EMERGENCY SIGNAL SWITCH
IGNITION SWITCH
IGN. KEY WARNING SWITCH
W/S/W
E77B-14401-G
E0VB-14489-L
D8BB-14489-DA
D8BB-14489-AA
E3EB-14A459-EA
15 R/Y (16) D
13 R/BK (16) D
12 LG/BK (16) D
3 LG/W (18) D
3A LG/W (20) D
9 LG/O (18) D
511 LG (18) D
385 W/R (18) D
44 LB (18) D
2 W/LB (18) D
2A W/LB (20) D
5 O/LB (18) D
30 BK/LG (24) L
297 BK/LG (12) D
262 BR/PK (12) D
16 R/LG (12) D
32 R/LB (18) D
208A GY (20) D
977 P/W (20) D
687 GY/Y (10) D
37A Y (12) D
37C Y (12) D
158 BK/PK (20) D
941 BK/W (20) D
58 W (16) D
63 R (16) D
57R BK (12) D
32 R/LB (18) D
9 LG/O (18) D
5 O/LB (18) D
977 P/W (20) D
3 LG/W (18) D
12 LG/BK (16) D
13 R/BK (16) D

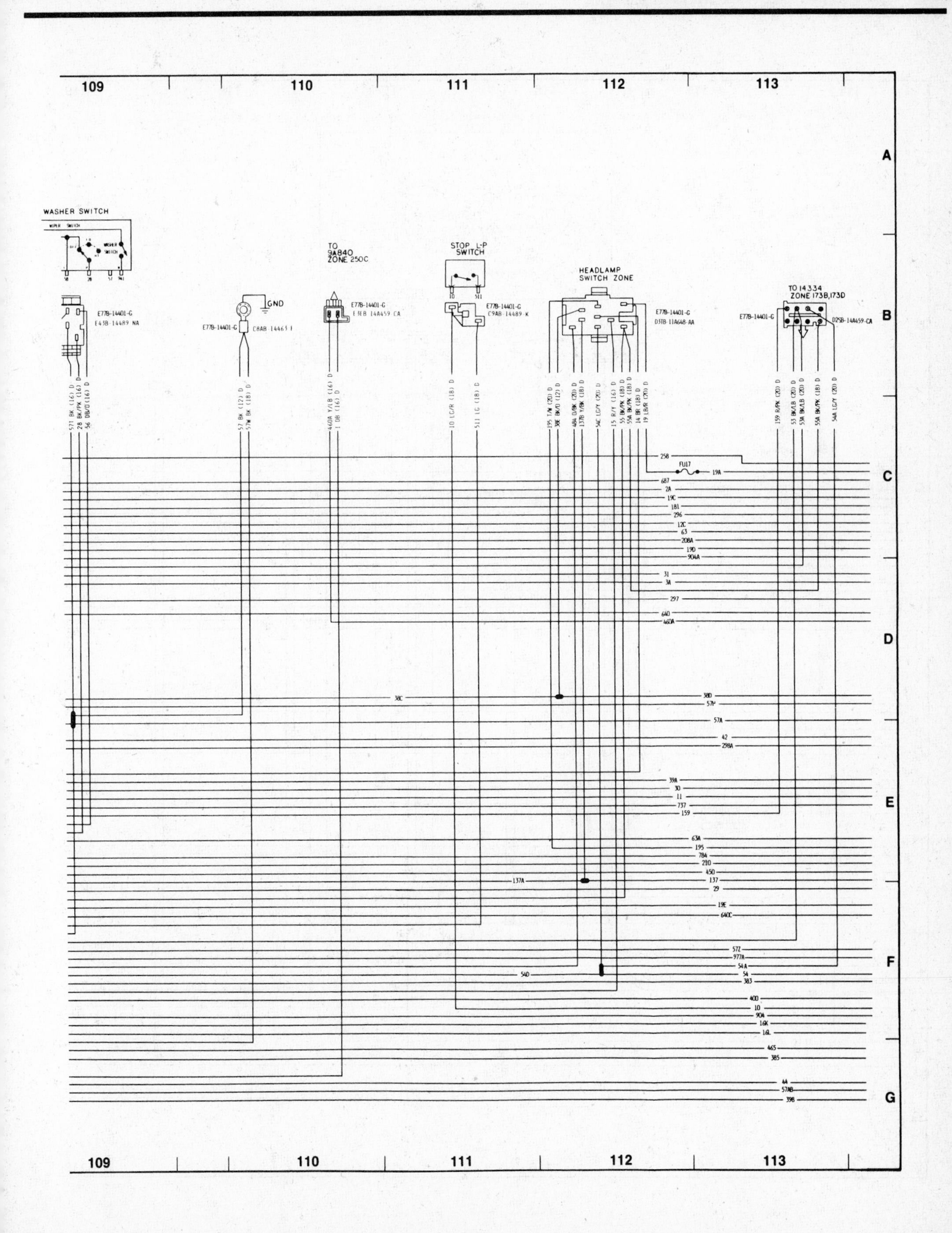
WASHER SWITCH
TO
9A840
ZONE 250C
STOP L-P
SWITCH
HEADLAMP
SWITCH ZONE
TO 14334
ZONE 173B,173D
GND
109
110
111
112
113
A
B
C
D
E
F
G

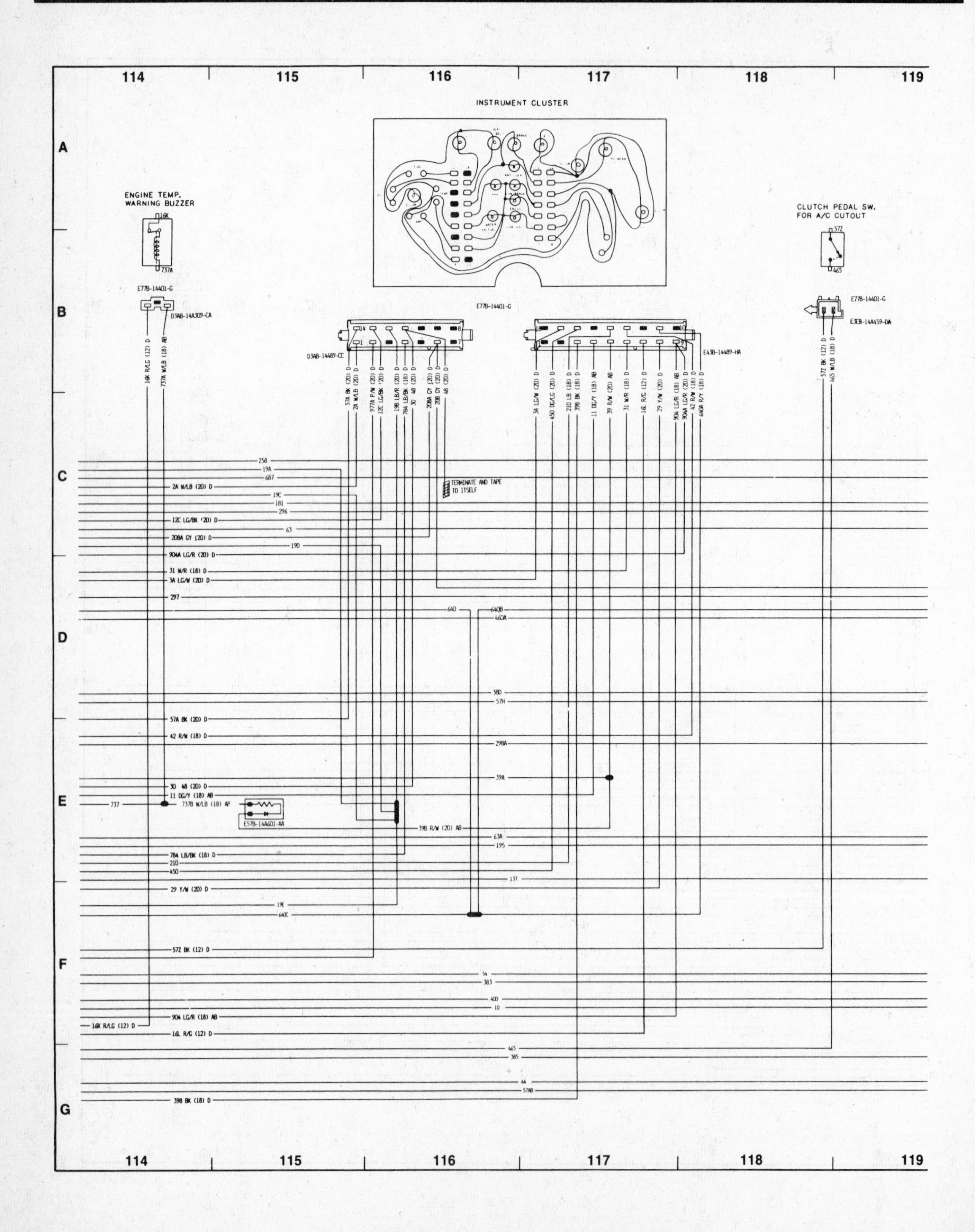
INSTRUMENT CLUSTER
ENGINE TEMP. WARNING BUZZER
CLUTCH PEDAL SW. FOR A/C CUTOUT
TERMINATE AND TAPE TO ITSELF
114
115
116
117
118
119
A
B
C
D
E
F
G

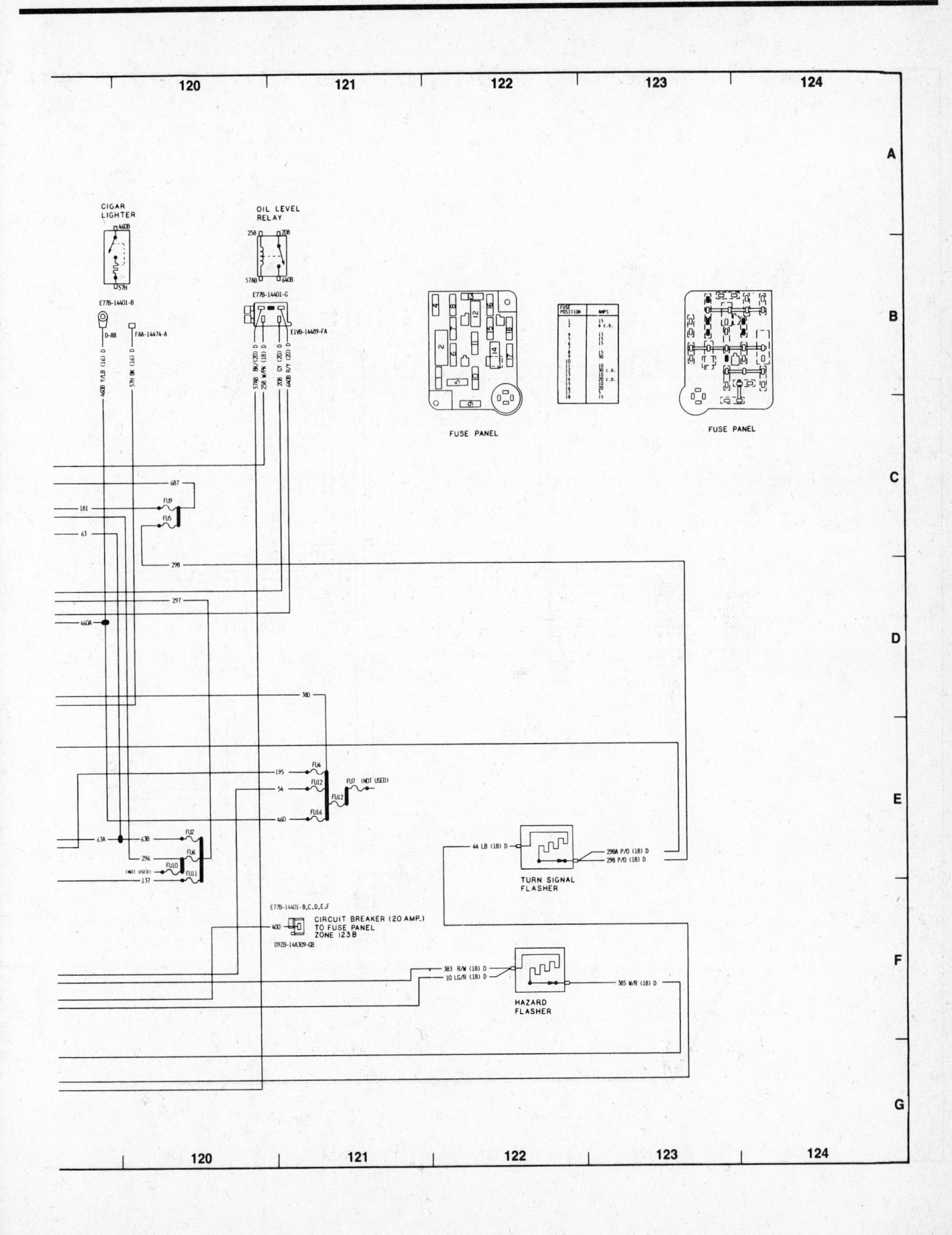

120
121
122
123
124
CIGAR LIGHTER
OIL LEVEL RELAY
FUSE PANEL
FUSE PANEL
TURN SIGNAL FLASHER
HAZARD FLASHER
CIRCUIT BREAKER (20 AMP.) TO FUSE PANEL ZONE 123B
A
B
C
D
E
F
G

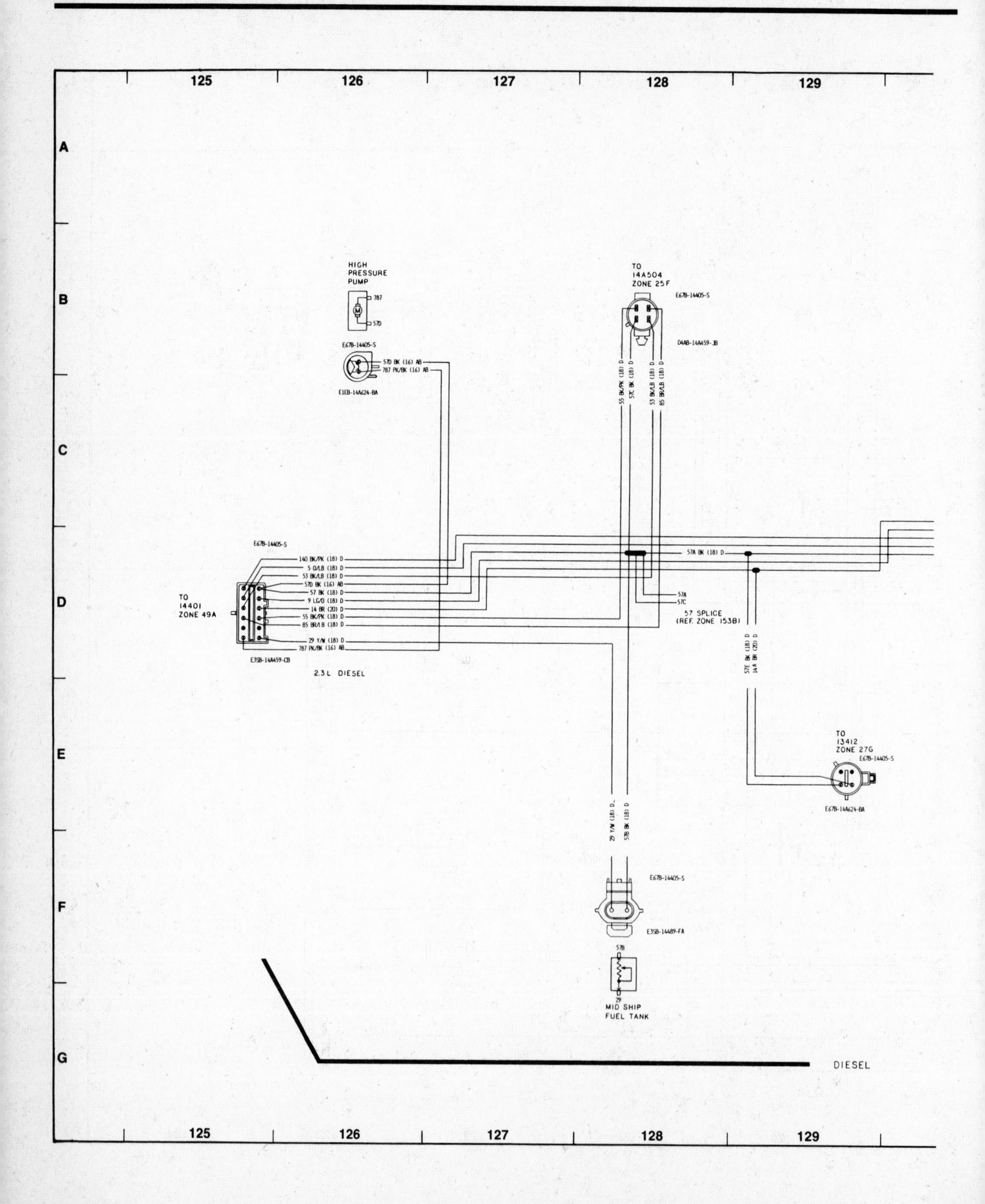
125
126
127
128
129
A
B
C
D
E
F
G
HIGH PRESSURE PUMP
E67B-14405-S
E1EB-14A624-BA
570 BK (16) AB
787 PK/BK (16) AB
TO 14A504 ZONE 25F
D4AB-14A459-JB
55 BK/PK (18) D
57C BK (18) D
53 BK/LB (18) D
85 BR/LB (18) D
TO 14401 ZONE 49A
140 BK/PK (18) D
5 O/LB (18) D
57D BK (16) AB
57 BK (18) D
9 LG/O (18) D
14 BR (20) D
29 Y/W (18) D
E35B-14A459-CB
2.3L DIESEL
57A BK (18) D
57A
57C
57 SPLICE (REF. ZONE 153B)
57E BK (18) D
14A BR (20) D
TO 13412 ZONE 27G
E67B-14A624-BA
29 Y/W (18) D
57B BK (18) D
E35B-14489-FA
57B
MID SHIP FUEL TANK
DIESEL

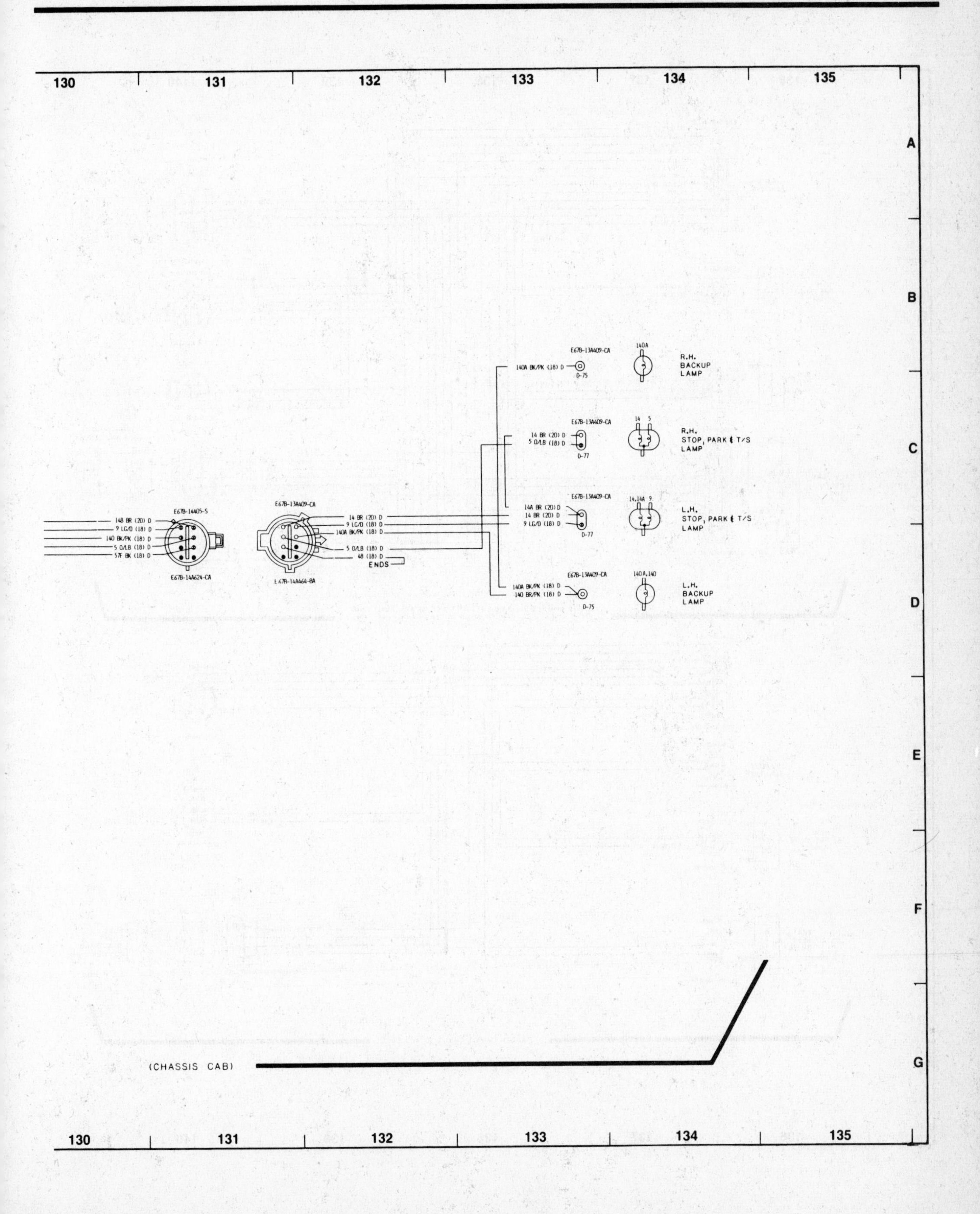
130
131
132
133
134
135
A
B
C
D
E
F
G
14B BR (20) D
9 LG/O (18) D
140 BK/PK (18) D
5 O/LB (18) D
57F BK (18) D
E67B-14405-S
E67B-14A624-CA
E67B-13A409-CA
14 BR (20) D
9 LG/O (18) D
140A BK/PK (18) D
5 O/LB (18) D
48 (18) D
ENDS
E67B-14A464-BA
E67B-13A409-CA
140A BK/PK (18) D
D-75
140A
R.H.
BACKUP
LAMP
E67B-13A409-CA
14 BR (20) D
5 O/LB (18) D
D-77
14 5
R.H.
STOP, PARK & T/S
LAMP
E67B-13A409-CA
14A BR (20) D
14 BR (20) D
9 LG/O (18) D
D-77
14,14A 9
L.H.
STOP, PARK & T/S
LAMP
E67B-13A409-CA
140A BK/PK (18) D
140 BR/PK (18) D
D-75
140A,140
L.H.
BACKUP
LAMP
(CHASSIS CAB)

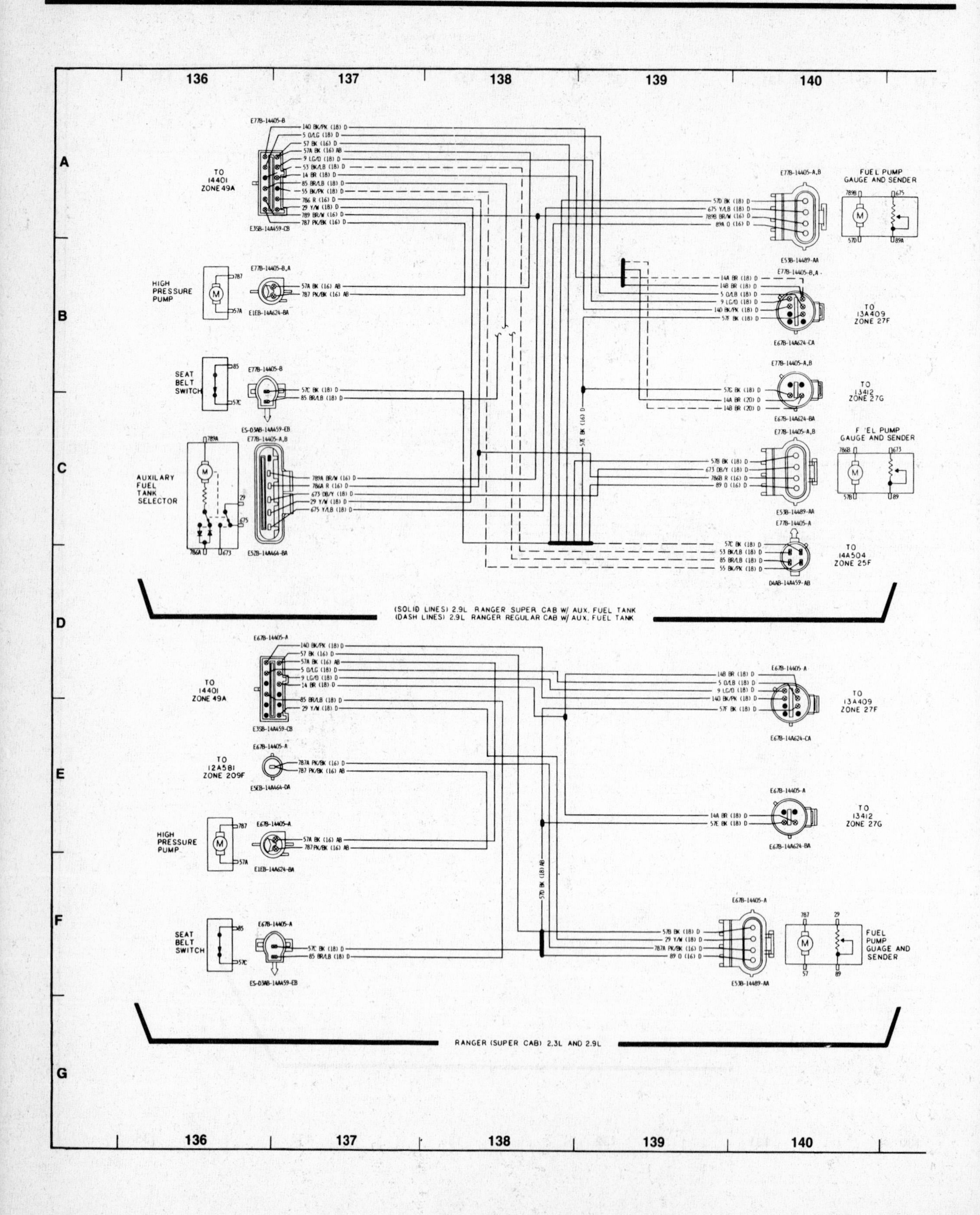
136
137
138
139
140
A
B
C
D
E
F
G
TO 14401 ZONE 49A
FUEL PUMP GAUGE AND SENDER
HIGH PRESSURE PUMP
TO 13A409 ZONE 27F
SEAT BELT SWITCH
TO 13412 ZONE 27G
AUXILARY FUEL TANK SELECTOR
TO 14A504 ZONE 25F
(SOLID LINES) 2.9L RANGER SUPER CAB W/ AUX. FUEL TANK
(DASH LINES) 2.9L RANGER REGULAR CAB W/ AUX. FUEL TANK
TO 12A581 ZONE 209F
FUEL PUMP GUAGE AND SENDER
RANGER (SUPER CAB) 2.3L AND 2.9L

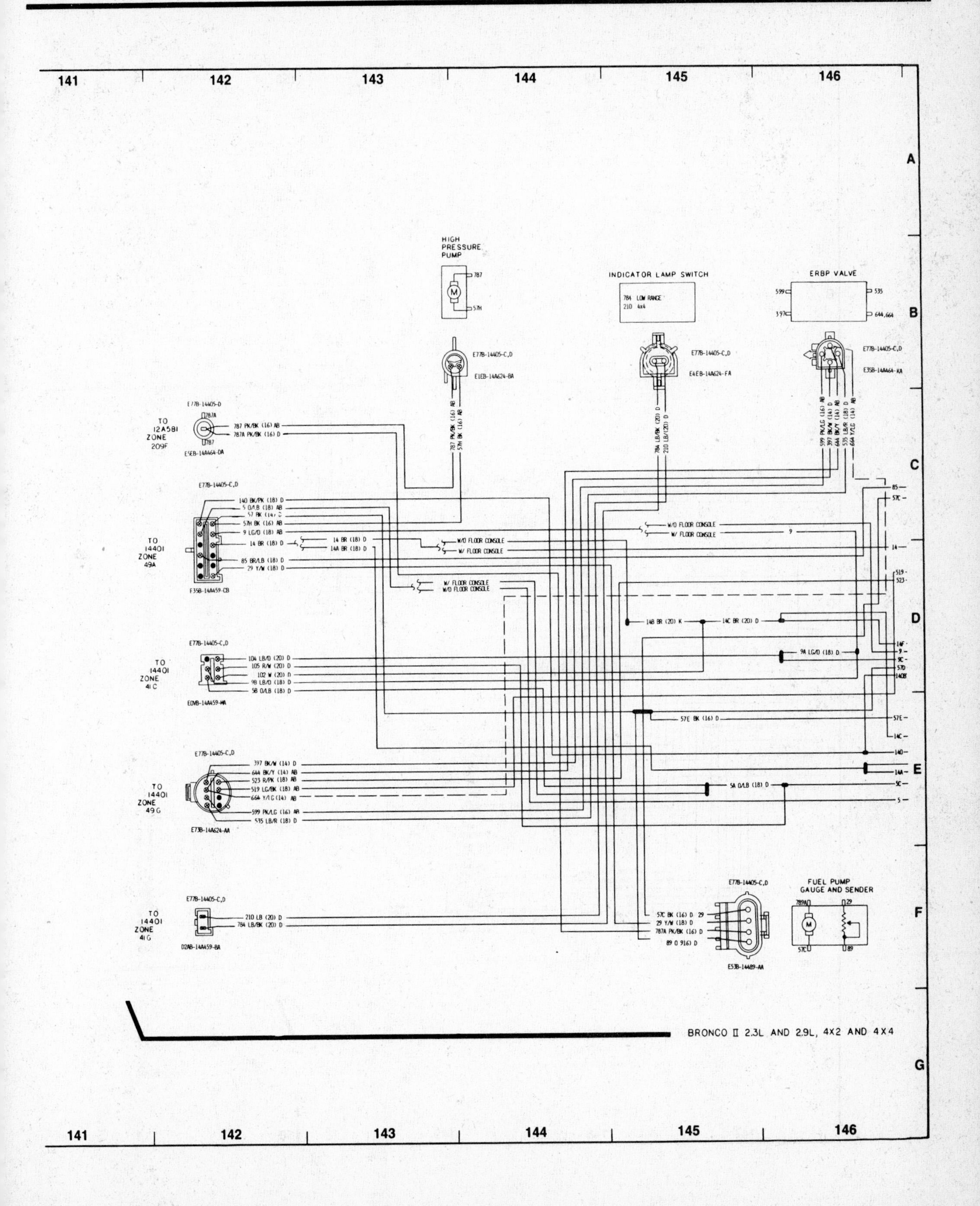
HIGH PRESSURE PUMP
INDICATOR LAMP SWITCH
784 LOW RANGE
210 4x4
ERBP VALVE
TO 12A581 ZONE 209F
TO 14401 ZONE 49A
TO 14401 ZONE 41C
TO 14401 ZONE 49G
TO 14401 ZONE 41G
W/O FLOOR CONSOLE
W/ FLOOR CONSOLE
FUEL PUMP GAUGE AND SENDER
BRONCO II 2.3L AND 2.9L, 4X2 AND 4X4

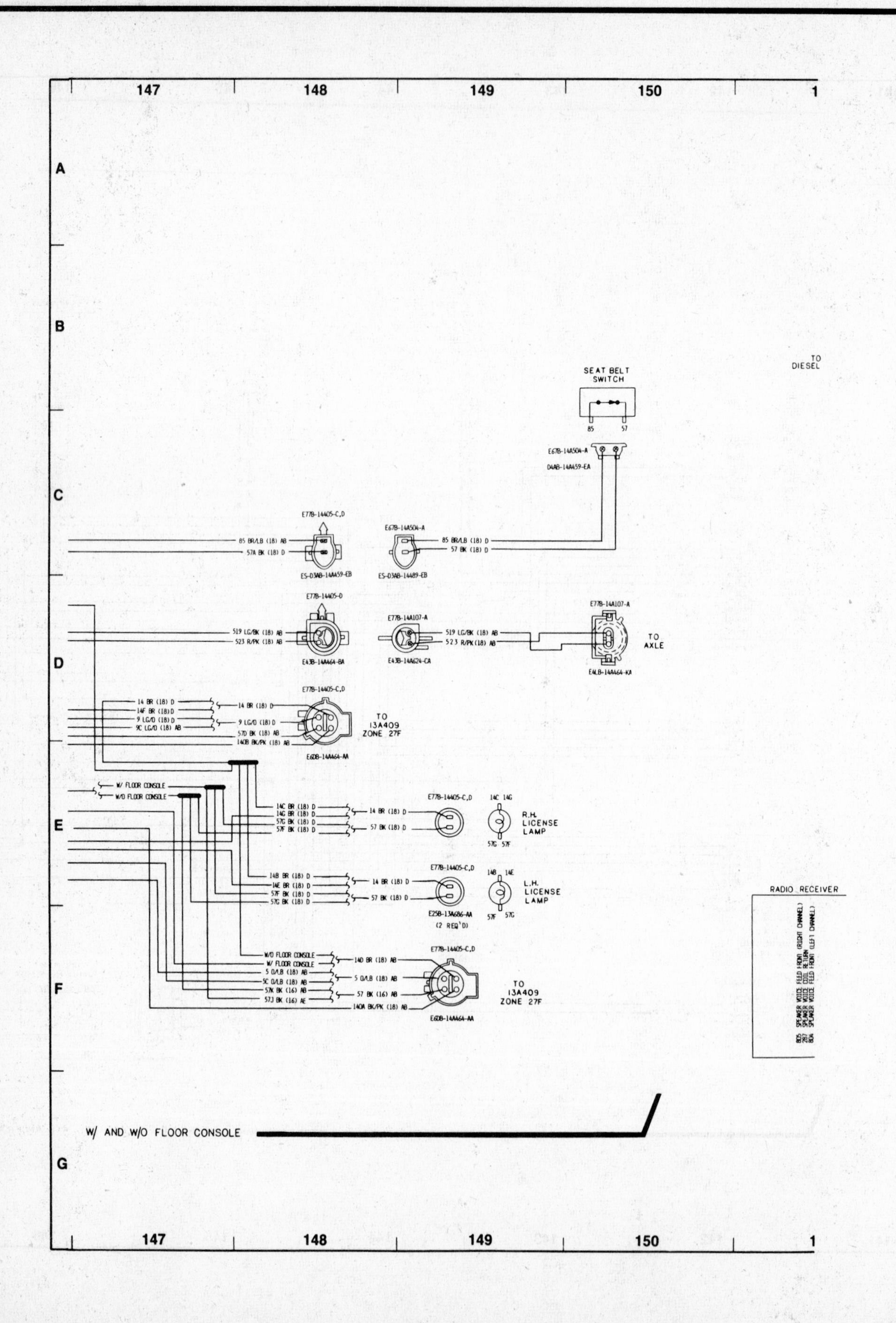
147
148
149
150
1
A
B
C
D
E
F
G
SEAT BELT SWITCH
85
57
TO DIESEL
E67B-14A504-A
D4AB-14A459-EA
E77B-14405-C,D
85 BR/LB (18) AB
57A BK (18) D
ES-D3AB-14A459-EB
E67B-14A504-A
85 BR/LB (18) D
57 BK (18) D
ES-D3AB-14489-EB
E77B-14405-D
519 LG/BK (18) AB
523 R/PK (18) AB
E43B-14A464-BA
E77B-14A107-A
519 LG/BK (18) AB
523 R/PK (18) AB
E43B-14A624-CA
E77B-14A107-A
TO AXLE
E4LB-14A464-KA
E77B-14405-C,D
14 BR (18) D
14F BR (18) D
9 LG/O (18) D
9C LG/O (18) AB
14 BR (18) D
9 LG/O (18) D
57D BK (18) AB
140B BK/PK (18) AB
TO 13A409 ZONE 27F
E6DB-14A464-AA
W/ FLOOR CONSOLE
W/O FLOOR CONSOLE
14C BR (18) D
14G BR (18) D
57G BK (18) D
57F BK (18) D
14 BR (18) D
57 BK (18) D
E77B-14405-C,D
14C 14G
57G 57F
R.H. LICENSE LAMP
14B BR (18) D
14E BR (18) D
57F BK (18) D
57G BK (18) D
14 BR (18) D
57 BK (18) D
E77B-14405-C,D
E25B-13A686-AA
(2 REQ'D)
14B 14E
57F 57G
L.H. LICENSE LAMP
RADIO RECEIVER
805 SPEAKER VOICE FEED FRONT (RIGHT CHANNEL)
287 SPEAKER VOICE COIL RETURN
804 SPEAKER VOICE FEED FRONT (LEFT CHANNEL)
W/O FLOOR CONSOLE
W/ FLOOR CONSOLE
5 O/LB (18) AB
5C O/LB (18) AB
57K BK (16) AB
57J BK (16) AE
140D BR (18) AB
5 O/LB (18) AB
57 BK (16) AB
140A BK/PK (18) AB
E77B-14405-C,D
E6DB-14A464-AA
TO 13A409 ZONE 27F
W/ AND W/O FLOOR CONSOLE

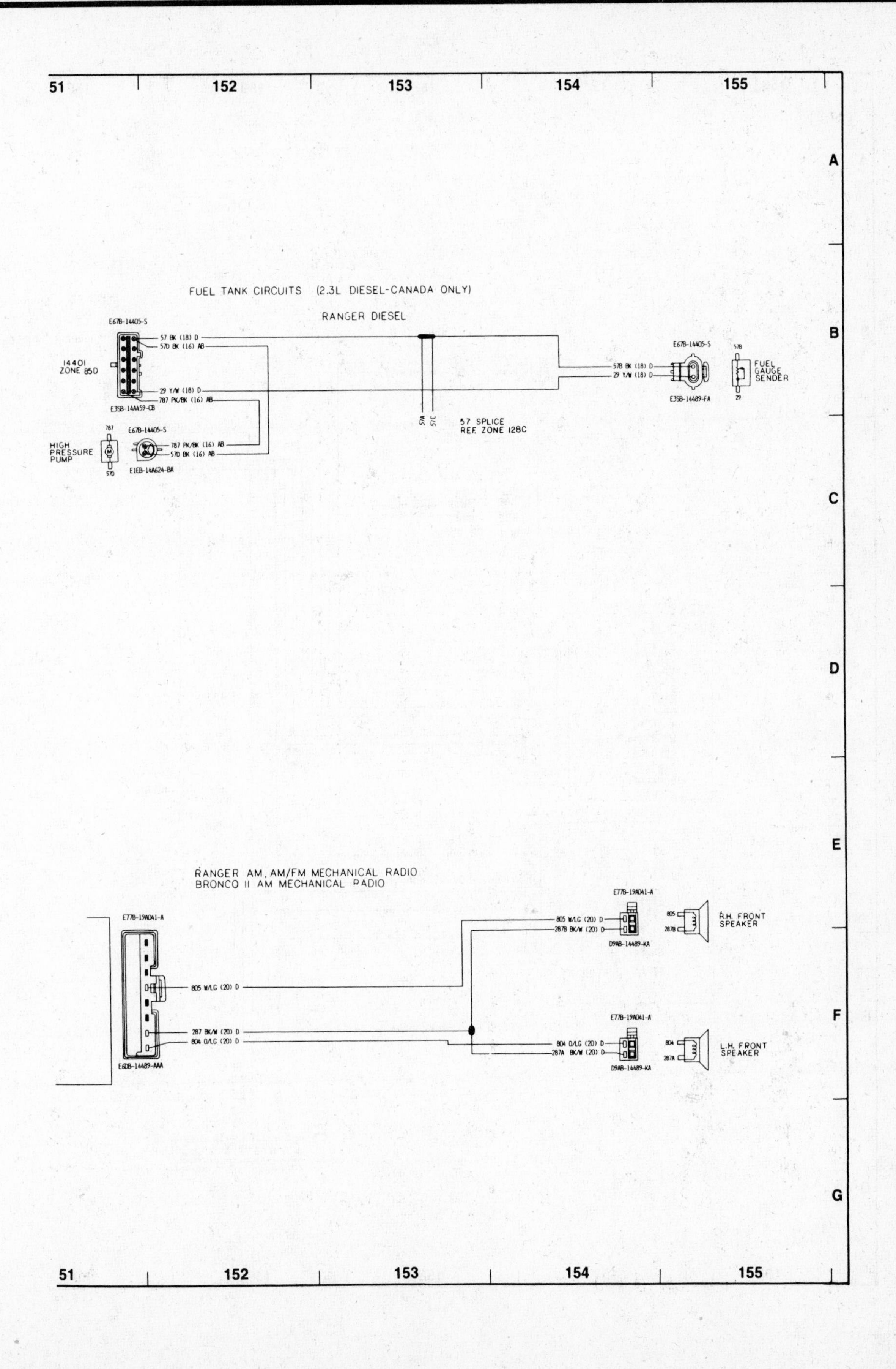
51
152
153
154
155
A
B
C
D
E
F
G
FUEL TANK CIRCUITS (2.3L DIESEL-CANADA ONLY)
RANGER DIESEL
E67B-14405-S
57 BK (18) D
570 BK (16) AB
14401
ZONE 85D
29 Y/W (18) D
787 PK/BK (16) AB
E3SB-14A459-CB
787
E67B-14405-S
HIGH
PRESSURE
PUMP
787 PK/BK (16) AB
570 BK (16) AB
570
E1EB-14A624-BA
57A
57C
57 SPLICE
REF. ZONE 128C
E67B-14405-S
57B BK (18) D
29 Y/W (18) D
E3SB-14489-FA
57B
29
FUEL
GAUGE
SENDER
RANGER AM, AM/FM MECHANICAL RADIO
BRONCO II AM MECHANICAL RADIO
E77B-19A041-A
805 W/LG (20) D
287 BK/W (20) D
804 O/LG (20) D
E6DB-14489-AAA
E77B-19A041-A
805 W/LG (20) D
287B BK/W (20) D
D9AB-14489-KA
805
287B
R.H. FRONT
SPEAKER
E77B-19A041-A
804 O/LG (20) D
287A BK/W (20) D
D9AB-14489-KA
804
287A
L.H. FRONT
SPEAKER

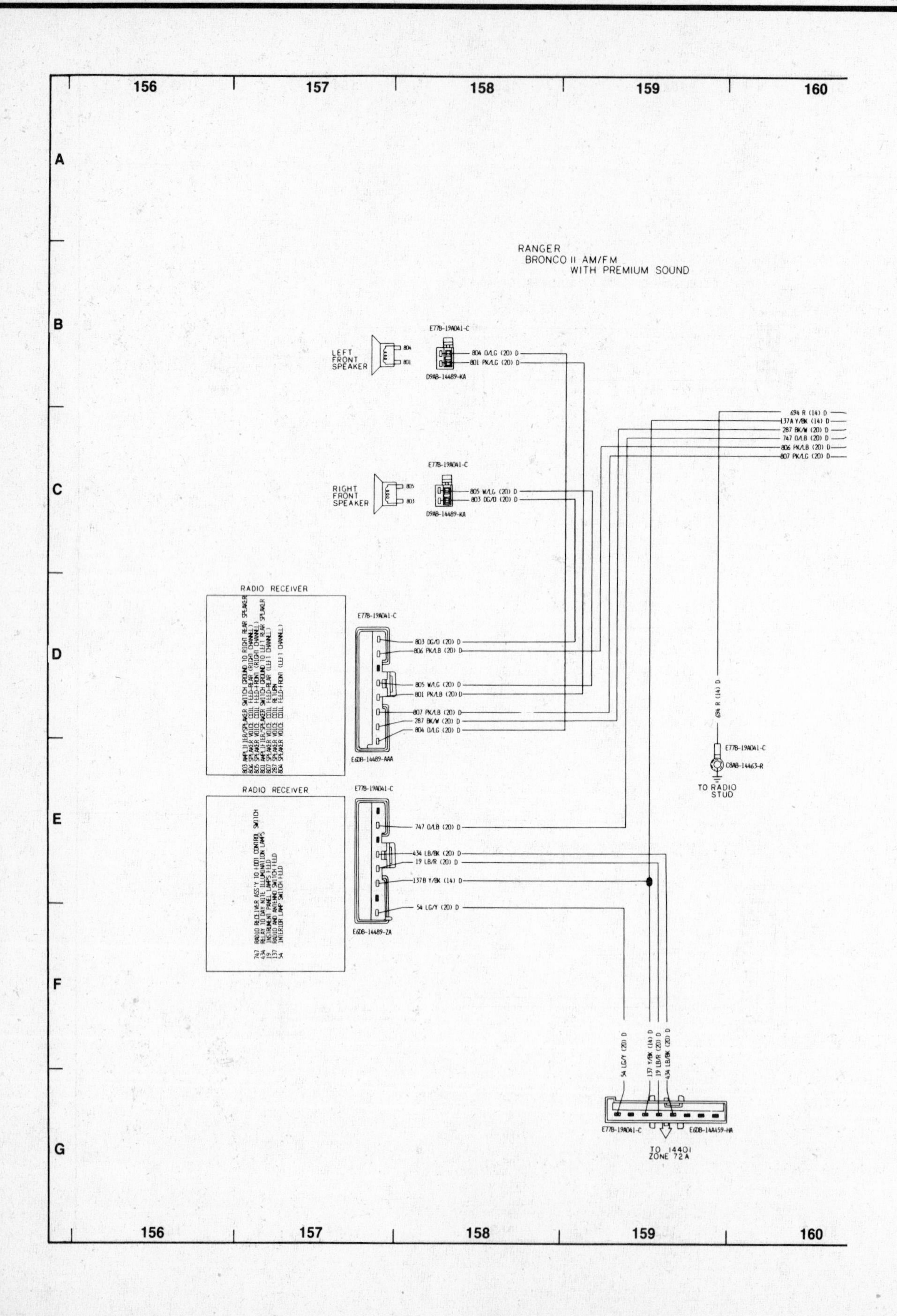
156
157
158
159
160
A
B
C
D
E
F
G
RANGER
BRONCO II AM/FM
WITH PREMIUM SOUND
LEFT FRONT SPEAKER
804
801
E77B-19A0A1-C
804 O/LG (20) D
801 PK/LG (20) D
D9AB-14489-KA
RIGHT FRONT SPEAKER
805
803
E77B-19A0A1-C
805 W/LG (20) D
803 DG/O (20) D
D9AB-14489-KA
694 R (14) D
137A Y/BK (14) D
287 BK/W (20) D
747 O/LB (20) D
806 PK/LB (20) D
807 PK/LG (20) D
RADIO RECEIVER
803 AMPLIFIER/SPEAKER SWITCH GROUND TO RIGHT REAR SPEAKER
806 SPEAKER VOICE COIL FEED-REAR (RIGHT CHANNEL)
805 SPEAKER VOICE COIL FEED-FRONT (RIGHT CHANNEL)
801 AMPLIFIER/SPEAKER SWITCH GROUND TO LEFT REAR SPEAKER
807 SPEAKER VOICE COIL FEED-REAR (LEFT CHANNEL)
287 SPEAKER VOICE COIL RETURN
804 SPEAKER VOICE COIL FEED-FRONT (LEFT CHANNEL)
E77B-19A041-C
803 DG/O (20) D
806 PK/LB (20) D
805 W/LG (20) D
801 PK/LB (20) D
807 PK/LB (20) D
287 BK/W (20) D
804 O/LG (20) D
E6DB-14489-AAA
RADIO RECEIVER
747 RADIO RECEIVER ASS'Y TO FOOT CONTROL SWITCH
434 RELAY TO DAY NITE ILLUMINATION LAMPS
19 INSTRUMENT PANEL LAMPS FEED
137 RADIO AND ANTENNO SWITCH FEED
54 INTERIOR LAMP SWITCH FEED
E77B-19A041-C
747 O/LB (20) D
434 LB/BK (20) D
19 LB/R (20) D
137B Y/BK (14) D
54 LG/Y (20) D
E6DB-14489-ZA
694 R (14) D
E77B-19A041-C
C8AB-14463-R
TO RADIO STUD
54 LG/Y (20) D
137 Y/BK (14) D
19 LB/R (20) D
434 LB/BK (20) D
E77B-19A041-C
E6DB-14A459-HA
TO 14401
ZONE 72A

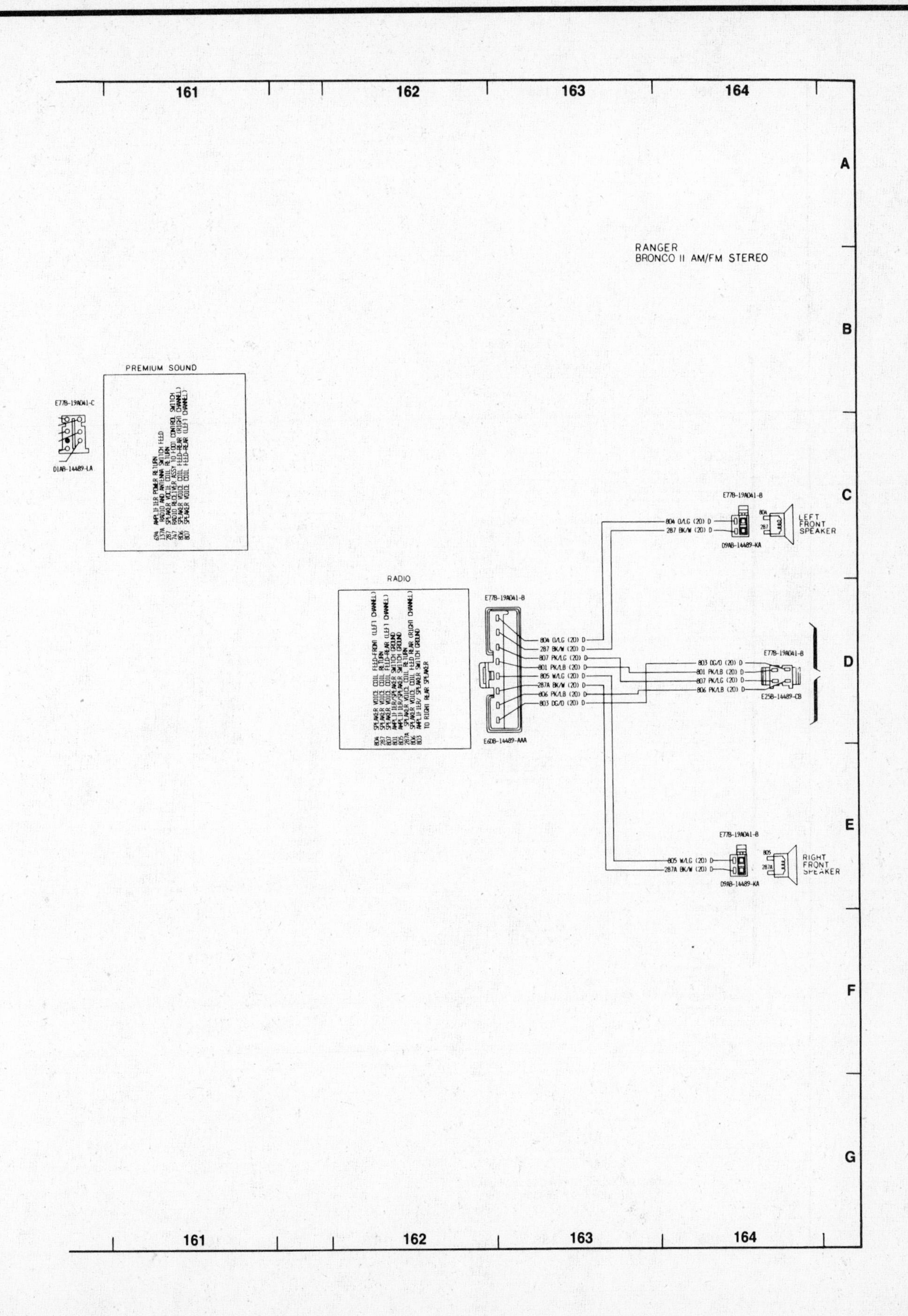
161
162
163
164
A
B
C
D
E
F
G
RANGER
BRONCO II AM/FM STEREO
PREMIUM SOUND
E77B-19A041-C
D1AB-14489-LA
694 AMPLIFIER POWER RETURN
137A RADIO AND ANTENNA SWITCH FEED
287 SPEAKER VOCIE COIL RETURN
747 RADIO RECEIVER ASSY TO FOOT CONTROL SWITCH
806 SPEAKER VOICE COIL FEED-REAR (RIGHT CHANNEL)
807 SPEAKER VOICE COIL FEED-REAR (LEFT CHANNEL)
RADIO
804 SPEAKER VOICE COIL FEED-FRONT (LEFT CHANNEL)
287 SPEAKER VOICE COIL RETURN
807 SPEAKER VOICE COIL FEED-REAR (LEFT CHANNEL)
801 AMPLIFIER/SPEAKER SWITCH GROUND
805 AMPLIFIER/SPEAKER SWITCH GROUND
287A SPEAKER VOICE COIL RETURN
806 SPEAKER VOICE COIL FEED REAR (RIGHT CHANNEL)
803 AMPLIFIER/ SPEAKER SWITCH GROUND
TO RIGHT REAR SPEAKER
E77B-19A041-B
E6DB-14489-AAA
804 O/LG (20) D
287 BK/W (20) D
807 PK/LG (20) D
801 PK/LB (20) D
805 W/LG (20) D
287A BK/W (20) D
806 PK/LB (20) D
803 DG/O (20) D
LEFT
FRONT
SPEAKER
D9AB-14489-KA
E25B-14489-CB
RIGHT
FRONT
SPEAKER
805
287A

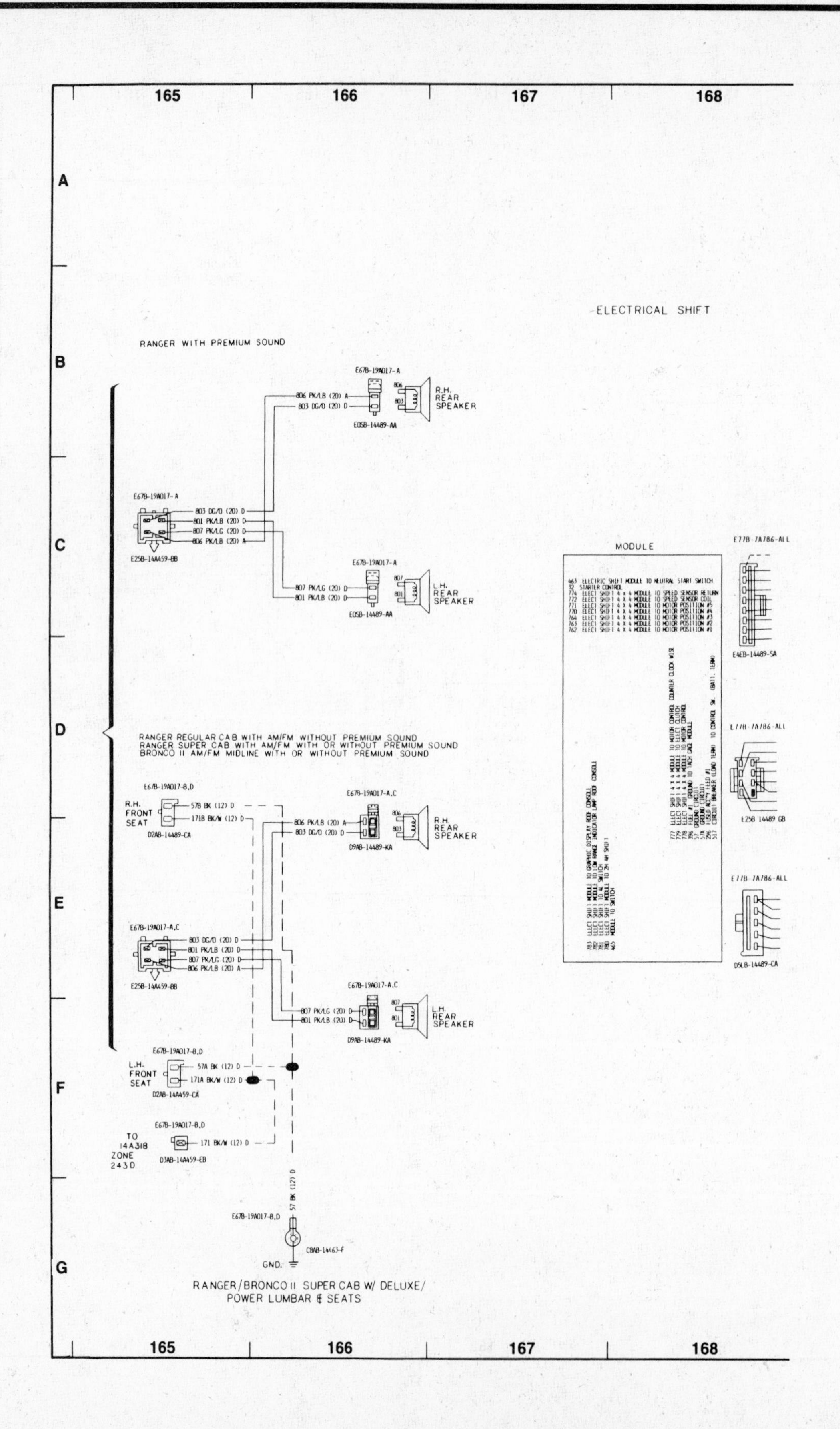

165
166
167
168
RANGER WITH PREMIUM SOUND
R.H. REAR SPEAKER
L.H. REAR SPEAKER
RANGER REGULAR CAB WITH AM/FM WITHOUT PREMIUM SOUND
RANGER SUPER CAB WITH AM/FM WITH OR WITHOUT PREMIUM SOUND
BRONCO II AM/FM MIDLINE WITH OR WITHOUT PREMIUM SOUND
R.H. FRONT SEAT
L.H. FRONT SEAT
TO 14A318 ZONE 243D
GND.
RANGER/BRONCO II SUPER CAB W/ DELUXE/ POWER LUMBAR & SEATS
ELECTRICAL SHIFT
MODULE

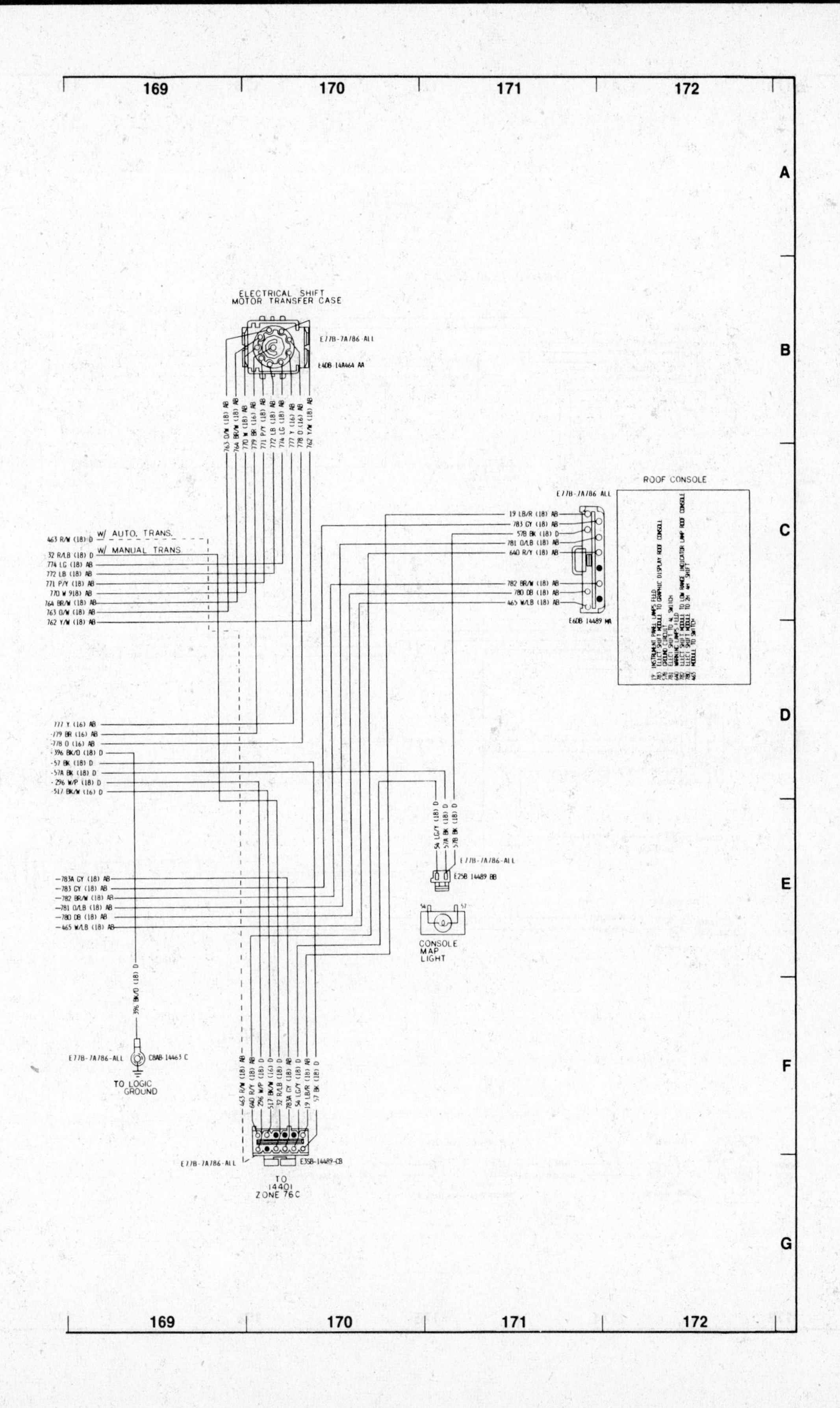
169
170
171
172
A
B
C
D
E
F
G
ELECTRICAL SHIFT MOTOR TRANSFER CASE
ROOF CONSOLE
W/ AUTO. TRANS.
W/ MANUAL TRANS.
CONSOLE MAP LIGHT
TO LOGIC GROUND
TO 14401 ZONE 76C

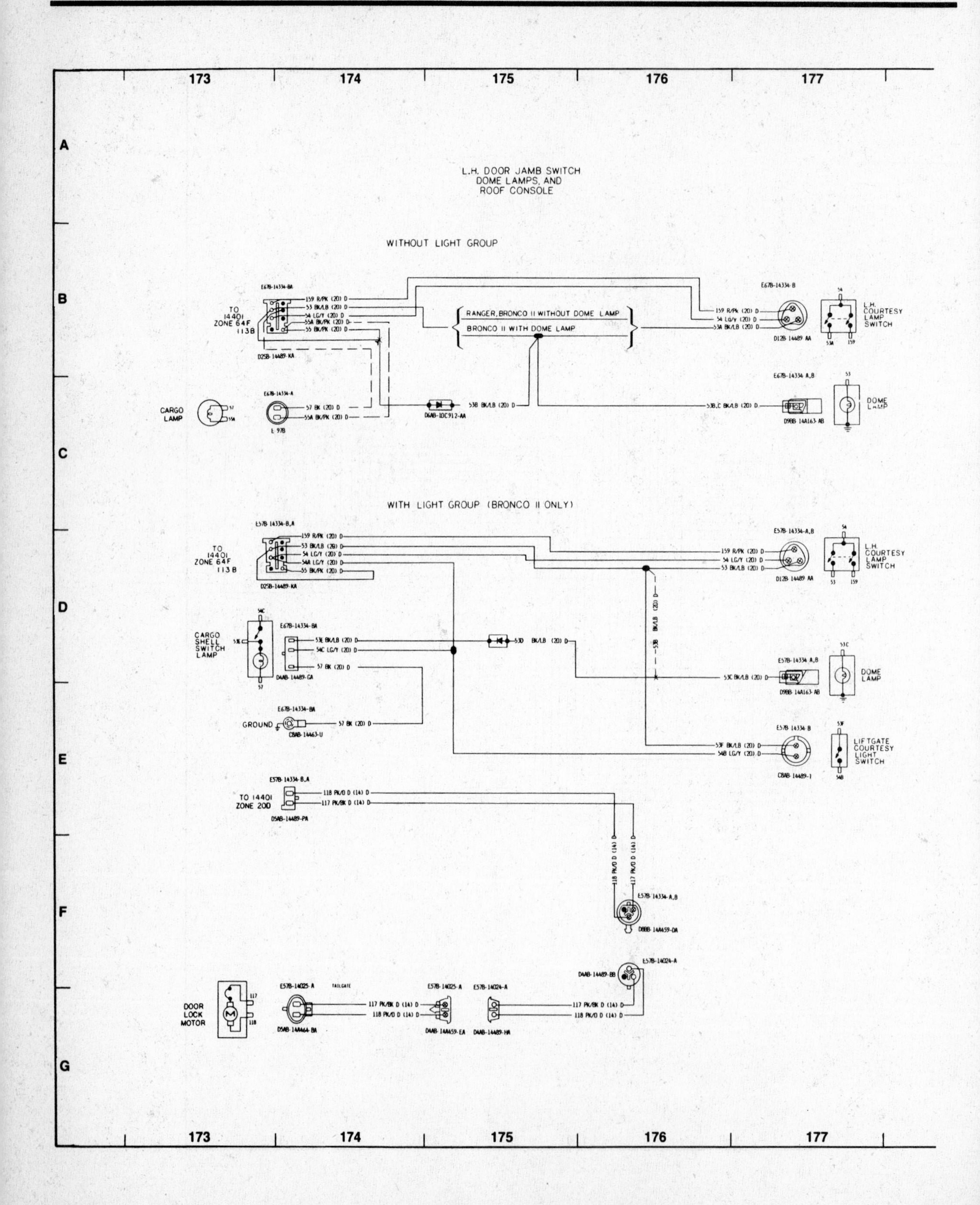
173
174
175
176
177
A
B
C
D
E
F
G
L.H. DOOR JAMB SWITCH
DOME LAMPS, AND
ROOF CONSOLE
WITHOUT LIGHT GROUP
TO 14401 ZONE 64F 113B
RANGER, BRONCO II WITHOUT DOME LAMP
BRONCO II WITH DOME LAMP
L.H. COURTESY LAMP SWITCH
CARGO LAMP
DOME LAMP
WITH LIGHT GROUP (BRONCO II ONLY)
TO 14401 ZONE 64F 113B
L.H. COURTESY LAMP SWITCH
CARGO SHELL SWITCH LAMP
DOME LAMP
GROUND
LIFTGATE COURTESY LIGHT SWITCH
TO 14401 ZONE 20D
DOOR LOCK MOTOR
TAILGATE

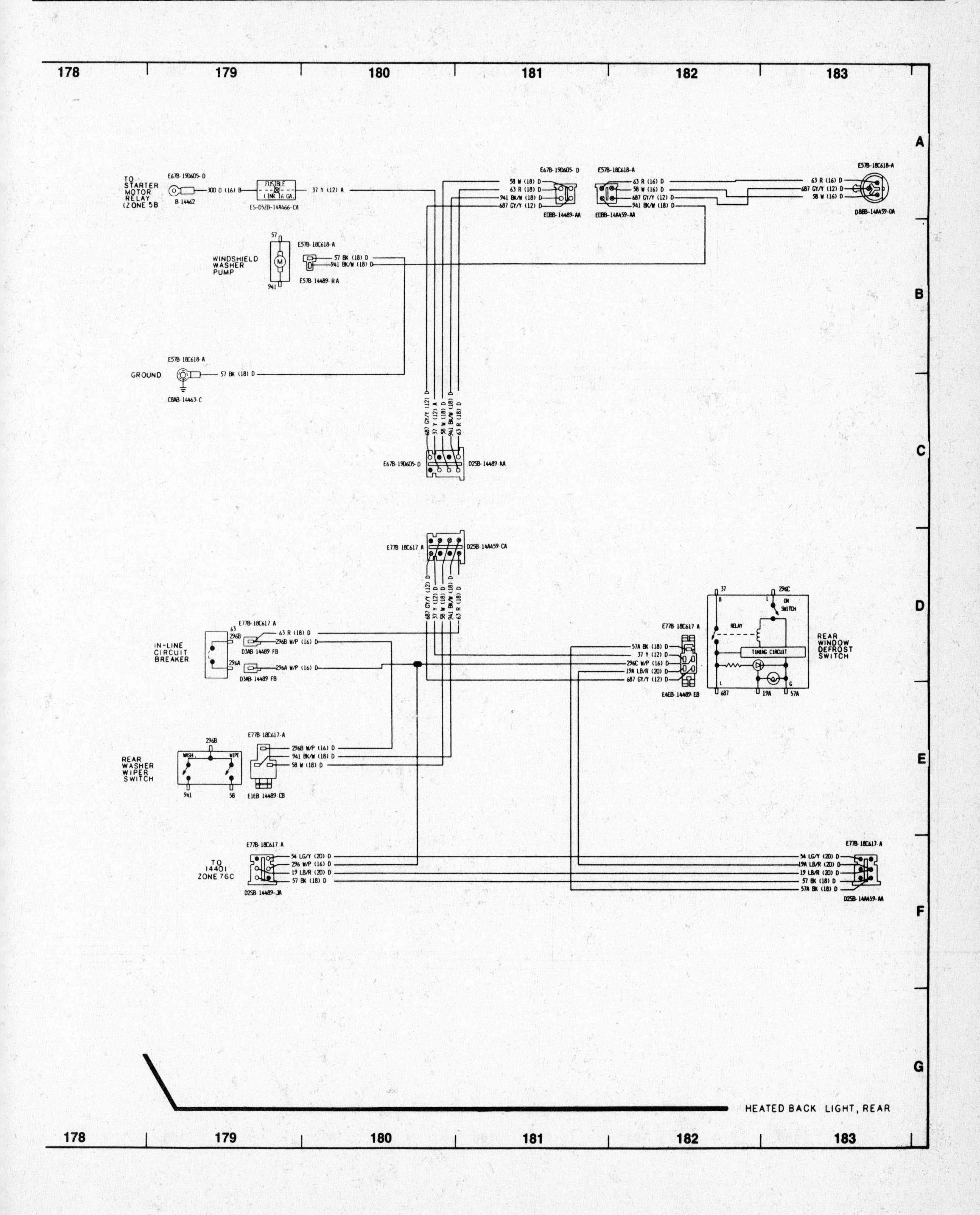

178
179
180
181
182
183
A
B
C
D
E
F
G
TO STARTER MOTOR RELAY (ZONE 5B
WINDSHIELD WASHER PUMP
GROUND
IN-LINE CIRCUIT BREAKER
REAR WASHER WIPER SWITCH
TO 14401 ZONE 76C
REAR WINDOW DEFROST SWITCH
FUSIBLE LINK 16 GA.
RELAY
TIMING CIRCUIT
ON SWITCH
HEATED BACK LIGHT, REAR

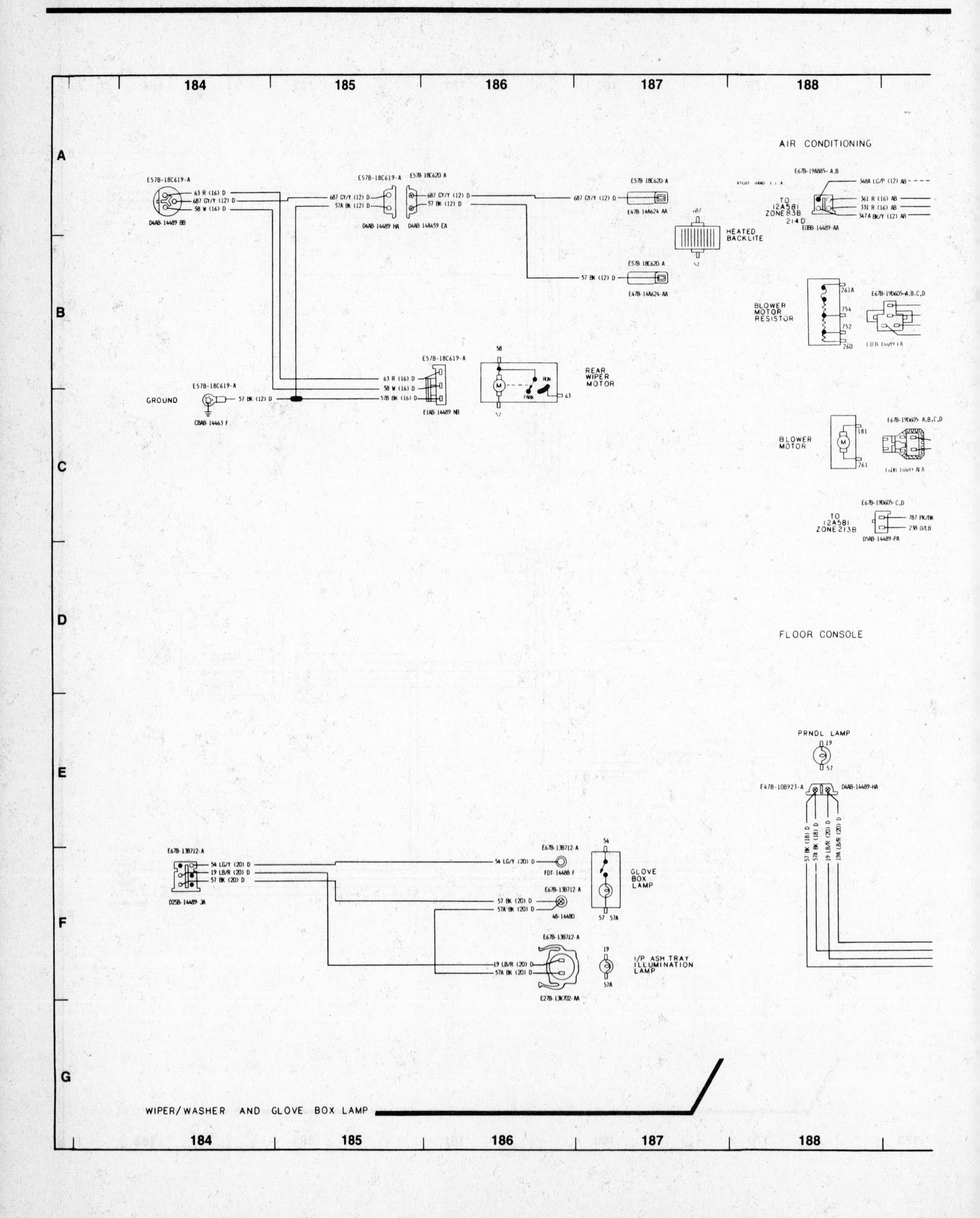
184
185
186
187
188
AIR CONDITIONING
HEATED BACKLITE
BLOWER MOTOR RESISTOR
BLOWER MOTOR
REAR WIPER MOTOR
GROUND
FLOOR CONSOLE
PRNDL LAMP
GLOVE BOX LAMP
I/P ASH TRAY ILLUMINATION LAMP
WIPER/WASHER AND GLOVE BOX LAMP

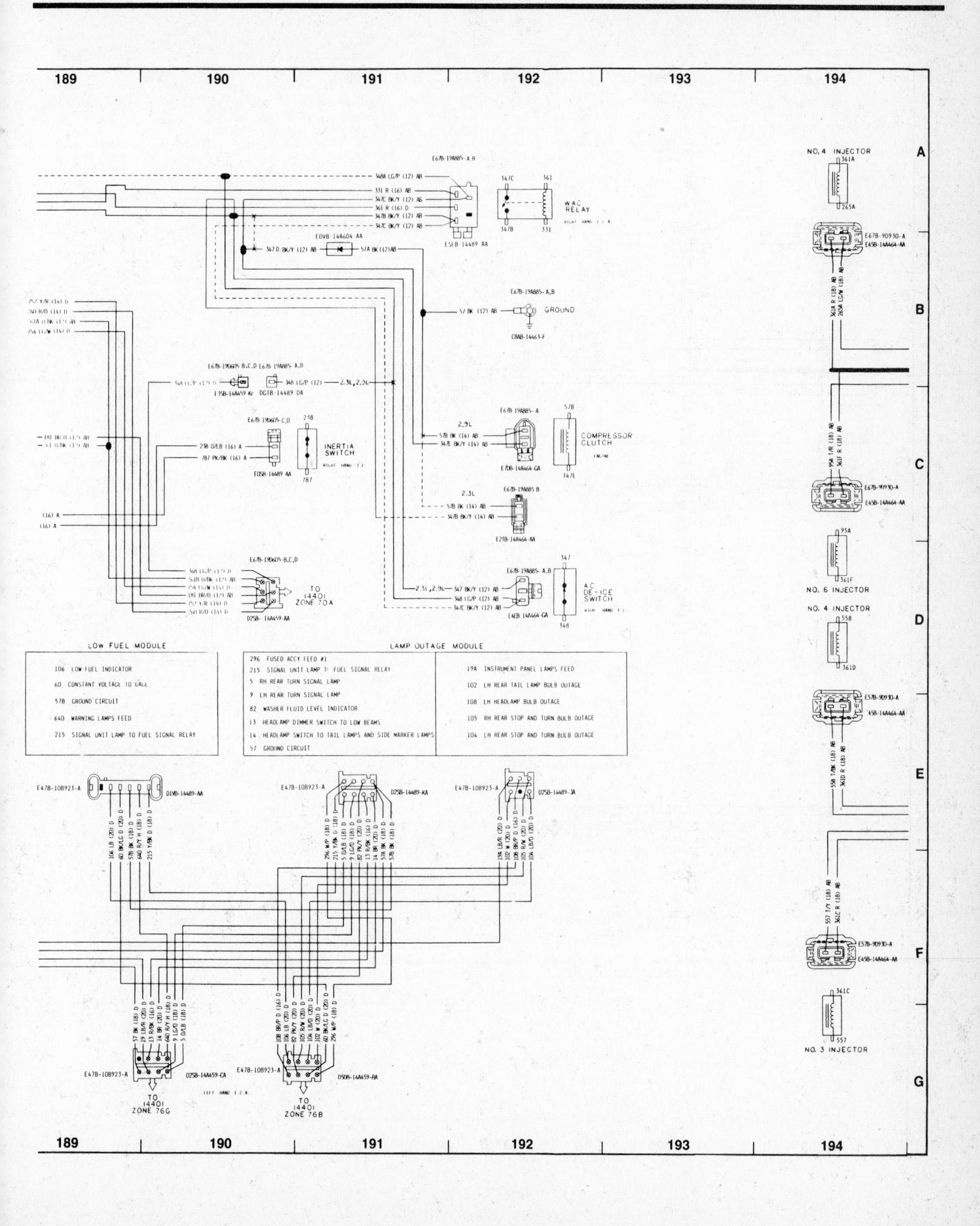
189
190
191
192
193
194
A
B
C
D
E
F
G
NO. 4 INJECTOR
E67B-9D930-A
E45B-14A464-AA
WAC RELAY
GROUND
C8AB-14463-F
COMPRESSOR CLUTCH
INERTIA SWITCH
E0SB-14489 AA
AC DE-ICE SWITCH
TO 14401 ZONE 70A
D2SB-14A459-AA
NO. 6 INJECTOR
NO. 4 INJECTOR
E57B-9D930-A
NO. 3 INJECTOR
LOW FUEL MODULE
106 LOW FUEL INDICATOR
60 CONSTANT VOLTAGE TO GAGE
57B GROUND CIRCUIT
640 WARNING LAMPS FEED
215 SIGNAL UNIT LAMP TO FUEL SIGNAL RELAY
LAMP OUTAGE MODULE
296 FUSED ACCY FEED #1
215 SIGNAL UNIT LAMP TO FUEL SIGNAL RELAY
5 RH REAR TURN SIGNAL LAMP
9 LH REAR TURN SIGNAL LAMP
82 WASHER FLUID LEVEL INDICATOR
13 HEADLAMP DIMMER SWITCH TO LOW BEAMS
14 HEADLAMP SWITCH TO TAIL LAMPS AND SIDE MARKER LAMPS
57 GROUND CIRCUIT
19A INSTRUMENT PANEL LAMPS FEED
102 LH REAR TAIL LAMP BULB OUTAGE
108 LH HEADLAMP BULB OUTAGE
105 RH REAR STOP AND TURN BULB OUTAGE
104 LH REAR STOP AND TURN BULB OUTAGE
E47B-10B923-A
D1VB-14489-AA
D2SB-14489-KA
D2SB-14489-JA
D2SB-14A459-CA
D50B-14A459-BA
TO 14401 ZONE 76G
TO 14401 ZONE 76B

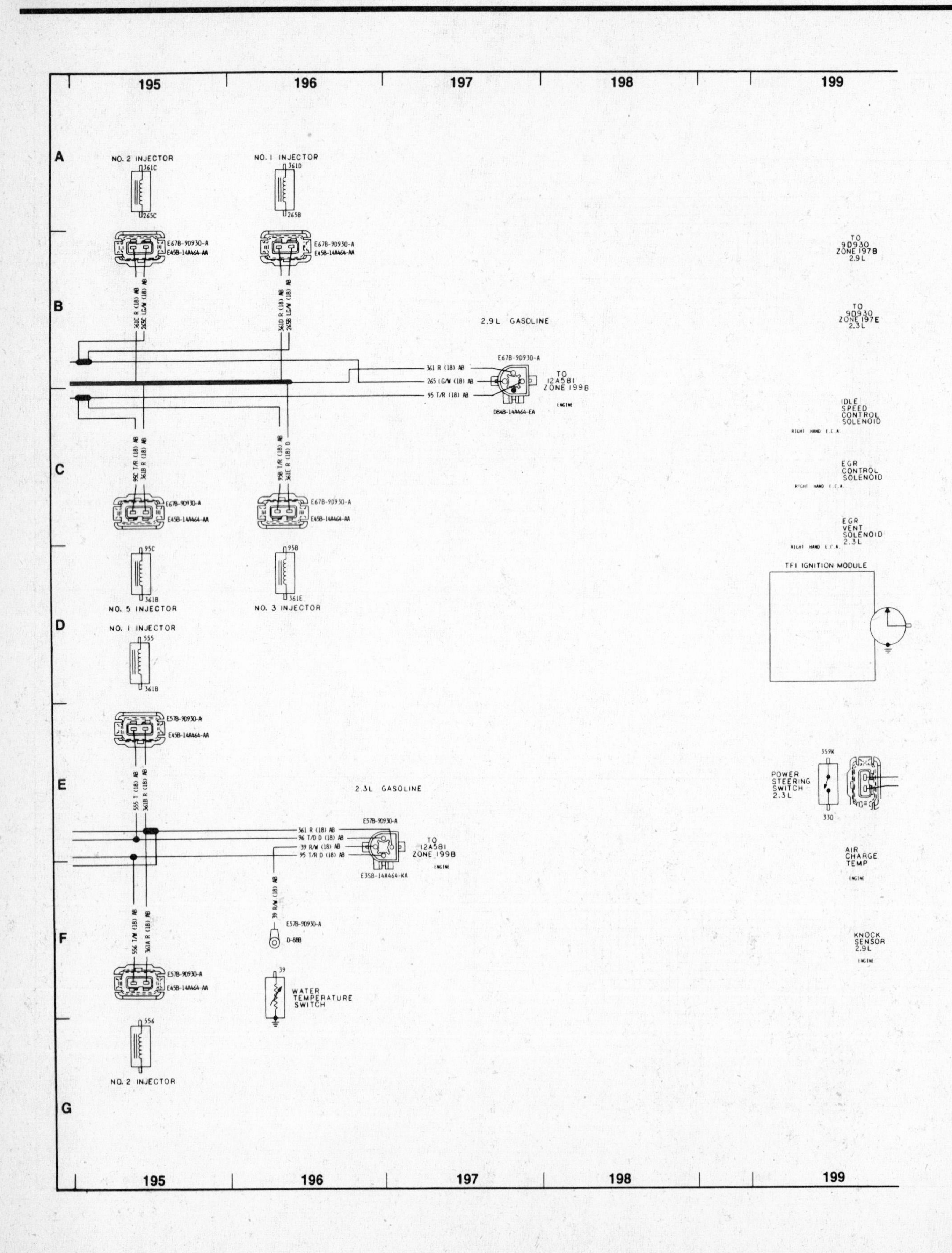
195
196
197
198
199
A
B
C
D
E
F
G
NO. 2 INJECTOR
NO. 1 INJECTOR
2.9L GASOLINE
TO 12A581 ZONE 199B
TO 9D930 ZONE 197B 2.9L
TO 9D930 ZONE 197E 2.3L
IDLE SPEED CONTROL SOLENOID
EGR CONTROL SOLENOID
EGR VENT SOLENOID 2.3L
NO. 5 INJECTOR
NO. 3 INJECTOR
TFI IGNITION MODULE
POWER STEERING SWITCH 2.3L
2.3L GASOLINE
AIR CHARGE TEMP
KNOCK SENSOR 2.9L
WATER TEMPERATURE SWITCH

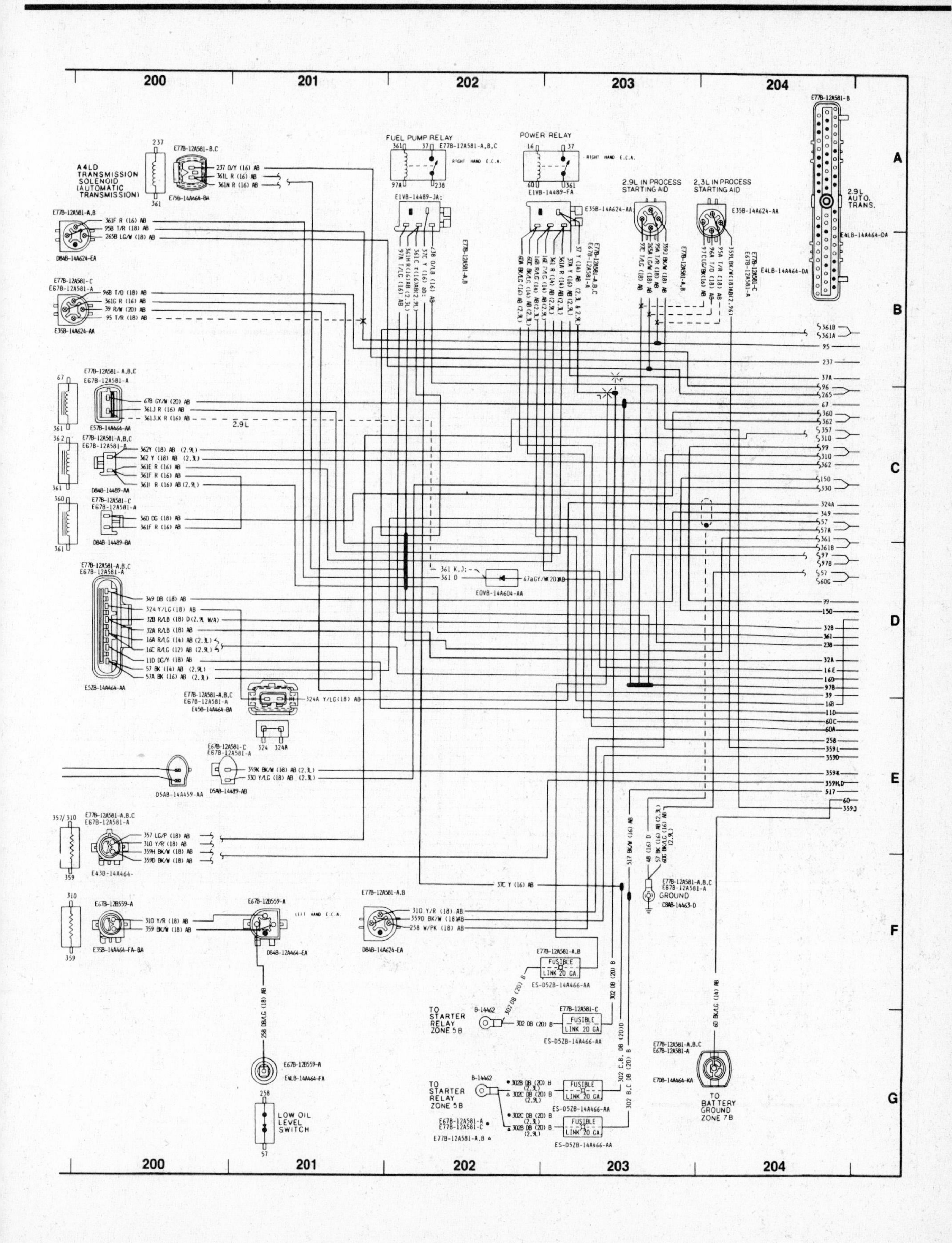
200
201
202
203
204
A4LD TRANSMISSION SOLENOID (AUTOMATIC TRANSMISSION)
FUEL PUMP RELAY
POWER RELAY
2.9L IN PROCESS STARTING AID
2.3L IN PROCESS STARTING AID
2.9L AUTO. TRANS.
A
B
C
D
E
F
G
TO STARTER RELAY ZONE 5B
FUSIBLE LINK 20 GA
GROUND
LOW OIL LEVEL SWITCH
TO BATTERY GROUND ZONE 7B

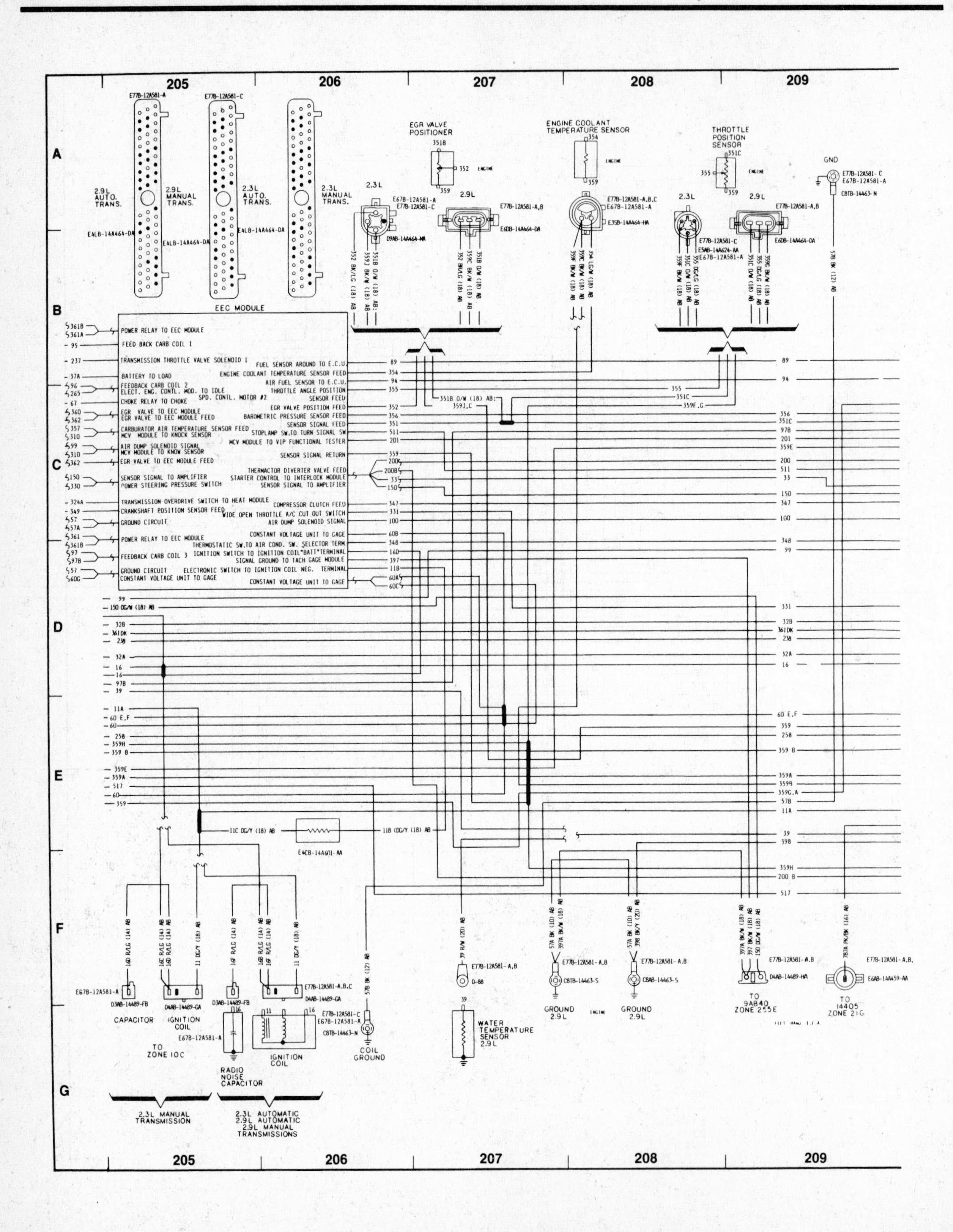
EEC MODULE
EGR VALVE POSITIONER
ENGINE COOLANT TEMPERATURE SENSOR
THROTTLE POSITION SENSOR
GND
2.9L AUTO. TRANS.
2.9L MANUAL TRANS.
2.3L AUTO. TRANS.
2.3L MANUAL TRANS.
POWER RELAY TO EEC MODULE
FEED BACK CARB COIL 1
TRANSMISSION THROTTLE VALVE SOLENOID 1
BATTERY TO LOAD
CHOKE RELAY TO CHOKE
FUEL SENSOR AROUND TO E.C.U.
ENGINE COOLANT TEMPERATURE SENSOR FEED
AIR FUEL SENSOR TO E.C.U.
THROTTLE ANGLE POSITION
SENSOR FEED
EGR VALVE POSITION FEED
BAROMETRIC PRESSURE SENSOR FEED
SENSOR SIGNAL FEED
STOPLAMP SW.TO TURN SIGNAL SW
MCV MODULE TO VIP FUNCTIONAL TESTER
SENSOR SIGNAL RETURN
THERMACTOR DIVERTER VALVE FEED
STARTER CONTROL TO INTERLOCK MODULE
SENSOR SIGNAL TO AMPLIFIER
POWER STEERING PRESSURE SWITCH
TRANSMISSION OVERDRIVE SWITCH TO HEAT MODULE
COMPRESSOR CLUTCH FEED
CRANKSHAFT POSITION SENSOR FEED
WIDE OPEN THROTTLE A/C CUT OUT SWITCH
AIR DUMP SOLENOID SIGNAL
GROUND CIRCUIT
CONSTANT VOLTAGE UNIT TO GAGE
THERMOSTATIC SW.TO AIR COND. SW. SELECTOR TERM
FEEDBACK CARB COIL 3
IGNITION SWITCH TO IGNITION COIL"BATT"TERMINAL
SIGNAL GROUND TO TACH GAGE MODULE
ELECTRONIC SWITCH TO IGNITION COIL NEG. TERMINAL
CAPACITOR
IGNITION COIL
TO ZONE 10C
RADIO NOISE CAPACITOR
COIL GROUND
WATER TEMPERATURE SENSOR 2.9L
GROUND 2.9L
TO 9A840 ZONE 255E
TO 14405 ZONE 21G
2.3L MANUAL TRANSMISSION
2.3L AUTOMATIC
2.9L AUTOMATIC
2.9L MANUAL
TRANSMISSIONS

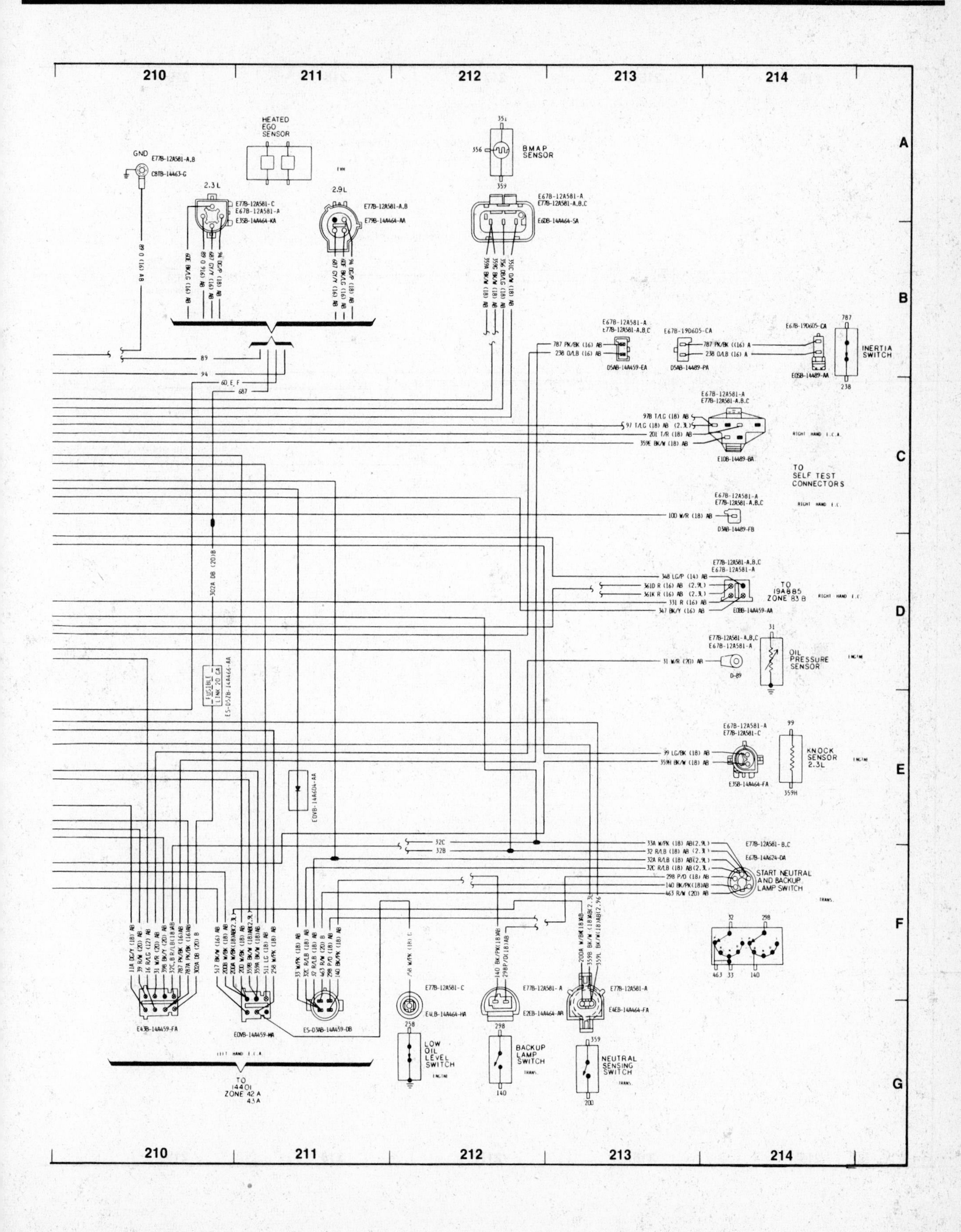

210
211
212
213
214
HEATED EGO SENSOR
BMAP SENSOR
GND
2.3 L
2.9 L
INERTIA SWITCH
TO SELF TEST CONNECTORS
TO 19A885 ZONE 83 B
OIL PRESSURE SENSOR
KNOCK SENSOR 2.3L
START NEUTRAL AND BACKUP LAMP SWITCH
FUSIBLE LINK 20 GA
LOW OIL LEVEL SWITCH
BACKUP LAMP SWITCH
NEUTRAL SENSING SWITCH
TO 14401 ZONE 42 A 43 A
A
B
C
D
E
F
G

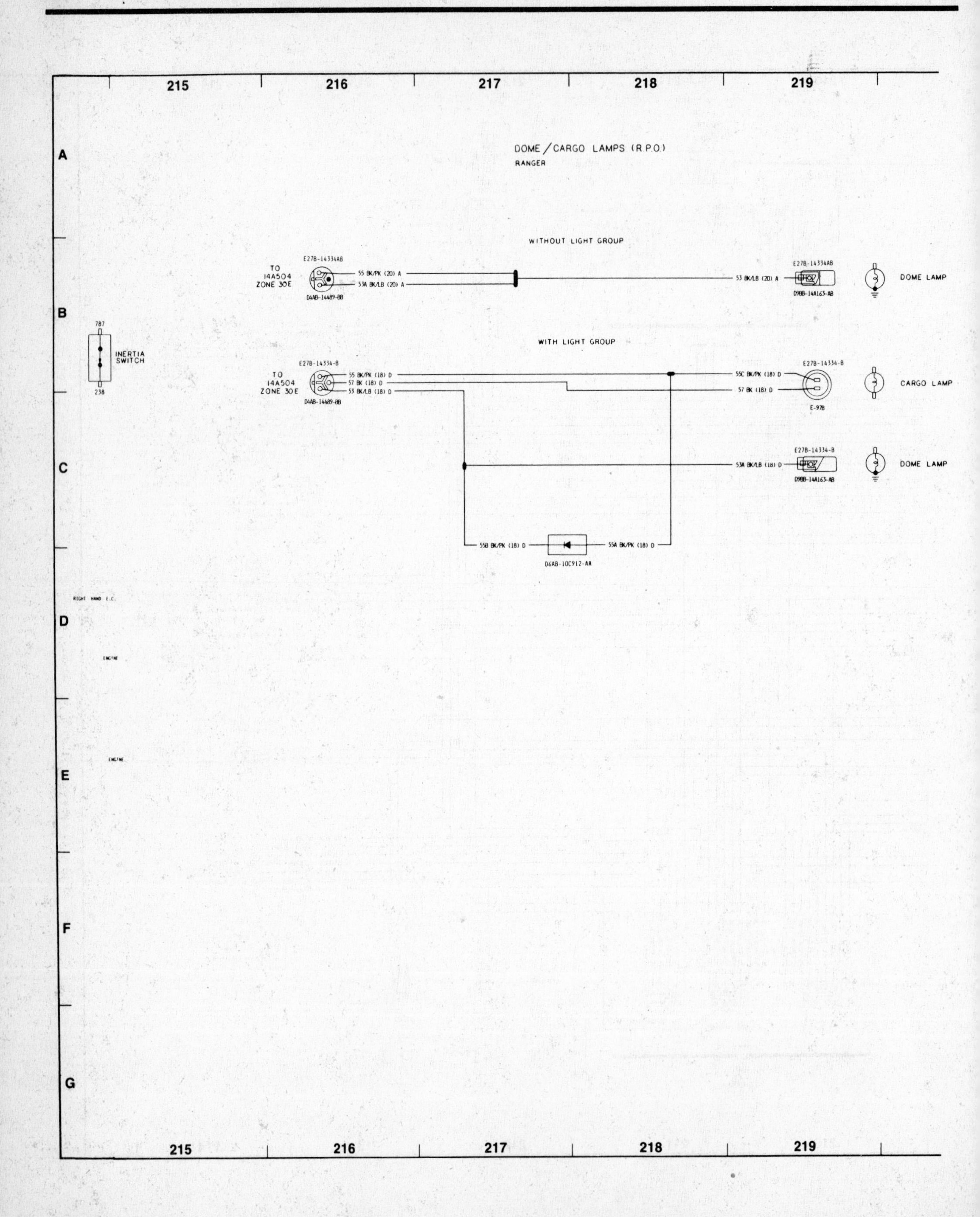
215
216
217
218
219
A
B
C
D
E
F
G
DOME / CARGO LAMPS (R.P.O.)
RANGER
WITHOUT LIGHT GROUP
TO
14A504
ZONE 30E
E27B-14334AB
D4AB-14489-BB
55 BK/PK (20) A
53A BK/LB (20) A
53 BK/LB (20) A
E27B-14334AB
D9BB-14A163-AB
DOME LAMP
787
INERTIA
SWITCH
238
WITH LIGHT GROUP
TO
14A504
ZONE 30E
E27B-14334-B
D4AB-14489-BB
55 BK/PK (18) D
57 BK (18) D
53 BK/LB (18) D
55C BK/PK (18) D
57 BK (18) D
E27B-14334-B
E-97B
CARGO LAMP
53A BK/LB (18) D
E27B-14334-B
D9BB-14A163-AB
DOME LAMP
55B BK/PK (18) D
55A BK/PK (18) D
D6AB-10C912-AA

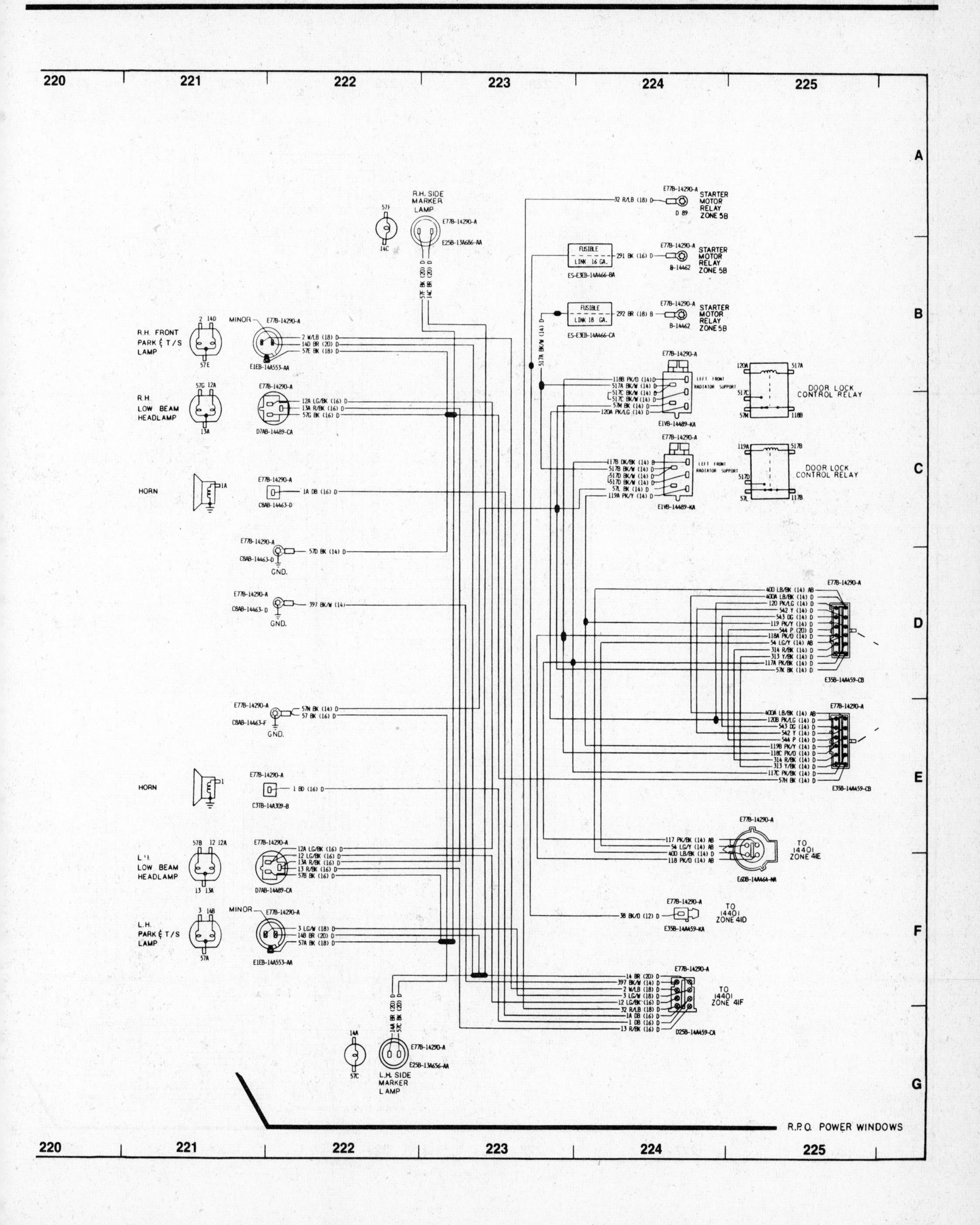

220
221
222
223
224
225
A
B
C
D
E
F
G
R.H. SIDE MARKER LAMP
STARTER MOTOR RELAY ZONE 5B
FUSIBLE LINK 16 GA.
FUSIBLE LINK 18 GA.
R.H. FRONT PARK & T/S LAMP
MINOR
R.H. LOW BEAM HEADLAMP
DOOR LOCK CONTROL RELAY
HORN
GND.
TO 14401 ZONE 41E
TO 14401 ZONE 41D
TO 14401 ZONE 41F
L.H. LOW BEAM HEADLAMP
L.H. PARK & T/S LAMP
L.H. SIDE MARKER LAMP
R.P.O. POWER WINDOWS

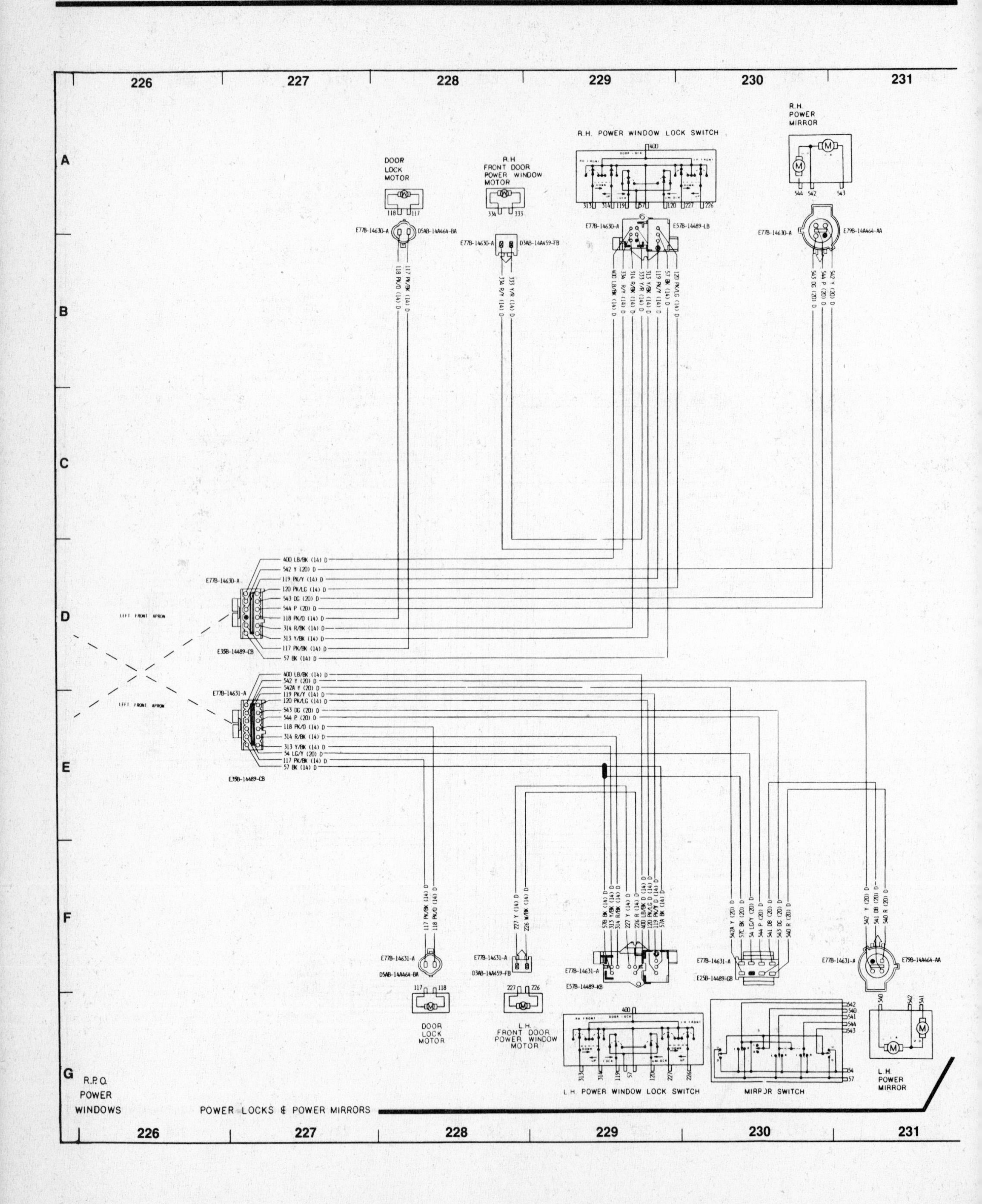

R.H. POWER WINDOW LOCK SWITCH
R.H. POWER MIRROR
DOOR LOCK MOTOR
R.H. FRONT DOOR POWER WINDOW MOTOR
LEFT FRONT APRON
L.H. FRONT DOOR POWER WINDOW MOTOR
L.H. POWER WINDOW LOCK SWITCH
MIRROR SWITCH
L.H. POWER MIRROR
R.P.O. POWER WINDOWS
POWER LOCKS & POWER MIRRORS

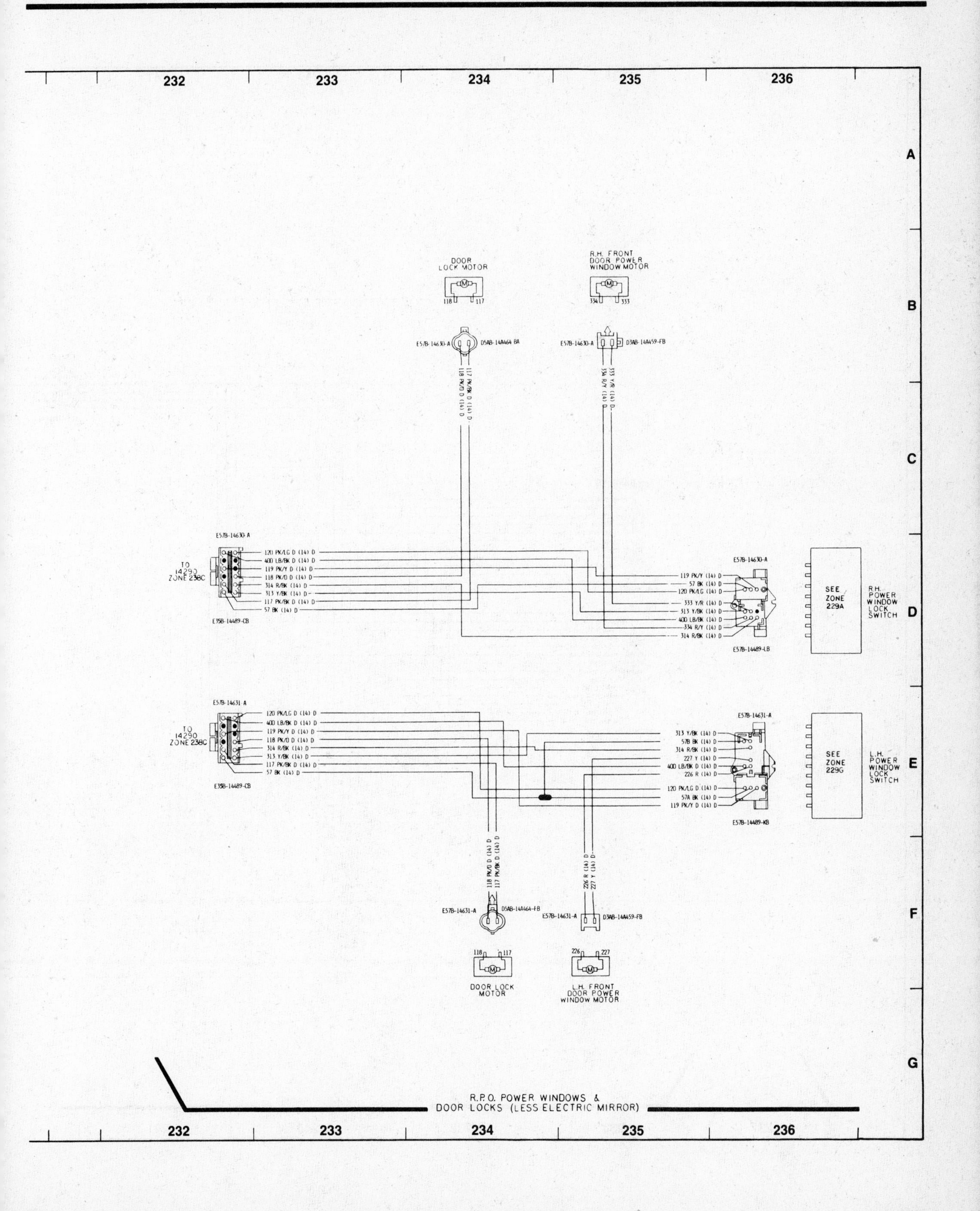

232
233
234
235
236
A
B
C
D
E
F
G
DOOR LOCK MOTOR
R.H. FRONT DOOR POWER WINDOW MOTOR
E57B-14630-A
D5AB-14A464 BA
D3AB-14A459-FB
TO 14290 ZONE 238C
120 PK/LG D (14) D
400 LB/BK D (14) D
119 PK/Y D (14) D
118 PK/O D (14) D
314 R/BK (14) D
313 Y/BK (14) D
117 PK/BK D (14) D
57 BK (14) D
E35B-14489-CB
119 PK/Y (14) D
57 BK (14) D
120 PK/LG (14) D
333 Y/R (14) D
313 Y/BK (14) D
400 LB/BK (14) D
334 R/Y (14) D
314 R/BK (14) D
E57B-14489-LB
SEE ZONE 229A
R.H. POWER WINDOW LOCK SWITCH
E57B-14631-A
313 Y/BK (14) D
57B BK (14) D
314 R/BK (14) D
227 Y (14) D
400 LB/BK D (14) D
226 R (14) D
120 PK/LG D (14) D
57A BK (14) D
119 PK/Y D (14) D
E57B-14489-KB
SEE ZONE 229G
L.H. POWER WINDOW LOCK SWITCH
D5AB-14A464-FB
D3AB-14A459-FB
DOOR LOCK MOTOR
L.H. FRONT DOOR POWER WINDOW MOTOR
R.P.O. POWER WINDOWS & DOOR LOCKS (LESS ELECTRIC MIRROR)

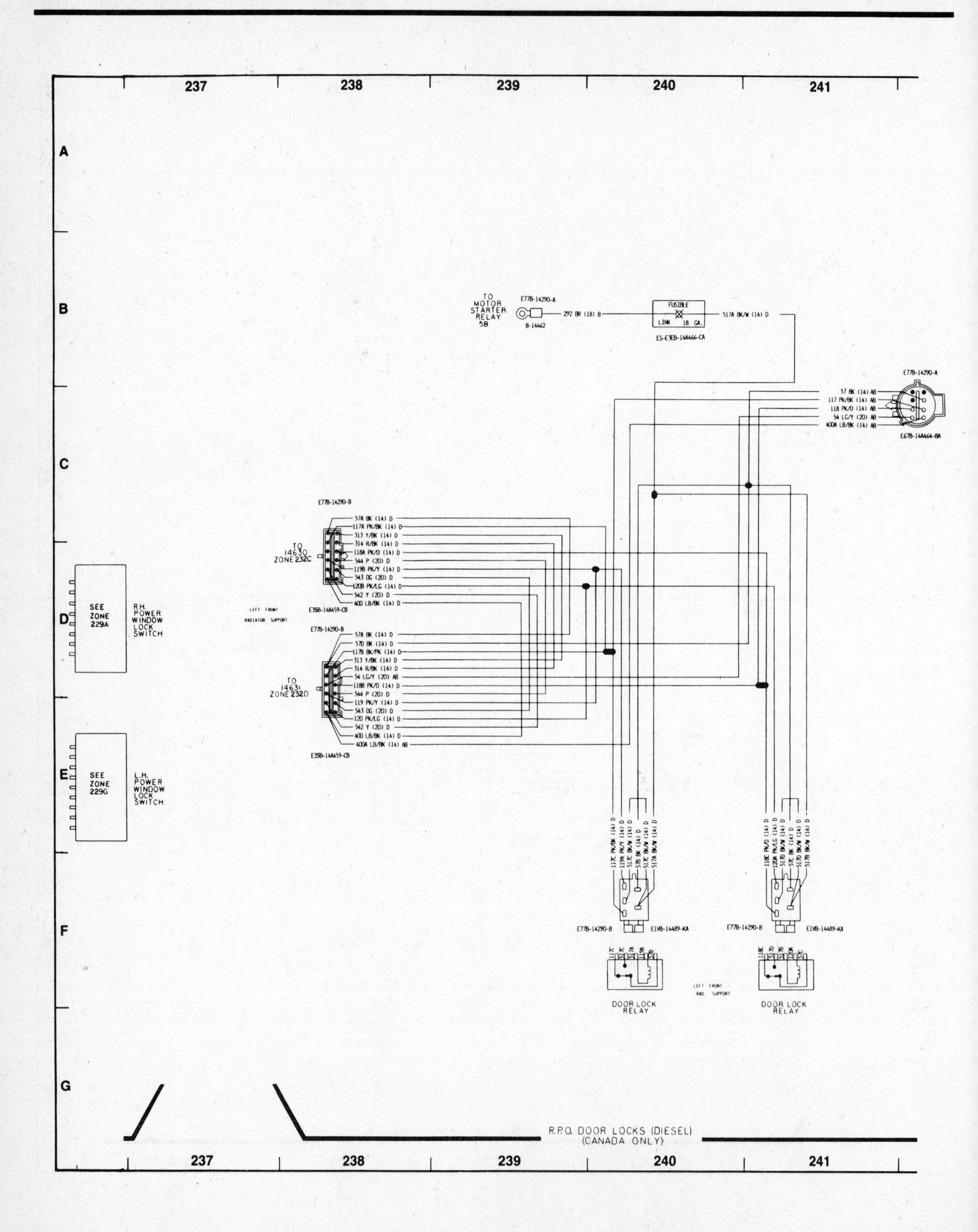

237
238
239
240
241
TO MOTOR STARTER RELAY 5B
E77B-14290-A
B-14462
292 BR (18) B
FUSIBLE LINK 18 GA.
ES-E3EB-14A466-CA
517A BK/W (14) D
E77B-14290-A
57 BK (14) AB
117 PK/BK (14) AB
118 PK/O (14) AB
54 LG/Y (20) AB
400A LB/BK (14) AB
E67B-14A464-BA
E77B-14290-B
TO 14630 ZONE 232C
57A BK (14) D
117A PK/BK (14) D
313 Y/BK (14) D
314 R/BK (14) D
118A PK/O (14) D
544 P (20) D
119B PK/Y (14) D
543 DG (20) D
120B PK/LG (14) D
542 Y (20) D
400 LB/BK (14) D
E35B-14A459-CB
SEE ZONE 229A
R.H. POWER WINDOW LOCK SWITCH
E77B-14290-B
TO 14631 ZONE 232D
57A BK (14) D
57D BK (14) D
117B BK/PK (14) D
313 Y/BK (14) D
314 R/BK (14) D
54 LG/Y (20) AB
118B PK/O (14) D
544 P (20) D
119 PK/Y (14) D
543 DG (20) D
120 PK/LG (14) D
542 Y (20) D
400 LB/BK (14) D
400A LB/BK (14) AB
E35B-14A459-CB
SEE ZONE 229G
L.H. POWER WINDOW LOCK SWITCH
E77B-14290-B
E1VB-14489-KA
E77B-14290-B
E1VB-14489-KA
DOOR LOCK RELAY
DOOR LOCK RELAY
R.P.O. DOOR LOCKS (DIESEL) (CANADA ONLY)

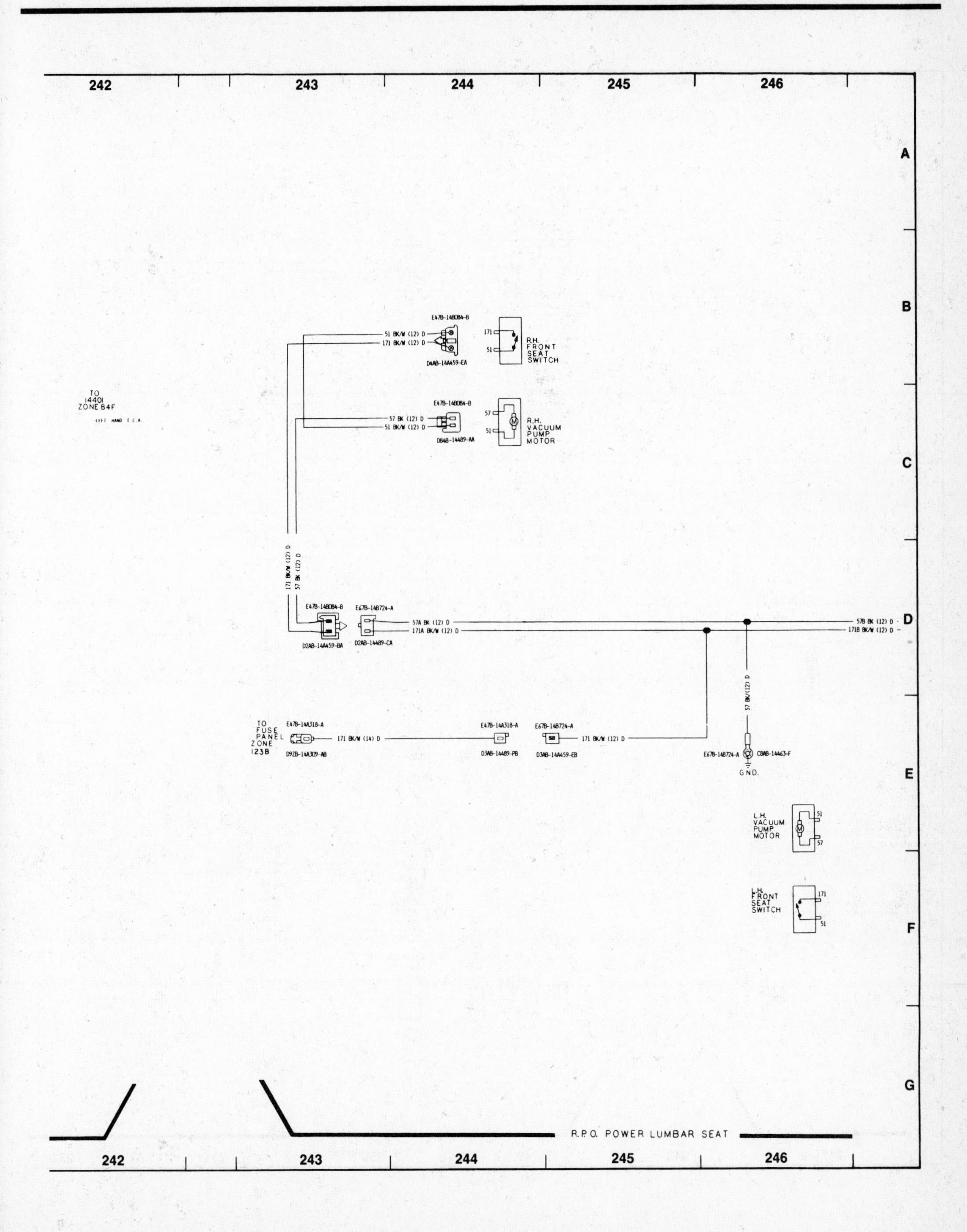
242
243
244
245
246
A
B
C
D
E
F
G
TO
14401
ZONE 84F
E47B-14B084-B
51 BK/W (12) D
171 BK/W (12) D
D4AB-14A459-EA
171
51
R.H.
FRONT
SEAT
SWITCH
E47B-14B084-B
57 BK (12) D
51 BK/W (12) D
D8AB-14489-AA
57
51
R.H.
VACUUM
PUMP
MOTOR
171 BK/W (12) D
57 BK (12) D
E47B-14B084-B
D2AB-14A459-BA
E67B-14B724-A
D2AB-14489-CA
57A BK (12) D
171A BK/W (12) D
57B BK (12) D
171B BK/W (12) D
57 BK/(12) D
TO
FUSE
PANEL
ZONE
123B
E47B-14A318-A
D9ZB-14A309-AB
171 BK/W (14) D
E47B-14A318-A
D3AB-14489-PB
E67B-14B724-A
D3AB-14A459-EB
171 BK/W (12) D
E67B-14B724-A
C8AB-14463-F
GND.
L.H.
VACUUM
PUMP
MOTOR
51
57
L.H.
FRONT
SEAT
SWITCH
171
51
R.P.O. POWER LUMBAR SEAT

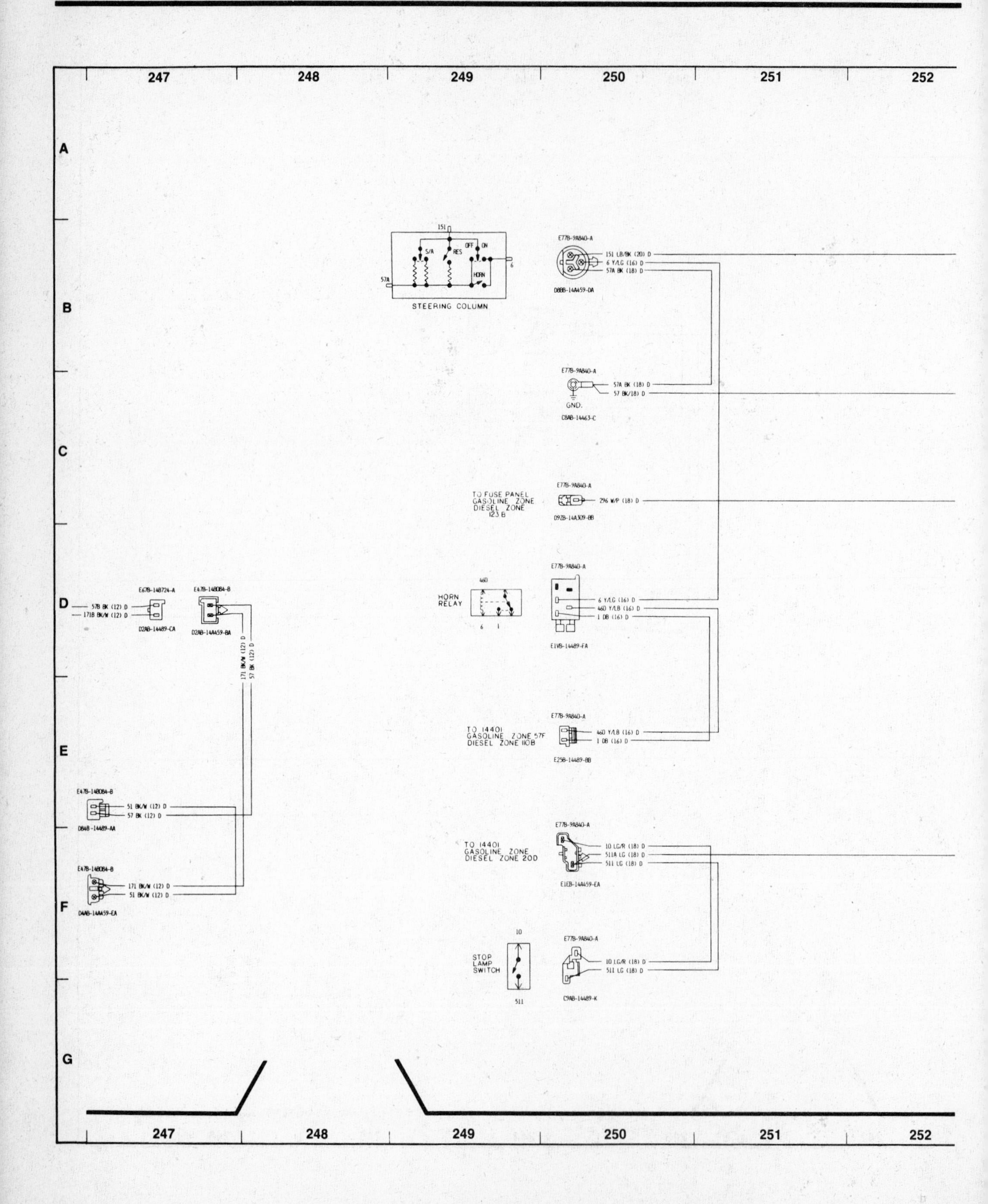
247
248
249
250
251
252
A
B
C
D
E
F
G
151
S/A
RES
OFF
ON
6
HORN
57A
STEERING COLUMN
E77B-9A840-A
151 LB/BK (20) D
6 Y/LG (16) D
57A BK (18) D
D8BB-14A459-DA
E77B-9A840-A
57A BK (18) D
57 BK/18) D
GND.
C8AB-14463-C
TO FUSE PANEL
GASOLINE ZONE
DIESEL ZONE
123 B
E77B-9A840-A
296 W/P (18) D
D9ZB-14A309-BB
460
HORN
RELAY
6
1
E77B-9A840-A
6 Y/LG (16) D
460 Y/LB (16) D
1 DB (16) D
E1VB-14489-FA
TO 14401
GASOLINE ZONE 57F
DIESEL ZONE 110B
E77B-9A840-A
460 Y/LB (16) D
1 DB (16) D
E25B-14489-BB
TO 14401
GASOLINE ZONE
DIESEL ZONE 20D
E77B-9A840-A
10 LG/R (18) D
511A LG (18) D
511 LG (18) D
E1EB-14A459-EA
10
STOP
LAMP
SWITCH
511
E77B-9A840-A
10 LG/R (18) D
511 LG (18) D
C9AB-14489-K
57B BK (12) D
171B BK/W (12) D
E67B-14B724-A
D2AB-14489-CA
E47B-14B084-B
D2AB-14A459-BA
171 BK/W (12) D
57 BK (12) D
E47B-14B084-B
51 BK/W (12) D
57 BK (12) D
D84B-14489-AA
E47B-14B084-B
171 BK/W (12) D
51 BK/W (12) D
D4AB-14A459-EA

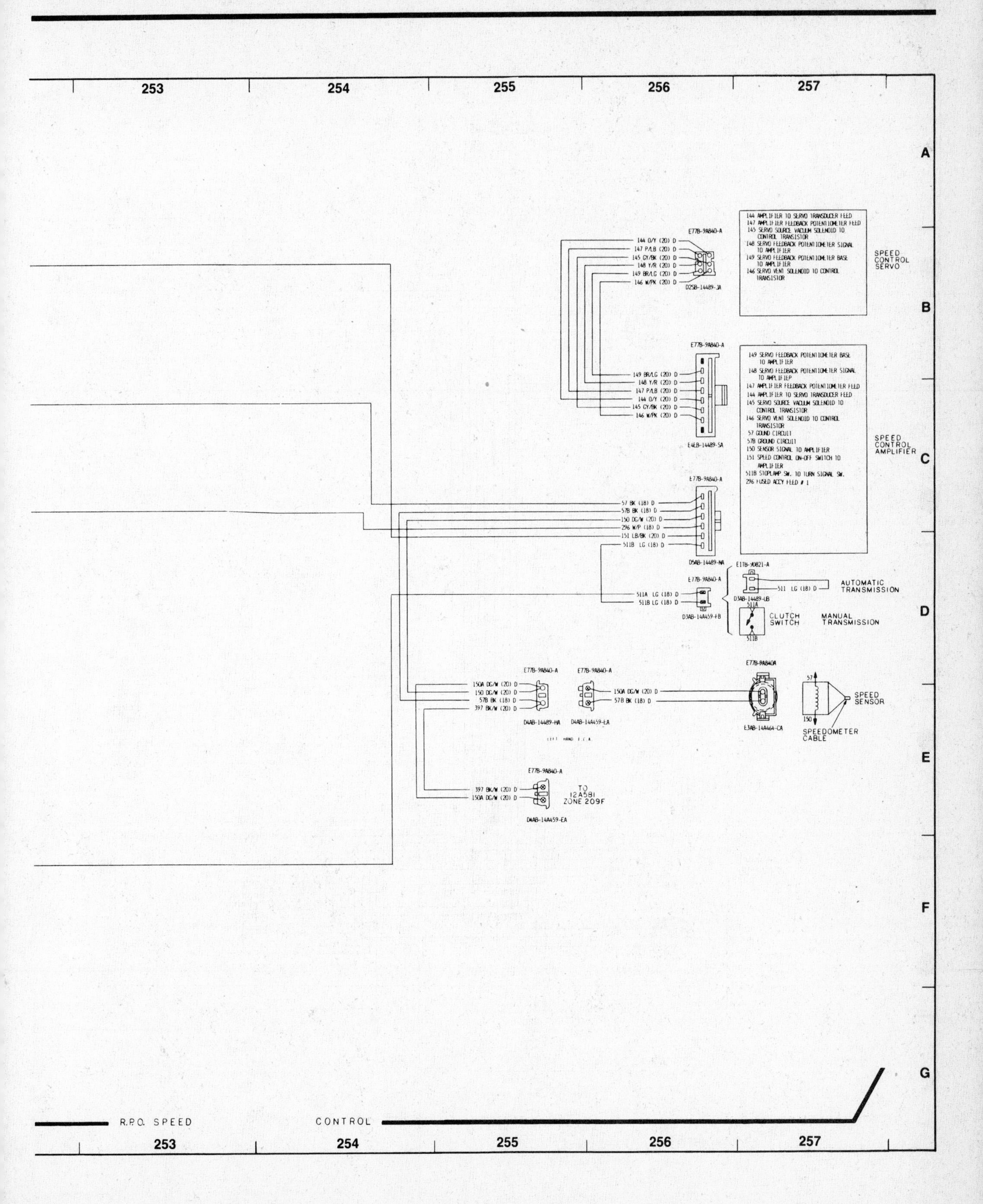
253
254
255
256
257
A
B
C
D
E
F
G
E77B-9A840-A
144 O/Y (20) D
147 P/LB (20) D
145 GY/BK (20) D
148 Y/R (20) D
149 BR/LG (20) D
146 W/PK (20) D
D2SB-14489-JA
144 AMPLIFIER TO SERVO TRANSDUCER FEED
147 AMPLIFIER FEEDBACK POTENTIOMETER FEED
145 SERVO SOURCE VACUUM SOLENOID TO CONTROL TRANSISTOR
148 SERVO FEEDBACK POTENTIOMETER SIGNAL TO AMPLIFIER
149 SERVO FEEDBACK POTENTIOMETER BASE TO AMPLIFIER
146 SERVO VENT SOLENOID TO CONTROL TRANSISTOR
SPEED CONTROL SERVO
E77B-9A840-A
149 BR/LG (20) D
148 Y/R (20) D
147 P/LB (20) D
144 O/Y (20) D
145 GY/BK (20) D
146 W/PK (20) D
E4EB-14489-SA
149 SERVO FEEDBACK POTENTIOMETER BASE TO AMPLIFIER
148 SERVO FEEDBACK POTENTIOMETER SIGNAL TO AMPLIFIEP
147 AMPLIFIER FEEDBACK POTENTIOMETER FEED
144 AMPLIFIER TO SERVO TRANSDUCER FEED
145 SERVO SOURCE VACUUM SOLENOID TO CONTROL TRANSISTOR
146 SERVO VENT SOLENOID TO CONTROL TRANSISTOR
57 GOUND CIRCUIT
57B GROUND CIRCUIT
150 SENSOR SIGNAL TO AMPLIFIER
151 SPEED CONTROL ON-OFF SWITCH TO AMPLIFIER
511B STOPLAMP SW. TO TURN SIGNAL SW.
296 FUSED ACCY FEED # 1
SPEED CONTROL AMPLIFIER
E77B-9A840-A
57 BK (18) D
57B BK (18) D
150 DG/W (20) D
296 W/P (18) D
151 LB/BK (20) D
511B LG (18) D
D5AB-14489-NA
E1TB-9D821-A
E77B-9A840-A
511 LG (18) D
AUTOMATIC TRANSMISSION
511A LG (18) D
511B LG (18) D
D3AB-14A459-FB
D3AB-14489-LB
511A
CLUTCH SWITCH
MANUAL TRANSMISSION
511B
E77B-9A840-A
E77B-9A840-A
150A DG/W (20) D
150 DG/W (20) D
57B BK (18) D
397 BK/W (20) D
150A DG/W (20) D
57B BK (18) D
D4AB-14489-HA
D4AB-14A459-EA
LEFT HAND E.C.A.
E77B-9A840A
E3AB-14A464-CA
57
150
SPEED SENSOR
SPEEDOMETER CABLE
E77B-9A840-A
397 BK/W (20) D
150A DG/W (20) D
TO 12A581 ZONE 209F
D4AB-14A459-EA
R.P.O. SPEED CONTROL

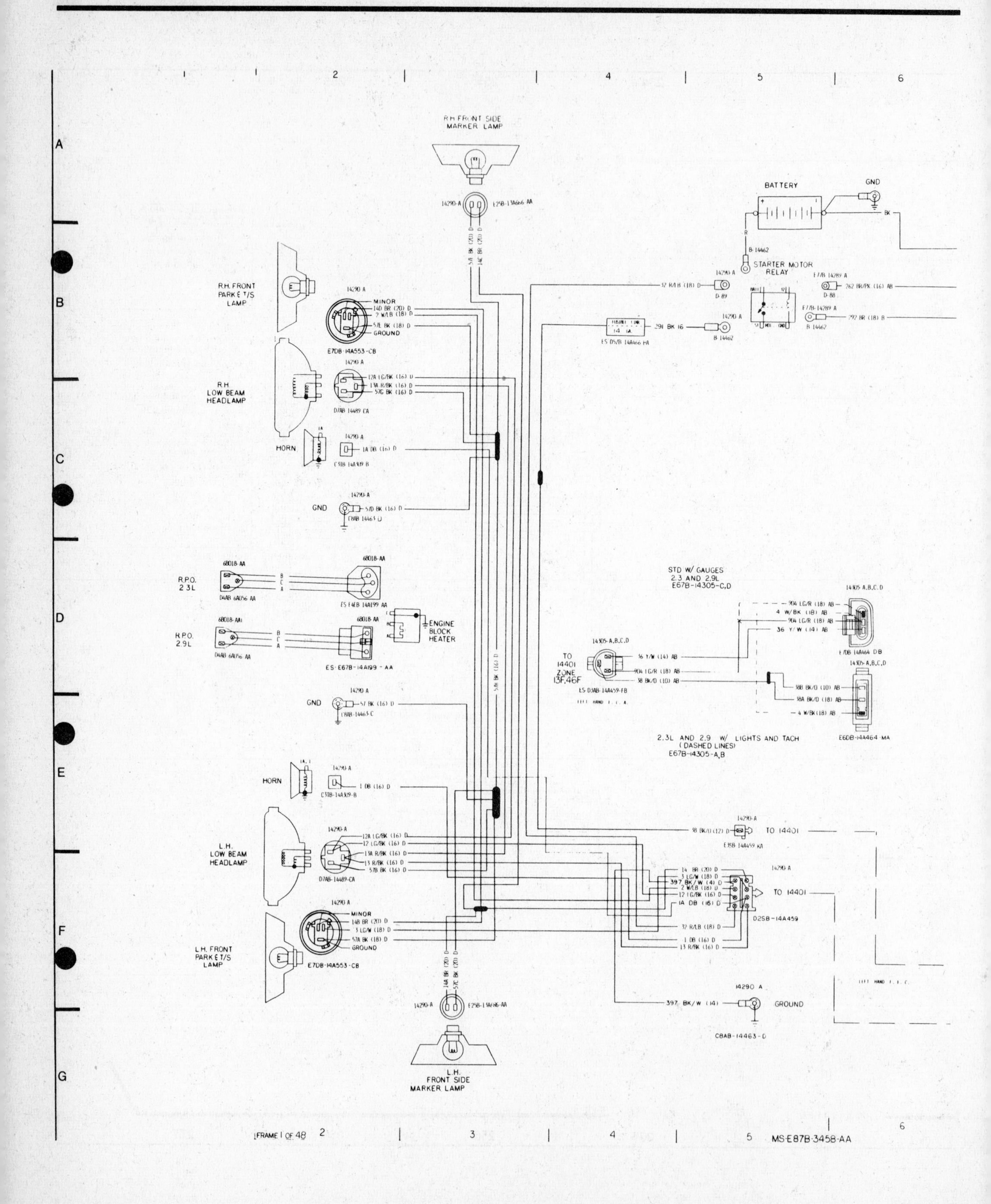
R.H. FRONT SIDE MARKER LAMP
BATTERY
GND
STARTER MOTOR RELAY
R.H. FRONT PARK & T/S LAMP
R.H. LOW BEAM HEADLAMP
HORN
GND
R.P.O. 2.3L
R.P.O. 2.9L
ENGINE BLOCK HEATER
GND
STD W/ GAUGES 2.3 AND 2.9L E67B-14305-C,D
TO 14401 ZONE 13F,46F
2.3L AND 2.9 W/ LIGHTS AND TACH (DASHED LINES) E67B-14305-A,B
HORN
L.H. LOW BEAM HEADLAMP
TO 14401
TO 14401
L.H. FRONT PARK & T/S LAMP
GROUND
L.H. FRONT SIDE MARKER LAMP
FRAME 1 OF 48
MS-E87B-3458-AA

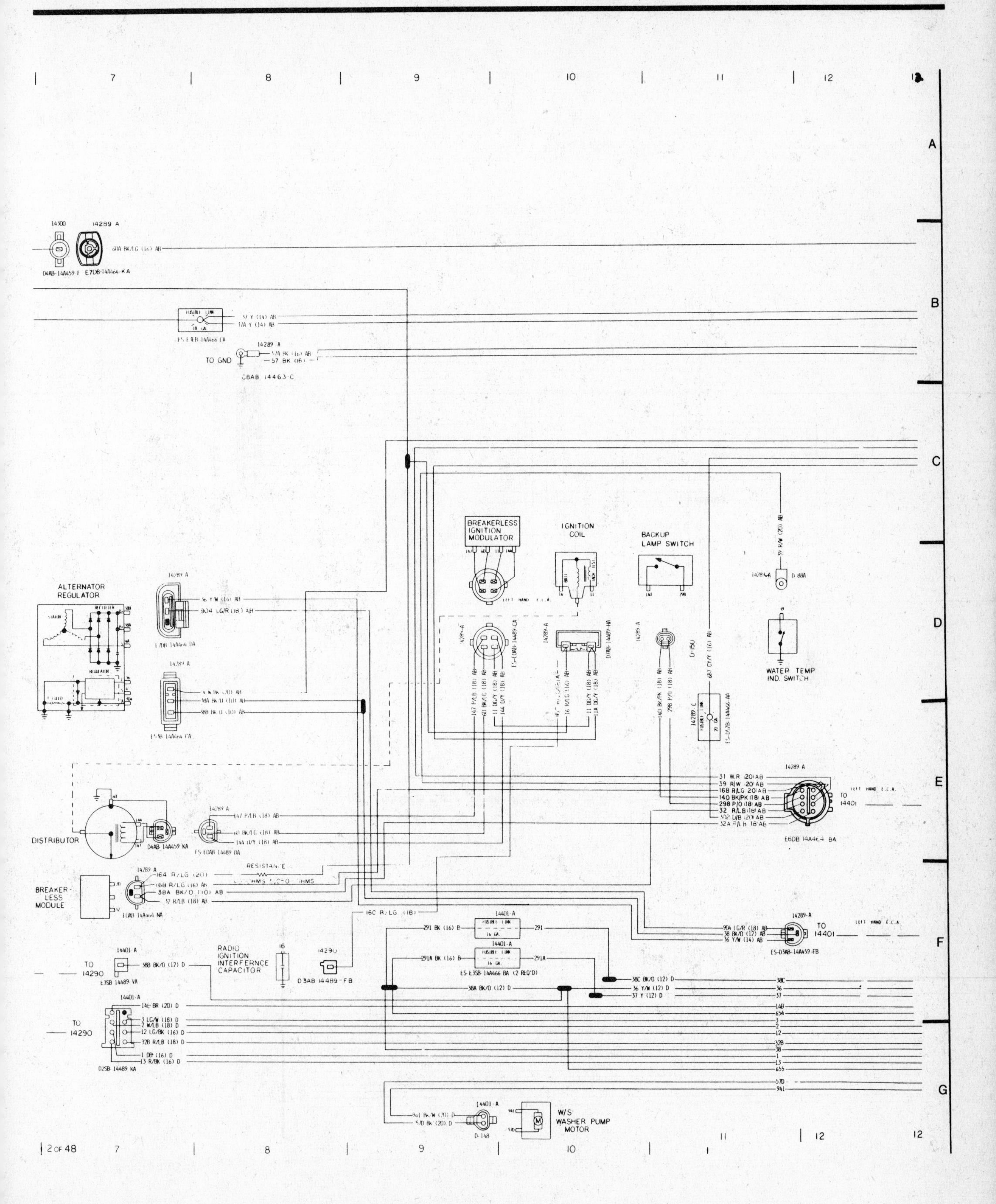
BREAKERLESS IGNITION MODULATOR
IGNITION COIL
BACKUP LAMP SWITCH
ALTERNATOR REGULATOR
DISTRIBUTOR
BREAKER-LESS MODULE
WATER TEMP IND. SWITCH
RADIO IGNITION INTERFERNCE CAPACITOR
W/S WASHER PUMP MOTOR
TO GND
TO 14290
TO 14401
2 OF 48

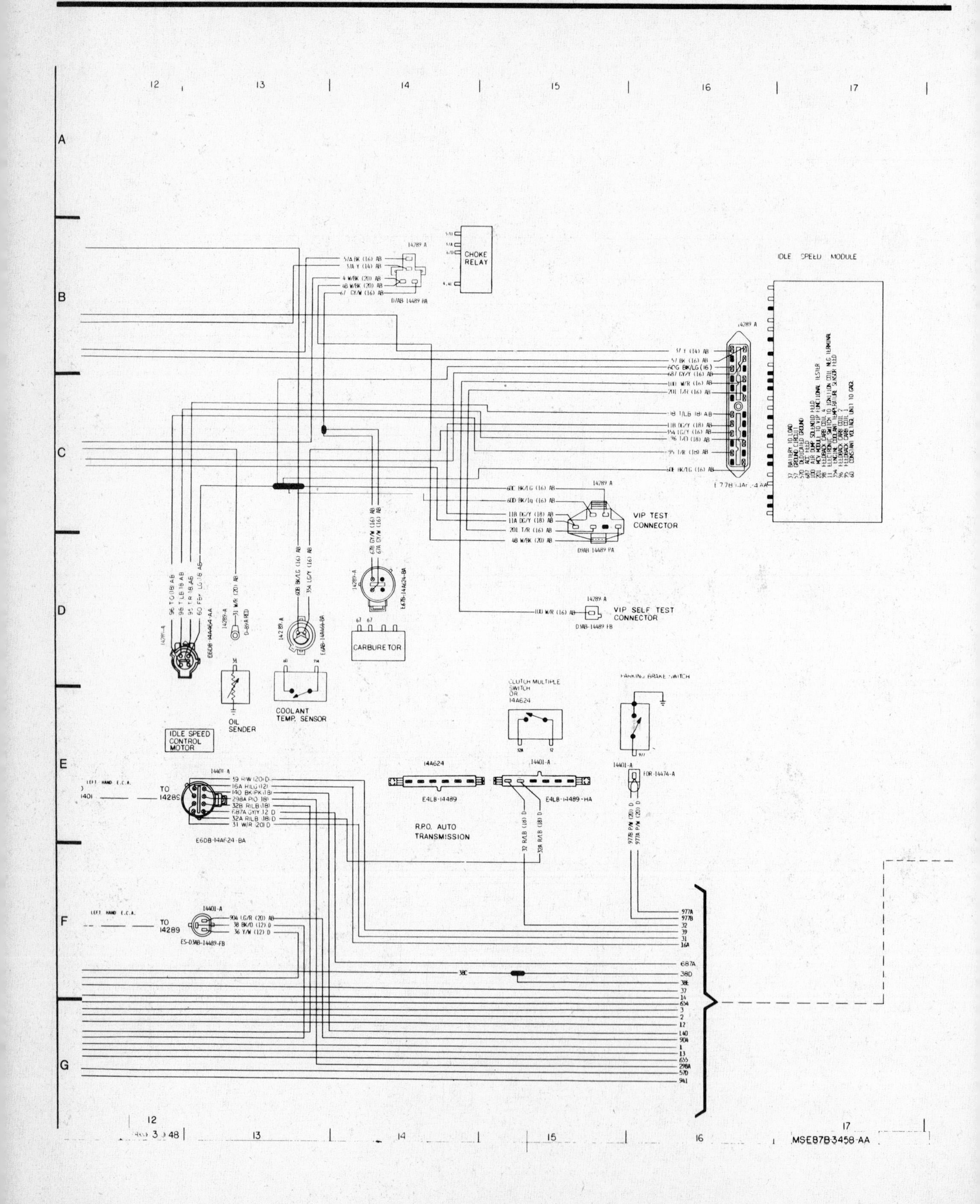

CHOKE RELAY
IDLE SPEED MODULE
VIP TEST CONNECTOR
VIP SELF TEST CONNECTOR
CARBURETOR
OIL SENDER
COOLANT TEMP. SENSOR
IDLE SPEED CONTROL MOTOR
CLUTCH MULTIPLE SWITCH OR 14A624
PARKING BRAKE SWITCH
R.P.O. AUTO TRANSMISSION
LEFT HAND E.C.A.
TO 14289
MSE87B-3458-AA

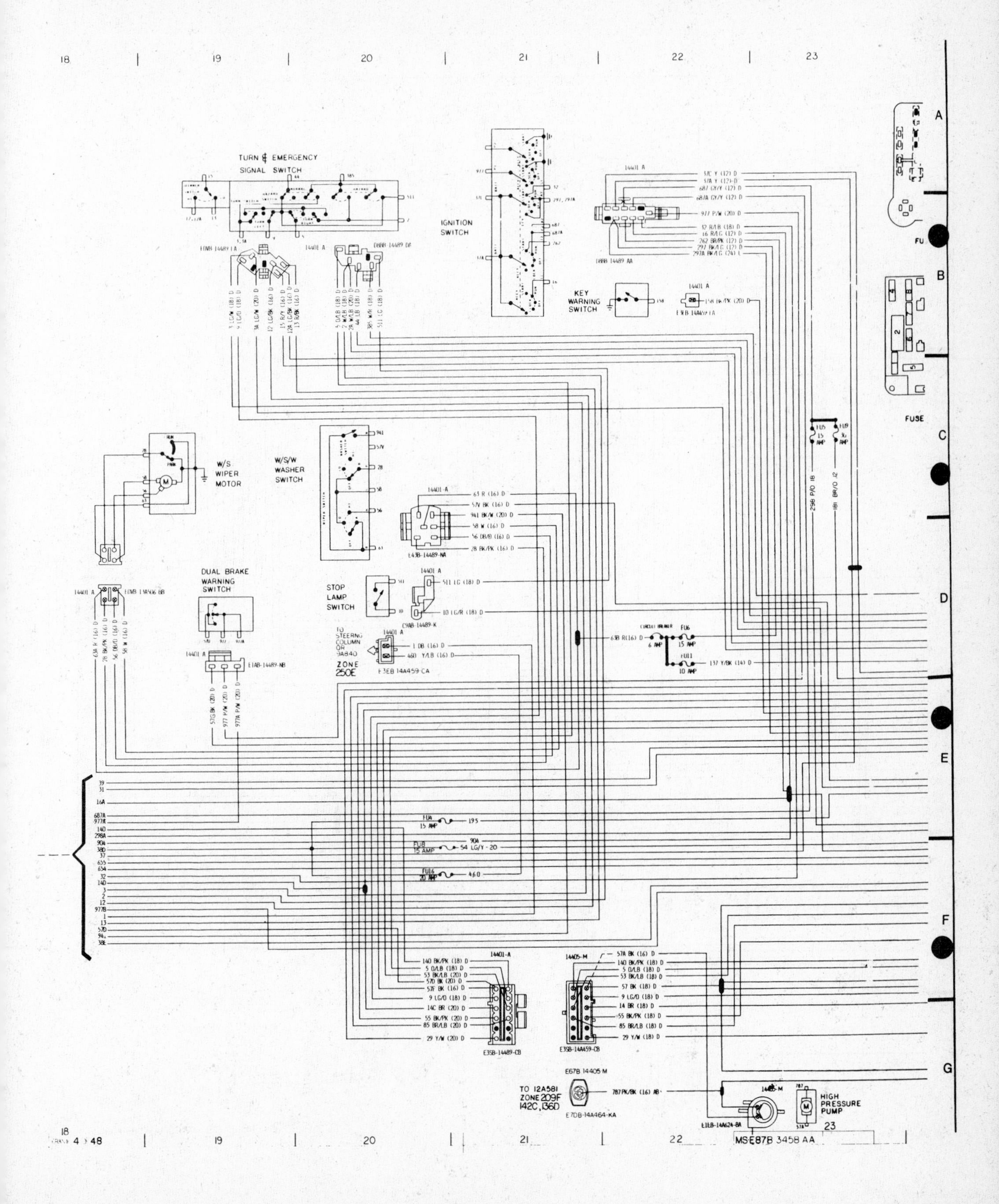

TURN & EMERGENCY SIGNAL SWITCH
IGNITION SWITCH
KEY WARNING SWITCH
W/S WIPER MOTOR
W/S/W WASHER SWITCH
DUAL BRAKE WARNING SWITCH
STOP LAMP SWITCH
TO STEERING COLUMN OR 9A840 ZONE 250E
TO 12A581 ZONE 209F 142C, 136D
HIGH PRESSURE PUMP
FUSE
MSE87B 3458 AA

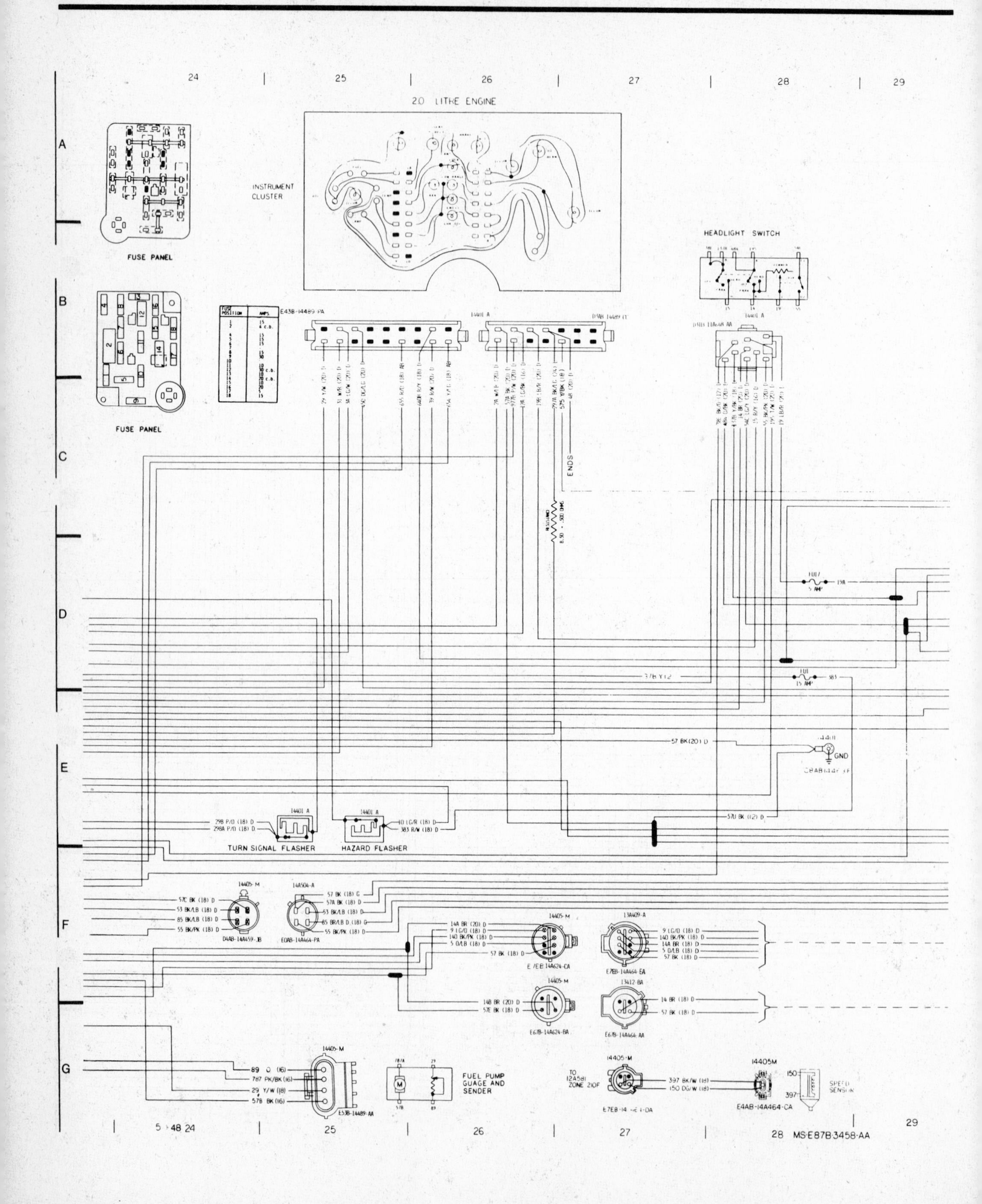

20 LITRE ENGINE
INSTRUMENT CLUSTER
FUSE PANEL
FUSE PANEL
HEADLIGHT SWITCH
ENDS
TURN SIGNAL FLASHER
HAZARD FLASHER
FUEL PUMP GUAGE AND SENDER
SPEED SENSOR
GND
MS-E87B-3458-AA

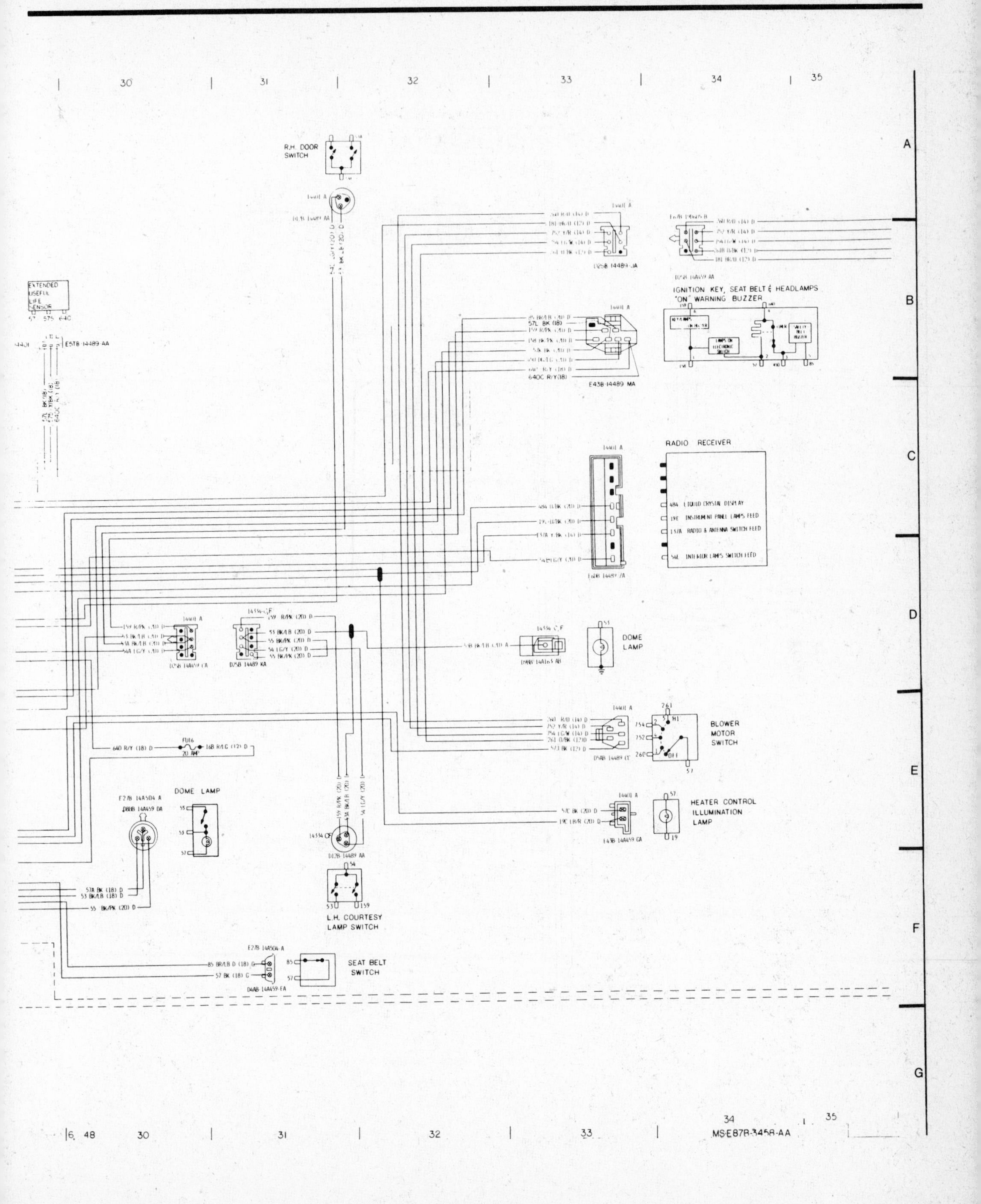
R.H. DOOR SWITCH
EXTENDED USEFUL LIFE SENSOR
IGNITION KEY, SEAT BELT & HEADLAMPS "ON" WARNING BUZZER
RADIO RECEIVER
DOME LAMP
BLOWER MOTOR SWITCH
HEATER CONTROL ILLUMINATION LAMP
L.H. COURTESY LAMP SWITCH
SEAT BELT SWITCH
MS-E87B-3458-AA

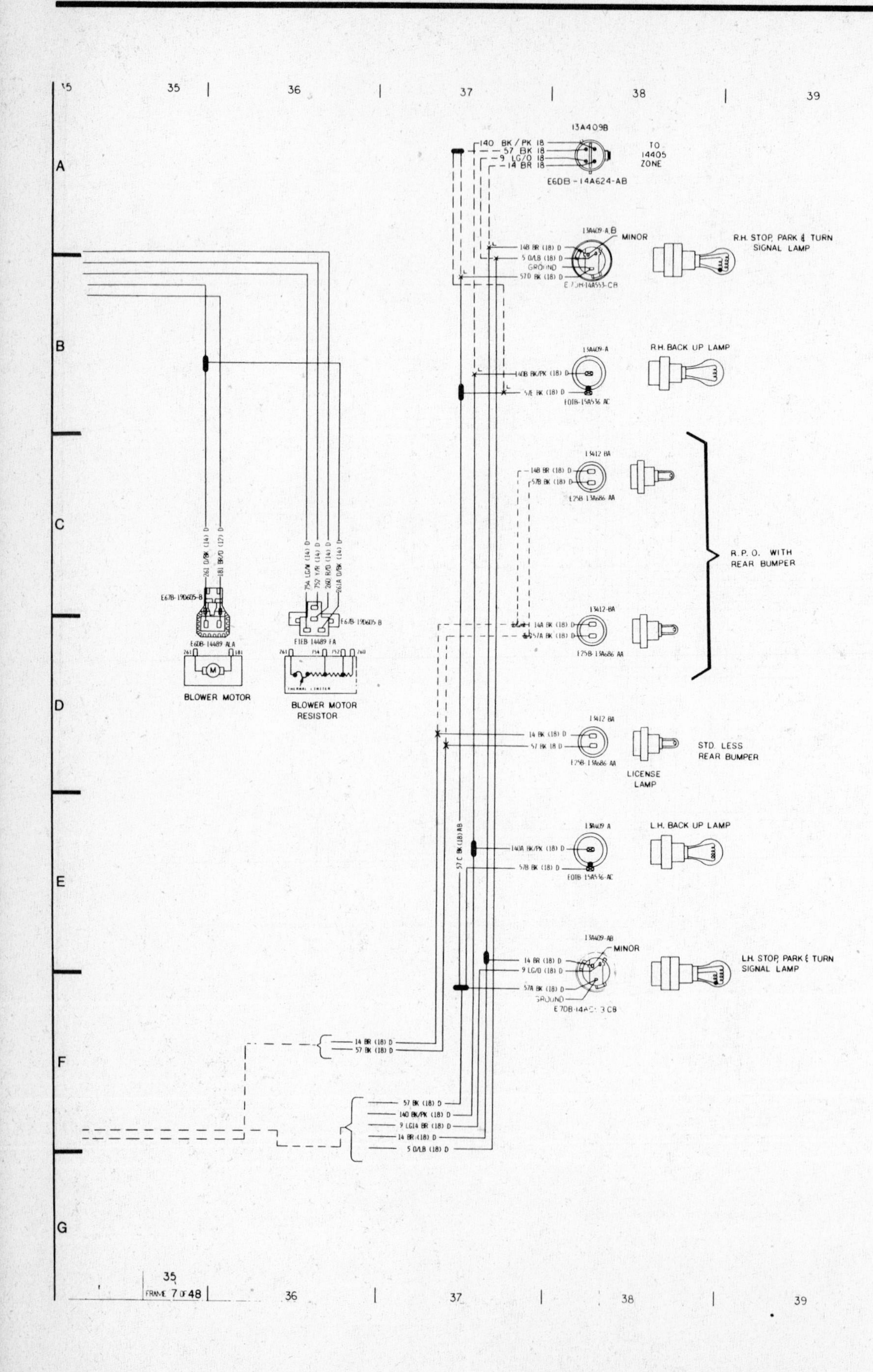
13A409B
TO 14405 ZONE
E6DB-14A624-AB
MINOR
R.H. STOP, PARK & TURN SIGNAL LAMP
R.H. BACK UP LAMP
R.P.O. WITH REAR BUMPER
STD. LESS REAR BUMPER
LICENSE LAMP
L.H. BACK UP LAMP
L.H. STOP, PARK & TURN SIGNAL LAMP
BLOWER MOTOR
BLOWER MOTOR RESISTOR
FRAME 7 OF 48

A

B

C

D

E

F

G

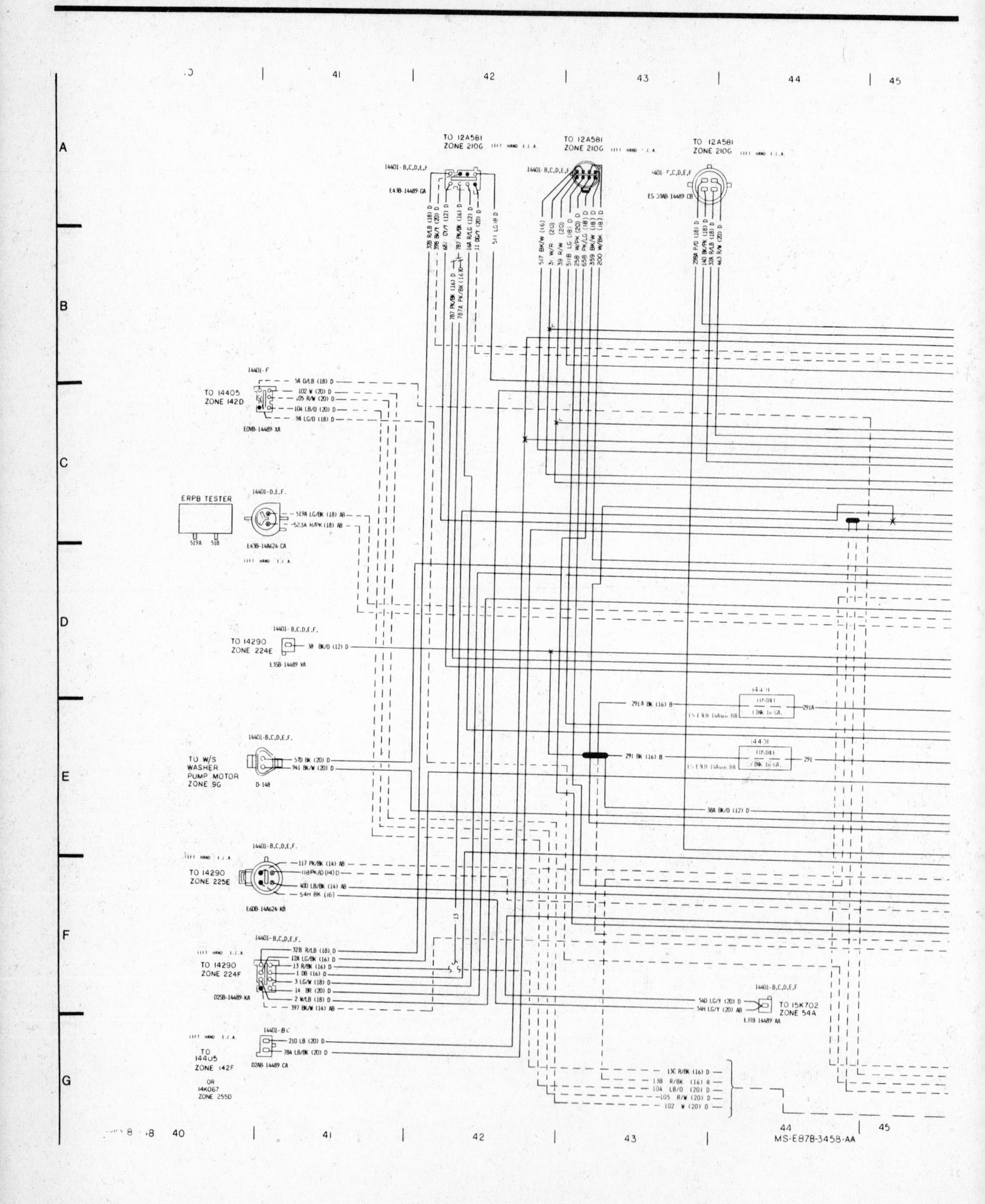
TO 12A581
ZONE 210G
TO 12A581
ZONE 210G
TO 12A581
ZONE 210G
TO 14405
ZONE 142D
ERPB TESTER
TO 14290
ZONE 224E
TO W/S
WASHER
PUMP MOTOR
ZONE 9G
TO 14290
ZONE 225E
TO 14290
ZONE 224F
TO
14405
ZONE 142F
OR
14K067
ZONE 255D
TO 15K702
ZONE 54A
MS-E87B-3458-AA

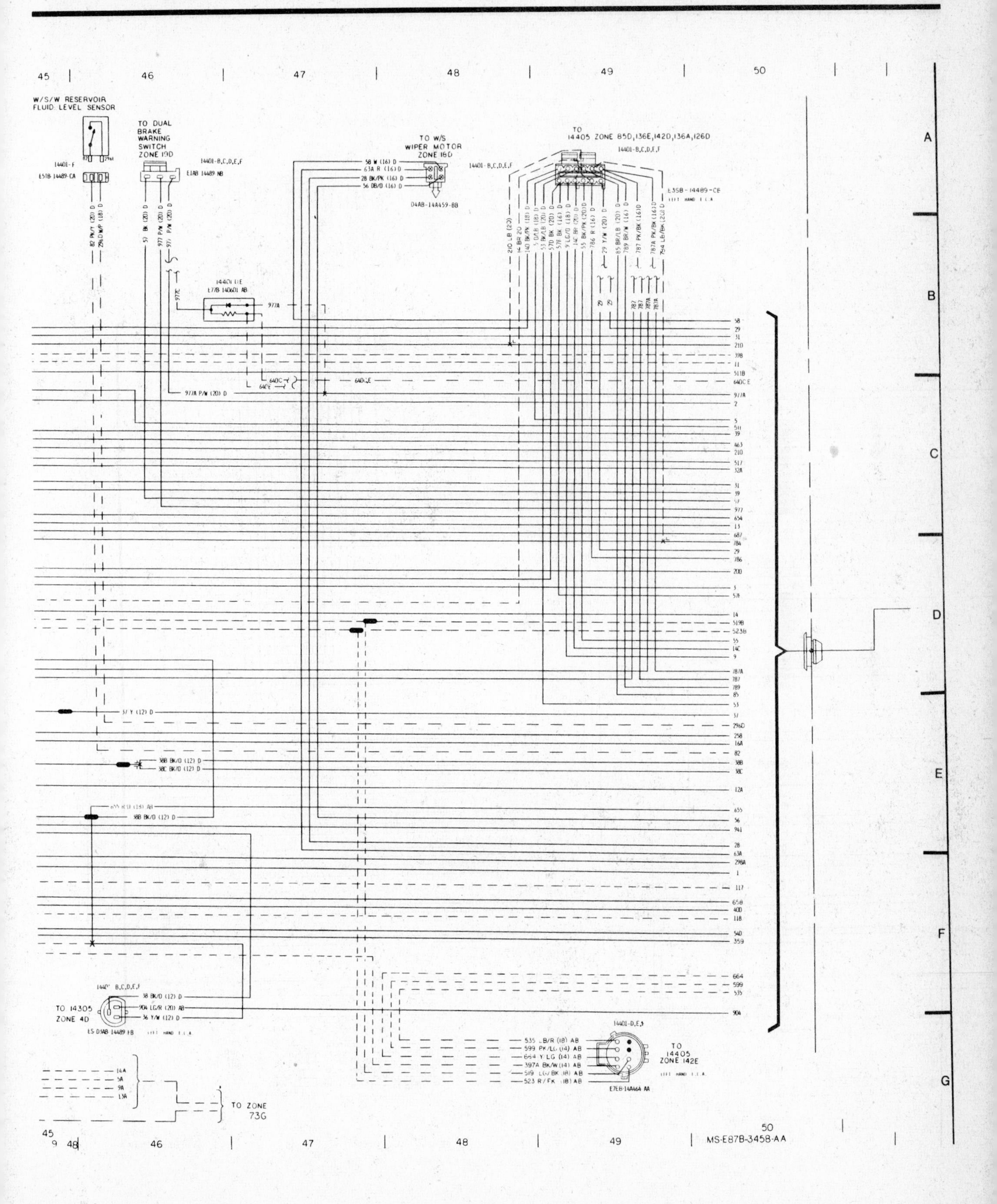
W/S/W RESERVOIR FLUID LEVEL SENSOR
TO DUAL BRAKE WARNING SWITCH ZONE 19D
TO W/S WIPER MOTOR ZONE 16D
TO 14405 ZONE 85D,136E,142D,136A,126D
TO 14305 ZONE 4D
TO 14405 ZONE 142E
TO ZONE 73G
MS-E87B-3458-AA

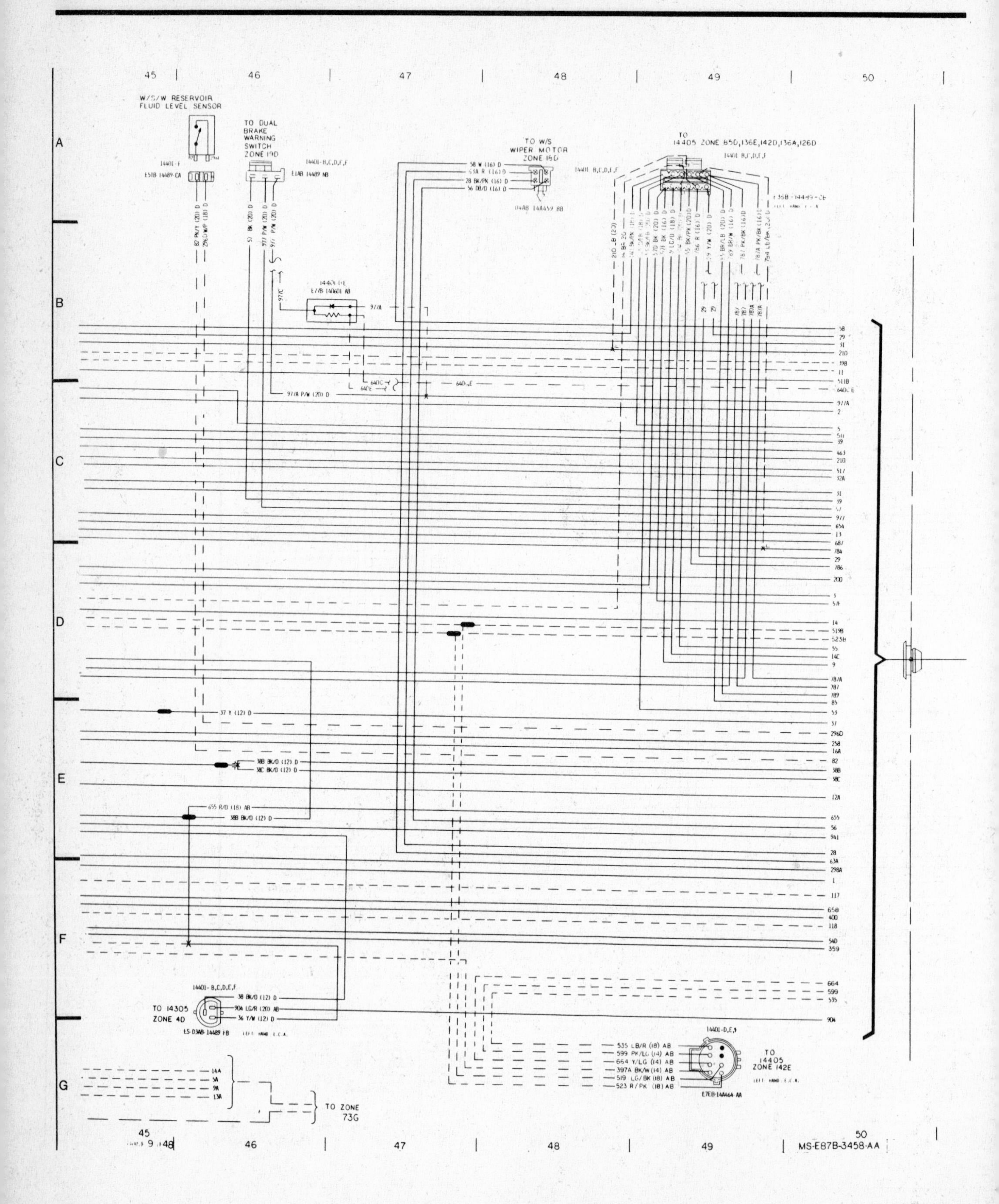
W/S/W RESERVOIR FLUID LEVEL SENSOR
TO DUAL BRAKE WARNING SWITCH ZONE 19D
TO W/S WIPER MOTOR ZONE 16D
TO 14405 ZONE 85D,136E,142D,136A,126D
TO 14305 ZONE 4D
TO 14405 ZONE 142E
TO ZONE 73G
535 LB/R (18) AB
599 PK/LG (14) AB
664 Y/LG (14) AB
397A BK/W (14) AB
519 LG/BK (18) AB
523 R/PK (18) AB
37 Y (12) D
38B BK/O (12) D
38C BK/O (12) D
655 R/O (18) AB
38 BK/O (12) D
904 LG/R (20) AB
36 Y/W (12) D
58 W (16) D
63A R (16) D
28 BK/PK (16) D
56 DB/O (16) D
97/A P/W (20) D
MS-E87B-3458-AA

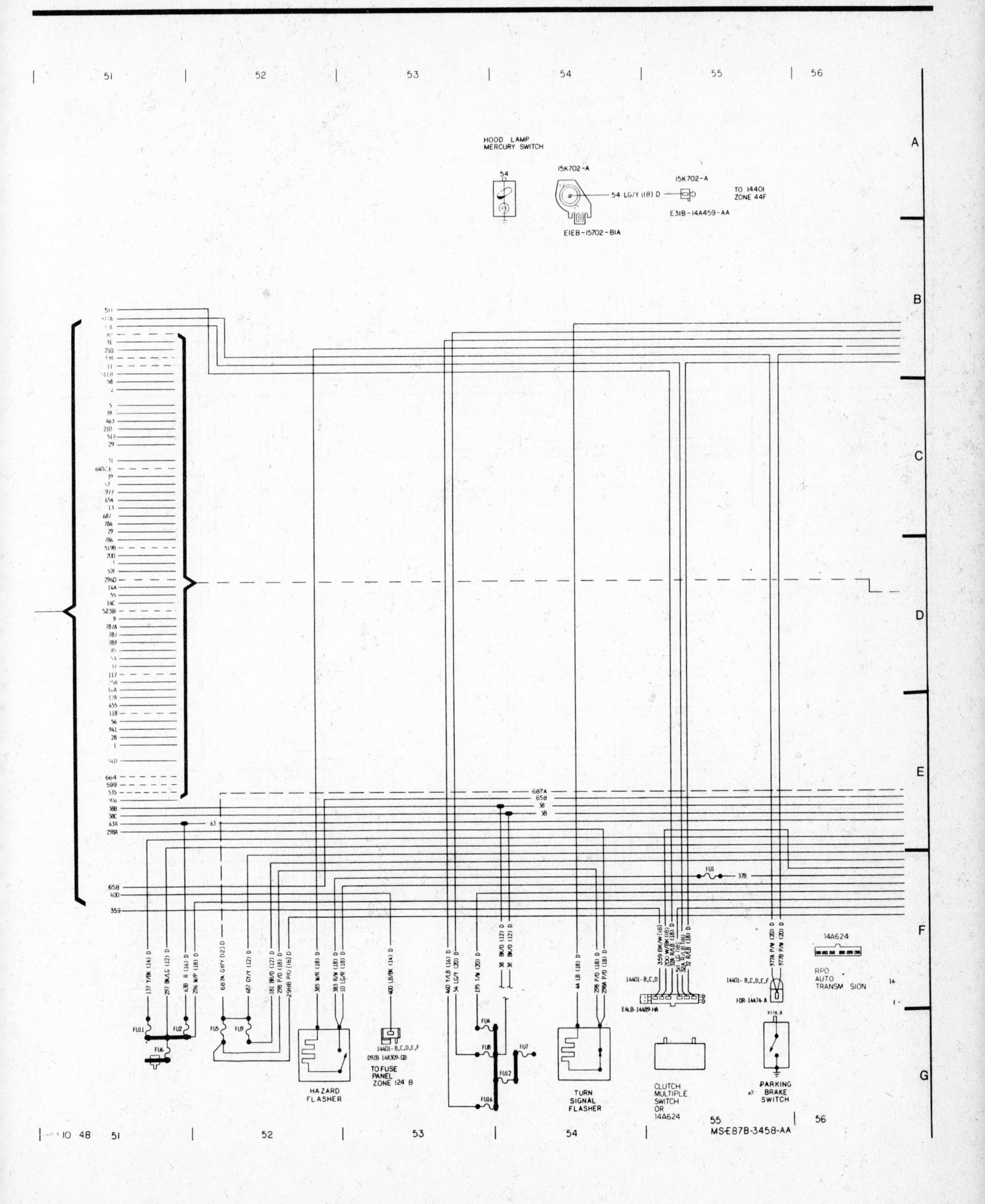
HOOD LAMP
MERCURY SWITCH
15K702-A
54 LG/Y (18) D
TO 14401
ZONE 44F
E31B-14A459-AA
E1EB-15702-B1A
687A
658
38
38
63
FU1
378
14A624
RPO
AUTO
TRANSMISSION
14401-B,C,D
14401-B,C,D,E,F
HAZARD
FLASHER
TO FUSE
PANEL
ZONE 124 B
TURN
SIGNAL
FLASHER
CLUTCH
MULTIPLE
SWITCH
OR
14A624
PARKING
BRAKE
SWITCH
MS-E87B-3458-AA

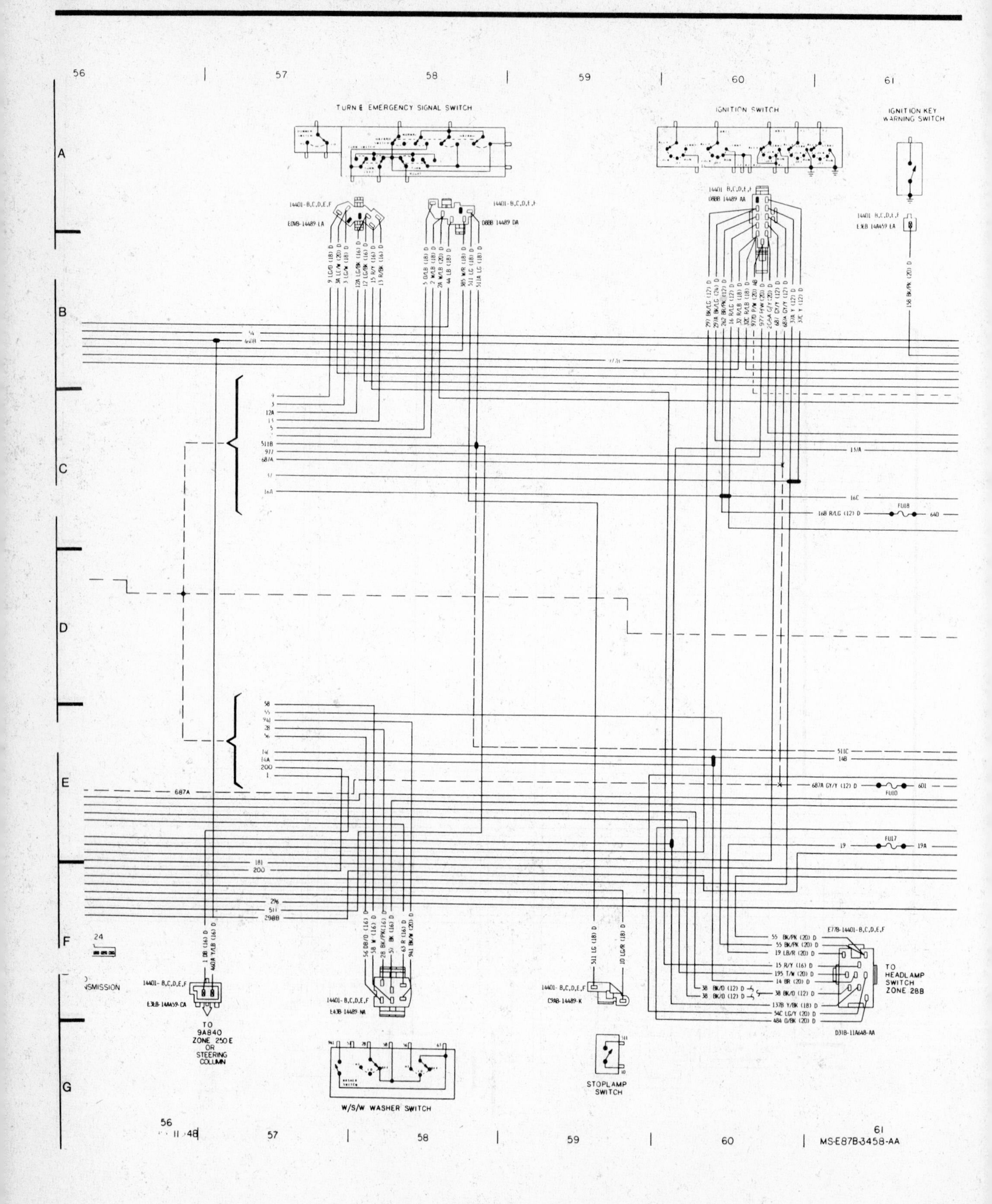

TURN & EMERGENCY SIGNAL SWITCH
IGNITION SWITCH
IGNITION KEY WARNING SWITCH
TO 9A840 ZONE 250 E OR STEERING COLUMN
W/S/W WASHER SWITCH
STOPLAMP SWITCH
TO HEADLAMP SWITCH ZONE 28B
MS-E87B-3458-AA

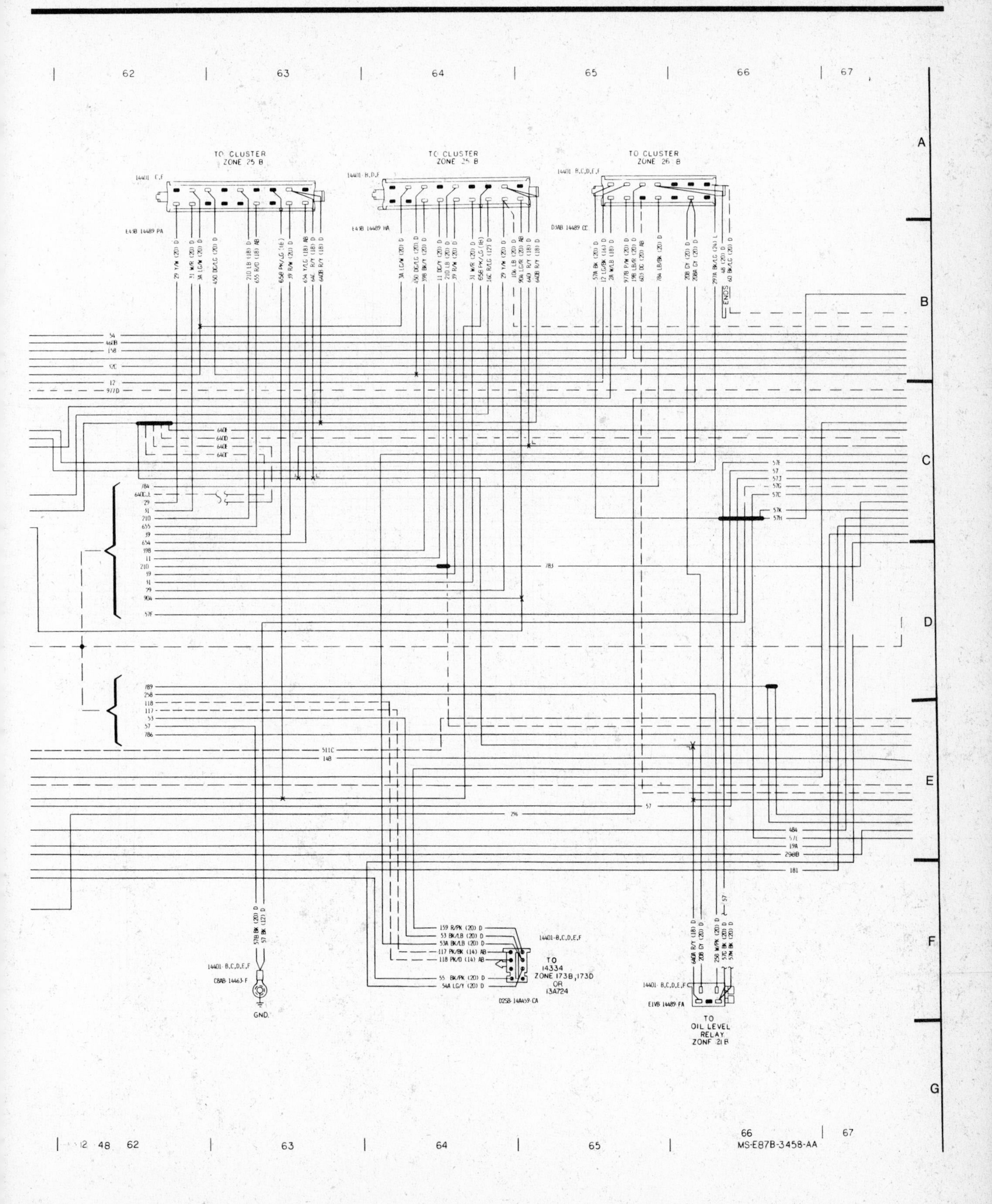
TO CLUSTER
ZONE 25 B
TO CLUSTER
ZONE 25 B
TO CLUSTER
ZONE 26 B
TO
14334
ZONE 173B, 173D
OR
13A724
GND.
TO
OIL LEVEL
RELAY
ZONE 21B
MS-E87B-3458-AA

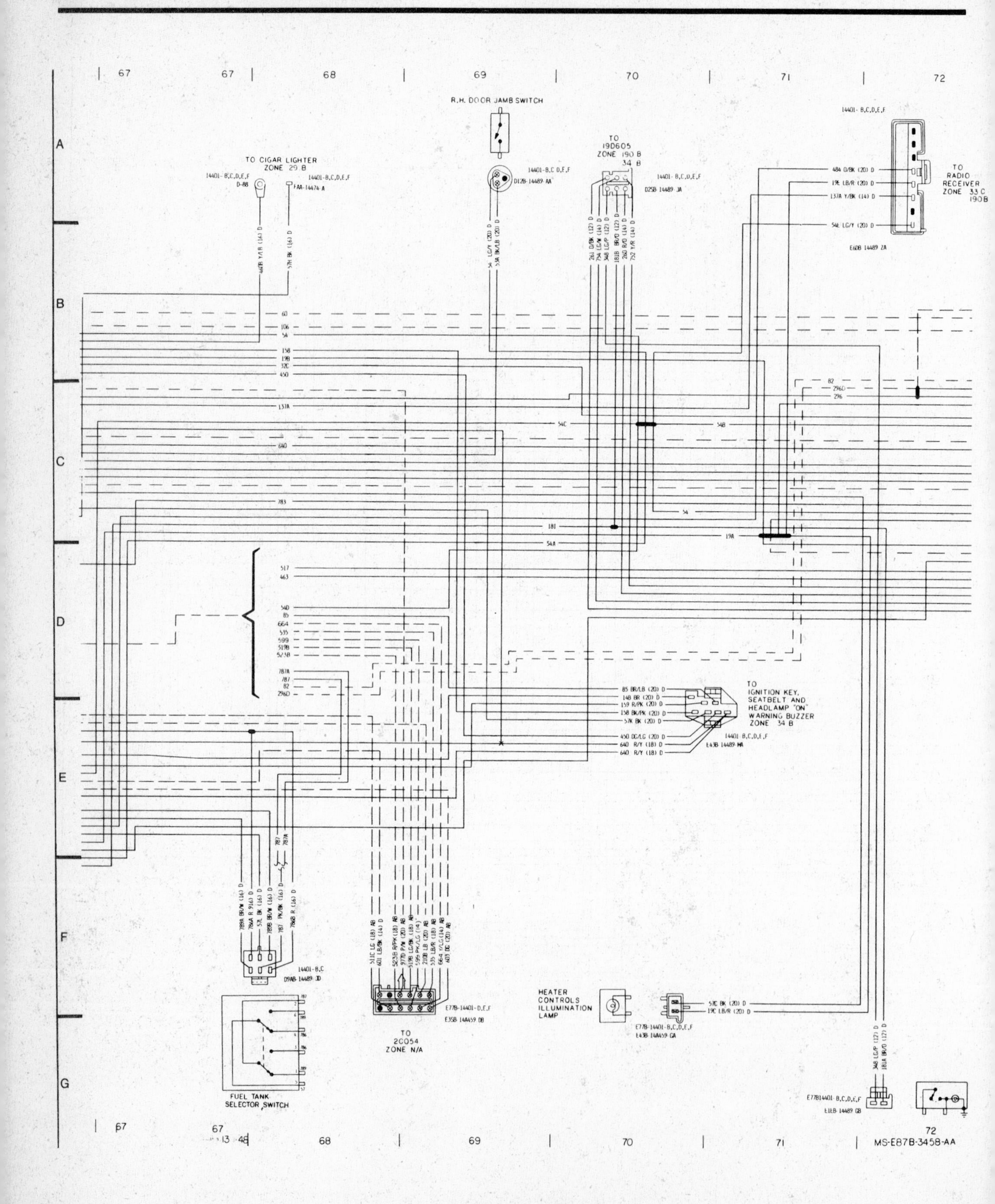

R.H. DOOR JAMB SWITCH
TO CIGAR LIGHTER ZONE 29 B
TO 19D605 ZONE 190 B 34 B
TO RADIO RECEIVER ZONE 33 C 190 B
TO IGNITION KEY, SEATBELT AND HEADLAMP "ON" WARNING BUZZER ZONE 34 B
HEATER CONTROLS ILLUMINATION LAMP
TO 2C054 ZONE N/A
FUEL TANK SELECTOR SWITCH
MS-E87B-3458-AA

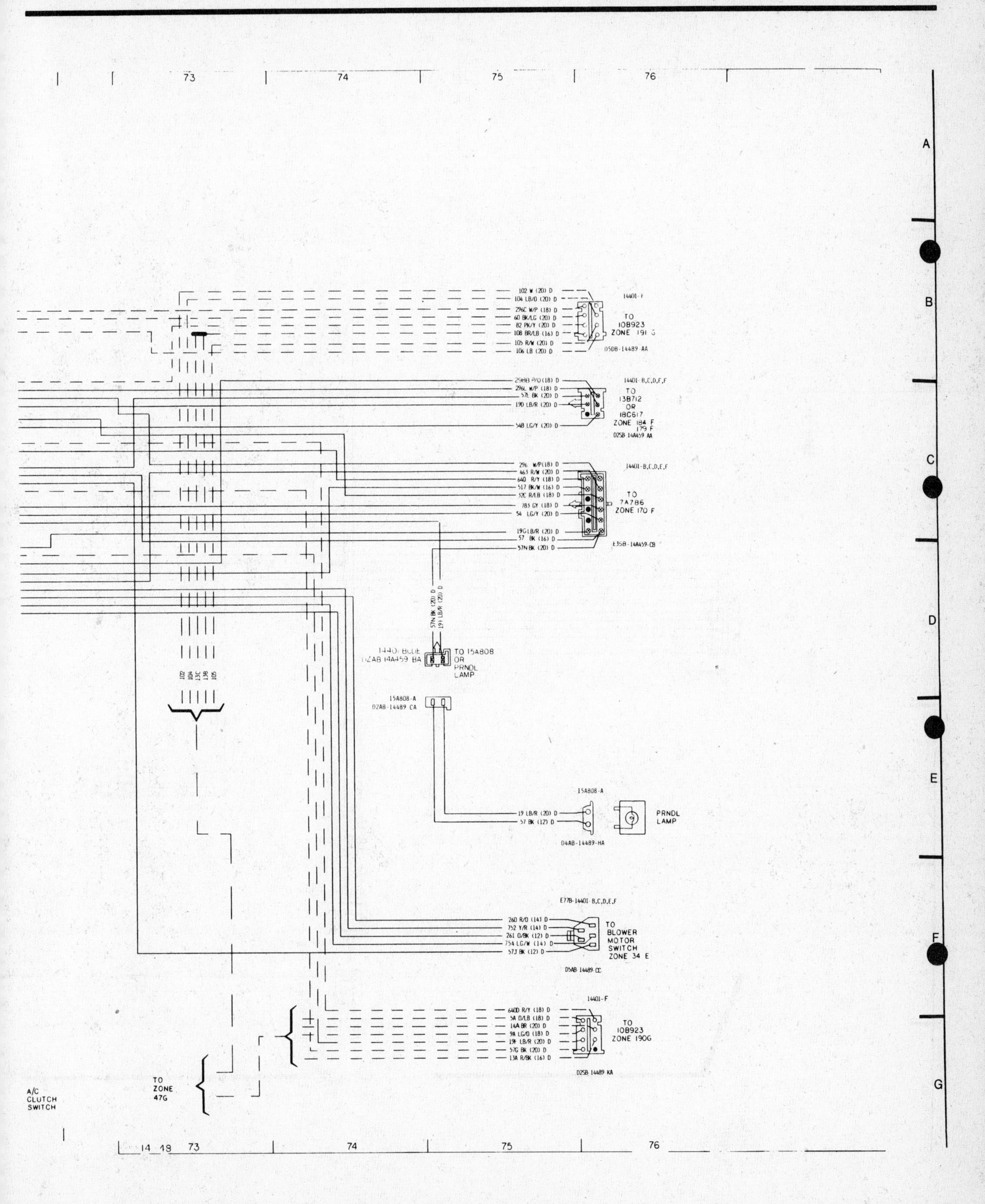
73
74
75
76
A
B
C
D
E
F
G
102 W (20) D
104 LB/O (20) D
296C W/P (18) D
60 BK/LG (20) D
82 PK/Y (20) D
108 BR/LB (16) D
105 R/W (20) D
106 LB (20) D
14401-F
TO
10B923
ZONE 191 G
D5DB-14489-AA
298B P/O (18) D
296L W/P (18) D
57E BK (20) D
190 LB/R (20) D
54B LG/Y (20) D
14401-B,C,D,F,F
TO
13B712
OR
18C617
ZONE 184 F
179 F
D2SB 14A459 AA
296 W/P (18) D
463 R/W (20) D
640 R/Y (18) D
517 BK/W (16) D
32C R/LB (18) D
783 GY (18) D
54 LG/Y (20) D
19G LB/R (20) D
57 BK (16) D
57N BK (20) D
14401-B,C,D,E,F
TO
7A786
ZONE 170 F
E35B-14A459-CB
57N BK (20) D
19+ LB/R (20) D
D2AB 14A459 BA
TO 15A808
OR
PRNDL
LAMP
15A808-A
D2AB-14489 CA
15A808-A
19 LB/R (20) D
57 BK (12) D
PRNDL
LAMP
D4AB-14489-HA
E77B-14401-B,C,D,E,F
260 R/O (14) D
752 Y/R (14) D
261 O/BK (12) D
754 LG/W (14) D
57J BK (12) D
TO
BLOWER
MOTOR
SWITCH
ZONE 34 E
D5AB 14489 CC
14401-F
640D R/Y (18) D
5A O/LB (18) D
14A BR (20) D
9A LG/O (18) D
19+ LB/R (20) D
57G BK (20) D
13A R/BK (16) D
TO
10B923
ZONE 190G
D2SB 14489 KA
A/C
CLUTCH
SWITCH
TO
ZONE
47G
14 48

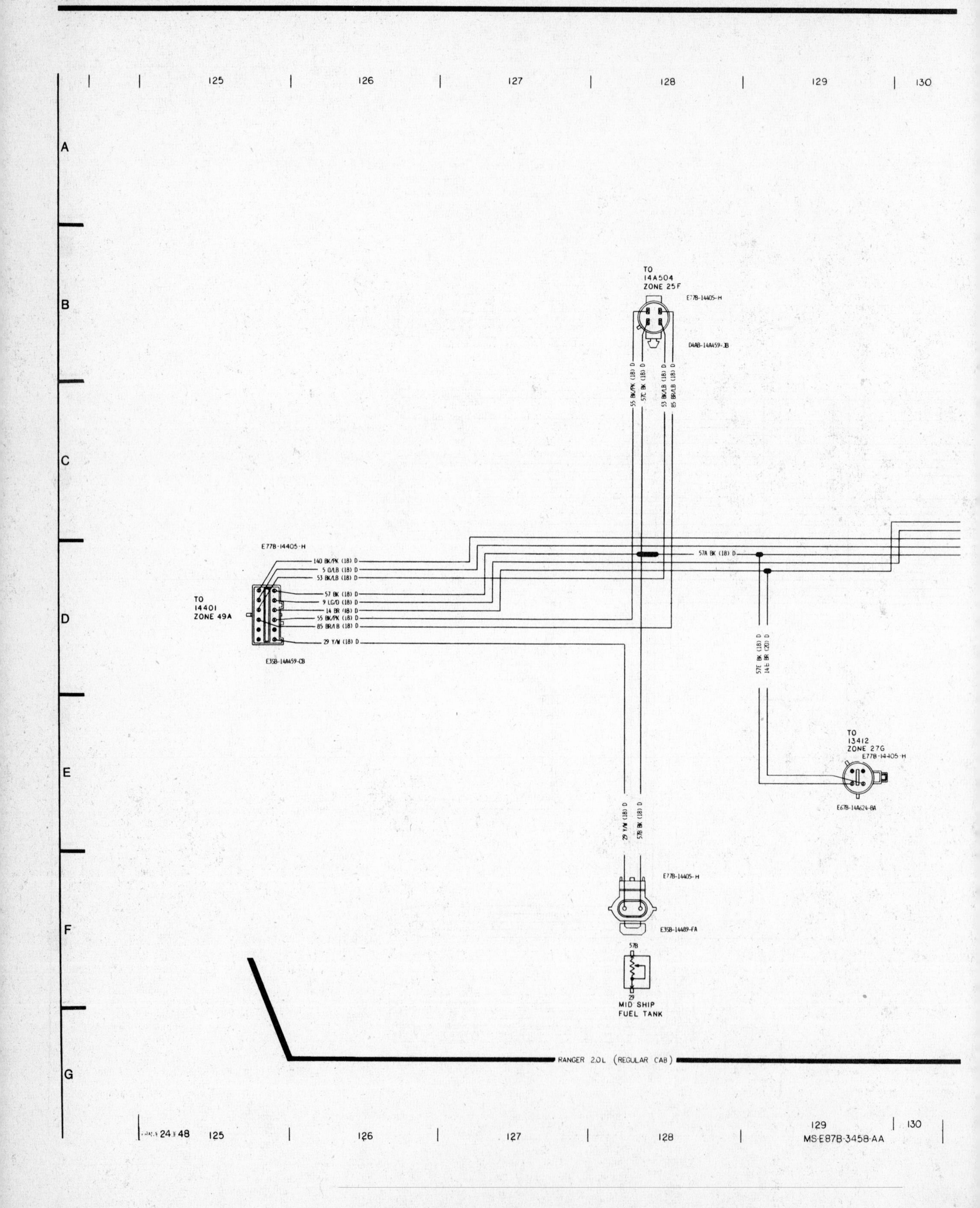
TO
14A504
ZONE 25F
E77B-14405-H
D4AB-14A459-JB
55 BK/PK (18) D
57C BK (18) D
53 BK/LB (18) D
85 BR/LB (18) D
E77B-14405-H
140 BK/PK (18) D
5 O/LB (18) D
53 BK/LB (18) D
57 BK (18) D
9 LG/O (18) D
14 BR (18) D
55 BK/PK (18) D
85 BR/LB (18) D
29 Y/W (18) D
TO
14401
ZONE 49A
E35B-14A459-CB
57A BK (18) D
57E BK (18) D
14E BR (20) D
TO
13412
ZONE 27G
E77B-14405-H
E67B-14A624-BA
29 Y/W (18) D
57B BK (18) D
E77B-14405-H
E35B-14489-FA
57B
29
MID SHIP
FUEL TANK
RANGER 2.0L (REGULAR CAB)
MS-E87B-3458-AA

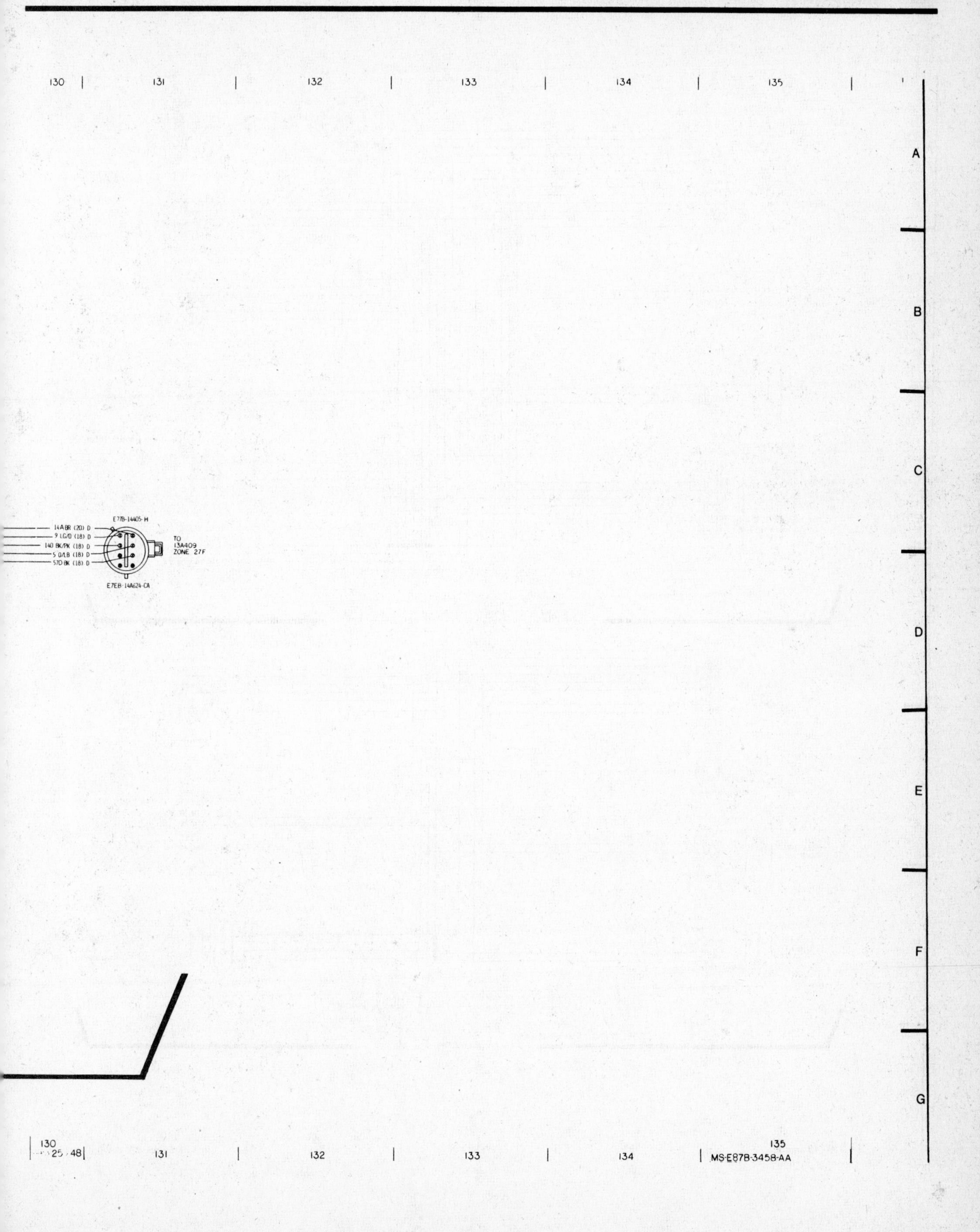
130
131
132
133
134
135
A
B
C
D
E
F
G
E7TB-14405-H
14A BR (20) D
9 LG/O (18) D
140 BK/PK (18) D
5 O/LB (18) D
57D BK (18) D
TO
13A409
ZONE 27F
E7EB-14A624-CA
MS-E87B-3458-AA

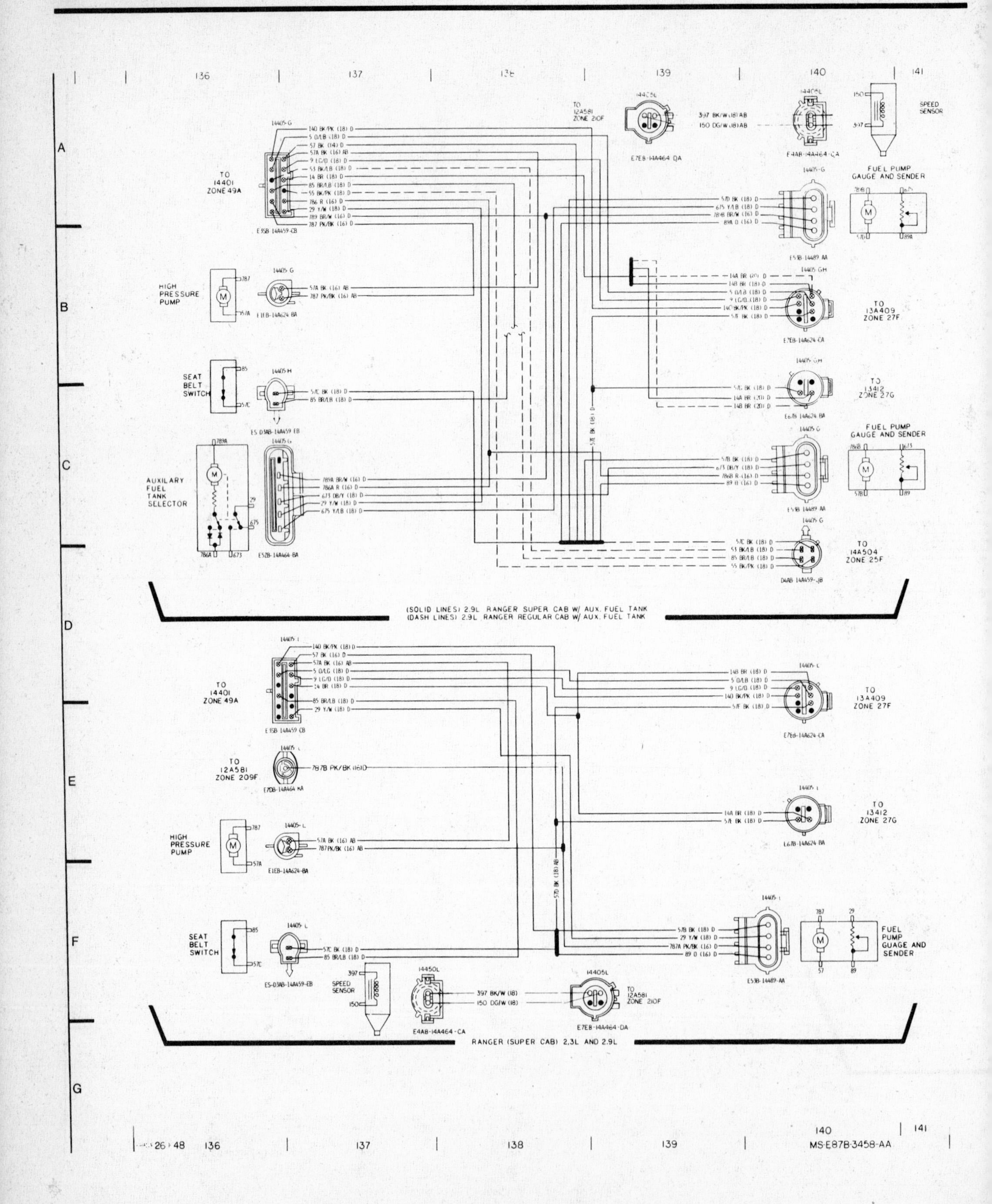

(SOLID LINES) 2.9L RANGER SUPER CAB W/ AUX. FUEL TANK
(DASH LINES) 2.9L RANGER REGULAR CAB W/ AUX. FUEL TANK
RANGER (SUPER CAB) 2.3L AND 2.9L
HIGH PRESSURE PUMP
SEAT BELT SWITCH
AUXILARY FUEL TANK SELECTOR
FUEL PUMP GAUGE AND SENDER
SPEED SENSOR
TO 14401 ZONE 49A
TO 13A409 ZONE 27F
TO 13412 ZONE 27G
TO 14A504 ZONE 25F
TO 12A581 ZONE 210F
TO 12A581 ZONE 209F
MS-E87B-3458-AA

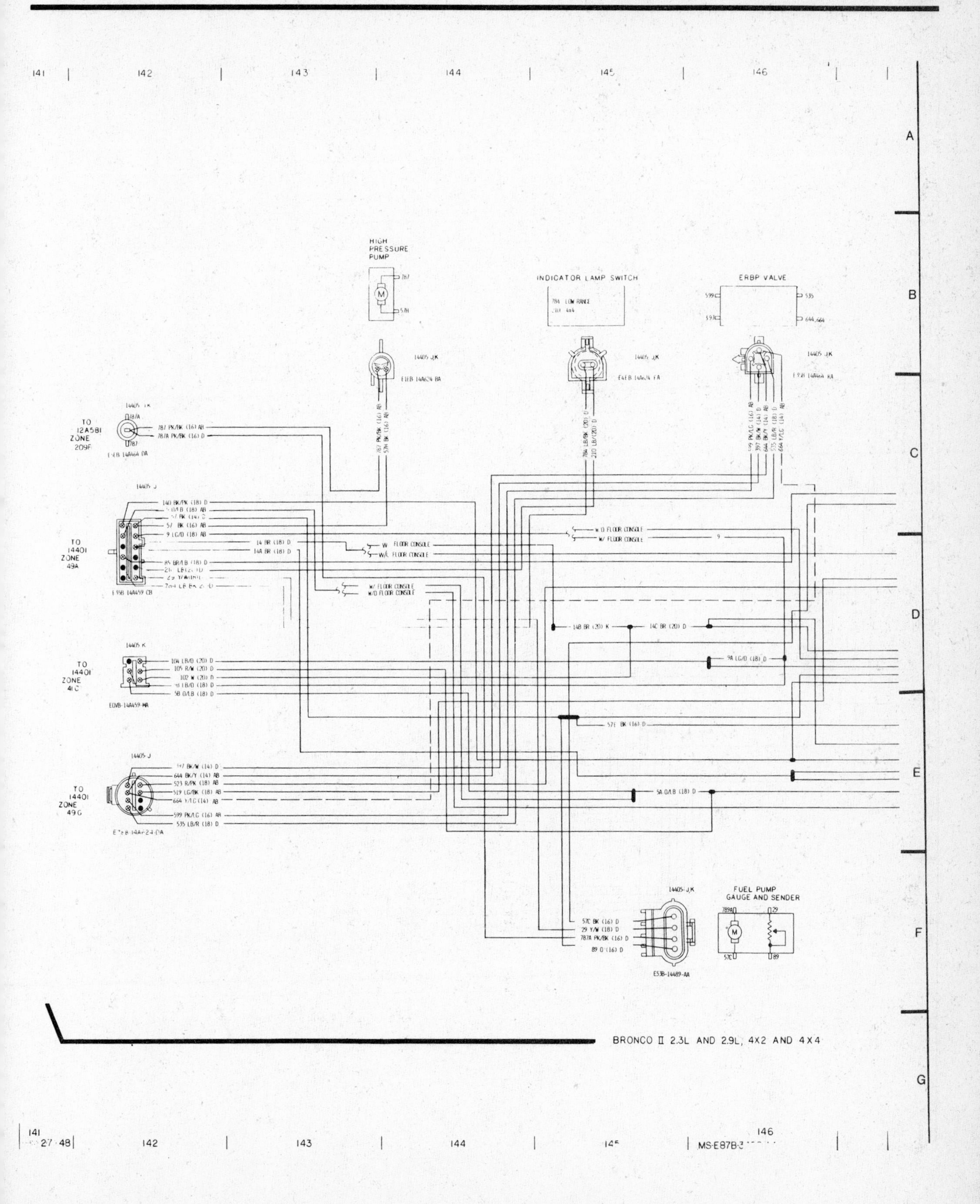
HIGH PRESSURE PUMP
INDICATOR LAMP SWITCH
ERBP VALVE
FUEL PUMP GAUGE AND SENDER
TO 12A581 ZONE 209F
TO 14401 ZONE 49A
TO 14401 ZONE 41C
TO 14401 ZONE 49G
W/ FLOOR CONSOLE
W/O FLOOR CONSOLE
BRONCO II 2.3L AND 2.9L, 4X2 AND 4X4

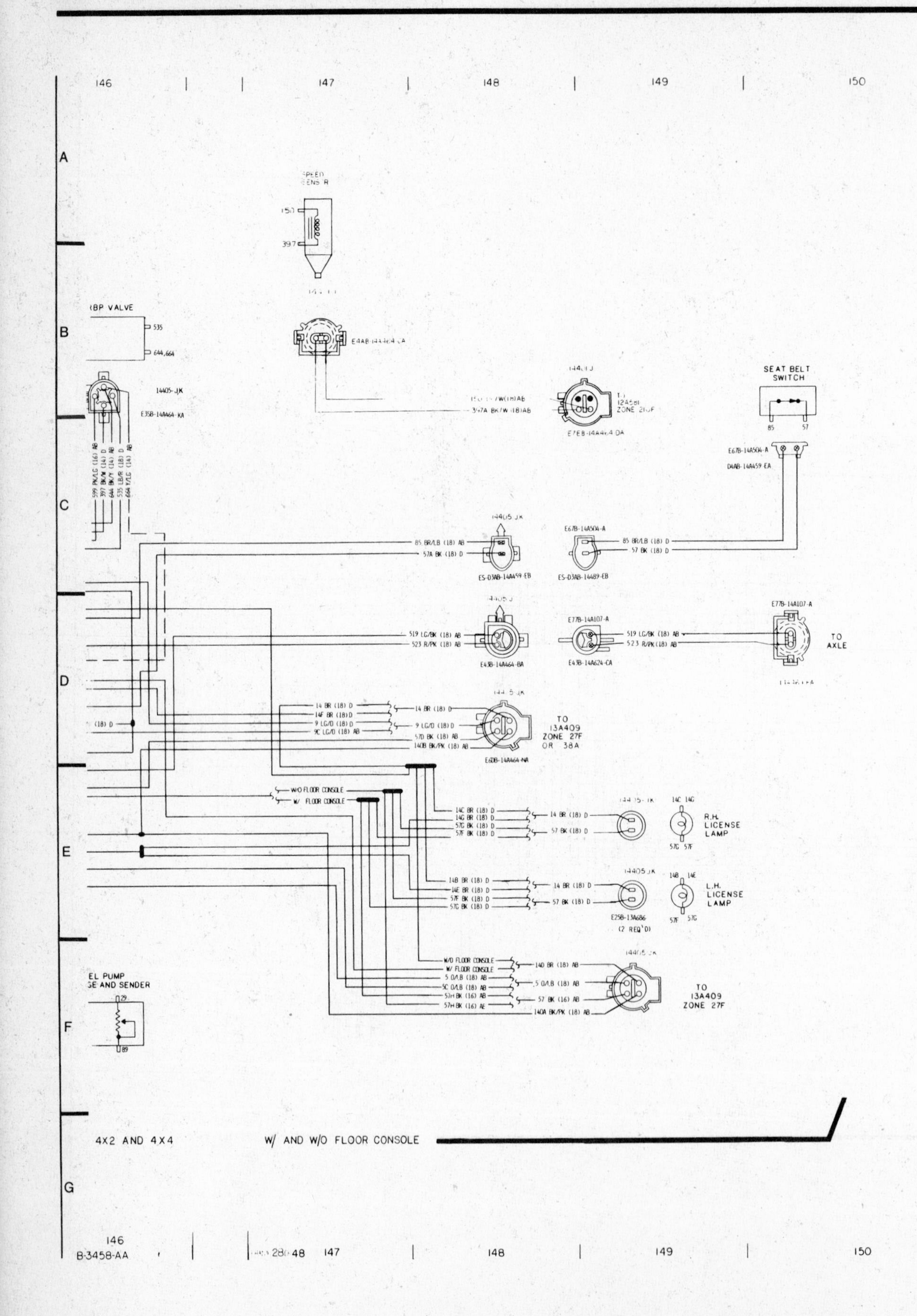

146
147
148
149
150
A
B
C
D
E
F
G
SEAT BELT SWITCH
TO AXLE
TO 13A409 ZONE 27F OR 38A
R.H. LICENSE LAMP
L.H. LICENSE LAMP
TO 13A409 ZONE 27F
W/O FLOOR CONSOLE
W/ FLOOR CONSOLE
4X2 AND 4X4
W/ AND W/O FLOOR CONSOLE
B-3458-AA

A

B

C

D

E

F

G

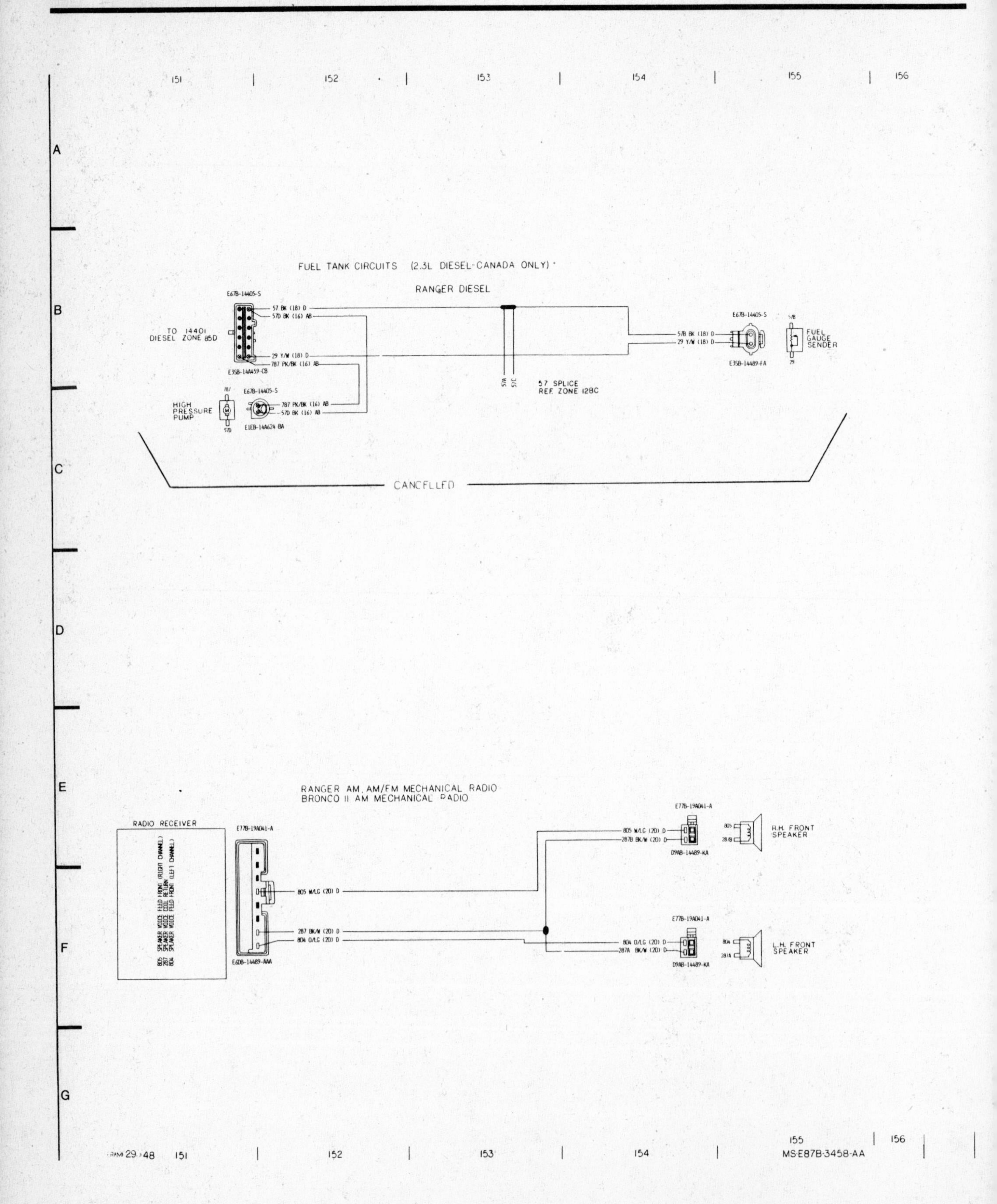

151
152
153
154
155
156
A
B
C
D
E
F
G
FUEL TANK CIRCUITS (2.3L DIESEL-CANADA ONLY)
RANGER DIESEL
TO 14401
DIESEL ZONE 85D
57 SPLICE
REF. ZONE 128C
HIGH
PRESSURE
PUMP
FUEL
GAUGE
SENDER
CANCELLED
RANGER AM, AM/FM MECHANICAL RADIO
BRONCO II AM MECHANICAL RADIO
RADIO RECEIVER
R.H. FRONT
SPEAKER
L.H. FRONT
SPEAKER
MS-E87B-3458-AA

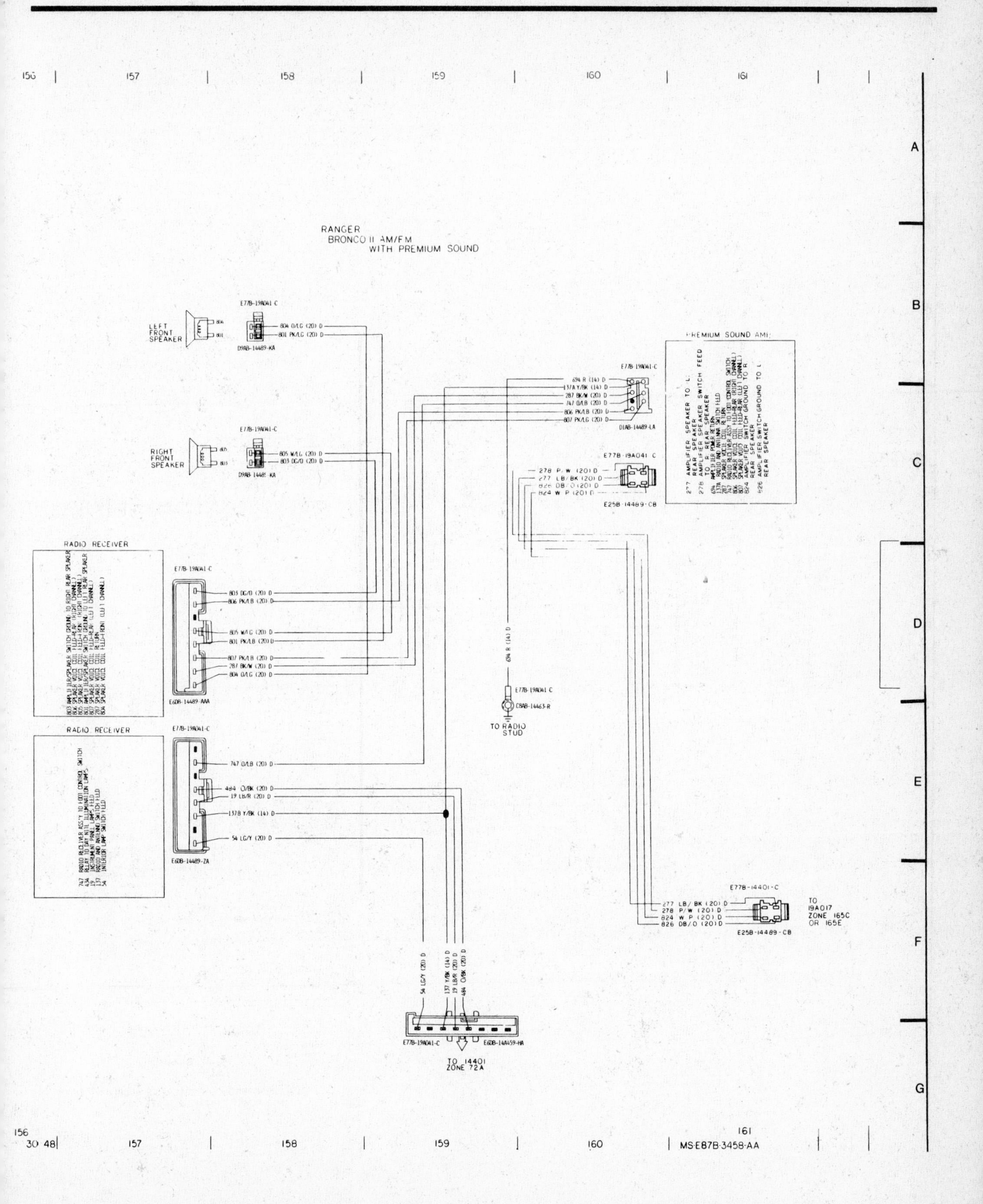
RANGER
BRONCO II AM/FM
WITH PREMIUM SOUND
LEFT FRONT SPEAKER
RIGHT FRONT SPEAKER
RADIO RECEIVER
PREMIUM SOUND AMP
TO RADIO STUD
TO 14401 ZONE 72A
TO 19A017 ZONE 165C OR 165E
MS-E87B-3458-AA

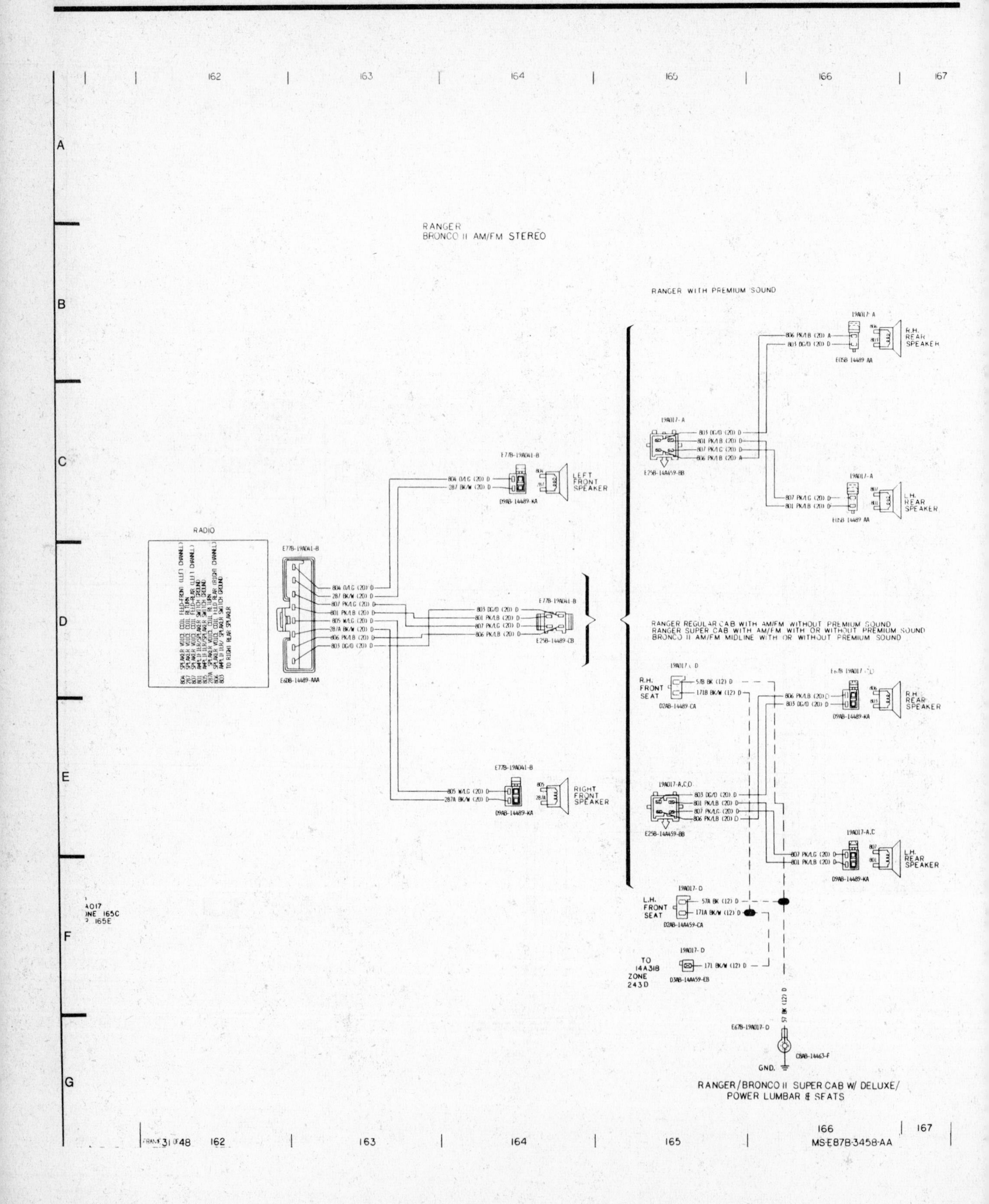
RANGER
BRONCO II AM/FM STEREO
RANGER WITH PREMIUM SOUND
RADIO
LEFT FRONT SPEAKER
RIGHT FRONT SPEAKER
R.H. REAR SPEAKER
L.H. REAR SPEAKER
RANGER REGULAR CAB WITH AM/FM WITHOUT PREMIUM SOUND
RANGER SUPER CAB WITH AM/FM WITH OR WITHOUT PREMIUM SOUND
BRONCO II AM/FM MIDLINE WITH OR WITHOUT PREMIUM SOUND
R.H. FRONT SEAT
L.H. FRONT SEAT
TO 14A318 ZONE 243D
GND.
RANGER/BRONCO II SUPER CAB W/ DELUXE/
POWER LUMBAR & SEATS
MS-E87B-3458-AA

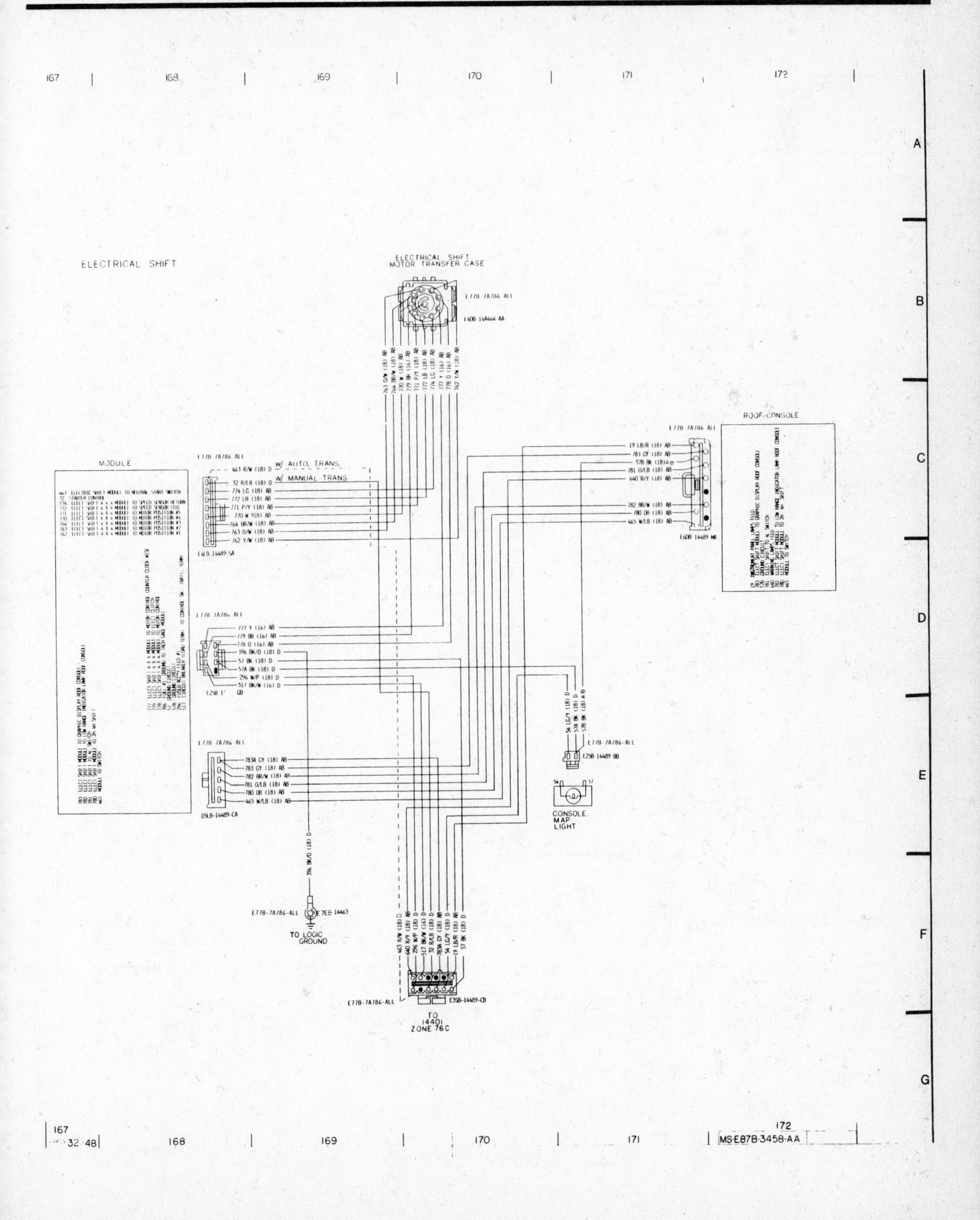
ELECTRICAL SHIFT
ELECTRICAL SHIFT MOTOR TRANSFER CASE
MODULE
ROOF CONSOLE
W/ AUTO. TRANS.
W/ MANUAL TRANS.
CONSOLE MAP LIGHT
TO LOGIC GROUND
TO 14401 ZONE 76C
MS-E87B-3458-AA

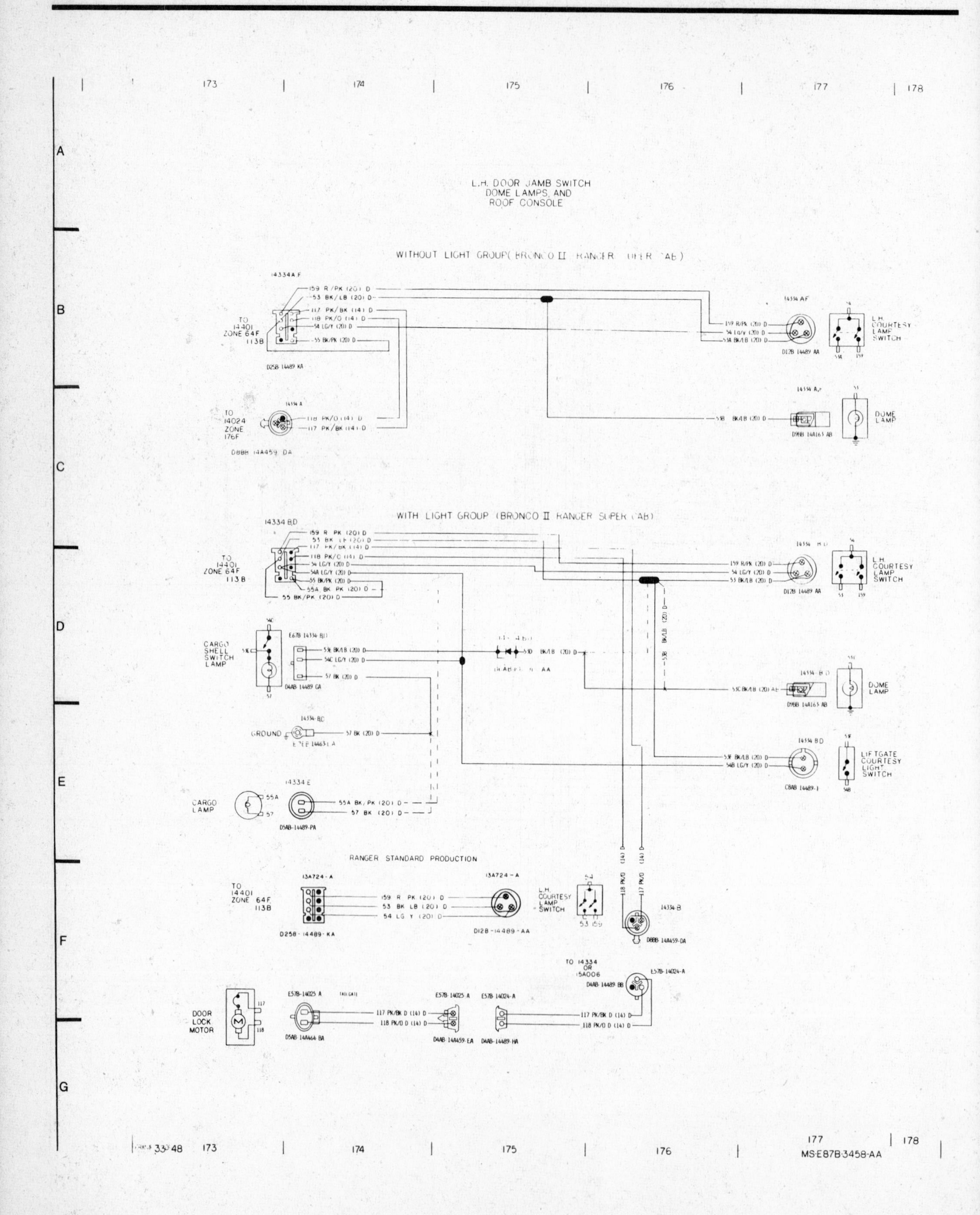

L.H. DOOR JAMB SWITCH
DOME LAMPS, AND
ROOF CONSOLE
WITHOUT LIGHT GROUP
WITH LIGHT GROUP (BRONCO II RANGER SUPER CAB)
L.H. COURTESY LAMP SWITCH
DOME LAMP
CARGO SHELL SWITCH LAMP
GROUND
CARGO LAMP
LIFTGATE COURTESY LIGHT SWITCH
RANGER STANDARD PRODUCTION
DOOR LOCK MOTOR
MS-E87B-3458-AA

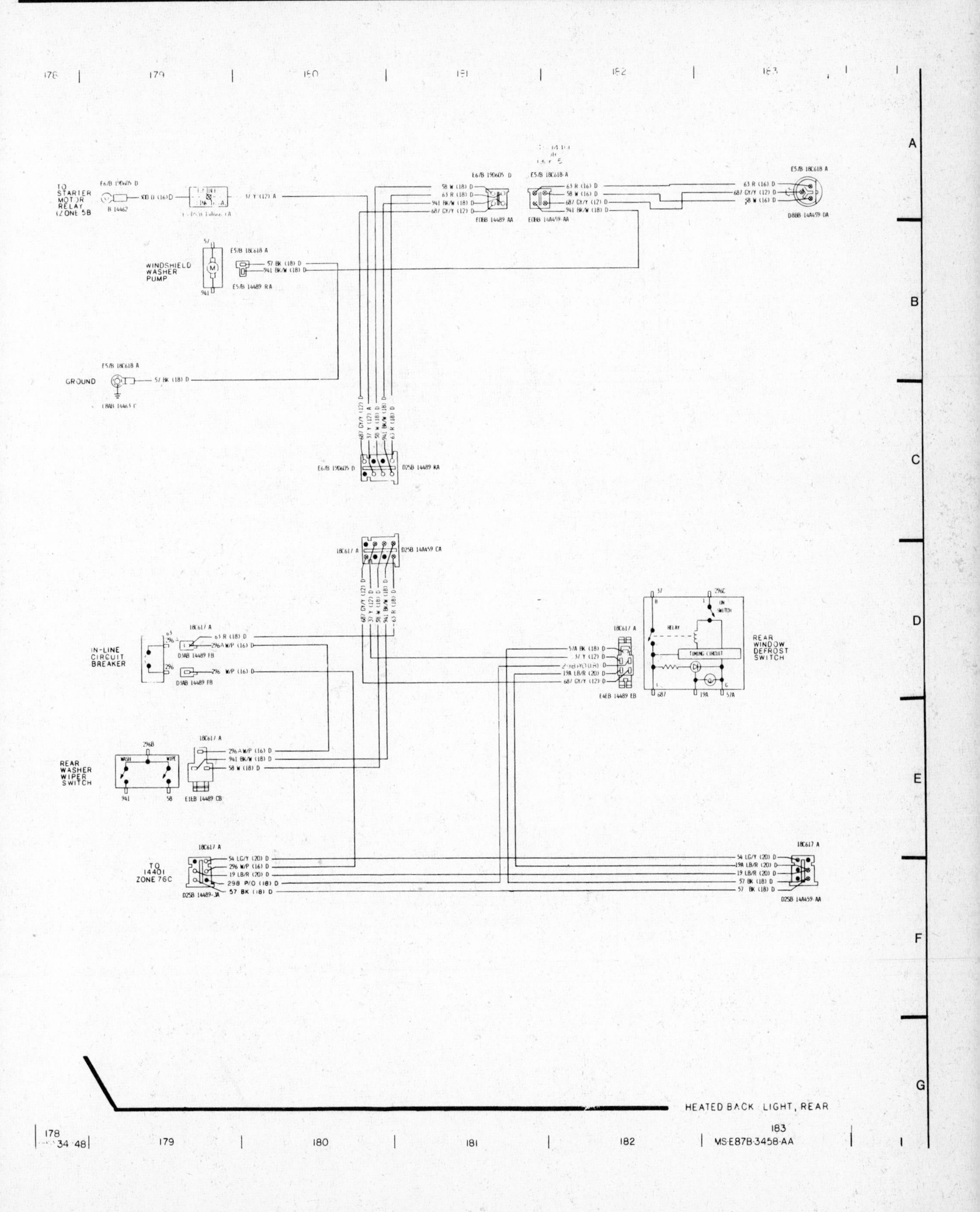

TO STARTER MOTOR RELAY (ZONE 5B
WINDSHIELD WASHER PUMP
GROUND
IN-LINE CIRCUIT BREAKER
REAR WASHER WIPER SWITCH
TO 14401 ZONE 76C
REAR WINDOW DEFROST SWITCH
TIMING CIRCUIT
RELAY
HEATED BACK LIGHT, REAR
MS E87B 3458 AA

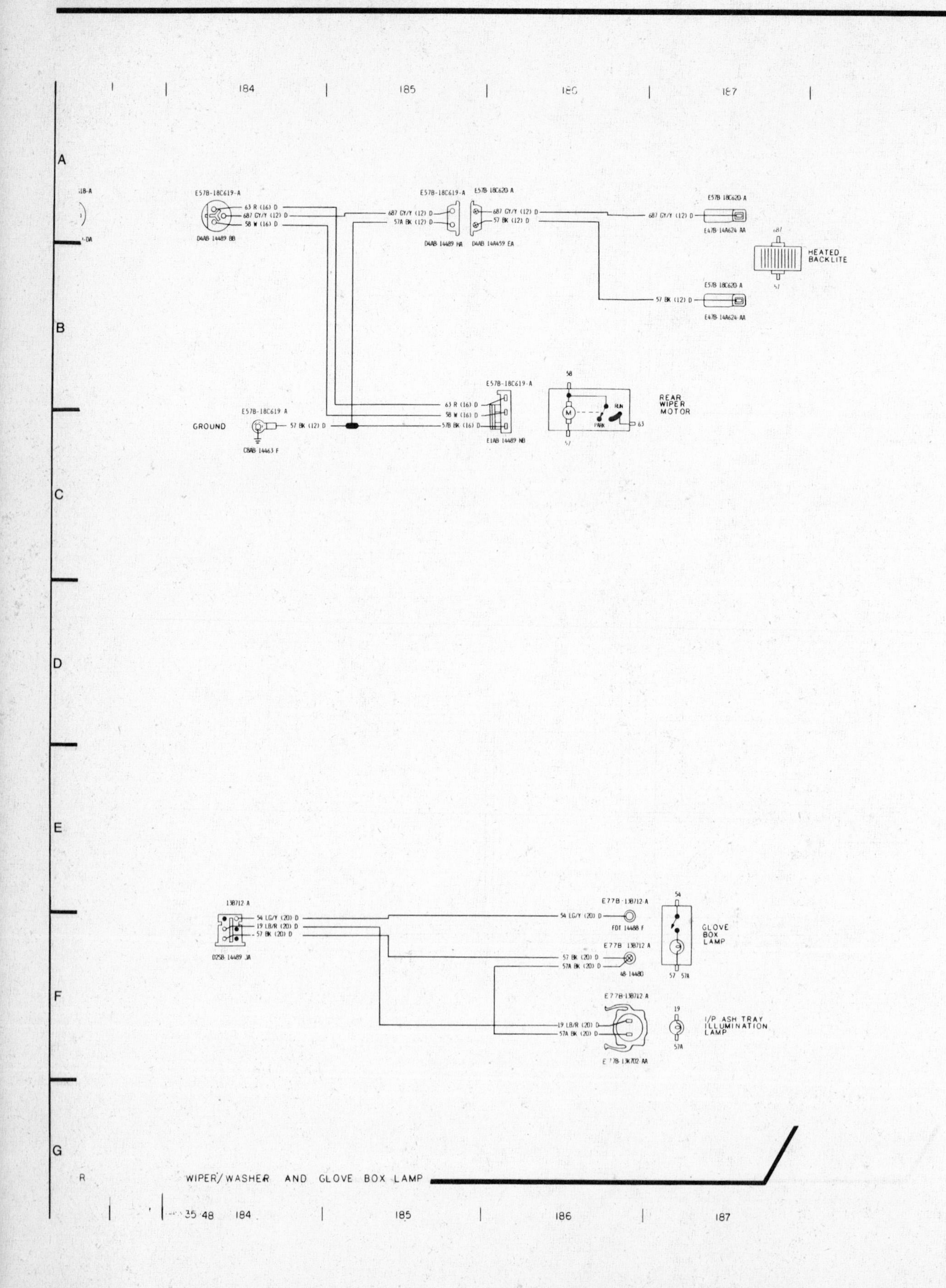
184
185
186
187
A
B
C
D
E
F
G
E57B-18C619-A
63 R (16) D
687 GY/Y (12) D
58 W (16) D
D4AB 14489 BB
E57B-18C619-A
E57B 18C620-A
687 GY/Y (12) D
57A BK (12) D
687 GY/Y (12) D
57 BK (12) D
D4AB 14489 HA
D4AB 14A459 EA
E57B 18C620-A
687 GY/Y (12) D
E47B 14A624 AA
687
HEATED
BACKLITE
57
E57B 18C620-A
57 BK (12) D
E47B 14A624-AA
E57B-18C619-A
63 R (16) D
58 W (16) D
57B BK (16) D
E1AB 14489 NB
58
RUN
PARK
63
57
REAR
WIPER
MOTOR
GROUND
E57B-18C619 A
57 BK (12) D
C8AB 14463 F
13B712 A
54 LG/Y (20) D
19 LB/R (20) D
57 BK (20) D
D2SB 14489 JA
E77B 13B712 A
54 LG/Y (20) D
FDT 14488 F
E77B 13B712 A
57 BK (20) D
57A BK (20) D
48 14480
54
57 57A
GLOVE
BOX
LAMP
E77B 13B712 A
19 LB/R (20) D
57A BK (20) D
E77B 13K702 AA
19
57A
I/P ASH TRAY
ILLUMINATION
LAMP
WIPER/WASHER AND GLOVE BOX LAMP
184
185
186
187

A
B
C
D
E
F
G

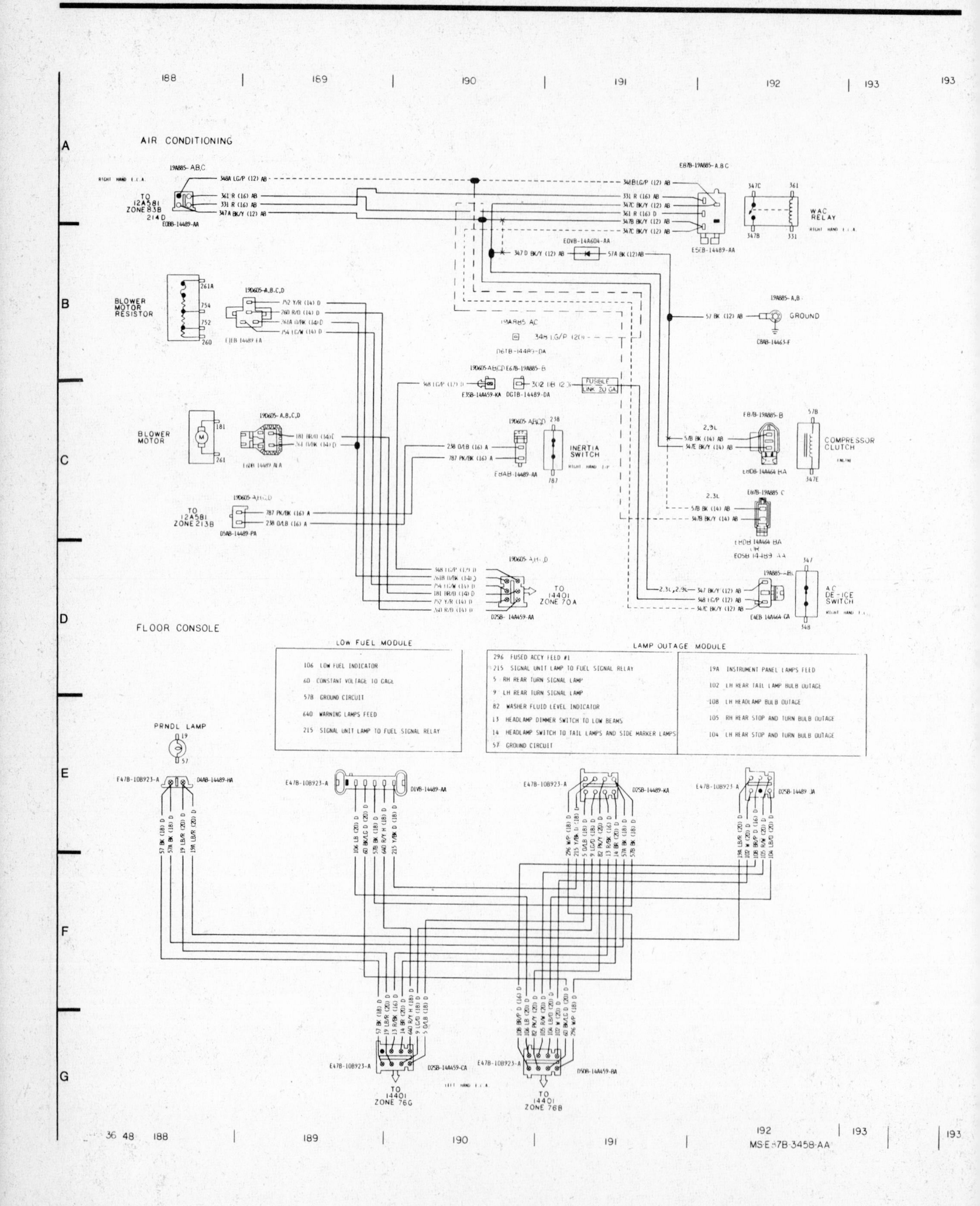

AIR CONDITIONING
BLOWER MOTOR RESISTOR
BLOWER MOTOR
TO 12A581 ZONE 83B
TO 12A581 ZONE 213B
INERTIA SWITCH
COMPRESSOR CLUTCH
WAC RELAY
GROUND
A/C DE-ICE SWITCH
TO 14401 ZONE 70A
FLOOR CONSOLE
LOW FUEL MODULE
106 LOW FUEL INDICATOR
60 CONSTANT VOLTAGE TO GAGE
57B GROUND CIRCUIT
640 WARNING LAMPS FEED
215 SIGNAL UNIT LAMP TO FUEL SIGNAL RELAY
LAMP OUTAGE MODULE
296 FUSED ACCY FEED #1
215 SIGNAL UNIT LAMP TO FUEL SIGNAL RELAY
5 RH REAR TURN SIGNAL LAMP
9 LH REAR TURN SIGNAL LAMP
82 WASHER FLUID LEVEL INDICATOR
13 HEADLAMP DIMMER SWITCH TO LOW BEAMS
14 HEADLAMP SWITCH TO TAIL LAMPS AND SIDE MARKER LAMPS
57 GROUND CIRCUIT
19A INSTRUMENT PANEL LAMPS FEED
102 LH REAR TAIL LAMP BULB OUTAGE
108 LH HEADLAMP BULB OUTAGE
105 RH REAR STOP AND TURN BULB OUTAGE
104 LH REAR STOP AND TURN BULB OUTAGE
PRNDL LAMP
TO 14401 ZONE 76G
TO 14401 ZONE 76B
MS-E87B-3458-AA

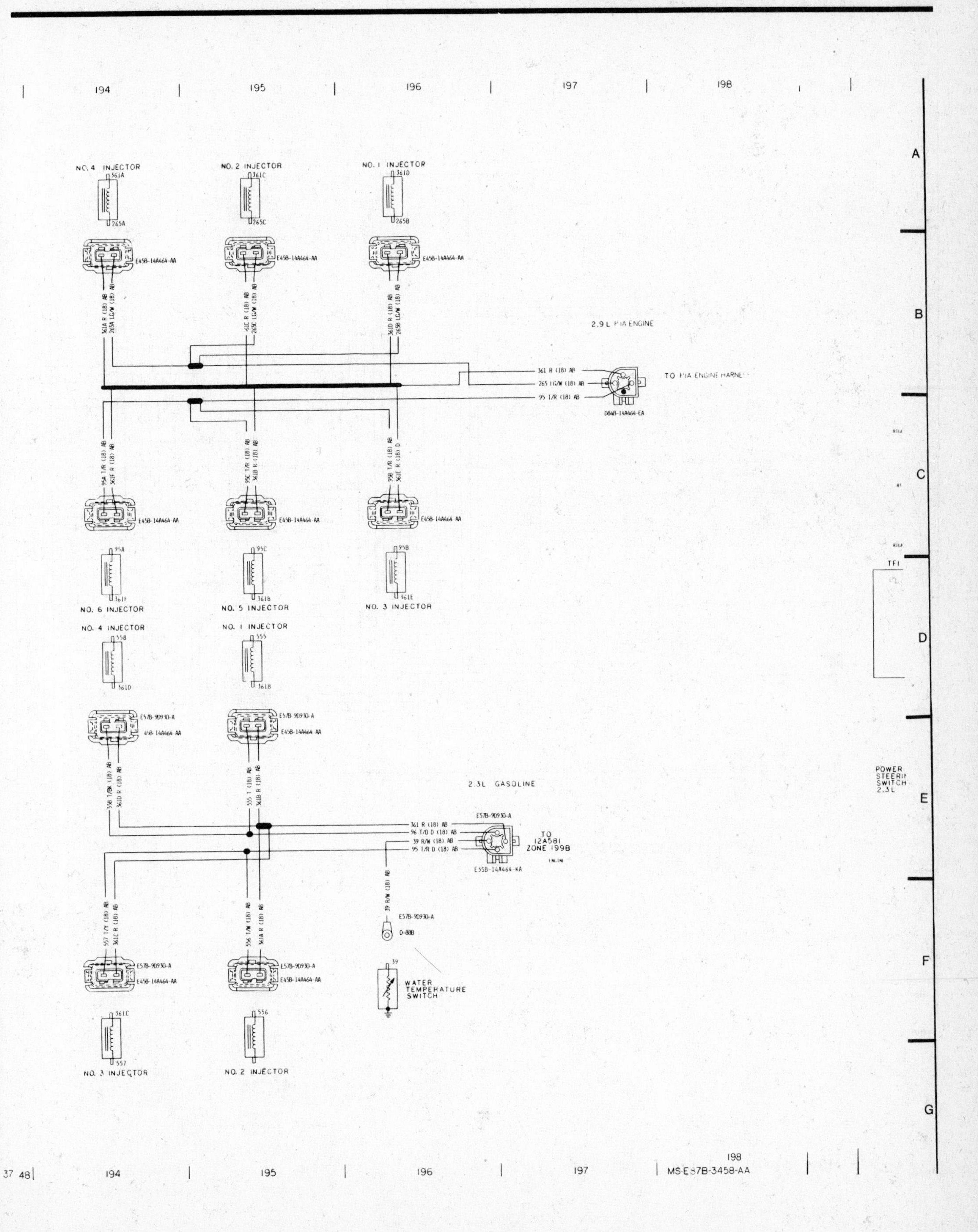
NO. 4 INJECTOR
NO. 2 INJECTOR
NO. 1 INJECTOR
E45B-14A464-AA
2.9 L PFIA ENGINE
TO PFIA ENGINE HARNESS
361 R (18) AB
265 LG/W (18) AB
95 T/R (18) AB
D84B-14A464-EA
NO. 6 INJECTOR
NO. 5 INJECTOR
NO. 3 INJECTOR
2.3L GASOLINE
E57B-9D930-A
TO 12A581 ZONE 199B
E35B-14A464-KA
D-88B
WATER TEMPERATURE SWITCH
POWER STEERING SWITCH 2.3L
TFI
MS-E87B-3458-AA

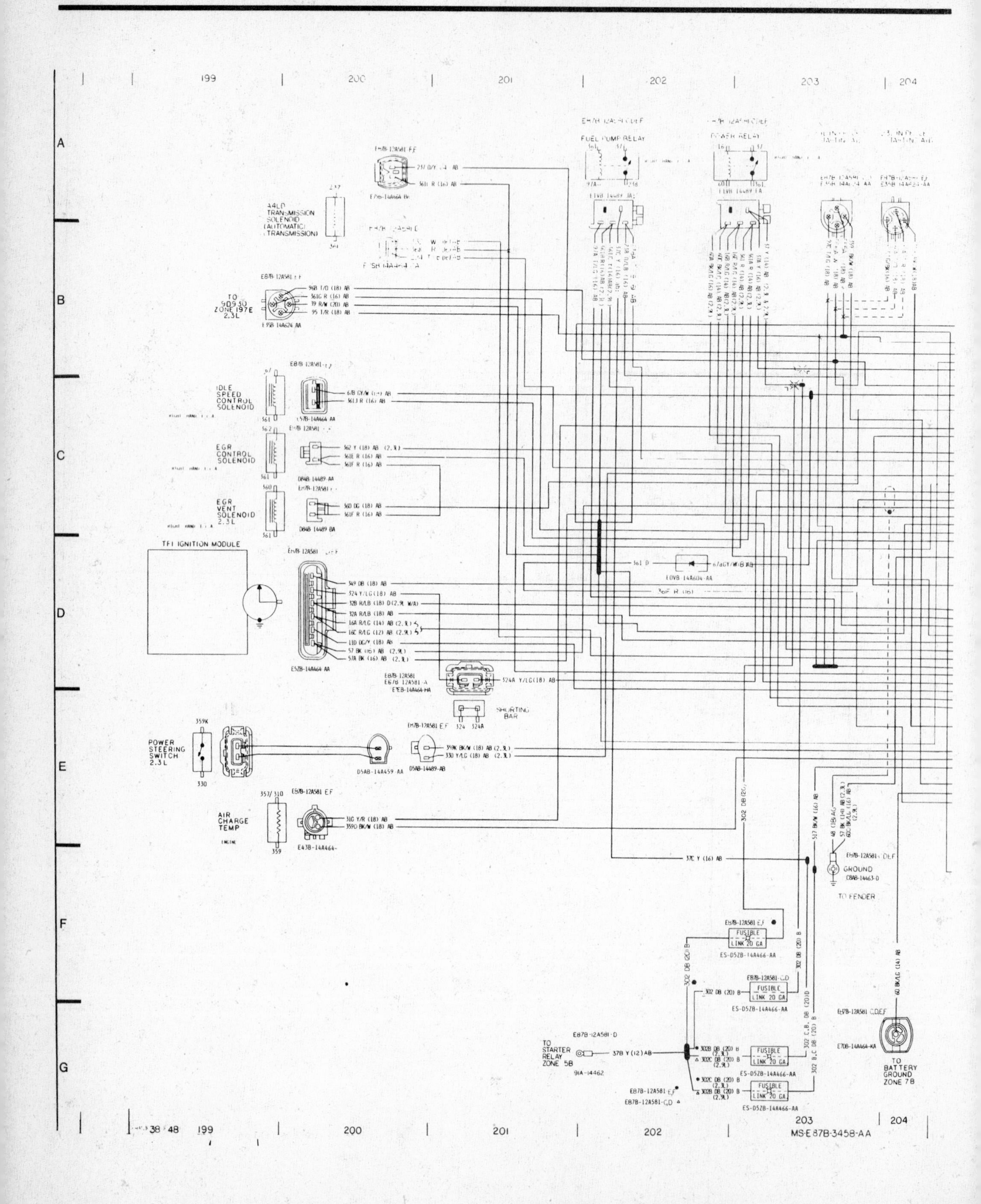
FUEL PUMP RELAY
POWER RELAY
A4LD TRANSMISSION SOLENOID (AUTOMATIC TRANSMISSION)
TO 9D930 ZONE 197E 2.3L
IDLE SPEED CONTROL SOLENOID
EGR CONTROL SOLENOID
EGR VENT SOLENOID 2.3L
TFI IGNITION MODULE
SHORTING BAR
POWER STEERING SWITCH 2.3L
AIR CHARGE TEMP
GROUND
TO FENDER
FUSIBLE LINK 20 GA
TO STARTER RELAY ZONE 5B
TO BATTERY GROUND ZONE 7B
MS-E87B-3458-AA

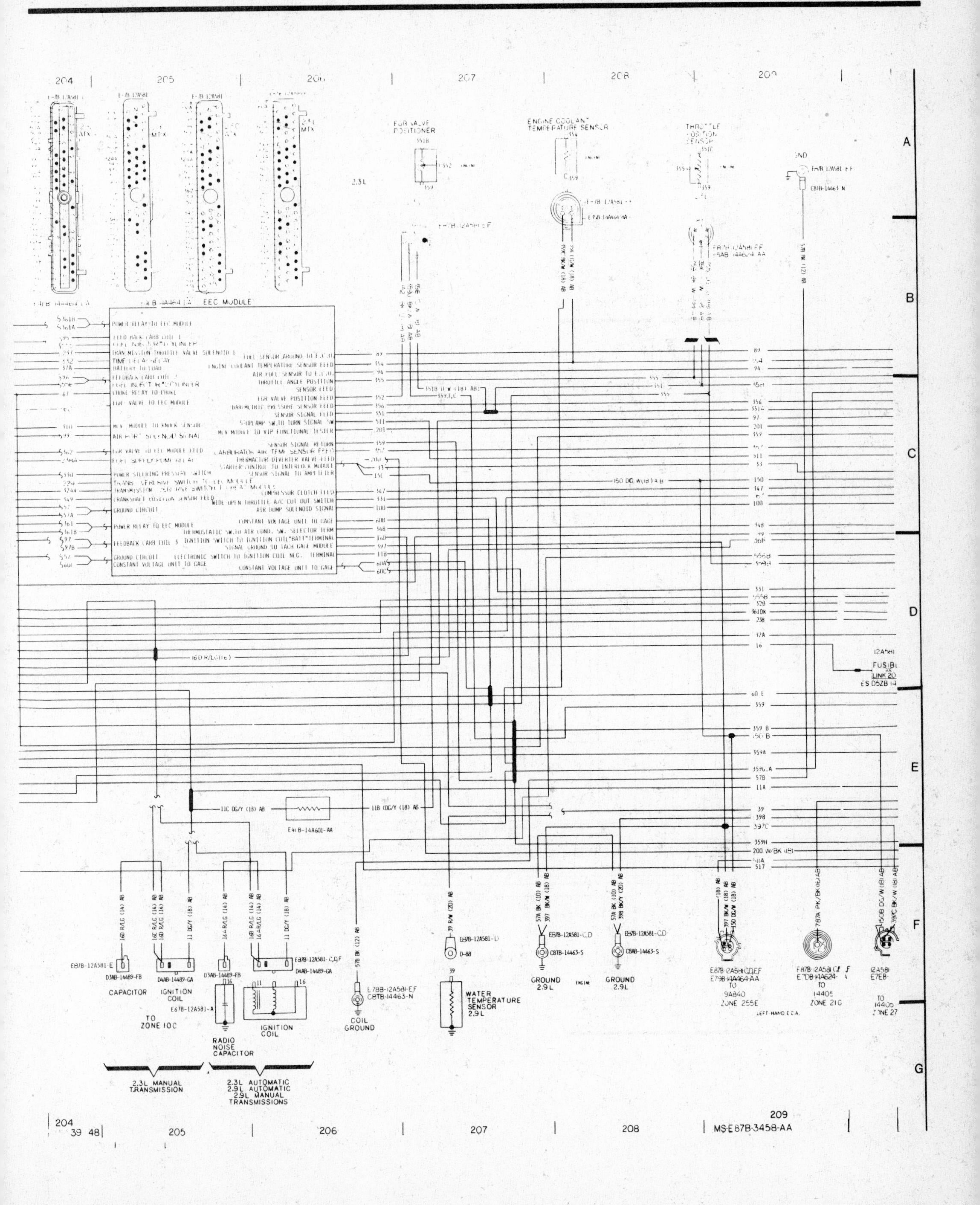
EEC MODULE
EGR VALVE POSITIONER
ENGINE COOLANT TEMPERATURE SENSOR
THROTTLE POSITION SENSOR
FUSIBLE LINK 20
CAPACITOR
IGNITION COIL
TO ZONE 10C
RADIO NOISE CAPACITOR
IGNITION COIL
COIL GROUND
WATER TEMPERATURE SENSOR 2.9 L
GROUND 2.9 L
GROUND 2.9 L
2.3L MANUAL TRANSMISSION
2.3L AUTOMATIC
2.9L AUTOMATIC
2.9L MANUAL
TRANSMISSIONS
MS-E87B-3458-AA

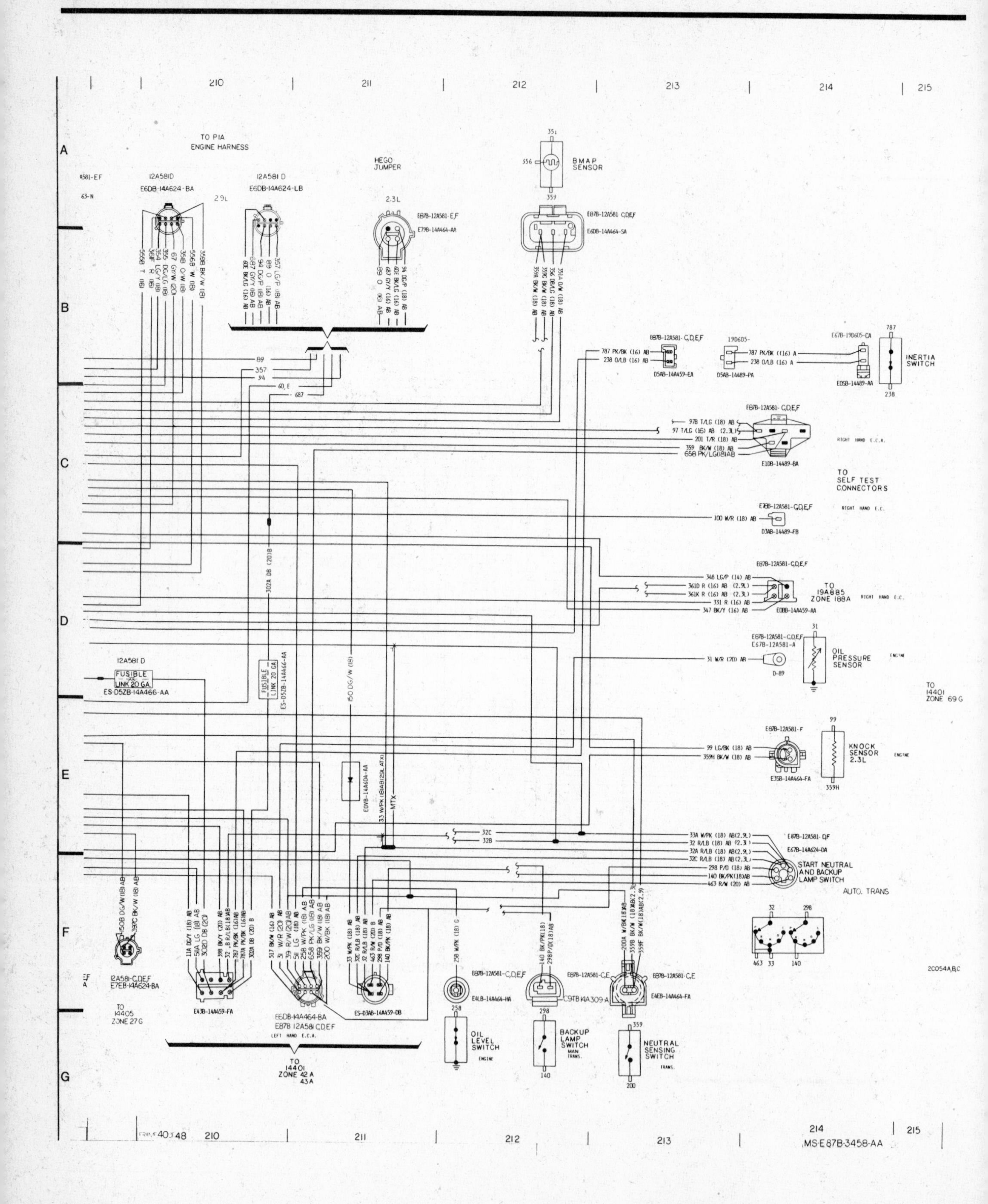

TO PIA ENGINE HARNESS
HEGO JUMPER
BMAP SENSOR
INERTIA SWITCH
TO SELF TEST CONNECTORS
TO 19A885 ZONE 188A
OIL PRESSURE SENSOR
TO 14401 ZONE 69 G
KNOCK SENSOR 2.3L
START NEUTRAL AND BACKUP LAMP SWITCH
AUTO. TRANS
FUSIBLE LINK 20 GA.
TO 14405 ZONE 27 G
TO 14401 ZONE 42 A 43A
OIL LEVEL SWITCH
BACKUP LAMP SWITCH
NEUTRAL SENSING SWITCH
MS-E87B-3458-AA

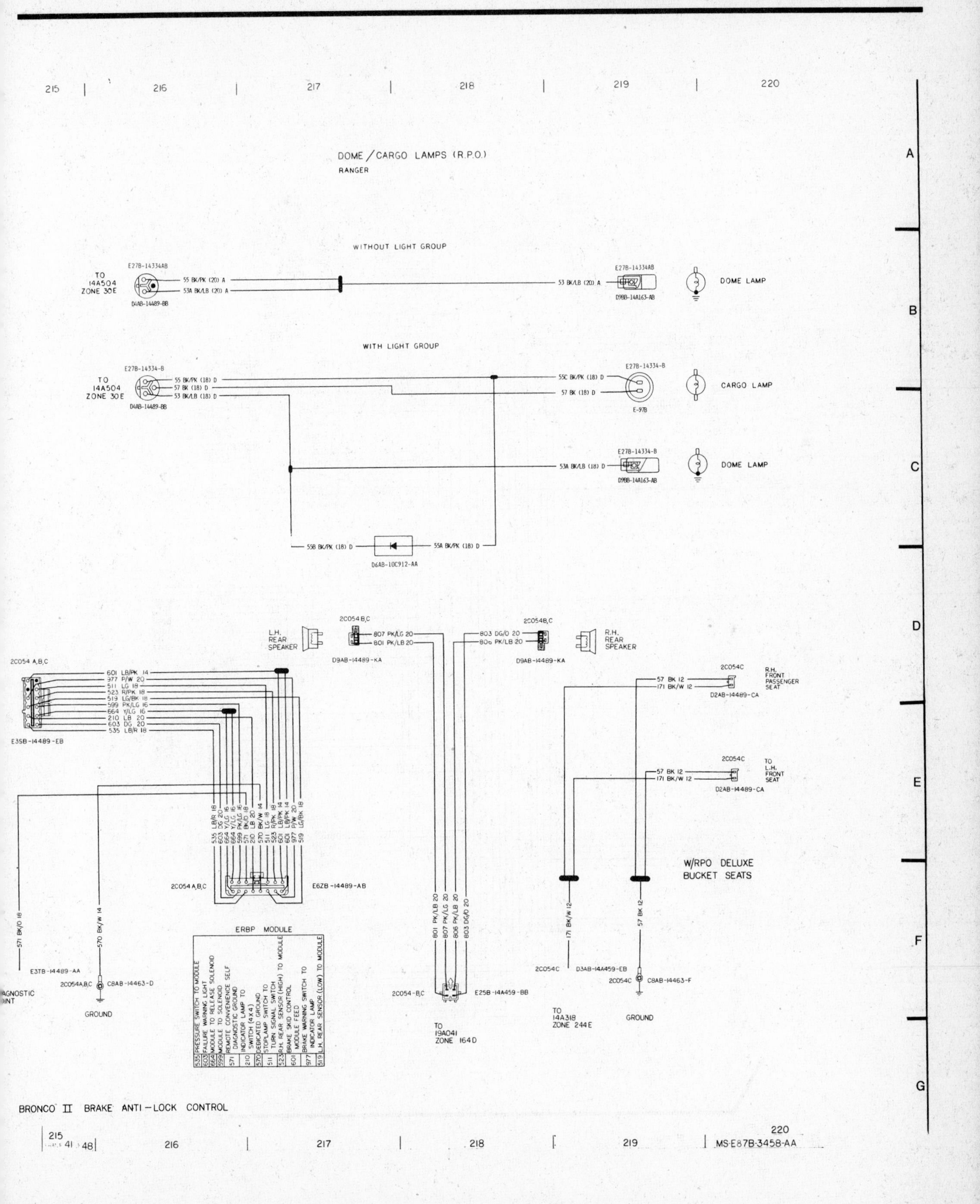

215
216
217
218
219
220
A
B
C
D
E
F
G
DOME / CARGO LAMPS (R.P.O.)
RANGER
WITHOUT LIGHT GROUP
TO
14A504
ZONE 30E
E27B-14334AB
55 BK/PK (20) A
53A BK/LB (20) A
D4AB-14489-BB
53 BK/LB (20) A
D9BB-14A163-AB
DOME LAMP
WITH LIGHT GROUP
E27B-14334-B
55 BK/PK (18) D
57 BK (18) D
53 BK/LB (18) D
55C BK/PK (18) D
57 BK (18) D
E-97B
CARGO LAMP
53A BK/LB (18) D
DOME LAMP
55B BK/PK (18) D
55A BK/PK (18) D
D6AB-10C912-AA
L.H.
REAR
SPEAKER
2C054 B,C
807 PK/LG 20
801 PK/LB 20
D9AB-14489-KA
2C054B,C
803 DG/O 20
806 PK/LB 20
R.H.
REAR
SPEAKER
2C054 A,B,C
601 LB/PK 14
977 P/W 20
511 LG 18
523 R/PK 18
519 LG/BK 18
599 PK/LG 16
664 Y/LG 16
210 LB 20
603 DG 20
535 LB/R 18
E35B-14489-EB
2C054C
57 BK 12
171 BK/W 12
R.H.
FRONT
PASSENGER
SEAT
D2AB-14489-CA
TO
L.H.
FRONT
SEAT
W/RPO DELUXE
BUCKET SEATS
E6ZB-14489-AB
571 BK/O 18
570 BK/W 14
E3TB-14489-AA
C8AB-14463-D
GROUND
ERBP MODULE
535 PRESSURE SWITCH TO MODULE
603 FAILURE WARNING LIGHT
664 MODULE TO RELEASE SOLENOID
599 MODULE TO SOLENOID
571 REMOTE CONVENIENCE SELF DIAGNOSTIC GROUND
210 INDICATOR LAMP TO SWITCH (4X4)
570 DEDICATED GROUND
511 STOPLAMP SWITCH TO TURN SIGNAL SWITCH
523 R.H. REAR SENSOR (HIGH) TO MODULE
601 BRAKE SKID CONTROL MODULE FEED
977 BRAKE WARNING SWITCH TO INDICATOR LAMP
519 L.H. REAR SENSOR (LOW) TO MODULE
801 PK/LB 20
807 PK/LG 20
806 PK/LB 20
803 DG/O 20
2C054-B,C
E25B-14A459-BB
TO
19A041
ZONE 164D
171 BK/W 12
57 BK 12
D3AB-14A459-EB
C8AB-14463-F
TO
14A318
ZONE 244E
GROUND
BRONCO II BRAKE ANTI-LOCK CONTROL
MS-E87B-3458-AA

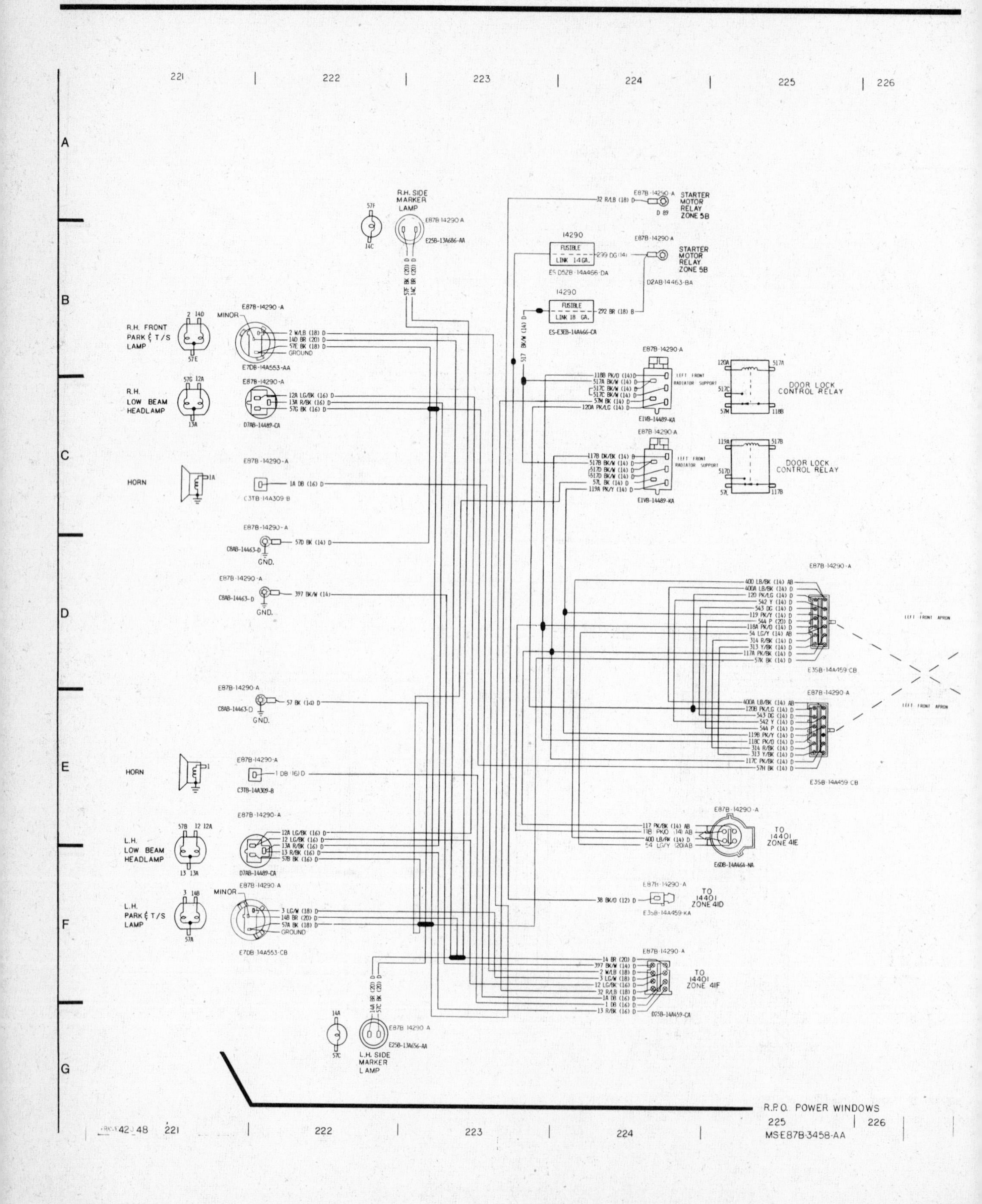
R.H. SIDE MARKER LAMP
STARTER MOTOR RELAY ZONE 5B
FUSIBLE LINK 14 GA.
FUSIBLE LINK 18 GA.
R.H. FRONT PARK & T/S LAMP
R.H. LOW BEAM HEADLAMP
DOOR LOCK CONTROL RELAY
HORN
GND.
L.H. LOW BEAM HEADLAMP
L.H. PARK & T/S LAMP
TO 14401 ZONE 41E
TO 14401 ZONE 41D
TO 14401 ZONE 41F
L.H. SIDE MARKER LAMP
R.P.O. POWER WINDOWS
MS-E87B-3458-AA

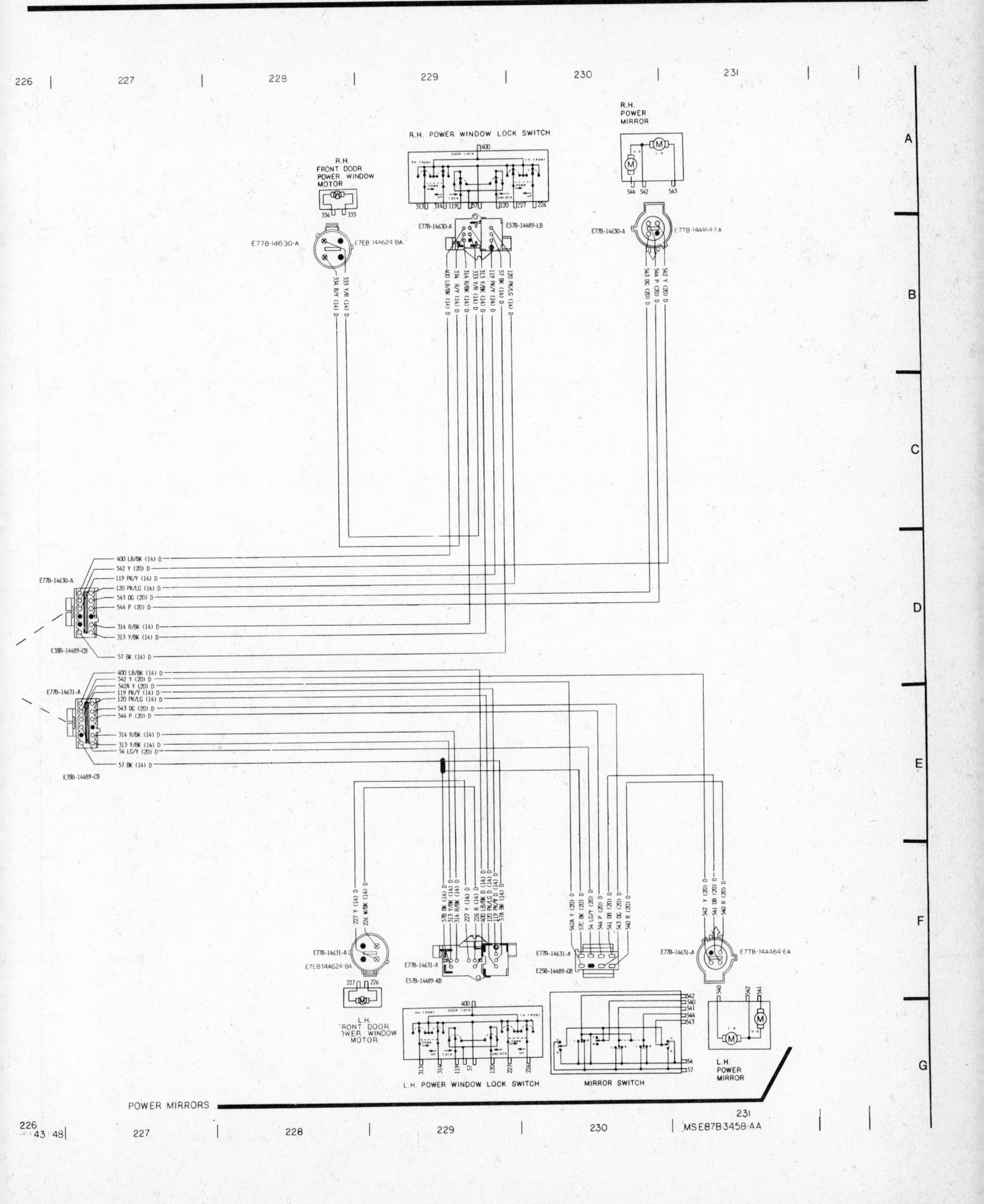
R.H. POWER WINDOW LOCK SWITCH
R.H. FRONT DOOR POWER WINDOW MOTOR
R.H. POWER MIRROR
L.H. FRONT DOOR POWER WINDOW MOTOR
L.H. POWER WINDOW LOCK SWITCH
MIRROR SWITCH
L.H. POWER MIRROR
POWER MIRRORS
MSE87B3458-AA

237 238 239 240 241 242

A

B

C

D

E

F

G

237 45 48 238 239 240 241 242 MSE87B-3458-AA

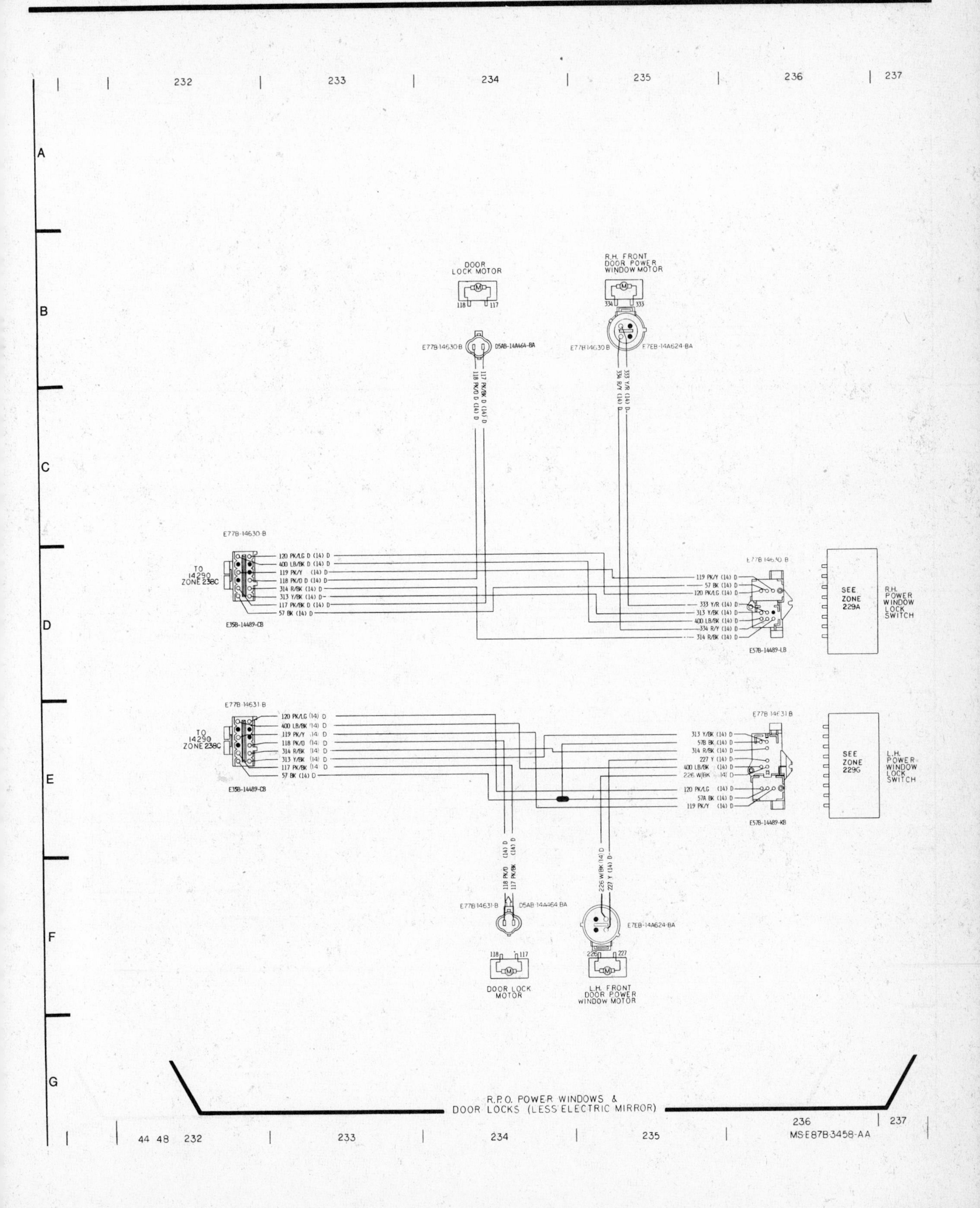
232
233
234
235
236
237
A
B
C
D
E
F
G
DOOR LOCK MOTOR
118
117
E77B-14630-B
D5AB-14A464-BA
117 PK/BK D (14) D
118 PK/O D (14) D
R.H. FRONT DOOR POWER WINDOW MOTOR
334
333
E77B-14630-B
E7EB-14A624-BA
333 Y/R (14) D
334 R/Y (14) D
E77B-14630-B
TO 14290 ZONE 238C
120 PK/LG D (14) D
400 LB/BK D (14) D
119 PK/Y (14) D
118 PK/O D (14) D
314 R/BK (14) D
313 Y/BK (14) D
117 PK/BK D (14) D
57 BK (14) D
E35B-14489-CB
E77B-14630-B
119 PK/Y (14) D
57 BK (14) D
120 PK/LG (14) D
333 Y/R (14) D
313 Y/BK (14) D
400 LB/BK (14) D
334 R/Y (14) D
314 R/BK (14) D
E57B-14489-LB
SEE ZONE 229A
R.H. POWER WINDOW LOCK SWITCH
E77B-14631-B
TO 14290 ZONE 238C
120 PK/LG (14) D
400 LB/BK (14) D
119 PK/Y (14) D
118 PK/O (14) D
314 R/BK (14) D
313 Y/BK (14) D
117 PK/BK (14) D
57 BK (14) D
E35B-14489-CB
E77B-14631-B
313 Y/BK (14) D
57B BK (14) D
314 R/BK (14) D
227 Y (14) D
400 LB/BK (14) D
226 W/BK (14) D
120 PK/LG (14) D
57A BK (14) D
119 PK/Y (14) D
E57B-14489-KB
SEE ZONE 229G
L.H. POWER WINDOW LOCK SWITCH
118 PK/O (14) D
117 PK/BK (14) D
E77B-14631-B
D5AB-14A464-BA
118
117
DOOR LOCK MOTOR
226 W/BK (14) D
227 Y (14) D
E7EB-14A624-BA
226
227
L.H. FRONT DOOR POWER WINDOW MOTOR
R.P.O. POWER WINDOWS & DOOR LOCKS (LESS ELECTRIC MIRROR)
44 48
MS-E87B3458-AA

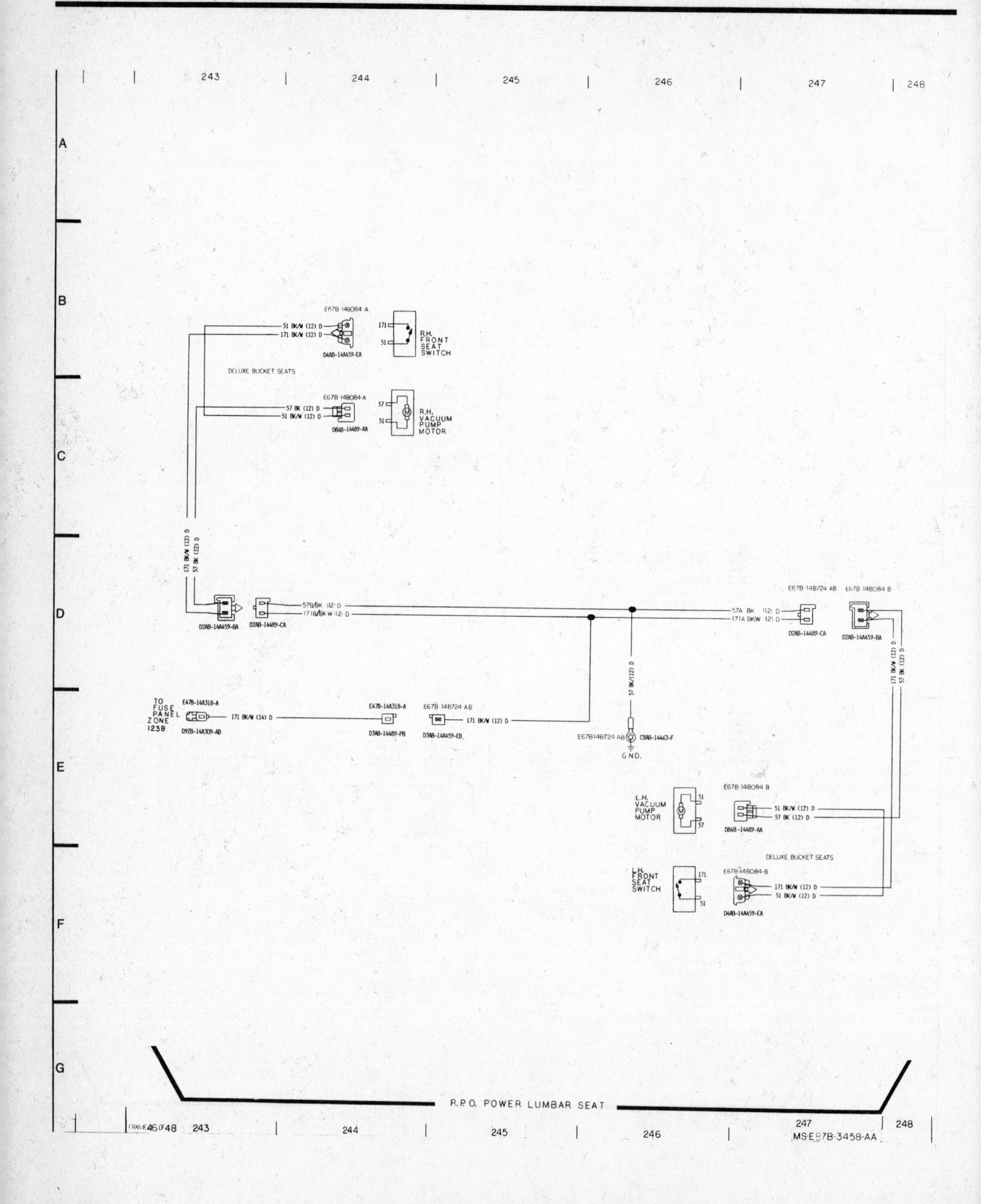
243
244
245
246
247
248
A
B
C
D
E
F
G
E67B-14B084-A
51 BK/W (12) D
171 BK/W (12) D
D4AB-14A459-EA
R.H. FRONT SEAT SWITCH
DELUXE BUCKET SEATS
E67B-14B084-A
57 BK (12) D
51 BK/W (12) D
D8AB-14489-AA
R.H. VACUUM PUMP MOTOR
171 BK/W (12) D
57 BK (12) D
D2AB-14A459-BA
D2AB-14489-CA
57B/BK (12) D
171B/BK W (12) D
E67B-14B724 AB
E67B 14B084 B
57A BK (12) D
171A BK/W (12) D
D2AB-14489-CA
D2AB-14A459-BA
171 BK/W (12) D
57 BK (12) D
57 BK/(12) D
TO FUSE PANEL ZONE 123B
E47B-14A318-A
D9ZB-14A309-AB
171 BK/W (14) D
E47B-14A318-A
D3AB-14489-PB
E67B 14B724 AB
D3AB-14A459-EB
171 BK/W (12) D
E67B14B724 AB
C8AB-14463-F
GND.
L.H. VACUUM PUMP MOTOR
E67B 14B084 B
51 BK/W (12) D
57 BK (12) D
D8AB-14489-AA
DELUXE BUCKET SEATS
L.H. FRONT SEAT SWITCH
E67B-14B084-B
171 BK/W (12) D
51 BK/W (12) D
D4AB-14A459-EA
R.P.O. POWER LUMBAR SEAT
FRAME 46 OF 48
MS-E87B-3458-AA

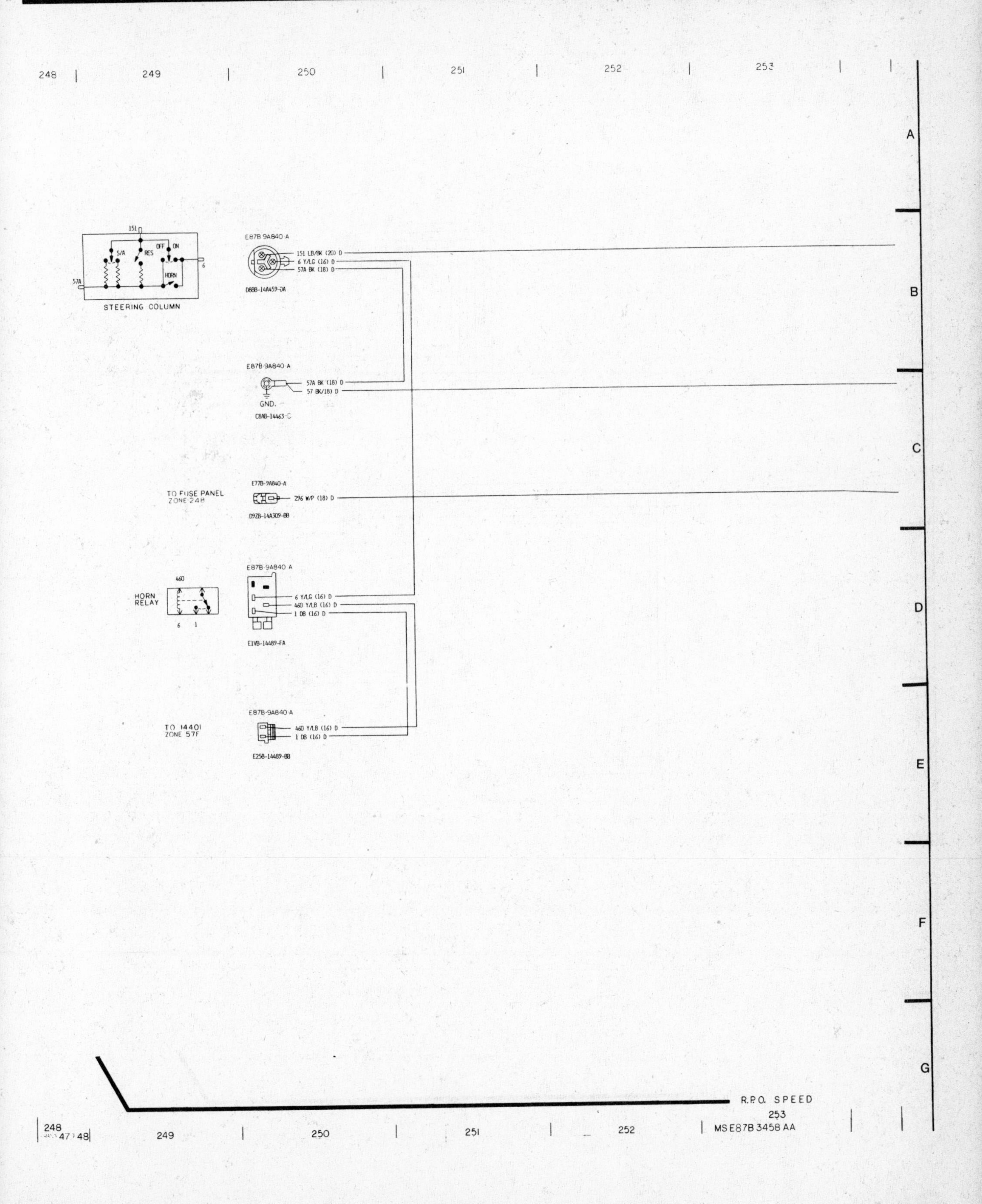
248
249
250
251
252
253
A
B
C
D
E
F
G
151
S/A
RES
OFF
ON
HORN
57A
6
STEERING COLUMN
E87B 9A840 A
151 LB/BK (20) D
6 Y/LG (16) D
57A BK (18) D
D8BB-14A459-DA
E87B 9A840 A
57A BK (18) D
57 BK/18) D
GND.
C8AB-14463-C
E77B-9A840-A
TO FUSE PANEL
ZONE 248
296 W/P (18) D
D9ZB-14A309-BB
E87B 9A840 A
460
HORN
RELAY
6 Y/LG (16) D
460 Y/LB (16) D
1 DB (16) D
E1VB-14489-FA
E87B-9A840-A
TO 14401
ZONE 57F
460 Y/LB (16) D
1 DB (16) D
E25B-14489-BB
R.P.O. SPEED
MS E87B 3458 AA

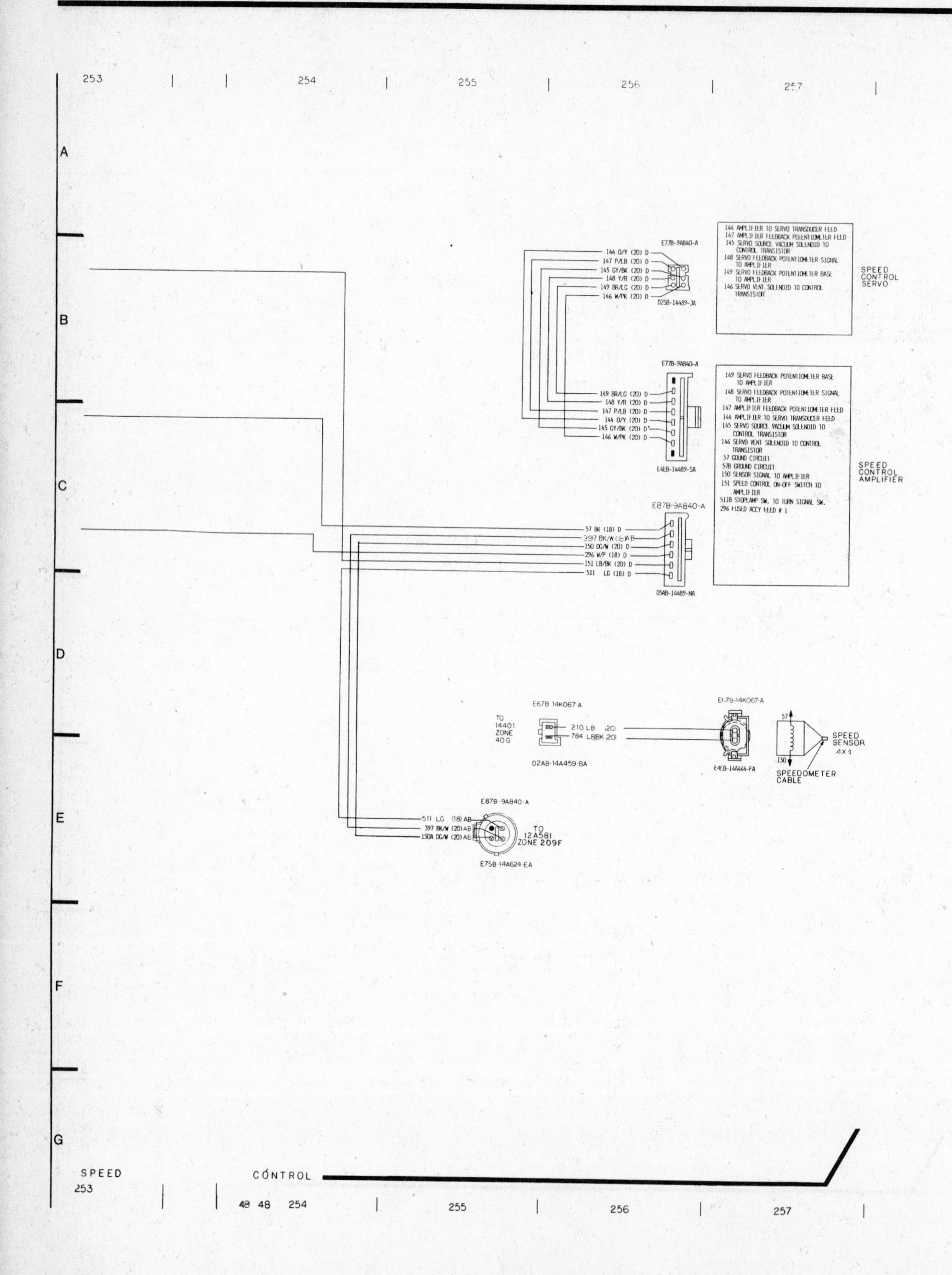
253
254
255
256
257
A
B
C
D
E
F
G
E77B-9A840-A
144 O/Y (20) D
147 P/LB (20) D
145 GY/BK (20) D
148 Y/R (20) D
149 BR/LG (20) D
146 W/PK (20) D
D2SB-14489-JA
144 AMPLIFIER 10 SERVO 1RANSDUCER FEED
147 AMPLIFIER FEEDBACK POTENTIOMETER FEED
145 SERVO SOURCE VACUUM SOLENOID 10 CONTROL TRANSISTOR
148 SERVO FEEDBACK POTENTIOMETER SIGNAL 10 AMPLIFIER
149 SERVO FEEDBACK POTENTIOMETER BASE 10 AMPLIFIER
146 SERVO VENT SOLENOID 10 CONTROL TRANSISTOR
SPEED CONTROL SERVO
E77B-9A840-A
149 BR/LG (20) D
148 Y/R (20) D
147 P/LB (20) D
144 O/Y (20) D
145 GY/BK (20) D
146 W/PK (20) D
E4EB-14489-SA
149 SERVO FEEDBACK POTENTIOMETER BASE 10 AMPLIFIER
148 SERVO FEEDBACK POTENTIOMETER SIGNAL 10 AMPLIFIER
147 AMPLIFIER FEEDBACK POTENTIOMETER FEED
144 AMPLIFIER 10 SERVO TRANSDUCER FEED
145 SERVO SOURCE VACUUM SOLENOID 10 CONTROL TRANSISTOR
146 SERVO VENT SOLENOID 10 CONTROL TRANSISTOR
57 GOUND CIRCUIT
57B GROUND CIRCUIT
150 SENSOR SIGNAL 10 AMPLIFIER
151 SPEED CONTROL ON-OFF SWITCH 10 AMPLIFIER
511B STOPLAMP SW. 10 TURN SIGNAL SW.
296 FUSED ACCY FEED # 1
SPEED CONTROL AMPLIFIER
E87B-9A840-A
57 BK (18) D
397 BK/W (18) AB
150 DG/W (20) D
296 W/P (18) D
151 LB/BK (20) D
511 LG (18) D
D5AB-14489-NA
E67B 14K067 A
TO 14401 ZONE 40 G
210 LB (20)
784 LB/BK (20)
D2AB-14A459-BA
E67B-14K067-A
E4EB-14A464-FA
57
150
SPEED SENSOR 4X4
SPEEDOMETER CABLE
E87B-9A840-A
511 LG (18) AB
397 BK/W (20) AB
150A DG/W (20) AB
TO 12A581 ZONE 209F
E7SB-14A624-EA
SPEED CONTROL
253
48 48 254
255
256
257

A

B

C

D

E

F

G

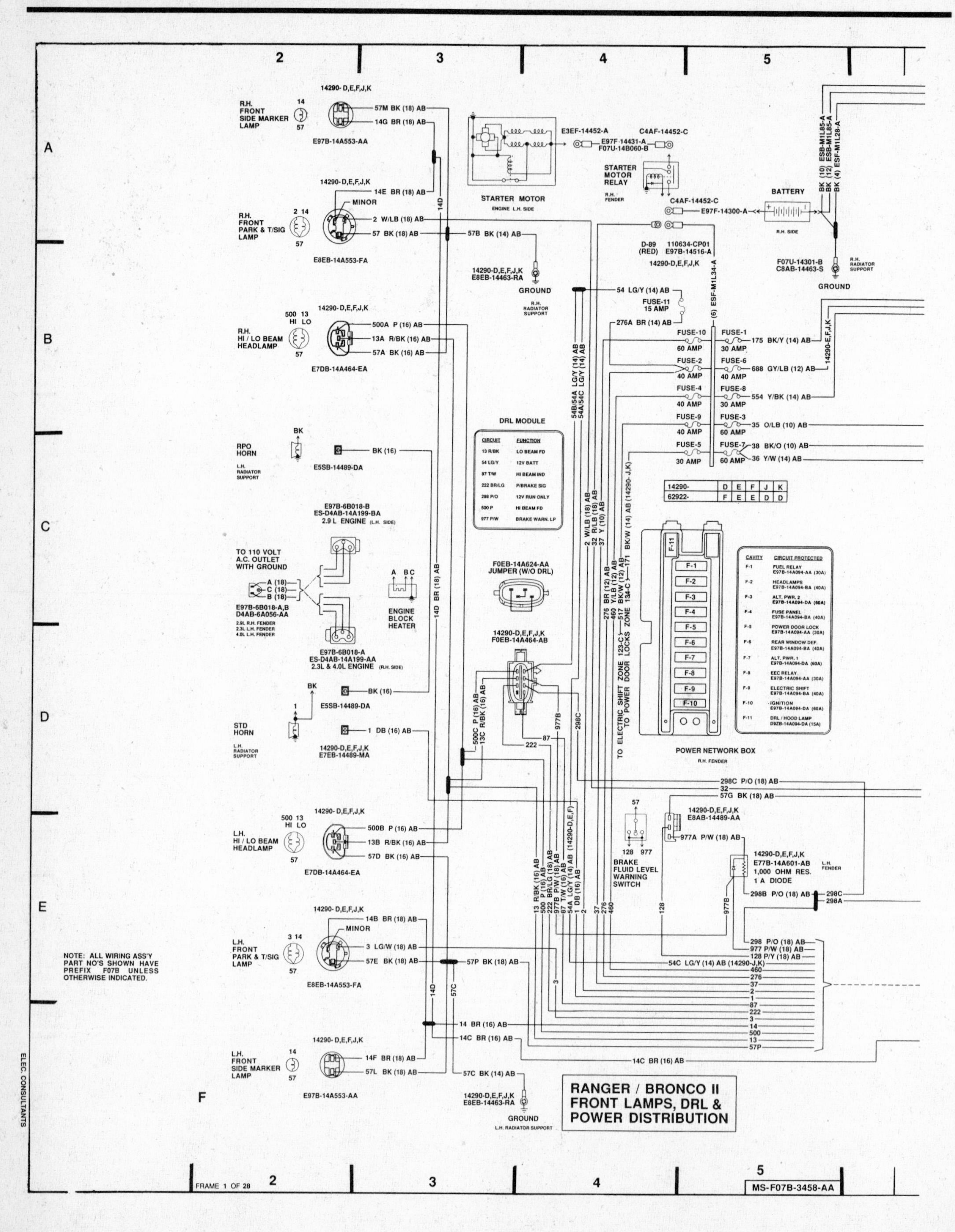
RANGER / BRONCO II
FRONT LAMPS, DRL &
POWER DISTRIBUTION
NOTE: ALL WIRING ASS'Y PART NO'S SHOWN HAVE PREFIX F07B UNLESS OTHERWISE INDICATED.
R.H. FRONT SIDE MARKER LAMP
R.H. FRONT PARK & T/SIG LAMP
R.H. HI / LO BEAM HEADLAMP
RPO HORN
STD HORN
L.H. HI / LO BEAM HEADLAMP
L.H. FRONT PARK & T/SIG LAMP
L.H. FRONT SIDE MARKER LAMP
STARTER MOTOR
STARTER MOTOR RELAY
BATTERY
GROUND
DRL MODULE
ENGINE BLOCK HEATER
TO 110 VOLT A.C. OUTLET WITH GROUND
F0EB-14A624-AA JUMPER (W/O DRL)
POWER NETWORK BOX
BRAKE FLUID LEVEL WARNING SWITCH
TO ELECTRIC SHIFT ZONE 123-C TO POWER DOOR LOCKS ZONE 134-C
FRAME 1 OF 28
MS-F07B-3458-AA
ELEC. CONSULTANTS

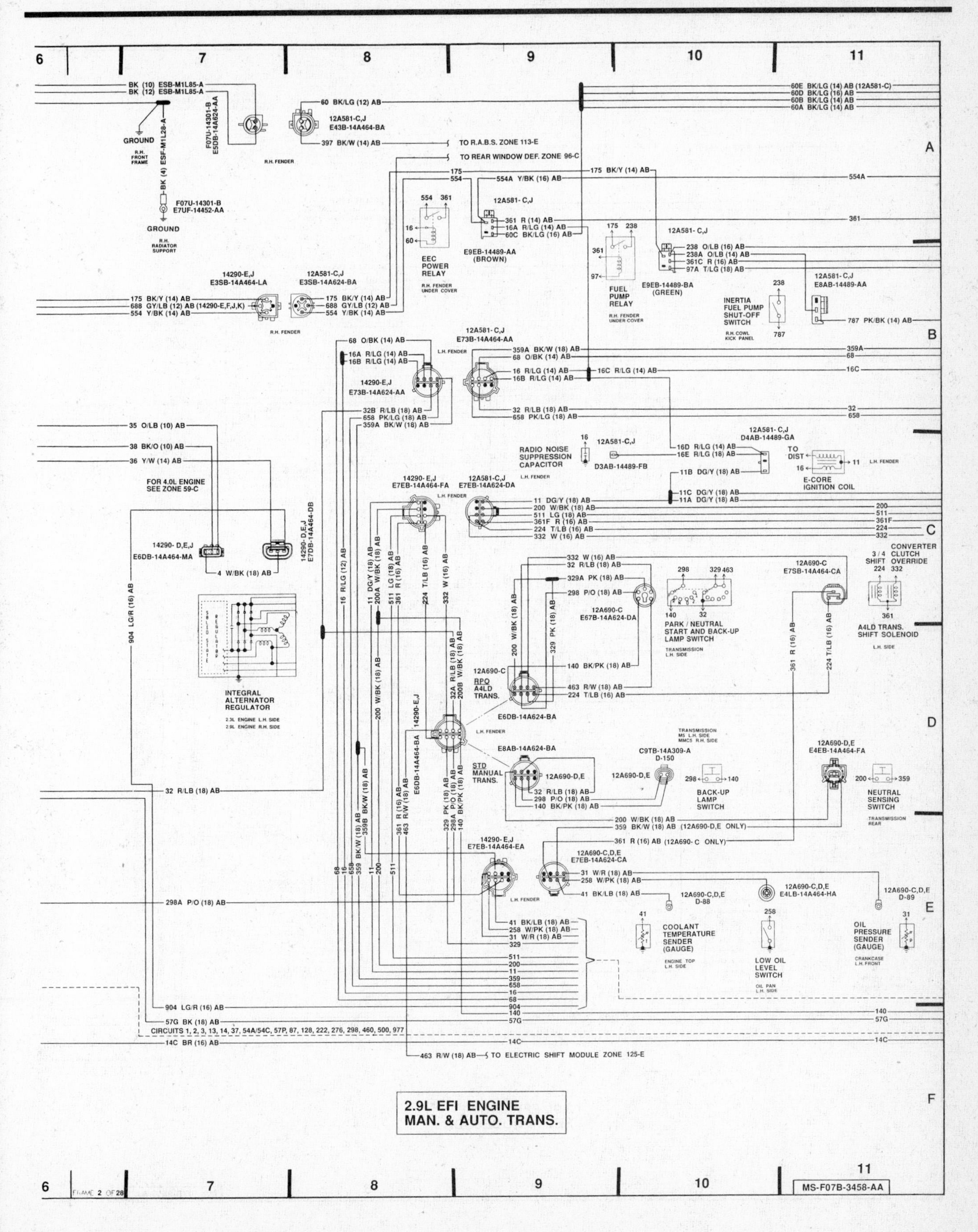
2.9L EFI ENGINE
MAN. & AUTO. TRANS.
MS-F07B-3458-AA

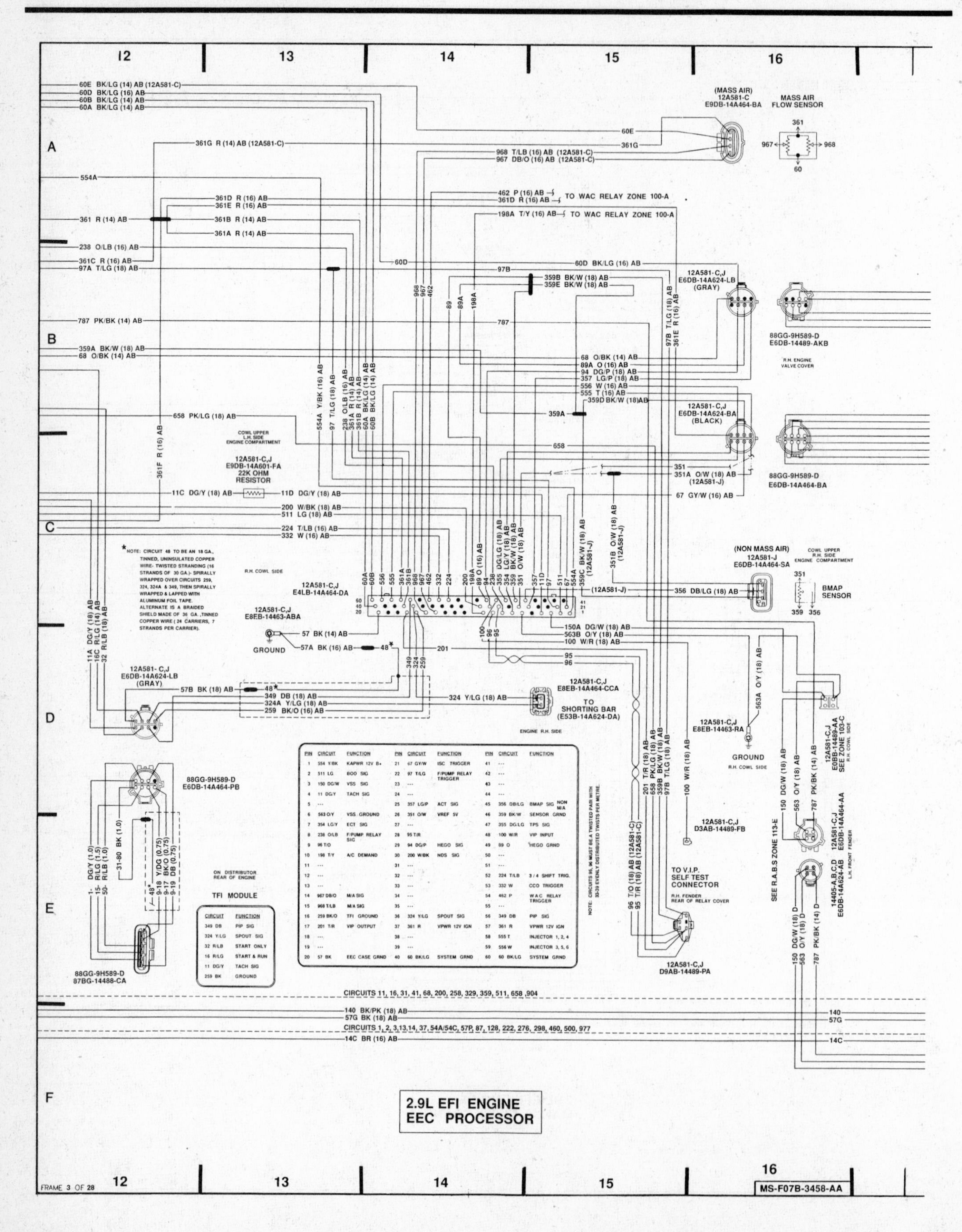
12
13
14
15
16
A
B
C
D
E
F
60E BK/LG (14) AB (12A581-C)
60D BK/LG (16) AB
60B BK/LG (14) AB
60A BK/LG (14) AB
361G R (14) AB (12A581-C)
554A
361D R (16) AB
361E R (16) AB
361 R (14) AB
361B R (14) AB
361A R (14) AB
238 O/LB (16) AB
361C R (16) AB
97A T/LG (18) AB
787 PK/BK (14) AB
359A BK/W (18) AB
68 O/BK (14) AB
658 PK/LG (18) AB
(MASS AIR)
12A581-C
E9DB-14A464-BA
MASS AIR FLOW SENSOR
968 T/LB (16) AB (12A581-C)
967 DB/O (16) AB (12A581-C)
462 P (16) AB
361D R (16) AB
TO WAC RELAY ZONE 100-A
198A T/Y (16) AB
TO WAC RELAY ZONE 100-A
60D BK/LG (16) AB
359B BK/W (18) AB
359E BK/W (18) AB
12A581-C,J
E6DB-14A624-LB
(GRAY)
88GG-9H589-D
E6DB-14489-AKB
R.H. ENGINE VALVE COVER
68 O/BK (14) AB
89A O (16) AB
94 DG/P (18) AB
357 LG/P (18) AB
556 W (16) AB
555 T (16) AB
359D BK/W (18) AB
12A581-C,J
E6DB-14A624-BA
(BLACK)
COWL UPPER L.H. SIDE ENGINE COMPARTMENT
12A581-C,J
E9DB-14A601-FA
22K OHM RESISTOR
351
351A O/W (18) AB
(12A581-J)
88GG-9H589-D
E6DB-14A464-BA
67 GY/W (16) AB
11C DG/Y (18) AB
11D DG/Y (18) AB
200 W/BK (18) AB
511 LG (18) AB
224 T/LB (16) AB
332 W (16) AB
(NON MASS AIR)
12A581-J
E6DB-14A464-SA
COWL UPPER R.H. SIDE ENGINE COMPARTMENT
BMAP SENSOR
*NOTE: CIRCUIT 48 TO BE AN 18 GA., TINNED, UNINSULATED COPPER WIRE- TWISTED STRANDING (16 STRANDS OF 30 GA.)- SPIRALLY WRAPPED OVER CIRCUITS 259, 324, 324A & 349, THEN SPIRALLY WRAPPED & LAPPED WITH ALUMINUM FOIL TAPE. ALTERNATE IS A BRAIDED SHIELD MADE OF 36 GA. TINNED COPPER WIRE (24 CARRIERS, 7 STRANDS PER CARRIER).
R.H. COWL SIDE
12A581-C,J
E4LB-14A464-DA
12A581-C,J
E8EB-14463-ABA
57 BK (14) AB
57A BK (16) AB
GROUND
356 DB/LG (18) AB
150A DG/W (18) AB
563B O/Y (18) AB
100 W/R (18) AB
12A581- C,J
E6DB-14A624-LB
(GRAY)
57B BK (18) AB
349 DB (18) AB
324A Y/LG (18) AB
259 BK/O (16) AB
324 Y/LG (18) AB
12A581-C,J
E8EB-14A464-CCA
TO SHORTING BAR
(E53B-14A624-DA)
ENGINE R.H. SIDE
12A581-C,J
E8EB-14463-RA
GROUND
R.H. COWL SIDE
88GG-9H589-D
E6DB-14A464-PB
ON DISTRIBUTOR REAR OF ENGINE
TFI MODULE
CIRCUIT FUNCTION
349 DB PIP SIG
324 Y/LG SPOUT SIG
32 R/LB START ONLY
16 R/LG START & RUN
11 DG/Y TACH SIG
259 BK GROUND
88GG-9H589-D
87BG-14488-CA
PIN CIRCUIT FUNCTION PIN CIRCUIT FUNCTION PIN CIRCUIT FUNCTION
1 554 Y/BK KAPWR 12V B+ 21 67 GY/W ISC TRIGGER 41 ---
2 511 LG BOO SIG 22 97 T/LG F/PUMP RELAY TRIGGER 42 ---
3 150 DG/W VSS SIG 23 --- 43 ---
4 11 DG/Y TACH SIG 24 --- 44 ---
5 --- 25 357 LG/P ACT SIG 45 356 DB/LG BMAP SIG NON M/A
6 563 O/Y VSS GROUND 26 351 O/W VREF 5V 46 359 BK/W SENSOR GRND
7 354 LG/Y ECT SIG 27 --- 47 355 DG/LG TPS SIG
8 238 O/LB F/PUMP RELAY SIG 28 95 T/R 48 100 W/R VIP INPUT
9 96 T/O 29 94 DG/P HEGO SIG 49 89 O HEGO GRND
10 198 T/Y A/C DEMAND 30 200 W/BK NDS SIG 50 ---
11 --- 31 --- 51 ---
12 --- 32 --- 52 224 T/LB 3/4 SHIFT TRIG.
13 --- 33 --- 53 332 W CCO TRIGGER
14 967 DB/O M/A SIG 34 --- 54 462 P WAC RELAY TRIGGER
15 968 T/LB M/A SIG 35 --- 55 ---
16 259 BK/O TFI GROUND 36 324 Y/LG SPOUT SIG 56 349 DB PIP SIG
17 201 T/R VIP OUTPUT 37 361 R VPWR 12V IGN 57 361 R VPWR 12V IGN
18 --- 38 --- 58 555 T INJECTOR 1, 2, 4
19 --- 39 --- 59 556 W INJECTOR 3, 5, 6
20 57 BK EEC CASE GRND 40 60 BK/LG SYSTEM GRND 60 60 BK/LG SYSTEM GRND
NOTE: CIRCUITS 95, 96 MUST BE A TWISTED PAIR WITH 33-39 EVENLY DISTRIBUTED TWISTS PER METRE.
12A581-C,J
D3AB-14489-FB
TO V.I.P. SELF TEST CONNECTOR
R.H. FENDER REAR OF RELAY COVER
12A581-C,J
D9AB-14489-PA
12A581-C,J
E0BB-14489-AA
SEE ZONE 103-C
SEE R.A.B.S ZONE 113-E
14405-A,B,C,D
E6DB-14A624-AB
12A581-C,J
E6DB-14A464-AA
L.H. FRONT FENDER
CIRCUITS 11, 16, 31, 41, 68, 200, 258, 329, 359, 511, 658, 904
140 BK/PK (18) AB
57G BK (18) AB
CIRCUITS 1, 2, 3, 13, 14, 37, 54A/54C, 57P, 87, 128, 222, 276, 298, 460, 500, 977
14C BR (16) AB
2.9L EFI ENGINE
EEC PROCESSOR
FRAME 3 OF 28
MS-F07B-3458-AA

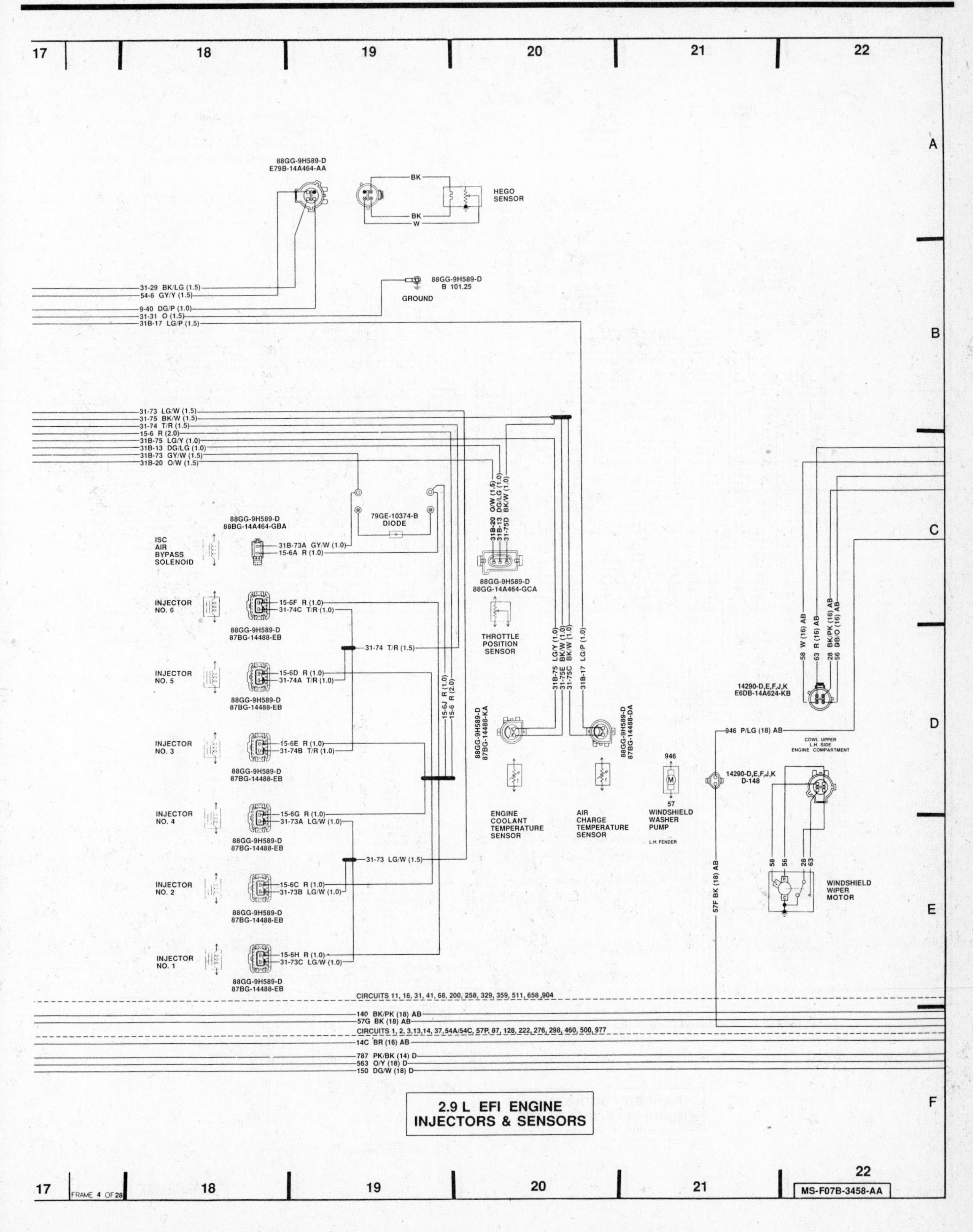

2.9 L EFI ENGINE
INJECTORS & SENSORS

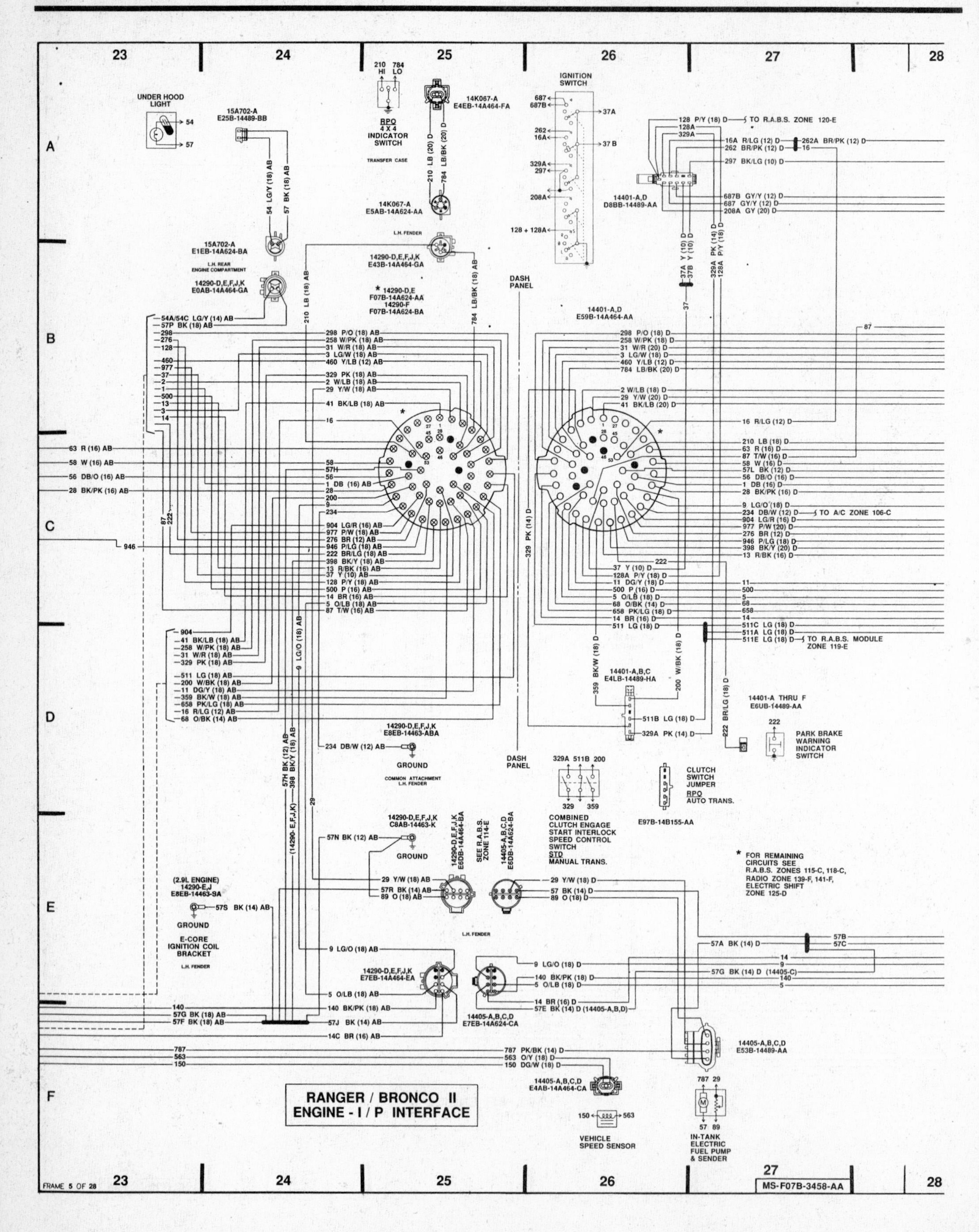
UNDER HOOD LIGHT
15A702-A
E25B-14489-BB
15A702-A
E1EB-14A624-BA
L.H. REAR ENGINE COMPARTMENT
14290-D,E,F,J,K
E0AB-14A464-GA
210 HI
784 LO
RPO
4 X 4
INDICATOR SWITCH
TRANSFER CASE
14K067-A
E4EB-14A464-FA
14K067-A
E5AB-14A624-AA
L.H. FENDER
14290-D,E,F,J,K
E43B-14A464-GA
* 14290-D,E
F07B-14A624-AA
14290-F
F07B-14A624-BA
IGNITION SWITCH
14401-A,D
D8BB-14489-AA
128 P/Y (18) D TO R.A.B.S. ZONE 120-E
DASH PANEL
14401-A,D
E59B-14A464-AA
9 LG/O (18) D TO A/C ZONE 106-C
511E LG (18) D TO R.A.B.S. MODULE ZONE 119-E
14401-A,B,C
E4LB-14489-HA
14401-A THRU F
E6UB-14489-AA
PARK BRAKE WARNING INDICATOR SWITCH
14290-D,E,F,J,K
E8EB-14463-ABA
GROUND
COMMON ATTACHMENT L.H. FENDER
DASH PANEL
COMBINED CLUTCH ENGAGE START INTERLOCK SPEED CONTROL SWITCH
STD
MANUAL TRANS.
CLUTCH SWITCH JUMPER
RPO
AUTO TRANS.
E97B-14B155-AA
14290-D,E,F,J,K
C8AB-14463-K
GROUND
SEE R.A.B.S. ZONE 114-E
* FOR REMAINING CIRCUITS SEE R.A.B.S. ZONES 115-C, 118-C, RADIO ZONE 139-F, 141-F, ELECTRIC SHIFT ZONE 125-D
(2.9L ENGINE)
14290-E,J
E8EB-14463-SA
GROUND
E-CORE IGNITION COIL BRACKET
L.H. FENDER
L.H. FENDER
14290-D,E,F,J,K
E7EB-14A464-EA
14405-A,B,C,D
E7EB-14A624-CA
14405-A,B,C,D
E53B-14489-AA
14405-A,B,C,D
E4AB-14A464-CA
VEHICLE SPEED SENSOR
IN-TANK ELECTRIC FUEL PUMP & SENDER
RANGER / BRONCO II
ENGINE - I / P INTERFACE
FRAME 5 OF 28
MS-F07B-3458-AA

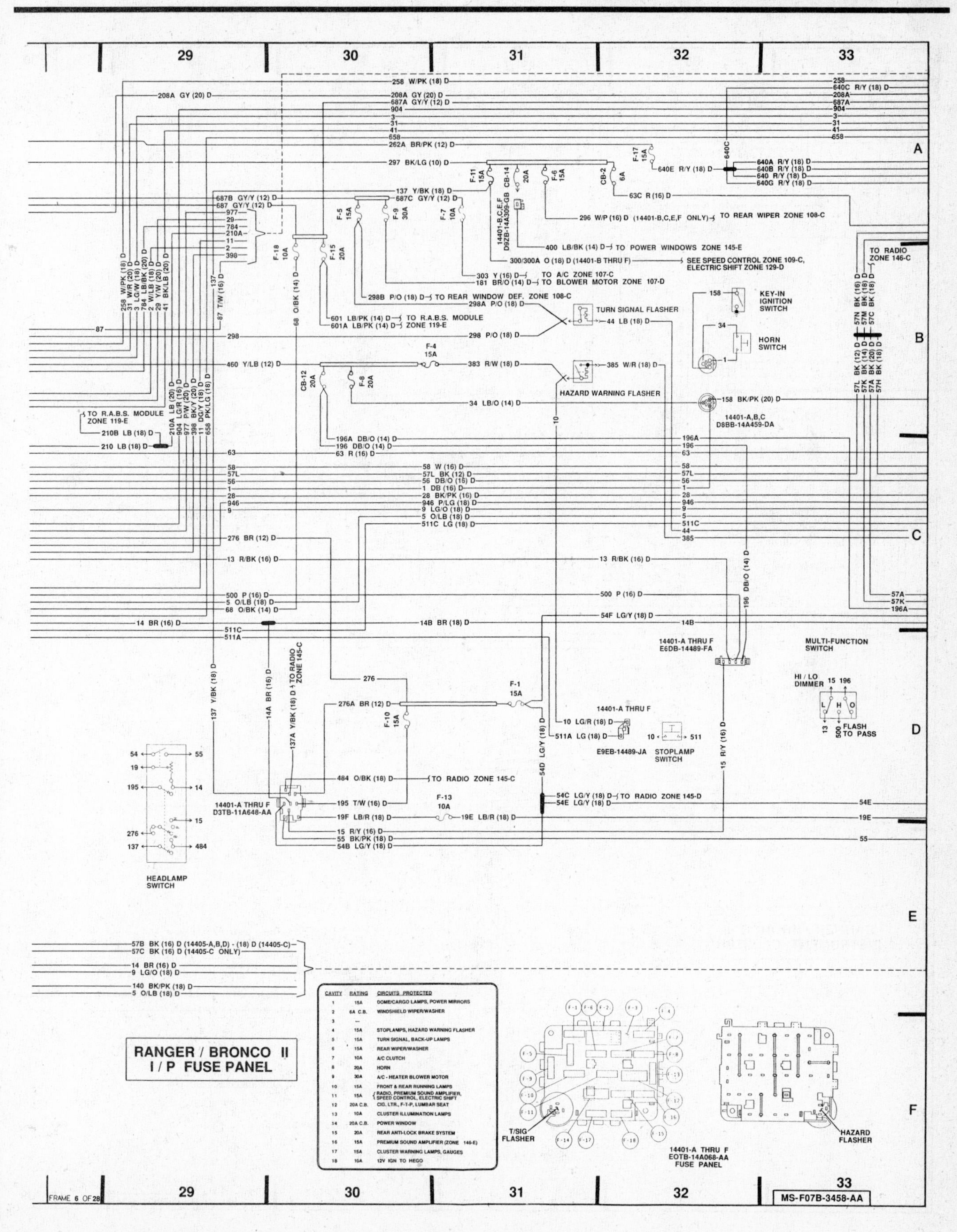

CAVITY	RATING	CIRCUITS PROTECTED
1	15A	DOME/CARGO LAMPS, POWER MIRRORS
2	6A C.B.	WINDSHIELD WIPER/WASHER
3	—	
4	15A	STOPLAMPS, HAZARD WARNING FLASHER
5	15A	TURN SIGNAL, BACK-UP LAMPS
6	15A	REAR WIPER/WASHER
7	10A	A/C CLUTCH
8	20A	HORN
9	30A	A/C - HEATER BLOWER MOTOR
10	15A	FRONT & REAR RUNNING LAMPS
11	15A	RADIO, PREMIUM SOUND AMPLIFIER, SPEED CONTROL, ELECTRIC SHIFT
12	20A C.B.	CIG. LTR., F-T-P, LUMBAR SEAT
13	10A	CLUSTER ILLUMINATION LAMPS
14	20A C.B.	POWER WINDOW
15	20A	REAR ANTI-LOCK BRAKE SYSTEM
16	15A	PREMIUM SOUND AMPLIFIER (ZONE 146-E)
17	15A	CLUSTER WARNING LAMPS, GAUGES
18	10A	12V IGN TO HEGO

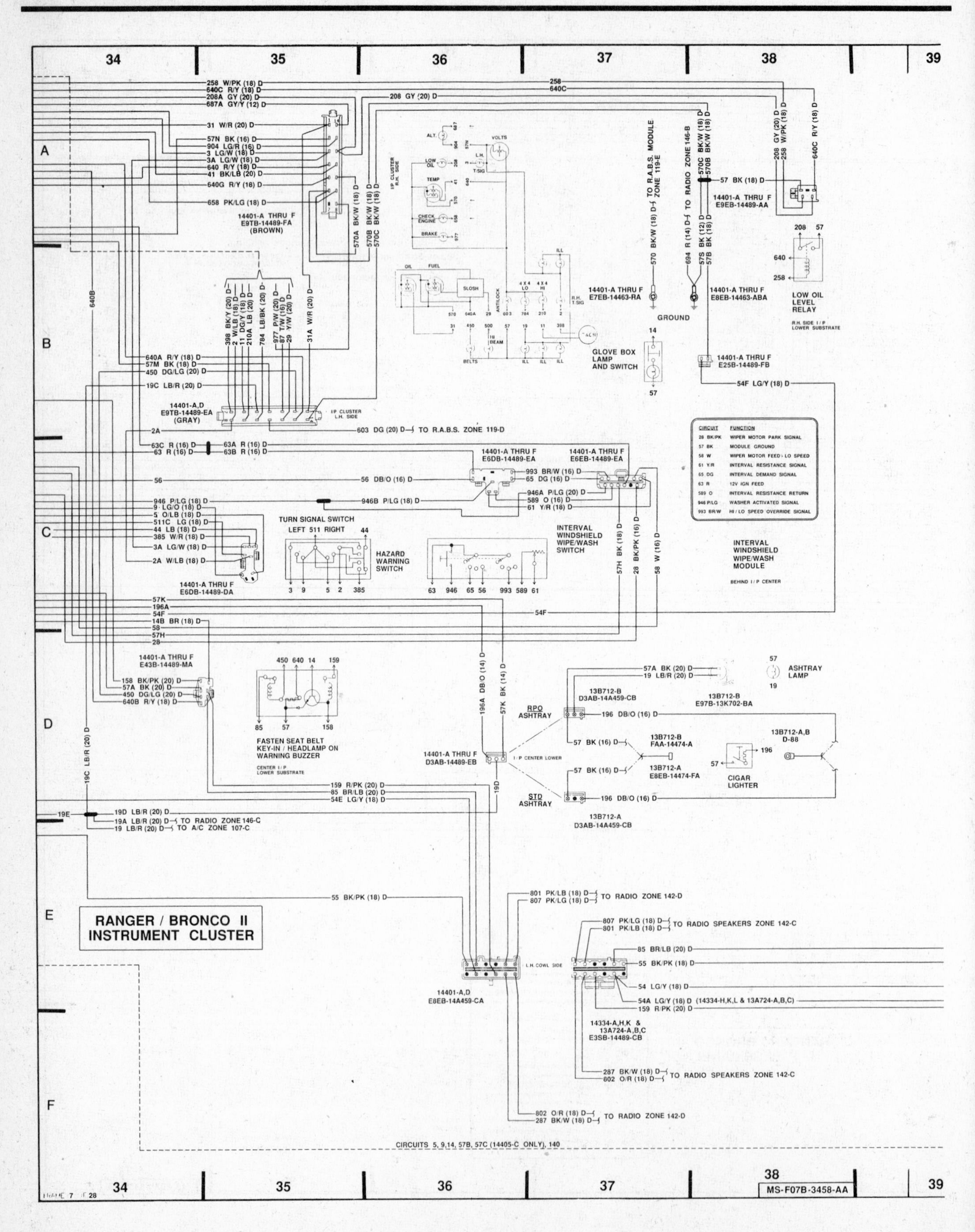
34
35
36
37
38
39
A
B
C
D
E
F
258 W/PK (18) D
640C R/Y (18) D
208A GY (20) D
687A GY/Y (12) D
208 GY (20) D
31 W/R (20) D
57N BK (16) D
904 LG/R (16) D
3 LG/W (18) D
3A LG/W (18) D
640 R/Y (18) D
41 BK/LB (20) D
640G R/Y (18) D
658 PK/LG (18) D
14401-A THRU F
E9TB-14489-FA
(BROWN)
570A BK/W (18) D
570B BK/W (18) D
570C BK/W (18) D
I/P CLUSTER R.H. SIDE
ALT.
VOLTS
LOW OIL
TEMP
CHECK ENGINE
BRAKE
L.H.
T/SIG
OIL
FUEL
SLOSH
ANTILOCK
4 X 4 LO
4 X 4 HI
R.H. T/SIG
ILL
HI BEAM
BELTS
TO R.A.B.S. MODULE ZONE 119-E
570 BK/W (18) D
TO RADIO ZONE 146-B
694 R (14) D
57S BK (12) D
57B BK (18) D
14401-A THRU F
E7EB-14463-RA
14401-A THRU F
E8EB-14463-ABA
GROUND
57 BK (18) D
14401-A THRU F
E9EB-14489-AA
208 GY (20) D
258 W/PK (18) D
640C R/Y (18) D
LOW OIL LEVEL RELAY
R.H. SIDE I/P LOWER SUBSTRATE
GLOVE BOX LAMP AND SWITCH
14401-A THRU F
E25B-14489-FB
54F LG/Y (18) D
640B
398 BK/Y (20) D
2 W/LB (18) D
11 DG/Y (18) D
210A LB (20) D
784 LB/BK (20) D
977 P/W (20) D
87 T/W (16) D
29 Y/W (20) D
31A W/R (20) D
640A R/Y (18) D
57M BK (18) D
450 DG/LG (20) D
19C LB/R (20) D
14401-A,D
E9TB-14489-EA
(GRAY)
I/P CLUSTER L.H. SIDE
2A
603 DG (20) D TO R.A.B.S. ZONE 119-D
63C R (16) D
63 R (16) D
63A R (16) D
63B R (16) D
14401-A THRU F
E6DB-14489-EA
14401-A THRU F
E6EB-14489-EA
56
56 DB/O (16) D
993 BR/W (16) D
65 DG (16) D
946A P/LG (20) D
589 O (16) D
61 Y/R (18) D
946 P/LG (18) D
946B P/LG (18) D
9 LG/O (18) D
5 O/LB (18) D
511C LG (18) D
44 LB (18) D
385 W/R (18) D
3A LG/W (18) D
2A W/LB (18) D
14401-A THRU F
E6DB-14489-DA
TURN SIGNAL SWITCH
LEFT 511 RIGHT
HAZARD WARNING SWITCH
INTERVAL WINDSHIELD WIPE/WASH SWITCH
57H BK (18) D
28 BK/PK (16) D
58 W (16) D
CIRCUIT FUNCTION
28 BK/PK WIPER MOTOR PARK SIGNAL
57 BK MODULE GROUND
58 W WIPER MOTOR FEED - LO SPEED
61 Y/R INTERVAL RESISTANCE SIGNAL
65 DG INTERVAL DEMAND SIGNAL
63 R 12V IGN FEED
589 O INTERVAL RESISTANCE RETURN
946 P/LG WASHER ACTIVATED SIGNAL
993 BR/W HI / LO SPEED OVERRIDE SIGNAL
INTERVAL WINDSHIELD WIPE/WASH MODULE
BEHIND I/P CENTER
57K
196A
54F
14B BR (18) D
58
57H
28
14401-A THRU F
E43B-14489-MA
158 BK/PK (20) D
57A BK (20) D
450 DG/LG (20) D
640B R/Y (18) D
FASTEN SEAT BELT KEY-IN / HEADLAMP ON WARNING BUZZER
CENTER I/P LOWER SUBSTRATE
196A DB/O (14) D
57K BK (14) D
57A BK (20) D
19 LB/R (20) D
ASHTRAY LAMP
13B712-B
D3AB-14A459-CB
13B712-B
E97B-13K702-BA
RPO ASHTRAY
196 DB/O (16) D
57 BK (16) D
13B712-B
FAA-14474-A
13B712-A
E8EB-14474-FA
13B712-A,B
D-88
CIGAR LIGHTER
14401-A THRU F
D3AB-14489-EB
I/P CENTER LOWER
STD ASHTRAY
13B712-A
D3AB-14A459-CB
19C LB/R (20) D
159 R/PK (20) D
85 BR/LB (20) D
54E LG/Y (18) D
19D
19E
19D LB/R (20) D
19A LB/R (20) D TO RADIO ZONE 146-C
19 LB/R (20) D TO A/C ZONE 107-C
55 BK/PK (18) D
801 PK/LB (18) D TO RADIO ZONE 142-D
807 PK/LG (18) D
807 PK/LG (18) D TO RADIO SPEAKERS ZONE 142-C
801 PK/LB (18) D
RANGER / BRONCO II INSTRUMENT CLUSTER
L.H. COWL SIDE
85 BR/LB (20) D
55 BK/PK (18) D
54 LG/Y (18) D
54A LG/Y (18) D (14334-H,K,L & 13A724-A,B,C)
159 R/PK (20) D
14401-A,D
E8EB-14A459-CA
14334-A,H,K &
13A724-A,B,C
E3SB-14489-CB
287 BK/W (18) D TO RADIO SPEAKERS ZONE 142-C
802 O/R (18) D
802 O/R (18) D TO RADIO ZONE 142-D
287 BK/W (18) D
CIRCUITS 5, 9, 14, 57B, 57C (14405-C ONLY), 140
MS-F07B-3458-AA

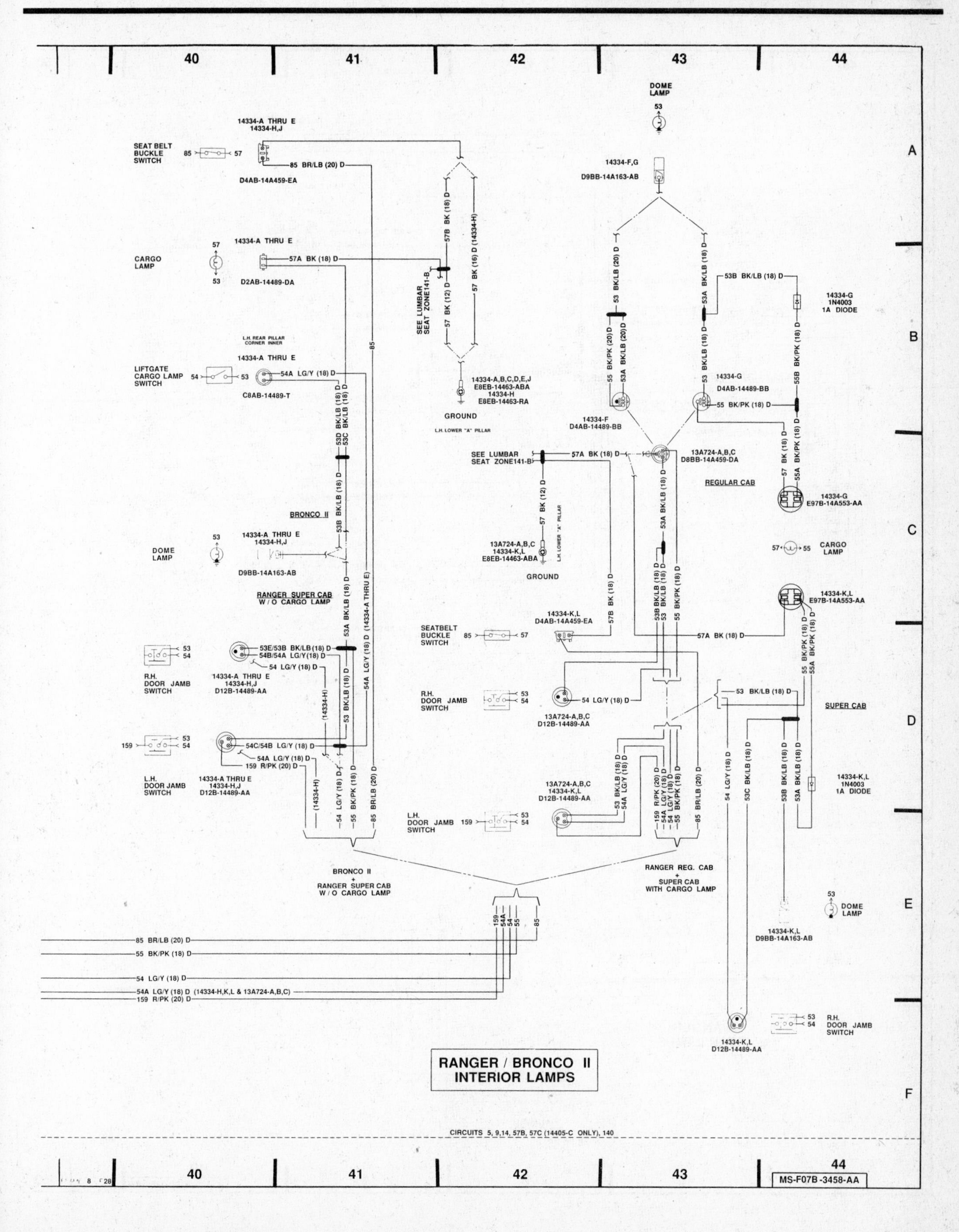
RANGER / BRONCO II
INTERIOR LAMPS
DOME LAMP
SEAT BELT BUCKLE SWITCH
CARGO LAMP
LIFTGATE CARGO LAMP SWITCH
GROUND
BRONCO II
RANGER SUPER CAB W/O CARGO LAMP
REGULAR CAB
SUPER CAB
R.H. DOOR JAMB SWITCH
L.H. DOOR JAMB SWITCH
SEE LUMBAR SEAT ZONE141-B
RANGER REG. CAB + SUPER CAB WITH CARGO LAMP
1N4003 1A DIODE
CIRCUITS 5, 9,14, 57B, 57C (14405-C ONLY), 140
MS-F07B-3458-AA

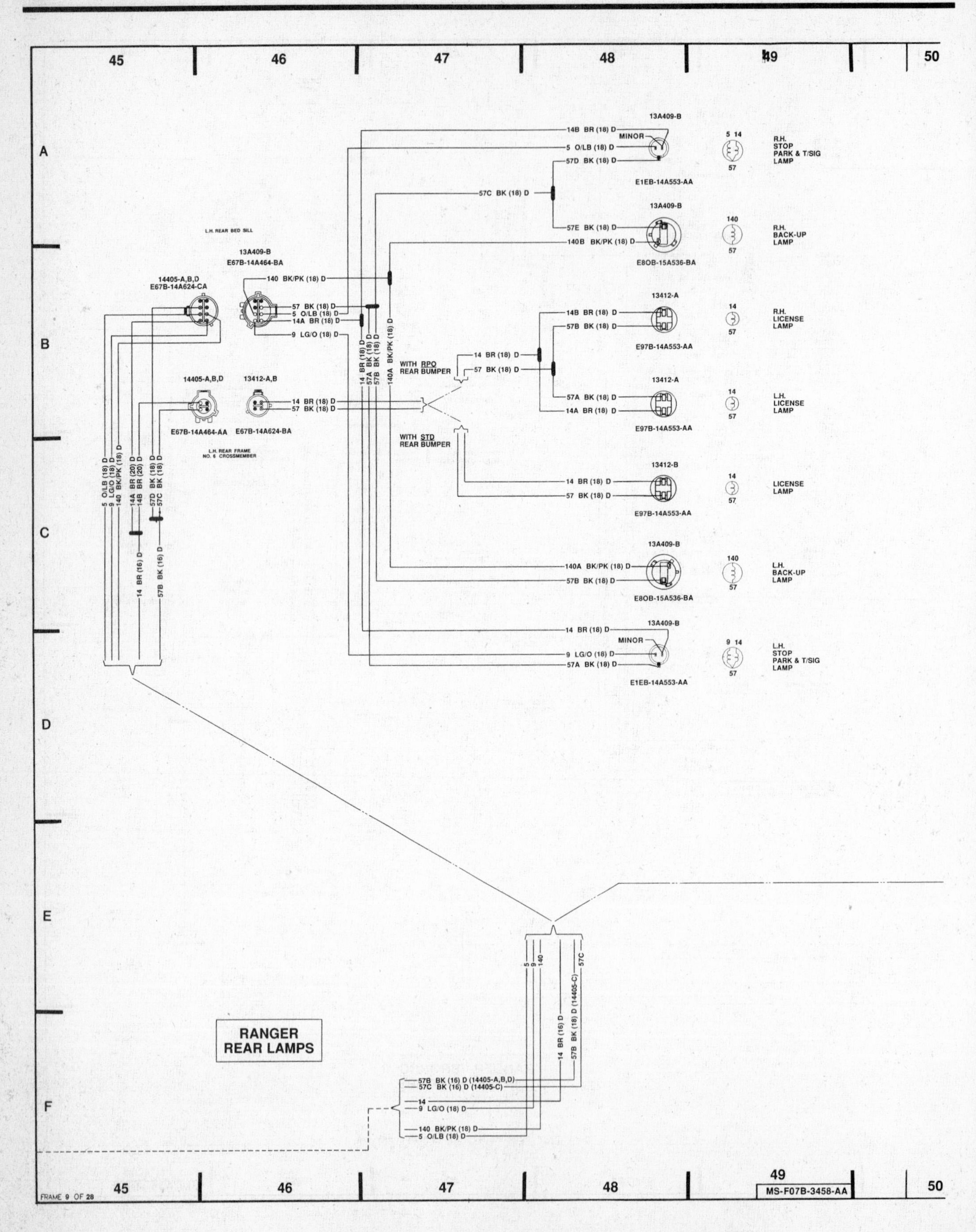
RANGER
REAR LAMPS
13A409-B
14B BR (18) D
MINOR
5 O/LB (18) D
57D BK (18) D
E1EB-14A553-AA
R.H.
STOP
PARK & T/SIG
LAMP
57C BK (18) D
57E BK (18) D
140B BK/PK (18) D
E8OB-15A536-BA
R.H.
BACK-UP
LAMP
L.H. REAR BED SILL
13A409-B
E67B-14A464-BA
14405-A,B,D
E67B-14A624-CA
140 BK/PK (18) D
57 BK (18) D
5 O/LB (18) D
14A BR (18) D
9 LG/O (18) D
13412-A
14B BR (18) D
57B BK (18) D
E97B-14A553-AA
R.H.
LICENSE
LAMP
WITH RPO
REAR BUMPER
14 BR (18) D
57 BK (18) D
14405-A,B,D
13412-A,B
E67B-14A464-AA
E67B-14A624-BA
L.H. REAR FRAME
NO. 6 CROSSMEMBER
57A BK (18) D
14A BR (18) D
L.H.
LICENSE
LAMP
WITH STD
REAR BUMPER
13412-B
14 BR (18) D
57 BK (18) D
LICENSE
LAMP
140A BK/PK (18) D
57B BK (18) D
L.H.
BACK-UP
LAMP
14 BR (18) D
9 LG/O (18) D
57A BK (18) D
L.H.
STOP
PARK & T/SIG
LAMP
57B BK (16) D (14405-A,B,D)
57C BK (16) D (14405-C)
140 BK/PK (18) D
5 O/LB (18) D
14 BR (16) D
57B BK (18) D (14405-C)
FRAME 9 OF 28
MS-F07B-3458-AA

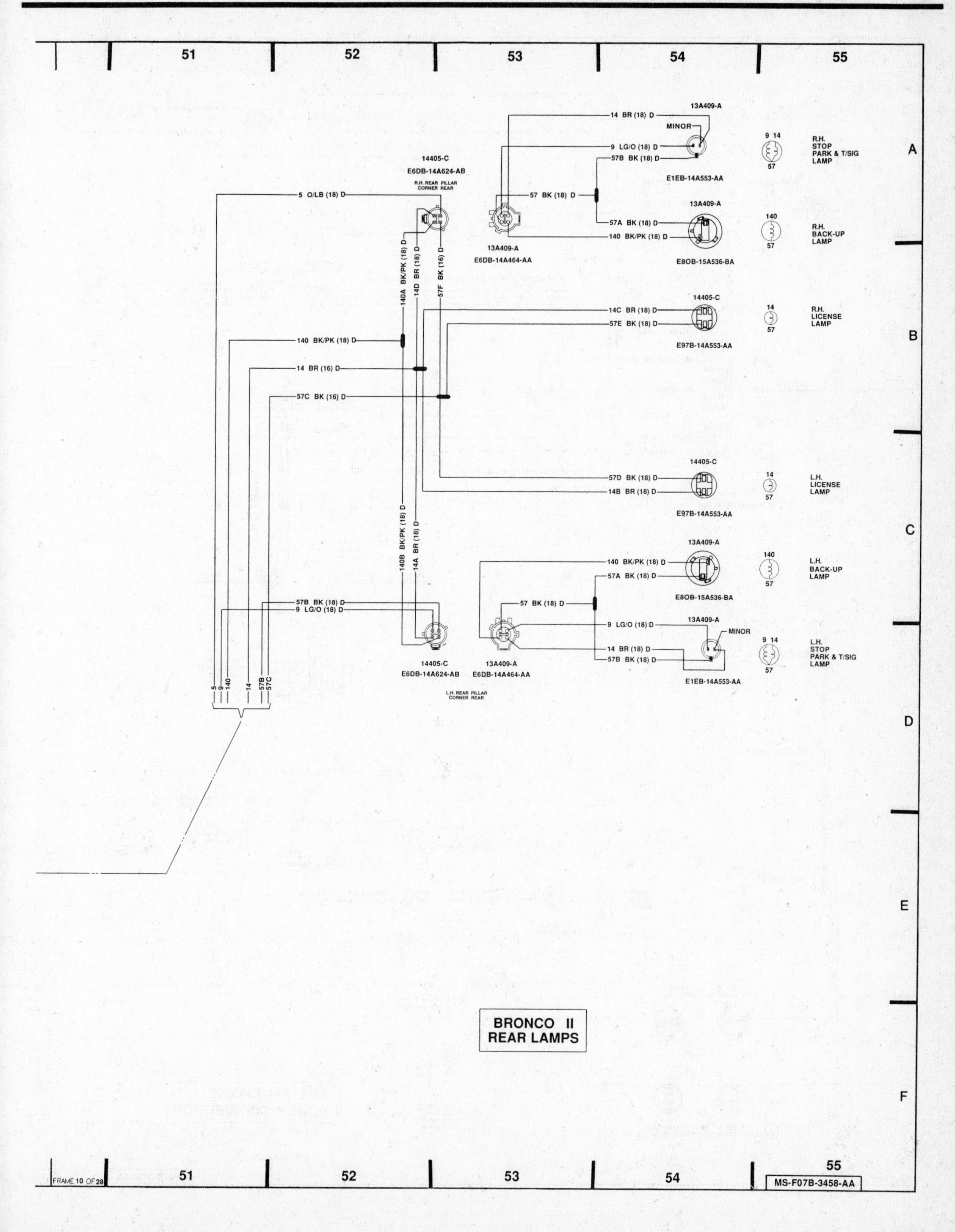
51
52
53
54
55
A
B
C
D
E
F
13A409-A
14 BR (18) D
MINOR
9 LG/O (18) D
57B BK (18) D
E1EB-14A553-AA
9 14
57
R.H.
STOP
PARK & T/SIG
LAMP
14405-C
E6DB-14A624-AB
R.H. REAR PILLAR
CORNER REAR
5 O/LB (18) D
57 BK (18) D
13A409-A
57A BK (18) D
140 BK/PK (18) D
13A409-A
E6DB-14A464-AA
E8OB-15A536-BA
140
R.H.
BACK-UP
LAMP
140A BK/PK (18) D
14D BR (18) D
57F BK (16) D
14405-C
14C BR (18) D
57E BK (18) D
E97B-14A553-AA
R.H.
LICENSE
LAMP
140 BK/PK (18) D
14 BR (16) D
57C BK (16) D
14405-C
57D BK (18) D
14B BR (18) D
E97B-14A553-AA
L.H.
LICENSE
LAMP
13A409-A
140 BK/PK (18) D
57A BK (18) D
E8OB-15A536-BA
L.H.
BACK-UP
LAMP
140B BK/PK (18) D
14A BR (18) D
57B BK (18) D
9 LG/O (18) D
57 BK (18) D
9 LG/O (18) D
13A409-A
MINOR
14 BR (18) D
57B BK (18) D
L.H.
STOP
PARK & T/SIG
LAMP
14405-C
E6DB-14A624-AB
13A409-A
E6DB-14A464-AA
E1EB-14A553-AA
L.H. REAR PILLAR
CORNER REAR
5
9
140
14
57B
57C
BRONCO II
REAR LAMPS
FRAME 10 OF 28
MS-F07B-3458-AA

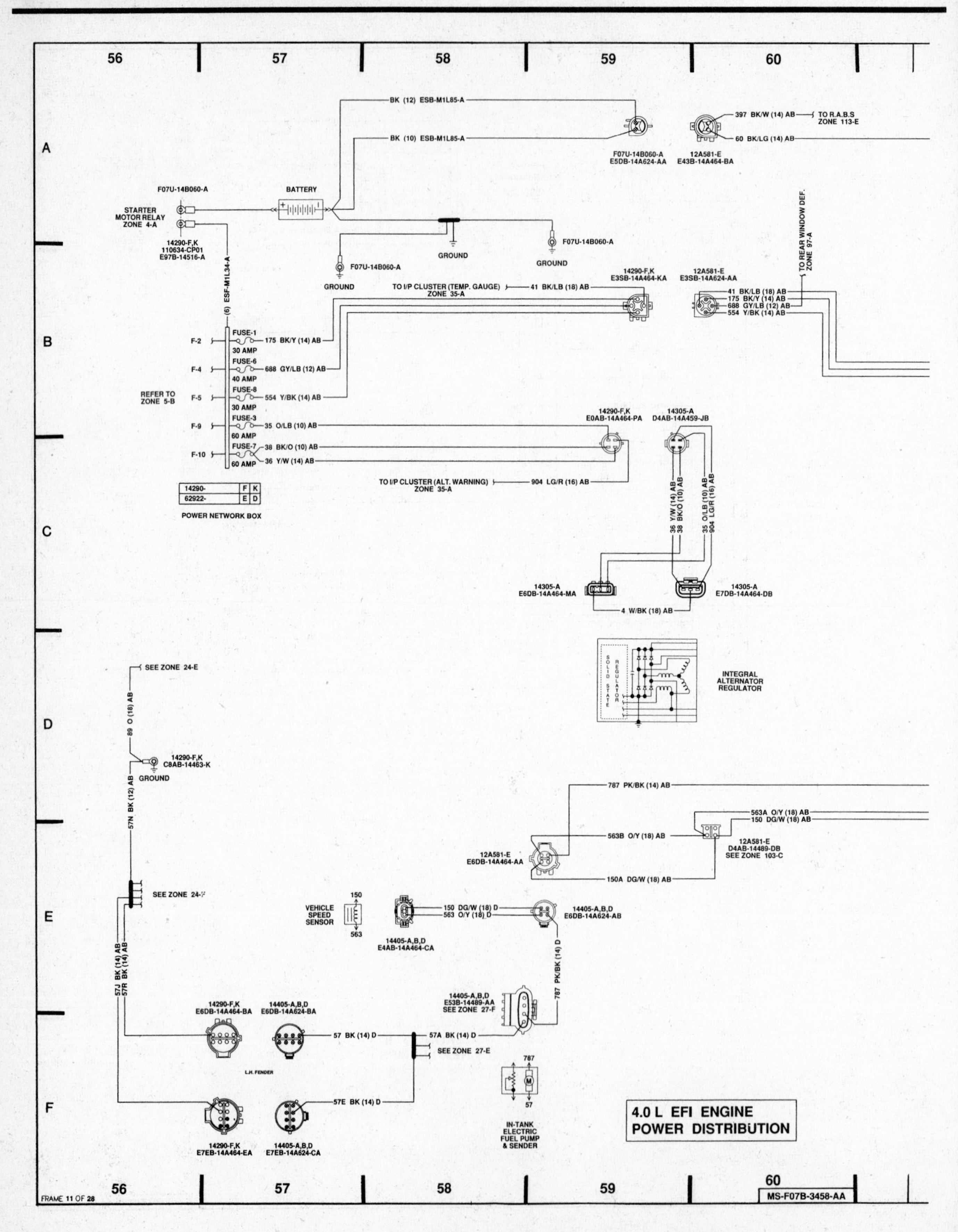

56
57
58
59
60
BK (12) ESB-M1L85-A
BK (10) ESB-M1L85-A
397 BK/W (14) AB
TO R.A.B.S ZONE 113-E
60 BK/LG (14) AB
F07U-14B060-A
E5DB-14A624-AA
12A581-E
E43B-14A464-BA
F07U-14B060-A
BATTERY
STARTER MOTOR RELAY ZONE 4-A
14290-F,K
110634-CP01
E97B-14516-A
GROUND
TO REAR WINDOW DEF. ZONE 97-A
ESF-M1L34-A
TO I/P CLUSTER (TEMP. GAUGE) ZONE 35-A
41 BK/LB (18) AB
14290-F,K
E3SB-14A464-KA
12A581-E
E3SB-14A624-AA
175 BK/Y (14) AB
688 GY/LB (12) AB
554 Y/BK (14) AB
FUSE-1
30 AMP
FUSE-6
40 AMP
FUSE-8
30 AMP
FUSE-3
60 AMP
FUSE-7
60 AMP
REFER TO ZONE 5-B
35 O/LB (10) AB
38 BK/O (10) AB
36 Y/W (14) AB
14290-F,K
E0AB-14A464-PA
14305-A
D4AB-14A459-JB
TO I/P CLUSTER (ALT. WARNING) ZONE 35-A
904 LG/R (16) AB
POWER NETWORK BOX
14305-A
E6DB-14A464-MA
14305-A
E7DB-14A464-DB
4 W/BK (18) AB
SOLID STATE REGULATOR
INTEGRAL ALTERNATOR REGULATOR
SEE ZONE 24-E
89 O (18) AB
14290-F,K
C8AB-14463-K
GROUND
57N BK (12) AB
787 PK/BK (14) AB
563A O/Y (18) AB
150 DG/W (18) AB
563B O/Y (18) AB
12A581-E
D4AB-14489-DB
SEE ZONE 103-C
12A581-E
E6DB-14A464-AA
150A DG/W (18) AB
VEHICLE SPEED SENSOR
150 DG/W (18) D
563 O/Y (18) D
14405-A,B,D
E6DB-14A624-AB
14405-A,B,D
E4AB-14A464-CA
57J BK (14) AB
57R BK (14) AB
PK/BK (14) D
14405-A,B,D
E53B-14489-AA
SEE ZONE 27-F
14290-F,K
E6DB-14A464-BA
14405-A,B,D
E6DB-14A624-BA
57 BK (14) D
57A BK (14) D
SEE ZONE 27-E
L.H. FENDER
57E BK (14) D
IN-TANK ELECTRIC FUEL PUMP & SENDER
14290-F,K
E7EB-14A464-EA
14405-A,B,D
E7EB-14A624-CA
4.0 L EFI ENGINE POWER DISTRIBUTION
FRAME 11 OF 28
MS-F07B-3458-AA

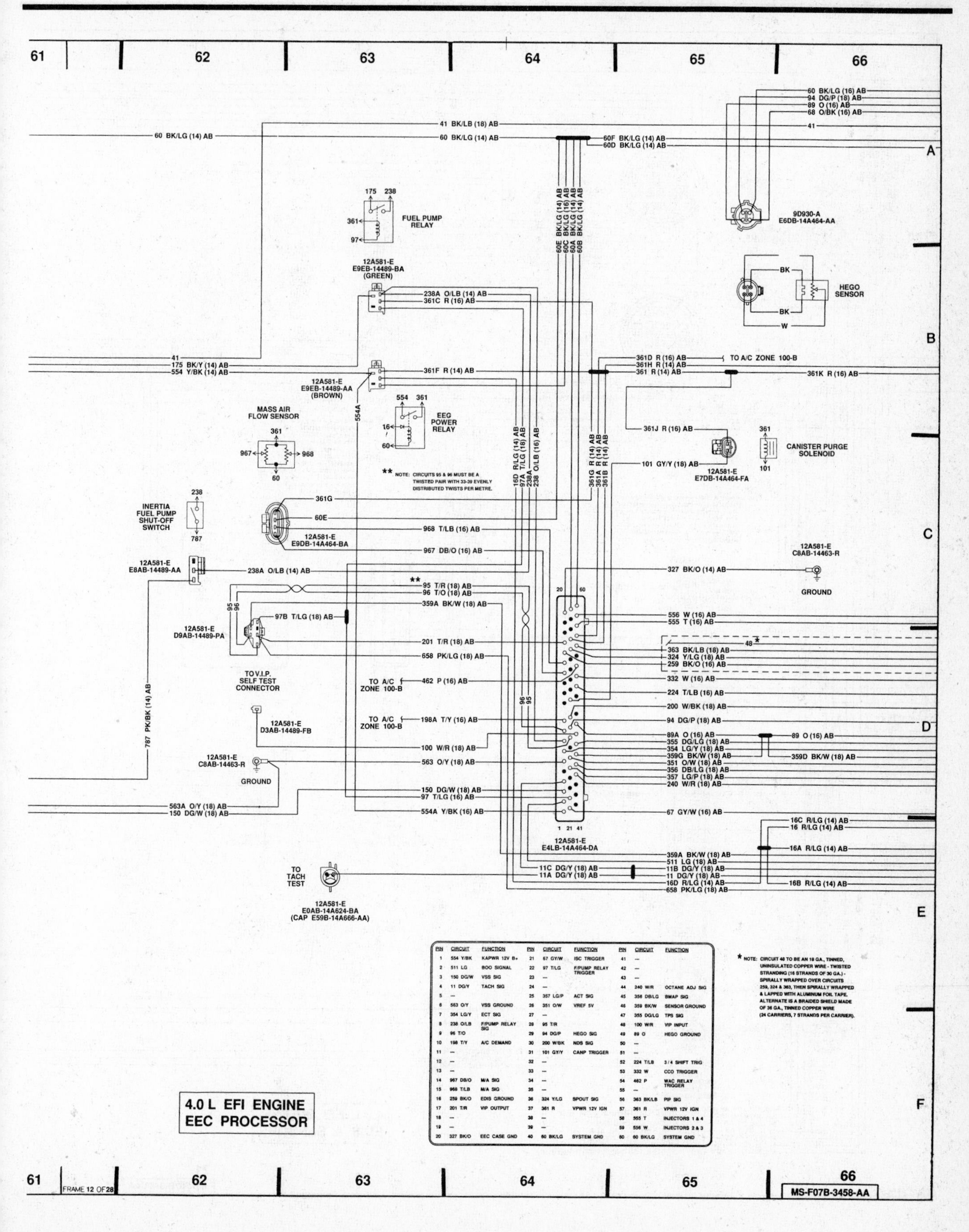

PIN	CIRCUIT	FUNCTION	PIN	CIRCUIT	FUNCTION	PIN	CIRCUIT	FUNCTION
1	554 Y/BK	KAPWR 12V B+	21	67 GY/W	ISC TRIGGER	41	—	
2	511 LG	BOO SIGNAL	22	97 T/LG	F/PUMP RELAY TRIGGER	42	—	
3	150 DG/W	VSS SIG	23	—		43	—	
4	11 DG/Y	TACH SIG	24	—		44	240 W/R	OCTANE ADJ SIG
5	—		25	357 LG/P	ACT SIG	45	356 DB/LG	BMAP SIG
6	563 O/Y	VSS GROUND	26	351 O/W	VREF 5V	46	359 BK/W	SENSOR GROUND
7	354 LG/Y	ECT SIG	27	—		47	355 DG/LG	TPS SIG
8	238 O/LB	F/PUMP RELAY SIG	28	95 T/R		48	100 W/R	VIP INPUT
9	96 T/O		29	94 DG/P	HEGO SIG	49	89 O	HEGO GROUND
10	198 T/Y	A/C DEMAND	30	200 W/BK	NDS SIG	50	—	
11	—		31	101 GY/Y	CANP TRIGGER	51	—	
12	—		32	—		52	224 T/LB	3 / 4 SHIFT TRIG
13	—		33	—		53	332 W	CCO TRIGGER
14	967 DB/O	M/A SIG	34	—		54	462 P	WAC RELAY TRIGGER
15	968 T/LB	M/A SIG	35	—		55	—	
16	259 BK/O	EDIS GROUND	36	324 Y/LG	SPOUT SIG	56	363 BK/LB	PIP SIG
17	201 T/R	VIP OUTPUT	37	361 R	VPWR 12V IGN	57	361 R	VPWR 12V IGN
18	—		38	—		58	555 T	INJECTORS 1 & 4
19	—		39	—		59	556 W	INJECTORS 2 & 3
20	327 BK/O	EEC CASE GND	40	60 BK/LG	SYSTEM GND	60	60 BK/LG	SYSTEM GND

* NOTE: CIRCUIT 48 TO BE AN 18 GA., TINNED, UNINSULATED COPPER WIRE - TWISTED STRANDING (16 STRANDS OF 30 GA.) - SPIRALLY WRAPPED OVER CIRCUITS 259, 324 & 363, THEN SPIRALLY WRAPPED & LAPPED WITH ALUMINUM FOIL TAPE. ALTERNATE IS A BRAIDED SHIELD MADE OF 36 GA., TINNED COPPER WIRE (24 CARRIERS, 7 STRANDS PER CARRIER).

4.0 L EFI ENGINE EEC PROCESSOR

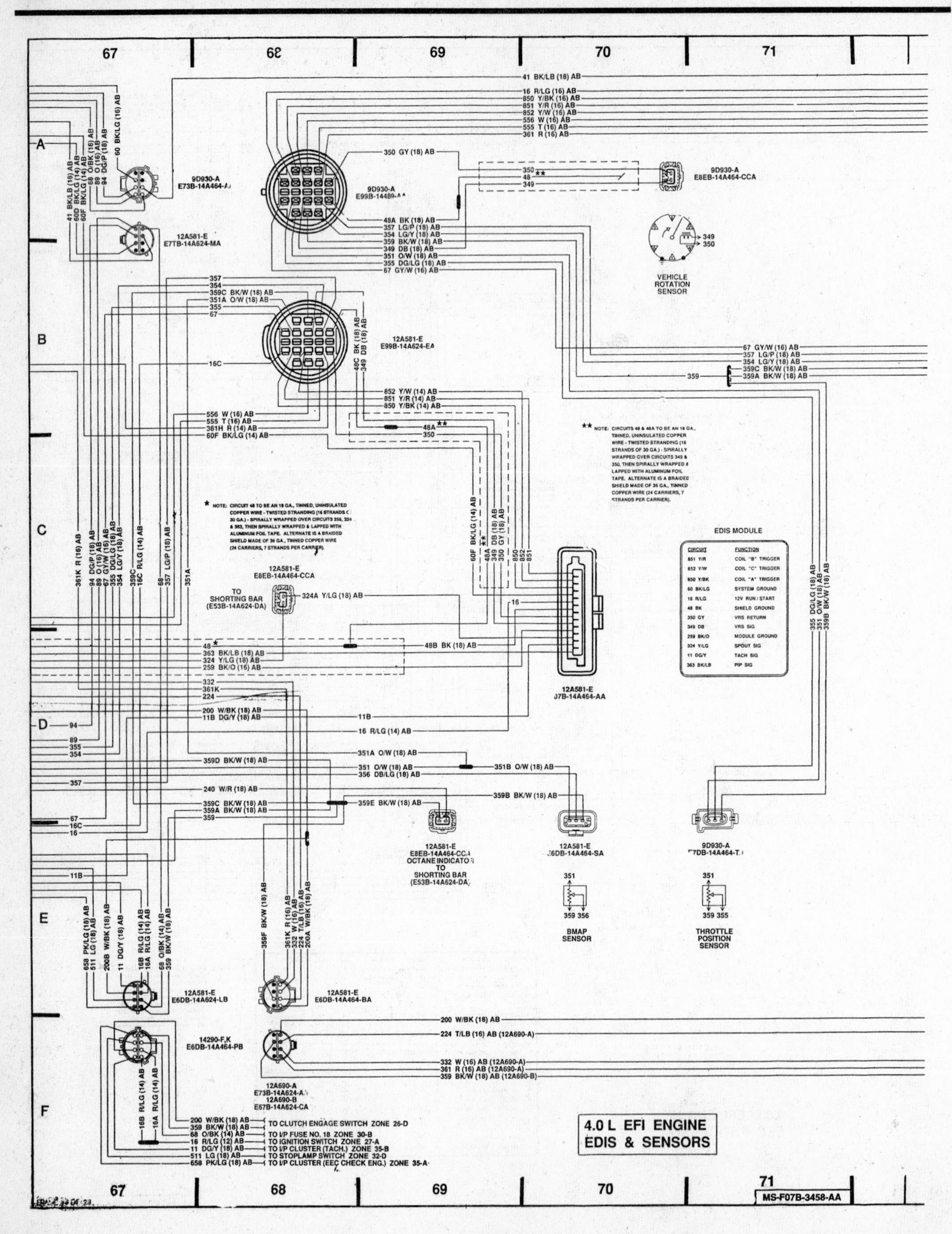

4.0 L EFI ENGINE
EDIS & SENSORS
EDIS MODULE
VEHICLE ROTATION SENSOR
BMAP SENSOR
THROTTLE POSITION SENSOR
TO CLUTCH ENGAGE SWITCH ZONE 26-D
TO I/P FUSE NO. 18 ZONE 30-B
TO IGNITION SWITCH ZONE 27-A
TO I/P CLUSTER (TACH.) ZONE 35-B
TO STOPLAMP SWITCH ZONE 32-D
TO I/P CLUSTER (EEC CHECK ENG.) ZONE 35-A
MS-F07B-3458-AA

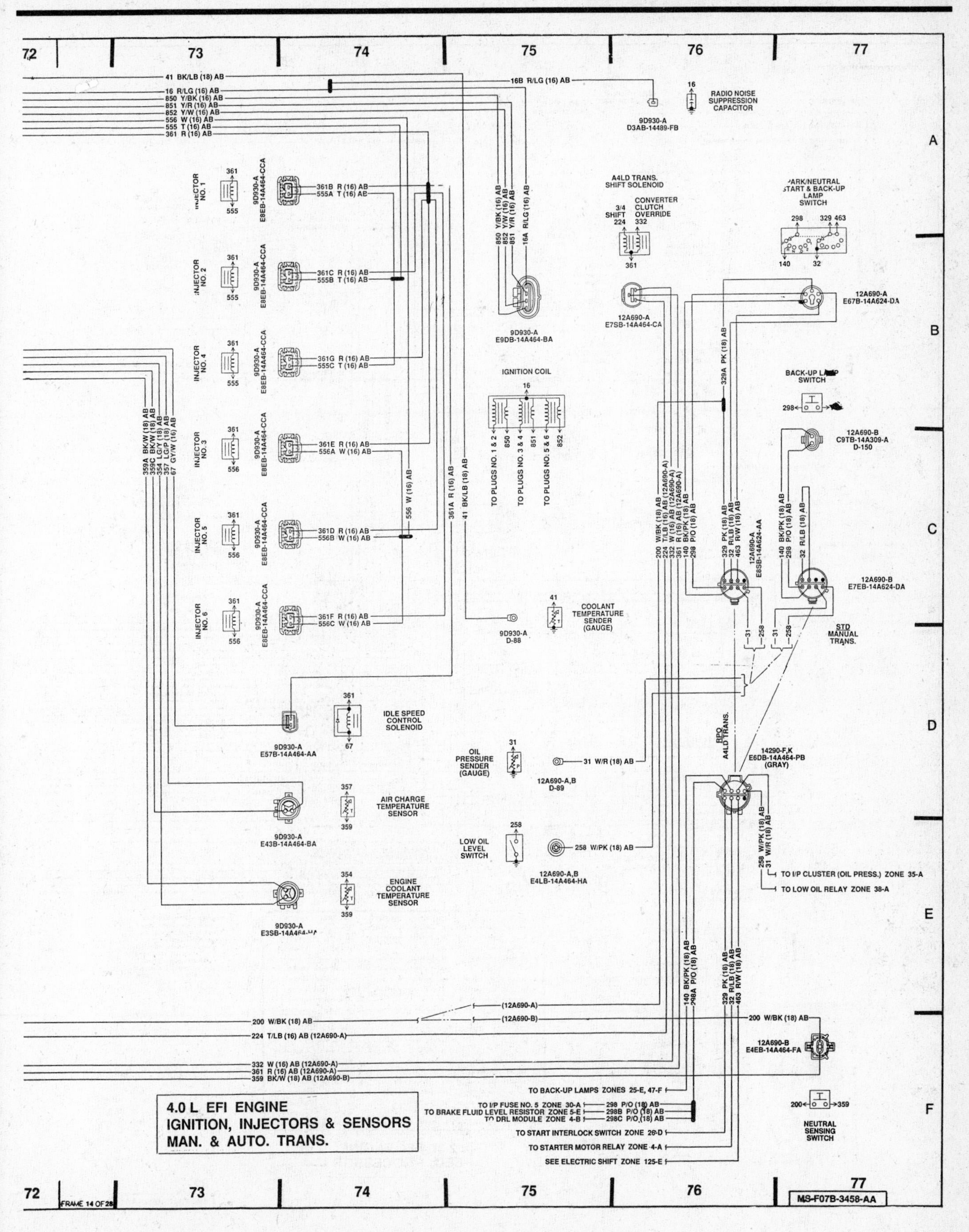
4.0 L EFI ENGINE
IGNITION, INJECTORS & SENSORS
MAN. & AUTO. TRANS.
INJECTOR NO. 1
INJECTOR NO. 2
INJECTOR NO. 4
INJECTOR NO. 3
INJECTOR NO. 5
INJECTOR NO. 6
IDLE SPEED CONTROL SOLENOID
AIR CHARGE TEMPERATURE SENSOR
ENGINE COOLANT TEMPERATURE SENSOR
IGNITION COIL
RADIO NOISE SUPPRESSION CAPACITOR
A4LD TRANS. SHIFT SOLENOID
PARK/NEUTRAL START & BACK-UP LAMP SWITCH
BACK-UP LAMP SWITCH
COOLANT TEMPERATURE SENDER (GAUGE)
OIL PRESSURE SENDER (GAUGE)
LOW OIL LEVEL SWITCH
STD MANUAL TRANS.
NEUTRAL SENSING SWITCH
TO BACK-UP LAMPS ZONES 25-E, 47-F
TO I/P FUSE NO. 5 ZONE 30-A
TO BRAKE FLUID LEVEL RESISTOR ZONE 5-E
TO DRL MODULE ZONE 4-B
TO START INTERLOCK SWITCH ZONE 26-D
TO STARTER MOTOR RELAY ZONE 4-A
SEE ELECTRIC SHIFT ZONE 125-E
TO I/P CLUSTER (OIL PRESS.) ZONE 35-A
TO LOW OIL RELAY ZONE 38-A
FRAME 14 OF 28
MS-F07B-3458-AA

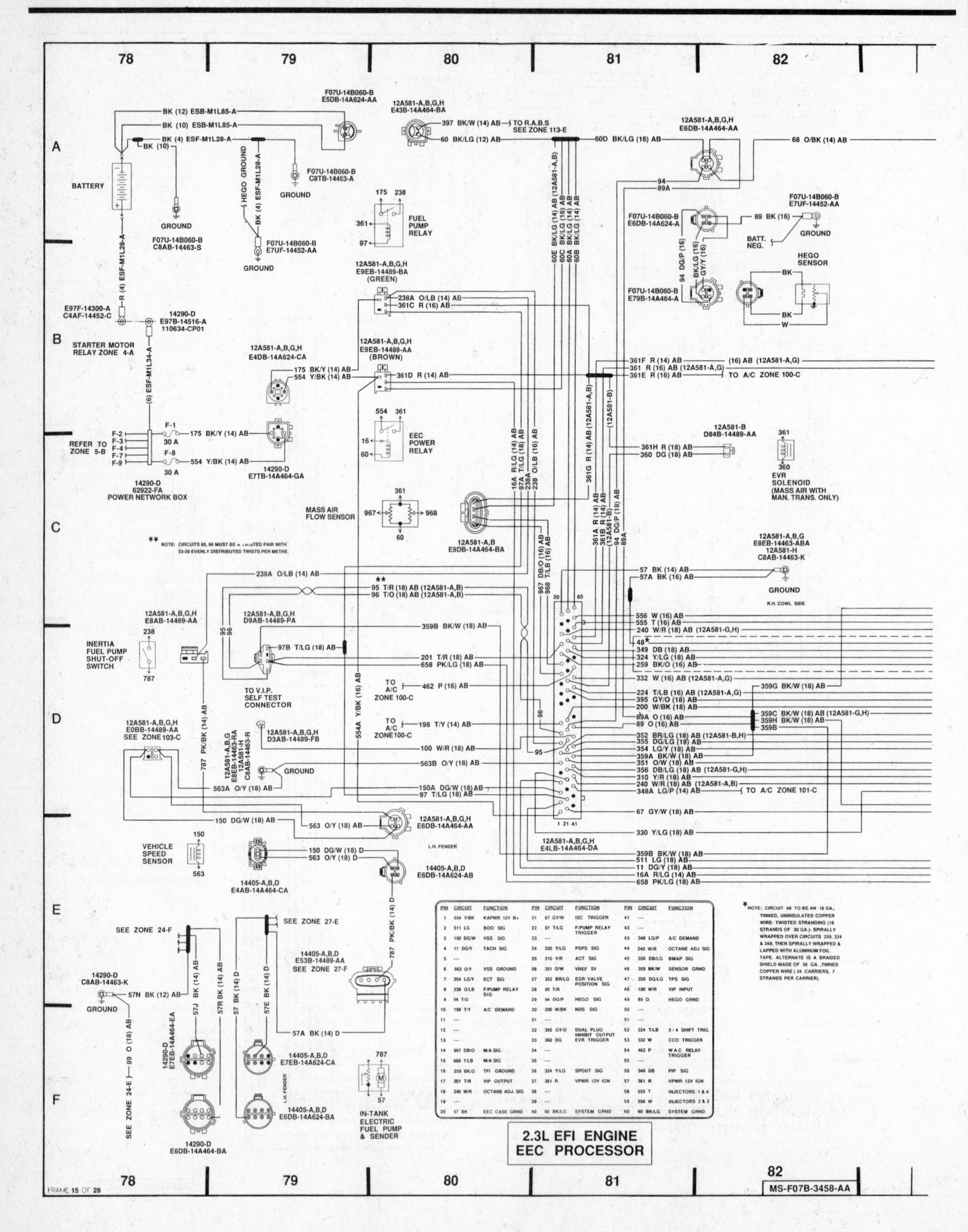

PIN	CIRCUIT	FUNCTION	PIN	CIRCUIT	FUNCTION	PIN	CIRCUIT	FUNCTION
1	554 Y/BK	KAPWR 12V B+	21	67 GY/W	ISC TRIGGER	41	---	
2	511 LG	BOO SIG	22	97 T/LG	F/PUMP RELAY TRIGGER	42	---	
3	150 DG/W	VSS SIG	23	---		43	348 LG/P	A/C DEMAND
4	11 DG/Y	TACH SIG	24	330 Y/LG	PSPS SIG	44	240 W/R	OCTANE ADJ SIG
5	---		25	310 Y/R	ACT SIG	45	356 DB/LG	BMAP SIG
6	563 O/Y	VSS GROUND	26	351 O/W	VREF 5V	46	359 BK/W	SENSOR GRND
7	354 LG/Y	ECT SIG	27	352 BR/LG	EGR VALVE POSITION SIG	47	355 DG/LG	TPS SIG
8	238 O/LB	F/PUMP RELAY SIG	28	95 T/R		48	100 W/R	VIP INPUT
9	96 T/O		29	94 DG/P	HEGO SIG	49	89 O	HEGO GRND
10	198 T/Y	A/C DEMAND	30	200 W/BK	NDS SIG	50	---	
11	---		31	---		51	---	
12	---		32	395 GY/O	DUAL PLUG INHIBIT OUTPUT	52	224 T/LB	3 / 4 SHIFT TRIG.
13	---		33	360 DG	EVR TRIGGER	53	332 W	CCO TRIGGER
14	967 DB/O	M/A SIG	34	---		54	462 P	W A C RELAY TRIGGER
15	968 T/LB	M/A SIG	35	---		55	---	
16	259 BK/O	TFI GROUND	36	324 Y/LG	SPOUT SIG	56	349 DB	PIP SIG
17	201 T/R	VIP OUTPUT	37	361 R	VPWR 12V IGN	57	361 R	VPWR 12V IGN
18	240 W/R	OCTANE ADJ. SIG	38	---		58	555 T	INJECTORS 1 & 4
19	---		39	---		59	556 W	INJECTORS 2 & 3
20	57 BK	EEC CASE GRND	40	60 BK/LG	SYSTEM GRND	60	60 BK/LG	SYSTEM GRND

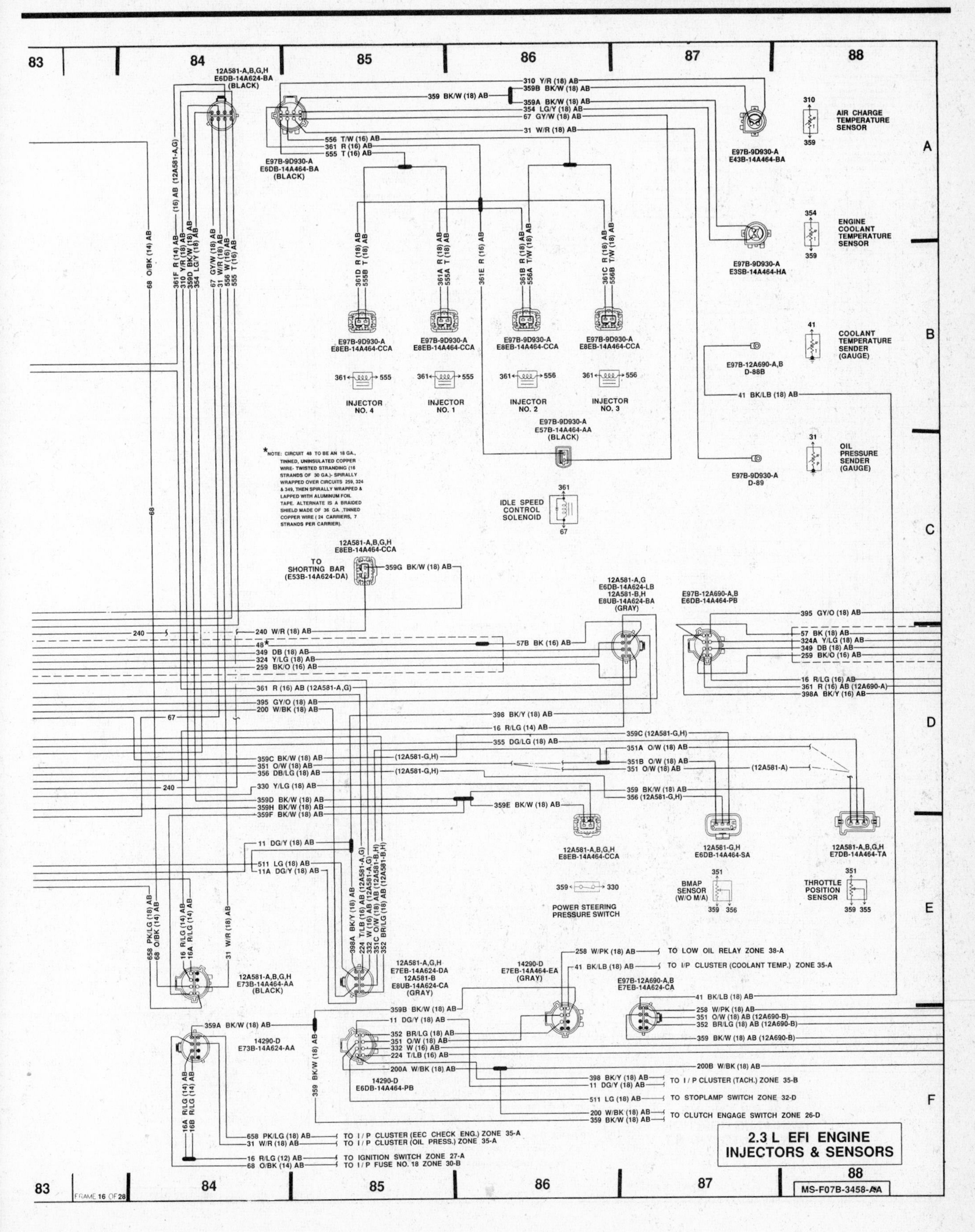

2.3 L EFI ENGINE
INJECTORS & SENSORS
AIR CHARGE TEMPERATURE SENSOR
ENGINE COOLANT TEMPERATURE SENSOR
COOLANT TEMPERATURE SENDER (GAUGE)
OIL PRESSURE SENDER (GAUGE)
INJECTOR NO. 4
INJECTOR NO. 1
INJECTOR NO. 2
INJECTOR NO. 3
IDLE SPEED CONTROL SOLENOID
TO SHORTING BAR (E53B-14A624-DA)
POWER STEERING PRESSURE SWITCH
BMAP SENSOR (W/O M/A)
THROTTLE POSITION SENSOR
TO LOW OIL RELAY ZONE 38-A
TO I/P CLUSTER (COOLANT TEMP.) ZONE 35-A
TO I / P CLUSTER (TACH.) ZONE 35-B
TO STOPLAMP SWITCH ZONE 32-D
TO CLUTCH ENGAGE SWITCH ZONE 26-D
TO I / P CLUSTER (EEC CHECK ENG.) ZONE 35-A
TO I / P CLUSTER (OIL PRESS.) ZONE 35-A
TO IGNITION SWITCH ZONE 27-A
TO I / P FUSE NO. 18 ZONE 30-B
*NOTE: CIRCUIT 48 TO BE AN 18 GA., TINNED, UNINSULATED COPPER WIRE- TWISTED STRANDING (16 STRANDS OF 30 GA.)- SPIRALLY WRAPPED OVER CIRCUITS 259, 324 & 349, THEN SPIRALLY WRAPPED & LAPPED WITH ALUMINUM FOIL TAPE. ALTERNATE IS A BRAIDED SHIELD MADE OF 36 GA. TINNED COPPER WIRE (24 CARRIERS, 7 STRANDS PER CARRIER).
FRAME 16 OF 28
MS-F07B-3458-AA

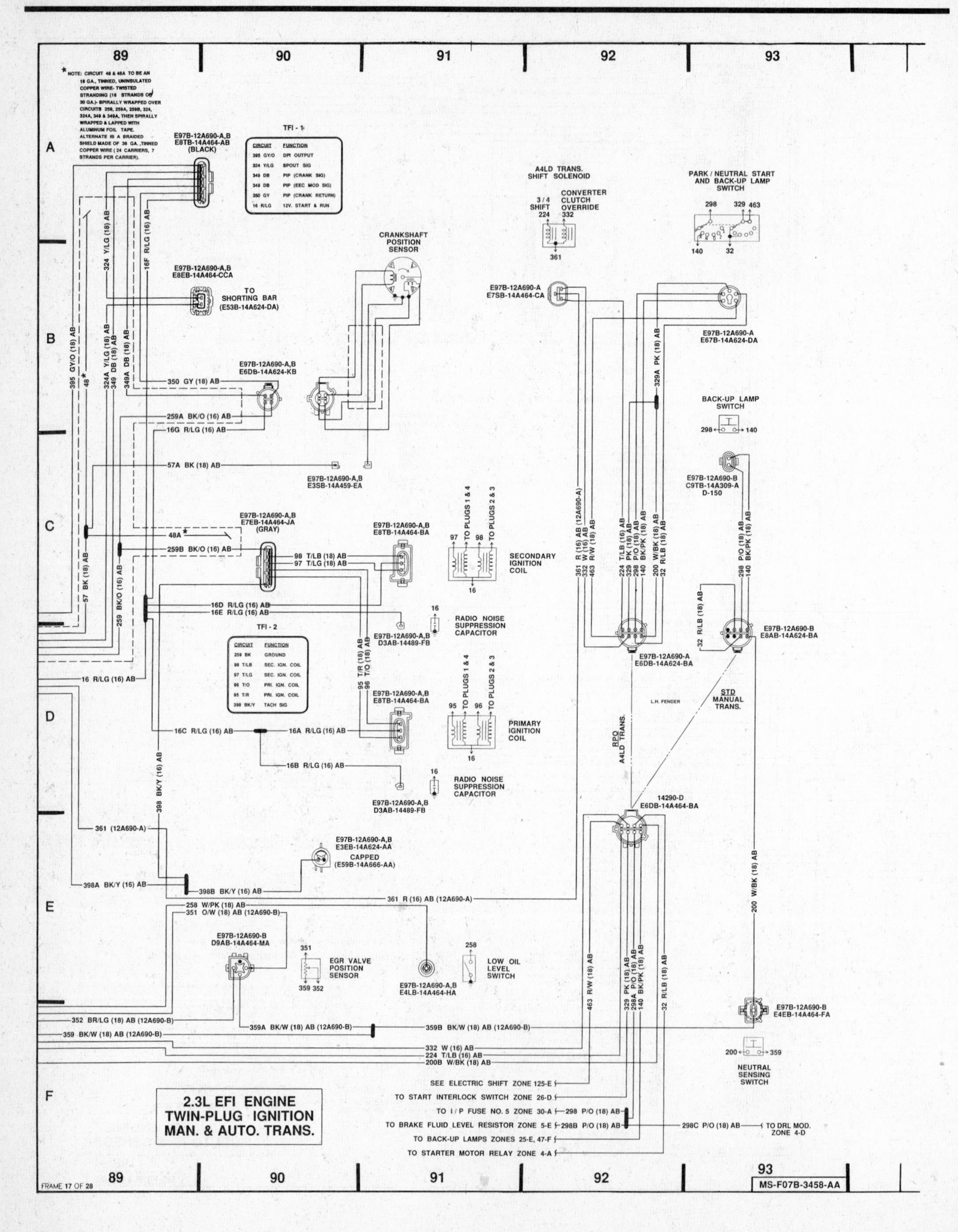
2.3L EFI ENGINE
TWIN-PLUG IGNITION
MAN. & AUTO. TRANS.
FRAME 17 OF 28
MS-F07B-3458-AA

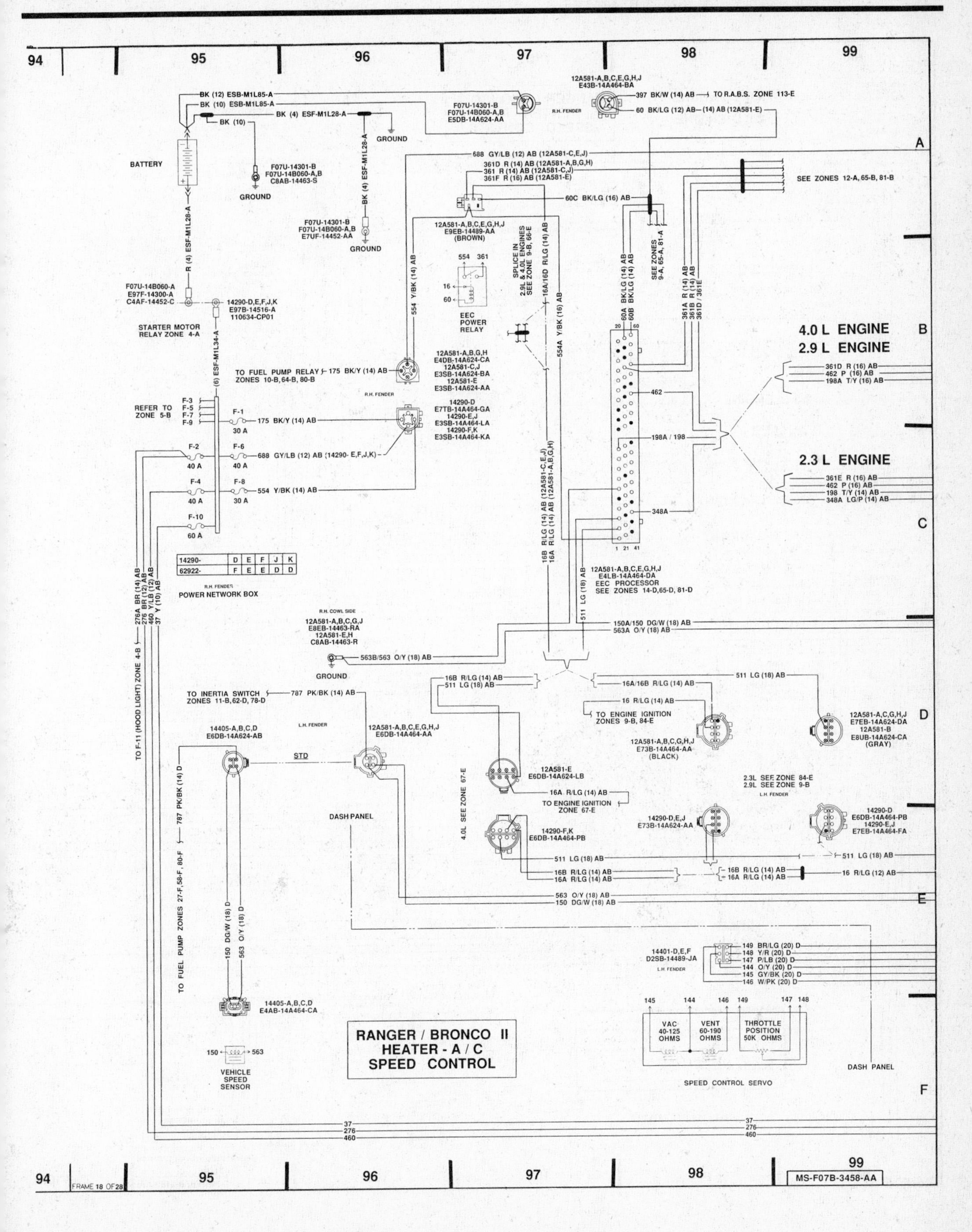
RANGER / BRONCO II
HEATER - A / C
SPEED CONTROL
BATTERY
GROUND
STARTER MOTOR RELAY ZONE 4-A
POWER NETWORK BOX
EEC POWER RELAY
4.0 L ENGINE
2.9 L ENGINE
2.3 L ENGINE
EEC PROCESSOR SEE ZONES 14-D, 65-D, 81-D
TO FUEL PUMP RELAY ZONES 10-B, 64-B, 80-B
TO INERTIA SWITCH ZONES 11-B, 62-D, 78-D
TO ENGINE IGNITION ZONES 9-B, 84-E
TO ENGINE IGNITION ZONE 67-E
TO R.A.B.S. ZONE 113-E
SEE ZONES 12-A, 65-B, 81-B
VEHICLE SPEED SENSOR
SPEED CONTROL SERVO
VAC 40-125 OHMS
VENT 60-190 OHMS
THROTTLE POSITION 50K OHMS
DASH PANEL
TO F-11 (HOOD LIGHT) ZONE 4-B
TO FUEL PUMP ZONES 27-F, 58-F, 80-F
MS-F07B-3458-AA
FRAME 18 OF 28

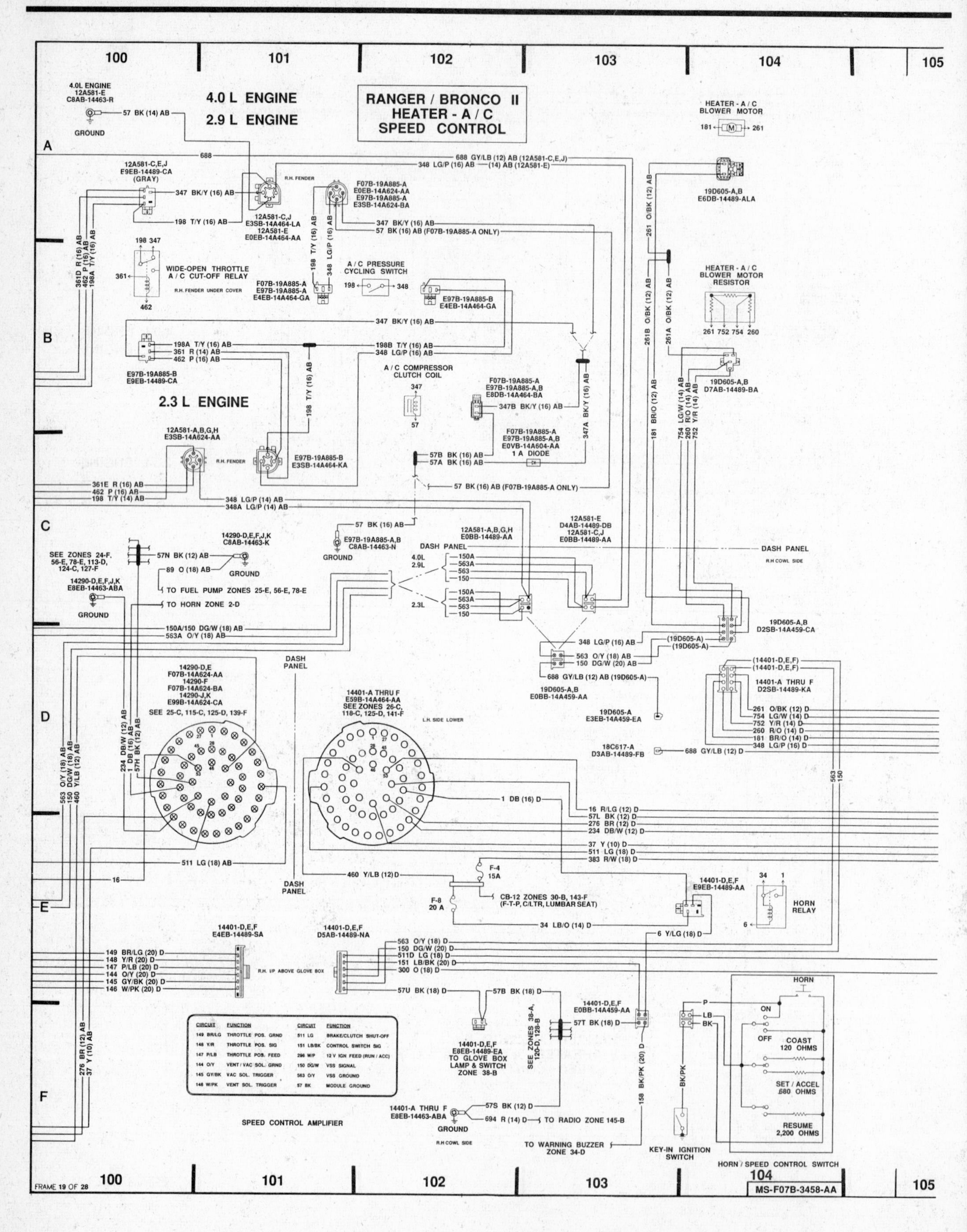

CIRCUIT	FUNCTION	CIRCUIT	FUNCTION
149 BR/LG	THROTTLE POS. GRND	511 LG	BRAKE/CLUTCH SHUT-OFF
148 Y/R	THROTTLE POS. SIG	151 LB/BK	CONTROL SWITCH SIG
147 P/LB	THROTTLE POS. FEED	296 W/P	12 V IGN FEED (RUN / ACC)
144 O/Y	VENT / VAC SOL. GRND	150 DG/W	VSS SIGNAL
145 GY/BK	VAC SOL. TRIGGER	563 O/Y	VSS GROUND
146 W/PK	VENT SOL. TRIGGER	57 BK	MODULE GROUND

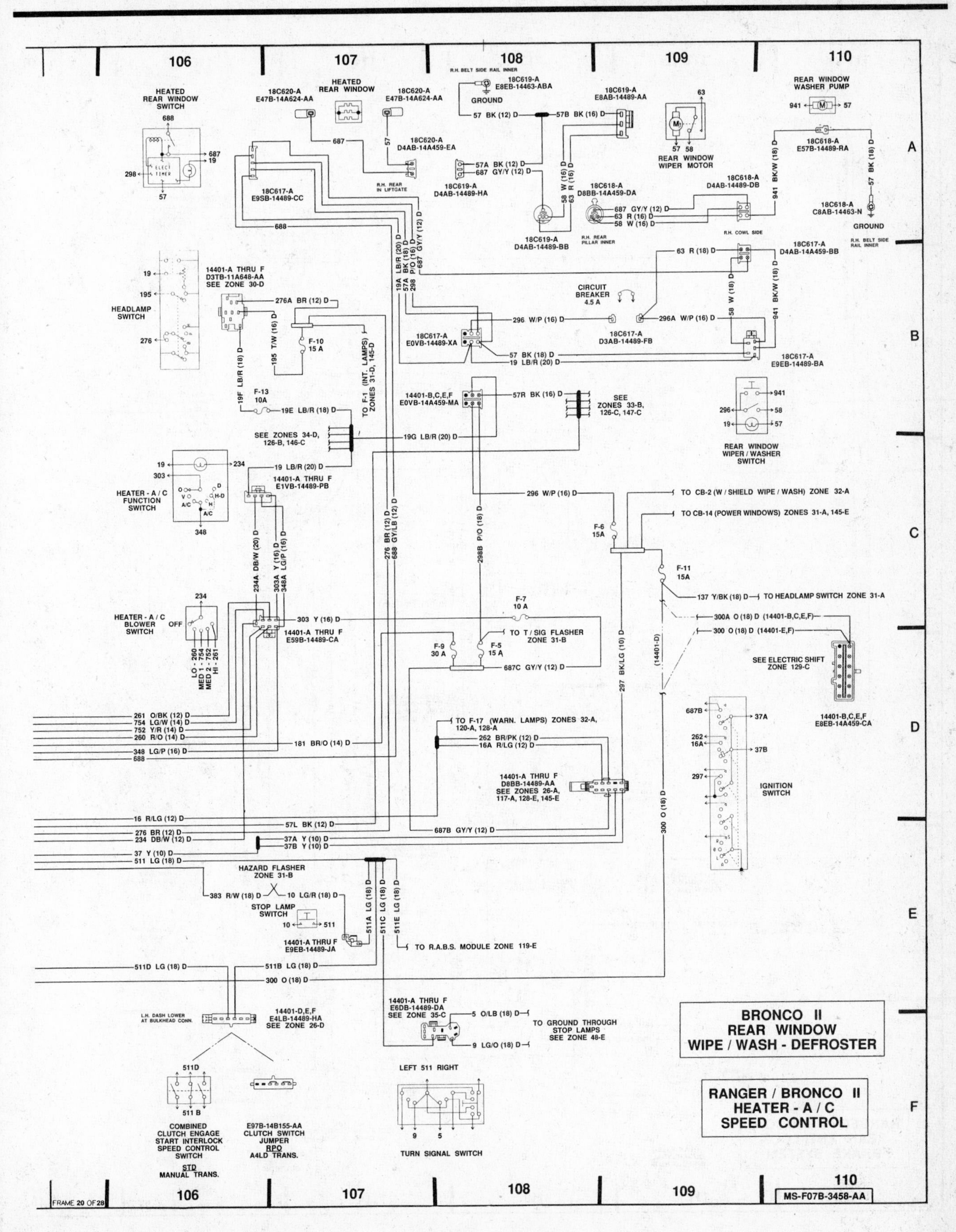
BRONCO II
REAR WINDOW
WIPE / WASH - DEFROSTER
RANGER / BRONCO II
HEATER - A / C
SPEED CONTROL
MS-F07B-3458-AA

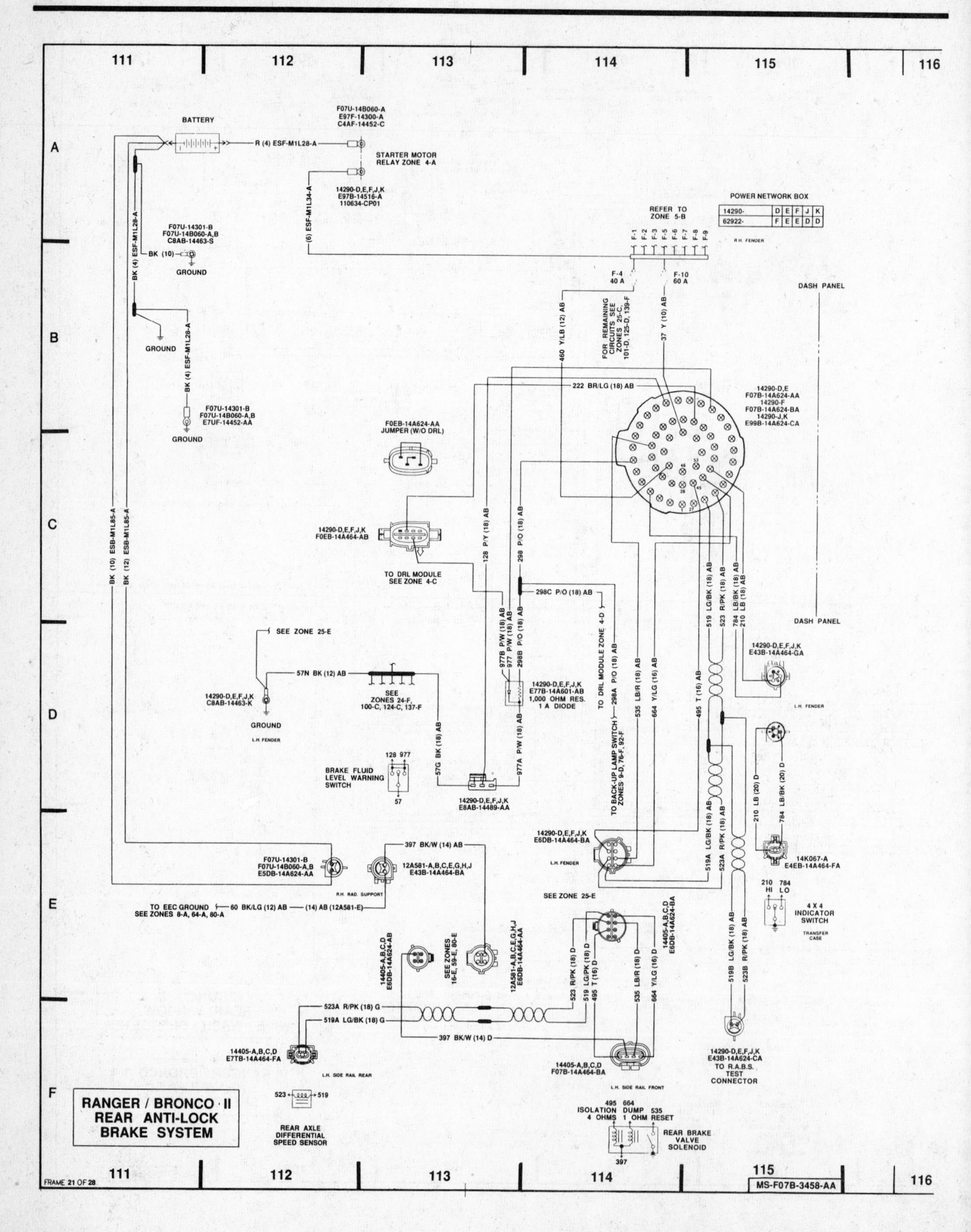

BATTERY
STARTER MOTOR RELAY ZONE 4-A
GROUND
POWER NETWORK BOX
REFER TO ZONE 5-B
DASH PANEL
F0EB-14A624-AA JUMPER (W/O DRL)
TO DRL MODULE SEE ZONE 4-C
SEE ZONE 25-E
BRAKE FLUID LEVEL WARNING SWITCH
TO EEC GROUND SEE ZONES 8-A, 64-A, 80-A
4 X 4 INDICATOR SWITCH
TO R.A.B.S. TEST CONNECTOR
REAR AXLE DIFFERENTIAL SPEED SENSOR
ISOLATION 4 OHMS
DUMP 1 OHM
RESET
REAR BRAKE VALVE SOLENOID
RANGER / BRONCO II REAR ANTI-LOCK BRAKE SYSTEM
FRAME 21 OF 28
MS-F07B-3458-AA

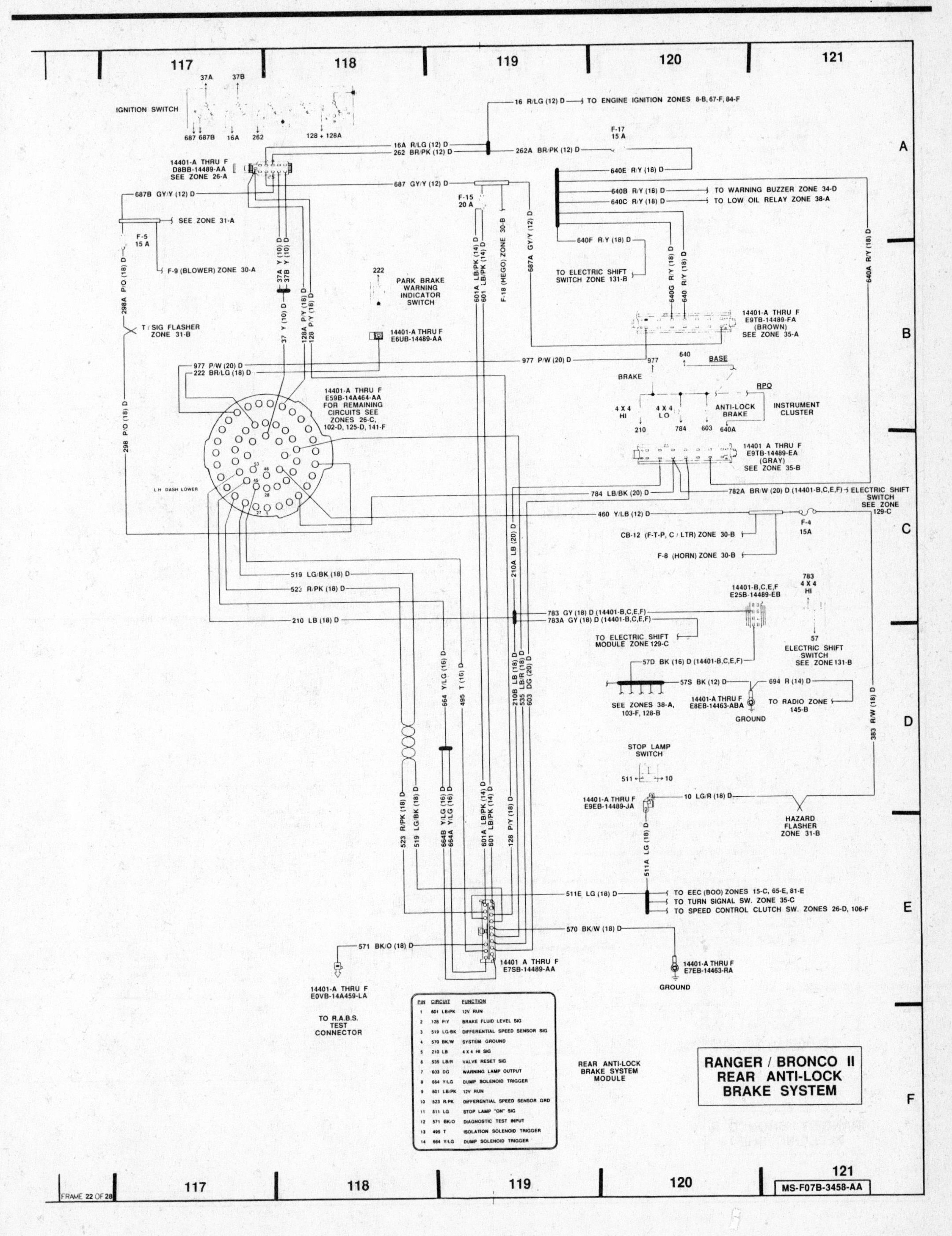

PIN	CIRCUIT	FUNCTION
1	601 LB/PK	12V RUN
2	128 P/Y	BRAKE FLUID LEVEL SIG
3	519 LG/BK	DIFFERENTIAL SPEED SENSOR SIG
4	570 BK/W	SYSTEM GROUND
5	210 LB	4 X 4 HI SIG
6	535 LB/R	VALVE RESET SIG
7	603 DG	WARNING LAMP OUTPUT
8	664 Y/LG	DUMP SOLENOID TRIGGER
9	601 LB/PK	12V RUN
10	523 R/PK	DIFFERENTIAL SPEED SENSOR GRD
11	511 LG	STOP LAMP "ON" SIG
12	571 BK/O	DIAGNOSTIC TEST INPUT
13	495 T	ISOLATION SOLENOID TRIGGER
14	664 Y/LG	DUMP SOLENOID TRIGGER

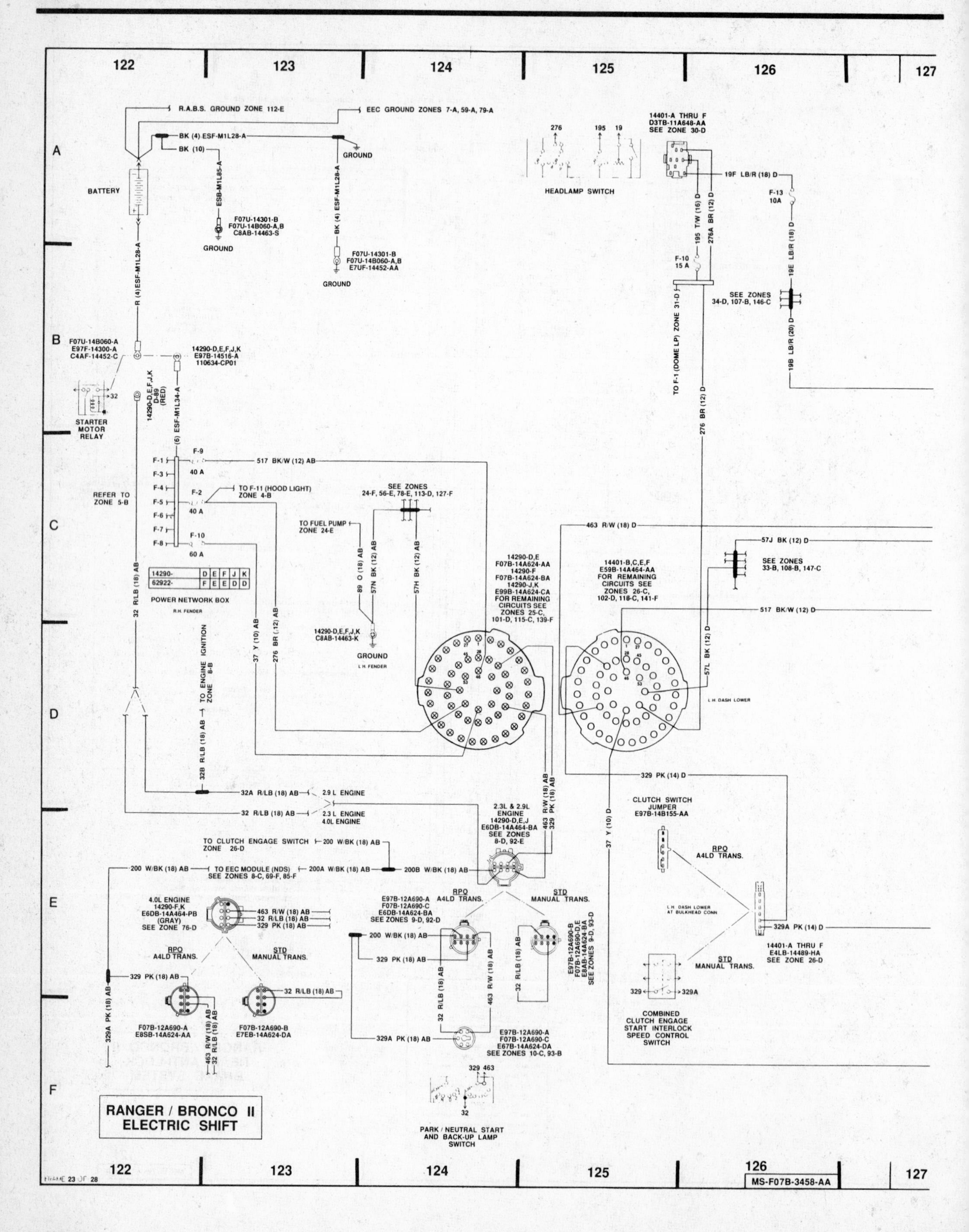
122
123
124
125
126
127
R.A.B.S. GROUND ZONE 112-E
EEC GROUND ZONES 7-A, 59-A, 79-A
BK (4) ESF-M1L28-A
BK (10)
GROUND
BATTERY
ESB-M1L85-A
BK (4) ESF-M1L28-A
F07U-14301-B
F07U-14B060-A,B
C8AB-14463-S
GROUND
F07U-14301-B
F07U-14B060-A,B
E7UF-14452-AA
GROUND
HEADLAMP SWITCH
14401-A THRU F
D3TB-11A648-AA
SEE ZONE 30-D
19F LB/R (18) D
F-13
10A
195 T/W (16) D
276A BR (12) D
F-10
15 A
19E LB/R (18) D
SEE ZONES
34-D, 107-B, 146-C
19B LB/R (20) D
TO F-1 (DOME LP) ZONE 31-D
276 BR (12) D
R (4) ESF-M1L28-A
F07U-14B060-A
E97F-14300-A
C4AF-14452-C
14290-D,E,F,J,K
E97B-14516-A
110634-CP01
STARTER
MOTOR
RELAY
14290-D,E,F,J,K
D-89
(RED)
(6) ESF-M1L34-A
F-9
40 A
517 BK/W (12) AB
TO F-11 (HOOD LIGHT)
ZONE 4-B
F-2
40 A
F-10
60 A
REFER TO
ZONE 5-B
SEE ZONES
24-F, 56-E, 78-E, 113-D, 127-F
TO FUEL PUMP
ZONE 24-E
89 O (18) AB
57N BK (12) AB
57H BK (12) AB
463 R/W (18) D
57J BK (12) D
SEE ZONES
33-B, 108-B, 147-C
517 BK/W (12) D
POWER NETWORK BOX
R.H. FENDER
32 R/LB (18) AB
37 Y (10) AB
276 BR (12) AB
14290-D,E,F,J,K
C8AB-14463-K
GROUND
L.H. FENDER
14290-D,E
F07B-14A624-AA
14290-F
F07B-14A624-BA
14290-J,K
E99B-14A624-CA
FOR REMAINING
CIRCUITS SEE
ZONES 25-C,
101-D, 115-C, 139-F
14401-B,C,E,F
E59B-14A464-AA
FOR REMAINING
CIRCUITS SEE
ZONES 26-C,
102-D, 118-C, 141-F
57L BK (12) D
L.H. DASH LOWER
TO ENGINE IGNITION ZONE 8-B
32B R/LB (18) AB
32A R/LB (18) AB
2.9 L ENGINE
32 R/LB (18) AB
2.3 L ENGINE
4.0L ENGINE
329 PK (14) D
CLUTCH SWITCH
JUMPER
E97B-14B155-AA
2.3L & 2.9L
ENGINE
14290-D,E,J
E6DB-14A464-BA
SEE ZONES
8-D, 92-E
463 R/W (18) AB
329 PK (18) AB
37 Y (10) D
TO CLUTCH ENGAGE SWITCH
ZONE 26-D
200 W/BK (18) AB
200 W/BK (18) AB
TO EEC MODULE (NDS)
SEE ZONES 8-C, 69-F, 85-F
200A W/BK (18) AB
200B W/BK (18) AB
RPO
A4LD TRANS.
4.0L ENGINE
14290-F,K
E6DB-14A464-PB
(GRAY)
SEE ZONE 76-D
463 R/W (18) AB
32 R/LB (18) AB
329 PK (18) AB
RPO
A4LD TRANS.
STD
MANUAL TRANS.
E97B-12A690-A
F07B-12A690-C
E6DB-14A624-BA
SEE ZONES 9-D, 92-D
RPO
A4LD TRANS.
STD
MANUAL TRANS.
L.H. DASH LOWER
AT BULKHEAD CONN
329A PK (14) D
14401-A THRU F
E4LB-14489-HA
SEE ZONE 26-D
STD
MANUAL TRANS.
200 W/BK (18) AB
329 PK (18) AB
32 R/LB (18) AB
463 R/W (18) AB
32 R/LB (18) AB
E97B-12A690-B
F07B-12A690-D,E
E8SB-14A624-BA
SEE ZONES 9-D, 93-D
329 PK (18) AB
32 R/LB (18) AB
329A PK (18) AB
463 R/W (18) AB
32 R/LB (18) AB
F07B-12A690-A
E8SB-14A624-AA
F07B-12A690-B
E7EB-14A624-DA
329A PK (18) AB
E97B-12A690-A
F07B-12A690-C
E67B-14A624-DA
SEE ZONES 10-C, 93-B
329
329A
COMBINED
CLUTCH ENGAGE
START INTERLOCK
SPEED CONTROL
SWITCH
329 463
32
RANGER / BRONCO II
ELECTRIC SHIFT
PARK / NEUTRAL START
AND BACK-UP LAMP
SWITCH
FRAME 23 OF 28
MS-F07B-3458-AA

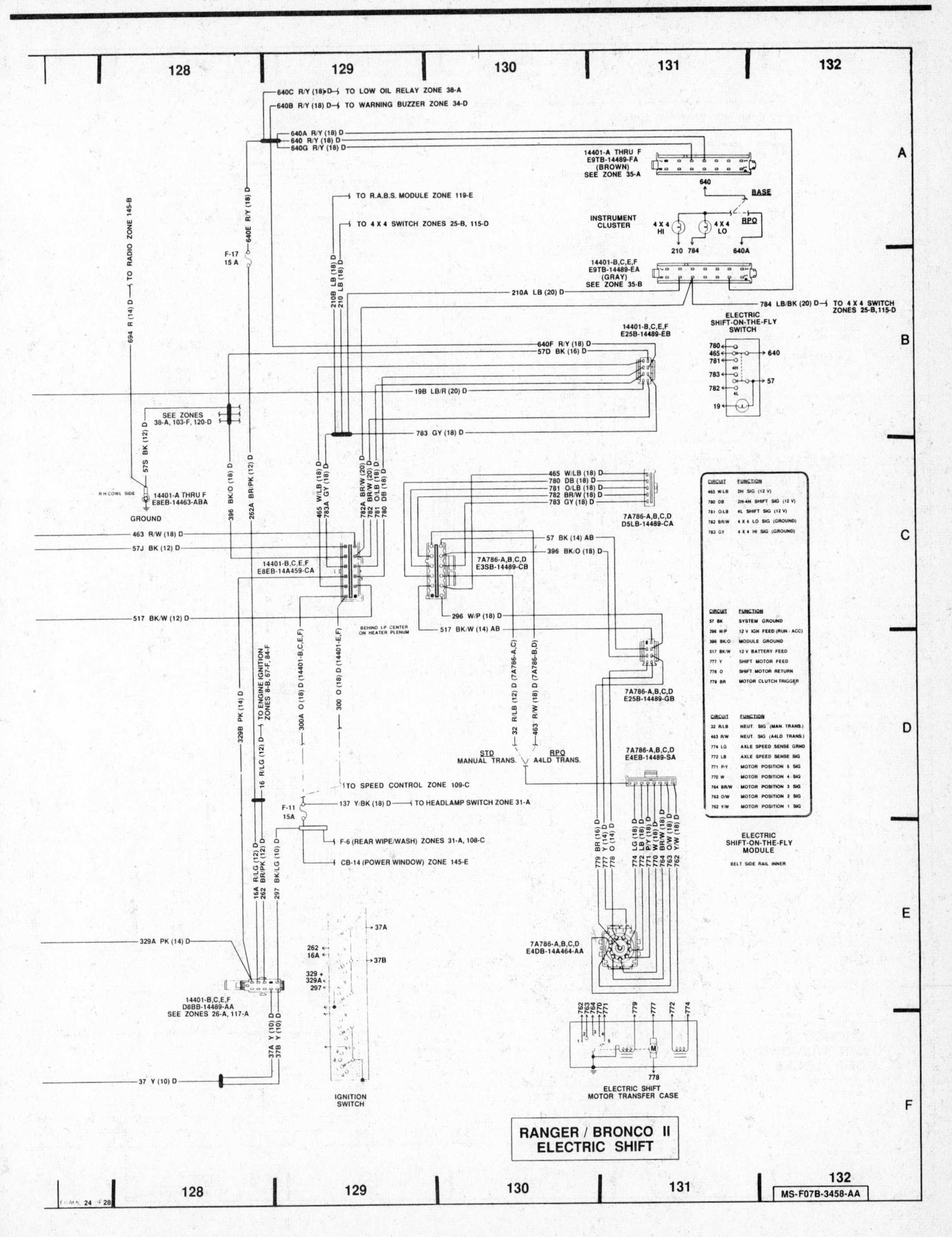

RANGER / BRONCO II
ELECTRIC SHIFT
MS-F07B-3458-AA

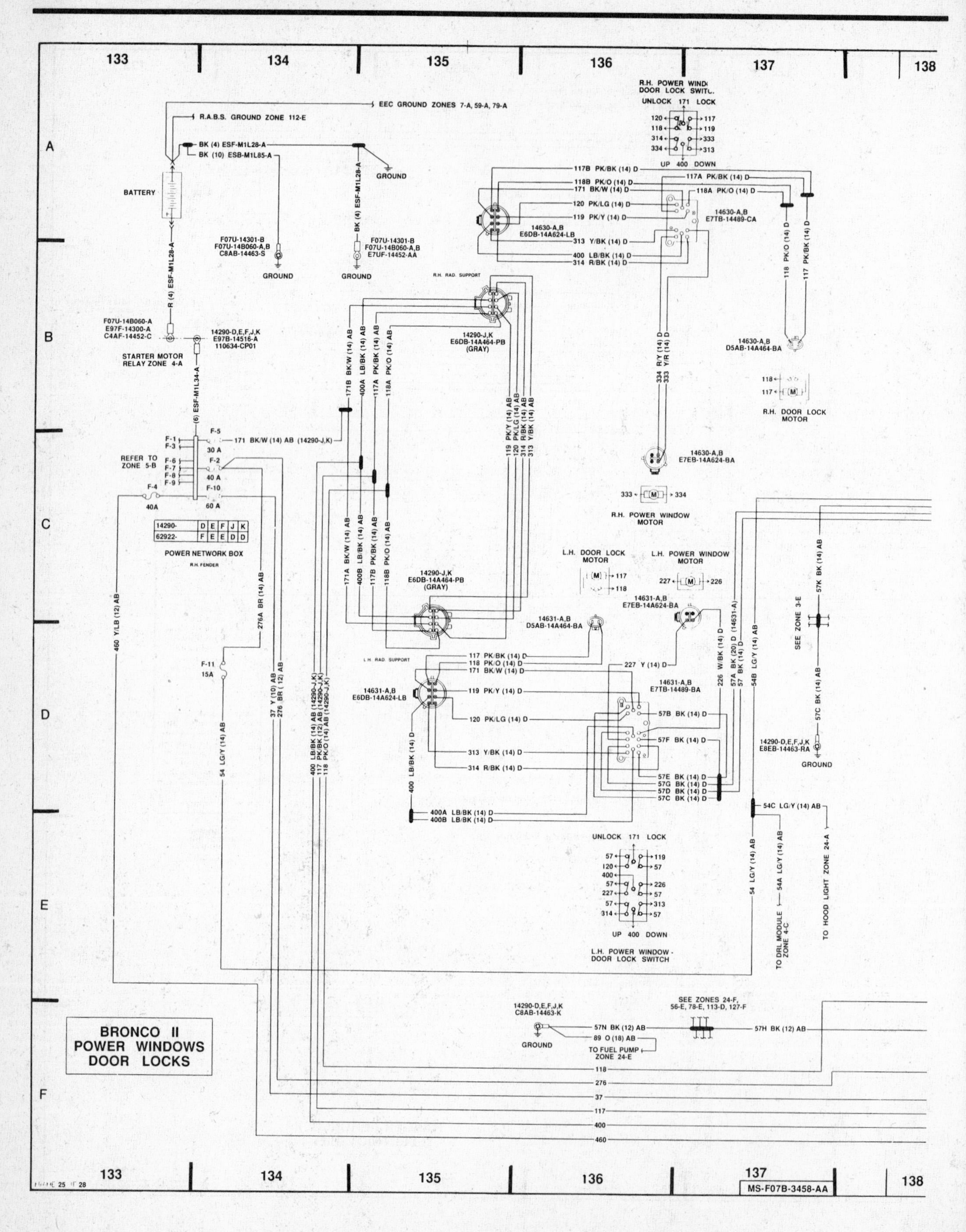
BRONCO II
POWER WINDOWS
DOOR LOCKS
R.H. POWER WINDOW DOOR LOCK SWITCH
L.H. POWER WINDOW - DOOR LOCK SWITCH
R.H. DOOR LOCK MOTOR
R.H. POWER WINDOW MOTOR
L.H. DOOR LOCK MOTOR
L.H. POWER WINDOW MOTOR
BATTERY
POWER NETWORK BOX
STARTER MOTOR RELAY ZONE 4-A
GROUND
MS-F07B-3458-AA

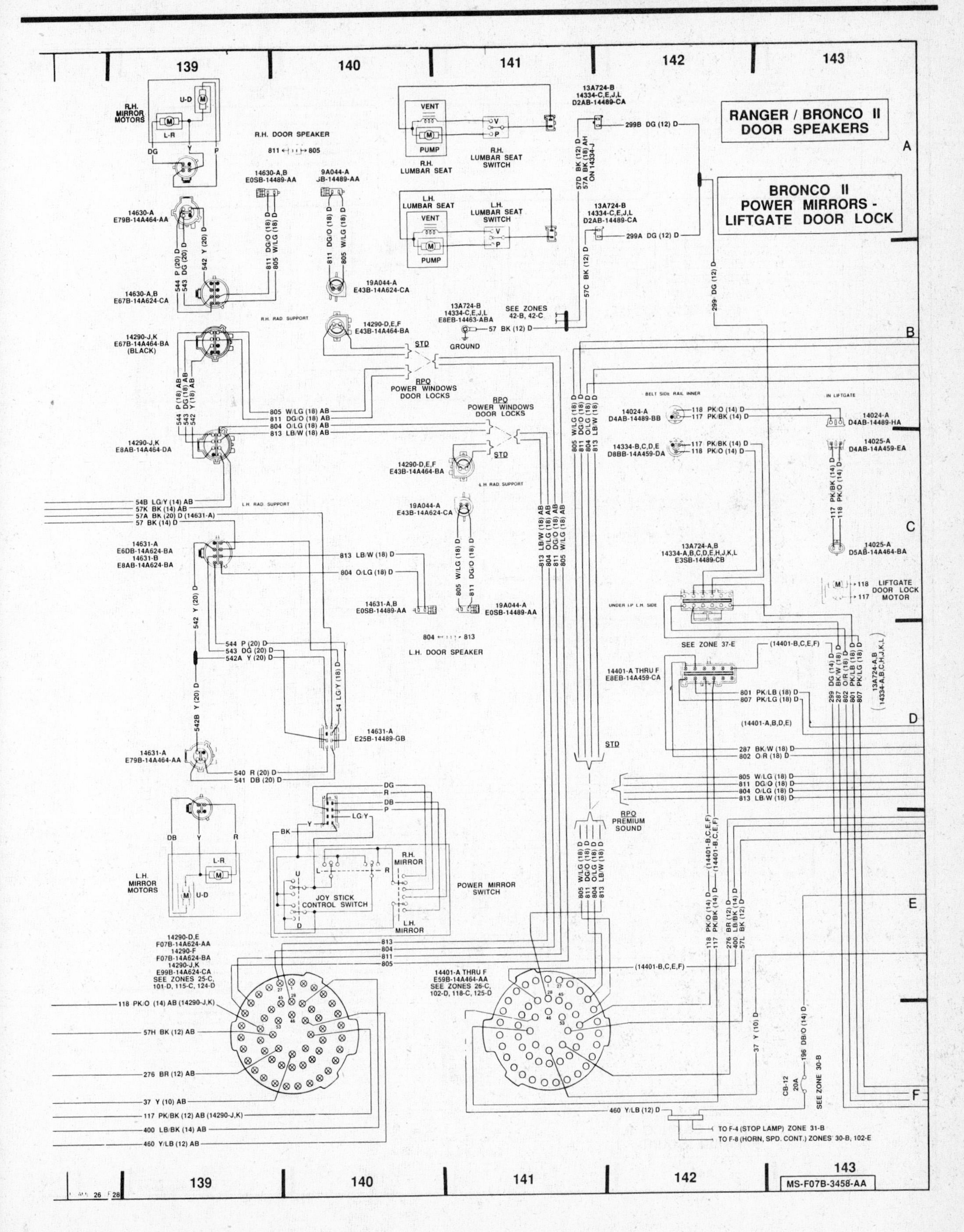
RANGER / BRONCO II DOOR SPEAKERS
BRONCO II POWER MIRRORS - LIFTGATE DOOR LOCK
R.H. MIRROR MOTORS
R.H. DOOR SPEAKER
R.H. LUMBAR SEAT
R.H. LUMBAR SEAT SWITCH
L.H. LUMBAR SEAT
L.H. LUMBAR SEAT SWITCH
VENT
PUMP
GROUND
RPO POWER WINDOWS DOOR LOCKS
STD
L.H. DOOR SPEAKER
LIFTGATE DOOR LOCK MOTOR
RPO PREMIUM SOUND
L.H. MIRROR MOTORS
JOY STICK CONTROL SWITCH
R.H. MIRROR
L.H. MIRROR
POWER MIRROR SWITCH
TO F-4 (STOP LAMP) ZONE 31-B
TO F-8 (HORN, SPD. CONT.) ZONES 30-B, 102-E
MS-F07B-3458-AA

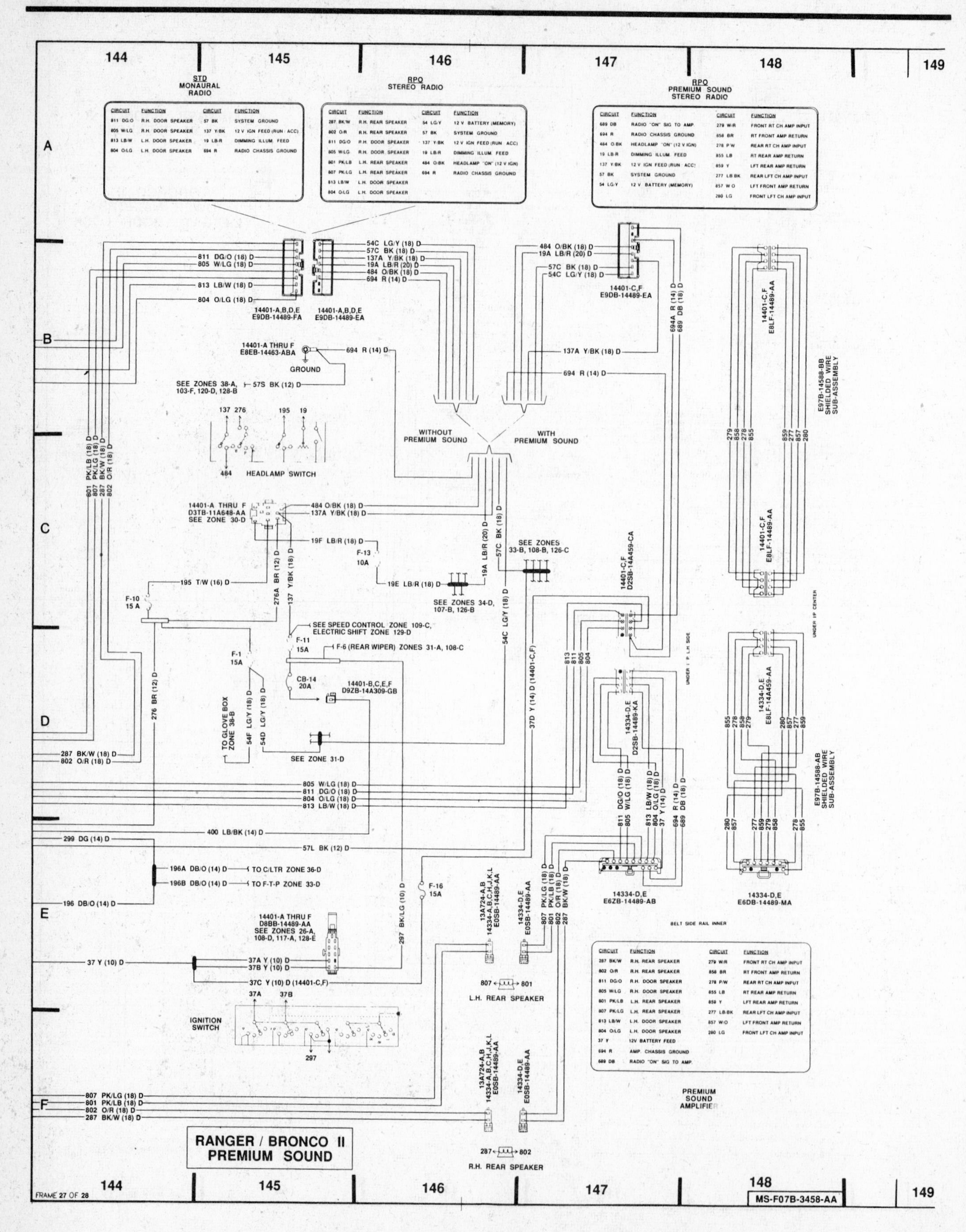
144
145
146
147
148
149
STD
MONAURAL
RADIO
CIRCUIT FUNCTION CIRCUIT FUNCTION
811 DG/O R.H. DOOR SPEAKER 57 BK SYSTEM GROUND
805 W/LG R.H. DOOR SPEAKER 137 Y/BK 12 V IGN FEED (RUN / ACC)
813 LB/W L.H. DOOR SPEAKER 19 LB/R DIMMING ILLUM. FEED
804 O/LG L.H. DOOR SPEAKER 694 R RADIO CHASSIS GROUND
RPO
STEREO RADIO
CIRCUIT FUNCTION CIRCUIT FUNCTION
287 BK/W R.H. REAR SPEAKER 54 LG/Y 12 V BATTERY (MEMORY)
802 O/R R.H. REAR SPEAKER 57 BK SYSTEM GROUND
811 DG/O R.H. DOOR SPEAKER 137 Y/BK 12 V IGN FEED (RUN ACC)
805 W/LG R.H. DOOR SPEAKER 19 LB/R DIMMING ILLUM. FEED
801 PK/LB L.H. REAR SPEAKER 484 O/BK HEADLAMP "ON" (12 V IGN)
807 PK/LG L.H. REAR SPEAKER 694 R RADIO CHASSIS GROUND
813 LB/W L.H. DOOR SPEAKER
804 O/LG L.H. DOOR SPEAKER
RPO
PREMIUM SOUND
STEREO RADIO
CIRCUIT FUNCTION CIRCUIT FUNCTION
689 DB RADIO "ON" SIG TO AMP 279 W/R FRONT RT CH AMP INPUT
694 R RADIO CHASSIS GROUND 858 BR RT FRONT AMP RETURN
484 O/BK HEADLAMP "ON" (12 V IGN) 278 P/W REAR RT CH AMP INPUT
19 LB/R DIMMING ILLUM. FEED 855 LB RT REAR AMP RETURN
137 Y/BK 12 V IGN FEED (RUN ACC) 859 Y LFT REAR AMP RETURN
57 BK SYSTEM GROUND 277 LB/BK REAR LFT CH AMP INPUT
54 LG/Y 12 V BATTERY (MEMORY) 857 W/O LFT FRONT AMP RETURN
280 LG FRONT LFT CH AMP INPUT
A
B
C
D
E
F
811 DG/O (18) D
805 W/LG (18) D
813 LB/W (18) D
804 O/LG (18) D
54C LG/Y (18) D
57C BK (18) D
137A Y/BK (18) D
19A LB/R (20) D
484 O/BK (18) D
694 R (14) D
14401-A,B,D,E
E9DB-14489-FA
14401-A,B,D,E
E9DB-14489-EA
484 O/BK (18) D
19A LB/R (20) D
57C BK (18) D
54C LG/Y (18) D
14401-C,F
E9DB-14489-EA
694A R (14) D
689 DB (18) D
14401-A THRU F
E8EB-14463-ABA
694 R (14) D
GROUND
SEE ZONES 38-A, 103-F, 120-D, 128-B
57S BK (12) D
137A Y/BK (18) D
694 R (14) D
137 276
195 19
484
HEADLAMP SWITCH
WITHOUT
PREMIUM SOUND
WITH
PREMIUM SOUND
14401-C,F
E8LF-14489-AA
E97B-14588-BB
SHIELDED WIRE
SUB-ASSEMBLY
279 858 278 855
859 277 857 280
801 PK/LB (18) D
807 PK/LG (18) D
287 BK/W (18) D
802 O/R (18) D
14401-A THRU F
D3TB-11A648-AA
SEE ZONE 30-D
484 O/BK (18) D
137A Y/BK (18) D
19F LB/R (18) D
F-13
10A
19E LB/R (18) D
19A LB/R (20) D
57C BK (18) D
SEE ZONES
33-B, 108-B, 126-C
SEE ZONES 34-D,
107-B, 126-B
14401-C,F
D2SB-14A459-CA
14401-C,F
E8LF-14489-AA
195 T/W (16) D
F-10
15 A
276A BR (12) D
137 Y/BK (18) D
SEE SPEED CONTROL ZONE 109-C,
ELECTRIC SHIFT ZONE 129-D
F-11
15A
F-6 (REAR WIPER) ZONES 31-A, 108-C
F-1
15A
CB-14
20A
14401-B,C,E,F
D9ZB-14A309-GB
54C LG/Y (18) D
37D Y (14) D (14401-C,F)
813
811
805
804
UNDER I. P. LH SIDE
UNDER IP CENTER
276 BR (12) D
TO GLOVE BOX
ZONE 38-B
54F LG/Y (18) D
54D LG/Y (18) D
SEE ZONE 31-D
14334-D,E
D2SB-14489-KA
14334-D,E
E8LF-14A459-AA
855 278 858 279
280 857 277 859
287 BK/W (18) D
802 O/R (18) D
805 W/LG (18) D
811 DG/O (18) D
804 O/LG (18) D
813 LB/W (18) D
811 DG/O (18) D
805 W/LG (18) D
813 LB/W (18) D
804 O/LG (18) D
37 Y (14) D
694 R (14) D
689 DB (18) D
E97B-14588-AB
SHIELDED WIRE
SUB-ASSEMBLY
280 857
277 859 279 858
278 855
400 LB/BK (14) D
299 DG (14) D
57L BK (12) D
196A DB/O (14) D
TO C/LTR ZONE 36-D
196B DB/O (14) D
TO F-T-P ZONE 33-D
196 DB/O (14) D
807 PK/LG (18) D
801 PK/LB (18) D
802 O/R (18) D
287 BK/W (18) D
14334-D,E
E6ZB-14489-AB
14334-D,E
E6DB-14489-MA
F-16
15A
297 BK/LG (10) D
13A724-A,B
14334-A,B,C,H,J,K,L
E0SB-14489-AA
14334-D,E
E0SB-14489-AA
BELT SIDE RAIL INNER
14401-A THRU F
D8BB-14489-AA
SEE ZONES 26-A,
108-D, 117-A, 128-E
37 Y (10) D
37A Y (10) D
37B Y (10) D
37C Y (10) D (14401-C,F)
807
801
L.H. REAR SPEAKER
37A
37B
IGNITION
SWITCH
297
CIRCUIT FUNCTION CIRCUIT FUNCTION
287 BK/W R.H. REAR SPEAKER 279 W/R FRONT RT CH AMP INPUT
802 O/R R.H. REAR SPEAKER 858 BR RT FRONT AMP RETURN
811 DG/O R.H. DOOR SPEAKER 278 P/W REAR RT CH AMP INPUT
805 W/LG R.H. DOOR SPEAKER 855 LB RT REAR AMP RETURN
801 PK/LB L.H. REAR SPEAKER 859 Y LFT REAR AMP RETURN
807 PK/LG L.H. REAR SPEAKER 277 LB/BK REAR LFT CH AMP INPUT
813 LB/W L.H. DOOR SPEAKER 857 W/O LFT FRONT AMP RETURN
804 O/LG L.H. DOOR SPEAKER 280 LG FRONT LFT CH AMP INPUT
37 Y 12V BATTERY FEED
694 R AMP. CHASSIS GROUND
689 DB RADIO "ON" SIG TO AMP.
807 PK/LG (18) D
801 PK/LB (18) D
802 O/R (18) D
287 BK/W (18) D
13A724-A,B
14334-A,B,C,H,J,K,L
E0SB-14489-AA
14334-D,E
E0SB-14489-AA
PREMIUM
SOUND
AMPLIFIER
RANGER / BRONCO II
PREMIUM SOUND
287
802
R.H. REAR SPEAKER
FRAME 27 OF 28
MS-F07B-3458-AA

150 151 152 153 154

A

B

C

D

E

F

FRAME 28 OF 28

150 151 152 153 154

MS-F07B-3458-AA

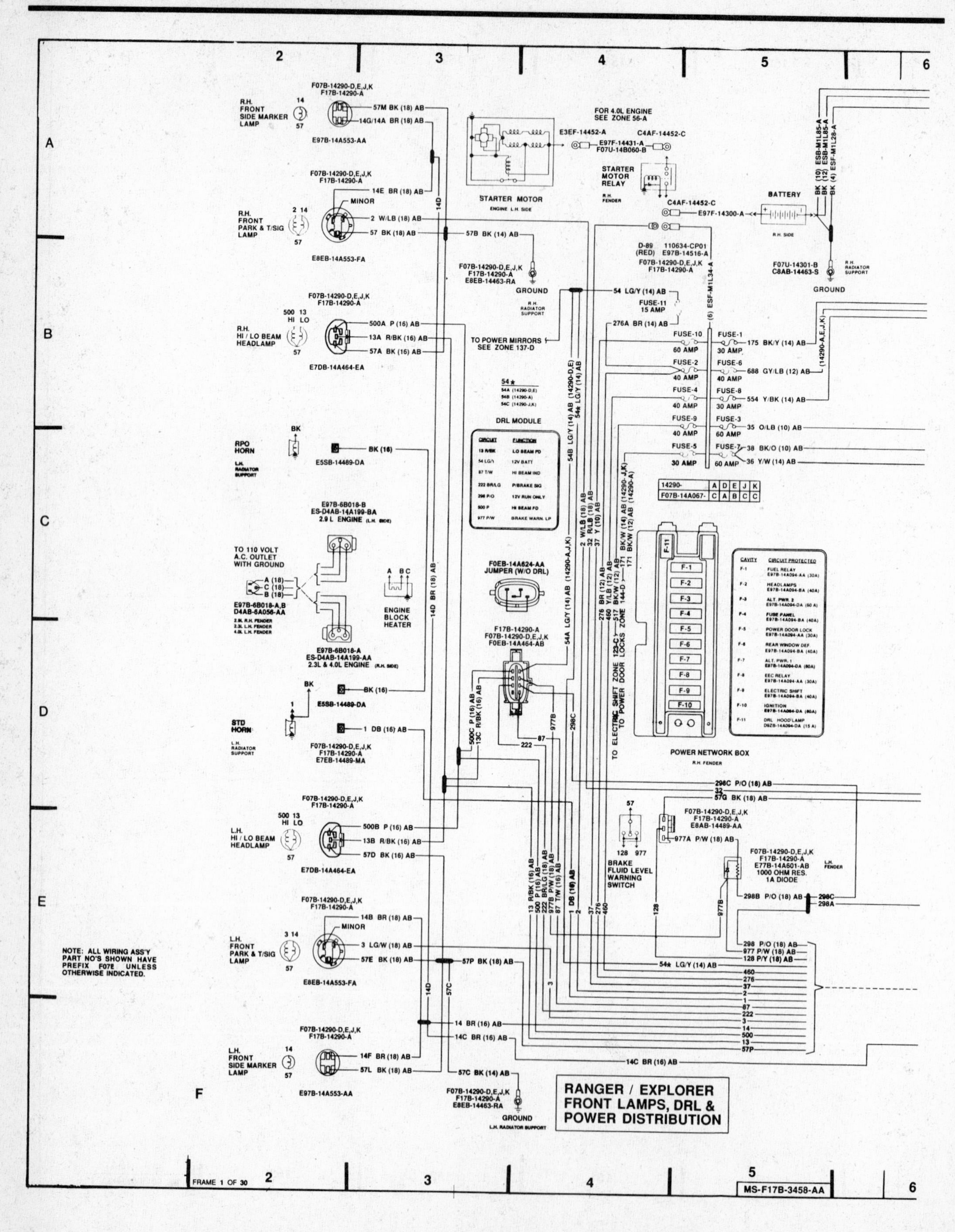
R.H. FRONT SIDE MARKER LAMP
F07B-14290-D,E,J,K
F17B-14290-A
57M BK (18) AB
14G/14A BR (18) AB
E97B-14A553-AA
R.H. FRONT PARK & T/SIG LAMP
14E BR (18) AB
MINOR
2 W/LB (18) AB
57 BK (18) AB
E8EB-14A553-FA
57B BK (14) AB
F07B-14290-D,E,J,K
F17B-14290-A
E8EB-14463-RA
GROUND
R.H. RADIATOR SUPPORT
R.H. HI / LO BEAM HEADLAMP
500A P (16) AB
13A R/BK (16) AB
57A BK (16) AB
E7DB-14A464-EA
TO POWER MIRRORS SEE ZONE 137-D
RPO HORN
L.H. RADIATOR SUPPORT
BK (16)
E5SB-14489-DA
E97B-6B018-B
ES-D4AB-14A199-BA
2.9 L ENGINE (L.H. SIDE)
TO 110 VOLT A.C. OUTLET WITH GROUND
A (18)
C (18)
B (18)
E97B-6B018-A,B
D4AB-6A056-AA
2.9L R.H. FENDER
2.3L L.H. FENDER
4.0L L.H. FENDER
ENGINE BLOCK HEATER
E97B-6B018-A
ES-D4AB-14A199-AA
2.3L & 4.0L ENGINE (R.H. SIDE)
STD HORN
1 DB (16) AB
E5SB-14489-DA
F07B-14290-D,E,J,K
F17B-14290-A
E7EB-14489-MA
STARTER MOTOR
ENGINE L.H. SIDE
FOR 4.0L ENGINE SEE ZONE 56-A
E3EF-14452-A
E97F-14431-A
F07U-14B060-B
C4AF-14452-C
STARTER MOTOR RELAY
R.H. FENDER
C4AF-14452-C
E97F-14300-A
BATTERY
R.H. SIDE
ESB-M1L85-A
ESB-M1L85-A
ESF-M1L28-A
BK (10)
BK (12)
BK (4)
F07U-14301-B
C8AB-14463-S
R.H. RADIATOR SUPPORT
GROUND
D-89 (RED)
110634-CP01
E97B-14516-A
F07B-14290-D,E,J,K
F17B-14290-A
54 LG/Y (14) AB
FUSE-11 15 AMP
276A BR (14) AB
(6) ESF-M1L34-A
FUSE-10 60 AMP
FUSE-1 30 AMP
175 BK/Y (14) AB
FUSE-2 40 AMP
FUSE-6 40 AMP
688 GY/LB (12) AB
FUSE-4 40 AMP
FUSE-8 30 AMP
554 Y/BK (14) AB
FUSE-9 40 AMP
FUSE-3 60 AMP
35 O/LB (10) AB
FUSE-5 30 AMP
FUSE-7 60 AMP
38 BK/O (10) AB
36 Y/W (14) AB
(14290-A,E,J,K)
54★
54A (14290-D,E)
54B (14290-A)
54C (14290-J,K)
DRL MODULE
CIRCUIT | FUNCTION
13 R/BK | LO BEAM FD
54 LG/Y | 12V BATT
87 T/W | HI BEAM IND
222 BR/LG | P/BRAKE SIG
298 P/O | 12V RUN ONLY
500 P | HI BEAM FD
977 P/W | BRAKE WARN. LP
F0EB-14A624-AA
JUMPER (W/O DRL)
F17B-14290-A
F07B-14290-D,E,J,K
F0EB-14A464-AB
14290- | A | D | E | J | K
F07B-14A067- | C | A | B | C | C
CAVITY | CIRCUIT PROTECTED
F-1 | FUEL RELAY E97B-14A094-AA (30A)
F-2 | HEADLAMPS E97B-14A094-BA (40A)
F-3 | ALT. PWR. 2 E97B-14A094-DA (60 A)
F-4 | FUSE PANEL E97B-14A094-BA (40A)
F-5 | POWER DOOR LOCK E97B-14A094-AA (30A)
F-6 | REAR WINDOW DEF. E97B-14A094-BA (40A)
F-7 | ALT. PWR. 1 E97B-14A094-DA (60A)
F-8 | EEC RELAY E97B-14A094-AA (30A)
F-9 | ELECTRIC SHIFT E97B-14A094-BA (40A)
F-10 | IGNITION E97B-14A094-DA (60A)
F-11 | DRL HOOD LAMP D9ZB-14A094-DA (15 A)
POWER NETWORK BOX
R.H. FENDER
TO ELECTRIC SHIFT ZONE 123-C
TO POWER DOOR LOCKS ZONE 144-D
298C P/O (18) AB
57G BK (18) AB
BRAKE FLUID LEVEL WARNING SWITCH
F07B-14290-D,E,J,K
F17B-14290-A
E8AB-14489-AA
977A P/W (18) AB
F07B-14290-D,E,J,K
F17B-14290-A
E77B-14A601-AB
1000 OHM RES.
1A DIODE
L.H. FENDER
298B P/O (18) AB
L.H. HI / LO BEAM HEADLAMP
500B P (16) AB
13B R/BK (16) AB
57D BK (16) AB
E7DB-14A464-EA
L.H. FRONT PARK & T/SIG LAMP
14B BR (18) AB
3 LG/W (18) AB
57E BK (18) AB
57P BK (18) AB
E8EB-14A553-FA
NOTE: ALL WIRING ASS'Y PART NO'S SHOWN HAVE PREFIX F07E UNLESS OTHERWISE INDICATED.
298 P/O (18) AB
977 P/W (18) AB
128 P/Y (18) AB
54★ LG/Y (14) AB
14 BR (16) AB
14C BR (16) AB
L.H. FRONT SIDE MARKER LAMP
14F BR (18) AB
57L BK (18) AB
E97B-14A553-AA
57C BK (14) AB
F07B-14290-D,E,J,K
F17B-14290-A
E8EB-14463-RA
GROUND
L.H. RADIATOR SUPPORT
RANGER / EXPLORER FRONT LAMPS, DRL & POWER DISTRIBUTION
FRAME 1 OF 30
MS-F17B-3458-AA

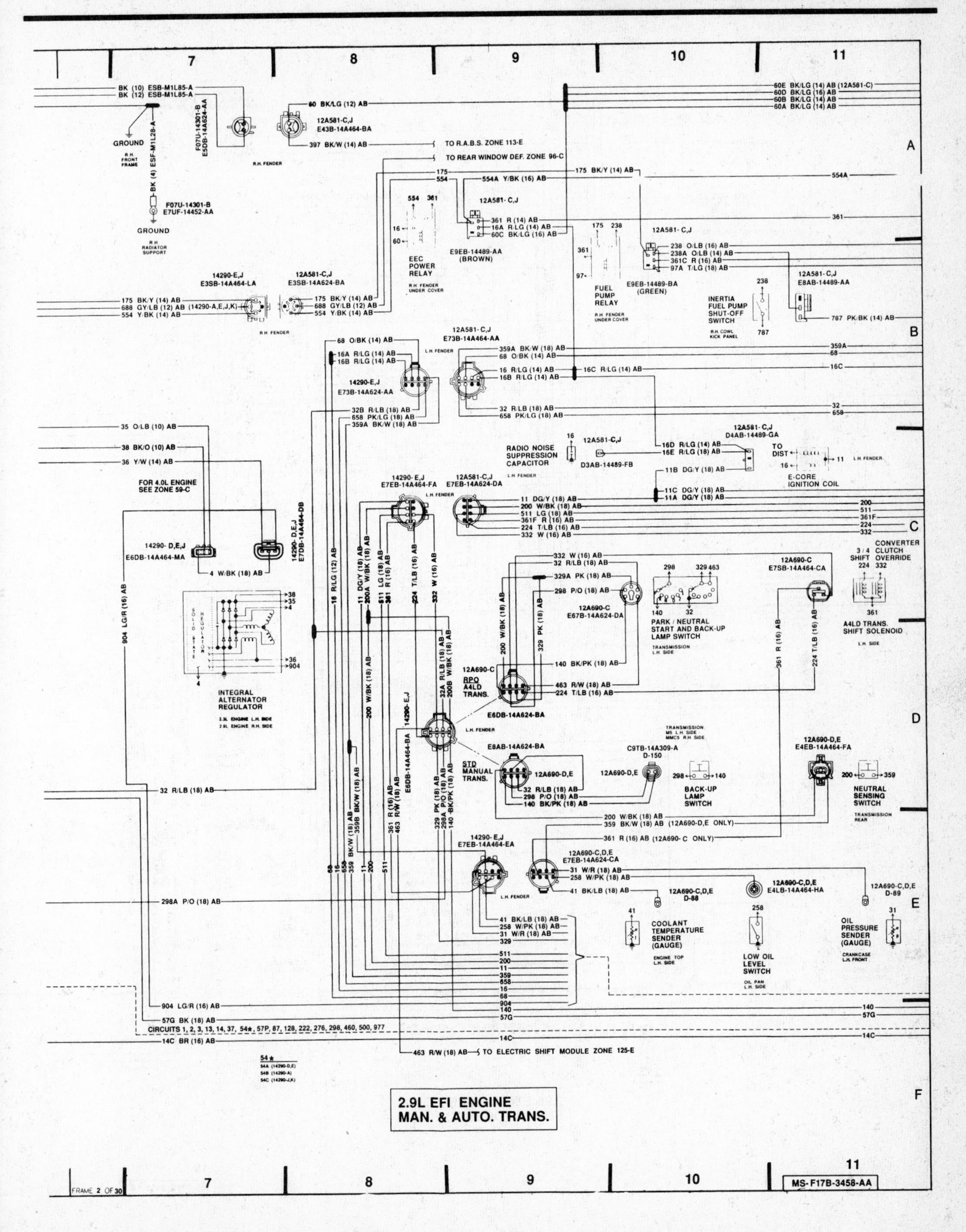

2.9L EFI ENGINE
MAN. & AUTO. TRANS.

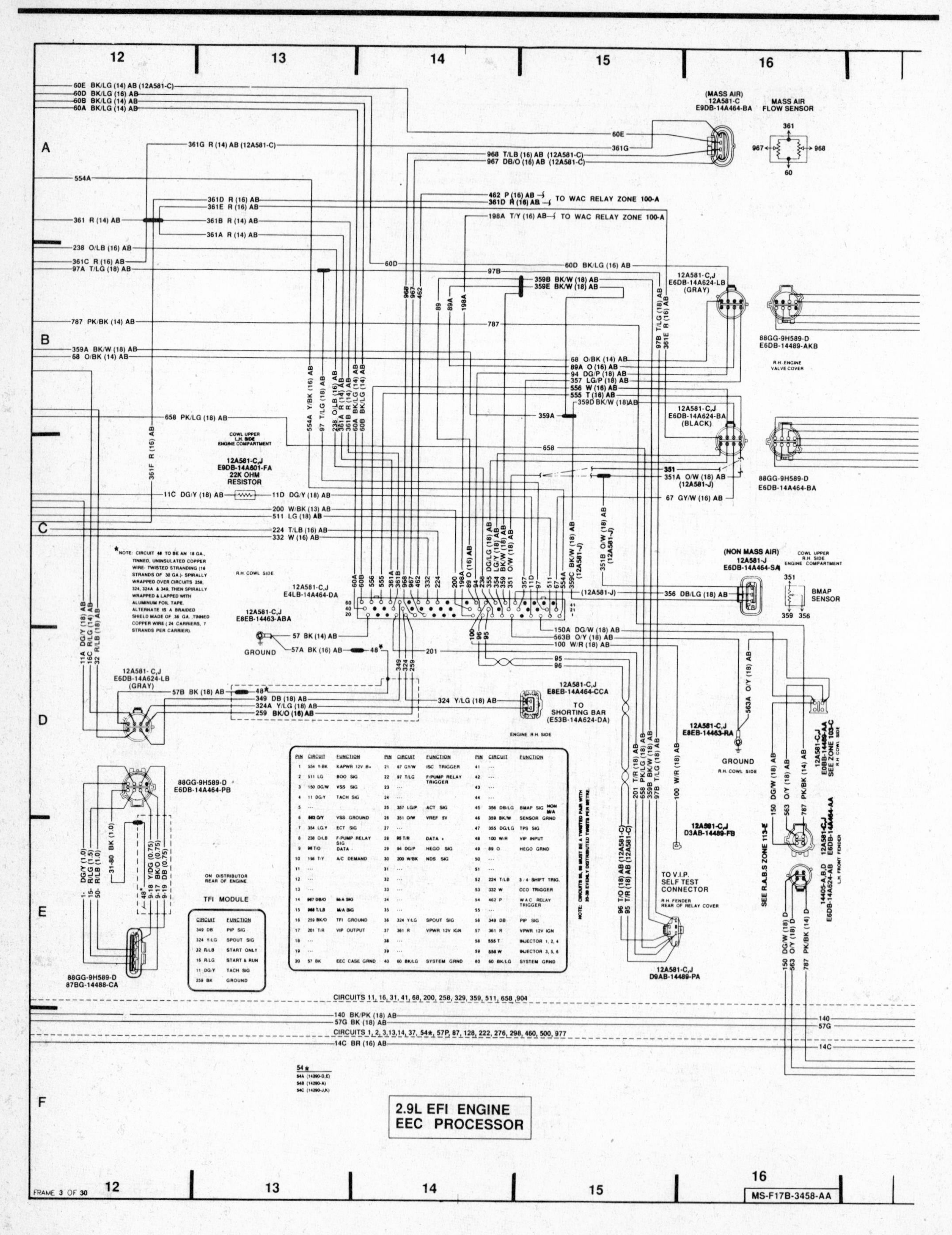

*NOTE: CIRCUIT 48 TO BE AN 18 GA., TINNED, UNINSULATED COPPER WIRE- TWISTED STRANDING (16 STRANDS OF 30 GA.)- SPIRALLY WRAPPED OVER CIRCUITS 259, 324, 324A & 349, THEN SPIRALLY WRAPPED & LAPPED WITH ALUMINUM FOIL TAPE. ALTERNATE IS A BRAIDED SHIELD MADE OF 36 GA., TINNED COPPER WIRE (24 CARRIERS, 7 STRANDS PER CARRIER).

PIN	CIRCUIT	FUNCTION	PIN	CIRCUIT	FUNCTION	PIN	CIRCUIT	FUNCTION
1	554 Y/BK	KAPWR 12V B+	21	67 GY/W	ISC TRIGGER	41	---	
2	511 LG	BOO SIG	22	97 T/LG	F/PUMP RELAY TRIGGER	42	---	
3	150 DG/W	VSS SIG	23	---		43	---	
4	11 DG/Y	TACH SIG	24	---		44	---	
5	---		25	357 LG/P	ACT SIG	45	356 DB/LG	BMAP SIG NON M/A
6	563 O/Y	VSS GROUND	26	351 O/W	VREF 5V	46	359 BK/W	SENSOR GRND
7	354 LG/Y	ECT SIG	27	---		47	355 DG/LG	TPS SIG
8	238 O/LB	F/PUMP RELAY SIG	28	95 T/R	DATA +	48	100 W/R	VIP INPUT
9	96 T/O	DATA -	29	94 DG/P	HEGO SIG	49	89 O	HEGO GRND
10	198 T/Y	A/C DEMAND	30	200 W/BK	NDS SIG	50	---	
11	---		31	---		51	---	
12	---		32	---		52	224 T/LB	3/4 SHIFT TRIG.
13	---		33	---		53	332 W	CCO TRIGGER
14	967 DB/O	M/A SIG	34	---		54	462 P	W A C RELAY TRIGGER
15	968 T/LB	M/A SIG	35	---		55	---	
16	259 BK/O	TFI GROUND	36	324 Y/LG	SPOUT SIG	56	349 DB	PIP SIG
17	201 T/R	VIP OUTPUT	37	361 R	VPWR 12V IGN	57	361 R	VPWR 12V IGN
18	---		38	---		58	555 T	INJECTOR 1, 2, 4
19	---		39	---		59	556 W	INJECTOR 3, 5, 6
20	57 BK	EEC CASE GRND	40	60 BK/LG	SYSTEM GRND	60	60 BK/LG	SYSTEM GRND

CIRCUIT	FUNCTION
349 DB	PIP SIG
324 Y/LG	SPOUT SIG
32 R/LB	START ONLY
16 R/LG	START & RUN
11 DG/Y	TACH SIG
259 BK	GROUND

54★
54A (14290-D,E)
54B (14290-A)
54C (14290-J,K)

2.9L EFI ENGINE EEC PROCESSOR

FRAME 3 OF 30

MS-F17B-3458-AA

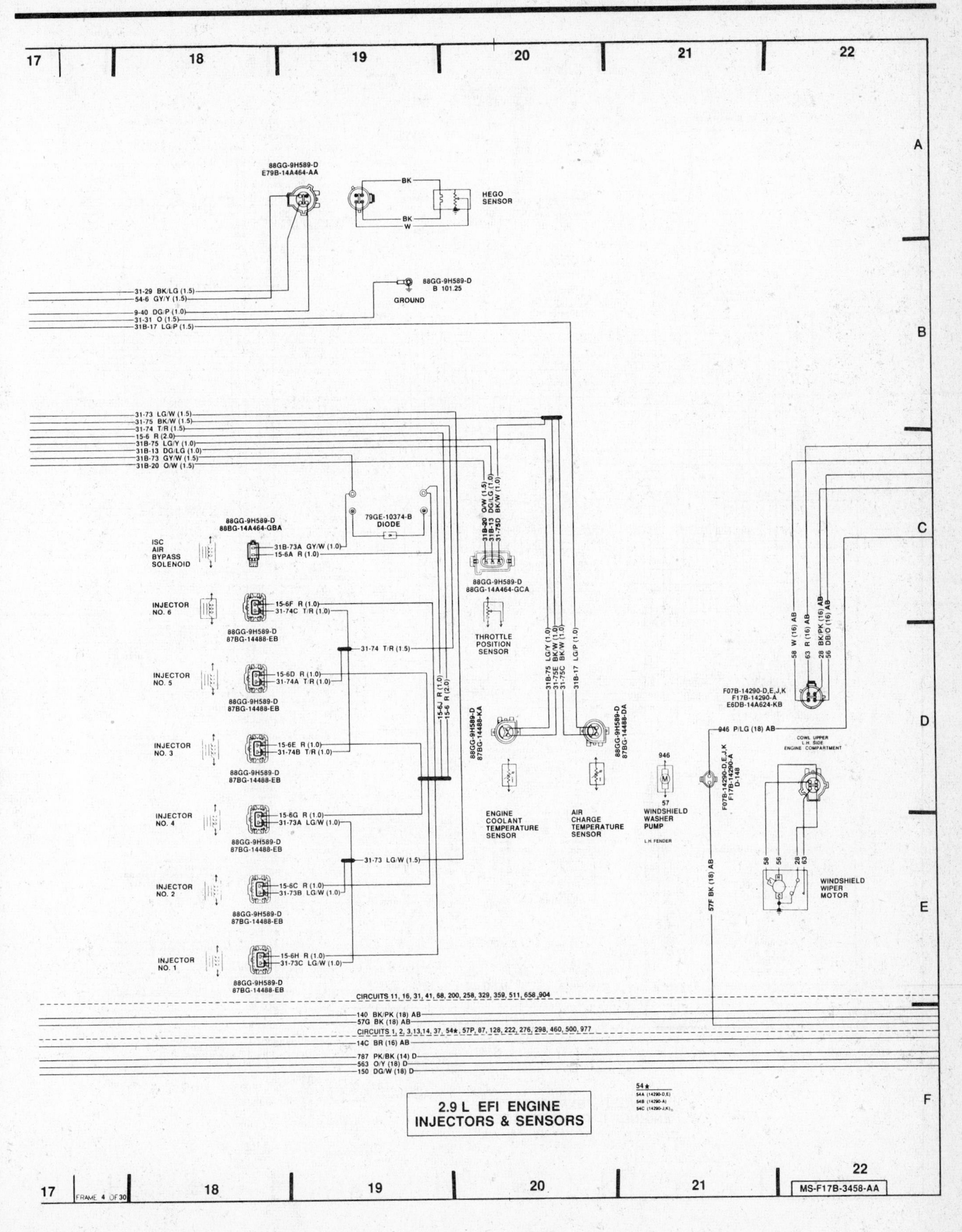

88GG-9H589-D
E79B-14A464-AA
BK
HEGO SENSOR
BK
W
88GG-9H589-D
B 101.25
GROUND
31-29 BK/LG (1.5)
54-6 GY/Y (1.5)
9-40 DG/P (1.0)
31-31 O (1.5)
31B-17 LG/P (1.5)
31-73 LG/W (1.5)
31-75 BK/W (1.5)
31-74 T/R (1.5)
15-6 R (2.0)
31B-75 LG/Y (1.0)
31B-13 DG/LG (1.0)
31B-73 GY/W (1.5)
31B-20 O/W (1.5)
79GE-10374-B DIODE
88GG-9H589-D
88BG-14A464-GBA
ISC AIR BYPASS SOLENOID
31B-73A GY/W (1.0)
15-6A R (1.0)
88GG-9H589-D
88GG-14A464-GCA
THROTTLE POSITION SENSOR
INJECTOR NO. 6
15-6F R (1.0)
31-74C T/R (1.0)
88GG-9H589-D
87BG-14488-EB
31-74 T/R (1.5)
INJECTOR NO. 5
15-6D R (1.0)
31-74A T/R (1.0)
INJECTOR NO. 3
15-6E R (1.0)
31-74B T/R (1.0)
INJECTOR NO. 4
15-6G R (1.0)
31-73A LG/W (1.0)
31-73 LG/W (1.5)
INJECTOR NO. 2
15-6C R (1.0)
31-73B LG/W (1.0)
INJECTOR NO. 1
15-6H R (1.0)
31-73C LG/W (1.0)
15-6J R (1.0)
15-6 R (2.0)
88GG-9H589-D
87BG-14488-KA
88GG-9H589-D
87BG-14488-DA
ENGINE COOLANT TEMPERATURE SENSOR
AIR CHARGE TEMPERATURE SENSOR
946
57
WINDSHIELD WASHER PUMP
L.H. FENDER
946 P/LG (18) AB
COWL UPPER L.H. SIDE ENGINE COMPARTMENT
F07B-14290-D,E,J,K
F17B-14290-A
E6DB-14A624-KB
58 W (16) AB
63 R (16) AB
28 BK/PK (16) AB
56 DB/O (16) AB
57F BK (18) AB
WINDSHIELD WIPER MOTOR
CIRCUITS 11, 16, 31, 41, 88, 200, 258, 329, 359, 511, 658, 904
140 BK/PK (18) AB
57G BK (18) AB
CIRCUITS 1, 2, 3, 13, 14, 37, 54★, 57P, 87, 128, 222, 276, 298, 460, 500, 977
14C BR (16) AB
787 PK/BK (14) D
563 O/Y (18) D
150 DG/W (18) D
2.9 L EFI ENGINE INJECTORS & SENSORS
FRAME 4 OF 30
MS-F17B-3458-AA

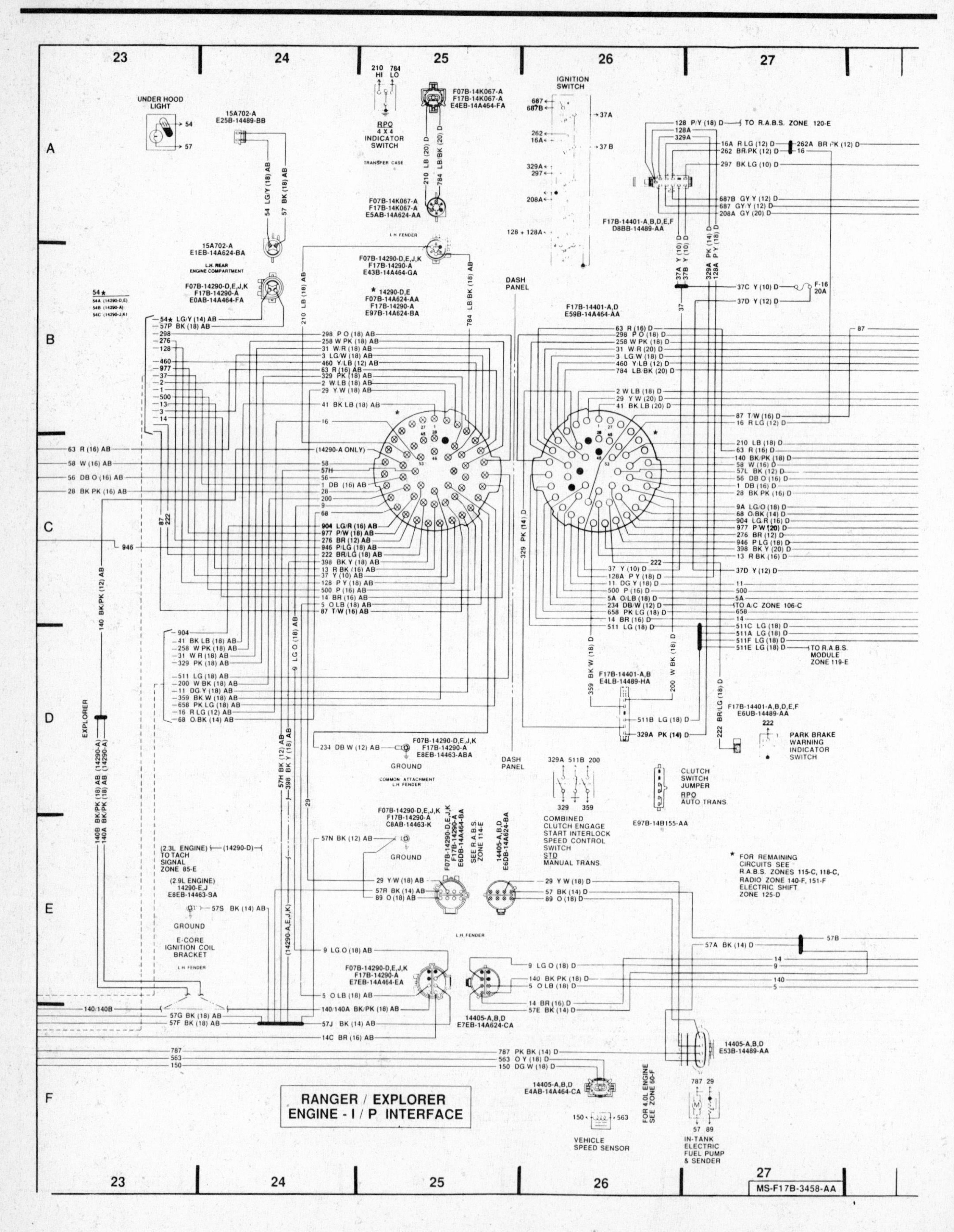
RANGER / EXPLORER
ENGINE - I / P INTERFACE
UNDER HOOD LIGHT
IGNITION SWITCH
RPO 4 X 4 INDICATOR SWITCH
DASH PANEL
GROUND
PARK BRAKE WARNING INDICATOR SWITCH
CLUTCH SWITCH JUMPER
RPO AUTO TRANS.
COMBINED CLUTCH ENGAGE START INTERLOCK SPEED CONTROL SWITCH
STD MANUAL TRANS.
* FOR REMAINING CIRCUITS SEE R.A.B.S. ZONES 115-C, 118-C, RADIO ZONE 140-F, 151-F ELECTRIC SHIFT ZONE 125-D
E-CORE IGNITION COIL BRACKET
VEHICLE SPEED SENSOR
IN-TANK ELECTRIC FUEL PUMP & SENDER
EXPLORER
MS-F17B-3458-AA

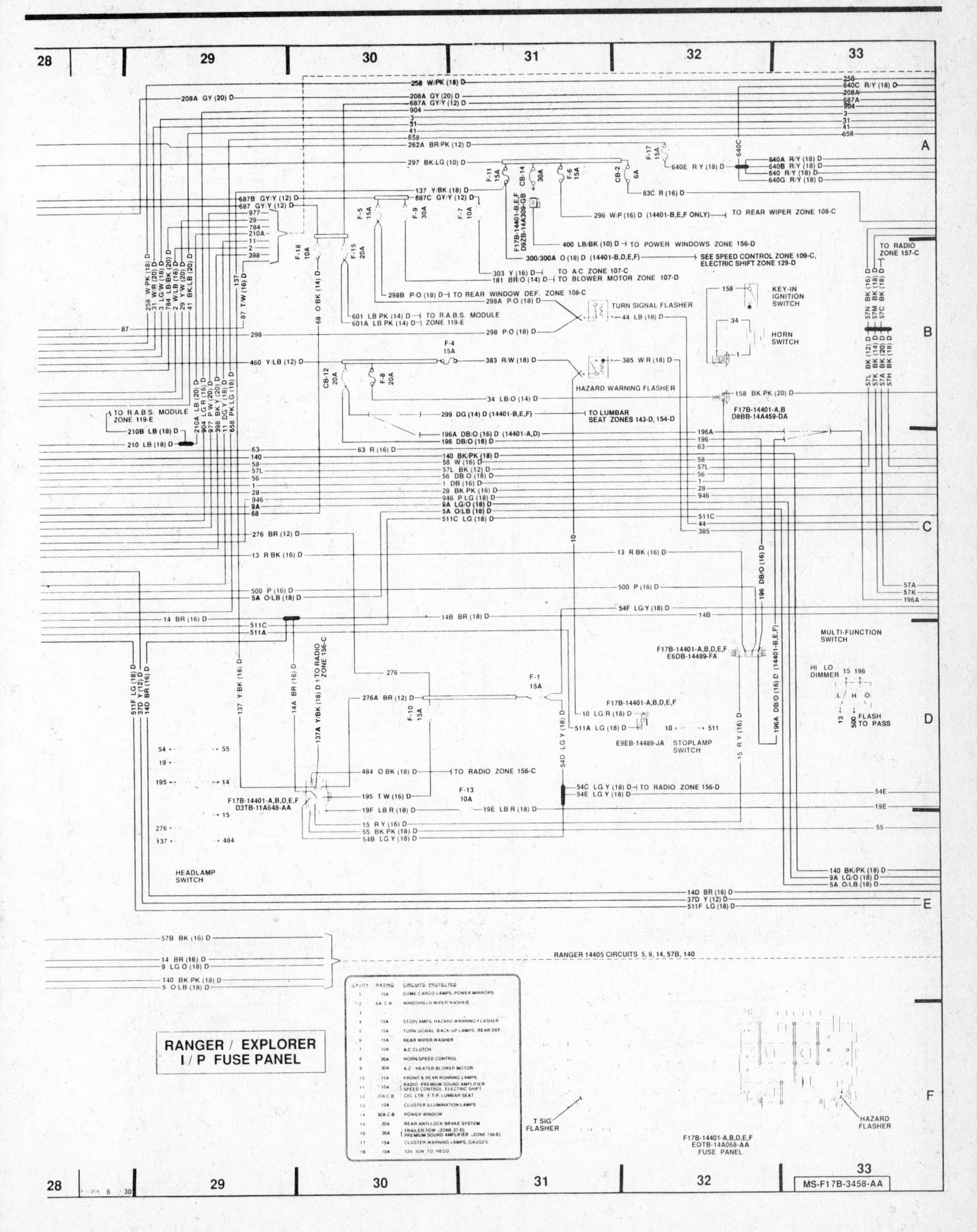

CAVITY	RATING	CIRCUITS PROTECTED
1	15A	DOME CARGO LAMPS, POWER MIRRORS
2	6A C.B.	WINDSHIELD WIPER WASHER
3	—	
4	15A	STOPLAMPS, HAZARD WARNING FLASHER
5	15A	TURN SIGNAL, BACK UP LAMPS, REAR DEF
6	15A	REAR WIPER WASHER
7	10A	A/C CLUTCH
8	20A	HORN/SPEED CONTROL
9	30A	A/C - HEATER BLOWER MOTOR
10	15A	FRONT & REAR RUNNING LAMPS
11	15A	RADIO PREMIUM SOUND AMPLIFIER SPEED CONTROL, ELECTRIC SHIFT
12	20A C.B.	CIG LTR, F-T-P, LUMBAR SEAT
13	10A	CLUSTER ILLUMINATION LAMPS
14	30A C.B.	POWER WINDOW
15	20A	REAR ANTI-LOCK BRAKE SYSTEM
16	20A	TRAILER TOW (ZONE 27-D) PREMIUM SOUND AMPLIFIER (ZONE 155-E)
17	15A	CLUSTER WARNING LAMPS, GAUGES
18	10A	12V IGN TO HEGO

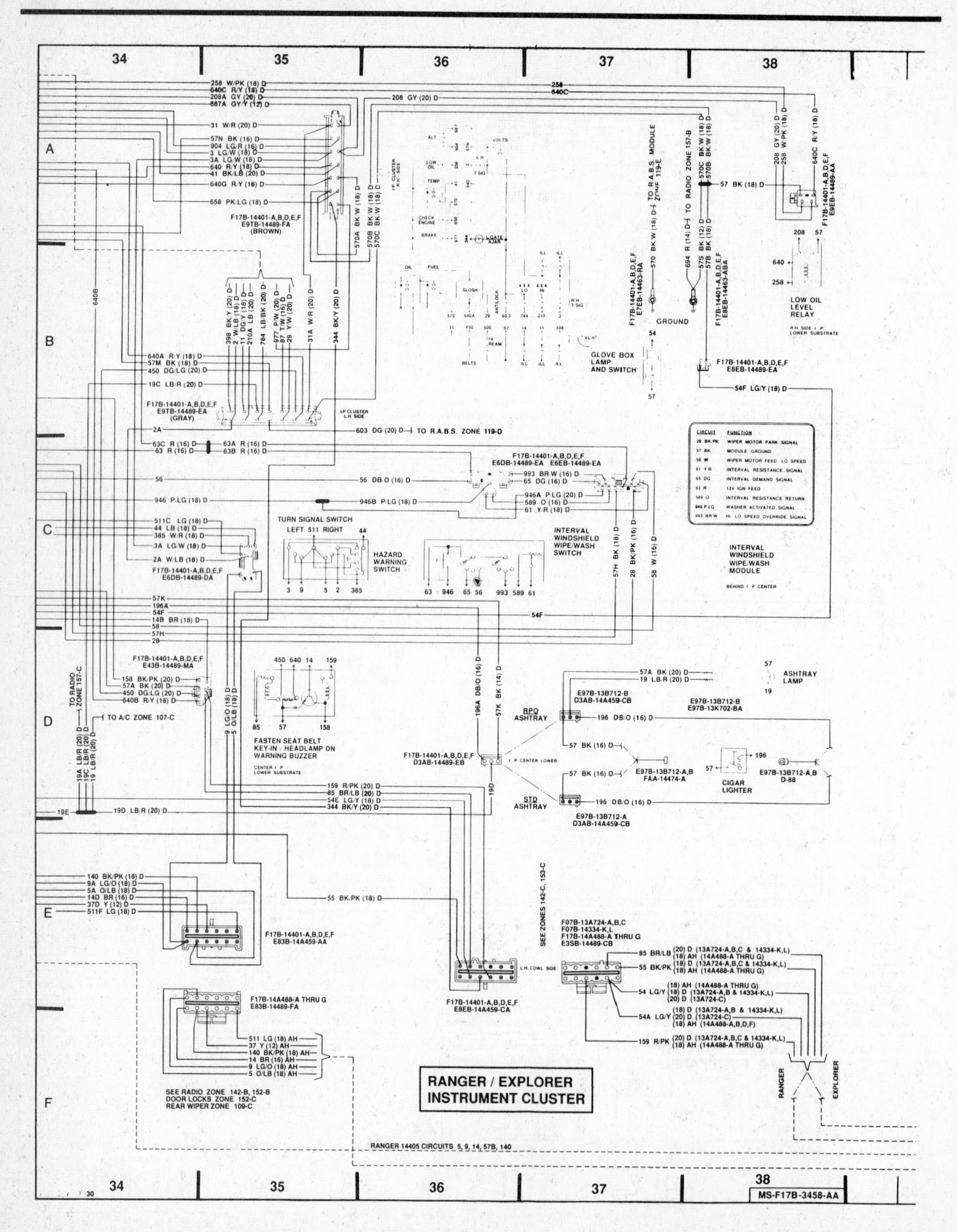
RANGER / EXPLORER
INSTRUMENT CLUSTER
TURN SIGNAL SWITCH
HAZARD WARNING SWITCH
INTERVAL WINDSHIELD WIPE/WASH SWITCH
INTERVAL WINDSHIELD WIPE/WASH MODULE
FASTEN SEAT BELT KEY-IN / HEADLAMP ON WARNING BUZZER
GLOVE BOX LAMP AND SWITCH
LOW OIL LEVEL RELAY
ASHTRAY LAMP
CIGAR LIGHTER
RPO ASHTRAY
STD ASHTRAY
GROUND
RANGER 14405 CIRCUITS 5, 9, 14, 57B, 140
SEE RADIO ZONE 142-B, 152-B
DOOR LOCKS ZONE 152-C
REAR WIPER ZONE 109-C
MS-F17B-3458-AA

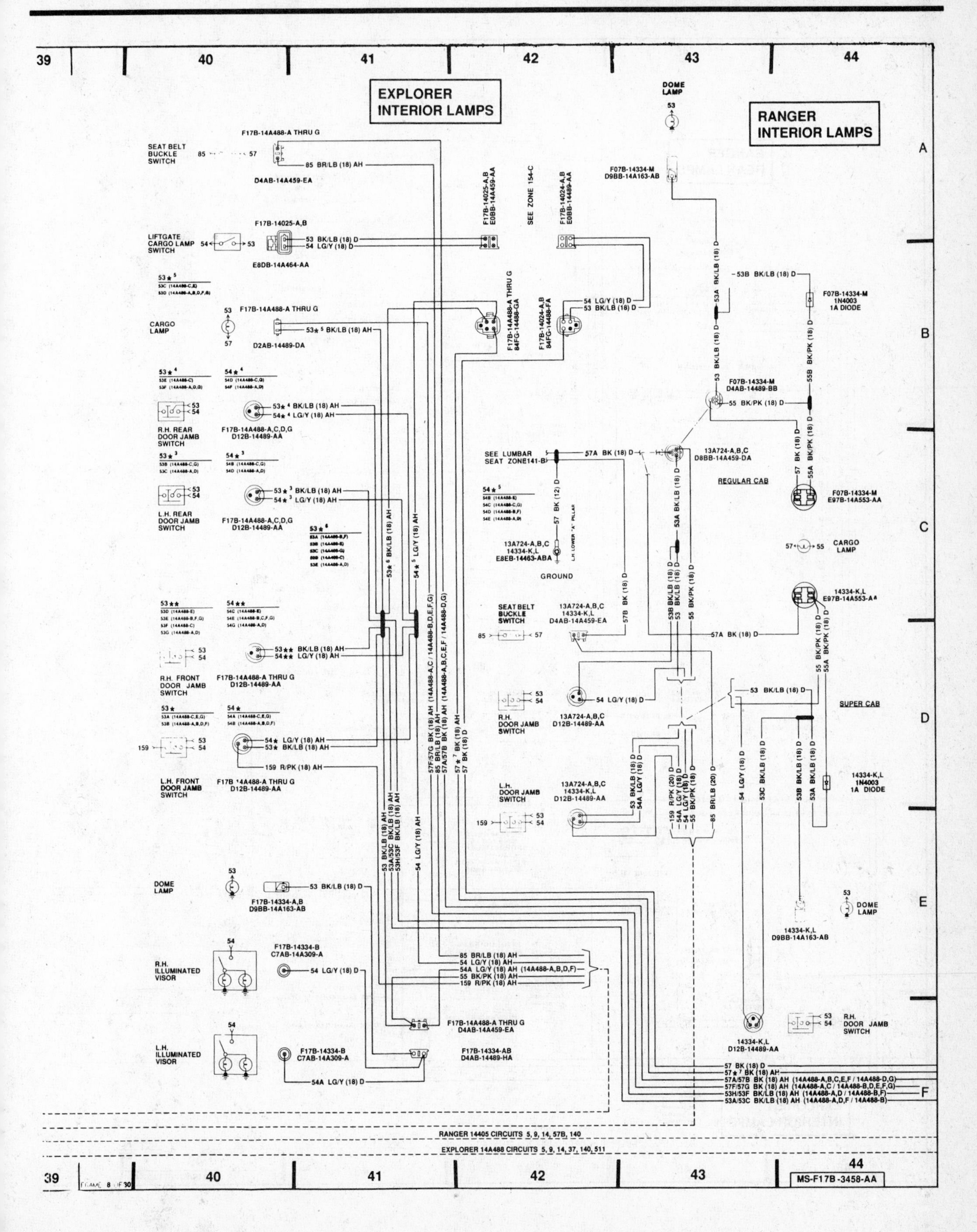
EXPLORER INTERIOR LAMPS
RANGER INTERIOR LAMPS
SEAT BELT BUCKLE SWITCH
LIFTGATE CARGO LAMP SWITCH
CARGO LAMP
R.H. REAR DOOR JAMB SWITCH
L.H. REAR DOOR JAMB SWITCH
R.H. FRONT DOOR JAMB SWITCH
L.H. FRONT DOOR JAMB SWITCH
DOME LAMP
R.H. ILLUMINATED VISOR
L.H. ILLUMINATED VISOR
SEE ZONE 154-C
SEE LUMBAR SEAT ZONE141-B
REGULAR CAB
SUPER CAB
GROUND
R.H. DOOR JAMB SWITCH
L.H. DOOR JAMB SWITCH
RANGER 14405 CIRCUITS 5, 9, 14, 57B, 140
EXPLORER 14A488 CIRCUITS 5, 9, 14, 37, 140, 511
MS-F17B-3458-AA

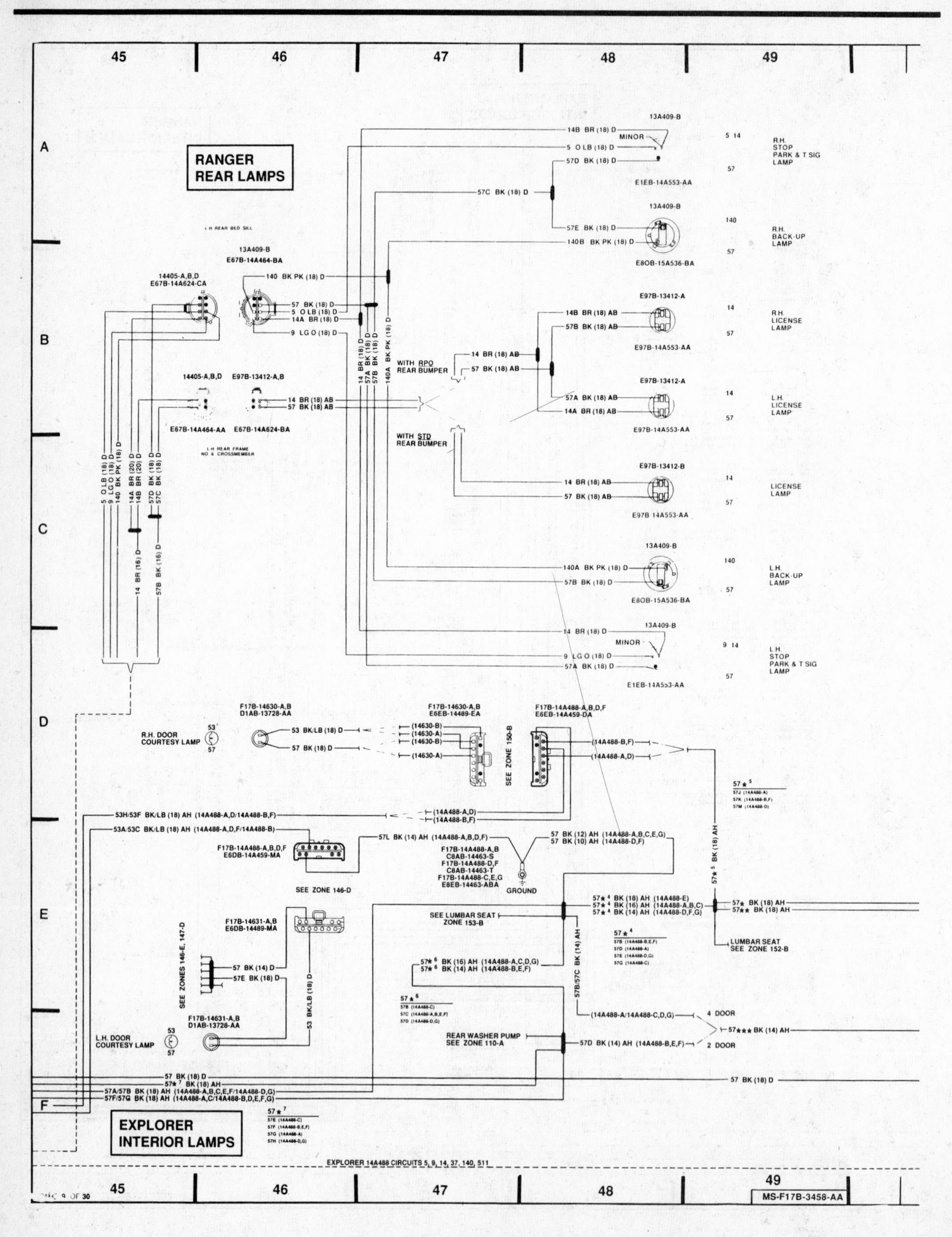

RANGER
REAR LAMPS
EXPLORER
INTERIOR LAMPS
R.H. STOP PARK & T SIG LAMP
R.H. BACK-UP LAMP
R.H. LICENSE LAMP
L.H. LICENSE LAMP
LICENSE LAMP
L.H. BACK-UP LAMP
L.H. STOP PARK & T SIG LAMP
WITH RPO REAR BUMPER
WITH STD REAR BUMPER
R.H. DOOR COURTESY LAMP
L.H. DOOR COURTESY LAMP
GROUND
SEE ZONE 146-D
SEE ZONE 150-B
SEE LUMBAR SEAT ZONE 153-B
LUMBAR SEAT SEE ZONE 152-B
REAR WASHER PUMP SEE ZONE 110-A
4 DOOR
2 DOOR
SEE ZONES 146-E, 147-D
EXPLORER 14A488 CIRCUITS 5, 9, 14, 37, 140, 511
MS-F17B-3458-AA

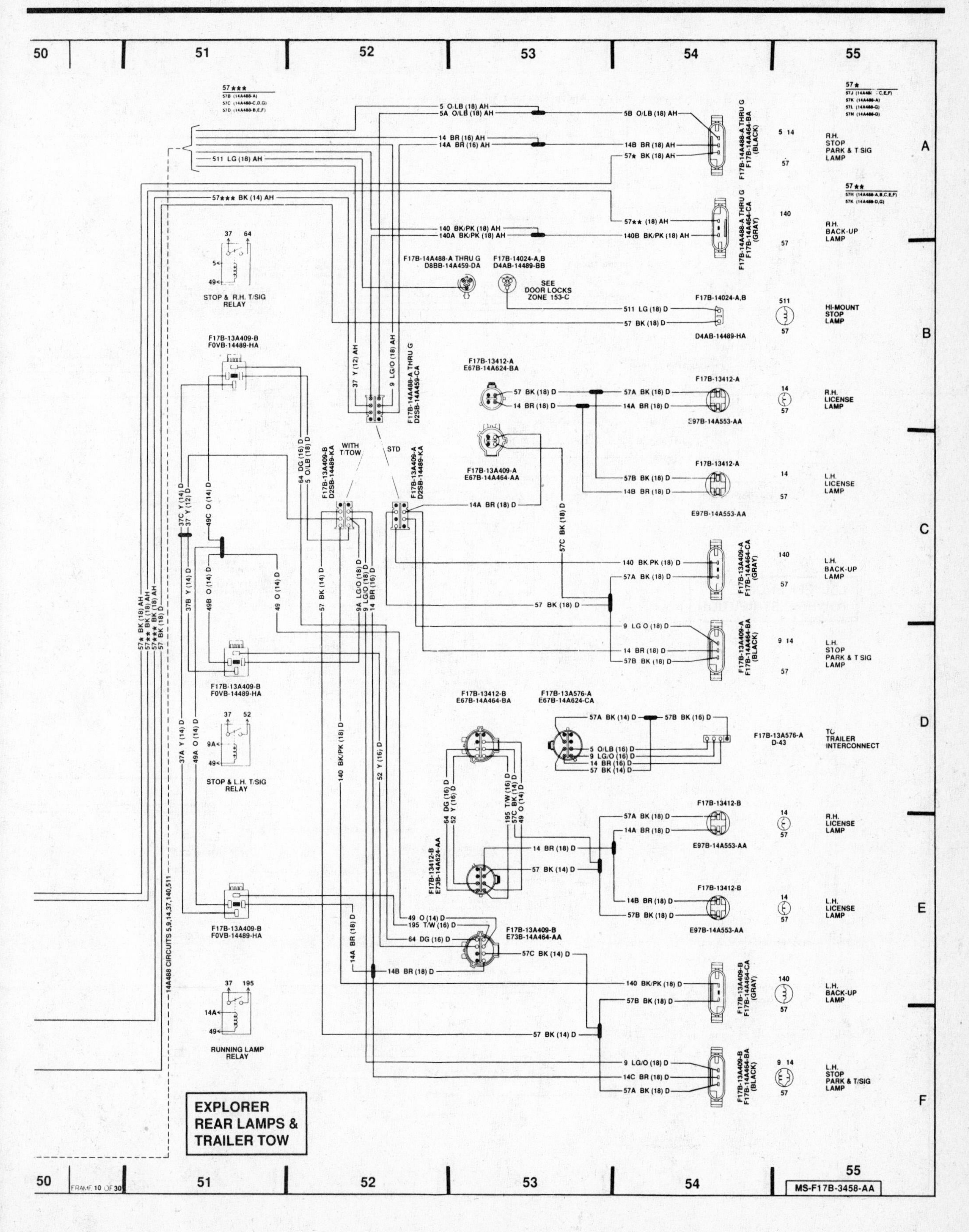
EXPLORER
REAR LAMPS &
TRAILER TOW
R.H. STOP PARK & T SIG LAMP
R.H. BACK-UP LAMP
HI-MOUNT STOP LAMP
R.H. LICENSE LAMP
L.H. LICENSE LAMP
L.H. BACK-UP LAMP
L.H. STOP PARK & T SIG LAMP
TO TRAILER INTERCONNECT
STOP & R.H. T/SIG RELAY
STOP & L.H. T/SIG RELAY
RUNNING LAMP RELAY
SEE DOOR LOCKS ZONE 153-C
14A488 CIRCUITS 5,9,14,37,140,511
FRAME 10 OF 30
MS-F17B-3458-AA

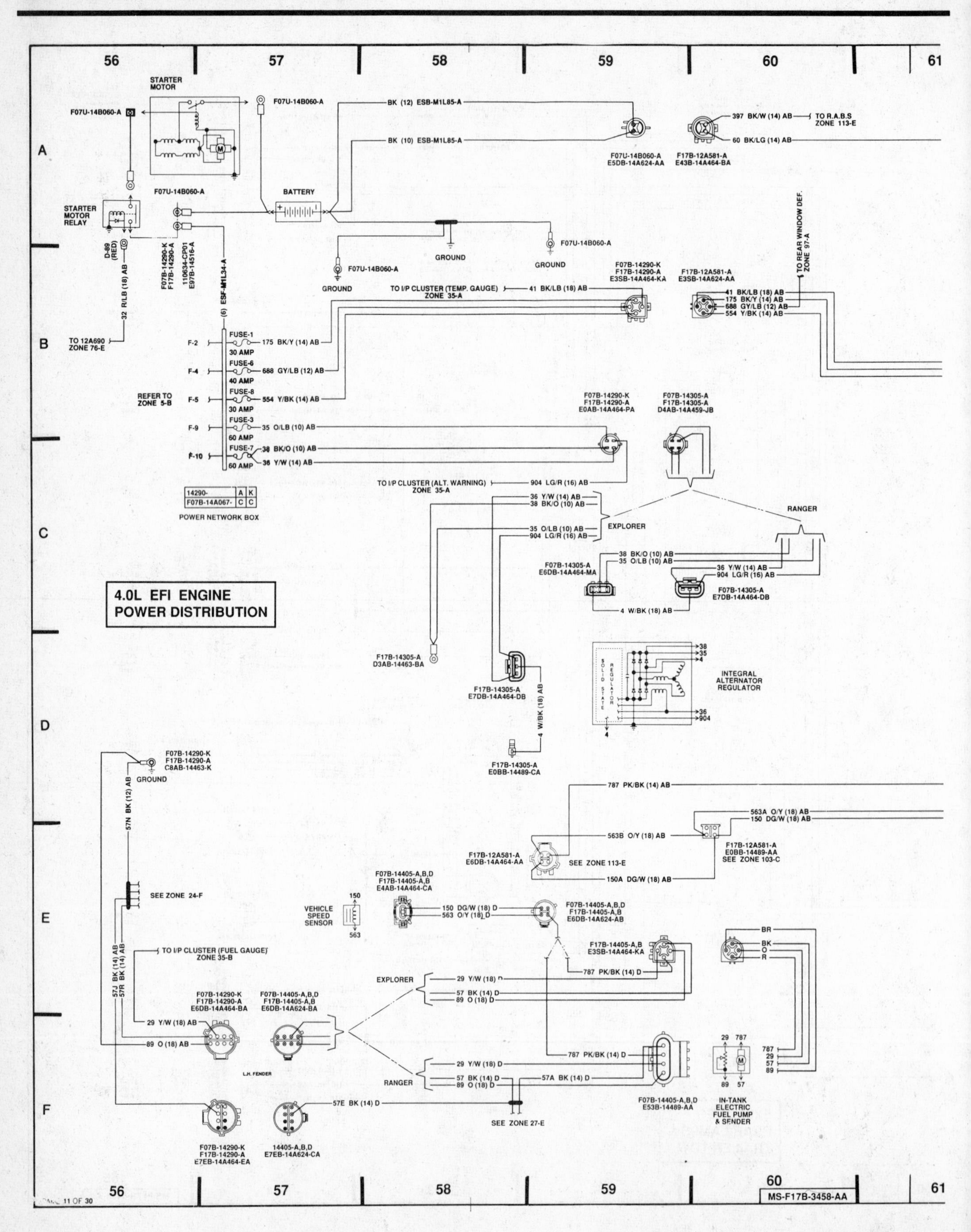
4.0L EFI ENGINE
POWER DISTRIBUTION
STARTER MOTOR
F07U-14B060-A
BK (12) ESB-M1L85-A
BK (10) ESB-M1L85-A
F07U-14B060-A
E5DB-14A624-AA
F17B-12A581-A
E43B-14A464-BA
397 BK/W (14) AB
TO R.A.B.S ZONE 113-E
60 BK/LG (14) AB
STARTER MOTOR RELAY
BATTERY
GROUND
D-89 (RED)
32 R/LB (18) AB
F07B-14290-K
F17B-14290-A
110634-CP01
E97B-14516-A
(6) ESF-M1L34-A
TO REAR WINDOW DEF. ZONE 97-A
TO I/P CLUSTER (TEMP. GAUGE) ZONE 35-A
41 BK/LB (18) AB
E3SB-14A464-KA
F17B-12A581-A
E3SB-14A624-AA
175 BK/Y (14) AB
688 GY/LB (12) AB
554 Y/BK (14) AB
TO 12A690 ZONE 76-E
FUSE-1 30 AMP
FUSE-6 40 AMP
FUSE-8 30 AMP
FUSE-3 60 AMP
FUSE-7 60 AMP
F-2
F-4
F-5
F-9
F-10
REFER TO ZONE 5-B
35 O/LB (10) AB
38 BK/O (10) AB
36 Y/W (14) AB
F07B-14290-K
F17B-14290-A
E0AB-14A464-PA
F07B-14305-A
F17B-14305-A
D4AB-14A459-JB
TO I/P CLUSTER (ALT. WARNING) ZONE 35-A
904 LG/R (16) AB
14290- A K
F07B-14A067- C C
POWER NETWORK BOX
EXPLORER
RANGER
F07B-14305-A
E6DB-14A464-MA
F07B-14305-A
E7DB-14A464-DB
4 W/BK (18) AB
F17B-14305-A
D3AB-14463-BA
F17B-14305-A
E7DB-14A464-DB
INTEGRAL ALTERNATOR REGULATOR
SOLID STATE REGULATOR
F17B-14305-A
E0BB-14489-CA
F07B-14290-K
F17B-14290-A
C8AB-14463-K
GROUND
57N BK (12) AB
787 PK/BK (14) AB
563A O/Y (18) AB
150 DG/W (18) AB
563B O/Y (18) AB
F17B-12A581-A
E6DB-14A464-AA
SEE ZONE 113-E
F17B-12A581-A
E0BB-14489-AA
SEE ZONE 103-C
150A DG/W (18) AB
SEE ZONE 24-F
F07B-14405-A,B,D
F17B-14405-A,B
E4AB-14A464-CA
VEHICLE SPEED SENSOR
150 DG/W (18) D
563 O/Y (18) D
F07B-14405-A,B,D
F17B-14405-A,B
E6DB-14A624-AB
TO I/P CLUSTER (FUEL GAUGE) ZONE 35-B
F17B-14405-A,B
E3SB-14A464-KA
787 PK/BK (14) D
57J BK (14) AB
57R BK (14) AB
F07B-14290-K
F17B-14290-A
E6DB-14A464-BA
F07B-14405-A,B,D
F17B-14405-A,B
E6DB-14A624-BA
29 Y/W (18) AB
89 O (18) AB
29 Y/W (18) D
57 BK (14) D
89 O (18) D
57A BK (14) D
L.H. FENDER
57E BK (14) D
SEE ZONE 27-E
F07B-14405-A,B,D
E53B-14489-AA
IN-TANK ELECTRIC FUEL PUMP & SENDER
F07B-14290-K
F17B-14290-A
E7EB-14A464-EA
14405-A,B,D
E7EB-14A624-CA
MS-F17B-3458-AA

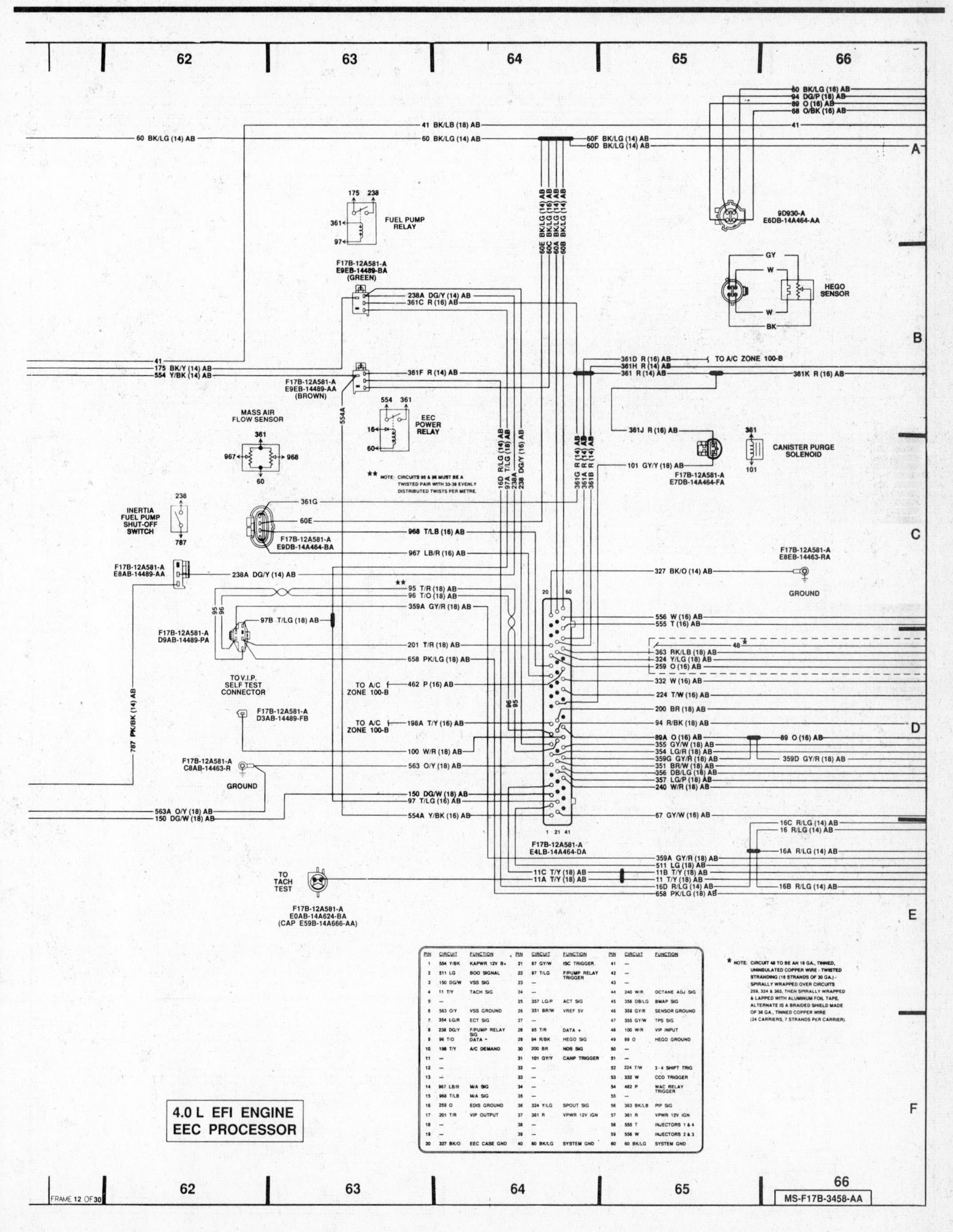

PIN	CIRCUIT	FUNCTION	PIN	CIRCUIT	FUNCTION	PIN	CIRCUIT	FUNCTION
1	554 Y/BK	KAPWR 12V B+	21	67 GY/W	ISC TRIGGER	41	—	
2	511 LG	BOO SIGNAL	22	97 T/LG	F/PUMP RELAY TRIGGER	42	—	
3	150 DG/W	VSS SIG	23	—		43	—	
4	11 T/Y	TACH SIG	24	—		44	240 W/R	OCTANE ADJ SIG
5	—		25	357 LG/P	ACT SIG	45	356 DB/LG	BMAP SIG
6	563 O/Y	VSS GROUND	26	351 BR/W	VREF 5V	46	359 GY/R	SENSOR GROUND
7	354 LG/R	ECT SIG	27	—		47	355 GY/W	TPS SIG
8	238 DG/Y	F/PUMP RELAY SIG	28	95 T/R	DATA +	48	100 W/R	VIP INPUT
9	96 T/O	DATA -	29	94 R/BK	HEGO SIG	49	89 O	HEGO GROUND
10	198 T/Y	A/C DEMAND	30	200 BR	NDS SIG	50	—	
11	—		31	101 GY/Y	CANP TRIGGER	51	—	
12	—		32	—		52	224 T/W	3/4 SHIFT TRIG
13	—		33	—		53	332 W	CCO TRIGGER
14	967 LB/R	M/A SIG	34	—		54	462 P	WAC RELAY TRIGGER
15	968 T/LB	M/A SIG	35	—		55	—	
16	259 O	EDIS GROUND	36	324 Y/LG	SPOUT SIG	56	363 BK/LB	PIP SIG
17	201 T/R	VIP OUTPUT	37	361 R	VPWR 12V IGN	57	361 R	VPWR 12V IGN
18	—		38	—		58	555 T	INJECTORS 1 & 4
19	—		39	—		59	556 W	INJECTORS 2 & 3
20	327 BK/O	EEC CASE GND	40	60 BK/LG	SYSTEM GND	60	60 BK/LG	SYSTEM GND

* NOTE: CIRCUIT 48 TO BE AN 18 GA., TINNED, UNINSULATED COPPER WIRE - TWISTED STRANDING (16 STRANDS OF 30 GA.) - SPIRALLY WRAPPED OVER CIRCUITS 259, 324 & 363, THEN SPIRALLY WRAPPED & LAPPED WITH ALUMINUM FOIL TAPE. ALTERNATE IS A BRAIDED SHIELD MADE OF 36 GA., TINNED COPPER WIRE (24 CARRIERS, 7 STRANDS PER CARRIER).

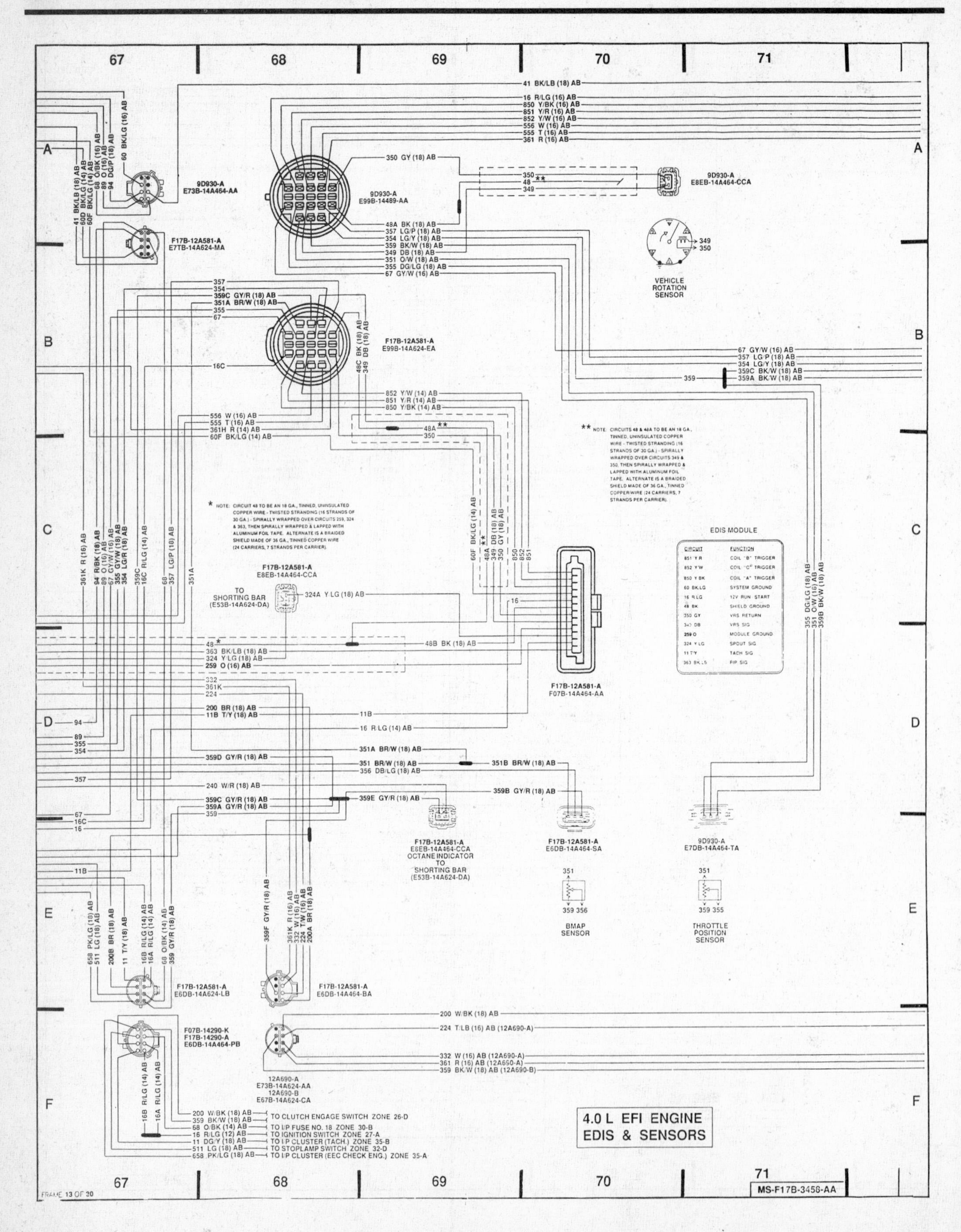

4.0 L EFI ENGINE
EDIS & SENSORS
EDIS MODULE
VEHICLE ROTATION SENSOR
BMAP SENSOR
THROTTLE POSITION SENSOR
MS-F17B-3458-AA
FRAME 13 OF 30

7 Drive Train

QUICK REFERENCE INDEX

GENERAL INDEX

UNDERSTANDING THE MANUAL TRANSMISSION

Because of the way an internal combustion engine breathes, it can produce torque, or twisting force, only within a narrow speed range. Most modern, overhead valve engines must turn at about 2,500 rpm to produce their peak torque. By 4,500 rpm they are producing so little torque that continued increases in engine speed produce no power increases.

The torque peak on overhead camshaft engines is, generally, much higher, but much narrower.

The manual transmission and clutch are employed to vary the relationship between engine speed and the speed of the wheels so that adequate engine power can be produced under all circumstances. The clutch allows engine torque to be applied to the transmission input shaft gradually, due to mechanical slippage. The van can, consequently, be started smoothly from a full stop.

The transmission changes the ratio between the rotating speeds of the engine and the wheels by the use of gears. 4-speed or 5-speed transmissions are most common. The lower gears allow full engine power to be applied to the rear wheels during acceleration at low speeds.

The transmission contains a mainshaft which passes all the way through the transmission, from the clutch to the driveshaft. This shaft is separated at one point, so that front and rear portions can turn at different speeds.

Power is transmitted by a countershaft in the lower gears and reverse. The gears of the countershaft mesh with gears on the mainshaft, allowing power to be carried from one to the other. All the countershaft gears are integral with that shaft, while several of the mainshaft gears can either rotate independently of the shaft or be locked to it. Shifting from one gear to the next causes one of the gears to be freed from rotating with the shaft and locks another to it. Gears are locked and unlocked by internal dog clutches which slide between the center of the gear and the shaft. The forward gears usually employ synchronizers; friction members which smoothly bring gear and shaft to the same speed before the toothed dog clutches are engaged.

MANUAL TRANSMISSION

Identification

All manual transmissions are equipped with an identification tag which must be used when servicing the unit. The 4-speed manual transmissions are designated code X and the 5-speed overdrive transmissions are code 5 or M/D.

The 4-speed manual transmission and the Mitsubishi 5-speed overdrive transmission service tags are located toward the right front side of the transmission case. The Mazda M5OD 5-speed overdrive transmission service tag is located on the left hand side of the transmission case.

Some early Ranger and Bronco II are equipped with a 4-speed manual transmission. The 4-speed manual transmission used on the 2.2L diesel engines and gasoline engines are fully synchronized with all gears, except reverse, in constant mesh. All forward speed gears are helically cut for quiet operation. The reverse gear and reverse idler gear are spur-cut. Gear shifting is directly control by a floor shift mechanism, which is built into the extension housing. When fitted to the diesel engine, the transmission features an integral clutch housing and gear case and a separate extension housing.

The Mazda (Toyo Kogyo) 5-speed overdrive transmission is fully synchronized with all gears, except reverse gear, which is in constant mesh. All forward speed gears are helically cut for quiet operation. The reverse gear and reverse idler gear are spur-cut. Gear shifting is directly control by a floor shift mechanism, which is built into the extension housing. The 5-speed overdrive transmission used on the gasoline engines is very similar to the 4-speed transmission used on the Ranger gasoline engine. When fitted to the diesel engine, the 5-speed overdrive

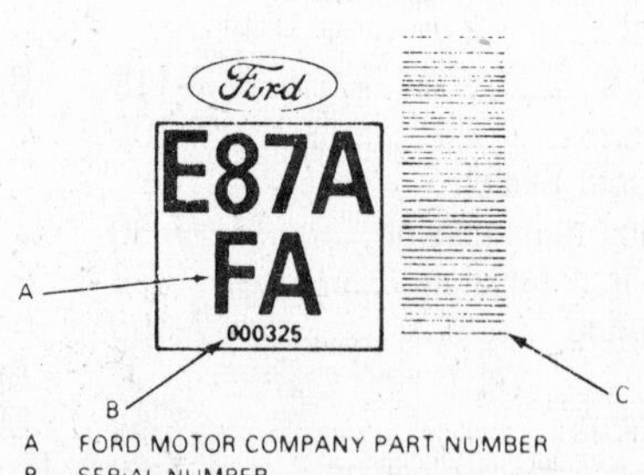

Transmission service tag – Mazda M5OD

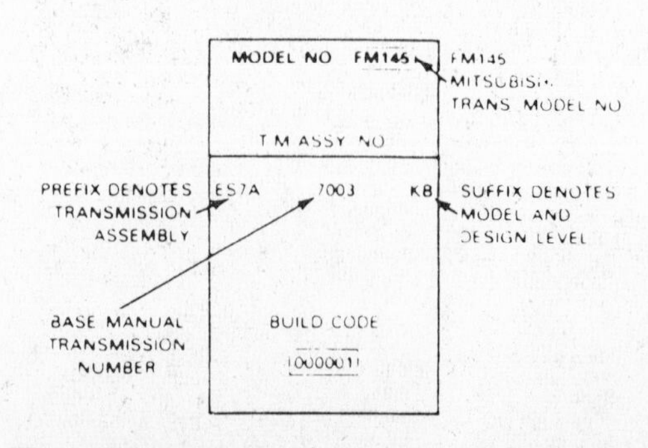

Transmission service tag – Mitsubishi

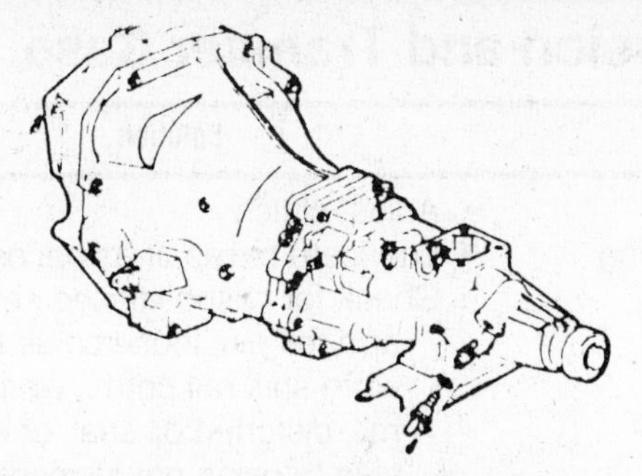

4-speed manual transmission assembly — diesel engine

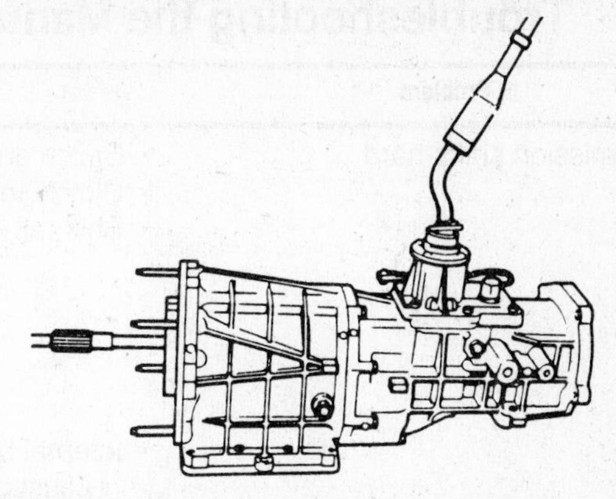

Mitsubishi FM146 5-speed transmission assembly

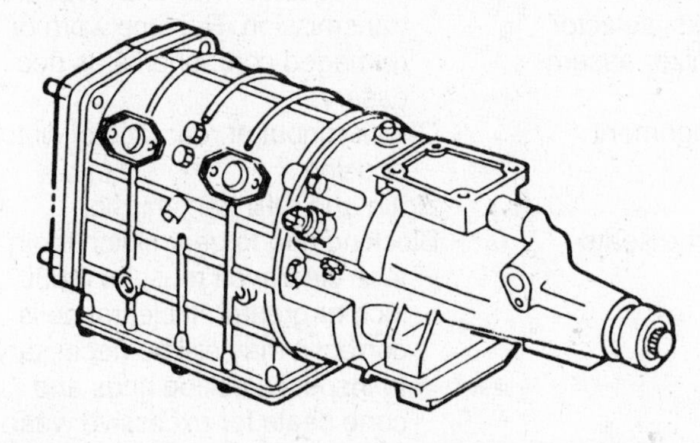

4-speed manual transmission assembly — gasoline engine

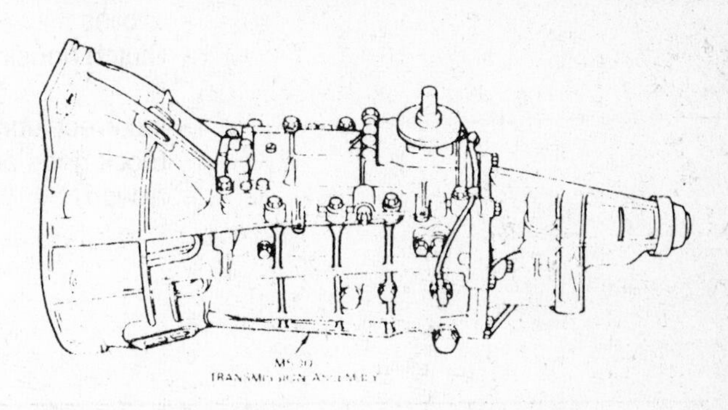

Mazda M50D 5-speed transmission assembly

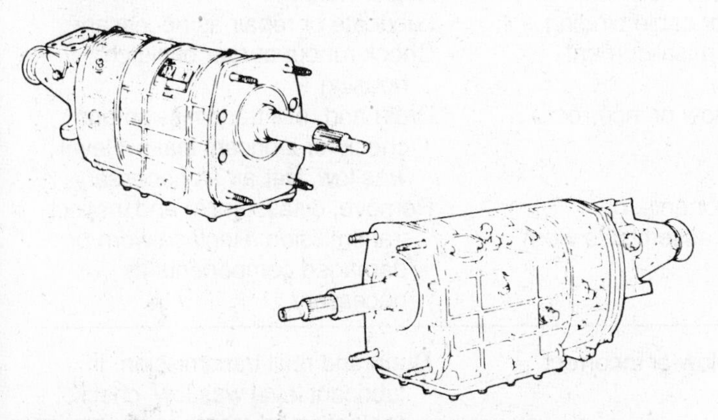

Mazda (Toyo Kogyo) 5-speed transmission assembly

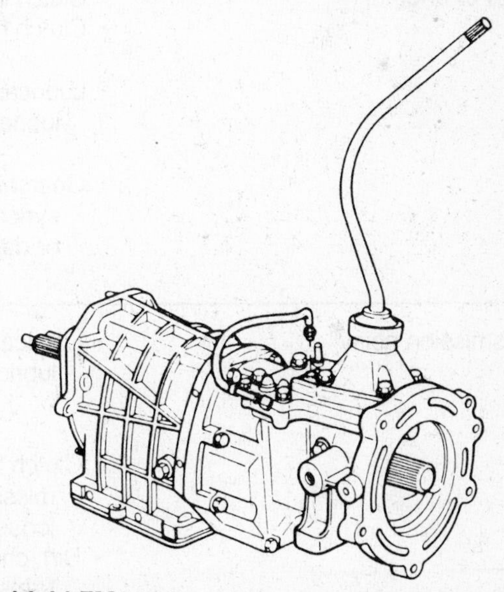

Mitsubishi FM145 5-speed transmission assembly

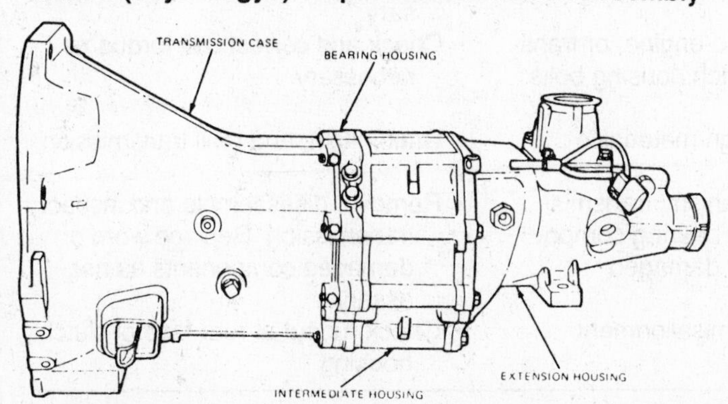

5-speed manual transmission assembly — diesel engine

transmission features a bearing housing, intermediate housing and extension housing.

The Mazda M5OD 5-speed overdrive transmission is a top shift, fully synchronized transmission. The M5OD transmission comes in both a M5OD-R1 and a ZF M5OD-HD version. It's main case, top cover and extension housing are constructed of aluminum alloy.

The Mitsubishi 5-speed overdrive transmissions are fully synchronized in all forward gears and reverse. The Mitsubishi transmission includes the FM132, FM145 and the FM146 ver-

Troubleshooting the Manual Transmission and Transfer Case

Problem	Cause	Solution
Transmission shifts hard	· Clutch adjustment incorrect	· Adjust clutch
	· Clutch linkage or cable binding	· Lubricate or repair as necessary
	· Shift rail binding	· Check for mispositioned selector arm roll pin, loose cover bolts, worn shift rail bores, worn shift rail, distorted oil seal, or extension housing not aligned with case. Repair as necessary.
	· Internal bind in transmission caused by shift forks, selector plates, or synchronizer assemblies	· Remove, dissemble and inspect transmission. Replace worn or damaged components as necessary.
	· Clutch housing misalignment	· Check runout at rear face of clutch housing
	· Incorrect lubricant	· Drain and refill transmission
	· Block rings and/or cone seats worn	· Blocking ring to gear clutch tooth face clearance must be 0.030 inch or greater. If clearance is correct it may still be necessary to inspect blocking rings and cone seats for excessive wear. Repair as necessary.
Gear clash when shifting from one gear to another	· Clutch adjustment incorrect	· Adjust clutch
	· Clutch linkage or cable binding	· Lubricate or repair as necessary
	· Clutch housing misalignment	· Check runout at rear of clutch housing
	· Lubricant level low or incorrect lubricant	· Drain and refill transmission and check for lubricant leaks if level was low. Repair as necessary.
	· Gearshift components, or synchronizer assemblies worn or damaged	· Remove, disassemble and inspect transmission. Replace worn or damaged components as necessary.
Transmission noisy	· Lubricant level low or incorrect lubricant	· Drain and refill transmission. If lubricant level was low, check for leaks and repair as necessary.
	· Clutch housing-to-engine, or transmission-to-clutch housing bolts loose	· Check and correct bolt torque as necessary
	· Dirt, chips, foreign material in transmission	· Drain, flush, and refill transmission
	· Gearshift mechanism, transmission gears, or bearing components worn or damaged	· Remove, disassemble and inspect transmission. Replace worn or damaged components as necessary.
	· Clutch housing misalignment	· Check runout at rear face of clutch housing

Troubleshooting the Manual Transmission and Transfer Case (cont.)

Problem	Cause	Solution
Jumps out of gear	• Clutch housing misalignment	• Check runout at rear face of clutch housing
	• Gearshift lever loose	• Check lever for worn fork. Tighten loose attaching bolts.
	• Offset lever nylon insert worn or lever attaching nut loose	• Remove gearshift lever and check for loose offset lever nut or worn insert. Repair or replace as necessary.
	• Gearshift mechanism, shift forks, selector plates, interlock plate, selector arm, shift rail, detent plugs, springs or shift cover worn or damaged	• Remove, disassemble and inspect transmission cover assembly. Replace worn or damaged components as necessary.
	• Clutch shaft or roller bearings worn or damaged	• Replace clutch shaft or roller bearings as necessary
Jumps out of gear (cont.)	• Gear teeth worn or tapered, synchronizer assemblies worn or damaged, excessive end play caused by worn thrust washers or output shaft gears	• Remove, disassemble, and inspect transmission. Replace worn or damaged components as necessary.
	• Pilot bushing worn	• Replace pilot bushing
Will not shift into one gear	• Gearshift selector plates, interlock plate, or selector arm, worn, damaged, or incorrectly assembled	• Remove, disassemble, and inspect transmission cover assembly. Repair or replace components as necessary.
	• Shift rail detent plunger worn, spring broken, or plug loose	• Tighten plug or replace worn or damaged components as necessary
	• Gearshift lever worn or damaged	• Replace gearshift lever
	• Synchronizer sleeves or hubs, damaged or worn	• Remove, disassemble and inspect transmission. Replace worn or damaged components.
Locked in one gear—cannot be shifted out	• Shift rail(s) worn or broken, shifter fork bent, setscrew loose, center detent plug missing or worn	• Inspect and replace worn or damaged parts
	• Broken gear teeth on countershaft gear, clutch shaft, or reverse idler gear	• Inspect and replace damaged part
	Gearshift lever broken or worn, shift mechanism in cover incorrectly assembled or broken, worn damaged gear train components	• Disassemble transmission. Replace damaged parts or assemble correctly.

Troubleshooting the Manual Transmission and Transfer Case (cont.)

Problem	Cause	Solution
Transfer case difficult to shift or will not shift into desired range	• Vehicle speed too great to permit shifting	• Stop vehicle and shift into desired range. Or reduce speed to 3–4 km/h (2–3 mph) before attempting to shift.
	• If vehicle was operated for extended period in 4H mode on dry paved surface, driveline torque load may cause difficult shifting	• Stop vehicle, shift transmission to neutral, shift transfer case to 2H mode and operate vehicle in 2H on dry paved surfaces
	• Transfer case external shift linkage binding	• Lubricate or repair or replace linkage, or tighten loose components as necessary
	• Insufficient or incorrect lubricant	• Drain and refill to edge of fill hole
	• Internal components binding, worn, or damaged	• Disassemble unit and replace worn or damaged components as necessary
Transfer case noisy in all drive modes	• Insufficient or incorrect lubricant	• Drain and refill to edge of fill hole Check for leaks and repair if necessary. Note: If unit is still noisy after drain and refill, disassembly and inspection may be required to locate source of noise.
Noisy in—or jumps out of four wheel drive low range	• Transfer case not completely engaged in 4L position	• Stop vehicle, shift transfer case in Neutral, then shift back into 4L position
	• Shift linkage loose or binding	• Tighten, lubricate, or repair linkage as necessary
	• Shift fork cracked, inserts worn, or fork is binding on shift rail	• Disassemble unit and repair as necessary
Lubricant leaking from output shaft seals or from vent	• Transfer case overfilled • Vent closed or restricted	• Drain to correct level • Clear or replace vent if necessary
Lubricant leaking from output shaft seals or from vent (cont.)	• Output shaft seals damaged or installed incorrectly	• Replace seals. Be sure seal lip faces interior of case when installed. Also be sure yoke seal surfaces are not scored or nicked. Remove scores, nicks with fine sandpaper or replace yoke(s) if necessary.
Abnormal tire wear	• Extended operation on dry hard surface (paved) roads in 4H range	• Operate in 2H on hard surface (paved) roads

sions. These transmissions are equipped with top mounted shifter, which operates shift rails through a set of shift forks. The shift forks operates the synchronizer sleeves and allow shifts through 1-2, 3-4 and overdrive-reverse. A shift interlock system, located in the side of the transmission case, prevents the shift rails from engaging 2 gears at the same time.

Adjustments

SHIFTER AND LINKAGE ADJUSTMENTS

Both the 4-speed and 5-speed transmissions are directly controlled with a floor shift mechanism built into the transmission extension housing. There are no adjustments necessary on these transmissions.

Shift Handle

REMOVAL AND INSTALLATION

1. Disconnect the negative battery cable.

NOTE: Do not remove the shift ball, unless the shift ball or boot is to be replaced. Otherwise, remove the shift ball, boot and lever as an assembly.

2. Remove the plastic shift pattern insert from the shift ball.
3. Heat the shift ball to 140–180°F (60–82°C), using a heat gun.
4. Position a block of wood beneath the shift ball and carefully hammer the ball from the lever. Be careful not to damage the finish on the shift lever.
5. Place the gearshift lever in **N** position.
6. Remove the rubber boot retainer screws.
7. Remove the shift lever-to-extension housing/transfer case adapter housing retaining bolts. Pull the gearshift lever straight up and away from the gearshift lever retainer.

To install:

8. Prior to installing the shift lever, lubricate the shift lever ball stud, using C1AZ-19590-B (ESA-M1C75-B) or equivalent.
9. Fit the shift lever into place and install the retaining bolts. Tighten and torque the retaining bolts 20–27 ft. lbs. (28–36 Nm).
10. Install the rubber boot and retaining screws.
11. If the shift ball was removed, heat the shift ball to 140°–180°F (60°–82°C), using a heat gun. Tap the ball onto the shift lever, using a $^{7}/_{16}$ in. (11mm will work) socket and mallet.
12. Place the shift lever in **N** position. Then, align the shift pattern plastic insert with the vehicle centerline and install it to the shift ball.

Neutral Sensing Switch

All manual transmission vehicles are equipped with a neutral sensing switch. The neutral sensing switch signals the vehicle on-board computer, which allows the vhicle to strt only when the transmission is in **N**.

REMOVAL AND INSTALLATION

1. Disconnect the negative battery cable.
2. Raise and support the vehicle safely.
3. Place the transmission in any position other than **N**.
4. Clean the area around the switch, then remove the switch.

To install:

5. Install the switch and tighten 8–12 ft. lbs (11–16Nm).
6. Reconnect the harness connector to the switch.

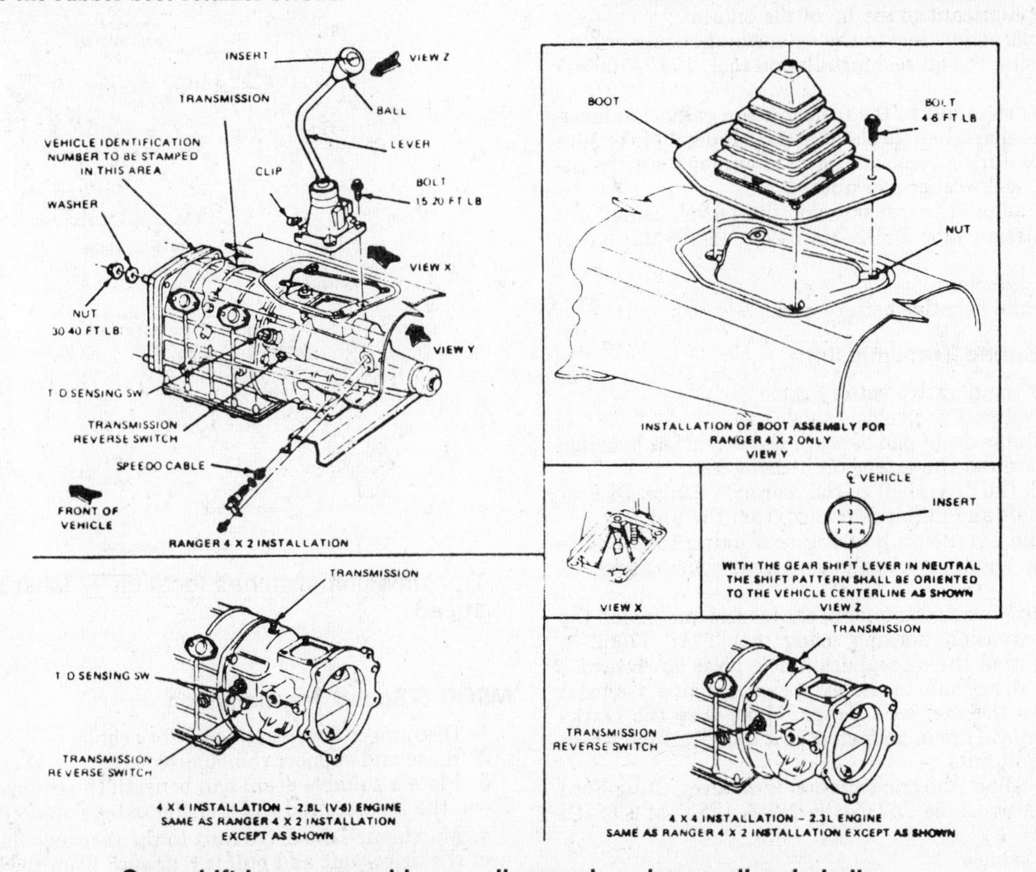

Gear shift lever assembly, gasoline engine shown, diesel similar

7. Lower the vehicle.
8. Reconnect the negative battery cable.

Back-up Lamp Switch

REMOVAL AND INSTALLATION

1. Disconnect the negative battery cable.
2. Raise and support the vehicle safely.
3. Place the transmission in any position other than **R** or **N**.
4. Clean the area around the switch, then remove the switch.

To install:

5. Install the switch and tighten 8–12 ft. lbs (11–16Nm).
6. Reconnect the harness connector to the switch.
7. Lower the vehicle.
8. Reconnect the negative battery cable.

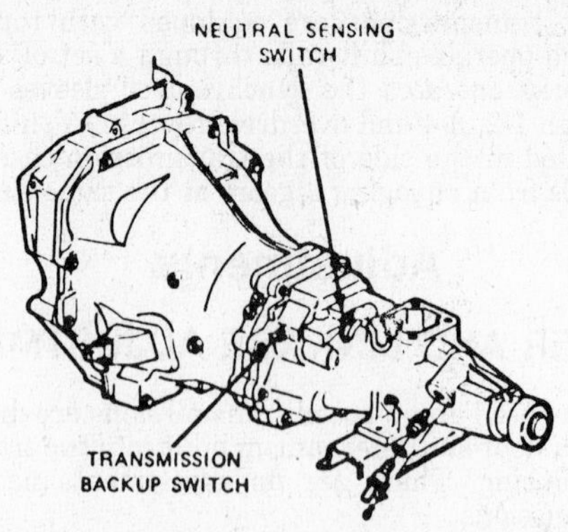

Transmission switches location – 4-speed (diesel) shown

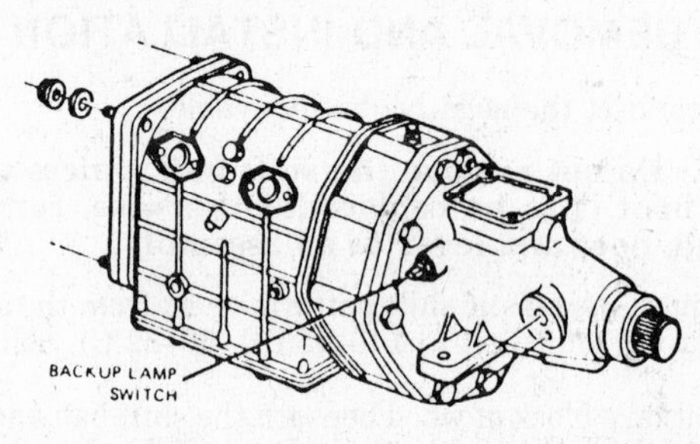

Transmission switch location – Mazda (Toyo Kogyo) 5-speed

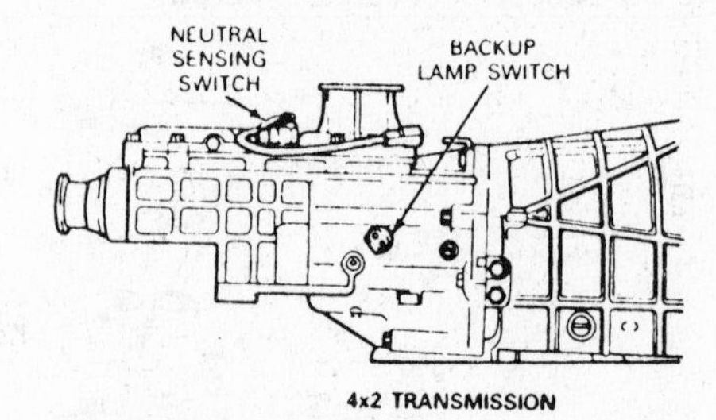

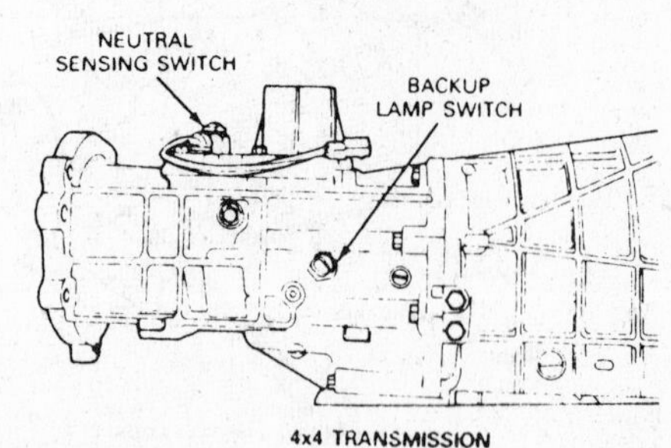

Transmission switches location – Mitsubishi 5-speed

Extension Housing Seal

REMOVAL AND INSTALLATION

4-Speed Transmission and 5-speeed

1. Disconnect the negative battery cable.
2. Raise and support the vehicle safely.
3. Place a suitable drain pan beneath the extension housing. Clean the area around the extension housing seal.
4. Matchmark the driveshaft to the rear axle flange. Disconnect the driveshaft and pull it rearward from the transmission.
5. Using an oil seal removal tool, T71P–7657–A or equivalent, remove the extension housing seal.

To install:

6. Apply gear lubricant to the lip of the oil seal.
7. Fit the replacement seal to the extension housing and install the seal, using the oil seal installation tool, T71P–7095–A or equivalent.
8. Install the driveshaft to the transmission extension housing. Connect the driveshaft to the rear axle flange. Make sure the marks made during removal are in alignment. Fit the attaching washer, lockwasher and nuts.
9. Check and adjust the transmission fluid level, using Ford manual transmission lube D8DZ–19C547–A (ESP–M2C83–C) or equivalent.
10. Lower the vehicle.
11. Reconnect the negative battery cable.

Mitsubishi 5-Speed Transmission

1. Disconnect the negative battery cable.
2. Raise and support the vehicle safely.
3. Place a suitable drain pan beneath the extension housing. Clean the area around the extension housing seal.
4. Matchmark the driveshaft to the rear axle flange. Disconnect the driveshaft and pull it rearward from the unit.
5. Remove the extension housing seal using tool T74P–77248–A or equivalent, remove the extension housing seal.

To install:

6. Lubricate the inside diameter of the oil seal and install the seal into the extension housing using tool T74P–77052–A. Check to ensure that the oil seal drain hole faces downward.
7. Install the driveshaft to the extension housing. Connect the driveshaft to the rear axle flange. Make sure the marks made during removal are in alignment. Fit the attaching washer, lockwasher and nuts.
8. Check and adjust the transmission fluid level, using Ford manual transmission lube D8DZ–19C547–A (ESP–M2C83–C) or equivalent.
9. Lower the vehicle.
10. Reconnect the negative battery cable.

M5OD 5-Speed Transmission

1. Disconnect the negative battery cable.
2. Raise and support the vehicle safely.
3. Place a suitable drain pan beneath the extension housing. Clean the area around the transfer extension housing seal.
4. Matchmark the driveshaft to the rear axle flange. Disconnect the driveshaft and pull it rearward from the unit.
5. Remove the extension housing seal using tool T74P–

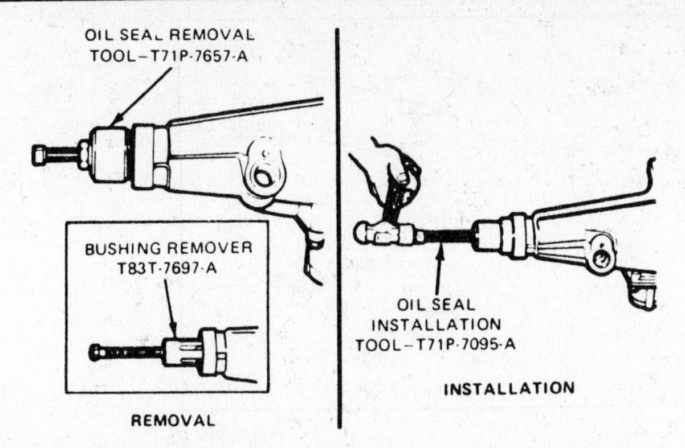

Extension housing oil seal, removal and installation – 4-Speed and Toyo Kogyo transmissions

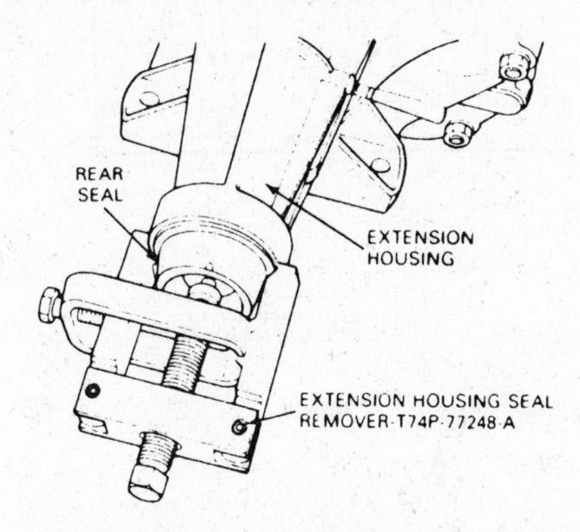

Extension housing oil seal removal – Mitsubishi 2WD transmission

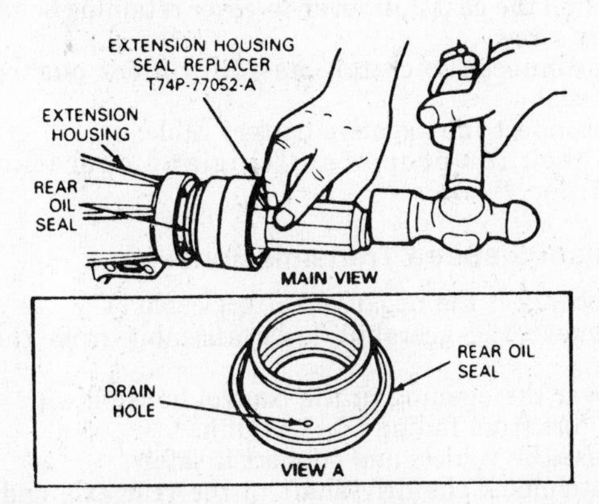

Extension housing oil seal installation – Mitsubishi 2WD and M50D 2WD transmissions

77248–A or equivalent, remove the extension housing seal.

To install:

6. Lubricate the inside diameter of the oil seal and install the seal into the extension housing using tool T61L–7657–A. Check to ensure that the oil seal drain hole faces downward.
7. Install the driveshaft to the extension housing. Connect the driveshaft to the rear axle flange. Make sure the marks made during removal are in alignment. Fit the attaching washer, lockwasher and nuts.
8. Check and adjust the transmission fluid level, using Ford manual transmission lube D8DZ–19C547–A (ESP–M2C83–C) or equivalent.
9. Lower the vehicle.
10. Reconnect the negative battery cable.

Transmission

REMOVAL AND INSTALLATION

4-Speed Transmission And Mazda (Toyo Kogyo) 5-Speed Transmission

1. Disconnect the negative battery cable.
2. Remove the gearshift lever, shim and bushing from the unit. Cover the opening in the housing with a cloth to prevent dirt from falling into the unit.
3. Disconnect the clutch master cylinder pushrod from the clutch pedal.
4. Raise the vehicle and support it safely.
5. Matchmark the driveshaft to the rear axle flange. Disconnect the driveshaft and pull it rearward from the transmission. Plug the extension housing to prevent leakage.
6. Remove the dust shield and slave cylinder from the clutch housing. Secure it to the side, using a piece of wire.
7. Disconnect the starter motor and back-up lamp wiring harness.
8. Remove the starter motor and speedometer cable.
9. Place a wood block on a service jack and position the jack under the engine oil pan.
10. Position a transmission jack under the transmission assembly.
11. On 4 × 4 vehicles, remove the transfer case.
12. Remove the transmission-to-engine retaining bolts and washers.
13. Remove the transmission mount and damper retaining bolts/nuts.
14. Remove the crossmember-to-frame retaining nuts and remove the crossmember from the vehicle.
15. Carefully lower the transmission jack, while working the transmission off the dowel pins. Slide the transmission rearward and remove it from the vehicle.

To install:

16. Check that the mating surfaces of the clutch housing, engine rear and dowel holes are free of burrs, dirt and paint.
17. Place the transmission on the transmission jack. Place the transmission under the vehicle, then raise it into position. Align the input shaft splines with the clutch disc splines and work the transmission forward into the locating dowels.
18. Install the transmission-to-engine retaining bolts and washers. Tighten and torque the retaining bolts 28–38 ft. lbs. (38–52Nm). Remove the transmission jack.

NOTE: Always place flat washers between the retaining bolts to avoid damaging the aluminum surface.

19. Install the starter motor. Tighten and torque the retaining bolts 15–20 ft. lbs. (20–27Nm).
20. Raise the engine and install the crossmember, transmission mount and damper. Torque the retaining to specifications.
21. Remove the service jack from supporting the engine.
22. On 4WD vehicles, install the transfer case.
23. Install the driveshaft to the transmission extension housing.
24. Connect the driveshaft to the rear axle flange. Make sure the marks made during removal are in alignment. Fit the attaching washer, lockwasher and nuts. Torque the driveshaft-to-flange retaining nuts to 61–87 ft. lbs. on all except the Ranger w/ 4WD, which gets a torque of 41–55 ft. lbs.
25. Install the dust shield and slave cylinder to the clutch housing. Torque the slave cylinder-to-clutch housing bolts 15–20 ft. lbs. (21–27Nm) and the dust shield bolts 5–10 ft. lbs. (7–13Nm).

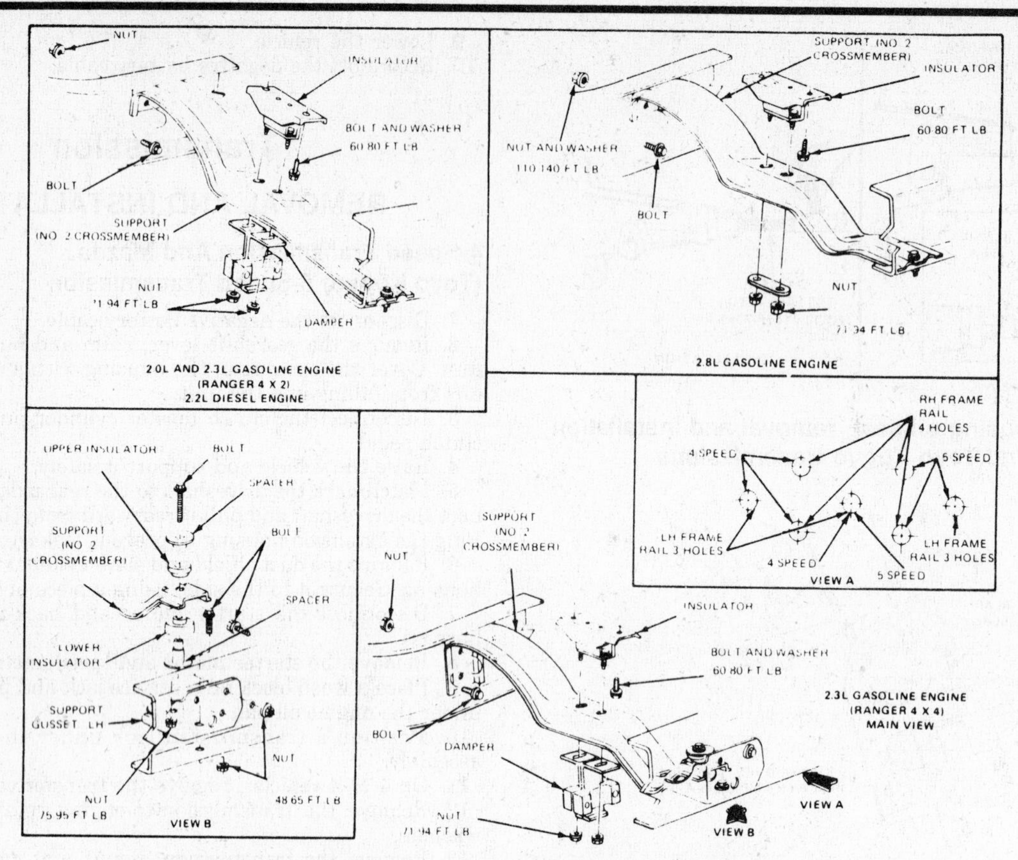

Crossmember installation – 4-speed gasoline engines

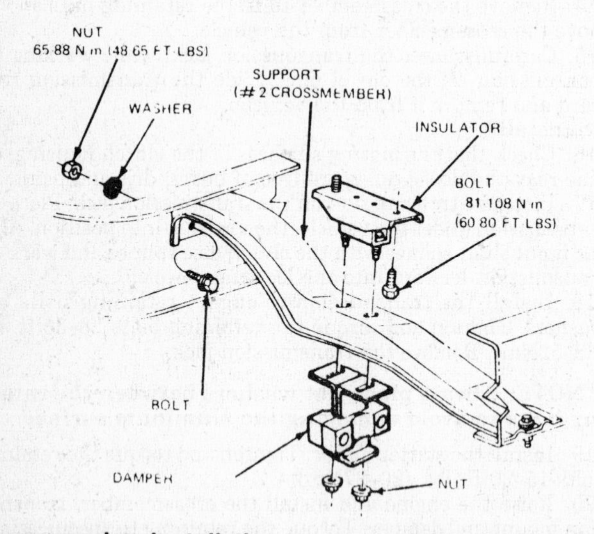

Crossmember installation – 4-speed diesel engines

26. Reconnect the starter motor and back-up lamp wiring harness.
27. Install the speedometer cable.
28. Check and adjust the transmission fluid level, using Ford manual transmission lube D8DZ–19C547–A (ESP–M2C83–C) or equivalent.
29. Lower the vehicle.
30. Remove the cloth and avoid getting dirt in the unit. Install the gearshift lever, shim and bushing into the gearshift lever retainer.
31. Install the gearshift lever-to-cover retaining bolts and boot retaining screws.
32. Reconnect the clutch master cylinder pushrod to the clutch pedal.
33. Reconnect the negative battery cable.
34. Check for proper shifting and operation of the transmission.

Mitsubishi 5-speed Transmissions

1. Disconnect the negative battery cable.
2. Remove the gearshift lever assembly from the control housing.
3. Cover the opening in the control housing with a cloth to prevent dirt from falling into the unit.
4. Raise the vehicle and support it safely.
5. Matchmark the driveshaft to the rear axle and transfer case flange, as required. Disconnect the rear driveshaft at both the rear axle and transfer case flanges.
6. Disconnect the forward driveshaft from the front axle and remove by sliding forward. Install a suitable plug in the transfer case adapter to prevent fluid leakage.
7. Disconnect the hydraulic fluid line from the clutch slave cylinder.
8. Disconnect the speedometer from the transfer case/extension housing.
9. Disconnect the starter motor cable, the back-up lamp switch wire and the neutral safety switch wire.
10. Place a wood block on a service jack and position the jack under the engine oil pan.
11. Support the transfer case using the proper equipment. Carefully remove the transfer case from the vehicle.
12. Remove the starter. Place a transmission jack under the transmission.

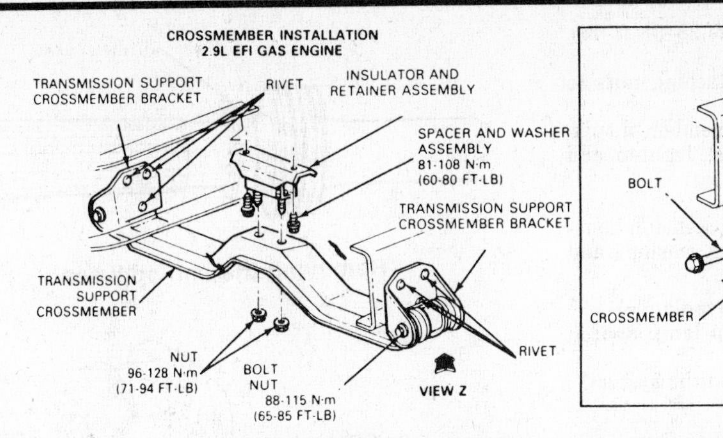

Crossmember installation – 2.9L EFI gasoline engine

13. Remove the transmission-to-engine retaining bolts and washers.
14. Remove the nuts and bolts attaching the transmission mount and damper to the crossmember.
15. Remove the nuts and bolts attaching the crossmember to the frame side rails and remove the crossmember.
16. Lower the engine jack. Work the clutch housing off the locating dowels and slide the clutch housing and the transmission rearward until the input shaft clears the clutch disc.
17. Remove the transmission from the vehicle.
18. Remove the clutch housing from the transmission.

To install:

19. Install the clutch housing to the transmission.
20. Check that the mating surfaces of the clutch housing, engine rear and dowel holes are free of burrs, dirt and paint.
21. Place the transmission on the transmission jack. Position the transmission under the vehicle, then raise it into position. Align the input shaft splines with the clutch disc splines and work the transmission forward into the locating dowels.
22. Install the transmission-to-engine retaining bolts and washers. Tighten and torque the retaining bolts 28–38 ft. lbs. (38–52Nm). Remove the transmission jack.

NOTE: Always place flat washers between the retaining bolts to avoid damaging the aluminum surface.

23. Install the starter motor. Tighten the attaching nuts to 15–20 ft. lbs. (21–27 Nm).
24. Raise the engine and install the rear crossmember, insulator and damper and attaching nuts and bolts. Tighten and torque the bolts to specification.
25. Install the transfer case.
26. Install the rear driveshaft in the transfer case adapter and attach it to the rear axle flange. Make sure the marks made during removal are in alignment. Install the attaching nuts and bolts. Torque circular flange bolts to 61–87 ft. lbs.; all others to 41–55 ft. lbs.
27. Install the front driveshaft in the transfer case adapter and attach it to the front axle flange. Make sure the marks made during removal are in alignment. Install the attaching nuts and bolts. Torque the nuts to 50 ft. lbs.
28. Connect the starter cable, back-up lamp switch wire, shift indicator wire and the neutral safety switch.
29. Connect the clutch hydraulic line to slave cylinder on the input shaft. Bleed the hydraulic clutch system.
30. Install the speedometer cable.
31. Remove the fill plug and check the fluid level. The fluid level should be level with the bottom of the fill hole. Ford manual transmission lube SAE 80W, D8DZ-19C547-A (ESP-M2C83-C) or equivalent is recommended.
32. Lower the vehicle.
33. Remove the cloth over the transfer case adapter opening. Avoid getting dirt in the adapter.
34. Install the gearshift lever assembly in the control housing. Make sure the ball on the lever is in the socket in the unit. Install the attaching bolts and tighten to 6–10 ft. lbs. (8–14 Nm).
35. Install the boot cover and bolts.
36. Reconnect the negative battery cable.
37. Check for proper shifting and operation of the transmission.

M5OD 5-speed Transmissions

1. Disconnect the negative battery cable.
2. Remove the gearshift lever assembly from the control housing.
3. Cover the opening in the control housing with a cloth to prevent dirt from falling into the unit.
4. Raise the vehicle and support it safely.
5. Matchmark the driveshaft to the rear axle flange. Pull the driveshaft rearward and disconnect it from the transmission.
6. Disconnect the clutch hydraulic line a the clutch housing. Plug the lines.
7. Disconnect the speedometer from the transfer case/extension housing.
8. Disconnect the starter motor and back-up lamp switch harness connector.
9. Place a wood block on a service jack and position the jack under the engine oil pan.
10. On 4WD vehicles, remove the transfer case from the vehicle.
11. Remove the starter motor.
12. Position a transmission jack, under the transmission.
13. Remove the transmission-to-engine retaining bolts and washers.
14. Remove the nuts and bolts attaching the transmission mount and damper to the crossmember.
15. Remove the nuts and bolts attaching the crossmember to the frame side rails and remove the crossmember.
16. Lower the engine jack. Work the clutch housing off the locating dowels and slide the clutch housing and the transmission rearward until the input shaft clears the clutch disc.
17. Remove the transmission from the vehicle.

To install:

18. Check that the mating surfaces of the clutch housing, engine rear and dowel holes are free of burrs, dirt and paint.
19. Place the transmission on the transmission jack. Position the transmission under the vehicle, then raise it into position. Align the input shaft splines with the clutch disc splines and work the transmission forward into the locating dowels.
20. Install the transmission-to-engine retaining bolts and

washers. Tighten and torque the retaining bolts 28–38 ft. lbs. (38–52Nm). Remove the transmission jack.

21. Install the starter motor. Tighten the attaching nuts to 15–20 ft. lbs. (21–27 Nm).
22. Raise the engine and install the rear crossmember, insulator and damper and attaching nuts and bolts. Tighten and torque the bolts to specification.
23. On 4WD vehicles, install the transfer case.
24. Insert the driveshaft nto the transmission extension housing and install the center bearing attaching nuts, washers and lockwashers.
25. Connect the driveshaft to the rear axle drive flange.
26. Connect the starter motor and back-up lamp switch connectors.
27. Connect the hydraulic clutch line and bleed the system.
28. Install the speedometer cable.
29. Check and adjust the fluid level.
30. Lower the vehicle.
33. Install the gearshift lever assembly. Install the boot cover and bolts.
34. Reconnect the negative battery cable.
35. Check for proper shifting and operation of the transmission.

4-Speed Transmission Overhaul (Gasoline Engine)

DISASSEMBLY

Before disassembly, clean the exterior of the transmission assembly before any attempt is made to disassemble it, in order to prevent dirt or other foreign material from entering the transmission assembly and damaging its internal parts. If steam cleaning is done to the exterior of the transmission, immediate disassembly should be done to avoid rusting, caused by condensation forming on the internal parts.

1. Mount the transmission in a suitable holding fixture.
2. Remove the bell housing-to-transmission case retaining nuts. Remove the bell housing and gasket from the case.
3. Remove the drain plug. Drain the lubricant into a suitable container.
4. Clean the metal filings from the drain plug, then reinstall it to the case.
5. Place the transmission in **N** position.
6. Remove the gearshift lever retainer-to-extension housing retaining bolts.
7. Remove the extension housing-to-case retaining bolts.
8. Remove the control lever by raising it toward the left side and sliding it toward the rear of the transmission.
9. Carefully slide the extension housing from the case. Be careful not to damage the oil seal.
10. If the control lever is being replaced, remove the control lever retaining bolt and remove the control lever.
11. Remove the back-up lamp switch from the transmission, if required.
12. Remove the anti-spill seal from the output shaft and discard. A seal is not necessary durnig reassembly.
13. Remove the snapring that retains the speedometer drive gear to the mainshaft. Slide the gear off the mainshaft and remove the lock ball.
14. Alternately loosen the transmission case cover bolts. Remove the case cover and gasket.
15. Remove the spring cap bolts, noting their location. Then, with the use of a magnet, remove the detent springs and detent balls.
16. To access the shift fork roll pins, remove the retaining bolts that secures the 2 blind covers of the transmission case. Remove the blind covers and gaskets.

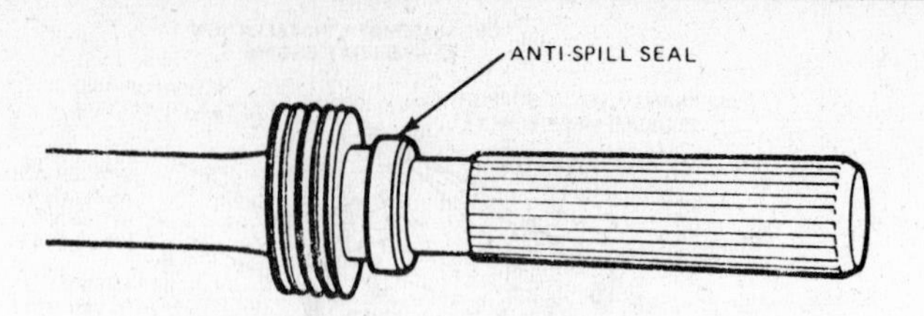

Removing the anti-spill seal

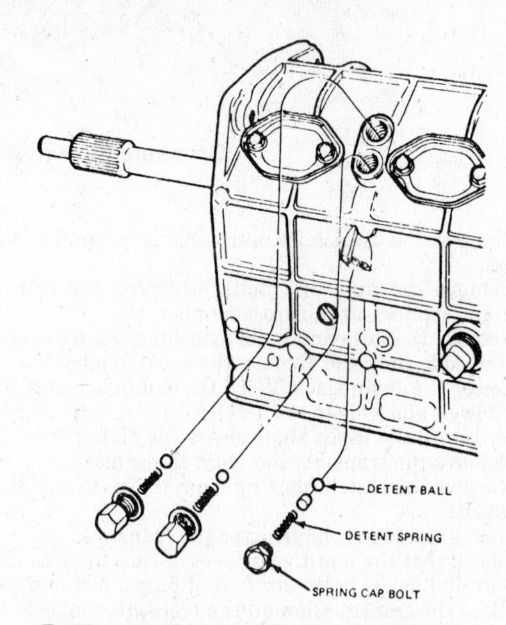

Detent springs and balls, location

17. Slide the reverse shift fork shaft assembly and reverse idler gear out of the transmission case.
18. To access the roll pin in the 3rd–4th gear shift fork, shift the transmission into 4th gear. With a drift, drive the roll pin from the 3rd–4th fork assembly. Slide the 3rd–4th shift fork shaft out the rear of the transmission case.
19. Remove the roll pin from the 1st–2nd shift fork. Then, slide the 1st–2nd shift fork shaft assembly out the rear of the transmission case. Remove both interlock pins.
20. To lock the gears, reinstall the reverse idler gear. Then, install the synchronizer ring holder and countershaft spacer between the 4th speed synchronizer ring and synchromesh gear on the mainshaft. Shift the transmission into 2nd gear to lock the mainshaft assembly from rotating.
21. Straighten the lockwasher. Then, using the locknut adapter and wrench, remove the locknut and washer that secures the reverse idler gear. Slide the reverse idler gear from the mainshaft and remove the key from the mainshaft.
22. Remove the reverse idler gear.
23. Remove the snapring from the rear end of the countershaft and slide the countershaft reverse gear off the countershaft.
24. Remove the bearing retainer together with the reverse idler gear shaft.
25. Remove the countershaft rear bearing, by install the bearing puller, puller rings, remover tube and forcing screw. Make certain the jaws of the puller is behind the front bearing retainer ring in the 2 recessed areas of the case.

NOTE: The retainer ring may need to be turned to position the split in the retainer ring midway between the recessed areas, before the puller can be installed.

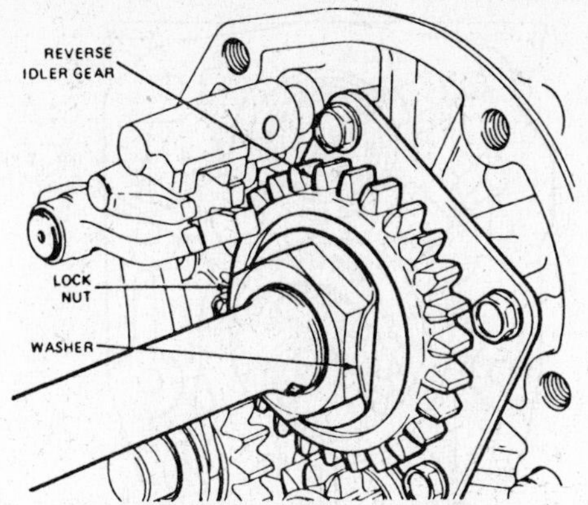

Reverse idler gear installation

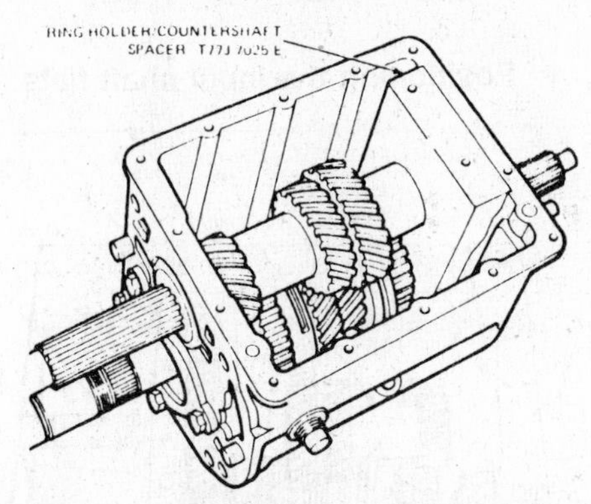

Synchronizer ring and countershaft holding tool, installation

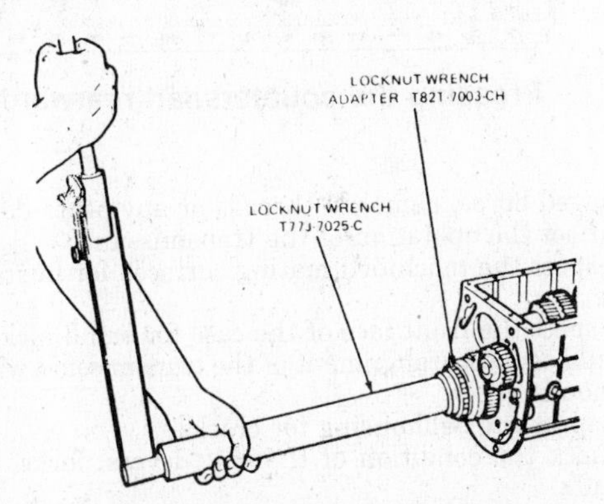

Reverse idler gear locknut, removal

26. Turn the forcing screw clockwise to remove the bearing.
27. Remove the mainshaft rear bearing, by install the puller, puller rings, remover and replacer tube and forcing screw. Make certain the jaws of the puller is behind the rear mainshaft bearing retainer ring in the 2 recessed areas of the case.

NOTE: The retainer ring may need to be turned to position the split in the ring midway between the recessed areas, before the puller can be installed.

28. Turn the forcing screw clockwise to remove the bearing.
29. Remove the shim and spacer from behind the mainshaft rear bearing.
30. To remove the front cover, install two 10mm × 1.5 nuts on the 4 studs and draw the studs out of the case. Remove the bolts

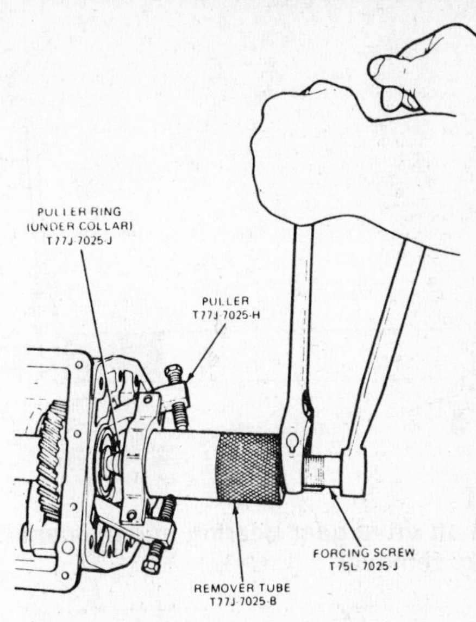

Countershaft bearing removal

and front cover. Save the shim which is on the inside of the cover.
31. Remove the snapring from the input shaft.
32. Remove the mainshaft drive gear bearing, by install the puller, puller rings, remover and replacer tube and forcing screw. Make certain the jaws of the puller is behind the rear mainshaft drive gear bearing retainer ring in the 2 recessed areas of the case.

NOTE: The retainer ring may need to be turned to position the split in the ring midway between the recessed areas, before the puller can be installed.

33. Turn the forcing screw clockwise to remove the bearing.
34. Rotate both shift forks so that the main geartrain will fall to the bottom of the case. Remove the shift forks.
35. Rotate the input shaft so that 1 of the 2 flats on the input shaft face upward.
36. Insert the synchronizer ring holder and countershaft spacer between the 1st gear on the countershaft and the rear of the case. Then, remove the snapring from the front of the countershaft.
37. Install the forcing screw, press frame and press frame adapter against the countershaft assembly.
38. Turn the forcing screw clockwise to press the countershaft rearward. Press the countershaft, approximately $^3/_{16}$ in. (5mm), until it contacts the synchronizer ring holder and countershaft spacer.
39. Remove the countershaft front bearing, by install the bearing puller, puller rings, remover tube and forcing screw. Make certain the jaws of the puller is behind the front bearing retainer ring in 2 recessed areas of the case.

NOTE: The retainer ring may need to be turned to position the split in the ring midway between the recessed areas, before the puller can be installed.

40. Turn the forcing screw clockwise to remove the bearing.
41. Remove the shim from behind the countershaft front bearing.
42. Remove the input shaft from the transmission case. Remove the synchronizer ring and caged bearing from the main driveshaft.
43. Remove the countershaft from the transmission case.
44. Press the inner race of the countershaft center bearing from the countershaft, using the axle bearing seal plate and pinion bearing cone remover.

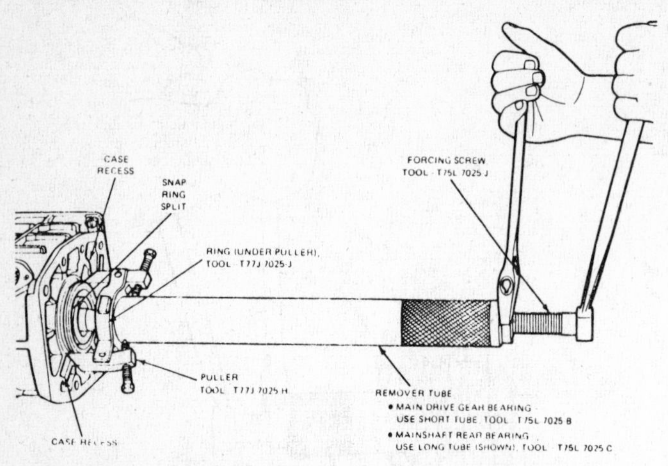

Mainshaft drive gear bearing and mainshaft rear bearing, removal

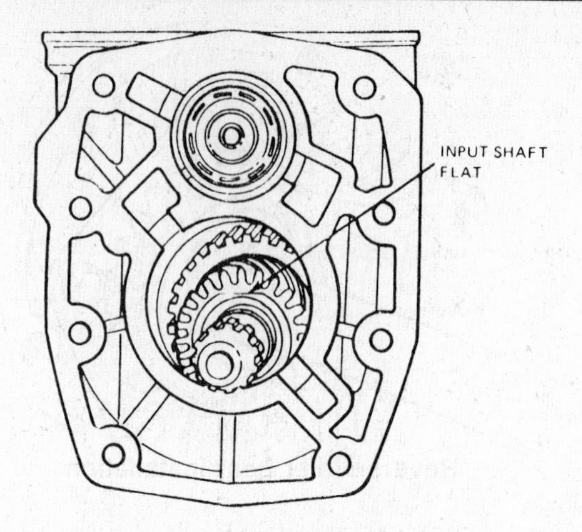

Positioning the input shaft flats

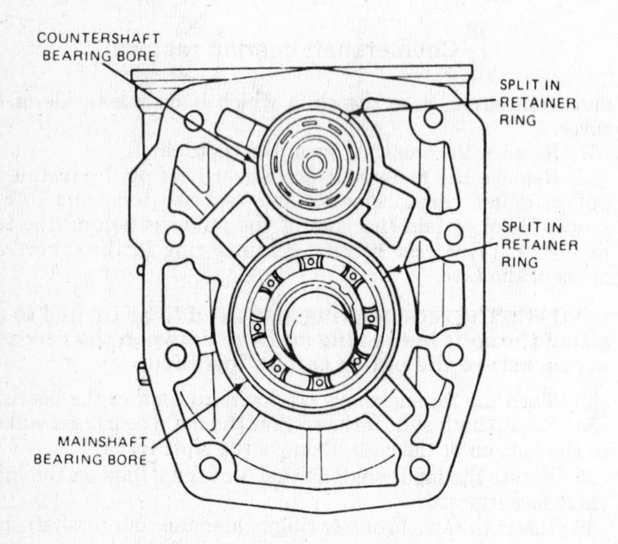

Retainer ring installation

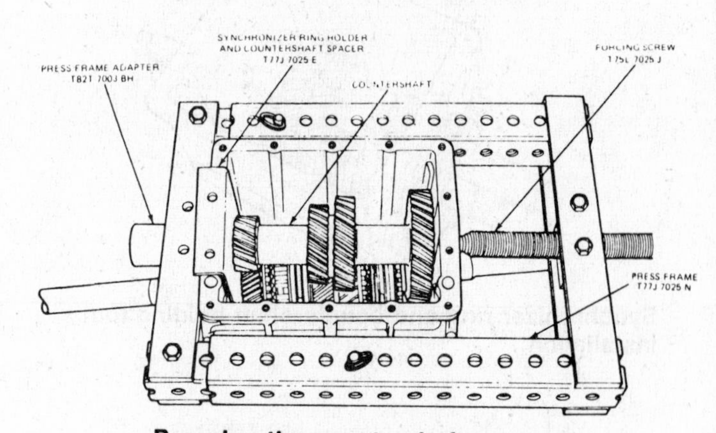

Pressing the countershaft rearward

45. Remove the mainshaft and gear assembly from the transmission case.
46. Remove the snapring from the front of the mainshaft.
47. Slide the 3rd–4th clutch hub and sleeve assembly, 3rd synchronizer ring and 3rd gear off the front of the mainshaft.

NOTE: Do not mix the synchronizer rings.

48. Slide the thrust washer, 1st gear and gear sleeve off the rear mainshaft. Using a press, remove the bushing from the 1st gear.
49. Remove the 1st–2nd clutch hub and sleeve assembly from the mainshaft.

CLEANING AND INSPECTION

1. Wash all parts except the ball bearings and seals, in a suitable cleaning solvent.
2. Dry all parts with compressed air.
3. Do not clean, wash or soak the transmission seals in cleaning solvent.
4. Lubricate the bearings and wrap them in a clean, lint free cloth until ready for use.
5. Inspect the transmission case and housing for cracks, worn or damaged bores, damaged threads or any other damage that could affect the operation of the transmission.
6. Inspect the machined mating surfaces for burrs, nicks or damage.
7. Inspect the front face of the case for small nicks or burrs that could cause misalignment of the transmission with the flywheel housing.
8. Inspect the bellhousing for cracks.
9. Check the condition of the shift levers, forks, shift rails and shafts.
10. Replace bearings that are broken, worn or rough.
11. Check the contact surface of the shift fork shaft with the control lever for wear with a feeler gauge. The clearance should be less that 0.8mm.
12. Check the wear of the synchronizer rings, by fitting the synchronizer ring evenly to the gear cone and measuring the clearance between the side faces of the synchronizer ring and the gear with a feeler gauge. If less than 0.8mm, replace the synchronizer ring.
13. Check the clearance between the shift fork and the clutch sleeve. The clearance should be less that 0.5mm.

NOTE: Before beginning the assembly procedure, the following 3 measurement checks must be performed:

Mainshaft Thrust Play

Check the mainshaft thrust play by measuring the depth of the mainshaft bearing bore in the transmission rear cage by us-

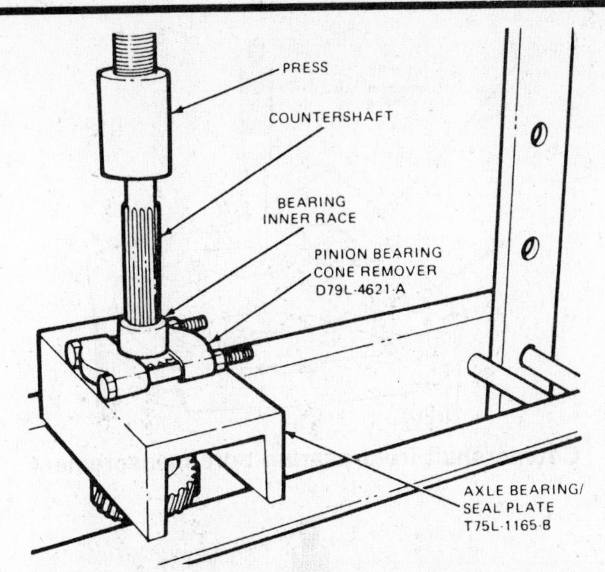

Removing the countershaft bearing inner race

ing a depth micrometer. Measure the mainshaft rear bearing height. The difference between the 2 measurements indicates the required thickness of the adjusting shim. The standard thrust play is 0–0.1mm. Adjusting shims are available in 0.1mm and 0.3mm sizes.

Countershaft Thrust Play

Check the countershaft thrust play by measuring the depth of the countershaft front bearing bore in the transmission case by using a depth micrometer. Measure the countershaft front bearing height. The difference between the 2 measurements indicates the required thickness of the adjusting shim. The standard thrust play is 0–0.1mm. Adjusting shims are available in 0.1mm and 0.3mm sizes.

Mainshaft Bearing Clearance

Check the mainshaft bearing clearance by measuring the depth of the bearing bore in the clutch adapter plate with a depth micrometer. Make sure the micrometer is on the second step of the plate. Measure the bearing height. The difference between the 2 measurements indicates the required thickness of the adjusting shim. The standard clearance is 0–0.1mm. If an adjusting shim is required, select a shim to bring the clearance to within specifications.

NOTE: As each part is assembled, coat the part with gear lubricant.

ASSEMBLY

1. Lubricate and assemble the 1st–2nd synchromesh mechanism by installing the clutch hub to the sleeve. Place the 3 synchronizer keys into the clutch hub key slots and install the key springs to the clutch hub.

NOTE: When installing the key springs, the open end tab of the springs should be inserted into the hub holes with the springs turned in the same direction. This will keep the spring tension on each key uniform.

2. Assemble the 3rd–4th synchromesh mechanisms in the same manner as the 1st–2nd synchromech mechanism.
3. Fit the synchronizer ring on the 3rd gear and slide the 3rd gear to the front of the mainshaft with the synchronizer ring toward the front.

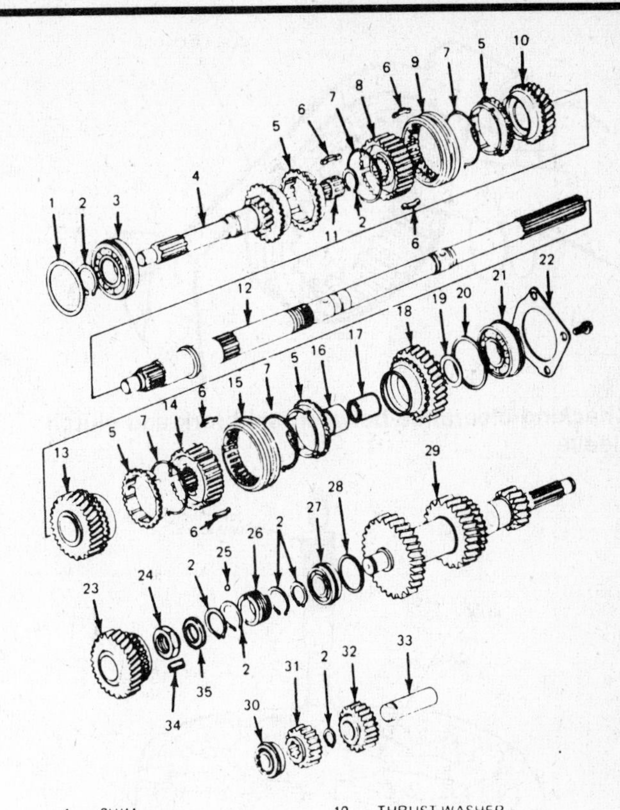

1. SHIM
2. SNAP RING
3. MAIN DRIVE SHAFT BEARING
4. MAIN DRIVE SHAFT GEAR
5. SYNCHRONIZER RING
6. SYNCHRONIZER KEY
7. SYNCHRONIZER KEY SPRING
8. 3RD-AND-4TH CLUTCH HUB
9. CLUTCH SLEEVE
10. 3RD GEAR
11. CAGED BEARING
12. MAIN SHAFT
13. 2ND GEAR
14. 1ST-AND-2ND CLUTCH HUB
15. CLUTCH SLEEVE
16. FIRST GEAR SLEEVE
17. BUSHING
18. 1ST GEAR
19. THRUST WASHER
20. SHIM
21. MAIN SHAFT FRONT BEARING
22. BEARING COVER
23. REVERSE GEAR
24. MAIN SHAFT LOCK NUT
25. LOCK BALL
26. SPEEDOMETER DRIVE GEAR
27. COUNTER SHAFT FRONT BEARING
28. SHIM
29. COUNTER SHAFT
30. COUNTER SHAFT REAR BEARING
31. COUNTER REVERSE GEAR
32. REVERSE IDLER GEAR
33. IDLER GEAR SHAFT
34. KEY
35. LOCKWASHER

Gears and shafts assembly, exploded view

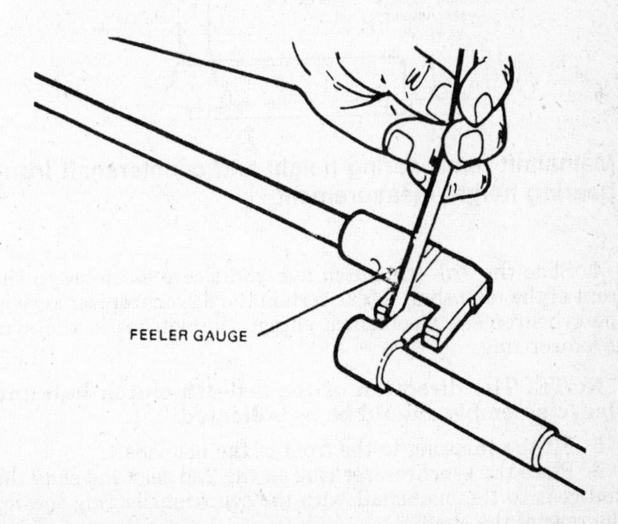

Checking clearance between the shift fork shaft and control lever

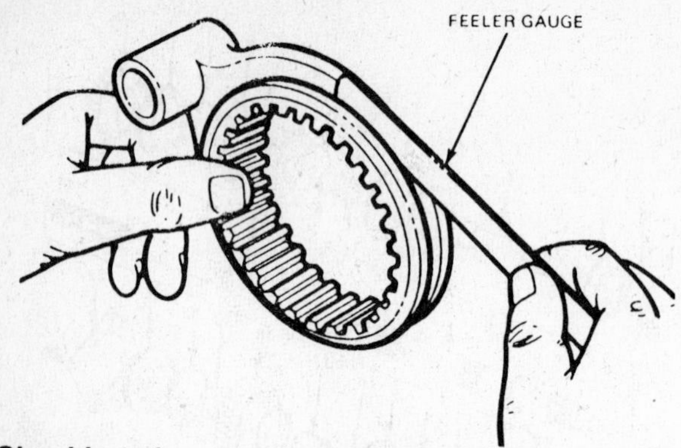

Checking clearance between shift fork and clutch sleeve

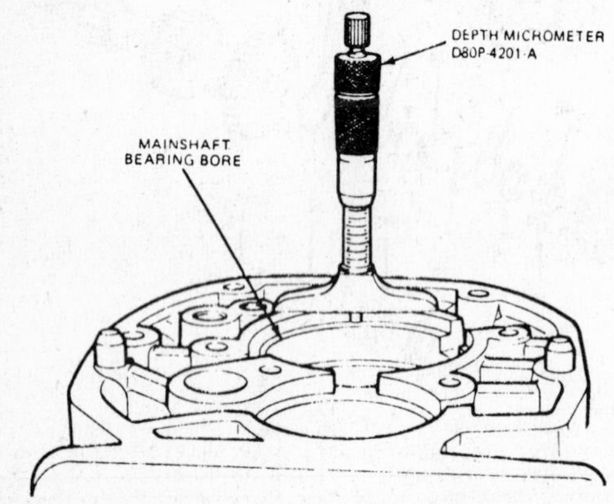

Mainshaft bearing bore depth measurement

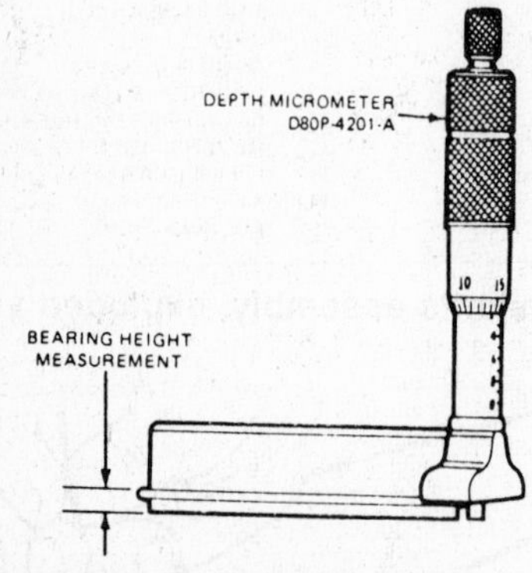

Mainshaft rear bearing height and countershaft front bearing height measurement

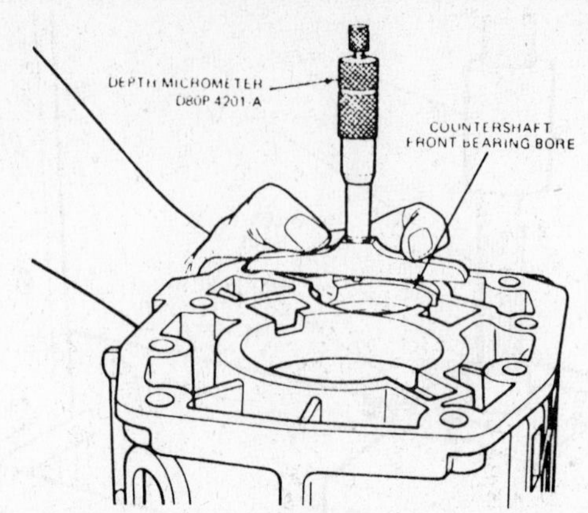

Countershaft front bearing bore measurement

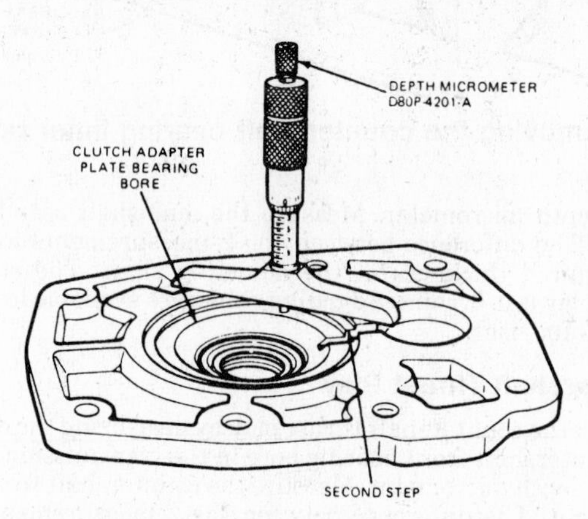

Clutch adapter plate bore measurement

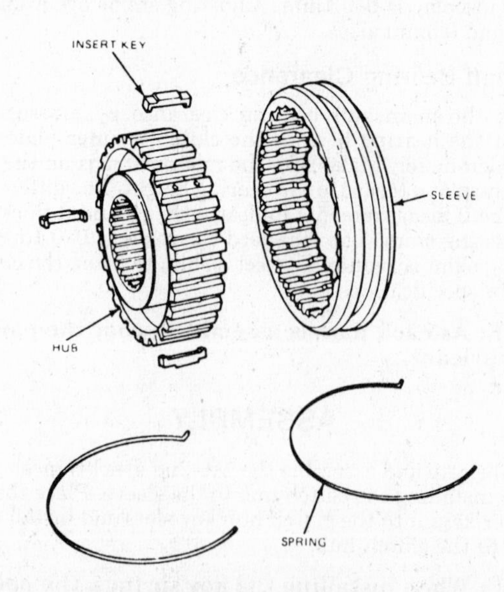

Synchronizer mechanism

4. Slide the 3rd–4th clutch hub and sleeve assembly to the front of the mainshaft. Make certain the 3 synchronizer keys in the synchromesh mechanism engage the notches in the synchronizer ring.

NOTE: The direction of the 3rd–4th clutch hub and sleeve assembly should be as indicated.

5. Fit the snapring to the front of the mainshaft.
6. Place the synchronizer ring on the 2nd gear and slide the 2nd gear to the mainshaft with the synchronizer ring toward the rear of the shaft.
7. Fit the 1st–2nd clutch hub and sleeve assembly to the mainshaft with the oil groove of the clutch hub toward the front of the mainshaft. Make certain the 3 synchronizer keys in the synchromesh mechanism engage the notches in the 2nd synchronizer ring.
8. Fit the 1st gear sleeve to the mainshaft. Using a press, press the bushing in the 1st gear.
9. Place the synchronizer ring on the 1st gear and slide the

1st gear onto the mainshaft with the synchronizer ring facing the front of the shaft. Rotate the 1st gear until it engage the 3 notches in the synchronizer ring with the synchronizer keys.

10. Fit the original thrust washer on the mainshaft.
11. Position the mainshaft and gears assembly in the case.
12. Position the caged bearing in the front end of the mainshaft.
13. Place the synchronizer ring on the input shaft (4th gear) and install the input shaft to the front end of the mainshaft. Make certain the 3 synchronizer keys in the 3rd–4th synchromesh mechanism engage the notches in the synchronizer ring.
14. Position the 1st–2nd shift fork and 3rd–4th shift fork in the groove of the clutch hub and sleeve assembly.
15. Press the inner race of the countershaft rear bearing onto the countershaft, using the center bearing replacer T77J–7025–K or equivalent.
16. Position the countershaft gear in the case. Make certain the countershaft gear engages each gear of the mainshaft assembly.
17. Install the correct shim in the mainshaft rear bearing bore. This shim was determine by the mainshaft thrust play measurement.
18. Position the main drive gear bearing and the mainshaft rear bearing into the proper bearing bores. Make certain the synchronizer and shifter forks have not been moved out of position.
19. Install the dummy bearing replacer, mainshaft front bearing replacer, replacer tube, press frame adapter and press frame. Position the synchronizer ring holder and countershaft spacer between the mainshaft drive gear and the synchronizer ring. Turn the forcing screw on the press frame until both bearings are properly seated.
20. Install the main drive gear bearing snapring.
21. Fit the correct shim in the countershaft front bearing bore. This shim was determine by the countershaft thrust play measurement.
22. Position the countershaft front and rear bearings in the bores and install the tools. Turn the forcing screw until the bearing is properly seated. Use the rear bearing as a pilot.
23. Secure the countershaft front bearing with the snapring.
24. Install the bearing retainer together with the reverse idler gear shaft to the transmission case and tighten the attaching bolts.
25. Slide the counter reverse gear onto the countershaft with the chamfer to the rear, then fit the snapring to secure the counter reverse gear.
26. Fit the key to the mainshaft.
27. Slide the reverse gear and lock washer onto the mainshaft, with the tab facing outward and the chamfer on teeth toward the rear. Install a new locknut and finger tighten.
28. Lock the transmission mainshaft, by shifting into 2nd gear and reverse gear. Tighten the locknut 145–203 ft. lbs. (197–275Nm).
29. Place the 3rd–4th clutch sleeve in 3rd gear, using the synchronizer ring holder and countershaft spacer, T77J–7003–CH or equivalent.
30. Using a feeler gauge, check the clearance between the synchronizer key and the exposed edge of the synchronizer ring. If the measurement is greater than 2.0mm, the synchronizer key can pop out of position. To correct this condition, change the selective thrust washer between the mainshaft rear bearing and the 1st gear. Thrust washer are available in 2.5mm, 3.0mm and 3.5mm.
31. Recheck the clearance. If the clearance is within specifications, bend the tab on the lockwasher.
32. From the rear of the case, slide the 1st–2nd shift fork shaft assembly into place. Install the roll pin. Secure the 1st–2nd shift fork to the fork shaft by staking the roll pin.

NOTE: Always use a new roll pins.

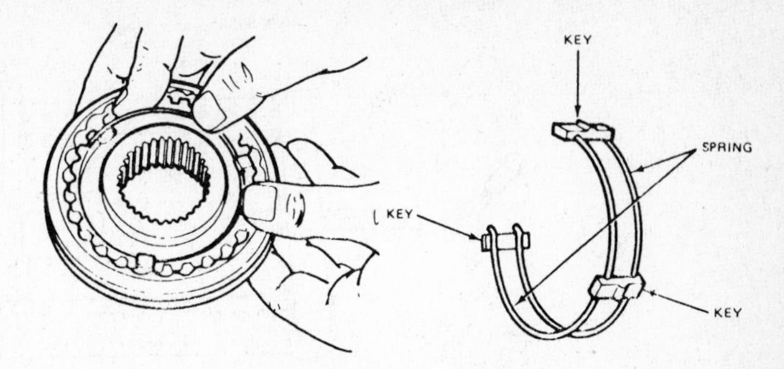

Key springs, installation

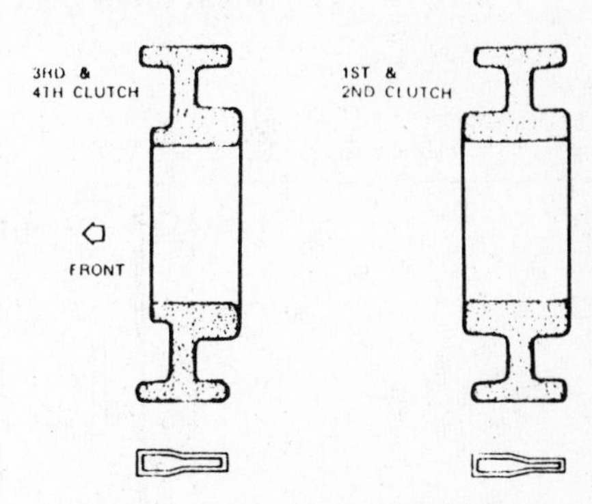

Clutch hub assembly direction

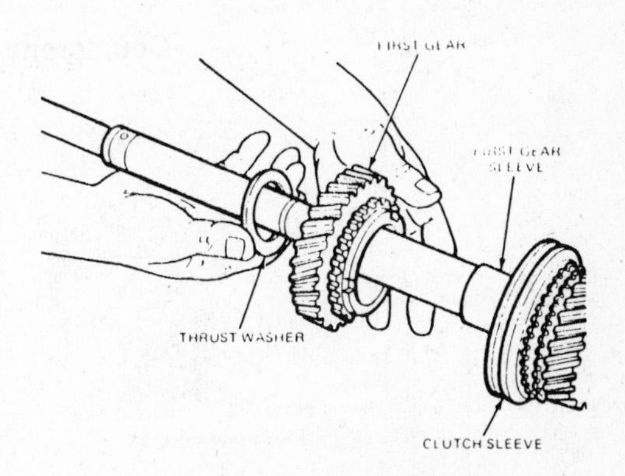

1st gear installation

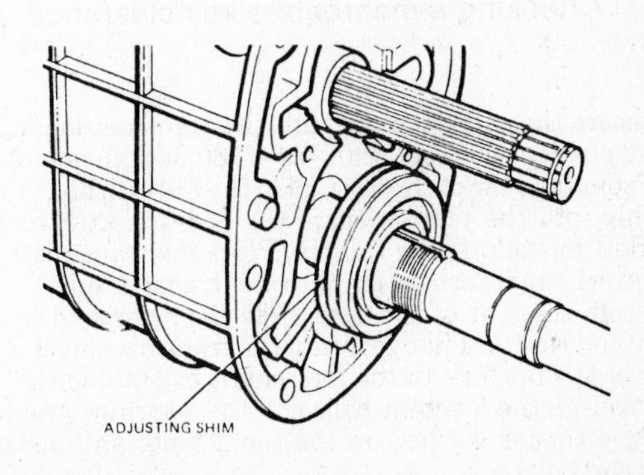

Mainshaft thrust play adjusting shim, installation

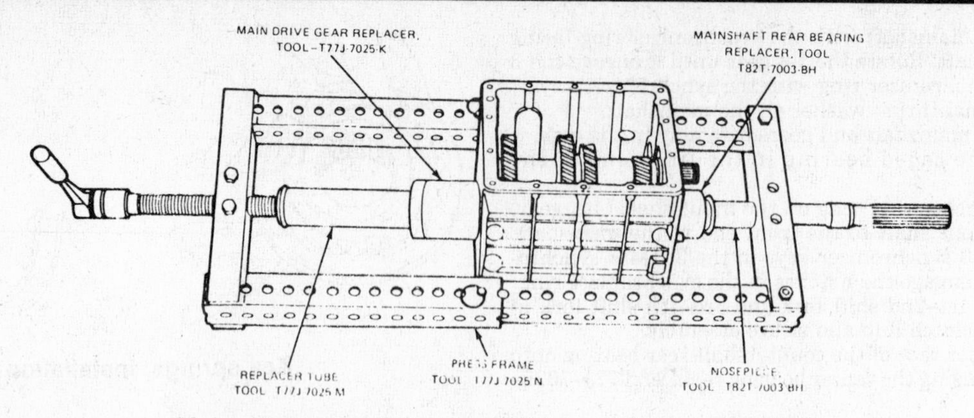

Main drive gear and mainshaft rear bearing installation

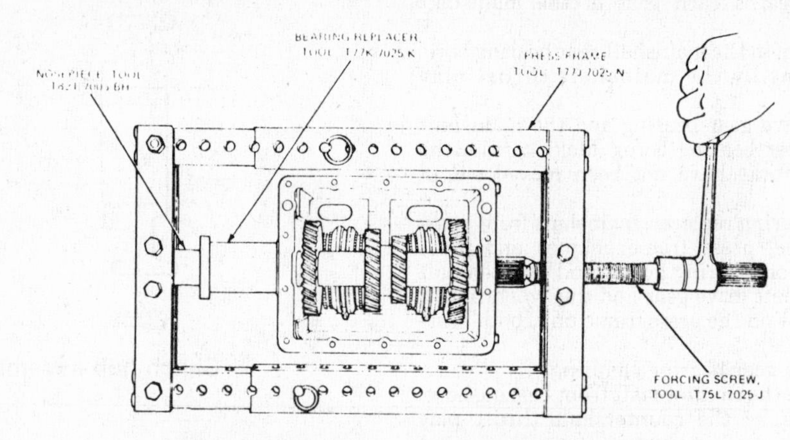

Countershaft front bearing installation

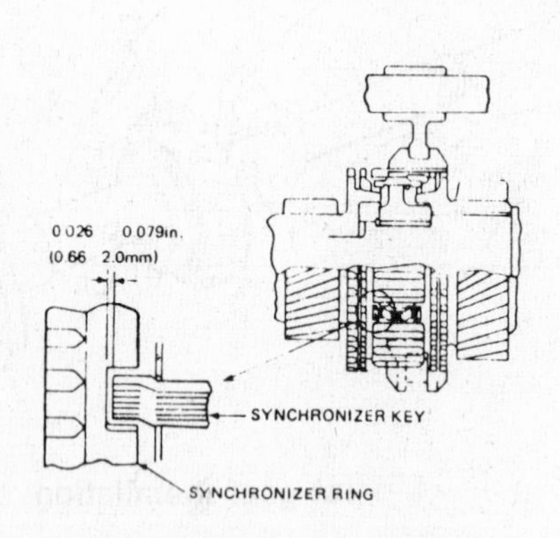

Checking synchronizer key clearance

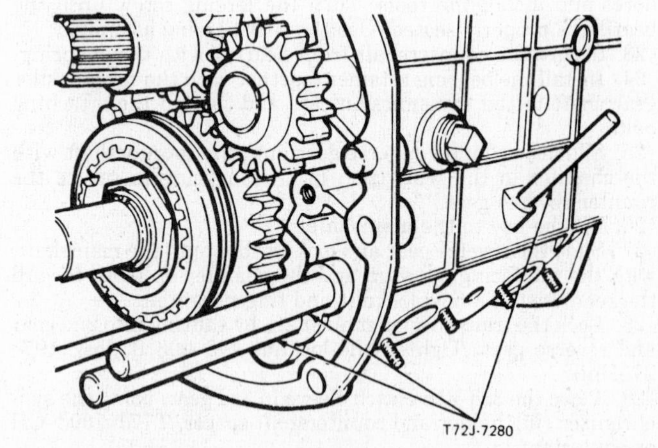

Shifter hardware installation

33. Insert the interlock pin into the transmission, using the lockout pin replacer tool set T72J–7280 or equivalent.

34. From the rear of the case, slide the 3rd–4th shift fork shaft assembly into the place. Secure the 3rd–4th shift fork to the fork shaft by staking the roll pin. Place the transmission in **N**.

35. Insert the interlock pin into the transmission.

36. From the rear of the case, slide the reverse fork shaft assembly and reverse idler gear into the transmission case. Secure the reverse shift fork to the fork shaft, by staking the roll pin.

37. Position the 3 detent balls and the 3 springs into the case. Place the copper washer on the top 2 bolts and install the 3 spring cap bolts.

38. Install the 2 blind covers and gaskets. Tighten the retaining bolts 23–34 ft. lbs. (32–45Nm).

39. Install the lock ball, speedometer drive gear and snapring onto the mainshaft.

40. Apply a thin coat of sealant E2AZ–19562–A or equivalent to the contact surfaces of the transmission case and extension housing.

41. Install the extension housing with the gearshift control lever end laid down to the left as far as it will go. Tighten the retaining bolts 41–59 ft. lbs. (56–79Nm).

NOTE: The lower 2 bolts must be coated with locite or equivalent.

42. If removed, fit the speedometer driven gear to the extension housing and tighten the retaining bolt.
43. Check for proper operation of the gearshift control lever.
44. Install the transmission case cover and gasket, with the drain plug to the rear. Install and tighten the cover retaining bolts 23–34 ft. lbs.
45. Fit the gearshift lever retainer and gasket to the extension housing. Tighten the retaining bolts.
46. Fit the correct shim on the second step of the clutch adapter plate. This shim was determine by the mainshaft bearing clearance measurement.
47. Apply a thin coat of sealant E2AZ–19562–A or equivalent to the clutch adapter plate, then install the clutch adapter plate to the transmission.
48. Remove the filler plug and add 3.0 pts. (1.4L) or Ford Manual Transmission Lube, D8DZ–19547–A or equivalent. Install the filler plug and tighten 18–29 ft. lbs. (25–39Nm).

4-Speed Transmission Overhaul (Diesel Engine)

DISASSEMBLY

Before disassembly, clean the exterior of the transmission assembly before any attempt is made to disassemble it, in order to prevent dirt or other foreign material from entering the transmission assembly and damaging its internal parts. If steam cleaning is done to the exterior of the transmission, immediate disassembly should be done to avoid rusting, caused by condensation forming on the internal parts.

1. Mount the transmission in a suitable holding fixture T57L–500–B or equivalent.
2. Remove the drain plug and drain the lubricant into a suitable container.
3. Remove the fork and release bearing from the clutch housing.
4. Remove the front cover-to-case retaining bolts and remove the front cover.
5. Remove the snapring from the input shaft.
6. Remove the outer snapring on the input shaft bearing. Install the bearing collet tool, remover tube and the forcing screw on the input shaft front bearing. Slide the bearing collet sleeve over the remover tube and bearing collet, then turn the forcing screw to remove the input shaft bearing.
7. Remove the extension housing-to-transmission case retaining bolts. Slide the extension housing off the mainshaft, with the control lever end laid down and to the left as far as it will go.
8. Remove the control lever end-to-control rod retaining bolts and remove the control lever end and rod from the extension housing.
9. Remove the speedometer driven gear assembly from the extension housing.
10. Remove the back-up lamp switch and neutral sensing switch.
11. Remove the snapring that secures the speedometer drive gear to the mainshaft. Slide the speedometer drive gear off the mainshaft and remove the lock ball.
12. Install the bearing pusher tool over the countershaft front bearing. Turn the forcing screw to force the countershaft, together with the countershaft front bearing from the transmission housing.
13. Slide the bearing housing and gear shaft assembly from the transmission housing.
14. Remove the 3 spring cap bolts, 3 springs and shift locking balls. The reverse spring is shortest.

NOTE: The spring-loaded lower ball will pop out.

SPECIAL SERVICE TOOLS

Number	Description
T75L-1165-B	Axle Bearing Seal Plate
D80P-4201-A	Depth Micrometer
D79L-4621-A	Pinion Bearing Remover
T82T-7003-BH	Press Frame Adapter
T82T-7003-CH	Locknut Wrench Adapter
T82T-7003-DH	Mainshaft Front Bearing Replacer
T75L-7025-B	Remover and Replacer Tube — Short Tube
T77J-7025-B	Remover Tube — Short Tube
T75L-7025-C	Remover and Replacer Tube — Long Tube
T75L-7025-J	Forcing Screw
T75L-7025-Q	Replacer Dummy Bearing
T77J-7025-C	Locknut Wrench
T77J-7025-E	Synchronizer Ring Holder and Countershaft Spacer
T77J-7025-G	Bell Housing Seal Replacer
T77J-7025-H	Puller
T77J-7025-J	Puller Ring
T77J-7025-K	Center Bearing Replacer
T77J-7025-L	Countershaft Front Bearing Replacer
T77J-7025-M	Mainshaft Front Bearing Replacer
T77J-7025-N	Portable Press Frame
T71P-7095-A	Oil Seal Installation Tool
T71P-7657-A	Oil Seal Removal Tool

TRANSMISSION ASSEMBLY TORQUES

Attachment	Torque	
	Ft-Lbs	N·m
Cap — Shift Rail Detent Spring	29-43	40-58
Nut — Main Shaft Gear Retaining	145-203	197-275
Pivot — Clutch Release Lever	23-34	32-46
Plug — Interlock Pin Bore	7.5-11.0	11-14
Plug — Drain	29-43	40-58
Plug — Filler	18-29	25-39
Switch — Back-Up Lamp	22-29	30-39
Nut Bolt Size		
6 mm	5-7.5	7-11
8 mm	12-17	17-23
10 mm	23-34	32-45
12 mm	41-59	56-79

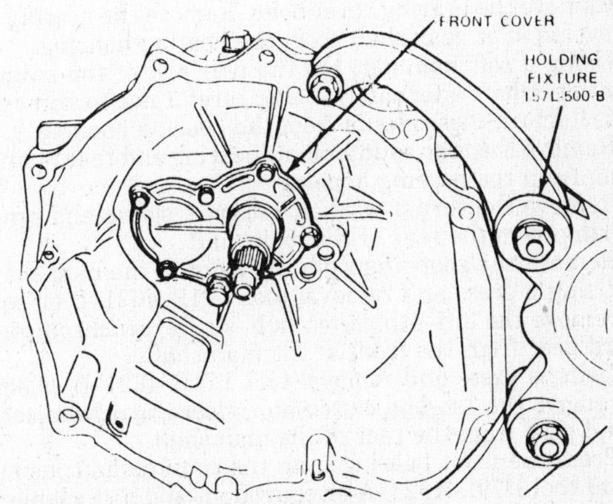

Front cover removal

15. Remove the reverse shift rod and shift fork assembly and reverse gear from the bearing housing.
16. Remove the roll pins that retains the shift forks to the rods. Push each of the shift rods rearward through the fork and bearing housing, then remove the shift rods and forks.

NOTE: Mark the 3rd–4th and 1st–2nd shift forks before removal to simplify installation.

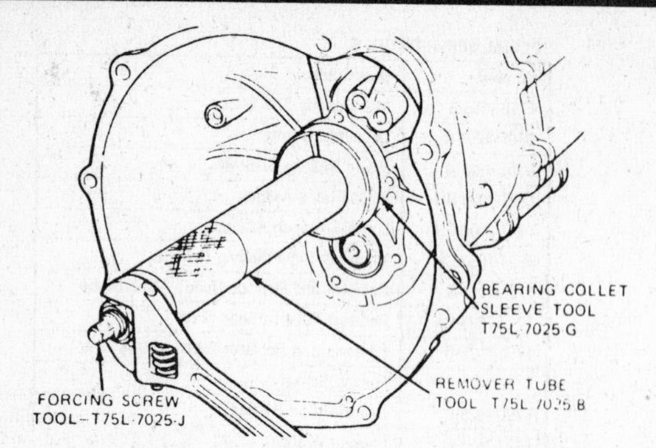

Removing input shaft front bearing

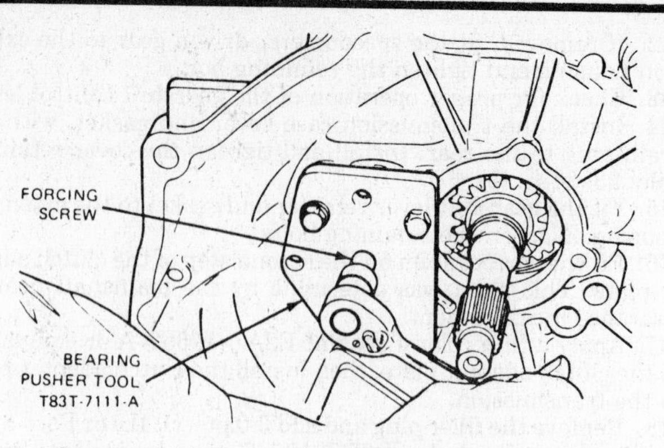

Countershaft and front bearing removal

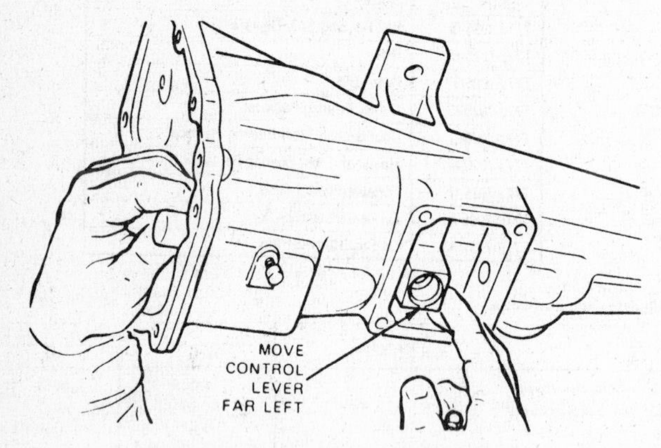

Shift control lever removal

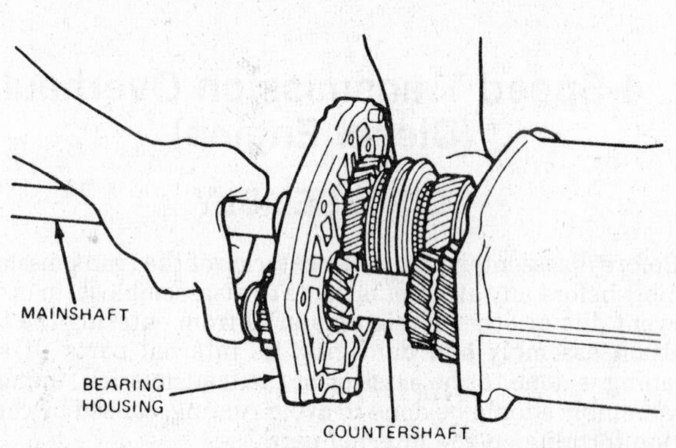

Bearing holder and gear shaft assembly removal

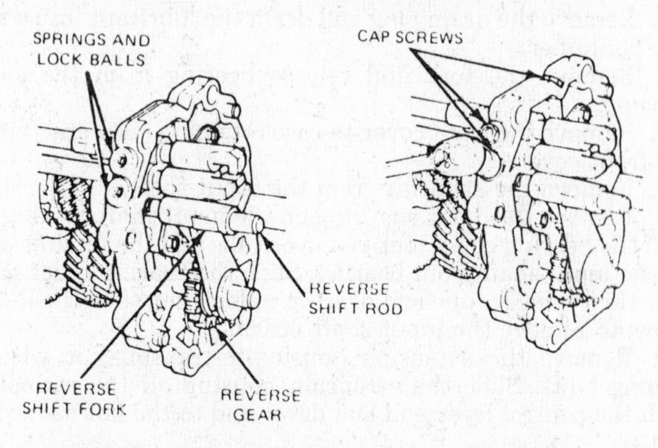

Shift lock balls and springs removal

17. Remove the lower reverse shift rod locking ball and spring and the interlock pins from the bearing housing.
18. Straighten the tab of the lockwasher. Lock the transmission synchronizer into any 2 gears and remove the mainshaft locknut.
19. Remove the snapring from the rear end of the countershaft and slide off the counter reverse gear.
20. Remove the bearing cover bolts. Remove the bearing cover and reverse idler gear shaft from the bearing housing.
21. Using a soft hammer, tap the rear end of the mainshaft and countershaft alternately, being careful not to damage the shafts. Remove these shafts from the bearing housing.
22. Remove the rear countershaft bearing and rear mainshaft bearing from the bearing housing.
23. Remove the thrust washer, 1st gear, sleeve and synchronizer ring from the rear of the mainshaft.
24. Remove the snapring from the front of the mainshaft.
25. Using a press and removal tool T71P–4621–B or equivalent, remove the 3rd–4th clutch hub, sleeve, synchronizer ring and 3rd gear from the front of the mainshaft.
26. Using a press and removal tool T71P–4621–B or equivalent, remove the 1st–2nd clutch hub, sleeve, synchronizer ring and 2nd gear from the rear of the mainshaft.
27. Press the front bearing from the countershaft, using the removal tool D79L–4621–A or equivalent and a stock piece.

CLEANING AND INSPECTION

1. Wash all parts except the ball bearings and seals, in a suitable cleaning solvent.
2. Dry all parts with compressed air.
3. Do not clean, wash or soak the transmission seals in cleaning solvent.
4. Lubricate the bearings and wrap them in a clean, lint free cloth until ready for use.
5. Inspect the transmission case and housing for cracks, worn or damaged bores, damaged threads or any other damage that could affect the operation of the transmission.
6. Inspect the machined mating surfaces for burrs, nicks or damage.
7. Inspect the front face of the case for small nicks or burrs that could cause misalignment of the transmission with the flywheel housing.
8. Inspect the bellhousing for cracks.
9. Check the condition of the shift levers, forks, shift rails and shafts.
10. Replace bearings that are broken, worn or rough.
11. Check the contact surface of the shift fork shaft with the control lever for wear with a feeler gauge. The clearance should be less that 0.8mm.
12. Check the wear of the synchronizer rings, by fitting the synchronizer ring evenly to the gear cone and measuring the

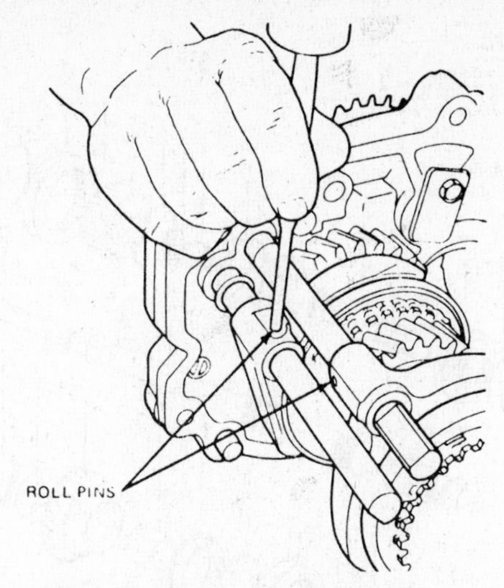

Shift fork roll pins removal

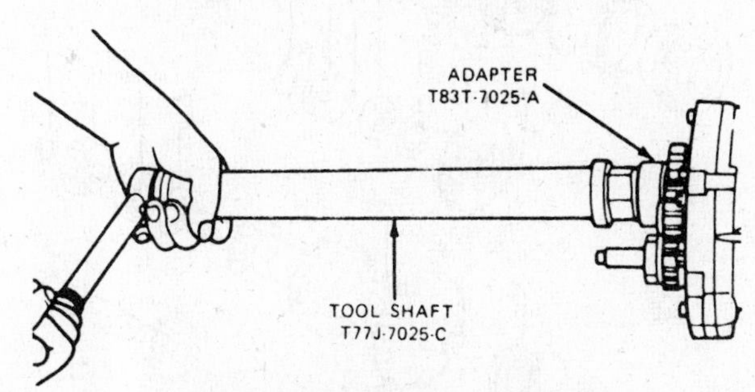

Mainshaft locknut removal

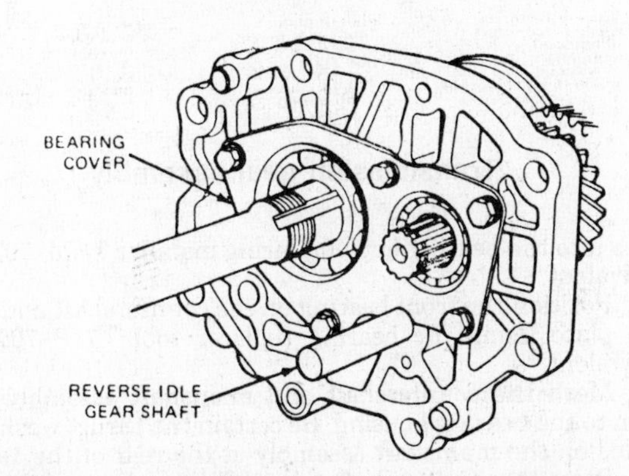

Bearing cover and reverse idler gear removal

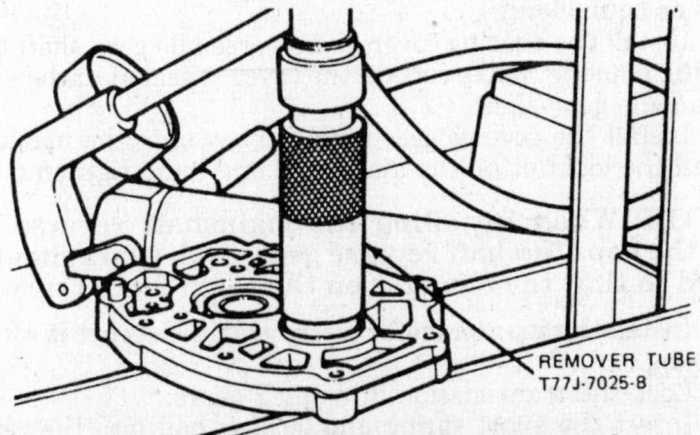

Rear countershaft bearing removal

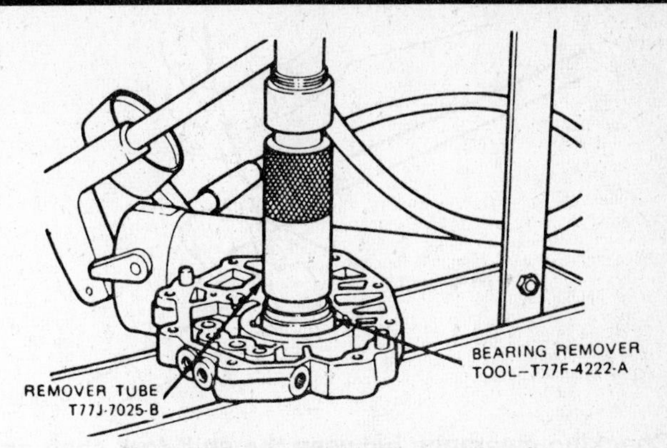

Rear mainshaft bearing removal

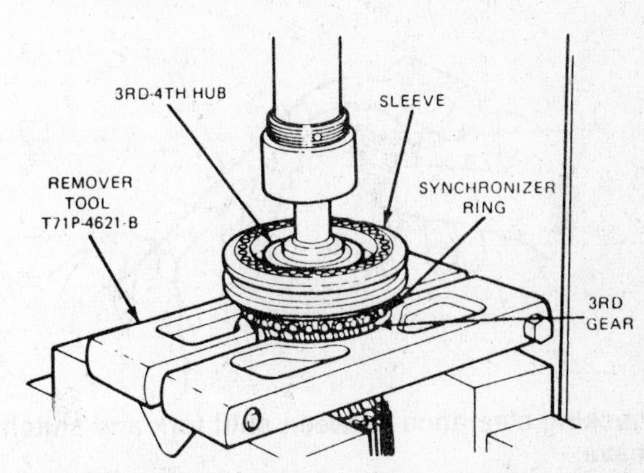

3rd–4th clutch hub and 3rd gear assembly removal

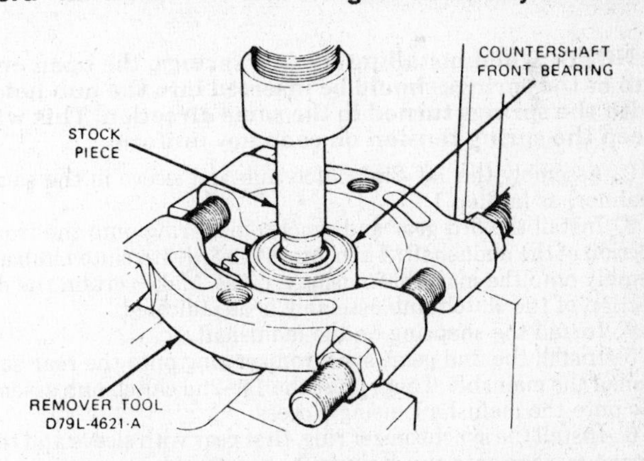

Countershaft front bearing removal

clearance between the side faces of the synchronizer ring and the gear with a feeler gauge. If less than 0.8mm, replace the synchronizer ring.

13. Check the clearance between the shift fork and the clutch sleeve. The clearance should be less that 0.5mm.

ASSEMBLY

1. Lubricate and assemble the 3rd–4th clutch by installing the clutch hub and synchronizer into the sleeve. Place the 3 synchronizer keys into the clutch hub key slots and install the key springs to the clutch hub.

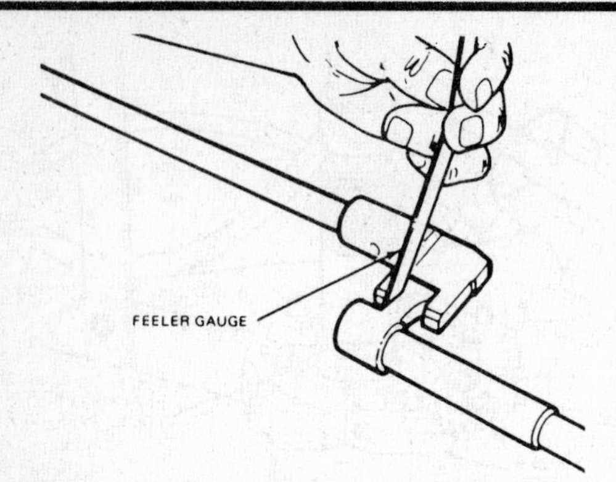

Checking clearance between the shift fork shaft and control lever

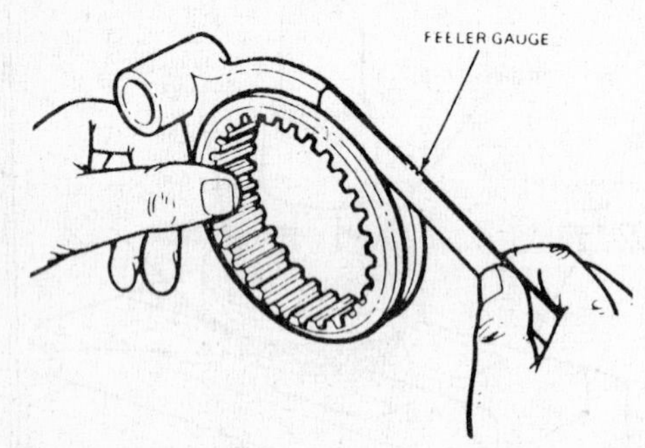

Checking clearance between shift fork and clutch sleeve

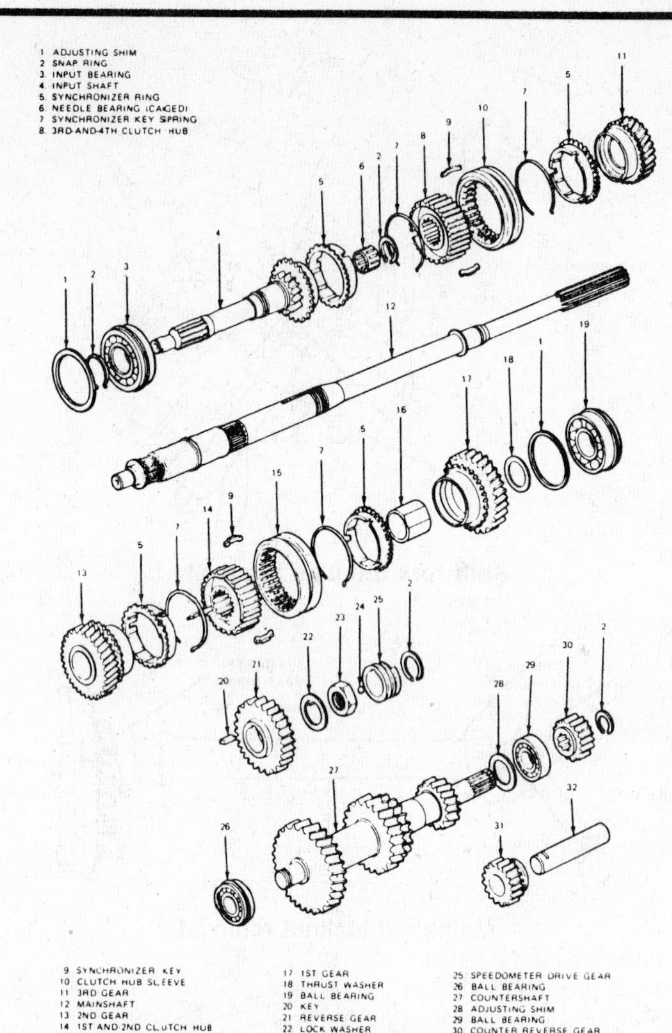

Transmission gear assembly

NOTE: When installing the key springs, the open end tab of the springs should be inserted into the hub holes with the springs turned in the same direction. This will keep the spring tension on each key uniform.

2. Assembly the 1st–2nd clutch hub and sleeve in the same manner, as in Step 1.
3. Install the 3rd gear and synchronizer ring onto the front section of the mainshaft. Then, press the 3rd–4th clutch hub assembly onto the mainshaft, using a press. Make certain the direction of the clutch hub assembly is as indicated.
4. Install the snapring on the mainshaft.
5. Install the 2nd gear, synchronizer ring onto the rear section of the mainshft. Then, press the 1st–2nd clutch hub assembly onto the mainshaft, using a press.
6. Install the synchronizer ring, first gear with sleeve and the thrust washer onto the mainshaft.
7. Install the input shaft and the needle roller bearing to the mainshaft.
8. Check the countershaft rear bearing clearance, by measuring the depth (A) of the countershaft bearing bore in the bearing housing with a depth micrometer. Then, measure the countershaft bearing height (B). The difference between the 2 measurements indicates the required shim thickness. The clearance should be less that 0.1mm. Adjusting shims are available in thickness of 0.1mm and 0.3mm.
9. Check the mainshaft bearing clearance in the same manner as for the countershaft rear bearing clearance. The clearance should be less that 0.1mm. Adjusting shims are available in thickness of 0.1mm and 0.3mm.
10. Position the proper shim on the countershaft rear bearing and press into the bearing housing, using installer T77J–7025–B or equivalent.
11. Position the proper shim on the mainshaft bearing and press into the bearing housing, using installer T77J–7025–K or equivalent.
12. Position the front bearing on the countershaft and press it into place using the bearing replacer tool T71P–7025–A or equivalent.
13. Mesh the countershaft and mainshaft assembly and fit them to the bearing housing. Be certain the thrust washer is installed on the mainshaft assembly at the rear of the 1st gear.
14. Press the countershaft assembly into place, while holding the mainshaft assembly. Use the bearing replacer tool, T71P–7025 or equivalent.
15. Install the bearing cover and reverse idle gear shaft to the bearing housing. Make certain the cover is seated in the groove on the idle gear shaft.
16. Install the reverse gear with the key onto the mainshaft. Install the locknut on the mainshaft and hand-tighten.

NOTE: When installing the mainshaft reverse gear and the countershaft reverse gear, both gears should be fitted so that the chamfer on the teeth faces rearward.

17. Install the countershaft reverse gear and secure it with the snapring.
18. Lock the transmission into any 2 gears.
19. Insert the short spring and locking ball into the reverse bore of the bearing housing.

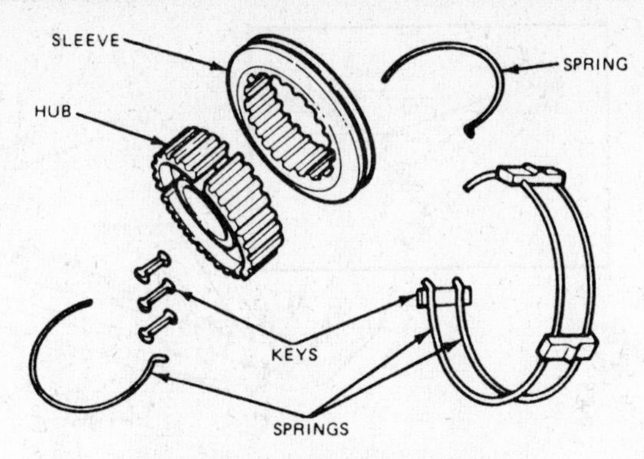

3rd–4th clutch assembly

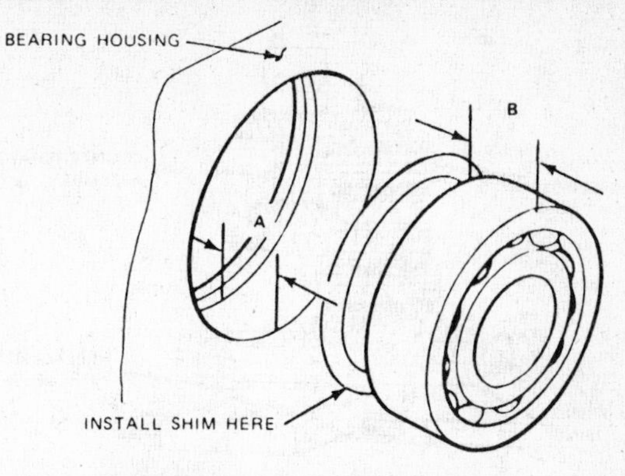

Mainshaft bearing clearance

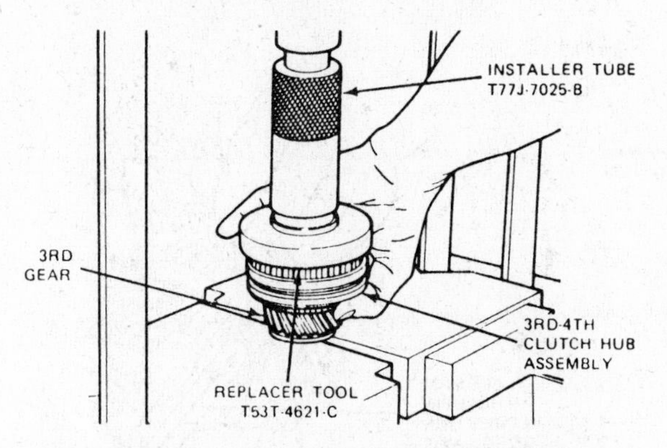

3rd–4th clutch hub assembly, installation

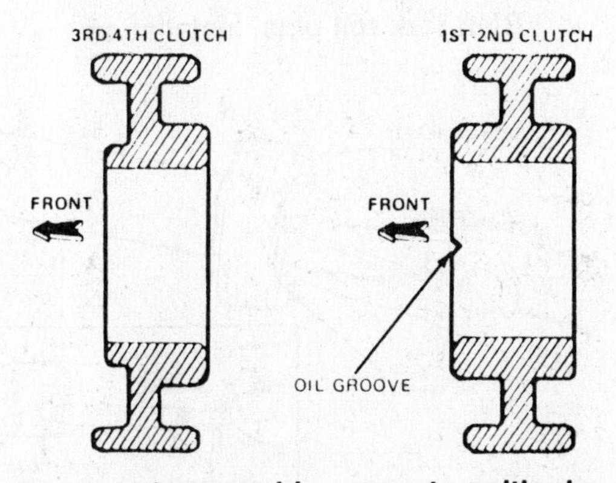

Clutch hub assembly, correct positioning

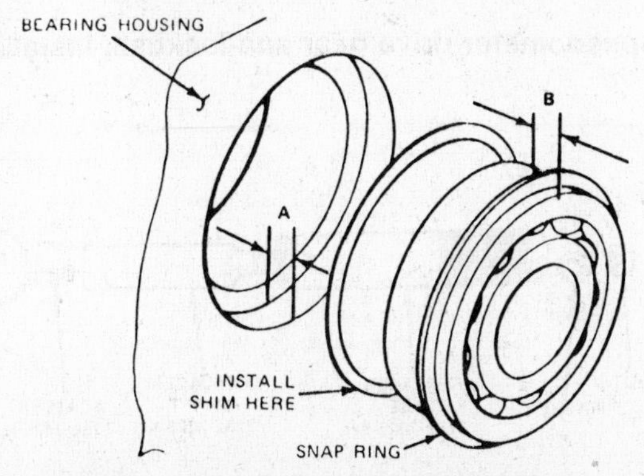

Countershaft rear bearing clearance

20. Hold the locking ball down with a punch or suitable tool. Then, install the reverse shift rod and shift lever assembly with the reverse idle gear.
21. Using the dummy shift rails, install each shift fork rod and interlock pins.
22. Install the 1st–2nd shift fork and 3rd–4th shift fork to their respective clutch sleeves.
23. Align the roll pin holes of each shift fork and rod. Then, install the new roll pins.
24. Install the shift locking balls and springs into their respective positions and install the spring cap bolt.

NOTE: The short spring and ball are installed in the reverse bore.

25. Apply a thin coat of silicone sealer on both contacting surfaces of the bearing housing. Then, install the bearing housing assembly to the transmission case.
26. Temporarily attach the bearing housing to the transmission with 2 top and bottom bolts. Then, tighten the extension housing retaining bolts allowing the countershaft front bearing to be position in the bore.

NOTE: It may be necessary to remove the shift rod bore plugs form the bell housing to align the shift rods. After installation of the bearing housing assembly is completed, reinstall the plugs using silicone sealer.

27. Using the adapter and tool shaft, tighten the mainshaft locknut to 116–174 ft. lbs. (160–240Nm). Then, bend the tab on the lockwasher over the locknut.
28. Install the speedometer drive gear with the lock ball onto the mainshaft and secure it with a snapring.
29. With the outer snapring in place on the main driveshaft front bearing, place the bearing, shim 389117–2S and adapter tool over the input shaft. Thread the replacer shaft onto the adapter tool and install the replacer tube over the replacer shaft. Then, install the nut and washer on the forcing screw.
30. Check that all tool assembly is aligned, then slowly tighten the nut until the adapter is secure on the input shaft.
31. Tighten the nut on the forcing screw until the bearing outer snapring is seated against the housing. Remove the bearing installation tools.
32. Install the input shaft snapring.
33. Install the speedometer driven gear assembly to the extension housing and attach the bolt and lock plate.
34. Insert the shift control lever from the front side of the extension housing and attach the control lever end to the control lever. Tighten the attaching bolts 20–25 ft. lbs. (28–34Nm).
35. Install the back-up lamp switch and neutral sensing switch to the extension housing and tighten the switches to 20–25 ft. lbs. (28–34Nm).
36. Remove the bolts that was previously used to temporarily hold the bearing housing.

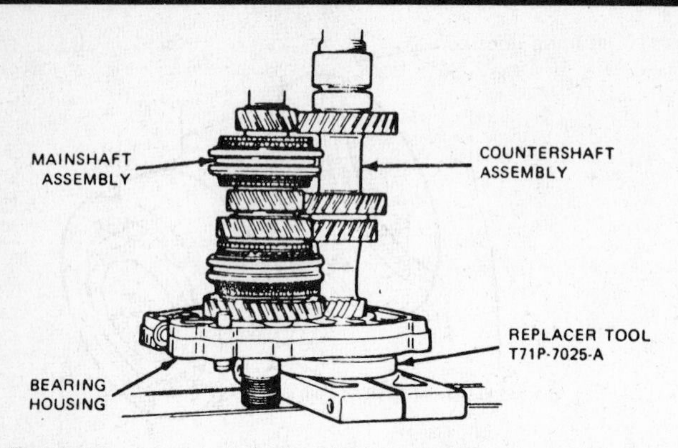

Pressing countershaft assembly into the bearing housing

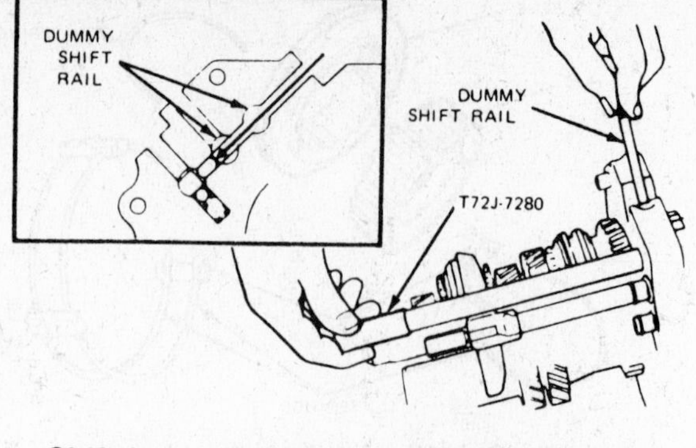

Shift fork rod interlock pins, installation

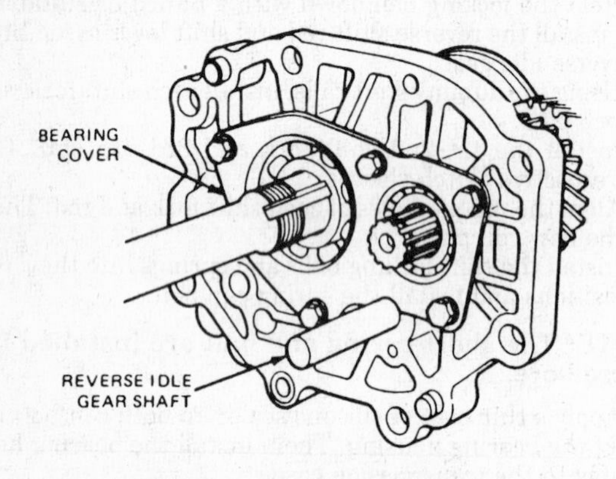

Bearing cover installation

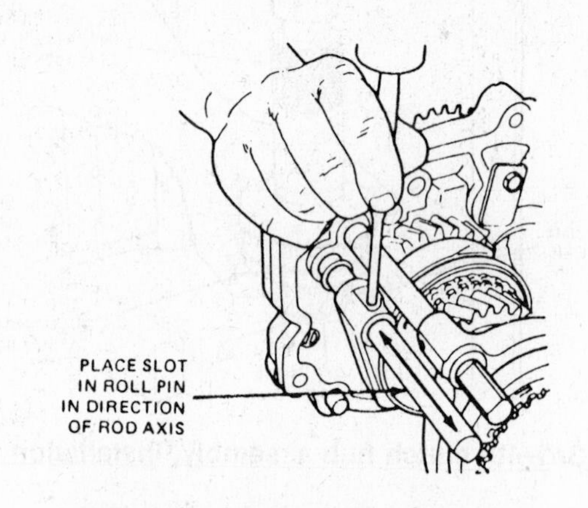

Shift fork roll pins, installation

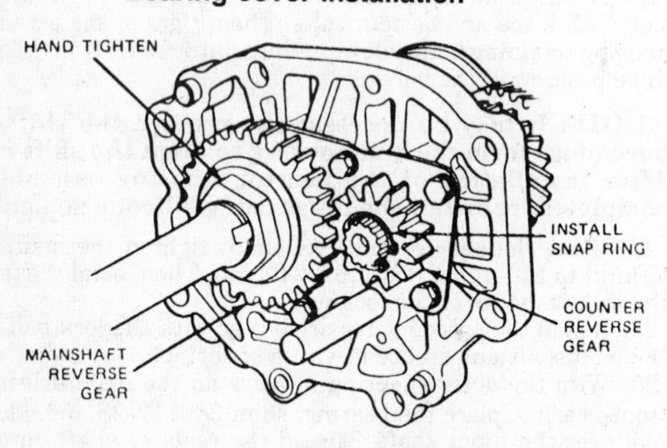

Reverse gear installation

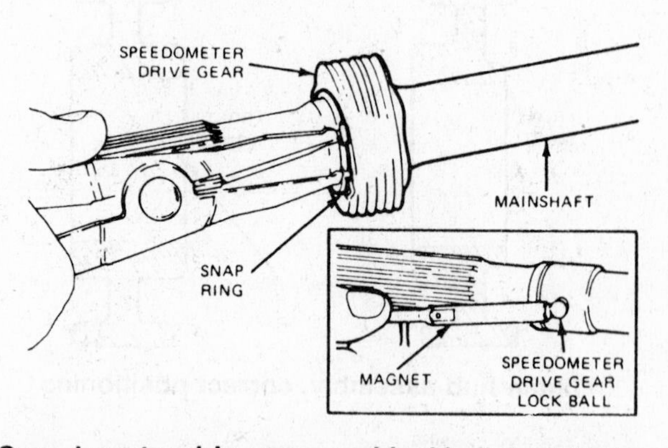

Speedometer drive gear and lockball, installation

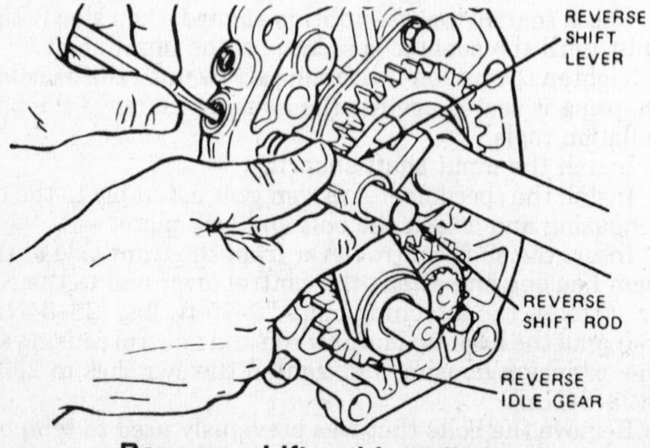

Reverse shift rod installation

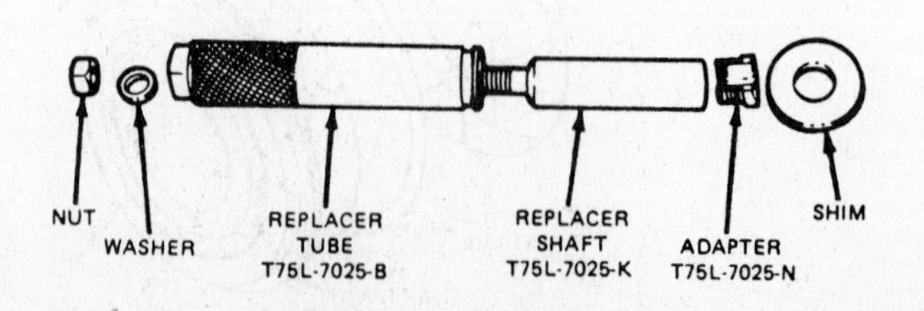

Input shaft front bearing installation tool

37. Apply a thin coat of silicone sealer on the contacting surface of the bearing housing and extension housing. Then, install the extension housing to the bearing housing, with the control lever laid down to the left as far as it will go.

38. Install the extension housing retaining bolts and tighten to 41–59 ft. lbs. (56–79Nm). Check that the control rod operates properly.

39. Install the gearshift lever retainer and gasket to the extension housing. Tighten the retaining bolts to 41–59 ft. lbs. (56–79Nm). Install the gearshift lever, if required.

40. Insall a new oil seal in the front cover, using installer tool T71P-7050–A or equivalent.

41. Apply gear lubricant to the lip of the oil seal and install the front cover to the transmission case. Check the front cover clearance. If necessary, the clearance can be adjusted by inserting an adjusting shim of 0.15mm.

NOTE: When the front cover is installed, the clearance between the bearing outer race and the front cover should be less than 0.1mm.

42. Install the release bearing and release fork.

Mazda (Toyo Kogyo) 5-Speed Transmission

DISASSEMBLY

Before disassembly, clean the exterior of the transmission assembly before any attempt is made to disassemble it, in order to prevent dirt or other foreign material from entering the transmission assembly and damaging its internal parts. If steam cleaning is done to the exterior of the transmission, immediate disassembly should be done to avoid rusting, caused by condensation forming on the internal parts.

1. Remove the nuts attaching the bellhousing to the transmission case. Remove the bellhousing and gasket.
2. Remove the drain plug and drain the lubricant from the transmission into a suitable container. Clean the metal filings from the magnet of the drain plug, if necessary. Reinstall the drain plug.
3. Secure the transmission in a suitable holding fixture.
4. Place the transmission in **N**.
5. Remove the speedometer sleeve and driven gear assembly from the extension housing.
6. Remove the bolts and nuts attaching the extension housing to the transmission case.

NOTE: There are 2 longer outer bolts and 1 shorter center (bottom) bolt used.

7. Raise the control lever to the left and slide it toward the rear of the transmission.
8. Slide the extension housing off of the mainshaft.
9. Pull the control lever and rod out of the front end of the extension housing.
10. If required, remove the backup lamp switch from the extension housing.
11. Remove the anti-spill seal from the mainshaft and discard it. A seal is not necessary for assembly.
12. Remove the snapring that secures the speedometer drive gear to the mainshaft. Slide the drive gear off the mainshaft and remove the lockball.

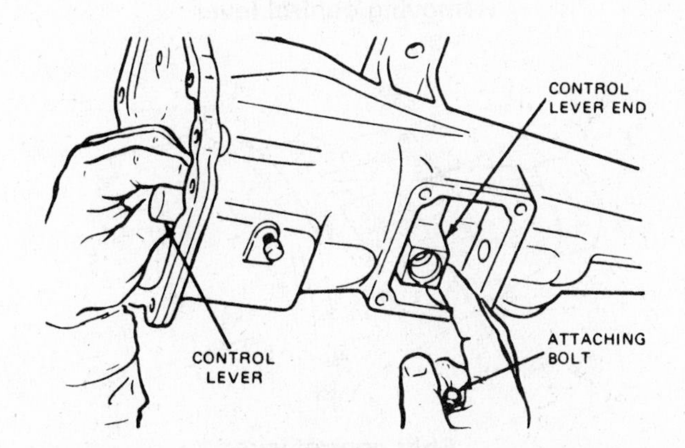

Shift control lever and lever end, installation

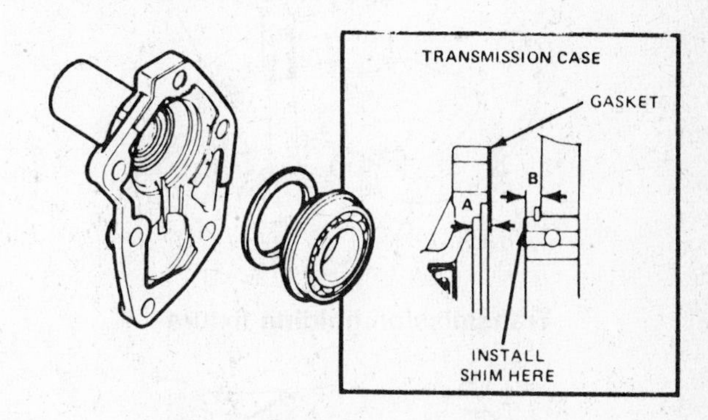

Checking front cover clearance

SPECIFICATIONS

SPECIAL SERVICE TOOLS

Tool Number	Description	Tool Number	Description
T50T-100-A	Impact Slide Hammer	T75L-7025-K	Shaft Sleeve Replacer
T57L-500-B	Bench Mounted Holding Fixture	T75L-7025-N	Shaft Collar
T75P-3504-G	Bearing Race	T75L-7025-Q	Dummy Bearing
D80P-4201-A	Depth Micrometer	T77J-7025-B	Remover Tube (5-1 2")
T77F-4222-A	Diff. Bearing Cup Replacer	T77J-7025-C	Locknut Wrench
D79L-4621-A	Pinion Bearing Cone Remover	T77J-7025-F	Staking Tool
T53T-4621-C	Pinion Bearing Cone Replacer	T77J-7025-K	Center Bearing Replacer
T57L-4621-B	Pinion Bearing Cone Replacer	T77J-7025-L	Countershaft Front Bearing Replacer
T62F-4621-A	Pinion Bearing Cone Replacer	T83T-7025-A	Locknut Wrench Adapter
T71P-4621-B	Pinion Bearing Cone Replacer	T71P-7050-A	Input Shaft Seal Replacer
T71P-7025-A	Output Shaft Bearing Replacer	T71P-7095-A	Extension Housing Seal Remover
T75L-7025-B	Remover Replacer Tube (Short)	T83T-7111-A	Countershaft Remover
T75L-7025-E	Bearing Collet	T72J-7280	Dummy Shift Rails (2) & Interlock Pins
T75L-7025-G	Bearing Collet Sleeve	T83T-7697-A	Extension Housing Bushing Remover
T75L-7025-J	Forcing Screw		

TRANSMISSION ASSEMBLY TORQUES

Attachment	Torque Ft-Lbs	Torque N·m
Cap — Shift Rail Detent Spring	29-43	40-58
Nut — Main Shaft Gear Retaining	116-174	160-240
Pivot — Clutch Release Lever	23-34	32-46
Plug — Drain	29-43	40-58
Plug — Filler	18-29	25-39
Switch — Back-Up Lamp	22-29	30-39
Bolt — Control Lever End to Control Lever	20-25	28-34
Nut/Bolt Size		
6mm	5-7.5	7-11
8mm	12-17	17-23
10mm	23-34	32-45
12mm	41-59	56-79

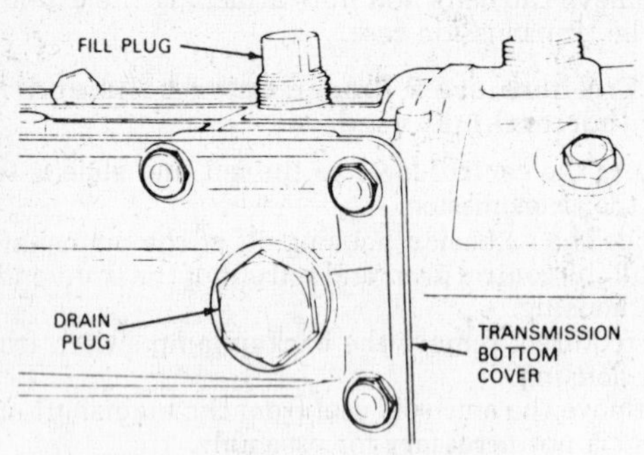

Transmission drain plug

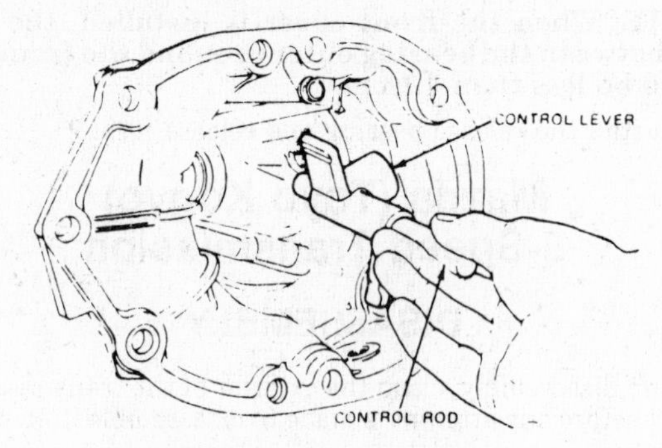

Removing control lever

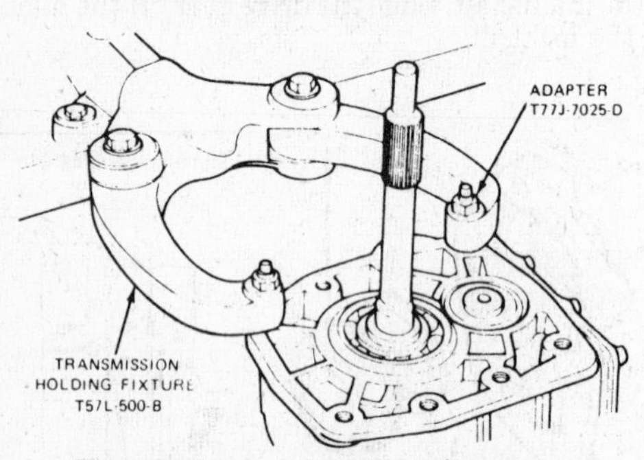

Transmission holding fixture

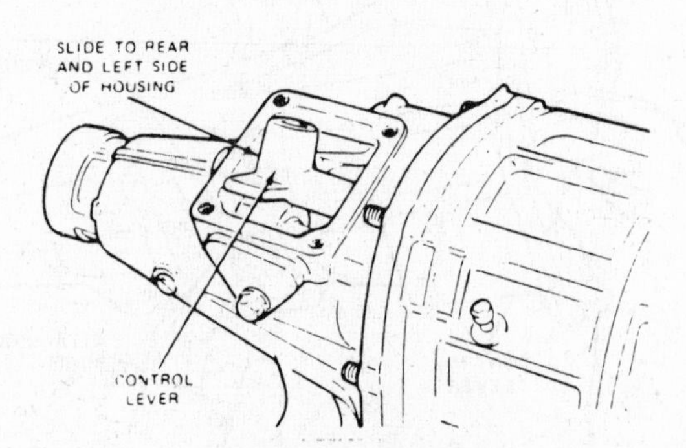

Shift control lever

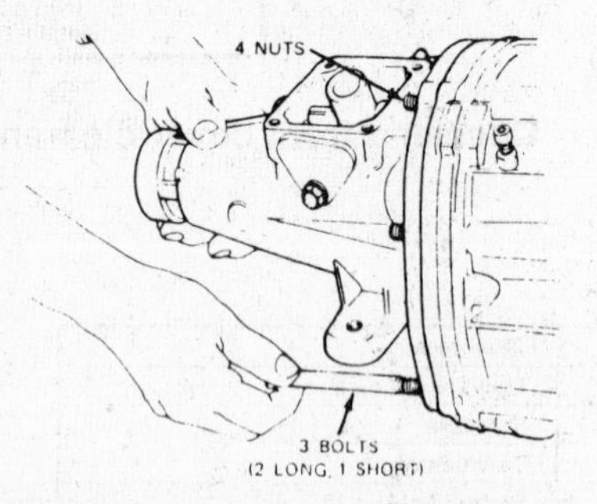

Extension housing mounting

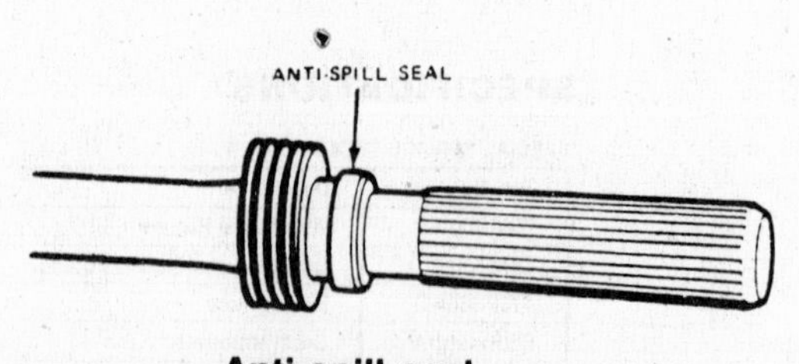

Anti-spill seal

13. Evenly loosen the bolts securing the transmission case cover to the transmission case. Remove the cover and gasket.
14. Mark the shift rails and forks to aid during transmission assembly. Remove the roll pins attaching the shift rod ends to the shift rod and remove the shift rod ends.
15. Pry the bearing housing away from the transmission case without damaging the housing or case. Slide the bearing housing off of the mainshaft.
16. Remove the snapring and washer retaining the mainshaft rear bearing to the mainshaft.
17. Using the proper tools, remove the mainshaft rear bearing.
18. Remove the snapring from the rear end of the countershaft and using the proper tools, remove the countershaft rear bearing.
19. Remove the counter 5th gear and spacer from the rear of the countershaft.

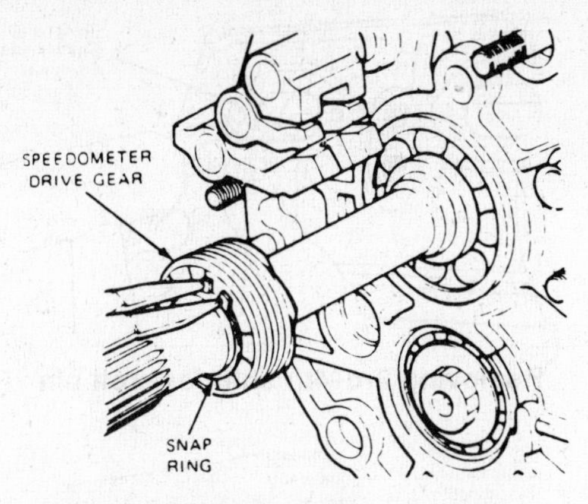

Removing speedometer drive gear

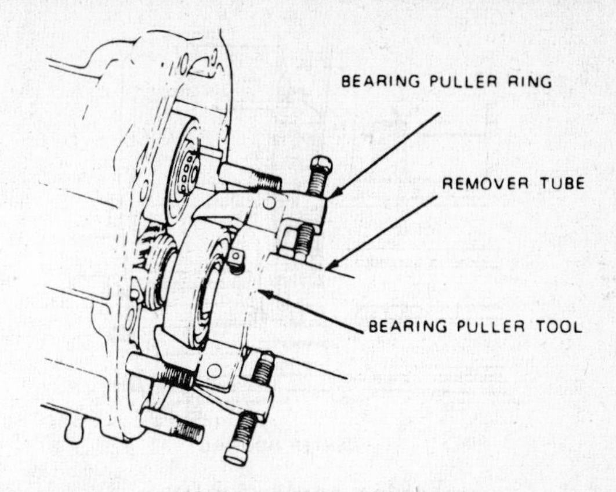

Installing mainshaft rear bearing puller tools

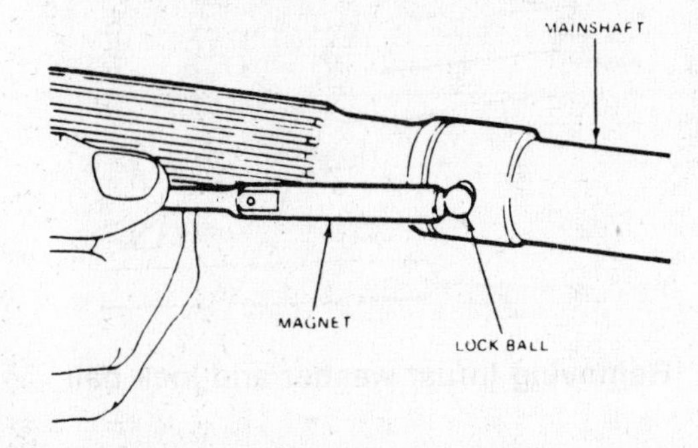

Removing speedometer drive gear lock ball

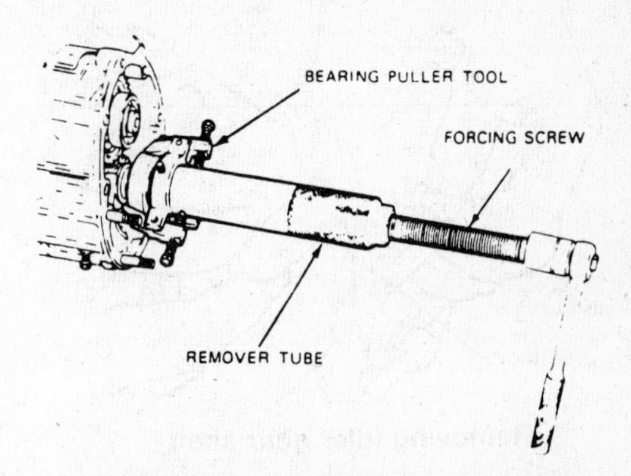

Removing mainshaft rear bearing

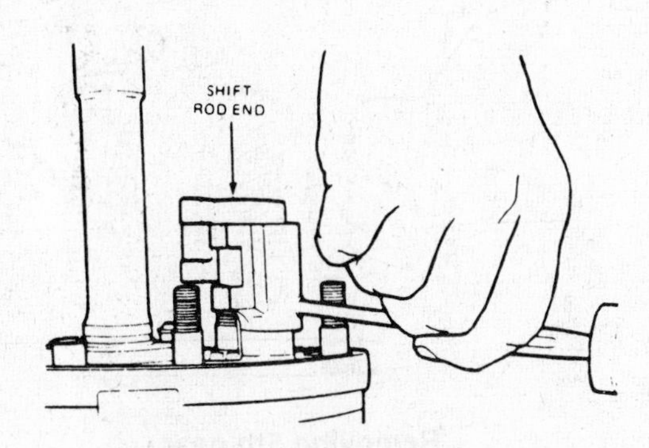

Removing shift rod end

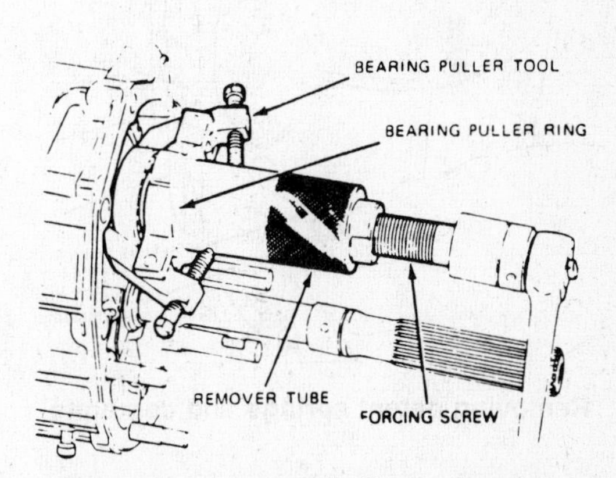

Removing countershaft rear bearing

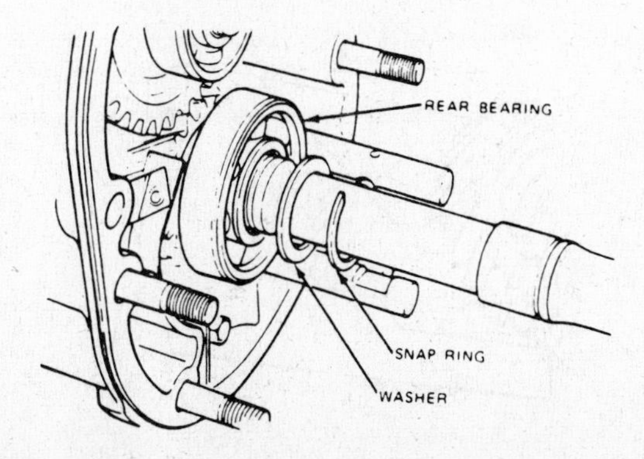

Mainshaft snapring and washer removal

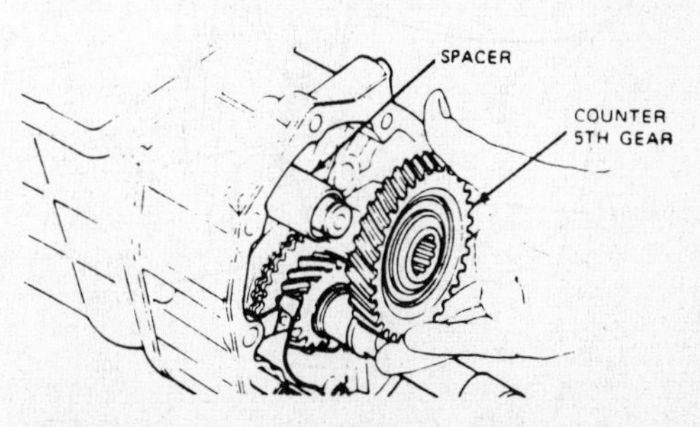

Removing counter 5th gear and spacer

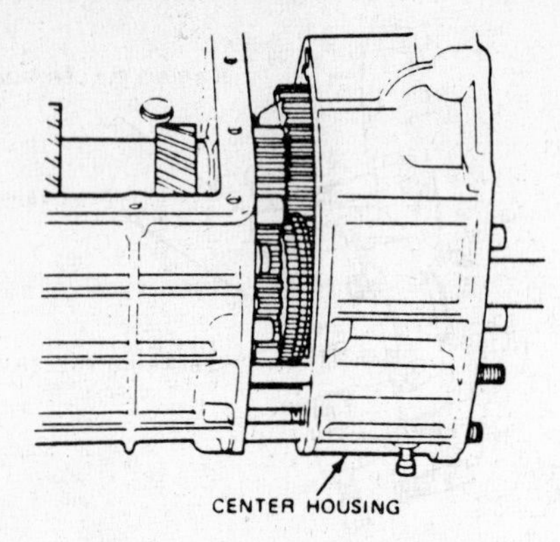

removing center housing

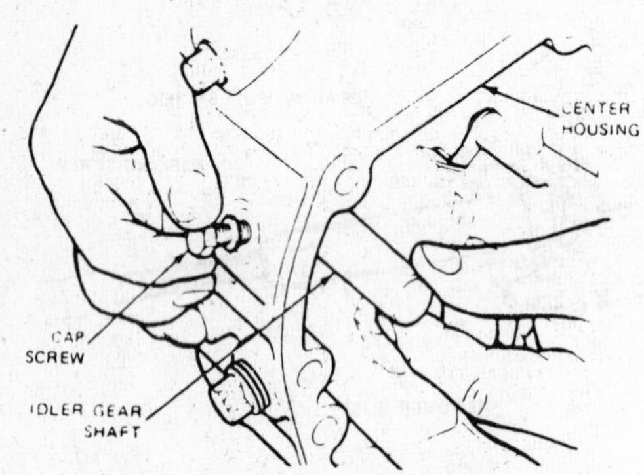

Removing idler gear shaft

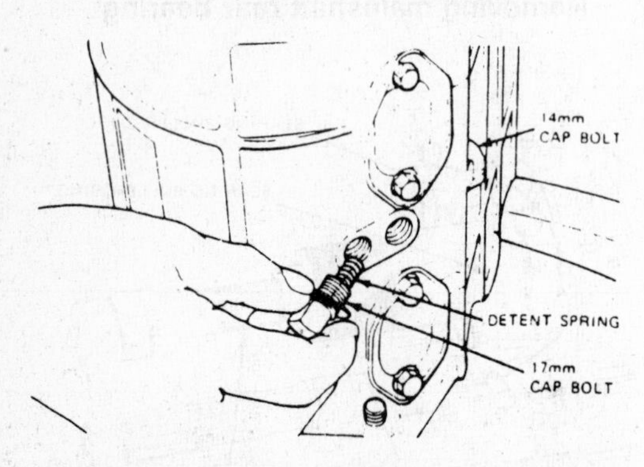

Removing detent springs and cap bolts

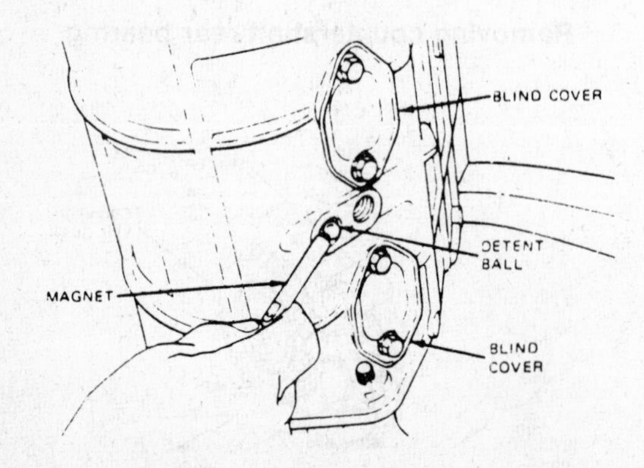

Removing detent balls

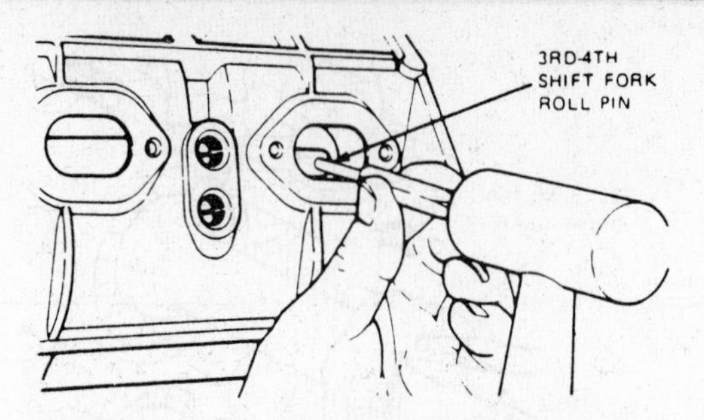

Removing 3rd–4th shift fork roll pin

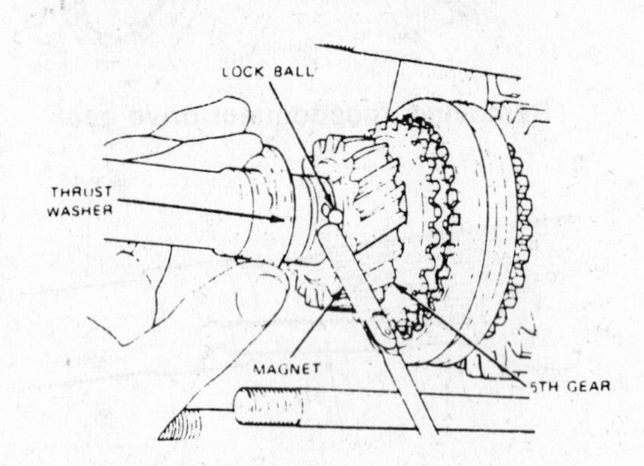

Removing thrust washer and lock ball

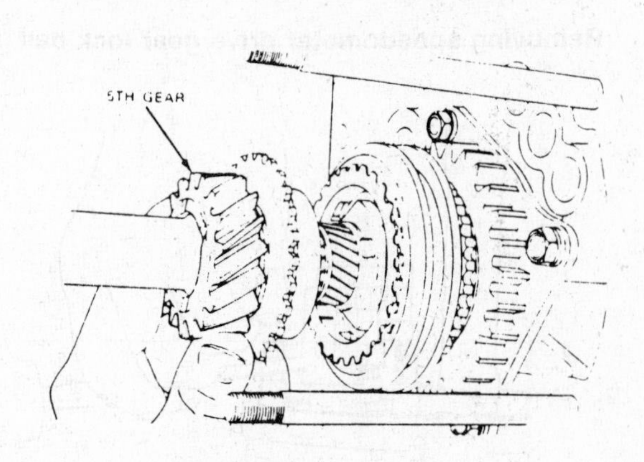

Removing 5th gear

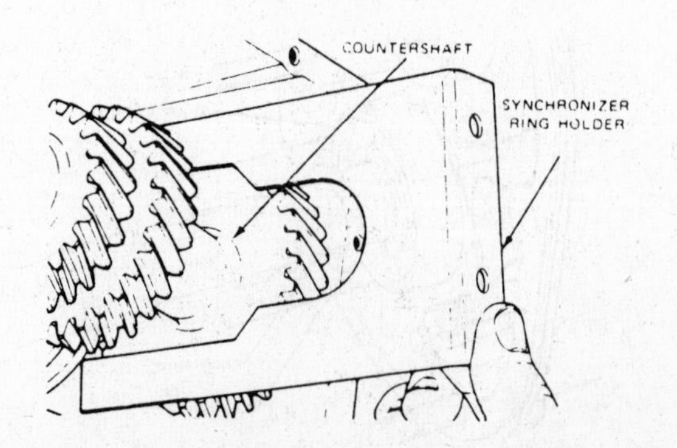

Installing synchronizer ring holder

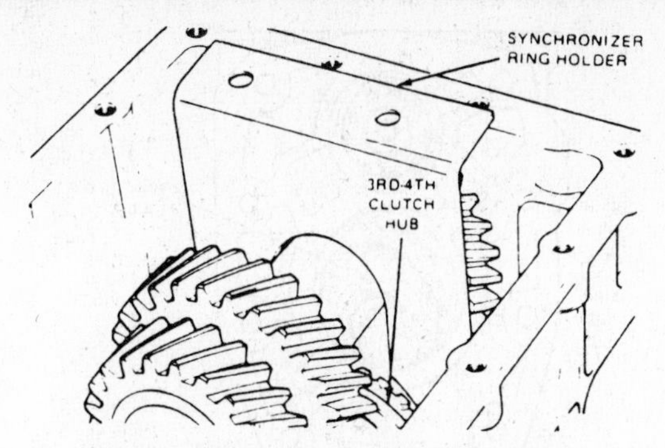

Synchronizer ring holder installed

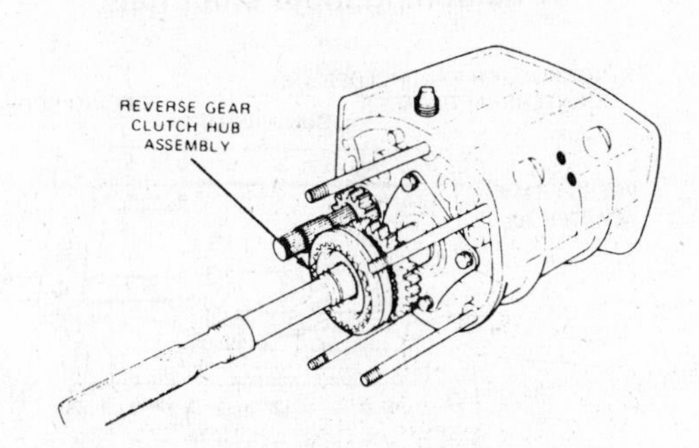

Removing reverse gear and clutch hub

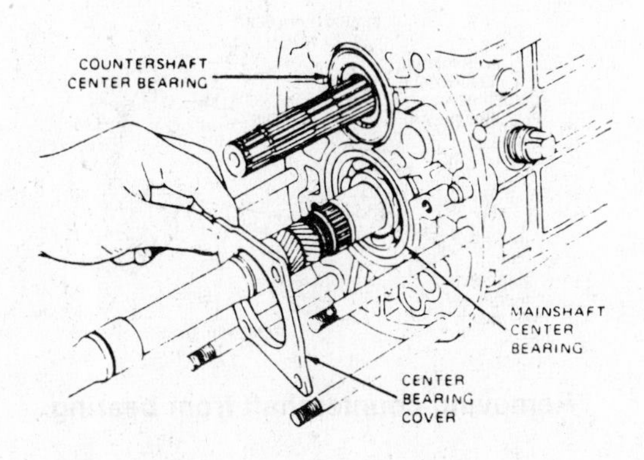

Removing center bearing cover

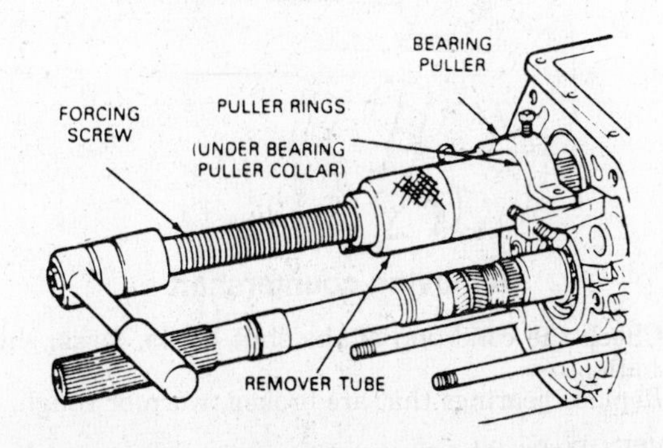

Removing countershaft center bearing

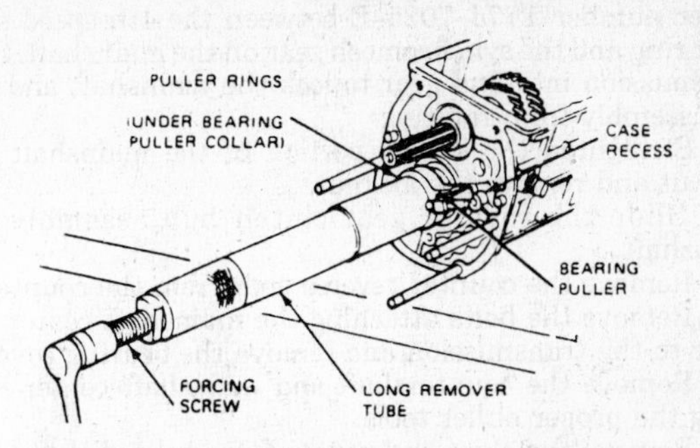

Removing mainshaft center bearing

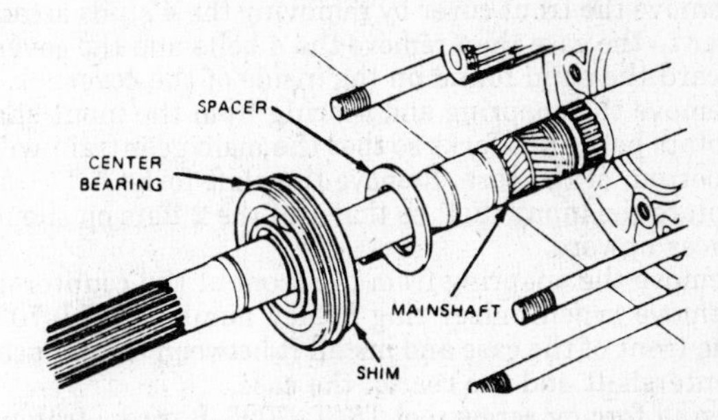

Mainshaft center bearing shim and spacer

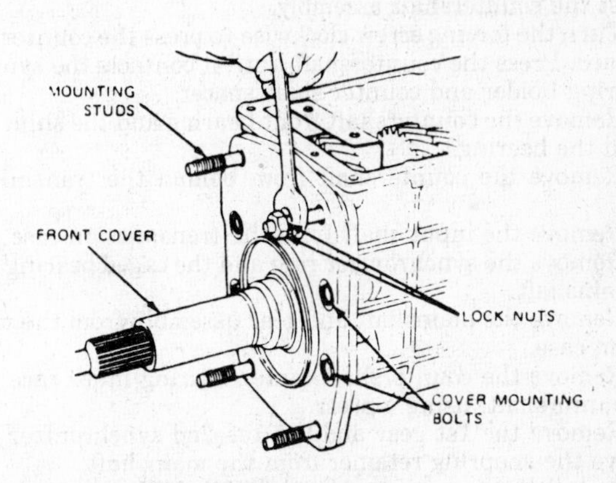

Front cover removal

20. Remove the center housing. Remove the reverse idler gear and 2 spacers with the housing.
21. Remove the screw from the center housing and remove the idler gear shaft.
22. Remove the 3 spring bolts, 2 on the case upper portion and 1 on the case side. Remove the detent springs and balls from the transmission case.
23. Remove the 4 bolts attaching the blind covers to the transmission case and remove the blind covers and gaskets.
24. Remove the roll pin from the 5th–reverse shift fork and slide the shift fork shaft out of the transmission case.
25. Shift the transmission into 4th gear. This will provide adequate space to drive out the 3rd–4th shift fork roll pin. Slide the 3rd–4th shift fork shaft out of the rear of the transmission case.
26. Remove the roll pin from the 1st–2nd shift fork and slide the shift fork shaft assembly out of the rear of the transmission case. Remove both interlock pins.
27. Remove the snapring that secures the 5th gear to the mainshaft.
28. Remove the thrust washer and lock ball, 5th gear and synchronizer ring from the rear of the mainshaft.
29. Install the synchronizer ring holder and countershaft

spacer number T77J–7025–E between the 4th speed synchronizer ring and the synchromesh gear on the mainshaft. Shift the transmission into 2nd gear to lock the mainshaft and prevent the assembly from rotating.

30. Straighten the staked portion of the mainshaft bearing locknut and remove the locknut.
31. Slide the reverse gear clutch hub assembly off the mainshaft.
32. Remove the counter reverse gear from the countershaft.
33. Remove the bolts attaching the mainshaft center bearing cover to the transmission and remove the bearing cover.
34. Remove the countershaft and mainshaft center bearings using the proper puller tools.
35. Remove the shim and spacer from behind the mainshaft rear bearing along with the bearing.
36. Remove the front cover by removing the 4 studs attaching the cover to the case then remove the 4 bolts and the cover. Do not discard the shim found on the inside of the cover.
37. Remove the snapring and bearing from the input shaft.
38. Rotate both shift forks so that the main gear train will fall to the bottom of the case. Remove the shift forks.
39. Rotate the input shaft so that 1 of the 2 flats on the input shaft faces upward.
40. Remove the snapring from the front of the countershaft.
41. Remove synchronizer ring holder number T77J–7025–E from the front of the case and install it between the 1st gear on the countershaft and the rear of the case.
42. Install forcing screw tool T84T–7025–B, press frame tool T77J–7025–N and press frame adapter tool T82T–7003–BH against the countershaft assembly.
43. Turn the forcing screw clockwise to press the countershaft rearward. Press the countershaft until it contacts the synchronizer ring holder and counter shaft spacer.
44. Remove the countershaft front bearing and the shim from behind the bearing.
45. Remove the countershaft from behind the transmission case.
46. Remove the input shaft from the transmission case.
47. Remove the synchronizer ring and the caged bearing from the mainshaft.
48. Remove the mainshaft and gear assembly from the transmission case.
49. Remove the countershaft center bearing inner race from the countershaft, using a press.
50. Remove the 1st gear and the 1st–2nd synchronizer ring. Remove the snapring retainer from the mainshaft.
51. Install bearing remover tool T71P–4621–B between the 2nd and 3rd gear.
52. Press the mainshaft out of the 3rd gear and the 3rd–4th clutch hub sleeve.
53. Press the 1st and 2nd clutch hub and sleeve assembly and the 1st gear sleeve from the mainshaft.

Cleaning and Inspection

1. Wash all parts except the ball bearings and seals, in a suitable cleaning solvent.
2. Dry all parts with compressed air.
3. Do not clean, wash or soak the transmission seals in cleaning solvent.
4. Lubricate the bearings and wrap them in a clean, lint free cloth until ready for use.
5. Inspect the transmission case and housing for cracks, worn or damaged bores, damaged threads or any other damage that could affect the operation of the transmission.
6. Inspect the machined mating surfaces for burrs, nicks or damage.
7. Inspect the front face of the case for small nicks or burrs that could cause misalignment of the transmission with the flywheel housing.
8. Inspect the bellhousing for cracks.

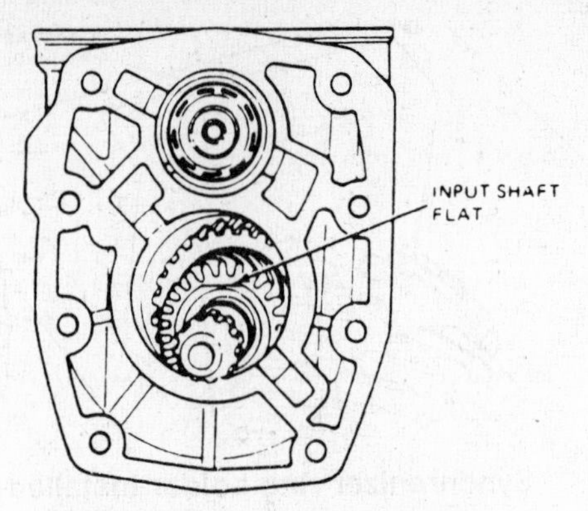

Positioning input shaft flats

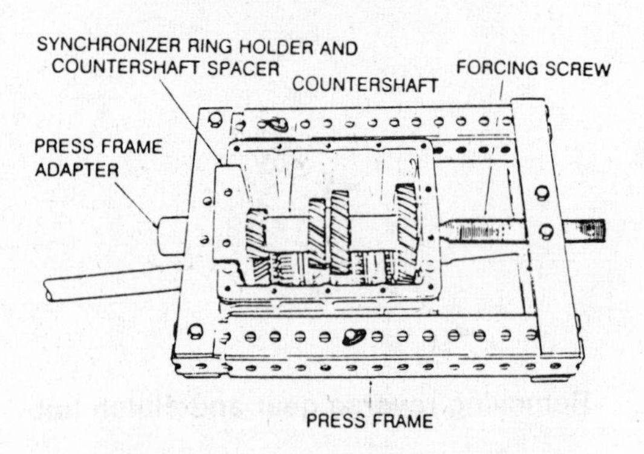

Pressing countershaft rearward

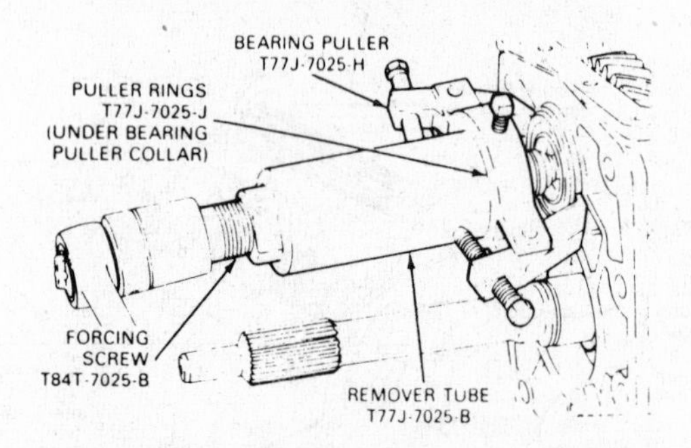

Removing countershaft front bearing

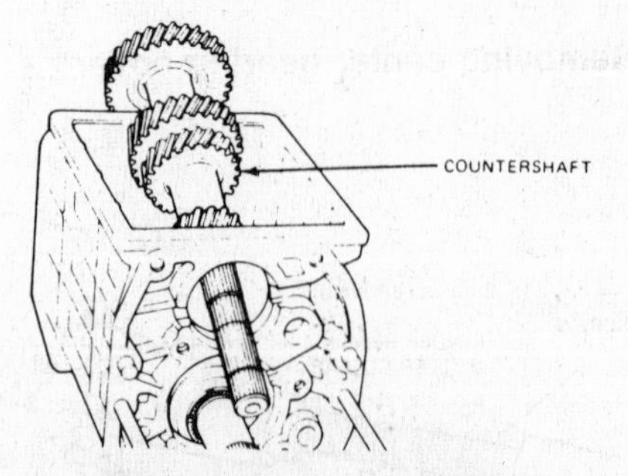

Removing countershaft

9. Check the condition of the shift levers, forks, shift rails and shafts.
10. Replace bearings that are broken, worn or rough.

NOTE: Before beginning the assembly procedure, the following 3 measurement checks must be performed:

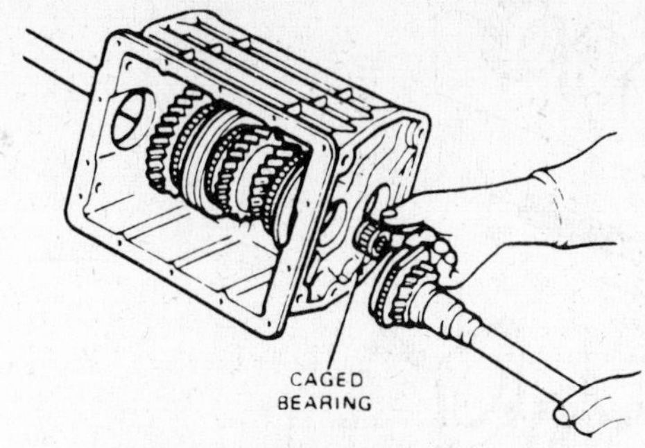

Input shaft removal

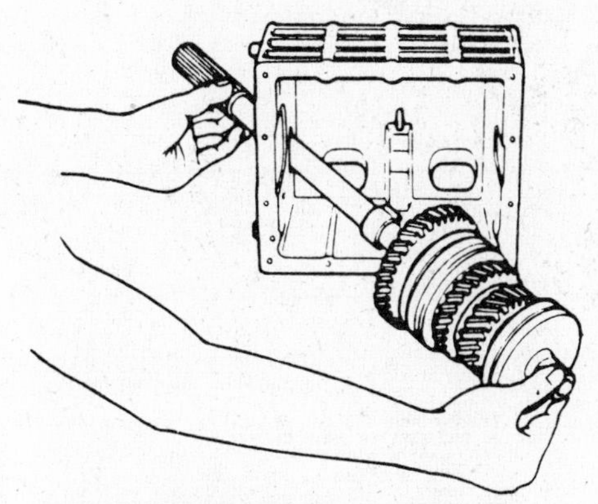

Mainshaft and gear removal

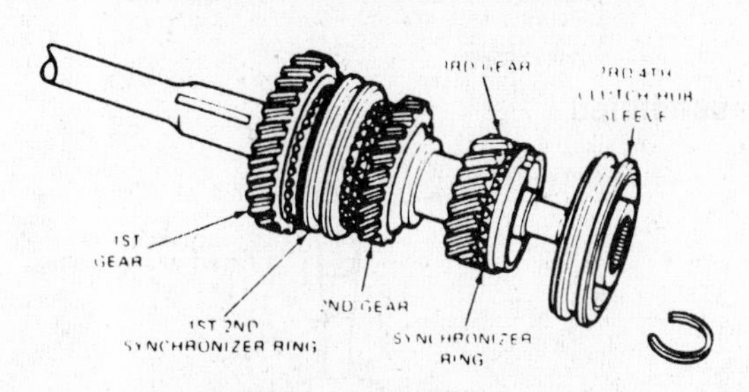

Mainshaft and gear assembly

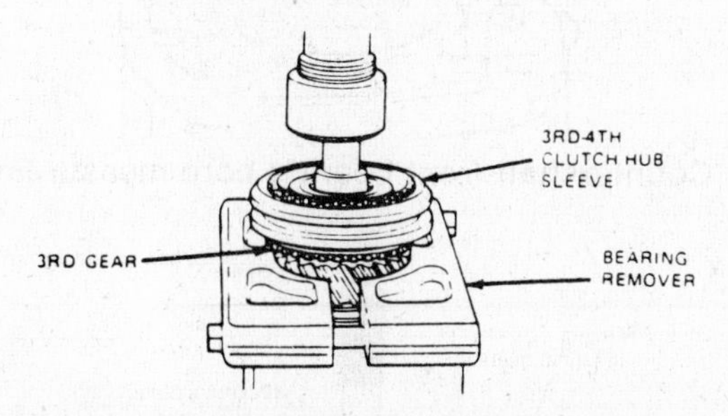

3rd gear and 3rd–4th clutch hub sleeve, removal

Mainshaft Thrust Play

Check the mainshaft thrust play by measuring the depth of the mainshaft bearing bore in the transmission rear cage by using a depth micrometer. Measure the mainshaft rear bearing height. The difference between the 2 measurements indicates the required thickness of the adjusting shim. The standard

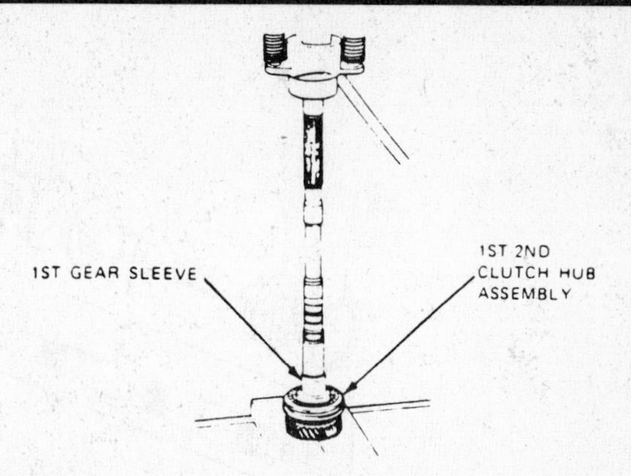

Removing 1st–2nd clutch hub assembly from mainshaft

thrust play is 0–0.1mm. Adjusting shims are available in 0.1mm and 0.3mm sizes.

Countershaft Thrust Play

Check the countershaft thrust play by measuring the depth of the countershaft front bearing bore in the transmission case by using a depth micrometer. Measure the countershaft front bearing height. The difference between the 2 measurements indicates the required thickness of the adjusting shim. The standard thrust play is 0–0.1mm. Adjusting shims are available in 0.1mm and 0.3mm sizes.

Mainshaft Bearing Clearance

Check the mainshaft bearing clearance by measuring the depth of the bearing bore in the clutch adapter plate with a depth micrometer. Make sure the micrometer is on the second step of the plate. Measure the bearing height. The difference between the 2 measurements indicates the required thickness of the adjusting shim. The standard clearance is 0–0.1mm. If an adjusting shim is required, select a shim to bring the clearance to within specifications.

ASSEMBLY

1. Lubricate and assemble the 1st–2nd synchromesh mechanism and the 3rd–4th synchromesh mechanism by installing the clutch hub to the sleeve. Install the 3 synchronizer keys into the clutch hub key slots and install the key springs to the clutch hub.
2. Install the synchronizer ring on the 2nd gear and install the 2nd gear to the mainshaft with the synchronizer ring toward the rear of the shaft.
3. Slide the 1st–2nd clutch hub and sleeve assembly to the mainshaft with the oil grooves of the clutch hub toward the front of the mainshaft.

NOTE: Make sure that the 3 synchronizer keys in the synchromesh mechanism engage the notches in the 2nd synchronizer ring.

4. Insert the 1st gear sleeve on the mainshaft.
5. Install the synchronizer ring on the 3rd gear along with the caged roller bearing and slide the 3rd gear to the front of the mainshaft with the synchronizer ring toward the front.
6. Press the 3rd–4th clutch hub and sleeve assembly to the front of the mainshaft.

NOTE: Make sure that the 3 synchronizer keys in the synchromesh mechanism engage the notches in the synchronizer ring.

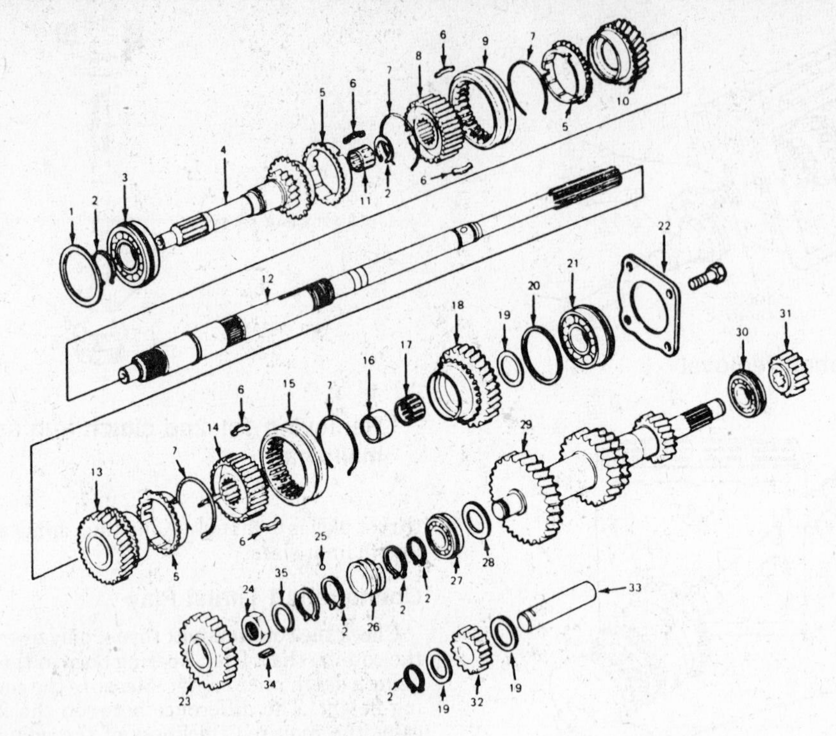

1. SHIM
2. SNAP RING
3. MAIN DRIVESHAFT BEARING
4. MAIN DRIVESHAFT GEAR
5. SYNCHRONIZER RING
6. SYNCHRONIZER KEY
7. SYNCHRONIZER KEY SPRING
8. 3RD-AND-4TH CLUTCH HUB
9. CLUTCH SLEEVE
10. 3RD GEAR
11. CAGED BEARING
12. MAINSHAFT
13. 2ND GEAR
14. 1ST AND 2ND CLUTCH HUB
15. CLUTCH SLEEVE
16. 1ST GEAR SLEEVE
17. BEARING
18. 1ST GEAR
19. THRUST WASHER
20. SHIM
21. MAINSHAFT FRONT BEARING
22. BEARING COVER
23. REVERSE GEAR
24. MAINSHAFT LOCK NUT
25. LOCK BALL
26. SPEEDOMETER DRIVE GEAR
27. COUNTERSHAFT FRONT BEARING
28. SHIM
29. COUNTERSHAFT
30. COUNTERSHAFT REAR BEARING
31. COUNTER REVERSE GEAR
32. REVERSE IDLER GEAR
33. IDLER GEAR SHAFT
34. KEY
35. LOCKWASHER

Shafts and gears disassembled

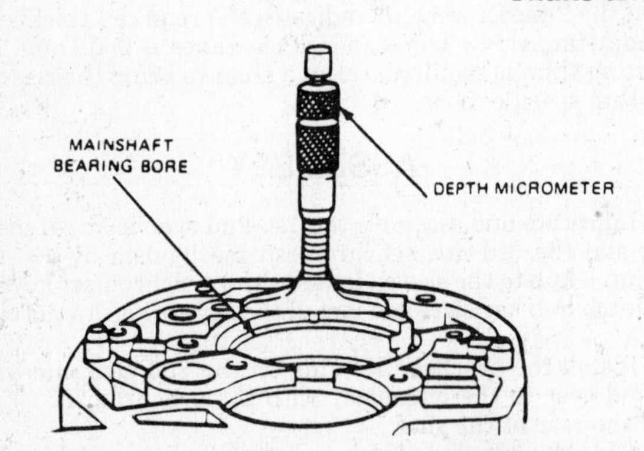

Mainshaft bearing bore depth measurement

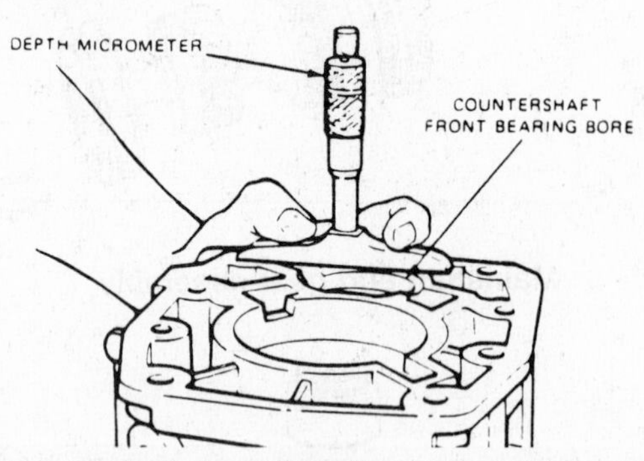

Countershaft front bearing bore measurement

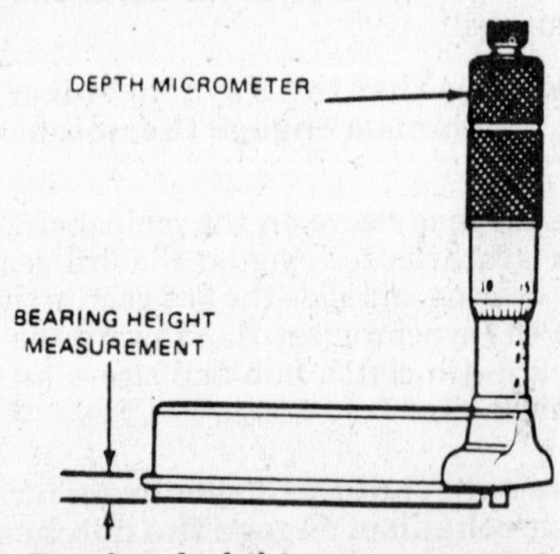

Bearing height measurement

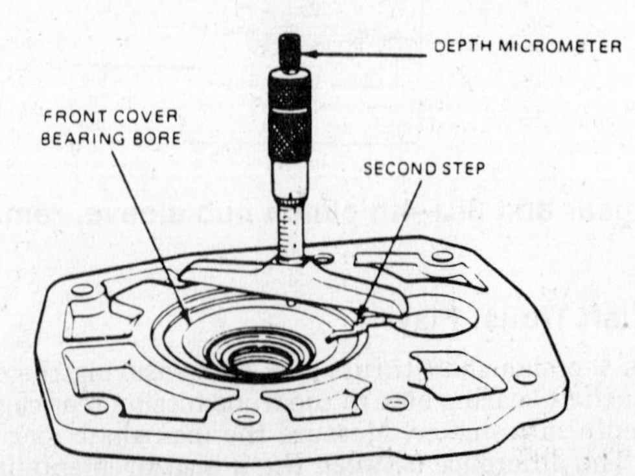

Clutch adapter plate bore measurement

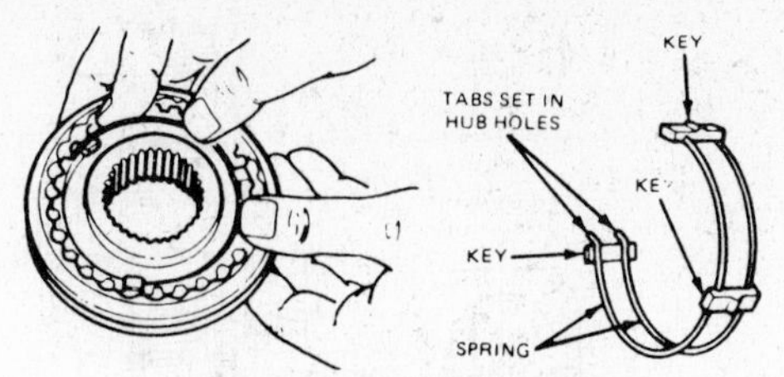

Installing key springs

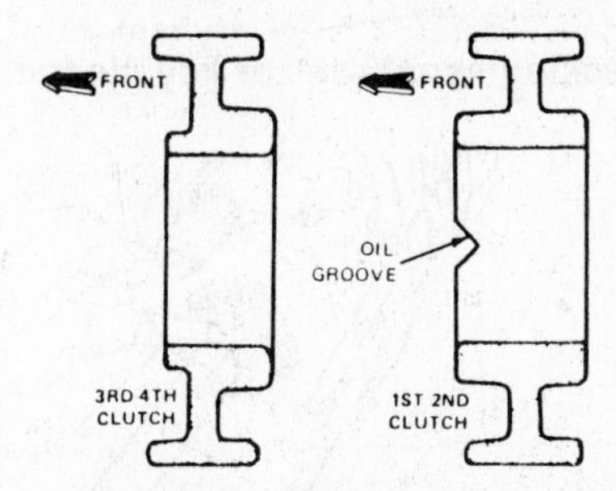

Clutch hub assembly direction

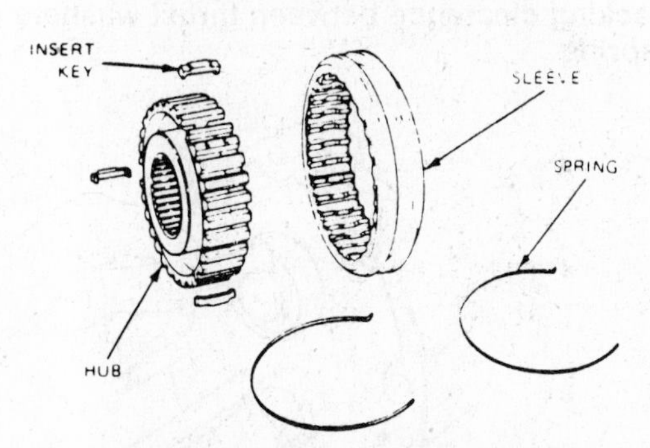

Synchromesh mechanism

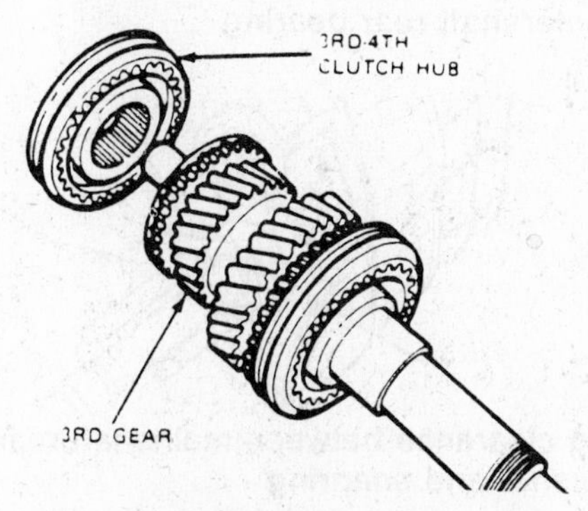

3rd–4th clutch hub installation

7. Install the snapring to the front of the mainshaft.
8. Slide the needle bearing for the 1st gear onto the mainshaft.
9. Install the synchronizer ring on the 1st gear. Slide the 1st gear onto the mainshaft with the synchronizer ring facing the front of the shaft. Rotate the 1st gear as necessary, to engage the 3 notches in the synchronizer ring with the synchronizer keys.
10. Install the original thrust washer to the mainshaft.
11. Install the mainshaft and gear assembly in the case.

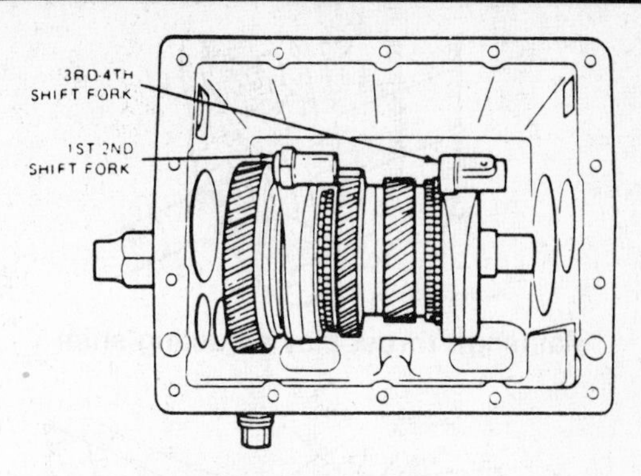

Shift fork installation

12. Install the 1st–2nd and 3rd–4th shift forks in the groove of the clutch hub and sleeve assembly.
13. Install the caged bearing in the front end of the mainshaft.
14. Install the synchronizer ring on the input shaft and install the input shaft to the front end of the mainshaft.

NOTE: Make sure that the 3 synchronizer keys in the 3rd–4th synchromesh mechanism engage the notches in the synchronizer ring.

15. Press the inner race of the countershaft rear bearing onto the countershaft.
16. Install the countershaft gear in the case, making sure that the countershaft gear engages each gear of the mainshaft assembly.
17. Install the shim on the mainshaft center bearing.
18. Install the input shaft bearing and the mainshaft center bearing in their bearing bores. Make sure that the synchronizer and shifter forks have not been moved out of position.
19. Install the input shaft bearing snapring.
20. Install the shim in the countershaft front bearing bore.
21. Install the countershaft front and center bearings in the bores. Properly seat the bearing.
22. Install the snapring to secure the countershaft front bearing.
23. Install the bearing cover to the transmission case and tighten the attaching bolts to 41–59 ft. lbs. (56–79 Nm).
24. Install the reverse idler gear and shaft with a spacer on each side of the shaft.
25. Slide the counter reverse gear (chamfer side forward) and spacer onto the countershaft.
26. Slide the thrust washer, reverse gear, caged roller bearings and clutch hub assembly onto the mainshaft. Install a new locknut hand tight.
27. Shift into 2nd gear and reverse gear to lock the rotation of the mainshaft. tighten the locknut to 115–172 ft. lbs. (156–233 Nm).
28. Install the 4th–3rd clutch sleeve in the 3rd gear.
29. If new synchronizers have been installed, check the clearance between the synchronizer key and the exposed edge of the synchronizer ring with a feeler gauge. If the measurement is greater than 2.0mm, the synchronizer key can pop out of position. To correct this, change the thrust washer between the mainshaft center bearing and the 1st gear. Thrust washers are available in 2.5mm, 3.0mm and 3.5mm.
30. Position the 5th synchronizer ring on the 5th gear. Slide the 5th gear onto the mainshaft with the synchronizer ring toward the front of the shaft. Rotate the 5th gear as necessary, to engage the 3 notches in the synchronizer ring with the synchronizer keys in the reverse and clutch hub assembly.
31. Install the lock ball and the thrust washer on the rear of the 5th gear.
32. Install the snapring on the rear of the thrust washer.

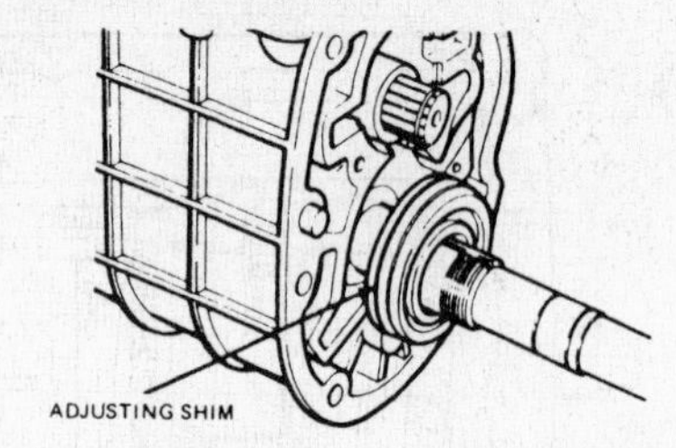

Mainshaft thrust play adjusting shim

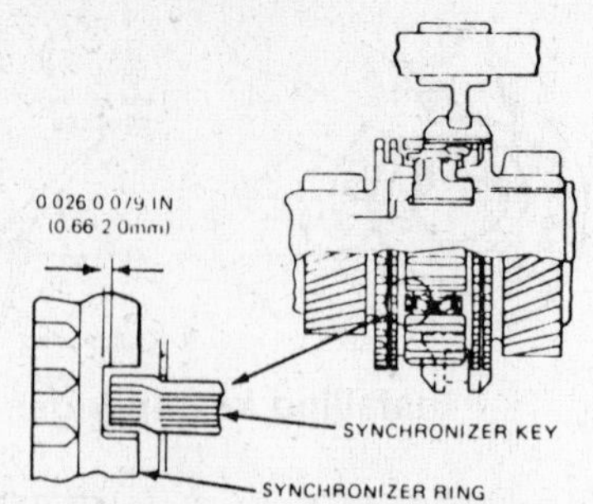

Checking synchronizer key clearance

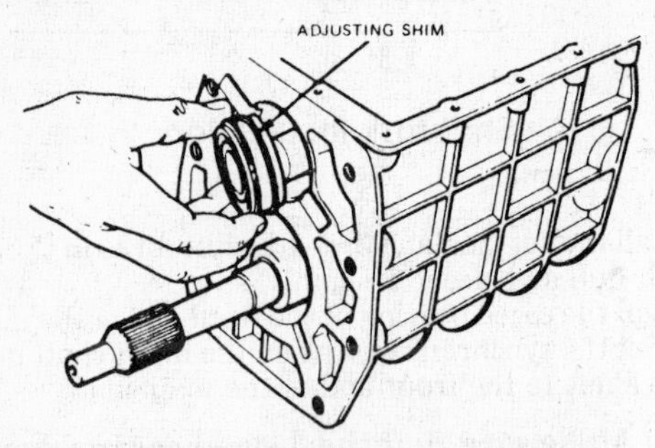

Countershaft play adjusting shim

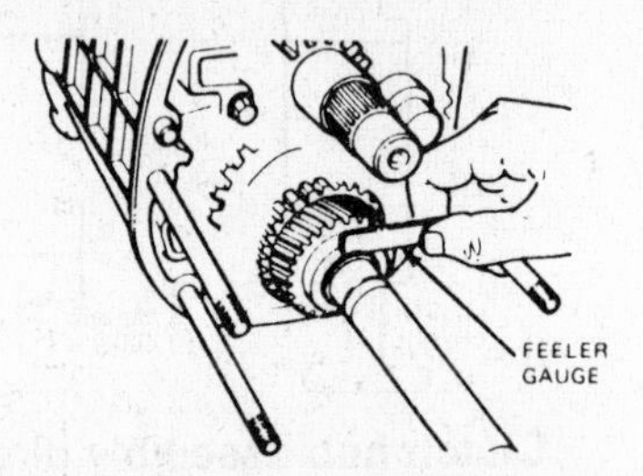

Checking clearance between thrust washers and snapring

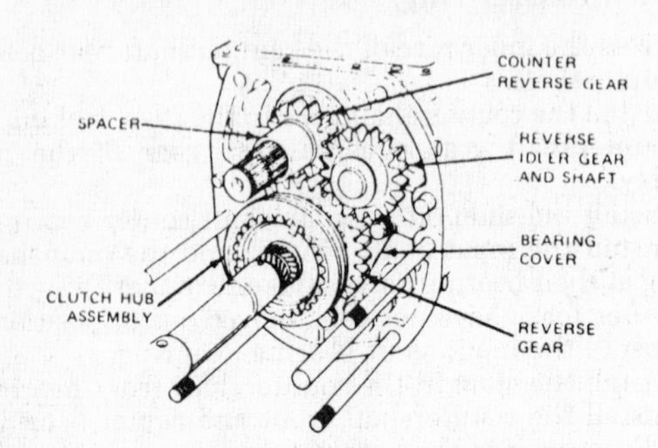

Installing reverse gears

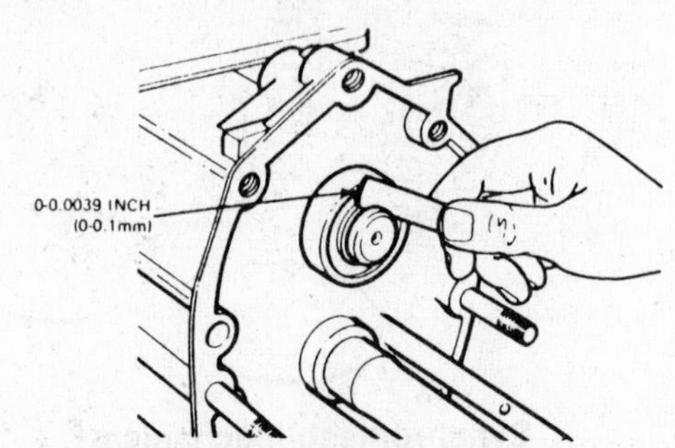

Checking clearance between snapring and countershaft rear bearing

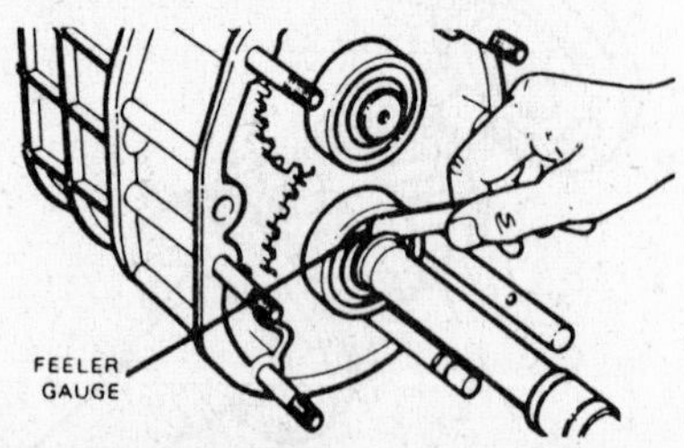

Checking clearance between mainshaft rear bearing thrust washer and snapring

Check the clearance between the thrust washer and the snapring. If the clearance is not within 0.1–0.3mm, select the proper size thrust washer to bring the clearance within specifications. Thrust washers are available as follows:

- 6.0mm
- 6.2mm
- 6.4mm
- 6.5mm
- 6.6mm
- 6.7mm
- 6.8mm
- 7.0mm
- 7.2mm

33. Slide the 1st–2nd shift fork shaft assembly into the case. Secure the shift fork to the shaft with a roll pin.
34. Insert the interlock pin into the transmission.
35. Shift the transmission into 4th gear. Slide the 3rd–4th shift fork shaft into the case, from the rear of the case. Secure the 3rd–4th shift fork to the shaft with a roll pin.
36. Shift the synchronizer hub into 5th gear. Install the reverse and 5th fork on the clutch hub and slide the reverse–5th fork shaft into the case.
37. Install the 2 blind covers and gaskets. Tighten the attaching bolts to 23–34 ft. lbs. (32–45 Nm).
38. Install the 3 detent balls and springs into the case and install the spring bolts.
39. Install the center housing on the case. Align the reverse idler gear shaft boss with the center housing attaching bolt boss. Tighten the idler shaft bolt to 41–59 ft. lbs. (56–79 Nm).
40. Slide the counter 5th gear to the countershaft.
41. Install the countershaft rear bearing on the countershaft and press it into position.
42. Install the thrust washer and snapring to the rear of the countershaft rear bearing. Check the clearance between the thrust washer and the snapring. If the clearance is not within 0–0.15mm, select the proper size thrust washer to bring the clearance within specifications. Thrust washers are available in 1.9mm, 2.0mm, 2.1mm and 2.2mm.
43. Install the mainshaft rear bearing on the mainshaft.
44. Install the thrust washer and snapring to the rear of the

Special Tools

Tool Number	Description
T57L-500-B	Bench mounted holding fixture
T75L-1165-B	Axle bearing seal plate
D80P-4201-A	Depth micrometer
D79L-4621-A	Pinion bearing cone remover
T53T-4621-C	Pinion bearing cone replacer
T57L-4621-B	Pinion bearing cone replacer
T62F-4621-A	Pinion bearing cone replacer
T71P-4621-B	Pinion bearing cone replacer
T82T-7003-BH	Adjustable press frame adapter
T82T-7003-DH	Mainshaft front bearing replacer
T75L-7025-B	Remover/replacer tube (short)
T75L-7025-C	Remover and replacer tube (long)
T84T-7025-B	Forcing screw
T75L-7025-Q	Dummy bearing
T77J-7025-B	Remover tube (5½ in.)

Tool Number	Description
T77J-7025-C	Locknut wrench
T77J-7025-D	Bench holding fixture adapters
T77J-7025-E	Synchronizer ring holder and counter shaft spacer
T77J-7025-F	Staking tool
T77J-7025-H	Bearing puller
T77J-7025-J	Puller ring
T77J-7025-K	Center bearing replacer
T77J-7025-L	Countershaft front bearing replacer
T77J-7025-M	Mainshaft front bearing replacer
T77J-7025-N	Adjustable press frame
T71P-7095-A	Extension housing seal replacer
T72J-7280	Dummy shift rails (2) and interlock pins
T72J-7697	Extension housing bushing remover
T71P-7657-A	Extension housing seal remover
T72J-7697-A	Extension housing bushing replacer

mainshaft rear bearing. Check the clearance between the thrust washer and the snapring. If the clearance is not within 0–0.10mm, select the proper size thrust washer to bring the clearance within specifications. Thrust washers are available in 2.0mm, 2.15mm and 2.30mm.

45. Install the bearing housing on the center housing.
46. Install each shift fork shaft end onto the proper shift fork shaft and secure with roll pins.
47. Install the lock ball, speedometer drive gear and snapring onto the mainshaft.
48. If removed, install the control lever and rod in the extension housing.
49. Install the extension housing in the bearing housing with the gearshift control lever end laid down to the left as far as it will go. Tighten the attaching bolts and nut to 60–80 ft. lbs. (82–108 Nm).
50. Install the speedometer driven gear assembly to the extension housing and secure it with a bolt.
51. Check to ensure that the gearshift control lever operates properly.
52. Install the transmission case cover gasket and the cover with the drain plug to the rear. Tighten the bolts to 23–34 ft. lbs. (32–45 Nm).
53. Install the shim on the 2nd step of the front cover.
54. Install the front cover to the transmission case and tighten the 4 bolts and 4 studs.
55. Fill with 3.6 pints (1.7L) of API GL4 or GL5 SAE 80W90 gear lubricant. Install the filler plugs and tighten to 18–29 ft. lbs. (25–39 Nm).

Torque Specifications

	ft. lbs.	Nm
Shift rail detent spring cap	29–43	40–58
Mainshaft gear retaining nut	145–203	197–275
Clutch release level pivot	23–34	32–46
Interlock pin bore plug	7.5–11	11–14
Drain plug	29–43	40–58
Filler plug	18–29	25–39
Backup lamp switch	22–29	30–39
Shift indicator switch	22–29	30–39

Mitsubishi 5-speed Transmission Overhaul

DISASSEMBLY

Before disassembly, clean the exterior of the transmission assembly before any attempt is made to disassemble it, in order to prevent dirt or other foreign material from entering the transmission assembly and damaging its internal parts. If steam cleaning is done to the exterior of the transmission, immediate disassembly should be done to avoid rusting, caused by condensation forming on the internal parts.

1. Remove the transmission. Make sure the transmission is in the **N** position.
2. Remove the nuts retaining the clutch housing to the transmission and remove the housing. If not removed, remove the clutch slave cylinder from the input shaft.
3. Remove the back-up lamp switch and shift indicator switch from the transmission.
4. Drain the fluid from the transmission by removing the drain plug from the pan.
5. Remove the bolts retaining the pan to the case and remove the pan. Remove and discard the gasket. Remove all traces of the gasket from the mating surfaces of the pan and case.
6. On the FM145, remove the bolts retaining the cover to the transfer case adapter and remove the cover (with stopper bracket inside). Remove and discard the gasket. Clean all traces of gasket material from the mating surfaces of the adapter and cover.
7. On the FM146, remove the bolts retaining the control housing to the transfer case adapter (4WD models) or extension housing (2WD models). Remove the control housing (with reverse lockout assembly attached). Remove and discard the rub-

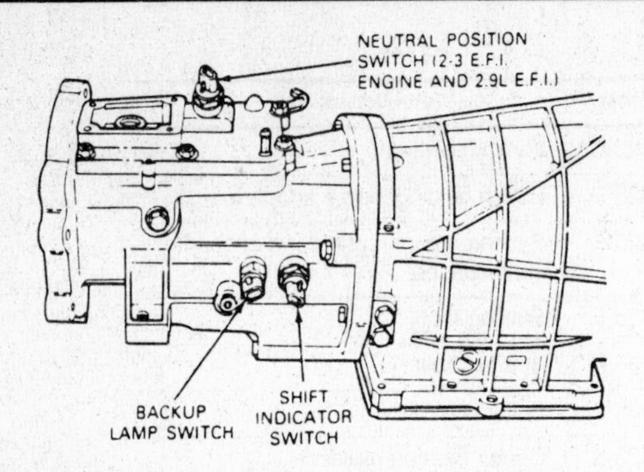

Back-up lamp, neutral safety and shift indicator switches – location

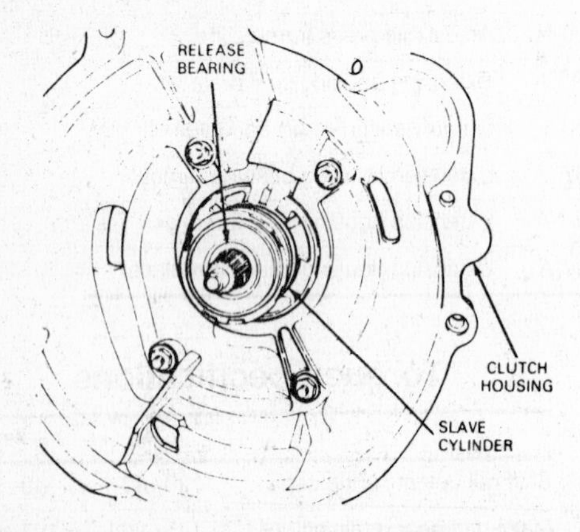

Clutch housing assembly, exploded view

ber seal. Clean the mating surfaces of the adapter and control housing.

8. On 4WD models, remove the detent spring and ball from the adapter housing.

9. On the FM145, remove the 3 shift gate roll pin access plugs (2 on the side, 1 on the bottom) with a 6mm allen-head wrench. On the FM146 and FM132, remove the 2 shift gate roll pin access plugs (1 on the side, 1 on the bottom) with a 6mm allen-head wrench.

10. Shift the transmission into the appropriate gear to align the roll pin with the access hole. Using a punch, drive the roll pins from the shift gates through the access holes.

11. From the right side of the adapter (4WD models) or control housing (2WD models), remove the bolt, spring and neutral return plunger. Note that the plunger has a slot in the center for the detent ball.

12. From the left side of the adapter/control housing, remove the bolt, spring and neutral return plunger.

13. Lift the gate selector lever out of the shift gates. Move the lever to the rear of the adapter as far as it will go. This will allow clearance to remove the adapter/control housing from the case.

14. Remove the bolts retaining the transfer case adapter/control housing to the transmission case. Note that 3 different bolts lengths (35mm, 55mm and 110mm) are used to retain the case to the adapter. Mark the bolt holes accordingly.

15. Remove the adapter/control housing from the case. Make sure the shift gates do not bind in the adapter during removal. Rotate the gates on the rails as required. Remove and discard the gasket. Clean all traces of gasket material from the mating surfaces of the case and adapter.

Note: On 2WD models, it may be necessary to remove the rear seal from the extension housing.

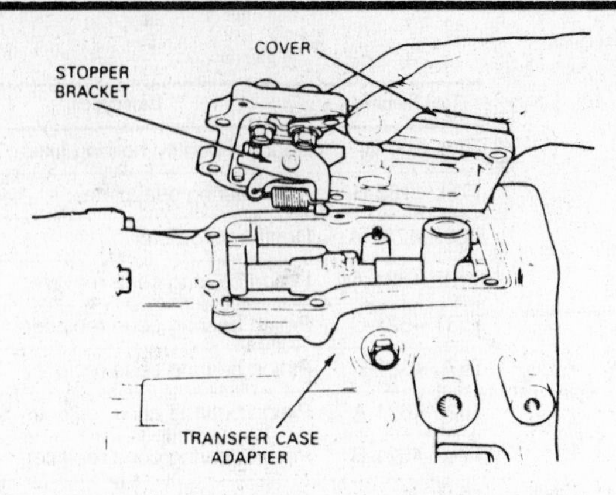

Removing transfer cover and gasket – FM145

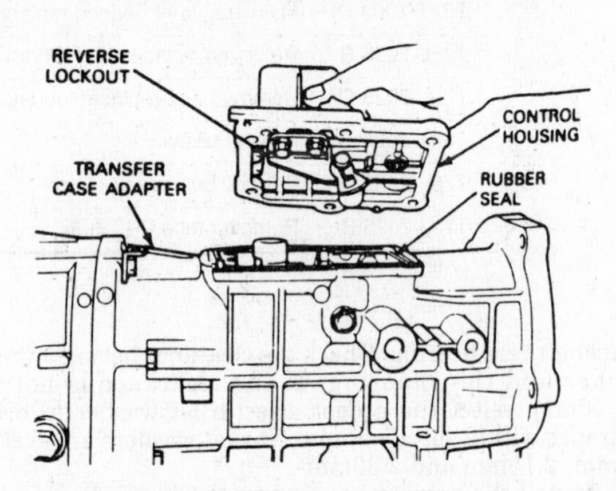

Removing control housing and rubber seal – FM146

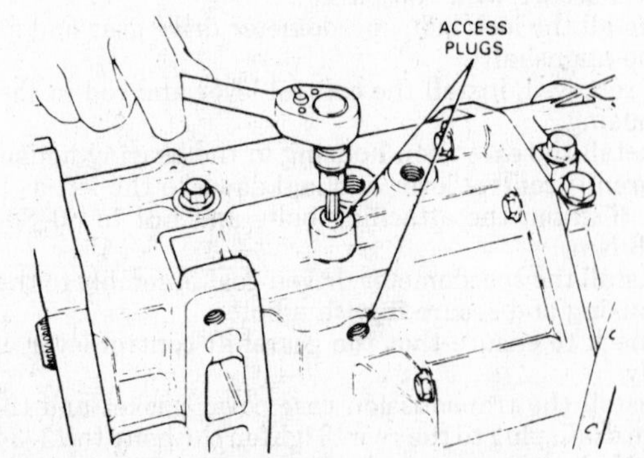

Removing shaft gate roll pin access plugs – FM145 shown

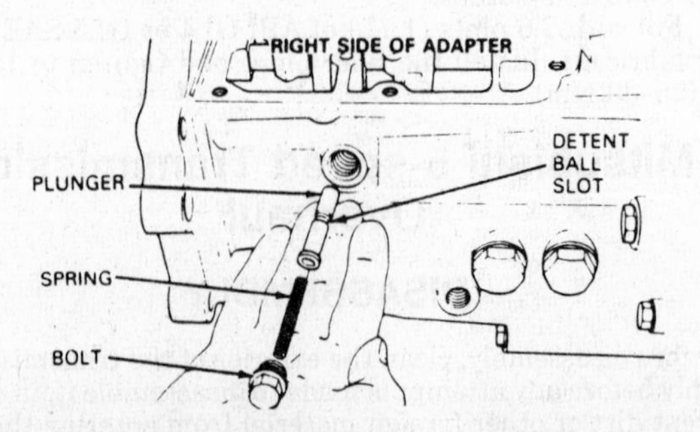

Removing neutral return plunger, right side – 4WD model shown, 2WD similar

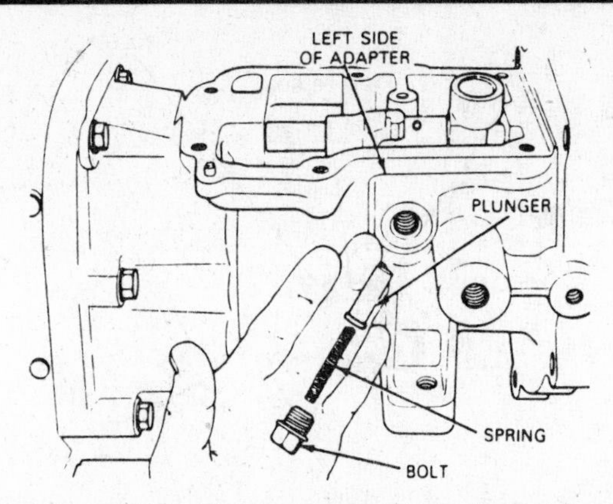

Removing neutral return plunger, left side

16. Identify each shift rail and gate. Remove the gates from the rails.
17. From inside the case, drive out the roll pins retaining the 1st–2nd and 3rd–4th shift forks to the rails. Remove the 1st–2nd shift fork.
18. Drive out the overdrive–reverse shift fork roll pin.

NOTE: The roll pin in the switch actuator does not need to be removed for transmission disassembly.

19. Remove the set screw on the top of the case and remove the poppet spring and steel ball. Remove the 2 bolts on the side of the case. Remove the poppet springs and steel balls.
20. Remove the overdrive–reverse shift rail and the 3rd–4th shift rail from the case. Remove the overdrive–reverse shift fork. When the 2 shift rails are removed, the interlock pins can be removed from the case.

NOTE: The 1st–2nd shift rail is unable to be removed at this time.

21. On 4WD models, remove the oil seal sleeve from the output shaft, using a puller. Do not reuse the oil seal sleeve, once removed.
22. On 2WD models, remove the snapring, speedometer drive gear and steel ball.
23. Unstake the locknuts on the mainshaft and countershaft using mainshaft locknut staking tool, T77J–7025–F or equivalent.
24. Double engage the transmission in 2 gears to lockup the transmission. This is done by engaging 2 of the synchronizers. This is necessary to remove the locknuts.
25. Remove the countershaft locknut with a 30mm socket. Discard the locknut.
26. Remove the mainshaft locknut with mainshaft locknut wrench, T77J–7025–C or equivalent. Discard the locknut.
27. Pull the rear bearing off the mainshaft using tube T77J–7025–B, forcing screw, T84T–7025–B, puller, T77J–7025–H, hand puller ring, T77J–7025–J or equivalents. Remove and discard the bearing.
28. Remove the spacer and lock ball from the mainshaft.
29. Remove the counter-overdrive gear and ball bearing from the countershaft by installing the jaws of puller, T77F–4220–B1 behind the gear. While removing the gear, remove the 1st–2nd shift rail from the case.
30. Remove the 1st–2nd and 3rd–4th shift forks from the case.
31. Remove the overdrive gear, needle bearing, spacer and synchronizer ring from the mainshaft.
32. Remove the overdrive synchronizer sleeve from the synchronizer hub on the mainshaft.

NOTE: Do not lose the 3 keys and 2 springs in the hub. A spring is located on each side of the hub.

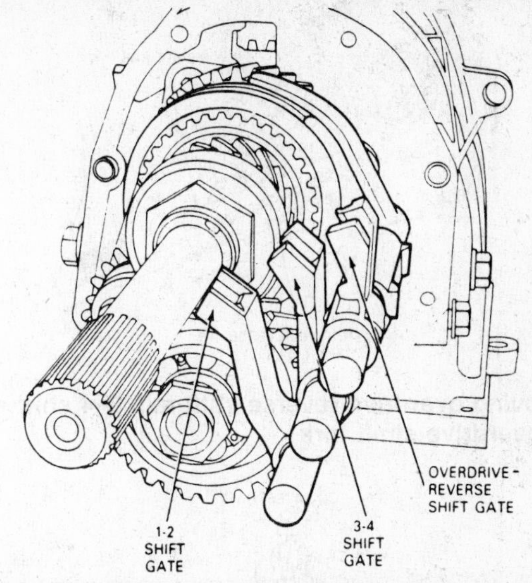

Correct location of 1–2, 3–4 and overdrive shift gates

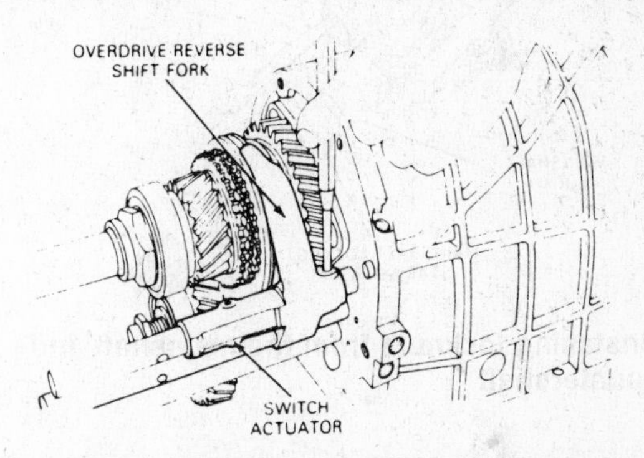

Removing 1–2, 3–4 and overdrive shift forks

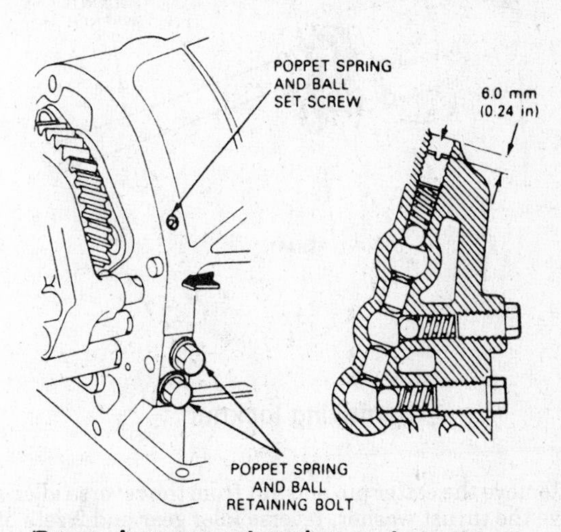

Removing poppet spring and steel ball

33. Pull the overdrive synchronizer hub and overdrive gear bearing sleeve from the mainshaft using bearing puller, T77J–7025–H, puller ring, T77J–7025–J, tube, T75L–7025–B and forcing screw, T84T–7025–B or equivalents.
34. Slide the reverse gear and needle bearing assembly off the mainshaft.
35. Slide the counter/reverse gear and distance spacer off the countershaft.

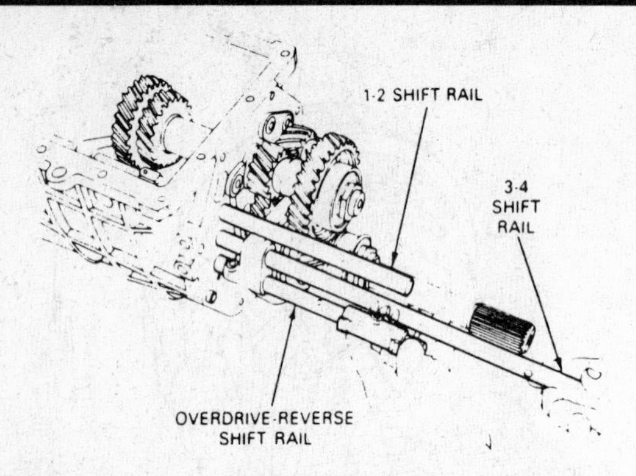

Removing overdrive reverse shift rail, 3–4 shift rail and overdrive shaft fork

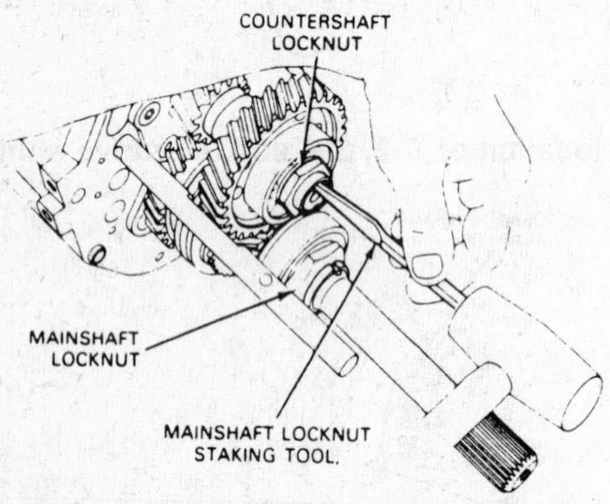

Unstaking locknuts from the mainshaft and countershaft

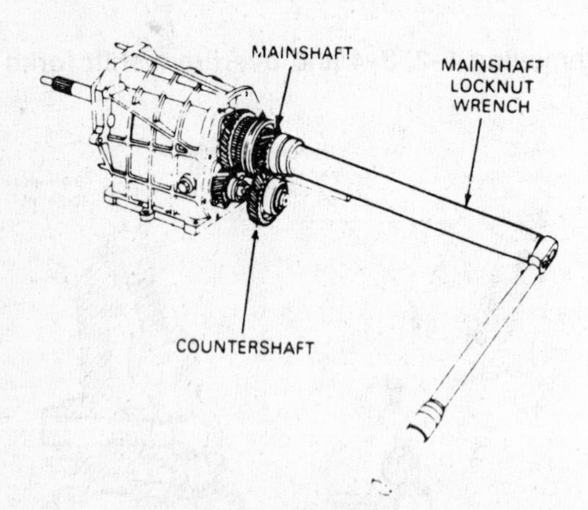

Removing locknut

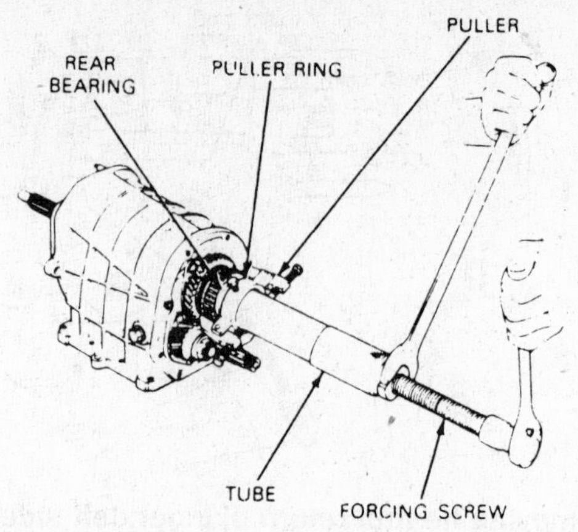

Removing rear bearing from mainshaft

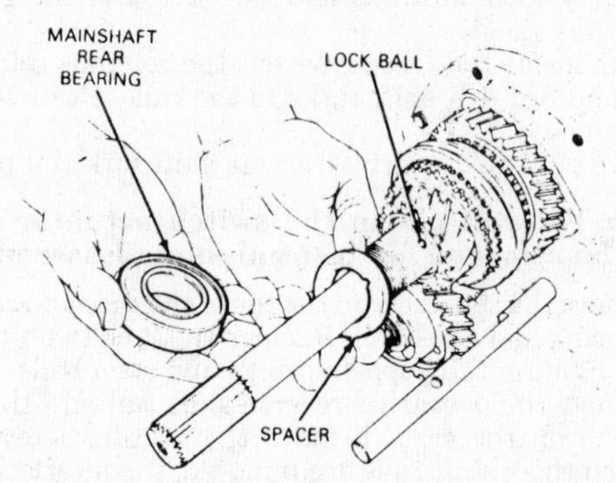

Removing lock ball and spacer from mainshaft

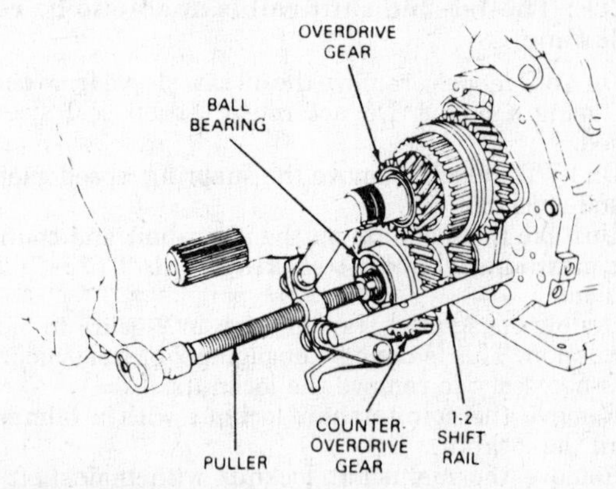

Removing counter-overdrive gear and ball bearing from countershaft

36. Remove the cotter pin and nut from the reverse idler shaft. Remove the thrust washer, reverse idler gear and 2 sets of needle bearings from the shaft.
37. Remove the 6mm allen-head bolts that attach the mainshaft rear bearing retainer to the case and remove the retainer. Remove and discard the gasket. Clean any traces of gasket material from the mating surfaces of the case and retainer.
38. Remove the 6mm allen-head bolts that retain the reverse idler gearshaft assembly to the case.
39. Pull the reverse-idler gearshaft assembly out of the case using slide hammer, T50T–100–A and reverse idler gearshaft remover, T85T–7140–A or equivalents.
40. Use a double nut procedure to remove the studs that retain the input shaft front bearing retainer to the case. Remove the bolts that attach the retainer to the case.
41. Remove the input shaft front bearing retainer from the case. Remove and discard the gasket. Clean all traces of gasket material from the mating surfaces of the case and retainer.
42. Remove the selective shim from inside the input shaft front bearing retainer.

WARNING: Do not discard the selective shim!

43. Remove the small selective snapring that retains the input shaft to the bearing.

WARNING: Do not discard the selective snapring!

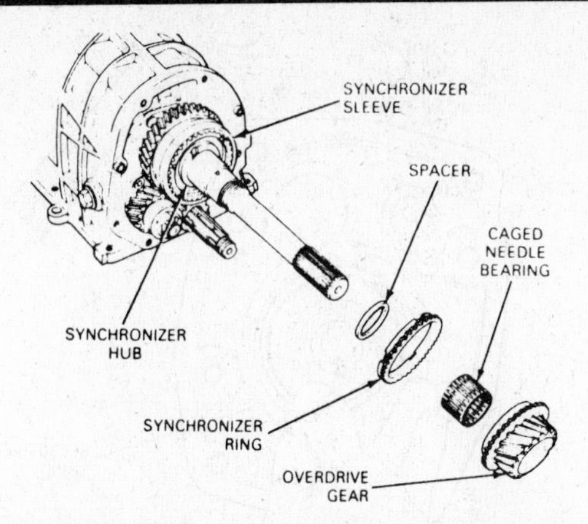

Synchronizer sleeve and related components

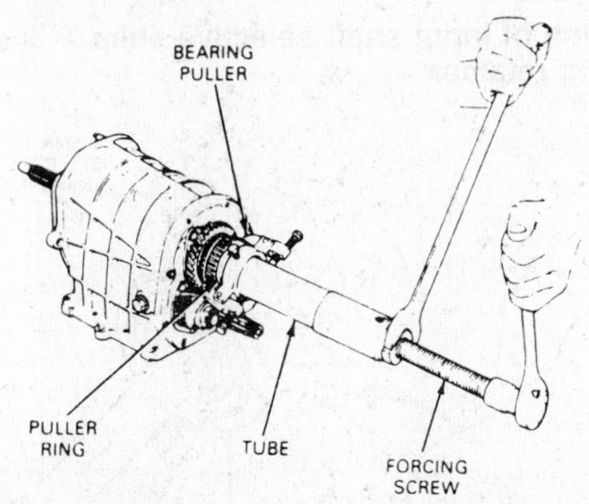

Removing synchronizer hub and overdrive gear bearing sleeve

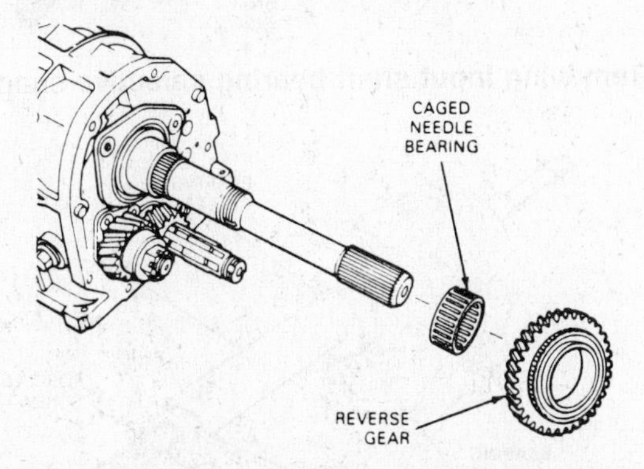

Removing reverse gear and needle bearing assembly

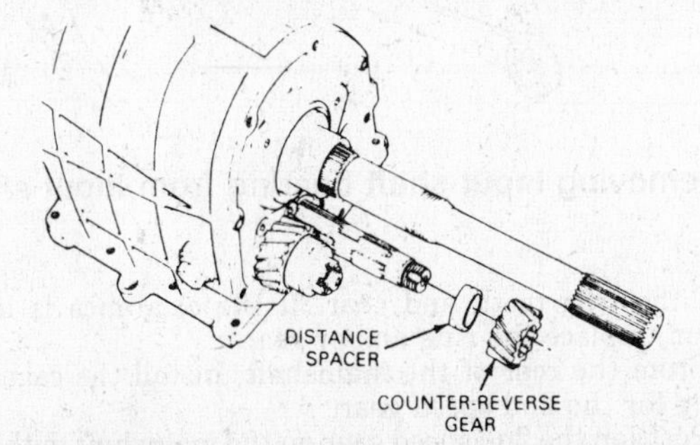

Removing counter-reverse gear and distance spacer from countershaft

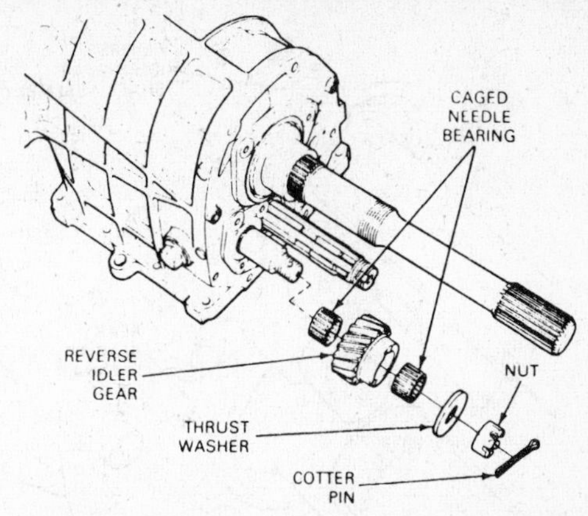

Removing reverse idler gear and needle bearing from shaft

44. Remove the large snapring that retains the input shaft bearing to the case.
45. Remove the bearing from the input shaft and case using tube, T75L–7025–B, bearing collets, T75L–7025–D, bearing collet sleeve, T75L–7025–G and forcing screw, T84T–7025–B. Remove and discard the bearing.
46. Rotate the input shaft so the flats on the shaft face the countershaft, providing clearance to remove the input shaft. Remove the input shaft. The output shaft (mainshaft) may have to be pulled to the rear of the case. Remove the small caged needle bearing from the inside of the input gear.
47. Remove the snapring from the mainshaft (output shaft) outer bearing race.
48. Remove the outer mainshaft bearing race, ball bearing and bearing sleeve from the case using tube, T75L–7025–B, mainshaft bearing collet remover, T85T–7065–A, bearing collet sleeve, T77F–7025–C and forcing screw, T84T–7025–B. Discard the outer bearing race and ball bearing.

NOTE: The inner race of the front bearing will remain on the mainshaft.

49. Remove the countershaft front spacer and bearing race.
50. Remove the countershaft from the case. The mainshaft assembly may have to be moved slightly to the side to allow clearance for countershaft removal.
51. Remove the mainshaft assembly from the case.

Mainshaft Disassembly and Assembly

1. Remove the selective snapring that retains the 3rd–4th synchronizer assembly to the mainshaft. A new snapring will be used in the assembly.
2. Remove the 3rd–4th synchronizer assembly (hub, sleeve, spring and keys), synchronizer ring, 3rd speed gear and caged needle bearing from the front of the mainshaft. Note the position of the synchronizer hub and sleeve during disassembly.
3. Position the mainshaft assembly in a press so the 2nd speed gear is supported by the press bed. Press the mainshaft down and out from the 1st–2nd gear assembly.
4. Separate the inner ball bearing, bearing sleeve, 1st speed gear, caged needle bearing, 1st–2nd synchronizer assembly (hub, sleeve, rings and keys), 2nd speed gear and caged needle bearing. Note the direction of the 1st–2nd synchronizer hub and sleeve during disassembly. Discard the inner ball bearing.

To Assembly:

Prior to assembly, lubricate all components with standard transmission lubricant, SAE 80W, D8DZ–19C547–A (ESP–M2C83–C) or equivalent.

5. Check the clearance between the synchronizer rings and gears. Install the ring on the gear and insert a feeler gauge be-

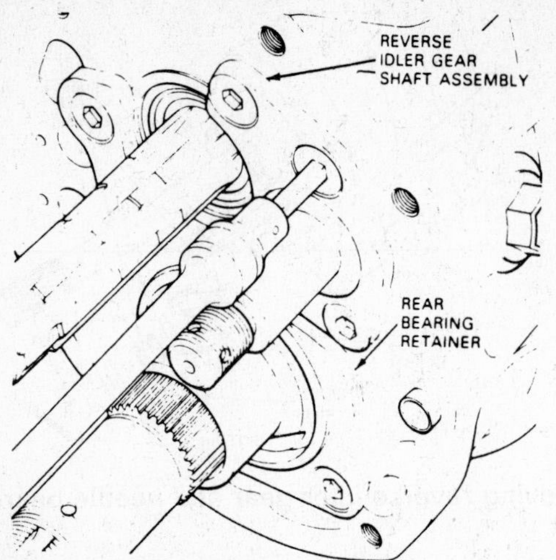

Removing mainshaft bearing retainer and reverse idler gear shaft assembly

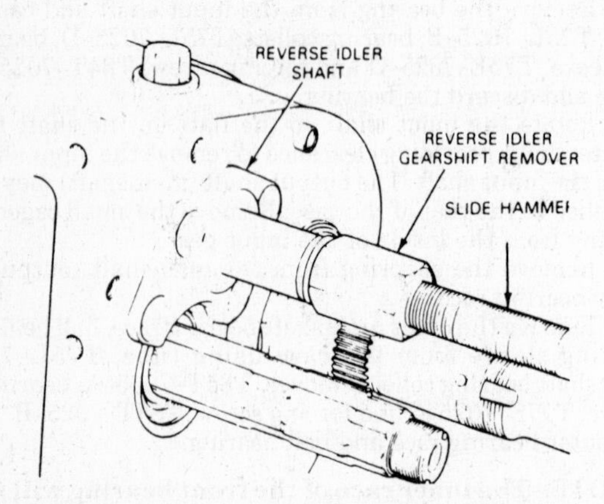

Removing reverse idler gear shaft from case

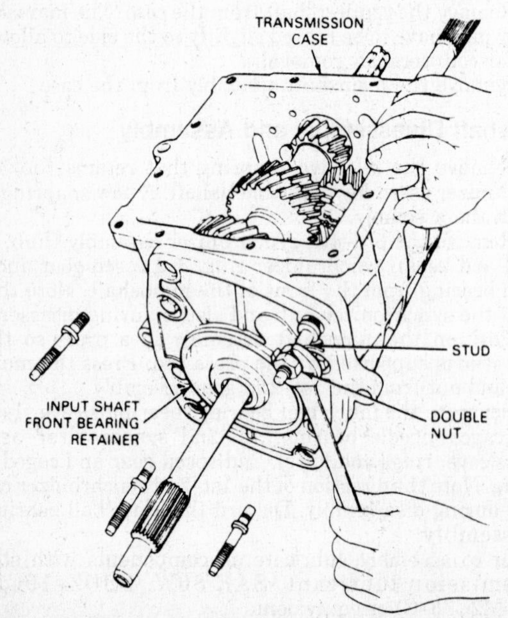

Removing input shaft front bearing retainer from case

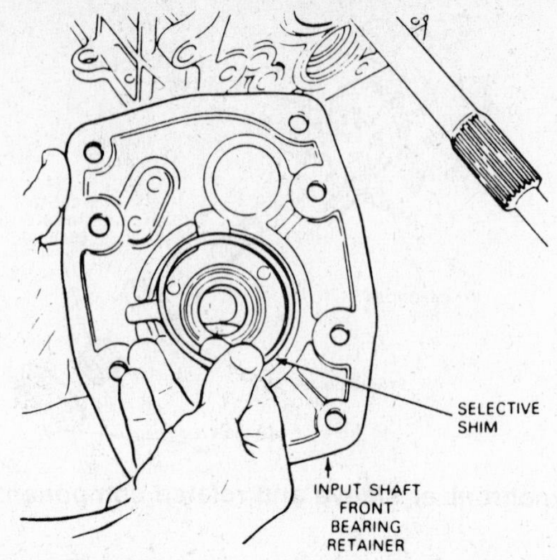

Location of input shaft selective shim – inside front bearing retainer

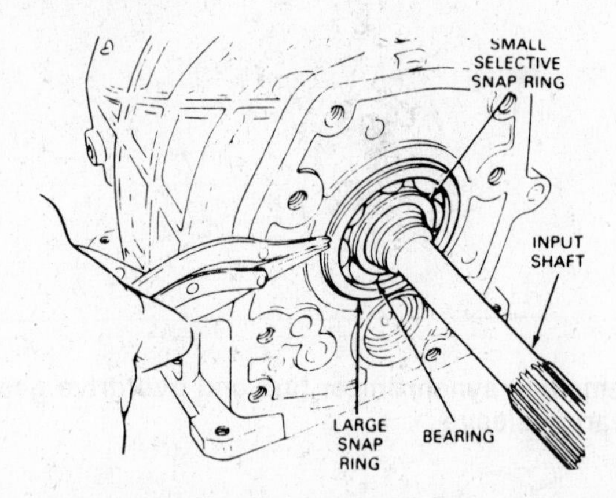

Removing input shaft bearing selective snapring

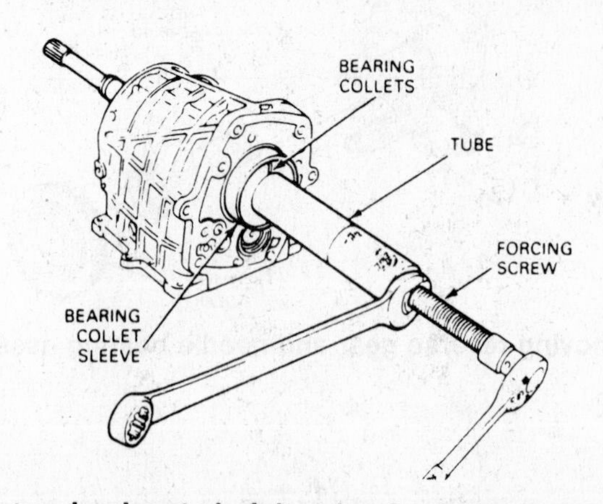

Removing input shaft bearing from input shaft

tween the ring teeth and gear. If the clearance is less than 0.23mm, replace the ring and/or gear.

6. From the rear of the mainshaft, install the caged needle bearing for the 2nd speed gear.

7. Position the 2nd speed gear on the mainshaft with the synchronizer ring surface facing the rear of the shaft.

8. Install the syncrhonizer ring on the 2nd speed gear.

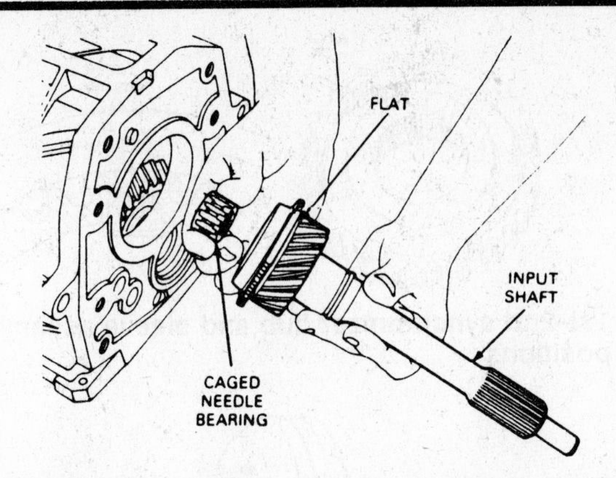

Removing input shaft and needle bearing from case

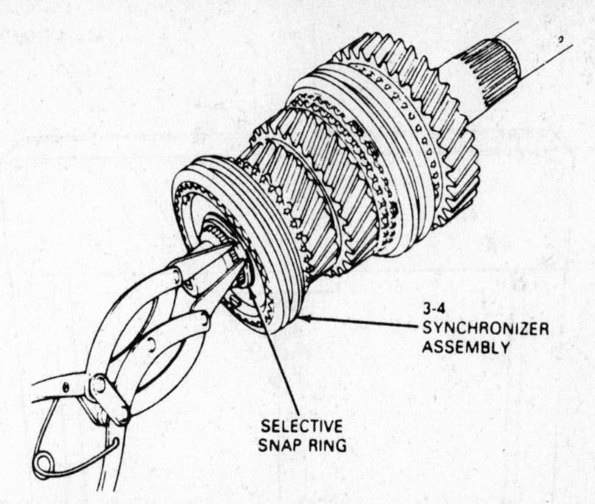

Removing selective snapring from the 3–4 synchronizer assembly

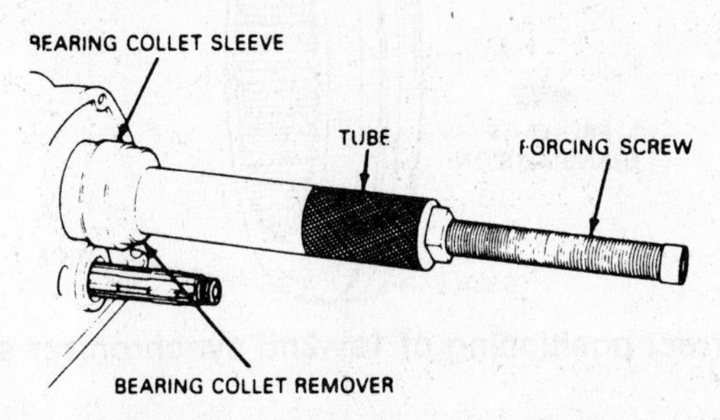

Removing mainshaft bearing race and sleeve from case

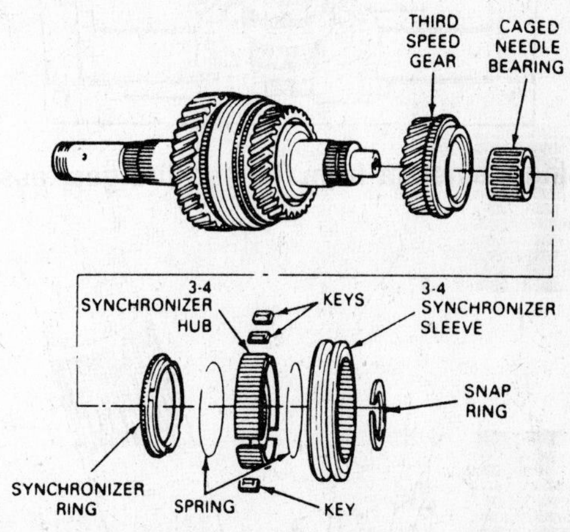

Removing 3–4 synchronizer assembly from mainshaft

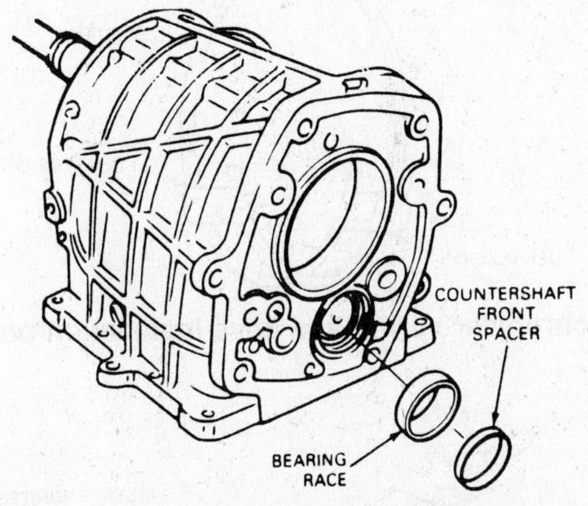

Removing countershaft front spacer and bearing race from case

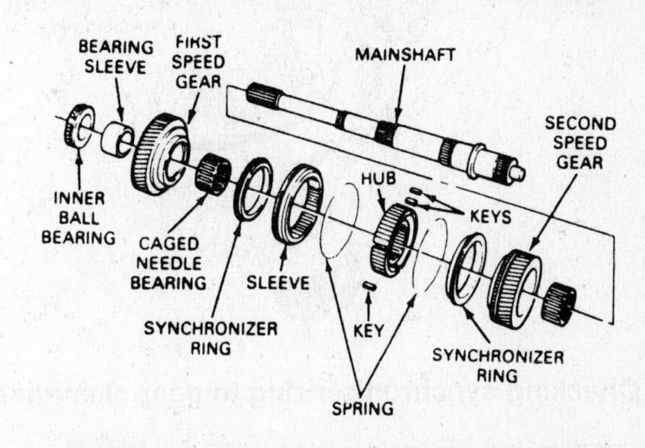

Mainshaft components, exploded view

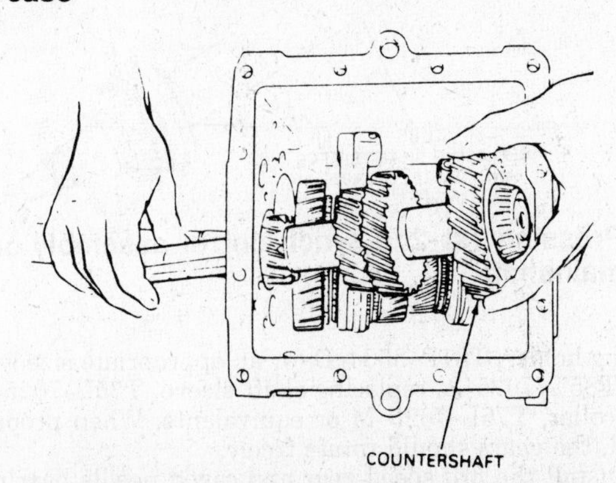

Countershaft and mainshaft removal

9. Position the 1st–2nd synchronizer assembly on the rear of the mainshaft, making sure of the following:

a. The splines of the mainshaft and synchronizer are properly aligned.

b. The rear of the 1st–2nd hub is identified by a ridge machined on the rear surface. The ridge must face the front of the mainshaft.

c. The synchronizer sleeve has a tooth missing at 6 positions. Assemble the hub to the sleeve so the single tooth between the 2 missing portions will touch the synchronizer key.

d. The synchronizer keys and springs are properly installed. The open ends of the spring do not face each other.

10. Press the 1st–2nd synchronizer assembly in position on

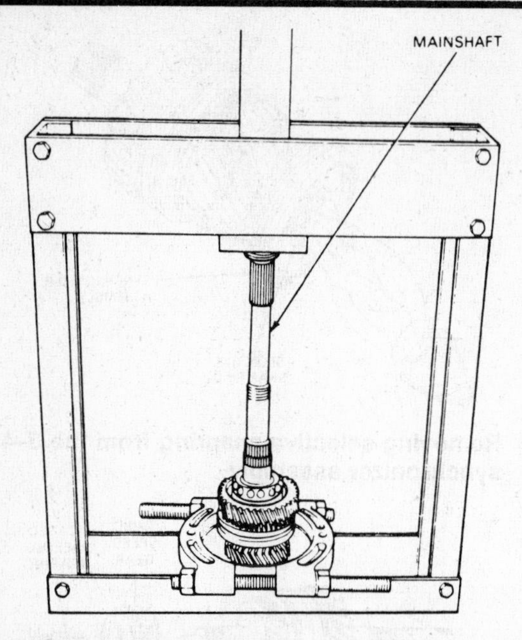

Pressing mainshaft from the 1st–2nd gear assembly

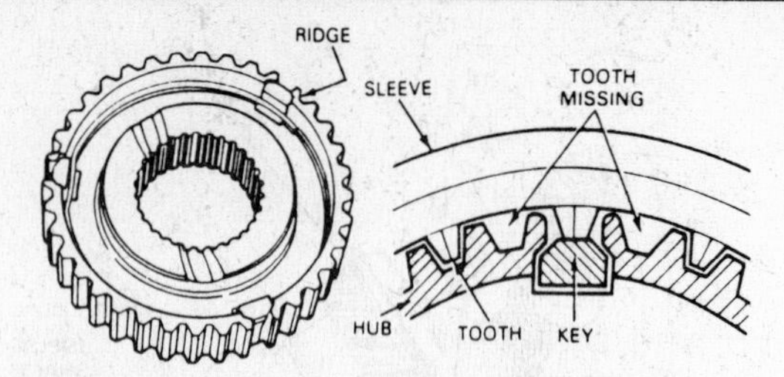

1st–2nd synchronizer hub and sleeve assembly positions

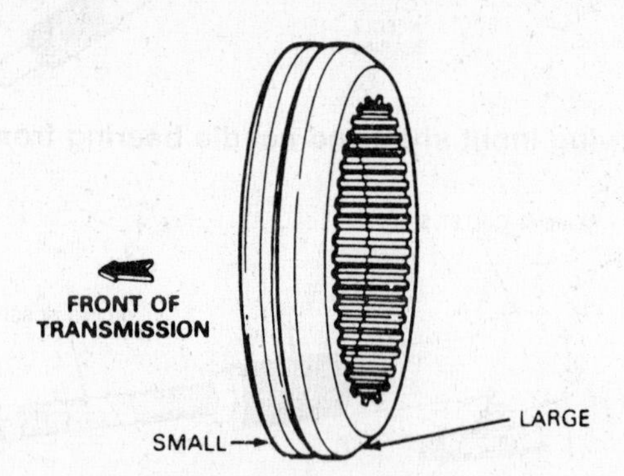

Correct positioning of 1st–2nd synchronizer sleeve

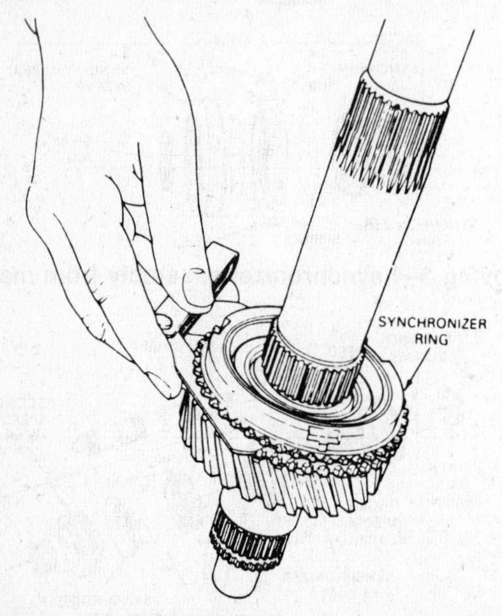

Checking synchronizer ring to gear clearance

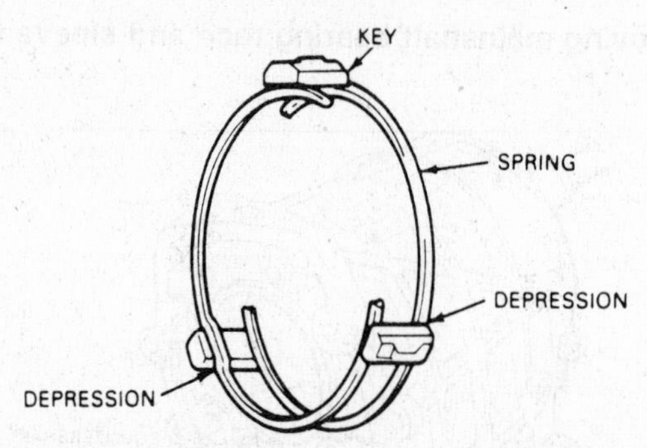

Synchronizer spring and keys installation position

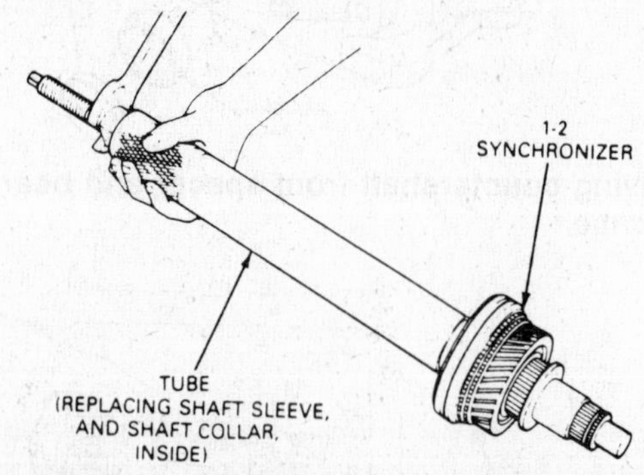

Pressing 1st–2nd synchronizer assembly on mainshaft

the mainshaft using replacing shaft sleeve tool, T75L–7025–K, shaft collar, T75L–7025–M and tube, T75L–7025–C or equivalents. If properly installed, the 2nd speed gear should rotate freely.

11. Position the 1st gear bearing sleeve on the mainshaft. Press the sleeve on the shaft using replacing shaft sleeve, T75L–7025–K, shaft collar, T75L–7025–M, rack bushing holder, T81P–3504–D (or an appropriate washer) and tube, T75L–7025–C or equivalents. When properly installed, the sleeve should be against the synchronizer hub. Make sure the gears rotate freely.

12. Install the synchronizer ring on the 1st–2nd synchronizer assembly.

13. Install the caged needle bearing and 1st speed gear.

14. Slide the inner ball in position on the mainshaft.

15. Press the inner ball bearing on the mainshaft using rack bushing holder, T81P–3504–D (or an appropriate size washer), tube, T85T–7025–A, replacing shaft sleeve, T75L–7025–K and shaft collar, T75L–7025–M or equivalents. When properly installed, the gears should rotate freely.

16. Install the 3rd speed gear and caged needle bearing over the front of the mainshaft.

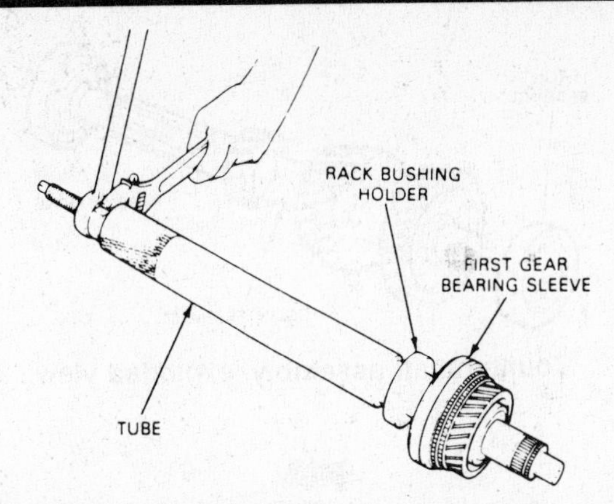

Pressing 1st gear bearing sleeve on mainshaft

17. Install the synchronizer ring against the 3rd speed gear.
18. Make sure the 3rd–4th synchronizer assembly is properly installed. Be sure of the following:
  a. The splines of the mainshaft and synchronizer are properly aligned.
  b. The small diameter boss of the hub faces the front of the mainshaft.
  c. The small bevel angle of the sleeve faces the front of the mainshaft.
  d. The synchronizer sleeve has a tooth missing at 6 positions. Assemble the hub to the sleeve so the single tooth between the 2 missing portions will touch the synchronizer key.
  e. The synchronizer springs and keys are properly installed. The open ends of each spring do not face each other.
19. Install the 3rd–4th synchronizer assembly on the front of the mainshaft.
20. Install a new selective snapring that retains the 3rd–4th synchronizer assembly to the mainshaft. Select the thickest snapring that fits in the groove.

Countershaft Disassembly and Assembly

1. Press the front and rear bearing off the countershaft, using bearing splitter, D84L–1123–A or equivalent. Remove and discard the bearings.
2. Press the new bearing on the countershaft, using a press and countershaft bearing replacer.

Input Shaft Disassembly and Assembly

1. Position bearing splitter, D84L–1123–A or equivalent, behind the bearing and press the input shaft out of the bearing. Discard the bearing.
2. Position a new bearing on the input shaft and press the bearing onto the shaft using tube, T75L–7025–B, replacing shaft sleeve, T75L–7025–K, shaft collar, T75L–7025–M and rack bushing holder, T81P–3504–D or equivalents.

Cleaning and Inspection

1. Wash all parts except the ball bearings and seals, in a suitable cleaning solvent.
2. Dry all parts with compressed air.
3. Do not clean, wash or soak the transmission seals in cleaning solvent.
4. Lubricate the bearings and wrap them in a clean, lint free cloth until ready for use.
5. Inspect the transmission case and housing for cracks, worn or damaged bores, damaged threads or any other damage that could affect the operation of the transmission.
6. Inspect the machined mating surfaces for burrs, nicks or damage.

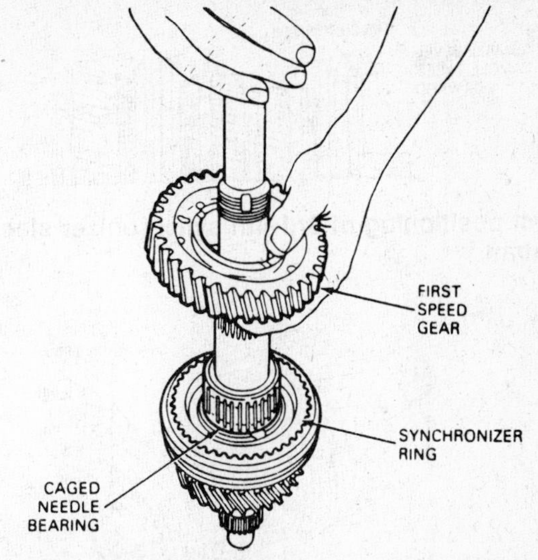

1st–2nd synchronizer assembly and needle bearing to mainshaft installation

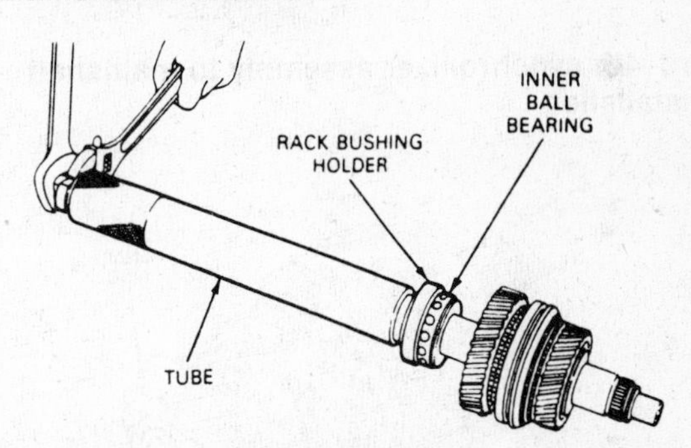

Pressing the inner ball bearing on mainshaft

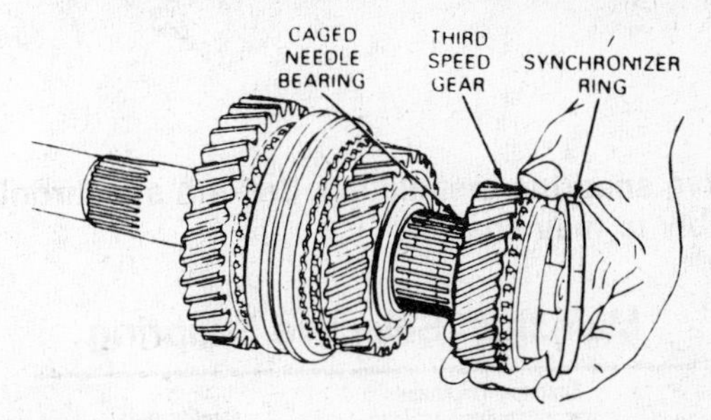

Installing 3rd gear, caged needle bearing and ring on mainshaft

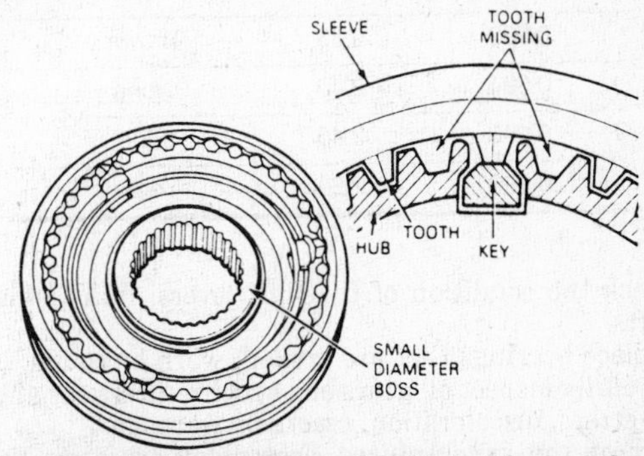

Correct installation position for 3rd–4th synchronizer hub and sleeve

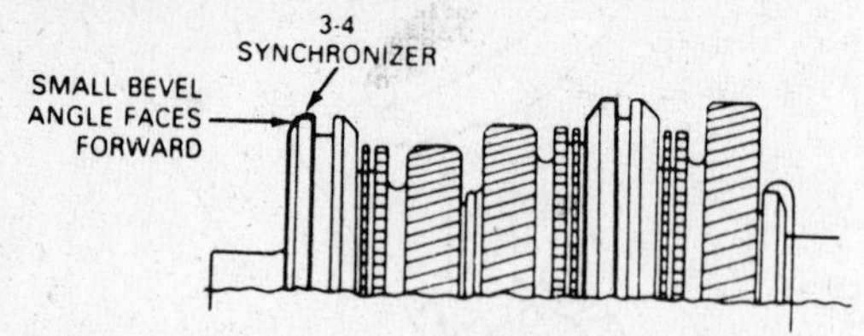

Correct positioning of 3rd–4th synchronizer sleeve to mainshaft

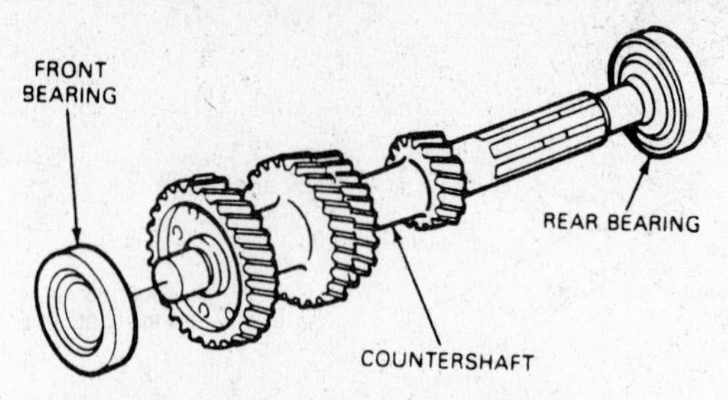

Countershaft assembly, exploded view

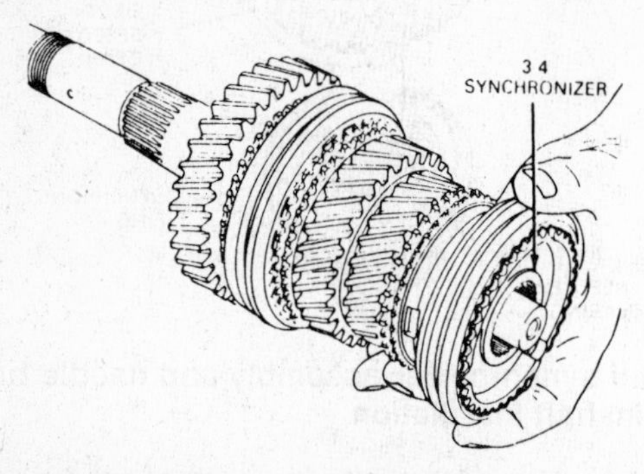

3rd–4th synchronizer assembly to mainshaft installation

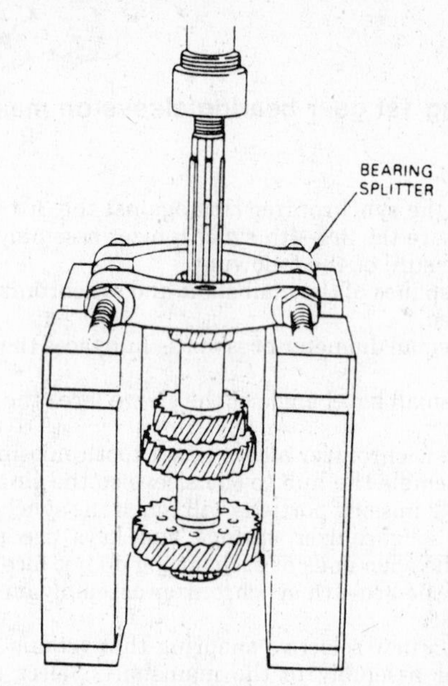

Pressing front and rear bearing from countershaft

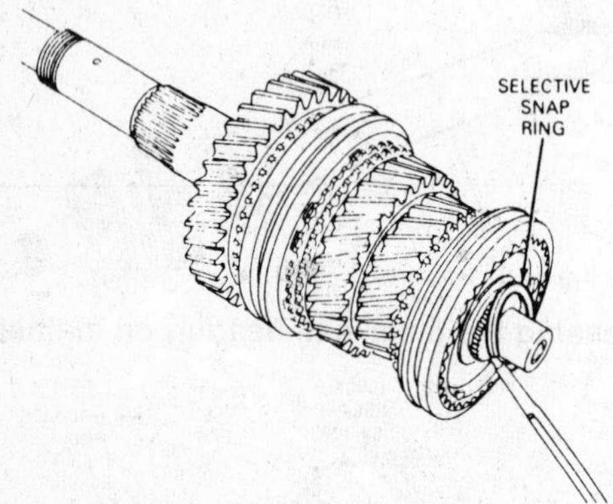

Selective snapring installation, 3rd–4th synchronizer assembly to mainshaft

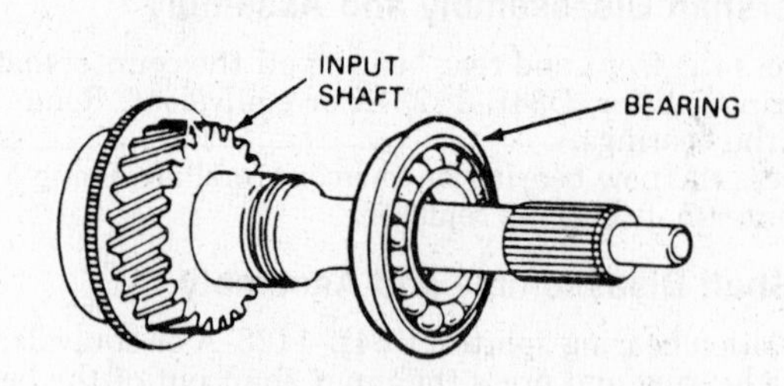

Input shaft, exploded view

Mainshaft Selective Snapring

Snapring Thickness		Identification Color
in.	mm	
0.091	2.30	White
0.093	2.35	Brown
0.094	2.40	None
0.096	2.45	Blue
0.098	2.50	Yellow

7. Check the condition of the shift levers, forks, shift rails and shafts.
8. Replace bearings that are broken, worn or rough.
9. Carefully inspect all gears and synchronizers for any signs of wear, stress, discoloration, cracks or warpage.
10. Inspect the synchronizer sleeves for excessive wear or distortion.
11. Inspect all roller and needle bearings.

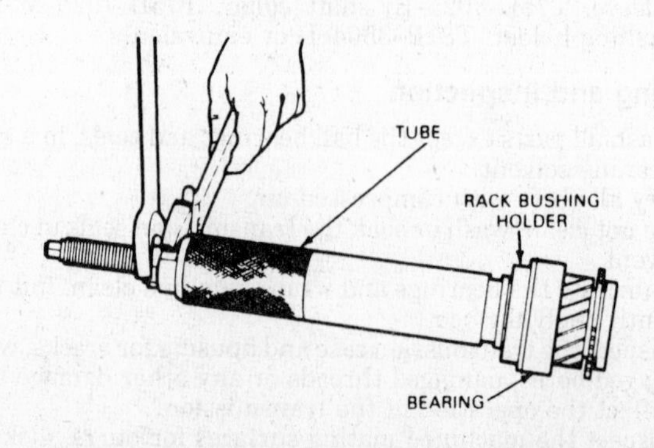

Pressing new bearing to input shaft

TRANSMISSION ASSEMBLY

1. Install the mainshaft assembly in the case.
2. Choose and install a new selective snapring in front of the input shaft bearing. Select the thickest snapring that will fit in the groove.
3. Install the small caged needle bearing inside the input gear. Install the synchronizer ring on the input shaft. Check the clearance between the ring and gear. If the clearance is less than 0.23mm, replace the ring and/or input shaft.
4. Install the synchronizer ring and input shaft in the case. Rotate the input shaft so the flats face the countershaft to provide installation clearance.

NOTE: It may be necessary to tap the input shaft into position with a brass hammer.

5. Install a new snapring on the outer bearing race. The longer portion of the race must be installed in the case.
6. Slide the outer ball bearing on the mainshaft. Press the bearing on the mainshaft and in the race using tube, T85T-7025-A, replacing shaft sleeve, T75L-7-25-K and shaft collar, T75L-7025-M or equivalent. When pressed in position, all gears and shafts must rotate freely.
7. Install the 3rd-4th shift fork into its synchronizer sleeve. The roll pin boss on the fork must face to the rear.
8. Install the countershaft into the case. It may be necessary to move the mainshaft to one side in order for the countershaft to be easily inserted.
9. Install the 1st-2nd shift fork. The roll pin boss must face toward the 3rd/4th shift fork.

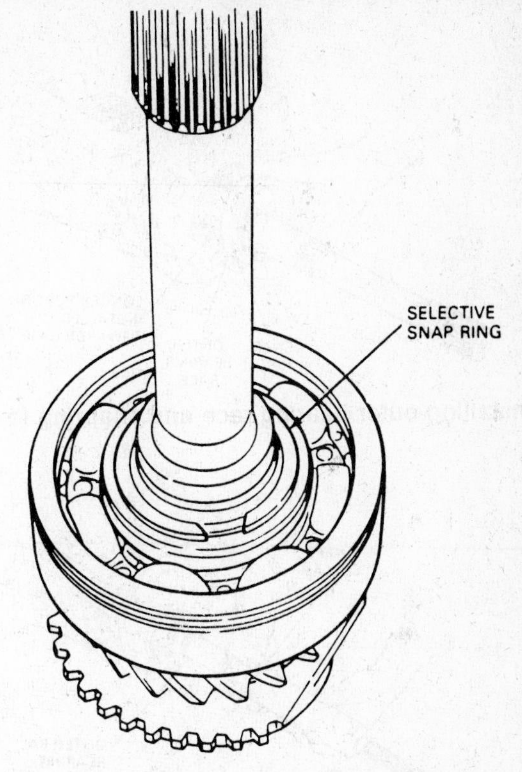

Input shaft selective snapring installation

Input Shaft Selective Snapring

Snapring Thickness		Identification Color
in.	mm	
0.085	2.15	Blue
0.087	2.22	None
0.090	2.29	Brown
0.093	2.36	White

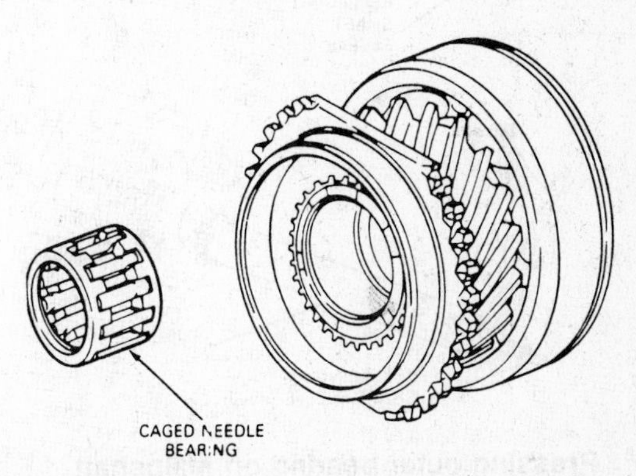

Synchronizer ring and caged needle bearing assembly

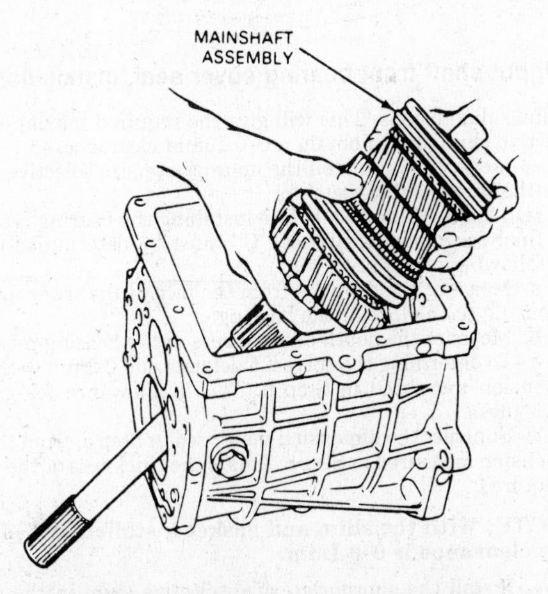

Mainshaft assembly to case installation

10. If removed, drive a new seal into the input shaft front bearing retainer using seal installer, T85T-7011-A and driver handle, T80T-4000-W or equivalents.
11. Install the large snapring that retains the input shaft bearing to the case.
12. On 1983-88 models, check the input shaft front bearing retainer-to-bearing clearance as follows:
    a. With the retainer selective shim removed, use a depth micrometer to measure the distance between the top machined surface to the spacer surface (second landing). Record the reading.
    b. Bottom the input shaft bearing so the snapring is flush against the case.
    c. Using a depth micrometer, measure the distance from the top of the outer front bearing race to the machined surface of the case.
    d. Subtract the distance of the bearing-to-case from the re-

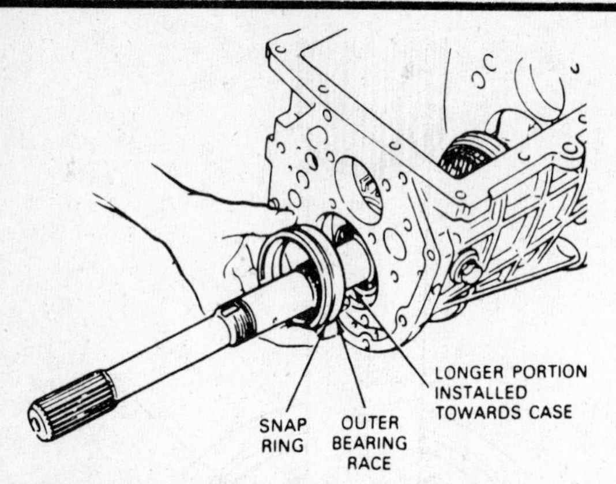

Installing outer bearing race and snapring to case

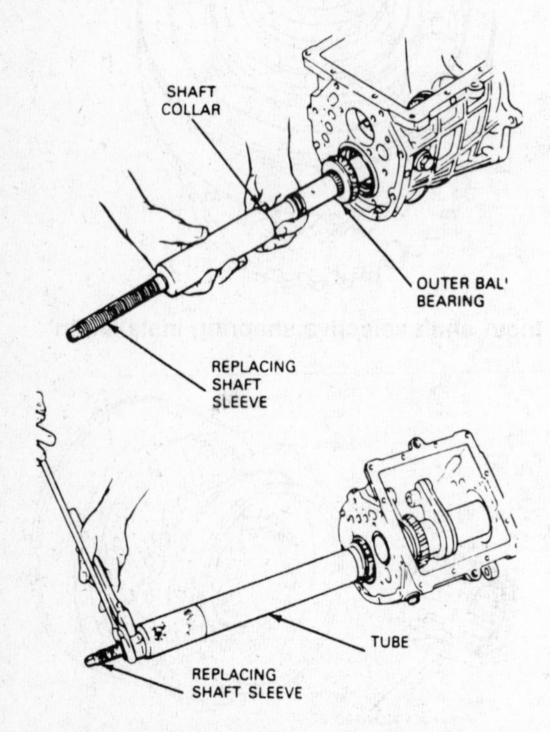

Pressing outer bearing on mainshaft

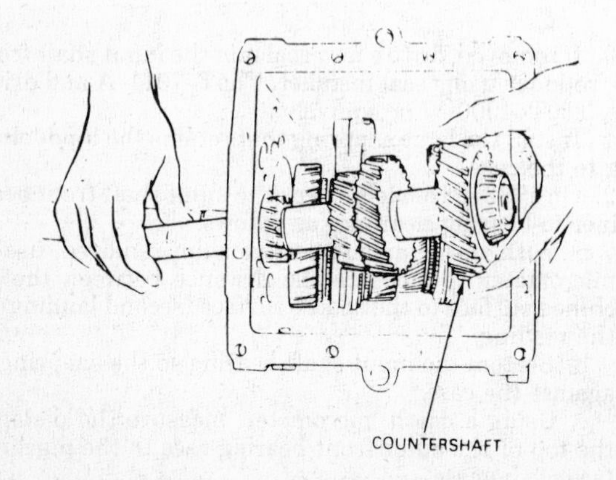

Countershaft to case installation

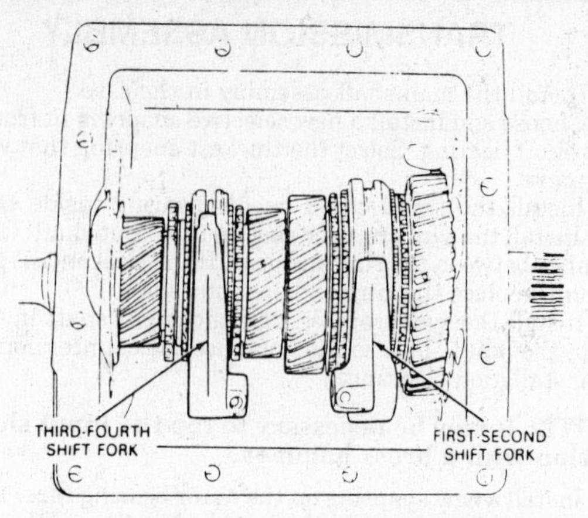

1st–2nd shift fork, proper positioning

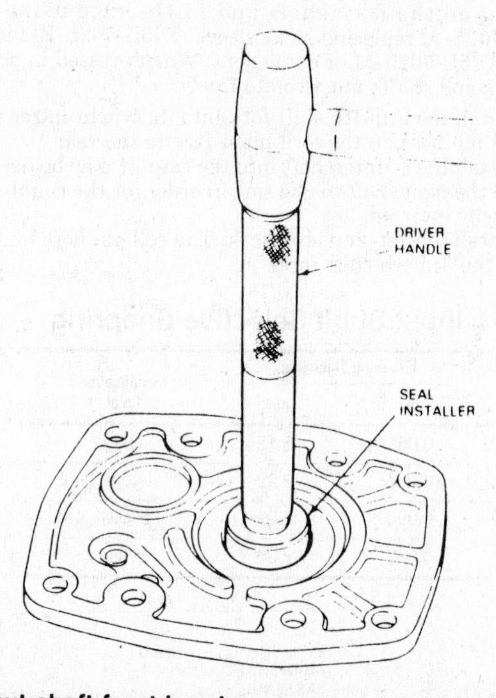

Input shaft front bearing cover seal, installation

tainer dimensions. This will give the required maximum selective shim size to obtain a 0–0.10mm clearance.

e. Measure and install the appropriate size selective shim in the front bearing retainer.

13. On 1989–91 models, before installing the bearing retainer the final assembled clearance "C" must be determined using the following procedure:

a. Measure the distance from the face of the transmission housing to the face of the bearing.

b. Measure the depth of the recess in the bearing retainer.

c. To determine the correct thickness add 0.2mm to the dimension measured in Step b. This is allowance for gasket thickness.

d. Subtract the dimension measured in Step a, from the dimension measured in Step c. This is the thickness of the shim required.

NOTE: With the shim and gasket installed, the allowable clearance is 0–0.1mm.

e. Install the appropriate size selective shim in the front bearing retainer.

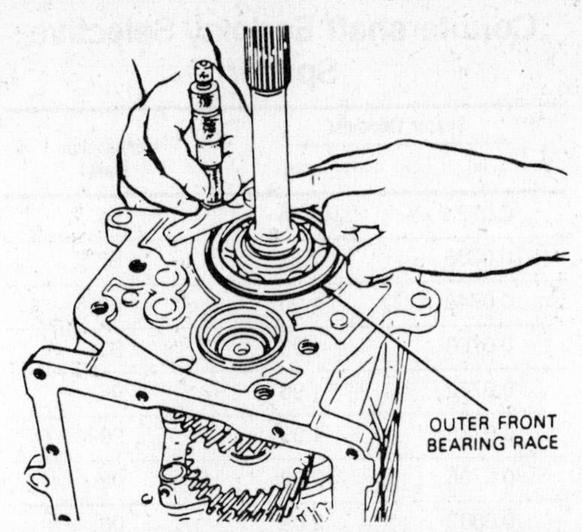

Using a depth micrometer to determine correct selective shim, for bearing retainer – 1983-88

Input Shaft Bearing Retainer-to-Bearing Selective Shim

Shim Thickness in.	Shim Thickness mm	Identification Color
0.033	0.84	Black
0.037	0.93	None
0.040	1.02	Red
0.044	1.11	White
0.047	1.20	Yellow
0.051	1.29	Blue
0.054	1.38	Green

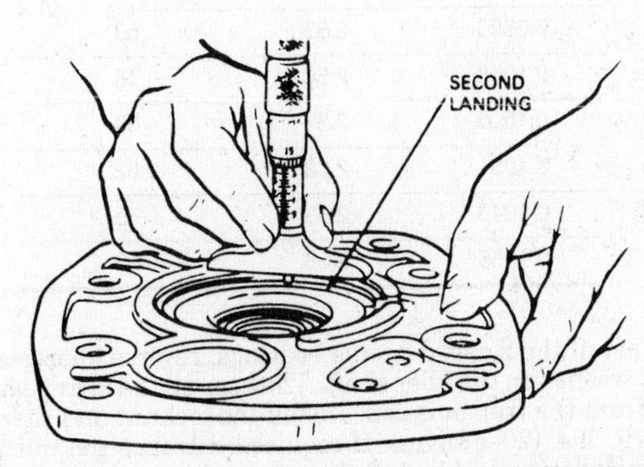

Checking input shaft front bearing retainer-to-bearing clearance – 1983–88 models

14. Install the countershaft front outer bearing race and non-selective spacer. Install the countershaft rear outer bearing race.
15. Install a new gasket between the front bearing retainer and case. Position the retainer on the case (with selective shim installed). Install the bolts and studs and tighten to 22–30 ft. lbs. (30–41 Nm).
16. Check and adjust the countershaft endplay as follows:
a. Place the transmission so the rear of the mainshaft and

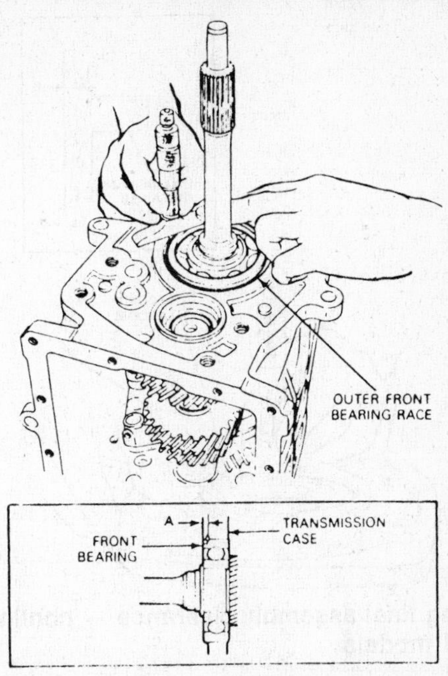

Checking input shaft front bearing retainer-to-bearing clearance – continued – 1989–91

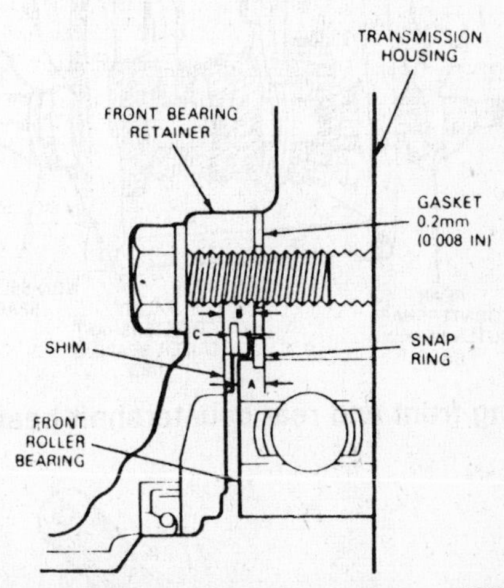

Checking final assembly clearance – 1989–91 models

countershaft face upward. Install the countershaft rear selective spacer.
b. Force the countershaft downward so it bottoms against the front bearing retainer.
c. Place a straight edge across the rear countershaft selective spacer in the case.
d. Try to turn the spacer. If the spacer turns lightly, replace the spacer with the next larger size. Install a spacer so the clearance between the spacer and straight edge is 0–0.05mm. Install the correct size spacer over the countershaft rear bearing cup.
17. Install the rear bearing retainer on the case with the 6mm bolts. Tighten to 11–16 ft. lbs. (15–21 Nm).

NOTE: Be sure the spacer installed in the previous Step does not fall out of place when installing the rear bearing retainer.

18. Position the reverse idler gear shaft on the case. Install the

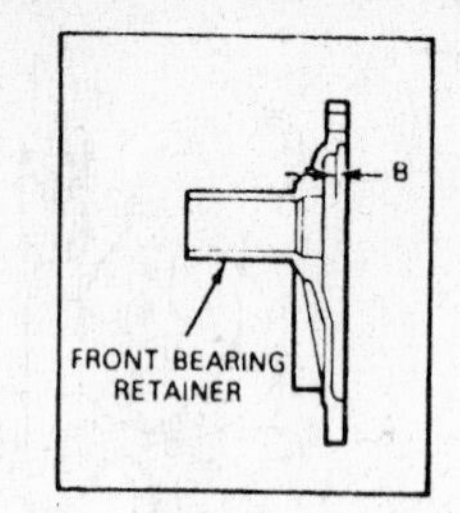

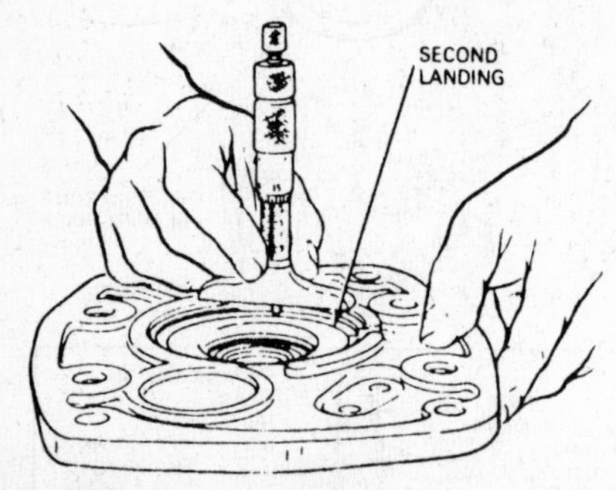

Checking final assembly clearance – continued – 1989–91 models

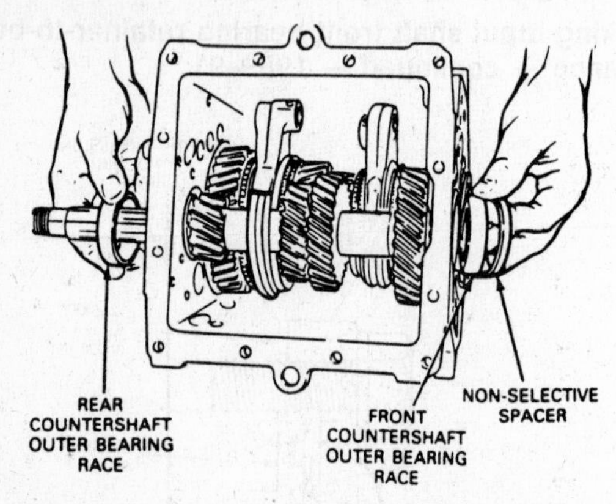

Installing front and rear countershaft bearings

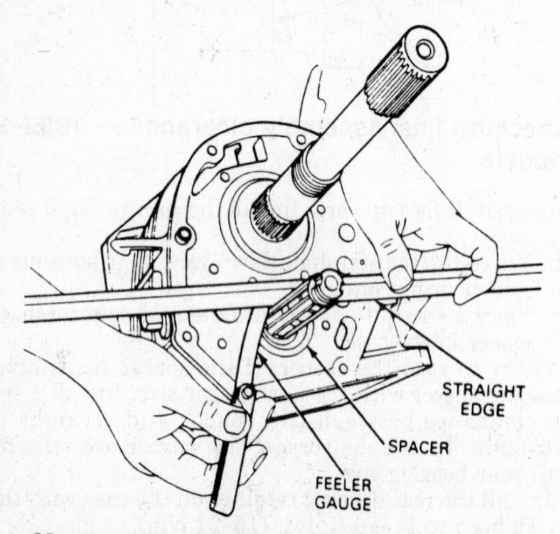

Measuring countershaft endplay

6mm allen-head bolts to act as a pilot. Install reverse idler gear shaft remover, T85T-7140-A on the shaft and drive the assembly into place. Tighten the bolts to 11–16 ft. lbs. (15–21 Nm).

Countershaft Endplay Selective Spacer

Spacer Thickness in.	Spacer Thickness mm	Identification Mark
0.0724	1.84	84
0.0736	1.87	87
0.0748	1.90	90
0.0760	1.93	93
0.0772	1.96	96
0.0783	1.99	99
0.0795	2.02	02
0.0807	2.05	06
0.0819	2.08	08
0.0831	2.11	11
0.0843	2.14	14
0.0854	2.17	17
0.0866	2.20	20
0.0878	2.23	23
0.0890	2.26	26
0.0902	2.29	29
0.0913	2.32	32
0.0925	2.35	35
0.0937	2.38	38
0.0949	2.41	41
0.0961	2.44	44
0.0972	2.47	47
0.0984	2.50	50
0.0996	2.53	53
0.1008	2.56	56
0.1020	2.59	59
0.1031	2.62	62
0.1043	2.65	65
0.1055	2.68	68

19. Install the 2 caged needle bearings, reverse idler gear and thrust washer on the idler shaft. The boss on the idler gear faces away from the transmission. Install the locknut and tighten to 15–42 ft. lbs. (20–58 Nm). If required, advance the nut to the next castellation and install the cotter pin.

NOTE: If required, cut 1 end of the cotter pin when it is bent over to prevent interference with the counter-overdrive gear.

20. Install the spacers and counter/reverse gear on the countershaft.
21. Press the reverse gear sleeve on the mainshaft using tube, T85T-7025-A, shaft sleeve replacer, T75L-7025-K, shaft collar T75L-7025-M and forcing screw, T84T-7025-B or equivalents.
22. Install the caged needle bearing and reverse gear on the mainshaft.
23. Assemble the overdrive synchronizer hub and sleeve as follows:

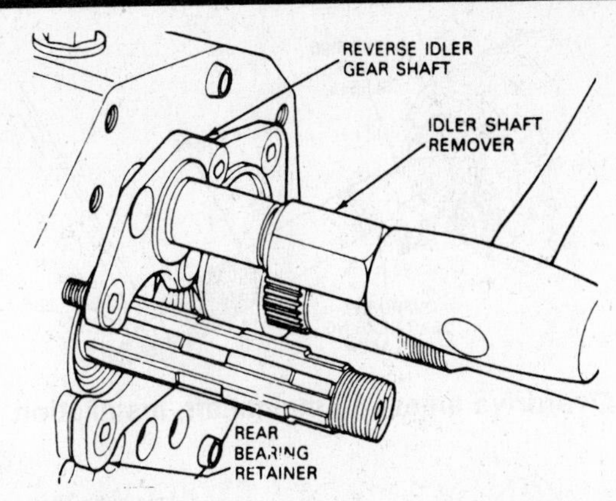

Rear bearing retainer and reverse idler shaft, installation

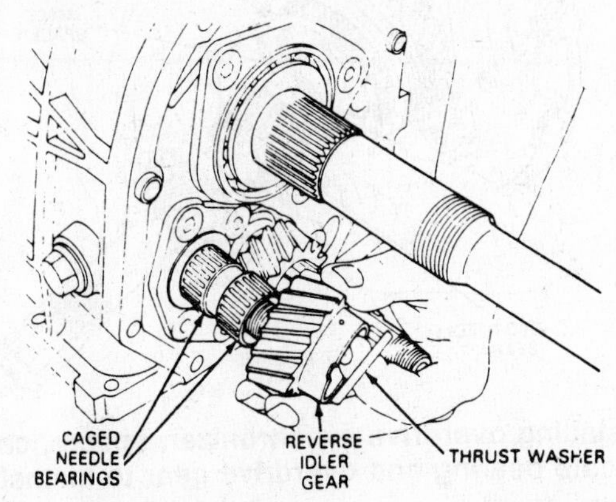

Installing reverse idler gear, caged needle bearing and thrust washer

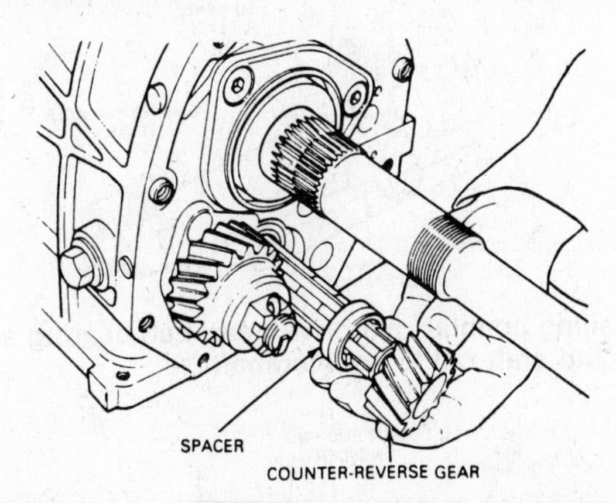

Counter-reverse gear to countershaft, installation

a. Install the hub in the sleeve. The recessed boss on the sleeve must face the front of the transmission. The large boss on the hub must face the front of the transmission.

b. When installing hub in the sleeve and the 3 keys, make sure that the single tooth between the 2 spaces will touch the key. Install the springs so the open ends do not face each other.

24. Install the overdrive synchronizer on the mainshaft. The recessed boss of the sleeve must face the front of the transmission.

25. Press the sleeve of the overdrive gear on the mainshaft using tube, T75L–7025–B, shaft sleeve replacer, T75L–7025–K, shaft collar, T75L–7025–M and overdrive gear bearing replacer, T85T–7061–A or equivalents.

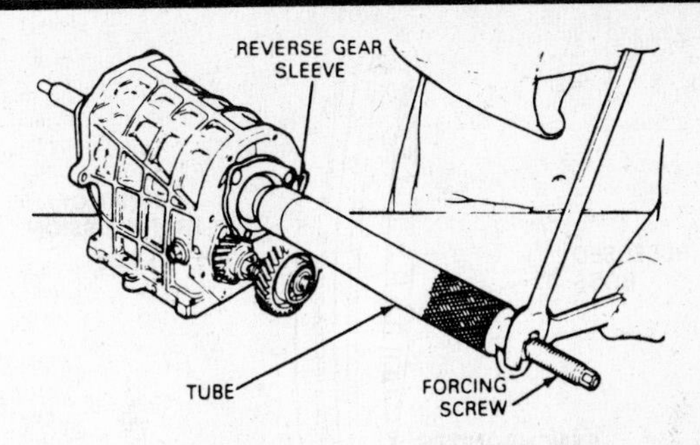

Reverse gear sleeve to mainshaft, installation

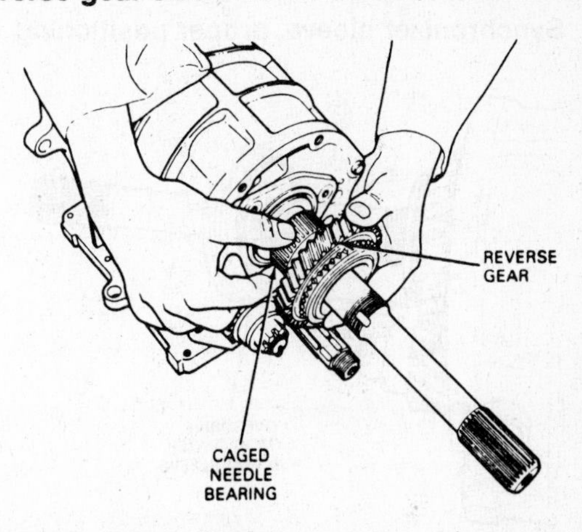

Installing caged needle bearing and reverse gear to shaft

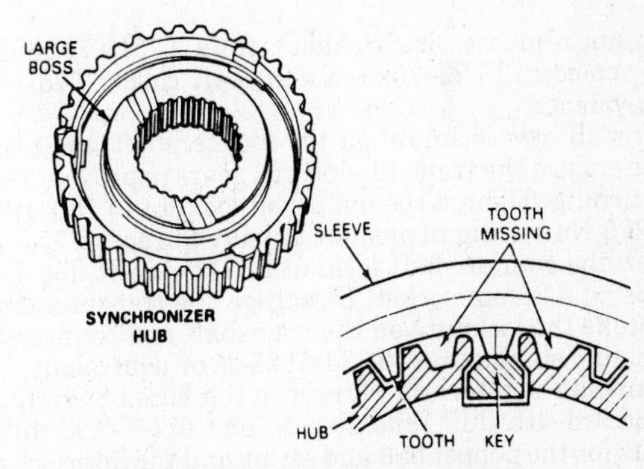

Hub and key, proper positioning

26. Install the ring on the overdrive synchronizer.

27. Slide the small spacer, caged needle bearing and overdrive gear on the mainshaft. Check the clearance between the overdrive gear and synchronizer ring. If the clearance is less than 0.23mm, replace the ring and/or overdrive gear.

28. Install the counter-overdrive gear and ball bearing onto the countershaft along with the 1st–2nd shift rail. Seat the bearing into position using countershaft bearing replacer collect, T85T–7121–A, rear countershaft bearing installer adapter, T85T–7111–A and remover and replacer tube, T77J–7025–B or equivalents. Make sure the rail engages the forks.

29. Install the lock ball and spacer on the mainshaft.

30. Place the rear bearing over the mainshaft and press the bearing in position using rack bushing holder, T81P–3504–D

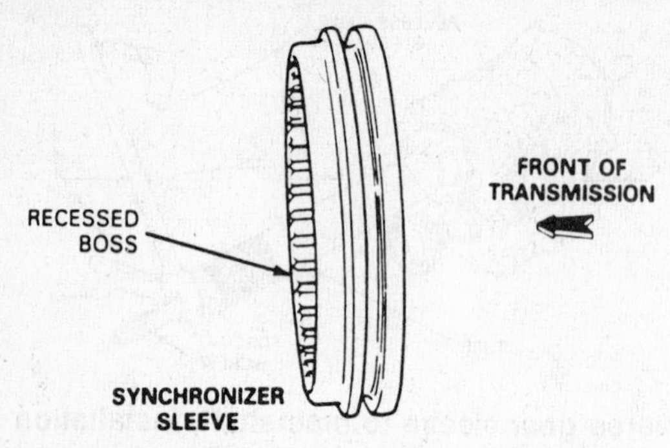

Synchronizer sleeve, proper positioning

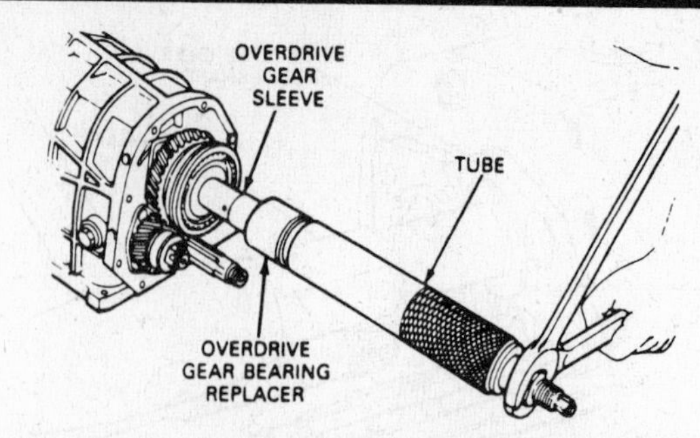

Overdrive sleeve to mainshaft, installation

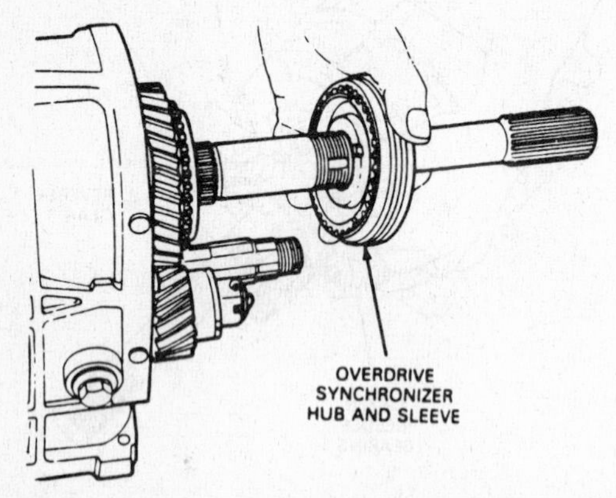

Overdrive synchronizer to mainshaft, installation

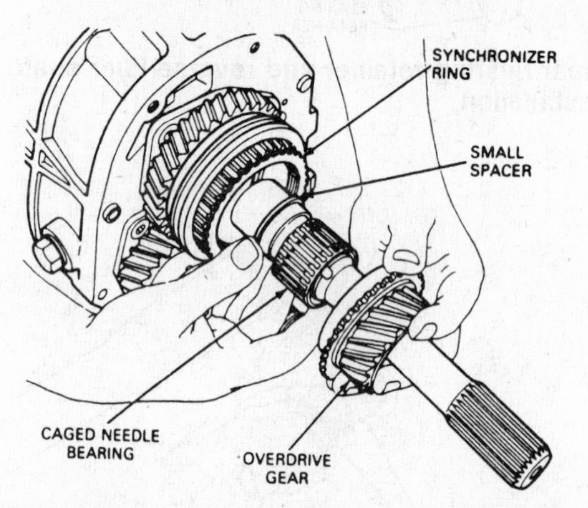

Installing overdrive synchronizer, spacer, caged needle bearing and overdrive gear to mainshaft

(or an appropriate size washer), tube T75L–7025–B, shaft sleeve replacer, T75L–7025–K and shaft collar, T75L–7025–M or equivalents.

31. Install new locknuts on the countershaft and mainshaft. Double engage the transmission in 2 gears to prevent the shafts from turning. Tighten the mainshaft locknut to 180–195 ft. lbs. (245–265 Nm) using mainshaft locknut wrench, T77J–7025–C. Tighten the countershaft locknut to 115–135 ft. lbs. (157–186 Nm) using a 30mm socket. Disengage the transmission.
32. Stake the locknuts on the mainshaft and countershaft using locknut staking tool, T77J–7025–F or equivalent.
33. Install the interlock plunger in the bore between the 1st–2nd and 3rd–4th shift rails. Reposition the 1st–2nd shift rail so the flats for the poppet ball and spring and the interlock plunger are in the correct position. Make sure the roll pin holes for the shift forks are in alignment.
34. Install the overdrive–reverse shift fork on the synchronizer sleeve. Slide the 3rd–4th shift rail through the overdrive–reverse shift fork into the case and into the 3rd–4th shift fork inside the case. Position the shift rail flats to accept the poppet balls and interlock plunger. Insert the interlock plunger in the bore between the 3rd–4th shift rail and overdrive–reverse shift rail. Make sure the roll pin holes in the fork are in alignment.
35. Insert the overdrive–reverse shift rail so it engages the forks in the case. Make sure the roll pin holes in the fork and rail are in alignment.
36. Insert the poppet ball and spring in the overdrive–reverse (upper) bore in the case. The small end of the spring should be installed toward the ball. Install the set screw and tighten until the set screw head is 6mm below the top of the bore.
37. Insert a poppet spring and poppet ball in the 3rd–4th and 1st–2nd bore (side 2 bores in the case). The small end of the spring must face towards the ball. Install and tighten the bolts.
38. Install the roll pins in the 1st–2nd and 3rd–4th shift forks.

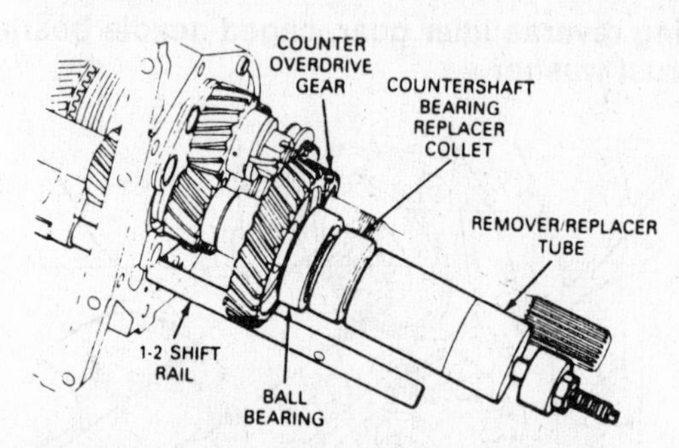

Installing counter-overdrive gear, ball bearing and 1st–2nd shift rail to countershaft

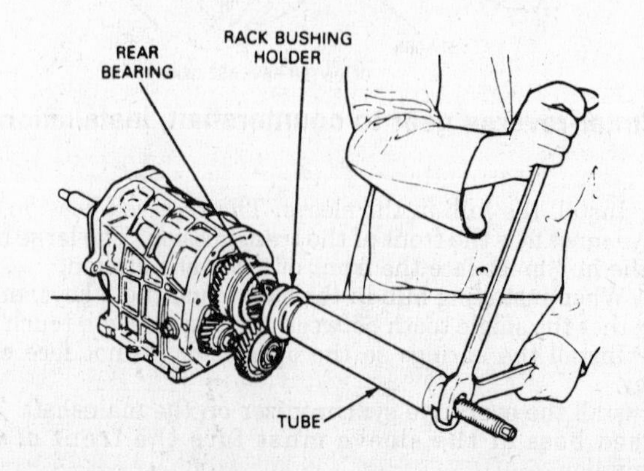

Lock ball and spacer to mainshaft, installation

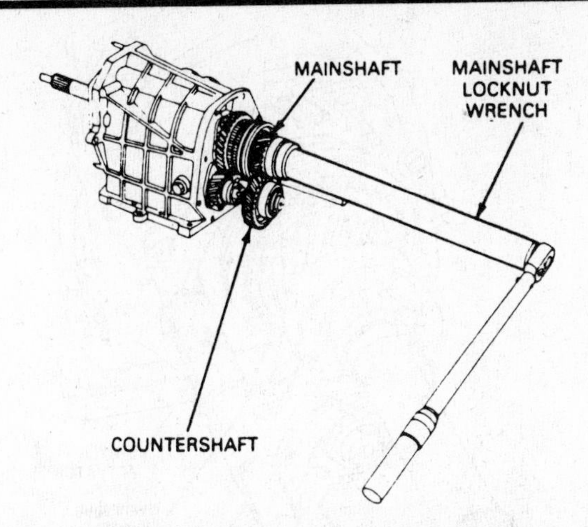

Tightening countershaft and mainshaft locknuts

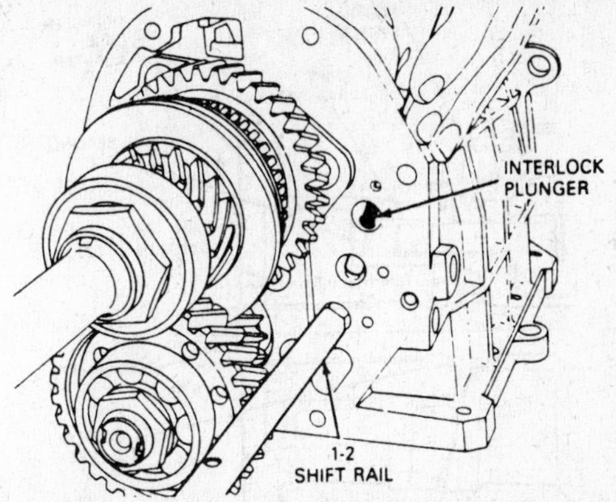

Installing interlock plunger and checking the alignment of the shift forks

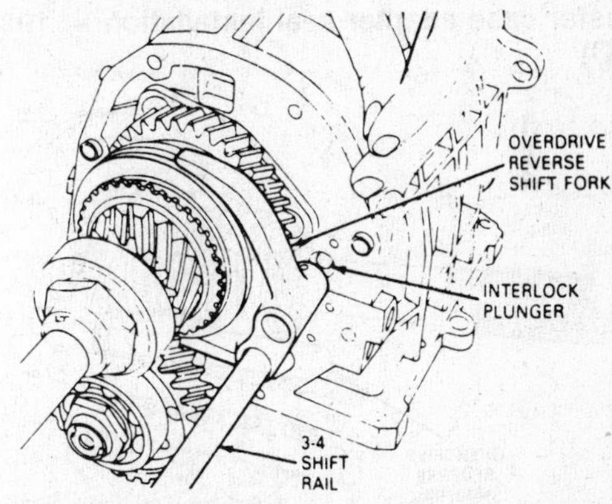

Positioning shift rail flats to accept the poppet balls and interlock plunger

39. Install the shift gates on the appropriate shift rails. Move the 1st–2nd gate and 3rd–4th gate to the rear of the rail.
40. Position a new gasket between the transmission case and the transfer case adapter/extension housing. Make sure the selector arm is out of the gates and the change shifter is at the rear of the adapter. Position the adapter on the case making sure the shift gates clear the adapter. Make sure the shift rails and rear bearings line up with the bores in the adapter.
41. Install the 3 sizes of bolts (35mm, 55mm and 110mm) in the appropriate holes in the adapter. Tighten the bolts to 11–16 ft. lbs. (15–21 Nm).
42. On 1989–91 4WD models, lubricate the inside diameter of the oil seal and position the seal around the output shaft, into the transfer case adapter. Then, press the seal into place.
43. Install the neutral return plungers, springs and bolts in the adapter/control housing. The longer plunger with the slot for the detent ball is installed on the right side of the adapter.
44. Position the shaft gates so the roll pin holes in the gates and rails are in alignment. Install the roll pins through the access holes. Install the access hole plugs.
45. Position the pan and new gasket on the case. Install the bolts and tighten to 11–16 ft. lbs. (15–21 Nm). Do not overtighten. Install the drain plug and tighten to 25–32 ft. lbs. (35–44 Nm).
46. On 4WD models, insert the plunger detent ball and spring in the hole above the neutral return plunger in the adapter case.
47. On the FM145, make sure the stopper bracket assembly on the cover for the transfer case adapter moves smoothly. Position a new gasket on the adapter and install the housing cover. Install and tighten to bolts to 11–16 ft. lbs. (15–21 Nm).
48. On the FM146, make sure the reverse lockout assembly on the control housing moves smoothly. Position a new seal on the adapter/extension housing and install the control housing assembly. Install and tighten the retaining bolts to 11–16 ft. lbs. (15–21 Nm).
49. Install the back-up lamp switch and the shift indicator lamp switch in the adapter/extension housing.
50. On 2WD models, install the rear oil seal into the extension housing.
51. Remove the fill plug and fill the transmission to the bottom of the fill hole with standard transmission lube (SAE 80W) D8DZ-19C547-A (ESP-M2C83-C) or equivalent. The fluid capacity is 2.4 quarts (1.3L). Install the fill plug and tighten to 22–25 ft. lbs. (30–34 Nm).
52. Position the clutch slave cylinder on the input shaft. Position the clutch housing on the transmission case and install and tighten the nuts.

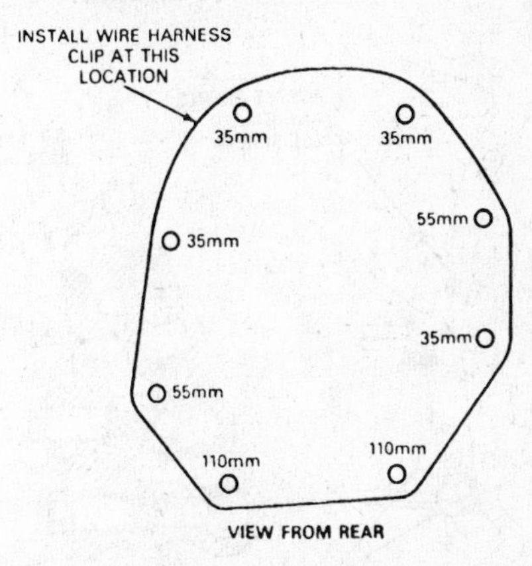

Adapter or extension housing, appropriate bolts pattern

M5OD Transmission Overhaul

DISASSEMBLY

Before disassembly, clean the exterior of the transmission assembly before any attempt is made to disassemble it, in order to prevent dirt or other foreign material from entering the transmission assembly and damaging its internal parts. If steam cleaning is done to the exterior of the transmission, immediate

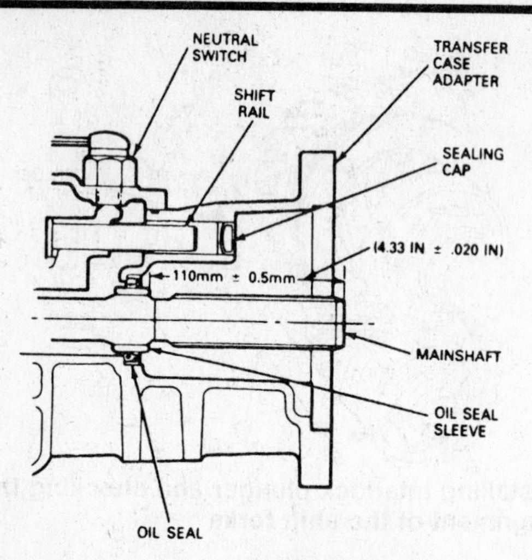

Transfer case adapter seal installation — 1989–91 (4WD)

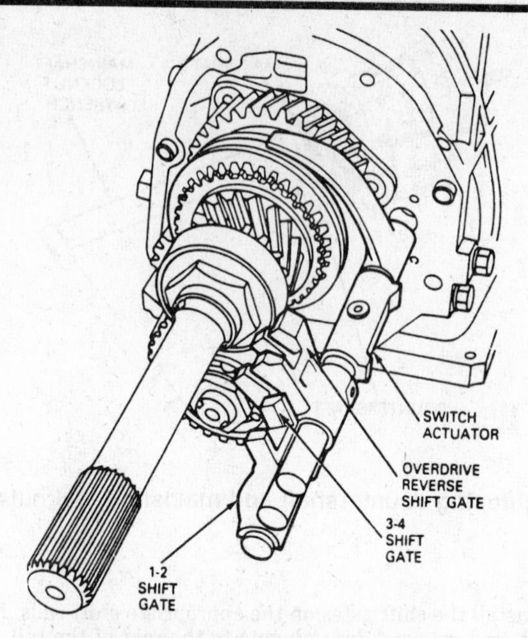

Installing gates on correct rails

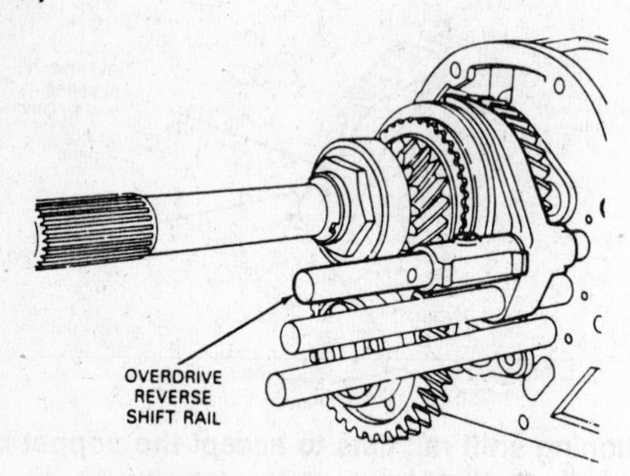

Overdrive shift rails, installation

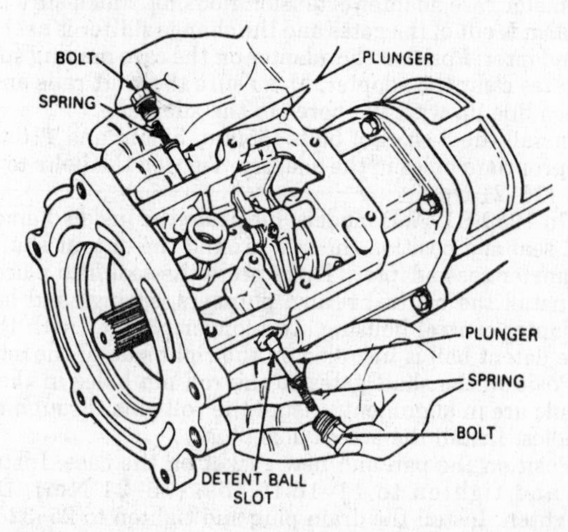

Installing return plungers and springs — right and left sides

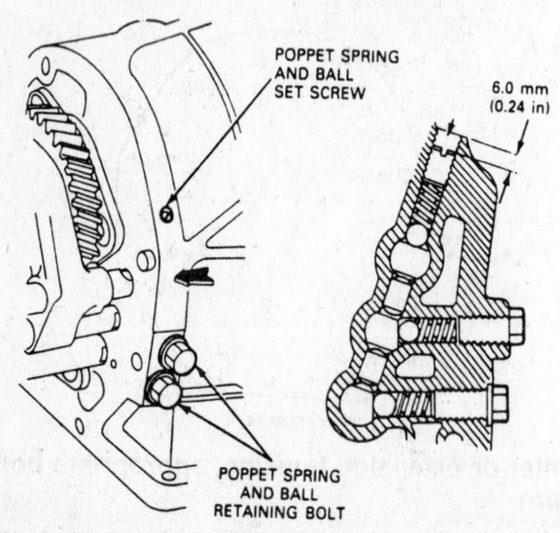

Installing the poppet balls and springs in overdrive-reverse upper bore of case

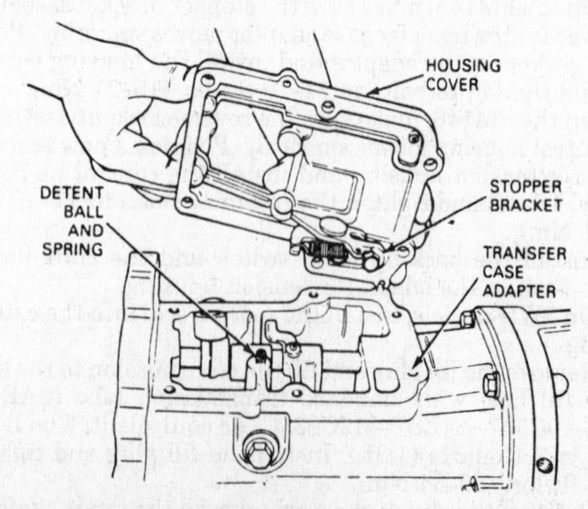

Installing stopper bracket and cover on transfer case adapter — FM145 shown

disassembly should be done to avoid rusting, caused by condensation forming on the internal parts.

1. Remove the transmission drain plug and drain the transmission fluid into a suitable drain pan.

2. Remove the shift lever and dust boot is necessary. Remove the 10 top cover assembly retaining bolts. Remove the top cover assembly.

Special Tools

Number	Description	Number	Description
T50T-100-A	Impact slide hammer	T75L-7025-K	Replacing shaft sleeve
D84L-1123-A	Bearing splitter	T75L-7025-M	Shaft collar
T81P-3504-D	Rack bushing holder	T77F-7025-C	Bearing collet sleeve
T80T-4000-W	Driver handle	T77J-7025-B	Tube
D80P-4201-A	Depth micrometer	T77J-7025-C	Mainshaft locknut wrench
D82L-4201-C	Metric depth micrometer	T77J-7025-F	Mainshaft locknut staking tool
D83L-4201-A	Straight edge	T77J-7025-H	Puller
T77F-4220-B1	Puller	T77J-7025-J	Puller ring
T85T-7011-A	Seal installer	T85T-7025-A	Tube
T75L-7025-B	Tube	T85T-7061-A	Overdrive gear bearing replacer
T75L-7025-C	Tube	T85T-7065-A	Mainshaft bearing collet remover
T75L-7025-D	Bearing collet	T85T-7121-A	Countershaft bearing replacer
T75L-7025-G	Bearing collet sleeve	T85T-7140-A	Reverse idler gearshaft replacer
T84T-7025-B	Forcing screw		

3. Remove the 9 extension housing retaining bolts. Pry gently at the locations provided on the extension housing and transmission case. Remove the extension housing from the case.

NOTE: On the 2WD vehicles, if it is necessary to remove the rubber seal from the extension housing, the extension housing must be installed on the transmission case. Remove the extension housing rear seal using a suitable seal removal tool.

4. On 2WD vehicles, if necessary, remove rear oil passage from the extension housing. Remove and discard anti-spill oil seal from the output shaft. Remove the speedometer drive gear and seal.

NOTE: For reference during assembly, observe and record speedometer drive gear color. Depending upon application, 1 or 2 different speedometer drive gear may be installed. It will be color coded either green or white. Speedometer drive gear colors and corresponding part numbers are as follows, white E8TZ–17285–B and green E8TZ–17285–C.

5. Lock the transmission into 1st/3rd gears. Using countershaft locknut staking tool T–77J–7025–F or equivalent, release the staked areas securing the output shaft and countershaft locknut. The staked areas of locknuts must be fully released or damage to shaft threads will result.
6. Remove and discard the countershaft rear bearing locknut. Remove the countershaft bearing and thrust washer.
7. Using the mainshaft locknut wrench tool Y–88T–7025–A and remover tube tool T–75L–7025–B or equivalents, remove and discard output shaft locknut. Remove the reverse idler shaft fixing bolt. Remove the reverse idler gear assembly by grasping and pulling rearward.
8. Remove the output shaft rear bearing from the output shaft using remover/replacer tube tool T75L–7025–B, TOD forcing screw tool T84T–7025–B, bearing puller tool T77J–7025–H and puller ring tool T77J–7025–J or equivalents.
9. Using a suitable brass drift and hammer, drive the reverse gear from the output shaft. Remove the sleeve from the output shaft. Remove the counter reverse gear with the 2 needle bearings and reverse synchronizer ring.
10. Remove the thrust washer and split washer from the countershaft. Remove the 5th/reverse shift rod fixing bolt.

Torque Specifications

Description	ft. lb.	Nm
Clutch housing to engine	28–38	38–51
Clutch housing to transmission	30–40	41–54
Countershaft locknut	115–137	157–186
Damper to insulator on cross-member	71–94	97–127
Drain plug	25–32	35–44
Fill plug	22–25	30–34
Front bearing retainer to case	22–30	30–41
Housing cover to transfer case adapter	11–16	15–21
Insulator to transmission	60–80	81–108
Crossmember to side rail bracket	65–85	88–115
Mainshaft locknut	180–195	245–265
Pan to case	11–16	15–21
Rear bearing to case	22–30	30–41
Reverse idler gear nut	15–42	20–58
Reverse idler gearshaft assembly to case	11–16	15–21
Shift lever assembly to transfer case adapter	6–10	8–14
Starter motor to clutch housing	15–20	21–27
Stud to front retainer and case	22–30	30–41

11. Remove the 5th/reverse synchronizer hub and sleeve as an assembly. Remove the 5th/reverse shift fork and rod. Do not separate the steel ball and spring (remove from the shift fork groove) unless necessary.
12. Remove the 5th gear synchronizer ring. Remove the 5th/reverse counter lever lockplate retaining bolt and inner circlip. Remove the counter lever assembly from the transmis-

sion case. Do not remove the Torx® nut retaining the counter lever pin at this time.

13. Remove the 5th gear (counter) with needle bearing. Remove the 5th gear from the output shaft using a bearing collet sleeve for an 89mm bearing collet tool T75L–7025–G, remover/replacer tube tool T85T–7025–A, TOD forcing screw tool T84T–7025–B and gear removal collet tool T88T–7061–A or equivalents.

14. Remove the 5th gear sleeve and position ball, TOD forcing screw tool T84T–7025–B, countershaft 5th gear sleeve puller tool T84T–7025–J, gear removal collet tool T88T–7025–J1 and remover/replacer tube tool T77–7025–B or equivalents.

15. Remove the 6 center bearing cover retaining bolts. Remove the center bearing cover. For reference during assembly, observe that the reference arrow in the middle of the center bearing cover points upward. Observe the flanged side of the center bearing cover faces inward.

16. Remove the 6 front bearing cover attaching bolts. Remove the front bearing cover by threading 2 of the originally installed retaining bolts into the front bearing cover service bolt locations (9 o'clock and 3 o'clock). Alternately tighten bolts until the front bearing cover can be lifted away by hand. Remove and discard the front bearing cover oil baffle. The bolt threaded into the service bolt locations will bottom out and lift the front bearing cover away from the transmission case. Do not remove the plastic scoop ring from the input shaft at this time.

17. Remove the oil trough retaining bolt and oil trough from the upper transmission case. Pull the input shaft forward and remove the input bearing outer race. Pull the output shaft rearward.

18. Pull input shaft forward and separate it from the output shaft. Incline output shaft upward and lift it from the transmission case. Remove the input shaft from the transmission case.

19. Remove the countershaft bearing outer races (front and center) by moving the countershaft forward and rearward. Pull the countershaft rearward enough to permit the tool clearance behind the front countershaft bearing. Using bearing race puller tool T88T–7120–A and slide hammer tool T50T–100–A or equivalents, remove the front countershaft bearing. Tap the bearing gently during removal. A forceful blow can cause damage to the bearing or transmission case.

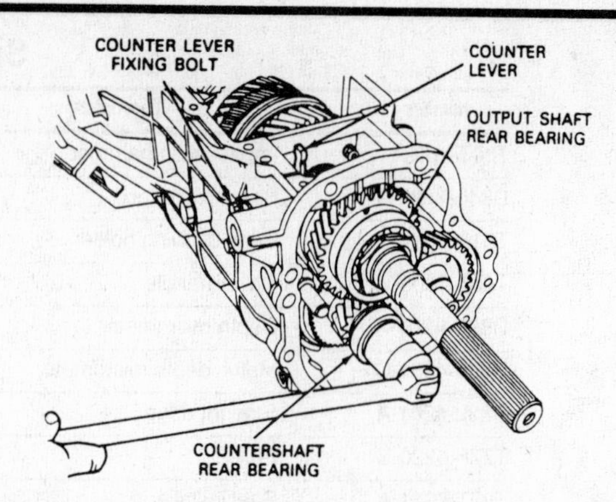

Removing countershaft bearing and thrust washer

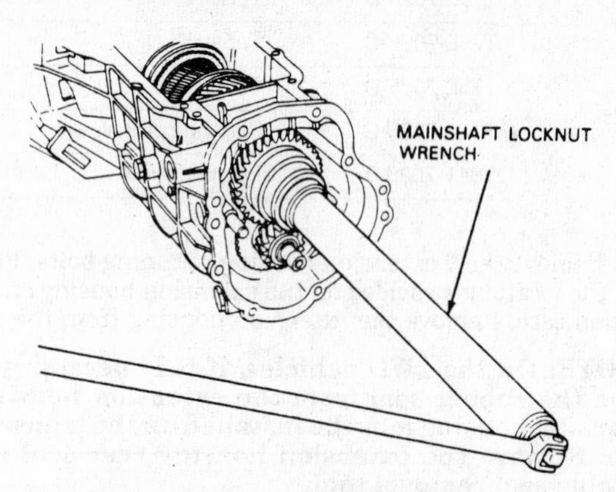

Removing output shaft locknut

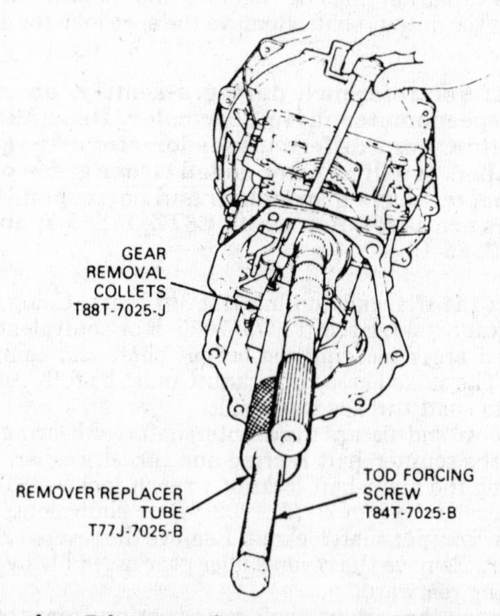

Removing 5th gear sleeve and positioning ball

CLEANING

During overhaul, all components of the transmission (except bearing assemblies) should be thoroughly cleaned with solvent and dried with air pressure prior to inspection and reassembly.

1. Clean the bearing assemblies as follows:

NOTE: Proper cleaning of bearings is of utmost importance. Bearings should always be cleaned separately from other parts.

a. Soak all bearing assemblies in clean solvent or fuel oil. Bearings should never be cleaned in a hot solution tank.

b. Slush bearings in solvent until all old lubricant is loosened. Hold races so that bearings will not rotate; then clean bearings with a soft bristled brush until all dirt has been removed. Remove loose particles of dirt by tapping bearing flat against a block of wood.

c. Rinse bearings in clean solvent; then blow bearings dry with air pressure.

WARNING: Do not spin bearings while drying!

d. After drying, rotate each bearing slowly while examining balls or rollers for roughness, damage, or excessive wear. Replace all bearings that are not in first class condition.

NOTE: After cleaning and inspecting bearings, lubricate generously with recommended lubricant, then wrap each bearing in clean paper until ready for reassembly.

2. Remove all portions of old gaskets from parts, using a stiff brush or scraper.

INSPECTION

1. Inspect all parts for discoloration or warpage.
2. Examine all gears and splines for chipped, worn, broken or

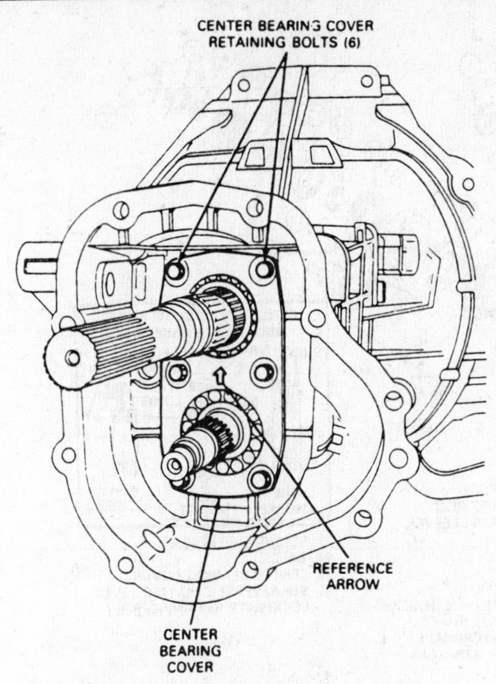

Removing center bearing cover

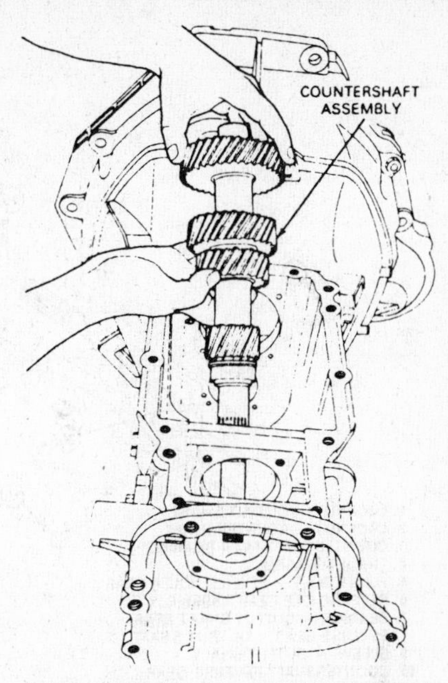

Removing countershaft assembly

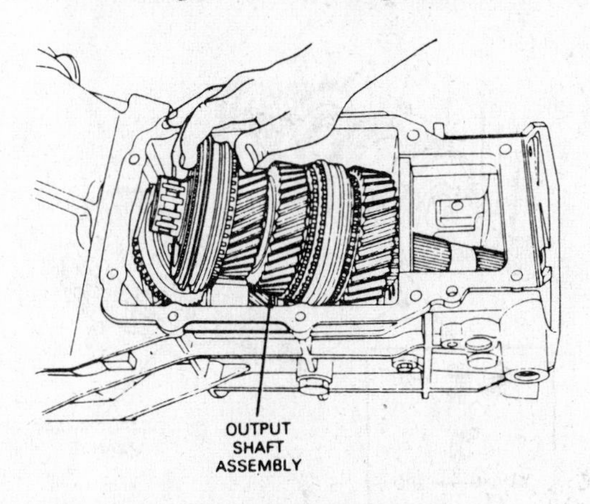

Removing output shaft assembly

nicked teeth. Small nicks or burrs may be removed with a fine abrasive stone.

3. Inspect the breather assembly to make sure that it is open and not damaged.
4. Check all threaded parts for damaged, stripped, or crossed threads.
5. Replace all gaskets, oil seals and snaprings.
6. Inspect housings, retainers and covers for cracks or other damage. Replace the damaged parts.
7. Inspect keys and keyways for condition and fit.
8. Inspect shift forks for wear, distortion or any other damage.
9. Check detent ball springs for free length, compressed length, distortion or collapsed coils.
10. Check bearing fit on their respective shafts and in their bores or cups. Inspect bearings, shafts and cups for wear.

NOTE: If either bearings or cups are worn or damaged, it is advisable to replace both parts.

11. Inspect all bearing rollers or balls for pitting or galling.
12. Examine detent balls for corrosion or brinneling. If shift bar detents show wear, replace them.
13. Inspect the synchronizer ring for wear. To check the wear of the synchronizer ring, fit the synchronizer ring evenly to the gear cone. Measure the clearance between the side faces of the synchronizer ring and gear with a feeler gauge. If the clearance is less than 0.8mm, replace the synchronizer ring or gear.
14. Check the contact surfaces of the shift fork and clutch hub sleeve for evidence of wear or damage. Measure from the shift fork to the clutch hub sleeve. If the clearance exceeds 0.8mm, replace the shift fork/clutch hub sleeve.
    a. The standard clearance for the 1st/2nd and 3rd/4th shift fork/clutch hub sleeves is 0.08–0.35mm. The maximum is 0.8mm.
    b. The standard clearance for the 5th/reverse shift fork/clutch hub sleeves is 0.08–0.38mm. The maximum is 0.8mm.
15. Replace all worn or damaged parts. When assembling the transmission, coat all moving parts with recommended lubricant.

Input Shaft Disassembly and Assembly

1. Remove and discard the plastic scoop ring.
2. Press the tappered roller bearing from the input shaft using bearing cone remover tool T71P–4621–B or equivalent.

INSPECTION

1. Clean and inspect the input shaft and gear assemblies.
2. Replace any and all worn or damaged components, as necessary.

ASSEMBLY

1. Install the input shaft tappered roller bearing onto the input shaft using a suitable press and bearing cone replacer tool T88T–7025–B or equivalent.
2. Install a plastic scoop ring onto the input shaft. Manually rotate the ring clockwise to ensure that the input oil holes properly engage the scoop ring. A click should be heard as the scoop ring notches align with the input shaft oil holes.

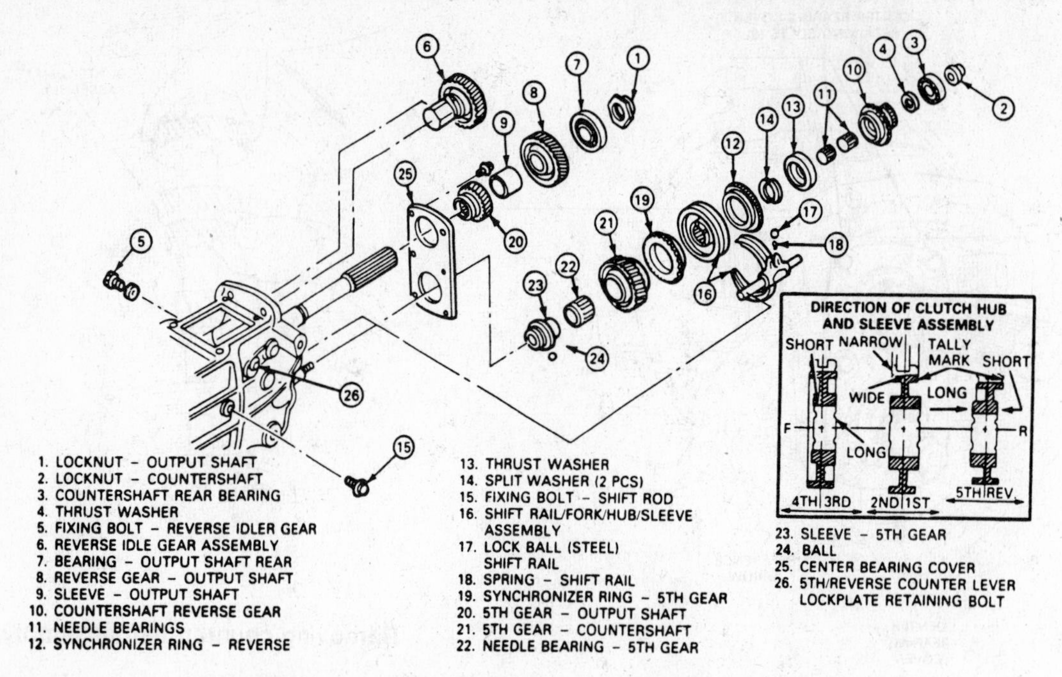

Rear housing disassembled — exploded view

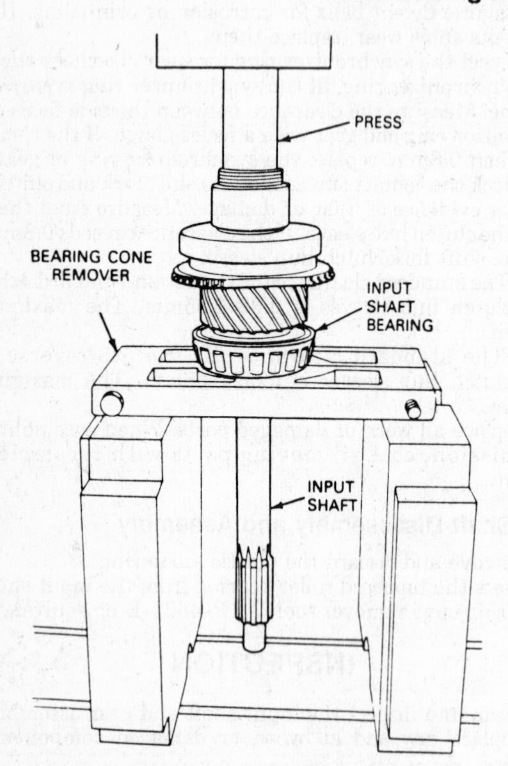

Input shaft bearing installation

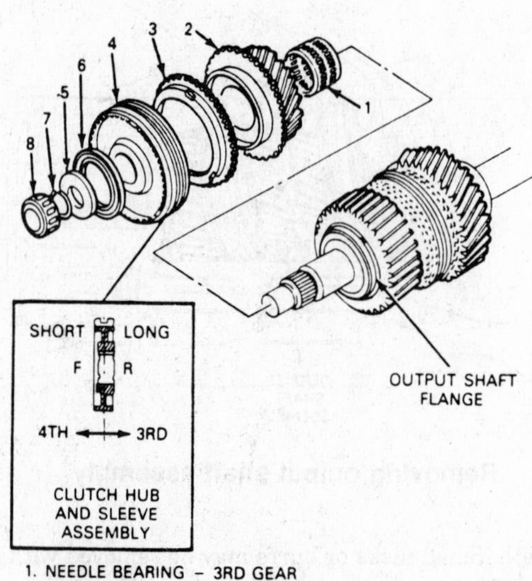

1. NEEDLE BEARING – 3RD GEAR
2. THIRD GEAR
3. SYNCHRONIZER RING – 3RD GEAR
4. CLUTCH HUB AND SLEEVE ASSEMBLY – 3RD/4TH
5. SPACER
6. NEEDLE BEARING (PLAIN)
7. RETAINING RING
8. ROLLER BEARING – PILOT BEARING

Disassembling output shaft

Output Shaft Disassembly and Assembly

1. Pull back and separate the 3rd gear and the 2nd gear from the output shaft flange.
2. Remove the pilot bearing (needle roller), snapring, needle bearing and spacer from the front (short side of the flange) of the output shaft.
3. Position the front (short side of the flange) of the output shaft so that it faces upward. Lift off the following components as an assembly.
   a. Clutch hub and sleeve assembly (3rd/4th).
   b. Synchronizer ring (3rd).
   c. 3rd gear.
   d. Needle bearing.
4. Position the output shaft with the rear end (long side of the flange) facing upward. Position the output shaft into the press with the press cradle contacting the lower part of 2nd gear.

NOTE: Ensure that the output shaft flange does not contact or ride up onto the press cradle. Improper positioning of the output shaft can cause component damage.

5. Press off the following components as a unit, center bearing, 1st gear sleeve, 1st gear needle bearing, 1st/2nd clutch hub and sleeve assembly, 1st/2nd synchronizer rings, 2nd gear and the needle bearing using bearing replacer tool T53T–4621–B and bearing cone replacer tool T88T–7025–B or equivalents.

INSPECTION

1. Clean and inspect the output shaft and gear assemblies.
2. Replace any and all worn or damaged components as necessary.
3. Check the output shaft for run-out by mounting the shaft between V-blocks and applying a dial indicator tool to several places along the shaft.
4. The standard reading of the dial indicator for the run-out should be less than 0.05mm. If the run-out exceeds the specifications, replace the mainshaft.
5. Replace the input shaft if the splines are damaged. If the needle bearing surface in the bore of the bearing is worn or rough, or if the cone surface is damaged, replace the shaft.

ASSEMBLY

During assembly apply the recommended transmission fluid to all rotating or sliding parts.

1. Position the output shaft so that the rear end (long side of the flange) faces upward. Install the following components in the order listed as follows:
   a. 2nd gear needle bearing.
   b. 2nd gear.
   c. 2nd gear synchronizer rings.
   d. 1st/2nd clutch hub and sleeve assembly.
   e. 1st gear synchronizer rings.
   f. 1st gear needle bearing.
   g. 1st gear.
   h. 1st gear sleeve.
   i. Center bearing (inner).

NOTE: To install the components onto the output shaft, press the components into position using bearing replacer tool T53T–4621–B and bearing plate tool T75L–1165–B or equivalents.

2. Ensure that the center bearing race is installed into the transmission case.
3. When installing the 1st/2nd clutch hub and sleeve, ensure that the smaller width of the sleeve faces 2nd gear (front) side. Ensure that the reference marks face the rear of the transmission, they reference the synchronizer key installation position.
4. Install the center bearing to the output shaft.
5. Position the output flange so that the front (short side) of the output shaft flange faces upward. Install the 3rd gear needle bearing, 3rd gear and 3rd gear synchronizer ring.
6. Install the 3rd/4th clutch hub and sleeve as follows:
   a. Mate the clutch hub synchronizer key groove with reference mark on the clutch hub sleeve. The mark should face rearward.
   b. Install the longer flange on the clutch hub sleeve toward 3rd gear (rear) side.

NOTE: The front and rear sides of the clutch hub are the same except for the reference mark on one side.

7. Install the spacer, needle bearing (install the rollers upward), retaining ring and pilot bearing (roller). Install the original retaining ring. Check the clutch hub endplay using a feeler gauge.

Output Shaft Retaining Ring

Part Number	Thickness in.	Thickness mm
E8TZ-7030-A	0.059	1.50
E8TZ-7030-B	0.061	1.55
E8TZ-7030-C	0.0629	1.60
E8TZ-7030-D	0.0649	1.65
E8TZ-7030-E	0.0669	1.70
E8TZ-7030-F	0.0688	1.75
E8TZ-7030-G	0.0708	1.80
E8TZ-7030-H	0.0728	1.85
E8TZ-7030-J	0.0748	1.90
E8TZ-7030-K	0.0767	1.95

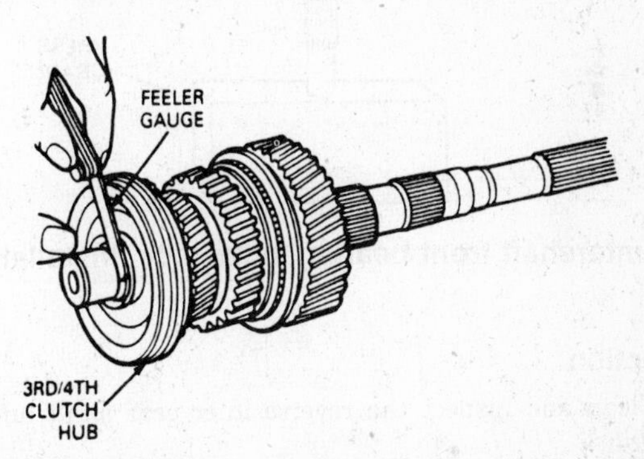

Measuring endplay at 3rd–4th clutch hub on output shaft

8. If necessary, adjust the 3rd/4th clutch hub endplay to 0–0.05mm by selecting a required retaining ring.

Countershaft Disassembly

1. Using a suitable press and bearing cone remover tool T71P–4621–B or equivalent, remove the countershaft center bearing inner race.
2. Using a press and bearing spliter tool D84L–1123–A or equivalent, remove the countershaft front bearing inner race.

Inspection

1. Clean and inspect the output shaft and gear assemblies.
2. Replace any and all worn or damaged components, as necessary.

Assembly

1. Using a suitable press, a press plate and bearing replacer tool T53T–4621–B or equivalent, install the center bearing inner race.
2. Using a suitable press, a press plate and bearing replacer tool T53T–4621–B or equivalent, install the countershaft front bearing inner race.

Reverse Idler Gear Shaft Disassembly

1. Remove the retaining ring, spacer, idler gear, needle bearing and thrust washer from the reverse idler gear shaft.

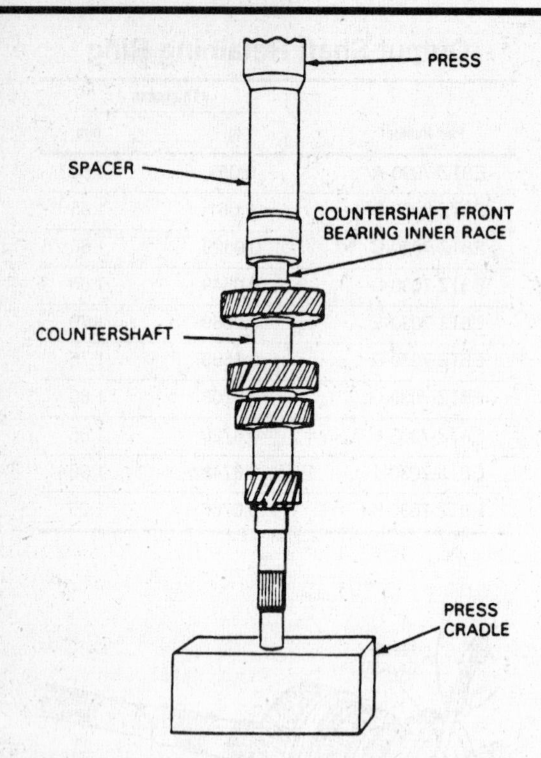

Countershaft front bearing inner race, installation

Inspection

1. Clean and inspect the reverse idler gear shaft and gear assemblies.
2. Replace any and all worn or damaged components, as necessary.

Assembly

1. Install the thrust washer onto the reverse idler gear shaft. Ensure that the tab on the thrust washer mates with the groove on the reverse idler shaft to prevent rotation of the thrust washer.
2. Install the needle bearing, idler gear and spacer.
3. Install the original retaining ring onto the reverse idler gear shaft. Insert a feeler gauge between the retaining ring and reverse idler gear to measure the reverse idler gear endplay.
4. Using the proper size retaining rings, adjust the endplay to 0.1–0.2mm.

Top Cover Disassembly

1. If necessary, remove the dust boot and shift lever from the top cover. Remove the 3 dust cover screws and remove the dust cover.

NOTE: For reference during assembly, notice that the grooves in the bushing align with the slots in the lower shift lever pivot ball. Notice that the notch in the lower shift lever faces toward the front of the transmission.

2. Position the top cover assembly into a suitable holding fixture.
3. Remove the backup lamp switch from the top cover. Remove the backup lamp switch pin from the groove in the top cover. There is only 1 type of backup lamp pin used.
4. Using a $^{5}/_{32}$ in. (4mm) drift punch, remove the spring pins retaining the shift forks to the shift rail. Discard the original spring pins.
5. Ensure that the 5th/reverse shift rail is in the fully forward position. Remove the spring pin from the end of the 5th/reverse rail.
6. Remove the 3 rubber plugs blocking the shift rod service bores.

NOTE: Perform the following shift rail removal procedures with great care. Cover the lock ball bores and friction device and spring seats with a clean cloth held firmly in place during shift rail removal. Failure to firmly cover the lock ball bores and friction device can result in component loss when the ball/friction device and spring forcefully leave their installed positions. Be sure to wear safety goggles while performing the shift rail removal procedure.

7. Remove the 1st/2nd shift rail from the top cover through the service bore. If necessary, insert a $^{5}/_{16}$ in. (8mm) drift punch through the spring pin bore and gently rock the shift rail from side to side while maintaining rearward pressure.
8. Remove the 3rd/4th shift rail from the top cover through the service bore. If necessary, insert a $^{5}/_{16}$ in. (8mm) drift punch through the spring pin bore and gently rock the shift rail from side to side while maintaining rearward pressure.
9. Remove the 5th/reverse cam lockout plate retaining bolts and remove the 5th/reverse cam lockout plate.

Reverse Idler Gear Shaft Retaining Ring

Part Number	Thickness in.	Thickness mm
E8TZ-7156-F	0.059	1.5
E8TZ-7156-E	0.0629	1.6
E8TZ-7156-D	0.0669	1.7
E8TZ-7156-C	0.0708	1.8
E8TZ-7156-B	0.0748	1.9

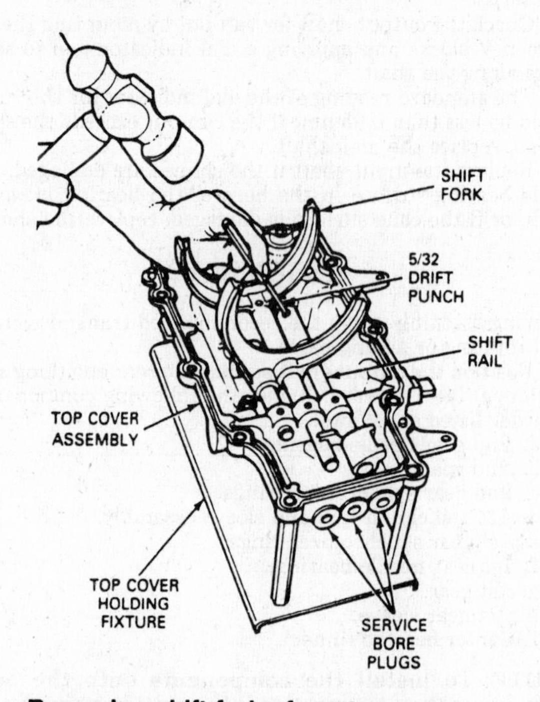

Removing shift forks from top cover

Inspection

1. Clean and inspect the shift rail and shift fork assemblies.
2. Replace any and all worn or damaged components, as necessary.

Assembly

1. With the top cover in a suitable holding fixture. Position the 5th/reverse cam lockout plate to the top cover. Install the 5th/reverse cam lockout plate retaining bolts and torque them to 6–7 ft. lbs. (8–10mm).
2. Position the 3rd/4th shift rail into the top cover through the service bore. If necessary, insert a 5/16 in. (8mm) drift punch through the spring bore and gently rock the shift rail from side to side while maintaining forward pressure. Position the detent ball and spring into the top cover spring seats.
3. Compress the detent ball and spring assembly using a suitable tool and push the shift rail into position over the detent ball. Position the friction device and spring into the top cover seats. Compress the fiction device and spring assembly using a suitable tool and push the shift rail into position over the friction device. Install the spring pins, retaining the shift rail to top cover. Install the spring retaining 3rd/4th shift fork to the shift rail.
4. Position the 1st/2nd shift rail into the top cover through the service bore. If necessary, insert a 5/16 in. (8mm) drift punch through the spring bore and gently rock the shift rail from side to side while maintaining forward pressure. Position the detent ball and spring into the top cover spring seats.
5. Compress the detent ball and spring assembly using a suitable tool and push the shift rail into position over the detent ball. Position the friction device and spring into the top cover seats. Compress the fiction device and spring assembly using a suitable tool and push the shift rail into position over the friction device. Install the spring pins, retaining the shift rail to top cover. Install the spring retaining 1st/2nd shift fork to the shift rail.
6. Position the 5th/reverse shift rail into the top cover through the service bore. If necessary, insert a 5/16 in. (8mm) drift punch through the spring bore and gently rock the shift rail from side to side while maintaining forward pressure. Position the detent ball and spring into the top cover spring seats. Compress the detent ball and spring assembly using a suitable tool and push the shift rail into position over the detent ball. Position the friction device and spring into the top cover seats.
7. Compress the fiction device and spring assembly using a suitable tool and push the shift rail into position over the friction device. Install the spring pins, retaining the shift rail to top cover. Install the spring retaining 5th/reverse shift fork to the shift rail.
8. Install the rubber plugs into the service bores. Install the interlock pins into the 1st/2nd and 3rd/4th shift rails. Ensure that the large and small interlock pins are installed into their original positions.

NOTE: Improper installation of the interlock pins will prevent activation of the backup lamp switch.

9. Apply a sealant to the backup lamp switch. Install the switch to the top cover and torque the switch to 18–26 ft. lbs. (25–35 Nm).
10. Position the lower shift lever and dust cover assembly to the top cover. Install and tighten the 3 retaining screws.

TRANSMISSION ASSEMBLY

1. Position the countershaft into the transmission case through the top opening.
2. Position the input shaft into the transmission case through the top opening. Be sure that the needle roller bearing is installed into the input shaft.

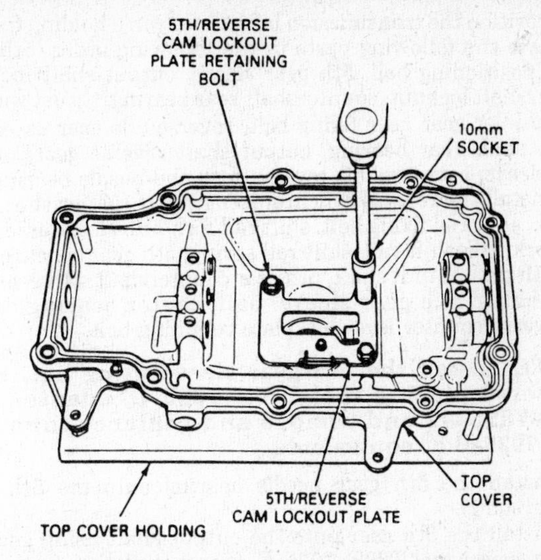

5th-reverse cam lockout plate installation

3. Position the output shaft assembly into the transmission case. Mate the input shaft and output shaft assemblies by positioning them at an upward angle and setting them together. Be sure that 4th gear synchronizer ring is installed at this time.
4. Install the output shaft center bearing outer race using a brass drift. Seat the center bearing outer races.
5. Install the countershaft center bearing. Be sure that the center bearing outer races are squarely position in their bores.
6. Position the center bearing cover to the transmission case with reference arrow pointing upward. Install and tighten the center bearing cover retaining bolts and torque them to 14–19 ft. lbs. Be sure that all center bearing cover retaining bolt heads are marked with an **8**.
7. Position the transmission vertically (input shaft and clutch housing facing upward). Be sure that the input shaft front bearing outer race is squarely position in the bore. If removed install the front cover oil seal using front cover seal installer tool T77J–7025–G or equivalent.
8. Install the countershaft front bearing by hand.

NOTE: If any related parts (such as outputshaft, bearing, etc.) have been replaced, measure the dimensions of the height of the input shaft bearing outer race above the transmission front bearing cover mating surface. Depth of the front cover outer race bore (input shaft). Depth of the countershaft front bearing race (transmission case to the front cover mating surface). Depth of the front cover outer race bore (countershaft). After measuring all dimensions, select bearing shim to maintain the endplay within specified limits.

9. Remove any sealant residue remaining on the mating surfaces of the transmission and front cover. To prevent damage to the oil seal lip during assembly, tape the input shaft splines along their entire length.
10. Apply a thin coat of oil to the front cover oil seal lip. Position the bearing shim and baffle plate into the front cover (install the shim with groove showing). Install the spacer to transmission case countershaft front bearing bore. If necessary apply a sufficient quantity of petroleum jelly to the shim, bearing cover and oil baffle to retain them in position during assembly.
11. Apply a 1/8 in. (3mm) bead of RTV sealant to the front cover and front cover retaining bolt threads. Install the front bearing cover to the transmission case. Install and torque the front bearing cover retaining bolts to 9–17 ft. lbs. (12–22mm). Be sure the front bearing cover retaining bolt heads are marked with a **6**.

12. Position the transmission horizontally in a holding fixture. Assemble the following parts in the following order, 5th gear sleeve positioning ball, 5th gear sleeve, output shaft locknut, countershaft locknut, countershaft rear bearing, thrust washer, reverse idler gear gear fixing bolt, reverse idle gear assembly, output shaft rear bearing, output shaft reverse gear, output shaft sleeve, countershaft reverse gear and needle bearing.

13. Install the reverse synchronizer ring, thrust washer, split washer, shift rod fixing bolt, shift rail/fork/hub/sleeve assembly, steel lock ball, shift rail, shift rail spring, 5th gear synchronizer ring, 5th gear output shaft, 5th gear countershaft, 5th gear needle bearing, 5th gear sleeve, ball, center bearing cover, 5th/reverse counter lever lockplate retaining bolt.

NOTE: Install the 5th gear sleeve using nut, shaft adapter replacing tool T75L-7025-L, adapter tool T88T-7025-J2 and remove and replacer tube tool T75L-7025-B or equivalents.

14. Install the 5th gear needle bearing onto the 5th gear (countershaft).

15. Install the 5th gear onto the output shaft using gear installing spacer tool T88T-7025-F, gear installation spacer tool T88T-7025-G, shaft adapter tool T75L-7025-P, shaft adapter screw tool T75L-7025-K, remover/replacer tube tool T75L-7025-B (2WD vehicles), or remover/replacer tube tool T85T-7025-A (4WD vehicles) nut and washer or equivalents. Be sure that the long flange on the 5th gear faces forward.

NOTE: On the 2WD vehicles, to install the 5th gear assembly, perform the process in 2 steps. First install the gear installing spacer tool T88T-7025-F, when the tool bottoms out add gear installing spacer tool T88T-7025-G and press the 5th gear assembly the rest of the way into position. On the 4WD vehicles, installation of the 5th gear assembly is similar to the 2WD installation except that remover/replacer tube tool T85T-7025-A and TOD bearing remover/replacer adapter tool T84T-7025-A are used.

16. Position the counter lever assembly to the transmission and install the thrust washer and retaining ring. Apply sealant to the counter lever fixing bolt threads. Install the counter lever fixing bolt and torque it to 6-7 ft. lbs. (8-10 Nm).

17. If removed, position the 5th/reverse shift fork and shift rail to the top cover. Insert the 5th/reverse shift rail through the top cover bore and 5th/reverse shift fork. Install the spring and detent ball to lower part of the rod.

18. Assemble the 5th/reverse synchronizer hub, sleeve and 5th gear synchronizer ring to the 5th/reverse shift fork and rod assembly. Be sure to install the longer flange (on the 5th/reverse hub, sleeve and synchonizer assembly) toward the front of the transmission. The reference mark on the synchronizer sleeve must be installed toward the reverse gear side.

19. Install the 5th/reverse shift fork and shift rail assembly (including 5th/reverse synchronizer hub, sleeve and 5th gear synchronizer ring) to countershaft. Mate the shift fork gate to the 5th/reverse counter lever end. Install the 5th/reverse fork and shift rail assembly with threaded fixing bolt bores (in rail and transmission case) aligned with each other.

NOTE: For ease of assembly, position the 5th/reverse shift fork into the rearmost threaded bore of the 3 detent positions. Return the shift fork to the neutral gear position after installation.

20. Apply a suitable sealant to the 5th/reverse shift rail fixing bolt threads. Install the 5th/reverse shift rail fixing bolt to the transmission case and torque it to 16-22 ft. lbs. (20-30 Nm).

21. Apply a suitable sealant to the oil passage retaining bolt. Position the oil passage to transmission case and install the retaining bolt. Torque the oil passage retaining bolt to 6-7 ft. lbs. (8-9 Nm).

22. Install the split washer and thrust washer onto the countershaft.

NOTE: If the clutch hub and or counter reverse gear have been replaced, new split washers must be selected to maintain endplay within specified limits. Measure the endplay using a feeler gauge. Be sure the spilt washers are a matched set of identical thickness.

23. Install the reverse synchronizer ring and needle bearings onto the counter reverse gear. Install the counter reverse gear and needle bearing onto the countershaft as an assembly. Install the thrust washer to the countershaft.

24. Press the thrust washer forward against the shoulder on the countershaft. Maintain forward pressure against the thrust washer and insert a feeler gauge between the thrust washer and counter reverse gear. Using the proper size thrust washer, bring the counter reverse gear endplay into specifications. The counter reverse gear endplay should be 0.25-0.35mm.

25. Temporarily install a suitable spacer (inner bore larger than 21mm, outer bore smaller than 36mm, 15-20mm over all length) in place of the countershaft bearing. Loosely install the countershaft locknut to retain the components.

NOTE: The installation of a suitable spacer prevents the thrust washers from slipping off the shaft and avoids interference with the reverse idler gears.

25. Install the reverse idler gear assembly. Apply a suitable sealant to the threads of the reverse idler gear fixing bolt. Torque the bolt to 58-86 ft. lbs. (79-116 Nm).

27. Drive the sleeve and reverse gear assembly onto the output shaft using gear installing spacer tool T88T-7025-G, shaft adapter tool T75L-7025-P, shaft adapter screw tool T75L-7025-K, remover/replacer tube tool T75L-7025-B (2WD vehicles), or remover/replacer tube tool T85T-7025-A (4WD vehicles) nut and washer or equivalents. Install the reverse gear with the longer flange facing rearward.

28. Install the output shaft rear bearing using gear installing spacer T88T-7025-G, shaft adapter tool T75L-7025-P, shaft adapter screw tool T75L-7025-K, remover/replacer tube tool T75L-7025-B (2WD vehicles), or remover/replacer tube tool T85T-7025-A (4WD vehicles) nut and washer or equivalents. Remove the temporary spacer from the countershaft.

29. Install the countershaft rear bearing by hand. Tightening the shaft locknuts without fully seating the bearing can cause damage to the output shaft threads.

30. Lock the transmission into 1st/3rd gears. Install a new output and countershaft locknuts. Torque the output shaft locknut to 160-200 ft. lbs. (216-274 Nm) and the countershaft locknut to 94-144 ft. lbs. (128-196 Nm).

NOTE: Always install new output and countershaft locknuts when assembling the transmission. Locknuts unstaked during disassembly cannot be reused.

31. Stake the locknuts to the bottom shaft groove using countershaft locknut staking tool T77J-7025-F or equivalent.

32. Install the speedometer drive gear and steel ball to the output shaft. Install the snapring retaining the speedometer drive gear to the output shaft (2WD vehicles).

NOTE: The speedometer drive gear contains 3 detents into which the steel drive ball can be installed. The steel drive ball can be installed into any of the 3 detents.

For reference during assembly, observe and record speedometer drive gear color. Depending upon application, 1 or 2 different speedometer drive gear may be installed. It will be color coded either green or white. Speedometer drive gear colors and corresponding part numbers are as follows, white E8TZ-17285-B and green E8TZ-17285-C.

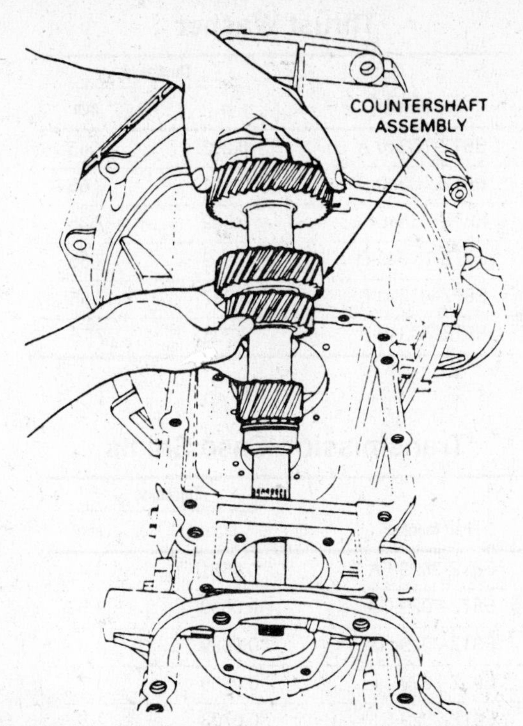

Installing countershaft assembly into transmission case

33. Remove any sealant residue from the mating surfaces of the transmission case and extension housing. Apply a ⅛ in. (3mm) bead of RTV sealant or equivalent to the transmission case.

NOTE: The extension housing bushing cannot be serviced. If the bushing requires service, the extension housing must be replaced as a unit.

34. Position the extension housing to the transmission case and install the extension housing retaining bolts. Torque the bolts to 24–34 ft. lbs. (32–46 Nm). Place the synchronizers into the neutral gear position. Be sure that the shift forks on the top cover assembly are in the neutral position.
35. Position the top cover on the transmission case. Carefully engage the shift forks with the synchronizers.
36. Apply a suitable sealant to the 2 rearmost top cover retaining bolts and install them into the top cover rear retaining bolt locations. Install the remaining top cover retaining bolts (no sealant) and torque them to 12–16 ft. lbs. (16–22 Nm).

NOTE: Do not apply sealant to the top cover or transmission mating surfaces. If necessary, apply a small quantity of grease to the sealing gasket to retain the gasket in position during assembly.

37. If removed, install the rear oil seal into the extension housing using extension housing seal replacer tool T61L–7657–A or equivalent. Be sure that the oil seal drain hole faces downward.
38. Refill the transmission with the recommended lubricant to the proper level. Install the transmission drain plug and torque it to 29–43 ft. lbs. (40–58 Nm).

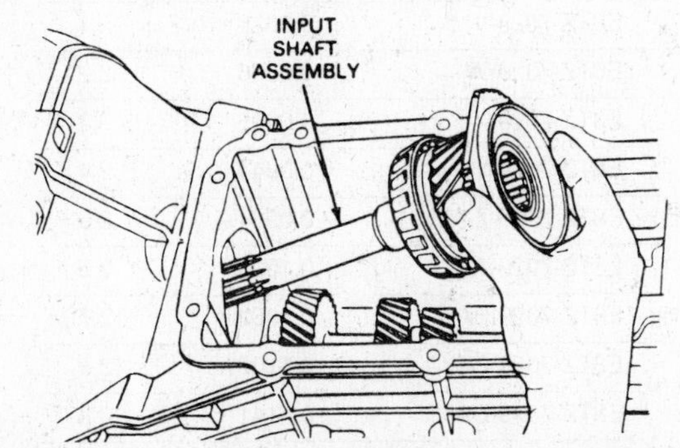

Installing input shaft assembly into transmission case

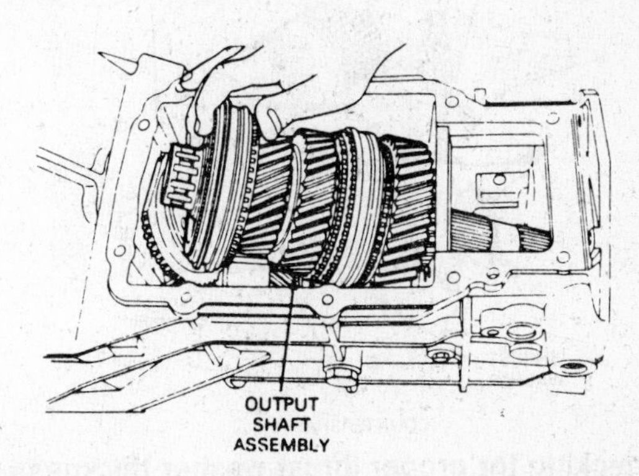

Installing output shaft assembly into transmission case

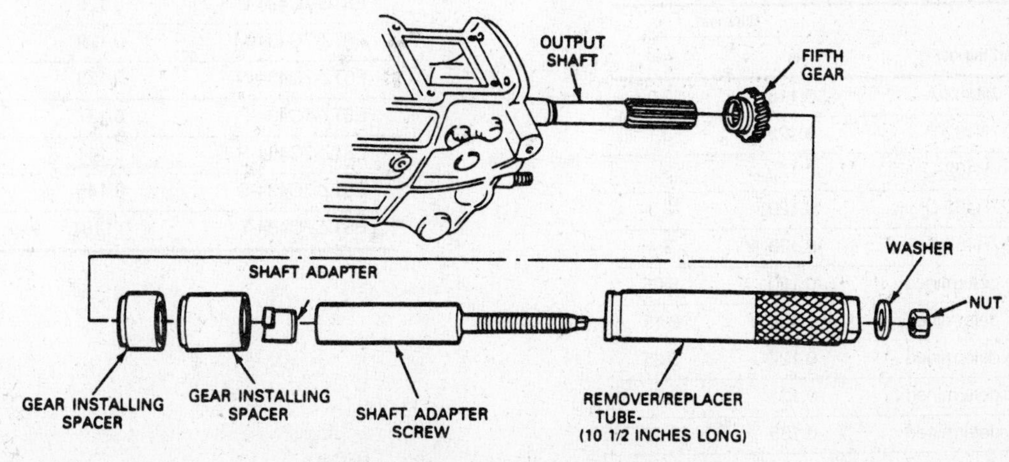

Installing 5th gear onto output shaft

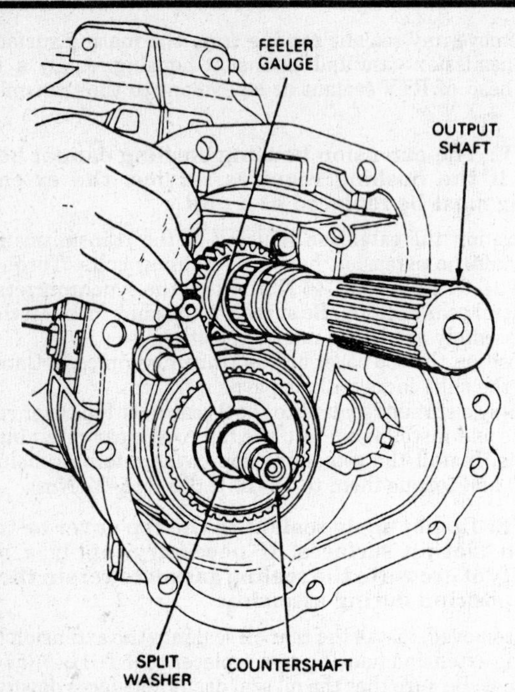

Measuring counter reverse gear endplay

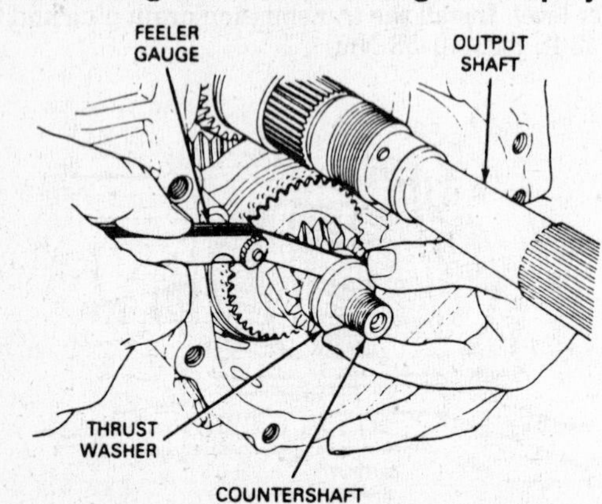

Checking for proper thrust washer thickness

Counter Reverse Gear Split Washer

Part Number	Thickness in.	Thickness mm
E8TZ-7R482-A	0.118	3.0
E8TZ-7R482-B	0.122	3.1
E8TZ-7R482-C	0.125	3.2
E8TZ-7R482-D	0.129	3.3
E8TZ-7R482-E	0.133	3.4
To be determined	0.120	3.05
To be determined	0.124	3.15
To be determined	0.127	3.25
To be determined	0.131	3.35
To be determined	0.135	3.45
To be determined	0.137	3.50

Thrust Washer

Part Number	Thickness in.	Thickness mm
E8TZ-7C340-A	0.293	7.45
E8TZ-7C340-B	0.301	7.65
E8TZ-7C340-C	0.309	7.85
E8TZ-7C340-D	0.289	7.35
E8TZ-7C340-E	0.297	7.55
E8TZ-7C340-F	0.305	7.75

Transmission Case Shims

Part Number	Thickness in.	Thickness mm
E8TZ-7029-FA	0.0551	1.4
E8TZ-7029-GA	0.0590	1.5
E8TZ-7029-Ha	0.0629	1.6
E8TZ-7029-Ja	0.0669	1.7
E8TZ-7029-S	0.0708	1.8
E8TZ-7029-T	0.0748	1.9
E8TZ-7029-U	0.0787	2.0
E8TZ-7029-V	0.0826	2.1
E8TZ-7029-W	0.0866	2.2
E8TZ-7029-X	0.0905	2.3
E8TZ-7029-Y	0.0944	2.4
E8TZ-7029-Z	0.0984	2.5
E8TZ-7029-AA	0.1023	2.6
E8TZ-7029-BA	0.1062	2.7
E8TZ-7029-CA	0.1102	2.8
E8TZ-7029-DA	0.1141	2.9
E8TZ-7029-EA	0.1181	3.0
E8TZ-7C434-K	0.122	3.1
E8TZ-7C434-L	0.125	3.2
E8TZ-7C434-M	0.129	3.3
E8TZ-7C434-N	0.133	3.4
E8TZ-7C434-P	0.137	3.5
E8TZ-7C434-R	0.141	3.6
E8TZ-7C434-S	0.145	3.7
E8TZ-7C434-T	0.1181	3.0

Torque Specifications

Description	ft. lbs.	Nm
Output shaft locknut	160–203	216–274
Countershaft locknut	94–144	128–196
Extension housing retaining bolts	24–34	32–46
Reverse idler shaft fixing bolt	58–86	79–116
Center bearing cover	14–19	18–26
Front bearing cover	12–17	16–22
Fifth/reverse cam lockout plate	6–7	8–10
Dust boot	6–8	8–11
Top cover	12–16	16–22
Filler plug	29–43	40–58
Front oil passage	6–8	8–10
Counter lever shaft fixing bolt	6–8	8–10
Rock plate	6–7	8–10
Drain plug	29–43	40–58
Backup lamp switch	18–26	25–35
Neutral switch (if equipped)	18–26	25–35
Rear oil passage (extension housing)	5–7	8–9
5th/reverse shift rail fixing bolt	16–22	20–30

Special Tools

Number	Description
T74P-77248-A	Extension housing seal remover
T77J-7025-F	Countershaft locknut staking tool
T88T-7025-A	Mainshaft locknut wrench
T77J-7025-B	Remover/replacer tube
T75L-7025-B	Remover/replacer tube
T84T-7025-B	TOD forcing screw

Special Tools

Number	Description
T77L-7025-H	Bearing puller
T77J-7025-L	Puller ring
T75L-7025-G	Bearing collet sleeve for 3.5 in. bearing collets
T85T-7025-A	Remover/replacer tube
T84T-7025-A	TOD bearing remover/replacer adapter
T88T-7061-A	Gear removal collet
T88T-7120-A	Bearing race puller
T50T-100-A	Slide hammer
T53T-4621-B	Bearing replacer
T88T-7025-B	Bearing cone replacer
T75L-1165-B	Bearing plate
T71P-4621-B	Bearing cone remover
D84L-1123-A	Bearing splitter
T88T-7025-C	Top cover holding fixture
D84L-7000-B	Roll pin punch set
T88T-7025-F	Gear installing spacer
T88T-7025-G	Gear installing spacer
T75T-7025-P	Shaft adapter
T75L-7025-K	Shaft adapter screw
T61L-7657-A	Extension housing seal replacer
T00L-4201-C	Dial indicator
T77J-7025-G	Front cover seal installer
T75L-7025-L	Shaft adapter—replacing
T88T-7025-J2	Adapter
T88T-7025-J	Countershaft fifth gear sleeve puller
T88T-7025-J1	Gear removal collets
D82L-4201-C	Depth micrometer

CLUTCH

Understanding the Clutch

The purpose of the clutch is to disconnect and connect engine power from the transmission. A car at rest requires a lot of engine torque to get all that weight moving. An internal combustion engine does not develop a high starting torque (unlike steam engines), so it must be allowed to operate without any load until it builds up enough torque to move the car. Torque increases with engine rpm. The clutch allows the engine to build up torque by physically disconnecting the engine from the transmission, relieving the engine of any load or resistance. The transfer of engine power to the transmission (the load) must be smooth and gradual; if it weren't, drive line components would wear out or break quickly. This gradual power transfer is made possible by gradually releasing the clutch pedal. The clutch disc and pressure plate are the connecting link between the engine and transmission. When the clutch pedal is released, the disc and plate contact each other (clutch engagement), physically joining the engine and transmission. When the pedal is pushed in, the disc and plate separate (the clutch is disengaged), disconnecting the engine from the transmission.

The clutch assembly consists of the flywheel, the clutch disc, the clutch pressure plate, the throw-out bearing and fork, the actuating linkage and the pedal. The flywheel and clutch pressure plate (driving members) are connected to the engine crankshaft and rotate with it. The clutch disc is located between the flywheel and pressure plate, and splined to the transmission shaft. A driving member is one that is attached to the engine and transfers engine power to a driven member (clutch disc) on the transmission shaft. A driving member (pressure plate) rotates (drives) a driven member (clutch disc) on contact and, in so doing, turns the transmission shaft. There is a circular diaphragm spring within the pressure plate cover (transmission side). In a relaxed state (when the clutch pedal is fully released), this spring is convex; that it, it is dished outward toward the transmission. Pushing in the clutch pedal actuates an attached linkage rod. Connected to the other end of this rod is the throw-out bearing fork. The throw-out bearing is attached to the fork. When the clutch pedal is depressed, the clutch linkage pushes the fork and bearing forward to contact the diaphragm spring of the pressure plate. The outer edges of the spring are secured to the pressure plate and are pivoted on rings so that when the center of the spring is compressed by the throw-out bearing, the outer edges bow outward and, by so doing, pull the pressure plate in the same direction – away from the clutch disc. This action separates the disc from the plate, disengaging the clutch and allowing the transmission to be shifted into another gear. A coil type clutch return spring attached to the clutch pedal arm permits full release of the pedal. Releasing the pedal pulls the throw-out bearing away from the diaphragm spring resulting in a reversal of spring position. As bearing pressure is gradually released from the spring center, the outer edges of the spring bow outward, pushing the pressure plate into closer contact with the clutch disc. As the disc and plate move closer together, friction between the two increases and slippage is reduced until, when full spring pressure is applied (by fully releasing the pedal), The speed of the disc and plate are the same. This stops all slipping, creating a direct connection between the plate and disc which results in the transfer of power from the engine to the transmission. The clutch disc is now rotating with the pressure plate at engine speed and, because it is splined to the transmission shaft, the shaft now turns at the same engine speed. Understanding clutch operation can be rather difficult at first; if you're still confused after reading this, consider the following analogy. The action of the diaphragm spring can be compared to that of an oil can bottom. The bottom of an oil can is shaped very much like the clutch diaphragm spring and pushing in on the can bottom and then releasing it produces a similar effect. As mentioned earlier, the clutch pedal return spring permits full release of the pedal and reduces linkage slack due to wear. As the linkage wears, clutch free-pedal travel will increase and free-travel will decrease as the clutch wears. Free-travel is actually throw-out bearing lash.

The diaphragm spring type clutches used are available in two different designs: flat diaphragm springs or bent spring. The bent fingers are bent back to create a centrifugal boost ensuring quick re-engagement at higher engine speeds. This design enables pressure plate load to increase as the clutch disc wears and makes low pedal effort possible even with a heavy duty clutch. The throw-out bearing used with the bent finger design is 1¼ in. (31.75mm) long and is shorter than the bearing used with the flat finger design. These bearings are not interchangeable. If the longer bearing is used with the bent finger clutch, free-pedal travel will not exist. This results in clutch slippage and rapid wear.

The transmission varies the gear ratio between the engine and rear wheels. It can be shifted to change engine speed as driving conditions and loads change. The transmission allows disengaging and reversing power from the engine to the wheels.

Adjustments

NOTE: The clutch system is hydraulically activated. The hydraulic clutch system locates the clutch pedal and provides automatic clutch adjustment. No adjustment of the clutch linkage or pedal position is required.

Clutch Pedal

REMOVAL AND INSTALLATION

1. Disconnect the negative battery cable.
2. Remove the lockpin and remove the master cylinder pushrod from the clutch pedal. Remove the bushing.
3. Remove the retainer clip and remove the clutch pedal and shaft assembly from the pedal support bracket.
4. Remove the bolts securing the parking brake mechanism and position it out of the way (Explorer Only).

NOTE: It may be necessary to remove the left side kick panel to facilitate clutch pedal removal.

5. Before installing the clutch pedal and shaft assembly, inspect the shaft bushings and replace if necessary.

To install:

6. Clean and lubricate the clutch pedal shaft bushings lightly with SAE 30 engine oil. Install the clutch pedal and shaft assembly into place.
7. Install the retainer clip.
8. Install the parking brake mechanism (Explorer Only).
9. Install the bushing and master cylinder pushrod to the clutch pedal. Install the lockpin.
10. If removed, install the left side kick panel.
11. Reconnect the negative battery cable.

Clutch/Starter Interlock Switch

REMOVAL AND INSTALLATION

1983–84

1. Disconnect the negative battery cable.
2. Disconnect the connector at the switch by flexing the retaining tab on the switch housing and withdraw the connector.
3. Remove the clip that retains the switch to the clutch pedal.
4. Remove the screw that retains the switch to the support bracket and remove the switch.

To install:

5. Fit the replacement switch into place and install the switch-to-bracket retaining screw.
6. Install the retaining clip.
7. Reconnect the switch connector.
8. Reconnect the negative battery cable.

ADJUSTMENT

1. If the clip is out of position on the rod, remove both halves of the clip.
2. Position the clip closer to the switch and snap the clips together on the rod.
3. Depress the clutch pedal to the floor to adjust the switch.

Clutch Interlock 3-function Switch

The clutch interlock 3-function switch provides the 3 following functions:

Troubleshooting Basic Clutch Problems

Problem	Cause
Excessive clutch noise	Throwout bearing noises are more audible at the lower end of pedal travel. The usual causes are: · Riding the clutch · Too little pedal free-play · Lack of bearing lubrication A bad clutch shaft pilot bearing will make a high pitched squeal, when the clutch is disengaged and the transmission is in gear or within the first 2" of pedal travel. The bearing must be replaced. Noise from the clutch linkage is a clicking or snapping that can be heard or felt as the pedal is moved completely up or down. This usually requires lubrication. Transmitted engine noises are amplified by the clutch housing and heard in the passenger compartment. They are usually the result of insufficient pedal free-play and can be changed by manipulating the clutch pedal.
Clutch slips (the car does not move as it should when the clutch is engaged)	This is usually most noticeable when pulling away from a standing start. A severe test is to start the engine, apply the brakes, shift into high gear and SLOWLY release the clutch pedal. A healthy clutch will stall the engine. If it slips it may be due to: · A worn pressure plate or clutch plate · Oil soaked clutch plate · Insufficient pedal free-play
Clutch drags or fails to release	The clutch disc and some transmission gears spin briefly after clutch disengagement. Under normal conditions in average temperatures, 3 seconds is maximum spin-time. Failure to release properly can be caused by: · Too light transmission lubricant or low lubricant level · Improperly adjusted clutch linkage
Low clutch life	Low clutch life is usually a result of poor driving habits or heavy duty use. Riding the clutch, pulling heavy loads, holding the car on a grade with the clutch instead of the brakes and rapid clutch engagement all contribute to low clutch life.

- It requires the clutch pedal to be depressed to the floor in order to start the engine.
- If cuts off the speed control system when the clutch pedal is depressed.
- It provides a fuel control signal to the EEC system.

REMOVAL AND INSTALLATION

1986–91

1. Disconnect the negative battery cable.
2. Disconnect the connector at the switch by flexing the retaining tab on the switch housing and withdraw the connector.
3. Rotate the switch ½ turn to expose the plastic retainer.
4. Push the tabs together to allow the retainer to slide rearward and separate from the switch.
5. Remove the switch from the pushrod.

To install:

6. Fit the switch to the master cylinder pushrod.
7. Install the plastic retainer.
8. Rotate the switch into the position to attach the clip. Reconnect the switch connector.
9. Reconnect the negative battery cable.

Clutch Hydraulic System

The hydraulic clutch system operates much like a hydraulic brake system. When you push down (disengage) the clutch pedal, the mechanical clutch pedal movement is converted into hydraulic fluid movement, which is then converted back into mechanical movement by the slave cylinder to actuate the clutch release lever.

The system consists of a combination clutch fluid reservoir/master cylinder assembly, a slave cylinder mounted on the bellhousing, and connecting tubing.

Fluid level is checked at the master cylinder reservoir. The hydraulic clutch system continually remains in adjustment, like a hydraulic disc brake system, so not clutch linkage or pedal adjustment is necessary.

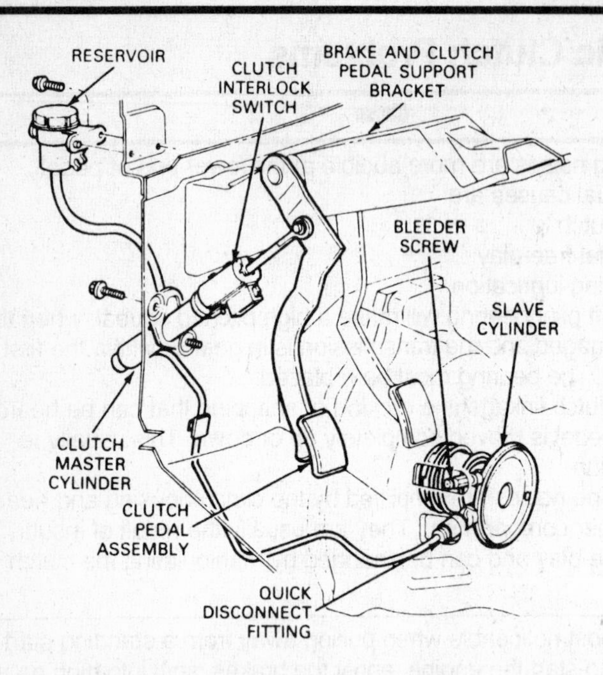

Typical hydraulic clutch system

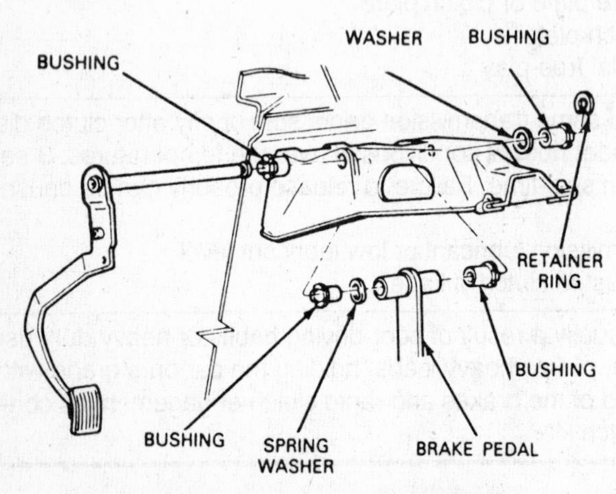

Clutch pedal removal and installation

Master Cylinder and Reservoir

REMOVAL AND INSTALLATION

1983–84

NOTE: The clutch hydraulic system is serviced as a complete unit; it has been bled of air and filled with fluid. Individual components of the system are not available separately.

1. Remove the lock pin and disconnect the master cylinder push rod from the clutch pedal.

2a. 4–2.0L,2.3L gasoline engines: Remove the bolt attaching the dust shield to the clutch housing and remove the dust shield. Push the slave cylinder rearward to disengage from the recess in the housing lugs, then slide outward to remove.

2b. 6–2.8L gasoline engine and 4–2.2L diesel engine: Remove the bolts attaching the slave cylinder to the clutch housing and remove the slave cylinder. Disengage the push rod from the release lever as the cylinder is removed. Retain the push rod to release lever plastic bearing inserts.

3. Remove the two bolts attaching the master cylinder to the firewall.

4. Remove the two bolts attaching the fluid reservoir to the cowl access cover.

5. Remove the master cylinder from the opening in the firewall and remove the hydraulic system assembly upward and out of the engine compartment.

To install:

6. Position the hydraulic system downward into the engine compartment. The slave to master cylinder tube routing is to be above the brake tubes and below the steering column shaft.

NOTE: On 6–2.8L vehicles, the tube must lay on top of the clutch housing.

7. Insert the master cylinder push rod through the opening in the firewall, position the master cylinder on the firewall, and install the attaching bolts. Torque the bolts to 15–20 ft. lbs.

8. Position the fluid reservoir on the cowl opening cover and install the attaching bolts. Torque the attaching bolts to 15–20 ft. lbs.

9. Install the slave cylinder by pushing the slave cylinder push rod into the cylinder, engage the push rod and plastic bearing inserts into the release lever, and attach the cylinder to the clutch housing.

NOTE: With a new system, the slave cylinder contains a shipping strap that pre-positions the push rod for installation, and also provides a bearing insert. Following installation of the slave cylinder, the first actuation of

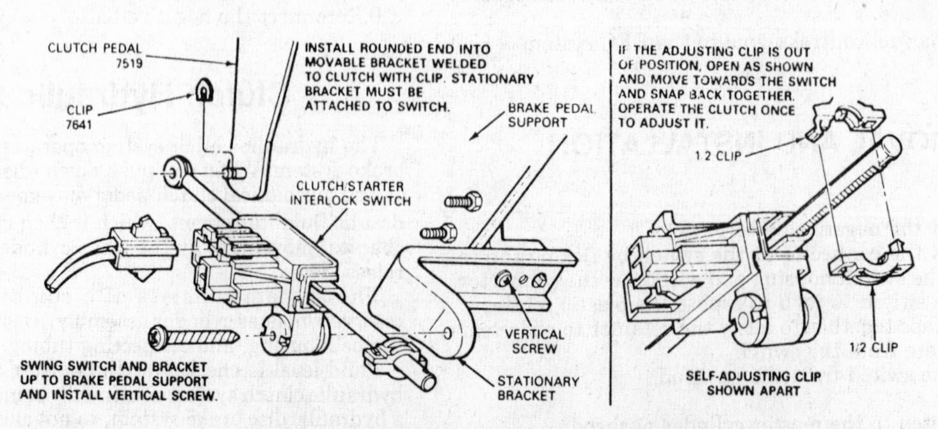

Clutch/starter interlock switch assembly

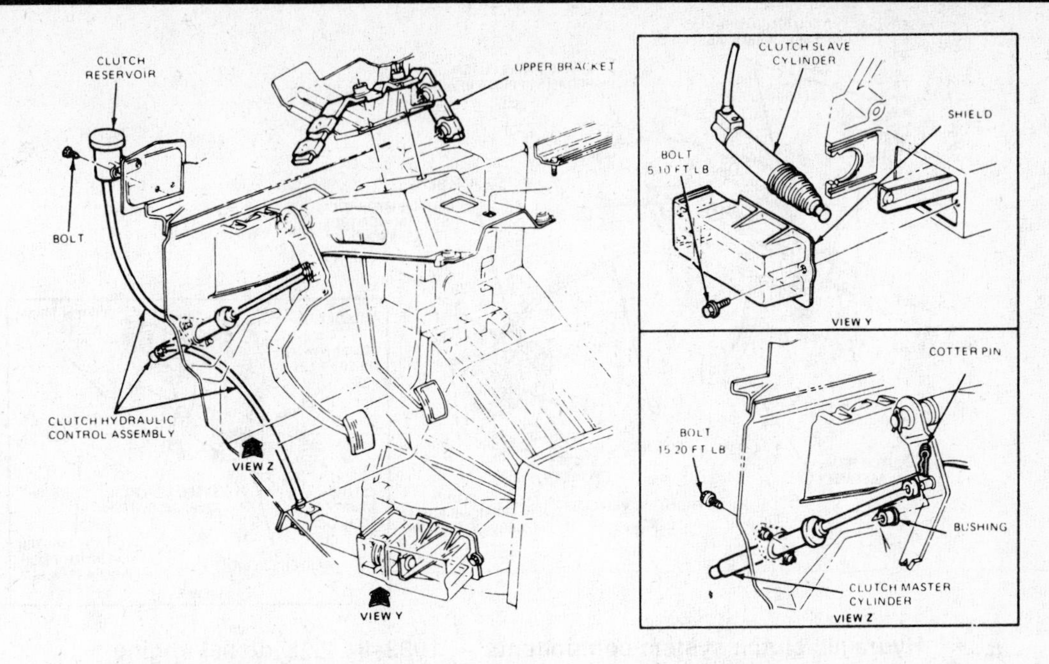

Hydraulic clutch system components — 1983–84 2.0L and 2.3L engines

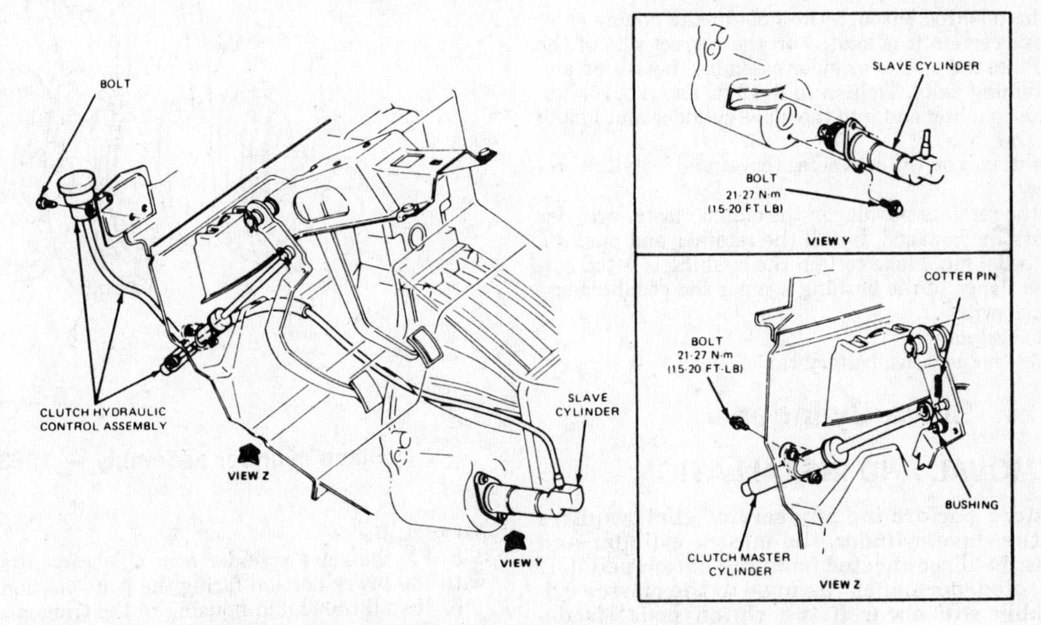

Hydraulic clutch system components — 1984 2.8L engine

the clutch pedal will break the shipping strap and give normal system operation.

10a. 4–2.0L,2.3L engines: Snap the dust shield into position. Install the retaining bolt and tighten to 5–10 ft. lbs.

10b. 6–2.8L gasoline and 4–2.2L diesel engines: Install the bolts attaching the slave cylinder to the clutch housing and torque them to 15–20 ft. lbs.

11. Clean and apply a light film of oil to the master cylinder push rod bushing and install the bushing and push rod to the clutch pedal. Retain with the lock pin.

12. Depress the clutch pedal at least 10 times to verify smooth operation and proper clutch release.

1985–91

1. Disconnect the negative battery cable.
2. Disconnect the clutch master cylinder pushrod from the clutch pedal.
3. Remove the switch from the master cylinder assembly, if equipped.
4. Remove the screw retaining the fluid reservoir to the cowl access cover.
5. Disconnect the tube from the slave cylinder and plug both openings.
6. Remove the bolts retaining the clutch master cylinder to the dash panel and remove the clutch master cylinder assembly.

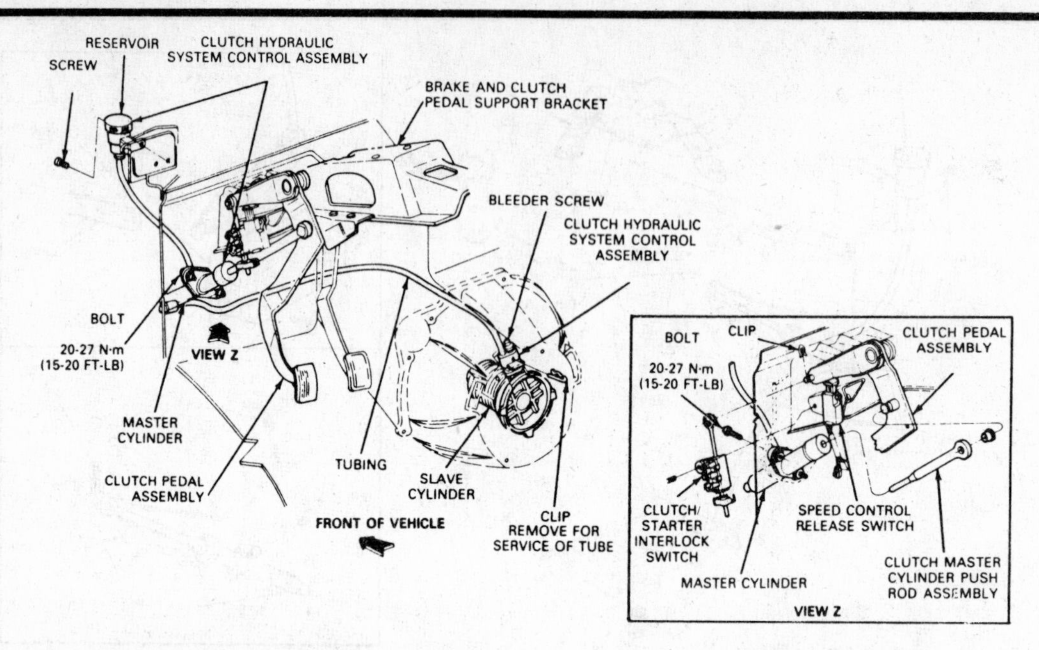

Hydraulic clutch system components — 1983–84 2.2L diesel engine

To install:

7. Install the pushrod through the hole in the engine compartment. Make certain it is located on the correct side of the clutch pedal. Place the master cylinder assembly in position and install the retaining bolts. Tighten to 8–12 ft. lbs. (11–16Nm).
8. Insert the coupling end into the slave cylinder and install the tube into the clips.
9. Fit the reservoir on the cowl access cover and install the retaining screws.
10. Replace the retainer bushing in the clutch master cylinder pushrod if worn or damaged. Install the retainer and pushrod on the clutch pedal pin. Make certain the bushing is fitted correctly with the flange of the bushing against the pedal blade.
11. Install the switch.
12. Bleed the system.
13. Reconnect the negative battery cable.

Slave Cylinder

REMOVAL AND INSTALLATION

NOTE: Before performing any service that requires removal of the slave cylinder, the master cylinder and pushrod must be disconnected from the clutch pedal. If not disconnected, permanent damage to the master cylinder assembly will occur if the clutch pedal is depressed while the slave cylinder is disconnected.

1983–84

The clutch hydraulic system is serviced as a complete unit; it has been bled of air and filled with fluid. Individual components of the system are not available separately. Refer to Clutch Master Cylinder and Reservoir removal and installation.

1985

1. Disconnect the negative battery cable.
2. Remove the transmission assembly.
3. Remove the clutch housing-to-transmission retaining nuts and remove the housing assembly.
4. Remove the slave cylinder from the transmission input shaft.

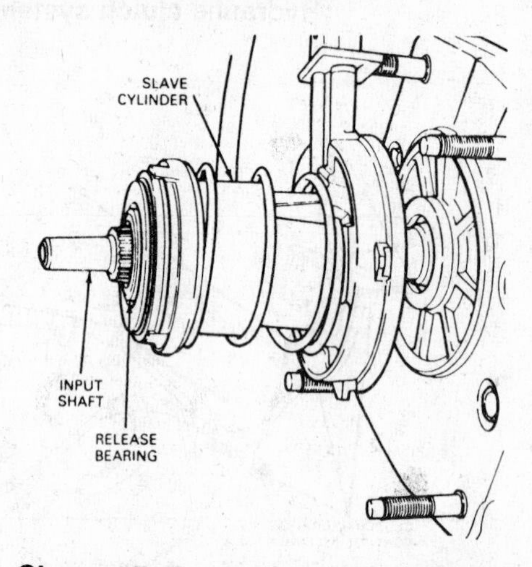

Slave cylinder assembly — 1983–85

To install:

5. Fit the slave cylinder over the transmission input shaft with the tower portion facing the transmission.
6. Install the clutch housing to the transmission. Make certain the slave cylinder is properly located in the notches of the clutch housing.
7. Install the transmission.
8. Bleed the hydraulic system.

1986–91

1. Disconnect the negative battery cable.
2. Disconnect the coupling at the transmission, using the clutch coupling removal tool T88T–70522–A or equivalent. Slide the white plastic sleeve toward the slave cylinder while applying a slight tug on the tube.
3. Remove the transmission assembly.

NOTE: On the 2.9L (4WD) vehicles, the clutch housing must be removed with the transmission assembly.

4. Remove the slave cylinder-to-transmission retaining bolts.

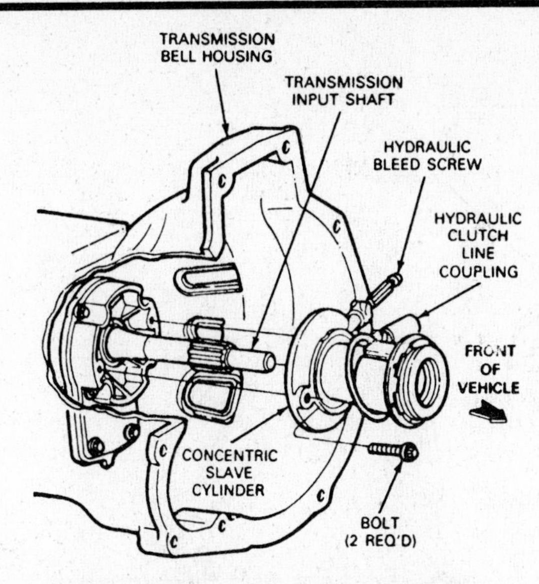

Slave cylinder assembly — 1986–91

5. Remove the slave cylinder from the transmission input shaft.

To install:

6. Fit the slave cylinder over the transmission input shaft with the bleed screws and coupling facing the left side of the transmission.
7. Install the slave cylinder retaining bolts. Torque to 13–19 ft. lbs. (18–26Nm).
8. Install the transmission.
9. Reconnect the coupling to the slave cylinder.
10. Bleed the system.
11. Reconnect the negative battery cable.

Bleeding the System

The following procedure is recommended for bleeding a hydraulic system installed on the vehicle. The largest portion of the filling is carried out by gravity. It is recommended that the original clutch tube with quick connect be replaced when servicing the hydraulic system because air can be trapped in the quick connect and prevent complete bleeding of the system. The replacement tube does not include a quick connect.

1. Clean the dirt and grease from the dust cap.
2. Remove the cap and diaphragm and fill th reservoir to the top with approved brake fluid C6AZ–19542–AA or BA, (ESA–M6C25–A) or equivalent.

NOTE: To keep brake fluid from entering the clutch housing, route a suitable rubber tube of appropriate inside diameter from the bleed screw to a container.

3. Loosen the bleed screw, located in the slave cylinder body, next to the inlet connection. Fluid will now begin to move from the master cylinder down the tube to the slave cylinder.

NOTE: The reservoir must be kept full at all time during the bleeding operation, to ensure no additional air enters the system.

4. Notice the bleed screw outlet. When the slave is full, a steady stream of fluid comes from the slave outlet. Tighten the bleed screw.
5. Depress the clutch pedal to the floor and hold for 1–2 seconds. Release the pedal as rapidly as possible. The pedal must be released completely. Pause for 1–2 seconds. Repeat 10 times.
6. Check the fluid level in the reservoir. The fluid should be level with the step when the diaphragm is removed.
7. Repeat Step 5 and 6 five times. Replace the reservoir diaphragm and cap.
8. Hold the pedal to the floor, crack open the bleed screw to allow any additional air to escape. Close the bleed screw, then release the pedal.
9. Check the fluid in the reservoir. The hydraulic system should now be fully bled and should release the clutch.
10. Check the vehicle by starting, pushing the clutch pedal to the floor and selecting reverse gear. There should be no grating of gears. If there is, and the hydraulic system still contains air, repeat the bleeding procedure from Step 5.

Driven Disc and Pressure Plate

CAUTION

The clutch driven disc contains asbestos, which has been determined to be a cancer causing agent. Never clean clutch surfaces with compressed air! Avoid inhaling any dust from any clutch surface! When cleaning clutch surface, use a commercially available brake cleaning fluid.

REMOVAL AND INSTALLATION

1983–84

1. Disconnect the negative battery cable.
2. Disconnect the clutch master cylinder from the clutch pedal and remove.
3. Raise the vehicle and support it safely.
4. Remove the dust shield from the clutch housing. Disconnect the hydraulic clutch linkage from the housing and release lever.
5. Remove the starter.
6. Remove the transmission from the vehicle.
7. Mark the assembled position of the pressure plate and cover the flywheel, to aid during re-assembly.
8. Loosen the pressure plate and cover attaching bolts evenly until the pressure plate springs are expanded, and remove the bolts.
9. Remove the pressure plate and cover assembly and the clutch disc from the flywheel. Remove the pilot bearing only for replacement.

To install:

10. Position the clutch disc on the flywheel so that the Clutch Alignment Shaft D79T–7550–A or equivalent can enter the clutch pilot bearing and align the disc.
11. When installing the original pressure plate and cover assembly, align the assembly and flywheel according to the marks made during the removal operations. Position the pressure plate and cover assembly on the flywheel, align the pressure plate and disc, and install the retaining bolts that fasten the assembly to the flywheel. Torque the bolts to 15–24 ft. lbs., and remove the clutch disc alignment tool.
12. Install the transmission into the vehicle.
13. Install the hydraulic clutch linkage on the housing in position with the release lever. Install the dust shield and install the starter.
14. Lower the vehicle and connect the clutch hydraulic system master cylinder to the clutch pedal. Check the clutch for proper operation.
15. Reconnect the negative battery cable.

1985–88

1. Disconnect the negative battery cable.
2. Disconnect the clutch hydraulic system master cylinder from the clutch pedal and remove.
3. Raise the vehicle and support it safely.
4. Remove the starter.
5. Disconnect the hydraulic coupling at the transmission.

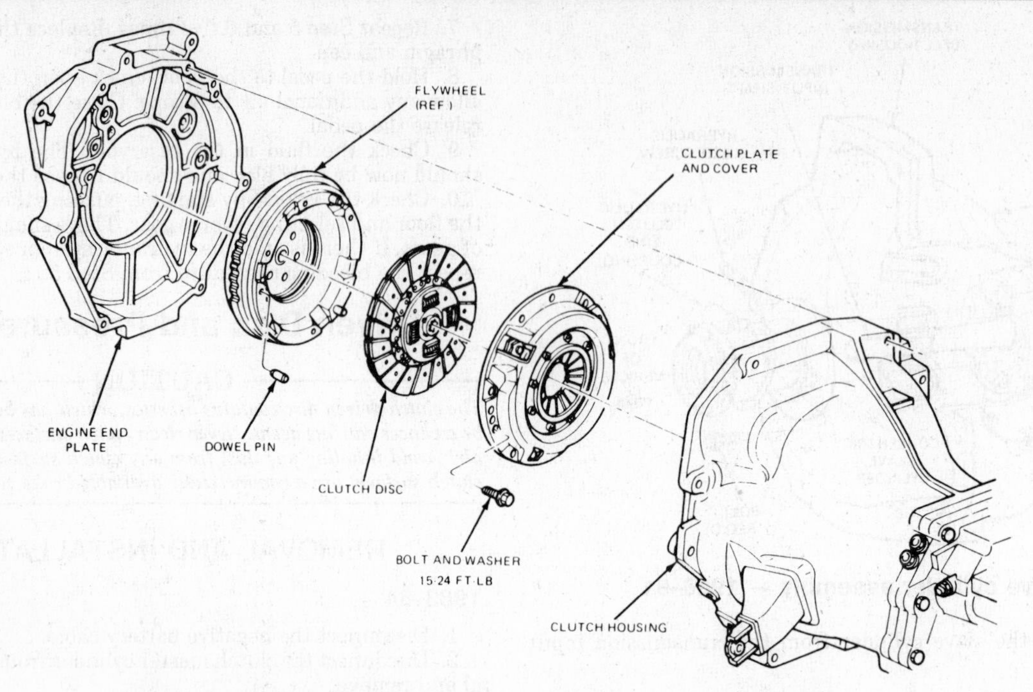

Clutch housing installation — 1983–84 2.2L I-4 diesel engine

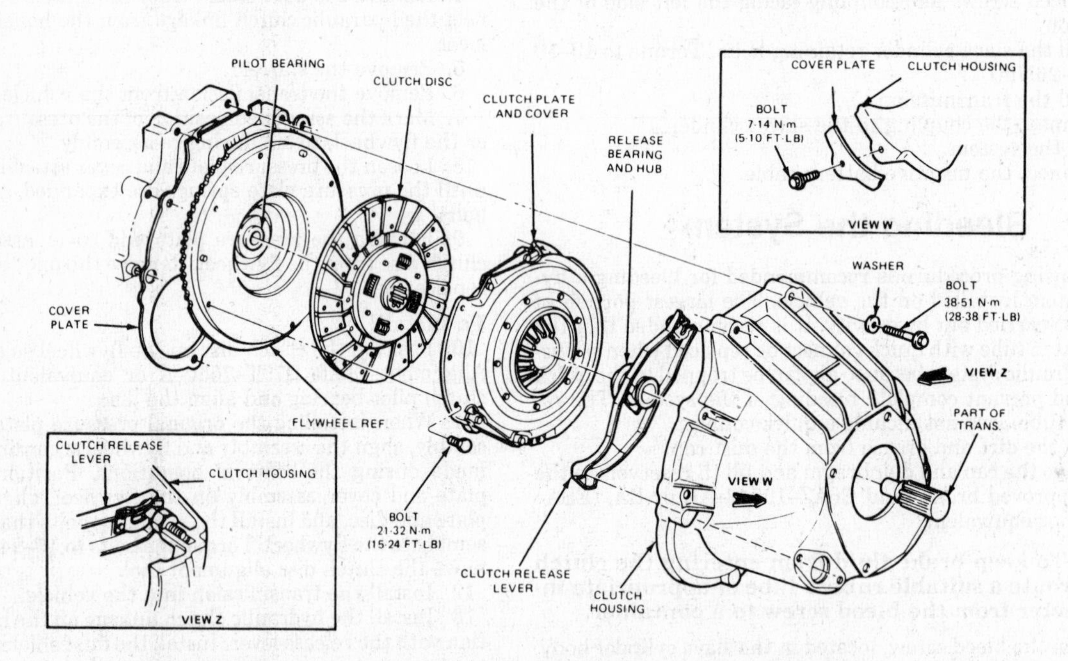

Clutch housing installation — 1983–84 2.8L V-6 gasoline engine

NOTE: Clean the area around the hose and slave cylinder to prevent fluid contamination.

6. Remove the transmission from the vehicle.
7. Mark the assembled position of the pressure plate and cover the flywheel, to aid during re-assembly.
8. Loosen the pressure plate and cover attaching bolts evenly until the pressure plate springs are expanded, and remove the bolts.
9. Remove the pressure plate and cover assembly and the clutch disc from the flywheel. Remove the pilot bearing only for replacement.

To install:

10. Position the clutch disc on the flywheel so that the Clutch Alignment Shaft Tool T74P–7137–K or equivalent can enter the clutch pilot bearing and align the disc.
11. When reinstalling the original pressure plate and cover assembly, align the assembly and flywheel according to the marks made during the removal operations. Position the pressure

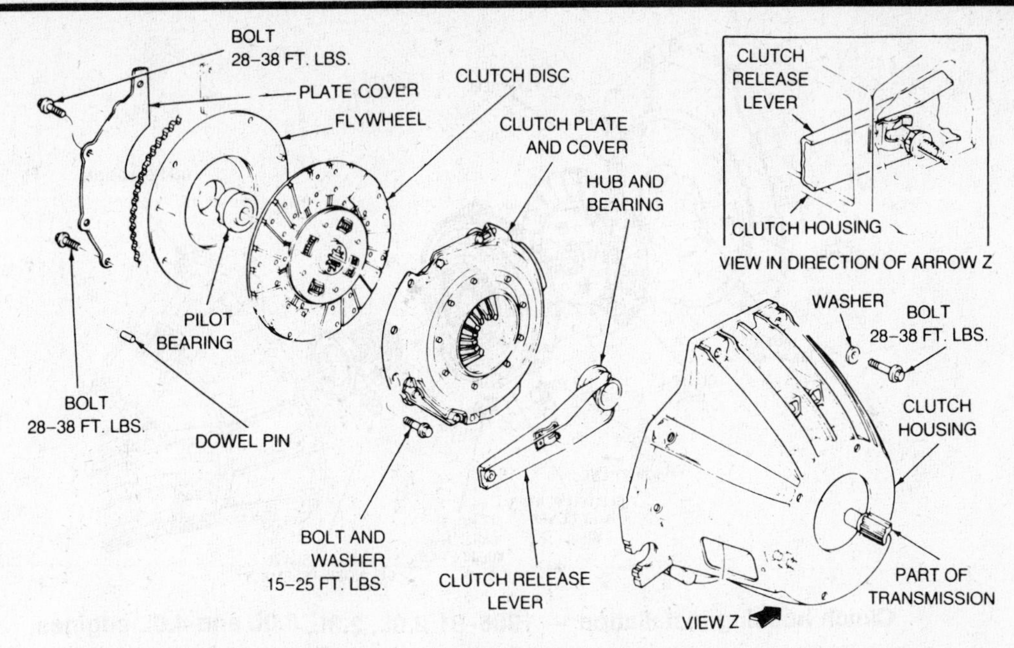

Clutch housing installation – 1983–84 2.0L and 2.3L I-4 gasoline engines

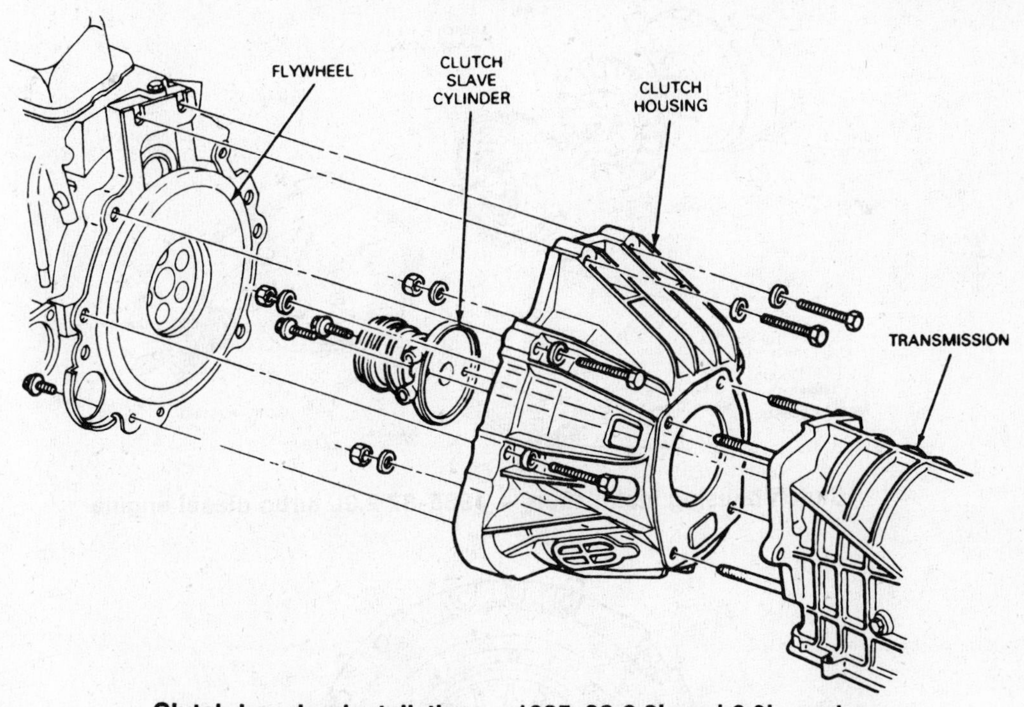

Clutch housing installation – 1985–88 2.8L and 2.9L engines

plate and cover assembly on the flywheel, align the pressure plate and disc, and install the retaining bolts that fasten the assembly to the flywheel. Tighten the bolts to 15–25 ft.lbs. (21–35 Nm) in the proper sequence. Remove the clutch disc pilot tool.

12. Install the transmission into the vehicle.

13. Connect the coupling by pushing the male coupling into the slave cylinder.

14. Connect the hydraulic clutch master cylinder pushrod to the clutch pedal.

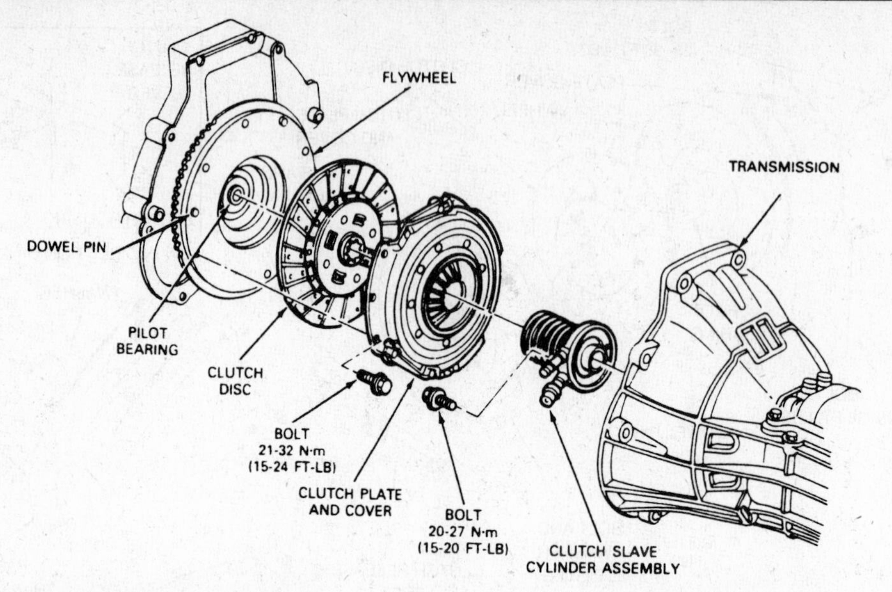

Clutch housing installation – 1985–91 2.0L, 2.3L, 3.0L and 4.0L engines

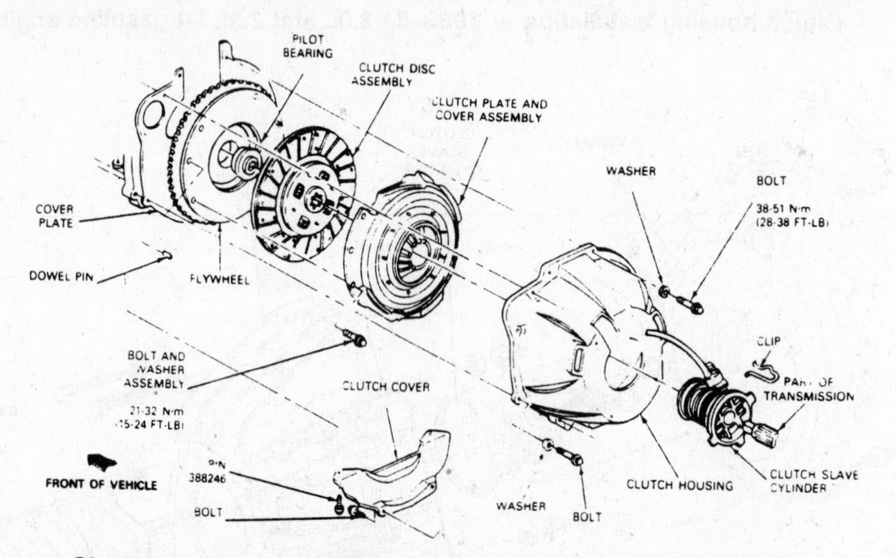

Clutch housing installation – 1985–87 2.3L turbo diesel engine

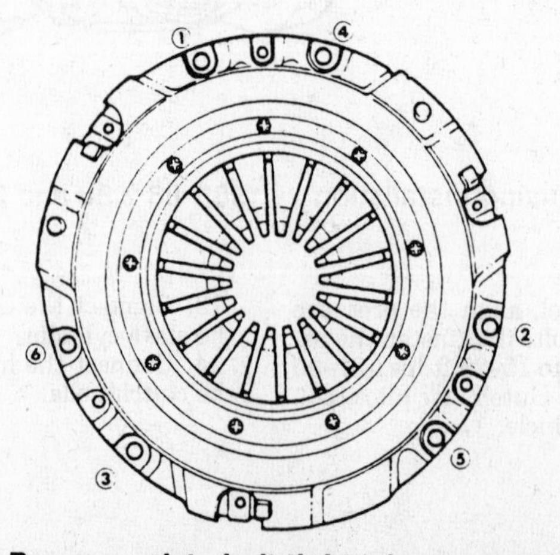

Pressure plate bolt tightening sequence

AUTOMATIC TRANSMISSION

Understanding Automatic Transmissions

The automatic transmission allows engine torque and power to be transmitted to the rear wheels within a narrow range of engine operating speeds. The transmission will allow the engine to turn fast enough to produce plenty of power and torque at very low speeds, while keeping it at a sensible rpm at high vehicle speeds. The transmission performs this job entirely without driver assistance. The transmission uses a light fluid as the medium for the transmission of power. This fluid also works in the operation of various hydraulic control circuits and as a lubricant. Because the transmission fluid performs all of these three functions, trouble within the unit can easily travel from one part to another. For this reason, and because of the complexity and unusual operating principles of the transmission, a very sound understanding of the basic principles of operation will simplify troubleshooting.

THE TORQUE CONVERTER

The torque converter replaces the conventional clutch. It has three functions:

1. It allows the engine to idle with the vehicle at a standstill, even with the transmission in gear.
2. It allows the transmission to shift from range to range smoothly, without requiring that the driver close the throttle during the shift.
3. It multiplies engine torque to an increasing extent as vehicle speed drops and throttle opening is increased. This has the effect of making the transmission more responsive and reduces the amount of shifting required.

The torque converter is a metal case which is shaped like a sphere that has been flattened on opposite sides. It is bolted to the rear end of the engine's crankshaft. Generally, the entire metal case rotates at engine speed and serves as the engine's flywheel.

The case contains three sets of blades. One set is attached directly to the case. This set forms the torus or pump. Another set is directly connected to the output shaft, and forms the turbine. The third set is mounted on a hub which, in turn, is mounted on a stationary shaft through a one-way clutch. This third set is known as the stator.

A pump, which is driven by the converter hub at engine speed, keeps the torque converter full of transmission fluid at all times. Fluid flows continuously through the unit to provide cooling.

Under low speed acceleration, the torque converter functions as follows:

The torus is turning faster than the turbine. It picks up fluid at the center of the converter and, through centrifugal force, slings it outward. Since the outer edge of the converter moves faster than the portions at the center, the fluid picks up speed.

The fluid then enters the outer edge of the turbine blades. It then travels back toward the center of the converter case along the turbine blades. In impinging upon the turbine blades, the fluid loses the energy picked up in the torus.

If the fluid were now to immediately be returned directly into the torus, both halves of the converter would have to turn at approximately the same speed at all times, and torque input and output would both be the same.

In flowing through the torus and turbine, the fluid picks up two types of flow, or flow in two separate directions. It flows through the turbine blades, and it spins with the engine. The stator, whose blades are stationary when the vehicle is being accelerated at low speeds, converts one type of flow into another. Instead of allowing the fluid to flow straight back into the torus,

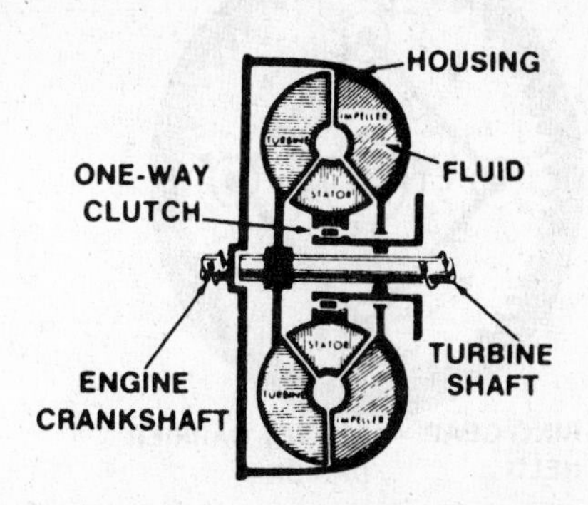

The torque converter housing is rotated by the engine's crankshaft and turns the impeller. The impeller spins the turbine, which gives motion to the turbine shaft, driving gear

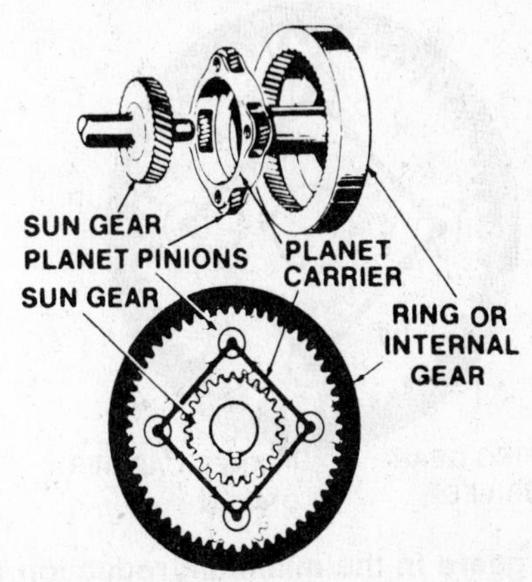

Planetary gears are similar to manual transmission gears but are composed of 3 parts

the stator's curved blades turn the fluid almost 90° toward the direction of rotation of the engine. Thus the fluid does not flow as fast toward the torus, but is already spinning when the torus picks it up. This has the effect of allowing the torus to turn much faster than the turbine. This difference in speed may be compared to the difference in speed between the smaller and larger gears in any gear train. The result is that engine power output is higher, and engine torque is multiplied.

As the speed of the turbine increases, the fluid spins faster and faster in the direction of engine rotation. As a result, the ability of the stator to redirect the fluid flow is reduced. Under cruising conditions, the stator is eventually forced to rotate on its one-way clutch in the direction of engine rotation. Under these conditions, the torque converter begins to behave almost like a solid shaft, with the torus and turbine speeds being almost equal.

THE PLANETARY GEARBOX

The ability of the torque converter to multiply engine torque is limited. Also, the unit tends to be more efficient when the turbine is rotating at relatively high speeds. Therefore, a planetary gearbox is used to carry the power output of the turbine to the driveshaft.

Planetary gears function very similarly to conventional transmission gears. However, their construction is different in that

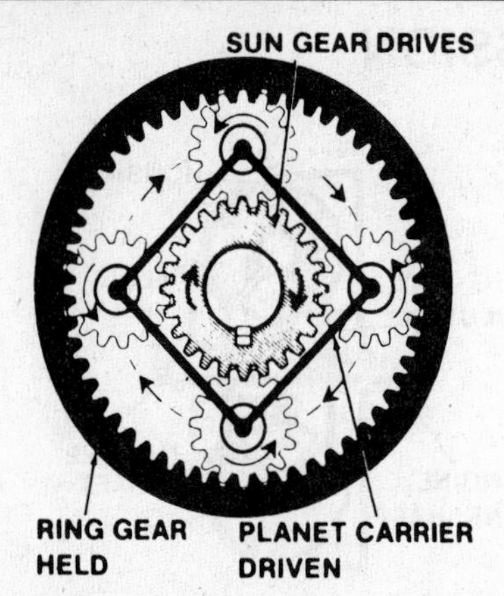

Planetary gears in the maximum reduction (low) range. The ring gear is held and a lower gear ratio is obtained

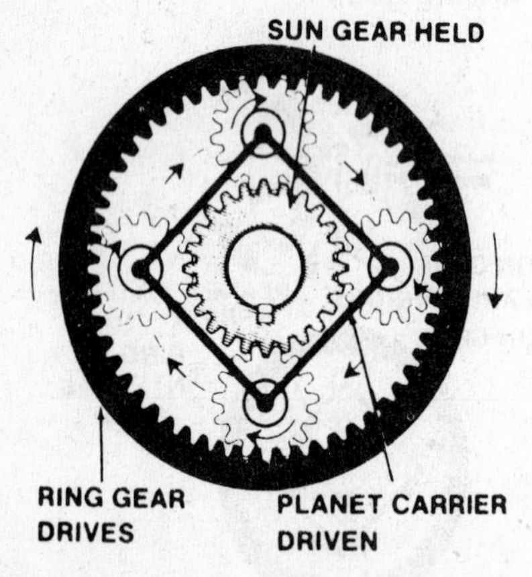

Planetary gears in the minimum reduction (drive) range. The ring gear is allowed to revolve, producing a higher gear ratio

three elements make up one gear system, and, in that all three elements are different from one another. The three elements are: an outer gear that is shaped like a hoop, with teeth cut into the inner surface; a sun gear, mounted on a shaft and located at the very center of the outer gear; and a set of three planet gears, held by pins in a ring-like planet carrier, meshing with both the sun gear and the outer gear. Either the outer gear or the sun gear may be held stationary, providing more than one possible torque multiplication factor for each set of gears. Also, if all three gears are forced to rotate at the same speed, the gearset forms, in effect, a solid shaft.

Most modern automatics use the planetary gears to provide either a single reduction ratio of about 1.8:1, or two reduction gears: a low of about 2.5:1, and an intermediate of about 1.5:1. Bands and clutches are used to hold various portions of the gearsets to the transmission case or to the shaft on which they are mounted. Shifting is accomplished, then, by changing the portion of each planetary gearset which is held to the transmission case or to the shaft.

THE SERVOS AND ACCUMULATORS

The servos are hydraulic pistons and cylinders. They resemble the hydraulic actuators used on many familiar machines, such as bulldozers. Hydraulic fluid enters the cylinder, under

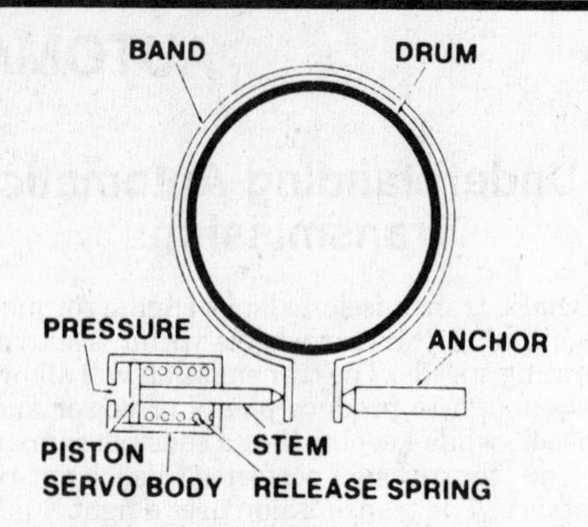

Servos, operated by hydraulic pressure, are used to apply or release the bands

pressure, and forces the piston to move to engage the band or clutches.

The accumulators are used to cushion the engagement of the servos. The transmission fluid must pass through the accumulator on the way to the servo. The accumulator housing contains a thin piston which is sprung away from the discharge passage of the accumulator. When fluid passes through the accumulator on the way to the servo, it must move the piston against spring pressure, and this action smooths out the action of the servo.

THE HYDRAULIC CONTROL SYSTEM

The hydraulic pressure used to operate the servos comes from the main transmission oil pump. This fluid is channeled to the various servos through the shift valves. There is generally a manual shift valve which is operated by the transmission selector lever and an automatic shift valve for each automatic upshift the transmission provides: i.e., 2-speed automatics have a low/high shift valve, while 3-speeds have a 1–2 valve, and a 2–3 valve.

There are two pressures which effect the operation of these valves. One is the governor pressure which is affected by vehicle speed. The other is the modulator pressure which is affected by intake manifold vacuum or throttle position. Governor pressure rises with an increase in vehicle speed, and modulator pressure rises as the throttle is opened wider. By responding to these two pressures, the shift valves cause the upshift points to be delayed with increased throttle opening to make the best use of the engine's power output.

Most transmissions also make use of an auxiliary circuit for downshifting. This circuit may be actuated by the throttle linkage or the vacuum line which actuates the modulator, or by a cable or solenoid. It applies pressure to a special downshift surface on the shift valve or valves.

The transmission modulator also governs the line pressure, used to actuate the servos. In this way, the clutches and bands will be actuated with a force matching the torque output of the engine.

Identification

There are 3 optional automatic transmissions used in the Ford Ranger/Bronco II/Explorer. They may be identified by checking the transmission code on the Safety Standard Certification Label attached to the driver's side door post, in the space marked **Trans.** The transmissions can also be identified by a tag attached to the lower left hand extension attaching bolt. The transmission codes are as follows:

- C3 transmission is code V
- C5 transmission is code W
- A4LD (4-speed overdrive) transmission is code T

Troubleshooting Basic Automatic Transmission Problems

Problem	Cause	Solution
Fluid leakage	• Defective pan gasket • Loose filler tube • Loose extension housing to transmission case • Converter housing area leakage	• Replace gasket or tighten pan bolts • Tighten tube nut • Tighten bolts • Have transmission checked professionally
Fluid flows out the oil filler tube	• High fluid level • Breather vent clogged • Clogged oil filter or screen • Internal fluid leakage	• Check and correct fluid level • Open breather vent • Replace filter or clean screen (change fluid also) • Have transmission checked professionally
Transmission overheats (this is usually accompanied by a strong burned odor to the fluid)	• Low fluid level • Fluid cooler lines clogged • Heavy pulling or hauling with insufficient cooling • Faulty oil pump, internal slippage	• Check and correct fluid level • Drain and refill transmission. If this doesn't cure the problem, have cooler lines cleared or replaced. • Install a transmission oil cooler • Have transmission checked professionally
Buzzing or whining noise	• Low fluid level • Defective torque converter, scored gears	• Check and correct fluid level • Have transmission checked professionally
No forward or reverse gears or slippage in one or more gears	• Low fluid level • Defective vacuum or linkage controls, internal clutch or band failure	• Check and correct fluid level • Have unit checked professionally
Delayed or erratic shift	• Low fluid level • Broken vacuum lines • Internal malfunction	• Check and correct fluid level • Repair or replace lines • Have transmission checked professionally

Lockup Torque Converter Service Diagnosis

Problem	Cause	Solution
No lockup	• Faulty oil pump • Sticking governor valve • Valve body malfunction (a) Stuck switch valve (b) Stuck lockup valve (c) Stuck fail-safe valve • Failed locking clutch • Leaking turbine hub seal • Faulty input shaft or seal ring	• Replace oil pump • Repair or replace as necessary • Repair or replace valve body or its internal components as necessary • Replace torque converter • Replace torque converter • Repair or replace as necessary

Lockup Torque Converter Service Diagnosis

Problem	Cause	Solution
Will not unlock	• Sticking governor valve • Valve body malfunction (a) Stuck switch valve (b) Stuck lockup valve (c) Stuck fail-safe valve	• Repair or replace as necessary • Repair or replace valve body or its internal components as necessary
Stays locked up at too low a speed in direct	• Sticking governor valve • Valve body malfunction (a) Stuck switch valve (b) Stuck lockup valve (c) Stuck fail-safe valve	• Repair or replace as necessary • Repair or replace valve body or its internal components as necessary
Locks up or drags in low or second	• Faulty oil pump • Valve body malfunction (a) Stuck switch valve (b) Stuck fail-safe valve	• Replace oil pump • Repair or replace valve body or its internal components as necessary
Sluggish or stalls in reverse	• Faulty oil pump • Plugged cooler, cooler lines or fittings • Valve body malfunction (a) Stuck switch valve (b) Faulty input shaft or seal ring	• Replace oil pump as necessary • Flush or replace cooler and flush lines and fittings • Repair or replace valve body or its internal components as necessary
Loud chatter during lockup engagement (cold)	• Faulty torque converter • Failed locking clutch • Leaking turbine hub seal	• Replace torque converter • Replace torque converter • Replace torque converter
Vibration or shudder during lockup engagement	• Faulty oil pump • Valve body malfunction • Faulty torque converter • Engine needs tune-up	• Repair or replace oil pump as necessary • Repair or replace valve body or its internal components as necessary • Replace torque converter • Tune engine
Vibration after lockup engagement	• Faulty torque converter • Exhaust system strikes underbody • Engine needs tune-up • Throttle linkage misadjusted	• Replace torque converter • Align exhaust system • Tune engine • Adjust throttle linkage
Vibration when revved in neutral Overheating: oil blows out of dip stick tube or pump seal	• Torque converter out of balance • Plugged cooler, cooler lines or fittings • Stuck switch valve	• Replace torque converter • Flush or replace cooler and flush lines and fittings • Repair switch valve in valve body or replace valve body
Shudder after lockup engagement	• Faulty oil pump • Plugged cooler, cooler lines or fittings • Valve body malfunction	• Replace oil pump • Flush or replace cooler and flush lines and fittings • Repair or replace valve body or its internal components as necessary

Lockup Torque Converter Service Diagnosis

Problem	Cause	Solution
	• Faulty torque converter • Fail locking clutch • Exhaust system strikes underbody • Engine needs tune-up • Throttle linkage misadjusted	• Replace torque converter • Replace torque converter • Align exhaust system • Tune engine • Adjust throttle linkage

Transmission Fluid Indications

The appearance and odor of the transmission fluid can give valuable clues to the overall condition of the transmission. Always note the appearance of the fluid when you check the fluid level or change the fluid. Rub a small amount of fluid between your fingers to feel for grit and smell the fluid on the dipstick.

If the fluid appears:	It indicates:
Clear and red colored	• Normal operation
Discolored (extremely dark red or brownish) or smells burned	• Band or clutch pack failure, usually caused by an overheated transmission. Hauling very heavy loads with insufficient power or failure to change the fluid, often result in overheating. Do not confuse this appearance with newer fluids that have a darker red color and a strong odor (though not a burned odor).
Foamy or aerated (light in color and full of bubbles)	• The level is too high (gear train is churning oil) • An internal air leak (air is mixing with the fluid). Have the transmission checked professionally.
Solid residue in the fluid	• Defective bands, clutch pack or bearings. Bits of band material or metal abrasives are clinging to the dipstick. Have the transmission checked professionally.
Varnish coating on the dipstick	• The transmission fluid is overheating

Fluid Pan and Filter

REMOVAL AND INSTALLATION

NOTE: For the C5 Automatic Transmission (Code W) use fluid that meets Ford Specification ESP–M2C166–H (Type H) or equivalent. The C3 Automatic Transmission (Code V) and the A4LD Automatic Transmission (Code T) use Dexron II® or equivalent automatic transmission fluid.

1. Disconnect the negative battery cable.
2. Raise and support the vehicle safely.
3. Loosen the transmission pan attaching bolts to drain the fluid from the transmission.
4. When all the fluid has drained from the transmission, remove and thoroughly clean the pan and screen. Discard the pan gasket.

To install:

5. Place a new gasket on the pan, and install the pan on the transmission.
6. Add 3 quarts of fluid to the transmission through the filler tuber.
7. Check the fluid level. Adjust, if required.

NOTE: If it is necessary to perform a complete drain and refill, it will be necessary to remove the remaining fluid from the torque converter and the cooler lines.

To drain the torque converter:

a. Remove the converter housing lower cover.
b. Rotate the torque converter until the drain plug comes into view.
c. Remove the drain plug and allow the transmission fluid to drain.
d. Flush the cooler lines completely.

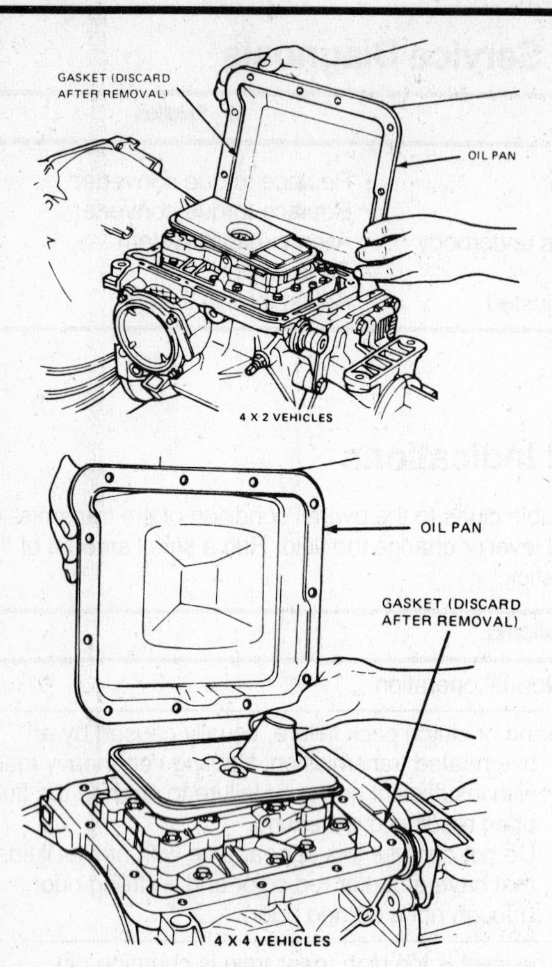

C-5 automatic transmission, sectional view

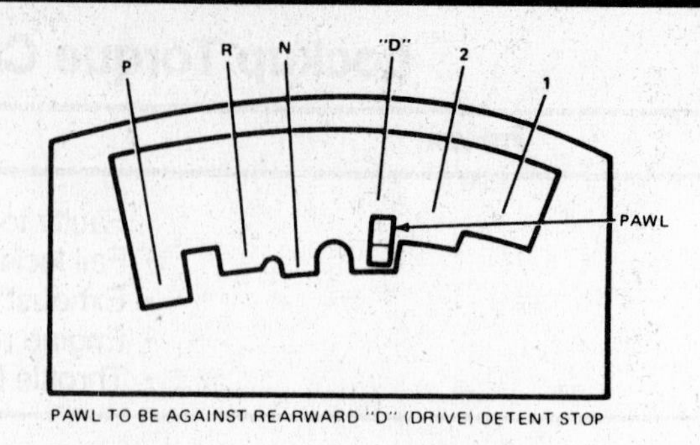

Pawl Positioning for linkage adjustment – C-3 and C-5 transmission

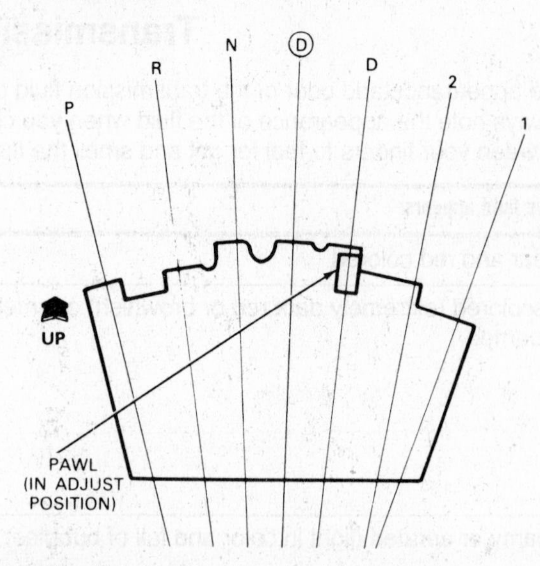

Pawl Positioning for linkage adjustment – A4LD transmission

Adjustments

SHIFT/MANUAL LINKAGE

Before the linkage is adjusted, be sure the engine idle speed and anti-stall dashpot are properly adjusted.

1. Position the transmission selector control lever in **D** position and loosen the trunion bolt. On 4ALD transmission, do not use the overdrive **D** position.

NOTE: Make sure that the shift lever detent pawl is held against the rearward Drive detent stop during the linkage adjustment procedure.

2. Position the transmission manual lever in the **D** position, by moving the bellcrank lever all the way rearward, then forward 3 detents.
3. With the transmission selector lever/manual lever in the **D** position, apply light forward pressure to the shifter control tower arm while tightening the trunion bolt to 12–23 ft. lbs. Forward pressure on the shifter lower arm will ensure correct positioning within the **D** detent as noted in Step 1.

After adjustment, check for Park engagement. The control lever must move to the right when engaged in Park. Check the transmission control lever in all detent positions with the engine running to ensure correct detent transmission action. Readjust if necessary.

KICKDOWN ROD ADJUSTMENT

C3 and C5 Transmissions

NOTE: The engine should be at operating temperature whenever kickdown rod adjustments are made.

1. Place a 6 lb. weight on the kickdown lever.
2. Rotate the throttle to the wide open position.
3. Insert a 0.06 in. (1.5mm) feeler gauge between the throttle lever and the adjusting screw.
4. Rotate the adjusting screw until it makes contact with the feeler gauge. Remove the feeler gauge.
5. After the adjustment has been made, a clearance of 0.001–0.008 in. (0.025–0.203mm) is acceptable.
6. Remove the weight from the kickdown lever.

KICKDOWN CABLE

A4LD transmission

The kickdown cable is self-adjusting over a tolerance range of 1 in. (25mm). If the cable requires readjustment, reset the by depressing the semi-circular metal tab on the self-adjuster mechanism and pulling the cable forward (toward the front of the vehicle) to the "Zero" position setting. The cable will then automatically readjust to the proper length when kicked down.

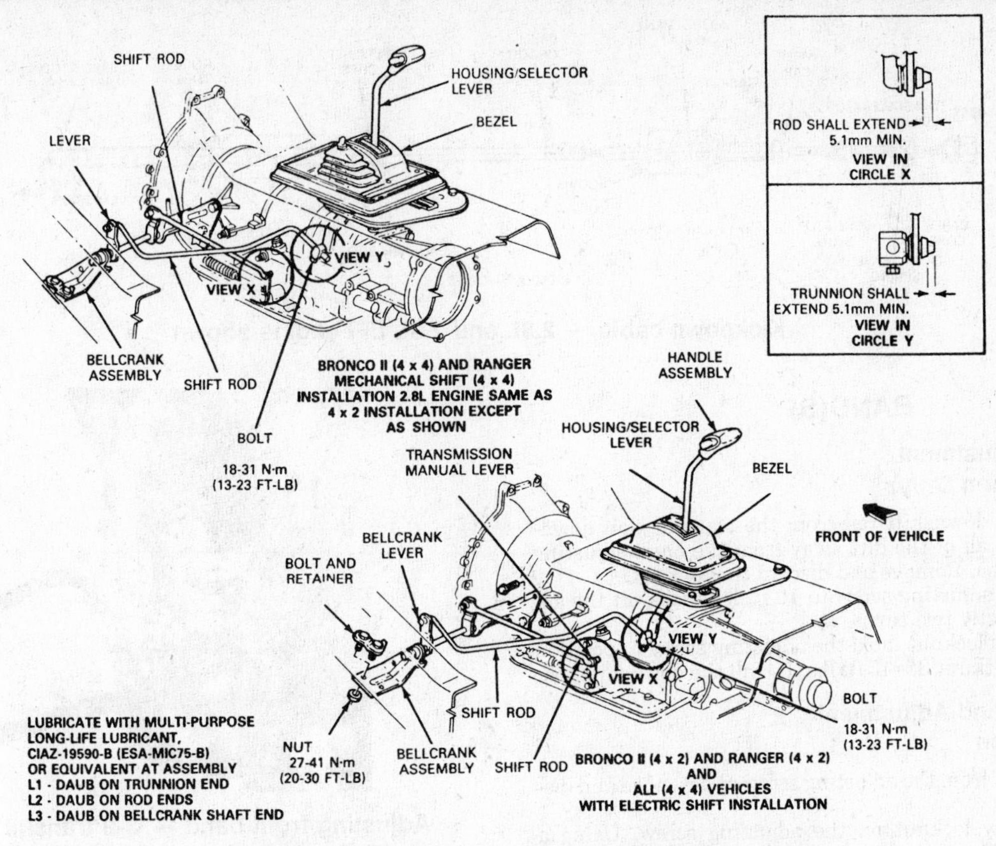

Shift control linkage — A4LD transmission shown

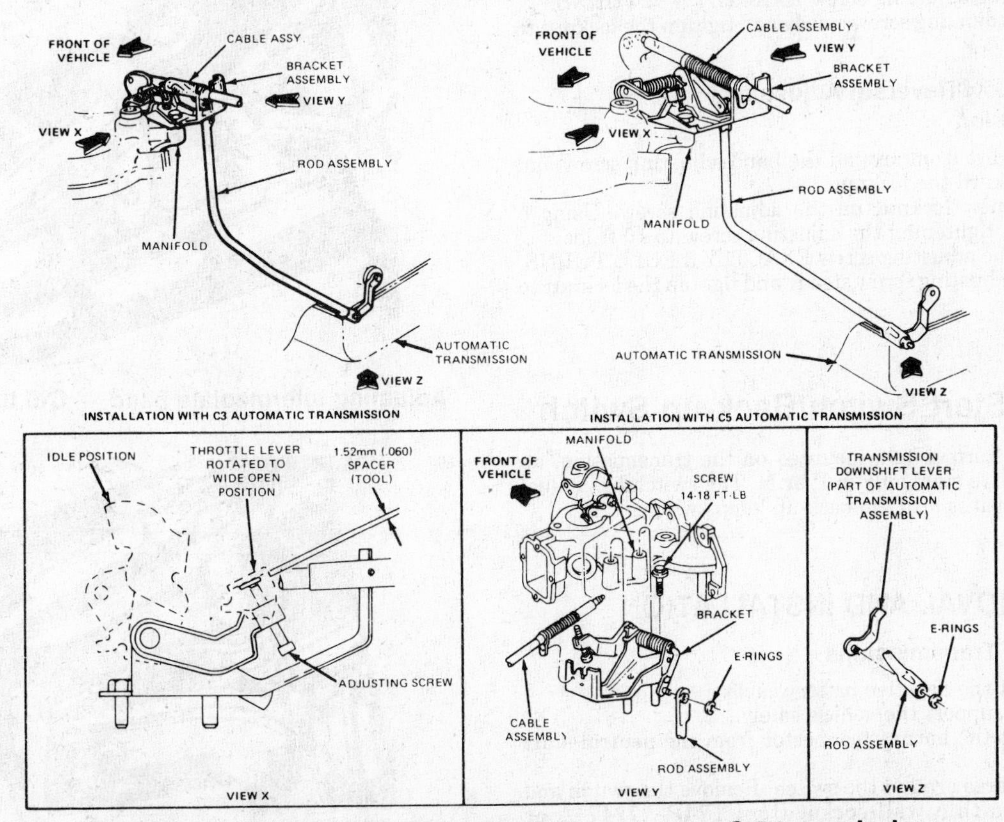

Kickdown rod installation and adjustment — C-3 and C-5 transmissions

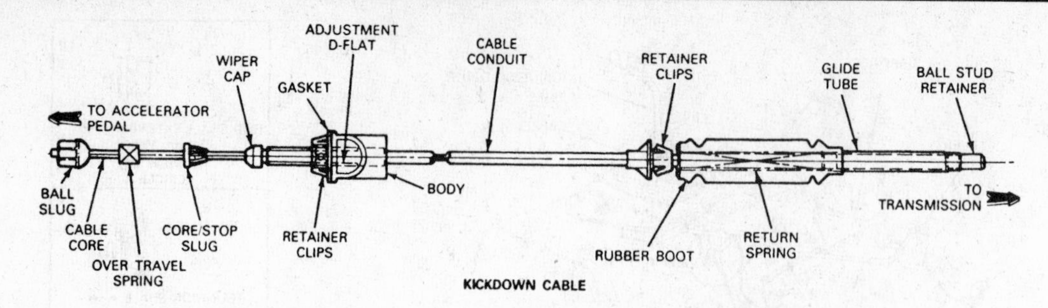

Kickdown cable – 2.3L and 2.9L EFI engine shown

BAND(S)

Front Band Adjustment (C3 Transmission Only)

1. Remove the downshift rod from the transmission downshift lever. Clean all of the dirt away from the band adjusting nut and screw area. Remove and discard the locknut.
2. Tighten the adjusting screw to 10 ft. lbs. Back off the adjusting screw exactly two turns.
3. Install a new locknut, hold the adjusting screw in position and tighten the locknut 35–45 ft. lbs. Install the downshift rod.

Intermediate Band Adjustment C5 Transmission

1. Clean all dirt from the adjusting screw and remove and discard the locknut.
2. Install a new locknut on the adjusting screw. Using a torque wrench, tighten the adjusting screw to 10 ft. lbs.
3. Back off the adjusting screw EXACTLY 4 ¼ TURNS.
4. Hold the adjusting screw steady and tighten the locknut to 40 ft. lbs.

Rear Band (Low-Reverse) Adjustment C5 Transmission

1. Clean all dirt from around the band adjusting screw and remove and discard the locknut.
2. Install a new locknut on the adjusting screw. Using a torque wrench, tightening the adjusting screw to 10 ft.lbs.
3. Back off the adjusting screw EXACTLY 3 FULL TURNS.
4. Hold the adjusting screw steady and tighten the locknut to 40 ft. lbs.

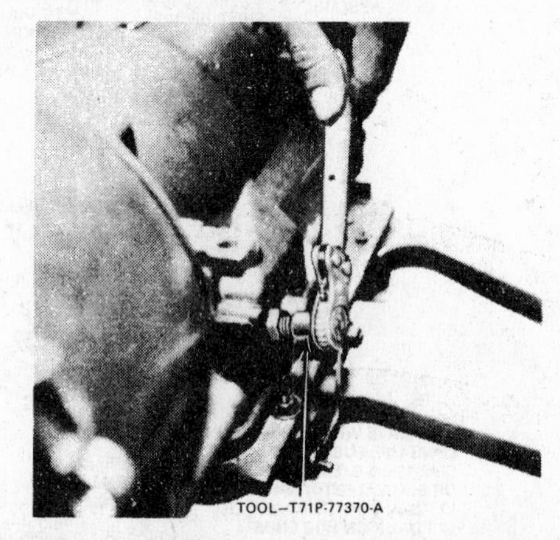

Adjusting front band – C-3 transmission

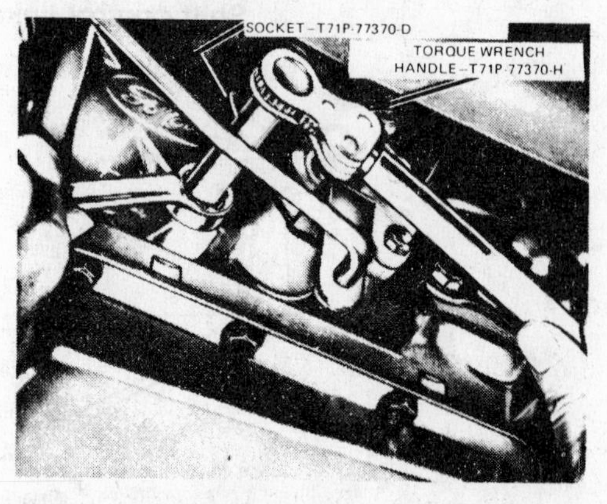

Adjusting intermediate band – C-5 transmission

Neutral Start Switch/Back-up Switch

The neutral start switch, mounted on the transmission, allows the vehicle to start only in **P** or **N**. The switch has a dual purpose, in that it is also the back-up lamp switch.

REMOVAL AND INSTALLATION

C3 and A4LD Transmissions

1. Disconnect the negative battery cable.
2. Raise and support the vehicle safely.
3. Disconnect the harness connector from the neutral start switch.
4. Clean the area around the switch. Remove the switch and O-ring, using a thin wall socket (tool T74P–77247–A or equivalent).

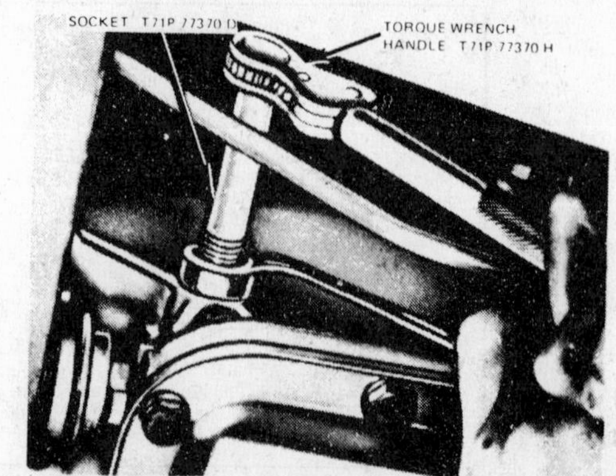

Adjusting low-reverse band – C-5 transmission

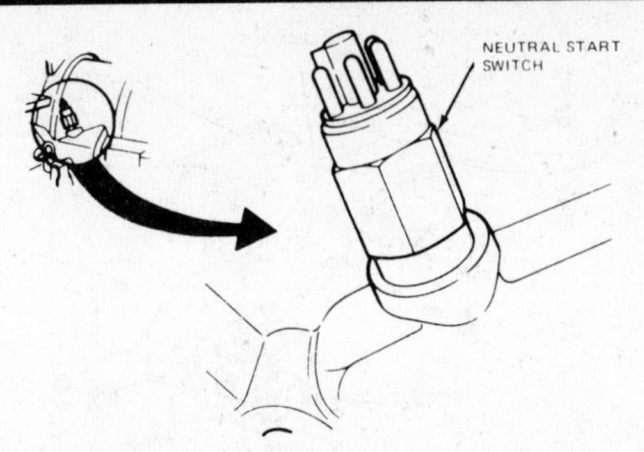

Neutral start switch installation – C-3 and A4LD transmission

To install:

5. Fit a new O-ring to the switch. Install the switch.
6. Reconnect the harness connector to the switch.
7. Lower the vehicle.
8. Reconnect the negative battery cable.
9. Check the operation of the switch, with the parking brake engaged. The engine should only start in **N** or **P**. The back-up lamps should come ON only in **R**.

C5 Transmission

1. Disconnect the negative battery cable.
2. Raise and support the vehicle safely.
3. Remove the downshift linkage rod from the transmission downshift lever.
4. Apply rust penetrant to the outer lever attaching nut to prevent breaking the inner lever shaft. Remove the transmission downshift outer lever attaching nut and lever.
5. Remove the two neutral start switch retaining bolts.
6. Disconnect the multiple wire connector.
7. Remove the neutral start switch from the transmission.

To install:

8. Install the neutral start switch on the transmission. Install the two retaining bolts.
9. Adjust the neutral safety switch following the above procedure.
10. Install the outer downshift lever and retaining nut, and tighten the nut. Install the downshift linkage rod with the retaining clips.
11. Connect the wire multiple connector. Check the operation of the switch. The engine should start only with the transmission selector lever in **N** or **P**.

ADJUSTMENT

C5 Transmission

1. With the automatic transmission linkage properly adjusted, loosen the two switch attaching bolts.
2. Place the transmission selector lever in neutral. Rotate the switch and insert the gauge pin (No. 43, $^{3}/_{32}$ in. drill shank end) fully into the gauge pin holes of the switch. The gauge pin has to be inserted a full ½ in. (13mm) into the three holes of the switch. Move the switch as necessary to allow the drill to rest against the case.
3. Tighten the two switch attaching bolts 55–75 inch lbs. Remove the drill from the switch.
4. Check the operation of the switch. The engine should start only with the transmission selector lever in **N** or **P**.

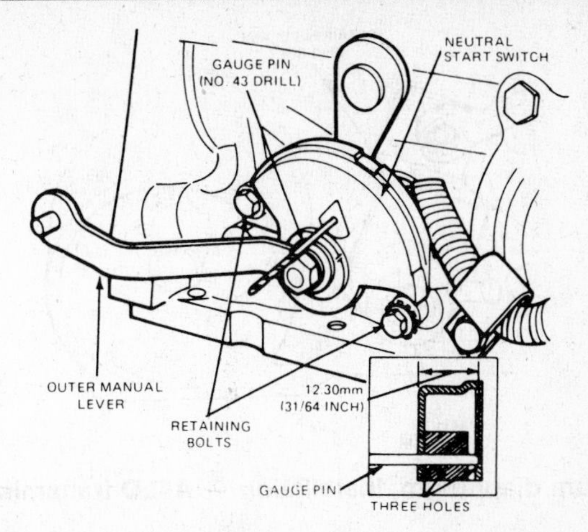

Adjusting neutral safety switch – C-5 transmission

Vacuum Diaphragm

REMOVAL AND INSTALLATION

A4LD Transmission

1. Disconnect the negative battery cable.
2. Raise and support the vehicle safely.
3. Disconnect the hose from the vacuum diaphragm.
4. Remove the vacuum diaphragm retaining clamp bolt and clamp. Do not pry on the clamp.
5. Pull the vacuum diaphragm from the transmission case and remove the vacuum diaphragm control rod from the transmission case.

To install:

6. Install the vacuum diaphragm control rod from the transmission case.
7. Push the vacuum diaphragm into the case and secure it with the clamp and bolt. Tighten to 80–106 inch lbs. (9–12Nm).
8. Fit the vacuum hose to the diaphragm.
9. Lower the vehicle.
10. Reconnect the negative battery cable.

Extension Housing Seal

REMOVAL AND INSTALLATION

1. Disconnect the negative battery cable.
2. Raise and support the vehicle safely.
3. Matchmark the driveshaft end yoke and rear axle companion flange to assure proper positioning during assembly. Remove the driveshaft.
4. Remove the oil seal from the extension housing, using seal remover T71P–7657–A or equivalent.

To install:

Before install the replacement seal, inspect the sealing surface of the universal joint yoke for scores. If scoring is found, replace the yoke.

5. Install the new seal, using seal installer T74P–77052–A or equivalent. Coat the inside diameter at the end of the rubber boot portion of the seal with long-life lubricant (C1AZ–19590–BA or equivalent).
6. Align the matchmarks and install the driveshaft.
7. Lower the vehicle.
8. Reconnect the negative battery cable.

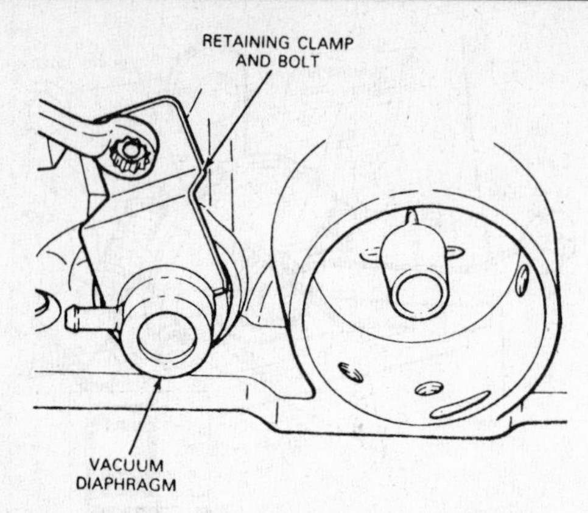

Vacuum diaphragm, installation – A4LD transmission

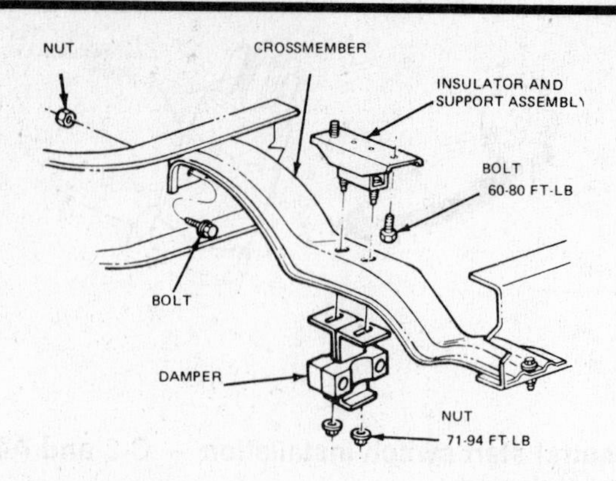

Crossmember removal and installation – C-3 transmission

Transmission

REMOVAL AND INSTALLATION

C3 Transmission

1. Raise the vehicle and safely support on jackstands. Place a drain pan under the transmission fluid pan. Starting at the rear of the pan and working toward the front, loosen the attaching bolts and allow the fluid to drain. Then remove all of the pan attaching bolts except two at the front, to to allow the fluid to further drain. After all the fluid has drained, install two bolts on the rear side of the pan to temporarily hold it in place.
2. Remove the converter drain plug access cover and adapter plate bolts from the lower end of the converter housing.
3. Remove the 4 flywheel to converter attaching nuts. Crank the engine to turn the converter to gain access to the nuts, using a wrench on the crankshaft pulley attaching bolt.On belt driven overheads camshaft engines, never turn the engine backwards.
4. Crank the engine until the converter drain plug is accessible and remove the plug. Place a drain pan under the converter to catch the fluid. After all the fluid has been drained from the converter, reinstall the plug and tighten to 20–30 ft. lbs. Remove the driveshaft. Install cover, plastic bag etc. over end of extension housing.
5. Remove the speedometer cable from the extension housing. Disconnect the shift rod at the transmission manual lever.Disconnect the downshift rod at the transmission downshift lever.
6. Remove the starter-to-converter housing attaching bolts and position the starter out of the way.
7. Disconnect the neutral safety switch wires from the switch. Remove the vacuum line from the transmission vacuum modulator.
8. Position a suitable jack under the transmission and raise it slightly.
9. Remove the engine rear support-to-crossmember bolts. Remove the crossmember-to-frame side support attaching bolts and remove the crossmember insulator and support and damper.
10. Lower the jack under the transmission and allow the transmission to hang.
11. Position a jack to the front of the engine and raise the engine to gain access to the two upper converter housing-to-engine attaching bolts.
12. Disconnect the oil cooler lines at the transmission. Plug all openings to keep out dirt.
13. Remove the lower converter housing-to-engine attaching bolts. Remove the transmission filler tube.
14. Secure the transmission to the jack with a safety chain.

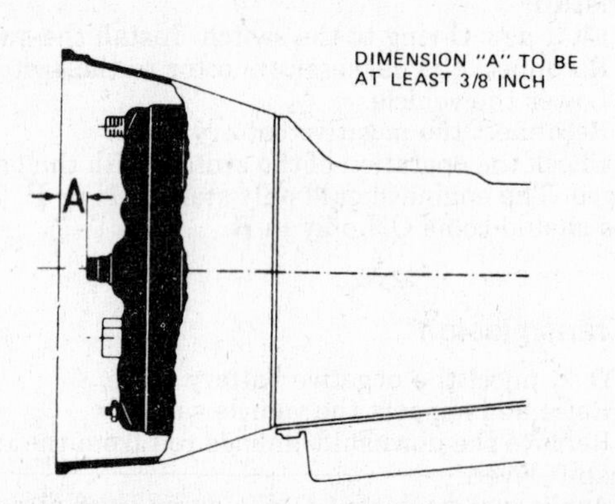

Positioning of converter hub to bell housing flange – C-3 transmission

15. Remove the two upper converter housing-to-engine attaching bolts. Move the transmission to the rear and down to remove it from under the vehicle.

To install:

16. Position the converter to the transmission making sure the converter hub is fully engaged in the pump. With the converter properly installed, place the transmission on the jack and secure with safety chain.
17. Rotate the converter so the drive studs and drain plug are in alignment with their holes in the flywheel. With the transmission mounted on a transmission jack, move the converter and transmission assembly forward into position being careful not to damage the flywheel and the converter pilot.

WARNING: During this move, to avoid damage, do not allow the transmission to get into a nosed down position as this will cause the converter to move forward and disengage from the pump gear. The converter must rest squarely against the flywheel. This indicates that the converter pilot is not binding in the engine crankshaft.

18. Install the two upper converter housing-to-engine attaching bolts and tighten to 28–38 ft. lbs.
19. Remove the safety chain from the transmission. Insert the filler tube in the stub tube and secure it to the cylinder block with the attaching bolt. Tighten the bolt to 28–38 ft. lbs. If the stub tube is loosened or dislodged, it should be replaced. Install the oil cooler lines in the retaining clip at the cylinder block. Connect the lines to the transmission case.
20. Remove the jack supporting the front of the engine. Raise the transmission. Position the crossmember, insulator and sup-

port and damper to the frame side supports and install the attaching bolts. Tighten the bolt to 60–80 ft. lbs.

21. Lower the transmission and install the rear engine support-to-crossmember nut. Tighten the bolt to 60–80 ft. lbs.

22. Remove the transmission jack. Install the vacuum hose on the transmission vacuum unit. Install the vacuum line into the retaining clip.

23. Connect the neutral safety switch plug to the switch. Install the starter and tighten the attaching bolts to 15–20 ft. lbs.

24. Install the 4 flywheel-to-converter attaching nuts.

When assembling the flywheel to the converter, first install the attaching nuts and tighten to 20–34 ft. lbs.

25. Install the converter drain plug access cover and adaptor plate bolts. Tighten the bolts to 12–16 ft. lbs.

26. Connect the muffler inlet pipe to the exhaust manifold.

27. Connect the transmission shift rod to the manual lever. Connect the downshift rod to the downshift lever.

28. Connect the speedometer cable to the extension housing. Install the driveshaft. Tighten the companion flange U-bolt attaching nuts to 70–95 ft.lbs.

29. Adjust the manual and downshift linkage as required.

30. Lower the vehicle. Fill the transmission to the correct level with the specified fluid.

31. Start the engine and shift the transmission to all ranges, then recheck the fluid level.

C5 Transmission
2WD Vehicles

1. Raise the vehicle and safely support on jackstands. Place the drain pan under the transmission fluid pan. Starting at the rear of the pan and working toward the front, loosen the attaching bolts and allow the fluid to drain. Finally remove all of the pan attaching bolts except two at the front, to allow the fluid to further drain. With fluid drained, install two bolts on the rear side of the pan to temporarily hold it in place.

2. Remove the converter drain plug access cover from the lower end of the converter housing.

3. Remove the converter-to-flywheel attaching nuts. Place a wrench on the crankshaft pulley attaching bolt to turn the converter to gain access to the nuts.

4. Place a drain pan under the converter to catch the fluid. With the wrench on the crankshaft pulley attaching bolt, turn the converter to gain access to the converter drain plug and remove the plug. After the fluid has been drained, reinstall the plug.

5. Disconnect the driveshaft from the rear axle and slide shaft rearward from the transmission. Install a suitable cover or plug in the extension housing to prevent fluid leakage. Mark the rear driveshaft yoke and axle flange so they can be installed in their original position.

6. Disconnect the cable from the terminal on the starter motor. Remove the three attaching bolts and remove the starter motor. Disconnect the neutral start switch wires at the plug connector.

7. Remove the rear mount-to-crossmember attaching nuts and the two crossmember-to-frame attaching bolts. Remove the right and left gusset.

8. Remove the two engine rear insulator-to-extension housing attaching bolts.

9. Disconnect the TV linkage rod from the transmission TV lever. Disconnect the manual rod from the transmission manual lever at the transmission.

10. Remove the two bolts securing the bell-crank bracket to the converter housing.

11. Raise the transmission with a suitable jack to provide clearance to remover the crossmember. Remove the rear mount from the crossmember and remove the crossmember from the side supports. Lower the transmission to gain access to the oil cooler lines. Disconnect each oil line from the fittings on the transmission.

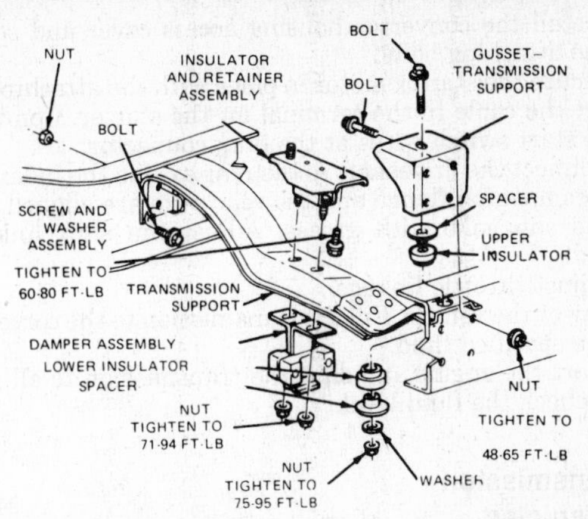

Crossmember removal and installation – C-5 transmission

12. Disconnect the speedometer cable from the extension housing.

13. Remove the bolt that secures the transmission fluid filler tube to the cylinder block. Lift the filler tube and the dipstick from the transmission.

14. Secure the transmission to the jack with the chain. Remove the converter housing-to-cylinder block attaching bolts.

15. Carefully move the transmission and converter assembly away from the engine and, at the same time, lower the jack to clear the underside of the vehicle.

To install:

16. Tighten the converter drain plug to specifications. Position the converter on the transmission, making sure the converter drive flats are fully engaged in the pump gear by rotating the converter.

17. With the converter properly installed, place the transmission on the jack. Secure the transmission to the jack with a chain.

18. Rotate the converter until the studs and drain plug are in alignment with the holes in the flywheel. Move the converter and transmission assembly forward into position, using care not to damage the flywheel and the converter pilot. The converter must rest squarely against the flywheel. This indicates that the converter pilot is not binding in the engine crankshaft.

19. Install and tighten the converter housing-to-engine attaching bolts to specification.

20. Remove the safety chain from around the transmission.

21. Install the new O-ring on the lower end of the transmission filler tube. Insert the tube in the transmission case and secure the tube to the engine with the attaching bolt.

22. Connect the speedometer cable to the extension housing.

23. Connect the oil cooler lines to the right side of transmission case.

24. Secure the engine rear support to the extension housing and tighten the bolts to specification.

25. Position the crossmember on the side supports. Lower the transmission and remove the jack. Secure the crossmember to the side supports with the attaching bolts.

26. Position the damper assembly over the engine rear support studs. (The painted face of the damper is facing forward when installed in the vehicle.) Secure the rear engine support to the crossmember.

27. Position the bellcrank to the converter housing and install the two attaching bolts.

28. Connect the TV linkage rod to the transmission TV lever. Connect the manual linkage rod to the manual lever at the transmission.

29. Secure the converter-to-flywheel attaching nuts and tighten them to specification.

30. Install the converter housing access cover and secure it with the attaching bolts.
31. Secure the starter motor in place with the attaching bolts. Connect the cable to the terminal on the starter. Connect the neutral start switch wires at the plug connector.
32. Connect the driveshaft to the rear axle so the index marks on the companion flange and the rear yoke are aligned. Lubricate the slip yoke with grease. Adjust the shift linkage as required.
33. Adjust throttle linkage.
34. Lower the vehicle. Fill the transmission to the correct level with the specified fluid.
33. Start the engine and shift the transmission to all ranges, then recheck the fluid level.

C5 Transmission
4WD Vehicles

1. Remove the bolt securing the fluid filler tube to the engine valve cover bracket.
2. Raise the vehicle on a hoist and support with jackstands.
3. Place a drain pan under the transmission fluid pan.
4. Starting at the rear of the pan and working towards the front, loosen the attaching bolts and allow the fluid to drain. Finally, remove all of the pan attaching bolts except two at the front, to allow the fluid to drain further. With fluid drained, install two bolts on the rear side of the pan to temporarily hold it in place.
5. Remove the converter drain plug access cover from the lower end of the converter housing.
6. Remove the converter-to-flywheel attaching nuts.
7. Place a wrench on the crankshaft pulley attaching bolt to turn the converter to gain access to the nuts.
8. Place a drain pan under the converter to catch the fluid. With the wrench on the crankshaft pulley attaching bolt, turn the converter to gain access to the converter drain plug and remove the plug. After the fluid has been drained, reinstall the plug.
9. Disconnect the cable from the terminal at the starter motor. Remove the three attaching bolts and remove the starter motor. Disconnect the neutral safety switch wires at the plug connector.
10. Remove the two engine rear insulator-to-extension housing attaching bolts.
11. Disconnect the TV linkage rod from the transmission TV lever. Disconnect the manual rod from the transmission manual lever at the transmission. Disconnect the downshift and manual linkage rods from the levers on the transmission.
12. Remove the vacuum hose from the vacuum diaphragm unit. Remove the vacuum line from the retaining clip.
13. Remove the two bolts securing the bellcrank bracket to the converter housing.
14. Remove the transfer case. Refer to Transfer Case in this chapter for the correct procedure.
15. Raise the transmission with a transmission jack to provide clearance to remove the crossmember. Remove the rear mount from the crossmember and remove the crossmember from the side supports.
16. Lower the transmission to gain access to the oil cooler lines.
17. Disconnect each oil line from the fittings on the transmission.
18. Disconnect the speedometer cable from the extension housing.
19. Secure the transmission to the jack with the chain.
20. Secure the converter housing-to-cylinder block attaching bolts.
21. Carefully move the transmission and converter assembly away from the engine and, at the same time, lower the jack to clear the underside of the vehicle.
22. Remove the converter and mount the transmission in a holding fixture.

To install:

23. Tighten the converter drain plug to 15–28 ft. lbs.
24. Position the converter on the transmission, making sure the converter drive flats are fully engaged in the pump gear by rotating the converter.
25. With the converter properly installed, place the transmission on the jack. Secure the transmission to the jack with a chain.
26. Rotate the converter until the studs and drain plug are in alignment with the holes in the flywheel.
27. Move the converter and transmission assembly forward into position using care not to damage the flywheel and the converter pilot. The converter must rest squarely against the flywheel. This indicates that the converter pilot is not binding in the engine crankshaft.
28. Install and tighten the converter housing-to-engine attaching bolts to 22–32 inch lbs.
29. Remove the safety chain from around the transmission.
30. Install a new O-ring on the lower end of the transmission filler tube. Insert the tube in the transmission case.
31. Connect the speedometer cable to the extension housing.
32. Connect the oil cooler lines to the right of the transmission case.
33. Position the crossmember on the side supports.
34. Position the rear mount insulator on the crossmember and install the attaching bolts and nuts.
35. Install the transfer case. Refer to Transfer Case in this chapter.
36. Secure the engine rear support to the extension housing and tighten the bolts.
37. Lower the transmission and remove the jack.
38. Secure the crossmember to the side supports with the attaching bolts and tighten the bolts.
39. Position the bellcrank to the converter housing and install the two attaching bolts.
40. Connect the downshift and manual linkage rods to their respective levers on the transmission.
41. Connect the vacuum line to the vacuum diaphragm making sure that the line is in the retaining clip.
42. Secure the converter-to-flywheel attaching nuts and tighten them to 20–34 ft. lbs.
43. Install the converter housing access cover and secure it with the attaching bolts.
44. Secure the starter motor in place with the attaching bolts. Connect the cable to the terminal on the starter. Connect the neutral safety switch wires at the plug connector.
45. Adjust the shift linkage as required. Refer to shift linkage adjustment, shown earlier in this chapter.
46. Remove the jackstands and lower the vehicle.
47. Position the transmission fluid filler tube to the valve cover bracket and secure with the attaching bolt.
48. Fill the transmission to the correct level with the specified fluid.
49. Start the engine and shift the transmission to all ranges, then recheck the fluid level.

A4LD Transmission

1. Disconnect the negative battery cable.
2. Raise the vehicle and support it safely.
3. Position a drain pan under the transmission pan.
4. Starting at the rear, loosen, but do not remove the pan bolts.
5. Loosen the pan from the transmission and allow the fluid to drain gradually.
6. Remove all of the pan bolts except 2 at the front or rear and allow the fluid to continue draining.
7. Remove the converter access cover from the lower right side of the converter housing on the 3.0L engine. Remove the

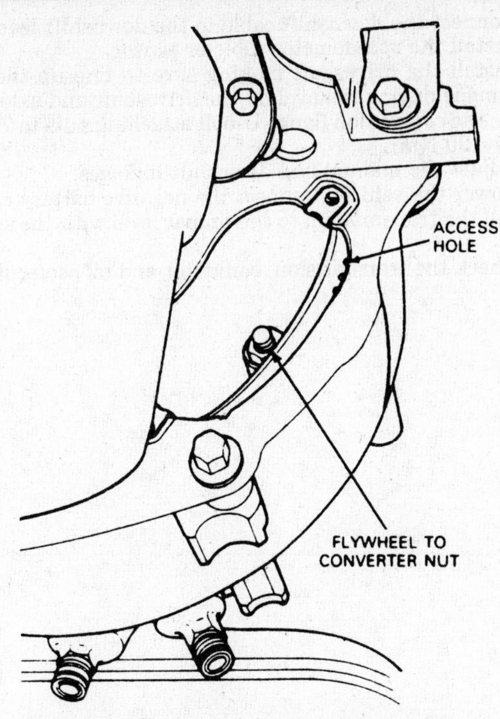

Torque converter nut access hole

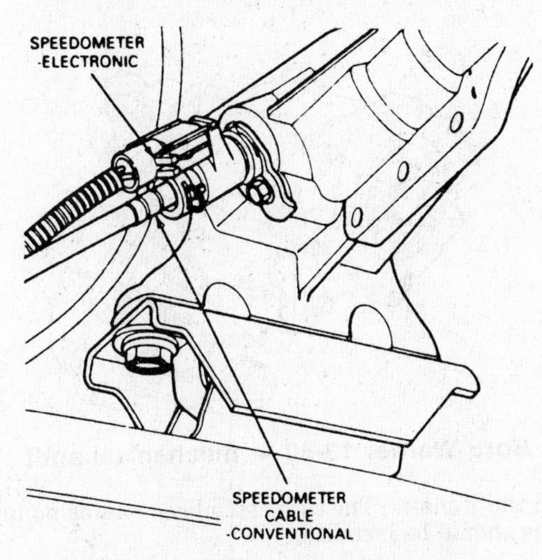

Speedometer cable connection

cover from the bottom of the engine oil pan on the 2.3L engine. Remove a bolt on the access cover of the 2.9L engine and swing the cover open. Remove the access cover and adapter plate bolts from the lower left side of the converter housing on all other applications.

8. Remove the flywheel to converter attaching nuts. Use a socket and breaker bar on the crankshaft pulley attaching bolt. Rotate the pulley clockwise as viewed from the front to gain access to each of the nuts.

NOTE: On belt driven overhead cam engines, never rotate the pulley in a counterclockwise direction as viewed from the front.

9. Scribe a mark indexing the driveshaft to the rear axle flange. Remove the driveshaft.
10. Remove the speedometer cable from the extension housing.
11. Disconnect the shift rod or cable at the transmission manual lever and retainer bracket.
12. Disconnect the downshift cable from the downshift lever. Depress the tab on the retainer and remove the kickdown cable from the bracket.
13. Disconnect the neutral start switch wires, converter clutch solenoid and the 3–4 shift soleniod connector.
14. Remove the starter mounting bolts and the ground cable. Remove the starter.
15. Remove the vacuum line from the transmission vacuum modulator.
16. Remove the filler tube from the transmission.
17. Position a transmission jack under the transmission and raise it slightly.
18. Remove the engine rear support to crossmember bolts.
19. Remove the crossmember to frame side support attaching nuts and bolts. Remove the crossmember.
20. Remove the converter housing to engine bolts.
21. Slightly lower the jack to gain access to the oil cooler lines. Disconnect the oil cooler lines at the transmission. Plug all openings to keep dirt and contamination out.
22. Move the transmission to the rear so it disengages from the dowel pins and the converter is disengaged from the flywheel. Lower the transmission from the vehicle.
23. Remove the torque converter from the transmission.

NOTE: If the transmission is to be removed for a period of time, support the engine with a safety stand and wood block.

To install:

Proper installation of the converter requires full engagement of the converter hub in the pump gear. To accomplish this, the converter must be pushed and at the same time rotated through what feels like 2 notches or bumps. When fully installed, rotation of the converter will usually result in a clicking noise heard, caused by the converter surface touching the housing to case bolts.

This should not be a concern, but an indication of proper converter installation since, when the converter is attached to the engine flywheel, it will be pulled slightly forward away from the bolt heads. Besides the clicking sound, the converter should rotate freely with no binding.

For reference, a properly installed converter will have a distance from the converter pilot nose from face to converter housing outer face of 11–14mm.

1. Install the converter on the transmission.
2. With the converter properly installed, position the transmission on the jack.
3. Rotate the converter so that the drive studs are in alignment with the holes in the flywheel.
4. Move the converter and transmission assembly forward into position, being careful not to damage the flywheel and converter pilot. The converter housing is piloted into position by the dowels in the rear of the engine block.

NOTE: During this move, to avoid damage, do not allow the transmission to get into a nose down position as this will cause the converter to move forward and disengage from the pump gear.

5. Install the converter housing to engine attaching bolts and tighten to 28–38 ft. lbs. (38–51 Nm). The 2 longer bolts are located at the dowel holes.
6. Remove the jack supporting the engine.
7. Raise the transmission. Position the crossmember to the frame side supports. Install the attaching bolts and tighten to 20–30 ft. lbs. (27–41 Nm).
8. Lower the transmission and install the rear engine to crossmember nut and tighten to 60–80 ft. lbs. (82–108 Nm). Remove the transmission jack.
9. Install the filler tube in the transmission.

10. Install the oil cooler lines in the retaining clip at the cylinder block. Connect the lines to the transmission case.
11. Install the vacuum hose on the transmission vacuum unit. Install the vacuum line into the retaining clip.
12. Connect the neutral start switch plug to the neutral start switch. Connect the converter clutch solenoid wires and the 3–4 shift solenoid wires.
13. Install the starter and tighten the bolts to 15–20 ft. lbs. (20–27 Nm).
14. Install the flywheel to converter attaching nuts and tighten to 20–34 ft. lbs. (27–46 Nm).
15. Connect the muffler inlet pipe to the exhaust manifold.
16. Connect the transmission shift rod or cable to the manual lever.
17. Connect the downshift cable to the downshift lever.
18. Install the speedometer cable or sensor.
19. Install the driveshaft making sure to line up the scribe marks made during removal on the driveshaft and axle flange. Tighten the companion flange U-bolt attaching nuts to 70–95 ft. lbs. (95–130 Nm).
20. Adjust the manual and downshift linkages.
21. Lower the vehicle. Connect the negative battery cable.
22. Fill the transmission to the proper level with the specified fluid.
23. Check the transmission, converter and oil cooler lines for leaks.

TRANSFER CASE

Identification

There are 4 transfer cases used on the Ranger/Bronco II/Explorer. There are 2 versions of the 13-50 (Mechanical shift and Electronic shift), the 13-54 and the 13-59.

The Borg Warner 13-50, mechanical shift transfer case, is a 3-piece aluminum part time unit. It transfers power from the transmission to the rear axle and when actuated, also the front drive axle. The unit is lubricated by a positive displacement oil pump that channels oil flow through drilled holes in the rear output shaft. The pump turns with the rear output shaft and allows towing of the vehicle at maximum legal road speeds for extended distances without disconnecting the front and/or rear driveshaft.

The Borg Warner 13-50, electronic shift transfer case, transfers power from the transmission to the rear axle and also the front drive axle, when electronically actuated.

This system consists of a pushbutton control, an electronic control module, an electric shift motor with an integral shift position sensor and a speed sensor.

The electric shift motor, mounted externally at the rear of the transfer case, drives a rotary helical cam. The cam moves the 2WD–4WD shift fork and 4H–4L reduction shift fork to the selected vehicle drive position.

The Borg Warner 13-54 is a 3-piece aluminum part time transfer case. It transfers power from the transmission to the rear axle and when actuated, also the front drive axle. The unit is lubricated by a positive displacement oil pump that channels oil flow through drilled holes in the rear output shaft. The pump turns with the rear output shaft and allows towing of the vehicle at maximum legal road speeds for extended distances without disconecting the front and/or rear driveshaft.

The Borg Warner 13-59 transfer drive case is a 3-piece aluminum assembly. It consists of a front mounting adapter, a front case and a rear cover that transmits power from the transmission to the rear axle. The input shaft and the output shaft are connected together by a coupling sleeve providing direct drive between the 2 shafts. The case assembly contains no lubricant and none should be installed.

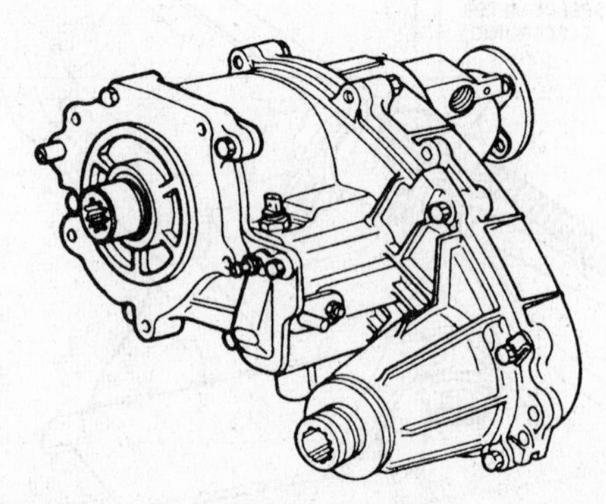

Borg-Warner 13-50 – mechanical shift

Adjustments

Manual Shift 13-50 and 13-54

The following procedure should be used, if a partial or incomplete engagement of the transfer case shift lever detent is experienced or if the control assembly requires removal.

1. Disconnect the negative battery cable.
2. Raise the shift boot to expose the top surface of the cam plates.
3. Loosen the 1 large and 1 small bolt, approximately 1 turn. Move the transfer case shift lever to the **4L** position (lever down).
4. Move the cam plate rearward until the bottom chamfered corner of the neutral lug just contacts the forward right edge of the shift lever.
5. Hold the cam plate in this position and torque the larger bolt first to 70–90 ft. lbs. (95–122 Nm) and torque the smaller bolt to 31–42 ft. lbs. (42–57 Nm).

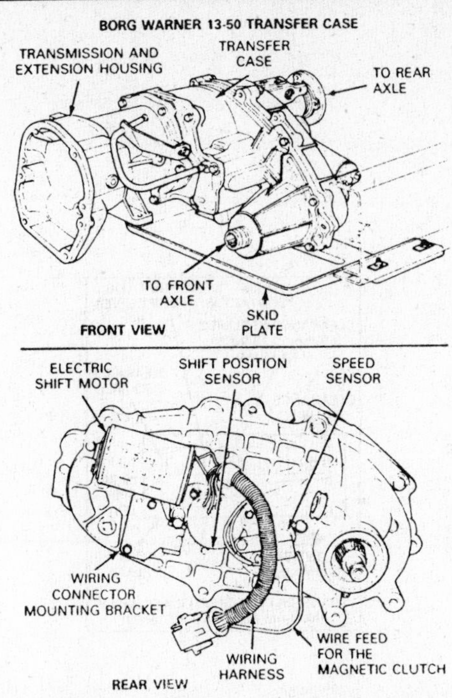

Borg-Warner 13-50 – electronic shift

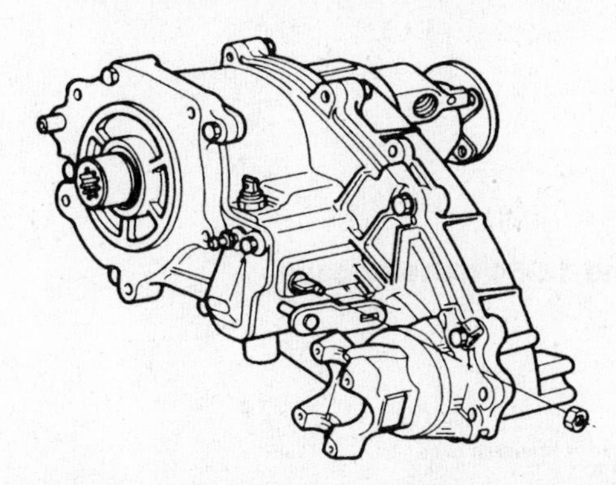

Borg-Warner 13-54 – mechanical shift

6. Move the transfer case in cab shift lever to all shift positions to check for positive engagement. There should be a clearance between the shift lever and the cam plate in the **2H** front and **4H** rear (clearance not to exceed 3.3mm) and **4L** shift positions.
7. Install the shift boot assembly.
8. Reconnect the negative battery cable.

Borg Warner 13-59

The Borg Warner 13–59 does not require adjustment.

Transfer Case

REMOVAL AND INSTALLATION

Borg Warner 13-50 Manual Shift

CAUTION

The catalytic converter is located beside the transfer case. Be careful when working around the catalytic converter because of the extremely high temperatures generated by the converter!

1. Disconnect the negative battery cable.
2. Raise the vehicle and support it safely.
3. Remove the skid plate from frame, if so equipped.
4. Place a drain pan under transfer case, remove the drain plug and drain the fluid from the transfer case.
5. Disconnect the 4-wheel drive indicator switch wire connector at the transfer case.
6. Disconnect the front driveshaft from the axle input yoke.
7. Loosen the clamp retaining the front driveshaft boot to the transfer case, and pull the driveshaft and front boot assembly out of the transfer case front output shaft.
8. Disconnect the rear driveshaft from the transfer case output shaft yoke.
9. Disconnect the speedometer drive gear from the transfer case rear cover.
10. Disconnect the vent hose from the control lever.
11. Loosen or remove the large bolt and the small bolt retaining the shifter to the extension housing. Pull on the control lever until the bushing slides off the transfer case shift lever pin. If necessary, unscrew the shift lever from the control lever.
12. Remove the heat shield from the transfer case.
13. Support the transfer case with a transmission jack.
14. Remove the five bolts retaining the transfer case to the transmission and the extension housing.

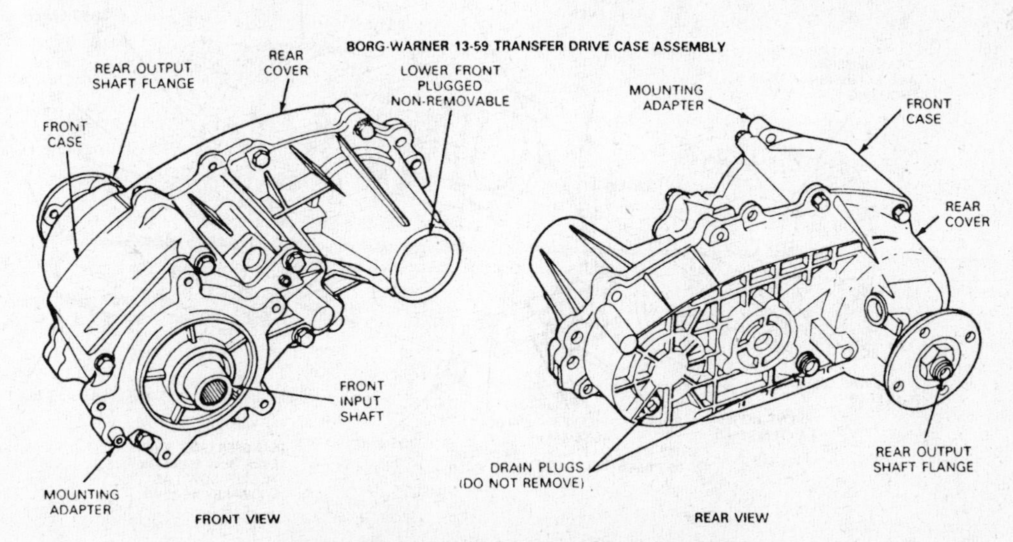

Borg-Warner 13-59 transfer drive case

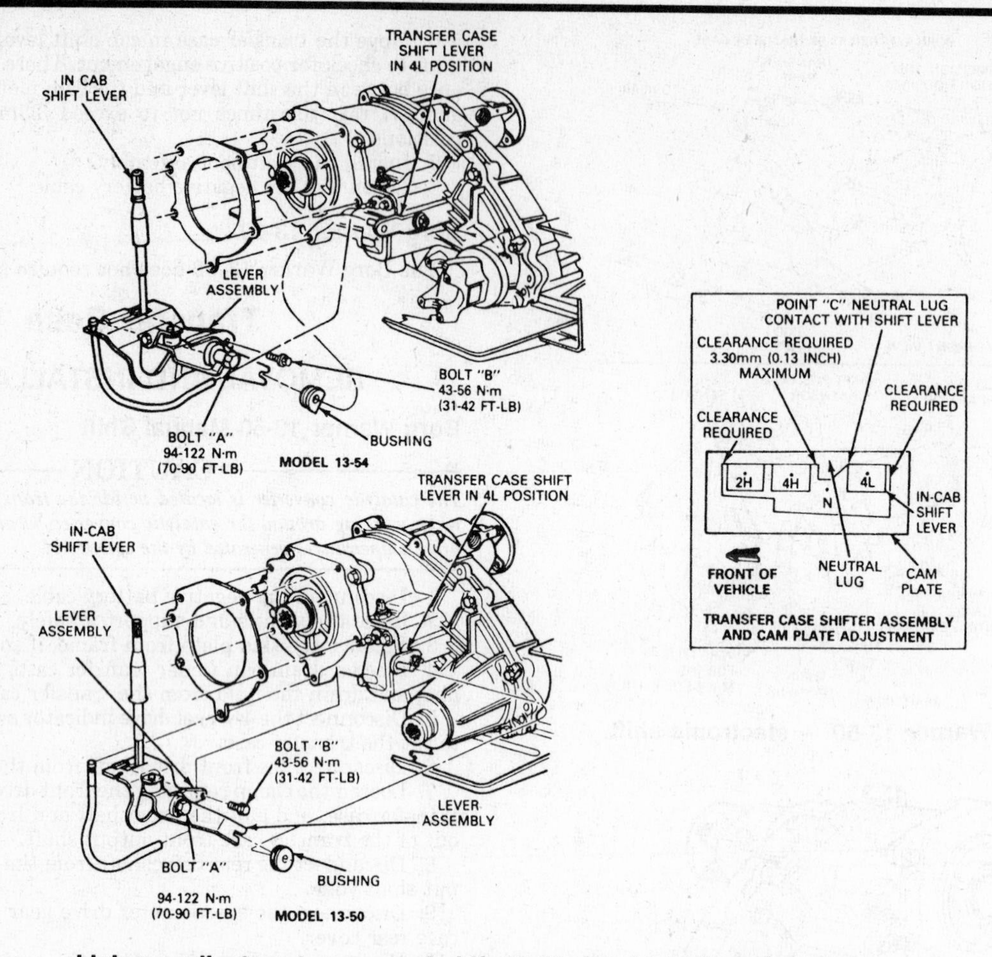

Linkage adjustment – manual shift 13-50 and 13-54 transfer case

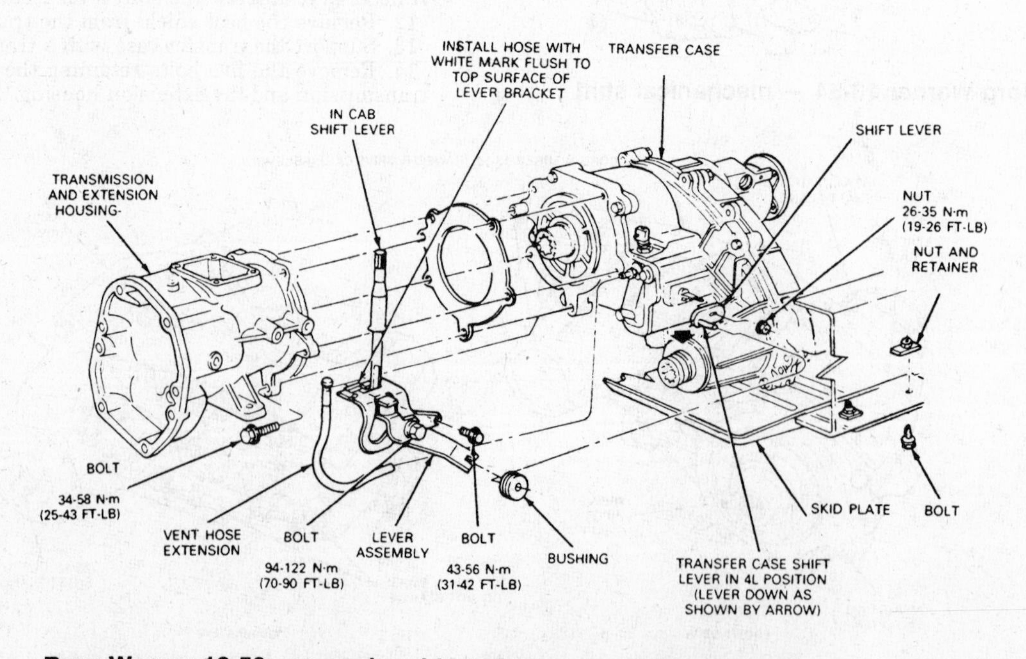

Borg-Warner 13-50, removal and installation – mechanical shift

15. Slide the transfer case rearward off the transmission output shaft and lower the transfer case from the vehicle. Remove the gasket from between the transfer case and extension housing.

To install:

16. Place a new gasket between the transfer case and the extension housing.
17. Raise the transfer case with the transmission jack so that the transmission output shaft aligns with the splined transfer case input shaft. Slide the transfer case forward onto the transmission output shaft and onto the dowel pin. Install the five bolts retaining the transfer case to the extension housing. Torque the bolts to 25–35 ft. lbs. in the sequence shown in the diagram.
18. Remove the transmission jack from the transfer case.
19. Install the heat shield on the transfer case. Torque the bolts to 27–37 ft. lbs.
20. Move the control lever until the bushing is in position over the transfer case shift lever pin. Install and hand start the attaching bolts. First, tighten the large bolt retaining the shifter to the extension housing to 70–90 ft. lbs. then the small bolt to 31–42 ft. lbs.

NOTE: Always tighten the large bolt retaining the shifter to the extension housing before tightening the small bolt.

21. Install the vent assembly so the white marking on the hose is in position in the notch in the shifter.

NOTE: The upper end of the vent hose should be 2 in. (51mm) above the top of the shifter and positioned inside of the shift lever boot.

22. Connect the speedometer drive gear to the transfer case rear cover. Torque the screw to 20–25 inch lbs.
23. Connect the rear driveshaft to the transfer case output shaft yoke. Torque the bolts to 12–15 ft. lbs.
24. Clean the transfer case front output shaft female splines. Apply 5–8 grams of Lubriplate® or equivalent to the splines. Insert the front driveshaft male spline.
25. Connect the front driveshaft to the axle input yoke. Torque the bolts to 12–15 ft.lbs.
26. Push the driveshaft boot to engage the external groove on the transfer case front output shaft. Secure with a clamp.
27. Connect the 4-wheel drive indicator switch wire connector at the transfer case.
28. Install the drain plug and torque to 14–22 ft. lbs. Remove the fill plug and install 3 U.S. pints of DEXRON® II, automatic transmission fluid. Install fill plug and torque to 14–22 ft. lbs.
29. Install the skid plate to the frame. Torque the nuts and bolts to 22–30 ft. lbs.
30. Remove the jackstands and lower the vehicle from the hoist.
31. Reconnect the negative battery cable.

Borg Warner 13-50 Electric Shift

CAUTION

The catalytic converter is located beside the transfer case. Be careful when working around the catalytic converter because of the extremely high temperatures generated by the converter.

1. Disconnect the negative battery cable.
2. Raise the vehicle and support it safely.
3. If so equipped, remove the skid plate from frame.
4. Place a drain pan under transfer case, remove the drain plug and drain fluid from the transfer case.
5. Remove the wire connector from the feed wire harness at the rear of the transfer case. Be sure to squeeze the locking tabs, then pull the connectors apart.
6. Disconnect the front driveshaft from the axle input yoke.
7. Loosen the clamp retaining the front driveshaft boot to the transfer case, and pull the driveshaft and front boot assembly out of the transfer case front output shaft.
8. Disconnect the rear driveshaft from the transfer case output shaft yoke.
9. Disconnect the speedometer driven gear from the transfer case rear cover.
10. Disconnect the vent hose from the control lever.
11. Loosen or remove the large bolt and the small bolt retaining the shifter to the extension housing. Pull on the control lever until the bushing slides off the transfer case shift lever pin. If necessary, unscrew the shift lever from the control lever.
12. Remove the heat shield from the transfer case.
13. Support the transfer case with a transmission jack.
14. Remove the 5 bolts retaining the transfer case to the transmission and the extension housing.
15. Slide the transfer case rearward off the transmission output shaft and lower the transfer case from the vehicle. Remove the gasket from between the transfer case and extension housing.

To install:

16. Install the heat shield onto the transfer case and place a new gasket between the transfer case and adapter.
17. Raise the transfer case with a suitable transmission jack or equivalent, raise it high enough so that the transmission output shaft aligns with the splined transfer case input shaft.
18. Slide the transfer case forward on to the transmission output shaft and onto the dowel pin. Install transfer case retaining bolts and torque them to 26–43 ft. lbs. (35–58 Nm).
19. Connect the rear driveshaft to the rear output shaft yoke and torque the retaining bolts to 20–28 ft. lbs. (27–38 Nm). Attach the shift rod to the transfer case shift lever and transfer case control rod and attach with retaining rings.
20. Connect the speedometer driven gear to the transfer case. Connect the 4WD indicator switch wire connector at the transfer case.
21. Connect the front driveshaft to the front output yoke and torque the yoke nut to 8–15 ft. lbs. (11–20 Nm). Attach the heat shield to the engine mounting bracket and mounting lug on the transfer case.
22. Install the skid plate to the frame. Connect the the wire connectors on the rear of the transfer case, making sure the retaining tabs lock.
23. Fill the transfer case with 6.5 pints of Dexron®II transmission fluid or equivalent. Torque the fill plug to 14–22 ft. lbs. (19–32 Nm). Lower the vehicle. Start the engine and check the transfer case for correct operation. Stop the engine and check the fluid level, add as necessary.
24. Reconnect the negative battery cable.

Borg Warner 13-54 Mechanical Shift

CAUTION

The catalytic converter is located beside the transfer case. Be careful when working around the catalytic converter because of the extremely high temperatures generated by the converter!

1. Disconnect the negative battery cable.
2. Raise the vehicle and support it safely.
3. Remove the skid plate from frame, if so equipped.
4. Remove the damper from the transfer case, if so equipped.
5. Place a drain pan under transfer case, remove the drain plug and drain the fluid from the transfer case.
6. Disconnect the 4WD drive indicator switch wire connector at the transfer case.
7. Disconnect the front driveshaft from the transfer case output shaft flange and wire the shaft out of the way.
8. Disconnect the rear driveshaft from the transfer case output shaft flange and wire the shaft out of the way.
9. Disconnect the speedometer drive gear from the transfer case rear cover.

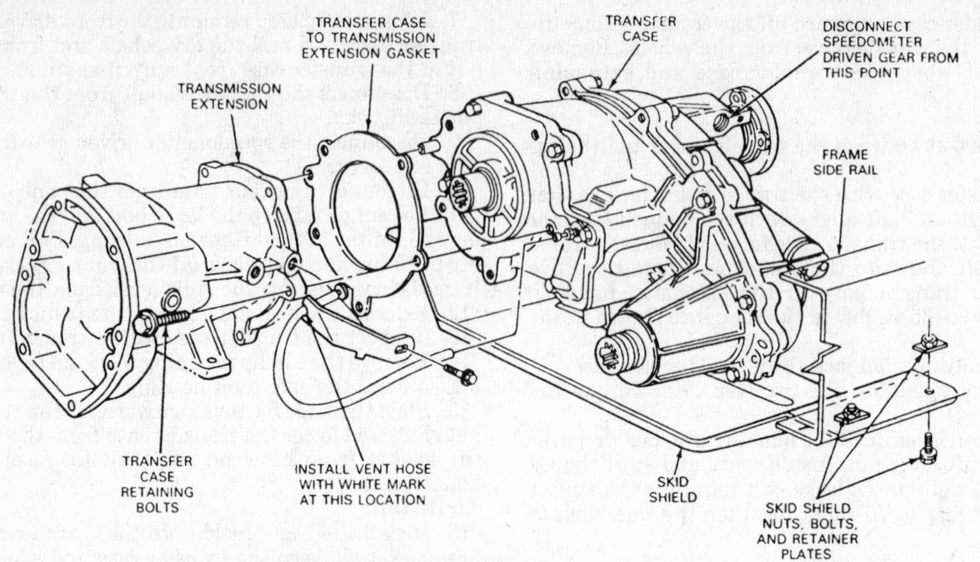

Borg-Warner 13-50, removal and installation – electrical shift

10. Disconnect the vent hose from the control lever.
11. Remove the shift lever retaining nut and remove the lever.
12. Remove the bolts that retains the shifter to the extension housing. Note the size and location of the bolts to aid during installation. Remove the lever assembly and bushing.
13. Support the transfer case with a suitable jack and remove the 5 bolts retaining the transfer case to the transmission and the extension housing.
14. Slide the transfer case rearward off the transmission output shaft and lower the transfer case from the vehicle. Remove and discard the gasket between the transfer case and extension housing.

To install:

15. Fit a new gasket to the front mounting surface of the transfer case assembly. It may be necessary to use a small daub of sealant to hold the gasket down.
16. Raise the transfer case with the transmission jack so that the transmission output shaft aligns with the splined transfer case input shaft. Slide the transfer case forward onto the transmission output shaft and onto the dowel pin. Install the five bolts retaining the transfer case to the extension housing. Torque the bolts to 25–35 ft. lbs. (34–48Nm) in the sequence shown in the diagram.
17. Remove the transmission jack from the transfer case.
18. Install and adjust the shifter, as required.

NOTE: Always tighten the large bolt retaining the shifter to the extension housing before tightening the small bolt.

19. Install the vent assembly so the white marking on the hose is in position in the notch in the shifter.

NOTE: The upper end of the vent hose should be ¾ in. (19mm) above the top of the shifter and positioned inside of the shift lever boot.

20. Connect the speedometer drive gear to the transfer case rear cover. Torque the screw to 20–25 inch lbs. (2.3–2.8Nm).
21. Connect the rear driveshaft to the transfer case output shaft yoke. Torque the bolts to 61–87 ft. lbs. (83–118Nm).
22. Connect the front driveshaft to the transfer case output shaft yoke. Torque the bolts to 12–16 ft.lbs. (16–22Nm).
23. Connect the 4WD indicator switch wire connector at the transfer case.
24. Install the drain plug and torque to 14–22 ft. lbs. Remove the fill plug and install 3 U.S. pints of DEXRON® II, automatic transmission fluid. Install fill plug and torque to 14–22 ft. lbs.

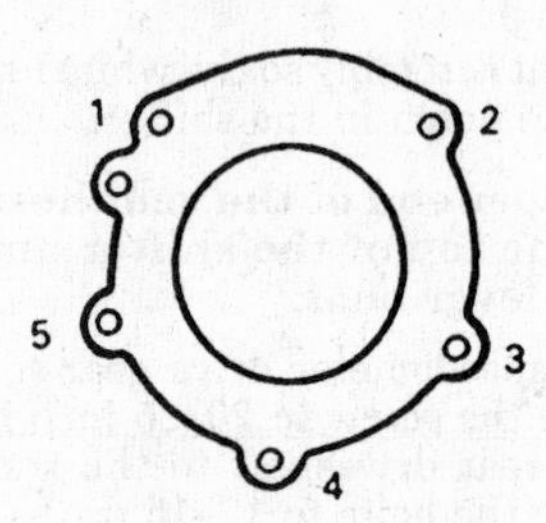

Case to extension – bolt torque sequence

25. Install the damper to the transfer case, if so equipped. Using new damper bolts, tighten the bolts to 25–35 ft. lbs. (34–48Nm).
26. Install the skid plate to the frame. Torque the nuts and bolts to 15–20 ft. lbs. (20–27Nm).
27. Lower the vehicle.
28. Reconnect the negative battery cable.

Borg Warner 13-59

CAUTION

The catalytic converter is located beside the heat shield. Be careful when working around the converter because of the extremely high temperatures generated by the converter!

1. Disconnect the negative battery cable.
2. Raise the vehicle and support safely.
3. Disconnect the rear driveshaft from the transfer case output shaft flange.
4. Disconnect the speedometer driven gear from the transfer case rear cover.
5. Support the transfer case with a transmission jack.
6. Remove the 5 bolts retaining the transfer case to the transmission and the extension housing.
7. Slide the transfer case rearward off the transmission output shaft and lower the transfer case from the vehicle. Remove the gasket from between the transfer case and extension housing.

To Install:

8. Place a new gasket between the transfer case and adapter.

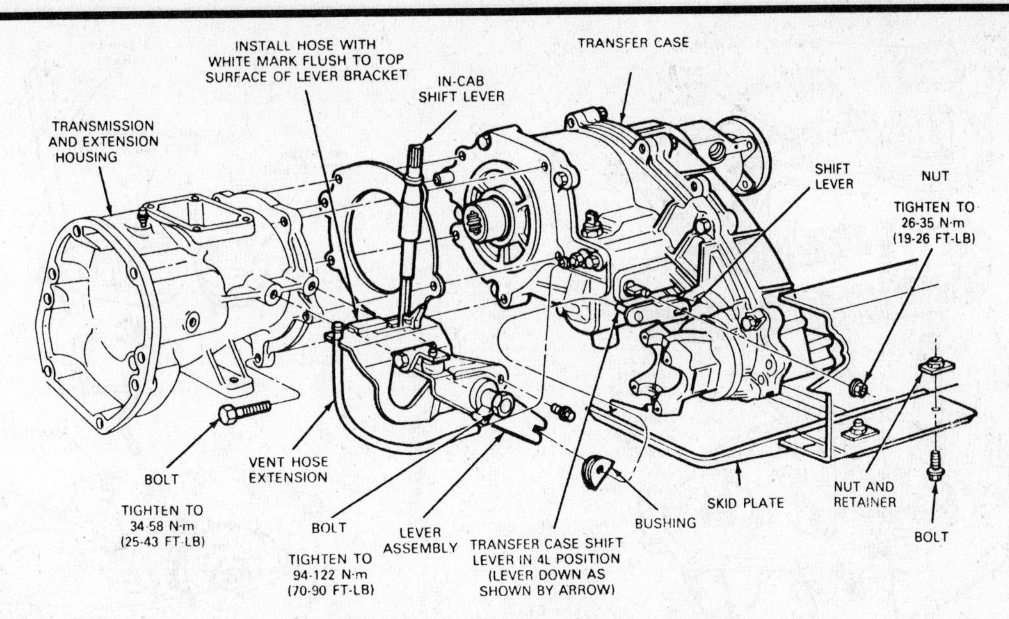

Borg-Warner 13-54, removal and installation

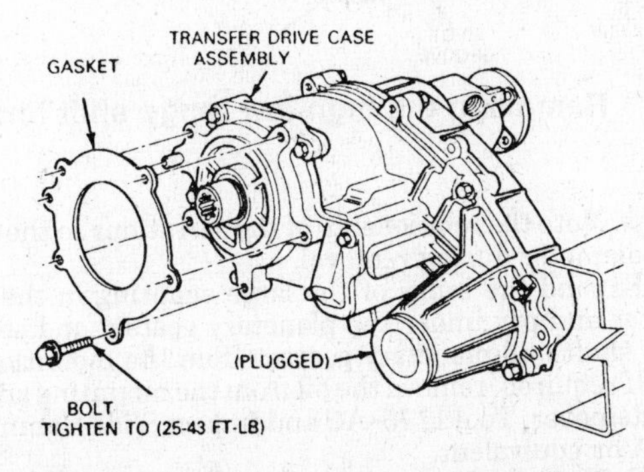

Borg-Warner 13-59, removal and installation

9. Raise the transfer case with a suitable transmission jack or equivalent, raise it high enough so that the transmission output shaft aligns with the splined transfer case input shaft.
10. Slide the transfer case forward on to the transmission output shaft and onto the dowel pin. Install transfer case retaining bolts and torque them to 25–43 ft. lbs.
11. Remove the transmission jack from the transfer case.
12. Connect the speedometer cable assembly to the transfer case rear cover. Tighten the screw to 20–25 inch lbs.
13. Connect the rear driveshaft to the output shaft flange and torque the yoke nut to 61–87 ft. lbs. (83–118 Nm).
14. Lower the vehicle.
15. Reconnect the negative battery cable.

13-50 Overhaul

The manual and electronic shift units are essemtially the same internally. Any differences between the two will be noted. The following procedures apply to both.

SHIFT LEVER REMOVAL

NOTE: Remove the shift ball only if the shift ball, boot or lever is to be replaced. If the ball, boot or lever is not being replaced, remove the ball, boot and lever as an assembly.

1. Remove the plastic insert from the shift ball. Warm the ball with a heat gun to 60–87°C (140–180°F) and knock the ball off the lever with a block of wood and a hammer. Be careful not to damage the finish on the shift lever.
2. Remove the rubber boot and floor pan cover.
3. Disconnect the vent hose from the control lever.
4. Unscrew the shift lever from the control lever.
5. Remove the bolts retaining the shifter to the extension housing. Remove the control lever and bushings.

TRANSFER CASE DISASSEMBLY

1. Remove the transfer case from the vehicle.
2. Remove the transfer case drain plug with a ⅜ in. drive ratchet and drain the fluid.
3. Remove the 4WD indicator switch and the breather vent.
4. Remove the rear output shaft yoke by removing the 30mm nut, steel washer and rubber seal from the output shaft.
5. Remove the bolts which retain the front case to the rear cover. Insert a ½ in. drive breaker bar between the three pry bosses and separate the front case from the rear cover. Remove all traces of RTV gasket sealant from the mating surfaces of the front case and rear cover.

WARNING: When removing RTV sealant, take care not to damage the mating surface of the aluminum case!

6. If the speedometer drive gear or ball bearing assembly is to be replaced, first, drive out the output shaft oil seal from either the inside of the rear cover with a brass drift and hammer or from the outside by bending and pulling on the curved-up lip of the oil seal. Remove and discard the oil seal. Remove the speedometer drive gear assembly (gear, clip and spacer). Note that the round end of the speedometer gear clip faces the inside of the rear cover.
7. Remove the internal snapring that retains the rear output shaft ball bearing in the bore. From the outside of the case, drive out the ball bearing with Output Shaft Bearing Replacer, T83T–7025–B and Drive Handle, T80T–4000–W or equivalent.
8. If required, remove the front output shaft caged needle bearing from the rear cover with Puller Collet, D80L–100–S and Impact Slide Hammer, T50T–100–A or equivalent.
9. Remove the 2W–4W shift fork spring from the boss in the rear cover.
10. Remove the shift collar hub from the output shaft. Remove the 2W–4W lock-up assembly and the 2W–4W shift fork togeth-

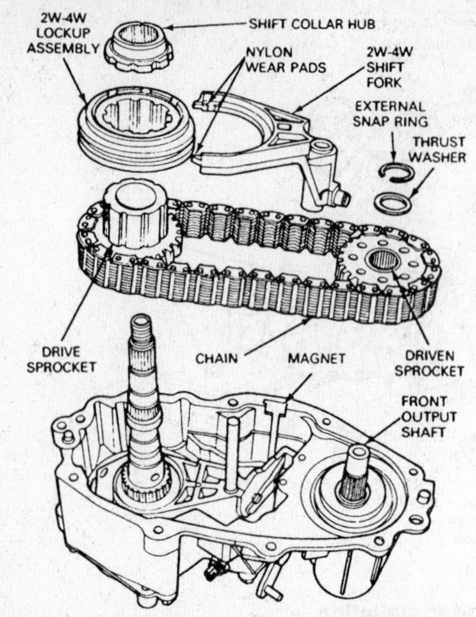

Removing shift collar hub from output shaft

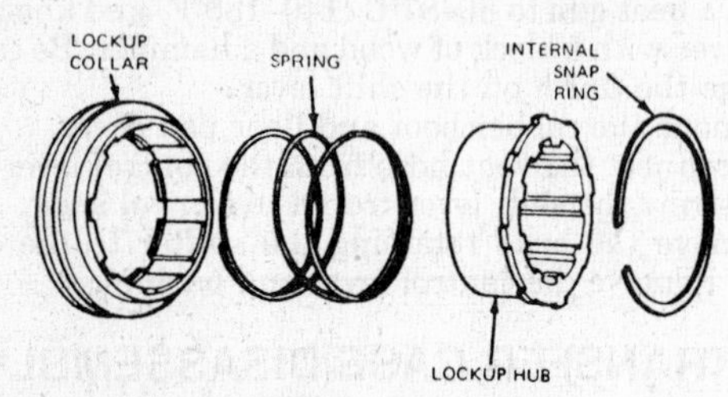

Disassembling the 2W–4W lock-up assembly

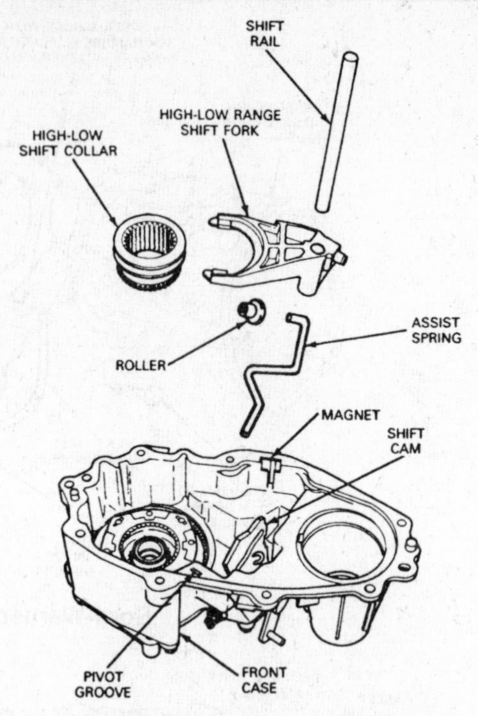

Removing the high-low range shift fork

er as an assembly. Remove the 2W–4W fork from the 2W–4W lock-up assembly. If required, remove the pin, roller, retainer assembly by pressing it out of the 2W–4W shift fork.

11. If required to disassemble the 2W–4W lock-up assembly, remove the internal snapring and pull the lock-up hub and spring from the lock-up collar.

12. Remove the external snapring and thrust washer that retains the drive sprocket to the front output shaft.

13. Remove the chain, driven sprocket and drive sprocket as an assembly.

14. Remove the collector magnet from the notch in the front case bottom.

15. Remove the output shaft and oil pump as an assembly.

16. If required to disassemble the oil pump, remove the bolts from the body. Note the position and markings of the front cover, body, pins, spring, rear cover, and pump retainer as removed.

17. Pull out the shift rail.

18. Slip the high-low range shift fork out of the inside track of the shift cam and remove the high-low shift fork and high-low shift collar together. If required, remove the pin, roller retainer assembly by pressing it out of the high-low range shift fork.

19. Remove the high-low shift hub from out of the planetary gearset in the front case.

20. Push and pull out the anchor end of the assist spring from the locking post in the front case half. Remove the assist spring and roller out of the shift cam.

21. Turn the front case over and remove the bolts retaining the mounting adapter to the front case. Remove the mounting adapter, input shaft and planetary gearset as an assembly.

22. If required, remove the ring gear from the front case using a press. Note the relationship of the serrations to the chamfered pilot diameter during removal.

23. Expand the tangs of the large snapring in the mounting adapter and pry under the planetary gearset and separate the input shaft and planetary gearset from the mounting adapter.

24. If required, remove the oil from the mounting adapter with Seal Remover, Tool 1175–AC and Impact Slide Hammer, T50T–100–A or equivalent.

25. Remove the internal snapring from the planetary carrier and separate the planetary gearset from the input shaft assembly.

26. Remove the external snapring from the input shaft. Place the input shaft assembly in a press and remove the ball bearing from the input shaft using Bearing Splitter, D79L–4621–A or equivalent. Remove the thrust washer, thrust plate and sun gear off the input shaft.

27. Move the shift lever by hand until the shift cam is in the FOUR WHEEL HIGH detent position (4WH) and mark a line on the outside of the front case using the side of the shift lever and a grease pencil.

28. Remove the 2 head set screws from the front case and from the shift cam.

29. Turn the front case over and remove the external clip. Pry the shift lever out of the front case and shift cam.

WARNING: Do not pound on the external clip during removal! Removal of 4WD indicator switch will ease removal of the shift lever and shift cam assembly.

30. Remove the O-ring from the second groove in the shift lever shaft.

31. Remove the detent plunger and compression spring from the inside of the front case.

32. Remove the internal snapring and remove the ball bearing retainer from the front case by tapping on the face of the front output shaft and U-joint assembly with a plastic hammer. Remove the internal snapring and drive the ball bearing out of the bearing retainer using Output Shaft Bearing Replacer, T83T–7025–B and Driver Handle, T80T–4000–W or equivalent.

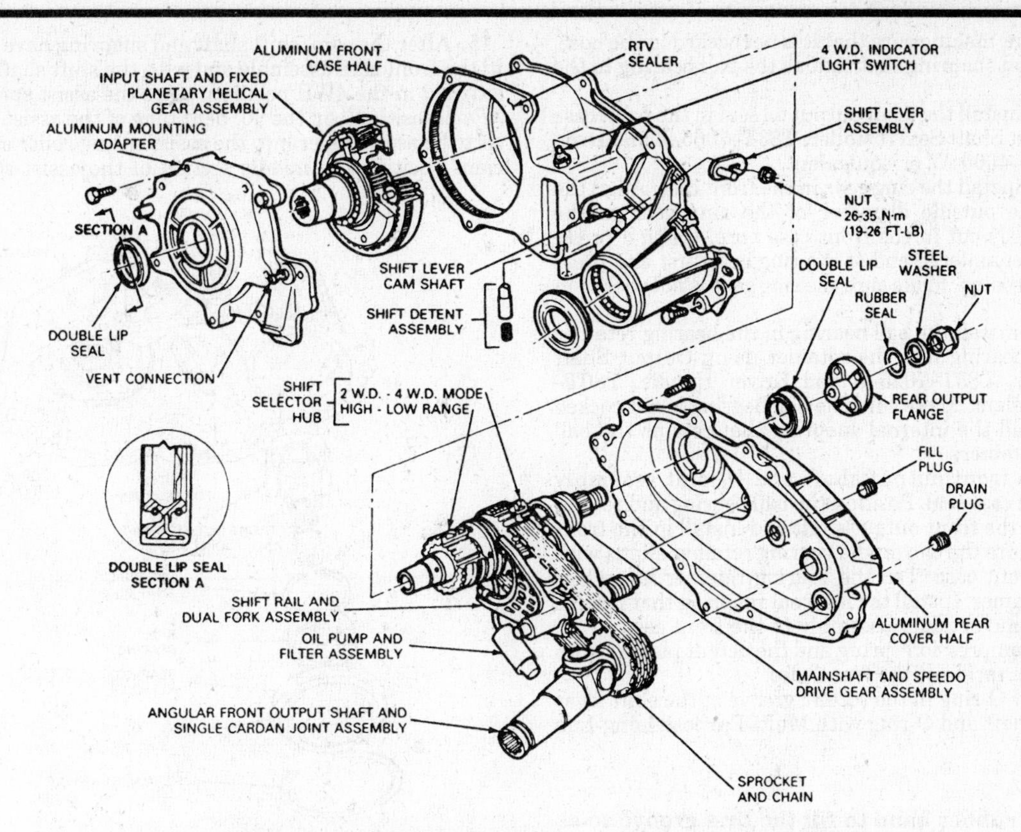

Transfer case disassemble – 13-50 mechanical shift

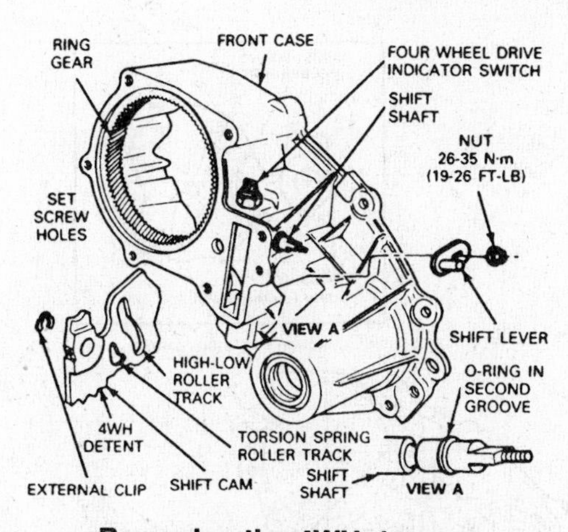

Removing the 4WH detent

NOTE: The clip is required to prevent the bearing retainer from rotating. Do not discard the clip.

33. Remove the front output shaft and U-joint assembly from the front case. If required, remove the oil seal with Seal Remover, Tool–1175–AC and Impact Slide Hammer, T50T–100–A or equivalent. If required, remove the internal snapring and drive the ball bearing out of the front case bore using Output Shaft Replacer, T83T–7025–B and Driver Handle, T80T–4000–W or equivalent.

34. If required, place the front output shaft and U-joint assembly in a vise, being careful not to damage the assembly. Use copper or wood vise jaws.

35. Remove the internal snaprings that retain the bearings in the shaft.

36. Position the U-Joint Tool, T74P–4635–C or equivalent, over the shaft ears and press the bearing out. If the bearing cannot be pressed all the way out, remove it with vise grip or channel lock pliers.

37. Re-position the U-joint tool on the spider in order to remove the opposite bearing.

38. Repeat the above procedure until all bearings are removed.

ASSEMBLY

Before assembly, lubricate all parts with DEXRON®II, Automatic Transmission Fluid.

1. If removed, start a new bearing into an end of the shaft ear. Support the output shaft in a vise equipped with copper or wood jaws, in order not to damage the shaft.

2. Position the spider into the bearing and press the bearing below the snapring groove using U-joint Tool, T74P–4635–C or equivalent.

3. Remove the tool and install a new internal snapring on the groove.

4. Start a new bearing into the opposite side of the shaft ear and using the tool, press the bearing until the opposite bearing contacts the snapring.

5. Remove the tool and install a new internal snapring in the groove.

6. Re-position the front output shaft assembly and install the other two bearings in the same manner.

7. Check the U-joint for freedom of movement. If a binding condition occurs due to misalignment during the installation procedure, tap the ears of both shafts sharply to relieve the bind. Do not install the front output shaft assembly if the U-joint shows any sign of binding.

8. If removed, drive the ball bearing into the front output case bore using Output Shaft Bearing Replacer, T83T–7025–B and Drive Handle, T80T–4000–W or equivalent. Drive the ball

bearing in straight, making sure that it is not cocked in the bore. Install the internal snapring that retains the ball bearing to the front case.

9. If removed, install the front output oil seal in the front case bore using Output Shaft Seal Installer, T83T–7065–B and Driver Handle, T80T–4000–W or equivalent.

10. If removed, install the ring gear in the front case. Align the serrations on the outside diameter of the ring gear to the serations previously cut in the front case bore. Using a press, start the piloted chamfered end of the ring gear first and press in until it is fully seated. Make sure the ring gear is not cocked in the bore.

11. If removed, install the ball bearing in the bearing retainer bore. Drive the bearing into the retainer using Output Shaft Bearing Replacer, T83T–7025–B and Driver Handle, T80T–4000–W or equivalent. Make sure the ball bearing is not cocked in the bore. Install the internal snapring that retains the ball bearing to the retainer.

12. Install the front output shaft and U-joint assembly through the front case seal. Position the ball bearing and retainer assembly over the front output shaft and install in the front case bore. Make sure the clip on the bearing retainer aligns with the slot in the front case. Tap the bearing retainer into place with a plastic hammer. Install the internal snapring that retains the ball bearing and retainer assembly to the front case.

13. Install the compression spring and the detent plunger into the bore from the inside of the front case.

14. Install a new O-ring in the second groove of the shift lever shaft. Coat the shaft and O-ring with Multi-Purpose Long-Life Lubricant.

NOTE: Use a rubber band to fill the first groove so as not to cut the O-ring. Discard the rubber band.

15. After the cam, shift shaft and snapring have been installed in the front case assembly and with the shift shaft assembly positioned in the 4WH position, place the assist spring roller into "Part-position" on the 90° bent tang of the assist spring and insert the assist roller into the assist spring/roller slot of the shift cam. Position the middle section of the assist spring into the

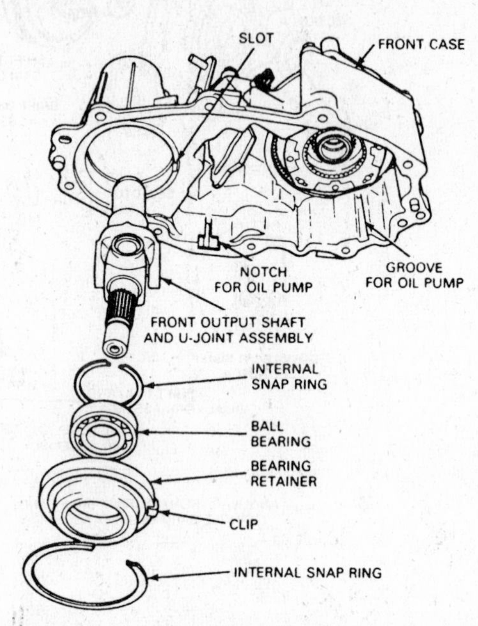

Installing front output shaft and U-joint assembly

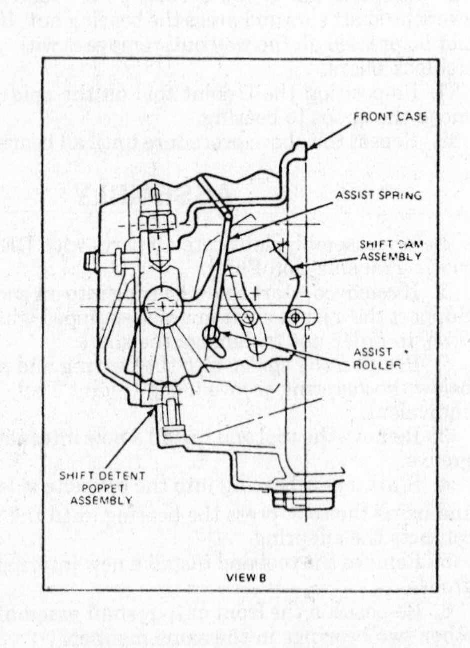

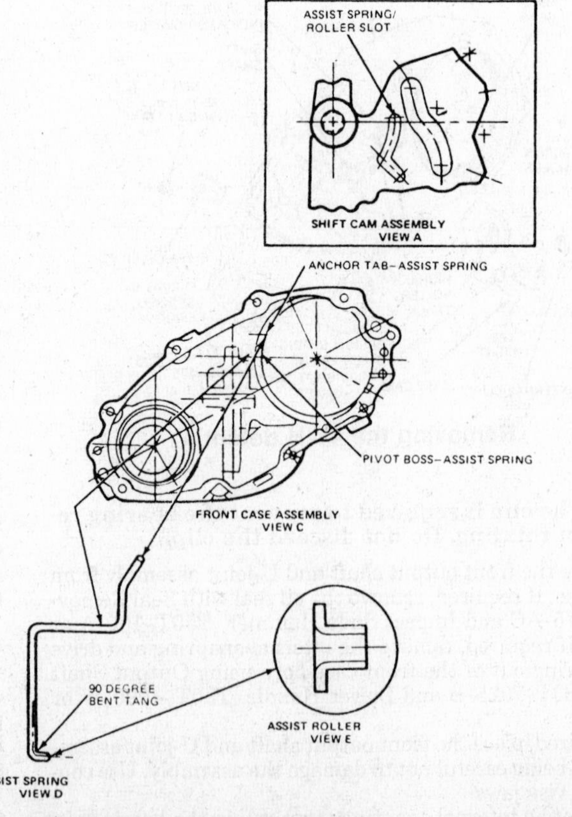

Assist roller and assist spring/roller slot, installation

groove of the front case pivot boss and push to lock the upper end of the assist spring behind the front case spring anchor tab.

16. Install the head set screws in the front case and in the shift cam. Tighten the screws to 6.8–9.5 Nm (5–7 ft. lbs.). Make sure the set screw in the front case is in the first groove of the shift lever shaft and not bottomed against the shaft itself. The shift lever should be able to move freely to all detent positions.

17. Slide the sun gear, thrust plate, thrust washer, and press the ball bearing over the input shaft. Install the external ssnapring to the input shaft.

NOTE: The sun gear recessed face and the snapring groove on the ball bearing outer race should be toward the rear of the transfer case. The stepped face of the thrust washer should face towards the ball bearing.

18. Install the planetary gear set to the sun gear and input shaft assembly. Install the internal snapring to the planetary carrier.

19. Drive the oil seal into the bore of the mounting adapter with Input Shaft Seal Installer, T83–T–7065–A and Driver Handle, T80T–4000–W or equivalent.

20. Place the tanged snapring in the mounting adapter groove. Position the input shaft and planetary gearset in the mounting adapter and push inward until the planetary assembly and input shaft assembly are seated in the adapter. When properly seated, the tanged snapring will snap into place. Check installation by holding the mounting adapter by hand and tapping the face of the input shaft against a wooden block to ensure that the snapring is engaged.

21. Remove all traces of RTV gasket sealant from the mating surfaces of the front case and mounting adapter. Install a bead of RTV gasket sealant on the surface of the front case.

22. Position the mounting adapter on the front case. Install the retaining bolts and tighten to 31–41 Nm (25–30 ft. lbs.).

24. If removed, install a new pin, roller and retainer assembly to the high-low shift fork.

25. Install the high-low shift fork and shift collar on the output shaft together. Slip the high-low shift fork roller bushing into the high-low roller track of the shift cam.

NOTE: Make sure the nylon wear pads are installed on the shift fork. Make sure the dot on the pad is installed in the fork holes.

26. Install the shift rail through the high-low fork and make sure the shift rail is seated in the bore in the front case.

27. Place the oil pump cover with the word TOP facing the front of the front case. Install the 2 pump pins with the flats facing toward the rear cover with the spring between the pins and place the assembly in the oil pump bore in the output shaft. Place the oil pump body and pick-up tube over the shaft and make sure the pins are riding against the inside of the pump body. Place the oil pump rear cover with the word **TOP REAR** facing the rear of the front case. The word TOP on the front cover and the rear cover should be on the same side. Install the pump retainer so the tabs face the front of the transfer case, the 4 bolts and rotate the output shaft while tightening the bolts to prevent the pump from binding. Tighten the bolts to 35–75 ft. lbs. (4.3–9.2Nm).

NOTE: Prime the pump through the oil filler pick-up tube while turning the output shaft.

Coat all pump part prior to assembly using Mercon® Automatic Transmission Fluid or equivalent.

The output shaft must turn freely within the oil pump. If binding occurs, loosen the 4 bolts and retighten again.

28. Install the output shaft and oil pump assembly in the input shaft. Make certain the external splines of the output shaft engage the internal splines of the high-low shift collar. Make certain the oil pump retainer and the oil filter leg are in the groove and notch of the front case.

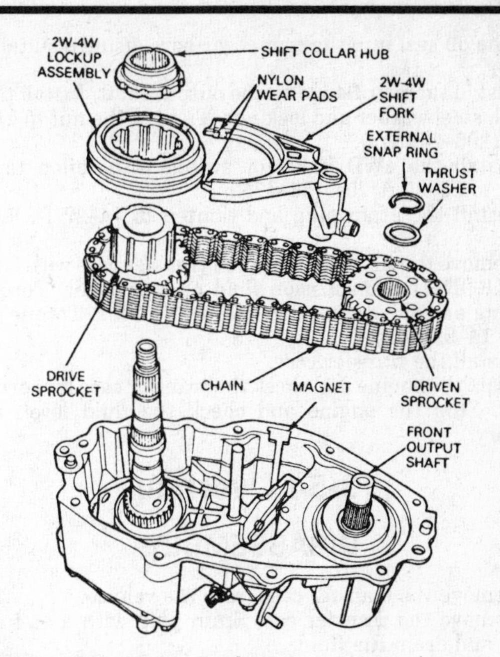

Installing chain, drive sprocket and driven sprocket assembly

29. Install the collector magnet in the notch in the front case.

30. Install the chain, drive sprocket and driven sprocket as an assembly over the shaft. Install the thrust washer on the front output shaft and install a new external snapring over the thrust washer to retain the driven sprocket.

31. If disassembled, assemble the 2W–4W shift fork to the 2W–4W lock up assembly. Install the spring in the lock up collar. Place the lock up hub over the spring and engage the lock up hub in the notches in the lock up collar. Retain the lock up hub to the lock up collar with an internal snapring.

32. Install the 2W–4W shift fork and lock-up assembly on the output shaft.

33. Install the shift collar hub to the output shaft.

34. If removed, drive the gaged needle bearing into the rear cover bore with the needle bearing replacer tool T83T–7127–A and driver handled T80T–4000–W or equivalent.

35. If removed, install the ball bearing in the rear cover bore. Drive the bearing into the rear cover bore with output shaft bearing replacer tool No. T83T–7025–B and driver handled T80–4000–W or equivalent. Make sure the ball bearing is not cocked in the bore. Install the internal snapring that retains the ball bearing to the rear cover.

36. Prior to the final assembly of the rear cover to the front case half, the transfer case shift lever assembly should be shifted into the 4H detent position to assure positioning of the shift rail to the rear cover.

37. Coat the mating surface of the front case with a bead of Non-Acid Cure Silicone Rubber E7TZ–19562–A or equivalent.

38. Install the 2W–4W shift fork spring to the shift rail and shift fork with spring mounted in a vertical position.

39. Position the rear cover so that the spring boss engages the 2W–4W shift fork spring and shift rail. Install the nine bolts, starting with the bolts on the rear cover and torque the bolts to 23–30 ft. lbs. (31–41Nm).

NOTE: If the rear cover assembly does not seat properly, move the rear cover up and down slightly to permit the end of the shift rail to enter the shift rail hole in the rear cover boss.

40. Install the speedometer drive gear into the rear cover bore.

Drive the oil seal into the rear cover bore, using a suitable seal installer.

41. Install the rear flange on the output shaft. Install the rubber seal, steel washer and lock nut. Tighten the nut to 150–180 ft. lbs. (203–244Nm).
42. Install the 4WD indicator switch with teflon tape and tighten to 25–35 ft. lbs. (34–47Nm).
43. Install the drain plug and tighten to 14–22 ft. lbs. (19–30Nm).
44. Remove the fill plug. Refill the transfer case with 3.0 pints of DEXRON®II transmission fluid or equivalent. Torque the level plug and the drain plugs to 14–22 ft. lbs. Torque the fill plug to 14–22 ft. lbs.
45. Install the transfer case.
46. Start the engine and check the transfer case for correct operation. Stop the engine and check the fluid level, add as necessary.

13-54 Overhaul

DISASSEMBLY

1. Remove the transfer case from the vehicle.
2. Remove the transfer case drain plug with a 3/8 in. drive ratchet and drain the fluid.
3. Remove the 4WD indicator switch and the breather vent.
4. Using a 30mm thin wall socket, remove the rear output nut, washer, rubber seal and yoke.
5. Remove the bolts which retain the front case to the rear cover. Insert a 1/2 in. drive breaker bar between the three pry bosses and separate the front case from the rear cover. Remove all traces of RTV gasket sealant from the mating surfaces of the front case and rear cover.

WARNING: When removing RTV sealant, take care not to damage the mating surface of the aluminum case!

6. If the speedometer drive gear or ball bearing assembly is to be replaced, first, drive out the output shaft oil seal from either the inside of the rear cover with a brass drift and hammer or from the outside by bending and pulling on the curved-up lip of the oil seal. Remove and discard the oil seal.
7. Remove the internal snapring that retains the rear output shaft ball bearing in the bore. From the outside of the case, drive out the ball bearing using D80L–630–A, step plate with a suitable drift.
8. If required, remove the front output shaft caged needle bearing from the rear cover with Puller Collet, D80L–100–S and Impact Slide Hammer, T50T–100–A or equivalent.
9. Remove the 2W–4W shift fork spring from the boss on the 2W–4W shift fork.
10. Remove the shift collar hub from the output shaft. Remove the 2W–4W lock-up assembly and the 2W–4W shift fork together as an assembly. Remove the 2W–4W fork from the 2W–4W lock-up assembly.
11. If required to disassemble the 2W–4W lock-up assembly, remove the internal snapring and pull the lock-up hub and spring from the lock-up collar.
12. Remove the external snapring and thrust washer that retains the drive sprocket to the front output shaft.
13. Remove the chain, driven sprocket and drive sprocket as an assembly.
14. Remove the collector magnet from the notch in the front case bottom.
15. Remove the output shaft and oil pump as an assembly.
16. If required to disassemble the oil pump, remove the bolts from the body. Note the position and markings of the front cover, body, pins, spring, rear cover, and pump retainer as removed.
17. Pull out the shift rail.
18. Slip the high-low range shift fork out of the inside track of the shift cam and remove the high-low shift fork and high-low shift collar together. If required, remove the pin, roller retainer assembly by pressing it out of the high-low range shift fork.
19. Push and pull out the anchor end of the assist spring from

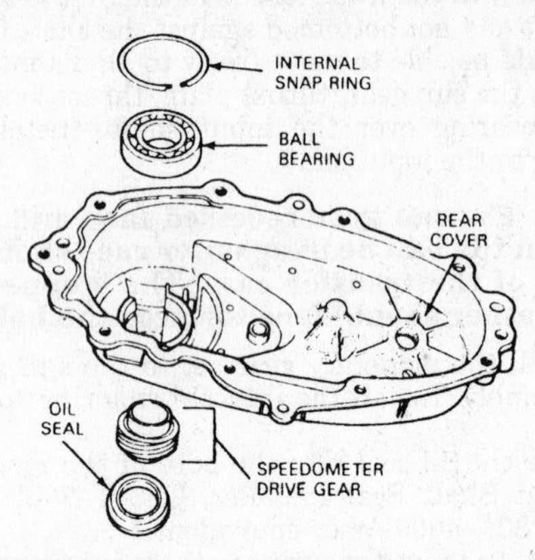

Removing the rear output shaft ball bearing

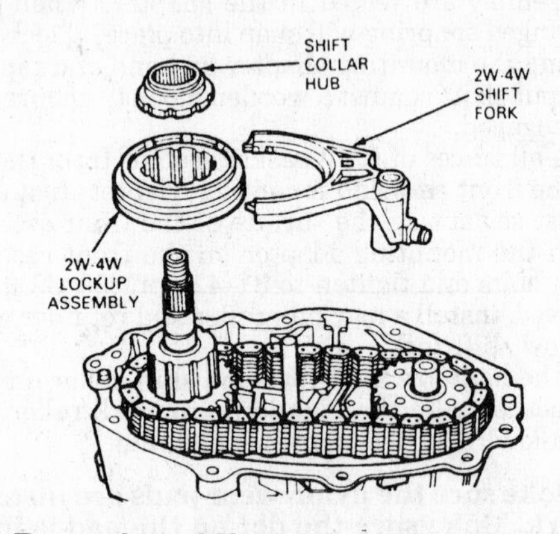

Removing shift collar from output shaft

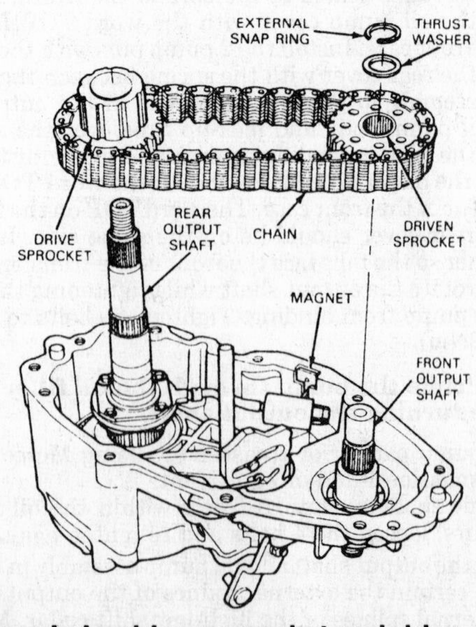

Removing chain, driven sprocket and drive sprocket assembly

the locking post in the front case half. Remove the assist spring and roller out of the shift cam.

20. Turn the front case over and remove the bolts retaining the mounting adapter to the front case. Remove the mounting adapter, input shaft and planetary gearset as an assembly.

21. If required, remove the ring gear from the front case using a press. Note the relationship of the serrations to the chamfered pilot diameter during removal.

22. Expand the tangs of the large snapring in the mounting adapter. With the input shaft against a bench, push the adapter down and slide the adapter off the ball bearing. Lift the input shaft and planetary gearset from the adapter.

23. If required, remove the oil from the mounting adapter by prying and pulling on the curved-up lip of the oil seal.

24. Remove the internal snapring from the planetary carrier and separate the planetary gearset from the input shaft assembly.

25. Remove the external snapring from the input shaft. Place the input shaft assembly in a press and remove the ball bearing from the input shaft using Bearing Splitter, D79L–4621–A or equivalent. Remove the thrust washer, thrust plate and sun gear off the input shaft.

26. Move the shift lever by hand down to the 4L position, then up 2 detents to the 4H position. Mark a line on the outside of the front case using the side of the shift lever and a grease pencil.

27. Remove the 2 head set screws from the front case and from the shift cam.

28. Turn the front case over and remove the external clip. Pry the shift shaft out of the front case and shift cam. Do not pound on the external clip during removal.

NOTE: The shifter lever and cam shaft do not have to be disassembled unless the parks have to be replaced.

29. Remove the O-ring from the second groove in the shift lever shaft.

30. Remove the detent plunger and compression spring from the inside of the front case.

31. If required, remove the front output shaft oil seal by prying and pulling on the curved-up lip of the oil seal.

32. If required, remove the internal snapring and drive the ball bearing out of the front case bore, using step plate D80L–630–A or equivalent, and a drift.

ASSEMBLY

Before assembly, lubricate all parts with Motorcraft® Automatic Transmission Fluid, XT-2-QDX or equivalent.

1. If removed, drive the ball bearing into the front output case bore, using Output Shaft Bearing Replacer, T83T–7025–B and Driver Handle, T80T–4000–W or equivalent. Make sure the ball bearing is not cocked in the bore. Install the internal snapring that retains the ball bearing to the front case.

2. If removed, install the front output oil seal in the front case bore, using Output Shaft Seal Installer, T83T–7065–B and Driver Handle, T80T–4000–W or equivalent.

3. If removed, install the ring gear in the front case. Align the serrations on the outside diameter of the ring gear to the serrations previously cut in the front case bore. Using a press, start the piloted chamfered end of the ring gear first and press in until it is fully seated. Make sure the ring gear is not cocked in the bore.

4. Install the compression spring and the detent plunger into the bore from the inside of the front case.

5. Install a new O-ring in the second groove of the shift lever shaft. Coat the shaft and O-ring with Multi-Purpose Long-Life Lubricant.

NOTE: Use a rubber band to fill the first groove so as not to cut the O-ring. Discard the rubber band.

6. After the cam, shift shaft and snapring have been installed

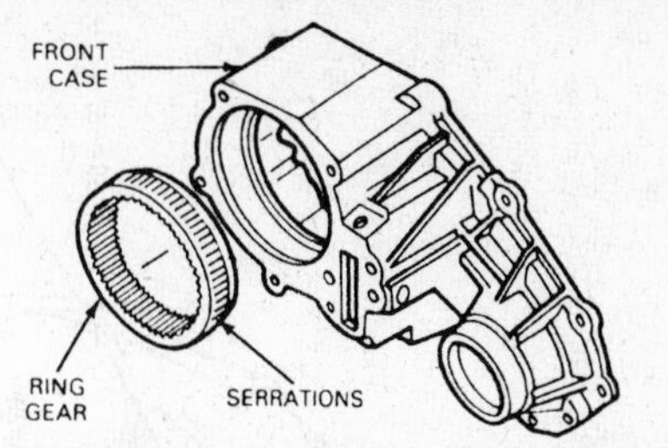

Installing ring gear to front case

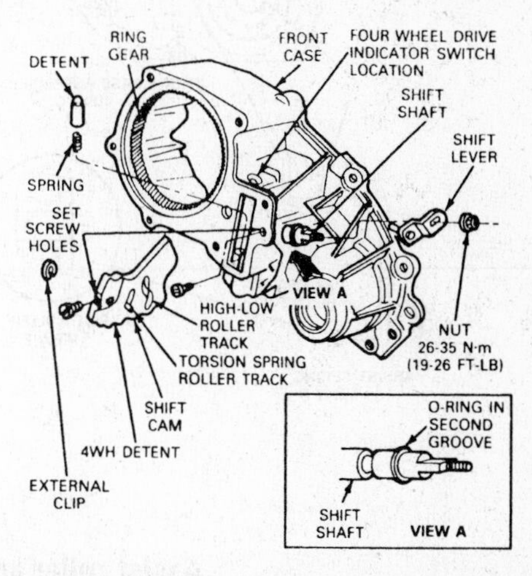

Installing compression spring and detent plunger

in the front case assembly and with the shift shaft assembly positioned in the 4WH position, place the assist spring roller into "Part-position" on the 90° bent tang of the assist spring and insert the assist roller into the assist spring/roller slot of the shift cam. Position the middle section of the assist spring into the groove of the front case pivot boss and push to lock the upper end of the assist spring behind the front case spring anchor tab.

7. Install the head set screws in the front case and in the shift cam. Tighten the screws to 6.8–9.5 Nm (5–7 ft. lbs.). Make sure the set screw in the front case is in the first groove of the shift lever shaft and not bottomed against the shaft itself. The shift lever should be able to move freely to all detent positions.

8. Slide the sun gear, thrust plate, thrust washer, and press the ball bearing over the input shaft. Install the external ssnapring to the input shaft.

NOTE: The sun gear recessed face and the snapring groove on the ball bearing outer race should be toward the rear of the transfer case. The stepped face of the thrust washer should face towards the ball bearing.

9. Install the planetary gear set to the sun gear and input shaft assembly. Install the internal snapring to the planetary carrier.

10. Drive the oil seal into the bore of the mounting adapter with Input Shaft Seal Installer, T83–T–7065–A and Driver Handle, T80T–4000–W or equivalent.

11. Place the tanged snapring in the mounting adapter groove. Position the input shaft and planetary gearset in the mounting adapter and push inward until the planetary assembly and input shaft assembly are seated in the adapter. When properly seated, the tanged snapring will snap into place. Check installation by holding the mounting adapter by hand and tapping the face of the input shaft against a wooden block to ensure that the snapring is engaged.

12. Remove all traces of RTV gasket sealant from the mating

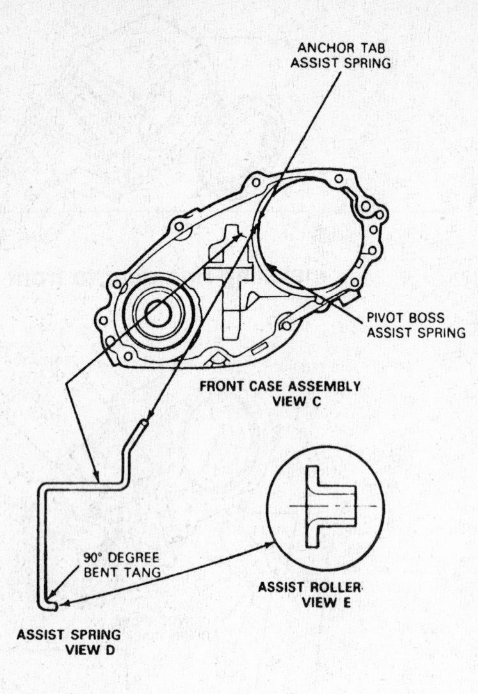

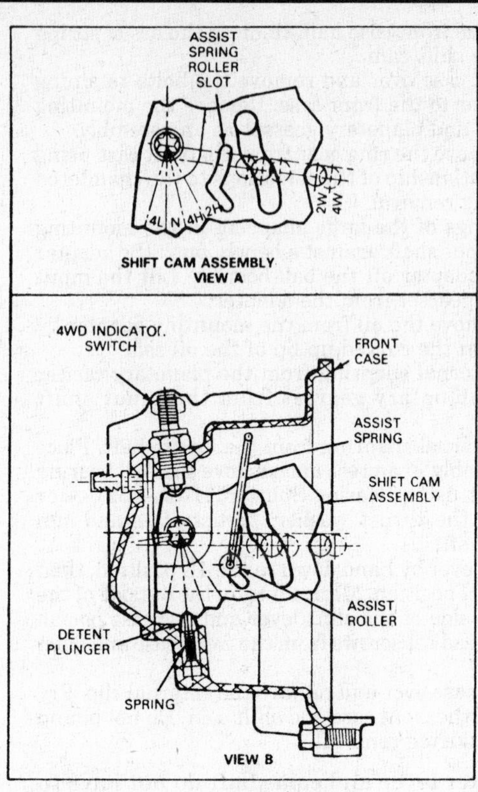

Assist roller and assist spring/roller slot, installation

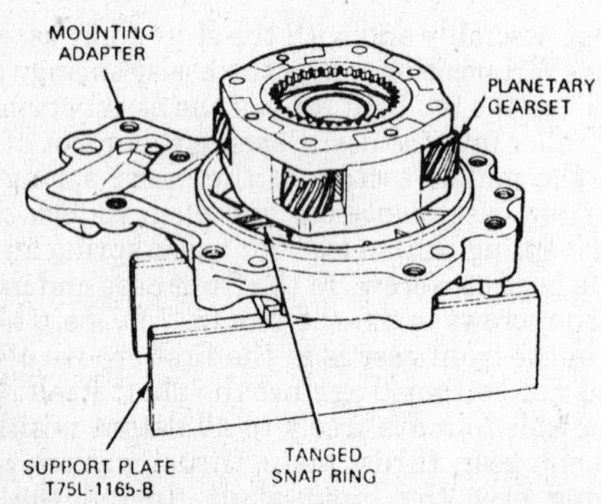

Installing planetary assembly and input shaft assembly in adapter

surfaces of the front case and mounting adapter. Install a bead of RTV gasket sealant on the surface of the front case.

13. Position the mounting adapter on the front case. Install the retaining bolts and tighten to 31–41 Nm (25–30 ft. lbs.).

14. If removed, install a new pin, roller and retainer assembly to the high-low shift fork.

NOTE: NOTE:

Make sure the nylon wear pads are installed on the shift fork. Make sure the dot on the pad is installed in the fork holes.

15. Install the high-low shift fork and shift collar on the output shaft together. Slip the high-low shift fork roller bushing into the high-low roller track of the shift cam.

16. Install the shift rail through the high-low fork and make sure the shift rail is seated in the bore in the front case. Using the shift lever, position the high-low shift fork in the 4H position.

17. Place the oil pump cover with the word TOP facing the front of the front case. Install the 2 pump pins with the flats facing toward the rear cover with the spring between the pins and place the assembly in the oil pump bore in the output shaft. Place the oil pump body and pick-up tube over the shaft and make sure the pins are riding against the inside of the pump body. Place the oil pump rear cover with the word **TOP REAR** facing the rear of the front case. The word TOP on the front cover and the rear cover should be on the same side. Install the pump retainer so the tabs face the front of the transfer case, the 4 bolts and rotate the output shaft while tightening the bolts to prevent the pump from binding. Tighten the bolts to 35–75 ft. lbs. (4.3–9.2Nm).

NOTE: Prime the pump through the oil filler pick-up tube while turning the output shaft.

Coat all pump part prior to assembly using Mercon® Automatic Transmission Fluid or equivalent.

The output shaft must turn freely within the oil pump. If binding occurs, loosen the 4 bolts and retighten again.

18. Install the output shaft and oil pump assembly in the input shaft. Make certain the external splines of the output shaft engage the internal splines of the high-low shift collar. Make certain the oil pump retainer and the oil filter leg are in the groove and notch of the front case.

19. Install the collector magnet in the notch in the front case.

20. Install the chain, drive sprocket and driven sprocket as an assembly over the shaft. Install the thrust washer on the front output shaft and install a new external snapring over the thrust washer to retain the driven sprocket.

21. If disassembled, assemble the 2W–4W shift fork to the 2W–4W lock up assembly. Install the spring in the lock up collar. Place the lock up hub over the spring and engage the lock up hub in the notches in the lock up collar. Retain the lock up hub to the lock up collar with an internal snapring.

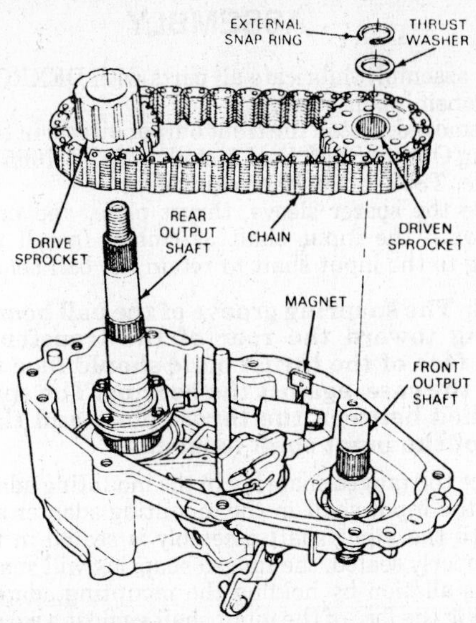

Installing the chain, drive sprocket and driven sprocket assembly

22. Install the 2W–4W shift fork and lock-up assembly on the output shaft and onto the shift rail.
23. Install the shift collar hub to the output shaft.
24. If removed, drive the gaged needle bearing into the rear cover bore with the needle bearing replacer tool T83T–7127–A and driver handled T80T–4000–W or equivalent.
25. If removed, install the ball bearing in the rear cover bore. Drive the bearing into the rear cover bore with output shaft bearing replacer tool No. T83T–7025–B and driver handled T80–4000–W or equivalent. Make sure the ball bearing is not cocked in the bore. Install the internal snapring that retains the ball bearing to the rear cover.
26. Prior to the final assembly of the rear cover to the front case half, the transfer case shift lever assembly should be shifted into the 4H detent position to assure positioning of the shift rail to the rear cover.
27. Coat the mating surface of the front case with a bead of Non-Acid Cure Silicone Rubber E7TZ–19562–A or equivalent.
28. Install the 2W–4W shift fork spring to the shift rail and shift fork with spring mounted in a vertical position.
29. Position the rear cover so that the spring boss engages the 2W–4W shift fork spring and shift rail. Install the nine bolts, starting with the bolts on the rear cover and torque the bolts to 23–30 ft. lbs. (31–41Nm).

NOTE: If the rear cover assembly does not seat properly, move the rear cover up and down slightly to permit the end of the shift rail to enter the shift rail hole in the rear cover boss.

30. Install the speedometer drive gear into the rear cover bore. Drive the oil seal into the rear cover bore, using a suitable seal installer.
31. Install the rear flange on the output shaft. Install the rubber seal, steel washer and lock nut. Tighten the nut to 150–180 ft. lbs. (203–244Nm).
32. Install the front yoke, rubber seal, steel washer and locknut to the front output shaft. Tighten the nut to 150–180 ft. lbs. (203–244Nm).
33. Install the 4WD indicator switch with teflon tape and tighten to 25–35 ft. lbs. (34–47Nm).
34. Install the drain plug and tighten to 14–22 ft. lbs. (19–30Nm).
35. Remove the fill plug. Refill the transfer case with 2½ pints

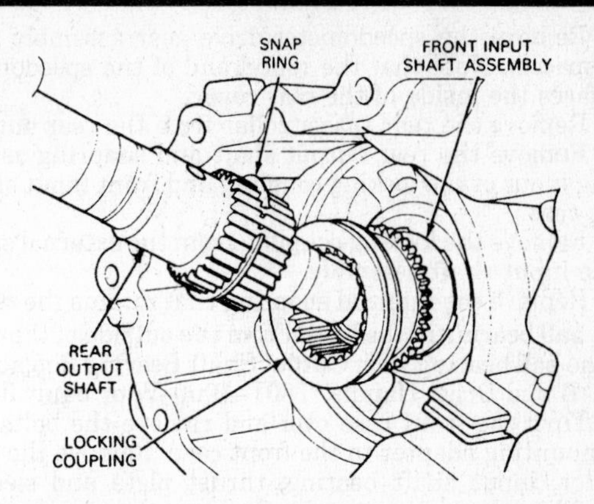

Removing rear output shaft and snapring

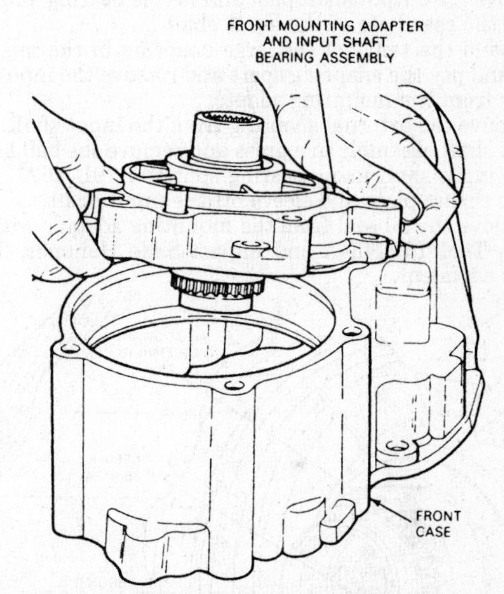

Removing mounting adapter and input shaft bearing

(1.2L) of Motorcraft Mercon® Automatic Transmission fluid or equivalent. Torque the level plug and the drain plugs to 14–22 ft. lbs. Torque the fill plug to 14–22 ft. lbs.
36. Install the transfer case.
37. Start the engine and check the transfer case for correct operation. Stop the engine and check the fluid level, add as necessary.

13-59 Overhaul

Since this unit is more of a spacer than an operating mechanism, there is no real overhaul. It can, however, be disassembled.

DISASSEMBLY

1. Remove the transfer case from the vehicle.
2. Remove the rear output shaft locknut, using a 30mm thin wall socket. Remove the nut, steel washer and rubber seal from the output shaft.
3. Remove the rear output shaft oil seal from the rear cover using Tool T74P–77248–A and T50T–100–A or equivalent.
4. Remove the bolts which retain the front case to the rear cover. Insert a ½ in. drive breaker bar between the three pry bosses and separate the front case from the rear cover.

5. Remove the speedometer drive gear assembly (gear, clip and spacer). Note that the round end of the speedometer gear clip faces the inside of the rear cover.
6. Remove the rear spacer collar from the rear output shaft.
7. Remove the rear output shaft and snapring assembly by lifting it out of the locking coupling and front input shaft in the front case.
8. Remove the locking coupling from the external splined end of the input shaft assembly.
9. Remove the internal snapring that retains the rear output shaft ball bearing in the bore. From the outside of the case, drive out the ball bearing with Output Shaft Bearing Replacer, T83T-7025-B and Drive Handle, T80T-4000-W or equivalent.
10. Turn the front case over and remove the bolts retaining the mounting adapter to the front case. Remove the mounting adapter, input shaft bearing, thrust plate and sleeve as an assembly.
11. Place the mounting adapter in a suitable holding fixture and remove the output shaft pilot and needle bearing and bushing from the rear bore of the input shaft.
12. Expand the tangs of the large snapring in the mounting adapter and pry the snapring apart and remove the input shaft assembly from the mounting adapter.
13. Remove the external snapring from the input shaft. Place the input shaft assembly in a press and remove the ball bearing from the input shaft using bearing splitter D79L-4621-A. Remove the thrust plate and sleeve off the input shaft.
14. Remove the oil seal from the mounting adapter with Seal Remover, Tool T77248-A and Impact Slide Hammer, T50T-100-A or eqivalent.

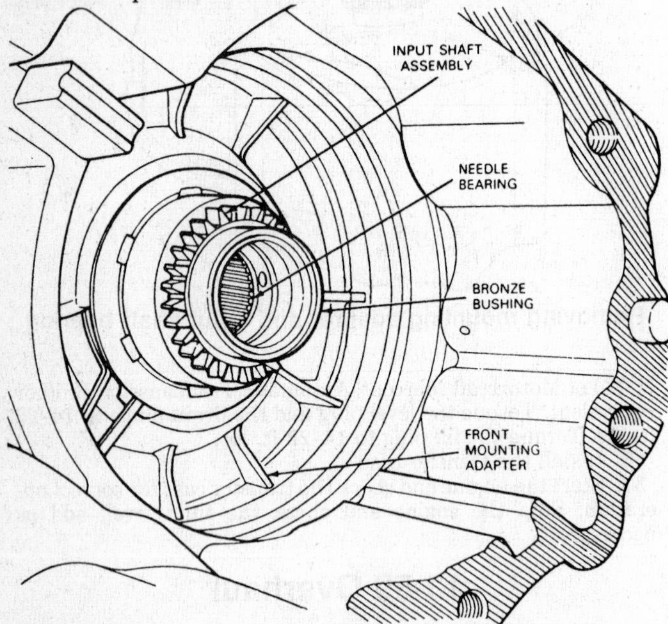

Removing output shaft pilot needle bearing and bushing

ASSEMBLY

Before assembly, lubricate all parts with DEXRON®II, Automatic Transmission Fluid.

1. If removed, install the front output oil seal in the front case bore using Output Shaft Seal Installer, T83T-7065-B and Driver Handle, T80T-4000-W or equivalent.
2. Slide the spacer sleeve, thrust plate, and press the ball bearing onto the input shaft assembly. Install the external ssnapring to the input shaft to retain the ball bearing.

NOTE: The snapring groove of the ball bearing should be facing toward the rear of the transfer case. The stepped face of the thrust plate should face toward the front of the case, against the bearing. The spacer sleeve is installed between the thrust plate and the external splines of the input shaft assembly.

3. Place the tanged snapring in the mounting adapter groove. Position the input shaft in the mounting adapter and push inward until the input shaft assembly is seated in the adapter. When properly seated, the tanged snapring will snap into place. Check installation by holding the mounting adapter by hand and tapping the face of the input shaft against a wooden block to ensure that the snapring is engaged.
4. Position the mounting adapter on the front case. Install six bolts and tighten to 31-41 Nm (25-30 ft. lbs.).
5. Install the locking coupling onto the external splines of the input shaft assembly.
6. Install the output shaft, and snapring assembly into the locking coupling and the input shaft. Make sure the external splines of the output shaft engauge the internal splines of the locking coupling and the input shaft.
7. Install the spacer hub onto the output shaft assembly with the square teeth side facing toward the front of the transfer case.
8. If removed, install the ball bearing in the rear cover bore. Drive the bearing into the rear cover bore with output shaft bearing replacer tool No. T83T-7025-B and driver handled T80-4000-W or equivalent. Make sure the ball bearing is not cocked in the bore. Install the internal snapring that retains the ball bearing to the rear cover.
9. Install the speedometer drive gear assembly into the rear cover bore with round end of the speedometer gear clip facing towards the inside of the rear cover. Pack the speedometer gear cavity in the rear cover with long life lubricant. Drive the oil seal into the rear cover bore with output shaft seal installer tool No. T83T-7065-B and driver handle # T80T-4000-W or equivalent.
10. Position the rear cover on the front case. Install the nine bolts, starting with the bolts on the rear cover and torque the bolts to 23-30 ft. lbs.
11. Install the rear output shaft flange. Coat the faces of the yoke nuts and output shaft threads with a suitable thread sealer. Torque the yoke nuts to 150-180 ft. lbs.
12. Install the transfer case to the vehicle.

NOTE: Do not fill the transfer case with any type of lubricant

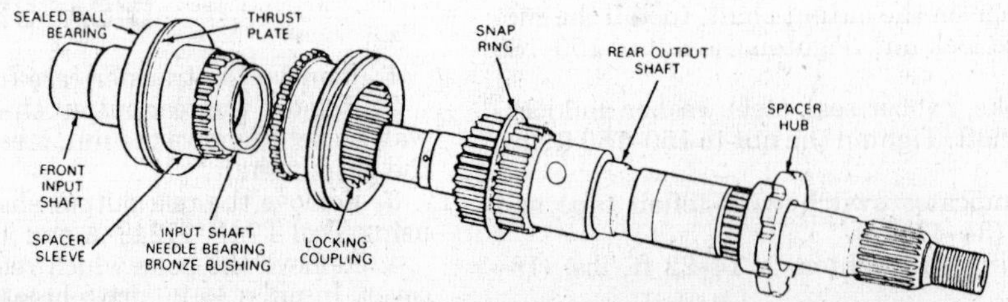

Input shaft assembly, locking coupling and output shaft assembly

DRIVELINE

General Description

The driveshaft is a steel tubular or aluminum shaft which is used to transfer the torque from the engine, through the transmission output shaft, to the differential in the axle, which in turn transmits torque to the wheels.

The splined slip yoke and transmission output shaft permit the driveshaft to move forward and rearward as the axle moves up and down. This provides smooth performance during vehicle operation.

The front driveshaft connects the power flow from the transfer case to the front drive axle.

DRIVELINE

Troubleshooting Basic Driveshaft and Rear Axle Problems

When abnormal vibrations or noises are detected in the driveshaft area, this chart can be used to help diagnose possible causes. Remember that other components such as wheels, tires, rear axle and suspension can also produce similar conditions.

BASIC DRIVESHAFT PROBLEMS

Problem	Cause	Solution
Shudder as car accelerates from stop or low speed	• Loose U-joint • Defective center bearing	• Replace U-joint • Replace center bearing
Loud clunk in driveshaft when shifting gears	• Worn U-joints	• Replace U-joints
Roughness or vibration at any speed	• Out-of-balance, bent or dented driveshaft • Worn U-joints • U-joint clamp bolts loose	• Balance or replace driveshaft • Replace U-joints • Tighten U-joint clamp bolts
Squeaking noise at low speeds	• Lack of U-joint lubrication	• Lubricate U-joint; if problem persists, replace U-joint
Knock or clicking noise	• U-joint or driveshaft hitting frame tunnel • Worn CV joint	• Correct overloaded condition • Replace CV joint

BASIC REAR AXLE PROBLEMS

First, determine when the noise is most noticeable.

Drive Noise—Produced under vehicle acceleration.

Coast Noise—Produced while the car coast with a closed throttle.

Float Noise—Occurs while maintaining constant car speed (just enough to keep speed constant) on a level road.

Road Noise

Brick or rough surfaced concrete roads produce noises that seem to come from the rear axle. Road noise is usually identical in Drive or Coast and driving on a different type of road will tell whether the road is the problem.

Tire Noise

Tire noises are often mistaken for rear axle problems. Snow treads or unevenly worn tires produce vibrations seeming to originate elsewhere. Temporarily inflating the tire to 40 lbs will significantly alter tire noise, but will have no effect on rear axle noises (which normally cease below about 30 mph).

Engine/Transmission Noise

Determine at what speed the noise is more pronounced, then stop the car in a quiet place. With the transmission in Neutral, run the engine through speeds corresponding to road speeds where the noise was noticed. Noises produced with the car standing still are coming from the engine or transmission.

Front Wheel Bearings

While holding the car speed steady, lightly apply the foot brake; this will often decease bearing noise, as some of the load is taken from the bearing.

Rear Axle Noises

Eliminating other possible sources can narrow the cause to the rear axle, which normally produces noise from worn gears or bearings. Gear noises tend to peak in a narrow speed range, while bearing noises will usually vary in pitch with engine speeds.

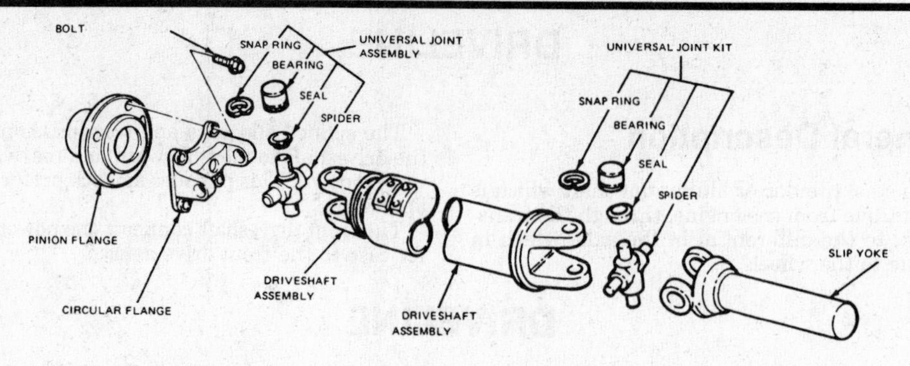

Rear driveshaft assembly – Single Cardan Type U- joint

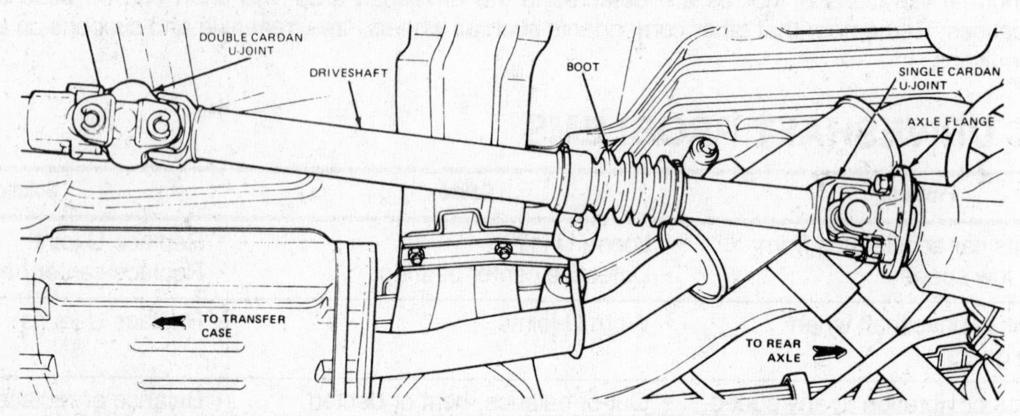

Rear driveshaft assembly – with Single Cardan and Double Cardan Type U-joint

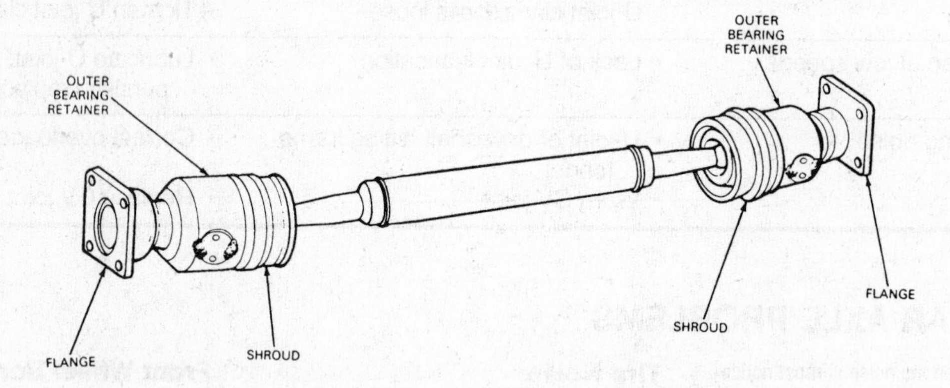

Rear driveshaft assembly – CV (Constant Velocity) Type U-joint

Some vehicles may be equipped with a Double Cardan type driveshaft. This driveshaft incorporates 2 U-joints, a centering socket yoke and a center slip at the transfer case end of each shaft. A single U-joint is used at the axle end of the shaft.

The Constant Velocity (CV) type U-joint allows the driveline angel to be adjusted according to the up-and-down movement of the vehicle without disturbing the power flow. The CV U-joint is composed of an outer bearing retainer and flange, sprig, cap, circlip, inner bearing assembly and wire ring. The inner bearing assembly is composed of a bearing cage, 6 ball bearings and an inner race.

The driveshafts used on Ranger/Bronco II/Explorer may be 1 of 3 types. They are as follows:

- Front and rear driveshaft – Single Cardan type U-joint
- Front and rear driveshaft – Double Cardan type U-joint
- Rear driveshaft – CV (Constant Velocity) type U-joint

Front Driveshaft Single Cardan Type

REMOVAL AND INSTALLATION

1. Disconnect the negative battery cable.
2. Raise and support the vehicle safely.

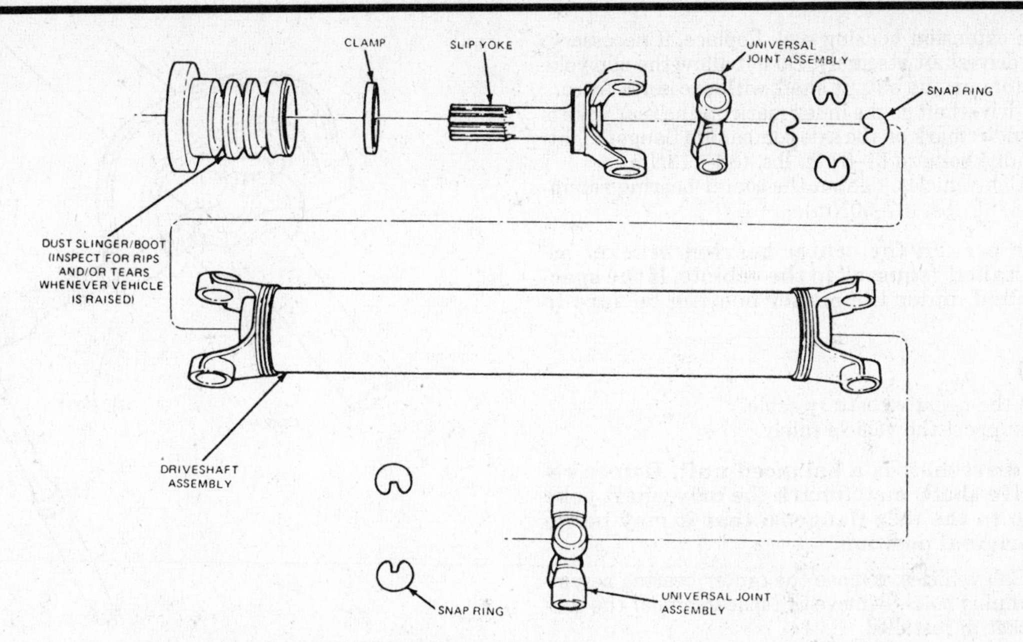

Front driveshaft and U-joint, exploded view – Single Cardan Type

NOTE: The driveshaft is a balanced unit. Before removing the drive shaft, matchmark the driveshaft in relationship to the end yoke so that it may be installed in its original position.

3. Using a shop cloth or gloves, pull back on the dust slinger to remove the boot from the transfer case slip yoke.
4. Remove the bolts and straps that retains the driveshaft to the front driving axle yoke. Remove the U-joint assembly from the front driving axle yoke.
5. Slide the splined yoke assembly out of the transfer case and remove the driveshaft assembly.
6. Inspect the boot for rips or tears. Inspect the stud yoke splines for wear or damage. Replace any damage parts.

To install:

7. Apply a light coating of Multi-purpose Long-Life lubricant C1AZ–19490–B or equivalent, to the yoke splines and the edge of the inner diameter of the rubber boot.
8. Slide the driveshaft into the transfer case front output yoke assembly. Make certain the wide tooth splines on the slip yoke are indexed to the output yoke in the transfer case.
9. Position the U-joint assembly in the front drive axle yoke in its original position. Install the retaining bolts and straps. Tighten the bolts to 10–15 ft. lbs. (14–20Nm).
10. Firmly press the dust slinger until the boot is felt to engage the output yoke in the transfer case.
11. Lower the vehicle.
12. Connect the negative battery cable.

NOTE: If replacement of the dust slinger/boot is necessary. Use the following procedure.

DUST SLINGER/BOOT REPLACEMENT

1. Remove the boot clamp using cutter pliers and discard the clamp. Remove the boot from the stud yoke.
2. Install a new dust slinger/boot on the stud yoke making certain the boot is seated in the groove in the yoke.
3. Install a new clamp on the boot. Position the clamp tabs in the slots so each tab fits into a slot. Then, crimp the clamp securely using a pair of clamp pliers T63P–9171–A or equivalent. Do not crimp to the point where the clamp damage the boot.

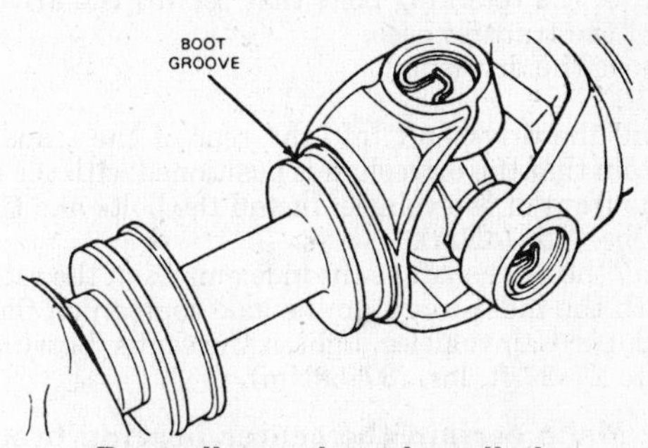

Dust slinger, boot installation

Rear Driveshaft Single Cardan Type

REMOVAL AND INSTALLATION

Except Ranger 4WD

1. Disconnect the negative battery cable.
2. Raise and support the vehicle safely.

NOTE: The driveshaft is a balanced unit. Before removing the drive shaft, matchmark the driveshaft yoke in relationship to the axle flange so that it may be installed in its original position.

3. On Super Cab vehicles, remove the center bearing assembly-to-frame retaining bolts. Remove the spacers under the center bearing bracket, if installed.
4. Remove the retaining bolts and disconnect the driveshaft from the axle companion flange. Pull the driveshaft rearward until the slip yoke clears the transmission extension housing and seal. Plug the extension housing to prevent lubricant leakage.

To install:

5. Lubricate the slip yoke splines with Multi-purpose Long-Life lubricant C1AZ–19490–B or equivalent. Remove the plug from the extension housing.

6. Inspect the extension housing seal. Replace, if necessary.
7. Install the driveshaft assembly. Do not allow the slip yoke assembly to bottom on the output shaft with excessive force.
8. Install the driveshaft so the index mark on the rear yoke is in line with the index mark on the axle companion flange. Tighten all cicular flange bolts to 61–87 ft. lbs. (83–118Nm).
9. On Super Cab vehicles, tighten the center bearing retaining bolts to 27–37 ft. lbs. (37–50Nm).

NOTE: Make certain the center bearing bracket assembly is reinstalled "square" to the vehicle. If the spacers were installed under the center bearing be sure to reinstall them.

Ranger (4WD)

1. Disconnect the negative battery cable.
2. Raise and support the vehicle safely.

NOTE: The driveshaft is a balanced unit. Before removing the drive shaft, matchmark the driveshaft yoke in relationship to the axle flange so that it may be installed in its original position.

3. On Super Cab vehicles, remove the center bearing assembly-to-frame retaining bolts. Remove the spacers under the center bearing bracket, if installed.
4. Remove the retaining bolts and disconnect the driveshaft from the axle companion flange.
5. Remove the retaining bolts that retains the driveshaft to the rear of the transfer case.
6. Remove the driveshaft.

To install:

7. Install the driveshaft into the rear of the transfer case. Make certain that the driveshaft is positioned with the slip yoke toward the front of the vehicle. Install the bolts and tighten to 41–55 ft. lbs. (55–74Nm).
8. Install the driveshaft so the index mark on the rear yoke is in line with the index mark on the axle companion flange.
9. On Super Cab vehicles, tighten the center bearing retaining bolts to 27–37 ft. lbs. (37–50Nm).

NOTE: Make certain the center bearing bracket assembly is reinstalled "square" to the vehicle. If the spacers were installed under the center bearing be sure to reinstall them.

DISASSEMBLY AND ASSEMBLY

2WD

1. Prior to disassembly, mark the position of the driveshaft components relative to the driveshaft tube. All components must be re-assembled in the same relationship to maintain proper balance.
2. Place the driveshaft on a suitable workbench.
3. Remove the snaprings that retain the bearing cups.
4. Position the U-joint removal tool, T74P–4635–C or equivalent, on the slip yoke and press out the bearing. If the bearing cup cannot be pressed all the way out of the slip yoke, remove it with vise grip or channel lock pliers.
5. Reposition the tool 180° to press on the spider and remove the remaining bearing cup from the opposite side.
6. On 2WD vehicles, remove the slip yoke form the spider. On 4WD vehicles, remove the spider from the driveshaft.
7. Remove the remaining bearing cups and spiders from the driveshaft in the same manner.
8. Clean the yoke area at each end of the driveshaft.
9. Four Wheel Drive (4WD) vehicles only:
   a. Remove the clamps on the driveshaft boot seal. Discard the clamp.
   b. Note the orientation of the slip yoke to the driveshaft tube for installation during assembly. Mark the position of the slip yoke to the driveshaft tube.
   c. Carefully pull the slip yoke from the driveshaft. Be careful not to damage the boot seal.
   d. Clean and inspect the spline area of the driveshaft.

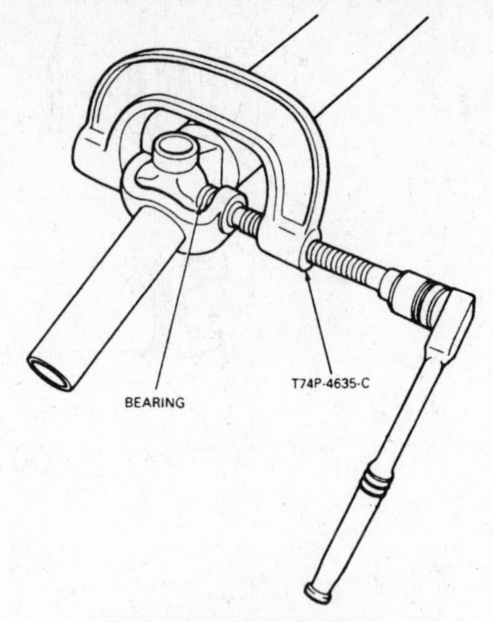

U-joint bearing removal

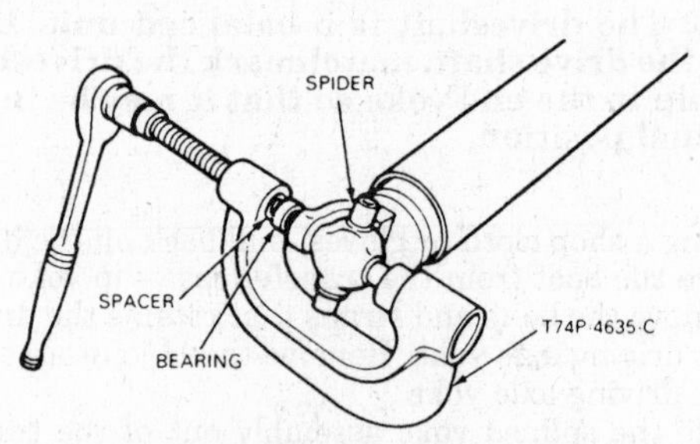

U-joint bearing installation

To assemble:

10. Four Wheel Drive (4WD) vehicles only:
   a. Lubricate the driveshaft slip splines with Multi-purpose Long-Life lubricant C1AZ–19490–B or equivalent.
   b. With the boot loosely installed on the driveshaft tube, install the slip yoke into the driveshaft splines in their original orientation.
   c. Using new clamps, install the driveshaft boot in its original position.
11. Start a new bearing cup into the yoke at the rear of the driveshaft.
12. Position the new spider in the rear yoke and press the bearing cup ¼ in. (6mm) below the yoke surface, using spacer.
13. Remove the tool and install a new snapring.
14. Start a new bearing cup into the opposite side of the yoke.
15. Position the U-joint tool and press on the bearing cup until the opposite bearing cup contacts the snapring.
16. Remove the tool and install a new snapring. It may be necessary to grind the surface of the snapring to permit easier entry.
17. Reposition the driveshaft and install the remaining bearing cups and spider in the same manner.
18. Check the universal joints for freedom of movement. If

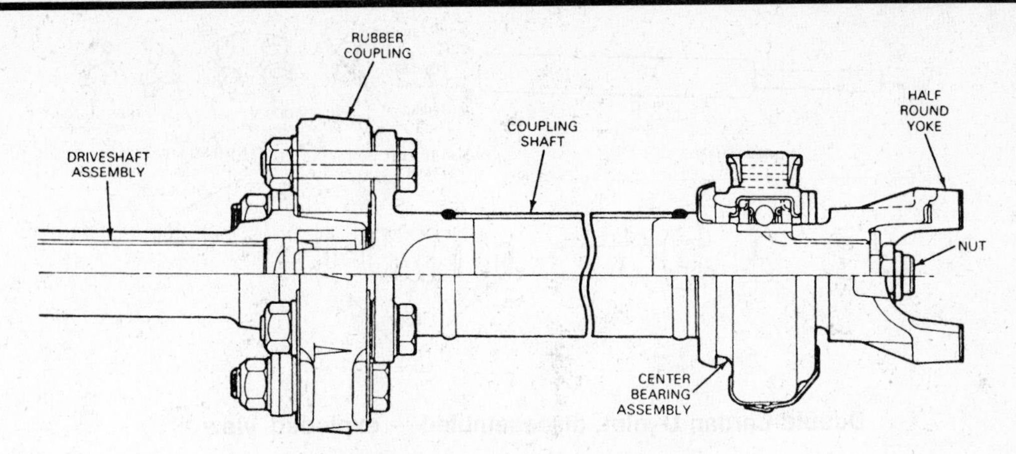

Driveshaft and coupling assembly

binding has resulted from misalignment during assembly, a sharp rap on the yokes with a brass hammer will seat the bearing needles. Be sure to support the shaft end during this procedure and do not strike the bearings themselves. Make certain the universal joints are free to rotate easily without binding before installing the driveshaft.

19. Lubricate the U-joint assemblies with Multi-purpose Long-Life lubricant C1AZ–19490–B or equivalent.

Center Bearing Single Cardan Type

REPLACEMENT

Super Cab

1. Remove the driveshaft from the vehicle.
2. Separate the driveshaft from the coupling shaft maintaining proper orientation.
3. Remove the nut retaining the half round yoke to the coupling shaft and remove the yoke.
4. Check the center bearing support for wear by rotating the outer area while holding the coupling shaft. If any wear or roughness is evident, replace the bearing.
5. Inspect the rubber insulator for evidence of hardeness, cracking or deterioration. Replace if damaged in any way.
6. Re-install the coupling shaft yoke.

NOTE: Be sure the yoke is re-installed on the coupling shaft in the same orientation as it was originally installed. The orientation is critical so that proper driveshaft balance and U-joint phasing is maintained. Tighten the retaining nut to 100–120 ft. lbs. (135–162Nm).

7. Re-assemble the driveshaft to the coupling shaft, maintaining proper orientation.

Front and Rear Driveshaft Double Cardan Type

REMOVAL AND INSTALLATION

1. Disconnect the negative battery cable.
2. Raise and support the vehicle safely.

NOTE: The driveshaft is a balanced unit. Before removing the drive shaft, matchmark the driveshaft in relationship to the axle flange so that it may be installed in its original position.

3. Remove the bolts retaining the flange to the transfer case. Disconnect the U-joint from the flange at the transfer case.
4. Remove the bolts retaining the flange to the rear axle. Disconnect the U-joint from the flange at the rear axle.
5. Remove the driveshaft.

To install:

6. Position the single U-joint end of the driveshaft to the rear axle and install the retaining bolts. Tighten the bolts to 61–87 ft. lbs. (83–118Nm).
7. Position the double Cardan U-joint to the transfer case and install the retaining bolts. Tighten the bolts to 12–16 ft. lbs. (17–22Nm).
8. Lower the vehicle.
9. Re-connect the negative battery cable.

DISASSEMBLY AND ASSEMBLY

1. Place the driveshaft on a suitable workbench.
2. Matchmark the positions of the spiders, the center yoke and the centering socket yoke as related to the stud yoke which is welded to the front of the driveshaft tube.

NOTE: The spiders must be assembled with the bosses in their original position to provide proper clearance.

3. Remove the snaprings that secure the bearings in the front of the center yoke.
4. Position the U-joint tool, T74P–4635–C or equivalent, on the center yoke. Thread the tool clockwise until the bearing protrudes approximately ⅜ in. (10mm) out of the yoke.
5. Position the bearing in a vice and tap on the center yoke to free it from the bearing. Lift the 2 bearing cups from the spider.
6. Re-position the tool on the yoke and move the remaining bearing in the opposite direction so that it protrudes approximately ⅜ in. (10mm) out of the yoke.
7. Position the bearing in a vice. Tap on the center yoke to free it from the bearing. Remove the spider from the center yoke.
8. Pull the centering socket yoke off the center stud. Remove the rubber seal from the centering ball stud.
9. remove the snaprings frm the center yoke and frm the driveshaft yoke.
10. Position the tool on the driveshaft yoke and press the bearing outward until the inside of the center yoke almost contacts the slinger ring at the front of the driveshaft yoke. Pressing

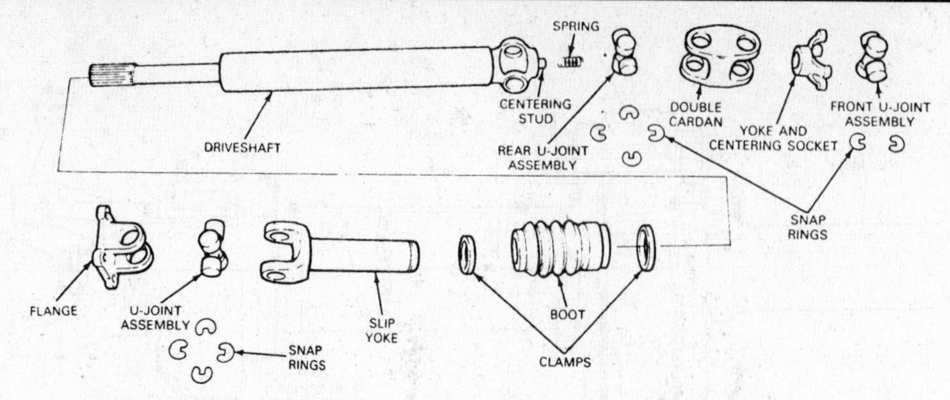

Double Cardan U-joint, disassembled — exploded view

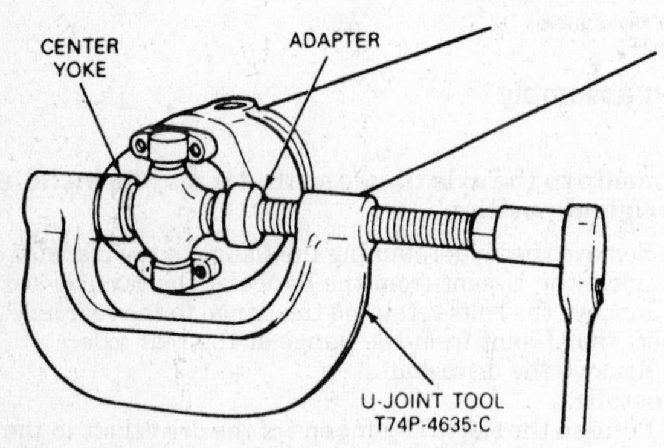

Pressing bearing from center yoke

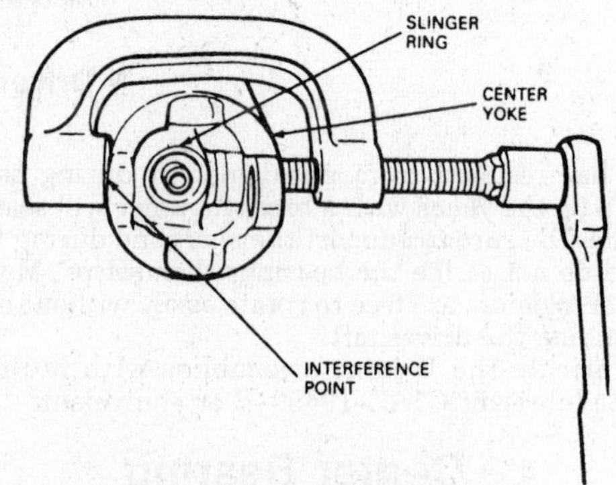

Center yoke interference point

beyound this point can distort the slinger ring interference point.

11. Clamp the exposed end of the bearing in a vice and drive on the center yoke with a soft-faced hammer to free it from the bearing.
12. Reposition the tool and press on the spider to remove the opposite bearing.
13. Remove the center yoke from the spider. Remove the spider form the driveshaft yoke.
14. Clean all serviceable parts in cleaning solvent. If using a repair kit, install all of the parts supplied in the kit.
15. Remove the clamps on the driveshaft boot seal. Discard the clamps.
16. Note the orientation of the slip yoke to the driveshaft tube for installation during assembly. Mark the position of the slip yoke to the driveshaft tube.
17. Carefully pull the slip yoke from the driveshfat. Be careful not to damage the boot seal.
18. Clean and inspect the spline area of the driveshaft.

To assemble:

19. Lubricate the driveshaft slip splines with Multi-purpose Long-Life lubricant C1AZ–19490–B or equivalent.
20. With the boot loosely installed on the driveshaft tube, install the slip yoke into the driveshaft splines in their original orientation.
21. Uning new clamps, install the driveshaft boot in its original position.
22. To assemble the double Cardan joint, position the spider in the driveshaft yoke. Make certain the spider bosses (or lubrication plugs on kits) will be in the same position as originally installed. Press in the bearing using the U-joint tool. Then, install the snaprings.
23. Pack the socket relief and the ball with Multi-purpose Long-Life lubricant C1AZ–19490–B or equivalent, then position the center yoke over the spider ends and press in the bearing. Install the snaprings.
24. Install a new seal on the centering ball stud. Position the centering socket yoke on the stud.
25. Place the front spider in the center yoke. Make certain the spider bosses (or lubrication plugs on kits) are properly positioned.
26. With the spider loosely positioned on the center stop, seat the first pair of bearings into the centering socket yoke. Then, press the second pair into the centering yoke. Install the snaprings.
27. Apply pressure on the centering socket yoke and install the remaining bearing cup.
28. If a kit was used, lubricate the U-joint through the grease fitting, using Multi-purpose Long-Life lubricant C1AZ–19490–B or equivalent.

Rear Driveshaft Constant Velocity (CV) Type U-joint

REMOVAL AND INSTALLATION

1. Disconnect the negative battery cable.
2. Raise and support the vehicle safely.

NOTE: The driveshaft is a balanced unit. Before removing the drive shaft, matchmark the driveshaft in relationship to the flange on the transfer case and the flange on the rear axle so that it may be installed in its original position.

3. Remove the bolts retaining the driveshaft to the transfer case.

4. Remove the bolts retaining the driveshaft to the rear axle flange.
5. Remove the driveshaft.

To install:

6. Position the driveshaft to the rear axle flange so that the marks made previously are line up. Install and tighten the retaining bolts to 61–87 ft. lbs. (83–118Nm).
7. Position the driveshaft to the transfer case flange so that the marks made previously are line up. Install and tighten the retaining bolts to 61–87 ft. lbs. (83–118Nm).
8. Lower the vehicle.
9. Reconnect the negative battery cable.

DISASSEMBLY AND ASSEMBLY

1. Place the driveshaft on a suitable workbench.

NOTE: The CV joint components are matched. Extreme care should be take not to mix or substitute components.

2. Remove the clamp retaining the shroud to the outer bearing race and flange assembly.
3. Carefully tap the shroud lightly with a blunt tool and remove the shroud. Be careful not to damage the shroud, dust boot or outer bearing race and flange assembly.
4. Peel the boot upward and away from the outer bearing race and flange assembly.
5. Remove the wire ring that retains the inner race to the outer race.
6. Remove the inner race and shaft assembly from the outer race and flange assembly. Remove the cap and spring from inside the outer retainer.
7. Remove the circlip retaining the inner race assembly to the shaft, using snapring pliers. Discard the clip and remove the inner race assembly.
8. If required, remove the clamp retaining the boot to the shaft and remove the boot.
9. Carefully pry the ball bearings from the cage. Be careful not to scratch or damage the cage, race or ball bearings.
10. Rotate the inner race to align with the cage windows and remove the inner race through the wider end of the cage.

To assemble:

12. Install the inner bearing race in the bearing cage. Install the race through the large end of the cage with the counterbore facing the large end of the cage.
13. Push the race to the top of the cage and rotate the race until all the ball slots are aligned with the windows. This will lock the race to the top of the cage.
14. With the bearing cage and inner race properly aligned, install the ball berings. The bearings can be pressed through the bearing cage with the heel of the hand. Repeat this step until the remaining ball bearings are installed.
15. If removed, install a new dust boot on the shaft, using a new clamp. Make certain the boot is seated i its groove.

NOTE: The clamp is a fixed diameter push-on metal ring.

16. Install the inner bearing assembly on the shaft. Make certain the circlip is exposed.
17. Install a new circlip on the shaft. Do not over-expand or twist the circlip during installation.
18. Install the spring and cap in the outer bearing retainer and flange.
19. Fill the outer bearing retainer with 3 oz. of Constant Velocity Joint Grease, D8RZ–19590–A or equivalent.
20. Insert the inner race and shaft assembly in the outer bearing retainer and flange.
21. Push the inner race down until the wire spring groove is visible and install the wire ring.
22. Fill the top of the outer bearing retainer with Constant Ve-

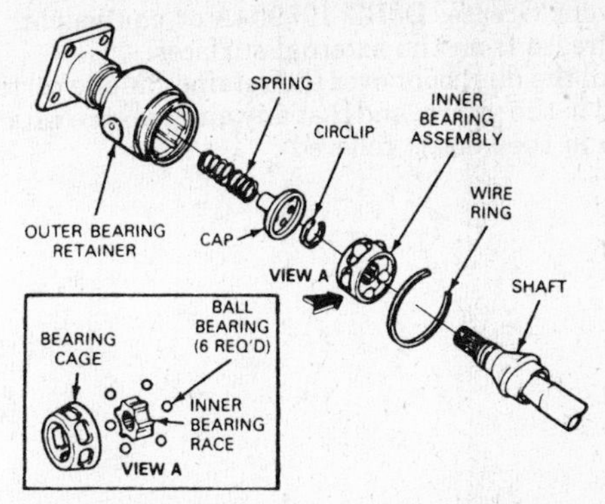

CV U-joint assembly, exploded view

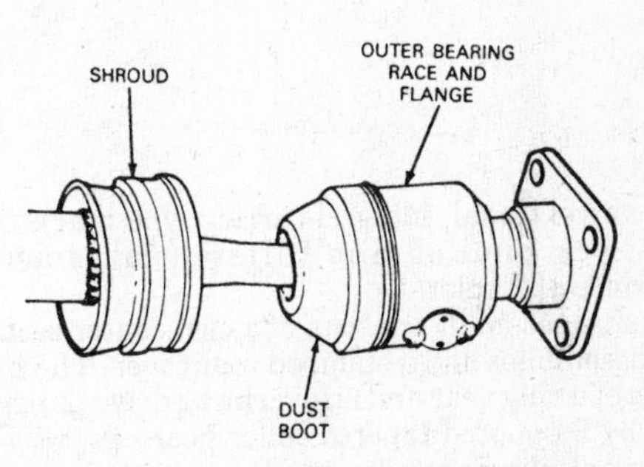

Removing shroud from outer bearing race and flange

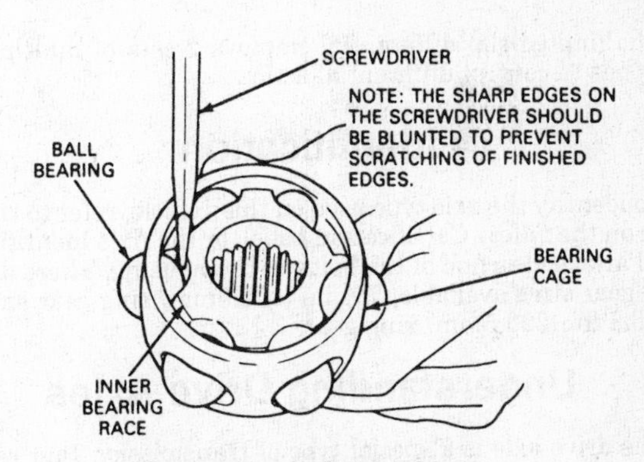

Removing ball bearings

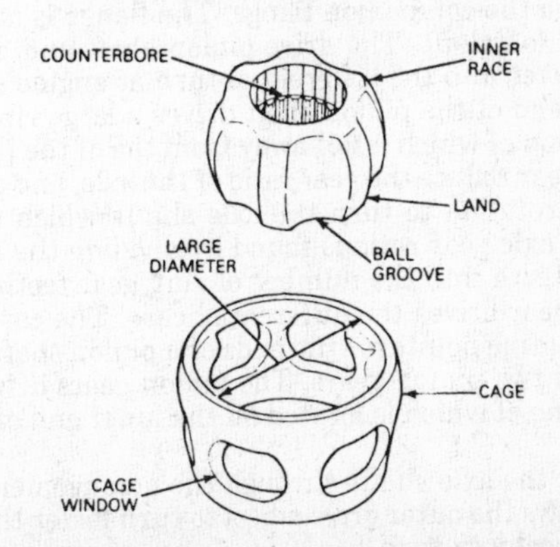

Assembling inner bearing race to cage

locity Joint Grease, D8RZ–19590–A or equivalent. Remove all excess grease from the external surfaces.

23. Pull the dust boot over the retainer. Make certain the boot is seated in the groove and that any air pressure which may have built up in the boot is relieved.

NOTE: Insert a dulled screwdriver blade between the boot and outer bearing retainer and allow the trapped air to escape from the boot.

24. Install the shroud over the boot and retainer and install the clamp.

REAR AXLE

A Ford conventional, integral-carrier type rear axle or Ford Traction-Lock Limited Slip Differential is used on the Ranger/Bronco II/Explorer.

The housing assembly consists of a cast center section with 2 steel tube assemblies and a stamped rear cover. The hypoid gear set consists of a ring gear and an overhung drive pinion, which is supported by 2 opposed tapered roller bearings.

The differential case is a 1 piece design with 2 openings to allow for assembly of the internal components and lubricant flow.

The limited-slip axle assembly, except for the differential case and its internal components, is identical to the conventional axle.

The limited-slip differential employs 2 sets of multiple disc clutches to control differential action.

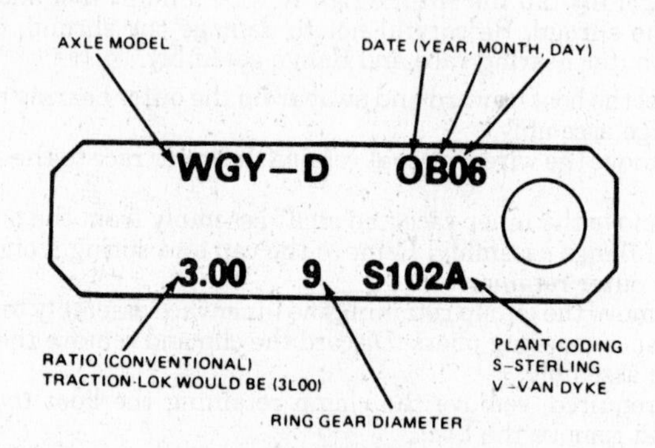

Axle identification tag

Identification

To identify the axle type used on this vehicle, refer to the axle code on the Safety Certification Label, or the Axle Identification label attached to one of the bolts on the housing. There are two ring gear sizes available; 7½ in. (190.5mm) ring gear axle and the 8.8 in. (223.5mm) ring gear.

Understanding Drive Axles

The drive axle is a special type of transmission that reduces the speed of the drive from the engine and transmission and divides the power to the wheels. Power enters the axle from the driveshaft via the companion flange. The flange is mounted on the drive pinion shaft. The drive pinion shaft and gear which carry the power into the differential turn at engine speed. The gear on the end of the pinion shaft drives a large ring gear the axis of rotation of which is 90° away from the of the pinion. The pinion and gear reduce the gear ratio of the axle, and change the direction of rotation to turn the axle shafts which drive both wheels. The axle gear ratio is found by dividing the number of pinion gear teeth into the number of ring gear teeth.

The ring gear drives the differential case. The case provides the two mounting points for the ends of a pinion shaft on which are mounted two pinion gears. The pinion gears drive the two side gears, one of which is located on the inner end of each axle shaft.

By driving the axle shafts through the arrangement, the differential allows the outer drive wheel to turn faster than the inner drive wheel in a turn.

The main drive pinion and the side bearings, which bear the weight of the differential case, are shimmed to provide proper bearing preload, and to position the pinion and ring gears properly.

WARNING: The proper adjustment of the relationship of the ring and pinion gears is critical. It should be attempted only by those with extensive equipment and/or experience.

Limited-slip differentials include clutches which tend to link each axle shaft to the differential case. Clutches may be engaged either by spring action or by pressure produced by the torque on the axles during a turn. During turning on a dry pavement, the effects of the clutches are overcome, and each wheel turns at the required speed. When slippage occurs at either wheel, however, the clutches will transmit some of the power to the wheel which has the greater amount of traction. Because of the presence of clutches, limited-slip units require a special lubricant.

Determining Axle Ratio

The drive axle is said to have a certain axle ratio. This number (usually a whole number and a decimal fraction) is actually a comparison of the number of gear teeth on the ring gear and the pinion gear. For example, a 4.11 rear means that theoretically, there are 4.11 teeth on the ring gear and one tooth on the pinion gear or, put another way, the driveshaft must turn 4.11 times to turn the wheels once. Actually, on a 4.11 rear, there might be 37 teeth on the ring gear and 9 teeth on the pinion gear. By dividing the number of teeth on the pinion gear into the number of teeth

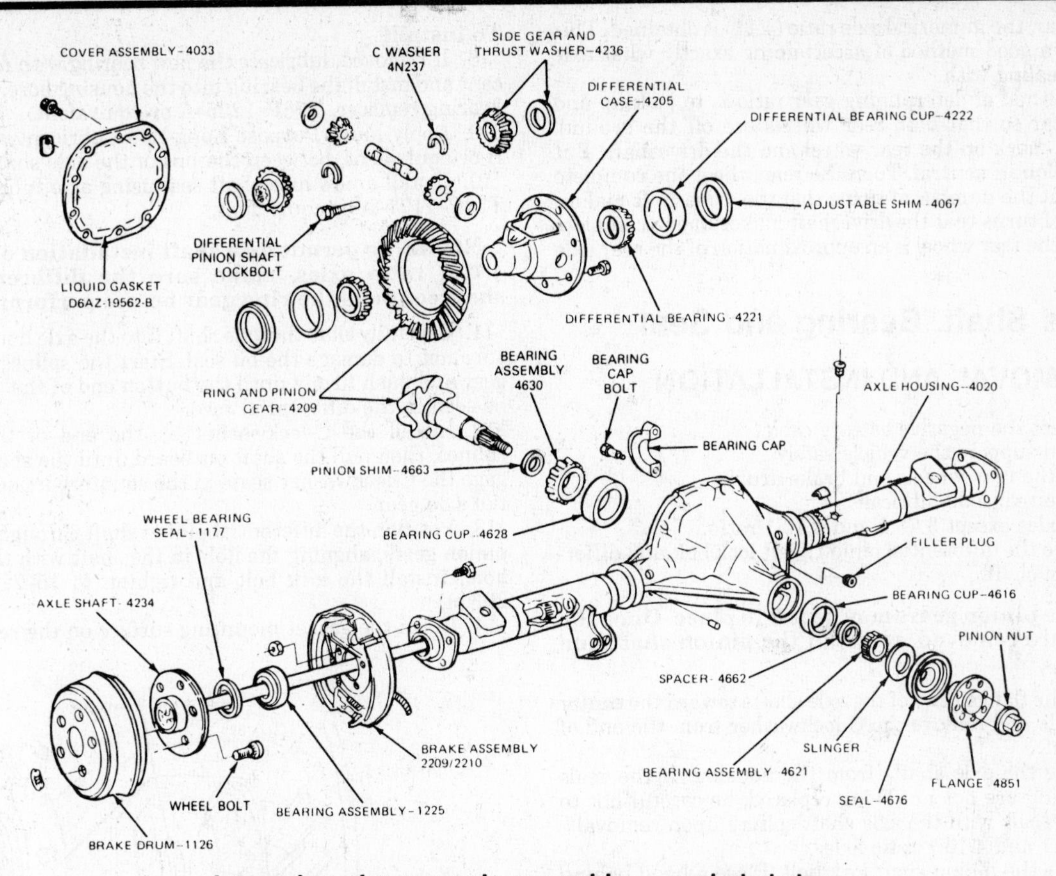

Integral carrier rear axle assembly — exploded view

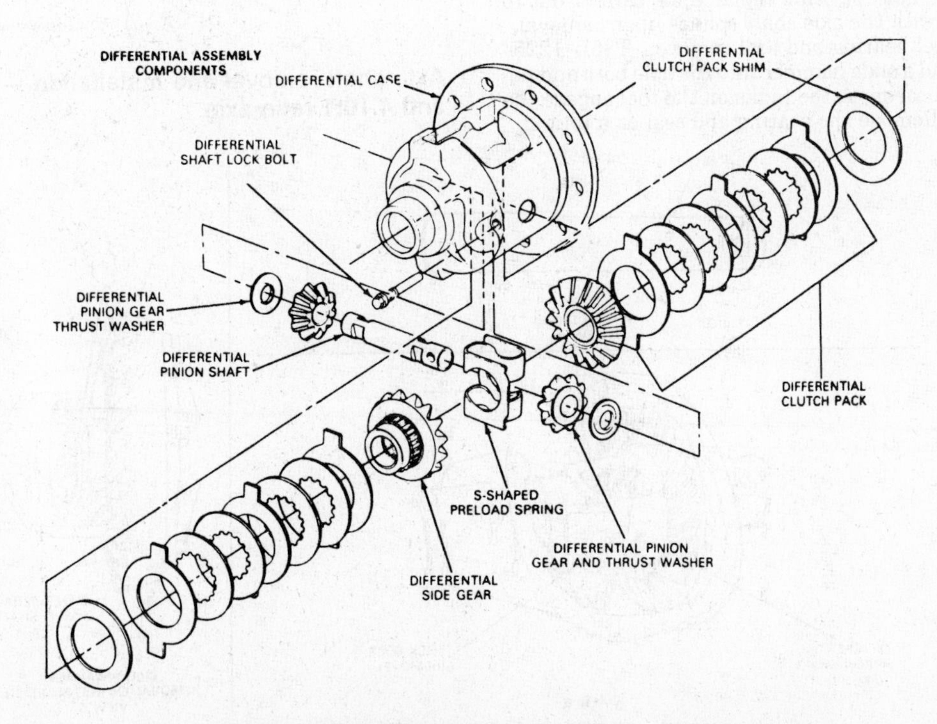

Limited slip differential assembly — exploded view

on the ring gear, the numerical axle ratio (4.11) is obtained. This also provides a good method of ascertaining exactly what axle ratio one is dealing with.

Another method of determining gear ratio is to jack up and support the car so that both rear wheels are off the ground. Make a chalk mark on the rear wheel and the driveshaft. Put the transmission in neutral. Turn the rear wheel one complete turn and count the number of turns that the driveshaft makes. The number of turns that the driveshaft makes in one complete revolution of the rear wheel is an approximation of the rear axle ratio.

Axle Shaft, Bearing and Seal

REMOVAL AND INSTALLATION

1. Disconnect the negative battery cable.
2. Raise and support the vehicle safely.
3. Remove the rear wheels and brake drums.
4. Drain the rear axle lubricant.
5. For all axles except 3.73:1 and 4.10:1 ratio.
   a. Remove the differential pinion shaft lock bolt and differential pinion shaft.

NOTE: The pinion gears may be left in place. Once the axle shafts are removed, reinstall the pinion shaft and lock bolt.

   b. Push the flanged end of the axle shafts toward the center of the vehicle and remove the C-lockwasher from the end of the axle shaft.
   c. Remove the axle shafts from the housing. If the seals and/or bearing are not not being replaced, be careful not to damage the seals with the axle shaft splines upon removal.
6. For 3.73:1 and 4.10:1 ratio axles.
   a. Remove the pinion shaft lock bolt. Place a hand behind the differential case and push out the pinion shaft until the step contacts the ring gear.
   b. Remove the C-lockwasher from the axle shafts.
   c. Remove the axle shafts from the housing. If the seals and/or bearing are not not being replaced, be careful not to damage the seals with the axle shaft splines upon removal.
7. Insert the wheel bearing and seal remover, T85L-1225-AH or equivalent, and a slide hammer into the axle bore and position it behind the bearing so the tanks on the tool engage the bearing outer race. Remove the bearing and seal as a unit.

To install:

8. If removed, lubricate the new bearing with rear axle lubricant and install the bearing into the housing bore. Use axle tube bearing replacer, T78P-1225-A or equivalent.
9. Apply Multi-Purpose Long-Life Lubricant, C1AZ-19590-B or equivalent, between the lips of the axle shaft seal.
10. Install a new axle shaft seal using axle tube seal replacer T78P-1177-A or equivalent.

NOTE: To permit axle shaft installation on 3.73:1 and 4.10:1 ratio axles, make sure the differential pinion shaft contacts the ring gear before performing Step 11.

11. Carefully slide the axle shaft into the axle housing, making sure not to damage the oil seal. Start the splines into the side gear and push firmly until the button end of the axle shaft can be seen in the differential case.
12. Install the C-lockwasher on the end of the axle shaft splines, then pull the shaft outboard until the shaft splines engage the C-lockwasher seats in the counterbore of the differential side gear.
13. Position the differential pinion shaft through the case and pinion gears, aligning the hole in the shaft with the lock screw hole. Install the lock bolt and tighten to 15-22 ft. lbs. (21-29Nm).
14. Clean the gasket mounting surface on the rear axle hous-

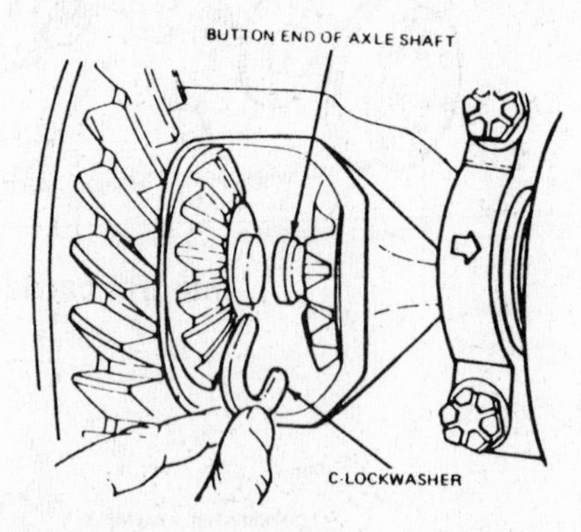

Axle shaft removal and installation — except 3.73:1 and 4.10:1 ratio axle

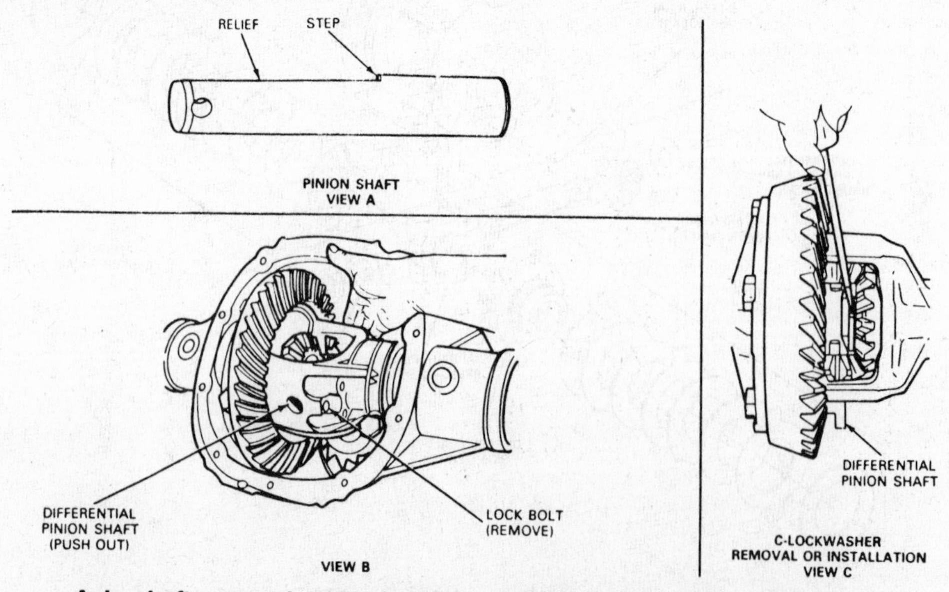

Axle shaft removal and installation — 3.73:1 and 4.10:1 ratio axle

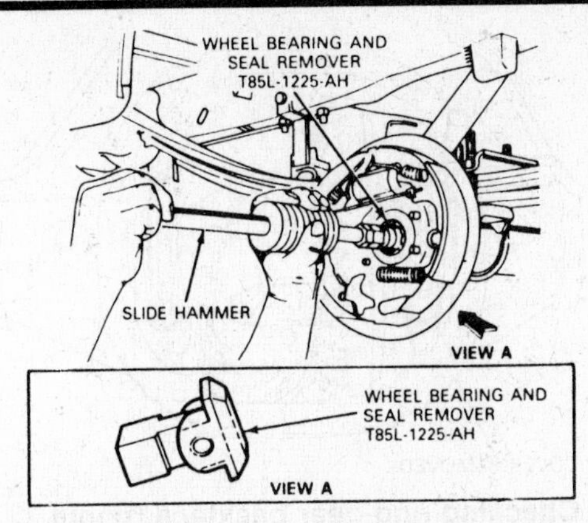

Axle shaft seal and wheel bearing removal

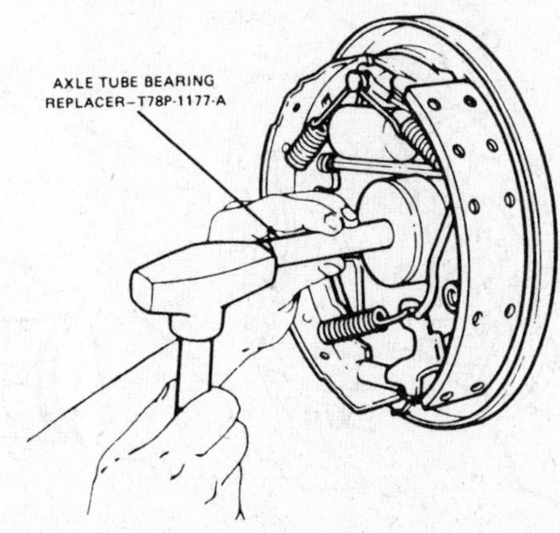

Wheel bearing and axle shaft seal installation

ing and cover. Apply a continuous bead of Silicone Rubber Sealant ESE-M4G195-A or equivalent to the carrier casting face.

15. Install the cover and tighten the retaining bolts to 25–35 ft. lbs. (20–34Nm).

NOTE: The cover assembly must be installed within 15 minutes of application of the silicone sealant.

16. Add lubricant until it is ¼ in. (6mm) below the bottom of the filler hole in the running position. Install the filler plug and tighten to 15–30 ft. lbs. (20–41Nm).

Pinion Oil Seal

REMOVAL AND INSTALLATION

1. Disconnect the negative battery cable.
2. Raise and support the vehicle safely. Allow the axle to drop to rebound position for working clearance.
3. Remove the rear wheels and brake drums. No drag must be present on the axle.
4. Mark the companion flanges and U-joints for correct reinstallation position.
5. Remove the driveshaft.
6. Using an inch pound torque wrench and socket on the pinion yoke nut measure the amount of torque needed to maintain differential rotation through several clockwise revolutions. Record the measurement.
7. Use a suitable tool to hold the companion flange. Remove the pinion nut.
8. Place a drain pan under the differential. Clean the area around the seal and mark the yoke-to-pinion relation.
9. Use a 2-jawed puller to remove the companion flange.
10. Remove the seal with a small prybar.

To install:

11. Thoroughly clean the oil seal bore.

NOTE: If you are not absolutely certain of the proper seal installation depth, the proper seal driver must be used. If the seal is misaligned or damaged during installation, it must be removed and a new seal installed.

12. Drive the new seal into place with a seal driver such as T83T-4676-A. Coat the seal lip with clean, waterproof wheel bearing grease.
13. Coat the splines with a small amount of wheel bearing grease and install the yoke, aligning the matchmarks. Never hammer the yoke onto the pinion!
14. Install a new nut on the pinion.
15. Hold the yoke with a holding tool. Tighten the pinion nut, taking frequent turning torque readings until the original preload reading is attained. If the original preload reading, that you noted before disassembly, is lower than the specified reading of 8–14 inch lbs. for used bearings; 16–29 inch lbs. for new bearings, keep tightening the pinion nut until the specified reading is reached. If the original preload reading is higher than the specified values, torque the nut just until the original reading is reached.

WARNING: Under no circumstances should the nut be backed off to reduce the preload reading! If the preload is exceeded, the yoke and bearing must be removed and a new collapsible spacer must be installed. The entire process of preload adjustment must be repeated.

15. Install the driveshaft using the matchmarks. Torque the nuts to 15 ft. lbs.
16. Lower the vehicle.
17. Reconnect the negative battery cable.

Axle Housing

REMOVAL AND INSTALLATION

1. Disconnect the negative battery cable.
2. Raise and support the vehicle safely.
3. Matchmark and disconnect the driveshaft at the axle.
4. Remove the wheels and brake drums.
5. Disengage the brake line from the clips that retain the line to the housing.
6. Disconnect the vent tube from the housing.
7. Remove the axle shafts.
8. Remove the brake backing plate from the housing and support them with wire. Do not disconnect the brake line.
9. Disconnect each rear shock absorber from the mounting bracket stud on the housing.
10. Lower the axle slightly to reduce some of the spring tension. At each rear spring, remove the spring clip (U-bolt) nuts, spring clips and spring seat caps.
11. Remove the housing from under the vehicle.

To Install:

12. Position the axle housing under the rear springs. Install the spring clips (U-bolts), spring seat clamps and nuts. Tighten the spring clamps evenly to 115 ft. lbs.
13. If a new axle housing is being installed, remove the bolts that attach the brake backing plate and bearing retainer from the old housing flanges. Position the bolts in the new housing flanges to hold the brake backing plates in position. Torque the bolts to 40 ft. lbs.
14. Install the axle shafts.
15. Connect the vent tube to the housing.
16. Position the brake line to the housing and secure it with the retaining clips.

17. Raise the axle housing and springs enough to allow connecting the rear shock absorbers to the mounting bracket studs on the housing. Torque the nuts to 60 ft. lbs.
18. Connect the driveshaft to the axle. Torque the nuts to 8–15 ft. lbs.
19. Install the brake drums and wheels.
20. Lower the vehicle.
21. Reconnect the negative battery cable.

Differential Case

BEFORE DISASSEMBLY

NOTE: The differential case assembly and drive pinion should be inspected before they are removed from the carrier casting. These inspections can find the cause of the complaint and determine the resolution.

1. Disconnect the negative battery cable.
2. Raise and support the vehicle safely.
3. Place a suitable drain pan beneath the differential case. Clean all dirt from the area of the carrier cover. Loosen the differential case retaining bolts and allow the lubricant to drain from the carrier housing. Remove the carrier case cover.
4. Wipe the lubricant from the internal parts and visually inspect the parts for wear and/or damage.
5. Rotate the gears to see is there is an roughness which would indicate damaged bearings or gears.
6. Check the ring gear teeth for signs of scoring, abnormal wear or nicks/chips.
7. Set up a dial indicator and check the ring gear backlash and ring gear backface runout. Backlash should be 0.008–0.015 in. (0.20–0.38mm) and backface runout should be under 0.004 in. (0.10mm).
8. A contact pattern is not an acceptable guide to check for noise. Proper gear set assembly must be checked using the pinion depth gauge, tool T79P–4020–A or equivalent, which shows the correct pinion shim required to assure acceptable running condition.

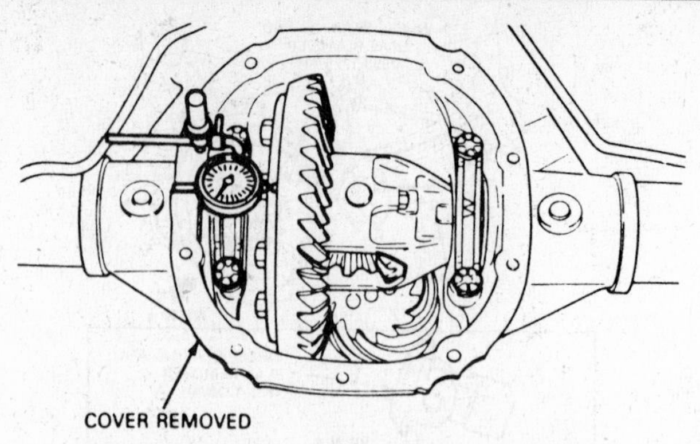

Checking ring gear backface runout

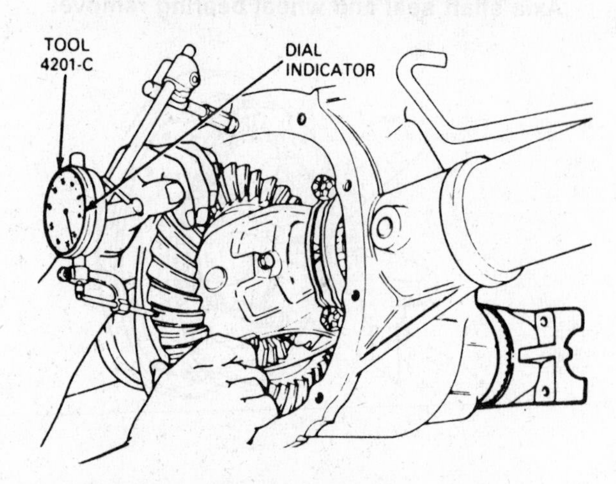

Checking ring gear backlash runout

REMOVAL AND INSTALLATION

1. Mark 1 differential bearing cap to help position the caps properly during assembly.

NOTE: Right and left bearing caps must not be interchanged. Note the scribe marks or triangles on the bearing caps. When reassembled the scribe marks or triangles must be pointing in the same direction as before removal.

2. Loosen the differential bearing cap bolts and bearing caps.
3. Pry the differential case, bearing cups and shims out until they are loose in the bearing caps. Remove the bearing caps and remove the differential assembly for the carrier.

NOTE: When using the pry bar, place a wood block between the pry bar and the axle housing to protect the casting face from damage.

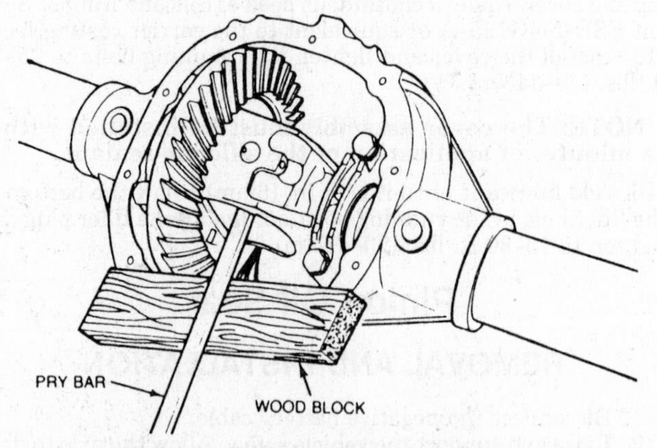

Differential assembly, removal

Differential Case
Integral Carrier Type Differential

DISASSEMBLY

1. Remove and discard the ring gear-to-differential case retaining bolts.
2. Lightly tap the ring gear to remove it from the differential case.
3. Install the differential assembly with the bearing cups and shims. Tighten the bearing cup bolts to 70–85 ft. lbs. (95–115Nm).
4. Check the runout of the differential case flange with a dial indicator. The maximum runout should not exceed 0.003 in. (0.08mm). If the runout exceeds specification, the ring gear is true and the trouble is due to either a damaged case or bearings.
5. Remove the differential case bearings, using remover tool, T7OP–4221–A or equivalent.

ASSEMBLY

1. Press new differential bearings on the case hubs, using re-

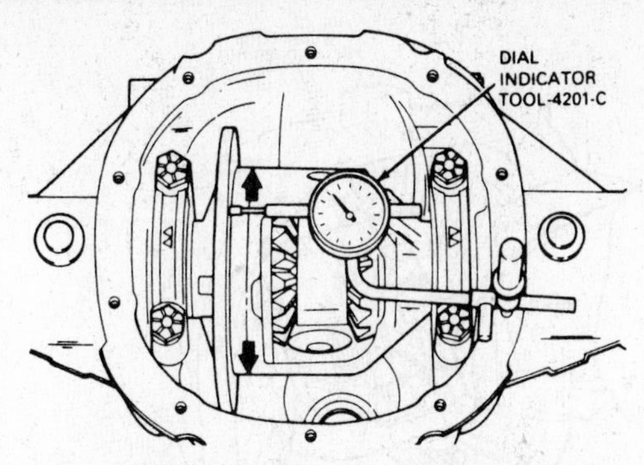

Checking runout of the differential case

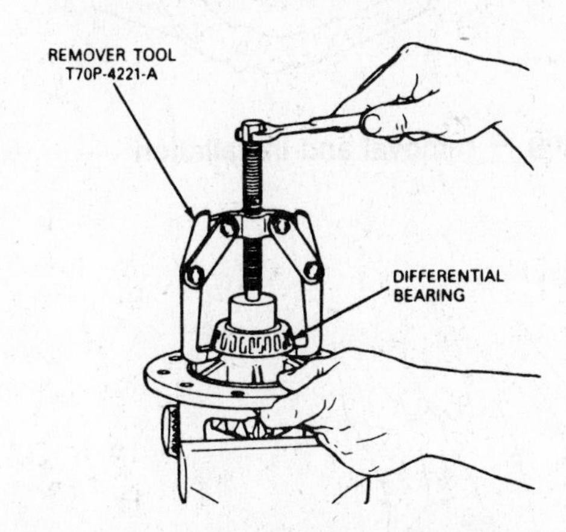

Removing differential case bearing

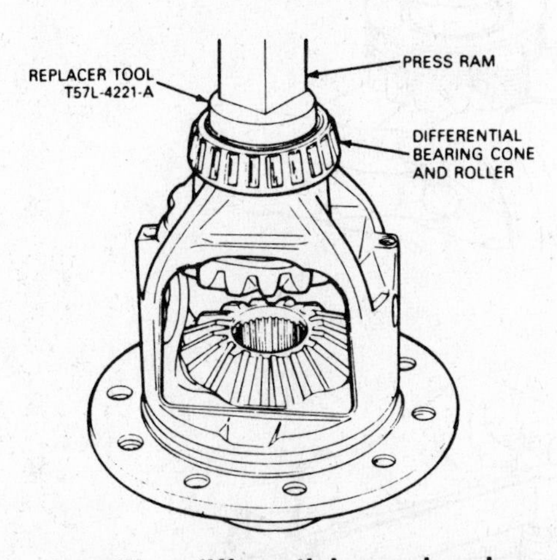

Installing differential case bearing

placing tool, T57L–4221–A or equivalent. Press against the bearing cone only.

2. Check the case runout again with the new bearings. If the runout is now within specification, use the new bearings for assembly. If the runout is still excessive, the case is damaged and should be replaced.

Traction-Lock Limited Slip Differential

OPERATIONAL CHECK

A limited-slip differential can be checked for proper operation without removing the differential from the axle housing as follows:

1. Disconnect the negative battery cable.
2. Place the transmission in **N** position.
3. Raise 1 rear wheel and remove the wheel cover.
4. Install torque tool T59L–4204–A or equivalent on the axle shaft flange studs.
5. Using a torque wrench, (torque wrench with a capacity of at least 200 ft. lbs.), rotate the axle shaft. The break-away torque required to start rotation should be at least 20 ft. lbs. (27Nm). The initial break-away torque may be higher that the continuous turning torque, but this is normal. The axle shaft should turn with even pressure throughout the check without slipping or binding. If the torque reading is less than specified, check the differential for proper adjustment.

NOTE: A vehicle equipped with a limited slip differential will always have both wheel driving. If only 1 wheel is raised off the floor and the rear axle is driven by the engine, the wheel on the floor could drive the vehicle off the stand or jack.

Adjustment

1. Disconnect the negative battery cable.
2. Raise and support the vehicle safely.
3. Remove the rear wheels and brake drums.
4. Remove the cover from the carrier casting face and drain the lubricant.
5. Perform an inspection before removing the differential case.
6. Working through the cover opening, remove the pinion shaft lock bolt and remove the pinion shaft.
7. Push the axle shafts inward until the C-locks at the end of the shafts are clear of the side gear recess.
8. Remove the C-locks and pull the axle shafts completely out of the housing.

NOTE: Care should be taken not to damage the axle seals when removing the axle shafts from the housing.

9. Drive the S-shaped preload spring half-way out of the differential case, using a suitable drift. Rotate the case 180°.
10. Hold the S-shaped preload spring with a pair of pliers and tap the spring until it is removed fro the differential.

NOTE: Be careful when removing the S-shaped spring, due to the spring tension.

11. Rotate the pinion gears until the gears can be removed fro the differential.
12. Remove and tag the gears, clutch pack and shims from both side. Inspect the clutch pack for wears. Replace, if necessary.

NOTE: Do not use cleaning solvents on the clutch plate friction surfaces. Wipe them clean only.

13. Install tool T84P–4946–A or equivalent on each of the side gear clutch packs without the shim. Tighten to 60 ft. lbs. (6.7Nm). Using a feeler blade, select the thickest blade that will enter between the tool and the clutch pack. This reading will be the thickness of the new shim.

NOTE: Be sure to lubricate friction plates with the proper hypoid gear lubricant prior to reassembly.

To install:

14. Install the clutch pack and new shim into the cavity in the differential case. Repeat this step for both sides.
15. Place the pinion gears and thrust washers 180° apart on the side gears. Rotate the tool until the pinion gears are aligned with the pinion shaft holes.
16. Hold the S-shaped preload spring up to the differential case window and with a soft-faced mallet, hammer the spring into position.

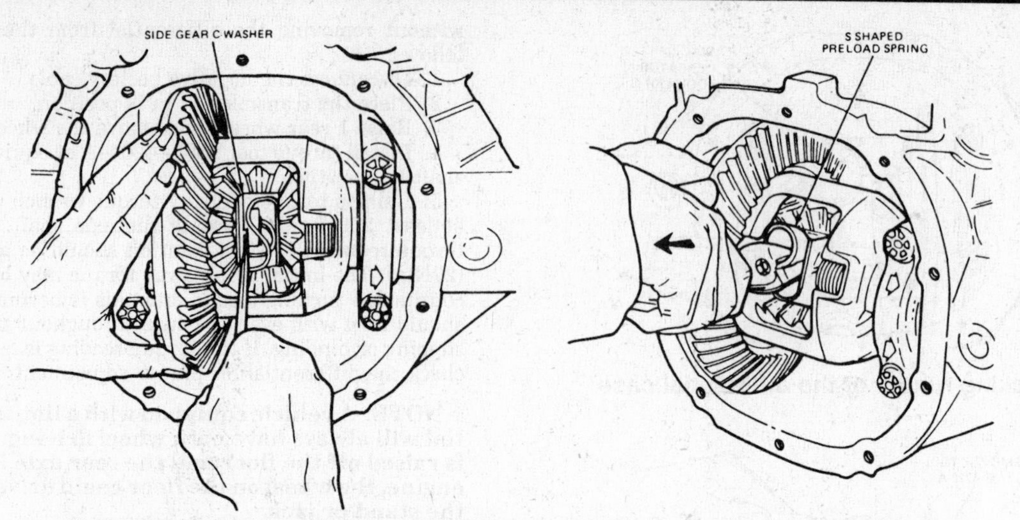

Side gear C-locks and S-shaped preload spring – removal and installation

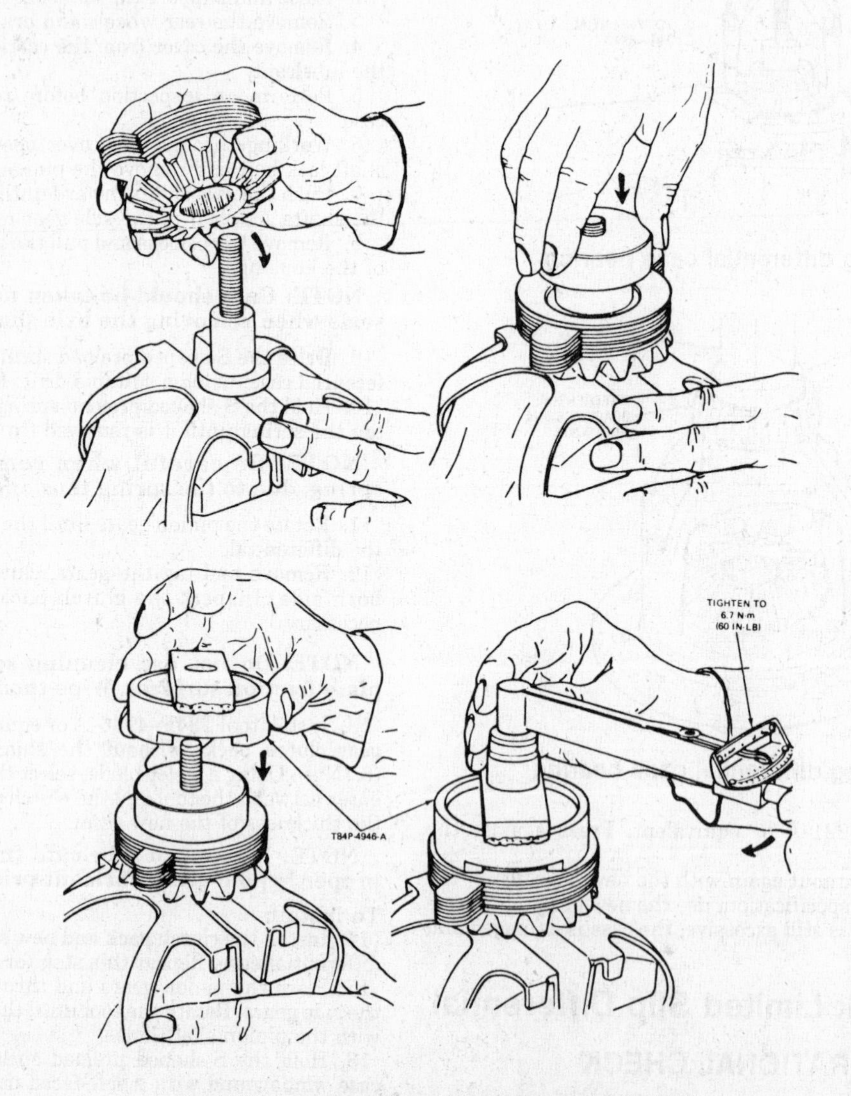

Installation of side gear clutch pack

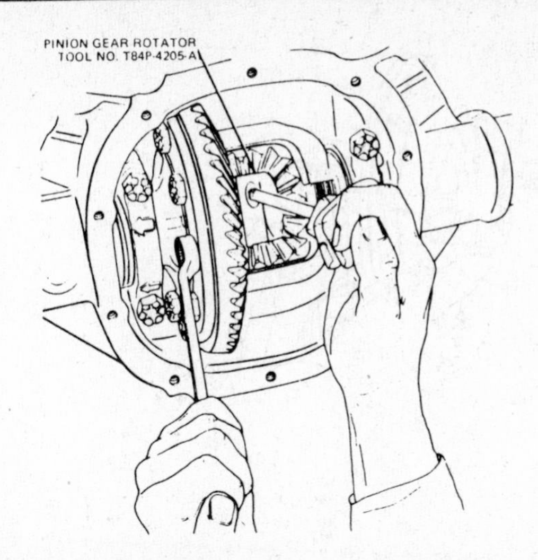

Differential pinion gears and thrust washers – removal and installation

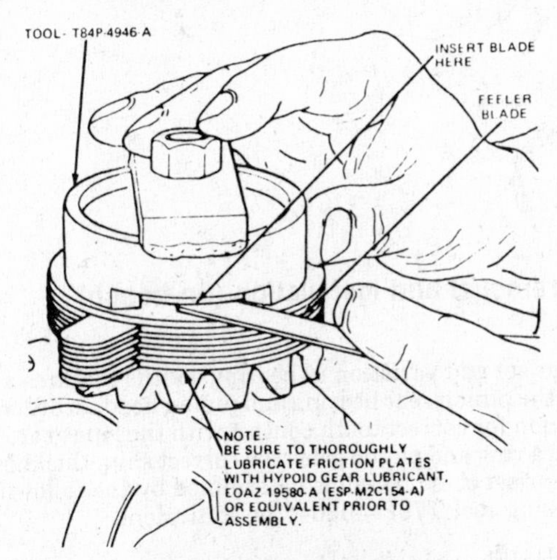

Feeler blade gauging

17. Install the axle shafts and C-locks into position and push the axle shaft outboard as far as possible.
18. Install the pinion shaft and pinion shaft lock bolt. Tighten the bolt to 15–30 ft. lbs. (20–40Nm).
19. Install the rear brake drums and wheels. Perform the traction-lock operational check to insure that the unit is within specification.
20. Apply silicone sealant, D6AZ–19562–B or equivalent, to the rear cover. Install the cover. Tighten the retaining bolts 25–35 ft. lbs. (34–47Nm).
21. Fill the unit to the bottom of the fill hole with the axle in the running position. Use 5.0 pints (2.6L) of Hypoid Gear Lubricant (EOAZ–19580–A) or equivalent.

Differential Case Limited Slip Differential

DISASSEMBLY

1. Remove and discard the ring gear-to-differential case retaining bolts.
2. Lightly tap the ring gear to remove it from the differential case.
3. Remove the differential pinion shaft lock bolt and remove the pinion shaft.
4. Drive out the S-shaped preload spring, using a suitable drift.

NOTE: Be careful when removing the S-shaped spring, due to the spring tension.

5. Rotate the pinion gears until the gears and thrust washers can be removed.
6. Remove and tag the gears, clutch pack and shims from both side. Inspect the clutch pack for wears. Replace, if necessary.

NOTE: Do not use cleaning solvents on the clutch plate friction surfaces. Wipe them clean only.

Clutch Pack Shim Selection

1. Assemble the clutch pack on the side gear. No shim is required at this point. However, all plates must be thoroughly prelubricated with Hypoid Gear Lubricant (EOAZ–19580–A) or equivalent.
2. Assemble tool T84P–4946–A or equivalent on the side gear clutch pack.
3. Using a feeler gauge tool, select the thickest feeler blade that will enter between the tool and the clutch pack. This reading will be the shim required for that clutch pack.

NOTE: Do not mix the clutches or shims.

4. Repeat Steps 1–3 for the opposite clutch pack.

ASSEMBLY

1. Prior to assembly, lubricate all parts with Hypoid Gear Lubricant (EOAZ–19580–A) or equivalent.
2. Mount the differential case in a soft jaw vise and place the clutch packs and side gears in their proper cavities in the differential case.
3. Place the pinion gears and thrust washers on the side gears. Install tool T84P–4946–A or equivalent in the differential case.
4. Rotate the pinion gears until the bores in the gears are aligned with the pinion shaft holes in the differential case. Remove the tool from the differential case.
5. Hold the S-shaped preload spring up to the differential case window and with a soft-faced mallet, hammer the spring into position.
6. Install the pinion shaft and lock bolt. Do not tighten the lock bolt at this point.
7. Prior to installation of the locking differential into a vehicle, a bench torque check must be made. With the locker tools T59L–4204–A or equivalent, check the torque required to rotate 1 side gear while the other is held stationary.
8. The initial break-away torque, if original clutch plates are used, should be no less that 20 ft. lbs. (27Nm). If new clutch plates are used, the break-away torque should be from 150–250 ft. lbs. (135–338Nm). The rotating torque required to keep the side gear turning with new clutch plates may fluctuate.
9. Clean the tapped holes in the ring gear with a suitable solvent. If the new bolts to be used show a green coating over approximately ½ in. (13mm) of the threaded area, use as is. If not coated, apply a small amount of threadlock and sealer, E0AZ–19554–B or equivalent. Tighten the bolts 70–85 ft. lbs. (95–115Nm).
10. Install the differential case and ring gear.

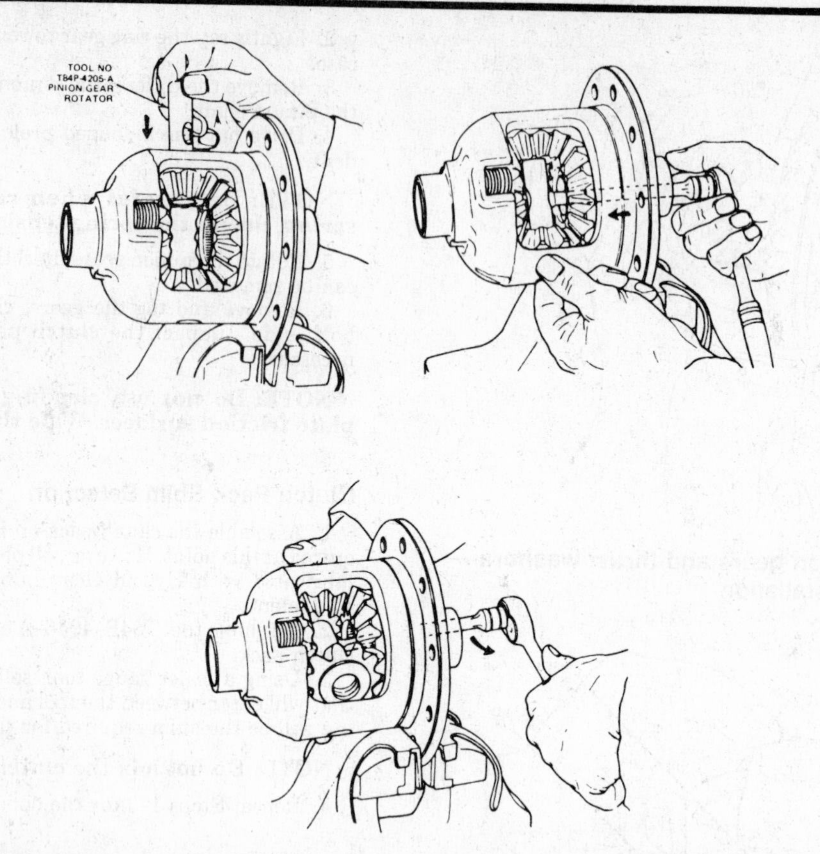

Differential pinion gears and thrust washers – removal and installation (on bench)

Drive Pinion

REMOVAL

1. Disconnect the negative battery cable.
2. Raise and support the vehicle safely.
3. Remove the differential case.
4. Mark the companion flanges and U-joints for correct reinstallation position.
5. Remove the driveshaft.
6. Use a suitable tool to hold the companion flange. Remove the pinion nut.
7. Use a 2-jawed puller to remove the companion flange. With a soft-faced hammer, drive the pinion out of the front bearing cone and remove it through the rear of the carrier casting.
8. Remove the drive pinion oil seal, front pinion bearing cone, roller and slinger from the carrier.

DISASSEMBLY AND ASSEMBLY

1. To remove the pinion rear bearing cone, use pinion bearing cone replacer, T719–4621–B or equivalent. Measure the shim found under the bearing cone with a micrometer. Record the thickness of the shim.
2. Determine the drive pinion depth following the procedure preceding the heading "Drive Pinion Depth Adjustment".
3. Place the selected shim(s) on the pinion shaft and press the pinion bearing until firmly seated on the shaft.

DRIVE PINION DEPTH ADJUSTMENT

Individual differences in machining the carrier casting and the gear set and variation in bearing widths requires a shim between the pinion rear bearing and pinion head, in order to locate the pinion for correct tooth contact with the ring gear. When replacing a ring and pinion gear, the correct shim thickness for the new gear set to be install is determined by the following procedure using tool T79P–4020–A or equivalent.

1. Place the rear pinion bearing over the aligning disc and insert it into the pinion bearing cup of the carrier. Place the front bearing into the front bearing cup and assemble the tool handle into the screw and tighten to 20 ft. lbs. (27Nm).

NOTE: The gauge block must be offset to obtain an accurate reading.

2. Center the gauge tube into the differential bearing bore. Install the bearing caps and tighten the bolts to 70–85 ft. lbs. (95–115Nm).

- Make sure that the gauge handle adapter screw, aligning adapter, gauge disc and gauge block assembly are securely mounted between the front and rear bearing. Recheck tool handle torque prior to gauging to ensure that the bearings are properly seated. This can affect final shim selection when improperly assembled. Clean the bearing cups and differential pedestal surfaces thoroughly. Apply only light oil film on the bearing assemblies prior to gauging.
- Gauge block should then be rotated several half turns to ensure rollers are properly seated in the bearings cups. Rotational torque on the gauge assembly should be 20 inch lbs., with new bearings. Final position should be approximately 45° in line with gauge tube high point. This area should be utilized for pinion shim selection. Selection of pinion shim with gauge block not lined up with tube high point will cause improper shim selection and may result in axle noise.

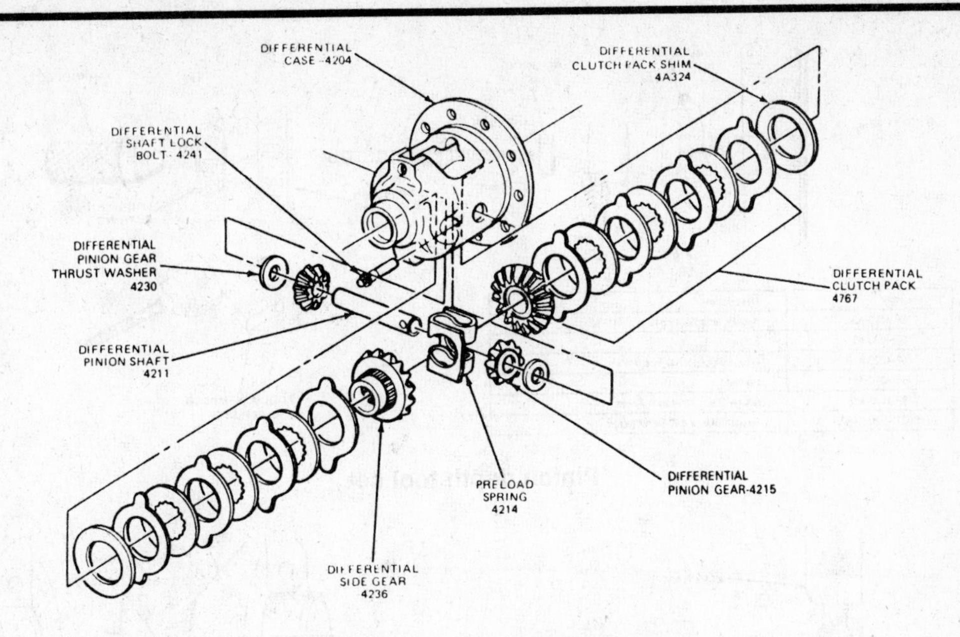

Differential assembly components

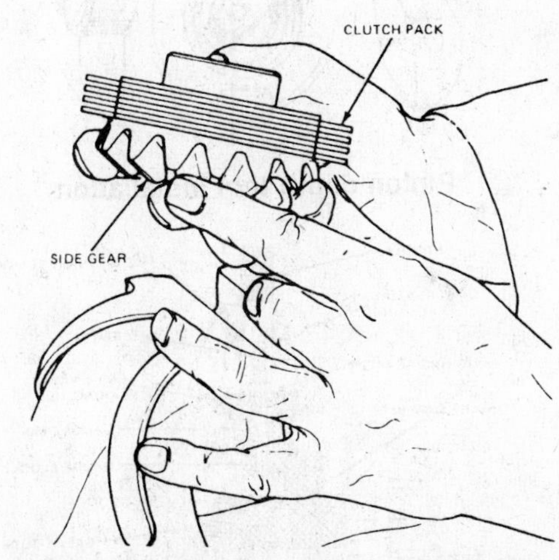

Clutch pack on side gear – installation

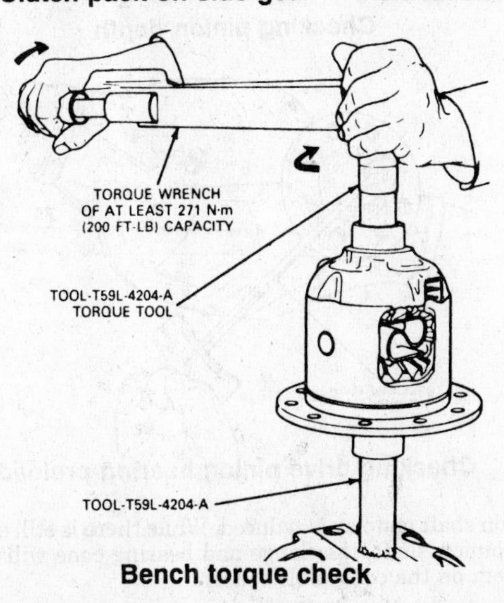

Bench torque check

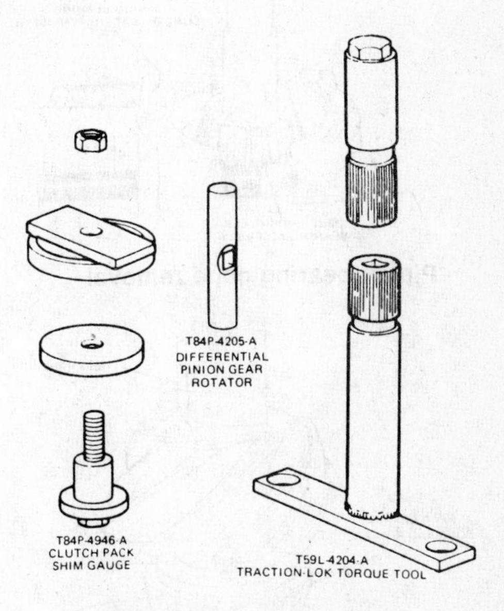

Special service tools

- Utilize pinion shims as the gauge for shim selection. This will minimize errors in attempting to stack feeler gauge stock together or simple addition errors in calculating correct shim thickness. Shims must be flat. Do not use dirty, bent, nicked or mutilated shims as a gauge.
- It is important to utilize a light drag on the shim for the correct selection. Do not attempt to force the shim between the gauge block and the gauge tube. This will minimize selection of a shim thicker than required which results in a deep tooth contact in final assembly for integral axles.
- If the pinion has a plus (+) marking, subtract this amount from the feeler gauge measurement. If the pinion has a minus (−) marking, add this amount to the feeler gauge measurement.

INSTALLATION

1. Install the pinion front bearing and slinger.

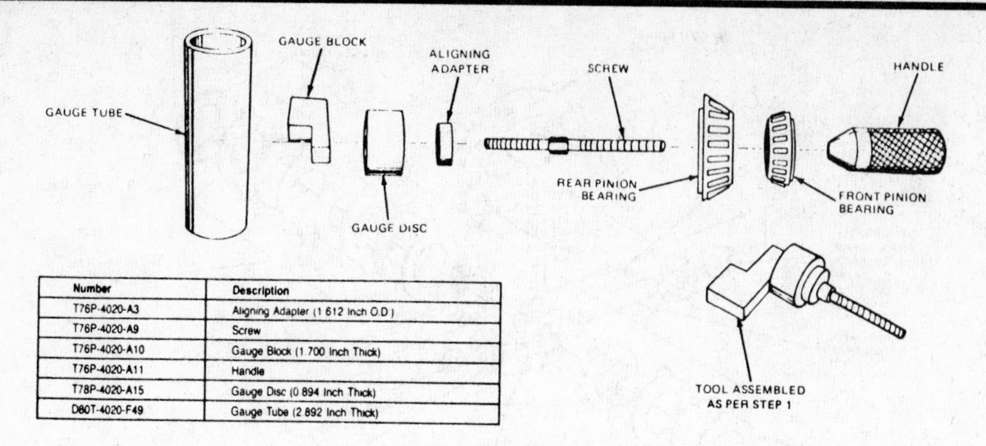

Number	Description
T76P-4020-A3	Aligning Adapter (1 612 Inch O.D)
T76P-4020-A9	Screw
T76P-4020-A10	Gauge Block (1.700 Inch Thick)
T76P-4020-A11	Handle
T78P-4020-A15	Gauge Disc (0.894 Inch Thick)
D80T-4020-F49	Gauge Tube (2.892 Inch Thick)

Pinion depth tool set

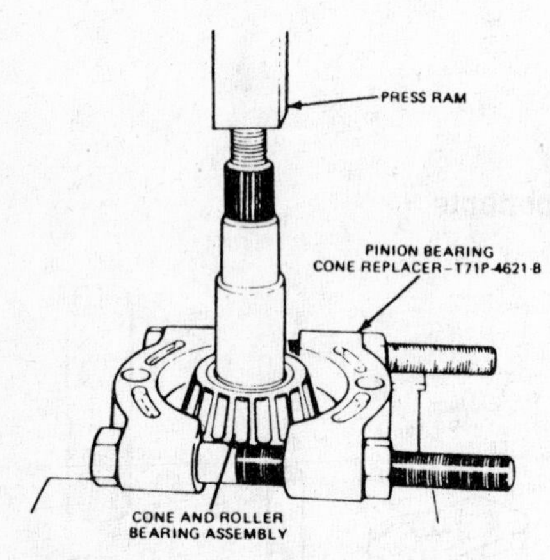

Pinion bearing cone removal

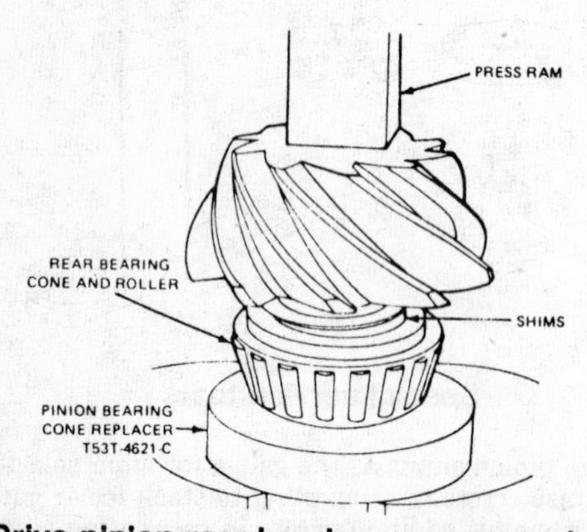

Drive pinion rear bearing cone assembly

2. Apply Multi-Purpose Long-Life Lubricant, C1AZ–19590–B or equivalent, between the lips of the pinion seal and install the pinion seal.

3. Install the companion flange into the seal and hold it firmly against the pinion front bearing cone. From the rear of the carrier casting, insert the pinion shaft, with a new spacer, into the flange.

4. Install a new pinion nut. Hold the flange with the companion flange holding tool, T78P–4851–A or equivalent. Tighten the pinion nut. As the nut is tighten, the pinion shaft is pulled into the front bearing cone and into the flange.

a. As the pinion shaft is pulled into the front bearing cone, pinion shaft endplay is reduced. While there is still endplay in the pinion shaft, the flange and bearing cone will be felt to bottom on the collapsible spacer.

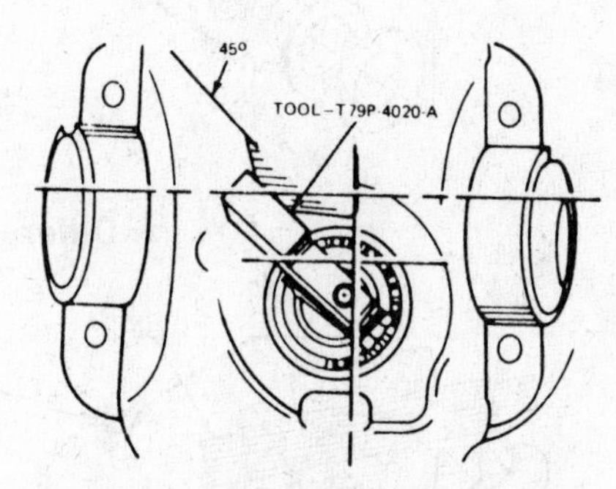

Pinion depth tool installation

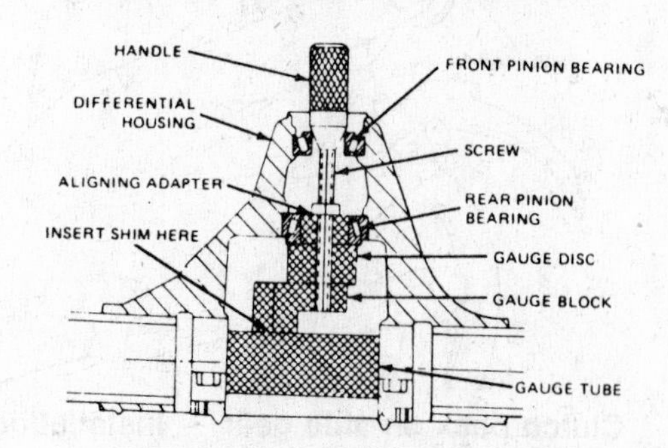

Checking pinion depth

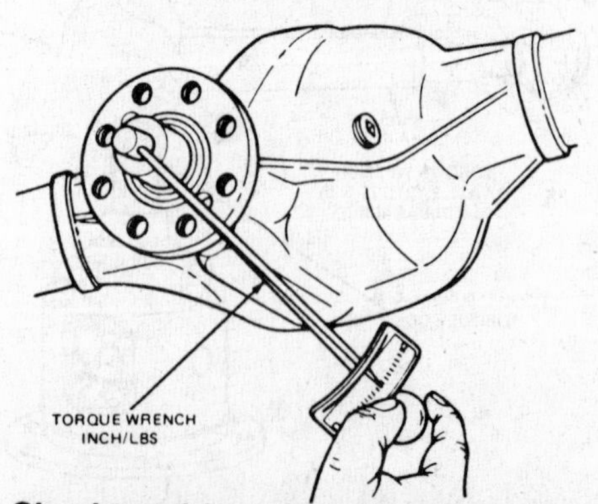

Checking drive pinion bearing preload

b. From this point, a much greater torque must be applied to turn the pinion nut, since the spacer must be collapsed. Very slowly, tighten the nut, but check the pinion shaft endplay often to see that the pinion bearing preload does not exceed the limits.

c. If the pinion nut is tightened to the point that the pinion bearing preload exceeds the limits, the pinion shaft must be removed and a new collapsible spacer installed.

NOTE: Do not decrease the preload by loosing the pinion nut. This will remove the compression between the pinion front and rear bearing cones and the collapsible spacer. It may also permit the front bearing cone to turn on the pinion shaft.

5. As soon as there is a preload on the bearings, turn the pinion shaft in both directions several times to set the bearing rollers.
6. Adjust the bearing preload to 8–14 inch lbs. (for original bearings) or 16–29 inch lbs. (for new bearings). Measure the preload with an inch-pound torque wrench.

FRONT DRIVE AXLE

Identification

The Dana Model 28 and Dana Model 35 front drive axles are used on the Ranger/Bronco II/Explorer. A code of F07A will appear on the axle identification tag if the vehicle is equipped with the Dana 35 axle and a code of E87A if equipped with the Dana 28 axle. The axles may be identified by a part number stamped on the left carrier arm between the fill plug and axle end or by using the Safety Certification Label attached to the door latch edge of the driver's side door.

The Dana 28 and Dana 35 are of the integral carrier housing hypoid-gear type, in which the centerline of the drive pinion is mounted above the centerline of the ring gear. The differential carrier is made of aluminum and mounted directly to the left axle arm assembly. The right hand axle shaft stub shaft is retained in the carrier by a C-clip in the differential case. The cover on the front of the carrier housing is integral with the left hand axle arm assembly.

Spindle, Right and Left Shaft and Joint Assembly

REMOVAL AND INSTALLATION

1. Disconnect the negative battery cable.
2. Raise and support the vehicle safely. Remove the wheel and tire assembly.
3. Remove the disc brake calipers and support the caliper on the vehicle's frame rail.
4. Remove the hub locks, wheel bearings and lock nuts.
5. Remove the hub and rotor. Remove the outer wheel bearing cone.
6. Remove the grease seal from the rotor with the seal remover tool, 1175–AC or equivalent, and a slide hammer. Discard the seal.
7. Remove the inner wheel bearing.
8. Remove the inner and outer bearing cups from the rotor with bearing cup puller tool, D78P–1225–B or equivalent.
9. Remove the nuts retaining the spindle to the steering knuckle. Tap the spindle with a plastic or rawhide hammer to jar the spindle from the knuckle. Remove the splash shield.
10. From the right side of the vehicle, remove the shaft and joint assembly by pulling the assembly out of the carrier.
11. From the right side of the carrier, remove and discard the keystone clamp from the shaft and joint assembly and the stub shaft. Slide the rubber boot onto the stub shaft and pull the shaft and joint assembly from the splines of the stub shaft.
12. Place the spindle in a vise on the second step of the spindle. Wrap a shop towel around the spindle or use a brass-jawed vise to protect the spindle.
13. Remove the oil seal and needle bearing from the spindle with a slide hammer and seal remover tool, 1175–AC or equivalent.
14. If required, remove the seal from the shaft, by driving off with a hammer.

To install:

15. Clean all dirt and grease from the spindle bearing bore. Bearing bore must be free from nicks and burrs.
16. Place the bearing in the bore with the manufacturer's identification facing outward. Drive the bearing into the bore using spindle bearing replacer tool, T83T–3123–A and drive handle T80T–4000–W or equivalent.
17. Install the grease seal in the bearing bore with the lip side of the seal facing towards the tool. Drive the seal in the bore using spindle bearing replacer tool, T83T–3123–A and drive handle T80T–4000–W or equivalent. Coat the bearing seal lip with Multi-Purpose Long Life Lubricant C1AZ–19590–B or equivalent.
18. If remove, install a new shaft seal. Place the shaft in a press and install the seal with seal installer tool, T83T–3132–A.
19. From the right side of the carrier, install the rubber boot and new keystone clamps on the stub shaft slip yoke. Since the splines on the shaft are phased, there is only 1 way to assemble the right shaft and joint assembly into the slip yoke. Align the missing spline in the slip yoke barrel with the gapless male spline on the shaft and joint assembly. Slide the right shaft and joint assembly into the slip yoke making sure the splines are

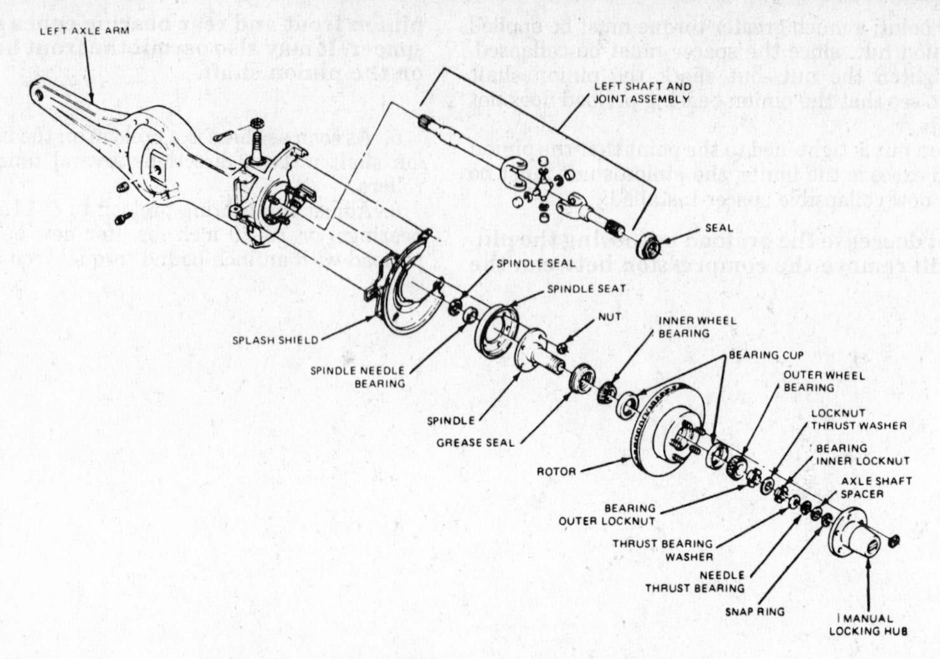

Spindle and left shaft and joint installation

fully engaged. Slide the boot over the assembly and crimp the keystone clamp.

20. From the left side of the carrier, slide the shaft and joint assembly through the knuckle and engage the splines on the shaft in the carrier.
21. Install the splash shield and spindle onto the steering knuckle. Install and tighten the spindle nuts to 35–45 ft. lbs. (47–61Nm).
22. Drive the bearing cups into the rotor, using bearing cup replacer T73T–4222–B and drive handle T80T–4000–W or equivalent.
23. Pack the inner and outer wheel bearings and the lip of the oil seal with Multi-Purpose Long Life Lubricant C1AZ–19590–B or equivalent.
24. Place the inner wheel bearing in the inner cup. Drive the grease seal into the bore with hub seal replacer tool, T83T–1175–B and drive handle T80T–4000–W or equivalent. Coat the bearing seal lip with Multi-Purpose Long Life Lubricant C1AZ–19590–B or equivalent.
25. Install the rotor on the spindle. Install the outer wheel bearing into the cup.

NOTE: Make certain the grease seal lip totally encircles the spindle.

26. Install the wheel bearing, locknut, thrust bearing, snapring and locking hubs.
27. Install the disc brake calipers. Install the wheel and tire assembly.
28. Lower the vehicle.
29. Reconnect the negative battery cable.

Right Hand Slip Yoke and Stub Shaft Assembly, Carrier, Carrier Oil Seal and Bearing

REMOVAL AND INSTALLATION

1. Disconnect the negative battery cable.
2. Raise and support the vehicle safely.

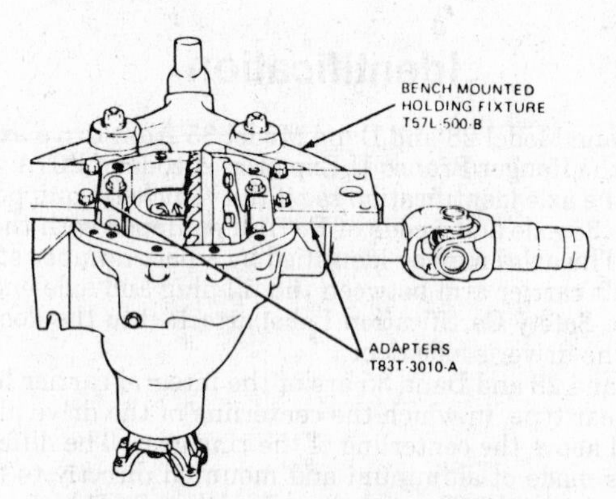

Installing carrier in holding fixture

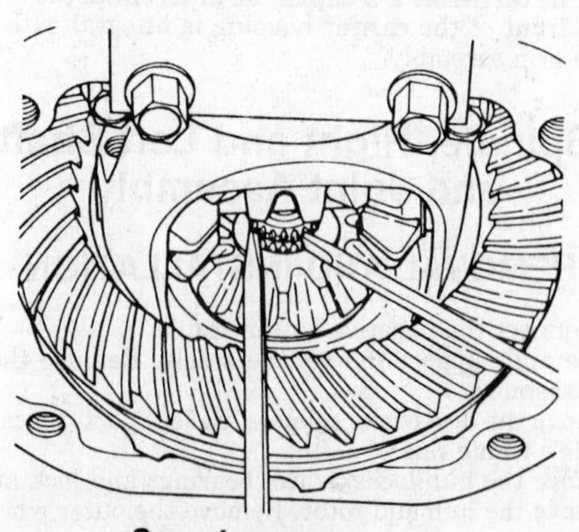

Snapring removal

3. Remove the nuts and U-bolts connecting the driveshaft to the yoke. Disconnect the driveshaft from the yoke. Wire the driveshaft aside.
4. Remove the spindles, the left and right shaft and U-joint assemblies.

5. Support the carrier with a suitable jack and remove the bolts retaining the carrier to the support arm. Separate the carrier from the support arm and drain the lubricant from the carrier. Remove the carrier from the vehicle.
6. Place the carrier in a holding fixture, T57L–500–B and adapter T83T–3010–A or equivalent.
7. Rotate the slip yoke and shaft assembly so the open side of the snapring is exposed. Remove the snapring from the shaft.
8. Remove the slip yoke and shaft assembly from the carrier.
9. Remove the oil seal and caged needle bearings at the same time, using slide hammer (T50T–100–A) and collet (D80L–100–A) or equivalent. Discard the seal and needle bearing.

To install:

10. Check that the bearing bore is free from nicks and burrs. Install a new caged needle bearing on the needle bearing replacer tool, T83T–1244–A or equivalent, with the manufacturer's name and part number facing outward towards the tool. Drive the needle bearing until it is seated in the bore.
11. Coat the seal with Multi-Purpose Long Life Lubricant C1AZ–19590–B or equivalent. Drive the seal into the carrier using needle bearing replacer tool, T83T–1244–A or equivalent.
12. Install the slip yoke and shaft assembly into the carrier so the grooves in the shaft is visible in the differential case.
13. Install the snapring in the groove in the shaft. Force the snapring into position. Do not tap the center of the snapring. This may damage the snapring.
14. Clean all traces of gasket sealant from the surfaces of the carrier and support arm and make sure the surfaces are free from dirt and oil.
15. Apply a bead of RTV sealant to the surface of the carrier. Position the carrier on a suitable jack and install it into position on the support arm, using guide pins to align. Install the retaining bolts and hand tighten. Then, tighten the bolts in a clockwise or counterclockwise pattern to 40–50 ft. lbs. (54–68Nm).
16. Install the shear bolt retaining the carrier to the axle arm and tighten to 75–95 ft. lbs. (102–129Nm).
17. Install both spindles, the left and right shaft and joint assemblies.
18. Connect the driveshaft to the yoke. Install the nuts and U-bolts and tighten to 8–15 ft. lbs. (11–20Nm).
19. Lower the vehicle.
20. Reconnect the negative battery cable.

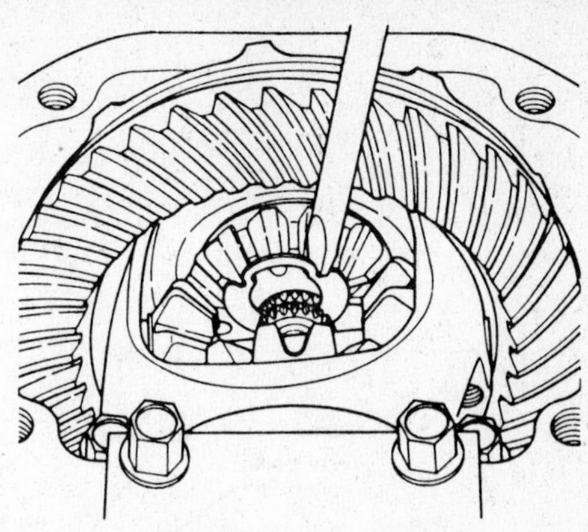
Snapring installation

Axle Housing

REMOVAL AND INSTALLATION

Ranger/Bronco II (Dana 28)

1. Disconnect the negative battery cable.
2. Raise and support the vehicle safely. Remove the wheel and tire assembly.

NOTE: Before removing the driveshaft from the front axle yoke, mark the yoke and driveshaft so that they can be reassembled in the same relative position, thus eliminating driveshaft imbalance.

3. Disconnect the driveshaft from the front axle yoke.
4. Remove the disc brake calipers and support the caliper on the vehicle's frame rail.
5. Remove the cotter pin and nut retaining the steering linkage to the spindle. Disconnect the linkage from the spindle.

NOTE: The axle arm assembly must be supported on the jack throughout spring removal and installation and must not be permitted to hang by the brake hose. If the length of the brake hose is not sufficient to provide adequate clearance for the removal and installation of the spring, the caliper must be removed.

6. Remove the bolt and nut and disconnect the shock absorber from the radius arm bracket.
7. Remove the stud and bolts that connect the radius arm bracket and radius arm to the axle arm. Remove the bracket and radius arm.
8. Remove the pivot bolt securing the right handle axle arm assembly to the crossmember. Remove the keystone clamps securing the axle shaft boot from the axle shaft slip yoke and axle shaft and slide the rubber boot over. Disconnect the right driveshaft from the slip yoke assembly. Lower the jack and remove the right axle arm assembly.
9. Position another jack under the differential housing. Remove the bolt that connects the left axle arm to the crosssmember. Lower the jacks and remove the left axle arm assembly.

To install:

10. Position the under the left axle arm assembly. Raise the axle arm until the arm is in position in the left pivot bracket. Install the nut and bolt and tighten to 120–150 ft. lbs. (163–203Nm).

NOTE: Do not remove the jack from under the differential housing at this time.

11. Place new keystone clamps for the axle shaft boot on the axle shaft assembly. Position the right axle arm on a jack and raise the right axle arm so the right driveshaft slides onto the slip yoke stub shaft and the axle arm is in position in the right pivot bracket. Install the nut and bolt and tighten to 120–150 ft. lbs. (163–203Nm).

NOTE: Do not remove the jack from the right axle arm at this time.

12. Position the radius arm and front bracket on the axle arms. Install a new stud and nut on the top of the axle and radius arm assembly and tighten to 160–220 ft. lbs. (217–298Nm). Install the bolts in the front of the bracket and tighten to 27–37 ft. lbs. (37–50Nm).
13. Install the seat, spacer retainer and coil spring on the stud and nut. Raise the jack to compress the coil spring. Install the nut and tighten to 70–100 ft. lbs. (95–135Nm).
14. Connect the shock absorber to the axle arm assembly. Install the nut and tighten to 42–72 ft. lbs. (57–97Nm).
15. Connect the tie rod ball joint to the spindle. Install the nut and tighten to 50–75 ft. lbs. (68–101Nm).
16. Lower the jacks from the axle arms.
17. Install the disc brake calipers. Install the wheel and tire assembly. Install the lug nuts and tighten to 85–115 ft. lbs. (115–155Nm).
18. Connect the front output shaft to the front axle yoke. Install the U-bolts and tighten to 8–15 ft. lbs. (11–20Nm).
19. Remove the jacks and lower the vehicle.
20. Reconnect the negative battery cable.

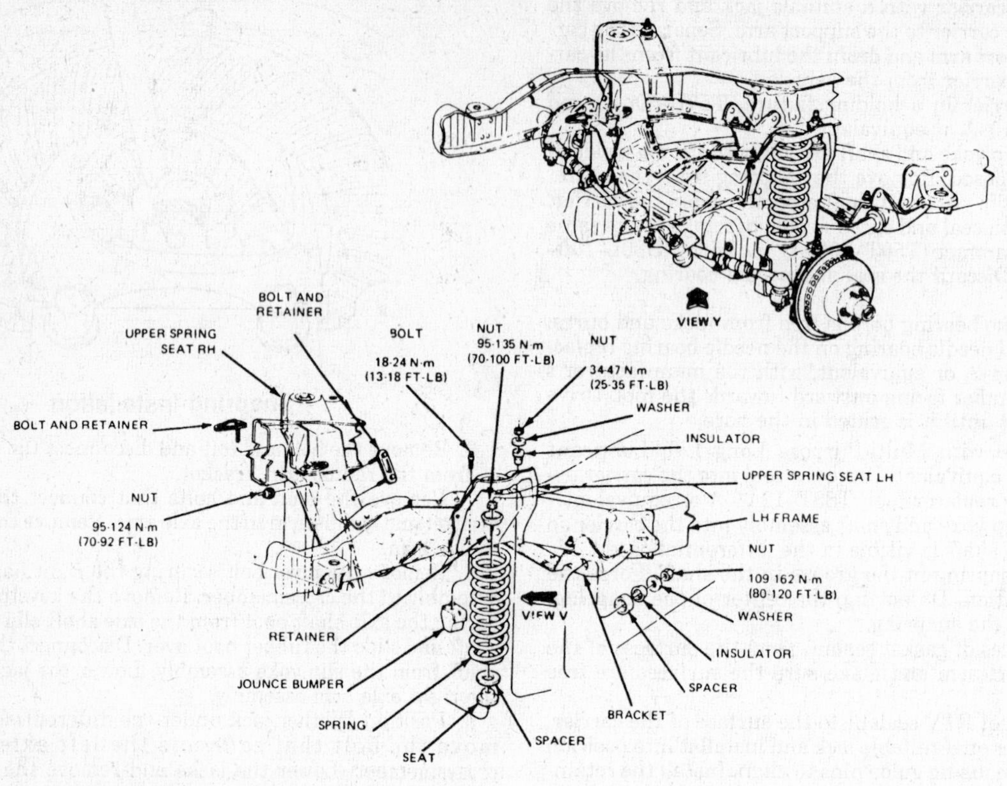

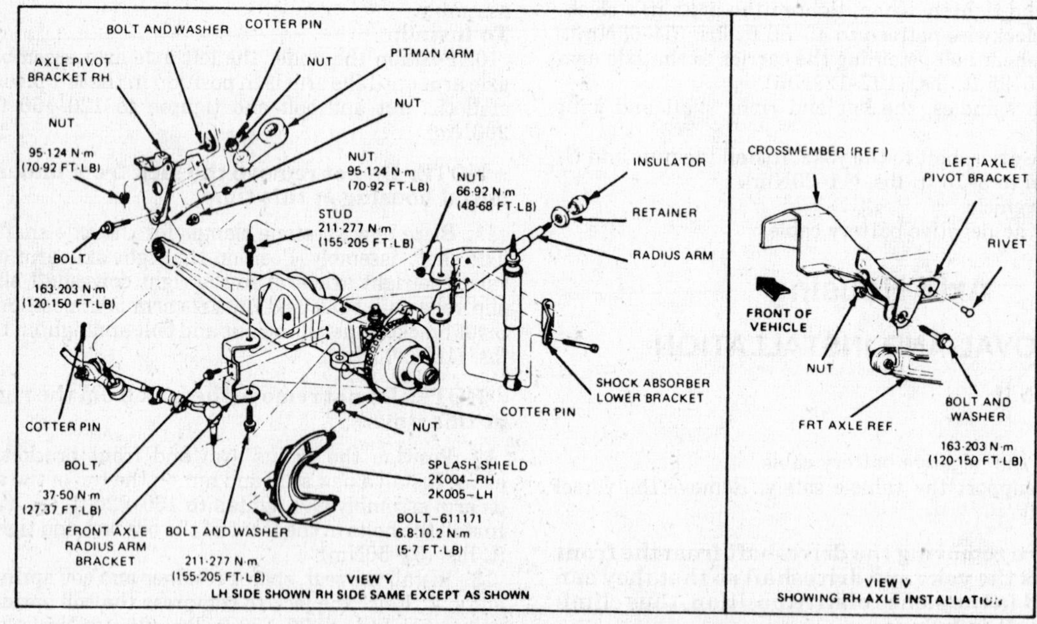

Front drive axle installation – Ranger/Bronco II

Explorer (Dana 35)

1. Disconnect the negative battery cable.
2. Raise and support the vehicle safely. Remove the wheel and tire assembly.
3. Remove the spindle, shaft and joint assembly.

NOTE: Before removing the driveshaft from the front axle yoke, mark the yoke and driveshaft so that they can be reassembled in the same relative position, thus eliminating driveshaft imbalance.

4. Disconnect the driveshaft from the front axle yoke.
5. Remove the cotter pin and nut retaining the steering linkage to the spindle. Disconnect the linkage from the spindle.
6. Remove the left stabilizer bar link lower bolt. Remove the link from the radius arm bracket.
7. Position a jack under the left axle arm assembly ad slightly compress the coil spring.
8. Remove the shock absorber lower nut and disconnect the shock absorber from the radius arm bracket.
9. Remove the nut which retains the lower portion of the

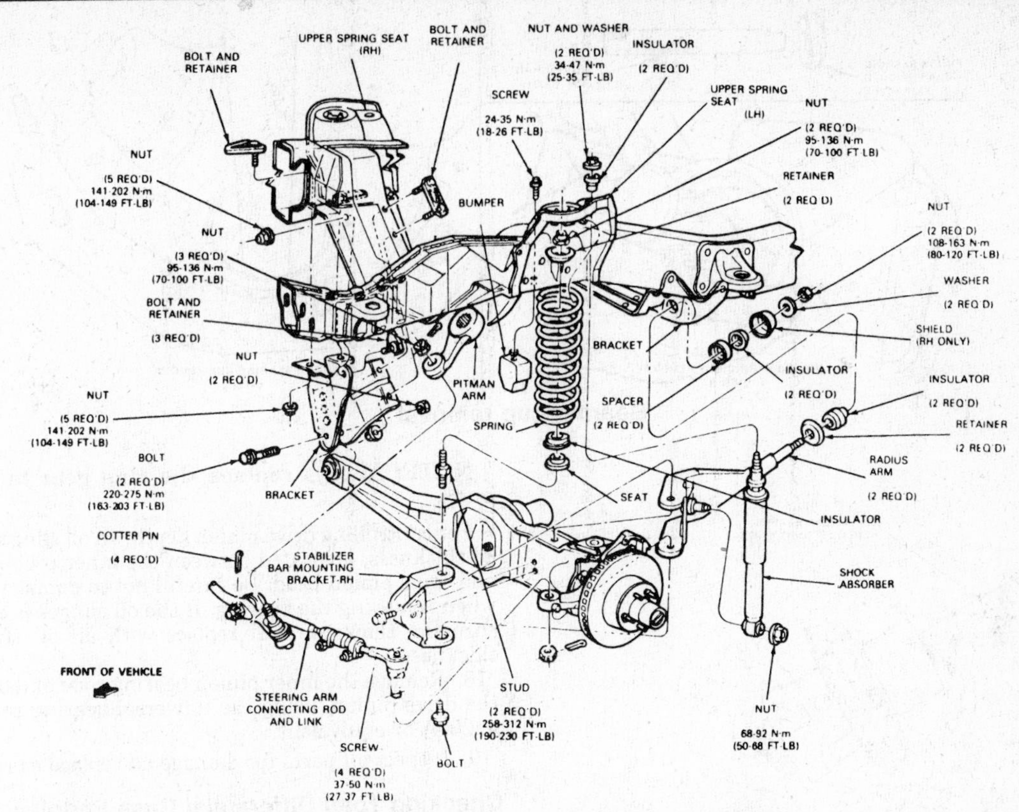

Front drive axle installation – Explorer

spring to the axle arm. Slowly lower the jack and remove the coil spring, spacer, seat and stud.

NOTE: The axle arm assembly must be supported on the jack throughout spring removal and installation and must not be permitted to hang by the brake hose. If the length of the brake hose is not sufficient to provide adequate clearance for the removal and installation of the spring, the caliper must be removed.

10. Remove the stud and bolts that connect the radius arm bracket and radius arm to the axle arm. Remove the bracket and radius arm.
11. Position another jack under the differential housing. Remove the bolt that connects the left axle arm to the axle pivot bracket. Lower the jacks and remove the left axle arm assembly.

To install:

12. Position the under the left axle arm assembly. Raise the axle arm until the arm is in position in the left pivot bracket. Install the nut and bolt and tighten to 120–150 ft. lbs. (163–203Nm).

NOTE: Do not remove the jack from the axle arm at this time.

13. Position the radius arm and front bracket on the axle arms. Install a new stud and nut on the top of the axle and radius arm assembly and tighten to 190–230 ft. lbs. (258–311Nm). Install the bolts in the front of the bracket and tighten to 27–37 ft. lbs. (37–50Nm).
14. Install the seat, spacer retainer and coil spring on the stud and nut. Raise the jack to compress the coil spring. Install the nut and tighten to 70–100 ft. lbs. (95–135Nm).
15. Connect the shock absorber to the radius arm. Install the nut and tighten to 42–72 ft. lbs. (57–97Nm).
16. Connect the tie rod ball joint to the knuckle. Install the nut and tighten to 50–75 ft. lbs. (68–101Nm). Install the stabilizer bar mounting bracket and tighten to 203–240 ft. lbs. (275–325Nm).
17. Connect the front driveshaft shaft to the front axle yoke. Install the U-bolts and tighten the nuts to 8–15 ft. lbs. (11–20Nm).

NOTE: Reassemble the yoke and driveshaft to the marks made during disassembly.

18. Install the spindle, shaft and joint assemblies.
19. Remove the jacks and lower the vehicle.
20. Reconnect the negative battery cable.

Carrier Assembly

DISASSEMBLY

1. Remove the left hand axle arm assembly.
2. Remove the carrier from the axle arm.
3. Place the carrier in a holding fixture, T57L–500–B and adapter T83T–3010–A or equivalent. Clean the gasket surface.
4. Remove the bearing caps.

NOTE: Notice the matched numbers or letters stamped on the cap and carrier in the horizontal and vertical position. These numbers or letters must be matched upon assembly.

5. Remove the differential case from the carrier. It necessary, carefully pry the differential case from the carrier using a pry bar.
6. Remove and tag the bearing cups to indicate which side of the carrier they were removed.
7. Turn the nose of the carrier up. Hold the end yoke with

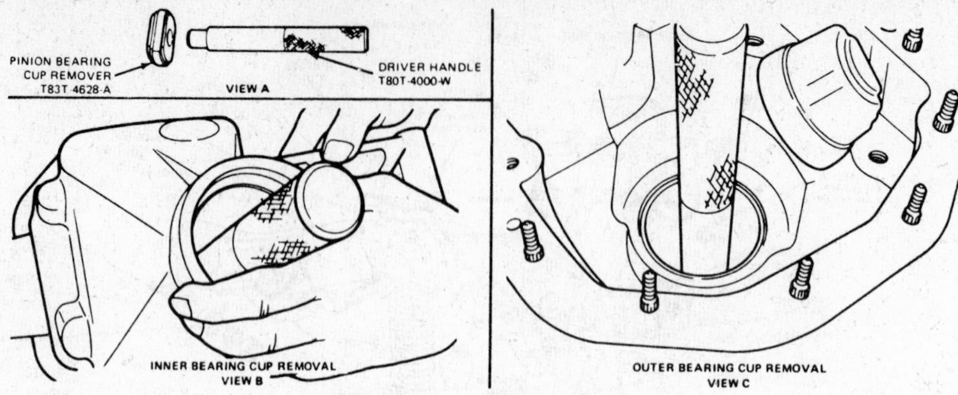

Bearing cup removal

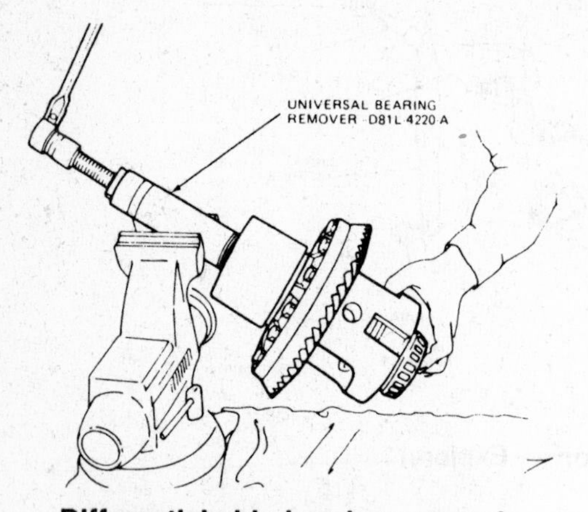

Differential side bearing removal

holding tool, T78P–4851–A or equivalent and remove the pinion nut and washer from the pinion shaft.

8. Remove the yoke, using removal tool T65L–4851–B or equivalent.

9. Lightly tap on the pinion to remove it. Catch the pinion to prevent damage to the pinion.

10. Remove the drive pinion oil seal, using tool T50T–100–A or equivalent. Discard the seal.

11. Remove the outer pinion bearing cone, oil slinger and collapsible spacer.

12. Remove the inner pinion bearing cup, using removal tool, T83T–4628–A or equivalent, and drive handle.

NOTE: An oil baffle is located between the inner bearing cup and carrier bore. Be careful not to damage the oil baffle when removing the inner bearing cup. If the baffle is damage, measure the thickness and replace with a baffle of equal thickness.

13. Remove the oil baffle from the inner bearing cup bore.

14. Turn the nose of the carrier down and remove the outer pinion bearing cup.

15. Remove the differential case bearings and shims from the case. Place step plate D80L–4220–A or equivalent under the bearing to protect the bearing. Turn the case over and remove the other bearing.

16. Wire the shims, bearing cup and cone together and identify from which side of the differential case they were removed.

17. Place a few shop towels over the vise to prevent damage to the ring gear when it is removed. Place the differential case in a vice. Remove the ring gear retaining bolts and lightly tap the ring gear from the case.

NOTE: Always replace the ring gear bolts with new ones.

For controlling drive pinion depth, an oil slinger with a selected thickness, is located between the inner pinion bearing cone and drive pinion head. Be careful not to damage the oil slinger when removing the bearing. If the oil slinger is damaged, measure the thickness and replace with an oil slinger of equal thickness.

18. Remove the inner pinion bearing cone and oil slinger from the drive pinion, using the universal bearing remover, D81L–4220–A or equivalent.

19. Inspect all parts for damage ad replace as required.

Checking Total Differential Case Endplay

1. Attach the ring gear to the differential case using new bolts. Tighten the bolts alternately and evenly to 50–60 ft. lbs. (68–81Nm).

2. Clean the trunnions on the differential and install the master differential bearings, T83T–4222–A or equivalent, onto the differential case. Remove all burrs and nicks from the hubs so master bearings rotate freely.

3. Place the differential case into the carrier, without the pinion. The differential case should move freely in the carrier. Mount a dial indicator against the differential case flange and locate the tip of the indicator on the flat surface of 1 ring gear bolt. Force the differential case toward the dial indicator as far as possible and zero the dial indicator with force still applied.

NOTE: The dial indicator should have a minimum of 0.200 in. (5mm) travel.

4. Force the differential case away from the dial indicator as far as it will go. Repeat this procedure until the same reading is obtained. Record the dial indicator reading. This reading indicates the amount of shims needed behind the differential side bearings to take up total clearance between the differential bearing and case.

NOTE: This reading will be used under pinion and ring gear backlash.

5. Remove the differential case from the carrier. Do not remove the master differential bearings at this time.

Checking Pinion Depth

Pinion depth is controlled by the selective oil slinger, located between the drive pinion head and the inner bearing cone and an oil baffle located between the inner bearing cup and carrier bore.

In the Pinion Depth Adjustment, the size of the selected thickness oil slinger and the oil baffle controls the pinion position. Increasing the oil slinger thickness moves the pinion to-

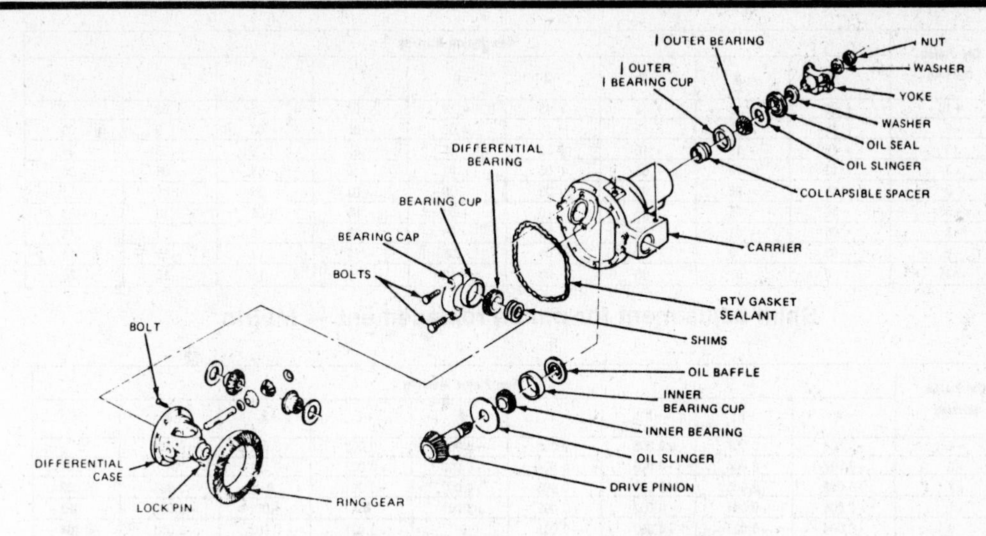

Carrier assembly — exploded view

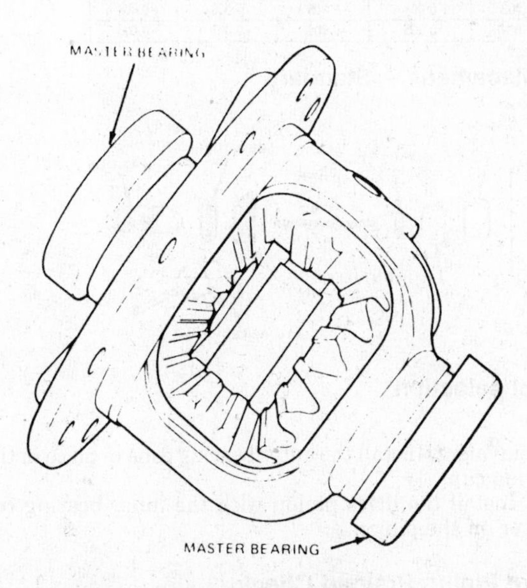

Master bearing installation

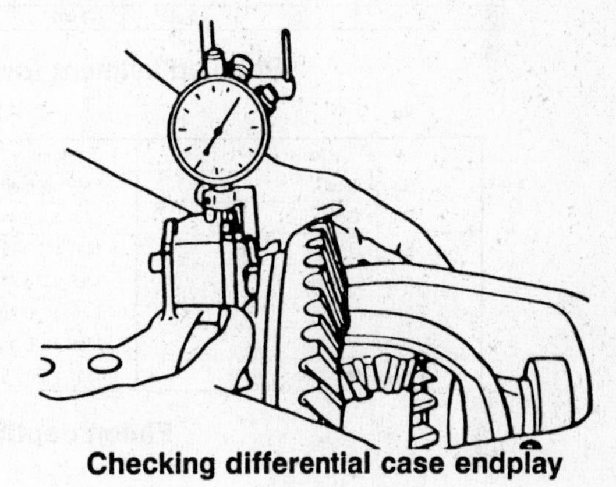

Checking differential case endplay

ward the ring gear. Decreasing the oil slinger moves the pinion away from the ring gear.

Ring gears and pinions are supplied in matched sets, with standard or metric markings. The bottom of each drive pinion is marked with a plus (+), a minus (−) or a zero (0) figure. These figures indicate the position for each gear set. The position is determined by the thickness of the baffle between the inner pinion bearing cup and the carrier bearing bore and the thickness of the selective oil slinger between the pinion head and the inner pinion bearing. Any pinion depth change is made by changing the thickness of the selective oil slinger.

1. Install the oil baffle i the inner bearing cup bore in the carrier.
2. Install the inner and outer pinion cups with forcing screw T75T–1176–A and pinion bearing cup replacers, T71P–4616–A or equivalent.
3. Place a new rear pinion bearing over the proper aligning adapter and insert into the pinion bearing retainer assembly. Place the front pinion bearing into the bearing cup in the carrier and assemble the handle onto the screw and hand tighten. Note the ³⁄₈ in. square drive in the handle to be used for obtaining the proper pinion bearing preload.

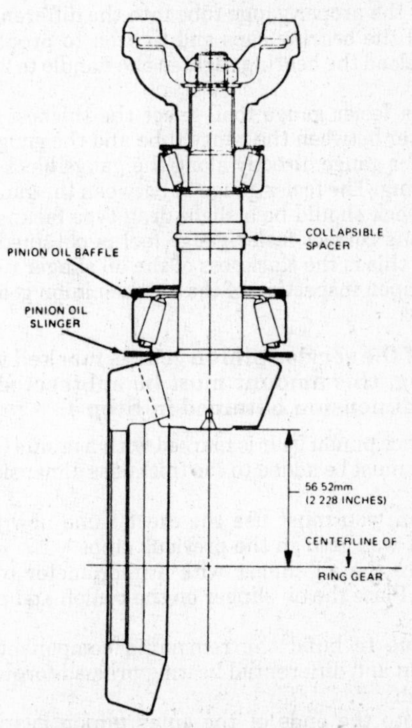

Ring gear and tooth contact

Old Pinion Marking	New Pinion Marking								
	−10	−8	−5	−3	0	+3	+5	+8	+10
+10	+.20	+.18	+.15	+.13	+.10	+.08	+.05	+.03	0
+8	+.18	+.15	+.13	+.10	+.08	+.05	+.03	0	−.03
+5	+.15	+.13	+.10	+.08	+.05	+.03	0	−.03	−.05
+3	+.13	+.10	+.08	+.05	+.03	0	−.03	−.05	−.08
0	+.10	+.08	+.05	+.03	0	−.03	−.05	−.08	−.10
−3	+.08	+.05	+.03	0	−.03	−.05	−.08	−.10	−.13
−5	+.05	+.03	0	−.03	−.05	−.08	−.10	−.13	−.15
−8	+.03	0	−.03	−.05	−.08	−.10	−.13	−.15	−.18
−10	0	−.03	−.05	−.08	−.10	−.13	−.15	−.18	−.20

Shim adjustment for pinion replacement — Metric

Old Pinion Marking	New Pinion Marking								
	−4	−3	−2	−1	0	+1	+2	+3	+4
+4	+0.008	+0.007	+0.006	+0.005	+0.004	+0.003	+0.002	+0.001	0
+3	+0.007	+0.006	+0.005	+0.004	+0.003	+0.002	+0.001	0	−0.001
+2	+0.006	+0.005	+0.004	+0.003	+0.002	+0.001	0	−0.001	−0.002
+1	+0.005	+0.004	+0.003	+0.002	+0.001	0	−0.001	−0.002	−0.003
0	+0.004	+0.003	+0.002	+0.001	0	−0.001	−0.002	−0.003	−0.004
−1	+0.003	+0.002	+0.001	0	−0.001	−0.002	−0.003	−0.004	−0.005
−2	+0.002	+0.001	0	−0.001	−0.002	−0.003	−0.004	−0.005	−0.006
−3	+0.001	0	−0.001	−0.002	−0.003	−0.004	−0.005	−0.006	−0.007
−4	0	−0.001	−0.002	−0.003	−0.004	−0.005	−0.006	−0.007	−0.008

Shim adjustment for pinion replacement — Standard

Tool	Number
Gauge Tube	T76P-4020-A7
Gauge Block	T76P-4020-A10
Gauge Disc	T83T-4020-F57
Aligning Adapter	T76P-4020-A1
Screw	T76P-4020-A9
Handle	T76P-4020-A11
Final Check Gauge Block	T83T-4020-F58

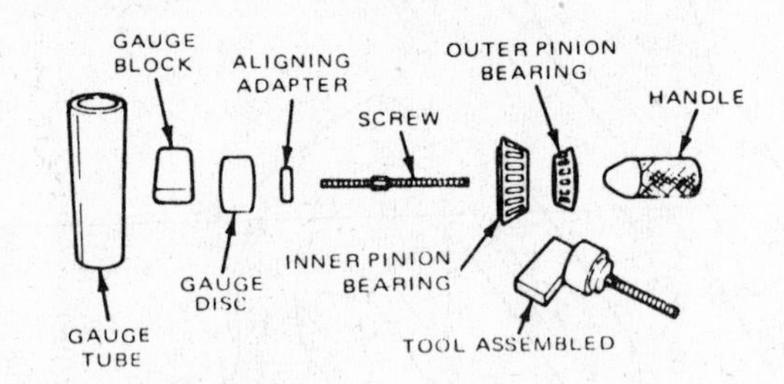

Pinion depth gauge tool selection

4. Center the proper gauge tube into the differential bearing bore. Install the bearing caps and tighten to proper specifications. To preload the bearing, tighten the handle to 20–40 ft. lbs. (2.3–4.5Nm).

5. Using a feeler gauge tool, select the thickest feeler shim that will enter between the gauge tube and the gauge block. Insert the feeler gauge directly along the gauge block to insure a correct reading. The feeler gauge fit between the gauge tube and the gauge block should be a slight drag-type feeling.

6. After the correct feeler gauge feel is obtained, check the reading and this is the thickness of the oil slinger required providing that upon inspection of the service pinion gear, there are no markings.

NOTE: If the service pinion gear is marked with a plus (+) reading, this amount must be subtracted from the thickness dimension obtained in Step 4.

If the service pinion gear is marked with a minus (−) reading, this amount must be added to the thickness dimension obtained in Step 4.

In addition, you must use the exact same new rear pinion bearing that was used in the previous steps.

7. Measure the oil slinger with a micrometer to verify the slinger size. Place the oil slinger on the pinion and press on the bearing.

8. Continue to build the remaining components with the proper pinion and differential bearing preload torques and ring gear backlash.

9. Lubricate the ends of the outer pinion bearings rollers with with Multi-Purpose Long-Life Lubricant, C1AZ–19590–B or equivalent. Install the outer bearing cone in place in the outer bearing cup.

10. Install the drive pinion with the inner bearing cone and slinger on the pinion.

Drive Pinion Preload Check and Final Depth Check

1. Install the pinion into the carrier.
2. Install the outer pinion bearing cone and oil slinger.
3. Assemble the end yoke, washer, deflector and slinger on the pinion shaft and align with special tool, T80T–4000–G and T78P–4851–A or equivalent, companion flange holder to seat the yoke. Install a new pinion nut.
4. Tighten the pinion nut until a rotating torque of 10 inch lbs. (1.1Nm) is obtained. Rotate the pinion several times to seat the bearing.
5. Install the gauge tube, T76P–4020–A7 or equivalent, in the carrier. Install the bearing caps and tighten to 35–40 ft. lbs. (48–54Nm).
6. Insert the final check gage block, T83T–4020–F58 or equivalent, on the top of the pinion button under the gauge tube. Place the thumb on the gauge block to make sure the block is level. Insert feeler gauges between the guage tube and final check gauge block until a slight drag is felt. The reading should be 0.50mm added to the drive pinion etching, which could be plus (+) or minus (−) with a tolerance ± 0.050mm.
7. With the drive pinion at the correct depth, remove the yoke, nut and washer and outer bearing cone.
8. Install a new collapsible spacer and outer bearing cone.

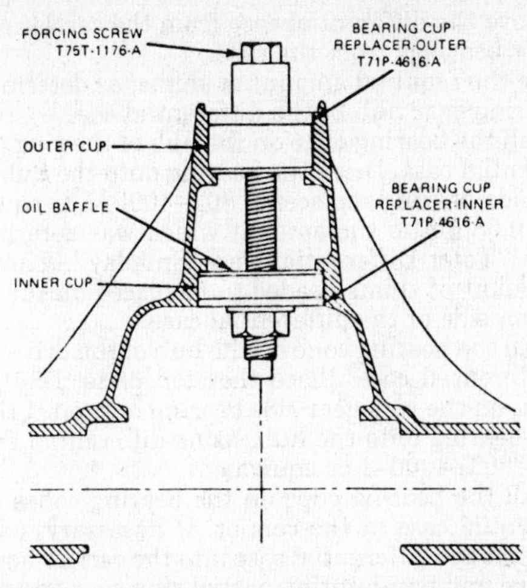

Pinion inner and outer cup installation

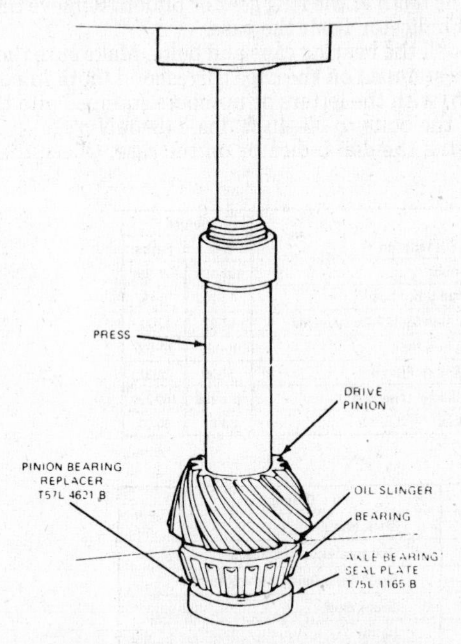

Drive pinion bearing installation

NOTE: Never reuse a collapsible spacer. Upon removal, always discard the spacer and install a new spacer when assembling.

9. Coat the oil seal with Hypoid Gear Lubricant C6AZ–19580–E or equivalent. Install the drive pinion oil seal. After installation, make sure the garter spring did not pop out. If the garter spring pops out, move and replace the seal.
10. Install the yoke, washer and nut. Tighten the nut to 175 ft. lbs. (237Nm).
11. Place an inch-pound torque wrench on the pinion nut. Check the pinion rotational torque. Rotational torque should be 15–35 inch lbs. (1.7–4.0Nm).
12. If the reading is more than 15 inch lbs. the collapsible spacer has been compressed too far. The spacer must be removed and a new collapsible spacer installed.

NOTE: Never tighten the pinion nut more that 225 ft. lbs. (305Nm) or the collapsible spacer will be compressed too far.

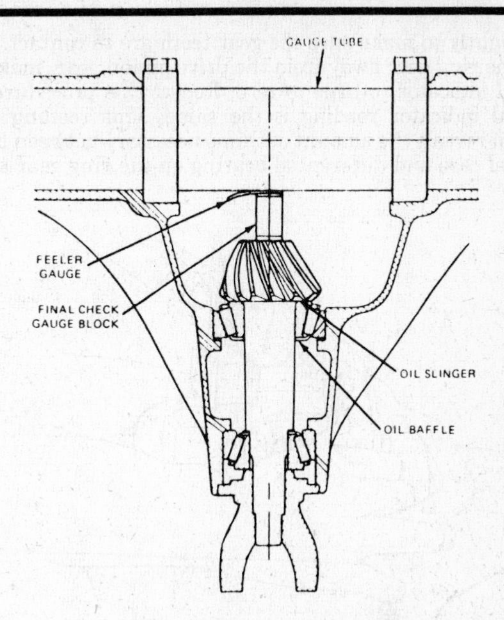

Final pinion depth check

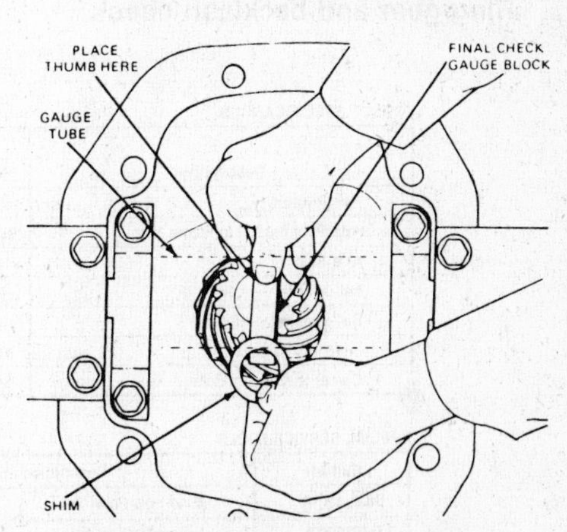

Checking final pinion depth

Checking Pinion and Ring Gear Backlash

Pinion and ring gear backlash is controlled by the preloaded shims located between the differential bearings and the differential case. This adjustment is described under the heading "Carrier Assemble".

CARRIER ASSEMBLY

Before assembling the differential case to the carrier, pinion and ring gear backlash must be adjusted.

Pinion and Ring Gear Backlash

Pinion and ring gear backlash is controlled by the preloaded shims located between the differential bearings and the differential case.

1. Install the differential case (with ring gear completely assembled to the differential case and master bearings installed) into the carrier.
2. Force the differential case away from the drive pinion gear, until it is completely seated against the bore face of the carrier. Position a dial indicator so the indicator tip rests on a differential case bolt. Zero the dial indicator.
3. Froce the ring gear against the pinion gear. Rock the ring

gear slightly to make sure the gear teeth are in contact. Then, force the ring gear away from the drive pinion gear, make sure the dial indicator returns to zero. Repeat this procedure until the dial indicator reading is the same. This reading minus 0.15mm reveals the amount of shims necessary between the differential case and differential bearing on the ring gear side.

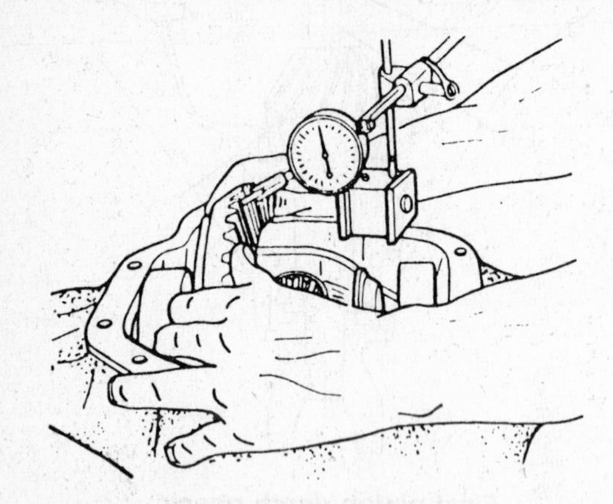

Ring gear and backlash check

4. Remove the differential case from the carrier and remove the master bearings from the case.

5. Place the required amount of shims, as determine in Step 3, on the ring gear hub of the differential case.

6. Install the bearing cone on the hub of the ring gear side of the differential case. Drive the bearing onto the hub, using differential side bearing replacer, T80T–4000–J or equivalent.

7. Add 0.08mm to the amount which was determine in the procedure "Total Differential Case Endplay". This is the required amount of shims needed to be placed on the hub of the drive pinion side of the differential case.

8. Install the bearing cone on the hub of the drive pinion side of the differential case. Place the step plate, D80L–630–5 or equivalent, on the ring gear side bearing to protect the bearing. Drive the bearing onto the hub, using differential side bearing replacer, T80T–4000–J or equivalent.

9. Install the bearing cups on the bearing cones and install the differential case in the carrier. If necessary, use a plastic mallet to seat the differential case into the carrier housing bore. With partial and non-hunting/partial ring gear and pinion sets, align the marks on the ring gear and drive pinion. Be careful not to nick the teeth of the ring gear or pinion. Remove the spreader and dial indicator from the case.

10. Install the bearing caps and bolts. Make sure the letters or numbers stamped on the caps correspond (both in position and direction) with the letters or numbers stamped into the carrier. Tighten the bolts to 35–40 ft. lbs. (48–54Nm).

11. Install the dial indicator on the case. Check the ring gear

TORQUE SPECIFICATIONS

Description	Torque N·m	Torque Ft-Lbs	Description	Torque N·m	Torque Ft-Lbs
Axle Pivot Bolt	163-203	120-150	Carrier Shear Bolt	102-129	75-95
Axle Pivot Bracket to Frame Nut	95-124	70-92	Front Driveshaft U-Bolt Nuts	11-20	8-15
Axle Stud	211-277	155-205	Lower Shock Absorber to Radius Arm Nut	57-97	42-72
Ball Joint Nut — Lower	109	80	Lower Spring Seat Nut	95-135	70-100
Ball Joint Nut — Upper	150	110	Radius Arm Bracket Front Bolt	37-50	27-37
Bearing Cap Bolts	48-54	35-40	Radius Arm Bracket Lower Bolt	217-298	160-220
Carrier to Axle Arm Bolts	54-68	40-50	Ring Gear Bolts	68-81	50-60

SPECIAL SERVICE TOOLS

Number	Description	Number	Description
D80L-100-A	Blind Hole Puller Set	T80T-4000-W	Driver Handle
T50T-100-A	Impact Slide Hammer	TOOL-4000-E	Differential Housing Spreader
T57L-500-B	Bench Mounted Holding Fixture	T76P-4020-A1	Pinion Depth Gauge Aligning Adapter
D80L-630-5	Step Plate Adapter	T76P-4020-A10	Gauge Block
T77F-1102-A	Bearing Cup Puller	T76P-4020-A11	Handle
T75L-1165-B	Axle Bearing/Seal Plate	T76P-4020-A7	Gauge Tube 2.563 O.D.
T83T-1175-B	Hub Seal Replacer	T76P-4020-A9	Screw
TOOL-1175-AC	Seal Remover	T83T-4020-F57	Pinion Depth Gauge Disc
T75T-1176-A	Threaded Drawbar	T83T-4020-F58	Final Check Gauge Block
D84T-1197-A	Four-Prong Spindle Nut Spanner Wrench	D78P-4201-B	Dial Indicator Magnetic Base
D78P-1225-B	Two Jaw Bearing Cup Puller	TOOL-4201-C	Dial Indicator with Bracketry
T83T-1244-A	Needle Bearing Replacer	D81L-4220-A	Universal Bearing Puller
T82T-3006-A1	Bushing Replacer	T73T-4222-B	Differential Bearing Cone Replacer
T83T-3006-A	Bushing Shell Flaring Tool	T83T-4222-A	Dummy Bearings
D79T-3010-BE	Forcing Screw (Part of D79T-3010-A)	T71P-4616-A	Pinion Bearing Cups Replacer
T71T-3010-R	Pinion Seal Replacer	T53T-4621-C	Pinion Bearing Cone Replacer
T80T-3010-A	Part of 4WD Ball Joint Set	T57L-4621-B	Pinion Bearing Cone Replacer
T80T-3010-A3	Part of 4WS Ball Joint Set	T83T-4628-A	Pinion Bearing Cup Driver
T83T-3010-A	Bench Mount/Housing Spreader Adapter	T74P-4635-C	C-Frame
T83T-3050-A	Ball Joint Tool Adapter	T65L-4851-B	Companion Flange Remover
T83T-3123-A	Spindle Bearing Seal Replacer	T78P-4851-A	Companion Flange Holder
T83T-3132-A	Spindle/Axle Seal Replacer	T83T-4851-A	Companion Flange Replacer
T64P-3590-F	Pitman Arm Remover	T78P-5638-A1	Forcing Screw
T80T-4000-B	Differential Spreader Adapters	T78P-5638-A4	Receiving Cup
T80T-4000-G	Companion Flange Replacer	T80T-5638-A2	Pivot Bushing Remover
T80T-4000-J	Differential Bearing Cone Replacer	T63P-9171-A	Keystone Clamp Pliers

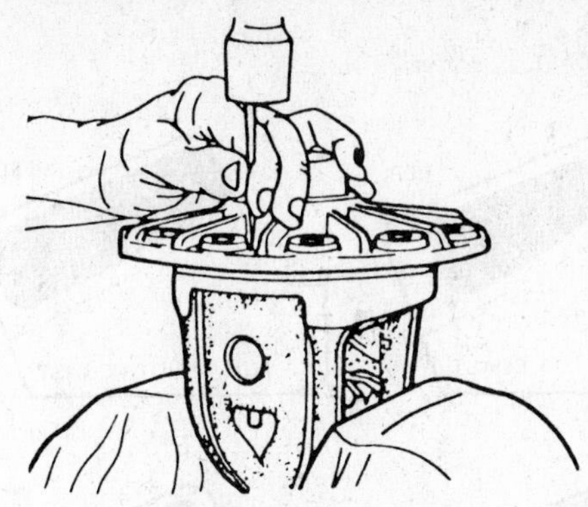
Removing lock pin

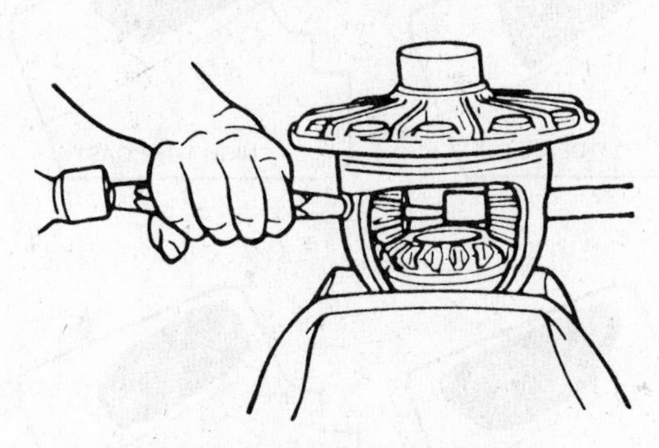
Removing pinion mate shaft

and pinion backlash at 3 equally spaced points on the ring gear. Backlash tolerance is 0.01–0.025mm and cannot vary more than ± 0.08mm among the 3 points.

NOTE: If the backlash is high, the ring gear must be moved closer to the pinion, by moving shims to the ring gear side from the opposite side. If the backlash is low, the ring gear must be moved away from the pinion by moving shims from the ring gear side to the opposite side.

11. Apply a bead of RTV sealant, ⅛–¼ in. (3mm × 6mm) high by a ¼–½ in. (6mm × 13mm) wide, on the mating surfaces of the carrier mounting face support arm.

NOTE: Allow 1 hour of curing time after the axle carrier is assembled to the axle arm before installing lubricant and operating the vehicle.

12. Mount the differential assembly to the left hand axle arm, using 2 guide pins. Install the retaining bolts. Tighten 1 bolt and then the other bolt which is directly opposite. Then, tighten all bolts in a clockwise direction to 40–50 ft. lbs. (54–68Nm).
13. Assemble the carrier shear bolt and nut. Tighten to 75–95 ft. lbs. (102–129Nm).
14. Fill the carrier with 17 ounces (0.50L) of Hypoid Gear Lubricant C6AZ–19580–E (ESW–M2C105) or equivalent.

Differential Case

DISASSEMBLY

1. Place the differential case in a vise.
2. Drive out the lock pin that retains the pinion mate shaft to the case. Remove the pinion mate shaft with a drift.
3. Rotate the pinion mate gears and side gears until the pinion mate gears turn to the windows of the case. Remove the pinion mate gears and spherical washers.
4. Lift the side gears and thrust washers from the case.
5. Insert lock pin. Peen some metal of the case over the pin to lock it in place.

INSPECTION

Inspect the case and parts for sign of wear. If wear exists on all parts, replace the entire differential case assembly. If one gear shows signs of wear, replace both pinion mate and side gears as a set.

ASSEMBLY

1. Lubricate all parts with Multi-Purpose Long-Life Lubricant, C1AZ–19590–B or equivalent.
2. Hold the side gears in place in the case with 1 hand and with install the pinion mate gears and spherical washers with the other hand. Rotate the side gears and pinion mate gears until the holes i the washers and pinion mate gears line up exactly with the holes in the case.
3. Insert the pinion mate shaft in the case. Make certain the lock pin hole in the shaft lines up with the lockpin holes in the case.

NOISE DIAGNOSIS

The Noise Is	Most Probably Produced By
• Identical under Drive or Coast	• Road surface, tires or front wheel bearings
• Different depending on road surface	• Road surface or tires
• Lower as the car speed is lowered	• Tires
• Similar with car standing or moving	• Engine or transmission
• A vibration	• Unbalanced tires, rear wheel bearing, unbalanced driveshaft or worn U-joint
• A knock or click about every 2 tire revolutions	• Rear wheel bearing
• Most pronounced on turns	• Damaged differential gears
• A steady low-pitched whirring or scraping, starting at low speeds	• Damaged or worn pinion bearing
• A chattering vibration on turns	• Wrong differential lubricant or worn clutch plates (limited slip rear axle)
• Noticed only in Drive, Coast or Float conditions	• Worn ring gear and/or pinion gear

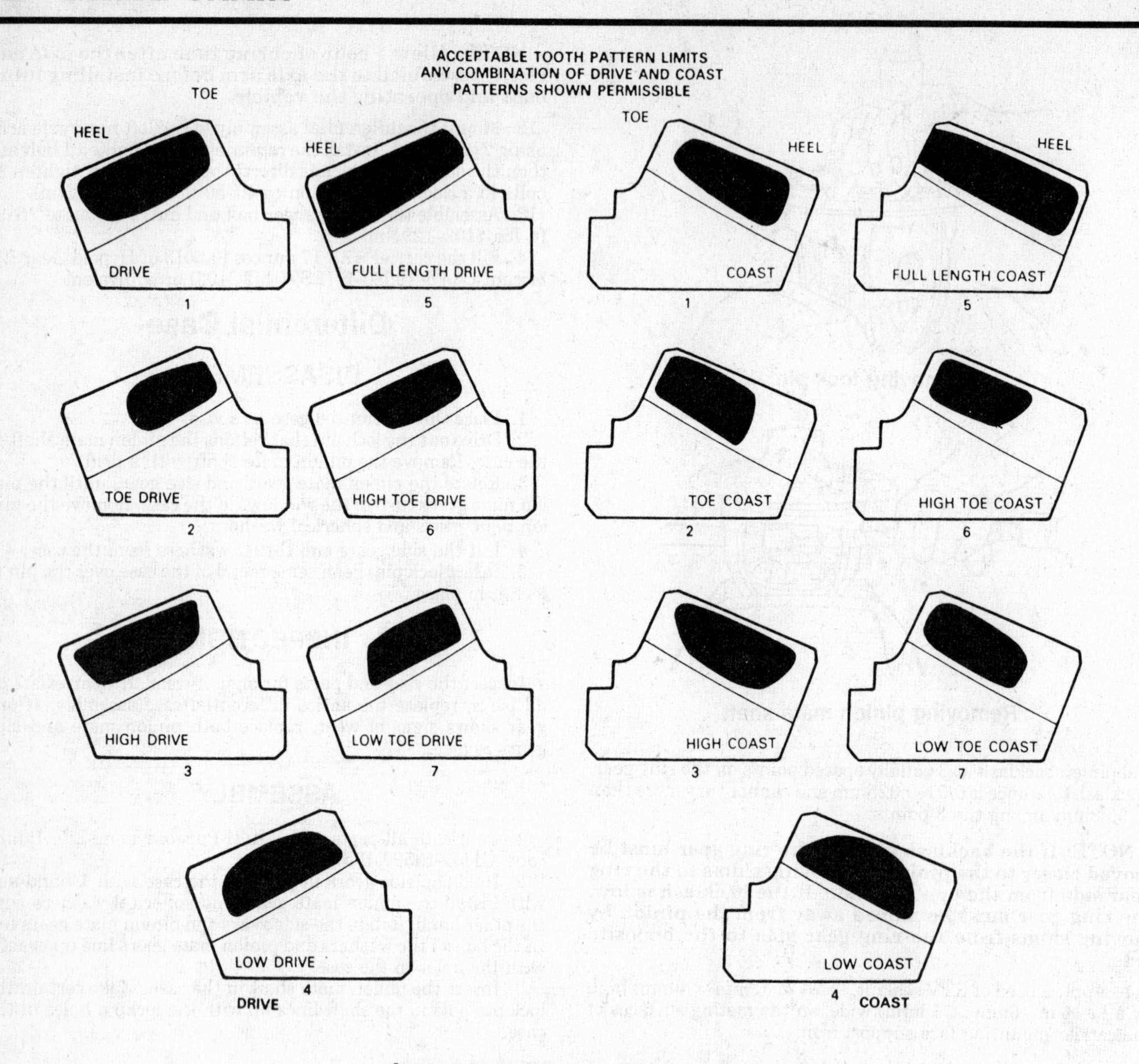

Acceptable gear tooth pattern limits

8 Suspension and Steering

QUICK REFERENCE INDEX

GENERAL INDEX

Troubleshooting Basic Steering and Suspension Problems

Problem	Cause	Solution
Hard steering (steering wheel is hard to turn)	• Low or uneven tire pressure • Loose power steering pump drive belt • Low or incorrect power steering fluid • Incorrect front end alignment • Defective power steering pump • Bent or poorly lubricated front end parts	• Inflate tires to correct pressure • Adjust belt • Add fluid as necessary • Have front end alignment checked/adjusted • Check pump • Lubricate and/or replace defective parts
Loose steering (too much play in the steering wheel)	• Loose wheel bearings • Loose or worn steering linkage • Faulty shocks • Worn ball joints	• Adjust wheel bearings • Replace worn parts • Replace shocks • Replace ball joints
Car veers or wanders (car pulls to one side with hands off the steering wheel)	• Incorrect tire pressure • Improper front end alignment • Loose wheel bearings • Loose or bent front end components • Faulty shocks	• Inflate tires to correct pressure • Have front end alignment checked/adjusted • Adjust wheel bearings • Replace worn components • Replace shocks
Wheel oscillation or vibration transmitted through steering wheel	• Improper tire pressures • Tires out of balance • Loose wheel bearings • Improper front end alignment • Worn or bent front end components	• Inflate tires to correct pressure • Have tires balanced • Adjust wheel bearings • Have front end alignment checked/adjusted • Replace worn parts
Uneven tire wear	• Incorrect tire pressure • Front end out of alignment • Tires out of balance	• Inflate tires to correct pressure • Have front end alignment checked/adjusted • Have tires balanced

WHEELS

Front Wheels

REMOVAL AND INSTALLATION

1. Set the parking brake. Block the diagonally opposite wheel.
2. On vehicles with automatic transmission position the selector lever in **PARK**.
3. On vehicles with manual transmission position the selector lever in **NEUTRAL**.
4. As necessary, remove the hubcap or wheel cover. Loosen the lug nuts, but do not remove them.
5. Raise the vehicle until the wheel and tire assembly clears the floor. Properly support the vehicle.
6. Remove the lug nuts. Remove the tire and wheel assembly from its mounting.

To install:

7. Position the wheel and tire assembly on its mounting.

CAUTION

Whenever a wheel is installed be sure to remove any corrosion, dirt or foreign material that may be present on the mounting surfaces of the hub, drum or rotor that contacts the wheel.

Installing wheels without proper metal to metal contact at the wheel mounting surfaces can cause the wheel lug nuts to loosen and could allow the wheel to come off while the vehicle is in motion!

8. Install the lug nuts. Be sure that the cone end of the lug nut faces inward.
9. With the lug nuts loosley installed, turn the wheel until one nut is at the top of the bolt circle. Tighten the lug nut until snug.

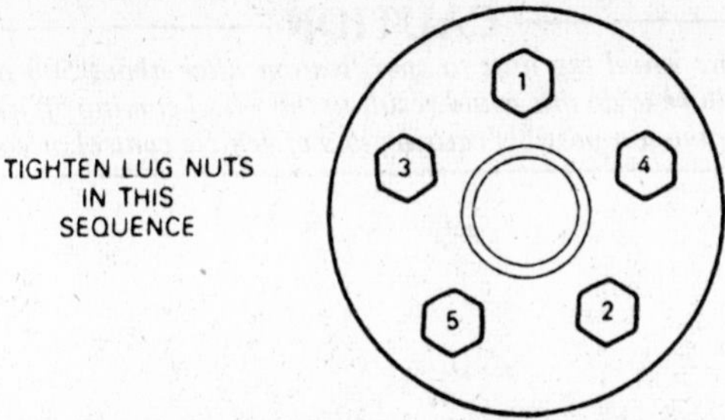

Lug nut torque sequence – 2WD and 4WD vehicles

10. In a criss cross manner tighten the remaining lug nuts until snug in order to minimize runout.
11. Lower the vehicle. Torque the lug nuts to 100 ft. lbs. in the proper sequence

CAUTION

Retighten the wheel lug nuts to specification after about 500 miles of driving. Failure to do this could result in the wheel coming off while the vehicle is in motion possibly causing loss of vehicle control or collision.

INSPECTION

Replace wheels if they are bent, cracked, leaking air or heavily rusted or if the lug nuts often become loose. Do not use bent wheels that have been straightened or do not use inner tubes in leaking wheels. Do not replace wheels with used wheels. Wheels that have been straightened or are leaking air or are used may have structural damage and could fail without warning.

Front Wheel Lug Nut Studs

REMOVAL AND INSTALLATION

1. Raise and support the vehicle safely.
2. Remove the tire and wheel assembly.
3. Remove the disc brake rotor. Be sure to properly support the brake caliper to avoid damage to the brake line hose.
4. Position the disc brake rotor in a press so that press ram pressure is not directly exerted on the disc brake rotor surface.
5. Using the proper press stock, press the lug stud from the disc brake rotor. Discard the lug stud. Remove the disc brake rotor from its mounting in the press.

To install:

6. Position a new lug stud in the disc brake rotor hole. Align the serrations of the new stud with the serration marks from the old lug stud.
7. Using a hammer tap the lug stud until the serrations on the stud are started in the hole. Be sure that the lug stud is not installed in an off centered position.
8. Reposition the disc brake rotor in the press so that the rotor is supported on the wheel mounting flange. Be sure to allow enough clearance for the stud to pass through the hole.
9. Do not apply ram pressure directly to the the rotor surface. Using the proper press stock, press the lug stud in position until the stud is flush against the inner surface of the disc brake rotor hub.
10. Install the disc brake rotor. Reposition the brake caliper. Install the tire and wheel assembly. Lower the vehicle.

CAUTION

Retighten the wheel lug nuts to specification after about 500 miles of driving. Failure to do this could result in the wheel coming off while the vehicle is in motion possibly causing loss of vehicle control or collision.

Rear Wheels

REMOVAL AND INSTALLATION

1. Set the parking brake. Block the diagonally opposite wheel.
2. On vehicles with automatic transmission position the selector lever in **PARK**.
3. On vehicles with manual transmission position the selector lever in **NEUTRAL**.
4. As necessary, remove the hubcap or wheel cover. Loosen the lug nuts, but do not remove them.
5. Raise the vehicle until the wheel and tire assembly clears the floor. Properly support the vehicle.
6. Remove the lug nuts. Remove the tire and wheel assembly from its mounting.

To install:

7. Position the wheel and tire assembly on its mounting.

CAUTION

Whenever a wheel is installed be sure to remove any corrosion, dirt or foreign material that may be present on the mounting surfaces of the hub, drum or rotor that contacts the wheel. Installing wheels without proper metal to metal contact at the wheel mounting surfaces can cause the wheel lug nuts to loosen and could allow the wheel to come off while the vehicle is in motion.

8. Install the lug nuts. Be sure that the cone end of the lug nut faces inward.
9. With the lug nuts loosley installed, turn the wheel until one nut is at the top of the bolt circle. Tighten the lug nut until snug.
10. In a criss cross manner tighten the remaining lug nuts until snug in order to minimize runout.
11. Lower the vehicle. Torque the lug nuts to 100 ft. lbs. in the proper sequence

CAUTION

Retighten the wheel lug nuts to specification after about 500 miles of driving. Failure to do this could result in the wheel coming off while the vehicle is in motion possibly causing loss of vehicle control or collision.

INSPECTION

Replace wheels if they are bent, cracked, leaking air or heavily rusted or if the lug nuts often become loose. Do not use bent wheels that have been straightened or do not use inner tubes in leaking wheels. Do not replace wheels with used wheels. Wheels that have been straightened or are leaking air or are used may have structural damage and could fail without warning.

Rear Wheel Lug Nut Studs

REMOVAL AND INSTALLATION

1. Raise and support the vehicle safely.
2. Remove the tire and wheel assembly.
3. Remove the brake drum.
4. Using wheel stud removal tool T74P–3044–A1 or equivalent, press the lug stud from its seat in the hub.

WARNING: Never use a hammer to remove the lug stud, as damage to the hub or bearing may result.

To install:

5. Insert the new lug stud in the hole in the hub. Rotate the stud slowly to assure the serrations are aligned with those made by the old lug nut stud.
6. Place 4 flat washers over the outside end of the lug nut stud and thread the wheel lug nut with the flat washer side against the washers.

7. Tighten the wheel nut until the stud head seats against the back side of the hub. Do not use air tools as the serrations may be stripped from the stud.

8. Remove the wheel lug nut and washers. Install the brake drum. Install the tire and wheel assembly. Lower the vehicle.

CAUTION

Retighten the wheel lug nuts to specification after about 500 miles of driving. Failure to do this could result in the wheel coming off while the vehicle is in motion possibly causing loss of vehicle control or collision.

2-WHEEL DRIVE FRONT SUSPENSION

Coil Springs

REMOVAL AND INSTALLATION

1. Raise the front of the vehicle and place jackstands under the frame and a jack under the axle.

WARNING: The axle must not be permitted to hang by the brake hose. If the length of the brake hoses is not sufficient to provide adequate clearance for removal and installation of the spring, the disc brake caliper must be removed from the spindle. A Strut Spring Compressor, T81P–5310–A or equivalent may be used to compress the spring sufficiently, so that the caliper does not have to be removed. After removal, the caliper must be placed on the frame or otherwise supported to prevent suspending the caliper from the caliper hose.

These precautions are absolutely necessary to prevent serious damage to the tube portion of the caliper hose assembly!

2. Disconnect the shock absorber at the lower shock stud. Remove the nut securing the lower retainer to spring seat. Remove the lower retainer.

3. Lower the axle as far as it will go without stretching the brake hose and tube assembly. The axle should now be unsupported without hanging by the brake hose. If not, then either remove the caliper or use Strut Spring Compressor Tool, T81P–5310–A or equivalent. Remove the spring.

4. If there is a lot of slack in the brake hose assembly, a pry bar can be used to lift the spring over the bolt that passes through the lower spring seat.

5. Rotate the spring so the built-in retainer on the upper spring seat is cleared.

6. Remove the spring from the vehicle.

To install:

7. If removed, install the bolt in the axle arm and install the nut all the way down. Install the spring lower seat and lower insulator. On the Bronco II and Explorer, also install the stabilizer bar mounting bracket and spring spacer.

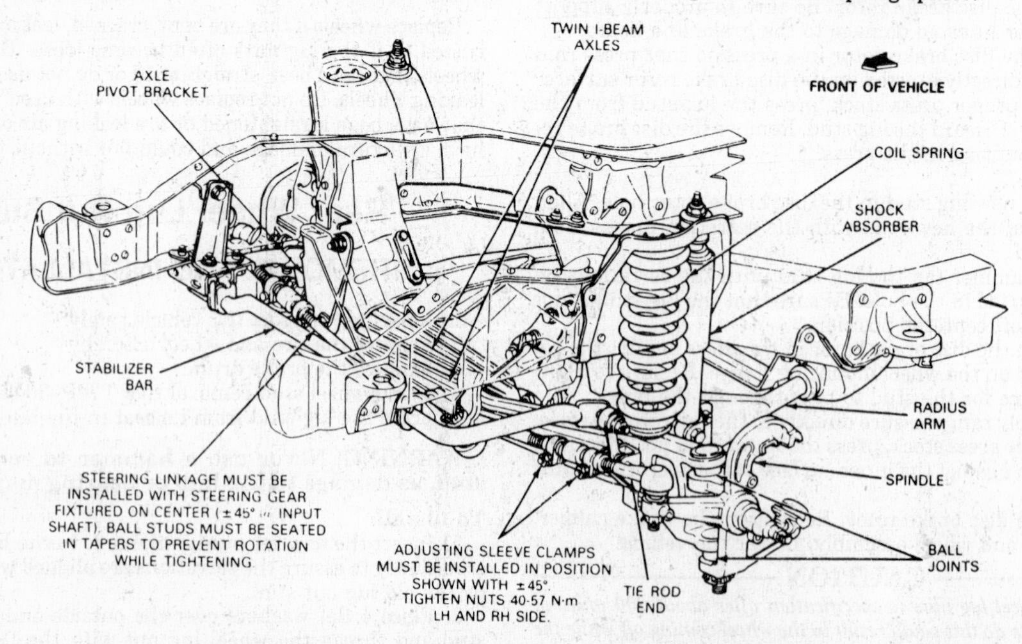

2WD Bronco II and Explorer front suspension assembly

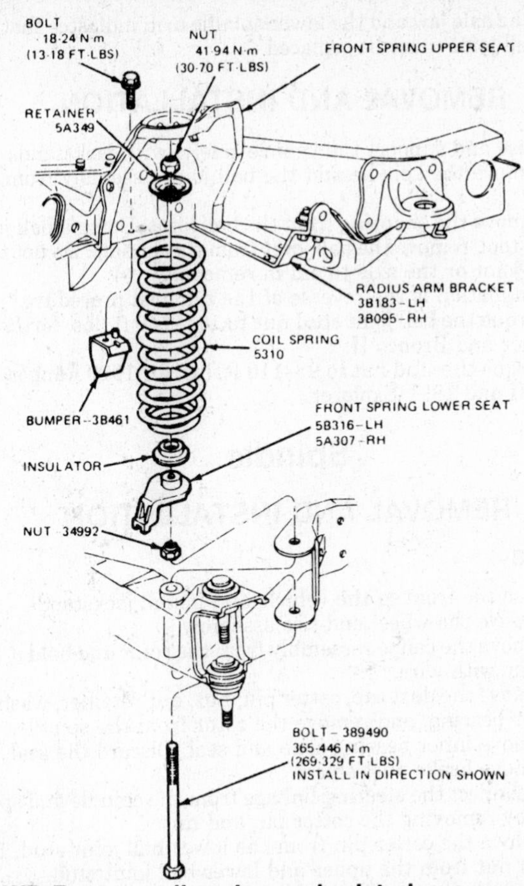

2WD Ranger coil spring and related components

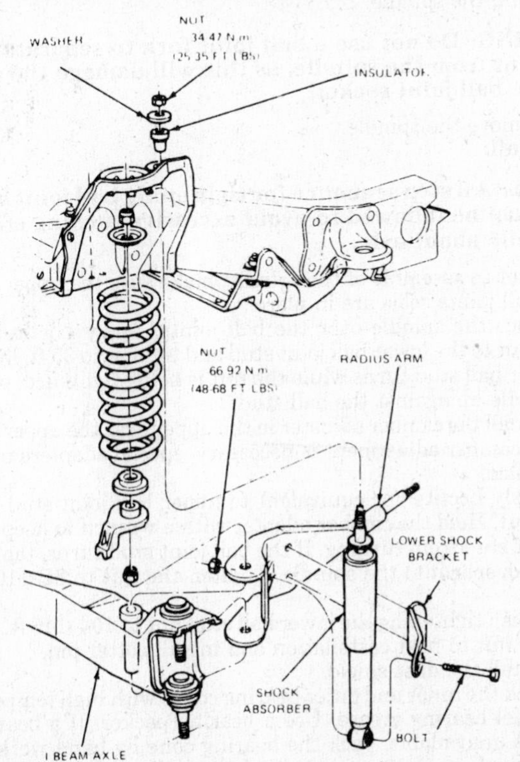

2WD Bronco II and Explorer coil spring and related components

8. With the axle in the lowest position, install the top of the spring in the upper seat. Rotate the spring into position.
9. Lift the lower end of the spring over the bolt.
10. Raise the axle slowly until the spring is seated in the lower spring upper seat. Install the lower retainer and nut.
11. Connect the shock absorber to the lower shock stud.
12. Remove the jack and jackstands and lower vehicle.

Shock Absorbers

REMOVAL AND INSTALLATION

NOTE: Low pressure gas shocks are charged with Nitrogen gas. Do not attempt to open, puncture or apply heat to them. Prior to installing a new shock absorber, hold it upright and extend it fully. Invert it and fully compress and extend it at least 3 times. This will bleed trapped air.

1. Raise the vehicle, as required to provide additional access and remove the bolt and nut attaching the shock absorber to the lower bracket on the radius arm.
2. Remove the nut, washer and insulator from the shock absorber at the frame bracket and remove the shock absorber.
3. Position the washer and insulator on the shock absorber rod and position the shock absorber to the frame bracket.
4. Position the insulator and washer on the shock absorber rod and install the attaching nut loosely.
5. Position the shock absorber to the lower bracket and install the attaching bolt and nut loosely.
6. Tighten the lower attaching bolts to 40–63 ft. lbs., and the upper attaching bolts to 25–35 ft. lbs.

TESTING

1. Visually check the shock absorbers for the presence of fluid leakage. A thin film of fluid is acceptable. Anything more than that means that the shock absorber must be replaced.
2. Disconnect the lower end of the shock absorber. Compress and extend the shock fully as fast as possible. If the action is not smooth in both directions, or there is no pressure resistance, replace the shock absorber. Shock absorbers should be replaced in pairs. In the case of relatively new shock absorbers, where one has failed, that one, alone, may be replaced.

Upper Ball Joint

INSPECTION

1. Check and adjust the front wheel bearings. Raise the vehicle and position a jackstand under the I-beam axle beneath the coil spring.
2. Have a helper grasp the lower edge of the tire and move the wheel assembly in and out.
3. While the wheel is being moved, observe the upper spindle arm and the upper part of the axle jaw.
4. A $^1/_{32}$ in. (0.8mm) or greater movement between the upper part of the axle jaw and the upper spindle arm indicates that the upper ball joint must be replaced

REMOVAL AND INSTALLATION

1. Raise and support the vehicle safely, with jackstands.
2. Remove the spindle and the ball joint assembly from the vehicle.
3. Remove the snapring from the ball joint. Using a ball joint removal tool, remove the ball joint from the spindle. Do not heat the ball joint or the axle to aid in removal.
4. Installation is the reverse of the removal procedure. Torque the ball joint stud nut to 85–110 ft. lbs. for 1983–89 Ranger and Bronco II.

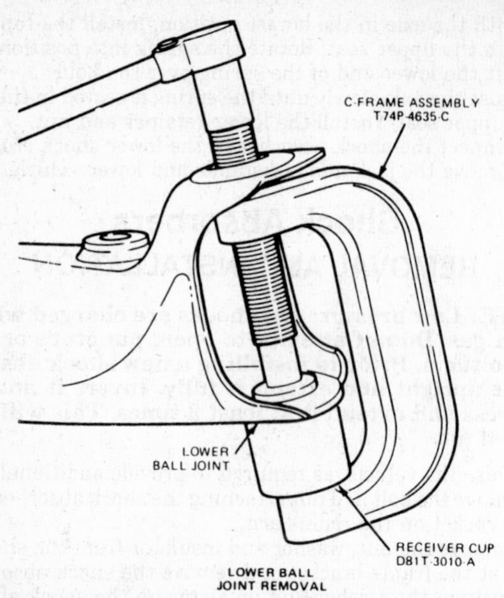

2WD Ranger, Bronco II and Explorer ball joint removal

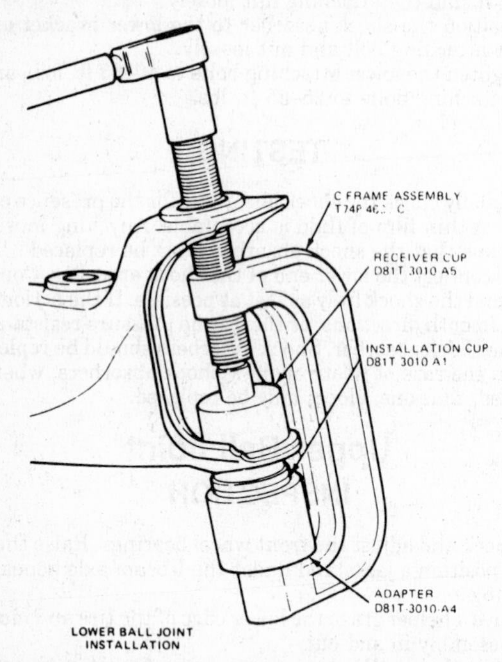

2WD Ranger, Bronco II and Explorer ball joint installation

Lower Ball Joint

INSPECTION

1. Check and adjust the front wheel bearings. Raise the vehicle and position a jackstand under the I-beam axle beneath the coil spring.
2. Have a helper grasp the upper edge of the tire and move the wheel assembly in and out.
3. While the wheel is being moved, observe the lower spindle arm and the lower part of the axle jaw.
4. A $^1/_{32}$ in. (0.8mm) or greater movement between the lower part of the axle jaw and the lower spindle arm indicates that the lower ball joint must be replaced

REMOVAL AND INSTALLATION

1. Raise and support the vehicle safely, with jackstands.
2. Remove the spindle and the ball joint assembly from the vehicle.
3. Remove the snapring from the ball joint. Using a ball joint removal tool, remove the ball joint from the spindle. Do not heat the ball joint or the axle to aid in removal.
4. Installation is the reverse of the removal procedure.
5. Torque the ball joint stud nut to 104–146 ft. lbs. for 1983–88 Ranger and Bronco II.
6. Torque the stud nut to 95–110 ft. lbs. for 1989 Ranger and Bronco II and 1991 Explorer.

Spindle

REMOVAL AND INSTALLATION

1983–88

1. Raise the front of the vehicle and install jackstands.
2. Remove the wheel and tire assembly.
3. Remove the caliper assembly from the rotor and hold it out of the way with wire.
4. Remove the dust cap, cotter pin, nut, nut retainer, washer, and outer bearing, and remove the rotor from the spindle.
5. Remove inner bearing cone and seal. Discard the seal.
6. Remove brake dust shield.
7. Disconnect the steering linkage from the spindle and spindle arm by removing the cotter pin and nut.
8. Remove the cotter pin from the lower ball joint stud. Remove the nut from the upper and lower ball joint stud.
9. Strike the lower side of the spindle to pop the ball joints loose from the spindle.

WARNING: Do not use a ball joint fork to separate the ball joint from the spindle, as this will damage the seal and the ball joint socket!

10. Remove the spindle.

To install:

NOTE: A 3 step sequence for tightening ball joint stud nuts must be followed to avoid excessive turning effort of spindle about axle.

11. Prior to assembly of the spindle, make sure the upper and lower ball joints seals are in place.
12. Place the spindle over the ball joints. Apply Loctite® or equivalent to the lower ball joint stud and tighten to 35 ft. lbs. If the lower ball stud turns while the nut is being tightened, push the spindle up against the ball stud.
13. Install the camber adjuster in the upper over the upper ball joint. If camber adjustment is necessary, special adapters must be installed.
14. Apply Loctite® or equivalent to upper ball joint stud and install nut. Hold the camber adapter with a wrench to keep the ball joint stud from turning. If the ball joint stud turns, tap the adapter deeper into the spindle. Tighten the nut to 85–110 ft. lbs.
15. Finish tightening the lower ball stud nut to 104–146 ft. lbs. Advance nut to next castellation and install cotter pin.
16. Install the dust shield.
17. Pack the inner and outer bearing cones with high temperature wheel bearing grease. Use a bearing packer. If a bearing packer is unavailable, pack the bearing cone by hand working the grease through the cage behind the rollers.
18. Install the inner bearing cone and seal. Install the hub and rotor on the spindle.

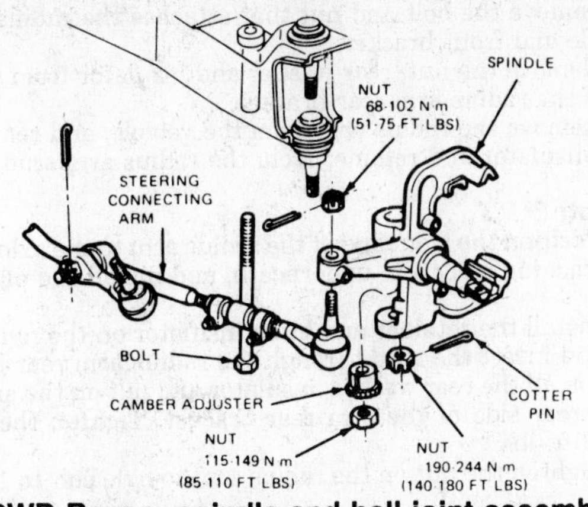

2WD Ranger spindle and ball joint assembly

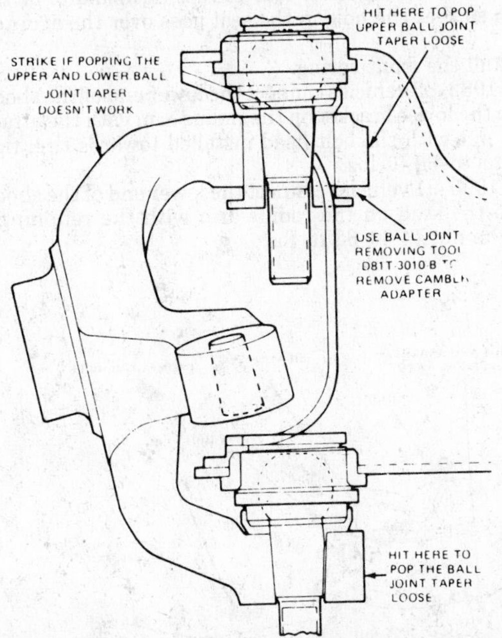

2WD Ranger, Bronco II and Explorer spindle assembly cross-section

19. Install the outer bearing cone, washer, and nut. Adjust bearing endplay and install the cotter pin and dust cap.
20. Install the caliper.
21. Connect the steering linkage to the spindle. Tighten the nut to 51–75 ft. lbs. and advance the nut as required for installation of the cotter pin.
22. Install the wheel and tire assembly. Lower the vehicle. Check, and if necessary, adjust the toe setting.

1989–91

1. Raise the front of the vehicle and install jackstands.
2. Remove the wheel and tire assembly.
3. Remove the caliper assembly from the rotor and hold it out of the way with wire.
4. Remove the dust cap, cotter pin, nut, nut retainer, washer, and outer bearing, and remove the rotor from the spindle.
5. Remove inner bearing cone and seal. Discard the seal.
6. Remove brake dust shield.
7. Disconnect the steering linkage from the spindle and spindle arm by removing the cotter pin and nut.
8. With Tie Rod removal tool 3290–D or equivalent remove the tie rod end from the spindle arm.
9. Remove the cotter pin and the castellated nut from the lower ball joint stud.
10. Remove the axle clamp bolt from the axle. Remove the camber adjuster from the upper ball joint stud and axle beam.
11. Strike the area inside the top of the axle to pop the lower ball joint loose from the axle beam.

WARNING: Do not use a ball joint fork to separate the ball joint from the spindle, as this will damage the seal and the ball joint socket!

12. Remove the spindle and the ball joint assembly from the axle.

To install:

NOTE: A 3 step sequence for tightening ball joint stud nuts must be followed to avoid excessive turning effort of spindle about axle.

13. Prior to assembly of the spindle, make sure the upper and lower ball joints seals are in place.
14. Place the spindle and the ball joint assembly into the axle.
15. Install the camber adjuster in the upper over the upper ball joint. If camber adjustment is necessary, special adapters must be installed.
16. Tighten the lower ball joint stud to 104–146 ft. lbs. for the

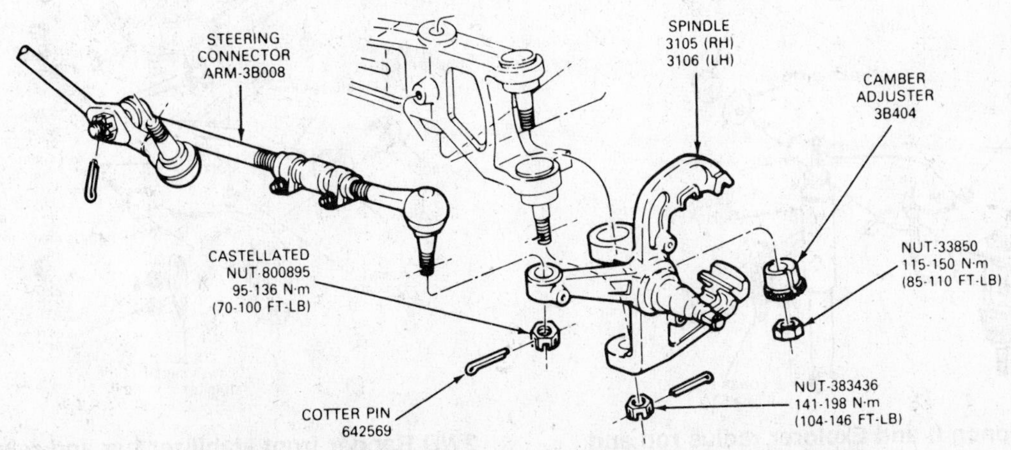

2WD Bronco II and Explorer spindle and ball joint assembly

Ranger and 95–110 ft. lbs. for the Bronco II and Explorer. Continue tightening the castellated nut until it lines up with the hole in the ball joint stud. Install the cotter pin. Install the dust shield.

17. Pack the inner and outer bearing cones with high temperature wheel bearing grease. Use a bearing packer. If a bearing packer is unavailable, pack the bearing cone by hand working the grease through the cage behind the rollers.

18. Install the inner bearing cone and seal. Install the hub and rotor on the spindle.

19. Install the outer bearing cone, washer, and nut. Adjust bearing endplay and install the cotter pin and dust cap.

20. Install the caliper.

21. Connect the steering linkage to the spindle. Tighten the nut to 52–74 ft. lbs. and advance the nut as required for installation of the cotter pin.

22. Install the wheel and tire assembly. Lower the vehicle. Check, and if necessary, adjust the toe setting.

Radius Arm

REMOVAL AND INSTALLATION

1. Raise the front of the vehicle, place jackstands under the frame. Place a jack under the axle.

WARNING: The axle must be supported on the jack throughout spring removal and installation, and must not be permitted to hang by the brake hose. If the length of the brake hose is not sufficient to provide adequate clearance for removal and installation of the spring, the disc brake caliper must be removed from the spindle. After removal, the caliper must be placed on the frame or otherwise supported to prevent suspending the the caliper from the caliper hose. These precautions are absolutely necessary to prevent serious damage to the tube portion of the caliper hose assembly.

2. Disconnect the lower end of the shock absorber from the shock lower bracket (bolt and nut).
3. Remove the front spring. Loosen the axle pivot bolt.
4. Remove the spring lower seat from the radius arm, and then remove the bolt and nut that attaches the radius arm to the axle and front bracket.
5. Remove the nut, rear washer and insulator from the rear side of the radius arm rear bracket.
6. Remove the radius arm from the vehicle, and remove the inner insulator and retainer from the radius arm stud.

To install:

7. Position the front end of the radius arm to the axle. Install the attaching bolt from underneath, and install the nut finger tight.
8. Install the retainer and inner insulator on the radius arm stud and insert the stud through the radius arm rear bracket.
9. Install the rear washer, insulator and nut on the arm stud at the rear side of the arm rear bracket. Tighten the nut to 81–120 ft. lbs.
10. Tighten the nut on the radius arm-to-axle bolt to 160–220 ft. lbs.
11. Install the spring lower seat and spring insulator on the radius arm so that the hole in the seat goes over the arm-to-axle bolt.
12. Install the front spring.
13. On 1983–89 vehicles connect the lower end of the shock absorber to the lower bracket on the radius arm with the attaching bolt and nut with the bolt head installed towards tire, tighten the nut to 48–68 ft. lbs.
14. On 1990–91 vehicles connect the lower end of the shock absorber to the stud on the radius arm with the retaining nut. Torque the nut to 40–63 ft. lbs.

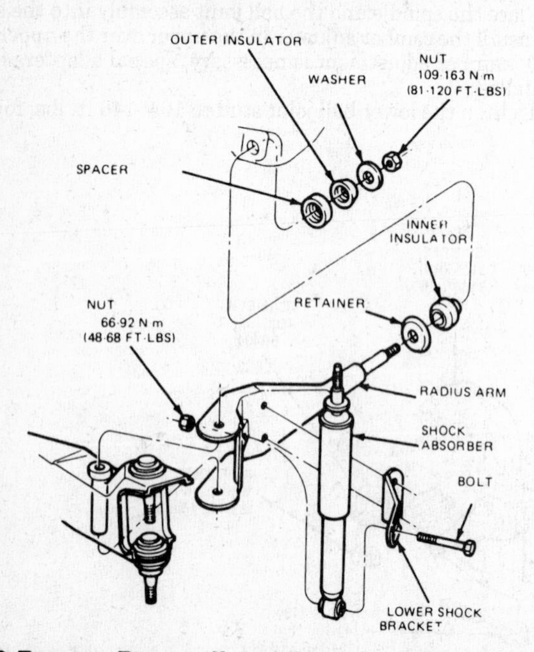

2WD Ranger, Bronco II and Explorer radius rod and related components

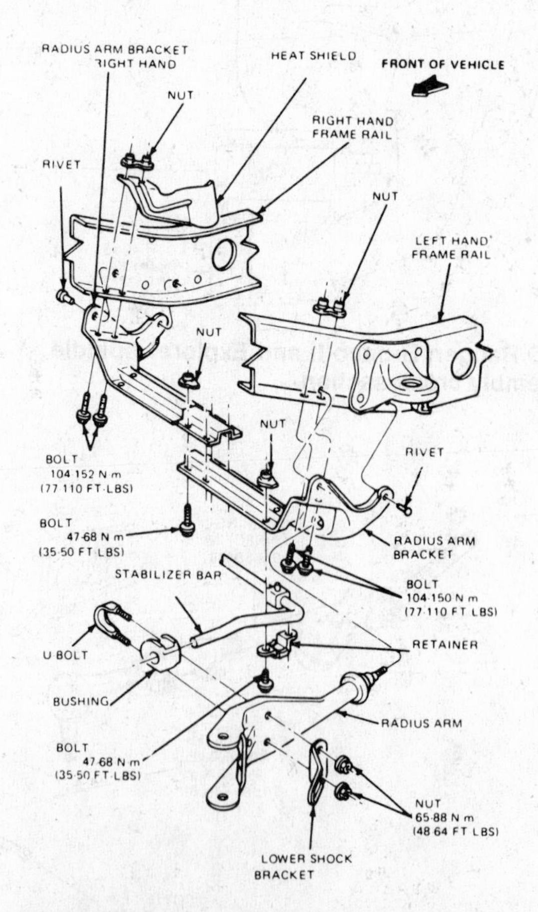

2WD Ranger front stabilizer bar and related components

Stabilizer Bar

REMOVAL AND INSTALLATION

Ranger

1983–89

1. As required, raise and support the vehicle safely.
2. Remove the nuts and U-bolts retaining the lower shock bracket/stabilizer bar bushing to radius arm.
3. Remove retainers and remove the stabilizer bar and bushing.

To install:

4. Place stabilizer bar in position on the radius arm and bracket.
5. Install retainers and U-bolts. Tighten retainer bolt to 35–50 ft. lbs. Tighten U-bolt nuts to 48–64 ft. lbs.

1990–91

1. As required, raise and support the vehicle safely.
2. Remove the nuts and bolts retaining the stabilizer bar to the end links.
3. Remove the retainers and the stabilizer bar and bushings from the vehicle.

To install:

4. Position the stabilizer bar to the axles and brackets.
5. Install the retainer and the end link bolts.
6. Torque the retainer bolts to 35–50 ft. lbs. Torque the end link nuts to 30–40 ft. lbs.

Bronco II and Explorer

1. As required, raise and support the vehicle safely.
2. Remove the nuts and washer and disconnect the stabilizer link assembly from the front I-beam axle.
3. Remove the mounting bolts and remove the stabilizer bar retainers from the stabilizer bar assembly.
4. Remove the stabilizer bar from the vehicle.

To install:

5. Place stabilizer bar in position on the frame mounting brackets.
6. Install retainers and tighten retainer bolt to 30–50 ft. lbs. If removed, install the stabilizer bar link assembly to the stabilizer bar. Install the nut and washer and tighten to 40–60 ft. lbs. on Bronco II and 30–40 ft. lbs. on Explorer.

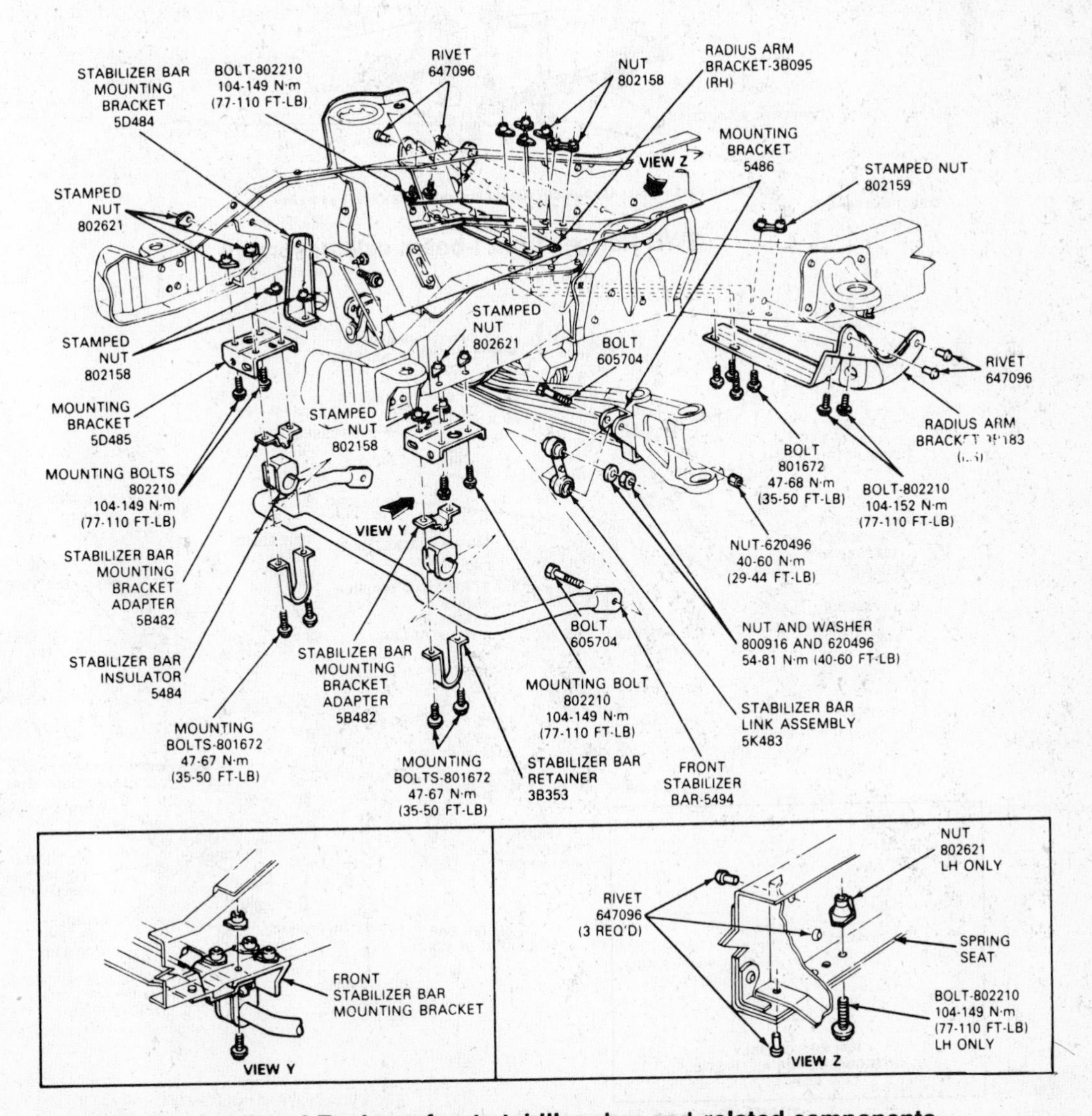

2WD Bronco II and Explorer front stabilizer bar and related components

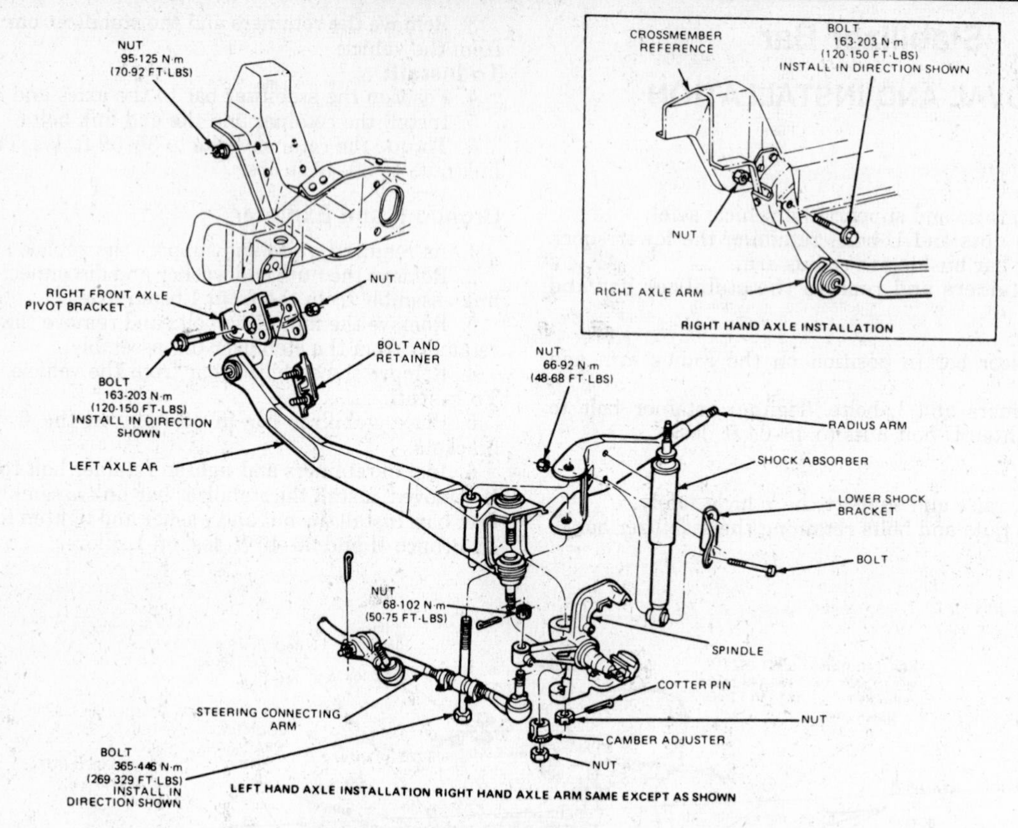

2WD Ranger front I-beam axle assembly

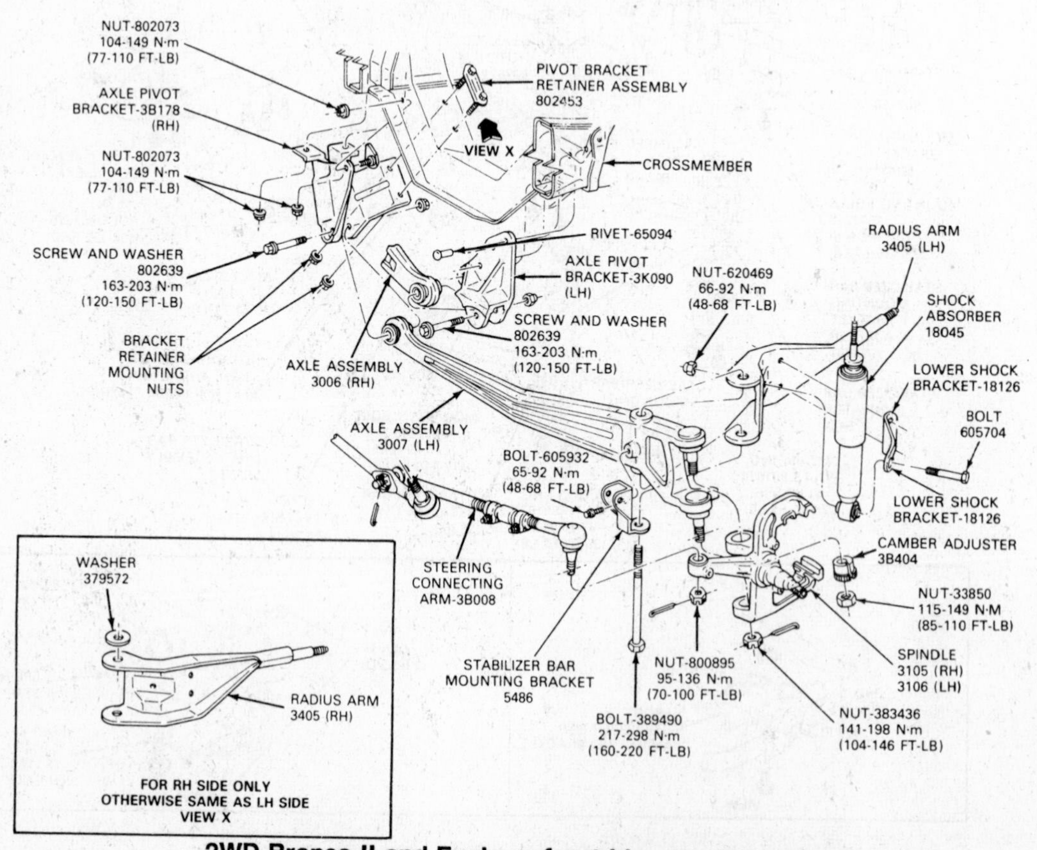

2WD Bronco II and Explorer front I-beam axle assembly

7. Position the stabilizer bar link in the I-beam mounting bracket. Install the bolt and tighten to 30–44 ft. lbs.

I-Beam Axle

REMOVAL AND INSTALLATION

1. Raise and safely support the vehicle. Remove the front wheel spindle. Remove the front spring. On 1989–91 vehicles, remove the front stabilizer bar, if equipped.
2. Remove the spring lower seat from the radius arm, and then remove the bolt and nut that attaches the stabilizer bar bracket, if equipped on 1989–91 vehicles and the radius arm to the (I-Beam) front axle.
3. If equipped, on 1983–88 vehicles disconnect the stabilizer bar from the front I-beam axle. Remove the axle-to-frame pivot bracket bolt and nut.
4. To install, position the axle to the frame pivot bracket and install the bolt and nut finger tight.
5. Position the opposite end of the axle to the radius arm, install the attaching bolt from underneath through the bracket, the radius arm,, and the axle. Install the nut and tighten to 120–150 ft. lbs.
6. Install the spring lower seat on the radius arm so that the hole in the seat indexes over the arm-to-axle bolt.
7. Install the front spring.

NOTE: Lower the vehicle on its wheels or properly support the vehicle at the front springs before tightening the axle pivot bolt and nut.

8. Tighten the axle-to-frame pivot bracket bolt to 120–150 ft. lbs.
9. Install the front wheel spindle.

Front Wheel Bearings

REPLACEMENT

1. Raise and support the vehicle safely. Remove the tire and wheel assembly from the hub and rotor.
2. Remove the caliper from its mounting and position it to the side with mechanics wire in order to prevent damage to the brake line hose.
3. Remove the grease cap from the hub. Remove the cotter pin, retainer, adjusting nut and flatwasher from the spindle.
4. Remove the outer bearing cone and roller assembly from the hub. Remove the hub and rotor from the spindle.
5. Using seal removal tool 1175–AC or equivalent remove and discard the grease seal. Remove the inner bearing cone and roller assembly from the hub.
6. Clean the inner and outer bearing assemblies in solvent. Inspect the bearings and the cones for wear and damage. Replace defective parts, as required.
7. If the cups are worn or damaged, remove them with front hub remover tool T81P–1104–C and tool T77F–1102–A or equivalent.
8. Wipe the old grease from the spindle. Check the spindle for excessive wear or damage. Replace defective parts, as required.

To install:

9. If the inner and outer cups were removed, use bearing driver handle tool T80–4000–W or equivalent and replace the cups. Be sure to seat the cups properly in the hub.
10. Use a bearing packer tool and properly repack the wheel bearings with the proper grade and type grease. If a bearing packer is not available work as much of the grease as possible between the rollers and cages. Also, grease the cone surfaces.
11. Position the inner bearing cone and roller assembly in the inner cup. A light film of grease should be included between the lips of the new grease retainer (seal).
12. Install the retainer using the proper installer tool. Be sure that the rtainer is properly seated.
13. Install the hub and rotor assembly onto the spindle. Keep the hub centered on the spindle to prevent damage to the spindle and the retainer.
14. Install the outer bearing cone and roller assembly and flatwasher on the spindle. Install the adjusting nut. Adjust the wheel bearings.
15. Install the retainer, a new cotter pin and the grease cap. Install the caliper.
16. Lower the vehicle and tighten the lug nuts to 100 ft. lbs. Before driving the vehicle pump the brake pedal several times to restore normal brake pedal travel.

CAUTION

Retighten the wheel lug nuts to specification after about 500 miles of driving. Failure to do this could result in the wheel coming off while the vehicle is in motion possibly causing loss of vehicle control or collision.

ADJUSTMENT

1. Raise and support the vehicle safely. Remove the wheel cover. Remove the grease cap from the hub.
2. Wipe the excess grease from the end of the spindle. Remove the cotter pin and retainer. Discard the cotter pin.
3. Loosen the adjusting nut 3 turns.

CAUTION

Obtain running clearance between the disc brake rotor surface and shoe linings by rocking the entire wheel assembly in and out several times in order to push the caliper and brake pads away from the rotor. An alternate method to obtain proper running clearance is to tap lightly on the caliper housing. Be sure not to tap on any other area that may damage the disc brake rotor or the brake lining surfaces. Do not pry on the phenolic caliper piston. The running clearance must be maintained throughout the adjustment procedure. If proper clearance cannot be maintained, the caliper must be removed from its mounting.

4. While rotating the wheel assembly, tighten the adjusting nut to 17–25 ft. lbs. in order to seat the bearings. Loosen the ad-

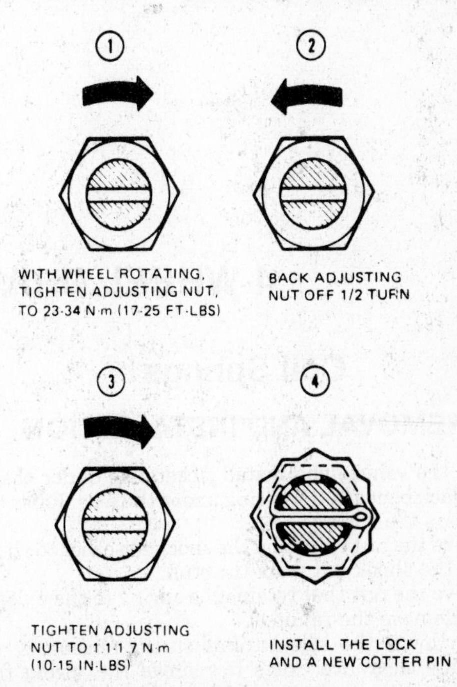

2WD Ranger, Bronco II and Explorer wheel bearing adjustment procedure

justing nut a half turn. Retighten the adjusting nut 18–20 inch lbs.

5. Place the retainer on the adjusting nut. The castellations on the retainer must be in alignment with the cotter pin holes in the spindle. Once this is accomplished install a new cotter pin and bend the ends to insure its being locked in place.

6. Check for proper wheel rotation. If correct, install the grease cap and wheel cover. If rotation is noisy or rough recheck your work and correct as required.

7. Lower the vehicle and tighten the lug nuts to 100 ft. lbs., if the wheel was removed. Before driving the vehicle pump the brake pedal several times to restore normal brake pedal travel.

CAUTION

If the wheel was removed, retighten the wheel lug nuts to specification after about 500 miles of driving. Failure to do this could result in the wheel coming off while the vehicle is in motion possibly causing loss of vehicle control or collision.

Front End Alignment

CASTER AND CAMBER

If you should start to notice abnormal tire wear patterns and handling (steering wheel is hard to return to straight ahead position after negotiating a turn on pavement), and misalignment of caster and camber are suspected, make the following checks:

1. Check the air pressure in all the tires. Make sure that the pressures agree with those specified for the tires and vehicle being checked.

2. Raise the front of the vehicle off the ground and support it safely. Grasp each front tire at the front and rear, and push the wheel inward and outward. If any free-play is noticed, adjust the wheel bearings.

NOTE: There is supposed to be a very, very small amount of free-play present where the wheel bearings are concerned. Replace the bearings if they are worn or damaged.

3. Check all steering linkage for wear or maladjustment. Adjust and/or replace all worn parts.

4. Check the steering gear mounting bolts and tighten if necessary.

5. Rotate each front wheel slowly, and observe the amount of lateral or side runout. If the wheel runout exceeds ⅛ in. (3mm), replace the wheel or install the wheel on the rear.

6. Inspect the radius arms to be sure they are not bent or damaged. Inspect the bushings at the radius arm-to-axle attachment and radius arm-to-frame attachment points for wear or looseness. Repair or replace parts as required.

Caster is the number of degrees of backward (positive) or forward (negative) tilt of the spindle or the line connecting the ball joint centers. Camber is the number of degrees the top of the wheel tilts outward (positive) or inward (negative) from a vertical plane.

Before checking caster or camber, perform the toe alignment check. Using alignment equipment known to be accurate and following the equipment manufacturer's instructions, measure and record the caster angle and the camber angle of both front wheels.

If the caster and camber measurements exceed the maximum variances, inspect for damaged front suspension components. Replace as required.

NOTE: Twin-I-Beam axles are not to be bent or twisted to correct caster or camber readings.

Both caster and camber adjustments are possible with service adjusters. These service adjusters are available in ½, 1 and 1½ degree increments. On of these adjusters is used to adjust both caster and camber.

4-WHEEL DRIVE FRONT SUSPENSION

Coil Springs

REMOVAL AND INSTALLATION

1. Raise the vehicle and install jackstands under the frame. Position a jack beneath the spring under the axle. Raise the jack and compress the spring.

2. Remove the nut retaining the shock absorber to the radius arm. Slide the shock out from the stud.

3. Remove the nut that retains the spring to the axle and radius arm. Remove the retainer.

4. Slowly lower the axle until all spring tension is released and adaquate clearance exists to remove the spring from its mounting.

5. Remove the spring by rotating the upper coil out of the tabs in the upper spring seat. Remove the spacer and the seat.

WARNING: The axle must be supported on the jack throughout spring removal and installation, and must not be permitted to hang by the brake hose. If the length of the brake hose is not sufficient to provide adequate clearance for removal and installation of the spring, the disc brake caliper must be removed from the spindle. After removal, the caliper must be placed on the frame or otherwise supported to prevent suspending the caliper from the brake line hose.

These precautions are absolutely necessary to prevent serious damage to the tube portion of the caliper hose assembly!

6. If required, remove the stud from the axle assembly.

To install:

7. If removed, install the stud on the axle and torque to 190–230 ft. lbs. Install the lower seat and spacer over the stud.

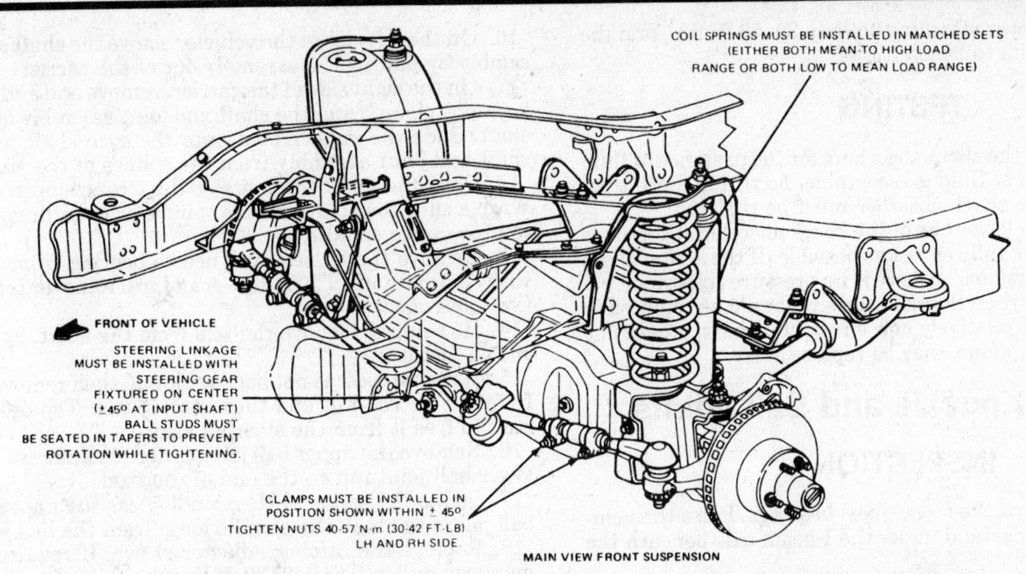

4WD Ranger, Bronco II and Explorer front suspension assembly

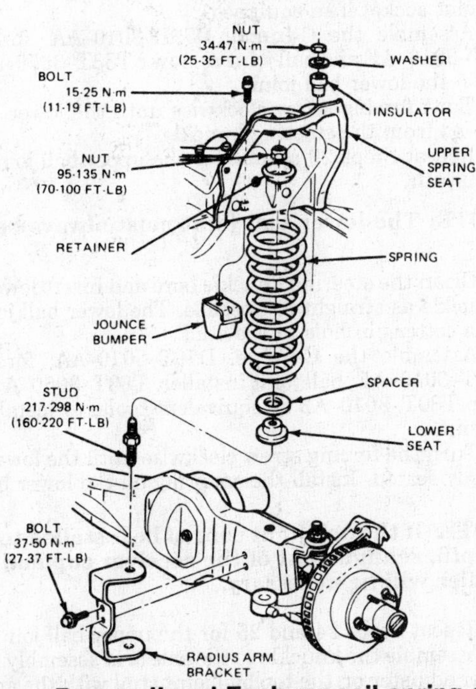

4WD Ranger, Bronco II and Explorer coil spring and related components

8. Place the spring in position and slowly raise the front axle. Ensure springs are positioned correctly in the upper spring seats.

9. Position the spring lower retainer over the stud and lower seat and torque the attaching nut to 70–100 ft. lbs.

10. Position the shock absorber to the lower stud and install the attaching nut. Tighten the nut to 41–63 ft. lbs. Lower the vehicle.

Shock Absorbers

REMOVAL AND INSTALLATION

NOTE: Low pressure gas shocks are charged with Nitrogen gas. Do not attempt to open, puncture or apply heat to them. Prior to installing a new shock absorber, hold it upright and extend it fully. Invert it and fully compress and extend it at least 3 times. This will bleed trapped air.

1. Raise the vehicle, as required to provide additional access and remove the bolt and nut attaching the shock absorber to the lower bracket on the radius arm.

2. Remove the nut, washer and insulator from the shock absorber at the frame bracket and remove the shock absorber.

To install:

3. Position the washer and insulator on the shock absorber rod and position the shock absorber to the frame bracket.

4. Position the insulator and washer on the shock absorber rod and install the attaching nut loosely.

5. Position the shock absorber to the lower bracket and install the attaching bolt and nut loosely.

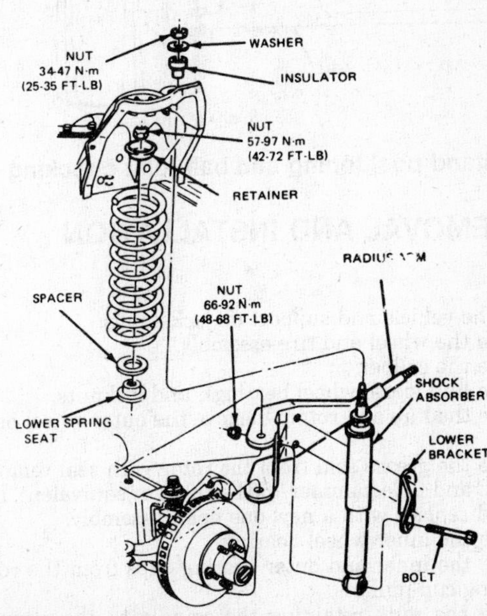

4WD Ranger, Bronco II and Explorer shock absorber and related components

6. Tighten the lower attaching bolts to 39–53 ft. lbs., and the upper attaching bolts to 25–35 ft. lbs.

TESTING

1. Visually check the shock absorbers for the presence of fluid leakage. A thin film of fluid is acceptable. Anything more than that means that the shock absorber must be replaced.
2. Disconnect the lower end of the shock absorber. Compress and extend the shock fully as fast as possible. If the action is not smooth in both directions, or there is no pressure resistance, replace the shock absorber. Shock absorbers should be replaced in pairs. In the case of relatively new shock absorbers, where one has failed, that one, alone, may be replaced.

Steering Knuckle and Ball Joints

INSPECTION

1. Check and adjust the front wheel bearings. Raise the vehicle and position a jackstand under the I-beam axle beneath the coil spring.
2. Have a helper grasp the lower edge of the tire and move the wheel assembly in and out.
3. While the wheel is being moved, observe the lower spindle arm and the lower part of the axle jaw.
4. A $^1/_{32}$ in. (0.8mm) or greater movement between the lower part of the axle jaw and the lower spindle arm indicates that the lower ball joint must be replaced
5. To check the upper ball joints, while the wheel is being moved, observe the upper spindle arm and the upper part of the axle jaw.
4. A $^1/_{32}$ in. (0.8mm) or greater movement between the upper part of the axle jaw and the upper spindle arm indicates that the upper ball joint must be replaced

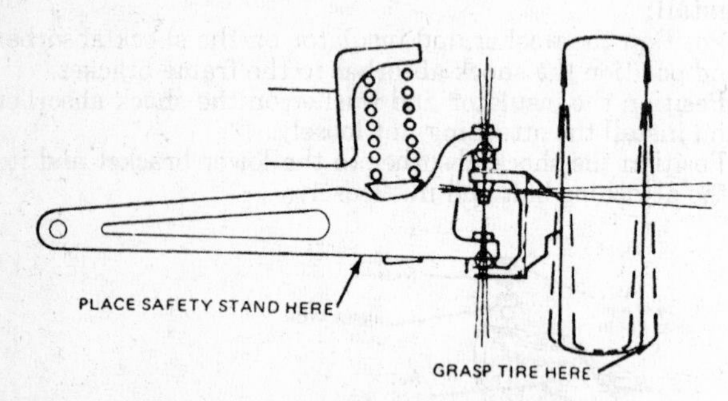

Safety stand positioning and ball joint checking

REMOVAL AND INSTALLATION

1983–89

1. Raise the vehicle and support on jackstands.
2. Remove the wheel and tire assembly.
3. Remove the caliper.
4. Remove hub locks, wheel bearings, and locknuts.
5. Remove the hub and rotor. Remove the outer wheel bearing cone.
6. Remove the grease seal from the rotor with seal remover tool 1175–AC and slide hammer 750T–100–A or equivalent. Discard seal and replace with a new one upon assembly.
7. Remove the inner wheel bearing.
8. Remove the inner and outer bearing cups from the rotor with a bearing cup puller.
9. Remove the nuts retaining the spindle to the steering knuckle. Tap the spindle with a plastic or rawhide hammer to jar the spindle from the knuckle. Remove the splash shield.
10. On the left side of the vehicle remove the shaft and joint assembly by pulling the assembly out of the carrier.
11. On the right side of the carrier, remove and and discard the keystone clamp from the shaft and joint assembly and the stub shaft. Slide the rubber boot onto the stub shaft and pull the shaft and joint assembly from the splines of the stub shaft.
12. Place the spindle in a vise on the second step of the spindle. Wrap a shop towel around the spindle or use a brass-jawed vise to protect the spindle.
13. Remove the oil seal and needle bearing from the spindle with slide hammer T50T–100–A and seal remover tool–1175–A–C or equivalent.
14. If required, remove the seal from the shaft, by driving off with a hammer.
15. If the tie rod has not been removed, then remove cotter pin from the tie rod nut and then remove nut. Tap on the tie rod stud to free it from the steering arm.
16. Remove the upper ball joint cotter pin and nut. Loosen the lower ball joint nut to the end of the stud.
17. Strike the inside of the spindle near the upper and lower ball joints to break the spindle loose from the ball joint studs.
18. Remove the camber adjuster sleeve. If required, use pitman arm puller, T64P–3590–F or equivalent to remove the adjuster out of the spindle. Remove the lower ball joint nut.
19. Place knuckle in vise and remove snapring from bottom ball joint socket if so equipped.
20. Assemble the C-frame, D79D–3010–AA, forcing screw, D79T–3010–AE and ball joint remover T83T–3050–A or equivalent on the lower ball joint.
21. Turn forcing screw clockwise until the lower ball joint is removed from the steering knuckle.
22. Repeat Steps 20 and 21 for the upper ball joint.

To install:

NOTE: The lower ball joint must always be installed first.

23. Clean the steering knuckle bore and insert lower ball joint in knuckle as straight as possible. The lower ball joint doesn't have a cotter pin hole in the stud.
24. Assemble the C-frame, D79T–3010–AA, forcing screw, D790T–3010–AE, ball joint installer, T83T–3050–A and receiver cup T80T–3010–A3 or equivalent tools, to install the lower ball joint.
25. Turn the forcing screw clockwise until the lower ball joint is firmly seated. Install the snapring on the lower ball joint.

NOTE: If the ball joint cannot be installed to the proper depth, realignment of the receiver cup and ball joint installer will be necessary.

26. Repeat Steps 24 and 25 for the upper ball joint.
27. Assemble the knuckle to the axle arm assembly. Install the camber adjuster on the top ball joint stud with the arrow pointing outboard for POSITIVE CAMBER and the arrow pointing inboard for NEGATIVE CAMBER and ZERO camber bushings will not have an arrow and may be rotated in either direction as long as the lugs on the yoke engage the slots in the bushing.

CAUTION

The following torque sequence must be followed exactly when securing the spindle. Excessive spindle turning effort may result in reduced steering returnability if this procedure is not followed.

28. Install a new nut on the bottom ball joint stud and tighten to 40 ft. lbs.
29. Install a new nut on the top ball stud and tighten to 85–100 ft. lbs., then advance nut until castellation aligns with cotter pin hole and install cotter pin.
30. Finish tightening the lower nut to 95–110 ft. lbs.

NOTE: The camber adjuster will seat itself into the spindle at a predetermined position during the tightening sequence. Do not attempt to adjust this position.

31. Clean all dirt and grease from the spindle bearing bore. Bearing bores must be free from nicks and burrs.
32. Place the bearing in the fore with the manufacturer's identification facing outward. Drive the bearing into the bore using spindle replacer, T83T-3123-A and driver handle T80T-4000-W or equivalent.
33. Install the grease seal in the bearing bore with the lip side of the seal facing towards the tool. Drive the seal in the bore with spindle bearing replacer, T83T-3123-A and driver handle T80-4000-W or equivalent. Coat the bearing seal lip with Lubriplate®.
34. If removed, install a new shaft seal. Place the shaft in a press, and install the seal with spindle/axle seal installer, T83T-3132-A, or equivalent.
35. On the right side of the carrier, install the rubber boot and new keystone clamps on the stub slip yoke. Since the splines on the shaft are phased, there is only one way to assemble the right shaft and joint assembly into the slip yoke. Align the missing spline in the slip yoke barrel with the gap less male spline on the shaft and joint assembly. Slide the right shaft and joint assembly into the slip yoke making sure the splines are fully engaged. Slide the boot over the assembly and crimp the keystone clamp using keystone clamp pliers, T63P-9171-A or equivalent.
36. On the left side of the carrier slide the shaft and joint assembly through the knuckle and engage the splines on the shaft in the carrier.
37. Install the splash shield and spindle onto the steering knuckle. Install and tighten the spindle nuts to 35–45 ft. lbs.
38. Drive the bearing cups into the rotor using bearing cup replacer T73T-4222-B and driver handle, T80T-4000-W or equivalent.
39. Pack the inner and outer wheel bearings and the lip of the

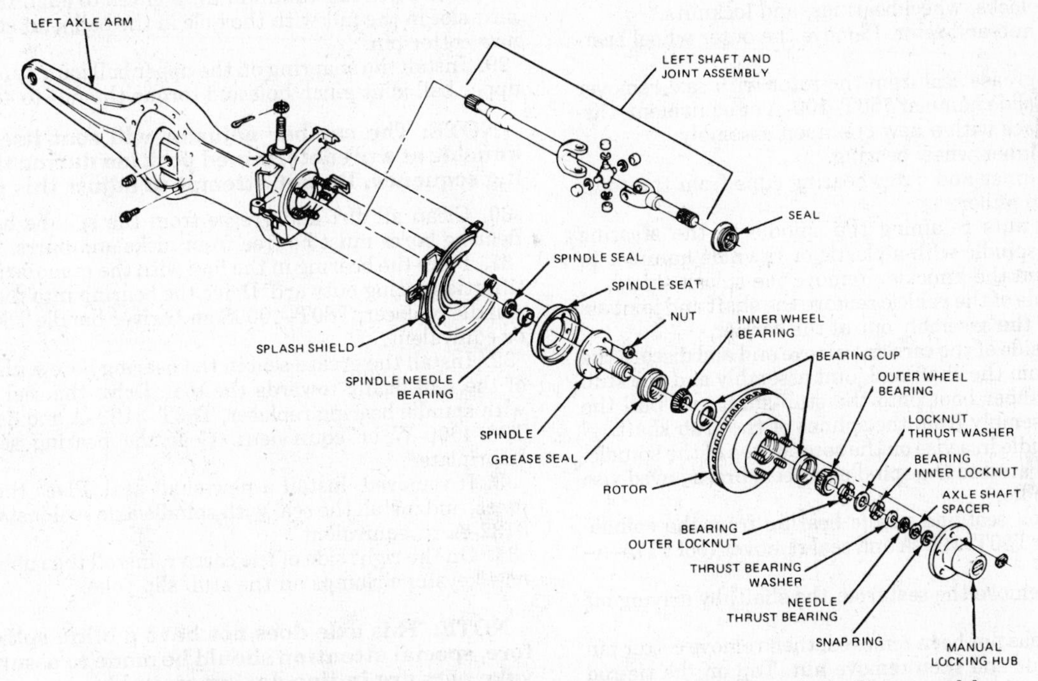

4WD Ranger, Bronco II and Explorer spindle and left hand shaft assembly

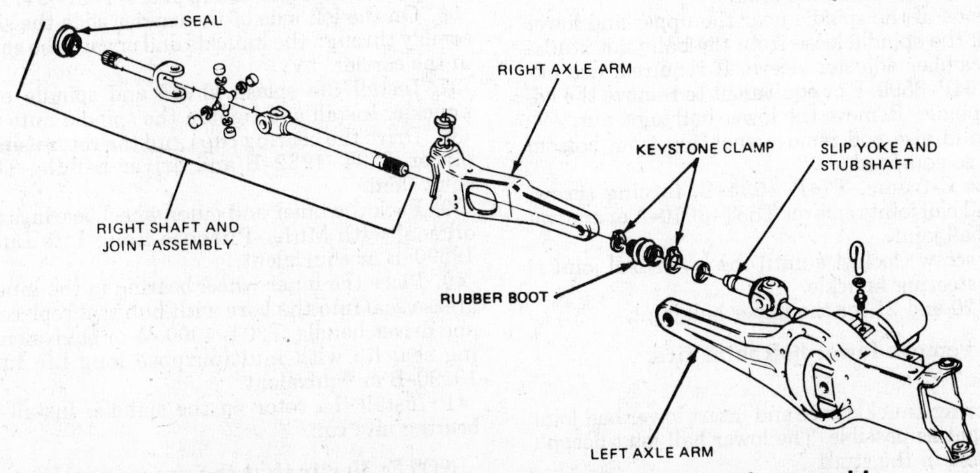

4WD Ranger, Bronco II and Explorer right hand shaft assembly

oil seal with Multi-Purpose Long-Life Lubricant, C1AZ-19590-B or equivalent.

40. Place the inner wheel bearing in the inner cup. Drive the grease seal into the bore with hub seal replacer, T83T-1175-B and driver handle, T80T-4000-W or equivalent. Coat the bearing seal lip with multipurpose long life lubricant, C1AZ-19590-B or equivalent.

41. Install the rotor on the spindle. Install the outer wheel bearing into cup.

NOTE: Verify that the grease seal lip totally encircles the spindle.

42. Install the wheel bearing, locknut, thrust bearing, snapring, and locking hubs.

1990-91

1. Raise the vehicle and support on jackstands.
2. Remove the wheel and tire assembly.
3. Remove the caliper.
4. Remove hub locks, wheel bearings, and locknuts.
5. Remove the hub and rotor. Remove the outer wheel bearing cone.
6. Remove the grease seal from the rotor with seal remover tool 1175-AC and slide hammer 750T-100-A or equivalent. Discard seal and replace with a new one upon assembly.
7. Remove the inner wheel bearing.
8. Remove the inner and outer bearing cups from the rotor with a bearing cup puller.
9. Remove the nuts retaining the spindle to the steering knuckle. Tap the spindle with a plastic or rawhide hammer to jar the spindle from the knuckle. Remove the splash shield.
10. On the left side of the vehicle remove the shaft and joint assembly by pulling the assembly out of the carrier.
11. On the right side of the carrier, remove and and discard the keystone clamp from the shaft and joint assembly and the stub shaft. Slide the rubber boot onto the stub shaft and pull the shaft and joint assembly from the splines of the stub shaft.
12. Place the spindle in a vise on the second step of the spindle. Wrap a shop towel around the spindle or use a brass-jawed vise to protect the spindle.
13. Remove the oil seal and needle bearing from the spindle with slide hammer T50T-100-A and seal remover tool 1175-A-C or equivalent.
14. If required, remove the seal from the shaft, by driving off with a hammer.
15. If the tie rod has not been removed, then remove cotter pin from the tie rod nut and then remove nut. Tap on the tie rod stud to free it from the steering arm.
16. Remove the upper ball joint cotter pin and nut. Loosen the lower ball joint nut to the end of the stud.
17. Strike the inside of the spindle near the upper and lower ball joints to break the spindle loose from the ball joint studs.
18. Remove the camber adjuster sleeve. If required, use pitman arm puller, T64P-3590-F or equivalent to remove the adjuster out of the spindle. Remove the lower ball joint nut.
19. Place knuckle in vise and remove snapring from bottom ball joint socket if so equipped.
20. Assemble the C-frame, T74P-4635-C, forcing screw, D79T-3010-AE and ball joint remover T83T-3050-A or equivalent on the lower ball joint.
21. Turn forcing screw clockwise until the lower ball joint is removed from the steering knuckle.
22. Repeat Steps 20 and 21 for the upper ball joint.

NOTE: Always remove lower ball joint first

To install:

23. Clean the steering knuckle bore and insert lower ball joint in knuckle as straight as possible. The lower ball joint doesn't have a cotter pin hole in the stud.
24. Assemble the C-frame, T74P-4635-C, forcing screw, D790T-3010-AE, ball joint installer, T83T-3050-A and receiver cup T80T-3010-A3 or equivalent tools, to install the lower ball joint.
25. Turn the forcing screw clockwise until the lower ball joint is firmly seated. Install the snapring on the lower ball joint.

NOTE: If the ball joint cannot be installed to the proper depth, realignment of the receiver cup and ball joint installer will be necessary.

26. Repeat Steps 24 and 25 for the upper ball joint.
27. Install the camber adjuster into the support arm. Position the slot in its original position.

CAUTION

The following torque sequence must be followed exactly when securing the spindle. Excessive spindle turning effort may result in reduced steering returnability if this procedure is not followed.

28. Install a new nut on the bottom of the ball joint stud and torque to 90 ft. lbs. (minimum). Tighten to align the nut to the next slot in the nut with the hole in the ball joint stud. Install a new cotter pin.
29. Install the snapring on the upper ball joint stud. Install the upper ball joint pinch bolt and torque the nut to 48-65 ft. lbs.

NOTE: The camber adjuster will seat itself into the knuckle at a predetermined position during the tightening sequence. Do not attempt to adjust this position.

30. Clean all dirt and grease from the spindle bearing bore. Bearing bores must be free from nicks and burrs.
31. Place the bearing in the fore with the manufacturer's identification facing outward. Drive the bearing into the bore using spindle replacer, T80T-4000S and driver handle T80T-4000-W or equivalent.
32. Install the grease seal in the bearing bore with the lip side of the seal facing towards the tool. Drive the seal in the bore with spindle bearing replacer, T83T-3123-A and driver handle T80-4000-W or equivalent. Coat the bearing seal lip with Lubriplate®.
33. If removed, install a new shaft seal. Place the shaft in a press, and install the seal with spindle/axle seal installer, T83T-3132-A, or equivalent.
34. On the right side of the carrier, install the rubber boot and new keystone clamps on the stub slip yoke.

NOTE: This axle does not have a blind spline. Therefore, special attention should be made to assure that the yoke ears are in line during assembly.

35. Slide the boot over the assembly and crimp the keystone clamp using keystone clamp pliers, T63P-9171-A or equivalent.
36. On the left side of the carrier slide the shaft and joint assembly through the knuckle and engage the splines on the shaft in the carrier.
37. Install the splash shield and spindle onto the steering knuckle. Install and tighten the spindle nuts to 40-50 ft. lbs.
38. Drive the bearing cups into the rotor using bearing cup replacer T73T-4222-B and driver handle, T80T-4000-W or equivalent.
39. Pack the inner and outer wheel bearings and the lip of the oil seal with Multi-Purpose Long-Life Lubricant, C1AZ-19590-B or equivalent.
40. Place the inner wheel bearing in the inner cup. Drive the grease seal into the bore with hub seal replacer, T80T-4000-T and driver handle, T80T-4000-W or equivalent. Coat the bearing seal lip with multipurpose long life lubricant, C1AZ-19590-B or equivalent.
41. Install the rotor on the spindle. Install the outer wheel bearing into cup.

NOTE: Verify that the grease seal lip totally encircles the spindle.

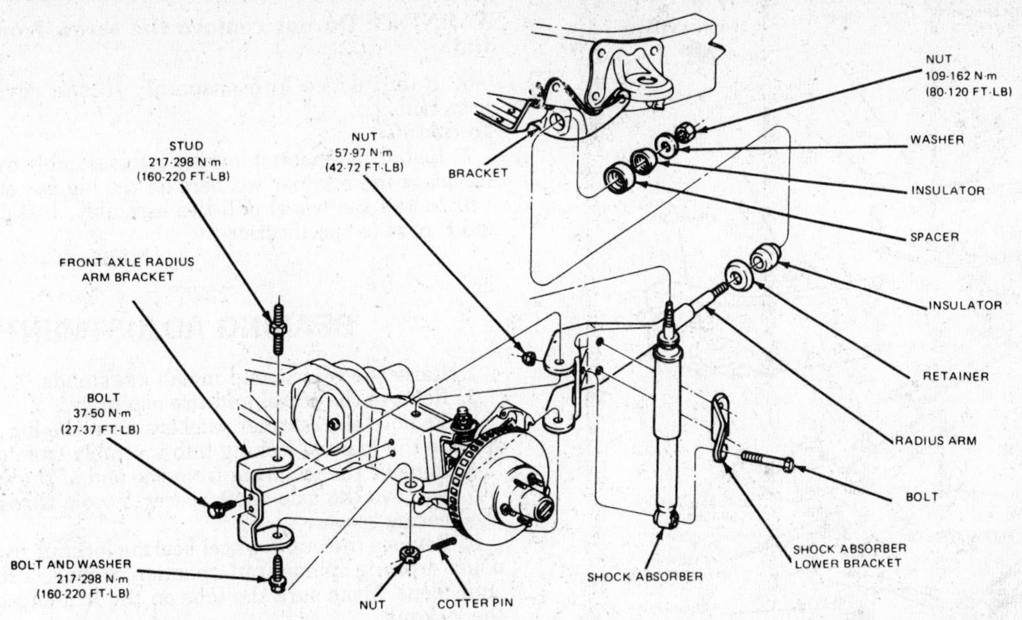

4WD Ranger, Bronco II and Explorer radius rod and related components

42. Install the wheel bearing, locknut, thrust bearing, snapring, and locking hubs.

Radius Arm

REMOVAL AND INSTALLATION

1. Raise the front of the vehicle, place jackstands under the frame. Place a jack under the axle.

WARNING: The axle must be supported on the jack throughout spring removal and installation, and must not be permitted to hang by the brake hose. If the length of the brake hose is not sufficient to provide adequate clearance for removal and installation of the spring, the disc brake caliper must be removed from the spindle. After removal, the caliper must be placed on the frame or otherwise supported to prevent suspending the the caliper from the caliper hose.

These precautions are absolutely necessary to prevent serious damage to the tube portion of the caliper hose assembly.

2. Disconnect the lower end of the shock absorber from the shock lower bracket on 1983–89 vehicles and the lower stud on 1990–91 vehicles. Remove the front spring from the vehicle.
3. Remove the spring lower seat and stud from the radius arm. Remove the bolts that attach the radius arm to the axle and front bracket.
4. Remove the nut, rear washer and insulator from the rear side of the radius arm rear bracket.
5. Remove the radius arm from the vehicle. Remove the inner insulator and retainer from the radius arm stud.

To install:

6. Position the front end of the radius arm from bracket to axle. Install the retaining bolts and stud in the bracket finger tight.
7. Install the retainer and inner insulator on the radius arm stud and insert the stud through the radius arm rear bracket.
8. Install the rear washer, insulator and nut on the arm stud at the rear side of the arm rear bracket. Tighten the nut to 80–120 ft. lbs.
9. Tighten the stud to 190–230 ft. lbs. Tighten the front bracket to axle bolts to 37–50 ft. lbs. and the lower bolt and washer to 190–230 ft. lbs.
10. Install the spring lower seat and spring insulator on the radius arm so that the hole in the seat goes over the arm to axle bolt. Tighten the axle pivot bolt to 120–150 ft. lbs.
11. Install the front spring. Connect the lower end of the shock absorber to the lower bracket on 1983–89 vehicles and to the stud 1990–91 vehicles of the radius arm and torque the retaining nut to 42–72 ft. lbs. on 1983–90 vehicles and 39–53 ft. lbs on 1991 vehicles.

Stabilizer Bar

REMOVAL AND INSTALLATION

1983–88

1. As required, raise and support the vehicle safely. Remove the nuts and U-bolts retaining the lower shock bracket/stabilizer bar bushing to the radius arm.
2. Remove the retainers and remove the stabilizer bar and bushing.
3. Place the stabilizer bar in position on the radius arm and bracket.
4. Install the retainers and U-bolts. Tighten the retainer bolts to 35–50 ft. lbs. Tighten the U-bolt nuts to 48–68 ft. lbs.

1989–91

1. As required, raise and support the vehicle safely. Remove the bolts and the retainers from the center and right hand end of the stabilizer bar.
2. Remove the nut, bolt and washer retaining the stabilizer bar to the stabilizer link.
3. Remove the stabilizer bar and bushings from the vehicle.
4. Installation is the reverse of the removal procedure. Tighten the retainer bolts to 35–50 ft. lbs. Tighten the stabilizer bar to link nut to 30–44 ft. lbs.

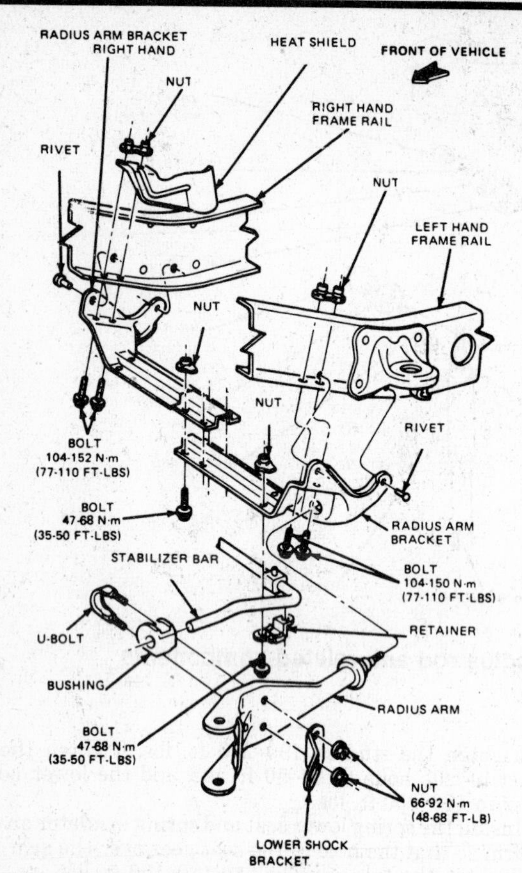

4WD Ranger, Bronco II and Explorer front stabilizer bar and related components

Manual Locking Hubs

REMOVAL AND INSTALLATION

1. Raise the vehicle and install jackstands.
2. Remove the wheel and tire assembly.
3. Remove the retainer washers from the lug nut studs and remove the manual locking hub assembly.
4. To remove the internal hub lock assembly from the outer body assembly, remove the outer lock ring seated in the hub body groove.
5. The internal assembly, spring and clutch gear will now slide out of the hub body.

WARNING: Do not remove the screw from the plastic dial!

6. Rebuild the hub assembly in the reverse order of disassembly.

To install:

7. Install the manual locking hub assembly over the spindle and place the retainer washers on the lug nut studs.
8. Install the wheel and tire assembly. Install the lug nuts and torque to specification.

BEARING ADJUSTMENT

1. Raise the vehicle and install jackstands.
2. Remove the wheel and tire assembly.
3. Remove the retainer washers from the lug nut studs and remove the manual locking hub assembly from the spindle.
4. Remove the snapring from the end of the spindle shaft.
5. Remove the axle shaft spacer, needle thrust bearing and the bearing spacer.
6. Remove the outer wheel bearing locknut from the spindle using 4 prong spindle nut spanner wrench, T86T–1197–A or equivalent. Make sure the tabs on the tool engage the slots in the locknut.
7. Remove the locknut washer from the spindle.
8. Loosen the inner wheel bearing locknut using 4 prong spindle nut spanner wrench, tool T83T–1197–A for 1983–89 vehicles and tool T86T–1197–A for 1990–91 vehicles or equivalent. Make sure that the tabs on the tool engage the slots in the locknut and that the slot in the tool is over the pin on the locknut.
9. Tighten the inner locknut to 35 ft. lbs. to seat the bearings.
10. Spin the rotor and back off the inner locknut ¼ turn. Install the lockwasher on the spindle. Retighten the inner locknut to 16 inch lbs. It may be necessary to turn the inner locknut slightly so that the pin on the locknut aligns with the closest hole in the lockwasher.
11. Install the outer wheel bearing locknut using 4 prong spindle nut spanner wrench, tool T83T–1197–A for 1983–89 vehicles and tool T86T–1197–A for 1990–91 vehicles or equivalent. Tighten locknut to 150 ft. lbs.
12. Install the bearing thrust spacer and needle thrust bearing, as required. Install the axle shaft spacer.
13. Clip the snapring onto the end of the spindle.
14. Install the manual hub assembly over the spindle. Install the retainer washers.
15. Install the wheel and tire assembly. Install and torque lug nuts to specification.
16. Check the endplay of the wheel and tire assembly on the spindle. Endplay should be 0.001–0.003 in. (0.025–0.076mm). On 1990–91 vehicles the maximum torque to rotate the hub should be 25 inch lbs.

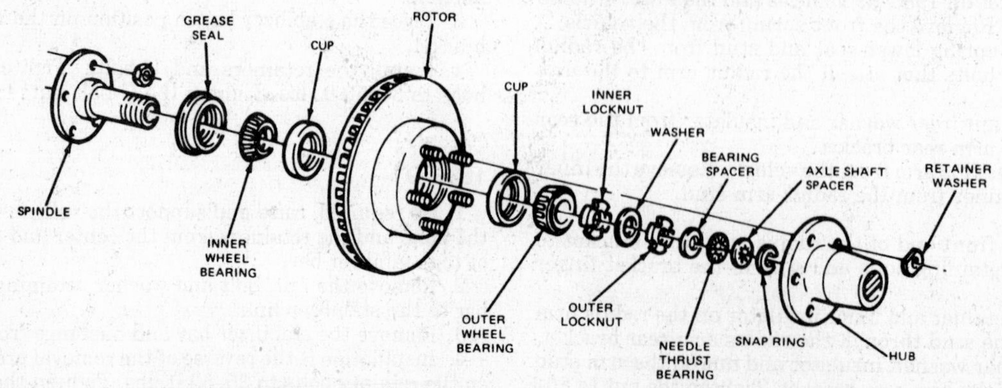

Ranger, Bronco II and Explorer manual locking hub assembly

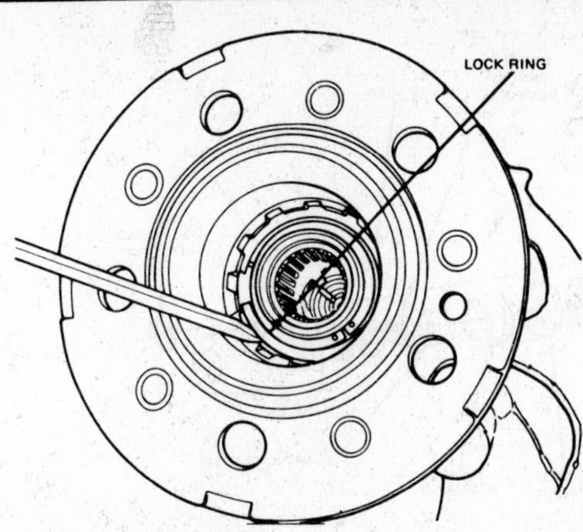

Ranger, Bronco II and Explorer manual locking hub outer ring removal

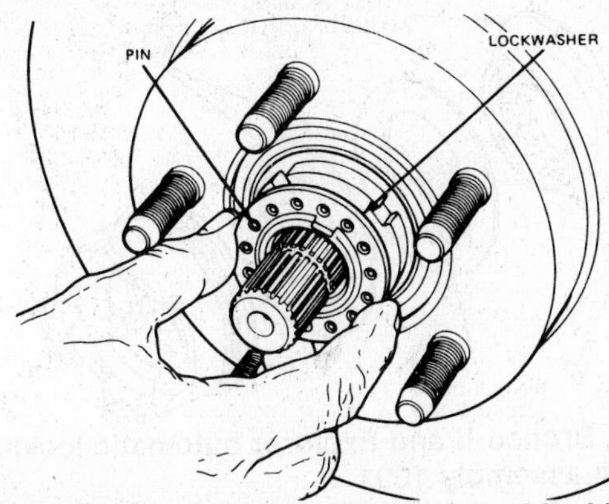

Ranger, Bronco II and Explorer manual locking hub lockwasher installation

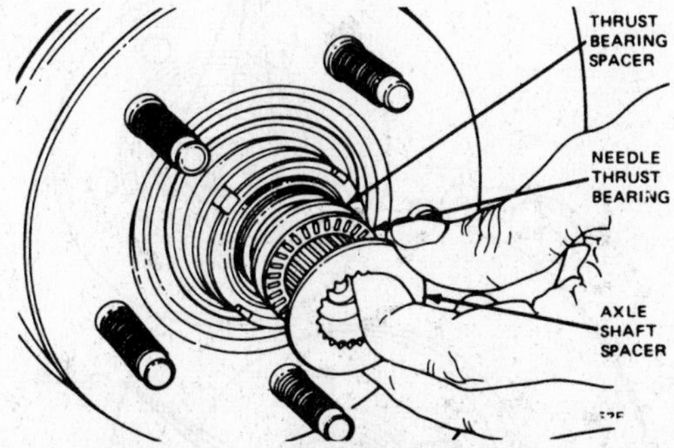

Ranger, Bronco II and Explorer manual locking hub thrust bearing and spacer removal

BEARING REPLACEMENT

1. Raise the vehicle and install jackstands.
2. Remove the wheel and tire assembly.
3. Remove the retainer washers and remove the manual locking hub assembly.
4. Remove the caliper and wire it to the side using mechanics wire.
5. Remove the snapring from the end of the spindle shaft.
6. Remove the axle shaft spacer, needle thrust bearing and the bearing spacer.
7. Remove the outer wheel bearing locknut from the spindle using 4 prong spindle nut spanner wrench, T86T–1197–A or equivalent. Make sure the tabs on the tool engage the slots in the locknut.
8. Remove the locknut washer from the spindle.
9. Remove the inner wheel bearing adjusting nut using 4 prong spindle nut spanner wrench, tool T83T–1197–A for 1983–89 vehicles and tool T86T–1197–A for 1990–91 vehicles or equivalent. Make sure that the tabs on the tool engage the slots in the locknut and that the slot in the tool is over the pin on the locknut.
10. Remove the disc brake rotor and the hub assembly. Remove the outer wheel bearing cone assembly.
11. Remove the grease seal from the rotor with seal remover tool 1175–AC and slide hammer 750T–100–A or equivalent. Discard seal and replace with a new one upon assembly.
12. Remove the inner wheel bearing.
13. Inspect the bearing cups for pits or cracks. If necessary, remove them with internal puller tool D80L–943–A and slide hammer 750T–100–A. or equivalent.

NOTE: If new cups are installed, install new cone and roller assemblies.

To install:

14. Lubricate the bearings with disc brake wheel bearing grease. Clean all old grease from the hub. Pack the cones and rollers. If a bearing packer is not available, work as much lubricant as possible between the rollers and the cages.
15. If bearing cups are to be installed, position cups in rotor and drive in place with bearing cup tool T73T–4222–B and driver handle T80T–4000–W.
16. Position the inner bearing in the inner cup in the rotor. Install the grease seal by driving in place with hub seal replacer tool T83T–1175–B and Driver Handle T80T–4000–W.
17. Carefully install the rotor onto the spindle. Install the outer wheel bearing in the rotor.
18. Install the inner adjusting nut with the pin facing out. Tighten the inner adjusting nut to 35 ft. lbs. to seat the bearings.
19. Spin the rotor and back off the inner nut ¼ turn. Retighten the inner nut to 16 inch lbs. Install the locking washer. It may be necessary to turn the inner nut slightly so that the pin on the nut aligns with the closest hole in the lockwasher.
20. Install the outer wheel bearing locknut using 4 prong spindle nut spanner wrench, tool T83T–1197–A for 1983–89 vehicles and tool T86T–1197–A for 1990–91 vehicles or equivalent. Tighten locknut to 150 ft. lbs.
21. Install the bearing thrust spacer and needle thrust bearing, as required. Install the axle shaft spacer.
22. Clip the snapring onto the end of the spindle.
23. Install the caliper. Install the locking hub assembly.
24. Install the wheel assembly. Lower the vehicle.

Automatic Locking Hubs

REMOVAL AND INSTALLATION

NOTE: The following procedures also include the bearing adjustment procedure.

1983–90

1. Raise the vehicle and install jackstands.
2. Remove the wheel and tire assembly.
3. Remove the retainer washers from the lug nut studs and remove the automatic locking hub assembly from the spindle.

To install:

4. Install the automatic locking hub assembly over the spindle by lining up the 3 legs in the hub assembly with 3 pockets in the cam assembly. Install the retainer washers.
5. Install the wheel and tire assembly. Torque the lug nuts to specification.

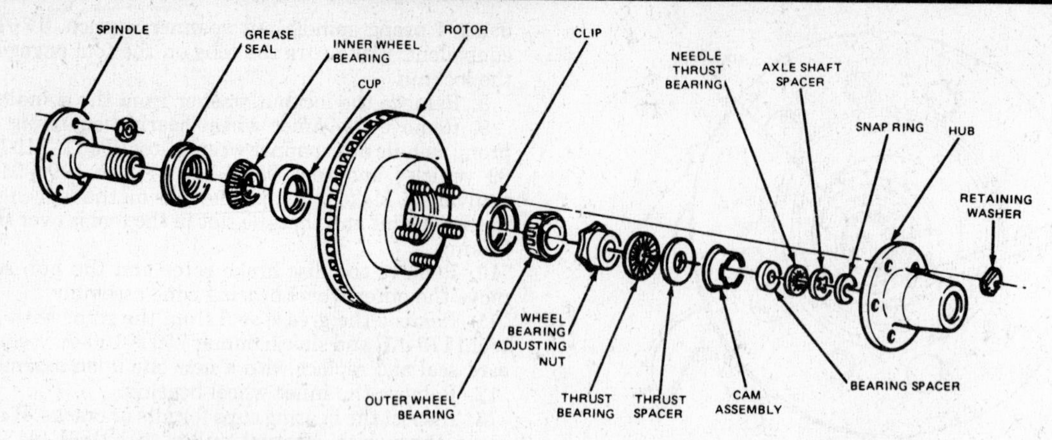

Ranger, Bronco II and Explorer automatic locking hub assembly

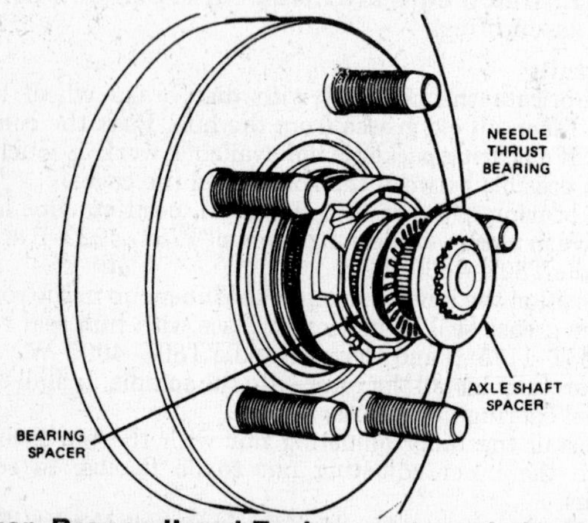

Ranger, Bronco II and Explorer automatic locking hub thrust bearing removal

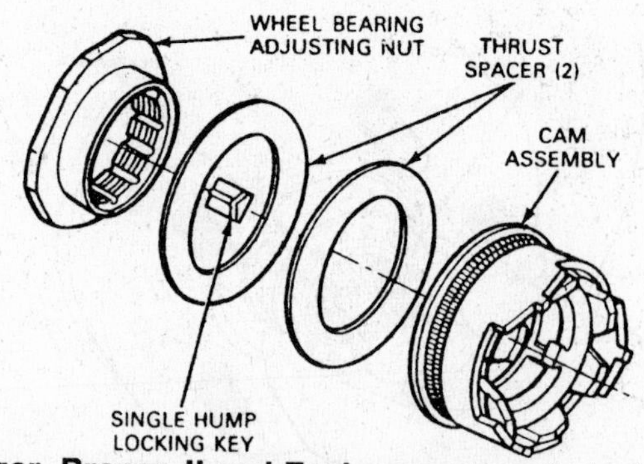

Ranger, Bronco II and Explorer automatic locking hub cam assembly 1991

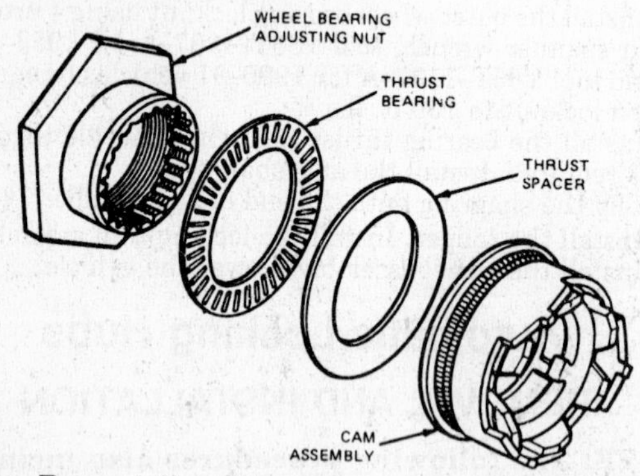

Ranger, Bronco II and Explorer automatic locking hub cam assembly 1983–90

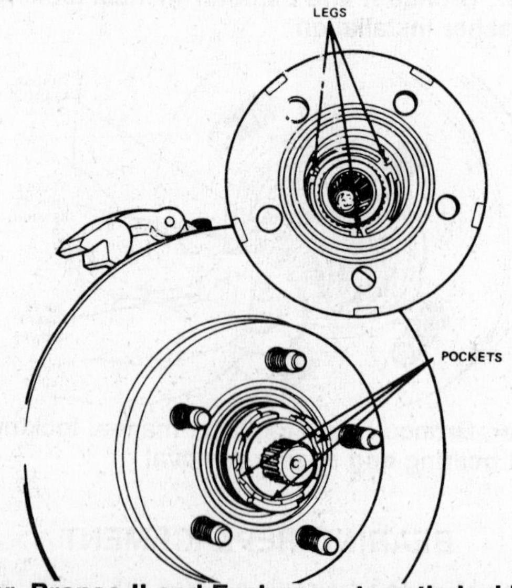

Ranger, Bronco II and Explorer automatic locking hub pocket location

1991

1. Raise the vehicle and install jackstands.
2. Remove the wheel and tire assembly.
3. Remove the retainer washers from the lug nut studs and remove the automatic locking hub assembly from the spindle.

To install:

4. Install the automatic locking hub assembly over the spindle by lining up the 3 legs in the hub assembly with the 3 pockets in the cam asembly. Install the retainer washers.
5. Install the wheel and tire assembly. Torque the lug nuts to specification.

BEARING REPLACEMENT

1. Raise the vehicle and install jackstands.

2. Remove the wheel and tire assembly.
3. Remove the retainer washers and remove the automatic locking hub assembly.
4. Remove the caliper and wire it to the side using mechanics wire.
5. Remove the snapring from the end of the spindle shaft.
6. Remove the axle shaft spacer, needle thrust bearing and the bearing spacer.
7. Remove the outer wheel bearing locknut from the spindle using 4 prong spindle nut spanner wrench, T86T–1197–A or equivalent. Make sure the tabs on the tool engage the slots in the locknut.
8. Remove the locknut washer from the spindle.
9. Remove the inner wheel bearing adjusting nut using 4 prong spindle nut spanner wrench, tool T83T–1197–A for 1983–89 vehicles and tool T86T–1197–A for 1990–91 vehicles or equivalent. Make sure that the tabs on the tool engage the slots in the locknut and that the slot in the tool is over the pin on the locknut.
10. Remove the disc brake rotor and the hub assembly. Remove the outer wheel bearing cone assembly.
11. Remove the grease seal from the rotor with seal remover tool 1175–AC and slide hammer 750T–100–A or equivalent. Discard seal and replace with a new one upon assembly.
12. Remove the inner wheel bearing.
13. Inspect the bearing cups for pits or cracks. If necessary, remove them with internal puller tool D80L–943–A and slide hammer 750T–100–A. or equivalent.

NOTE: If new cups are installed, install new cone and roller assemblies.

To install:
14. Lubricate the bearings with disc brake wheel bearing grease. Clean all old grease from the hub. Pack the cones and rollers. If a bearing packer is not available, work as much lubricant as possible between the rollers and the cages.
15. If bearing cups are to be installed, position cups in rotor and drive in place with bearing cup tool T73T–4222–B and driver handle T80T–4000–W.
16. Position the inner bearing in the inner cup in the rotor. Install the grease seal by driving in place with hub seal replacer tool T83T–1175–B and Driver Handle T80T–4000–W.
17. Carefully install the rotor onto the spindle. Install the outer wheel bearing in the rotor.
18. Install the inner adjusting nut, with the pin facing out. Tighten the inner locknut to 35 ft. lbs. to seat the bearings.
19. Spin the rotor and back off the inner locknut ¼ turn (90°). Retighten the inner locknut to 16 inch lbs. Install the locking washer. It may be necessary to turn the inner locknut slightly so that the pin on the locknut aligns with the closest hole in the lockwasher.
20. Install the outer wheel bearing locknut using 4 prong spindle nut spanner wrench, tool T83T–1197–A for 1983–89 vehicles and tool T86T–1197–A for 1990–91 vehicles or equivalent. Tighten locknut to 150 ft. lbs.
21. Install the bearing thrust spacer and needle thrust bearing, as required. Install the axle shaft spacer.
22. Clip the snapring onto the end of the spindle.
23. Install the caliper. Install the locking hub assembly.
24. Install the wheel. Lower the vehicle.

Front End Alignment

CASTER AND CAMBER

If you should start to notice abnormal tire wear patterns and handling (steering wheel is hard to return to straight ahead position after negotiating a turn on pavement), and misalignment of caster and camber are suspected, make the following checks:
1. Check the air pressure in all the tires. Make sure that the pressures agree with those specified for the tires and vehicle being checked.
2. Raise the front of the vehicle off the ground and support it safely. Grasp each front tire at the front and rear, and push the wheel inward and outward. If any free-play is noticed, adjust the wheel bearings.

NOTE: There is supposed to be a very, very small amount of free-play present where the wheel bearings are concerned. Replace the bearings if they are worn or damaged.

3. Check all steering linkage for wear or maladjustment. Adjust and/or replace all worn parts.
4. Check the steering gear mounting bolts and tighten if necessary.
5. Rotate each front wheel slowly, and observe the amount of lateral or side runout. If the wheel runout exceeds ⅛ in. (3mm), replace the wheel or install the wheel on the rear.
6. Inspect the radius arms to be sure they are not bent or damaged. Inspect the bushings at the radius arm-to-axle attachment and radius arm-to-frame attachment points for wear or looseness. Repair or replace parts as required.

Caster is the number of degrees of backward (positive) or forward (negative) tilt of the spindle or the line connecting the ball joint centers. Camber is the number of degrees the top of the wheel tilts outward (positive) or inward (negative) from a vertical plane.

Before checking caster or camber, perform the toe alignment check. Using alignment equipment known to be accurate and following the equipment manufacturer's instructions, measure and record the caster angle and the camber angle of both front wheels.

If the caster and camber measurements exceed the maximum variances, inspect for damaged front suspension components. Replace as required.

Both caster and camber adjustments are possible with service adjusters. These service adjusters are available in ½, 1 and 1½ degree increments. On of these adjusters is used to adjust both caster and camber.

WHEEL ALIGNMENT

Year	Model	Ride Height Min.	Ride Height Max.	Caster Min.	Caster Max.	Camber Min.	Camber Max.	Toe-in (inches)	Toe-in (Degrees)
1983	Ranger 2WD	2 3/4	3 1/4	4 1/2	7	−1	1	1/32	1/16
		3 1/4	3 1/2	4	6 1/2	−1/2	1 3/4	1/32	1/16
		3 1/2	4	3 3/8	5 7/8	0	2 3/8	1/32	1 1/16
		4	4 1/4	2 5/8	5 1/8	3/4	3	1/32	1/16
		4 1/4	4 3/4	2	4 1/2	1 1/2	3 3/4	1/32	1/16
	Ranger 4WD	2 3/4	3 1/4	5	8	−1	1/2	1/32	1/16
		3 1/4	3 1/2	4	7	0	1 1/2	1/32	1/16
		3 1/2	4	3	6	1/2	2	1/32	1/16
		4	4 1/4	2 1/2	5 1/2	1 1/4	2 3/4	1/32	1/16
		4 1/4	4 3/4	1 3/4	5	2	3 3/4	1/32	1/16
1984	Bronco II	2 3/4	3	5 1/2	8 1/2	−2	−1/2	1/32	1/16
		3 1/4	3 1/2	4	7	−1	1/2	1/32	1/16
		3 1/2	3 3/4	3	6	0	1 1/2	1/32	1/16
		4	4 1/4	2	5	1	2 1/2	1/32	1/16
		4 1/4	4 3/4	1	4	2	3 1/2	1/32	1/16
	Ranger 2WD (w/forged axle)	3 1/4	3 1/2	5 1/4	8 1/4	−2	−1/2	1/32	1/16
		3 1/2	3 3/4	4 1/2	7 1/2	−1 5/8	1/8	1/32	1/16
		3 3/4	4	3 1/2	6 1/2	−1/2	1	1/32	1/16
		4	4 1/4	2	6	1/4	1 3/4	1/32	1/16
		4 1/2	4 3/4	1 7/8	4 7/8	1 1/4	2 3/4	1/32	1/16
	Ranger 2WD (w/stamped axle)	3	3 1/4	5 1/4	8 1/4	−2	−1/2	1/32	1/16
		3 1/4	3 1/2	4 1/2	7 1/2	−1 5/8	1/8	1/32	1/16
		3 1/2	3 3/4	3 1/2	6 1/2	−1/2	1	1/32	1/16
		3 3/4	4	3	6	1/4	1 3/4	1/32	1/16
		4 1/4	4 1/2	1 7/8	4 7/8	1 1/4	2 3/4	1/32	1/16
	Ranger 4WD	2 3/4	3	5 1/2	8 1/2	−2	−1/2	1/32	1/16
		3 1/4	3 1/2	4	7	−1	1/2	1/32	1/16
		3 1/2	3 3/4	3	6	0	1 1/2	1/32	1/16
		4	4 1/4	2	5	1	2 1/2	1/32	1/16
		4 1/4	4 3/4	1	4	2	3 1/2	1/32	1/16
1985	Bronco II	2 3/4	3	5 1/2	8 1/2	−2	−1/2	1/32	1/16
		3 1/4	3 1/2	4	7	−1	1/2	1/32	1/16
		3 1/2	3 3/4	3	6	0	1 1/2	1/32	1/16
		4	4 1/4	2	5	1	2 1/2	1/32	1/16
		4 1/4	4 3/4	1	4	2	3 1/2	1/32	1/16
	Ranger 2WD (w/forged axle)	3 1/4	3 1/2	5 1/4	8 1/4	−2	−1/2	1/32	1/16
		3 1/2	3 3/4	4 1/2	7 1/2	−1 5/8	1/8	1/32	1/16
		3 3/4	4	3 1/2	6 1/2	−1/2	1	1/32	1/16
		4	4 1/4	2	6	1/4	1 3/4	1/32	1/16
		4 1/2	4 3/4	1 7/8	4 7/8	1 1/4	2 3/4	1/32	1/16
	Ranger 2WD (w/stamped axle)	3	3 1/4	5 1/4	8 1/4	−2	−1/2	1/32	1/16
		3 1/4	3 1/2	4 1/2	7 1/2	−1 5/8	1/8	1/32	1/16
		3 1/2	3 3/4	3 1/2	6 1/2	−1/2	1	1/32	1/16
		3 3/4	4	3	6	1/4	1 3/4	1/32	1/16
		4 1/4	4 1/2	1 7/8	4 7/8	1 1/4	2 3/4	1/32	1/16
	Ranger 4WD	2 3/4	3	5 1/2	8 1/2	−2	−1/2	1/32	1/16
		3 1/4	3 1/2	4	7	−1	1/2	1/32	1/16
		3 1/2	3 3/4	3	6	0	1 1/2	1/32	1/16
		4	4 1/4	2	5	1	2 1/2	1/32	1/16
		4 1/4	4 3/4	1	4	2	3 1/2	1/32	1/16

WHEEL ALIGNMENT

Year	Model	Ride Height		Caster		Camber		Toe-in	Toe-in
		Min.	Max.	Min.	Max.	Min.	Max.	(inches)	(Degrees)
1986	Bronco II	2 3/4	3	5 1/2	8 1/2	−2	−1/2	1/32	1/16
		3 1/4	3 1/2	4	7	−1	1/2	1/32	1/16
		3 1/2	3 3/4	3	6	0	1 1/2	1/32	1/16
		4	4 1/4	2	5	1	2 1/2	1/32	1/16
		4 1/4	4 3/4	1	4	2	3 1/2	1/32	1/16
	Ranger 2WD (w/forged axle)	3 1/4	3 1/2	5 1/14	8 1/4	−2	−1/2	1/32	1/16
		3 1/2	3 3/4	4 1/2	7 1/2	−1 5/8	1/8	1/32	1/16
		3 3/4	4	3 1/2	6 1/2	−1/2	1	1/32	1/16
		4	4 1/4	2	6	1/4	1 3/4	1/32	1/16
		4 1/2	4 3/4	1 7/8	4 7/8	1 1/4	2 3/4	1/32	1/16
	Ranger 2WD (w/stamped axle)	3	3 1/4	5 1/4	8 1/4	−2	−1/2	1/32	1/16
		3 1/4	3 1/2	4 1/2	7 1/2	−1 5/8	1/8	1/32	1/16
		3 1/2	3 3/4	3 1/2	6 1/2	−1/2	1	1/32	1/16
		3 3/4	4	3	6	1/4	1 3/4	1/32	1/16
		4 1/4	4 1/2	1 7/8	4 7/8	1 1/4	2 3/4	1/32	1/16
	Ranger 4WD	2 3/4	3	5 1/2	8 1/2	−2	−1/2	1/32	1/16
		3 1/4	3 1/2	4	7	−1	1/2	1/32	1/16
		3 1/2	3 3/4	3	6	0	1 1/2	1/32	1/16
		4	4 1/4	2	5	1	2 1/2	1/32	1/16
		4 1/4	4 3/4	1	4	2	3 1/2	1/32	1/16
1987	Bronco II	2 3/4	3	5 1/2	8 1/2	−2	−1/2	1/32	1/16
		3 1/4	3 1/2	4	7	−1	1/2	1/32	1/16
		3 1/2	3 3/4	3	6	0	1 1/2	1/32	1/16
		4	4 1/4	2	5	1	2 1/2	1/32	1/16
		4 1/4	4 3/4	1	4	2	3 1/2	1/32	1/16
	Ranger 2WD (w/forged axle)	3 1/4	3 1/2	5 1/4	8 1/4	−2	−1/2	1/32	1/16
		3 1/2	3 3/4	4 1/2	7 1/2	−1 5/8	1/8	1/32	1/16
		3 3/4	4	3 1/2	6 1/2	−1/2	1	1/32	1/16
		4	4 1/4	2	6	1/4	1 3/4	1/32	1/16
		4 1/2	4 3/4	1 7/8	4 7/8	1 1/4	2 3/4	1/32	1/16
	Ranger 2WD (w/stamped axle)	3	3 1/4	5 1/4	8 1/4	−2	−1/2	1/32	1/16
		3 1/4	3 1/2	4 1/2	7 1/2	−1 5/8	1/8	1/32	1/16
		3 1/2	3 3/4	3 1/2	6 1/2	−1/2	1	1/32	1/16
		3 3/4	4	3	6	1/4	1 3/4	1/32	1/16
		4 1/4	4 1/2	1 7/8	4 7/8	1 1/4	2 3/4	1/32	1/16
	Ranger 4WD (exc. STX)	2 3/4	3	5 1/2	8 1/2	−2	−1/2	1/32	1/16
		3 1/4	3 1/2	4	7	−1	1/2	1/32	1/16
		3 1/2	3 3/4	3	6	0	1 1/2	1/32	1/16
		4	4 1/4	2	5	1	2 1/2	1/32	1/16
		4 1/4	4 3/4	1	4	2	3 1/2	1/32	1/16
	Ranger 4WD (STX)	4 1/4	4 1/2	5 1/2	8 1/2	−2	−1/2	1/32	1/16
		4 1/2	5	4	7	−1	1/2	1/32	1/16
		5	5 1/4	3	6	0	1 1/2	1/32	1/16
		5 1/4	5 3/4	2	5	1	2 1/2	1/32	1/16
		5 3/4	6	1	4	2	3 1/2	1/32	1/16
1988	Bronco II	2 3/4	3	5 1/2	8 1/2	−2	−1/2	1/32	1/16
		3 1/4	3 1/2	4	7	−1	1/2	1/32	1/16
		3 1/2	3 3/4	3	6	0	1 1/2	1/32	1/16
		4	4 1/4	2	5	1	2 1/2	1/32	1/16
		4 1/4	4 3/4	1	4	2	3 1/2	1/32	1/16

WHEEL ALIGNMENT

Year	Model	Ride Height Min.	Ride Height Max.	Caster Min.	Caster Max.	Camber Min.	Camber Max.	Toe-in (inches)	Toe-in (Degrees)
1988	Ranger 2WD	3 1/4	3 1/2	5 1/4	8 1/4	−2	−1/2	1/32	1/16
	(w/forged	3 1/2	3 3/4	4 1/2	7 1/2	−1 5/8	1/8	1/32	1/16
	axle)	3 3/4	4	3 1/2	6 1/2	−1/2	1	1/32	1/16
		4	4 1/4	2	6	1/4	1 3/4	1/32	1/16
		4 1/2	4 3/4	1 7/8	4 7/8	1 1/4	2 3/4	1/32	1/16
	Ranger 2WD	3	3 1/4	5 1/4	8 1/4	−2	−1/2	1/32	1/16
	(w/stamped	3 1/4	3 1/2	4 1/2	7 1/2	−1 5/8	1/8	1/32	1/16
	axle)	3 1/2	3 3/4	3 1/2	6 1/2	−1/2	1	1/32	1/16
		3 3/4	4	3	6	1/4	1 3/4	1/32	1/16
		4 1/4	4 1/2	1 7/8	4 7/8	1 1/4	2 3/4	1/32	1/16
	Ranger 4WD	2 3/4	3	5 1/2	8 1/2	−2	−1/2	1/32	1/16
	(exc. STX)	3 1/4	3 1/2	4	7	−1	1/2	1/32	1/16
		3 1/2	3 3/4	3	6	0	1 1/2	1/32	1/16
		4	4 1/4	2	5	1	2 1/2	1/32	1/16
		4 1/4	4 3/4	1	4	2	3 1/2	1/32	1/16
	Ranger 4WD	4 1/4	4 1/2	5 1/2	8 1/2	−2	−1/2	1/32	1/16
	(STX)	4 1/2	5	4	7	−1	1/2	1/32	1/16
		5	5 1/4	3	6	0	1 1/2	1/32	1/16
		5 1/4	5 3/4	2	5	1	2 1/2	1/32	1/16
		5 3/4	6	1	4	2	3 1/2	1/32	1/16
1989	Bronco II	2 3/4	3	5 1/2	8 1/2	−2	−1/2	1/32	1/16
		3 1/4	3 1/2	4	7	−1	1/2	1/32	1/16
		3 1/2	3 3/4	3	6	0	1 1/2	1/32	1/16
		4	4 1/4	2	5	1	2 1/2	1/32	1/16
		4 1/4	4 3/4	1	4	2	3 1/2	1/32	1/16
	Ranger 2WD	3 1/4	3 1/2	5 1/4	8 1/4	−2	−1/2	1/32	1/16
	(w/forged	3 1/2	3 3/4	4 1/2	7 1/2	−1 5/8	1/8	1/32	1/16
	axle	3 3/4	4	3 1/2	6 1/2	−1/2	1	1/32	1/16
		4	4 1/4	2	6	1/4	1 3/4	1/32	1/16
		4 1/2	4 3/4	1 7/8	4 7/8	1 1/4	2 3/4	1/32	1/16
	Ranger 2WD	3	3 1/4	5 1/4	8 1/4	−2	−1/2	1/32	1/16
	(w/stamped	3 1/4	3 1/2	4 1/2	7 1/2	−1 5/8	1/8	1/32	1/16
	axle)	3 1/2	3 3/4	3 1/2	6 1/2	−1/2	1	1/32	1/16
		3 3/4	4	3	6	1/4	1 3/4	1/32	1/16
		4 1/4	4 1/2	1 7/8	4 7/8	1 1/4	2 3/4	1/32	1/16
	Ranger 4WD	2 3/4	3	5 1/2	8 1/2	−2	−1/2	1/32	1/16
	(exc. STX)	3 1/4	3 1/2	4	7	−1	1/2	1/32	1/16
		3 1/2	3 3/4	3	6	0	1 1/2	1/32	1/16
		4	4 1/4	2	5	1	2 1/2	1/32	1/16
		4 1/4	4 3/4	1	4	2	3 1/2	1/32	1/16
	Ranger 4WD	4 1/4	4 1/2	5 1/2	8 1/2	−2	−1/2	1/32	1/16
	(STX)	4 1/2	5	4	7	−1	1/2	1/32	1/16
		5	5 1/4	3	6	0	1 1/2	1/32	1/16
		5 1/4	5 3/4	2	5	1	2 1/2	1/32	1/16
		5 3/4	6	1	4	2	3 1/2	1/32	1/16
1990	Bronco II	2 3/4	3	5 1/2	8 1/2	−2	−1/2	1/32	1/16
		3 1/4	3 1/2	4	7	−1	1/2	1/32	1/16
		3 1/2	3 3/4	3	6	0	1 1/2	1/32	1/16
		4	4 1/4	2	5	1	2 1/2	1/32	1/16
		4 1/4	4 3/4	1	4	2	3 1/2	1/32	1/16

WHEEL ALIGNMENT

Year	Model	Ride Height Min.	Ride Height Max.	Caster Min.	Caster Max.	Camber Min.	Camber Max.	Toe-in (inches)	Toe-in (Degrees)
1990	Ranger 2WD	3 1/4	3 1/2	5 1/4	8 1/4	-2	-1/2	1/32	1/16
	(w/forged	3 1/2	3 3/4	4 1/2	7 1/2	-1 5/8	1/8	1/32	1/16
	axle)	3 3/4	4	3 1/2	6 1/2	-1/2	1	1/32	1/16
		4	4 1/4	2	6	1/4	1 3/4	1/32	1/16
		4 1/2	4 3/4	1 7/8	4 7/8	1 1/4	2 3/4	1/32	1/16
	Ranger 2WD	3	3 1/4	5 1/4	8 1/4	-2	-1/2	1/32	1/16
	(w/stamped	3 1/4	3 1/2	4 1/2	7 1/2	-1 5/8	1/8	1/32	1/16
	axle)	3 1/2	3 3/4	3 1/2	6 1/2	-1/2	1	1/32	1/16
		3 3/4	4	3	6	1/4	1 3/4	1/32	1/16
		4 1/4	4 1/2	1 7/8	4 7/8	1 1/4	2 3/4	1/32	1/16
	Ranger 4WD	2 3/4	3	5 1/2	8 1/2	-2	-1/2	1/32	1/16
	(exc. STX)	3 1/4	3 1/2	4	7	-1	1/2	1/32	1/16
		3 1/2	3 3/4	3	6	0	1 1/2	1/32	1/16
		4	4 1/4	2	5	1	2 1/2	1/32	1/16
		4 1/4	4 3/4	1	4	2	3 1/2	1/32	1/16
	Ranger 4WD	4 1/4	4 1/2	5 1/2	8 1/2	-2	-1/2	1/32	1/16
	(STX)	4 1/2	5	4	7	-1	1/2	1/32	1/16
		5	5 1/4	3	6	0	1 1/2	1/32	1/16
		5 1/4	5 3/4	2	5	1	2 1/2	1/32	1/16
		5 3/4	6	1	4	2	3 1/2	1/32	1/16
1991	Explorer	2 3/4	3	5 1/2	8 1/2	-2	-1/2	1/32	1/16
		3 1/4	3 1/2	4	7	-1	1/2	1/32	1/16
		3 1/2	3 3/4	3	6	0	1 1/2	1/32	1/16
		4	4 1/4	2	5	1	2 1/2	1/32	1/16
		4 1/4	4 3/4	1	4	2	3 1/2	1/32	1/16
	Ranger 2WD	3 1/4	3 1/2	5 1/4	8 1/4	-2	-1/2	1/32	1/16
	(w/forged	3 1/2	3 3/4	4 1/2	7 1/2	-1 5/8	1/8	1/32	1/16
	axle)	3 3/4	4	3 1/2	6 1/2	-1/2	1	1/32	1/16
		4	4 1/4	2	6	1/4	1 3/4	1/32	1/16
		4 1/2	4 3/4	1 7/8	4 7/8	1 1/4	2 3/4	1/32	1/16
	Ranger 2WD	3	3 1/4	5 1/4	8 1/4	-2	-1/2	1/32	1/16
	(w/stamped	3 1/4	3 1/2	4 1/2	7 1/2	-1 5/8	1/8	1/32	1/16
	axle)	3 1/2	3 3/4	3 1/2	6 1/2	-1/2	1	1/32	1/16
		3 3/4	4	3	6	1/4	1 3/4	1/32	1/16
		4 1/4	4 1/2	1 7/8	4 7/8	1 1/4	2 3/4	1/32	1/16
	Ranger 4WD	2 3/4	3	5 1/2	8 1/2	-2	-1/2	1/32	1/16
	(exc. STX)	3 1/4	3 1/2	4	7	-1	1/2	1/32	1/16
		3 1/2	3 3/4	3	6	0	1 1/2	1/32	1/16
		4	4 1/4	2	5	1	2 1/2	1/32	1/16
		4 1/4	4 3/4	1	4	2	3 1/2	1/32	1/16
	Ranger 4WD	4 1/4	4 1/2	5 1/2	8 1/2	-2	-1/2	1/32	1/16
	(STX)	4 1/2	5	4	7	-1	1/2	1/32	1/16
		5	5 1/4	3	6	0	1 1/2	1/32	1/16
		5 1/4	5 3/4	2	5	1	2 1/2	1/32	1/16
		5 3/4	6	1	4	2	3 1/2	1/32	1/16

REAR SUSPENSION

Leaf Springs

REMOVAL AND INSTALLATION

1. Raise the vehicle and install jackstands under the frame. The vehicle must be supported in such a way that the rear axle hangs free with the tires still touching the ground.
2. Remove the nuts from the spring U-bolts and drive the U-bolts from the U-bolt plate.
3. Remove the spring to bracket nut and bolt at the front of the spring.
4. Remove the shackle upper and lower nuts and bolts at the rear of the spring.
5. Remove the spring and shackle assembly from the rear shackle bracket.

To install:

6. Position the spring in the shackle. Install the upper shackle spring bolt and nut with the bolt head facing outward.
7. Position the front end of the spring in the bracket and install the bolt and nut.
8. Position the shackle in the rear bracket and install the nut and bolt.
9. Position the spring on top of the axle with the spring tie bolt centered in the hole provided in the seat.
10. Lower the vehicle to the floor. Torque the spring U-bolt nuts to 65–75 ft. lbs. Torque the front spring bolt to 75–115 ft. lbs. Torque the rear shackle nuts and bolts to 75–115 ft. lbs.

Shock Absorbers

REMOVAL AND INSTALLATION

1. Raise the vehicle and position jackstands under the axle or wheel, in order to take the load off of the shock absorber.
2. Remove the shock absorber lower retaining nut and bolt. Swing the lower end free of the mounting bracket on the axle housing.

To install:

3. Remove the retaining nut(s) from the upper shock absorber mounting mounting.

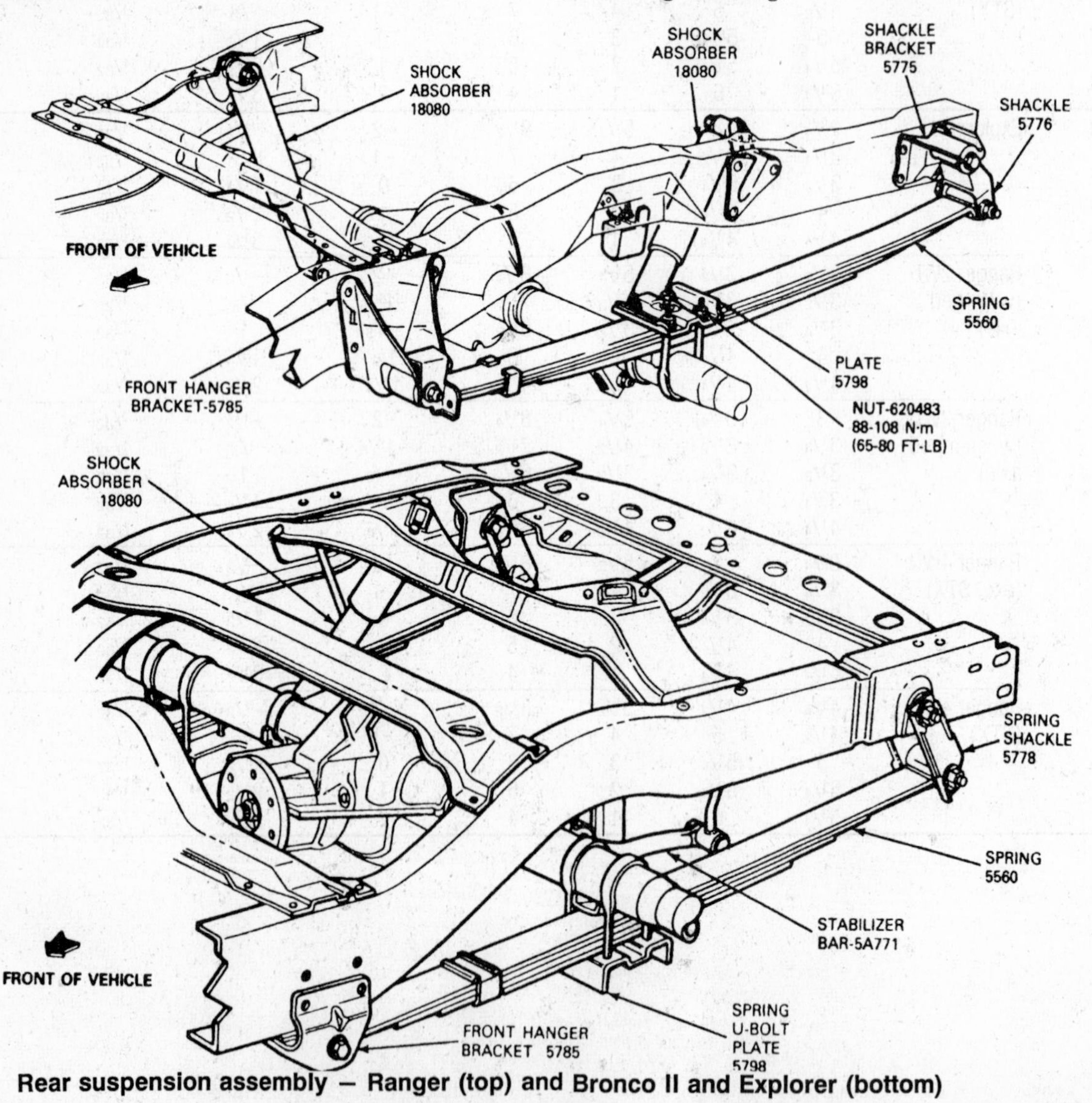

Rear suspension assembly – Ranger (top) and Bronco II and Explorer (bottom)

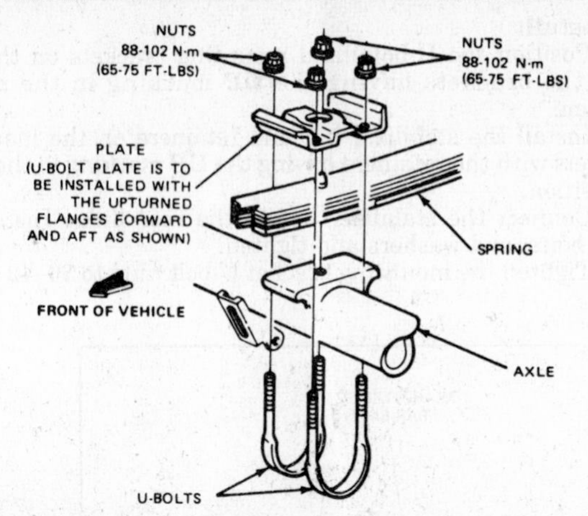

Ranger, Bronco II and Explorer U-bolt and spring plate assembly

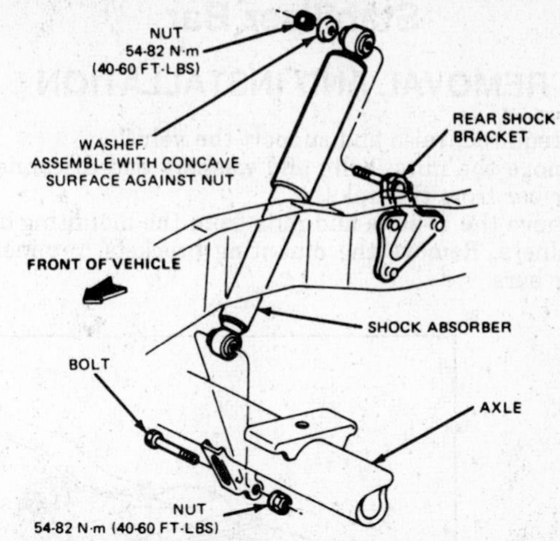

1983–91 Ranger and 1983–90 Bronco II rear shock absorber assembly

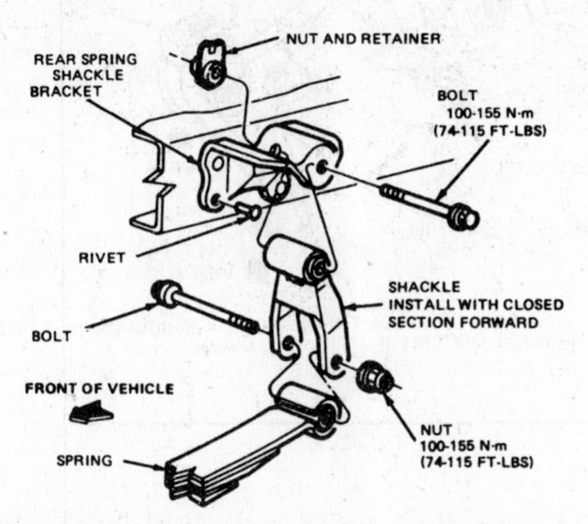

Ranger, Bronco II and Explorer spring to rear shackle assembly

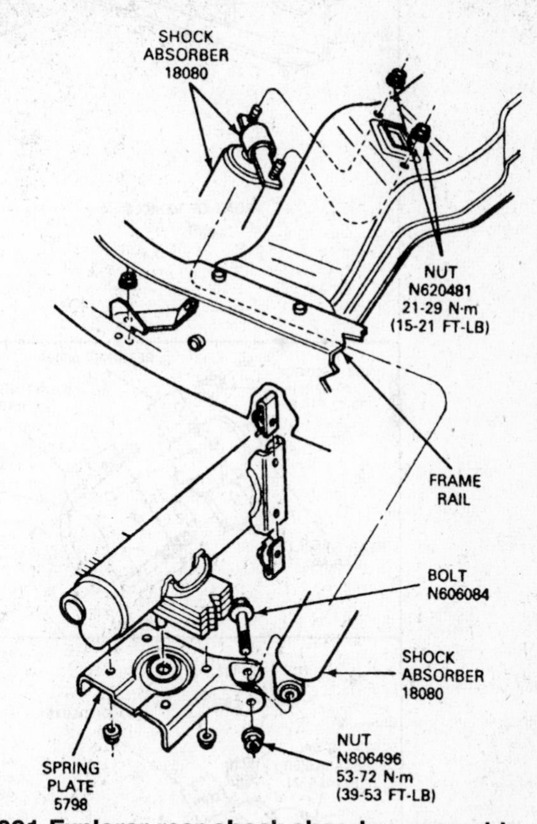

1991 Explorer rear shock absorber assembly

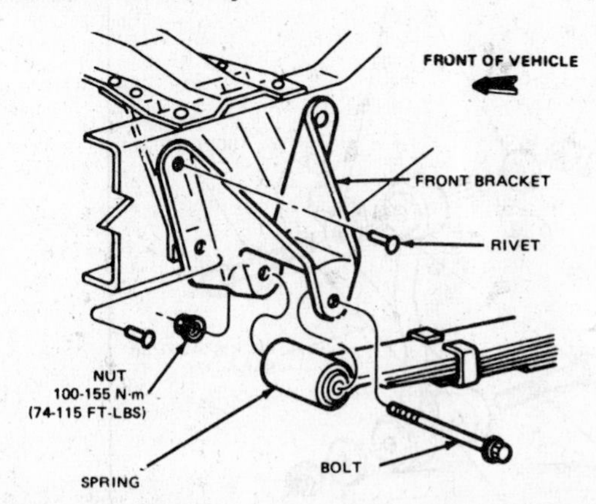

Ranger, Bronco II and Explorer spring to front bracket assembly

4. Remove the shock absorber from the vehicle.
5. Installation is the reverse of the removal procedure. Torque the lower shock absorber retaining bolt to 39–53 ft. lbs.
6. On the Ranger and Bronco II torque the upper shock absorber mounting nut to 39–53 ft. lbs. On the Explorer torque the upper shock absorber retaining nuts to 15–21 ft. lbs.

TESTING

1. Visually check the shock absorbers for the presence of fluid leakage. A thin film of fluid is acceptable. Anything more than that means that the shock absorber must be replaced.
2. Disconnect the lower end of the shock absorber. Compress and extend the shock fully as fast as possible. If the action is not smooth in both directions, or there is no pressure resistance, replace the shock absorber. Shock absorbers should be replaced in pairs. In the case of relatively new shock absorbers, where one has failed, that one, alone, may be replaced.

Stabilizer Bar

REMOVAL AND INSTALLATION

1. As required, raise and support the vehicle.
2. Remove the nuts, bolts and washers and disconnect the stabilizer bar from the links.
3. Remove the U-bolts and nuts from the mounting bracket and retainers. Remove the mounting brackets, retainers and stabilizer bars.

To install:

4. Position the U-bolts and mounting brackets on the axle with the brackets having the **UP** marking in the proper position.
5. Install the stabilizer bar and retainers on the mounting brackets with the retainers having the **UP** marking in the proper position.
6. Connect the stabilizer bar to the rear links. Install the nuts, bolts, and washers and tighten.
7. Tighten the mounting bracket U-bolt nuts to 30–42 ft. lbs.

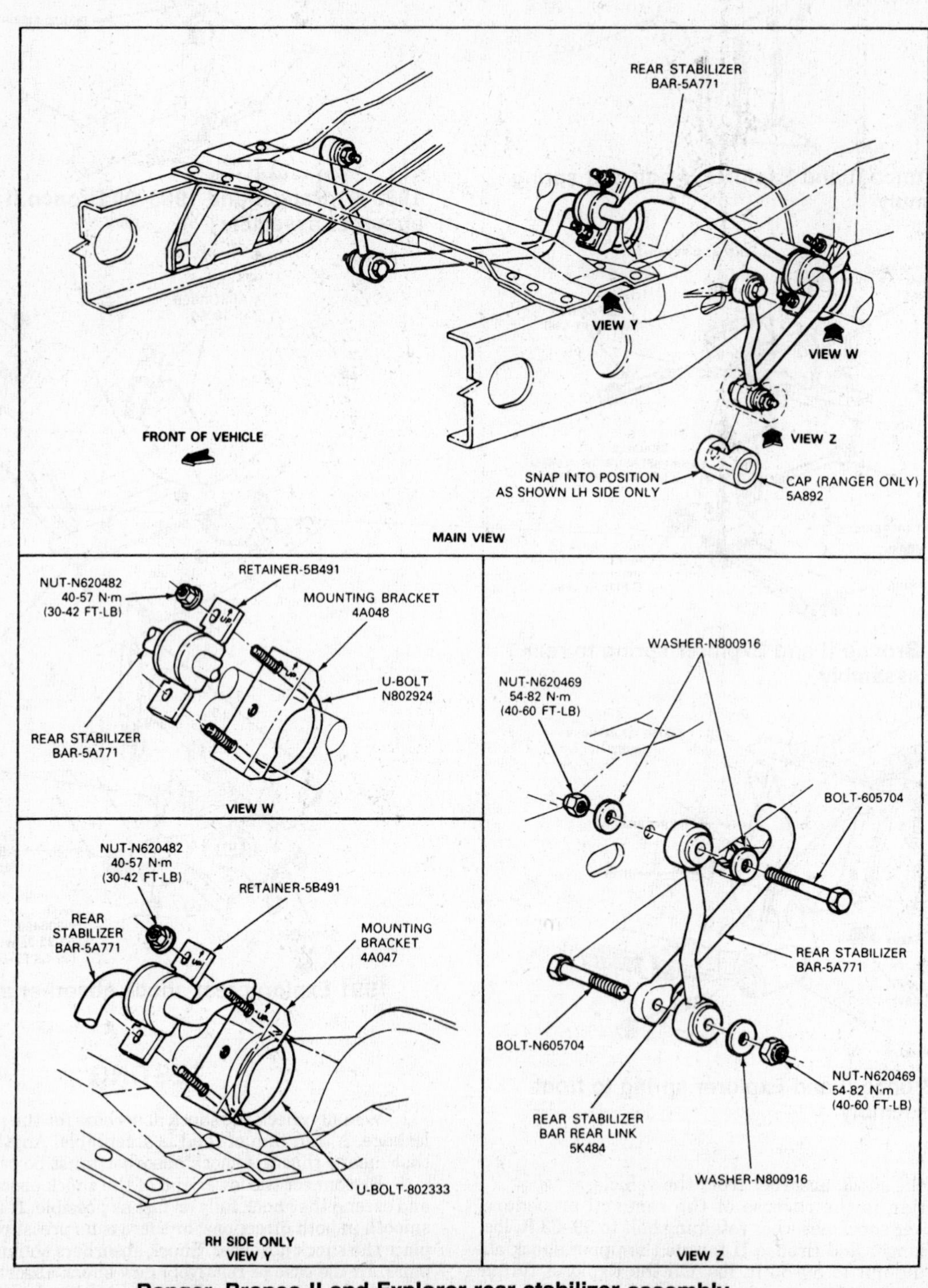

Ranger, Bronco II and Explorer rear stabilizer assembly

STEERING

Troubleshooting the Steering Column

Problem	Cause	Solution
Will not lock	• Lockbolt spring broken or defective	• Replace lock bolt spring
High effort (required to turn ignition key and lock cylinder)	• Lock cylinder defective • Ignition switch defective • Rack preload spring broken or deformed • Burr on lock sector, lock rack, housing, support or remote rod coupling • Bent sector shaft • Defective lock rack • Remote rod bent, deformed • Ignition switch mounting bracket bent • Distorted coupling slot in lock rack (tilt column)	• Replace lock cylinder • Replace ignition switch • Replace preload spring • Remove burr • Replace shaft • Replace lock rack • Replace rod • Straighten or replace • Replace lock rack
Will stick in "start"	• Remote rod deformed • Ignition switch mounting bracket bent	• Straighten or replace • Straighten or replace
Key cannot be removed in "off-lock"	• Ignition switch is not adjusted correctly • Defective lock cylinder	• Adjust switch • Replace lock cylinder
Lock cylinder can be removed without depressing retainer	• Lock cylinder with defective retainer • Burr over retainer slot in housing cover or on cylinder retainer	• Replace lock cylinder • Remove burr
High effort on lock cylinder between "off" and "off-lock"	• Distorted lock rack • Burr on tang of shift gate (automatic column) • Gearshift linkage not adjusted	• Replace lock rack • Remove burr • Adjust linkage
Noise in column	• One click when in "off-lock" position and the steering wheel is moved (all except automatic column) • Coupling bolts not tightened • Lack of grease on bearings or bearing surfaces • Upper shaft bearing worn or broken • Lower shaft bearing worn or broken • Column not correctly aligned • Coupling pulled apart • Broken coupling lower joint • Steering shaft snap ring not seated	• Normal—lock bolt is seating • Tighten pinch bolts • Lubricate with chassis grease • Replace bearing assembly • Replace bearing. Check shaft and replace if scored. • Align column • Replace coupling • Repair or replace joint and align column • Replace ring. Check for proper seating in groove.

Troubleshooting the Steering Column (cont.)

Problem	Cause	Solution
Noise in column	• Shroud loose on shift bowl. Housing loose on jacket—will be noticed with ignition in "off-lock" and when torque is applied to steering wheel.	• Position shroud over lugs on shift bowl. Tighten mounting screws.
High steering shaft effort	• Column misaligned • Defective upper or lower bearing • Tight steering shaft universal joint • Flash on I.D. of shift tube at plastic joint (tilt column only) • Upper or lower bearing seized	• Align column • Replace as required • Repair or replace • Replace shift tube • Replace bearings
Lash in mounted column assembly	• Column mounting bracket bolts loose • Broken weld nuts on column jacket • Column capsule bracket sheared	• Tighten bolts • Replace column jacket • Replace bracket assembly
Lash in mounted column assembly (cont.)	• Column bracket to column jacket mounting bolts loose • Loose lock shoes in housing (tilt column only) • Loose pivot pins (tilt column only) • Loose lock shoe pin (tilt column only) • Loose support screws (tilt column only)	• Tighten to specified torque • Replace shoes • Replace pivot pins and support • Replace pin and housing • Tighten screws
Housing loose (tilt column only)	• Excessive clearance between holes in support or housing and pivot pin diameters • Housing support-screws loose	• Replace pivot pins and support • Tighten screws
Steering wheel loose—every other tilt position (tilt column only)	• Loose fit between lock shoe and lock shoe pivot pin	• Replace lock shoes and pivot pin
Steering column not locking in any tilt position (tilt column only)	• Lock shoe seized on pivot pin • Lock shoe grooves have burrs or are filled with foreign material • Lock shoe springs weak or broken	• Replace lock shoes and pin • Clean or replace lock shoes • Replace springs
Noise when tilting column (tilt column only)	• Upper tilt bumpers worn • Tilt spring rubbing in housing	• Replace tilt bumper • Lubricate with chassis grease
One click when in "off-lock" position and the steering wheel is moved	• Seating of lock bolt	• None. Click is normal characteristic sound produced by lock bolt as it seats.
High shift effort (automatic and tilt column only)	• Column not correctly aligned • Lower bearing not aligned correctly • Lack of grease on seal or lower bearing areas	• Align column • Assemble correctly • Lubricate with chassis grease

Troubleshooting the Steering Column (cont.)

Problem	Cause	Solution
Improper transmission shifting—automatic and tilt column only	• Sheared shift tube joint • Improper transmission gearshift linkage adjustment • Loose lower shift lever	• Replace shift tube • Adjust linkage • Replace shift tube

Troubleshooting the Ignition Switch

Problem	Cause	Solution
Ignition switch electrically inoperative	• Loose or defective switch connector • Feed wire open (fusible link) • Defective ignition switch	• Tighten or replace connector • Repair or replace • Replace ignition switch
Engine will not crank	• Ignition switch not adjusted properly	• Adjust switch
Ignition switch wil not actuate mechanically	• Defective ignition switch • Defective lock sector • Defective remote rod	• Replace switch • Replace lock sector • Replace remote rod
Ignition switch cannot be adjusted correctly	• Remote rod deformed	• Repair, straighten or replace

Troubleshooting the Turn Signal Switch

Problem	Cause	Solution
Turn signal will not cancel	• Loose switch mounting screws • Switch or anchor bosses broken • Broken, missing or out of position detent, or cancelling spring	• Tighten screws • Replace switch • Reposition springs or replace switch as required
Turn signal difficult to operate	• Turn signal lever loose • Switch yoke broken or distorted • Loose or misplaced springs • Foreign parts and/or materials in switch • Switch mounted loosely	• Tighten mounting screws • Replace switch • Reposition springs or replace switch • Remove foreign parts and/or material • Tighten mounting screws
Turn signal will not indicate lane change	• Broken lane change pressure pad or spring hanger • Broken, missing or misplaced lane change spring • Jammed wires	• Replace switch • Replace or reposition as required • Loosen mounting screws, reposition wires and retighten screws

Troubleshooting the Turn Signal Switch (cont.)

Problem	Cause	Solution
Turn signal will not stay in turn position	• Foreign material or loose parts impeding movement of switch yoke • Defective switch	• Remove material and/or parts • Replace switch
Hazard switch cannot be pulled out	• Foreign material between hazard support cancelling leg and yoke	• Remove foreign material. No foreign material impeding function of hazard switch—replace turn signal switch.
No turn signal lights	• Inoperative turn signal flasher • Defective or blown fuse • Loose chassis to column harness connector • Disconnect column to chassis connector. Connect new switch to chassis and operate switch by hand. If vehicle lights now operate normally, signal switch is inoperative • If vehicle lights do not operate, check chassis wiring for opens, grounds, etc.	• Replace turn signal flasher • Replace fuse • Connect securely • Replace signal switch • Repair chassis wiring as required
Instrument panel turn indicator lights on but not flashing	• Burned out or damaged front or rear turn signal bulb • If vehicle lights do not operate, check light sockets for high resistance connections, the chassis wiring for opens, grounds, etc. • Inoperative flasher • Loose chassis to column harness connection • Inoperative turn signal switch • To determine if turn signal switch is defective, substitute new switch into circuit and operate switch by hand. If the vehicle's lights operate normally, signal switch is inoperative.	• Replace bulb • Repair chassis wiring as required • Replace flasher • Connect securely • Replace turn signal switch • Replace turn signal switch
Stop light not on when turn indicated	• Loose column to chassis connection • Disconnect column to chassis connector. Connect new switch into system without removing old.	• Connect securely • Replace signal switch

Troubleshooting the Turn Signal Switch (cont.)

Problem	Cause	Solution
Stop light not on when turn indicated (cont.)	Operate switch by hand. If brake lights work with switch in the turn position, signal switch is defective.	
	• If brake lights do not work, check connector to stop light sockets for grounds, opens, etc.	• Repair connector to stop light circuits using service manual as guide
Turn indicator panel lights not flashing	• Burned out bulbs	• Replace bulbs
	• High resistance to ground at bulb socket	• Replace socket
	• Opens, ground in wiring harness from front turn signal bulb socket to indicator lights	• Locate and repair as required
Turn signal lights flash very slowly	• High resistance ground at light sockets	• Repair high resistance grounds at light sockets
	• Incorrect capacity turn signal flasher or bulb	• Replace turn signal flasher or bulb
	• If flashing rate is still extremely slow, check chassis wiring harness from the connector to light sockets for high resistance	• Locate and repair as required
	• Loose chassis to column harness connection	• Connect securely
	• Disconnect column to chassis connector. Connect new switch into system without removing old. Operate switch by hand. If flashing occurs at normal rate, the signal switch is defective.	• Replace turn signal switch
Hazard signal lights will not flash—turn signal functions normally	• Blow fuse	• Replace fuse
	• Inoperative hazard warning flasher	• Replace hazard warning flasher in fuse panel
	• Loose chassis-to-column harness connection	• Conect securely
	• Disconnect column to chassis connector. Connect new switch into system without removing old. Depress the hazard warning lights. If they now work normally, turn signal switch is defective.	• Replace turn signal switch
	• If lights do not flash, check wiring harness "K" lead for open between hazard flasher and connector. If open, fuse block is defective	• Repair or replace brown wire or connector as required

Steering Wheel

REMOVAL AND INSTALLATION

1983–87

1. Disconnect the negative battery cable. Remove the steering wheel hub cover by removing the screws from the spokes and lifting the steering wheel hub cover. On the deluxe wheel, pop the hub emblem off. On sport wheels, unscrew the hub emblem.
2. Disconnect the horn switch wires by pulling the spoke terminal from the blade connectors. On vehicles equipped with speed control, squeeze or pinch the **J** clip ground wire terminal firmly and pull it out of the hole in the steering wheel. Do not pull the ground terminal out of the threaded hole without squeezing the terminal clip to relieve the spring retention of the terminal in the threaded hole.
3. Remove the horn switch assembly and disconnect the horn and speed control wire, if equipped.
4. Remove the steering wheel attaching nut.
5. Using a steering wheel puller, remove the steering wheel from the upper steering shaft. Do not use a knock-off type steering wheel puller or strike the end of the steering column upper shaft with a hammer. This could cause damage to the steering shaft bearing.

To install:

6. Position the steering wheel on the end of the steering wheel shaft. Align the mark and the flats on the steering wheel with the mark and the flats on the shaft, assuring that the straight ahead steering wheel position corresponds to the straight ahead position of the front wheels.
7. Install the wheel nut. Tighten the nut to 30–40 ft. lbs.
8. Install the horn switch assembly and connect the horn and speed control wire, if equipped.
9. Install the cover or trim emblem.
10. Check the steering column for proper operation.

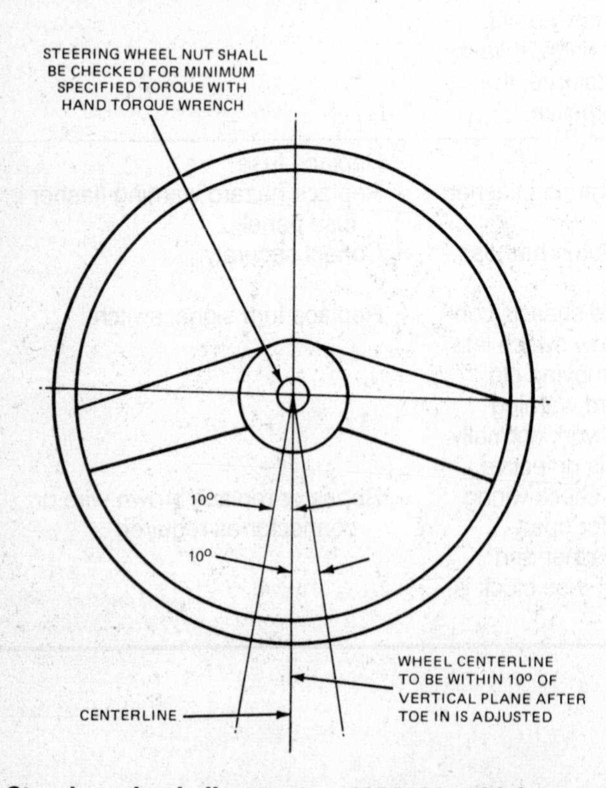

Steering wheel alignment – 1989–91 vehicles

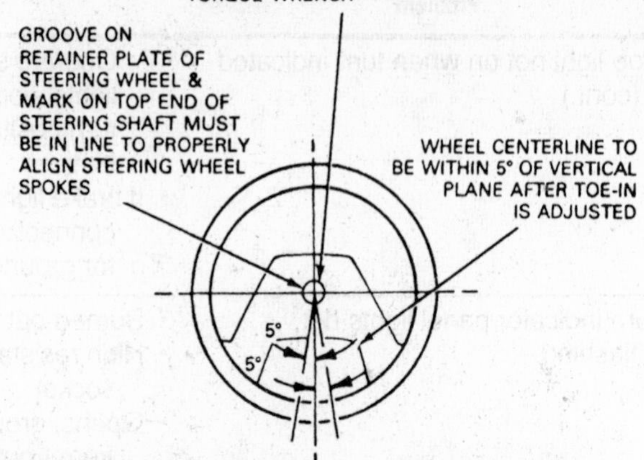

Steering wheel alignment – 1983–88 vehicles

1988–91

1. Disconnect the negative battery cable.
2. From the underside of the steering wheel, remove the screws that hold the steering wheel pad to the steering wheel spokes.
3. Lift up the steering wheel pad and disconnect the horn wires from the steering wheel pad by pulling the spade terminal from the blade connectors.
4. Remove the steering wheel pad. Remove the bolt from the steering shaft.
5. Using the proper steering wheel removal tool, remove the steering wheel from the steering column.

WARNING: Do not hammer on the steering wheel or the steering shaft or use a knock off type steering wheel puller as damage to the steering column will occur.

6. Installation is the reverse of the removal procedure. Be sure that the steering wheel is properly aligned before installin the lock bolt. Torque the steering wheel lock bolt to 23–33 ft. lbs.

Combination Switch

REMOVAL AND INSTALLATION

1983–88

WARNING: The corrugated outer tube steering shaft column upper support bracket assembly and shrouds affect energy absorption on impact. It is absolutely necessary to handle these components with care when performing any service operation!

1. Disconnect the negative battery cable. For tilt column only, remove the upper extension shroud by squeezing it at the 6 and 12 o'clock positions and popping it free of the retaining plate at the 3 o'clock position.
2. Remove the 2 trim shroud halves by removing the 2 attaching screws.
3. Remove the turn signal switch lever by grasping the lever and by using a pulling and twisting motion of the hand while pulling the lever straight out from the switch.
4. Peel back the foam sight shield from the turn signal switch.
5. Disconnect the 2 turn signal switch electrical connectors.
6. Remove the 2 self-tapping screws attaching the turn signal switch to the lock cylinder housing. Disengage the switch from the housing.

To install:

7. Align the turn signal switch mounting holes with the corresponding holes in the lock cylinder housing, and install 2 self-tapping screws.
8. Stick the foam sight shield to the turn signal switch.
9. Install the turn signal switch lever into the switch manually, by aligning the key on the lever with the keyway in the switch and by pushing the lever toward the switch to full engagement.
10. Install the 2 turn signal switch electrical connectors to full engagement.
11. Install the steering column trim shrouds.

1989–91

1. Disconnect the negative battery cable. Remove the steering wheel.
2. On vehicles equipped with tilt wheel, remove the tilt lever.
3. On vehicles equipped with tilt wheel, remove the steering column collar by pressing on the collar from the top and bottom while removing the collar.
4. Remove the instrument panel trim cover retaining screws. Remove the trim cover.
5. Remove the 2 screws from the bottom of the steering column shroud. Remove the bottom half of the shroud by pulling the shroud down and toward the rear of the vehicle.
6. If the vehicle is equipped with automatic transmission, move the shift lever as required to aid in removal of the shroud. Lift the top half of the shroud from the column.
7. If the vehicle is equipped with automatic transmission, disconnect the selector indicator actuation cable bv removing the screw from the column casting and the plastic plug at the end of the cable.
8. To remove the plastic plug from the shift lever socket casting push on the nose of the plug until the head clears the casting and pull the plug from the casting.
9. Remove the plastic clip that retains the combination switch wiring to the steering column bracket.
10. Remove the 2 self taping screws that retain the combination switch to the steering column casting. Disengage the switch from the casting.
11. Disconnect the 3 electrical connectors, using caution not to damage the locking tabs. Be sure not to damage the PNDRL cable.
12. Installation is the reverse of the removal procedure. Torque the combination switch retaining screws to 18–27 inch lbs.

Ignition Switch

REMOVAL AND INSTALLATION

1983–88

1. Rotate the lock cylinder key to the Lock position. Disconnect the negative battery cable from the battery.
2. For tilt column only, remove the upper extension shroud by squeezing it at the 6 and 12 o'clock positions and popping it free of the retaining plate at the 3 o'clock position.
3. Remove the 2 trim shroud halves by removing the 2 attaching screws.
4. Disconnect the ignition switch electrical connector.
5. Drill out the break-off head bolts connecting the switch to the lock cylinder housing by using a ⅛ in. (3mm) drill bit.
6. Remove the 2 bolts, using an Ex-3 Easy-out® tool or equivalent.
7. Disengage the ignition switch from the actuator pin.

To install:

8. Rotate the ignition key to the **RUN** position, approximately 90° clockwise from **LOCK**.
9. Install the replacement switch by aligning the holes on the switch casting base with the holes in the lock cylinder housing. Note that the replacement switch is provided in the RUN position. Minor movement of the lock cylinder to align the actuator pin with the **U** shaped slot in the switch carrier may be required.
10. Install the new break-off head bolts and tighten until heads shear off (approximately 35–50 inch lbs.).
11. Connect the electrical connector to the ignition switch.
12. Connect the negative battery cable to the battery terminal. Check the ignition switch for proper operation in all modes.
13. Install the steering column trim shrouds.

1989–91

1. Disconnect the negative battery cable.
2. Remove the steering wheel.
3. As necessary, remove all under dash panels in order to gain access to the ignition switch.
4. As necessary, lower the steering column to gain working clearance.
5. Disconnect the ignition switch electrical connectors.
6. Remove the ignition switch retaining screws from the studs. Disengage the ignition switch from switch rod. Remove the switch from the vehicle.

To install:

7. Position the lock cylinder in the **LOCK** position.
8. To set the switch, position a wire in the opening in the outer surface of the switch through its positions until the wire drops down into the slot.

NOTE: The slot is in the bottom of the switch where the rod must be inserted to allow full movement through the switch positions.

9. Position the ignition switch on the column studs and over the actuating rod. Torque the retaining nuts to 3.3–5.3 ft. lbs.
10. Remove the wire from the slot in the housing. Continue the installation in the reverse order of the removal procedure.

Ignition Lock Cylinder Assembly

REMOVAL AND INSTALLATION

1983–88

1. Disconnect the negative battery cable.
2. Remove the trim shroud. Remove the electrical connector from the key warning switch.
3. Turn the lock cylinder to the **RUN** position.
4. Place a ⅛ in. (3mm) diameter pin or small drift punch in the hole located at 4 o'clock and 1¼ in. (31.75mm) from the outer edge of the lock cylinder housing. Depress the retaining pin, and pull out the lock cylinder.

To install:

5. Prior to installation of the lock cylinder, lubricate the cylinder cavity, including the drive gear, with Lubriplate® or equivalent.
6. To install the lock cylinder, turn the lock cylinder to the RUN position, depress the retaining pin, and insert it into the lock cylinder housing. Assure that the cylinder is fully seated and aligned into the interlocking washer before turning the key to the **OFF** position. This action will permit the cylinder retaining pin to extend into the hole in the lock cylinder housing.
7. Using the ignition key, rotate the lock cylinder to ensure correct mechanical operation in all positions. Install the electrical connector onto the key warning switch.
8. Connect the battery ground cable.
9. Check for proper ignition functions and verify that the column is locked in the **LOCK** position.
10. Install the trim shrouds.

1989–91

1. Disconnect the negative battery cable. Remove the steering wheel.

2. On vehicles equipped with tilt wheel, remove the tilt lever.
3. On vehicles equipped with tilt wheel, remove the steering column collar by pressing on the collar from the top and bottom while removing the collar.
4. Remove the instrument panel trim cover retaining screws. Remove the trim cover.
5. Remove the 2 screws from the bottom of the steering column shroud. Remove the bottom half of the shroud by pulling the shroud down and toward the rear of the vehicle.
6. If the vehicle is equipped with automatic transmission, move the shift lever as required to aid in removal of the shroud. Lift the top half of the shroud from the column.
7. Turn the lock cylinder with the ignition key in it to the **ON** position. On vehicles equipped with automatic transmission be sure that the selector lever is in the **PARK** position.
8. Push down on the lock cylinder retaining pin with a ⅛ in. (3mm) diameter wire pin or small punch. Pull the lock cylinder from the column housing. Disconnect the lock cylinder wiring plug from the horn brush wiring connector.

To install:

9. Prior to installation of the lock cylinder, lubricate the cylinder cavity, including the drive gear, with Lubriplate® or equivalent.
10. To install the lock cylinder, turn the lock cylinder to the **ON** position, depress the retaining pin. Insert the lock cylinder housing into its housing in the flange casting. Be sure that the tab at the end of the cylinder aligns with the slot in the ignition drive gear.
11. Turn the key to the **OFF** position. This action will permit the cylinder retaining pin to extend into the cylinder casting housing hole.
12. Using the ignition key rotate the lock cylinder to ensure correct mechanical operation in all positions. Connect the key warning wire plug.
13. Install the steering column lower shroud. Install the steering wheel.
14. Check for proper vehicle operation in **PARK** and **NEUTRAL**. Also be sure that the start circuit cannot be actuated in **DRIVE** or **REVERSE**.

Steering Column

REMOVAL AND INSTALLATION

1983–88

1. Disconnect the negative battery terminal. Remove the bolt that retains the column steering shaft to the lower steering shaft assembly. Disengage the shaft.
2. Remove the steering wheel.
3. For tilt column only, remove the upper extension shroud by squeezing it at the 6 and 12 o'clock positions and popping it free of the retaining plate at the 3 o'clock position. Remove the 2 trim shroud halves by removing the 2 attaching screws.
4. Remove the steering column cover directly under the steering column on the instrument panel.
5. Disconnect all electrical connections. Loosen the 2 bolts that retain the column to the brake and clutch pedal support bracket. Do not remove these bolts.
6. Remove the 3 screws that retain the steering column toe plate lower seal to the dash.
7. Remove the 2 bolts that retain the column to the brake and clutch pedal support bracket.
8. Lower the steering column and carefully pull it out of the vehicle.

To install:

9. Installation is the reverse of the removal procedure. Be sure to torque the column retaining bolts to 15–22 ft. lbs.
10. Torque the steering column shaft assembly bolt to 19–28 ft. lbs. Check the steering column for proper alignment.

1989–91

1. Disconnect the negative battery cable. Set the parking brake. If equipped with automatic transmission position the selector lever in **NEUTRAL**.
2. Remove the bolt that holds the intermediate shaft to the steering column shaft.
3. Using the proper tool, compress the intermediate shaft until it is clear of the steering column shaft.
4. Remove the nuts from the studs and remove the shift cable bracket from the steering column bracket, if the vehicle is equipped with automatic transmission.
5. If the vehicle is equipped with automatic transmission disconnect the shift cable from the column lever. Remove the steering wheel.

NOTE: If the vehicle is equipped with tilt wheel, be sure that the steering wheel is in the full up position before removing it.

6. On vehicles equipped with tilt wheel, remove the tilt lever.
7. On vehicles equipped with tilt wheel, remove the steering column collar by pressing on the collar from the top and bottom while removing the collar.
8. Remove the instrument panel trim cover retaining screws. Remove the trim cover.
9. Remove the 2 screws from the bottom of the steering column shroud. Remove the bottom half of the shroud by pulling the shroud down and toward the rear of the vehicle.
10. If the vehicle is equipped with automatic transmission, move the shift lever as required to aid in removal of the shroud. Lift the top half of the shroud from the column.
11. If the vehicle is equipped with automatic transmission, disconnect the selector indicator actuation cable by removing the screw from the column casting and the plastic plug at the end of the cable.
12. To remove the plastic plug from the shift lever socket casting push on the nose of the plug until the head clears the casting and pull the plug from the casting.
13. Remove the plastic clip that retains the combination switch wiring to the steering column bracket.
14. Remove the 2 screws form the multi function switch. Remove the switch from the column and leave the wiring connectors attached to the switch. Position the switch and the wiring out of the way.
15. Disconnect the key warning buzzar wire from the horn brush wire. Remove the screw thet holds the horn brush connector to the column. Remove the connector.
16. Remove the 5 screws that hold the toe plate to the dash panel. Loosen the toe plate clamp bolt.
17. Support the steering column assembly. Remove the bolts that hold the breakaway bracket to the pedal support bracket.
18. Pry apart the locking tabs and disconnect the ignition switch wiring harness. Carefully remove the steering from the vehicle.
19. If the vehicle is equipped with automatic transmission, remove the shift cable bracket from the column.

To install:

20. If the vehicle is equipped with automatic transmission, install the shift cable bracket from the column. Torque the cable bracket bolts to 13–22 ft. lbs.
21. Position the steering column in the hole in the vehicle floor.
22. Connect the ignition switch wiring harness to the column connector. Install the column breakaway bolts, but do not tighten.
23. Install and torque the toe plate bolts to 5–8 ft. lbs. Torque the toe plate clamp bolt to 6–13 ft. lbs.
24. Torque the steering column breakaway bolts to 19–27 ft. lbs.
25. Continue the installation in the reverse order of the removal procedure. Torque the steering column shaft to the intermediate shaft U-joint pinch bolt to 25–35 ft. lbs.

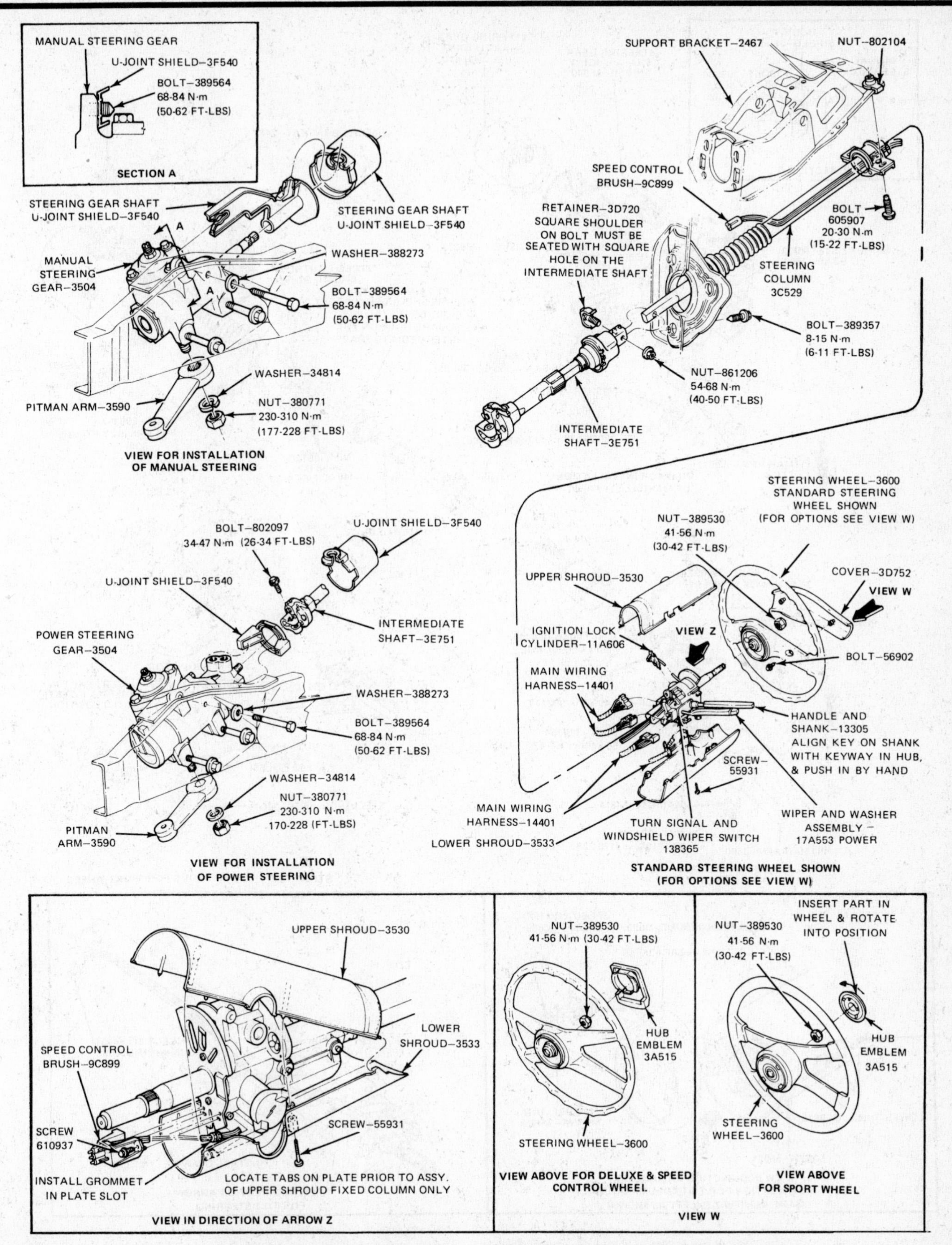

Steering column assembly – 1983–88 vehicles without tilt wheel

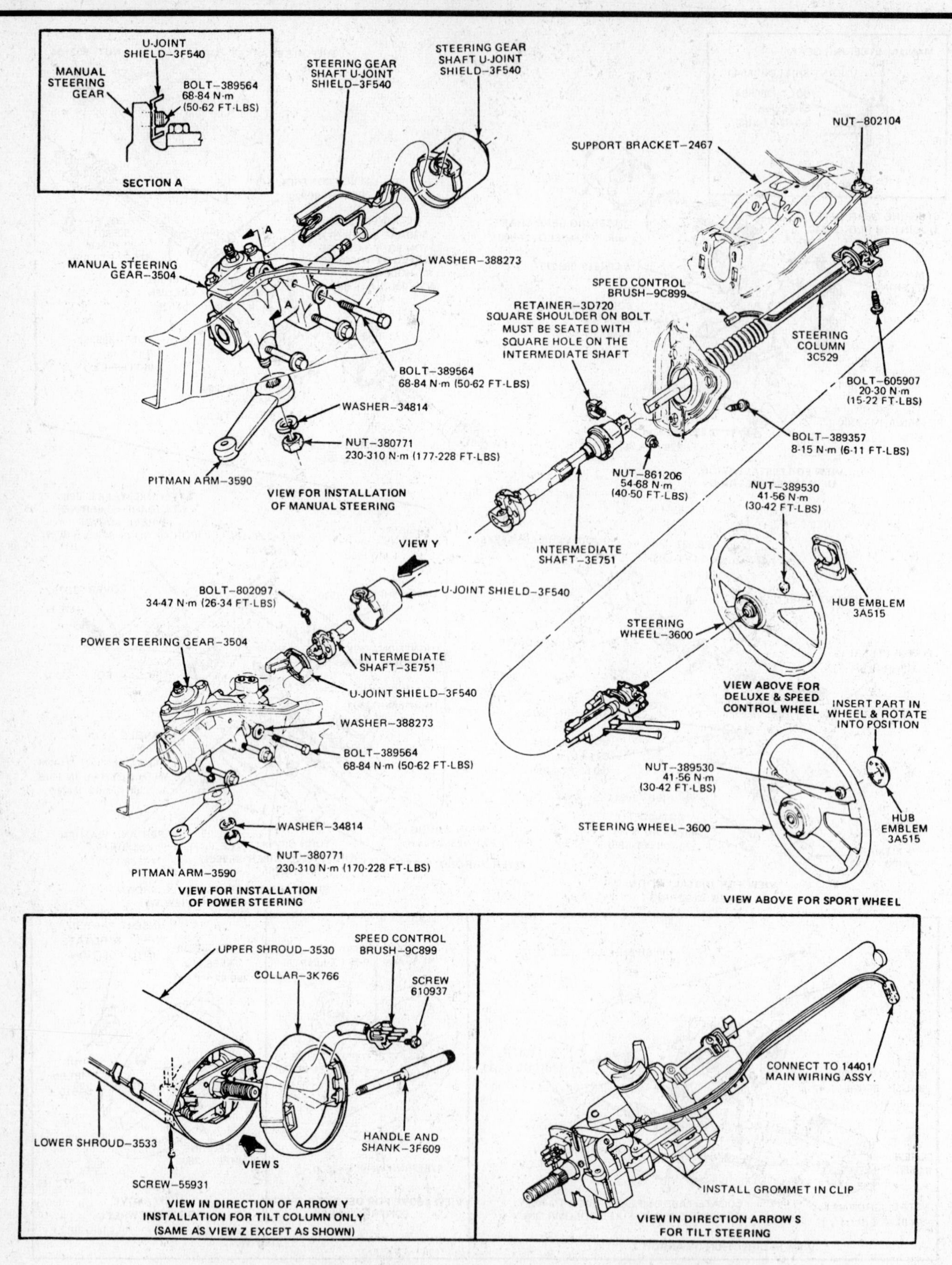

Steering column assembly – 1983–88 vehicles with tilt wheel

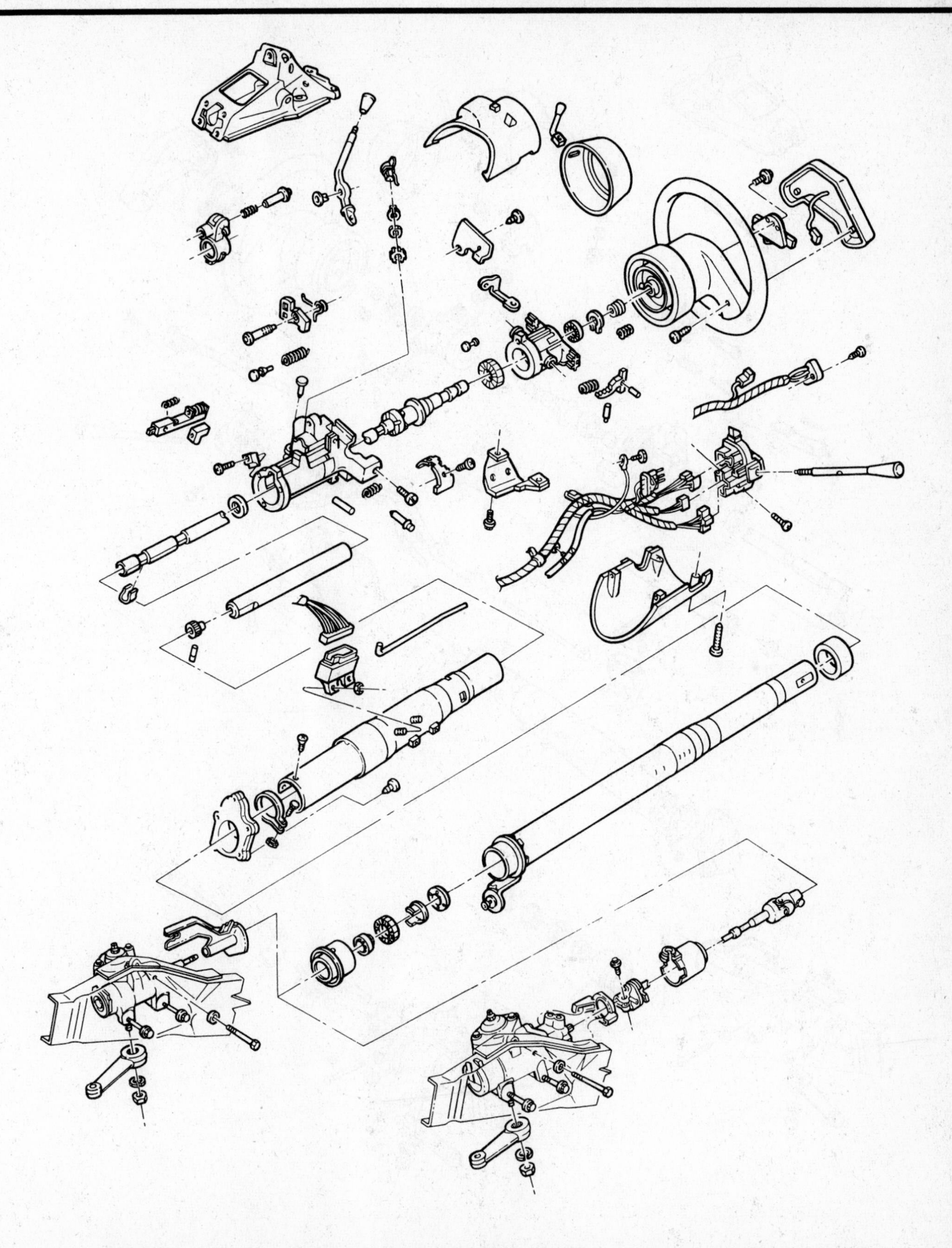

Steering column assembly – 1989–91 vehicles with tilt wheel and automatic transmission

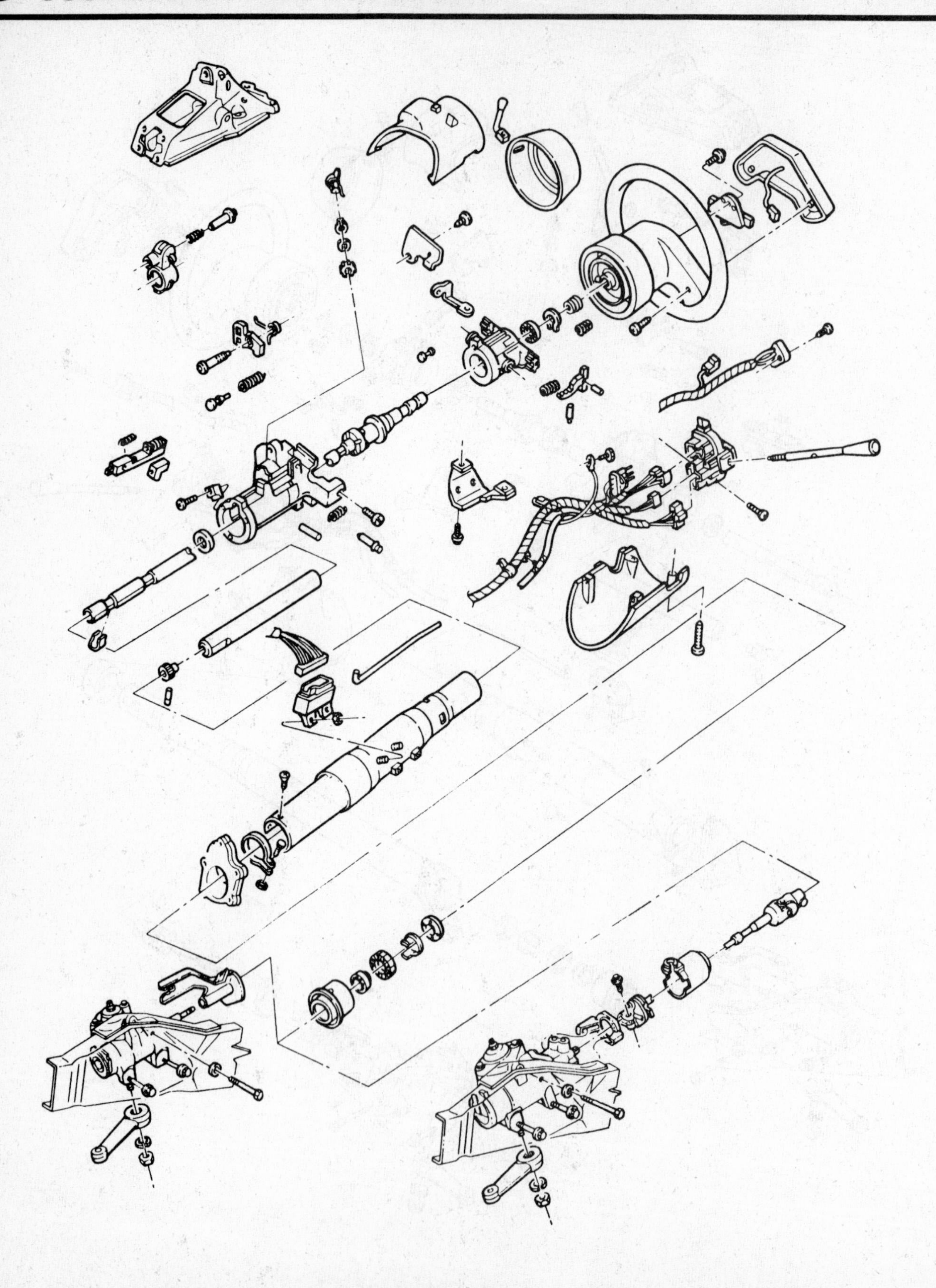

Steering column assembly — 1989–91 vehicles with tilt wheel and manual transmission

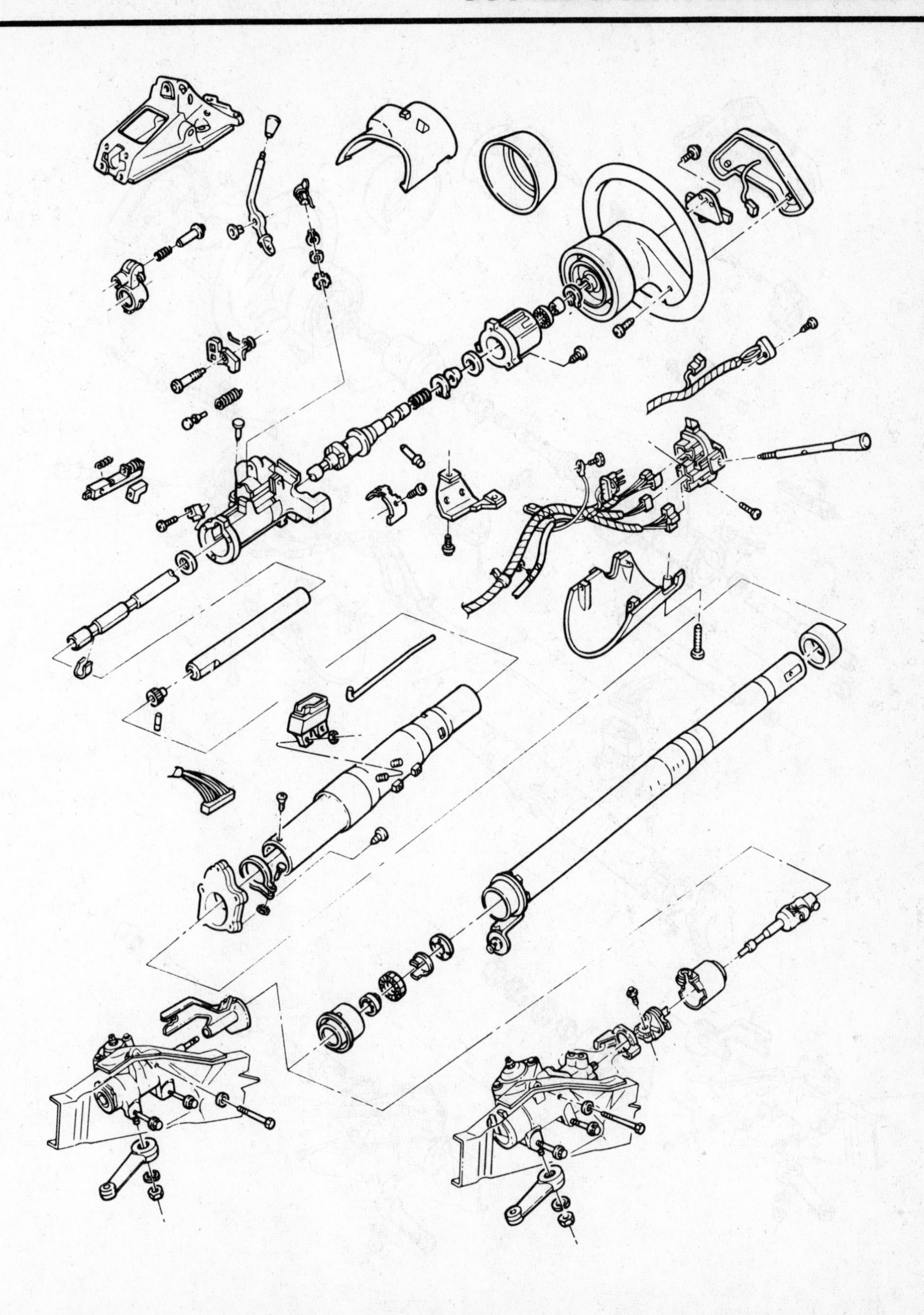

Steering column assembly – 1989–91 vehicles without tilt wheel and automatic transmission

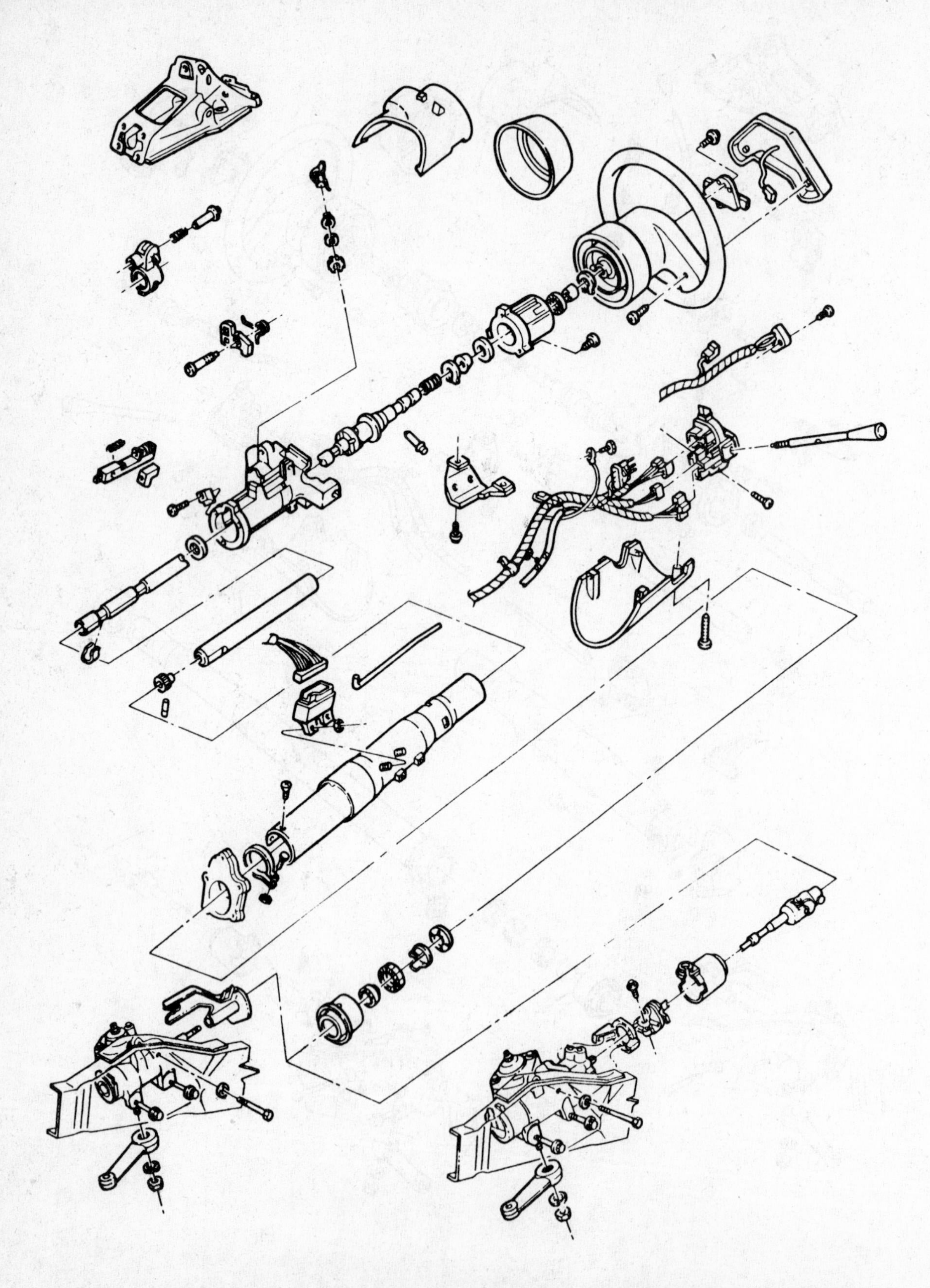

Steering column assembly – 1989–91 vehicles without tilt wheel and manual transmission

Steering Linkage

REMOVAL AND INSTALLATION

Pitman Arm

1. As required, raise and safely support the vehicle using jackstands.
2. Remove the cotter pin and nut from the drag link ball stud at the pitman arm.
3. Remove the drag link ball stud from the pitman arm using pitman arm removal tool T64P–3590–F or equivalent.
4. Remove the pitman arm retaining nut and washer. Remove the pitman arm from the steering gear sector shaft using tool T64P–3590–F or equivalent.

To install:

5. Installation is the reverse of the removal procedure. Torque the pitman arm attaching washer and nut to 170–230 ft. lbs. Torque the drag link ball stud nut to 50–70 ft. lbs. and install a new cotter pin.
6. Check and adjust front end alignment, as required.

Tie Rod

1. Raise and support the vehicle using jackstands. Be sure that the front wheels are in the straight ahead position.
2. Remove the nut and cotter pin from the ball stud on the drag link. Remove the ball stud from the drag link using pitman arm removal tool T64P–3590–F or equivalent.
3. Loosen the bolts on the tie rod adjusting sleeve. Be sure to count and record the number of turns it takes to remove the tie rod from the tie rod adjusting sleeve. Remove the tie rod from the vehicle.

To install:

4. Install the tie rod in the tie rod sleeve in the same number of turns it took to remove it. Torque the tie rod adjusting sleeve nuts to 30–42 ft. lbs.
5. Be sure that the sdjusting sleeve clamps are pointed down ± 45°. Tighten the tie rod ball stud to drag link retaining bolt to 50–75 ft. lbs. Install a new cotter pin.
6. Check and adjust front end alignment, as required.

Tie Rod Ends

1. Raise and support the vehicle using jackstands. Be sure that the front wheels are in the straight ahead position.
2. Remove the nut and cotter pin from the ball stud on the drag link. Remove the ball stud from the drag link using pitman arm removal tool T64P–3590–F or equivalent.
3. Loosen the bolts on the tie rod adjusting sleeve. Be sure to count and record the number of turns it takes to remove the sleeve from the ball stud.

To install:

4. Install the adjusting sleeve on the tie rod ball stud in the same number of turns it took to remove it. Loosley assemble the ball stud in the spindle arm.
5. Torque the retaining nuts to 30–42 ft. lbs. Be sure that the sdjusting sleeve clamps are pointed down ± 45°.
6. With the vehicle wheels in the straight ahead position install and torque the nut to 50–75 ft. lbs. Install a new cotter pin.
7. Check and adjust front end alignment, as required.

Drag Link

1. Raise and support the vehicle using jackstands. Be sure that the front wheels are in the straight ahead position.
2. Remove the nuts and cotter pins from the ball stud at the pitman arm and steering tie rod. Remove the ball studs from the linkage using pitman arm removal tool T64P–3590–F or equivalent.
3. Loosen the bolts on the drag link adjusting sleeve. Be sure to count and record the number of turns it takes to remove the drag link.

To install:

4. Install the drag link in the same number of turns it took to remove it. Tighten the adjusting sleeve nuts to 30–42 ft. lbs. Be sure that the sdjusting sleeve clamps are pointed down ± 45°.
5. Position the drag link ball stud in the pitman arm. Position

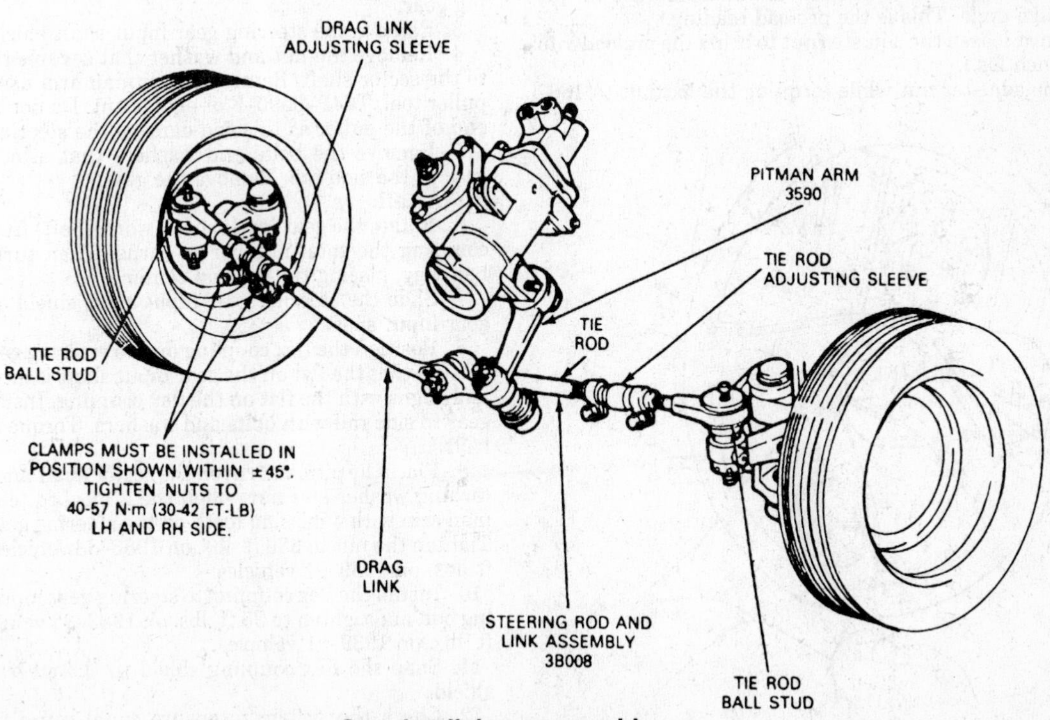

Steering linkage assembly

the steering tie rod ball stud in the drag link. With the vehicle wheels in the straight ahead position install and torque the nuts to 50–75 ft. lbs. Install a new cotter pin.

6. Check and adjust front end alignment, as required.

Manual Steering Gear

ADJUSTMENTS

Preload and Meshload Check

1. Raise and support the front of the vehicle on jackstands.
2. Disconnect the the pitman arm at the ball stud.
3. Lubricate the wormshaft seal with a drop of automatic transmission fluid.
4. Remove the horn pad from the steering wheel.
5. Turn the steering wheel slowly to one stop.
6. Using an inch-pound torque wrench on the steering wheel nut, check the amount of torque needed to rotate the steering wheel through a 1½ turn cycle. The preload should be 2–6 inch lbs. If correct, proceed with the rest of the Steps. If not perform preload and mesh adjustment.
7. Rotate the steering wheel from stop-to-stop, counting the total number of turns. Using that figure, center the steering wheel (½ the total turns).
8. Using the inch-pound torque wrench, rotate the steering wheel 90° to either side of center, noting the highest torque reading over center. The meshload should be 4–10 inch lbs., or at least 2 inch lbs. more than the preload figure.

Preload and Meshload Adjustment

1. Remove the steering gear from the vehicle.
2. Torque the sector cover bolts on the gear to 32–40 ft. lbs.
3. Loosen the preload adjuster nut and tighten the worm bearing adjuster nut until all endplay has been removed. Lubricate the wormshaft seal with a few drops of automatic transmission fluid.
4. Using an 11/16 in., 12-point socket and an inch lbs. torque wrench, carefully turn the wormshaft all the way to the right.
5. Turn the shaft back to the left and measure the torque over a 1½ turn cycle. This is the preload reading.
6. Tighten or loosen the adjuster nut to bring the preload into range (5–6 inch lbs.).
7. Hold the adjuster nut while torquing the locknut to 166–187 ft. lbs.

Manual steering gear — preload measurement

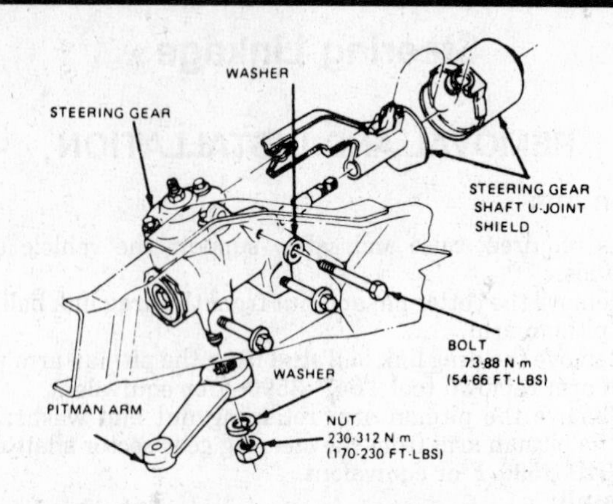

Manual steering gear and related components

8. Rotate the wormshaft stop-to-stop counting the total number of turns and center the shaft (½ the total turns).
9. Using the torque wrench and socket, measure the torque required to turn the shaft 90° to either side of center.
10. Turn the sector shaft adjusting screw as needed to bring the meshload torque within the 9–11 inch lbs. range, or at least 4 inch lbs. higher than the preload torque.
11. Hold the adjusting screw while tightening the locknut to 14–25 ft. lbs.
12. Install the gear.

REMOVAL AND INSTALLATION

1. Raise and safely support the vehicle using jackstands. Disengage the flex coupling shield from the steering gear input shaft shield and slide it up the intermediate shaft.
2. Remove the bolt that retains the flex coupling to the steering gear.
3. Remove the steering gear input shaft shield.
4. Remove the nut and washer that secures the pitman arm to the sector shaft. Remove the pitman arm using pitman arm puller tool, T64P–3590–F or equivalent. Do not hammer on the end of the puller as this can damage the steering gear.
5. Remove the bolts and washers that attach the steering gear to the side rail. Remove the gear.

To install:

6. Rotate the gear input shaft (wormshaft) from stop to stop, counting the total number of turns. Then turn back exactly half-way, placing the gear on center.
7. Slide the steering gear input shaft shield on the steering gear input shaft.
8. Position the flex coupling on the steering gear input shaft. Ensure that the flat on the gear input shaft is facing straight up and aligns with the flat on the flex coupling. Install the steering gear to side rail with bolts and washers. Torque the bolts to 66 ft. lbs.
9. Place the pitman arm on the sector shaft and install the attaching washer and nut. Align the 2 blocked teeth on the Pitman arm with 4 missing teeth on the steering gear sector shaft. Tighten the nut to 230 ft. lbs. on 1983–88 vehicles and 170–230 ft. lbs. on 1989–91 vehicles.
10. Install the flex coupling to steering gear input shaft attaching bolt and tighten to 35 ft. lbs. on 1983–88 vehicles and 50–62 ft. lbs. on 1989–91 vehicles.
11. Snap the flex coupling shield to the steering gear input shield.
12. Check the system to ensure equal turns from center to each lock position.

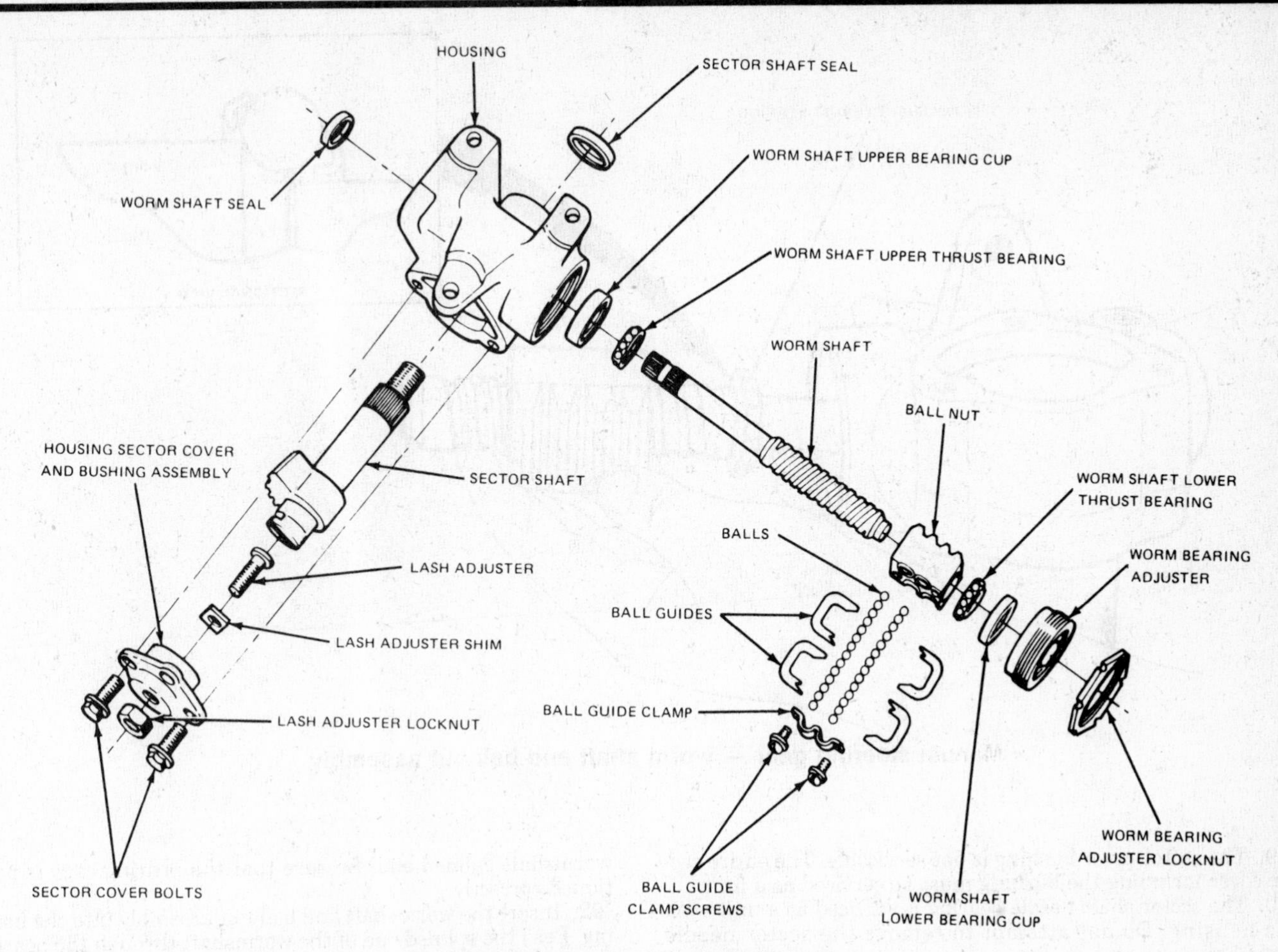

Manual steering gear — exploded view

OVERHAUL

1. Remove the steering gear from the vehicle. Position the assembly in a suitable holding fixture.
2. Rotate the wormshaft from stop to stop, counting the total number of turns. Then turn back exactly half way, placing the gear on center.
3. Remove the sector adjusting cover bolts. Remove the sector shaft along with the cover. Remove the cover from the shaft by turning the screw clockwise. Be sure to keep the shim with the screw.
4. Loosen the worm bearing adjuster locknut and remove the adjuster plug and the wormshaft thrust bearing.
5. Carefully pull the wormshaft and the ball nut assembly from the housing and remove the upper thrust bearing. Avoid damage to the return guides by keeping the ball nut from running down to either end of the worm.
6. Pry out the sector shaft and wormshaft seals and discard them.

NOTE: Because the individual parts of the manual steering gear are not servicable do not disassemble parts. If the worm gear does not rotate freely in the ball nut, replace the entire assembly. Avoid damage to the return guides by keeping the ball nut from running down to either end of the worm.

7. Remove the adjuster cup plug using tool T58L–101B or equivalent and a slide hammer.
8. Remove the bearing cup from the housing using a hammer and a suitable size bearing driver or socket.

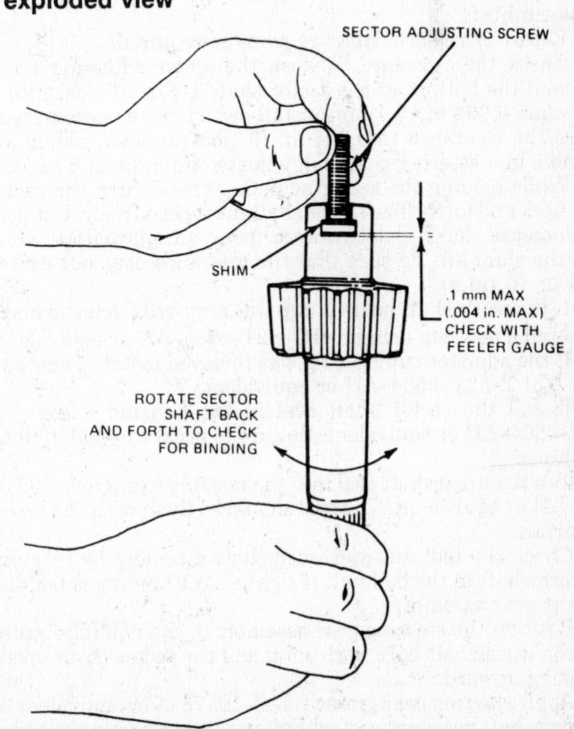

Manual steering gear — sector shaft inspection

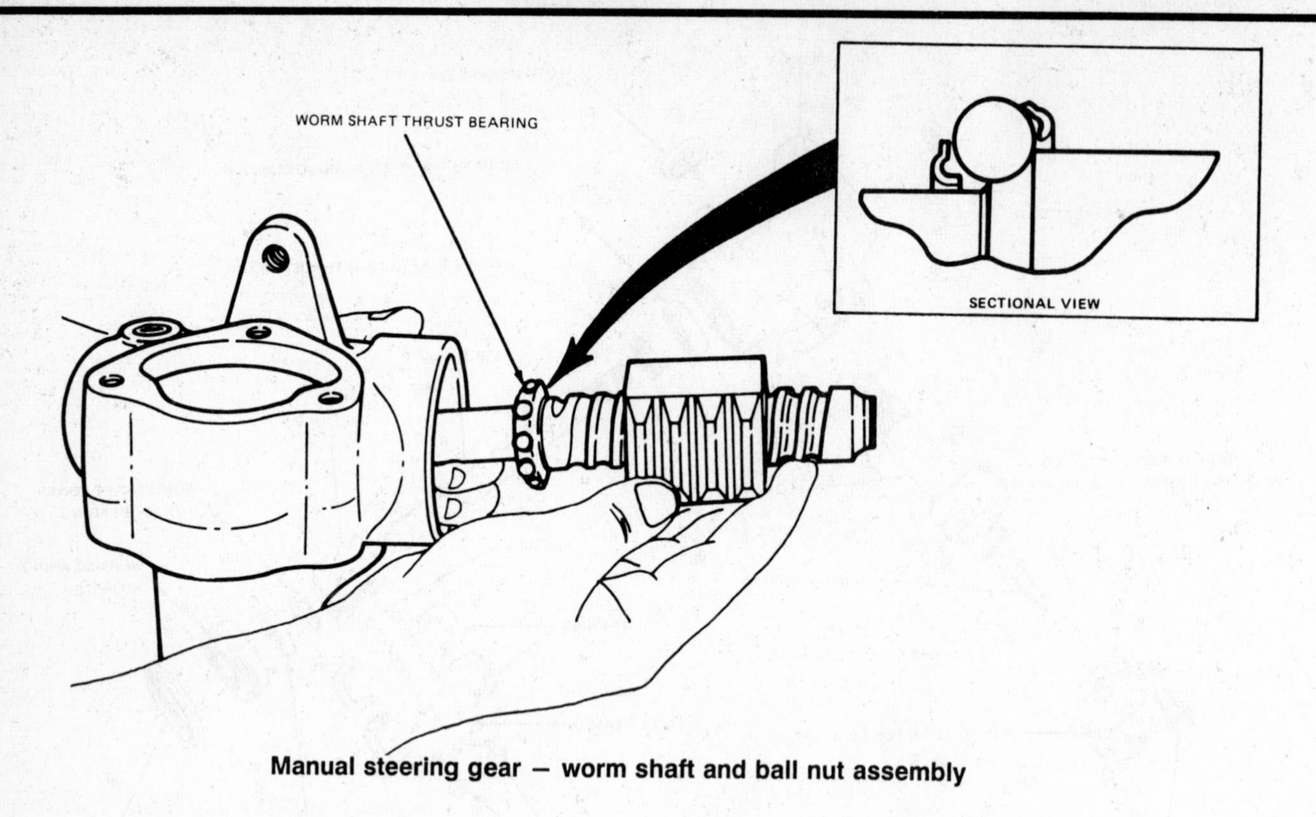

Manual steering gear — worm shaft and ball nut assembly

9. The sector cover bushing is not servicable. The entire sector cover including the bushing must be serviced as a unit.
10. The sector shaft needle bearing is serviced as a unit with the housing. Do not attempt to remove the sector needle assembly.

To assemble:

11. Repair or replace defective parts as required.
12. Check the clearance between the sector adjusting screw head and the bottom of the sector shaft T slot. If clearance is more than 0.004 in. (0.1mm) install new shims as necessary to reduce the clearance to 0.004 in. (0.1mm) or less. Shims are available in a steering gear lash adjuster shim kit in 5 sizes.
13. While holding the sector adjusting screw, turn the sector shaft back and forth. The sector shaft must turn freely. If it does not, increase the T slot clearance using an appropriate shim from the shim kit. Be sure that the clearance does not exceed 0.004 in. (0.1mm).
14. If the wormshaft bearing cup was removed from the housing install a new cup using tool T-82T-3504-AH or equivalent.
15. If the adjuster cup bearing was removed install a new cup using tool T-82T-3504-AH or equivalent.
16. Install the sector shaft seal in the housing using tool T82T-3504-AH or equivalent. Be sure to press the seal in until it bottoms.
17. Tap the wormshaft seal into the housing using tool T82T-3504-AH or equivalent. The seal should be flush with the housing surface.
18. Check the ball nut and wormshaft assembly by rotating the wormshaft in the ball nut. If tightness or binding is felt discard the gear assembly.
19. Position the steering gear assembly in the holding fixture with the wormshaft bore horizontal and the sector cover opening facing upward.
20. Apply steering gear grease C3AZ-19578-A or equivalent to the wormshaft bearings, sector shaft needle bearing in the housing and the sector cover bushing.
21. Slip one of the wormshaft thrust bearings over the wormshaft splined end. Be sure that the bearing cage is positioned correctly.
22. Insert the wormshaft and ball nut assembly into the housing. Feed the splined end of the wormshaft through the bearing cup and seal. Place the remaining wormshaft thrust bearing in the adjuster plug bearing cup.
23. Install the adjuster plug and locknut into the housing opening. Be careful to guide the wormshaft end into the bearing until nearly all of the endplay has been removed from the wormshaft.
24. Position the sector adjusting screw and shim into the sector shaft slot. Check the clearance between the sector adjusting screw head and the bottom of the sector shaft T slot. If clearance is more than 0.004 in. (0.1mm) install new shims as necessary to reduce the clearance to 0.004 in. (0.1mm) or less. Shims are available in a steering gear lash adjuster shim kit in 5 sizes.
25. Lubricate the steering gear assembly using 10.2 oz. (weight) of gear lubricant C3AZ-19578-A or equivalent. Rotate the wormshaft untilthe ball nut is near the end of its travel. Pack as much grease into the housing as possible without loosing it out at the sector shaft opening.
26. Rotate the wormshaft to move the ball nut near the other end of its travels and pack more grease into the housing. Rotate the wormshaft until the ball nut is in the center of its travel.
27. Insert the sector shaft assembly containing the adjuster screwand shim into the housing so that the center tooth of the sector shaft enters the center rack tooth space in the ball nut. Rotating the ball nut teeth slightly up will aid in alignment of the gear teeth and installation of the sector shaft.
28. Apply a ⅛ in. (3mm) bead of sealant D6AZ-19562-AA or equivalent to the mating surfaces of the sector cover and housing.
29. Wait about 5 minutes and engage the sector adjuster screw with the tapped hole in the center of the sector cover by turning the screw counterclockwise. Contine to turn the screw until the sector cover is flush with the housing.
30. Install the sector cover to housing retaining bolts. Do not

torque the bolts to 32–40 ft. lbs. unless there is a lash between the sector shaft and the wormshaft. Lash can be obtained by turning the screw counterclockwise.

31. Make the required steering gear adjustments. Install the steering gear in the vehicle.

Power Steering Gear

ADJUSTMENTS

Meshload

1. As required, raise and support the vehicle using jackstands.
2. Disconnect the pitman arm from the sector shaft using tool T64P–3590–F or equivalent.
3. Disconnect and cap the fluid return line at the reservoir return line pipe.
4. Place the end of the return line in a clean container and turn the steering wheel from stop to stop several times to discharge the fluid from the gear. Discard the used fluid.
5. Turn the steering gear 45° from the right stop.
6. Remove the steering wheel hub cover. Attach an inch lb. torque wrench to the steering wheel nut and determine the torque required to rotate the shaft slowly about ⅛ turn from the 45° position toward center.
7. Turn the steering wheel back to center and determine the torque required to rotate the shaft back and forth across the center position.
8. Specification for vehicles under 5000 miles is 12–24 inch lbs. If the vehicle has over 5000 miles, reset the meshload measured while rocking the input shaft over center is less than 10 inch lbs. greater than torque 45° from the right stop.
9. If reset is required loosen the adjuster locknut and turn the sector shaft adjuster screw until the reading is the specified value greater than the torque at 45° from the stop. Hold the sector shaft screw in place and tighten the locknut.

REMOVAL AND INSTALLATION

1. Disconnect the pressure and return lines from the steering gear. Plug the lines and the ports in the gear to prevent entry of dirt.
2. Remove the upper and lower steering gear shaft U-joint shield from the flex coupling. Remove the bolts that secure the flex coupling to the steering gear and to the column steering shaft assembly.
3. Raise the vehicle and remove the pitman arm attaching nut and washer.
4. Remove the pitman arm from the sector shaft using tool T64P–3590–F. Remove the tool from the pitman arm. Do not damage the seals.
5. Support the steering gear, and remove the steering gear attaching bolts.
6. Work the steering gear free of the flex coupling. Remove the steering gear from the vehicle.

To install:

7. Install the lower U-joint shield onto the steering gear lugs. Slide the upper U-joint shield into place on the steering shaft assembly.
8. Slide the flex coupling into place on the steering shaft assembly. Turn the steering wheel so that the spokes are in the horizontal position. Center the steering gear input shaft.
9. Slide the steering gear input shaft into the flex coupling and into place on the frame side rail. Install the attaching bolts and tighten to 50–62 ft. lbs. Tighten the flex coupling bolt 26–34 ft. lbs.
10. Be sure the wheels are in the straight ahead position, then install the pitman arm on the sector shaft. Install the pitman arm attaching washer and nut. Tighten nut to 170–230 ft. lbs.
11. Connect and tighten the pressure and the return lines to the steering gear.
12. Disconnect the coil wire. Fill the reservoir. Turn on the ignition and turn the steering wheel from left to right to distribute the fluid.
13. Recheck fluid level and add fluid, if necessary. Connect the coil wire, start the engine and turn the steering wheel from side to side. Inspect for fluid leaks.

OVERHAUL

1. Remove the steering gear from the vehicle. Hold the unit upside down and drain the fluid into a suitable container. Position the assembly in a holding fixture.
2. Remove the nut from the sector shaft adjusting screw. Turn the input shaft to either stop and then back 2 turns to the center of the gear. The indexing flat on the input shaft spline should face upward.

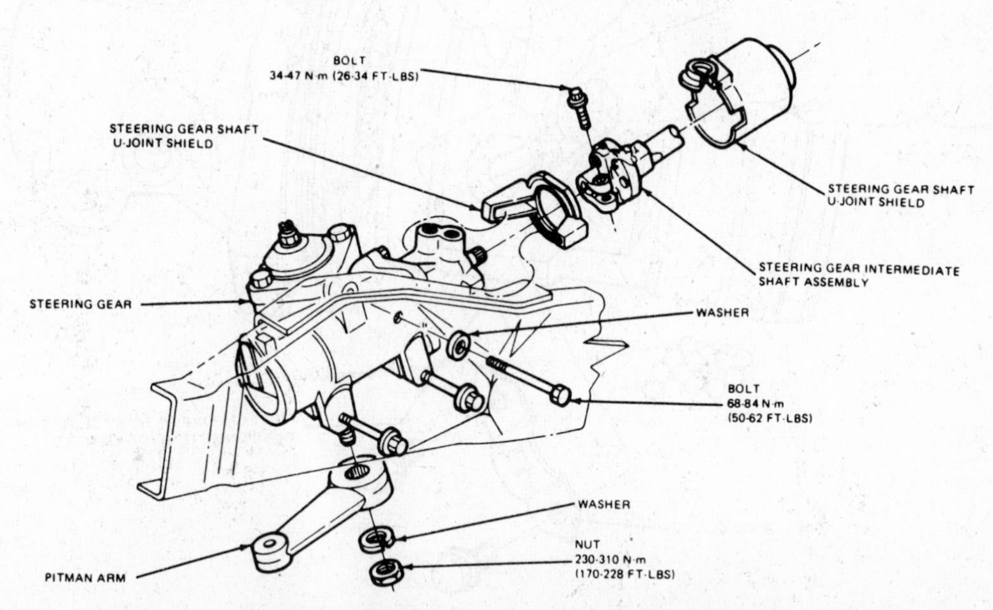

Power steering gear and related components

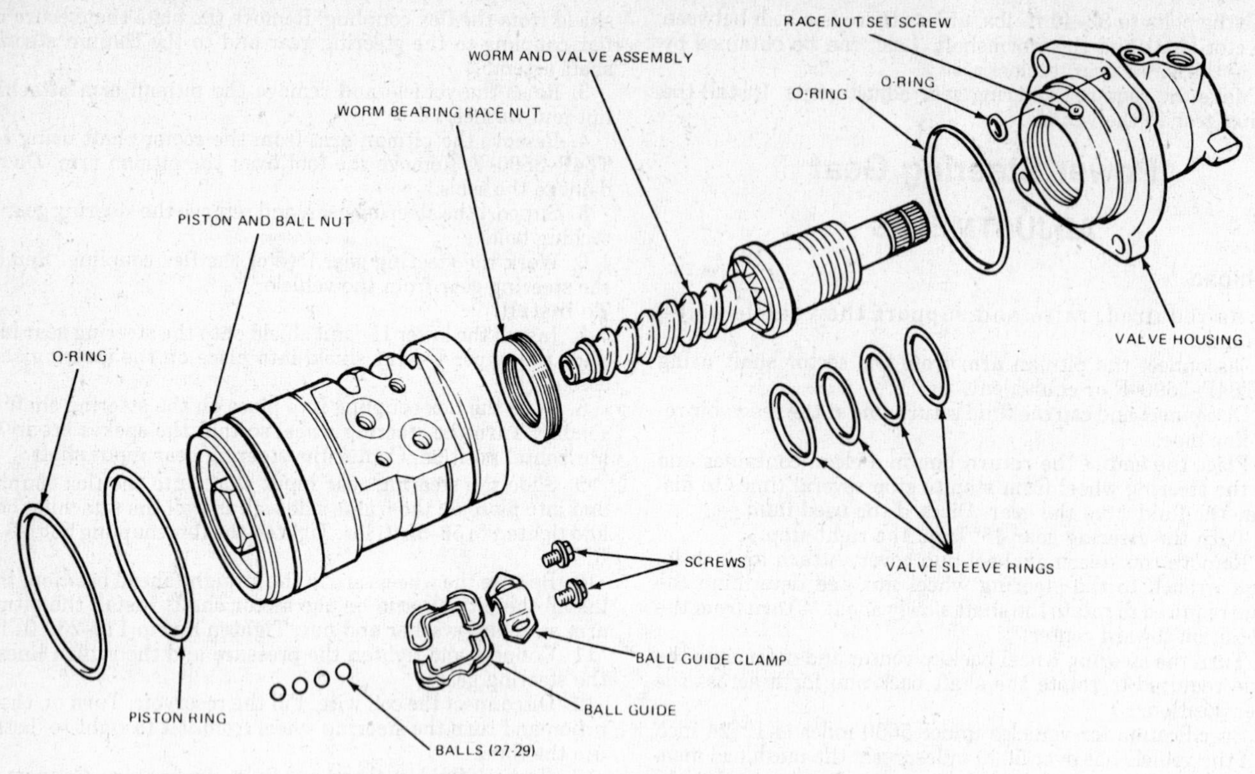

Power steering gear — ball nut housing assembly

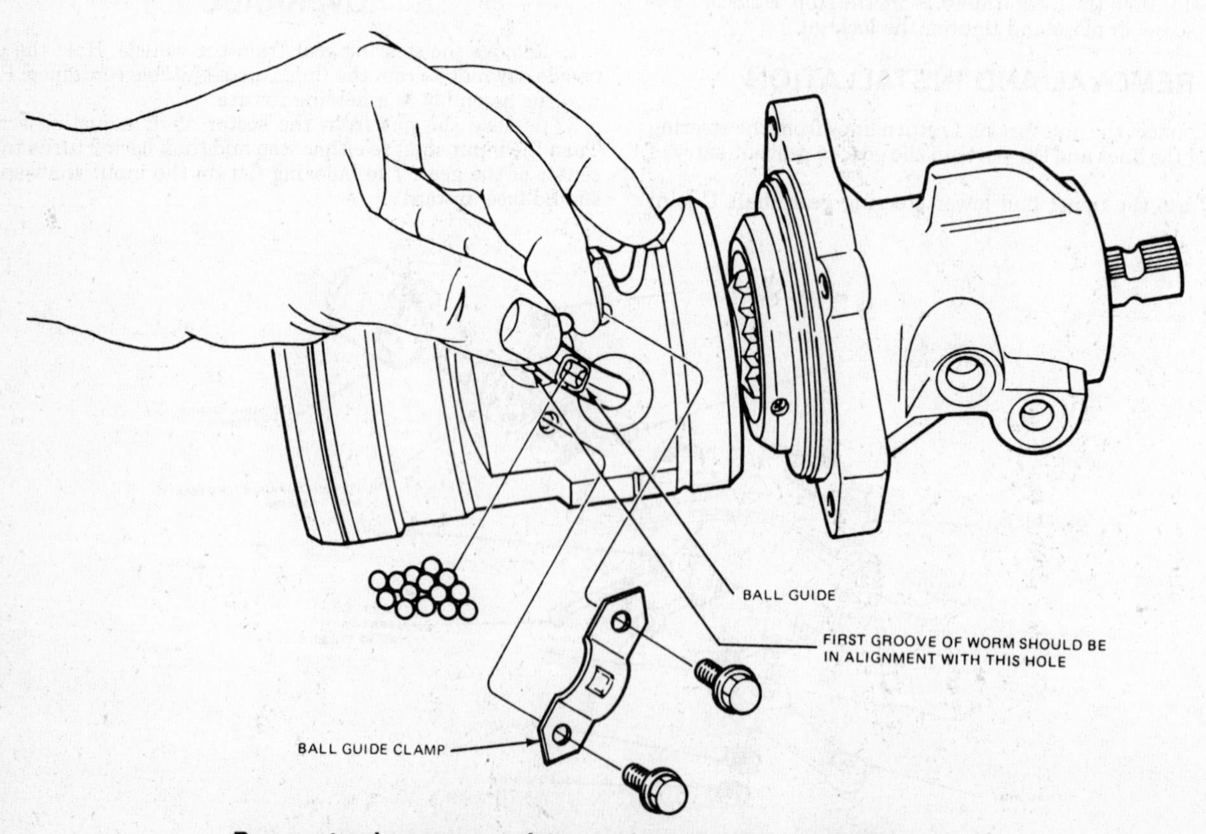

Power steering gear — piston and worm shaft assembly

3. Remove the sector cover retaining bolts. Tap the lower end of the sector shaft and loosen it. Lift the cover and shaft from the housing as an assembly. Discard the O-ring.

4. Turn the sector shaft cover counterclockwise and remove it from the sector shaft adjuster screw.

5. Remove the valve retaining bolts and identification tag. Lift the valve housing off the steering gear housing. Remove the valve and piston assembly and control valve gasket. Discard the gasket.

NOTE: If the housing or valve sleeve seals need to be replaced proceed to Step 11. If the sector shaft seals are to be replaced proceed to the steering gear housing section. The balls need to be removed if the valve sleeve rings are to be replaced.

6. With the piston held so that the ball guide faces up, remove the ball guide clamp screws and the ball guide clamp. With a finger over the opening in the ball guide turn the piston so that the ball guide faces down over a clean container. Let the guide tubes drop into the container.

7. Rotate the input shaft from stop to stop until all the balls fall from the piston into the container.

8. The valve assembly can now be removed from the piston. Inspect the piston bore to be sure that all the balls have been removed.

9. Install the valve body assembly in a suitable holding fixture. Loosen the allen head race nut screw from the valve housing. Remove the worm bearing race nut using tool T66P-3553-B or equivalent.

10. Carefully slide the input shaft, worm and valve assembly out of the valve housing.

To assemble:

11. Check all parts and repair or replace them as required. Apply a light coat of power steering fluid to all O-rings and teflon seals.

12. Carefully install the wormshaft and valve in the housing. Install the worm bearing race nut in the housing and torque the nut to 55–90 ft. lbs.

13. Install the allen head race nut set screw through the valve housing and torque the nut to 15–25 inch lbs.

14. Place the power cylinder piston on the bench with the ball guide holes facing upward. Insert the wormshaft into the piston so that the first groove is in line with the hole nearest the center of the piston.

15. Place the ball guides in the piston. Turning the wormshaft counterclockwise install the balls. If all the balls have not been installed upon reaching the left stop, rotate the input shaft in one direction, then in the other direction while inserting the remaining ball.

16. Secure the guides in the ball nut with the clamp. Torque

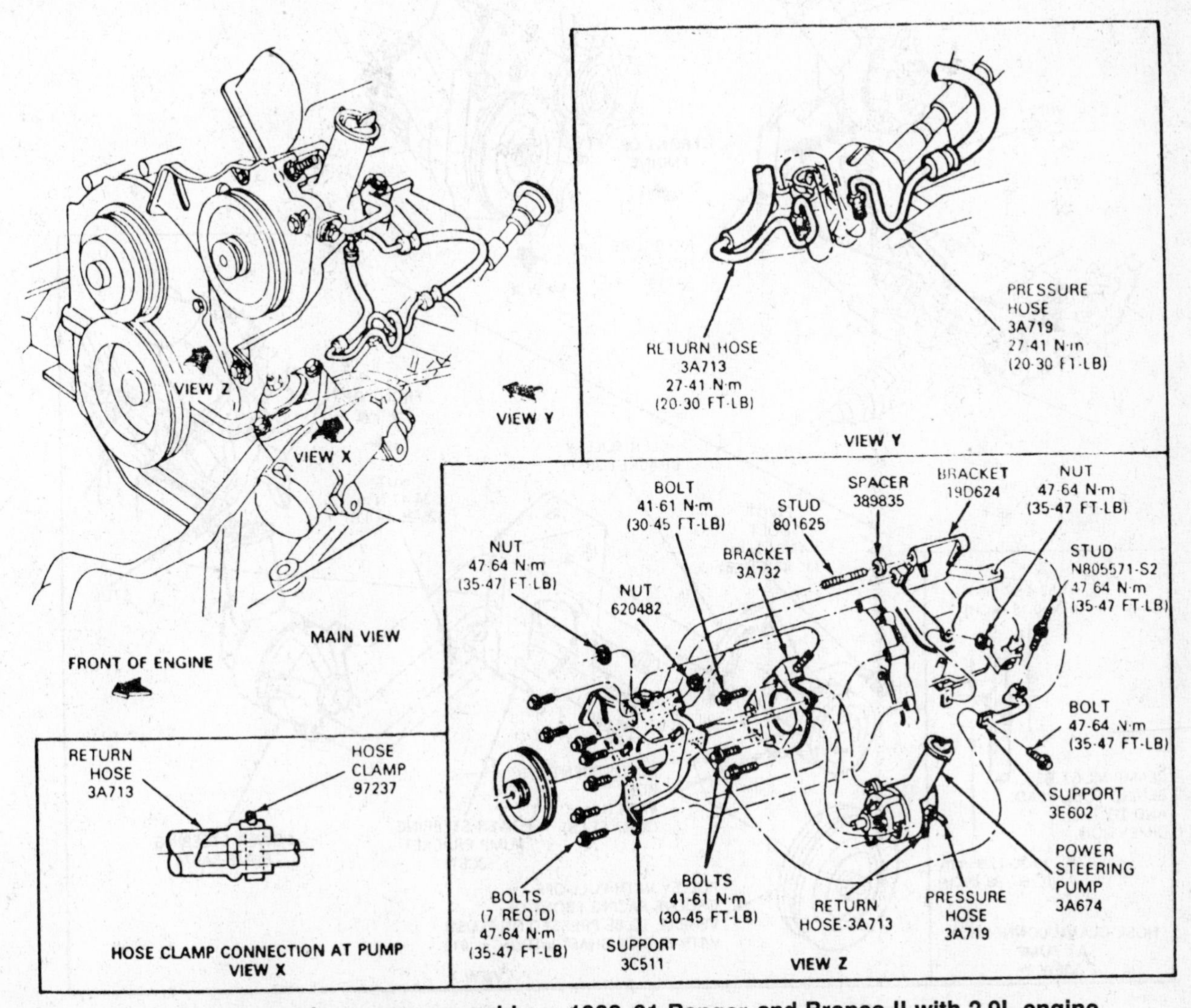

Power steering pump assembly — 1988–91 Ranger and Bronco II with 2.9L engine

the screws to 47–70 inch lbs. Apply petroleum jelly or equivalent to the teflon seal on the piston.

17. Place a new control valve O-ring on the valve housing. Slide the piston and valve into the gear housing. Be careful not to damage the piston ring.

18. Align the oil passage in the valve housing with the passage in the gear housing. Place a new O-ring onto the oil passage hole of the gear housing.

19. Install the identification tag onto the housing. Install, but do not tighten the retaining bolts.

20. Rotate the ball nut so that the teeth are in the same place as the sector teeth. Tighten the valve housing retaining bolts to 30–45 ft. lbs.

21. Position the sector shaft cover O-ring in the steering gear housing. Turn the input shaft to center the piston.

22. Apply petroleum jelly or equivalent to the sector shaft journal and position the sector shaft and cover assembly in the gear housing. Install the sector shaft cover retaining bolts and torque them to 55–70 ft. lbs.

23. Adjust the steering gear preload. Install the gear in the vehicle. Fill the power steering gear pump with the proper grade and type fluid.

Power Steering Pump

REMOVAL AND INSTALLATION

1983–88

1. Disconnect the negative battery cable.
2. Remove some power steering fluid from the reservoir by disconnecting the fluid return line hose at the reservoir. Drain the fluid into a container and discard it.
3. Remove the pressure hose from the pump. If equipped, disconnect the power steering pump pressure switch.
4. On the 2.0L and 2.3L engines, loosen the alternator pivot bolt and the adjusting bolt to remove belt tension.
5. On the 2.8L and 2.9L engines, loosen the adjusting nut and the slider bolts on the pump support to slacken the belt tension.
6. On the 2.2L diesel engine, loosen the adjustment bolt and the pivot bolt on the idler pulley to slacken belt tension.
7. Remove the drive belt from the pulley.
8. Install power steering pump pulley removal tool T69L-10300-B or equivalent. Hold the pump and rotate the tool counterclockwise to remove the pulley. Do not apply in and out pres-

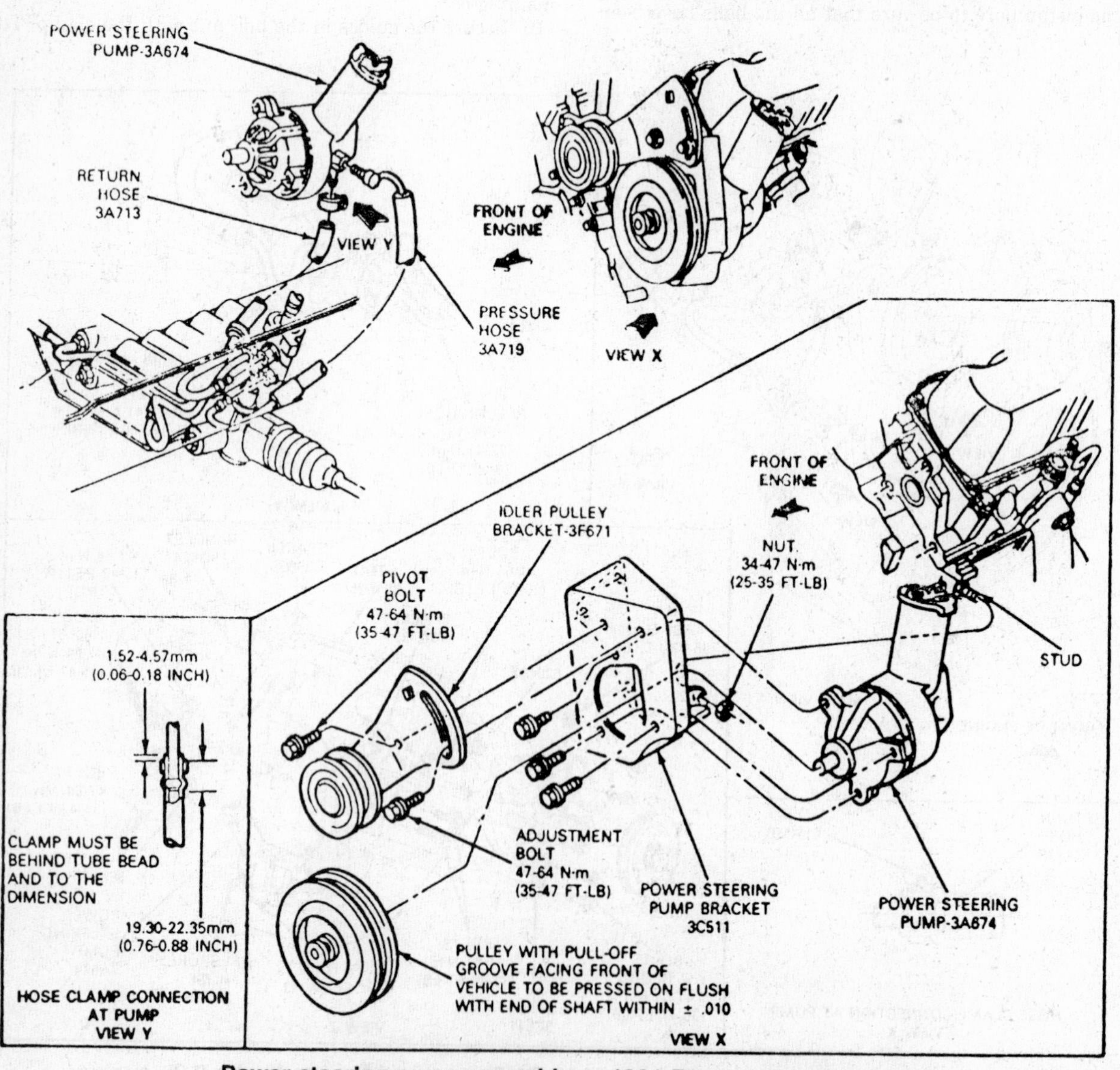

Power steering pump assembly — 1991 Ranger with 3.0L engine

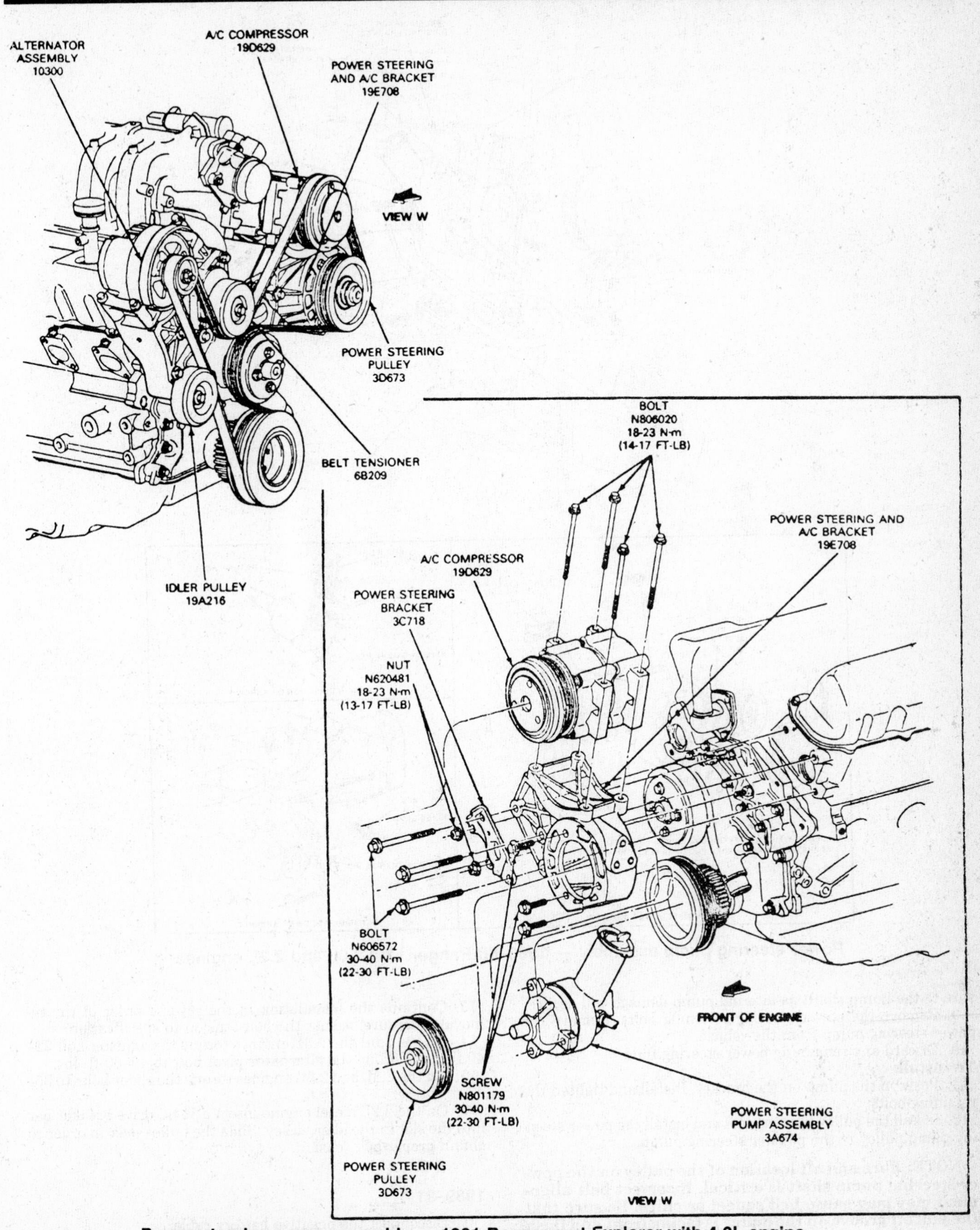

Power steering pump assembly – 1991 Ranger and Explorer with 4.0L engine

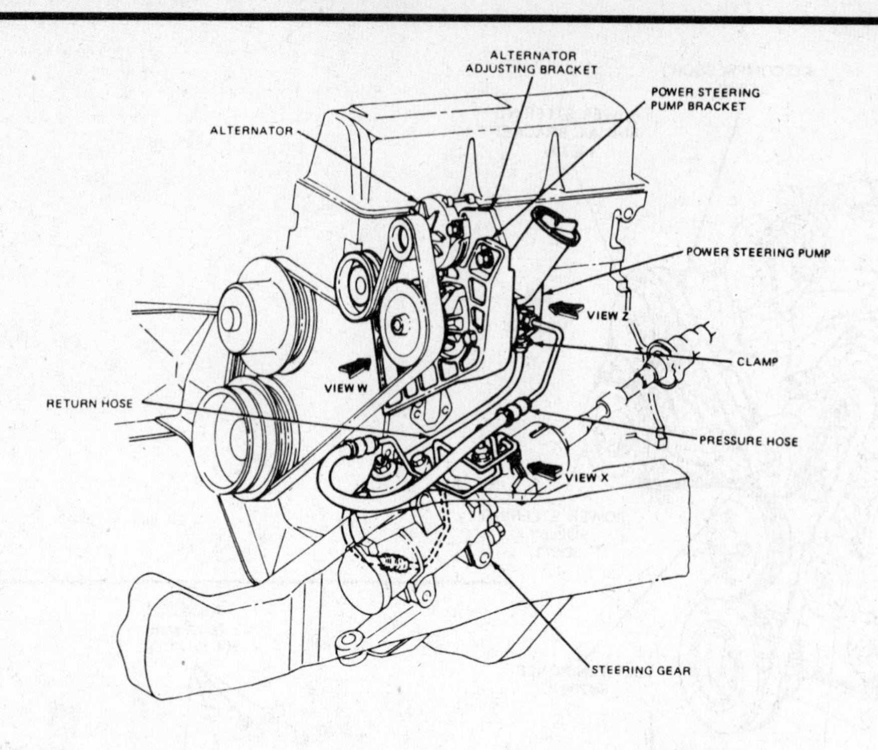

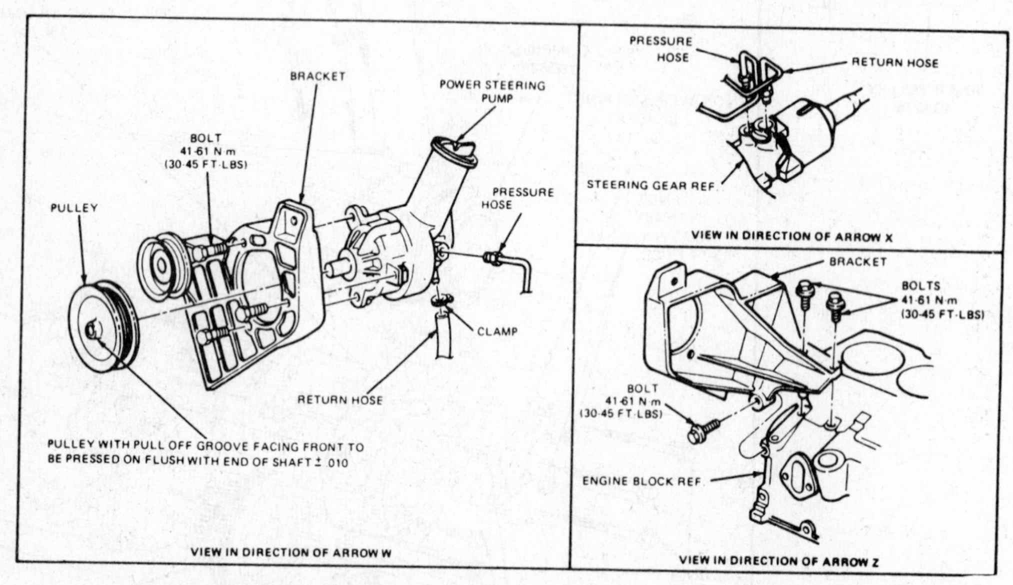

Power steering pump assembly – 1983–88 Ranger with 2.0L and 2.3L engines

sure to the pump shaft, as internal pump damage will occur.

9. Remove the power steering retaining bolts. Remove the power steering pump from the vehicle.
10. Discard any remaining power steering fluid.

To install:

11. Position the pump on the bracket. Install and tighten the retaining bolts.
12. Install the pulley removal tool and install the power steering pump pulley to the powwer steering pump.

NOTE: Fore and aft location of the pulley on the power steering pump shaft is critical. Incorrect belt alignment may may cause belt squeel or chirp. Be sure that the pull off groove on the pulley is facing front and flush with the end of the shaft ± 0.010 in. (0.25mm).

13. Continue the installation in the reverse order of the removal procedure. Adjust the belt tension to specification.
14. On the 2.0L and 2.3L engines torque the adjuster bolt 22–40 ft. lbs. Torque the alternator pivot bolt to 40–50 ft. lbs.
15. On the 2.8L and 2.9L engines torque the slider bolts to 35–47 ft. lbs.
16. On the 2,2L diesel engine insert a ½ in. drive breaker bar into the slot in the idler pulley. Slide the pulley over in order to obtain proper belt tension.

1989–91

1. Disconnect the negative battery cable.
2. Remove some power steering fluid from the reservoir by

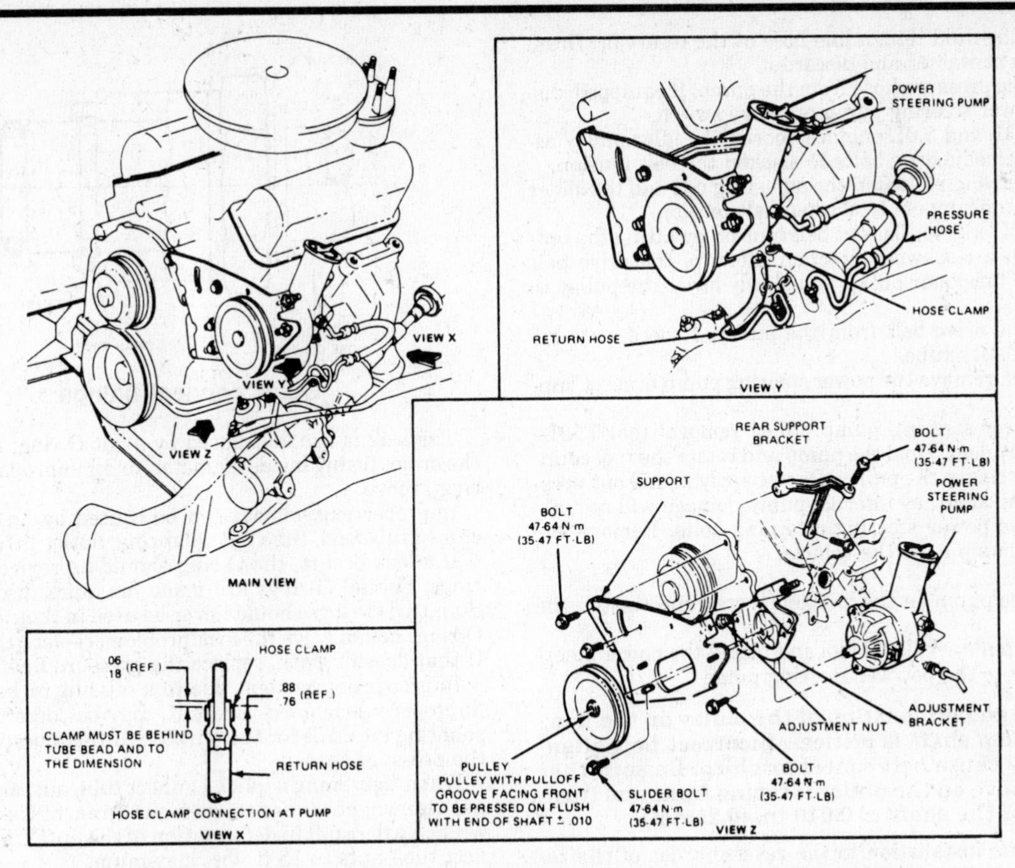

Power steering pump assembly – 1984–88 Bronco II with 2.8L engines

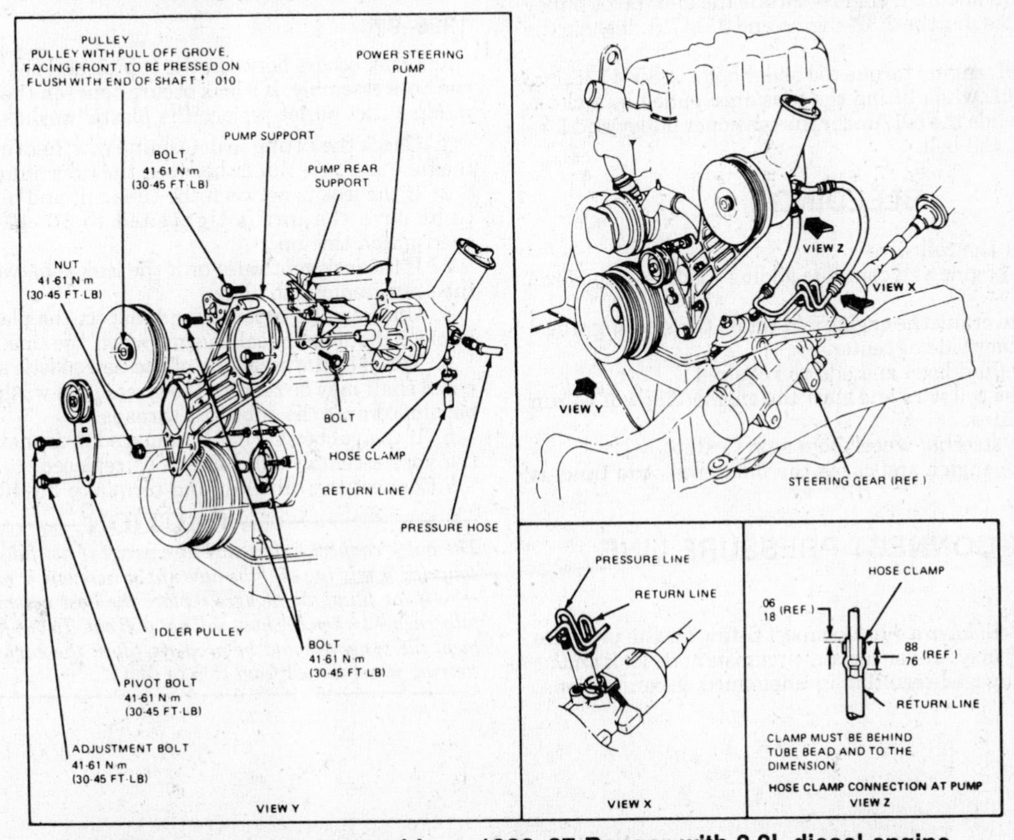

Power steering pump assembly – 1983–87 Ranger with 2.2L diesel engine

disconnecting the fluid return line hose at the reservoir. Drain the fluid into a container and discard it.

3. Remove the pressure hose from the pump. If equipped, disconnect the power steering pump pressure switch.
4. On the 2.3L and 3.0L engines, loosen the idler pulley assembly pivot and adjusting bolts to slacken the belt tension.
5. On the 2.9L engine loosen the adjusting nut and the slider bolts on the pump support to slacken belt tension.
6. On the 4.0L engine, slacken belt tension by lifting the tensioner pulley in a clockwise direction. Remove the drive belt from under the tensioner pulley and slowly lower the pulley to its stop.
7. Remove the drive belt from the pulley. If necessary, remove the oil dipstick tube.
8. If equipped, remove the power steering pump bracket support brace.
9. Install power steering pump pulley removal tool T69L–10300–B or equivalent. Hold the pump and rotate the tool counterclockwise to remove the pulley. Do not apply in and out pressure to the pump shaft, as internal pump damage will occur.
10. Remove the power steering retaining bolts. Remove the power steering pump from the vehicle.

To install:

11. Position the pump on the bracket. Install and tighten the retaining bolts.
12. Install the pulley removal tool and install the power steering pump pulley to the powwer steering pump.

NOTE: Fore and aft location of the pulley on the power steering pump shaft is critical. Incorrect belt alignment may may cause belt squeel or chirp. Be sure that the pull off groove on the pulley is facing front and flush with the end of the shaft ± 0.010 in. (0.25mm).

13. Continue the installation in the reverse order of the removal procedure. Adjust the belt tension to specification.
14. On the 2.3L and 3.0L engines torque the idler pivot pulley bolts 30–40 ft. lbs. for the 2.3L engine and 35–47 ft. lbs. for the 3.0L engine
15. On the 2.9L engine torque the slider bolts to 35–47 ft. lbs.
16. On the 4.0L while lifting the tensioner pulley in a clockwise direction, slide the belt under the tensioner pulley and lower the pulley to the belt.

BLEEDING

1. Disconnect the coil wire.
2. Crank the engine and continue adding fluid until the level stabilizes.
3. Continue to crank the engine and rotate the steering wheel about 30° to either side of center.
4. Check the fluid level and add as required.
5. Connect the coil wire and start the engine. Allow it to run for several minutes.
6. Rotate the steering wheel from stop to stop.
7. Shut of the engine and check the fluid level. Add fluid as necessary.

QUICK-CONNECT PRESSURE LINE

1983–89

Some pumps will have a quick-connect fitting for the pressure line. This fitting may, under certain circumstances, leak and/or be improperly engaged resulting in unplanned disconnection.

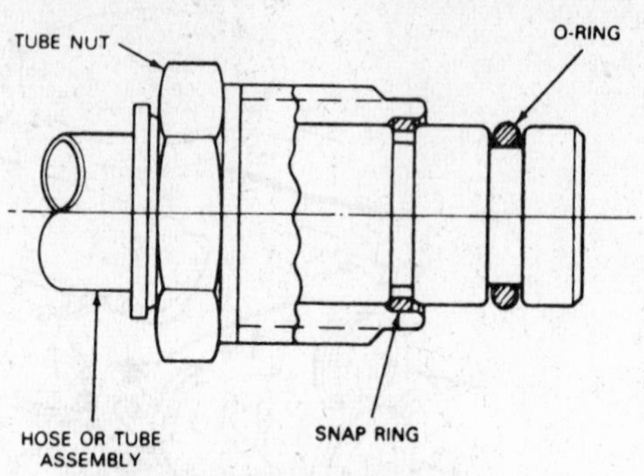

Quick connect fittings

The leak is usually caused by a cut O-ring, imperfections in the outlet fitting inside diameter, or an improperly machined O-ring groove.

Improper engagement can be caused by an improperly machined tube end, tube nut, snapring, outlet fitting or gear port.

If a leak occurs, the O-ring should be replaced with new O-rings. Special O-rings are made for quick-disconnect fittings. Standard O-rings should never be used in their place. If the new O-rings do not solve the leak problem, replace the outlet fitting. If that doesn't work, replace the pressure line.

Improper engagement due to a missing or bent snapring, or improperly machined tube nut, may be corrected with a Ford snapring kit made for the purpose. If that doesn't work, replace the pressure hose.

When tightening a quick-connect tube nut, always use a tube nut wrench; never use an open-end wrench! Use of an open-end wrench will result in deformation of the nut! Tighten quick-connect tube nuts to 15 ft. lbs. maximum.

Swivel and/or endplay of quick-connect fittings is normal.

1990–91

If a leak occurs between the tubing and the tube nut, replace the hose assembly. If a leak occurs between the tube nut and the pump outlet outlet replace the plastic washer.

1. Check the fitting to determine whether the leak is between the tube and tube nut or between the tube nut and pump outlet.
2. If the leak is between the tube nut and pump outlet check to be sure the nut is tightened to 30–40 ft. lbs. Do not overtighten this nut.
3. If the leak continues or if the leak is between the tube and tube nut, remove the line.
4. Unscrew the tube nut and inspect the plastic seal washer. Replace the plastic seal washer when the line is removed.
5. To aid in the assembly of the new plastic seal washer, a tapered shaft may be required to stretch the washer so that it may be slipped over the tube nut threads.
6. If the rubber O-ring is damaged it cannot be serviced and the hose assembly will have to be replaced.
7.Connect the tube nut and torque to 30–40 ft. lbs.

CAUTION

The quick connect fitting may disensgage if not fully assembled, if the snapring is missing or if the tube nut or hose end is not machined properly. If the fitting disengages replace the hose assembly. The fitting is fully engaged when the hose will not pull out. To test for positive engagement the system should be properly filled, the engine started and the steering wheel turned from stop to stop.

Troubleshooting the Manual Steering Gear

Problem	Cause	Solution
Hard or erratic steering	• Incorrect tire pressure	• Inflate tires to recommended pressures
	• Insufficient or incorrect lubrication	• Lubricate as required (refer to Maintenance Section)
	• Suspension, or steering linkage parts damaged or misaligned	• Repair or replace parts as necessary
	• Improper front wheel alignment	• Adjust incorrect wheel alignment angles
	• Incorrect steering gear adjustment	• Adjust steering gear
	• Sagging springs	• Replace springs
Play or looseness in steering	• Steering wheel loose	• Inspect shaft spines and repair as necessary. Tighten attaching nut and stake in place.
	• Steering linkage or attaching parts loose or worn	• Tighten, adjust, or replace faulty components
	• Pitman arm loose	• Inspect shaft splines and repair as necessary. Tighten attaching nut and stake in place
	• Steering gear attaching bolts loose	• Tighten bolts
	• Loose or worn wheel bearings	• Adjust or replace bearings
	• Steering gear adjustment incorrect or parts badly worn	• Adjust gear or replace defective parts
Wheel shimmy or tramp	• Improper tire pressure	• Inflate tires to recommended pressures
	• Wheels, tires, or brake rotors out-of-balance or out-of-round	• Inspect and replace or balance parts
	• Inoperative, worn, or loose shock absorbers or mounting parts	• Repair or replace shocks or mountings
	• Loose or worn steering or suspension parts	• Tighten or replace as necessary
	• Loose or worn wheel bearings	• Adjust or replace bearings
	• Incorrect steering gear adjustments	• Adjust steering gear
	• Incorrect front wheel alignment	• Correct front wheel alignment
Tire wear	• Improper tire pressure	• Inflate tires to recommended pressures
	• Failure to rotate tires	• Rotate tires
	• Brakes grabbing	• Adjust or repair brakes
	• Incorrect front wheel alignment	• Align incorrect angles
	• Broken or damaged steering and suspension parts	• Repair or replace defective parts
	• Wheel runout	• Replace faulty wheel
	• Excessive speed on turns	• Make driver aware of conditions

Troubleshooting the Manual Steering Gear

Problem	Cause	Solution
Vehicle leads to one side	· Improper tire pressures · Front tires with uneven tread depth, wear pattern, or different cord design (i.e., one bias ply and one belted or radial tire on front wheels) · Incorrect front wheel alignment · Brakes dragging · Pulling due to uneven tire construction	· Inflate tires to recommended pressures · Install tires of same cord construction and reasonably even tread depth, design, and wear pattern · Align incorrect angles · Adjust or repair brakes · Replace faulty tire

Troubleshooting the Power Steering Gear

Problem	Cause	Solution
Hissing noise in steering gear	· There is some noise in all power steering systems. One of the most common is a hissing sound most evident at standstill parking. There is no relationship between this noise and performance of the steering. Hiss may be expected when steering wheel is at end of travel or when slowly turning at standstill.	· Slight hiss is normal and in no way affects steering. Do not replace valve unless hiss is extremely objectionable. A replacement valve will also exhibit slight noise and is not always a cure. Investigate clearance around flexible coupling rivets. Be sure steering shaft and gear are aligned so flexible coupling rotates in a flat plane and is not distorted as shaft rotates. Any metal-to-metal contacts through flexible coupling will transmit valve hiss into passenger compartment through the steering column.
Rattle or chuckle noise in steering gear	· Gear loose on frame · Steering linkage looseness · Pressure hose touching other parts of car · Loose pitman shaft over center adjustment **NOTE:** A slight rattle may occur on turns because of increased clearance off the "high point." This is normal and clearance must not be reduced below specified limits to eliminate this slight rattle. · Loose pitman arm	· Check gear-to-frame mounting screws. · Check linkage pivot points for wear. Replace if necessary. · Adjust hose position. Do not bend tubing by hand. · Adjust to specifications · Tighten pitman arm nut to specifications

Troubleshooting the Power Steering Gear (cont.)

Problem	Cause	Solution
Squawk noise in steering gear when turning or recovering from a turn	• Damper O-ring on valve spool cut	• Replace damper O-ring
Poor return of steering wheel to center	• Tires not properly inflated • Lack of lubrication in linkage and ball joints • Lower coupling flange rubbing against steering gear adjuster plug • Steering gear to column misalignment • Improper front wheel alignment • Steering linkage binding • Ball joints binding • Steering wheel rubbing against housing • Tight or frozen steering shaft bearings • Sticking or plugged valve spool • Steering gear adjustments over specifications • Kink in return hose	• Inflate to specified pressure • Lube linkage and ball joints • Loosen pinch bolt and assemble properly • Align steering column • Check and adjust as necessary • Replace pivots • Replace ball joints • Align housing • Replace bearings • Remove and clean or replace valve • Check adjustment with gear out of car. Adjust as required. • Replace hose
Car leads to one side or the other (keep in mind road condition and wind. Test car in both directions on flat road)	• Front end misaligned • Unbalanced steering gear valve **NOTE:** If this is cause, steering effort will be very light in direction of lead and normal or heavier in opposite direction	• Adjust to specifications • Replace valve
Momentary increase in effort when turning wheel fast to right or left	• Low oil level • Pump belt slipping • High internal leakage	• Add power steering fluid as required • Tighten or replace belt • Check pump pressure. (See pressure test)
Steering wheel surges or jerks when turning with engine running especially during parking	• Low oil level • Loose pump belt • Steering linkage hitting engine oil pan at full turn • Insufficient pump pressure • Pump flow control valve sticking	• Fill as required • Adjust tension to specification • Correct clearance • Check pump pressure. (See pressure test). Replace relief valve if defective. • Inspect for varnish or damage, replace if necessary

Troubleshooting the Power Steering Gear (cont.)

Problem	Cause	Solution
Excessive wheel kickback or loose steering	• Air in system • Steering gear loose on frame • Steering linkage joints worn enough to be loose • Worn poppet valve • Loose thrust bearing preload adjustment • Excessive overcenter lash	• Add oil to pump reservoir and bleed by operating steering. Check hose connectors for proper torque and adjust as required. • Tighten attaching screws to specified torque • Replace loose pivots • Replace poppet valve • Adjust to specification with gear out of vehicle • Adjust to specification with gear out of car
Hard steering or lack of assist	• Loose pump belt • Low oil level **NOTE:** Low oil level will also result in excessive pump noise • Steering gear to column misalignment • Lower coupling flange rubbing against steering gear adjuster plug • Tires not properly inflated	• Adjust belt tension to specification • Fill to proper level. If excessively low, check all lines and joints for evidence of external leakage. Tighten loose connectors. • Align steering column • Loosen pinch bolt and assemble properly • Inflate to recommended pressure
Foamy milky power steering fluid, low fluid level and possible low pressure	• Air in the fluid, and loss of fluid due to internal pump leakage causing overflow	• Check for leak and correct. Bleed system. Extremely cold temperatures will cause system aeriation should the oil level be low. If oil level is correct and pump still foams, remove pump from vehicle and separate reservoir from housing. Check welsh plug and housing for cracks. If plug is loose or housing is cracked, replace housing.
Low pressure due to steering pump	• Flow control valve stuck or inoperative • Pressure plate not flat against cam ring	• Remove burrs or dirt or replace. Flush system. • Correct
Low pressure due to steering gear	• Pressure loss in cylinder due to worn piston ring or badly worn housing bore • Leakage at valve rings, valve body-to-worm seal	• Remove gear from car for disassembly and inspection of ring and housing bore • Remove gear from car for disassembly and replace seals

Troubleshooting the Power Steering Pump

Problem	Cause	Solution
Chirp noise in steering pump	• Loose belt	• Adjust belt tension to specification
Belt squeal (particularly noticeable at full wheel travel and stand still parking)	• Loose belt	• Adjust belt tension to specification
Growl noise in steering pump	• Excessive back pressure in hoses or steering gear caused by restriction	• Locate restriction and correct. Replace part if necessary.
Growl noise in steering pump (particularly noticeable at stand still parking)	• Scored pressure plates, thrust plate or rotor • Extreme wear of cam ring	• Replace parts and flush system • Replace parts
Groan noise in steering pump	• Low oil level • Air in the oil. Poor pressure hose connection.	• Fill reservoir to proper level • Tighten connector to specified torque. Bleed system by operating steering from right to left—full turn.
Rattle noise in steering pump	• Vanes not installed properly • Vanes sticking in rotor slots	• Install properly • Free up by removing burrs, varnish, or dirt
Swish noise in steering pump	• Defective flow control valve	• Replace part
Whine noise in steering pump	• Pump shaft bearing scored	• Replace housing and shaft. Flush system.
Hard steering or lack of assist	• Loose pump belt • Low oil level in reservoir **NOTE:** Low oil level will also result in excessive pump noise • Steering gear to column misalignment • Lower coupling flange rubbing against steering gear adjuster plug • Tires not properly inflated	• Adjust belt tension to specification • Fill to proper level. If excessively low, check all lines and joints for evidence of external leakage. Tighten loose connectors. • Align steering column • Loosen pinch bolt and assemble properly • Inflate to recommended pressure
Foaming milky power steering fluid, low fluid level and possible low pressure	• Air in the fluid, and loss of fluid due to internal pump leakage causing overflow	• Check for leaks and correct. Bleed system. Extremely cold temperatures will cause system aeriation should the oil level be low. If oil level is correct and pump still foams, remove pump from vehicle and separate reservoir from body. Check welsh plug and body for cracks. If plug is loose or body is cracked, replace body.

Troubleshooting the Power Steering Pump (cont.)

Problem	Cause	Solution
Low pump pressure	· Flow control valve stuck or inoperative · Pressure plate not flat against cam ring	· Remove burrs or dirt or replace. Flush system. · Correct
Momentary increase in effort when turning wheel fast to right or left	· Low oil level in pump · Pump belt slipping · High internal leakage	· Add power steering fluid as required · Tighten or replace belt · Check pump pressure. (See pressure test)
Steering wheel surges or jerks when turning with engine running especially during parking	· Low oil level · Loose pump belt · Steering linkage hitting engine oil pan at full turn · Insufficient pump pressure	· Fill as required · Adjust tension to specification · Correct clearance · Check pump pressure. (See pressure test). Replace flow control valve if defective.
Steering wheel surges or jerks when turning with engine running especially during parking (cont.)	· Sticking flow control valve	· Inspect for varnish or damage, replace if necessary
Excessive wheel kickback or loose steering	· Air in system	· Add oil to pump reservoir and bleed by operating steering. Check hose connectors for proper torque and adjust as required.
Low pump pressure	· Extreme wear of cam ring · Scored pressure plate, thrust plate, or rotor · Vanes not installed properly · Vanes sticking in rotor slots · Cracked or broken thrust or pressure plate	· Replace parts. Flush system. · Replace parts. Flush system. · Install properly · Freeup by removing burrs, varnish, or dirt · Replace part

9 Brakes

QUICK REFERENCE INDEX

GENERAL INDEX

BASIC OPERATING PRINCIPLES

BRAKE SPECIFICATIONS

		Master Cylinder	Caliper	Wheel Cylinder Bore		Rotor	Rotor Minimum	Rotor Maximum	Brake Drum Diameter		Machined Oversize	
Year	Model	Bore	Bore	Front	Rear	Diameter	Thickness	Run-out	Front	Rear	Front	Rear
1983	Ranger	0.9375	2.597	—	0.750	10.28①	0.810	0.003	—	9.00②	—	9.060
1984	Bronco II	0.9375	2.597	—	0.750	10.86①	0.810	0.003	—	9.00②	—	9.060
	Ranger	0.9375	2.597	—	0.750	10.28①	0.810	0.003	—	9.00②	—	9.060
1985	Bronco II	0.9375	2.597	—	0.750	10.86①	0.810	0.003	—	9.00②	—	9.060
	Ranger	0.9375	2.597	—	0.750	10.28①	0.810	0.003	—	9.00②	—	9.060
1986	Bronco II	0.9375	2.597	—	0.750	10.28①	0.810	0.003	—	9.00②	—	9.060
	Ranger	0.9375	2.597	—	0.750	10.28①	0.810	0.003	—	9.00②	—	9.060
1987	Bronco II	0.9375	2.597	—	0.750	10.28①	0.810	0.003	—	9.00②	—	9.060
	Ranger	0.9375	2.597	—	0.750	10.28①	0.810	0.003	—	9.00②	—	9.060
1988	Bronco II	0.9375	2.597	—	0.750	10.28①	0.810	0.003	—	9.00②	—	9.060
	Ranger	0.9375	2.597	—	0.750	10.28①	0.810	0.003	—	9.00②	—	9.060
1989	Bronco II	0.9375	2.597	—	0.750	10.28①	0.810	0.003	—	9.00②	—	9.060
	Ranger	0.9375	2.597	—	0.750	10.28①	0.810	0.003	—	9.00②	—	9.060
1990	Bronco II	0.9375	2.597	—	0.750	10.28①	0.810	0.003	—	9.00②	—	9.060
	Ranger	0.9375	2.597	—	0.750	10.28①	0.810	0.003	—	9.00②	—	9.060
1991	Explorer	0.9375	2.597	—	0.750	10.28①	0.810	0.003	—	9.00②	—	9.060
	Ranger	0.9375	2.597	—	0.750	10.28①	0.810	0.003	—	9.00②	—	9.060

① 4WD: 10.86
② Optional 10" rear brakes
NOTE: Always use specifications as a service guide for brake system components

Hydraulic systems are used to actuate the brakes of all automobiles. The system transports the power required to force the frictional surfaces of the braking system together from the pedal to the individual brake units at each wheel. A hydraulic system is used for 2 reasons.

First, fluid under pressure can be carried to all parts of an automobile by small pipes and flexible hoses without taking up a significant amount of room or posing routing problems.

Second, a great mechanical advantage can be given to the brake pedal end of the system, and the foot pressure required to actuate the brakes can be reduced by making the surface area of the master cylinder pistons smaller than that of any of the pistons in the wheel cylinders or calipers.

The master cylinder consists of a fluid reservoir and a double cylinder and piston assembly. Double type master cylinders are designed to separate the front and rear braking systems hydraulically in case of a leak.

Steel lines carry the brake fluid to a point on the vehicle's frame near each of the vehicle's wheels. The fluid is then carried to the calipers and wheel cylinders by flexible tubes in order to allow for suspension and steering movements.

In drum brake systems, each wheel cylinder contains 2 pistons, one at either end, which push outward in opposite directions.

In disc brake systems, the cylinders are part of the calipers. One cylinder in each caliper is used to force the brake pads against the disc.

All pistons employ some type of seal, usually made of rubber, to minimize fluid leakage. A rubber dust boot seals the outer end of the cylinder against dust and dirt. The boot fits around the outer end of the piston on disc brake calipers, and around the brake actuating rod on wheel cylinders.

The hydraulic system operates as follows: When at rest, the entire system, from the piston(s) in the master cylinder to those in the wheel cylinders or calipers, is full of brake fluid. Upon application of the brake pedal, fluid trapped in front of the master cylinder piston(s) is forced through the lines to the wheel cylinders. Here, it forces the pistons outward, in the case of drum brakes, and inward toward the disc, in the case of disc brakes. The motion of the pistons is opposed by return springs mounted outside the cylinders in drum brakes, and by spring seals, in disc brakes.

Upon release of the brake pedal, a spring located inside the master cylinder immediately returns the master cylinder pistons to the normal position. The pistons contain check valves and the master cylinder has compensating ports drilled in it. These are uncovered as the pistons reach their normal position. The piston check valves allow fluid to flow toward the wheel cylinders or calipers as the pistons withdraw. Then, as the return springs force the brake pads or shoes into the released position, the excess fluid reservoir through the compensating ports. It is during the time the pedal is in the released position that any fluid that has leaked out of the system will be replaced through the compensating ports.

Dual circuit master cylinders employ 2 pistons, located one behind the other, in the same cylinder. The primary piston is actuated directly by mechanical linkage from the brake pedal through the power booster. The secondary piston is actuated by fluid trapped between the 2 pistons. If a leak develops in front of the secondary piston, it moves forward until it bottoms against the front of the master cylinder, and the fluid trapped between the pistons will operate the rear brakes. If the rear brakes develop a leak, the primary piston will move forward until direct contact with the secondary piston takes place, and it will force the secondary piston to actuate the front brakes. In either case, the brake pedal moves farther when the brakes are applied, and less braking power is available.

All dual circuit systems use a switch to warn the driver when only half of the brake system is operational. This switch is located in a valve body which is mounted on the firewall or the frame below the master cylinder. A hydraulic piston receives pressure from both circuits, each circuit's pressure being applied to one end of the piston. When the pressures are in balance, the piston remains stationary. When one circuit has a leak, however, the greater pressure in that circuit during application of the brakes will push the piston to one side, closing the switch and activating the brake warning light.

In disc brake systems, this valve body also contains a metering valve and, in some cases, a proportioning valve. The metering valve keeps pressure from traveling to the disc brakes on the front wheels until the brake shoes on the rear wheels have contacted the drums, ensuring that the front brakes will never be used alone. The proportioning valve controls the pressure to the rear brakes to lessen the chance of rear wheel lock-up during very hard braking.

Warning lights may be tested by depressing the brake pedal and holding it while opening one of the wheel cylinder bleeder screws. If this does not cause the light to go on, substitute a new lamp, make continuity checks, and, finally, replace the switch as necessary.

The hydraulic system may be checked for leaks by applying pressure to the pedal gradually and steadily. If the pedal sinks very slowly to the floor, the system has a leak. This is not to be confused with a springy or spongy feel due to the compression of air within the lines. If the system leaks, there will be a gradual change in the position of the pedal with a constant pressure.

Check for leaks along all lines and at wheel cylinders. If no external leaks are apparent, the problem is inside the master cylinder.

Disc Brakes

BASIC OPERATING PRINCIPLES

Instead of the traditional expanding brakes that press outward against a circular drum, disc brake systems utilize a disc (rotor) with brake pads positioned on either side of it. Braking effect is achieved in a manner similar to the way you would squeeze a spinning phonograph record between your fingers. The disc (rotor) is a casting with cooling fins between the 2 braking surfaces. This enables air to circulate between the braking surfaces making them less sensitive to heat buildup and more resistant to fade. Dirt and water do not affect braking action since contaminants are thrown off by the centrifugal action of the rotor or scraped off the by the pads. Also, the equal clamping action of the 2 brake pads tends to ensure uniform, straight line stops. Disc brakes are inherently self-adjusting.

There are 3 general types of disc brake:

1. A fixed caliper.
2. A floating caliper.
3. A sliding caliper.

The fixed caliper design uses 2 pistons mounted on either side of the rotor (in each side of the caliper). The caliper is mounted rigidly and does not move.

The sliding and floating designs are quite similar. In fact, these 2 types are often lumped together. In both designs, the pad on the inside of the rotor is moved into contact with the rotor by hydraulic force. The caliper, which is not held in a fixed position, moves slightly, bringing the outside pad into contact with the rotor. There are various methods of attaching floating calipers. Some pivot at the bottom or top, and some slide on mounting bolts. In any event, the end result is the same.

Drum Brakes

BASIC OPERATING PRINCIPLES

Drum brakes employ 2 brake shoes mounted on a stationary backing plate. These shoes are positioned inside a circular drum which rotates with the wheel assembly. The shoes are held in place by springs. This allows them to slide toward the drums (when they are applied) while keeping the linings and drums in alignment. The shoes are actuated by a wheel cylinder which is mounted at the top of the backing plate. When the brakes are applied, hydraulic pressure forces the wheel cylinder's actuating links outward. Since these links bear directly against the top of the brake shoes, the tops of the shoes are then forced against the inner side of the drum. This action forces the bottoms of the 2 shoes to contact the brake drum by rotating the entire assembly slightly (known as servo action). When pressure within the wheel cylinder is relaxed, return springs pull the shoes back away from the drum.

Most modern drum brakes are designed to self-adjust themselves during application when the vehicle is moving in reverse. This motion causes both shoes to rotate very slightly with the drum, rocking an adjusting lever, thereby causing rotation of the adjusting screw.

Power Boosters

Power brakes operate just as non-power brake systems except in the actuation of the master cylinder pistons. A vacuum diaphragm is located on the front of the master cylinder and assists the driver in applying the brakes, reducing both the effort and travel he must put into moving the brake pedal.

The vacuum diaphragm housing is connected to the intake manifold by a vacuum hose. A check valve is placed at the point where the hose enters the diaphragm housing, so that during periods of low manifold vacuum brake assist vacuum will not be lost.

Depressing the brake pedal closes off the vacuum source and allows atmospheric pressure to enter on one side of the diaphragm. This causes the master cylinder pistons to move and apply the brakes. When the brake pedal is released, vacuum is applied to both sides of the diaphragm, and return springs return the diaphragm and master cylinder pistons to the released position. If the vacuum fails, the brake pedal rod will butt against the end of the master cylinder actuating rod, and direct mechanical application will occur as the pedal is depressed.

The hydraulic and mechanical problems that apply to conventional brake systems also apply to power brakes, and should be checked for if the tests below do not reveal the problem.

Test for a system vacuum leak as described below:

1. Operate the engine at idle without touching the brake pedal for at least one minute.
2. Turn off the engine, and wait one minute.
3. Test for the presence of assist vacuum by depressing the brake pedal and releasing it several times. Light application will produce less and less pedal travel, if vacuum was present. If there is no vacuum, air is leaking into the system somewhere.

Test for system operation as follows:

1. Pump the brake pedal (with engine off) until the supply vacuum is entirely gone.
2. Put a light, steady pressure on the pedal.
3. Start the engine, and operate it at idle. If the system is operating, the brake pedal should fall toward the floor if constant pressure is maintained on the pedal.

Power brake systems may be tested for hydraulic leaks just as ordinary systems are tested.

BRAKE SYSTEM

CAUTION

Clean, high quality brake fluid is essential to the safe and proper operation of the brake system. You should always buy the highest quality brake fluid that is available. If the brake fluid becomes contaminated, drain and flush the system and fill the master cylinder with new fluid.

Never reuse any brake fluid. Any brake fluid that is removed from the system should be discarded.

Adjustment

The drum brakes are self-adjusting and require a manual adjustment only after the brake shoes have been replaced.

NOTE: Disc brakes are not adjustable.

To adjust the rear brakes with drums installed, follow the procedure given below:

Troubleshooting the Brake System

Problem	Cause	Solution
Low brake pedal (excessive pedal travel required for braking action.)	• Excessive clearance between rear linings and drums caused by inoperative automatic adjusters	• Make 10 to 15 alternate forward and reverse brake stops to adjust brakes. If brake pedal does not come up, repair or replace adjuster parts as necessary.
	• Worn rear brakelining	• Inspect and replace lining if worn beyond minimum thickness specification
	• Bent, distorted brakeshoes, front or rear	• Replace brakeshoes in axle sets
	• Air in hydraulic system	• Remove air from system. Refer to Brake Bleeding.
Low brake pedal (pedal may go to floor with steady pressure applied.)	• Fluid leak in hydraulic system	• Fill master cylinder to fill line; have helper apply brakes and check calipers, wheel cylinders, differential valve tubes, hoses and fittings for leaks. Repair or replace as necessary.
	• Air in hydraulic system	• Remove air from system. Refer to Brake Bleeding.
	• Incorrect or non-recommended brake fluid (fluid evaporates at below normal temp).	• Flush hydraulic system with clean brake fluid. Refill with correct-type fluid.
	• Master cylinder piston seals worn, or master cylinder bore is scored, worn or corroded	• Repair or replace master cylinder
Low brake pedal (pedal goes to floor on first application—o.k. on subsequent applications.)	• Disc brake pads sticking on abutment surfaces of anchor plate. Caused by a build-up of dirt, rust, or corrosion on abutment surfaces	• Clean abutment surfaces
Fading brake pedal (pedal height decreases with steady pressure applied.)	• Fluid leak in hydraulic system	• Fill master cylinder reservoirs to fill mark, have helper apply brakes, check calipers, wheel cylinders, differential valve, tubes, hoses, and fittings for fluid leaks. Repair or replace parts as necessary.
	• Master cylinder piston seals worn, or master cylinder bore is scored, worn or corroded	• Repair or replace master cylinder

Troubleshooting the Brake System (cont.)

Problem	Cause	Solution
Decreasing brake pedal travel (pedal travel required for braking action decreases and may be accompanied by a hard pedal.)	• Caliper or wheel cylinder pistons sticking or seized	• Repair or replace the calipers, or wheel cylinders
	• Master cylinder compensator ports blocked (preventing fluid return to reservoirs) or pistons sticking or seized in master cylinder bore	• Repair or replace the master cylinder
	• Power brake unit binding internally	• Test unit according to the following procedure: (a) Shift transmission into neutral and start engine (b) Increase engine speed to 1500 rpm, close throttle and fully depress brake pedal (c) Slow release brake pedal and stop engine (d) Have helper remove vacuum check valve and hose from power unit. Observe for backward movement of brake pedal. (e) If the pedal moves backward, the power unit has an internal bind—replace power unit
Grabbing brakes (severe reaction to brake pedal pressure.)	• Brakelining(s) contaminated by grease or brake fluid	• Determine and correct cause of contamination and replace brakeshoes in axle sets
	• Parking brake cables incorrectly adjusted or seized	• Adjust cables. Replace seized cables.
	• Incorrect brakelining or lining loose on brakeshoes	• Replace brakeshoes in axle sets
	• Caliper anchor plate bolts loose	• Tighten bolts
	• Rear brakeshoes binding on support plate ledges	• Clean and lubricate ledges. Replace support plate(s) if ledges are deeply grooved. Do not attempt to smooth ledges by grinding.
	• Incorrect or missing power brake reaction disc	• Install correct disc
	• Rear brake support plates loose	• Tighten mounting bolts
Spongy brake pedal (pedal has abnormally soft, springy, spongy feel when depressed.)	• Air in hydraulic system	• Remove air from system. Refer to Brake Bleeding.
	• Brakeshoes bent or distorted	• Replace brakeshoes
	• Brakelining not yet seated with drums and rotors	• Burnish brakes
	• Rear drum brakes not properly adjusted	• Adjust brakes

Troubleshooting the Brake System (cont.)

Problem	Cause	Solution
Hard brake pedal (excessive pedal pressure required to stop vehicle. May be accompanied by brake fade.)	• Loose or leaking power brake unit vacuum hose	• Tighten connections or replace leaking hose
	• Incorrect or poor quality brake-lining	• Replace with lining in axle sets
	• Bent, broken, distorted brakeshoes	• Replace brakeshoes
	• Calipers binding or dragging on mounting pins. Rear brakeshoes dragging on support plate.	• Replace mounting pins and bushings. Clean rust or burrs from rear brake support plate ledges and lubricate ledges with molydisulfide grease. **NOTE:** If ledges are deeply grooved or scored, do not attempt to sand or grind them smooth—replace support plate.
	• Caliper, wheel cylinder, or master cylinder pistons sticking or seized	• Repair or replace parts as necessary
	• Power brake unit vacuum check valve malfunction	• Test valve according to the following procedure: (a) Start engine, increase engine speed to 1500 rpm, close throttle and immediately stop engine (b) Wait at least 90 seconds then depress brake pedal (c) If brakes are not vacuum assisted for 2 or more applications, check valve is faulty
	• Power brake unit has internal bind	• Test unit according to the following procedure: (a) With engine stopped, apply brakes several times to exhaust all vacuum in system (b) Shift transmission into neutral, depress brake pedal and start engine (c) If pedal height decreases with foot pressure and less pressure is required to hold pedal in applied position, power unit vacuum system is operating normally. Test power unit. If power unit exhibits a bind condition, replace the power unit.

Troubleshooting the Brake System (cont.)

Problem	Cause	Solution
Hard brake pedal (excessive pedal pressure required to stop vehicle. May be accompanied by brake fade.)	• Master cylinder compensator ports (at bottom of reservoirs) blocked by dirt, scale, rust, or have small burrs (blocked ports prevent fluid return to reservoirs).	• Repair or replace master cylinder **CAUTION:** Do not attempt to clean blocked ports with wire, pencils, or similar implements. Use compressed air only.
	• Brake hoses, tubes, fittings clogged or restricted	• Use compressed air to check or unclog parts. Replace any damaged parts.
	• Brake fluid contaminated with improper fluids (motor oil, transmission fluid, causing rubber components to swell and stick in bores	• Replace all rubber components, combination valve and hoses. Flush entire brake system with DOT 3 brake fluid or equivalent.
	• Low engine vacuum	• Adjust or repair engine
Dragging brakes (slow or incomplete release of brakes)	• Brake pedal binding at pivot	• Loosen and lubricate
	• Power brake unit has internal bind	• Inspect for internal bind. Replace unit if internal bind exists.
	• Parking brake cables incorrrectly adjusted or seized	• Adjust cables. Replace seized cables.
	• Rear brakeshoe return springs weak or broken	• Replace return springs. Replace brakeshoe if necessary in axle sets.
	• Automatic adjusters malfunctioning	• Repair or replace adjuster parts as required
	• Caliper, wheel cylinder or master cylinder pistons sticking or seized	• Repair or replace parts as necessary
	• Master cylinder compensating ports blocked (fluid does not return to reservoirs).	• Use compressed air to clear ports. Do not use wire, pencils, or similar objects to open blocked ports.
Vehicle moves to one side when brakes are applied	• Incorrect front tire pressure	• Inflate to recommended cold (reduced load) inflation pressure
	• Worn or damaged wheel bearings	• Replace worn or damaged bearings
	• Brakelining on one side contaminated	• Determine and correct cause of contamination and replace brakelining in axle sets
	• Brakeshoes on one side bent, distorted, or lining loose on shoe	• Replace brakeshoes in axle sets
	• Support plate bent or loose on one side	• Tighten or replace support plate
	• Brakelining not yet seated with drums or rotors	• Burnish brakelining
	• Caliper anchor plate loose on one side	• Tighten anchor plate bolts
	• Caliper piston sticking or seized	• Repair or replace caliper
	• Brakelinings water soaked	• Drive vehicle with brakes lightly applied to dry linings
	• Loose suspension component attaching or mounting bolts	• Tighten suspension bolts. Replace worn suspension components.
	• Brake combination valve failure	• Replace combination valve

Troubleshooting the Brake System (cont.)

Problem	Cause	Solution
Chatter or shudder when brakes are applied (pedal pulsation and roughness may also occur.)	• Brakeshoes distorted, bent, contaminated, or worn • Caliper anchor plate or support plate loose • Excessive thickness variation of rotor(s)	• Replace brakeshoes in axle sets • Tighten mounting bolts • Refinish or replace rotors in axle sets
Noisy brakes (squealing, clicking, scraping sound when brakes are applied.)	• Bent, broken, distorted brakeshoes • Excessive rust on outer edge of rotor braking surface	• Replace brakeshoes in axle sets • Remove rust
Noisy brakes (squealing, clicking, scraping sound when brakes are applied.) (cont.)	• Brakelining worn out—shoes contacting drum of rotor • Broken or loose holdown or return springs • Rough or dry drum brake support plate ledges • Cracked, grooved, or scored rotor(s) or drum(s) • Incorrect brakelining and/or shoes (front or rear).	• Replace brakeshoes and lining in axle sets. Refinish or replace drums or rotors. • Replace parts as necessary • Lubricate support plate ledges • Replace rotor(s) or drum(s). Replace brakeshoes and lining in axle sets if necessary. • Install specified shoe and lining assemblies
Pulsating brake pedal	• Out of round drums or excessive lateral runout in disc brake rotor(s)	• Refinish or replace drums, re-index rotors or replace

1. Raise the vehicle and support it with safety stands.
2. Remove the rubber plug from the adjusting slot on the backing plate.
3. Turn the adjusting screw using a Brake Shoe Adjustment Tool or equivalent inside the hole to expand the brake shoes until they drag against the brake drum and lock the drum.
4. Insert a small screwdriver or piece of firm wire (coat hanger wire) into the adjusting slot and push the automatic adjusting lever out and free of the starwheel on the adjusting screw and hold it there.
5. Engage the topmost tooth possible on the starwheel with the brake adjusting spoon. Move the end of the adjusting spoon upward to move the adjusting screw starwheel downward and contract the adjusting screw. Back off the adjusting screw starwheel until the wheel spins FREELY with a minimum of drag about 10 to 12 notches. Keep track of the number of turns that the starwheel is backed off, or the number of strokes taken with the brake adjusting spoon.
6. Repeat this operation for the other side. When backing off the brakes on the other side, the starwheel adjuster must be backed off the same number of turns to prevent side-to-side brake pull.
7. When all drum brakes are adjusted, remove the safety stands and lower the vehicle and make several stops while backing the vehicle, to equalize the brakes at all of the wheels.
8. Road test the vehicle. PERFORM THE ROAD TEST ONLY WHEN THE BRAKES WILL APPLY AND THE VEHICLE CAN BE STOPPED SAFELY!

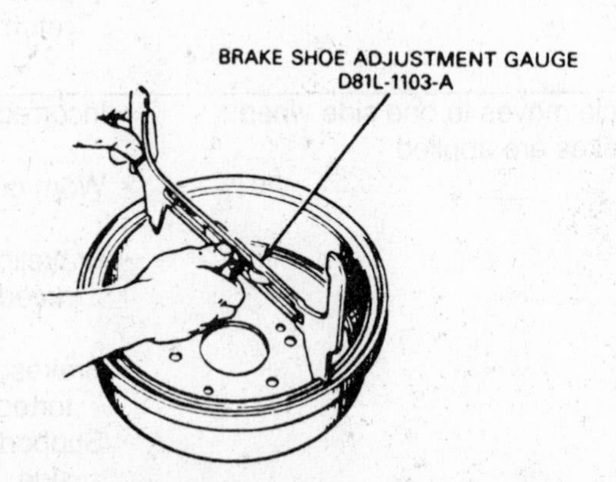

Brake shoe adjustment gauge – Step 1

Brake Light Switch

REMOVAL AND INSTALLATION

1. Lift the locking tab on the switch connector and disconnect the wiring.
2. Remove the hairpin retainer, slide the stoplamp switch, pushrod and nylon washer off of the pedal. Remove the washer, then the switch by sliding it up or down.

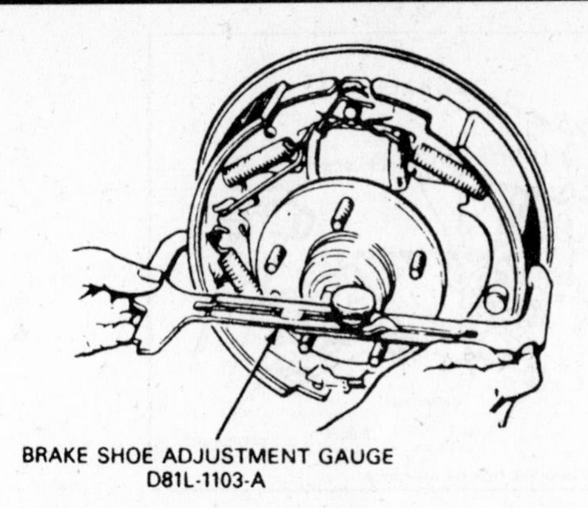

Brake shoe adjustment gauge – Step 2

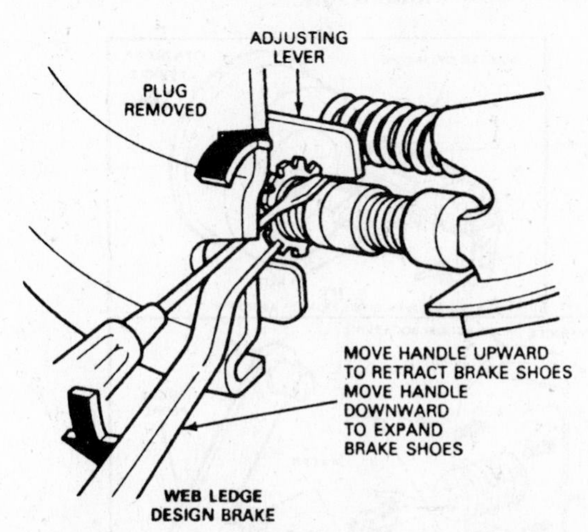

Brake shoe adjustment gauge – Step 3

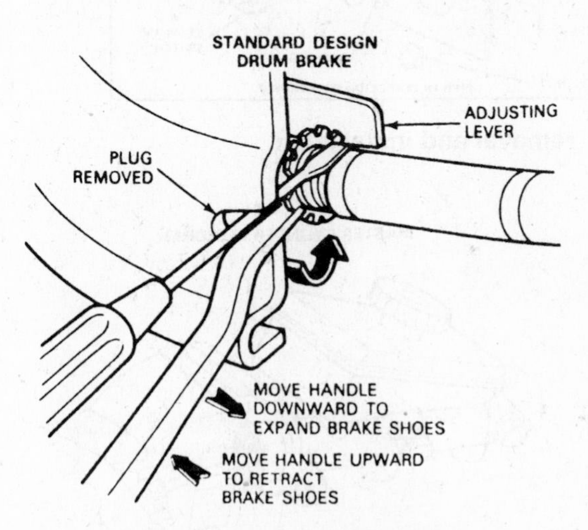

Adjusting rear brake shoes

NOTE: On some vehicles equipped with speed control, the spacer washer is replaced by the dump valve adapter washer.

3. To install the switch, position it so that the U-shaped side is nearest the pedal and directly over/under the pin.

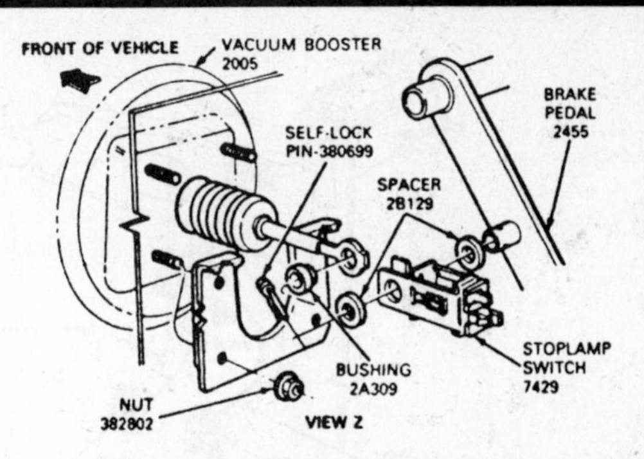

Stoplight switch mounting

4. Slide the switch up or down, trapping the master cylinder pushrod and bushing between the switch side plates.
5. Push the switch and pushrod assembly firmly towards the brake pedal arm. Assemble the outside white plastic washer to the pin and install the hairpin retainer.

NOTE: Don't substitute any other type of retainer. Use only the Ford specified hairpin retainer.

6. Assemble the connector on the switch.
7. Check stoplamp operation.

NOTE: Make sure that the stoplamp switch wiring has sufficient travel during a full pedal stroke.

Master Cylinder

REMOVAL AND INSTALLATION

Manual Brake System

1. Working from inside the cab below the instrument panel, disconnect the wires from the stop lamp switch.
2. Remove the retaining nut, shoulder bolt and spacers, securing the master cylinder push rod to the brake pedal assembly. Remove the stop lamp switch from the pedal.
3. Disconnect the brake hydraulic system lines (always use correct tool, a Line wrench) from the master cylinder.
4. Remove the master cylinder-to-dash panel retaining nuts, and remove the master cylinder.
5. Remove the boot from the master cylinder push rod.

To install:

6. Place the master cylinder assembly on the dash panel in the engine compartment and install the retaining bolts. Tighten the bolts to 13–25 ft. lbs.
7. Connect the hydraulic brake system lines to the master cylinder.
8. Secure the push rod to the brake pedal assembly using the shoulder bolt. Make sure the bushings and spacers are installed properly. Install self-locking nut.
9. Connect the wires to the stop lamp switch. Bleed the brake system as described in this Section. Centralize the differential valve. Fill the dual master cylinder reservoirs with DOT 3 brake fluid to within ¼ in. (6mm) of the top. Install the gasket and reservoir cover. Roadtest the vehicle.

Power Brake System

1. With the engine turned off, push the brake pedal down to expel vacuum from the brake booster system.
2. Disconnect the hydraulic lines (use correct tool, a Line Wrench) from the brake master cylinder.
3. Remove the brake booster-to-master cylinder retaining nuts and lock washers. Remove the master cylinder from the brake booster.

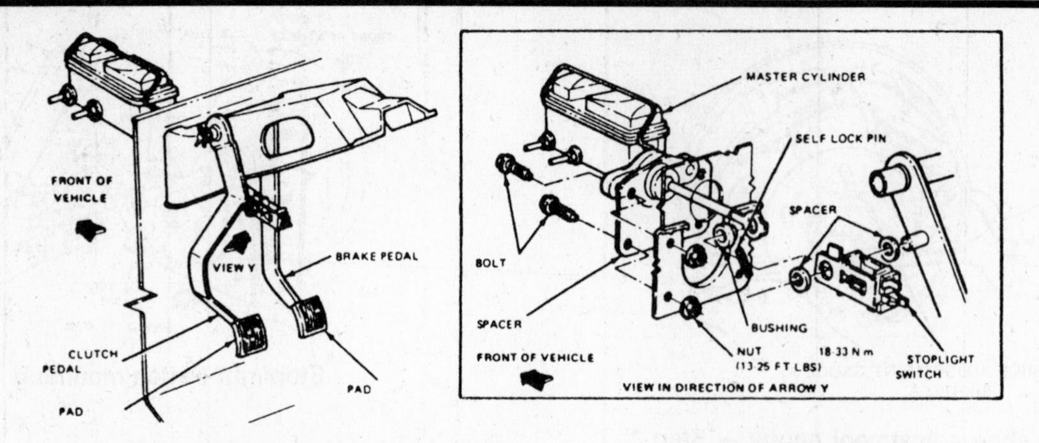

Manual brake master cylinder removal and installation

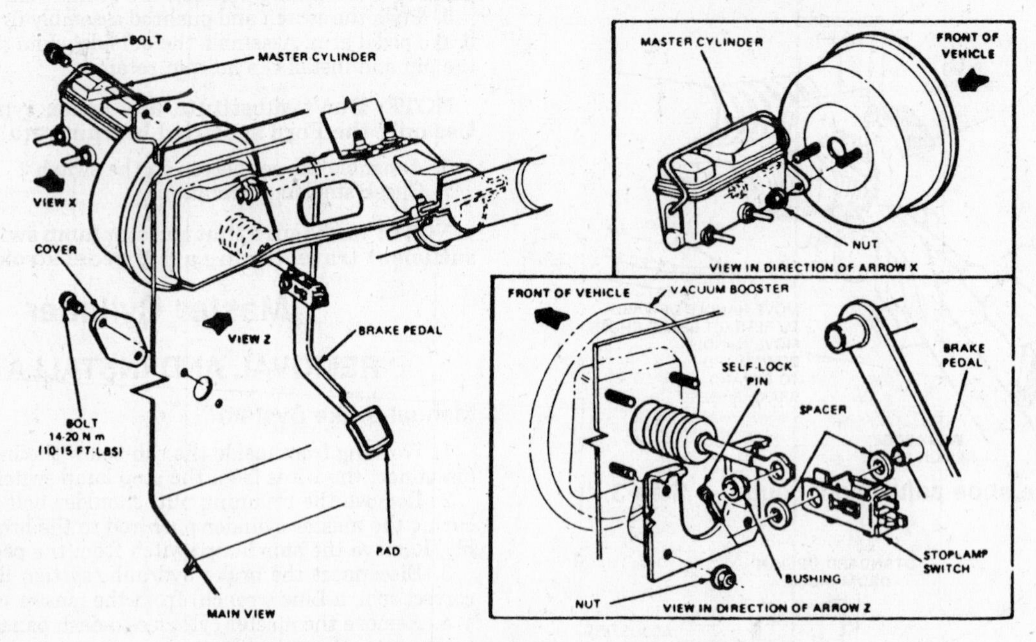

Power brake master cylinder removal and installation

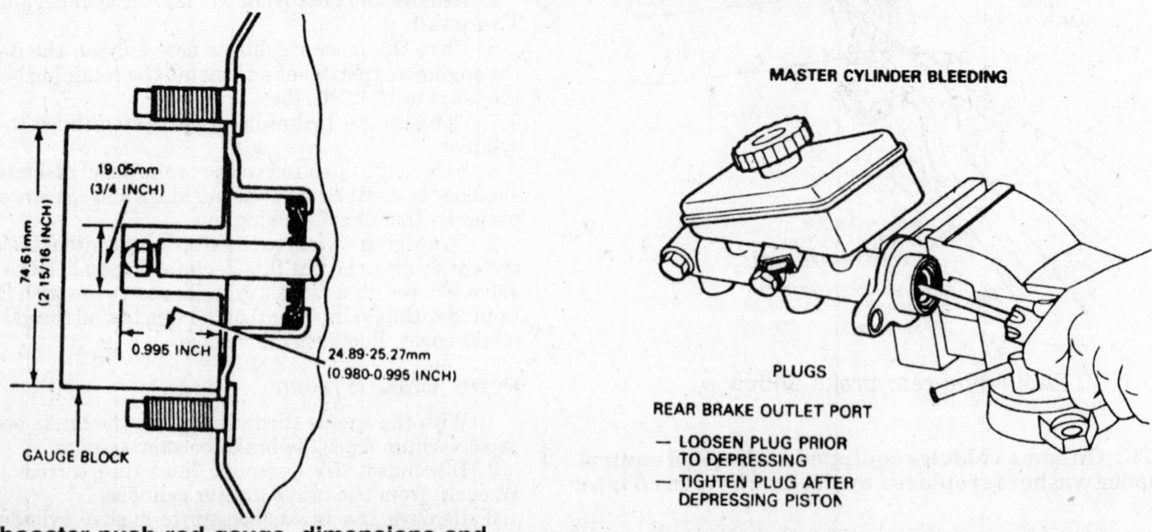

Bendix booster push rod gauge, dimensions and adjustment

Master cylinder bleeding procedure

To install:

4. Before installing the master cylinder, check the distance from the outer end of the booster assembly push rod to the front face of the brake booster assembly. Turn the push rod adjusting screw in or out as required to obtain the length shown. Refer to illustration in this Section.
5. Position the master cylinder assembly over the booster push rod and onto the 2 studs on the booster assembly. Install the attaching nuts and lockwashers and tighten to 13–25 ft. lbs.
6. Connect the hydraulic brake system lines to the master cylinder.
7. Bleed the hydraulic brake system (refer to procedure in this Section). Centralize the differential valve. Then, fill the dual master cylinder reservoirs with DOT 3 brake fluid to within ¼ in. (6mm) of the top. Install the gasket and reservoir cover. Roadtest the vehicle for proper operation.

When replacing the the master cylinder it is best to BENCH BLEED the master cylinder before installing it to the vehicle. Mount the master cylinder into a vise or suitable equivalent (do not damage the cylinder). Fill the cylinder to the correct level with the specified fluid. Block off all the outer brake line holes but one, then, using a long tool such as rod position it in the cylinder to actuate the brake master cylinder. Pump (push tool in and out) the brake master cylinder 3 or 4 times till brake fluid is release out and no air is in the brake fluid. Repeat this procedure until all brake fluid is released out of every hole and no air is expelled.

OVERHAUL

NOTE: Use this service procedure and exploded view diagrams as a guide for overhaul of the master cylinder assembly. If in doubt about overhaul condition or service procedure replace the complete assembly with a new master cylinder assembly.

The most important thing to remember when rebuilding the master cylinder is cleanliness. Work in clean surroundings with clean tools and clean cloths or paper for drying purposes. Have plenty of clean alcohol and brake fluid on hand to clean and lubricate the internal components. There are service repair kits available for overhauling the master cylinder.

1. Clean the outside of the master cylinder and remove the filler cap and gasket (diaphragm). Pour out any fluid that remains in the cylinder reservoir. Do not use any fluids other than brake fluid or alcohol to clean the master cylinder.
2. Unscrew the piston stop from the bottom of the cylinder body. Remove the O-ring seal from the piston stop. Discard the seal.
3. Remove the pushrod boot, if so equipped, from the groove at the rear of the master cylinder.
4. Remove the snapring retaining the primary and secondary piston assemblies within the cylinder body.
5. Remove the pushrod (if so equipped) and primary piston assembly from the master cylinder. Discard the piston assembly, including the boot (if so equipped).
6. Apply an air hose to the rear brake outlet port of the cylinder body and carefully blow the secondary piston out of the cylinder body.
7. Remove the return spring, spring retainer, cup protector, and cups from the secondary piston. Discard the cup protector and cups.
8. Clean all of the remaining parts in clean isopropyl alcohol and inspect the parts for chipping, excessive wear or damage. Replace them as required.

NOTE: When using a master cylinder repair kit, install all the parts supplied in the kit.

9. Check all recesses, reopenings and internal passages to be sure they are open and free from foreign matter. Use compressed air to blow out dirt and cleaning solvent remaining after the parts have been cleaned in the alcohol. Place all the parts on a clean pan, lint-free cloth, or paper to dry.
10. Dip all the parts, except the cylinder body, in clean brake fluid.
11. Assemble the 2 secondary cups, back-to-back, in the grooves near the end of the secondary piston.
12. Install the secondary piston assembly in the master cylinder.
13. Install a new O-ring on the piston stop, and start the stop into the cylinder body.
14. Position the boot, snapring and pushrod retainer on the pushrod. Make sure the pushrod retainer is seated securely on the ball end of the rod. Seat the pushrod in the primary piston assembly.

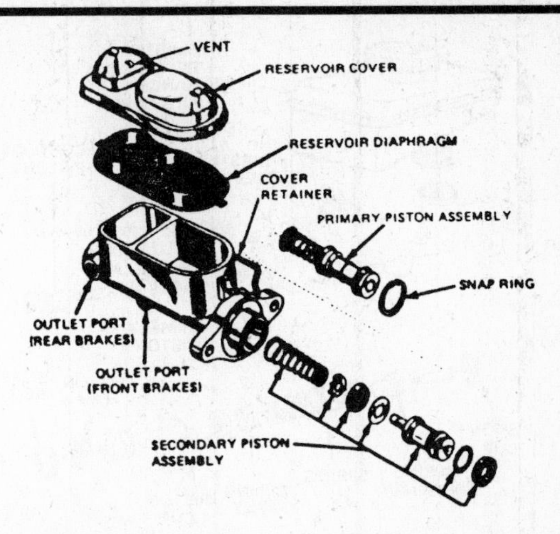

Exploded view of the master cylinder assembly

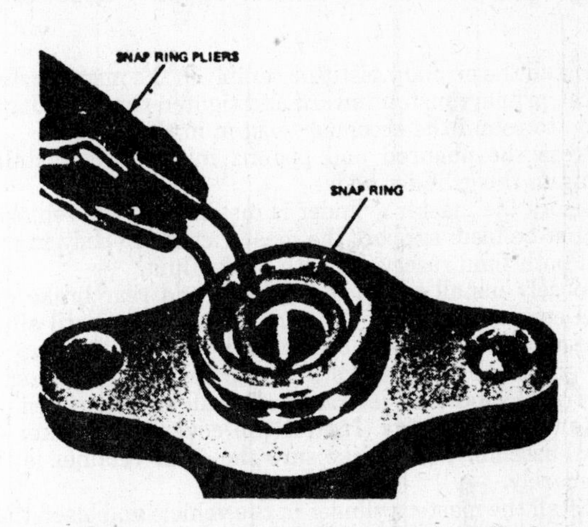

Snapring removal of the master cylinder

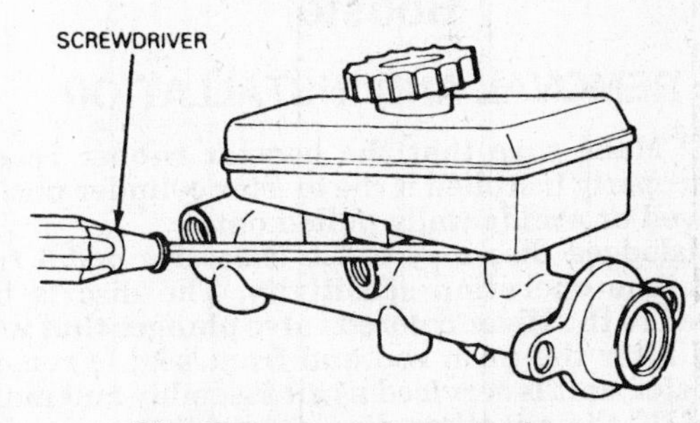

Plastic reservoir assembly

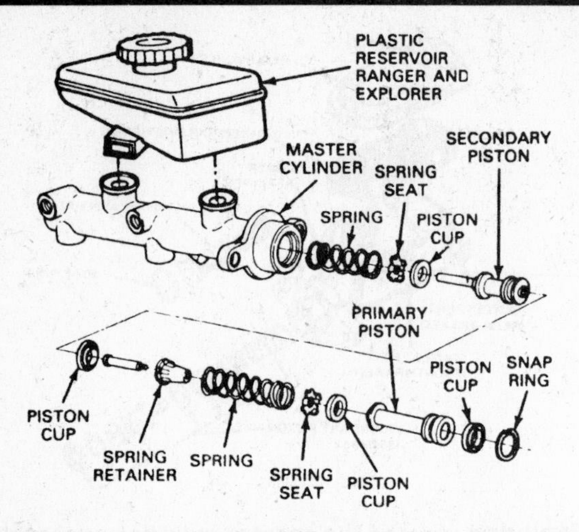

Exploded view of the master cylinder assembly

15. Install the primary piston assembly in the master cylinder. Push the primary piston inward and tighten the secondary piston stop to retain the secondary piston in the bore.
16. Press the pushrod and pistons inward and install the snapring in the cylinder body.
17. Before the master cylinder is installed on the vehicle, the unit must be bled: support the master cylinder body in a vise, and fill both fluid reservoirs with brake fluid.
18. Loosely install plugs in the front and rear brake outlet bores. Depress the primary piston several times until air bubbles cease to appear in the brake fluid.
19. Tighten the plugs and attempt to depress the piston. The piston travel should be restricted after all air is expelled.
20. Remove the plugs. Install the cover and gasket (diaphragm) assembly, and make sure the cover retainer is tightened securely.
21. Install the master cylinder in the vehicle and bleed the hydraulic system.

Booster

REMOVAL AND INSTALLATION

NOTE: Make sure that the booster rubber reaction disc is properly installed if the master cylinder push rod is removed or accidentally pulled out.

A dislodged disc may cause excessive pedal travel and extreme operation sensitivity. The disc is black compared to the silver colored valve plunger that will be exposed after the push rod and front seal is removed. The booster unit is serviced as an assembly and must be replaced if the reaction disc cannot be properly installed and aligned, or if it cannot be located within the unit itself.

1. Disconnect the stop lamp switch wiring to prevent running the battery down.
2. Support the master cylinder from the underside with a prop.
3. Remove the master cylinder-to-booster retaining nuts.
4. Loosen the clamp that secures the manifold vacuum hose to the booster check valve, and remove the hose. Remove the booster check valve.
5. Pull the master cylinder off the booster and leave it supported by the prop, far enough away to allow removal of the booster assembly.
6. From inside the cab on vehicles equipped with push rod mounted stop lamp switch, remove the retaining pin and slide the stop lamp switch, push rod, spacers and bushing off the brake pedal arm.
7. From the engine compartment remove the bolts that attach the booster to the dash panel.

To install:

8. Mount the booster assembly on the engine side of the dash panel by sliding the bracket mounting bolts and valve operating rod in through the holes in the dash panel.

NOTE: Make certain that the booster push rod is positioned on the correct side of the master cylinder to install onto the push pin prior to tightening the booster assembly to the dash.

9. From inside the cab, install the booster mounting bracket-to-dash panel retaining nuts.
10. Position the master cylinder on the booster assembly, install the retaining nuts, and remove the prop from underneath the master cylinder.

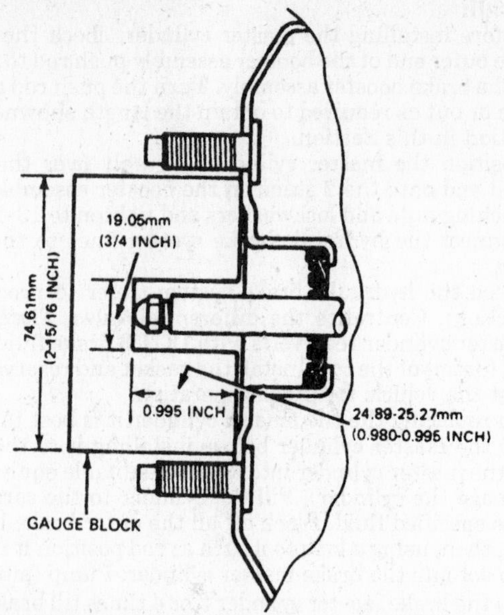

Bendix booster push rod gauge, dimensions and adjustment procedure

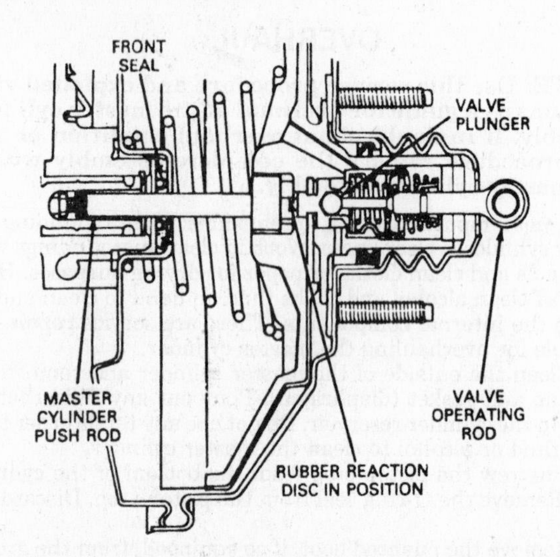

Cutaway view of brake booster assembly

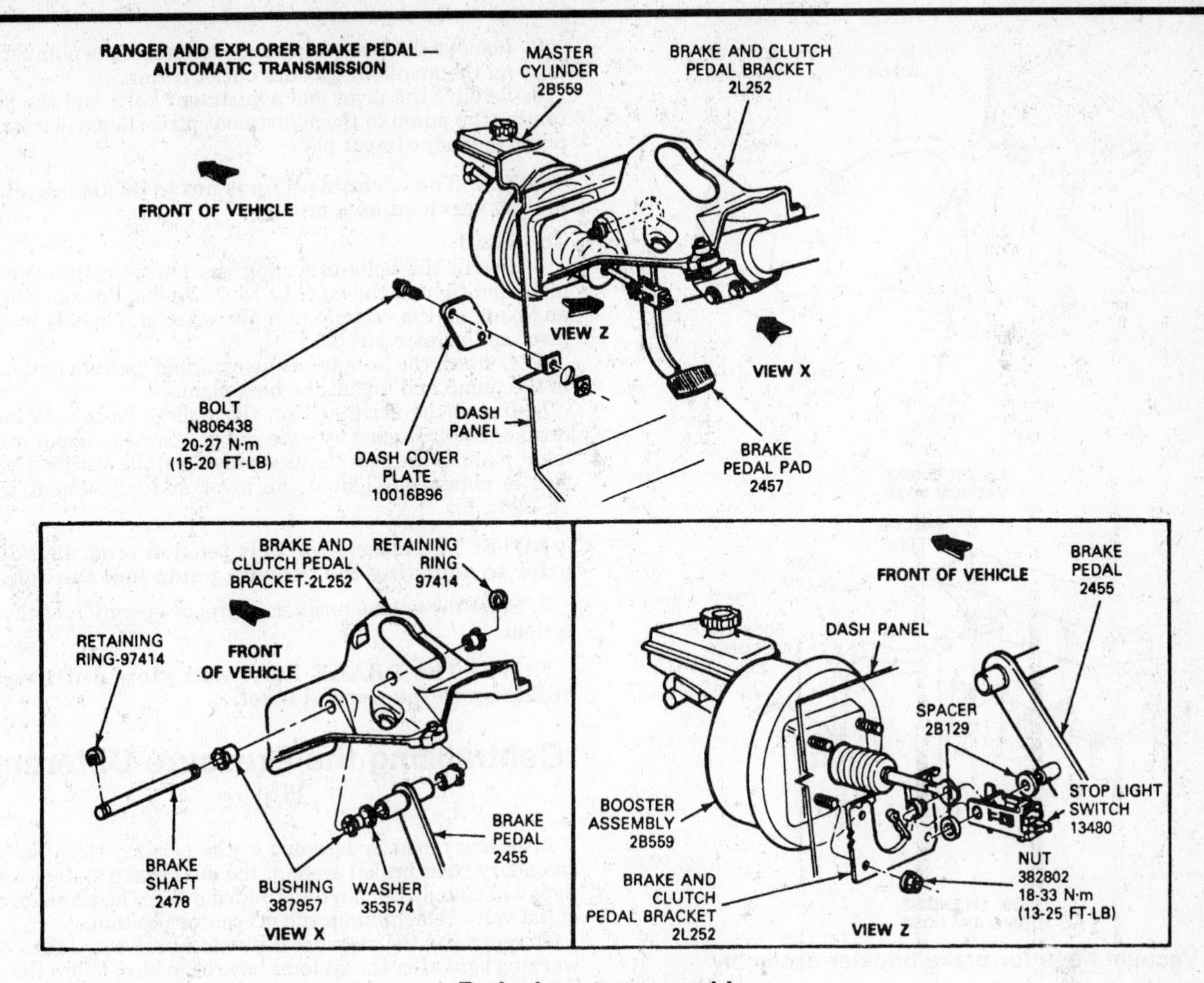

Brake booster assembly

11. Install the booster check valve. Connect the manifold vacuum hose to the booster check valve and secure with the clamp.
12. From inside the cab on vehicles equipped with push rod mounted stop lamp switch, install the bushing and position the switch on the end of the push rod. Then install the switch and rod on the pedal arm, along with spacers on each side, and secure with the retaining pin.
13. Connect the stop lamp switch wiring.
14. Start the engine and check brake operation.

Brake Hoses and Lines

HYDRAULIC BRAKE LINE CHECK

The hydraulic brake lines and brake linings are to be inspected at the recommended intervals in the maintenance schedule. Follow the steel tubing from the master cylinder to the flexible hose fitting at each wheel. If a section of the tubing is found to be damaged, replace the entire section with tubing of the same type (steel, not copper), size, shape, and length. When installing a new section of brake tubing, flush clean brake fluid or denatured alcohol through to remove any dirt or foreign material from the line. Be sure to flare both ends to provide sound, leakproof connections (replacement brake lines can purchased already made up at local parts store). When bending the tubing to fit the underbody contours, be careful not to kink or crack the line.

Check the flexible brake hoses that connect the steel tubing to each wheel cylinder. Replace the hose if it shows any signs of softening, cracking, or other damage. When installing a new front brake hose, position the hose to avoid contact with other chassis parts. Place a new copper gasket over the hose fitting and thread the hose assembly into the front wheel cylinder. A new rear brake hose must be positioned clear of the exhaust pipe or shock absorber. Thread the hose into the rear brake tube connector. When installing either a new front or rear brake hose, engage the opposite end of the hose to the bracket on the frame. Install the horseshoe type retaining clip and connect the tube to the hose with the tube fitting nut.

Always bleed the system after hose or line replacement. Before bleeding, make sure that the master cylinder is topped up with high temperature, extra heavy duty fluid of at least SAE 70R3 quality.

Vacuum Pump

Unlike gasoline engines, diesel engines have little vacuum available to power brake booster systems. The diesel is thus equipped with a vacuum pump, which is driven by a single belt off of the alternator. This pump is located on the top right side of the engine on the 2.2L and on the bottom left side of the engine on the 2.3L diesel engine.

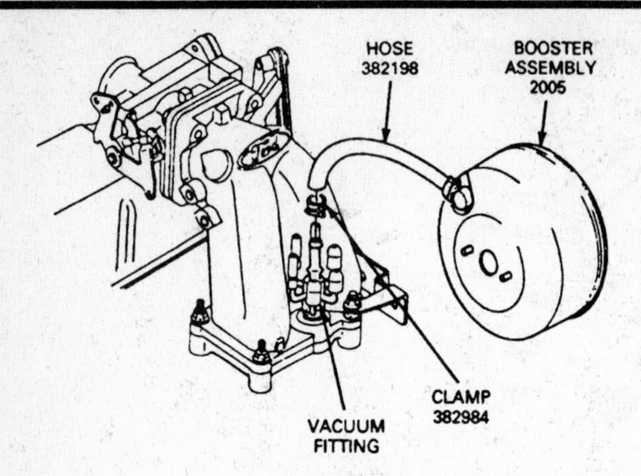

2.3L EFI ENGINE
VACUUM HOSE

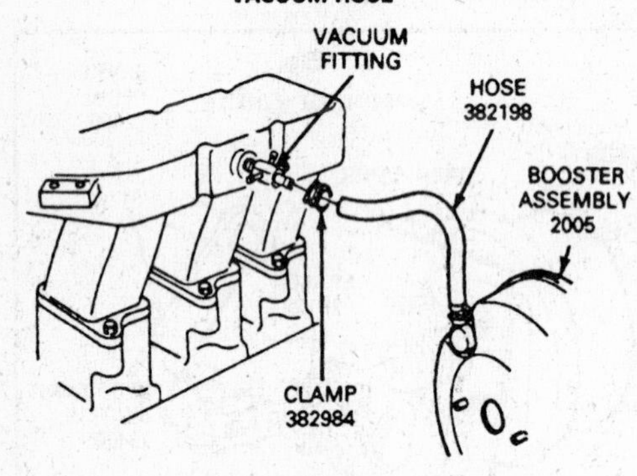

2.9L EFI ENGINE
VACUUM HOSE

Vacuum hose for brake booster assembly

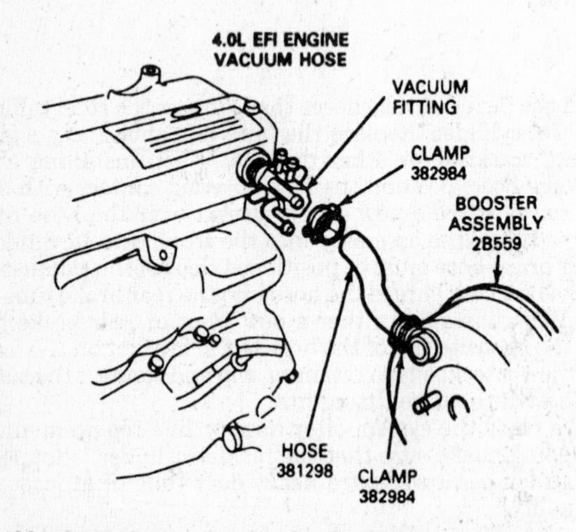

Vacuum hose for brake booster assembly

REMOVAL AND INSTALLATION

2.2L And 2.3L Diesel Engine

1. Loosen the vacuum pump adjustment bolt and the pivot bolt. Slide the pump downward and remove the drive belt from the pulley.

NOTE: If the vacuum pump drive belt is to be replaced, the alternator drive belt must be removed.

2. Remove the hose clamp and disconnect the pump from the hose on the manifold vacuum outlet fitting.
3. Remove the pivot and adjustment bolts and the bolts retaining the pump to the adjustment plate. Remove the vacuum pump and adjustment plate.

NOTE: The vacuum pump is not to be disassembled. It is only serviced as a unit.

To install:

4. Install the bolts attaching the pump to the adjustment plate and tighten the bolts to 15–20 ft. lbs. Position the pump and plate on the vacuum pump bracket and loosely install the pivot and adjustment bolts.
5. Connect the hose from the manifold vacuum outlet fitting to the pump and install the hose clamp.
6. Install the drive belt on the pulley. Place a 3/8 in. drive breaker bar or ratchet into the slot on the vacuum pump adjustment plate. Lift up on the assembly until the specified belt tension is obtained. Tighten the pivot and adjustment bolts to 15–20 ft. lbs.

NOTE: The alternator belt tension must be adjusted prior to adjusting the vacuum pump belt tension.

7. Start the engine and verify proper operation of the brake system.

NOTE: The BRAKE light will glow until vacuum builds up to the normal level.

Centralizing the Pressure Differential Valve

After any repair or bleeding of the primary (front brake) or secondary (rear brake) system, the dual brake system warning light will usually remain illuminated due to the pressure differential valve remaining in the off-center position.

To centralize the pressure differential valve and turn off the warning light after the systems have been bled, follow the procedure below.

1. Turn the ignition switch to the ACC or ON position.
2. Check the fluid level in the master cylinder reservoirs and fill them to within 1/4 in. (6mm) of the top with brake fluid, if necessary.
3. Depress the brake pedal and the piston should center itself causing the brake warning light to go out.
4. Turn the ignition switch to the OFF position.
5. Before driving the vehicle, check the operation of the brakes and be sure that a firm pedal is obtained.

Bleeding The Brakes

When any part of the hydraulic system has been disconnected for repair or replacement, air may get into the lines and cause spongy pedal action (because air can be compressed and brake fluid cannot). To correct this condition, it is necessary to bleed the hydraulic system after it has been properly connected to be sure all air is expelled from the brake cylinders and lines.

When bleeding the brake system, bleed one brake cylinder at a time, beginning at the cylinder with the longest hydraulic line (farthest from the master cylinder) first. ALWAYS Keep the master cylinder reservoir filled with brake fluid during the bleeding operation. Never use brake fluid that has been drained from the hydraulic system, no matter how clean it is.

It will be necessary to centralize the pressure differential value after a brake system failure has been corrected and the hydraulic system has been bled.

The primary and secondary hydraulic brake systems are individual systems and are bled separately. During the entire bleed-

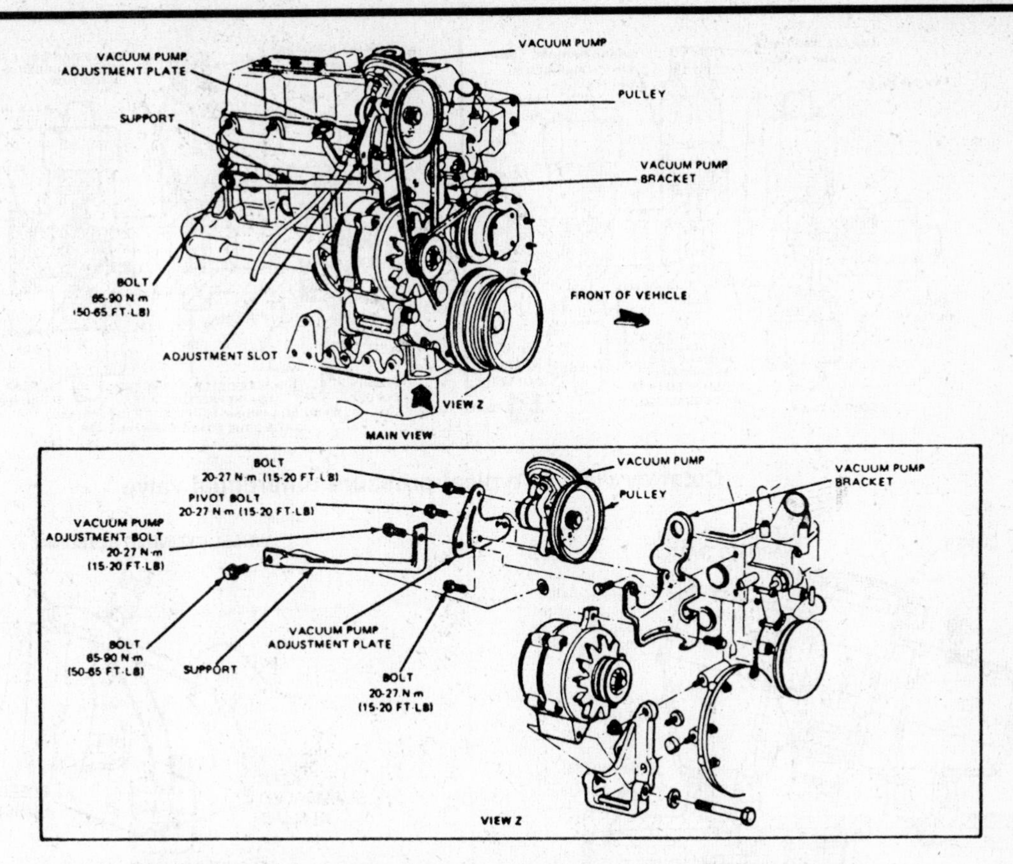

Vacuum pump removal and installation – 2.2L diesel engine

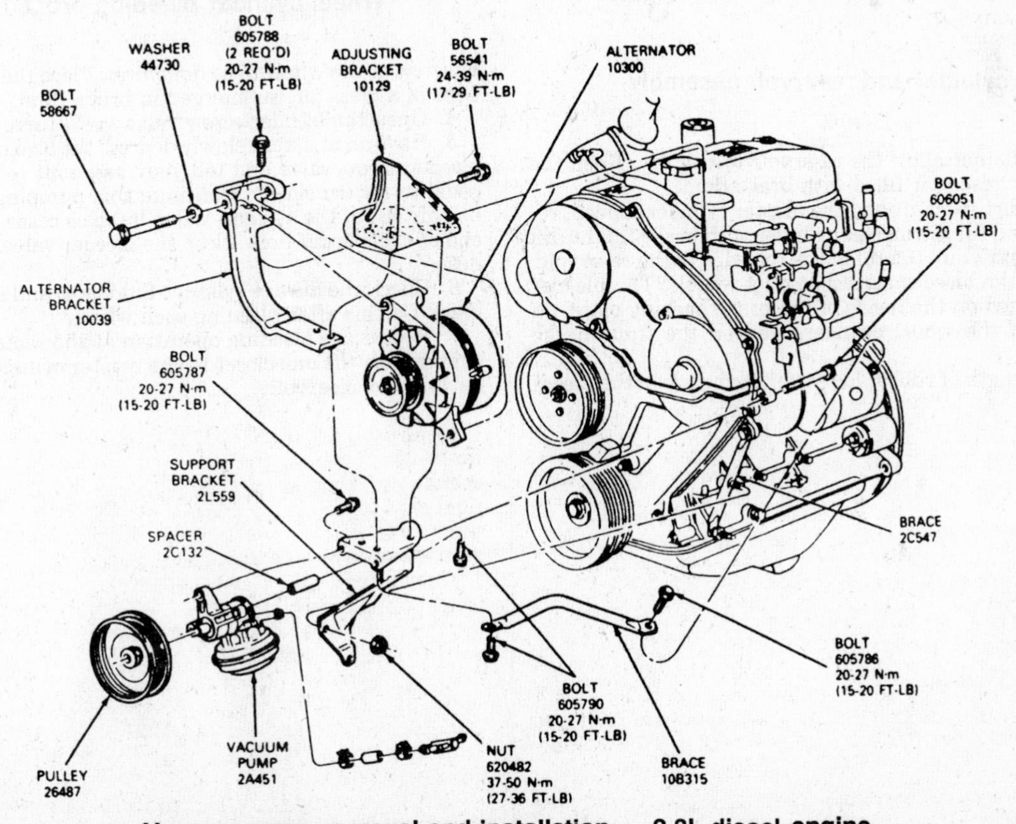

Vacuum pump removal and installation – 2.3L diesel engine

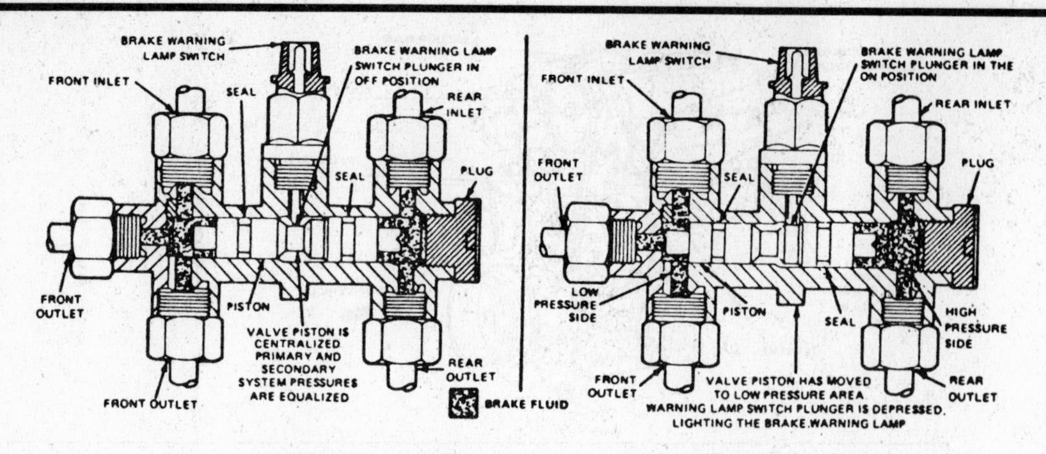

Cutaway view of typical pressure differential valve

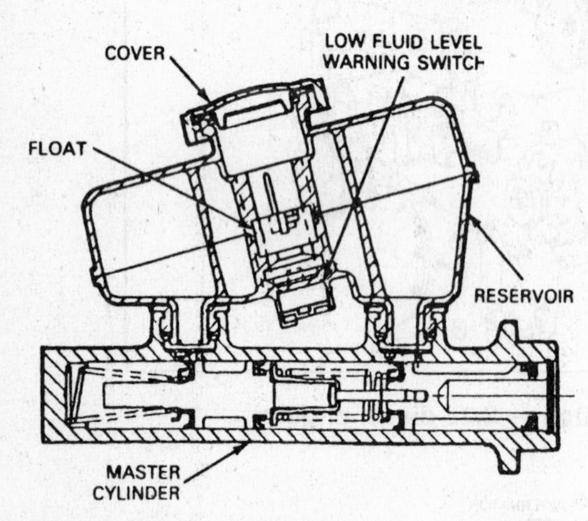

Master cylinder and reservoir assembly

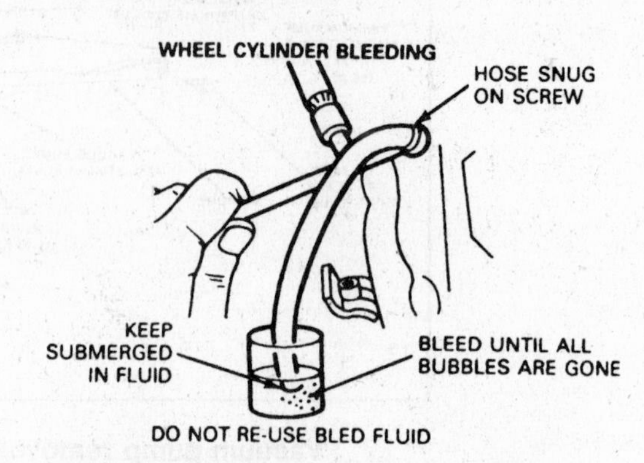

Wheel cylinder bleeding procedure

ing operation, do not allow the reservoir to run dry. Keep the master cylinder reservoir filled with brake fluid.

1. Clean all dirt from around the master cylinder fill cap, remove the cap and fill the master cylinder with brake fluid until the level is within ¼ in. (6mm) of the top edge of the reservoir.
2. Clean off the bleeder screws at all 4 wheels. The bleeder screws are located on the inside of the brake backing plate, on the backside of the wheel cylinders and on the front brake calipers.
3. Attach a length of rubber hose over the nozzle of the bleeder screw at the wheel to be done first. Place the other end of the hose in a glass jar, submerged in brake fluid.
4. Open the bleeder screw valve ½–¾ turn.
5. Have an assistant slowly depress the brake pedal. Close the bleeder screw valve and tell your assistant to allow the brake pedal to return slowly. Continue this pumping action to force any air out of the system. When bubbles cease to appear at the end of the bleeder hose, close the bleeder valve and remove the hose.
6. Check the master cylinder fluid level and add fluid accordingly. Do this after bleeding each wheel.
7. Repeat the bleeding operation at the remaining 3 wheels, ending with the one closet to the master cylinder. Fill the master cylinder reservoir.

FRONT DISC BRAKES

Pads

INSPECTION

Replace the front pads when the pad thickness is at the minimum thickness recommended by the Ford Motor Co. which is $^{1}/_{32}$ in. (0.8mm), or at the minimum allowed by the applicable state or local motor vehicle inspection code. Pad thickness may be checked by removing the wheel and looking through the inspection port in the caliper assembly.

REMOVAL AND INSTALLATION

NOTE: Always replace all disc pad assemblies on an axle. Never service one wheel only.

1. To avoid fluid overflow when the caliper piston is pressed into the caliper cylinder bores, siphon or dip part of the brake fluid out of the larger master cylinder reservoir (connected to the front disc brakes). Discard the removed fluid.
2. Raise the vehicle and install jackstands. Remove a front wheel and tire assembly.
3. Place an 8 in. (203mm) C-clamp on the caliper and tighten the clamp to bottom the caliper piston in the cylinder bore. Remove the clamp.

NOTE: Do not use a screwdriver or similar tool to pry piston away from the rotor.

4. There are 3 types of caliper pins used: a single tang type, a double tang type and a split-shell type. The pin removal process is dependent upon how the pin is installed (bolt head direction). Remove the upper caliper pin first.

NOTE: On some applications, the pin may be retained by a nut and Torx® head bolt (except the split-shell type).

5. If the bolt head is on the outside of the caliper, use the following procedure:
   a. From the inner side of the caliper, tap the bolt within the caliper pin until the bolt head on the outer side of the caliper shows a separation between the bolt head and the caliper pin.
   b. Using a hacksaw or bolt cutter, remove the bolt head from the bolt.
   c. Depress the tab on the bolt head end of the upper caliper pin with a screwdriver, while tapping on the pin with a hammer. Continue tapping until the tab is depressed by the V-slot.
   d. Place one end of a punch, ½ in. (13mm) or smaller, against the end of the caliper pin and drive the caliper pin out of the caliper toward the inside of the vehicle. Do not use a screwdriver or other edged tool to help drive out the caliper pin as the V-grooves may be damage.

WARNING: Never reuse caliper pins. Always install new pins whenever a caliper is removed.

6. If the nut end of the bolt is on the outside of the caliper, use the following procedure:
   a. Remove the nut from the bolt.
   b. Depress the lead tang on the end of the upper caliper pin with a screwdriver while tapping on the pin with a hammer. Continue tapping until the lead tang is depressed by the V-slot.
   c. Place one end of a punch, ½ in. (13mm) or smaller, against the end of the caliper pin and drive the caliper pin out of the caliper toward the inside of the vehicle. Do not use a

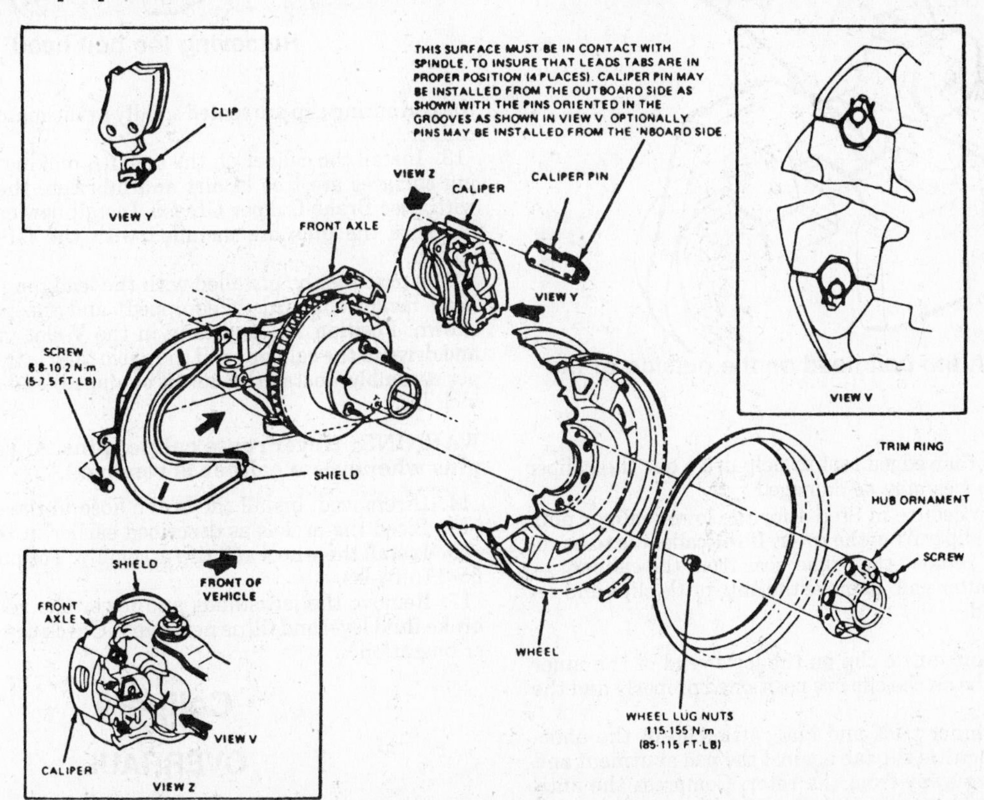

Front disc brake assembly

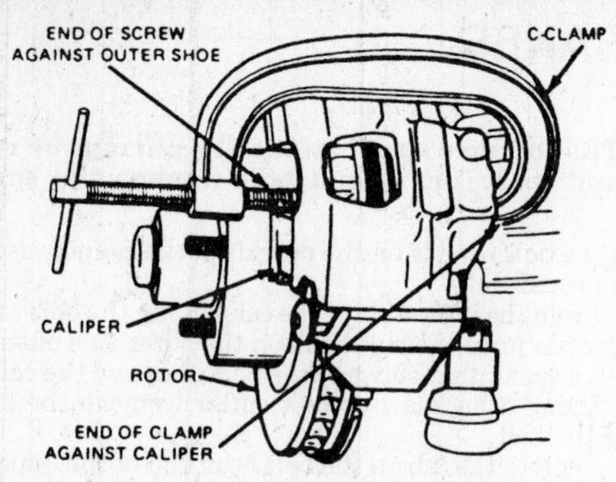

Bottoming the caliper piston

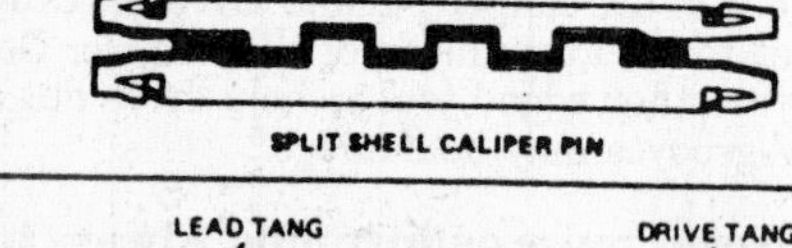

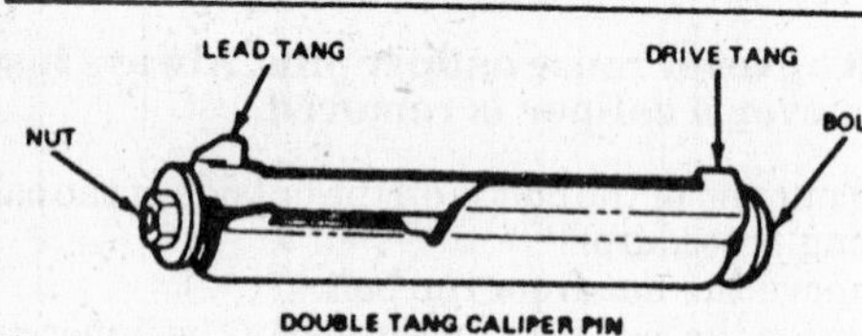

Caliper pins

Caliper pin with the bolt head on the outside of the caliper

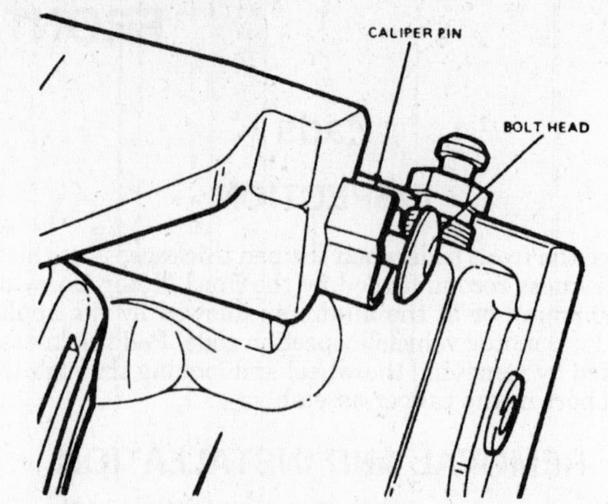

Separation between the bolt head and the caliper pin

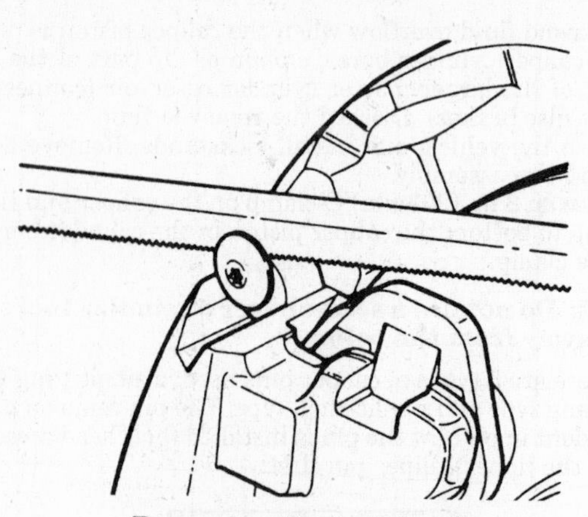

Removing the bolt head

screwdriver or other edged tool to help drive out the caliper pin as the V-grooves may be damaged.

7. Repeat the procedure in Step 4 for the lower caliper pin.
8. Remove the caliper from the rotor. If the caliper is to be removed for service, remove the brake hose from the caliper.
9. Remove the outer pad. Remove the anti-rattle clips and remove the inner pad.

To install:

10. Place a new anti-rattle clip on the lower end of the inner pad. Be sure the tabs on the clip are positioned properly and the clip is fully seated.
11. Position the inner pads and anti-rattle clip in the abutment with the anti-rattle clip tab against the pad abutment and the loop-type spring away from the rotor. Compress the anti-rattle clip and slide the upper end of the pad in position.
12. Install the outer pad, making sure the torque buttons on the pad spring clip are seated solidly in the matching holes in the caliper.
13. Install the caliper on the spindle, making sure the mounting surfaces are free of dirt and lubricate the caliper grooves with Disc Brake Caliper Grease. Install new caliper pins, making sure the pins are installed with the fang in position as shown.

The pin must be installed with the lead tang in first, the bolt head facing outward (if equipped) and the pin positioned as shown. Position the lead tang in the V-slot mounting surface and drive in the caliper until the drive tang is flush with the caliper assembly. Install the nut (if equipped) and tighten to 32–47 inch lbs.

WARNING: Never reuse caliper pins. Always install new pins whenever a caliper is removed.

14. If removed, install the brake hose to the caliper.
15. Bleed the brakes as described earlier in this Section.
16. Install the wheel and tire assembly. Torque the lug nuts to 85–115 ft. lbs.
17. Remove the jackstands and lower the vehicle. Check the brake fluid level and fill as necessary. Check the brakes for proper operation.

Calipers

OVERHAUL

1. For caliper removal, see the above procedure. Disconnect the brake hose.

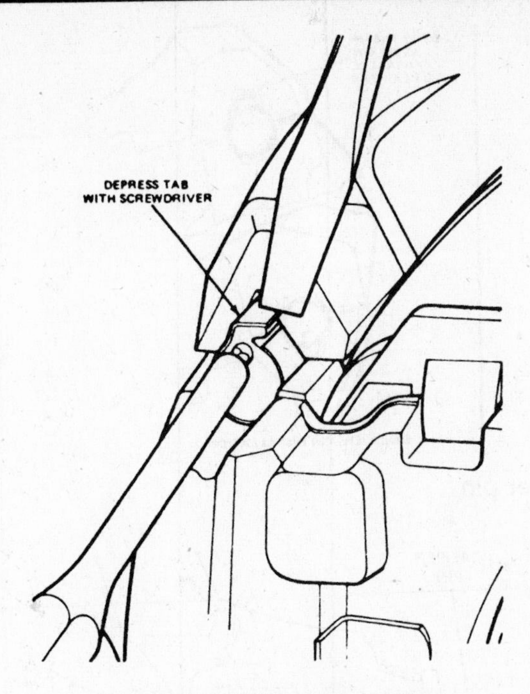

Removing the caliper pin

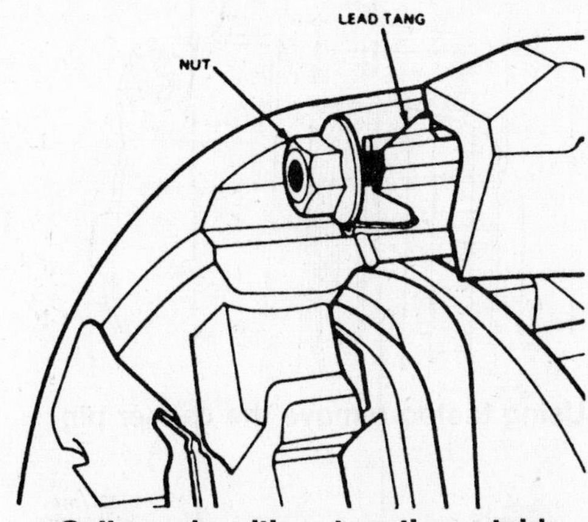

Caliper pin with nut on the outside

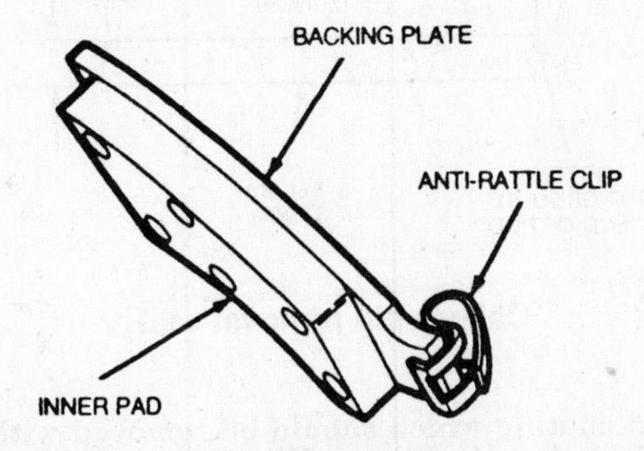

Anti-rattle clip installed on the inner pad

2. Clean the exterior of the caliper with denatured alcohol.
3. Remove the plug from the caliper inlet port and drain the fluid.
4. Air pressure is necessary to remove the piston. When a source of compressed air is found, such as a shop or gas station, apply air to the inlet port slowly and carefully until the piston pops out of its bore.

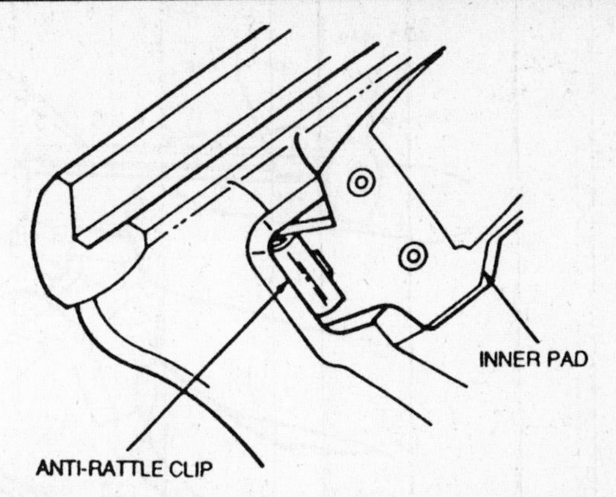

Installing the inner pad and the anti-rattle clip into the caliper

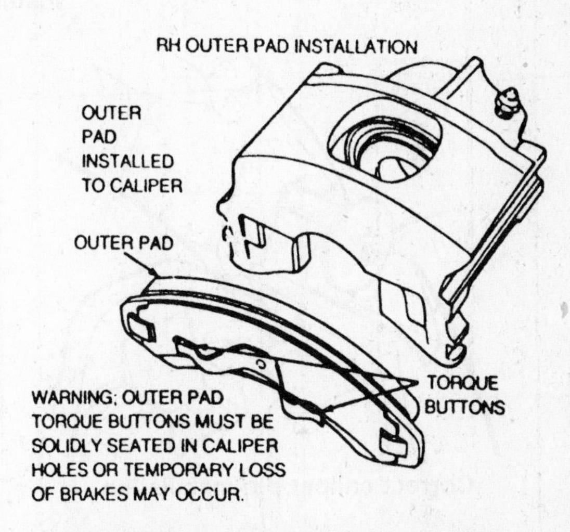

Installing the outer pad in the caliper

CAUTION

If high pressure air is applied, the piston will pop out with considerable force and cause damage or injury.

5. If the piston jams, release the air pressure and tap sharply on the piston end with a soft hammer. Reapply air pressure.
6. When the piston is out, remove the boot from the piston and the seal from the bore.
7. Clean the housing and piston with denatured alcohol. Dry with compressed air.
8. Lubricate the new piston seal, boot and piston with clean brake fluid, and assemble them in the caliper.
9. The dust boot can be worked in with the fingers and the piston should be pressed straight in until it bottoms. Be careful to avoid cocking the piston in the bore.
10. A C-clamp may be necessary to bottom the piston.
11. Install the caliper using the procedure given in the pad and caliper replacement procedure above.

Rotor (Disc)

REMOVAL AND INSTALLATION

1. Jack up the front of the vehicle and support on jackstands.

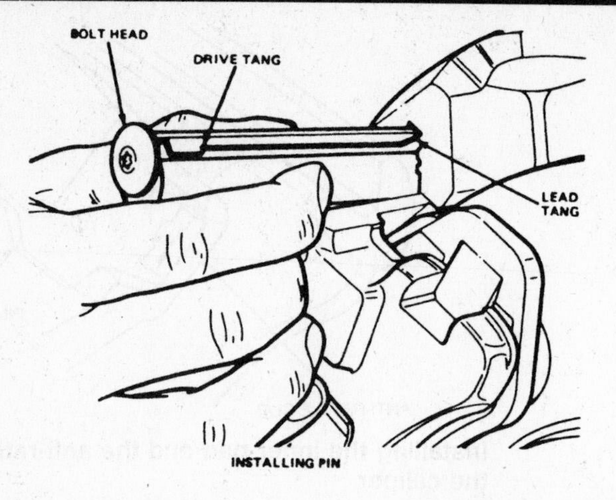

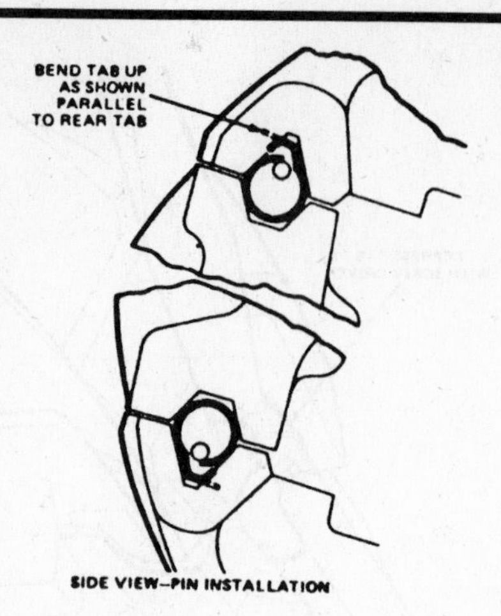

Installing the caliper pin

Correct caliper pin installation

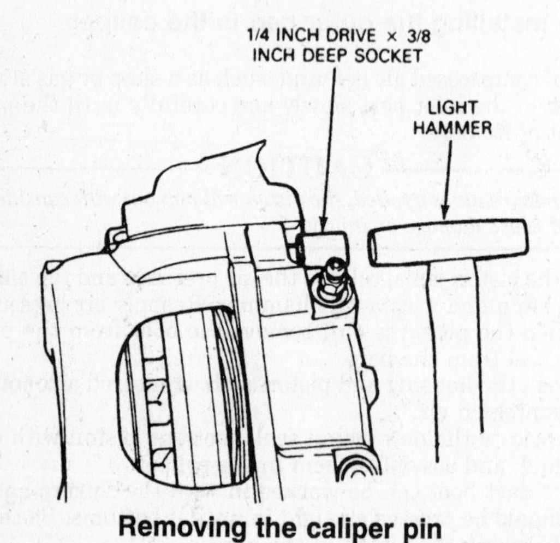

Removing the caliper pin

CALIPER PIN

Using tool to remove the caliper pin

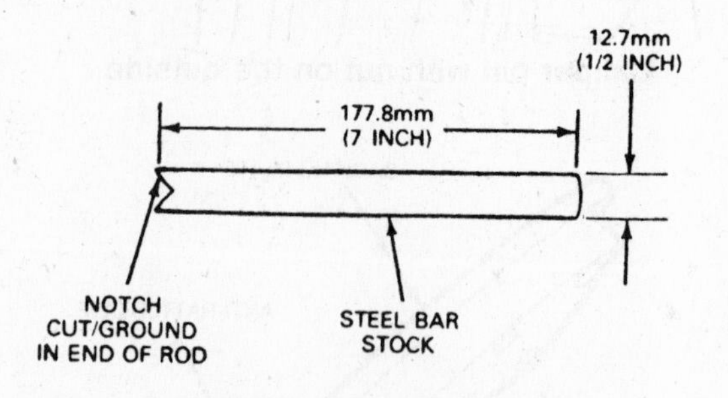

Caliper pin removal tool

2. Remove the wheel and tire.
3. Remove the caliper assembly as described earlier in this Section.
4. Follow the procedure given under hub and wheel bearing removal in Section 7 for models with manual and automatic locking hubs.

NOTE: New rotor assemblies come protected with an anti-rust coating which should be removed with denatured alcohol or degreaser. New hubs must be packed with EP wheel bearing grease. If the old rotors are to be reused, check them for cracks, grooves or waviness. Rotors that aren't too badly scored or grooved can be resurfaced by most automotive shops. Minimum rotor thickness should be 0.81 in. (20.5mm). If refinishing exceeds that, the rotor will have to be replaced.

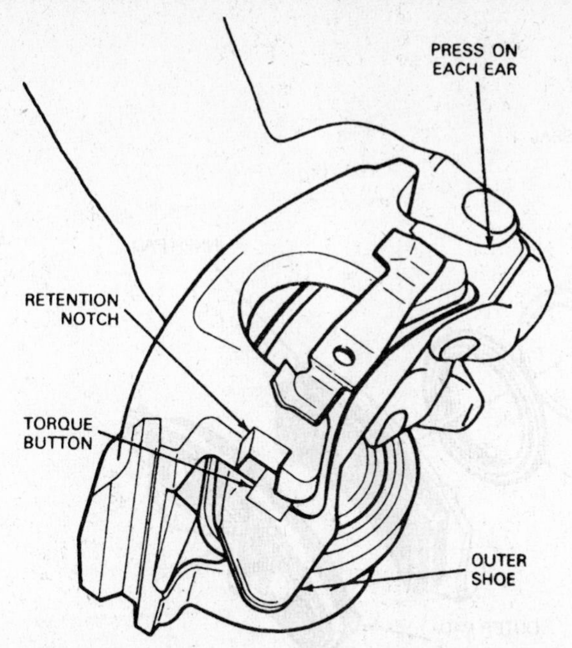

Removing the outer shoe from the caliper assembly

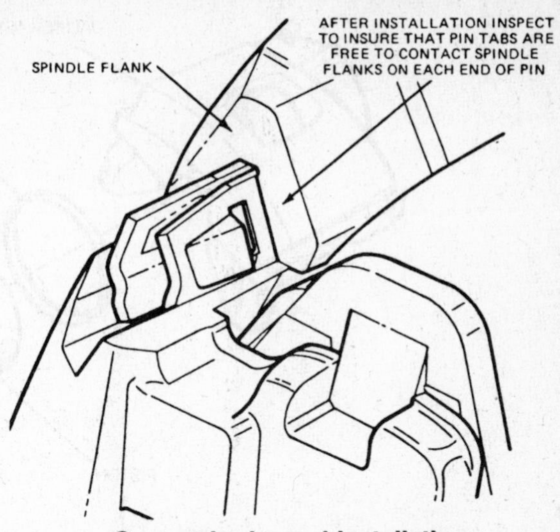

Correct brake pad installation

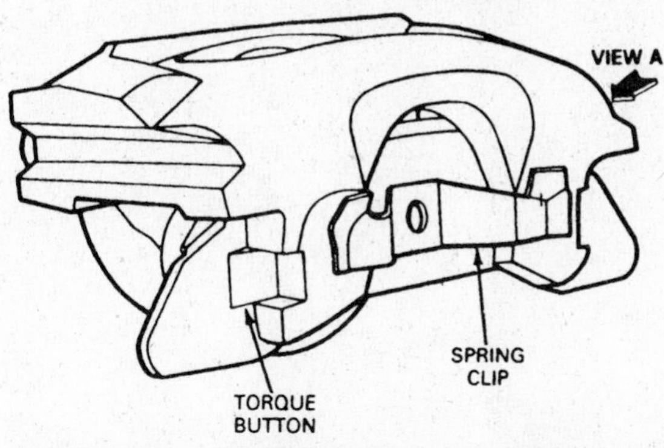

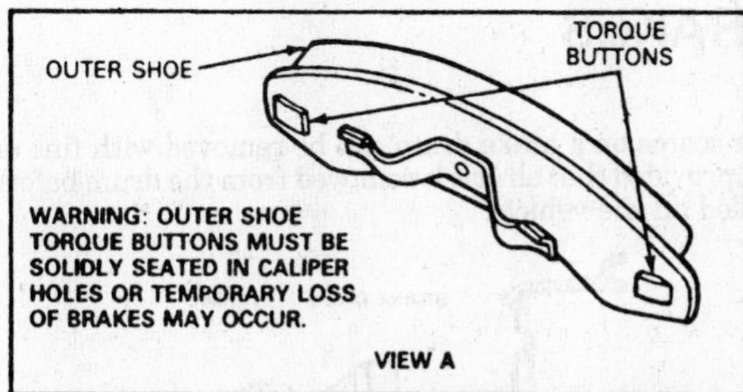

Correct brake shoe installation in caliper

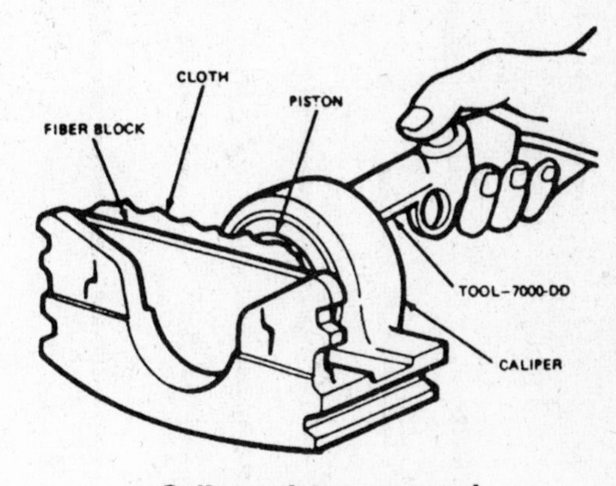

Caliper piston removal

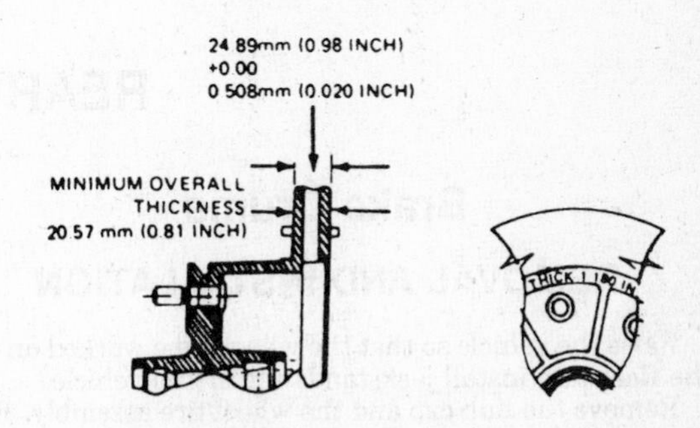

Disc brake service limits

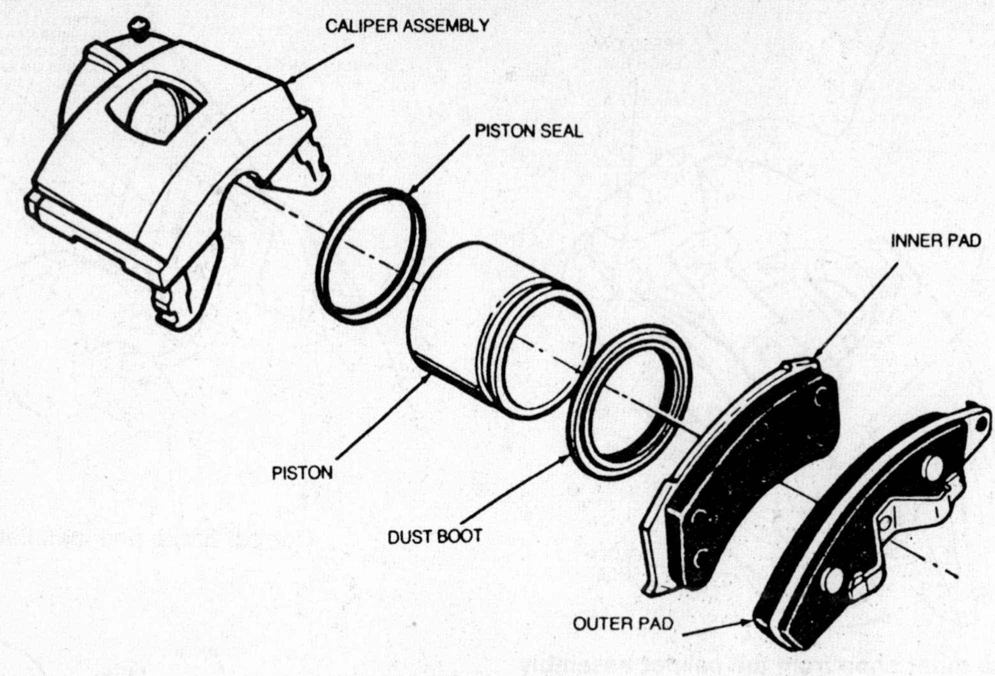

Caliper assembly

REAR DRUM BRAKES

Brake Drums

REMOVAL AND INSTALLATION

1. Raise the vehicle so that the wheel to be worked on is clear of the floor and install jackstands under the vehicle.
2. Remove the hub cap and the wheel/tire assembly. Remove the 3 retaining nuts and remove the brake drum. It may be necessary to back off the brake shoe adjustment in order to remove the brake drum. This is because the drum might be grooved or worn from being in service for an extended period of time.
3. Before installing a new brake drum, be sure and remove any protective coating with carburetor degreaser.
4. Install the brake drum in the reverse order of removal and adjusts the brakes.

INSPECTION

After the brake drum has been removed from the vehicle, it should be inspected for runout, severe scoring cracks, and the proper inside diameter.

Minor scores on a brake drum can be removed with fine emery cloth, provided that all grit is removed from the drum before it is installed on the vehicle.

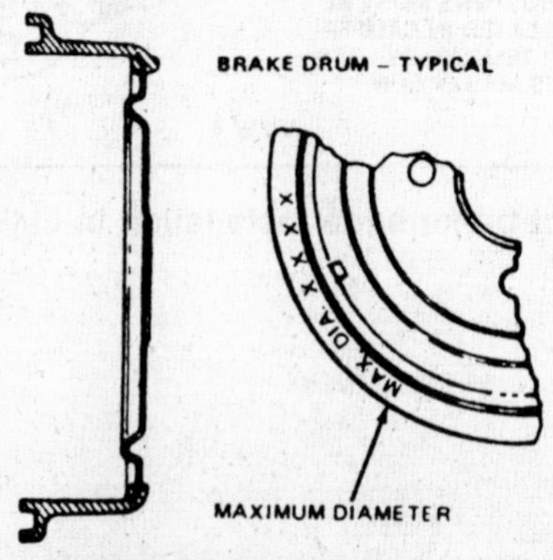

Rear brake drum maximum inside surface diameter marking location

A badly scored, rough, or out-of-round (runout) drum can be ground or turned on a brake drum lathe. Do not remove any more material from the drum than is necessary to provide a smooth surface for the brake shoe to contact. The maximum diameter of the braking surface is shown on the inside of each brake drum. Brake drums that exceed the maximum braking surface diameter shown on the brake drum, either through wear or refinishing, must be replaced. This is because after the outside wall of the brake drum reaches a certain thickness (thinner than the original thickness) the drum loses its ability to dissipate the heat created by the friction between the brake drum and the brake shoes, when the brakes are applied. Also the brake drum will have more tendency to warp and/or crack.

The maximum braking surface diameter specification, which is shown on each drum, allows for a 0.060 in. (1.5mm) machining cut over the original nominal drum diameter plus 0.030 in. (0.76mm) additional wear before reaching the diameter where the drum must be discarded. Use a brake drum micrometer to measure the inside diameter of the brake drums.

Brake Shoes

REMOVAL AND INSTALLATION

1. Raise and support the vehicle and remove the wheel and brake drum from the wheel to be worked on.

NOTE: If you have never replaced the brakes on a car before and you are not too familiar with the procedures involved, only disassemble and assemble one side at a time, leaving the other side intact as a reference during reassembly.

2. Install a clamp over the ends of the wheel cylinder to prevent the pistons of the wheel cylinder from coming out, causing loss of fluid.
3. Contract the brake shoes by pulling the self-adjusting lever away from the starwheel adjustment screw and turn the starwheel up and back until the pivot nut is drawn onto the starwheel as far as it will come.
4. Pull the adjusting lever, cable and automatic adjuster spring down and toward the rear to unhook the pivot hook from the large hole in the secondary shoe web. Do not attempt to pry the pivot hook from the hole.
5. Remove the automatic adjuster spring and the adjusting lever.
6. Remove the secondary shoe-to-anchor spring with a brake tool. (Brake tools are very common and are available at auto parts stores). Remove the primary shoe-to-anchor spring and unhook the cable anchor. Remove the anchor pin plate.
7. Remove the cable guide from the secondary shoe.
8. Remove the shoe hold-down springs, shoes, adjusting screw, pivot nut, and socket. Note the color of each hold-down spring for assembly. To remove the hold-down springs, reach being the brake backing plate and place one finger on the end of one of the brake hold-down spring mounting pins. Using a pair of pliers, grasp the washer-type retainer on top of the hold-down spring that corresponds to the pin that you are holding. Push down on the pliers and turn them 90° to align the slot in the washer with the head on the spring mounting pin. Remove the spring and washer retainer and repeat this operation on the hold-down spring on the other shoe.
9. Remove the parking brake link and spring. Disconnect the parking brake cable from the parking brake lever.
10. After removing the rear brake secondary shoe, disassemble the parking brake lever from the shoe by removing the retaining clip and spring washer.

To assemble and install the brake shoes:

11. Assemble the parking brake lever to the secondary shoe and secure it with the spring washer and retaining clip.

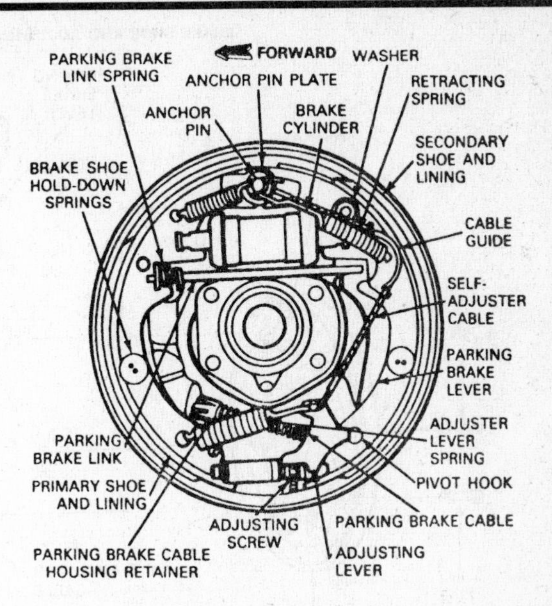

10-INCH REAR BRAKE (LEFT SIDE)

REAR BRAKE (LEFT SIDE)

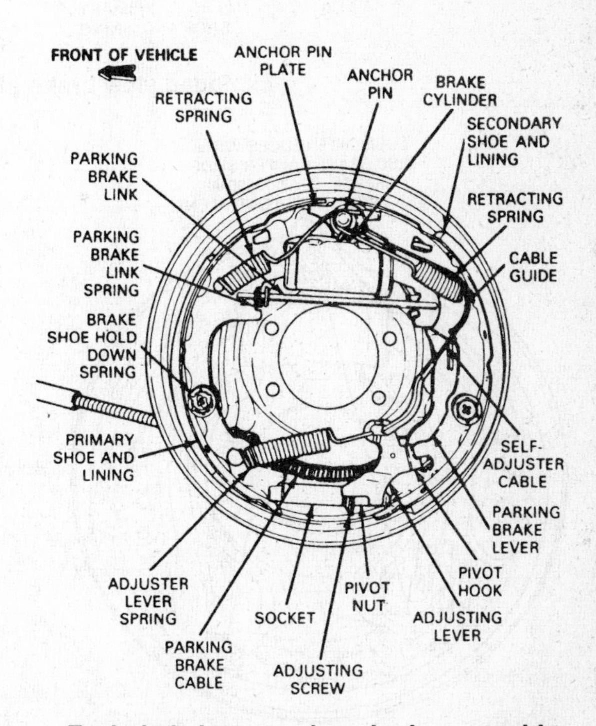

Exploded view rear drum brake assembly

12. Apply a light coating of Lubriplate® at the points where the brake shoes contact the backing plate.
13. Position the brake shoes on the backing plate, and install the hold-down spring pins, springs, and spring washer-type retainers. Install the parking brake link, spring and washer. Connect the parking brake cable to the parking brake lever.
14. Install the anchor pin plate, and place the cable anchor over the anchor pin with the crimped side toward the backing plate.
15. Install the primary shoe-to-anchor spring with the brake tool.

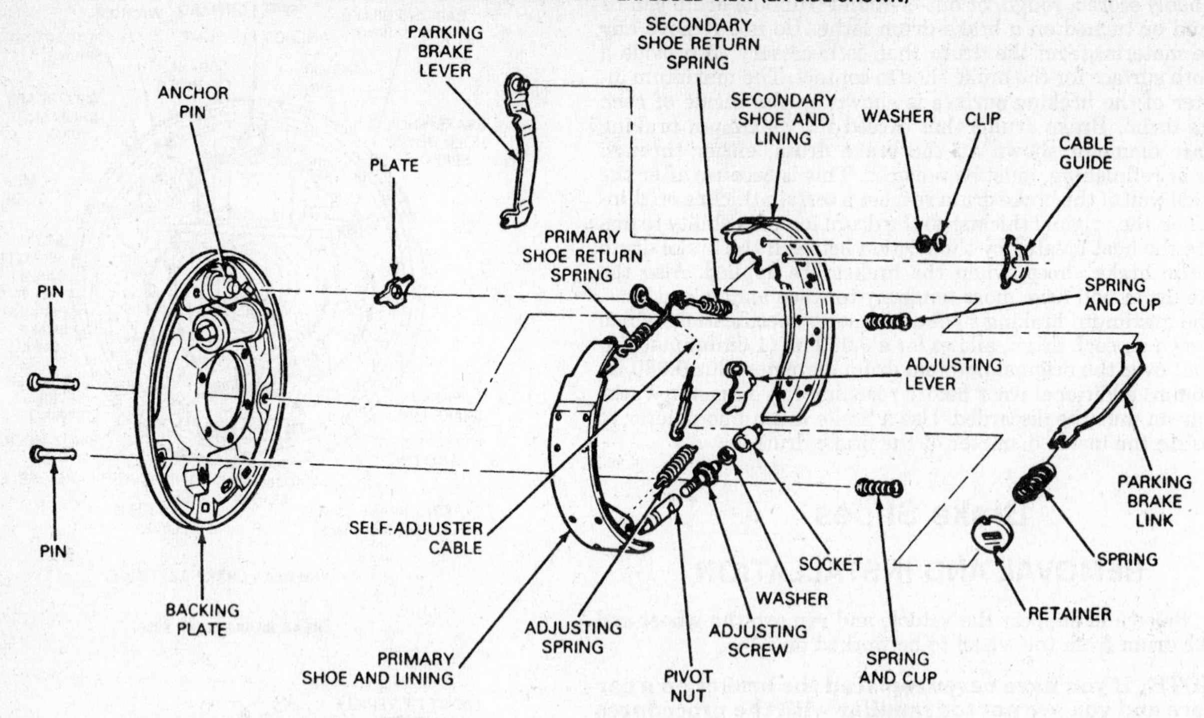

Exploded view brake shoe and adjusting screw assembly

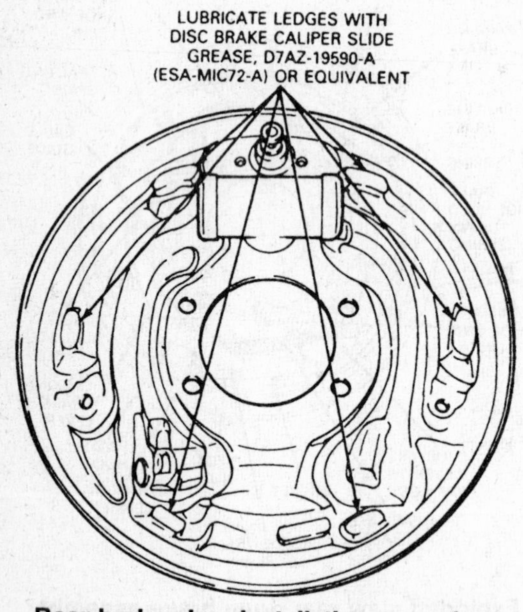

Rear brake shoe installation procedures

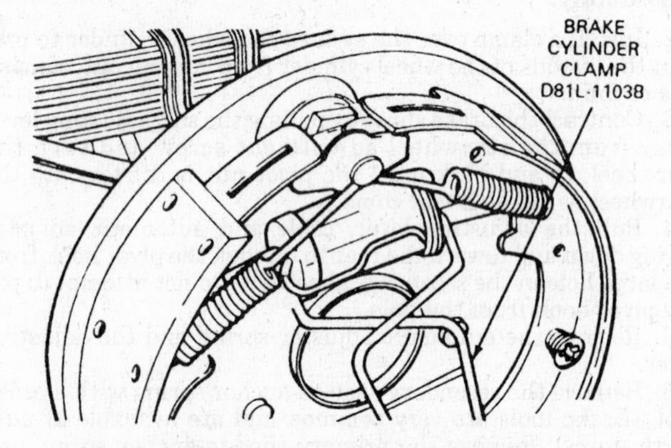

Rear brake shoe removal procedure

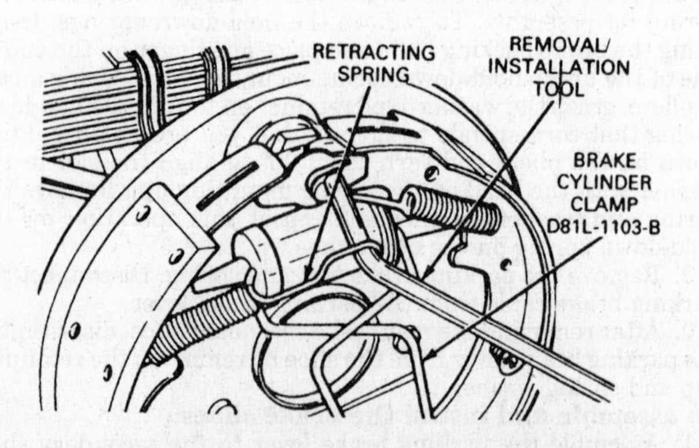

Rear brake shoe removal procedure

16. Install the cable guide on the secondary shoe web with the flanged holes fitted into the hole in the secondary shoe web. Thread the cable around the cable guide groove.
17. Install the secondary shoe-to-anchor (long) spring. Be sure that the cable end is not cocked or binding on the anchor pin when installed. All of the parts should be flat on the anchor pin. Remove the wheel cylinder piston clamp.
18. Apply Lubriplate® to the threads and the socket end of the adjusting starwheel screw. Turn the adjusting screw into the adjusting pivot nut to the limit of the threads and then back off ½ turn.

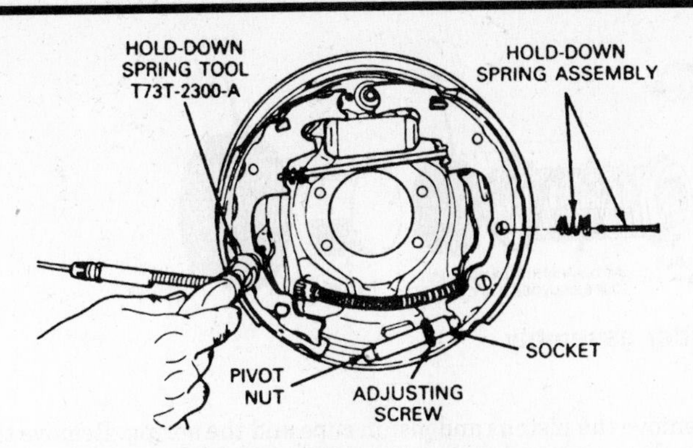

Rear brake shoe removal procedure

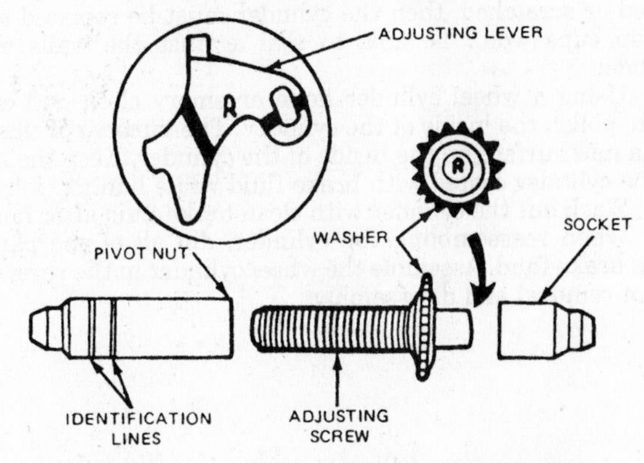

Exploded view adjusting screw assembly

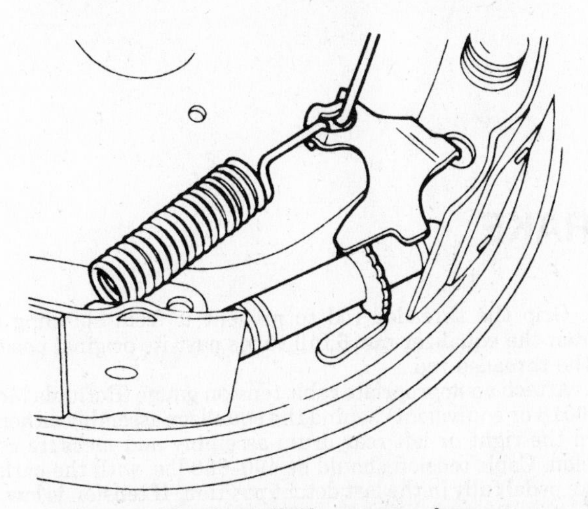

Rear brake installation procedure

NOTE: Interchanging the brake shoe adjusting screw assemblies from one side of the vehicle to the other would cause the brake shoes to retract rather than expand each time the automatic adjusting mechanism operated. To prevent this, the socket end of the adjusting screw is stamped with an R or an L for RIGHT or LEFT. The adjusting pivot nuts can be distinguished by the number of lines machined around the body of the nut; one line indicates left hand nut and 2 lines indicates a right hand nut.

19. Place the adjusting socket on the screw and install this as-

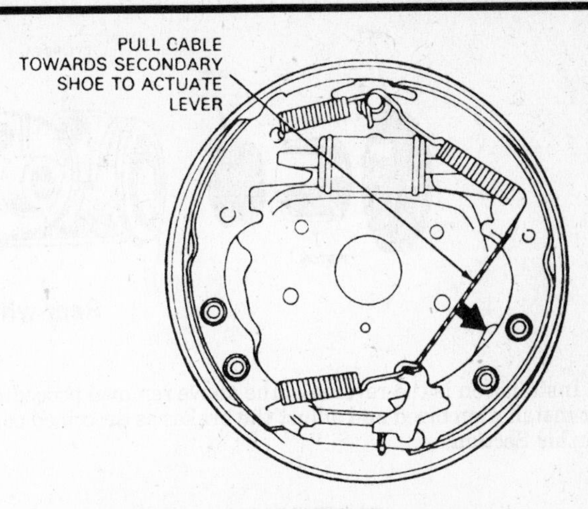

Rear brake installation procedure

sembly between the shoe ends with the adjusting screw nearest to the secondary shoe.

20. Place the cable hook into the hole in the adjusting lever from the backing plate side. The adjusting levers are stamped with an **R** (right) or an **L** (left) to indicate their installation on the right or left hand brake assembly.

21. Position the hooked end of the adjuster spring in the primary shoe web and connect the loop end of the spring to the adjuster lever hole.

22. Pull the adjuster lever, cable and automatic adjuster spring down toward the rear to engage the pivot hook in the large hole in the secondary shoe web.

23. After installation, check the action of the adjuster by pulling the section of the cable between the cable guide and the adjusting lever toward the secondary shoe web far enough to lift the lever past a tooth on the adjusting screw starwheel. The lever should snap into position behind the next tooth, and release of the cable should cause the adjuster spring to return the lever to its original position. This return action of the lever will turn the adjusting screw starwheel one tooth. The lever should contact the adjusting screw starwheel one tooth above the center line of the adjusting screw.

If the automatic adjusting mechanism does not perform properly, check the following:

1. Check the cable end fittings. The cable ends should fill or extend slightly beyond the crimped section of the fittings. If this is not the case, replace the cable.
2. Check the cable guide for damage. The cable groove should be parallel to the shoe web, and the body of the guide should lie flat against the web. Replace the cable guide if this is not so.
3. Check the pivot hook on the lever. The hook surfaces should be square with the body on the lever for proper pivoting. Repair or replace the hook as necessary.
4. Make sure that the adjusting screw starwheel is properly seated in the notch in the shoe web.

Wheel Cylinders

REMOVAL AND INSTALLATION

1. To remove the wheel cylinder, jack up the vehicle and remove the wheel, hub, and drum.
2. Disconnect the brake line at the fitting on the brake backing plate.
3. Remove the brake assemblies.
4. Remove the screws that hold the wheel cylinder to the backing plate and remove the wheel cylinder from the vehicle.

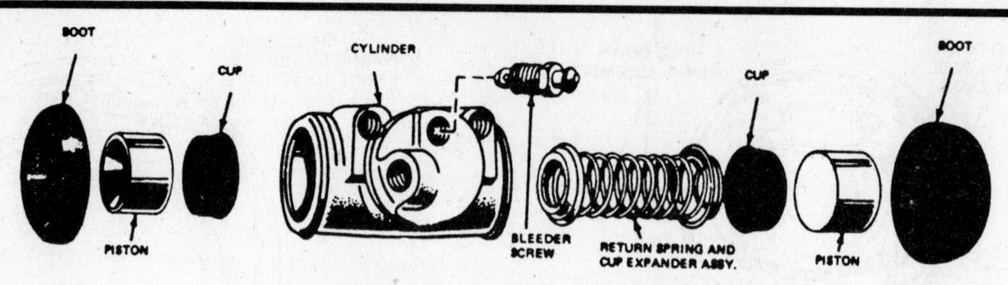

Rear wheel cylinder assembly

5. Installation is the reverse of the above removal procedure. After installation bleed and adjust the brakes as described earlier in this Section.

OVERHAUL

Wheel cylinder rebuilding kits are available for reconditioning wheel cylinders. The kits usually contain new cup springs, cylinder cups, and in some, new boots. The most important factor to keep in mind when rebuilding wheel cylinders is cleanliness. Keep all dirt away from the wheel cylinders when you are reassembling them.

1. Remove the wheel cylinder as described earlier.
2. Remove the rubber dust covers on the ends of the cylinder. Remove the pistons and piston cups and the spring. Remove the bleeder screw and make sure that it is not plugged.
3. Discard all of the parts that the rebuilding kit will replace.
4. Examine the inside of the cylinder. If it is severely rusted, pitted or scratched, then the cylinder must be replaced as the piston cups won't be able to seal against the walls of the cylinder.
5. Using a wheel cylinder hone or emery cloth and crocus cloth, polish the inside of the cylinder. The purpose of this is to put a new surface on the inside of the cylinder. Keep the inside of the cylinder coated with brake fluid while honing.
6. Wash out the cylinder with clean brake fluid after honing.
7. When reassembling the cylinder, dip all of the parts in clean brake fluid. Assemble the wheel cylinder in the reverse order of removal and disassembly.

PARKING BRAKE

Cable

ADJUSTMENT

Pre-Tension Procedure

NOTE: This procedure is to be used when a new Tension Limiter has been installed.

1. Depress the parking brake pedal.
2. Grip the Tension Limiter Bracket to prevent it from spinning and tighten the equalizer nut 2½ in. (63.5mm) up the rod.
3. Check to make sure the cinch strap has slipped less than 1⅜ in. (35mm) remaining.

Final Adjustment

NOTE: This procedure is to be used to remove the slack from the system if a new Tension Limiter has not been installed.

1. Make sure the brake drums are cold for correct adjustment.
2. Position the parking brake pedal to the fully depressed position.
3. Grip the threaded rod to prevent it from spinning and tighten the equalizer nut 6 full turns past its original position on the threaded rod.
4. Attach an appropriate cable tension gauge (Rotunda Model 21–0018 or equivalent) behind the equalizer assembly either toward the right or left rear drum assembly and measure cable tension. Cable tension should be 400–600 lbs. with the parking brake pedal fully in the last detent position. If tension is low, repeat Steps 2 and 3.
5. Release parking brake and check for rear wheel drag. The

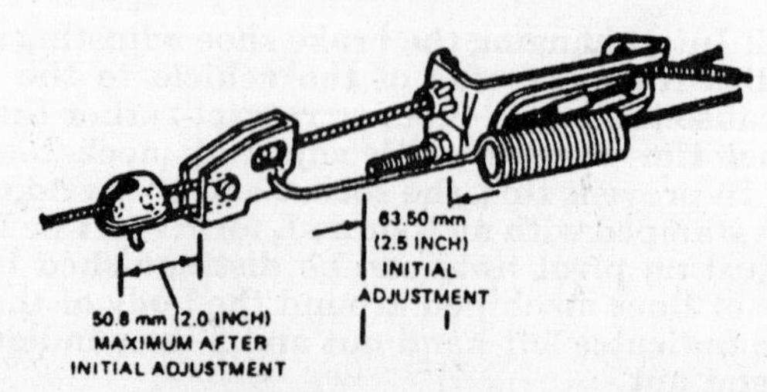

Pre-tension adjustment — parking brake assembly

cables should be tight enough to provide full application of the rear brake shoes, when the parking brake lever or foot pedal is placed in the fully applied position, yet loose enough to ensure complete release of the brake shoes when the lever is in the released position.

NOTE: The Tension Limiter will reset the parking brake tension any time the system is disconnected provided the distance between the bracket and the cinch strap hook is reduced during adjustment. When the cinch strap contacts the bracket, the system tension will increase significantly and over tensioning may result. If all available adjustment travel has been used, the tension limiter must be replaced.

REMOVAL AND INSTALLATION

Equalizer-To-Control Cable

1. Raise the vehicle on a hoist and support on jackstands. Back off the equalizer nut and remove slug of front cable from the tension limiter.
2. Remove the parking brake cable from the bracket.
3. Remove the jackstands and lower the vehicle. Remove the forward ball end of the parking brake cable from the control assembly clevis.
4. Remove the cable from the control assembly.
5. Using a fishing line wire leader or cord attached to the control lever end of the cable, remove the cable from the vehicle.

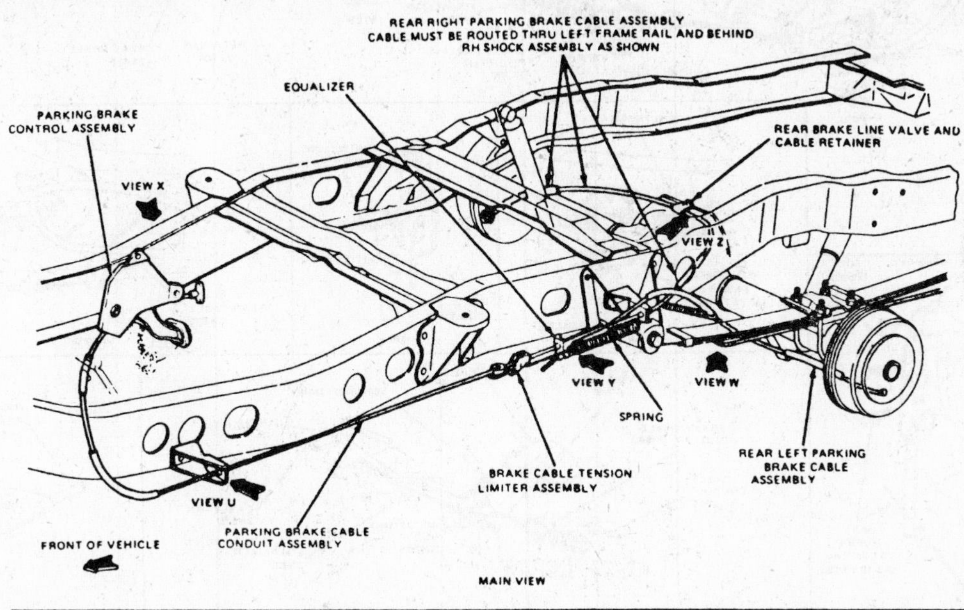

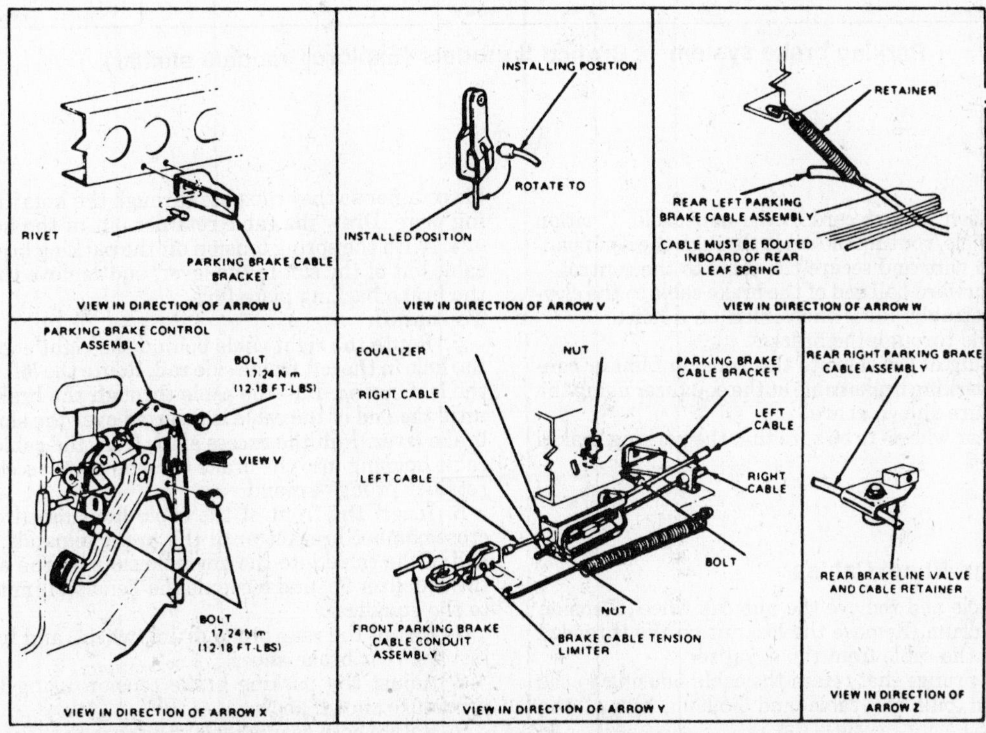

Parking brake system — Ranger models (Explorer models similar)

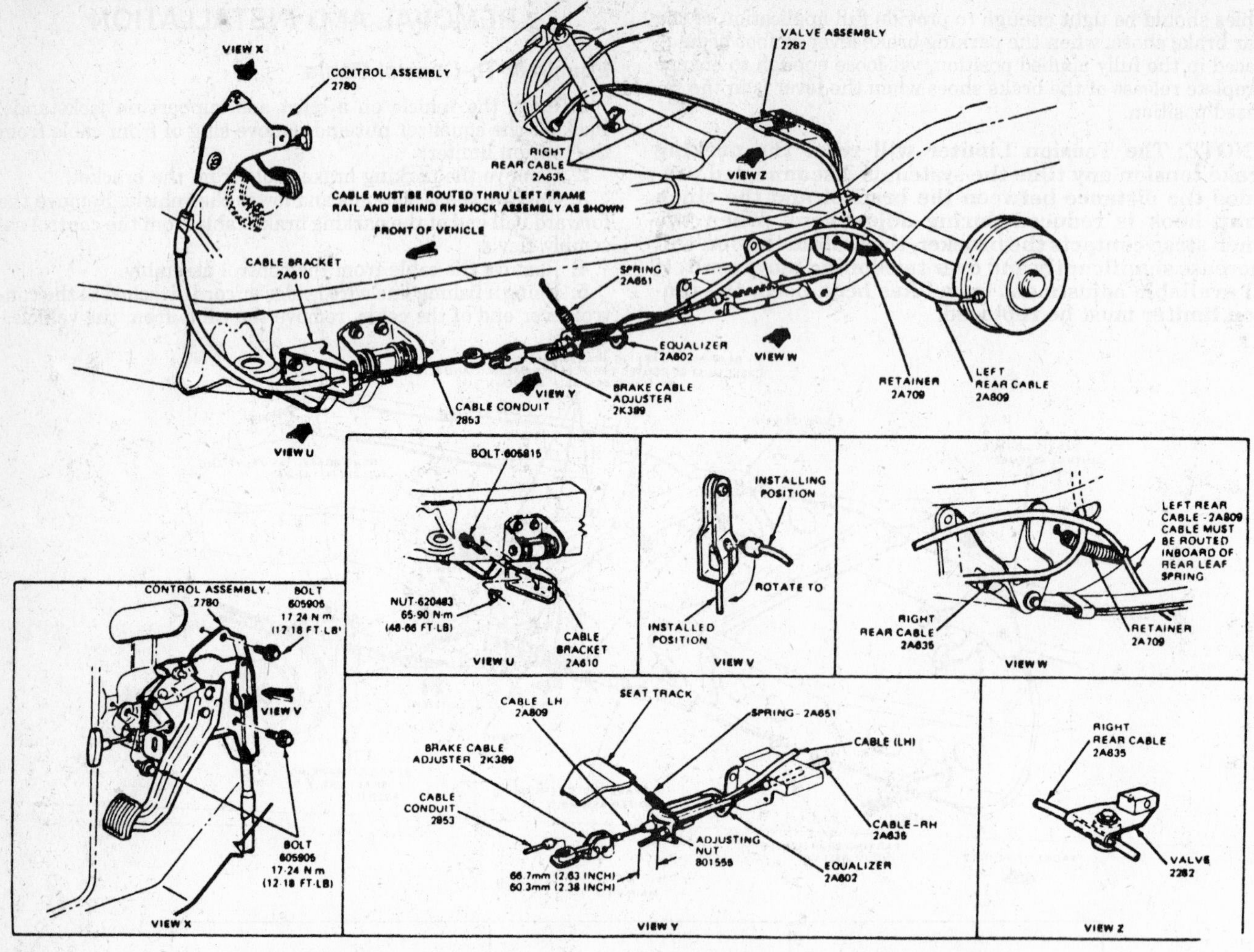

Parking brake system — Bronco II models (Explorer models similar)

To install:

6. Transfer the fish wire or cord to the new cable. Position the cable in the vehicle, routing the cable through the dash panel. Remove the fish wire and secure the cable to the control.
7. Connect the forward ball end of the brake cable to the clevis of the control assembly. Raise the vehicle on a hoist.
8. Route the cable through the bracket.
9. Connect the slug of the cable to the Tension Limiter connector. Adjust the parking brake cable at the equalizer using the appropriate procedure shown above.
10. Rotate both rear wheels to be sure that the parking brakes are not dragging.

Equalizer-To-Rear Wheel Cables

1. Raise the vehicle and remove the hub cap wheel, Tension Limiter and brake drum. Remove the locknut on the threaded rod and disconnect the cable from the equalizer.
2. Compress the prongs that retain the cable housing to the frame bracket, and pull the cable and housing out of the bracket.
3. Working on the wheel side, compress the prongs on the cable retainer so they can pass through the hole in the brake backing plate. Draw the cable retainer out of the hole.
4. With the spring tension off the parking brake lever, lift the cable out of the slot in the lever, and remove the cable through the brake backing plate hole.

To install:

5. Route the right cable behind the right shock and through the hole in the left frame side rail. Route the left cable inboard of the leaf spring. Pull the cable through the brake backing plate until the end of the cable is inserted over the slot in the parking brake lever. Pull the excess slack from the cable and insert the cable housing into the brake backing plate access hole until the retainer prongs expand.
6. Insert the front of the cable housing through the frame crossmember bracket until the prong expands. Insert the ball end of the cable into the key hole slots on the equalizer, rotate the equalizer 90° and recouple the Tension Limiter threaded rod to the equalizer.
7. Install the rear brake drum, wheel, and hub cap, and adjust the rear brake shoes.
8. Adjust the parking brake tension using the appropriate procedure shown above.
9. Rotate both rear wheels to be sure that the parking brakes are not dragging.

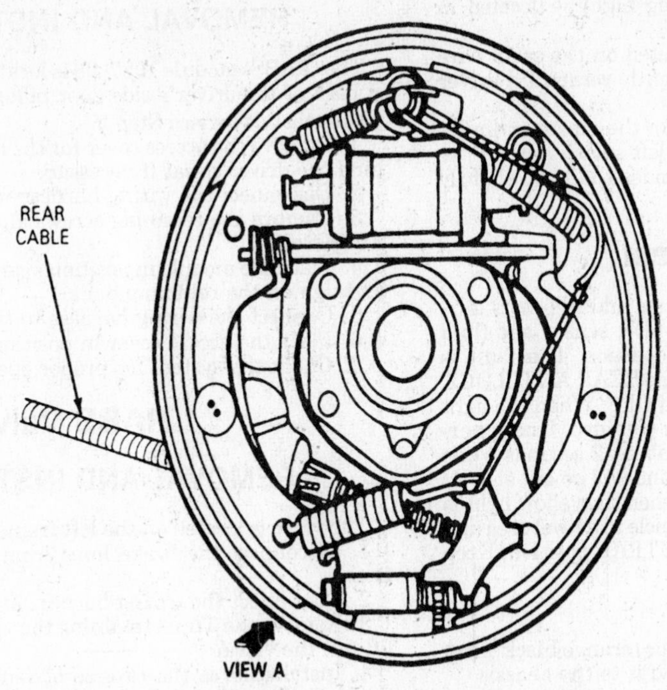

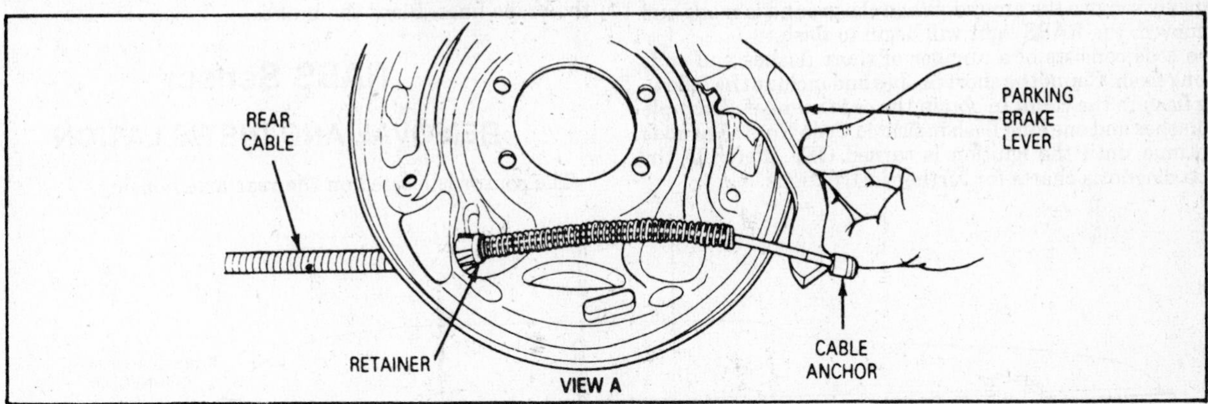

Parking brake cable installation

REAR ANTI-LOCK BRAKE SYSTEM (RABS)

Component Location

The RABS consists of the following components:

1. RABS module (1989–91) – located in the dash under the IP panel mount on a brace, center of panel area. The (RABS) module for 1987–88 model year is located behind an access cover in back of the driver's side door pillar.

2. Dual Solenoid Electro-Hydraulic Valve is located 5 in. (127mm) rearward of the No. 1 crossmember on the inboard side of the LH frame rail.

3. Speed Sensor and Excitor Ring – located in the rear differential housing.
4. Yellow REAR ANTI-LOCK Warning Light – located in the instrument cluster.
5. RABS Diagnostic Connector – located on the main wire bundle inside of cab under the dash, slightly rerarward driver side.
6. Diode/Resistoe Element – located in the master cylinder fluid level sensor wiring harness on the left splash apron.
7. Sensor test Connector – located under the hood slightly rearward the washer bottle.

System Self-Test

The RABS module preforms system tests and self-tests during startup and normal operation. The valve, sensor and fluid level circuits are monitored for proper opertion. If a fault is found, the RABS will be deactivated and the REAR ANTI LOCK light will be lit until the ignition is turned OFF. When the light is lit, the diagnostic flashout code may be obtained. Under normal operation, the light will stay on for about 2 seconds while the ignition switch is in the ON position and will go out shortly after. A flash code may be obtained only when the yellow light is ON. Before reading the code, drive the vehicle to a level area and place the shift lever in the PARK or NEUTRAL position. Keep the vehicle ignition ON.

TO OBTAIN THE FLASH CODE:
1. Locate the RABS diagnostic connector (orange/black wire) and attach a jumper wire to it and ground it to the chassis.
2. Quickly remove the ground. When the ground is made and then removed, the RABS light will begin to flash.
3. The code consists of a number of short flashes and ends with a long flash. Count the short flashes and include the following long flash in the count to obtain the code number. Example 3 short flashes and one long flash indicated Code No. 4. The code will continue until the ignition is turned OFF. Refer to the flashcode diagnosis charts for further instructions.

Computer (RABS) Module

REMOVAL AND INSTALLATION

The (RABS) module 1987–88 is located behind an access cover in back of the driver's side door pillar. On the 1989–91 model years start at service Step 2.
1. Remove the access cover for the module from the pillar behind the driver's seat if necessary.
2. Disconnect the wiring harness to the module.
3. Remove the retaining screws and remove the module.

To install:
4. Place the module in position against the door pillar. Install and tighten the retaining bolts.
5. Connect the wiring harness to the module.
6. Place the access cover in position.
7. Check the system for proper operation.

RABS Valve

REMOVAL AND INSTALLATION

The valve is located on the left frame rail above the rear axle.
1. Disconnect the brake lines from the valve and plug the lines.
2. Disconnect the wiring harness at the valve.
3. Remove the 3 nuts retaining the valve to the frame rail and lift out the valve.
4. Installation is the reverse of removal. Don't overtighten the brake lines. Bleed the brakes.

RABS Sensor

REMOVAL AND INSTALLATION

The sensor is located on the rear axle housing.

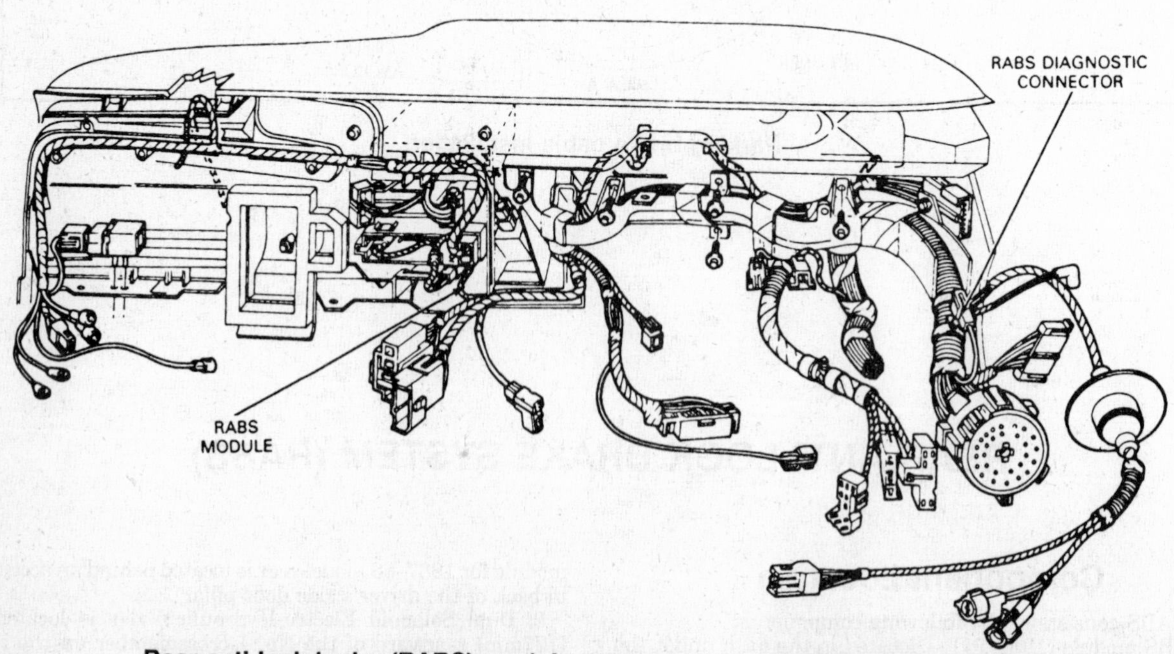

Rear anti-lock brake (RABS) module removal and installation 1989–91 year

1. Remove the sensor holddown bolt.
2. Remove the sensor.
3. Carefully clean the axle surface to keep dirt from entering the housing.
4. If a new sensor is being installed, lubricate the O-ring with clean engine oil. Carefully push the sensor into the housing aligning the mounting flange hole with the threaded hole in the housing. Torque the holddown bolt to 30 ft. lbs.

If the old sensor is being installed, clean it thoroughly and install a new O-ring coated with clean engine oil.

Exciter Ring

The ring is located on the differential case inside the axle housing. Once it is pressed of the case it cannot be reused. This job should be left to a qualified service technician.

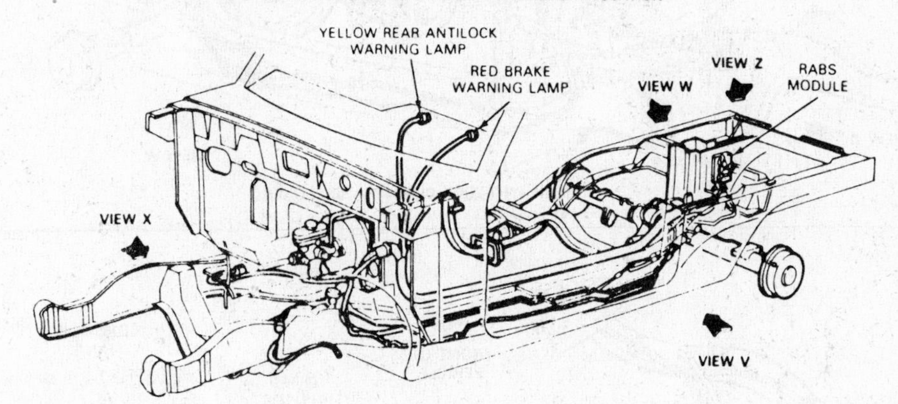

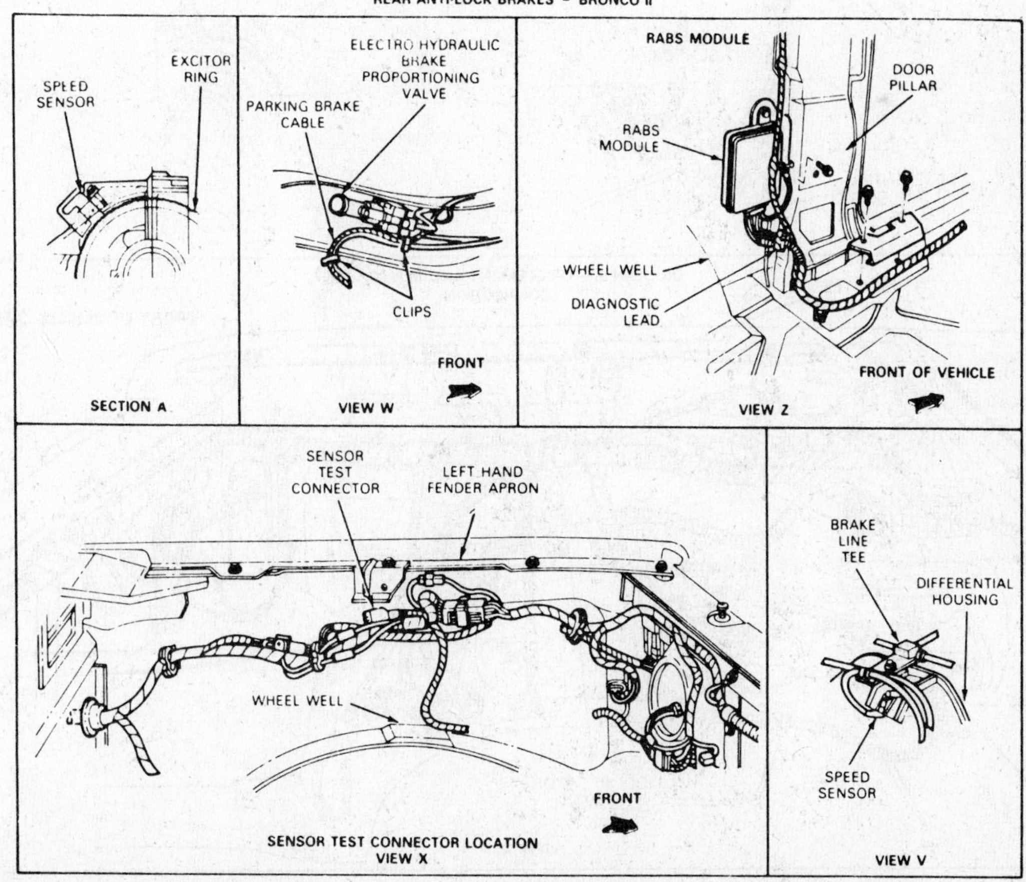

Rear anti-lock brake (RABS) system component locations 1987–88 year

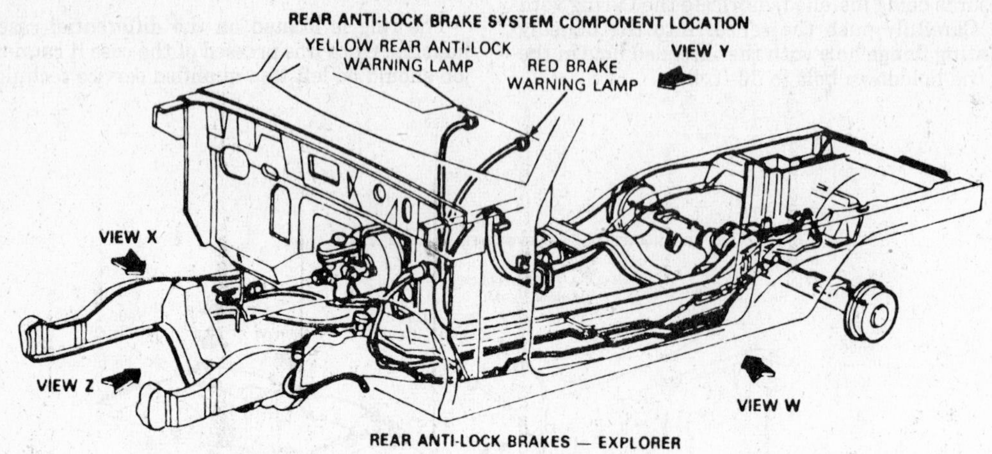

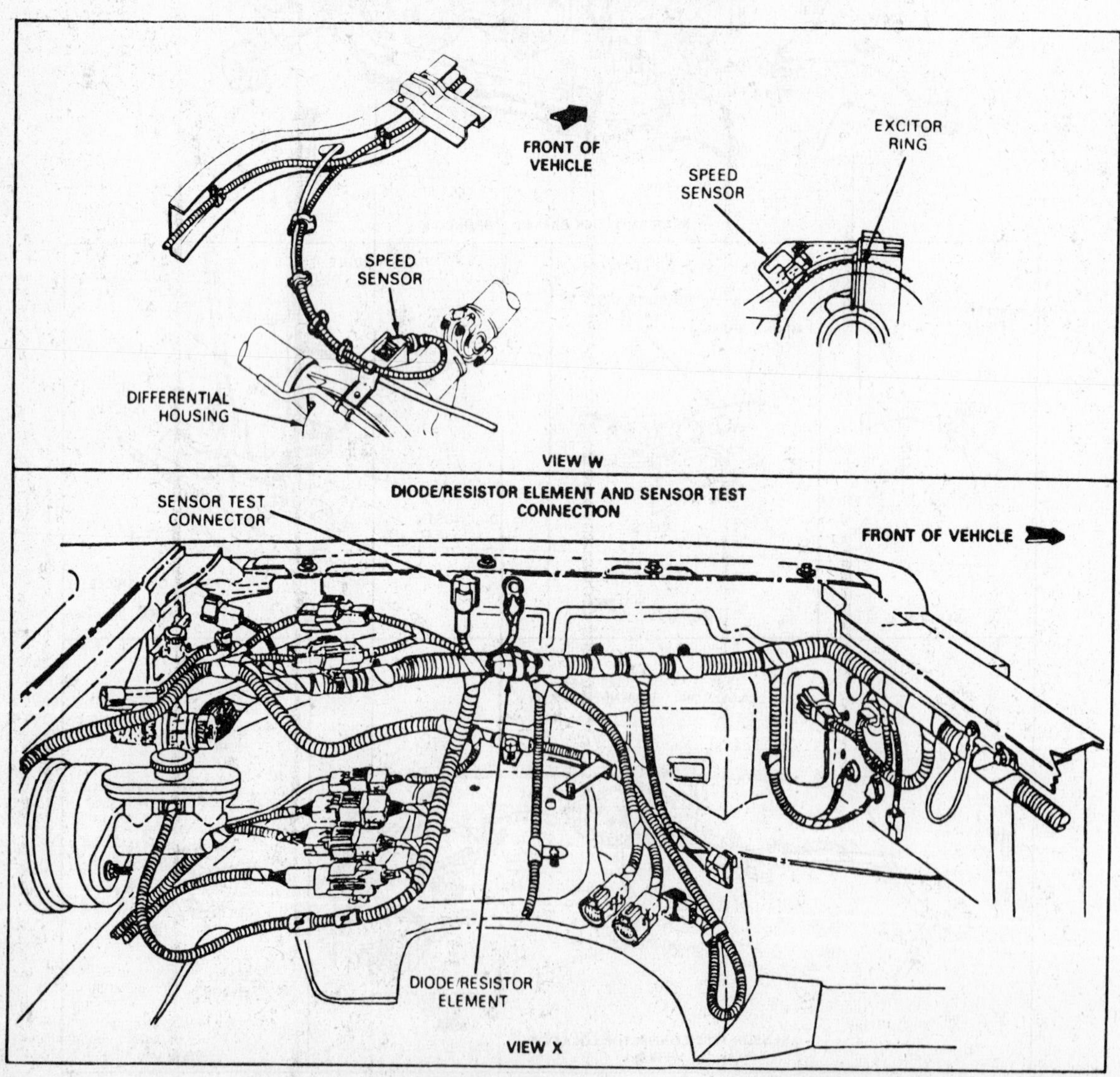

Rear anti-lock brake (RABS) system component locations 1989–91 year

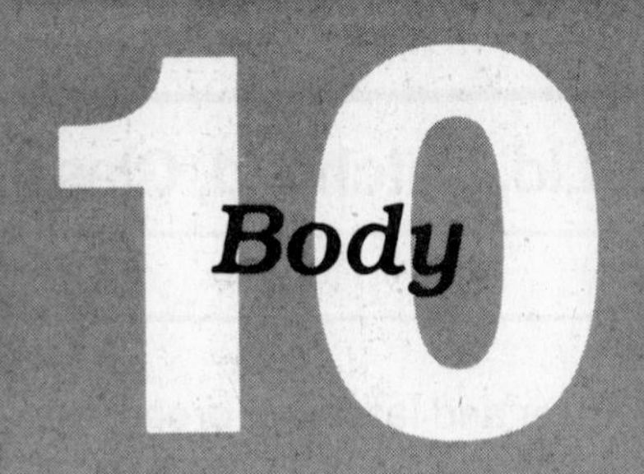

QUICK REFERENCE INDEX

GENERAL INDEX

Hood, Trunk Lid, Hatch Lid, Glass and Doors

Problem	Possible Cause	Correction
HOOD/TRUNK/HATCH LID		
Improper closure.	• Striker and latch not properly aligned.	• Adjust the alignment.
Difficulty locking and unlocking.	• Striker and latch not properly aligned.	• Adjust the alignment.
Uneven clearance with body panels.	• Incorrectly installed hood or trunk lid.	• Adjust the alignment.
WINDOW/WINDSHIELD GLASS		
Water leak through windshield	• Defective seal. • Defective body flange.	• Fill sealant • Correct.
Water leak through door window glass.	• Incorrect window glass installation. • Gap at upper window frame.	• Adjust position. • Adjust position.
Water leak through quarter window.	• Defective seal. • Defective body flange.	• Replace seal. • Correct.
Water leak through rear window.	• Defective seal. • Defective body flange.	• Replace seal. • Correct.
FRONT/REAR DOORS		
Door window malfunction.	• Incorrect window glass installation. • Damaged or faulty regulator.	• Adjust position. • Correct or replace.
Water leak through door edge.	• Cracked or faulty weatherstrip.	• Replace.
Water leak from door center.	• Drain hole clogged. • Inadequate waterproof skeet contact or damage.	• Remove foreign objects. • Correct or replace.
Door hard to open.	• Incorrect latch or striker adjustment.	• Adjust.
Door does not open or close completely.	• Incorrect door installation. • Defective door check strap. • Door check strap and hinge require grease.	• Adjust position. • Correct or replace. • Apply grease.
Uneven gap between door and body.	• Incorrect door installation.	• Adjust position.
Wind noise around door.	• Improperly installed weatherstrip. • Improper clearance between door glass and door weatherstrip. • Deformed door.	• Repair or replace. • Adjust. • Repair or replace.

EXTERIOR

Doors

REMOVAL AND INSTALLATION

1. If replacing the door with a stripped unit, remove all usable components from the old door.
2. If removing the rear door (Explorer), remove the scuff plate and center panel trim panel. Remove the upper and lower hinge access cover plate if equipped, and matchmark the hinge-to-body and hinge-to-door location. Support the door either on jackstands or have somebody hold it for you.
3. Remove the lower hinge-to-door bolts.
4. Remove the upper hinge-to-door bolts and lift the door off the hinges.
5. If the hinges are being replaced, remove them from the door pillar.

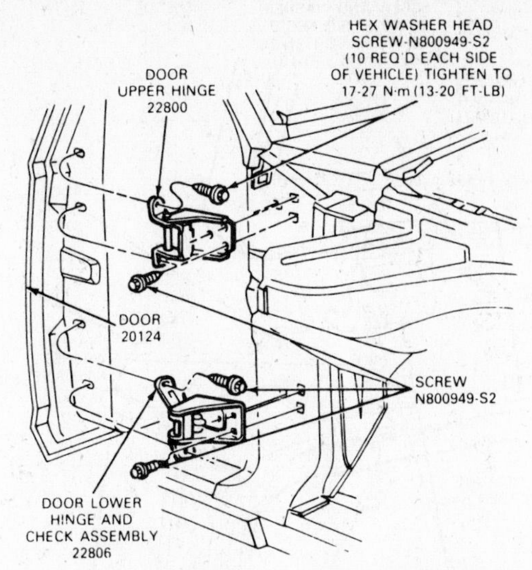

Door hinge mounting – Ranger and Bronco II

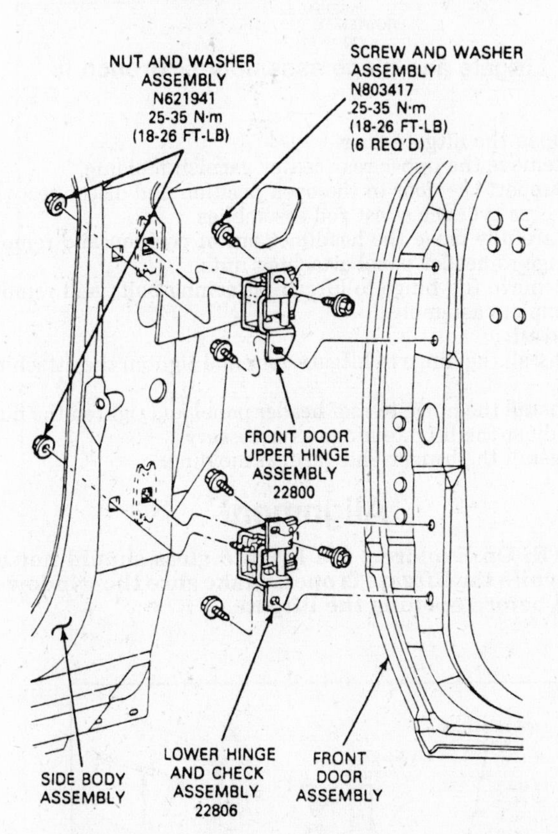

Front door hinge mounting – Explorer

To install:

6. If they were removed, install the hinges in the same position as before removal. Tighten the bolts finger tight.
7. Position the door on the hinges and install the bolts finger tight.
8. Install all previously removed hardware to the new door before adjustments are made.
9. Install the trim panel and scuff plate if they were removed.
10. Adjust the door and tighten all retaining bolts.
11. Install the hinge access cover plates.

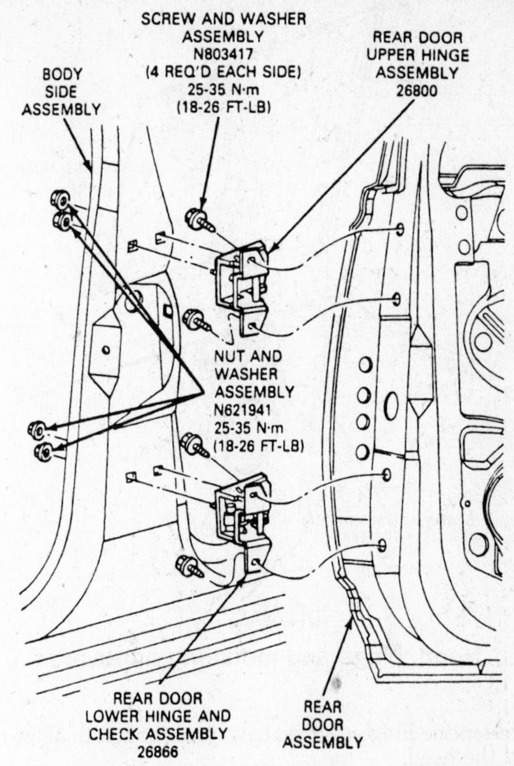

Rear door hinge mounting – Explorer

ADJUSTMENT

NOTE: Loosen the hinge-to-door bolts for lateral adjustment only. Loosen the hinge-to-body bolts for both lateral and vertical adjustments.

1. Determine which hinge bolts are to be loosened and back them out just enough to allow movement.
2. To move the door safely, use a padded pry bar. When the door is in the proper position, tighten the bolts to specification and check the door operation. There should be no binding or interference when the door is closed and opened.
3. Door closing adjustment can also be affected by the position of the lock striker plate. Loosen the striker plate bolts and move the striker plate just enough to permit proper closing and locking of the door.

Hood

REMOVAL AND INSTALLATION

NOTE: It is highly recommended that you have at least one assistant helping during this operation.

1. Open the hood.
3. Matchmark the hood-to-hinge position.
2. Have you're assistant(s) support the weight of the hood.
4. Remove the hood-to-hinge bolts and lift the hood off of the hinges.
5. Installation is the reverse of the removal. Loosely install the hood and align the matchmarks. Torque all bolts to 62–97 inch lbs.

ALIGNMENT

1. Open the hood and matchmark the hinge and latch positions.

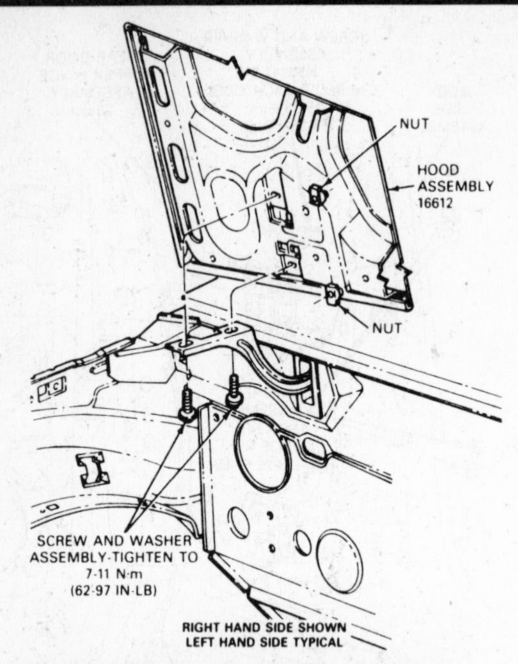

Hood, hinge and mounting hardware

2. Loosen the hinge-to-hood bolts just enough to allow movement of the hood.
3. Move the hood as required to obtain the proper fit and alignment between the hood and all adjoining body panels. Tighten the bolts to 62–97 inch lbs. when satisfactorily aligned.
4. Loosen the 2 latch attaching bolts.
5. Move the latch from side-to-side to align the latch with the striker. Torque the latch bolts.
6. Lubricate the latch and hinges and check the hood fit several times.

Liftgate

REMOVAL AND INSTALLATION

Bronco II and Explorer

NOTE: On Explorer, the liftgate glass should not be open while the liftgate is open. Make sure the window is closed before opening the liftgate.

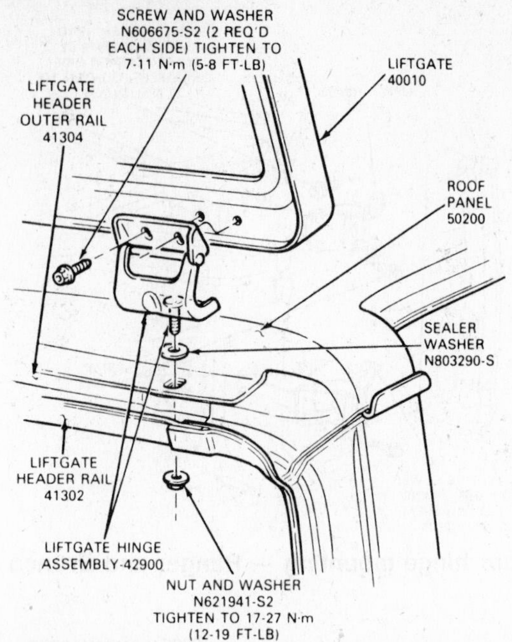

Liftgate and hinge assembly – Bronco II

1. Open the liftgate door.
2. Remove the upper rear center garnish molding.
3. Support the door in the open position and disconnect the liftgate gas cylinder assist rod assemblies.
4. Carefully move the headliner out of position and remove the hinge-to-header panel attaching nuts.
5. Remove the hinge-to-liftgate attaching bolts and remove the complete assembly.

To install:

6. Install the hinge to liftgate door and tighten the attaching bolts.
7. Install the hinge to roof header panel and tighten the nut.
8. Adjust the liftgate hinge as necessary.
9. Install the header and garnish molding.

Alignment

NOTE: On Explorer, the liftgate glass should not be open while the liftgate is open. Make sure the window is closed before opening the liftgate.

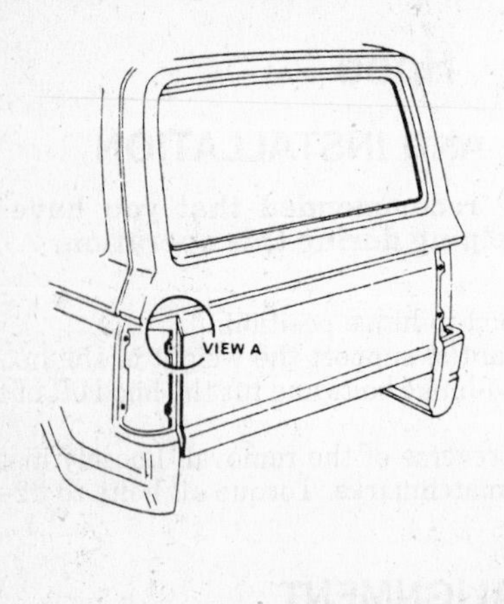

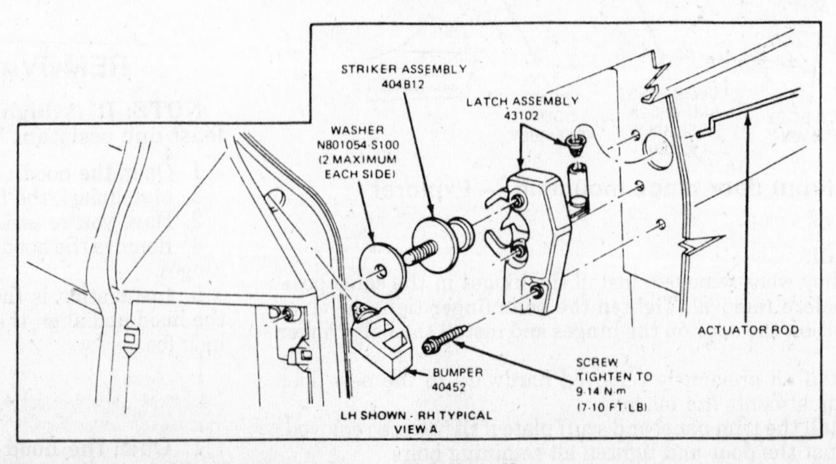

Liftgate striker and latch assembly – Bronco II

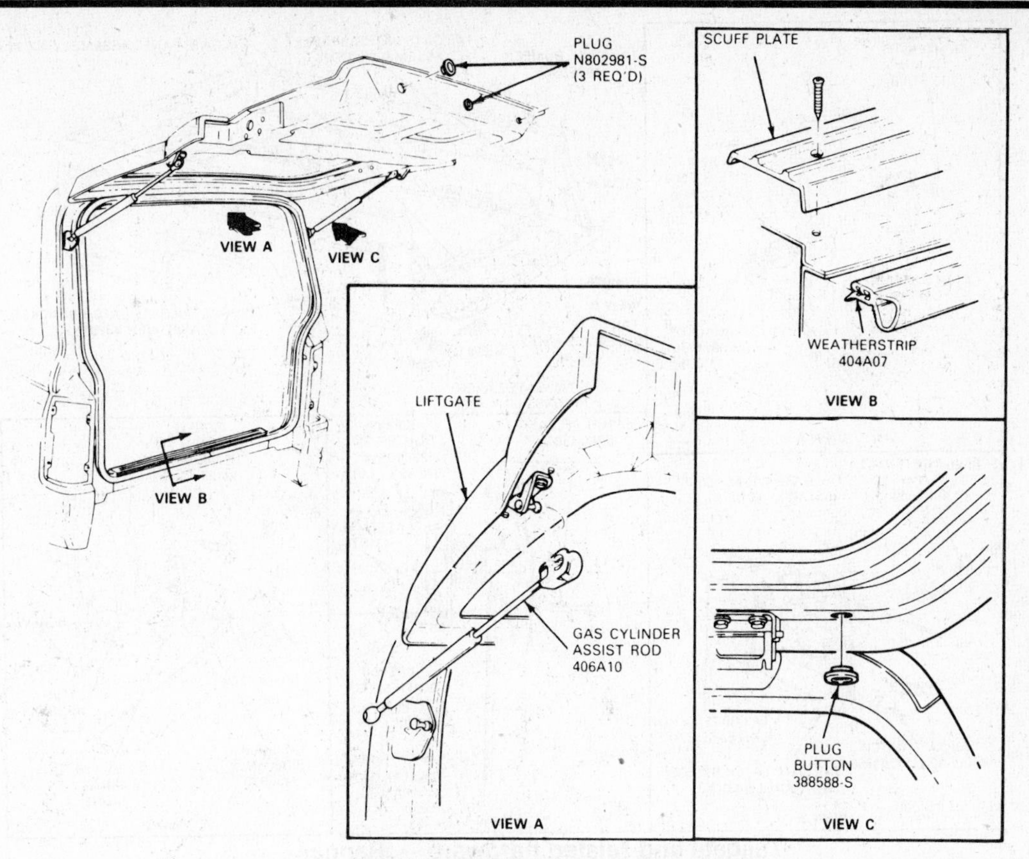

Liftgate gas cylinder assist rods

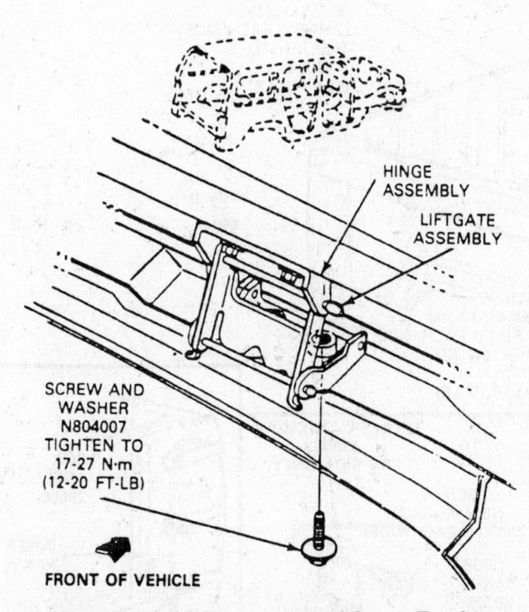

Liftgate and hinge assembly – Explorer

The liftgate can be adjusted slightly in or out and side to side by loosening the hinge-to-header nut or bolt. Some up and down adjustment can be accomplished by loosening the hinge bolts on the liftgate and moving the gate up or down. The liftgate should be adjusted for even and parallel fit with adjoining panels.

Tailgate

REMOVAL AND INSTALLATION

Ranger

1. Remove the tailgate support strap at the pillar T-head pivot.
2. Lift off the tailgate at the right hinge.
3. Pull off the left hinge.
4. Transfer all necessary hardware to the new tailgate if necessary.
5. Installation is the reverse of removal.

Front and Rear Bumpers

REMOVAL AND INSTALLATION

1. Support the bumper. Disconnect electrical pigtails, if applicable.
2. Remove the nuts and bolts attaching the bumper brackets to the frame. Once the bumper is removed from the truck, remove the brackets from the bumper.
3. Remove the valance panel and rubstrip from the bumper as required.
4. Installation is the reverse of removal. Use a leveling tool to ensure a level installation before tightening the bolts.
5. Support the bumper and torque the bracket-to-frame bolts to specifications.

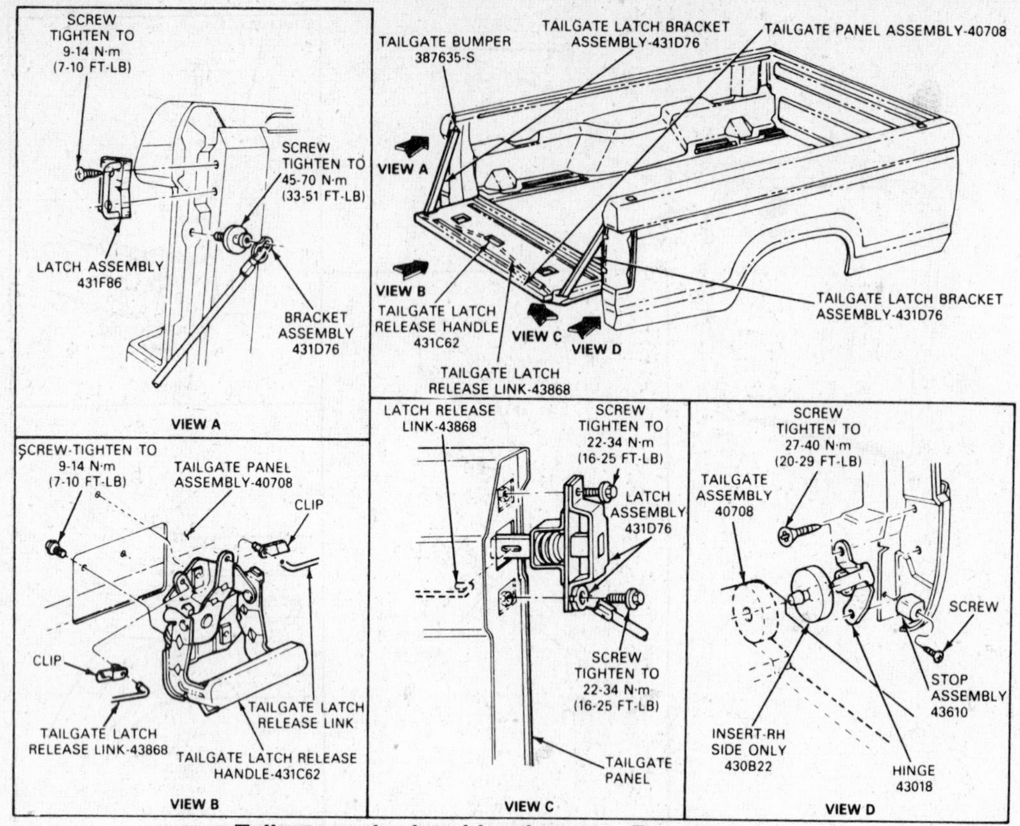

Tailgate and related hardware — Ranger

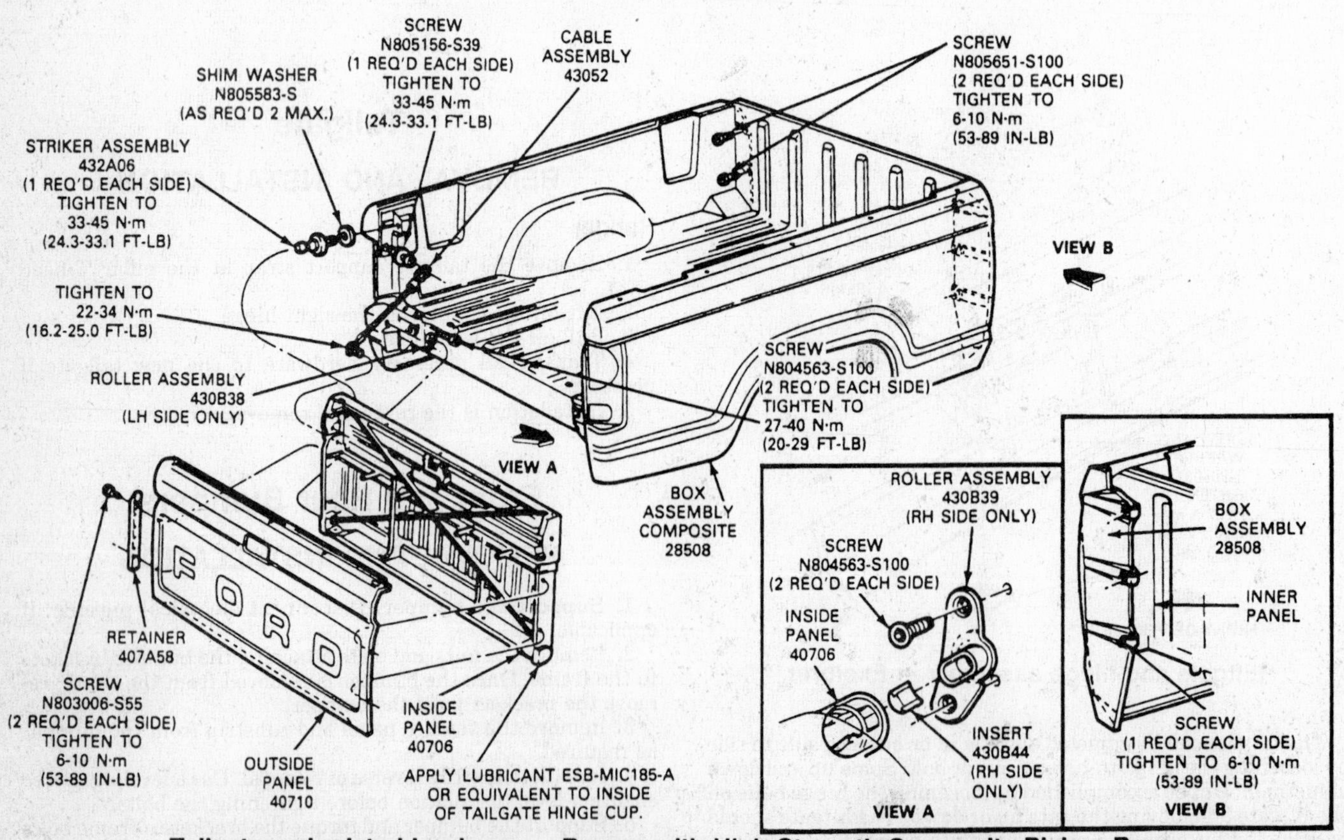

Tailgate and related hardware — Ranger with High Strength Composite Pickup Box

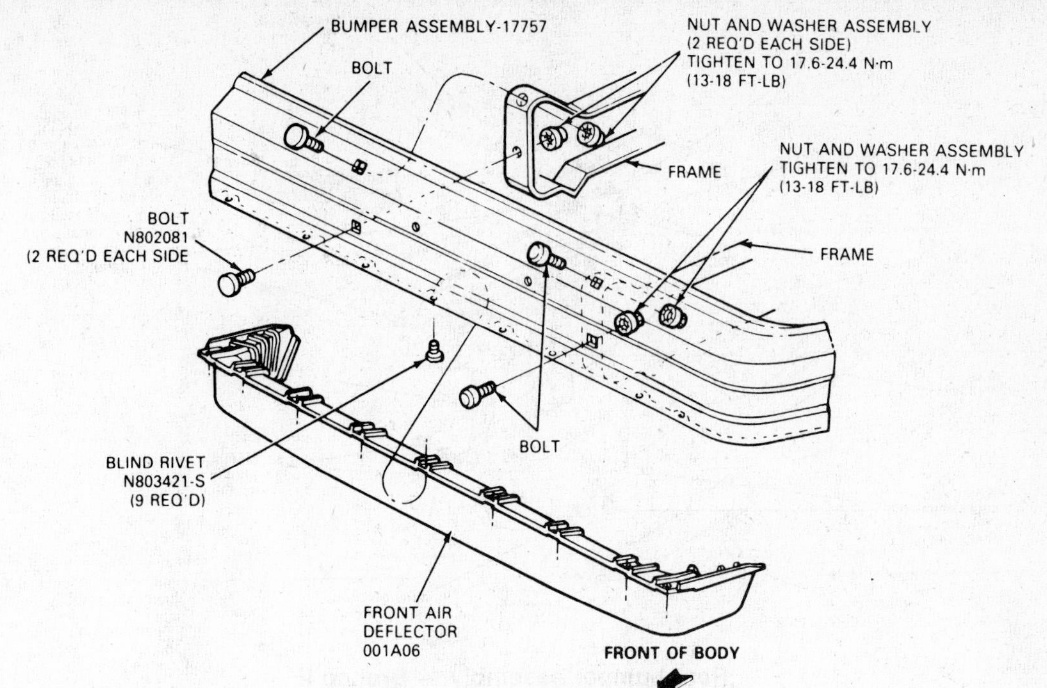

Front bumper assembly – Bronco II

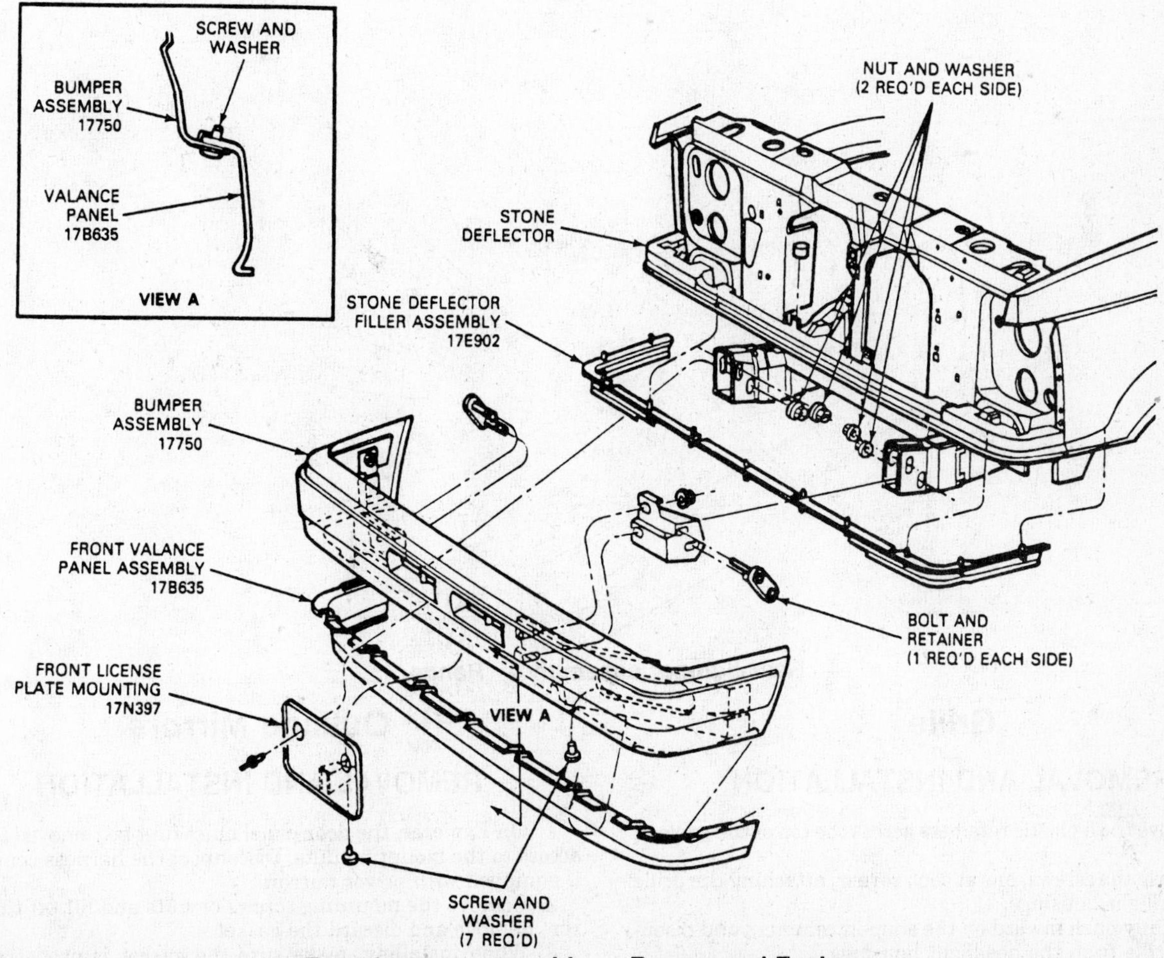

Front bumper assembly – Ranger and Explorer

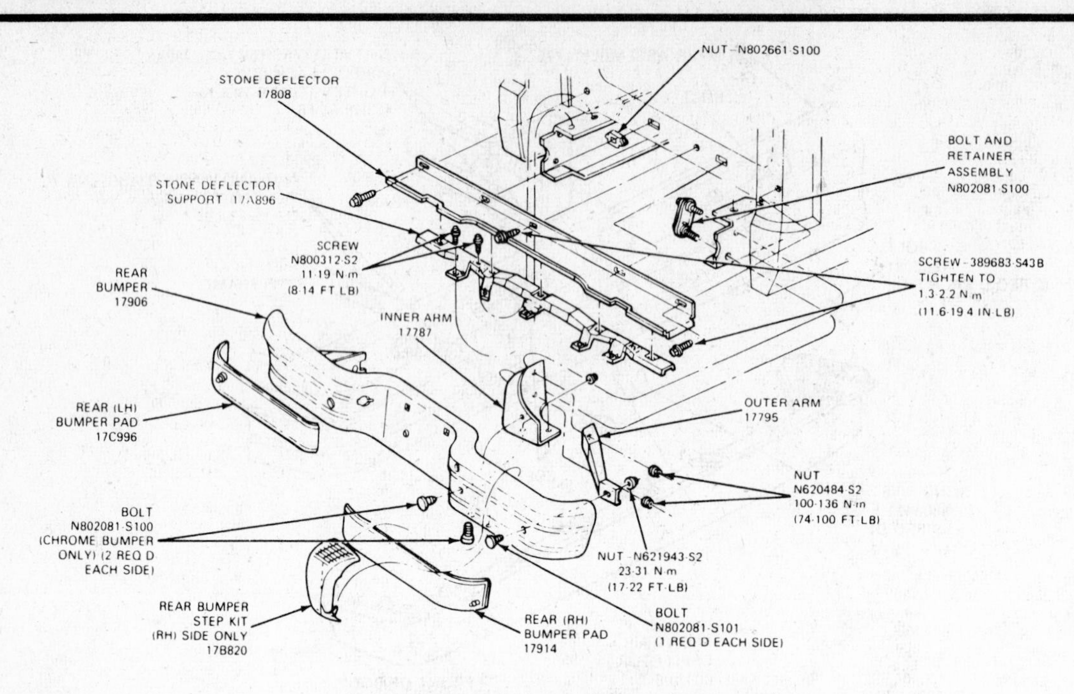

Rear bumper assembly — Bronco II

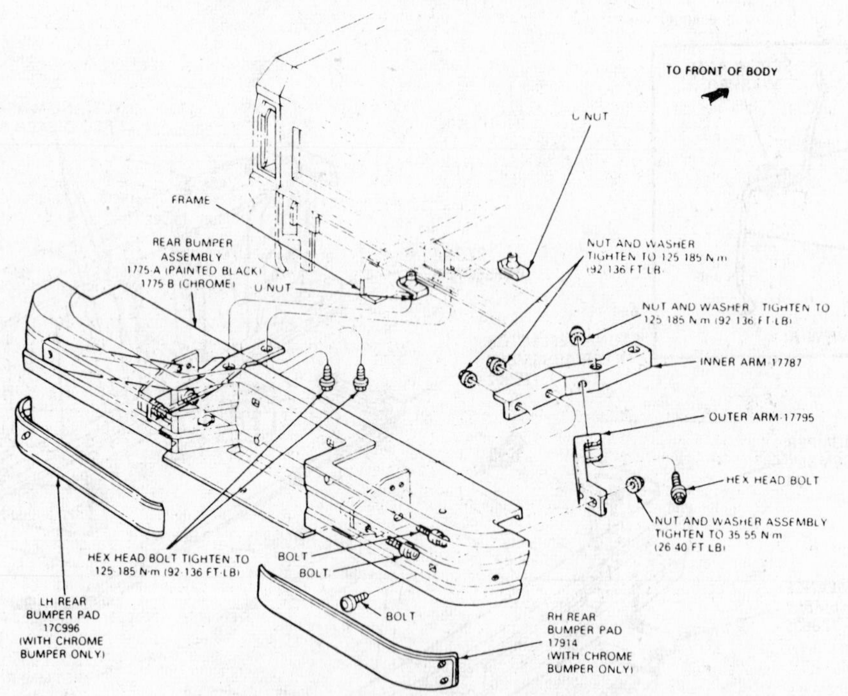

Rear bumper assembly — Ranger

Grille

REMOVAL AND INSTALLATION

1. Remove the 5 plastic retainers across the top of the grille, if equipped.
2. Remove the screws, one at each corner, attaching the grille to the headlight housings.
3. Carefully push inward on the snap-in retainers and disengage the grille from the headlight housings.
4. Installation is the reverse of removal.

Outside Mirrors

REMOVAL AND INSTALLATION

1. On Explorer, the door panel must first be removed to gain access to the mounting nuts. Disconnect the harness connector if equipped with power mirrors.
2. Remove the mounting screws or nuts and lift off the mirror. Remove and discard the gasket.
3. When installing, make sure the gasket is properly positioned before tightening the screws.

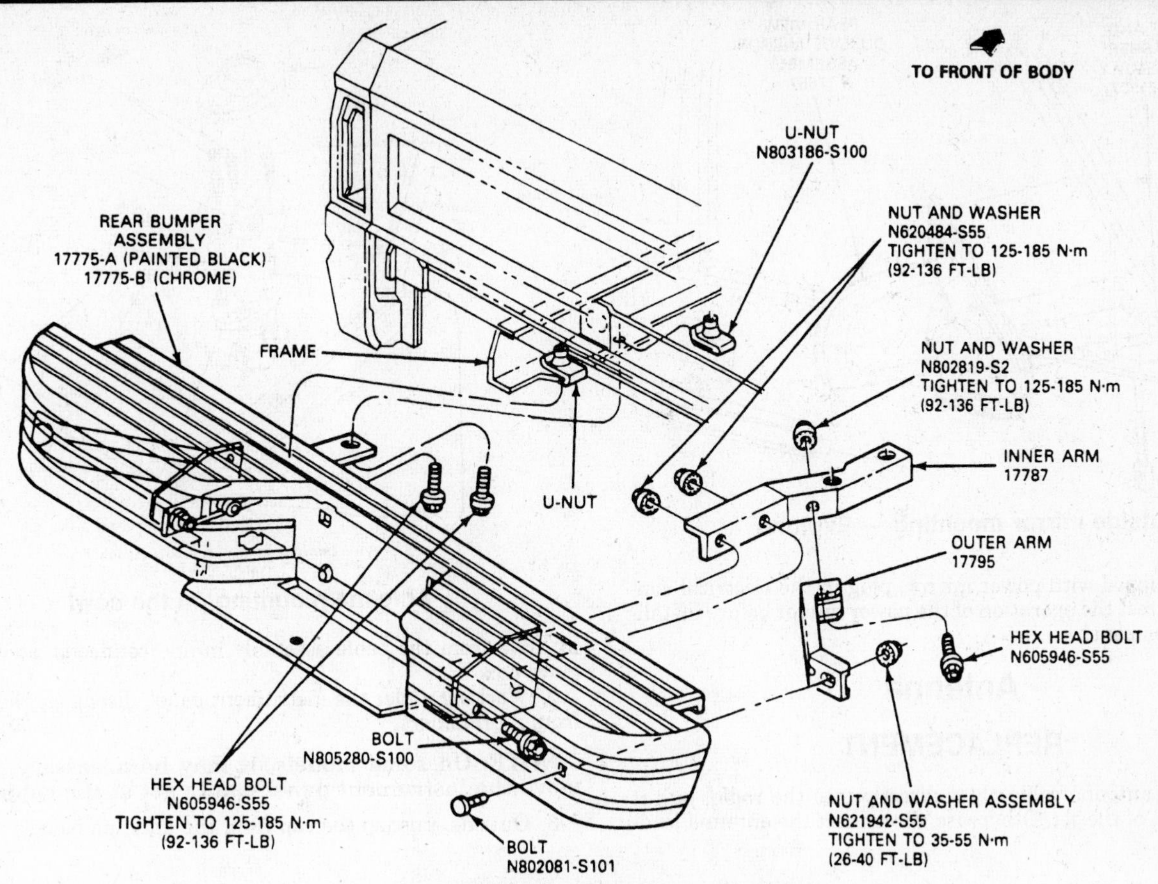

Rear bumper assembly — Explorer

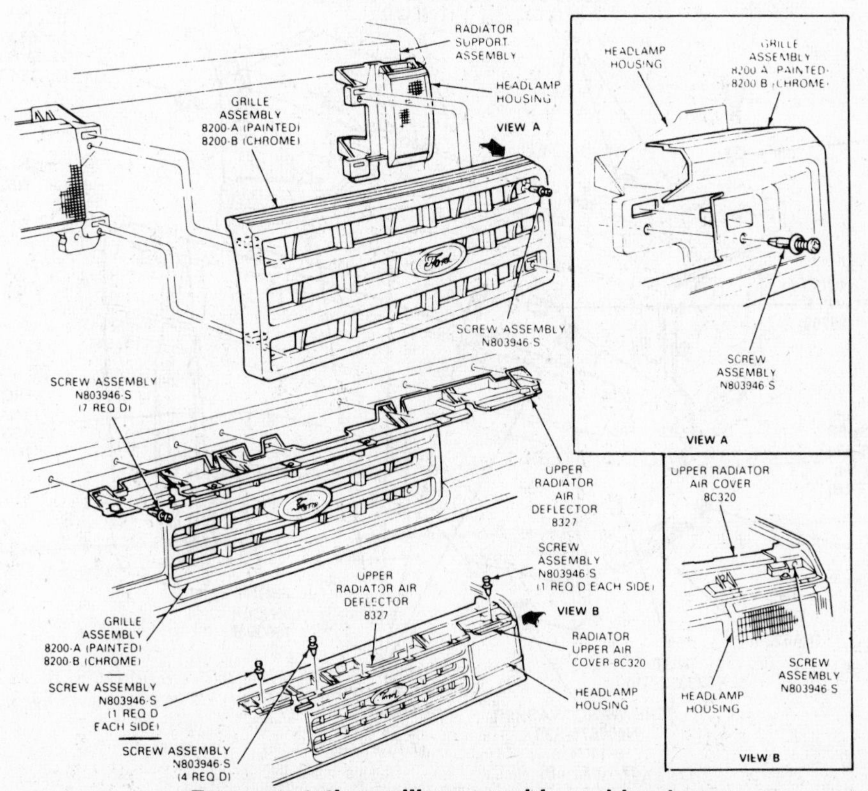

Representative grille assembly and hardware

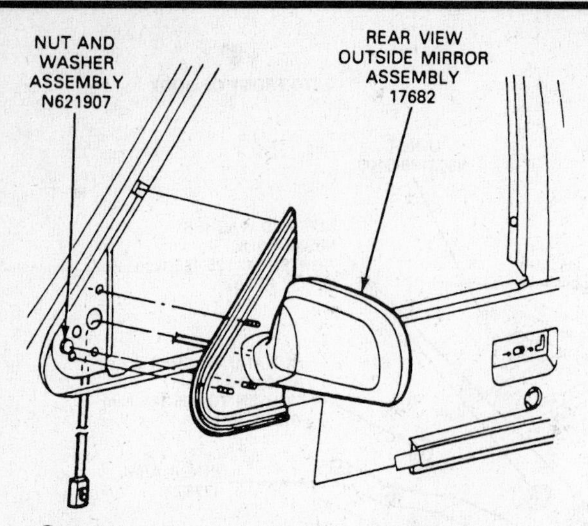

Outside mirror mounting – Explorer

4. If equipped with power mirror, plug in the electrical connector and test the operation of the power mirror before installing the door panel.

Antenna

REPLACEMENT

1. If the antenna cable plugs directly into the radio, pull it straight out of the set. Otherwise, disconnect the antenna lead-in cable from the cable assembly in-line connector above the glove box.

2. Working under the instrument panel, disengage the cable from its retainers.

NOTE: On some models, it may be necessary to remove the instrument panel pad to get at the cable.

3. Outside, unsnap the cap from the antenna base.

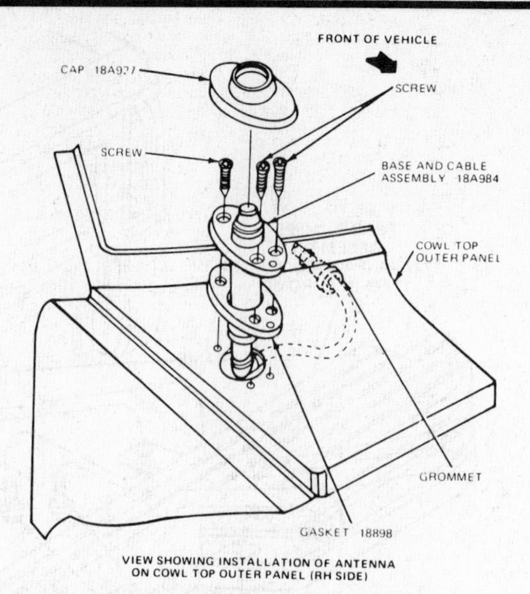

Antenna mounting on the cowl

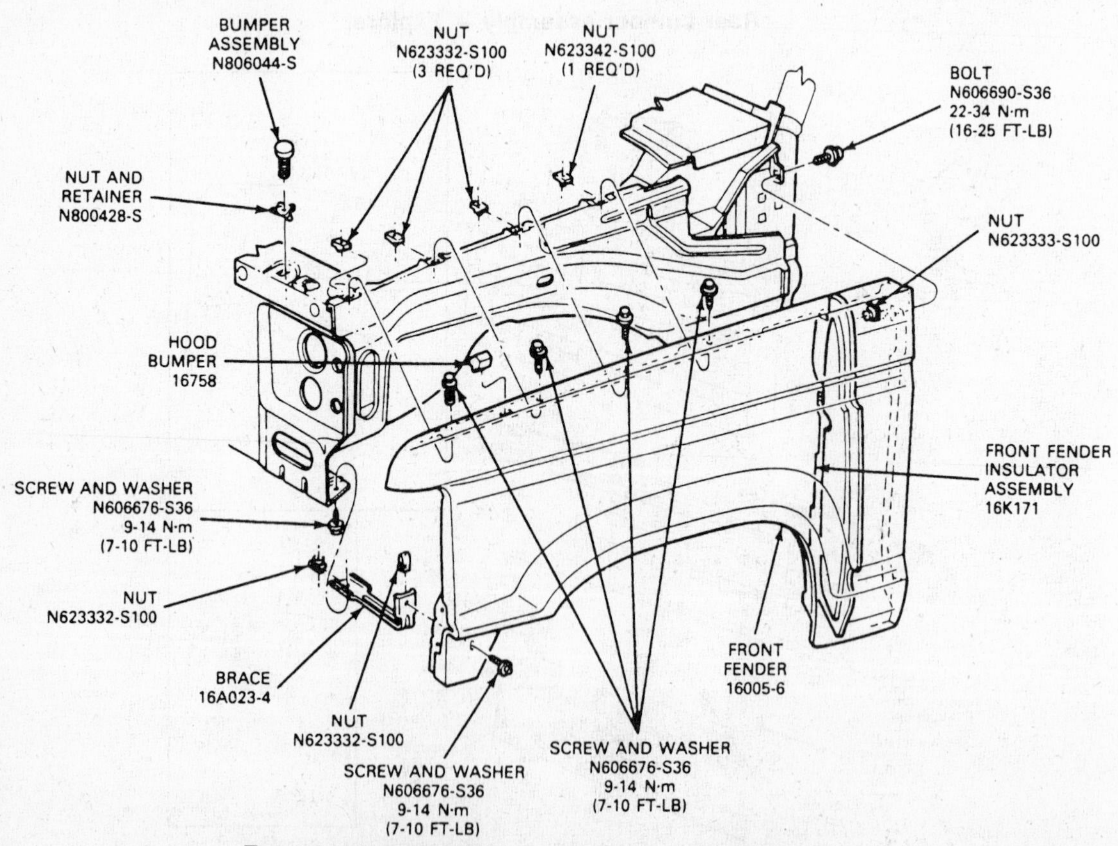

Front fender assembly – Ranger and Bronco II

4. Remove the 3 screws and lift off the antenna, pulling the cable with it, carefully.
5. Remove and discard the gasket.

To install:

6. Place the gasket in position on the cowl panel.
7. Insert the antenna cable through the hole and seat the antenna base on the cowl. Secure with the 3 screws.
8. Position the cap over the antenna base and snap it into place.
9. Route the cable in exactly the same position as before removal behind the instrument panel.
10. Connect the cable to the radio or in-line connector.

Fenders

REMOVAL AND INSTALLATION

Ranger and Bronco II

1. Clean out all dirt from the fender attaching hardware and lubricate them to ease removal.
2. Remove the grille.
3. Remove the bolt attaching the rear lower end of the fender to the lower corner of the cab.
4. From inside the cab, remove the bolt attaching the rear end of the fender to the cowl.
5. Remove the screws around the wheel opening attaching the fender apron.
6. Remove the bolts along the top of the apron attaching the fender.
7. Remove the bolt attaching the brace to radiator support assembly.
8. Remove the fender and remove the brace from the fender.

To install:

9. Position the nuts, retainers and brace on the fender.
10. Position the fender on the apron and loosely install the apron retaining bolts.
11. Loosely install the bolt from inside the cab attaching the rear end of the fender to the cowl.
12. Loosely install the bolt attaching the rear lower end of the fender to the lower corner of the cab.
13. Loosely install the bolt attaching the brace to the radiator support assembly.
14. Loosely install the 4 bolts along the fender inner body attaching the fender.
15. Adjust the position of the fender and tighten all mounting bolts. Torque the front and top bolts to 10 ft. lbs. and rear bottom bolts to 21 ft. lbs.
16. Install the grille.
17. Install the apron screws around the wheel opening.

Explorer

1. Clean out all dirt from the fender attaching hardware and lubricate them to ease removal.
2. Remove the headlight door and side marker lamp.

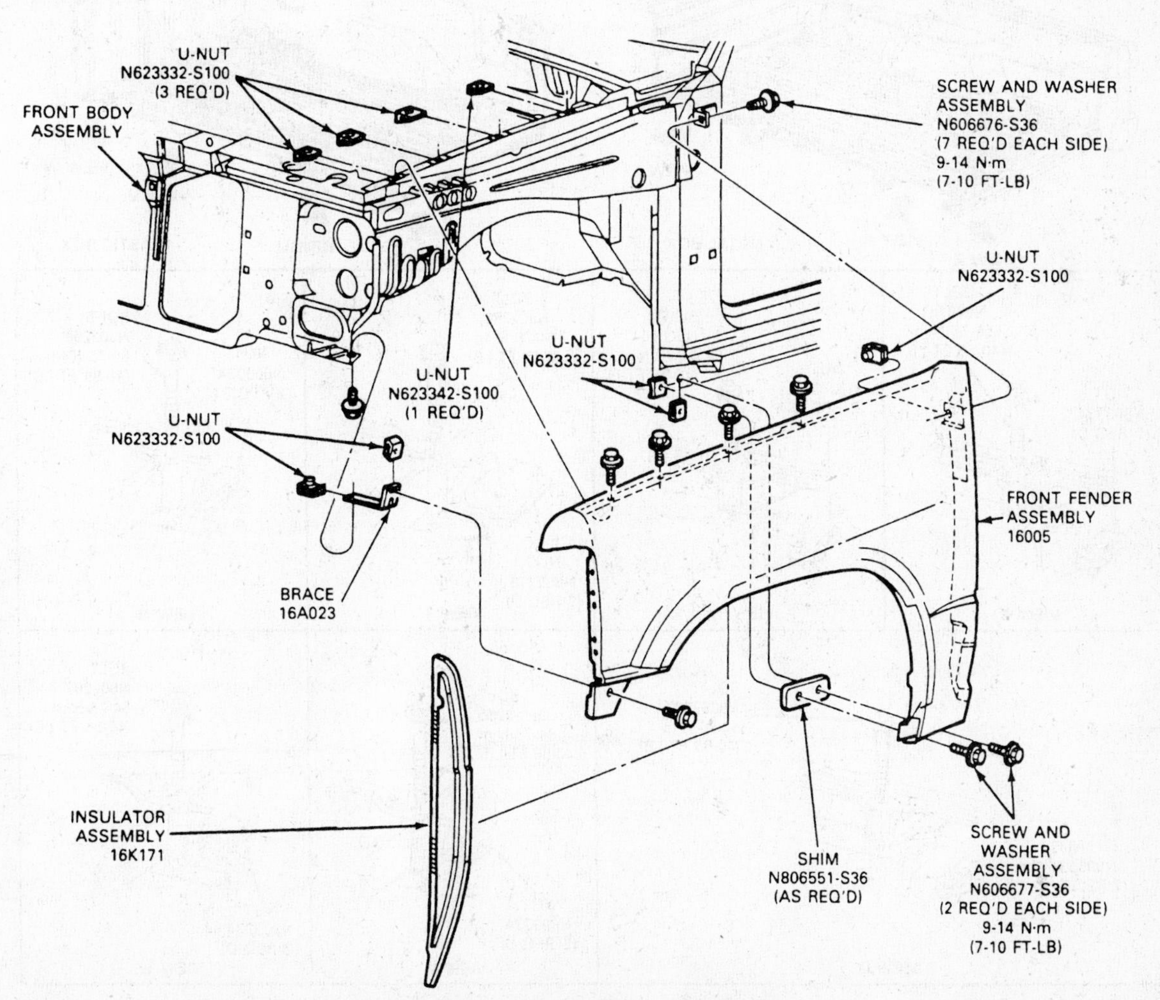

Front fender assembly — Explorer

3. Remove both bolts attaching the headlight assembly to the fender.
4. From inside the passenger compartment, remove the bolt attaching the rear end of the fender to the cowl.
5. Remove the 4 bolts along the top of the fender inner body attaching the fender.
6. Raise the truck to a convenient height.
7. Remove the bolt attaching the brace to the radiator support assembly.
8. Remove the 4 screws attaching the fender apron to the fender.
9. Remove the 2 bolts attaching the fender to the rocker panel.
10. Lower the vehicle and remove the fender. Remove the brace from the fender.

To install:

11. Position the nuts, retainers and brace on the fender.
10. Install the fender and raise the truck.
12. Install the 2 bolts attaching the fender to the rocker panel.
13. Install the 4 screws attaching the fender apron to the fender.
14. Install the bolt attaching the brace to the radiator support. Lower the truck.
15. Install the 4 bolts along the top of the fender inner body attaching the fender.
16. Install the bolt attaching the rear end of the fender to the cowl.
17. Install both bolts attaching the headlight assembly to the fender.
18. Install the headlight door and side marker lamp.

Pickup Truck Bed

REMOVAL AND INSTALLATION

Ranger

1. Clean out all dirt from the bed's attaching hardware and lubricate them to ease removal.
2. Remove all bolts attaching the bed to the frame.
3. Lift the bed off of the frame.
4. Installation is the reverse of removal.
5. Be sure to use new bolts and other mounting hardware when installing. Replacement parts should be of the same specification as the original parts; i.e. do not use Grade 5 bolts where Grade 8 bolts were originally used.

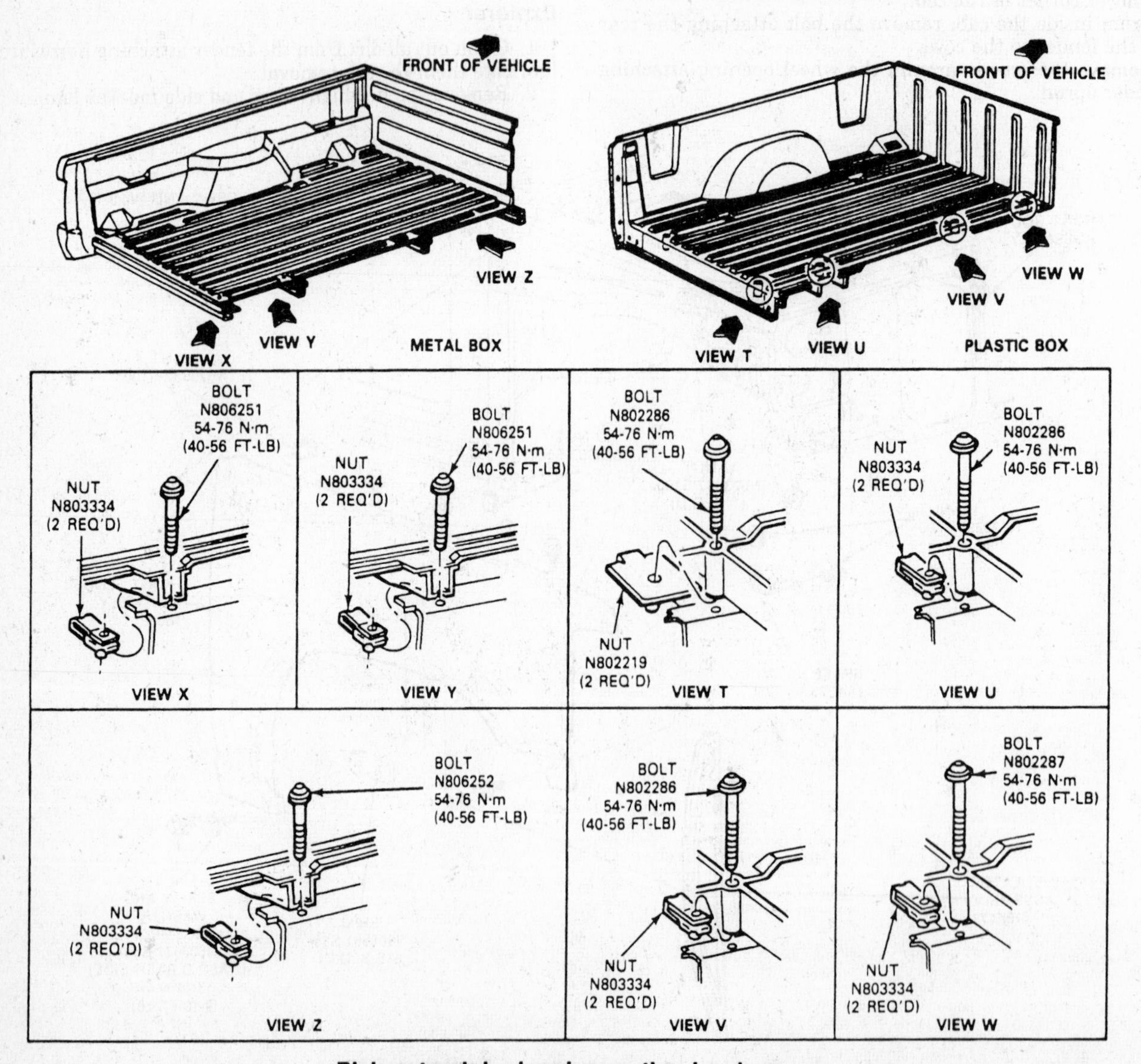

Pickup truck bed and mounting hardware

Chassis and Cab Mount Bushings

REMOVAL AND INSTALLATION

Use the accompanying illustrations as a guide.

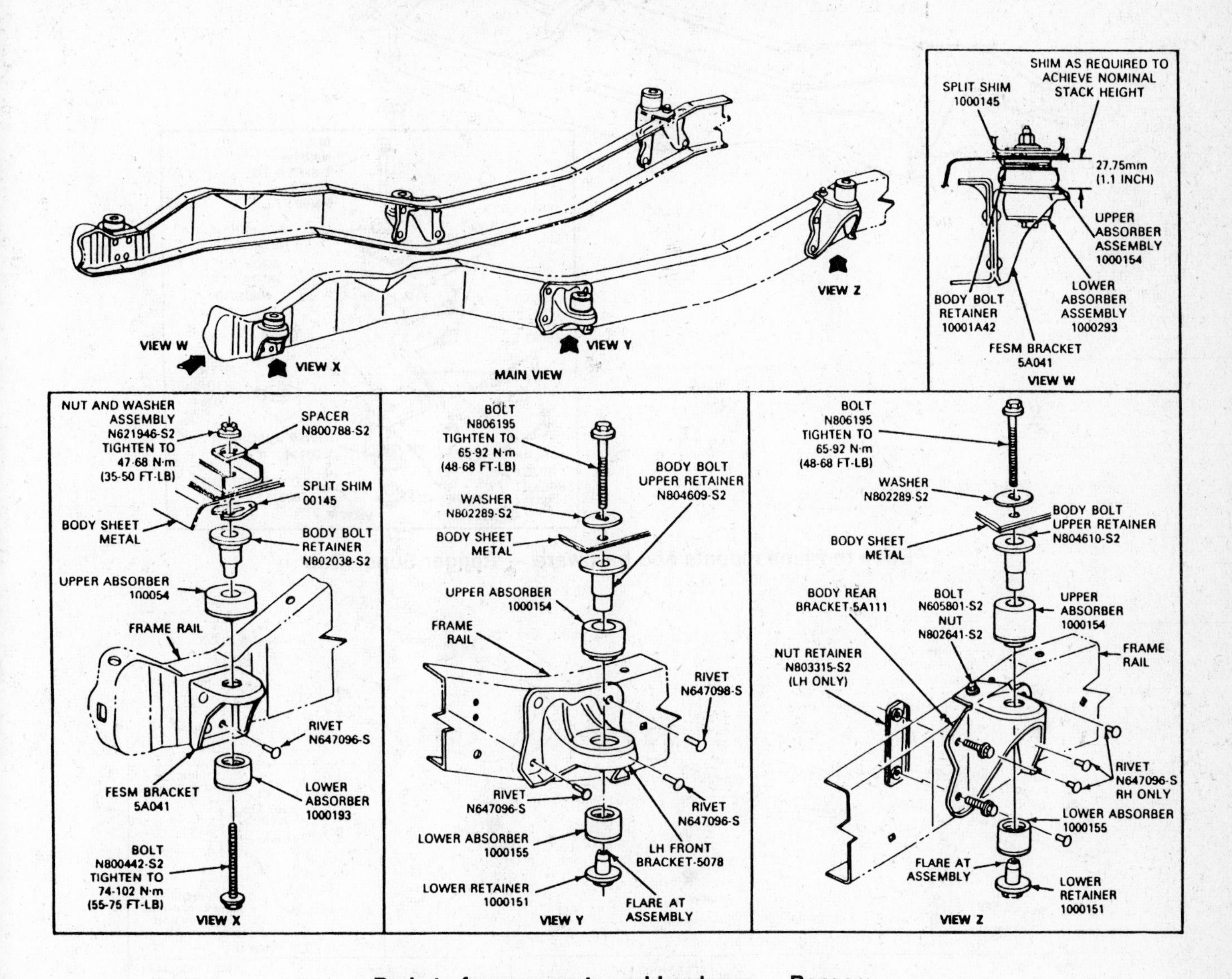

Body to frame mounts and hardware — Ranger

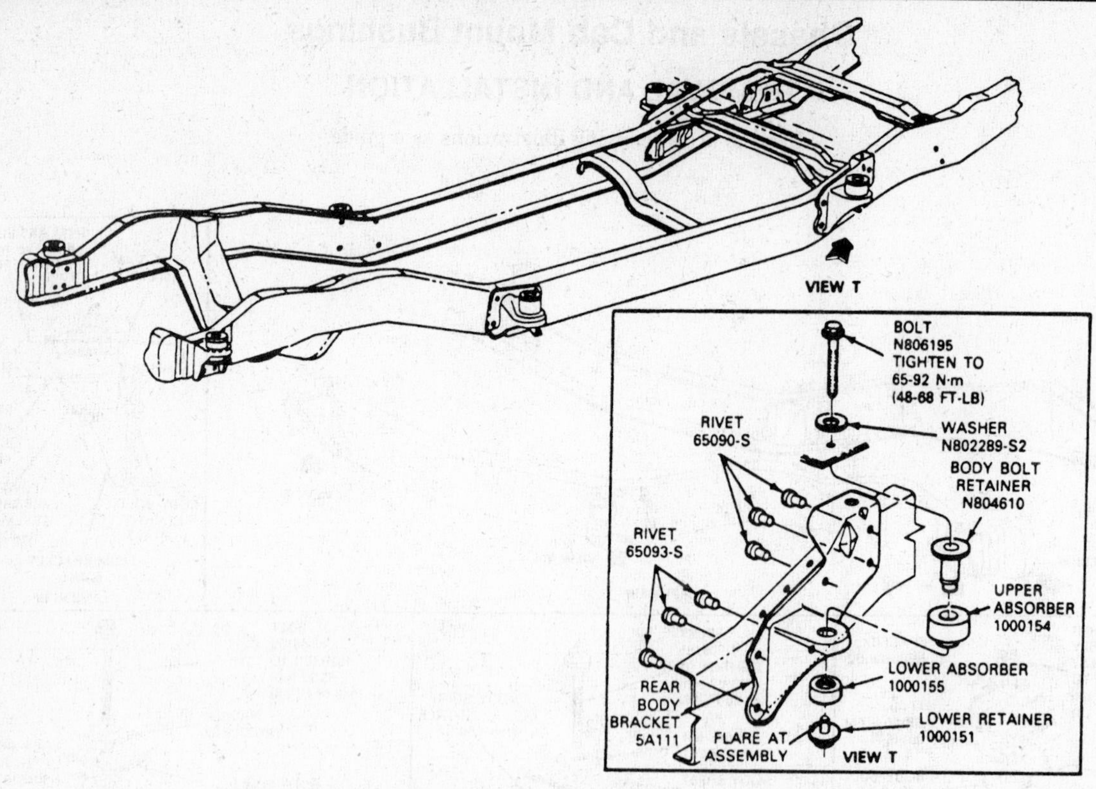

Body to frame mounts and hardware — Ranger Super Cab

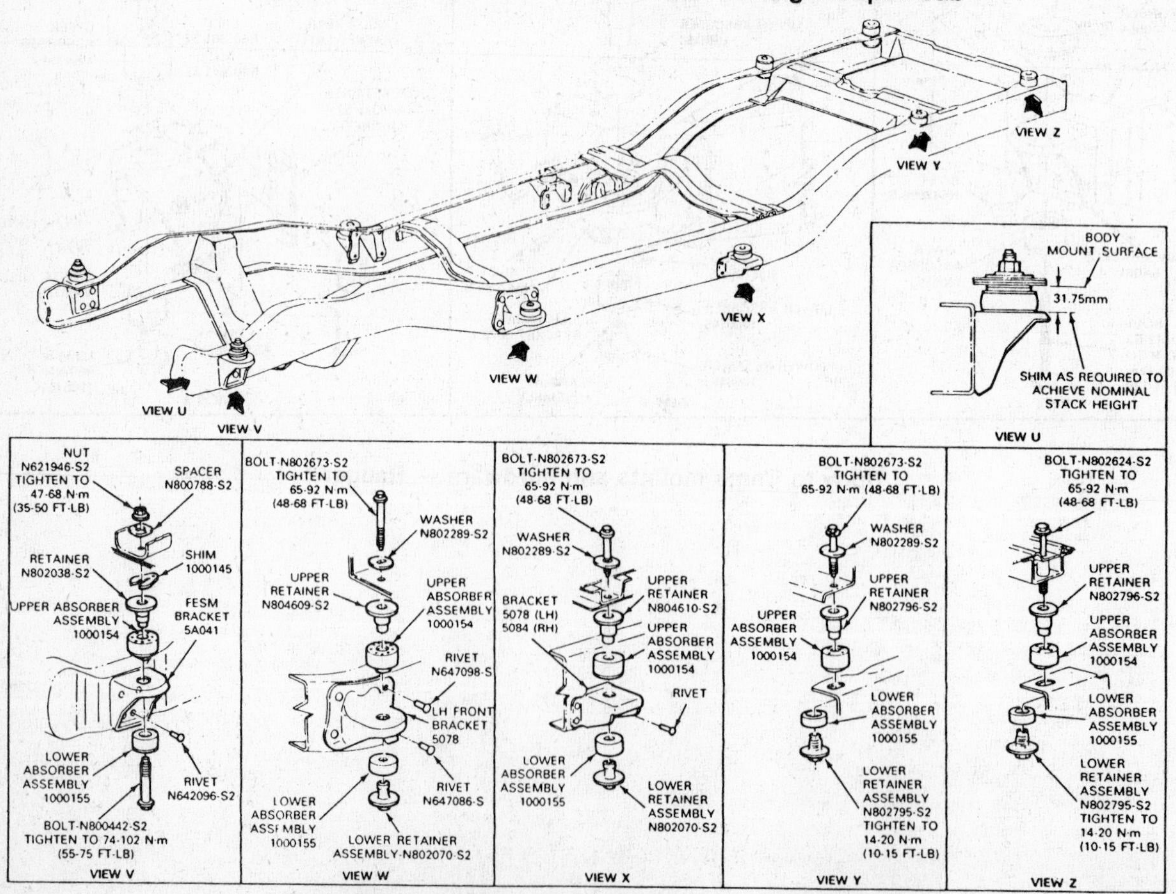

Body to frame mounts and hardware — Bronco II

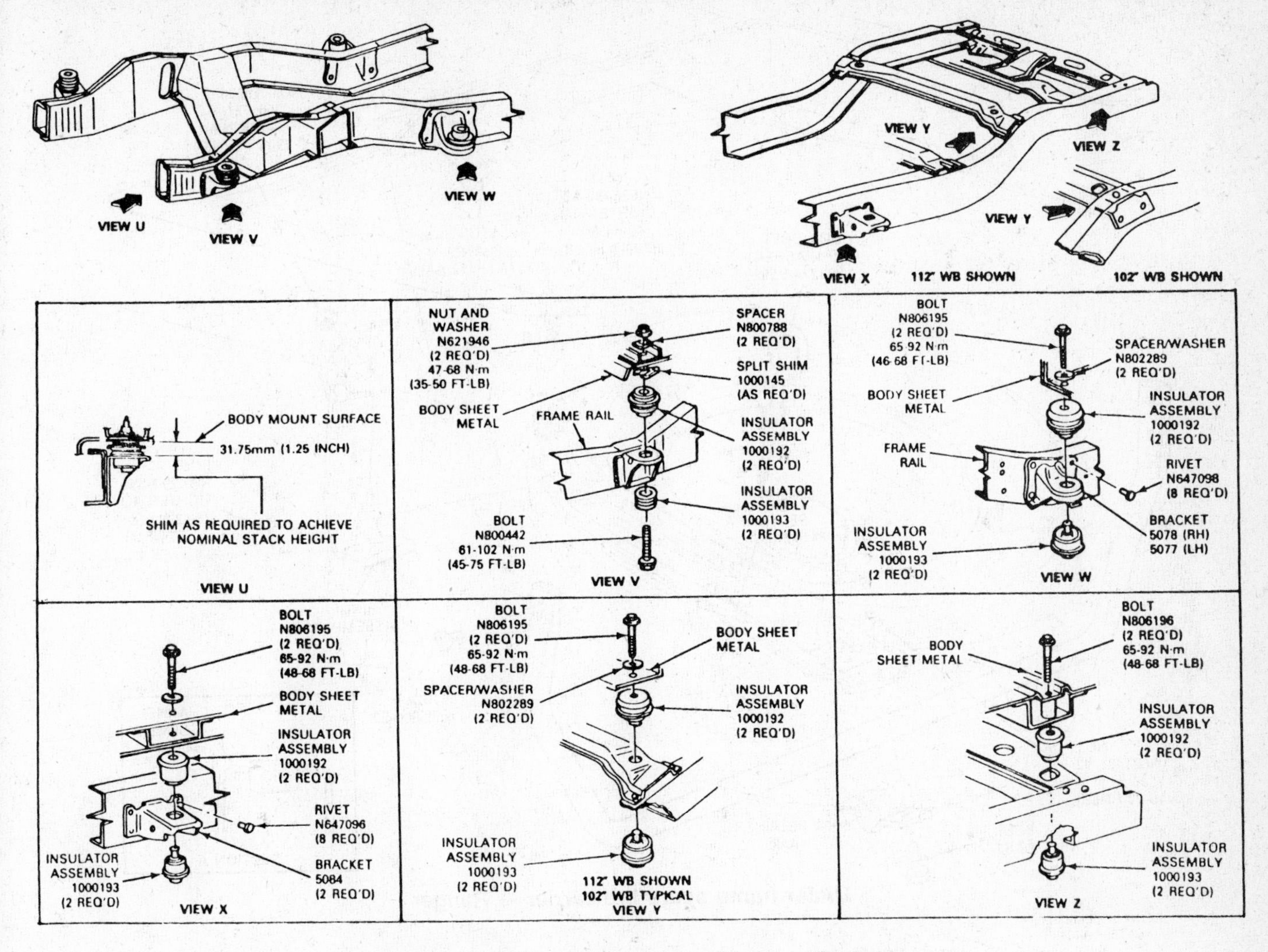

Body to frame mounts and hardware – Explorer

Spare Tire Carrier

REMOVAL AND INSTALLATION

Ranger

1. Remove the spare tire from the carrier.
2. Remove the bolt attaching the hinge assembly to the channel.
3. Remove the stud plate and the nuts attaching the hinge to the rear crossmember.
4. Remove the eye bolt from the channel.
5. Remove the bolts attaching the support assembly to the frame and remove the support.
6. Installation is the reverse of removal.

Bronco II

1. Remove the spare tire from the carrier.
2. Remove the bolts from the center of the bracket assembly.
3. Support the weight of the carrier and remove the bolts attaching the pivoting portion of the carrier to the body.
4. Remove the carrier from the vehicle.
5. Installation is the reverse of removal.

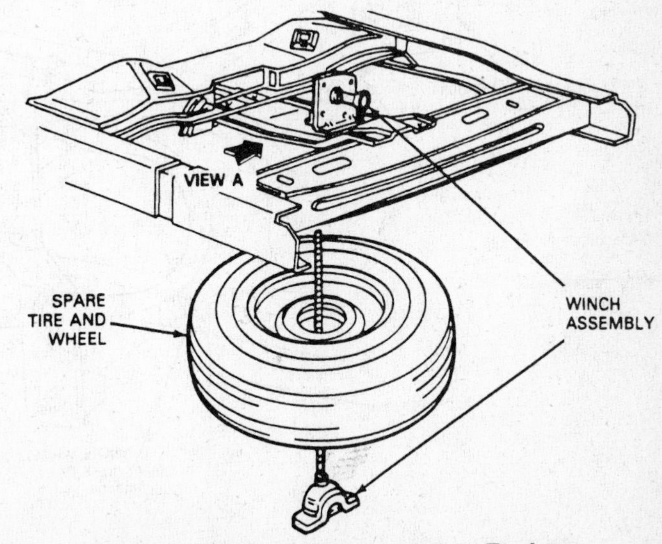

Under frame spare tire carrier – Explorer

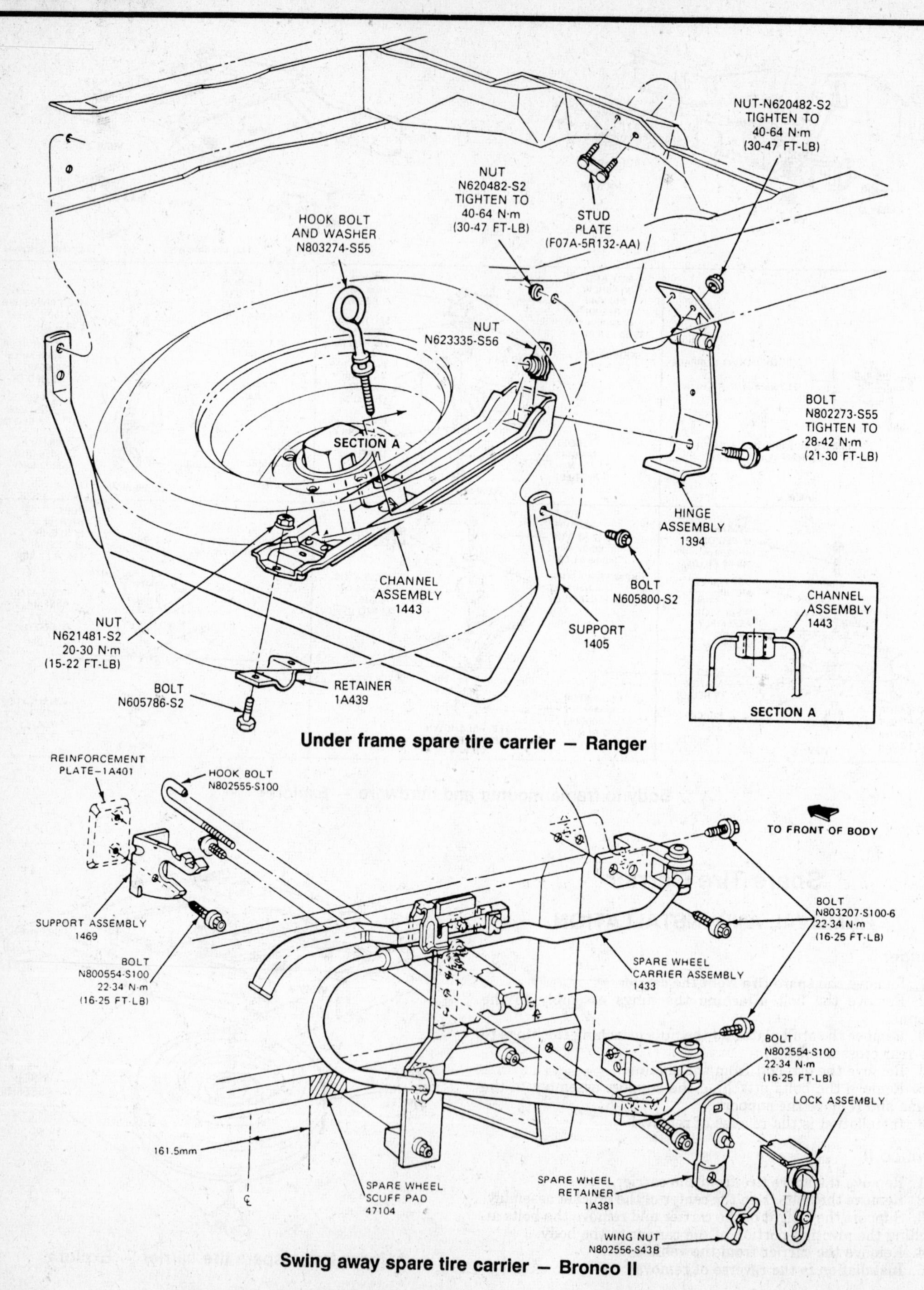

Under frame spare tire carrier — Ranger

Swing away spare tire carrier — Bronco II

Explorer

1. Lower the spare tire and work the lift cup through the wheel to release the cable.
2. Remove the 2 screws attaching the winch to the support and remove the winch assembly from the underside of the truck.

To install:

3. Position the winch assembly on its support and install the screws.
4. Insert the cable end through the center of the wheel and raise into position. Raise until the clutch overrides – an audible snap will be felt.

INTERIOR

Door Panels

REMOVAL AND INSTALLATION

Ranger and Bronco II

1. Open the window. Remove the armrest.
2. Remove the door handle screw and pull off the handle.
3. If equipped with manual windows, remove the window regulator handle screw and pull off the handle.

If equipped with power windows, remove the power window switch housing.

4. If equipped with manual door locks, remove the door lock control.

If equipped with power door locks, remove the power door lock switch housing.

5. If equipped with electric outside rear view mirrors, remove the power mirror switch housing.
6. Using a flat wood spatula, insert it carefully behind the panel and slide it along to find the push-pins. When you encoun-

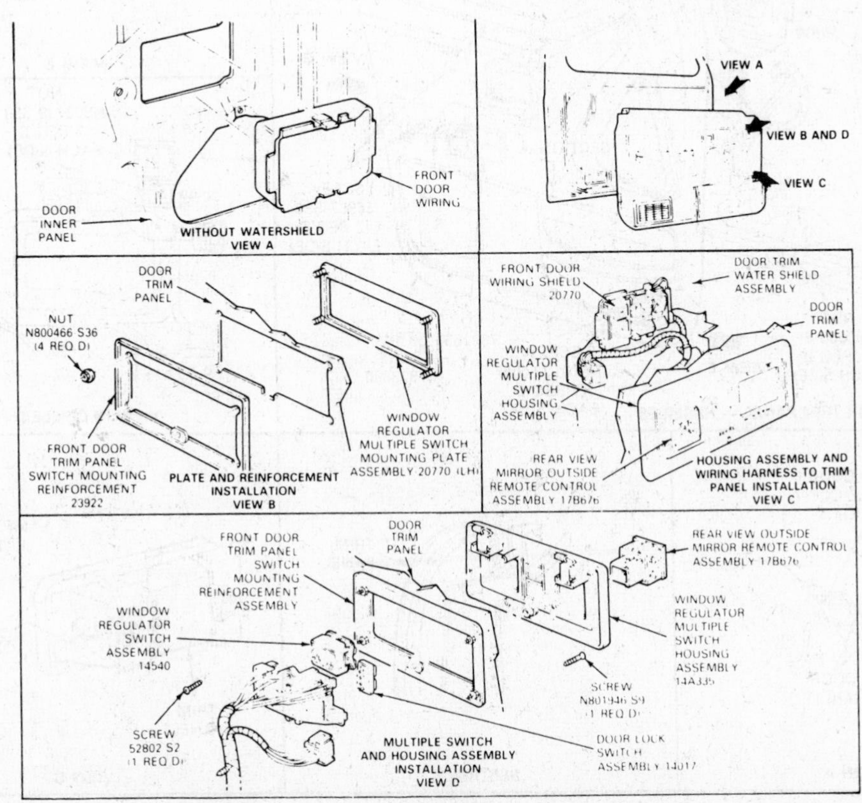

Door trim panel — Ranger and Bronco II

ter a pin, pry the pin outward. Do this until all the pins are out. NEVER PULL ON THE PANEL TO REMOVE THE PINS!

7. Installation is the reverse of removal.

Explorer

1. Open the window. Remove the 2 screws retaining the trim panel located above the door handle.
2. Remove the rim cup behind the door handle using a small prying tool. Retention nibs will flex for ease of removal.
3. If equipped with power accessories, use the notch at the lower end of the plate and pry the plate off. Remove the plate from the trim panel and pull the wiring harness from behind the panel. Disconnect the harness from the switches.
4. Using a flat wood spatula, insert it carefully behind the panel and slide it along to find the push-pins. When you encounter a pin, pry the pin outward. Do this until all the pins are out. NEVER PULL ON THE PANEL TO REMOVE THE PINS!
5. Lift slightly to disengage the panel from the flange at the top of the door.
6. Disconnect the door courtesy lamp and remove the panel completely. Replace any damaged or bent attaching clips.
7. Installation is the reverse of removal.

Interior Trim Panels

REMOVAL AND INSTALLATION

If you're going to remove major pieces of interior trim, it's best to heat the interior of the truck first; this makes the plastic soften slightly and will help prevent possible breakage.

Many of the attaching screws and pins are hidden by plugs, caps, etc. or are the attaching point for clips or brackets. The plugs and caps can be removed with light prying action. Don't force anything—if the trim isn't coming out, it's because you've missed a screw somewhere. Most attaching hardware is located around the perimeter or near the center of the piece. Don't forget to reinstall the ornamental plug or cap that hides the screw when the part is back in place.

Also, some moldings cannot be removed without first removing the adjacent molding. It may seem like more work than necessary, but if you force something that isn't meant to come out yet, something is bound to break. If seat belts are in the way, refer to that section for proper fastener torque values.

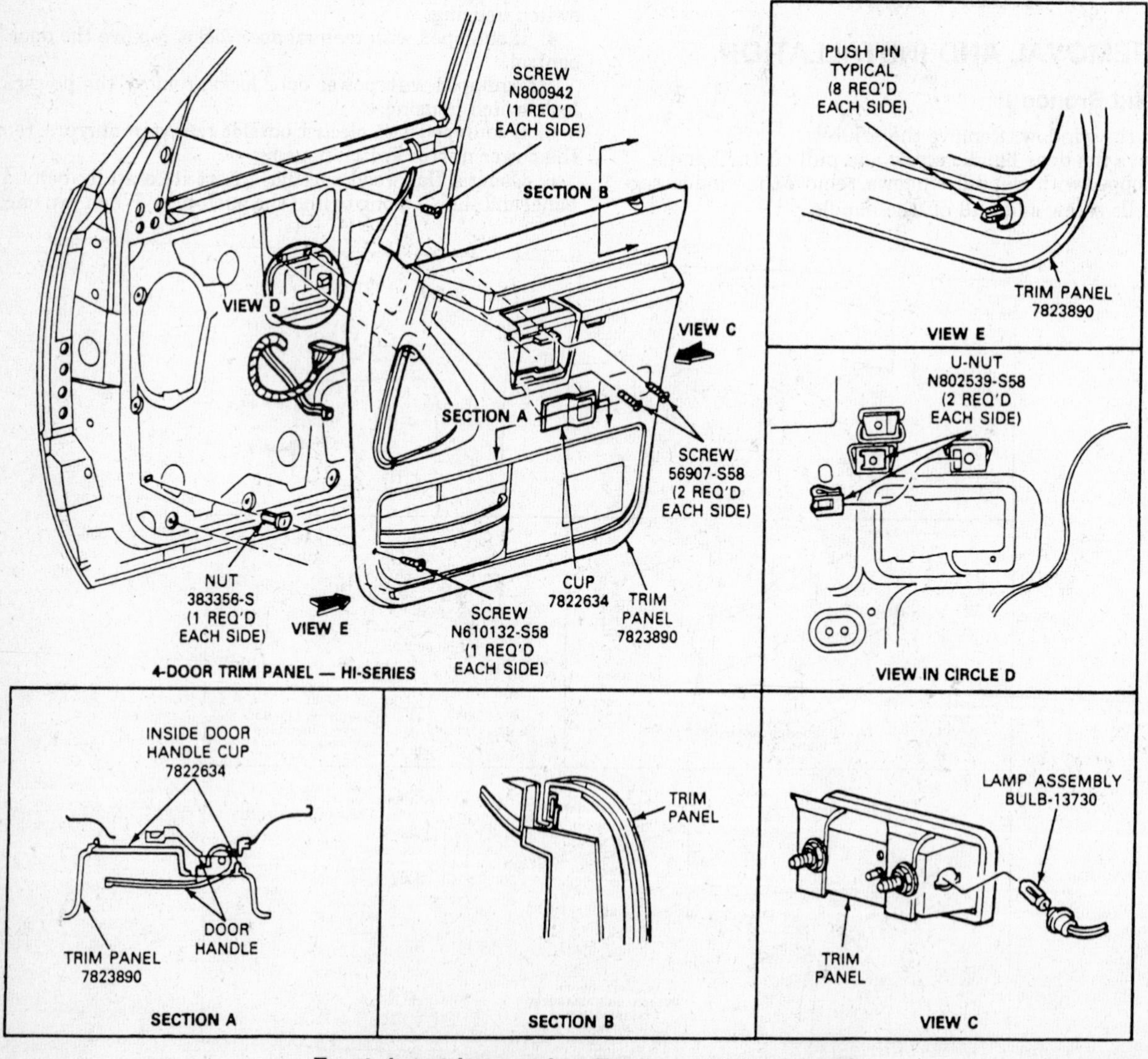

Front door trim panel – Explorer; Hi line shown

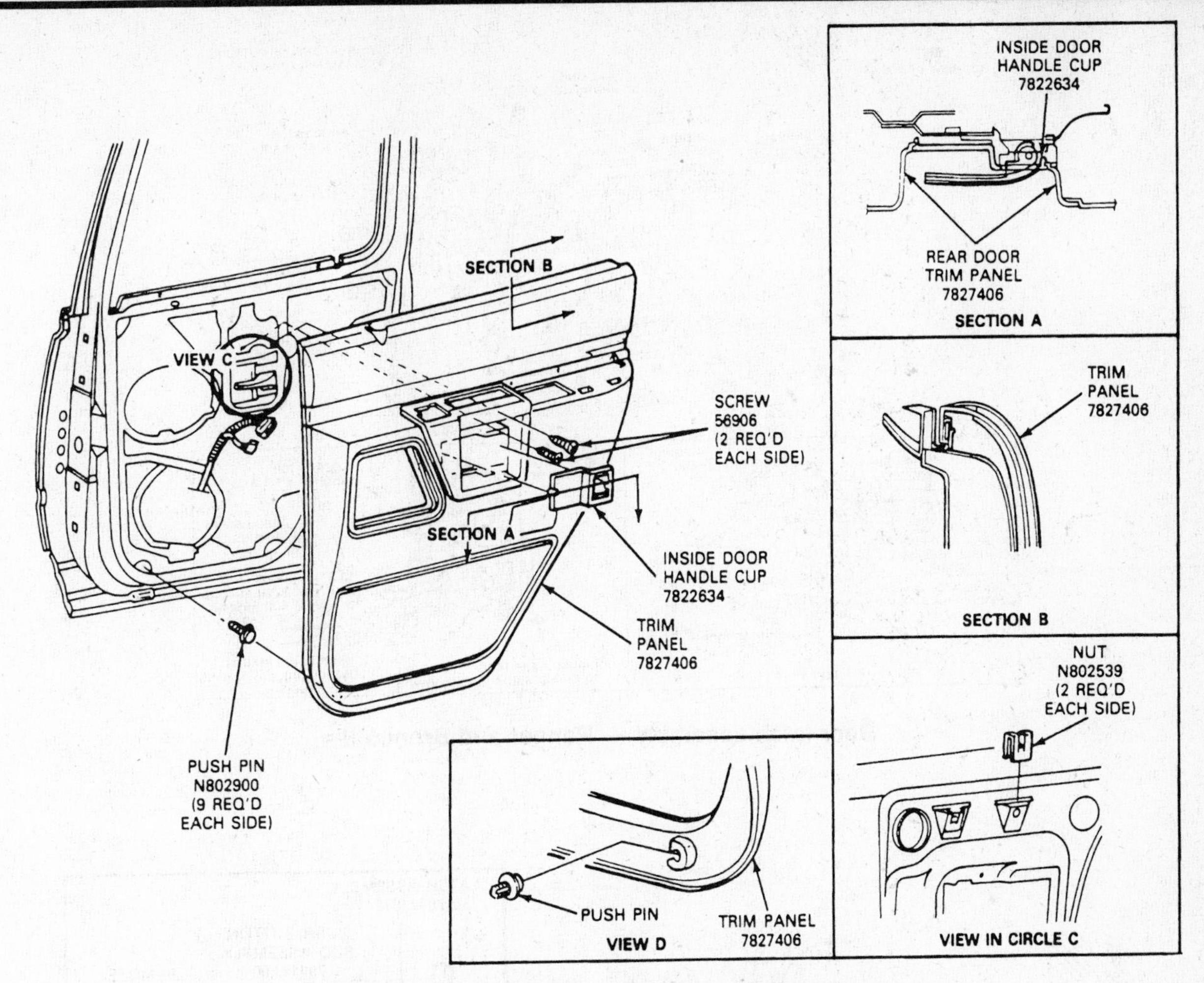

Rear door trim panel – Explorer; Hi line shown

Door Locks

REMOVAL AND INSTALLATION

Door Latch

1. Remove the door trim panel and watershield.
2. Disconnect the rods from the handle and lock cylinder, and from the remote control assembly.
3. Remove the latch assembly attaching screws and remove the latch from the door.
4. Installation is the reverse of removal.

Door Lock Cylinder

1. Open the window.
2. Remove the trim panel and watershield.
3. Disconnect the actuating rod from the lock control link clip.
4. Slide the retainer away from the lock cylinder.
5. Remove the cylinder from the door.
6. Use a new gasket when installing to ensure a watertight fit.
7. Lubricate the cylinder with suitable oil recommended for this application.

Door Glass and Regulator

REMOVAL AND INSTALLATION

Glass

1. Remove the door trim panel and speaker if applicable.
2. Remove the screw from the division bar. Remove the inside belt weaterstrip(s) if equipped.
3. Remove the 2 vent window attaching screws from the front edge of the door.
4. Lower the glass and pull the glass out of the run retainer near the vent window division bar, just enough to allow the removal of the vent window, if equipped.
5. Push the front edge of the glass downward and remove the rear glass run retainer from the door.
6. If equipped with retaining rivets, remove them carefully. Otherwise, remove the glass from the channel using Glass and Channel Removal Tool 2900, made by the Sommer and Mala Glass Machine Co. of Chicago, ILL., or its equivalent. Remove the glass through the belt opening if possible.

To install:

7. Install the glass spacer and retainer into the retention holes.

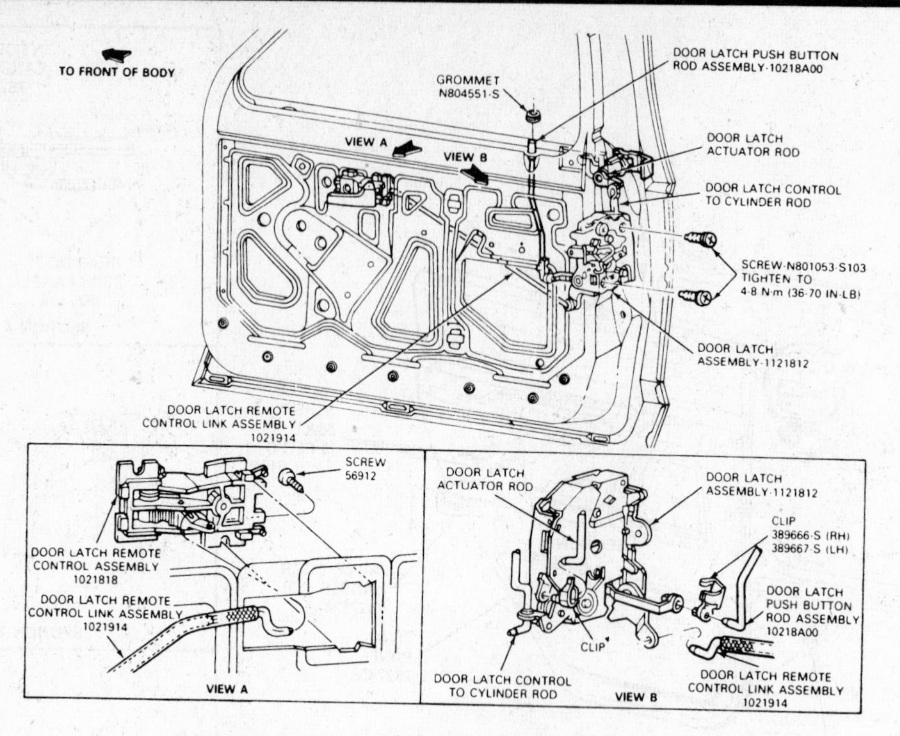

Door latch assembly – Ranger and Bronco II

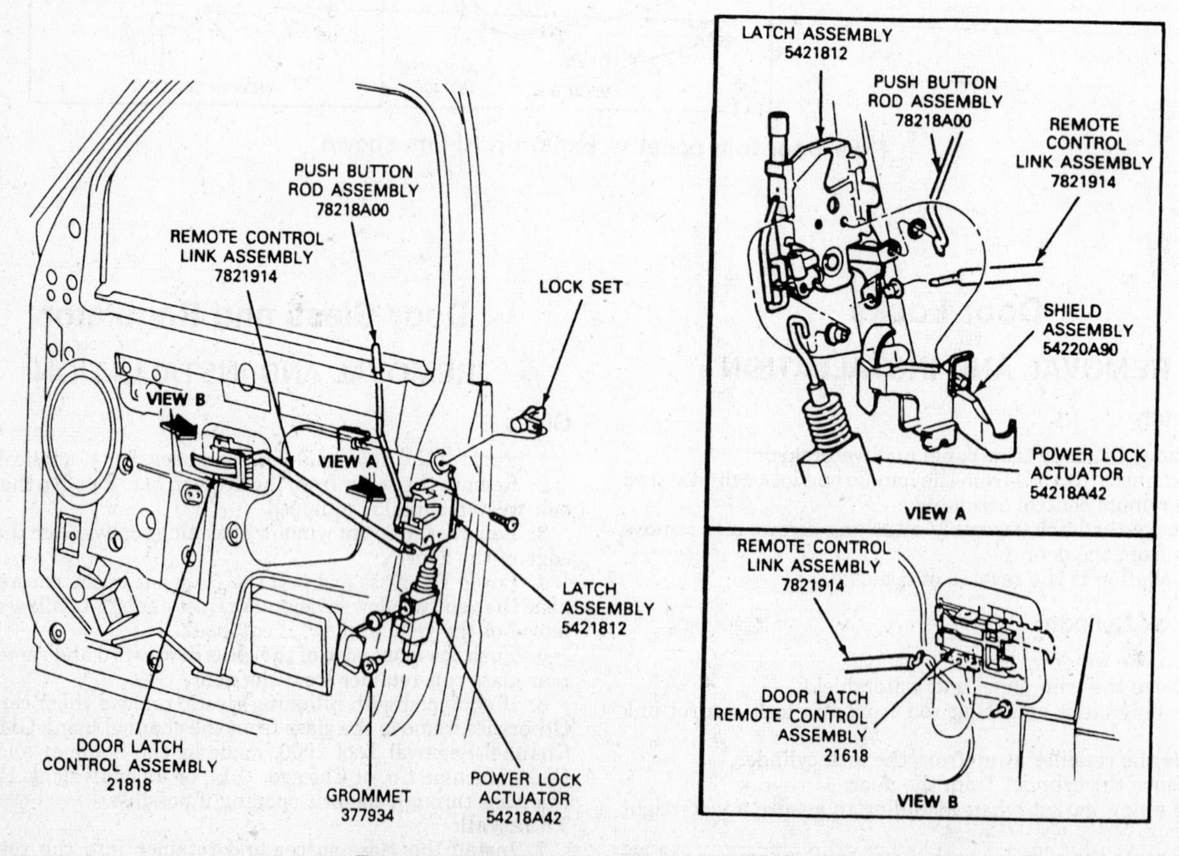

Front door latch assembly – Explorer

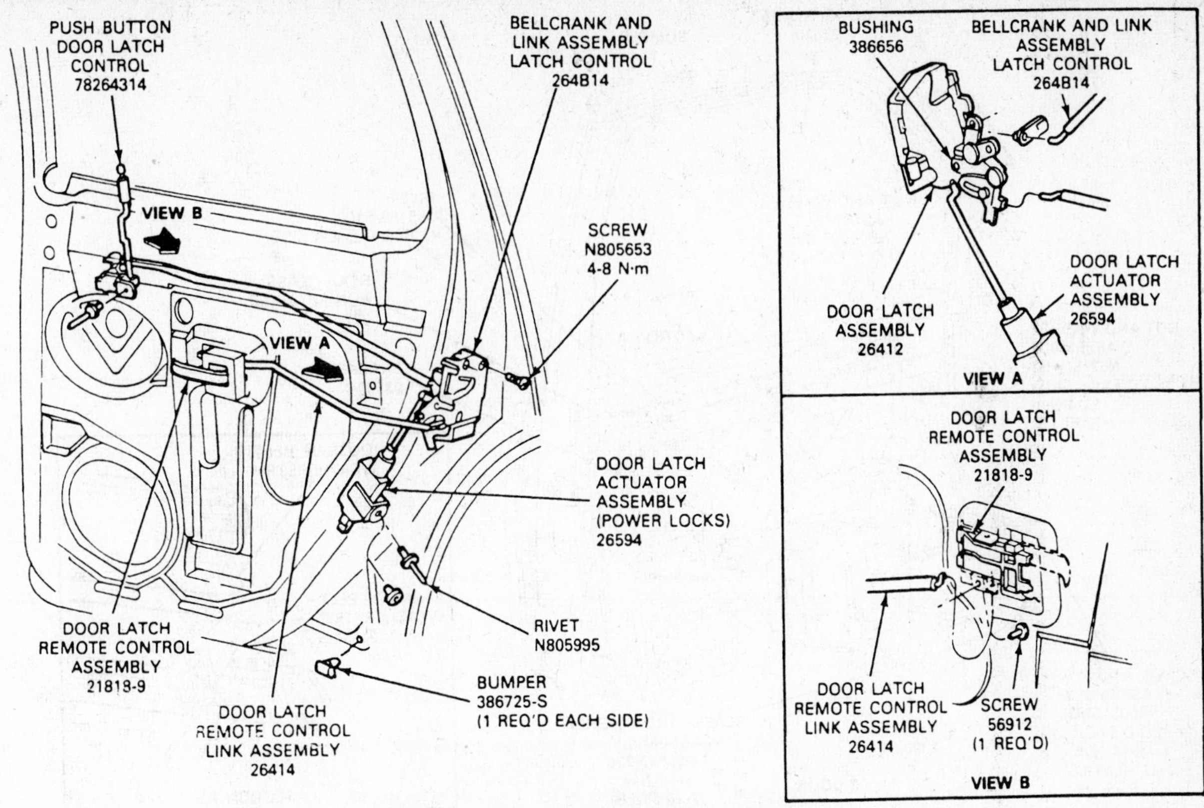

Rear door latch assembly – Explorer

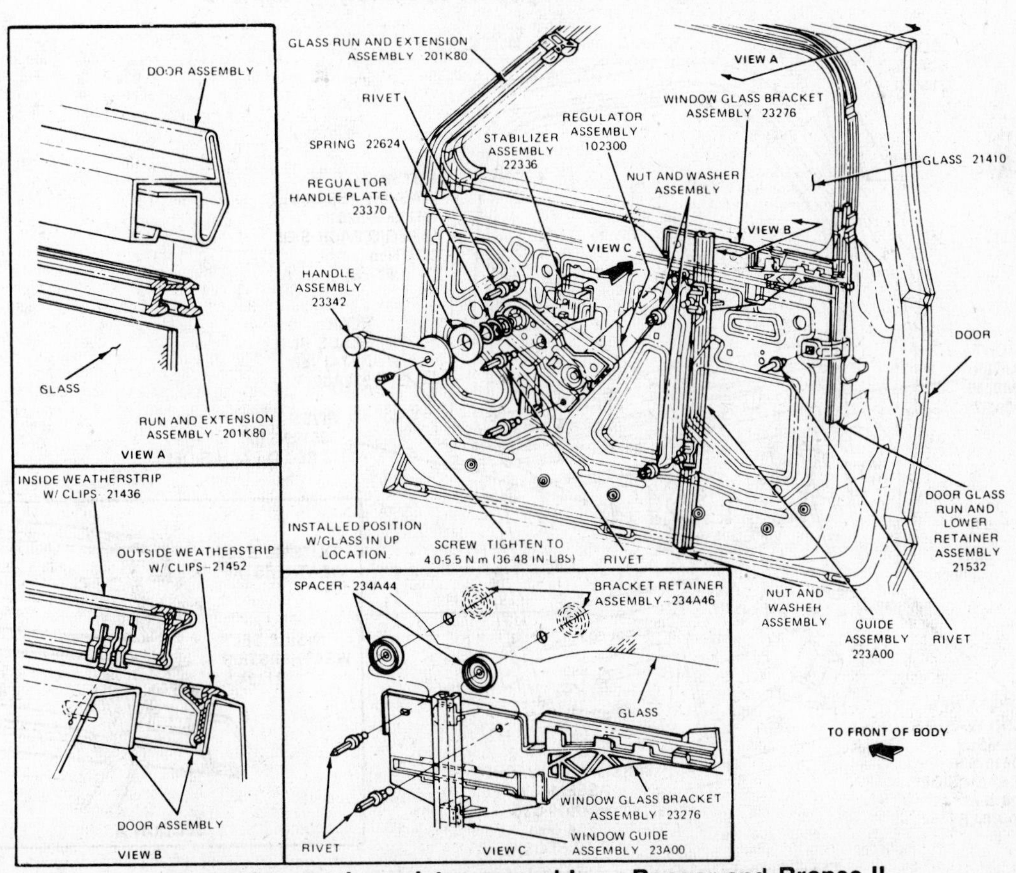

Door glass and regulator assembly – Ranger and Bronco II

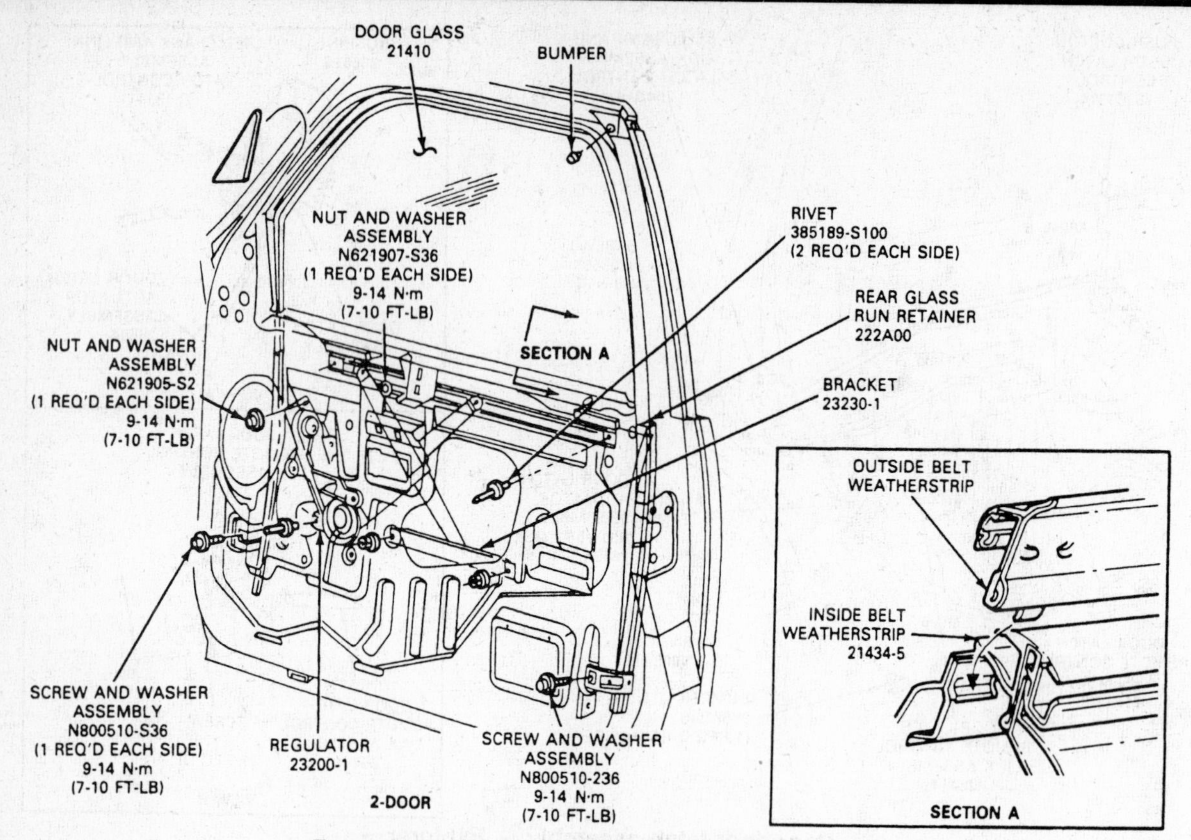

Front door glass and regulator assembly – Explorer

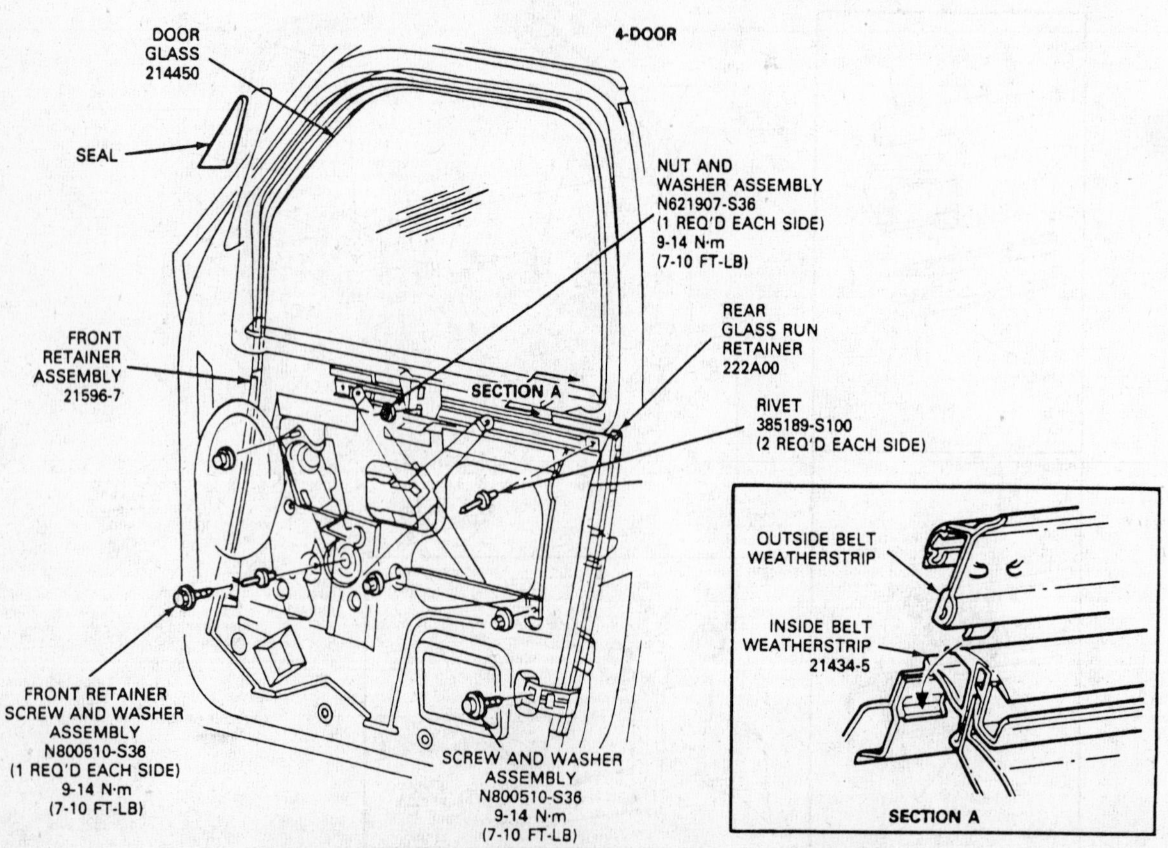

Rear door glass and regulator assembly – Explorer

8. Install the glass into the door, position on the bracket and align the retaining holes.
9. Carefully install the retaining rivets or equivalent.
10. Raise the glass to the full closed position.
11. Install the rear glass run retainer and glass run. Install the inside belt weatherstrip(s).
12. Check for smooth operation before installing the trim panel.

Regulator

EXCEPT FRONT DOOR – EXPLORER

1. Remove the door trim panel. If equipped with power windows, disconnect the wire from the regulator
2. Support the glass in the full UP position or remove completely.
3. Remove the window guide and glass bracket if equipped.
4. Remove the center pins from the regulator attaching rivets.
5. Drill out the regulator attaching rivets using a ¼ in. (6mm) drill bit.
6. Disengage the regulator arm from the glass bracket and remove the regulator.
7. Installation is the reverse of removal. ¼ in.-20 × ½ in. bolts and nuts may be used in place of the rivets to attach the regulator.

FRONT DOOR – EXPLORER

1. Remove the door trim panel and watershield.
2. Remove the inside door belt weatherstrip and glass stabilizer.
3. Remove the door glass.

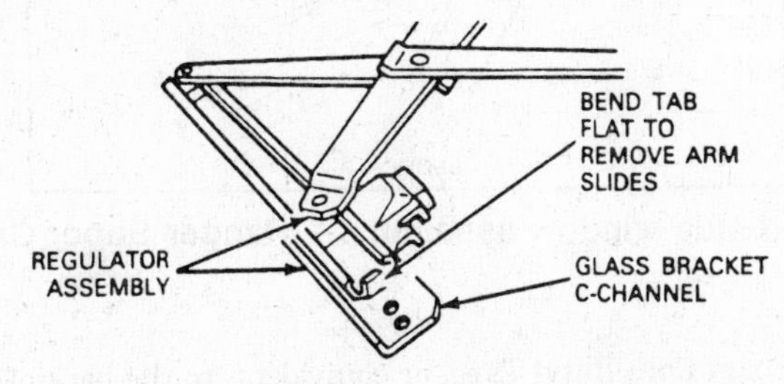

Regulator assembly – Explorer front door

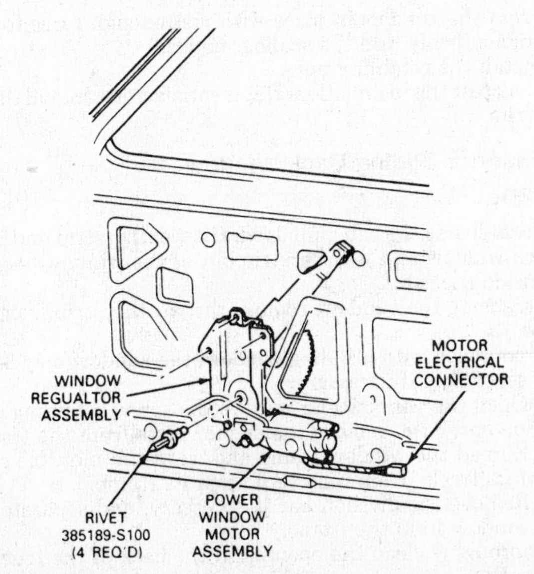

Power window motor and regulator assembly – Ranger and Bronco II

4. Remove the 2 nuts attaching the equalizer bracket.
5. Remove the rivets attaching the regulator base plate to the door.
6. Remove the regulator and glass bracket as an assembly from the door and transfer to a workbench.
7. Carefully bend the tab flat in order to remove the air slides from the glass bracket C-channel.
8. Install new regulator arm plastic guides into the C-channel and bend the tab back 90°. If the tab is broken or cracked, replace the glass bracket assembly. Make sure the rubber bumper is installed properly on the new glass bracket, if applicable.

CAUTION

If the regulator counterbalance spring is to be removed, make sure the regulator arms are in a fixed position prior to removal. This will prevent possible injury when the C-spring unwinds.

To install:

9. Assemble the glass bracket and regulator assembly.
10. Install the assembly in the door. Set the regulator base plate to the door using the base plate locator tab as a guide.
11. Attach the regulator to the door using new rivets. ¼ in.-20 × ½ in. bolts and nuts may be used in place of the rivets to attach the regulator.
12. Install the equalizer bracket, door belt weatherstrip and glass stabilizer.
13. Install the glass and check for smooth operation before installing the door trim panel.

Electric Window Motor

REMOVAL AND INSTALLATION

Ranger and Bronco II

1. Disconnect the negative battery cable.
2. Open the window. Remove the trim panel and watershield and support the window.
3. Disconnect the power window motor connector.
4. There may be a drill dimple in the door panel, opposite the concealed motor retaining bolt. Drill out the dimple to gain access to the bolt. Be careful to avoid damage to the wires. Remove the motor mounting bolts and remove the motor and regulator assembly.
5. Separate the motor and drive from the regulator on a workbench.
6. Installation is the reverse of removal.
7. Check for smooth operation before installing the trim panel.

Explorer

1. Raise the window fully if possible. If not, you will have to support the window during this procedure. Disconnect the battery ground.
2. Remove the door trim panel.
3. Disconnect the window motor wiring harness.
4. There may be a drill dimple in the door panel, opposite the concealed motor retaining bolt. Drill out the dimple to gain access to the bolt. Be careful to avoid damage to the wires.
5. Remove the motor mounting bolts (front door) or rivets(rear door).
6. Push the motor towards the outside of the door to disengage it from the gears. You'll have to support the window glass once the motor is disengaged.
7. Remove the motor from the door.
8. Installation is the reverse of removal. To avoid rusting in the drilled areas, prime and paint the exposed metal, or, cover the holes with waterproof body tape. Make sure that the motor works properly before installing the trim panel.

Windshield Glass and Bronco II Side Body Windows

REMOVAL AND INSTALLATION

NOTE: Successful windshield or fixed window replacement takes a lot of practice and proper tools and equipment. Although the following procedure may appear to be easy, it is very difficult in reality for an inexperienced glazier. It is highly recommended that you have a professional shop replace your windshield or window. Read through the entire procedure first; if you're going to attempt it, have an assistant help you.

1. Remove the windshield wiper arms and blades.
2. Remove all moldings, glass stops and retainers.
3. Remove the rear view mirror from inside the passenger compartment.
4. Cut the urethane seal from around the perimeter of the windshield or fixed window using a 3-foot length of single strand music wire, special Dual-Handle Knife D8 1P-42006-B or an equivalent method.
5. Remove the windshield or window from the truck.
6. Remove any excess sealing material from the sheet metal and repair the surface as required.

To install:

7. Use a clean brush and apply urethane metal primer to the windshield or window opening flange. Allow at least 30 minutes for the primer to dry.
8. Thoroughly clean and dry the windshield or window; pay particular attention to the outer edges.
9. Use a clean brush and apply primer to the outer edge of the inside of the windshield or window perimeter. Allow 5 minutes drying time.
10. Apply the foam tape along the bottom of the windshield from the glass edge on the inside of the windshield around the entire perimeter.
11. Properly align the windshield or window to the body of the truck. Place the windshield or window into the opening evenly. Install vinyl windshield molding. Matchmark the position of the windshield or window to the body.
12. Remove the windshield or window and place on a workbench.
13. Apply an even bead of urethane sealer around the entire windshield or window edge using an air pressure cartridge gun. The bead should be about ½ in. (13mm) high and wide. Do not allow more than 15 minutes drying time.
14. Install the windshield or window to its opening, paying attention to the alignment marks.
15. Install the windshield wipers.
16. Leak check the windshield or window. If satisfactory, install the moldings.

Stationary Glass

REMOVAL AND INSTALLATION

Fixed Side Window

RANGER SUPER CAB AND EXPLORER

1. Remove the interior trim around the window.
2. Remove the nuts from inside window assembly and remove the molding.
3. Remove the glass by pushing on it with enough force to separate the butyl seal.
4. Clean all traces of the original seal from the window opening and repair the sheet metal as required.

To install:

5. If the replacement window is not complete with sealer tape, then apply a continuous strip of $^5/_{16}$ × $^5/_{16}$ in. (8mm × 8mm) Foam Core Butyl Tape, or equivalent, to the back of the window. The ends must meet at the bottom and overlap 1–2 in. (25–50mm).
6. Press the window in place with just enough force to seat the window firmly into the sealing material.
7. Install the retaining nuts.
8. Leak test the installation. If it is satisfactory, install the interior trim.

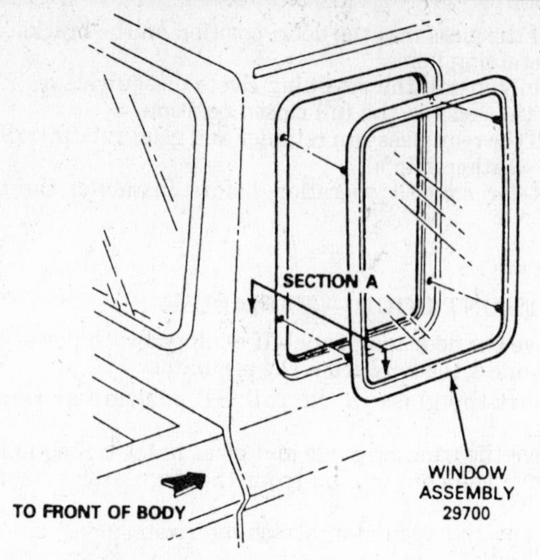

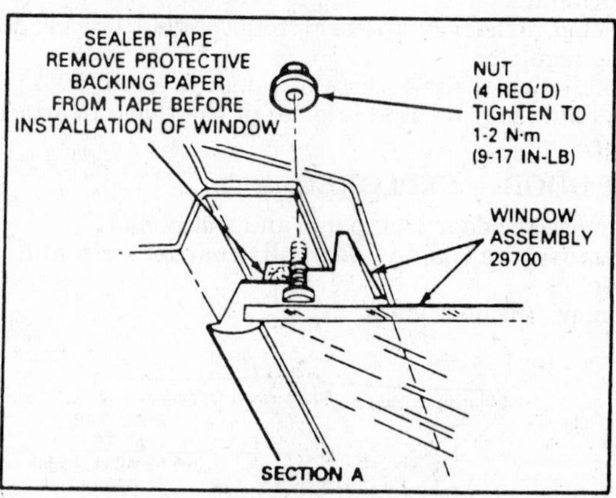

Fixed side window assembly – Ranger Super Cab

Stationary or Sliding Back Window

RANGER

1. From inside the cab, pull down the weatherstrip and push the back window and weatherstrip out of the window opening from inside the cab.
2. If reusing the window, remove the weatherstrip from the window.
3. If equipped with a sliding window, the window may be removed separately if necessary:
   a. Open the window and remove the screw retaining each division bar. Also, remove the anchor plate from the track.
   b. Spread the window frame and work the movable glass out of its track. Then remove it from its frame.
   c. Remove the division bar if necessary, and separate the rear window from the frame.
4. Thoroughly clean the opening in the back of the truck.

To install:

5. Assemble the sliding window to the back window if necesaray. Lubricate all seals to ease assembly.

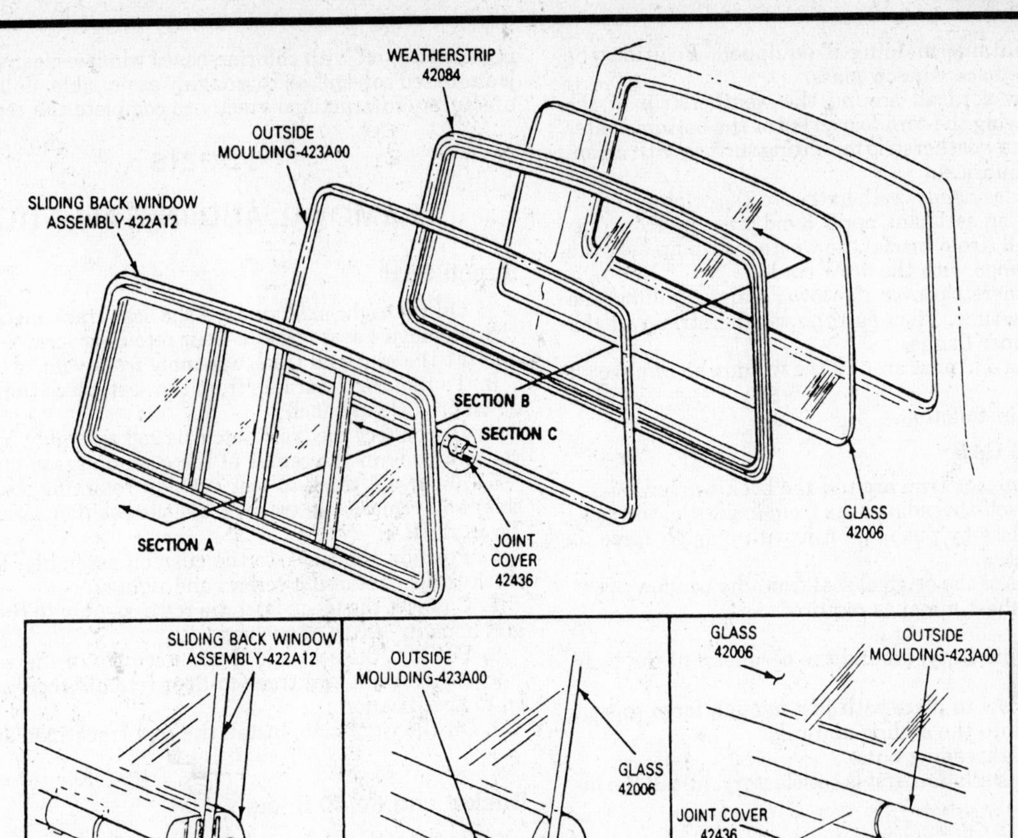

Back window assembly – Ranger

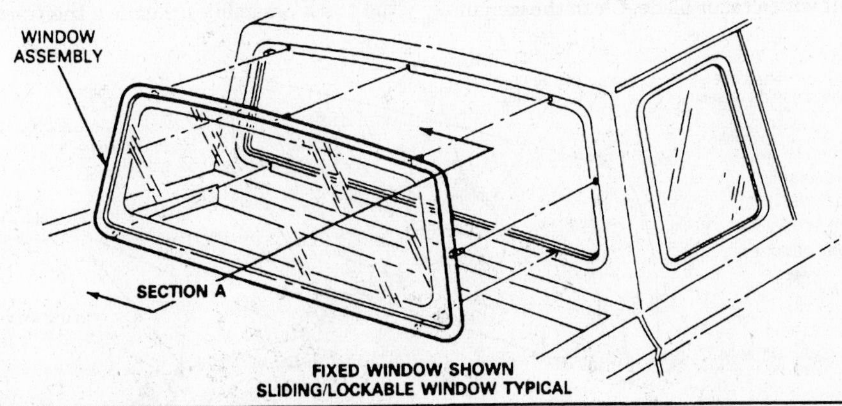

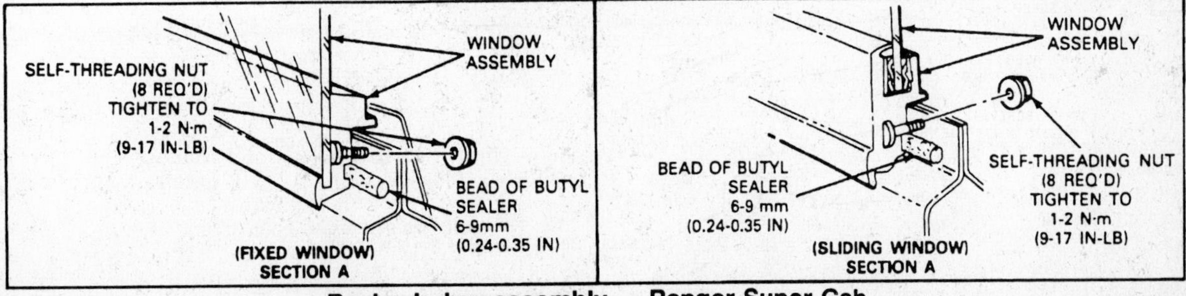

Back window assembly – Ranger Super Cab

6. Install the outside molding if equipped. Position the weatherstrip to the back window glass.
7. Install a draw cord all around the weatherstrip in the flange crevice, allowing the cord to overlap at the bottom center of the glass. Coat the weatherstrip mounting surface with an appropriate Rubber Lubricant.
8. Position the glass and weatherstrip to the window opening. With the aid of an assistant, apply hand pressure from outside the cab and pull (from inside) the weatherstrip lip over the window opening flange with the draw cord.
9. Pull the weatherstrip over the lower flange, pulling one end of the cord at a time. Then pull the weatherstrip over the side flanges and upper flange.
10. Clean the glass and area around the window to remove all excess sealer.
11. Leak test the installation.

RANGER SUPER CAB

1. Remove the interior trim around the back window.
2. Remove the 8 self-threading nuts from inside the window.
3. Remove the glass by pushing on it with enough force to separate the butyl seal.
4. Clean all traces of the original seal from the window opening and repair the sheet metal as required.

To install:

5. Apply new Liquid Butyl around the perimeter of the body recess.
6. Press the window in place with just enough force to seat the window firmly into the sealing material.
7. Install the self-threading nuts.
8. Leak test the installation. If it is satisfactory, install the interior trim.

Inside Rear View Mirror

The mirror is held in place with a single setscrew. Loosen the screw and lift the mirror off. Don't forget to unplug the electrical connector if the truck has an electric Day/Night mirror.

Repair kits for damaged mirrors are available and most auto parts stores. The most important part of the repair is the beginning. Mark the outside of the windshield to locate the pad, then scrape the old adhesive off with a razor blade. Clean the remaining adhesive off with chlorine-based window cleaner (not petroleum-based solvent) as thoroughly as possible. Follow the manufacturers instructions exactly to complete the repair..

Seats

REMOVAL AND INSTALLATION

Bench Seat

1. On the right side, remove the seat track insulator.
2. Remove 4 seat track-to-floor retaining screws (2 each side) and lift the seat and track assembly from vehicle.
3. To remove the tracks from the seat, place the seat upside-down on a clean bench.
4. Disconnect the track latch tie rod assembly from latch lever and hook in the center of the cushion assembly.
5. Remove 4 track-to-seat cushion retaining screws (2 each side) and remove the tracks from the cushion assembly.

To install:

6. Position the track to the cushion assembly. Install the 4 track-to-seat retaining screws and tighten.
7. Connect the track latch tie rod assembly to the latch lever and hook in center of cushion.
8. Position the seat and track assembly in the vehicle.
9. Install the 4 seat track-to-floor retaining screws and tighten to specification.
10. On the right side, install the seat track insulator.

Bucket and 60/40 Seats

1. Remove the seat track insulator (Ranger passenger seat only).
2. Remove the 4 seat track-to-floorpan screws (2 each side) and lift the seat and track assembly from the vehicle.
3. To remove the seat tracks from the seat cushion, position the seat upside down on a clean bench.
4. Disconnect the latch tie rod assembly and assist spring from the tracks.
5. Remove 4 track-to-seat cushion screws (2 each side) from the track assemblies. Remove the tracks from seat cushion.

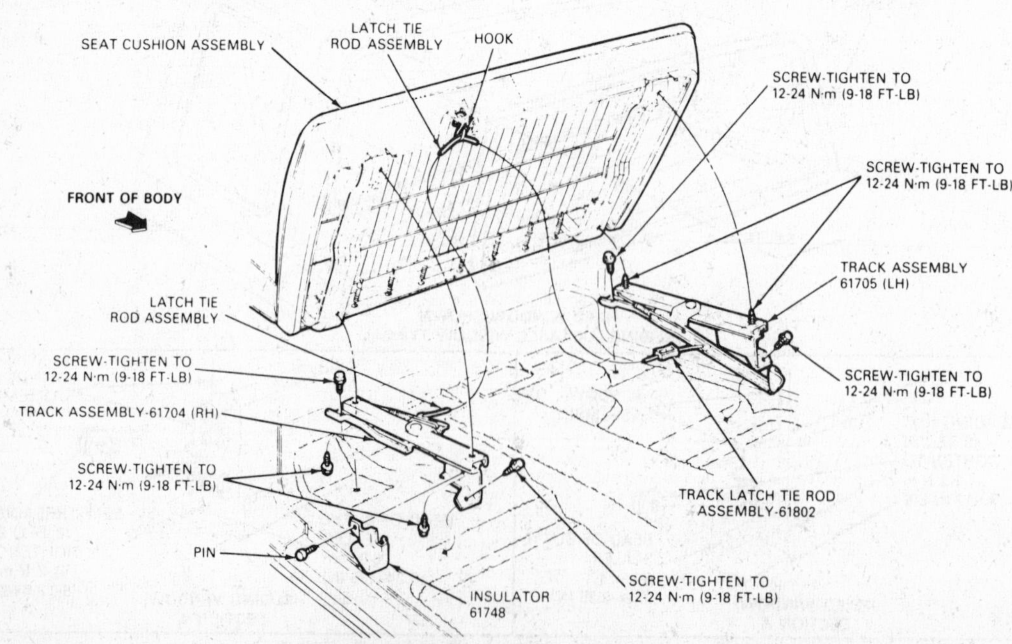

Bench seat and track — Ranger

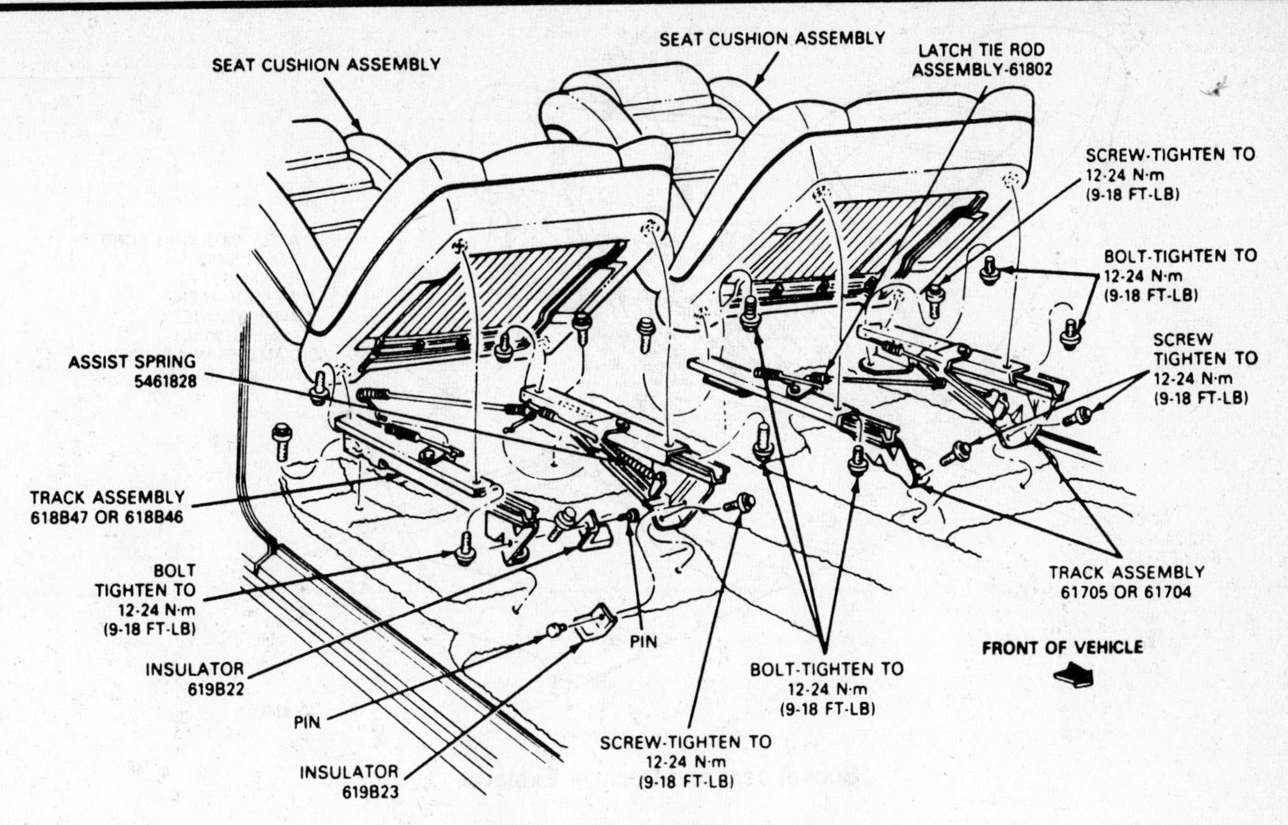

Bucket seats and tracks – Ranger

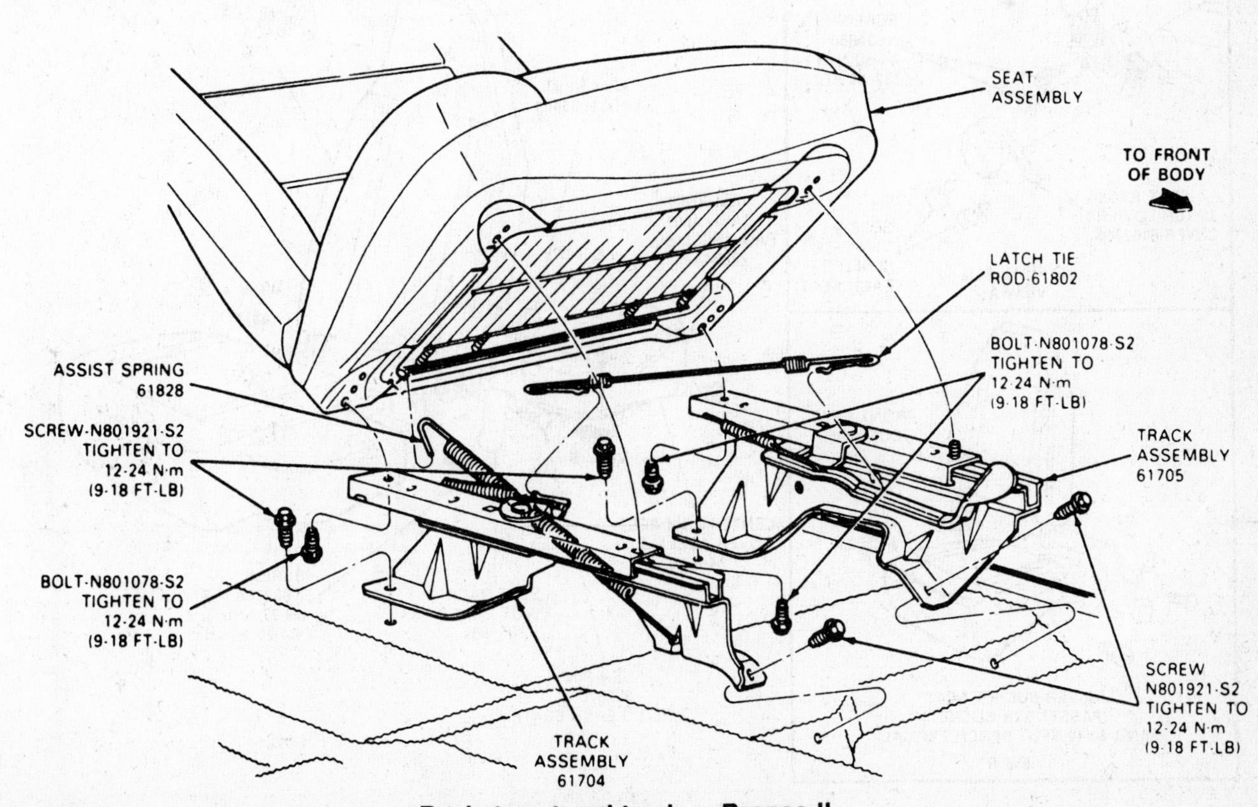

Bucket seat and track – Bronco II

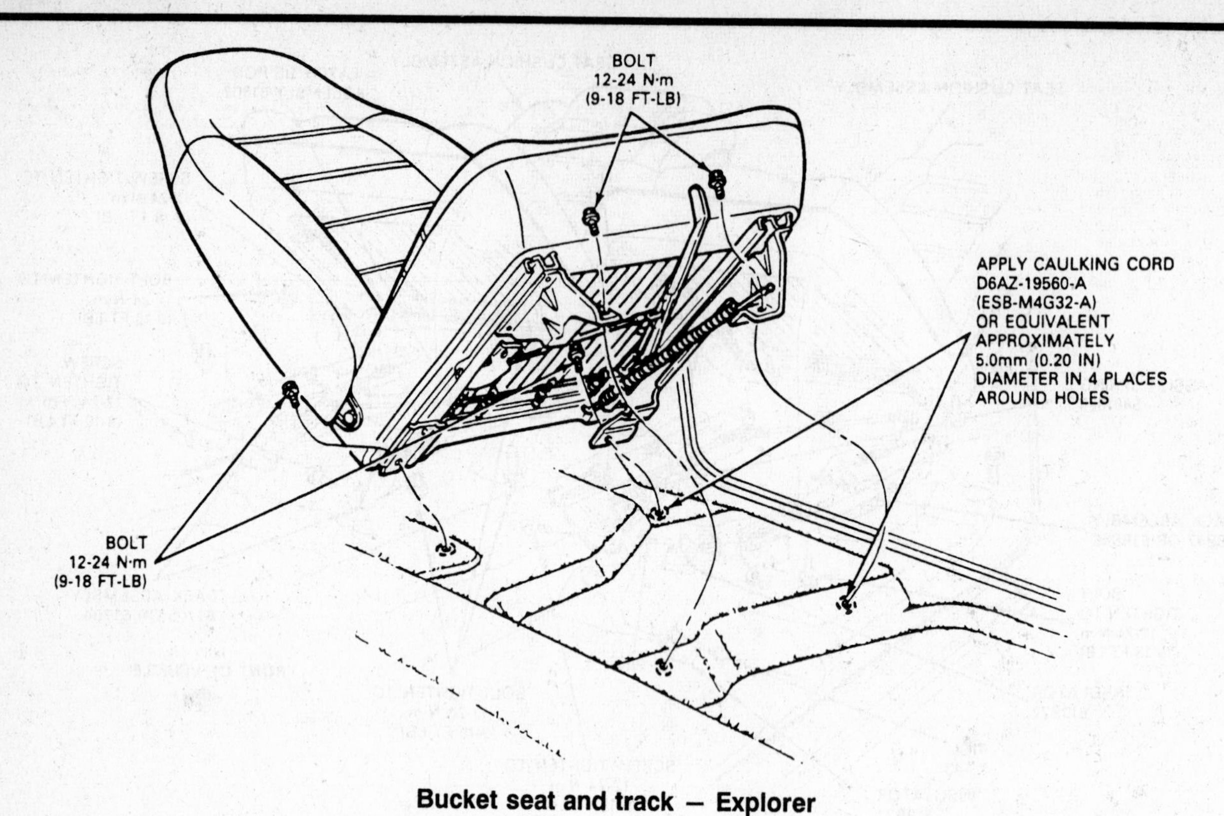

Bucket seat and track — Explorer

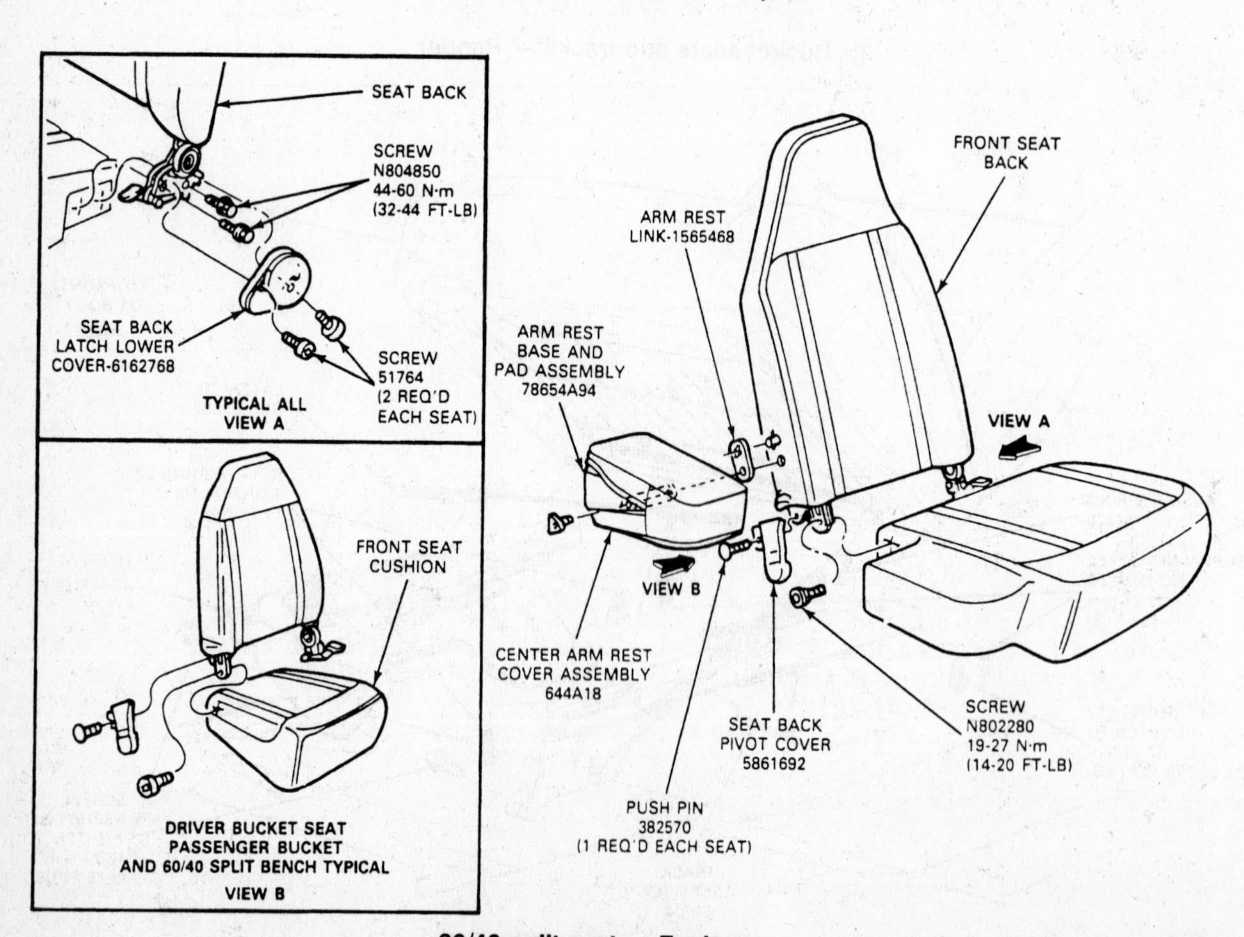

60/40 split seat — Explorer

To install:

6. Position the tracks to the seat cushion. Install the 4 track-to-seat cushion screws (2 each side) and tighten.
7. Connect the latch tie rod assembly and assist spring to the tracks.
8. Position the seat and track assembly in the vehicle.
9. Install 4 track-to-floorpan screws and tighten to specification.
10. Install the seat track insulators (Ranger passenger seat only).

Seat Belt Systems

REMOVAL AND INSTALLATION

1. Remove the retaining bolt from the retractor assembly if applicable.
2. Lift the carpet as required, and remove the retaining bolt securing the assembly to the floorpan or side panel.
3. Remove the bolt from the strap guide, if necessary, and remove the assembly from the vehicle.
4. Installation is the reverse of removal. Refer to the proper illustration for proper torque values.

Headliner

REMOVAL AND INSTALLATION

Ranger

1. Remove both sun visors.
2. Remove the assist handles if equipped.
3. Remove the upper windshield garnish molding.
4. Remove the front pillar moldings.
5. Remove the coat hooks.
6. Remove the lower back window garnish molding.
7. Remove the rear corner trim panels.
8. Remove the dome lamp lens and lamp assembly.
9. Carefully remove the headliner from the truck.

To install:

10. If replacing the headliner, preshape the replacement one along the score marks to about the same shape as the original.
11. Install the headliner to the roof of the truck.
12. Install the dome lamp assembly.
13. Install the rear corner trim panels and related items.
14. Install the lower back window garnish molding.
15. Install the coat hooks.

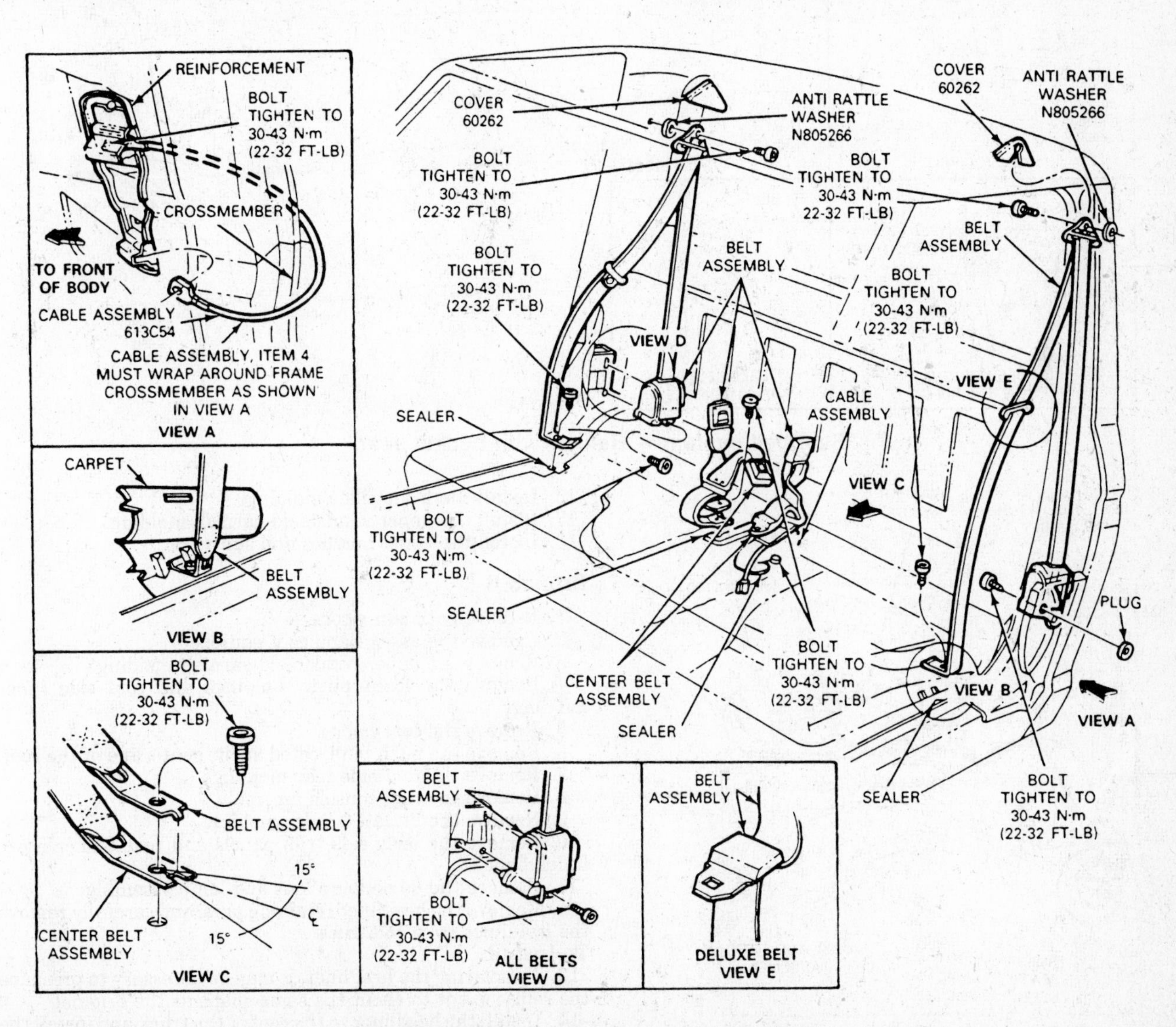

Seat belt system — Ranger with bench seat

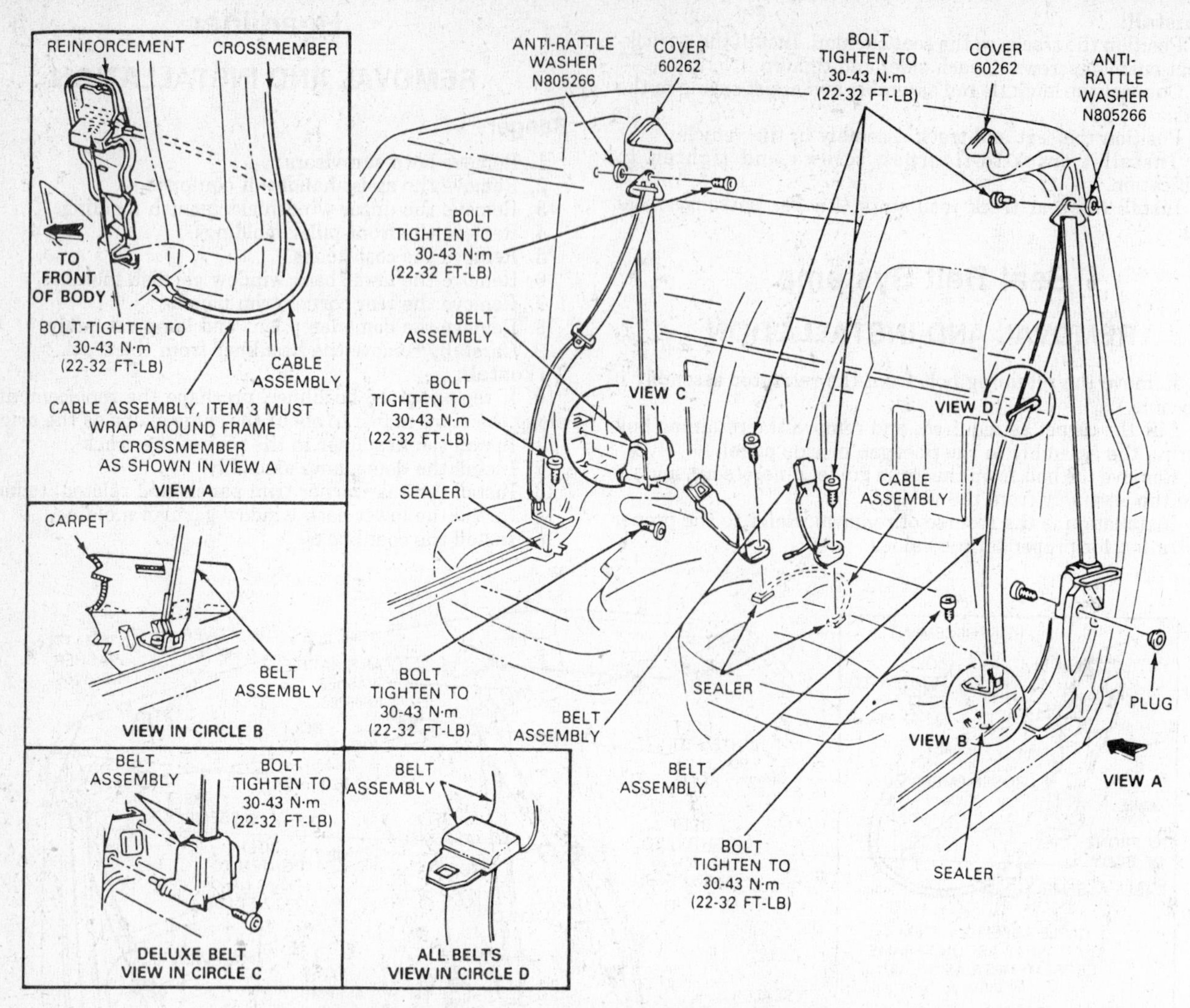

Seat belt system – Ranger with bucket seat

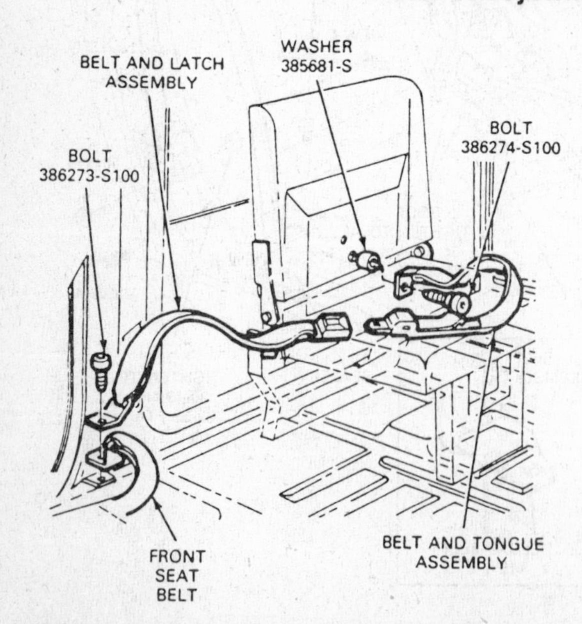

Rear seat belt system – Ranger Super Cab

16. Install the front pillar moldings.
17. Install the upper windshield garnish molding.
18. Install the assist handles and sun visors.

Bronco II

1. Remove both sun visors.
2. Remove the assist handles if equipped.
3. Remove the upper windshield garnish molding.
4. Remove the front pillar moldings and roof side inner moldings.
5. Remove the coat hooks.
6. Remove the push pin located at the centerline of the roof.
7. Remove the roof side rear moldings.
8. Remove the roof console assembly.
9. Remove the liftgate header molding.
10. Remove the body side trim panels and rear corner inner finish panels.
11. Remove the dome lamp lens and lamp assembly.
12. Remove the headliner retaining pins, and carefully remove the headliner from the truck.

To install:

13. If replacing the headliner, it may be necessary to preshape the replacement to about the same shape as the original.
14. Install the headliner to the roof of the truck and install the retaining pins.

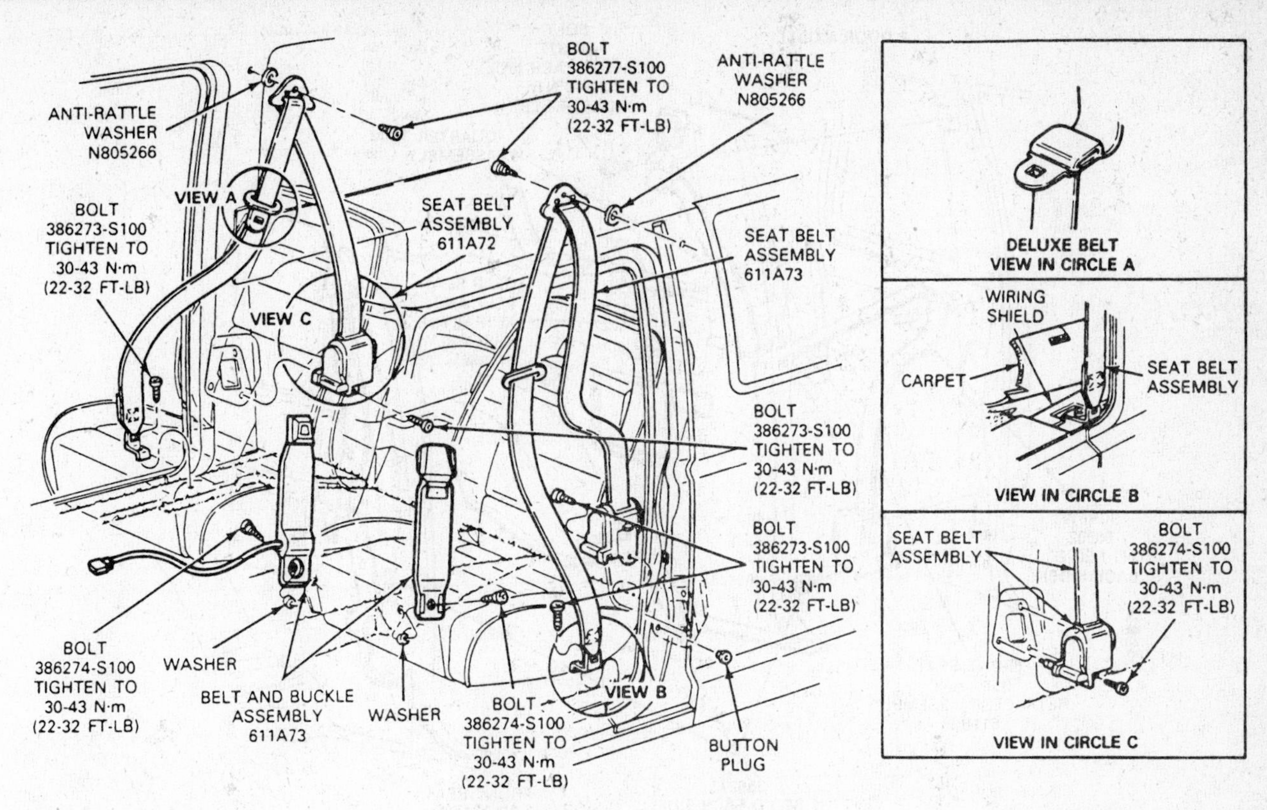

Seat belt system – Bronco II

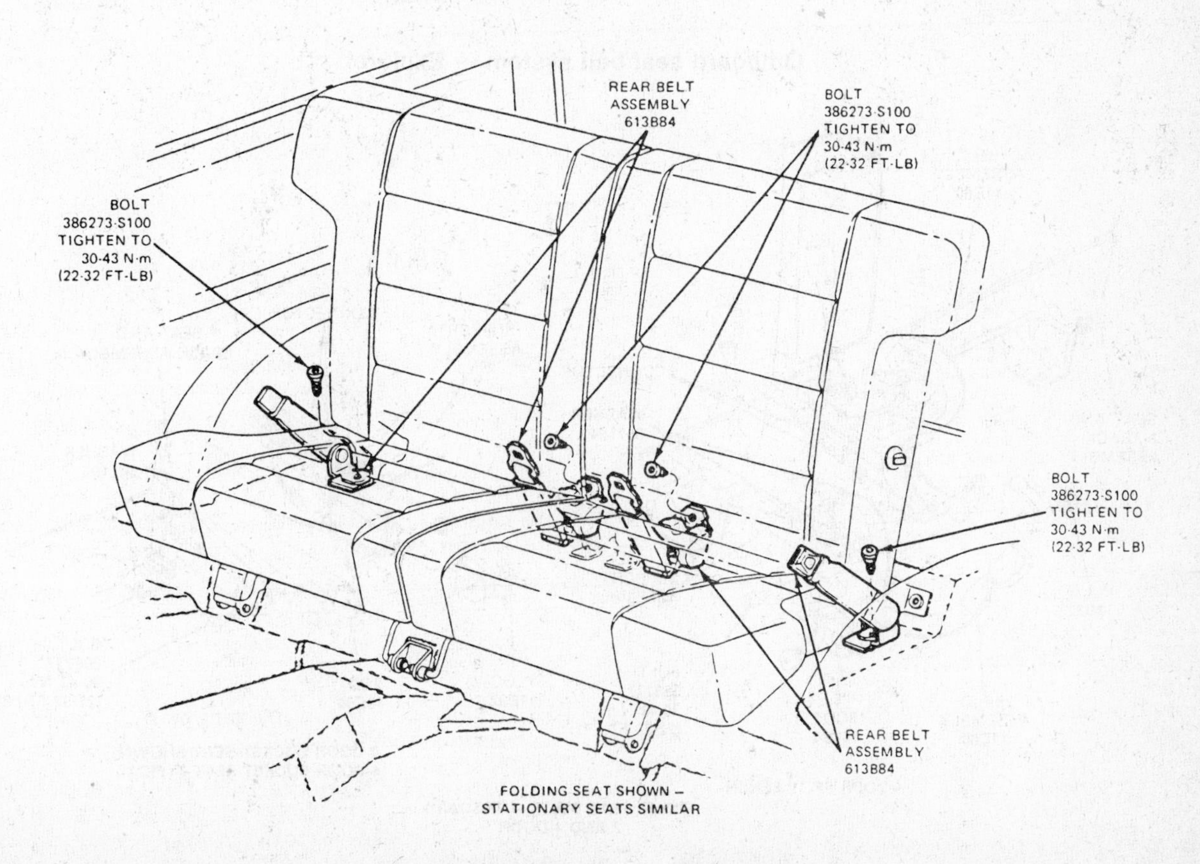

Rear seat belt system – Bronco II

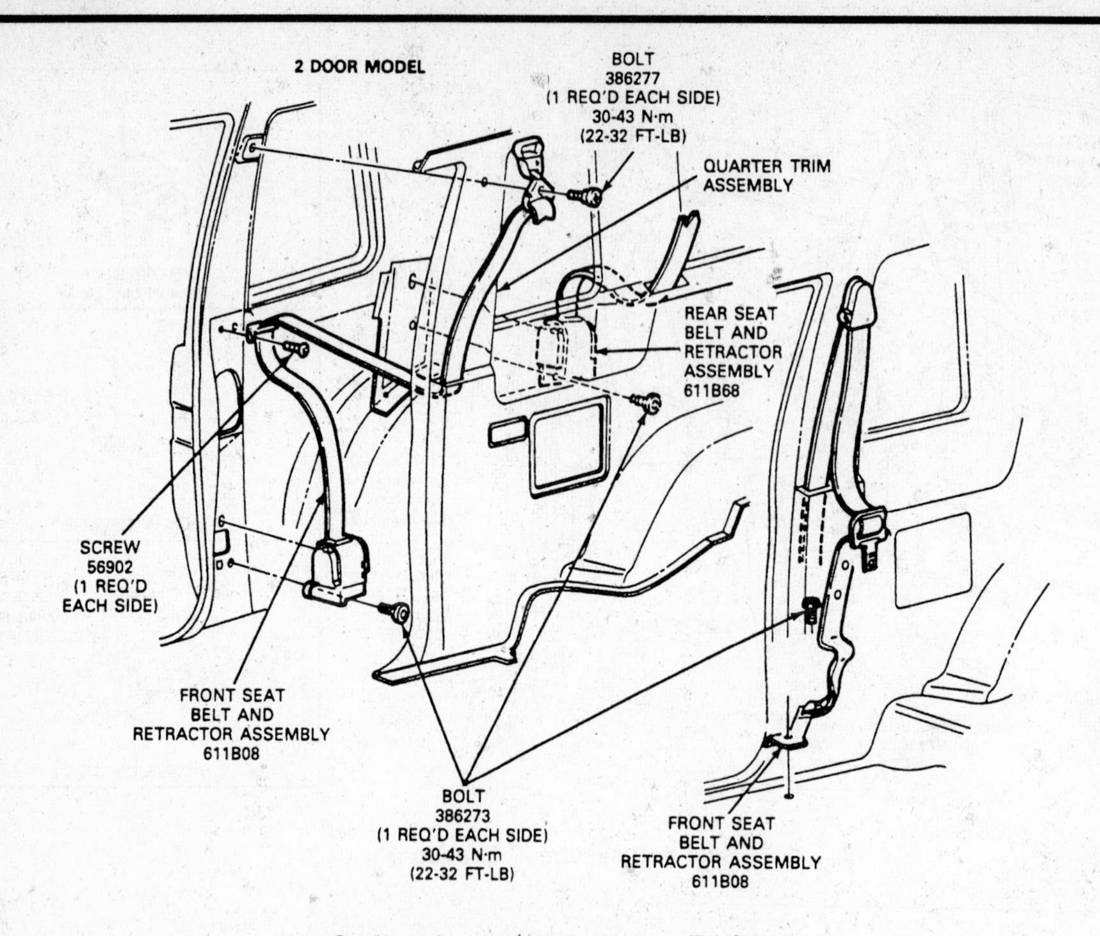

Outboard seat belt system – Explorer

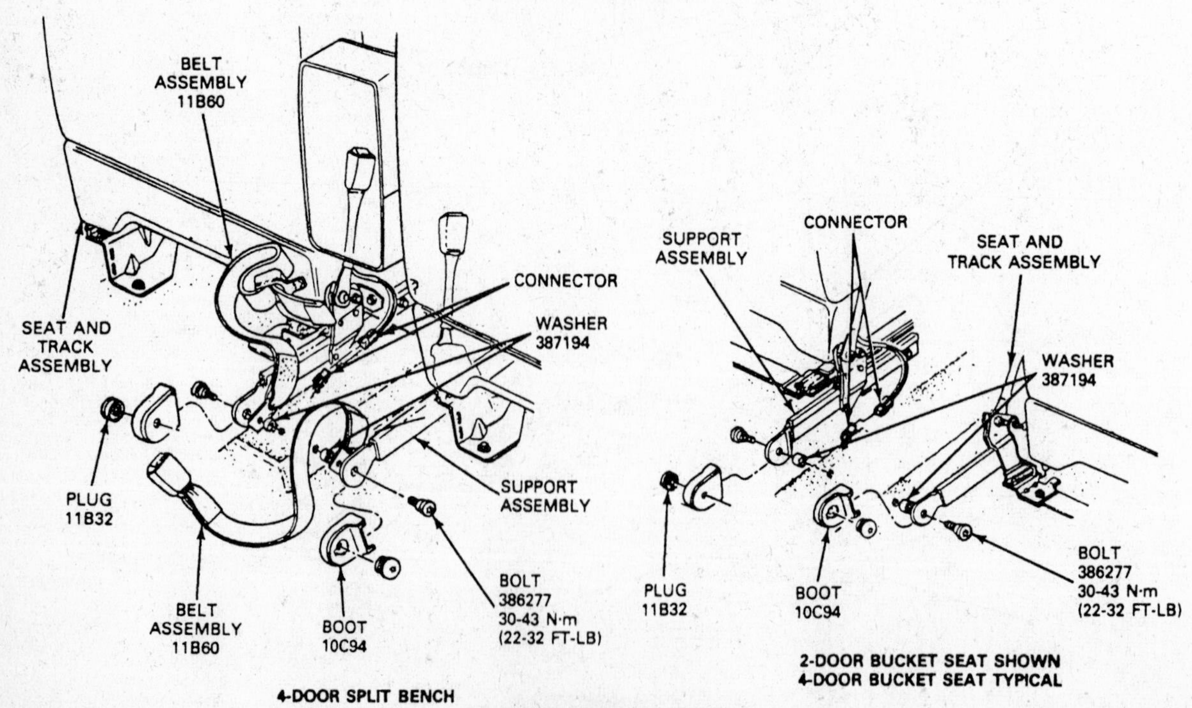

Center seat belt system – Explorer

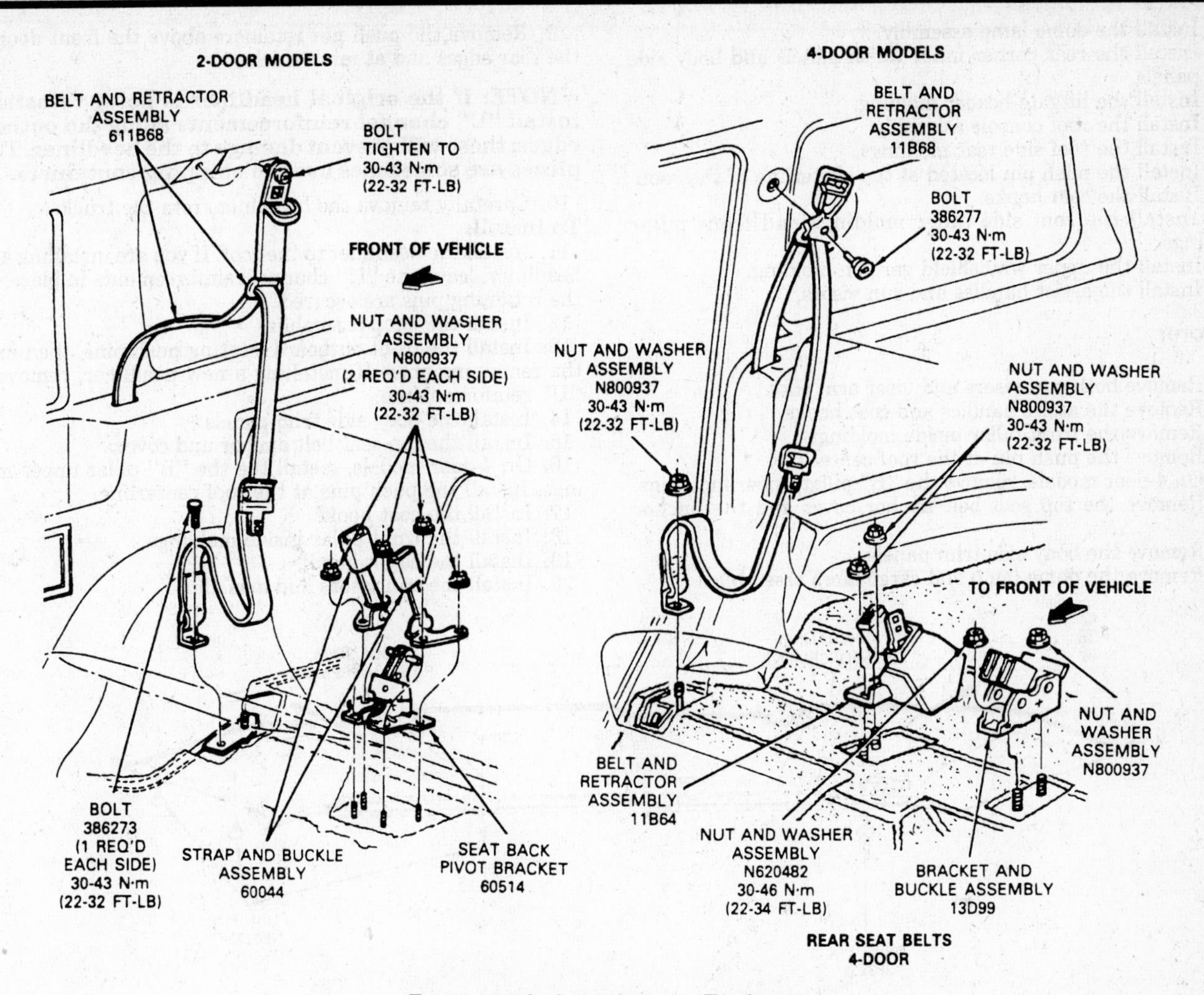

Rear seat belt system – Explorer

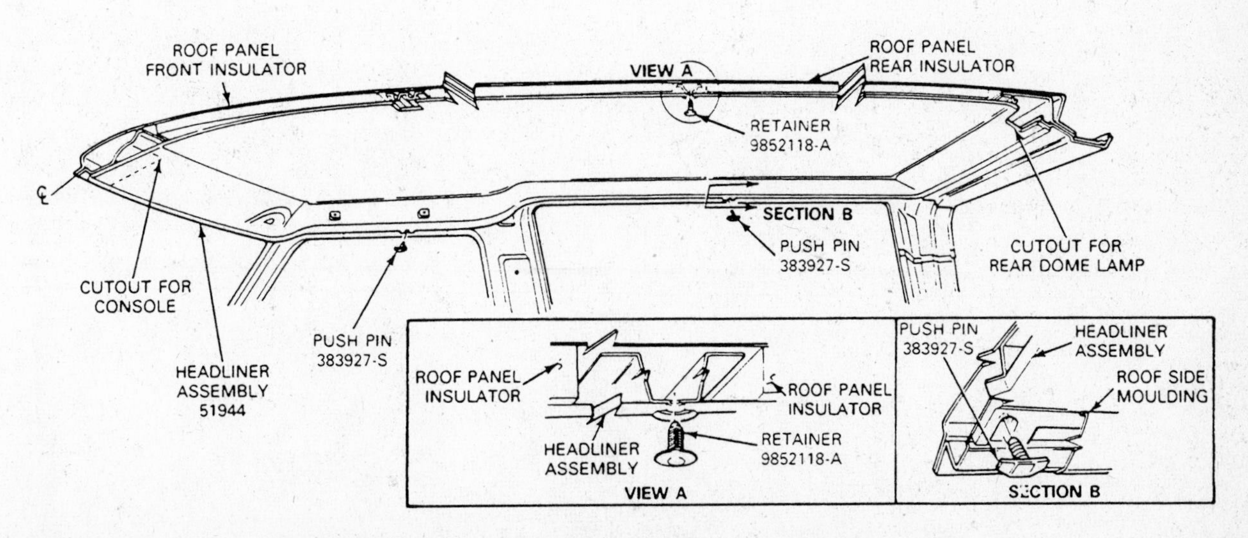

Headliner with close-ups of retaining pins – Bronco II

15. Install the dome lamp assembly.
16. Install the rear corner inner finish panels and body side trim panels.
17. Install the liftgate header molding.
18. Install the roof console assembly.
19. Install the roof side rear moldings.
20. Install the push pin located at the centerline of the roof.
21. Install the coat hooks.
22. Install the roof side inner moldings and front pillar moldings.
23. Install the upper windshield garnish molding.
24. Install the assist handles and sun visors.

Explorer

1. Remove both sun visors and visor arm clips.
2. Remove the assist handles and coat hooks.
3. Remove the front pillar inside moldings
4. Remove the push pin at the roof centerline.
5. On 4-door models, remove the "B" pillar upper moldings.
6. Remove the top seat belt anchor cover and the anchor itself.
7. Remove the body side trim panels.
8. Remove the dome lamp and cargo lamp assemblies.
9. Remove the push pin retainers above the front doors, at the rear edges and at mid-vehicle.

NOTE: If the original headliner is being reinstalled, install "U" channel reinforcements along the outboard edges; these will prevent damage to the headliner. These pieces are sometimes used in shipping containers.

10. Carefully remove the headliner from the truck.

To install:

11. Install the headliner to the roof. If you are installing a new headliner, leave the "U" channel reinforcements in place until the retaining pins are secured.
12. Install the lamp assemblies.
13. Install the front outboard locating push pins, then install the remaining ones. If installing a new headliner, remove the "U" reinforcements.
14. Install the body side trim panels.
15. Install the top seat belt anchor and cover.
16. On 4-door models, install the the "B" pillar upper moldings. Install the push pins at the roof centerline.
17. Install the coat hooks.
18. Install the front pillar inside moldings.
19. Install the assist handles.
20. Install the sun visors and arm clips.

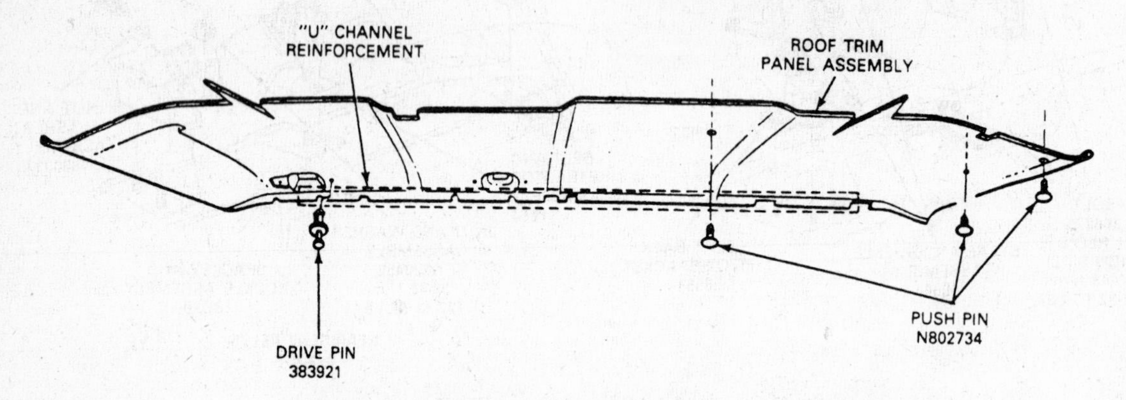

Headliner with retaining pins shown — Explorer

Glossary

AIR/FUEL RATIO: The ratio of air to gasoline by weight in the fuel mixture drawn into the engine.

AIR INJECTION: One method of reducing harmful exhaust emissions by injecting air into each of the exhaust ports of an engine. The fresh air entering the hot exhaust manifold causes any remaining fuel to be burned before it can exit the tailpipe.

ALTERNATOR: A device used for converting mechanical energy into electrical energy.

AMMETER: An instrument, calibrated in amperes, used to measure the flow of an electrical current in a circuit. Ammeters are always connected in series with the circuit being tested.

AMPERE: The rate of flow of electrical current present when one volt of electrical pressure is applied against one ohm of electrical resistance.

ANALOG COMPUTER: Any microprocessor that uses similar (analogous) electrical signals to make its calculations.

ARMATURE: A laminated, soft iron core wrapped by a wire that converts electrical energy to mechanical energy as in a motor or relay. When rotated in a magnetic field, it changes mechanical energy into electrical energy as in a generator.

ATMOSPHERIC PRESSURE: The pressure on the Earth's surface caused by the weight of the air in the atmosphere. At sea level, this pressure is 14.7 psi at 32°F (101 kPa at 0°C).

ATOMIZATION: The breaking down of a liquid into a fine mist that can be suspended in air.

AXIAL PLAY: Movement parallel to a shaft or bearing bore.

BACKFIRE: The sudden combustion of gases in the intake or exhaust system that results in a loud explosion.

BACKLASH: The clearance or play between two parts, such as meshed gears.

BACKPRESSURE: Restrictions in the exhaust system that slow the exit of exhaust gases from the combustion chamber.

BAKELITE: A heat resistant, plastic insulator material commonly used in printed circuit boards and transistorized components.

BALL BEARING: A bearing made up of hardened inner and outer races between which hardened steel balls roll.

BALLAST RESISTOR: A resistor in the primary ignition circuit that lowers voltage after the engine is started to reduce wear on ignition components.

BEARING: A friction reducing, supportive device usually located between a stationary part and a moving part.

BIMETAL TEMPERATURE SENSOR: Any sensor or switch made of two dissimilar types of metal that bend when heated or cooled due to the different expansion rates of the alloys. These types of sensors usually function as an on/off switch.

BLOWBY: Combustion gases, composed of water vapor and unburned fuel, that leak past the piston rings into the crankcase during normal engine operation. These gases are removed by the PCV system to prevent the buildup of harmful acids in the crankcase.

BRAKE PAD: A brake shoe and lining assembly used with disc brakes.

BRAKE SHOE: The backing for the brake lining. The term is, however, usually applied to the assembly of the brake backing and lining.

BUSHING: A liner, usually removable, for a bearing; an antifriction liner used in place of a bearing.

BYPASS: System used to bypass ballast resistor during engine cranking to increase voltage supplied to the coil.

CALIPER: A hydraulically activated device in a disc brake system, which is mounted straddling the brake rotor (disc). The caliper contains at least one piston and two brake pads. Hydraulic pressure on the piston(s) forces the pads against the rotor.

CAMSHAFT: A shaft in the engine on which are the lobes (cams) which operate the valves. The camshaft is driven by the crankshaft, via a belt, chain or gears, at one half the crankshaft speed.

CAPACITOR: A device which stores an electrical charge.

CARBON MONOXIDE (CO): A colorless, odorless gas given off as a normal byproduct of combustion. It is poisonous and extremely dangerous in confined areas, building up slowly to toxic levels without warning if adequate ventilation is not available.

CARBURETOR: A device, usually mounted on the intake manifold of an engine, which mixes the air and fuel in the proper proportion to allow even combustion.

CATALYTIC CONVERTER: A device installed in the exhaust system, like a muffler, that converts harmful byproducts of combustion into carbon dioxide and water vapor by means of a heat-producing chemical reaction.

CENTRIFUGAL ADVANCE: A mechanical method of advancing the spark timing by using flyweights in the distributor that react to centrifugal force generated by the distributor shaft rotation.

CHECK VALVE: Any one-way valve installed to permit the flow of air, fuel or vacuum in one direction only.

GLOSSARY

CHOKE: A device, usually a moveable valve, placed in the intake path of a carburetor to restrict the flow of air.

CIRCUIT: Any unbroken path through which an electrical current can flow. Also used to describe fuel flow in some instances.

CIRCUIT BREAKER: A switch which protects an electrical circuit from overload by opening the circuit when the current flow exceeds a predetermined level. Some circuit breakers must be reset manually, while most reset automatically

COIL (IGNITION): A transformer in the ignition circuit which steps up the voltage provided to the spark plugs.

COMBINATION MANIFOLD: An assembly which includes both the intake and exhaust manifolds in one casting.

COMBINATION VALVE: A device used in some fuel systems that routes fuel vapors to a charcoal storage canister instead of venting them into the atmosphere. The valve relieves fuel tank pressure and allows fresh air into the tank as the fuel level drops to prevent a vapor lock situation.

COMPRESSION RATIO: The comparison of the total volume of the cylinder and combustion chamber with the piston at BDC and the piston at TDC.

CONDENSER: 1. An electrical device which acts to store an electrical charge, preventing voltage surges.
2. A radiator-like device in the air conditioning system in which refrigerant gas condenses into a liquid, giving off heat.

CONDUCTOR: Any material through which an electrical current can be transmitted easily.

CONTINUITY: Continuous or complete circuit. Can be checked with an ohmmeter.

COUNTERSHAFT: An intermediate shaft which is rotated by a mainshaft and transmits, in turn, that rotation to a working part.

CRANKCASE: The lower part of an engine in which the crankshaft and related parts operate.

CRANKSHAFT: The main driving shaft of an engine which receives reciprocating motion from the pistons and converts it to rotary motion.

CYLINDER: In an engine, the round hole in the engine block in which the piston(s) ride.

CYLINDER BLOCK: The main structural member of an engine in which is found the cylinders, crankshaft and other principal parts.

CYLINDER HEAD: The detachable portion of the engine, fastened, usually, to the top of the cylinder block, containing all or most of the combustion chambers. On overhead valve engines, it contains the valves and their operating parts. On overhead cam engines, it contains the camshaft as well.

DEAD CENTER: The extreme top or bottom of the piston stroke.

DETONATION: An unwanted explosion of the air/fuel mixture in the combustion chamber caused by excess heat and compression, advanced timing, or an overly lean mixture. Also referred to as "ping".

DIAPHRAGM: A thin, flexible wall separating two cavities, such as in a vacuum advance unit.

DIESELING: A condition in which hot spots in the combustion chamber cause the engine to run on after the key is turned off.

DIFFERENTIAL: A geared assembly which allows the transmission of motion between drive axles, giving one axle the ability to turn faster than the other.

DIODE: An electrical device that will allow current to flow in one direction only.

DISC BRAKE: A hydraulic braking assembly consisting of a brake disc, or rotor, mounted on an axle, and a caliper assembly containing, usually two brake pads which are activated by hydraulic pressure. The pads are forced against the sides of the disc, creating friction which slows the vehicle.

DISTRIBUTOR: A mechanically driven device on an engine which is responsible for electrically firing the spark plug at a predetermined point of the piston stroke.

DOWEL PIN: A pin, inserted in mating holes in two different parts allowing those parts to maintain a fixed relationship.

DRUM BRAKE: A braking system which consists of two brake shoes and one or two wheel cylinders, mounted on a fixed backing plate, and a brake drum, mounted on an axle, which revolves around the assembly. Hydraulic action applied to the wheel cylinders forces the shoes outward against the drum, creating friction, slowing the vehicle.

DWELL: The rate, measured in degrees of shaft rotation, at which an electrical circuit cycles on and off.

ELECTRONIC CONTROL UNIT (ECU): Ignition module, module, amplifier or igniter. See Module for definition.

ELECTRONIC IGNITION: A system in which the timing and firing of the spark plugs is controlled by an electronic control unit, usually called a module. These systems have no points or condenser.

ENDPLAY: The measured amount of axial movement in a shaft.

ENGINE: A device that converts heat into mechanical energy.

EXHAUST MANIFOLD: A set of cast passages or pipes which conduct exhaust gases from the engine.

FEELER GAUGE: A blade, usually metal, of precisely predetermined thickness, used to measure the clearance between two parts. These blades usually are available in sets of assorted thicknesses.

F-Head: An engine configuration in which the intake valves are in the cylinder head, while the camshaft and exhaust valves are located in the cylinder block. The camshaft operates the intake valves via lifters and pushrods, while it operates the exhaust valves directly.

FIRING ORDER: The order in which combustion occurs in the cylinders of an engine. Also the order in which spark is distributed to the plugs by the distributor.

FLATHEAD: An engine configuration in which the camshaft and all the valves are located in the cylinder block.

FLOODING: The presence of too much fuel in the intake manifold and combustion chamber which prevents the air/fuel mixture from firing, thereby causing a no-start situation.

FLYWHEEL: A disc shaped part bolted to the rear end of the crankshaft. Around the outer perimeter is affixed the ring gear. The starter drive engages the ring gear, turning the flywheel, which rotates the crankshaft, imparting the initial starting motion to the engine.

FOOT POUND (ft.lb. or sometimes, ft. lbs.): The amount of energy or work needed to raise an item weighing one pound, a distance of one foot.

FUSE: A protective device in a circuit which prevents circuit overload by breaking the circuit when a specific amperage is present. The device is constructed around a strip or wire of a lower amperage rating than the circuit it is designed to protect. When an amperage higher than that stamped on the fuse is present in the circuit, the strip or wire melts, opening the circuit.

GEAR RATIO: The ratio between the number of teeth on meshing gears.

GENERATOR: A device which converts mechanical energy into electrical energy.

HEAT RANGE: The measure of a spark plug's ability to dissipate heat from its firing end. The higher the heat range, the hotter the plug fires.

HUB: The center part of a wheel or gear.

HYDROCARBON (HC): Any chemical compound made up of hydrogen and carbon. A major pollutant formed by the engine as a byproduct of combustion.

HYDROMETER: An instrument used to measure the specific gravity of a solution.

INCH POUND (in.lb. or sometimes, in. lbs.): One twelfth of a foot pound.

INDUCTION: A means of transferring electrical energy in the form of a magnetic field. Principle used in the ignition coil to increase voltage.

INJECTION PUMP: A device, usually mechanically operated, which meters and delivers fuel under pressure to the fuel injector.

INJECTOR: A device which receives metered fuel under relatively low pressure and is activated to inject the fuel into the engine under relatively high pressure at a predetermined time.

INPUT SHAFT: The shaft to which torque is applied, usually carrying the driving gear or gears.

INTAKE MANIFOLD: A casting of passages or pipes used to conduct air or a fuel/air mixture to the cylinders.

JOURNAL: The bearing surface within which a shaft operates.

KEY: A small block usually fitted in a notch between a shaft and a hub to prevent slippage of the two parts.

MANIFOLD: A casting of passages or set of pipes which connect the cylinders to an inlet or outlet source.

MANIFOLD VACUUM: Low pressure in an engine intake manifold formed just below the throttle plates. Manifold vacuum is highest at idle and drops under acceleration.

MASTER CYLINDER: The primary fluid pressurizing device in a hydraulic system. In automotive use, it is found in brake and hydraulic clutch systems and is pedal activated, either directly or, in a power brake system, through the power booster.

MODULE: Electronic control unit, amplifier or igniter of solid state or integrated design which controls the current flow in the ignition primary circuit based on input from the pick-up coil. When the module opens the primary circuit, the high secondary voltage is induced in the coil.

NEEDLE BEARING: A bearing which consists of a number (usually a large number) of long, thin rollers.

OHM: (Ω) The unit used to measure the resistance of conductor to electrical flow. One ohm is the amount of resistance that limits current flow to one ampere in a circuit with one volt of pressure.

OHMMETER: An instrument used for measuring the resistance, in ohms, in an electrical circuit.

OUTPUT SHAFT: The shaft which transmits torque from a device, such as a transmission.

OVERDRIVE: A gear assembly which produces more shaft revolutions than that transmitted to it.

OVERHEAD CAMSHAFT (OHC): An engine configuration in which the camshaft is mounted on top of the cylinder head and operates the valve either directly or by means of rocker arms.

OVERHEAD VALVE (OHV): An engine configuration in which all of the valves are located in the cylinder head and the camshaft is located in the cylinder block. The camshaft operates the valves via lifters and pushrods.

OXIDES OF NITROGEN (NOx): Chemical compounds of nitrogen produced as a byproduct of combustion. They combine with hydrocarbons to produce smog.

OXYGEN SENSOR: Used with the feedback system to sense the presence of oxygen in the exhaust gas and signal the computer which can reference the voltage signal to an air/fuel ratio.

PINION: The smaller of two meshing gears.

PISTON RING: An open ended ring which fits into a groove on the outer diameter of the piston. Its chief function is to form a seal between the piston and cylinder wall. Most automotive pistons have three rings: two for compression sealing; one for oil sealing.

PRELOAD: A predetermined load placed on a bearing during assembly or by adjustment.

PRIMARY CIRCUIT: Is the low voltage side of the ignition system which consists of the ignition switch, ballast resistor or resistance wire, bypass, coil, electronic control unit and pick-up coil as well as the connecting wires and harnesses.

PRESS FIT: The mating of two parts under pressure, due to the inner diameter of one being smaller than the outer diameter of the other, or vice versa; an interference fit.

RACE: The surface on the inner or outer ring of a bearing on which the balls, needles or rollers move.

REGULATOR: A device which maintains the amperage and/or voltage levels of a circuit at predetermined values.

RELAY: A switch which automatically opens and/or closes a circuit.

RESISTANCE: The opposition to the flow of current through a circuit or electrical device, and is measured in ohms. Resistance is equal to the voltage divided by the amperage.

RESISTOR: A device, usually made of wire, which offers a preset amount of resistance in an electrical circuit.

RING GEAR: The name given to a ring-shaped gear attached to a differential case, or affixed to a flywheel or as part a planetary gear set.

ROLLER BEARING: A bearing made up of hardened inner and outer races between which hardened steel rollers move.

ROTOR: 1. The disc-shaped part of a disc brake assembly, upon which the brake pads bear; also called, brake disc.
2. The device mounted atop the distributor shaft, which passes current to the distributor cap tower contacts.

SECONDARY CIRCUIT: The high voltage side of the ignition system, usually above 20,000 volts. The secondary includes the ignition coil, coil wire, distributor cap and rotor, spark plug wires and spark plugs.

SENDING UNIT: A mechanical, electrical, hydraulic or electromagnetic device which transmits information to a gauge.

SENSOR: Any device designed to measure engine operating conditions or ambient pressures and temperatures. Usually electronic in nature and designed to send a voltage signal to an on-board computer, some sensors may operate as a simple on/off switch or they may provide a variable voltage signal (like a potentiometer) as conditions or measured parameters change.

SHIM: Spacers of precise, predetermined thickness used between parts to establish a proper working relationship.

SLAVE CYLINDER: In automotive use, a device in the hydraulic clutch system which is activated by hydraulic force, disengaging the clutch.

SOLENOID: A coil used to produce a magnetic field, the effect of which is produce work.

SPARK PLUG: A device screwed into the combustion chamber of a spark ignition engine. The basic construction is a conductive core inside of a ceramic insulator, mounted in an outer conductive base. An electrical charge from the spark plug wire travels along the conductive core and jumps a preset air gap to a grounding point or points at the end of the conductive base. The resultant spark ignites the fuel/air mixture in the combustion chamber.

SPLINES: Ridges machined or cast onto the outer diameter of a shaft or inner diameter of a bore to enable parts to mate without rotation.

TACHOMETER: A device used to measure the rotary speed of an engine, shaft, gear, etc., usually in rotations per minute.

THERMOSTAT: A valve, located in the cooling system of an engine, which is closed when cold and opens gradually in response to engine heating, controlling the temperature of the coolant and rate of coolant flow.

TOP DEAD CENTER (TDC): The point at which the piston reaches the top of its travel on the compression stroke.

TORQUE: The twisting force applied to an object.

TORQUE CONVERTER: A turbine used to transmit power from a driving member to a driven member via hydraulic action, providing changes in drive ratio and torque. In automotive use, it links the driveplate at the rear of the engine to the automatic transmission.

TRANSDUCER: A device used to change a force into an electrical signal.

TRANSISTOR: A semi-conductor component which can be actuated by a small voltage to perform an electrical switching function.

TUNE-UP: A regular maintenance function, usually associated with the replacement and adjustment of parts and components in the electrical and fuel systems of a vehicle for the purpose of attaining optimum performance.

TURBOCHARGER: An exhaust driven pump which compresses intake air and forces it into the combustion chambers at higher than atmospheric pressures. The increased air pressure allows more fuel to be burned and results in increased horsepower being produced.

VACUUM ADVANCE: A device which advances the ignition timing in response to increased engine vacuum.

VACUUM GAUGE: An instrument used to measure the presence of vacuum in a chamber.

VALVE: A device which control the pressure, direction of flow or rate of flow of a liquid or gas.

VALVE CLEARANCE: The measured gap between the end of the valve stem and the rocker arm, cam lobe or follower that activates the valve.

VISCOSITY: The rating of a liquid's internal resistance to flow.

VOLTMETER: An instrument used for measuring electrical force in units called volts. Voltmeters are always connected parallel with the circuit being tested.

WHEEL CYLINDER: Found in the automotive drum brake assembly, it is a device, actuated by hydraulic pressure, which, through internal pistons, pushes the brake shoes outward against the drums.